WILEY PLUS

HELPING TEACHERS
AND STUDENTS
SUCCEED TOGETHER

WILEY

Principles of

ANATOMY & PHYSIOLOGY

13th Edition

Volume 1—Organization, Support and Movement, and Control Systems of the Human Body

International Student Version

Gerard J. Tortora
Bergen Community College

Bryan Derrickson
Valencia Community College

WILEY
John Wiley & Sons, Inc.

Contributing Subject Matter Expert: Tara D. Foss
Cover image from © Max Delson Martins Santos
Cover Design by Wendy Lai

Founded in 1807, John Wiley & Sons, Inc. has been a valued source of knowledge and understanding for more than 200 years, helping people around the world meet their needs and fulfill their aspirations. Our company is built on a foundation of principles that include responsibility to the communities we serve and where we live and work. In 2008, we launched a Corporate Citizenship Initiative, a global effort to address the environmental, social, economic, and ethical challenges we face in our business. Among the issues we are addressing are carbon impact, paper specifications and procurement, ethical conduct within our business and among our vendors, and community and charitable support. For more information, please visit our website: www.wiley.com/go/citizenship.

ISBN: 978-0-470-64608-3

Printed in Asia

10 9 8 7 6 5 4 3 2 1

HELPING TEACHERS AND STUDENTS SUCCEED TOGETHER

An anatomy and physiology course can be the gateway to a gratifying career in a whole host of health-related professions. It can also be an incredible challenge. Through years of collaboration with students and instructors alike, we have come to intimately understand not only the material but also the evolving dynamics of teaching and learning A&P. So with every new edition, it's our goal to find new ways to help instructors teach more easily and effectively and students to learn in a way that sticks.

We believe we bring together experience and innovation like no one else, offering a unique solution for A&P designed to help instructors and students succeed together. From constantly evolving animations and visualizations to design based on optimal learning to lessons firmly grounded in learning outcomes, everything is designed with the goal of helping instructors like you teach in a way that inspires confidence and resilience in students and better learning outcomes.

The thirteenth edition of ***Principles of Anatomy and Physiology,*** integrated with ***WileyPLUS,*** builds students' confidence; it takes the guesswork out of studying by providing students with a clear roadmap (one that tells them what to do, how to do it, and if they did it right). Students will take more initiative, so instructors can have greater impact.

Principles of Anatomy and Physiology 13e continues to offer a balanced presentation of content under the umbrella of our primary and unifying theme of homeostasis, supported by relevant discussions of disruptions to homeostasis. In addition, years of student feedback have convinced us that readers learn anatomy and physiology more readily when they remain mindful of the relationship between structure and function. As a writing team—an anatomist and a physiologist—our very different specializations offer practical advantages in fine-tuning the balance between anatomy and physiology.

On the following pages students will discover the tips and tools needed to make the most of their study time using the integrated text and media. Instructors will gain an overview of the changes to this edition and of the resources available to create dynamic classroom experiences as well as build meaningful assessment opportunities. Both students and instructors will be interested in the outstanding resources available to seamlessly link laboratory activity with lecture presentation and study time.

NOTES TO STUDENTS

The challenges of learning anatomy and physiology can be complex and time-consuming. This textbook and *WileyPLUS for Anatomy and Physiology* have been carefully designed to maximize your study time by simplifying the choices you make in deciding what to study and how to study it, and in assessing your understanding of the content.

Anatomy and Physiology Is a Visual Science

Studying the figures in this book is as important as reading the narrative. The tools described here will help you understand the concepts being presented in any figure and ensure that you get the most out of the visuals.

❶ **LEGEND.** Read this first. It explains what the figure is about.

❷ **KEY CONCEPT STATEMENT.** Indicated by a "key" icon, this reveals a basic idea portrayed in the figure.

❸ **ORIENTATION DIAGRAM.** Added to many figures, this small diagram helps you understand the perspective from which you are viewing a particular piece of anatomical art.

❹ **FIGURE QUESTION.** Found at the bottom of each figure and accompanied by a "question mark" icon, this serves as a self-check to help you understand the material as you go along.

❺ **FUNCTIONS BOX.** Included with selected figures, these provide a brief summary of the functions of the anatomical structure or system depicted.

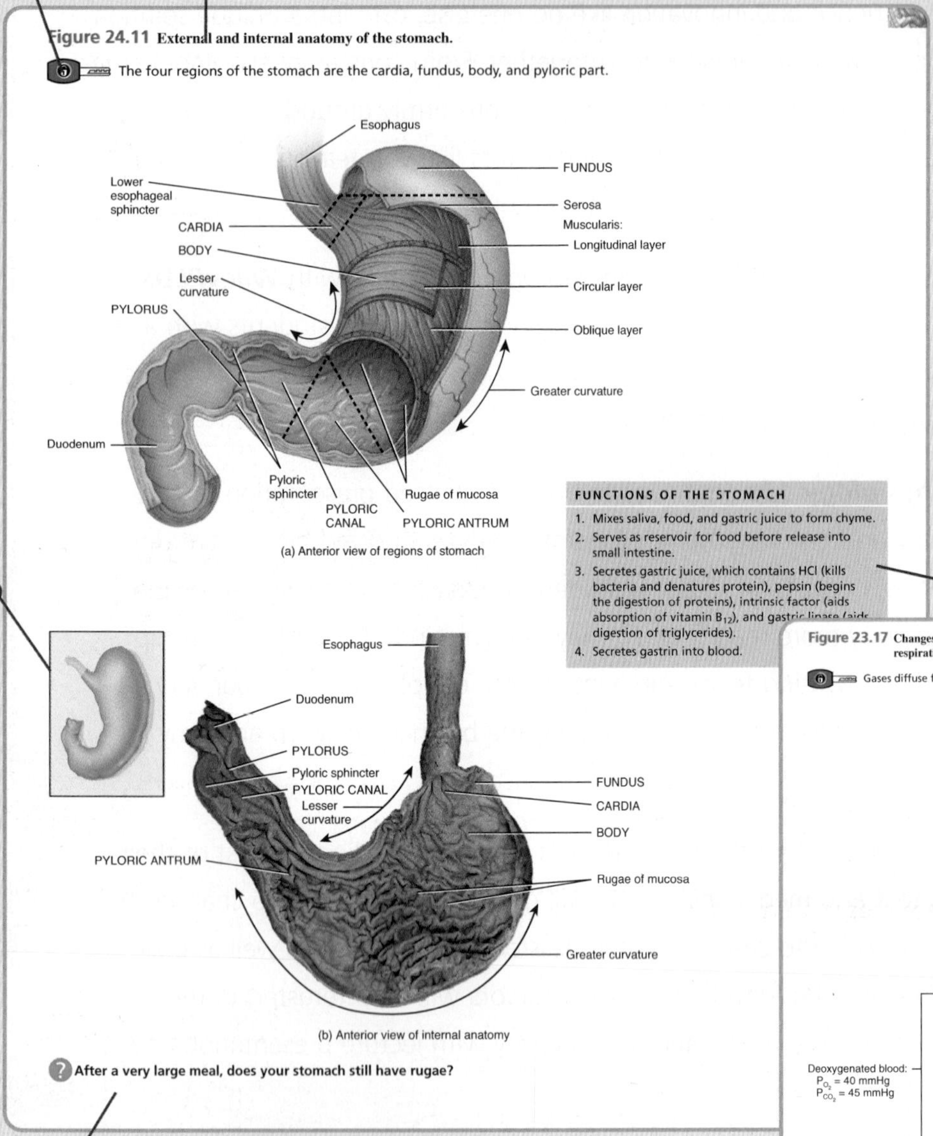

Figure 24.11 External and internal anatomy of the stomach.
The four regions of the stomach are the cardia, fundus, body, and pyloric part.
Esophagus
Lower esophageal sphincter
CARDIA
BODY
Lesser curvature
PYLORUS
Duodenum
Pyloric sphincter
PYLORIC CANAL
PYLORIC ANTRUM
Rugae of mucosa
FUNDUS
Serosa
Muscularis:
Longitudinal layer
Circular layer
Oblique layer
Greater curvature
(a) Anterior view of regions of stomach
FUNCTIONS OF THE STOMACH
1. Mixes saliva, food, and gastric juice to form chyme.
2. Serves as reservoir for food before release into small intestine.
3. Secretes gastric juice, which contains HCl (kills bacteria and denatures protein), pepsin (begins the digestion of proteins), intrinsic factor (aids absorption of vitamin B12), and gastric lipase (aids digestion of triglycerides).
4. Secretes gastrin into blood.
Esophagus
Duodenum
PYLORUS
Pyloric sphincter
PYLORIC CANAL
Lesser curvature
PYLORIC ANTRUM
FUNDUS
CARDIA
BODY
Rugae of mucosa
Greater curvature
(b) Anterior view of internal anatomy
After a very large meal, does your stomach still have rugae?

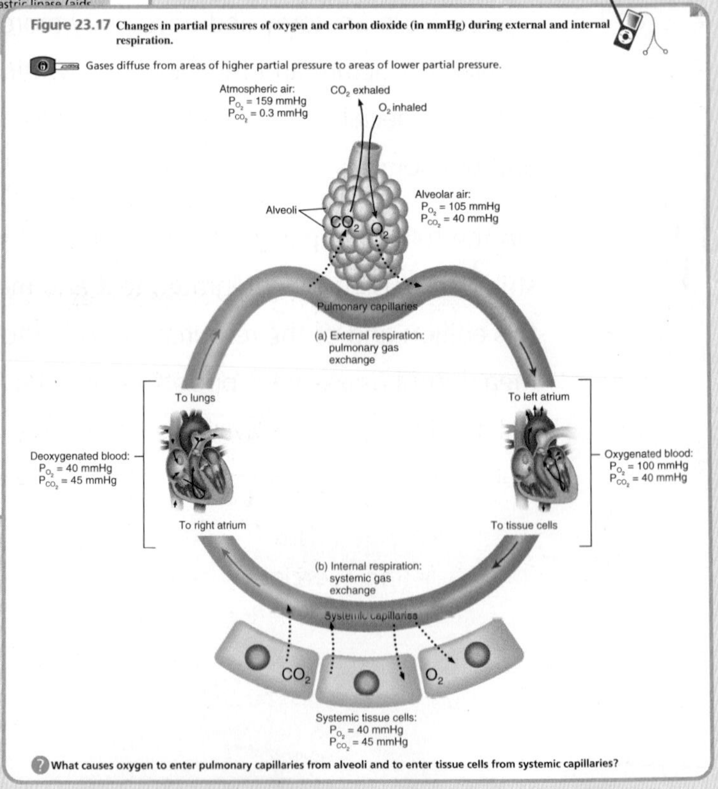

Figure 23.17 Changes in partial pressures of oxygen and carbon dioxide (in mmHg) during external and internal respiration.
Gases diffuse from areas of higher partial pressure to areas of lower partial pressure.
Atmospheric air: P_{O_2} = 159 mmHg P_{CO_2} = 0.3 mmHg
CO_2 exhaled
O_2 inhaled
Alveoli
Alveolar air: P_{O_2} = 105 mmHg P_{CO_2} = 40 mmHg
Pulmonary capillaries
(a) External respiration: pulmonary gas exchange
To lungs
To left atrium
Deoxygenated blood: P_{O_2} = 40 mmHg P_{CO_2} = 45 mmHg
Oxygenated blood: P_{O_2} = 100 mmHg P_{CO_2} = 40 mmHg
To right atrium
To tissue cells
(b) Internal respiration: systemic gas exchange
Systemic capillaries
Systemic tissue cells: P_{O_2} = 40 mmHg P_{CO_2} = 45 mmHg
What causes oxygen to enter pulmonary capillaries from alveoli and to enter tissue cells from systemic capillaries?

❻ **MP3 DOWNLOADS.** In each chapter you will find that several illustrations are marked with an icon that looks like an iPod. This indicates that an audio file that narrates and discusses the important elements of that particular illustration is available. You can access these downloads on the student companion website or within *WileyPLUS.*

NOTES TO STUDENTS

Studying physiology requires an understanding of the sequence of processes. Correlation of sequential processes in text and art is achieved through the use of special numbered lists in the narrative that correspond to numbered segments in the accompanying figure. This approach is used extensively throughout the book to lend clarity to the flow of complex processes.

Physiology of Hearing

The following events are involved in hearing (Figure 17.22):

1. The auricle directs sound waves into the external auditory canal.
2. When sound waves strike the tympanic membrane, the alternating waves of high and low pressure in the air cause the tympanic membrane to vibrate back and forth. The tympanic membrane vibrates slowly in response to low-frequency (low-pitched) sounds and rapidly in response to high-frequency (high-pitched) sounds.
3. The central area of the tympanic membrane connects to the malleus, which vibrates along with the tympanic membrane. This vibration is transmitted from the malleus to the incus and then to the stapes.
4. As the stapes moves back and forth, its oval-shaped footplate, which is attached via a ligament to the circumference of the oval window, vibrates in the oval window. The vibrations at the oval window are about 20 times more vigorous than the tympanic membrane because the auditory ossicles efficiently transmit small vibrations spread over a large surface area (the tympanic membrane) into larger vibrations at a smaller surface (the oval window).
5. The movement of the stapes at the oval window sets up fluid pressure waves in the perilymph of the cochlea. As the oval window bulges inward, it pushes on the perilymph of the scala vestibuli.
6. Pressure waves are transmitted from the scala vestibuli to the scala tympani and eventually to the round window, causing it to bulge outward into the middle ear. (See 9 in the figure.)
7. The pressure waves travel through the perilymph of the scala vestibuli, then the vestibular membrane, and then move into the endolymph inside the cochlear duct.
8. The pressure waves in the endolymph cause the basilar membrane to vibrate, which moves the hair cells of the spiral organ against the tectorial membrane. This leads to bending of the stereocilia and ultimately to the generation of nerve impulses in first-order neurons in cochlear nerve fibers.
9. Sound waves of various frequencies cause certain regions of the basilar membrane to vibrate more intensely than other regions. Each segment of the basilar membrane is "tuned" for

Figure 17.22 Events in the stimulation of auditory receptors in the right ear. The numbers correspond to the events listed in the text. The cochlea has been uncoiled to more easily visualize the transmission of sound waves and their distortion of the vestibular and basilar membranes of the cochlear duct.

Hair cells of the spiral organ (organ of Corti) convert a mechanical vibration (stimulus) into an electrical signal (receptor potential).

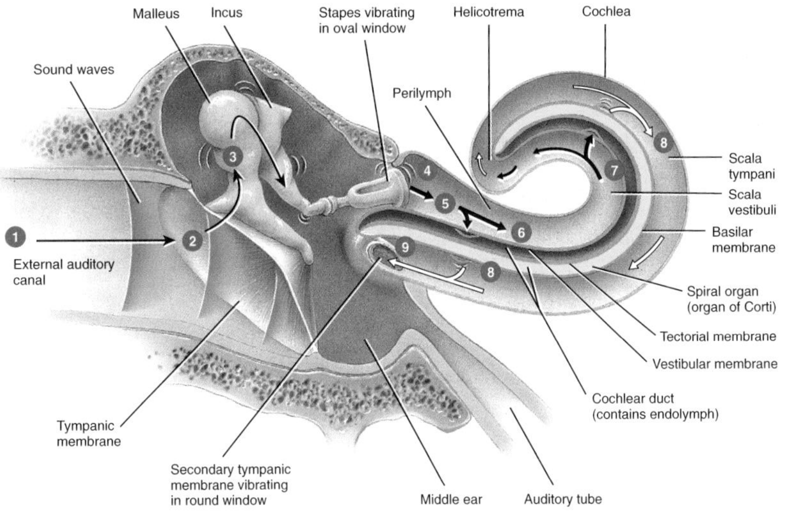

? Which part of the basilar membrane vibrates most vigorously in response to high-frequency (high-pitched) sounds?

WILEY PLUS There are many visual resources within *WileyPLUS,* in addition to the art from your text. These visual aids can help you master the topic you are studying. Examples closely integrated with the reading material include *animations*, *cadaver video clips*, and *Real Anatomy Views*. *Anatomy Drill and Practice* lets you test your knowledge of structures with simple-to-use drag-and-drop labeling exercises or fill-in-the-blank labeling. You can drill and practice on these activities using illustrations from the text, cadaver photographs, histology micrographs, or lab models.

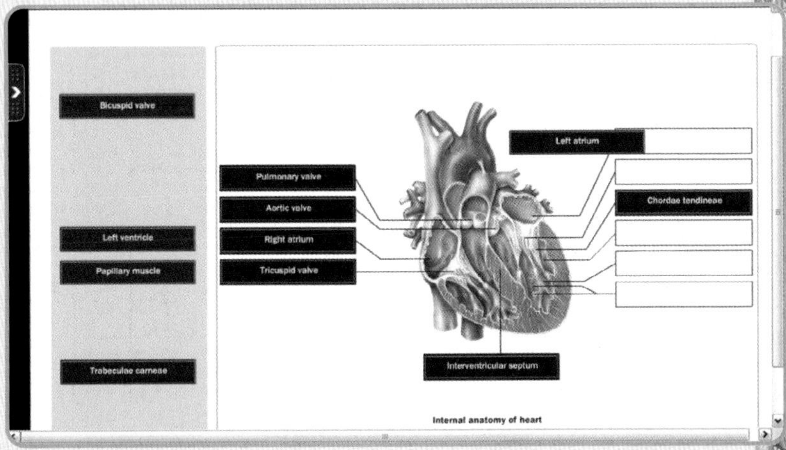

NOTES TO STUDENTS

Exhibits Organize Complex Anatomy into Manageable Modules

Many topics in this text have been organized into **Exhibits** that bring together all of the information and elements that you need to learn the complex terminology, anatomy, and the relevance of the anatomy into a simple-to-navigate content module. You will find Exhibits for tissues, bones, joints, skeletal muscles, nerves, and blood vessels. Most exhibits include the following:

1. Objective to focus your study.
2. Overview narrative of the structure(s).
3. Table summarizing key features of the structure(s).
4. Illustrations and photographs.
5. Checkpoint Question to assess your understanding.
6. Clinical Connection to provide relevance for learning the details.

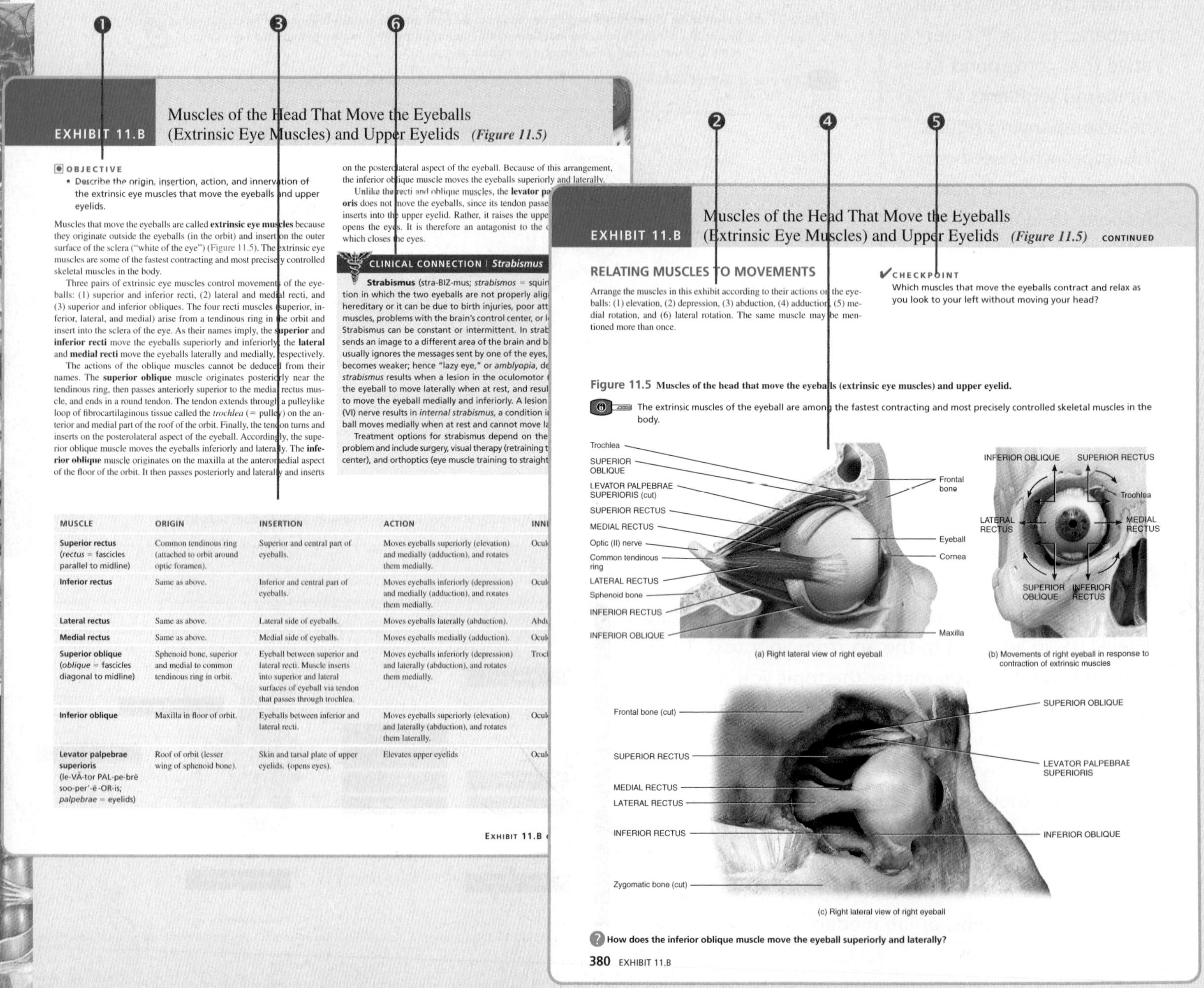

EXHIBIT 11.B Muscles of the Head That Move the Eyeballs (Extrinsic Eye Muscles) and Upper Eyelids *(Figure 11.5)*

OBJECTIVE

- Describe the origin, insertion, action, and innervation of the extrinsic eye muscles that move the eyeballs and upper eyelids.

Muscles that move the eyeballs are called **extrinsic eye muscles** because they originate outside the eyeballs (in the orbit) and insert on the outer surface of the sclera ("white of the eye") (Figure 11.5). The extrinsic eye muscles are some of the fastest contracting and most precisely controlled skeletal muscles in the body.

Three pairs of extrinsic eye muscles control movements of the eyeballs: (1) superior and inferior recti, (2) lateral and medial recti, and (3) superior and inferior obliques. The four recti muscles (superior, inferior, lateral, and medial) arise from a tendinous ring in the orbit and insert into the sclera of the eye. As their names imply, the **superior** and **inferior recti** move the eyeballs superiorly and inferiorly; the **lateral** and **medial recti** move the eyeballs laterally and medially, respectively.

The actions of the oblique muscles cannot be deduced from their names. The **superior oblique** muscle originates posteriorly near the tendinous ring, then passes anteriorly superior to the medial rectus muscle, and ends in a round tendon. The tendon extends through a pulleylike loop of fibrocartilaginous tissue called the *trochlea* (= pulley) on the anterior and medial part of the roof of the orbit. Finally, the tendon turns and inserts on the posterolateral aspect of the eyeball. Accordingly, the superior oblique muscle moves the eyeballs inferiorly and laterally. The **inferior oblique** muscle originates on the maxilla at the anteromedial aspect of the floor of the orbit. It then passes posteriorly and laterally and inserts on the posterolateral aspect of the eyeball. Because of this arrangement, the inferior oblique muscle moves the eyeballs superiorly and laterally.

Unlike the recti and oblique muscles, the **levator pa[...]oris** does not move the eyeballs, since its tendon passe[...] inserts into the upper eyelid. Rather, it raises the uppe[...] opens the eyes. It is therefore an antagonist to the [...] which closes the eyes.

CLINICAL CONNECTION | *Strabismus*

Strabismus (stra-BIZ-mus; *strabismos* = squin[...]) tion in which the two eyeballs are not properly alig[...] hereditary or it can be due to birth injuries, poor att[...] muscles, problems with the brain's control center, or l[...] Strabismus can be constant or intermittent. In strab[...] sends an image to a different area of the brain and b[...] usually ignores the messages sent by one of the eyes, [...] becomes weaker; hence "lazy eye," or *amblyopia*, de[...] *strabismus* results when a lesion in the oculomotor [...] the eyeball to move laterally when at rest, and resul[...] to move the eyeball medially and inferiorly. A lesion [...] (VI) nerve results in *internal strabismus*, a condition i[...] ball moves medially when at rest and cannot move la[...]

Treatment options for strabismus depend on the [...] problem and include surgery, visual therapy (retraining t[...] center), and orthoptics (eye muscle training to straight[...]

MUSCLE	ORIGIN	INSERTION	ACTION	INN[...]
Superior rectus (*rectus* = fascicles parallel to midline)	Common tendinous ring (attached to orbit around optic foramen).	Superior and central part of eyeballs.	Moves eyeballs superiorly (elevation) and medially (adduction), and rotates them medially.	Ocul[...]
Inferior rectus	Same as above.	Inferior and central part of eyeballs.	Moves eyeballs inferiorly (depression) and medially (adduction), and rotates them medially.	Ocul[...]
Lateral rectus	Same as above.	Lateral side of eyeballs.	Moves eyeballs laterally (abduction).	Abdu[...]
Medial rectus	Same as above.	Medial side of eyeballs.	Moves eyeballs medially (adduction).	Ocul[...]
Superior oblique (*oblique* = fascicles diagonal to midline)	Sphenoid bone, superior and medial to common tendinous ring in orbit.	Eyeball between superior and lateral recti. Muscle inserts into superior and lateral surfaces of eyeball via tendon that passes through trochlea.	Moves eyeballs inferiorly (depression) and laterally (abduction), and rotates them medially.	Troc[...]
Inferior oblique	Maxilla in floor of orbit.	Eyeballs between inferior and lateral recti.	Moves eyeballs superiorly (elevation) and laterally (abduction), and rotates them laterally.	Ocul[...]
Levator palpebrae superioris (le-VĀ-tor PAL-pe-brē soo-per'-ē-OR-is; *palpebrae* = eyelids)	Roof of orbit (lesser wing of sphenoid bone).	Skin and tarsal plate of upper eyelids. (opens eyes).	Elevates upper eyelids	Ocul[...]

EXHIBIT 11.B [...]

EXHIBIT 11.B Muscles of the Head That Move the Eyeballs (Extrinsic Eye Muscles) and Upper Eyelids *(Figure 11.5)* CONTINUED

RELATING MUSCLES TO MOVEMENTS

Arrange the muscles in this exhibit according to their actions on the eyeballs: (1) elevation, (2) depression, (3) abduction, (4) adduction, (5) medial rotation, and (6) lateral rotation. The same muscle may be mentioned more than once.

✔CHECKPOINT

Which muscles that move the eyeballs contract and relax as you look to your left without moving your head?

Figure 11.5 **Muscles of the head that move the eyeballs (extrinsic eye muscles) and upper eyelid.**

The extrinsic muscles of the eyeball are among the fastest contracting and most precisely controlled skeletal muscles in the body.

(a) Right lateral view of right eyeball

(b) Movements of right eyeball in response to contraction of extrinsic muscles

(c) Right lateral view of right eyeball

How does the inferior oblique muscle move the eyeball superiorly and laterally?

380 EXHIBIT 11.B

Clinical Discussions Make Your Study Relevant

The relevance of the anatomy and physiology that you are studying is best understood when you make the connection between normal structure and function and what happens when the body doesn't work the way it should. Throughout the chapters of the text you will find **Clinical Connections** that introduce you to interesting clinical perspectives related to the text discussion. In addition, at the end of each body system chapter you will find the **Disorders: Homeostatic Imbalances** section, which includes concise discussions of major diseases and disorders. These provide answers to many of your questions about medical problems. The **Medical Terminology** section that follows includes selected terms dealing with both normal and pathological conditions.

CLINICAL CONNECTION | *Arthroplasty*

Joints that have been severely damaged by diseases such as arthritis, or by injury, may be replaced surgically with artificial joints in a procedure referred to as **arthroplasty** (AR-thrō-plas′-tē; *arthr-*= joint; *plasty*=plastic repair of). Although most joints in the body can be repaired by arthroplasty, the ones most commonly replaced are the hips, knees, and shoulders. About 400,000 hip replacements and 300,000 knee replacements are performed annually in the United States. During the procedure, the ends of the damaged bones are removed and metal, ceramic, or plastic components are fixed in place. The goals of arthroplasty are to relieve pain and increase range of motion.

Partial hip replacements involve only the femur. **Total hip replacements** involve both the acetabulum and head of the femur (Figures A–C). The damaged portions of the acetabulum and the head of the femur are replaced by prefabricated prostheses (artificial devices). The acetabulum is shaped to accept the new socket, the head of the femur is removed, and the center of the femur is shaped to fit the femoral component. The acetabular component consists of a plastic such as polyethylene, and the femoral component is composed of a metal such as cobalt-chrome, titanium alloys, or stainless steel. These materials are designed to withstand a high degree of stress and to prevent a response by the immune system. Once the appropriate acetabular and femoral components are selected, they are attached to the healthy portion of bone with acrylic cement, which forms an interlocking mechanical bond.

Knee replacements are actually a resurfacing of cartilage and, like hip replacements, may be partial or total. In a **partial knee replacement (PKR)**, also called a **unicompartmental knee replacement**, only one side of the knee joint is replaced. Once the damaged cartilage is removed from the distal end of the femur, the femur is reshaped and a metal femoral component is cemented in place. Then the damaged cartilage from the proximal end of the tibia is removed, along with the meniscus. The tibia is reshaped and fitted with a plastic tibial component that is cemented into place. If the posterior surface of the patella is badly damaged, the patella is replaced with a plastic patellar component.

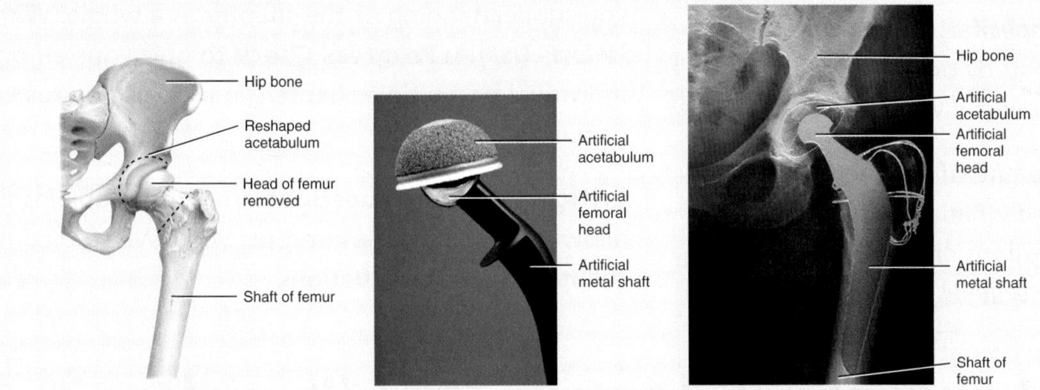
Hip bone
Reshaped acetabulum
Head of femur removed
Shaft of femur
Artificial acetabulum
Artificial femoral head
Artificial metal shaft
Hip bone
Artificial acetabulum
Artificial femoral head
Artificial metal shaft
Shaft of femur

(A) Preparation for total hip replacement (B) Components of an artificial hip joint prior to implantation (C) Radiograph of an artificial hip joint

WILEY PLUS

WileyPLUS offers you opportunities for even further Clinical Connections with animated and interactive case studies that relate specifically to one body system. Look for these under additional chapter resources as an interesting and engaging break from traditional study routines.

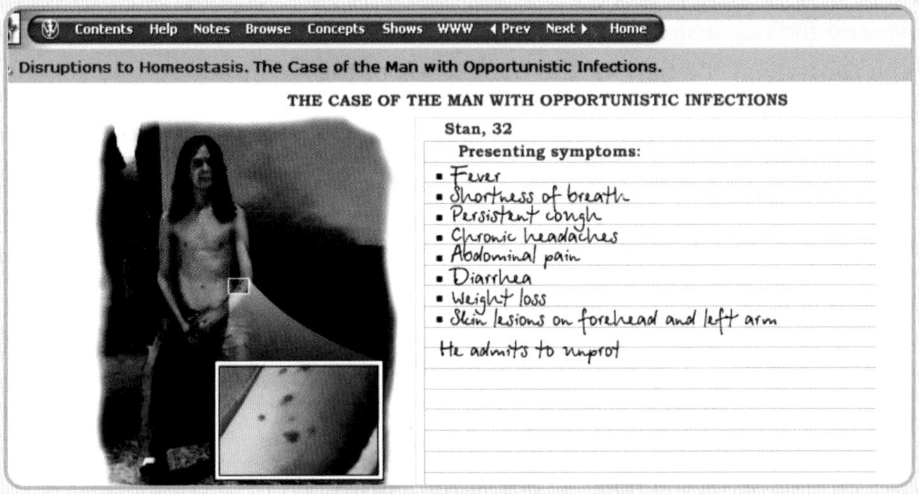
Contents Help Notes Browse Concepts Shows WWW ◂ Prev Next ▸ Home
Disruptions to Homeostasis. The Case of the Man with Opportunistic Infections.
THE CASE OF THE MAN WITH OPPORTUNISTIC INFECTIONS
Stan, 32
Presenting symptoms:
Fever
Shortness of breath
Persistent cough
Chronic headaches
Abdominal pain
Diarrhea
Weight loss
Skin lesions on forehead and left arm
He admits to unprot

NOTES TO STUDENTS

Chapter Resources Help You Focus and Review

Your book has a variety of special features that will make your time studying anatomy and physiology a more rewarding experience. These have been developed based on feedback from students—like you—who have used previous editions of the text. Their effectiveness is even further enhanced within *WileyPLUS for Anatomy and Physiology.*

Chapter Introductions set the stage for the content to come. Each chapter starts with a succinct overview of the particular system's role in maintaining homeostasis in your body, followed by an introduction to the chapter content. This opening page concludes with a question that always begins with "Did you ever wonder…?" These questions will capture your interest and encourage you to find the answer in the chapter material to come.

Objectives at the start of each section help you focus on what is important as you read. All of the content within *WileyPLUS* is tagged to these specific learning objectives so that you can organize your study or review what is still not clear in simple, more meaningful ways.

Checkpoint Questions at the end of each section help you assess if you have absorbed what you have read. Take time to review these questions or answer them within the Practice section of each *WileyPLUS* concept module, where your answers will automatically be graded to let you know where you stand.

Mnemonics are a memory aid that can be particularly helpful when learning specific anatomical features. Mnemonics are included throughout the text—some displayed in figures, tables, or Exhibits, and some included within the text discussion. We encourage you not only to use the mnemonics provided, but also to create your own to help you learn the multitude of terms involved in your study of human anatomy.

Chapter Review and Resource Summary is a helpful table at the end of chapters that offers you a concise summary of the important concepts from the chapter and links each section to the media resources available in *WileyPLUS for Anatomy and Physiology.*

Self-Quiz Questions give you an opportunity to evaluate your understanding of the chapter as a whole. Within *WileyPLUS,* use **Progress Check** to quiz yourself on individual or multiple chapters in preparation for exams or quizzes.

Critical Thinking Questions are word problems that allow you to apply the concepts you have studied in the chapter to specific situations.

Mastering the Language of Anatomy and Physiology

Throughout the text we have included **Pronunciations** and, sometimes, **Word Roots** for many terms that may be new to you. These appear in parentheses immediately following the new words. The pronunciations are repeated in the Glossary at the back of the book. Look at the words carefully and say them out loud several times. Learning to pronounce a new word will help you remember it and make it a useful part of your medical vocabulary. Take a few minutes to read the Pronunciation Key, found at the beginning of the Glossary at the end of this text, so it will be familiar as you encounter new words.

To provide more assistance in learning the language of anatomy, a full **Glossary** of terms with phonetic pronunciations appears at the end of the book. The basic building blocks of medical terminology—**Combining Forms, Word Roots, Prefixes, and Suffixes**—are listed inside the back cover, accompanied by **Eponyms,** traditional terms that include reference to a person's name, along with the current terminology.

WileyPLUS houses help for you in building your new language skills as well. The **Audio Glossary,** which is always available to you, lets you hear all these new, unfamiliar terms pronounced. Throughout the e-text, these terms can be clicked on and heard pronounced as you read. In addition, you can use the helpful **Mastering Vocabulary** program, which creates electronic flashcards for you of the key terms within each chapter for practice, as well as take a self-quiz specifically on the terms introduced in each chapter.

NOTES TO INSTRUCTORS

As active teachers of the course, we recognize both the rewards and challenges in providing a strong foundation for understanding the complexities of the human body. We believe that teaching goes beyond just sharing information. *How* we share information makes all the difference—especially, if as we do, you have an increasingly diverse population of students with varying learning abilities. As we revised this text we focused on those areas that we knew we could enhance to provide greater impact in terms of better learning outcomes. Feedback from many of you, as well as from the students we interact with in our own classrooms, guided us in ensuring that the revisions to the text, along with the powerful new *WileyPLUS for Anatomy and Physiology,* support the needs and challenges you face day to day in your own classrooms.

We focused on several key areas for revision: enhancing the all-important visuals, both drawings and photographs; increasing the use of Exhibits that provide a focused and functional organization of detailed content; adding some new and revising many of the tables to increase their effectiveness; updating and adding clinical material that helps students relate what they are learning to their desired career goals and the world around them; and making narrative changes aimed at increasing student engagement with—and comprehension of—the material.

The Art of Anatomy and Physiology

Illustrations throughout the text have been refined. The color palette for the skulls in Chapter 7, and for the brain and spinal cord throughout the text, has been adjusted for greater impact. Illustrations in each chapter have been revised and updated to provide greater clarity and more saturated colors. Particular emphasis was placed on revised drawings of joints, muscles, and blood vessels.

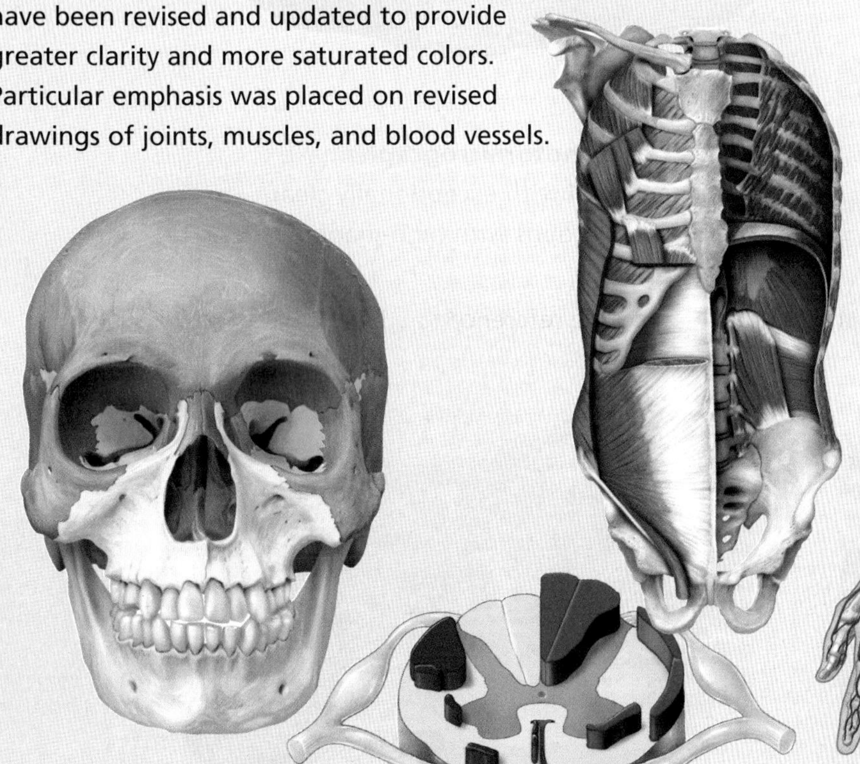

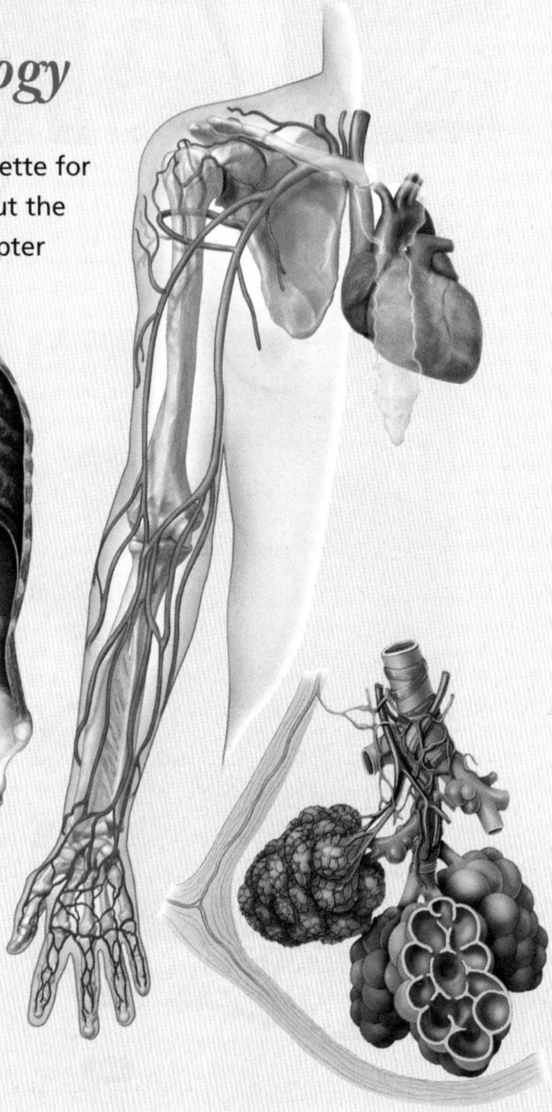

NOTES TO INSTRUCTORS

Cadaver photographs are included throughout the text to help students relate the content to real-life images. These are often paired with diagrams to help make the connections. Most of the meticulous dissections and outstanding photography come from Mark Nielsen's lab at the University of Utah.

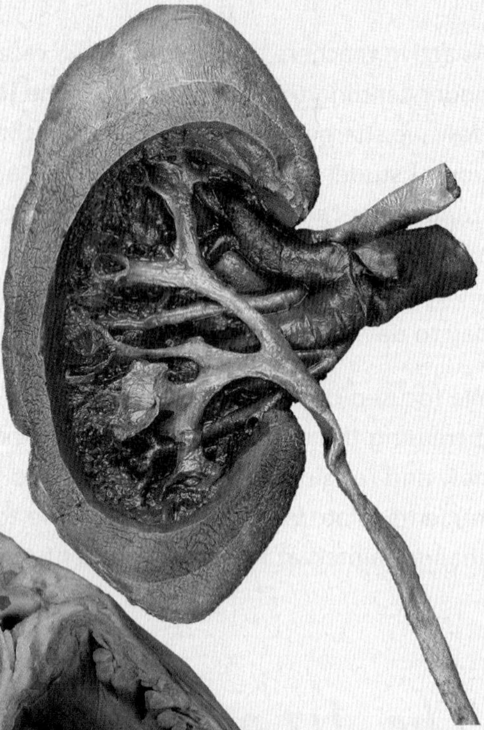

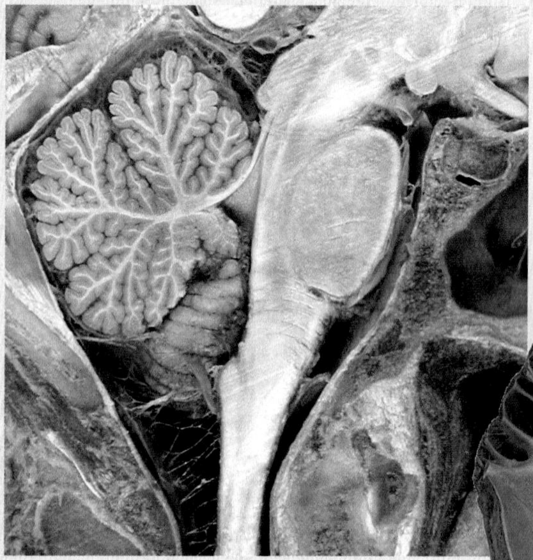

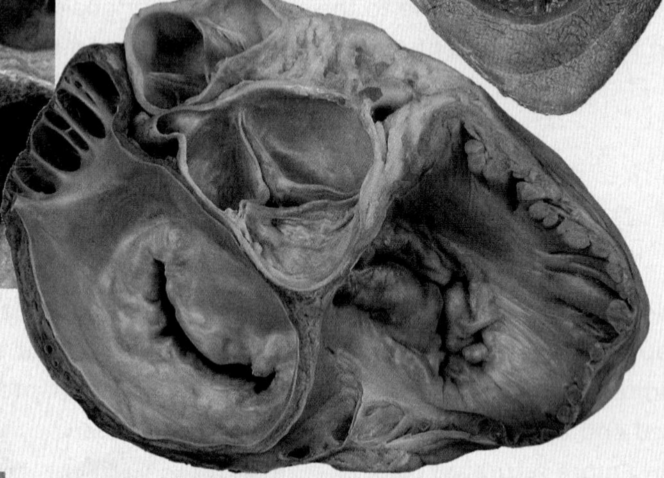

Most tissue **Photomicrographs** have been replaced with exceptionally clear photomicrographs with high-magnification blowups.

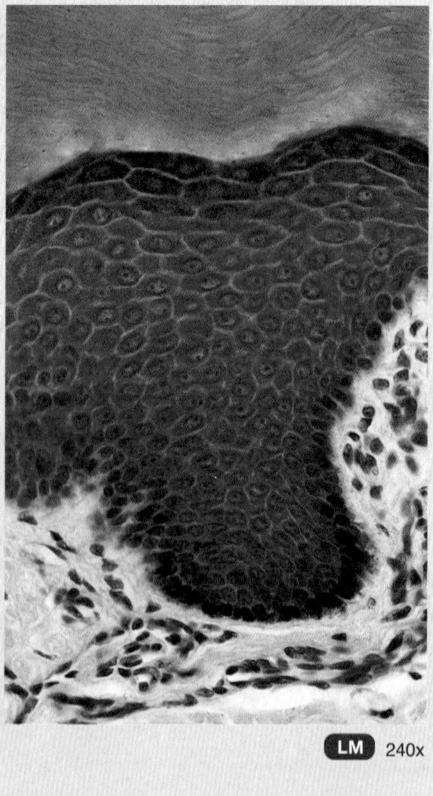

LM 240x

LM 630x

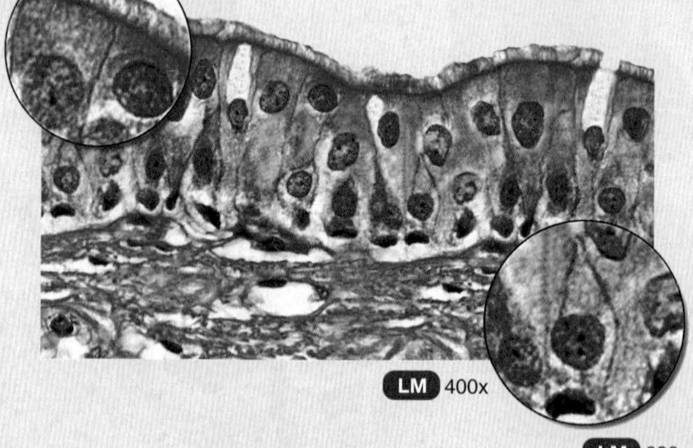

LM 400x

LM 630x

NOTES TO INSTRUCTORS

Exhibits and Tables

The use of the pedagogically designed **Exhibits** has been expanded to include the axial and appendicular skeletons, as well as cranial nerves, providing students with simplified presentations of complex content.

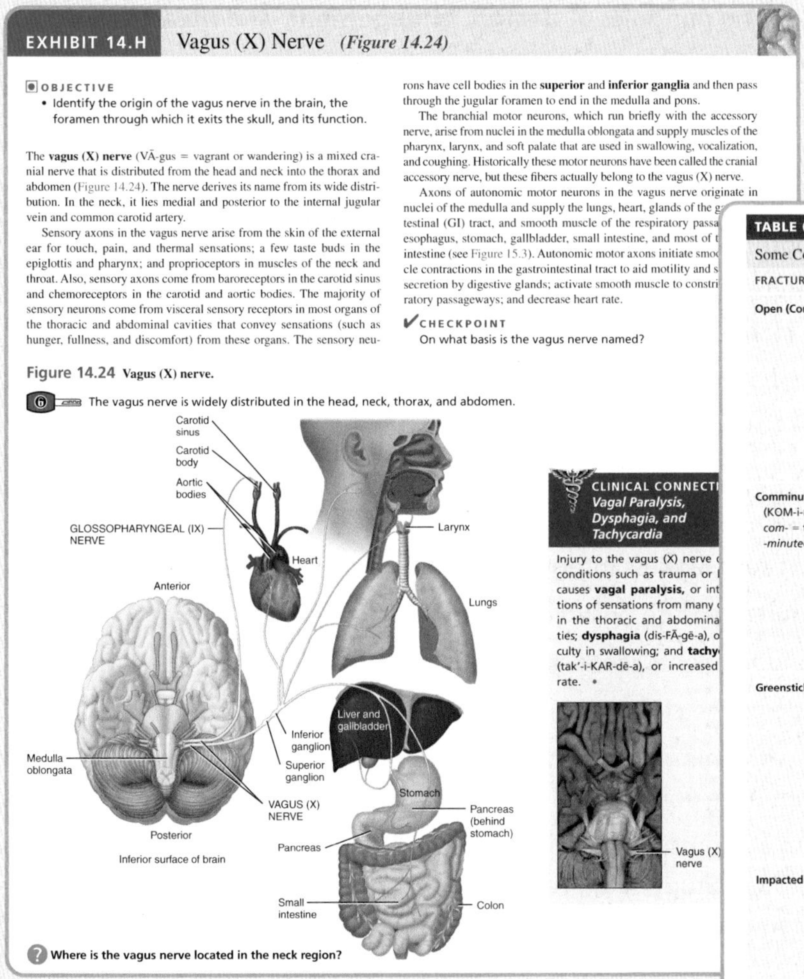

EXHIBIT 14.H Vagus (X) Nerve *(Figure 14.24)*

OBJECTIVE

- Identify the origin of the vagus nerve in the brain, the foramen through which it exits the skull, and its function.

The **vagus (X) nerve** (VĀ-gus = vagrant or wandering) is a mixed cranial nerve that is distributed from the head and neck into the thorax and abdomen (Figure 14.24). The nerve derives its name from its wide distribution. In the neck, it lies medial and posterior to the internal jugular vein and common carotid artery.

Sensory axons in the vagus nerve arise from the skin of the external ear for touch, pain, and thermal sensations; a few taste buds in the epiglottis and pharynx; and proprioceptors in muscles of the neck and throat. Also, sensory axons come from baroreceptors in the carotid sinus and chemoreceptors in the carotid and aortic bodies. The majority of sensory neurons come from visceral sensory receptors in most organs of the thoracic and abdominal cavities that convey sensations (such as hunger, fullness, and discomfort) from these organs. The sensory neurons have cell bodies in the **superior** and **inferior ganglia** and then pass through the jugular foramen to end in the medulla and pons.

The branchial motor neurons, which run briefly with the accessory nerve, arise from nuclei in the medulla oblongata and supply muscles of the pharynx, larynx, and soft palate that are used in swallowing, vocalization, and coughing. Historically these motor neurons have been called the cranial accessory nerve, but these fibers actually belong to the vagus (X) nerve.

Axons of autonomic motor neurons in the vagus nerve originate in nuclei of the medulla and supply the lungs, heart, glands of the g[...] testinal (GI) tract, and smooth muscle of the respiratory passa[...] esophagus, stomach, gallbladder, small intestine, and most of t[...] intestine (see Figure 15.3). Autonomic motor axons initiate smo[...] cle contractions in the gastrointestinal tract to aid motility and s[...] secretion by digestive glands; activate smooth muscle to constri[...] ratory passageways; and decrease heart rate.

CHECKPOINT

On what basis is the vagus nerve named?

Figure 14.24 **Vagus (X) nerve.**

The vagus nerve is widely distributed in the head, neck, thorax, and abdomen.

Where is the vagus nerve located in the neck region?

New **Tables,** including Skin Glands, Common Bone Fractures, Summary of the Levels of Organization within a Skeletal Muscle, and Summary of the Respiratory System, have been added, in addition to refinement of many of the existing tables with either new illustrations or rewritten text.

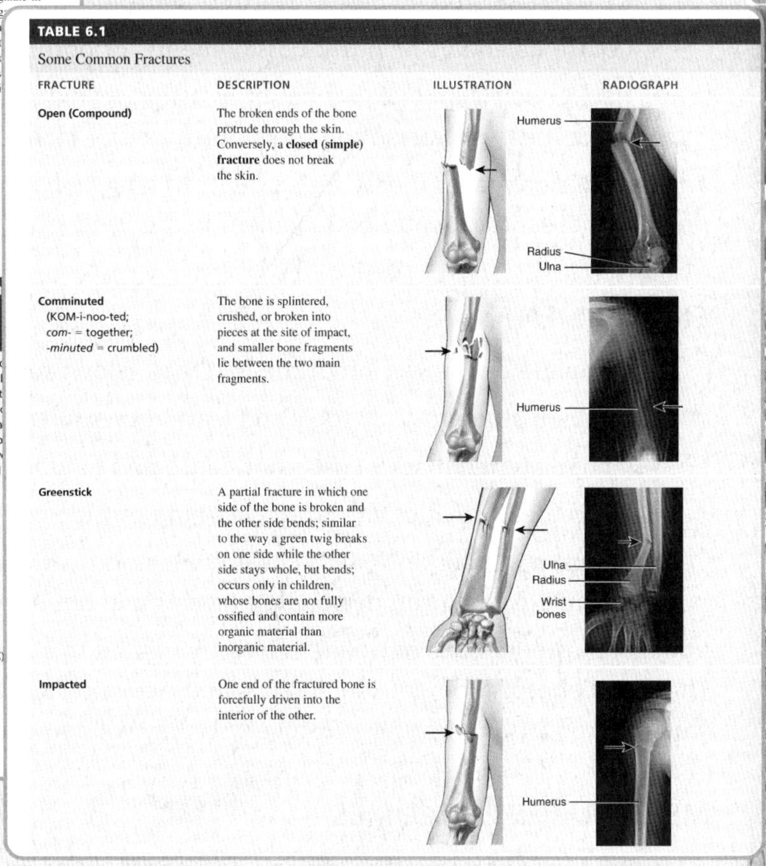

TABLE 6.1

Some Common Fractures

FRACTURE	DESCRIPTION	ILLUSTRATION	RADIOGRAPH
Open (Compound)	The broken ends of the bone protrude through the skin. Conversely, a **closed (simple) fracture** does not break the skin.		
Comminuted (KOM-i-noo-ted; *com-* = together; *-minuted* = crumbled)	The bone is splintered, crushed, or broken into pieces at the site of impact, and smaller bone fragments lie between the two main fragments.		
Greenstick	A partial fracture in which one side of the bone is broken and the other side bends; similar to the way a green twig breaks on one side while the other side stays whole, but bends; occurs only in children, whose bones are not fully ossified and contain more organic material than inorganic material.		
Impacted	One end of the fractured bone is forcefully driven into the interior of the other.		

Clinical Connections

Your students are fascinated by the **Clinical Connections** to the normal anatomy and physiology that they are learning. You'll find that the text is liberally peppered with engaging discussions of a wide variety of clinical scenarios from disease coverage to tests and procedures. As always, we have updated all of the Clinical Connections and Disorders: Homeostatic Imbalances sections to reflect the most current information. We have added several new Clinical Connections, such as a feature on fibromyalgia, to the text. A complete reference list of the Clinical Connections within each chapter follows the Table of Contents.

NOTES TO INSTRUCTORS

WileyPLUS *and You*

WileyPLUS for Anatomy and Physiology is an innovative, research-based online environment designed for both effective teaching and learning. Utilizing *WileyPLUS* in your course provides your students with an accessible, affordable, and active learning platform and gives you tools and resources to efficiently build presentations for a dynamic classroom experience and to create and manage effective assessment strategies. The underlying principles of **design, engagement,** and **measurable outcomes** provide the foundation for this powerful, new release of *WileyPLUS.*

DESIGN

- New research-based design helps students manage their time better and develop better study skills.
- Course Calendars help track assignments for both students and teachers.
- New Course Plan makes it easier to assign reading, activities, and assessment. Simple drag-and-drop tools make it easy to assign the course plan as-is or in any way that best reflects your course syllabus.

The new design makes it easy for students to know *what* it is they need to do, boosting their confidence and preparing them for greater engagement in class and lab.

ENGAGEMENT

- Complete online version of the textbook allows for seamless integration of all content.
- Relevant student study tools and learning resources ensure positive learning outcomes.
- Immediate feedback boosts confidence and helps students see a return on investment for each study session.
- Precreated activities encourage learning outside of the classroom.
- Course materials, including PowerPoint stacks with animations and Wiley's Visual Library for Anatomy and Physiology, help you personalize lessons and optimize your time.

Concept mastery in this discipline is directly related to students keeping up with the work and not falling behind. The new Concept Modules, Animations and Activities, Self Study, and Progress Checks in *WileyPLUS* will ensure that students know how to study effectively so they will remain engaged and stay on task.

MEASURABLE OUTCOMES

- Progress Check enables students to hone in on areas of weakness for increased success.
- Self-assessment and remediation for all Learning Objectives let students know exactly how their efforts have paid off.
- Instant reports monitor trends in class performance, use of course materials, and student progress toward learning objectives.

With new detailed reporting capabilities, students will know that they are doing it right. With increased confidence, motivation is sustained so students stay on task—success will follow.

...se contact your Wiley representative for details about these and other resources or visit our website at ...wileyplus.com.

RESOURCES FOR INTEGRATING LABORATORY EXPERIENCES

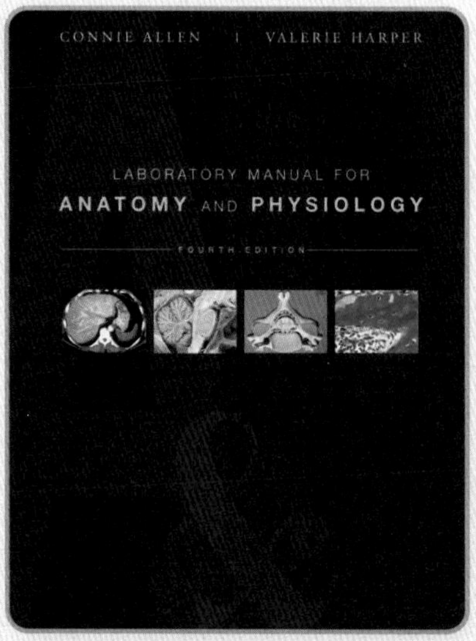

Laboratory Manual for Anatomy and Physiology, 4e

Connie Allen and Valerie Harper

Newly revised, ***Laboratory Manual for Anatomy and Physiology*** with ***WileyPLUS 5.0*** engages your students in active learning and focuses on the most important concepts in A&P. Exercises reflect the multiple ways in which students learn and provide guidance for anatomical exploration and application of critical thinking to analyzing physiological processes. A concise narrative, self-contained exercises that include a wide variety of activities and question types, and two types of lab reports for each exercise keep students focused on the task at hand. Depending on your needs, a *Cat Dissection Manual* or *Fetal Pig Dissection Manual* accompanies the main text. Rich media within *WileyPLUS* further enhance the student experience and include dissection videos, animations, and illustrated drill and practice exercises with illustrations, micrographs, cadaver photos, and popular lab models. Each lab text comes with access to *PowerPhys 2.0.*

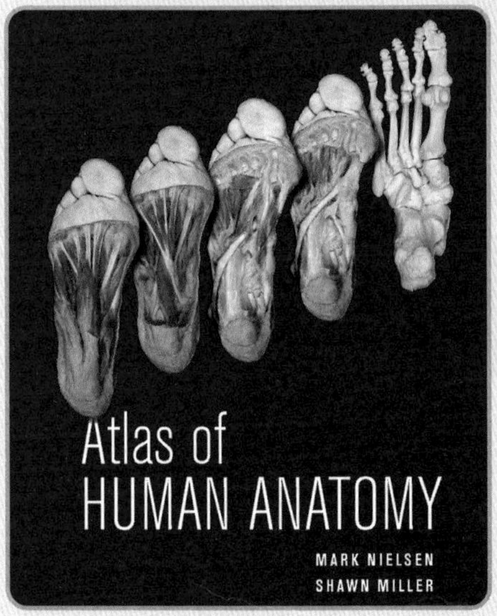

Atlas of Human Anatomy, 1e

Mark Nielsen and Shawn Miller

This new atlas filled with outstanding photographs of meticulously executed dissections of the human body has been developed to be a strong teaching and learning solution, not just a catalog of photographs. Organized around body systems, each chapter includes a narrative overview of the body system and is then followed with detailed photographs that accurately and realistically represent the anatomical structures. Histology is included. *Atlas of Human Anatomy* will work well in your laboratories, as a study companion to your textbook, and as a print companion to the *Real Anatomy* DVD.

Photographic Atlas of the Human Body, 2e

Gerard J. Tortora

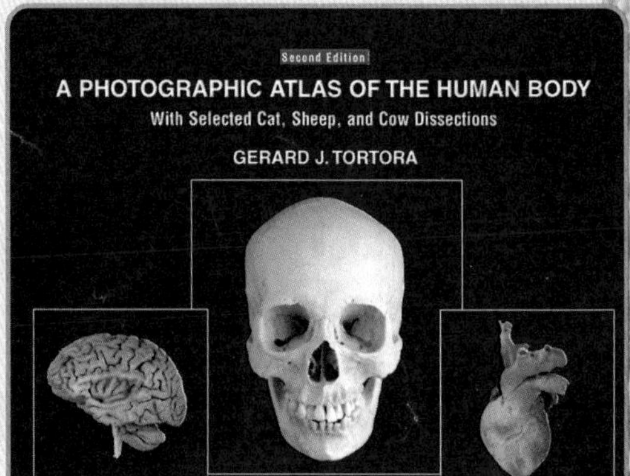

Like the new atlas from Nielsen and Miller, this popular atlas is also systemic in its approach to the photographic review of the human body. In addition to the excellent cadaver photographs and micrographs, this atlas also contains selected cat and sheep heart dissections. The high-quality imagery can be used in the classroom, in the laboratory, or for study and review.

RESOURCES FOR INTEGRATING LABORATORY EXPERIENCES

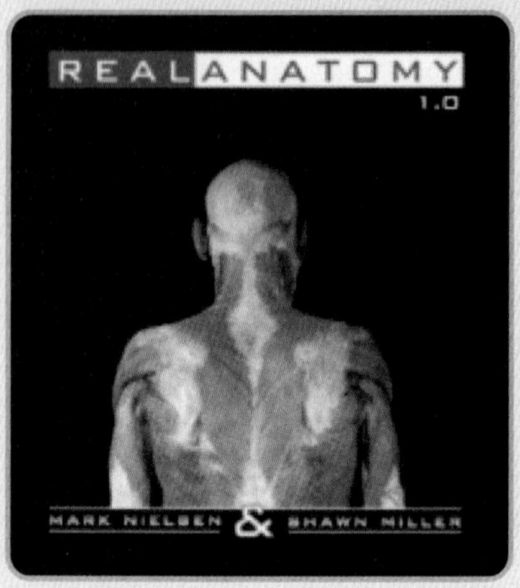

Real Anatomy

Mark Nielsen and Shawn Miller

Real Anatomy is 3-D imaging software that allows you to dissect through multiple layers of a three-dimensional real human body to study and learn the anatomical structures of all body systems.

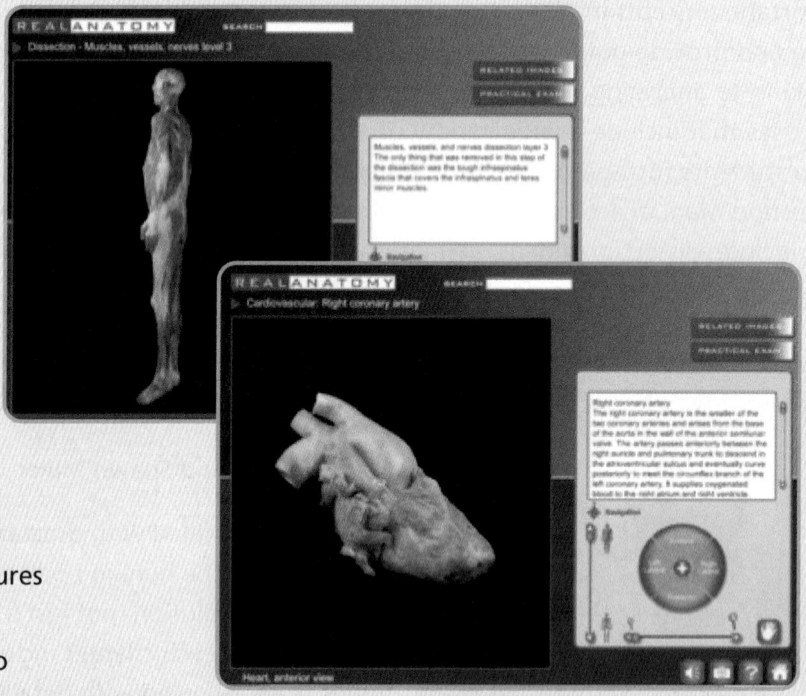

- Dissect through up to 40 layers of the body and discover the relationships of the structures to the whole.
- Rotate the body, as well as major organs, to view the image from multiple perspectives.
- Use a built-in zoom feature to get a closer look at detail.
- A unique approach to highlighting and labeling structures does not obscure the real anatomy in view.

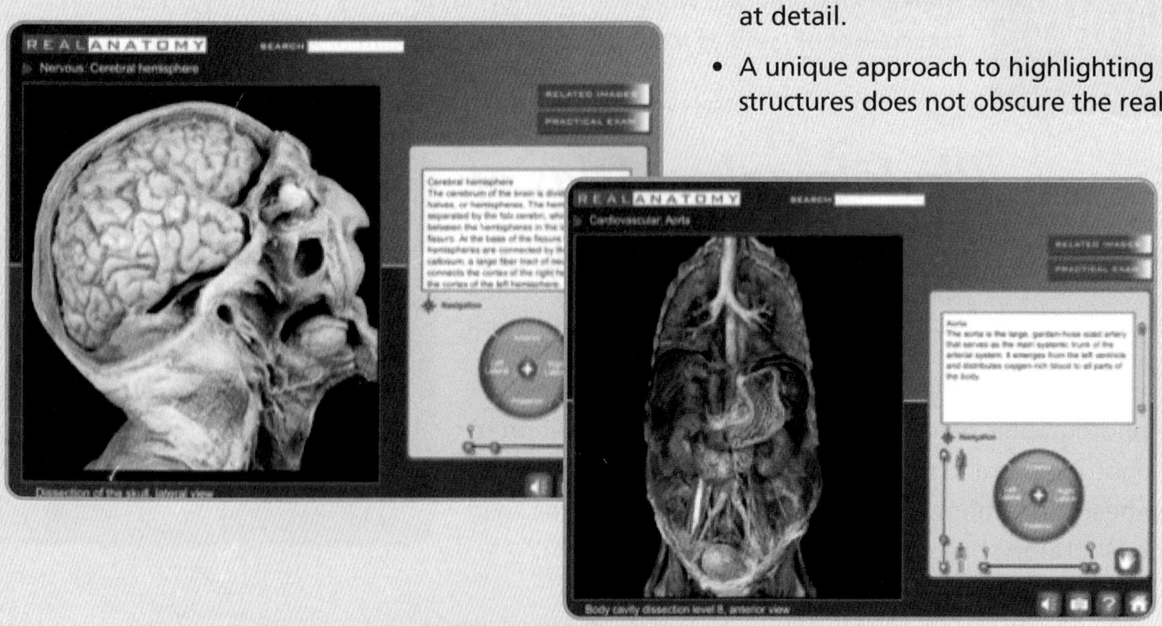

RESOURCES FOR INTEGRATING LABORATORY EXPERIENCES

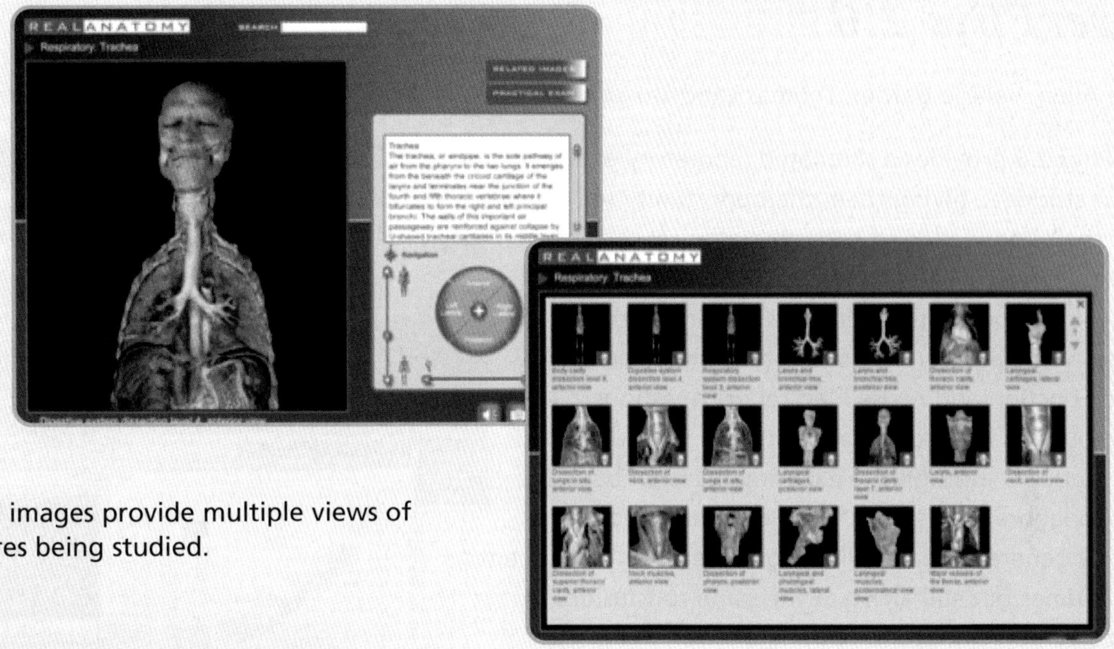

- Related images provide multiple views of structures being studied.

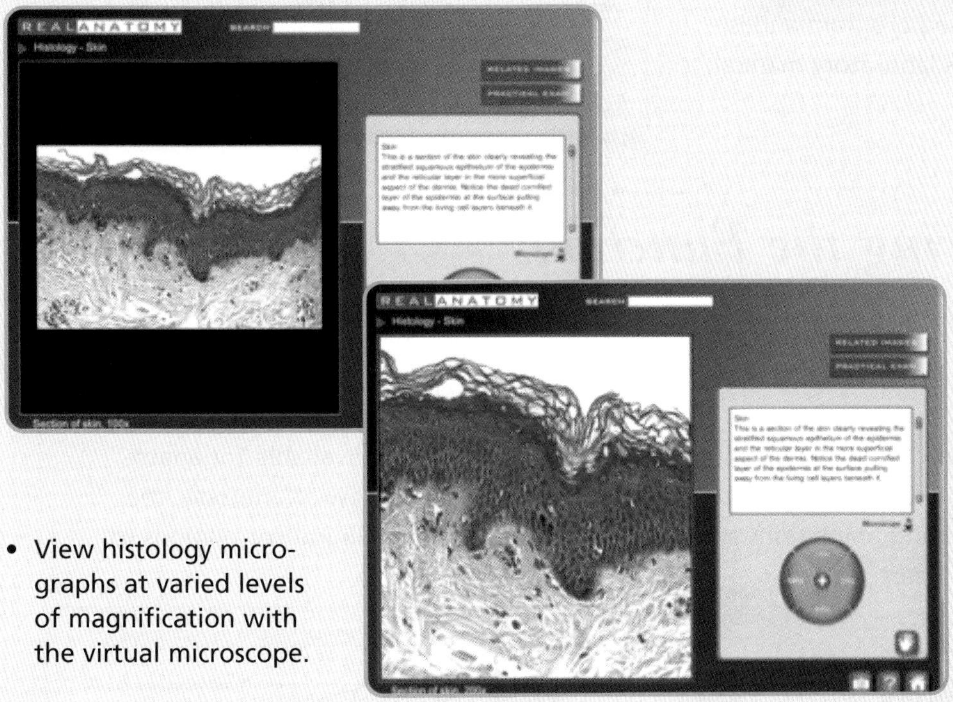

- View histology micrographs at varied levels of magnification with the virtual microscope.

- Snapshots of any image can be saved for use in PowerPoints, quizzes, or handouts.

- Audio pronunciation of all labeled structures is readily available.

Virtual Dissection—100% Real

REAL ANATOMY

RESOURCES FOR INTEGRATING LABORATORY EXPERIENCES

PowerPhys 2.0

Connie Allen, Valerie Harper, Thomas Lancraft, and Yuri Ivlev

PowerPhys 2.0 provides a simulated laboratory experience for students, giving them the opportunity to review their knowledge of core physiological concepts, predict outcomes of an experiment, collect data, analyze it, and report on their findings. This revised edition features a new activity on Homeostatic Imbalance of Thyroid Function and revised lab report questions throughout. An easy-to-use and intuitive interface guides students through the experiments from basic review to laboratory reports. All experiments contain randomly generated data, allowing students to experiment multiple times but still arrive at the same conclusions. A perfect addition to distant learning or hybrid courses, *PowerPhys 2.0* is a stand-alone web-based program and is fully integrated with Allen and Harper's laboratory manual.

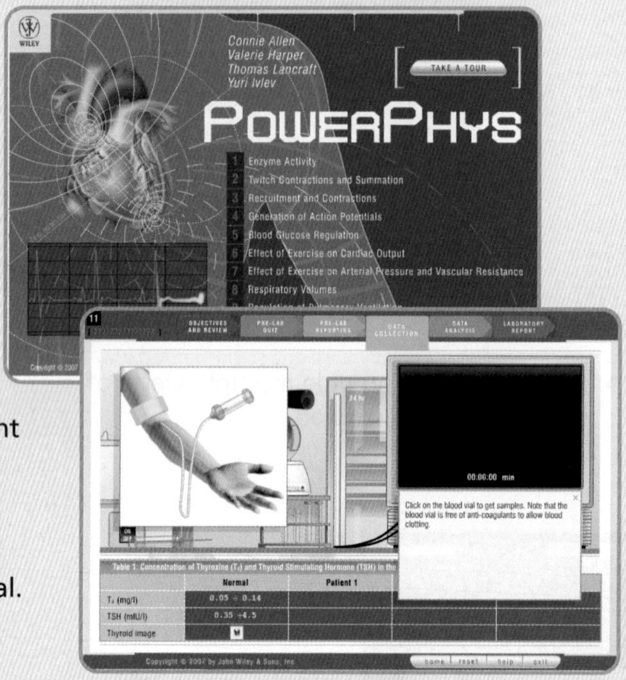

Interactions: Exploring the Functions of the Human Body 3.0

Thomas Lancraft and Frances Frierson

Interactions 3.0 is the most complete program of interactive animations and activities available for anatomy and physiology. A series of modules encompassing all body systems focuses on a review of anatomy, the examination of physiological processes using animations and interactive exercises, and clinical correlations to enhance student understanding. At the heart of ***Interactions*** is a focus on core principles—*homeostasis, communication, energy flow, fluid flow,* and *boundaries*—that underscore the key relationships between structure and function as well as interrelationships between systems. It is the reinforcement of these fundamental organizing principles that sets this series apart from others. ***Interactions*** is available on DVD, web-based, or fully integrated within *WileyPLUS.*

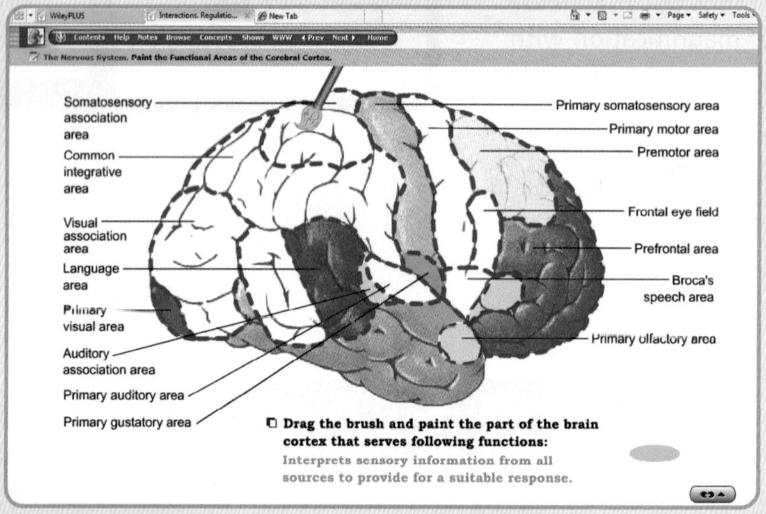

ACKNOWLEDGMENTS

We wish to especially thank several academic colleagues for their helpful contributions to this edition. Creating and implementing the integration of this text with *WileyPLUS for Anatomy and Physiology* was only possible because of the expertise and fine work of the following group of people. We are so very grateful to you:

Sarah Bales
Moraine Valley Community College

Celina Bellanceau
University of South Florida

Curtis DeFriez
Weber State University

Alan Erickson
South Dakota State University

Gibril Fadika
Hampton University

Pamela Fouche
Walters State Community College

Sophia Garcia
Tarrant County College–Trinity River

Clare Hays
Metropolitan State College of Denver

Jason Hunt
Brigham Young University–Idaho

Judy Learn
North Seattle Community College

Jerri K. Lindsey
Tarrant County College

Todd Miller
Hunter College

Erin Morrey
Georgia Perimeter College

Gus Pita
Hunter College

Susan Puglisi
Norwalk Community College

Saeed Rahmanian
Roane State Community College

Lori A. Smith
American River College

Randall Tracy
Worcester State College

Jay Zimmer
South Florida Community College

We are also very grateful to our colleagues who have reviewed the manuscript or participated in focus groups and offered numerous suggestions for improvement:

Charles J. Biggers
University of Memphis

Gladys Bolding
Georgia Perimeter College

Lois Borek
Georgia State University

Betsy Brantley
Lansing Community College

Arthur R. Buckley
University of Cincinnati

Alex Cheroske
Mesa Community College

Robert Comegys
Old Dominion University

Curtis DeFriez
Weber State University

William Dunscombe
Union County College

Heather Dy
Long Beach Community College

Christine Ross Earls
Fairfield University

Angela Edwards
University of South Carolina Allendale

Sharon Ellerton
Queensborough Community College

David Evans
Pennsylvania College of Technology

Gibril Fadika
Hampton University

Sandy Garrett
Texas Woman's University

Michael Harman
Lone Star College

Jane Horlings
Saddleback College

Barbara Hunnicutt
Seminole State College

Jason Hunt
Brigham Young University–Idaho

Alexander T. Imholtz
Prince George's Community College

Amy E. Jetton
Middle Tennessee State University

Becky Keck
Clemson University

Marc LaBella
Ocean County College

Ellen Lathrop-Davis
Community College of Baltimore County

Billy Bob Long
Del Mar College

Wayne M. Mason
Western Kentucky University

Karen McLellan
Indiana Purdue University Fort Wayne

Marie McMahon
Miramar College

Erin Morrey
Georgia Perimeter College

Maria Oehler
Florida State College at Jacksonville

Betsy Ott
Tyler Junior College

Gilbert Pitts
Austin Peay State University

Saeed Rahmanian
Roane State Community College

Terrence J. Ravine
University of South Alabama

Philip D. Reynolds
Troy University

John Roufaiel
SUNY Rockland Community College

Kelly Sexton
North Lake College

Colleen Sinclair
Towson University

Lori A. Smith
American River College

Nora Stevens
Portland Community College

Leo B. Stouder
Broward College

Dennis Strete
McLennan Community College

Peter Susan
Trident Technical College

Jared Taglialatela
Clayton State College

Bonnie J. Tarricone
Ivy Tech Community College

Heather Walker
Clemson University

Janice Webster
Ivy Tech Community College

Delores Wenzel
University of Georgia

Matthew A. Williamson
Georgia Southern University

Finally, our hats are off to everyone at Wiley. We enjoy collaborating with this enthusiastic, dedicated, and talented team of publishing professionals. Our thanks to the entire team: Bonnie Roesch, Executive Editor; Mary Berry and Karen Trost, Developmental Editors; Lorraina Raccuia, Project Editor; Lauren Morris, Program Assistant; Suzanne Ingrao, Outside Production Editor; Hilary Newman, Photo Manager; Claudia Volano, Illustration Coordinator; Anna Melhorn, Senior Illustration Editor; Madelyn Lesure, Senior Designer; Laura Ierardi, LCI Design; Linda Muriello, Senior Media Editor; and Clay Stone, Executive Marketing Manager.

Gerard J. Tortora
Department of Science and Health, S229
Bergen Community College
400 Paramus Road
Paramus, NJ 07652

Bryan Derrickson
Science Department
Valencia Community College
1800 S. Kirkman Rd.
Orlando, FL 32811

ABOUT THE AUTHORS

Courtesy of Heidi Chung.

Gerard J. Tortora is Professor of Biology and former Biology Coordinator at Bergen Community College in Paramus, New Jersey, where he teaches human anatomy and physiology as well as microbiology. He received his bachelor's degree in biology from Fairleigh Dickinson University and his master's degree in science education from Montclair State College. He is a member of many professional organizations, including the Human Anatomy and Physiology Society (HAPS), American Society of Microbiology (ASM), American Association for the Advancement of Science (AAAS), National Education Association (NEA), and Metropolitan Association of College and University Biologists (MACUB).

Above all, Jerry is devoted to his students and their aspirations. In recognition of this commitment, Jerry was the recipient of MACUB's 1992 President's Memorial Award. In 1996, he received a National Institute for Staff and Organizational Development (NISOD) excellence award from the University of Texas and was selected to represent Bergen Community College in a campaign to increase awareness of the contributions of community colleges to higher education.

Jerry is the author of several best-selling science textbooks and laboratory manuals, a calling that often requires an additional 40 hours per week beyond his teaching responsibilities. Nevertheless, he still makes time for four or five weekly aerobic workouts that include biking and running. He also enjoys attending college basketball and professional hockey games and performances at the Metropolitan Opera House.

To my mother, Angelina M. Tortora.
(August 20, 1913–August 14, 2010).
Her love, guidance, faith, support, and example continue to be the cornerstone of my personal and professional life. ***G.J.T.***

Bryan Derrickson is Professor of Biology at Valencia Community College in Orlando, Florida, where he teaches human anatomy and physiology as well as general biology and human sexuality. He received his bachelor's degree in biology from Morehouse College and his doctorate in cell biology from Duke University. Bryan's study at Duke was in the Physiology Division within the Department of Cell Biology, so while his degree is in cell biology, his training focused on physiology. At Valencia, he frequently serves on faculty hiring committees. He has served as a member of the Faculty Senate, which is the governing body of the college, and as a member of the Teaching and Learning Academy, which sets the standards for the acquisition of tenure by faculty members. Nationally, he is a member of the Human Anatomy and Physiology Society (HAPS) and the National Association of Biology Teachers (NABT).

Bryan has always wanted to teach. Inspired by several biology professors while in college, he decided to pursue physiology with an eye to teaching at the college level. He is completely dedicated to the success of his students. He particularly enjoys the challenges of his diverse student population, in terms of their age, ethnicity, and academic ability, and he finds being able to reach all of them, despite their differences, a rewarding experience. His students continually recognize Bryan's efforts and care by nominating him for a campus award known as the "Valencia Professor Who Makes Valencia a Better Place to Start." Bryan has received this award three times.

To my family: Rosalind, Hurley, Cherie, and Robb.
Your support and motivation have been invaluable. ***B.H.D.***

BRIEF CONTENTS

VOLUME 1

BRIEF CONTENTS

VOLUME 2

VOLUME 1 CONTENTS

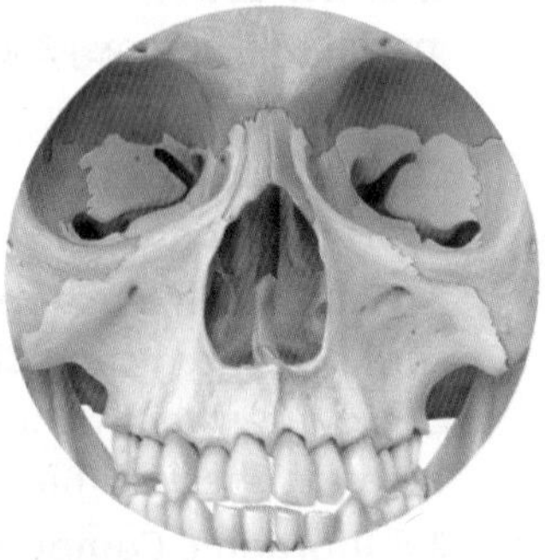

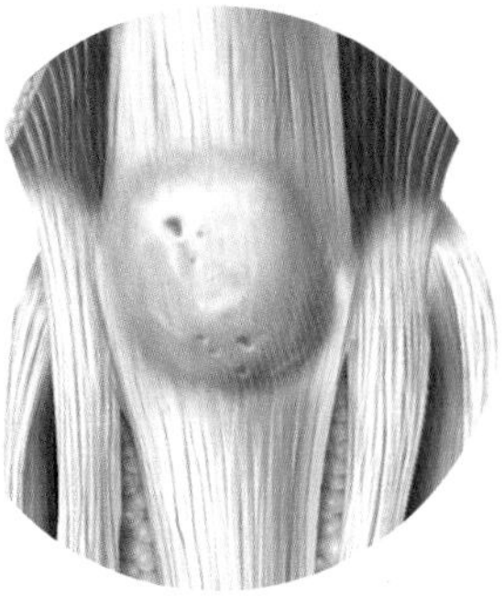

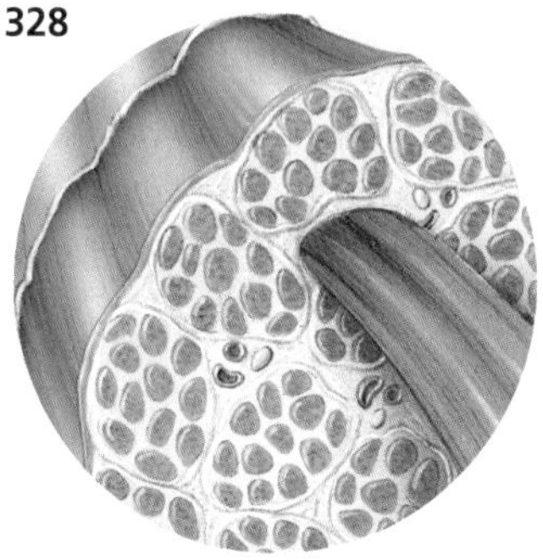

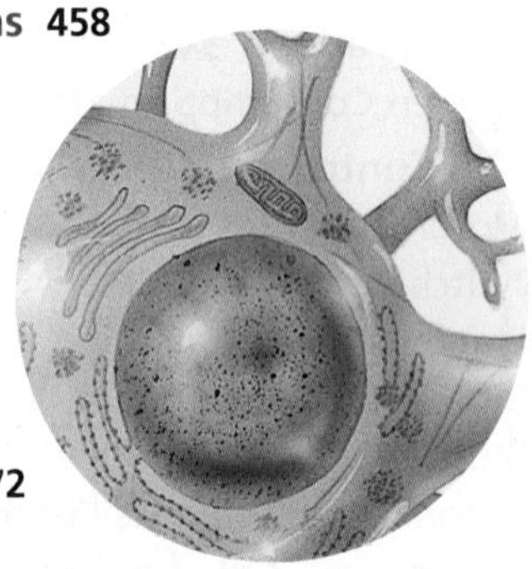

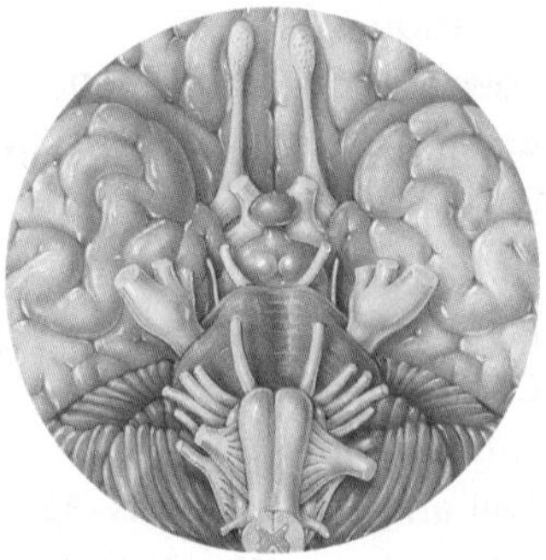

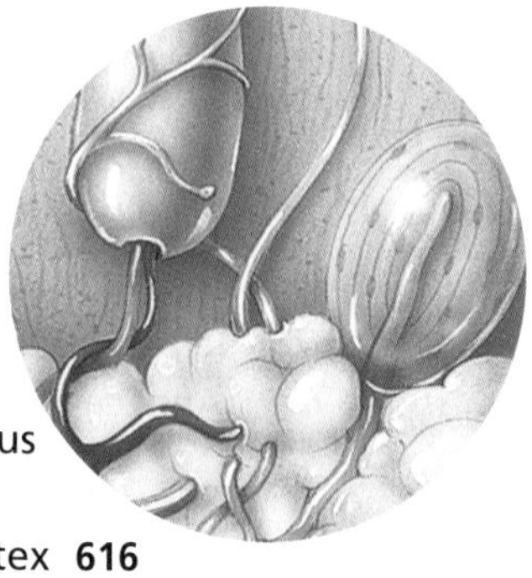

CLINICAL CONNECTIONS AND DISORDERS

1 AN INTRODUCTION TO THE HUMAN BODY

THE HUMAN BODY AND HOMEOSTASIS *Humans have many ways to maintain homeostasis, the state of relative stability of the body's internal environment. Disruptions to homeostasis often set in motion corrective cycles, called feedback systems, that help restore the conditions needed for health and life.*

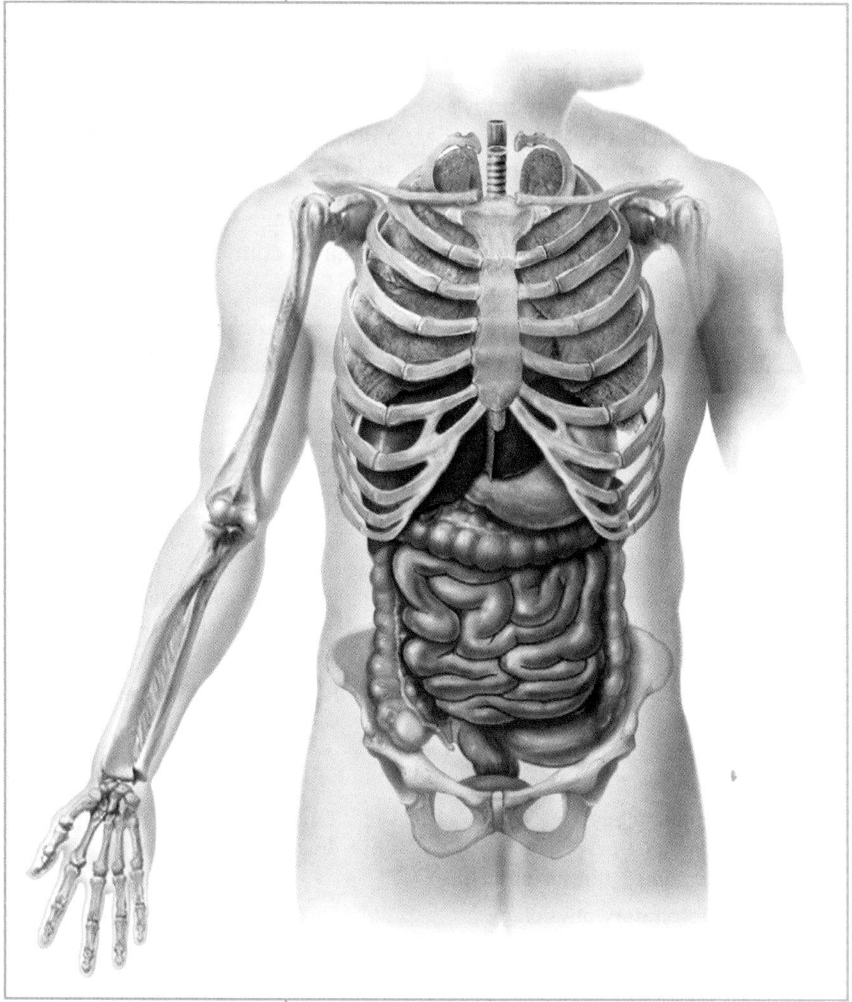

Our fascinating journey through the human body begins with an overview of the meanings of anatomy and physiology, followed by a discussion of the organization of the human body and the properties that it shares with all living things. Next, you will discover how the body regulates its own internal environment; this unceasing process, called homeostasis, is a major theme in every chapter of this book. Finally, we introduce the basic vocabulary that will help you speak about the body in a way that is understood by scientists and health-care professionals alike.

? ***Did you ever wonder why an autopsy is performed?***

1.1 ANATOMY AND PHYSIOLOGY DEFINED

OBJECTIVE

- Define anatomy and physiology, and name several subspecialties of these sciences.

Two branches of science—anatomy and physiology—provide the foundation for understanding the body's parts and functions. **Anatomy** (a-NAT-ō-mē; *ana-* = up; *-tomy* = process of cutting) is the science of body *structures* and the relationships among them. It was first studied by **dissection** (dis-SEK-shun; *dis-* = apart; *-section* = act of cutting), the careful cutting apart of body structures to study their relationships. Today, a variety of imaging techniques (see Table 1.3) also contribute to the advancement of anatomical knowledge. Whereas anatomy deals with structures of the body, **physiology** (fiz′-ē-OL-ō-jē; *physio-* = nature; *-logy* = study of) is the science of body *functions*—how the body parts work. Table 1.1 describes several subspecialties of anatomy and physiology.

Because structure and function are so closely related, you will learn about the human body by studying its anatomy and physiology together. The structure of a part of the body often reflects its functions. For example, the bones of the skull join tightly to form a rigid case that protects the brain. The bones of the fingers are more loosely joined to allow a variety of movements. The walls of the air sacs in the lungs are very thin, permitting rapid movement of inhaled oxygen into the blood. The lining of the urinary bladder is much thicker to prevent the escape of urine into the pelvic cavity, yet its construction allows for considerable stretching.

✔ CHECKPOINT

1. What body function might a respiratory therapist strive to improve? What structures are involved?
2. Give your own example of how the structure of a part of the body is related to its function.

1.2 LEVELS OF STRUCTURAL ORGANIZATION AND BODY SYSTEMS

OBJECTIVES

- Describe the body's six levels of structural organization.
- List the 11 systems of the human body, representative organs present in each, and their general functions.

The levels of organization of a language—letters, words, sentences, paragraphs, and so on—can be compared to the levels of

TABLE 1.1

Selected Subspecialties of Anatomy and Physiology

SUBSPECIALTIES OF ANATOMY	STUDY OF	SUBSPECIALTIES OF PHYSIOLOGY	STUDY OF
Embryology (em′-brē-OL-ō-jē; *embry-* = embryo; *-logy* = study of)	The first eight weeks of development after fertilization of a human egg.	**Neurophysiology** (NOOR-ō-fiz-ē-ol′-ō-jē; *neuro-* = nerve)	Functional properties of nerve cells.
Developmental biology	The complete development of an individual from fertilization to death.	**Endocrinology** (en′-dō-kri-NOL-ō-jē; *endo-* = within; *-crin* = secretion)	Hormones (chemical regulators in the blood) and how they control body functions.
Cell biology	Cellular structure and functions.	**Cardiovascular physiology** (kar-dē-ō-VAS-kū-lar; *cardi-* = heart; *-vascular* = blood vessels)	Functions of the heart and blood vessels.
Histology (hiss′-TOL-ō-jē; *hist-* = tissue)	Microscopic structure of tissues.		
Gross anatomy	Structures that can be examined without a microscope.	**Immunology** (im′-ū-NOL-ō-jē; *immun-* = not susceptible)	The body's defenses against disease-causing agents.
Systemic anatomy	Structure of specific systems of the body such as the nervous or respiratory systems.	**Respiratory physiology** (RES-pir-a-to′-rē; *respira-* = to breathe)	Functions of the air passageways and lungs.
Regional anatomy	Specific regions of the body such as the head or chest.	**Renal physiology** (RĒ-nal; *ren-* = kidney)	Functions of the kidneys.
Surface anatomy	Surface markings of the body to understand internal anatomy through visualization and palpation (gentle touch).	**Exercise physiology**	Changes in cell and organ functions due to muscular activity.
Radiographic anatomy (rā′-dē-ō-GRAF-ik; *radio-* = ray; *-graphic* = to write)	Body structures that can be visualized with x-rays.	**Pathophysiology** (PATH-ō-fiz-ē-ol′-ō-jē)	Functional changes associated with disease and aging.
Pathological anatomy (path′-ō-LOJ-i-kal; *path-* = disease)	Structural changes (gross to microscopic) associated with disease.		

organization of the human body. Your exploration of the human body will extend from atoms and molecules to the whole person. From the smallest to the largest, six levels of organization will help you to understand anatomy and physiology: the chemical, cellular, tissue, organ, system, and organismal levels of organization (Figure 1.1).

1 **Chemical level.** This very basic level can be compared to the *letters of the alphabet* and includes **atoms,** the smallest units of matter that participate in chemical reactions, and **molecules,** two or more atoms joined together. Certain atoms, such as carbon (C), hydrogen (H), oxygen (O), nitrogen (N), phosphorus (P), calcium (Ca), and sulfur (S), are essential for maintaining life. Two familiar molecules found in the body are deoxyribonucleic acid (DNA), the genetic material passed from one generation to the next, and glucose, commonly known as blood sugar. Chapters 2 and 25 focus on the chemical level of organization.

Figure 1.1 Levels of structural organization in the human body.

The levels of structural organization are chemical, cellular, tissue, organ, system, and organismal.

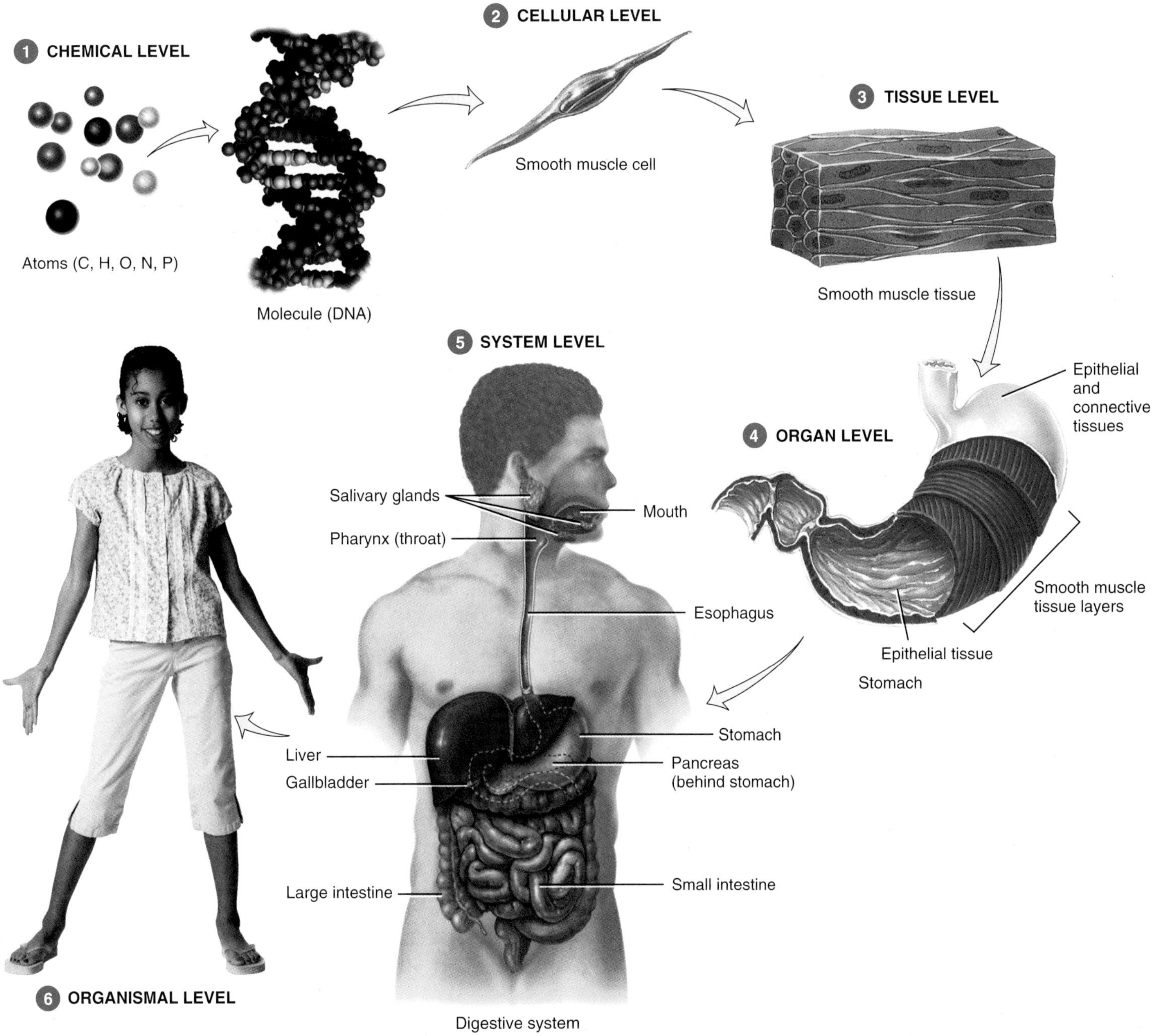
1 CHEMICAL LEVEL
Atoms (C, H, O, N, P)
Molecule (DNA)
2 CELLULAR LEVEL
Smooth muscle cell
3 TISSUE LEVEL
Smooth muscle tissue
4 ORGAN LEVEL
Epithelial and connective tissues
Smooth muscle tissue layers
Epithelial tissue
Stomach
5 SYSTEM LEVEL
Salivary glands
Pharynx (throat)
Mouth
Esophagus
Stomach
Liver
Gallbladder
Pancreas (behind stomach)
Large intestine
Small intestine
Digestive system
6 ORGANISMAL LEVEL

Which level of structural organization is composed of two or more different types of tissues that work together to perform a specific function?

2 **Cellular level.** Molecules combine to form **cells,** the basic structural and functional units of an organism that are composed of chemicals. Just as *words* are the smallest elements of language that make sense, cells are the smallest living units in the human body. Among the many kinds of cells in your body are muscle cells, nerve cells, and epithelial cells. Figure 1.1 shows a smooth muscle cell, one of the three types of muscle cells in the body. The cellular level of organization is the focus of Chapter 3.

3 **Tissue level. Tissues** are groups of cells and the materials surrounding them that work together to perform a particular function, similar to the way words are put together to form *sentences*. There are just four basic types of tissues in your body: epithelial tissue, connective tissue, muscular tissue, and nervous tissue. *Epithelial tissue* covers body surfaces, lines hollow organs and cavities, and forms glands. *Connective tissue* connects, supports, and protects body organs while distributing blood vessels to other tissues. *Muscular tissue* contracts to make body parts move and generates heat. *Nervous tissue* carries information from one part of the body to another through nerve impulses. Chapter 4 describes the tissue level of organization in greater detail. Shown in Figure 1.1 is smooth muscle tissue, which consists of tightly packed smooth muscle cells.

4 **Organ level.** At the organ level different types of tissues are joined together. Similar to the relationship between sentences and *paragraphs,* **organs** are structures that are composed of two or more different types of tissues; they have specific functions and usually have recognizable shapes. Examples of organs are the stomach, skin, bones, heart, liver, lungs, and brain. Figure 1.1 shows how several tissues make up the stomach. The stomach's outer covering is a layer of epithelial tissue and connective tissue that reduces friction when the stomach moves and rubs against other organs. Underneath are three layers of a type of muscular tissue called *smooth muscle tissue,* which contracts to churn and mix food and then push it into the next digestive organ, the small intestine. The innermost lining is an *epithelial tissue layer* that produces fluid and chemicals responsible for digestion in the stomach.

5 **System level.** A **system** (or *chapter* in our language analogy) consists of related organs (paragraphs) with a common function. An example of the system level, also called the *organ-system level,* is the digestive system, which breaks down and absorbs food. Its organs include the mouth, salivary glands, pharynx (throat), esophagus (food tube), stomach, small intestine, large intestine, liver, gallbladder, and pancreas. Sometimes an organ is part of more than one system. The pancreas, for example, is part of both the digestive system and the hormone-producing endocrine system.

6 **Organismal level.** An **organism** (OR-ga-nizm), any living individual, can be compared to a *book* in our analogy. All the parts of the human body functioning together constitute the total organism.

In the chapters that follow, you will study the anatomy and physiology of the body systems. Table 1.2 lists the components and introduces the functions of these systems. You will also discover that all body systems influence one another. As you study

TABLE 1.2

The Eleven Systems of the Human Body

INTEGUMENTARY SYSTEM (CHAPTER 5)

Components: **Skin** and associated structures, such as **hair, fingernails** and **toenails, sweat glands,** and **oil glands.**

Functions: Protects body; helps regulate body temperature; eliminates some wastes; helps make vitamin D; detects sensations such as touch, pain, warmth, and cold.

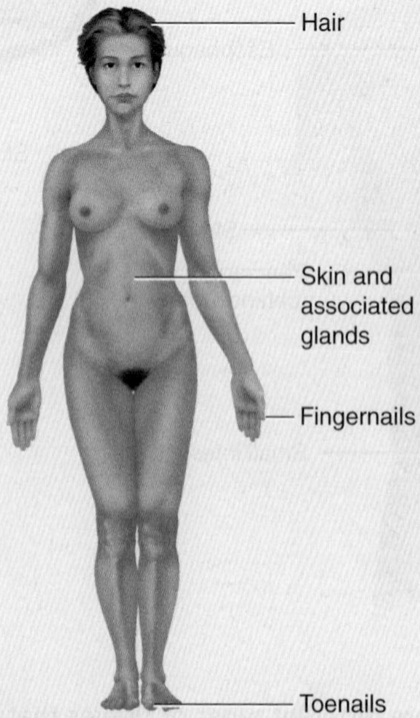
Hair
Skin and associated glands
Fingernails
Toenails

SKELETAL SYSTEM (CHAPTERS 6–9)

Components: **Bones** and **joints** of the body and their associated **cartilages.**

Functions: Supports and protects body; provides surface area for muscle attachments; aids body movements; houses cells that produce blood cells; stores minerals and lipids (fats).

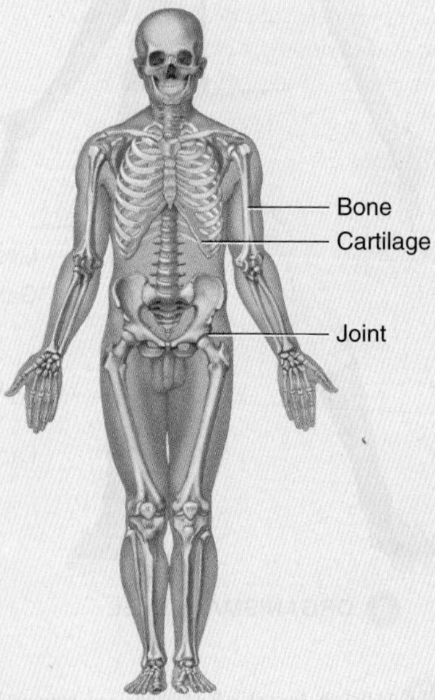
Bone
Cartilage
Joint

each of the body systems in more detail, you will discover how they work together to maintain health, provide protection from disease, and allow for reproduction of the human species.

CLINICAL CONNECTION | *Noninvasive Diagnostic Techniques*

Health-care professionals and students of anatomy and physiology commonly use several noninvasive diagnostic techniques to assess certain aspects of body structure and function. A **noninvasive diagnostic technique** is one that does not involve insertion of an instrument or device through the skin or a body opening. In **inspection,** the examiner observes the body for any changes that deviate from normal. For example, a physician may examine the mouth cavity for evidence of disease. Following inspection, one or more additional techniques may be employed. In **palpation** (pal-PĀ-shun; *palp-* = gently touching) the examiner feels body surfaces with the hands. An example is palpating the abdomen to detect enlarged or tender internal organs or abnormal masses. In **auscultation** (aws-kul-TĀ-shun; *auscult-* = listening) the examiner listens to body sounds to evaluate the functioning of certain organs, often using a stethoscope to amplify the sounds. An example is auscultation of the lungs during breathing to check for crackling sounds associated with abnormal fluid accumulation. In **percussion** (pur-KUSH-un; *percus-* = beat through) the examiner taps on the body surface with the fingertips and listens to the resulting echo. For example, percussion may reveal the abnormal presence of fluid in the lungs or air in the intestines. It may also provide information about the size, consistency, and position of an underlying structure. An understanding of anatomy is important for the effective application of most of these diagnostic techniques. •

✔ CHECKPOINT

3. Define the following terms: atom, molecule, cell, tissue, organ, system, and organism.
4. At what levels of organization would an exercise physiologist study the human body? *(Hint: Refer to Table 1.1.)*
5. Referring to Table 1.2, which body systems help eliminate wastes?

1.3 CHARACTERISTICS OF THE LIVING HUMAN ORGANISM

OBJECTIVES

- Define the important life processes of the human body.
- Define homeostasis and explain its relationship to interstitial fluid.

Basic Life Processes

Certain processes distinguish organisms, or living things, from nonliving things. Following are the six most important life processes of the human body:

1. **Metabolism** (me-TAB-ō-lizm) is the sum of all the chemical processes that occur in the body. One phase of metabolism is **catabolism** (ka-TAB-ō-lizm; *catabol-* = throwing down; *-ism* = a condition), the breakdown of complex chemical substances into simpler components. The other phase of

MUSCULAR SYSTEM (CHAPTERS 10, 11)

Components: Specifically, **skeletal muscle** tissue—muscle usually attached to bones (other muscle tissues include smooth and cardiac).

Functions: Participates in body movements, such as walking; maintains posture; produces heat.

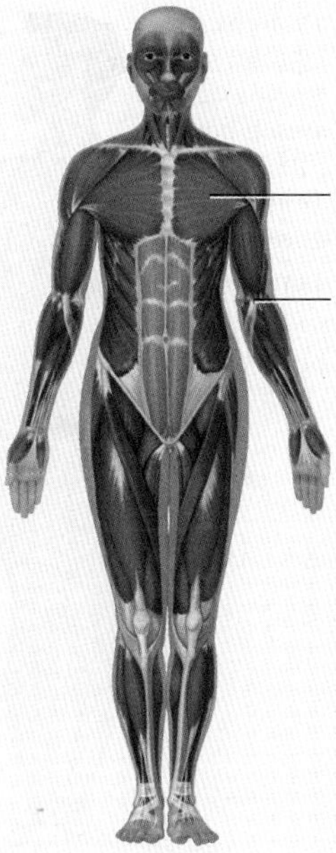

NERVOUS SYSTEM (CHAPTERS 12–17)

Components: **Brain, spinal cord, nerves,** and special sense organs, such as **eyes** and **ears.**

Functions: Generates action potentials (nerve impulses) to regulate body activities; detects changes in body's internal and external environments, interprets changes, and responds by causing muscular contractions or glandular secretions.

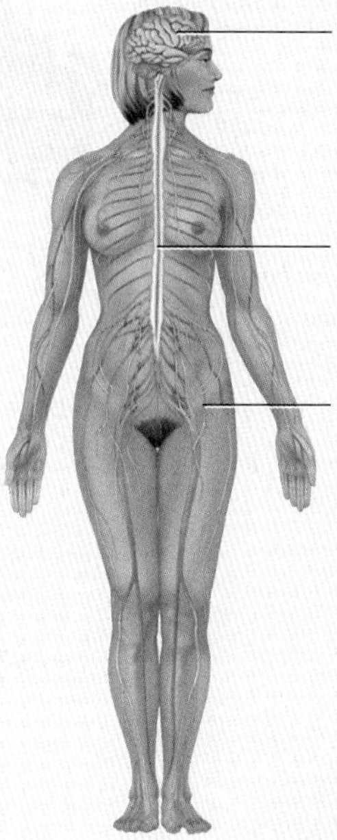

TABLE 1.2 CONTINUES

TABLE 1.2 CONTINUED

The Eleven Systems of the Human Body

ENDOCRINE SYSTEM (CHAPTER 18)

Components: Hormone-producing glands (**pineal gland, hypothalamus, pituitary gland, thymus, thyroid gland, parathyroid glands, adrenal glands, pancreas, ovaries,** and **testes**) and hormone-producing cells in several other organs.

Functions: Regulates body activities by releasing hormones (chemical messengers transported in blood from endocrine gland or tissue to target organ).

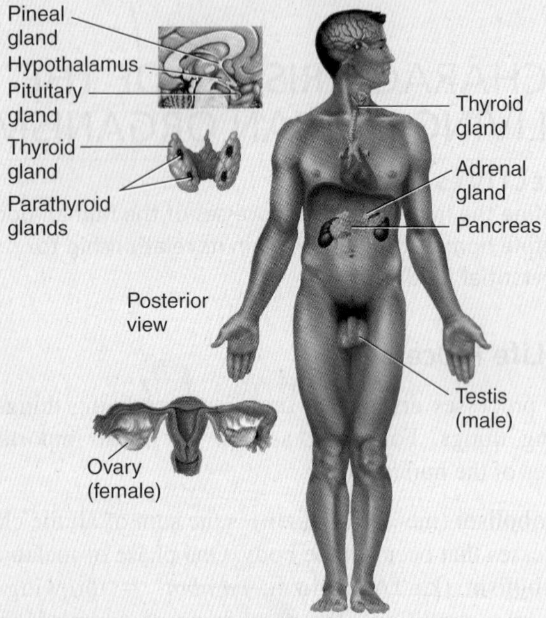

CARDIOVASCULAR SYSTEM (CHAPTERS 19–21)

Components: **Blood, heart,** and **blood vessels**.

Functions: Heart pumps blood through blood vessels; blood carries oxygen and nutrients to cells and carbon dioxide and wastes away from cells and helps regulate acid–base balance, temperature, and water content of body fluids; blood components help defend against disease and repair damaged blood vessels.

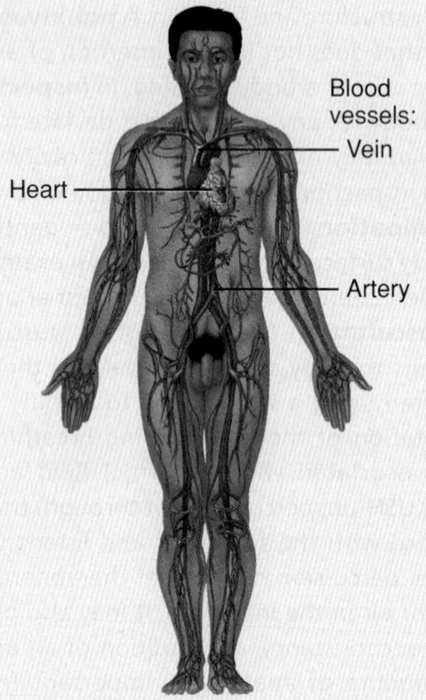

LYMPHATIC SYSTEM AND IMMUNITY (CHAPTER 22)

Components: **Lymphatic fluid** and **vessels; spleen, thymus, lymph nodes,** and **tonsils;** cells that carry out immune responses (**B cells, T cells,** and others).

Functions: Returns proteins and fluid to blood; carries lipids from gastrointestinal tract to blood; contains sites of maturation and proliferation of B cells and T cells that protect against disease-causing microbes.

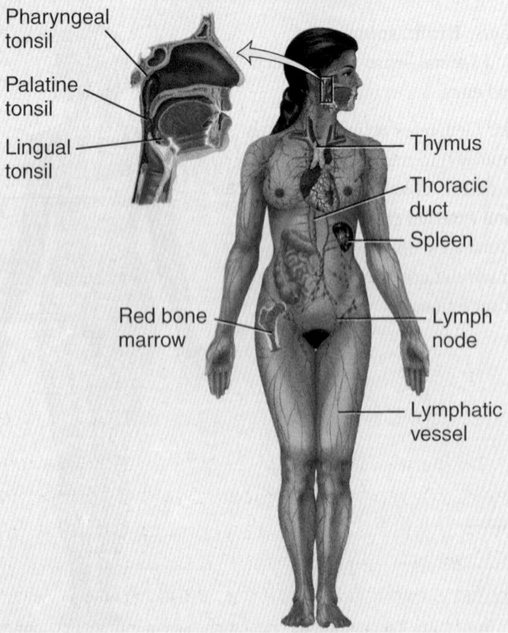

RESPIRATORY SYSTEM (CHAPTER 23)

Components: **Lungs** and air passageways such as the **pharynx (throat), larynx (voice box), trachea (windpipe),** and **bronchial tubes** leading into and out of lungs.

Functions: Transfers oxygen from inhaled air to blood and carbon dioxide from blood to exhaled air; helps regulate acid–base balance of body fluids; air flowing out of lungs through vocal cords produces sounds.

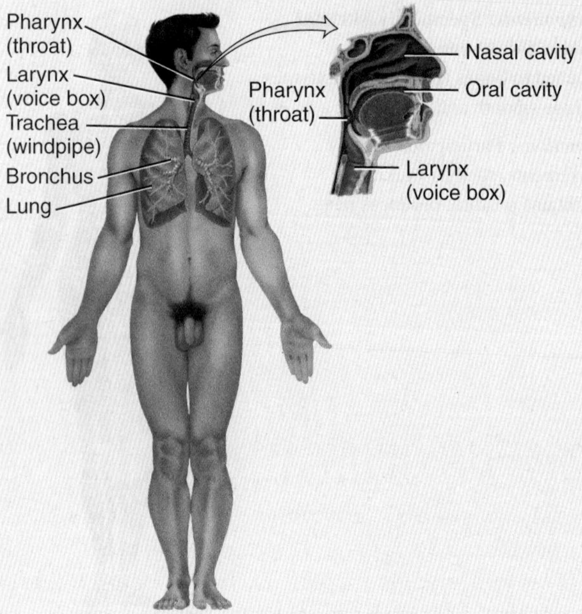

DIGESTIVE SYSTEM (CHAPTER 24)

Components: Organs of gastrointestinal tract, a long tube that includes the **mouth, pharynx** (throat), **esophagus** (food tube), **stomach, small** and **large intestines,** and **anus;** also includes accessory organs that assist in digestive processes, such as **salivary glands, liver, gallbladder,** and **pancreas.**

Functions: Achieves physical and chemical breakdown of food; absorbs nutrients; eliminates solid wastes.

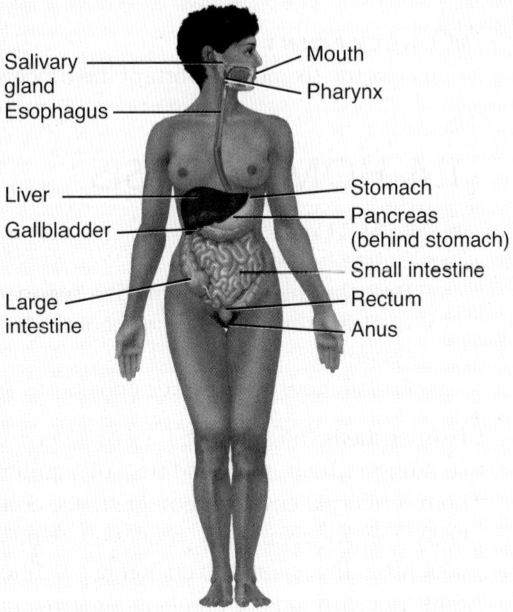
Salivary gland
Esophagus
Mouth
Pharynx
Liver
Gallbladder
Stomach
Pancreas (behind stomach)
Small intestine
Large intestine
Rectum
Anus

URINARY SYSTEM (CHAPTER 26)

Components: **Kidneys, ureters, urinary bladder,** and **urethra.**

Functions: Produces, stores, and eliminates urine; eliminates wastes and regulates volume and chemical composition of blood; helps maintain the acid–base balance of body fluids; maintains body's mineral balance; helps regulate production of red blood cells.

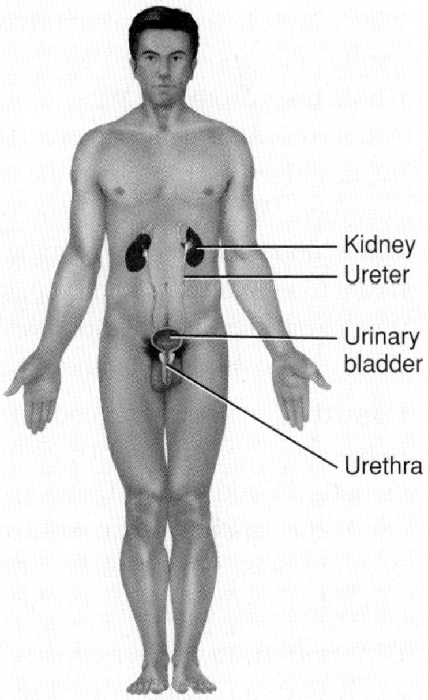
Kidney
Ureter
Urinary bladder
Urethra

REPRODUCTIVE SYSTEMS (CHAPTER 28)

Components: **Gonads** (**testes** in males and **ovaries** in females) and associated organs (**uterine tubes, uterus, vagina,** and **mammary glands** in females and **epididymides, ductus deferens, seminal vesicles, prostate,** and **penis** in males).

Functions: Gonads produce gametes (sperm or oocytes) that unite to form a new organism; gonads also release hormones that regulate reproduction and other body processes; associated organs transport and store gametes; mammary glands produce milk.

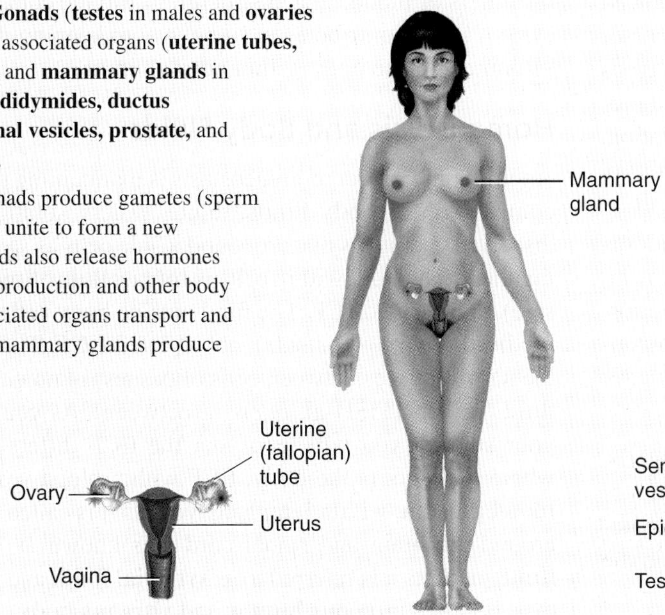
Mammary gland
Uterine (fallopian) tube
Ovary
Uterus
Vagina

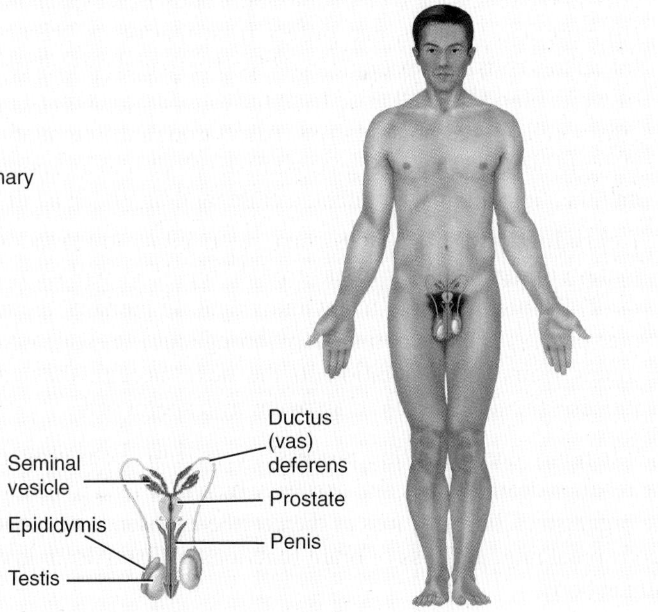
Ductus (vas) deferens
Seminal vesicle
Prostate
Epididymis
Penis
Testis

metabolism is **anabolism** (a-NAB-ō-lizm; *anabol-* = a raising up), the building up of complex chemical substances from smaller, simpler components. For example, digestive processes catabolize (split) proteins in food into amino acids. These amino acids are then used to anabolize (build) new proteins that make up body structures such as muscles and bones.

2. **Responsiveness** is the body's ability to detect and respond to changes. For example, an increase in body temperature during a fever represents a change in the internal environment (within the body), and turning your head toward the sound of squealing brakes is a response to a change in the external environment (outside the body) to prepare the body for a potential threat. Different

cells in the body respond to environmental changes in characteristic ways. Nerve cells respond by generating electrical signals known as nerve impulses (action potentials). Muscle cells respond by contracting, which generates force to move body parts.

3. **Movement** includes motion of the whole body, individual organs, single cells, and even tiny structures inside cells. For example, the coordinated action of leg muscles moves your whole body from one place to another when you walk or run. After you eat a meal that contains fats, your gallbladder contracts and releases bile into the gastrointestinal tract to help digest them. When a body tissue is damaged or infected, certain white blood cells move from the bloodstream into the affected tissue to help clean up and repair the area. Inside the cell, various parts, such as secretory vesicles (see Figure 3.20), move from one position to another to carry out their functions.
4. **Growth** is an increase in body size that results from an increase in the size of existing cells, an increase in the number of cells, or both. In addition, a tissue sometimes increases in size because the amount of material between cells increases. In a growing bone, for example, mineral deposits accumulate between bone cells, causing the bone to grow in length and width.
5. **Differentiation** (dif′-er-en-shē-Ā-shun) is the development of a cell from an unspecialized to a specialized state. Such precursor cells, which can divide and give rise to cells that undergo differentiation, are known as **stem cells.** As you will see later in the text, each type of cell in the body has a specialized structure and function that differs from that of its precursor (ancestor) cells. For example, red blood cells and several types of white blood cells all arise from the same unspecialized precursor cells in red bone marrow. Also through differentiation, a single fertilized human egg (ovum) develops into an embryo, and then into a fetus, an infant, a child, and finally an adult.
6. **Reproduction** (rē-prō-DUK-shun) refers either to (1) the formation of new cells for tissue growth, repair, or replacement, or (2) the production of a new individual. In humans, the former process occurs continuously throughout life, which continues from one generation to the next through the latter process, the fertilization of an ovum by a sperm cell.

When any one of the life processes ceases to occur properly, the result is death of cells and tissues, which may lead to death of the organism. Clinically, loss of the heartbeat, absence of spontaneous breathing, and loss of brain functions indicate death in the human body.

CLINICAL CONNECTION | *Autopsy*

An **autopsy** (AW-top-sē = seeing with one's own eyes) or *necropsy* is a postmortem (after death) examination of the body and dissection of its internal organs to confirm or determine the cause of death. An autopsy can uncover the existence of diseases not detected during life, determine the extent of injuries, and explain how those injuries may have contributed to a person's death. It also may provide more information about a disease, assist in the accumulation of statistical data, and educate health-care students. Moreover, an autopsy can reveal conditions that may affect offspring or siblings (such as congenital heart defects). Sometimes an autopsy is legally required, such as during a criminal investigation. It may also be useful in resolving disputes between beneficiaries and insurance companies about the cause of death. •

✔CHECKPOINT

6. List the six most important life processes in the human body.

1.4 HOMEOSTASIS

OBJECTIVES

- Define homeostasis.
- Describe the components of a feedback system.
- Contrast the operation of negative and positive feedback systems.
- Explain how homeostatic imbalances are related to disorders.

Homeostasis (hō′-mē-ō-STĀ-sis; *homeo-* = sameness; *-stasis* = standing still) is the condition of equilibrium (balance) in the body's internal environment due to the constant interaction of the body's many regulatory processes. Homeostasis is a dynamic condition. In response to changing conditions, the body's equilibrium can shift among points in a narrow range that is compatible with maintaining life. For example, the level of glucose in blood normally stays between 70 and 110 milligrams of glucose per 100 milliliters of blood.* Each structure, from the cellular level to the system level, contributes in some way to keeping the internal environment of the body within normal limits.

Homeostasis and Body Fluids

An important aspect of homeostasis is maintaining the volume and composition of **body fluids,** dilute, watery solutions containing dissolved chemicals that are found inside cells as well as surrounding them. The fluid within cells is **intracellular fluid** (*intra-* = inside), abbreviated **ICF.** The fluid outside body cells is **extracellular fluid** (*extra-* = outside), abbreviated **ECF.** The ECF that fills the narrow spaces between cells of tissues is known as **interstitial fluid** (in′-ter-STISH-al; *inter-* = between). As you progress with your studies, you will learn that the ECF differs depending on where it occurs in the body: ECF within blood vessels is termed **blood plasma,** within lymphatic vessels it is called **lymph,** in and around the brain and spinal cord it is known as **cerebrospinal fluid,** in joints it is referred to as **synovial fluid,** and the ECF of the eyes is called **aqueous humor** and **vitreous body.**

The proper functioning of body cells depends on precise regulation of the composition of the interstitial fluid surrounding them. Because of this, interstitial fluid is often called the body's *internal environment.* The composition of interstitial fluid changes as substances move back and forth between it and blood plasma. Such exchange of materials occurs across the thin walls of the smallest blood vessels in the body, the *blood capillaries.* This movement in both directions across capillary walls provides needed materials,

*Appendix A describes metric measurements.

such as glucose, oxygen, ions, and so on, to tissue cells. It also removes wastes, such as carbon dioxide, from interstitial fluid.

Control of Homeostasis

Homeostasis in the human body is continually being disturbed. Some disruptions come from the external environment in the form of physical insults such as the intense heat of a hot summer day or a lack of enough oxygen for that two-mile run. Other disruptions originate in the internal environment, such as a blood glucose level that falls too low when you skip breakfast. Homeostatic imbalances may also occur due to psychological stresses in our social environment—the demands of work and school, for example. In most cases the disruption of homeostasis is mild and temporary, and the responses of body cells quickly restore balance in the internal environment. However, in some cases the disruption of homeostasis may be intense and prolonged, as in poisoning, overexposure to temperature extremes, severe infection, or major surgery.

Fortunately, the body has many regulating systems that can usually bring the internal environment back into balance. Most often, the nervous system and the endocrine system, working together or independently, provide the needed corrective measures. The nervous system regulates homeostasis by sending electrical signals known as *nerve impulses (action potentials)* to organs that can counteract changes from the balanced state. The endocrine system includes many glands that secrete messenger molecules called *hormones* into the blood. Nerve impulses typically cause rapid changes, but hormones usually work more slowly. Both means of regulation, however, work toward the same end, usually through negative feedback systems.

Feedback Systems

The body can regulate its internal environment through many feedback systems. A **feedback system** or *feedback loop* is a cycle of events in which the status of a body condition is monitored, evaluated, changed, remonitored, reevaluated, and so on. Each monitored variable, such as body temperature, blood pressure, or blood glucose level, is termed a *controlled condition.* Any disruption that changes a controlled condition is called a *stimulus.* A feedback system includes three basic components: a receptor, a control center, and an effector (Figure 1.2).

1. A **receptor** is a body structure that monitors changes in a controlled condition and sends input to a control center. This pathway is called an *afferent pathway* (AF-er-ent; *af-* = toward; *-ferrent* = carried), since the information flows *toward* the control center. Typically, the *input* is in the form of nerve impulses or chemical signals. For example, certain nerve endings in the skin sense temperature and can detect changes, such as a dramatic drop in temperature.
2. A **control center** in the body, for example, the brain, sets the range of values within which a controlled condition should be maintained (set point), evaluates the input it receives from receptors, and generates output commands when they are needed. *Output* from the control center typically occurs as nerve impulses, or hormones or other chemical signals. This pathway is called an *efferent pathway* (EF-er-ent; *ef-* = away from), since the information flows *away from* the control center. In our skin temperature example, the brain acts as the control center, receiving nerve impulses from the skin receptors and generating nerve impulses as output.
3. An **effector** (e-FEK-tor) is a body structure that receives output from the control center and produces a *response* or effect that changes the controlled condition. Nearly every organ or tissue in the body can behave as an effector. When your body temperature drops sharply, your brain (control center) sends nerve impulses

Figure 1.2 Operation of a feedback system. The dashed return arrow symbolizes negative feedback.

6 The three basic components of a feedback system are the receptor, control center, and effector.

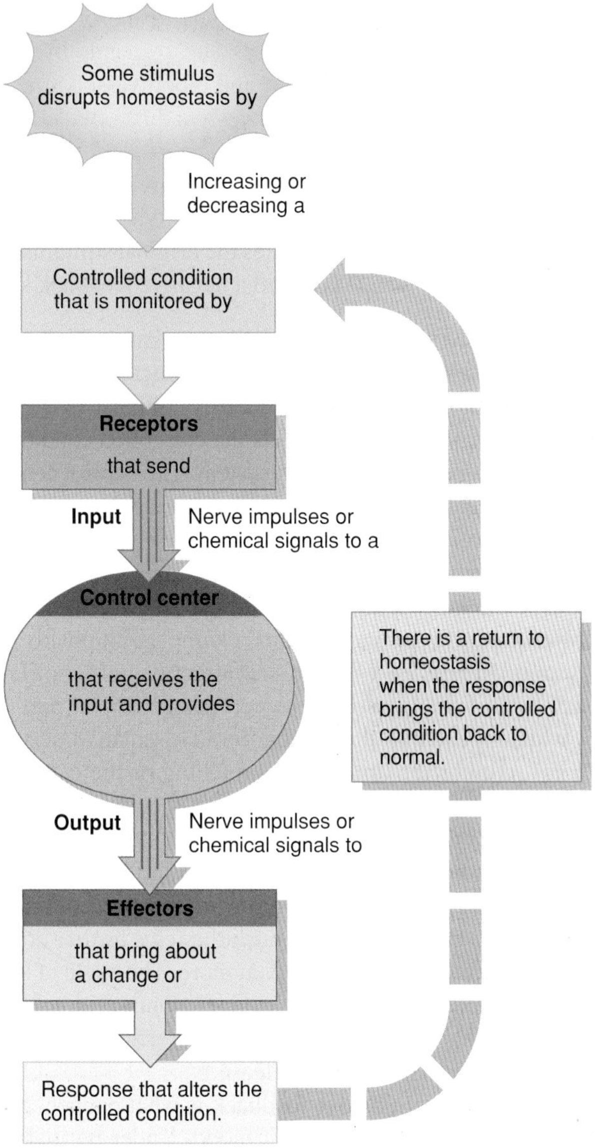
Some stimulus disrupts homeostasis by
Increasing or decreasing a
Controlled condition that is monitored by
Receptors
that send
Input
Nerve impulses or chemical signals to a
Control center
that receives the input and provides
There is a return to homeostasis when the response brings the controlled condition back to normal.
Output
Nerve impulses or chemical signals to
Effectors
that bring about a change or
Response that alters the controlled condition.

? What is the main difference between negative and positive feedback systems?

(output) to your skeletal muscles (effectors). The result is shivering, which generates heat and raises your body temperature.

A group of receptors and effectors communicating with their control center forms a feedback system that can regulate a controlled condition in the body's internal environment. In a feedback system, the response of the system "feeds back" information to change the controlled condition in some way, either negating it (negative feedback) or enhancing it (positive feedback).

Negative Feedback Systems A **negative feedback system** *reverses* a change in a controlled condition. Consider the regulation of blood pressure. Blood pressure (BP) is the force exerted by blood as it presses against the walls of blood vessels. When the heart beats faster or harder, BP increases. If some internal or external stimulus causes blood pressure (controlled condition) to rise, the following sequence of events occurs (Figure 1.3). *Baroreceptors* (the receptors), pressure-sensitive nerve cells located in the walls of certain blood vessels, detect the higher pressure. The baroreceptors send nerve impulses (input) to the brain (control center), which interprets the impulses and responds by sending nerve impulses (output) to the heart and blood vessels (the effectors). Heart rate decreases and blood vessels dilate (widen), which cause BP to decrease (response). This sequence of events quickly returns the controlled condition—blood pressure—to normal, and homeostasis is restored. Notice that the activity of the effector causes BP to drop, a result that negates the original stimulus (an increase in BP). This is why it is called a negative feedback system.

Positive Feedback Systems Unlike a negative feedback system, a **positive feedback system** tends to *strengthen* or *reinforce* a change in one of the body's controlled conditions. In a positive feedback system, the response affects the controlled condition differently than in a negative feedback system. The control center still provides commands to an effector, but this time the effector produces a physiological response that adds to or *reinforces* the initial change in the controlled condition. The action of a positive feedback system continues until it is interrupted by some mechanism.

Normal childbirth provides a good example of a positive feedback system (Figure 1.4). The first contractions of labor (stimulus) push part of the fetus into the cervix, the lowest part of the uterus, which opens into the vagina. Stretch-sensitive nerve cells (receptors) monitor the amount of stretching of the cervix (controlled condition). As stretching increases, they send more nerve impulses (input) to the brain (control center), which in turn releases the hormone oxytocin (output) into the blood. Oxytocin causes muscles in the wall of the uterus (effector) to contract even more forcefully. The contractions push the fetus farther down the uterus, which stretches the cervix even more. The cycle of stretching, hormone release, and ever-stronger contractions is interrupted only by the birth of the baby. Then, stretching of the cervix ceases and oxytocin is no longer released.

Another example of positive feedback is what happens to your body when you lose a great deal of blood. Under normal conditions, the heart pumps blood under sufficient pressure to body cells to provide them with oxygen and nutrients to maintain homeostasis. Upon severe blood loss, blood pressure drops and blood cells

Figure 1.3 Homeostatic regulation of blood pressure by a negative feedback system. Note that the response is fed back into the system, and the system continues to lower blood pressure until there is a return to normal blood pressure (homeostasis).

If the response reverses the stimulus, a system is operating by negative feedback.

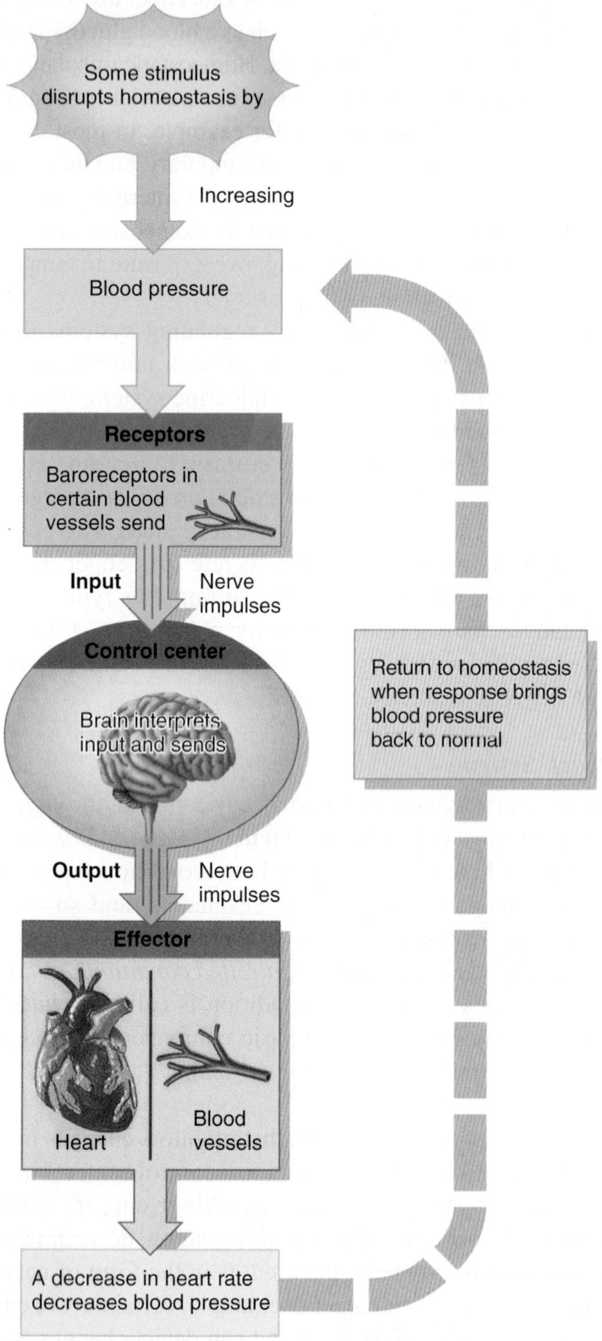

What would happen to heart rate if some stimulus caused blood pressure to decrease? Would this occur by way of positive or negative feedback?

Figure 1.4 Positive feedback control of labor contractions during birth of a baby. The solid return arrow symbolizes positive feedback.

6 If the response enhances or intensifies the stimulus, a system is operating by positive feedback.

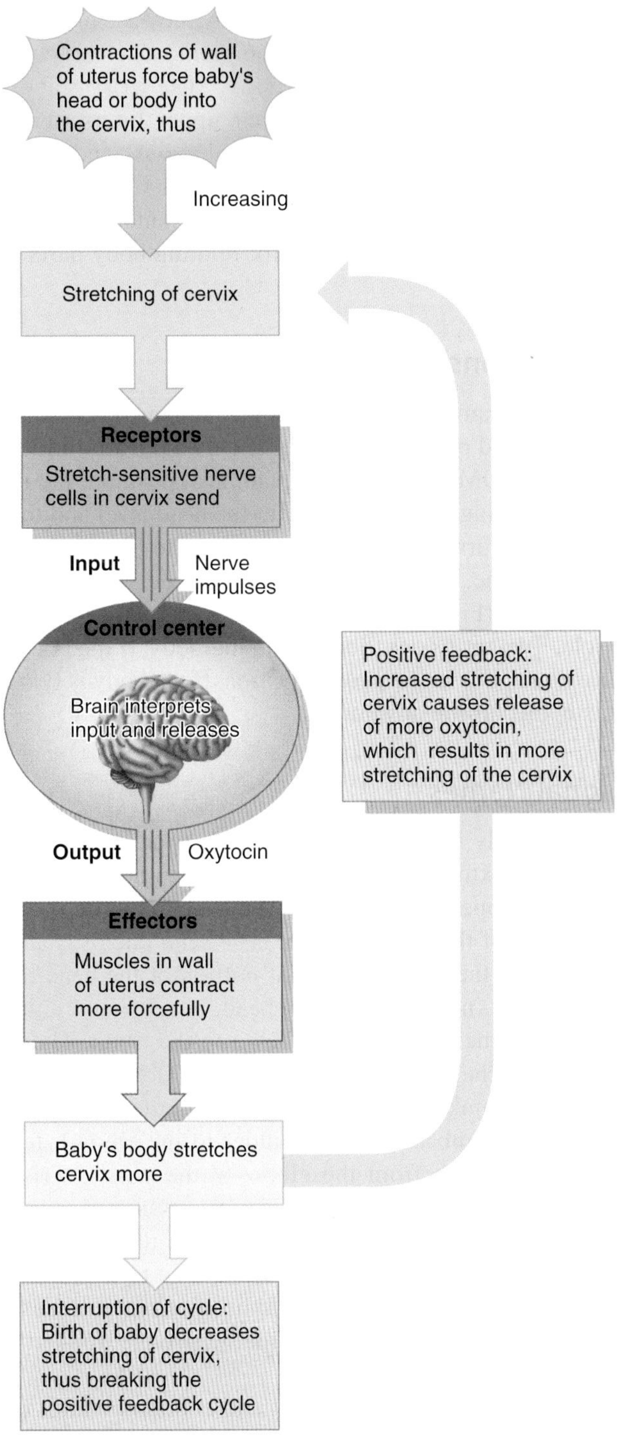

Contractions of wall of uterus force baby's head or body into the cervix, thus
Increasing
Stretching of cervix
Receptors
Stretch-sensitive nerve cells in cervix send
Input
Nerve impulses
Control center
Brain interprets input and releases
Output
Oxytocin
Effectors
Muscles in wall of uterus contract more forcefully
Baby's body stretches cervix more
Interruption of cycle: Birth of baby decreases stretching of cervix, thus breaking the positive feedback cycle
Positive feedback: Increased stretching of cervix causes release of more oxytocin, which results in more stretching of the cervix

Why do positive feedback systems that are part of a normal physiological response include some mechanism that terminates the system?

(including heart cells) receive less oxygen and function less efficiently. If the blood loss continues, heart cells become weaker, the pumping action of the heart decreases further, and blood pressure continues to fall. This is an example of a positive feedback cycle that has serious consequences and may even lead to death if there is no medical intervention. As you will see in Chapter 19, blood clotting is also an example of a positive feedback system.

These examples suggest some important differences between positive and negative feedback systems. Because a positive feedback system continually reinforces a change in a controlled condition, some event outside the system must shut it off. If the action of a positive feedback system is not stopped, it can "run away" and may even produce life-threatening conditions in the body. The action of a negative feedback system, by contrast, slows and then stops as the controlled condition returns to its normal state. Usually, positive feedback systems reinforce conditions that do not happen very often, and negative feedback systems regulate conditions in the body that remain fairly stable over long periods.

Homeostatic Imbalances

You've seen homeostasis defined as a condition in which the body's internal environment remains relatively stable. The body's ability to maintain homeostasis gives it tremendous healing power and a remarkable resistance to abuse. The physiological processes responsible for maintaining homeostasis are in large part also responsible for your good health.

For most people, lifelong good health is not something that happens effortlessly. The many factors in this balance called health include the following:

- The environment and your own behavior.
- Your genetic makeup.
- The air you breathe, the food you eat, and even the thoughts you think.

The way you live your life can either support or interfere with your body's ability to maintain homeostasis and recover from the inevitable stresses life throws your way.

Many diseases are the result of years of poor health behavior that interferes with the body's natural drive to maintain homeostasis. An obvious example is smoking-related illness. Smoking tobacco exposes sensitive lung tissue to a multitude of chemicals that cause cancer and damage the lung's ability to repair itself. Because diseases such as emphysema and lung cancer are difficult to treat and are very rarely cured, it is much wiser to quit smoking—or never start—than to hope a doctor can "fix" you once you are diagnosed with a lung disease. Developing a lifestyle that works with, rather than against, your body's homeostatic processes helps you maximize your personal potential for optimal health and well-being.

As long as all of the body's controlled conditions remain within certain narrow limits, body cells function efficiently, homeostasis is maintained, and the body stays healthy. Should one or more components of the body lose their ability to contribute to homeostasis, however, the normal balance among all of the body's processes may be disturbed. If the homeostatic imbalance is moderate, a disorder or disease may occur; if it is severe, death may result.

A **disorder** is any abnormality of structure or function. **Disease** is a more specific term for an illness characterized by a recognizable set of signs and symptoms. A *local disease* affects one part or a limited region of the body (for example, a sinus infection); a *systemic disease* affects either the entire body or several parts of it (for example, influenza). Diseases alter body structures and functions in characteristic ways. A person with a disease may experience **symptoms,** *subjective* changes in body functions that are not apparent to an observer. Examples of symptoms are headache, nausea, and anxiety. *Objective* changes that a clinician can observe and measure are called **signs.** Signs of disease can be either anatomical, such as swelling or a rash, or physiological, such as fever, high blood pressure, or paralysis.

The science that deals with why, when, and where diseases occur and how they are transmitted among individuals in a community is known as **epidemiology** (ep′-i-dē-mē-OL-ō-jē; *epi-* = upon; *-demi* = people). **Pharmacology** (far′-ma-KOL-ō-jē; *pharmac-* = drug) is the science that deals with the effects and uses of drugs in the treatment of disease.

CLINICAL CONNECTION | *Diagnosis of Disease*

Diagnosis (di′-ag-NŌ-sis; *dia-* = through; *-gnosis* = knowledge) is the science and skill of distinguishing one disorder or disease from another. The patient's symptoms and signs, his or her medical history, a physical exam, and laboratory tests provide the basis for making a diagnosis. Taking a *medical history* consists of collecting information about events that might be related to a patient's illness. These include the chief complaint (primary reason for seeking medical attention), history of present illness, past medical problems, family medical problems, social history, and review of symptoms. A *physical examination* is an orderly evaluation of the body and its functions. This process includes the noninvasive techniques of inspection, palpation, auscultation, and percussion that you learned about earlier in the chapter, along with measurement of vital signs (temperature, pulse, respiratory rate, and blood pressure), and sometimes laboratory tests. •

✔CHECKPOINT

7. Describe the locations of intracellular fluid, extracellular fluid, interstitial fluid, and blood plasma.
8. Why is interstitial fluid called the internal environment of the body?
9. What types of disturbances can act as stimuli that initiate a feedback system?
10. Define receptor, control center, and effector.
11. What is the difference between symptoms and signs of a disease? Give examples of each.

1.5 BASIC ANATOMICAL TERMINOLOGY

OBJECTIVES

- Describe the anatomical position.
- Relate the anatomical names and the corresponding common names for various regions of the human body.
- Define the anatomical planes, anatomical sections, and directional terms used to describe the human body.
- Outline the major body cavities, the organs they contain, and their associated linings.

Scientists and health-care professionals use a common language of special terms when referring to body structures and their functions. The language of anatomy they use has precisely defined meanings that allow us to communicate clearly and precisely. For example, is it correct to say, "The wrist is above the fingers"? This might be true if your upper limbs (described shortly) are at your sides. But if you hold your hands up above your head, your fingers would be above your wrists. To prevent this kind of confusion, anatomists use a standard anatomical position and a special vocabulary for relating body parts to one another.

Body Positions

Descriptions of any region or part of the human body assume that it is in a standard position of reference called the **anatomical position** (an′-a-TOM-i-kal). In the anatomical position, the subject stands erect facing the observer, with the head level and the eyes facing directly forward. The feet are flat on the floor and directed forward, and the upper limbs are at the sides with the palms turned forward (Figure 1.5). In the anatomical position, the body is upright. Two terms describe a reclining body. If the body is lying face down, it is in the **prone** position. If the body is lying face up, it is in the **supine** position.

Regional Names

The human body is divided into several major regions that can be identified externally. The principal regions are the head, neck, trunk, upper limbs, and lower limbs (Figure 1.5). The **head** consists of the skull and face. The *skull* encloses and protects the brain; the *face* is the front portion of the head that includes the eyes, nose, mouth, forehead, cheeks, and chin. The **neck** supports the head and attaches it to the trunk. The **trunk** consists of the chest, abdomen, and pelvis. Each **upper limb** attaches to the trunk and consists of the shoulder, armpit, arm (portion of the limb from the shoulder to the elbow), forearm (portion of the limb from the elbow to the wrist), wrist, and hand. Each **lower limb** also attaches to the trunk and consists of the buttock, thigh (portion of the limb from the buttock to the knee), leg (portion of the limb from the knee to the ankle), ankle, and foot. The *groin* is the area on the front surface of the body marked by a crease on each side, where the trunk attaches to the thighs.

Figure 1.5 shows the anatomical and common names of major parts of the body. The anatomical term in the form of an adjective for each part appears first and is followed in parentheses by the corresponding common name. For example, if you receive a tetanus shot in your *gluteal region,* it is an injection in your *buttock*. Because the anatomical term for a body part usually is based on a Greek or Latin word, it may look different

Figure 1.5 **The anatomical position.** The anatomical names and corresponding common names (in parentheses) are indicated for specific body regions. For example, the cephalic region is the head.

In the anatomical position, the subject stands erect facing the observer with the head level and the eyes facing forward. The feet are flat on the floor and directed forward, and the upper limbs are at the sides with the palms facing forward.

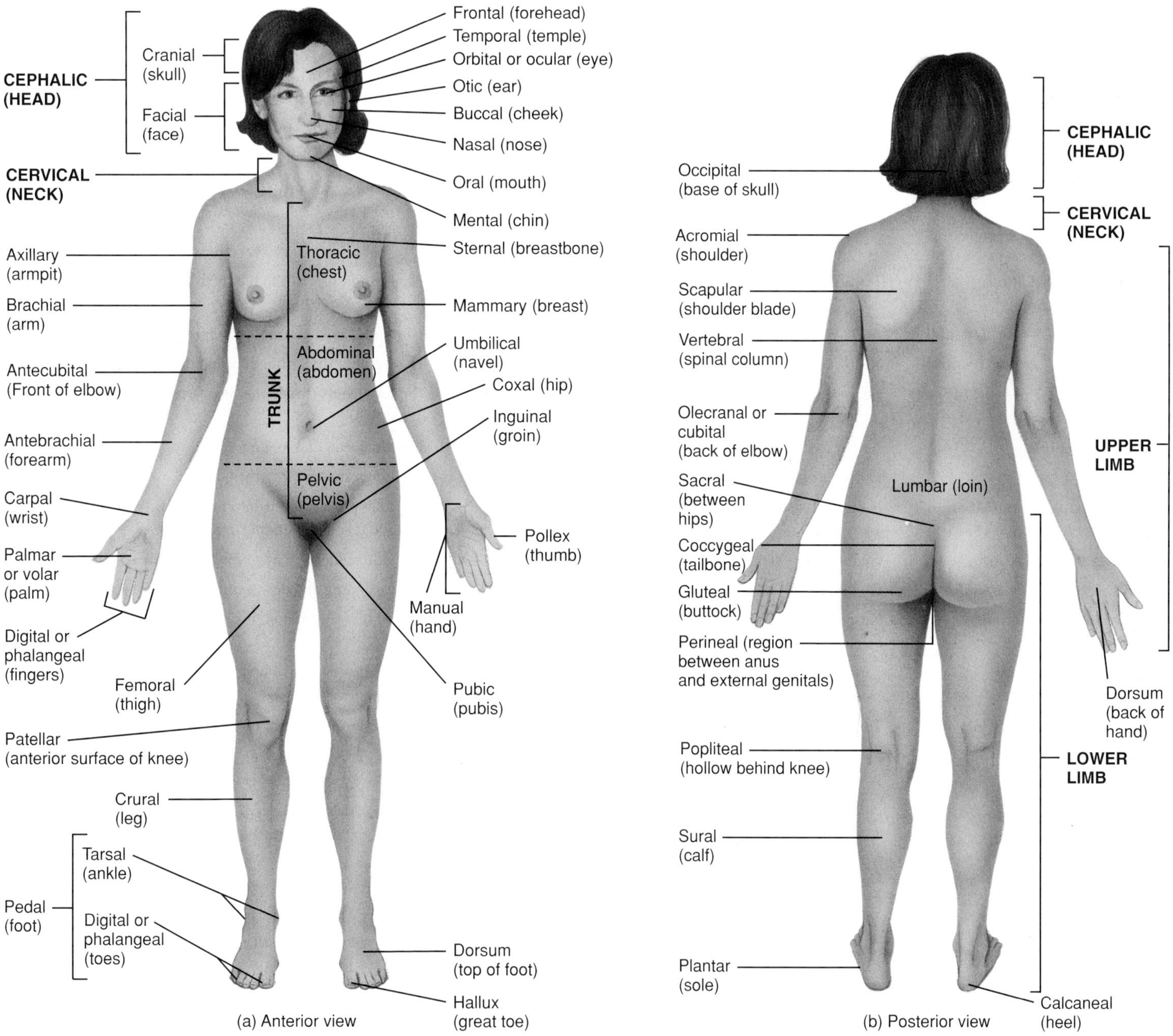

CEPHALIC (HEAD)
Cranial (skull)
Facial (face)
CERVICAL (NECK)
Frontal (forehead)
Temporal (temple)
Orbital or ocular (eye)
Otic (ear)
Buccal (cheek)
Nasal (nose)
Oral (mouth)
Mental (chin)
Sternal (breastbone)
Thoracic (chest)
Axillary (armpit)
Brachial (arm)
Mammary (breast)
Abdominal (abdomen)
Umbilical (navel)
Antecubital (Front of elbow)
Coxal (hip)
TRUNK
Inguinal (groin)
Antebrachial (forearm)
Pelvic (pelvis)
Carpal (wrist)
Pollex (thumb)
Palmar or volar (palm)
Manual (hand)
Digital or phalangeal (fingers)
Femoral (thigh)
Pubic (pubis)
Patellar (anterior surface of knee)
Crural (leg)
Tarsal (ankle)
Pedal (foot)
Digital or phalangeal (toes)
Dorsum (top of foot)
Hallux (great toe)
(a) Anterior view
CEPHALIC (HEAD)
Occipital (base of skull)
CERVICAL (NECK)
Acromial (shoulder)
Scapular (shoulder blade)
Vertebral (spinal column)
Olecranal or cubital (back of elbow)
UPPER LIMB
Lumbar (loin)
Sacral (between hips)
Coccygeal (tailbone)
Gluteal (buttock)
Perineal (region between anus and external genitals)
Dorsum (back of hand)
LOWER LIMB
Popliteal (hollow behind knee)
Sural (calf)
Plantar (sole)
Calcaneal (heel)
(b) Posterior view

What is the usefulness of defining one standard anatomical position?

from the common name for the same part or area. For example, the Latin word for *axilla* (ak-SIL-a) is the common name armpit. Thus, the axillary nerve is one of the nerves passing within the armpit. You will learn more about the Greek and Latin word roots of anatomical and physiological terms as you read this book.

Directional Terms

To locate various body structures, anatomists use specific **directional terms,** words that describe the position of one body part relative to another. Several directional terms are grouped in pairs that have opposite meanings, such as anterior (front) and posterior (back). Exhibit 1.A and Figure 1.6 present the main directional terms.

EXHIBIT 1.A Directional Terms *(Figure 1.6)*

OBJECTIVE

- Define each directional term used to describe the human body.

OVERVIEW

Most of the directional terms used to describe the relationship of one part of the body to another can be grouped into pairs that have opposite meanings. For example, **superior** means toward the upper part of the body, and **inferior** means toward the lower part of the body. It is important to understand that directional terms have relative meanings; they make sense only when used to describe the position of one structure relative to another. For example, your knee is superior to your ankle, even though both are located in the inferior half of the body. Study the directional terms below and the example of how each is used. As you read the examples, look at Figure 1.6 to see the location of each structure.

✔CHECKPOINT

Which directional terms can be used to specify the relationships between (1) the elbow and the shoulder, (2) the left and right shoulders, (3) the sternum and the humerus, and (4) the heart and the diaphragm?

DIRECTIONAL TERM	DEFINITION	EXAMPLE OF USE
Superior (soo′-PĒR-ē-or) **(cephalic** or **cranial)**	Toward the head, or the upper part of a structure.	The heart is superior to the liver.
Inferior (in′-FĒR-ē-or) **(caudal)**	Away from the head, or the lower part of a structure.	The stomach is inferior to the lungs.
Anterior (an-TĒR-ē-or) **(ventral)***	Nearer to or at the front of the body.	The sternum (breastbone) is anterior to the heart.
Posterior (pos-TĒR-ē-or) **(dorsal)**	Nearer to or at the back of the body.	The esophagus (food tube) is posterior to the trachea (windpipe).
Medial (MĒ-dē-al)	Nearer to the midline.†	The ulna is medial to the radius.
Lateral (LAT-er-al)	Farther from the midline.	The lungs are lateral to the heart.
Intermediate (in′-ter-MĒ-dē-at)	Between two structures.	The transverse colon is intermediate between the ascending and descending colons.
Ipsilateral (ip-si-LAT-er-al)	On the same side of the body as another structure.	The gallbladder and ascending colon are ipsilateral.
Contralateral (CON-tra-lat-er-al)	On the opposite side of the body from another structure.	The ascending and descending colons are contralateral.
Proximal (PROK-si-mal)	Nearer to the attachment of a limb to the trunk; nearer to the origination of a structure.	The humerus (arm bone) is proximal to the radius.
Distal (DIS-tal)	Farther from the attachment of a limb to the trunk; farther from the origination of a structure.	The phalanges (finger bones) are distal to the carpals (wrist bones).
Superficial (soo′-per-FISH-al) **(external)**	Toward or on the surface of the body.	The ribs are superficial to the lungs.
Deep (internal)	Away from the surface of the body.	The ribs are deep to the skin of the chest and back.

*Note that the terms *anterior* and *ventral* mean the same thing in humans. However, in four-legged animals *ventral* refers to the belly side and is therefore *inferior.* Similarly, the terms *posterior* and *dorsal* mean the same thing in humans, but in four-legged animals *dorsal* refers to the back side and is therefore *superior.*

†Recall that the **midline** is an imaginary vertical line that divides the body into equal right and left sides.

Figure 1.6 Directional terms.

Directional terms precisely locate various parts of the body relative to one another.

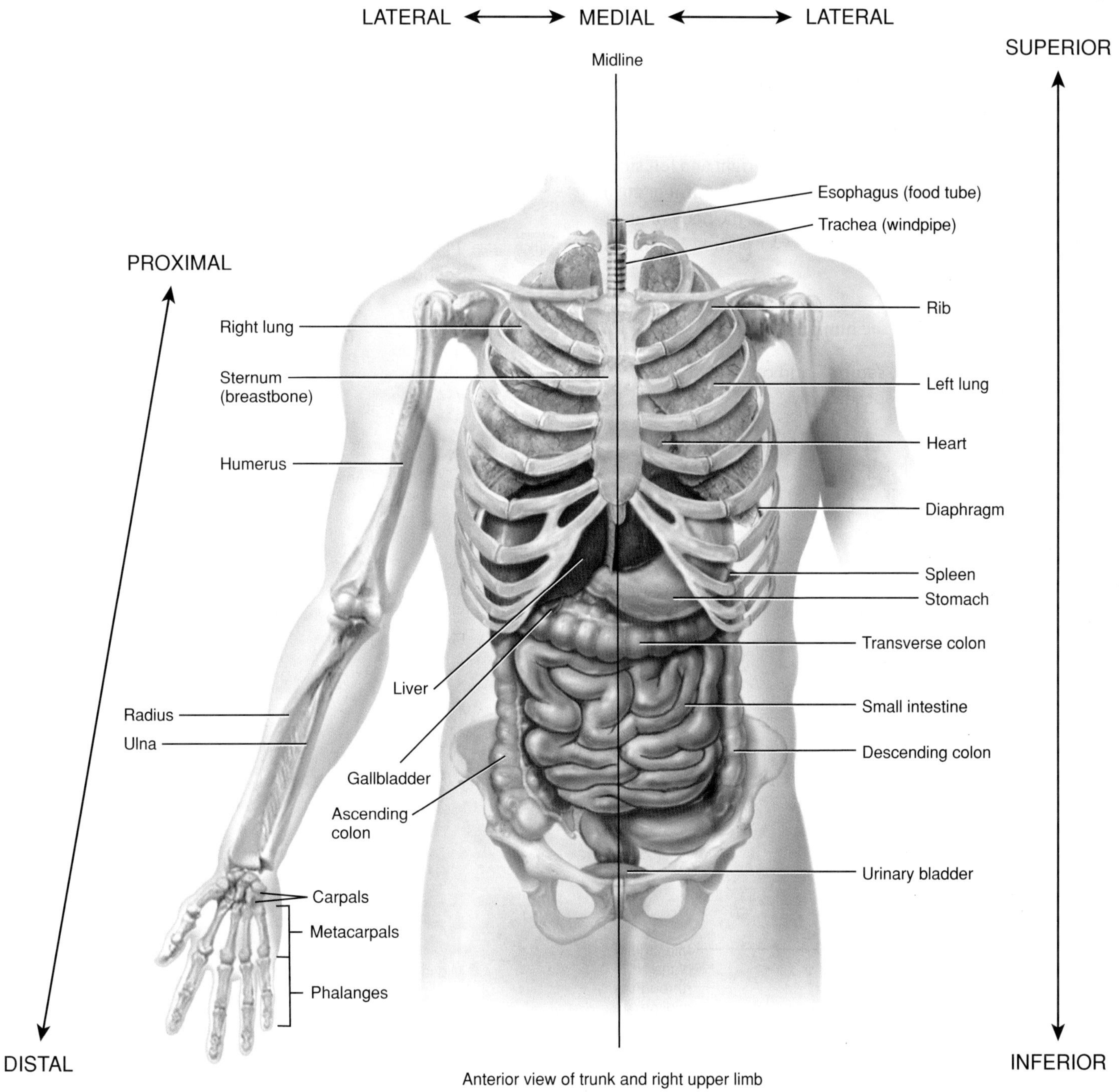

LATERAL
MEDIAL
LATERAL
Midline
SUPERIOR
PROXIMAL
Esophagus (food tube)
Trachea (windpipe)
Right lung
Rib
Sternum (breastbone)
Left lung
Heart
Humerus
Diaphragm
Spleen
Stomach
Transverse colon
Liver
Radius
Small intestine
Ulna
Descending colon
Gallbladder
Ascending colon
Urinary bladder
Carpals
Metacarpals
Phalanges
DISTAL
INFERIOR

Anterior view of trunk and right upper limb

Is the radius proximal to the humerus? Is the esophagus anterior to the trachea? Are the ribs superficial to the lungs? Is the urinary bladder medial to the ascending colon? Is the sternum lateral to the descending colon?

Planes and Sections

You will also study parts of the body relative to **planes,** imaginary flat surfaces that pass through the body parts (Figure 1.7). A **sagittal plane** (SAJ-i-tal; *sagitt-* = arrow) is a vertical plane that divides the body or an organ into right and left sides. More specifically, when such a plane passes through the midline of the body or an organ and divides it into *equal* right and left sides, it is called a **midsagittal plane** or a **median plane.** The **midline** is an imaginary vertical line that divides the body into equal left and right sides. If the sagittal plane does not pass through the midline but instead divides the body or an organ into *unequal* right and left sides, it is called a **parasagittal plane** (*para-* = near). A **frontal** or **coronal plane** (kō-RŌ-nal; *corona* = crown) divides the body or an organ into anterior (front) and posterior (back) portions. A **transverse plane** divides the body or an organ into superior (upper) and inferior (lower) portions. Other names for a transverse plane are a **cross-sectional** or **horizontal plane.** Sagittal, frontal, and transverse planes are all at right angles to one another. An **oblique plane** (ō-BLĒK), by contrast, passes through the body or an organ at an oblique angle (any angle other than a 90-degree angle).

Figure 1.7 Planes through the human body.

Frontal, transverse, sagittal, and oblique planes divide the body in specific ways.

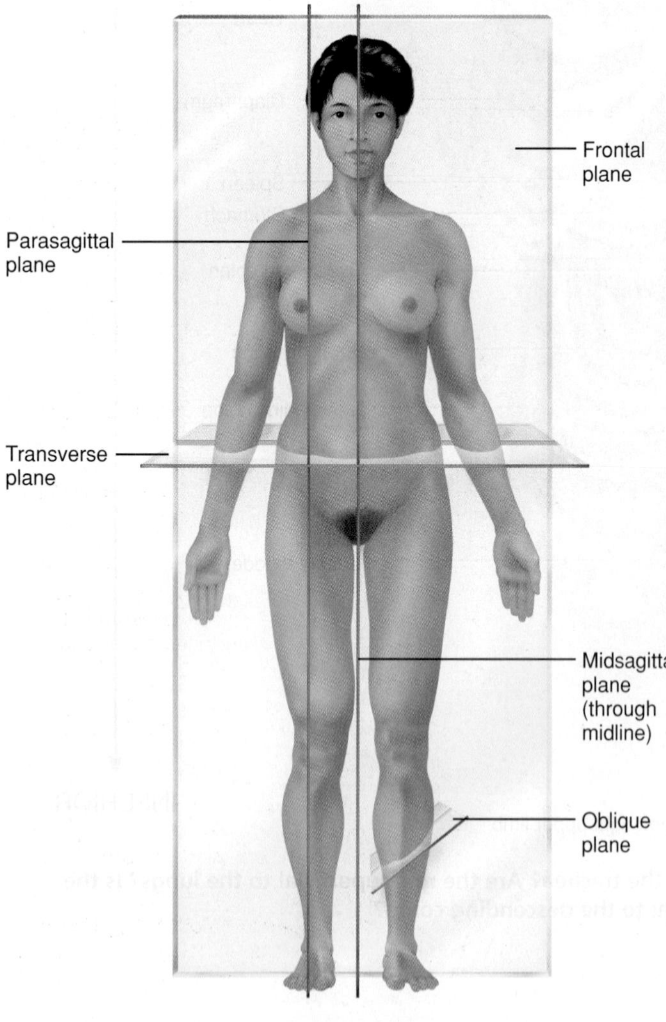

Anterior view

Which plane divides the heart into anterior and posterior portions?

Figure 1.8 Planes and sections through different parts of the brain. The diagrams (left) show the planes, and the photographs (right) show the resulting sections. Note: The arrows in the diagrams indicate the direction from which each section is viewed. This aid is used throughout the book to indicate viewing perspectives.

Planes divide the body in various ways to produce sections.

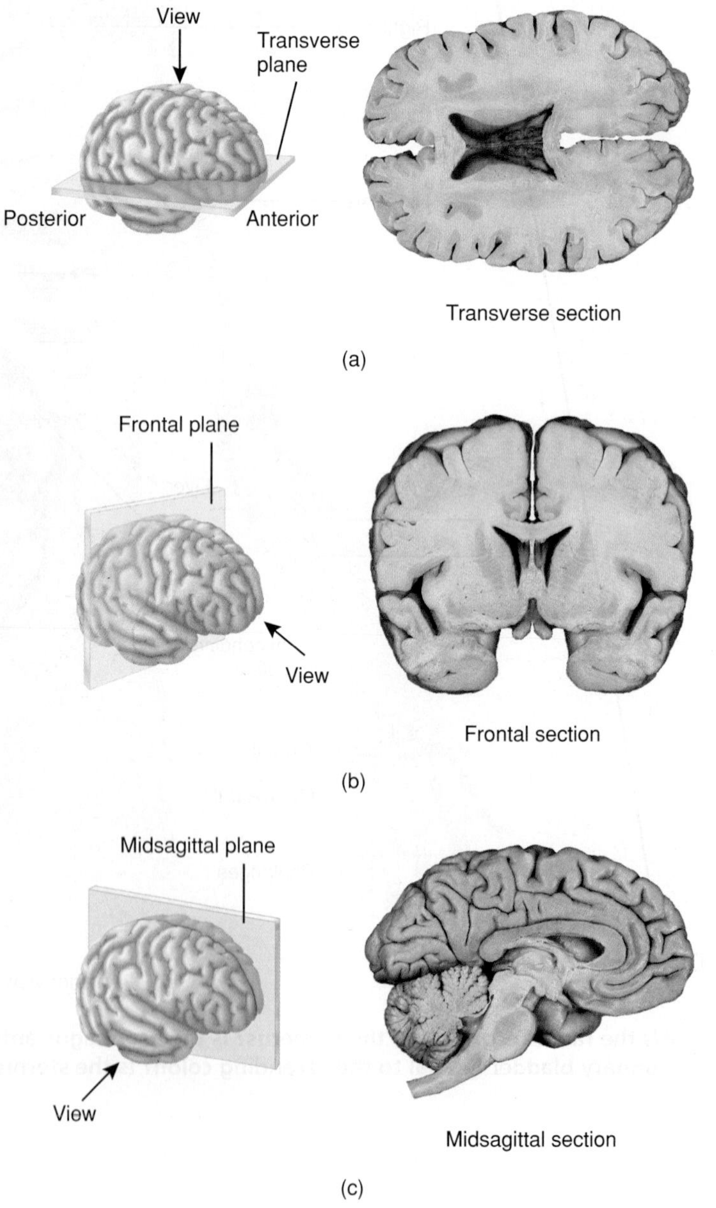

Which plane divides the brain into unequal right and left portions?

When you study a body region, you often view it in section. A **section** is a cut of the body or one of its organs made along one of the planes just described. It is important to know the plane of the section so you can understand the anatomical relationship of one part to another. Figure 1.8a–c indicates how three different sections—*transverse, frontal,* and *midsagittal*—provide different views of the brain.

Body Cavities

Body cavities are spaces within the body that help protect, separate, and support internal organs. Bones, muscles, ligaments, and other structures separate the various body cavities from one another. Here we discuss several body cavities (Figure 1.9).

The cranial bones form a hollow space of the head called the **cranial cavity** (KRĀ-nē-al), which contains the brain. The bones of the vertebral column (backbone) form the **vertebral (spinal) canal** (VER-te-bral), which contains the spinal cord. The cranial cavity and vertebral canal are continuous with one another. Three layers of protective tissue, the **meninges** (me-NIN-jēz), and a shock-absorbing fluid surround the brain and spinal cord.

The major body cavities of the trunk are the thoracic and abdominopelvic cavities. The **thoracic cavity** (thor-AS-ik; *thorac-* = chest) or chest cavity (Figure 1.10) is formed by the ribs, the muscles of the chest, the sternum (breastbone), and the thoracic portion of the vertebral column. Within the thoracic cavity are the **pericardial cavity** (per′-i-KAR-dē-al; *peri-* = around; *-cardial* = heart), a fluid-filled space that surrounds the heart, and two fluid-filled spaces called **pleural cavities** (PLOOR-al; *pleur-* = rib or side), one around each lung. The central part of the thoracic cavity is an anatomical region called the **mediastinum** (mē′-dē-as-TĪ-num; *media-* = middle; *-stinum* = partition). It is between the lungs, extending from the sternum to the vertebral column and from the first rib to the diaphragm (Figure 1.10a, b). The mediastinum

Figure 1.9 Body cavities. The dashed line in (a) indicates the border between the abdominal and pelvic cavities.

The major cavities of the trunk are the thoracic and abdominopelvic cavities.

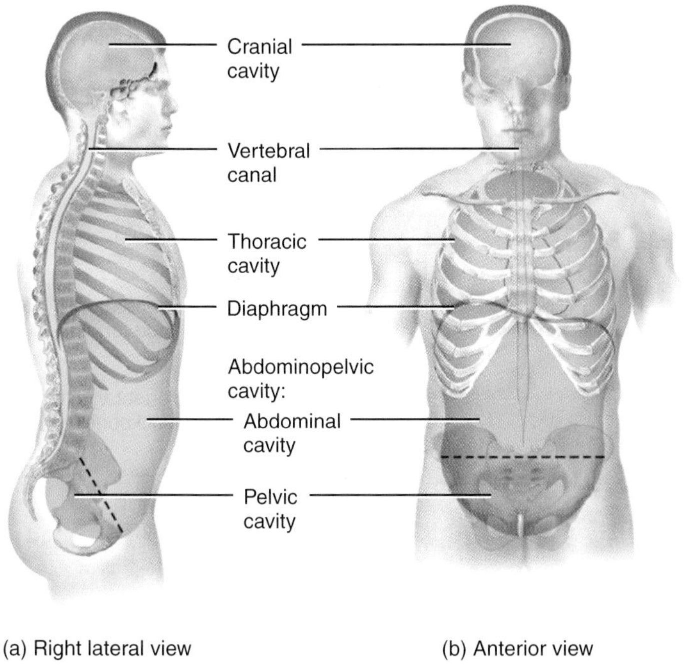

CAVITY	COMMENTS
Cranial cavity	Formed by cranial bones and contains brain.
Vertebral canal	Formed by vertebral column and contains spinal cord and the beginnings of spinal nerves.
Thoracic cavity*	Chest cavity; contains pleural and pericardial cavities and mediastinum.
Pleural cavity	Each surrounds a lung; the serous membrane of each pleural cavity is the pleura.
Pericardial cavity	Surrounds the heart; the serous membrane of the pericardial cavity is the pericardium.
Mediastinum	Central portion of thoracic cavity between the lungs; extends from sternum to vertebral column and from first rib to diaphragm; contains heart, thymus, esophagus, trachea, and several large blood vessels.
Abdominopelvic cavity	Subdivided into abdominal and pelvic cavities.
Abdominal cavity	Contains stomach, spleen, liver, gallbladder, small intestine, and most of large intestine; the serous membrane of the abdominal cavity is the peritoneum.
Pelvic cavity	Contains urinary bladder, portions of large intestine, and internal organs of reproduction.

* See Figure 1.10 for details of the thoracic cavity.

In which cavities are the following organs located: urinary bladder, stomach, heart, small intestine, lungs, internal female reproductive organs, thymus, spleen, liver? Use the following symbols for your responses: T = thoracic cavity, A = abdominal cavity, or P = pelvic cavity.

Figure 1.10 The thoracic cavity. The dashed lines indicate the borders of the mediastinum. Note: When transverse sections are viewed inferiorly (from below), the anterior aspect of the body appears on top and the left side of the body appears on the right side of the illustration.

The thoracic cavity contains three smaller cavities and the mediastinum.

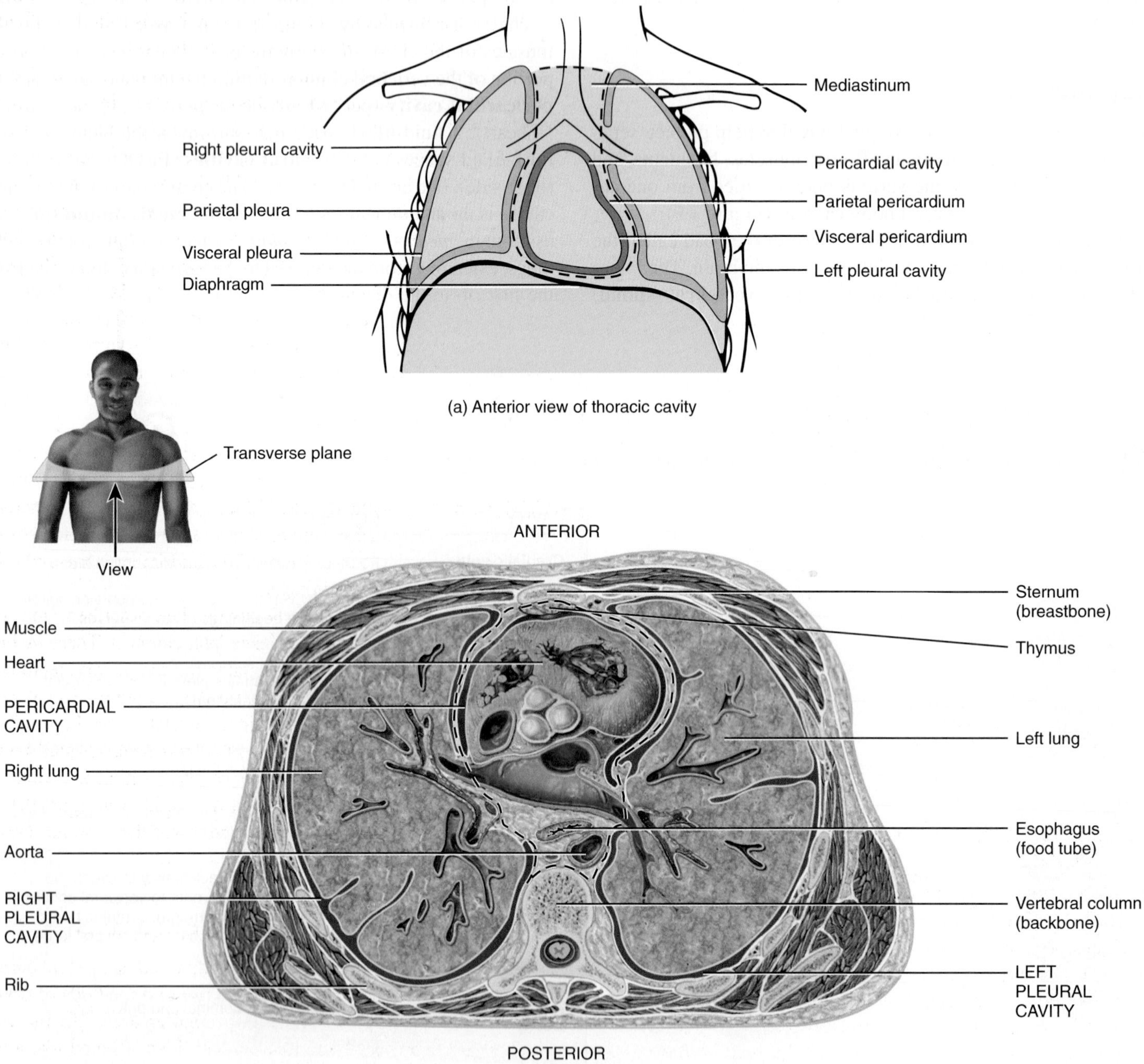

(a) Anterior view of thoracic cavity

(b) Inferior view of transverse section of thoracic cavity

What is the name of the cavity that surrounds the heart? Which cavities surround the lungs?

contains all thoracic organs except the lungs themselves. Among the structures in the mediastinum are the heart, esophagus, trachea, thymus, and several large blood vessels that enter and exit the heart. The **diaphragm** (DĪ-a-fram = partition or wall) is a dome-shaped muscle that separates the thoracic cavity from the abdominopelvic cavity.

The **abdominopelvic cavity** (ab-dom′-i-nō-PEL-vik; see Figure 1.9) extends from the diaphragm to the groin and is encircled

Figure 1.11 The abdominopelvic cavity. The dashed lower line shows the approximate boundary between the abdominal and pelvic cavities.

The abdominopelvic cavity extends from the diaphragm to the groin.

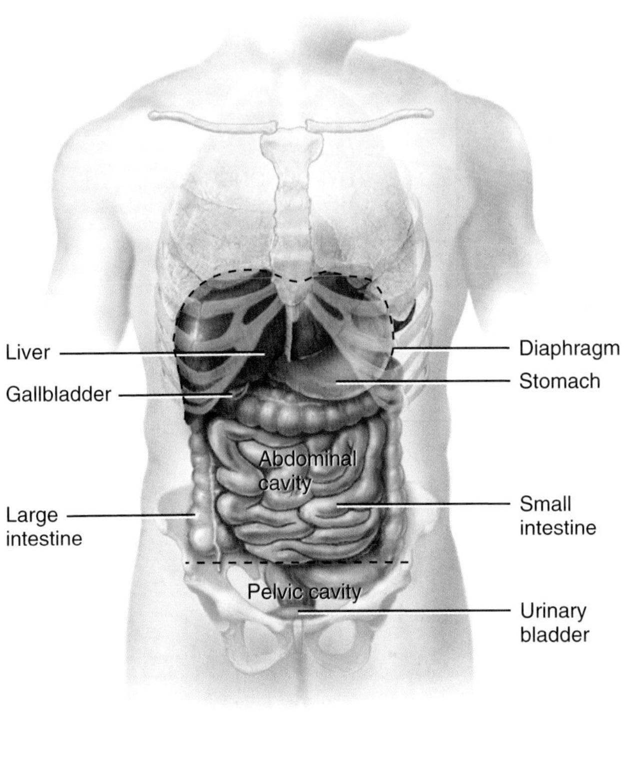
Liver
Gallbladder
Diaphragm
Stomach
Abdominal cavity
Large intestine
Small intestine
Pelvic cavity
Urinary bladder
Anterior view

To which body systems do the organs shown here within the abdominal and pelvic cavities belong? (*Hint:* Refer to Table 1.2.)

by the abdominal muscular wall and the bones and muscles of the pelvis. As the name suggests, the abdominopelvic cavity is divided into two portions, even though no wall separates them (Figure 1.11). The superior portion, the **abdominal cavity** (ab-DOM-i-nal; *abdomin-* = belly), contains the stomach, spleen, liver, gallbladder, small intestine, and most of the large intestine. The inferior portion, the **pelvic cavity** (PEL-vik; *pelv-* = basin), contains the urinary bladder, portions of the large intestine, and internal organs of the reproductive system. Organs inside the thoracic and abdominopelvic cavities are called **viscera** (VIS-er-a).

Thoracic and Abdominal Cavity Membranes

A **membrane** is a thin, pliable tissue that covers, lines, partitions, or connects structures. One example is a slippery, double-layered membrane called a **serous membrane** (SĒR-us), which covers the viscera within the thoracic and abdominal cavities and also lines the walls of the thorax and abdomen. The parts of a serous membrane are (1) the *parietal layer* (pa-RĪ-e-tal), a thin epithelium that lines the walls of the cavities, and (2) the *visceral layer* (VIS-er-al), a thin epithelium that covers and adheres to the viscera within the cavities. A small amount of lubricating fluid (serous fluid) between the two layers reduces friction, allowing the viscera to slide somewhat during movements, such as when the lungs inflate and deflate during breathing.

The serous membrane of the pleural cavities is called the **pleura** (PLOO-ra). The *visceral pleura* clings to the surface of the lungs, and the anterior part of the *parietal pleura* lines the chest wall, covering the superior surface of the diaphragm (see Figure 1.10a). In between is the *pleural cavity,* filled with a small amount of lubricating fluid (see Figure 1.10). The serous membrane of the pericardial cavity is the **pericardium** (per′-i-KAR-dē-um). The *visceral pericardium* covers the surface of the heart; the *parietal pericardium* lines the chest wall. Between them is the *pericardial cavity,* filled with a small amount of lubricating fluid (see Figure 1.10). The **peritoneum** (per-i-tō-NĒ-um) is the serous membrane of the abdominal cavity. The *visceral peritoneum* covers the abdominal viscera, and the *parietal peritoneum* lines the abdominal wall, covering the inferior surface of the diaphragm. Between them is the *peritoneal cavity,* which contains a small amount of lubricating fluid. Most abdominal organs are located in the peritoneal cavity. Some are located between the parietal peritoneum and the posterior abdominal wall. Such organs are said to be *retroperitoneal* (re′-trō-per-i-tō-NĒ-al; *retro-* = behind). The kidneys, adrenal glands, pancreas, duodenum of the small intestine, ascending and descending colons of the large intestine, and portions of the abdominal aorta and inferior vena cava are retroperitoneal.

In addition to the major body cavities just described, you will also learn about other body cavities in later chapters. These include the *oral (mouth) cavity,* which contains the tongue and teeth (see Figure 24.5); the *nasal cavity* in the nose (see Figure 23.1); the *orbital cavities (orbits),* which contain the eyeballs (see Figure 7.3); the *middle ear cavities (middle ears),* which contain small bones (see Figure 17.4); and the *synovial cavities,* which are found in freely movable joints and contain synovial fluid (see Figure 9.3).

A summary of the major body cavities and their membranes is presented in the table included in Figure 1.9.

Abdominopelvic Regions and Quadrants

To describe the location of the many abdominal and pelvic organs more easily, anatomists and clinicians use two methods of dividing the abdominopelvic cavity into smaller areas. In the first method, two horizontal and two vertical lines, aligned like a tic-tac-toe grid, partition this cavity into nine **abdominopelvic regions** (Figure 1.12a,b). The top horizontal line, the *subcostal line* (*sub-* = under; *costal* = rib), is drawn just inferior to the rib cage, across the inferior portion of the stomach; the bottom horizontal line, the *transtubercular line* (trans-too-BER-kū-lar), is drawn just inferior to the tops of the hip bones. Two vertical lines, the left and right *midclavicular lines* (mid-kla-VIK-ū-lar), are drawn through the midpoints of the clavicles (collar bones), just medial to the nipples. The four lines divide the abdominopelvic cavity into a larger middle section and smaller left and right sections.

Figure 1.12 Regions and quadrants of the abdominopelvic cavity.

The nine-region designation is used for anatomical studies; the quadrant designation is used to locate the site of pain, tumors, or some other abnormality.

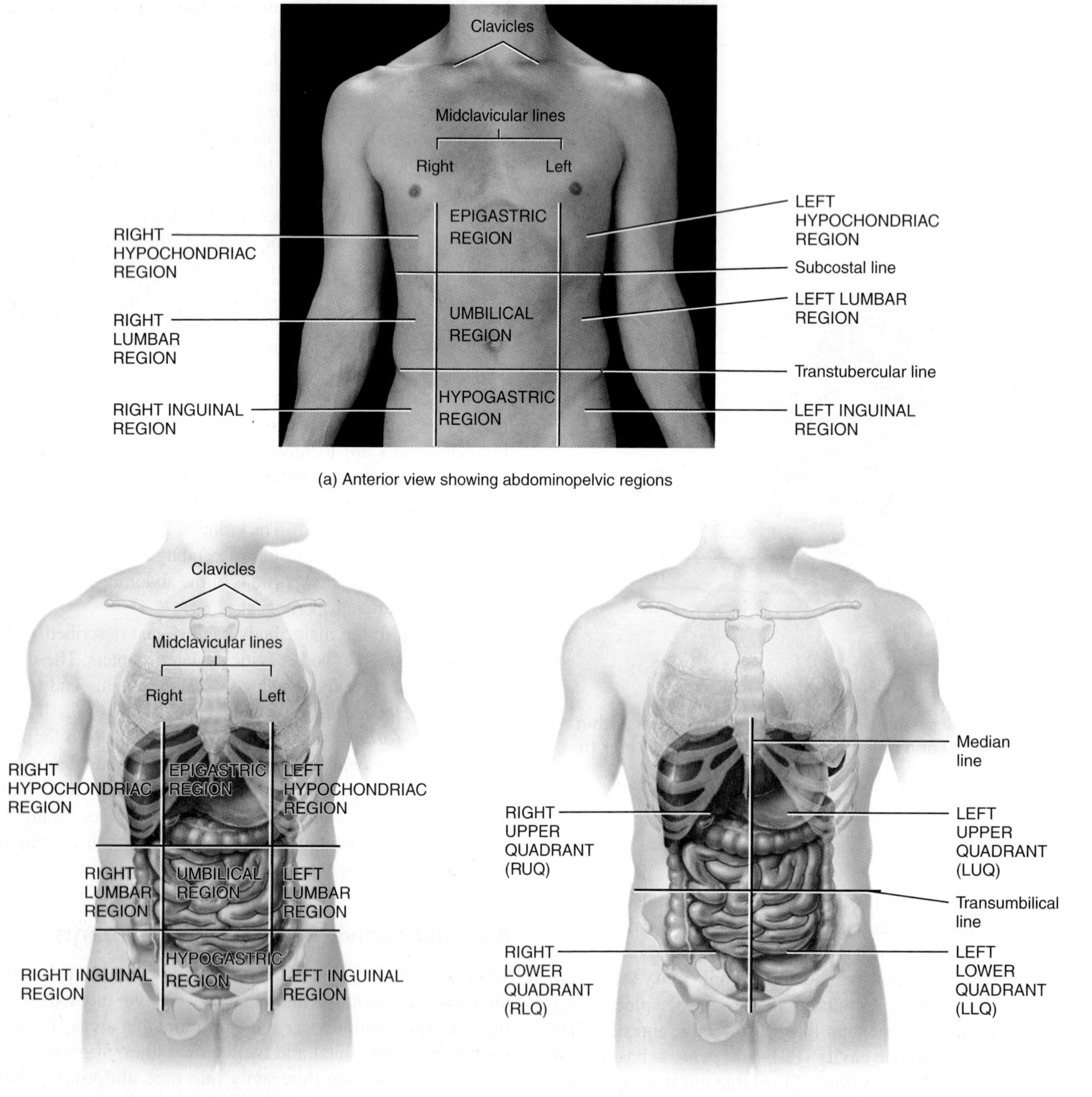

(a) Anterior view showing abdominopelvic regions

(b) Anterior view showing location of abdominopelvic regions

(c) Anterior view showing location of abdominopelvic quadrants

In which abdominopelvic region is each of the following found: most of the liver, ascending colon, urinary bladder, and most of the small intestine? In which abdominopelvic quadrant would pain from appendicitis (inflammation of the appendix) be felt?

The names of the nine abdominopelvic regions are **right hypochondriac** (hī′-pō-KON-drē-ak), **epigastric** (ep-i-GAS-trik), **left hypochondriac, right lumbar, umbilical** (um-BIL-i-kal) **left lumbar, right inguinal (iliac)** (IN-gwi-nal), **hypogastric (pubic),** and **left inguinal (iliac).**

The second method is simpler and divides the abdominopelvic cavity into **quadrants** (KWOD-rantz; *quad-* = one-fourth), as shown in Figure 1.12c. In this method, a midsagittal line (the *median line*) and a transverse line (the *transumbilical line*) are passed through the **umbilicus** (um-BI-li-kus; *umbilic-* = navel) or *belly button.* The names of the abdominopelvic quadrants are **right upper quadrant (RUQ), left upper quadrant (LUQ), right lower quadrant (RLQ),** and **left lower quadrant (LLQ).** The nine-region division is more widely used for anatomical studies, and quadrants are more commonly used by clinicians for describing the site of abdominopelvic pain, a tumor, or another abnormality.

✔CHECKPOINT

12. Locate each region shown in Figure 1.5 on your own body, and then identify it by its anatomical name and the corresponding common name.
13. What structures separate the various body cavities from one another?
14. Locate the nine abdominopelvic regions and the four abdominopelvic quadrants on yourself, and list some of the organs found in each.

1.6 MEDICAL IMAGING

OBJECTIVE

- Describe the principles and importance of medical imaging procedures in the evaluation of organ functions and the diagnosis of disease.

Medical imaging refers to techniques and procedures used to create images of the human body. Various types of medical imaging allow visualization of structures inside our bodies and are increasingly helpful for precise diagnosis of a wide range of anatomical and physiological disorders. The grandparent of all medical imaging techniques is conventional radiography (x-rays), in medical use since the late 1940s. The newer imaging technologies not only contribute to diagnosis of disease, but they also are advancing our understanding of normal anatomy and physiology. Table 1.3 describes some commonly used medical imaging techniques. Other imaging methods, such as cardiac catheterization, will be discussed in later chapters.

✔CHECKPOINT

15. Which forms of medical imaging would be used to show a blockage in an artery of the heart?
16. Of the medical imaging techniques outlined in Table 1.3, which one best reveals the physiology of a structure?
17. Which medical imaging technique would you use to determine whether a bone was broken?

TABLE 1.3

Common Medical Imaging Procedures

RADIOGRAPHY

Procedure: A single barrage of x-rays passes through the body, producing an image of interior structures on x-ray–sensitive film. The resulting two-dimensional image is a *radiograph* (RĀ-dē-ō-graf′), commonly called an x-ray.

Comments: Relatively inexpensive, quick, and simple to perform; usually provides sufficient information for diagnosis. X-rays do not easily pass through dense structures, so bones appear white. Hollow structures, such as the lungs, appear black. Structures of intermediate density, such as skin, fat, and muscle, appear as varying shades of gray. At low doses, x-rays are useful for examining soft tissues such as the breast **(mammography)** and for determining bone density **(bone densitometry).**

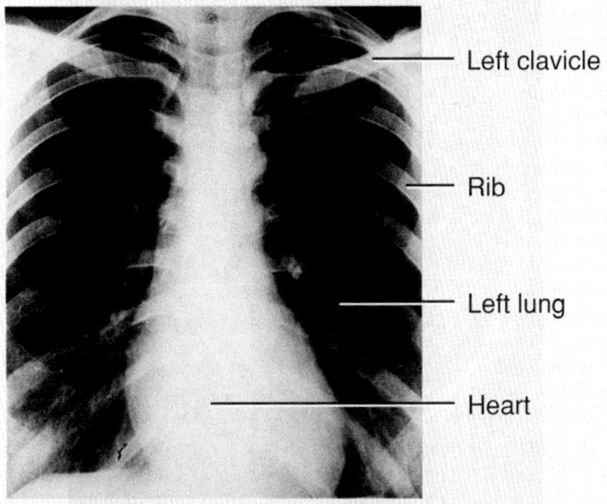

Left clavicle
Rib
Left lung
Heart

Radiograph of the thorax in anterior view

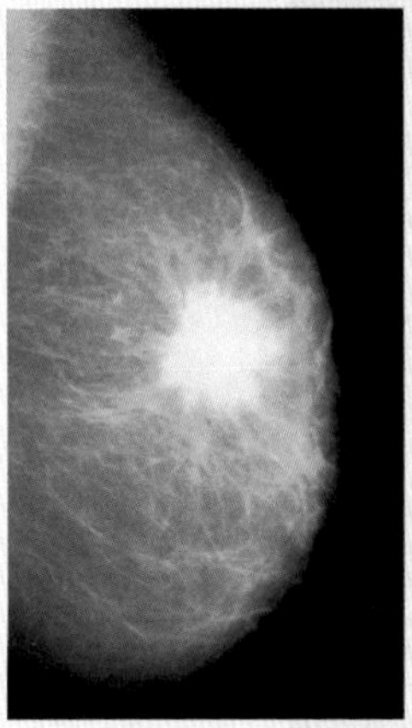
Mammogram of a female breast showing a cancerous tumor (white mass with uneven border)

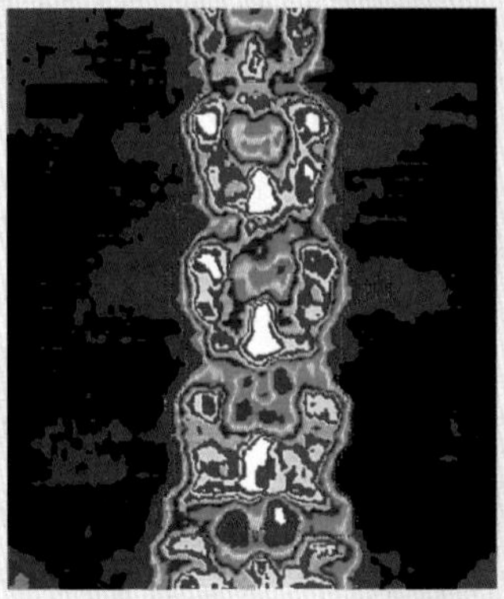
Bone densitometry scan of the lumbar spine in anterior view

TABLE 1.3 CONTINUES

TABLE 1.3 CONTINUED

Common Medical Imaging Procedures

RADIOGRAPHY (CONTINUED)

It is necessary to use a substance called a contrast medium to make hollow or fluid-filled structures visible (appear white) in radiographs. X-rays make structures that contain contrast media appear white. The medium may be introduced by injection, orally, or rectally, depending on the structure to be imaged. Contrast x-rays are used to image blood vessels **(angiography),** the urinary system **(intravenous urography),** and the gastrointestinal tract **(barium contrast x-ray).**

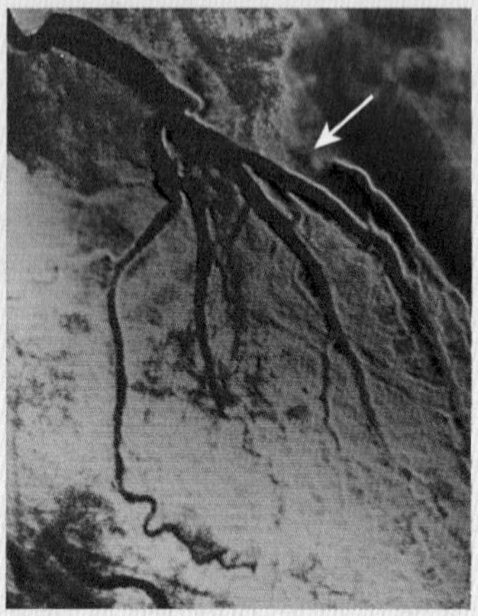

Angiogram of an adult human heart showing a blockage in a coronary artery (arrow)

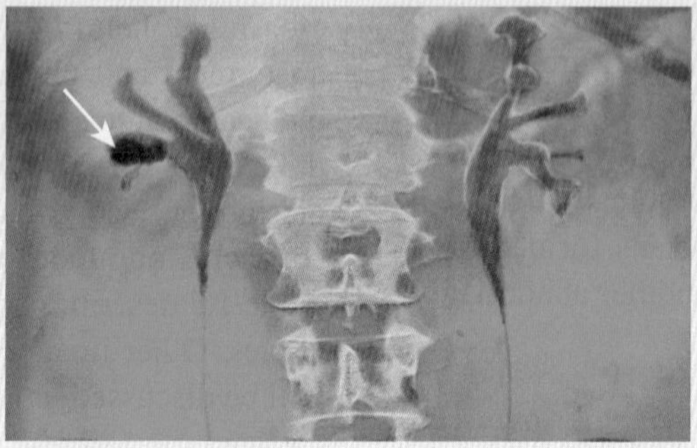

Intravenous urogram showing a kidney stone (arrow) in the right kidney

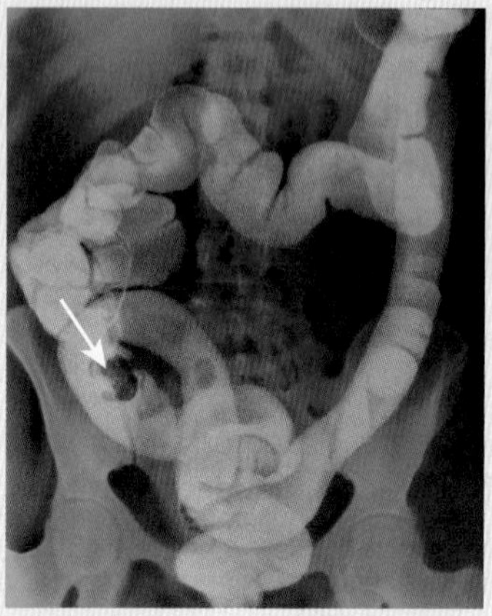

Barium contrast x-ray showing a cancer of the ascending colon (arrow)

MAGNETIC RESONANCE IMAGING (MRI)

Procedure: The body is exposed to a high-energy magnetic field, which causes protons (small positive particles within atoms, such as hydrogen) in body fluids and tissues to arrange themselves in relation to the field. Then a pulse of radio waves "reads" these ion patterns, and a color-coded image is assembled on a video monitor. The result is a two- or three-dimensional blueprint of cellular chemistry.

Comments: Relatively safe, but cannot be used on patients with metal in their bodies. Shows fine details for soft tissues but not for bones. Most useful for differentiating between normal and abnormal tissues. Used to detect tumors and artery-clogging fatty plaques; reveal brain abnormalities; measure blood flow; and detect a variety of musculoskeletal, liver, and kidney disorders.

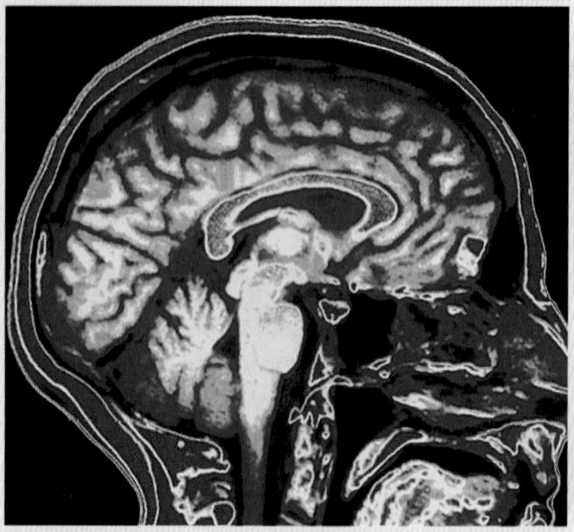

Magnetic resonance image of the brain in sagittal section

DIFFUSION TENSOR IMAGING (DTI)

Procedure: **Diffusion tensor imaging (DTI)** is a variation of MRI that tracks the movement of water molecules along the length of axons, long processes of nerve cells that form the white matter in the brain. The white matter is organized into bundles of axons called tracts that connect one part of the brain with another. DTI provides images of the brain's white matter.

Comments: Used in the diagnosis of addictions, epilepsy, brain tumors, traumatic brain injury, stroke, multiple sclerosis, and neurodegenerative diseases.

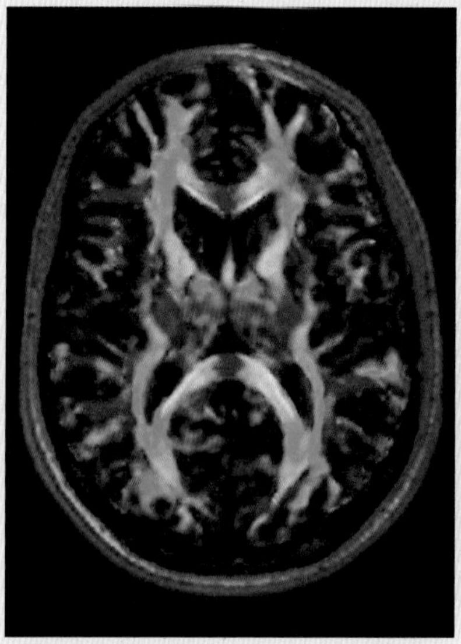

Diffusion tensor image of transverse section of the brain showing white matter (various colors)

COMPUTED TOMOGRAPHY (CT)

[formerly called computerized axial tomography (CAT) scanning]

Procedure: In this form of computer-assisted radiography, an x-ray beam traces an arc at multiple angles around a section of the body. The resulting transverse section of the body, called a *CT scan,* is shown on a video monitor.

Comments: Visualizes soft tissues and organs with much more detail than conventional radiographs. Differing tissue densities show up as various shades of gray. Multiple scans can be assembled to build three-dimensional views of structures (described next). Whole-body CT scanning typically targets the torso and appears to provide the most benefit in screening for lung cancers, coronary artery disease, and kidney cancers.

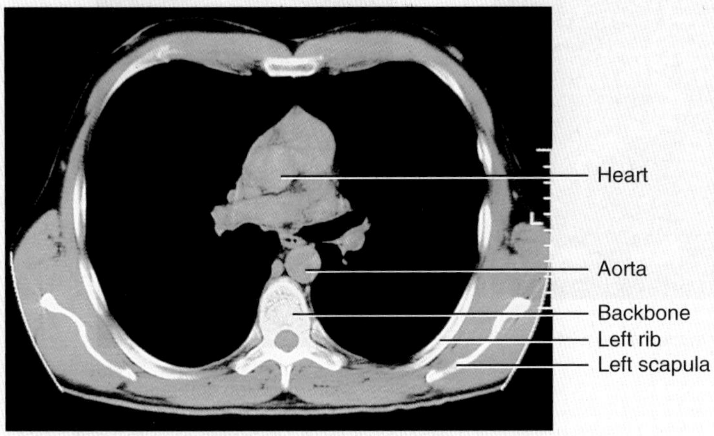

Computed tomography scan of the thorax in inferior view

CORONARY (CARDIAC) COMPUTED TOMOGRAPHY ANGIOGRAPHY (CCTA) SCAN

Procedure: In this form of computer-assisted radiography, an iodine-containing contrast medium is injected into a vein and a beta-blocker is given to decrease heart rate. Then, numerous x-ray beams trace an arc around the heart and a scanner detects the x-ray beams and transmits them to a computer, which transforms the information into a three-dimensional image of the coronary blood vessels on a monitor. The image produced is called a *CCTA scan* and can be generated in less than 20 seconds.

Comments: Used primarily to determine if there are any coronary artery blockages (for example, atherosclerotic plaque or calcium) that may require an intervention such as angioplasty or stent. The CCTA scan can be rotated, enlarged, and moved at any angle. The procedure can take thousands of images of the heart within the time of a single heartbeat, so it provides a great amount of detail about the heart's structure and function.

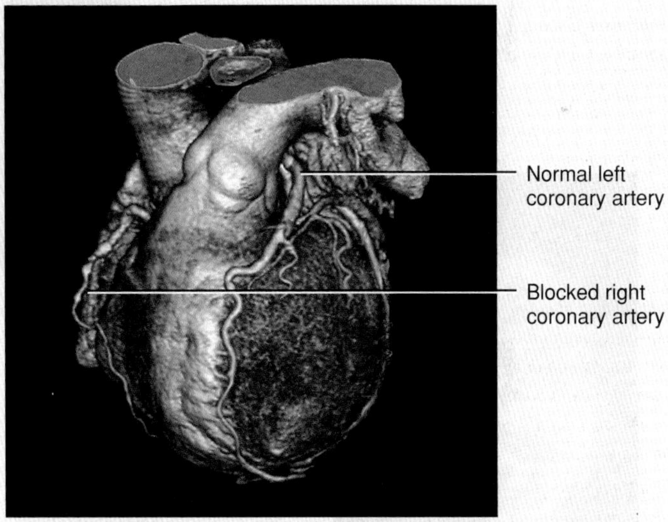

CCTA scan of coronary arteries

ULTRASOUND SCANNING

Procedure: High-frequency sound waves produced by a handheld wand reflect off body tissues and are detected by the same instrument. The image, which may be still or moving, is called a *sonogram* (SON-ō-gram) and is shown on a video monitor.

Comments: Safe, noninvasive, painless, and uses no dyes. Most commonly used to visualize the fetus during pregnancy. Also used to observe the size, location, and actions of organs and blood flow through blood vessels **(Doppler ultrasound).**

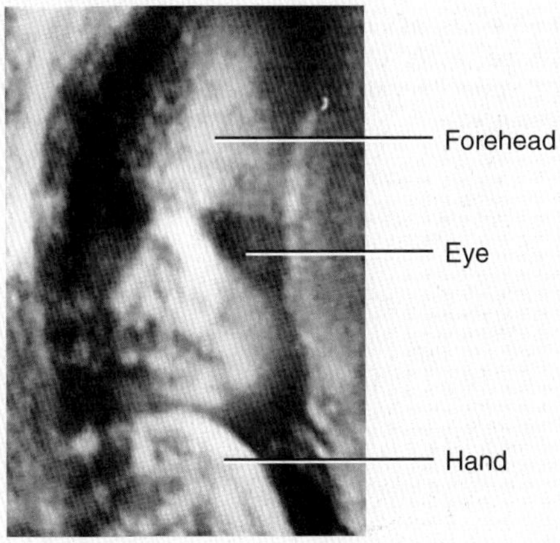

Sonogram of a fetus (Courtesy of Andrew Joseph Tortora and Damaris Soler)

TABLE 1.3 CONTINUED

Common Medical Imaging Procedures

POSITRON EMISSION TOMOGRAPHY (PET)

Procedure: A substance that emits positrons (positively charged particles) is injected into the body, where it is taken up by tissues. The collision of positrons with negatively charged electrons in body tissues produces gamma rays (similar to x-rays) that are detected by gamma cameras positioned around the subject. A computer receives signals from the gamma cameras and constructs a *PET scan* image, displayed in color on a video monitor. The PET scan shows where the injected substance is being used in the body. In the PET scan image shown here, the black and blue colors indicate minimal activity; the red, orange, yellow, and white colors indicate areas of increasingly greater activity.

Comments: Used to study the physiology of body structures, such as metabolism in the brain or heart.

ANTERIOR

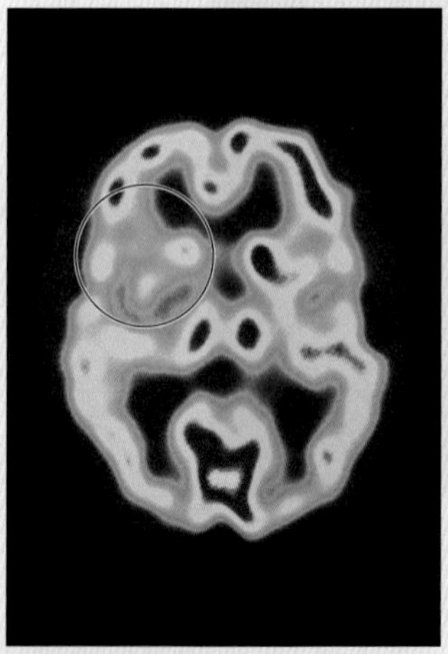

POSTERIOR

Positron emission tomography scan of a transverse section of the brain (circled area at upper left indicates where a stroke has occurred)

RADIONUCLIDE SCANNING

Procedure: A *radionuclide* (radioactive substance) is introduced intravenously into the body and carried by the blood to the tissue to be imaged. Gamma rays emitted by the radionuclide are detected by a gamma camera outside the subject, and the data are fed into a computer. The computer constructs a *radionuclide image* and displays it in color on a video monitor. Areas of intense color take up a lot of the radionuclide and represent high tissue activity; areas of less intense color take up smaller amounts of the radionuclide and represent low tissue activity. **Single-photon-emission computerized tomography (SPECT) scanning** is a specialized type of radionuclide scanning that is especially useful for studying the brain, heart, lungs, and liver.

Comments: Used to study activity of a tissue or organ, such as searching for malignant tumors in body tissue or scars that may interfere with heart muscle activity.

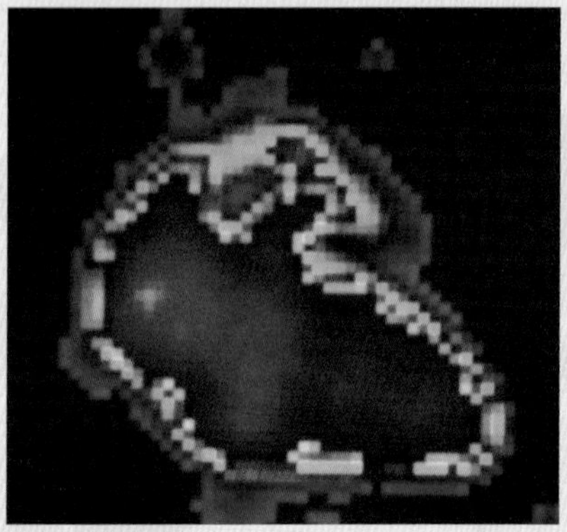

Radionuclide (nuclear) scan of a normal human heart

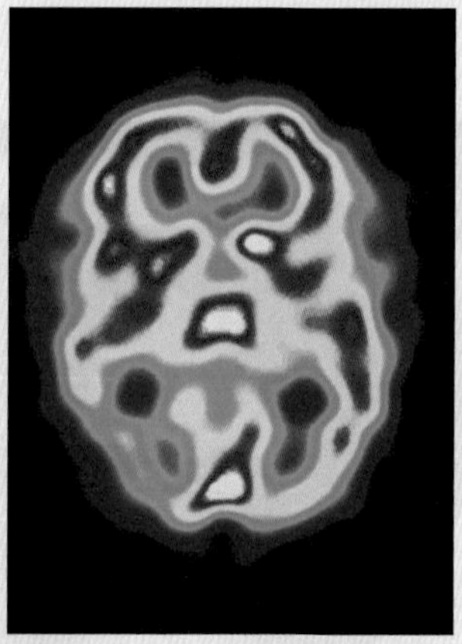

Single-photon-emission computerized tomography (SPECT) scan of a transverse section of the brain (green area at lower left indicates a migraine attack)

ENDOSCOPY

Procedure: Endoscopy involves the visual examination of the inside of body organs or cavities using a lighted instrument with lenses called an *endoscope*. The image is viewed through an eyepiece on the endoscope or projected onto a monitor.

Comments: Examples include *colonoscopy* (used to examine the interior of the colon, which is part of the large intestine), *laparoscopy* (used to examine the organs within the abdominopelvic cavity), and *arthroscopy* (used to examine the interior of a joint, usually the knee).

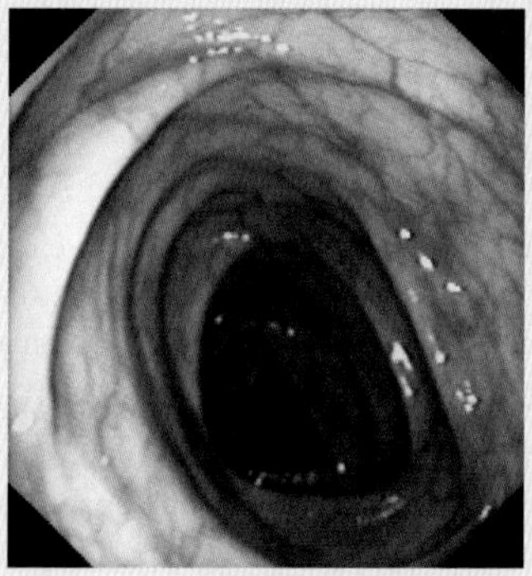

Interior view of the colon as shown by colonoscopy

CHAPTER REVIEW AND RESOURCE SUMMARY

WILEY PLUS

Review

1.1 Anatomy and Physiology Defined

1. Anatomy is the science of body structures and the relationships among structures; physiology is the science of body functions.
2. Dissection is the careful cutting apart of body structures to study their relationships.
3. Some subspecialties of anatomy are embryology, developmental biology, cell biology, histology, gross anatomy, systemic anatomy, regional anatomy, surface anatomy, radiographic anatomy, and pathological anatomy (see Table 1.1).
4. Some subspecialties of physiology are neurophysiology, endocrinology, cardiovascular physiology, immunology, respiratory physiology, renal physiology, exercise physiology, and pathophysiology (see Table 1.1).

1.2 Levels of Structural Organization and Body Systems

1. The human body consists of six levels of structural organization: chemical, cellular, tissue, organ, system, and organismal.
2. Cells are the basic structural and functional living units of an organism and the smallest living units in the human body.
3. Tissues are groups of cells and the materials surrounding them that work together to perform a particular function.
4. Organs are composed of two or more different types of tissues; they have specific functions and usually have recognizable shapes.
5. Systems consist of related organs that have a common function.
6. An organism is any living individual.
7. Table 1.2 introduces the 11 systems of the human organism: the integumentary, skeletal, muscular, nervous, endocrine, cardiovascular, lymphatic, respiratory, digestive, urinary, and reproductive systems.

Resource

Anatomy Overview - The Integumentary System
Anatomy Overview - The Skeletal System
Anatomy Overview - The Muscular System
Anatomy Overview - The Nervous System
Anatomy Overview - The Endocrine System
Anatomy Overview - The Cardiovascular System
Anatomy Overview - The Lymphatic and Immune Systems
Anatomy Overview - The Respiratory System
Anatomy Overview - The Digestive System
Anatomy Overview - The Urinary System
Anatomy Overview - The Reproductive Systems
Exercise - Concentrate on Systemic Functions
Exercise - Find the System Outsiders

1.3 Characteristics of the Living Human Organism

1. All organisms carry on certain processes that distinguish them from nonliving things.
2. Among the life processes in humans are metabolism, responsiveness, movement, growth, differentiation, and reproduction.

Resource

Animation - Communication, Regulation, and Homeostasis

1.4 Homeostasis

1. Homeostasis is a condition of equilibrium in the body's internal environment produced by the interplay of all the body's regulatory processes.
2. Body fluids are dilute, watery solutions. Intracellular fluid (ICF) is inside cells, and extracellular fluid (ECF) is outside cells. Plasma is the ECF within blood vessels. Interstitial fluid is the ECF that fills spaces between tissue cells; because it surrounds all body cells, interstitial fluid is called the body's internal environment.
3. Disruptions of homeostasis come from external and internal stimuli and psychological stresses. When disruption of homeostasis is mild and temporary, responses of body cells quickly restore balance in the internal environment. If disruption is extreme, regulation of homeostasis may fail.
4. Most often, the nervous and endocrine systems acting together or separately regulate homeostasis. The nervous system detects body changes and sends nerve impulses to counteract changes in controlled conditions. The endocrine system regulates homeostasis by secreting hormones.
5. Feedback systems include three components: (1) Receptors monitor changes in a controlled condition and send input to a control center (afferent pathway). (2) The control center sets the value (set point) at which a controlled condition should be maintained, evaluates the input it receives from receptors (efferent pathway), and generates output commands when they are needed. (3) Effectors receive output from the control center and produce a response (effect) that alters the controlled condition.
6. If a response reverses the original stimulus, the system is operating by negative feedback. If a response enhances the original stimulus, the system is operating by positive feedback.
7. One example of negative feedback is the regulation of blood pressure. If a stimulus causes blood pressure (controlled condition) to rise, baroreceptors (pressure-sensitive nerve cells, the receptors) in blood vessels send impulses (input) to the brain (control center). The brain sends impulses (output) to the heart (effector). As a result, heart rate decreases (response) and blood pressure decreases to normal (restoration of homeostasis).

Resource

Animation - Communication, Regulation, and Homeostasis
Animation - Homeostatic Relationships
Animation - Negative Feedback Control of Blood Pressure
Animation - Negative Feedback Control of Temperature
Animation - Positive Feedback Control of Labor
Exercise - System Contributions to Body Homeostasis
Concepts and Connections - Negative Feedback Loop
Concepts and Connections - Regulation and Communication

Review | **Resource**

8. One example of positive feedback occurs during the birth of a baby. When labor begins, the cervix of the uterus is stretched (stimulus), and stretch-sensitive nerve cells in the cervix (receptors) send nerve impulses (input) to the brain (control center). The brain responds by releasing oxytocin (output), which stimulates the uterus (effector) to contract more forcefully (response). Movement of the fetus further stretches the cervix, more oxytocin is released, and even more forceful contractions occur. The cycle is broken with the birth of the baby.
9. Disruptions of homeostasis—homeostatic imbalances—can lead to disorders, diseases, and even death. A disorder is a general term for any abnormality of structure or function. A disease is an illness with a definite set of signs and symptoms.
10. Symptoms are subjective changes in body functions that are not apparent to an observer; signs are objective changes that can be observed and measured.

1.5 Basic Anatomical Terminology

1. Descriptions of any region of the body assume the body is in the anatomical position, in which the subject stands erect facing the observer, with the head level and the eyes facing directly forward. The feet are flat on the floor and directed forward, and the upper limbs are at the sides, with the palms turned forward. A body lying facedown is prone; a body lying faceup is supine.
2. Regional names are terms given to specific regions of the body. The principal regions are the head, neck, trunk, upper limbs, and lower limbs. Within the regions, specific body parts have anatomical names and corresponding common names. Examples are thoracic (chest), nasal (nose), and carpal (wrist).
3. Directional terms indicate the relationship of one part of the body to another. Exhibit 1.A summarizes commonly used directional terms.
4. Planes are imaginary flat surfaces that are used to divide the body or organs to visualize interior structures. A midsagittal plane divides the body or an organ into *equal* right and left sides. A parasagittal plane divides the body or an organ into *unequal* right and left sides. A frontal plane divides the body or an organ into anterior and posterior portions. A transverse plane divides the body or an organ into superior and inferior portions. An oblique plane passes through the body or an organ at an oblique angle.
5. Sections are cuts of the body or its organs made along a plane. They are named according to the plane along which the cut is made and include transverse, frontal, and sagittal sections.
6. Figure 1.9 summarizes body cavities and their membranes. Body cavities are spaces in the body that help protect, separate, and support internal organs. The cranial cavity contains the brain, and the vertebral canal contains the spinal cord. The meninges are protective tissues that line the cranial cavity and vertebral canal. The diaphragm separates the thoracic cavity from the abdominopelvic cavity. Viscera are organs within the thoracic and abdominopelvic cavities. A serous membrane lines the wall of the cavity and adheres to the viscera.
7. The thoracic cavity is subdivided into three smaller cavities: a pericardial cavity, which contains the heart, and two pleural cavities, each of which contains a lung. The central part of the thoracic cavity is an anatomical region called the mediastinum. It is located between the pleural cavities, extending from the sternum to the vertebral column and from the first rib to the diaphragm. It contains all thoracic viscera except the lungs.
8. The abdominopelvic cavity is divided into a superior abdominal and an inferior pelvic cavity. Viscera of the abdominal cavity include the stomach, spleen, liver, gallbladder, small intestine, and most of the large intestine. Viscera of the pelvic cavity include the urinary bladder, portions of the large intestine, and internal organs of the reproductive system.
9. Serous membranes line the walls of the thoracic and abdominal cavities and cover the organs within them. They include the pleura, associated with the lungs; the pericardium, associated with the heart; and the peritoneum, associated with the abdominal cavity.
10. To describe the location of organs more easily, the abdominopelvic cavity is divided into nine regions: right hypochondriac, epigastric, left hypochondriac, right lumbar, umbilical, left lumbar, right inguinal (iliac), hypogastric (pubic), and left inguinal (iliac). To locate the site of an abdominopelvic abnormality in clinical studies, the abdominopelvic cavity is divided into quadrants: right upper quadrant (RUQ), left upper quadrant (LUQ), right lower quadrant (RLQ), and left lower quadrant (LLQ).

Anatomy Overview - Serous Membrane
Figure 1.5 - The Anatomical Position
Figure 1.6 - Directional Terms
Figure 1.9 - Body Cavities
Figure 1.10 - The Thoracic Cavity

1.6 Medical Imaging

1. Medical imaging refers to techniques and procedures used to create images of the human body. They allow visualization of internal structures to diagnose abnormal anatomy and deviations from normal physiology.
2. Table 1.3 summarizes and illustrates several medical imaging techniques.

SELF-QUIZ QUESTIONS

Fill in the blanks in the following statements.

1. A(n) ______ is a group of similar cells and their surrounding materials performing specific functions.

2. The sum of all of the body's chemical processes is ______. It consists of two parts: the phase that builds up new substances is ______, and the phase that breaks down substances is ______.

3. The fluid located within cells is the ______, whereas the fluid located outside of the cells is ______.

Indicate whether the following statements are true or false.

4. In a positive feedback system, the response enhances or intensifies the original stimulus.

5. A person lying face down would be in the supine position.

6. The highest level of structural organization is the system level.

Choose the one best answer to the following questions.

7. A plane that separates the body into unequal right and left sides is a
(a) transverse plane (b) frontal plane
(c) midsagittal plane (d) coronal plane
(e) parasagittal plane

8. Midway through a 5-mile workout, a runner begins to sweat profusely. The sweat glands producing the sweat would be considered which part of a feedback loop?
(a) controlled condition (b) receptors (c) stimulus
(d) effectors (e) control center

9. An unspecialized stem cell becomes a brain cell during fetal development. This is an example of
(a) differentiation (b) growth (c) organization
(d) responsiveness (e) homeostasis.

10. A radiography technician needs to x-ray a growth on the urinary bladder. To accomplish this, the camera must be positioned on the ____ region.
(a) left inguinal (b) epigastric (c) hypogastric
(d) right inguinal (e) umbilical

11. Which of the following would not be associated with the thoracic cavity? (1) pericardium, (2) mediastinum, (3) peritoneum, (4) pleura,
(a) 2 and 3 (b) 2 (c) 3
(d) 1 and 4 (e) 3 and 4

12. Match the following common names and anatomical descriptive adjectives:

____ (a) axillary	(1) skull
____ (b) inguinal	(2) eye
____ (c) cervical	(3) cheek
____ (d) cranial	(4) armpit
____ (e) oral	(5) arm
____ (f) brachial	(6) groin
____ (g) orbital	(7) buttock
____ (h) gluteal	(8) neck
____ (i) buccal	(9) mouth
____ (j) coxal	(10) hip

13. Choose the term that best fits the blank in each statement. Some answers may be used more than once.

____ (a) Your eyes are _____ to your chin.	(1) superior
____ (b) Your skin is _____ to your heart.	(2) inferior
____ (c) Your right shoulder is _____ and _____ from your umbilicus (belly button).	(3) anterior
____ (d) In the anatomical position, your thumb is _____.	(4) posterior
____ (e) Your buttocks are _____.	(5) medial
____ (f) Your right foot and right hand are _____.	(6) lateral
____ (g) Your knee is _____ between your thigh and toes.	(7) intermediate
____ (h) Your lungs are _____ to your spinal column.	(8) ipsilateral
____ (i) Your breastbone is _____ to your chin.	(9) contralateral
____ (j) Your calf is _____ to your heel.	(10) proximal
	(11) distal
	(12) superficial
	(13) deep

14. Match the following cavities to their definitions:

____ (a) a fluid-filled space that surrounds the heart	(1) cranial cavity
____ (b) the cavity that contains the brain	(2) vertebral canal
____ (c) a cavity formed by the ribs, muscles of the chest, sternum, and part of the vertebral column	(3) thoracic cavity
____ (d) a cavity that contains the stomach, spleen, liver, gallbladder, small intestine, and most of the large intestine	(4) pericardial cavity
____ (e) fluid-filled space that surrounds a lung	(5) pleural cavity
____ (f) the cavity that contains the urinary bladder, part of the large intestine, and the organs of the reproductive system	(6) abdominal cavity
____ (g) the canal that contains the spinal cord	(7) pelvic cavity

15. Match the following systems with their functions:

____ (a) nervous system	(1) regulates body activities through hormones (chemicals) transported in the blood to various target organs of the body
____ (b) endocrine system	(2) produces gametes; releases hormones from gonads
____ (c) urinary system	(3) protects against disease; returns fluids to blood
____ (d) cardiovascular system	(4) protects body by forming a barrier to the outside environment; helps regulate body temperature
____ (e) muscular system	(5) transports oxygen and nutrients to cells; protects against disease; carries wastes away from cells
____ (f) respiratory system	(6) regulates body activities through action potentials (nerve impulses); receives sensory information; interprets and responds to the information
____ (g) digestive system	(7) carries out the physical and chemical breakdown of food and absorption of nutrients
____ (h) skeletal system	(8) transfers oxygen and carbon dioxide between air and blood
____ (i) integumentary system	(9) supports and protects the body; provides internal framework; provides a place for muscle attachment
____ (j) lymphatic system and immunity	(10) powers movements of the body and stabilizes body position
____ (k) reproductive system	(11) eliminates wastes; regulates the volume and chemical composition of blood

CRITICAL THINKING QUESTIONS

1. You are studying for your first anatomy and physiology exam and want to know which areas of your brain are working hardest as you study. Your classmate suggests that you could have a computed tomography (CT) scan done to assess your brain activity. Would this be the best way to determine brain activity levels?

2. There is much interest in using stem cells to help in the treatment of diseases such as type I diabetes, which is due to a malfunction of some of the normal cells in the pancreas. What would make stem cells useful in disease treatment?

3. On her first anatomy and physiology exam, Heather defined homeostasis as "the condition in which the body approaches room temperature and stays there." Do you agree with Heather's definition?

? ANSWERS TO FIGURE QUESTIONS

1.1 Organs are composed of two or more different types of tissues that work together to perform a specific function.

1.2 The difference between negative and positive feedback systems is that in negative feedback systems, the response reverses the original stimulus, but in positive feedback systems, the response enhances the original stimulus.

1.3 When something causes blood pressure to decrease, then heart rate increases due to operation of this negative feedback system.

1.4 Because positive feedback systems continually intensify or reinforce the original stimulus, some mechanism is needed to end the response.

1.5 Having one standard anatomical position allows directional terms to be clearly defined so that any body part can be described in relation to any other part.

1.6 No, the radius is distal to the humerus. No, the esophagus is posterior to the trachea. Yes, the ribs are superficial to the lungs. Yes, the urinary bladder is medial to the ascending colon. No, the sternum is medial to the descending colon.

1.7 The frontal plane divides the heart into anterior and posterior portions.

1.8 The parasagittal plane (not shown in the figure) divides the brain into unequal right and left portions.

1.9 Urinary bladder = P, stomach = A, heart = T, small intestine = A, lungs = T, internal female reproductive organs = P, thymus = T, spleen = A, liver = A.

1.10 The pericardial cavity surrounds the heart, and the pleural cavities surround the lungs.

1.11 The illustrated abdominal cavity organs all belong to the digestive system (liver, gallbladder, stomach, small intestine, and most of the large intestine). Illustrated pelvic cavity organs belong to the urinary system (the urinary bladder) and the digestive system (part of the large intestine).

1.12 The liver is mostly in the epigastric region; the ascending colon is in the right lumbar region; the urinary bladder is in the hypogastric region; most of the small intestine is in the umbilical region. The pain associated with appendicitis would be felt in the right lower quadrant (RLQ).

2 THE CHEMICAL LEVEL OF ORGANIZATION

CHEMISTRY AND HOMEOSTASIS *Maintaining the proper assortment and quantity of thousands of different chemicals in your body, and monitoring the interactions of these chemicals with one another, are two important aspects of homeostasis.*

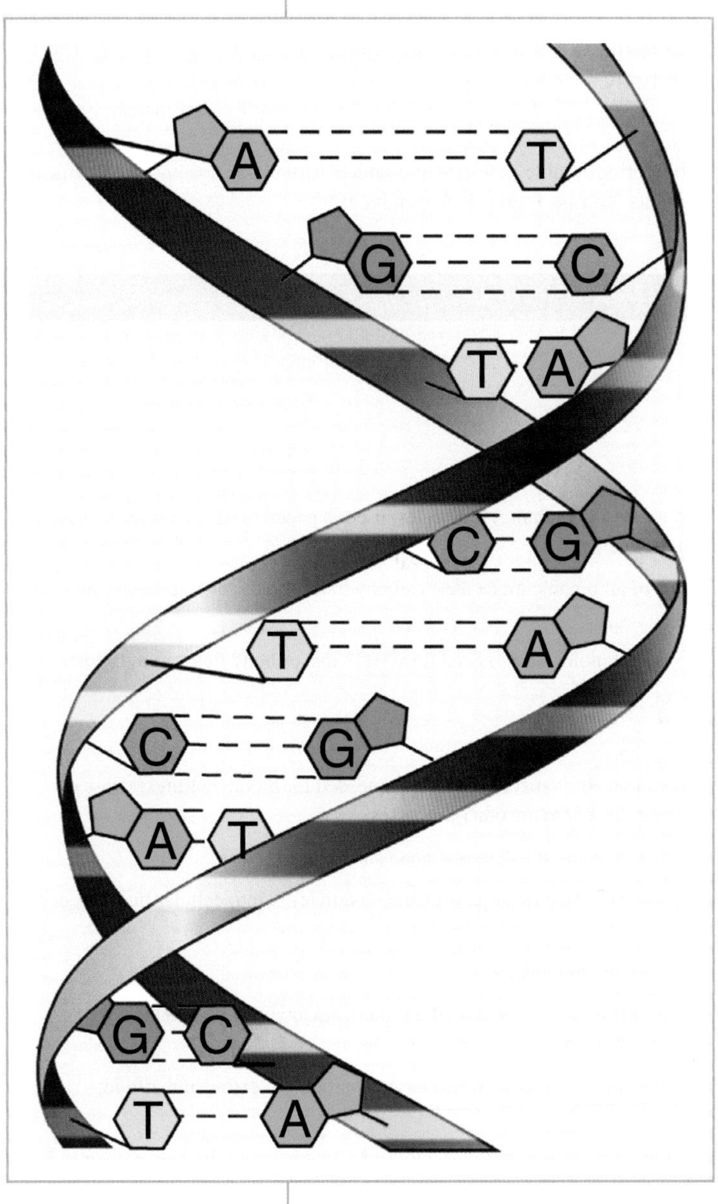

You learned in Chapter 1 that the chemical level of organization, the lowest level of structural organization, consists of atoms and molecules. These letters of the anatomical alphabet ultimately combine to form body organs and systems of astonishing size and complexity. In this chapter, we consider how atoms bond together to form molecules, and how atoms and molecules release or store energy in processes known as chemical reactions. You will also learn about the vital importance of water, which accounts for nearly two-thirds of your body weight, in chemical reactions and the maintenance of homeostasis. Finally, we present several groups of molecules whose unique properties contribute to assembly of your body's structures and to powering the processes that enable you to live.

Chemistry (KEM-is-trē) is the science of the structure and interactions of matter. All living and nonliving things consist of **matter,** which is anything that occupies space and has **mass.** Mass is the amount of matter in any object, which does not change. *Weight,* the force of gravity acting on matter, does change. When objects are farther from Earth, the pull of gravity is weaker; this is why the weight of an astronaut is close to zero in outer space.

? *Did you ever wonder how fatty acids relate to health and disease?*

2.1 HOW MATTER IS ORGANIZED

OBJECTIVES

- Identify the main chemical elements of the human body.
- Describe the structures of atoms, ions, molecules, free radicals, and compounds.

Chemical Elements

Matter exists in three states: solid, liquid, and gas. *Solids,* such as bones and teeth, are compact and have a definite shape and volume. *Liquids,* such as blood plasma, have a definite volume and assume the shape of their container. *Gases,* like oxygen and carbon dioxide, have neither a definite shape nor volume. *All* forms of matter—both living and nonliving—are made up of a limited number of building blocks called **chemical elements.** Each element is a substance that cannot be split into a simpler substance by ordinary chemical means. Scientists now recognize 117 elements. Of these, 92 occur naturally on Earth. The rest have been produced from the natural elements using particle accelerators or nuclear reactors. Each element is designated by a **chemical symbol,** one or two letters of the element's name in English, Latin, or another language. Examples of chemical symbols are H for hydrogen, C for carbon, O for oxygen, N for nitrogen, Ca for calcium, and Na for sodium (*natrium* = sodium).*

Twenty-six different chemical elements normally are present in your body. Just four elements, called the **major elements,** constitute about 96% of the body's mass: oxygen, carbon, hydrogen, and nitrogen. Eight others, the **lesser elements,** contribute about 3.6% to the body's mass: calcium, phosphorus (P), potassium (K), sulfur (S), sodium, chlorine (Cl), magnesium (Mg), and iron (Fe). An additional 14 elements—the **trace elements**—are present in tiny amounts. Together, they account for the remaining about 0.4% of the body's mass. Several trace elements have important functions in the body. For example, iodine is needed to make thyroid hormones. The functions of some trace elements are unknown. Table 2.1 lists the main chemical elements of the human body.

*The periodic table of elements, which lists all of the known chemical elements, can be found in Appendix B.

TABLE 2.1

Main Chemical Elements in the Body

CHEMICAL ELEMENT (SYMBOL)	% OF TOTAL BODY MASS	SIGNIFICANCE
MAJOR ELEMENTS	(about 96)	
Oxygen (O)	65.0	Part of water and many organic (carbon-containing) molecules; used to generate ATP, a molecule used by cells to temporarily store chemical energy.
Carbon (C)	18.5	Forms backbone chains and rings of all organic molecules: carbohydrates, lipids (fats), proteins, and nucleic acids (DNA and RNA).
Hydrogen (H)	9.5	Constituent of water and most organic molecules; ionized form (H^+) makes body fluids more acidic.
Nitrogen (N)	3.2	Component of all proteins and nucleic acids.
LESSER ELEMENTS	(about 3.6)	
Calcium (Ca)	1.5	Contributes to hardness of bones and teeth; ionized form (Ca^{2+}) needed for blood clotting, release of some hormones, contraction of muscle, and many other processes.
Phosphorus (P)	1.0	Component of nucleic acids and ATP; required for normal bone and tooth structure.
Potassium (K)	0.35	Ionized form (K^+) is the most plentiful cation (positively charged particle) in intracellular fluid; needed to generate action potentials.
Sulfur (S)	0.25	Component of some vitamins and many proteins.
Sodium (Na)	0.2	Ionized form (Na^+) is the most plentiful cation in extracellular fluid; essential for maintaining water balance; needed to generate action potentials.
Chlorine (Cl)	0.2	Ionized form (Cl^-) is the most plentiful anion (negatively charged particle) in extracellular fluid; essential for maintaining water balance.
Magnesium (Mg)	0.1	Ionized form (Mg^{2+}) needed for action of many enzymes, molecules that increase the rate of chemical reactions in organisms.
Iron (Fe)	0.005	Ionized forms (Fe^{2+} and Fe^{3+}) are part of hemoglobin (oxygen-carrying protein in red blood cells) and some enzymes.
TRACE ELEMENTS	(about 0.4)	Aluminum (Al), boron (B), chromium (Cr), cobalt (Co), copper (Cu), fluorine (F), iodine (I), manganese (Mn), molybdenum (Mo), selenium (Se), silicon (Si), tin (Sn), vanadium (V), and zinc (Zn).

Structure of Atoms

Each element is made up of **atoms,** the smallest units of matter that retain the properties and characteristics of the element. Atoms are extremely small. Two hundred thousand of the largest atoms would fit on the period at the end of this sentence. Hydrogen atoms, the smallest atoms, have a diameter less than 0.1 nanometer (0.1×10^{-9} m = 0.0000000001 m), and the largest atoms are only five times larger.

Dozens of different **subatomic particles** compose individual atoms. However, only three types of subatomic particles are important for understanding the chemical reactions in the human body: protons, neutrons, and electrons (Figure 2.1). The dense central core of an atom is its **nucleus.** Within the nucleus are positively charged **protons (p^+)** and uncharged (neutral) **neutrons (n^0)**. The tiny, negatively charged **electrons (e^-)** move about in a large space surrounding the nucleus. They do not follow a fixed path or orbit but instead form a negatively charged "cloud" that envelops the nucleus (Figure 2.1a).

Even though their exact positions cannot be predicted, specific groups of electrons are most likely to move about within certain regions around the nucleus. These regions, called **electron shells,** are depicted as simple circles around the nucleus. Because each electron shell can hold a specific number of electrons, the electron shell model best conveys this aspect of atomic structure (Figure 2.1b). The first electron shell (nearest the nucleus) never holds more than 2 electrons. The second shell holds a maximum of 8 electrons, and the third can hold up to 18 electrons. The electron shells fill with electrons in a specific order, beginning with the first shell. For example, notice in Figure 2.2 that sodium (Na), which has 11 electrons total, contains 2 electrons in the first shell, 8 electrons in the second shell, and 1 electron in the third shell. The most massive element present in the human body is iodine, which has a total of 53 electrons: 2 in the first shell, 8 in the second shell, 18 in the third shell, 18 in the fourth shell, and 7 in the fifth shell.

The number of electrons in an atom of an element always equals the number of protons. Because each electron and proton carries one charge, the negatively charged electrons and the positively charged protons balance each other. Thus, each atom is electrically neutral; its total charge is zero.

Figure 2.1 Two representations of the structure of an atom. Electrons move about the nucleus, which contains neutrons and protons. (a) In the electron cloud model of an atom, the shading represents the chance of finding an electron in regions outside the nucleus. (b) In the electron shell model, filled circles represent individual electrons, which are grouped into concentric circles according to the shells they occupy. Both models depict a carbon atom, with six protons, six neutrons, and six electrons.

An atom is the smallest unit of matter that retains the properties and characteristics of its element.

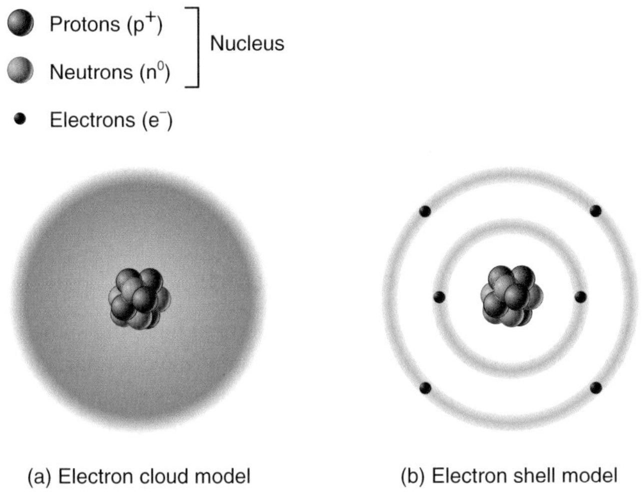
Protons (p^+)
Neutrons (n^0)
Nucleus
Electrons (e^-)
(a) Electron cloud model
(b) Electron shell model

How are the electrons of carbon distributed between the first and second electron shells?

Atomic Number and Mass Number

The *number of protons* in the nucleus of an atom is an atom's **atomic number.** Figure 2.2 shows that atoms of different elements have different atomic numbers because they have different numbers of protons. For example, oxygen has an atomic number of 8 because its nucleus has 8 protons, and sodium has an atomic number of 11 because its nucleus has 11 protons.

The **mass number** of an atom is the sum of its protons and neutrons. Because sodium has 11 protons and 12 neutrons, its mass number is 23 (Figure 2.2). Although all atoms of one element have the same number of protons, they may have different numbers of neutrons and thus different mass numbers. **Isotopes** are atoms of an element that have different numbers of neutrons and therefore different mass numbers. In a sample of oxygen, for example, most atoms have 8 neutrons, and a few have 9 or 10, but all have 8 protons and 8 electrons. Most isotopes are stable, which means that their nuclear structure does not change over time. The stable isotopes of oxygen are designated ^{16}O, ^{17}O, and ^{18}O (or O-16, O-17, and O-18). As you may already have determined, the numbers indicate the mass number of each isotope. As you will discover shortly, the number of electrons of an atom determines its chemical properties. Although the isotopes of an element have different numbers of neutrons, they have identical chemical properties because they have the same number of electrons.

Certain isotopes called **radioactive isotopes** are unstable; their nuclei decay (spontaneously change) into a stable configuration. Examples are H-3, C-14, O-15, and O-19. As they decay, these atoms emit radiation—either subatomic particles or packets of energy—and in the process often transform into a different element. For example, the radioactive isotope of carbon, C-14, decays to N-14. The decay of a radioisotope may be as fast as a fraction of a second or as slow as millions of years. The **half-life** of an isotope is the time required for half of the radioactive atoms in a sample of that isotope to decay into a more stable form. The half-life of C-14, which is used to determine the age of organic samples, is about 5730 years; the half-life of I-131, an important clinical tool, is 8 days.

Figure 2.2 Atomic structures of several stable atoms.

The atoms of different elements have different atomic numbers because they have different numbers of protons.

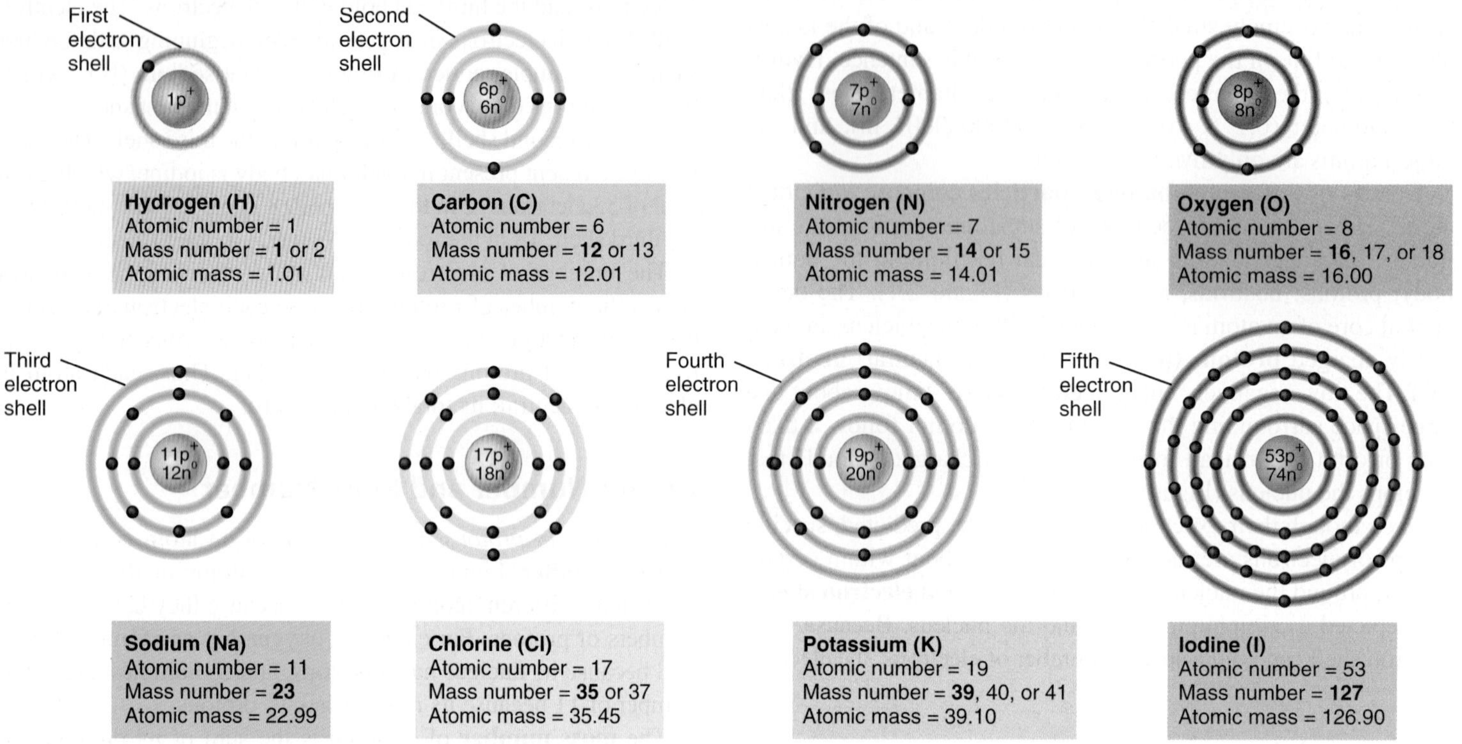

Atomic number = number of protons in an atom
Mass number = number of protons and neutrons in an atom (boldface indicates most common isotope)
Atomic mass = average mass of all stable atoms of a given element in daltons

Which four of these elements are present most abundantly in living organisms?

CLINICAL CONNECTION | *Harmful and Beneficial Effects of Radiation*

Radioactive isotopes may have either harmful or helpful effects. Their radiations can break apart molecules, posing a serious threat to the human body by producing tissue damage and/or causing various types of cancer. Although the decay of naturally occurring radioactive isotopes typically releases just a small amount of radiation into the environment, localized accumulations can occur. Radon-222, a colorless and odorless gas that is a naturally occurring radioactive breakdown product of uranium, may seep out of the soil and accumulate in buildings. It is not only associated with many cases of lung cancer in smokers but has also been implicated in many cases of lung cancer in nonsmokers. Beneficial effects of certain radioisotopes include their use in medical imaging procedures to diagnose and treat certain disorders. Some radioisotopes can be used as **tracers** to follow the movement of certain substances through the body. Thallium-201 is used to monitor blood flow through the heart during an exercise stress test. Iodine-131 is used to detect cancer of the thyroid gland and to assess its size and activity, and may also be used to destroy part of an overactive thyroid gland. Cesium-137 is used to treat advanced cervical cancer, and iridium-192 is used to treat prostate cancer. •

Atomic Mass

The standard unit for measuring the mass of atoms and their subatomic particles is a **dalton,** also known as an *atomic mass unit (amu).* A neutron has a mass of 1.008 daltons, and a proton has a mass of 1.007 daltons. The mass of an electron, at 0.0005 dalton, is almost 2000 times smaller than the mass of a neutron or proton. The **atomic mass** (also called the *atomic weight*) of an element is the average mass of all its naturally occurring isotopes. Typically, the atomic mass of an element is close to the mass number of its most abundant isotope.

Ions, Molecules, and Compounds

As we discussed, atoms of the same element have the same number of protons. The atoms of each element have a characteristic way of losing, gaining, or sharing their electrons when interacting with other atoms to achieve stability. The way that electrons behave enables atoms in the body to exist in electrically charged forms called ions, or to join with each other into complex combinations called molecules. If an atom either *gives up* or *gains* electrons, it becomes an ion. An **ion** is an atom that has a positive or negative charge because it has unequal numbers of protons and electrons. *Ionization* is the process of giving up or gaining electrons. An ion of an atom is symbolized by writing its chemical symbol followed by the number of its positive (+) or negative (–) charges. Thus, Ca^{2+} stands for a calcium ion that has two positive charges because it has lost two electrons.

When two or more atoms *share* electrons, the resulting combination is called a **molecule** (MOL-e-kūl). A *molecular formula* indicates the elements and the number of atoms of each element that make up a molecule. A molecule may consist of two atoms of the same kind, such as an oxygen molecule (Figure 2.3a). The molecular formula for a molecule of oxygen is O_2. The subscript 2 indicates that the molecule contains two atoms of oxygen. Two or more different kinds of atoms may also form a molecule, as in a water molecule (H_2O). In H_2O one atom of oxygen shares electrons with two atoms of hydrogen.

A **compound** is a substance that contains atoms of two or more different elements. Most of the atoms in the body are joined into compounds. Water (H_2O) and sodium chloride (NaCl), common table salt, are compounds. However, a molecule of oxygen (O_2) is not a compound because it consists of atoms of only one element.

A **free radical** is an atom or group of atoms with an unpaired electron in the outermost shell. A common example is superoxide, which is formed by the addition of an electron to an oxygen molecule (Figure 2.3b). Having an unpaired electron makes a free radical unstable, highly reactive, and destructive to nearby molecules. Free radicals become stable by either giving up their unpaired electron to, or taking on an electron from, another molecule. In so doing, free radicals may break apart important body molecules.

CLINICAL CONNECTION | *Free Radicals and Antioxidants*

There are several sources of free radicals, including exposure to ultraviolet radiation in sunlight, exposure to x-rays, and some reactions that occur during normal metabolic processes. Certain harmful substances, such as carbon tetrachloride (a solvent used in dry cleaning), also give rise to free radicals when they participate in metabolic reactions in the body. Among the many disorders, diseases, and conditions linked to oxygen-derived free radicals are cancer, atherosclerosis, Alzheimer disease, emphysema, diabetes mellitus, cataracts, macular degeneration, rheumatoid arthritis, and deterioration associated with aging. Consuming more **antioxidants**—substances that inactivate oxygen-derived free radicals—is thought to slow the pace of damage caused by free radicals. Important dietary antioxidants include selenium, zinc, beta-carotene, and vitamins C and E. Red, blue, or purple fruits and vegetables contain high levels of antioxidants. •

✔ **CHECKPOINT**

1. List the names and chemical symbols of the 12 most abundant chemical elements in the human body.
2. What are the atomic number, mass number, and atomic mass of carbon? How are they related?
3. Define isotopes and free radicals.

Figure 2.3 Atomic structures of an oxygen molecule and a superoxide free radical.

A free radical has an unpaired electron in its outermost electron shell.

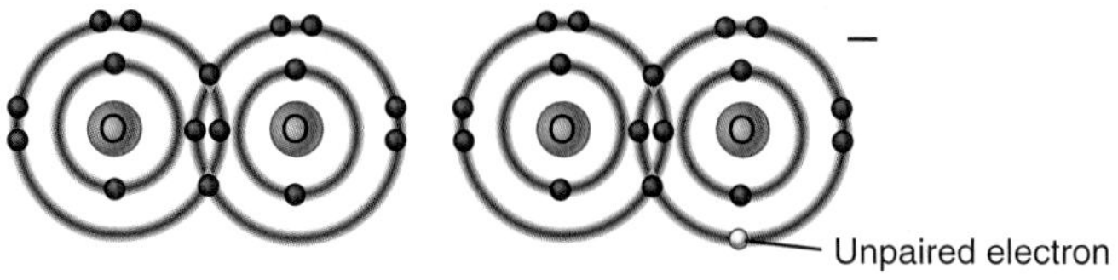

(a) Oxygen molecule (O_2) (b) Superoxide free radical (O_2^-)

What substances in the body can inactivate oxygen-derived free radicals?

2.2 CHEMICAL BONDS

OBJECTIVES

- Describe how valence electrons form chemical bonds.
- Distinguish among ionic, covalent, and hydrogen bonds.

The forces that hold together the atoms of a molecule or a compound are **chemical bonds.** The likelihood that an atom will form a chemical bond with another atom depends on the number of electrons in its outermost shell, also called the **valence shell.** An atom with a valence shell holding eight electrons is *chemically stable,* which means it is unlikely to form chemical bonds with other atoms. Neon, for example, has eight electrons in its valence shell, and for this reason it does not bond easily with other atoms. The valence shell of hydrogen and helium is the first electron shell, which holds a maximum of two electrons. Because helium has two valence electrons, it too is stable and seldom bonds with other atoms. Hydrogen, on the other hand, has only one valence electron (see Figure 2.2), so it binds readily with other atoms.

The atoms of most biologically important elements do not have eight electrons in their valence shells. Under the right conditions, two or more atoms can interact in ways that produce a chemically stable arrangement of eight valence electrons for each atom. This chemical principle, called the **octet rule** (*octet* = set of eight), helps explain why atoms interact in predictable ways. One atom is more likely to interact with another atom if doing so will leave both with eight valence electrons. For this to happen, an atom either empties its partially filled valence shell, fills it with donated electrons, or shares electrons with other atoms. The way that valence electrons are distributed determines what kind of chemical bond results. We will consider three types of chemical bonds: ionic bonds, covalent bonds, and hydrogen bonds.

Ionic Bonds

As you have already learned, when atoms lose or gain one or more valence electrons, ions are formed. Positively and negatively charged ions are attracted to one another—opposites attract. The force of attraction that holds together ions with opposite charges is an **ionic bond.** Consider sodium and chlorine atoms, the components of common table salt. Sodium has one valence electron (Figure 2.4a). If sodium *loses* this electron, it is left with the eight electrons in its second shell, which becomes the valence shell. As a result, however, the total number of protons (11) exceeds the number of electrons (10). Thus, the sodium atom has become a **cation** (KAT-ī-on), or positively charged ion. A sodium ion has a charge of 1+ and is written Na^+. By contrast, chlorine has seven valence electrons (Figure 2.4b). If chlorine *gains* an electron from a neighboring atom, it will have a complete octet in its third electron shell. After gaining an electron, the total number of electrons (18) exceeds the number of protons (17), and the chlorine atom has become an **anion** (AN-ī-on), a negatively charged ion. The ionic form of chlorine is called a *chloride* ion. It has a charge of 1− and is written Cl^-. When an atom of sodium donates its sole valence electron to an atom of chlorine, the resulting positive and negative charges pull both ions tightly together, forming an ionic bond (Figure 2.4c). The resulting compound is sodium chloride, written NaCl.

In general, ionic compounds exist as solids, with an orderly, repeating arrangement of the ions, as in a crystal of NaCl (Figure 2.4d). A crystal of NaCl may be large or small—the total number of ions can vary—but the ratio of Na^+ to Cl^- is always 1:1. In the body, ionic bonds are found mainly in teeth and bones, where they give great strength to these important structural tissues. An ionic compound that breaks apart into positive and negative ions in solution is called an **electrolyte** (e-LEK-trō-līt). Most ions in the body are dissolved in body fluids as electrolytes, so named because their solutions can conduct an electric current. (In Chapter 27 we will discuss the chemistry and importance of electrolytes.) Table 2.2 lists the names and symbols of common ions in the body.

TABLE 2.2

Common Ions in the Body

CATIONS		ANIONS	
NAME	SYMBOL	NAME	SYMBOL
Hydrogen ion	H^+	Fluoride ion	F^-
Sodium ion	Na^+	Chloride ion	Cl^-
Potassium ion	K^+	Iodide ion	I^-
Ammonium ion	NH_4^+	Hydroxide ion	OH^-
Magnesium ion	Mg^{2+}	Bicarbonate ion	HCO_3^-
Calcium ion	Ca^{2+}	Oxide ion	O^{2-}
Iron(II) ion	Fe^{2+}	Sulfate ion	SO_4^{2-}
Iron(III) ion	Fe^{3+}	Phosphate ion	PO_4^{3-}

Figure 2.4 Ions and ionic bond formation. (a) A sodium atom can have a complete octet of electrons in its outermost shell by losing one electron. (b) A chlorine atom can have a complete octet by gaining one electron. (c) An ionic bond may form between oppositely charged ions. (d) In a crystal of NaCl, each Na^+ is surrounded by six Cl^-. In (a), (b), and (c), the electron that is lost or accepted is colored red.

An ionic bond is the force of attraction that holds together oppositely charged ions.

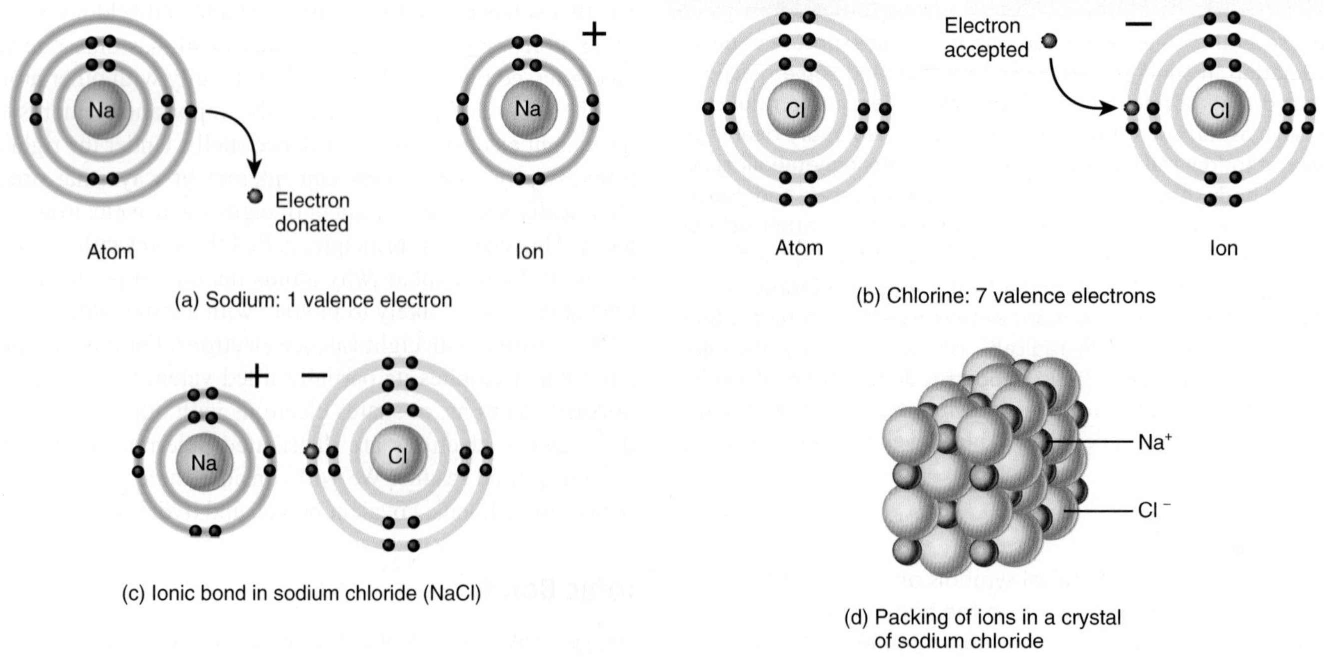

What are cations and anions?

Covalent Bonds

When a **covalent bond** forms, two or more atoms *share* electrons rather than gaining or losing them. Atoms form a covalently bonded molecule by sharing one, two, or three pairs of valence electrons. The larger the number of electron pairs shared between two atoms, the stronger the covalent bond. Covalent bonds may form between atoms of the same element or between atoms of different elements. They are the most common chemical bonds in the body, and the compounds that result from them form most of the body's structures.

A **single covalent bond** results when two atoms share one electron pair. For example, a molecule of hydrogen forms when two hydrogen atoms share their single valence electrons (Figure 2.5a), which allows both atoms to have a full valence shell at

Figure 2.5 Covalent bond formation. The red electrons are shared equally in (a)–(d) and unequally in (e). In writing the structural formula of a covalently bonded molecule, each straight line between the chemical symbols for two atoms denotes a pair of shared electrons. In molecular formulas, the number of atoms in each molecule is noted by subscripts.

6 In a covalent bond, two atoms share one, two, or three pairs of valence electrons.

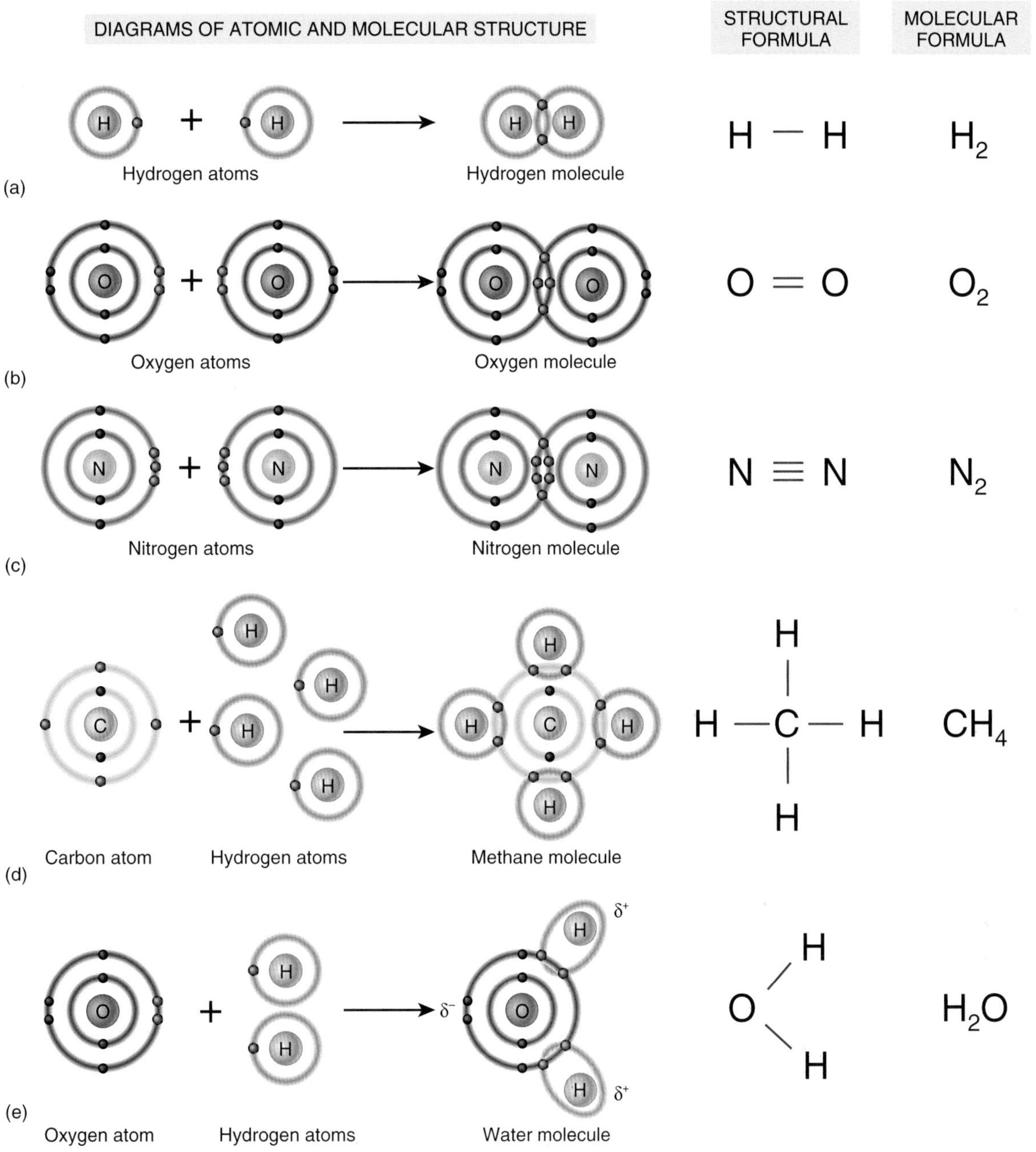

? What is the principal difference between an ionic bond and a covalent bond?

least part of the time. A **double covalent bond** results when two atoms share two pairs of electrons, as happens in an oxygen molecule (Figure 2.5b). A **triple covalent bond** occurs when two atoms share three pairs of electrons, as in a molecule of nitrogen (Figure 2.5c). Notice in the *structural formulas* for covalently bonded molecules in Figure 2.5 that the number of lines between the chemical symbols for two atoms indicates whether the bond is a single (—), double (═), or triple (≡) covalent bond.

The same principles of covalent bonding that apply to atoms of the same element also apply to covalent bonds between atoms of different elements. The gas methane (CH_4) contains covalent bonds formed between the atoms of two different elements, one carbon and four hydrogens (Figure 2.5d). The valence shell of the carbon atom can hold eight electrons but has only four of its own. The single electron shell of a hydrogen atom can hold two electrons, but each hydrogen atom has only one of its own. A methane molecule contains four separate single covalent bonds. Each hydrogen atom shares one pair of electrons with the carbon atom.

In some covalent bonds, two atoms share the electrons equally—one atom does not attract the shared electrons more strongly than the other atom. This type of bond is a **nonpolar covalent bond.** The bonds between two identical atoms are always nonpolar covalent bonds (Figure 2.5a–c). The bonds between carbon and hydrogen atoms are also nonpolar, such as the four C—H bonds in a methane molecule (Figure 2.5d).

In a **polar covalent bond,** the sharing of electrons between two atoms is unequal—the nucleus of one atom attracts the shared electrons more strongly than the nucleus of the other atom. When polar covalent bonds form, the resulting molecule has a partial negative charge near the atom that attracts electrons more strongly. This atom has greater **electronegativity,** the power to attract electrons to itself. At least one other atom in the molecule then will have a partial positive charge. The partial charges are indicated by a lowercase Greek delta with a minus or plus sign: δ^- or δ^+. A very important example of a polar covalent bond in living systems is the bond between oxygen and hydrogen in a molecule of water (Figure 2.5e); in this molecule, the nucleus of the oxygen atom attracts the electrons more strongly than do the nuclei of the hydrogen atoms, so the oxygen atom is said to have greater electronegativity. Later in the chapter, we will see how polar covalent bonds allow water to dissolve many molecules that are important to life. Bonds between nitrogen and hydrogen and those between oxygen and carbon are also polar bonds.

Hydrogen Bonds

The polar covalent bonds that form between hydrogen atoms and other atoms can give rise to a third type of chemical bond, a hydrogen bond (Figure 2.6). A **hydrogen bond** forms when a hydrogen atom with a partial positive charge (δ^+) attracts the partial negative charge (δ^-) of neighboring electronegative atoms, most often larger oxygen or nitrogen atoms. Thus, hydrogen bonds result from attraction of oppositely charged parts of molecules rather than from sharing of electrons as in covalent bonds, or the loss or gain of electrons as in ionic bonds. Hydrogen bonds are weak compared to ionic and covalent bonds. Thus, they cannot bind atoms into molecules. However, hydrogen bonds do establish important links between molecules or between different parts of a large molecule, such as a protein or nucleic acid (both discussed later in this chapter).

The hydrogen bonds that link neighboring water molecules give water considerable *cohesion,* the tendency of like particles to stay together. The cohesion of water molecules creates a very high **surface tension,** a measure of the difficulty of stretching or breaking the surface of a liquid. At the boundary between water and air, water's surface tension is very high because the water molecules are much more attracted to one another than they are attracted to molecules in the air. This is readily seen when a spider walks on water or a leaf floats on water. The influence of water's surface tension on the body can be seen in the way it increases the work required for breathing. A thin film of watery fluid coats the air sacs of the lungs. So, each inhalation must have enough force to overcome the opposing effect of surface tension as the air sacs stretch and enlarge when taking in air.

Even though single hydrogen bonds are weak, very large molecules may contain thousands of these bonds. Acting collectively, hydrogen bonds provide considerable strength and stability and help determine the three-dimensional shape of large molecules. As you will see later in this chapter, a large molecule's shape determines how it functions.

Figure 2.6 Hydrogen bonding among water molecules. Each water molecule forms hydrogen bonds (indicated by dotted lines) with three to four neighboring water molecules.

Hydrogen bonds occur because hydrogen atoms in one water molecule are attracted to the partial negative charge of the oxygen atom in another water molecule.

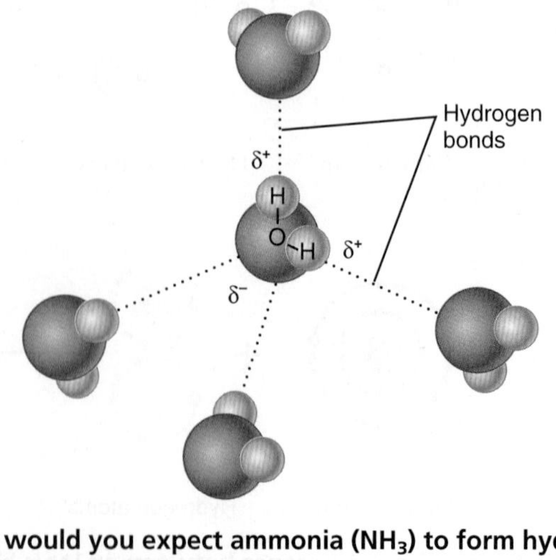

Why would you expect ammonia (NH_3) to form hydrogen bonds with water molecules?

✔CHECKPOINT

4. Which electron shell is the valence shell of an atom, and what is its significance?
5. Compare the properties of ionic, covalent, and hydrogen bonds.
6. What information is conveyed when you write the molecular or structural formula for a molecule?

2.3 CHEMICAL REACTIONS

OBJECTIVES

- Define a chemical reaction.
- Describe the various forms of energy.
- Compare exergonic and endergonic chemical reactions.
- Describe the role of activation energy and catalysts in chemical reactions.
- Describe synthesis, decomposition, exchange, and reversible reactions.

A **chemical reaction** occurs when new bonds form or old bonds break between atoms. Chemical reactions are the foundation of all life processes, and as we have seen, the interactions of valence electrons are the basis of all chemical reactions. Consider how hydrogen and oxygen molecules react to form water molecules (Figure 2.7). The starting substances—two H_2 and one O_2—are known as the **reactants.** The ending substances—two molecules of H_2O—are the **products.** The arrow in the figure indicates the direction in which the reaction proceeds. In a chemical reaction, the total mass of the reactants equals the total mass of the products. Thus, the number of atoms of each element is the same before and after the reaction. However, because the atoms are rearranged, the reactants and products have different chemical properties. Through thousands of different chemical reactions, body structures are built and body functions are carried out. The term **metabolism** refers to all the chemical reactions occurring in the body.

Figure 2.7 The chemical reaction between two hydrogen molecules (H_2) and one oxygen molecule (O_2) to form two molecules of water (H_2O). Note that the reaction occurs by breaking old bonds and making new bonds.

The number of atoms of each element is the same before and after a chemical reaction.

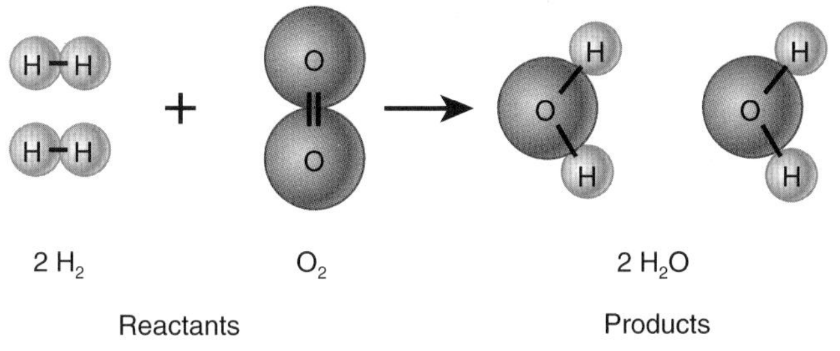

H–H
H–H
O
O
H
O
H
H
O
H
2 H2
O2
2 H2O
Reactants
Products

Why does this reaction require two molecules of H_2?

Forms of Energy and Chemical Reactions

Each chemical reaction involves energy changes. **Energy** (*en-* = in; *-ergy* = work) is the capacity to do work. Two principal forms of energy are **potential energy,** energy stored by matter due to its position, and **kinetic energy,** the energy associated with matter in motion. For example, the energy stored in water behind a dam or in a person poised to jump down some steps is potential energy. When the gates of the dam are opened or the person jumps, potential energy is converted into kinetic energy. **Chemical energy** is a form of potential energy that is stored in the bonds of compounds and molecules. The total amount of energy present at the beginning and end of a chemical reaction is the same. Although energy can be neither created nor destroyed, it may be converted from one form to another. This principle is known as the **law of conservation of energy.** For example, some of the chemical energy in the foods we eat is eventually converted into various forms of kinetic energy, such as mechanical energy used to walk and talk. Conversion of energy from one form to another generally releases heat, some of which is used to maintain normal body temperature.

Energy Transfer in Chemical Reactions

Chemical bonds represent stored chemical energy, and chemical reactions occur when new bonds are formed or old bonds are broken between atoms. The *overall reaction* may either release energy or absorb energy. **Exergonic reactions** (*ex-* = out) release more energy than they absorb. By contrast, **endergonic reactions** (*end-* = within) absorb more energy than they release.

A key feature of the body's metabolism is the coupling of exergonic reactions and endergonic reactions. Energy released from an exergonic reaction often is used to drive an endergonic one. In general, exergonic reactions occur as nutrients, such as glucose, are broken down. Some of the energy released may be trapped in the covalent bonds of adenosine triphosphate (ATP), which we describe more fully later in this chapter. If a molecule of glucose is completely broken down, the chemical energy in its bonds can be used to produce as many as 38 molecules of ATP. The energy transferred to the ATP molecules is then used to drive endergonic reactions needed to build body structures, such as muscles and bones. The energy in ATP is also used to do the mechanical work involved in the contraction of muscle or the movement of substances into or out of cells.

Activation Energy

Because particles of matter such as atoms, ions, and molecules have kinetic energy, they are continuously moving and colliding with one another. A sufficiently forceful collision can disrupt the movement of valence electrons, causing an existing chemical bond to break or a new one to form. The collision energy needed to break the chemical bonds of the reactants is called

activation energy of the reaction (Figure 2.8). This initial energy "investment" is needed to start a reaction. The reactants must absorb enough energy for their chemical bonds to become unstable and their valence electrons to form new combinations. Then, as new bonds form, energy is released to the surroundings.

Both the concentration of particles and the temperature influence the chance that a collision will occur and cause a chemical reaction.

- ***Concentration.*** The more particles of matter present in a confined space, the greater the chance that they will collide (think of people crowding into a subway car at rush hour). The concentration of particles increases when more are added to a given space or when the pressure on the space increases, which forces the particles closer together so that they collide more often.
- ***Temperature.*** As temperature rises, particles of matter move about more rapidly. Thus, the higher the temperature of matter, the more forcefully particles will collide, and the greater the chance that a collision will produce a reaction.

Catalysts

As we have seen, chemical reactions occur when chemical bonds break or form after atoms, ions, or molecules collide with one another. Body temperature and the concentrations of molecules in body fluids, however, are far too low for most chemical reactions to occur rapidly enough to maintain life. Raising the temperature and the number of reacting particles of matter in the body could increase the frequency of collisions and thus increase the rate of chemical reactions, but doing so could also damage or kill the body's cells.

Substances called catalysts solve this problem. **Catalysts** are chemical compounds that speed up chemical reactions by lowering the activation energy needed for a reaction to occur (Figure 2.9). The most important catalysts in the body are enzymes, which we will discuss later in this chapter.

A catalyst does not alter the difference in potential energy between the reactants and the products. Rather, it lowers the amount of energy needed to start the reaction.

For chemical reactions to occur, some particles of matter—especially large molecules—not only must collide with sufficient force, but they must hit one another at precise spots. A catalyst helps to properly orient the colliding particles. Thus, they interact at the spots that make the reaction happen. Although the action of a catalyst helps to speed up a chemical reaction, the catalyst itself is unchanged at the end of the reaction. A single catalyst molecule can assist one chemical reaction after another.

Types of Chemical Reactions

After a chemical reaction takes place, the atoms of the reactants are rearranged to yield products with new chemical properties.

Figure 2.8 Activation energy.

Activation energy is the energy needed to break chemical bonds in the reactant molecules so a reaction can start.

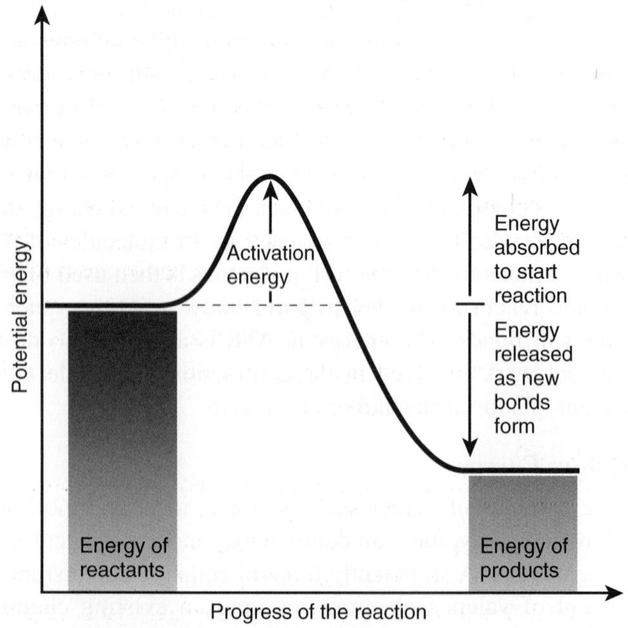

Why is the reaction illustrated here exergonic?

Figure 2.9 Comparison of energy needed for a chemical reaction to proceed with a catalyst (blue curve) and without a catalyst (red curve).

Catalysts speed up chemical reactions by lowering the activation energy.

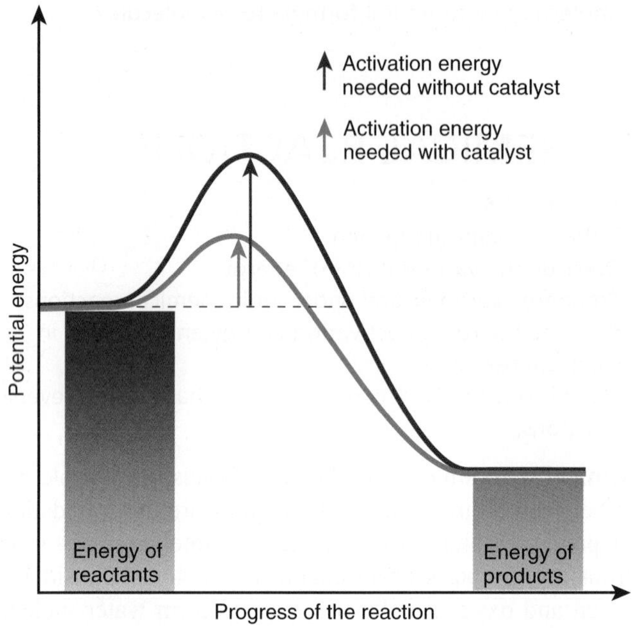

Does a catalyst change the potential energies of the products and reactants?

In this section we will look at the types of chemical reactions common to all living cells. Once you have learned them, you will be able to understand the chemical reactions so important to the operation of the human body that are discussed throughout the book.

Synthesis Reactions—Anabolism

When two or more atoms, ions, or molecules combine to form new and larger molecules, the processes are called **synthesis reactions.** The word *synthesis* means "to put together." A synthesis reaction can be expressed as follows:

$$\underset{\text{Atom, ion, or molecule A}}{A} + \underset{\text{Atom, ion, or molecule B}}{B} \xrightarrow{\text{Combine to form}} \underset{\text{New molecule AB}}{AB}$$

One example of a synthesis reaction is the reaction between two hydrogen molecules and one oxygen molecule to form two molecules of water (see Figure 2.7). Another example of a synthesis reaction is the formation of ammonia from nitrogen and hydrogen:

$$\underset{\text{One nitrogen molecule}}{N_2} + \underset{\text{Three hydrogen molecules}}{3H_2} \xrightarrow{\text{Combine to form}} \underset{\text{Two ammonia molecules}}{2NH_3}$$

All the synthesis reactions that occur in your body are collectively referred to as **anabolism** (a-NAB-ō-lizm). Overall, anabolic reactions are usually endergonic because they absorb more energy than they release. Combining simple molecules like amino acids (discussed shortly) to form large molecules such as proteins is an example of anabolism.

Decomposition Reactions—Catabolism

Decomposition reactions split up large molecules into smaller atoms, ions, or molecules. A decomposition reaction is expressed as follows:

$$\underset{\text{Molecule AB}}{AB} \xrightarrow{\text{Breaks down into}} \underset{\text{Atom, ion, or molecule A}}{A} + \underset{\text{Atom, ion, or molecule B}}{B}$$

The decomposition reactions that occur in your body are collectively referred to as **catabolism** (ka-TAB-ō-lizm). Overall, catabolic reactions are usually exergonic because they release more energy than they absorb. For instance, the series of reactions that break down glucose to pyruvic acid, with the net production of two molecules of ATP, are important catabolic reactions in the body. These reactions will be discussed in Chapter 25.

Exchange Reactions

Many reactions in the body are **exchange reactions;** they consist of both synthesis and decomposition reactions. One type of exchange reaction works like this:

$$AB + CD \longrightarrow AD + BC$$

The bonds between A and B and between C and D break (decomposition), and new bonds then form (synthesis) between A and D and between B and C. An example of an exchange reaction is

$$\underset{\text{Hydrochloric acid}}{HCl} + \underset{\text{Sodium bicarbonate}}{NaHCO_3} \longrightarrow \underset{\text{Carbonic acid}}{H_2CO_3} + \underset{\text{Sodium chloride}}{NaCl}$$

Notice that the ions in both compounds have "switched partners": The hydrogen ion (H^+) from HCl has combined with the bicarbonate ion (HCO_3^-) from $NaHCO_3$, and the sodium ion (Na^+) from $NaHCO_3$ has combined with the chloride ion (Cl^-) from HCl.

Reversible Reactions

Some chemical reactions proceed in only one direction, from reactants to products, as previously indicated by the single arrows. Other chemical reactions may be reversible. In a **reversible reaction,** the products can revert to the original reactants. A reversible reaction is indicated by two half-arrows pointing in opposite directions:

$$AB \underset{\text{Combines to form}}{\overset{\text{Breaks down into}}{\rightleftharpoons}} A + B$$

Some reactions are reversible only under special conditions:

$$AB \underset{\text{Heat}}{\overset{\text{Water}}{\rightleftharpoons}} A + B$$

In that case, whatever is written above or below the arrows indicates the condition needed for the reaction to occur. In these reactions, AB breaks down into A and B only when water is added, and A and B react to produce AB only when heat is applied. Many reversible reactions in the body require catalysts called enzymes. Often, different enzymes guide the reactions in opposite directions.

Oxidation–Reduction Reactions

You will learn in Chapter 25 that chemical reactions called oxidation–reduction reactions are essential to life, since they are the reactions that break down food molecules to produce energy. These reactions are concerned with the transfer of electrons between atoms and molecules. **Oxidation** refers to the loss of electrons, and in the process the oxidized substance releases energy. **Reduction** refers to the gain of electrons, and in the process the reduced substance gains energy. **Oxidation–reduction reactions** are always parallel; when one substance is oxidized, another is reduced at the same time. When a food molecule, such as glucose, is oxidized, the energy produced is used by a cell to carry out its various functions.

✔CHECKPOINT

7. What is the relationship between reactants and products in a chemical reaction?
8. Compare potential energy and kinetic energy.
9. How do catalysts affect activation energy?
10. How are anabolism and catabolism related to synthesis and decomposition reactions, respectively?
11. Why are oxidation–reduction reactions important?

2.4 INORGANIC COMPOUNDS AND SOLUTIONS

OBJECTIVES

- Describe the properties of water and those of inorganic acids, bases, and salts.
- Distinguish among solutions, colloids, and suspensions.
- Define pH and explain the role of buffer systems in homeostasis.

Most of the chemicals in your body exist in the form of compounds. Biologists and chemists divide these compounds into two principal classes: inorganic compounds and organic compounds. **Inorganic compounds** usually lack carbon and are structurally simple. Their molecules also have only a few atoms and cannot be used by cells to perform complicated biological functions. They include water and many salts, acids, and bases. Inorganic compounds may have either ionic or covalent bonds. Water makes up 55–60% of a lean adult's total body mass; all other inorganic compounds combined add 1–2%. Examples of inorganic compounds that contain carbon are carbon dioxide (CO_2), bicarbonate ion (HCO_3^-), and carbonic acid (H_2CO_3). **Organic compounds** always contain carbon, usually contain hydrogen, and always have covalent bonds. Most are large molecules and many are made up of long chains of carbon atoms. Organic compounds make up the remaining 38–43% of the human body.

Water

Water is the most important and abundant inorganic compound in all living systems. Although you might be able to survive for weeks without food, without water you would die in a matter of days. Nearly all the body's chemical reactions occur in a watery medium. Water has many properties that make it such an indispensable compound for life. We have already mentioned the most important property of water, its polarity—the uneven sharing of valence electrons that confers a partial negative charge near the one oxygen atom and two partial positive charges near the two hydrogen atoms in a water molecule (see Figure 2.5e). This property alone makes water an excellent solvent for other ionic or polar substances, gives water molecules cohesion (the tendency to stick together), and allows water to resist temperature changes.

Water as a Solvent

In medieval times people searched in vain for a "universal solvent," a substance that would dissolve all other materials. They found nothing that worked as well as water. Although it is the most versatile solvent known, water is not the universal solvent sought by medieval alchemists. If it were, no container could hold it because it would dissolve all potential containers! What exactly is a solvent? In a **solution,** a substance called the **solvent** dissolves another substance called the **solute.** Usually there is more solvent than solute in a solution. For example, your sweat is a dilute solution of water (the solvent) plus small amounts of salts (the solutes).

The versatility of water as a solvent for ionized or polar substances is due to its polar covalent bonds and its bent shape, which allows each water molecule to interact with several neighboring ions or molecules. Solutes that are charged or contain polar covalent bonds are **hydrophilic** (*hydro-* = water; *-philic* = loving), which means they dissolve easily in water. Common examples of hydrophilic solutes are sugar and salt. Molecules that contain mainly nonpolar covalent bonds, by contrast, are **hydrophobic** (*-phobic* = fearing). They are not very water soluble. Examples of hydrophobic compounds include animal fats and vegetable oils.

To understand the dissolving power of water, consider what happens when a crystal of a salt such as sodium chloride (NaCl) is placed in water (Figure 2.10). The electronegative oxygen

Figure 2.10 How polar water molecules dissolve salts and polar substances. When a crystal of sodium chloride is placed in water, the slightly negative oxygen end (red) of water molecules is attracted to the positive sodium ions (Na^+), and the slightly positive hydrogen portions (gray) of water molecules are attracted to the negative chloride ions (Cl^-). In addition to dissolving sodium chloride, water also causes it to dissociate, or separate into charged particles, which is discussed shortly.

Water is a versatile solvent because its polar covalent bonds, in which electrons are shared unequally, create positive and negative regions.

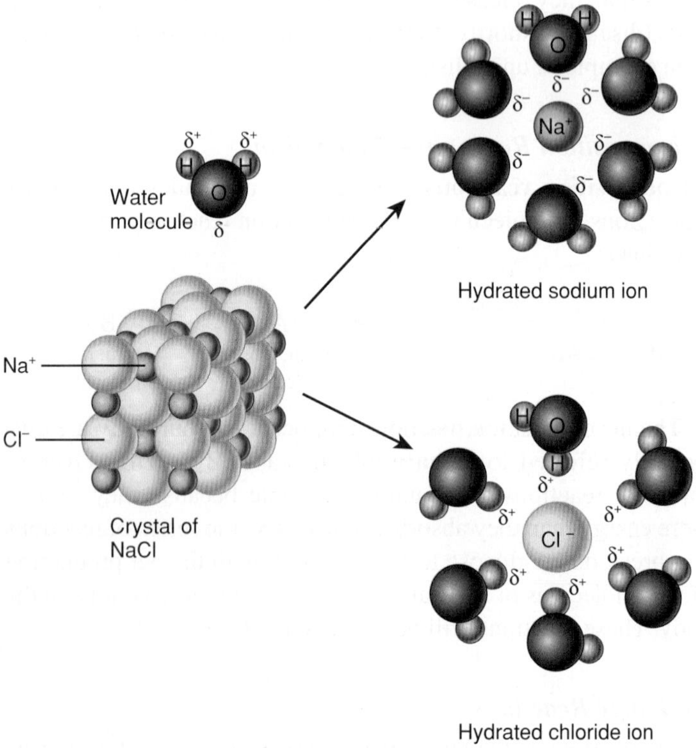

Table sugar (sucrose) easily dissolves in water but is not an electrolyte. Is it likely that all the covalent bonds between atoms in table sugar are nonpolar bonds? Why or why not?

atom in water molecules attracts the sodium ions (Na^+), and the electropositive hydrogen atoms in water molecules attract the chloride ions (Cl^-). Soon, water molecules surround and separate Na^+ and Cl^- ions from each other at the surface of the crystal, breaking the ionic bonds that held NaCl together. The water molecules surrounding the ions also lessen the chance that Na^+ and Cl^- will come together and re-form an ionic bond.

The ability of water to form solutions is essential to health and survival. Because water can dissolve so many different substances, it is an ideal medium for metabolic reactions. Water enables dissolved reactants to collide and form products. Water also dissolves waste products, which allows them to be flushed out of the body in the urine.

Water in Chemical Reactions

Water serves as the medium for most chemical reactions in the body and participates as a reactant or product in certain reactions. During digestion, for example, decomposition reactions break down large nutrient molecules into smaller molecules by the addition of water molecules. This type of reaction is called **hydrolysis** (hī-DROL-i-sis; *-lysis* = to loosen or break apart). Hydrolysis reactions enable dietary nutrients to be absorbed into the body. By contrast, when two smaller molecules join to form a larger molecule in a **dehydration synthesis reaction** (*de-* = from, down, or out; *hydra-* = water), a water molecule is one of the products formed. As you will see later in the chapter, such reactions occur during synthesis of proteins and other large molecules (for example, see Figure 2.21).

Thermal Properties of Water

In comparison to most substances, water can absorb or release a relatively large amount of heat with only a modest change in its own temperature. For this reason, water is said to have a high *heat capacity*. The reason for this property is the large number of hydrogen bonds in water. As water absorbs heat energy, some of the energy is used to break hydrogen bonds. Less energy is then left over to increase the motion of water molecules, which would increase the water's temperature. The high heat capacity of water is the reason it is used in automobile radiators; it cools the engine by absorbing heat without its own temperature rising to an unacceptably high level. The large amount of water in the body has a similar effect: It lessens the impact of environmental temperature changes, helping to maintain the homeostasis of body temperature.

Water also requires a large amount of heat to change from a liquid to a gas. Its *heat of vaporization* is high. As water evaporates from the surface of the skin, it removes a large quantity of heat, providing an important cooling mechanism.

Water as a Lubricant

Water is a major component of mucus and other lubricating fluids throughout the body. Lubrication is especially necessary in the chest (pleural and pericardial cavities) and abdomen (peritoneal cavity), where internal organs touch and slide over one another. It is also needed at joints, where bones, ligaments, and tendons rub against one another. Inside the gastrointestinal tract, mucus and other watery secretions moisten foods, which aids their smooth passage through the digestive system.

Solutions, Colloids, and Suspensions

A **mixture** is a combination of elements or compounds that are physically blended together but not bound by chemical bonds. For example, the air you are breathing is a mixture of gases that includes nitrogen, oxygen, argon, and carbon dioxide. Three common liquid mixtures are solutions, colloids, and suspensions.

Once mixed together, solutes in a solution remain evenly dispersed among the solvent molecules. Because the solute particles in a solution are very small, a solution looks clear and transparent.

A **colloid** differs from a solution mainly because of the size of its particles. The solute particles in a colloid are large enough to scatter light, just as water droplets in fog scatter light from a car's headlight beams. For this reason, colloids usually appear translucent or opaque. Milk is an example of a liquid that is both a colloid and a solution: The large milk proteins make it a colloid, whereas calcium salts, milk sugar (lactose), ions, and other small particles are in solution.

The solutes in both solutions and colloids do not settle out and accumulate on the bottom of the container. In a **suspension,** by contrast, the suspended material may mix with the liquid or suspending medium for some time, but eventually it will settle out. Blood is an example of a suspension. When freshly drawn from the body, blood has an even, reddish color. After blood sits for a while in a test tube, red blood cells settle out of the suspension and drift to the bottom of the tube (see Figure 19.1a). The upper layer, the liquid portion of blood, appears pale yellow and is called blood plasma. Blood plasma is both a solution of ions and other small solutes and a colloid due to the presence of larger plasma proteins.

The **concentration** of a solution may be expressed in several ways. One common way is by a mass per volume **percentage,** which gives the relative mass of a solute found in a given volume of solution. For example, you may have seen the following on the label of a bottle of wine: "Alcohol 14.1% by volume." Another way expresses concentration in units of **moles per liter (mol/L),** which relate to the total number of molecules in a given volume of solution. A **mole** is the amount of any substance that has a mass in grams equal to the sum of the atomic masses of all its atoms. For example, 1 mole of the element chlorine (atomic mass = 35.45) is 35.45 grams and 1 mole of the salt sodium chloride (NaCl) is 58.44 grams (22.99 for Na + 35.45 for Cl). Just as a dozen always means 12 of something, a mole of anything has the same number of particles: 6.023×10^{23}. This huge number is called *Avogadro's number.* Thus, measurements of substances that are stated in moles tell us about the numbers of atoms, ions, or molecules present. This is important when chemical reactions are occurring because each reaction requires a set number of

TABLE 2.3

Percentage and Molarity

DEFINITION	EXAMPLE
Percentage (mass per volume) Number of grams of a substance per 100 milliliters (mL) of solution	To make a 10% NaCl solution, take 10 g of NaCl and add enough water to make a total of 100 mL of solution.
Molarity: 5 moles (mol) per liter A 1 molar (1 M) solution = 1 mole of a solute in 1 liter of solution	To make a 1 molar (1 M) solution of NaCl, dissolve 1 mole of NaCl (58.44 g) in enough water to make a total of 1 liter of solution.

atoms of specific elements. Table 2.3 describes these ways of expressing concentration.

Inorganic Acids, Bases, and Salts

When inorganic acids, bases, or salts dissolve in water, they **dissociate** (dis′-sō-sē-ĀT); that is, they separate into ions and become surrounded by water molecules. An **acid** (Figure 2.11a) is a substance that dissociates into one or more **hydrogen ions (H^+)** and one or more anions. Because H^+ is a single proton with one positive charge, an acid is also referred to as a **proton donor.** A **base,** by contrast (Figure 2.11b), removes H^+ from a solution and is therefore a **proton acceptor.** Many bases dissociate into one or more **hydroxide ions (OH^-)** and one or more cations.

A **salt,** when dissolved in water, dissociates into cations and anions, neither of which is H^+ or OH^- (Figure 2.11c). In the body, salts such as potassium chloride are electrolytes that are important for carrying electrical currents (ions flowing from one place to another), especially in nerve and muscular tissues. The ions of salts also provide many essential chemical elements in intracellular and extracellular fluids such as blood, lymph, and the interstitial fluid of tissues.

Figure 2.11 Dissociation of inorganic acids, bases, and salts.

Dissociation is the separation of inorganic acids, bases, and salts into ions in a solution.

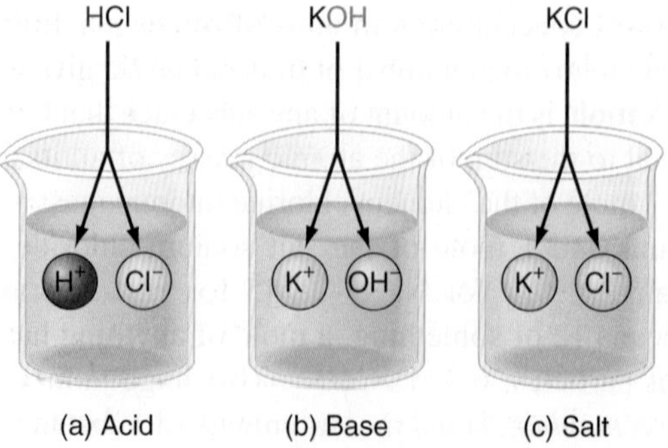
HCl
KOH
KCl
H+
Cl−
K+
OH−
K+
Cl−
(a) Acid
(b) Base
(c) Salt

? The compound $CaCO_3$ (calcium carbonate) dissociates into a calcium ion (Ca^{2+}) and a carbonate ion (CO_3^{2-}). Is it an acid, a base, or a salt? What about H_2SO_4, which dissociates into two H^+ and one SO_4^{2-}?

Acids and bases react with one another to form salts. For example, the reaction of hydrochloric acid (HCl) and potassium hydroxide (KOH), a base, produces the salt potassium chloride (KCl) and water (H_2O). This exchange reaction can be written as follows:

$$\underset{\text{Acid}}{HCl} + \underset{\text{Base}}{KOH} \longrightarrow \underset{\text{Dissociated ions}}{H^+ + Cl^- + K^+ + OH^-} \longrightarrow \underset{\text{Salt}}{KCl} + \underset{\text{Water}}{H_2O}$$

Acid–Base Balance: The Concept of pH

To ensure homeostasis, intracellular and extracellular fluids must contain almost balanced quantities of acids and bases. The more hydrogen ions (H^+) dissolved in a solution, the more acidic the solution; the more hydroxide ions (OH^-), the more basic (alkaline) the solution. The chemical reactions that take place in the body are very sensitive to even small changes in the acidity or alkalinity of the body fluids in which they occur. Any departure from the narrow limits of normal H^+ and OH^- concentrations greatly disrupts body functions.

A solution's acidity or alkalinity is expressed on the **pH scale,** which extends from 0 to 14 (Figure 2.12). This scale is based on the concentration of H^+ in moles per liter. A pH of 7 means that a solution contains one ten-millionth (0.0000001) of a mole of hydrogen ions per liter. The number 0.0000001 is written as 1×10^{-7} in scientific notation, which indicates that the number is 1 with the decimal point moved seven places to the left. To convert this value to pH, the negative exponent (–7) is changed to a positive number (7). A solution with a H^+ concentration of 0.0001 (10^{-4}) moles per liter has a pH of 4; a solution with a H^+ concentration of 0.000000001 (10^{-9}) moles per liter has a pH of 9; and so on. It is important to realize that a change of one whole number on the pH scale represents a *tenfold* change in the number of H^+. A pH of 6 denotes 10 times more H^+ than a pH of 7, and a pH of 8 indicates 10 times fewer H^+ than a pH of 7 and 100 times fewer H^+ than a pH of 6.

The midpoint of the pH scale is 7, where the concentrations of H^+ and OH^- are equal. A substance with a pH of 7, such as pure water, is neutral. A solution that has more H^+ than OH^- is an **acidic solution** and has a pH below 7. A solution that has more OH^- than H^+ is a **basic (alkaline) solution** and has a pH above 7.

Maintaining pH: Buffer Systems

Although the pH of body fluids may differ, as we have discussed, the normal limits for each fluid are quite narrow. Table 2.4 shows the pH values for certain body fluids along with those of some common substances outside the body. Homeostatic mechanisms maintain the pH of blood between 7.35 and 7.45, which is slightly more basic than pure water. You will learn in Chapter 27 that if the

Figure 2.12 **The pH scale.** A pH below 7 indicates an acidic solution—more H^+ than OH^-. A pH above 7 indicates a basic (alkaline) solution; that is, there are more OH^- than H^+.

The lower the numerical value of the pH, the more acidic is the solution because the H^+ concentration becomes progressively greater. The higher the pH, the more basic the solution.

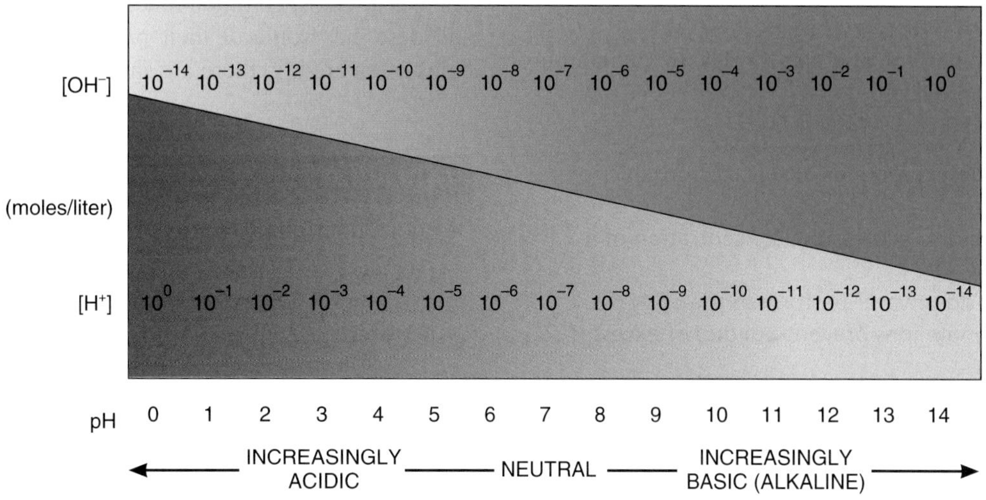

At pH 7 (neutrality), the concentrations of H^+ and OH^- are equal (10^{-7} mol/liter). What are the concentrations of H^+ and OH^- at pH 6? Which pH is more acidic, 6.82 or 6.91? Which pH is closer to neutral, 8.41 or 5.59?

TABLE 2.4

pH Values of Selected Substances

SUBSTANCE*	pH VALUE
• Gastric juice (found in the stomach)	1.2–3.0
Lemon juice	2.3
Vinegar	3.0
Carbonated soft drink	3.0–3.5
Orange juice	3.5
• Vaginal fluid	3.5–4.5
Tomato juice	4.2
Coffee	5.0
• Urine	4.6–8.0
• Saliva	6.35–6.85
Milk	6.8
Distilled (pure) water	7.0
• Blood	7.35–7.45
• Semen (fluid containing sperm)	7.20–7.60
• Cerebrospinal fluid (fluid associated with nervous system)	7.4
• Pancreatic juice (digestive juice of the pancreas)	7.1–8.2
• Bile (liver secretion that aids fat digestion)	7.6–8.6
Milk of magnesia	10.5
Lye (sodium hydroxide)	14.0

*Bullets (•) denote substances in the human body.

pH of blood falls below 7.35, a condition called *acidosis* occurs, and if the pH rises above 7.45, it results in a condition called *alkalosis;* both conditions can seriously compromise homeostasis. Saliva is slightly acidic, and semen is slightly basic. Because the kidneys help remove excess acid from the body, urine can be quite acidic.

Even though strong acids and bases are continually taken into and formed by the body, the pH of fluids inside and outside cells remains almost constant. One important reason is the presence of **buffer systems,** which function to convert strong acids or bases into weak acids or bases. Strong acids (or bases) ionize easily and contribute many H^+ (or OH^-) to a solution. Therefore, they can change pH drastically, which can disrupt the body's metabolism. Weak acids (or bases) do not ionize as much and contribute fewer H^+ (or OH^-). Hence, they have less effect on the pH. The chemical compounds that can convert strong acids or bases into weak ones are called **buffers.** They do so by removing or adding protons (H^+).

One important buffer system in the body is the **carbonic acid–bicarbonate buffer system.** Carbonic acid (H_2CO_3) can act as a weak acid, and the bicarbonate ion (HCO_3^-) can act as a weak base. Hence, this buffer system can compensate for either an excess or a shortage of H^+. For example, if there is an excess of H^+ (an acidic condition), HCO_3^- can function as a weak base and remove the excess H^+, as follows:

$$\underset{\text{Hydrogen ion}}{H^+} + \underset{\text{Bicarbonate ion (weak base)}}{HCO_3^-} \longrightarrow \underset{\text{Carbonic acid}}{H_2CO_3}$$

If there is a shortage of H^+ (an alkaline condition), by contrast, H_2CO_3 can function as a weak acid and provide needed H^+ as follows:

$$\underset{\text{Carbonic acid (weak acid)}}{H_2CO_3} \longrightarrow \underset{\text{Hydrogen ion}}{H^+} + \underset{\text{Bicarbonate ion}}{HCO_3^-}$$

Chapter 27 describes buffers and their roles in maintaining acid–base balance in more detail.

✔ CHECKPOINT

12. How do inorganic compounds differ from organic compounds?
13. Describe two ways to express the concentration of a solution.
14. What functions does water perform in the body?
15. How do bicarbonate ions prevent buildup of excess H^+?

2.5 ORGANIC COMPOUNDS

OBJECTIVES

- Describe the functional groups of organic molecules.
- Identify the building blocks and functions of carbohydrates, lipids, and proteins.
- Describe the structure and functions of deoxyribonucleic acid (DNA), ribonucleic acid (RNA), and adenosine triphosphate (ATP).

Many organic molecules are relatively large and have unique characteristics that allow them to carry out complex functions. Important categories of organic compounds include carbohydrates, lipids, proteins, nucleic acids, and adenosine triphosphate (ATP).

Carbon and Its Functional Groups

Carbon has several properties that make it particularly useful to living organisms. For one thing, it can form bonds with one to thousands of other carbon atoms to produce large molecules that can have many different shapes. Due to this property of carbon, the body can build many different organic compounds, each of which has a unique structure and function. Moreover, the large size of most carbon-containing molecules and the fact that some do not dissolve easily in water make them useful materials for building body structures.

Organic compounds are usually held together by covalent bonds. Carbon has four electrons in its outermost (valence) shell. It can bond covalently with a variety of atoms, including other carbon atoms, to form rings and straight or branched chains. Other elements that most often bond with carbon in organic compounds are hydrogen, oxygen, and nitrogen. Sulfur and phosphorus are also present in organic compounds. The other elements listed in Table 2.1 are present in a smaller number of organic compounds.

The chain of carbon atoms in an organic molecule is called the **carbon skeleton.** Many of the carbons are bonded to hydrogen atoms, yielding a **hydrocarbon.** Also attached to the carbon skeleton are distinctive **functional groups,** other atoms or molecules bound to the hydrocarbon skeleton. Each type of functional group has a specific arrangement of atoms that confers characteristic chemical properties on the organic molecule attached to it. Table 2.5 lists the most common functional groups of organic molecules and describes some of their properties. Because organic molecules often are big, there are shorthand methods for representing

TABLE 2.5

Major Functional Groups of Organic Molecules

NAME AND STRUCTURAL FORMULA*	OCCURRENCE AND SIGNIFICANCE
Hydroxyl R—O—H	*Alcohols* contain an —OH group, which is polar and hydrophilic due to its electronegative O atom. Molecules with many —OH groups dissolve easily in water.
Sulfhydryl R—S—H	*Thiols* have an —SH group, which is polar and hydrophilic due to its electronegative S atom. Certain amino acids (for example, cysteine) contain —SH groups, which help stabilize the shape of proteins.
Carbonyl R—C(=O)—R	*Ketones* contain a carbonyl group within the carbon skeleton. The carbonyl group is polar and hydrophilic due to its electronegative O atom.
or R—C(=O)—H	*Aldehydes* have a carbonyl group at the end of the carbon skeleton.
Carboxyl R—C(=O)—OH or R—C(=O)—O^-	*Carboxylic acids* contain a carboxyl group at the end of the carbon skeleton. All amino acids have a —COOH group at one end. The negatively charged form predominates at the pH of body cells and is hydrophilic.
Ester R—C(=O)—O—R	*Esters* predominate in dietary fats and oils and also occur in our body as triglycerides. Aspirin is an ester of salicylic acid, a pain-relieving molecule found in the bark of the willow tree.
Phosphate R—O—P(=O)(O^-)—O^-	*Phosphates* contain a phosphate group (—PO_4^{2-}), which is very hydrophilic due to the dual negative charges. An important example is adenosine triphosphate (ATP), which transfers chemical energy between organic molecules during chemical reactions.
Amino R—N(H)—H or R—N^+(H)(H)—H	*Amines* have an —NH_2 group, which can act as a base and pick up a hydrogen ion, giving the amino group a positive charge. At the pH of body fluids, most amino groups have a charge of 1+. All amino acids have an amino group at one end.

*R = variable group.

Figure 2.13 **Alternative ways to write the structural formula for glucose.**

In standard shorthand, carbon atoms are understood to be at locations where two bond lines intersect, and single hydrogen atoms are not indicated.

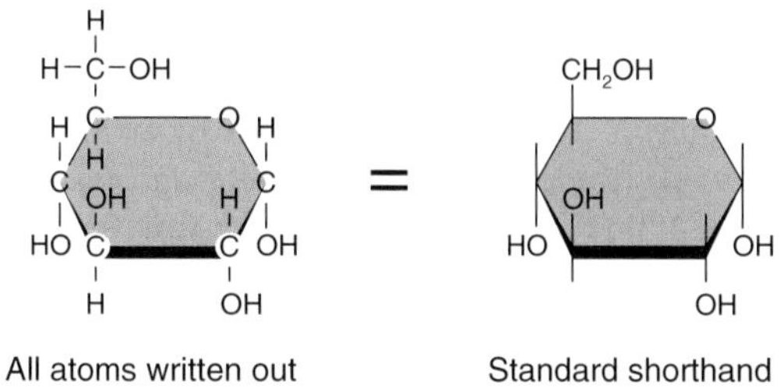
All atoms written out
Standard shorthand

How many hydroxyl groups does a molecule of glucose have? How many carbon atoms are part of glucose's carbon skeleton?

their structural formulas. Figure 2.13 shows two ways to indicate the structure of the sugar glucose, a molecule with a ring-shaped carbon skeleton that has several hydroxyl groups attached.

Small organic molecules can combine into very large molecules that are called **macromolecules** (*macro-* = large). Macromolecules are usually **polymers** (*poly-* = many; *-mers* = parts). A polymer is a large molecule formed by the covalent bonding of many identical or similar small building-block molecules called **monomers** (*mono-* = one). Usually, the reaction that joins two monomers is a dehydration synthesis. In this type of reaction, a hydrogen atom is removed from one monomer and a hydroxyl group is removed from the other to form a molecule of water (see Figure 2.15a). Macromolecules such as carbohydrates, lipids, proteins, and nucleic acids are assembled in cells via dehydration synthesis reactions.

Molecules that have the same molecular formula but different structures are called **isomers** (Ī-so-merz; *iso-* = equal or the same). For example, the molecular formulas for the sugars glucose and fructose are both $C_6H_{12}O_6$. The individual atoms, however, are positioned differently along the carbon skeleton (see Figure 2.15a), giving the sugars different chemical properties.

Carbohydrates

Carbohydrates include sugars, glycogen, starches, and cellulose. Even though they are a large and diverse group of organic compounds and have several functions, carbohydrates represent only 2–3% of your total body mass. In humans and animals, carbohydrates function mainly as a source of chemical energy for generating ATP needed to drive metabolic reactions. Only a few carbohydrates are used for building structural units. One example is deoxyribose, a type of sugar that is a building block of deoxyribonucleic acid (DNA), the molecule that carries inherited genetic information.

Carbon, hydrogen, and oxygen are the elements found in carbohydrates. The ratio of hydrogen to oxygen atoms is usually

TABLE 2.6

Major Carbohydrate Groups

TYPE OF CARBOHYDRATE	EXAMPLES
Monosaccharides (simple sugars that contain from 3 to 7 carbon atoms)	Glucose (the main blood sugar). Fructose (found in fruits). Galactose (in milk sugar). Deoxyribose (in DNA). Ribose (in RNA).
Disaccharides (simple sugars formed from the combination of two monosaccharides by dehydration synthesis)	Sucrose (table sugar) = glucose + fructose. Lactose (milk sugar) = glucose + galactose. Maltose = glucose + glucose.
Polysaccharides (from tens to hundreds of monosaccharides joined by dehydration synthesis)	Glycogen (stored form of carbohydrates in animals). Starch (stored form of carbohydrates in plants and main carbohydrates in food). Cellulose (part of cell walls in plants that cannot be digested by humans but aids movement of food through intestines).

2:1, the same as in water. Although there are exceptions, carbohydrates generally contain one water molecule for each carbon atom. This is the reason they are called carbohydrates, which means "watered carbon." The three major groups of carbohydrates, based on their sizes, are monosaccharides, disaccharides, and polysaccharides (Table 2.6).

Monosaccharides and Disaccharides: The Simple Sugars

Monosaccharides and disaccharides are known as **simple sugars.** The monomers of carbohydrates, **monosaccharides** (mon′-ō-SAK-a-rīds; *sacchar-* = sugar), contain from three to seven carbon atoms. They are designated by names ending in "-ose" with a prefix that indicates the number of carbon atoms. For example, monosaccharides with three carbons are called *trioses* (*tri-* = three). There are also *tetroses* (four-carbon sugars), *pentoses* (five-carbon sugars), *hexoses* (six-carbon sugars), and *heptoses* (seven-carbon sugars). Examples of pentoses and hexoses are illustrated in Figure 2.14. Cells throughout the body break down the hexose glucose to produce ATP.

A **disaccharide** (dī-SAK-a-rīd; *di-* = two) is a molecule formed from the combination of two monosaccharides by dehydration synthesis (Figure 2.15). For example, molecules of the monosaccharides glucose and fructose combine to form a molecule of the disaccharide sucrose (table sugar), as shown in Figure 2.15a. Glucose and fructose are isomers. As you learned earlier in the chapter, isomers have the same molecular formula, but the relative positions of the oxygen and carbon atoms are different, causing the compounds to have different chemical properties. Notice that the formula for sucrose is $C_{12}H_{22}O_{11}$, not $C_{12}H_{24}O_{12}$, because a molecule of water is removed as the two monosaccharides are joined.

Figure 2.14 Monosaccharides. The structural formulas of selected monosaccharides are shown.

Monosaccharides are the monomers used to build carbohydrates.

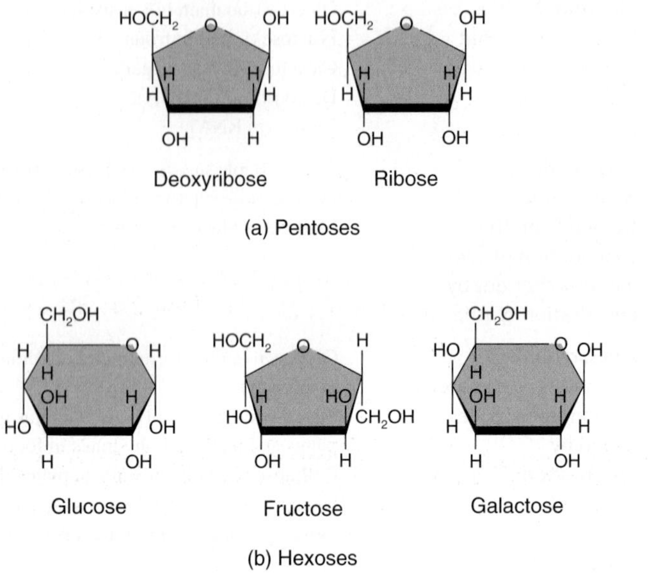

Which of these monosaccharides are hexoses?

Disaccharides can also be split into smaller, simpler molecules by hydrolysis. A molecule of sucrose, for example, may be hydrolyzed into its components, glucose and fructose, by the addition of water. Figure 2.15a also illustrates this reaction.

CLINICAL CONNECTION | *Artificial Sweeteners*

Some individuals use **artificial sweeteners** to limit their sugar consumption for medical reasons, while others do so to avoid calories that might result in weight gain. Examples of artificial sweeteners include aspartame (trade names NutraSweet® and Equal®), saccharin (Sweet 'N Low®), and sucralose (Splenda®). Aspartame is 200 times sweeter than sucrose and it adds essentially no calories to the diet because only small amounts of it are used to produce a sweet taste. Saccharin is about 400 times sweeter than sucrose, and sucralose is 600 times sweeter than sucrose. Both saccharin and sucralose have zero calories because they pass through the body without being metabolized. Artificial sweeteners are also used as sugar substitutes because they do not cause tooth decay. In fact, studies have shown that using artificial sweeteners in the diet helps reduce the incidence of dental cavities. •

Polysaccharides

The third major group of carbohydrates is the **polysaccharides** (pol′-ē-SAK-a-rīds). Each polysaccharide molecule contains tens or hundreds of monosaccharides joined through dehydration synthesis

Figure 2.15 Disaccharides. (a) The structural and molecular formulas for the monosaccharides glucose and fructose and the disaccharide sucrose. In dehydration synthesis (read from left to right), two smaller molecules, glucose and fructose, are joined to form a larger molecule of sucrose. Note the loss of a water molecule. In hydrolysis (read from right to left), the addition of a water molecule to the larger sucrose molecule breaks the disaccharide into two smaller molecules, glucose and fructose. Shown in (b) and (c) are the structural formulas of the disaccharides lactose and maltose, respectively.

A disaccharide consists of two monosaccharides that have combined by dehydration synthesis.

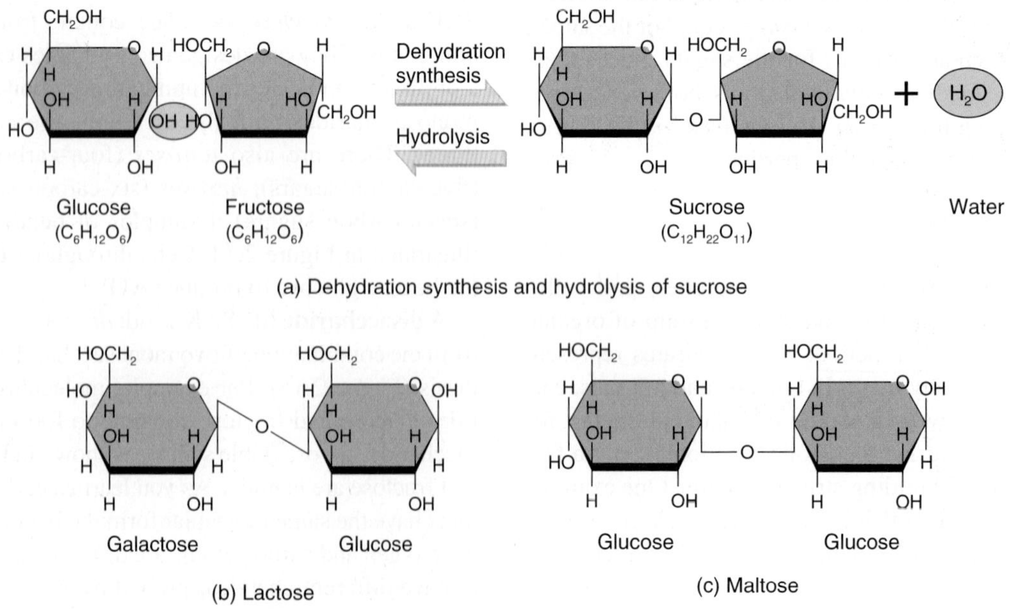

How many carbon atoms are there in fructose? In sucrose?

reactions. Unlike simple sugars, polysaccharides usually are insoluble in water and do not taste sweet. The main polysaccharide in the human body is **glycogen,** which is made entirely of glucose monomers linked to one another in branching chains (Figure 2.16). A limited amount of carbohydrates is stored as glycogen in the liver and skeletal muscles. **Starches** are polysaccharides formed from glucose by plants. They are found in foods such as pasta and potatoes and are the major carbohydrates in the diet. Like disaccharides, polysaccharides such as glycogen and starches can be broken down into monosaccharides through hydrolysis reactions. For example, when the blood glucose level falls, liver cells break down glycogen into glucose and release it into the blood, making it available to body cells, which break it down to synthesize ATP. **Cellulose** is a polysaccharide formed from glucose by plants that cannot be digested by humans but does provide bulk to help eliminate feces.

Lipids

A second important group of organic compounds is **lipids** (*lip-* = fat). Lipids make up 18–25% of body mass in lean adults. Like carbohydrates, lipids contain carbon, hydrogen, and oxygen. Unlike carbohydrates, they do not have a 2:1 ratio of hydrogen to oxygen. The proportion of electronegative oxygen atoms in lipids is usually smaller than in carbohydrates, so there are fewer polar covalent bonds. As a result, most lipids are insoluble in polar solvents such as water; they are *hydrophobic.* Because they are hydrophobic, only the smallest lipids (some fatty acids) can dissolve in watery blood plasma. To become more soluble in blood plasma, other lipid molecules join with hydrophilic protein molecules. The resulting lipid/protein complexes are termed **lipoproteins.** Lipoproteins are soluble because the proteins are on the outside and the lipids are on the inside.

The diverse lipid family includes fatty acids, triglycerides (fats and oils), phospholipids (lipids that contain phosphorus), steroids (lipids that contain rings of carbon atoms), eicosanoids (20-carbon lipids), and a variety of other lipids, including fat-soluble vitamins (vitamins A, D, E, and K) and lipoproteins. Table 2.7 introduces the various types of lipids and highlights their roles in the human body.

Figure 2.16 Part of a glycogen molecule, the main polysaccharide in the human body.

Glycogen is made up of glucose monomers and is the stored form of carbohydrate in the human body.

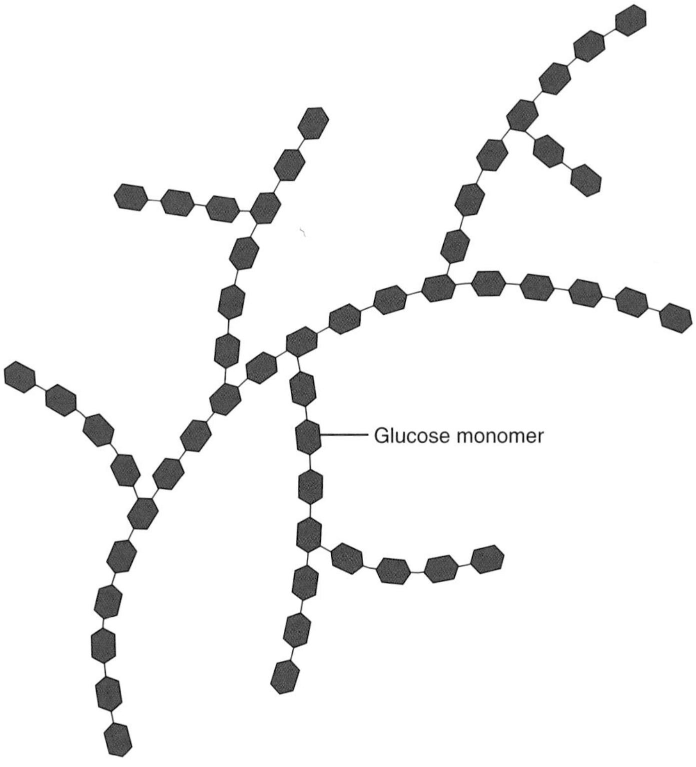

Which body cells store glycogen?

TABLE 2.7

Types of Lipids in the Body

TYPE OF LIPID	FUNCTIONS
Fatty acids	Used to synthesize triglycerides and phospholipids or catabolized to generate adenosine triphosphate (ATP).
Triglycerides *(fats and oils)*	Protection, insulation, energy storage.
Phospholipids	Major lipid component of cell membranes.
Steroids	
Cholesterol	Minor component of all animal cell membranes; precursor of bile salts, vitamin D, and steroid hormones.
Bile salts	Needed for digestion and absorption of dietary lipids.
Vitamin D	Helps regulate calcium level in body; needed for bone growth and repair.
Adrenocortical hormones	Help regulate metabolism, resistance to stress, and salt and water balance.
Sex hormones	Stimulate reproductive functions and sexual characteristics.
Eicosanoids *(prostaglandins and leukotrienes)*	Have diverse effects on modifying responses to hormones, blood clotting, inflammation, immunity, stomach acid secretion, airway diameter, lipid breakdown, and smooth muscle contraction.
Other lipids	
Carotenes	Needed for synthesis of vitamin A (used to make visual pigments in eye); function as antioxidants.
Vitamin E	Promotes wound healing, prevents tissue scarring, contributes to normal structure and function of nervous system, and functions as antioxidant.
Vitamin K	Required for synthesis of blood-clotting proteins.
Lipoproteins	Transport lipids in blood, carry triglycerides and cholesterol to tissues, and remove excess cholesterol from blood.

Fatty Acids

Among the simplest lipids are the **fatty acids,** which are used to synthesize triglycerides and phospholipids. Fatty acids can also be catabolized to generate adenosine triphosphate (ATP). A fatty acid consists of a carboxyl group and a hydrocarbon chain (Figure 2.17a). Fatty acids can be either saturated or unsaturated. A **saturated fatty acid** contains only *single covalent bonds* between the carbon atoms of the hydrocarbon chain. Because they lack double bonds, each carbon atom of the hydrocarbon chain is *saturated with hydrogen atoms* (see, for example, palmitic acid in Figure 2.17a). An **unsaturated fatty acid** contains one or more *double covalent bonds* between the carbon atoms of the hydrocarbon chain.

Figure 2.17 **Fatty acid structure and triglyceride synthesis.** Shown in (a) are the structures of a saturated fatty acid and an unsaturated fatty acid. Each time a glycerol and a fatty acid are joined in dehydration synthesis (b), a molecule of water is removed. An ester linkage joins the glycerol to each of the three molecules of fatty acids, which vary in length and in number and location of double bonds between carbon atoms (C═C). Shown here (c) is a triglyceride molecule that contains two saturated fatty acids and a monounsaturated fatty acid. The kink (bend) in the oleic acid occurs at the double bond.

One glycerol and three fatty acids are the building blocks of triglycerides.

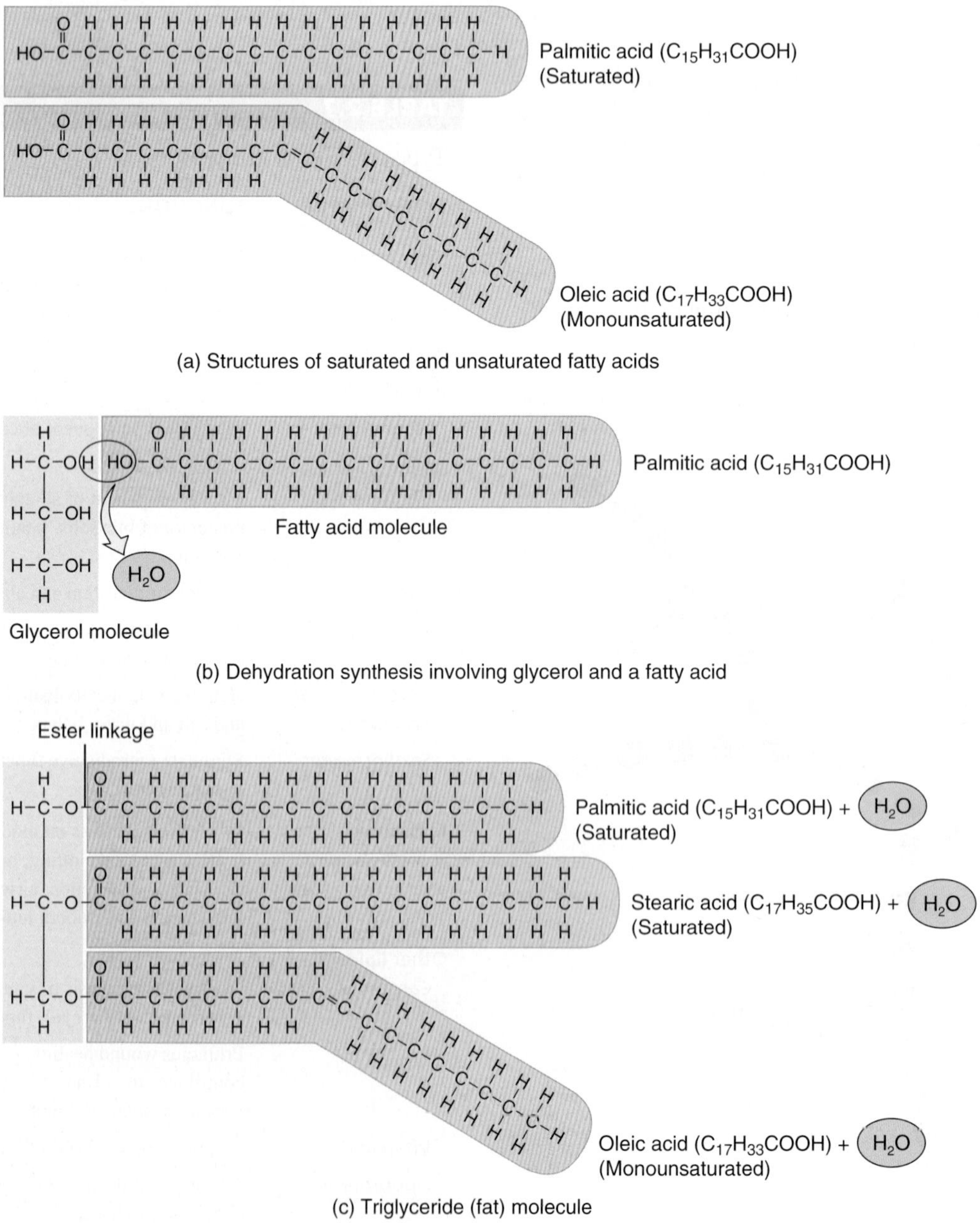

Does the oxygen in the water molecule removed during dehydration synthesis come from the glycerol or from a fatty acid?

Thus, the fatty acid is not completely saturated with hydrogen atoms (see, for example, oleic acid in Figure 2.17a). The unsaturated fatty acid has a kink (bend) at the site of the double bond. If the fatty acid has just one double bond in the hydrocarbon chain, it is *monounsaturated* and it has just one kink. If a fatty acid has more than one double bond in the hydrocarbon chain, it is *polyunsaturated* and it contains more than one kink.

Triglycerides

The most plentiful lipids in your body and in your diet are the **triglycerides** (trī-GLI-ser-īds; *tri-* = three), also known as **triacylglycerols.** A triglyceride consists of two types of building blocks, a single glycerol molecule and three fatty acid molecules. A three-carbon **glycerol** molecule forms the backbone of a triglyceride (Figure 2.17b, c). Three fatty acids are attached by dehydration synthesis reactions, one to each carbon of the glycerol backbone. The chemical bond formed where each water molecule is removed is an *ester linkage* (see Table 2.5). The reverse reaction, hydrolysis, breaks down a single molecule of a triglyceride into three fatty acids and glycerol.

Triglycerides can be either solids or liquids at room temperature. A **fat** is a triglyceride that is a solid at room temperature. The fatty acids of a fat are mostly saturated. Because these saturated fatty acids lack double bonds in their hydrocarbon chains, they can closely pack together and solidify at room temperature. A fat that mainly consists of saturated fatty acids is called a **saturated fat.** Although saturated fats occur mostly in meats (especially red meats) and nonskim dairy products (whole milk, cheese, and butter), they are also found in a few plant products, such as cocoa butter, palm oil, and coconut oil. Diets that contain large amounts of saturated fats are associated with disorders such as heart disease and colorectal cancer.

An **oil** is a triglyceride that is a liquid at room temperature. The fatty acids of an oil are mostly unsaturated. Recall that unsaturated fatty acids contain one or more double bonds in their hydrocarbon chains. The kinks at the sites of the double bonds prevent the unsaturated fatty acids of an oil from closely packing together and solidifying. The fatty acids of an oil can be either monounsaturated or polyunsaturated. **Monounsaturated fats** contain triglycerides that mostly consist of monounsaturated fatty acids. Olive oil, peanut oil, canola oil, most nuts, and avocados are rich in triglycerides with monounsaturated fatty acids. **Polyunsaturated fats** contain triglycerides that mostly consist of polyunsaturated fatty acids. Corn oil, safflower oil, sunflower oil, soybean oil, and fatty fish (salmon, tuna, and mackerel) contain a high percentage of polyunsaturated fatty acids. Both monounsaturated and polyunsaturated fats are believed to decrease the risk of heart disease.

Triglycerides are the body's most highly concentrated form of chemical energy. Triglycerides provide more than twice as much energy per gram as do carbohydrates and proteins. Our capacity to store triglycerides in adipose (fat) tissue is unlimited for all practical purposes. Excess dietary carbohydrates, proteins, fats, and oils all have the same fate: They are deposited in adipose tissue as triglycerides.

CLINICAL CONNECTION | *Fatty Acids in Health and Disease*

As its name implies, a group of fatty acids called **essential fatty acids (EFAs)** is essential to human health. However, they cannot be made by the human body and must be obtained from foods or supplements. Among the more important EFAs are *omega-3 fatty acids, omega-6 fatty acids,* and cis-*fatty acids.*

Omega-3 and omega-6 fatty acids are polyunsaturated fatty acids that are believed to work together to promote health. They may have a protective effect against heart disease and stroke by lowering total cholesterol, raising HDL (high-density lipoproteins or "good cholesterol") and lowering LDL (low-density lipoproteins or "bad cholesterol"). In addition, omega-3 and omega-6 fatty acids decrease bone loss by increasing calcium utilization by the body; reduce symptoms of arthritis due to inflammation; promote wound healing; improve certain skin disorders (psoriasis, eczema, and acne); and improve mental functions. Primary sources of omega-3 fatty acids include flaxseed, fatty fish, oils that have large amounts of polyunsaturated fatty acids, fish oils, and walnuts. Primary sources of omega-6 fatty acids include most processed foods (cereals, breads, white rice), eggs, baked goods, oils with large amounts of polyunsaturated fatty acids, and meats (especially organ meats, such as liver).

Note in Figure 2.17a that the hydrogen atoms on either side of the double bond in oleic acid are on the same side of the unsaturated fatty acid. Such an unsaturated fatty acid is called a *cis*-fatty acid. *Cis*-fatty acids are nutritionally beneficial unsaturated fatty acids that are used by the body to produce hormonelike regulators and cell membranes. However, when *cis*-fatty acids are heated, pressurized, and combined with a catalyst (usually nickel) in a process called *hydrogenation,* they are changed to unhealthy *trans*-fatty acids. In *trans*-fatty acids the hydrogen atoms are on opposite sides of the double bond of an unsaturated fatty acid. Hydrogenation is used by manufacturers to make vegetable oils solid at room temperature and less likely to turn rancid. Hydrogenated or *trans*-fatty acids are common in commercially baked goods (crackers, cakes, and cookies), salty snack foods, some margarines, and fried foods (donuts and french fries). When oil is used for frying and if the oil is reused (like in fast food french fry machines), *cis*-fatty acids are converted to *trans*-fatty acids. If a product label contains the words hydrogenated or partially hydrogenated, then the product contains *trans*-fatty acids. Among the adverse effects of *trans*-fatty acids are an increase in total cholesterol, a decrease in HDL, an increase in LDL, and an increase in triglycerides. These effects, which can increase the risk of heart disease and other cardiovascular diseases, are similar to those caused by saturated fats. •

Phospholipids

Like triglycerides, **phospholipids** have a glycerol backbone and two fatty acid chains attached to the first two carbons. In the third position, however, a phosphate group (PO_4^{3-}) links a small charged group that usually contains nitrogen (N) to the backbone (Figure 2.18). This portion of the molecule (the "head") is polar and can form hydrogen bonds with water molecules. The two fatty acids (the "tails"), by contrast, are nonpolar and can interact only with other lipids. Molecules that have both polar and nonpolar parts are said to be **amphipathic** (am-fē-PATH-ik;

Figure 2.18 **Phospholipids.** (a) In the synthesis of phospholipids, two fatty acids attach to the first two carbons of the glycerol backbone. A phosphate group links a small charged group to the third carbon in glycerol. In (b), the circle represents the polar head region, and the two wavy lines represent the two nonpolar tails. Double bonds in the fatty acid hydrocarbon chain often form kinks in the tail.

Phospholipids are amphipathic molecules, having both polar and nonpolar regions.

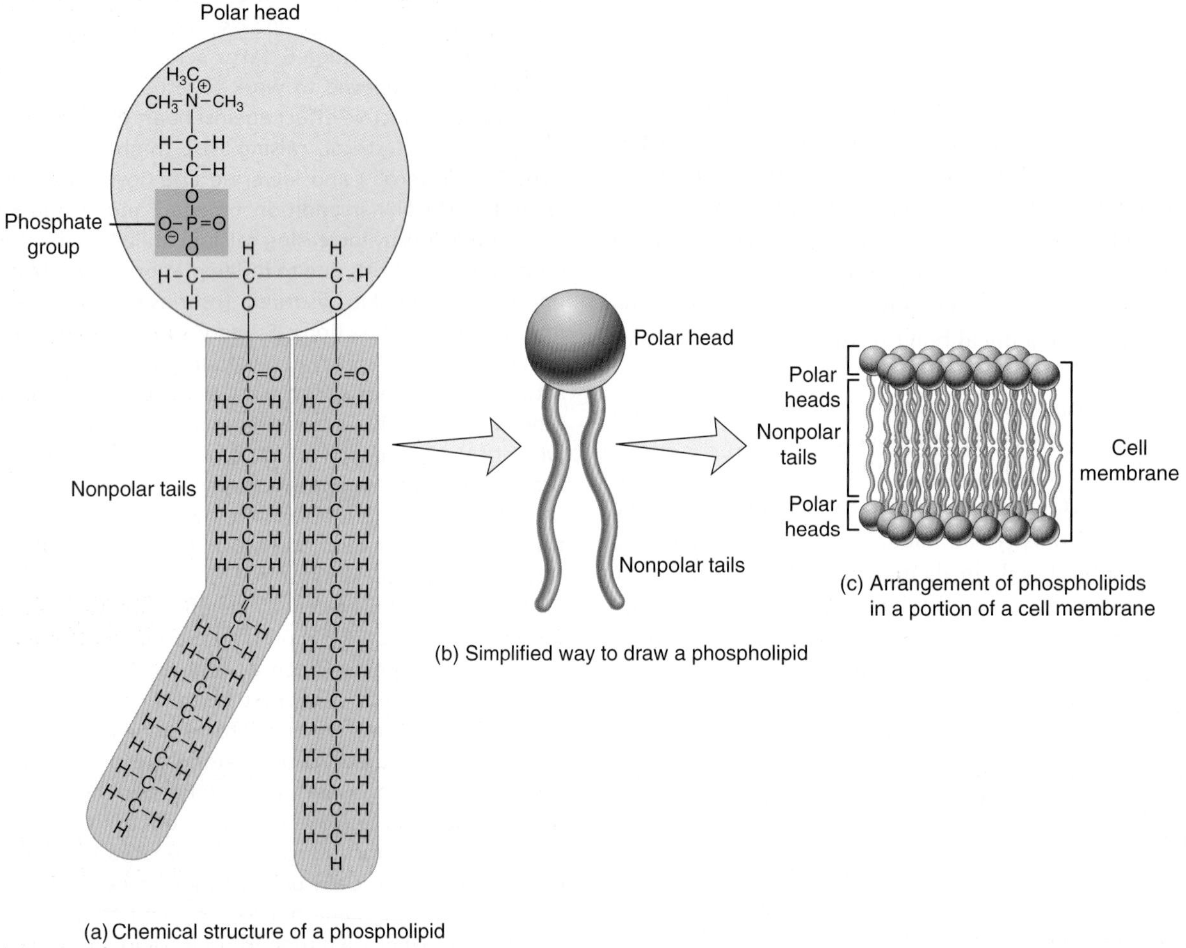

Which portion of a phospholipid is hydrophilic, and which portion is hydrophobic?

amphi- = on both sides; *-pathic* = feeling). Amphipathic phospholipids line up tail-to-tail in a double row to make up much of the membrane that surrounds each cell (Figure 2.18c).

Steroids

The structure of **steroids** differs considerably from that of the triglycerides. Steroids have four rings of carbon atoms (colored gold in Figure 2.19). Body cells synthesize other steroids from cholesterol (Figure 2.19a), which has a large nonpolar region consisting of the four rings and a hydrocarbon tail. In the body, the commonly encountered steroids, such as cholesterol, estrogens, testosterone, cortisol, bile salts, and vitamin D, are known as **sterols** because they also have at least one hydroxyl (alcohol) group (—OH). The polar hydroxyl groups make sterols weakly amphipathic. Cholesterol is needed for cell membrane structure; estrogens and testosterone are required for regulating sexual functions; cortisol is necessary for maintaining normal blood sugar levels; bile salts are needed for lipid digestion and absorption; and vitamin D is related to bone growth. In Chapter 10, we will discuss the use of anabolic steroids by athletes to increase muscle size, strength, and endurance.

Other Lipids

Eicosanoids (ī-KŌ-sa-noyds; *eicosan-* = twenty) are lipids derived from a 20-carbon fatty acid called arachidonic acid. The two principal subclasses of eicosanoids are the **prostaglandins** (pros′-ta-GLAN-dins) and the **leukotrienes** (loo′-kō-TRĪ-ēnz). Prostaglandins have a wide variety of functions. They modify responses to hormones, contribute to the inflammatory response (Chapter 22), prevent stomach ulcers, dilate (enlarge) airways to the lungs, regulate body temperature, and influence formation of blood clots, to name just a few. Leukotrienes participate in allergic and inflammatory responses.

Other lipids also include fat-soluble vitamins such as beta-carotenes (the yellow-orange pigments in egg yolk, carrots, and

Figure 2.19 Steroids. All steroids have four rings of carbon atoms. The individual rings are designated by the letters A, B, C, and D.

Cholesterol, which is synthesized in the liver, is the starting material for synthesis of other steroids in the body.

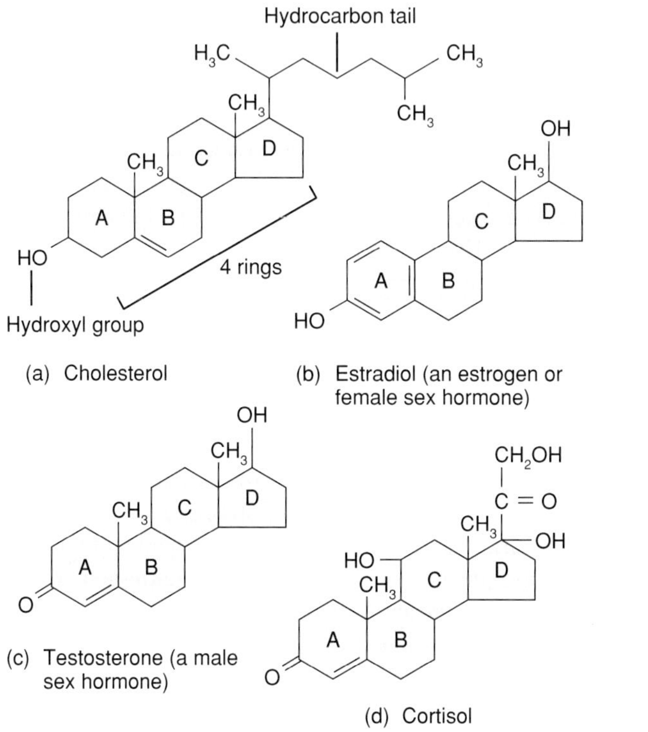

How is the structure of estradiol different from that of testosterone?

tomatoes that are converted to vitamin A); vitamins D, E, and K; and lipoproteins.

✔CHECKPOINT

16. How are carbohydrates classified?
17. How are dehydration synthesis and hydrolysis reactions related?
18. What is the importance to the body of triglycerides, phospholipids, steroids, lipoproteins, and eicosanoids?
19. Distinguish among saturated, monounsaturated, and polyunsaturated fats.

Proteins

Proteins are large molecules that contain carbon, hydrogen, oxygen, and nitrogen. Some proteins also contain sulfur. A normal, lean adult body is 12–18% protein. Much more complex in structure than carbohydrates or lipids, proteins have many roles in the body and are largely responsible for the structure of body tissues. Enzymes are proteins that speed up most biochemical reactions. Other proteins work as "motors" to drive muscle contraction. Antibodies are proteins that defend against invading microbes. Some hormones that regulate homeostasis also are proteins. Table 2.8 describes several important functions of proteins.

Amino Acids and Polypeptides

The monomers of proteins are **amino acids** (a-MĒ-nō). Each of the 20 different amino acids has a hydrogen (H) atom and three important functional groups attached to a central carbon atom (Figure 2.20a): (1) an amino group (—NH_2), (2) an acidic carboxyl group (—COOH), and (3) a side chain (R group). At the normal pH of body fluids, both the amino group and the carboxyl group are ionized (Figure 2.20b). The different side chains give each amino acid its distinctive chemical identity (Figure 2.20c).

A protein is synthesized in stepwise fashion—one amino acid is joined to a second, a third is then added to the first two, and so on. The covalent bond joining each pair of amino acids is a **peptide bond.** It always forms between the carbon of the carboxyl group (—COOH) of one amino acid and the nitrogen of the amino group (—NH_2) of another. As the peptide bond is formed, a molecule of water is removed (Figure 2.21), making this a

TABLE 2.8

Functions of Proteins

TYPE OF PROTEIN	FUNCTIONS
Structural	Form structural framework of various parts of body. *Examples:* collagen in bone and other connective tissues; keratin in skin, hair, and fingernails.
Regulatory	Function as hormones that regulate various physiological processes; control growth and development; as neurotransmitters, mediate responses of nervous system. *Examples:* the hormone insulin (regulates blood glucose level); the neurotransmitter known as substance P (mediates sensation of pain in nervous system).
Contractile	Allow shortening of muscle cells, which produces movement. *Examples:* myosin; actin.
Immunological	Aid responses that protect body against foreign substances and invading pathogens. *Examples:* antibodies; interleukins.
Transport	Carry vital substances throughout body. *Example:* hemoglobin (transports most oxygen and some carbon dioxide in blood).
Catalytic	Act as enzymes that regulate biochemical reactions. *Examples:* salivary amylase; sucrase; ATPase.

Figure 2.20 Amino acids. (a) In keeping with their name, amino acids have an amino group (shaded blue) and a carboxyl (acid) group (shaded red). The side chain (R group) is different in each amino acid. (b) At pH close to 7, both the amino group and the carboxyl group are ionized. (c) Glycine is the simplest amino acid; the side chain is a single H atom. Cysteine is one of two amino acids that contain sulfur (S). The side chain in tyrosine contains a six-carbon ring. Lysine has a second amino group at the end of its side chain.

Body proteins contain 20 different amino acids, each of which has a unique side chain.

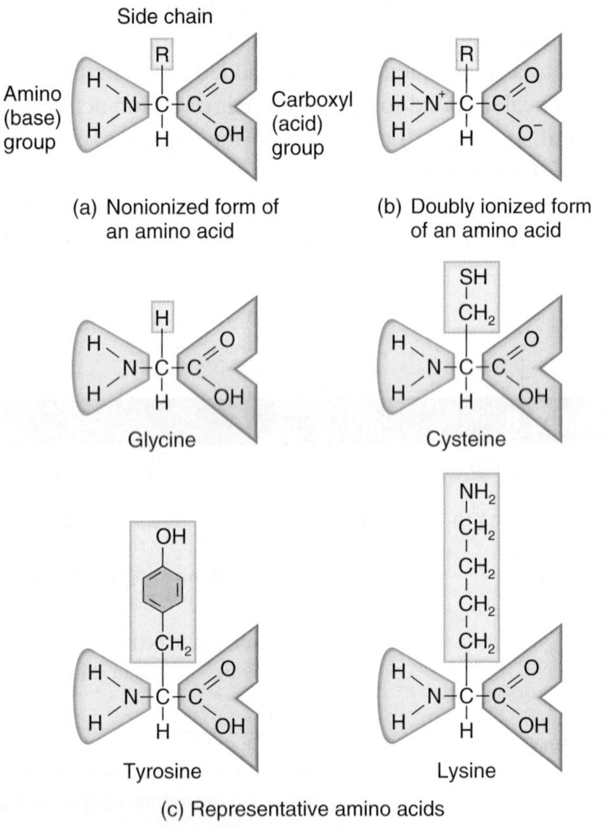
Side chain
Amino (base) group
Carboxyl (acid) group
(a) Nonionized form of an amino acid
(b) Doubly ionized form of an amino acid
Glycine
Cysteine
Tyrosine
Lysine
(c) Representative amino acids

In an amino acid, what is the minimum number of carbon atoms? Of nitrogen atoms?

dehydration synthesis reaction. Breaking a peptide bond, as occurs during digestion of dietary proteins, is a hydrolysis reaction (Figure 2.21).

When two amino acids combine, a **dipeptide** results. Adding another amino acid to a dipeptide produces a **tripeptide.** Further additions of amino acids result in the formation of a chainlike **peptide** (4–9 amino acids) or **polypeptide** (10–2000 or more amino acids). Small proteins may consist of a single polypeptide chain with as few as 50 amino acids. Larger proteins have hundreds or thousands of amino acids and may consist of two or more polypeptide chains folded together.

Because each variation in the number or sequence of amino acids can produce a different protein, a great variety of proteins is possible. The situation is similar to using an alphabet of 20 letters to form words. Each different amino acid is like a letter, and their various combinations give rise to a seemingly endless diversity of words (peptides, polypeptides, and proteins).

Levels of Structural Organization in Proteins

Proteins exhibit four levels of structural organization. The **primary structure** is the unique sequence of amino acids that are linked by covalent peptide bonds to form a polypeptide chain (Figure 2.22a). A protein's primary structure is genetically determined, and any changes in a protein's amino acid sequence can have serious consequences for body cells. In **sickle-cell disease,** for example, a nonpolar amino acid (valine) replaces a polar amino acid (glutamate) through two mutations in the oxygen-carrying protein hemoglobin. This change of amino acids diminishes hemoglobin's water solubility. As a result, the altered hemoglobin tends to form crystals inside red blood cells, producing deformed, sickle-shaped cells that cannot properly squeeze through narrow blood vessels. The symptoms and treatment of sickle-cell disease are discussed in Disorders: Homeostatic Imbalances in Chapter 19.

The **secondary structure** of a protein is the repeated twisting or folding of neighboring amino acids in the polypeptide chain (Figure 2.22b). Two common secondary structures are *alpha helixes* (clockwise spirals) and *beta pleated sheets.* The secondary structure of a protein is stabilized by hydrogen bonds, which form at regular intervals along the polypeptide backbone.

Figure 2.21 Formation of a peptide bond between two amino acids during dehydration synthesis. In this example, glycine is joined to alanine, forming a dipeptide (read from left to right). Breaking a peptide bond occurs via hydrolysis (read from right to left).

Amino acids are the monomers used to build proteins.

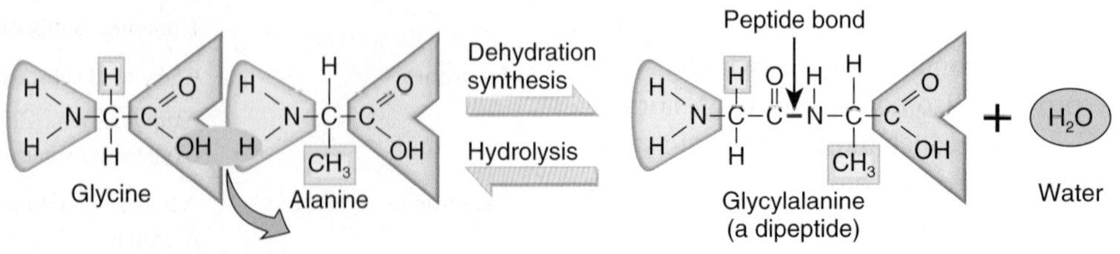
Glycine
Alanine
Dehydration synthesis
Hydrolysis
Peptide bond
Glycylalanine (a dipeptide)
Water

What type of reaction takes place during catabolism of proteins?

Figure 2.22 Levels of structural organization in proteins. (a) The primary structure is the sequence of amino acids in the polypeptide. (b) Common secondary structures include alpha helixes and beta pleated sheets. For simplicity, the amino acid side groups are not shown here. (c) The tertiary structure is the overall folding pattern that produces a distinctive, three-dimensional shape. (d) The quaternary structure in a protein is the arrangement of two or more polypeptide chains relative to one another.

6 The unique shape of each protein permits it to carry out specific functions.

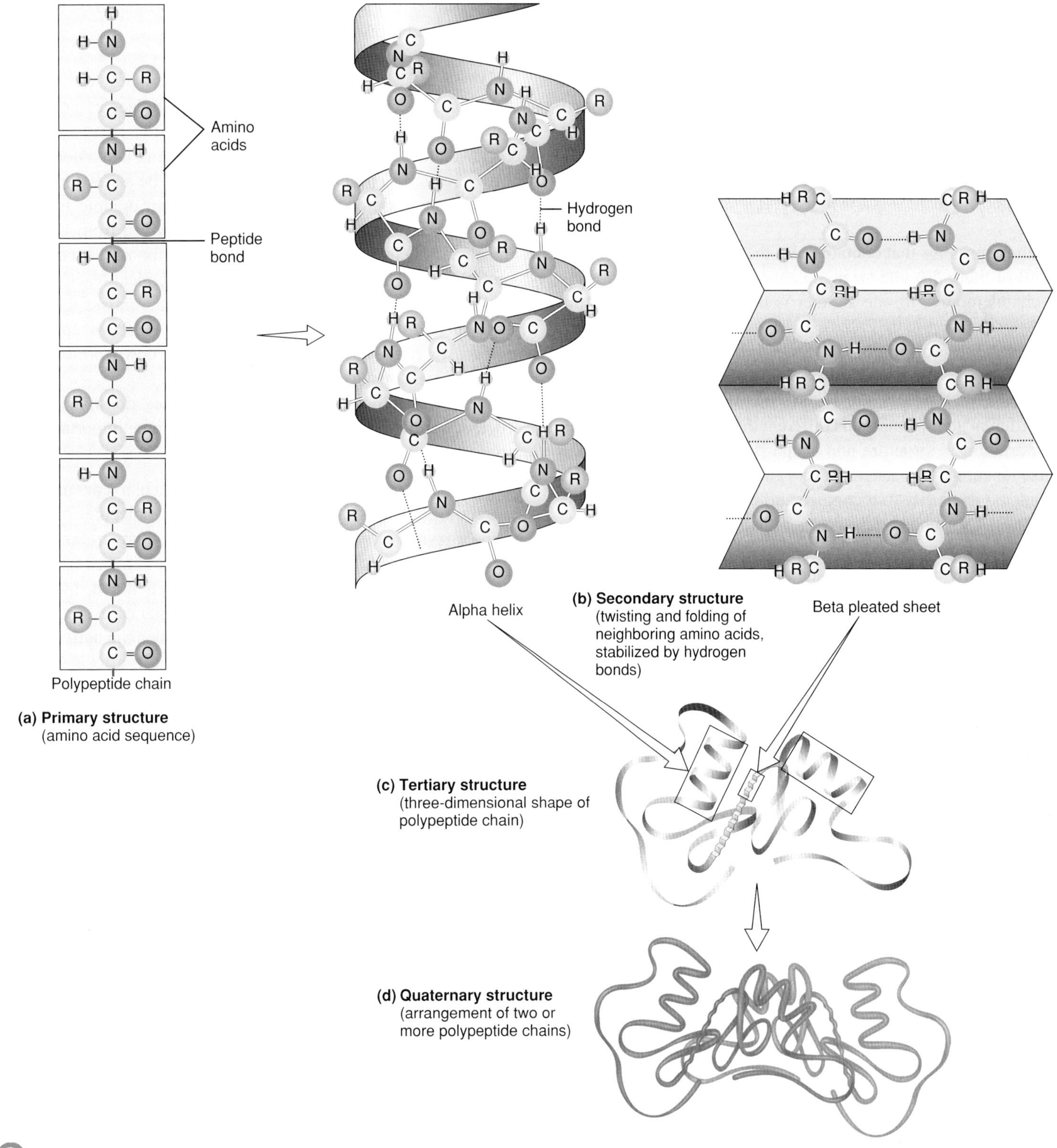

? Do all proteins have a quaternary structure?

The **tertiary structure** (TUR-shē-er′-ē) refers to the three-dimensional shape of a polypeptide chain. Each protein has a unique tertiary structure that determines how it will function. The tertiary folding pattern may allow amino acids at opposite ends of the chain to be close neighbors (Figure 2.22c). Several types of bonds can contribute to a protein's tertiary structure. The strongest but least common bonds, S—S covalent bonds called *disulfide bridges,* form between the sulfhydryl groups of two monomers of the amino acid cysteine. Many weak bonds—hydrogen bonds, ionic bonds, and hydrophobic interactions—also help determine the folding pattern. Some parts of a polypeptide are attracted to water (hydrophilic), and other parts are repelled by it (hydrophobic). Because most proteins in our body exist in watery surroundings, the folding process places most amino acids with hydrophobic side chains in the central core, away from the protein's surface. Often, helper molecules known as *chaperones* aid the folding process.

In those proteins that contain more than one polypeptide chain (not all of them do), the arrangement of the individual polypeptide chains relative to one another is the **quaternary structure** (KWA-ter-ner′-ē; Figure 2.22d). The bonds that hold polypeptide chains together are similar to those that maintain the tertiary structure.

Proteins vary tremendously in structure. Different proteins have different architectures and different three-dimensional shapes. This variation in structure and shape is directly related to their diverse functions. In practically every case, the function of a protein depends on its ability to recognize and bind to some other molecule. Thus, a hormone binds to a specific protein on a cell in order to alter its function, and an antibody protein binds to a foreign substance (antigen) that has invaded the body. A protein's unique shape permits it to interact with other molecules to carry out a specific function.

On the basis of overall shape, proteins are classified as fibrous or globular. **Fibrous proteins** are insoluble in water and their polypeptide chains form long strands that are parallel to each other. Fibrous proteins have many structural functions. Examples include *collagen* (strengthens bones, ligaments, and tendons), *elastin* (provides stretch in skin, blood vessels, and lung tissue), *keratin* (forms structure of hair and nails and waterproofs the skin), *dystrophin* (reinforces parts of muscle cells), *fibrin* (forms blood clots), and *actin* and *myosin* (are involved in contraction of muscle cells, division in all cells, and transport of substances within cells). **Globular proteins** are more or less soluble in water and their polypeptide chains are spherical (globular) in shape. Globular proteins have metabolic functions. Examples include *enzymes,* which function as catalysts; *antibodies* and *complement proteins,* which help protect us against disease; *hemoglobin,* which transports oxygen; *lipoproteins,* which transport lipids and cholesterol; *albumins,* which help regulate blood pH; *membrane proteins,* which transport substances into and out of cells; and some *hormones* such as *insulin,* which helps regulate blood sugar level.

Homeostatic mechanisms maintain the temperature and chemical composition of body fluids, which allow body proteins to keep their proper three-dimensional shapes. If a protein encounters an altered environment, it may unravel and lose its characteristic shape (secondary, tertiary, and quaternary structure). This process is called **denaturation.** Denatured proteins are no longer functional. Although in some cases denaturation can be reversed, a frying egg is a common example of permanent denaturation. In a raw egg the soluble egg-white protein (albumin) is a clear, viscous fluid. When heat is applied to the egg, the protein denatures, becomes insoluble, and turns white.

Enzymes

In living cells, most catalysts are protein molecules called **enzymes** (EN-zīms). Some enzymes consist of two parts—a protein portion, called the **apoenzyme** (ā′-pō-EN-zīm), and a nonprotein portion, called a **cofactor.** The cofactor may be a metal ion (such as iron, magnesium, zinc, or calcium) or an organic molecule called a *coenzyme.* Coenzymes often are derived from vitamins. The names of enzymes usually end in the suffix *-ase.* All enzymes can be grouped according to the types of chemical reactions they catalyze. For example, *oxidases* add oxygen, *kinases* add phosphate, *dehydrogenases* remove hydrogen, *ATPases* split ATP, *anhydrases* remove water, *proteases* break down proteins, and *lipases* break down triglycerides.

Enzymes catalyze specific reactions. They do so with great efficiency and with many built-in controls. Three important properties of enzymes are as follows:

1. ***Enzymes are highly specific.*** Each particular enzyme binds only to specific **substrates**—the reactant molecules on which the enzyme acts. Of the more than 1000 known enzymes in your body, each has a characteristic three-dimensional shape with a specific surface configuration, which allows it to recognize and bind to certain substrates. In some cases, the part of the enzyme that catalyzes the reaction, called the **active site,** is thought to fit the substrate like a key fits in a lock. In other cases the active site changes its shape to fit snugly around the substrate once the substrate enters the active site. This change in shape is known as an *induced fit.*

   Not only is an enzyme matched to a particular substrate; it also catalyzes a specific reaction. From among the large number of diverse molecules in a cell, an enzyme must recognize the correct substrate and then take it apart or merge it with another substrate to form one or more specific products.
2. ***Enzymes are very efficient.*** Under optimal conditions; enzymes can catalyze reactions at rates that are from 100 million to 10 billion times more rapid than those of similar reactions occurring without enzymes. The number of substrate molecules that a single enzyme molecule can convert to product molecules in one second is generally between 1 and 10,000 and can be as high as 600,000.
3. ***Enzymes are subject to a variety of cellular controls.*** Their rate of synthesis and their concentration at any given time are under the control of a cell's genes. Substances within the cell may either enhance or inhibit the activity of a given enzyme. Many enzymes have both active and inactive forms in

cells. The rate at which the inactive form becomes active or vice versa is determined by the chemical environment inside the cell.

Enzymes lower the activation energy of a chemical reaction by decreasing the "randomness" of the collisions between molecules. They also help bring the substrates together in the proper orientation so that the reaction can occur. Figure 2.23 depicts how an enzyme works:

1. The substrates make contact with the active site on the surface of the enzyme molecule, forming a temporary intermediate compound called the **enzyme–substrate complex.** In this reaction the two substrate molecules are sucrose (a disaccharide) and water.
2. The substrate molecules are transformed by the rearrangement of existing atoms, the breakdown of the substrate molecule, or the combination of several substrate molecules into the products of the reaction. Here the products are two monosaccharides: glucose and fructose.
3. After the reaction is completed and the reaction products move away from the enzyme, the unchanged enzyme is free to attach to other substrate molecules.

Sometimes a single enzyme may catalyze a reversible reaction in either direction, depending on the relative amounts of the substrates and products. For example, the enzyme *carbonic anhydrase* catalyzes the following reversible reaction:

$$\underset{\text{Carbon dioxide}}{CO_2} + \underset{\text{Water}}{H_2O} \xrightleftharpoons{\text{Carbonic anhydrase}} \underset{\text{Carbonic acid}}{H_2CO_3}$$

During exercise, when more CO_2 is produced and released into the blood, the reaction flows to the right, increasing the amount of carbonic acid in the blood. Then, as you exhale CO_2, its level in the blood falls and the reaction flows to the left, converting carbonic acid to CO_2 and H_2O.

Figure 2.23 How an enzyme works.

An enzyme speeds up a chemical reaction without being altered or consumed.

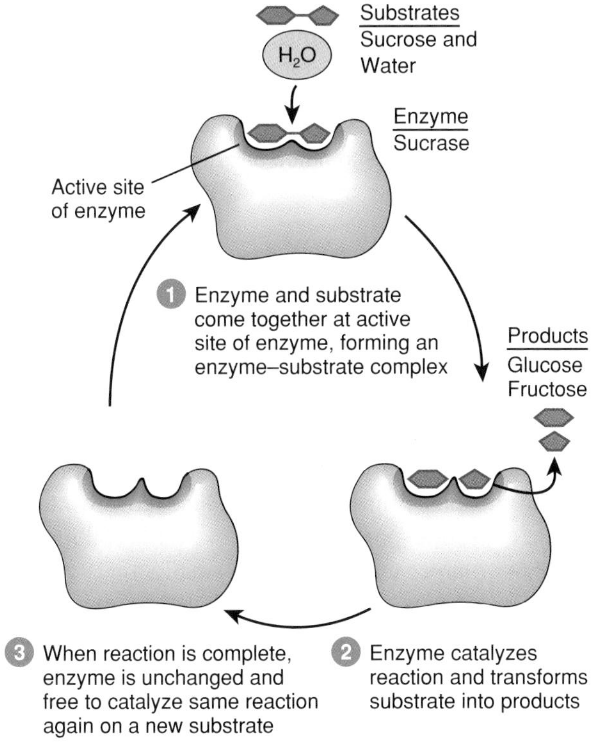

Substrates
Sucrose and Water
H2O
Enzyme
Sucrase
Active site of enzyme
1 Enzyme and substrate come together at active site of enzyme, forming an enzyme–substrate complex
Products
Glucose
Fructose
3 When reaction is complete, enzyme is unchanged and free to catalyze same reaction again on a new substrate
2 Enzyme catalyzes reaction and transforms substrate into products

(a) Mechanism of enzyme action

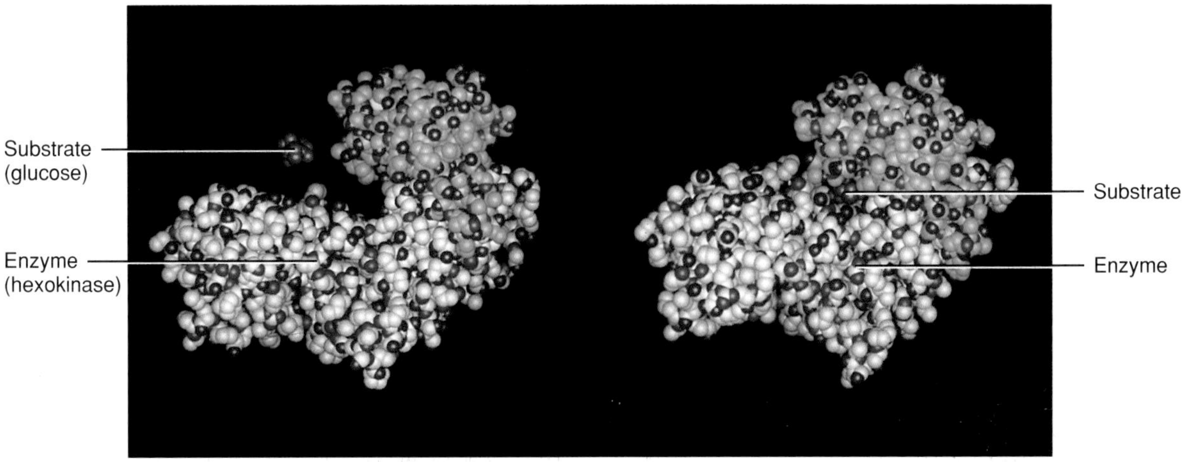

Substrate (glucose)
Enzyme (hexokinase)
Substrate
Enzyme

(b) Molecular model of enzyme and substrate uncombined (left) and enzyme–substrate complex (right)

? Why is it that sucrase cannot catalyze the formation of sucrose from glucose and fructose?

Nucleic Acids: Deoxyribonucleic Acid (DNA) and Ribonucleic Acid (RNA)

Nucleic acids (noo-KLĒ-ic), so named because they were first discovered in the nuclei of cells, are huge organic molecules that contain carbon, hydrogen, oxygen, nitrogen, and phosphorus. Nucleic acids are of two varieties. The first, **deoxyribonucleic acid (DNA)** (dē-ok′-sē-rī-bō-nū-KLĒ-ik), forms the inherited genetic material inside each human cell. In humans, each **gene** (JĒN) is a segment of a DNA molecule. Our genes determine the traits we inherit, and by controlling protein synthesis they regulate most of the activities that take place in body cells throughout our lives. When a cell divides, its hereditary information passes on to the next generation of cells. **Ribonucleic acid (RNA),** the second type of nucleic acid, relays instructions from the genes to guide each cell's synthesis of proteins from amino acids.

A nucleic acid is a chain of repeating monomers called **nucleotides.** Each nucleotide of DNA consists of three parts (Figure 2.24a):

1. ***Nitrogenous base.*** DNA contains four different nitrogenous bases, which contain atoms of C, H, O, and N. In DNA the four **nitrogenous bases** are adenine (A), thymine (T), cytosine (C), and guanine (G). Adenine and guanine are larger, double-ring bases called **purines** (PŪR-ēnz); thymine and cytosine are smaller, single-ring bases called **pyrimidines** (pī-RIM-i-dēnz). The nucleotides are named according to the base that is present. For instance, a nucleotide containing thymine is called a thymine nucleotide, one containing adenine is called an adenine nucleotide, and so on.
2. ***Pentose sugar.*** A five-carbon sugar called **deoxyribose** attaches to each base in DNA.
3. ***Phosphate group.*** Phosphate groups (PO_4^{3-}) alternate with pentose sugars to form the "backbone" of a DNA strand; the bases project inward from the backbone chain (Figure 2.24b).

In 1953, F.H.C. Crick of Great Britain and J.D. Watson, a young American scientist, published a brief paper describing how these three components might be arranged in DNA. Their insights into data gathered by others led them to construct a model so elegant and simple that the scientific world immediately knew it was correct! In the Watson–Crick **double helix** model, DNA resembles a spiral ladder (Figure 2.24b). Two strands of alternating phosphate groups and deoxyribose sugars form the uprights of the ladder. Paired bases, held together by hydrogen bonds, form the rungs. Because adenine always pairs with thymine, and cytosine always pairs with guanine, if you know the sequence of bases in one strand of DNA, you can predict the sequence on the complementary (second) strand. Each time DNA is copied, as when living cells divide to increase their number, the two strands unwind. Each strand serves as the template or mold on which to construct a new second strand. Any change that occurs in the base sequence of a DNA strand is called a *mutation.* Some mutations can result in the death of a cell, cause cancer, or produce genetic defects in future generations.

RNA, the second variety of nucleic acid, differs from DNA in several respects. In humans, RNA is single-stranded. The sugar in the RNA nucleotide is the pentose **ribose,** and RNA contains the pyrimidine base uracil (U) instead of thymine. Cells contain three different kinds of RNA: messenger RNA, ribosomal RNA, and transfer RNA. Each has a specific role to perform in carrying out the instructions coded in DNA (see Figure 3.29).

CLINICAL CONNECTION | *DNA Fingerprinting*

A technique called **DNA fingerprinting** is used in research and in courts of law to ascertain whether a person's DNA matches the DNA obtained from samples or pieces of legal evidence such as blood stains or hairs. In each person, certain DNA segments contain base sequences that are repeated several times. Both the number of repeat copies in one region and the number of regions subject to repeat are different from one person to another. DNA fingerprinting can be done with minute quantities of DNA—for example, from a single strand of hair, a drop of semen, or a spot of blood. It also can be used to identify a crime victim or a child's biological parents and even to determine whether two people have a common ancestor. •

Adenosine Triphosphate

Adenosine triphosphate (a-DEN-ō-sēn trī-FOS-fāt) or **ATP** is the "energy currency" of living systems (Figure 2.25). ATP transfers the energy liberated in exergonic catabolic reactions to power cellular activities that require energy (endergonic reactions). Among these cellular activities are muscular contractions, movement of chromosomes during cell division, movement of structures within cells, transport of substances across cell membranes, and synthesis of larger molecules from smaller ones. As its name implies, ATP consists of three phosphate groups attached to adenosine, a unit composed of adenine and the five-carbon sugar ribose.

When a water molecule is added to ATP, the third phosphate group (PO_4^{3-}), symbolized by Ⓟ in the following discussion, is removed, and the overall reaction liberates energy. The enzyme that catalyzes the hydrolysis of ATP is called *ATPase.* Removal of the third phosphate group produces a molecule called **adenosine diphosphate (ADP)** in the following reaction:

$$\underset{\text{Adenosine triphosphate}}{\text{ATP}} + \underset{\text{Water}}{H_2O} \xrightarrow{\textit{ATPase}} \underset{\text{Adenosine diphosphate}}{\text{ADP}} + \underset{\text{Phosphate group}}{\textcircled{P}} + \underset{\text{Energy}}{\text{E}}$$

As noted previously, the energy supplied by the catabolism of ATP into ADP is constantly being used by the cell. As the supply of ATP at any given time is limited, a mechanism exists to replenish it: The enzyme *ATP synthase* catalyzes the addition of a phosphate group to ADP in the following reaction:

$$\underset{\text{Adenosine diphosphate}}{\text{ADP}} + \underset{\text{Phosphate group}}{\textcircled{P}} + \underset{\text{Energy}}{\text{E}} \xrightarrow{\textit{ATP synthase}} \underset{\text{Adenosine triphosphate}}{\text{ATP}} + \underset{\text{Water}}{H_2O}$$

Where does the cell get the energy required to produce ATP? The energy needed to attach a phosphate group to ADP is supplied

Figure 2.24 **DNA molecule.** (a) A nucleotide consists of a base, a pentose sugar, and a phosphate group. (b) The paired bases project toward the center of the double helix. The structure is stabilized by hydrogen bonds (dotted lines) between each base pair. There are two hydrogen bonds between adenine and thymine and three between cytosine and guanine.

Nucleotides are the monomers of nucleic acids.

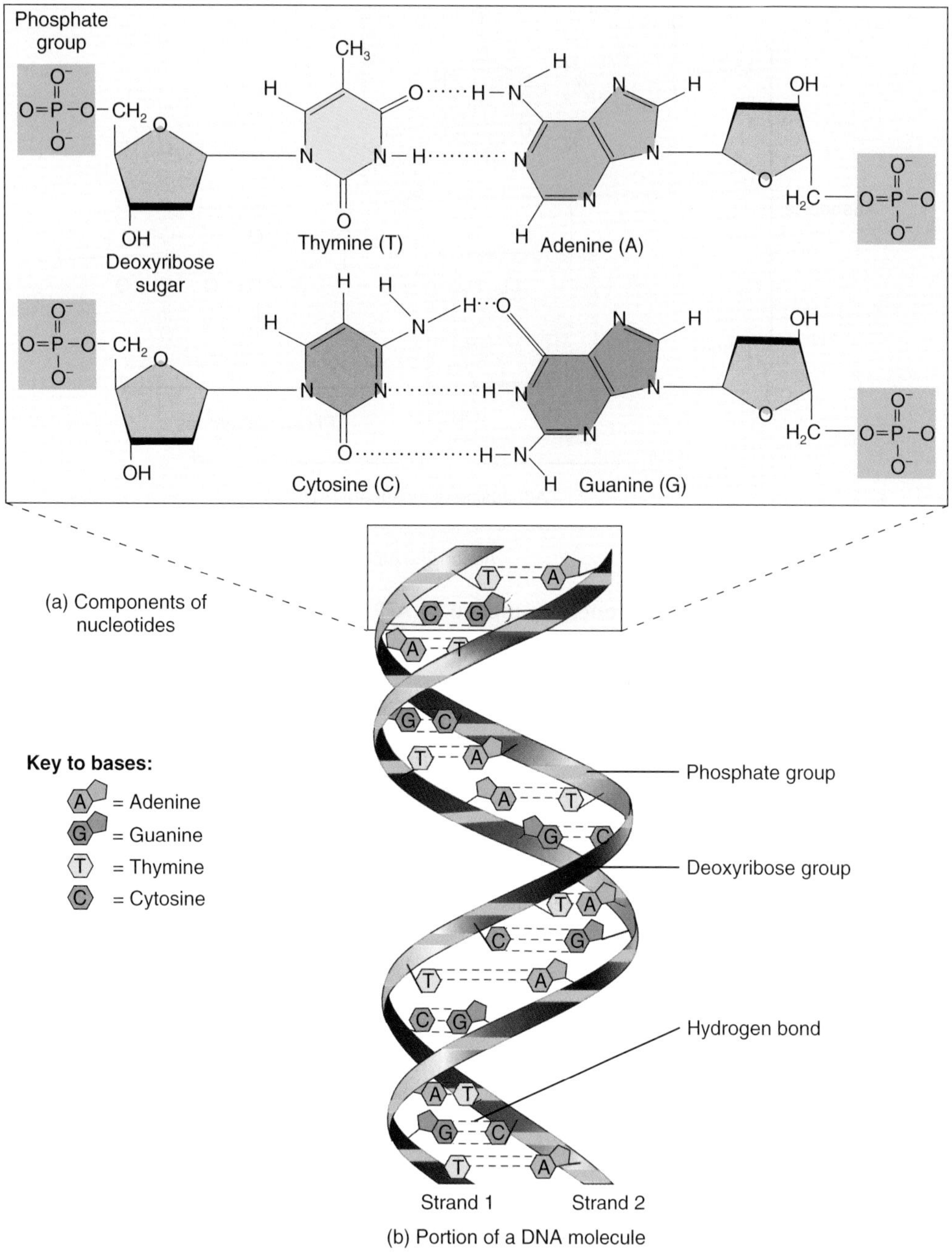
Phosphate group
Deoxyribose sugar
Thymine (T)
Adenine (A)
Cytosine (C)
Guanine (G)
(a) Components of nucleotides
Key to bases:
= Adenine
= Guanine
= Thymine
= Cytosine
Phosphate group
Deoxyribose group
Hydrogen bond
Strand 1
Strand 2
(b) Portion of a DNA molecule

Which bases always pair with each other?

Figure 2.25 Structures of ATP and ADP. "Squiggles" (~) indicate the two phosphate bonds that can be used to transfer energy. Energy transfer typically involves hydrolysis of the last phosphate bond of ATP.

ATP transfers chemical energy to power cellular activities.

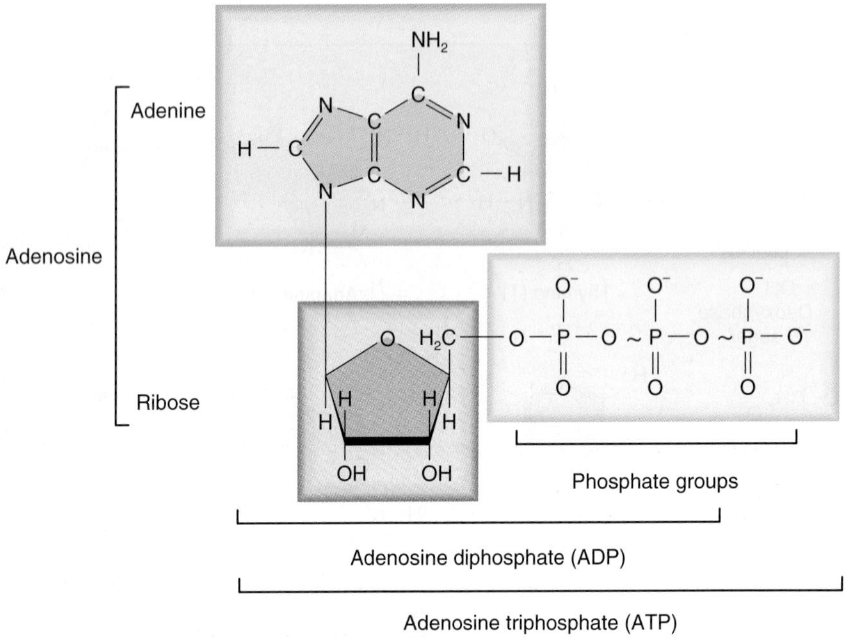

Adenine
Adenosine
Ribose
Phosphate groups
Adenosine diphosphate (ADP)
Adenosine triphosphate (ATP)

What are some cellular activities that depend on energy supplied by ATP?

mainly by the catabolism of glucose in a process called cellular respiration. Cellular respiration has two phases, anaerobic and aerobic:

1. ***Anaerobic phase.*** In a series of reactions that do not require oxygen, glucose is partially broken down by a series of catabolic reactions into pyruvic acid. Each glucose molecule that is converted into a pyruvic acid molecule yields two molecules of ATP.
2. ***Aerobic phase.*** In the presence of oxygen, glucose is completely broken down into carbon dioxide and water. These reactions generate heat and 36 or 38 ATP molecules.

Chapters 10 and 25 cover the details of cellular respiration.

In Chapter 1, you learned that the human body comprises various levels of organization; this chapter has just showed you the alphabet of atoms and molecules that is the basis for the language of the body. Now that you have an understanding of the chemistry of the human body, you are ready to form words; in Chapter 3 you will see how atoms and molecules are organized to form structures of cells and perform the activities of cells that contribute to homeostasis.

CHECKPOINT

20. Define a protein. What is a peptide bond?
21. Outline the levels of structural organization in proteins.
22. Distinguish between fibrous and globular proteins in terms of structure and function.
23. How do DNA and RNA differ?
24. In the reaction catalyzed by ATP synthase, what are the substrates and products? Is this an exergonic or endergonic reaction?

CHAPTER REVIEW AND RESOURCE SUMMARY

WILEY PLUS

Review

2.1 How Matter Is Organized

1. All forms of matter are composed of chemical elements.
2. Oxygen, carbon, hydrogen, and nitrogen make up about 96% of body mass.
3. Each element is made up of small units called atoms. Atoms consist of a nucleus, which contains protons and neutrons, plus electrons that move about the nucleus in regions called electron shells.
4. The number of protons (the atomic number) distinguishes the atoms of one element from those of another element.
5. The mass number of an atom is the sum of its protons and neutrons.

Resource

Anatomy Overview - Common Biomolecules
Animation - Atomic Structure and the Basis of Bonds

Review | **Resource**

6. Different atoms of an element that have the same number of protons but different numbers of neutrons are called isotopes. Radioactive isotopes are unstable and decay.
7. The atomic mass of an element is the average mass of all naturally occurring isotopes of that element.
8. An atom that *gives up* or *gains* electrons becomes an ion—an atom that has a positive or negative charge because it has unequal numbers of protons and electrons. Positively charged ions are cations; negatively charged ions are anions.
9. If two atoms share electrons, a molecule is formed. Compounds contain atoms of two or more elements.
10. A free radical is an atom or group of atoms with an unpaired electron in its outermost shell. A common example is superoxide, an anion which is formed by the addition of an electron to an oxygen molecule.

2.2 Chemical Bonds

1. Forces of attraction called chemical bonds hold atoms together. These bonds result from gaining, losing, or sharing electrons in the valence shell.
2. Most atoms become stable when they have an octet of eight electrons in their valence (outermost) electron shell.
3. When the force of attraction between ions of opposite charge holds them together, an ionic bond has formed.
4. In a covalent bond, atoms share pairs of valence electrons. Covalent bonds may be single, double, or triple and either nonpolar or polar.
5. An atom of hydrogen that forms a polar covalent bond with an oxygen atom or a nitrogen atom may also form a weaker bond, called a hydrogen bond, with an electronegative atom. The polar covalent bond causes the hydrogen atom to have a partial positive charge (δ^+) that attracts the partial negative charge (δ^-) of neighboring electronegative atoms, often oxygen or nitrogen.

Resources:
- Animation - Chemical Bonding
- Figure 2.6 - Hydrogen Bonding among Water Molecules
- Exercise - Bond Boulevard
- Concepts and Connections - Chemical Bonds

2.3 Chemical Reactions

1. When atoms combine with or break apart from other atoms, a chemical reaction occurs. The starting substances are the reactants, and the ending ones are the products.
2. Energy, the capacity to do work, is of two principal kinds: potential (stored) energy and kinetic energy (energy of motion).
3. Endergonic reactions require energy; exergonic reactions release energy. ATP couples endergonic and exergonic reactions.
4. The initial energy investment needed to start a reaction is the activation energy. Reactions are more likely when the concentrations and the temperatures of the reacting particles are higher.
5. Catalysts accelerate chemical reactions by lowering the activation energy. Most catalysts in living organisms are protein molecules called enzymes.
6. Synthesis reactions involve the combination of reactants to produce larger molecules. The reactions are anabolic and usually endergonic.
7. In decomposition reactions, a substance is broken down into smaller molecules. The reactions are catabolic and usually exergonic.
8. Exchange reactions involve the replacement of one atom or atoms by another atom or atoms.
9. In reversible reactions, end products can revert to the original reactants.

Resources:
- Animation - Types of Reactions and Equilibrium
- Exercise - Reaction Race

2.4 Inorganic Compounds and Solutions

1. Inorganic compounds usually are small and usually lack carbon. Organic substances always contain carbon, usually contain hydrogen, and always have covalent bonds.
2. Water is the most abundant substance in the body. It is an excellent solvent and suspending medium, participates in hydrolysis and dehydration synthesis reactions, and serves as a lubricant. Because of its many hydrogen bonds, water molecules are cohesive, which causes a high surface tension. Water also has a high capacity for absorbing heat and a high heat of vaporization.
3. Inorganic acids, bases, and salts dissociate into ions in water. An acid ionizes into hydrogen ions (H^+) and anions and is a proton donor; many bases ionize into cations and hydroxide ions (OH^-), and all are proton acceptors. A salt ionizes into neither H^+ nor OH^-.
4. Mixtures are combinations of elements or compounds that are physically blended together but are not bound by chemical bonds. Solutions, colloids, and suspensions are mixtures with different properties.

Resources:
- Anatomy Overview - Common Biomolecules: Water
- Anatomy Overview- Common Biomolecules: Blood Gases
- Anatomy Overview - Common Biomolecules: Electrolytes
- Animation - Acids and Bases
- Animation - Water and Fluid Flow
- Exercise - Destination: Acid–Base Balance

Review

5. Two ways to express the concentration of a solution are percentage (mass per volume), expressed in grams per 100 mL of a solution, and moles per liter. A mole (abbreviated mol) is the amount in grams of any substance that has a mass equal to the combined atomic mass of all its atoms.
6. The pH of body fluids must remain fairly constant for the body to maintain homeostasis. On the pH scale, 7 represents neutrality. Values below 7 indicate acidic solutions, and values above 7 indicate alkaline solutions. Normal blood pH is 7.35–7.45.
7. Buffer systems remove or add protons (H^+) to help maintain pH homeostasis.
8. One important buffer system is the carbonic acid–bicarbonate buffer system. The bicarbonate ion (HCO_3^-) acts as a weak base and removes excess H^+, and carbonic acid (H_2CO_3) acts as a weak acid and adds H^+.

2.5 Organic Compounds

1. Carbon, with its four valence electrons, bonds covalently with other carbon atoms to form large molecules of many different shapes. Attached to the carbon skeletons of organic molecules are functional groups that confer distinctive chemical properties.
2. Small organic molecules are joined together to form larger molecules by dehydration synthesis reactions in which a molecule of water is removed. In the reverse process, called hydrolysis, large molecules are broken down into smaller ones by the addition of water.
3. Carbohydrates provide most of the chemical energy needed to generate ATP. They may be monosaccharides, disaccharides, or polysaccharides.
4. Lipids are a diverse group of compounds that include fatty acids, triglycerides (fats and oils), phospholipids, steroids, and eicosanoids. Triglycerides protect, insulate, provide energy, and are stored. Phospholipids are important cell membrane components. Steroids are important in cell membrane structure, regulating sexual functions, maintaining normal blood sugar level, aiding lipid digestion and absorption, and helping bone growth. Eicosanoids (prostaglandins and leukotrienes) modify hormone responses, contribute to inflammation, dilate airways, and regulate body temperature.
5. Proteins are constructed from amino acids. They give structure to the body, regulate processes, provide protection, help muscles contract, transport substances, and serve as enzymes. Levels of structural organization among proteins include primary, secondary, tertiary, and (sometimes) quaternary. Variations in protein structure and shape are related to their diverse functions.
6. Deoxyribonucleic acid (DNA) and ribonucleic acid (RNA) are nucleic acids consisting of nitrogenous bases, five-carbon (pentose) sugars, and phosphate groups. DNA is a double helix and is the primary chemical in genes. RNA takes part in protein synthesis.
7. Adenosine triphosphate (ATP) is the principal energy-transferring molecule in living systems. When it transfers energy to an endergonic reaction, it is decomposed to adenosine diphosphate (ADP) and a phosphate group. ATP is synthesized from ADP and a phosphate group using the energy supplied by various decomposition reactions, particularly those of glucose.

Resource

Anatomy Overview - Carbohydrates
Anatomy Overview - Lipids
Anatomy Overview - Proteins
Anatomy Overview - Nucleic Acids
Animation: Enzyme Functions and ATP Parts 1 & 2
Figure 2.21 - Formation of a Peptide Bond between Two Amino Acids

Figure 2.23 - How an Enzyme Works
Exercise - Enzyme Anticipation
Concepts and Connections - Organic Molecules

SELF-QUIZ QUESTIONS

Fill in the blanks in the following statements.

1. An atom with a mass number of 18 that contains 10 neutrons would have an atomic number of ____.
2. Matter exists in three forms: ____, ____, and ____.
3. The building blocks of carbohydrates are the monomers ____ while the building blocks of proteins are the monomers ____.

Indicate whether the following statements are true or false.

4. The elements that compose most of the body's mass are carbon, hydrogen, oxygen, and nitrogen.
5. Ionic bonds are created when atoms share electrons in the valence shell.
6. Human blood has a normal pH between 7.35 and 7.45 and is considered slightly alkaline.

Choose the one best answer to the following questions.

7. Which of the following would be considered a compound? (1) $C_6H_{12}O_6$, (2) O_2, (3) Fe, (4) H_2, (5) CH_4.
(a) all are compounds (b) 1, 2, 4, and 5 (c) 1 and 5
(d) 2 and 4 (e) 3
8. The monosaccharides glucose and fructose combine to form the disaccharide sucrose by a process known as
(a) dehydration synthesis (b) hydrolysis
(c) decomposition (d) hydrogen bonding
(e) ionization.

9. Which of the following is *not* a function of proteins?
(a) provide structural framework
(b) bring about contraction
(c) transport materials throughout the body
(d) store energy
(e) regulate many physiological processes

10. Which of the following organic compounds are classified as lipids? (1) polysaccharides, (2) triglycerides, (3) steroids, (4) enzymes, (5) eicosanoids.
(a) 1, 2, and 4 (b) 2, 3, and 5 (c) 2 and 5
(d) 2, 3, 4, and 5 (e) 2 and 3

11. A compound dissociates in water and forms a cation other than H^+ and an anion other than OH^-. This substance most likely is a(n)
(a) acid (b) base (c) enzyme
(d) buffer (e) salt.

12. Which of the following statements regarding ATP are *true*? (1) ATP is the energy currency for the cell. (2) The energy supplied by the hydrolysis of ATP is constantly being used by cells. (3) Energy is required to produce ATP. (4) The production of ATP involves both aerobic and anaerobic phases. (5) The process of producing energy in the form of ATP is termed the law of conservation of energy.
(a) 1, 2, 3, and 4 (b) 1, 2, 3, and 5 (c) 2, 4, and 5
(d) 1, 2, and 4 (e) 3, 4, and 5

13. During the course of analyzing an unknown chemical, a chemist determines that the chemical is composed of carbon, hydrogen, and oxygen in the proportion of 1 carbon to 2 hydrogens to 1 oxygen. The chemical is probably
(a) an amino acid (b) DNA
(c) a triglyceride (d) a protein
(e) a monosaccharide.

14. Match the following reactions with the term that describes them:

____ (a) $H_2 + Cl_2 \longrightarrow 2HCl$	(1) synthesis reaction
____ (b) $3\ NaOH + H_3PO_4 \longrightarrow Na_3PO_4 + 3\ H_2O$	(2) exchange reaction
____ (c) $CaCO_3 + CO_2 + H_2O \longrightarrow Ca(HCO_3)_2$	(3) decomposition reaction
____ (d) $NH_3 + H_2O \rightleftharpoons NH_4^+ + OH^-$	(4) reversible reaction
____ (e) $C_{12}H_{22}O_{11} + H_2O \longrightarrow C_6H_{12}O_6 + C_6H_{12}O_6$	

15. Match the following:

____ (a) an abundant polar covalent molecule that serves as a solvent, has a high heat capacity, creates a high surface tension, and serves as a lubricant	(1) acid
____ (b) a substance that dissociates into one or more hydrogen ions and one or more anions	(2) free radical
____ (c) a substance that dissociates into cations and anions, neither of which is a hydrogen ion or a hydroxide ion	(3) base
____ (d) a proton acceptor	(4) buffer
____ (e) a measure of hydrogen ion concentration	(5) enzyme
____ (f) a chemical compound that can convert strong acids and bases into weak ones	(6) ion
____ (g) a catalyst for chemical reactions that is specific, efficient, and under cellular control	(7) pH
____ (h) a single-stranded compound that contains a five-carbon sugar, and the bases adenine, cytosine, guanine, and uracil	(8) salt
____ (i) a compound that functions to temporarily store and then transfer energy liberated in exergonic reactions to cellular activities that require energy	(9) RNA
____ (j) a double-stranded compound that contains a five-carbon sugar, the bases adenine, thymine, cytosine, and guanine, and the body's genetic material	(10) ATP
____ (k) a charged atom	(11) water
____ (l) a charged atom with an unpaired electron in its outermost shell	(12) DNA

CRITICAL THINKING QUESTIONS

1. Your best friend has decided to begin frying his breakfast eggs in margarine instead of butter because he has heard that eating butter is bad for his heart. Has he made a wise choice? Are there other alternatives?

2. A 4-month-old baby is admitted to the hospital with a temperature of 102°F (38.9°C). Why is it critical to treat the fever as quickly as possible?

3. During chemistry lab, Maria places sucrose (table sugar) in a glass beaker, adds water, and stirs. As the table sugar disappears, she loudly proclaims that she has chemically broken down the sucrose into fructose and glucose. Is Maria's chemical analysis correct?

? ANSWERS TO FIGURE QUESTIONS

2.1 In carbon, the first shell contains two electrons and the second shell contains four electrons.

2.2 The four most plentiful elements in living organisms are oxygen, carbon, hydrogen, and nitrogen.

2.3 Antioxidants such as selenium, zinc, beta-carotene, vitamin C, and vitamin E can inactivate free radicals derived from oxygen.

2.4 A cation is a positively charged ion; an anion is a negatively charged ion.

2.5 An ionic bond involves the *loss* and *gain* of electrons; a covalent bond involves the *sharing* of pairs of electrons.

2.6 The N atom in ammonia is electronegative. Because it attracts electrons more strongly than do the H atoms, the nitrogen end of ammonia acquires a slight negative charge, allowing H atoms in water molecules (or in other ammonia molecules) to form hydrogen bonds with it. Likewise, O atoms in water molecules can form hydrogen bonds with H atoms in ammonia molecules.

2.7 The number of hydrogen atoms in the reactants must equal the number in the products—in this case, four hydrogen atoms total. Put another way, two molecules of H_2 are needed to react with each molecule of O_2 so that the number of H atoms and O atoms in the reactants is the same as the number of H atoms and O atoms in the products.

2.8 This reaction is exergonic because the reactants have more potential energy than the products.

2.9 No. A catalyst does not change the potential energies of the products and reactants; it only lowers the activation energy needed to get the reaction going.

2.10 No. Because sugar easily dissolves in a polar solvent (water), you can correctly predict that it has several polar covalent bonds.

2.11 $CaCO_3$ is a salt, and H_2SO_4 is an acid.

2.12 At pH = 6, $[H^+] = 10^{-6}$ mol/liter and $[OH^-] = 10^{-8}$ mol/liter. A pH of 6.82 is more acidic than a pH of 6.91. Both pH = 8.41 and pH = 5.59 are 1.41 pH units from neutral (pH = 7).

2.13 Glucose has five —OH groups and six carbon atoms.

2.14 Hexoses are six-carbon sugars; examples include glucose, fructose, and galactose.

2.15 There are 6 carbons in fructose and 12 in sucrose.

2.16 Cells in the liver and in skeletal muscle store glycogen.

2.17 The oxygen in the water molecule comes from a fatty acid.

2.18 The polar head is hydrophilic, and the nonpolar tails are hydrophobic.

2.19 The only differences between estradiol and testosterone are the number of double bonds and the types of functional groups attached to ring A.

2.20 An amino acid has a minimum of two carbon atoms and one nitrogen atom.

2.21 Hydrolysis occurs during catabolism of proteins.

2.22 No. Proteins consisting of a single polypeptide chain do not have a quaternary structure.

2.23 Sucrase has specificity for the sucrose molecule and thus would not "recognize" glucose and fructose.

2.24 In DNA, thymine always pairs with adenine, and cytosine always pairs with guanine.

2.25 Cellular activities that depend on energy supplied by ATP include muscular contractions, movement of chromosomes, transport of substances across cell membranes, and synthesis (anabolic) reactions.

3 THE CELLULAR LEVEL OF ORGANIZATION

CELLS AND HOMEOSTASIS *Cells carry out a multitude of functions that help each system contribute to the homeostasis of the entire body. At the same time, all cells share key structures and functions that support their intense activity.*

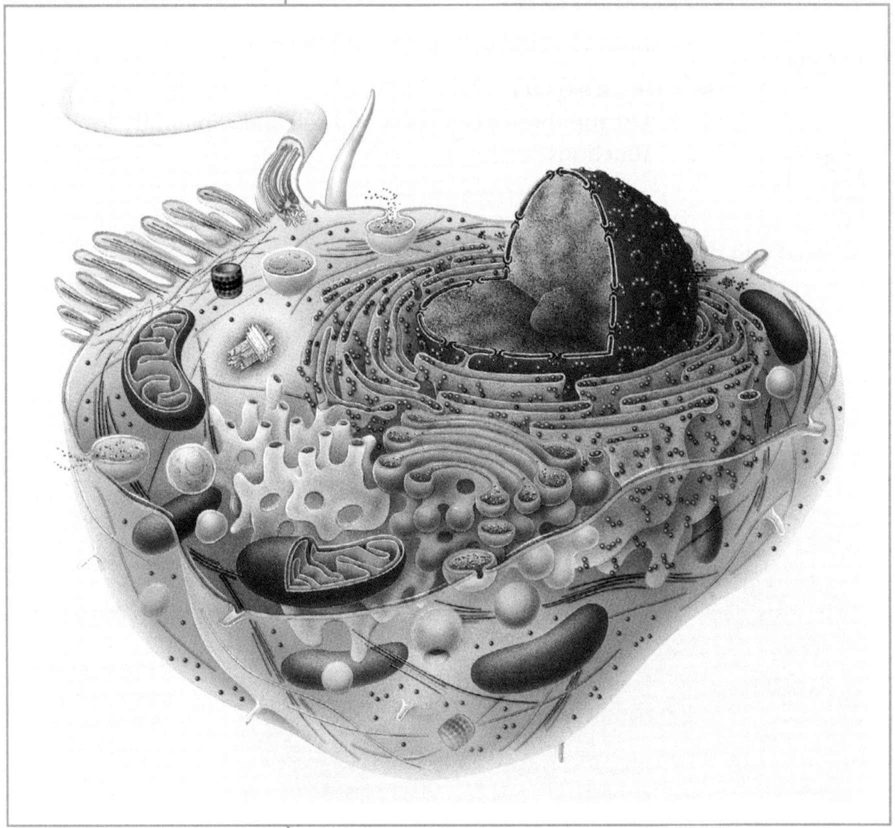

In the previous chapter you learned about the atoms and molecules that compose the alphabet of the language of the human body. These are combined into about 200 different types of words called **cells**—living structural and functional units enclosed by a membrane. All cells arise from existing cells by the process of **cell division,** in which one cell divides into two identical cells. Different types of cells fulfill unique roles that support homeostasis and contribute to the many functional capabilities of the human organism. **Cell biology** or *cytology* is the study of cellular structure and function. As you study the various parts of a cell and their relationships to one another, you will learn that cell structure and function are intimately related. In this chapter, you will learn that cells carry out a dazzling array of chemical reactions to create and maintain life processes—in part, by isolating specific types of chemical reactions within specialized cellular structures.

? ***Did you ever wonder why cancer is so difficult to treat?***

3.1 PARTS OF A CELL

OBJECTIVE

- Name and describe the three main parts of a cell.

Figure 3.1 provides an overview of the typical structures found in body cells. Most cells have many of the structures shown in this diagram. For ease of study, we divide the cell into three main parts: plasma membrane, cytoplasm, and nucleus.

1. The **plasma membrane** forms the cell's flexible outer surface, separating the cell's *internal environment* (everything inside the cell) from the *external environment* (everything outside the cell). It is a selective barrier that regulates the flow of materials into and out of a cell. This selectivity helps establish and maintain the appropriate environment for normal cellular activities. The plasma membrane also plays a key role in communication among cells and between cells and their external environment.
2. The **cytoplasm** (SĪ-tō-plasm; *-plasm* = formed or molded) consists of all the cellular contents between the plasma membrane and the nucleus. This compartment has two components: cytosol and organelles. **Cytosol** (SĪ-tō-sol), the fluid portion of cytoplasm, contains water, dissolved solutes, and suspended particles. Within the cytosol are several different types of **organelles** (or-gan-ELZ = little organs). Each type of organelle has a characteristic shape and specific functions. Examples include the cytoskeleton, ribosomes, endoplasmic reticulum, Golgi complex, lysosomes, peroxisomes, and mitochondria.
3. The **nucleus** (NOO-klē-us = nut kernel) is a large organelle that houses most of a cell's DNA. Within the nucleus, each **chromosome** (KRŌ-mō-sōm; *chromo-* = colored), a single molecule of DNA associated with several proteins, contains thousands of hereditary units called **genes** that control most aspects of cellular structure and function.

✔ CHECKPOINT

1. List the three main parts of a cell and explain their functions.

Figure 3.1 Typical structures found in body cells.

The cell is the basic living, structural, and functional unit of the body.

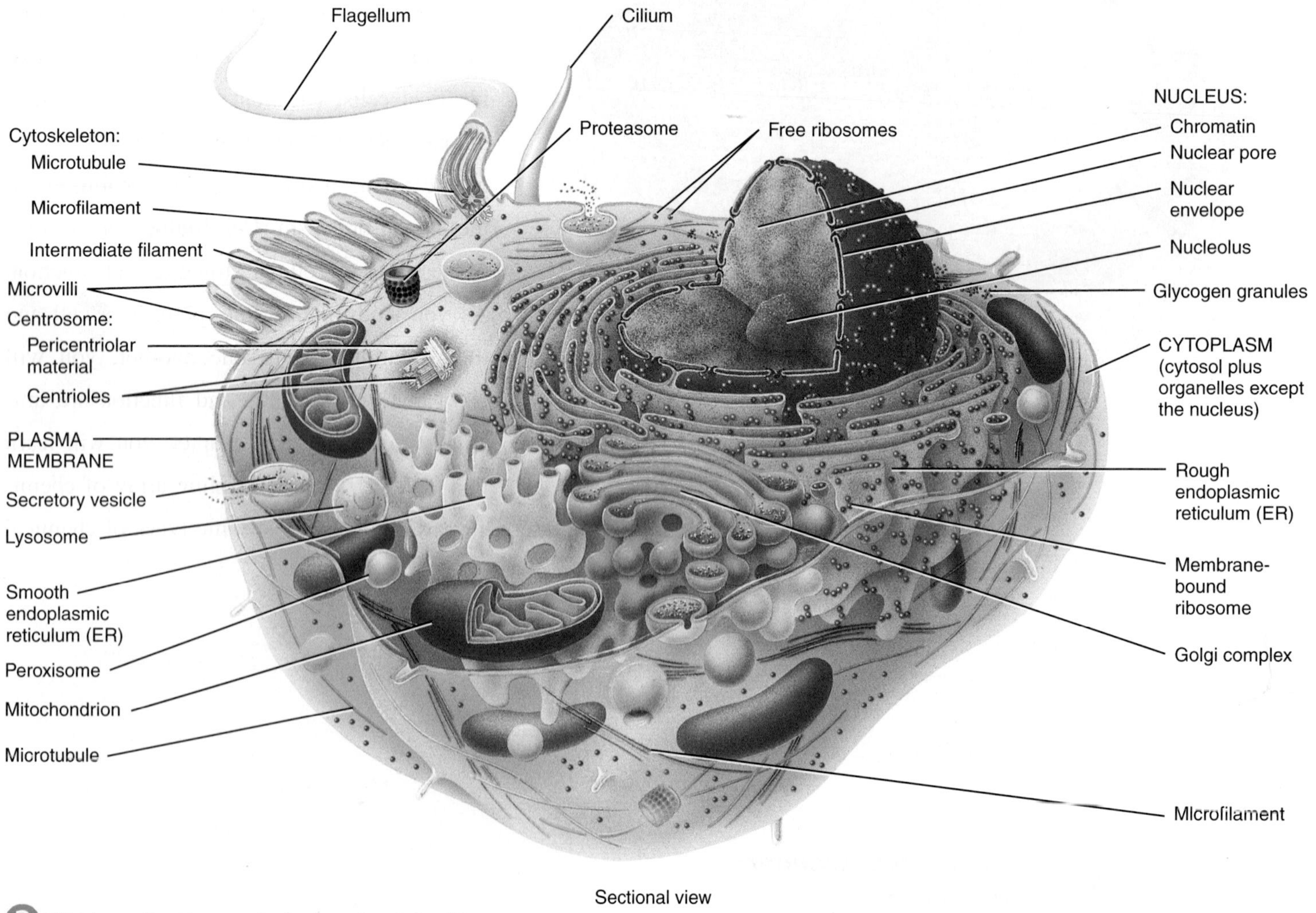

? **What are the three principal parts of a cell?**

3.2 THE PLASMA MEMBRANE

OBJECTIVES

- Distinguish between cytoplasm and cytosol.
- Explain the concept of selective permeability.
- Define the electrochemical gradient and describe its components.

The **plasma membrane,** a flexible yet sturdy barrier that surrounds and contains the cytoplasm of a cell, is best described by using a structural model called the **fluid mosaic model.** According to this model, the molecular arrangement of the plasma membrane resembles a continually moving sea of fluid lipids that contains a mosaic of many different proteins (Figure 3.2). Some proteins float freely like icebergs in the lipid sea, whereas others are anchored at specific locations like islands. The membrane lipids allow passage of several types of lipid-soluble molecules but act as a barrier to the entry or exit of charged or polar substances. Some of the proteins in the plasma membrane allow movement of polar molecules and ions into and out of the cell. Other proteins can act as signal receptors or as molecules that link the plasma membrane to intracellular or extracellular proteins.

Structure of the Plasma Membrane

The Lipid Bilayer

The basic structural framework of the plasma membrane is the **lipid bilayer,** two back-to-back layers made up of three types of lipid molecules—phospholipids, cholesterol, and glycolipids (Figure 3.2). About 75% of the membrane lipids are **phospholipids,** lipids that contain phosphorus. Present in smaller amounts are **cholesterol** (about 20%), a steroid with an attached —OH (hydroxyl) group, and various **glycolipids** (about 5%), lipids with attached carbohydrate groups.

The bilayer arrangement occurs because the lipids are **amphipathic** (am-fē-PATH-ik) molecules, which means that they have both polar and nonpolar parts. In phospholipids (see Figure 2.18), the polar part is the phosphate-containing "head," which is *hydrophilic* (*hydro-* = water; *-philic* = loving). The nonpolar parts are the two long fatty acid "tails," which are *hydrophobic* (*-phobic* = fearing) hydrocarbon chains. Because "like seeks like," the phospholipid molecules orient themselves in the bilayer with their hydrophilic heads facing outward. In this way, the heads face a watery fluid on either side—cytosol on the inside and extracellular fluid on the outside. The hydrophobic fatty acid tails in each half

Figure 3.2 The fluid mosaic arrangement of lipids and proteins in the plasma membrane.

Membranes are fluid structures because the lipids and many of the proteins are free to rotate and move sideways in their own half of the bilayer.

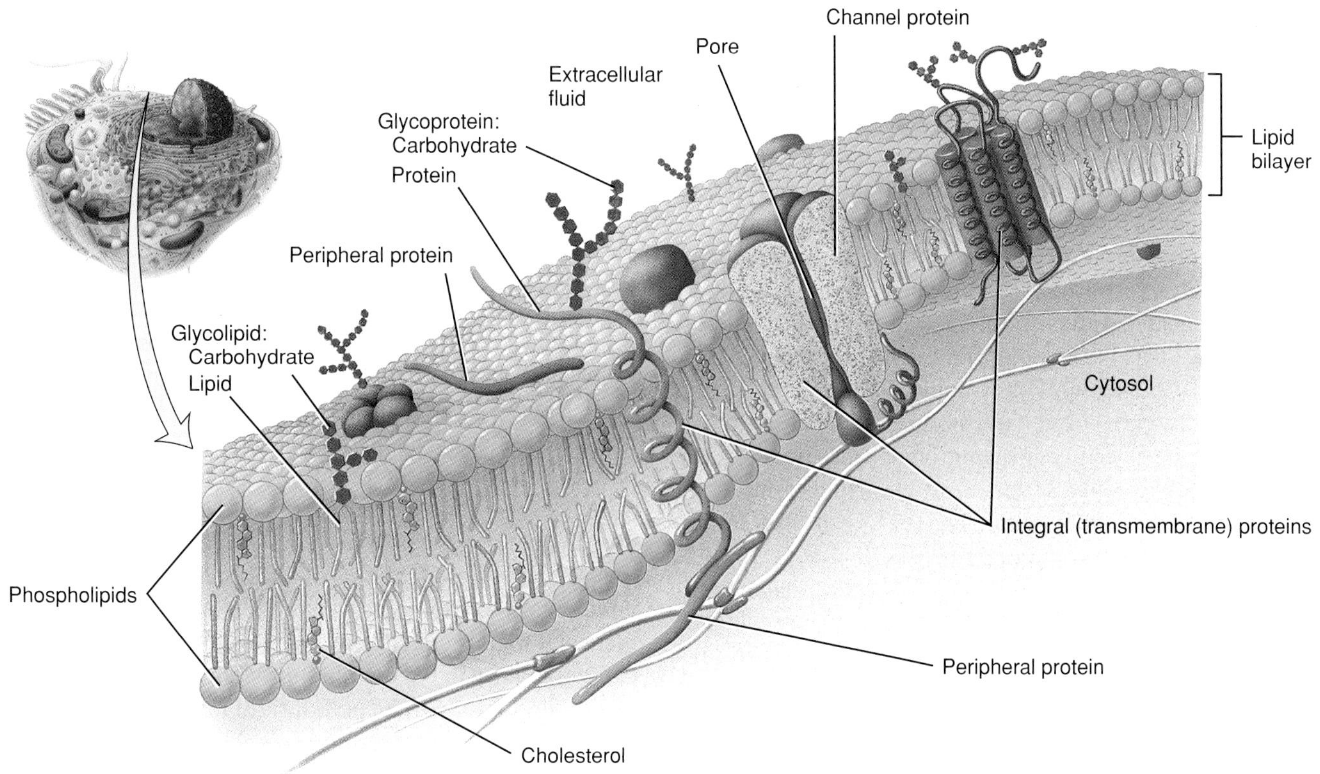

What is the glycocalyx?

of the bilayer point toward one another, forming a nonpolar, hydrophobic region in the membrane's interior.

Cholesterol molecules are weakly amphipathic (see Figure 2.19a) and are interspersed among the other lipids in both layers of the membrane. The tiny —OH group is the only polar region of cholesterol, and it forms hydrogen bonds with the polar heads of phospholipids and glycolipids. The stiff steroid rings and hydrocarbon tail of cholesterol are nonpolar; they fit among the fatty acid tails of the phospholipids and glycolipids. The carbohydrate groups of glycolipids form a polar "head"; their fatty acid "tails" are nonpolar. Glycolipids appear only in the membrane layer that faces the extracellular fluid, which is one reason the two sides of the bilayer are asymmetric, or different.

Arrangement of Membrane Proteins

Membrane proteins are classified as integral or peripheral according to whether they are firmly embedded in the membrane (Figure 3.2). **Integral proteins** extend into or through the lipid bilayer among the fatty acid tails and are firmly embedded in it. Most integral proteins are **transmembrane proteins,** which means that they span the entire lipid bilayer and protrude into both the cytosol and extracellular fluid. A few integral proteins are tightly attached to one side of the bilayer by covalent bonding to fatty acids. Like membrane lipids, integral membrane proteins are amphipathic. Their hydrophilic regions protrude into either the watery extracllular fluid or the cytosol, and their hydrophobic regions extend among the fatty acid tails.

As their name implies, **peripheral proteins** (pe-RIF-er-al) are not as firmly embedded in the membrane. They are attached to the polar heads of membrane lipids or to integral proteins at the inner or outer surface of the membrane.

Many integral proteins are **glycoproteins,** proteins with carbohydrate groups attached to the ends that protrude into the extracellular fluid. The carbohydrates are *oligosaccharides* (*oligo-* = few; *-saccharides* = sugars), chains of 2 to 60 monosaccharides that may be straight or branched. The carbohydrate portions of glycolipids and glycoproteins form an extensive sugary coat called the **glycocalyx** (glī-kō-KĀL-iks). The pattern of carbohydrates in the glycocalyx varies from one cell to another. Therefore, the glycocalyx acts like a molecular "signature" that enables cells to recognize one another. For example, a white blood cell's ability to detect a "foreign" glycocalyx is one basis of the immune response that helps us destroy invading organisms. In addition, the glycocalyx enables cells to adhere to one another in some tissues and protects cells from being digested by enzymes in the extracellular fluid. The hydrophilic properties of the glycocalyx attract a film of fluid to the surface of many cells. This action makes red blood cells slippery as they flow through narrow blood vessels and protects cells that line the airways and the gastrointestinal tract from drying out.

Functions of Membrane Proteins

Generally, the types of lipids in cellular membranes vary only slightly. In contrast, the membranes of different cells and various intracellular organelles have remarkably different assortments of proteins that determine many of the membrane's functions (Figure 3.3).

- Some integral proteins form **ion channels,** *pores* or holes that specific ions, such as potassium ions (K^+), can flow through to get into or out of the cell. Most ion channels are *selective;* they allow only a single type of ion to pass through.
- Other integral proteins act as **carriers,** selectively moving a polar substance or ion from one side of the membrane to the other. Carriers are also known as *transporters.*
- Integral proteins called **receptors** serve as cellular recognition sites. Each type of receptor recognizes and binds a specific type of molecule. For instance, insulin receptors bind the hormone insulin. A specific molecule that binds to a receptor is called a **ligand** (LĪ-gand; *liga* = tied) of that receptor.
- Some integral proteins are **enzymes** that catalyze specific chemical reactions at the inside or outside surface of the cell.
- Integral proteins may also serve as **linkers** that anchor proteins in the plasma membranes of neighboring cells to one another or to protein filaments inside and outside the cell. Peripheral proteins also serve as enzymes and linkers.
- Membrane glycoproteins and glycolipids often serve as **cell-identity markers.** They may enable a cell to (1) recognize other cells of the same kind during tissue formation or (2) recognize and respond to potentially dangerous foreign cells. The ABO blood type markers are one example of cell-identity markers. When you receive a blood transfusion, the blood type must be compatible with your own, or red blood cells may clump together.

In addition, peripheral proteins help support the plasma membrane, anchor integral proteins, and participate in mechanical activities such as moving materials and organelles within cells, changing cell shape in dividing and muscle cells, and attaching cells to one another.

Membrane Fluidity

Membranes are fluid structures; that is, most of the membrane lipids and many of the membrane proteins easily rotate and move sideways in their own half of the bilayer. Neighboring lipid molecules exchange places about 10 million times per second and may wander completely around a cell in only a few minutes! Membrane fluidity depends both on the number of double bonds in the fatty acid tails of the lipids that make up the bilayer, and on the amount of cholesterol present. Each double bond puts a "kink" in the fatty acid tail (see Figure 2.18), which increases membrane fluidity by preventing lipid molecules from packing tightly in the membrane. Membrane fluidity is an excellent compromise for the cell; a rigid membrane would lack mobility, and a completely fluid membrane would lack the structural organization and mechanical support required by the cell. Membrane fluidity allows interactions to occur within the plasma membrane, such as the assembly of membrane proteins. It also

Figure 3.3 Functions of membrane proteins.

Membrane proteins largely reflect the functions a cell can perform.

Extracellular fluid | Plasma membrane | Cytosol

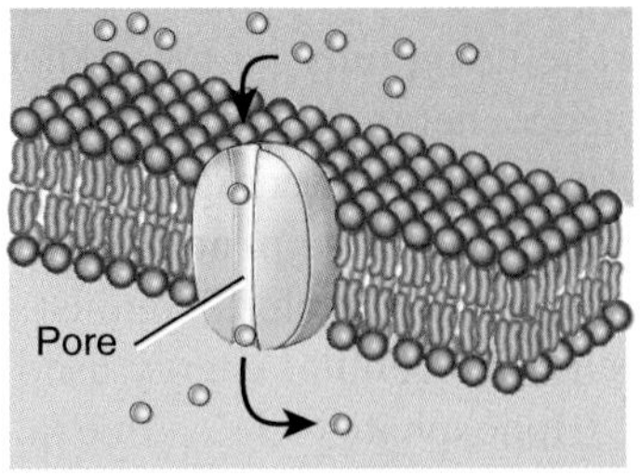

Ion channel (integral)
Allows specific ion (○) to move through water-filled pore. Most plasma membranes include specific channels for several common ions.

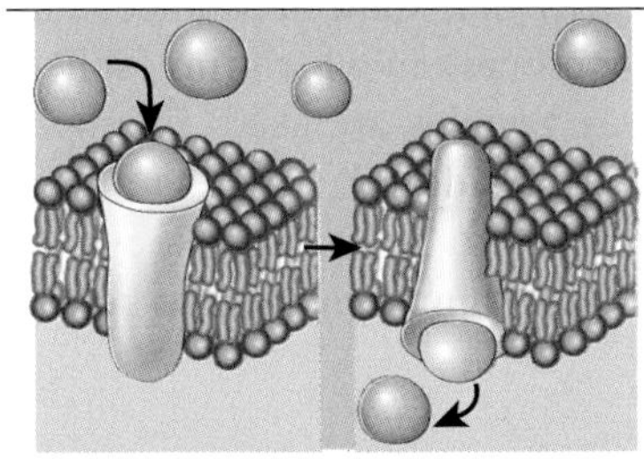

Carrier (integral)
Transports specific substances (○) across membrane by changing shape. For example, amino acids, needed to synthesize new proteins, enter body cells via carriers. Carrier proteins are also known as *transporters*.

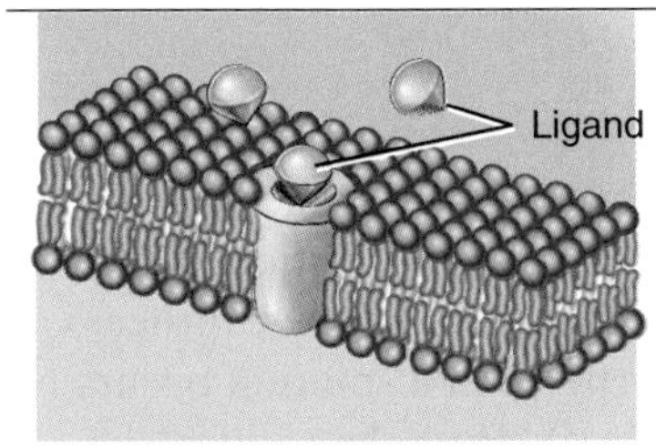

Receptor (integral)
Recognizes specific ligand (▽) and alters cell's function in some way. For example, antidiuretic hormone binds to receptors in the kidneys and changes the water permeability of certain plasma membranes.

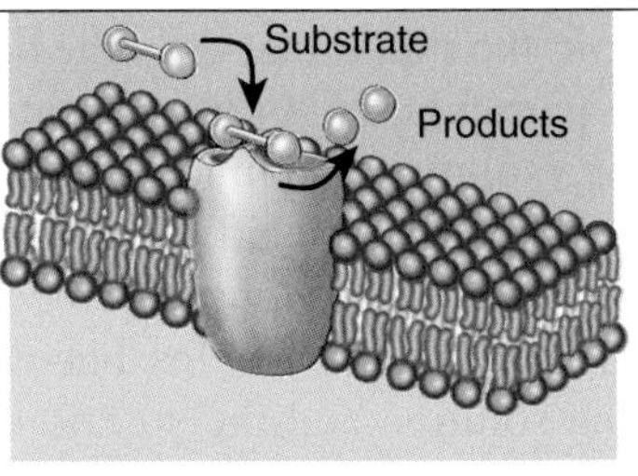

Enzyme (integral and peripheral)
Catalyzes reaction inside or outside cell (depending on which direction the active site faces). For example, lactase protruding from epithelial cells lining your small intestine splits the disaccharide lactose in the milk you drink.

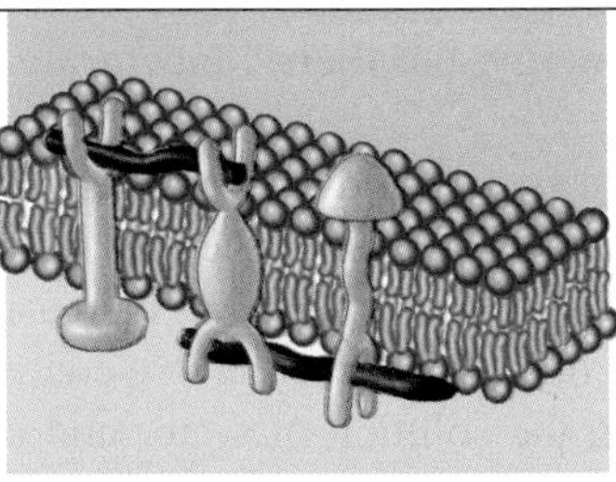

Linker (integral and peripheral)
Anchors filaments inside and outside the plasma membrane, providing structural stability and shape for the cell. May also participate in movement of the cell or link two cells together.

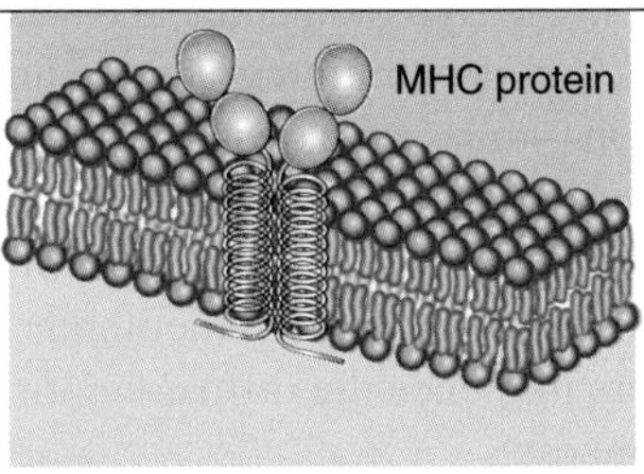

Cell identity marker (glycoprotein)
Distinguishes your cells from anyone else's (unless you are an identical twin). An important class of such markers are the major histocompatibility (MHC) proteins.

? When stimulating a cell, the hormone insulin first binds to a protein in the plasma membrane. This action best represents which membrane protein function?

enables the movement of the membrane components responsible for cellular processes such as cell movement, growth, division, and secretion, and the formation of cellular junctions. Fluidity allows the lipid bilayer to self-seal if torn or punctured. When a needle is pushed through a plasma membrane and pulled out, the puncture site seals spontaneously, and the cell does not burst. This property of the lipid bilayer allows a procedure called intracytoplasmic sperm injection to help infertile couples conceive a child; scientists can fertilize an oocyte by injecting a sperm cell through a tiny syringe. It also permits removal and replacement of a cell's nucleus in cloning experiments, such as the one that created Dolly, the famous cloned sheep.

Despite the great mobility of membrane lipids and proteins in their own half of the bilayer, they seldom flip-flop from one half of the bilayer to the other, because it is difficult for hydrophilic parts of membrane molecules to pass through the hydrophobic core of the membrane. This difficulty contributes to the asymmetry of the membrane bilayer.

Because of the way it forms hydrogen bonds with neighboring phospholipid and glycolipid heads and fills the space between bent fatty acid tails, cholesterol makes the lipid bilayer stronger but less fluid at normal body temperature. At low temperatures, cholesterol has the opposite effect—it increases membrane fluidity.

Membrane Permeability

The term *permeable* means that a structure permits the passage of substances through it, while *impermeable* means that a structure does not permit the passage of substances through it. The permeability of the plasma membrane to different substances varies. Plasma membranes permit some substances to pass more readily than others. This property of membranes is termed **selective permeability** (per′-mē-a-BIL-i-tē).

The lipid bilayer portion of the membrane is permeable to nonpolar, uncharged molecules, such as oxygen, carbon dioxide, and steroids, but is impermeable to ions and large, uncharged polar molecules such as glucose. It is also *slightly* permeable to small, uncharged polar molecules such as water and urea, a waste product from the breakdown of amino acids. The slight permeability to water and urea is an unexpected property since they are polar molecules. These two small molecules are thought to pass through the lipid bilayer in the following way. As the fatty acid tails of membrane phospholipids and glycolipids randomly move about, small gaps briefly appear in the hydrophobic environment of the membrane's interior. Water and urea molecules are small enough to move from one gap to another until they have crossed the membrane.

Transmembrane proteins that act as channels and carriers increase the plasma membrane's permeability to a variety of ions and uncharged polar molecules that, unlike water and urea molecules, cannot cross the lipid bilayer unassisted. Channels and carriers are very selective. Each one helps a specific molecule or ion to cross the membrane. Macromolecules, such as proteins, are so large that they are unable to pass across the plasma membrane except by endocytosis and exocytosis (discussed later in this chapter).

Gradients across the Plasma Membrane

The selective permeability of the plasma membrane allows a living cell to maintain different concentrations of certain substances on either side of the plasma membrane. A **concentration gradient** is a difference in the concentration of a chemical from one place to another, such as from the inside to the outside of the plasma membrane. Many ions and molecules are more concentrated in either the cytosol or the extracellular fluid. For instance, oxygen molecules and sodium ions (Na^+) are more concentrated in the extracellular fluid than in the cytosol; the opposite is true of carbon dioxide molecules and potassium ions (K^+).

The plasma membrane also creates a difference in the distribution of positively and negatively charged ions between the two sides of the plasma membrane. Typically, the inner surface of the plasma membrane is more negatively charged and the outer surface is more positively charged. A difference in electrical charges between two regions constitutes an **electrical gradient.** Because it occurs across the plasma membrane, this charge difference is termed the **membrane potential.**

As you will see shortly, the concentration gradient and electrical gradient are important because they help move substances across the plasma membrane. In many cases a substance will move across a plasma membrane *down its concentration gradient.* That is to say, a substance will move "downhill," from where it is more concentrated to where it is less concentrated, to reach equilibrium. Similarly, a positively charged substance will tend to move toward a negatively charged area, and a negatively charged substance will tend to move toward a positively charged area. The combined influence of the concentration gradient and the electrical gradient on movement of a particular ion is referred to as its **electrochemical gradient.**

✔CHECKPOINT

2. How do hydrophobic and hydrophilic regions govern the arrangement of membrane lipids in a bilayer?
3. What substances can and cannot diffuse through the lipid bilayer?
4. "The proteins present in a plasma membrane determine the functions that a membrane can perform." Is this statement true or false? Explain your answer.
5. How does cholesterol affect membrane fluidity?
6. Why are membranes said to have selective permeability?
7. What factors contribute to an electrochemical gradient?

3.3 TRANSPORT ACROSS THE PLASMA MEMBRANE

OBJECTIVE

- Describe the processes that transport substances across the plasma membrane.

Transport of materials across the plasma membrane is essential to the life of a cell. Certain substances must move into the cell to support metabolic reactions. Other substances that have been produced by the cell for export or as cellular waste products must move out of the cell.

Substances generally move across cellular membranes via transport processes that can be classified as passive or active, depending on whether they require cellular energy. In **passive processes,** a substance moves down its concentration or electrical gradient to cross the membrane using only its own kinetic energy (energy of motion). Kinetic energy is intrinsic to the particles that are moving. There is no input of energy from the cell. An example is simple diffusion. In **active processes,** cellular energy is used to drive the substance "uphill" against its concentration or electrical gradient. The cellular energy used is usually in the form of adenosine triphosphate (ATP). An example is active transport. Another way that some substances may enter and leave cells is an active process in which tiny, spherical membrane sacs referred to as **vesicles** are used. Examples include endocytosis, in which vesicles detach from the plasma membrane while bringing materials into a cell, and exocytosis, the merging of vesicles with the plasma membrane to release materials from the cell.

Passive Processes

The Principle of Diffusion

Learning why materials diffuse across membranes requires an understanding of how diffusion occurs in a solution. **Diffusion** (di-FŪ-zhun; *diffus-* = spreading) is a passive process in which the random mixing of particles in a solution occurs because of the particles' kinetic energy. Both the *solutes,* the dissolved substances, and the *solvent,* the liquid that does the dissolving, undergo diffusion. If a particular solute is present in high concentration in one area of a solution and in low concentration in another area, solute molecules will diffuse toward the area of lower concentration—they move *down their concentration gradient.* After some time, the particles become evenly distributed throughout the solution and the solution is said to be at equilibrium. The particles continue to move about randomly due to their kinetic energy, but their concentrations do not change.

For example, when you place a crystal of dye in a water-filled container (Figure 3.4), the color is most intense in the area closest to the dye because its concentration is higher there. At increasing distances, the color is lighter and lighter because the dye concentration is lower. Some time later, the solution of water and dye will have a uniform color, because the dye molecules and water molecules have diffused down their concentration gradients until they are evenly mixed in solution—they are at equilibrium (ē-kwi-LIB-rē-um).

In this simple example, no membrane was involved. Substances may also diffuse through a membrane, if the membrane is permeable to them. Several factors influence the diffusion rate of substances across plasma membranes:

- ***Steepness of the concentration gradient.*** The greater the difference in concentration between the two sides of the membrane, the higher is the rate of diffusion. When charged particles are diffusing, the steepness of the electrochemical gradient determines the diffusion rate across the membrane.

Figure 3.4 **Principle of diffusion.** At the beginning of our experiment, a crystal of dye placed in a cylinder of water dissolves (a) and then diffuses from the region of higher dye concentration to regions of lower dye concentration (b). At equilibrium (c), the dye concentration is uniform throughout, although random movement continues.

In diffusion, a substance moves down its concentration gradient.

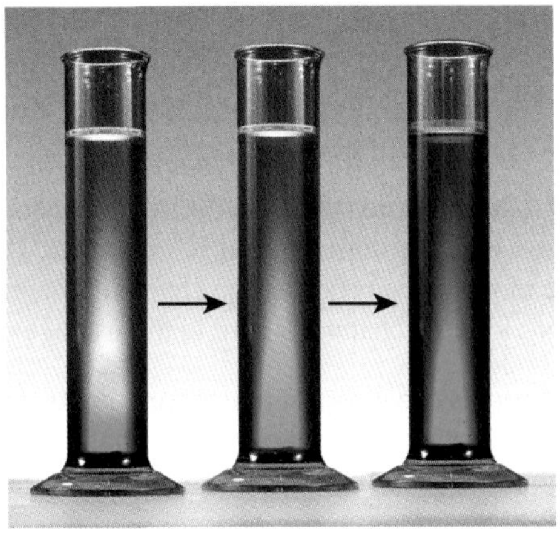

How would having a fever affect body processes that involve diffusion?

- ***Temperature.*** The higher the temperature, the faster the rate of diffusion. All of the body's diffusion processes occur more rapidly in a person with a fever.
- ***Mass of the diffusing substance.*** The larger the mass of the diffusing particle, the slower its diffusion rate. Smaller molecules diffuse more rapidly than larger ones.
- ***Surface area.*** The larger the membrane surface area available for diffusion, the faster is the diffusion rate. For example, the air sacs of the lungs have a large surface area available for diffusion of oxygen from the air into the blood. Some lung diseases, such as emphysema, reduce the surface area. This slows the rate of oxygen diffusion and makes breathing more difficult.
- ***Diffusion distance.*** The greater the distance over which diffusion must occur, the longer it takes. Diffusion across a plasma membrane takes only a fraction of a second because the membrane is so thin. In pneumonia, fluid collects in the lungs; the additional fluid increases the diffusion distance because oxygen must move through both the built-up fluid and the membrane to reach the bloodstream.

Now that you have a basic understanding of the nature of diffusion, we will consider three types of diffusion: simple diffusion, facilitated diffusion, and osmosis.

Simple Diffusion

Simple diffusion is a passive process in which substances move freely through the lipid bilayer of the plasma membranes of cells without the help of membrane transport proteins (Figure 3.5). Nonpolar, hydrophobic molecules move across the lipid bilayer through the process of simple diffusion. Such molecules include oxygen, carbon dioxide, and nitrogen gases; fatty acids; steroids; and fat-soluble vitamins (A, D, E, and K). Small, uncharged polar

Figure 3.5 **Simple diffusion, channel-mediated facilitated diffusion, and carrier-mediated facilitated diffusion.**

In simple diffusion, a substance moves across the lipid bilayer of the plasma membrane without the help of membrane transport proteins. In facilitated diffusion, a substance moves across the lipid bilayer aided by a channel protein or a carrier protein.

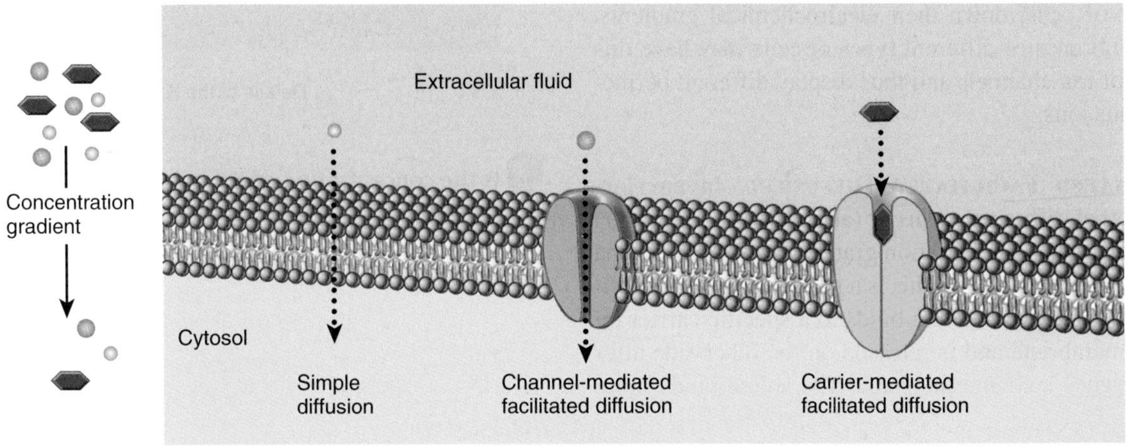

What types of molecules move across the lipid bilayer of the plasma membrane via simple diffusion?

molecules such as water, urea, and small alcohols also pass through the lipid bilayer by simple diffusion. Simple diffusion through the lipid bilayer is important in the movement of oxygen and carbon dioxide between blood and body cells, and between blood and air within the lungs during breathing. It also is the route for absorption of some nutrients and excretion of some wastes by body cells.

Facilitated Diffusion

Solutes that are too polar or highly charged to move through the lipid bilayer by simple diffusion can cross the plasma membrane by a passive process called **facilitated diffusion.** In this process, an integral membrane protein assists a specific substance across the membrane. The integral membrane protein can be either a membrane channel or a carrier.

CHANNEL-MEDIATED FACILITATED DIFFUSION In **channel-mediated facilitated diffusion,** a solute moves down its concentration gradient across the lipid bilayer through a membrane channel (Figure 3.5). Most membrane channels are *ion channels,* integral transmembrane proteins that allow passage of small, inorganic ions that are too hydrophilic to penetrate the nonpolar interior of the lipid bilayer. Each ion can diffuse across the membrane only at certain sites. In typical plasma membranes, the most numerous ion channels are selective for K^+ (potassium ions) or Cl^- (chloride ions); fewer channels are available for Na^+ (sodium ions) or Ca^{2+} (calcium ions). Diffusion of ions through channels is generally slower than free diffusion through the lipid bilayer because channels occupy a smaller fraction of the membrane's total surface area than lipids. Still, facilitated diffusion through channels is a very fast process: More than a million potassium ions can flow through a K^+ channel in one second!

A channel is said to be *gated* when part of the channel protein acts as a "plug" or "gate," changing shape in one way to open the pore and in another way to close it (Figure 3.6). Some gated channels randomly alternate between the open and closed positions; others are regulated by chemical or electrical changes inside and outside the cell. When the gates of a channel are open, ions diffuse into or out of cells, down their electrochemical gradients. The plasma membranes of different types of cells may have different numbers of ion channels and thus display different permeabilities to various ions.

CARRIER-MEDIATED FACILITATED DIFFUSION In **carrier-mediated facilitated diffusion,** a *carrier* (also called a *transporter*) moves a solute down its concentration gradient across the plasma membrane (see Figure 3.5). Since this is a passive process, no cellular energy is required. The solute binds to a specific carrier on one side of the membrane and is released on the other side after the carrier undergoes a change in shape. The solute binds more often to the carrier on the side of the membrane with a higher concentration of solute. Once the concentration is the same on both sides of the membrane, solute molecules bind to the carrier on the cytosolic side and move out to the extracellular fluid as rapidly as they bind to the carrier on the extracellular side and move into the cytosol. The rate of carrier-mediated facilitated diffusion (how quickly it occurs) is determined by the steepness of the concentration gradient across the membrane.

The number of carriers available in a plasma membrane places an upper limit, called the *transport maximum,* on the rate at which facilitated diffusion can occur. Once all of the carriers are occupied, the transport maximum is reached, and a further increase in the concentration gradient does not increase the rate of facilitated diffusion. Thus, much like a completely saturated sponge can absorb no more water, the process of carrier-mediated facilitated diffusion exhibits *saturation.*

Figure 3.6 Channel-mediated facilitated diffusion of potassium ions (K^+) through a gated K^+ channel. A gated channel is one in which a portion of the channel protein acts as a gate to open or close the channel's pore to the passage of ions.

Channels are integral membrane proteins that allow specific, small, inorganic ions to pass across the membrane by facilitated diffusion.

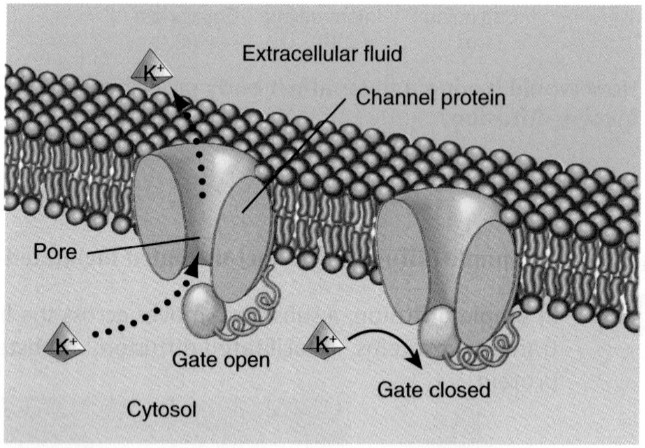

Details of the K^+ channel

Is the concentration of K^+ in body cells higher in the cytosol or in the extracellular fluid?

Substances that move across the plasma membrane by carrier-mediated facilitated diffusion include glucose, fructose, galactose, and some vitamins. Glucose, the body's preferred energy source for making ATP, enters many body cells by carrier-mediated facilitated diffusion as follows (Figure 3.7):

1. Glucose binds to a specific type of carrier protein called the *glucose transporter* (GluT) on the outside surface of the membrane.
2. As the transporter undergoes a change in shape, glucose passes through the membrane.
3. The transporter releases glucose on the other side of the membrane.

The selective permeability of the plasma membrane is often regulated to achieve homeostasis. For instance, the hormone insulin, via the action of the insulin receptor, promotes the insertion of many copies of glucose transporters into the plasma membranes of certain cells. Thus, the effect of insulin is to elevate the transport maximum for facilitated diffusion of glucose into cells. With more glucose transporters available, body cells can pick up glucose from the blood more rapidly. An inability to produce or utilize insulin is called diabetes mellitus (Chapter 18).

Figure 3.7 Carrier-mediated facilitated diffusion of glucose across a plasma membrane. The carrier protein binds to glucose in the extracellular fluid and releases it into the cytosol.

Carriers are integral membrane proteins that undergo changes in shape in order to move substances across the membrane by facilitated diffusion.

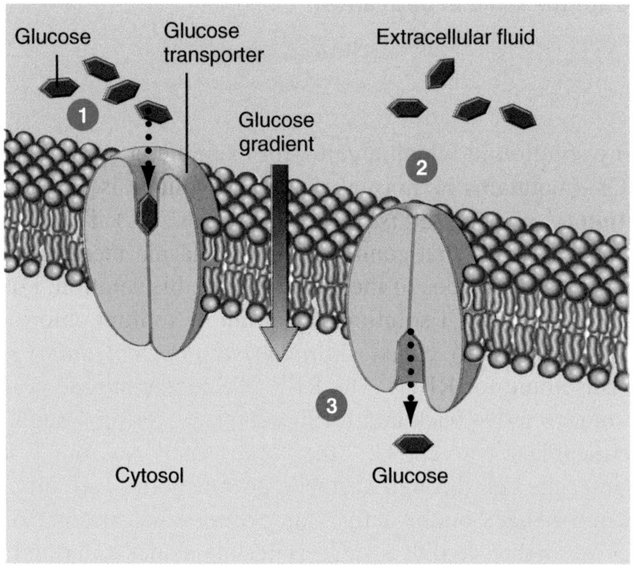
Glucose
Glucose transporter
Extracellular fluid
1
Glucose gradient
2
3
Cytosol
Glucose

Does insulin alter glucose transport by facilitated diffusion?

Osmosis

Osmosis (oz-MŌ-sis) is a type of diffusion in which there is net movement of a solvent through a selectively permeable membrane. Like the other types of diffusion, osmosis is a passive process. In living systems, the solvent is water, which moves by osmosis across plasma membranes from an area of *higher water concentration* to an area of *lower water concentration*. Another way to understand this idea is to consider the solute concentration: In osmosis, water moves through a selectively permeable membrane from an area of *lower solute concentration* to an area of *higher solute concentration*. During osmosis, water molecules pass through a plasma membrane in two ways: (1) by moving between neighboring phospholipid molecules in the lipid bilayer via simple diffusion, as previously described, and (2) by moving through **aquaporins** (ak-wa-POR-ins; *aqua-* = water), integral membrane proteins that function as water channels.

Osmosis occurs only when a membrane is permeable to water but is not permeable to certain solutes. A simple experiment can demonstrate osmosis. Consider a U-shaped tube in which a selectively permeable membrane separates the left and right arms of the tube. A volume of pure water is poured into the left arm, and the same volume of a solution containing a solute that cannot pass through the membrane is poured into the right arm (Figure 3.8a). Because the *water* concentration is higher on the left and lower on the right, net movement of water molecules—osmosis—occurs from left to right, so that the water is moving down its concentration gradient. At the same time, the membrane prevents diffusion of the solute from the right arm into the left arm. As a result, the volume of water in the left arm decreases, and the volume of solution in the right arm increases (Figure 3.8b).

You might think that osmosis would continue until no water remained on the left side, but this is *not* what happens. In this experiment, the higher the column of solution in the right arm becomes, the more pressure it exerts on its side of the membrane. Pressure exerted in this way by a liquid, known as **hydrostatic pressure,** forces water molecules to move back into the left arm. Equilibrium is reached when just as many water molecules move from right to left due to the hydrostatic pressure as move from left to right due to osmosis (Figure 3.8b).

To further complicate matters, the solution with the impermeable solute also exerts a force, called the **osmotic pressure.** The osmotic pressure of a solution is proportional to the concentration of the solute particles that cannot cross the membrane—the higher the solute concentration, the higher the solution's osmotic

Figure 3.8 **Principle of osmosis.** Water molecules move through the selectively permeable membrane; the solute molecules in the right arm cannot pass through the membrane. (a) As the experiment starts, water molecules move from the left arm into the right arm, down the water concentration gradient. (b) After some time, the volume of water in the left arm has decreased and the volume of solution in the right arm has increased. At equilibrium, there is no net osmosis: Hydrostatic pressure forces just as many water molecules to move from right to left as osmosis forces water molecules to move from left to right. (c) If pressure is applied to the solution in the right arm, the starting conditions can be restored. This pressure, which stops osmosis, is equal to the osmotic pressure.

Osmosis is the movement of water molecules through a selectively permeable membrane.

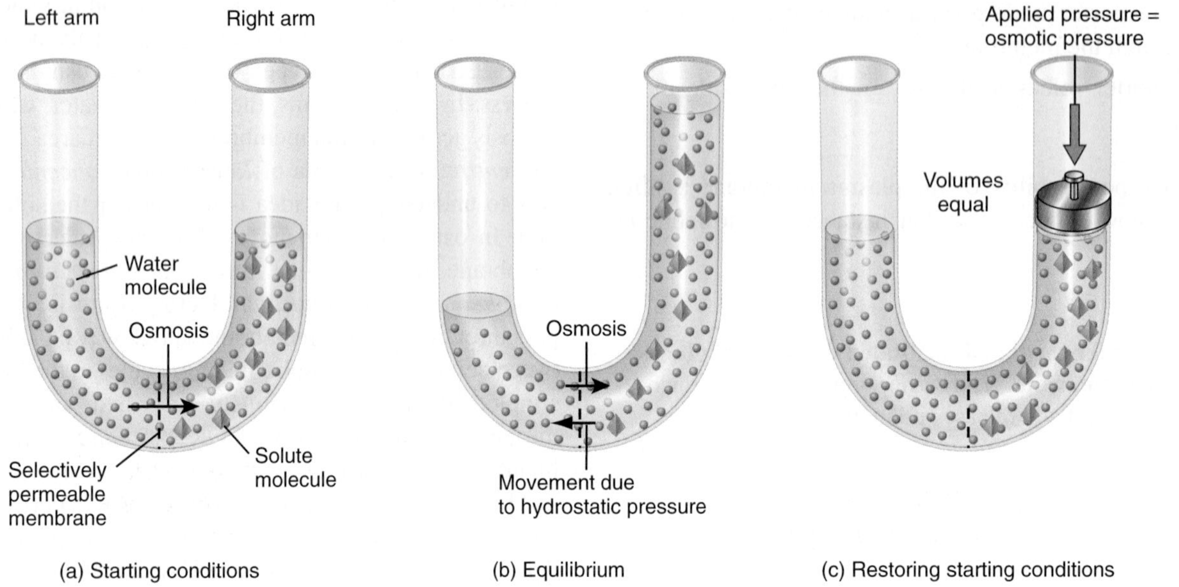

Will the fluid level in the right arm rise until the water concentrations are the same in both arms?

pressure. Consider what would happen if a piston were used to apply more pressure to the fluid in the right arm of the tube in Figure 3.8. With enough pressure, the volume of fluid in each arm could be restored to the starting volume, and the concentration of solute in the right arm would be the same as it was at the beginning of the experiment (Figure 3.8c). The amount of pressure needed to restore the starting condition equals the osmotic pressure. So, in our experiment osmotic pressure is the pressure needed to stop the movement of water from the left tube into the right tube. Notice that the osmotic pressure of a solution does not produce the movement of water during osmosis. Rather it is the pressure that would *prevent* such water movement.

Normally, the osmotic pressure of the cytosol is the same as the osmotic pressure of the interstitial fluid outside cells. Because the osmotic pressure on both sides of the plasma membrane (which is selectively permeable) is the same, cell volume remains relatively constant. When body cells are placed in a solution having a different osmotic pressure than cytosol, however, the shape and volume of the cells change. As water moves by osmosis into or out of the cells, their volume increases or decreases. A solution's **tonicity** (tō-NIS-i-tē; *tonic* = tension) is a measure of the solution's ability to change the volume of cells by altering their water content.

Any solution in which a cell—for example, a red blood cell (RBC)—maintains its normal shape and volume is an **isotonic solution** (ī′-sō-TON-ik; *iso-* = same) (Figure 3.9). The concentrations of solutes that cannot cross the plasma membrane are the same on both sides of the membrane in this solution. For instance, a 0.9% NaCl solution (0.9 gram of sodium chloride in 100 mL of solution), called a *normal (physiological) saline solution,* is isotonic for RBCs. The RBC plasma membrane permits the water to move back and forth, but it behaves as though it is impermeable to Na^+ and Cl^-, the solutes. (Any Na^+ or Cl^- ions that enter the cell through channels or transporters are immediately moved back out by active transport or other means.) When RBCs are bathed in 0.9% NaCl, water molecules enter and exit at the same rate, allowing the RBCs to keep their normal shape and volume.

A different situation results if RBCs are placed in a **hypotonic solution** (hī′-pō-TON-ik; *hypo-* = less than), a solution that has a *lower* concentration of solutes than the cytosol inside the RBCs (Figure 3.9). In this case, water molecules enter the cells faster than they leave, causing the RBCs to swell and eventually to burst. The rupture of RBCs in this manner is called **hemolysis** (hē-MOL-i-sis; *hemo-* = blood; *-lysis* = to loosen or split apart); the rupture of other types of cells due to placement in a hypotonic solution is referred to simply as **lysis.** Pure water is very hypotonic and causes rapid hemolysis.

A **hypertonic solution** (hī′-per-TON-ik; *hyper-* = greater than) has a *higher* concentration of solutes than does the cytosol inside RBCs (Figure 3.9). One example of a hypertonic solution is a 2% NaCl solution. In such a solution, water molecules move out of the cells faster than they enter, causing the cells to shrink. Such shrinkage of cells is called **crenation** (kre-NĀ-shun).

CLINICAL CONNECTION | *Medical Uses of Isotonic, Hypertonic, and Hypotonic Solutions*

RBCs and other body cells may be damaged or destroyed if exposed to hypertonic or hypotonic solutions. For this reason, most **intravenous (IV) solutions,** liquids infused into the blood of a vein, are isotonic. Examples are isotonic saline (0.9% NaCl) and D5W, which stands for dextrose 5% in water. Sometimes infusion of a hypertonic solution such as mannitol is useful to treat patients who have *cerebral edema,* excess interstitial fluid in the brain. Infusion of such a solution relieves fluid overload by causing osmosis of water from interstitial fluid into the blood. The kidneys then excrete the excess water from the blood into the urine. Hypotonic solutions, given either orally or through an IV, can be used to treat people who are dehydrated. The water in the hypotonic solution moves from the blood into interstitial fluid and then into body cells to rehydrate them. Water and most sports drinks that you consume to "rehydrate" after a workout are hypotonic relative to your body cells. •

✔CHECKPOINT

8. What factors can increase the rate of diffusion?
9. How does simple diffusion compare with facilitated diffusion?
10. What is osmotic pressure?
11. Distinguish among isotonic, hypotonic, and hypertonic solutions.

Figure 3.9 Tonicity and its effects on red blood cells (RBCs). The arrows indicate the direction and degree of water movement into and out of the cells. One example of an isotonic solution for RBCs is 0.9% NaCl.

Cells placed in an isotonic solution maintain their shape because there is no net water movement into or out of the cells.

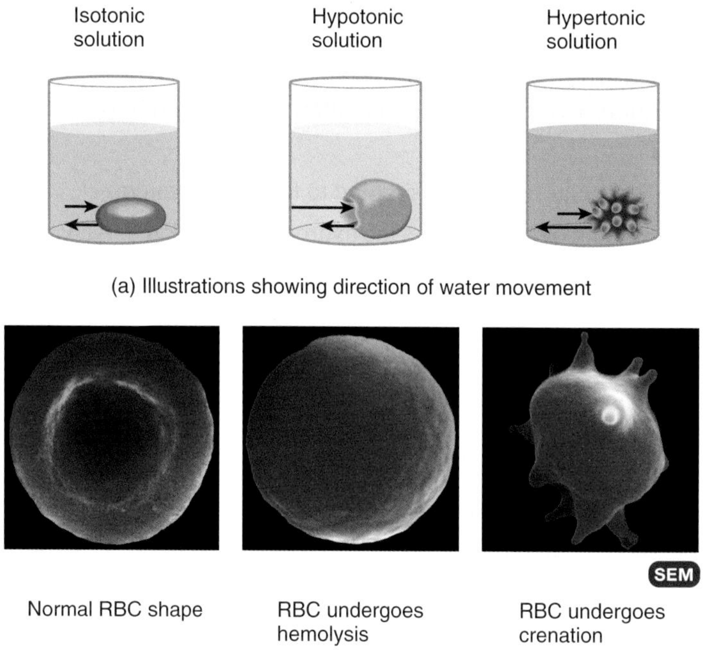

Isotonic solution
Hypotonic solution
Hypertonic solution
SEM
Normal RBC shape
RBC undergoes hemolysis
RBC undergoes crenation

(a) Illustrations showing direction of water movement

(b) Scanning electron micrographs (all 15,000x)

Will a 2% solution of NaCl cause hemolysis or crenation of RBCs? Why?

Active Processes

Active Transport

Some polar or charged solutes that must enter or leave body cells cannot cross the plasma membrane through any form of passive transport because they would need to move "uphill," *against* their concentration gradients. Such solutes may be able to cross the membrane by a process called **active transport.** Active transport is considered an active process because energy is required for carrier proteins to move solutes across the membrane against a concentration gradient. Two sources of cellular energy can be used to drive active transport: (1) Energy obtained from hydrolysis of adenosine triphosphate (ATP) is the source in *primary active transport;* (2) energy stored in an ionic concentration gradient is the source in *secondary active transport.* Like carrier-mediated facilitated diffusion, active transport processes exhibit a transport maximum and saturation. Solutes actively transported across the plasma membrane include several ions, such as Na^+, K^+, H^+, Ca^{2+}, I^- (iodide ions), and Cl^-; amino acids; and monosaccharides. (Note that some of these substances also cross the membrane via facilitated diffusion when the proper channel proteins or carriers are present.)

Primary Active Transport In **primary active transport,** energy derived from hydrolysis of ATP changes the shape of a carrier protein, which "pumps" a substance across a plasma membrane against its concentration gradient. Indeed, carrier proteins that mediate primary active transport are often called **pumps.** A typical body cell expends about 40% of the ATP it generates on primary active transport. Chemicals that turn off ATP production—for example, the poison cyanide—are lethal because they shut down active transport in cells throughout the body.

The most prevalent primary active transport mechanism expels sodium ions (Na^+) from cells and brings potassium ions (K^+) in. Because of the specific ions it moves, this carrier is called the **sodium-potassium pump.** Because a part of the sodium-potassium pump acts as an *ATPase,* an enzyme that hydrolyzes ATP, another name for this pump is **Na^+/K^+ ATPase.**

All cells have thousands of sodium-potassium pumps in their plasma membranes. These sodium-potassium pumps maintain a low concentration of Na^+ in the cytosol by pumping these ions into the extracellular fluid against the Na^+ concentration gradient. At the same time, the pumps move K^+ into cells against the K^+ concentration gradient. Because K^+ and Na^+ slowly leak back across the plasma membrane down their electrochemical gradients—through passive transport or secondary active transport—the sodium-potassium pumps must work nonstop to maintain a low concentration of Na^+ and a high concentration of K^+ in the cytosol.

Figure 3.10 depicts the operation of the sodium-potassium pump:

1. Three Na^+ in the cytosol bind to the pump protein.
2. Binding of Na^+ triggers the hydrolysis of ATP into ADP, a reaction that also attaches a phosphate group Ⓟ to the pump protein. This chemical reaction changes the shape of the pump protein, expelling the three Na^+ into the extracellular fluid. Now the shape of the pump protein favors binding of two K^+ in the extracellular fluid to the pump protein.
3. The binding of K^+ triggers release of the phosphate group from the pump protein. This reaction again causes the shape of the pump protein to change.
4. As the pump protein reverts to its original shape, it releases K^+ into the cytosol. At this point, the pump is again ready to bind three Na^+, and the cycle repeats.

The different concentrations of Na^+ and K^+ in cytosol and extracellular fluid are crucial for maintaining normal cell volume and for the ability of some cells to generate electrical signals such as action potentials. Recall that the tonicity of a solution is proportional to the concentration of its solute particles that cannot penetrate the membrane. Because sodium ions that diffuse into a cell or enter through secondary active transport are immediately pumped out, it is as if they never entered. In effect, sodium ions behave as if they cannot penetrate the membrane. Thus, sodium ions are an important contributor to the tonicity of the extracellular fluid. A similar condition holds for K^+ in the cytosol. By helping to maintain normal tonicity on each side of the plasma membrane, the sodium-potassium pumps ensure that cells neither shrink nor swell due to the movement of water by osmosis out of or into cells.

SECONDARY ACTIVE TRANSPORT In **secondary active transport,** the energy stored in a Na^+ or H^+ concentration gradient is used to drive other substances across the membrane against their own concentration gradients. Because a Na^+ or H^+ gradient is established by primary active transport, secondary active transport *indirectly* uses energy obtained from the hydrolysis of ATP.

The sodium-potassium pump maintains a steep concentration gradient of Na^+ across the plasma membrane. As a result, the sodium ions have stored or potential energy, just like water behind a dam. Accordingly, if there is a route for Na^+ to leak back in, some of the stored energy can be converted to kinetic energy (energy of motion) and used to transport other substances *against their concentration gradients.* In essence, secondary active transport proteins harness the energy in the Na^+ concentration gradient by providing routes for Na^+ to leak into cells. In secondary active transport, a carrier protein simultaneously binds to Na^+ and another substance and then changes its shape so that both substances cross the membrane at the same time. If these transporters move two substances in the same direction they are called **symporters** (sim-PORT-ers; *sym-* = same); **antiporters** (an′-tē-PORT-ers), in contrast, move two substances in opposite directions across the membrane (*anti-* = against).

Figure 3.10 The sodium-potassium pump (Na^+/K^+ATPase) expels sodium ions (Na^+) and brings potassium ions (K^+) into the cell.

Sodium-potassium pumps maintain a low intracellular concentration of sodium ions.

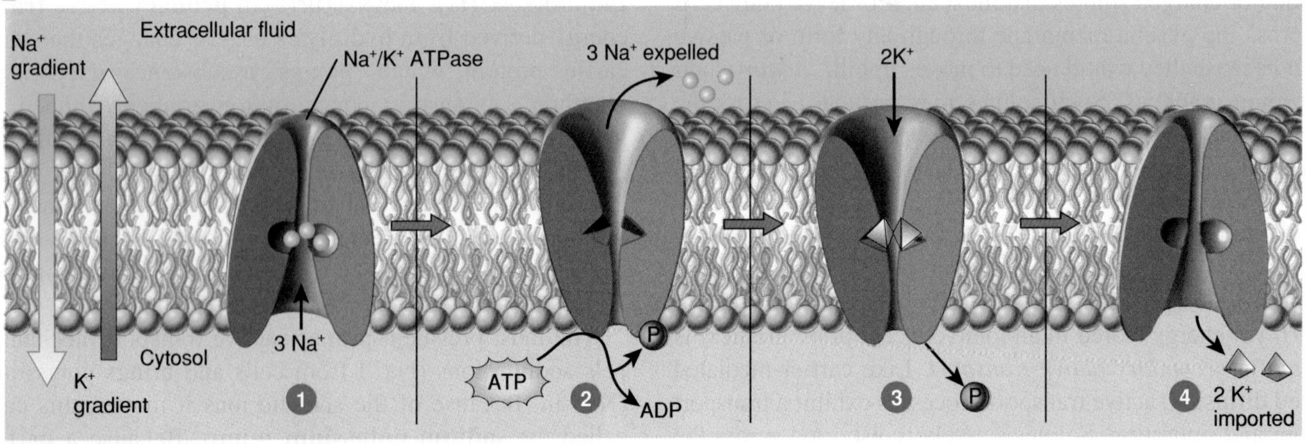

What is the role of ATP in the operation of this pump?

Plasma membranes contain several antiporters and symporters that are powered by the Na^+ gradient (Figure 3.11). For instance, the concentration of calcium ions (Ca^{2+}) is low in the cytosol because Na^+/Ca^{2+} antiporters eject calcium ions. Likewise, Na^+/H^+ antiporters help regulate the cytosol's pH (H^+ concentration) by expelling excess H^+. By contrast, dietary glucose and amino acids are absorbed into cells that line the small intestine by Na^+/glucose and Na^+/amino acid symporters (Figure 3.11b). In each case, sodium ions are moving down their concentration gradient while the other solutes move "uphill," against their concentration gradients. Keep in mind that all these symporters and antiporters can do their job because the sodium-potassium pumps maintain a low concentration of Na^+ in the cytosol.

CLINICAL CONNECTION | *Digitalis Increases Ca^{2+} in Heart Muscle Cells*

Digitalis often is given to patients with *heart failure,* a condition of weakened pumping action by the heart. Digitalis exerts its effect by slowing the action of the sodium-potassium pumps, which lets more Na^+ accumulate inside heart muscle cells. The result is a decreased Na^+ concentration gradient across the plasma membrane, which causes the Na^+/Ca^{2+} antiporters to slow down. As a result, more Ca^{2+} remains inside heart muscle cells. The slight increase in the level of Ca^{2+} in the cytosol of heart muscle cells increases the force of their contractions and thus strengthens the force of the heartbeat. •

Transport in Vesicles

A **vesicle** (VES-i-kul = little blister or bladder), as noted earlier, is a small, spherical sac. As you will learn later in this chapter, a variety of substances are transported in vesicles from one structure to another within cells. Vesicles also import materials from and release materials into extracellular fluid. During **endocytosis** (en′-dō-sī-TŌ-sis; *endo-* = within), materials move into a cell in a vesicle formed from the plasma membrane. In **exocytosis** (ek′-sō-sī-TŌ-sis; *exo-* = out), materials move out of a cell by the fusion with the plasma membrane of vesicles formed inside the cell. Both endocytosis and exocytosis require energy supplied by ATP. Thus, transport in vesicles is an active process.

ENDOCYTOSIS Here we consider three types of endocytosis: receptor-mediated endocytosis, phagocytosis, and bulk-phase endocytosis. **Receptor-mediated endocytosis** is a highly selective type of endocytosis by which cells take up specific ligands. (Recall that ligands are molecules that bind to specific receptors.) A vesicle forms after a receptor protein in the plasma membrane recognizes and binds to a particular particle in the extracellular fluid. For instance, cells take up cholesterol-containing low-density lipoproteins (LDLs), transferrin (an iron-transporting protein in the blood), some vitamins, antibodies, and certain hormones by receptor-mediated endocytosis. Receptor-mediated endocytosis of LDLs (and other ligands) occurs as follows (Figure 3.12):

1. ***Binding.*** On the extracellular side of the plasma membrane, an LDL particle that contains cholesterol binds to a specific receptor in the plasma membrane to form a receptor–LDL complex. The receptors are integral membrane proteins that are concentrated in regions of the plasma membrane called *clathrin-coated pits.* Here, a protein called *clathrin* attaches to the membrane on its cytoplasmic

Figure 3.11 Secondary active transport mechanisms. (a) Antiporters carry two substances across the membrane in opposite directions. (b) Symporters carry two substances across the membrane in the same direction.

Secondary active transport mechanisms use the energy stored in an ionic concentration gradient (here, for Na^+). Because primary active transport pumps that hydrolyze ATP maintain the gradient, secondary active transport mechanisms consume ATP indirectly.

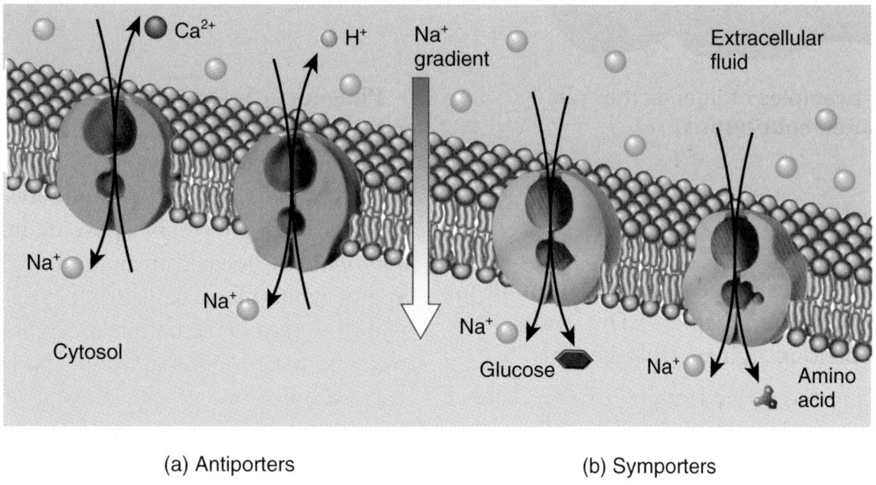

What is the main difference between primary and secondary active transport mechanisms?

Figure 3.12 Receptor-mediated endocytosis of a low-density lipoprotein (LDL) particle.

Receptor-mediated endocytosis imports materials that are needed by cells.

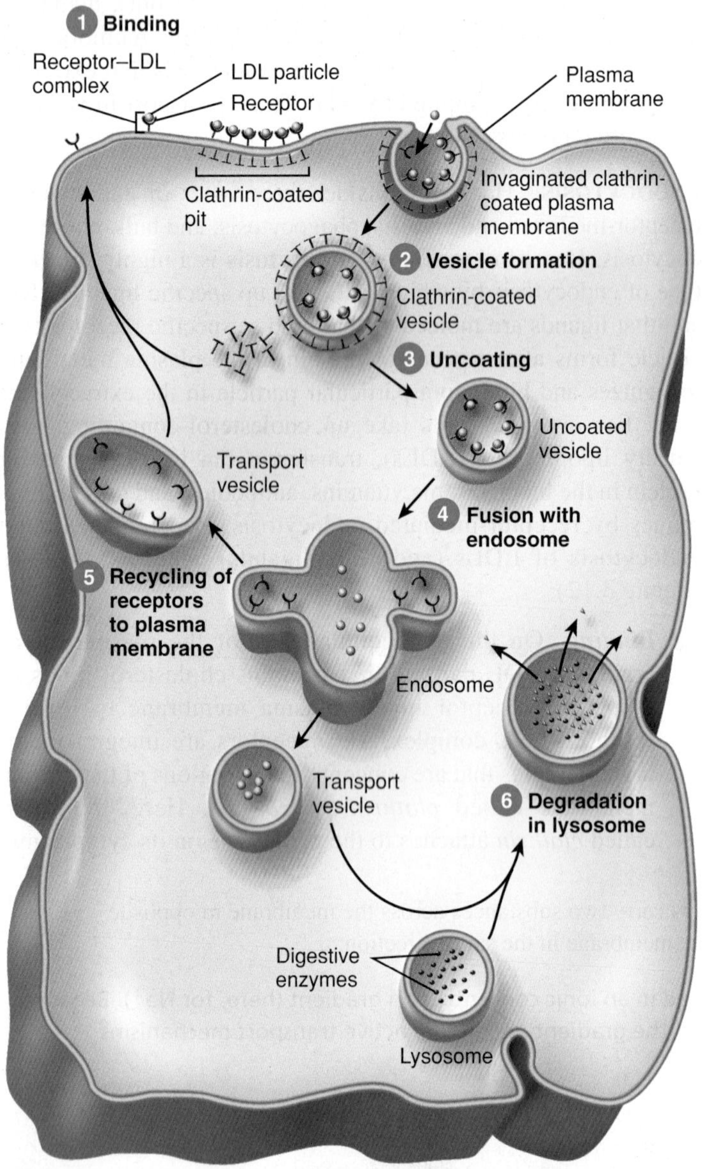

1 Binding
Receptor–LDL complex
LDL particle
Receptor
Plasma membrane
Clathrin-coated pit
Invaginated clathrin-coated plasma membrane
2 Vesicle formation
Clathrin-coated vesicle
3 Uncoating
Uncoated vesicle
Transport vesicle
4 Fusion with endosome
5 Recycling of receptors to plasma membrane
Endosome
Transport vesicle
6 Degradation in lysosome
Digestive enzymes
Lysosome

What are several other examples of ligands that can undergo receptor-mediated endocytosis?

side. Many clathrin molecules come together, forming a basketlike structure around the receptor–LDL complexes that causes the membrane to invaginate (fold inward).

2 ***Vesicle formation.*** The invaginated edges of the membrane around the clathrin-coated pit fuse, and a small piece of the membrane pinches off. The resulting vesicle, known as a *clathrin-coated vesicle,* contains the receptor–LDL complexes.

3 ***Uncoating.*** Almost immediately after it is formed, the clathrin-coated vesicle loses its clathrin coat to become an *uncoated vesicle.* Clathrin molecules either return to the inner surface of the plasma membrane or help form coats on other vesicles inside the cell.

4 ***Fusion with endosome.*** The uncoated vesicle quickly fuses with a vesicle known as an *endosome.* Within an endosome, the LDL particles separate from their receptors.

5 ***Recycling of receptors to plasma membrane.*** Most of the receptors accumulate in elongated protrusions of the endosome (the arms of the cross-shaped vesicle at the center of the figure). These pinch off, forming transport vesicles that return the receptors to the plasma membrane. An LDL receptor is returned to the plasma membrane about 10 minutes after it enters a cell.

6 ***Degradation in lysosomes.*** Other transport vesicles, which contain the LDL particles, bud off the endosome and soon fuse with a *lysosome*. Lysosomes contain many digestive enzymes. Certain enzymes break down the large protein and lipid molecules of the LDL particle into amino acids, fatty acids, and cholesterol. These smaller molecules then leave the lysosome. The cell uses cholesterol for rebuilding its membranes and for synthesis of steroids, such as estrogen. Fatty acids and amino acids can be used for ATP production or to build other molecules needed by the cell.

CLINICAL CONNECTION | *Viruses and Receptor-Mediated Endocytosis*

Although receptor-mediated endocytosis normally imports needed materials, some viruses are able to use this mechanism to enter and infect body cells. For example, the human immunodeficiency virus (HIV), which causes acquired immunodeficiency syndrome (AIDS), can attach to a receptor called CD4. This receptor is present in the plasma membrane of white blood cells called helper T cells. After binding to CD4, HIV enters the helper T cell via receptor-mediated endocytosis. •

Phagocytosis (fag′-ō-sī-TŌ-sis; *phago-* = to eat) is a form of endocytosis in which the cell engulfs large solid particles, such as worn-out cells, whole bacteria, or viruses (Figure 3.13). Only a few body cells, termed **phagocytes** (FAG-ō-sīts), are able to carry out phagocytosis. Two main types of phagocytes are *macrophages,* located in many body tissues, and *neutrophils,* a type of white blood cell. Phagocytosis begins when the particle binds to a plasma membrane receptor on the phagocyte, causing it to extend **pseudopods** (SOO-dō-pods; *pseudo-* = false; *-pods* = feet), projections of its plasma membrane and cytoplasm. Pseudopods surround the particle outside the cell, and the membranes fuse to form a vesicle called a *phagosome,* which enters the cytoplasm. The phagosome fuses with one or more lysosomes, and lysosomal enzymes break down the ingested material.

Figure 3.13 Phagocytosis. Pseudopods surround a particle, and the membranes fuse to form a phagosome.

Phagocytosis is a vital defense mechanism that helps protect the body from disease.

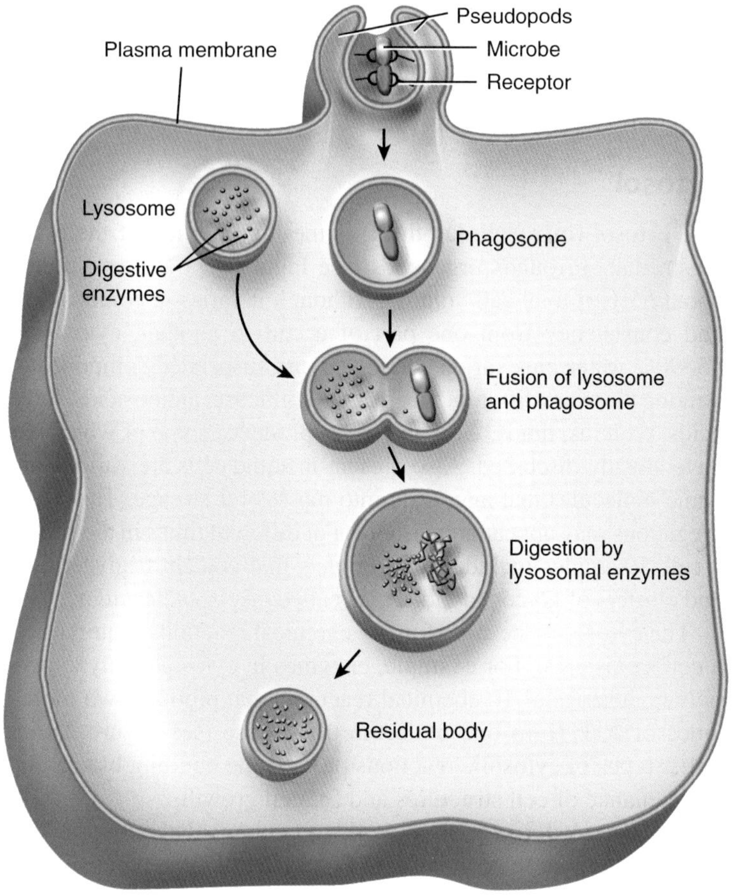

(a) Diagram of the process

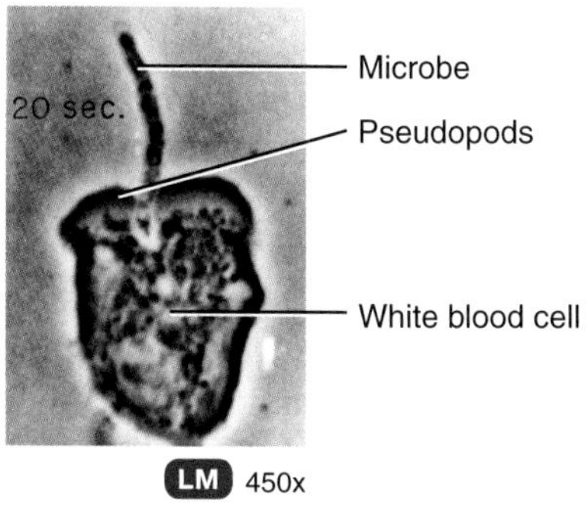

(b) White blood cell engulfs microbe

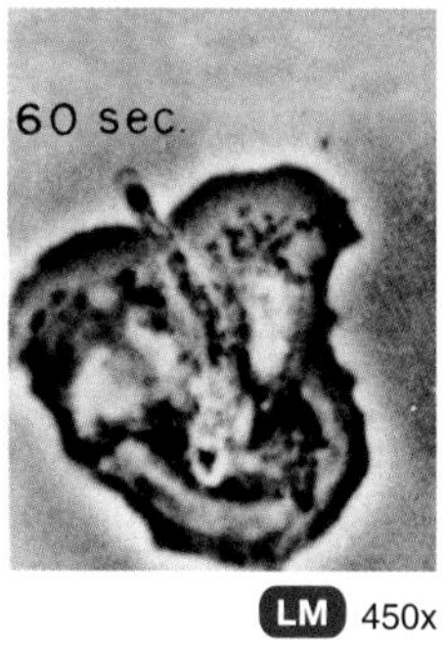

(c) White blood cell destroys microbe

What triggers pseudopod formation?

In most cases, any undigested materials in the phagosome remain indefinitely in a vesicle called a *residual body.* The residual bodies are then either secreted by the cell via exocytosis or they remain stored in the cell as lipofuscin granules.

CLINICAL CONNECTION | *Phagocytosis and Microbes*

The process of phagocytosis is a vital defense mechanism that helps protect the body from disease. Through phagocytosis, macrophages dispose of invading microbes and billions of aged, worn-out red blood cells every day; neutrophils also help rid the body of invading microbes. **Pus** is a mixture of dead neutrophils, macrophages, and tissue cells and fluid in an infected wound. •

Most body cells carry out **bulk-phase endocytosis,** also called **pinocytosis** (pi-nō-sī-TŌ-sis; *pino-* = to drink), a form of endocytosis in which tiny droplets of extracellular fluid are taken up (Figure 3.14). No receptor proteins are involved; all solutes dissolved in the extracellular fluid are brought into the cell. During bulk-phase endocytosis, the plasma membrane folds inward and forms a vesicle containing a droplet of extracellular fluid. The vesicle detaches or "pinches off" from the plasma membrane and

Figure 3.14 Bulk-phase endocytosis. The plasma membrane folds inward, forming a vesicle.

Most body cells carry out bulk-phase endocytosis, the nonselective uptake of tiny droplets of extracellular fluid.

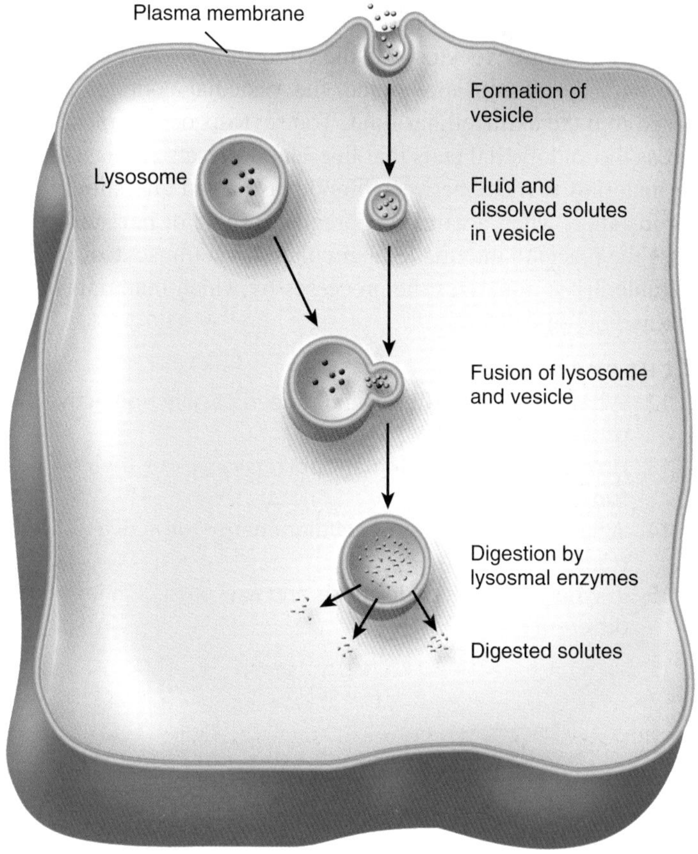

How do receptor-mediated endocytosis and phagocytosis differ from bulk-phase endocytosis?

enters the cytosol. Within the cell, the vesicle fuses with a lysosome, where enzymes degrade the engulfed solutes. The resulting smaller molecules, such as amino acids and fatty acids, leave the lysosome to be used elsewhere in the cell. Bulk-phase endocytosis occurs in most cells, especially absorptive cells in the intestines and kidneys.

Exocytosis In contrast with endocytosis, which brings materials into a cell, **exocytosis** releases materials from a cell. All cells carry out exocytosis, but it is especially important in two types of cells: (1) secretory cells that liberate digestive enzymes, hormones, mucus, or other secretions and (2) nerve cells that release substances called *neurotransmitters* (see Figure 12.23). In some cases, wastes are also released by exocytosis. During exocytosis, membrane-enclosed vesicles called *secretory vesicles* form inside the cell, fuse with the plasma membrane, and release their contents into the extracellular fluid.

Segments of the plasma membrane lost through endocytosis are recovered or recycled by exocytosis. The balance between endocytosis and exocytosis keeps the surface area of a cell's plasma membrane relatively constant. Membrane exchange is quite extensive in certain cells. In your pancreas, for example, the cells that secrete digestive enzymes can recycle an amount of plasma membrane equal to the cell's entire surface area in 90 minutes.

Transcytosis Transport in vesicles may also be used to successively move a substance into, across, and out of a cell. In this active process, called **transcytosis** (tranz′-sī-TŌ-sis), vesicles undergo endocytosis on one side of a cell, move across the cell, and then undergo exocytosis on the opposite side. As the vesicles fuse with the plasma membrane, the vesicular contents are released into the extracellular fluid. Transcytosis occurs most often across the endothelial cells that line blood vessels and is a means for materials to move between blood plasma and interstitial fluid. For instance, when a woman is pregnant, some of her antibodies cross the placenta into the fetal circulation via transcytosis.

Table 3.1 summarizes the processes by which materials move into and out of cells.

✔ **CHECKPOINT**

12. What is the key difference between passive and active processes?
13. How do symporters and antiporters carry out their functions?
14. What are the sources of cellular energy for active transport?
15. In what ways are endocytosis and exocytosis similar and different?

3.4 CYTOPLASM

OBJECTIVE

- Describe the structure and function of cytoplasm, cytosol, and organelles.

Cytoplasm consists of all the cellular contents between the plasma membrane and the nucleus, and has two components: (1) the cytosol and (2) organelles, tiny structures that perform different functions in the cell.

Cytosol

The **cytosol (intracellular fluid)** is the fluid portion of the cytoplasm that surrounds organelles (see Figure 3.1) and constitutes about 55% of total cell volume. Although it varies in composition and consistency from one part of a cell to another, cytosol is 75–90% water plus various dissolved and suspended components. Among these are different types of ions, glucose, amino acids, fatty acids, proteins, lipids, ATP, and waste products, some of which we have already discussed. Also present in some cells are various organic molecules that aggregate into masses for storage. These aggregations may appear and disappear at different times in the life of a cell. Examples include *lipid droplets* that contain triglycerides, and clusters of glycogen molecules called *glycogen granules.*

The cytosol is the site of many chemical reactions required for a cell's existence. For example, enzymes in cytosol catalyze *glycolysis,* a series of 10 chemical reactions that produce two molecules of ATP from one molecule of glucose (see Figure 25.4). Other types of cytosolic reactions provide the building blocks for maintenance of cell structures and for cell growth.

The **cytoskeleton** is a network of protein filaments that extends throughout the cytosol (see Figure 3.1). Three types of filamentous proteins contribute to the cytoskeleton's structure, as well as the structure of other organelles. In the order of their increasing diameter, these structures are microfilaments, intermediate filaments, and microtubules.

Microfilaments **Microfilaments** (mī-krō-FIL-a-ments) are the thinnest elements of the cytoskeleton. They are composed of the proteins *actin* and *myosin* and are most prevalent at the edge of a cell (Figure 3.15a). Microfilaments have two general functions: They help generate movement and provide mechanical support. With respect to movement, microfilaments are involved in muscle contraction, cell division, and cell locomotion, such as occurs during the migration of embryonic cells during development, the invasion of tissues by white blood cells to fight infection, or the migration of skin cells during wound healing.

Microfilaments provide much of the mechanical support that is responsible for the basic strength and shapes of cells. They anchor the cytoskeleton to integral proteins in the plasma membrane. Microfilaments also provide mechanical support for cell extensions called **microvilli** (mī-krō-VIL-ī; *micro-* = small; *-villi* = tufts of hair; singular is *microvillus*), nonmotile, microscopic fingerlike

TABLE 3.1

Transport of Materials into and out of Cells

TRANSPORT PROCESS	DESCRIPTION	SUBSTANCES TRANSPORTED
PASSIVE PROCESSES	Movement of substances down a concentration gradient until equilibrium is reached; do not require cellular energy in the form of ATP.	
Diffusion	Movement of molecules or ions down a concentration gradient due to their kinetic energy until they reach equilibrium.	
Simple diffusion	Passive movement of a substance down its concentration gradient through the lipid bilayer of the plasma membrane without the help of membrane transport proteins.	Nonpolar, hydrophobic solutes: oxygen, carbon dioxide, and nitrogen gases; fatty acids; steroids; and fat-soluble vitamins. Polar molecules such as water, urea, and small alcohols.
Facilitated diffusion	Passive movement of a substance down its concentration gradient through the lipid bilayer by transmembrane proteins that function as channels or carriers.	Polar or charged solutes: glucose; fructose; galactose; some vitamins; and ions such as K^+, Cl^-, Na^+, and Ca^{2+}.
Osmosis	Passive movement of water molecules across a selectively permeable membrane from an area of higher to lower water concentration until equilibrium is reached.	Solvent: water in living systems.
ACTIVE PROCESSES	Movement of substances against a concentration gradient; requires cellular energy in the form of ATP.	
Active Transport	Active process in which a cell expends energy to move a substance across the membrane against its concentration gradient by transmembrane proteins that function as carriers.	Polar or charged solutes.
Primary active transport	Active process in which a substance moves across the membrane against its concentration gradient by pumps (carriers) that use energy supplied by hydrolysis of ATP.	Na^+, K^+, Ca^{2+}, H^+, I^-, Cl^-, and other ions.
Secondary active transport	Coupled active transport of two substances across the membrane using energy supplied by a Na^+ or H^+ concentration gradient maintained by primary active transport pumps. Antiporters move Na^+ (or H^+) and another substance in opposite directions across the membrane; symporters move Na^+ (or H^+) and another substance in the same direction across the membrane.	Antiport: Ca^{2+}, H^+ out of cells. Symport: glucose, amino acids into cells.
Transport in Vesicles	Active process in which substances move into or out of cells in vesicles that bud from plasma membrane; requires energy supplied by ATP.	
Endocytosis	Movement of substances into a cell in vesicles.	
Receptor-mediated endocytosis	Ligand–receptor complexes trigger infolding of a clathrin-coated pit that forms a vesicle containing ligands.	Ligands: transferrin, low-density lipoproteins (LDLs), some vitamins, certain hormones, and antibodies.
Phagocytosis	"Cell eating"; movement of a solid particle into a cell after pseudopods engulf it to form a phagosome.	Bacteria, viruses, and aged or dead cells.
Bulk-phase endocytosis	"Cell drinking"; movement of extracellular fluid into a cell by infolding of plasma membrane to form a vesicle.	Solutes in extracellular fluid.
Exocytosis	Movement of substances out of a cell in secretory vesicles that fuse with the plasma membrane and release their contents into the extracellular fluid.	Neurotransmitters, hormones, and digestive enzymes.
Transcytosis	Movement of a substance through a cell as a result of endocytosis on one side and exocytosis on the opposite side.	Substances, such as antibodies, across endothelial cells. This is a common route for substances to pass between blood plasma and interstitial fluid.

Figure 3.15 **Cytoskeleton.**

The cytoskeleton is a network of three types of protein filaments—microfilaments, intermediate filaments, and microtubules—that extend throughout the cytoplasm.

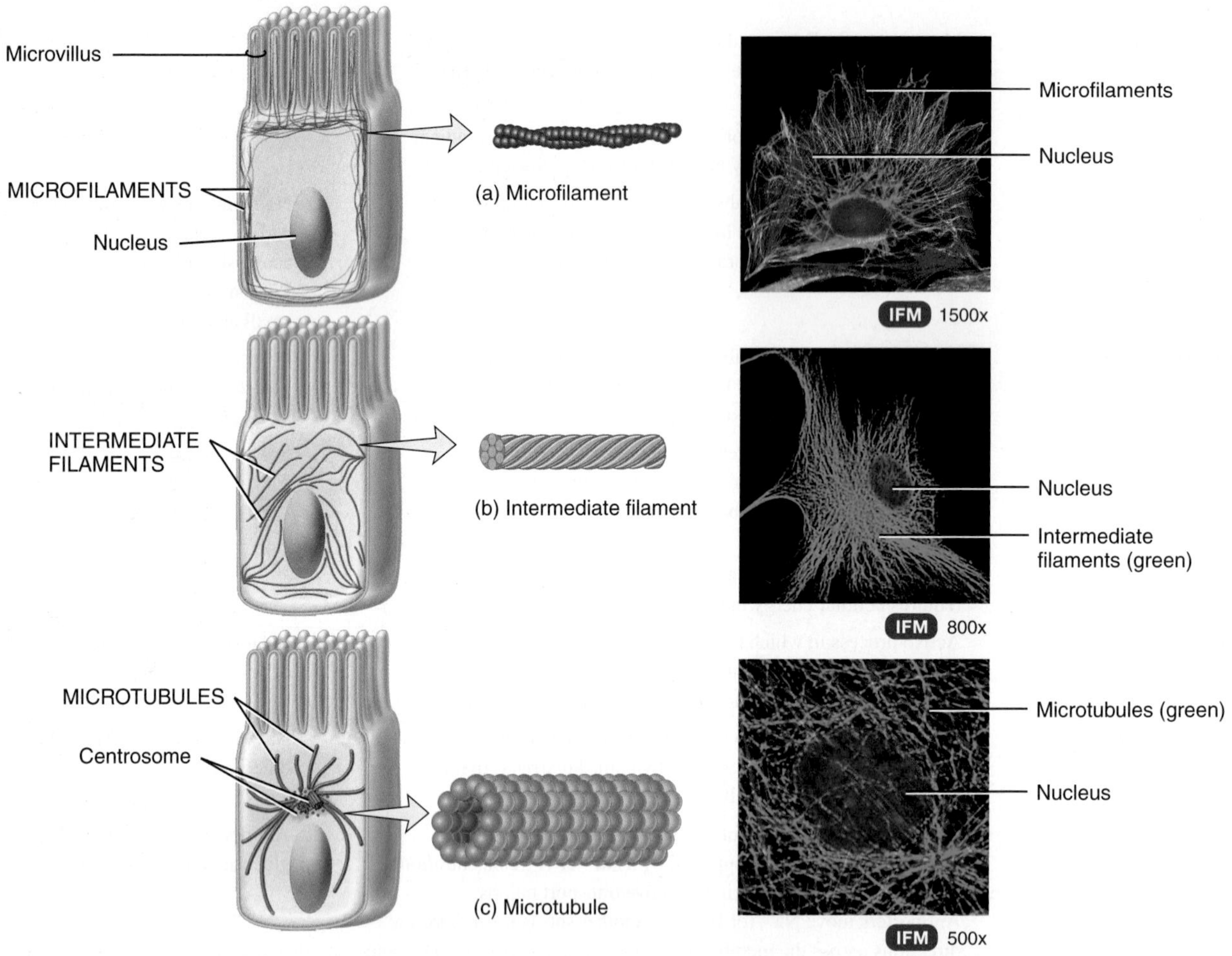
Microvillus
MICROFILAMENTS
Nucleus
(a) Microfilament
Microfilaments
Nucleus
IFM 1500x
INTERMEDIATE FILAMENTS
(b) Intermediate filament
Nucleus
Intermediate filaments (green)
IFM 800x
MICROTUBULES
Centrosome
(c) Microtubule
Microtubules (green)
Nucleus
IFM 500x

FUNCTIONS OF THE CYTOSKELETON

1. Serves as a scaffold that helps to determine a cell's shape and to organize the cellular contents.
2. Aids movement of organelles within the cell, of chromosomes during cell division, and of whole cells such as phagocytes.

Which cytoskeletal component helps form the structure of centrioles, cilia, and flagella?

projections of the plasma membrane. Within each microvillus is a core of parallel microfilaments that supports it. Because they greatly increase the surface area of the cell, microvilli are abundant on cells involved in absorption, such as the epithelial cells that line the small intestine.

INTERMEDIATE FILAMENTS As their name suggests, **intermediate filaments** are thicker than microfilaments but thinner than microtubules (Figure 3.15b). Several different proteins can compose intermediate filaments, which are exceptionally strong. They are found in parts of cells subject to mechanical stress; they help stabilize the position of organelles such as the nucleus and help attach cells to one another.

MICROTUBULES **Microtubules** (mī-krō-TOO-būls′) are the largest of the cytoskeletal components and are long, unbranched hollow tubes composed mainly of the protein *tubulin.* The assembly of microtubules begins in an organelle called the centrosome (discussed shortly). The microtubules grow outward from the centrosome toward the periphery of the cell (Figure 3.15c). Microtubules help determine cell shape. They also function in the movement of organelles such as secretory vesicles, of chromosomes during cell division, and of specialized cell projections, such as cilia and flagella.

Organelles

As noted earlier, **organelles** are specialized structures within the cell that have characteristic shapes, and they perform specific functions in cellular growth, maintenance, and reproduction. Despite the many chemical reactions going on in a cell at any given time, there is little interference among reactions because they are confined to different organelles. Each type of organelle has its own set of enzymes that carry out specific reactions, and serves as a functional compartment for specific biochemical processes. The numbers and types of organelles vary in different cells, depending on the cell's function. Although they have different functions, organelles often cooperate to maintain homeostasis. Even though the nucleus is a large organelle, it is discussed in a separate section because of its special importance in directing the life of a cell.

Centrosome

The **centrosome** (SEN-trō-sōm), located near the nucleus, consists of two components: a pair of centrioles and pericentriolar material (Figure 3.16a). The two **centrioles** (SEN-trē-ōls) are cylindrical structures, each composed of nine clusters of three microtubules (triplets) arranged in a circular pattern (Figure 3.16b). The long axis of one centriole is at a right angle to the long axis of the other (Figure 3.16c). Surrounding the centrioles is **pericentriolar material** (per′-ē-sen′-trē-Ō-lar), which contains hundreds of ring-shaped complexes composed of the protein *tubulin.* These tubulin complexes are the organizing centers for growth of the mitotic spindle, which plays a critical role in cell division, and for microtubule formation in nondividing cells. During cell division, centrosomes replicate so that succeeding generations of cells have the capacity for cell division.

Cilia and Flagella

Microtubules are the dominant components of cilia and flagella, which are motile projections of the cell surface. **Cilia** (SIL-ē-a = eyelashes; singular is *cilium*) are numerous, short, hairlike projections that extend from the surface of the cell (see Figures 3.1 and 3.17b). Each cilium contains a core of 20 microtubules surrounded by plasma membrane (Figure 3.17a). The microtubules are arranged such that one pair in the center is surrounded by nine clusters of two fused microtubules (doublets). Each cilium is anchored to a *basal body* just below the surface of the plasma membrane. A basal body is similar in structure to a centriole and functions in initiating the assembly of cilia and flagella.

A cilium displays an oarlike pattern of beating; it is relatively stiff during the power stroke (oar digging into the water), but more flexible during the recovery stroke (oar moving above the water preparing for a new stroke) (Figure 3.17d). The coordinated movement of many cilia on the surface of a cell causes the steady movement of fluid along the cell's surface. Many cells of the respiratory tract, for example, have hundreds of cilia that help sweep foreign particles trapped in mucus away from the lungs. In cystic fibrosis, the extremely thick mucous secretions that are produced interfere with ciliary action and the normal functions of the respiratory tract.

Figure 3.16 Centrosome.

Located near the nucleus, the centrosome consists of a pair of centrioles and pericentriolar material.

FUNCTIONS OF CENTROSOMES

1. The pericentriolar material of the centrosome contains tubulins that build microtubules in nondividing cells.
2. The pericentriolar material of the centrosome forms the mitotic spindle during cell division.

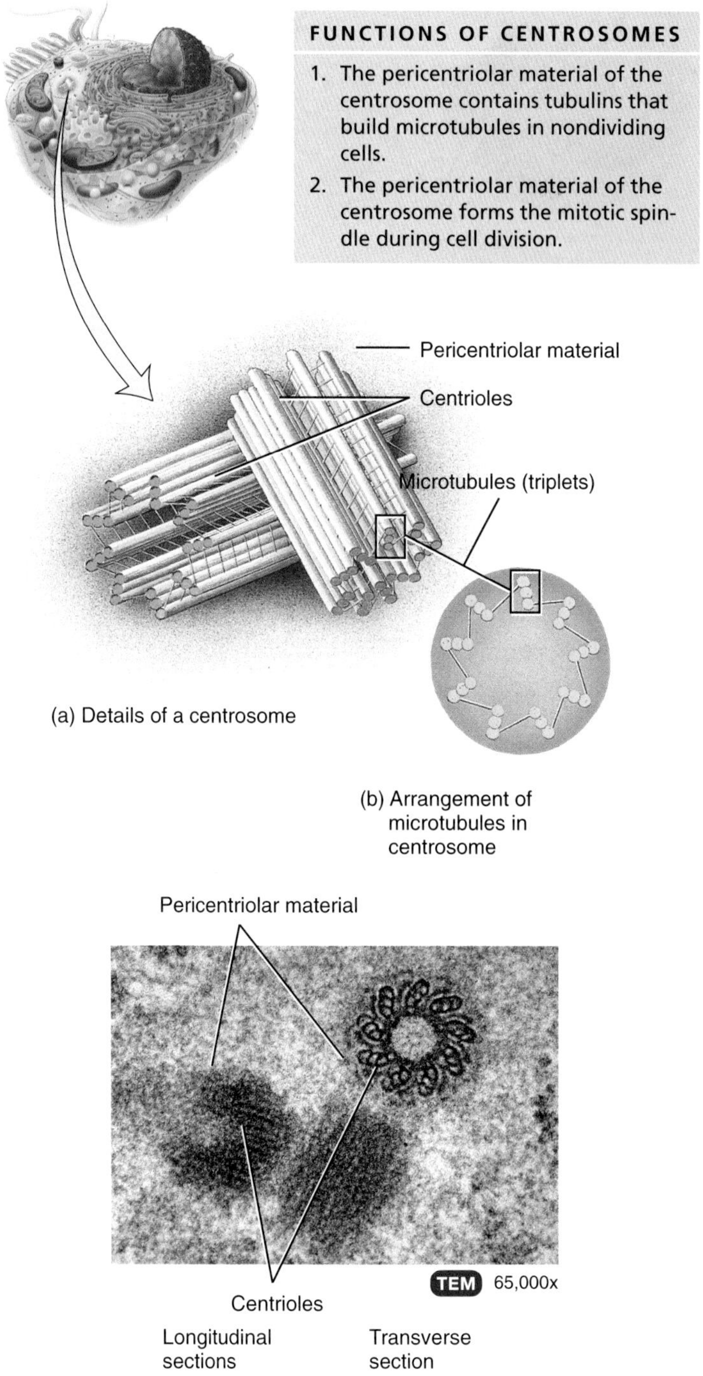

(a) Details of a centrosome

(b) Arrangement of microtubules in centrosome

(c) Centrioles

If you observed that a cell did not have a centrosome, what could you predict about its capacity for cell division?

Figure 3.17 Cilia and flagella.

A cilium contains a core of microtubules with one pair in the center surrounded by nine clusters of doublet microtubules.

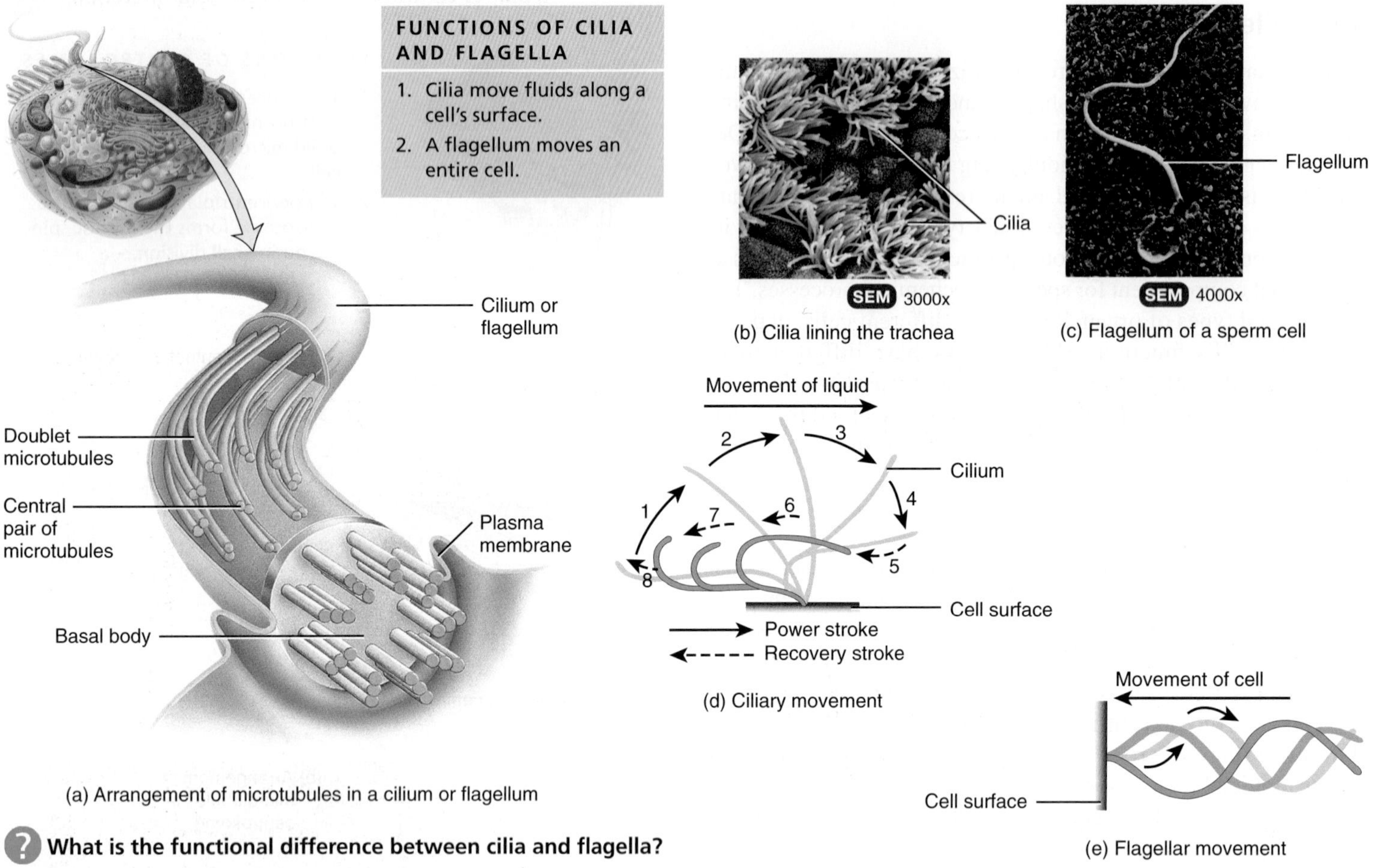
FUNCTIONS OF CILIA AND FLAGELLA
1. Cilia move fluids along a cell's surface.
2. A flagellum moves an entire cell.
Cilium or flagellum
Doublet microtubules
Central pair of microtubules
Plasma membrane
Basal body
(a) Arrangement of microtubules in a cilium or flagellum
Cilia
SEM 3000x
(b) Cilia lining the trachea
Flagellum
SEM 4000x
(c) Flagellum of a sperm cell
Movement of liquid
Cilium
1
2
3
4
5
6
7
8
Cell surface
Power stroke
Recovery stroke
(d) Ciliary movement
Movement of cell
Cell surface
(e) Flagellar movement

What is the functional difference between cilia and flagella?

> **CLINICAL CONNECTION | *Cilia and Smoking***
>
> The movement of cilia is also paralyzed by nicotine in cigarette smoke. For this reason, smokers cough often to remove foreign particles from their airways. Cells that line the uterine (fallopian) tubes also have cilia that sweep oocytes (egg cells) toward the uterus, and females who smoke have an increased risk of ectopic (outside the uterus) pregnancy. •

Flagella (fla-JEL-a = whip; singular is *flagellum*) are similar in structure to cilia but are typically much longer. Flagella usually move an entire cell. A flagellum generates forward motion along its axis by rapidly wiggling in a wavelike pattern (Figure 3.17e). The only example of a flagellum in the human body is a sperm cell's tail, which propels the sperm toward the oocyte in the uterine tube (Figure 3.17c).

Ribosomes

Ribosomes (RĪ-bō-sōms; *-somes* = bodies) are the sites of protein synthesis. The name of these tiny structures reflects their high content of one type of ribonucleic acid, **ribosomal RNA (rRNA),** but each one also includes more than 50 proteins. Structurally, a ribosome consists of two subunits, one about half the size of the other (Figure 3.18). The large and small subunits are made separately in the nucleolus, a spherical body inside the nucleus. Once produced, the large and small subunits exit the nucleus separately, then come together in the cytoplasm.

Some ribosomes are attached to the outer surface of the nuclear membrane and to an extensively folded membrane called the endoplasmic reticulum. These ribosomes synthesize proteins destined for specific organelles, for insertion in the plasma membrane, or for export from the cell. Other ribosomes are "free" or unattached to other cytoplasmic structures. Free ribosomes synthesize proteins used in the cytosol. Ribosomes are also located within mitochondria, where they synthesize mitochondrial proteins.

Endoplasmic Reticulum

The **endoplasmic reticulum** (en′-dō-PLAS-mik re-TIK-ū-lum; *-plasmic* = cytoplasm; *reticulum* = network) or **ER** is a network of membranes in the form of flattened sacs or tubules (Figure 3.19). The ER extends from the nuclear envelope (membrane around the

Figure 3.18 Ribosomes.

Ribosomes are the sites of protein synthesis.

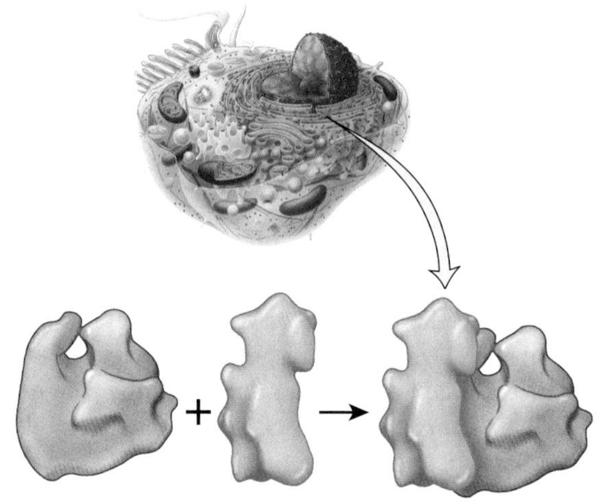
Large subunit
Small subunit
Complete functional ribosome

Details of ribosomal subunits

FUNCTIONS OF RIBOSOMES
1. Ribosomes associated with endoplasmic reticulum synthesize proteins destined for insertion in the plasma membrane or secretion from the cell.
2. Free ribosomes synthesize proteins used in the cytosol.

Where are subunits of ribosomes synthesized and assembled?

Figure 3.19 Endoplasmic reticulum.

The endoplasmic reticulum is a network of membrane-enclosed sacs or tubules that extend throughout the cytoplasm and connect to the nuclear envelope.

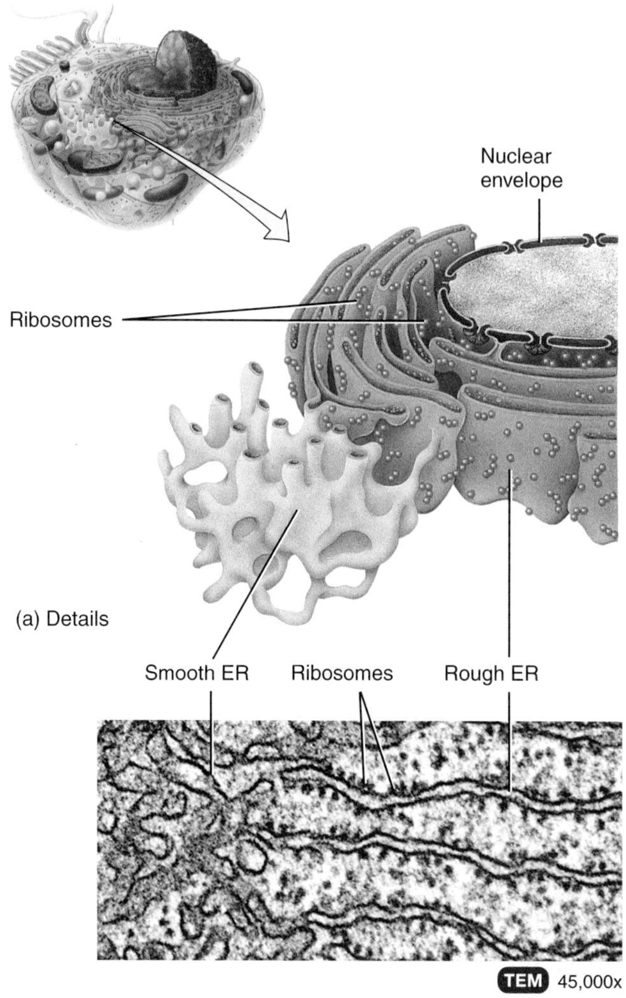
Nuclear envelope
Ribosomes
Smooth ER
Ribosomes
Rough ER
TEM 45,000x

(a) Details

(b) Transverse section

FUNCTIONS OF ENDOPLASMIC RETICULUM
1. Rough ER synthesizes glycoproteins and phospholipids that are transferred into cellular organelles, inserted into the plasma membrane, or secreted during exocytosis.
2. Smooth ER synthesizes fatty acids and steroids, such as estrogens and testosterone; inactivates or detoxifies drugs and other potentially harmful substances; removes the phosphate group from glucose-6-phosphate; and stores and releases calcium ions that trigger contraction in muscle cells.

What are the structural and functional differences between rough and smooth ER?

nucleus), to which it is connected, throughout the cytoplasm. The ER is so extensive that it constitutes more than half of the membranous surfaces within the cytoplasm of most cells.

Cells contain two distinct forms of ER, which differ in structure and function. **Rough ER** is continuous with the nuclear membrane and usually is folded into a series of flattened sacs. The outer surface of rough ER is studded with ribosomes, the sites of protein synthesis. Proteins synthesized by ribosomes attached to rough ER enter spaces within the ER for processing and sorting. In some cases, enzymes attach the proteins to carbohydrates to form glycoproteins. In other cases, enzymes attach the proteins to phospholipids, also synthesized by rough ER. These molecules (glycoproteins and phospholipids) may be incorporated into the membranes of organelles, inserted into the plasma membrane, or secreted via exocytosis. Thus rough ER produces secretory proteins, membrane proteins, and many organellar proteins.

Smooth ER extends from the rough ER to form a network of membrane tubules (Figure 3.19). Unlike rough ER, smooth ER does not have ribosomes on the outer surfaces of its membrane. However, smooth ER contains unique enzymes that make it functionally more diverse than rough ER. Because it lacks ribosomes, smooth ER does not synthesize proteins, but it does synthesize fatty acids and steroids, such as estrogens and testosterone. In liver cells, enzymes of the smooth ER help release glucose into the bloodstream and inactivate or detoxify lipid-soluble drugs or potentially harmful substances, such as alcohol, pesticides, and *carcinogens* (cancer-causing agents). In liver, kidney, and intestinal cells a smooth ER enzyme removes the phosphate group from glucose-6-phosphate, which allows the "free" glucose to enter the bloodstream. In muscle cells, the calcium ions (Ca^{2+}) that trigger contraction are released from the sarcoplasmic reticulum, a form of smooth ER.

CLINICAL CONNECTION | *Smooth ER and Drug Tolerance*

One of the functions of smooth ER, as noted earlier, is to detoxify certain drugs. Individuals who repeatedly take such drugs, such as the sedative phenobarbital, develop changes in the smooth ER in their liver cells. Prolonged administration of phenobarbital results in increased tolerance to the drug; the same dose no longer produces the same degree of sedation. With repeated exposure to the drug, the amount of smooth ER and its enzymes increases to protect the cell from its toxic effects. As the amount of smooth ER increases, higher and higher dosages of the drug are needed to achieve the original effect. This could result in an increased possibility of overdose and increased drug dependence. •

Golgi Complex

Most of the proteins synthesized by ribosomes attached to rough ER are ultimately transported to other regions of the cell. The first step in the transport pathway is through an organelle called the **Golgi complex** (GOL-jē). It consists of 3 to 20 **cisternae** (sis-TER-nē = cavities; singular is *cisterna*), small, flattened membranous sacs with bulging edges that resemble a stack of pita bread (Figure 3.20). The cisternae are often curved, giving the Golgi complex a cuplike shape. Most cells have several Golgi complexes, and Golgi complexes are more extensive in cells that secrete proteins, a clue to the organelle's role in the cell.

The cisternae at the opposite ends of a Golgi complex differ from each other in size, shape, and enzymatic activity. The convex **entry** or ***cis*** **face** is a cisterna that faces the rough ER. The concave **exit** or ***trans*** **face** is a cisterna that faces the plasma membrane. Sacs between the entry and exit faces are called **medial cisternae.** Transport vesicles (described shortly) from the ER merge to form the entry face. From the entry face, the cisternae are thought to mature, in turn becoming medial and then exit cisternae.

Figure 3.20 Golgi complex.

The opposite faces of a Golgi complex differ in size, shape, content, and enzymatic activities.

FUNCTIONS OF GOLGI COMPLEX

1. Modifies, sorts, packages, and transports proteins received from the rough ER.
2. Forms secretory vesicles that discharge processed proteins via exocytosis into extracellular fluid; forms membrane vesicles that ferry new molecules to the plasma membrane; forms transport vesicles that carry molecules to other organelles, such as lysosomes.

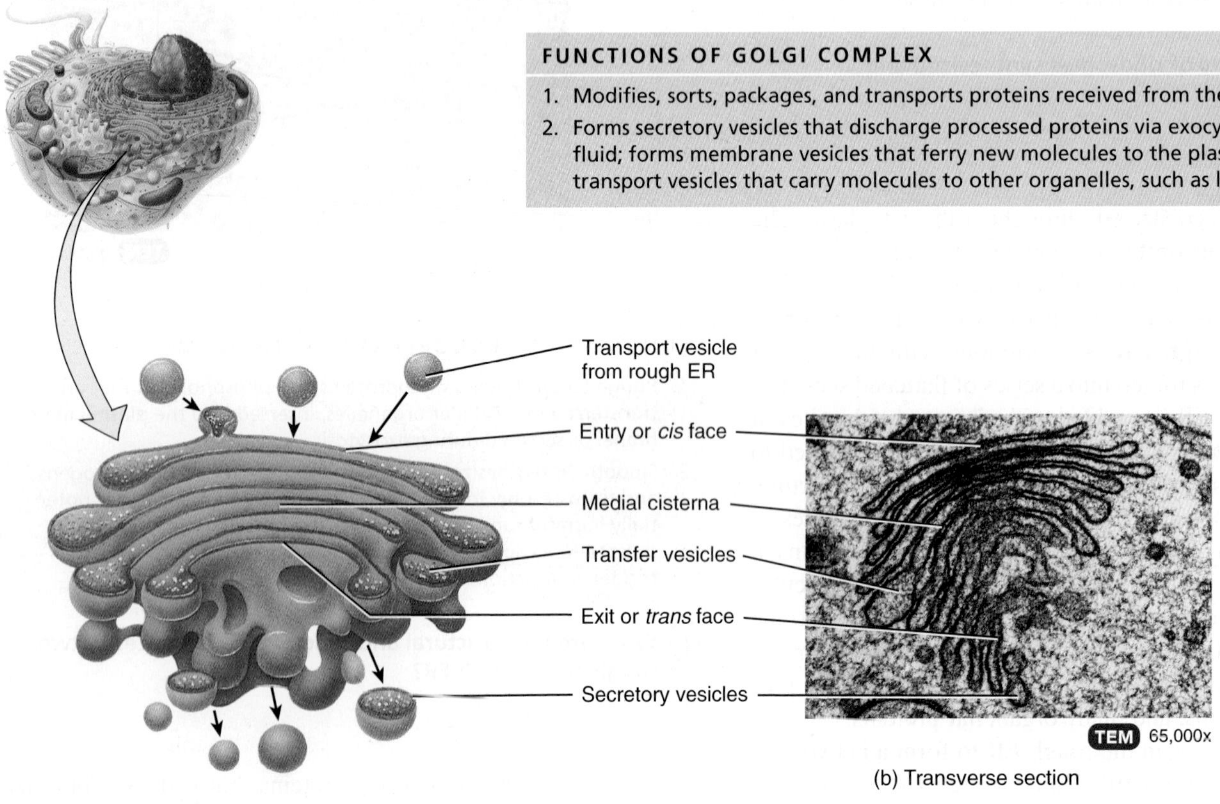
Transport vesicle from rough ER
Entry or *cis* face
Medial cisterna
Transfer vesicles
Exit or *trans* face
Secretory vesicles
TEM 65,000x

(a) Details

(b) Transverse section

How do the entry and exit faces differ in function?

Different enzymes in the entry, medial, and exit cisternae of the Golgi complex permit each of these areas to modify, sort, and package proteins into vesicles for transport to different destinations. The entry face receives and modifies proteins produced by the rough ER. The medial cisternae add carbohydrates to proteins to form glycoproteins and lipids to proteins to form lipoproteins. The exit face modifies the molecules further and then sorts and packages them for transport to their destinations.

Proteins arriving at, passing through, and exiting the Golgi complex do so through maturation of the cisternae and exchanges that occur via transfer vesicles (Figure 3.21):

1. Proteins synthesized by ribosomes on the rough ER are surrounded by a piece of the ER membrane, which eventually buds from the membrane surface to form transport vesicles.
2. Transport vesicles move toward the entry face of the Golgi complex.
3. Fusion of several transport vesicles creates the entry face of the Golgi complex and releases proteins into its lumen (space).
4. The proteins move from the entry face into one or more medial cisternae. Enzymes in the medial cisternae modify the proteins to form glycoproteins, glycolipids, and lipoproteins. **Transfer vesicles** that bud from the edges of the cisternae move specific enzymes back toward the entry face and move some partially modified proteins toward the exit face.
5. The products of the medial cisternae move into the lumen of the exit face.
6. Within the exit face cisterna, the products are further modified and are sorted and packaged.
7. Some of the processed proteins leave the exit face and are stored in **secretory vesicles.** These vesicles deliver the proteins to the plasma membrane, where they are discharged by exocytosis into the extracellular fluid. For example, certain pancreatic cells release the hormone insulin in this way.
8. Other processed proteins leave the exit face in **membrane vesicles** that deliver their contents to the plasma membrane for incorporation into the membrane. In doing so, the Golgi complex adds new segments of plasma membrane as existing segments are lost and modifies the number and distribution of membrane molecules.
9. Finally, some processed proteins leave the exit face in transport vesicles that will carry the proteins to another cellular destination. For instance, transport vesicles carry digestive enzymes to lysosomes; the structure and functions of these important organelles are discussed next.

Figure 3.21 Processing and packaging of proteins by the Golgi complex.

6 All proteins exported from the cell are processed in the Golgi complex.

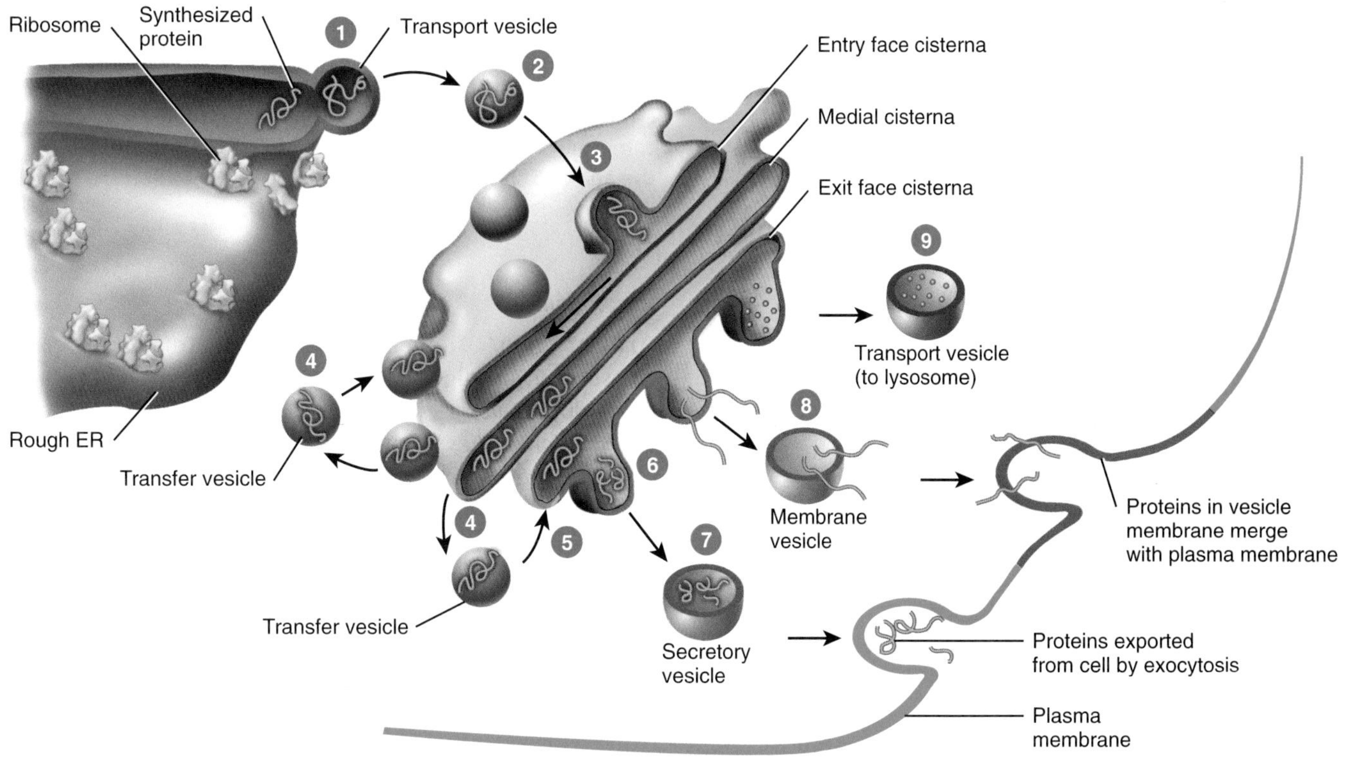

What are the three general destinations for proteins that leave the Golgi complex?

Lysosomes

Lysosomes (LĪ-sō-sōms; *lyso-* = dissolving; *-somes* = bodies) are membrane-enclosed vesicles that form from the Golgi complex (Figure 3.22). They can contain as many as 60 kinds of powerful digestive and hydrolytic enzymes that can break down a wide variety of molecules once lysosomes fuse with vesicles formed during endocytosis. Because lysosomal enzymes work best at an acidic pH, the lysosomal membrane includes active transport pumps that import hydrogen ions (H^+). Thus, the lysosomal interior has a pH of 5, which is 100 times more acidic than the pH of the cytosol (pH 7). The lysosomal membrane also includes transporters that move the final products of digestion, such as glucose, fatty acids, and amino acids, into the cytosol.

Figure 3.22 Lysosomes.

Lysosomes contain several types of powerful digestive enzymes.

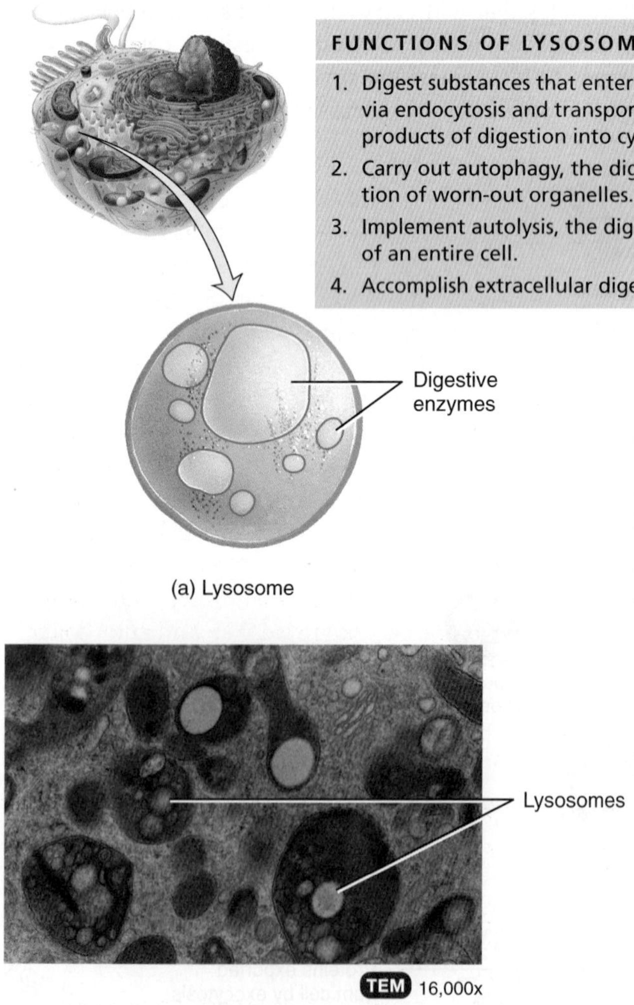

FUNCTIONS OF LYSOSOMES

1. Digest substances that enter a cell via endocytosis and transport final products of digestion into cytosol.
2. Carry out autophagy, the digestion of worn-out organelles.
3. Implement autolysis, the digestion of an entire cell.
4. Accomplish extracellular digestion.

(a) Lysosome

(b) Several lysosomes

What is the name of the process by which worn-out organelles are digested by lysosomes?

Lysosomal enzymes also help recycle worn-out cell structures. A lysosome can engulf another organelle, digest it, and return the digested components to the cytosol for reuse. In this way, old organelles are continually replaced. The process by which entire worn-out organelles are digested is called **autophagy** (aw-TOF-a-jē; *auto-* = self; *-phagy* = eating). In autophagy, the organelle to be digested is enclosed by a membrane derived from the ER to create a vesicle called an **autophagosome** (aw-tō-FĀ-gō-sōm); the vesicle then fuses with a lysosome. In this way, a human liver cell, for example, recycles about half of its cytoplasmic contents every week. Autophagy is also involved in cellular differentiation, control of growth, tissue remodeling, adaptation to adverse environments, and cell defense. Lysosomal enzymes may also destroy the entire cell that contains them, a process known as **autolysis** (aw-TOL-i-sis). Autolysis occurs in some pathological conditions and also is responsible for the tissue deterioration that occurs immediately after death.

As we just discussed, most lysosomal enzymes act within a cell. However, some operate in extracellular digestion. One example occurs during fertilization. The head of a sperm cell releases lysosomal enzymes that aid its penetration of the oocyte by dissolving its protective coating in a process called the acrosomal reaction (see Section 29.1).

CLINICAL CONNECTION | *Tay-Sachs Disease*

Some disorders are caused by faulty or absent lysosomal enzymes. For instance, **Tay-Sachs disease** (TĀ-SAKS), which most often affects children of Ashkenazi (eastern European Jewish) descent, is an inherited condition characterized by the absence of a single lysosomal enzyme called Hex A. This enzyme normally breaks down a membrane glycolipid called ganglioside G_{M2} that is especially prevalent in nerve cells. As the excess ganglioside G_{M2} accumulates, the nerve cells function less efficiently. Children with Tay-Sachs disease typically experience seizures and muscle rigidity. They gradually become blind, demented, and uncoordinated and usually die before the age of 5. Tests can now reveal whether an adult is a carrier of the defective gene. •

Peroxisomes

Another group of organelles similar in structure to lysosomes, but smaller, are the **peroxisomes** (pe-ROKS-i-sōms; *peroxi-* = peroxide; *-somes* = bodies; see Figure 3.1). Peroxisomes, also called *microbodies,* contain several *oxidases,* enzymes that can oxidize (remove hydrogen atoms from) various organic substances. For instance, amino acids and fatty acids are oxidized in peroxisomes as part of normal metabolism. In addition, enzymes in peroxisomes oxidize toxic substances, such as alcohol. Thus, peroxisomes are very abundant in the liver, where detoxification of alcohol and other damaging substances occurs. A by-product of the oxidation reactions is hydrogen peroxide (H_2O_2), a potentially toxic compound, and associated free radicals such as superoxide. However, peroxisomes also contain the enzyme *catalase,* which decomposes H_2O_2. Because production and degradation of H_2O_2 occur within the same organelle, peroxisomes protect other parts of the cell from the toxic effects of H_2O_2. Peroxisomes also contain enzymes that destroy superoxide. Without peroxisomes, byproducts of metabolism could accumulate inside a cell and result in cellular death. Peroxisomes can self-replicate. New peroxisomes may form from preexisting ones by enlarging and dividing. They may also form by a process in which components accumulate at a given site in the cell and then assemble into a peroxisome.

Proteasomes

As you have just learned, lysosomes degrade proteins delivered to them in vesicles. Cytosolic proteins also require disposal at certain times in the life of a cell. Continuous destruction of unneeded, damaged, or faulty proteins is the function of tiny barrel-shaped structures consisting of four stacked rings of proteins around a central core called **proteasomes** (PRŌ-tē-a-sōms = protein bodies). For example, proteins that are part of metabolic pathways need to be degraded after they have accomplished their function. Such protein destruction plays a part in negative feedback by halting a pathway once the appropriate response has been achieved. A typical body cell contains many thousands of proteasomes, in both the cytosol and the nucleus. Discovered only recently because they are far too small to discern under the light microscope and do not show up well in electron micrographs, proteasomes were so named because they contain myriad *proteases,* enzymes that cut proteins into small peptides. Once the enzymes of a proteasome have chopped up a protein into smaller chunks, other enzymes then break down the peptides into amino acids, which can be recycled into new proteins.

CLINICAL CONNECTION | *Proteasomes and Disease*

Some diseases could result from failure of proteasomes to degrade abnormal proteins. For example, clumps of misfolded proteins accumulate in brain cells of people with Parkinson disease and Alzheimer disease. Discovering why the proteasomes fail to clear these abnormal proteins is a goal of ongoing research. •

Mitochondria

Because they generate most of the ATP through aerobic (oxygen-requiring) respiration, **mitochondria** (mī-tō-KON-drē-a; *mito-* = thread; *-chondria* = granules; singular is *mitochondrion*) are referred to as the "powerhouses" of the cell. A cell may have as few as a hundred or as many as several thousand mitochondria, depending on its activity. Active cells, such as those found in the muscles, liver, and kidneys, which use ATP at a high rate, have a large number of mitochondria. For example, regular exercise can lead to an increase in the number of mitochondria in muscle cells, and this allows muscle cells to function more efficiently. Mitochondria are usually located within the cell where oxygen enters the cell or where the ATP is used, for example, among the contractile proteins in muscle cells.

A mitochondrion consists of an **outer mitochondrial membrane** and an **inner mitochondrial membrane** with a small fluid-filled space between them (Figure 3.23). Both membranes are similar in structure to the plasma membrane. The inner mitochondrial membrane contains a series of folds called **mitochondrial cristae** (KRIS-tē = ridges). The central fluid-filled cavity of a mitochondrion, enclosed by the inner mitochondrial membrane, is the **mitochondrial matrix.** The elaborate folds of the cristae provide an enormous surface area for the chemical reactions that are part of the aerobic phase of *cellular respiration,* the reactions that produce most of a cell's ATP (see Chapter 25). The enzymes that catalyze these reactions are located on the cristae and in the matrix of the mitochondria.

Mitochondria also play an important and early role in **apoptosis** (ap′-ōp-TŌ-sis or ap-ō-TŌ-sis = a falling off), the orderly, genetically programmed death of a cell. In response to stimuli such as large numbers of destructive free radicals, DNA damage, growth factor deprivation, or lack of oxygen and nutrients, certain chemicals are released from mitochondria following the formation of a pore in the outer mitochondrial membrane. One of the chemicals released into the cytosol of the cell is cytochrome *c,* which while inside the mitochondria is involved in aerobic cellular respiration. In the cytosol, however, cytochrome *c* and other substances initiate a cascade of activation of protein-digesting enzymes that bring about apoptosis.

Like peroxisomes, mitochondria self-replicate, a process that occurs during times of increased cellular energy demand or before cell division. Synthesis of some of the proteins needed for mitochondrial functions occurs on the ribosomes that are present in the mitochondrial matrix. Mitochondria even have their own DNA, in the form of multiple copies of a circular DNA molecule that contains 37 genes. These mitochondrial genes control the synthesis of 2 ribosomal RNAs, 22 transfer RNAs, and 13 proteins that build mitochondrial components.

Although the nucleus of each somatic cell contains genes from both your mother and your father, mitochondrial genes are inherited only from your mother. This is due to the fact that all mitochondria in a cell are descendants of those that were present in the oocyte (egg) during the fertilization process. The head of a sperm (the part that penetrates and fertilizes an oocyte) normally lacks most organelles, such as mitochondria, ribosomes, endoplasmic reticulum, and the Golgi complex, and any sperm mitochondria that do enter the oocyte are soon destroyed. Since all mitochondrial genes are inherited from the maternal parent, mitochondrial DNA can be used to trace maternal lineage (in other words, to determine whether two or more individuals are related through their mother's side of the family).

✔CHECKPOINT

16. What are some of the chemicals present in cytosol?
17. What is the function of cytosol?
18. Define an organelle.
19. Which organelles are surrounded by a membrane and which are not?
20. Which organelles contribute to synthesizing protein hormones and packaging them into secretory vesicles?
21. What happens on the cristae and in the matrix of mitochondria?

Figure 3.23 Mitochondria.

Within mitochondria, chemical reactions of aerobic cellular respiration generate ATP.

FUNCTIONS OF MITOCHONDRIA
1. Generate ATP through reactions of aerobic cellular respiration.
2. Play an important early role in apoptosis.

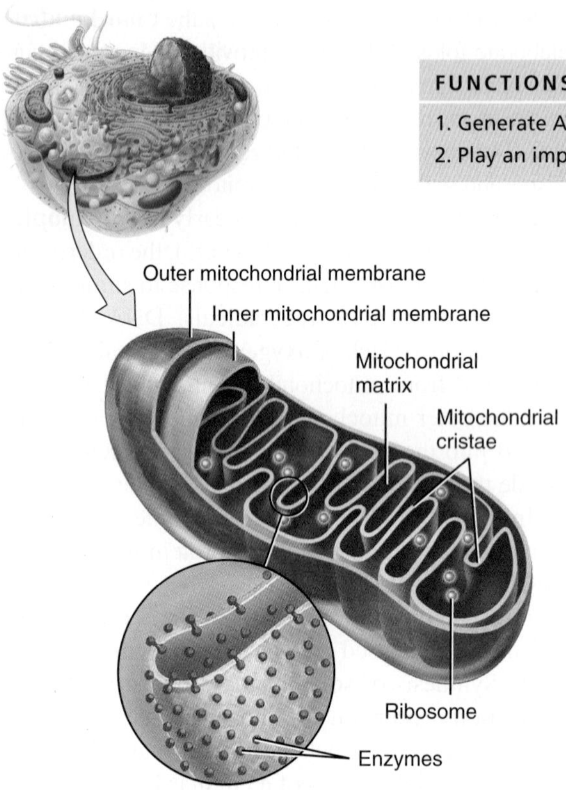

(a) Details

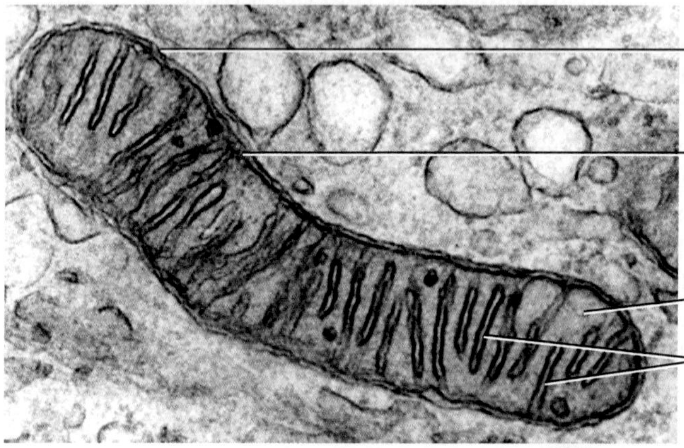

(b) Transverse section

How do the mitochondrial cristae contribute to its ATP-producing function?

3.5 NUCLEUS

OBJECTIVE

- Describe the structure and function of the nucleus.

The **nucleus** is a spherical or oval-shaped structure that usually is the most prominent feature of a cell (Figure 3.24). Most cells have a single nucleus, although some, such as mature red blood cells, have none. In contrast, skeletal muscle cells and a few other types of cells have multiple nuclei. A double membrane called the **nuclear envelope** separates the nucleus from the cytoplasm. Both layers of the nuclear envelope are lipid bilayers similar to the plasma membrane. The outer membrane of the nuclear envelope is continuous with rough ER and resembles it in structure. Many openings called **nuclear pores** extend through the nuclear envelope. Each nuclear pore consists of a circular arrangement of proteins surrounding a large central opening that is about 10 times wider than the pore of a channel protein in the plasma membrane.

Nuclear pores control the movement of substances between the nucleus and the cytoplasm. Small molecules and ions move through the pores passively by diffusion. Most large molecules, such as RNAs and proteins, cannot pass through the nuclear pores by diffusion. Instead, their passage involves an active transport process in which the molecules are recognized and selectively transported through the nuclear pore into or out of the nucleus. For example, proteins needed for nuclear functions move from the cytosol into the nucleus; newly formed RNA molecules move from the nucleus into the cytosol in this manner.

Inside the nucleus are one or more spherical bodies called **nucleoli** (noo′-KLĒ-ō-li; singular is *nucleolus*) that function in producing ribosomes. Each nucleolus is simply a cluster of protein, DNA, and RNA; it is not enclosed by a membrane. Nucleoli are the sites of synthesis of rRNA and assembly of rRNA and proteins into ribosomal subunits. Nucleoli are quite prominent in cells that synthesize large amounts of protein, such as muscle and liver cells. Nucleoli disperse and disappear during cell division and reorganize once new cells are formed.

Within the nucleus are most of the cell's hereditary units, called **genes,** which control cellular structure and direct cellular activities. Genes are arranged along **chromosomes** (KRŌ-mō-sōms;

Figure 3.24 Nucleus.

The nucleus contains most of the cell's genes, which are located on chromosomes.

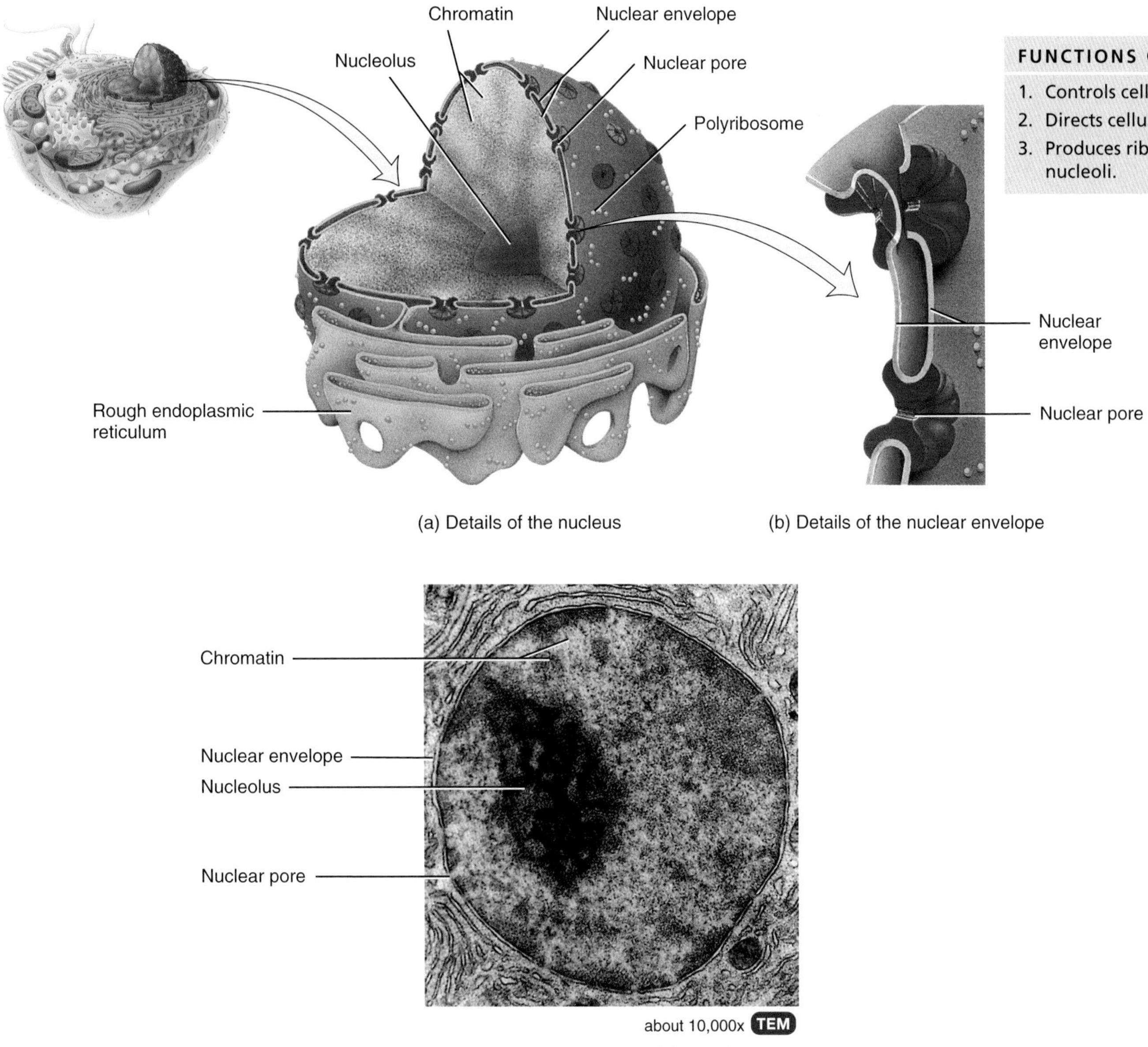
Chromatin
Nuclear envelope
Nucleolus
Nuclear pore
Polyribosome
Rough endoplasmic reticulum
Nuclear envelope
Nuclear pore
Chromatin
Nuclear envelope
Nucleolus
Nuclear pore
about 10,000x TEM

(a) Details of the nucleus

(b) Details of the nuclear envelope

(c) Transverse section of the nucleus

FUNCTIONS OF NUCLEUS

1. Controls cellular structure.
2. Directs cellular activities.
3. Produces ribosomes in nucleoli.

What is chromatin?

chromo- = colored). Human somatic (body) cells have 46 chromosomes, 23 inherited from each parent. Each chromosome is a long molecule of DNA that is coiled together with several proteins (Figure 3.25). This complex of DNA, proteins, and some RNA is called **chromatin** (KRŌ-ma-tin). The total genetic information carried in a cell or an organism is its **genome** (JĒ-nōm).

In cells that are not dividing, the chromatin appears as a diffuse, granular mass. Electron micrographs reveal that chromatin has a beads-on-a-string structure. Each bead is a **nucleosome** (NOO-klē-ō-sōm) that consists of double-stranded DNA wrapped twice around a core of eight proteins called **histones,** which help organize the coiling and folding of DNA. The string between the beads is called **linker DNA,** which holds adjacent nucleosomes together. In cells that are not dividing, another histone promotes coiling of nucleosomes into a larger-diameter **chromatin fiber,** which then folds into large loops. Just before cell division takes place, however, the DNA replicates (duplicates) and the loops condense even more, forming a pair of **chromatids** (KRŌ-ma-tids). As you will see shortly, during cell division a pair of chromatids constitutes a chromosome.

The main parts of a cell, their structure, and their functions are summarized in Table 3.2.

Figure 3.25 **Packing of DNA into a chromosome in a dividing cell.** When packing is complete, two identical DNA molecules and their histones form a pair of chromatids, which are held together by a centromere.

A chromosome is a highly coiled and folded DNA molecule that is combined with protein molecules.

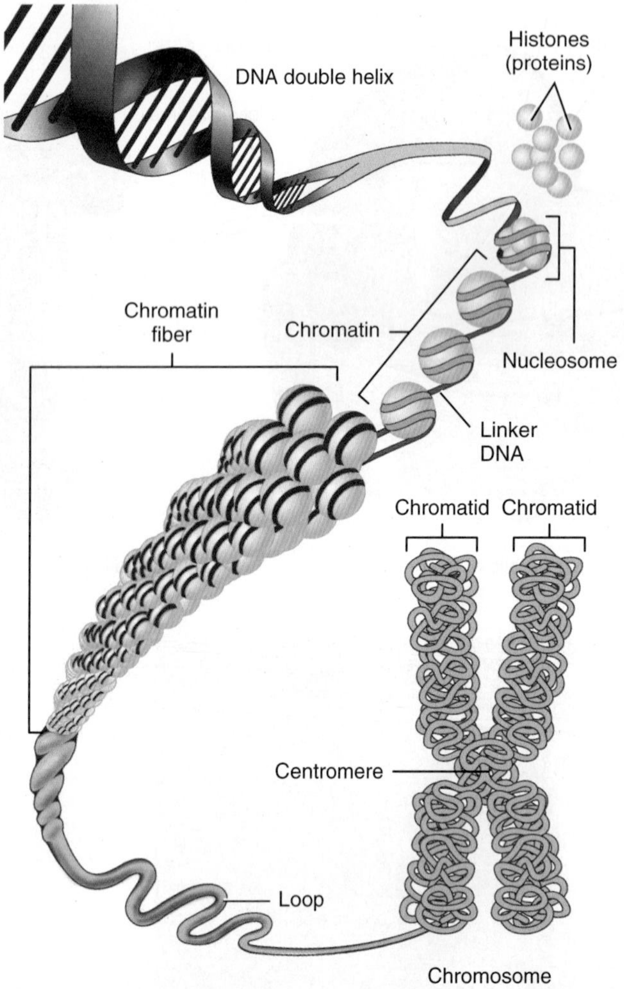

(a) Illustration

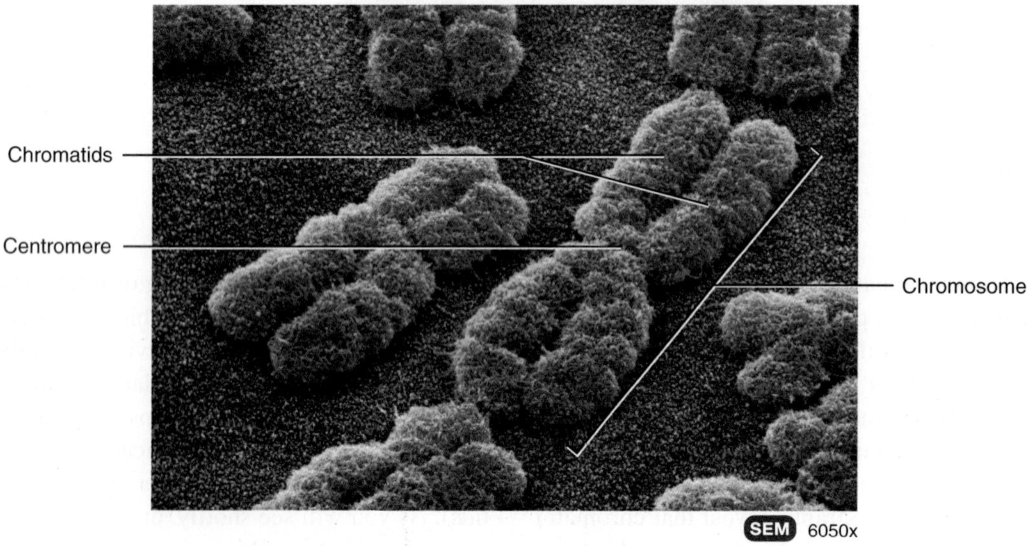

(b)

What are the components of a nucleosome?

TABLE 3.2

Cell Parts and Their Functions

PART	DESCRIPTION	FUNCTIONS
PLASMA MEMBRANE	Fluid mosaic lipid bilayer (phospholipids, cholesterol, and glycolipids) studded with proteins; surrounds cytoplasm.	Protects cellular contents; makes contact with other cells; contains channels, transporters, receptors, enzymes, cell-identity markers, and linker proteins; mediates entry and exit of substances.
CYTOPLASM	Cellular contents between plasma membrane and nucleus—cytosol and organelles.	Site of all intracellular activities except those occurring in the nucleus.
Cytosol	Composed of water, solutes, suspended particles, lipid droplets, and glycogen granules. Within the cytoplasm is the cytoskeleton, a network of microfilaments, intermediate filaments, and microtubules.	Fluid in which many of cell's metabolic reactions occur. Maintains shape and general organization of cellular contents; responsible for cellular movements.
Organelles	Specialized structures with characteristic shapes.	Each organelle has specific functions.
Centrosome	Pair of centrioles plus pericentriolar material.	The pericentriolar material contains tubulins, which are used for growth of the mitotic spindle and microtubule formation.
Cilia and flagella	Motile cell surface projections that contain 20 microtubules and a basal body.	Cilia: move fluids over cell's surface; flagella: move entire cell.
Ribosome	Composed of two subunits containing ribosomal RNA and proteins; may be free in cytosol or attached to rough ER.	Protein synthesis.
Endoplasmic reticulum (ER)	Membranous network of flattened sacs or tubules. Rough ER is covered by ribosomes and is attached to the nuclear envelope; smooth ER lacks ribosomes.	Rough ER: synthesizes glycoproteins and phospholipids that are transferred to cellular organelles, inserted into plasma membrane, or secreted during exocytosis; smooth ER: synthesizes fatty acids and steroids, inactivates or detoxifies drugs, removes phosphate group from glucose-6-phosphate, and stores and releases calcium ions in muscle cells.
Golgi complex	Consists of 3–20 flattened membranous sacs called cisternae; structurally and functionally divided into entry (*cis*) face, medial cisternae, and exit (*trans*) face.	Entry (*cis*) face accepts proteins from rough ER; medial cisternae form glycoproteins, glycolipids, and lipoproteins; exit (*trans*) face modifies molecules further, then sorts and packages them for transport to their destinations.
Lysosome	Vesicle formed from Golgi complex; contains digestive enzymes.	Fuses with and digests contents of endosomes, pinocytic vesicles, and phagosomes and transports final products of digestion into cytosol; digests worn-out organelles (autophagy), entire cells (autolysis), and extracellular materials.
Peroxisome	Vesicle containing oxidases (oxidative enzymes) and catalase (decomposes hydrogen peroxide); new peroxisomes bud from preexisting ones.	Oxidizes amino acids and fatty acids; detoxifies harmful substances, such as hydrogen peroxide and associated free radicals.
Proteasome	Tiny barrel-shaped structure that contains proteases (proteolytic enzymes).	Degrades unneeded, damaged, or faulty proteins by cutting them into small peptides.
Mitochondrion	Consists of an outer and an inner mitochondrial membrane, cristae, and matrix; new mitochondria form from preexisting ones.	Site of aerobic cellular respiration reactions that produce most of a cell's ATP. Plays an important early role in apoptosis.
NUCLEUS	Consists of a nuclear envelope with pores, nucleoli, and chromosomes, which exist as a tangled mass of chromatin in interphase cells.	Nuclear pores control the movement of substances between the nucleus and cytoplasm, nucleoli produce ribosomes, and chromosomes consist of genes that control cellular structure and direct cellular functions.

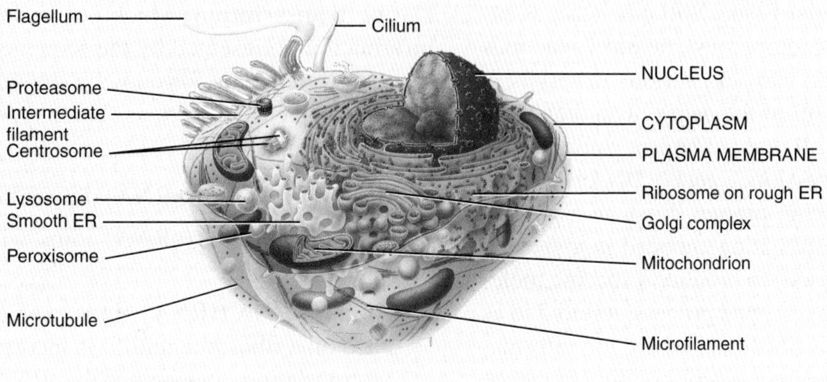

CLINICAL CONNECTION | *Genomics*

In the last decade of the twentieth century, the genomes of humans, mice, fruit flies, and more than 50 microbes were sequenced. As a result, research in the field of **genomics** (jē-NŌM-iks), the study of the relationships between the genome and the biological functions of an organism, has flourished. The Human Genome Project began in June 1990 as an effort to sequence all of the nearly 3.2 billion nucleotides of our genome, and was completed in April 2003. More than 99.9% of the nucleotide bases are identical in everyone. Less than 0.1% of our DNA (1 in each 1000 bases) accounts for inherited differences among humans. Surprisingly, at least half of the human genome consists of repeated sequences that do not code for proteins, so-called junk DNA. The average gene consists of 3000 nucleotides, but sizes vary greatly. The largest known human gene, with 2.4 million nucleotides, codes for the protein dystrophin. Scientists now know that the total number of genes in the human genome is about 30,000, far fewer than the 100,000 previously predicted to exist. Information regarding the human genome and how it is affected by the environment seeks to identify and discover the functions of the specific genes that play a role in genetic diseases. Genomic medicine also aims to design new drugs and to provide screening tests to enable physicians to provide more effective counseling and treatment for disorders with significant genetic components, such as hypertension (high blood pressure), obesity, diabetes, and cancer. •

✔CHECKPOINT

22. How do large particles enter and exit the nucleus?
23. Where are ribosomes produced?
24. How is DNA packed in the nucleus?

Figure 3.26 Overview of gene expression. Synthesis of a specific protein requires transcription of a gene's DNA into RNA and translation of RNA into a corresponding sequence of amino acids.

Transcription occurs in the nucleus; translation occurs in the cytoplasm.

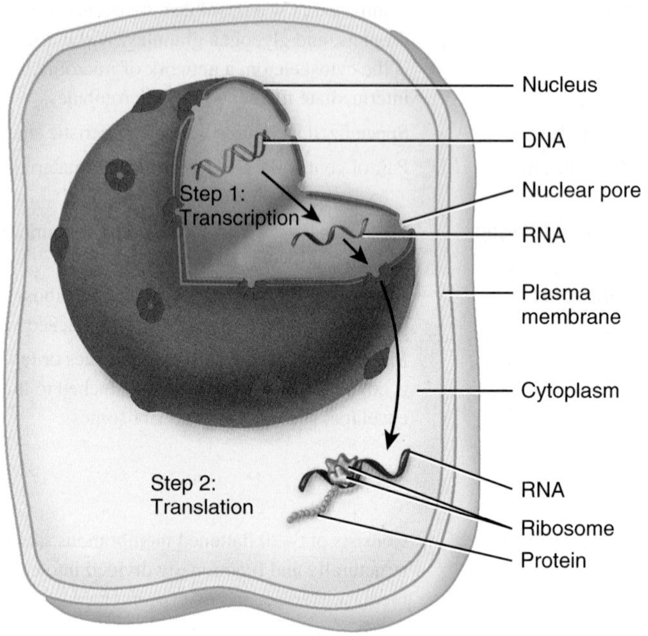

? **Why are proteins important in the life of a cell?**

3.6 PROTEIN SYNTHESIS

OBJECTIVE

- Describe the sequence of events in protein synthesis.

Although cells synthesize many chemicals to maintain homeostasis, much of the cellular machinery is devoted to synthesizing large numbers of diverse proteins. The proteins in turn determine the physical and chemical characteristics of cells and, therefore, of the organisms formed from them. Some proteins help assemble cellular structures such as the plasma membrane, the cytoskeleton, and other organelles. Others serve as hormones, antibodies, and contractile elements in muscular tissue. Still others act as enzymes, regulating the rates of the numerous chemical reactions that occur in cells, or transporters, carrying various materials in the blood. Just as genome means all of the genes in an organism, **proteome** (PRŌ-te-ōm) refers to all of an organism's proteins.

In the process called **gene expression,** a gene's DNA is used as a template for synthesis of a specific protein. First, in a process aptly named *transcription,* the information encoded in a specific region of DNA is *transcribed* (copied) to produce a specific molecule of RNA (ribonucleic acid). In a second process, referred to as translation, the RNA attaches to a ribosome, where the information contained in RNA is *translated* into a corresponding sequence of amino acids to form a new protein molecule (Figure 3.26).

DNA and RNA store genetic information as sets of three nucleotides. A sequence of three such nucleotides in DNA is called a **base triplet.** Each DNA base triplet is transcribed as a complementary sequence of three nucleotides, called a **codon.** A given codon specifies a particular amino acid. The **genetic code** is the set of rules that relate the base triplet sequence of DNA to the corresponding codons of RNA and the amino acids they specify.

Transcription

During **transcription,** which occurs in the nucleus, the genetic information represented by the sequence of base triplets in DNA serves as a template for copying the information into a complementary sequence of codons. Three types of RNA are made from the DNA template:

1. **Messenger RNA (mRNA)** directs the synthesis of a protein.
2. **Ribosomal RNA (rRNA)** joins with ribosomal proteins to make ribosomes.
3. **Transfer RNA (tRNA)** binds to an amino acid and holds it in place on a ribosome until it is incorporated into a protein during translation. One end of the tRNA carries a specific amino acid, and the opposite end consists of a triplet of nucleotides

called an **anticodon.** By pairing between complementary bases, the tRNA anticodon attaches to the mRNA codon. Each of the more than 20 different types of tRNA binds to only one of the 20 different amino acids.

The enzyme **RNA polymerase** (po-LIM-er-ās) catalyzes transcription of DNA. However, the enzyme must be instructed where to start the transcription process and where to end it. Only one of the two DNA strands serves as a template for RNA synthesis. The segment of DNA where transcription begins, a special nucleotide sequence called a **promoter,** is located near the beginning of a gene (Figure 3.27a). This is where RNA polymerase attaches to the DNA. During transcription, bases pair in a complementary manner: The bases cytosine (C), guanine (G), and thymine (T) in the DNA template pair with guanine, cytosine, and adenine (A), respectively, in the RNA strand (Figure 3.27b). However, adenine in the DNA template pairs with uracil (U), not thymine, in RNA:

A		U
T		A
G		C
	⟶	
C		G
A		U
T		A
Template DNA base sequence		Complementary RNA base sequence

Transcription of the DNA strand ends at another special nucleotide sequence called a **terminator,** which specifies the end of the gene (Figure 3.27a). When RNA polymerase reaches the terminator, the enzyme detaches from the transcribed RNA molecule and the DNA strand.

Not all parts of a gene actually code for parts of a protein. Regions within a gene called **introns** *do not* code for parts of proteins. They are located between regions called **exons** that *do* code for segments of a protein. Immediately after transcription, the transcript includes information from both introns and exons and is called **pre-mRNA.** The introns are removed from pre-mRNA by **small nuclear ribonucleoproteins** (snRNPs, pronounced "snurps"; Figure 3.27b). The snRNPs are enzymes that cut out the introns and splice together the exons. The resulting product is a functional mRNA molecule that passes through a pore in the nuclear envelope to reach the cytoplasm, where translation takes place.

Although the human genome contains around 30,000 genes, there are probably 500,000 to 1 million human proteins. How can so many proteins be coded for by so few genes? Part of the answer lies in **alternative splicing** of mRNA, a process in which the pre-mRNA transcribed from a gene is spliced in different ways to produce several different mRNAs. The different mRNAs are then translated into different proteins. In this way, one gene may code for 10 or more different proteins. In addition, chemical modifications are made to proteins after translation, for example, as proteins pass through the Golgi complex. Such chemical alterations can produce two or more different proteins from a single translation.

Figure 3.27 Transcription. DNA transcription begins at a promoter and ends at a terminator.

During transcription, the genetic information in DNA is copied to RNA.

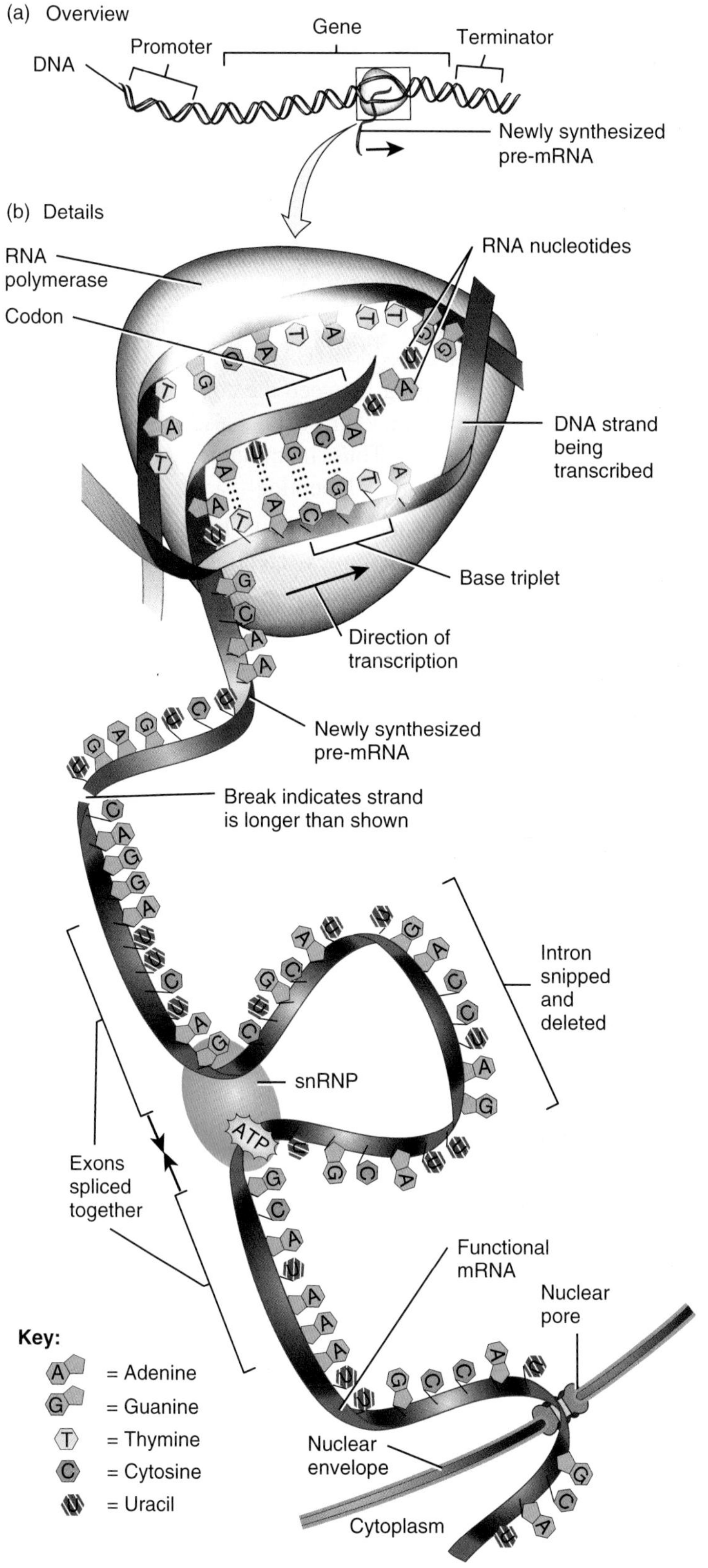

If the DNA template had the base sequence AGCT, what would be the mRNA base sequence, and what enzyme would catalyze DNA transcription?

Translation

In the process of **translation,** the nucleotide sequence in an mRNA molecule specifies the amino acid sequence of a protein. Ribosomes in the cytoplasm carry out translation. The small subunit of a ribosome has a *binding site* for mRNA; the large subunit has two binding sites for tRNA molecules, a *P site* and an *A site* (Figure 3.28). The first tRNA molecule bearing its specific amino acid attaches to mRNA at the P site. The A site holds the next tRNA molecule bearing its amino acid. Translation occurs in the following way (Figure 3.29):

1. An mRNA molecule binds to the small ribosomal subunit at the mRNA binding site. A special tRNA, called *initiator tRNA,* binds to the start codon (AUG) on mRNA, where translation begins. The tRNA anticodon (UAC) attaches to the mRNA codon (AUG) by pairing between the complementary bases. Besides being the start codon, AUG is also the codon for the amino acid methionine. Thus, methionine is always the first amino acid in a growing polypeptide.
2. Next, the large ribosomal subunit attaches to the small ribosomal subunit–mRNA complex, creating a functional ribosome. The initiator tRNA, with its amino acid (methionine), fits into the P site of the ribosome.
3. The anticodon of another tRNA with its attached amino acid pairs with the second mRNA codon at the A site of the ribosome.
4. A component of the large ribosomal subunit catalyzes the formation of a peptide bond between methionine, which separates from its tRNA at the P site, and the amino acid carried by the tRNA at the A site.

Figure 3.28 Translation. During translation, an mRNA molecule binds to a ribosome. Then, the mRNA nucleotide sequence specifies the amino acid sequence of a protein.

(6) Ribosomes have a binding site for mRNA and a P site and an A site for attachment of tRNAs.

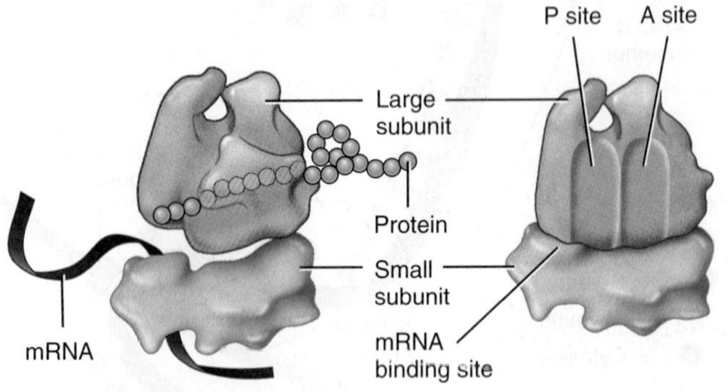

(a) Components of a ribosome and their relationship to mRNA and protein during translation

(b) Interior view of tRNA binding sites

? What roles do the P and A sites serve?

5. After peptide bond formation, the tRNA at the P site detaches from the ribosome, and the ribosome shifts the mRNA strand by one codon. The tRNA in the A site bearing the two-peptide protein shifts into the P site, allowing another tRNA with its amino acid to bind to a newly exposed codon at the A site. Steps 3 through 5 occur repeatedly, and the protein lengthens progressively.
6. Protein synthesis ends when the ribosome reaches a stop codon, which causes the completed protein to detach from the final tRNA. When the tRNA vacates the ribosome, the ribosome splits into its large and small subunits.

Protein synthesis progresses at a rate of about 15 peptide bonds per second. As the ribosome moves along the mRNA and before it completes synthesis of the whole protein, another ribosome may attach behind it and begin translation of the same mRNA strand. Several ribosomes attached to the same mRNA constitute a **polyribosome.** The simultaneous movement of several ribosomes along the same mRNA molecule permits the translation of one mRNA into several identical proteins at the same time.

CLINICAL CONNECTION | *Recombinant DNA*

Scientists have developed techniques for inserting genes from other organisms into a variety of host cells. Manipulating the cell in this way can cause the host organism to produce proteins it normally does not synthesize. Organisms so altered are called **recombinants** (rē-KOM-bi-nants), and their DNA—a combination of DNA from different sources—is called **recombinant DNA.** When recombinant DNA functions properly, the host will synthesize the protein specified by the new gene it has acquired. The technology that has arisen from the manipulation of genetic material is referred to as **genetic engineering.**

The practical applications of recombinant DNA technology are enormous. Strains of recombinant bacteria now produce large quantities of many important therapeutic substances, including *human growth hormone (hGH),* required for normal growth and metabolism; *insulin,* a hormone that helps regulate blood glucose level and is used by diabetics; *interferon (IFN),* an antiviral (and possibly anticancer) substance; and *erythropoietin (EPO),* a hormone that stimulates production of red blood cells. •

✔CHECKPOINT

25. What is meant by the term gene expression?
26. What is the difference between transcription and translation?

3.7 CELL DIVISION

OBJECTIVES

- Discuss the stages, events, and significance of somatic and reproductive cell division.
- Describe the signals that induce somatic cell division.

Most cells of the human body undergo **cell division,** the process by which cells reproduce themselves. The two types of cell division—somatic cell division and reproductive cell division—accomplish different goals for the organism.

Figure 3.29 Protein elongation and termination of protein synthesis during translation.

During protein synthesis the small and large ribosomal subunits join to form a functional ribosome. When the process is complete, they separate.

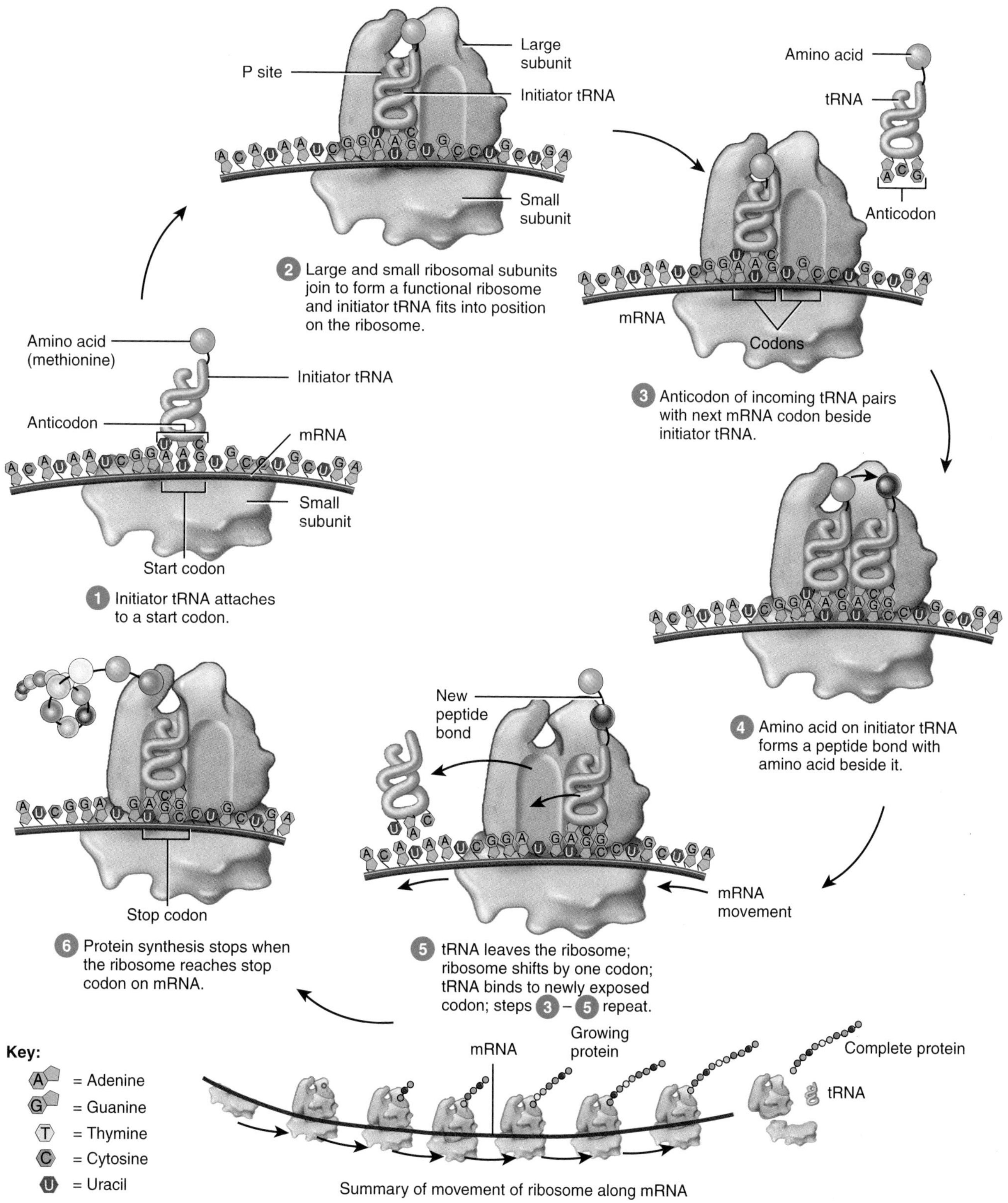

Large subunit
P site
Initiator tRNA
Small subunit
2 Large and small ribosomal subunits join to form a functional ribosome and initiator tRNA fits into position on the ribosome.
Amino acid
tRNA
Anticodon
mRNA
Codons
3 Anticodon of incoming tRNA pairs with next mRNA codon beside initiator tRNA.
Amino acid (methionine)
Initiator tRNA
Anticodon
mRNA
Small subunit
Start codon
1 Initiator tRNA attaches to a start codon.
4 Amino acid on initiator tRNA forms a peptide bond with amino acid beside it.
New peptide bond
mRNA movement
5 tRNA leaves the ribosome; ribosome shifts by one codon; tRNA binds to newly exposed codon; steps 3 – 5 repeat.
Stop codon
6 Protein synthesis stops when the ribosome reaches stop codon on mRNA.
Key:
= Adenine
= Guanine
= Thymine
= Cytosine
= Uracil
mRNA
Growing protein
Complete protein
tRNA
Summary of movement of ribosome along mRNA

What is the function of a stop codon?

A **somatic cell** (sō-MAT-ik; *soma* = body) is any cell of the body other than a germ cell. A **germ cell** is a gamete (sperm or oocyte) or any precursor cell destined to become a gamete. In **somatic cell division,** a cell undergoes a nuclear division called **mitosis** (mī-TŌ-sis; *mitos* = thread) and a cytoplasmic division called **cytokinesis** (sī-tō-ki-NĒ-sis; *cyto-* = cell; *-kinesis* = movement) to produce two genetically identical cells, each with the same number and kind of chromosomes as the original cell. Somatic cell division replaces dead or injured cells and adds new ones during tissue growth.

Reproductive cell division is the mechanism that produces gametes, the cells needed to form the next generation of sexually reproducing organisms. This process consists of a special two-step division called **meiosis,** in which the number of chromosomes in the nucleus is reduced by half.

Somatic Cell Division

The **cell cycle** is an orderly sequence of events in which a somatic cell duplicates its contents and divides in two. Some cells divide more than others. Human cells, such as those in the brain, stomach, and kidneys, contain 23 pairs of chromosomes, for a total of 46. One member of each pair is inherited from each parent. The two chromosomes that make up each pair are called **homologous chromosomes** (hō-MOL-ō-gus; *homo-* = same) or **homologs;** they contain similar genes arranged in the same (or almost the same) order. When examined under a light microscope, homologous chromosomes generally look very similar. The exception to this rule is one pair of chromosomes called the **sex chromosomes,** designated X and Y. In females the homologous pair of sex chromosomes consists of two large X chromosomes; in males the pair consists of an X and a much smaller Y chromosome. Because somatic cells contain two sets of chromosomes, they are called **diploid cells** (DIP-loyd; *dipl-* = double; *-oid* = form), symbolized **2n.**

When a cell reproduces, it must replicate (duplicate) all its chromosomes to pass its genes to the next generation of cells. The cell cycle consists of two major periods: interphase, when a cell is not dividing, and the mitotic (M) phase, when a cell is dividing (Figure 3.30).

Figure 3.30 **The cell cycle.** Not illustrated is cytokinesis (division of the cytoplasm), which usually occurs during late anaphase of the mitotic phase.

In a complete cell cycle, a starting cell duplicates its contents and divides into two identical cells.

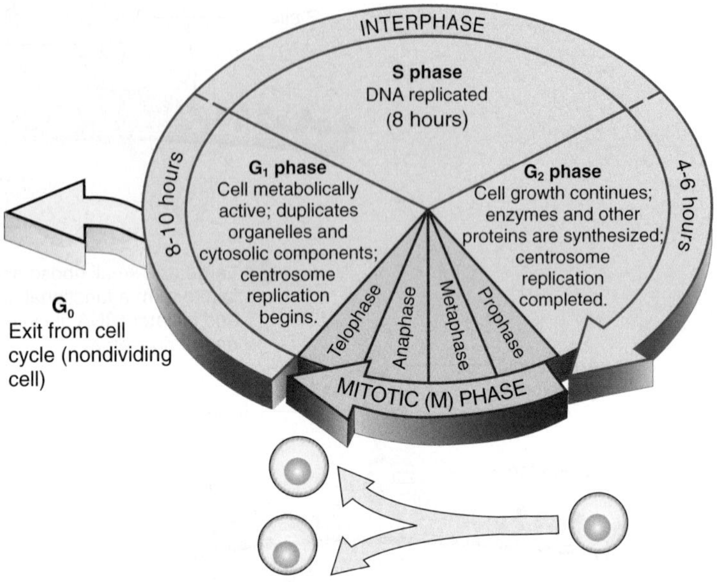

During which phase of the cell cycle does DNA replication occur?

Interphase

During **interphase** (IN-ter-fāz) the cell replicates its DNA through a process that will be described shortly. It also produces additional organelles and cytosolic components in anticipation of cell division. Interphase is a state of high metabolic activity; it is during this time that the cell does most of its growing. Interphase consists of three phases: G_1, S, and G_2 (Figure 3.30). The S stands for *synthesis* of DNA. Because the G phases are periods when there is no activity related to DNA duplication, they are thought of as *gaps* or interruptions in DNA duplication.

The **G_1 phase** is the interval between the mitotic phase and the S phase. During G_1, the cell is metabolically active; it replicates most of its organelles and cytosolic components but not its DNA. Replication of centrosomes also begins in the G_1 phase. Virtually all the cellular activities described in this chapter happen during G_1. For a cell with a total cell cycle time of 24 hours, G_1 lasts 8 to 10 hours. However, the duration of this phase is quite variable. It is very short in many embryonic cells or cancer cells. Cells that remain in G_1 for a very long time, perhaps destined never to divide again, are said to be in the **G_0 phase.** Most nerve cells are in the G_0 phase. Once a cell enters the S phase, however, it is committed to go through the rest of the cell cycle.

The **S phase,** the interval between G_1 and G_2, lasts about 8 hours. During the S phase, DNA replication occurs. As a result of DNA replication, the two identical cells formed during cell division later in the cell cycle will have the same genetic material. The **G_2 phase** is the interval between the S phase and the mitotic phase. It lasts 4 to 6 hours. During G_2, cell growth continues, enzymes and other proteins are synthesized in preparation for cell division, and replication of centrosomes is completed. When DNA replicates during the S phase, its helical structure partially uncoils, and the two strands separate at the points where hydrogen bonds connect base pairs (Figure 3.31). Each exposed base of the old DNA strand then pairs with the complementary base of a newly synthesized nucleotide. A new DNA strand takes shape as chemical bonds form between neighboring nucleotides. The uncoiling and complementary base pairing continues until each of the two original DNA strands is joined with a newly formed complementary DNA strand. The original DNA molecule has become two identical DNA molecules.

Figure 3.31 Replication of DNA. The two strands of the double helix separate by breaking the hydrogen bonds (shown as dotted lines) between nucleotides. New, complementary nucleotides attach at the proper sites, and a new strand of DNA is synthesized alongside each of the original strands. Arrows indicate hydrogen bonds forming again between pairs of bases.

Replication doubles the amount of DNA.

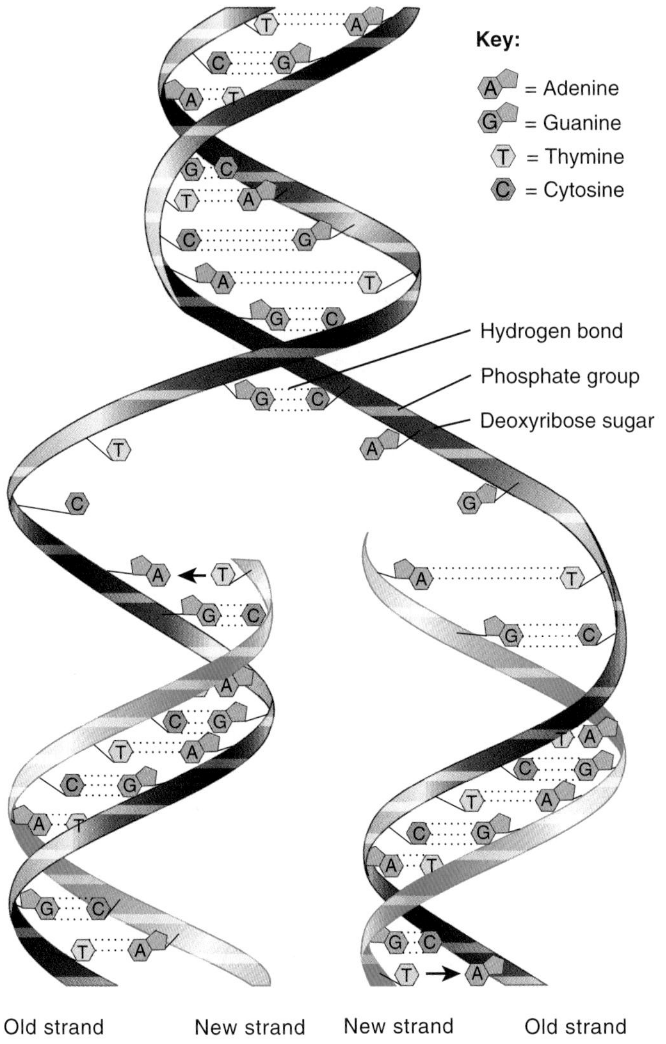

Key:
= Adenine
= Guanine
= Thymine
= Cytosine
Hydrogen bond
Phosphate group
Deoxyribose sugar
Old strand
New strand
New strand
Old strand

Why must DNA replication occur before cytokinesis in somatic cell division?

A microscopic view of a cell during interphase shows a clearly defined nuclear envelope, a nucleolus, and a tangled mass of chromatin (Figure 3.32a). Once a cell completes its activities during the G_1, S, and G_2 phases of interphase, the mitotic phase begins.

Mitotic Phase

The **mitotic (M) phase** of the cell cycle, which results in the formation of two identical cells, consists of a nuclear division (mitosis) and a cytoplasmic division (cytokinesis) to form two identical cells. The events that occur during mitosis and cytokinesis are plainly visible under a microscope because chromatin condenses into discrete chromosomes.

NUCLEAR DIVISION: MITOSIS **Mitosis,** as noted earlier, is the distribution of two sets of chromosomes into two separate nuclei. The process results in the *exact* partitioning of genetic information. For convenience, biologists divide the process into four stages: prophase, metaphase, anaphase, and telophase. However, mitosis is a continuous process; one stage merges seamlessly into the next.

1. **Prophase** (PRŌ-fāz). During early prophase, the chromatin fibers condense and shorten into chromosomes that are visible under the light microscope (Figure 3.32b). The condensation process may prevent entangling of the long DNA strands as they move during mitosis. Because longitudinal DNA replication took place during the S phase of interphase, each prophase chromosome consists of a pair of identical strands called *chromatids.* A constricted region called a **centromere** (SEN-trō-mēr) holds the chromatid pair together. At the outside of each centromere is a protein complex known as the **kinetochore** (ki-NET-ō-kor). Later in prophase, tubulins in the pericentriolar material of the centrosomes start to form the **mitotic spindle,** a football-shaped assembly of microtubules that attach to the kinetochore (Figure 3.32b). As the microtubules lengthen, they push the centrosomes to the poles (ends) of the cell so that the spindle extends from pole to pole. The mitotic spindle is responsible for the separation of chromatids to opposite poles of the cell. Then, the nucleolus disappears and the nuclear envelope breaks down.
2. **Metaphase** (MET-a-fāz). During metaphase, the microtubules of the mitotic spindle align the centromeres of the chromatid pairs at the exact center of the mitotic spindle (Figure 3.32c). This midpoint region is called the **metaphase plate.**
3. **Anaphase** (AN-a-fāz). During anaphase, the centromeres split, separating the two members of each chromatid pair, which move toward opposite poles of the cell (Figure 3.32d). Once separated, the chromatids are termed *chromosomes.* As the chromosomes are pulled by the microtubules of the mitotic spindle during anaphase, they appear V-shaped because the centromeres lead the way, dragging the trailing arms of the chromosomes toward the pole.
4. **Telophase** (TEL-ō-fāz). The final stage of mitosis, telophase, begins after chromosomal movement stops (Figure 3.32e). The identical sets of chromosomes, now at opposite poles of the cell, uncoil and revert to the threadlike chromatin form. A nuclear envelope forms around each chromatin mass, nucleoli reappear in the identical nuclei, and the mitotic spindle breaks up.

CYTOPLASMIC DIVISION: CYTOKINESIS As noted earlier, division of a cell's cytoplasm and organelles into two identical cells is called **cytokinesis.** This process usually begins in late anaphase with the formation of a **cleavage furrow,** a slight indentation of the plasma membrane, and is completed after telophase. The

Figure 3.32 Cell division: mitosis and cytokinesis. Begin the sequence at 1 at the top of the figure and read clockwise to complete the process.

In somatic cell division, a single starting cell divides to produce two identical diploid cells.

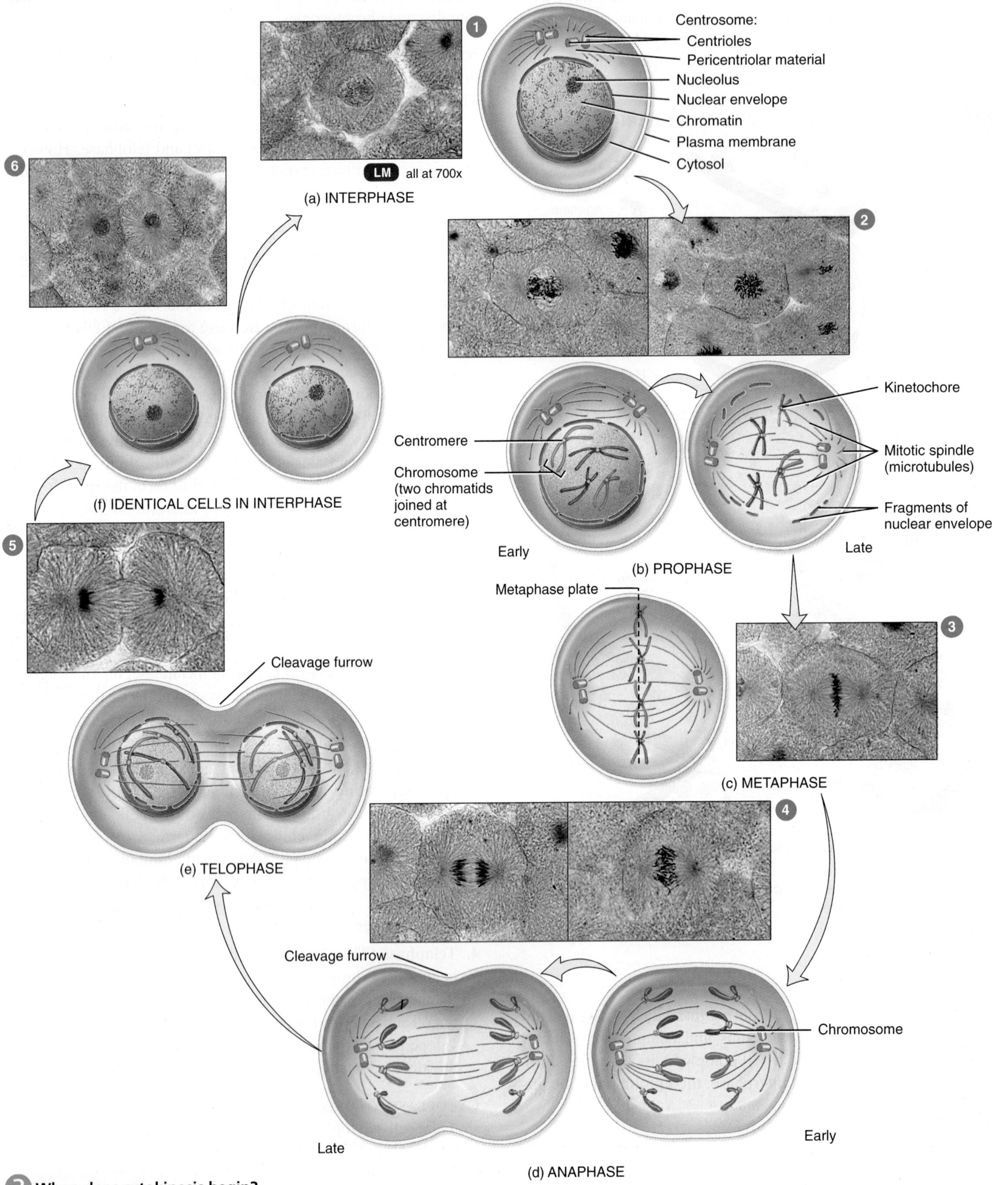

1
Centrosome:
Centrioles
Pericentriolar material
Nucleolus
Nuclear envelope
Chromatin
Plasma membrane
Cytosol
LM all at 700x
(a) INTERPHASE
2
Kinetochore
Centromere
Chromosome (two chromatids joined at centromere)
Mitotic spindle (microtubules)
Fragments of nuclear envelope
Early
Late
(b) PROPHASE
Metaphase plate
3
(c) METAPHASE
4
Cleavage furrow
Chromosome
Late
Early
(d) ANAPHASE
(e) TELOPHASE
Cleavage furrow
5
(f) IDENTICAL CELLS IN INTERPHASE
6

When does cytokinesis begin?

cleavage furrow usually appears midway between the centrosomes and extends around the periphery of the cell (Figure 3.32d, e). Actin microfilaments that lie just inside the plasma membrane form a *contractile ring* that pulls the plasma membrane progressively inward. The ring constricts the center of the cell, like tightening a belt around the waist, and ultimately pinches it in two. Because the plane of the cleavage furrow is always perpendicular to the mitotic spindle, the two sets of chromosomes end up in separate cells. When cytokinesis is complete, interphase begins (Figure 3.32f).

The sequence of events can be summarized as

$$G_1 \longrightarrow \text{S phase} \longrightarrow G_2 \text{ phase} \longrightarrow \text{mitosis} \longrightarrow \text{cytokinesis}$$

Table 3.3 summarizes the events of the cell cycle in somatic cells.

CLINICAL CONNECTION | *Mitotic Spindle and Cancer*

One of the distinguishing features of cancer cells is uncontrolled division, which results in the formation of a mass of cells called a *neoplasm* or *tumor*. One of the ways to treat cancer is by *chemotherapy*, the use of anticancer drugs. Some of these drugs stop cell division by inhibiting the formation of the mitotic spindle. Unfortunately, these types of anticancer drugs also kill all types of rapidly dividing cells in the body, causing side effects such as nausea, diarrhea, hair loss, fatigue, and decreased resistance to disease. •

Control of Cell Destiny

A cell has three possible destinies: (1) to remain alive and functioning without dividing, (2) to grow and divide, or (3) to die. Homeostasis is maintained when there is a balance between cell proliferation and cell death. The signals that tell a cell when to exist in the G_0 phase, when to divide, and when to die have been the subjects of intense and fruitful research in recent years.

Within a cell, there are enzymes called **cyclin-dependent protein kinases (Cdk's)** that can transfer a phosphate group from ATP to a protein to activate the protein; other enzymes can remove the phosphate group from the protein to deactivate it. The activation and deactivation of Cdk's at the appropriate time is crucial in the initiation and regulation of DNA replication, mitosis, and cytokinesis.

Switching the Cdk's on and off is the responsibility of cellular proteins called **cyclins** (SĪK-lins), so named because their levels rise and fall during the cell cycle. The joining of a specific cyclin and Cdk molecule triggers various events that control cell division.

The activation of specific cyclin–Cdk complexes is responsible for progression of a cell from G_1 to S to G_2 to mitosis in a specific order. If any step in the sequence is delayed, all subsequent steps are delayed in order to maintain the normal sequence. The levels of cyclins in the cell are very important in determining the timing and sequence of events in cell division. For example, the level of the cyclin that helps drive a cell from G_2 to mitosis rises throughout the G_1, S, and G_2 phases and into mitosis. The high level triggers mitosis, but toward the end of mitosis, the level declines rapidly and mitosis ends. Destruction of this cyclin, as well as others in the cell, is by proteasomes.

Cellular death is also regulated. Throughout the lifetime of an organism, certain cells undergo apoptosis, an orderly, genetically programmed death (see the discussion of mitochondria in Section 3.4). In apoptosis, a triggering agent from either outside or inside the cell causes "cell-suicide" genes to produce enzymes that damage the cell in several ways, including disruption of its cytoskeleton and nucleus. As a result, the cell shrinks and pulls away from neighboring cells. Although the plasma membrane

TABLE 3.3

Events of the Somatic Cell Cycle

PHASE	ACTIVITY
Interphase	Period between cell divisions; chromosomes not visible under light microscope.
G_1 phase	Metabolically active cell duplicates most of its organelles and cytosolic components; replication of chromosomes begins. (Cells that remain in the G_1 phase for a very long time, and possibly never divide again, are said to be in the G_0 phase.)
S phase	Replication of DNA and centrosomes.
G_2 phase	Cell growth, enzyme and protein synthesis continue; replication of centrosomes complete.
Mitotic phase	Parent cell produces identical cells with identical chromosomes; chromosomes visible under light microscope.
Mitosis	Nuclear division; distribution of two sets of chromosomes into separate nuclei.
Prophase	Chromatin fibers condense into paired chromatids; nucleolus and nuclear envelope disappear; each centrosome moves to an opposite pole of the cell.
Metaphase	Centromeres of chromatid pairs line up at metaphase plate.
Anaphase	Centromeres split; identical sets of chromosomes move to opposite poles of cell.
Telophase	Nuclear envelopes and nucleoli reappear; chromosomes resume chromatin form; mitotic spindle disappears.
Cytokinesis	Cytoplasmic division; contractile ring forms cleavage furrow around center of cell, dividing cytoplasm into separate and equal portions.

remains intact, the DNA within the nucleus fragments and the cytoplasm shrinks. Nearby phagocytes then ingest the dying cell via a complex process that involves a receptor protein in the plasma membrane of the phagocyte that binds to a lipid in the plasma membrane of the dying cell. Apoptosis removes unneeded cells during fetal development, such as the webbing between digits. It continues to occur after birth to regulate the number of cells in a tissue and eliminate potentially dangerous cells such as cancer cells.

Apoptosis is a normal type of cell death; in contrast, **necrosis** (ne-KRŌ-sis = death) is a pathological type of cell death that results from tissue injury. In necrosis, many adjacent cells swell, burst, and spill their cytoplasm into the interstitial fluid. The cellular debris usually stimulates an inflammatory response by the immune system, a process that does not occur in apoptosis.

CLINICAL CONNECTION | *Tumor-Suppressor Genes*

Abnormalities in genes that regulate the cell cycle or apoptosis are associated with many diseases. For example, damage to genes called **tumor-suppressor genes,** which produce proteins that normally inhibit cell division, causes some types of cancer. Loss or alteration of a tumor-suppressor gene called *p53* on chromosome 17 is the most common genetic change leading to a wide variety of tumors, including breast and colon cancers. The normal *p53* protein arrests cells in the G_1 phase, which prevents cell division. Normal *p53* protein also assists in repair of damaged DNA and induces apoptosis in the cells where DNA repair was not successful. For this reason, the *p53* gene is nicknamed "the guardian angel of the genome." •

Reproductive Cell Division

In the process called sexual reproduction, each new organism is the result of the union of two different gametes (fertilization), one produced by each parent. If gametes had the same number of chromosomes as somatic cells, the number of chromosomes would double at fertilization. **Meiosis** (mī-Ō-sis; *mei-* = lessening; *-osis* = condition of), the reproductive cell division that occurs in the gonads (ovaries and testes), produces gametes in which the number of chromosomes is reduced by half. As a result, gametes contain a single set of 23 chromosomes and thus are **haploid (*n*) cells** (HAP-loyd; *hapl-* = single). Fertilization restores the diploid number of chromosomes.

Meiosis

Unlike mitosis, which is complete after a single round, meiosis occurs in two successive stages: **meiosis I** and **meiosis II.** During the interphase that precedes meiosis I, the chromosomes of the diploid cell start to replicate. As a result of replication, each chromosome consists of two sister (genetically identical) chromatids, which are attached at their centromeres. This replication of chromosomes is similar to the one that precedes mitosis in somatic cell division.

MEIOSIS I Meiosis I, which begins once chromosomal replication is complete, consists of four phases: prophase I, metaphase I, anaphase I, and telophase I (Figure 3.33a). Prophase I is an extended phase in which the chromosomes shorten and thicken, the nuclear envelope and nucleoli disappear, and the mitotic spindle forms. Two events that are not seen in mitotic prophase occur during prophase I of meiosis (Figure 3.33b). First, the two sister chromatids of each pair of homologous chromosomes pair off, an event called **synapsis** (sin-AP-sis). The resulting four chromatids form a structure called a **tetrad.** Second, parts of the chromatids of two homologous chromosomes may be exchanged with one another. Such an exchange between parts of nonsister (genetically different) chromatids is called **crossing-over.** This process, among others, permits an exchange of genes between chromatids of homologous chromosomes. Due to crossing-over, the resulting cells are genetically unlike each other and genetically unlike the starting cell that produced them. Crossing-over results in *genetic recombination*—that is, the formation of new combinations of genes—and accounts for part of the great genetic variation among humans and other organisms that form gametes via meiosis.

In metaphase I, the tetrads formed by the homologous pairs of chromosomes line up along the metaphase plate of the cell, with homologous chromosomes side by side (Figure 3.33a). During anaphase I, the members of each homologous pair of chromosomes separate as they are pulled to opposite poles of the cell by the microtubules attached to the centromeres. The paired chromatids, held by a centromere, remain together. (Recall that during mitotic anaphase, the centromeres split and the sister chromatids separate.) Telophase I and cytokinesis of meiosis are similar to telophase and cytokinesis of mitosis. The net effect of meiosis I is that each resulting cell contains the haploid number of chromosomes because it contains only one member of each pair of the homologous chromosomes present in the starting cell.

MEIOSIS II The second stage of meiosis, meiosis II, also consists of four phases: prophase II, metaphase II, anaphase II, and telophase II (Figure 3.33a). These phases are similar to those that occur during mitosis; the centromeres split, and the sister chromatids separate and move toward opposite poles of the cell.

In summary, meiosis I begins with a diploid starting cell and ends with two cells, each with the haploid number of chromosomes. During meiosis II, each of the two haploid cells formed during meiosis I divides; the net result is four haploid gametes that are genetically different from the original diploid starting cell.

Figure 3.34 compares the events of meiosis and mitosis.

✔CHECKPOINT

27. Distinguish between somatic and reproductive cell division and explain the importance of each.
28. What is the significance of interphase?
29. Outline the major events of each stage of the mitotic phase of the cell cycle.
30. How are apoptosis and necrosis similar? How do they differ?
31. How are haploid and diploid cells different?
32. What are homologous chromosomes?

Figure 3.33 Meiosis, reproductive cell division. Details of each of the stages are discussed in the text.

In reproductive cell division, a single diploid starting cell undergoes meiosis I and meiosis II to produce four haploid gametes that are genetically different from the starting cell that produced them.

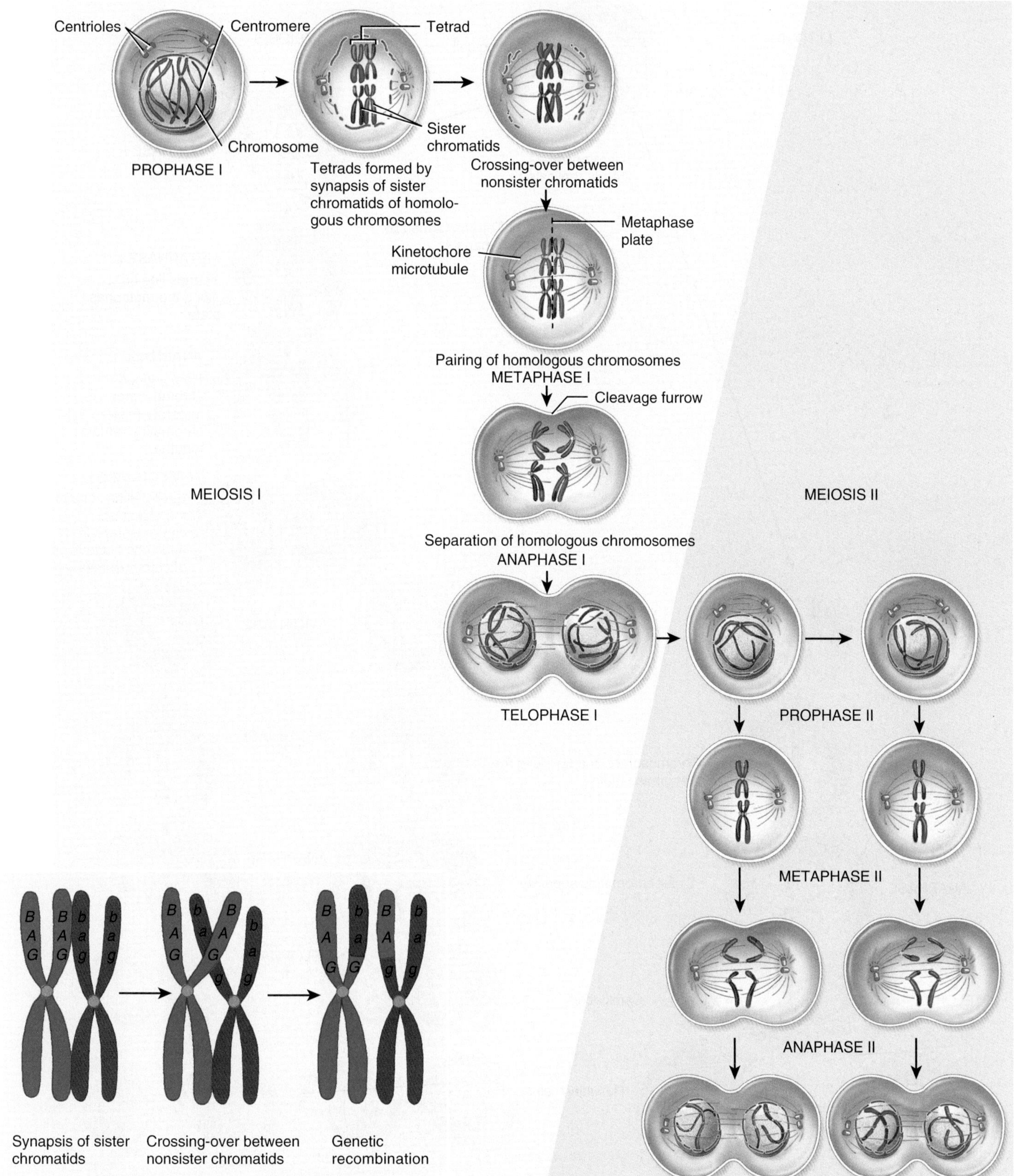

Centrioles
Centromere
Tetrad
Chromosome
Sister chromatids
PROPHASE I
Tetrads formed by synapsis of sister chromatids of homologous chromosomes
Crossing-over between nonsister chromatids
Metaphase plate
Kinetochore microtubule
Pairing of homologous chromosomes
METAPHASE I
Cleavage furrow
MEIOSIS I
MEIOSIS II
Separation of homologous chromosomes
ANAPHASE I
TELOPHASE I
PROPHASE II
METAPHASE II
ANAPHASE II
TELOPHASE II
B A G
b a g
Synapsis of sister chromatids
Crossing-over between nonsister chromatids
Genetic recombination

(b) Details of crossing-over during prophase I

(a) Stages of meiosis

How does crossing-over affect the genetic content of the haploid gametes?

Figure 3.34 Comparison between mitosis (left) and meiosis (right) in which the starting cell has two pairs of homologous chromosomes.

⑥ The phases of meiosis II and mitosis are similar.

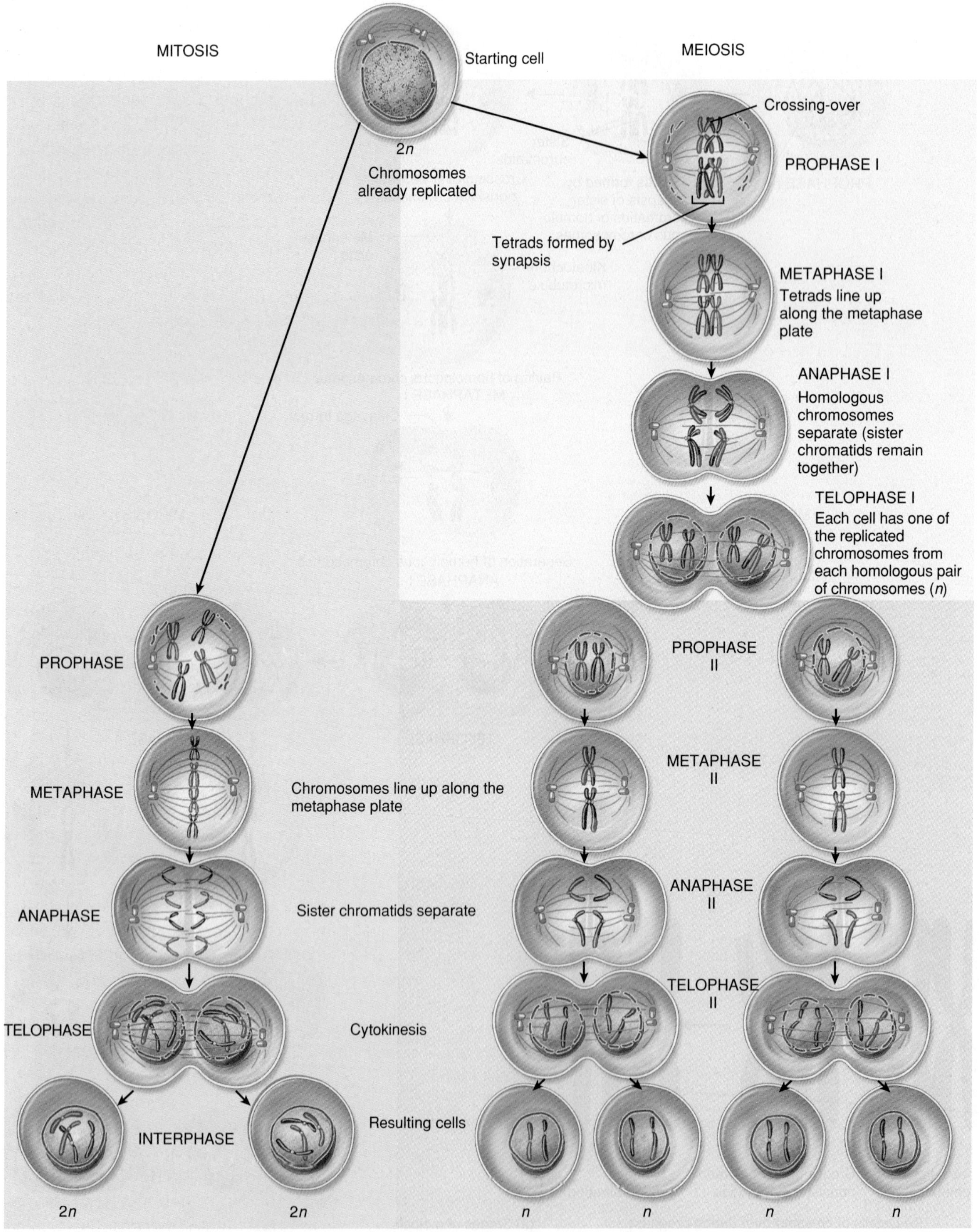

MITOSIS
Starting cell
MEIOSIS
2n
Chromosomes already replicated
Crossing-over
PROPHASE I
Tetrads formed by synapsis
METAPHASE I
Tetrads line up along the metaphase plate
ANAPHASE I
Homologous chromosomes separate (sister chromatids remain together)
TELOPHASE I
Each cell has one of the replicated chromosomes from each homologous pair of chromosomes (n)
PROPHASE
PROPHASE II
METAPHASE
Chromosomes line up along the metaphase plate
METAPHASE II
ANAPHASE
Sister chromatids separate
ANAPHASE II
TELOPHASE
Cytokinesis
TELOPHASE II
INTERPHASE
Resulting cells
2n
2n
n
n
n
n

Somatic cells with diploid number of chromosomes (not replicated)

Gametes with haploid number of chromosomes (not replicated)

How does anaphase I of meiosis differ from anaphase of mitosis?

3.8 CELLULAR DIVERSITY

OBJECTIVE

- Describe how cells differ in size and shape.

The body of an average human adult is composed of nearly 100 trillion cells. All of these cells can be classified into about 200 different cell types. Cells vary considerably in size. The sizes of cells are measured in units called *micrometers* (mī-KROM-i-ters). One micrometer (μm) is equal to 1 one-millionth of a meter, or 10^{-6} m (1/25,000 of an inch). High-powered microscopes are needed to see the smallest cells of the body. The largest cell, a single oocyte, has a diameter of about 140 μm and is barely visible to the unaided eye. A red blood cell has a diameter of 8 μm. To better visualize this, an average hair from the top of your head is approximately 100 μm in diameter.

The shapes of cells also vary considerably (Figure 3.35). They may be round, oval, flat, cube-shaped, column-shaped, elongated, star-shaped, cylindrical, or disc-shaped. A cell's shape is related to its function in the body. For example, a sperm cell has a long whiplike tail (flagellum) that it uses for locomotion. The disc shape of a red blood cell gives it a large surface area that enhances its ability to pass oxygen to other cells. The long, spindle shape of a relaxed smooth muscle cell shortens as it contracts. This change in shape allows groups of smooth muscle cells to narrow or widen the passage for blood flowing through blood vessels. In this way, they regulate blood flow through various tissues. Recall that some cells contain microvilli, which greatly increase their surface area. Microvilli are common in the epithelial cells that line the small intestine, where the large surface area speeds the absorption of digested food. Nerve cells have long extensions that permit them to conduct nerve impulses over great distances. As you will see in the following chapters, cellular diversity also permits organization of cells into more complex tissues and organs.

CHECKPOINT

33. How is cell shape related to function? Give several of your own examples.

Figure 3.35 Diverse shapes and sizes of human cells. The relative difference in size between the smallest and largest cells is actually much greater than shown here.

The nearly 100 trillion cells in an average adult can be classified into about 200 different cell types.

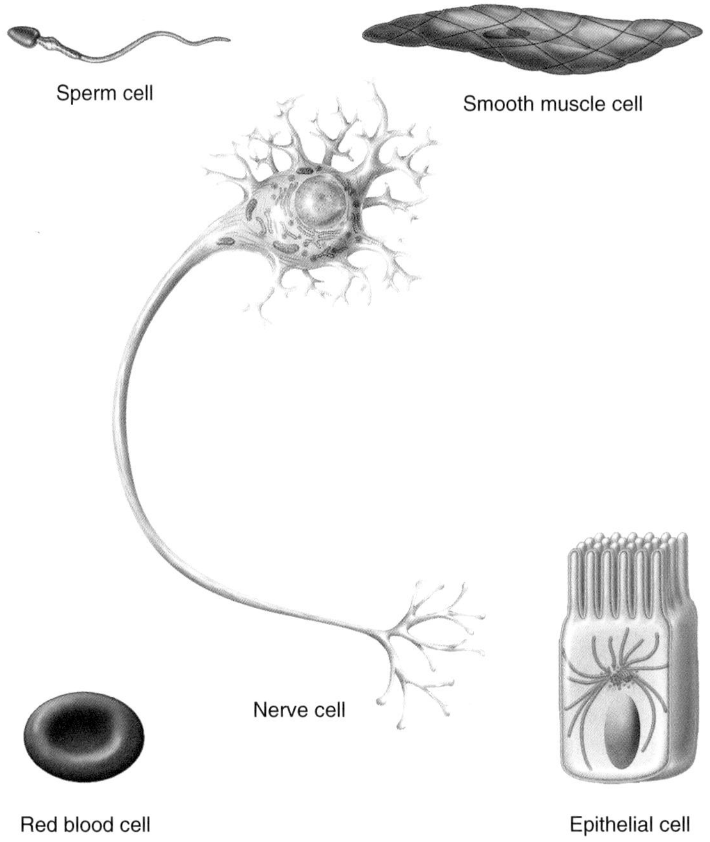
Sperm cell
Smooth muscle cell
Nerve cell
Red blood cell
Epithelial cell

Why are sperm the only body cells that need to have a flagellum?

3.9 AGING AND CELLS

OBJECTIVE

- Describe the cellular changes that occur with aging.

Aging is a normal process accompanied by a progressive alteration of the body's homeostatic adaptive responses. It produces observable changes in structure and function and increases vulnerability to environmental stress and disease. The specialized branch of medicine that deals with the medical problems and care of elderly persons is **geriatrics** (jer′-ē-AT-riks; *ger-* = old age; *-iatrics* = medicine). **Gerontology** (jer′-on-TOL-ō-jē) is the scientific study of the process and problems associated with aging.

Although many millions of new cells normally are produced each minute, several kinds of cells in the body—skeletal muscle cells and nerve cells—do not divide because they are arrested permanently in the G_0 phase (see the discussion of interphase earlier in the chapter). Experiments have shown that many other cell types have only a limited capability to divide. Normal cells grown outside the body divide only a certain number of times and then stop. These observations suggest that cessation of mitosis is a normal, genetically programmed event. According to this view, "aging genes" are part of the genetic blueprint at birth. These genes have an important function in normal cells, but their activities slow over time. They bring about aging by slowing down or halting processes vital to life.

Another aspect of aging involves **telomeres** (TĒ-lō-mērz), specific DNA sequences found only at the tips of each chromosome. These pieces of DNA protect the tips of chromosomes from erosion and from sticking to one another. However, in most normal body cells each cycle of cell division shortens the telomeres. Eventually, after many cycles of cell division, the telomeres can be completely gone and even some of the functional chromosomal material may be lost. These observations suggest that erosion of DNA from the tips of our chromosomes contributes greatly to aging and death of cells. It has recently been learned that individuals who experience high levels of stress have significantly shorter telomere length.

Glucose, the most abundant sugar in the body, plays a role in the aging process. It is haphazardly added to proteins inside and outside cells, forming irreversible cross-links between adjacent protein molecules. With advancing age, more cross-links form, which contributes to the stiffening and loss of elasticity that occur in aging tissues.

CLINICAL CONNECTION | *Free Radicals*

Free radicals produce oxidative damage in lipids, proteins, or nucleic acids by "stealing" an electron to accompany their unpaired electrons. Some effects are wrinkled skin, stiff joints, and hardened arteries. Normal metabolism—for example, aerobic cellular respiration in mitochondria—produces some free radicals. Others are present in air pollution, radiation, and certain foods we eat. Naturally occurring enzymes in peroxisomes and in the cytosol normally dispose of free radicals. Certain dietary substances, such as vitamin E, vitamin C, beta-carotene, zinc, and selenium, are referred to as *antioxidants* because they inhibit the formation of free radicals. •

Some theories of aging explain the process at the cellular level, while others concentrate on regulatory mechanisms operating within the entire organism. For example, the immune system may start to attack the body's own cells. This *autoimmune response* might be caused by changes in cell-identity markers at the surface of cells that cause antibodies to attach to and mark the cell for destruction. As changes in the proteins on the plasma membrane of cells increase, the autoimmune response intensifies, producing the well-known signs of aging. In the chapters that follow, we will discuss the effects of aging on each body system in sections similar to this one.

CLINICAL CONNECTION | *Progeria and Werner Syndrome*

Progeria (prō-JER-ē-a) is a disease characterized by normal development in the first year of life followed by rapid aging. It is caused by a genetic defect in which telomeres are considerably shorter than normal. Symptoms include dry and wrinkled skin, total baldness, and birdlike facial features. Death usually occurs around age 13.

Werner syndrome (VER-ner) is a rare, inherited disease that causes a rapid acceleration of aging, usually while the person is only in his or her twenties. It is characterized by wrinkling of the skin, graying of the hair and baldness, cataracts, muscular atrophy, and a tendency to develop diabetes mellitus, cancer, and cardiovascular disease. Most afflicted individuals die before age 50. Recently, the gene that causes Werner syndrome has been identified. Researchers hope to use the information to gain insight into the mechanisms of aging, as well as to help those suffering from the disorder. •

✔ **CHECKPOINT**

34. What is one reason that some tissues become stiffer as they age?

DISORDERS: HOMEOSTATIC IMBALANCES

Most chapters in the text are followed by concise discussions of major diseases and disorders that illustrate departures from normal homeostasis. These discussions provide answers to many questions that you might ask about medical problems.

Cancer

Cancer is a group of diseases characterized by uncontrolled or abnormal cell division. When cells in a part of the body divide without control, the excess tissue that develops is called a **tumor** or **neoplasm** (NĒ-ō-plazm; *neo-* = new). The study of tumors is called **oncology** (on-KOL-ō-jē; *onco-* = swelling or mass). Tumors may be cancerous and often fatal, or they may be harmless. A cancerous neoplasm is called a **malignant tumor** (ma-LIG-nant) or **malignancy.** One property of most malignant tumors is their ability to undergo **metastasis** (me-TAS-ta-sis), the spread of cancerous cells to other parts of the body. A **benign tumor** (be-NĪN) is a neoplasm that does not metastasize. An example is a wart. Most benign tumors may be removed surgically if they interfere with normal body function or become disfiguring. Some benign tumors can be inoperable and perhaps fatal.

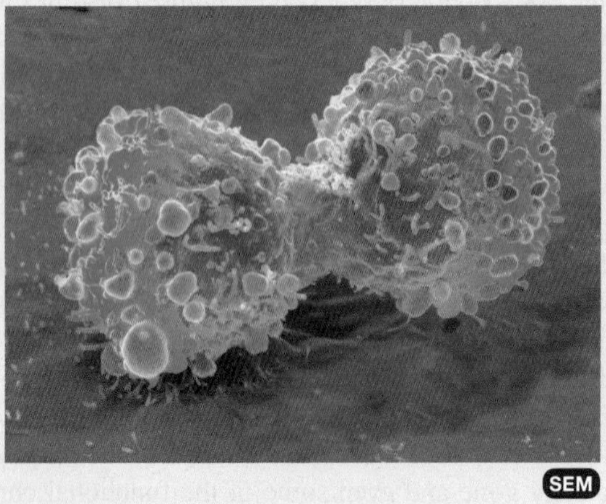

Lung cancer cell dividing

Types of Cancer

The name of a cancer is derived from the type of tissue in which it develops. Most human cancers are **carcinomas** (kar-si-NŌ-maz; *carcin* = cancer; *-omas* = tumors), malignant tumors that arise from epithelial cells. **Melanomas** (mel-a-NŌ-maz; *melan-* = black), for example, are cancerous growths of melanocytes, skin epithelial cells that produce the pigment melanin. **Sarcoma** (sar-KŌ-ma; *sarc-* = flesh) is a general term for any cancer arising from muscle cells or connective tissues. For example, **osteogenic sarcoma** (*osteo-* = bone; *-genic* = origin), the most frequent type of childhood cancer, destroys normal bone tissue. **Leukemia** (loo-KĒ-mē-a; *leuk-* = white; *-emia* = blood) is a cancer of blood-forming organs characterized by rapid growth of abnormal leukocytes (white blood cells). **Lymphoma** (lim-FŌ-ma) is a malignant disease of lymphatic tissue—for example, of lymph nodes.

Growth and Spread of Cancer

Cells of malignant tumors duplicate rapidly and continuously. As malignant cells invade surrounding tissues, they often trigger **angiogenesis** (an′-jē-ō-JEN-e-sis), the growth of new networks of blood vessels. Proteins that stimulate angiogenesis in tumors are called **tumor angiogenesis factors (TAFs).** The formation of new blood vessels can occur either by overproduction of TAFs or by the lack of naturally occurring angiogenesis inhibitors. As the cancer grows, it begins to compete with normal tissues for space and nutrients. Eventually, the normal tissue decreases in size and dies. Some malignant cells may detach from the initial (primary) tumor and invade a body cavity or enter the blood or lymph, then circulate to and invade other body tissues, establishing secondary tumors. Malignant cells resist the antitumor defenses of the body. The pain associated with cancer develops when the tumor presses on nerves or blocks a passageway in an organ so that secretions build up pressure, or as a result of dying tissue or organs.

Causes of Cancer

Several factors may trigger a normal cell to lose control and become cancerous. One cause is environmental agents: substances in the air we breathe, the water we drink, and the food we eat. A chemical agent or radiation that produces cancer is called a **carcinogen** (car-SIN-ō-jen). Carcinogens induce **mutations** (mū-TĀ-shuns), permanent changes in the DNA base sequence of a gene. The World Health Organization estimates that carcinogens are associated with 60–90% of all human cancers. Examples of carcinogens are hydrocarbons found in cigarette tar, radon gas from the earth, and ultraviolet (UV) radiation in sunlight.

Intensive research efforts are now directed toward studying cancer-causing genes, or **oncogenes** (ON-kō-jēnz). When inappropriately activated, these genes have the ability to transform a normal cell into a cancerous cell. Most oncogenes derive from normal genes called **proto-oncogenes** that regulate growth and development. The proto-oncogene undergoes some change that causes it (1) to be expressed inappropriately, (2) to make its products in excessive amounts, or (3) to make its products at the wrong time. Some oncogenes cause excessive production of growth factors, chemicals that stimulate cell growth. Others may trigger changes in a cell-surface receptor, causing it to send signals as though it were being activated by a growth factor. As a result, the growth pattern of the cell becomes abnormal.

Proto-oncogenes in every cell carry out normal cellular functions until a malignant change occurs. It appears that some proto-oncogenes are activated to oncogenes by mutations in which the DNA of the proto-oncogene is altered. Other proto-oncogenes are activated by a rearrangement of the chromosomes so that segments of DNA are exchanged. Rearrangement activates proto-oncogenes by placing them near genes that enhance their activity.

Some cancers have a viral origin. Viruses are tiny packages of nucleic acids, either RNA or DNA, that can reproduce only while inside the cells they infect. Some viruses, termed **oncogenic viruses,** cause cancer by stimulating abnormal proliferation of cells. For instance, the *human papillomavirus (HPV)* causes virtually all cervical cancers in women. The virus produces a protein that causes proteasomes to destroy p53, a protein that normally suppresses unregulated cell division. In the absence of this suppressor protein, cells proliferate without control.

Recent studies suggest that certain cancers may be linked to a cell having abnormal numbers of chromosomes. As a result, the cell could potentially have extra copies of oncogenes or too few copies of tumor-suppressor genes, which in either case could lead to uncontrolled cell proliferation. There is also some evidence suggesting that cancer may be caused by normal stem cells that develop into cancerous stem cells capable of forming malignant tumors.

Later in the book, we will discuss the process of inflammation, which is a defensive response to tissue damage. It appears that inflammation contributes to various steps in the development of cancer. Some evidence suggests that chronic inflammation stimulates the proliferation of mutated cells and enhances their survival, promotes angiogenesis, and contributes to invasion and metastasis of cancer cells. There is a clear relationship between certain chronic inflammatory conditions and the transformation of inflamed tissue into a malignant tissue. For example, chronic gastritis (inflammation of the stomach lining) and peptic ulcers may be a causative factor in 60–90% of stomach cancers. Chronic hepatitis (inflammation of the liver) and cirrhosis of the liver are believed to be responsible for about 80% of liver cancers. Colorectal cancer is ten times more likely to occur in patients with chronic inflammatory diseases of the colon, such as ulcerative colitis and Crohn's disease. And the relationship between asbestosis and silicosis, two chronic lung inflammatory conditions, and lung cancer has long been recognized. Chronic inflammation is also an underlying contributor to rheumatoid arthritis, Alzheimer disease, depression, schizophrenia, cardiovascular disease, and diabetes.

Carcinogenesis: A Multistep Process

Carcinogenesis (kar′-si-nō-JEN-e-sis) is a multistep process of cancer development in which as many as 10 distinct mutations may have to accumulate in a cell before it becomes cancerous. The progression of genetic changes leading to cancer is best understood for colon (colorectal) cancer. Such cancers, as well as lung and breast cancer, take years or decades to develop. In colon cancer, the tumor begins as an area of increased cell proliferation that results from one mutation. This growth then progresses to abnormal, but noncancerous, growths called adenomas. After two or three additional mutations, a mutation of the tumor-suppressor gene *p53* occurs and a carcinoma develops. The fact that so many mutations are needed for a cancer to develop indicates that cell growth is normally controlled with many sets of checks and balances. Thus, it is not surprising that a compromised immune system contributes significantly to carcinogenesis.

Treatment of Cancer

Many cancers are removed surgically. However, cancer that is widely distributed throughout the body or exists in organs with essential functions, such as the brain, which might be greatly harmed by surgery, may be treated with chemotherapy and radiation therapy instead. Sometimes surgery, chemotherapy, and radiation therapy are used in combination. Chemotherapy involves administering drugs that cause death of cancerous cells. Radiation therapy breaks chromosomes, thus blocking cell division. Because cancerous cells divide rapidly, they are more vulnerable to the destructive effects of chemotherapy and radiation therapy than are normal cells. Unfortunately for the patients, hair follicle cells, red bone marrow cells, and cells lining the gastrointestinal tract also are rapidly dividing. Hence, the side effects of chemotherapy and radiation therapy include hair loss due to death of hair follicle cells, vomiting and nausea due to death of cells lining the stomach and intestines, and susceptibility to infection due to slowed production of white blood cells in red bone marrow.

Treating cancer is difficult because it is not a single disease and because the cells in a single tumor population rarely behave all in the same way. Although most cancers are thought to derive from a single abnormal cell, by the time a tumor reaches a clinically detectable size, it may contain a diverse population of abnormal cells. For example, some cancerous cells metastasize readily, and others do not. Some are sensitive to chemotherapy drugs and some are drug-resistant. Because of differences in drug resistance, a single chemotherapeutic agent may destroy susceptible cells but permit resistant cells to proliferate.

Another potential treatment for cancer that is currently under development is *virotherapy,* the use of viruses to kill cancer cells. The viruses employed in this strategy are designed so that they specifically target cancer cells without affecting the healthy cells of the body. For example, proteins (such as antibodies) that specifically bind to receptors found only in cancer cells are attached to viruses. Once inside the body, the viruses bind to cancer cells and then infect them. The cancer cells are eventually killed once the viruses cause cellular lysis.

Researchers are also investigating the role of *metastasis regulatory genes* that control the ability of cancer cells to undergo metastasis. Scientists hope to develop therapeutic drugs that can manipulate these genes and, therefore, block metastasis of cancer cells.

MEDICAL TERMINOLOGY

Most chapters in this text are followed by a glossary of key medical terms that include both normal and pathological conditions. You should familiarize yourself with these terms because they will play an essential role in your medical vocabulary.

Some of these conditions, as well as ones discussed in the text, are referred to as local or systemic. A *local disease* is one that affects one part or a limited area of the body. A *systemic disease* affects the entire body or several parts.

Anaplasia (an′-a-PLĀ-zē-a; *an-* = not; *-plasia* = to shape) The loss of tissue differentiation and function that is characteristic of most malignancies.

Atrophy (AT-rō-fē; *a-* = without; *-trophy* = nourishment) A decrease in the size of cells, with a subsequent decrease in the size of the affected tissue or organ; wasting away.

Dysplasia (dis-PLĀ-zē-a; *dys-* = abnormal) Alteration in the size, shape, and organization of cells due to chronic irritation or inflammation; may progress to neoplasia (tumor formation, usually malignant) or revert to normal if the irritation is removed.

Hyperplasia (hī-per-PLĀ-zē-a; *hyper-* = over) Increase in the number of cells of a tissue due to an increase in the frequency of cell division.

Hypertrophy (hī-PER-trō-fē) Increase in the size of cells without cell division.

Metaplasia (met′-a-PLĀ-zē-a; *meta-* = change) The transformation of one type of cell into another.

Progeny (PROJ-e-nē; *pro-* = forward; *-geny* = production) Offspring or descendants.

Proteomics (prō′-tē-Ō-miks; *proteo-* = protein) The study of the proteome (all of an organism's proteins) in order to identify all the proteins produced; it involves determining the three-dimensional structure of proteins so that drugs can be designed to alter protein activity to help in the treatment and diagnosis of disease.

Tumor marker A substance introduced into circulation by tumor cells that indicates the presence of a tumor, as well as the specific type. Tumor markers may be used to screen, diagnose, make a prognosis, evaluate a response to treatment, and monitor for recurrence of cancer.

CHAPTER REVIEW AND RESOURCE SUMMARY

Review	Resource
Introduction 1. A cell is the basic, living, structural and functional unit of the body. 2. Cell biology is the scientific study of cellular structure and function.	
3.1 Parts of a Cell 1. Figure 3.1 provides an overview of the typical structures in body cells. 2. The principal parts of a cell are the plasma membrane; the cytoplasm, the cellular contents between the plasma membrane and nucleus; and the nucleus.	Anatomy Overview - Cell Structure and Function Figure 3.1 - Structures in Body Cells Concepts and Connections - Human Cell
3.2 The Plasma Membrane 1. The plasma membrane, which surrounds and contains the cytoplasm of a cell, is composed of proteins and lipids. 2. According to the fluid mosaic model, the membrane is a mosaic of proteins floating like icebergs in a lipid bilayer sea. 3. The lipid bilayer consists of two back-to-back layers of phospholipids, cholesterol, and glycolipids. The bilayer arrangement occurs because the lipids are amphipathic, having both polar and nonpolar parts. 4. Integral proteins extend into or through the lipid bilayer; peripheral proteins associate with membrane lipids or integral proteins at the inner or outer surface of the membrane.	Anatomy Overview - Plasma Membrane Structure Figure 3.2 - The Fluid Mosaic Arrangement of the Plasma Membrane Exercise - Paint a Cell Membrane Concepts and Connections - Membrane Functions

Review	Resource

5. Many integral proteins are glycoproteins, with sugar groups attached to the ends that face the extracellular fluid. Together with glycolipids, the glycoproteins form a glycocalyx on the extracellular surface of cells.
6. Membrane proteins have a variety of functions. Integral proteins are channels and carriers that help specific solutes cross the membrane; receptors that serve as cellular recognition sites; enzymes that catalyze specific chemical reactions; and linkers that anchor proteins in the plasma membranes to protein filaments inside and outside the cell. Peripheral proteins serve as enzymes and linkers; support the plasma membrane; anchor integral proteins; and participate in mechanical activities. Membrane glycoproteins function as cell-identity markers.
7. Membrane fluidity is greater when there are more double bonds in the fatty acid tails of the lipids that make up the bilayer. Cholesterol makes the lipid bilayer stronger but less fluid at normal body temperature. Its fluidity allows interactions to occur within the plasma membrane, enables the movement of membrane components, and permits the lipid bilayer to self-seal when torn or punctured.
8. The membrane's selective permeability permits some substances to pass more readily than others. The lipid bilayer is permeable to most nonpolar, uncharged molecules. It is impermeable to ions and charged or polar molecules other than water and urea. Channels and carriers increase the plasma membrane's permeability to small- and medium-sized polar and charged substances, including ions, that cannot cross the lipid bilayer.
9. The selective permeability of the plasma membrane supports the existence of concentration gradients, differences in the concentrations of chemicals between one side of the membrane and the other.

3.3 Transport across the Plasma Membrane

1. In passive processes, a substance moves down its concentration gradient across the membrane using its own kinetic energy of motion. In active processes, cellular energy is used to drive the substance "uphill" against its concentration gradient.
2. In diffusion, molecules or ions move from an area of higher concentration to an area of lower concentration until an equilibrium is reached. The rate of diffusion across a plasma membrane is affected by the steepness of the concentration gradient, temperature, mass of the diffusing substance, surface area available for diffusion, and the distance over which diffusion must occur.
3. Nonpolar, hydrophobic molecules such as oxygen, carbon dioxide, nitrogen, steroids, and fat-soluble vitamins (A, E, D, and K) plus small, polar, uncharged molecules such as water, urea, and small alcohols diffuse through the lipid bilayer of the plasma membrane via simple diffusion.
4. In channel-mediated facilitated diffusion, a solute moves down its concentration gradient across the lipid bilayer through a membrane channel. Examples include ion channels that allow specific ions such as K^+, Cl^-, Na^+, or Ca^{2+} (which are too hydrophilic to penetrate the membrane's nonpolar interior) to move across the plasma membrane. In carrier-mediated facilitated diffusion, a solute such as glucose binds to a specific carrier protein on one side of the membrane and is released on the other side after the carrier undergoes a change in shape.
5. Osmosis is a type of diffusion in which there is net movement of water through a selectively permeable membrane from an area of higher water concentration to an area of lower water concentration. In an isotonic solution, red blood cells maintain their normal shape; in a hypotonic solution, they undergo hemolysis; in a hypertonic solution, they undergo crenation.
6. Substances can cross the membrane against their concentration gradient by active transport. Actively transported substances include ions such as Na^+, K^+, H^+, Ca^{2+}, I^-, and Cl^-; amino acids; and monosaccharides. Two sources of energy drive active transport: Energy obtained from hydrolysis of ATP is the source in primary active transport, and energy stored in a Na^+ or H^+ concentration gradient is the source in secondary active transport. The most prevalent primary active transport pump is the sodium-potassium pump, also known as Na^+/K^+ ATPase. Secondary active transport mechanisms include both symporters and antiporters that are powered by either a Na^+ or H^+ concentration gradient. Symporters move two substances in the same direction across the membrane; antiporters move two substances in opposite directions.
7. In endocytosis, tiny vesicles detach from the plasma membrane to move materials across the membrane into a cell; in exocytosis, vesicles merge with the plasma membrane to move materials out of a cell. Receptor-mediated endocytosis is the selective uptake of large molecules and particles (ligands) that bind to specific receptors in membrane areas called clathrin-coated pits. In bulk-phase endocytosis (pinocytosis), the ingestion of extracellular fluid, a vesicle surrounds the fluid to take it into the cell.
8. Phagocytosis is the ingestion of solid particles. Some white blood cells destroy microbes that enter the body in this way.

Resource (3.3):

Animation - Membrane Functions
Animation - Transport across the Plasma Membrane
Figure 3.5 - Types of Diffusion
Figure 3.9 - Tonicity and Its Effects on Red Blood Cells
Figure 3.12 - Receptor-Mediated Endocytosis of a Low-Density Lipoprotein Particle
Exercise - Membrane Function Matchup
Exercise - Create a Transport Condition
Exercise - Osmosis—Move It!
Concepts and Connections - Membrane Transport Processes

Review | **Resource**

9. In transcytosis, vesicles undergo endocytosis on one side of a cell, move across the cell, and undergo exocytosis on the opposite side.

3.4 Cytoplasm

1. Cytoplasm—all the cellular contents within the plasma membrane except for the nucleus—consists of cytosol and organelles. Cytosol is the fluid portion of cytoplasm, containing water, ions, glucose, amino acids, fatty acids, proteins, lipids, ATP, and waste products. It is the site of many chemical reactions required for a cell's existence.
2. Components of the cytoskeleton, a network of several kinds of protein filaments that extend throughout the cytoplasm, include microfilaments, intermediate filaments, and microtubules. The cytoskeleton provides a structural framework for the cell and is responsible for cell movements.
3. Organelles are specialized structures with characteristic shapes that have specific functions. The centrosome consists of pericentriolar material and a pair of centrioles. The pericentriolar material organizes microtubules in nondividing cells and the mitotic spindle in dividing cells.
4. Cilia and flagella, motile projections of the cell surface, are formed by basal bodies. Cilia move fluid along the cell surface; flagella move an entire cell.
5. Ribosomes consist of two subunits made in the nucleus that are composed of ribosomal RNA and ribosomal proteins. They serve as sites of protein synthesis.
6. Endoplasmic reticulum (ER) is a network of membranes that form flattened sacs or tubules; it extends from the nuclear envelope throughout the cytoplasm. Rough ER is studded with ribosomes that synthesize proteins; the proteins then enter the space within the ER for processing and sorting. Rough ER produces secretory proteins, membrane proteins, and organelle proteins; forms glycoproteins; synthesizes phospholipids; and attaches proteins to phospholipids. Smooth ER lacks ribosomes. It synthesizes fatty acids and steroids; inactivates or detoxifies drugs and other potentially harmful substances; removes phosphate from glucose-6-phosphate; and releases calcium ions that trigger contraction in muscle cells.
7. The Golgi complex consists of flattened sacs called cisternae. The entry, medial, and exit regions of the Golgi complex contain different enzymes that permit each to modify, sort, and package proteins for transport in secretory vesicles, membrane vesicles, or transport vesicles to different cellular destinations.
8. Lysosomes are membrane-enclosed vesicles that contain digestive enzymes. Endosomes, phagosomes, and pinocytic vesicles deliver materials to lysosomes for degradation. Lysosomes function in digestion of worn-out organelles (autophagy), digestion of a host cell (autolysis), and extracellular digestion.
9. Peroxisomes contain oxidases that oxidize amino acids, fatty acids, and toxic substances; the hydrogen peroxide produced in the process is destroyed by catalase. The proteases contained in proteasomes, another kind of organelle, continually degrade unneeded, damaged, or faulty proteins by cutting them into small peptides.
10. Mitochondria consist of a smooth outer membrane, an inner membrane containing mitochondrial cristae, and a fluid-filled cavity called the mitochondrial matrix. These so-called powerhouses of the cell produce most of a cell's ATP and can play an important early role in apoptosis.

Resource:
Anatomy Overview - Cell Structure and Function
Figure 3.21 - Processing and Packaging of Proteins by the Golgi Complex
Exercise - Concentrate on Cellular Functions
Exercise - Target Practice
Exercise - Protein - Producing Processes
Exercise - Synthesize and Transport a Protein
Concepts and Connections - Human Cell
Concepts and Connections - Membranous Organelles

3.5 Nucleus

1. The nucleus consists of a double nuclear envelope; nuclear pores, which control the movement of substances between the nucleus and cytoplasm; nucleoli, which produce ribosomes; and genes arranged on chromosomes, which control cellular structure and direct cellular activities.
2. Human somatic cells have 46 chromosomes, 23 inherited from each parent. The total genetic information carried in a cell or an organism is its genome.

Resource:
Anatomy Overview - The Nucleus

3.6 Protein Synthesis

1. Cells make proteins by transcribing and translating the genetic information contained in DNA.
2. The genetic code is the set of rules that relates the base triplet sequences of DNA to the corresponding codons of RNA and the amino acids they specify.
3. In transcription, the genetic information in the sequence of base triplets in DNA serves as a template for copying the information into a complementary sequence of codons in messenger RNA. Transcription begins on DNA in a region called a promoter. Regions of DNA that code for protein synthesis are called exons; those that do not are called introns.

Resource:
Animation - Protein Synthesis
Figure 3.29 - Protein Elongation
Exercise - Synthesize and Transport a Protein
Exercise - Protein-Producing Processes
Concepts and Connections - Protein Synthesis

Review | **Resource**

4. Newly synthesized pre-mRNA is modified before leaving the nucleus.
5. In the process of translation, the nucleotide sequence of mRNA specifies the amino acid sequence of a protein. The mRNA binds to a ribosome, specific amino acids attach to tRNA, and anticodons of tRNA bind to codons of mRNA, bringing specific amino acids into position on a growing polypeptide. Translation begins at the start codon and ends at the stop codon.

3.7 Cell Division

1. Cell division, the process by which cells reproduce themselves, consists of nuclear division (mitosis or meiosis) and cytoplasmic division (cytokinesis). Cell division that replaces cells or adds new ones is called somatic cell division and involves mitosis and cytokinesis. Cell division that results in the production of gametes (sperm and ova) is called reproductive cell division and consists of meiosis and cytokinesis.
2. The cell cycle, an orderly sequence of events in which a somatic cell duplicates its contents and divides in two, consists of interphase and a mitotic phase. Human somatic cells contain 23 pairs of homologous chromosomes and are thus diploid ($2n$). Before the mitotic phase, the DNA molecules, or chromosomes, replicate themselves so that identical sets of chromosomes can be passed on to the next generation of cells.
3. A cell between divisions that is carrying on every life process except division is said to be in interphase, which consists of three phases: G_1, S, and G_2. During the G_1 phase, the cell replicates its organelles and cytosolic components, and centrosome replication begins; during the S phase, DNA replication occurs; during the G_2 phase, enzymes and other proteins are synthesized and centrosome replication is completed.
4. Mitosis is the splitting of the chromosomes and the distribution of two identical sets of chromosomes into separate and equal nuclei; it consists of prophase, metaphase, anaphase, and telophase.
5. In cytokinesis, which usually begins in late anaphase and ends once mitosis is complete, a cleavage furrow forms at the cell's metaphase plate and progresses inward, pinching in through the cell to form two separate portions of cytoplasm.
6. A cell can either remain alive and functioning without dividing, grow and divide, or die. The control of cell division depends on specific cyclin-dependent protein kinases and cyclins.
7. Apoptosis is normal, programmed cell death. It first occurs during embryological development and continues throughout the lifetime of an organism.
8. Certain genes regulate both cell division and apoptosis. Abnormalities in these genes are associated with a wide variety of diseases and disorders.
9. In sexual reproduction, each new organism is the result of the union of two different gametes, one from each parent. Gametes contain a single set of chromosomes (23) and thus are haploid (n).
10. Meiosis is the process that produces haploid gametes; it consists of two successive nuclear divisions, called meiosis I and meiosis II. During meiosis I, homologous chromosomes undergo synapsis (pairing) and crossing-over; the net result is two haploid cells that are genetically unlike each other and unlike the starting diploid parent cell that produced them. During meiosis II, two haploid cells divide to form four haploid cells.

Resource:
Animation - The Cell Cycle and Division Processes
Figure 3.34 - Comparison between Mitosis and Meiosis

Exercise - Mitosis Matchup
Exercise - Meiosis Matchup
Exercise - Cell Cycle Quiz

3.8 Cellular Diversity

1. The almost 200 different types of cells in the body vary considerably in size and shape.
2. The sizes of cells are measured in micrometers. One micrometer (μm) equals 10^{-6} m (1/25,000 of an inch). Cells in the body range from 8 μm to 140 μm in size.
3. A cell's shape is related to its function.

Resource:
Anatomy Overview - Cell Structure and Function
Concepts and Connections - Human Cell

3.9 Aging and Cells

1. Aging is a normal process accompanied by progressive alteration of the body's homeostatic adaptive responses.
2. Many theories of aging have been proposed, including genetically programmed cessation of cell division, buildup of free radicals, and an intensified autoimmune response.

SELF-QUIZ QUESTIONS

Fill in the blanks in the following statements.

1. The three major parts of the cell are the ______, ______, and ______.

2. Cell death that is genetically programmed is known as ______, while cell death which is due to tissue injury is known as ______.

3. ______ are special DNA sequences located at the ends of chromosomes and whose erosion contributes to cellular aging and death.

4. The mRNA base sequence that is complementary to the DNA base sequence ATC would be ______.

Indicate whether the following statements are true or false.

5. A small membrane surface area will increase the rate of diffusion across the cell membrane.

6. The cells created during meiosis are genetically different from the original cell.

7. An important and abundant active mechanism that helps maintain cellular tonicity is the Na^+/K^+ ATPase pump.

Choose the one best answer to the following questions.

8. If the concentration of solutes in the ECF and ICFs are equal, the cell is in a(n) ______ solution.
(a) hypertonic (b) hydrophobic (c) saturated
(d) hypotonic (e) isotonic

9. Which membrane protein is *incorrectly* matched with its function?
(a) receptor: allows recognition of specific molecules
(b) ion channel: allows passage of specific ions through the membrane
(c) carrier: allows cells to recognize each other and foreign cells
(d) linker: allows binding of one cell to another and provides stability and shape to a cell
(e) enzyme: catalyzes cellular reactions

10. Place the following steps in protein synthesis in the correct order.
(a) binding of anticodons of tRNA to codons of mRNA
(b) modification of newly synthesized pre-mRNA by snRNPs before leaving the nucleus and entering the cytoplasm
(c) attachment of RNA polymerase at promoter
(d) binding of mRNA to a ribosome's small subunit
(e) amino acids joined by peptide bonds
(f) joining of large and small ribosomal subunits to create a functional ribosome
(g) transcription of a segment of DNA onto mRNA
(h) detachment of protein from ribosome when ribosome reaches stop codon on mRNA
(i) detachment of RNA polymerase after reaching terminator
(j) attachment of specific amino acids to tRNA
(k) binding of initiator tRNA to start codon on mRNA

11. Which of the following organelles function primarily in decomposition reactions? (1) ribosomes, (2) proteasomes, (3) lysosomes, (4) centrosomes, (5) peroxisomes
(a) 2, 3, and 5 (b) 3 and 5 (c) 2, 4, and 5
(d) 1 and 4 (e) 2 and 5

12. Which of the following statements regarding the nucleus are true? (1) Nucleoli within the nucleus are the sites of ribosome synthesis. (2) The nucleus contains the cell's hereditary units. (3) The nuclear membrane is a solid, impermeable membrane. (4) Protein synthesis occurs within the nucleus. (5) In nondividing cells, DNA is found in the nucleus in the form of chromatin.
(a) 1, 2, and 3 (b) 1, 2, and 4 (c) 1, 2, and 5
(d) 2, 4, and 5 (e) 2, 3, and 4

13. Match the following:

____ (a) mitosis	(1) cytoplasmic division
____ (b) meiosis	(2) somatic cell division resulting in the formation of two identical cells
____ (c) prophase	(3) reproductive cell division that reduces the number of chromosomes by half
____ (d) metaphase	(4) stage of cell division when replication of DNA occurs
____ (e) anaphase	(5) stage when chromatin fibers condense and shorten to form chromosomes
____ (f) telophase	(6) stage when centromeres split and chromatids move to opposite poles of the cell
____ (g) cytokinesis	(7) stage when centromeres of chromatid pairs line up at the center of the mitotic spindle
____ (h) interphase	(8) stage when chromosomes uncoil and revert to chromatin

14. Match the following:

____ (a) cytoskeleton
____ (b) centrosome
____ (c) ribosomes
____ (d) rough ER
____ (e) smooth ER
____ (f) Golgi complex
____ (g) lysosomes
____ (h) peroxisomes
____ (i) mitochondria
____ (j) cilia
____ (k) flagellum
____ (l) proteasomes
____ (m) vesicles

(1) membrane-enclosed vesicles formed in the Golgi complex that contain strong hydrolytic and digestive enzymes
(2) network of protein filaments that extend throughout the cytoplasm, providing cellular shape, organization, and movement
(3) sites of protein synthesis
(4) contain enzymes that break apart unneeded, damaged, or faulty proteins into small peptides
(5) site where secretory proteins and membrane molecules are synthesized
(6) membrane-enclosed vesicles that contain enzymes that oxidize various organic substances
(7) short microtubular structures extending from the plasma membrane and involved in movement of materials along the cell's surface
(8) modifies, sorts, packages, and transports molecules synthesized in the rough ER
(9) an organizing center for growth of the mitotic spindle
(10) function in ATP generation
(11) functions in synthesizing fatty acids and steroids, helping liver cells release glucose into the bloodstream, and detoxification
(12) membrane-bound sacs that transport, transfer, or secrete proteins
(13) long microtubular structure extending from the plasma membrane and involved in movement of a cell

15. Match the following:

____ (a) diffusion
____ (b) osmosis
____ (c) facilitated diffusion
____ (d) primary active transport
____ (e) secondary active transport
____ (f) vesicular transport
____ (g) phagocytosis
____ (h) pinocytosis
____ (i) exocytosis
____ (j) receptor-mediated endocytosis
____ (k) transcytosis

(1) passive transport in which a solute binds to a specific carrier on one side of the membrane and is released on the other side
(2) movement of materials out of the cell by fusing of secretory vesicles with the plasma membrane
(3) the random mixing of particles in a solution due to the kinetic energy of the particles; substances move from high to low concentrations until equilibrium is reached
(4) transport of substances either into or out of the cell by means of small, spherical membranous sacs formed by budding off from existing membranes
(5) uses energy derived from hydrolysis of ATP to change the shape of a carrier protein, which "pumps" a substance across a cellular membrane against its concentration gradient
(6) vesicular movement involving endocytosis on one side of a cell and subsequent exocytosis on the opposite side of the cell
(7) type of endocytosis that involves the nonselective uptake of tiny droplets of extracellular fluid
(8) type of endocytosis in which large solid particles are taken in
(9) movement of water from an area of higher to an area of lower water concentration through a selectively permeable membrane
(10) process that allows a cell to take specific ligands from the ECF by forming vesicles
(11) indirectly uses energy obtained from the breakdown of ATP; involves symporters and antiporters

CRITICAL THINKING QUESTIONS

1. Mucin is a protein present in saliva and other secretions. When mixed with water, it becomes the slippery substance known as mucus. Trace the route taken by mucin through the cell, from its synthesis to its secretion, listing all the organelles and processes involved.
2. Sam does not consume alcohol, while his brother Sebastian regularly drinks large quantities of alcohol. If we could examine the liver cells of each of these brothers, would we see a difference in smooth ER and peroxisomes? Explain.
3. Marathon runners can become dehydrated due to the extreme physical activity. What types of fluids should they consume in order to rehydrate their cells?

? ANSWERS TO FIGURE QUESTIONS

3.1 The three main parts of a cell are the plasma membrane, cytoplasm, and nucleus.

3.2 The glycocalyx is the sugary coat on the extracellular surface of the plasma membrane. It is composed of the carbohydrate portions of membrane glycolipids and glycoproteins.

3.3 The membrane protein that binds to insulin acts as a receptor.

3.4 Because fever involves an increase in body temperature, the rates of all diffusion processes would increase.

3.5 Nonpolar, hydrophobic molecules (oxygen, carbon dioxide, and nitrogen gases; fatty acids; steroids; and fat-soluble vitamins) plus small, uncharged polar molecules (water, urea, and small alcohols) move across the lipid bilayer of the plasma membrane through the process of simple diffusion.

3.6 The concentration of K^+ is higher in the cytosol of body cells than in extracellular fluids.

3.7 Yes. Insulin promotes insertion of glucose transporter (GluT) in the plasma membrane, which increases cellular glucose uptake by carrier-mediated facilitated diffusion.

3.8 No. The water concentrations can never be the same in the two arms because the left arm contains pure water and the right arm contains a solution that is less than 100% water.

3.9 A 2% solution of NaCl will cause crenation of RBCs because it is hypertonic.

3.10 ATP adds a phosphate group to the pump protein, which changes the pump's three-dimensional shape. ATP transfers energy to power the pump.

3.11 In secondary active transport, hydrolysis of ATP is used indirectly to drive the activity of symporter or antiporter proteins; this reaction directly powers the pump protein in primary active transport.

3.12 Transferrin, vitamins, and hormones are other examples of ligands that can undergo receptor-mediated endocytosis.

3.13 The binding of particles to a plasma membrane receptor triggers pseudopod formation.

3.14 Receptor-mediated endocytosis and phagocytosis involve receptor proteins; bulk-phase endocytosis does not.

3.15 Microtubules help to form centrioles, cilia, and flagella.

3.16 A cell without a centrosome probably would not be able to undergo cell division.

3.17 Cilia move fluids across cell surfaces; flagella move an entire cell.

3.18 Large and small ribosomal subunits are synthesized separately in the nucleolus in the nucleus and then assembled in the cytoplasm.

3.19 Rough ER has attached ribosomes; smooth ER does not. Rough ER synthesizes proteins that will be exported from the cell; smooth ER is associated with lipid synthesis and other metabolic reactions.

3.20 The entry face receives and modifies proteins from rough ER; the exit face modifies, sorts, and packages molecules for transport to other destinations.

3.21 Some proteins are secreted from the cell by exocytosis, some are incorporated into the plasma membrane, and some occupy storage vesicles that become lysosomes.

3.22 Digestion of worn-out organelles by lysosomes is called autophagy.

3.23 Mitochondrial cristae increase the surface area available for chemical reactions and contain some of the enzymes needed for ATP production.

3.24 Chromatin is a complex of DNA, proteins, and some RNA.

3.25 A nucleosome is a double-stranded molecule of DNA wrapped twice around a core of eight histones (proteins).

3.26 Proteins determine the physical and chemical characteristics of cells.

3.27 The DNA base sequence AGCT would be transcribed into the mRNA base sequence UCGA by RNA polymerase.

3.28 The P site holds the tRNA attached to the growing polypeptide. The A site holds the tRNA carrying the next amino acid to be added to the growing polypeptide.

3.29 When a ribosome encounters a stop codon at the A site, it releases the completed protein from the final tRNA.

3.30 DNA replicates during the S phase of interphase of the cell cycle.

3.31 DNA replication must occur before cytokinesis so that each of the new cells will have a complete genome.

3.32 Cytokinesis usually starts in late anaphase.

3.33 The result of crossing-over is that four haploid gametes are genetically unlike each other and genetically unlike the starting cell that produced them.

3.34 During anaphase I of meiosis, the paired chromatids are held together by a centromere and do not separate. During anaphase of mitosis, the paired chromatids separate and the centromeres split.

3.35 Sperm, which use the flagella for locomotion, are the only body cells required to move considerable distances.

4 THE TISSUE LEVEL OF ORGANIZATION

TISSUES AND HOMEOSTASIS *The four basic types of tissues in the human body contribute to homeostasis by providing diverse functions including protection, support, communication among cells, and resistance to disease, to name just a few.*

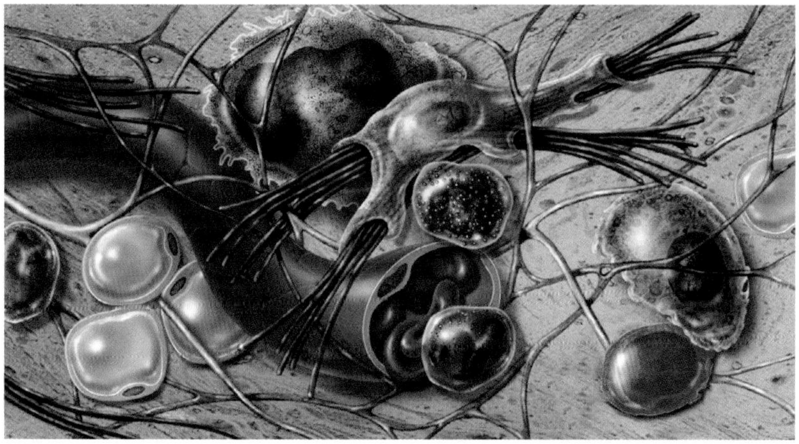

As you learned in Chapter 3, a cell is a complex collection of compartments, each of which carries out a host of biochemical reactions that make life possible. However, a cell seldom functions as an isolated unit in the body. Instead, cells usually work together in groups called tissues. A **tissue** is a group of cells that usually have a common origin in an embryo and function together to carry out specialized activities. The structure and properties of a specific tissue are influenced by factors such as the nature of the extracellular material that surrounds the tissue cells and the connections between the cells that compose the tissue. Tissues may be hard, semisolid, or even liquid in their consistency, a range exemplified by bone, fat, and blood. In addition, tissues vary tremendously with respect to the kinds of cells present, how the cells are arranged, and the types of fibers present, if any.

Histology (hiss′-TOL-ō-jē; *histo-* = tissue; *-logy* = study of) is the science that deals with the study of tissues. A **pathologist** (pa-THOL-ō-jist; *patho-* = disease) is a physician who specializes in laboratory studies of cells and tissues to help other physicians make accurate diagnoses. One of the principal functions of a pathologist is to examine tissues for any changes that might indicate disease.

? ***Did you ever wonder whether the complications of liposuction outweigh the benefits?***

4.1 TYPES OF TISSUES

OBJECTIVE

- Name the four basic types of tissues that make up the human body, and state the characteristics of each.

Body tissues can be classified into four basic types according to their structure and function (Figure 4.1):

1. **Epithelial tissues** cover body surfaces and line hollow organs, body cavities, and ducts; they also forms glands. This tissue allows the body to interact with both its internal and external environments.
2. **Connective tissues** protect and support the body and its organs. Various types of connective tissues bind organs together, store energy reserves as fat, and help provide the body with immunity to disease-causing organisms.
3. **Muscular tissues** are composed of cells specialized for contraction and generation of force. In the process, muscular tissues generate heat that warms the body.
4. **Nervous tissue** detects changes in a variety of conditions inside and outside the body and responds by generating electrical signals called nerve action potentials (nerve impulses) that activate muscular contractions and glandular secretions.

Epithelial tissues and most types of connective tissues, except cartilage, bone, and blood, are more general in nature and have a wide distribution in the body. These tissues are components of most body organs and have a wide range of structure and function. We will look at epithelial tissues and connective tissues in some detail in this chapter. The general features of bone tissue and blood will be introduced here, but their detailed discussion is presented in Chapters 6 and 19, respectively. Similarly, the structure and function of muscular tissues and nervous tissue are introduced here and examined in detail in Chapters 10 and 12, respectively.

Normally, most cells within a tissue remain anchored to other cells or structures. Only a few cells, such as phagocytes, move freely through the body, searching for invaders to destroy. However, many cells migrate extensively during the growth and development process before birth.

CLINICAL CONNECTION | *Biopsy*

A **biopsy** (BĪ-op-sē; *bio-* = life; *-opsy* = to view) is the removal of a sample of living tissue for microscopic examination. This procedure is used to help diagnose many disorders, especially cancer, and to discover the cause of unexplained infections and inflammations. Both normal and potentially diseased tissues are removed for purposes of comparison. Once the tissue samples are removed, either surgically or through a needle and syringe, they may be preserved, stained to highlight special properties, or cut into thin sections for microscopic observation. Sometimes a biopsy is conducted while a patient is anesthetized during surgery to help a physician determine the most appropriate treatment. For example, if a biopsy of thyroid tissue reveals malignant cells, the surgeon can proceed immediately with the most appropriate procedure. •

✔CHECKPOINT

1. Define a tissue.
2. What are the four basic types of human tissues?

4.2 CELL JUNCTIONS

OBJECTIVE

- Describe the structure and functions of the five main types of cell junctions.

Before looking more specifically at the types of tissues, we will first examine how cells are held together to form tissues. Most epithelial cells and some muscle and nerve cells are tightly joined into functional units. **Cell junctions** are contact points between the plasma membranes of tissue cells. Here we consider the five most important types of cell junctions: tight junctions, adherens junctions, desmosomes, hemidesmosomes, and gap junctions (Figure 4.2).

Figure 4.1 Types of tissues.

Each of the four types of tissues has different cells that vary in shape, structure, function, and distribution.

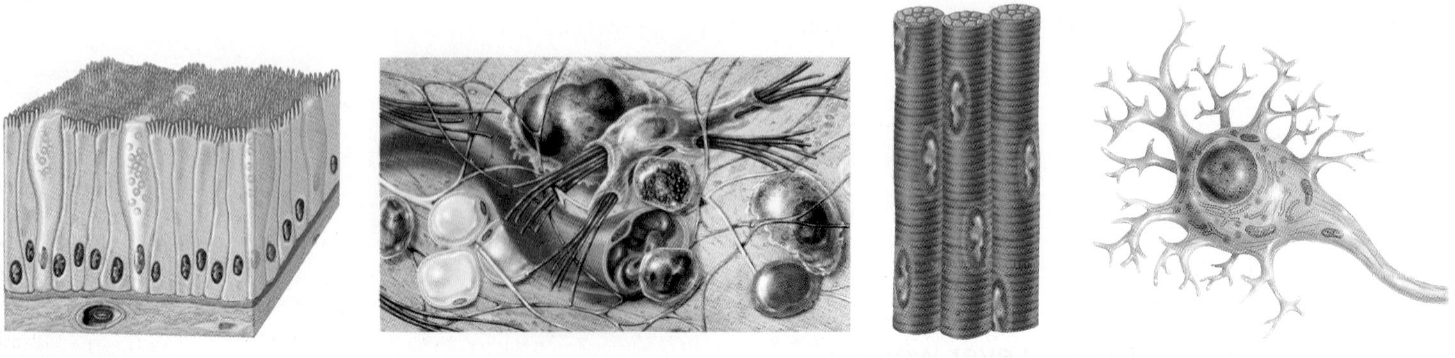

(a) Epithelial tissue (b) Connective tissue (c) Muscular tissue (d) Nervous tissue

? **What are some key differences in function among the four tissue types?**

Figure 4.2 Cell junctions.

Most epithelial cells and some muscle and nerve cells contain cell junctions.

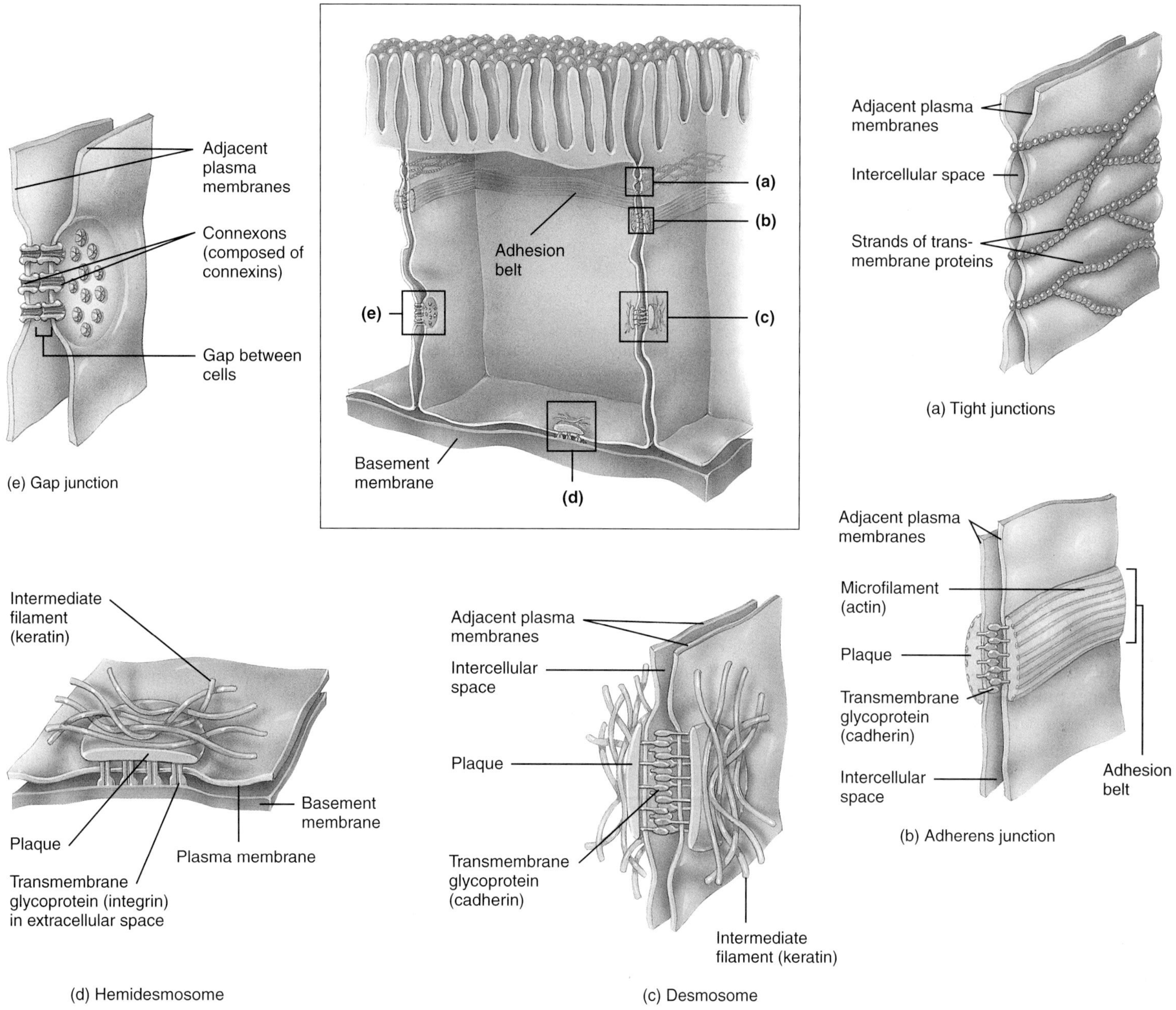

Adjacent plasma membranes
Connexons (composed of connexins)
Gap between cells
(e) Gap junction
Adhesion belt
(a)
(b)
(c)
(d)
(e)
Basement membrane
Adjacent plasma membranes
Intercellular space
Strands of trans-membrane proteins
(a) Tight junctions
Adjacent plasma membranes
Microfilament (actin)
Plaque
Transmembrane glycoprotein (cadherin)
Intercellular space
Adhesion belt
(b) Adherens junction
Intermediate filament (keratin)
Basement membrane
Plaque
Plasma membrane
Transmembrane glycoprotein (integrin) in extracellular space
(d) Hemidesmosome
Adjacent plasma membranes
Intercellular space
Plaque
Transmembrane glycoprotein (cadherin)
Intermediate filament (keratin)
(c) Desmosome

Which type of cell junction functions in communication between adjacent cells?

Tight Junctions

Tight junctions consist of weblike strands of transmembrane proteins that fuse together the outer surfaces of adjacent plasma membranes to seal off passageways between adjacent cells (Figure 4.2a). Cells of epithelial tissues that line the stomach, intestines, and urinary bladder have many tight junctions. They inhibit the passage of substances between cells and prevent the contents of these organs from leaking into the blood or surrounding tissues.

Adherens Junctions

Adherens junctions (ad-HER-ens) contain *plaque* (PLAK), a dense layer of proteins on the inside of the plasma membrane that attaches both to membrane proteins and to microfilaments of the cytoskeleton (Figure 4.2b). Transmembrane glycoproteins called **cadherins** join the cells. Each cadherin inserts into the plaque from the opposite side of the plasma membrane, partially crosses the intercellular space (the space between the cells), and connects to cadherins of an adjacent cell. In epithelial cells, adherens junctions often form extensive zones called **adhesion belts** because they encircle the cell similar to the way a belt encircles your waist. Adherens junctions help epithelial surfaces resist separation during various contractile activities, as when food moves through the intestines.

Desmosomes

Like adherens junctions, **desmosomes** (DEZ-mō-sōms; *desmo-* = band) contain plaque and have transmembrane glycoproteins (cadherins) that extend into the intercellular space between adjacent cell membranes and attach cells to one another (Figure 4.2c). However, unlike adherens junctions, the plaque of desmosomes does not attach to microfilaments. Instead, a desmosome plaque attaches to elements of the cytoskeleton known as intermediate filaments, which consist of the protein keratin. The intermediate filaments extend from desmosomes on one side of the cell across the cytosol to desmosomes on the opposite side of the cell. This structural arrangement contributes to the stability of the cells and tissue. These spot-weld-like junctions are common among the cells that make up the epidermis (the outermost layer of the skin) and among cardiac muscle cells in the heart. Desmosomes prevent epidermal cells from separating under tension and cardiac muscle cells from pulling apart during contraction.

Hemidesmosomes

Hemidesmosomes (*hemi-* = half) resemble desmosomes, but they do not link adjacent cells. The name arises from the fact that they look like half of a desmosome (Figure 4.2d). However, the transmembrane glycoproteins in hemidesmosomes are **integrins** rather than cadherins. On the inside of the plasma membrane, integrins attach to intermediate filaments made of the protein keratin. On the outside of the plasma membrane, the integrins attach to the protein *laminin,* which is present in the basement membrane (discussed shortly). Thus, hemidesmosomes anchor cells not to each other but to the basement membrane.

Gap Junctions

At **gap junctions,** membrane proteins called **connexins** form tiny fluid-filled tunnels called *connexons* that connect neighboring cells (Figure 4.2e). The plasma membranes of gap junctions are not fused together as in tight junctions but are separated by a very narrow intercellular gap (space). Through the connexons, ions and small molecules can diffuse from the cytosol of one cell to another, but the passage of large molecules such as vital intracellular proteins is prevented. The transfer of nutrients, and perhaps wastes, takes place through gap junctions in avascular tissues such as the lens and cornea of the eye. Gap junctions allow the cells in a tissue to communicate with one another. In a developing embryo, some of the chemical and electrical signals that regulate growth and cell differentiation travel via gap junctions. Gap junctions also enable nerve or muscle impulses to spread rapidly among cells, a process that is crucial for the normal operation of some parts of the nervous system and for the contraction of muscle in the heart, gastrointestinal tract, and uterus.

✔ **CHECKPOINT**

3. Which type of cell junction prevents the contents of organs from leaking into surrounding tissues?
4. Which types of cell junctions are found in epithelial tissues?

4.3 COMPARISON BETWEEN EPITHELIAL AND CONNECTIVE TISSUES

OBJECTIVE

- State the main differences between epithelial and connective tissues.

Before examining epithelial tissues and connective tissues in more detail, let's compare these two widely distributed tissues (Figure 4.3). Major structural differences between an epithelial tissue and a connective tissue are immediately obvious under a light microscope. The first obvious difference is the number of cells in relation to the extracellular matrix (the substance between cells). In an epithelial tissue many cells are tightly packed together with little or no extracellular matrix, whereas in a connective tissue a large amount of extracellular material separates cells that are usually widely scattered. The second obvious difference is that an epithelial tissue has no blood vessels, whereas most connective tissues have significant networks of blood vessels. Another key difference is that epithelial tissues almost always form surface layers and are not covered by another tissue. An exception is the epithelial lining of blood vessels where blood constantly passes over the epithelium. While these key structural distinctions account for some of the major functional differences between these tissue types, they also lead to a common bond. Because epithelial tissues lack blood vessels and form surfaces, they are always found immediately adjacent to blood-vessel-rich connective tissues, which enable them to make the exchanges with blood necessary for the delivery of oxygen and nutrients and the removal of wastes that are critical processes for their survival and function.

✔ **CHECKPOINT**

5. Why are epithelial and connective tissues found adjacent to each other?

4.4 EPITHELIAL TISSUES

OBJECTIVES

- Describe the general features of epithelial tissues.
- List the location, structure, and function of each different type of epithelial tissues.

An **epithelial tissue** (ep-i-THĒ-lē-al) or **epithelium** (plural is *epithelia*) consists of cells arranged in continuous sheets, in either

Figure 4.3 Comparison between epithelial tissues and connective tissues.

The ratio of cells to extracellular matrix is a major difference between epithelial tissues and connective tissues.

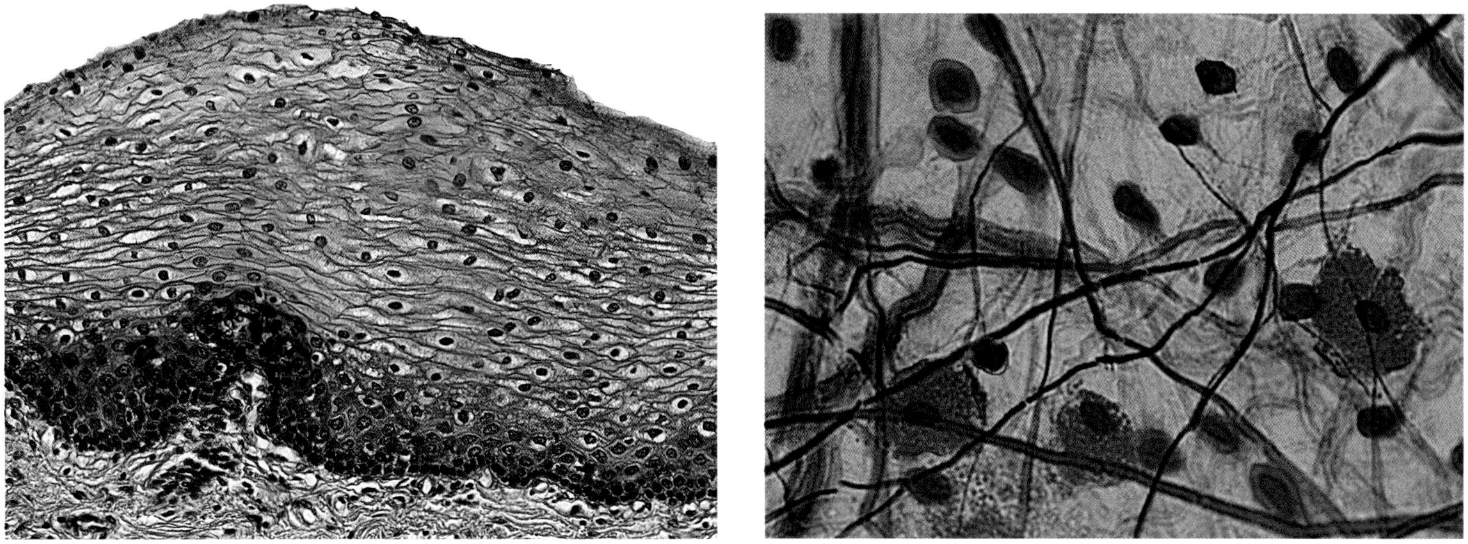

(a) Epithelial tissue with many cells tightly packed together and little to no extracellular matrix.

(b) Connective tissue with a few scattered cells surrounded by large amounts of extracellular matrix.

What relationship between epithelial tissues and connective tissues is important for the survival and function of epithelial tissues?

single or multiple layers. Because the cells are closely packed and are held tightly together by many cell junctions, there is little intercellular space between adjacent plasma membranes. Epithelial tissues form coverings and linings throughout the body. They are rarely covered by another tissue, so they always have a free surface. Epithelial tissues have three major functions. They serve as (1) selective barriers that limit or aid the transfer of substances into and out of the body; (2) secretory surfaces that release products produced by the cells onto their free surfaces; and (3) protective surfaces that resist the abrasive influences of the environment.

The various surfaces of epithelial cells often differ in structure and have specialized functions. The **apical (free) surface** of an epithelial cell faces the body surface, a body cavity, the lumen (interior space) of an internal organ, or a tubular duct that receives cell secretions (Figure 4.4). Apical surfaces may contain cilia or microvilli. The **lateral surfaces** of an epithelial cell, which face the adjacent cells on either side, may contain tight junctions, adherens junctions, desmosomes, and/or gap junctions. The **basal surface** of an epithelial cell is opposite the apical surface. The basal surfaces of the deepest layer of epithelial cells adhere to extracellular materials such as the basement membrane. Hemidesmosomes in the basal surfaces of the deepest layer of epithelial cells anchor the epithelium to the basement membrane (described next). In discussing epithelia with multiple layers, the term *apical layer* refers to the most superficial layer of cells, and the *basal layer* is the deepest layer of cells.

Figure 4.4 Surfaces of epithelial cells and the structure and location of the basement membrane.

The basement membrane is found between epithelial tissues and connective tissues.

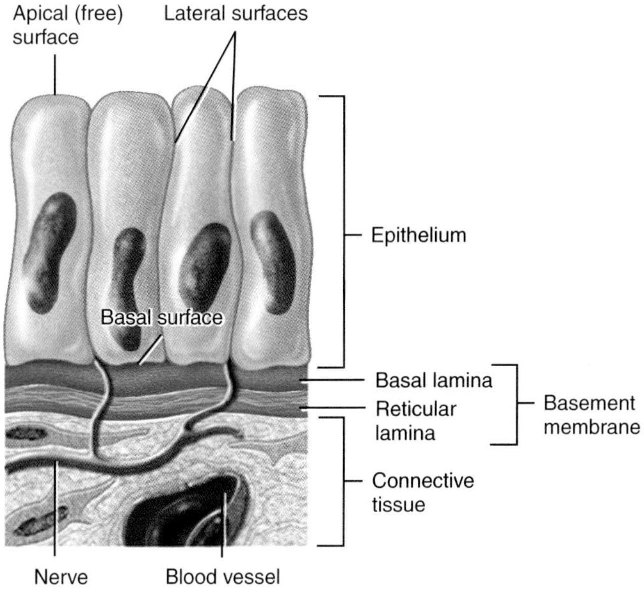

What are the functions of the basement membrane?

The **basement membrane** is a thin extracellular layer that commonly consists of two layers, the basal lamina and reticular lamina. The *basal lamina* (*lamina* = thin layer) is closer to—and secreted by—the epithelial cells. It contains proteins such as laminin and collagen (described shortly), as well as glycoproteins and proteoglycans (also described shortly). As you have already learned, the laminin molecules in the basal lamina adhere to integrins in hemidesmosomes and thus attach epithelial cells to the basement membrane (see Figure 4.2d). The *reticular lamina* is closer to the underlying connective tissue and contains proteins such as collagen produced by connective tissue cells called *fibroblasts* (see Figure 4.8). In addition to attaching to and supporting the overlying epithelial tissue, basement membranes have other functions. They form a surface along which epithelial cells migrate during growth or wound healing, restrict passage of larger molecules between epithelium and connective tissue, and participate in filtration of blood in the kidneys.

CLINICAL CONNECTION | *Basement Membranes and Disease*

Under certain conditions, basement membranes become markedly thickened, due to increased production of collagen and laminin. In untreated cases of diabetes mellitus, the basement membrane of small blood vessels (capillaries) thickens, especially in the eyes and kidneys. Because of this the blood vessels cannot function properly and blindness and kidney failure may result. •

Epithelial tissues have their own nerve supply, but as mentioned previously are **avascular** (*a-* = without; *-vascular* = vessel), relying on the blood vessels of the adjacent connective tissue to bring nutrients and remove wastes. Exchange of substances between epithelial tissues and connective tissues occurs by diffusion.

Because epithelial tissues form boundaries between the body's organs, or between the body and the external environment, they are repeatedly subjected to physical stress and injury. A high rate of cell division allows epithelial tissues to constantly renew and repair themselves by sloughing off dead or injured cells and replacing them with new ones. Epithelial tissues play many different roles in the body; the most important are protection, filtration, secretion, absorption, and excretion. In addition, epithelial tissues combine with nervous tissue to form special organs for smell, hearing, vision, and touch.

Epithelial tissues may be divided into two types. (1) **Covering and lining epithelium** forms the *outer* covering of the skin and some internal organs. It also forms the *inner* lining of blood vessels, ducts, and body cavities, and the interior of the respiratory, digestive, urinary, and reproductive systems. (2) **Glandular epithelium** makes up the secreting portion of glands such as the thyroid gland, adrenal glands, and sweat glands.

Classification of Epithelial Tissues

Types of covering and lining epithelial tissues are classified according to two characteristics: the arrangement of cells into layers and the shapes of the cells (Figure 4.5).

Figure 4.5 Cell shapes and arrangement of layers for covering and lining epithelium.

6 Cell shapes and arrangement of layers are the bases for classifying covering and lining epithelium.

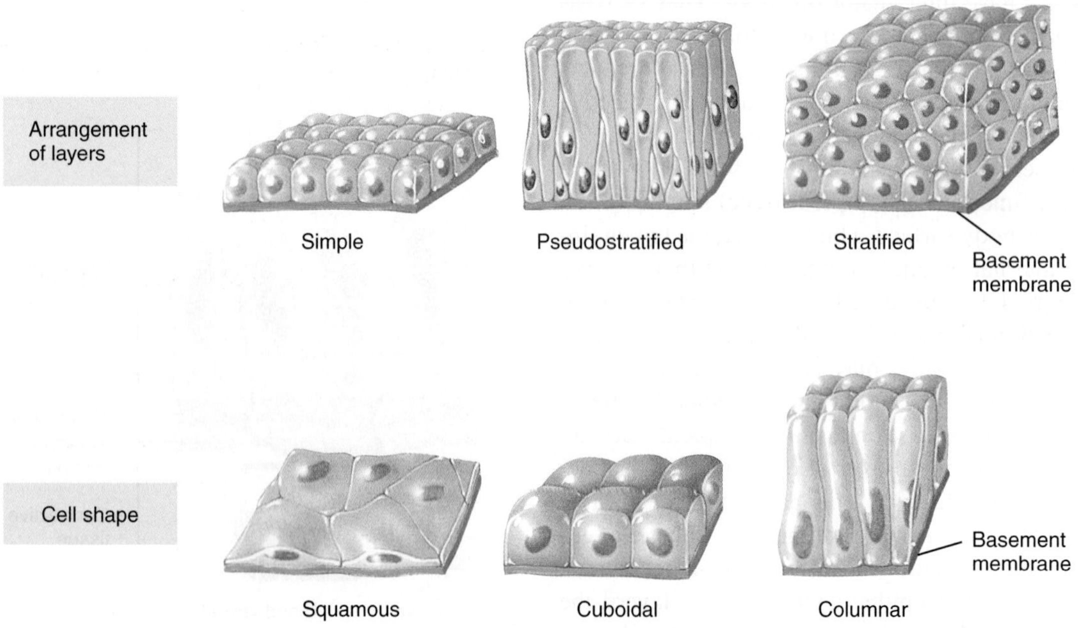
Arrangement of layers
Simple
Pseudostratified
Stratified
Basement membrane
Cell shape
Squamous
Cuboidal
Columnar
Basement membrane

? Which cell shape is best adapted for the rapid movement of substances from one cell to another?

1. ***Arrangement of cells in layers*** (Figure 4.5). The cells are arranged in one or more layers depending on function:
   a. *Simple epithelium* is a single layer of cells that functions in diffusion, osmosis, filtration, secretion, or absorption. **Secretion** is the production and release of substances such as mucus, sweat, or enzymes. **Absorption** is the intake of fluids or other substances such as digested food from the intestinal tract.
   b. *Pseudostratified epithelium* (*pseudo-* = false) appears to have multiple layers of cells because the cell nuclei lie at different levels and not all cells reach the apical surface, but it is actually a simple epithelium because all its cells rest on the basement membrane. Cells that do extend to the apical surface may contain cilia; others (goblet cells) secrete mucus.
   c. *Stratified epithelium* (*stratum* = layer) consists of two or more layers of cells that protect underlying tissues in locations where there is considerable wear and tear.
2. ***Cell shapes*** (Figure 4.5). Epithelial cells vary in shape depending on their function:
   a. *Squamous* cells (SKWĀ-mus = flat) are thin, which allows for the rapid passage of substances through them.
   b. *Cuboidal* cells are as tall as they are wide and are shaped like cubes or hexagons. They may have microvilli at their apical surface and function in either secretion or absorption.
   c. *Columnar* cells are much taller than they are wide, like columns, and protect underlying tissues. Their apical surfaces may have cilia or microvilli, and they often are specialized for secretion and absorption.
   d. *Transitional* cells change shape, from squamous to cuboidal and back, as organs such as the urinary bladder stretch (distend) to a larger size and then collapse to a smaller size.

When we combine the two characteristics (arrangements of layers and cell shapes), we come up with the following types of epithelial tissues:

I. Simple epithelium
   A. Simple squamous epithelium
   B. Simple cuboidal epithelium
   C. Simple columnar epithelium (nonciliated and ciliated)
   D. Pseudostratified columnar epithelium (nonciliated and ciliated)

II. Stratified epithelium
   A. Stratified squamous epithelium (keratinized, when surface cells are dead and become hardened, and nonkeratinized, when surface cells remain alive)*
   B. Stratified cuboidal epithelium*
   C. Stratified columnar epithelium*
   D. Transitional epithelium

We will now examine the important features of covering and lining epithelium.

Covering and Lining Epithelium

As noted earlier, covering and lining epithelium forms the outer covering of the skin and some internal organs. It also forms the inner lining of blood vessels, ducts, and body cavities, and the interior of the respiratory, digestive, urinary, and reproductive systems. Table 4.1 describes covering and lining epithelium in more detail. The discussion of each type consists of a photomicrograph, a corresponding diagram, and an inset that identifies a major location of the tissue in the body. Descriptions, locations, and functions of the tissues accompany each illustration.

CLINICAL CONNECTION | *Papanicolaou Test*

A **Papanicolaou test** (pa-pa-NI-kō-lō), also called a **Pap test** or **Pap smear,** involves collection and microscopic examination of epithelial cells that have been scraped off the apical layer of a tissue. A very common type of Pap test involves examining the cells from the nonkeratinized stratified squamous epithelium of the vagina and cervix (inferior portion) of the uterus. This type of Pap test is performed mainly to detect early changes in the cells of the female reproductive system that may indicate a precancerous condition or cancer. In performing a Pap smear, the cells are scraped from the tissue and then smeared on a microscope slide. The slides are then sent to a laboratory for analysis. Pap tests should be started within three years of the onset of sexual activity, or age 21, whichever comes first. Annual screening is recommended for females ages 21–30 and every 2–3 years for females age 30 or older following three consecutive negative Pap tests. •

*This classification is based on the shape of the cells at the apical surface.

TABLE 4.1

Epithelial Tissues: Covering and Lining Epithelium

A. SIMPLE SQUAMOUS EPITHELIUM

Description	Single layer of flat cells that resembles a tiled floor when viewed from apical surface; centrally located nucleus that is flattened and oval or spherical in shape.
Location	Most commonly (1) lines the cardiovascular and lymphatic system (heart, blood vessels, lymphatic vessel linings), where it is known as **endothelium** (en′-dō-THĒ-lē-um; *endo-* = within; *-thelium* = covering), and (2) forms the epithelial layer of serous membranes (peritoneum, pleura, pericardium), where it is called **mesothelium** (mez′-ō-THĒ-lē-um; *meso-* = middle). Also found in air sacs of lungs, glomerular (Bowman's) capsule of kidneys, inner surface of tympanic membrane (eardrum).
Function	Present at sites of filtration (such as blood filtration in kidneys) or diffusion (such as diffusion of oxygen into blood vessels of lungs) and at site of secretion in serous membranes. Not found in body areas subject to mechanical stress (wear and tear).

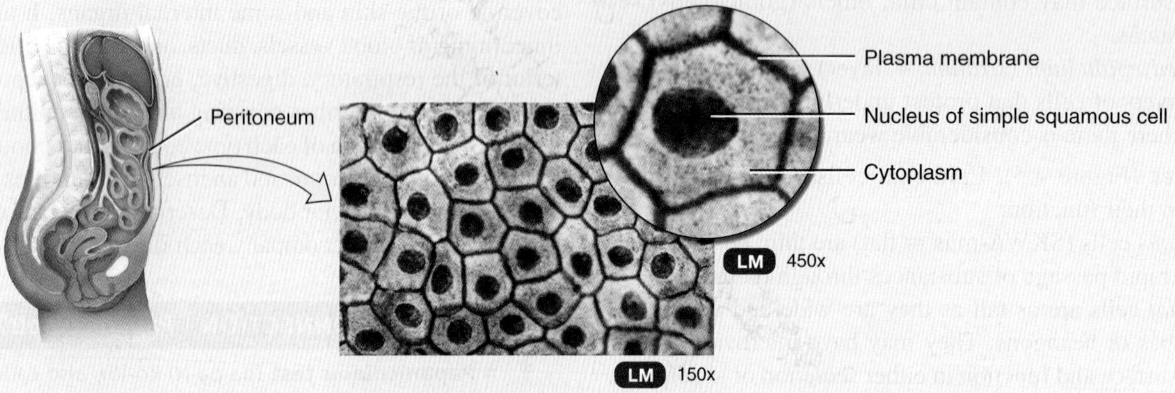
Peritoneum
Plasma membrane
Nucleus of simple squamous cell
Cytoplasm
LM 450x
LM 150x

Surface view of simple squamous epithelium of mesothelial lining of peritoneum

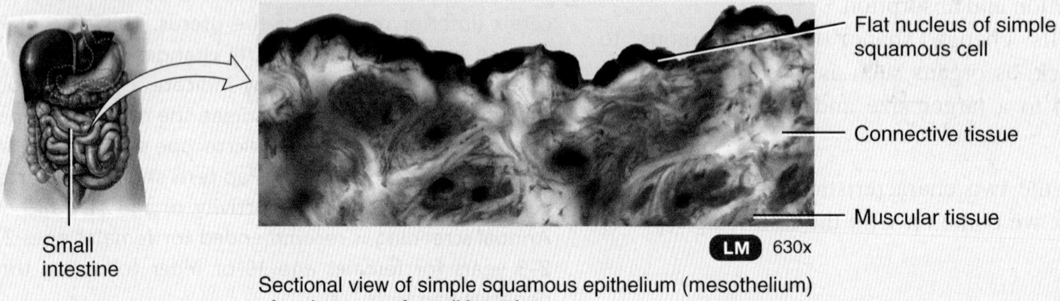
Small intestine
Flat nucleus of simple squamous cell
Connective tissue
Muscular tissue
LM 630x

Sectional view of simple squamous epithelium (mesothelium) of peritoneum of small intestine

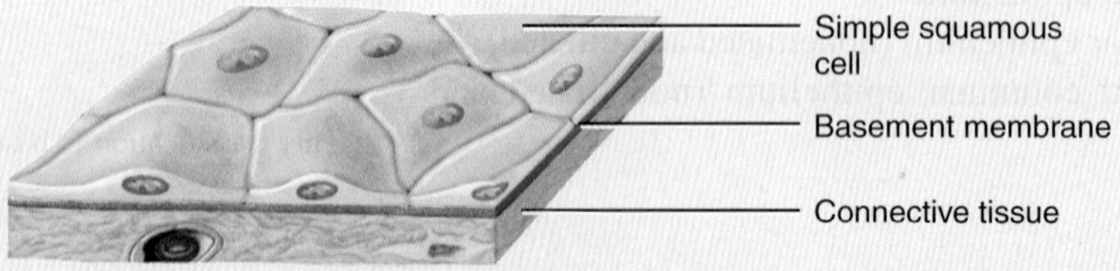
Simple squamous cell
Basement membrane
Connective tissue

Simple squamous epithelium

B. SIMPLE CUBOIDAL EPITHELIUM

Description Single layer of cube-shaped cells; round, centrally located nucleus. Cuboidal cell shape is obvious when tissue is sectioned and viewed from the side. (Note: Strictly cuboidal cells could not form small tubes; these cuboidal cells are more pie-shaped but still nearly as high as they are wide at the base.)

Location Covers surface of ovary; lines anterior surface of capsule of lens of the eye; forms pigmented epithelium at posterior surface of retina of the eye; lines kidney tubules and smaller ducts of many glands; makes up secreting portion of some glands, such as thyroid gland and ducts of some glands such as pancreas.

Function Secretion and absorption.

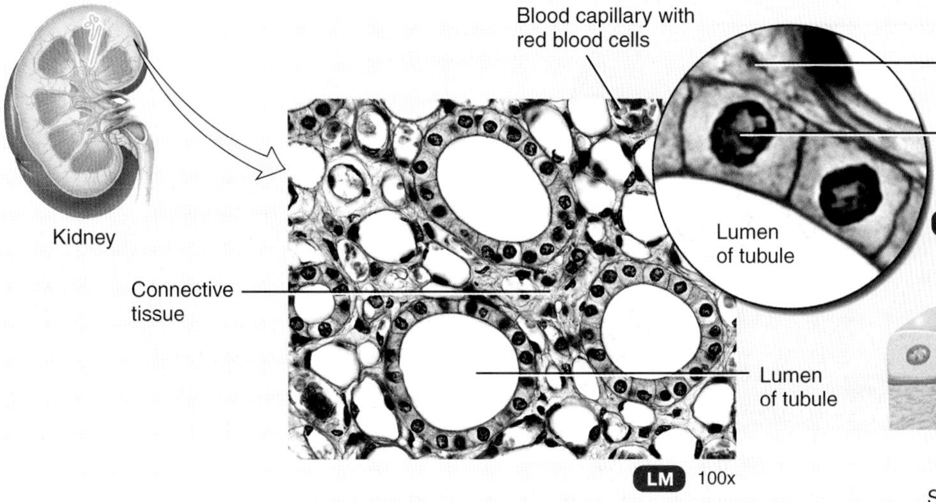

Sectional view of simple cuboidal epithelium of urinary tubules

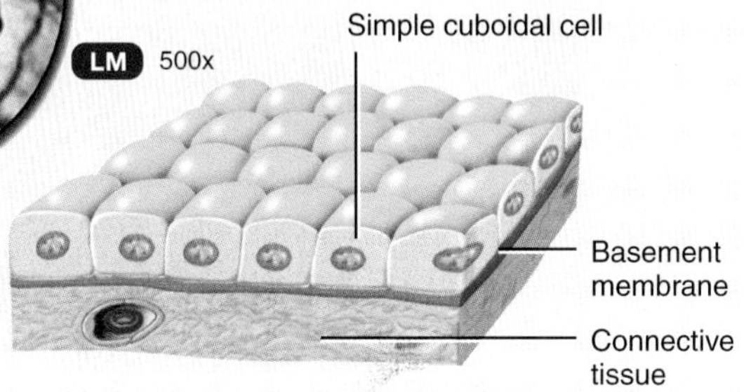

Simple cuboidal epithelium

C. NONCILIATED SIMPLE COLUMNAR EPITHELIUM

Description Single layer of nonciliated columnlike cells with oval nuclei near base of cells; contains (1) columnar epithelial cells with microvilli at apical surface and (2) goblet cells. **Microvilli,** fingerlike cytoplasmic projections, increase surface area of plasma membrane (see Figure 3.1), thus increasing cell's rate of absorption. **Goblet cells** are modified columnar epithelial cells that secrete mucus, a slightly sticky fluid, at their apical surfaces. Before release, mucus accumulates in upper portion of cell, causing it to bulge and making the whole cell resemble a goblet or wine glass.

Location Lines gastrointestinal tract (from stomach to anus), ducts of many glands, and gallbladder.

Function Secretion and absorption; larger columnar cells contain more organelles and thus are capable of higher level of secretion and absorption than are cuboidal cells. Secreted mucus lubricates linings of digestive, respiratory, and reproductive tracts, and most of urinary tract; helps prevent destruction of stomach lining by acidic gastric juice secreted by stomach.

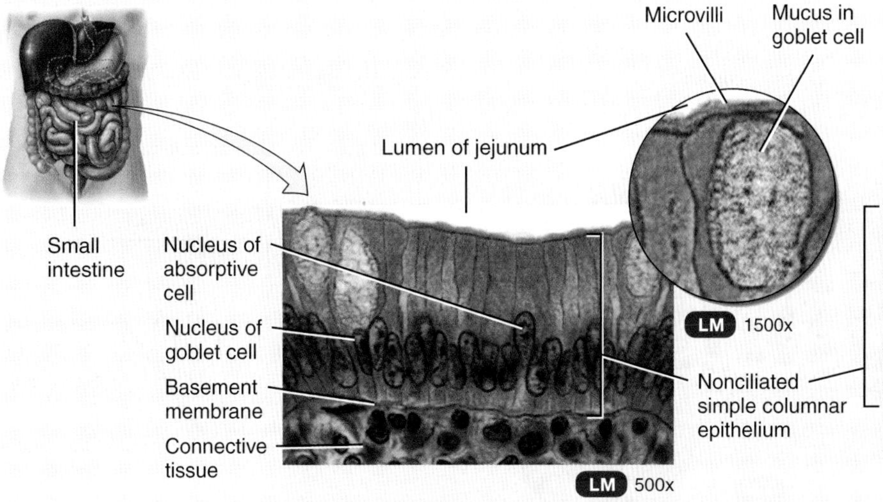

Sectional view of nonciliated simple columnar epithelium of lining of jejunum of small intestine

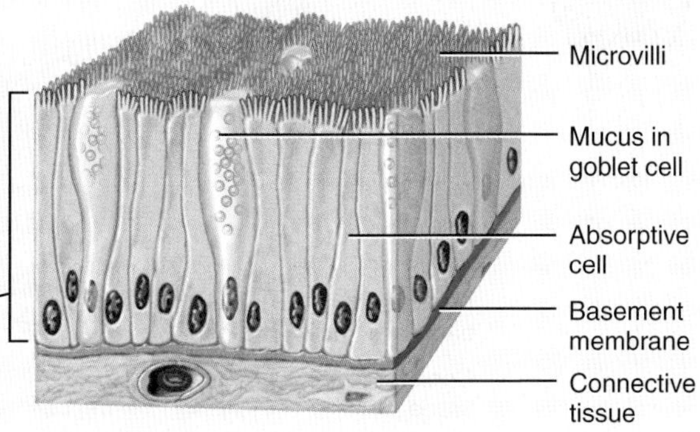

Nonciliated simple columnar epithelium

TABLE 4.1 CONTINUES

TABLE 4.1 CONTINUED

Epithelial Tissues: Covering and Lining Epithelium

D. CILIATED SIMPLE COLUMNAR EPITHELIUM

Description Single layer of ciliated columnlike cells with oval nuclei near base of cells. Goblet cells are usually interspersed among ciliated columnar epithelia.

Location Lines some bronchioles (small tubes) of respiratory tract, uterine (fallopian) tubes, uterus, some paranasal sinuses, central canal of spinal cord, and ventricles of brain.

Function Cilia beat in unison, moving mucus and foreign particles toward throat, where they can be coughed up and swallowed or spit out. Coughing and sneezing speed up movement of cilia and mucus. Cilia also help move oocytes expelled from ovaries through uterine (fallopian) tubes into uterus.

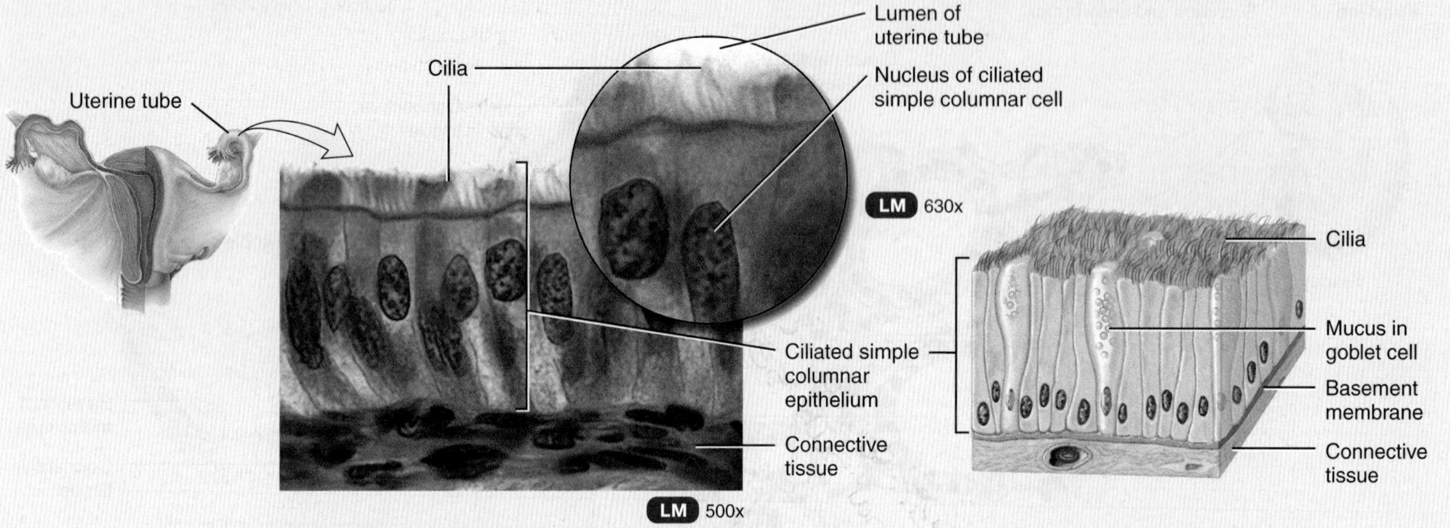

Sectional view of ciliated simple columnar epithelium of uterine tube

Ciliated simple columnar epithelium

E. PSEUDOSTRATIFIED COLUMNAR EPITHELIUM

Description Appears to have several layers because cell nuclei are at various levels. All cells are attached to basement membrane in a single layer, but some cells do not extend to apical surface. When viewed from side, these features give false impression of a multilayered tissue (thus the name pseudostratified; *pseudo* = false).

Pseudostratified ciliated columnar epithelium contains cells that extend to surface and secrete mucus (goblet cells) or bear cilia.

Pseudostratified nonciliated columnar epithelium contains cells without cilia and lacks goblet cells.

Location Ciliated variety lines airways of most of upper respiratory tract; nonciliated variety lines larger ducts of many glands, epididymis, and part of male urethra.

Function Ciliated variety secretes mucus that traps foreign particles, and cilia sweep away mucus for elimination from body; nonciliated variety functions in absorption and protection.

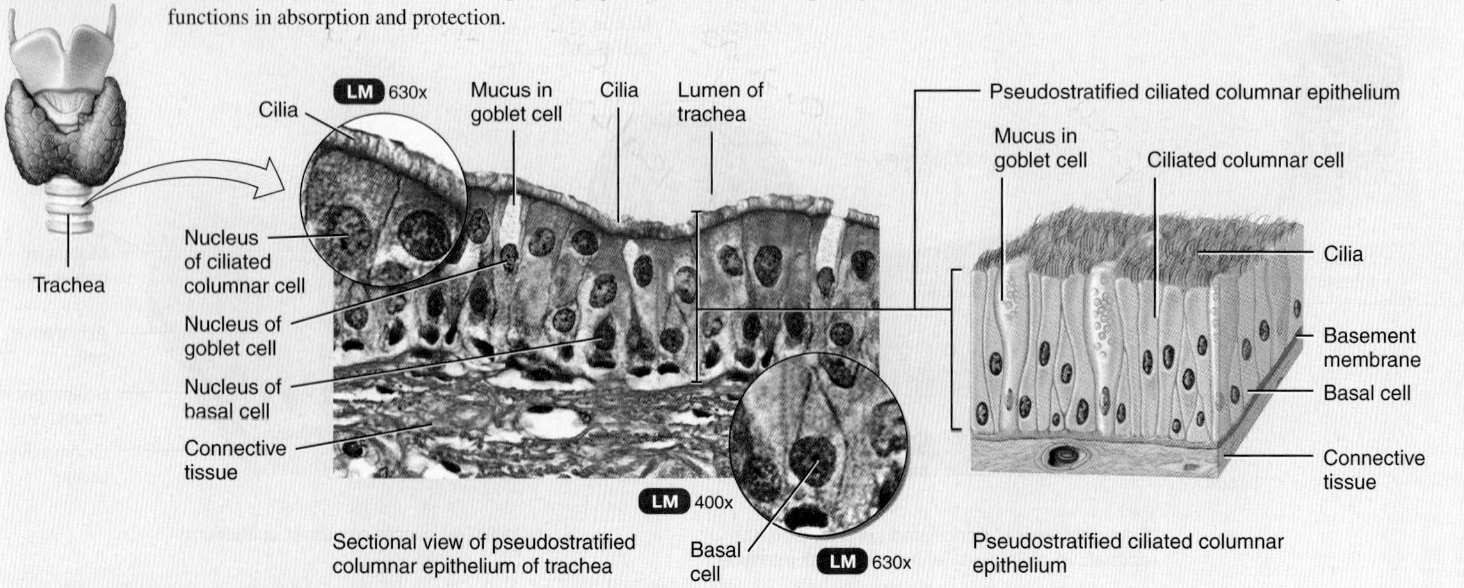

Sectional view of pseudostratified columnar epithelium of trachea

Pseudostratified ciliated columnar epithelium

F. STRATIFIED SQUAMOUS EPITHELIUM

Description Two or more layers of cells; cells in apical layer and several layers deep to it are squamous; cells in deeper layers vary from cuboidal to columnar. As basal cells divide, daughter cells arising from cell divisions push upward toward apical layer. As they move toward surface and away from blood supply in underlying connective tissue, they become dehydrated and less metabolically active. Tough proteins predominate as cytoplasm is reduced, and cells become tough, hard structures that eventually die. At apical layer, after dead cells lose cell junctions they are sloughed off, but they are replaced continuously as new cells emerge from basal cells.

Keratinized stratified squamous epithelium develops tough layer of keratin in apical layer of cells and several layers deep to it (see Figure 5.3). (**Keratin** is a tough, fibrous intracellular protein that helps protect skin and underlying tissues from heat, microbes, and chemicals.) Relative amount of keratin increases in cells as they move away from nutritive blood supply and organelles die.

Nonkeratinized stratified squamous epithelium does not contain large amounts of keratin in apical layer and several layers deep and is constantly moistened by mucus from salivary and mucous glands; organelles are not replaced.

Location Keratinized variety forms superficial layer of skin; nonkeratinized variety lines wet surfaces (lining of mouth, esophagus, part of epiglottis, part of pharynx, and vagina) and covers tongue.

Function Protection against abrasion, water loss, ultraviolet radiation, and foreign invasion. Both types form first line of defense against microbes.

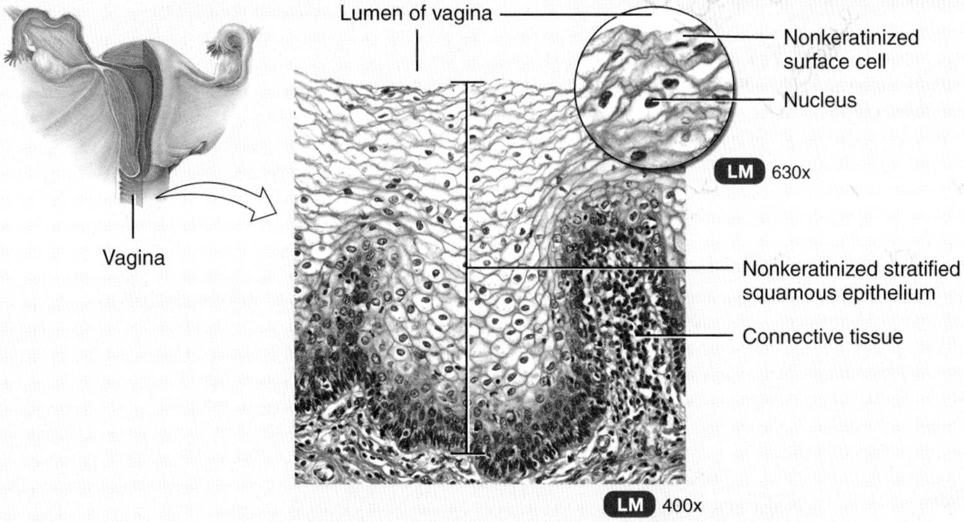
Lumen of vagina
Nonkeratinized surface cell
Nucleus
LM 630x
Vagina
Nonkeratinized stratified squamous epithelium
Connective tissue
LM 400x

Sectional view of nonkeratinized stratified squamous epithelium of lining of vagina

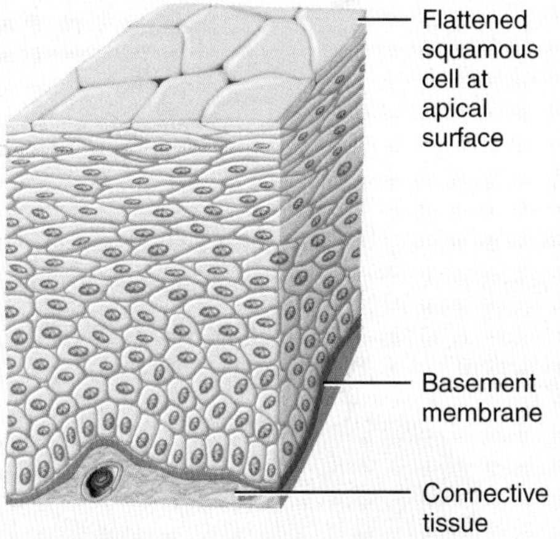
Flattened squamous cell at apical surface
Basement membrane
Connective tissue

Nonkeratinized stratified squamous epithelium

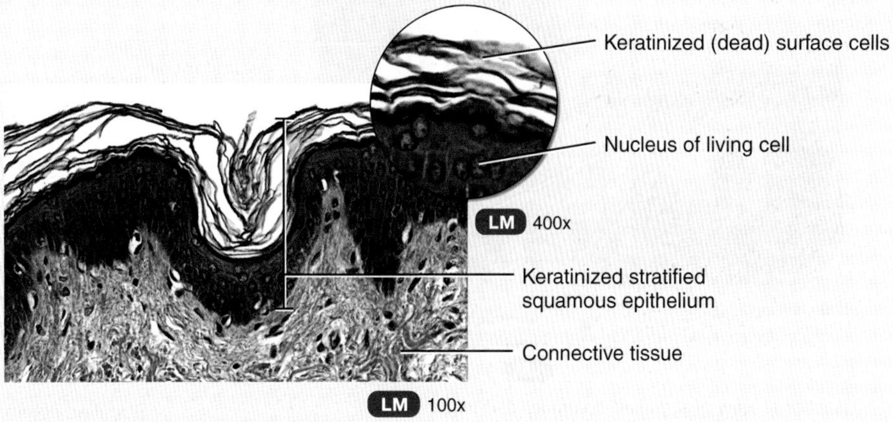
Keratinized (dead) surface cells
Nucleus of living cell
LM 400x
Keratinized stratified squamous epithelium
Connective tissue
LM 100x

Sectional view of keratinized stratified squamous epithelium of epidermis

TABLE 4.1 CONTINUES

TABLE 4.1 CONTINUED

Epithelial Tissues: Covering and Lining Epithelium

G. STRATIFIED CUBOIDAL EPITHELIUM

Description Two or more layers of cells; cells in apical layer are cube-shaped; fairly rare type.

Location Ducts of adult sweat glands and esophageal glands, part of male urethra.

Function Protection; limited secretion and absorption.

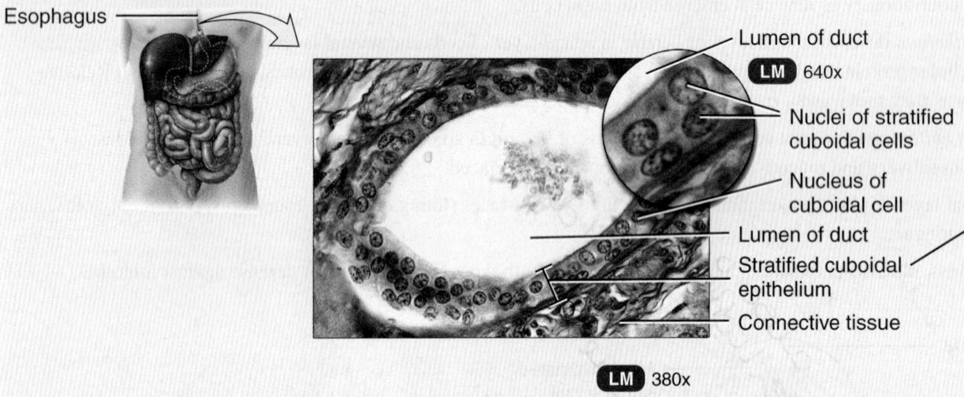
Esophagus
Lumen of duct
LM 640x
Nuclei of stratified cuboidal cells
Nucleus of cuboidal cell
Lumen of duct
Stratified cuboidal epithelium
Connective tissue
LM 380x

Sectional view of stratified cuboidal epithelium of the duct of an esophageal gland

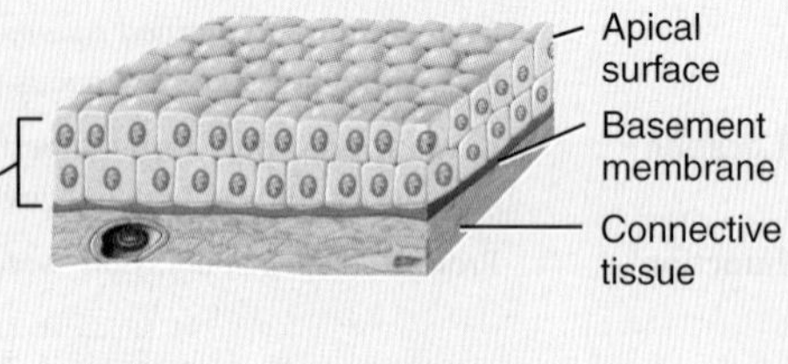
Apical surface
Basement membrane
Connective tissue

Stratified cuboidal epithelium

H. STRATIFIED COLUMNAR EPITHELIUM

Description Basal layers usually consist of shortened, irregularly shaped cells; only apical layer has columnar cells; uncommon.

Location Lines part of urethra; large excretory ducts of some glands, such as esophageal glands; small areas in anal mucous membrane; part of conjunctiva of eye.

Function Protection and secretion.

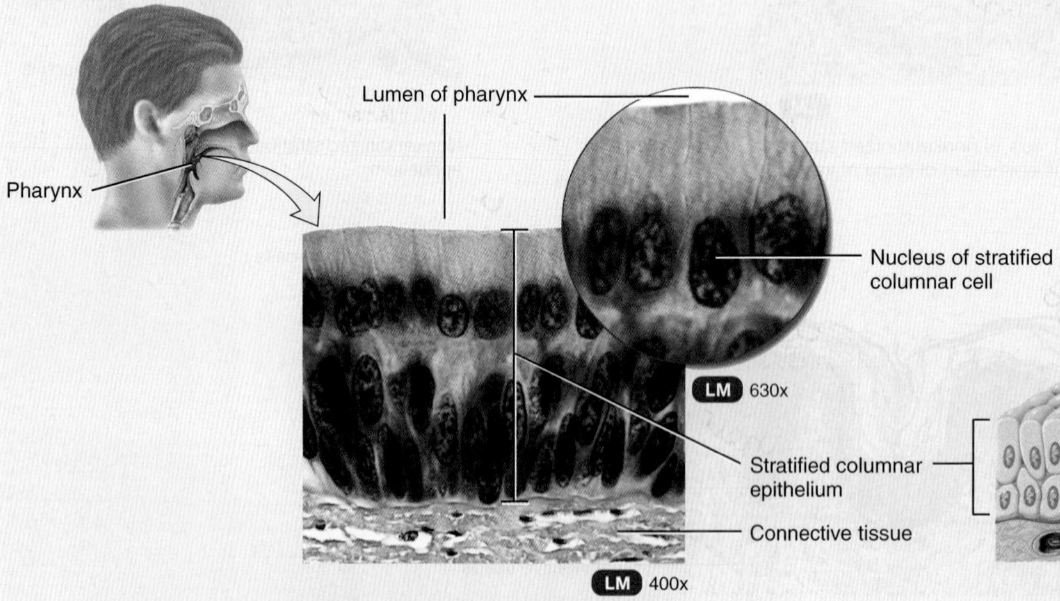
Pharynx
Lumen of pharynx
Nucleus of stratified columnar cell
LM 630x
Stratified columnar epithelium
Connective tissue
LM 400x

Sectional view of stratified columnar epithelium of lining of pharynx

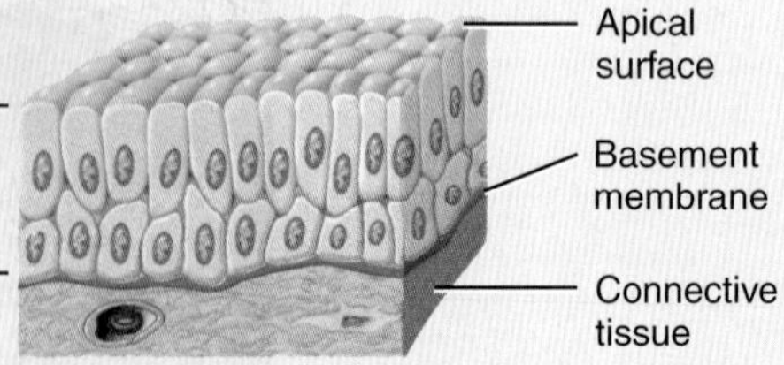
Apical surface
Basement membrane
Connective tissue

Stratified columnar epithelium

I. TRANSITIONAL EPITHELIUM

Description Variable appearance (transitional). In relaxed or unstretched state, looks like stratified cuboidal epithelium, except apical layer cells tend to be large and rounded. As tissue is stretched, cells become flatter, giving the appearance of stratified squamous epithelium. Multiple layers and elasticity make it ideal for lining hollow structures (urinary bladder) subject to expansion from within.

Location Lines urinary bladder and portions of ureters and urethra.

Function Allows urinary organs to stretch and maintain protective lining while holding variable amounts of fluid without rupturing.

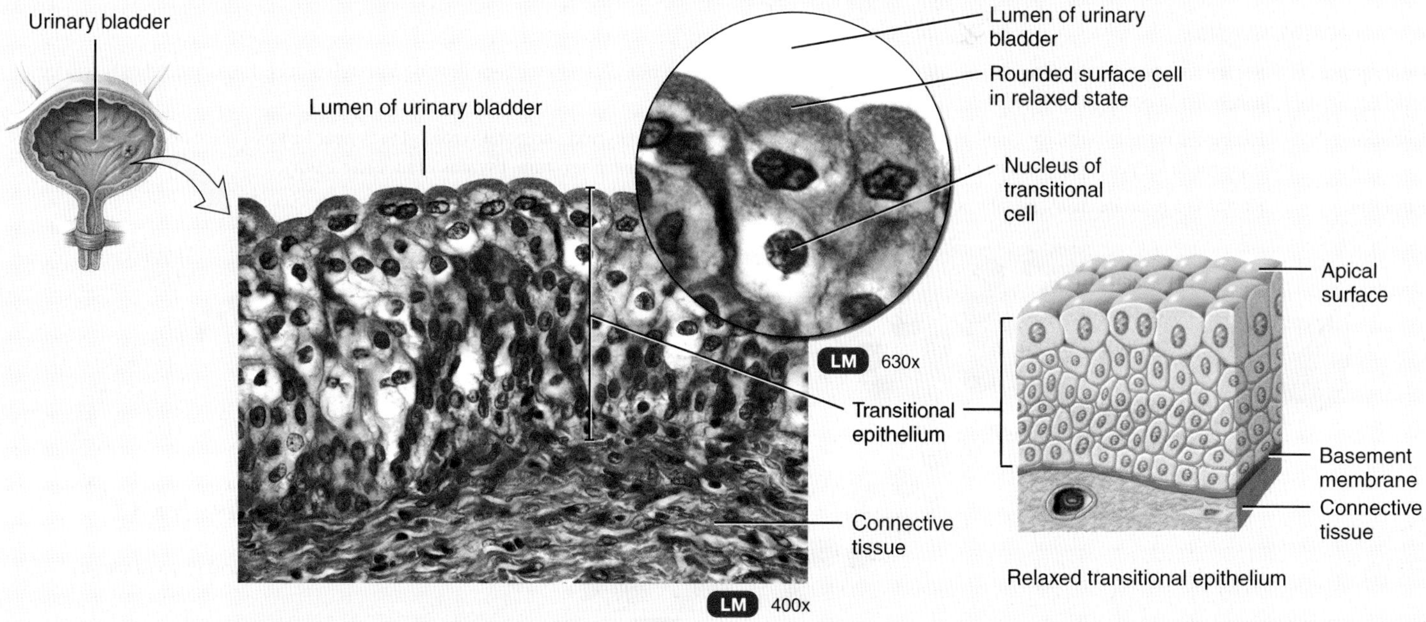

Sectional view of transitional epithelium of urinary bladder in relaxed (empty) state

Relaxed transitional epithelium

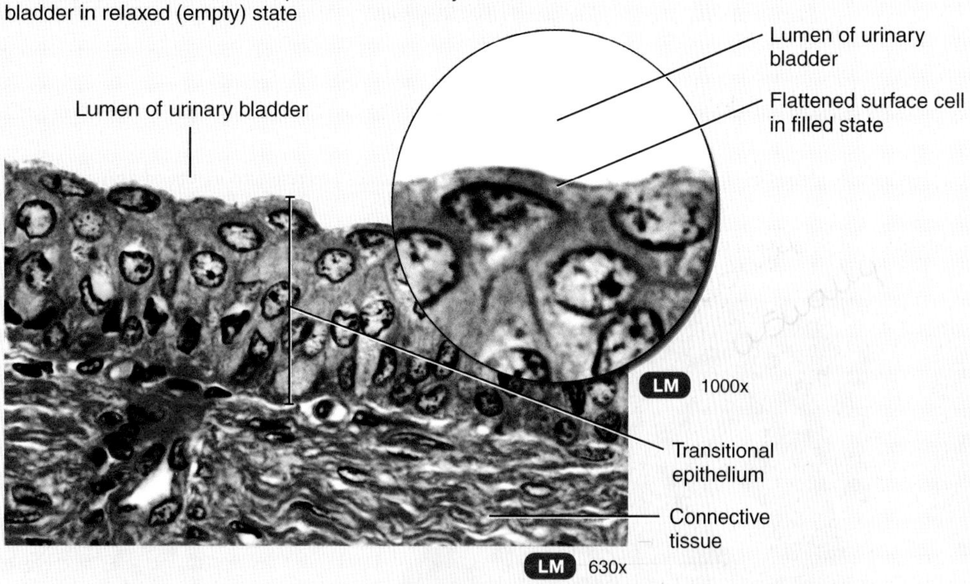

Sectional view of transitional epithelium of urinary bladder in filled state

Glandular Epithelium

The function of glandular epithelium is secretion, which is accomplished by glandular cells that often lie in clusters deep to the covering and lining epithelium. A **gland** may consist of a single cell or a group of cells that secrete substances into ducts (tubes), onto a surface, or into the blood. All glands of the body are classified as either endocrine or exocrine.

The secretions of **endocrine glands** (EN-dō-krin; *endo-* = inside; *-crine* = secretion; Table 4.2), called hormones, enter the

TABLE 4.2

Epithelial Tissues: Glandular Epithelium

A. ENDOCRINE GLANDS

Description	Secretions (*hormones*) enter interstitial fluid and diffuse directly into bloodstream without flowing through a duct. Endocrine glands will be described in detail in Chapter 18.
Location	Examples include pituitary gland at base of brain, pineal gland in brain, thyroid and parathyroid glands near larynx (voice box), adrenal glands superior to kidneys, pancreas near stomach, ovaries in pelvic cavity, testes in scrotum, thymus in thoracic cavity.
Function	Hormones regulate many metabolic and physiological activities to maintain homeostasis.

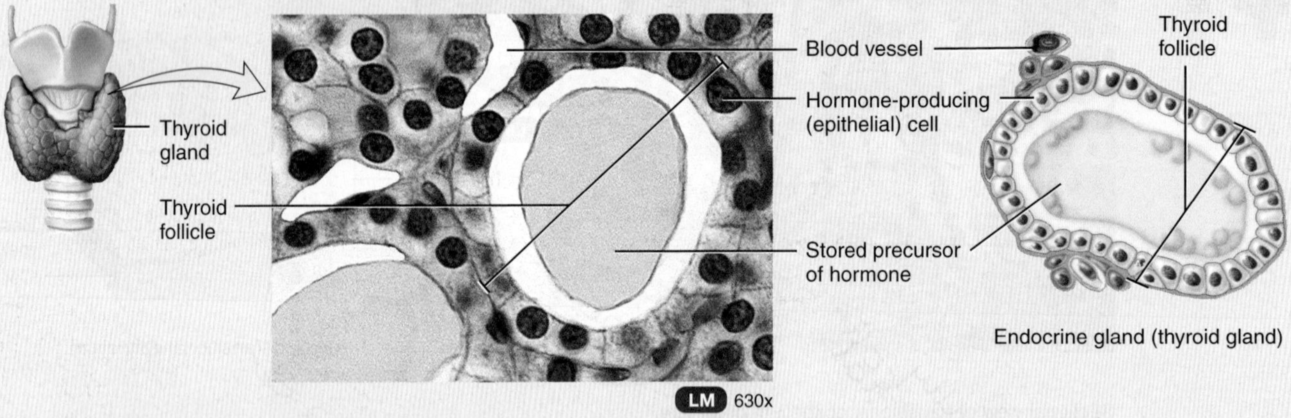

Sectional view of endocrine gland (thyroid gland)

Endocrine gland (thyroid gland)

B. EXOCRINE GLANDS

Description	Secretory products released into ducts that empty onto surface of a covering and lining epithelium, such as skin surface or lumen of hollow organ.
Location	Sweat, oil, and earwax glands of skin; digestive glands such as salivary glands (secrete into mouth cavity) and pancreas (secretes into small intestine).
Function	Produce substances such as sweat to help lower body temperature, oil, earwax, saliva, or digestive enzymes.

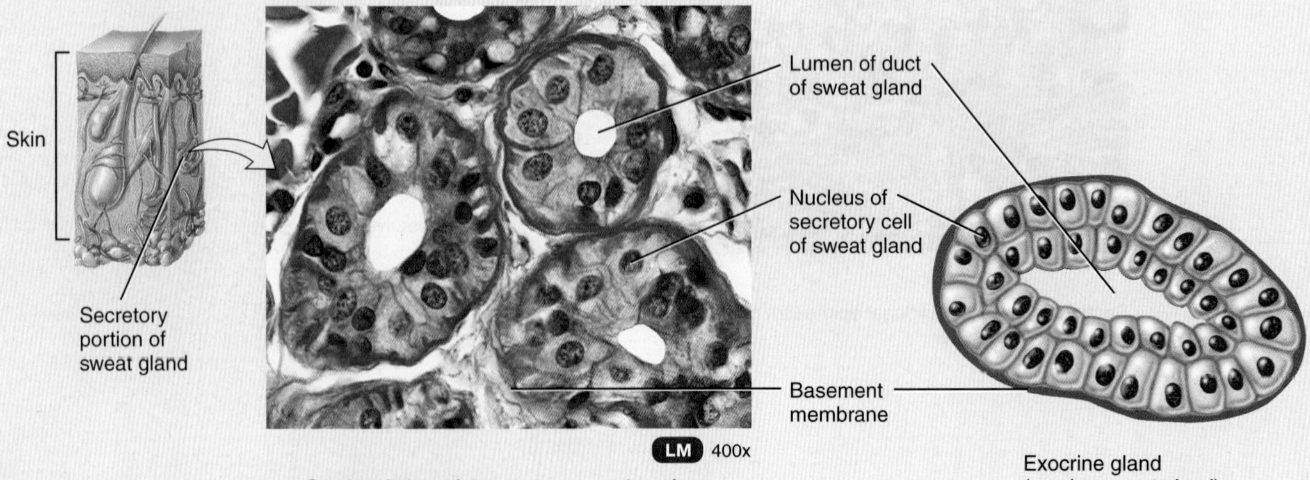

Sectional view of the secretory portion of an exocrine gland (eccrine sweat gland)

Exocrine gland (eccrine sweat gland)

interstitial fluid and then diffuse directly into the bloodstream without flowing through a duct. Endocrine glands will be described in detail in Chapter 18. Endocrine secretions have far-reaching effects because they are distributed throughout the body by the bloodstream.

Exocrine glands (EKS-ō-krin; *exo-* = outside; Table 4.2) secrete their products into ducts that empty onto the surface of a covering and lining epithelium such as the skin surface or the lumen of a hollow organ. The secretions of exocrine glands have limited effects and some of them would be harmful if they entered the bloodstream. As you will learn later in the text, some glands of the body, such as the pancreas, ovaries, and testes, are mixed glands that contain both endocrine and exocrine tissue.

Structural Classification of Exocrine Glands

Exocrine glands are classified as unicellular or multicellular. As the name implies, **unicellular glands** are single-celled glands. Goblet cells are important unicellular exocrine glands that secrete mucus directly onto the apical surface of a lining epithelium. Most exocrine glands are **multicellular glands,** composed of many cells that form a distinctive microscopic structure or macroscopic organ. Examples include sudoriferous (sweat), sebaceous (oil), and salivary glands.

Multicellular glands are categorized according to two criteria: (1) whether their ducts are branched or unbranched and (2) the shape of the secretory portions of the gland (Figure 4.6). If the duct of the gland does not branch, it is a **simple gland.** If the duct branches, it is a **compound gland.** Glands with tubular secretory parts are **tubular glands;** those with rounded secretory portions are **acinar glands** (AS-i-nar; *acin-* = berry), also called *alveolar glands.* **Tubuloacinar glands** have both tubular and more rounded secretory parts.

Combinations of these features are the criteria for the following structural classification scheme for multicellular exocrine glands:

I. Simple glands
   A. **Simple tubular.** Tubular secretory part is straight and attaches to a single unbranched duct. Example: glands in the large intestine.

Figure 4.6 **Multicellular exocrine glands.** Pink represents the secretory portion; lavender represents the duct.

Structural classification of multicellular exocrine glands is based on the branching pattern of the duct and the shape of the secreting portion.

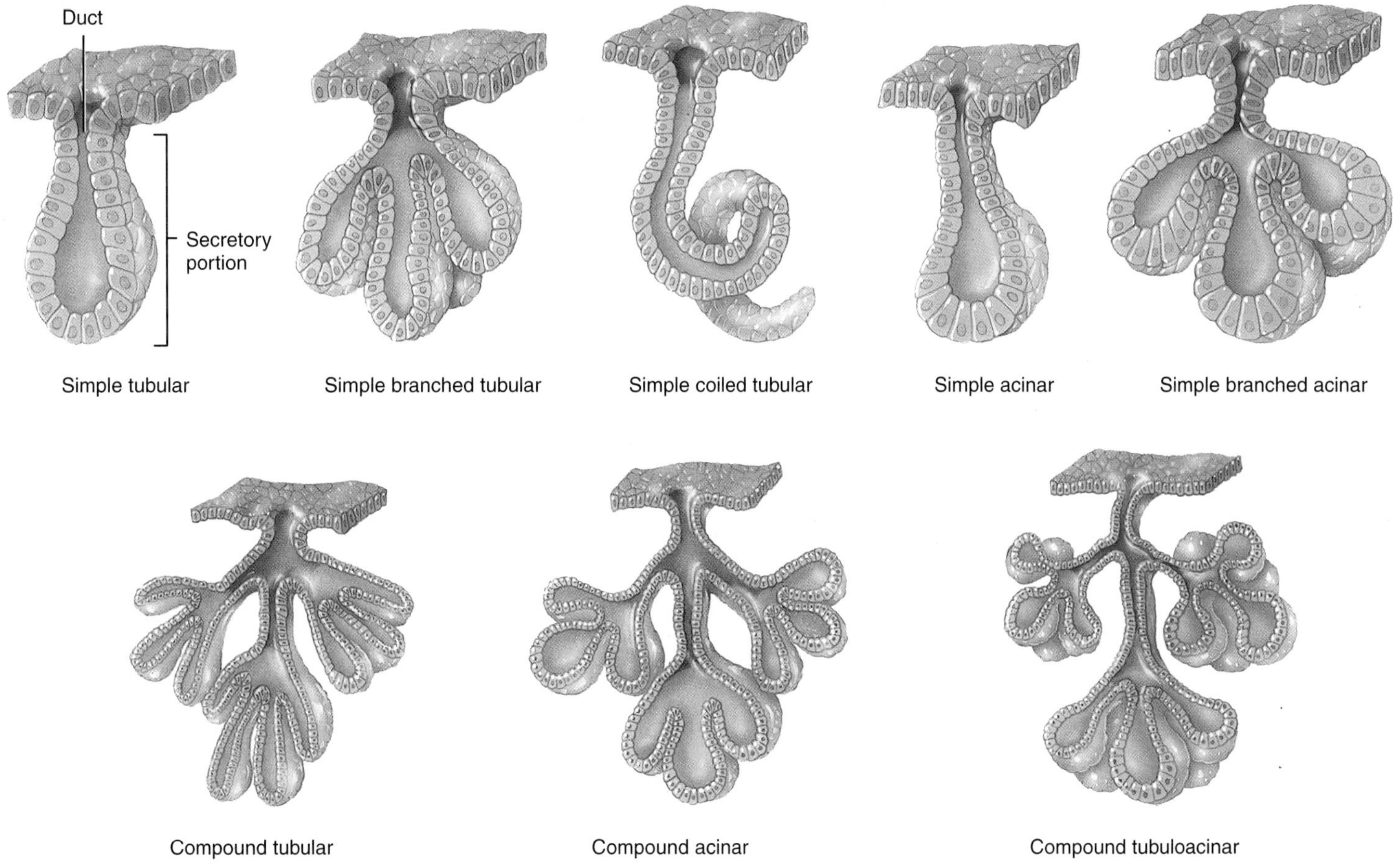

How do simple multicellular exocrine glands differ from compound ones?

B. **Simple branched tubular.** Tubular secretory part is branched and attaches to a single unbranched duct. Example: gastric glands.
C. **Simple coiled tubular.** Tubular secretory part is coiled and attaches to a single unbranched duct. Example: sweat glands.
D. **Simple acinar.** Secretory portion is rounded and attaches to a single unbranched duct. Example: glands of the penile urethra.
E. **Simple branched acinar.** Rounded secretory part is branched and attaches to a single unbranched duct. Example: sebaceous glands.

II. Compound glands
A. **Compound tubular.** Secretory portion is tubular and attaches to a branched duct. Example: bulbourethral (Cowper's) glands.
B. **Compound acinar.** Secretory portion is rounded and attaches to a branched duct. Example: mammary glands.
C. **Compound tubuloacinar.** Secretory portion is both tubular and rounded and attaches to a branched duct. Example: acinar glands of the pancreas.

Functional Classification of Exocrine Glands

The functional classification of exocrine glands is based on how their secretions are released. Each of these secretory processes begins with the endoplasmic reticulum and Golgi complex working together to form intracellular secretory vesicles that contain the secretory product. Secretions of **merocrine glands** (MER-ō-krin; *mero-* = a part) are synthesized on ribosomes attached to rough ER; processed, sorted, and packaged by the Golgi complex; and released from the cell in secretory vesicles via exocytosis (Figure 4.7a). Most exocrine glands of the body are merocrine glands. Examples include the salivary glands and pancreas. **Apocrine glands** (AP-ō-krin; *apo-* = from) accumulate their secretory product at the apical surface of the secreting cell. Then, that portion of the cell pinches off by exocytosis from the rest of the cell to release the secretion (Figure 4.7b). The remaining part of the cell repairs itself and repeats the process. Electron microscopy has recently confirmed that this is the mechanism of secretion of milk fats in the mammary glands. Recent evidence reveals that the sweat glands of the skin, named apocrine sweat glands after this mode of secretion, actually undergo merocrine secretion. The cells of **holocrine glands** (HŌ-lō-krin; *holo-* = entire) accumulate a secretory product in their cytosol. As the secretory cell matures, it ruptures and becomes the secretory product (Figure 4.7c). Because the cell ruptures in this mode of secretion, the secretion contains large amounts of lipids from the plasma membrane and intracellular membranes. The sloughed off cell is replaced by a new cell. One example of a holocrine gland is a sebaceous gland of the skin.

✔CHECKPOINT
6. Describe the various layering arrangements and cell shapes of epithelium.
7. What characteristics are common to all epithelial tissues?
8. How is the structure of the following epithelial tissues related to their functions: simple squamous, simple cuboidal, simple columnar (nonciliated and ciliated), pseudostratified columnar (ciliated and nonciliated), stratified squamous (keratinized and nonkeratinized), stratified cuboidal, stratified columnar, and transitional?
9. Where are endothelium and mesothelium located?
10. What is the difference between endocrine glands and exocrine glands? Name and give examples of the three functional classes of exocrine glands based on how their secretions are released.

4.5 CONNECTIVE TISSUES

OBJECTIVES
- Describe the general features of connective tissues.
- Describe the structure, location, and function of the various types of connective tissues.

Connective tissues are one of the most abundant and widely distributed tissues in the body. In their various forms, connective tissues have a variety of functions. They bind together, support, and strengthen other body tissues; protect and insulate internal organs; compartmentalize structures such as skeletal muscles; serve as the major transport system within the body (blood, a fluid connective tissue); are the primary locations of stored energy reserves (adipose, or fat, tissue); and are the main source of immune responses.

General Features of Connective Tissues

Connective tissues consist of two basic elements: extracellular matrix and cells. A connective tissue's **extracellular matrix** (MĀ-triks) is the material located between its widely spaced cells. The extracellular matrix consists of *protein fibers* and *ground substance,* the material between the cells and the fibers. The extracellular fibers are secreted by the connective tissue cells and account for many of the functional properties of the tissue in addition to controlling the surrounding watery environment via specific proteoglycan molecules (described shortly). The structure of the extracellular matrix determines much of the tissue's qualities. For instance, in cartilage, the extracellular matrix is firm but pliable. The extracellular matrix of bone, by contrast, is hard and inflexible.

Recall that, in contrast to epithelial tissues, connective tissues do not usually occur on body surfaces. Also, unlike epithelial tissues, connective tissues usually are highly vascular; that is, they have a rich blood supply. Exceptions include cartilage, which is avascular, and tendons, with a scanty blood supply. Except for cartilage, connective tissues, like epithelial tissues, are supplied with nerves.

Connective Tissue Cells

Embryonic cells called mesenchymal cells give rise to the cells of connective tissues. Each major type of connective tissue contains

Figure 4.7 Functional classification of multicellular exocrine glands.

The functional classification of exocrine glands is based on whether a secretion is a product of a cell or consists of an entire or a partial glandular cell.

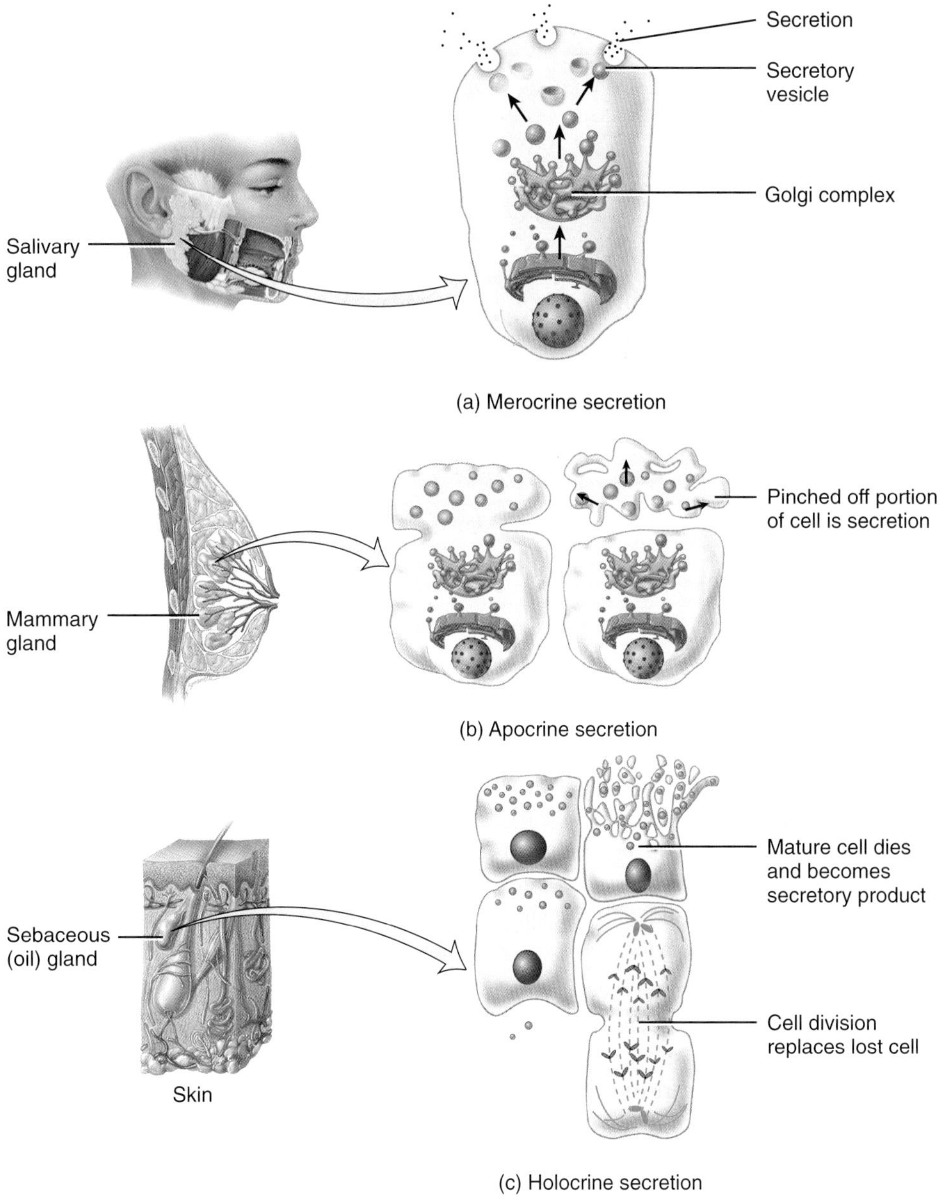

What class of glands are sebaceous (oil) glands? Salivary glands?

an immature class of cells with a name ending in *-blast,* which means "to bud or sprout." These immature cells are called *fibroblasts* in loose and dense connective tissue (described shortly), *chondroblasts* in cartilage, and *osteoblasts* in bone. Blast cells retain the capacity for cell division and secrete the extracellular matrix that is characteristic of the tissue. In cartilage and bone, once the extracellular matrix is produced, the immature cells differentiate into mature cells with names ending in *-cyte,* namely, chondrocytes and osteocytes. Mature cells have reduced capacities for cell division and extracellular matrix formation and are mostly involved in monitoring and maintaining the extracellular matrix.

The types of cells in connective tissues vary according to the type of tissue and include the following (Figure 4.8):

1. **Fibroblasts** (FĪ-brō-blasts; *fibro-* = fibers) are large, flat cells with branching processes. They are present in all the general connective tissues, and usually are the most numerous. Fibroblasts migrate through the connective tissues, secreting the fibers and certain components of the ground substance of the extracellular matrix.

Figure 4.8 Representative cells and fibers present in connective tissues.

Fibroblasts are usually the most numerous connective tissue cells.

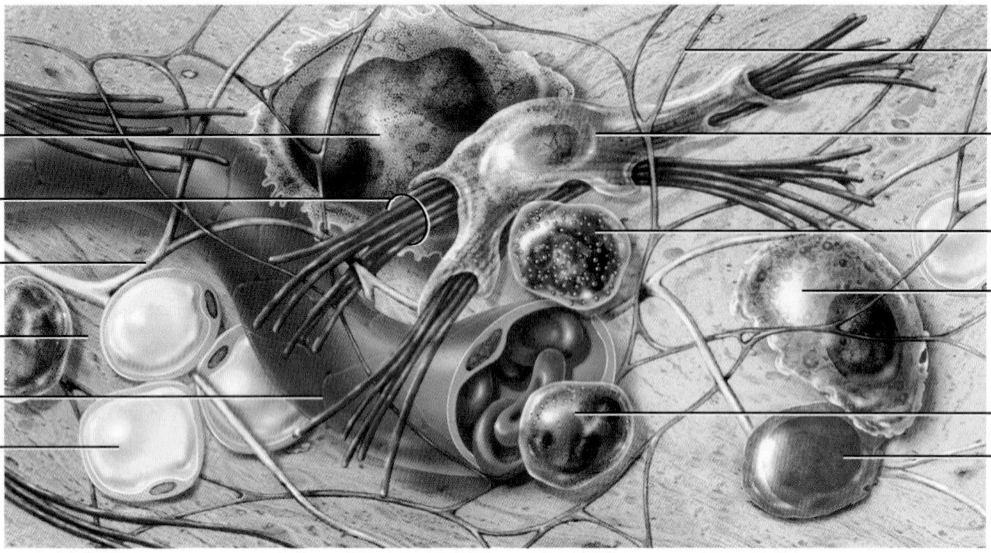

What is the function of fibroblasts?

2. **Macrophages** (MAK-rō-fā-jez; *macro-* = large; *-phages* = eaters) develop from *monocytes,* a type of white blood cell. Macrophages have an irregular shape with short branching projections and are capable of engulfing bacteria and cellular debris by phagocytosis. *Fixed macrophages* reside in a particular tissue; examples include alveolar macrophages in the lungs or splenic macrophages in the spleen. *Wandering macrophages* have the ability to move throughout the tissue and gather at sites of infection or inflammation to carry on phagocytosis.
3. **Plasma cells** are small cells that develop from a type of white blood cell called a *B lymphocyte*. Plasma cells secrete antibodies, proteins that attack or neutralize foreign substances in the body. Thus, plasma cells are an important part of the body's immune response. Although they are found in many places in the body, most plasma cells reside in connective tissues, especially in the gastrointestinal and respiratory tracts. They are also abundant in the salivary glands, lymph nodes, spleen, and red bone marrow.
4. **Mast cells** are abundant alongside the blood vessels that supply connective tissue. They produce histamine, a chemical that dilates small blood vessels as part of the inflammatory response, the body's reaction to injury or infection. In addition, researchers have recently discovered that mast cells can bind to, ingest, and kill bacteria.
5. **Adipocytes** (AD-i-pō-sīts), also called fat cells or *adipose* cells, are connective tissue cells that store triglycerides (fats). They are found deep to the skin and around organs such as the heart and kidneys.
6. **White blood cells** are not found in significant numbers in normal connective tissues. However, in response to certain conditions they migrate from blood into connective tissues. For example, *neutrophils* gather at sites of infection, and *eosinophils* migrate to sites of parasitic invasions and allergic responses.

Connective Tissue Extracellular Matrix

Each type of connective tissue has unique properties, based on the specific extracellular materials between the cells. The extracellular matrix consists of two major components: (1) the ground substance and (2) the fibers.

Ground Substance

As noted earlier, the **ground substance** is the component of a connective tissue between the cells and fibers. The ground substance may be fluid, semifluid, gelatinous, or calcified. It supports cells, binds them together, stores water, and provides a medium for exchange of substances between the blood and cells. It plays an active role in how tissues develop, migrate, proliferate, and change shape, and in how they carry out their metabolic functions.

Ground substance contains water and an assortment of large organic molecules, many of which are complex combinations of polysaccharides and proteins. The polysaccharides include hyaluronic acid, chondroitin sulfate, dermatan sulfate, and keratan sulfate. Collectively, they are referred to as **glycosaminoglycans** (glī-kōs-a-mē′-nō-GLĪ-kans) or **GAGs.** Except for hyaluronic acid, the GAGs are associated with proteins called **proteoglycans** (prō-tē-ō-GLĪ-kans). The proteoglycans form a core protein and the GAGs project from the protein like the bristles of a brush. One of the most important properties of GAGs is that they trap water, making the ground substance more jellylike.

Hyaluronic acid (hī′-a-loo-RON-ik) is a viscous, slippery substance that binds cells together, lubricates joints, and helps maintain the shape of the eyeballs. White blood cells, sperm cells, and some bacteria produce *hyaluronidase,* an enzyme that breaks apart hyaluronic acid, thus causing the ground substance of connective tissue to become more liquid. The ability to produce hyaluronidase helps white blood cells move more easily through connective tissues to reach sites of infection and aids penetration of an oocyte by a sperm cell during fertilization. It also accounts for the rapid spread of bacteria through connective tissues. **Chondroitin sulfate** (kon-DROY-tin) provides support and adhesiveness in cartilage, bone, skin, and blood vessels. The skin, tendons, blood vessels, and heart valves contain **dermatan sulfate;** bone, cartilage, and the cornea of the eye contain **keratan sulfate.** Also present in the ground substance are **adhesion proteins,** which are responsible for linking components of the ground substance to one another and to the surfaces of cells. The main adhesion protein of connective tissues is **fibronectin,** which binds to both collagen fibers (discussed shortly) and ground substance, linking them together. Fibronectin also attaches cells to the ground substance.

CLINICAL CONNECTION | *Chondroitin Sulfate, Glucosamine, and Joint Disease*

In recent years, **chondroitin sulfate** and **glucosamine** (a proteoglycan) have been used as nutritional supplements either alone or in combination to promote and maintain the structure and function of joint cartilage, to provide pain relief from osteoarthritis, and to reduce joint inflammation. Although these supplements have benefited some individuals with moderate to severe osteoarthritis, the benefit is minimal in lesser cases. More research is needed to determine how they act and why they help some people and not others. •

Fibers

Three types of **fibers** are embedded in the extracellular matrix between the cells: collagen fibers, elastic fibers, and reticular fibers (Figure 4.8). They function to strengthen and support connective tissues.

Collagen fibers (KOL-a-jen; *colla* = glue) are very strong and resist pulling forces (tension), but they are not stiff, which allows tissue flexibility. The properties of different types of collagen fibers vary from tissue to tissue. For example, the collagen fibers found in cartilage and bone form different associations with surrounding molecules. As a result of these associations, the collagen fibers in cartilage are surrounded by more water molecules than those in bone, which gives cartilage a more cushioning effect. Collagen fibers often occur in parallel bundles (see Table 4.5, dense regular connective tissue). The bundle arrangement adds great tensile strength to the tissue. Chemically, collagen fibers consist of the protein *collagen,* which is the most abundant protein in your body, representing about 25% of the total. Collagen fibers are found in most types of connective tissues, especially bone, cartilage, tendons (which attach muscle to bone), and ligaments (which attach bone to bone).

Elastic fibers, which are smaller in diameter than collagen fibers, branch and join together to form a fibrous network within a connective tissue. An elastic fiber consists of molecules of the protein *elastin* surrounded by a glycoprotein named *fibrillin,* which adds strength and stability. Because of their unique molecular structure, elastic fibers are strong but can be stretched up to 150% of their relaxed length without breaking. Equally important, elastic fibers have the ability to return to their original shape after being stretched, a property called *elasticity.* Elastic fibers are plentiful in skin, blood vessel walls, and lung tissue.

Reticular fibers (*reticul-* = net), consisting of *collagen* arranged in fine bundles with a coating of glycoprotein, provide support in the walls of blood vessels and form a network around the cells in some tissues, such as areolar connective tissue (a-RĒ-ō-lar; *areol-* = small space), adipose tissue, nerve fibers, and smooth muscle tissue. Produced by fibroblasts, reticular fibers are much thinner than collagen fibers and form branching networks. Like collagen fibers, reticular fibers provide support and strength. Reticular fibers are plentiful in reticular connective tissue, which forms the **stroma** (STRŌ-ma = bed or covering) or supporting framework of many soft organs, such as the spleen and lymph nodes. These fibers also help form the basement membrane.

CLINICAL CONNECTION | *Marfan Syndrome*

Marfan syndrome (MAR-fan) is an inherited disorder caused by a defective fibrillin gene. The result is abnormal development of elastic fibers. Tissues rich in elastic fibers are malformed or weakened. Structures affected most seriously are the covering layer of bones (periosteum), the ligament that suspends the lens of the eye, and the walls of the large arteries. People with Marfan syndrome tend to be tall and have disproportionately long arms, legs, fingers, and toes. A common symptom is blurred vision caused by displacement of the lens of the eye. The most life-threatening complication of Marfan syndrome is weakening of the aorta (the main artery that emerges from the heart), which can suddenly burst. •

Classification of Connective Tissues

Because of the diversity of cells and extracellular matrix and the differences in their relative proportions, the classification of connective tissues is not always clear-cut and several classifications exist. We offer the following classification scheme:

I. Embryonic connective tissues
   - A. Mesenchyme
   - B. Mucous connective tissue

II. Mature connective tissues
   - A. Loose connective tissues
     1. Areolar connective tissue
     2. Adipose tissue
     3. Reticular connective tissue
   - B. Dense connective tissues
     1. Dense regular connective tissue
     2. Dense irregular connective tissue
     3. Elastic connective tissue

C. Cartilage
  1. Hyaline cartilage
  2. Fibrocartilage
  3. Elastic cartilage
D. Bone tissue
E. Liquid connective tissue
  1. Blood tissue
  2. Lymph

Embryonic Connective Tissues

Note that our classification scheme has two major subclasses of connective tissues: embryonic and mature. **Embryonic connective tissues** are present primarily in the *embryo,* the developing human from fertilization through the first two months of pregnancy, and in the *fetus,* the developing human from the third month of pregnancy to birth (Table 4.3).

TABLE 4.3

Embryonic Connective Tissues

A. MESENCHYME

Description	Irregularly shaped mesenchymal cells embedded in semifluid ground substance that contains delicate reticular fibers.
Location	Almost exclusively under skin and along developing bones of embryo; some in adult connective tissue, especially along blood vessels.
Function	Forms almost all other types of connective tissue.

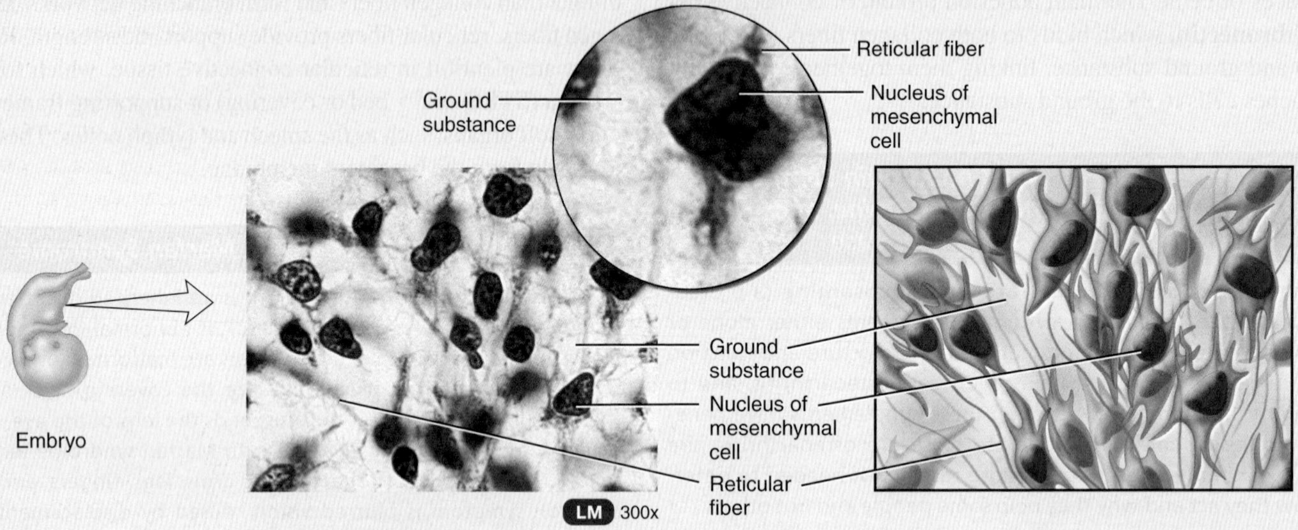
LM 1000x
Reticular fiber
Ground substance
Nucleus of mesenchymal cell
Embryo
Ground substance
Nucleus of mesenchymal cell
Reticular fiber
LM 300x

Sectional view of mesenchyme of a developing embryo

Mesenchyme

B. MUCOUS CONNECTIVE TISSUE

Description	Widely scattered fibroblasts embedded in viscous, jellylike ground substance that contains fine collagen fibers.
Location	Umbilical cord of fetus.
Function	Support.

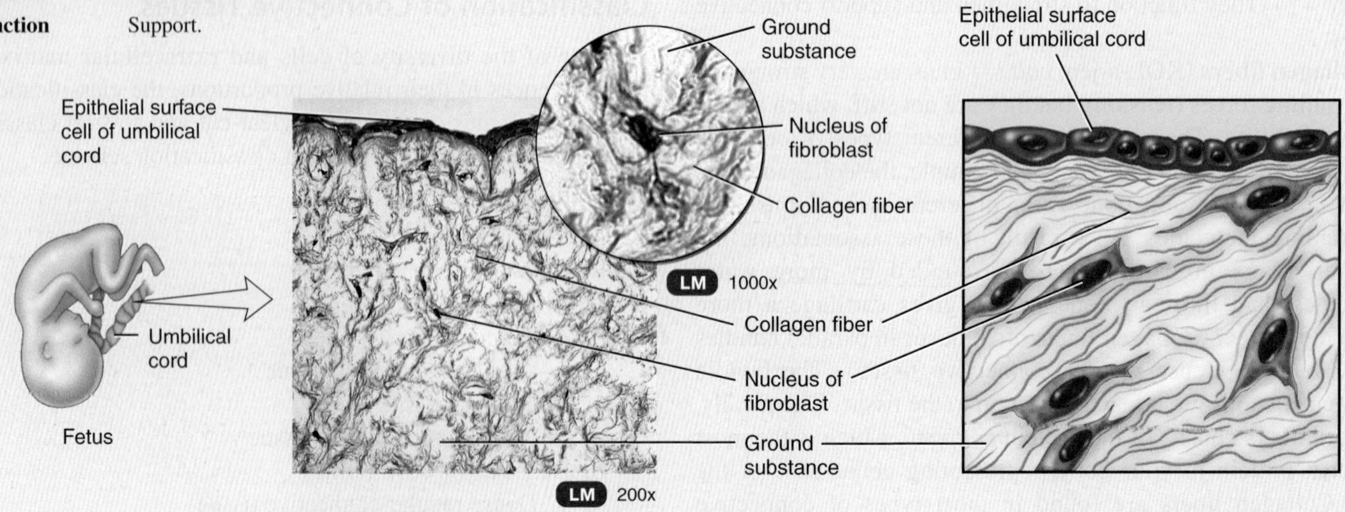
Ground substance
Epithelial surface cell of umbilical cord
Epithelial surface cell of umbilical cord
Nucleus of fibroblast
Collagen fiber
LM 1000x
Umbilical cord
Fetus
Collagen fiber
Nucleus of fibroblast
Ground substance
LM 200x

Sectional view of mucous connective tissue of the umbilical cord

Mucous connective tissue

Mature Connective Tissues

The second major subclass of connective tissues, **mature connective tissues,** are present in the newborn. Their cells arise primarily from mesenchyme. In the next section we explore the numerous types of mature connective tissues. The five types of mature connective tissues are (1) loose connective tissues, (2) dense connective tissues, (3) cartilage, (4) bone tissue, and (5) liquid connective tissue (blood tissue and lymph). We now examine each in detail.

Loose Connective Tissues

The fibers of **loose connective tissues** are *loosely* arranged between cells. The types of loose connective tissues are areolar connective tissue, adipose tissue, and reticular connective tissue (Table 4.4).

CLINICAL CONNECTION | *Liposuction*

A surgical procedure called **liposuction** (LIP-ō-suk′-shun; *lip-* = fat) or **suction lipectomy** (*-ectomy* = to cut out) involves suctioning out small amounts of adipose tissue from various areas of the body. After an incision is made in the skin, the fat is removed through a stainless steel tube, called a cannula, with the assistance of a powerful vacuum pressure unit that suctions out the fat. The technique can be used as a body-contouring procedure in regions such as the thighs, buttocks, arms, breasts, and abdomen, and to transfer fat to another area of the body. Postsurgical complications that may develop include fat that may enter blood vessels broken during the procedure and obstruct blood flow, infection, loss of feeling in the area, fluid depletion, injury to internal structures, and severe postoperative pain. •

Dense Connective Tissues

Dense connective tissues contain more fibers, which are thicker and more *densely* packed, but have considerably fewer cells than loose connective tissues. There are three types: dense regular connective tissue, dense irregular connective tissue, and elastic connective tissue (Table 4.5).

Cartilage

Cartilage (KAR-ti-lij) consists of a dense network of collagen fibers and elastic fibers firmly embedded in chondroitin sulfate, a gel-like component of the ground substance. Cartilage can endure considerably more stress than loose and dense connective tissues. The strength of cartilage is due to its collagen fibers, and its *resilience* (ability to assume its original shape after deformation) is due to chondroitin sulfate.

Like other connective tissues, cartilage has few cells and large quantities of extracellular matrix. It differs from other connective tissues, however, in not having nerves or blood vessels in its extracellular matrix. Interestingly, cartilage does not have a blood supply because it secretes an *antiangiogenesis* factor (an′-tē-an′-jē-ō-JEN-e-sis; *anti-* = against; *angio-* = vessel; *-genesis* = production), a substance that prevents blood vessel growth. Because of this property, antiangiogenesis factor is being studied as a possible cancer treatment. If cancer cells can be stopped from promoting new blood vessel growth, their rapid rate of cell division and expansion can be slowed or even halted.

TABLE 4.4

Mature Connective Tissues: Loose Connective Tissues

A. AREOLAR CONNECTIVE TISSUE

Description	One of the most widely distributed connective tissues; consists of fibers (collagen, elastic, reticular) arranged randomly and several kinds of cells (fibroblasts, macrophages, plasma cells, adipocytes, mast cells, and a few white blood cells) embedded in semifluid ground substance (hyaluronic acid, chondroitin sulfate, dermatan sulfate, and keratan sulfate).
Location	In and around nearly every body structure (thus, called "packing material" of the body): in subcutaneous layer deep to skin; papillary (superficial) region of dermis of skin; lamina propria of mucous membranes; around blood vessels, nerves, and body organs.
Function	Strength, elasticity, support.

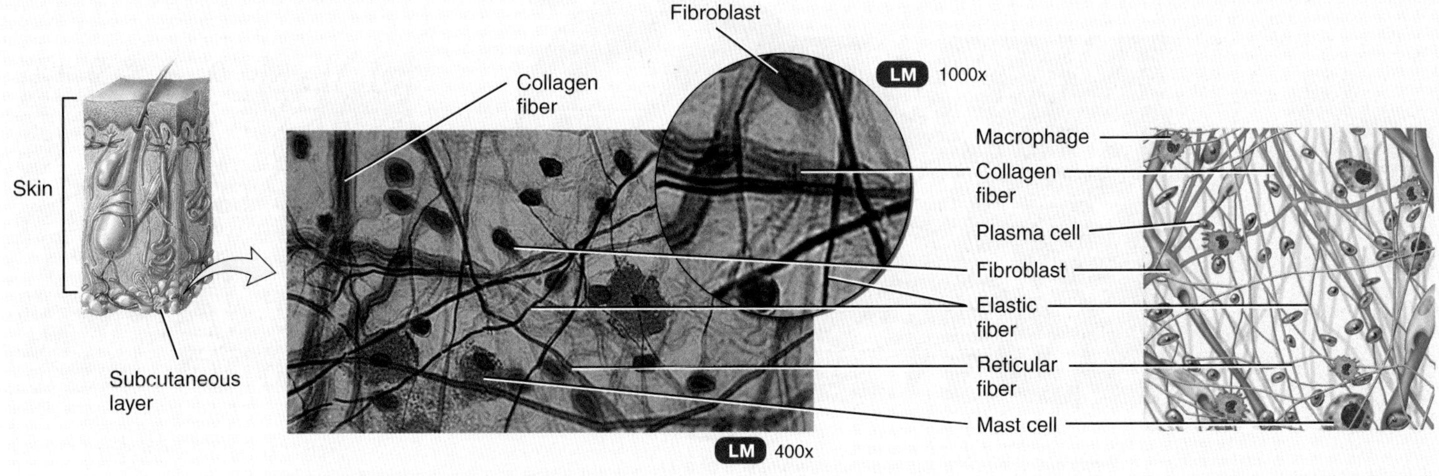

Sectional view of subcutaneous areolar connective tissue

Areolar connective tissue

TABLE 4.4 CONTINUES

TABLE 4.4 CONTINUED

Mature Connective Tissues: Loose Connective Tissues

B. ADIPOSE TISSUE

Description Has cells derived from fibroblasts (called **adipocytes**) that are specialized for storage of triglycerides (fats) as a large, centrally located droplet. Cell fills up with a single, large triglyceride droplet, and cytoplasm and nucleus are pushed to periphery of cell. With weight gain, amount of adipose tissue increases and new blood vessels form. Thus, an obese person has many more blood vessels than does a lean person, a situation that can cause high blood pressure, since the heart has to work harder. Most adipose tissue in adults is *white adipose tissue* (just described). *Brown adipose tissue* (BAT) is darker due to very rich blood supply and numerous pigmented mitochondria that participate in aerobic cellular respiration. BAT is widespread in the fetus and infant; adults have only small amounts.

Location Wherever areolar connective tissue is located: subcutaneous layer deep to skin, around heart and kidneys, yellow bone marrow, padding around joints and behind eyeball in eye socket.

Function Reduces heat loss through skin; serves as an energy reserve; supports and protects organs. In newborns, BAT generates heat to maintain proper body temperature.

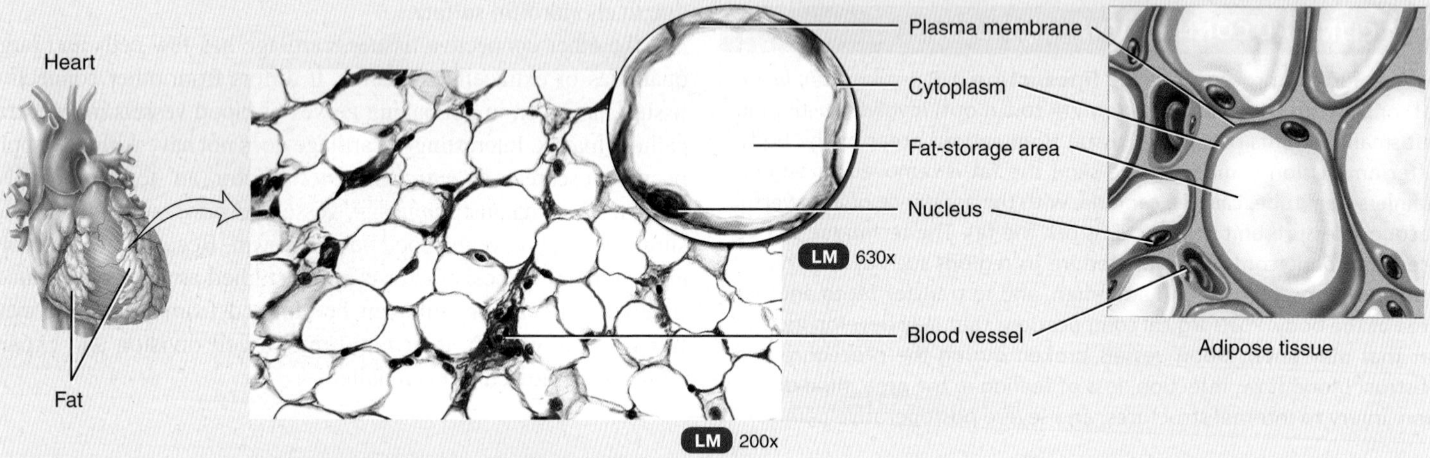

Sectional view of adipose tissue showing adipocytes of white fat and details of an adipocyte

Adipose tissue

C. RETICULAR CONNECTIVE TISSUE

Description Fine interlacing network of reticular fibers (thin form of collagen fiber) and reticular cells.

Location Stroma (supporting framework) of liver, spleen, lymph nodes; red bone marrow; reticular lamina of basement membrane; around blood vessels and muscles.

Function Forms stroma of organs; binds smooth muscle tissue cells; filters and removes worn-out blood cells in spleen and microbes in lymph nodes.

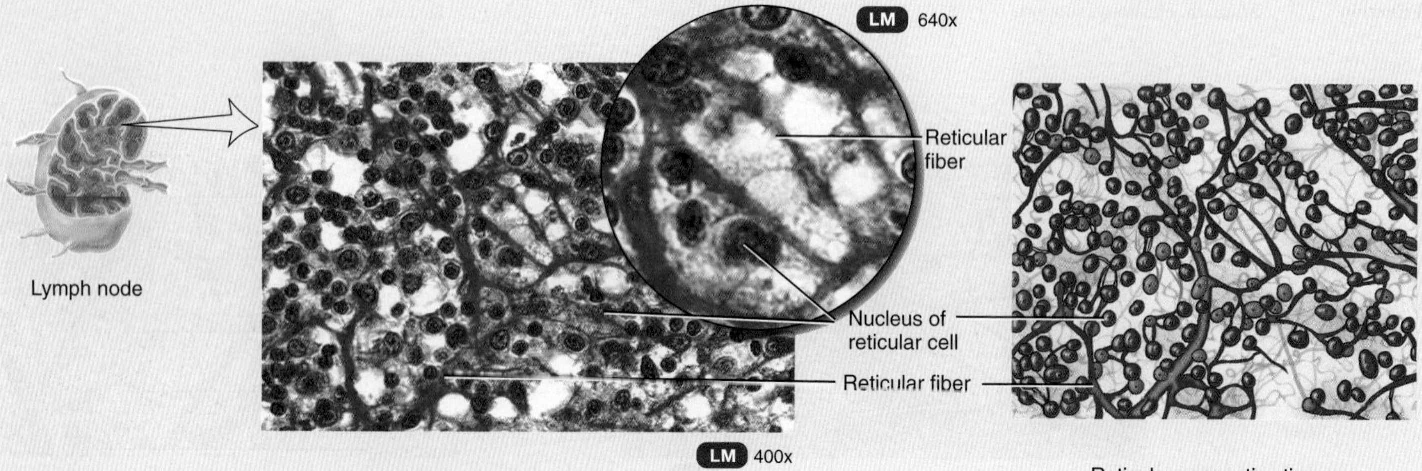

Sectional view of reticular connective tissue of a lymph node

Reticular connective tissue

TABLE 4.5

Mature Connective Tissues: Dense Connective Tissues

D. DENSE REGULAR CONNECTIVE TISSUE

Description Shiny white extracellular matrix; mainly collagen fibers *regularly* arranged in bundles with fibroblasts in rows between bundles. Collagen fibers are not living (are protein structures secreted by fibroblasts), so damaged tendons and ligaments heal very slowly.

Location Forms tendons (attach muscle to bone), most ligaments (attach bone to bone), and aponeuroses (sheetlike tendons that attach muscle to muscle or muscle to bone).

Function Provides strong attachment between various structures. Tissue structure withstands pulling (tension) along long axis of fibers.

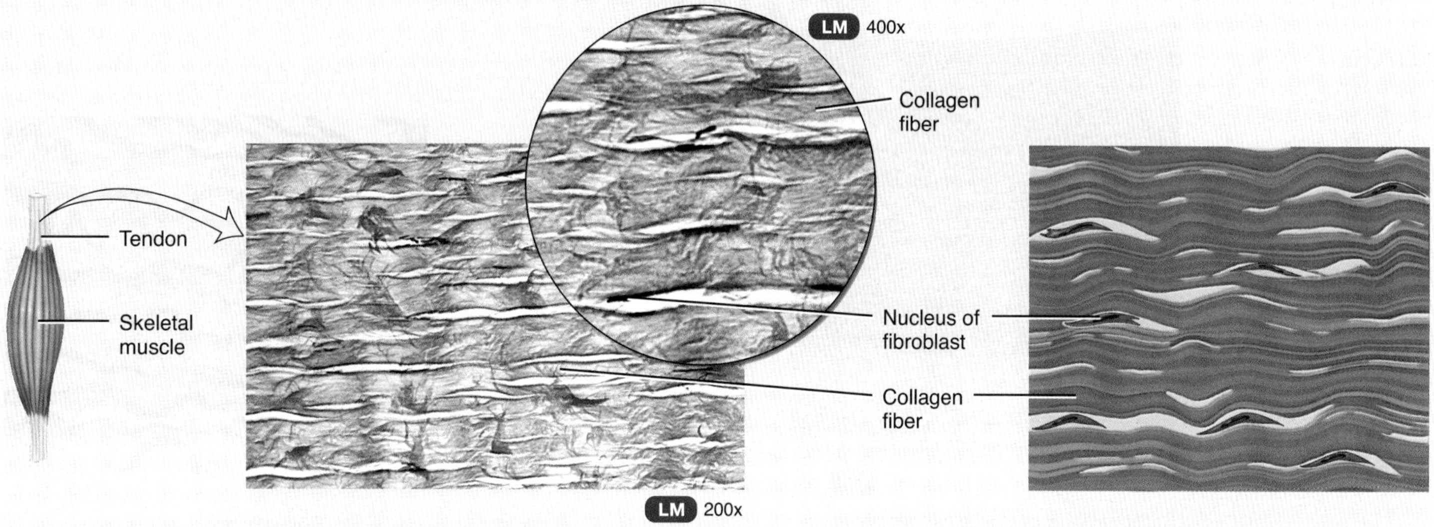

Sectional view of dense regular connective tissue of a tendon

Dense regular connective tissue

E. DENSE IRREGULAR CONNECTIVE TISSUE

Description Collagen fibers; usually *irregularly* arranged with a few fibroblasts.

Location Often occurs in sheets, such as fasciae (tissue beneath skin and around muscles and other organs), reticular (deeper) region of dermis of skin, fibrous pericardium of heart, periosteum of bone, perichondrium of cartilage, joint capsules, membrane capsules around various organs (kidneys, liver, testes, lymph nodes); also in heart valves.

Function Provides tensile (pulling) strength in many directions.

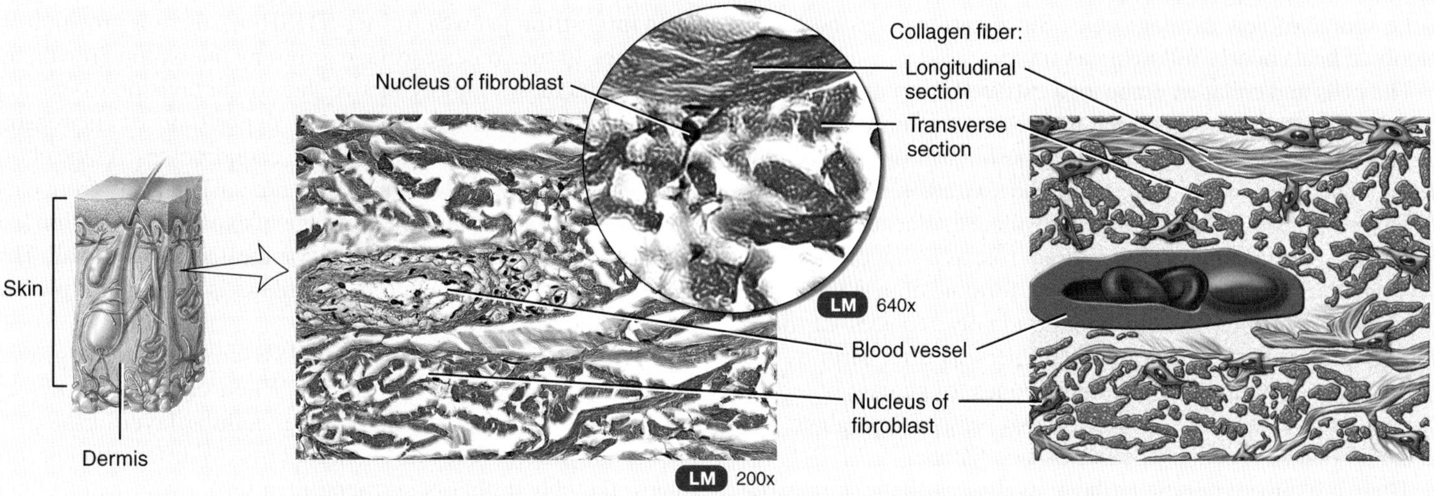

Sectional view of dense irregular connective tissue of reticular region of dermis

Dense irregular connective tissue

TABLE 4.5 CONTINUES

TABLE 4.5 CONTINUED

Mature Connective Tissues: Dense Connective Tissues

F. ELASTIC CONNECTIVE TISSUE

Description	Predominantly elastic fibers with fibroblasts between fibers; unstained tissue is yellowish.
Location	Lung tissue, walls of elastic arteries, trachea, bronchial tubes, true vocal cords, suspensory ligaments of penis, some ligaments between vertebrae.
Function	Allows stretching of various organs; is strong and can recoil to original shape after being stretched. Elasticity is important to normal functioning of lung tissue (recoils in exhaling) and elastic arteries (recoil between heartbeats to help maintain blood flow).

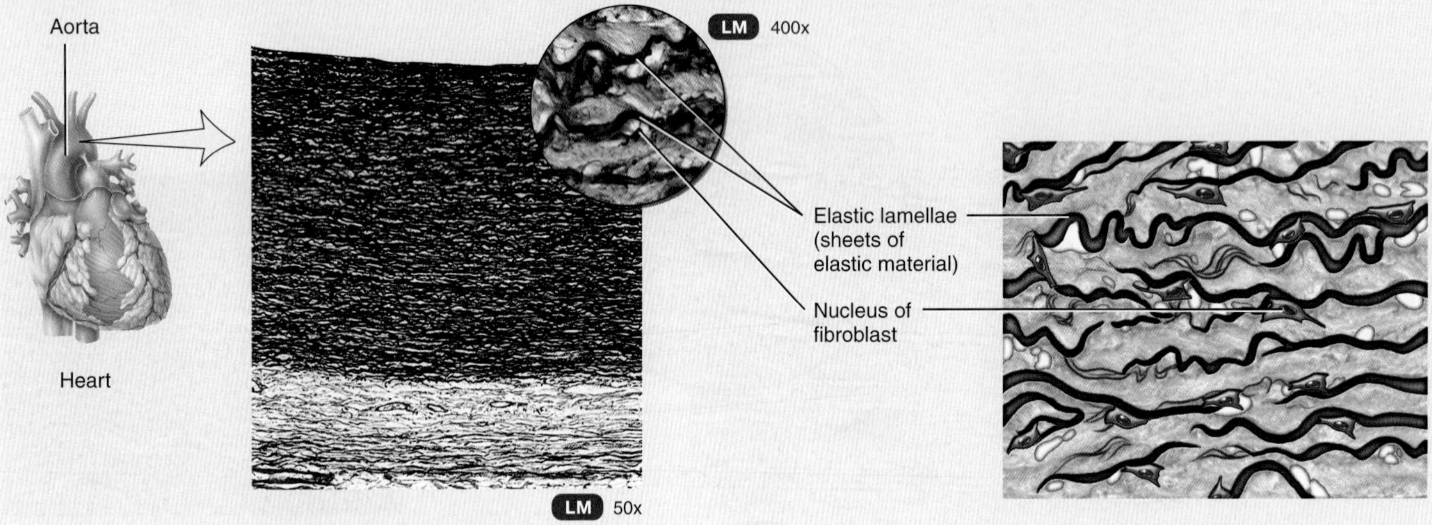

Sectional view of elastic connective tissue of aorta

Elastic connective tissue

The cells of mature cartilage, called **chondrocytes** (KON-drō-sīts; *chondro-* = cartilage), occur singly or in groups within spaces called **lacunae** (la-KOO-nē = little lakes; singular is *lacuna,* pronounced la-KOO-na) in the extracellular matrix. A covering of dense irregular connective tissue called the **perichondrium** (per′-i-KON-drē-um; *peri-* = around) surrounds the surface of most cartilage and contains blood vessels and nerves and is the source of new cartilage cells. Since cartilage has no blood supply, it heals poorly following an injury.

The cells and collagen-embedded extracellular matrix of cartilage form a strong, firm material that resists tension (stretching), compression (squeezing), and shear (pushing in opposite directions). The chondroitin sulfate in the extracellular matrix is largely responsible for cartilage's resilience. Because of these properties, cartilage plays an important role as a support tissue in the body. It is also a precursor to bone, forming almost the entire embryonic skeleton. Though bone gradually replaces cartilage during further development, cartilage persists after birth as the growth plates within bone that allow bones to increase in length during the growing years. Cartilage also persists throughout life as the lubricated articular surfaces of most joints.

There are three types of cartilage: hyaline cartilage, fibrocartilage, and elastic cartilage (Table 4.6).

Repair and Growth of Cartilage

Metabolically, cartilage is a relatively inactive tissue that grows slowly. When injured or inflamed, cartilage repair proceeds slowly, in large part because cartilage is avascular. Substances needed for repair and blood cells that participate in tissue repair must diffuse or migrate into the cartilage. The growth of cartilage follows two basic patterns: interstitial growth and appositional growth.

In **interstitial growth** (in′-ter-STISH-al), there is growth from within the tissue. When cartilage grows by interstitial growth, the cartilage increases rapidly in size due to the division of existing chondrocytes and the continuous deposition of increasing amounts of extracellular matrix by the chondrocytes. As the chondrocytes synthesize new matrix, they are pushed away from each other. These events cause the cartilage to expand from within like bread rising, which is the reason for the term *inter*stitial. This growth pattern occurs while the cartilage is young and pliable, during childhood and adolescence.

In **appositional growth** (ap-ō-ZISH-un-al), there is growth at the outer surface of the tissue. When cartilage grows by appositional growth, cells in the inner cellular layer of the perichondrium differentiate into chondroblasts. As differentiation continues, the chondroblasts surround themselves with extracellular matrix and become chondrocytes. As a result, matrix accumulates

TABLE 4.6

Mature Connective Tissues: Cartilage

A. HYALINE CARTILAGE

Description	Hyaline (*hyalinos* = glassy) cartilage contains a resilient gel as ground substance and appears in the body as a bluish-white, shiny substance (can stain pink or purple when prepared for microscopic examination). Fine collagen fibers are not visible with ordinary staining techniques; prominent chondrocytes are found in lacunae surrounded by perichondrium (exceptions: articular cartilage in joints and cartilage of epiphyseal plates, where bones lengthen during growth).
Location	Most abundant cartilage in body; at ends of long bones, anterior ends of ribs, nose, parts of larynx, trachea, bronchi, bronchial tubes, embryonic and fetal skeleton.
Function	Provides smooth surfaces for movement at joints, flexibility, and support; weakest type of cartilage.

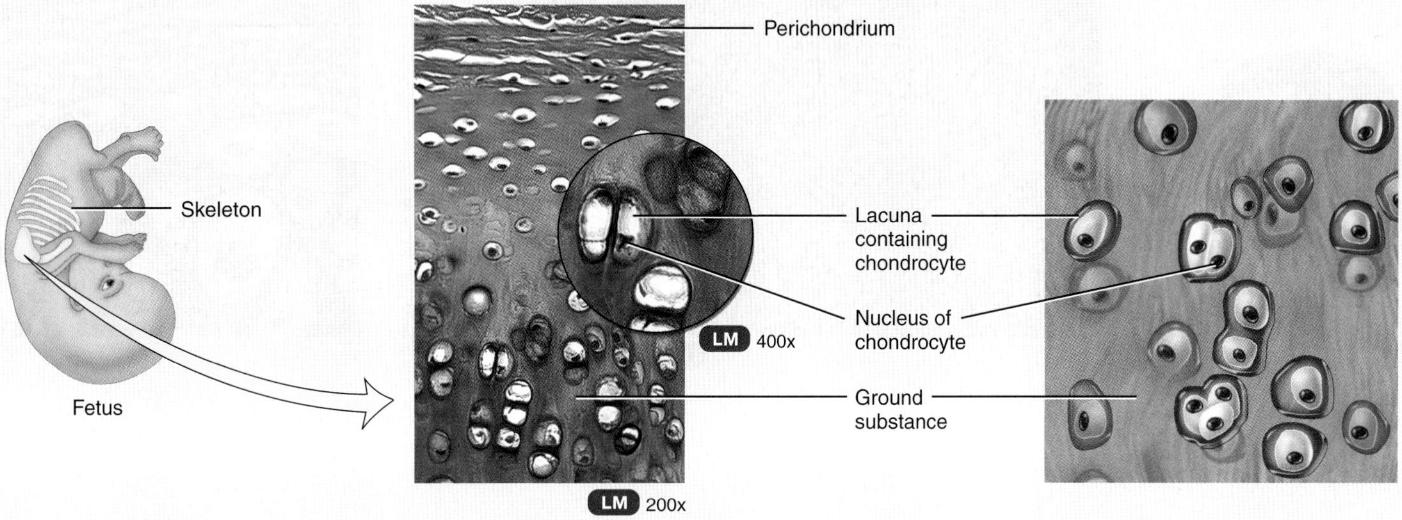

Sectional view of hyaline cartilage of a developing fetal bone

Hyaline cartilage

B. FIBROCARTILAGE

Description	Chondrocytes scattered among clearly visible thick bundles of collagen fibers within extracellular matrix; lacks perichondrium.
Location	Pubic symphysis (where hip bones join anteriorly), intervertebral discs, menisci (cartilage pads) of knee, portions of tendons that insert into cartilage.
Function	Support and joining structures together. Strength and rigidity make it the strongest type of cartilage.

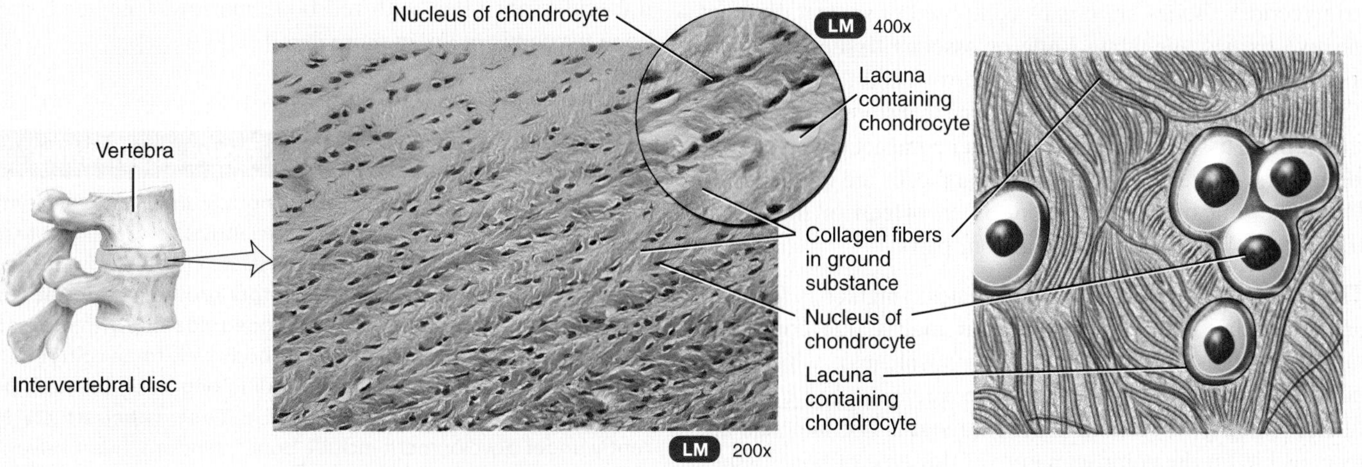

Sectional view of fibrocartilage of intervertebral disc

Fibrocartilage

TABLE 4.6 CONTINUES

TABLE 4.6 CONTINUED

Mature Connective Tissues: Cartilage

C. ELASTIC CARTILAGE

Description Chondrocytes in threadlike network of elastic fibers within extracellular matrix; perichondrium present.

Location Lid on top of larynx (epiglottis), part of external ear (auricle), auditory (eustachian) tubes.

Function Provides strength and elasticity; maintains shape of certain structures.

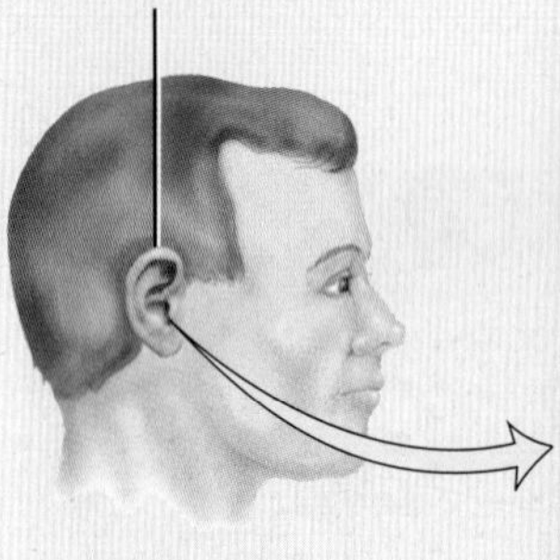

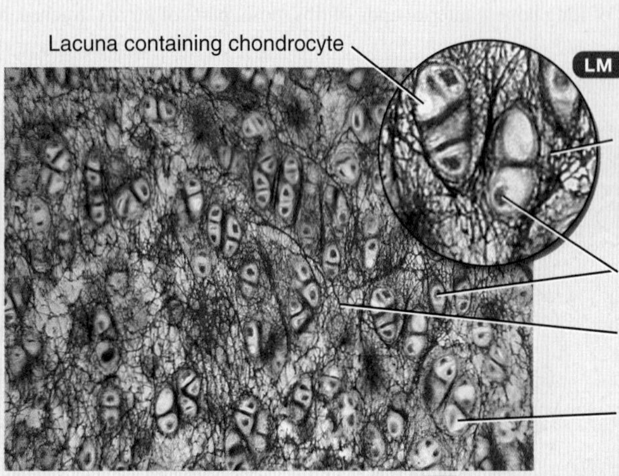

Sectional view of elastic cartilage of auricle of ear

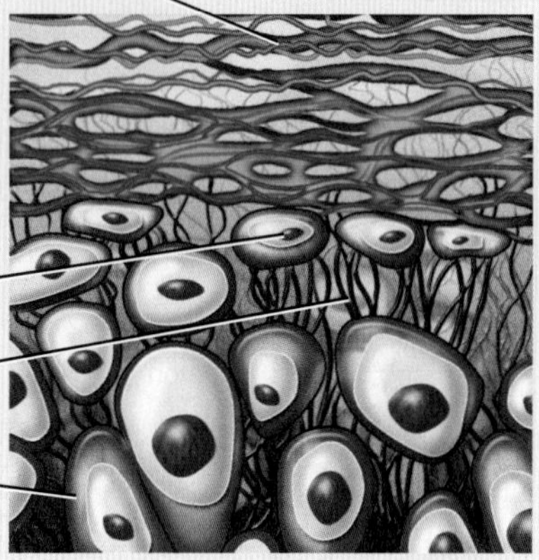

Elastic cartilage

beneath the perichondrium on the outer surface of the cartilage, causing it to grow in width. Appositional growth starts later than interstitial growth and continues through adolescence.

Bone Tissue

Cartilage, joints, and bones make up the skeletal system. The skeletal system supports soft tissues, protects delicate structures, and works with skeletal muscles to generate movement. Bones store calcium and phosphorus; house red bone marrow, which produces blood cells; and contain yellow bone marrow, a storage site for triglycerides. Bones are organs composed of several different connective tissues, including **bone** or **osseous tissue** (OS-ē-us), the periosteum, red and yellow bone marrow, and the endosteum (a membrane that lines a space within bone that stores yellow bone marrow). Bone tissue is classified as either compact or spongy, depending on how its extracellular matrix and cells are organized.

The basic unit of **compact bone** is an **osteon** or **haversian system** (Table 4.7). Each osteon has four parts:

1. The **lamellae** (la-MEL-lē = little plates; singular is *lamella*) are concentric rings of extracellular matrix that consist of mineral salts (mostly calcium and phosphates), which give bone its hardness and compressive strength, and collagen fibers, which give bone its tensile strength. The lamellae are responsible for the compact nature of this type of bone tissue.
2. **Lacunae** (la-KOO-nē; singular is *lacuna,* pronounced la-KOO-na) are small spaces between lamellae that contain mature bone cells called **osteocytes.**
3. Projecting from the lacunae are **canaliculi** (kan-a-LIK-ū-lī = little canals), networks of minute canals containing the processes of osteocytes. Canaliculi provide routes for nutrients to reach osteocytes and for wastes to leave them.
4. A **central (haversian) canal** contains blood vessels and nerves.

Spongy bone lacks osteons. Rather, it consists of columns of bone called **trabeculae** (tra-BEK-ū-lē = little beams), which contain lamellae, osteocytes, lacunae, and canaliculi. Spaces between trabeculae are filled with red bone marrow. Chapter 6 presents bone tissue histology in more detail.

CLINICAL CONNECTION | *Tissue Engineering*

Tissue engineering is a technology that combines synthetic material with cells and has allowed scientists to grow new tissues in the laboratory to replace damaged tissues in the body. Tissue engineers have already developed laboratory-grown versions of skin and cartilage using scaffolding beds of biodegradable synthetic materials or collagen as substrates that permit body cells to be cultured. As the cells divide and assemble, the scaffolding degrades; the new, permanent tissue is then implanted in the patient. Other structures currently under development include bones, tendons, heart valves, bone marrow, and intestines. Work is also under way to develop insulin-producing cells for diabetics, dopamine-producing cells for Parkinson disease patients, and even entire livers and kidneys. •

TABLE 4.7

Mature Connective Tissues: Bone Tissue

Description	Compact bone tissue consists of osteons (haversian systems) that contain lamellae, lacunae, osteocytes, canaliculi, and central (haversian) canals. By contrast, spongy bone tissue (see Figure 6.3a, b) consists of thin columns called trabeculae; spaces between trabeculae are filled with red bone marrow.
Location	Both compact and spongy bone tissue make up the various parts of bones of the body.
Function	Support, protection, storage; houses blood-forming tissue; serves as levers that act with muscle tissue to enable movement.

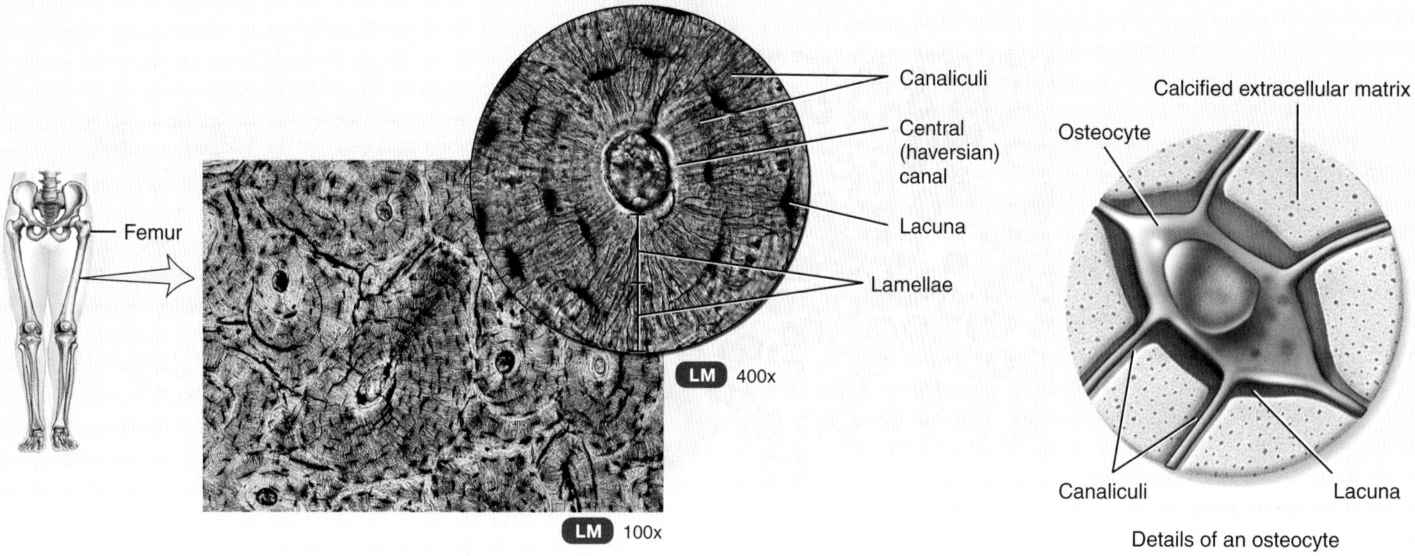
Femur
Canaliculi
Central (haversian) canal
Lacuna
Lamellae
LM 400x
LM 100x
Calcified extracellular matrix
Osteocyte
Canaliculi
Lacuna

Sectional view of several osteons (haversian systems) of femur (thigh bone)

Details of an osteocyte

Liquid Connective Tissue

Blood Tissue **Blood tissue** (or simply **blood**) is a connective tissue with a liquid extracellular matrix and formed elements. The extracellular matrix is called blood plasma. The **blood plasma** is a pale yellow fluid that consists mostly of water with a wide variety of dissolved substances—nutrients, wastes, enzymes, plasma proteins, hormones, respiratory gases, and ions. Suspended in the blood plasma are **formed elements**—red blood cells (erythrocytes), white blood cells (leukocytes), and platelets (thrombocytes) (Table 4.8). **Red blood cells** transport oxygen to body cells and remove some carbon dioxide from them. **White blood cells** are involved in phagocytosis, immunity, and allergic reactions. **Platelets** (PLĀT-lets) participate in blood clotting. The details of blood are considered in Chapter 19.

Lymph **Lymph** is the extracellular fluid that flows in lymphatic vessels. It is a connective tissue that consists of several types of cells in a clear liquid extracellular matrix that is similar to blood plasma but with much less protein. The composition of lymph varies from one part of the body to another. For example, lymph leaving lymph nodes includes many lymphocytes, a type of white blood cell, in contrast to lymph from the small intestine, which has a high content of newly absorbed dietary lipids. The details of lymph are considered in Chapter 22.

✔ **CHECKPOINT**

11. In what ways do connective tissues differ from epithelia?
12. What are the features of the cells, ground substance, and fibers that make up connective tissues?
13. How are connective tissues classified? List the various types.
14. Describe how the structures of the following connective tissues are related to their functions: areolar connective tissue, adipose tissue, reticular connective tissue, dense regular connective tissue, dense irregular connective tissue, elastic connective tissue, hyaline cartilage, fibrocartilage, elastic cartilage, bone tissue, blood tissue, and lymph.
15. What is the difference between interstitial and appositional growth of cartilage?

4.6 MEMBRANES

OBJECTIVES

- Define a membrane.
- Describe the classification of membranes.

Membranes are flat sheets of pliable tissue that cover or line a part of the body. The majority of membranes consist of an epithelial layer and an underlying connective tissue layer and are called **epithelial membranes.** The principal epithelial membranes of the

TABLE 4.8

Mature Connective Tissues: Blood

Description	Blood plasma and formed elements: red blood cells (erythrocytes), white blood cells (leukocytes), platelets (thrombocytes).
Location	Within blood vessels (arteries, arterioles, capillaries, venules, veins), within chambers of heart.
Function	Red blood cells: transport oxygen and some carbon dioxide; white blood cells: carry on phagocytosis and mediate allergic reactions and immune system responses; platelets: essential for blood clotting.

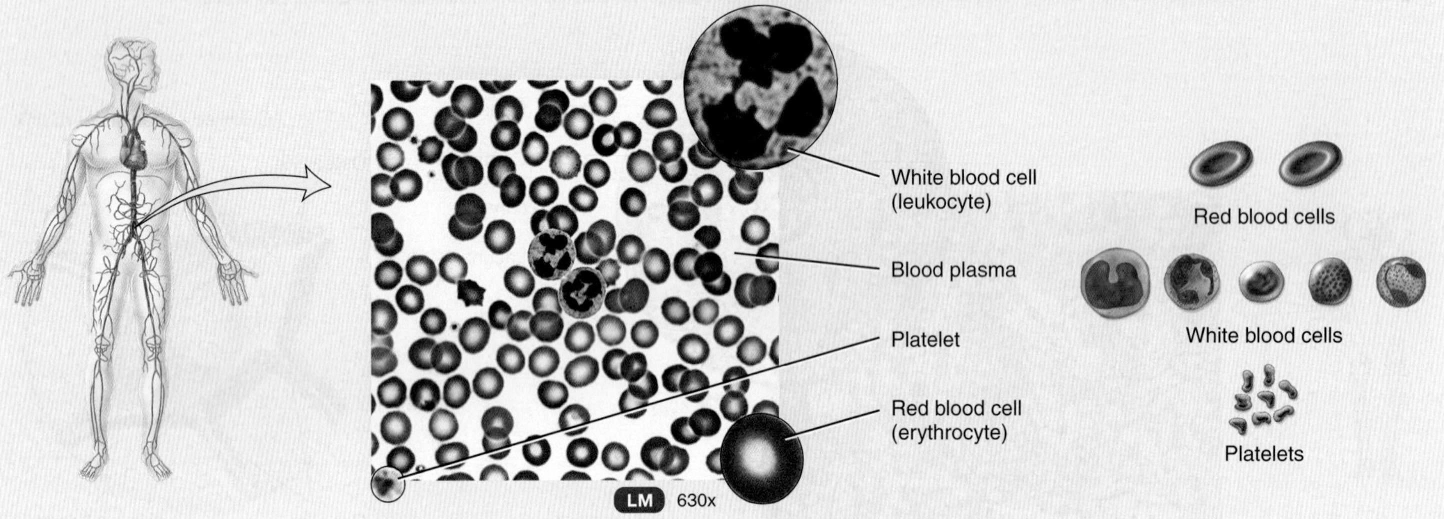
White blood cell
(leukocyte)
Blood plasma
Platelet
Red blood cell
(erythrocyte)
LM 630x
Red blood cells
White blood cells
Platelets

Blood in blood vessels

Blood smear (all enlargements are 1500x)

body are mucous membranes, serous membranes, and the cutaneous membrane, or skin. Another type of membrane, a synovial membrane, lines joints and contains connective tissue but no epithelium.

Epithelial Membranes

Mucous Membranes

A **mucous membrane** or **mucosa** (mū-KŌ-sa) lines a body cavity that opens directly to the exterior. Mucous membranes line the entire digestive, respiratory, and reproductive tracts, and much of the urinary tract. They consist of a lining layer of epithelium and an underlying layer of connective tissue (Figure 4.9a).

The epithelial layer of a mucous membrane is an important feature of the body's defense mechanisms because it is a barrier that microbes and other pathogens have difficulty penetrating. Usually, tight junctions connect the cells, so materials cannot leak in between them. Goblet cells and other cells of the epithelial layer of a mucous membrane secrete mucus, and this slippery fluid prevents the cavities from drying out. It also traps particles in the respiratory passageways and lubricates food as it moves through the gastrointestinal tract. In addition, the epithelial layer secretes some of the enzymes needed for digestion and is the site of food and fluid absorption in the gastrointestinal tract. The epithelia of mucous membranes vary greatly in different parts of the body. For example, the mucous membrane of the small intestine is nonciliated simple columnar epithelium, and the large airways to the lungs consist of pseudostratified ciliated columnar epithelium (see Table 4.1).

The connective tissue layer of a mucous membrane is areolar connective tissue and is called the **lamina propria** (LAM-i-na PRŌ-prē-a; *propria* = one's own), so named because it belongs to (is owned by) the mucous membrane. The lamina propria supports the epithelium, binds it to the underlying structures, allows some flexibility of the membrane, and affords some protection for underlying structures. It also holds blood vessels in place and is the vascular source for the overlying epithelium. Oxygen and nutrients diffuse from the lamina propria to the covering epithelium; carbon dioxide and wastes diffuse in the opposite direction.

Serous Membranes

A **serous membrane** (SĒR-us = watery) or **serosa** lines a body cavity that does not open directly to the exterior (thoracic or abdominal cavities), and it covers the organs that are within the cavity. Serous membranes consist of areolar connective tissue covered by mesothelium (simple squamous epithelium) (Figure 4.9b). You will recall from Chapter 1 that serous membranes have two layers: The layer attached to and lining the cavity wall is called the **parietal layer** (pa-RĪ-e-tal; *pariet-* = wall); the layer that covers and adheres to the organs within the cavity is the **visceral layer** (*viscer-* = body organ) (see Figure 1.10a). The mesothelium of a serous membrane secretes **serous fluid,** a watery lubricant that allows organs to glide easily over one another or to slide against the walls of cavities.

Recall from Chapter 1 that the serous membrane lining the thoracic cavity and covering the lungs is the **pleura.** The serous membrane lining the heart cavity and covering the heart is the

Figure 4.9 Membranes.

A membrane is a flat sheet of pliable tissues that covers or lines a part of the body.

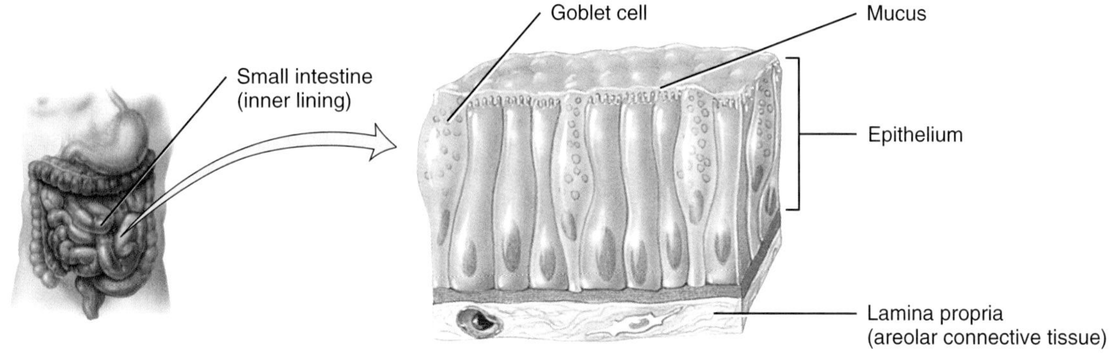
Goblet cell
Mucus
Small intestine
(inner lining)
Epithelium
Lamina propria
(areolar connective tissue)

(a) Mucous membrane

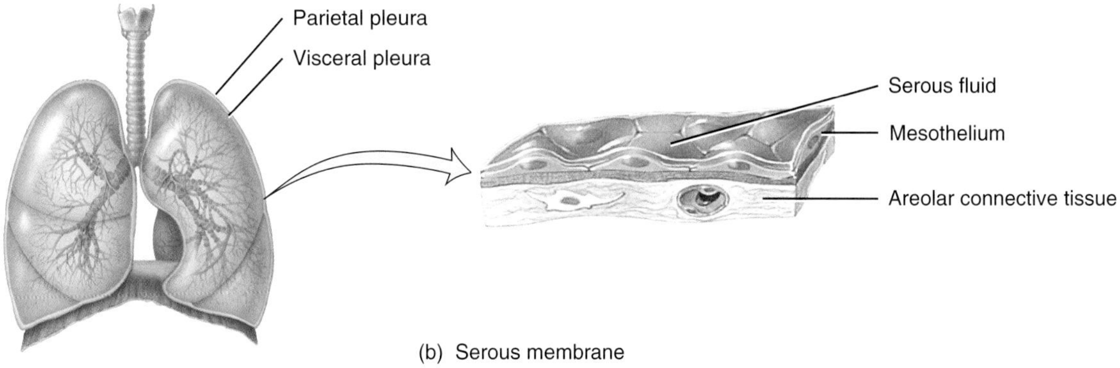
Parietal pleura
Visceral pleura
Serous fluid
Mesothelium
Areolar connective tissue

(b) Serous membrane

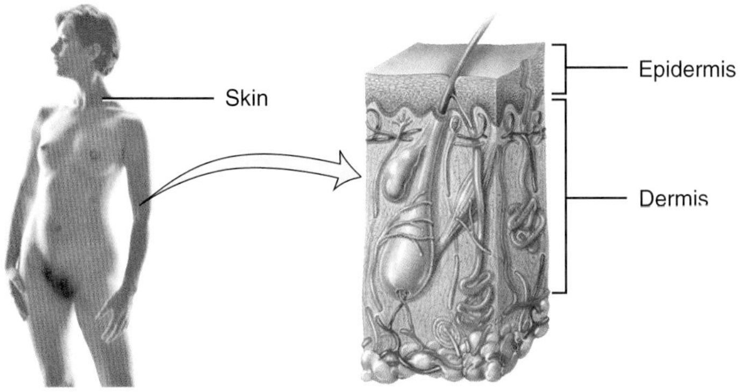
Skin
Epidermis
Dermis

(c) Skin (cutaneous membrane)

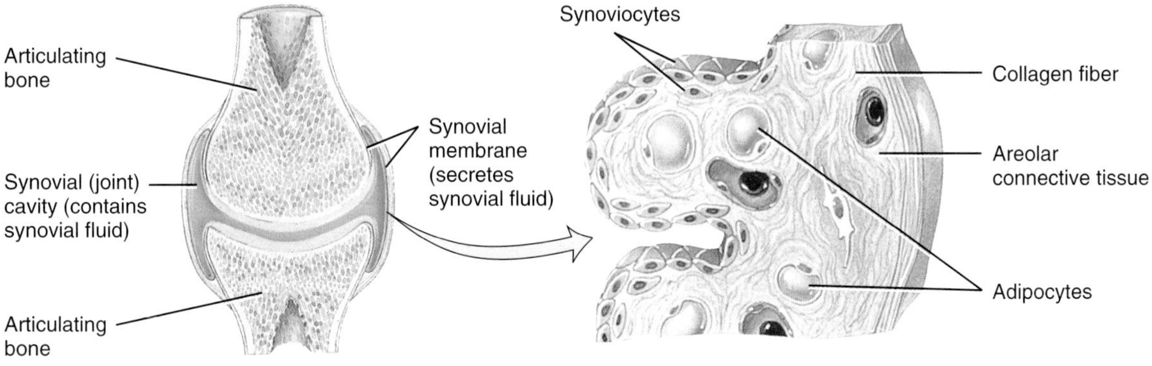
Articulating
bone
Synovial (joint)
cavity (contains
synovial fluid)
Articulating
bone
Synovial
membrane
(secretes
synovial fluid)
Synoviocytes
Collagen fiber
Areolar
connective tissue
Adipocytes

(d) Synovial membrane

What is an epithelial membrane?

pericardium. The serous membrane lining the abdominal cavity and covering the abdominal organs is the **peritoneum.**

Cutaneous Membrane

The **cutaneous membrane** or **skin** covers the entire surface of the body and consists of a superficial portion called the *epidermis* and a deeper portion called the *dermis* (Figure 4.9c). The epidermis consists of keratinized stratified squamous epithelium, which protects underlying tissues. The dermis consists of dense irregular connective tissue and areolar connective tissue. Details of the cutaneous membrane are presented in Chapter 5.

Synovial Membranes

Synovial membranes (si-NŌ-vē-al; *syn-* = together, referring here to a place where bones come together; *-ova* = egg, because of their resemblance to the slimy egg white of an uncooked egg) line the cavities of freely movable joints (joint cavities). Like serous membranes, synovial membranes line structures that do not open to the exterior. Unlike mucous, serous, and cutaneous membranes, they lack an epithelium and are therefore not epithelial membranes. Synovial membranes are composed of a discontinuous layer of cells called **synoviocytes** (si-NŌ-vē-ō-sīts), which are closer to the synovial cavity (space between the bones), and a layer of connective tissue (areolar and adipose) deep to the synoviocytes (Figure 4.9d). Synoviocytes secrete some of the components of synovial fluid. **Synovial fluid** lubricates and nourishes the cartilage covering the bones at movable joints and contains macrophages that remove microbes and debris from the joint cavity.

✔**CHECKPOINT**

16. Define the following kinds of membranes: mucous, serous, cutaneous, and synovial. How do they differ from one another?
17. Where is each type of membrane located in the body? What are their functions?

TABLE 4.9

Muscular Tissues

A. SKELETAL MUSCLE TISSUE

Description	Long, cylindrical, striated fibers (*striations* are alternating light and dark bands within fibers that are visible under a light microscope). Skeletal muscle fibers vary greatly in length, from a few centimeters in short muscles to 30–40 cm (about 12–16 in.) in longest muscles. A muscle fiber is a roughly cylindrical, multinucleated cell with nuclei at periphery. Skeletal muscle is considered *voluntary* because it can be made to contract or relax by conscious control.
Location	Usually attached to bones by tendons.
Function	Motion, posture, heat production, protection.

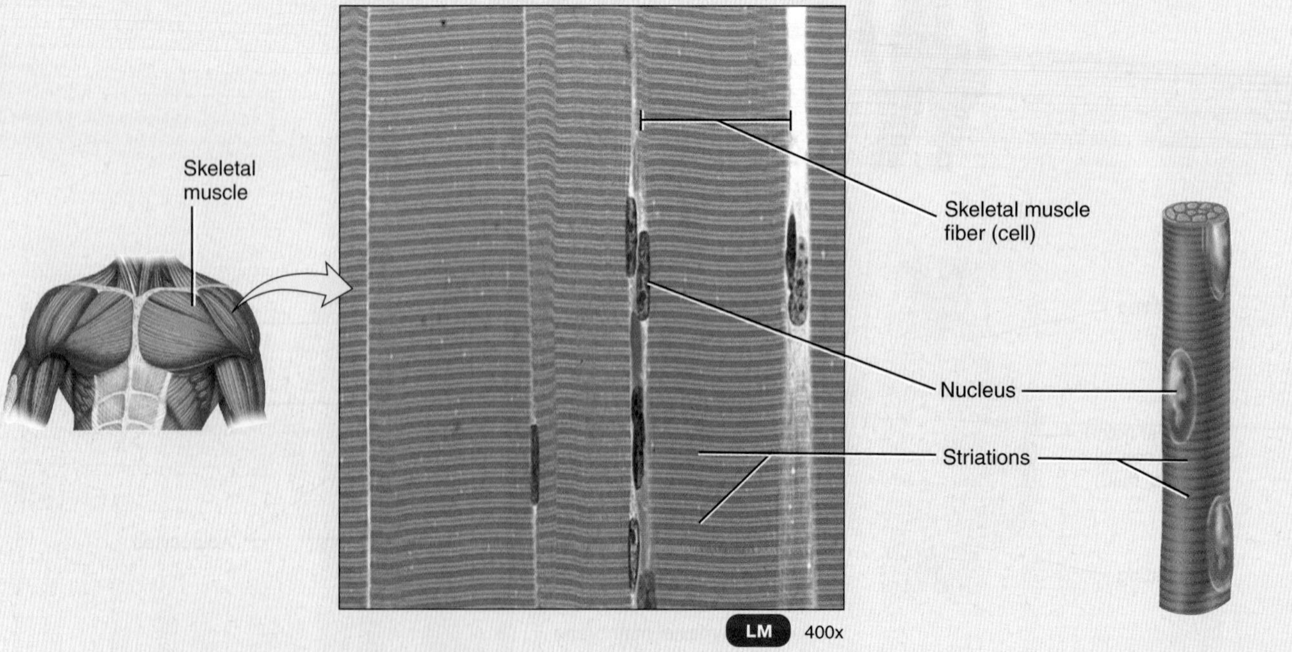
Skeletal muscle
Skeletal muscle fiber (cell)
Nucleus
Striations
LM 400x

Longitudinal section of skeletal muscle tissue

Skeletal muscle fiber

4.7 MUSCULAR TISSUES

OBJECTIVES

- Describe the general features of muscular tissues.
- Contrast the structure, location, and mode of control of skeletal, cardiac, and smooth muscle tissue.

Muscular tissues consist of elongated cells called *muscle fibers* or *myocytes* that can use ATP to generate force. As a result, muscular tissues produce body movements, maintain posture, and generate heat. They also provides protection. Based on their location and certain structural and functional features, muscular tissues are classified into three types: skeletal, cardiac, and smooth (Table 4.9).

Chapter 10 provides a more detailed discussion of muscular tissue.

✔CHECKPOINT

18. Which types of muscular tissues are striated? Which are smooth?
19. Which types of muscular tissues have gap junctions?

4.8 NERVOUS TISSUE

OBJECTIVE

- Describe the structural features and functions of nervous tissue.

Despite the awesome complexity of the nervous system, it consists of only two principal types of cells: neurons and neuroglia. **Neurons** (NOO-rons; *neuro-* = nerve), or nerve cells, are sensitive to various stimuli. They convert stimuli into electrical signals called **nerve action potentials (nerve impulses)** and conduct these action potentials to other neurons, to muscle tissue, or to glands. Most neurons consist of three basic parts: a cell body and two kinds of cell processes—dendrites and axons (Table 4.10). The **cell body** contains the nucleus and other organelles. **Dendrites** (*dendr-* = tree) are tapering, highly branched, and usually short cell processes (extensions). They are the major receiving or input portion of a neuron. The **axon** (*axo-* = axis) of a neuron is a single, thin, cylindrical process that may be very long. It is the

B. CARDIAC MUSCLE TISSUE

Description Branched, striated fibers with usually only one centrally located nucleus (occasionally two). Attach end to end by transverse thickenings of plasma membrane called *intercalated discs* (in-TER-ka-lāt-ed; *intercalate* = to insert between), which contain desmosomes and gap junctions. Desmosomes strengthen tissue and hold fibers together during vigorous contractions. Gap junctions provide route for quick conduction of electrical signals (muscle action potentials) throughout heart. *Involuntary* (not conscious) control.

Location Heart wall.

Function Pumps blood to all parts of body.

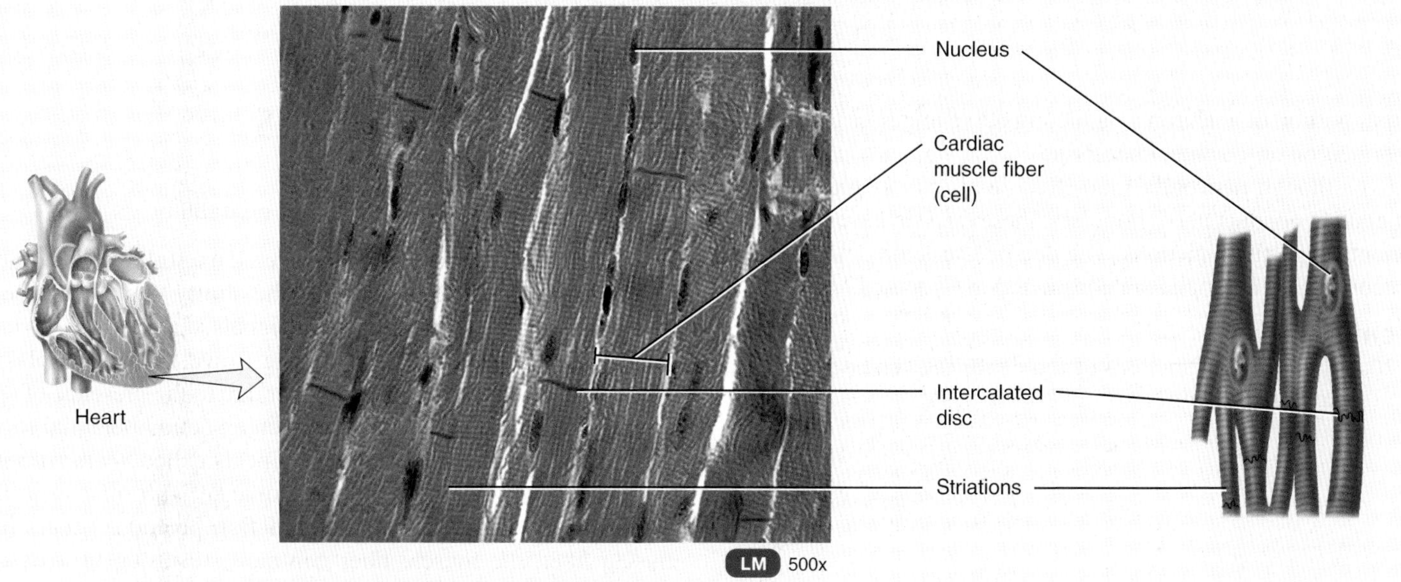
Heart
Nucleus
Cardiac
muscle fiber
(cell)
Intercalated
disc
Striations
LM 500x

Longitudinal section of cardiac muscle tissue

Cardiac muscle fibers

TABLE 4.9 CONTINUES

TABLE 4.9 CONTINUED

Muscular Tissues

C. SMOOTH MUSCLE TISSUE

Description	Fibers usually *involuntary,* nonstriated (lack striations, hence the term *smooth*). Smooth muscle fiber is a small spindle-shaped cell thickest in middle, tapering at each end, and containing a single, centrally located nucleus. Gap junctions connect many individual fibers in some smooth muscle tissues (for example, in wall of intestines). Can produce powerful contractions as many muscle fibers contract in unison. Where gap junctions are absent, such as iris of eye, smooth muscle fibers contract individually, like skeletal muscle fibers.
Location	Iris of eyes; walls of hollow internal structures such as blood vessels, airways to lungs, stomach, intestines, gallbladder, urinary bladder, and uterus.
Function	Motion (constriction of blood vessels and airways, propulsion of foods through gastrointestinal tract, contraction of urinary bladder and gallbladder).

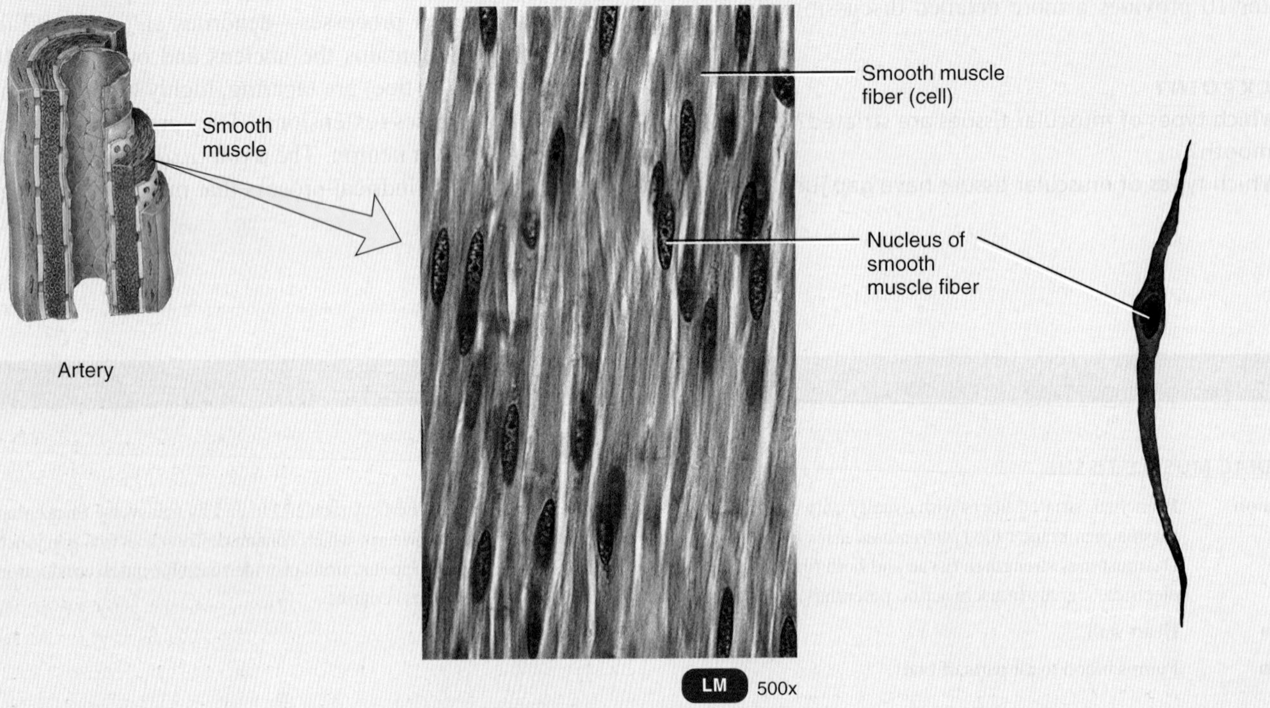

Longitudinal section of smooth muscle tissue

Smooth muscle fiber

output portion of a neuron, conducting nerve impulses toward another neuron or to some other tissue.

Even though **neuroglia** (noo-RŌG-lē-a; *-glia* = glue) do not generate or conduct nerve impulses, these cells do have many important supportive functions. The detailed structure and function of neurons and neuroglia are considered in Chapter 12.

✔CHECKPOINT

20. What are the functions of the dendrites, cell body, and axon of a neuron?

4.9 EXCITABLE CELLS

OBJECTIVE

- Explain the concept of electrical excitability.

Neurons and muscle fibers are considered **excitable cells** because they exhibit **electrical excitability,** the ability to respond to certain stimuli by producing electrical signals such as *action potentials.* Action potentials can propagate (travel) along the plasma membrane of a neuron or muscle fiber due to the presence of specific voltage-gated ion channels. When an action potential forms in a neuron, the neuron releases chemicals called *neurotransmitters,* which allow neurons to communicate with other neurons, muscle fibers, or glands. When an action potential occurs in a muscle fiber, the muscle fiber contracts, resulting in activities such as the movement of the limbs, the propulsion of food through the small intestine, and the movement of blood out of the

TABLE 4.10

Nervous Tissue

Description	(1) Neurons (nerve cells), which consist of cell body and processes extending from cell body (one to multiple dendrites and a single axon); and (2) neuroglia, which do not generate or conduct nerve impulses but have other important supporting functions.
Location	Nervous system.
Function	Exhibits sensitivity to various types of stimuli; converts stimuli into nerve impulses (action potentials); conducts nerve impulses to other neurons, muscle fibers, or glands.

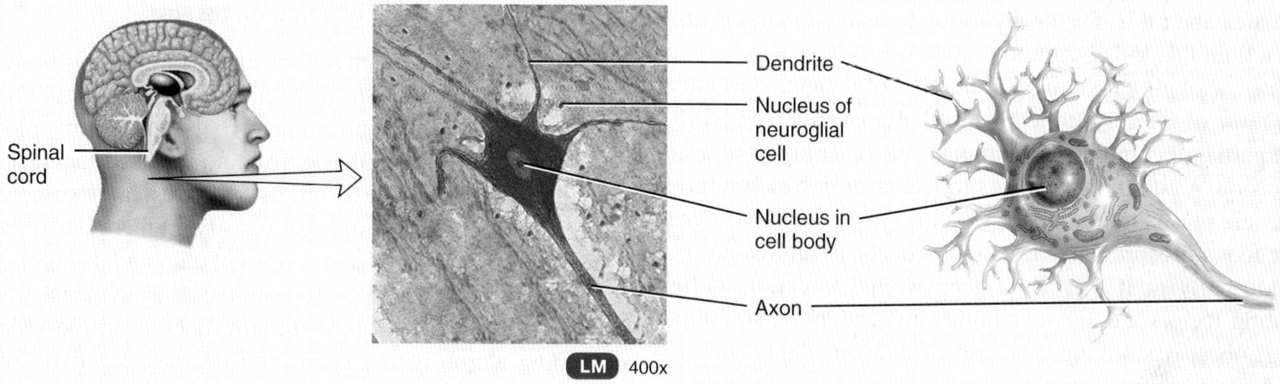
Spinal cord
Dendrite
Nucleus of neuroglial cell
Nucleus in cell body
Axon
LM 400x
Neuron of spinal cord

heart and into the blood vessels of the body. The muscle action potential and the nerve action potential are discussed in detail in Chapters 10 and 12, respectively.

✔CHECKPOINT

21. Why is electrical excitability important to neurons and muscle fibers?

4.10 TISSUE REPAIR: RESTORING HOMEOSTASIS

OBJECTIVE

- Describe the role of tissue repair in restoring homeostasis.

Tissue repair is the replacement of worn-out, damaged, or dead cells. New cells originate by cell division from the **stroma,** the supporting connective tissue, or from the **parenchyma** (pa-REN-ki-ma), cells that constitute the functioning part of the tissue or organ. In adults, each of the four basic tissue types (epithelial, connective, muscle, and nervous) has a different capacity for replenishing parenchymal cells lost by damage, disease, or other processes.

Epithelial cells, which endure considerable wear and tear (and even injury) in some locations, have a continuous capacity for renewal. In some cases, immature, undifferentiated cells called **stem cells** divide to replace lost or damaged cells. For example, stem cells reside in protected locations in the epithelia of the skin and gastrointestinal tract to replenish cells sloughed from the apical layer, and stem cells in red bone marrow continually provide new red and white blood cells and platelets. In other cases, mature, differentiated cells can undergo cell division; examples include hepatocytes (liver cells) and endothelial cells in blood vessels.

Some connective tissues also have a continuous capacity for renewal. One example is bone, which has an ample blood supply. Connective tissues such as cartilage can replenish cells much less readily, in part because of a smaller blood supply.

Muscular tissues have a relatively poor capacity for renewal of lost cells. Even though skeletal muscle tissue contains stem cells called *satellite cells,* they do not divide rapidly enough to replace extensively damaged muscle fibers. Cardiac muscle tissue lacks satellite cells, and existing cardiac muscle fibers do not undergo mitosis to form new cells. Recent evidence suggests that stem cells do migrate into the heart from the blood. There, they can differentiate and replace a limited number of cardiac muscle fibers and endothelial cells in heart blood vessels. Smooth muscle fibers can proliferate to some extent, but they do so much more slowly than the cells of epithelial or connective tissues.

Nervous tissue has the poorest capacity for renewal. Although experiments have revealed the presence of some stem cells in the brain, they normally do not undergo mitosis to replace damaged neurons. Discovering why this is so is a major goal of researchers who seek ways to repair nervous tissue damaged by injury or disease.

The restoration of an injured tissue or organ to normal structure and function depends entirely on whether parenchymal cells are active in the repair process. If parenchymal cells accomplish the repair, **tissue regeneration** is possible, and a near-perfect reconstruction of the injured tissue may occur. However, if fibroblasts of the stroma are active in the repair, the replacement tissue will be a new connective tissue. The fibroblasts synthesize collagen and other extracellular matrix materials that aggregate to form scar tissue, a process known as **fibrosis.** Because scar tissue is not specialized to perform the functions of the parenchymal tissue, the original function of the tissue or organ is impaired.

When tissue damage is extensive, as in large, open wounds, both the connective tissue stroma and the parenchymal cells are active in repair; fibroblasts divide rapidly, and new collagen fibers are manufactured to provide structural strength. Blood capillaries also sprout new buds to supply the healing tissue with the materials it needs. All these processes create an actively growing connective tissue called **granulation tissue** (gran-ū-LĀ-shun). This new tissue forms across a wound or surgical incision to provide a framework (stroma) that supports the epithelial cells that migrate into the open area and fill it. The newly formed granulation tissue also secretes a fluid that kills bacteria.

At times, a small but significant number of patients develop a complication of surgery called **wound dehiscence** (dē-HISS-ens), the partial or complete separation of the outer layers of a sutured incision. A common cause is surgical error in which sutures or staples are placed too far apart, too close to the incision edges, or under too much pressure. It can also occur if sutures are removed too early or if there is a deep wound infection. Other contributing factors are age, chemotherapy, coughing, straining, vomiting, obesity, smoking, and use of anticoagulants such as aspirin. A major complication of wound dehiscence is the protrusion of an organ through the open wound, especially the intestines. This can lead to peritonitis (inflammation of the peritoneum) and septic shock (shock that results from bacterial toxins due to vasodilation).

Three factors affect tissue repair: nutrition, blood circulation, and age. Nutrition is vital because the healing process places a great demand on the body's store of nutrients. Adequate protein in the diet is important because most of the structural components of a tissue are proteins. Several vitamins also play a direct role in wound healing and tissue repair. For example, vitamin C directly affects the normal production and maintenance of matrix materials, especially collagen, and strengthens and promotes the formation of new blood vessels. In a person with vitamin C deficiency, even superficial wounds fail to heal, and the walls of the blood vessels become fragile and are easily ruptured.

Proper blood circulation is essential to transport oxygen, nutrients, antibodies, and many defensive cells to the injured site. The blood also plays an important role in the removal of tissue fluid, bacteria, foreign bodies, and debris, elements that would otherwise interfere with healing. The third factor in tissue repair, age, is the topic of the next section.

CLINICAL CONNECTION | *Adhesions*

Scar tissue can form **adhesions** (ad-HĒ-zhuns; *adhaero* = to stick to), abnormal joining of tissues. Adhesions commonly form in the abdomen around a site of previous inflammation such as an inflamed appendix, and they can develop after surgery. Although adhesions do not always cause problems, they can decrease tissue flexibility, cause obstruction (such as in the intestine), and make a subsequent operation, such as a cesarean section (c-section), more difficult. In rare cases adhesions can result in infertility. An *adhesiotomy,* the surgical release of adhesions, may be required. •

✔CHECKPOINT

22. How are stromal and parenchymal repair of a tissue different?
23. What is the importance of granulation tissue?

4.11 AGING AND TISSUES

OBJECTIVE

- Describe the effects of aging on tissues.

In later chapters, the effects of aging on specific body systems will be considered. With respect to tissues, epithelial tissues get progressively thinner and connective tissues become more fragile with aging. This is evidenced by an increased incidence of skin and mucous membrane disorders, wrinkles, more susceptibility to bruises, increased loss of bone density, higher rates of bone fractures, and increased episodes of joint pain and disorders. There is also an effect of aging on muscle tissue as evidenced by loss of skeletal muscle mass and strength, decline in the efficiency of pumping action of the heart, and decreased activity of smooth muscle–containing organs, for example, organs of the gastrointestinal tract.

Generally, tissues heal faster and leave less obvious scars in the young than in the aged. In fact, surgery performed on fetuses leaves no scars at all. The younger body is generally in a better nutritional state, its tissues have a better blood supply, and its cells have a higher metabolic rate. Thus, its cells can synthesize needed materials and divide more quickly. The extracellular components of tissues also change with age. Glucose, the most abundant sugar in the body, plays a role in the aging process. As the body ages, glucose is haphazardly added to proteins inside and outside cells, forming irreversible cross-links between adjacent protein molecules. With advancing age, more cross-links form, which contributes to the stiffening and loss of elasticity that occur in aging tissues. Collagen fibers, responsible for the strength of tendons, increase in number and change in quality with aging. Changes in the collagen of arterial walls affect the flexibility of arteries as much as the fatty deposits associated with atherosclerosis (see Coronary Artery Disease in Disorders: Homeostatic Imbalances section of Chapter 20). Elastin, another extracellular component, is responsible for the elasticity of blood vessels and skin. It thickens, fragments, and acquires a greater affinity for calcium with age—changes that may also be associated with the development of atherosclerosis.

✔CHECKPOINT

24. What common changes occur in epithelial and connective tissues with aging?

DISORDERS: HOMEOSTATIC IMBALANCES

Disorders of epithelial tissues are mainly specific to individual organs, such as peptic ulcer disease (PUD), which erodes the epithelial lining of the stomach or small intestine. For this reason, epithelial disorders are described along with their relevant body systems throughout the text. The most prevalent disorders of connective tissues are **autoimmune diseases**—diseases in which antibodies produced by the immune system fail to distinguish what is foreign from what is self and attack the body's own tissues. One of the most common autoimmune disorders is rheumatoid arthritis, which attacks the synovial membranes of joints. Because connective tissues are one of the most abundant and widely distributed of the four main types of tissues, disorders related to them often affect multiple body systems. Common disorders of muscular tissues and nervous tissue are described at the ends of Chapters 10 and 12, respectively.

Sjögren's Syndrome

Sjögren's syndrome (SHO-grenz) is a common autoimmune disorder in which your own immune system attacks the lacrimal (tear) glands and salivary glands, especially, and causes inflammation and destruction of these exocrine glands. Signs include dryness of the eyes, mouth, nose, ears, skin, and vagina, and salivary gland enlargement. Systemic effects include fatigue, arthritis, difficulty in swallowing, pancreatitis (inflammation of the pancreas), pleuritis (inflammation of the pleurae of the lungs), and muscle and joint pain. The disorder affects females more than males by a ratio of 9 to 1. About 20% of older adults experience some signs of Sjögren's. Treatment is supportive and includes using artificial tears to moisten the eyes, sipping fluids, chewing sugarless gum, using a saliva substitute to moisten the mouth, and applying moisturizing creams for the skin. If symptoms or complications are severe, medications may be used. These include cyclosporine eyedrops, pilocarpine to increase saliva production, immunosuppressants, nonsteroidal anti-inflammatory drugs, and corticosteroids.

Systemic Lupus Erythematosus

Systemic lupus erythematosus (er-i-thē-ma-TŌ-sus), **SLE,** or simply lupus, is a chronic inflammatory disease of connective tissue occurring mostly in nonwhite women during their childbearing years. It is an autoimmune disease that can cause tissue damage in every body system. The disease, which can range from a mild condition in most patients to a rapidly fatal disease, is marked by periods of exacerbation and remission. The prevalence of SLE is about 1 in 2000, with females more likely to be afflicted than males by a ratio of 8 or 9 to 1.

Although the cause of SLE is unknown, genetic, environmental, and hormonal factors all have been implicated. The genetic component is suggested by studies of twins and family history. Environmental factors include viruses, bacteria, chemicals, drugs, exposure to excessive sunlight, and emotional stress. Sex hormones, such as estrogens, may also trigger SLE.

Signs and symptoms of SLE include painful joints, low-grade fever, fatigue, mouth ulcers, weight loss, enlarged lymph nodes and spleen, sensitivity to sunlight, rapid loss of large amounts of scalp hair, and anorexia. A distinguishing feature of lupus is an eruption across the bridge of the nose and cheeks called a "butterfly rash." Other skin lesions may occur, including blistering and ulceration. The erosive nature of some SLE skin lesions was thought to resemble the damage inflicted by the bite of a wolf—thus, the name *lupus* (= wolf). The most serious complications of the disease involve inflammation of the kidneys, liver, spleen, lungs, heart, brain, and gastrointestinal tract. Because there is no cure for SLE, treatment is supportive, including anti-inflammatory drugs such as aspirin, and immunosuppressive drugs.

MEDICAL TERMINOLOGY

Atrophy (AT-rō-fē; *a-* = without; *-trophy* = nourishment) A decrease in the size of cells, with a subsequent decrease in the size of the affected tissue or organ.

Hypertrophy (hī-PER-trō-fē; *hyper-* = above or excessive) Increase in the size of a tissue because its cells enlarge without undergoing cell division.

Tissue rejection An immune response of the body directed at foreign proteins in a transplanted tissue or organ; immunosuppressive drugs, such as cyclosporine, have largely overcome tissue rejection in heart-, kidney-, and liver-transplant patients.

Tissue transplantation The replacement of a diseased or injured tissue or organ. The most successful transplants involve use of a person's own tissues or those from an identical twin.

Xenotransplantation (zen'-ō-trans-plan-TĀ-shun; *xeno-* = strange, foreign) The replacement of a diseased or injured tissue or organ with cells or tissues from an animal. Porcine (from pigs) and bovine (from cows) heart valves are used for some heart-valve replacement surgeries.

CHAPTER REVIEW AND RESOURCE SUMMARY

WILEY PLUS

Review

4.1 Types of Tissues

1. A tissue is a group of cells, usually with a similar embryological origin, that is specialized for a particular function.
2. The various tissues of the body are classified into four basic types: epithelial, connective, muscular, and nervous tissue.

Resource

Anatomy Overview - Epithelial Tissues
Anatomy Overview - Connective Tissues
Anatomy Overview - Muscle Tissue
Anatomy Overview - Nervous Tissue
Concepts and Connections - Human Tissue

Review | **Resource**

4.2 Cell Junctions

1. Cell junctions are points of contact between adjacent plasma membranes.
2. Tight junctions form fluid-tight seals between cells; adherens junctions, desmosomes, and hemidesmosomes anchor cells to one another or to the basement membrane; and gap junctions permit electrical and chemical signals to pass between cells.

Resource:
Animation - Intracellular Junctions
Figure 4.2 - Cell Junctions
Exercise - Getting It Together

4.3 Comparison between Epithelial and Connective Tissues

1. Epithelial tissues have many cells tightly packed together and are avascular.
2. Connective tissues have relatively few cells with lots of extracellular material.

Resource:
Anatomy Overview - Epithelial Tissues
Anatomy Overview - Connective Tissues

4.4 Epithelial Tissues

1. The subtypes of epithelial tissues include covering and lining epithelia and glandular epithelia.
2. Epithelial tissues consist mostly of cells with little extracellular material between adjacent plasma membranes. The apical, lateral, and basal surfaces of epithelial cells are modified in various ways to carry out specific functions. Epithelium is arranged in sheets and attached to a basement membrane. Although epithelial tissues are avascular, they have a nerve supply. The high rate of cell division gives epithelial tissues a high capacity for renewal.
3. Epithelial layers can be simple (one layer), pseudostratified (appears to be several layers but is actually a single layer), or stratified (several layers). The cell shapes may be squamous (flat), cuboidal (cubelike), columnar (rectangular), or transitional (variable). The subtypes of epithelia include covering and lining epithelia and glandular epithelia.
4. Simple squamous epithelium, which consists of a single layer of flat cells (Table 4.1), is found in parts of the body where filtration or diffusion is a priority process. One type, called endothelium, lines the heart and blood vessels. Another type, called mesothelium, forms the serous membranes that line the thoracic and abdominopelvic cavities and cover the organs within them.
5. Simple cuboidal epithelium consists of a single layer of cube-shaped cells that function in secretion and absorption (Table 4.1). It is found covering the ovaries, in the kidneys and eyes, and lining some glandular ducts.
6. Nonciliated simple columnar epithelium, a single layer of nonciliated rectangular cells (Table 4.1), lines most of the gastrointestinal tract and contains specialized cells that perform absorption and secrete mucus. Ciliated simple columnar epithelium, which consists of a single layer of ciliated rectangular cells (Table 4.1), is found in a few portions of the upper respiratory tract, where it moves foreign particles trapped in mucus out of the respiratory tract. A ciliated variety of pseudostratified columnar epithelium (Table 4.1) contains goblet cells and lines most of the upper respiratory tract; a nonciliated variety has no goblet cells and lines ducts of many glands, the epididymis, and part of the male urethra. The ciliated variety moves mucus in the respiratory tract. The nonciliated variety functions in absorption and protection.
7. Stratified epithelium consists of several layers of cells: Cells of the apical layer of stratified squamous epithelium and several layers deep to it are flat (Table 4.1); a nonkeratinized variety lines the mouth, and a keratinized variety forms the epidermis, the most superficial layer of the skin. Cells at the apical layer of stratified cuboidal epithelium are cube-shaped (Table 4.1); found in adult sweat glands and in a portion of the male urethra, stratified cuboidal epithelium protects and provides limited secretion and absorption. Cells of the apical layer of stratified columnar epithelium have a columnar shape (Table 4.1); this type of stratified epithelium is found in a portion of the male urethra and in large excretory ducts of some glands, and functions in protection and secretion.
8. Transitional epithelium consists of several layers of cells whose appearance varies with the degree of stretching (Table 4.1). It lines the urinary bladder.
9. A gland is a single cell or a group of epithelial cells adapted for secretion. There are two types of glands: endocrine and exocrine. Endocrine glands secrete hormones into interstitial fluid and then into the blood (Table 4.2). Exocrine glands (mucous, sweat, oil, and digestive glands) secrete into ducts or directly onto a free surface (Table 4.2).
10. The structural classification of exocrine glands includes unicellular and multicellular glands. The functional classification of exocrine glands includes merocrine, apocrine, and holocrine glands.

Resource:
Anatomy Overview - Epithelial Tissues
Anatomy Overview - Epithelial Membranes
Figure 4.5 - Cell Shapes and Arrangement of Layers for Covering and Lining Epithelium

Review

4.5 Connective Tissues

1. Connective tissues, one of the most abundant body tissues, consist of relatively few cells and an abundant extracellular matrix of ground substance and protein fibers. They do not usually occur on free surfaces, they have a nerve supply (except for cartilage), and they are highly vascular (except for cartilage, tendons, and ligaments).
2. Cells in connective tissues are derived primarily from mesenchymal cells. Cell types include fibroblasts (secrete extracellular matrix), macrophages (perform phagocytosis), plasma cells (secrete antibodies), mast cells (produce histamine), adipocytes (store fat), and white blood cells (migrate from blood in response to infections).
3. The ground substance and fibers make up the extracellular matrix. The ground substance supports and binds cells together, provides a medium for the exchange of materials, stores water, and actively influences cell functions. Substances found in the ground substance include water and polysaccharides such as hyaluronic acid, chondroitin sulfate, dermatan sulfate, and keratan sulfate (glycosaminoglycans). Also present are proteoglycans and adhesion proteins.
4. The fibers in the extracellular matrix provide strength and support and are of three types: (a) Collagen fibers (composed of collagen) are found in large amounts in bone, tendons, and ligaments. (b) Elastic fibers (composed of elastin, fibrillin, and other glycoproteins) are found in skin, blood vessel walls, and lungs. (c) Reticular fibers (composed of collagen and glycoprotein) are found around fat cells, nerve fibers, and skeletal and smooth muscle cells.
5. The two major subclasses of connective tissues are embryonic connective tissues (found in the embryo and fetus) and mature connective tissues (present in the newborn). The embryonic connective tissues (summarized in Table 4.3) are mesenchyme, which forms almost all other connective tissues, and mucous connective tissue, found in the umbilical cord of the fetus, where it gives support. Mature connective tissues differentiate from mesenchyme and are subdivided into several types: loose or dense connective tissue, cartilage, bone tissue, and liquid connective tissue.
6. Loose connective tissue includes areolar connective tissue, adipose tissue, and reticular connective tissue. Areolar connective tissue consists of the three types of fibers (collagen, elastic, and reticular), several types of cells, and a semifluid ground substance (Table 4.4); it is found in the subcutaneous layer, in mucous membranes, and around blood vessels, nerves, and body organs. Adipose tissue consists of adipocytes, which store triglycerides (Table 4.4); it is found in the subcutaneous layer, around organs, and in yellow bone marrow. Brown adipose tissue (BAT) generates heat. Reticular connective tissue consists of reticular fibers and reticular cells and is found in the liver, spleen, and lymph nodes (Table 4.4).
7. Dense connective tissue includes dense regular connective tissue, dense irregular connective tissue, and elastic connective tissue. Dense regular connective tissue consists of parallel bundles of collagen fibers and fibroblasts (Table 4.5); it forms tendons, most ligaments, and aponeuroses. Dense irregular connective tissue usually consists of randomly arranged collagen fibers and a few fibroblasts (Table 4.5); it is found in fasciae, the dermis of skin, and membrane capsules around organs. Elastic connective tissue consists of branching elastic fibers and fibroblasts (Table 4.5) and is found in the walls of large arteries, lungs, trachea, and bronchial tubes.
8. Cartilage contains chondrocytes and has a rubbery extracellular matrix (chondroitin sulfate) containing collagen and elastic fibers. Hyaline cartilage, which consists of a gel-like ground substance and appears bluish white in the body, is found in the embryonic skeleton, at the ends of bones, in the nose, and in respiratory structures (Table 4.6); it is flexible, allows movement, provides support, and is usually surrounded by a perichondrium. Fibrocartilage is found in the pubic symphysis, intervertebral discs, and menisci (cartilage pads) of the knee joint (Table 4.6); it contains chondrocytes scattered among clearly visible bundles of collagen fibers. Elastic cartilage maintains the shape of organs such as the epiglottis of the larynx, auditory (eustachian) tubes, and external ear (Table 4.6); its chondrocytes are located within a threadlike network of elastic fibers, and it has a perichondrium.
9. Bone or osseous tissue consists of an extracellular matrix of mineral salts and collagen fibers that contribute to the hardness of bone, and osteocytes that are located in lacunae (Table 4.7). It supports and protects the body, provides a surface area for muscle attachment, helps the body move, stores minerals, and houses blood-forming tissue.
10. There are two types of liquid connective tissues: blood and lymph. Blood tissue consists of blood plasma and formed elements—red blood cells, white blood cells, and platelets (Table 4.8); its cells transport oxygen and carbon dioxide, carry on phagocytosis, participate in allergic reactions, provide immunity, and bring about blood clotting. Lymph, the extracellular fluid that flows in lymphatic vessels, is a clear fluid similar to blood plasma but with less protein.

Resource

Anatomy Overview - Connective Tissues
Figure 4.7 - Functional Classification of Multicellular Exocrine Glands
Figure 4.8 - Representative Cells and Fibers in Connective Tissues

Review	Resource
4.6 Membranes 1. An epithelial membrane consists of an epithelial layer overlying a connective tissue layer. Types include mucous, serous, and cutaneous membranes. 2. Mucous membranes line cavities that open to the exterior, such as the gastrointestinal tract. 3. Serous membranes line closed cavities (pleura, pericardium, peritoneum) and cover the organs in the cavities. These membranes consist of parietal and visceral layers. 4. The cutaneous membrane is the skin. It covers the entire body and consists of a superficial epidermis (epithelium) and a deep dermis (connective tissue). 5. Synovial membranes line joint cavities and consist of areolar connective tissue; they do not have an epithelial layer.	Anatomy Overview - Epithelial Membranes
4.7 Muscular Tissues 1. Muscular tissues consist of cells called muscle fibers or myocytes that are specialized for contraction. They provide motion, maintenance of posture, heat production, and protection. 2. Skeletal muscle tissue is attached to bones and is striated and voluntary (Table 4.9). 3. The action of cardiac muscle tissue, which forms most of the heart wall and is striated, is involuntary (Table 4.9). 4. Smooth muscle tissue is found in the walls of hollow internal structures (blood vessels and viscera) and is nonstriated and involuntary (Table 4.9).	Anatomy Overview - Muscle Tissue
4.8 Nervous Tissue 1. The nervous system is composed of neurons (nerve cells) and neuroglia (protective and supporting cells) (Table 4.10). 2. Neurons respond to stimuli by converting the stimuli into electrical signals called nerve action potentials (nerve impulses), and conducting nerve impulses to other cells. 3. Most neurons consist of a cell body and two types of processes, dendrites and axons.	Anatomy Overview - Nervous Tissue Exercise - Concentrate on Tissue Function Exercise - Name That Tissue Exercise - Tissue Identification
4.9 Excitable Cells 1. Electrical excitability is the ability to respond to certain stimuli by producing electrical signals such as action potentials. 2. Because neurons and muscle fibers exhibit electrical excitability, they are considered excitable cells.	Anatomy Overview - Nervous Tissue
4.10 Tissue Repair: Restoring Homeostasis 1. Tissue repair is the replacement of worn-out, damaged, or dead cells by healthy ones. 2. Stem cells may divide to replace lost or damaged cells. 3. If the injury is superficial, tissue repair involves parenchymal regeneration; if damage is extensive, granulation tissue is involved. 4. Good nutrition and blood circulation are vital to tissue repair.	
4.11 Aging and Tissues 1. Tissues heal faster and leave less obvious scars in the young than in the aged; surgery performed on fetuses leaves no scars. 2. The extracellular components of tissues, such as collagen and elastic fibers, also change with age.	

SELF-QUIZ QUESTIONS

Fill in the blanks in the following statements.

1. The four types of tissues are _____ , _____ , _____ , and _____ .
2. Epithelial tissue tends to be classified according to two criteria: _____ and _____.

Indicate whether the following statements are true or false.

3. Epithelial tissue cells have an apical surface at the top and are attached to a basement membrane at the bottom.
4. Connective tissue fibers that are arranged in bundles and lend strength and flexibility to a tissue are collagen fibers.

Choose the one best answer to the following questions.

5. Which of the following muscular tissues can be voluntarily controlled? (1) cardiac, (2), smooth, (3) skeletal.
   (a) 1, 2, and 3 (b) 2 (c) 1
   (d) 1 and 3 (e) 3
6. Which of the following tissues is avascular?
   (a) cardiac muscle (b) stratified squamous epithelial
   (c) compact bone (d) skeletal muscle
   (e) adipose
7. If the lining of an organ produces and releases mucus, which of the following cells would likely be found in the tissue lining the organ?
   (a) goblet cells (b) mast cells (c) macrophages
   (d) osteoblasts (e) fibroblasts
8. Why does damaged cartilage heal slowly?
   (a) Damaged cartilage undergoes fibrosis, which interferes with the movement of materials needed for repair.
   (b) Cartilage does not contain fibroblasts, which are needed to produce the fibers in cartilage tissue.
   (c) Cartilage is avascular, so materials needed for repair must diffuse from surrounding tissue.
   (d) Chondrocytes cannot be replaced once they are damaged.
   (e) Chondrocytes undergo mitosis slowly, which delays healing.
9. Which of the following is *true* concerning serous membranes?
   (a) A serous membrane lines a body part that opens directly to the body's exterior.
   (b) The parietal portion of a serous membrane attaches to the organ.
   (c) The visceral portion of a serous membrane attaches to a body cavity wall.
   (d) The serous membrane covering the heart is known as the peritoneum.
   (e) The serous membrane covering the lungs is known as the pleura.
10. The type of exocrine gland that forms its secretory product and simply releases it from the cell by exocytosis is the
    (a) apocrine gland (b) merocrine gland (c) holocrine gland
    (d) endocrine gland (e) tubular gland.
11. Tissue changes that occur with aging can be due to (1) cross-links forming between glucose and proteins, (2) a decrease in the amount of collagen fibers, (3) a decreased blood supply, (4) improper nutrition, (5) a higher cellular metabolic rate.
    (a) 1, 2, 3, 4, and 5 (b) 1, 2, 3, and 4 (c) 1 and 4
    (d) 1, 3, and 4 (e) 1, 2, and 3
12. What type of cell junction would be required for cells to communicate with one another?
    (a) adherens junction (b) desmosome (c) gap junction
    (d) tight junction (e) hemidesmosome
13. Match the following epithelial tissues to their descriptions:
    ____(a) contains a single layer of flat cells; found in the body where filtration (kidney) or diffusion (lungs) are priority processes
    ____(b) found in the superficial part of skin; provides protection from heat, microbes, and chemicals
    ____(c) contains cube-shaped cells functioning in secretion and absorption
    ____(d) lines the lower respiratory tract and uterine tubes; wavelike motion of cilia propels materials through the lumen
    ____(e) contains cells with microvilli and goblet cells; found in linings of the digestive, reproductive, and urinary tracts
    ____(f) found in the urinary bladder; contains cells that can change shape (stretch or relax)
    ____(g) contains cells that are all attached to the basement membrane, although some do not reach the surface; those cells that do extend to the surface secrete mucus or contain cilia
    ____(h) a fairly rare type of epithelium that has a mainly protective function

    (1) pseudostratified ciliated columnar epithelium
    (2) ciliated simple columnar epithelium
    (3) transitional epithelium
    (4) simple squamous epithelium
    (5) simple cuboidal epithelium
    (6) nonciliated simple columnar epithelium
    (7) stratified cuboidal epithelium
    (8) keratinized stratified squamous epithelium
14. For each of the following items, indicate the tissue type with which they are associated. Use **E** for epithelial tissues, **C** for connective tissues, **M** for muscle tissues, and **N** for nervous tissue.
    ____(a) bind, support
    ____(b) contains elongated cells that generate force
    ____(c) neuroglia
    ____(d) avascular
    ____(e) may contain fibroblasts
    ____(f) tightly packed cells
    ____(g) intercalated discs
    ____(h) goblet cells
    ____(i) contain extracellular matrix
    ____(j) striated
    ____(k) generate action potentials
    ____(l) cilia
    ____(m) ground substance
    ____(n) apical surface
    ____(o) excitable

15. Match the following connective tissues to their descriptions:
____(a) the tissue from which all other connective tissues eventually arise
____(b) connective tissue with a clear, liquid matrix that flows in lymphatic vessels
____(c) connective tissue consisting of several kinds of cells, containing all three fiber types randomly arranged, and found in the subcutaneous layer deep to the skin
____(d) a loose connective tissue specialized for triglyceride storage
____(e) tissue that contains reticular fibers and reticular cells and forms the stroma of certain organs such as the spleen
____(f) tissue with irregularly arranged collagen fibers found in the dermis of the skin
____(g) tissue found in the lungs that is strong and can recoil back to its original shape after being stretched
____(h) tissue that affords flexibility at joints and reduces joint friction
____(i) tissue that provides strength and rigidity and is the strongest of the three types of cartilage
____(j) bundles of collagen arranged in parallel patterns; compose tendons and ligaments
____(k) tissue that forms the internal framework of the body and works with skeletal muscle to generate movement
____(l) tissue that contains a network of elastic fibers, providing strength, elasticity, and maintenance of shape; located in the external ear
____(m) connective tissue with formed elements suspended in a liquid matrix called plasma

(1) blood
(2) fibrocartilage
(3) mesenchyme
(4) dense regular connective tissue
(5) lymph
(6) hyaline cartilage
(7) dense irregular connective tissue
(8) areolar connective tissue
(9) reticular connective tissue
(10) bone (osseous tissue)
(11) elastic connective tissue
(12) elastic cartilage
(13) adipose tissue

CRITICAL THINKING QUESTIONS

1. Imagine that you live 50 years in the future, and that you can custom-design a human to suit the environment. Your assignment is to customize the human's tissues so that the individual can survive on a large planet with gravity, a cold, dry climate, and a thin atmosphere. What adaptations would you incorporate into the structure and/or amount of tissues, and why?

2. You are entering a "Cutest Baby Contest" and have asked your colleagues to help you choose the most adorable picture of yourself as a baby. One of your colleagues rudely points out that you were quite chubby as an infant. You, however, are not offended. Explain to your colleague the benefit of that "baby fat."

3. You've been on a "bread-and-water" diet for three weeks and have noticed that a cut on your shin won't heal and bleeds easily. Why?

? ANSWERS TO FIGURE QUESTIONS

4.1 Epithelial tissues cover the body, line various structures, and form glands. Connective tissues protect, support, bind organs together, store energy, and help provide immunity. Muscular tissue contracts and generates force and heat. Nervous tissue detects changes in the environment and generates nerve impulses that activate muscular contraction and glandular secretion.

4.2 Gap junctions allow cellular communication via passage of electrical and chemical signals between adjacent cells.

4.3 Since epithelial tissues are avascular, they depend on the blood vessels in connective tissues for oxygen, nutrients, and waste disposal.

4.4 The basement membrane provides physical support for the epithelium and plays a part in growth and wound healing, restriction of molecule movement between tissues, and blood filtration in the kidneys.

4.5 Because the cells are so thin, substances move most rapidly through squamous cells.

4.6 Simple multicellular exocrine glands have a nonbranched duct; compound multicellular exocrine glands have a branched duct.

4.7 Sebaceous (oil) glands are holocrine glands, and salivary glands are merocrine glands.

4.8 Fibroblasts secrete the fibers and ground substance of the extracellular matrix.

4.9 An epithelial membrane is a membrane that consists of an epithelial layer and an underlying layer of connective tissue.

5 THE INTEGUMENTARY SYSTEM

THE INTEGUMENTARY SYSTEM AND HOMEOSTASIS *The integumentary system contributes to homeostasis by protecting the body and helping regulate body temperature. It also allows you to sense pleasurable, painful, and other stimuli in your external environment.*

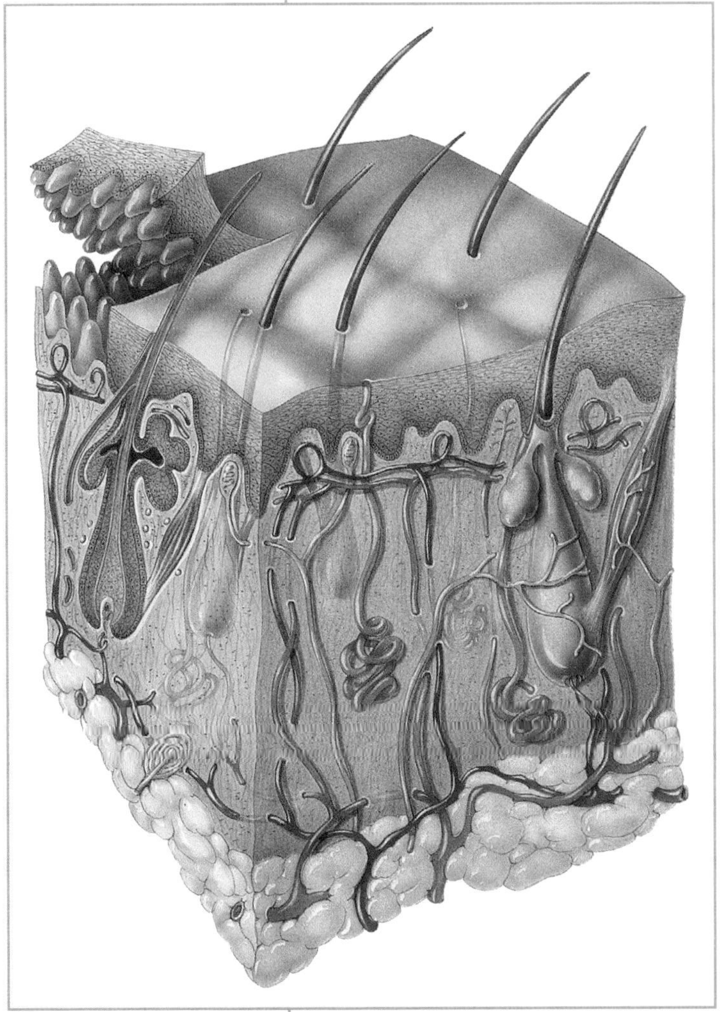

Recall from Chapter 1 that a system consists of a group of organs working together to perform specific activities. The **integumentary system** (in-teg-ū-MEN-tar-ē; *in-* = inward; *-tegere* = to cover) is composed of the skin, hair, oil and sweat glands, nails, and sensory receptors. The integumentary system helps maintain a constant body temperature, protects the body, and provides sensory information about the surrounding environment. Of all the body's organs, none is more easily inspected or more exposed to infection, disease, and injury than the skin. Although its location makes it vulnerable to damage from trauma, sunlight, microbes, and pollutants in the environment, the skin's protective features ward off such damage. Because of its visibility, skin reflects our emotions (frowning, blushing) and some aspects of normal physiology (such as sweating). Changes in skin color may also indicate homeostatic imbalances in the body. For example, the bluish skin color associated with hypoxia (oxygen deficiency at the tissue level) is one sign of heart failure as well as other disorders. Abnormal skin eruptions or rashes such as chickenpox, cold sores, or measles may reveal systemic infections or diseases of internal organs, while other conditions, such as warts, age spots, or pimples, may involve the skin alone. So important is the skin to self-image that many people spend a great deal of time and money to restore it to a more normal or youthful appearance. **Dermatology** (der′-ma-TOL-ō-jē; *dermato-* = skin; *-logy* = study of) is the medical specialty that deals with the diagnosis and treatment of integumentary system disorders.

? ***Did you ever wonder why it is so difficult to save the life of someone with extensive third-degree burns?***

5.1 STRUCTURE OF THE SKIN

OBJECTIVES

- Describe the layers of the epidermis and the cells that compose them.
- Compare the composition of the papillary and reticular regions of the dermis.
- Explain the basis for different skin colors.

The **skin,** also known as the **cutaneous membrane** (kū-TĀ-nē-us), covers the external surface of the body and is the largest organ of the body in both surface area and weight. In adults, the skin covers an area of about 2 square meters (22 square feet) and weighs 4.5–5 kg (10–11 lb), about 7% of total body weight. It ranges in thickness from 0.5 mm (0.02 in.) on the eyelids to 4.0 mm (0.16 in.) on the heels. Over most of the body it is 1–2 mm (0.04–0.08 in.) thick. The skin consists of two main parts (Figure 5.1). The superficial,

Figure 5.1 **Components of the integumentary system.** The skin consists of a superficial, thin epidermis and a deep, thicker dermis. Deep to the skin is the subcutaneous layer, which attaches the dermis to underlying fascia.

The integumentary system includes the skin, hair, oil and sweat glands, nails, and sensory receptors.

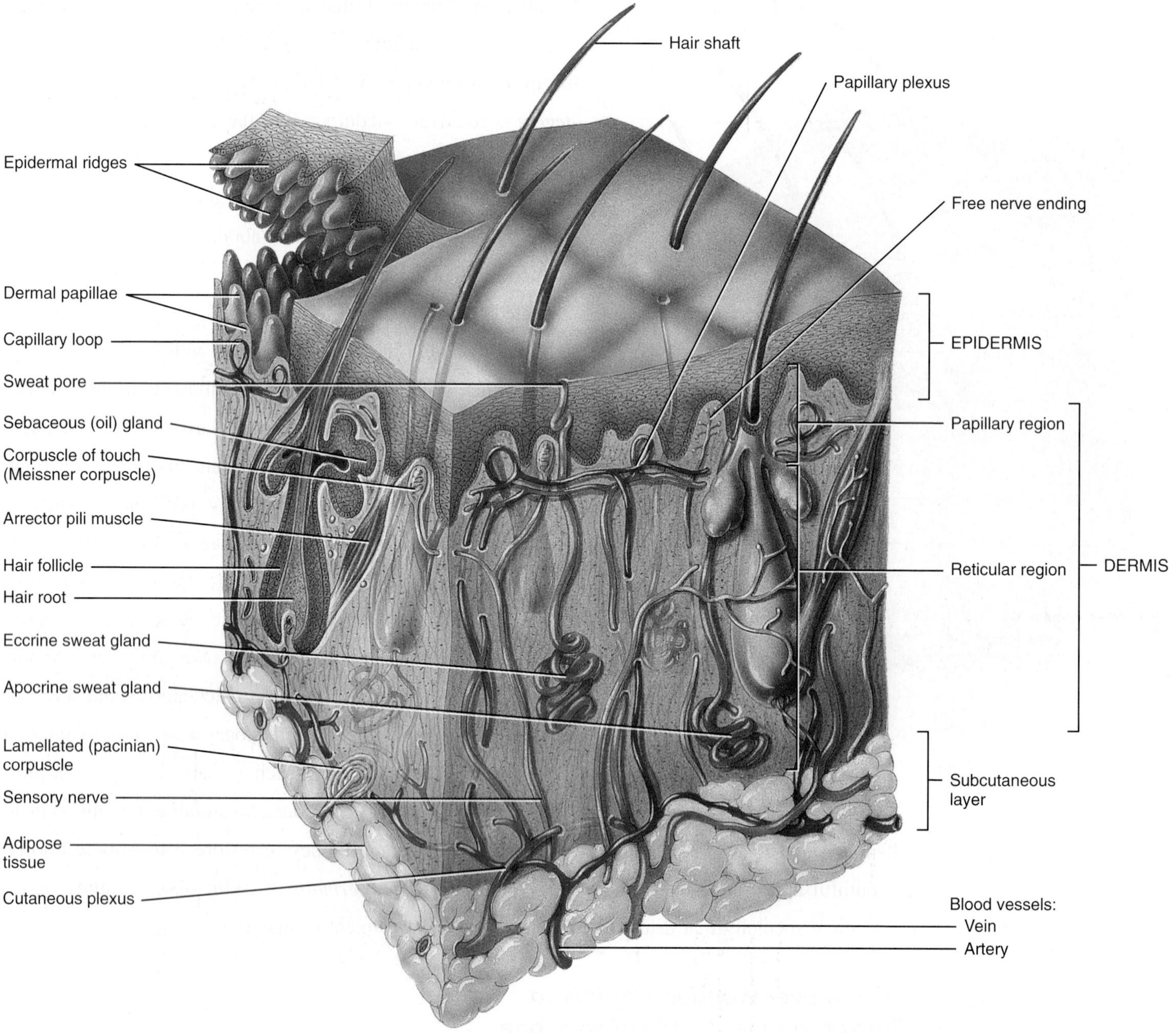

(a) Sectional view of skin and subcutaneous layer

thinner portion, which is composed of *epithelial tissue,* is the **epidermis** (ep′-i-DERM-is; *epi-* = above). The deeper, thicker *connective tissue* portion is the **dermis.** While the epidermis is avascular, the dermis is vascular. For this reason, if you cut the epidermis there is no bleeding, but if you cut the dermis there is bleeding.

Deep to the dermis, but not part of the skin, is the **subcutaneous (subQ) layer.** Also called the **hypodermis** (*hypo-* = below), this layer consists of areolar and adipose tissues. Fibers that extend from the dermis anchor the skin to the subcutaneous layer, which in turn attaches to underlying fascia, the connective tissue around muscles and bones. The subcutaneous layer serves as a storage depot for fat and contains large blood vessels that supply the skin. This region (and sometimes the dermis) also contains nerve endings called **pacinian (lamellated) corpuscles** (pa-SIN-ē-an) that are sensitive to pressure (Figure 5.1).

Epidermis

The **epidermis** is composed of keratinized stratified squamous epithelium. It contains four principal types of cells: keratinocytes, melanocytes, Langerhans cells, and Merkel cells

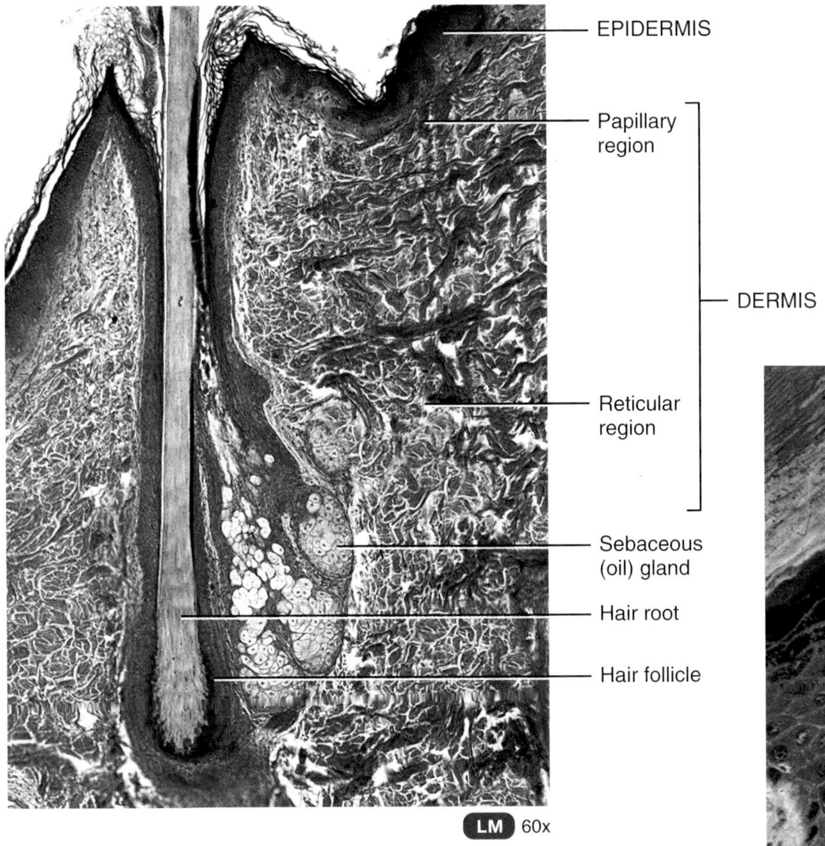

(b) Sectional view of skin

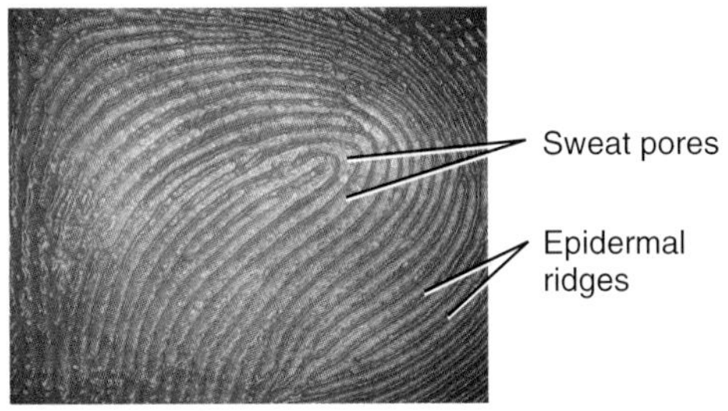

(c) Epidermal ridges and sweat pores

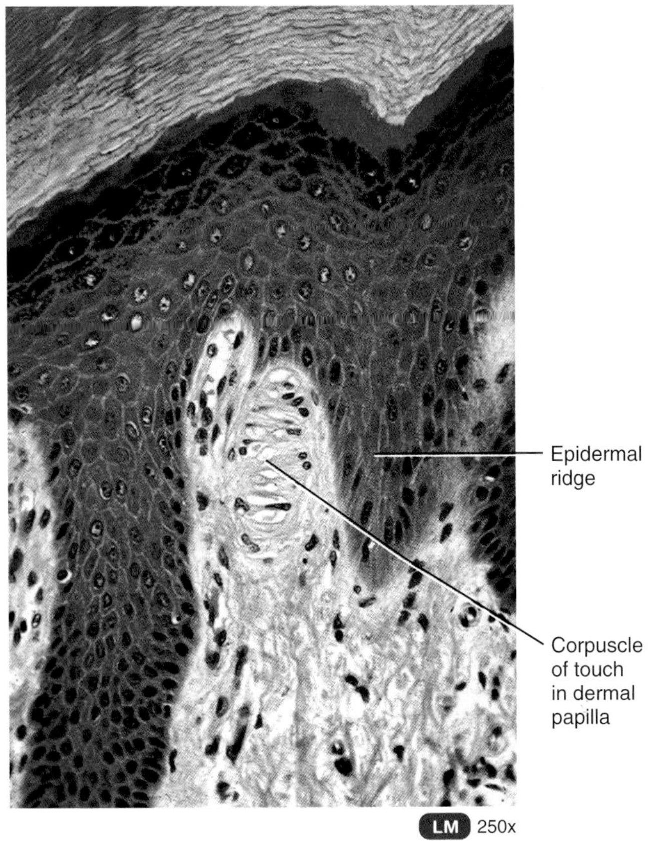

(d) Sectional view of dermal papillae and epidermal ridges

FUNCTIONS OF THE INTEGUMENTARY SYSTEM

1. Regulates body temperature.
2. Stores blood.
3. Protects body from external environment.
4. Detects cutaneous sensations.
5. Excretes and absorbs substances.
6. Synthesizes vitamin D.

? What types of tissues make up the epidermis and the dermis?

(Figure 5.2). About 90% of epidermal cells are **keratinocytes** (ker-a-TIN-ō-sīts′; *keratino-* = hornlike; *-cytes* = cells), which are arranged in four or five layers and produce the protein **keratin** (KER-a-tin) (Figure 5.2a). Recall from Chapter 4 that keratin is a tough, fibrous protein that helps protect the skin and underlying tissues from abrasions, heat, microbes, and chemicals. Keratinocytes also produce lamellar granules, which release a water-repellent sealant that decreases water entry and loss and inhibits the entry of foreign materials.

About 8% of the epidermal cells are **melanocytes** (MEL-a-nō-sīts′; *melano-* = black), which develop from the ectoderm of a developing embryo and produce the pigment melanin (Figure 5.2b). Their long, slender projections extend between the keratinocytes and transfer melanin granules to them. **Melanin** (MEL-a-nin) is a yellow-red or brown-black pigment that contributes to skin color and absorbs damaging ultraviolet (UV) light. Once inside keratinocytes, the melanin granules cluster to form a protective veil over the nucleus, on the side toward the skin surface. In this way, they shield the nuclear DNA from damage by UV light. Although their melanin granules effectively protect keratinocytes, melanocytes themselves are particularly susceptible to damage by UV light.

Figure 5.2 Cells in the epidermis. Besides keratinocytes, the epidermis contains melanocytes, which produce the pigment melanin; Langerhans cells, which participate in immune responses; and Merkel cells, which function in the sensation of touch.

Most of the epidermis consists of keratinocytes, which produce the protein keratin (protects underlying tissues) and lamellar granules (contain a waterproof sealant).

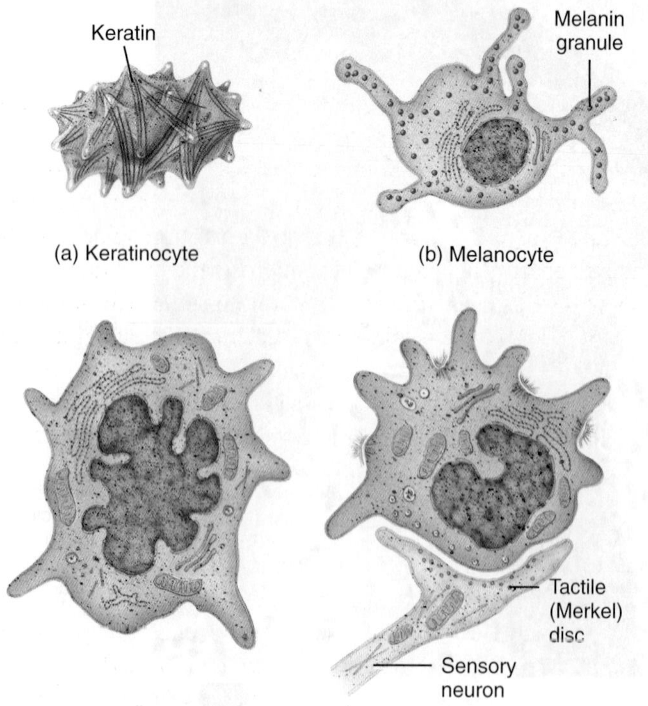

What is the function of melanin?

Langerhans cells (LANG-er-hans), also called *epidermal dendritic cells,* arise from red bone marrow and migrate to the epidermis (Figure 5.2c), where they constitute a small fraction of the epidermal cells. They participate in immune responses mounted against microbes that invade the skin, and are easily damaged by UV light. Their role in the immune response is to help other cells of the immune system recognize an invading microbe and destroy it.

Merkel cells (MER-kel) are the least numerous of the epidermal cells. They are located in the deepest layer of the epidermis, where they contact the flattened process of a sensory neuron (nerve cell), a structure called a **Merkel (tactile) disc** (Figure 5.2d). Merkel cells and their associated Merkel discs detect touch sensations.

Several distinct layers of keratinocytes in various stages of development form the epidermis (Figure 5.3). In most regions of the body the epidermis has four strata or layers—stratum basale, stratum spinosum, stratum granulosum, and a thin stratum corneum. This is called **thin skin.** Where exposure to friction is greatest, such as in the fingertips, palms, and soles, the epidermis has five layers—stratum basale, stratum spinosum, stratum granulosum, stratum lucidum, and a thick stratum corneum. This is called **thick skin.** The details of thin and thick skin are discussed later in the chapter.

Stratum Basale

The deepest layer of the epidermis is the **stratum basale** (ba-SA-lē; *basal* = base), composed of a single row of cuboidal or columnar keratinocytes. Some cells in this layer are *stem cells* that undergo cell division to continually produce new keratinocytes. The nuclei of keratinocytes in the stratum basale are large, and their cytoplasm contains many ribosomes, a small Golgi complex, a few mitochondria, and some rough endoplasmic reticulum. The cytoskeleton within keratinocytes of the stratum basale includes scattered intermediate filaments, called *keratin intermediate filaments (tonofilaments).* The keratin intermediate filaments form the tough protein keratin in its more superficial epidermal layers. Keratin protects the deeper layers from injury. Keratin intermediate filaments attach to desmosomes, which bind cells of the stratum basale to each other and to the cells of the adjacent stratum spinosum, and to hemidesmosomes, which bind the keratinocytes to the basement membrane positioned between the epidermis and the dermis. Melanocytes and Merkel cells with their associated Merkel discs are scattered among the keratinocytes of the basal layer. The stratum basale is also known as the **stratum germinativum** (jer′-mi-na-TĒ-vum; *germ-* = sprout) to indicate its role in forming new cells.

CLINICAL CONNECTION | *Skin Grafts*

New skin cannot regenerate if an injury destroys a large area of the stratum basale and its stem cells. Skin wounds of this magnitude require skin grafts in order to heal. A **skin graft** is the transfer of a

Figure 5.3 Layers of the epidermis.

The epidermis consists of keratinized stratified squamous epithelium.

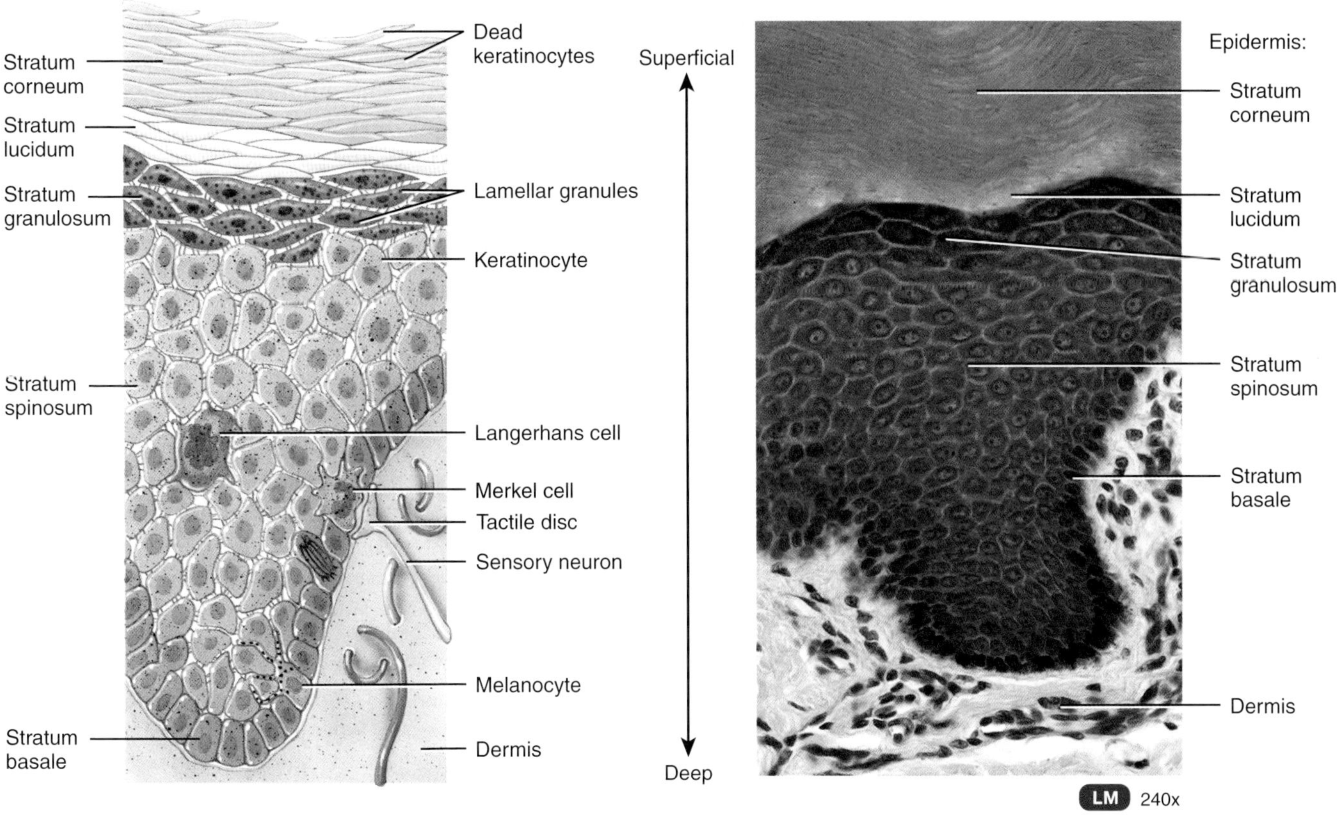

(a) Four principal cell types in epidermis

(b) Photomicrograph of a portion of the skin

Which epidermal layer includes stem cells that continually undergo cell division?

patch of healthy skin taken from a donor site to cover a wound. A skin graft is performed to protect against fluid loss and infection, to promote tissue healing, to reduce scar formation, to prevent loss of function, and for cosmetic reasons. To avoid tissue rejection, the transplanted skin is usually taken from the same individual (*autograft*) or an identical twin (*isograft*). If skin damage is so extensive that an autograft would cause harm, a self-donation procedure called *autologous skin transplantation* (aw-TOL-ō-gus) may be used. In this procedure, performed most often for severely burned patients, small amounts of an individual's epidermis are removed, and the keratinocytes are cultured in the laboratory to produce thin sheets of skin. The new skin is transplanted back to the patient so that it covers the burn wound and generates a permanent skin. Also available as skin grafts for wound coverage are products (Apligraft and Transite) grown in the laboratory from the foreskins of circumcised infants. •

Stratum Spinosum

Superficial to the stratum basale is the **stratum spinosum** (spi-NŌ-sum; *spinos-* = thornlike). This stratum mainly consists of numerous keratinocytes arranged in 8–10 layers. Cells in the more superficial layers become somewhat flattened. The keratinocytes in the stratum spinosum, which are produced by the stem cells in the basal layer, have the same organelles as cells of the stratum basale and some retain their ability to divide. The keratinocytes of this layer produce coarser bundles of keratin in intermediate filaments than those of the basal layer. Although they are rounded and larger in living tissue, cells of the stratum spinosum shrink and pull apart when prepared for microscopic examination so that they appear to be covered with thornlike spines (thus, the name) (Figure 5.3a). At each spinelike projection, bundles of keratin intermediate filaments insert into desmosomes, which tightly join the cells to one another. This arrangement provides both strength and flexibility to the skin. Langerhans cells and projections of melanocytes are also present in the stratum spinosum.

Stratum Granulosum

At about the middle of the epidermis, the **stratum granulosum** (gran-ū-LŌ-sum; *granulos-* = little grains) consists of three to five layers of flattened keratinocytes that are undergoing apoptosis. (Recall from Chapter 3 that *apoptosis* is an orderly, genetically programmed cell death in which the nucleus fragments before the cells die.) The nuclei and other organelles of these cells begin to degenerate as they move farther from their source of nutrition (the dermal blood vessels). Even though keratin intermediate filaments are no longer being produced by these cells, they become more apparent because the organelles in the cells are regressing. A distinctive feature of cells in this layer is the presence of darkly staining granules of a protein called **keratohyalin** (ker′-a-tō-HĪ-a-lin), which assembles keratin intermediate filaments into keratin. Also present in the keratinocytes are membrane-enclosed **lamellar granules** (la-MEL-ar), which fuse with the plasma membrane and release a lipid-rich secretion. This secretion is deposited in the spaces between cells of the stratum granulosum, stratum lucidum, and stratum corneum. The lipid-rich secretion acts as a water-repellent sealant, retarding loss and entry of water and entry of foreign materials. As their nuclei break down during apoptosis, the keratinocytes of the stratum granulosum can no longer carry on vital metabolic reactions, and they die. Thus, the stratum granulosum marks the transition between the deeper, metabolically active strata and the dead cells of the more superficial strata.

Stratum Lucidum

The **stratum lucidum** (LOO-si-dum; *lucid-* = clear) is present only in the thick skin of areas such as the fingertips, palms, and soles. It consists of four to six layers of flattened clear, dead keratinocytes that contain large amounts of keratin and thickened plasma membranes. This probably provides an additional level of toughness in this region of thick skin.

Stratum Corneum

The **stratum corneum** (KOR-nē-um; *corne-* = horn or horny) consists on average of 25 to 30 layers of flattened dead keratinocytes, but can range in thickness from a few cells in thin skin to 50 or more cell layers in thick skin. The cells are extremely thin, flat, plasma membrane–enclosed packages of keratin that no longer contain a nucleus or any internal organelles. They are the final product of the differentiation process of the keratinocytes. The cells within each layer overlap one another like the scales on the skin of a snake. Neighboring layers of cells also form strong connections with one another. The plasma membranes of adjacent cells are arranged in complex, wavy folds that fit together like pieces of a jigsaw puzzle to hold the layers together. In this outer stratum of the epidermis, cells are continuously shed and replaced by cells from the deeper strata. Its multiple layers of dead cells help the stratum corneum to protect deeper layers from injury and microbial invasion. Constant exposure of skin to friction stimulates increased cell production and keratin production that results in the formation of a **callus,** an abnormal thickening of the stratum corneum.

Keratinization and Growth of the Epidermis

Newly formed cells in the stratum basale are slowly pushed to the surface. As the cells move from one epidermal layer to the next, they accumulate more and more keratin, a process called **keratinization** (ker′-a-tin-i-ZĀ-shun). Then they undergo apoptosis. Eventually the keratinized cells slough off and are replaced by underlying cells that in turn become keratinized. The whole process by which cells form in the stratum basale, rise to the surface, become keratinized, and slough off takes about four to six weeks in an average epidermis of 0.1 mm (0.004 in.) thickness. Nutrients and oxygen diffuse to the avascular epidermis from blood vessels in the dermis. The epidermal cells of the stratum basale are closest to these blood vessels and receive most of the nutrients and oxygen. These cells are the most active metabolically and continuously undergo cell division to produce new keratinocytes. As the new keratinocytes are pushed farther from the blood supply by continuing cell division, the epidermal strata above the basale receive fewer nutrients and the cells become less active and eventually die. The rate of cell division in the stratum basale increases when the outer layers of the epidermis are stripped away, as occurs in abrasions and burns. The mechanisms that regulate this remarkable growth are not well understood, but hormonelike proteins such as **epidermal growth factor (EGF)** play a role. An excessive amount of keratinized cells shed from the skin of the scalp is called **dandruff.**

Table 5.1 summarizes the distinctive features of the epidermal strata.

CLINICAL CONNECTION | *Psoriasis*

Psoriasis (sō-RĪ-a-sis) is a common and chronic skin disorder in which keratinocytes divide and move more quickly than normal from the stratum basale to the stratum corneum. They are shed prematurely in as little as 7 to 10 days. The immature keratinocytes make an abnormal keratin, which forms flaky, silvery scales at the skin surface, most often on the knees, elbows, and scalp (dandruff). Effective treatments—various topical ointments and ultraviolet phototherapy—suppress cell division, decrease the rate of cell growth, or inhibit keratinization. •

TABLE 5.1

Summary of Epidermal Strata (see Figure 5.3)

STRATUM	DESCRIPTION
Basale	Deepest layer, composed of single row of cuboidal or columnar keratinocytes that contain scattered keratin intermediate filaments (tonofilaments); stem cells undergo cell division to produce new keratinocytes; melanocytes and Merkel cells associated with Merkel discs are scattered among keratinocytes.
Spinosum	Eight to ten rows of many-sided keratinocytes with bundles of keratin intermediate filaments; contains armlike processes of melanocytes and Langerhans cells.
Granulosum	Three to five rows of flattened keratinocytes, in which organelles are beginning to degenerate; cells contain the protein keratohyalin (converts keratin intermediate filaments into keratin) and lamellar granules (release lipid-rich, water-repellent secretion).
Lucidum	Present only in skin of fingertips, palms, and soles; consists of four to six rows of clear, flat, dead keratinocytes with large amounts of keratin.
Corneum	Twenty-five to thirty rows of dead, flat keratinocytes that contain mostly keratin.

Dermis

The second, deeper part of the skin, the **dermis,** is composed of dense irregular connective tissue containing collagen and elastic fibers. This woven network of fibers has great tensile strength (resists pulling or stretching forces). The dermis also has the ability to stretch and recoil easily. It is much thicker than the epidermis, and this thickness varies from region to region in the body, reaching its greatest thickness on the palms and soles. Leather, which we use for belts, shoes, baseball gloves, and basketballs, is the dried and treated dermis of other animals. The few cells present in the dermis include predominantly fibroblasts, with some macrophages, and a few adipocytes near its boundary with the subcutaneous layer. Blood vessels, nerves, glands, and hair follicles (epithelial invaginations of the epidermis) are embedded in the dermal layer. The dermis is essential to the survival of the epidermis, and these adjacent layers form many important structural and functional relations. Based on its tissue structure, the dermis can be divided into a thin superficial papillary region and a thick deeper reticular region.

The **papillary region** makes up about one-fifth of the thickness of the total layer (see Figure 5.1). It consists of thin collagen and fine elastic fibers. Its surface area is greatly increased by **dermal papillae** (pa-PIL-ē = nipples), small, nipple-shaped structures that project into the undersurface of the epidermis. All dermal papillae contain **capillary loops** (blood vessels). Some also contain tactile receptors called **Meissner corpuscles** (MĪS-ner) or *corpuscles of touch,* nerve endings that are sensitive to touch. Still other dermal papillae also contain **free nerve endings,** dendrites that lack any apparent structural specialization. Different free nerve endings initiate signals that give rise to sensations of warmth, coolness, pain, tickling, and itching.

The **reticular region** (*reticul-* = netlike), which is attached to the subcutaneous layer, consists of bundles of thick collagen fibers, scattered fibroblasts, and various wandering cells (such as macrophages). Some adipose cells can be present in the deepest part of the layer, along with some coarse elastic fibers (see Figure 5.1). The collagen fibers in the reticular region are arranged in a netlike manner and have a more regular arrangement than those in the papillary region. The more regular orientation of the large collagen fibers helps the skin resist stretching. Blood vessels, nerves, hair follicles, sebaceous (oil) glands, and sudoriferous (sweat) glands occupy the spaces between fibers.

The combination of collagen and elastic fibers in the reticular region provides the skin with strength, **extensibility** (ek-sten′-si-BIL-i-tē), the ability to stretch, and **elasticity** (e-las-TISS-i-tē), the ability to return to original shape after stretching. The extensibility of skin can be readily seen around joints and in pregnancy and obesity.

CLINICAL CONNECTION | *Stretch Marks*

Because of the collagenous, vascular structure of the dermis, **striae** (STRĪ-ē = streaks) or **stretch marks,** a form of internal scarring, can result from the internal damage to this layer that occurs when the skin is stretched too much. When the skin is overstretched, the lateral bonding between adjacent collagen fibers is disrupted and small dermal blood vessels rupture. This is why stretch marks initially appear as reddish streaks at these sites. Later, after scar tissue (which is poorly vascularized) forms at these sites of dermal breakdown, the stretch marks appear as silvery white streaks. Stretch marks often occur in the abdominal skin during pregnancy, on the skin of weight-lifters where the skin is stretched by a rapid increase in muscle mass, and in the stretched skin accompanying gross obesity. •

The surfaces of the palms, fingers, soles, and toes have a series of ridges and grooves. They appear either as straight lines or as a pattern of loops and whorls, as on the tips of the digits. These **epidermal ridges** are produced during the third month of fetal development as downward projections of the epidermis into the dermis between the dermal papillae of the papillary region (see Figure 5.1). The epidermal ridges create a strong bond between the epidermis and dermis in a region of high mechanical stress. The epidermal ridges also increase the surface area of the epidermis and thus increase the grip of the hand or foot by increasing friction. Finally, the epidermal ridges greatly increase surface area, which increases the number of Meissner corpuscles and thus increases tactile sensitivity. Because the ducts of sweat glands open on the tops of the epidermal ridges as sweat pores, the sweat and ridges form **fingerprints** (or **footprints**) on touching a smooth object. The epidermal ridge pattern is in part genetically determined and is unique for each individual. Even identical twins have different patterns. Normally, the ridge pattern does not change during life, except to enlarge, and thus can serve as the basis for identification. The study of the pattern of epidermal ridges is called **dermatoglyphics** (der′-ma-tō-GLIF-iks; *glyphe* = carved work).

In addition to forming epidermal ridges, the complex papillary surface of the dermis has other functional properties. The dermal papillae greatly increase the surface contact between the dermis and epidermis. This increased dermal contact surface, with its extensive network of small blood vessels, serves as an important source of nutrition for the overlying epidermis. Molecules diffuse from the small blood capillaries in the dermal papillae to the cells of the stratum basale, allowing the basal epithelial stem cells to divide and the keratinocytes to grow and develop. As keratinocytes push toward the surface and away from the dermal blood source, they are no longer able to obtain the nutrition they require, leading to the eventual breakdown of their organelles.

The dermal papillae fit together with the complementary epidermal ridge to form an extremely strong junction between the two layers. This jigsaw puzzle–like connection strengthens the skin against shearing forces (forces that laterally shift in relation to each other) that attempt to separate the epidermis from the dermis.

CLINICAL CONNECTION | *Tension Lines and Surgery*

In certain regions of the body, collagen fibers within the reticular region tend to orient more in one direction than another because of natural tension experienced by these regions of the skin resulting from bony projections, orientation of muscles, and movements of joints. **Tension lines (lines of cleavage)** in the skin indicate the predominant direction of underlying collagen fibers. Knowledge of tension lines is especially important to plastic surgeons. For example, a surgical incision running parallel to the collagen fibers will heal with only a fine scar. A surgical incision made across the rows of fibers disrupts the collagen, and the wound tends to gape open and heal in a broad, thick scar. •

Table 5.2 summarizes the structural features of the papillary and reticular regions of the dermis.

The Structural Basis of Skin Color

Melanin, hemoglobin, and carotene are three pigments that impart a wide variety of colors to skin. The amount of **melanin** causes the skin's color to vary from pale yellow to reddish-brown to black. The difference between the two forms of melanin, *pheomelanin* (fē-ō-MEL-a-nin) (yellow to red) and *eumelanin* (ū-MEL-a-nin) (brown to black), is most apparent in the hair. Melanocytes, the melanin-producing cells, are most plentiful in the epidermis of the penis, nipples of the breasts, area just around the nipples (areolae), face, and limbs. They are also present in mucous membranes. Because the *number* of melanocytes is about the same in all people, differences in skin color are due mainly to the *amount of pigment* the melanocytes produce and transfer to keratinocytes. In some people who are genetically predisposed, melanin accumulates in patches called *freckles.* Freckles typically are reddish or brown and tend to be more visible in the summer than the winter. As a person ages, *age (liver) spots* may develop. These flat blemishes have nothing to do with the liver. They look like freckles and range in color from light brown to black. Like freckles, age spots are accumulations of melanin. Age spots are darker than freckles and build up over time due to exposure to sunlight. Age spots do not fade away during the winter months and are more common in adults over forty. A round, flat, or raised area that represents a benign localized overgrowth of melanocytes and usually develops in childhood or adolescence is called a **nevus** (NĒ-vus), or a **mole.**

TABLE 5.2

Summary of Papillary and Reticular Regions of the Dermis (see Figure 5.1b)

REGION	DESCRIPTION
Papillary	Superficial portion of dermis (about one-fifth); consists of dense irregular connective tissue with thin collagen and fine elastic fibers; contains dermal ridges that house capillaries, Meissner corpuscles, and free nerve endings.
Reticular	Deeper portion of dermis (about four-fifths); consists of dense irregular connective tissue with bundles of thick collagen and some coarse elastic fibers. Spaces between fibers contain some adipose cells, hair follicles, nerves, sebaceous glands, and sudoriferous glands.

Melanocytes synthesize melanin from the amino acid *tyrosine* in the presence of an enzyme called *tyrosinase.* Synthesis occurs in an organelle called a **melanosome** (MEL-an-ō-sōm). Exposure to ultraviolet (UV) light increases the enzymatic activity within melanosomes and thus increases melanin production. Both the amount and darkness of melanin increase on UV exposure, which gives the skin a tanned appearance and helps protect the body against further UV radiation. Melanin absorbs UV radiation, prevents damage to DNA in epidermal cells, and neutralizes free radicals that form in the skin following damage by UV radiation. Thus, within limits, melanin serves a protective function. In response to DNA damage, melanin production increases. As you will see later, exposing the skin to a *small* amount of UV light is actually necessary for the skin to begin the process of vitamin D synthesis. However, repeatedly exposing the skin to a *large* amount of UV light may cause skin cancer. A tan is lost when the melanin-containing keratinocytes are shed from the stratum corneum.

CLINICAL CONNECTION | *Albinism and Vitiligo*

Albinism (AL-bin-izm; *albin-* = white) is the inherited inability of an individual to produce melanin. Most **albinos** (al-BĪ-nōs), people affected by albinism, have melanocytes that are unable to synthesize tyrosinase. Melanin is missing from their hair, eyes, and skin. This results in problems with vision and a tendency of the skin to burn easily on overexposure to sunlight.

In another condition, called **vitiligo** (vit-i-LĪ-gō), the partial or complete loss of melanocytes from patches of skin produces irregular white spots. The loss of melanocytes may be related to an immune system malfunction in which antibodies attack the melanocytes. •

Dark-skinned individuals have large amounts of melanin in the epidermis, so their skin color ranges from yellow to reddish-brown to black. Light-skinned individuals have little melanin in the epidermis. Thus, the epidermis appears translucent, and skin color ranges from pink to red depending on the oxygen content of the blood moving through capillaries in the dermis. The red color is due to **hemoglobin** (hē-mō-GLŌ-bin), the oxygen-carrying pigment in red blood cells.

Carotene (KAR-ō-tēn; *carot* = carrot) is a yellow-orange pigment that gives egg yolks and carrots their color. This precursor of vitamin A, which is used to synthesize pigments needed for vision, is stored in the stratum corneum and fatty areas of the dermis and subcutaneous layer in response to excessive dietary intake. In fact, so much carotene may be deposited in the skin after eating large amounts of carotene-rich foods that the skin actually turns orange, which is especially apparent in light-skinned individuals. Decreasing carotene intake eliminates the problem.

CLINICAL CONNECTION | *Skin Color as a Diagnostic Clue*

The color of skin and mucous membranes can provide clues for diagnosing certain conditions. When blood is not picking up an adequate amount of oxygen from the lungs, as in someone who has stopped breathing, the mucous membranes, nail beds, and skin appear bluish or **cyanotic** (sī-a-NOT-ik; *cyan-* = blue). **Jaundice** (JON-dis; *jaund-* = yellow) is due to a buildup of the yellow pigment bilirubin in the skin. This condition gives a yellowish appearance to the skin and the whites of the eyes, and usually indicates liver disease. **Erythema** (er-e-THĒ-ma; *eryth-* = red), redness of the skin, is caused by engorgement of capillaries in the dermis with blood due to skin injury, exposure to heat, infection, inflammation, or allergic reactions. **Pallor** (PAL-or), or paleness of the skin, may occur in conditions such as shock and anemia. All skin color changes are observed most readily in people with lighter-colored skin and may be more difficult to discern in people with darker skin. However, examination of the nail beds and gums can provide some information about circulation in individuals with darker skin. •

Tattooing and Body Piercing

Tattooing is a permanent coloration of the skin in which a foreign pigment is deposited with a needle into the dermis. It is believed that the practice originated in ancient Egypt between 4000 and 2000 B.C. Today, tattooing is performed in one form or another by nearly all peoples of the world, and it is estimated that about one in five U.S. college students has one or more tattoos. They are created by injecting ink with a needle that punctures the epidermis, moves between 50 and 3000 times per minute, and deposits the ink in the dermis. Since the dermis is stable (unlike the epidermis, which is shed about every four to six weeks), tattoos are permanent. However, they can fade over time due to exposure to sunlight, improper healing, picking scabs, and flushing away of ink particles by the lymphatic system. Tattoos can be removed by lasers, which use concentrated beams of light. In the procedure, which requires a series of treatments, the tattoo inks and pigments selectively absorb the high-intensity laser light without destroying normal surrounding skin tissue. The laser causes the tattoo to dissolve into small ink particles that are eventually removed by the immune system. Laser removal of tattoos involves a considerable investment in time and money and can be quite painful.

Body piercing, the insertion of jewelry through an artificial opening, is also an ancient practice employed by Egyptian pharaohs and Roman soldiers, and a current tradition among many Americans. Today it is estimated that about one in three U.S. college students has had a body piercing. For most piercing locations, the piercer cleans the skin with an antiseptic, retracts the skin with forceps, and pushes a needle through the skin. Then the jewelry is connected to the needle and pushed through the skin. Total healing can take up to a year. Among the sites that are pierced are the ears, nose, eyebrows, lips, tongue, nipples, navel, and genitals. Potential complications of body piercing are infections, allergic reactions, and anatomical damage (such as nerve damage or cartilage deformation). In addition, body piercing jewelry may interfere with certain medical procedures such as masks used for resuscitation, airway management procedures, urinary catheterization, radiographs, and delivery of a baby.

✔ **CHECKPOINT**

1. What structures are included in the integumentary system?
2. How does the process of keratinization occur?
3. What are the structural and functional differences between the epidermis and dermis?
4. How are epidermal ridges formed?
5. What are the three pigments in the skin and how do they contribute to skin color?
6. What is a tattoo? What are some potential problems associated with body piercing?

5.2 ACCESSORY STRUCTURES OF THE SKIN

OBJECTIVE

- Contrast the structure, distribution, and functions of hair, skin glands, and nails.

Accessory structures of the skin—hair, skin glands, and nails—develop from the embryonic epidermis. They have a host of important functions. For example, hair and nails protect the body, and sweat glands help regulate body temperature.

Hair

Hairs, or *pili* (PĪ-lī), are present on most skin surfaces except the palms, palmar surfaces of the fingers, the soles, and plantar surfaces of the feet. In adults, hair usually is most heavily distributed across the scalp, in the eyebrows, in the axillae (armpits), and around the external genitalia. Genetic and hormonal influences largely determine the thickness and the pattern of hair distribution.

Although the protection it offers is limited, hair on the head guards the scalp from injury and the sun's rays. It also decreases heat loss from the scalp. Eyebrows and eyelashes protect the eyes from foreign particles, similar to the way hair in the nostrils and in the external ear canal defends those structures. Touch receptors (hair root plexuses) associated with hair follicles are activated whenever a hair is moved even slightly. Thus, hairs also function in sensing light touch.

Anatomy of a Hair

Each hair is composed of columns of dead, keratinized epidermal cells bonded together by extracellular proteins. The **shaft** is the superficial portion of the hair, which projects above the surface of the skin (Figure 5.4a). The **root** is the portion of the hair deep to the shaft that penetrates into the dermis, and sometimes into the subcutaneous layer. The shaft and root of the hair both consist of three concentric layers of cells: medulla, cortex, and cuticle of the hair (Figure 5.4c, d). The inner *medulla,* which may be lacking in thinner hair, is composed of two or three rows of irregularly shaped cells that contain large amounts of pigment granules in

Figure 5.4 Hair.

Hairs are growths of epidermis composed of dead, keratinized epidermal cells.

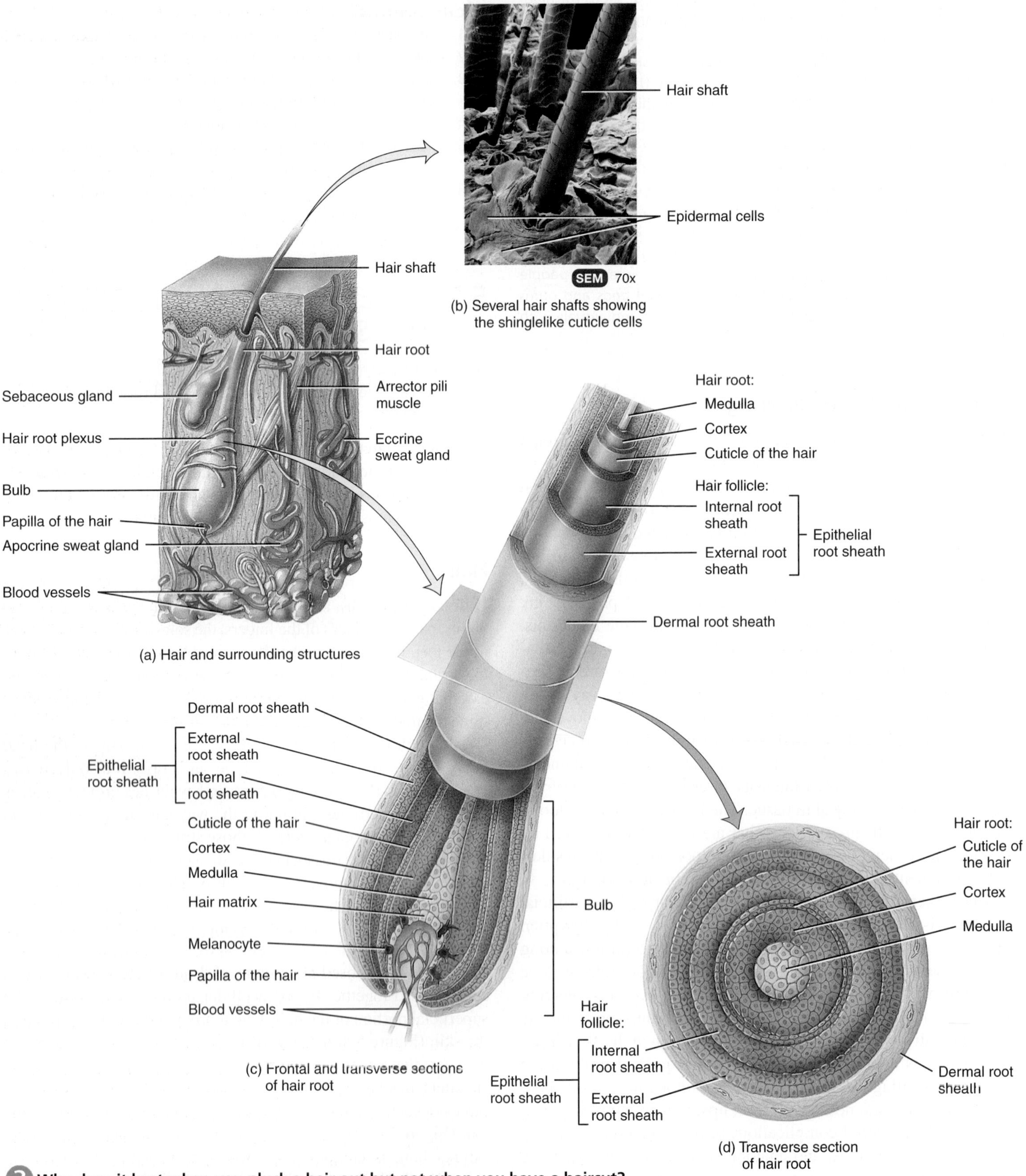
Hair shaft
Epidermal cells
SEM 70x
Hair shaft
Hair root
Sebaceous gland
Arrector pili muscle
Hair root plexus
Eccrine sweat gland
Bulb
Papilla of the hair
Apocrine sweat gland
Blood vessels
Hair root:
Medulla
Cortex
Cuticle of the hair
Hair follicle:
Internal root sheath
External root sheath
Epithelial root sheath
Dermal root sheath
Dermal root sheath
Epithelial root sheath
External root sheath
Internal root sheath
Cuticle of the hair
Cortex
Medulla
Hair matrix
Melanocyte
Papilla of the hair
Blood vessels
Bulb
Hair root:
Cuticle of the hair
Cortex
Medulla
Hair follicle:
Epithelial root sheath
Internal root sheath
External root sheath
Dermal root sheath

(a) Hair and surrounding structures

(b) Several hair shafts showing the shinglelike cuticle cells

(c) Frontal and transverse sections of hair root

(d) Transverse section of hair root

Why does it hurt when you pluck a hair out but not when you have a haircut?

dark hair, small amounts of pigment granules in gray hair, and a lack of pigment granules and the presence of air bubbles in white hair. The middle *cortex* forms the major part of the shaft and consists of elongated cells. The *cuticle of the hair,* the outermost layer, consists of a single layer of thin, flat cells that are the most heavily keratinized. Cuticle cells on the shaft are arranged like shingles on the side of a house, with their free edges pointing toward the end of the hair (Figure 5.4b).

Surrounding the root of the hair is the **hair follicle** (FOL-i-kul), which is made up of an external root sheath and an internal root sheath, together referred to as an **epithelial root sheath** (Figure 5.4c, d). The *external root sheath* is a downward continuation of the epidermis. The *internal root sheath* is produced by the matrix (described shortly) and forms a cellular tubular sheath of epithelium between the external root sheath and the hair. The dense dermis surrounding the hair follicle is called the **dermal root sheath.**

The base of each hair follicle and its surrounding dermal root sheath is an onion-shaped structure, the **bulb** (Figure 5.4c). This structure houses a nipple-shaped indentation, the **papilla of the hair,** which contains areolar connective tissue and many blood vessels that nourish the growing hair follicle. The bulb also contains a germinal layer of cells called the **hair matrix.** The hair matrix cells arise from the stratum basale, the site of cell division. Hence, hair matrix cells are responsible for the growth of existing hairs, and they produce new hairs when old hairs are shed. This replacement process occurs within the same follicle. Hair matrix cells also give rise to the cells of the internal root sheath.

CLINICAL CONNECTION | *Hair Removal*

A substance that removes hair is called a **depilatory** (de-PIL-a-tō-rē). It dissolves the protein in the hair shaft, turning it into a gelatinous mass that can be wiped away. Because the hair root is not affected, regrowth of the hair occurs. In **electrolysis,** an electric current is used to destroy the hair matrix so the hair cannot regrow. **Laser treatments** may also be used to remove hair. •

Sebaceous (oil) glands (discussed shortly) and a bundle of smooth muscle cells are also associated with hairs (Figure 5.4a). The smooth muscle is the **arrector pili** (a-REK-tor PĪ-lī; *arrect-* = to raise). It extends from the superficial dermis of the skin to the dermal root sheath around the side of the hair follicle. In its normal position, hair emerges at an angle to the surface of the skin. Under physiological or emotional stress, such as cold or fright, autonomic nerve endings stimulate the arrector pili muscles to contract, which pulls the hair shafts perpendicular to the skin surface. This action causes "goose bumps" or "gooseflesh" because the skin around the shaft forms slight elevations.

Surrounding each hair follicle are dendrites of neurons that form a **hair root plexus** (PLEK-sus), which is sensitive to touch (Figure 5.4a). The hair root plexuses generate nerve impulses if their hair shafts are moved.

Hair Growth

Each hair follicle goes through a growth cycle, which consists of a growth stage, a regression stage, and a resting stage. During the **growth stage,** cells of the hair matrix divide. As new cells from the hair matrix are added to the base of the hair root, existing cells of the hair root are pushed upward and the hair grows longer. While the cells of the hair are being pushed upward, they become keratinized and die. Following the growth stage is the **regression stage,** when the cells of the hair matrix stop dividing, the hair follicle atrophies (shrinks), and the hair stops growing. After the regression stage, the hair follicle enters a **resting stage.** Following the resting stage, a new growth cycle begins. The old hair root falls out or is pushed out of the hair follicle, and a new hair begins to grow in its place. Scalp hair is in the growth stage for 2 to 6 years, the regression stage for 2 to 3 weeks, and the resting stage for about 3 months. At any time, about 85% of scalp hairs are in the growth stage. Visible hair is dead, but until the hair is pushed out of its follicle by a new hair, portions of its root within the scalp are alive.

Normal hair loss in the adult scalp is about 70–100 hairs per day. Both the rate of growth and the replacement cycle may be altered by illness, radiation therapy, chemotherapy, age, genetics, gender, and severe emotional stress. Rapid weight-loss diets that severely restrict calories or protein increase hair loss. The rate of shedding also increases for three to four months after childbirth. **Alopecia** (al′-ō-PĒ-shē-a), the partial or complete lack of hair, may result from genetic factors, aging, endocrine disorders, chemotherapy, or skin disease.

CLINICAL CONNECTION | *Chemotherapy and Hair Loss*

Chemotherapy is the treatment of disease, usually cancer, by means of chemical substances or drugs. Chemotherapeutic agents interrupt the life cycle of rapidly dividing cancer cells. Unfortunately, the drugs also affect other rapidly dividing cells in the body, such as the hair matrix cells of a hair. It is for this reason that individuals undergoing chemotherapy experience hair loss. Since about 15% of the hair matrix cells of scalp hairs are in the resting stage, these cells are not affected by chemotherapy. Once chemotherapy is stopped, the hair matrix cells replace lost hair follicles and hair growth resumes. •

Types of Hairs

Hair follicles develop at about twelve weeks after fertilization. Usually by the fifth month of development, the follicles produce very fine, nonpigmented, downy hairs called **lanugo** (la-NOO-gō = wool or down) that cover the body of the fetus. Prior to birth, the lanugo of the eyebrows, eyelashes, and scalp are shed and replaced by long, coarse, heavily pigmented hairs called **terminal hairs.** The lanugo of the rest of the body are replaced by **vellus hairs** (VEL-us = fleece), commonly called "peach fuzz," which are short, fine, pale hairs that are barely visible to the naked eye. During childhood, vellus hairs cover most of the body except for the hairs of the eyebrows, eyelashes, and scalp, which are terminal hairs. In response to hormones (androgens) secreted at puberty, terminal hairs replace vellus hairs in the axillae (armpits) and pubic regions of boys and girls and they replace vellus hairs on the face, limbs, and chests of boys, which leads to the formation of a mustache, a beard, hairy arms and legs, and a hairy chest. During adulthood, about 95% of body hair on males is terminal hair and 5% is vellus hair; on females, about 35% of body hair is terminal hair and 65% is vellus hair.

Hair Color

The color of hair is due primarily to the amount and type of melanin in its keratinized cells. Melanin is synthesized by melanocytes scattered in the matrix of the bulb and passes into cells of the cortex and medulla of the hair (Figure 5.4c). Dark-colored hair contains mostly eumelanin (brown to black); blond and red hair contain variants of pheomelanin (yellow to red). Hair becomes gray because of a progressive decline in melanin production; gray hair contains only a few melanin granules. White hair results from the lack of melanin and the accumulation of air bubbles in the shaft.

Hair coloring is a process that adds or removes pigment. Temporary hair dyes coat the surface of a hair shaft and usually wash out within 2 or 3 shampoos. Semipermanent dyes penetrate the hair shaft moderately and do fade and wash out of hair after about 5 to 10 shampoos. Permanent hair dyes penetrate deeply into the hair shaft and don't wash out but are eventually lost as the hair grows out.

CLINICAL CONNECTION | *Hair and Hormones*

At puberty, when the testes begin secreting significant quantities of androgens (masculinizing sex hormones), males develop the typical male pattern of hair growth throughout the body, including a beard and a hairy chest. In females at puberty, the ovaries and the adrenal glands produce small quantities of androgens, which promote hair growth throughout the body including the axillae and pubic region. Occasionally, a tumor of the adrenal glands, testes, or ovaries produces an excessive amount of androgens. The result in females or prepubertal males is **hirsutism** (HER-soo-tizm; *hirsut-* = shaggy), excessive body hair or body hair in areas that usually are not hairy.

Surprisingly, androgens also must be present for occurrence of the most common form of baldness, **androgenic alopecia** (an′-drō-JEN-ik al′-ō-PĒ-shē-a) or **male-pattern baldness.** In genetically predisposed adults, androgens inhibit hair growth. In men, hair loss usually begins with a receding hairline followed by hair loss in the temples and crown. Women are more likely to have thinning of hair on top of the head. The first drug approved for enhancing scalp hair growth was minoxidil (Rogaine). It causes vasodilation (widening of blood vessels), thus increasing circulation; direct stimulation of hair follicle cells to pass into growth stage follicles; and inhibition of androgens. In about a third of the people who try it, minoxidil improves hair growth, causing scalp follicles to enlarge and lengthening the growth cycle. For many, however, the hair growth is meager. Minoxidil does not help people who already are bald. •

Skin Glands

Recall from Chapter 4 that glands are epithelial cells that secrete a substance. Several kinds of exocrine glands are associated with the skin: sebaceous (oil) glands, sudoriferous (sweat) glands, and ceruminous glands. Mammary glands, which are specialized sudoriferous glands that secrete milk, are discussed in Chapter 28 along with the female reproductive system.

Sebaceous Glands

Sebaceous glands (se-BĀ-shus; *sebace-* = greasy) or **oil glands** are simple, branched acinar (rounded) glands. With few exceptions, they are connected to hair follicles (see Figures 5.1 and 5.4a). The secreting portion of a sebaceous gland lies in the dermis and usually opens into the neck of a hair follicle. In some locations, such as the lips, glans penis, labia minora, and tarsal glands of the eyelids, sebaceous glands open directly onto the surface of the skin. Absent in the palms and soles, sebaceous glands are small in most areas of the trunk and limbs, but large in the skin of the breasts, face, neck, and superior chest.

Sebaceous glands secrete an oily substance called **sebum** (SĒ-bum), a mixture of triglycerides, cholesterol, proteins, and inorganic salts. Sebum coats the surface of hairs and helps keep them from drying and becoming brittle. Sebum also prevents excessive evaporation of water from the skin, keeps the skin soft and pliable, and inhibits the growth of some (but not all) bacteria.

CLINICAL CONNECTION | *Acne*

During childhood, sebaceous glands are relatively small and inactive. At puberty, androgens from the testes, ovaries, and adrenal glands stimulate sebaceous glands to grow in size and increase their production of sebum. **Acne** is an inflammation of sebaceous glands that usually begins at puberty, when the sebaceous glands are stimulated by androgens. Acne occurs predominantly in sebaceous follicles that have been colonized by bacteria, some of which thrive in the lipid-rich sebum. The infection may cause a cyst or sac of connective tissue cells to form, which can destroy and displace epidermal cells. This condition, called **cystic acne,** can permanently scar the epidermis. Treatment consists of gently washing the affected areas once or twice daily with a mild soap, topical antibiotics (such as clindamycin and erythromycin), topical drugs such as benzoyl peroxide or tretinoin, and oral antibiotics (such as tetracycline, minocycline, erythromycin, and isotretinoin). Contrary to popular belief, foods such as chocolate or fried foods do not cause or worsen acne. •

Sudoriferous Glands

There are three million to four million **sweat glands,** or **sudoriferous glands** (soo′-dor-IF-er-us; *sudor-* = sweat; *-ferous* = bearing) in the body. The cells of these glands release sweat, or perspiration, into hair follicles or onto the skin surface through pores. Sweat glands are divided into two main types, eccrine and apocrine, based on their structure and type of secretion.

Eccrine sweat glands (EK-rin; *eccrine* = secreting outwardly) are simple, coiled tubular glands that are much more common than apocrine sweat glands (see Figures 5.1 and 5.4a). They are distributed throughout the skin of most regions of the body, especially in the skin of the forehead, palms, and soles. Eccrine sweat glands are not present, however, in the margins of the lips, nail beds of the fingers and toes, glans penis, glans clitoris, labia minora, or eardrums. The secretory portion of eccrine sweat glands is located mostly in the deep dermis (sometimes in the upper subcutaneous layer). The excretory duct projects through the dermis and epidermis and ends as a pore at the surface of the epidermis (see Figure 5.1).

The sweat produced by eccrine sweat glands (about 600 mL per day) consists of water, ions (mostly Na^+ and Cl^-), urea, uric acid, ammonia, amino acids, glucose, and lactic acid. The main function of eccrine sweat glands is to help regulate body temperature through evaporation. As sweat evaporates, large quantities of heat energy leave the body surface. The homeostatic regulation of body temperature is known as **thermoregulation.** This role of eccrine sweat glands in helping the body to achieve thermoregulation is known as **thermoregulatory sweating.** During thermoregulatory sweating, sweat first forms on the forehead and scalp and then extends to the rest of the body, forming last on the palms and soles. Sweat that evaporates from the skin before it is perceived as moisture is termed **insensible perspiration** (*in-* = not). Sweat that is excreted in larger amounts and is seen as moisture on the skin is called **sensible perspiration.**

The sweat produced by eccrine sweat glands also plays a small role in eliminating wastes such as urea, uric acid, and ammonia from the body. However, the kidneys play more of a role in the excretion of these waste products from the body than eccrine sweat glands.

Eccrine sweat glands also release sweat in response to an emotional stress such as fear or embarrassment. This type of sweating is referred to as **emotional sweating** or a *cold sweat.* In contrast to thermoregulatory sweating, emotional sweating first occurs on the palms, soles, and axillae and then spreads to other areas of the body. As you will soon learn, apocrine sweat glands are also active during emotional sweating.

Apocrine sweat glands (AP-ō-krin; *apo-* = separated from) are also simple, coiled tubular glands (see Figures 5.1 and 5.4a). They are found mainly in the skin of the axilla (armpit), groin, areolae (pigmented areas around the nipples) of the breasts, and bearded regions of the face in adult males. These glands were once thought to release their secretions in an apocrine manner (see text coverage in Chapter 4 and Figure 4.7b)—by pinching off a portion of the cell. We now know, however, that their secretion is via exocytosis, which is characteristic of eccrine glands (see Figure 5.4a). Nevertheless, the term *apocrine* is still used. The secretory portion of these sweat glands is located in the lower dermis or upper subcutaneous layer, and the excretory duct opens into hair follicles (see Figure 5.1).

Compared to eccrine sweat, apocrine sweat appears milky or yellowish in color. Apocrine sweat contains the same components as eccrine sweat plus lipids and proteins. Sweat secreted from apocrine sweat glands is odorless. However, when apocrine sweat interacts with bacteria on the surface of the skin, the bacteria metabolize its components, causing apocrine sweat to have a musky odor that is often referred to as body odor. Eccrine sweat glands start to function soon after birth, but apocrine sweat glands do not begin to function until puberty.

Apocrine sweat glands, along with eccrine sweat glands, are active during emotional sweating. In addition, apocrine sweat glands secrete sweat during sexual activities. In contrast to eccrine sweat glands, apocrine sweat glands are not active during thermoregulatory sweating and, therefore, do not play a role in thermoregulation.

Ceruminous Glands

Modified sweat glands in the external ear, called **ceruminous glands** (se-RŪ-mi-nus; *cer-* = wax), produce a waxy lubricating secretion. The secretory portions of ceruminous glands lie in the subcutaneous layer, deep to sebaceous glands. Their excretory ducts open either directly onto the surface of the external auditory canal (ear canal) or into ducts of sebaceous glands. The combined secretion of the ceruminous and sebaceous glands is a yellowish material called **cerumen** (se-ROO-men), or earwax. Cerumen, together with hairs in the external auditory canal, provides a sticky barrier that impedes the entrance of foreign bodies and insects. Cerumen also waterproofs the canal and prevents bacteria and fungi from entering cells.

Table 5.3 presents a summary of skin glands.

CLINICAL CONNECTION | *Impacted Cerumen*

Some people produce an abnormally large amount of cerumen in the external auditory canal. If it accumulates until it becomes impacted (firmly wedged), sound waves may be prevented from reaching the eardrum. Treatments for **impacted cerumen** include periodic ear irrigation with enzymes to dissolve the wax and removal of wax with a blunt instrument by trained medical personnel. The use of cotton-tipped swabs or sharp objects is not recommended for this purpose because they may push the cerumen further into the external auditory canal and damage the eardrum. •

Nails

Nails are plates of tightly packed, hard, dead, keratinized epidermal cells that form a clear, solid covering over the dorsal surfaces of the distal portions of the digits. Each nail consists of a nail body, a free edge, and a nail root (Figure 5.5). The **nail body (plate)** is the visible portion of the nail. It is comparable to the stratum corneum of the general epidermis, with the exception that its flattened, keratinized cells fill with a harder type of keratin and the cells are not shed. Below the nail body is a region of epithelium and a deeper layer of dermis. Most of the nail body appears pink because of blood flowing through the capillaries in the underlying dermis. The **free edge** is the part of the nail body that may extend past the distal end of the digit. The free edge is white because there are no underlying capillaries. The **nail root** is the portion of the nail that is buried in a fold of skin. The whitish, crescent-shaped area of the proximal end of the nail body is called the **lunula** (LOO-noo-la = little moon). It appears whitish because the vascular tissue underneath does not show through due to a thickened region of epithelium in the area. Beneath the free edge is a thickened region of stratum corneum called the **hyponychium** (hī′-pō-NIK-ē-um; *hypo-* = below; *-onych* = nail), or **nail bed,** which secures the nail to the fingertip. The **eponychium** (ep′-ō-NIK-ē-um; *ep-* = above) or **cuticle** is a narrow band of epidermis that extends from and adheres to the margin (lateral border) of the nail wall. It occupies the proximal border of the nail and consists of stratum corneum.

The proximal portion of the epithelium deep to the nail root is the **nail matrix.** The superficial nail matrix cells divide mitotically

TABLE 5.3

Summary of Skin Glands (see Figures 5.1 and 5.4a)

FEATURE	SEBACEOUS (OIL) GLANDS	ECCRINE SWEAT GLANDS	APOCRINE SWEAT GLANDS	CERUMINOUS GLANDS
Distribution	Largely in lips, glans penis, labia minora, and tarsal glands; small in trunk and limbs; absent in palms and soles.	Throughout skin of most regions of body, especially skin of forehead, palms, and soles.	Skin of axillae, groin, areolae, bearded regions of face, clitoris, and labia minora.	External auditory canal.
Location of secretory portion	Dermis.	Mostly in deep dermis (sometimes in upper subcutaneous layer).	Mostly in deep dermis and upper subcutaneous layer.	Subcutaneous layer.
Termination of excretory duct	Mostly connected to hair follicles.	Surface of epidermis.	Hair follicle.	Surface of external auditory canal or into ducts of sebaceous glands.
Secretion	Sebum (mixture of triglycerides, cholesterol, proteins, and inorganic salts).	Perspiration, which consists of water, ions (Na^+, Cl^-), urea, uric acid, ammonia, amino acids, glucose, and lactic acid.	Perspiration, which consists of same components as eccrine sweat glands plus lipids and proteins.	Cerumen, a waxy material.
Functions	Prevent hairs from drying out, prevent water loss from skin, keep skin soft, inhibit growth of some bacteria.	Regulation of body temperature, waste removal, stimulated during emotional stress.	Stimulated during emotional stress and sexual excitement.	Impede entrance of foreign bodies and insects into external ear canal, waterproof canal, prevent microbes from entering cells.
Onset of function	Relatively inactive during childhood; activated during puberty.	Soon after birth.	Puberty.	Soon after birth.

Figure 5.5 **Nails.** Shown is a fingernail.

Nail cells arise by transformation of superficial cells of the nail matrix.

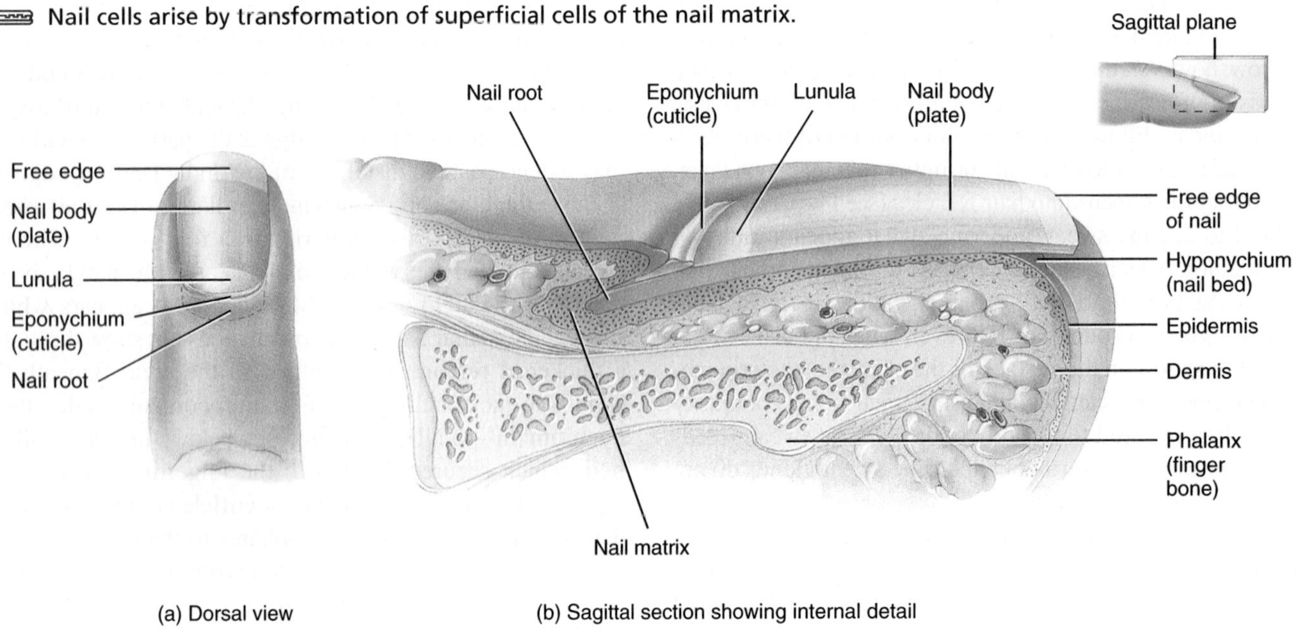

Why are nails so hard?

to produce new nail cells. Nail growth occurs by the transformation of superficial cells of the matrix into new nail cells. The growth rate of nails is determined by the rate of mitosis in matrix cells, which is influenced by factors such as a person's age, health, and nutritional status. Nail growth also varies according to the season, the time of day, and environmental temperature. The average growth in the length of fingernails is about 1 mm (0.04 in.) per week. The growth rate is somewhat slower in toenails. The longer the digit the faster the nail grows.

Nails have a variety of functions:

1. They protect the distal end of the digits.
2. They provide support and counterpressure to the palmar surface of the fingers to enhance touch perception and manipulation.
3. They allow us to grasp and manipulate small objects, and they can be used to scratch and groom the body in various ways.

✔CHECKPOINT

7. Describe the structure of a hair. What causes "goose bumps"?
8. Contrast the locations and functions of sebaceous (oil) glands, sudoriferous (sweat) glands, and ceruminous glands.
9. Describe the parts of a nail.

5.3 TYPES OF SKIN

OBJECTIVE

- Compare structural and functional differences in thin and thick skin.

Although the skin over the entire body is similar in structure, there are quite a few local variations related to thickness of the epidermis, strength, flexibility, degree of keratinization, distribution and type of hair, density and types of glands, pigmentation, vascularity (blood supply), and innervation (nerve supply). Two major types of skin are recognized on the basis of certain structural and functional properties: **thin (hairy) skin** and **thick (hairless) skin.** The greatest contributor to epidermal thickness is the increased number of layers in the stratum corneum. This arises in response to the greater mechanical stress in regions of thick skin.

Table 5.4 presents a comparison of the features of thin and thick skin.

✔CHECKPOINT

10. What criteria are used to distinguish thin and thick skin?

5.4 FUNCTIONS OF THE SKIN

OBJECTIVE

- Describe how the skin contributes to the regulation of body temperature, storage of blood, protection, sensation, excretion and absorption, and synthesis of vitamin D.

Now that you have a basic understanding of the structure of the skin, you can better appreciate its many functions, which were introduced at the beginning of this chapter. The numerous functions of the integumentary system (mainly the skin) include thermoregulation, storage of blood, protection, cutaneous sensations, excretion and absorption, and synthesis of vitamin D.

Thermoregulation

Recall that **thermoregulation** is the homeostatic regulation of body temperature. The skin contributes to thermoregulation in two ways: by liberating sweat at its surface and by adjusting the flow of blood in the dermis. In response to high environmental temperature or heat produced by exercise, sweat production from eccrine

TABLE 5.4

Comparison of Thin and Thick Skin

FEATURE	THIN SKIN	THICK SKIN
Distribution	All parts of body except areas such as palms, palmar surface of digits, and soles.	Areas such as palms, palmar surface of digits, and soles.
Epidermal thickness	0.10–0.15 mm (0.004–0.006 in.).	0.6–4.5 mm (0.024–0.18 in.), due mostly to a thicker stratum corneum.
Epidermal strata	Stratum lucidum essentially lacking; thinner strata spinosum and corneum.	Stratum lucidum present; thicker strata spinosum and corneum.
Epidermal ridges	Lacking due to poorly developed, fewer, and less-well-organized dermal papillae.	Present due to well-developed and more numerous dermal papillae organized in parallel rows.
Hair follicles and arrector pili muscles	Present.	Absent.
Sebaceous glands	Present.	Absent.
Sudoriferous glands	Fewer.	More numerous.
Sensory receptors	Sparser.	Denser.

sweat glands increases; the evaporation of sweat from the skin surface helps lower body temperature. In addition, blood vessels in the dermis of the skin dilate (become wider); consequently, more blood flows through the dermis, which increases the amount of heat loss from the body (see Figure 25.19). In response to low environmental temperature, production of sweat from eccrine sweat glands is decreased, which helps conserve heat. Also, the blood vessels in the dermis of the skin constrict (become narrow), which decreases blood flow through the skin and reduces heat loss from the body. And, skeletal muscle contractions generate body heat.

Blood Reservoir

The dermis houses an extensive network of blood vessels that carry 8–10% of the total blood flow in a resting adult. For this reason, the skin acts as a **blood reservoir.**

Protection

The skin provides **protection** to the body in various ways. Keratin protects underlying tissues from microbes, abrasion, heat, and chemicals, and the tightly interlocked keratinocytes resist invasion by microbes. Lipids released by lamellar granules inhibit evaporation of water from the skin surface, thus guarding against dehydration; they also retard entry of water across the skin surface during showers and swims. The oily sebum from the sebaceous glands keeps skin and hairs from drying out and contains *bactericidal chemicals* (substances that kill bacteria). The acidic pH of perspiration retards the growth of some microbes. The pigment melanin helps shield against the damaging effects of ultraviolet light. Two types of cells carry out protective functions that are immunological in nature. Epidermal Langerhans cells alert the immune system to the presence of potentially harmful microbial invaders by recognizing and processing them, and macrophages in the dermis phagocytize bacteria and viruses that manage to bypass the Langerhans cells of the epidermis.

Cutaneous Sensations

Cutaneous sensations are sensations that arise in the skin, including tactile sensations—touch, pressure, vibration, and tickling—as well as thermal sensations such as warmth and coolness. Another cutaneous sensation, pain, usually is an indication of impending or actual tissue damage. There is a wide variety of nerve endings and receptors distributed throughout the skin, including the tactile discs of the epidermis, the corpuscles of touch in the dermis, and hair root plexuses around each hair follicle. Chapter 16 provides more details on the topic of cutaneous sensations.

Excretion and Absorption

The skin normally has a small role in **excretion,** the elimination of substances from the body, and **absorption,** the passage of materials from the external environment into body cells. Despite the almost waterproof nature of the stratum corneum, about 400 mL of water evaporates through it daily. A sedentary person loses an additional 200 mL per day as sweat; a physically active person loses much more. Besides removing water and heat from the body, sweat also is the vehicle for excretion of small amounts of salts, carbon dioxide, and two organic molecules that result from the breakdown of proteins—ammonia and urea.

The absorption of water-soluble substances through the skin is negligible, but certain lipid-soluble materials do penetrate the skin. These include fat-soluble vitamins (A, D, E, and K), certain drugs, and the gases oxygen and carbon dioxide. Toxic materials that can be absorbed through the skin include organic solvents such as acetone (in some nail polish removers) and carbon tetrachloride (dry-cleaning fluid); salts of heavy metals such as lead, mercury, and arsenic; and the substances in poison ivy and poison oak. Since topical (applied to the skin) steroids, such as cortisone, are lipid-soluble, they move easily into the papillary region of the dermis. Here, they exert their anti-inflammatory properties by inhibiting histamine production by mast cells (recall that histamine contributes to inflammation). Certain drugs that are absorbed by the skin may be administered by applying adhesive patches to the skin.

CLINICAL CONNECTION | *Transdermal Drug Administration*

Most drugs are either absorbed into the body through the digestive system or injected into subcutaneous tissue or muscle. An alternative route, **transdermal (transcutaneous) drug administration,** enables a drug contained within an adhesive skin patch to pass across the epidermis and into the blood vessels of the dermis. The drug is released continuously at a controlled rate over a period of one to several days. This method of administration is especially useful for drugs that are quickly eliminated from the body because such drugs, if taken in other forms, would have to be taken quite frequently. Because the major barrier to penetration of most drugs is the stratum corneum, transdermal absorption is most rapid in regions of the skin where this layer is thin, such as the scrotum, face, and scalp. A growing number of drugs are available for transdermal administration, including nitroglycerin, for prevention of angina pectoris (chest pain associated with heart disease); scopolamine, for motion sickness; estradiol, used for estrogen-replacement therapy during menopause; ethinyl estradiol and norelgestromin in contraceptive patches; nicotine, used to help people stop smoking; and fentanyl, used to relieve severe pain in cancer patients. •

Synthesis of Vitamin D

Synthesis of vitamin D requires activation of a precursor molecule in the skin by ultraviolet (UV) rays in sunlight. Enzymes in the liver and kidneys then modify the activated molecule, finally producing *calcitriol,* the most active form of vitamin D. Calcitriol is a hormone that aids in the absorption of calcium from foods in the gastrointestinal tract into the blood. Only a small amount of exposure to UV light (about 10 to 15 minutes at least twice a week) is required for vitamin D synthesis. People who avoid sun exposure and individuals who live in colder, northern climates may require vitamin D supplements to avoid vitamin D deficiency. Most cells of the immune system have vitamin D receptors, and the cells activate vitamin D in response to an infection, especially a respiratory infection, such as influenza. Vitamin D is believed to enhance phagocytic activity, increase the production of antimicrobial substances in phagocytes, regulate immune functions, and help reduce inflammation.

✔**CHECKPOINT**

11. In what two ways does the skin help regulate body temperature?
12. How does the skin serve as a protective barrier?
13. What sensations arise from stimulation of neurons in the skin?
14. What types of molecules can penetrate the stratum corneum?

5.5 MAINTAINING HOMEOSTASIS: SKIN WOUND HEALING

OBJECTIVE

- Explain how epidermal wounds and deep wounds heal.

Skin damage sets in motion a sequence of events that repairs the skin to its normal (or near-normal) structure and function. Two kinds of wound-healing processes can occur, depending on the depth of the injury. Epidermal wound healing occurs following wounds that affect only the epidermis; deep wound healing occurs following wounds that penetrate the dermis.

Epidermal Wound Healing

Even though the central portion of an epidermal wound may extend to the dermis, the edges of the wound usually involve only slight damage to superficial epidermal cells. Common types of epidermal wounds include abrasions, in which a portion of skin has been scraped away, and minor burns.

In response to an epidermal injury, basal cells of the epidermis surrounding the wound break contact with the basement membrane. The cells then enlarge and migrate across the wound (Figure 5.6a). The cells appear to migrate as a sheet until advancing cells from opposite sides of the wound meet. When epidermal cells encounter one another, they stop migrating due to a cellular response called **contact inhibition.** Migration of the epidermal cells stops completely when each is finally in contact with other epidermal cells on all sides.

As the basal epidermal cells migrate, a hormone called *epidermal growth factor* stimulates basal stem cells to divide and replace the ones that have moved into the wound. The relocated basal epidermal cells divide to build new strata, thus thickening the new epidermis (Figure 5.6b).

Deep Wound Healing

Deep wound healing occurs when an injury extends to the dermis and subcutaneous layer. Because multiple tissue layers must be repaired, the healing process is more complex than in epidermal wound healing. In addition, because scar tissue is formed, the healed tissue loses some of its normal function. Deep wound healing occurs in four phases: an inflammatory phase, a migratory phase, a proliferative phase, and a maturation phase.

During the **inflammatory phase,** a blood clot forms in the wound and loosely unites the wound edges (Figure 5.6c). As its name implies, this phase of deep wound healing involves **inflammation,** a vascular and cellular response that helps eliminate microbes, foreign material, and dying tissue in preparation for repair. The vasodilation and increased permeability of blood vessels associated with inflammation enhance delivery of helpful cells. These include phagocytic white blood cells called neutrophils; monocytes, which develop into macrophages that

Figure 5.6 Skin wound healing.

In an epidermal wound, the injury is restricted to the epidermis; in a deep wound, the injury extends deep into the dermis.

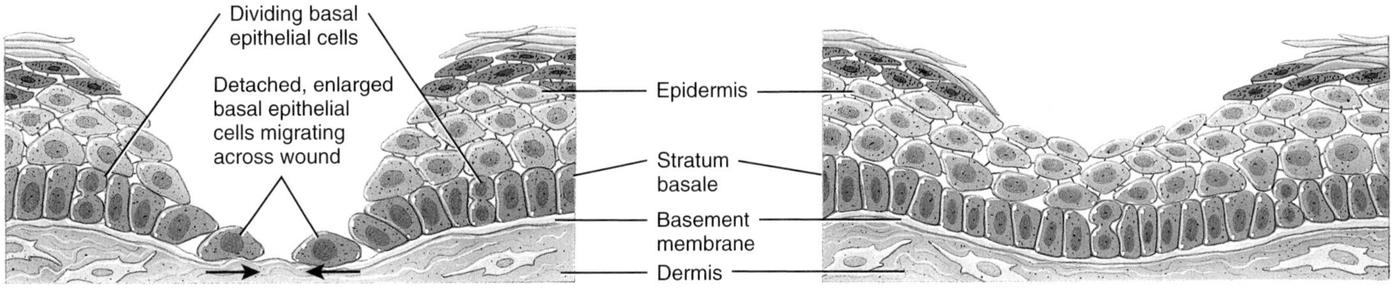

(a) Division of basal epithelial cells and migration across wound

(b) Thickening of epidermis

Epidermal wound healing

FIGURE 5.6 CONTINUES

FIGURE 5.6 CONTINUED

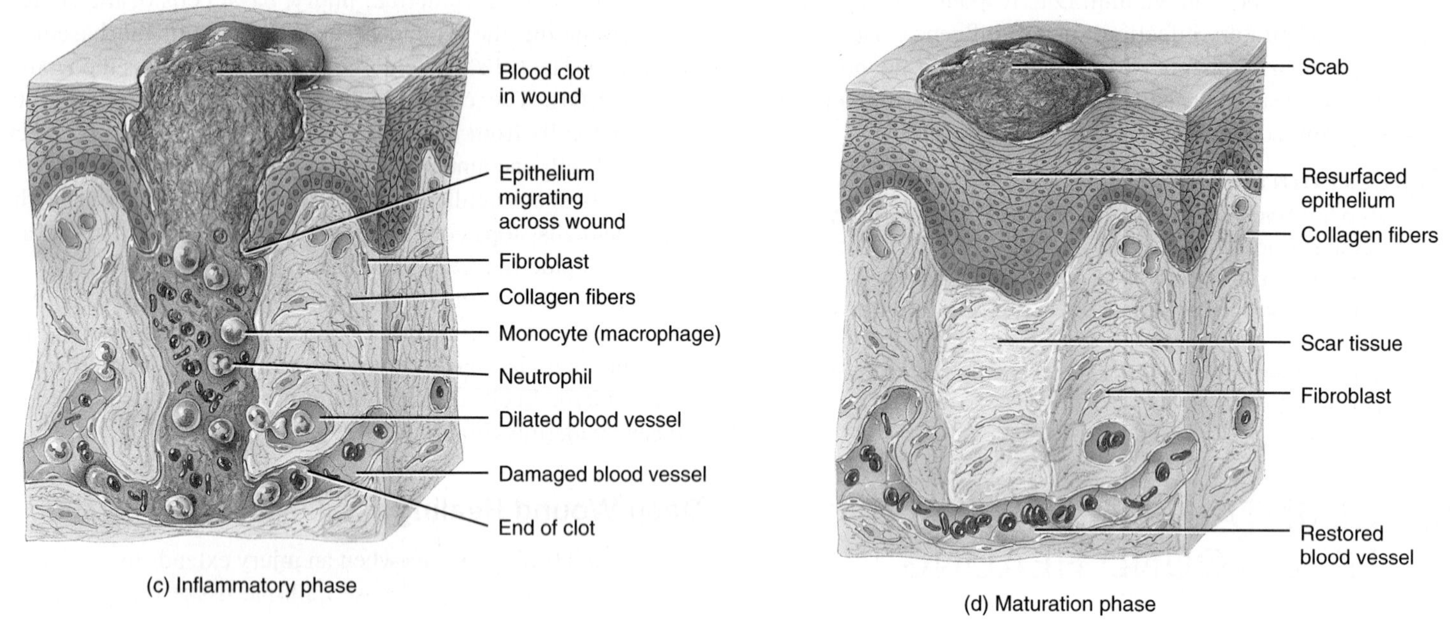
Blood clot in wound
Epithelium migrating across wound
Fibroblast
Collagen fibers
Monocyte (macrophage)
Neutrophil
Dilated blood vessel
Damaged blood vessel
End of clot
(c) Inflammatory phase
Scab
Resurfaced epithelium
Collagen fibers
Scar tissue
Fibroblast
Restored blood vessel
(d) Maturation phase

Deep wound healing

Would you expect an epidermal wound to bleed? Why or why not?

phagocytize microbes; and mesenchymal cells, which develop into fibroblasts.

The three phases that follow do the work of repairing the wound. In the **migratory phase,** the clot becomes a scab, and epithelial cells migrate beneath the scab to bridge the wound. Fibroblasts migrate along fibrin threads and begin synthesizing scar tissue (collagen fibers and glycoproteins), and damaged blood vessels begin to regrow. During this phase, the tissue filling the wound is called **granulation tissue.** The **proliferative phase** is characterized by extensive growth of epithelial cells beneath the scab, deposition by fibroblasts of collagen fibers in random patterns, and continued growth of blood vessels. Finally, during the **maturation phase,** the scab sloughs off once the epidermis has been restored to normal thickness. Collagen fibers become more organized, fibroblasts decrease in number, and blood vessels are restored to normal (Figure 5.6d).

The process of scar tissue formation is called **fibrosis.** Sometimes, so much scar tissue is formed during deep wound healing that a raised scar—one that is elevated above the normal epidermal surface—results. If such a scar remains within the boundaries of the original wound, it is a **hypertrophic scar.** If it extends beyond the boundaries into normal surrounding tissues, it is a **keloid scar,** also called a **cheloid scar.** Scar tissue differs from normal skin in that its collagen fibers are more densely arranged, it has decreased elasticity, it has fewer blood vessels, and it may or may not contain the same number of hairs, skin glands, or sensory structures as undamaged skin. Because of the arrangement of collagen fibers and the scarcity of blood vessels, scars usually are lighter in color than normal skin.

✔ CHECKPOINT

15. Why doesn't epidermal wound healing result in scar formation?

5.6 DEVELOPMENT OF THE INTEGUMENTARY SYSTEM

OBJECTIVE

- Describe the development of the epidermis, its accessory structures, and the dermis.

The *epidermis* is derived from the **ectoderm,** which covers the surface of the embryo. Initially, at about the fourth week after fertilization, the epidermis consists of only a single layer of ectodermal cells (Figure 5.7a). At the beginning of the seventh week the single layer, called the **basal layer,** divides and forms a superficial protected layer of flattened cells called the **periderm** (Figure 5.7b). The peridermal cells are continuously sloughed off, and by the fifth month of development secretions from sebaceous glands mix with them and hairs to form a fatty substance called **vernix caseosa** (VER-niks KĀ-sē-ō-sa; *vernix* = varnish; *caseosa* = cheese). This substance covers and protects the skin of the fetus from the constant exposure to the amniotic fluid in which it is bathed. In addition, the

Figure 5.7 Development of the integumentary system.

The epidermis develops from ectoderm, and the dermis develops from mesoderm.

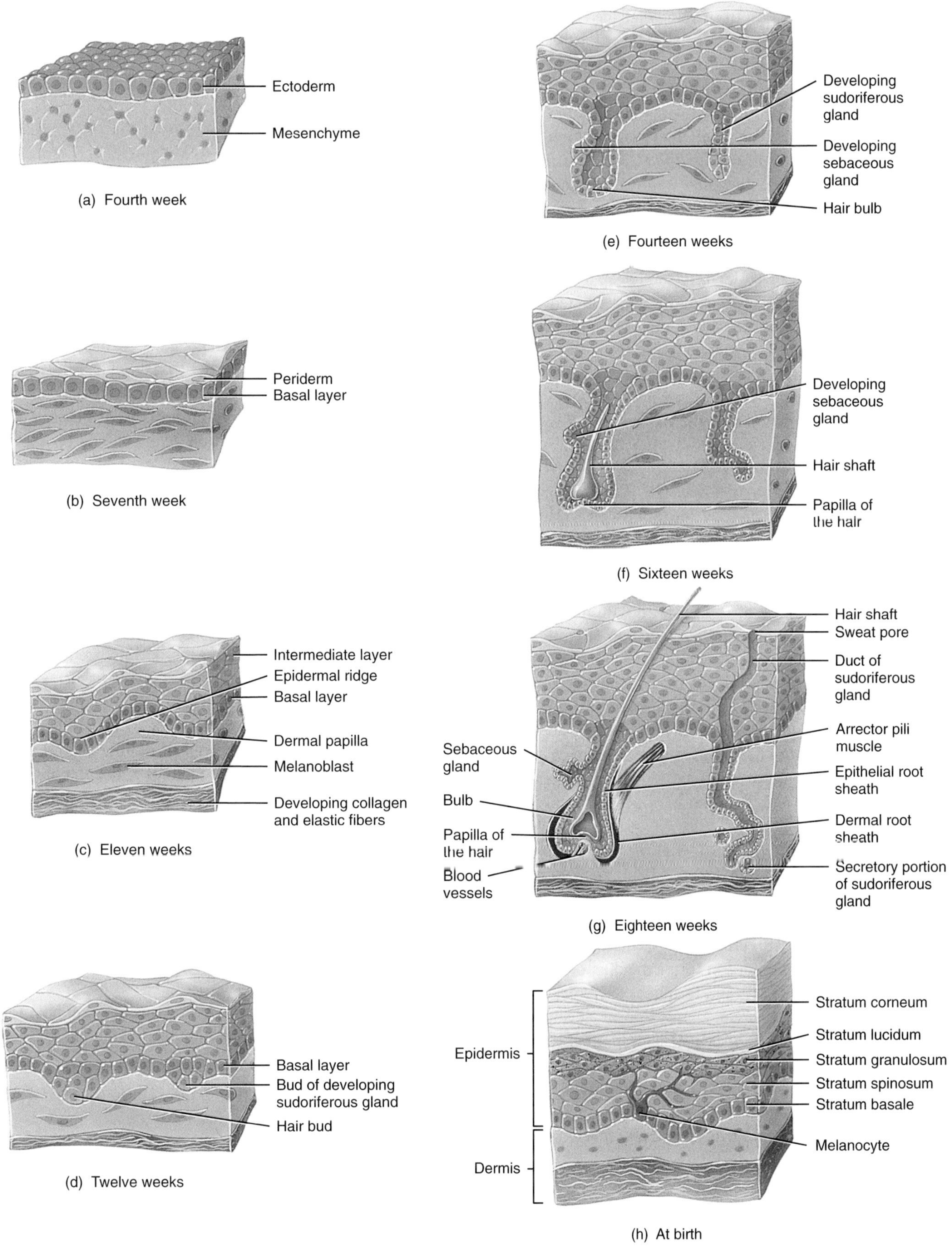

Ectoderm
Mesenchyme
(a) Fourth week
Periderm
Basal layer
(b) Seventh week
Intermediate layer
Epidermal ridge
Basal layer
Dermal papilla
Melanoblast
Developing collagen and elastic fibers
(c) Eleven weeks
Basal layer
Bud of developing sudoriferous gland
Hair bud
(d) Twelve weeks
Developing sudoriferous gland
Developing sebaceous gland
Hair bulb
(e) Fourteen weeks
Developing sebaceous gland
Hair shaft
Papilla of the hair
(f) Sixteen weeks
Hair shaft
Sweat pore
Duct of sudoriferous gland
Arrector pili muscle
Epithelial root sheath
Dermal root sheath
Secretory portion of sudoriferous gland
Sebaceous gland
Bulb
Papilla of the hair
Blood vessels
(g) Eighteen weeks
Epidermis
Dermis
Stratum corneum
Stratum lucidum
Stratum granulosum
Stratum spinosum
Stratum basale
Melanocyte
(h) At birth

What is the composition of vernix caseosa?

vernix caseosa facilitates the birth of the fetus because of its slippery nature and protects the skin from being damaged by the nails.

By about 11 weeks (Figure 5.7c), the basal layer forms an intermediate layer of cells. Proliferation of the basal cells eventually forms all layers of the epidermis, which are present at birth (Figure 5.7h). *Epidermal ridges* form along with the epidermal layers. By about the eleventh week, cells from the ectoderm migrate into the dermis and differentiate into *melanoblasts.* These cells soon enter the epidermis and differentiate into *melanocytes.* Later in the first trimester of pregnancy, *Langerhans cells,* which arise from red bone marrow, invade the epidermis. *Merkel cells* appear in the epidermis in the fourth to sixth months; their origin is unknown.

The *dermis* arises from **mesoderm** located deep to the surface ectoderm. The mesoderm gives rise to a loosely organized embryonic connective tissue called **mesenchyme** (MEZ-en-kīm; see Figure 5.7a). By 11 weeks, the mesenchymal cells differentiate into fibroblasts and begin to form collagen and elastic fibers. As the epidermal ridges form, parts of the superficial dermis project into the epidermis and develop into the *dermal papillae,* which contain capillary loops, corpuscles of touch, and free nerve endings (Figure 5.7c).

Hair follicles develop at about 12 weeks as downgrowths of the basal layer of the epidermis into the deeper dermis. The downgrowths are called **hair buds** (Figure 5.7d). As the hair buds penetrate deeper into the dermis, their distal ends become club-shaped and are called **hair bulbs** (Figure 5.7e). Invaginations of the hair bulbs, called **papillae of the hair,** fill with mesoderm in which blood vessels and nerve endings develop (Figure 5.7f). Cells in the center of a hair bulb develop into the *matrix,* which forms the *hair,* and the peripheral cells of the hair bulb form the *epithelial root sheath;* mesenchyme in the surrounding dermis develops into the *dermal root sheath* and *arrector pili muscle* (Figure 5.7g). By the fifth month, the hair follicles produce lanugo (delicate fetal hair; see Types of Hairs earlier in the chapter). It is produced first on the head and then on other parts of the body, and is usually shed prior to birth.

Most *sebaceous (oil) glands* develop as outgrowths from the sides of hair follicles at about four months and remain connected to the follicles (Figure 5.7e). Most *sudoriferous (sweat) glands* are derived from downgrowths **(buds)** of the stratum basale of the epidermis into the dermis (Figure 5.7d). As the buds penetrate into the dermis, the proximal portion forms the duct of the sweat gland and the distal portion coils and forms the secretory portion of the gland (Figure 5.7g). Sweat glands appear at about five months on the palms and soles and a little later in other regions.

Nails are developed at about 10 weeks. Initially they consist of a thick layer of epithelium called the **primary nail field.** The nail itself is keratinized epithelium and grows distally from its base. It is not until the ninth month that the nails actually reach the tips of the digits.

✔CHECKPOINT

16. Which structures develop as downgrowths of the stratum basale?

5.7 AGING AND THE INTEGUMENTARY SYSTEM

OBJECTIVE

- Describe the effects of aging on the integumentary system.

The pronounced effects of skin aging do not become noticeable until people reach their late forties. Most of the age-related changes occur in the proteins in the dermis. Collagen fibers in the dermis begin to decrease in number, stiffen, break apart, and disorganize into a shapeless, matted tangle. Elastic fibers lose some of their elasticity, thicken into clumps, and fray, an effect that is greatly accelerated in the skin of smokers. Fibroblasts, which produce both collagen and elastic fibers, decrease in number. As a result, the skin forms the characteristic crevices and furrows known as *wrinkles.*

With further aging, Langerhans cells dwindle in number and macrophages become less-efficient phagocytes, thus decreasing the skin's immune responsiveness. Moreover, decreased size of sebaceous glands leads to dry and broken skin that is more susceptible to infection. Production of sweat diminishes, which probably contributes to the increased incidence of heat stroke in the elderly. There is a decrease in the number of functioning melanocytes, resulting in gray hair and atypical skin pigmentation. Hair loss increases with aging as hair follicles stop producing hairs. About 25% of males begin to show signs of hair loss by age 30 and about two-thirds have significant hair loss by age 60. Both males and females develop pattern baldness. An increase in the size of some melanocytes produces pigmented blotching (age spots). Walls of blood vessels in the dermis become thicker and less permeable, and subcutaneous adipose tissue is lost. Aged skin (especially the dermis) is thinner than young skin, and the migration of cells from the basal layer to the epidermal surface slows considerably. With the onset of old age, skin heals poorly and becomes more susceptible to pathological conditions such as skin cancer and pressure sores. **Rosacea** (ro-ZĀ-shē-a = rosy) is a skin condition that affects mostly light-skinned adults between the ages of 30 and 60. It is characterized by redness, tiny pimples, and noticeable blood vessels, usually in the central area of the face.

Growth of nails and hair slows during the second and third decades of life. The nails also may become more brittle with age, often due to dehydration or repeated use of cuticle remover or nail polish.

Several cosmetic anti-aging treatments are available to diminish the effects of aging or sun-damaged skin. These include the following:

1. **Topical products** that bleach the skin to tone down blotches and blemishes (hydroquinone) or decrease fine wrinkles and roughness (retinoic acid).
2. **Microdermabrasion** (mī-krō-DER-ma-brā′-zhun; *mikros-* = small; *-derm-* = skin; *-abrasio* = to wear away), the use of tiny crystals under pressure to remove and vacuum the skin's surface cells to improve skin texture and reduce blemishes.

3. **Chemical peel,** the application of a mild acid (such as glycolic acid) to the skin to remove surface cells to improve skin texture and reduce blemishes.
4. **Laser resurfacing,** the use of a laser to clear up blood vessels near the skin surface, even out blotches and blemishes, and decrease fine wrinkles.
5. **Dermal fillers,** injections of collagen from cows, hyaluronic acid, calcium hydroxylapatite (Radiesse®), or polylactic acid (Sculptra®) that plumps up the skin to smooth out wrinkles and fill in furrows, such as those around the nose and mouth and between the eyebrows.
6. **Fat transplantation,** in which fat from one part of the body is injected into another location such as around the eyes.
7. **Botulinum toxin** or **Botox®**, a diluted version of a toxin that is injected into the skin to paralyze skeletal muscles that cause the skin to wrinkle.
8. **Radio frequency nonsurgical facelift,** the use of radio frequency emissions to tighten the deeper layers of the skin of the jowls, neck, and sagging eyebrows and eyelids.
9. **Facelift, browlift,** or **necklift,** invasive surgery in which loose skin and fat are removed surgically and the underlying connective tissue and muscle are tightened.

CLINICAL CONNECTION | *Sun Damage, Sunscreens, and Sunblocks*

Although basking in the warmth of the sun may feel good, it is not a healthy practice. There are two forms of ultraviolet radiation that affect the health of the skin. Longer-wavelength ultraviolet A (UVA) rays make up nearly 95% of the ultraviolet radiation that reaches the earth. UVA rays are not absorbed by the ozone layer. They penetrate the furthest into the skin, where they are absorbed by melanocytes and thus are involved in sun tanning. UVA rays also depress the immune system. Shorter-wavelength ultraviolet B (UVB) rays are partially absorbed by the ozone layer and do not penetrate the skin as deeply as UVA rays. UVB rays cause sunburn and are responsible for most of the tissue damage (production of oxygen free radicals which disrupt collagen and elastic fibers) that results in wrinkling and aging of the skin and cataract formation. Both UVA and UVB rays are thought to cause skin cancer. Long-term overexposure to sunlight results in dilated blood vessels, age spots, freckles, and changes in skin texture.

Exposure to ultraviolet radiation (either natural sunlight or the artificial light of a tanning booth) may also produce **photosensitivity,** a heightened reaction of the skin after consumption of certain medications or contact with certain substances. Photosensitivity is characterized by redness, itching, blistering, peeling, hives, and even shock. Among the medications or substances that may cause a photosensitivity reaction are certain antibiotics (tetracycline), nonsteroidal anti-inflammatory drugs (ibuprofen or naproxen), certain herbal supplements (St. John's wort), some birth control pills, some high blood pressure medications, some antihistamines, and certain artificial sweeteners, perfumes, aftershaves, lotions, detergents, and medicated cosmetics.

Self-tanning lotions (sunless tanners), topically applied substances, contain a color additive (dihydroxyacetone) that produces a tanned appearance by interacting with proteins in the skin.

Sunscreens are topically applied preparations that contain various chemical agents (such as benzophenone or one of its derivatives) that absorb UVB rays but let most of the UVA rays pass through.

Sunblocks are topically applied preparations that contain substances such as zinc oxide that reflect and scatter both UVB and UVA rays.

Both sunscreens and sunblocks are graded according to a *sun protection factor (SPF)* rating, which measures the level of protection they supposedly provide against UV rays. The higher the rating, presumably the greater the degree of protection. As a precautionary measure, individuals who plan to spend a significant amount of time in the sun should use a sunscreen or a sunblock with an SPF of 15 or higher. Although sunscreens protect against sunburn, there is considerable debate as to whether they actually protect against skin cancer. In fact, some studies suggest that sunscreens increase the incidence of skin cancer because of the false sense of security they provide. •

✔CHECKPOINT

17. What factors contribute to the susceptibility of aging skin to infection?

• • •

To appreciate the many ways that skin contributes to homeostasis of other body systems, examine *Focus on Homeostasis: The Integumentary System.* This feature is the first of 11, found at the end of selected chapters, that explain how the body system under consideration contributes to the homeostasis of all other body systems. Next, in Chapter 6, we will explore how bone tissue is formed and how bones are assembled into the skeletal system, which, like the skin, protects many of our internal organs.

Focus on Homeostasis

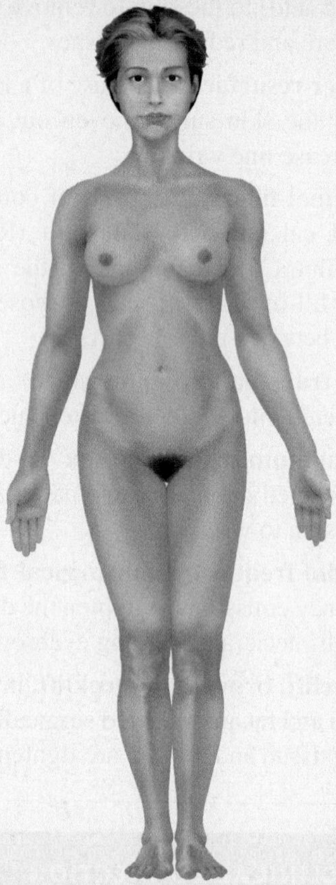

THE INTEGUMENTARY SYSTEM

BODY SYSTEM		CONTRIBUTION OF THE INTEGUMENTARY SYSTEM
For all body systems		Skin and hair provide barriers that protect all internal organs from damaging agents in external environment; sweat glands and skin blood vessels regulate body temperature, needed for proper functioning of other body systems.
Skeletal system		Skin helps activate vitamin D, needed for proper absorption of dietary calcium and phosphorus to build and maintain bones.
Muscular system		Skin helps provide calcium ions, needed for muscle contraction.
Nervous system	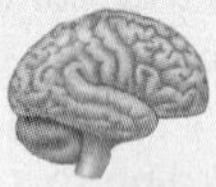	Nerve endings in skin and subcutaneous tissue provide input to brain for touch, pressure, thermal, and pain sensations.
Endocrine system		Keratinocytes in skin help activate vitamin D to calcitriol, a hormone that aids absorption of dietary calcium and phosphorus.
Cardiovascular system	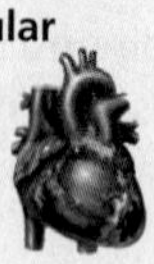	Local chemical changes in dermis cause widening and narrowing of skin blood vessels, which help adjust blood flow to skin.
Lymphatic system and immunity		Skin is "first line of defense" in immunity, providing mechanical barriers and chemical secretions that discourage penetration and growth of microbes; Langerhans cells in epidermis participate in immune responses by recognizing and processing foreign antigens; macrophages in dermis phagocytize microbes that penetrate skin surface.
Respiratory system	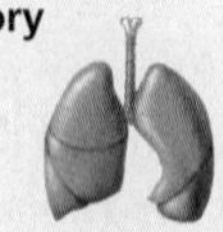	Hairs in nose filter dust particles from inhaled air; stimulation of pain nerve endings in skin may alter breathing rate.
Digestive system	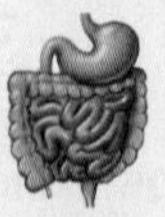	Skin helps activate vitamin D to the hormone calcitriol, which promotes absorption of dietary calcium and phosphorus in small intestine.
Urinary system	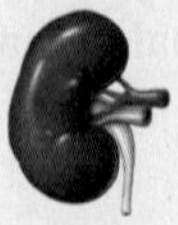	Kidney cells receive partially activated vitamin D hormone from skin and convert it to calcitriol; some waste products are excreted from body in sweat, contributing to excretion by urinary system.
Reproductive systems	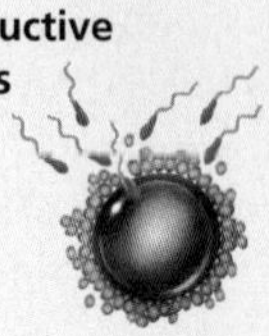	Nerve endings in skin and subcutaneous tissue respond to erotic stimuli, thereby contributing to sexual pleasure; suckling of a baby stimulates nerve endings in skin, leading to milk ejection; mammary glands (modified sweat glands) produce milk; skin stretches during pregnancy as fetus enlarges.

DISORDERS: HOMEOSTATIC IMBALANCES

Skin Cancer

Excessive exposure to ultraviolet radiation from the sun or tanning beds has caused virtually all of the one million cases of **skin cancer** diagnosed annually in the United States. One-half of all cancers in the United States are skin cancers. There are three common forms of skin cancer. **Basal cell carcinomas** account for about 78% of all skin cancers. The tumors arise from cells in the stratum basale of the epidermis and rarely metastasize. **Squamous cell carcinomas,** which account for about 20% of all skin cancers, arise from the stratum spinosum of the epidermis, and they have a variable tendency to metastasize. Basal and squamous cell carcinomas are together known as *nonmelanoma skin cancer.*

Malignant melanomas arise from melanocytes and account for about 2% of all skin cancers. The estimated lifetime risk of developing melanoma is now 1 in 75, double the risk only 20 years ago. In part, this increase is due to depletion of the ozone layer, which absorbs some UV light high in the atmosphere. But the main reason for the increase is that more people are spending more time in the sun and in tanning beds. Malignant melanomas metastasize rapidly and can kill a person within months of diagnosis.

The key to successful treatment of malignant melanoma is early detection. The early warning signs of malignant melanoma are identified by the acronym ABCDE (Figure 5.8). *A* is for *asymmetry;* malignant melanomas tend to lack symmetry. This means that they have irregular shapes, such as two very different looking halves. *B* is for *border;* malignant melanomas have irregular—notched, indented, scalloped, or indistinct—borders. *C* is for *color;* malignant melanomas have uneven coloration and may contain several colors. *D* is for *diameter;* ordinary moles typically are smaller than 6 mm (0.25 in.), about the size of a pencil eraser. *E* is for *evolving;* malignant melanomas change in size, shape, and color. Once a malignant melanoma has the characteristics of A, B, and C, it is usually larger than 6 mm.

Among the risk factors for skin cancer are the following:

1. ***Skin type.*** Individuals with light-colored skin who never tan but always burn are at high risk.
2. ***Sun exposure.*** People who live in areas with many days of sunlight per year and at high altitudes (where ultraviolet light is more intense) have a higher risk of developing skin cancer. Likewise, people who engage in outdoor occupations and those who have suffered three or more severe sunburns have a higher risk.
3. ***Family history.*** Skin cancer rates are higher in some families than in others.
4. ***Age.*** Older people are more prone to skin cancer owing to longer total exposure to sunlight.
5. ***Immunological status.*** Immunosuppressed individuals have a higher incidence of skin cancer.

Figure 5.8 **Comparison of a normal nevus (mole) and a malignant melanoma.**

Excessive exposure to ultraviolet radiation from the sun or tanning beds accounts for almost all cases of skin cancer.

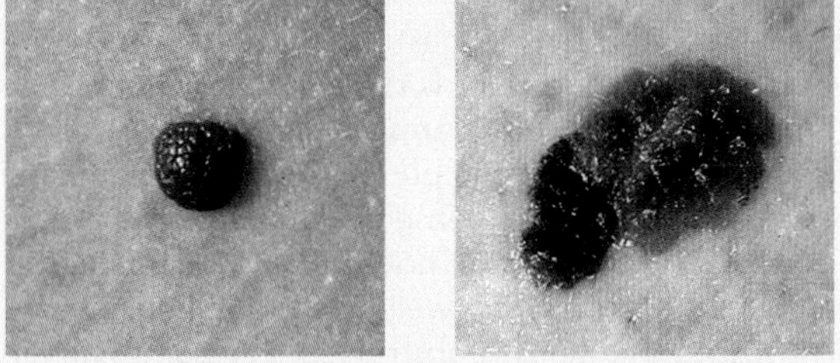

(a) Normal nevus (mole) (b) Malignant melanoma

? **Which is the most common type of skin cancer?**

Burns

A **burn** is tissue damage caused by excessive heat, electricity, radioactivity, or corrosive chemicals that denature (break down) proteins in the skin. Burns destroy some of the skin's important contributions to homeostasis—protection against microbial invasion and dehydration, and thermoregulation.

Burns are graded according to their severity. A *first-degree burn* involves only the epidermis (Figure 5.9a). It is characterized by mild pain and erythema (redness) but no blisters. Skin functions remain intact. Immediate flushing with cold water may lessen the pain and damage caused by a first-degree burn. Generally, healing of a first-degree burn will occur in 3 to 6 days and may be accompanied by flaking or peeling. One example of a first-degree burn is mild sunburn.

A *second-degree burn* destroys the epidermis and part of the dermis (Figure 5.9b). Some skin functions are lost. In a second-degree burn, redness, blister formation, edema, and pain result. In a blister the epidermis separates from the dermis due to the accumulation of tissue fluid between them. Associated structures, such as hair follicles, sebaceous glands, and sweat glands, usually are not injured. If there is no infection, second-degree burns heal without skin grafting in about 3 to 4 weeks, but scarring may result. First- and second-degree burns are collectively referred to as *partial-thickness burns.*

A *third-degree burn* or *full-thickness burn* destroys the epidermis, dermis, and subcutaneous layer (Figure 5.9c). Most skin functions are lost. Such burns vary in appearance from marble-white to mahogany colored to charred, dry wounds. There is marked edema, and the burned region is numb because sensory nerve endings have been destroyed. Regeneration occurs slowly, and much granulation tissue forms before being covered by epithelium. Skin grafting may be required to promote healing and to minimize scarring.

The injury to the skin tissues directly in contact with the damaging agent is the *local effect* of a burn. Generally, however, the *systemic effects* of a major burn are a greater threat to life. The systemic effects of a burn may include (1) a large loss of water, plasma, and plasma proteins, which causes shock; (2) bacterial infection; (3) reduced circulation of blood; (4) decreased production of urine; and (5) diminished immune responses.

The seriousness of a burn is determined by its depth and extent of area involved, as well as the person's age and general health. According to the American Burn Association's classification of burn injury, a major burn includes third-degree burns over 10% of body surface area; or second-degree burns over 25% of body surface area; or any third-degree burns on the face, hands, feet, or *perineum* (per-i-NĒ-um, which includes the anal and urogenital regions). When the burn

Figure 5.9 **Burns.**

A burn is tissue damage caused by agents that destroy the proteins in the skin.

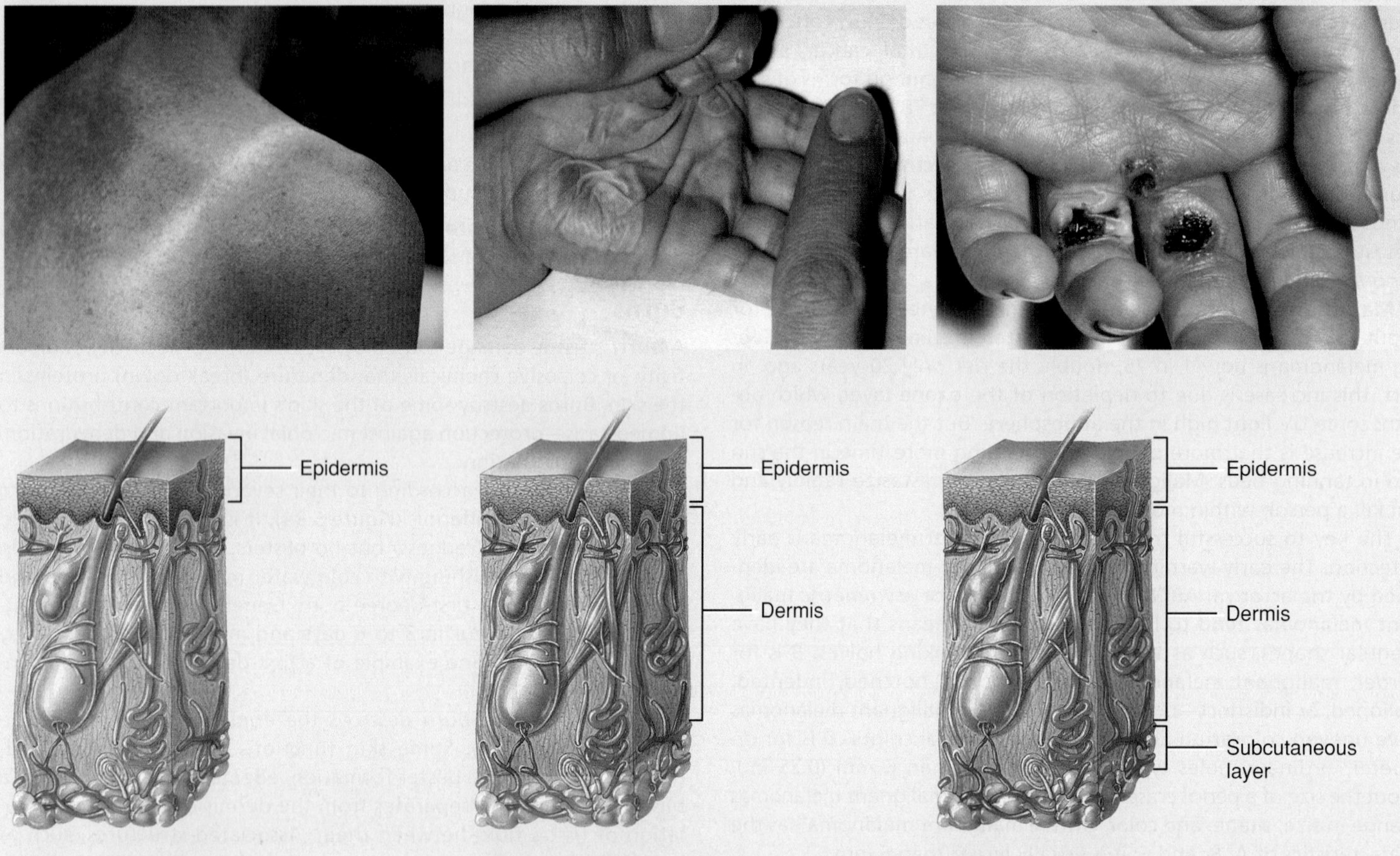

(a) First-degree burn (sunburn)

(b) Second-degree burn (note the blisters in the photograph above)

(c) Third-degree burn

What factors determine the seriousness of a burn?

area exceeds 70%, more than half the victims die. A quick means for estimating the surface area affected by a burn in an adult is the **rule of nines** (Figure 5.10):

1. Count 9% if both the anterior and posterior surfaces of the head and neck are affected.
2. Count 9% for both the anterior and posterior surfaces of each upper limb (total of 18% for both upper limbs).
3. Count four times nine, or 36%, for both the anterior and posterior surfaces of the trunk, including the buttocks.
4. Count 9% for the anterior and 9% for the posterior surfaces of each lower limb as far up as the buttocks (total of 36% for both lower limbs).
5. Count 1% for the perineum.

In severely burned patients with full-thickness or deep partial-thickness burns where there is not sufficient autograft, a tissue engineered product called Integra® Dermal Regeneration Template (DRT) is available. It is designed to promote organized regeneration of the dermis while providing a protective barrier against fluid loss and microbes. Integra® DRT consists of two layers, just like human skin. The bottom layer, called the matrix layer, is composed of bovine (cow) collagen and the carbohydrate glycosaminoglycan (GAG). It mimics the dermis, functions as an extracellular layer, and induces the body's own dermal cells to migrate into the area and regenerate a new dermis. The outer layer, called the silicone layer, consists of a thin layer of silicone that mimics the epidermis. Its role is to close the wound, control fluid loss, and serve as a protective barrier. Once the dermis has regenerated sufficiently (about three weeks), the silicone layer is removed and a thin sheet of the patient's own epidermal cells is applied.

Many people who have been burned in fires also inhale smoke. If the smoke is unusually hot or dense or if inhalation is prolonged, serious problems can develop. The hot smoke can damage the trachea (windpipe), causing its lining to swell. As the swelling narrows the trachea, airflow into the lungs is obstructed. Further, small airways inside the lungs can also narrow, producing wheezing or shortness of breath. A person who has inhaled smoke is given oxygen through a face mask, and a tube may be inserted into the trachea to assist breathing.

Figure 5.10 Rule-of-nines method for determining the extent of a burn. The percentages are the approximate proportions of the body surface area.

The rule of nines is a quick way for estimating the surface area affected by a burn in an adult.

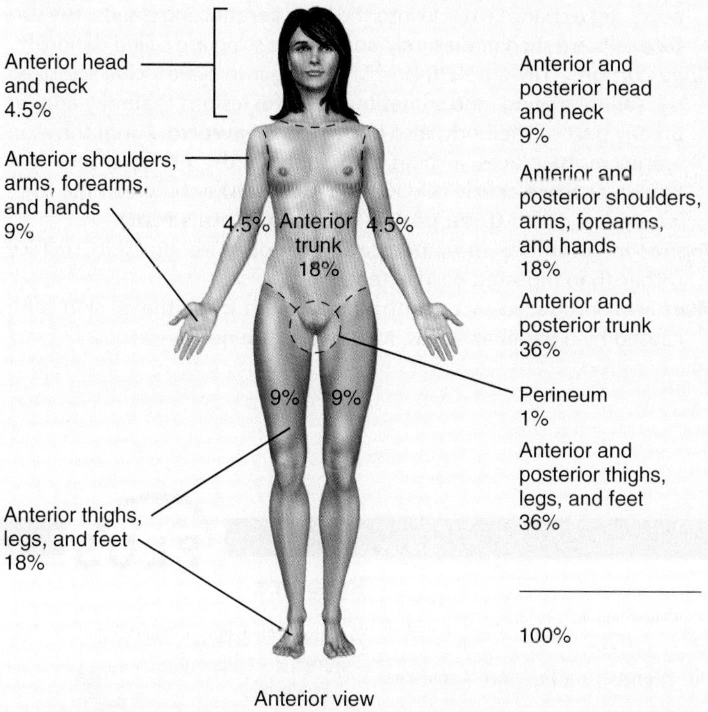
Anterior head and neck 4.5%
Anterior shoulders, arms, forearms, and hands 9%
4.5%
Anterior trunk 18%
4.5%
9%
9%
Anterior thighs, legs, and feet 18%
Anterior and posterior head and neck 9%
Anterior and posterior shoulders, arms, forearms, and hands 18%
Anterior and posterior trunk 36%
Perineum 1%
Anterior and posterior thighs, legs, and feet 36%
100%
Anterior view

What percentage of the body would be burned if only the anterior trunk and anterior left upper limb were involved?

Pressure Ulcers

Pressure ulcers, also known as *decubitus ulcers* (dē-KŪ-bi-tus) or *bedsores,* are caused by a constant deficiency of blood flow to tissues (Figure 5.11). Typically the affected tissue overlies a bony projection that has been subjected to prolonged pressure against an object such as a bed, cast, or splint. If the pressure is relieved in a few hours, redness occurs but no lasting tissue damage results. Blistering of the affected area may indicate superficial damage; a reddish-blue discoloration may indicate deep tissue damage. Prolonged pressure causes tissue ulceration. Small breaks in the epidermis become infected, and the sensitive subcutaneous layer and deeper tissues are damaged. Eventually, the tissue dies. Pressure ulcers occur most often in bedridden patients. With proper care, pressure ulcers are preventable, but they can develop very quickly in patients who are very old or very ill.

Figure 5.11 Pressure ulcer.

A pressure ulcer is a shedding of epithelium caused by a constant deficiency of blood flow to tissues.

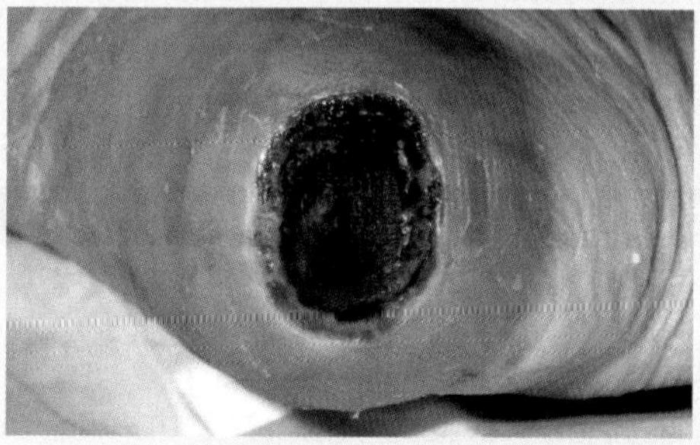

Pressure ulcer on heel

What parts of the body are usually affected by pressure ulcers?

MEDICAL TERMINOLOGY

Abrasion (a-BRĀ-shun; *ab-* = away; *-rasion* = scraped) An area where skin has been scraped away.

Blister A collection of serous fluid within the epidermis or between the epidermis and dermis, due to short-term but severe friction. The term **bulla** (BUL-a) refers to a large blister.

Callus (KAL-lus = hard skin) An area of hardened and thickened skin that is usually seen in palms and soles and is due to persistent pressure and friction.

Cold sore A lesion, usually in an oral mucous membrane, caused by Type 1 herpes simplex virus (HSV) transmitted by oral or respiratory routes. The virus remains dormant until triggered by factors such as ultraviolet light, hormonal changes, and emotional stress. Also called a **fever blister.**

Comedo (KOM-ē-dō = to eat up) A collection of sebaceous material and dead cells in the hair follicle and excretory duct of the sebaceous (oil) gland. Usually found over the face, chest, and back, and more commonly during adolescence. Also called a **blackhead.**

Contact dermatitis (der-ma-TĪ-tis; *dermat-* = skin; *-itis* = inflammation of) Inflammation of the skin characterized by redness, itching, and swelling and caused by exposure of the skin to chemicals that bring about an allergic reaction, such as poison ivy toxin.

Contusion (kon-TOO-shun; *contundere* = to bruise) Condition in which tissue deep to the skin is damaged, but the epidermis is not broken.

Corn A painful conical thickening of the stratum corneum of the epidermis found principally over toe joints and between the toes, often caused by friction or pressure. Corns may be hard or soft, depending on their location. Hard corns are usually found over toe joints, and soft corns are usually found between the fourth and fifth toes.

Cyst (SIST = sac containing fluid) A sac with a distinct connective tissue wall, containing a fluid or other material.

Eczema (EK-ze-ma; *ekzeo-* = to boil over) An inflammation of the skin characterized by patches of red, blistering, dry, extremely itchy skin. It occurs mostly in skin creases in the wrists, backs of the knees, and fronts of the elbows. It typically begins in infancy and many children outgrow the condition. The cause is unknown but is linked to genetics and allergies.

Frostbite Local destruction of skin and subcutaneous tissue on exposed surfaces as a result of extreme cold. In mild cases, the skin is blue and swollen and there is slight pain. In severe cases there is considerable swelling, some bleeding, no pain, and blistering. If untreated, gangrene may develop. Frostbite is treated by rapid rewarming.

Hemangioma (hē-man′-jē-Ō-ma; *hem-* = blood; *-angi-* = blood vessel; *-oma* = tumor) Localized benign tumor of the skin and subcutaneous layer that results from an abnormal increase in blood vessels. One type is a **portwine stain,** a flat, pink, red, or purple lesion present at birth, usually at the nape of the neck.

Hives Reddened elevated patches of skin that are often itchy. Most commonly caused by infections, physical trauma, medications, emotional stress, food additives, and certain food allergies. Also called **urticaria** (ūr-ti-KAR-ē-a).

Keloid (KĒ-loid; *kelis* = tumor) An elevated, irregular darkened area of excess scar tissue caused by collagen formation during healing. It extends beyond the original injury and is tender and frequently painful. It occurs in the dermis and underlying subcutaneous tissue, usually after trauma, surgery, a burn, or severe acne; more common in people of African descent.

Keratosis (ker′-a-TŌ-sis; *kera-* = horn) Formation of a hardened growth of epidermal tissue, such as *solar keratosis,* a premalignant lesion of the sun-exposed skin of the face and hands.

Laceration (las-er-Ā-shun; *lacer-* = torn) An irregular tear of the skin.

Lice Contagious arthropods that include two basic forms. **Head lice** are tiny, jumping arthropods that suck blood from the scalp. They lay eggs, called nits, and their saliva causes itching that may lead to complications. **Pubic lice** are tiny arthropods that do not jump; they look like miniature crabs.

Papule (PAP-ūl; *papula* = pimple) A small, round skin elevation less than 1 cm in diameter. One example is a pimple.

Pruritus (proo-RĪ-tus; *pruri-* = to itch) Itching, one of the most common dermatological disorders. It may be caused by skin disorders (infections), systemic disorders (cancer, kidney failure), psychogenic factors (emotional stress), or allergic reactions.

Psoriasis (sō-RĪ-a-sis) A common and chronic skin disorder in which keratinocytes divide and move more quickly than normal from the stratum basale to the stratum corneum. As a result, the surface cells never get a chance to cycle into the later keratinizing stages. The surface cells are shed immaturely and on the scalp are called dandruff.

Tinea corporis (TIN-ē-a KOR-po-ris) A fungal infection characterized by scaling, itching, and sometimes painful lesions that may appear on any part of the body, also known as **ringworm.** Fungi thrive in warm, moist places such as skin folds of the groin, where it is known as **tinea cruris** (KROO-ris) **(jock itch)** or between the toes, where it is called **tinea pedis** (PE-dis) **(athlete's foot).**

Topical In reference to a medication, applied to the skin surface rather than ingested or injected.

Wart Mass produced by uncontrolled growth of epithelial skin cells; caused by a papillomavirus. Most warts are noncancerous.

CHAPTER REVIEW AND RESOURCE SUMMARY

WILEY PLUS

Review	Resource
5.1 Structure of the Skin 1. The integumentary system consists of the skin, hair, oil and sweat glands, nails, and sensory receptors. 2. The skin is the largest organ of the body in surface area and weight. The principal parts of the skin are the epidermis (superficial) and dermis (deep). 3. The subcutaneous layer (hypodermis) is deep to the dermis and not part of the skin. It anchors the dermis to underlying tissues and organs, and it contains pacinian (lamellated) corpuscles. 4. The types of cells in the epidermis are keratinocytes, melanocytes, Langerhans cells, and Merkel cells. 5. The epidermal layers, from deep to superficial, are the stratum basale, stratum spinosum, stratum granulosum, stratum lucidum (in thick skin only), and stratum corneum (see Table 5.1). Stem cells in the stratum basale undergo continuous cell division, producing keratinocytes for the other layers. 6. The dermis is composed of dense irregular connective tissue containing collagen and elastic fibers. It is divided into papillary and reticular regions. The papillary region contains thin collagen and fine elastic fibers, dermal papillae, and Meissner corpuscles. The reticular region contains bundles of thick collagen and some coarse elastic fibers, fibroblasts and macrophages, adipose tissue, hair follicles, nerves, sebaceous (oil) glands, and sudoriferous (sweat) glands. (See Table 5.2.) 7. Epidermal ridges provide the basis for fingerprints and footprints. 8. The color of skin is due to melanin, carotene, and hemoglobin. 9. In tattooing, a pigment is deposited with a needle in the dermis. Body piercing is the insertion of jewelry through an artificial opening.	Anatomy Overview - The Skin Figure 5.2 - Cells in the Epidermis Figure 5.3 - Layers of the Epidermis
5.2 Accessory Structures of the Skin 1. Accessory structures of the skin—hair, skin glands, and nails—develop from the embryonic epidermis. 2. A hair consists of a shaft, most of which is superficial to the surface, a root that penetrates the dermis and sometimes the subcutaneous layer, and a hair follicle. 3. Associated with each hair follicle is a sebaceous (oil) gland, an arrector pili muscle, and a hair root plexus. 4. New hairs develop from division of hair matrix cells in the bulb; hair replacement and growth occur in a cyclical pattern consisting of growth, regression, and resting stages. 5. Hairs offer a limited amount of protection—from the sun, heat loss, and entry of foreign particles into the eyes, nose, and ears. They also function in sensing light touch.	Anatomy Overview - Hair Anatomy Overview - Nails Figure 5.4 - Hair

Review | **Resource**

6. Lanugo of the fetus is shed before birth. Most body hair on males is terminal (coarse, pigmented); most body hair on females is vellus (fine).
7. Sebaceous (oil) glands are usually connected to hair follicles; they are absent from the palms and soles. Sebaceous glands produce sebum, which moistens hairs and waterproofs the skin. Clogged sebaceous glands may produce acne.
8. There are two types of sudoriferous (sweat) glands: eccrine and apocrine. Eccrine sweat glands have an extensive distribution; their ducts terminate at pores at the surface of the epidermis. Eccrine sweat glands are involved in thermoregulation and waste removal and are stimulated during emotional stress. Apocrine sweat glands are limited to the skin of the axillae, groin, and areolae; their ducts open into hair follicles. Apocrine sweat glands are stimulated during emotional stress and sexual excitement. (See Table 5.3.)
9. Ceruminous glands are modified sudoriferous glands that secrete cerumen. They are found in the external auditory canal (ear canal).
10. Nails are hard, dead keratinized epidermal cells over the dorsal surfaces of the distal portions of the digits. The principal parts of a nail are the nail body, free edge, nail root, lunula, eponychium, and nail matrix. Cell division of the nail matrix cells produces new nails.

5.3 Types of Skin

1. Thin skin covers all parts of the body except for the palms, palmar surfaces of the digits, and the soles.
2. Thick skin covers the palms, palmar surfaces of the digits, and soles. (See Table 5.4.)

5.4 Functions of the Skin

1. Skin functions include body temperature regulation, blood storage, protection, sensation, excretion and absorption, and synthesis of vitamin D.
2. The skin participates in thermoregulation by liberating sweat at its surface and by adjusting the flow of blood in the dermis.
3. The skin provides physical, chemical, and biological barriers that help protect the body.
4. Cutaneous sensations include tactile sensations, thermal sensations, and pain.

Resource: Anatomy Overview - The Integument and Disease Resistance
Animation - Nonspecific Disease Resistance

5.5 Maintaining Homeostasis: Skin Wound Healing

1. In an epidermal wound, the central portion of the wound usually extends down to the dermis; the wound edges involve only superficial damage to the epidermal cells.
2. Epidermal wounds are repaired by enlargement and migration of basal cells, contact inhibition, and division of migrating and stationary basal cells.
3. During the inflammatory phase of deep wound healing, a blood clot unites the wound edges, epithelial cells migrate across the wound, vasodilation and increased permeability of blood vessels enhance delivery of phagocytes, and mesenchymal cells develop into fibroblasts.
4. During the migratory phase, fibroblasts migrate along fibrin threads and begin synthesizing collagen fibers and glycoproteins.
5. During the proliferative phase, epithelial cells grow extensively.
6. During the maturation phase, the scab sloughs off, the epidermis is restored to normal thickness, collagen fibers become more organized, fibroblasts begin to disappear, and blood vessels are restored to normal.

Resource: Figure 5.6 - Skin Wound Healing

5.6 Development of the Integumentary System

1. The epidermis develops from the embryonic ectoderm, and the accessory structures of the skin (hair, nails, and skin glands) are epidermal derivatives.
2. The dermis is derived from mesodermal cells.

5.7 Aging and the Integumentary System

1. Most effects of aging begin to occur when people reach their late forties.
2. Among the effects of aging are wrinkling, loss of subcutaneous adipose tissue, atrophy of sebaceous glands, and decrease in the number of melanocytes and Langerhans cells.

SELF-QUIZ QUESTIONS

Fill in the blanks in the following statements.

1. The epidermal layer that is found in thick skin but not in thin skin is the ____.
2. The most common sweat glands that release a watery secretion are ____ sweat glands; modified sweat glands in the ear are ____ glands; sweat glands located in the axillae, groin, areolae, and beards of males and that release a slightly viscous, lipid-rich secretion are ____ sweat glands.

Indicate whether the following statements are true or false.

3. An individual with a dark skin color has more melanocytes than a fair-skinned person.
4. In order to permanently prevent growth of an unwanted hair, you must destroy the hair matrix.

Choose the one best answer to the following questions.

5. The layer of the epidermis that contains stem cells undergoing mitosis is the
(a) stratum corneum. (b) stratum lucidum.
(c) stratum basale. (d) stratum spinosum.
(e) stratum granulosum.
6. The substance that helps promote mitosis in epidermal skin cells is
(a) keratohyalin. (b) melanin. (c) carotene.
(d) collagen. (e) epidermal growth factor.
7. Which of the following is *not* a function *of* skin?
(a) calcium production (b) vitamin D synthesis
(c) protection (d) excretion of wastes
(e) temperature regulation
8. To expose underlying tissues in the bottom of the foot, a foot surgeon must first cut through the skin. Place the following layers in the order that the scalpel would cut. (1) stratum lucidum, (2) stratum corneum, (3) stratum basale, (4) stratum granulosum, (5) stratum spinosum.
(a) 3, 5, 4, 1, 2 (b) 2, 1, 5, 4, 3 (c) 2, 1, 4, 5, 3
(d) 1, 3, 5, 4, 2 (e) 3, 4, 5, 1, 2
9. Aging of the skin can result in
(a) an increase in collagen and elastic fibers.
(b) a decrease in the activity of sebaceous glands.
(c) a thickening of the skin.
(d) an increased blood flow to the skin.
(e) an increase in toenail growth.
10. Which of the following is *not* true?
(a) Albinism is an inherited inability of melanocytes to produce melanin.
(b) Striae occurs when the dermis is overstretched to the point of tearing.
(c) In order to prevent excessive scarring, surgeons should cut parallel to the lines of cleavage.
(d) The papillary layer of the dermis is responsible for fingerprints.
(e) Much of the body's fat is located in the dermis of the skin.
11. A patient is brought into the emergency room suffering from a burn. The patient does not feel any pain at the burn site. Using a gentle pull on a hair, the examining physician can remove entire hair follicles from the patient's arm. This patient is suffering from what type of burn?
(a) third degree (b) second degree (c) first degree
(d) partial thickness (e) localized
12. Which of the following statements are *true?* (1) Nails are composed of tightly packed, hard, keratinized cells of the epidermis that form a clear, solid covering over the dorsal surface of the terminal end of digits. (2) The free edge of the nail is white due to the absence of capillaries. (3) Nails help us grasp and manipulate small objects. (4) Nails protect the ends of digits from trauma. (5) Nail color is due to a combination of melanin and carotene.
(a) 1, 2, and 3 (b) 1, 3, and 4 (c) 1, 2, 3, and 4
(d) 2, 3, and 4 (e) 1, 3, and 5
13. Match the following:

____ (a) produce the protein that helps protect the skin and underlying tissues from light, heat, microbes, and many chemicals	(1) Merkel cells
____ (b) produce a pigment that contributes to skin color and absorbs ultraviolet light	(2) callus
____ (c) cells that arise from red bone marrow, migrate to the epidermis, and participate in immune responses	(3) keratinocytes
____ (d) cells thought to function in the sensation of touch	(4) Langerhans cells
____ (e) located in the dermis, they function in the sensations of warmth, coolness, pain, itching, and tickling	(5) melanocytes
____ (f) smooth muscles associated with the hair follicles; when contracted, they pull the hair shafts perpendicular to the skin's surface	(6) free nerve endings
____ (g) an abnormal thickening of the epidermis	(7) sebaceous glands
____ (h) release a lipid-rich secretion that functions as a water-repellent sealant in the stratum granulosum	(8) lamellar granules
____ (i) pressure-sensitive cells found mostly in the subcutaneous layer	(9) pacinian (lamellated) corpuscles
____ (j) a fatty substance that covers and protects the skin of the fetus from the constant exposure to amniotic fluid	(10) vernix caseosa
____ (k) associated with hair follicles, these secrete an oily substance that helps prevent hair from becoming brittle, prevents evaporation of water from the skin's surface, and inhibits the growth of certain bacteria	(11) arrector pili

14. Match the following:

____ (a) deep region of the dermis composed primarily of dense irregular connective tissue
____ (b) composed of keratinized stratified squamous epithelial tissue
____ (c) not considered part of the skin, it contains areolar and adipose tissues and blood vessels; attaches skin to underlying tissues and organs
____ (d) superficial region of the dermis; composed of areolar connective tissue

(1) subcutaneous layer (hypodermis)
(2) papillary region
(3) reticular region
(4) epidermis

15. Match the following and place the phases of deep wound healing in the correct order:

____ (a) epithelial cells migrate under scab to bridge the wound; formation of granulation tissue
____ (b) sloughing of scab; reorganization of collagen fibers; blood vessels return to normal
____ (c) vasodilation and increased permeability of blood vessels to deliver cells involved in phagocytosis; clot formation
____ (d) extensive growth of epithelial cells beneath scab; random deposition of collagen fibers; continued growth of blood vessels

(1) proliferative phase
(2) inflammatory phase
(3) maturation phase
(4) migratory phase

CRITICAL THINKING QUESTIONS

1. The amount of dust that collects in a house with an assortment of dogs, cats, and people is truly amazing. A lot of these dust particles had a previous "life" as part of the home's living occupants. Where did the dust originate on the human body?

2. Josie reassures her mother that the tattoo she received at the tattoo parlor will eventually disappear. She knows this because she has learned in biology class that skin cells are shed every four weeks. Is Josie correct?

3. Six months ago, Chef Eduardo sliced through the end of his right thumbnail. Although the surrounding nail grows normally, this part of his nail remains split and doesn't seem to want to "heal." What has happened to cause this?

? ANSWERS TO FIGURE QUESTIONS

5.1 The epidermis is composed of epithelial tissue; the dermis is made up of connective tissue.

5.2 Melanin protects DNA of keratinocytes from the damaging effects of UV light.

5.3 The stratum basale is the layer of the epidermis with stem cells that continually undergo cell division.

5.4 Plucking a hair stimulates hair root plexuses in the dermis, some of which are sensitive to pain. Because the cells of a hair shaft are already dead and the hair shaft lacks nerves, cutting hair is not painful.

5.5 Nails are hard because they are composed of tightly packed, hard, keratinized epidermal cells.

5.6 Since the epidermis is avascular, an epidermal wound would not produce any bleeding.

5.7 Vernix caseosa consists of secretions from sebaceous glands, sloughed off peridermal cells, and hairs.

5.8 Basal cell carcinoma is the most common type of skin cancer.

5.9 The seriousness of a burn is determined by the depth and extent of the area involved, the individual's age, and general health.

5.10 About 22.5% of the body would be involved (4.5% [anterior arm] + 18% [anterior trunk]).

5.11 Pressure ulcers typically develop in tissues that overlie bony projections subjected to pressure, such as the shoulders, hips, buttocks, heels, and ankles.

6 THE SKELETAL SYSTEM: BONE TISSUE

BONE TISSUE AND HOMEOSTASIS *Bone tissue is continuously growing, remodeling, and repairing itself. It contributes to homeostasis of the body by providing support and protection, producing blood cells, and storing minerals and triglycerides.*

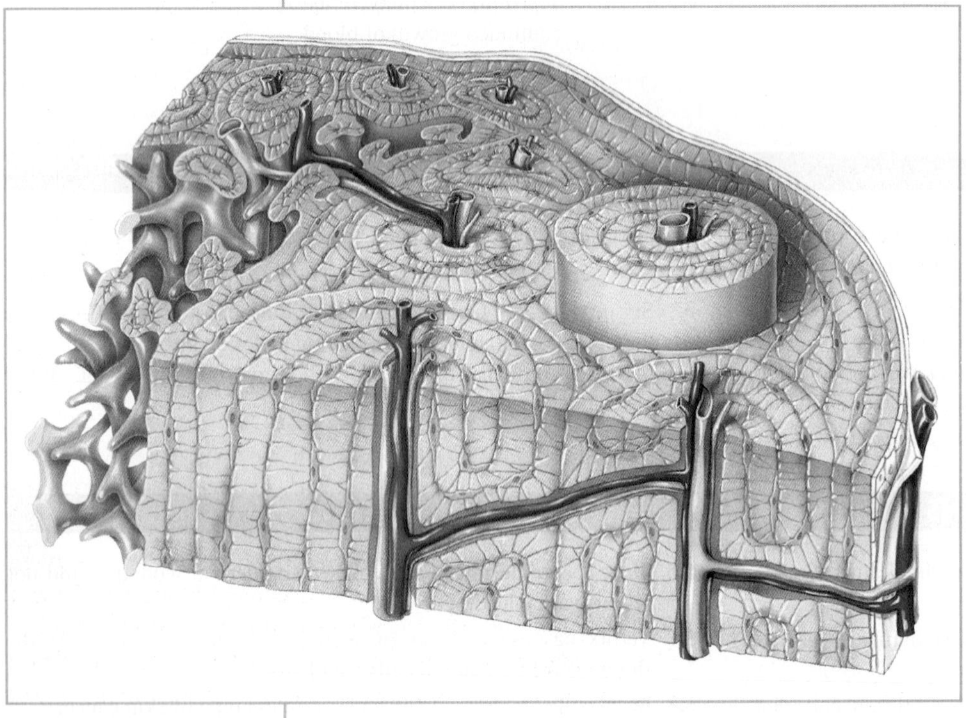

A bone is composed of several different tissues working together: bone or osseous tissue, cartilage, dense connective tissues, epithelium, adipose tissue, and nervous tissue. For this reason, each individual bone in your body is considered an organ. Bone tissue, a complex and dynamic living tissue, continually engages in a process called *remodeling*—the construction of new bone tissue and breaking down of old bone tissue. The entire framework of bones and their cartilages, along with ligaments and tendons, constitutes the **skeletal system.** In this chapter we will survey the various components of bones to help you understand how bones form and age, and how exercise affects their density and strength. The study of bone structure and the treatment of bone disorders is called **osteology** (os-tē-OL-ō-jē; *osteo-* = bone; *-logy* = study of).

? ***Did you ever wonder why more females than males are affected by osteoporosis?***

6.1 FUNCTIONS OF BONE AND THE SKELETAL SYSTEM

OBJECTIVE

- Describe the six main functions of the skeletal system.

Bone tissue makes up about 18% of the weight of the human body. The skeletal system performs several basic functions:

1. ***Support.*** The skeleton serves as the structural framework for the body by supporting soft tissues and providing attachment points for the tendons of most skeletal muscles.
2. ***Protection.*** The skeleton protects the most important internal organs from injury. For example, cranial bones protect the brain, vertebrae (backbones) protect the spinal cord, and the rib cage protects the heart and lungs.
3. ***Assistance in movement.*** Most skeletal muscles attach to bones; when they contract, they pull on bones to produce movement. This function is discussed in detail in Chapter 10.
4. ***Mineral homeostasis (storage and release).*** Bone tissue stores several minerals, especially calcium and phosphorus, which contribute to the strength of bone. Bone tissue stores about 99% of the body's calcium. On demand, bone releases minerals into the blood to maintain critical mineral balances (homeostasis) and to distribute the minerals to other parts of the body.
5. ***Blood cell production.*** Within certain bones, a connective tissue called **red bone marrow** produces red blood cells, white blood cells, and platelets, a process called **hemopoiesis** (hēm-ō-poy-Ē-sis; *hemo-* = blood; *-poiesis* = making). Red bone marrow consists of developing blood cells, adipocytes, fibroblasts, and macrophages within a network of reticular fibers. It is present in developing bones of the fetus and in some adult bones, such as the hip (pelvic) bones, ribs, sternum (breastbone), vertebrae (backbones), skull, and ends of the bones of the humerus (arm bone) and femur (thigh bone). In a newborn, all bone marrow is red and is involved in hemopoiesis. With increasing age, much of the bone marrow changes from red to yellow. Blood cell production is considered in detail in Section 19.2.
6. ***Triglyceride storage.*** **Yellow bone marrow** consists mainly of adipose cells, which store triglycerides. The stored triglycerides are a potential chemical energy reserve.

✔CHECKPOINT

1. How does the skeletal system function in support, protection, movement, and storage of minerals?
2. Describe the role of bones in blood cell production.
3. Which bones contain red bone marrow?
4. How do red bone marrow and yellow bone marrow differ in composition and function?

6.2 STRUCTURE OF BONE

OBJECTIVE

- Describe the structure and functions of each part of a long bone.

We will now examine the structure of bone at the macroscopic level. Macroscopic bone structure may be analyzed by considering the parts of a long bone, such as the humerus (the arm bone) shown in Figure 6.1a. A *long bone* is one that has greater length than width. A typical long bone consists of the following parts:

1. The **diaphysis** (dī-AF-i-sis = growing between) is the bone's shaft or body—the long, cylindrical, main portion of the bone.
2. The **epiphyses** (e-PIF-i-sēz = growing over; singular is *epiphysis*) are the proximal and distal ends of the bone.
3. The **metaphyses** (me-TAF-i-sēz; *meta-* = between; singular is *metaphysis*) are the regions between the diaphysis and the epiphyses. In a growing bone, each metaphysis contains an *epiphyseal (growth) plate* (ep′-i-FIZ-ē-al), a layer of hyaline cartilage that allows the diaphysis of the bone to grow in length (described later in the chapter). When a bone ceases to grow in length at about ages 18–21, the cartilage in the epiphyseal plate is replaced by bone; the resulting bony structure is known as the *epiphyseal line.*
4. The **articular cartilage** is a thin layer of hyaline cartilage covering the part of the epiphysis where the bone forms an articulation (joint) with another bone. Articular cartilage reduces friction and absorbs shock at freely movable joints. Because articular cartilage lacks a perichondrium and lacks blood vessels, repair of damage is limited.
5. The **periosteum** (per-ē-OS-tē-um; *peri-* = around) is a tough connective tissue sheath and its associated blood supply that surrounds the bone surface wherever it is not covered by articular cartilage. It is composed of an *outer fibrous layer* of dense irregular connective tissue and an *inner osteogenic layer* that consists of cells. Some of the cells enable bone to grow in thickness, but not in length. The periosteum also protects the bone, assists in fracture repair, helps nourish bone tissue, and serves as an attachment point for ligaments and tendons. The periosteum is attached to the underlying bone by **perforating (Sharpey's) fibers,** thick bundles of collagen that extend from the periosteum into the bone extracellular matrix.
6. The **medullary cavity** (MED-ū-lar-ē; *medulla-* = marrow, pith), or **marrow cavity,** is a hollow, cylindrical space within the diaphysis that contains fatty yellow bone marrow and numerous blood vessels in adults. This cavity minimizes the weight of the bone by reducing the dense bony material where it is least needed. The long bones' tubular design provides maximum strength with minimum weight.

Figure 6.1 **Parts of a long bone.** The spongy bone tissue of the epiphyses and metaphyses contains red bone marrow, and the medullary cavity of the diaphysis contains yellow bone marrow (in adults).

A long bone is covered by articular cartilage at the articular surfaces of its proximal and distal epiphyses and by periosteum around all other parts of the bone.

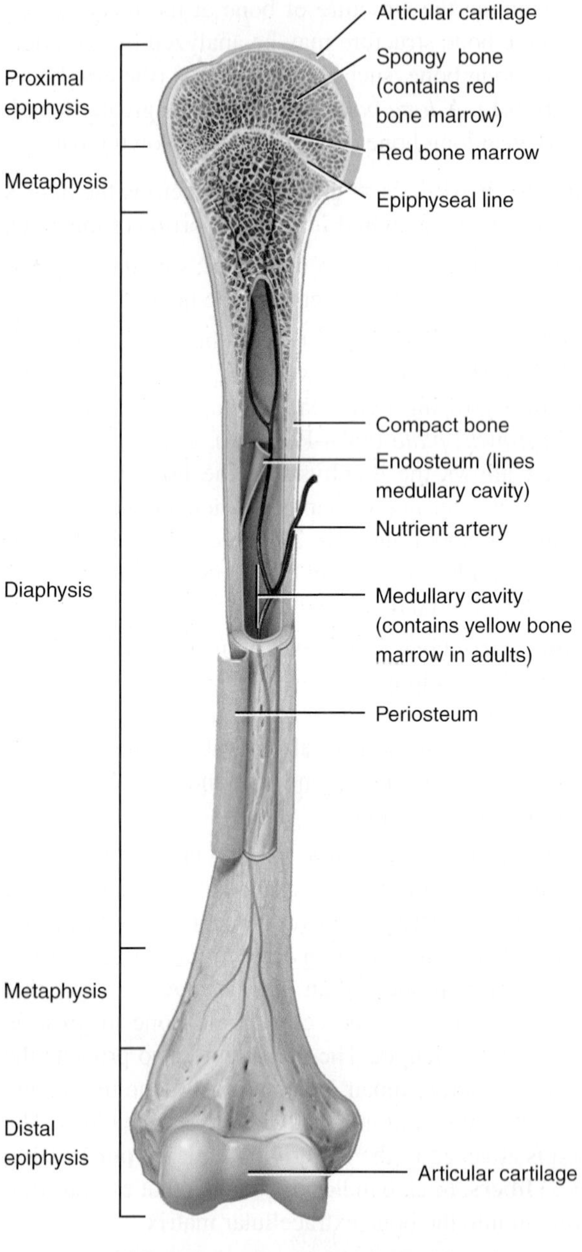

(a) Partially sectioned humerus (arm bone)

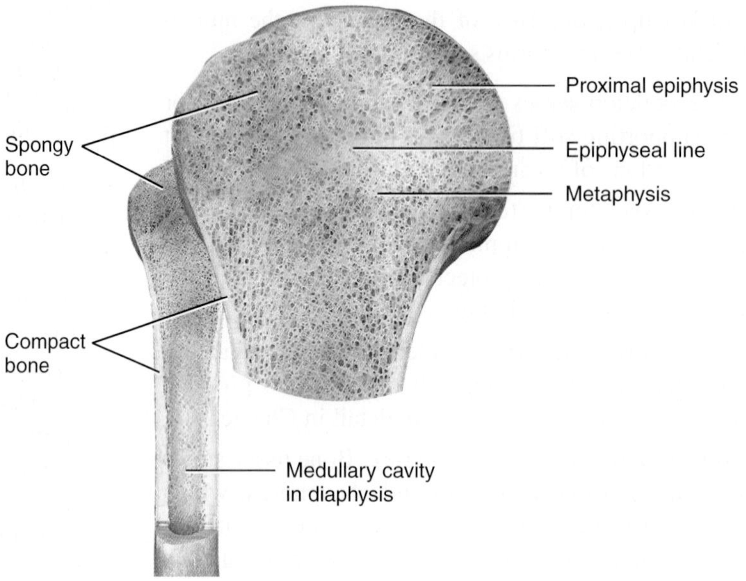

(b) Partially sectioned humerus

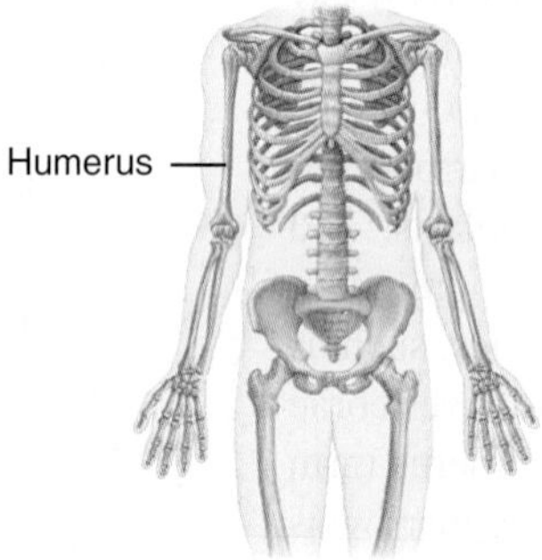

FUNCTIONS OF BONE TISSUE

1. Supports soft tissue and provides attachment for skeletal muscles.
2. Protects internal organs.
3. Assists in movement, along with skeletal muscles.
4. Stores and releases minerals.
5. Contains red bone marrow, which produces blood cells.
6. Contains yellow bone marrow, which stores triglycerides (fats).

What is the functional significance of the periosteum?

7. The **endosteum** (end-OS-tē-um; *endo-* = within) is a thin membrane that lines the medullary cavity. It contains a single layer of bone-forming cells and a small amount of connective tissue.

CHECKPOINT

5. Diagram the parts of a long bone, and list the functions of each part.

6.3 HISTOLOGY OF BONE TISSUE

OBJECTIVES

- Explain why bone tissue is classified as a connective tissue.
- Describe the cellular composition of bone tissue and the functions of each type of cell.
- Compare the structural and functional differences between compact and spongy bone tissue.

We will now examine the structure of bone at the microscopic level. Like other connective tissues, **bone,** or **osseous tissue** (OS-ē-us), contains an abundant extracellular matrix that surrounds widely separated cells. The extracellular matrix is about 15% water, 30% collagen fibers, and 55% crystallized mineral salts. The most abundant mineral salt is calcium phosphate [$Ca_3(PO_4)_2$]. It combines with another mineral salt, calcium hydroxide [$Ca(OH)_2$], to form crystals of **hydroxyapatite** [$Ca_{10}(PO_4)_6(OH)_2$] (hī-drok-sē-AP-a-tīt). As the crystals form, they combine with still other mineral salts, such as calcium carbonate ($CaCO_3$), and ions such as magnesium, fluoride, potassium, and sulfate. As these mineral salts are deposited in the framework formed by the collagen fibers of the extracellular matrix, they crystallize and the tissue hardens. This process, called **calcification** (kal′-si-fi-KĀ-shun), is initiated by bone-building cells called osteoblasts (described shortly).

It was once thought that calcification simply occurred when enough mineral salts were present to form crystals. We now know that the process requires the presence of collagen fibers. Mineral salts first begin to crystallize in the microscopic spaces between collagen fibers. After the spaces are filled, mineral crystals accumulate around the collagen fibers. The combination of crystallized salts and collagen fibers is responsible for the characteristics of bone.

Although a bone's *hardness* depends on the crystallized inorganic mineral salts, a bone's *flexibility* depends on its collagen fibers. Like reinforcing metal rods in concrete, collagen fibers and other organic molecules provide *tensile strength,* resistance to being stretched or torn apart. Soaking a bone in an acidic solution, such as vinegar, dissolves its mineral salts, causing the bone to become rubbery and flexible. As you will see shortly, when the need for particular minerals arises or as part of bone formation or breakdown, bone cells called osteoclasts secrete enzymes and acids that break down both the mineral salts and the collagen fibers of the extracellular matrix of bone.

Four types of cells are present in bone tissue: osteogenic cells, osteoblasts, osteocytes, and osteoclasts (Figure 6.2).

1. **Osteogenic cells** (os′-tē-ō-JEN-ik; *-genic* = producing) are unspecialized bone stem cells derived from mesenchyme, the tissue from which almost all connective tissues are formed. They are the only bone cells to undergo cell division; the resulting cells develop into osteoblasts. Osteogenic cells are found along the inner portion of the periosteum, in the endosteum, and in the canals within bone that contain blood vessels.
2. **Osteoblasts** (OS-tē-ō-blasts′; *-blasts* = buds or sprouts) are bone-building cells. They synthesize and secrete collagen fibers and other organic components needed to build the extracellular matrix of bone tissue, and they initiate calcification (described shortly). As osteoblasts surround themselves with extracellular matrix, they become trapped in their secretions and become osteocytes. (Note: The ending *-blast* in the name of a bone cell or any other connective tissue cell means that the cell secretes extracellular matrix.)
3. **Osteocytes** (OS-tē-ō-sīts′; *-cytes* = cells), mature bone cells, are the main cells in bone tissue and maintain its daily metabolism, such as the exchange of nutrients and wastes with the blood. Like osteoblasts, osteocytes do not undergo cell division.

Figure 6.2 Types of cells in bone tissue.

Osteogenic cells undergo cell division and develop into osteoblasts, which secrete bone extracellular matrix.

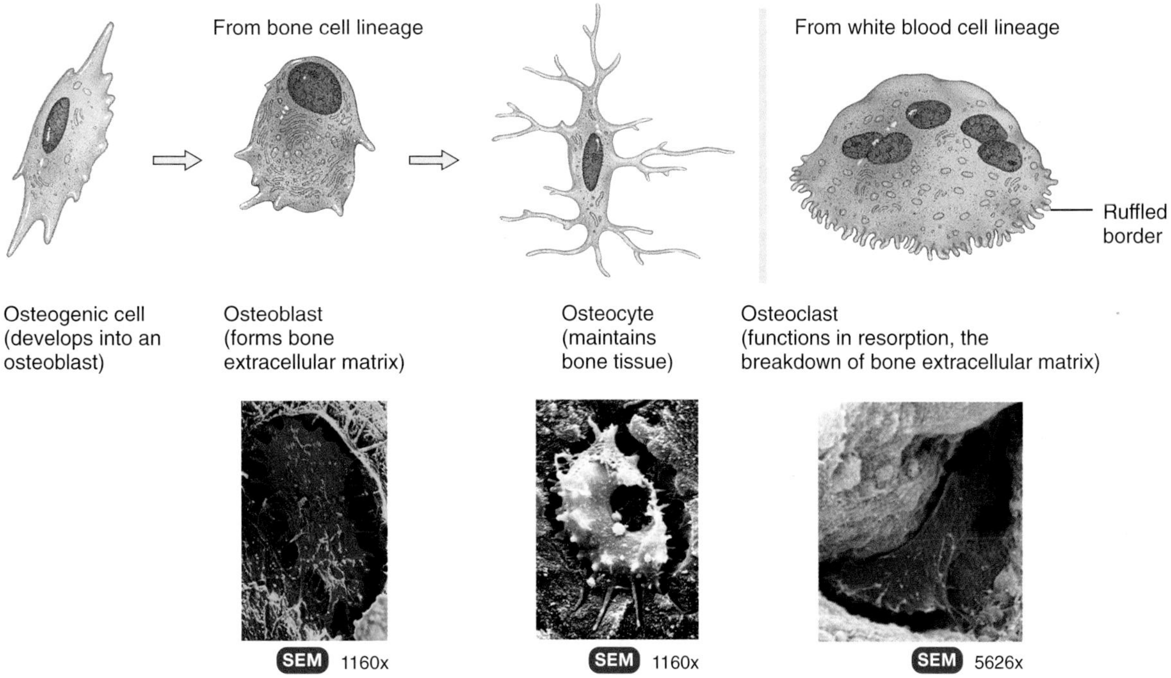
From bone cell lineage
From white blood cell lineage
Ruffled border
Osteogenic cell (develops into an osteoblast)
Osteoblast (forms bone extracellular matrix)
Osteocyte (maintains bone tissue)
Osteoclast (functions in resorption, the breakdown of bone extracellular matrix)
SEM 1160x
SEM 1160x
SEM 5626x

Dr. Richard Kessel and Randy Kardon/Tissues & Organs/Visuals Unlimited

Dr. Richard Kessel and Randy Kardon/Tissues & Organs/Visuals Unlimited

Why is bone resorption important?

(Note: The ending *-cyte* in the name of a bone cell or any other tissue cell means that the cell maintains the tissue.)

4. **Osteoclasts** (OS-tē-ō-klasts′; *-clast* = break) are huge cells derived from the fusion of as many as 50 monocytes (a type of white blood cell) and are concentrated in the endosteum. On the side of the cell that faces the bone surface, the osteoclast's plasma membrane is deeply folded into a *ruffled border*. Here the cell releases powerful lysosomal enzymes and acids that digest the protein and mineral components of the underlying extracellular bone matrix. This breakdown of bone extracellular matrix, termed **resorption** (rē-SORP-shun), is part of the normal development, maintenance, and repair of bone. (Note: The ending *-clast* means that the cell breaks down extracellular matrix.) As you will see later, in response to certain hormones, osteoclasts help regulate blood calcium level (see Section 6.7). They are also target cells for drug therapy used to treat osteoporosis (see Disorders: Homeostatic Imbalances at the end of this chapter).

A mnemonic that will help you remember the difference between the function of osteoblasts and osteoclasts is as follows: osteo**B**lasts **B**uild bone, while osteo**C**lasts **C**arve out bone.

Bone is not completely solid but has many small spaces between its cells and extracellular matrix components. Some spaces serve as channels for blood vessels that supply bone cells with nutrients. Other spaces act as storage areas for red bone marrow. Depending on the size and distribution of the spaces, the regions of a bone may be categorized as compact or spongy (see Figure 6.1). Overall, about 80% of the skeleton is compact bone and 20% is spongy bone.

Compact Bone Tissue

Compact bone tissue contains few spaces (Figure 6.3a) and is the strongest form of bone tissue. It is found beneath the periosteum of all bones and makes up the bulk of the diaphyses of long bones. Compact bone tissue provides protection and support and resists the stresses produced by weight and movement.

Compact bone tissue is composed of repeating structural units called **osteons,** or **haversian systems** (ha-VER-shan). Each osteon consists of concentric lamellae arranged around a **central (haversian) canal.** Resembling the growth rings of a tree, the **concentric lamellae** (LA-mel-ē) are circular plates of mineralized extracellular matrix of increasing diameter, surrounding a small network of blood vessels, lymphatics, and nerves located in the central canal (Figure 6.3a). These tube-like units of bone generally form a series of parallel cylinders that, in long bones, tend to run parallel to the long axis of the bone. Between the concentric lamellae are small spaces called **lacunae** (la-KOO-nē = little lakes; singular is *lacuna*), which contain osteocytes. Radiating in all directions from the lacunae are tiny **canaliculi** (kan-a-LIK-ū-lī = small channels), which are filled with extracellular fluid. Inside the canaliculi are slender fingerlike processes of osteocytes (see inset at right of Figure 6.3a). Neighboring osteocytes communicate via gap junctions (see Section 4.2). The canaliculi connect lacunae with one another and with the central canals, forming an intricate, miniature system of interconnected canals throughout the bone. This system provides many routes for nutrients and oxygen to reach the osteocytes and for the removal of wastes.

Osteons in compact bone tissue are aligned in the same direction and are parallel to the length of the diaphysis. As a result, the shaft of a long bone resists bending or fracturing even when considerable force is applied from either end. Compact bone tissue tends to be thickest in those parts of a bone where stresses are applied in relatively few directions. The lines of stress in a bone are not static. They change as a person learns to walk and in response to repeated strenuous physical activity, such as weight training. The lines of stress in a bone also can change because of fractures or physical deformity. Thus, the organization of osteons is not static but changes over time in response to the physical demands placed on the skeleton.

The areas between neighboring osteons contain lamellae called **interstitial lamellae** (in′-ter-STISH-al), which also have lacunae with osteocytes and canaliculi. Interstitial lamellae are fragments of older osteons that have been partially destroyed during bone rebuilding or growth.

Blood vessels, lymphatic vessels, and nerves from the periosteum penetrate the compact bone through transverse **perforating canals** or *Volkmann's canals* (FOLK-mans). The vessels and nerves of the perforating canals connect with those of the medullary cavity, periosteum, and central canals.

Arranged around the entire outer and inner circumference of the shaft of a long bone are lamellae called **circumferential lamellae** (ser′-kum-fer-EN-shē-al). They develop during initial bone formation. The circumferential lamellae directly deep to the periosteum are called *outer circumferential lamellae*. They are connected to the periosteum by **perforating** (*Sharpey's*) **fibers.** The circumferential lamellae that line the medullary cavity are called *inner circumferential lamellae* (Figure 6.3a).

Spongy Bone Tissue

In contrast to compact bone tissue, **spongy bone tissue,** also referred to as *trabecular* or *cancellous bone tissue,* does not contain osteons (Figure 6.3b, c). Spongy bone tissue is always located in the *interior* of a bone, protected by a covering of compact bone. It consists of lamellae that are arranged in an irregular pattern of thin columns called **trabeculae** (tra-BEK-ū-lē = little beams; singular is *trabecula*). Between the trabeculae are spaces that are visible to the unaided eye. These macroscopic spaces are filled with red bone marrow in bones that produce blood cells, and yellow bone marrow (adipose tissue) in other bones. Both types of bone marrow contain numerous small blood vessels that provide nourishment to the osteocytes. Each trabecula consists of concentric lamellae, osteocytes that lie in lacunae, and canaliculi that radiate outward from the lacunae.

Spongy bone tissue makes up most of the interior bone tissue of short, flat, sesamoid, and irregularly shaped bones. In long bones it forms the core of the epiphyses beneath the paper-thin layer of compact bone, and forms a variable narrow rim bordering the medullary cavity of the diaphysis. Spongy bone is always covered by a layer of compact bone for protection.

At first glance, the trabeculae of spongy bone tissue may appear to be less organized than the osteons of compact bone tissue.

Figure 6.3 Histology of compact and spongy bone. (a) Sections through the diaphysis of a long bone, from the surrounding periosteum on the right, to compact bone in the middle, to spongy bone and the medullary cavity on the left. The inset at the upper right shows an osteocyte in a lacuna. (b, c) Details of spongy bone. See Table 4.7 for a photomicrograph of compact bone tissue and Figure 6.11a for a scanning electron micrograph of spongy bone tissue.

Bone tissue is organized in concentric lamellae around a central (haversian) canal in compact bone and in irregularly arranged lamellae in the trabeculae in spongy bone.

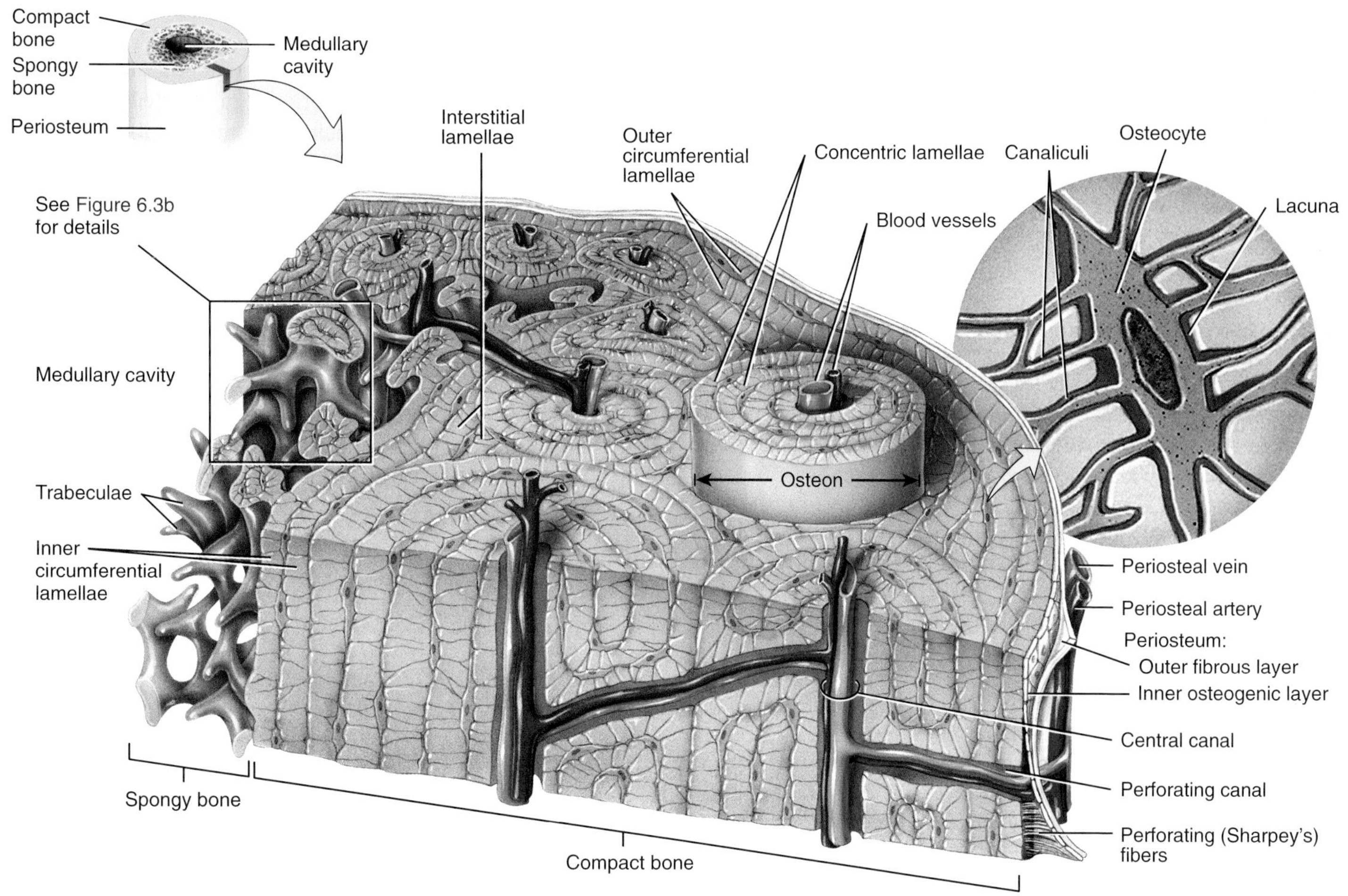

(a) Osteons (haversian systems) in compact bone and trabeculae in spongy bone

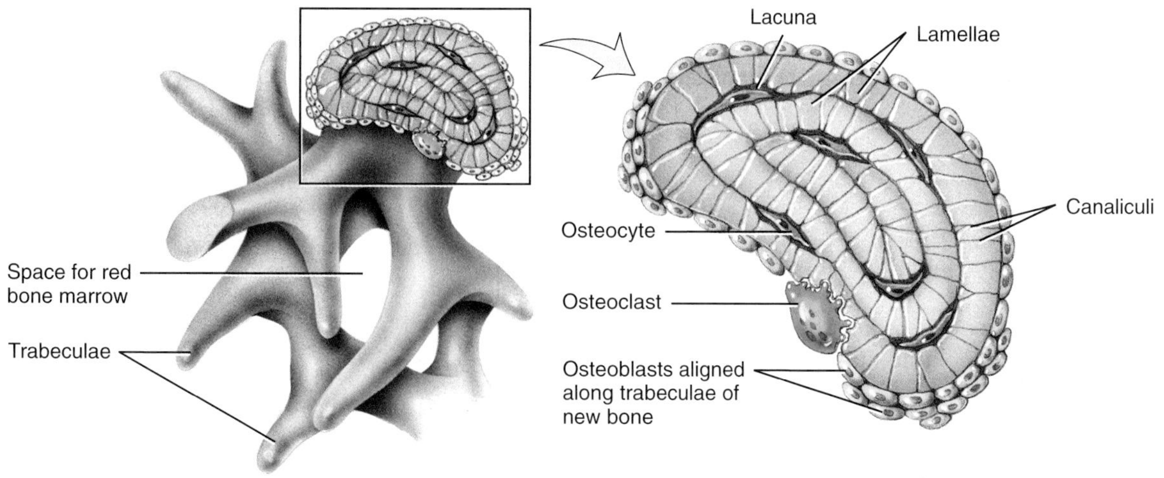

(b) Enlarged aspect of spongy bone trabeculae

(c) Details of a section of a trabecula

As people age, some central (haversian) canals may become blocked. What effect would this have on the surrounding osteocytes?

However, they are precisely oriented along lines of stress, a characteristic that helps bones resist stresses and transfer force without breaking. Spongy bone tissue tends to be located where bones are not heavily stressed or where stresses are applied from many directions. The trabeculae do not achieve their final arrangement until locomotion is completely learned. In fact, the arrangement can even be altered as lines of stress change due to a poorly healed fracture or a deformity.

Spongy bone tissue is different from compact bone tissue in two respects. First, spongy bone tissue is light, which reduces the overall weight of a bone. This reduction in weight allows the bone to move more readily when pulled by a skeletal muscle. Second, the trabeculae of spongy bone tissue support and protect the red bone marrow. Spongy bone in the hip bones, ribs, sternum (breastbone), vertebrae, and the proximal ends of the humerus and femur is the only site where red bone marrow is stored and, thus, the site where hemopoiesis (blood cell production) occurs in adults.

CLINICAL CONNECTION | *Bone Scan*

A **bone scan** is a diagnostic procedure that takes advantage of the fact that bone is living tissue. A small amount of a radioactive tracer compound that is readily absorbed by bone is injected intravenously. The degree of uptake of the tracer is related to the amount of blood flow to the bone. A scanning device (gamma camera) measures the radiation emitted from the bones, and the information is translated into a photograph that can be read like an x-ray on a monitor. Normal bone tissue is identified by a consistent gray color throughout because of its uniform uptake of the radioactive tracer. Darker or lighter areas may indicate bone abnormalities. Darker areas called "hot spots" are areas of increased metabolism that absorb more of the radioactive tracer due to increased blood flow. Hot spots may indicate bone cancer, abnormal healing of fractures, or abnormal bone growth. Lighter areas called "cold spots" are areas of decreased metabolism that absorb less of the radioactive tracer due to decreased blood flow. Cold spots may indicate problems such as degenerative bone disease, decalcified bone, fractures, bone infections, Paget's disease, and rheumatoid arthritis. A bone scan detects abnormalities 3 to 6 months sooner than standard x-ray procedures and exposes the patient to less radiation. A bone scan is the standard test for bone density screening, particularly important in screening for osteoporosis in females. •

✔CHECKPOINT

6. Why is bone considered a connective tissue?
7. What factors contribute to the hardness and tensile strength of bone?
8. List the four types of cells in bone tissue and their functions.
9. What is the composition of the extracellular matrix of bone tissue?
10. How are compact and spongy bone tissues different in microscopic appearance, location, and function?
11. What is a bone scan and how is it used clinically?

6.4 BLOOD AND NERVE SUPPLY OF BONE

OBJECTIVE

- Describe the blood and nerve supply of bone.

Bone is richly supplied with blood. Blood vessels, which are especially abundant in portions of bone containing red bone marrow, pass into bones from the periosteum. We will consider the blood supply of a long bone such as the mature tibia (shin bone) shown in Figure 6.4.

Periosteal arteries (per-ē-OS-tē-al), small arteries accompanied by nerves, enter the diaphysis through many perforating (Volkmann's) canals and supply the periosteum and outer part of the compact bone (see Figure 6.3a). Near the center of the diaphysis, a large **nutrient artery** passes through a hole in compact bone called the **nutrient foramen.** On entering the medullary cavity, the nutrient artery divides into proximal and distal branches that course toward each end of the bone. These branches supply both the inner part of compact bone tissue of the diaphysis and the

Figure 6.4 **Blood supply of a mature long bone.**

6 Bone is richly supplied with blood vessels.

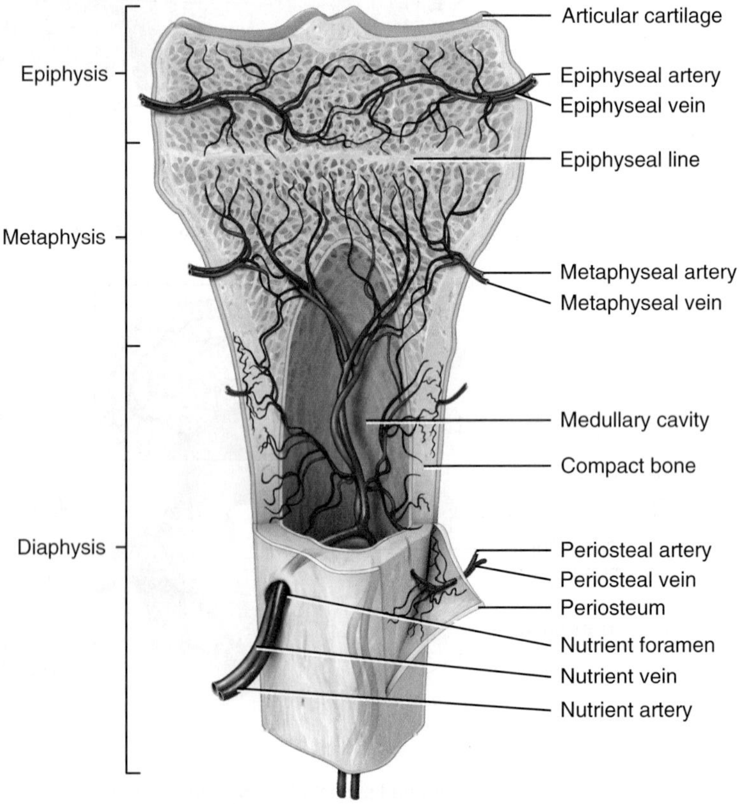
Epiphysis
Metaphysis
Diaphysis
Articular cartilage
Epiphyseal artery
Epiphyseal vein
Epiphyseal line
Metaphyseal artery
Metaphyseal vein
Medullary cavity
Compact bone
Periosteal artery
Periosteal vein
Periosteum
Nutrient foramen
Nutrient vein
Nutrient artery

Partially sectioned tibia (shin bone)

? Where do periosteal arteries enter bone tissue?

spongy bone tissue and red bone marrow as far as the epiphyseal plates (or lines). Some bones, like the tibia, have only one nutrient artery; others, like the femur (thigh bone), have several. The ends of long bones are supplied by the metaphyseal and epiphyseal arteries, which arise from arteries that supply the associated joint. The **metaphyseal arteries** (met-a-FIZ-ē-al) enter the metaphyses of a long bone and, together with the nutrient artery, supply the red bone marrow and bone tissue of the metaphyses. The **epiphyseal arteries** (ep′-i-FIZ-ē-al) enter the epiphyses of a long bone and supply the red bone marrow and bone tissue of the epiphyses.

Veins that carry blood away from long bones are evident in three places: (1) One or two **nutrient veins** accompany the nutrient artery and exit through the diaphysis; (2) numerous **epiphyseal veins** and **metaphyseal veins** accompany their respective arteries and exit through the epiphyses; and (3) many small **periosteal veins** accompany their respective arteries and exit through the periosteum.

Nerves accompany the blood vessels that supply bones. The periosteum is rich in sensory nerves, some of which carry pain sensations. These nerves are especially sensitive to tearing or tension, which explains the severe pain resulting from a fracture or a bone tumor. For the same reason there is some pain associated with a bone marrow needle biopsy. In this procedure, a needle is inserted into the middle of the bone to withdraw a sample of red bone marrow to examine it for conditions such as leukemias, metastatic neoplasms, lymphoma, Hodgkin's disease, and aplastic anemia. As the needle penetrates the periosteum, pain is felt. Once it passes through, there is little pain.

✔ **CHECKPOINT**

12. Explain the location and roles of the nutrient arteries, nutrient foramina, epiphyseal arteries, and periosteal arteries.
13. Which part of a bone contains sensory nerves associated with pain?
14. Describe one situation in which these sensory neurons are important.
15. How is a bone marrow needle biopsy performed? What conditions are diagnosed through this procedure?

6.5 BONE FORMATION

OBJECTIVES

- Describe the steps of intramembranous and endochondral ossification.
- Explain how bone grows in length and thickness.
- Describe the process involved in bone remodeling.

The process by which bone forms is called **ossification** (os′-i-fi-KĀ-shun; *ossi-* = bone; *-fication* = making) or **osteogenesis** (os′-tē-ō-JEN-e-sis). Bone formation occurs in four principal situations: (1) the initial formation of bones in an embryo and fetus, (2) the growth of bones during infancy, childhood, and adolescence until their adult sizes are reached, (3) the remodeling of bone (replacement of old bone by new bone tissue throughout life), and (4) the repair of fractures (breaks in bones) throughout life.

Initial Bone Formation in an Embryo and Fetus

We will first consider the initial formation of bone in an embryo and fetus. The embryonic "skeleton," initially composed of mesenchyme in the general shape of bones, is the site where cartilage formation and ossification occur during the sixth week of embryonic development. Bone formation follows one of two patterns.

The two methods of bone formation, which both involve the replacement of a preexisting connective tissue with bone, do not lead to differences in the structure of mature bones, but are simply different methods of bone development. In the first type of ossification, called **intramembranous ossification** (in′-tra-MEM-bra-nus; *intra-* = within; *-membran-* = membrane), bone forms directly within mesenchyme, which is arranged in sheetlike layers that resemble membranes. In the second type, **endochondral ossification** (en′-dō-KON-dral; *endo-* = within; *-chondral* = cartilage), bone forms within hyaline cartilage that develops from mesenchyme.

Intramembranous Ossification

Intramembranous ossification is the simpler of the two methods of bone formation. The flat bones of the skull, most of the facial bones, mandible (lower jawbone), and the medial part of the clavicle (collar bone) are formed in this way. Also, the "soft spots" that help the fetal skull pass through the birth canal later harden as they undergo intramembranous ossification, which occurs as follows (Figure 6.5):

1. ***Development of the ossification center.*** At the site where the bone will develop, specific chemical messages cause the mesenchymal cells to cluster together and differentiate, first into osteogenic cells and then into osteoblasts. The site of such a cluster is called an **ossification center.** Osteoblasts secrete the organic extracellular matrix of bone until they are surrounded by it.

Figure 6.5 Intramembranous ossification. Refer to this figure as you read the corresponding numbered paragraphs in the text. Illustrations 1 and 2 show a smaller field of vision at higher magnification than illustrations 3 and 4.

6 Intramembranous ossification involves the formation of bone within mesenchyme arranged in sheetlike layers that resemble membranes.

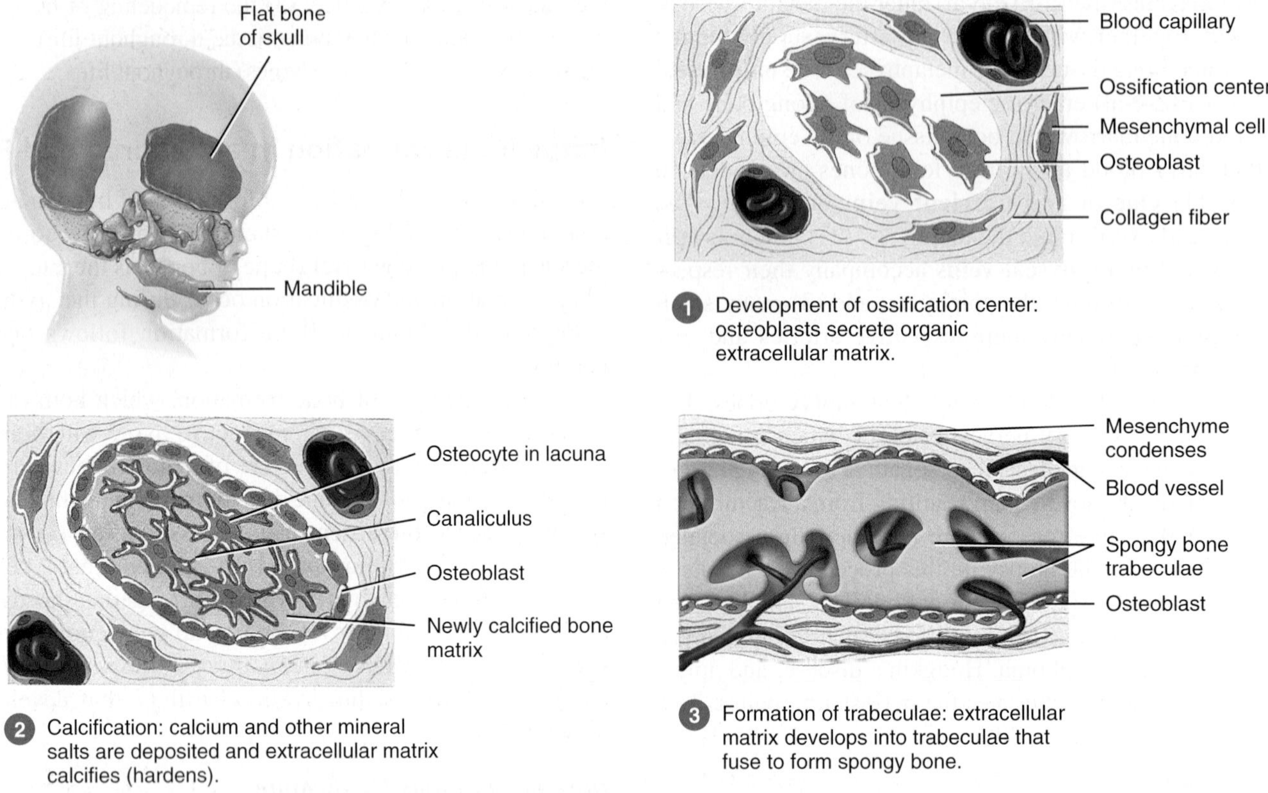
Flat bone of skull
Mandible
Blood capillary
Ossification center
Mesenchymal cell
Osteoblast
Collagen fiber
Osteocyte in lacuna
Canaliculus
Osteoblast
Newly calcified bone matrix
Mesenchyme condenses
Blood vessel
Spongy bone trabeculae
Osteoblast

1 Development of ossification center: osteoblasts secrete organic extracellular matrix.

2 Calcification: calcium and other mineral salts are deposited and extracellular matrix calcifies (hardens).

3 Formation of trabeculae: extracellular matrix develops into trabeculae that fuse to form spongy bone.

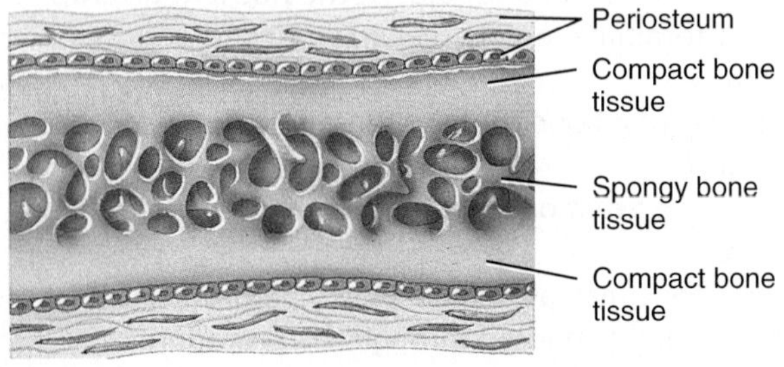
Periosteum
Compact bone tissue
Spongy bone tissue
Compact bone tissue

4 Development of the periosteum: mesenchyme at the periphery of the bone develops into the periosteum.

? Which bones of the body develop by intramembranous ossification?

2 ***Calcification.*** Next, the secretion of extracellular matrix stops, and the cells, now called osteocytes, lie in lacunae and extend their narrow cytoplasmic processes into canaliculi that radiate in all directions. Within a few days, calcium and other mineral salts are deposited and the extracellular matrix hardens or calcifies (calcification).

3 ***Formation of trabeculae.*** As the bone extracellular matrix forms, it develops into trabeculae that fuse with one another to form spongy bone around the network of blood vessels in the tissue. Connective tissue that is associated with the blood vessels in the trabeculae differentiates into red bone marrow.

4 ***Development of the periosteum.*** In conjunction with the formation of trabeculae, the mesenchyme condenses at the periphery of the bone and develops into the periosteum. Eventually, a thin layer of compact bone replaces the surface layers of the spongy bone, but spongy bone remains in the center. Much of the newly formed bone is remodeled (destroyed and reformed) as the bone is transformed into its adult size and shape.

Endochondral Ossification

The replacement of cartilage by bone is called endochondral ossification. Although most bones of the body are formed in this way, the process is best observed in a long bone. It proceeds as follows (Figure 6.6):

1 ***Development of the cartilage model.*** At the site where the bone is going to form, specific chemical messages cause the mesenchymal cells to crowd together in the general shape of the future bone, and then develop into chondroblasts. The chondroblasts secrete cartilage extracellular matrix, producing a **cartilage model** consisting of hyaline cartilage. A covering called the **perichondrium** (per′-i-KON-drē-um) develops around the cartilage model.

Figure 6.6 Endochondral ossification.

During endochondral ossification, bone gradually replaces a cartilage model.

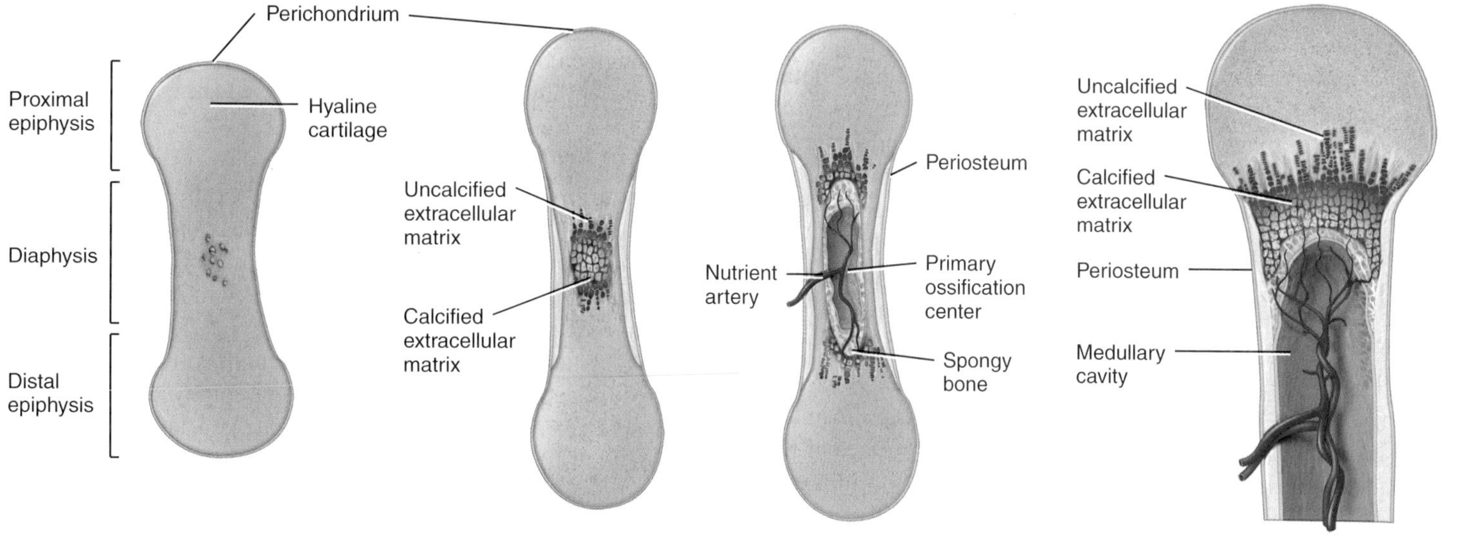

1. Development of cartilage model: mesenchymal cells develop into chondroblasts, which form the cartilage model.
2. Growth of cartilage model: growth occurs by cell division of chondrocytes.
3. Development of primary ossification center: in this region of the diaphysis, bone tissue replaces most of the cartilage.
4. Development of the medullary (marrow) cavity: bone breakdown by osteoclasts forms the medullary cavity.

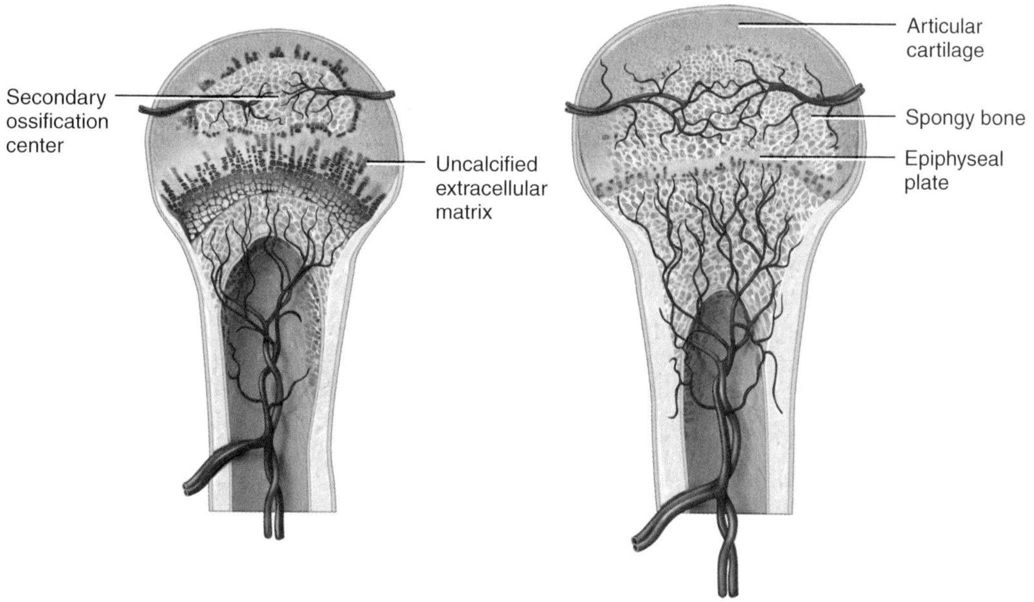

5. Development of secondary ossification centers: these occur in the epiphyses of the bone.
6. Formation of articular cartilage and epiphyseal plate: both structures consist of hyaline cartilage.

(a) Sequence of events

(b) Fifteen-week fetus. The red areas represent bones that are forming (calcified). Dark gray areas represent cartilage (uncalcified).

Where in the cartilage model do secondary ossification centers develop during endochondral ossification?

2 ***Growth of the cartilage model.*** Once chondroblasts become deeply buried in the cartilage extracellular matrix, they are called chondrocytes. The cartilage model grows in length by continual cell division of chondrocytes, accompanied by further secretion of the cartilage extracellular matrix. This type of cartilaginous growth, called **interstitial (endogenous) growth** (growth from within), results in an increase in length. In contrast, growth of the cartilage in thickness is due mainly

to the deposition of extracellular matrix material on the cartilage surface of the model by new chondroblasts that develop from the perichondrium. This process is called **appositional (exogenous) growth** (a-pō-ZISH-o-nal), meaning growth at the outer surface. Interstitial growth and appositional growth of cartilage are described in more detail in Section 4.5.

As the cartilage model continues to grow, chondrocytes in its mid-region hypertrophy (increase in size) and the surrounding cartilage extracellular matrix begins to calcify. Other chondrocytes within the calcifying cartilage die because nutrients can no longer diffuse quickly enough through the extracellular matrix. As these chondrocytes die, the spaces left behind by dead chondrocytes merge into small cavities called lacunae.

3 ***Development of the primary ossification center.*** Primary ossification proceeds *inward* from the external surface of the bone. A nutrient artery penetrates the perichondrium and the calcifying cartilage model through a nutrient foramen in the midregion of the cartilage model, stimulating osteogenic cells in the perichondrium to differentiate into osteoblasts. Once the perichondrium starts to form bone, it is known as the **periosteum.** Near the middle of the model, periosteal capillaries grow into the disintegrating calcified cartilage, inducing growth of a **primary ossification center,** a region where bone tissue will replace most of the cartilage. Osteoblasts then begin to deposit bone extracellular matrix over the remnants of calcified cartilage, forming spongy bone trabeculae. Primary ossification spreads from this central location toward both ends of the cartilage model.

4 ***Development of the medullary (marrow) cavity.*** As the primary ossification center grows toward the ends of the bone, osteoclasts break down some of the newly formed spongy bone trabeculae. This activity leaves a cavity, the medullary (marrow) cavity, in the diaphysis (shaft). Eventually, most of the wall of the diaphysis is replaced by compact bone.

5 ***Development of the secondary ossification centers.*** When branches of the epiphyseal artery enter the epiphyses, **secondary ossification centers** develop, usually around the time of birth. Bone formation is similar to what occurs in primary ossification centers. However, in the secondary ossification centers spongy bone remains in the interior of the epiphyses (no medullary cavities are formed here). In contrast to primary ossification, secondary ossification proceeds *outward* from the center of the epiphysis toward the outer surface of the bone.

6 ***Formation of articular cartilage and the epiphyseal (growth) plate.*** The hyaline cartilage that covers the epiphyses becomes the articular cartilage. Prior to adulthood, hyaline cartilage remains between the diaphysis and epiphysis as the epiphyseal (growth) plate, the region responsible for the lengthwise growth of long bones that you will learn about next.

Bone Growth during Infancy, Childhood, and Adolescence

During infancy, childhood, and adolescence, bones throughout the body grow in thickness by appositional growth, and long bones lengthen by the addition of bone material on the diaphyseal side of the epiphyseal plate by interstitial growth.

Growth in Length

The growth in length of long bones involves the following two major events: (1) interstitial growth of cartilage on the epiphyseal side of the epiphyseal plate and (2) replacement of cartilage on the diaphyseal side of the epiphyseal plate with bone by endochondral ossification.

To understand how a bone grows in length, you need to know some of the details of the structure of the epiphyseal plate. The **epiphyseal (growth) plate** (ep-i-FIZ-ē-al) is a layer of hyaline cartilage in the metaphysis of a growing bone that consists of four zones (Figure 6.7b):

1. ***Zone of resting cartilage.*** This layer is nearest the epiphysis and consists of small, scattered chondrocytes. The term "resting" is used because the cells do not function in bone growth. Rather, they anchor the epiphyseal plate to the epiphysis of the bone.
2. ***Zone of proliferating cartilage.*** Slightly larger chondrocytes in this zone are arranged like stacks of coins. These chondrocytes undergo interstitial growth as they divide and secrete extracellular matrix. The chondrocytes in this zone divide to replace those that die at the diaphyseal side of the epiphyseal plate.
3. ***Zone of hypertrophic cartilage*** (hī-per-TRŌ-fik). This layer consists of large, maturing chondrocytes arranged in columns.
4. ***Zone of calcified cartilage.*** The final zone of the epiphyseal plate is only a few cells thick and consists mostly of chondrocytes that are dead because the extracellular matrix around them has calcified. Osteoclasts dissolve the calcified cartilage, and osteoblasts and capillaries from the diaphysis invade the area. The osteoblasts lay down bone extracellular matrix, replacing the calcified cartilage by the process of endochondral ossification. Recall that endochondral ossification is the replacement of cartilage with bone. As a result, the zone of calcified cartilage becomes the "new diaphysis" that is firmly cemented to the rest of the diaphysis of the bone.

The activity of the epiphyseal plate is the only way that the diaphysis can increase in length. As a bone grows, chondrocytes proliferate on the epiphyseal side of the plate. New chondrocytes replace older ones, which are destroyed by calcification. Thus, the cartilage is replaced by bone on the diaphyseal side of the plate. In this way the thickness of the epiphyseal plate remains relatively constant, but the bone on the diaphyseal side increases

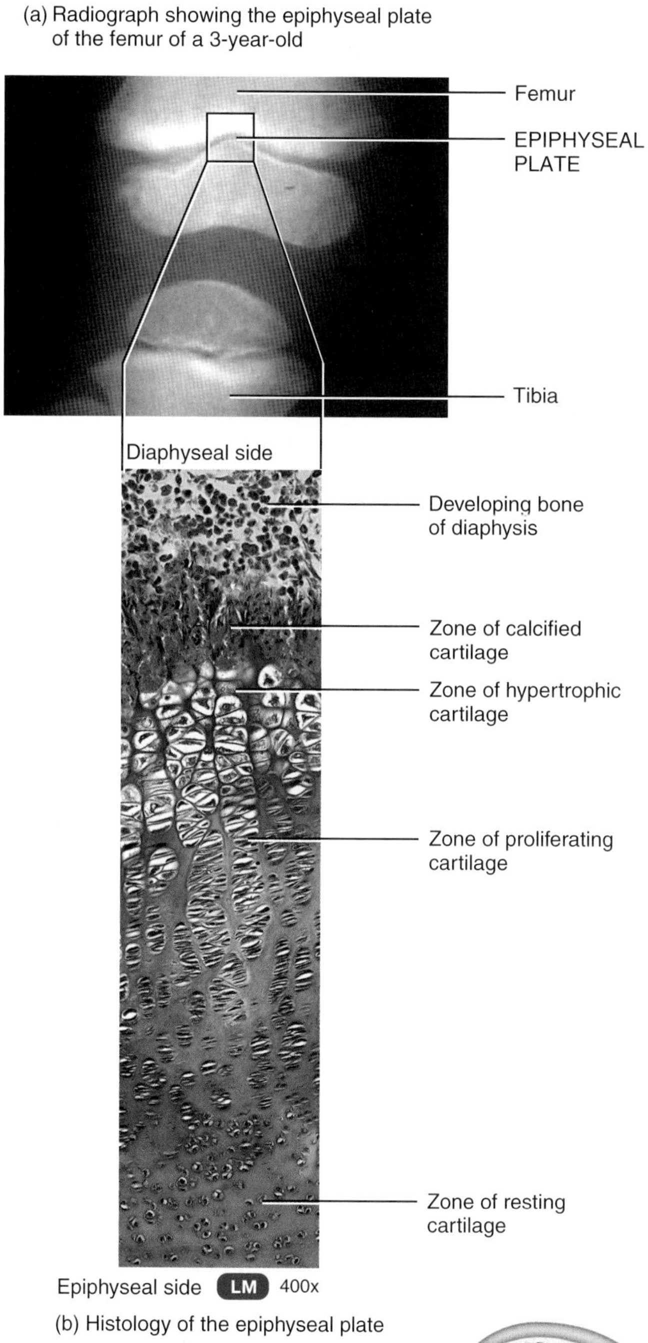
(a) Radiograph showing the epiphyseal plate of the femur of a 3-year-old
Femur
EPIPHYSEAL PLATE
Tibia
Diaphyseal side
Developing bone of diaphysis
Zone of calcified cartilage
Zone of hypertrophic cartilage
Zone of proliferating cartilage
Zone of resting cartilage
Epiphyseal side
LM 400x
(b) Histology of the epiphyseal plate

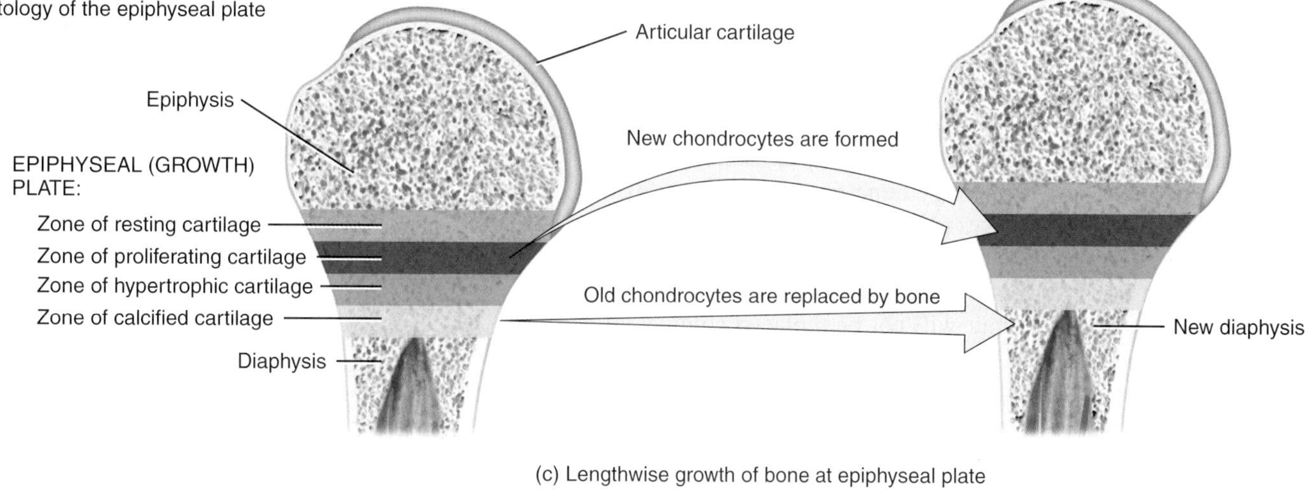
Articular cartilage
Epiphysis
EPIPHYSEAL (GROWTH) PLATE:
Zone of resting cartilage
Zone of proliferating cartilage
Zone of hypertrophic cartilage
Zone of calcified cartilage
Diaphysis
New chondrocytes are formed
Old chondrocytes are replaced by bone
New diaphysis
(c) Lengthwise growth of bone at epiphyseal plate

Figure 6.7 Epiphyseal (growth) plate. The epiphyseal (growth) plate appears as a dark band between whiter calcified areas in the radiograph (x-ray) shown in part (a).

The epiphyseal (growth) plate allows the diaphysis of a bone to increase in length.

How does the epiphyseal (growth) plate account for the lengthwise growth of the diaphysis?

in length (Figure 6.7c). If a bone fracture damages the epiphyseal plate, the fractured bone may be shorter than normal once adult stature is reached. This is because damage to cartilage, which is avascular, accelerates closure of the epiphyseal plate due to the cessation of cartilage cell division, thus inhibiting lengthwise growth of the bone.

When adolescence comes to an end (at about age 18 in females and age 21 in males), the epiphyseal plates close; that is, the epiphyseal cartilage cells stop dividing and bone replaces all remaining cartilage. The epiphyseal plate fades, leaving a bony structure called the **epiphyseal line.** With the appearance of the epiphyseal line, bone growth in length stops completely.

Closure of the epiphyseal plate is a gradual process and the degree to which it occurs is useful in determining bone age, predicting adult height, and establishing age at death from skeletal remains, especially in infants, children, and adolescents. For example, an open epiphyseal plate indicates a younger person, while a partially closed epiphyseal plate or a completely closed one indicates an older person. It should also be kept in mind that closure of the epiphyseal plate, on average, takes place 1–2 years earlier in females.

Growth in Thickness

Like cartilage, bone can grow in thickness (diameter) only by appositional growth (Figure 6.8a):

1. At the bone surface, periosteal cells differentiate into osteoblasts, which secrete the collagen fibers and other organic molecules that form bone extracellular matrix. The osteoblasts become surrounded by extracellular matrix and develop into osteocytes. This process forms bone ridges on either side of a periosteal blood vessel. The ridges slowly enlarge and create a groove for the periosteal blood vessel.
2. Eventually, the ridges fold together and fuse, and the groove becomes a tunnel that encloses the blood vessel. The former periosteum now becomes the endosteum that lines the tunnel.
3. Osteoblasts in the endosteum deposit bone extracellular matrix, forming new concentric lamellae. The formation of additional concentric lamellae proceeds inward toward the periosteal blood vessel. In this way, the tunnel fills in, and a new osteon is created.
4. As an osteon is forming, osteoblasts under the periosteum deposit new circumferential lamellae, further increasing the thickness of the bone. As additional periosteal blood vessels become enclosed as in step 1, the growth process continues.

Recall that as new bone tissue is being deposited on the outer surface of bone, the bone tissue lining the medullary cavity is destroyed by osteoclasts in the endosteum. In this way, the medullary cavity enlarges as the bone increases in thickness (Figure 6.8b).

Remodeling of Bone

Like skin, bone forms before birth but continually renews itself thereafter. **Bone remodeling** is the ongoing replacement of old bone tissue by new bone tissue. It involves **bone resorption,** the removal of minerals and collagen fibers from bone by osteoclasts, and **bone deposition,** the addition of minerals and collagen fibers to bone by osteoblasts. Thus, bone resorption results in the destruction of bone extracellular matrix, while bone deposition results in the formation of bone extracellular matrix. At any given time, about 5% of the total bone mass in the body is being remodeled. The renewal rate for compact bone tissue is about 4% per year, and for spongy bone tissue it is about 20% per year. Remodeling also takes place at different rates in different regions of the body. The distal portion of the femur is replaced about every four months. By contrast, bone in certain areas of the shaft of the femur will not be replaced completely during an individual's life. Even after bones have reached their adult shapes and sizes, old bone is continually destroyed and new bone is formed in its place. Remodeling also removes injured bone, replacing it with new bone tissue. Remodeling may be triggered by factors such as exercise, sedentary lifestyle, and changes in diet.

Remodeling has several other benefits. Since the strength of bone is related to the degree to which it is stressed, if newly formed bone is subjected to heavy loads, it will grow thicker and therefore be stronger than the old bone. Also, the shape of a bone can be altered for proper support based on the stress patterns experienced during the remodeling process. Finally, new bone is more resistant to fracture than old bone.

CLINICAL CONNECTION | *Remodeling and Orthodontics*

Orthodontics (or-thō-DON-tiks) is the branch of dentistry concerned with the prevention and correction of poorly aligned teeth. The movement of teeth by braces places a stress on the bone that forms the sockets that anchor the teeth. In response to this artificial stress, osteoclasts and osteoblasts remodel the sockets so that the teeth align properly. •

During the process of bone resorption, an osteoclast attaches tightly to the bone surface at the endosteum or periosteum and forms a leakproof seal at the edges of its ruffled border (see Figure 6.2). Then it releases protein-digesting lysosomal enzymes and several acids into the sealed pocket. The enzymes digest collagen fibers and other organic substances while the acids dissolve the bone minerals. Working together, several osteoclasts carve out a small tunnel in the old bone. The degraded bone proteins and extracellular matrix minerals, mainly calcium and phosphorus, enter an osteoclast by endocytosis, cross the cell in vesicles, and undergo exocytosis on the side opposite the ruffled border. Now in the interstitial fluid, the products of bone resorption diffuse into nearby blood capillaries. Once a small area of bone has been resorbed, osteoclasts depart and osteoblasts move in to rebuild the bone in that area.

CLINICAL CONNECTION | *Paget's Disease*

A delicate balance exists between the actions of osteoclasts and osteoblasts. Should too much new tissue be formed, the bones become abnormally thick and heavy. If too much mineral material is deposited in the bone, the surplus may form thick bumps, called *spurs,* on the bone that interfere with movement at joints. Excessive loss of calcium or tissue weakens the bones, and they may break, as occurs in osteoporosis, or they may become too flexible, as in rickets and osteomalacia. In **Paget's disease,** there is an excessive proliferation of osteoclasts so that bone resorption occurs faster than bone deposition. In response, osteoblasts attempt to compensate, but the new bone is weaker because it has a higher proportion of spongy to compact bone, mineralization is decreased, and the newly synthesized extracellular matrix contains abnormal proteins. The newly formed bone, especially that of the pelvis, limbs, lower vertebrae, and skull, becomes enlarged, hard, and brittle and fractures easily. •

Figure 6.8 **Bone growth in thickness.**

As new bone is deposited on the outer surface of bone by osteoblasts, the bone tissue lining the medullary cavity is destroyed by osteoclasts in the endosteum.

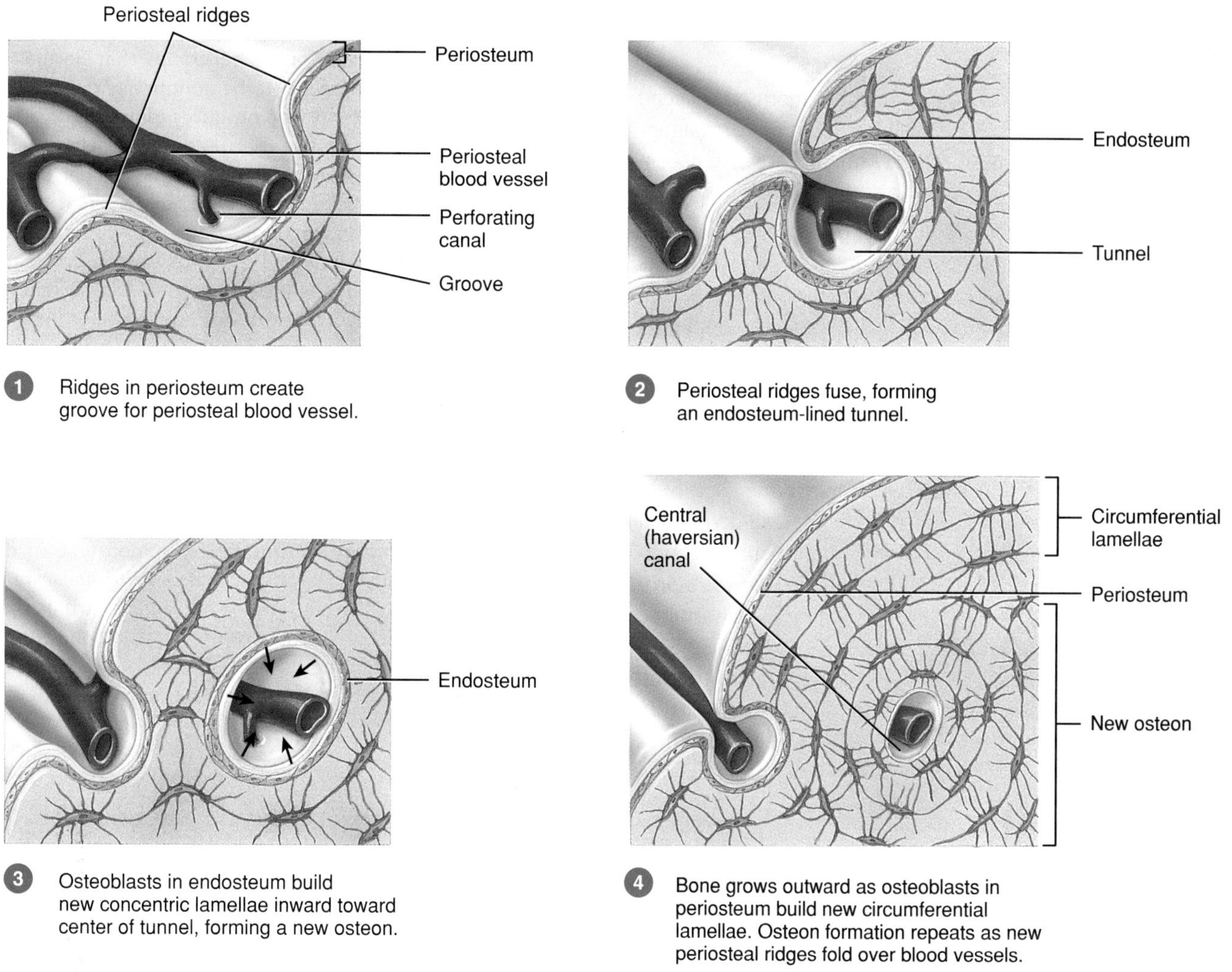
Periosteal ridges
Periosteum
Periosteal blood vessel
Perforating canal
Groove
1 Ridges in periosteum create groove for periosteal blood vessel.
Endosteum
Tunnel
2 Periosteal ridges fuse, forming an endosteum-lined tunnel.
Endosteum
3 Osteoblasts in endosteum build new concentric lamellae inward toward center of tunnel, forming a new osteon.
Central (haversian) canal
Circumferential lamellae
Periosteum
New osteon
4 Bone grows outward as osteoblasts in periosteum build new circumferential lamellae. Osteon formation repeats as new periosteal ridges fold over blood vessels.

(a) Microscopic details

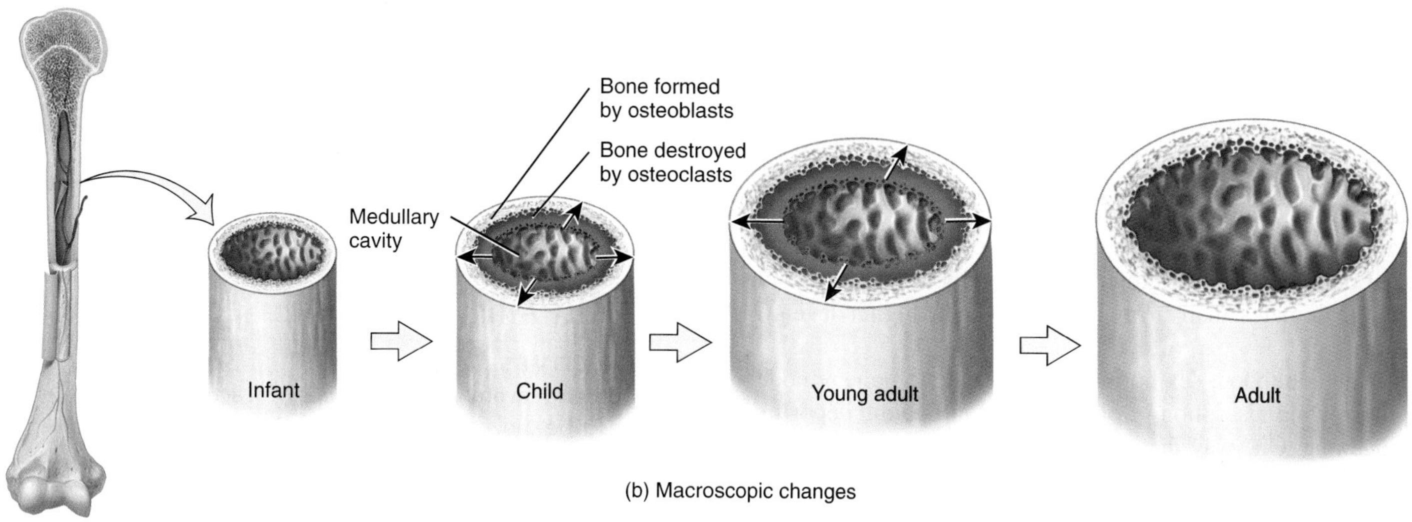
Bone formed by osteoblasts
Bone destroyed by osteoclasts
Medullary cavity
Infant
Child
Young adult
Adult

(b) Macroscopic changes

How does the medullary cavity enlarge during growth in thickness?

Factors Affecting Bone Growth and Bone Remodeling

Normal bone metabolism—growth in the young and bone remodeling in the adult—depends on several factors. These include adequate dietary intake of minerals and vitamins, as well as sufficient levels of several hormones.

1. ***Minerals.*** Large amounts of calcium and phosphorus are needed while bones are growing, as are smaller amounts of magnesium, fluoride, and manganese. These minerals are also necessary during bone remodeling.
2. ***Vitamins.*** Vitamin A stimulates activity of osteoblasts. Vitamin C is needed for synthesis of collagen, the main bone protein. As you will soon learn, vitamin D helps build bone by increasing the absorption of calcium from foods in the gastrointestinal tract into the blood. Vitamins K and B_{12} are also needed for synthesis of bone proteins.
3. ***Hormones.*** During childhood, the hormones most important to bone growth are the insulinlike growth factors (IGFs), which are produced by the liver and bone tissue (see Section 18.6). IGFs stimulate osteoblasts, promote cell division at the epiphyseal plate and in the periosteum, and enhance synthesis of the proteins needed to build new bone. IGFs are produced in response to the secretion of human growth hormone (hGH) from the anterior lobe of the pituitary gland (see Section 18.6). Thyroid hormones (T_3 and T_4) from the thyroid gland also promote bone growth by stimulating osteoblasts. In addition, the hormone insulin from the pancreas promotes bone growth by increasing the synthesis of bone proteins.

At puberty, the secretion of hormones known as sex hormones causes a dramatic effect on bone growth. The **sex hormones** include estrogens (produced by the ovaries) and androgens such as testosterone (produced by the testes). Although females have much higher levels of estrogens and males have higher levels of androgens, females also have low levels of androgens, and males have low levels of estrogens. The adrenal glands of both sexes produce androgens, and other tissues, such as adipose tissue, can convert androgens to estrogens. These hormones are responsible for increased osteoblast activity and synthesis of bone extracellular matrix and the sudden "growth spurt" that occurs during the teenage years. Estrogens also promote changes in the skeleton that are typical of females, such as widening of the pelvis. Ultimately sex hormones, especially estrogens in both sexes, shut down growth at epiphyseal (growth) plates, causing elongation of the bones to cease. Lengthwise growth of bones typically ends earlier in females than in males due to their higher levels of estrogens.

During adulthood, sex hormones contribute to bone remodeling by slowing resorption of old bone and promoting deposition of new bone. One way that estrogens slow resorption is by promoting apoptosis (programmed death) of osteoclasts. As you will see shortly, parathyroid hormone, calcitriol (the active form of vitamin D), and calcitonin are other hormones that can affect bone remodeling.

Moderate weight-bearing exercises maintain sufficient strain on bones to increase and maintain their density.

CLINICAL CONNECTION | *Hormonal Abnormalities that Affect Height*

Excessive or deficient secretion of hormones that normally control bone growth can cause a person to be abnormally tall or short. Oversecretion of hGH during childhood produces **giantism,** in which a person becomes much taller and heavier than normal. Undersecretion of hGH produces **pituitary dwarfism,** in which a person has short stature. (The usual adult height of a *dwarf* is under 4 feet 10 inches.) Although the head, trunk, and limbs of a pituitary dwarf are smaller than normal, they are proportionate. The condition can be treated medically with hGH until epiphyseal plate closure. Oversecretion of hGH during adulthood is called **acromegaly** (ak'-rō-MEG-a-lē). Although hGH cannot produce further lengthening of the long bones because the epiphyseal (growth) plates are already closed, the bones of the hands, feet, and jaws thicken and other tissues enlarge. In addition, the eyelids, lips, tongue, and nose enlarge, and the skin thickens and develops furrows, especially on the forehead and soles. **Achondroplasia** (ā-kon-dro-PLĀ-zē-a; *a-* = without; *-chondro-* = cartilage; *-plasia* = to mold) is an inherited condition in which the conversion of cartilage to bone is abnormal. It results in the most common type of dwarfism, called **achondroplastic dwarfism.** These individuals are typically about 4 feet tall as adults. They have an average-size trunk, short limbs, and a slightly enlarged head with a prominent forehead and flattened nose at the bridge. The condition is essentially untreatable, although some individuals opt for limb-lengthening surgery. •

✔CHECKPOINT

16. What are the major events of intramembranous ossification and endochondral ossification, and how are they different?
17. Describe the zones of the epiphyseal (growth) plate and their functions, and the significance of the epiphyseal line.
18. Explain how bone growth in length differs from bone growth in thickness.
19. How could the metaphyseal area of a bone help determine the age of a skeleton?
20. Define remodeling, and describe the roles of osteoblasts and osteoclasts in the process.
21. What factors affect bone growth and bone remodeling?

6.6 FRACTURE AND REPAIR OF BONE

OBJECTIVES

- Describe several common types of fractures.
- Describe the sequence of events involved in fracture repair.

A **fracture** (FRAK-choor) is any break in a bone. Fractures are named according to their severity, the shape or position of the fracture line, or even the physician who first described them. Some of the common types of fractures are shown in Table 6.1.

TABLE 6.1

Some Common Fractures

FRACTURE	DESCRIPTION	ILLUSTRATION	RADIOGRAPH
Open (Compound)	The broken ends of the bone protrude through the skin. Conversely, a **closed (simple) fracture** does not break the skin.	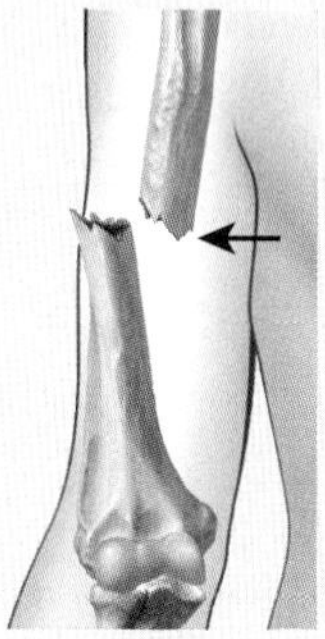	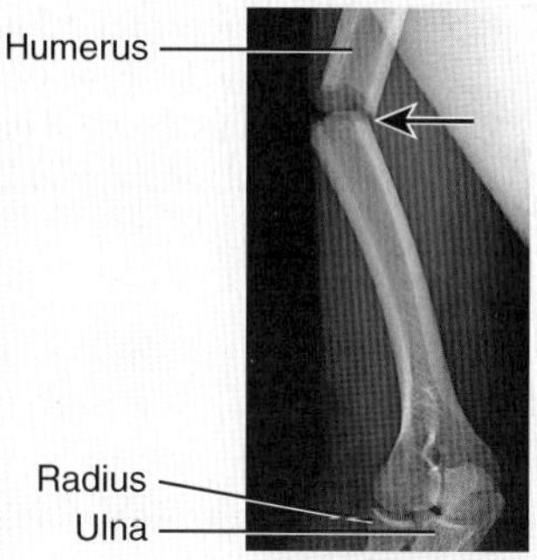
Comminuted (KOM-i-noo-ted; *com-* = together; *-minuted* = crumbled)	The bone is splintered, crushed, or broken into pieces at the site of impact, and smaller bone fragments lie between the two main fragments.	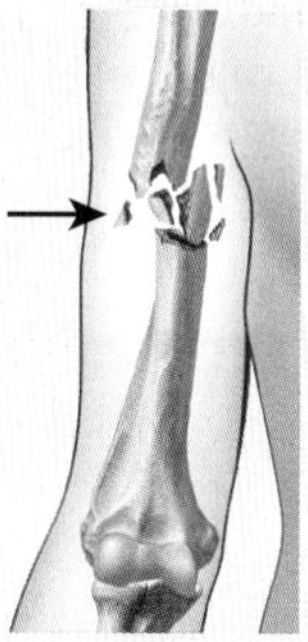	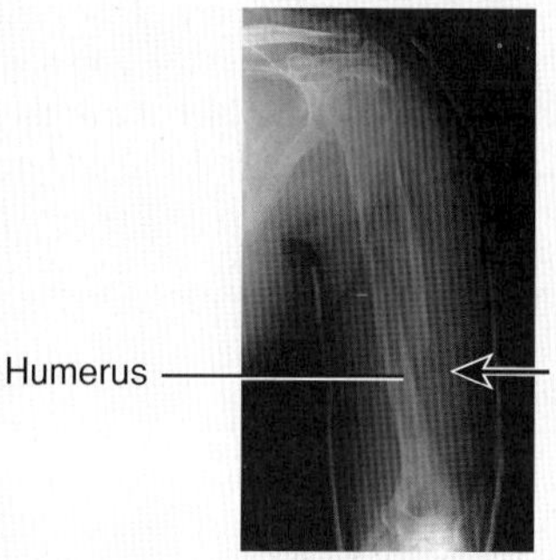
Greenstick	A partial fracture in which one side of the bone is broken and the other side bends; similar to the way a green twig breaks on one side while the other side stays whole, but bends; occurs only in children, whose bones are not fully ossified and contain more organic material than inorganic material.	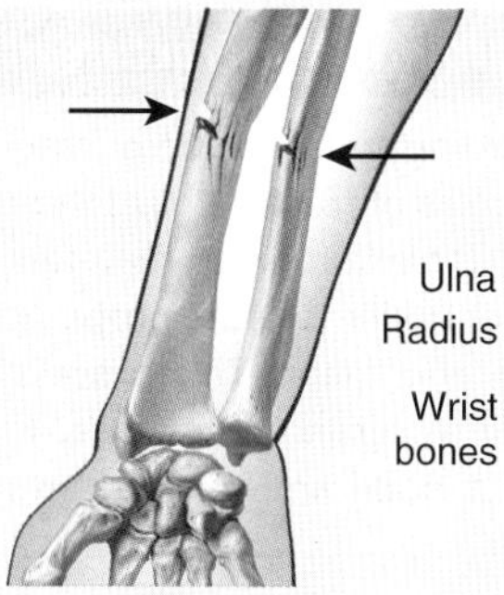	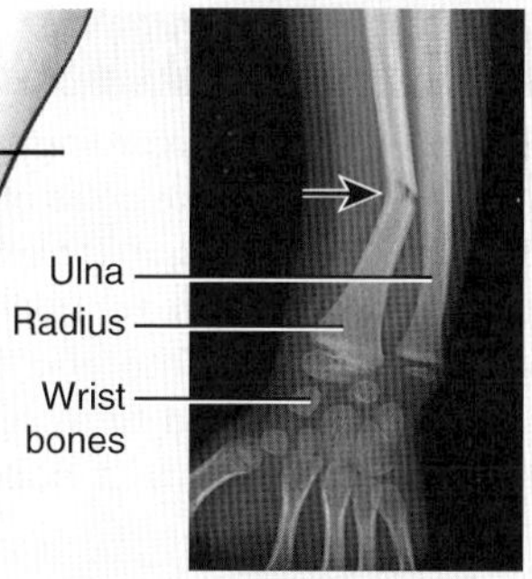
Impacted	One end of the fractured bone is forcefully driven into the interior of the other.	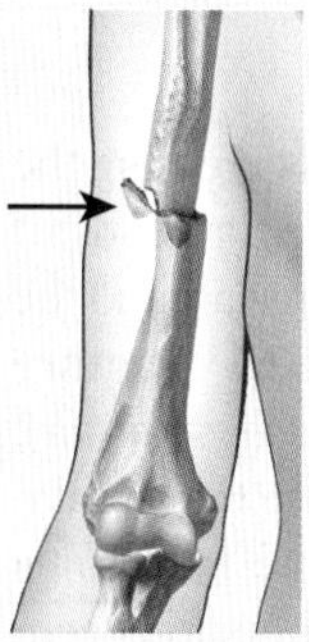	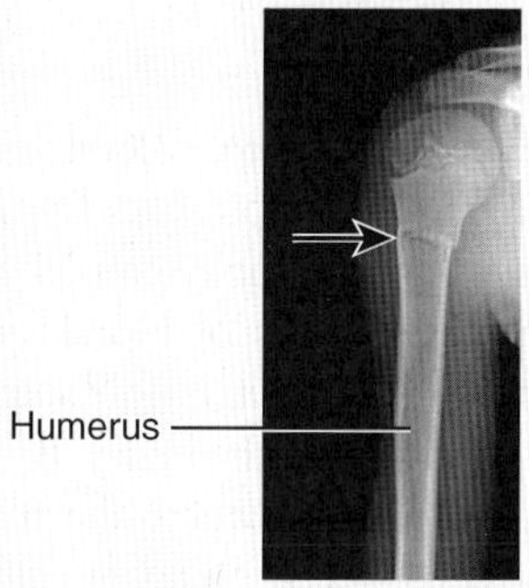

TABLE 6.1 CONTINUES

In some cases, a bone may fracture without visibly breaking. A **stress fracture** is a series of microscopic fissures in bone that forms without any evidence of injury to other tissues. In healthy adults, stress fractures result from repeated, strenuous activities such as running, jumping, or aerobic dancing. Stress fractures are quite painful and also result from disease processes that disrupt normal bone calcification, such as osteoporosis (discussed in Disorders: Homeostatic Imbalances at the end of this chapter). About 25% of stress fractures involve the tibia. Although standard x-ray images often fail to reveal the presence of stress fractures, they show up clearly in a bone scan.

TABLE 6.1 CONTINUED

Some Common Fractures

FRACTURE	DESCRIPTION	ILLUSTRATION	RADIOGRAPH
Pott	Fracture of the distal end of the lateral leg bone (fibula), with serious injury of the distal tibial articulation.	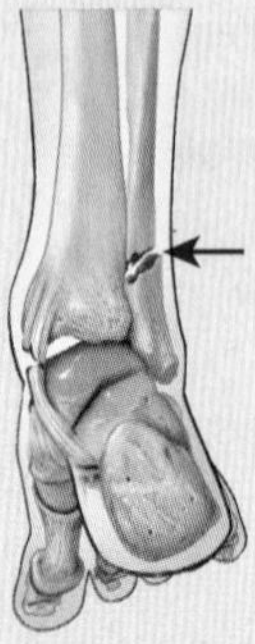	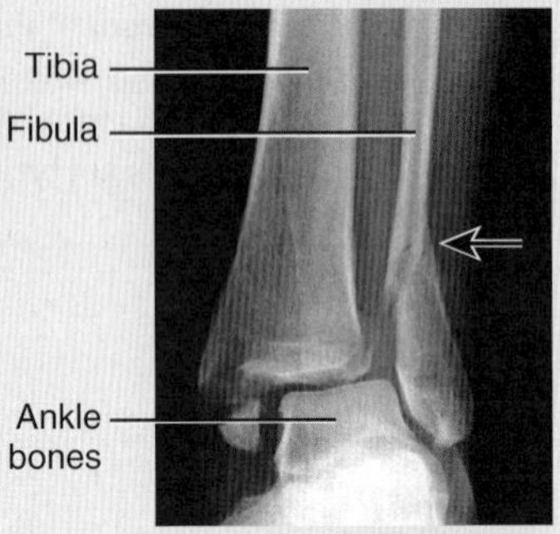 Tibia Fibula Ankle bones
Colles' (KOL-ēz)	Fracture of the distal end of the lateral forearm bone (radius) in which the distal fragment is displaced posteriorly.	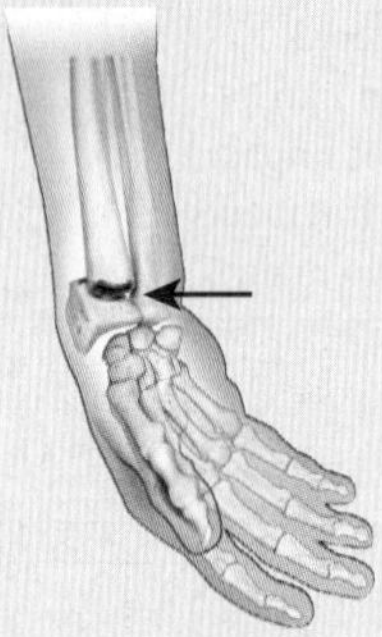	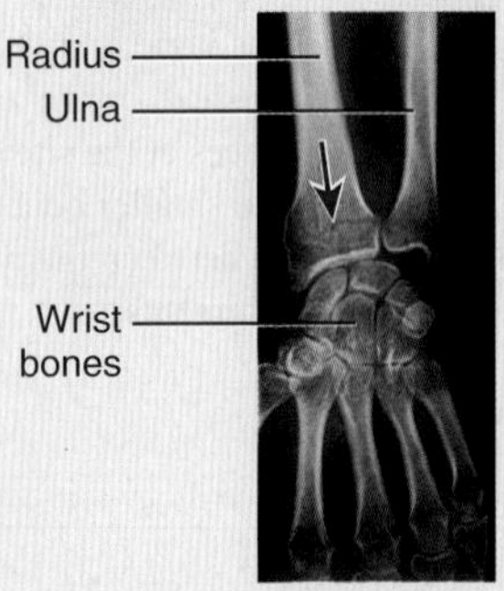 Radius Ulna Wrist bones

CLINICAL CONNECTION | ***Treatments for Fractures***

Treatments for fractures vary according to age, type of fracture, and the bone involved. The ultimate goals of fracture treatment are realignment of the bone fragments, immobilization to maintain realignment, and restoration of function. For bones to unite properly, the fractured ends must be brought into alignment. This process, called **reduction,** is commonly referred to as setting a fracture. In **closed reduction,** the fractured ends of a bone are brought into alignment by manual manipulation, and the skin remains intact. In **open reduction,** the fractured ends of a bone are brought into alignment by a surgical procedure using internal fixation devices such as screws, plates, pins, rods, and wires. Following reduction, a fractured bone may be kept immobilized by a cast, sling, splint, elastic bandage, external fixation device, or a combination of these devices. •

The repair of a bone fracture involves the following steps (Figure 6.9):

1. ***Formation of fracture hematoma.*** Blood vessels crossing the fracture line are broken. As blood leaks from the torn ends of the vessels, a mass of blood (usually clotted) forms around the site of the fracture. This mass of blood, called a **fracture hematoma** (hē′-ma-TŌ-ma; *hemat-* = blood; *-oma* = tumor), usually forms 6 to 8 hours after the injury. Because the circulation of blood stops at the site where the fracture hematoma forms, nearby bone cells die. Swelling and inflammation occur in response to dead bone cells, producing additional cellular debris. Phagocytes (neutrophils and macrophages) and osteoclasts begin to remove the dead or damaged tissue in and around the fracture hematoma. This stage may last up to several weeks.
2. ***Fibrocartilaginous callus formation.*** Fibroblasts from the periosteum invade the fracture site and produce collagen fibers. In addition, cells from the periosteum develop into chondroblasts and begin to produce fibrocartilage in this region. These events lead to the development of a **fibrocartilaginous (soft) callus** (fī-brō-kar-ti-LAJ-i-nus), a mass of repair tissue consisting of collagen fibers and cartilage that bridges the broken ends of the bone. Formation of the fibrocartilaginous callus takes about 3 weeks.
3. ***Bony callus formation.*** In areas closer to well-vascularized healthy bone tissue, osteogenic cells develop into osteoblasts, which begin to produce spongy bone trabeculae. The trabeculae join living and dead portions of the original bone fragments. In time, the fibrocartilage is converted to spongy bone, and the callus is then referred to as a **bony (hard) callus.** The bony callus lasts about 3 to 4 months.
4. ***Bone remodeling.*** The final phase of fracture repair is **bone remodeling** of the callus. Dead portions of the original fragments of broken bone are gradually resorbed by osteoclasts. Compact bone replaces spongy bone around the periphery of the fracture. Sometimes, the repair process is so thorough that the fracture line is undetectable, even in a radiograph (x-ray).

However, a thickened area on the surface of the bone remains as evidence of a healed fracture.

Although bone has a generous blood supply, healing sometimes takes months. The calcium and phosphorus needed to strengthen and harden new bone are deposited only gradually, and bone cells generally grow and reproduce slowly. The temporary disruption in their blood supply also helps explain the slowness of healing of severely fractured bones.

✔CHECKPOINT

22. List the types of fractures and outline the four steps involved in fracture repair.

Figure 6.9 Steps in repair of a bone fracture.

Bone heals more rapidly than cartilage because its blood supply is more plentiful.

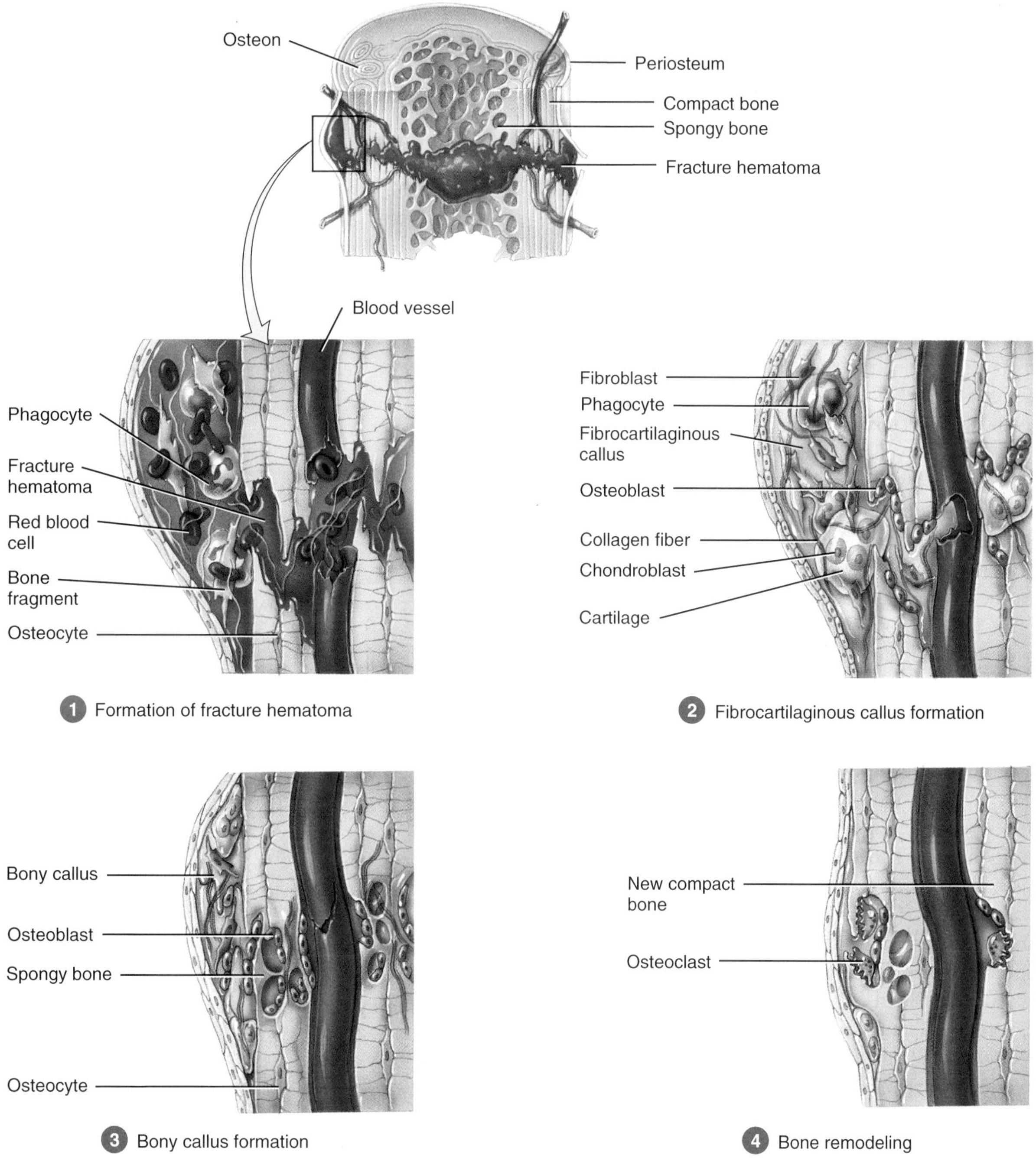

? Why does it sometimes take months for a fracture to heal?

6.7 BONE'S ROLE IN CALCIUM HOMEOSTASIS

OBJECTIVES

- Describe the importance of calcium in the body.
- Explain how blood calcium level is regulated.

Bone is the body's major calcium reservoir, storing 99% of total body calcium. One way to maintain the level of calcium in the blood is to control the rates of calcium resorption from bone into blood and of calcium deposition from blood into bone. Both nerve and muscle cells depend on a stable level of calcium ions (Ca^{2+}) in extracellular fluid to function properly. Blood clotting also requires Ca^{2+}. Also, many enzymes require Ca^{2+} as a cofactor (an additional substance needed for an enzymatic reaction to occur). For this reason, the blood plasma level of Ca^{2+} is very closely regulated between 9 and 11 mg/100 mL. Even small changes in Ca^{2+} concentration outside this range may prove fatal—the heart may stop (cardiac arrest) if the concentration goes too high, or breathing may cease (respiratory arrest) if the level falls too low. The role of bone in calcium homeostasis is to help "buffer" the blood Ca^{2+} level, releasing Ca^{2+} into blood plasma (using osteoclasts) when the level decreases, and absorbing Ca^{2+} (using osteoblasts) when the level rises.

Ca^{2+} exchange is regulated by hormones, the most important of which is **parathyroid hormone (PTH)** secreted by the parathyroid glands (see Figure 18.13). This hormone increases blood Ca^{2+} level. PTH secretion operates via a negative feedback system (Figure 6.10). If some stimulus causes the blood Ca^{2+} level to decrease, parathyroid gland cells (receptors) detect this change and increase their production of a molecule known as cyclic adenosine monophosphate (cyclic AMP). The gene for PTH within the nucleus of a parathyroid gland cell (the control center) detects the intracellular increase in cyclic AMP (the input). As a result, PTH synthesis speeds up, and more PTH (the output) is released into the blood. The presence of higher levels of PTH increases the number and activity of osteoclasts (effectors), which step up the pace of bone resorption. The resulting release of Ca^{2+} from bone into blood returns the blood Ca^{2+} level to normal.

PTH also acts on the kidneys (effectors) to decrease loss of Ca^{2+} in the urine, so more is retained in the blood. And PTH stimulates formation of **calcitriol** (the active form of vitamin D), a hormone that promotes absorption of calcium from foods in the gastrointestinal tract into the blood. Both of these actions also help elevate blood Ca^{2+} level.

Another hormone works to decrease blood Ca^{2+} level. When blood Ca^{2+} rises above normal, *parafollicular cells* in the thyroid gland secrete **calcitonin (CT)** (kal-si-TŌ-nin). CT inhibits activity of osteoclasts, speeds blood Ca^{2+} uptake by bone, and accelerates Ca^{2+} deposition into bones. The net result is that CT promotes bone formation and decreases blood Ca^{2+} level. Despite these effects, the role of CT in normal calcium homeostasis is uncertain because it can be completely absent without causing symptoms. Nevertheless, calcitonin harvested from salmon (Miacalcin®) is an effective drug for treating osteoporosis because it slows bone resorption.

Figure 18.14 summarizes the roles of parathyroid hormone, calcitriol, and calcitonin in regulation of blood Ca^{2+} level.

Figure 6.10 Negative feedback system for the regulation of blood calcium (Ca^{2+}) concentration. PTH = parathyroid hormone.

Release of calcium from bone matrix and retention of calcium by the kidneys are the two main ways that blood calcium level can be increased.

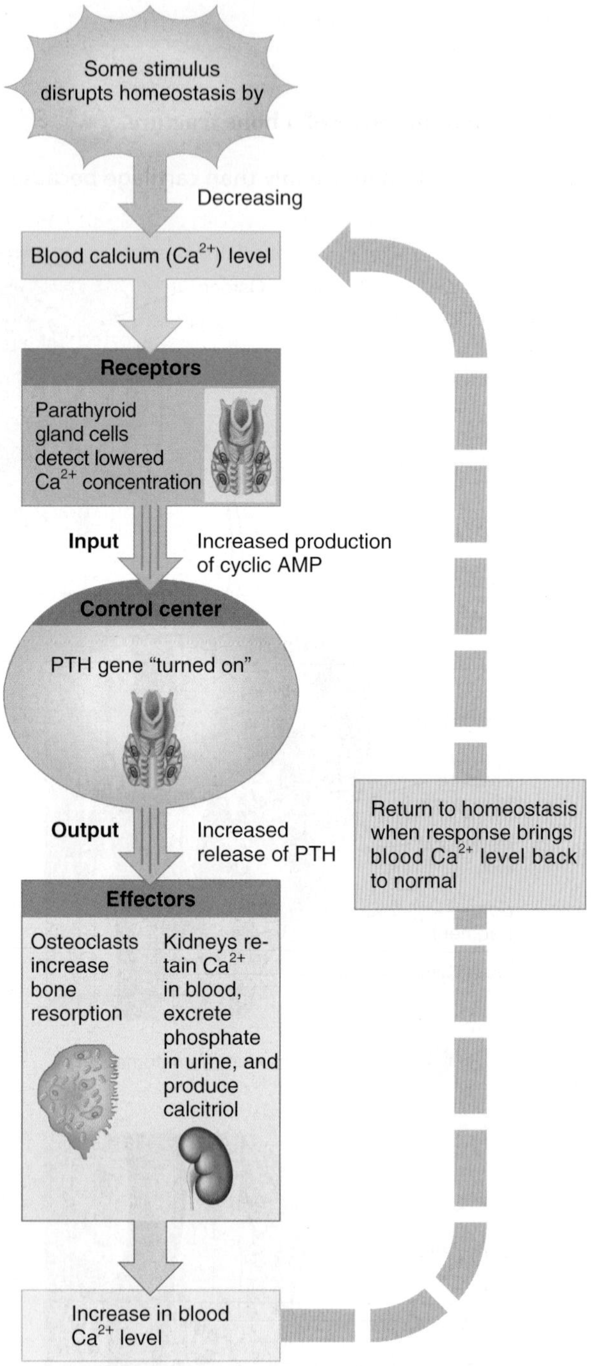

Some stimulus disrupts homeostasis by
Decreasing
Blood calcium (Ca2+) level
Receptors
Parathyroid gland cells detect lowered Ca2+ concentration
Input
Increased production of cyclic AMP
Control center
PTH gene "turned on"
Output
Increased release of PTH
Effectors
Osteoclasts increase bone resorption
Kidneys retain Ca2+ in blood, excrete phosphate in urine, and produce calcitriol
Increase in blood Ca2+ level
Return to homeostasis when response brings blood Ca2+ level back to normal

What body functions depend on proper levels of Ca^{2+}?

CHECKPOINT

23. How do hormones act on bone to regulate calcium homeostasis?

6.8 EXERCISE AND BONE TISSUE

OBJECTIVE

- Describe how exercise and mechanical stress affect bone tissue.

Within limits, bone tissue has the ability to alter its strength in response to changes in mechanical stress. When placed under stress, bone tissue becomes stronger through increased deposition of mineral salts and production of collagen fibers by osteoblasts. Without mechanical stress, bone does not remodel normally because bone resorption occurs more quickly than bone formation. Research has shown that high-impact intermittent strains more strongly influence bone deposition as compared with lower-impact constant strains. Therefore, running and jumping stimulate bone growth more dramatically than walking.

The main mechanical stresses on bone are those that result from the pull of skeletal muscles and the pull of gravity. If a person is bedridden or has a fractured bone in a cast, the strength of the unstressed bones diminishes because of the loss of bone minerals and decreased numbers of collagen fibers. Astronauts subjected to the microgravity of space also lose bone mass. In any of these cases, bone loss can be dramatic—as much as 1% per week. In contrast, the bones of athletes, which are repetitively and highly stressed, become notably thicker and stronger than those of astronauts or nonathletes. Weight-bearing activities, such as walking or moderate weight lifting, help build and retain bone mass. Adolescents and young adults should engage in regular weight-bearing exercise prior to the closure of the epiphyseal plates to help build total mass prior to its inevitable reduction with aging. However, people of all ages can and should strengthen their bones by engaging in weight-bearing exercise.

✔CHECKPOINT

24. How do mechanical stresses strengthen bone tissue?
25. Would children raised in space ever be able to return to Earth?
26. Why is it important to engage in weight-bearing exercises before the epiphyseal plates close?

6.9 AGING AND BONE TISSUE

OBJECTIVE

- Describe the effects of aging on bone tissue.

From birth through adolescence, more bone tissue is produced than is lost during bone remodeling. In young adults the rates of bone deposition and resorption are about the same. As the level of sex hormones diminishes during middle age, especially in women after menopause, a decrease in bone mass occurs because bone resorption by osteoclasts outpaces bone deposition by osteoblasts. In old age, loss of bone through resorption occurs more rapidly than bone gain. Because women's bones generally are smaller and less massive than men's bones to begin with, loss of bone mass in old age typically has a greater adverse effect in females. These factors contribute to the higher incidence of osteoporosis in females.

There are two principal effects of aging on bone tissue: loss of bone mass and brittleness. Loss of bone mass results from **demineralization** (dē-min′-er-al-i-ZĀ-shun), the loss of calcium and other minerals from bone extracellular matrix. This loss usually begins after age 30 in females, accelerates greatly around age 45 as levels of estrogens decrease, and continues until as much as 30% of the calcium in bones is lost by age 70. Once bone loss begins in females, about 8% of bone mass is lost every 10 years. In males, calcium loss typically does not begin until after age 60, and about 3% of bone mass is lost every 10 years. The loss of calcium from bones is one of the problems in osteoporosis (described shortly).

The second principal effect of aging on the skeletal system, brittleness, results from a decreased rate of protein synthesis. Recall that the organic part of bone extracellular matrix, mainly collagen fibers, gives bone its tensile strength. The loss of tensile strength causes the bones to become very brittle and susceptible to fracture. In some elderly people, collagen fiber synthesis slows, in part due to diminished production of human growth hormone. In addition to increasing the susceptibility to fractures, loss of bone mass also leads to deformity, pain, loss of height, and loss of teeth.

Table 6.2 summarizes the factors that influence bone metabolism.

✔CHECKPOINT

27. What is demineralization, and how does it affect the functioning of bone?
28. What changes occur in the organic part of bone extracellular matrix with aging?

TABLE 6.2

Summary of Factors That Affect Bone Growth

FACTOR	COMMENT
MINERALS	
Calcium and phosphorus	Make bone extracellular matrix hard.
Magnesium	Helps form bone extracellular matrix.
Fluoride	Helps strengthen bone extracellular matrix.
Manganese	Activates enzymes involved in synthesis of bone extracellular matrix.
VITAMINS	
Vitamin A	Needed for the activity of osteoblasts during remodeling of bone; deficiency stunts bone growth; toxic in high doses.
Vitamin C	Needed for synthesis of collagen, the main bone protein; deficiency leads to decreased collagen production, which slows down bone growth and delays repair of broken bones.
Vitamin D	Active form (calcitriol) is produced by the kidneys; helps build bone by increasing absorption of calcium from gastrointestinal tract into blood; deficiency causes faulty calcification and slows down bone growth; may reduce the risk of osteoporosis but is toxic if taken in high doses.
Vitamins K and B_{12}	Needed for synthesis of bone proteins; deficiency leads to abnormal protein production in bone extracellular matrix and decreased bone density.
HORMONES	
Human growth hormone (hGH)	Secreted by the anterior lobe of the pituitary gland; promotes general growth of all body tissues, including bone, mainly by stimulating production of insulinlike growth factors.
Insulinlike growth factors (IGFs)	Secreted by the liver, bones, and other tissues upon stimulation by human growth hormone; promotes normal bone growth by stimulating osteoblasts and by increasing the synthesis of proteins needed to build new bone.
Thyroid hormones (thyroxine and triiodothyronine)	Secreted by thyroid gland; promote normal bone growth by stimulating osteoblasts.
Insulin	Secreted by the pancreas; promotes normal bone growth by increasing the synthesis of bone proteins.
Sex hormones (estrogens and testosterone)	Secreted by the ovaries in women (estrogens) and by the testes in men (testosterone); stimulate osteoblasts and promote the sudden "growth spurt" that occurs during the teenage years; shut down growth at the epiphyseal plates around age 18–21, causing lengthwise growth of bone to end; contribute to bone remodeling during adulthood by slowing bone resorption by osteoclasts and promoting bone deposition by osteoblasts.
Parathyroid hormone (PTH)	Secreted by the parathyroid glands; promotes bone resorption by osteoclasts; enhances recovery of calcium ions from urine; promotes formation of the active form of vitamin D (calcitriol).
Calcitonin (CT)	Secreted by the thyroid gland; inhibits bone resorption by osteoclasts.
EXERCISE	
	Weight-bearing activities stimulate osteoblasts and, consequently, help build thicker, stronger bones and retard loss of bone mass that occurs as people age.
AGING	
	As the level of sex hormones diminishes during middle age to older adulthood, especially in women after menopause, bone resorption by osteoclasts outpaces bone deposition by osteoblasts, which leads to a decrease in bone mass and an increased risk of osteoporosis.

DISORDERS: HOMEOSTATIC IMBALANCES

Osteoporosis

Osteoporosis (os′-tē-ō-pō-RŌ-sis; *por* = passageway; *-osis* = condition), literally a condition of porous bones, affects 10 million people a year in the United States (Figure 6.11). In addition, 18 million people have low bone mass (*osteopenia*), which puts them at risk for osteoporosis. The basic problem is that bone resorption (breakdown) outpaces bone deposition (formation). In large part this is due to depletion of calcium from the body—more calcium is lost in urine, feces, and sweat than is absorbed from the diet. Bone mass becomes so depleted that bones fracture, often spontaneously, under the mechanical stresses of everyday living. For example, a hip fracture might result from simply sitting down too quickly. In the United States, osteoporosis causes more than one and a half million fractures a year, mainly in the hips, wrists, and vertebrae. Osteoporosis afflicts the entire skeletal system. In addition to fractures, osteoporosis causes shrinkage of vertebrae, height loss, hunched backs, and bone pain.

Osteoporosis primarily affects middle-aged and elderly people, 80 percent of them women. Older women suffer from osteoporosis more often than men for two reasons: (1) Women's bones are less massive than men's bones, and (2) production of estrogens in women declines dramatically at menopause, while production of the main androgen, testosterone, in older men wanes gradually and only slightly. Estrogens and testosterone stimulate osteoblast activity and synthesis of bone matrix. Besides gender, risk factors for developing osteoporosis include a family history of the disease, European or Asian ancestry, thin or small body build, an inactive lifestyle, cigarette smoking, a diet low in calcium and vitamin D, more than two alcoholic drinks a day, and the use of certain medications.

Figure 6.11 Comparison of spongy bone tissue from (a) a normal young adult and (b) a person with osteoporosis. Notice the weakened trabeculae in (b). Compact bone tissue is similarly affected by osteoporosis.

In osteoporosis, bone resorption outpaces bone formation, so bone mass decreases.

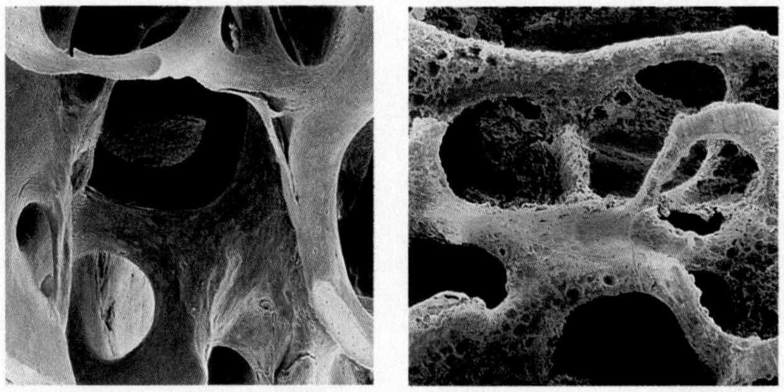

SEM 30x
(a) Normal bone

SEM 30x
(b) Osteoporotic bone

If you wanted to develop a drug to lessen the effects of osteoporosis, would you look for a chemical that inhibits the activity of osteoblasts or that of osteoclasts?

Osteoporosis is diagnosed by taking a family history and undergoing a *bone mineral density* (BMD) test. Performed like x-rays, BMD tests measure bone density. They can also be used to confirm a diagnosis of osteoporosis, determine the rate of bone loss, and monitor the effects of treatment.

Treatment options for osteoporosis are varied. With regard to nutrition, a diet high in calcium is important to reduce the risk of fractures. Vitamin D is necessary for the body to utilize calcium. In terms of exercise, regular performance of weight-bearing exercises has been shown to maintain and build bone mass. These exercises include walking, jogging, hiking, climbing stairs, playing tennis, and dancing. Resistance exercises, such as weight lifting, also build bone strength and muscle mass.

Medications used to treat osteoporosis are generally of two types: (1) **antireabsorptive drugs** slow down the progression of bone loss and (2) **bone-building drugs** promote increasing bone mass. Among the antireabsorptive drugs are (1) *bisphosphonates,* which inhibit osteoclasts (Fosamax®, Actonel®, Boniva®, and calcitonin); (2) *selective estrogen receptor modulators*, which mimic the effects of estrogens without unwanted side effects (Raloxifene®, Evista®); and (3) estrogen replacement therapy (ERT), which replaces estrogens lost during and after menopause (Premarin®), and hormone replacement therapy (HRT), which replaces estrogens and progesterone lost during and after menopause (Prempro®). ERT helps maintain and increase bone mass after menopause. Women on ERT have a slightly increased risk of stroke and blood clots. HRT also helps maintain and increase bone mass. Women on HRT have increased risks of heart disease, breast cancer, stroke, blood clots, and dementia.

Among the bone-building drugs is parathyroid hormone (PTH), which stimulates osteoblasts to produce new bone (Forteo®). Others are under development.

Rickets and Osteomalacia

Rickets and **osteomalacia** (os′-tē-ō-ma-LĀ-shē-a; *malacia* = softness) are two forms of the same disease that result from inadequate calcification of the extracellular bone matrix, usually caused by a vitamin D deficiency. Rickets is a disease of children in which the growing bones become "soft" or rubbery and are easily deformed. Because new bone formed at the epiphyseal (growth) plates fails to ossify, bowed legs and deformities of the skull, rib cage, and pelvis are common. Osteomalacia is the adult counterpart of rickets, sometimes called *adult rickets*. New bone formed during remodeling fails to calcify, and the person experiences varying degrees of pain and tenderness in bones, especially the hip and legs. Bone fractures also result from minor trauma. Prevention and treatment for rickets and osteomalacia consists of the administration of adequate vitamin D and exposure to moderate amounts of sunlight.

MEDICAL TERMINOLOGY

Osteoarthritis (os′-tē-ō-ar-THRĪ-tis; *arthr* = joint) The degeneration of articular cartilage such that the bony ends touch; the resulting friction of bone against bone worsens the condition. Usually associated with the elderly.

Osteomyelitis (os′-tē-ō-mī-e-LĪ-tis) An infection of bone characterized by high fever, sweating, chills, pain, nausea, pus formation, edema, and warmth over the affected bone and rigid overlying muscles. It is often caused by bacteria, usually *Staphylococcus aureus.* The bacteria may reach the bone from outside the body (through open fractures, penetrating wounds, or orthopedic surgical procedures); from other sites of infection in the body (abscessed teeth, burn infections, urinary tract infections, or upper respiratory infections) via the blood; and from adjacent soft tissue infections (as occurs in diabetes mellitus).

Osteopenia (os′-tē-ō-PĒ-nē-a; *penia* = poverty) Reduced bone mass due to a decrease in the rate of bone synthesis to a level too low to compensate for normal bone resorption; any decrease in bone mass below normal. An example is osteoporosis.

Osteosarcoma (os′-tē-ō-sar-KŌ-ma; *sarcoma* = connective tissue tumor) Bone cancer that primarily affects osteoblasts and occurs most often in teenagers during their growth spurt; the most common sites are the metaphyses of the thigh bone (femur), shin bone (tibia), and arm bone (humerus). Metastases occur most often in lungs; treatment consists of multidrug chemotherapy and removal of the malignant growth, or amputation of the limb.

CHAPTER REVIEW AND RESOURCE SUMMARY

WILEY PLUS

Review	Resource
Introduction 1. A bone is made up of several different tissues: bone or osseous tissue, cartilage, dense connective tissue, epithelium, adipose tissue, and nervous tissue. 2. The entire framework of bones and their cartilages constitutes the skeletal system.	
6.1 Functions of Bone and the Skeletal System 1. The skeletal system functions in support, protection, movement, mineral homeostasis, blood cell production, and triglyceride storage.	Anatomy Overview - Bone Structure and Tissues
6.2 Structure of Bone 1. Parts of a typical long bone are the diaphysis (shaft), proximal and distal epiphyses (ends), metaphyses, articular cartilage, periosteum, medullary (marrow) cavity, and endosteum.	Exercise - Growing Long Bone
6.3 Histology of Bone Tissue 1. Bone tissue consists of widely separated cells surrounded by large amounts of extracellular matrix. 2. The four principal types of cells in bone tissue are osteogenic cells, osteoblasts (bone-building cells), osteocytes (maintain daily activity of bone), and osteoclasts (bone-destroying cells). 3. The extracellular matrix of bone contains abundant mineral salts (mostly hydroxyapatite) and collagen fibers. 4. Compact bone tissue consists of osteons (haversian systems) with little space between them. 5. Compact bone tissue lies over spongy bone tissue in the epiphyses and makes up most of the bone tissue of the diaphysis. Functionally, compact bone tissue is the strongest form of bone and protects, supports, and resists stress. 6. Spongy bone tissue does not contain osteons. It consists of trabeculae surrounding many red bone marrow–filled spaces. 7. Spongy bone tissue forms most of the structure of short, flat, and irregular bones, and the interior of the epiphyses in long bones. Functionally, spongy bone tissue trabeculae offer resistance along lines of stress, support and protect red bone marrow, and make bones lighter for easier movement.	Anatomy Overview - Compact Bone Anatomy Overview - Spongy Bone Figure 6.3 - Histology of Compact and Spongy Bone Exercise - Concentrate on Bone Tissue
6.4 Blood and Nerve Supply of Bone 1. Long bones are supplied by periosteal, nutrient, metaphyseal, and epiphyseal arteries; veins accompany the arteries. 2. Nerves accompany blood vessels in bone; the periosteum is rich in sensory neurons.	
6.5 Bone Formation 1. The process by which bone forms, called ossification, occurs in four principal situations: (1) the initial formation of bones in an embryo and fetus; (2) the growth of bones during infancy, childhood, and adolescence until their adult sizes are reached; (3) the remodeling of bone (replacement of old bone by new bone tissue throughout life); and (4) the repair of fractures (breaks in bones) throughout life. 2. Bone development begins during the sixth or seventh week of embryonic development. The two types of ossification, intramembranous and endochondral, involve the replacement of a preexisting connective tissue with bone. Intramembranous ossification refers to bone formation directly within mesenchyme arranged in sheetlike layers that resemble membranes. Endochondral ossification refers to	Animation - Bone Formation Animation - Bone Elongation and Bone Widening Animation - Bone Remodeling Animation - Regulation of Blood Calcium Animation - Bone Dynamics and Tissue Animation - Regulation of Bone Growth

Review

bone formation within hyaline cartilage that develops from mesenchyme. The primary ossification center of a long bone is in the diaphysis. Cartilage degenerates, leaving cavities that merge to form the medullary cavity. Osteoblasts lay down bone. Next, ossification occurs in the epiphyses, where bone replaces cartilage, except for the epiphyseal (growth) plate.

3. The epiphyseal plate consists of four zones: zone of resting cartilage, zone of proliferating cartilage, zone of hypertrophic cartilage, and zone of calcified cartilage. Because of the cell division in the epiphyseal (growth) plate, the diaphysis of a bone increases in length.
4. Bone grows in thickness or diameter due to the addition of new bone tissue by periosteal osteoblasts around the outer surface of the bone (appositional growth).
5. Bone remodeling is an ongoing process in which osteoclasts carve out small tunnels in old bone tissue and then osteoblasts rebuild it.
6. In bone resorption, osteoclasts release enzymes and acids that degrade collagen fibers and dissolve mineral salts.
7. Dietary minerals (especially calcium and phosphorus) and vitamins (A, C, D, K, and B_{12}) are needed for bone growth and maintenance. Insulinlike growth factors (IGFs), human growth hormone, thyroid hormones, and insulin stimulate bone growth.
8. Sex hormones slow resorption of old bone and promote new bone deposition.

Resource

Figure 6.5 - Intramembranous Ossification
Figure 6.6 - Enchondral Ossification
Figure 6.7 - Epiphyseal (Growth) Plate
Exercise - Observe the Ossification
Exercise - Bone Growth Sequencing
Exercise - Concentrate on Bone Tissue
Exercise - Regulation Sequences

6.6 Fracture and Repair of Bone

1. A fracture is any break in a bone. Types of fractures include closed (simple), open (compound), comminuted, greenstick, impacted, stress, Pott's, and Colles'.
2. Fracture repair involves formation of a fracture hematoma, a fibrocartilaginous callus and a bony callus, and bone remodeling.

Animation - Bone Remodeling
Figure 6.9 - Steps in Repair of a Bone Fracture

6.7 Bone's Role in Calcium Homeostasis

1. Bone is the major reservoir for calcium in the body.
2. Parathyroid hormone (PTH) secreted by the parathyroid glands increases blood Ca^{2+} level. Calcitonin (CT) from the thyroid gland has the potential to decrease blood Ca^{2+} level. Vitamin D enhances absorption of calcium and phosphate and thus raises the blood levels of these substances.

Animation - Regulation of Blood Calcium
Concepts and Connections - Blood Calcium Regulation
Exercise - Calcium Homeostasis
Exercise - Regulation Sequencing

6.8 Exercise and Bone Tissue

1. Mechanical stress increases bone strength by increasing deposition of mineral salts and production of collagen fibers.
2. Removal of mechanical stress weakens bone through demineralization and collagen fiber reduction.

Animation - Response to Stress in Adult Bones

6.9 Aging and Bone Tissue

1. The principal effect of aging is demineralization, a loss of calcium from bones, which is due to reduced osteoblast activity.
2. Another effect is decreased production of extracellular matrix proteins (mostly collagen fibers), which makes bones more brittle and thus more susceptible to fracture.

SELF-QUIZ QUESTIONS

Fill in the blanks in the following statements.

1. Bone growth in length is called _____ growth, and bone growth in diameter (thickness) is called _____ growth.
2. The crystallized inorganic mineral salts in bone contribute to bone's _____, while the collagen fibers and other organic molecules provide bone with _____.

Indicate whether the following statements are true or false.

3. Bone resorption involves increased activity of osteoclasts.
4. The formation of bone from cartilage is known as endochondral ossification.
5. The growth of bone is controlled primarily by hormones.

Choose the one best answer to the following questions.

6. Place in order the steps involved in intramembranous ossification. (1) Bony matrices fuse to form trabeculae. (2) Clusters of osteoblasts form a center of ossification that secretes the organic extracellular matrix. (3) Spongy bone is replaced with compact bone on the bone's surface. (4) Periosteum develops on the bone's periphery. (5) The extracellular matrix hardens by deposition of calcium and mineral salts.
(a) 2, 4, 5, 1, 3 (b) 4, 3, 5, 1, 2 (c) 1, 2, 5, 4, 3
(d) 2, 5, 1, 4, 3 (e) 5, 1, 3, 4, 2

7. Place in order the steps involved in endochondral ossification. (1) Nutrient artery invades the perichondrium. (2) Osteoclasts create a marrow cavity. (3) Chondrocytes enlarge and calcify. (4) Secondary ossification centers appear at epiphyses. (5) Osteoblasts become active in the primary ossification center.
(a) 3, 1, 5, 2, 4 (b) 3, 1, 5, 4, 2 (c) 1, 3, 5, 2, 4
(d) 1, 2, 3, 5, 4 (e) 2, 5, 4, 3, 1

8. Spongy bone differs from compact bone because spongy bone
(a) is composed of numerous osteons (haversian systems).
(b) is found primarily in the diaphyses of long bones, and compact bone is found primarily in the epiphyses of long bones.
(c) contains osteons all aligned in the same direction along lines of stress.
(d) does not contain osteocytes contained in lacunae.
(e) is composed of trabeculae that are oriented along lines of stress.

9. A primary effect of weight-bearing exercise on bones is to
(a) provide oxygen for bone development.
(b) increase the demineralization of bone.
(c) maintain and increase bone mass.
(d) stimulate the release of sex hormones for bone growth.
(e) utilize the stored triglycerides from the yellow bone marrow.

10. Place in order the steps involved in the repair of a bone fracture. (1) Osteoblast production of trabeculae and bony callus formation; (2) formation of a hematoma at the site of fracture; (3) resorption of remaining bone fragments and remodeling of bone; (4) migration of fibroblasts to the fracture site; (5) bridging of broken ends of bones by a fibrocartilaginous callus.
(a) 2, 4, 5, 1, 3 (b) 2, 5, 4, 1, 3 (c) 1, 2, 5, 4, 3
(d) 2, 5, 1, 3, 4 (e) 5, 2, 4, 1, 3

11. Match the following:

____ (a) column-like layer of maturing chondrocytes
____ (b) layer of small, scattered chondrocytes anchoring the epiphyseal (growth) plate to the bone
____ (c) layer of actively dividing chondrocytes
____ (d) region of dead chondrocytes

(1) zone of hypertrophic cartilage
(2) zone of calcified cartilage
(3) zone of proliferating cartilage
(4) zone of resting cartilage

12. Match the following:

____ (a) decreases blood calcium levels by accelerating calcium deposition in bones and inhibiting osteoclasts
____ (b) required for collagen synthesis
____ (c) during childhood, it promotes growth at epiphyseal plate; production stimulated by human growth hormone
____ (d) involved in bone growth by increasing osteoblast activity; causes long bones to stop growing in length
____ (e) required for protein synthesis
____ (f) active form of vitamin D; raises blood calcium levels by increasing absorption of calcium from digestive tract
____ (g) raises blood calcium levels by increasing bone resorption

(1) PTH
(2) CT
(3) calcitriol
(4) insulinlike growth factors
(5) sex hormones
(6) vitamin C
(7) vitamin K

13. Match the following:

____ (a) space within the shaft of the bone that contains yellow bone marrow
____ (b) triglyceride storage tissue
____ (c) hemopoietic tissue
____ (d) thin layer of hyaline cartilage covering the ends of bones where they form a joint
____ (e) proximal and distal ends of bones
____ (f) the long, cylindrical main portion of the bone; the shaft
____ (g) in a growing bone, the region that contains the epiphyseal (growth) plate
____ (h) the tough covering that surrounds the bone surface wherever cartilage is not present
____ (i) a layer of hyaline cartilage in the area between the shaft and end of a growing bone
____ (j) membrane lining the medullary cavity
____ (k) a remnant of the active epiphyseal (growth) plate; a sign that the bone has stopped growing in length
____ (l) bundles of collagen fibers that attach periosteum to bone

(1) articular cartilage
(2) endosteum
(3) medullary cavity
(4) diaphysis
(5) epiphyses
(6) metaphysis
(7) periosteum
(8) red bone marrow
(9) yellow bone marrow
(10) perforating (Sharpey's) fibers
(11) epiphyseal line
(12) epiphyseal (growth) plate

14. Match the following:

____ (a) a broken bone in which one end of the fractured bone is driven into the other end
____ (b) a condition of porous bones characterized by decreased bone mass and increased susceptibility to fractures
____ (c) splintered bone, with smaller fragments lying between main fragments
____ (d) a broken bone that does not break through the skin
____ (e) a partial break in a bone in which one side of the bone is broken and the other side bends
____ (f) a broken bone that protrudes through the skin
____ (g) microscopic bone breaks resulting from inability to withstand repeated stressful impact
____ (h) a degeneration of articular cartilage allowing the bony ends to touch; worsens due to friction between the bones
____ (i) condition characterized by failure of new bone formed by remodeling to calcify in adults
____ (j) an infection of bone

(1) closed (simple) fracture
(2) open (compound) fracture
(3) impacted fracture
(4) greenstick fracture
(5) stress fracture
(6) comminuted fracture
(7) osteoporosis
(8) osteomalacia
(9) osteoarthritis
(10) osteomyelitis

15. Match the following:

____ (a) small spaces between lamellae that contain osteocytes
____ (b) tiny canals that penetrate compact bone; carry blood vessels, lymphatic vessels, and nerves from the periosteum
____ (c) areas between osteons; fragments of old osteons
____ (d) cells that secrete the components required to build bone
____ (e) microscopic unit of compact bone tissue
____ (f) interconnected, tiny canals filled with extracellular fluid; connect lacunae to each other and to the central canal
____ (g) canals that extend longitudinally through the bone and carry blood vessels and nerves to the central canal
____ (h) large cells derived from numerous monocytes; involved in resorption by release of lysosomal enzymes and acids
____ (i) irregular lattice of thin columns of bone found in spongy bone tissue
____ (j) rings of hard calcified matrix found just beneath the periosteum and lining the medullary cavity
____ (k) mature cells that maintain the daily metabolism of bone
____ (l) an opening in the shaft of the bone allowing an artery to pass into the bone
____ (m) unspecialized stem cells derived from mesenchyme

(1) osteogenic cells
(2) osteocytes
(3) osteon (haversian system)
(4) perforating (Volkmann's) canals
(5) circumferential lamellae
(6) osteoblasts
(7) trabeculae
(8) interstitial lamellae
(9) canaliculi
(10) osteoclasts
(11) nutrient foramen
(12) lacunae
(13) haversian (central) canals

CRITICAL THINKING QUESTIONS

1. Taryn is a high school senior who is undergoing a strenuous running regimen for several hours a day in order to qualify for her state high school track meet. Lately she has experienced intense pain in her right leg that is hindering her workouts. Her physician performs an examination of her right leg. The doctor doesn't notice any outward evidence of injury; he then orders a bone scan. What does her doctor suspect the problem is?

2. While playing basketball, nine-year-old Marcus fell and broke his left arm. The arm was placed in a cast and appeared to heal normally. As an adult, Marcus was puzzled because it seemed that his right arm is longer than his left arm. He measured both arms and he was correct—his right arm is longer! How would you explain to Marcus what happened?

3. Astronauts in space exercise as part of their daily routine, yet they still have problems with bone weakness after prolonged stays in space. Why does this happen?

? ANSWERS TO FIGURE QUESTIONS

6.1 The periosteum is essential for growth in bone thickness, bone repair, and bone nutrition. It also serves as a point of attachment for ligaments and tendons.

6.2 Bone resorption is necessary for the development, maintenance, and repair of bone.

6.3 The central (haversian) canals are the main blood supply to the osteocytes of an osteon (haversian system), so their blockage would lead to death of the osteocytes.

6.4 Periosteal arteries enter bone tissue through perforating (Volkmann's) canals.

6.5 Flat bones of the skull, most facial bones, the mandible (lower jawbone), and the medial part of the clavicle develop by intramembranous ossification.

6.6 Secondary ossification centers develop in the regions of the cartilage model that will give rise to the epiphyses.

6.7 The lengthwise growth of the diaphysis is caused by cell divisions in the zone of proliferating cartilage and replacement of the zone of calcified cartilage with bone (new diaphysis).

6.8 The medullary cavity enlarges by activity of the osteoclasts in the endosteum.

6.9 Healing of bone fractures can take months because calcium and phosphorus deposition is a slow process, and bone cells generally grow and reproduce slowly.

6.10 Heartbeat, respiration, nerve cell functioning, enzyme functioning, and blood clotting all depend on proper levels of calcium.

6.11 A drug that inhibits the activity of osteoclasts might lessen the effects of osteoporosis, since osteoclasts are responsible for bone resorption.

7 THE SKELETAL SYSTEM: THE AXIAL SKELETON

THE AXIAL SKELETON AND HOMEOSTASIS *The bones of the axial skeleton contribute to homeostasis by protecting many of the body's organs such as the brain, spinal cord, heart, and lungs. They are also important in support and calcium storage and release.*

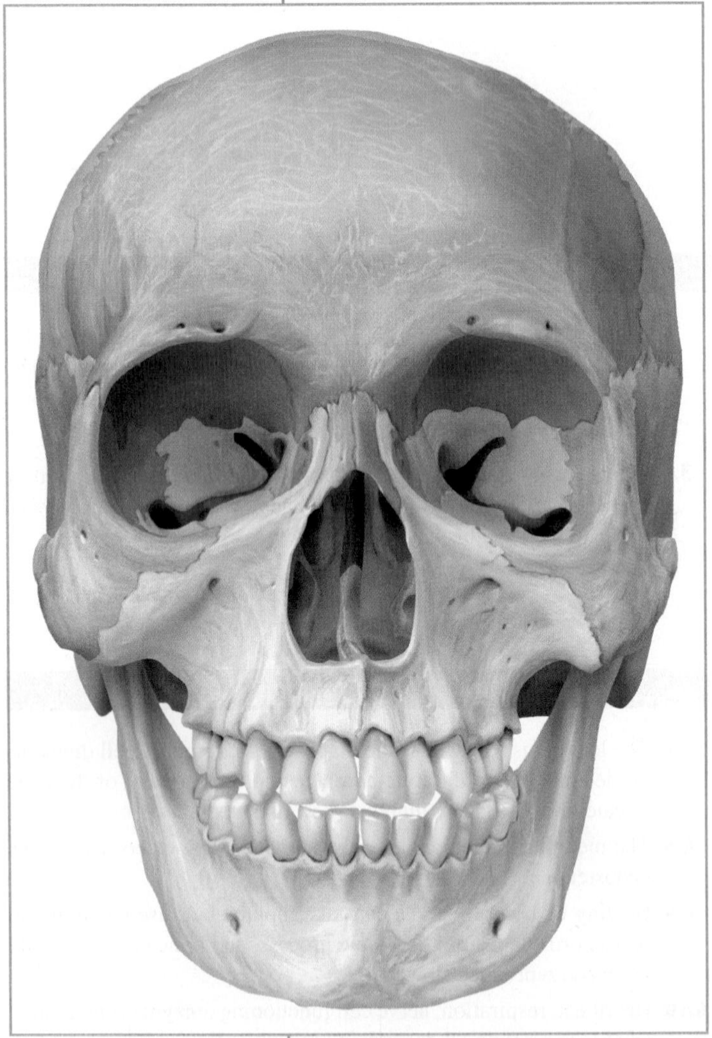

Without bones, you could not survive. You would be unable to perform movements such as walking or grasping, and the slightest blow to your head or chest could damage your brain or heart. Because the skeletal system forms the framework of the body, a familiarity with the names, shapes, and positions of individual bones will help you locate and name many other anatomical features. For example, the radial artery, the site where the pulse is usually taken, is named for its closeness to the radius, the lateral bone of the forearm. The ulnar nerve is named for its proximity to the ulna, the medial bone of the forearm. The frontal lobe of the brain lies deep to the frontal (forehead) bone. The tibialis anterior muscle lies along the anterior surface of the tibia (shin bone). Parts of certain bones also serve to locate structures within the skull and to outline the lungs, heart, and abdominal and pelvic organs.

Movements such as throwing a ball, biking, and walking require interactions between bones and muscles. To understand how muscles produce different movements, from high-fives to three-point shots, you will need to learn where the muscles attach on individual bones and what types of joints are involved. Together, the bones, muscles, and joints form an integrated system called the **musculoskeletal system.** The branch of medical science concerned with the prevention or correction of disorders of the musculoskeletal system is called **orthopedics** (or′-thō-PĒ-diks; *ortho-* = correct; *-pedi* = child).

? ***Did you ever wonder what causes people to become measurably shorter as they age?***

7.1 DIVISIONS OF THE SKELETAL SYSTEM

OBJECTIVE

- Describe how the skeleton is organized into axial and appendicular divisions.

The adult human skeleton consists of 206 named bones, most of which are paired, with one member of each pair on the right and left sides of the body. The skeletons of infants and children have more than 206 bones because some of their bones (sacrum and coccyx) fuse later in life. Examples are the hip bones and some bones (sacrum and coccyx) of the vertebral column (backbone).

Bones of the adult skeleton are grouped into two principal divisions: the **axial skeleton** and the **appendicular skeleton** (*appendic-* = to hang onto). Table 7.1 presents the 80 bones of the axial skeleton and the 126 bones of the appendicular skeleton. Figure 7.1 shows how both divisions join to form the complete skeleton (the bones of the axial skeleton are shown in blue). You can remember the names of the divisions if you think of the axial skeleton as consisting of the bones that lie around the longitudinal *axis* of the human body, an imaginary vertical line that runs through the body's center of gravity from the head to the space between the feet: skull bones, auditory ossicles (ear bones), hyoid bone (see Figure 7.5), ribs, sternum (breastbone), and bones of the vertebral column. The appendicular skeleton consists of the bones of the **upper** and **lower limbs (extremities** or **appendages)**, plus the bones forming the **girdles** that connect the limbs to the axial skeleton. Functionally, the auditory ossicles in the middle ear, which vibrate in response to sound waves that strike the eardrum, are not part of either the axial or appendicular skeleton, but they are grouped with the axial skeleton for convenience (see Chapter 17).

We will organize our study of the skeletal system around the two divisions of the skeleton, with emphasis on how the many bones of the body are interrelated. In this chapter we focus on the axial skeleton, looking first at the skull and then at the bones of the vertebral column and the chest. In Chapter 8 we explore the appendicular skeleton, examining in turn the bones of the pectoral (shoulder) girdle and upper limbs, and then the pelvic (hip) girdle and the lower limbs. Before we examine the axial skeleton, we direct your attention to some general characteristics of bones.

CHECKPOINT

1. On what basis is the skeleton grouped into the axial and appendicular divisions?

7.2 TYPES OF BONES

OBJECTIVE

- Classify bones based on their shape or location.

Almost all bones of the body can be classified into five main types based on shape: long, short, flat, irregular, and sesamoid

TABLE 7.1

The Bones of the Adult Skeletal System

DIVISION OF THE SKELETON	STRUCTURE	NUMBER OF BONES	DIVISION OF THE SKELETON	STRUCTURE	NUMBER OF BONES
Axial skeleton	**Skull**		**Appendicular skeleton**	**Pectoral (shoulder) girdles**	
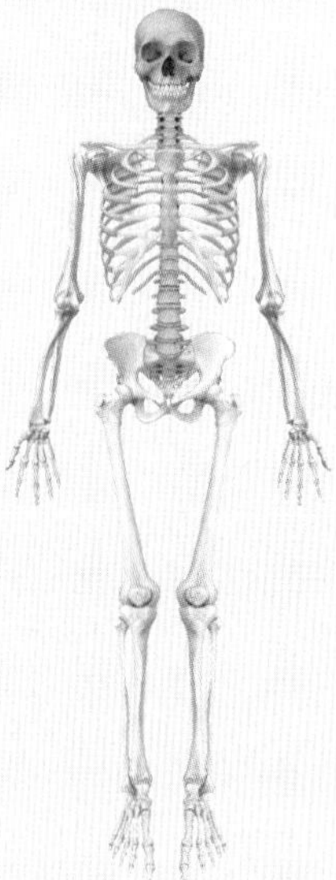	Cranium	8	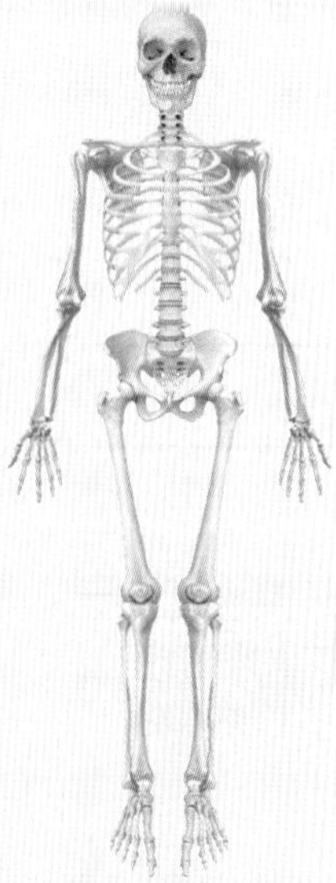	Clavicle	2
	Face	14		Scapula	2
	Hyoid	1		**Upper limbs**	
	Auditory ossicles	6		Humerus	2
	Vertebral column	26		Ulna	2
	Thorax			Radius	2
	Sternum	1		Carpals	16
	Ribs	24		Metacarpals	10
	Number of bones = 80			Phalanges	28
				Pelvic (hip) girdle	
				Hip, pelvic, or coxal bone	2
				Lower limbs	
				Femur	2
				Patella	2
				Fibula	2
				Tibia	2
				Tarsals	14
				Metatarsals	10
				Phalanges	28
				Number of bones = 126	
				Total bones in an adult skeleton = 206	

Figure 7.1 Divisions of the skeletal system. The axial skeleton is indicated in blue. (Note the position of the hyoid bone in Figure 7.5.)

The adult human skeleton consists of 206 bones grouped into two divisions: the axial skeleton and the appendicular skeleton.

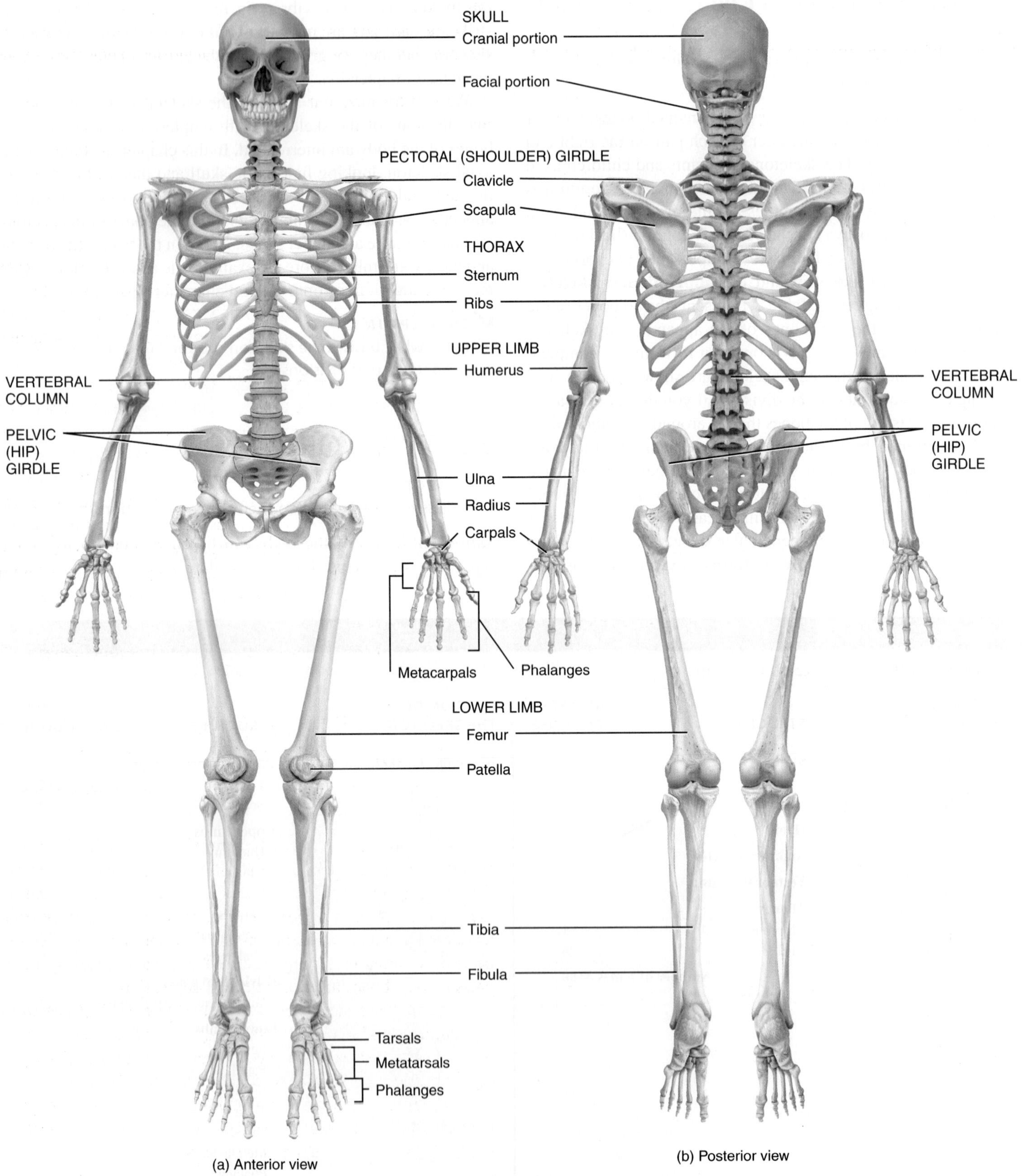

SKULL
Cranial portion
Facial portion
PECTORAL (SHOULDER) GIRDLE
Clavicle
Scapula
THORAX
Sternum
Ribs
UPPER LIMB
Humerus
VERTEBRAL COLUMN
PELVIC (HIP) GIRDLE
Ulna
Radius
Carpals
Metacarpals
Phalanges
LOWER LIMB
Femur
Patella
Tibia
Fibula
Tarsals
Metatarsals
Phalanges
(a) Anterior view
(b) Posterior view

Which of the following structures are part of the axial skeleton, and which are part of the appendicular skeleton? Skull, clavicle, vertebral column, shoulder girdle, humerus, pelvic girdle, and femur.

(Figure 7.2). As you learned in Chapter 6, **long bones** have greater length than width, consist of a shaft and a variable number of extremities or epiphyses (ends), and are slightly curved for strength. A curved bone absorbs the stress of the body's weight at several different points, so that it is evenly distributed. If bones were straight, the weight of the body would be unevenly distributed, and the bone would fracture more easily. Long bones consist mostly of *compact bone tissue* in their diaphyses but have considerable amounts of *spongy bone tissue* in their epiphyses. Long bones vary tremendously in size and include those in the femur (thigh bone), tibia and fibula (leg bones), humerus (arm bone), ulna and radius (forearm bones), and phalanges (finger and toe bones).

Short bones are somewhat cube-shaped and are nearly equal in length and width. They consist of spongy bone tissue except at the surface, which has a thin layer of compact bone tissue. Examples of short bones are most carpal (wrist) bones and most tarsal (ankle) bones.

Flat bones are generally thin and composed of two nearly parallel plates of compact bone tissue enclosing a layer of spongy bone tissue. Flat bones afford considerable protection and provide extensive areas for muscle attachment. Flat bones include the cranial bones, which protect the brain; the sternum (breastbone) and ribs, which protect organs in the thorax; and the scapulae (shoulder blades).

Irregular bones have complex shapes and cannot be grouped into any of the previous categories. They vary in the amount of spongy and compact bone present. Such bones include the vertebrae (backbones), hip bones, certain facial bones, and the calcaneus.

Sesamoid bones (SES-a-moyd = shaped like a sesame seed) develop in certain tendons where there is considerable friction, tension, and physical stress, such as the palms and soles. They may vary in number from person to person, are not always completely ossified, and typically measure only a few millimeters in diameter. Notable exceptions are the two patellae (kneecaps), large sesamoid bones located in the quadriceps femoris tendon (see Figure 11.20a) that are normally present in everyone. Functionally, sesamoid bones protect tendons from excessive wear and tear, and they often change the direction of pull of a tendon, which improves the mechanical advantage at a joint.

An additional type of bone is classified by location rather than shape. **Sutural bones** (SOO-chur-al; *sutur-* = seam) are small bones located in sutures (joints) between certain cranial bones (see Figure 7.6). Their number varies greatly from person to person.

Recall from Chapter 6 that in adults, red bone marrow is restricted to flat bones such as the ribs, sternum (breastbone), and skull; irregular bones such as vertebrae (backbones) and hip bones; long bones such as the proximal epiphyses of the femur (thigh bone) and humerus (arm bone); and some short bones.

CHECKPOINT

2. Give examples of long, short, flat, and irregular bones.

Figure 7.2 Types of bones based on shape. The bones are not drawn to scale.

The shapes of bones largely determine their functions.

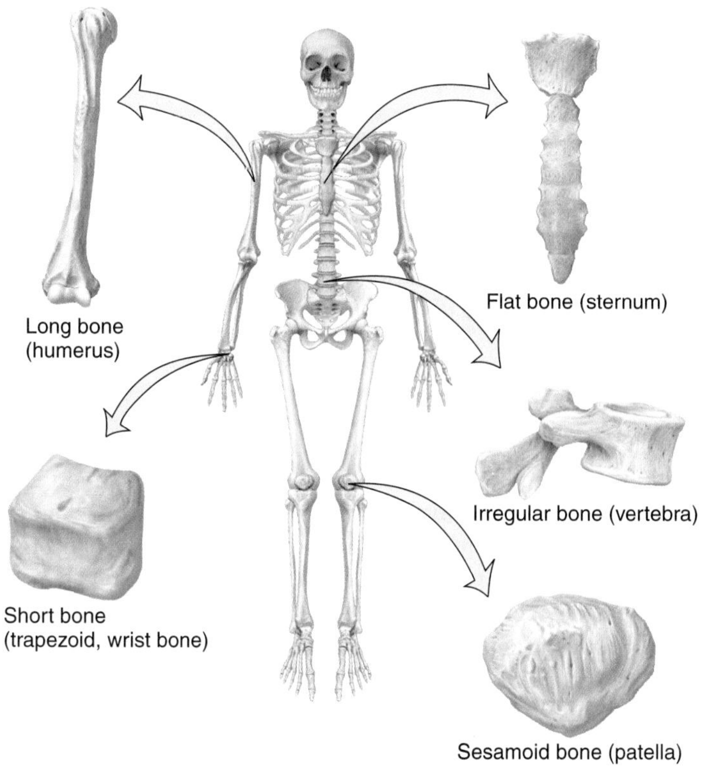
Long bone (humerus)
Flat bone (sternum)
Short bone (trapezoid, wrist bone)
Irregular bone (vertebra)
Sesamoid bone (patella)

Which type of bone primarily provides protection and a large surface area for muscle attachment?

7.3 BONE SURFACE MARKINGS

OBJECTIVE

- Describe the principal surface markings on bones and the functions of each.

Bones have characteristic **surface markings,** structural features adapted for specific functions. Most are not present at birth but develop in response to certain forces and are most prominent in the adult skeleton. In response to tension on a bone surface from tendons, ligaments, aponeuroses, and fasciae, new bone is deposited, resulting in raised or roughened areas. Conversely, compression on a bone surface results in a depression.

There are two major types of surface markings: (1) *depressions and openings,* which allow the passage of soft tissues (such as blood vessels, nerves, ligaments, and tendons) or form joints, and (2) *processes,* projections or outgrowths that either help form joints or serve as attachment points for connective tissue (such as ligaments and tendons). Table 7.2 describes the various surface markings and provides examples of each.

CHECKPOINT

3. What are surface markings? What are their general functions?

TABLE 7.2

Bone Surface Markings

MARKING	DESCRIPTION	EXAMPLE
DEPRESSIONS AND OPENINGS: SITES ALLOWING THE PASSAGE OF SOFT TISSUE (NERVES, BLOOD VESSELS, LIGAMENTS, TENDONS) OR FORMATION OF JOINTS		
Fissure (FISH-ur)	Narrow slit between adjacent parts of bones through which blood vessels or nerves pass.	Superior orbital fissure of sphenoid bone (Figure 7.12).
Foramen (fō-RĀ-men = hole; plural is *foramina*)	Opening through which blood vessels, nerves, or ligaments pass.	Optic foramen of sphenoid bone (Figure 7.12).
Fossa (FOS-a = trench; plural is *fossae,* FOS-ē)	Shallow depression.	Coronoid fossa of humerus (Figure 8.5a).
Sulcus (SUL-kus = groove; plural is *sulci,* SUL-sī)	Furrow along bone surface that accommodates blood vessel, nerve, or tendon.	Intertubercular sulcus of humerus (Figure 8.5a).
Meatus (mē-Ā-tus = passageway; plural is *meati,* mē-Ā-tī)	Tubelike opening.	External auditory meatus of temporal bone (Figure 7.4b).
PROCESSES: PROJECTIONS OR OUTGROWTHS ON BONE THAT FORM JOINTS OR ATTACHMENT POINTS FOR CONNECTIVE TISSUE, SUCH AS LIGAMENTS AND TENDONS		
Processes that form joints		
Condyle (KON-dīl; *condylus* = knuckle)	Large, round protuberance with a smooth articular surface at end of bone.	Lateral condyle of femur (Figure 8.13a).
Facet (FAS-et or fa-SET)	Smooth, flat, slightly concave or convex articular surface.	Superior articular facet of vertebra (Figure 7.18d).
Head	Usually rounded articular projection supported on neck (constricted portion) of bone.	Head of femur (Figure 8.13a).
Processes that form attachment points for connective tissue		
Crest	Prominent ridge or elongated projection.	Iliac crest of hip bone (Figure 8.10b).
Epicondyle (*epi-* = above)	Typically roughened projection above condyle.	Medial epicondyle of femur (Figure 8.13a).
Line (*linea*)	Long, narrow ridge or border (less prominent than crest).	Linea aspera of femur (Figure 8.13b).
Spinous process	Sharp, slender projection.	Spinous process of vertebra (Figure 7.17).
Trochanter (trō-KAN-ter)	Very large projection.	Greater trochanter of femur (Figure 8.13b).
Tubercle (TOO-ber-kul; *tuber-* = knob)	Variably sized rounded projection.	Greater tubercle of humerus (Figure 8.5a).
Tuberosity	Variably sized projection that has a rough, bumpy surface.	Ischial tuberosity of hip bone (Figure 8.10b).

7.4 SKULL

OBJECTIVES

- Name the cranial and facial bones and indicate whether they are paired or single.
- Describe the following special features of the skull: sutures, paranasal sinuses, and fontanels.

The **skull** is the bony framework of the head. It contains 22 bones (not counting the bones of the middle ears) and rests on the superior end of the vertebral column (backbone). The bones of the skull are grouped into two categories: cranial bones and facial bones. The **cranial bones** (*crani-* = brain case) form the cranial cavity, which encloses and protects the brain. The eight cranial bones are the frontal bone, two parietal bones, two temporal bones, the occipital bone, the sphenoid bone, and the ethmoid bone. Fourteen **facial bones** form the face: two nasal bones, two maxillae (or maxillas), two zygomatic bones, the mandible, two lacrimal bones, two palatine bones, two inferior nasal conchae, and the vomer. Exhibits 7.A through 7.G illustrate the bones of the skull from different views.

Cranial Bones—Frontal Bone *(Figure 7.3)*

OBJECTIVE

- Identify the location and surface features of the frontal bone.

The **frontal bone** forms the forehead (the anterior part of the cranium), the roofs of the *orbits* (eye sockets), and most of the anterior part of the cranial floor (Figure 7.3). Soon after birth, the left and right sides of the frontal bone are united by the *metopic suture,* which usually disappears between the ages of six and eight.

Note the *frontal squama,* a scalelike plate of bone that forms the forehead in the anterior view of the skull shown in Figure 7.3a. It gradually slopes inferiorly from the coronal suture, on the top of the skull, then angles abruptly and becomes almost vertical above the orbits. At the superior border of the orbits, the frontal bone thickens, forming the *supraorbital margin* (*supra-* = above; *-orbi* = circle). From this margin, the frontal bone extends posteriorly to form the roof of the orbit, which is part of the floor of the cranial cavity. Within the supraorbital margin, slightly

Figure 7.3 Anterior view of the skull.

The skull consists of cranial bones and facial bones.

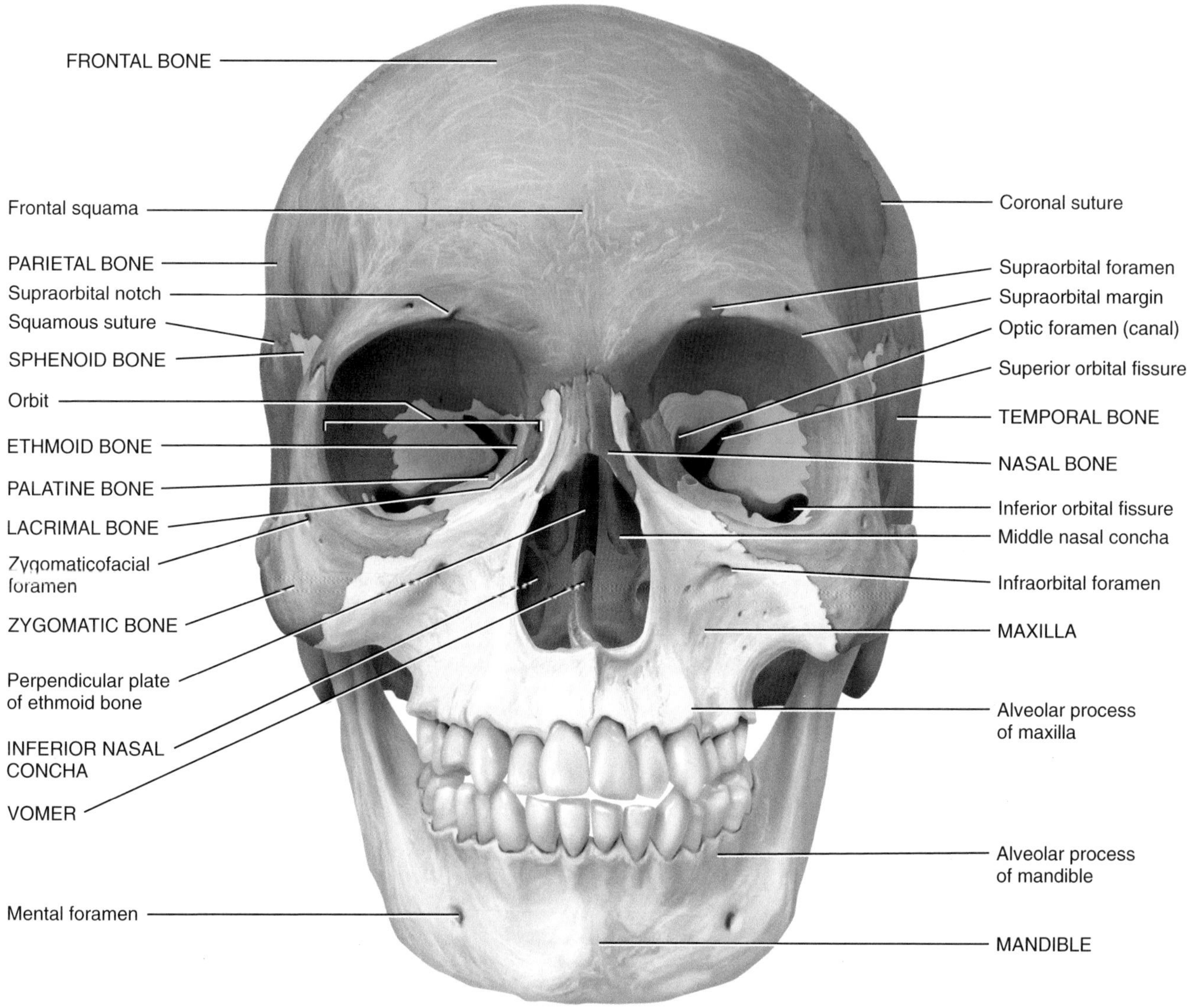

(a) Anterior view

EXHIBIT 7.A CONTINUES

medial to its midpoint, is a hole called the *supraorbital foramen.* Sometimes the foramen is incomplete and is called the *supraorbital notch.* As you read about each foramen associated with a cranial bone, refer to Table 7.3 to note which structures pass through it. The *frontal sinuses* lie deep to the frontal squama. Sinuses, or more technically paranasal sinuses, are mucous membrane–lined cavities within certain skull bones that will be discussed later.

✔CHECKPOINT

What structures pass through the supraorbital foramen?

CLINICAL CONNECTION | *Black Eye*

A **black eye** is a bruising around the eye, commonly due to an injury to the face, rather than an eye injury. In response to trauma, blood and other fluids accumulate in the space around the eye, causing the swelling and dark discoloration. One cause might be a blow to the sharp ridge just superior to the supraorbital margin that fractures the frontal bone, resulting in bleeding. Another is a blow to the nose. Certain surgical procedures (face lift, eyelid surgery, jaw surgery, or nasal surgery) can also result in black eyes. •

FIGURE 7.3 CONTINUED

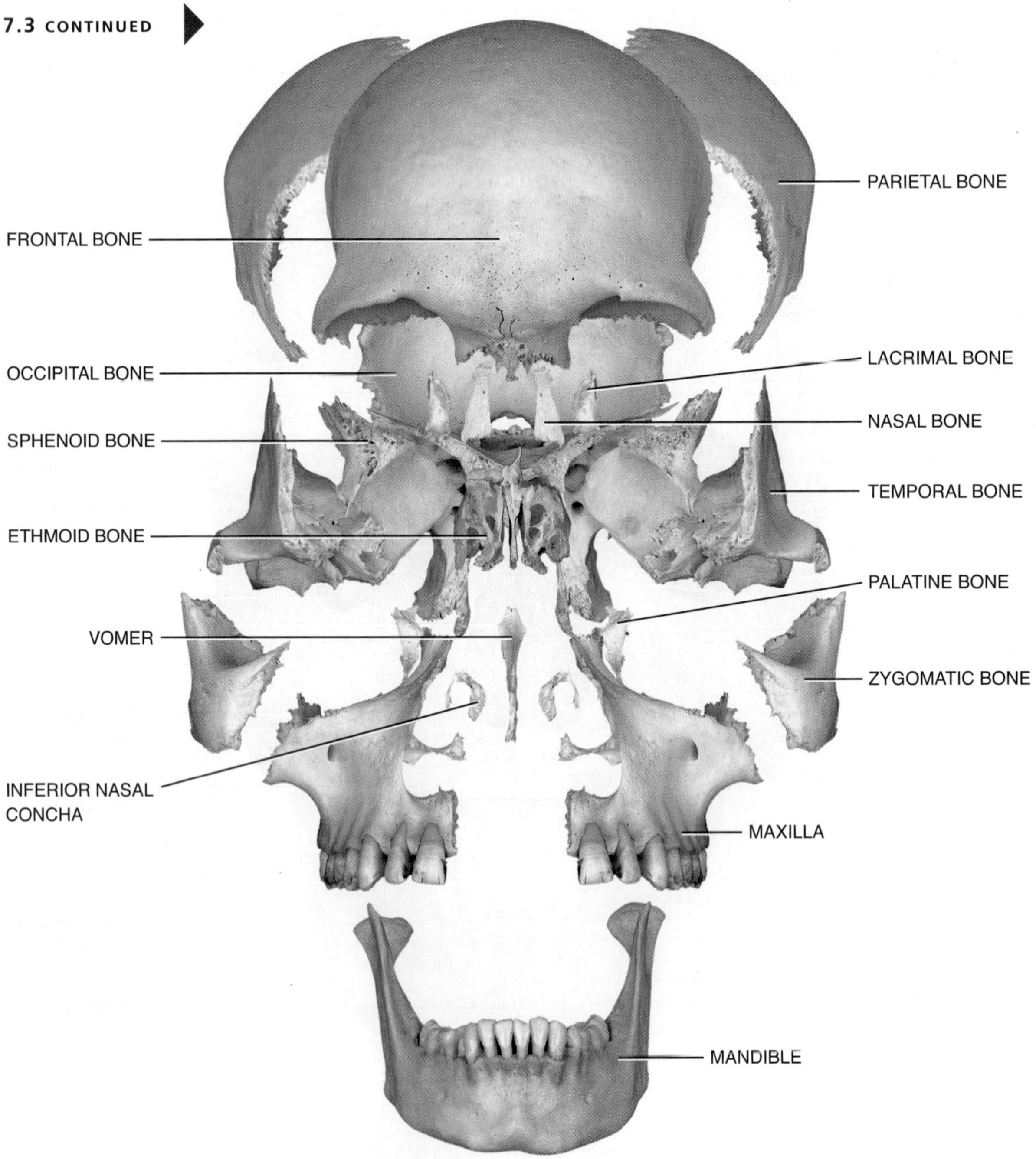

(b) Anterior view of disarticulated skull

? Which of the bones shown here are cranial bones?

EXHIBIT 7.B Cranial Bones–Parietal Bones *(Figure 7.4)*

OBJECTIVE

- Identify the location and surface features of the parietal bones.

The two **parietal bones** (pa-RĪ-e-tal; *pariet-* = wall) form the greater portion of the sides and roof of the cranial cavity (Figure 7.4). The internal surfaces of the parietal bones contain many protrusions and depressions that accommodate the blood vessels supplying the dura mater, the superficial connective tissue (meninx) covering of the brain.

✔CHECKPOINT

How do the parietal bones relate to the cranial cavity?

Figure 7.4 Superior and right lateral views of the skull.

The zygomatic arch is formed by the zygomatic process of the temporal bone and the temporal process of the zygomatic bone.

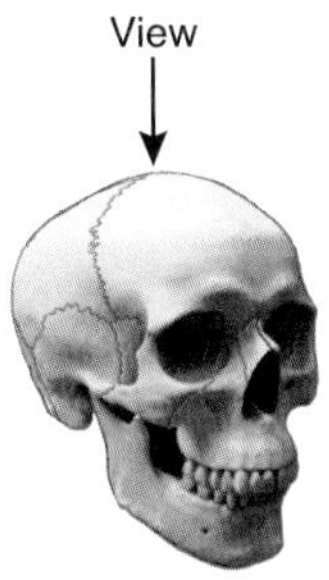
View

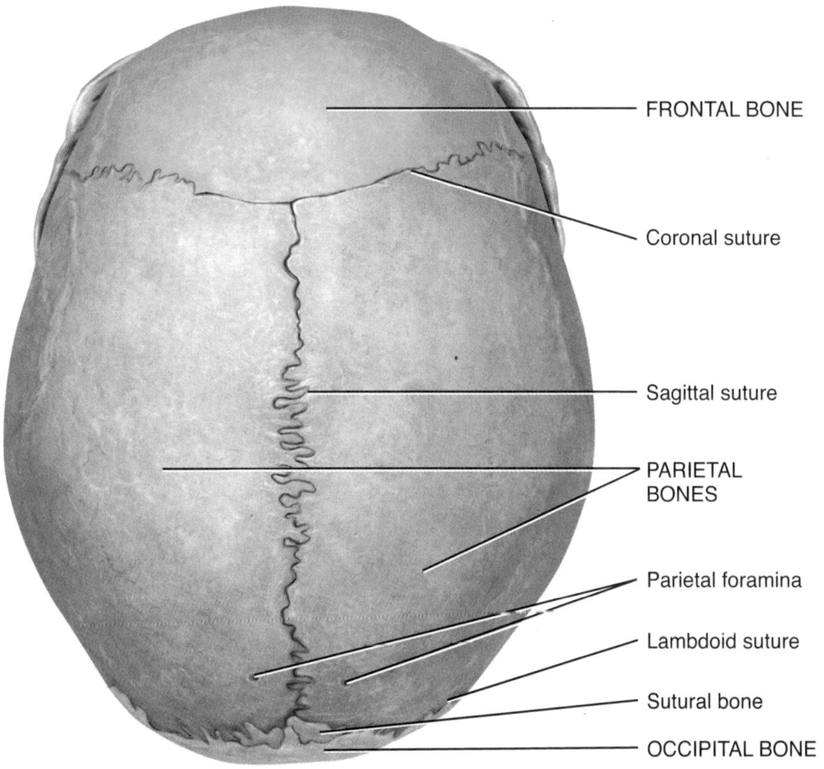
FRONTAL BONE
Coronal suture
Sagittal suture
PARIETAL BONES
Parietal foramina
Lambdoid suture
Sutural bone
OCCIPITAL BONE

(a) Superior view

EXHIBIT 7.B CONTINUES

FIGURE 7.4 CONTINUED

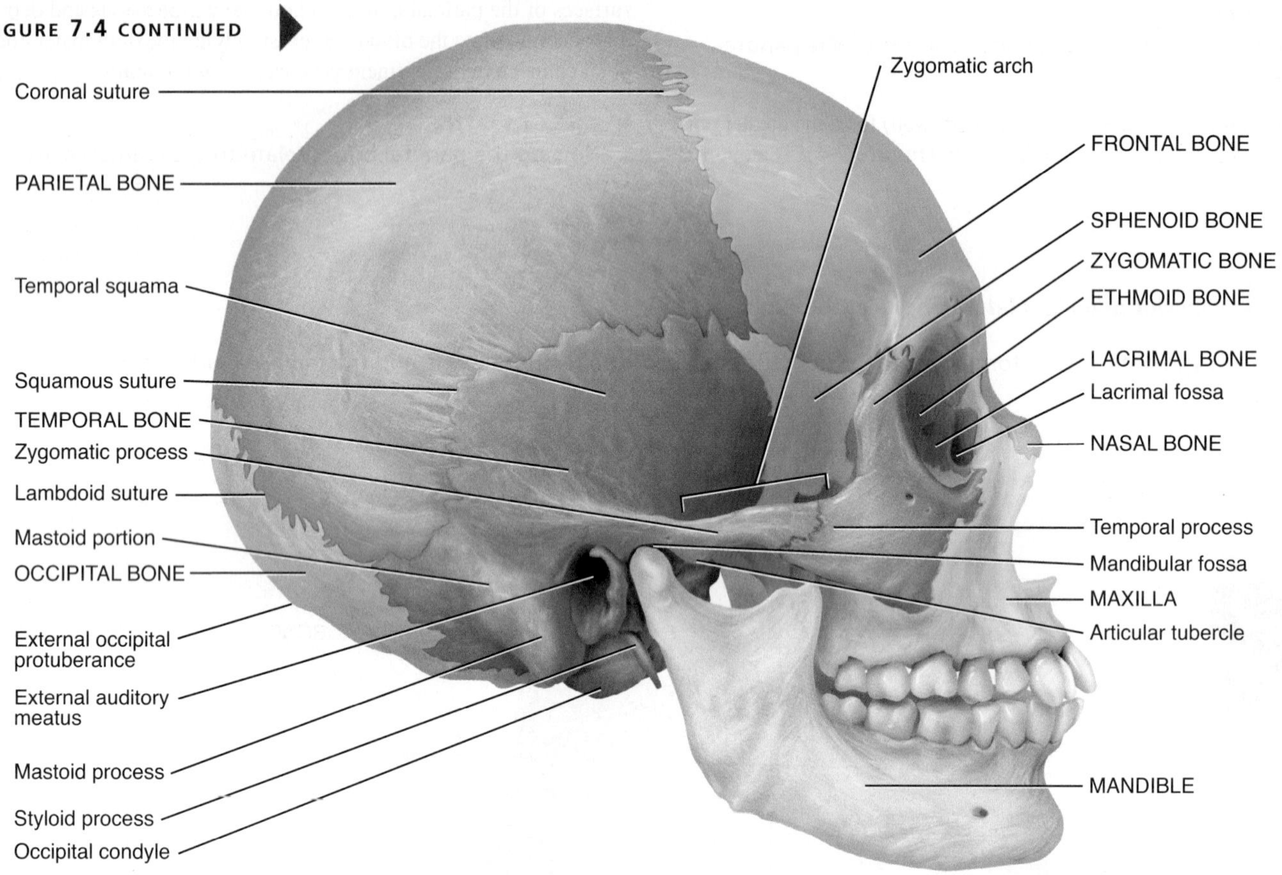
Coronal suture
PARIETAL BONE
Temporal squama
Squamous suture
TEMPORAL BONE
Zygomatic process
Lambdoid suture
Mastoid portion
OCCIPITAL BONE
External occipital protuberance
External auditory meatus
Mastoid process
Styloid process
Occipital condyle
Zygomatic arch
FRONTAL BONE
SPHENOID BONE
ZYGOMATIC BONE
ETHMOID BONE
LACRIMAL BONE
Lacrimal fossa
NASAL BONE
Temporal process
Mandibular fossa
MAXILLA
Articular tubercle
MANDIBLE

(b) Right lateral view

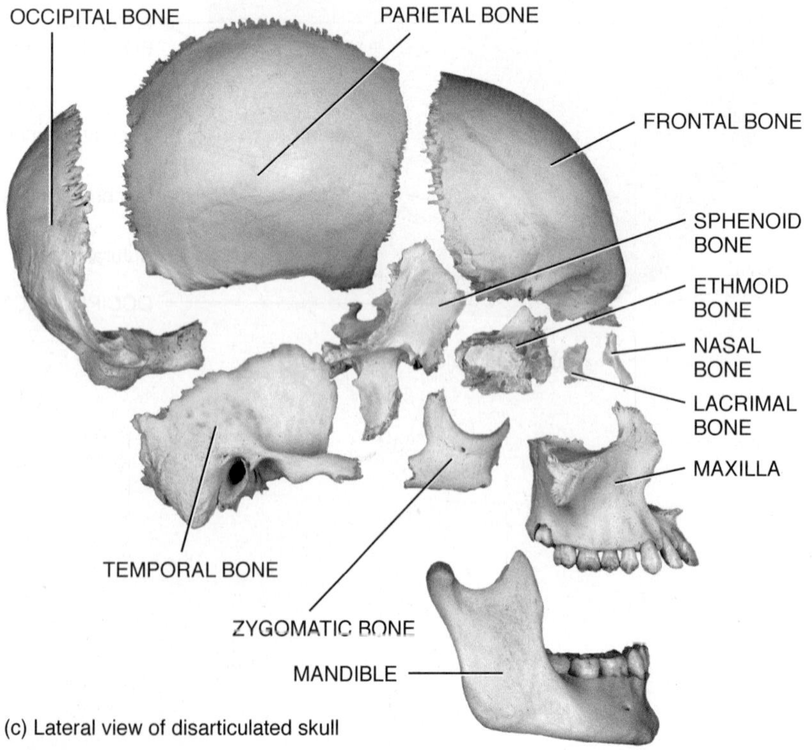
OCCIPITAL BONE
PARIETAL BONE
FRONTAL BONE
SPHENOID BONE
ETHMOID BONE
NASAL BONE
LACRIMAL BONE
MAXILLA
TEMPORAL BONE
ZYGOMATIC BONE
MANDIBLE

(c) Lateral view of disarticulated skull

? What major bones are joined by (1) the squamous suture, (2) the lambdoid suture, and (3) the coronal suture?

EXHIBIT 7.C Cranial Bones—Temporal Bones *(Figure 7.5)*

OBJECTIVE

- Identify the location and surface features of the temporal bones.

The paired **temporal bones** (*tempor-* = temple) form the inferior lateral aspects of the cranium and part of the cranial floor. (The terms *temporal* and *temple* are derived from the Latin word *tempus,* meaning "time," in reference to the graying of hair in the temple area, a sign of time's passing.) In Figure 7.3a, note the *temporal squama* (= scale), the thin, flat part of the temporal bone that forms the anterior and superior part of the *temple* (the region of the cranium around the ear). Projecting from the inferior portion of the temporal squama is the *zygomatic process,* which articulates (forms a joint) with the temporal process of the zygomatic (cheek) bone. Together, the zygomatic process of the temporal bone and the temporal process of the zygomatic bone form the *zygomatic arch.*

A socket called the *mandibular fossa* is located on the inferior posterior surface of the zygomatic process of each temporal bone. Anterior to the mandibular fossa is a rounded elevation, the *articular tubercle* (Figure 7.4b). The mandibular fossa and articular tubercle articulate with the mandible (lower jawbone) to form the *temporomandibular joint (TMJ).*

The *mastoid portion* (*mastoid* = breast-shaped; Figure 7.4b) of the temporal bone is located posterior and inferior to the *external auditory meatus* (*meatus* = passageway), or ear canal, which directs sound waves into the ear. In an adult, this portion of the bone contains several *mastoid air cells* that communicate with the hollow space of the middle ear. These tiny air-filled compartments are separated from the brain by thin bony partitions. Middle ear infections that go untreated can spread into the mastoid air cells, causing a painful inflammation called **mastoiditis** (mas′-toy-DĪ-tis).

The *mastoid process* is a rounded projection of the mastoid portion of the temporal bone posterior and inferior to the external auditory meatus. It is the point of attachment for several neck muscles. The *internal auditory meatus* (Figure 7.5) is the opening through which the facial (VII) nerve and vestibulocochlear (VIII) nerve pass. The *styloid process* (*styl-* = stake or pole) projects inferiorly from the inferior surface of the temporal bone and serves as a point of attachment for muscles and ligaments of the tongue and neck (see Figure 7.4b). Between the styloid process and the mastoid process is the *stylomastoid foramen,* through which the facial (VII) nerve and stylomastoid artery pass (see Figure 7.7).

At the floor of the cranial cavity (see Figure 7.8a) is the *petrous portion* (*petrous* = rock) of the temporal bone. This triangular part, located at the base of the skull between the sphenoid and occipital bones, houses the internal ear and the middle ear, structures involved in hearing and equilibrium (balance). It also contains the *carotid foramen,* through which the carotid artery passes (see Figure 7.7). Posterior to the carotid foramen and anterior to the occipital bone is the *jugular foramen,* a passageway for the jugular vein.

CHECKPOINT

What structures form the zygomatic arch?

Figure 7.5 Medial view of sagittal section of the skull. Although the hyoid bone is not part of the skull, it is included here for reference.

The cranial bones are the frontal, parietal, temporal, occipital, sphenoid, and ethmoid bones. The facial bones are the nasal bones, maxillae, zygomatic bones, lacrimal bones, palatine bones, inferior nasal conchae, mandible, and vomer.

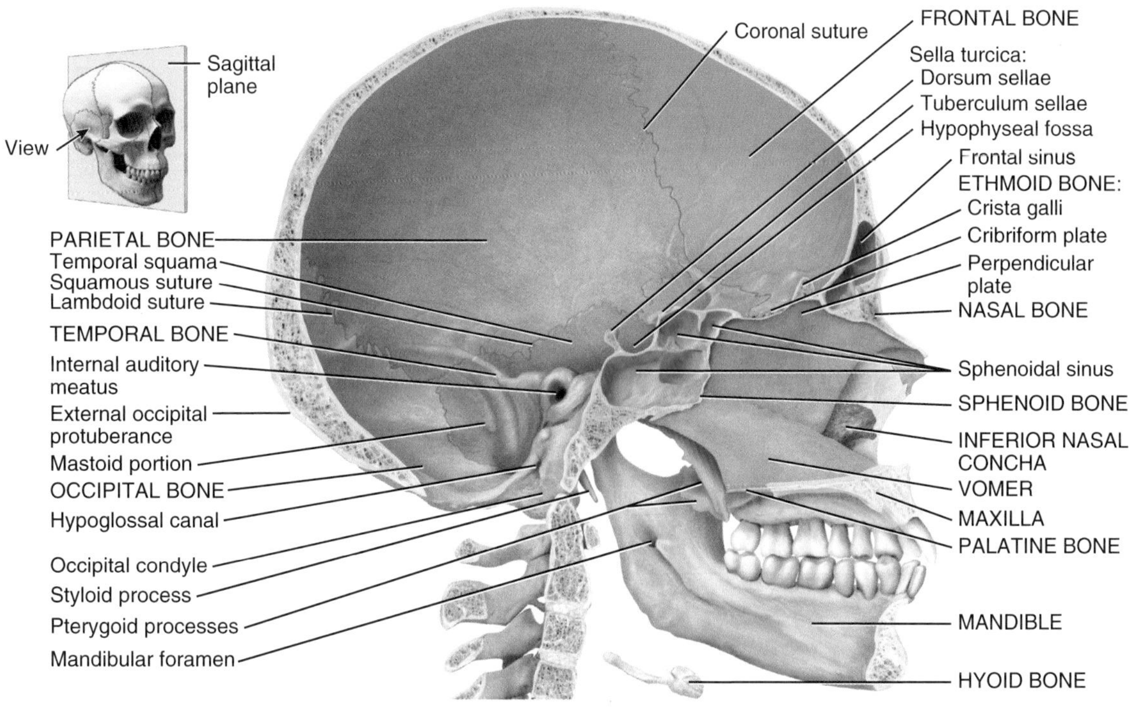

Medial view of sagittal section

With which bones does the temporal bone articulate?

Cranial Bones—Occipital Bone *(Figures 7.6 and 7.7)*

OBJECTIVE

- Identify the location and surface features of the occipital bone.

The **occipital bone** (ok-SIP-i-tal; *occipit-* = back of head) forms the posterior part and most of the base of the cranium (Figure 7.6; also see Figure 7.4). Also view the occipital bone and surrounding structures in the inferior view of the skull in Figure 7.7. The *foramen magnum* (= large hole) is in the inferior part of the bone. The medulla oblongata (inferior part of the brain) connects with the spinal cord within this foramen, and the vertebral and spinal arteries also pass through it along with the accessory (XI) nerve. The *occipital condyles,* oval processes with convex surfaces on either side of the foramen magnum (Figure 7.7), articulate with depressions on the first cervical vertebra (atlas) to form the *atlanto-occipital joint,* which allows you to nod your head "yes." Superior to each occipital condyle on the inferior surface of the skull is the *hypoglossal canal* (*hypo-* = under; *-glossal* = tongue). (See Figure 7.5.)

The *external occipital protuberance* is the most prominent midline projection on the posterior surface of the bone just above the foramen magnum. You may be able to feel this structure as a bump on the back

Figure 7.6 Posterior view of the skull. The sutures are exaggerated for emphasis.

The occipital bone forms most of the posterior and inferior portions of the cranium.

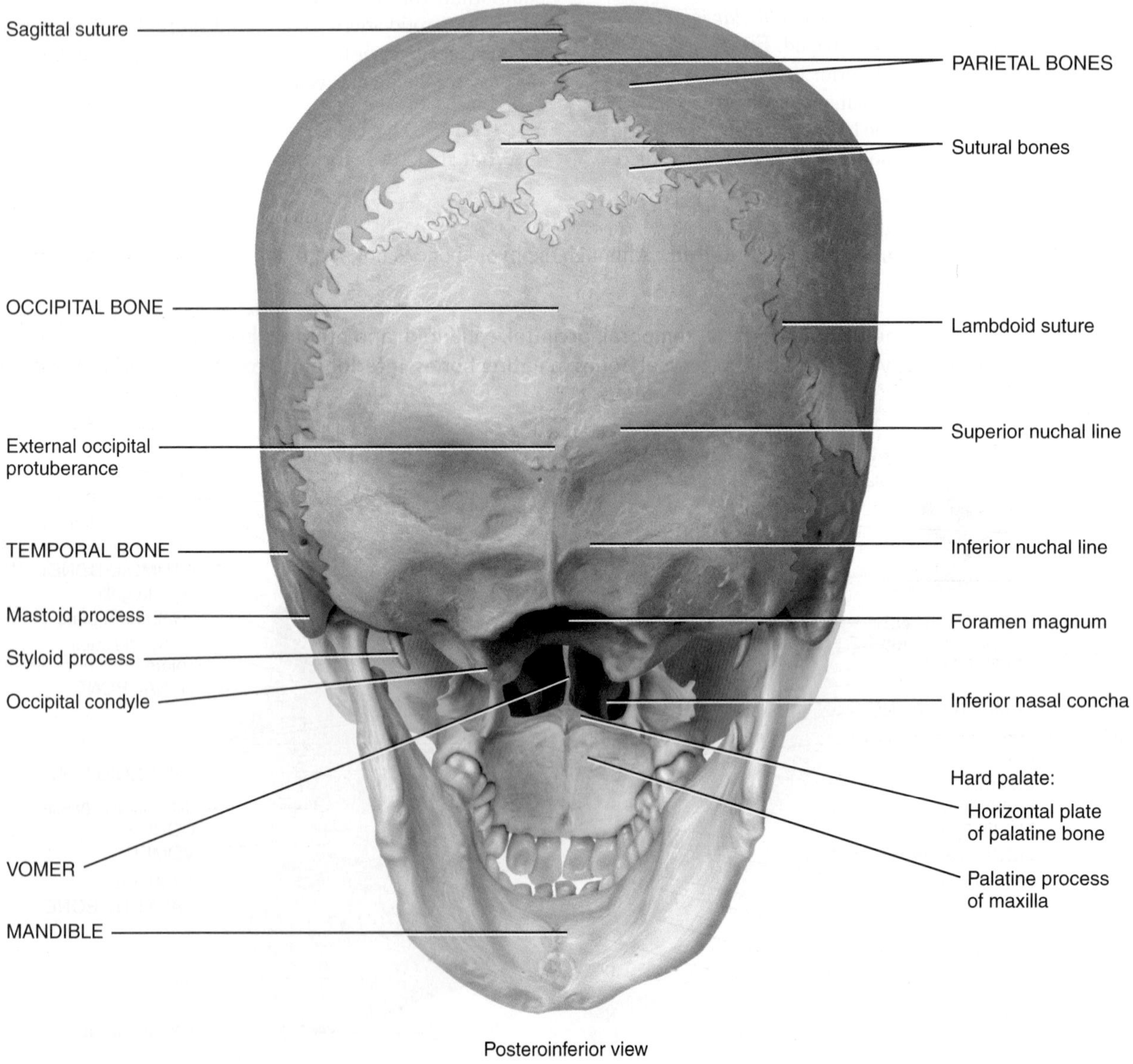

Which bones form the posterior, lateral portion of the cranium?

of your head, just above your neck. (See Figure 7.4b.) A large fibrous, elastic ligament, the *ligamentum nuchae* (*nucha-* = nape of neck), extends from the external occipital protuberance to the seventh cervical vertebra to help support the head. Extending laterally from the protuberance are two curved ridges, the *superior nuchal lines,* and below these are two *inferior nuchal lines,* which are areas of muscle attachment (Figure 7.7).

✔ CHECKPOINT

What structures pass through the hypoglossal canal?

Figure 7.7 Inferior view of the skull. The mandible (lower jawbone) has been removed.

The occipital condyles of the occipital bone articulate with the first cervical vertebra to form the atlanto-occipital joint.

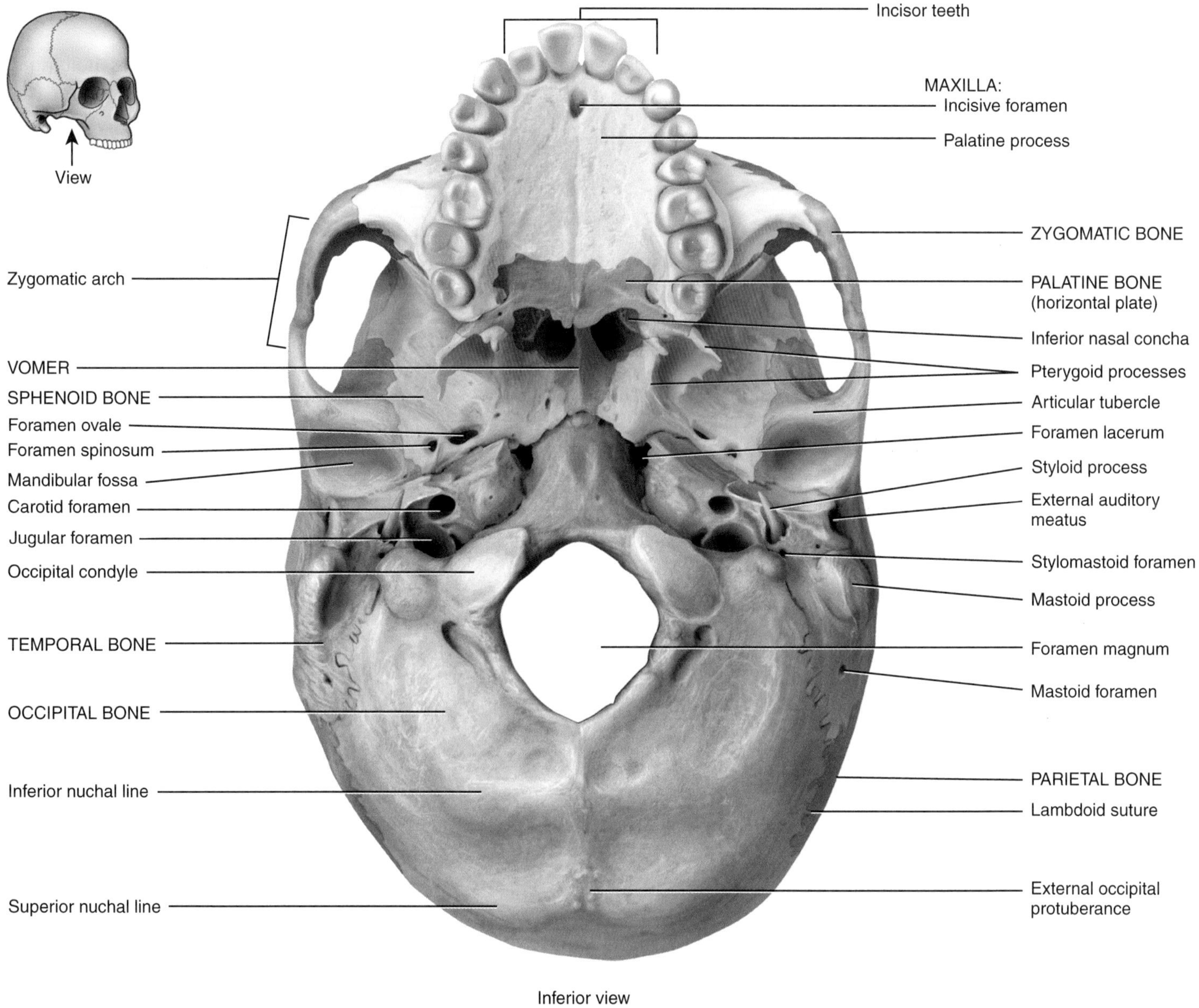

View
Incisor teeth
MAXILLA:
Incisive foramen
Palatine process
ZYGOMATIC BONE
PALATINE BONE
(horizontal plate)
Inferior nasal concha
Pterygoid processes
Articular tubercle
Foramen lacerum
Styloid process
External auditory
meatus
Stylomastoid foramen
Mastoid process
Foramen magnum
Mastoid foramen
PARIETAL BONE
Lambdoid suture
External occipital
protuberance
Zygomatic arch
VOMER
SPHENOID BONE
Foramen ovale
Foramen spinosum
Mandibular fossa
Carotid foramen
Jugular foramen
Occipital condyle
TEMPORAL BONE
OCCIPITAL BONE
Inferior nuchal line
Superior nuchal line

Inferior view

What organs of the nervous system join together within the foramen magnum?

Cranial Bones—Sphenoid Bone *(Figure 7.8)*

OBJECTIVE

- Identify the location and surface features of the sphenoid bone.

The **sphenoid bone** (SFĒ-noyd = wedge-shaped) lies at the middle part of the base of the skull (Figures 7.7 and 7.8). This bone is the keystone of the cranial floor because it articulates with all the other cranial bones, holding them together. View the floor of the cranium superiorly (Figure 7.8a) and note the sphenoid articulations. The sphenoid bone joins anteriorly with the frontal and ethmoid bones, laterally with the temporal bones, and posteriorly with the occipital bone. The sphenoid lies posterior and slightly superior to the nasal cavity and forms part of the floor, side walls, and rear wall of the orbit (see Figure 7.12).

The shape of the sphenoid resembles a butterfly with outstretched wings (Figure 7.8b). The *body* of the sphenoid is the hollowed cube-like medial portion between the ethmoid and occipital bones. The space inside the body is the *sphenoidal sinus,* which drains into the nasal cavity (see Figure 7.13). The *sella turcica* (SEL-a TUR-si-ka; *sella* = saddle; *turcica* = Turkish) is a bony saddle-shaped structure on the superior surface of the body of the sphenoid (Figure 7.8a). The anterior part of the sella turcica, which forms the horn of the saddle, is a ridge called the *tuberculum sellae.* The seat of the saddle is a depression, the *hypophyseal fossa* (hī-pō-FIZ-ē-al), which contains the pituitary gland. The posterior part of the sella turcica, which forms the back of the saddle, is another ridge called the *dorsum sellae.*

Figure 7.8 **Sphenoid bone.**

The sphenoid bone is called the keystone of the cranial floor because it articulates with all other cranial bones, holding them together.

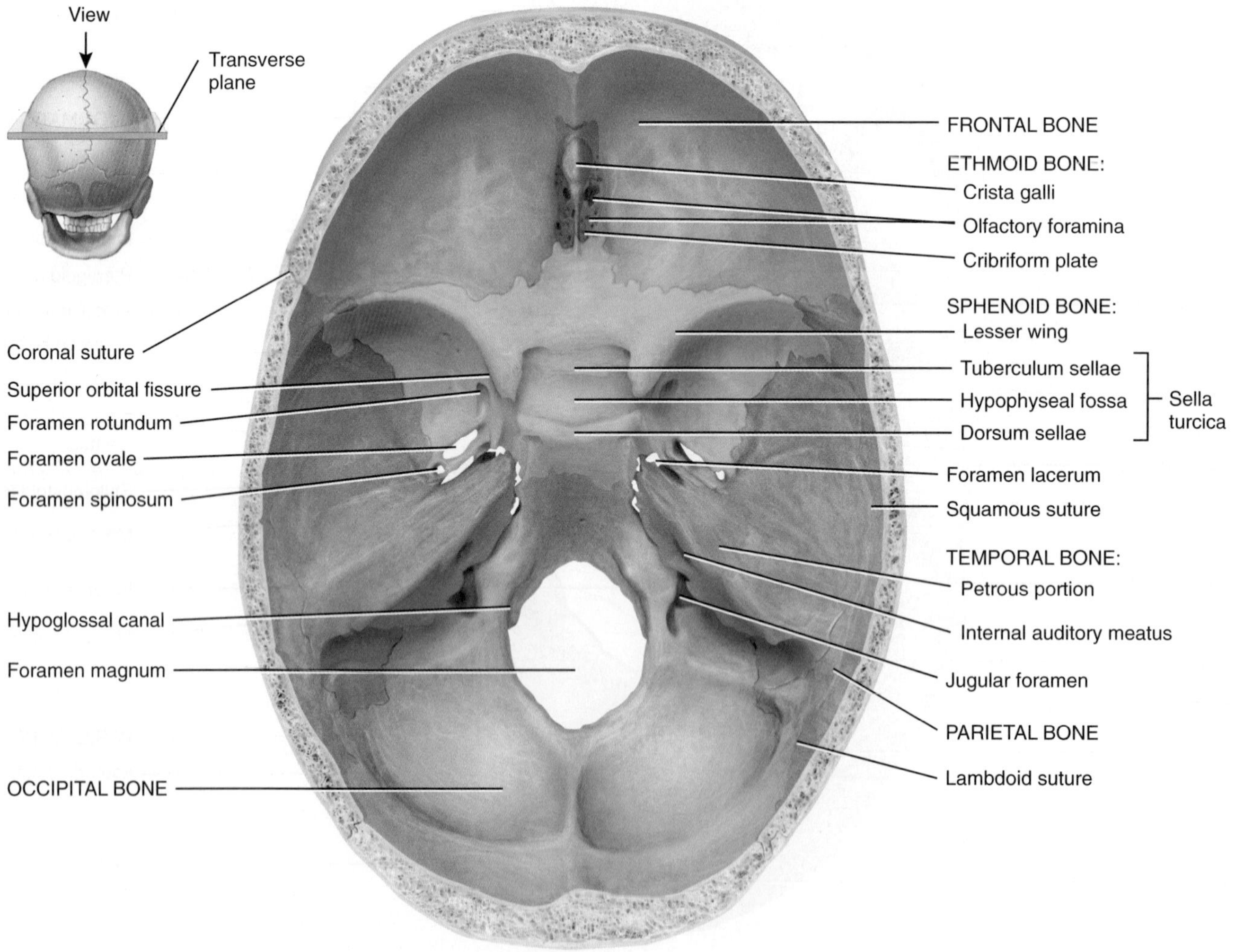

(a) Superior view of sphenoid bone in floor of cranium

The *greater wings* of the sphenoid project laterally from the body and form the anterolateral floor of the cranium. The greater wings also form part of the lateral wall of the skull just anterior to the temporal bone and can be viewed externally. The *lesser wings,* which are smaller, form a ridge of bone anterior and superior to the greater wings. They form part of the floor of the cranium and the posterior part of the orbit of the eye.

Between the body and lesser wing just anterior to the sella turcica is the *optic foramen* or *canal* (*optic* = eye), through which the optic (II) nerve and ophthalmic artery pass into the orbit. Lateral to the body between the greater and lesser wings is a triangular slit called the *superior orbital fissure.* This fissure may also be seen in the anterior view of the orbit in Figure 7.12. Blood vessels and cranial nerves pass through this fissure.

The *pterygoid processes* (TER-i-goyd = winglike) project inferiorly from the points where the body and greater wings of the sphenoid bone unite; they form the lateral posterior region of the nasal cavity (see Figures 7.7 and 7.8b). Some of the muscles that move the mandible attach to the pterygoid processes. At the base of the lateral pterygoid process in the greater wing is the *foramen ovale* (= oval hole). The *foramen lacerum* (= lacerated), covered in part by a layer of fibrocartilage in living subjects, is bounded anteriorly by the sphenoid bone and medially by the sphenoid and occipital bones. It transmits a branch of the ascending pharyngeal artery. Another foramen associated with the sphenoid bone is the *foramen rotundum* (= round hole) located at the junction of the anterior and medial parts of the sphenoid bone. The maxillary branch of the trigeminal (V) nerve passes through the foramen rotundum.

✔CHECKPOINT

Why is the sphenoid bone called the keystone of the cranial floor?

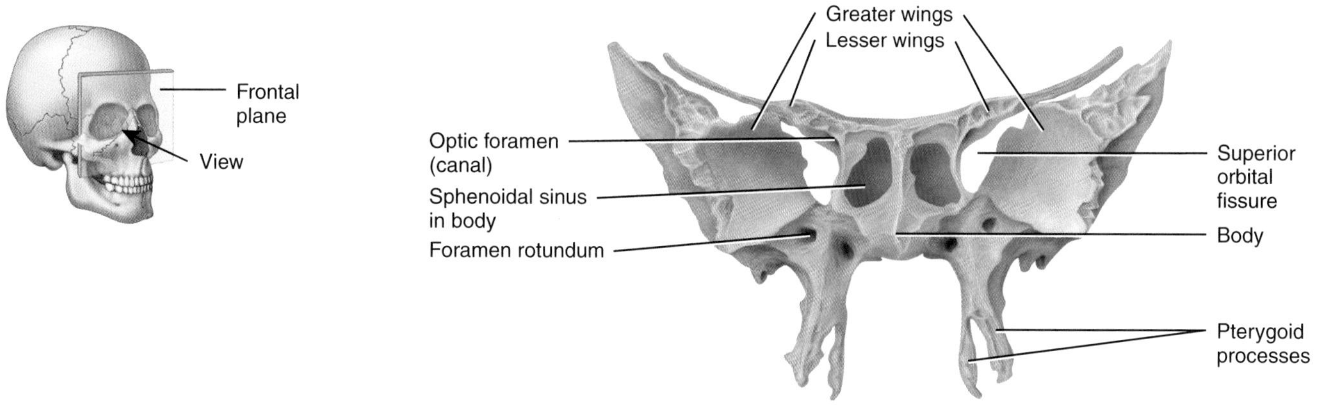

Frontal plane
View
Greater wings
Lesser wings
Optic foramen (canal)
Sphenoidal sinus in body
Foramen rotundum
Superior orbital fissure
Body
Pterygoid processes

(b) Anterior view of sphenoid bone

? Name the bones that articulate with the sphenoid bone, starting at the crista galli of the ethmoid bone and going in a clockwise direction.

Cranial Bones—Ethmoid Bone *(Figure 7.9)*

OBJECTIVE

- Identify the location and surface features of the ethmoid bone.

The **ethmoid bone** (ETH-moyd = like a sieve) is a delicate bone located in the anterior part of the cranial floor medial to the orbits and is sponge-like in appearance (Figure 7.9). It is anterior to the sphenoid and posterior to the nasal bones. The ethmoid bone forms (1) part of the anterior portion of the cranial floor; (2) the medial wall of the orbits; (3) the superior portion of the **nasal septum,** a partition that divides the nasal cavity into right and left sides; and (4) most of the superior sidewalls of the nasal cavity. The ethmoid bone is a major superior supporting structure of the nasal cavity and forms an extensive surface area in the nasal cavity.

The *cribriform plate* (*cribri-* = sieve) of the ethmoid bone lies in the anterior floor of the cranium and forms the roof of the nasal cavity. The cribriform plate contains the *olfactory foramina* (*olfact-* = to smell) through which the olfactory nerves pass. Projecting superiorly from the cribriform plate is a triangular process called the *crista galli* (*crista* = crest; *galli* = cock), which serves as a point of attachment for the falx cerebri, the membrane that separates the two sides of the brain. Projecting inferiorly from the cribriform plate is the *perpendicular plate,* which forms the superior portion of the nasal septum (see Figure 7.11).

The *lateral masses* of the ethmoid bone compose most of the wall between the nasal cavity and the orbits. They contain 3 to 18 air spaces called ethmoidal cells. The ethmoidal cells together form the *ethmoidal sinuses* (see Figure 7.13). The lateral masses contain two thin, scroll-

Figure 7.9 Ethmoid bone.

The ethmoid bone forms part of the anterior portion of the cranial floor, the medial wall of the orbits, the superior portions of the nasal septum, and most of the side walls of the nasal cavity.

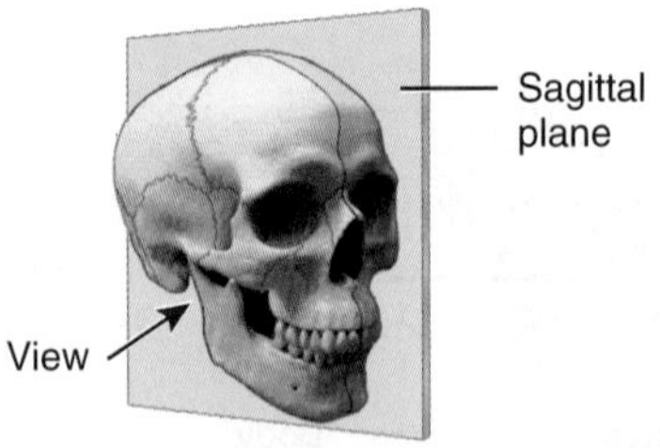
Sagittal plane
View

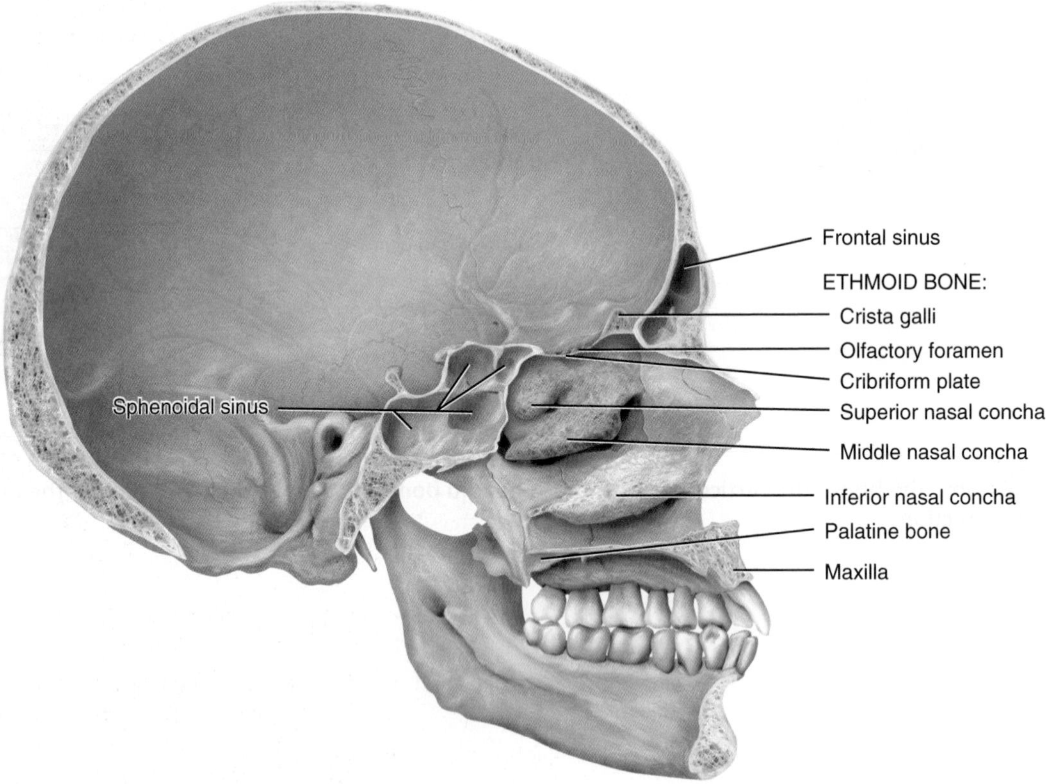
Frontal sinus
ETHMOID BONE:
Crista galli
Olfactory foramen
Cribriform plate
Superior nasal concha
Middle nasal concha
Inferior nasal concha
Palatine bone
Maxilla
Sphenoidal sinus

(a) Medial view of sagittal section

shaped projections lateral to the nasal septum. These are called the *superior nasal concha* (KON-ka = shell) or *turbinate* and the *middle nasal concha (turbinate)*. The plural form is *conchae* (KON-kē). A third pair of conchae, the inferior nasal conchae, are separate bones (discussed shortly). The conchae greatly increase the vascular and mucous membrane surface area in the nasal cavity, which warms and moistens (humidifies) inhaled air before it passes into the lungs. The conchae also cause inhaled air to swirl, and the result is that many inhaled particles become trapped in the mucus that lines the nasal cavity. This action of the conchae helps cleanse inhaled air before it passes into the rest of the respiratory passageways. The superior nasal conchae are near the olfactory foramina of the cribriform plate where the sensory receptors for olfaction (smell) terminate in the mucous membrane of the superior nasal conchae. Thus, they increase the surface area for the sense of smell.

✔ **CHECKPOINT**

The ethmoid bone forms which other cranial structures?

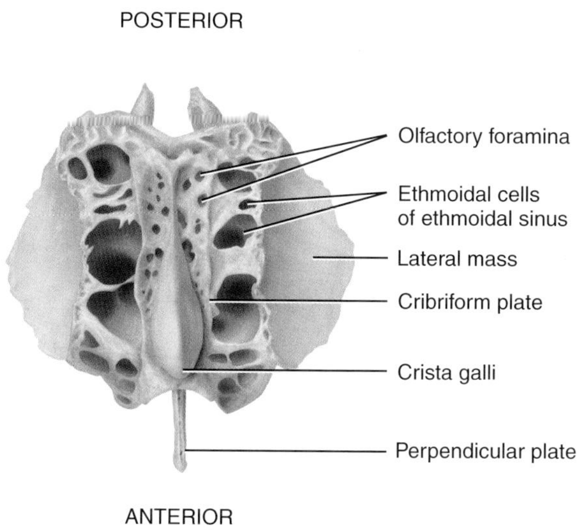
POSTERIOR
Olfactory foramina
Ethmoidal cells of ethmoidal sinus
Lateral mass
Cribriform plate
Crista galli
Perpendicular plate
ANTERIOR

(b) Superior view

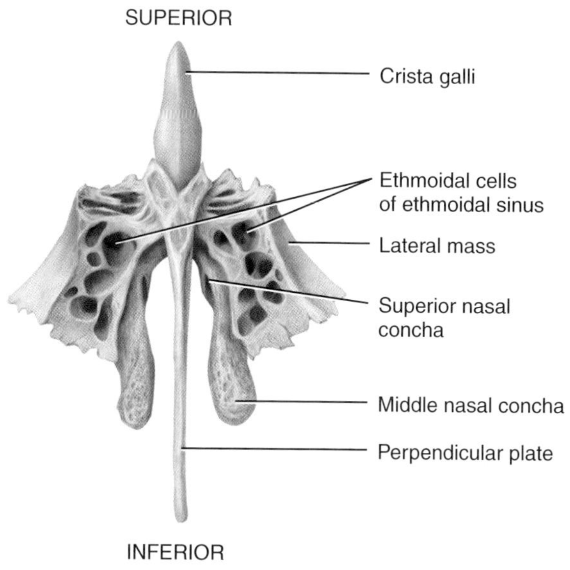
SUPERIOR
Crista galli
Ethmoidal cells of ethmoidal sinus
Lateral mass
Superior nasal concha
Middle nasal concha
Perpendicular plate
INFERIOR

(c) Anterior view

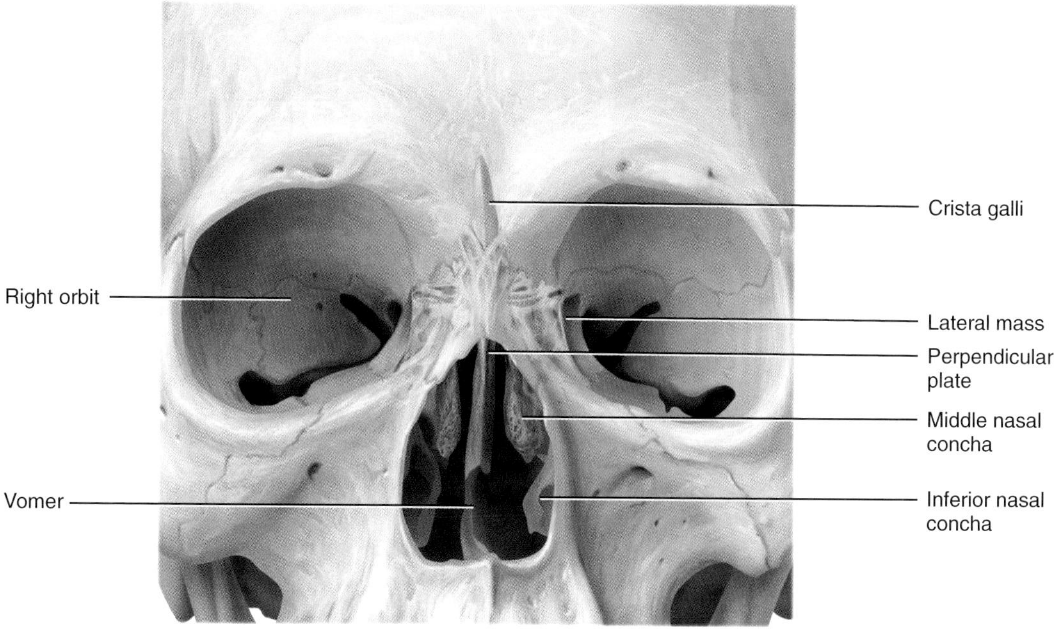
Crista galli
Right orbit
Lateral mass
Perpendicular plate
Middle nasal concha
Vomer
Inferior nasal concha

(d) Anterior view of position of ethmoid bone in skull (projected to the surface)

EXHIBIT 7.F CONTINUES ▶

FIGURE 7.9 CONTINUED

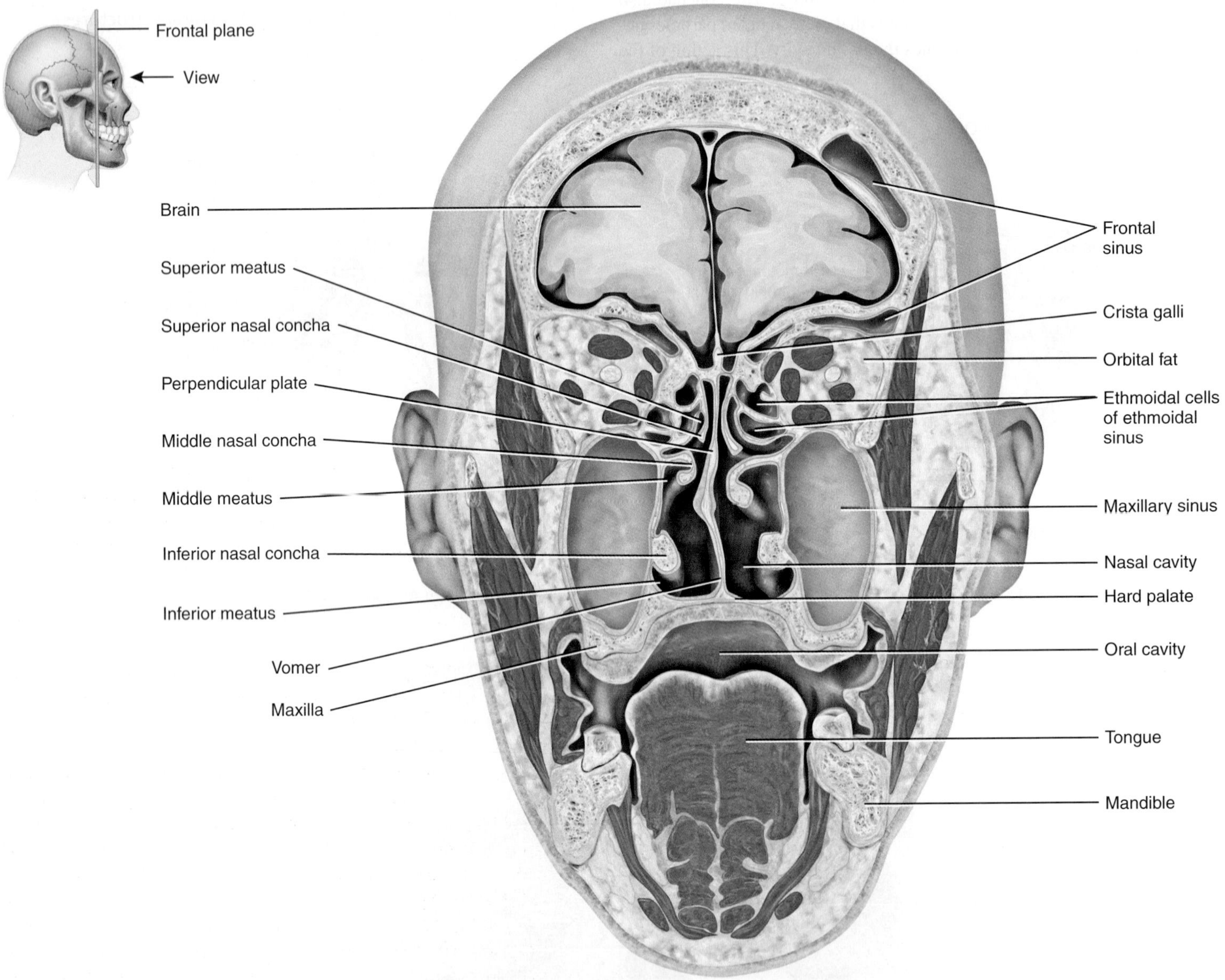

Frontal plane
View
Brain
Superior meatus
Superior nasal concha
Perpendicular plate
Middle nasal concha
Middle meatus
Inferior nasal concha
Inferior meatus
Vomer
Maxilla
Frontal sinus
Crista galli
Orbital fat
Ethmoidal cells of ethmoidal sinus
Maxillary sinus
Nasal cavity
Hard palate
Oral cavity
Tongue
Mandible

(e) Frontal section through the head

What part of the ethmoid bone forms the superior part of the nasal septum? The medial walls of the orbits?

Facial Bones—Nasal, Lacrimal, Palatine, Inferior Nasal Conchae, Vomer, Maxillae, Zygomatic, and Mandible *(Figure 7.10)*

OBJECTIVE

- Identify the location and surface features of the following bones: nasal, lacrimal, palatine, inferior nasal conchae, vomer, maxillae, zygomatic, and mandible.

The shape of the face changes dramatically during the first two years after birth. The brain and cranial bones expand, the first set of teeth form and erupt (emerge), and the paranasal sinuses increase in size. Growth of the face ceases at about 16 years of age. The 14 facial bones include two nasal bones, two maxillae (or maxillas), two zygomatic bones, the mandible, two lacrimal bones, two palatine bones, two inferior nasal conchae, and the vomer.

Nasal Bones

The paired **nasal bones** are small, flattened, rectangular-shaped bones that form the bridge of the nose (see Figure 7.3a). These small bones protect the upper entry to the nasal cavity and provide attachment for a couple of thin muscles of facial expression. For those of you who wear glasses, they are the bones that form the resting place for the bridge of the glasses. The major structural portion of the nose consists of cartilage.

Lacrimal Bones

The paired **lacrimal bones** (LAK-ri-mal; *lacrim-* = teardrops) are thin and roughly resemble a fingernail in size and shape (see Figures 7.3, 7.4b, and 7.12). These bones, the smallest bones of the face, are posterior and lateral to the nasal bones and form a part of the medial wall of each orbit. The lacrimal bones each contain a *lacrimal fossa,* a vertical tunnel formed with the maxilla, that houses the lacrimal sac, a structure that gathers tears and passes them into the nasal cavity (see Figure 7.12).

Palatine Bones

The two L-shaped **palatine bones** (PAL-a-tīn) form the posterior portion of the hard palate, part of the floor and lateral wall of the nasal cavity, and a small portion of the floors of the orbits (see Figures 7.7 and 7.12). The posterior portion of the hard palate is formed by the *horizontal plates* of the palatine bones (see Figures 7.6 and 7.7).

Inferior Nasal Conchae

The two **inferior nasal conchae,** which are inferior to the middle nasal conchae of the ethmoid bone, are separate bones, not part of the ethmoid bone (see Figures 7.3 and 7.9a). These scroll-like bones form a part of the inferior lateral wall of the nasal cavity and project into the nasal cavity. All three pairs of nasal conchae (superior, middle, and inferior) increase the surface area of the nasal cavity and help swirl and filter air before it passes into the lungs. However, only the superior nasal conchae of the ethmoid bone are involved in the sense of smell.

Vomer

The **vomer** (VŌ-mer = plowshare) is a roughly triangular bone on the floor of the nasal cavity that articulates superiorly with the perpendicular plate of the ethmoid bone and sphenoid bone and inferiorly with both the maxillae and palatine bones along the midline (see Figures 7.3a, 7.7, and 7.11). It forms the inferior portion of the bony nasal septum, the partition that divides the nasal cavity into right and left sides.

CLINICAL CONNECTION | *Cleft Palate and Cleft Lip*

Usually the palatine processes of the maxillary bones unite during weeks 10 to 12 of embryonic development. Failure to do so can result in one type of **cleft palate.** The condition may also involve incomplete fusion of the horizontal plates of the palatine bones (see Figure 7.7). Another form of this condition, called **cleft lip,** involves a split in the upper lip. Cleft lip and cleft palate often occur together. Depending on the extent and position of the cleft, speech and swallowing may be affected. In addition, children with cleft palate tend to have many ear infections, which can lead to hearing loss. Facial and oral surgeons recommend closure of cleft lip during the first few weeks following birth, and surgical results are excellent. Repair of cleft palate typically is completed between 12 and 18 months of age, ideally before the child begins to talk. Because the palate is important for pronouncing consonants, speech therapy may be required, and orthodontic therapy may be needed to align the teeth. Recent research strongly suggests that supplementation with folic acid (one of the B vitamins) during early pregnancy decreases the incidence of cleft palate and cleft lip. The mechanism behind this is not yet understood. •

Maxillae

The paired **maxillae** (mak-SIL-ē = jawbones; singular is *maxilla*) unite to form the upper jawbone. They articulate with every bone of the face except the mandible (lower jawbone) (see Figures 7.3a, 7.4b, and 7.7). The maxillae form part of the floors of the orbits, part of the lateral walls and floor of the nasal cavity, and most of the hard palate. The **hard palate** is the bony roof of the mouth, and is formed by the palatine processes of the maxillae and horizontal plates of the palatine bones. The hard palate separates the nasal cavity from the oral cavity.

Each maxilla contains a large *maxillary sinus* that empties into the nasal cavity (see Figure 7.13). The *alveolar process* (al-VĒ-ō-lar; *alveol-* = small cavity) of the maxilla is a ridgelike arch that contains the *alveoli* (sockets) for the maxillary (upper) teeth. The *palatine process* is a horizontal projection of the maxilla that forms the anterior three-quarters of the hard palate. The union and fusion of the maxillary bones normally is completed before birth. If this fusion fails, this condition is referred to as a cleft palate.

The *infraorbital foramen* (*infra-* = below; *-orbital* = orbit; see Figure 7.3a), an opening in the maxilla inferior to the orbit, allows passage of the infraorbital blood vessels and nerve, a branch of the maxillary division of the trigeminal (V) nerve. Another prominent foramen in the maxilla is the *incisive foramen* (= incisor teeth) just posterior to the incisor teeth (see Figure 7.7). It transmits branches of the greater palatine blood vessels and nasopalatine nerve. A final structure associated with the maxilla and sphenoid bone is the *inferior orbital fissure,* located between the greater wing of the sphenoid and the maxilla (see Figure 7.12).

Zygomatic Bones

The two **zygomatic bones** (*zygo-* = yokelike), commonly called cheekbones, form the prominences of the cheeks and part of the lateral wall and floor of each orbit (see Figure 7.12). They articulate with the frontal, maxilla, sphenoid, and temporal bones.

EXHIBIT 7.G CONTINUES ▶

The *temporal process* of the zygomatic bone projects posteriorly and articulates with the zygomatic process of the temporal bone to form the *zygomatic arch* (see Figure 7.4b).

Mandible

The **mandible** (*mand-* = to chew), or lower jawbone, is the largest, strongest facial bone (Figure 7.10). It is the only movable skull bone (other than the auditory ossicles, the small bones of the ear). In the lateral view, you can see that the mandible consists of a curved, horizontal portion, the *body,* and two perpendicular portions, the *rami* (RĀ-mī = branches; singular is *ramus*). The *angle* of the mandible is the area where each *ramus* meets the body. Each ramus has a posterior *condylar process* (KON-di-lar) that articulates with the mandibular fossa and articular tubercle of the temporal bone (see Figure 7.4b) to form the **temporomandibular joint (TMJ),** and an anterior *coronoid process* (KOR-ō-noyd) to which the temporalis muscle attaches. The depression between the coronoid and condylar processes is called the *mandibular notch.* The *alveolar process* is the ridgelike arch containing the *alveoli* (sockets) for the mandibular (lower) teeth.

The *mental foramen* (*ment-* = chin) is approximately inferior to the second premolar tooth. It is near this foramen that dentists reach the mental nerve when injecting anesthetics. Another foramen associated with the mandible is the *mandibular foramen* on the medial surface of each ramus, another site often used by dentists to inject anesthetics. The mandibular foramen is the beginning of the *mandibular canal,* which runs obliquely in the ramus and anteriorly to the body. Through the canal pass the inferior alveolar nerves and blood vessels, which are distributed to the mandibular teeth.

CLINICAL CONNECTION | *Temporomandibular Joint Syndrome*

One problem associated with the temporomandibular joint is **temporomandibular joint (TMJ) syndrome.** It is characterized by dull pain around the ear, tenderness of the jaw muscles, a clicking or popping noise when opening or closing the mouth, limited or abnormal opening of the mouth, headache, tooth sensitivity, and abnormal wearing of the teeth. TMJ syndrome can be caused by improperly aligned teeth, grinding or clenching the teeth, trauma to the head and neck, or arthritis. Treatments include application of moist heat or ice, limiting the diet to soft foods, administration of pain relievers such as aspirin, muscle retraining, use of a splint or bite plate to reduce clenching and teeth grinding (especially when worn at night), adjustment or reshaping of the teeth (orthodontic treatment), and surgery. •

✓CHECKPOINT

Which bones form the hard palate? Which bones form the nasal septum?

Figure 7.10 Mandible.

The mandible is the largest and strongest facial bone.

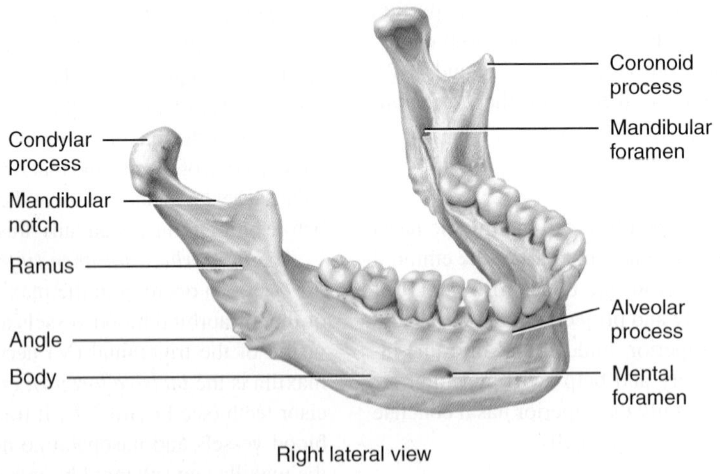

What is the distinctive functional feature of the mandible among all the skull bones?

General Features and Functions

Besides forming the large cranial cavity, the skull also forms several smaller cavities, including the nasal cavity and orbits (eye sockets), which open to the exterior. Certain skull bones also contain cavities called paranasal sinuses that are lined with mucous membranes and open into the nasal cavity. Also within the skull are small middle ear cavities in the temporal bones that house the structures that are involved in hearing and equilibrium (balance).

Other than the auditory ossicles (tiny bones involved in hearing), which are located within the temporal bones, the mandible is the only movable bone of the skull. Joints called sutures attach most of the skull bones together and are especially noticeable on the outer surface of the skull.

The skull has many surface markings, such as foramina (rounded passageways) and fissures (slit-like openings) through which blood vessels and nerves pass. You will learn the names of important skull bone surface markings as we describe each bone.

In addition to protecting the brain, the cranial bones stabilize the positions of the brain, blood vessels, lymphatic vessels, and nerves through the attachment of their inner surfaces to meninges (membranes). The outer surfaces of cranial bones provide large areas of attachment for muscles that move various parts of the head. The bones also provide attachment for some muscles that produce facial expressions such as the frown of concentration you wear when studying this book. The facial bones form the framework of the face and provide support for the entrances to the digestive and respiratory systems. Together, the cranial and facial bones protect and support the delicate special sense organs for vision, taste, smell, hearing, and equilibrium (balance).

Nasal Septum

The nasal cavity is a space inside the skull that is divided into right and left sides by a vertical partition called the **nasal septum,** which consists of bone and cartilage. The three components of the nasal septum are the vomer, septal cartilage, and the perpendicular plate of the ethmoid bone (Figure 7.11). The anterior border of the vomer articulates with the septal cartilage, which is hyaline cartilage, to form the anterior portion of the septum. The superior border of the vomer articulates with the perpendicular plate of the ethmoid bone to form the remainder of the nasal septum. The term "broken nose," in most cases, refers to damage to the septal cartilage rather than the nasal bones themselves.

CLINICAL CONNECTION | *Deviated Nasal Septum*

A **deviated nasal septum** is one that does not run along the midline of the nasal cavity. It deviates (bends) to one side. A blow to the nose can easily damage, or break, this delicate septum of bone and displace and damage the cartilage. Often, when a broken nasal septum heals, the bones and cartilage deviate to one side or the other. This deviated septum can block airflow into the constricted side of the nose, making it difficult to breathe through that half of the nasal cavity. The deviation usually occurs at the junction of the

Figure 7.11 Nasal septum.

The structures that form the nasal septum are the perpendicular plate of the ethmoid bone, the vomer, and septal cartilage.

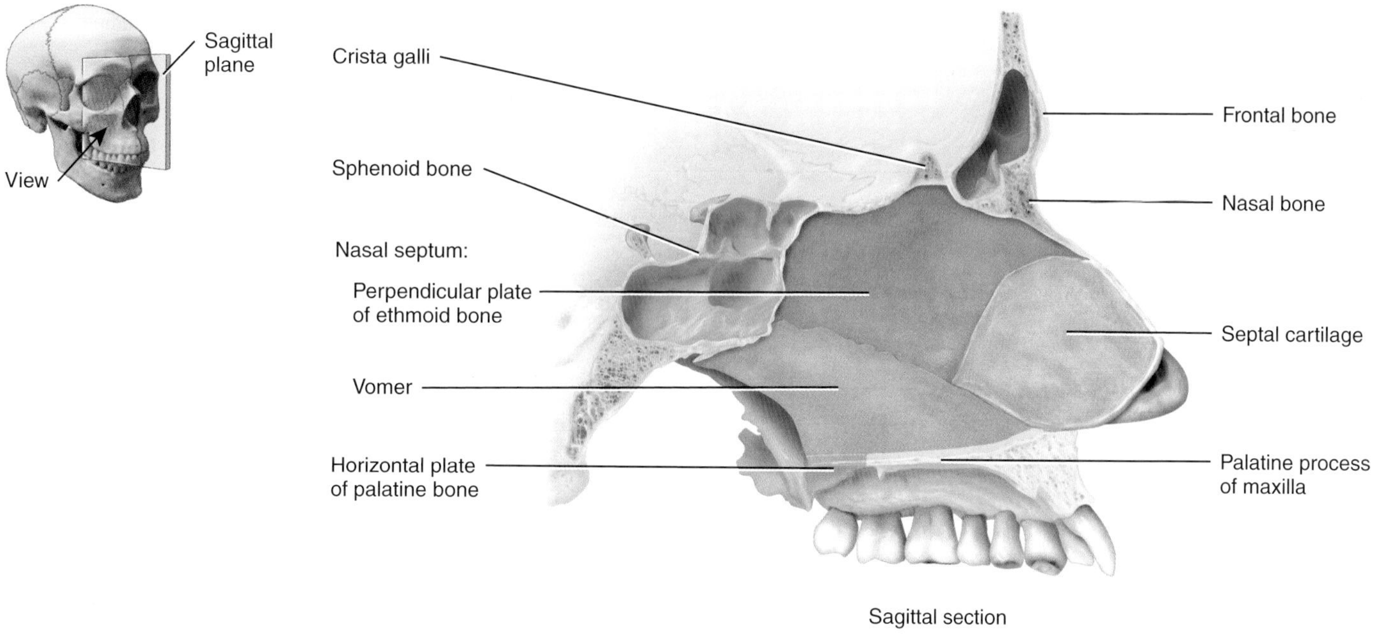
Sagittal plane
View
Crista galli
Sphenoid bone
Nasal septum:
Perpendicular plate of ethmoid bone
Vomer
Horizontal plate of palatine bone
Frontal bone
Nasal bone
Septal cartilage
Palatine process of maxilla
Sagittal section

What is the function of the nasal septum?

vomer bone with the septal cartilage. Septal deviations may also occur due to developmental abnormality. If the deviation is severe, it may block the nasal passageway entirely. Even a partial blockage may lead to infection. If inflammation occurs, it may cause nasal congestion, blockage of the paranasal sinus openings, chronic sinusitis, headache, and nosebleeds. The condition usually can be corrected or improved surgically. •

Orbits

Seven bones of the skull join to form each **orbit** (eye socket) or *orbital cavity,* which contains the eyeball and associated structures (Figure 7.12). The three cranial bones of the orbit are the frontal, sphenoid, and ethmoid; the four facial bones are the palatine, zygomatic, lacrimal, and maxilla. Each pyramid-shaped orbit has four regions that converge posteriorly:

1. Parts of the frontal and sphenoid bones comprise the *roof* of the orbit.
2. Parts of the zygomatic and sphenoid bones form the *lateral wall* of the orbit.
3. Parts of the maxilla, zygomatic, and palatine bones make up the *floor* of the orbit.
4. Parts of the maxilla, lacrimal, ethmoid, and sphenoid bones form the *medial wall* of the orbit.

Associated with each orbit are five openings:

1. The *optic foramen (canal)* is at the junction of the roof and medial wall.
2. The *superior orbital fissure* is at the superior lateral angle of the apex.
3. The *inferior orbital fissure* is at the junction of the lateral wall and floor.
4. The *supraorbital foramen* is on the medial side of the supraorbital margin of the frontal bone.
5. The *lacrimal fossa* is in the lacrimal bone.

Foramina

We mentioned most of the **foramina** (openings for blood vessels, nerves, or ligaments) of the skull in the descriptions of the cranial and facial bones that they penetrate. As preparation for studying other systems of the body, especially the nervous and cardiovascular systems, these foramina and the structures passing through them are listed in Table 7.3. For your convenience and for future reference, the foramina are listed alphabetically.

Figure 7.12 Details of the orbit (eye socket).

The orbit is a pyramid-shaped structure that contains the eyeball and associated structures.

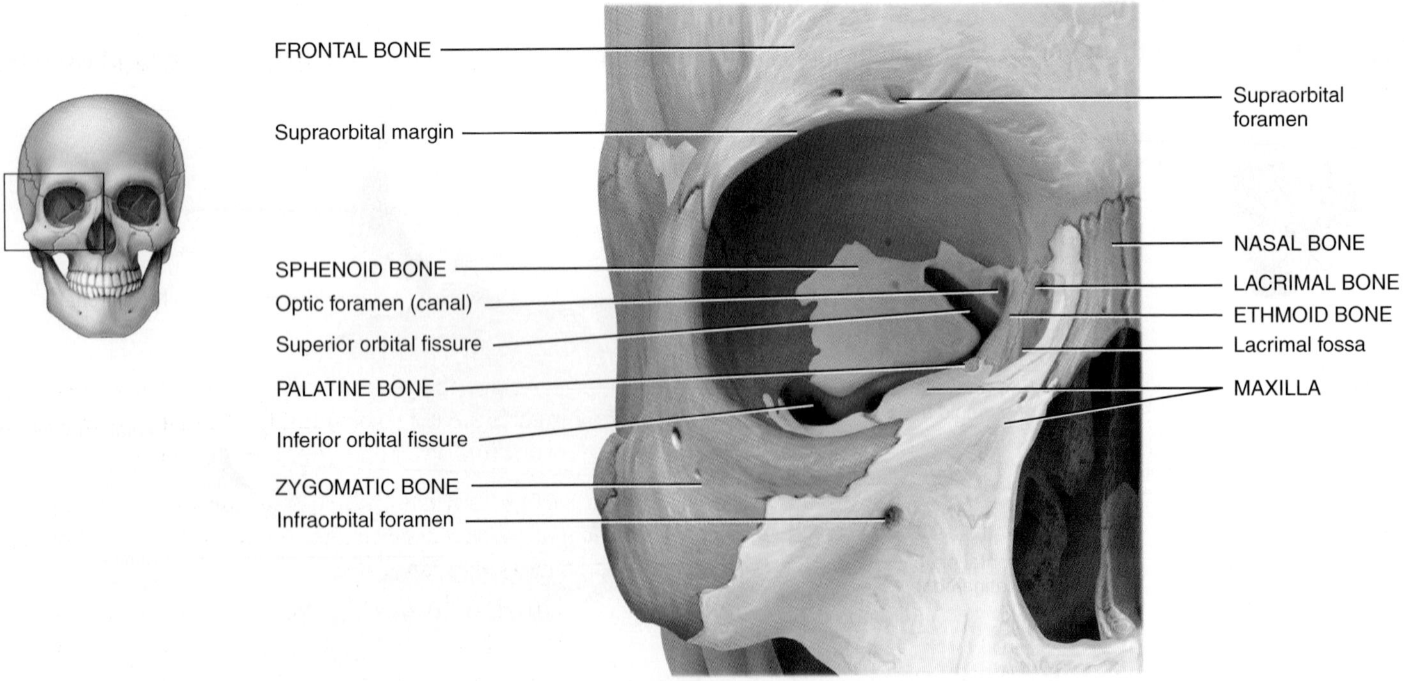

Anterior view showing the bones of the right orbit

Which seven bones form the orbit?

TABLE 7.3

Principal Foramina of the Skull

FORAMEN	LOCATION	STRUCTURES PASSING THROUGH*
Carotid (relating to carotid artery in neck)	Petrous portion of temporal bone (Figure 7.7).	Internal carotid artery, sympathetic nerves for eyes.
Hypoglossal (*hypo-* = under; *-glossus* = tongue)	Superior to base of occipital condyles (Figure 7.8a).	Cranial nerve XII (hypoglossal), branch of ascending pharyngeal artery.
Infraorbital (*infra-* = below)	Inferior to orbit in maxilla (Figure 7.12).	Infraorbital nerve and blood vessels, branch of maxillary division of cranial nerve V (trigeminal).
Jugular (*jugul-* = throat)	Posterior to carotid canal between petrous portion of temporal bone and occipital bone (Figure 7.8a).	Internal jugular vein; cranial nerves IX (glossopharyngeal), X (vagus), XI (accessory).
Lacerum (*lacerum* = lacerated)	Bounded anteriorly by sphenoid bone, posteriorly by petrous portion of temporal bone, medially by sphenoid and occipital bones (Figure 7.8a).	Branch of ascending pharyngeal artery.
Magnum (= large)	Occipital bone (Figure 7.7).	Medulla oblongata and its membranes (meninges), cranial nerve XI (accessory), vertebral and spinal arteries.
Mandibular (*mand-* = to chew)	Medial surface of ramus of mandible (Figure 7.10).	Inferior alveolar nerve and blood vessels.
Mastoid (= breast-shaped)	Posterior border of mastoid process of temporal bone (Figure 7.7).	Emissary vein to transverse sinus, branch of occipital artery to dura mater.
Mental (*ment-* = chin)	Inferior to second premolar tooth in mandible (Figure 7.10).	Mental nerve and vessels.
Olfactory (*olfact-* = to smell)	Cribriform plate of ethmoid bone (Figure 7.8a).	Cranial nerve I (olfactory).
Optic (= eye)	Between superior and inferior portions of small wing of sphenoid bone (Figure 7.12).	Cranial nerve II (optic), ophthalmic artery.
Ovale (= oval)	Greater wing of sphenoid bone (Figure 7.8a).	Mandibular branch of cranial nerve V (trigeminal).
Rotundum (= round)	Junction of anterior and medial parts of sphenoid bone (Figure 7.8a, b).	Maxillary branch of cranial nerve V (trigeminal).
Stylomastoid (*stylo-* = stake or pole)	Between styloid and mastoid processes of temporal bone (Figure 7.7).	Cranial nerve VII (facial), stylomastoid artery.
Supraorbital (*supra-* = above)	Supraorbital margin of orbit in frontal bone (Figure 7.12).	Supraorbital nerve and artery.

*The cranial nerves listed here are described in Table 14.4.

Unique Features of the Skull

The skull exhibits several unique features not seen in other bones of the body. These include sutures, paranasal sinuses, and fontanels.

Sutures

A **suture** (SOO-chur = seam) is an immovable joint in most cases in an adult skull that holds most skull bones together. Sutures in the skulls of infants and children, however, often are movable and function as important growth centers in the developing skull. The names of many sutures reflect the bones they unite. For example, the frontozygomatic suture is between the frontal bone and the zygomatic bone. Similarly, the sphenoparietal suture is between the sphenoid bone and the parietal bone. In other cases, however, the names of sutures are not so obvious. Of the many sutures found in the skull, we will identify only four prominent ones:

1. The **coronal suture** (KO-rō-nal; *coron-* = relating to the frontal or coronal plane) unites the frontal bone and both parietal bones (see Figure 7.4b).
2. The **sagittal suture** (SAJ-i-tal; *sagitt-* = arrow) unites the two parietal bones on the superior midline of the skull (see Figure 7.4a). The sagittal suture is so named because in the infant, before the bones of the skull are firmly united, the suture and the fontanels (soft spots) associated with it resemble an arrow.

3. The **lambdoid suture** (LAM-doyd) unites the two parietal bones to the occipital bone. This suture is so named because of its resemblance to the capital Greek letter lambda (Λ), as can be seen in Figure 7.6 (with the help of a little imagination). Sutural bones may occur within the sagittal and lambdoid sutures.
4. The two **squamous sutures** (SKWĀ-mus; *squam-* = flat, like the flat overlapping scales of a snake) unite the parietal and temporal bones on the lateral aspects of the skull (see Figure 7.4b).

Paranasal Sinuses

The **paranasal sinuses** (par′-a-NĀ-zal SĪ-nus-ez; *para-* = beside) are cavities within certain cranial and facial bones near the nasal cavity. They are most evident in a sagittal section of the skull (Figure 7.13). The paranasal sinuses are lined with mucous membranes that are continuous with the lining of the nasal cavity. Secretions produced by the mucous membranes of the paranasal sinuses drain into the lateral wall of the nasal cavity. Paranasal

Figure 7.13 Paranasal sinuses projected to the surface.

Paranasal sinuses are mucous membrane–lined spaces in the frontal, sphenoid, ethmoid, and maxillary bones that connect to the nasal cavity.

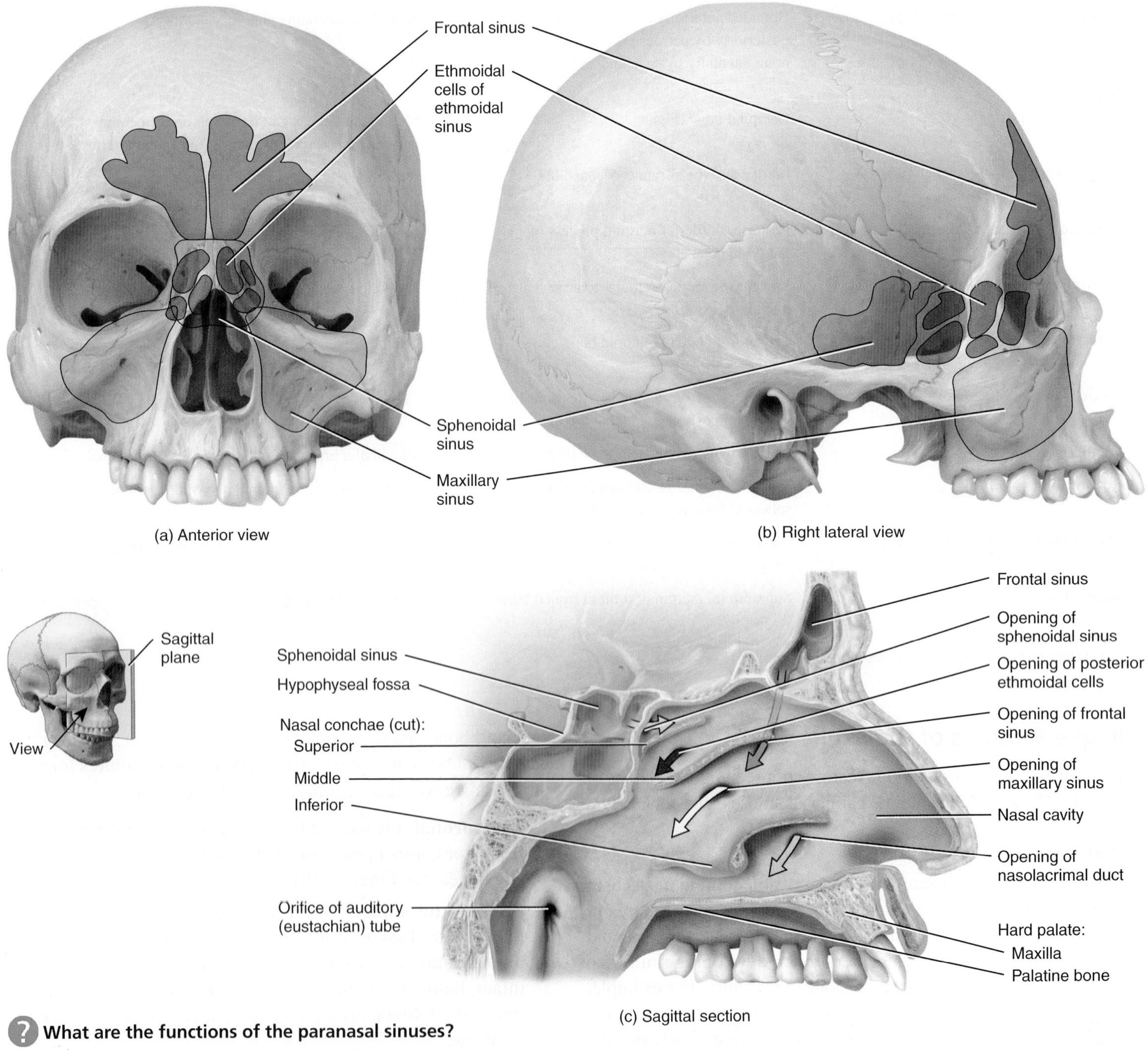

What are the functions of the paranasal sinuses?

sinuses are quite small or absent at birth, but increase in size during two periods of facial enlargement—during the eruption of the teeth and at the onset of puberty. They arise as outgrowths of the nasal mucosa that project into the surrounding bones. Skull bones containing the paranasal sinuses are the frontal, sphenoid, ethmoid, and maxillae. The paranasal sinuses allow the skull to increase in size without a change in the mass (weight) of the bone. The paranasal sinuses increase the surface area of the nasal mucosa, thus increasing the production of mucus to help moisten and cleanse inhaled air. In addition, the paranasal sinuses serve as resonating (echo) chambers within the skull that intensify and prolong sounds, thereby enhancing the quality of the voice. The influence of the paranasal sinuses on your voice becomes obvious when you have a cold; the passageways through which sound travels into and out of the paranasal sinuses become blocked by excess mucus production, changing the quality of your voice.

CLINICAL CONNECTION | *Sinusitis*

Sinusitis (sīn-ū-SĪ-tis) is an inflammation of the mucous membrane of one or more paranasal sinuses. It may be caused by a microbial infection (virus, bacterium, or fungus), allergic reactions, nasal polyps, or a severely deviated nasal septum. If the inflammation or an obstruction blocks the drainage of mucus into the nasal cavity, fluid pressure builds up in the paranasal sinuses, and a sinus headache may develop. Other symptoms may include nasal congestion, inability to smell, fever, and cough. Treatment options include decongestant sprays or drops, oral decongestants, nasal corticosteroids, antibiotics, analgesics to relieve pain, warm compresses, and surgery. •

Fontanels

The skull of a developing embryo consists of cartilage and mesenchyme arranged in thin plates around the developing brain. Gradually, ossification occurs, and bone slowly replaces most of the cartilage and mesenchyme. At birth, bone ossification is incomplete, and the mesenchyme-filled spaces become dense connective tissue regions between incompletely developed cranial bones called **fontanels** (fon-ta-NELZ = little fountains), commonly called "soft spots" (Figure 7.14). Fontanels are areas where unossified mesenchyme develops into the dense connective tissues of the skull. As bone formation continues after birth, the fontanels are eventually replaced with bone by intramembranous ossification, and the thin collagenous connective tissue junctions that remain between neighboring bones become the sutures. Functionally, the fontanels serve as spacers for the growth of neighboring skull bones and provide some flexibility to the fetal skull, allowing the skull to change shape as it passes through the birth canal and later permitting rapid growth of the brain during infancy. Although an infant may have many fontanels at birth, the form and location of six are fairly constant:

- The unpaired **anterior fontanel,** the largest fontanel, is located at the midline among the two parietal bones and the frontal bone, and is roughly diamond-shaped. It usually closes 18 to 24 months after birth.
- The unpaired **posterior fontanel** is located at the midline among the two parietal bones and the occipital bone. Because it is much smaller than the anterior fontanel, it generally closes about 2 months after birth.

Figure 7.14 Fontanels at birth.

6 Fontanels are mesenchyme-filled spaces between cranial bones that are present at birth.

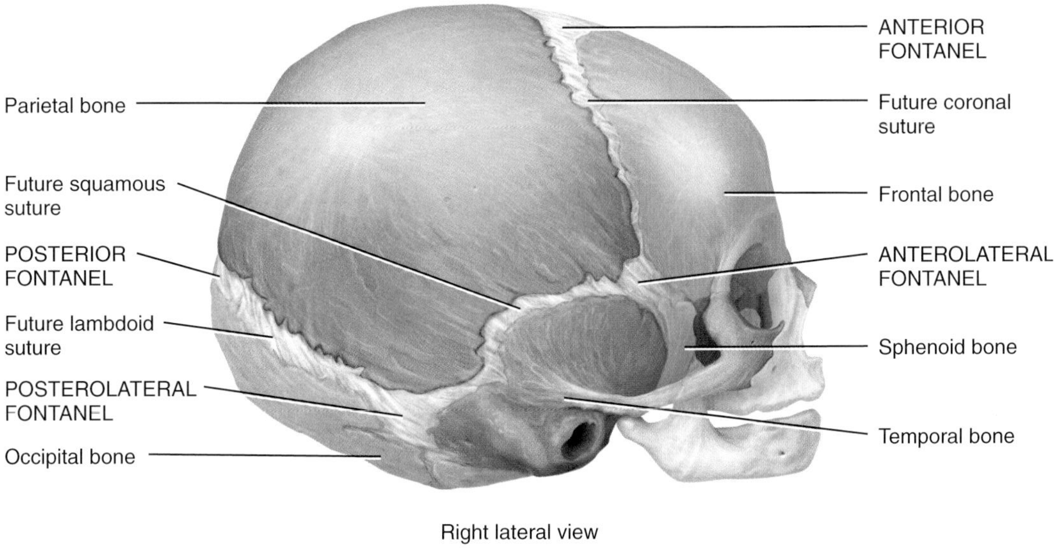

? Which fontanel is bordered by four different skull bones?

- The paired **anterolateral fontanels,** located laterally among the frontal, parietal, temporal, and sphenoid bones, are small and irregular in shape. Normally, they close about 3 months after birth.
- The paired **posterolateral fontanels,** located laterally among the parietal, occipital, and temporal bones, are irregularly shaped. They begin to close 1 to 2 months after birth, but closure is generally not complete until 12 months.

The amount of closure in fontanels helps a physician gauge the degree of brain development. In addition, the anterior fontanel serves as a landmark for withdrawal of blood for analysis from the superior sagittal sinus (a large midline vein within the covering tissues that surround the brain). (See Figure 21.24.)

✔CHECKPOINT

4. Describe the cavities within the skull and the nasal septum.
5. Which foramina and fissures are associated with the orbit?
6. What structures make up the nasal septum?
7. Define the following: foramen, suture, paranasal sinus, and fontanel.

7.5 HYOID BONE

OBJECTIVE

- Describe the relationship of the hyoid bone to the skull.

The single **hyoid bone** (= U-shaped) is a unique component of the axial skeleton because it does not articulate with any other bone. Rather, it is suspended from the styloid processes of the temporal bones by ligaments and muscles. Located in the anterior neck between the mandible and larynx (Figure 7.15a), the hyoid bone supports the tongue, providing attachment sites for some tongue muscles and for muscles of the neck and pharynx. The hyoid bone consists of a horizontal *body* and paired projections called the *lesser horns* and the *greater horns* (Figure 7.15b, c). Muscles and ligaments attach to the body and these paired projections.

The hyoid bone and the cartilages of the larynx and trachea are often fractured during strangulation. As a result, they are carefully examined at autopsy when manual strangulation is a suspected cause of death.

✔CHECKPOINT

8. What are the functions of the hyoid bone?

Figure 7.15 Hyoid bone.

The hyoid bone supports the tongue, providing attachment sites for muscles of the tongue, neck, and pharynx.

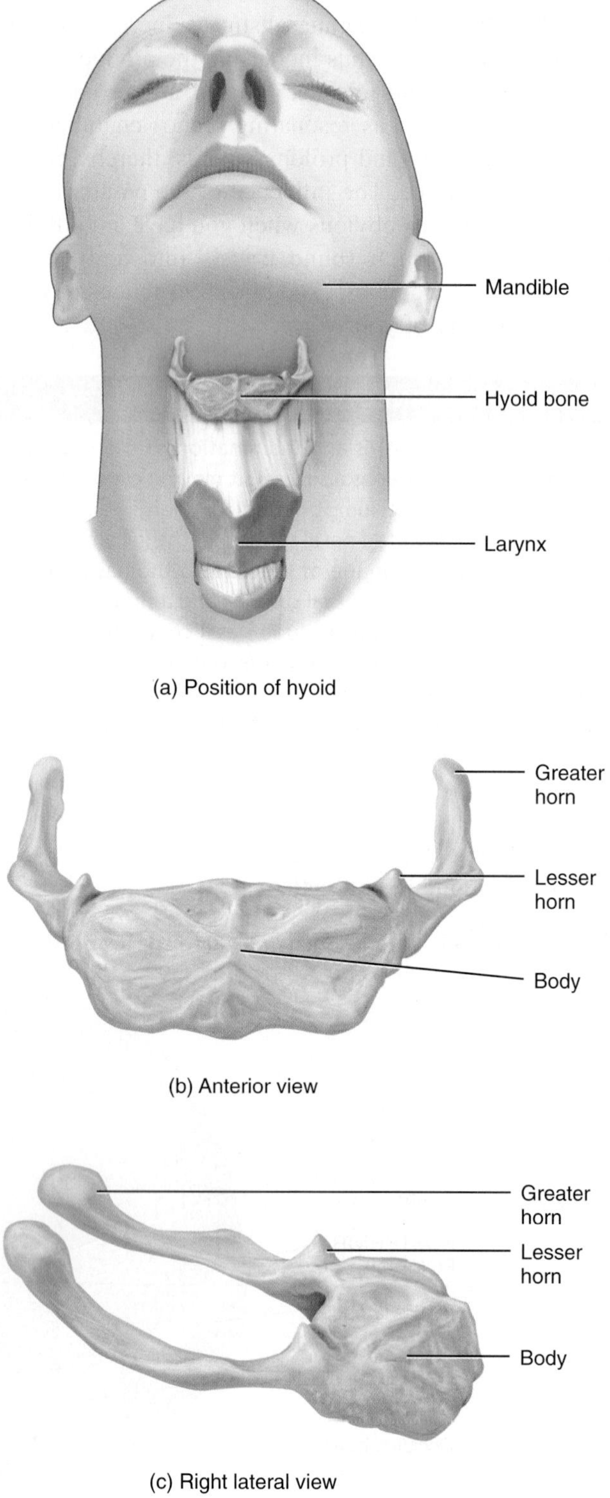
Mandible
Hyoid bone
Larynx
(a) Position of hyoid
Greater horn
Lesser horn
Body
(b) Anterior view
Greater horn
Lesser horn
Body
(c) Right lateral view

In what way is the hyoid bone different from all the other bones of the axial skeleton?

7.6 VERTEBRAL COLUMN

OBJECTIVES

- Identify the regions and normal curves of the vertebral column.
- Describe the structural and functional features of the bones in various regions of the vertebral column.

The **vertebral column,** also called the *spine, backbone,* or *spinal column,* makes up about two-fifths of your total height and is composed of a series of bones called **vertebrae** (VER-te-brē; singular is *vertebra*). The vertebral column, the sternum, and the ribs form the skeleton of the trunk of the body. The vertebral column consists of bone and connective tissue; the spinal cord that it surrounds and protects consists of nervous and connective tissues. At about 71 cm (28 in.) in an average adult male and about 61 cm (24 in.) in an average adult female, the vertebral column functions as a strong, flexible rod with elements that can move forward, backward, and sideways, and rotate. In addition to enclosing and protecting the spinal cord, it supports the head and serves as a point of attachment for the ribs, pelvic girdle, and muscles of the back and upper limbs.

The total number of vertebrae during early development is 33. As a child grows, several vertebrae in the sacral and coccygeal regions fuse. As a result, the adult vertebral column typically contains 26 vertebrae (Figure 7.16a). These are distributed as follows:

- 7 **cervical vertebrae** (*cervic-* = neck) are in the neck region.
- 12 **thoracic vertebrae** (*thorax* = chest) are posterior to the thoracic cavity.

Figure 7.16 Vertebral column. The numbers in parentheses in part (a) indicate the number of vertebrae in each region. In part (d), the relative size of the disc has been enlarged for emphasis.

The adult vertebral column typically contains 26 vertebrae.

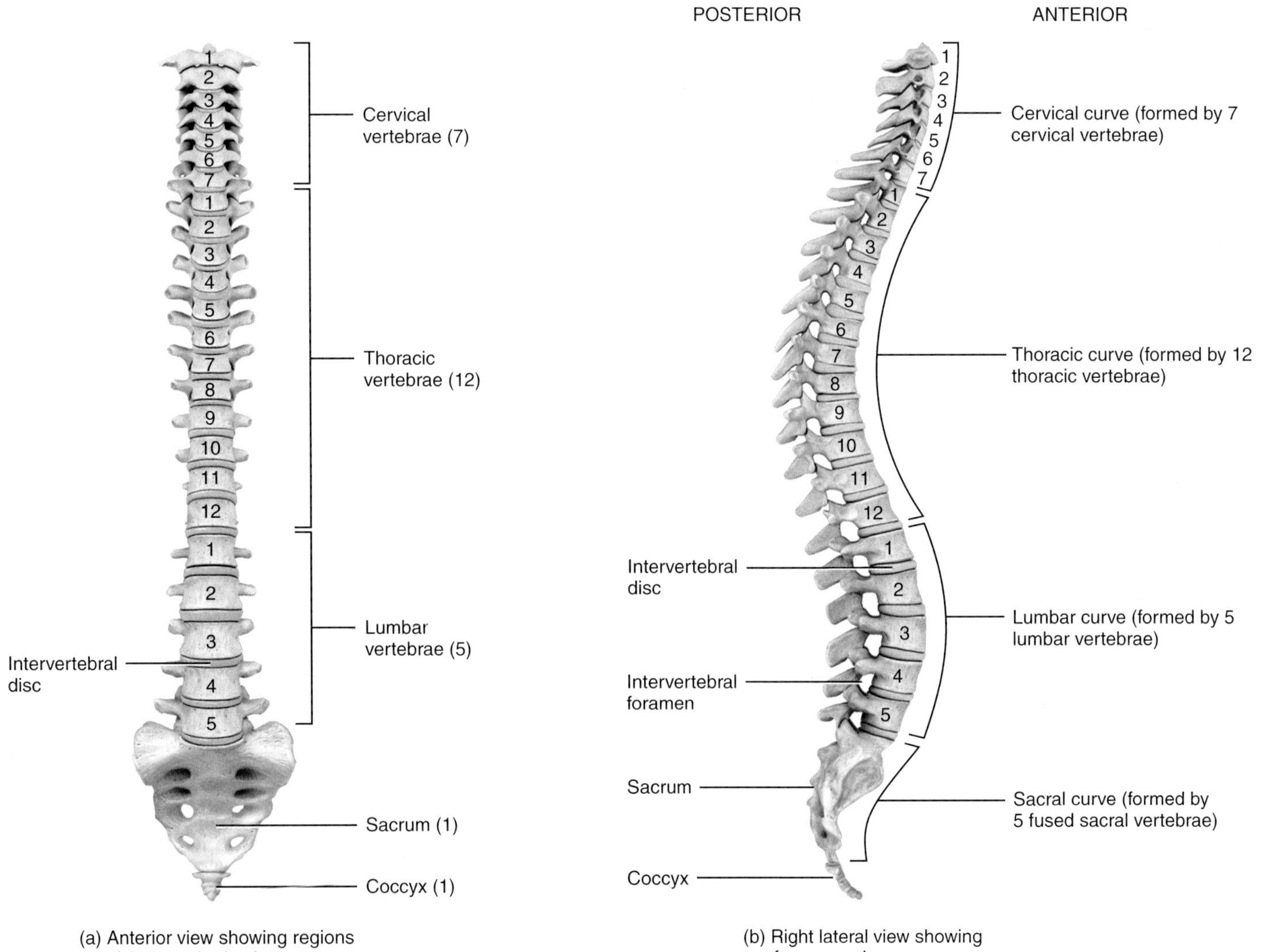

POSTERIOR
ANTERIOR
Cervical vertebrae (7)
Thoracic vertebrae (12)
Lumbar vertebrae (5)
Intervertebral disc
Sacrum (1)
Coccyx (1)
Cervical curve (formed by 7 cervical vertebrae)
Thoracic curve (formed by 12 thoracic vertebrae)
Lumbar curve (formed by 5 lumbar vertebrae)
Sacral curve (formed by 5 fused sacral vertebrae)
Intervertebral disc
Intervertebral foramen
Sacrum
Coccyx

(a) Anterior view showing regions of the vertebral column

(b) Right lateral view showing four normal curves

FIGURE 7.16 CONTINUES

FIGURE 7.16 CONTINUED

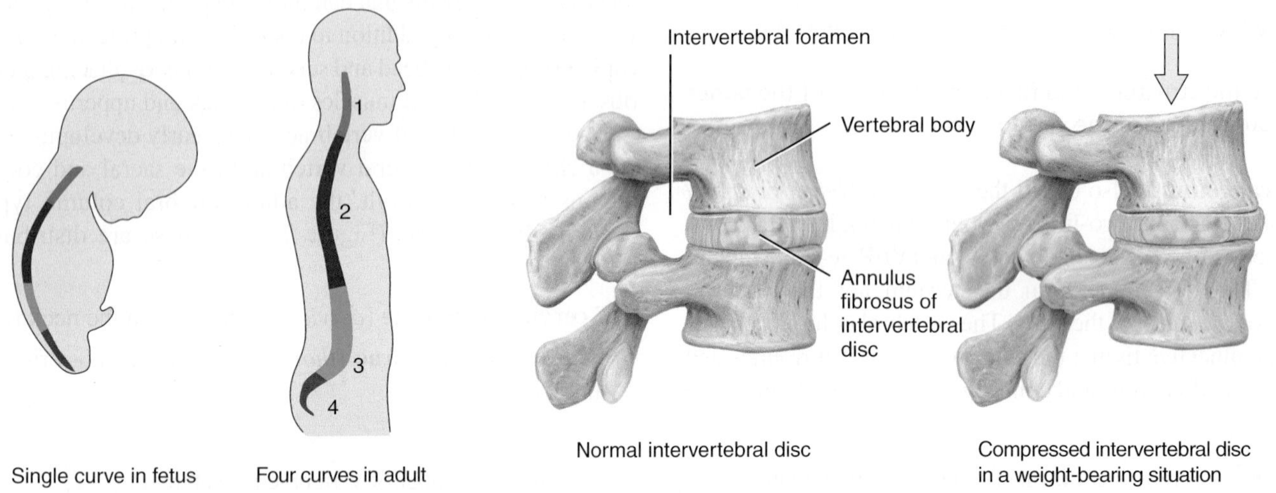
1
2
3
4
Single curve in fetus
Four curves in adult
Intervertebral foramen
Vertebral body
Annulus fibrosus of intervertebral disc
Normal intervertebral disc
Compressed intervertebral disc in a weight-bearing situation

(c) Fetal and adult curves

(d) Intervertebral disc

Which curves of the adult vertebral column are concave (relative to the anterior side of the body)?

- 5 **lumbar vertebrae** (*lumb-* = loin) support the lower back.
- 1 **sacrum** (SĀ-krum = sacred bone) consists of five fused **sacral vertebrae.**
- 1 **coccyx** (KOK-siks = cuckoo, because the shape resembles the bill of a cuckoo bird) usually consists of four fused **coccygeal vertebrae** (kok-SIJ-ē-al).

The cervical, thoracic, and lumbar vertebrae are movable, but the sacrum and coccyx are not. We will discuss each of these regions in detail shortly.

Normal Curves of the Vertebral Column

When viewed from the anterior or posterior, a normal adult vertebral column appears straight. But when viewed from the side, it shows four slight bends called **normal curves** (Figure 7.16b). Relative to the front of the body, the *cervical* and *lumbar curves* are convex (bulging out); the *thoracic* and *sacral curves* are concave (cupping in). The curves of the vertebral column increase its strength, help maintain balance in the upright position, absorb shocks during walking, and help protect the vertebrae from fracture.

The fetus has a single anteriorly concave curve throughout the length of the entire vertebral column (Figure 7.16c). At about the third month after birth, when an infant begins to hold its head erect, the anteriorly convex cervical curve develops. Later, when the child sits up, stands, and walks, the anteriorly convex lumbar curve develops. The thoracic and sacral curves are called *primary curves* because they retain the original curvature of the embryonic vertebral column. The cervical and lumbar curves are known as *secondary curves* because they begin to form later, several months after birth. All curves are fully developed by age 10. However, secondary curves may be progressively lost in old age.

Various conditions may exaggerate the normal curves of the vertebral column, or the column may acquire a lateral bend, resulting in **abnormal curves** of the vertebral column. Three such abnormal curves—kyphosis, lordosis, and scoliosis—are described in Disorders: Homeostatic Imbalances at the end of this chapter.

Intervertebral Discs

Intervertebral discs (in′-ter-VER-te-bral; *inter-* = between) are found between the bodies of adjacent vertebrae from the second cervical vertebra to the sacrum (Figure 7.16d) and account for about 25% of the height of the vertebral column. Each disc has an outer fibrous ring consisting of fibrocartilage called the *annulus fibrosus* (*annulus* = ringlike) and an inner soft, pulpy, highly elastic substance called the *nucleus pulposus* (*pulposus* = pulplike). The discs form strong joints, permit various movements of the vertebral column, and absorb vertical shock. Under compression, they flatten and broaden.

During the course of a day the discs compress and lose water from their cartilage so that we are a bit shorter at night. While we are sleeping there is less compression and rehydration occurs, so that we are taller when we awaken in the morning. With age, the nucleus pulposus hardens and becomes less elastic. Decrease in vertebral height with age results from bone loss in the vertebral bodies and not a decrease in thickness of the intervertebral discs.

Since intervertebral discs are avascular, the annulus fibrosus and nucleus pulposus rely on blood vessels from the bodies of vertebrae to obtain oxygen and nutrients and remove wastes.

Certain stretching exercises, such as yoga, decompress discs and increase general blood circulation, both of which speed up the uptake of oxygen and nutrients by discs and the removal of wastes.

Parts of a Typical Vertebra

Vertebrae in different regions of the spinal column vary in size, shape, and detail, but they are similar enough that we can discuss the structures (and the functions) of a typical vertebra (Figure 7.17). Vertebrae typically consist of a vertebral body, a vertebral arch, and several processes.

Vertebral Body

The **vertebral body,** the thick, disc-shaped anterior portion, is the weight-bearing part of a vertebra. Its superior and inferior surfaces are roughened for the attachment of cartilaginous intervertebral discs. The anterior and lateral surfaces contain nutrient foramina, openings through which blood vessels deliver nutrients and oxygen and remove carbon dioxide and wastes from bone tissue.

Vertebral Arch

Two short, thick processes, the *pedicles* (PED-i-kuls = little feet), project posteriorly from the vertebral body and then unite with the flat *laminae* (LAM-i-nē = thin layers) to form the **vertebral arch.** The vertebral arch extends posteriorly from the body of the vertebra; together, the vertebral body and the vertebral arch surround the spinal cord by forming the *vertebral foramen.* The vertebral foramen contains the spinal cord, adipose tissue, areolar connective tissue, and blood vessels. Collectively, the vertebral foramina of all vertebrae form the **vertebral (spinal) canal.** The pedicles exhibit superior and inferior indentations called *vertebral notches.* When the vertebral notches are stacked on top of one another, they form an opening between adjoining vertebrae on both sides of the column. Each opening, called an *intervertebral foramen,* permits the passage of a single spinal nerve carrying information to and from the spinal cord.

Processes

Seven **processes** arise from the vertebral arch. At the point where a lamina and pedicle join, a *transverse process* extends laterally

Figure 7.17 Structure of a typical vertebra, as illustrated by a thoracic vertebra. In part b, only one spinal nerve has been included, and it has been extended beyond the intervertebral foramen for clarity.

A vertebra consists of a vertebral body, a vertebral arch, and several processes.

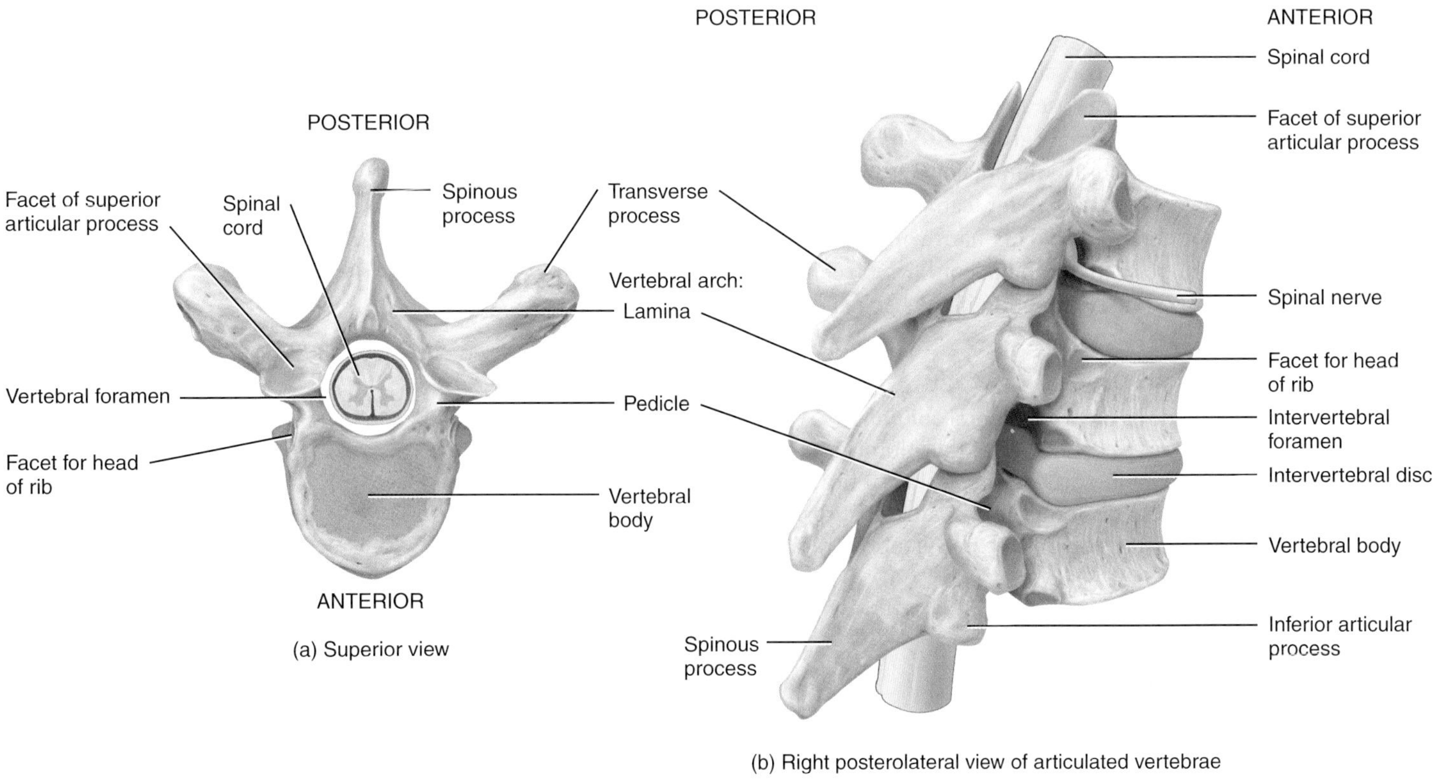

What are the functions of the vertebral and intervertebral foramina?

on each side. A single *spinous process (spine)* projects posteriorly from the junction of the laminae. These three processes serve as points of attachment for muscles. The remaining four processes form joints with other vertebrae above or below. The two *superior articular processes* of a vertebra articulate (form joints) with the two inferior articular processes of the vertebra immediately above them. In turn, the two *inferior articular processes* of that vertebra articulate with the two superior articular processes of the vertebra immediately below them, and so on. The articulating surfaces of the articular processes, which are referred to as *facets* (FAS-ets or fa-SETS = little faces), are covered with hyaline cartilage. The articulations formed between the vertebral bodies and articular facets of successive vertebrae are termed *intervertebral joints.*

Regions of the Vertebral Column

Exhibits 7.H through 7.K present the five regions of the vertebral column, beginning superiorly and moving inferiorly. The regions are the cervical, thoracic, lumbar, sacral, and coccygeal. Note that vertebrae in each region are numbered in sequence, from superior to inferior. When you actually view the bones of the vertebral column, you will notice that the transition from one region to the next is not abrupt but gradual, a feature that helps the vertebrae fit together.

Age-Related Changes in the Vertebral Column

With advancing age the vertebral column undergoes changes that are characteristic of the skeletal system in general. These changes include reduction in the mass and density of the bone along with a reduction in the collagen-to-mineral content within the bone, changes that make the bones more brittle and susceptible to damage. The articular surfaces, those surfaces where neighboring bones move against one another, lose their covering cartilage as they age; in their place rough bony growths form that lead to arthritic conditions. In the vertebral column, bony growths around the intervertebral discs, called *osteophytes,* can lead to a narrowing (stenosis) of the vertebral canal. This narrowing can lead to compression of spinal nerves and the spinal cord, which can manifest as pain and decreased muscle function in the back and lower limbs.

✔ **CHECKPOINT**

9. What are the functions of the vertebral column?
10. Describe the four curves of the vertebral column.
11. What are the three main parts of a typical vertebra?
12. What are the principal distinguishing characteristics of the bones of the various regions of the vertebral column?

7.7 THORAX

OBJECTIVE

- Identify the bones of the thorax and their functions.

The term **thorax** refers to the entire chest region. The skeletal part of the thorax, the **thoracic cage,** is a bony enclosure formed by the sternum, ribs and their costal cartilages, and the bodies of the thoracic vertebrae (see Exhibits 7.L and 7.M). The costal cartilages attach the ribs to the sternum. The thoracic cage is narrower at its superior end and broader at its inferior end and is flattened from front to back. It encloses and protects the organs in the thoracic and superior abdominal cavities, provides support for the bones of the upper limbs, and, as you will see in Chapter 23, plays a role in breathing.

✔ **CHECKPOINT**

13. What bones form the skeleton of the thorax?
14. What are the functions of the bones of the thorax?
15. How are ribs classified?

Vertebral Regions—Cervical Vertebrae *(Figure 7.18)*

OBJECTIVE

- Identify the location and surface features of the cervical vertebrae.

The bodies of the **cervical vertebrae** (C1–C7) are smaller than all other vertebrae except those that form the coccyx (Figure 7.18a). Their vertebral arches, however, are larger. All cervical vertebrae have three foramina: one vertebral foramen and two transverse foramina (Figure 7.18c). The *vertebral foramina* of cervical vertebrae are the largest in the spinal column because they house the cervical enlargement of the spinal cord. Each cervical transverse process contains a *transverse foramen* through which the vertebral artery and its accompanying vein and nerve fibers pass. The spinous processes of C2 through C6 are often *bifid*—that is, they branch into two small projections at the tips (Figure 7.18a, c).

The first two cervical vertebrae differ considerably from the others. The **atlas** (C1), named after the mythological Atlas who supported the world on his shoulders, is the first cervical vertebra inferior to the skull (Figure 7.18a, b). The atlas is a ring of bone with *anterior* and *posterior arches* and large *lateral masses*. It lacks a body and a spinous process. The superior surfaces of the lateral masses, called *superior articular facets*, are concave. They articulate with the occipital condyles of the occipital bone to form the paired *atlanto-occipital joints*. These articulations permit you to move your head to signify "yes." The inferior surfaces of the lateral masses, the *inferior articular facets,* articulate with the second cervical vertebra. The transverse processes and transverse foramina of the atlas are quite large.

The second cervical vertebra (C2), the **axis** (see Figure 7.18a, d, e), does have a vertebral body. A peglike process called the *dens* (= tooth) or *odontoid process* projects superiorly through the anterior portion of the vertebral foramen of the atlas. The dens makes a pivot on which the atlas and head rotate. This arrangement permits side-to-side movement of the head, as when you move your head to signify "no." The articulation formed between the anterior arch of the atlas and dens of the axis, and between their articular facets, is called the *atlanto-axial joint.* In some instances of trauma, the dens of the axis may be driven into the medulla oblongata of the brain. This type of injury is the usual cause of death from whiplash injuries.

The third through sixth cervical vertebrae (C3–C6), represented by the vertebra in Figure 7.18c, correspond to the structural pattern of the

Figure 7.18 **Cervical vertebrae.**

The cervical vertebrae are found in the neck region.

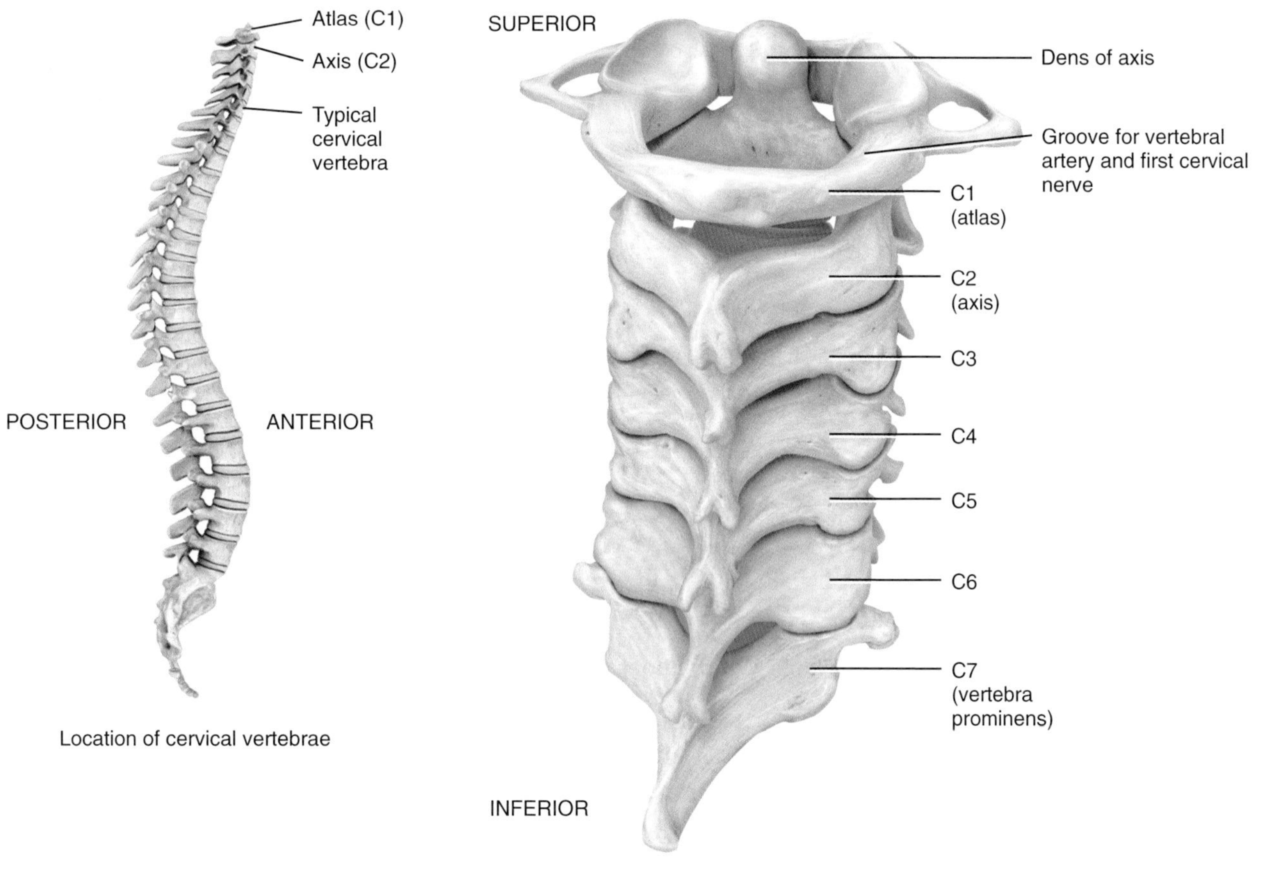

(a) Posterior view of articulated cervical vertebrae

EXHIBIT 7.H CONTINUES

typical cervical vertebra previously described. The seventh cervical vertebra (C7), called the *vertebra prominens,* is somewhat different (see Figure 7.18a). It has a large, nonbifid spinous process that may be seen and felt at the base of the neck, but otherwise is typical.

✔CHECKPOINT

How do the atlas and axis differ from the other cervical vertebrae?

FIGURE 7.18 CONTINUED

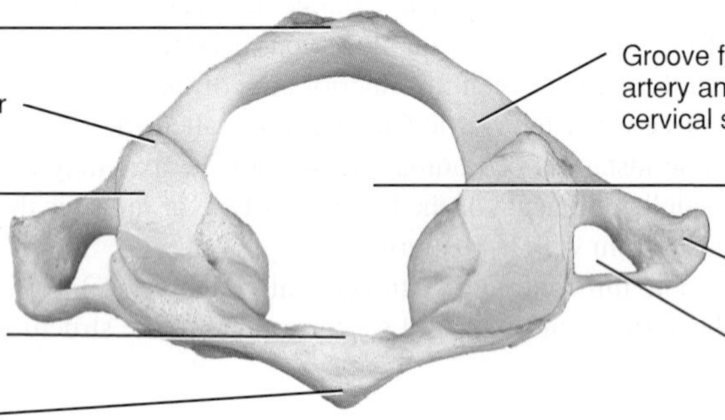

(b) Superior view of atlas (C1)

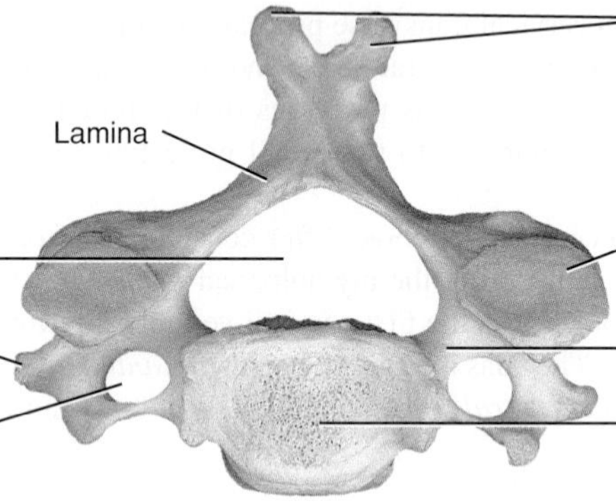

(c) Superior view of a typical cervical vertebra

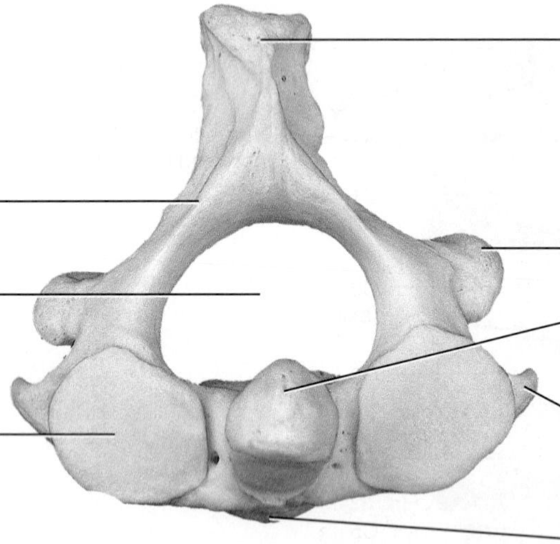

(d) Superior view of axis (C2)

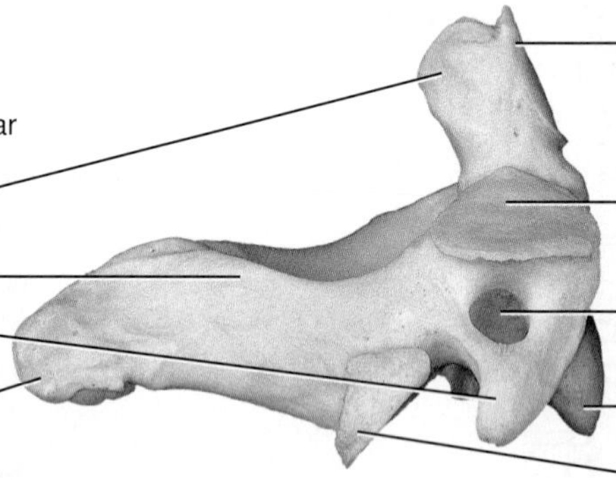

(e) Right lateral view of axis (C2)

Which joint permits you to move your head to signify "no"? Which bones are involved?

EXHIBIT 7.I Vertebral Regions—Thoracic Vertebrae *(Figure 7.19)*

OBJECTIVE

- Identify the location and surface features of the thoracic vertebrae.

Thoracic vertebrae (T1–T12; Figure 7.19) are considerably larger and stronger than cervical vertebrae. In addition, the spinous processes on T1 through T10 are long, laterally flattened, and directed inferiorly. In contrast, the spinous processes on T11 and T12 are shorter, broader, and directed more posteriorly. Compared to cervical vertebrae, thoracic vertebrae also have longer and larger transverse processes. They are easily identified by their *costal facets* (*cost-* = rib), which are articular surfaces for the ribs.

The feature of the thoracic vertebrae that distinguishes them from other vertebrae is that they articulate with the ribs. Except for T11 and T12, the transverse processes of thoracic vertebrae have costal facets that articulate with the *tubercles* of the ribs. Additionally, the vertebral bodies of thoracic vertebrae have articular surfaces that form articulations with the *heads* of the ribs (see Figure 7.23). The articular surfaces on the vertebral bodies are called either facets or demifacets. A *facet* is formed when the head of a rib articulates with the body of one vertebra. A *demifacet* is formed when the head of a rib articulates with two adjacent vertebral bodies. As you can see in Figure 7.19, on each side of the vertebral body T1 has a superior facet for the first rib and an inferior demifacet for the second rib. On each side of the vertebral body of T2–T8, there is a

Figure 7.19 Thoracic vertebrae.

The thoracic vertebrae are found in the chest region and articulate with the ribs.

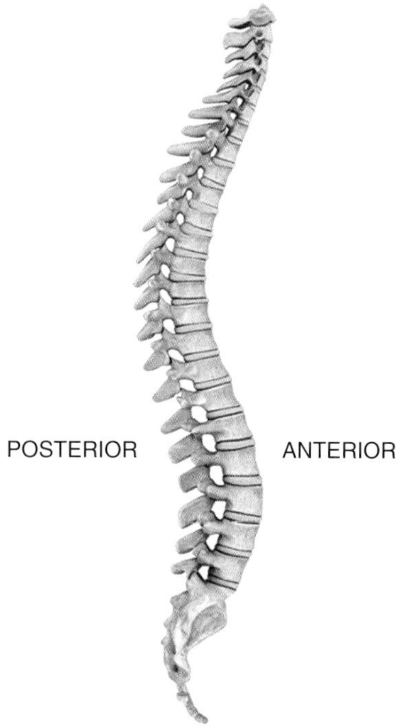
POSTERIOR
ANTERIOR

Location of thoracic vertebrae

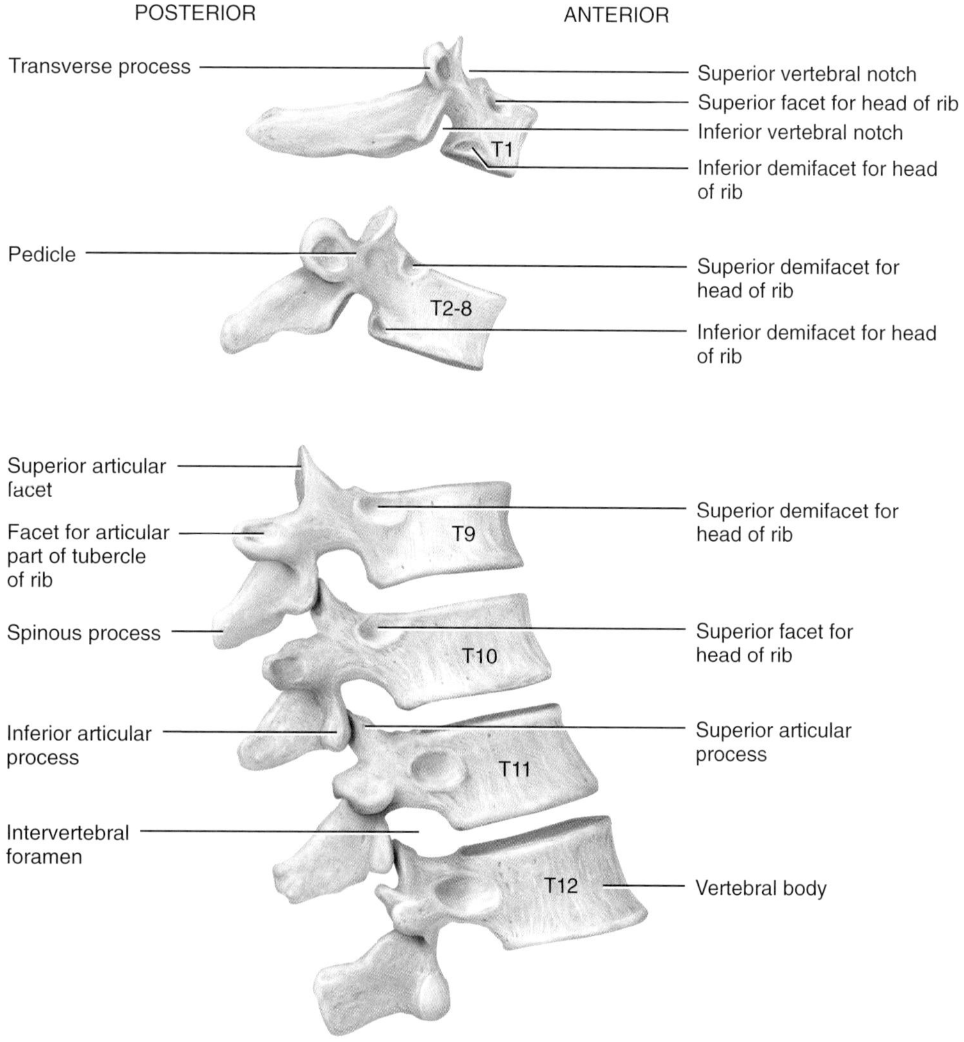
POSTERIOR
ANTERIOR
Transverse process
Superior vertebral notch
Superior facet for head of rib
Inferior vertebral notch
T1
Inferior demifacet for head of rib
Pedicle
Superior demifacet for head of rib
T2-8
Inferior demifacet for head of rib
Superior articular facet
Facet for articular part of tubercle of rib
T9
Superior demifacet for head of rib
Spinous process
T10
Superior facet for head of rib
Inferior articular process
T11
Superior articular process
Intervertebral foramen
T12
Vertebral body

(a) Right lateral view of several articulated thoracic vertebrae

EXHIBIT 7.I CONTINUES

superior demifacet and an inferior demifacet as ribs two through nine articulate with two vertebrae, and T10–T12 have a facet on each side of the vertebral body for ribs 10–12. These articulations between the thoracic vertebrae and ribs, called *vertebrocostal joints,* are distinguishing features of thoracic vertebrae. Movements of the thoracic region are limited by the attachment of the ribs to the sternum.

✔ **CHECKPOINT**

Describe several distinguishing features of thoracic vertebrae.

FIGURE 7.19 CONTINUED

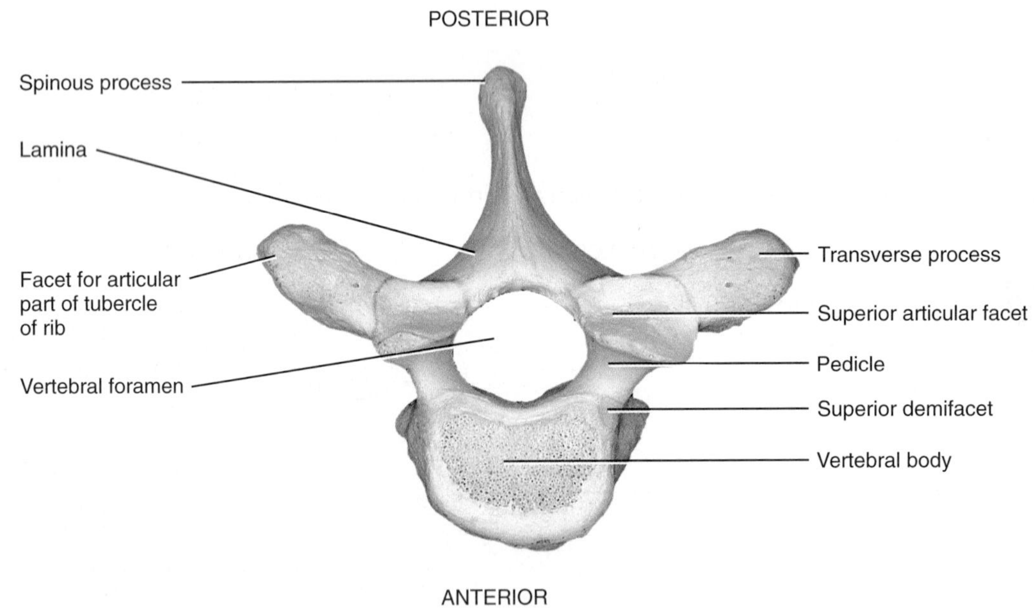
POSTERIOR
Spinous process
Lamina
Facet for articular part of tubercle of rib
Vertebral foramen
Transverse process
Superior articular facet
Pedicle
Superior demifacet
Vertebral body
ANTERIOR

(b) Superior view

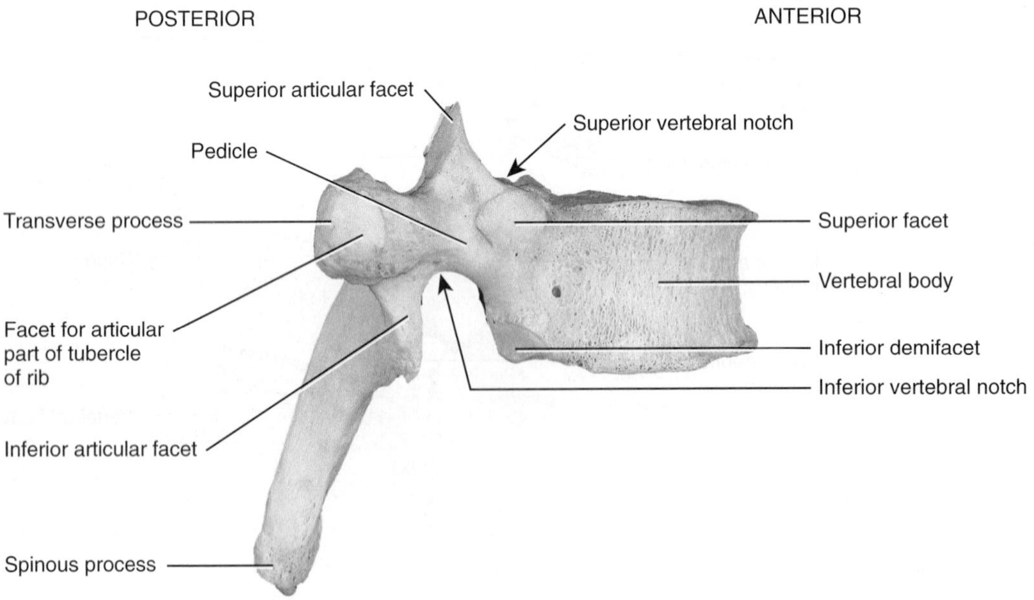
POSTERIOR
ANTERIOR
Superior articular facet
Superior vertebral notch
Pedicle
Transverse process
Superior facet
Vertebral body
Facet for articular part of tubercle of rib
Inferior demifacet
Inferior vertebral notch
Inferior articular facet
Spinous process

(c) Right lateral view

Which parts of the thoracic vertebrae articulate with the ribs?

Vertebral Regions—Lumbar Vertebrae

(Figure 7.20 and Table 7.4)

OBJECTIVE

- Identify the location and surface features of the lumbar vertebrae.

The **lumbar vertebrae** (L1–L5) are the largest and strongest of the unfused bones in the vertebral column (Figure 7.20) because the amount of body weight supported by the vertebrae increases toward the inferior end of the backbone. Their various projections are short and thick. The superior articular processes are directed medially instead of superiorly, and the inferior articular processes are directed laterally instead of inferiorly. The spinous processes are quadrilateral in shape, are thick and broad, and project nearly straight posteriorly. The spinous processes are well adapted for the attachment of the large back muscles.

Figure 7.20 Lumbar vertebrae.

Lumbar vertebrae are found in the lower back.

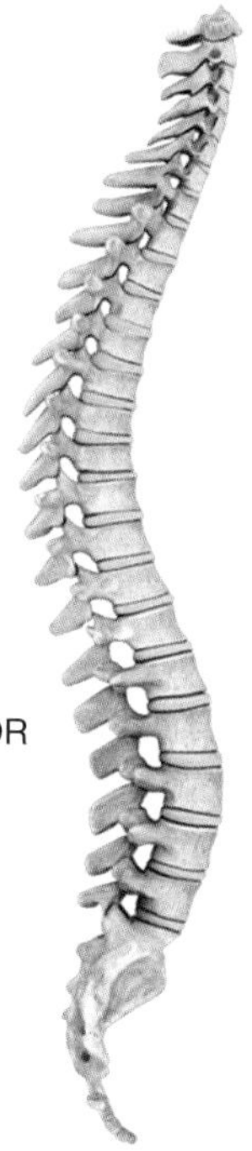

Location of lumbar vertebrae

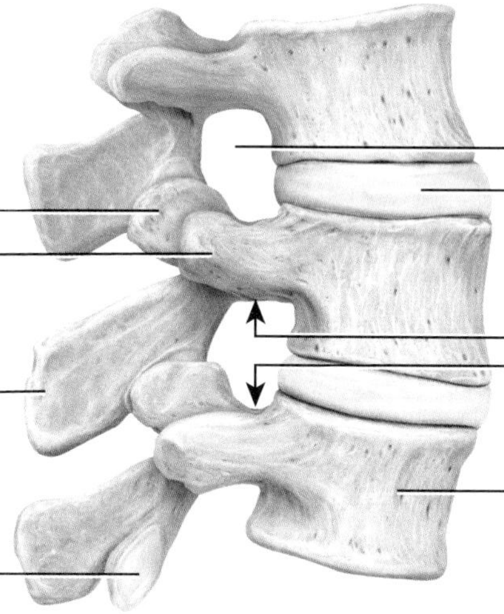

(a) Right lateral view of articulated lumbar vertebrae

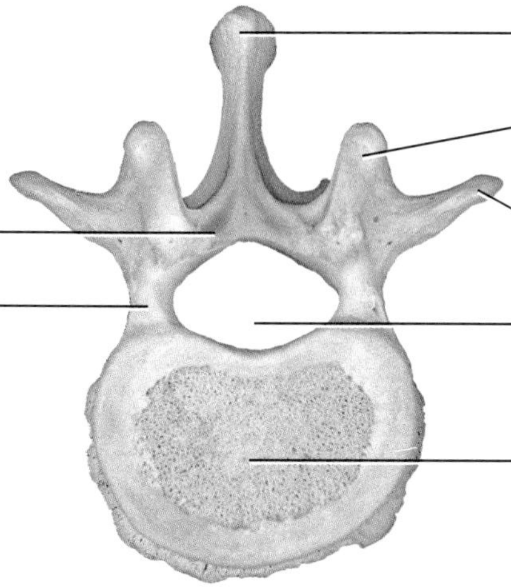

(b) Superior view

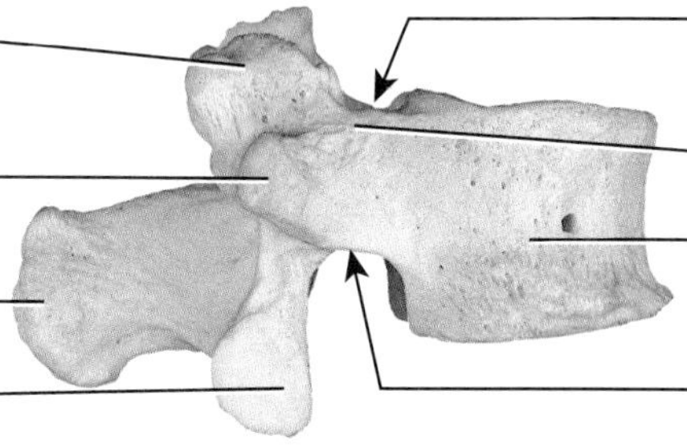

(c) Right lateral view

Why are the lumbar vertebrae the largest and strongest in the vertebral column?

EXHIBIT 7.J CONTINUES

Vertebral Regions—Lumbar Vertebrae

EXHIBIT 7.J *(Figure 7.20 and Table 7.4)* CONTINUED

A summary of the major structural differences among cervical, thoracic, and lumbar vertebrae is presented in Table 7.4.

✔**CHECKPOINT**

What are the distinguishing features of the lumbar vertebrae?

TABLE 7.4

Comparison of Major Structural Features of Cervical, Thoracic, and Lumbar Vertebrae

CHARACTERISTIC	CERVICAL	THORACIC	LUMBAR
Overall structure	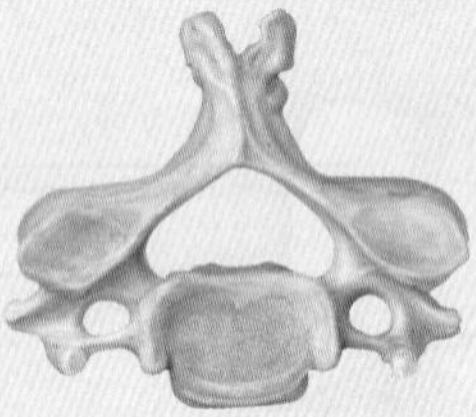	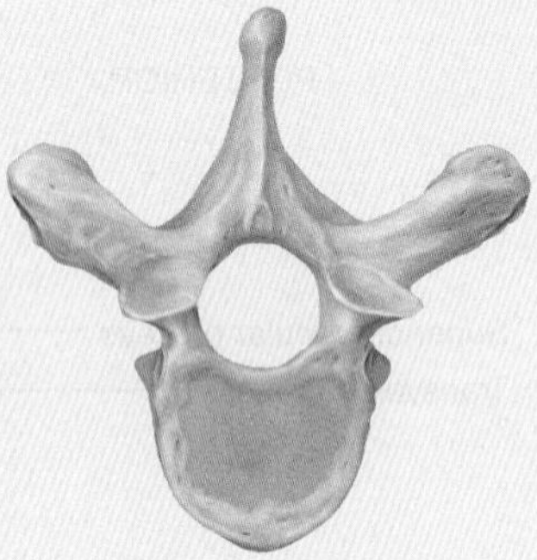	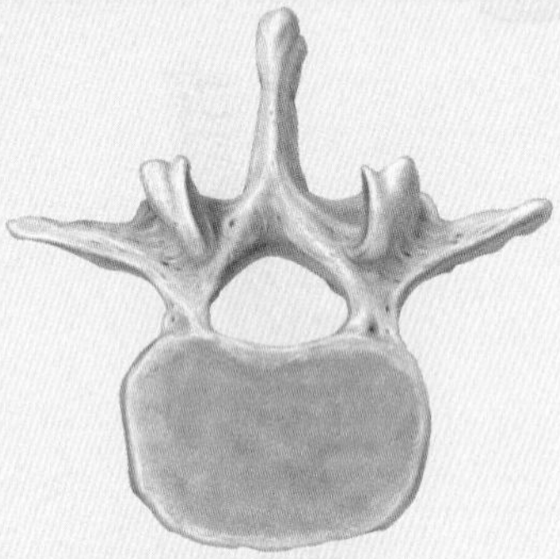
Size	Small.	Larger.	Largest.
Foramina	One vertebral and two transverse.	One vertebral.	One vertebral.
Spinous processes	Slender, often bifid (C2–C6).	Long, fairly thick (most project inferiorly).	Short, blunt (project posteriorly rather than inferiorly).
Transverse processes	Small.	Fairly large.	Large and blunt.
Articular facets for ribs	Absent.	Present.	Absent.
Direction of articular facets			
Superior	Posterosuperior.	Posterolateral.	Medial.
Inferior	Anteroinferior.	Anteromedial.	Lateral.
Size of intervertebral discs	Thick relative to size of vertebral bodies.	Thin relative to size of vertebral bodies.	Thickest.

Vertebral Regions—Sacral and Coccygeal Vertebrae *(Figure 7.21)*

OBJECTIVE

- Identify the location and surface features of the sacral and coccygeal vertebrae.

Sacrum

The **sacrum** (SĀ-krum) is a triangular bone formed by the union of five sacral vertebrae (S1–S5) (Figure 7.21a). The sacral vertebrae begin to fuse in individuals between 16 and 18 years of age, a process usually completed by age 30. Positioned at the posterior portion of the pelvic cavity medial to the two hip bones, the sacrum serves as a strong foundation for the pelvic girdle. The female sacrum is shorter, wider, and more curved between S2 and S3 than the male sacrum (see Table 8.1).

The concave anterior side of the sacrum faces the pelvic cavity. It is smooth and contains four *transverse lines (ridges)* that mark the joining of the sacral vertebral bodies (Figure 7.21a). At the ends of these lines are four pairs of *anterior sacral foramina*. The lateral portion of the superior surface of the sacrum contains a smooth surface called the *sacral ala* (ĀL-a = wing; plural is *alae,* ĀL-ē), which is formed by the fused transverse processes of the first sacral vertebra (S1).

The convex, posterior surface of the sacrum contains a *median sacral crest,* the fused spinous processes of the upper sacral vertebrae; a *lateral sacral crest,* the fused transverse processes of the sacral vertebrae; and four pairs of *posterior sacral foramina* (Figure 7.21b). These foramina connect with anterior sacral foramina to allow passage of nerves and blood vessels. The *sacral canal* is a continuation of the vertebral cavity. The laminae of the fifth sacral vertebra, and sometimes the fourth, fail to meet. This leaves an inferior entrance to the vertebral canal called the *sacral hiatus* (hī-Ā-tus = opening). On either side of the sacral hiatus is a *sacral cornu* (KOR-noo; *cornu* = horn; plural is *cornua,* KOR-noo-a), the inferior articular processes of the fifth sacral vertebra. They are connected by ligaments to the coccyx.

The narrow inferior portion of the sacrum is known as the *apex*. The broad superior portion of the sacrum is called the *base.* The anteriorly

Figure 7.21 Sacrum and coccyx.

The sacrum is formed by the union of five sacral vertebrae, and the coccyx is formed by the union of usually four coccygeal vertebrae.

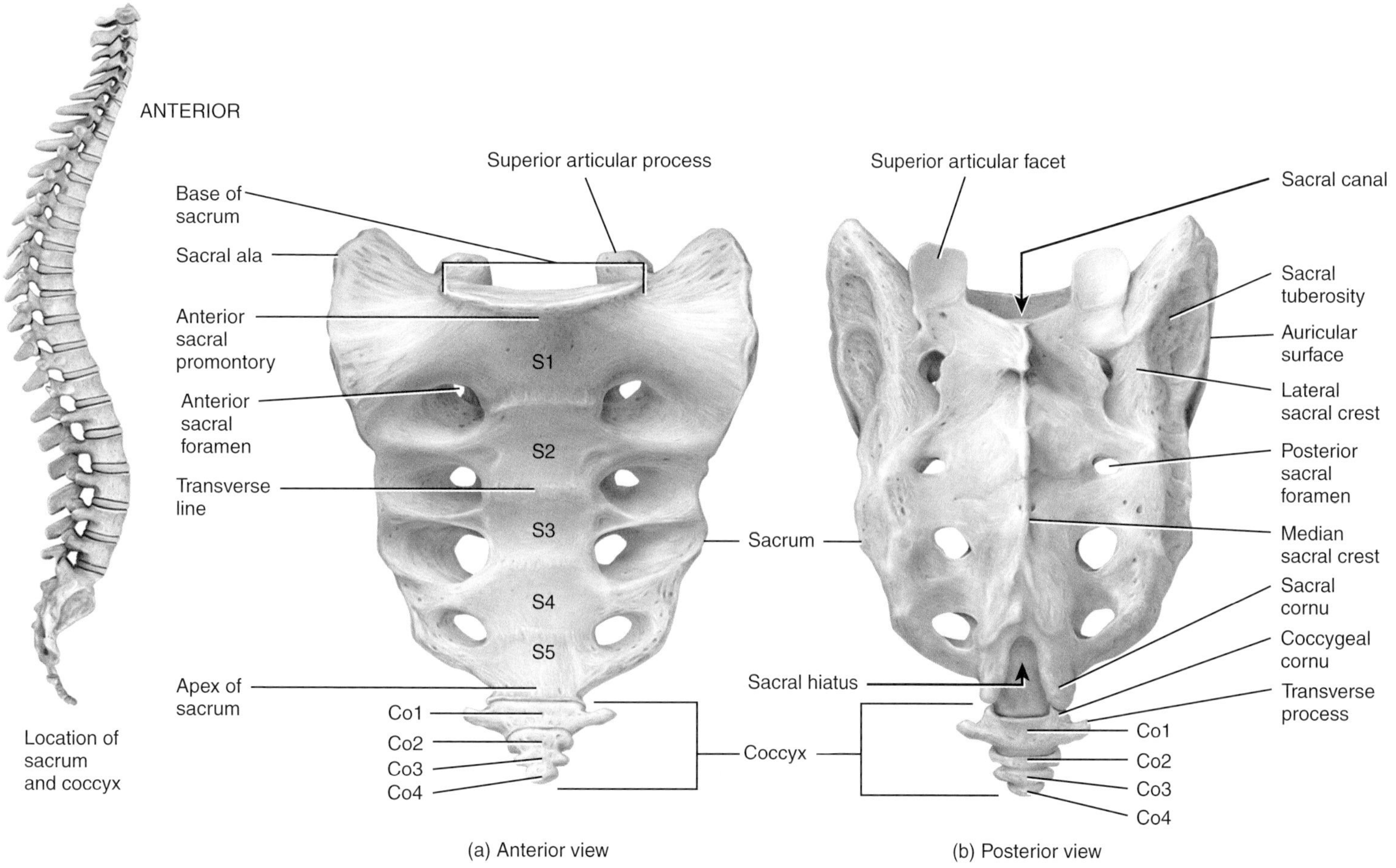

How many foramina pierce the sacrum, and what is their function?

EXHIBIT 7.K CONTINUES

projecting border of the base, called the *sacral promontory* (PROM-on-tō-rē), is one of the points used for measurements of the pelvis. On both lateral surfaces the sacrum has a large ear-shaped *auricular surface* that articulates with the ilium of each hip bone to form the *sacroiliac joint* (see Figure 8.9). Posterior to the auricular surface is a roughened surface, the *sacral tuberosity,* which contains depressions for the attachment of ligaments. The sacral tuberosity unites with the hip bones to form the sacroiliac joints. The *superior articular processes* of the sacrum articulate with the inferior articular processes of the fifth lumbar vertebra, and the base of the sacrum articulates with the body of the fifth lumbar vertebra to form the *lumbosacral joint.*

CLINICAL CONNECTION | *Caudal Anesthesia*

Anesthetic agents that act on the sacral and coccygeal nerves are sometimes injected through the sacral hiatus, a procedure called **caudal anesthesia.** The procedure is used most often to relieve pain during labor and to provide anesthesia to the perineal area. Because the sacral hiatus is between the sacral cornua, the cornua are important bony landmarks for locating the hiatus. Anesthetic agents may also be injected through the posterior sacral foramina. Since the injection site is inferior to the lowest portion of the spinal cord, there is little danger of damaging the cord. •

Coccyx

The **coccyx,** like the sacrum, is triangular in shape. It is formed by the fusion of usually four coccygeal vertebrae, indicated in Figure 7.21 as Co1–Co4. The coccygeal vertebrae fuse somewhat later than the sacral vertebrae, between the ages of 20 and 30. The dorsal surface of the body of the coccyx contains two long *coccygeal cornua* that are connected by ligaments to the sacral cornua. The coccygeal cornua are the pedicles and superior articular processes of the first coccygeal vertebra. They are on the lateral surfaces of the coccyx, formed by a series of *transverse processes;* the first pair are the largest. The coccyx articulates superiorly with the apex of the sacrum. In females, the coccyx points inferiorly to allow the passage of a baby during birth; in males, it points anteriorly (see Table 8.1).

✔CHECKPOINT

How many vertebrae fuse to form the sacrum and coccyx?

Thoracic Bones—Sternum *(Figure 7.22)*

OBJECTIVE

- Identify the location and surface features of the sternum.

Sternum

The **sternum,** or breastbone, is a flat, narrow bone located in the center of the anterior thoracic wall that measures about 15 cm (6 in.) in length and consists of three parts (Figure 7.22). The superior part is the **manubrium** (ma-NOO-brē-um = handlelike); the middle and largest part is the **body;** and the inferior, smallest part is the **xiphoid process** (ZĪ-foyd = sword-shaped). The segments of the sternum typically fuse by age 25, and the points of fusion are marked by transverse ridges.

The junction of the manubrium and body forms the *sternal angle.* The manubrium has a depression on its superior surface, the *suprasternal notch.* Lateral to the suprasternal notch are *clavicular notches* that articulate with the medial ends of the clavicles to form the *sternoclavicular joints.* The manubrium also articulates with the costal cartilages of the first and second ribs. The body of the sternum articulates directly or indirectly with the costal cartilages of the second through tenth ribs. The xiphoid process consists of hyaline cartilage during infancy and childhood and does not completely ossify until about age 40. No ribs are attached to it, but the xiphoid process provides attachment for some abdominal muscles. Incorrect positioning of the hands of a rescuer during cardiopulmonary resuscitation (CPR) may fracture the xiphoid process, driving it into internal organs. During thoracic surgery, the sternum may be split along the midline and the halves spread apart to allow surgeons access to structures in the thoracic cavity such as the thymus, heart, and great vessels of the heart. After surgery, the halves of the sternum are held together with wire sutures.

✔**CHECKPOINT**

What is the clinical significance of the xiphoid process?

Figure 7.22 **Skeleton of the thorax.**

The bones of the thorax enclose and protect organs in the thoracic cavity and in the superior abdominal cavity.

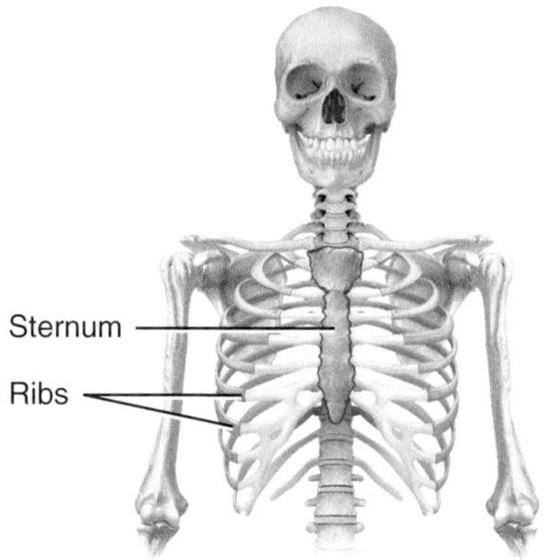

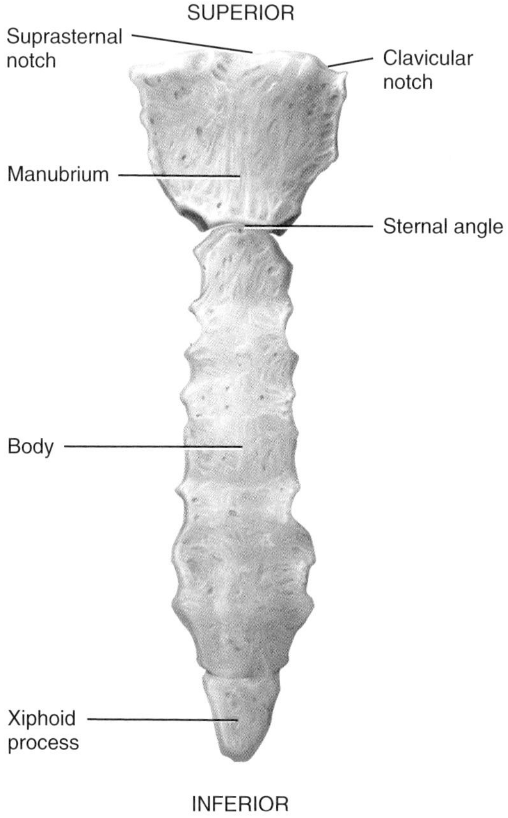

(a) Anterior view of sternum

EXHIBIT 7.L CONTINUES

FIGURE 7.22 CONTINUED

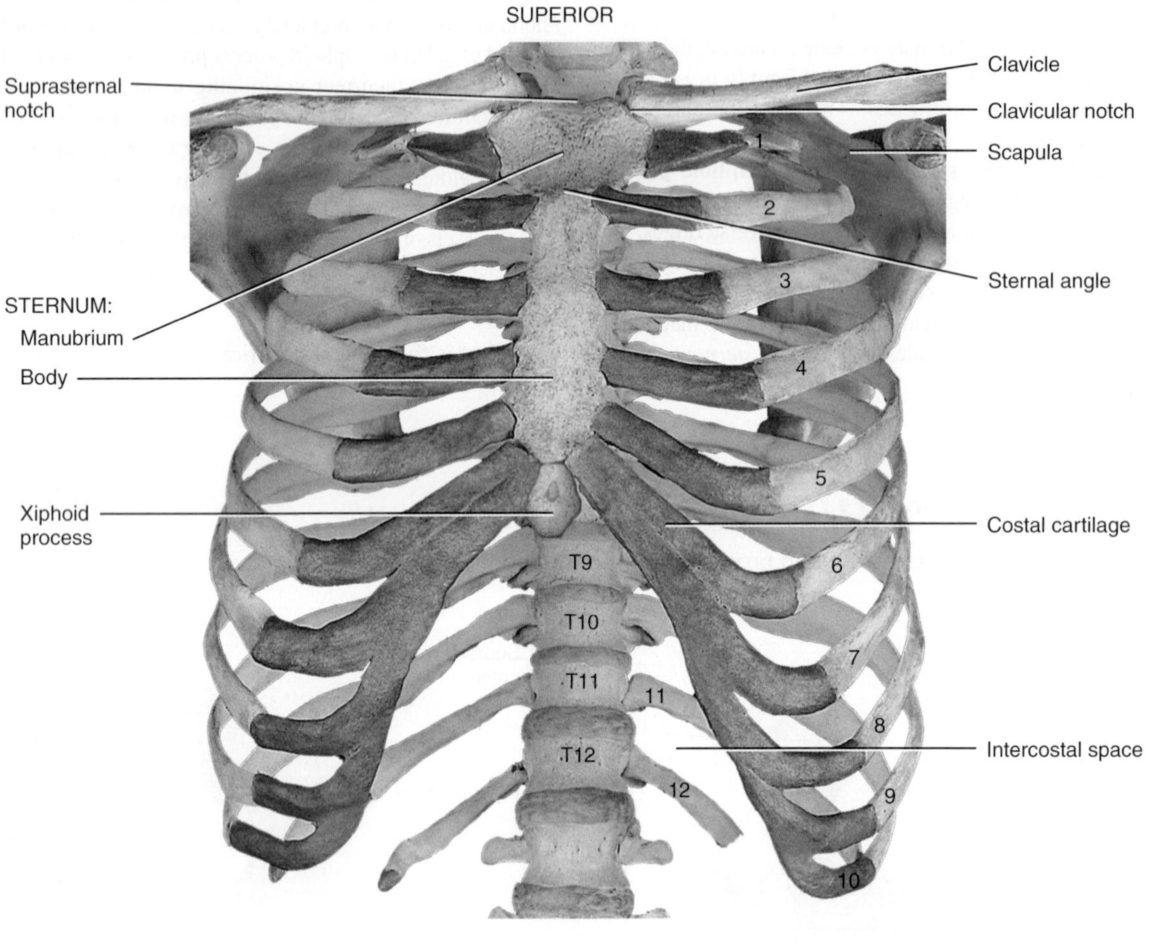
SUPERIOR
Suprasternal notch
Clavicle
Clavicular notch
Scapula
Sternal angle
STERNUM:
Manubrium
Body
Xiphoid process
Costal cartilage
Intercostal space
1
2
3
4
5
6
7
8
9
10
11
12
T9
T10
T11
T12
INFERIOR

(b) Anterior view of skeleton of thorax

With which ribs does the body of the sternum articulate?

Thoracic Bones—Ribs *(Figure 7.23)*

OBJECTIVE

- Identify the location and surface features of the ribs.

Twelve pairs of **ribs,** numbered 1–12 from superior to inferior, give structural support to the sides of the thoracic cavity (Figure 7.22b). The ribs increase in length from the first through seventh, and then decrease in length to the twelfth rib. Each rib articulates posteriorly with its corresponding thoracic vertebra.

The first through seventh pairs of ribs have a direct anterior attachment to the sternum by a strip of hyaline cartilage called *costal cartilage* (*cost-* = rib). The costal cartilages contribute to the elasticity of the thoracic cage and prevent various blows to the chest from fracturing the sternum and/or ribs. The ribs that have costal cartilages and attach directly to the sternum are called *true (vertebrosternal) ribs.* The articulations formed between the true ribs and the sternum are called *sternocostal joints.* The remaining five pairs of ribs are termed *false ribs* because their costal cartilages either attach indirectly to the sternum or do not attach to the sternum at all. The cartilages of the eighth, ninth, and tenth pairs of ribs attach to one another and then to the cartilages of the seventh pair of ribs. These false ribs are called *vertebrochondral ribs.* The eleventh and twelfth pairs of ribs are false ribs designated as *floating (vertebral) ribs* because the costal cartilages at their anterior ends do not attach to the sternum at all. These ribs attach only posteriorly to the thoracic vertebrae. Inflammation of one or more costal cartilages, called *costochondritis,* is characterized by local tenderness and pain in the anterior chest wall that may radiate. The symptoms mimic the chest pain associated with a heart attack (angina pectoris).

Figure 7.23a shows the parts of a typical (third through ninth) rib. The *head* is a projection at the posterior end of the rib that contains a pair of articular *facets* (superior and inferior). The facet of the head may fit either into a facet on the body of a single vertebra or into the demifacets of two adjoining vertebrae to form a *vertebrocostal joint.* The *neck* is a constricted portion of a rib just lateral to the head. A knoblike structure on the posterior surface, where the neck joins the body, is called a *tubercle*

Figure 7.23 The structure of ribs.

Each rib articulates posteriorly with its corresponding thoracic vertebra.

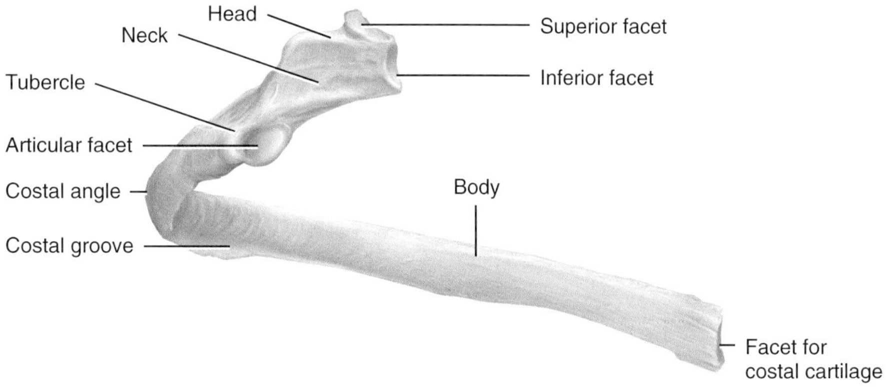

(a) Posterior view of left rib

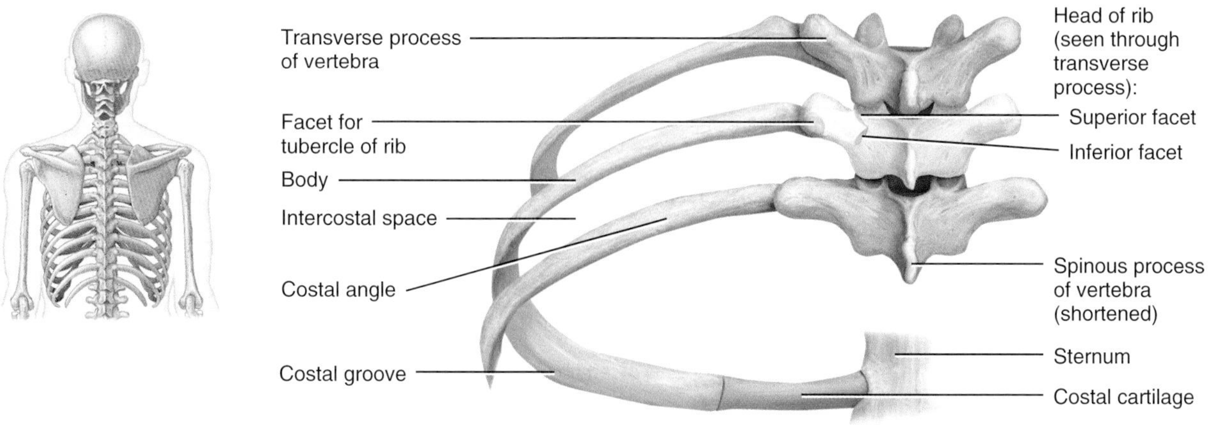

(b) Posterior view of left ribs articulated with thoracic vertebrae and sternum

EXHIBIT 7.M CONTINUES

(TOO-ber-kul). The *nonarticular part* of the tubercle attaches to the transverse process of a vertebra by a ligament (lateral costotransverse ligament). The *articular part* of the tubercle articulates with the facet of a transverse process of a vertebra (Figure 7.23c) to form vertebrocostal joints. The *body (shaft)* is the main part of the rib. A short distance beyond the tubercle, an abrupt change in the curvature of the shaft occurs. This point is called the *costal angle.* The inner surface of the rib has a *costal groove* that protects the intercostal blood vessels and a small nerve.

Spaces between ribs, called *intercostal spaces,* are occupied by intercostal muscles, blood vessels, and nerves. Surgical access to the lungs or other structures in the thoracic cavity is commonly obtained through an intercostal space. Special rib retractors are used to create a wide separation between ribs. The costal cartilages are sufficiently elastic in younger individuals to permit considerable bending without breaking.

In summary, the posterior portion of the rib connects to a thoracic vertebra by its head and the articular part of a tubercle. The facet of the head fits into either a facet on the body of one vertebra (T1 only) or into the demifacets of two adjoining vertebrae. The articular part of the tubercle articulates with the facet of the transverse process of the vertebra.

CLINICAL CONNECTION | *Rib Fractures, Dislocations, and Separations*

Rib fractures are the most common chest injuries. They usually result from direct blows, most often from impact with a steering wheel, falls, or crushing injuries to the chest. Ribs tend to break at the point where the greatest force is applied, but they may also break at their weakest point—the site of greatest curvature, just anterior to the costal angle. The middle ribs are the most commonly fractured. In some cases, fractured ribs may puncture the heart, great vessels of the heart, lungs, trachea, bronchi, esophagus, spleen, liver, and kidneys. Rib fractures are usually quite painful. Rib fractures are no longer bound with bandages because of the pneumonia that would result from lack of proper lung ventilation.

Dislocated ribs, which are common in body contact sports, involve displacement of a costal cartilage from the sternum, with resulting pain, especially during deep inhalations.

Separated ribs involve displacement of a rib and its costal cartilage; as a result, a rib may move superiorly, overriding the rib above and causing severe pain. •

✔ **CHECKPOINT**

How are ribs classified?

FIGURE 7.23 CONTINUED

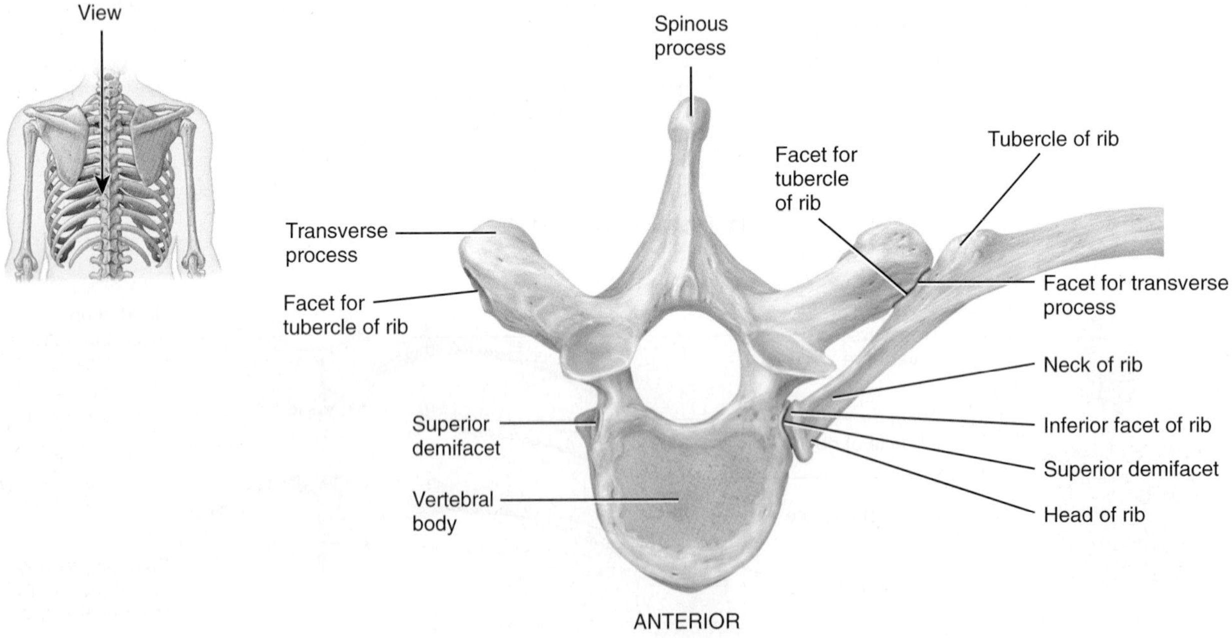

(c) Superior view of left rib articulated with thoracic vertebra

? How does a rib articulate with a thoracic vertebra?

DISORDERS: HOMEOSTATIC IMBALANCES

Herniated (Slipped) Disc

In their function as shock absorbers, intervertebral discs are constantly being compressed. If the anterior and posterior ligaments of the discs become injured or weakened, the pressure developed in the nucleus pulposus may be great enough to rupture the surrounding fibrocartilage (annulus fibrosus). If this occurs, the nucleus pulposus may herniate (protrude) posteriorly or into one of the adjacent vertebral bodies (Figure 7.24). This condition is called a **herniated (slipped) disc.** Because the lumbar region bears much of the weight of the body, and is the region of the most flexing and bending, herniated discs most often occur in the lumbar area.

Frequently, the nucleus pulposus slips posteriorly toward the spinal cord and spinal nerves. This movement exerts pressure on the spinal nerves, causing local weakness and acute pain. If the roots of the sciatic nerve, which passes from the spinal cord to the foot, are compressed, the pain radiates down the posterior thigh, through the calf, and occasionally into the foot. If pressure is exerted on the spinal cord itself, some of its neurons may be destroyed. Treatment options include bed rest, medications for pain, physical therapy and exercises, and *percutaneous endoscopic discectomy* (removal of disc material using a laser). A person with a herniated disc may also undergo a laminectomy, a procedure in which parts of the laminae of the vertebra and intervertebral disc are removed to relieve pressure on the nerves.

Figure 7.24 Herniated (slipped) disc.

Most often the nucleus pulposus herniates posteriorly.

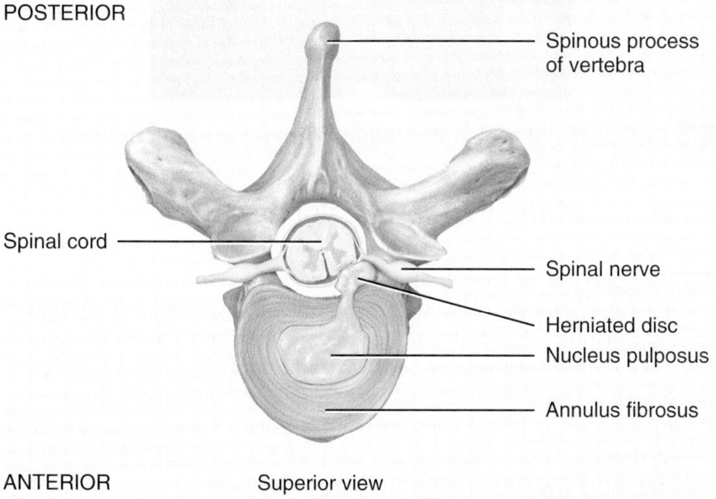
POSTERIOR
Spinous process of vertebra
Spinal cord
Spinal nerve
Herniated disc
Nucleus pulposus
Annulus fibrosus
ANTERIOR
Superior view

Why do most herniated discs occur in the lumbar region?

Abnormal Curves of the Vertebral Column

Various conditions may exaggerate the normal curves of the vertebral column, or the column may acquire a lateral bend, resulting in **abnormal curves** of the vertebral column.

Scoliosis (skō-lē-Ō-sis; *scolio-* = crooked), the most common of the abnormal curves, is a lateral bending of the vertebral column, usually in the thoracic region (Figure 7.25a). It may result from congenitally (present at birth) malformed vertebrae, chronic sciatica (pain in the lower back and lower limb), paralysis of muscles on one side of the vertebral column, poor posture, or one leg being shorter than the other.

Signs of scoliosis include uneven shoulders and waist, one shoulder blade more prominent than the other, one hip higher than the other, and leaning to one side. In severe scoliosis (a curve greater than 70 degrees), breathing is more difficult and the pumping action of the heart is less efficient. Chronic back pain and arthritis of the vertebral column may also develop. Treatment options include wearing a back brace, physical therapy, chiropractic care, and surgery (fusion of vertebrae and insertion of metal rods, hooks, and wires to reinforce the surgery).

Figure 7.25 Abnormal curves of the vertebral column.

An abnormal curve is the result of the exaggeration of a normal curve.

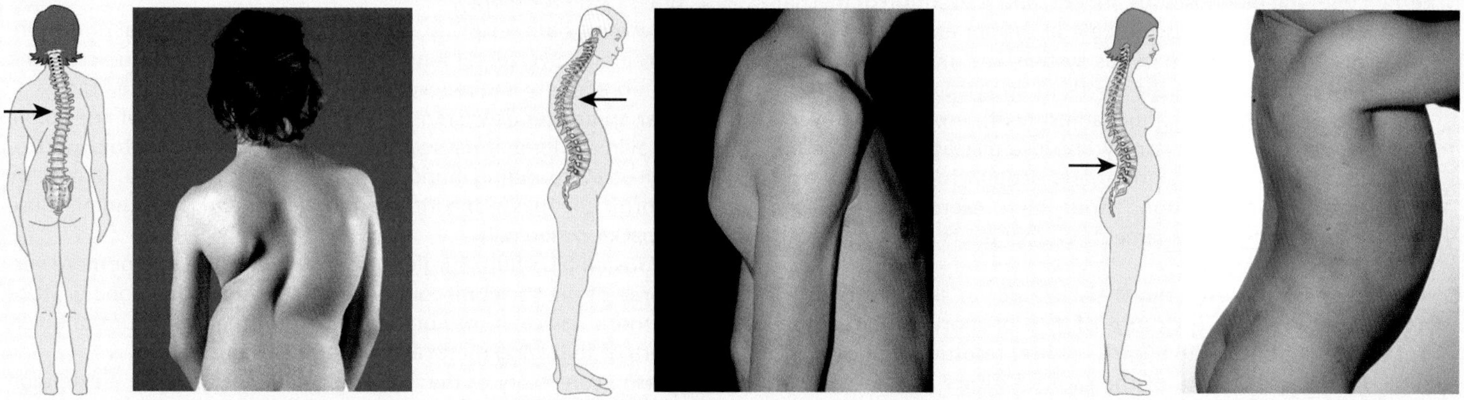

(a) Scoliosis (b) Kyphosis (c) Lordosis

Which abnormal curve is common in women with advanced osteoporosis?

Kyphosis (kī-FŌ-sis; *kyphoss-* = hump; *-osis* = condition) is an increase in the thoracic curve of the vertebral column that produces a "hunchback" look (Figure 7.25b). In tuberculosis of the spine, vertebral bodies may partially collapse, causing an acute angular bending of the vertebral column. In the elderly, degeneration of the intervertebral discs leads to kyphosis. Kyphosis may also be caused by rickets and poor posture. It is also common in females with advanced osteoporosis.

Lordosis (lor-DŌ-sis; *lord-* = bent backward), sometimes called *hollow back,* is an increase in the lumbar curve of the vertebral column (Figure 7.25c). It may result from increased weight of the abdomen as in pregnancy or extreme obesity, poor posture, rickets, osteoporosis, or tuberculosis of the spine.

Spina Bifida

Spina bifida (SPĪ-na BIF-i-da) is a congenital defect of the vertebral column in which laminae of L5 and/or S1 fail to develop normally and unite at the midline. The least serious form is called *spina bifida occulta.* It occurs in L5 or S1 and produces no symptoms. The only evidence of its presence is a small dimple with a tuft of hair in the overlying skin. Several types of spina bifida involve protrusion of meninges (membranes) and/or spinal cord through the defect in the laminae and are collectively termed *spina bifida cystica* because of the presence of a cystlike sac protruding from the backbone (Figure 7.26). If the sac contains the meninges from the spinal cord and cerebrospinal fluid, the condition is called *spina bifida with meningocele* (me-NING-gō-sēl). If the spinal cord and/or its nerve roots are in the sac, the condition is called *spina bifida with meningomyelocele* (me-ning-gō-MĪ-ē-lō-sēl). The larger the cyst and the number of neural structures it contains, the more serious the neurological problems. In severe cases, there may be partial or complete paralysis, partial or complete loss of urinary bladder and bowel control, and the absence of reflexes. An increased risk of spina bifida is associated with low levels of a B vitamin called folic acid during pregnancy. Spina bifida may be diagnosed prenatally by a test of the mother's blood for a substance produced by the fetus called alpha-fetoprotein, by sonography, or by amniocentesis (withdrawal of amniotic fluid for analysis).

Figure 7.26 **Spina bifida with meningomyelocele.**

Spina bifida is caused by a failure of laminae to unite at the midline.

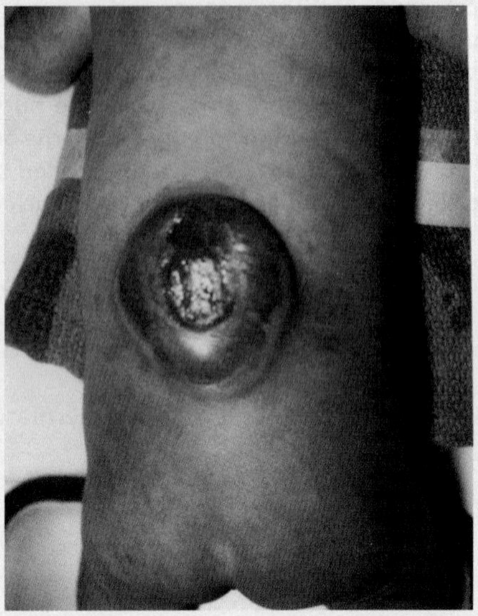

? **Deficiency of which B vitamin is linked to spina bifida?**

Fractures of the Vertebral Column

Fractures of the vertebral column often involve C1, C2, C4–T7, and T12–L2. Cervical or lumbar fractures usually result from a flexion–compression type of injury such as might be sustained in landing on the feet or buttocks after a fall or having a weight fall on the shoulders. Cervical vertebrae may be fractured or dislodged by a fall on the head with acute flexion of the neck, as might happen on diving into shallow water or being thrown from a horse. Spinal cord or spinal nerve damage may occur as a result of fractures of the vertebral column if the fractures compromise the foramina.

MEDICAL TERMINOLOGY

Chiropractic (kī-rō-PRAK-tik; *cheir-* = hand; *-praktikos* = efficient) A holistic health-care discipline that focuses on nerves, muscles, and bones. A ***chiropractor*** is a health-care professional who is concerned with the diagnoses, treatment, and prevention of mechanical disorders of the musculoskeletal system and the effects of these disorders on the nervous system and health in general. Treatment involves using the hands to apply specific force to adjust joints of the body (manual adjustment), especially the vertebral column. Chiropractors may also use massage, heat therapy, ultrasound, electric stimulation, and acupuncture. Chiropractors often provide information about diet, exercise, changes in lifestyle, and stress management. Chiropractors do not prescribe drugs or perform surgery.

Craniostenosis (krā-nē-ō-sten-Ō-sis; *cranio-* = skull; *-stenosis* = narrowing) Premature closure of one or more cranial sutures during the first 18 to 20 months of life, resulting in a distorted skull. Premature closure of the sagittal suture produces a long narrow skull; premature closure of the coronal suture results in a broad skull. Premature closure of all sutures restricts brain growth and development; surgery is necessary to prevent brain damage.

Craniotomy (krā-nē-OT-ō-mē; *cranio-* = skull; *-tome* = cutting) Surgical procedure in which part of the cranium is removed. It may be performed to remove a blood clot, a brain tumor, or a sample of brain tissue for biopsy.

Laminectomy (lam′-i-NEK-tō-mē; *lamina-* = layer) Surgical procedure to remove a vertebral lamina. It may be performed to access the vertebral cavity and relieve the symptoms of a herniated disc.

Lumbar spine stenosis (*sten-* = narrowed) Narrowing of the spinal canal in the lumbar part of the vertebral column, due to hypertrophy of surrounding bone or soft tissues. It may be caused by arthritic changes in the intervertebral discs and is a common cause of back and leg pain.

Spinal fusion (FŪ-zhun) Surgical procedure in which two or more vertebrae of the vertebral column are stabilized with a bone graft or synthetic device. It may be performed to treat a fracture of a vertebra or following removal of a herniated disc.

Whiplash injury Injury to the neck region due to severe hyperextension (backward tilting) of the head followed by severe hyperflexion (forward tilting) of the head, usually associated with a rear-end automobile collision. Symptoms are related to stretching and tearing of ligaments and muscles, vertebral fractures, and herniated vertebral discs.

CHAPTER REVIEW AND RESOURCE SUMMARY

Review | **Resource**

Introduction

1. Bones protect soft body parts and make movement possible; they also serve as landmarks for locating parts of other body systems.
2. The musculoskeletal system is composed of the bones, joints, and muscles working together.

7.1 Divisions of the Skeletal System (see Table 7.1)

1. The axial skeleton consists of bones arranged along the longitudinal axis. The parts of the axial skeleton are the skull, auditory ossicles (ear bones), hyoid bone, vertebral column, sternum, and ribs.
2. The appendicular skeleton consists of the bones of the girdles and the upper and lower limbs (extremities). The parts of the appendicular skeleton are the pectoral (shoulder) girdles, bones of the upper limbs, pelvic (hip) girdles, and bones of the lower limbs.

Resource: Anatomy Overview - The Axial Skeleton

7.2 Types of Bones

1. On the basis of shape, bones are classified as long, short, flat, irregular, or sesamoid. Sesamoid bones develop in tendons or ligaments.
2. Sutural bones are found within the sutures of some cranial bones.

7.3 Bone Surface Markings

1. Surface markings are structural features visible on the surfaces of bones.
2. Each marking—whether a depression, an opening, or a process—is structured for a specific function, such as joint formation, muscle attachment, or passage of nerves and blood vessels (see Table 7.2).

7.4 Skull

1. The 22 bones of the skull include cranial bones and facial bones.
2. The eight cranial bones are the frontal, parietal (2), temporal (2), occipital, sphenoid, and ethmoid.
3. The 14 facial bones are the nasal (2), maxillae (2), zygomatic (2), lacrimal (2), palatine (2), inferior nasal conchae (2), vomer, and mandible.
4. The nasal septum consists of the vomer, perpendicular plate of the ethmoid, and septal cartilage. The nasal septum divides the nasal cavity into left and right sides.
5. Seven skull bones form each of the orbits (eye sockets).
6. The foramina of the skull bones provide passages for nerves and blood vessels (see Table 7.3).
7. Sutures are immovable joints in adults that connect most bones of the skull. Examples are the coronal, sagittal, lambdoid, and squamous sutures.
8. Paranasal sinuses are cavities in bones of the skull that are connected to the nasal cavity. The frontal, sphenoid, and ethmoid bones and the maxillae contain paranasal sinuses.
9. Fontanels are mesenchyme-filled spaces between the cranial bones of fetuses and infants. The major fontanels are the anterior, posterior, anterolaterals (2), and posterolaterals (2). After birth, the fontanels fill in with bone and become sutures.

Resources:
Anatomy Overview - Facial Bones
Anatomy Overview - Fibrous Joints and Sutures
Figure 7.8 - Sphenoid Bone
Figure 7.12 - Details of the Orbit

7.5 Hyoid Bone

1. The hyoid bone is a U-shaped bone that does not articulate with any other bone.
2. It supports the tongue and provides attachment for some tongue muscles and for some muscles of the pharynx and neck.

7.6 Vertebral Column

1. The vertebral column, sternum, and ribs constitute the skeleton of the body's trunk.
2. The 26 bones of the adult vertebral column are the cervical vertebrae (7), the thoracic vertebrae (12), the lumbar vertebrae (5), the sacrum (5 fused vertebrae), and the coccyx (usually 4 fused vertebrae).
3. The adult vertebral column contains four normal curves (cervical, thoracic, lumbar, and sacral) that provide strength, support, and balance.
4. Each vertebra usually consists of a vertebral body, vertebral arch, and seven processes. Vertebrae in the different regions of the column vary in size, shape, and detail.

Resources:
Anatomy Overview - The Vertebral Column
Figure 7.16 - Vertebral Column
Table 7.4 - Comparison of Major Structural Features of Cervical, Thoracic, and Lumbar Vertebrae
Figure 7.21 - Sacrum and Coccyx

Review	Resource
7.7 Thorax 1. The thoracic skeleton consists of the sternum, ribs, costal cartilages, and thoracic vertebrae. 2. The thoracic cage protects vital organs in the chest area and upper abdomen.	Anatomy Overview - The Thorax Figure 7.22 - Skeleton of the Thorax Exercise - Bones, Bones, Bones

SELF-QUIZ QUESTIONS

Fill in the blanks in the following statements.

1. Membrane-filled spaces between cranial bones that enable the fetal skull to modify its size and shape for passage through the birth canal are called ______.

2. The hypophyseal fossa of the sella turcica of the sphenoid bone contains the ______.

3. The regions of the vertebral column that consist of fused vertebrae are the ______ and the ______.

Indicate whether the following statements are true or false.

4. The atlanto-occipital joints allow you to rotate the head, as in signifying "no."

5. Ribs that are not attached to the sternum are known as the true ribs.

Choose the one best answer to the following questions.

6. In which of the following bones are paranasal sinuses *not* found?
(a) frontal bone (b) sphenoid bone (c) lacrimal bones
(d) ethmoid bone (e) maxillae

7. Which of the following pairs are mismatched?
(a) mandible: only movable bone in the skull
(b) hyoid: bone that does not articulate with any other bone
(c) sacrum: supports lower back
(d) thoracic vertebrae: articulate with thoracic ribs posteriorly
(e) inferior nasal conchae: classified as facial bones

8. Which of the following bones are *not* paired?
(a) vomer (b) palatine (c) lacrimal
(d) maxilla (e) nasal

9. The suture located between a parietal and temporal bone is the
(a) lambdoid. (b) sagittal. (c) coronal.
(d) anterolateral. (e) squamous.

10. The primary vertebral curves that appear during fetal development are the (1) cervical curve, (2) thoracic curve, (3) lumbar curve, (4) coccyx curve, (5) sacral curve.
(a) 2 and 3 (b) 1 and 2 (c) 2 and 4
(d) 2 and 5 (e) 1 and 3

11. Which of the following are functions of the cranial bones? (1) protection of the brain; (2) attachment of muscles that move the head; (3) protection of the special sense organs; (4) attachment to the meninges; (5) attachment of muscles that produce facial expressions.
(a) 1, 2, and 5 (b) 1, 2, 4, and 5 (c) 2 and 5
(d) 1, 2, 3, and 5 (e) 1, 2, 3, 4, and 5

12. Match the following:

____ (a) prominent ridge or elongated projection	(1) foramen
____ (b) tubelike opening	(2) tuberosity
____ (c) large round protuberance at the end of a bone	(3) spinous process
____ (d) smooth, flat articular surface	(4) crest
____ (e) sharp, slender projection	(5) facet
____ (f) opening for passage of blood vessels, nerves, or ligaments	(6) fissure
____ (g) large, rounded, rough projection	(7) condyle
____ (h) shallow depression	(8) fossa
____ (i) narrow slit between adjacent parts of bones for passage of blood vessels or nerves	(9) meatus

13. Match the following:

____ (a) supraorbital foramen	(1) temporal bone
____ (b) temporomandibular joint	(2) sphenoid bone
____ (c) external auditory meatus	(3) cervical vertebrae
____ (d) foramen magnum	(4) ethmoid bone
____ (e) optic foramen	(5) articulation of mandibular fossa and articular tubercle of the temporal bone to the mandible
____ (f) cribriform plate	(6) occipital bone
____ (g) palatine process	(7) frontal bone
____ (h) ramus, body, and condylar process	(8) maxillae
____ (i) transverse foramina, bifid spinous processes	(9) mandible
____ (j) dens	(10) axis
____ (k) promontory	(11) sacrum
____ (l) costal cartilages	(12) sternum
____ (m) xiphoid process	(13) ribs

14. Match the following (the same answer may be used more than once):

____ (a) bones that have greater length than width and consist of a shaft and a variable number of extremities
____ (b) cube-shaped bones that are nearly equal in length and width
____ (c) bones that develop in certain tendons where there is considerable friction, tension, and physical stress
____ (d) small bones located within joints between certain cranial bones
____ (e) thin bones composed of two nearly parallel plates of compact bone enclosing a layer of spongy bone
____ (f) bones with complex shapes, including the vertebrae and some facial bones
____ (g) patella is an example
____ (h) bones that provide considerable protection and extensive areas for muscle attachment
____ (i) include femur, tibia, fibula, humerus, ulna, and radius
____ (j) include cranial bones, sternum, and ribs
____ (k) include almost all of the carpal (wrist) and tarsal (ankle) bones

(1) irregular bones
(2) long bones
(3) short bones
(4) flat bones
(5) sesamoid bones
(6) sutural bones

15. Match the following:

____ (a) forms the forehead
____ (b) form the inferior lateral aspects of the cranium and part of the cranial floor; contain zygomatic process and mastoid process
____ (c) forms part of the anterior portion of the cranial floor, medial wall of the orbits, superior portions of nasal septum, most of the side walls of the nasal cavity; is a major supporting structure of the nasal cavity
____ (d) form the prominence of the cheek and part of the lateral wall and floor of each orbit
____ (e) the largest, strongest facial bone; is the only movable skull bone
____ (f) a roughly triangular bone on the floor of the nasal cavity; one of the components of the nasal septum
____ (g) form greater portion of the sides and roof of the cranial cavity
____ (h) forms the posterior part and most of the base of the cranium; contains the foramen magnum
____ (i) called the keystone of the cranial floor; contains the sella turcica, optic foramen, and pterygoid processes
____ (j) form the bridge of the nose
____ (k) the smallest bones of the face; contain a vertical groove that houses a structure that gathers tears and passes them into the nasal cavity
____ (l) does not articulate with any other bone
____ (m) unite to form the upper jawbone and articulate with every bone of the face except the lower jawbone
____ (n) form the posterior part of the hard palate, part of the floor and lateral wall of the nasal cavity, and a small portion of the floors of the orbits
____ (o) scroll-like bones that form a part of the lateral walls of the nasal cavity; functions in the turbulent circulation and filtration of air

(1) temporal bones
(2) parietal bones
(3) frontal bone
(4) occipital bone
(5) sphenoid bone
(6) ethmoid bone
(7) nasal bones
(8) maxillae
(9) zygomatic bones
(10) lacrimal bones
(11) palatine bones
(12) vomer
(13) mandible
(14) inferior nasal conchae
(15) hyoid bone

CRITICAL THINKING QUESTIONS

1. Jimmy is in a car accident. He can't open his mouth and has been told that he suffers from the following: black eye, broken nose, broken cheek, broken upper jaw, damaged eye socket, and punctured lung. Describe *exactly* what structures have been affected by his car accident.

2. Bubba is a tug-of-war expert. He practices day and night by pulling on a rope attached to an 800-lb anchor. What kinds of changes would you expect that he develops in his bone structure?

3. A new mother brings her newborn infant home and has been told by her well-meaning friend not to wash the baby's hair for several months because the water and soap could "get through that soft area in the top of the head and cause brain damage." Explain to her why this is not true.

? ANSWERS TO FIGURE QUESTIONS

7.1 The skull and vertebral column are part of the axial skeleton. The clavicle, shoulder girdle, humerus, pelvic girdle, and femur are part of the appendicular skeleton.

7.2 Flat bones protect underlying organs and provide a large surface area for muscle attachment.

7.3 The frontal, parietal, sphenoid, ethmoid, temporal, and occipital bones are all cranial bones.

7.4 The parietal and temporal bones are joined by the squamous suture, the parietal and occipital bones are joined by the lambdoid suture, and the parietal and frontal bones are joined by the coronal suture.

7.5 The temporal bone articulates with the mandible and the parietal, sphenoid, zygomatic, and occipital bones.

7.6 The parietal bones form the posterior, lateral portion of the cranium.

7.7 The medulla oblongata of the brain connects with the spinal cord in the foramen magnum.

7.8 From the crista galli of the ethmoid bone, the sphenoid articulates with the frontal, parietal, temporal, occipital, temporal, parietal, and frontal bones, ending again at the crista galli of the ethmoid bone.

7.9 The perpendicular plate of the ethmoid bone forms the superior part of the nasal septum, and the lateral masses compose most of the medial walls of the orbits.

7.10 The mandible is the only movable skull bone, other than the auditory ossicles.

7.11 The nasal septum divides the nasal cavity into right and left sides.

7.12 Bones forming the orbit are the frontal, sphenoid, zygomatic, maxilla, lacrimal, ethmoid, and palatine.

7.13 The paranasal sinuses produce mucus and serve as resonating chambers for vocalizations.

7.14 The paired anterolateral fontanels are bordered by four different skull bones: the frontal, parietal, temporal, and sphenoid bones.

7.15 The hyoid bone does not articulate with any other bone.

7.16 The thoracic and sacral curves of the vertebral column are concave relative to the anterior of the body.

7.17 The vertebral foramina enclose the spinal cord; the intervertebral foramina provide spaces through which spinal nerves exit the vertebral column.

7.18 The atlas moving on the axis at the atlanto-axial point permits movement of the head to signify "no."

7.19 The facets and demifacets on the vertebral bodies of the thoracic vertebrae articulate with the heads of the ribs, and the facets on the transverse processes of these vertebrae articulate with the tubercles of the ribs.

7.20 The lumbar vertebrae are the largest and strongest in the body because the amount of weight supported by vertebrae increases toward the inferior end of the vertebral column.

7.21 There are four pairs of sacral foramina, for a total of eight. Each anterior sacral foramen joins a posterior sacral foramen at the intervertebral foramen. Nerves and blood vessels pass through these tunnels in the bone.

7.22 The body of the sternum articulates directly or indirectly with ribs 2–10.

7.23 The facet on the head of a rib fits into a facet on the body of a vertebra, and the articular part of the tubercle of a rib articulates with the facet of the transverse process of a vertebra.

7.24 Most herniated discs occur in the lumbar region because it bears most of the body weight and most flexing and bending occur there.

7.25 Kyphosis is common in individuals with advanced osteoporosis.

7.26 Deficiency of folic acid is associated with spina bifida.

8 THE SKELETAL SYSTEM: THE APPENDICULAR SKELETON

THE APPENDICULAR SKELETON AND HOMEOSTASIS *The bones of the appendicular skeleton contribute to homeostasis by providing attachment points and leverage for muscles, which aids body movements; by providing support and protection of internal organs, such as the reproductive organs; and by storing and releasing calcium.*

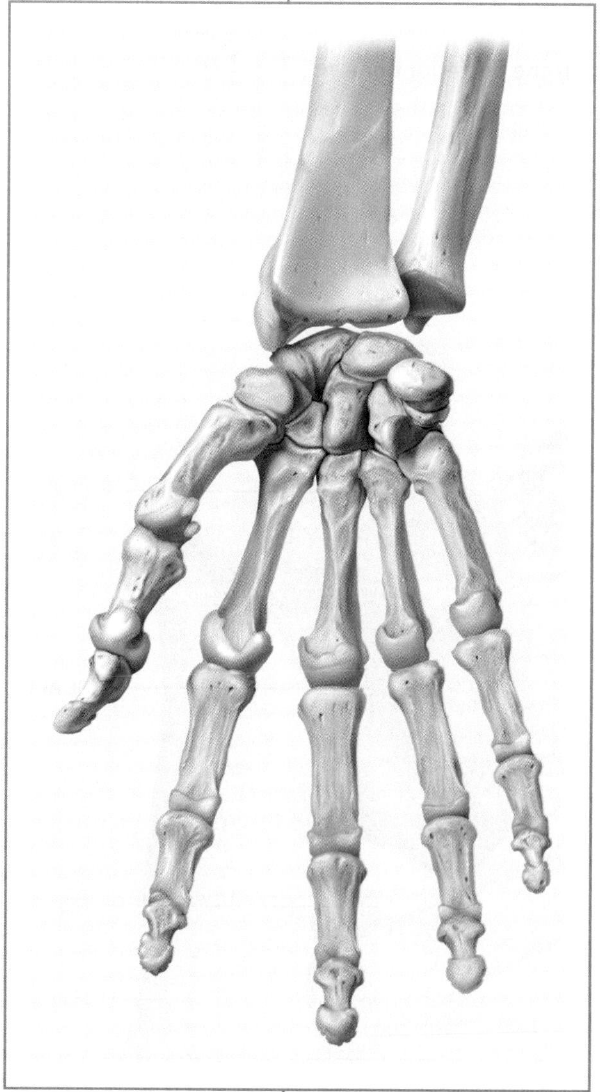

As noted in Chapter 7, the two main divisions of the skeletal system are the axial skeleton and the appendicular skeleton. As you learned in that chapter, the general function of the axial skeleton is the protection of internal organs; the primary function of the appendicular skeleton, the focus of this chapter, is movement. The appendicular skeleton includes the bones that make up the upper and lower limbs as well as the bones of the two girdles that attach the limbs to the axial skeleton. The bones of the appendicular skeleton are connected with one another and with skeletal muscles, permitting you to do things such as walk, write, use a computer, dance, swim, and play a musical instrument.

? ***Did you ever wonder what causes runner's knee?***

8.1 PECTORAL (SHOULDER) GIRDLE

OBJECTIVE

- Identify the bones of the pectoral (shoulder) girdle, their functions, and their principal markings.

The human body has two **pectoral** (PEK-tō-ral) or **shoulder girdles** that attach the bones of the upper limbs to the axial skeleton (Figure 8.1). Each of the two pectoral girdles consists of a clavicle and a scapula. The *clavicle* is the anterior bone and articulates with the manubrium of the sternum at the *sternoclavicular joint.* The scapula articulates with the clavicle at the *acromioclavicular joint* and with the humerus at the *glenohumeral (shoulder) joint.* The pectoral girdles do not articulate with the vertebral column and are held in position and stabilized by a group of large muscles that extend from the vertebral column and ribs to the scapula.

The bones of the pectoral girdle are discussed in Exhibits 8.A and 8.B.

CHECKPOINT

1. What is the function of the pectoral girdle?

Figure 8.1 Right pectoral (shoulder) girdle.

The clavicle is the anterior bone of the pectoral girdle, and the scapula is the posterior bone.

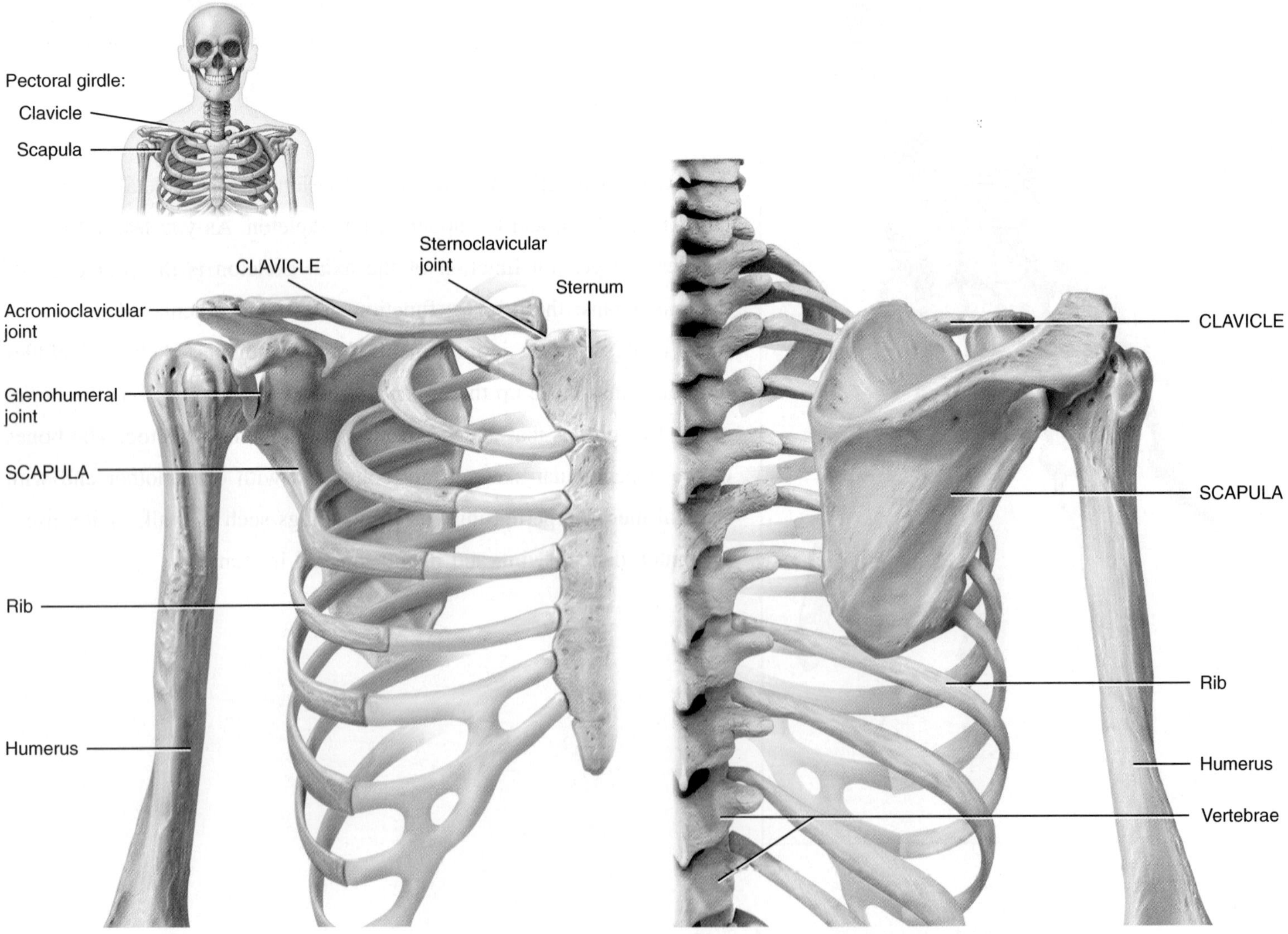
Pectoral girdle:
Clavicle
Scapula
CLAVICLE
Sternoclavicular joint
Sternum
Acromioclavicular joint
Glenohumeral joint
SCAPULA
Rib
Humerus
CLAVICLE
SCAPULA
Rib
Humerus
Vertebrae

(a) Anterior view of pectoral girdle

(b) Posterior view of pectoral girdle

What is the function of the pectoral girdle?

EXHIBIT 8.A Pectoral Girdle—Clavicle *(Figure 8.2)*

OBJECTIVE

- Describe the location and surface features of the clavicle.

Each slender, S-shaped **clavicle** (KLAV-i-kul = key), or *collarbone,* lies horizontally across the anterior part of the thorax superior to the first rib (Figure 8.2). It is subcutaneous (under the skin) and easily palpable along its length. The bone is S-shaped because the medial half is convex anteriorly (curves toward you when viewed in the anatomical position), and the lateral half is concave anteriorly (curves away from you). It is rougher and more curved in males.

The medial end, called the *sternal end,* is rounded and articulates with the manubrium of the sternum to form the *sternoclavicular joint.* The broad, flat, lateral end, the *acromial end* (a-KRŌ-mē-al), articulates with the acromion of the scapula to form the *acromioclavicular joint* (see Figure 8.1). The *conoid tubercle* (KŌ-noyd = conelike) on the inferior surface of the lateral end of the bone is a point of attachment for the conoid ligament, which attaches the clavicle and scapula. As its name implies, the *impression for the costoclavicular ligament* on the inferior surface of the sternal end is a point of attachment for the costoclavicular ligament (Figure 8.2b), which attaches the clavicle and first rib.

CLINICAL CONNECTION | *Fractured Clavicle*

The clavicle transmits mechanical force from the upper limb to the trunk. If the force transmitted to the clavicle is excessive, as when you fall on your outstretched arm, a **fractured clavicle** may result. A fractured clavicle may also result from a blow to the superior part of the anterior thorax, for example, as a result of an impact following an automobile accident. The clavicle is one of the most frequently broken bones in the body. Because the junction of the two curves of the clavicle is its weakest point, the clavicular midregion is the most frequent fracture site. Even in the absence of fracture, compression of the clavicle as a result of automobile accidents involving the use of shoulder harness seatbelts often causes damage to the brachial plexus (the network of nerves that enter the upper limb), which lies between the clavicle and the second rib. A fractured clavicle is usually treated with a figure-of-8 sling to keep the arm from moving outward. •

✔CHECKPOINT

Which joints are formed by the articulation of the clavicle with other bones? Which areas of the clavicle are involved in each joint?

Figure 8.2 Right clavicle.

The clavicle articulates medially with the manubrium of the sternum and laterally with the acromion of the scapula.

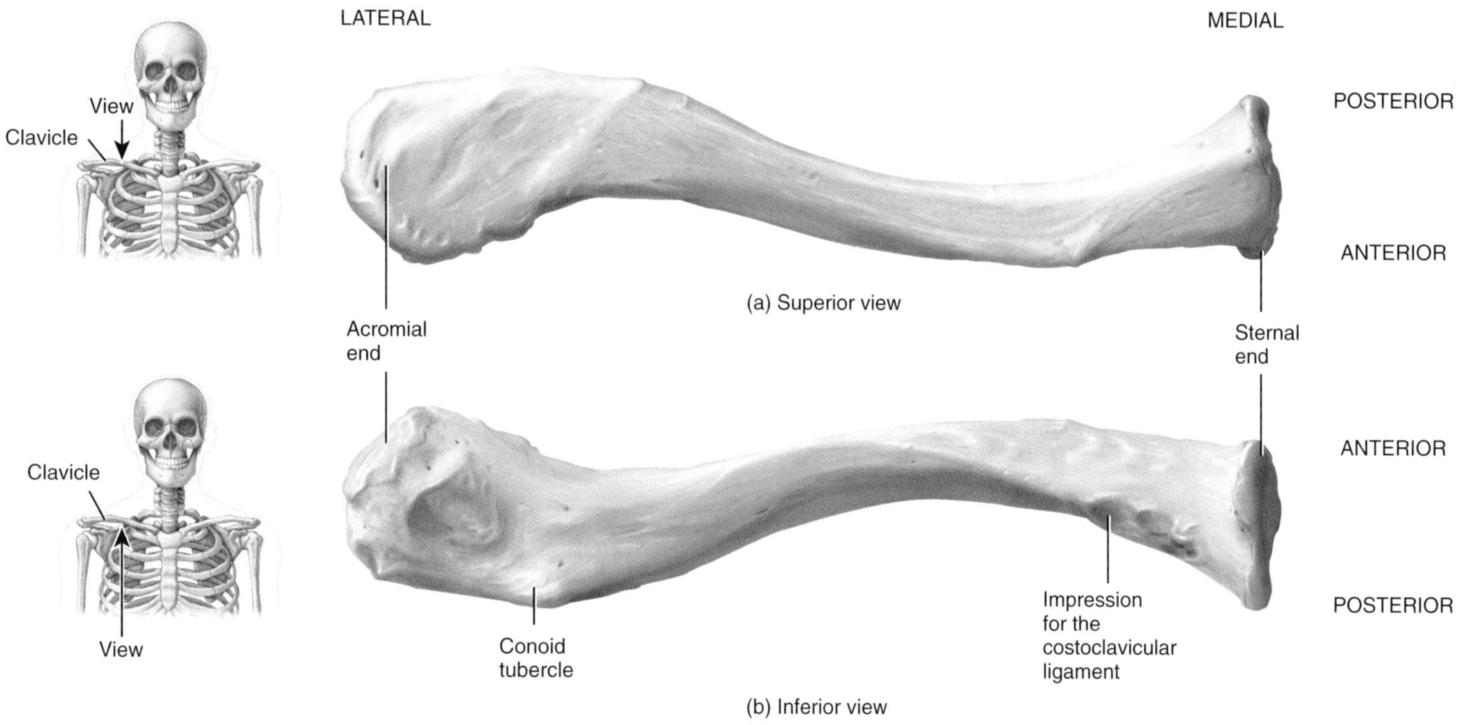

LATERAL
MEDIAL
View
Clavicle
POSTERIOR
ANTERIOR
(a) Superior view
Acromial end
Sternal end
Clavicle
ANTERIOR
POSTERIOR
View
Conoid tubercle
Impression for the costoclavicular ligament
(b) Inferior view

Which part of the clavicle is its weakest point?

EXHIBIT 8.B Pectoral Girdle—Scapula *(Figure 8.3)*

OBJECTIVE

- Describe the location and surface features of the scapula.

Each **scapula** (SCAP-ū-la; plural is *scapulae*), or *shoulder blade,* is a large, triangular, flat bone situated in the superior part of the posterior thorax between the levels of the second and seventh ribs (Figure 8.3).

A prominent ridge called the *spine* runs diagonally across the posterior surface of the scapula. The lateral end of the spine projects as a flattened, expanded process called the *acromion* (a-KRŌ-mē-on; *acrom-* = topmost; *-omos* = shoulder), easily felt as the high point of the shoulder. Tailors measure the length of the upper limb from the acromion. As noted earlier, the acromion articulates with the acromial end of the clavicle to form the *acromioclavicular joint*. Inferior to the acromion is a shallow depression, the *glenoid cavity,* that accepts the head of the humerus (arm bone) to form the *glenohumeral (shoulder) joint* (see Figure 8.1).

The thin edge of the scapula closer to the vertebral column is called the *medial (vertebral) border.* The thick edge of the scapula closer to the arm is called the *lateral (axillary) border.* The medial and lateral borders

Figure 8.3 **Right scapula (shoulder blade).**

The glenoid cavity of the scapula articulates with the head of the humerus to form the glenohumeral (shoulder) joint.

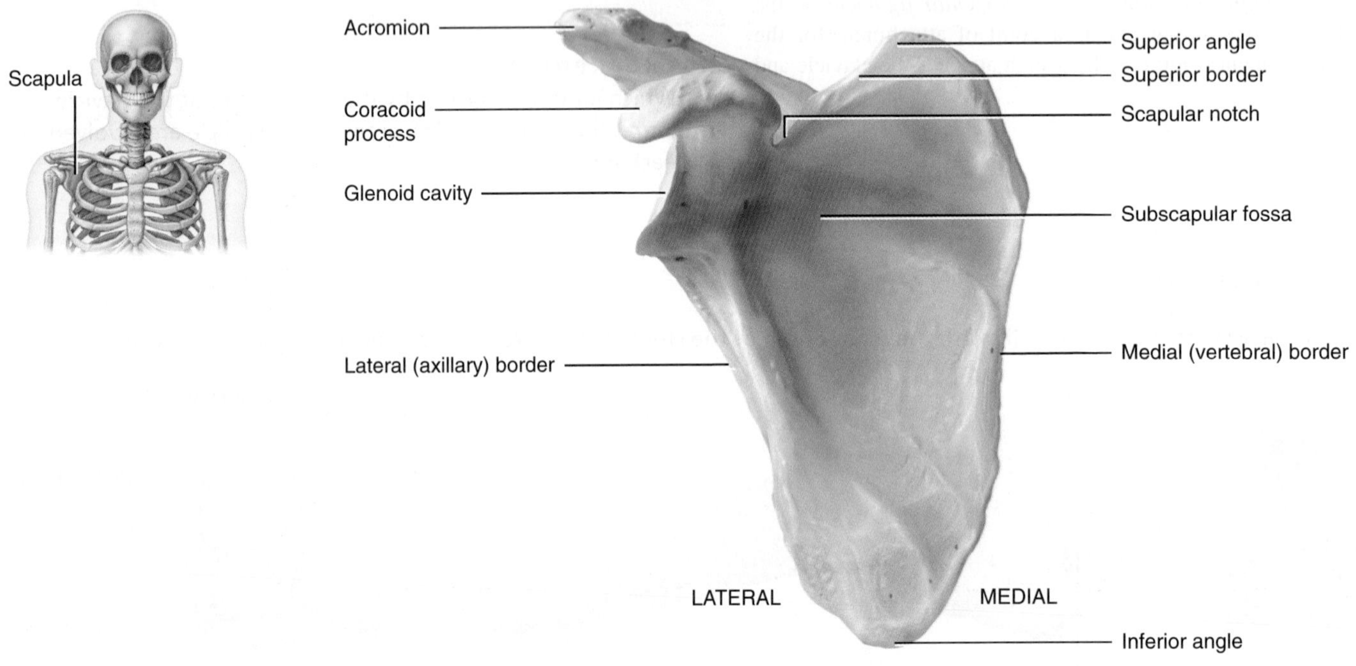
Scapula
Acromion
Coracoid process
Glenoid cavity
Lateral (axillary) border
Superior angle
Superior border
Scapular notch
Subscapular fossa
Medial (vertebral) border
LATERAL
MEDIAL
Inferior angle

(a) Anterior view

join at the *inferior angle.* The superior edge of the scapula, called the *superior border,* joins the medial border at the *superior angle.* The *scapular notch* is a prominent indentation along the superior border through which the suprascapular nerve passes.

At the lateral end of the superior border of the scapula is a projection of the anterior surface called the *coracoid process* (KOR-a-koyd = like a crow's beak), to which the tendons of muscles (pectoralis minor, coracobrachialis, and biceps brachii) and ligaments (coracoacromial, conoid, and trapezoid) attach. Superior and inferior to the spine on the posterior surface of the scapula are two fossae: The *supraspinous fossa* (sū-pra-SPĪ-nus) is a surface of attachment for the supraspinatus muscle of the shoulder, and the *infraspinous fossa* (in-fra-SPĪ-nus) serves as a surface of attachment for the infraspinatus muscle of the shoulder. On the anterior surface of the scapula is a slightly hollowed-out area called the *subscapular fossa,* a surface of attachment for the subscapularis muscle.

✔ **CHECKPOINT**

Which joints are formed by the scapula with other bones? What are the names of the parts of the scapula that form each joint?

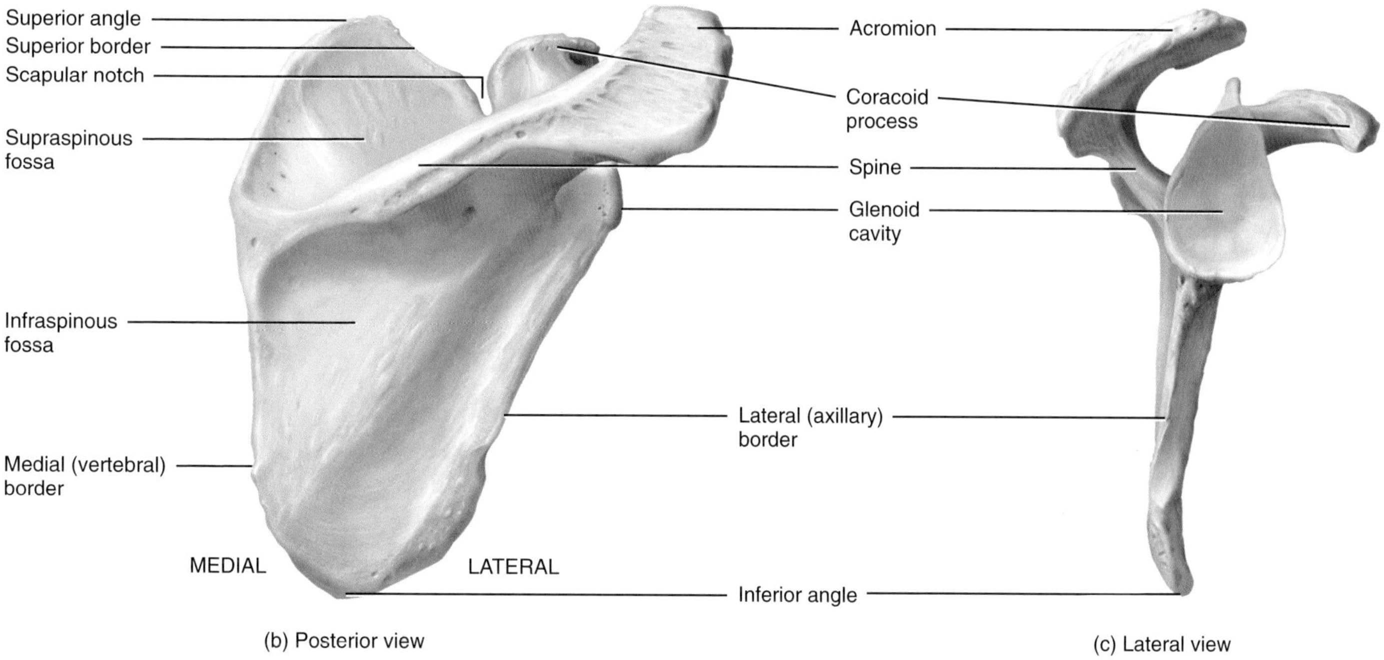
Superior angle
Superior border
Scapular notch
Supraspinous fossa
Infraspinous fossa
Medial (vertebral) border
MEDIAL
LATERAL
Acromion
Coracoid process
Spine
Glenoid cavity
Lateral (axillary) border
Inferior angle
(b) Posterior view
(c) Lateral view

Which part of the scapula forms the high point of the shoulder?

8.2 UPPER LIMB (EXTREMITY)

OBJECTIVE

- Identify the bones of the upper limb and their principal markings.

Each **upper limb (upper extremity)** has 30 bones in three locations—(1) the humerus in the arm; (2) the ulna and radius in the forearm; and (3) the 8 carpals in the carpus (wrist), the 5 metacarpals in the metacarpus (palm), and the 14 phalanges (bones of the digits) in the hand (Figure 8.4).

The bones of the upper limb are discussed in Exhibits 8.C through 8.E.

✓CHECKPOINT

2. Name the bones that form the upper limb, from proximal to distal.

Figure 8.4 Right upper limb.

6 Each upper limb includes a humerus, an ulna, a radius, carpals, metacarpals, and phalanges.

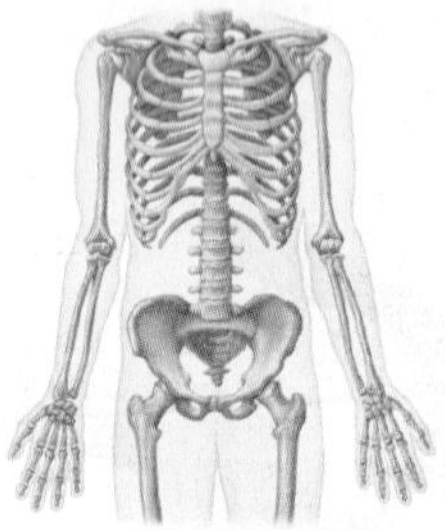

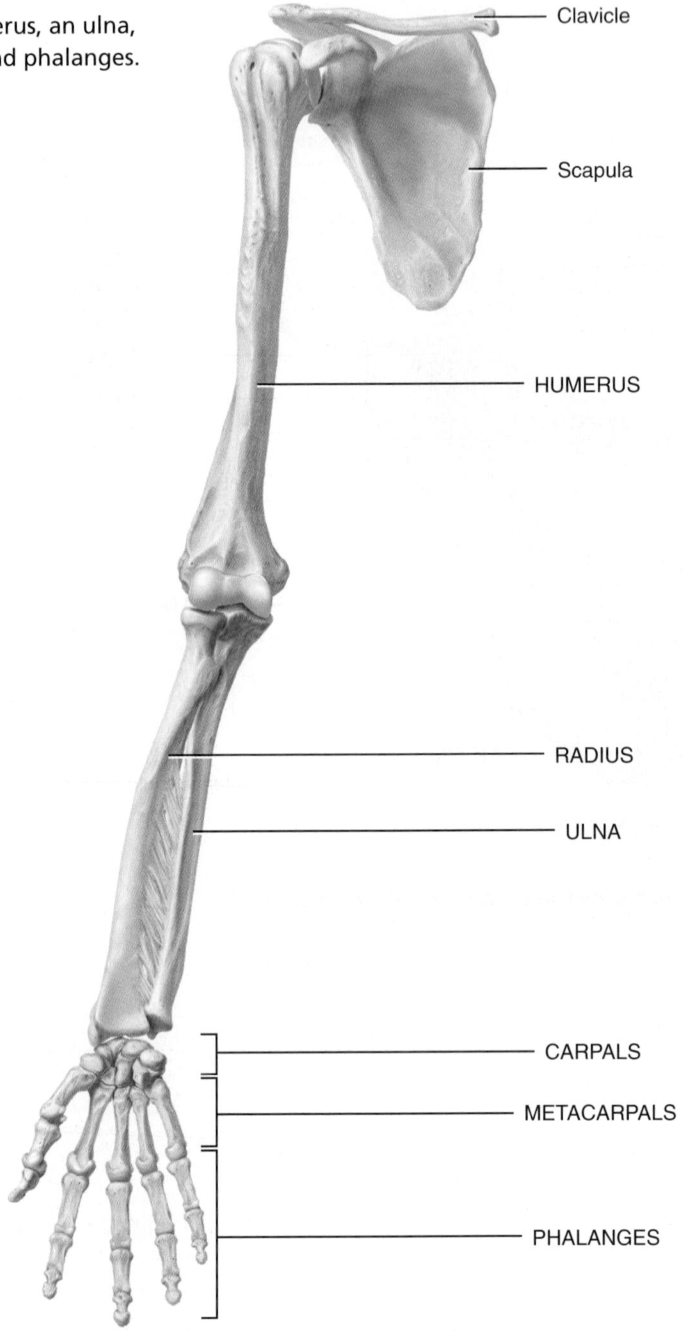
Clavicle
Scapula
HUMERUS
RADIUS
ULNA
CARPALS
METACARPALS
PHALANGES

Anterior view of free upper limb

? How many bones make up each upper limb?

Skeleton of the Arm—Humerus *(Figure 8.5)*

OBJECTIVE

- Identify the location and surface landmarks of the humerus.

The **humerus** (HŪ-mer-us), or arm bone, is the longest and largest bone of the upper limb (Figure 8.5). It articulates proximally with the scapula and distally with two bones, the ulna and the radius, to form the elbow joint.

The proximal end of the humerus features a rounded *head* that articulates with the glenoid cavity of the scapula to form the *glenohumeral (shoulder) joint.* Distal to the head is the *anatomical neck,* which is visible as an oblique groove. It is the former site of the epiphyseal (growth) plate in an adult humerus. The *greater tubercle* is a lateral projection distal to the anatomical neck. It is the most laterally palpable bony landmark of the shoulder region and is immediately inferior to the palpable

Figure 8.5 Right humerus in relation to the scapula, ulna, and radius.

The humerus is the longest and largest bone of the upper limb.

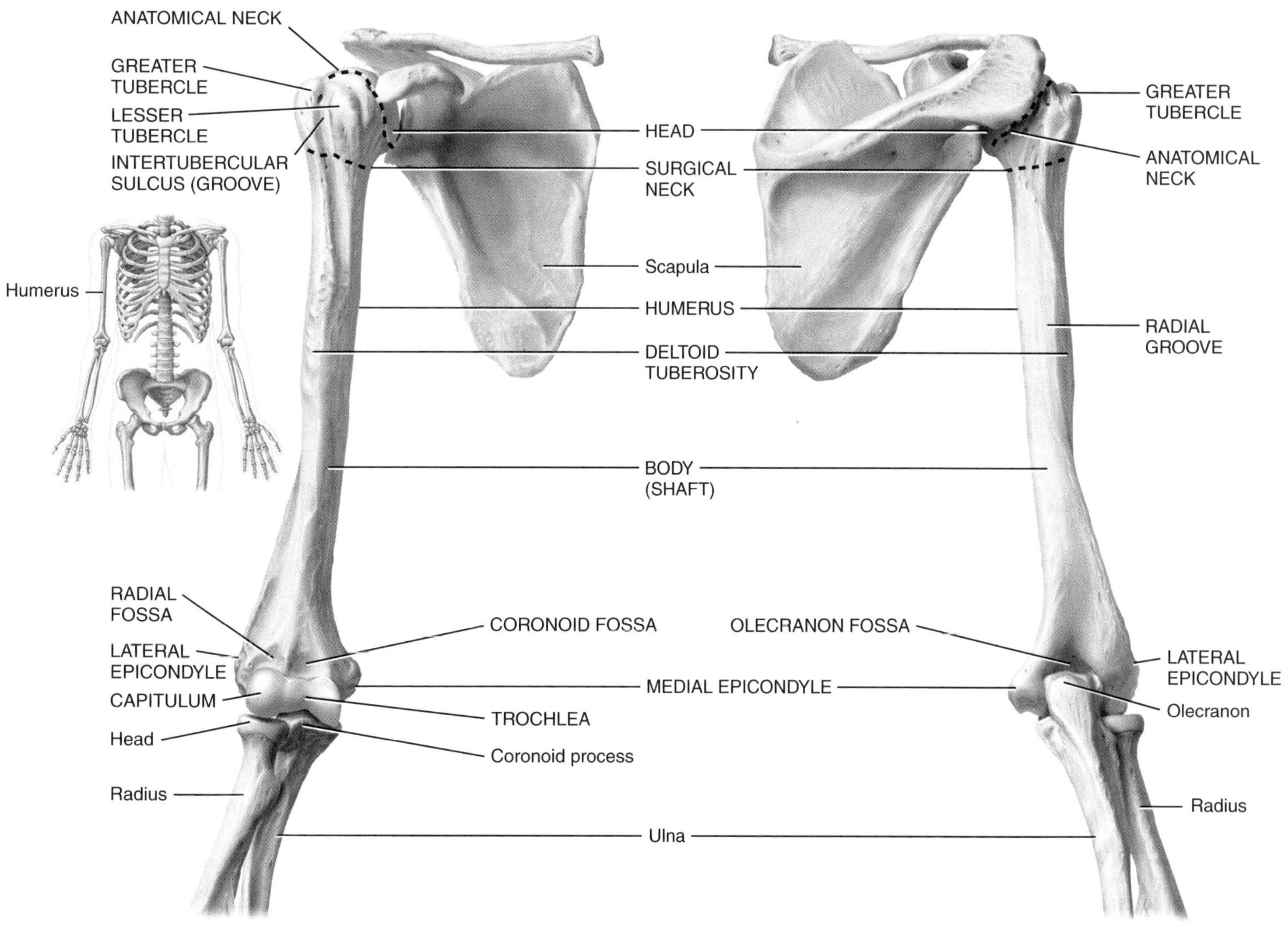

(a) Anterior view

(b) Posterior view

Which parts of the humerus articulate with the radius at the elbow? With the ulna at the elbow?

EXHIBIT 8.C CONTINUES

acromion of the scapula mentioned earlier. The *lesser tubercle* projects anteriorly. Between the two tubercles there is a groove named the *intertubercular sulcus*. The *surgical neck* is a constriction in the humerus just distal to the tubercles, where the head tapers to the shaft; it is so named because fractures often occur here.

The *body (shaft)* of the humerus is roughly cylindrical at its proximal end, but it gradually becomes triangular until it is flattened and broad at its distal end. Laterally, at the middle portion of the shaft, there is a roughened, V-shaped area called the *deltoid tuberosity.* This area serves as a point of attachment for the tendons of the deltoid muscle. On the posterior surface of the humerus is the *radial groove,* which runs along the deltoid tuberosity and contains the radial nerve.

Several prominent features are evident at the distal end of the humerus. The *capitulum* (ka-PIT-ū-lum; *capit-* = head) is a rounded knob on the lateral aspect of the bone that articulates with the head of the radius. The *radial fossa* is an anterior depression above the capitulum that articulates with the head of the radius when the forearm is flexed (bent). The *trochlea* (TROK-lē-a = pulley), located medial to the capitulum, is a spool-shaped surface that articulates with the trochlear notch of the ulna. The *coronoid fossa* (KOR-ō-noyd = crown-shaped) is an anterior depression that receives the coronoid process of the ulna when the forearm is flexed. The *olecranon fossa* (ō-LEK-ra-non = elbow) is a large posterior depression that receives the olecranon of the ulna when the forearm is extended (straightened). The *medial epicondyle* and *lateral epicondyle* are rough projections on either side of the distal end of the humerus to which the tendons of most muscles of the forearm are attached. The ulnar nerve may be palpated by rolling a finger over the skin surface above the posterior surface of the medial epicondyle. This nerve is the one that makes you feel a very severe pain when you hit your elbow, which for some reason is commonly referred to as the funnybone, even though this event is anything but funny.

✔ **CHECKPOINT**

Distinguish between the anatomical neck and the surgical neck of the humerus. Name the proximal and distal points formed by the humerus, and indicate which parts of the bones are involved.

EXHIBIT 8.D Skeleton of the Forearm—Ulna and Radius *(Figures 8.6 and 8.7)*

OBJECTIVE

- Identify the location and surface landmarks of the ulna and radius.

The **ulna** is located on the medial aspect (the little-finger side) of the forearm and is longer than the radius (Figure 8.6). You may find it convenient to use an aid called a *mnemonic device* (nē-MON-ik = memory) to learn new or unfamiliar information. One such mnemonic to help you remember the location of the ulna in relation to the hand is "p.u." (the **p**inky is on the **u**lna side).

At the proximal end of the ulna (Figure 8.6b) is the *olecranon,* which forms the prominence of the elbow. With the olecranon, an anterior projection called the *coronoid process* (Figure 8.6a) articulates with the trochlea of the humerus. The *trochlear notch* is a large curved area between the olecranon and coronoid process that forms part of the elbow joint (see Figure 8.7b). Lateral and inferior to the trochlear notch is a depression, the *radial notch,* which articulates with the head of the radius. Just inferior to the coronoid process is the *ulnar tuberosity,* to which the biceps brachii muscle attaches. The distal end of the ulna consists of a *head* that is separated from the wrist by a disc of fibrocartilage. A *styloid process* is located on the posterior side of the ulna's distal end. It provides attachment for the ulnar collateral ligament to the wrist.

The **radius** is the smaller bone of the forearm and is located on the lateral aspect (thumb side) of the forearm (Figure 8.6a). In contrast to the ulna, the radius is narrow at its proximal end and widens at its distal end.

The proximal end of the radius has a disc-shaped *head* that articulates with the capitulum of the humerus and the radial notch of the ulna. Inferior to the head is the constricted *neck.* A roughened area inferior to the neck on the anteromedial side, called the *radial tuberosity,* is a point of attachment for the tendons of the biceps brachii muscle. The shaft of the radius widens distally to form a *styloid process* on the lateral side, which can be felt proximal to the thumb. The styloid process provides attachment for the brachioradialis muscle and for attachment of the radial collateral ligament to the wrist. Fracture of the distal end of the radius is the most common fracture in adults older than 50 years.

The ulna and radius articulate with the humerus at the *elbow joint.* The articulation occurs in two places (Figure 8.7a, b): where the head of the radius articulates with the capitulum of the humerus, and where the trochlear notch of the ulna articulates with the trochlea of the humerus.

The ulna and the radius connect with one another at three sites. First, a broad, flat, fibrous connective tissue called the *interosseous membrane* (in-ter-OS-ē-us; *inter-* = between, *-osse* = bone) joins the shafts of the two bones (see Figure 8.6). This membrane also provides a site of attachment for some tendons of deep skeletal muscles of the forearm. The ulna and radius articulate directly at their proximal and distal ends (Figure 8.7b, c). Proximally, the head of the radius articulates with the ulna's radial notch. This articulation is the *proximal radioulnar joint.* Distally, the head of the ulna articulates with the *ulnar notch* of the radius. This articulation is the *distal radioulnar joint.* Finally, the distal end of the radius articulates with three bones of the wrist—the lunate, the scaphoid, and the triquetrum—to form the *radiocarpal (wrist) joint.*

✔ **CHECKPOINT**

How many joints are formed between the ulna and radius, what are their names, and what parts of the bones are involved?

Figure 8.6 Right ulna and radius in relation to the humerus and carpals.

In the forearm, the longer ulna is on the medial side, and the shorter radius is on the lateral side.

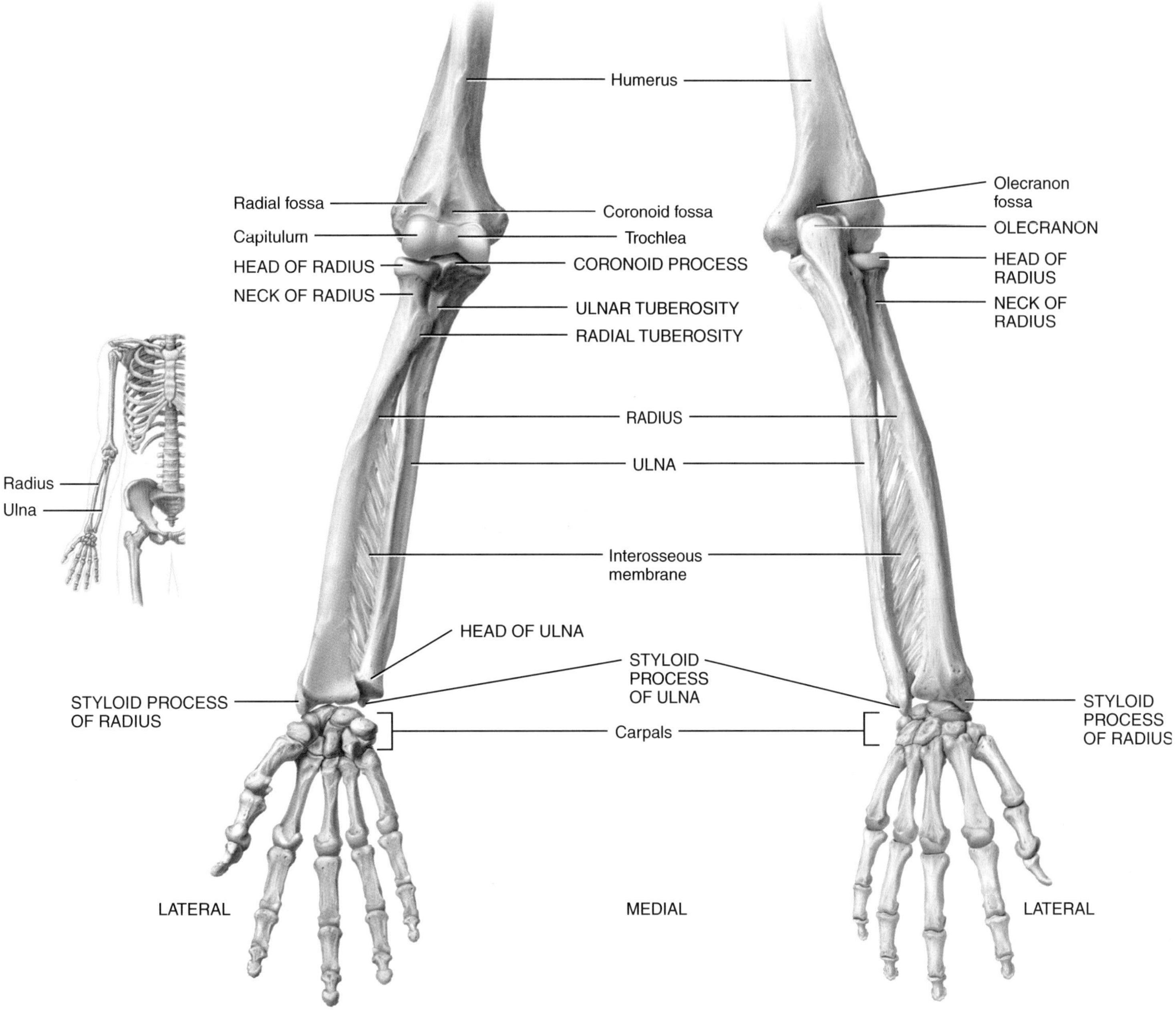

(a) Anterior view

(b) Posterior view

What part of the ulna is called the "elbow"?

EXHIBIT 8.D CONTINUES

Figure 8.7 Articulations formed by the ulna and radius. (a) Elbow joint. (b) Joint surfaces at proximal end of the ulna. (c) Joint surfaces at distal ends of radius and ulna.

The elbow joint is formed by two articulations: (1) the trochlear notch of the ulna with the trochlea of the humerus and (2) the head of the radius with the capitulum of the humerus.

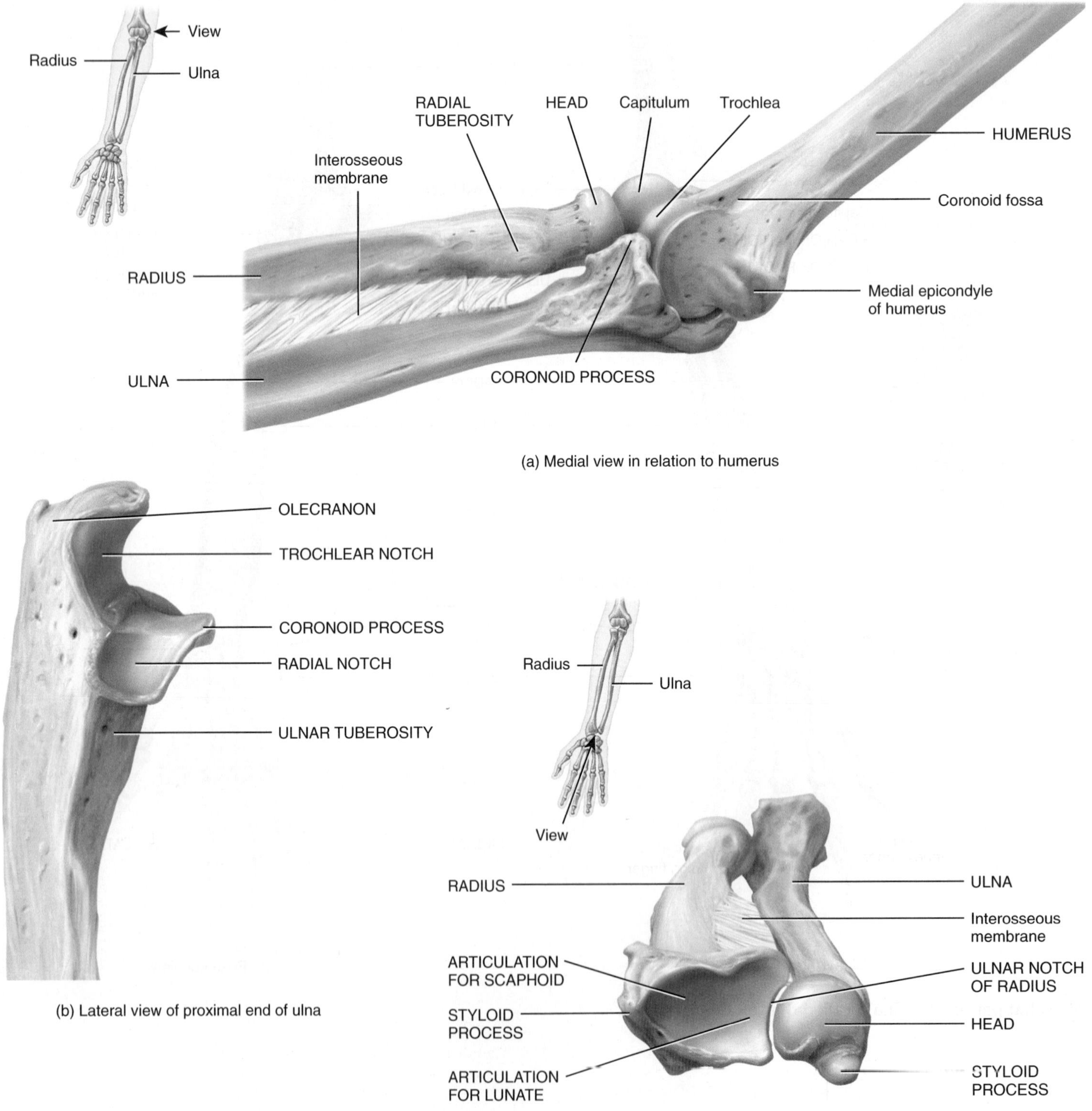

(a) Medial view in relation to humerus

(b) Lateral view of proximal end of ulna

(c) Inferior view of distal ends of radius and ulna

How many points of attachment are there between the radius and ulna?

Skeleton of the Hand—Carpals, Metacarpals, and Phalanges *(Figure 8.8)*

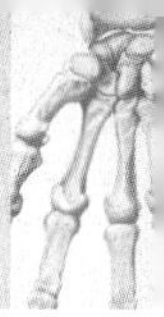

OBJECTIVE

- Identify the location and surface landmarks of the bones of the hand.

Carpals

The **carpus** (*wrist*) is the proximal region of the hand and consists of eight small bones, the **carpals,** joined to one another by ligaments (Figure 8.8). Articulations among carpal bones are called *intercarpal joints.* The carpals are arranged in two transverse rows of four bones each.

Figure 8.8 Right wrist and hand in relation to the ulna and radius.

The skeleton of the hand consists of the proximal carpals, the intermediate metacarpals, and the distal phalanges.

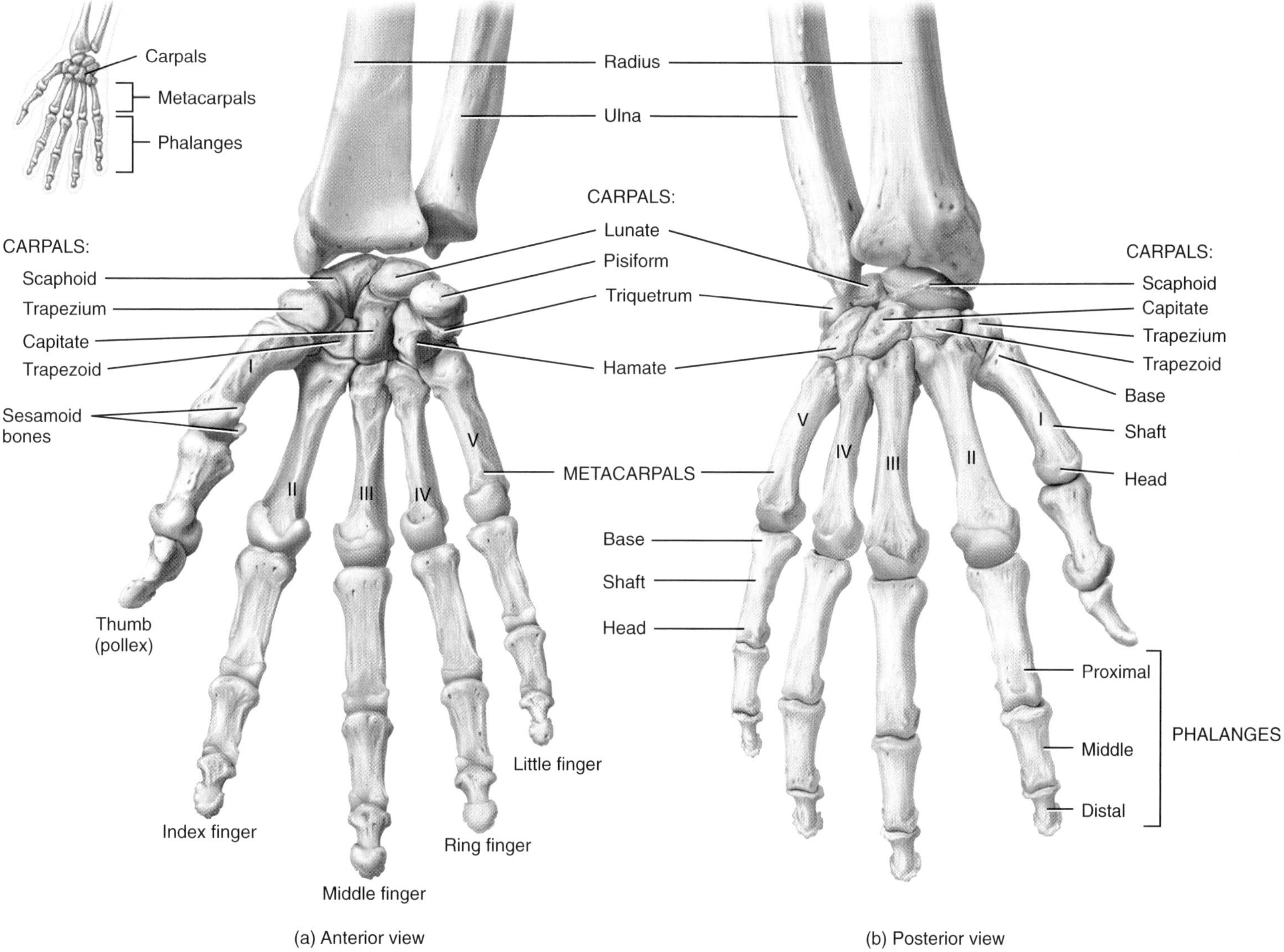

(a) Anterior view

(b) Posterior view

MNEMONIC for carpal bones*:

Stop **L**etting **T**hose **P**eople **T**ouch **T**he **C**adaver's **H**and.

Scaphoid	**L**unate	**T**riquetrum	**P**isiform	**T**rapezium	**T**rapezoid	**C**apitate	**H**amate
Proximal row				Distal row			
Lateral → Medial				Lateral → Medial			

* *Edward Tanner, University of Alabama, School of Medicine.*

Which is the most frequently fractured wrist bone?

EXHIBIT 8.E CONTINUES ▶

Skeleton of the Hand—Carpals, Metacarpals, and Phalanges *(Figure 8.8)* CONTINUED

Their names reflect their shapes. The carpals in the proximal row, from lateral to medial, are the

- **scaphoid** (SKAF-oyd = boatlike)
- **lunate** (LOO-nāt = moon-shaped)
- **triquetrum** (trī-KWĒ-trum = three-cornered)
- **pisiform** (PĪS-i-form = pea-shaped).

The proximal row of carpals articulates with the distal ends of the ulna and radius to form the *wrist joint*. The carpals in the distal row, from lateral to medial, are the

- **trapezium** (tra-PĒ-zē-um = four-sided figure with no two sides parallel)
- **trapezoid** (TRAP-e-zoyd = four-sided figure with two sides parallel)
- **capitate** (KAP-i-tāt = head-shaped)
- **hamate** (HAM-āt = hooked).

The capitate is the largest carpal bone; its rounded projection, the head, articulates with the lunate. The hamate is named for a large hook-shaped projection on its anterior surface. In about 70% of carpal fractures, only the scaphoid is broken. This is because the force of a fall on an outstretched hand is transmitted from the capitate through the scaphoid to the radius.

The anterior concave space formed by the pisiform and hamate (on the ulnar side), and the scaphoid and trapezium (on the radial side), with the rooflike covering of the *flexor retinaculum* (strong fibrous bands of connective tissue) is the **carpal tunnel.** The long flexor tendons of the digits and thumb and the median nerve pass through the carpal tunnel. Narrowing of the carpal tunnel, due to such factors as inflammation, may give rise to a condition called *carpal tunnel syndrome* (described in Clinical Connection: Carpal Tunnel Syndrome in Chapter 11).

There is a useful mnemonic for learning the names of the carpal bones provided in Figure 8.8. The first letter of the carpal bones from lateral to medial (proximal row, then distal row) corresponds to the first letter of each word in the mnemonic.

Metacarpals

The **metacarpus** (*meta-* = beyond), or *palm,* is the intermediate region of the hand and consists of five bones called **metacarpals.**

Each metacarpal bone consists of a proximal *base,* an intermediate *shaft,* and a distal *head* (Figure 8.8b). The metacarpal bones are numbered I to V (or 1–5), starting with the thumb, from lateral to medial. The bases articulate with the distal row of carpal bones to form the *carpometacarpal joints.* The heads articulate with the proximal phalanges to form the *metacarpophalangeal joints.* The heads of the metacarpals, commonly called "knuckles," are readily visible in a clenched fist.

CLINICAL CONNECTION | *Boxer's Fracture*

A **boxer's fracture** is a fracture of the fifth metacarpal, usually near the head of the bone. It frequently occurs after a person punches another person or an object, such as a wall. It is characterized by pain, swelling, and tenderness. There may also be a bump on the side of the hand. Treatment is either by casting or surgery, and the fracture usually heals in about 6 weeks. •

Phalanges

The **phalanges** (fa-LAN-jēz; *phalan-* = a battle line), or bones of the *digits,* make up the distal part of the hand. There are 14 phalanges in the five digits of each hand and, like the metacarpals, the digits are numbered I to V (or 1–5), beginning with the thumb, from lateral to medial. A single bone of a digit is referred to as a *phalanx* (FĀ-lanks).

Each phalanx consists of a proximal *base,* an intermediate *shaft,* and a distal *head.* The thumb (*pollex*) has two phalanges, and there are three phalanges in each of the other four digits. In order from the thumb, these other four digits are commonly referred to as the index finger, middle finger, ring finger, and little finger. The first row of phalanges, the *proximal row,* articulates with the metacarpal bones and second row of phalanges. The second row of phalanges, the *middle row,* articulates with the proximal row and the third row, called the *distal row.* The thumb has no middle phalanx. Joints between phalanges are called *interphalangeal joints.*

✔CHECKPOINT

Which is more distal, the base or the head of the carpals?
With which bones do the proximal phalanges articulate?

8.3 PELVIC (HIP) GIRDLE

OBJECTIVE

- Identify the bones of the pelvic girdle and their principal markings.

The **pelvic (hip) girdle** consists of the two **hip bones,** also called **coxal** (KOK-sal; *cox-* = hip) or **pelvic bones** or **os coxa** (Figure 8.9). The hip bones unite anteriorly at a joint called the **pubic symphysis** (PŪ-bik SIM-fi-sis). They unite posteriorly with the sacrum at the *sacroiliac joints.* The complete ring composed of the hip bones, pubic symphysis, and sacrum forms a deep, basin-like structure called the **bony pelvis** (*pelv-* = basin). The plural is *pelves* (PEL-vēz) or *pelvises.* Functionally, the bony pelvis provides a strong and stable support for the vertebral column and pelvic and lower abdominal organs. The pelvic girdle of the bony pelvis also connects the bones of the lower limbs to the axial skeleton.

Each of the two hip bones of a newborn consists of three bones separated by cartilage: a superior *ilium,* an inferior and anterior *pubis,* and an inferior and posterior *ischium.* By age 23, the three separate bones fuse together (Figure 8.10a). Although the hip bones function as single bones, anatomists commonly discuss each hip bone as three separate bones.

The bones of the pelvic girdle are discussed in Exhibit 8.F.

✔CHECKPOINT

3. Describe the distinguishing characteristics of the individual bones of the pelvic girdle.
4. Which bones form the acetabulum? What is its function?

Figure 8.9 **Bony pelvis.** Shown here is the female bony pelvis.

The hip bones unite anteriorly at the public symphysis and posteriorly at the sacrum to form the bony pelvis.

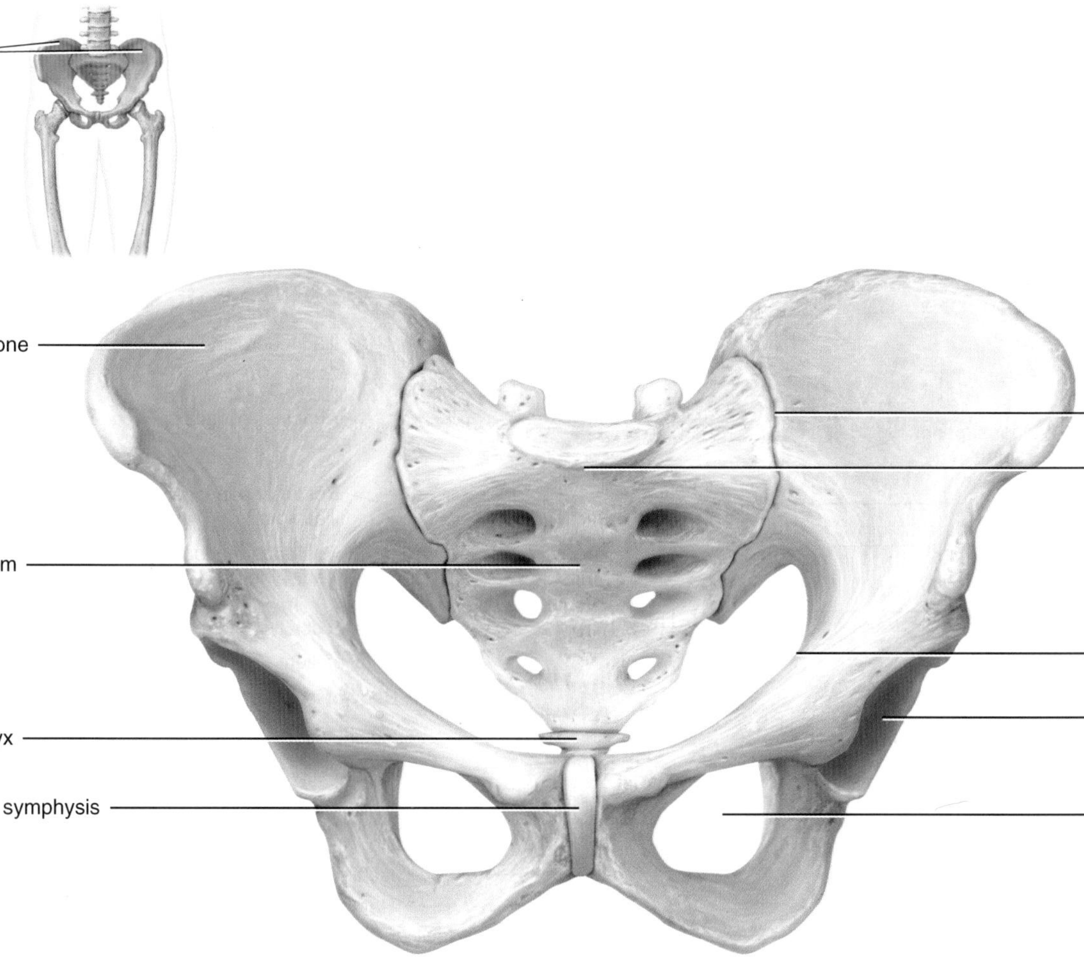
Pelvic (hip) girdle
Hip bone
Sacrum
Coccyx
Pubic symphysis
Sacroiliac joint
Sacral promontory
Pelvic brim
Acetabulum
Obturator foramen

Anterosuperior view of pelvic girdle

What are the functions of the bony pelvis?

Figure 8.10 Right hip bone. The lines of fusion of the ilium, ischium, and pubis depicted in part (a) are not always visible in an adult.

The acetabulum is the socket for the head of the femur, where the three parts of the hip bone converge and ossify.

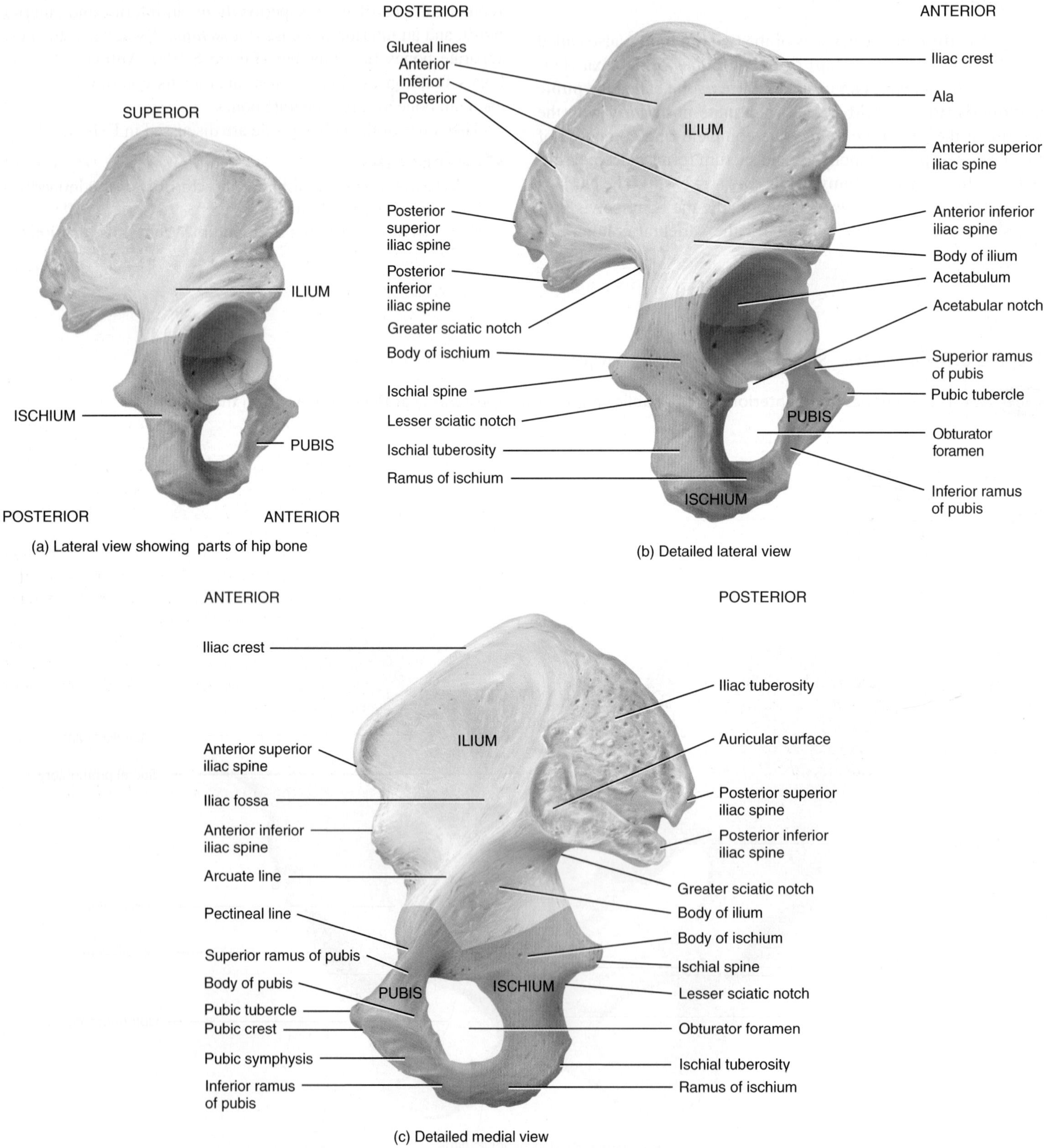

Which part of the hip bone articulates with the femur? With the sacrum?

EXHIBIT 8.F Bones of the Pelvic Girdle

OBJECTIVE

- Identify the locations and surface features of the three components of the hip bone.

Ilium

The **ilium** (IL-ē-um = flank), the largest of the three components of the hip bone (Figure 8.10b), is composed of a superior *ala* (= wing) and an inferior *body.* The body is one of the components of the *acetabulum,* the socket for the head of the femur. The superior border of the ilium, the *iliac crest,* ends anteriorly in a blunt *anterior superior iliac spine.* Bruising of the anterior superior iliac spine and associated soft tissues, such as occurs in body contact sports, is called a **hip pointer.** Below this spine is the *anterior inferior iliac spine.* Posteriorly, the iliac crest ends in a sharp *posterior superior iliac spine.* Below this spine is the *posterior inferior iliac spine.* The spines serve as points of attachment for the tendons of the muscles of the trunk, hip, and thighs. Below the posterior inferior iliac spine is the *greater sciatic notch* (sī-AT-ik), through which the sciatic nerve, the longest nerve in the body, passes.

The medial surface of the ilium contains the *iliac fossa,* a concavity where the tendon of the iliacus muscle attaches. Posterior to this fossa are the *iliac tuberosity,* a point of attachment for the sacroiliac ligament, and the *auricular surface* (*auric-* = ear-shaped), which articulates with the sacrum to form the *sacroiliac joint* (see Figure 8.9). Projecting anteriorly and inferiorly from the auricular surface is a ridge called the *arcuate line* (AR-kū-āt; *arc-* = bow).

The other conspicuous markings of the ilium are three arched lines on its lateral surface called the *posterior gluteal line* (*glut-* = buttock), the *anterior gluteal line,* and the *inferior gluteal line.* The gluteal muscles attach to the ilium between these lines.

Ischium

The **ischium** (IS-kē-um = hip), the inferior, posterior portion of the hip bone (Figure 8.10b, c), comprises a superior *body* and an inferior *ramus* (*ram-* = branch; plural is *rami*). The ramus is the portion of the ischium that fuses with the pubis. Features of the ischium include the prominent *ischial spine,* a *lesser sciatic notch* below the spine, and a rough and thickened *ischial tuberosity.* Because this prominent tuberosity is just deep to the skin, it commonly begins hurting after a relatively short time when you sit on a hard surface. Together, the ramus and the pubis surround the *obturator foramen* (OB-too-rā-tōr; *obtur-* = closed up), the largest foramen in the skeleton. The foramen is so named because, even though blood vessels and nerves pass through it, it is nearly completely closed by the fibrous *obturator membrane.*

Pubis

The **pubis** (PŪ-bis; plural is *pubes*), meaning pubic bone, is the anterior and inferior part of the hip bone (Figure 8.10b, c). A *superior ramus,* an *inferior ramus,* and a *body* between the rami make up the pubis. The anterior, superior border of the body is the *pubic crest,* and at its lateral end is a projection called the *pubic tubercle.* This tubercle is the beginning of a raised line, the *pectineal line* (pek-TIN-ē-al), which extends superiorly and laterally along the superior ramus to merge with the arcuate line of the ilium. These lines, as you will see shortly, are important landmarks for distinguishing the superior (false) and inferior (true) portions of the bony pelvis.

The *pubic symphysis* is the joint between the two pubes of the hip bones (see Figure 8.9). It consists of a disc of fibrocartilage. Inferior to this joint, the inferior rami of the two pubic bones converge to form the *pubic arch.* In the later stages of pregnancy, the hormone relaxin (produced by the ovaries and placenta) increases the flexibility of the pubic symphysis to ease delivery of the baby. Weakening of the joint, together with an already altered center of gravity due to an enlarged uterus, also changes the mother's gait during pregnancy.

The *acetabulum* (as-e-TAB-ū-lum = vinegar cup) is a deep fossa formed by the ilium, ischium, and pubis. It functions as the socket that accepts the rounded head of the femur. Together, the acetabulum and the femoral head form the *hip (coxal) joint.* On the inferior side of the acetabulum is a deep indentation, the *acetabular notch,* that forms a foramen through which blood vessels and nerves pass and serves as a point of attachment for ligaments of the femur (for example, the ligament of the head of the femur).

✔ **CHECKPOINT**

Why is the obturator foramen so named? Which joints are formed by the union of the hip bones with other bones?

8.4 FALSE AND TRUE PELVES

OBJECTIVES

- Distinguish between the false and true pelves.
- Explain why the false and true pelves are important clinically.

The bony pelvis is divided into superior and inferior portions by a boundary called the *pelvic brim* or *inlet* (Figure 8.11a). You can trace the pelvic brim by following the landmarks around parts of the hip bones to form the outline of an oblique plane. Beginning posteriorly at the *sacral promontory* of the sacrum, trace laterally and inferiorly along the *arcuate lines* of the ilium. Continue

Figure 8.11 **True and false pelves.** Shown here is the female pelvis. For simplicity, in part (a) the landmarks of the pelvic brim are shown only on the left side of the body, and the outline of the pelvic brim is shown only on the right side. The entire pelvic brim is shown in Table 8.1.

The true and false pelves are separated by the pelvic brim.

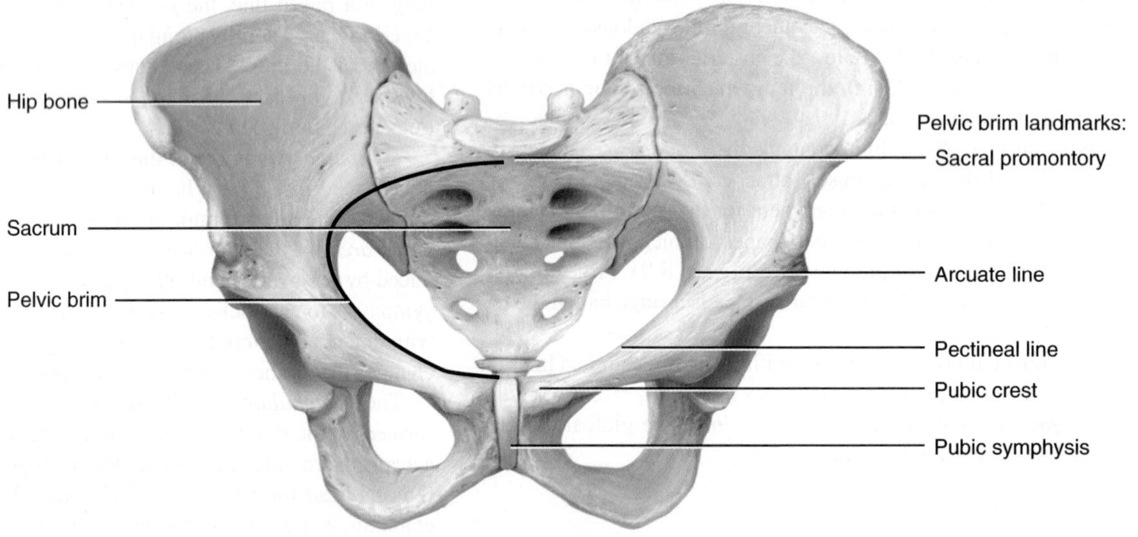
Hip bone
Sacrum
Pelvic brim
Pelvic brim landmarks:
Sacral promontory
Arcuate line
Pectineal line
Pubic crest
Pubic symphysis

(a) Anterior view of pelvic girdle

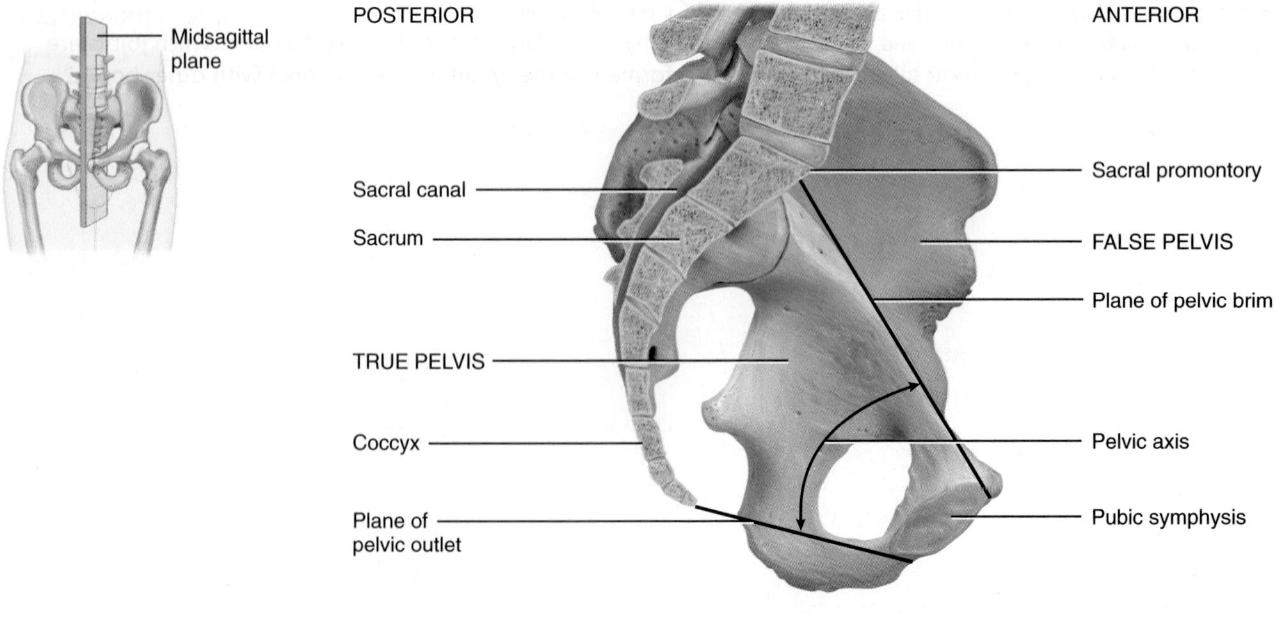
Midsagittal plane
POSTERIOR
ANTERIOR
Sacral canal
Sacrum
TRUE PELVIS
Coccyx
Plane of pelvic outlet
Sacral promontory
FALSE PELVIS
Plane of pelvic brim
Pelvic axis
Pubic symphysis

(b) Midsagittal section indicating locations of true and false pelves inferiorly along the *pectineal lines* of the pubis. Finally, trace anteriorly along the *pubic crest* to the superior portion of the *pubic symphysis*. Together, these points form an oblique plane that is higher in the back than in the front. The circumference of this plane is the pelvic brim.

The portion of the bony pelvis superior to the pelvic brim is referred to as the **false (greater) pelvis** (Figure 8.11b). It is bordered by the lumbar vertebrae posteriorly, the upper portions of the hip bones laterally, and the abdominal wall anteriorly. The space enclosed by the false pelvis is part of the lower abdomen; it contains the superior portion of the urinary bladder (when it is full) and the lower intestines in both genders and the uterus, ovaries, and uterine tubes of the female.

The portion of the bony pelvis inferior to the pelvic brim is the **true (lesser) pelvis** (Figure 8.11b). It has an inlet, an outlet, and a cavity. It is bounded by the sacrum and coccyx posteriorly, inferior portions of the ilium and ischium laterally, and the pubic bones anteriorly. The true pelvis surrounds the pelvic cavity, which was described in Chapter 1 (see Figure 1.9). The true pelvis contains the rectum and urinary bladder in both genders, the vagina and cervix of the uterus in females, and the prostate in males. The superior opening of the true pelvis, bordered by the pelvic brim, is called the *pelvic inlet;* the inferior opening of the true pelvis is the *pelvic outlet.* The *pelvic axis* is an imaginary line that curves through the true pelvis from the central point of the plane of the pelvic inlet to the central point of the plane of the pelvic outlet. During childbirth the pelvic axis is the route taken by the baby's head as it descends through the pelvis.

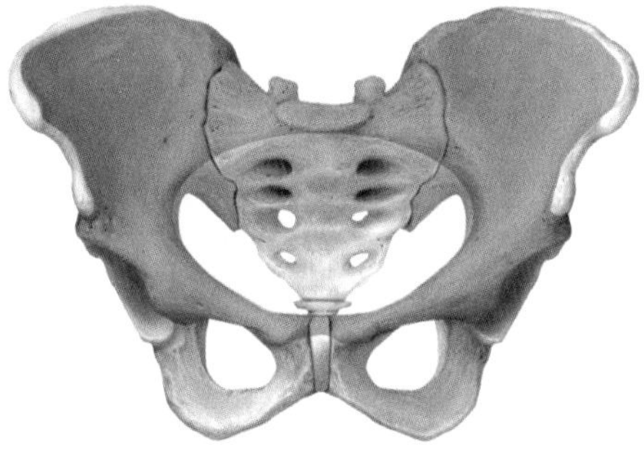

(c) Anterior view of false pelvis (pink)

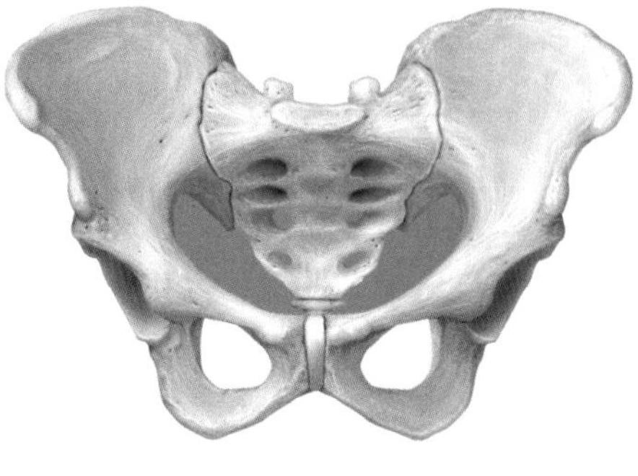

(d) Anterior view of true pelvis (blue)

What is the significance of the pelvic axis?

CLINICAL CONNECTION | *Pelvimetry*

Pelvimetry is the measurement of the size of the inlet and outlet of the birth canal, which may be done by ultrasonography or physical examination. Measurement of the pelvic cavity in pregnant females is important because the fetus must pass through the narrower opening of the pelvis at birth. A cesarean section is usually planned if it is determined that the pelvic cavity is too small to permit passage of the baby. •

CHECKPOINT

5. Why are the false and true pelves important clinically?

8.5 COMPARISON OF FEMALE AND MALE PELVES

OBJECTIVE

- Compare the principal differences between female and male pelves.

Generally, the bones of males are larger and heavier and possess larger surface markings than those of females of comparable age and physical stature. Sex-related differences in the features of bones are readily apparent when comparing the adult female and male pelves. Most of the structural differences in the pelves are adaptations to the requirements of pregnancy and childbirth. The female's pelvis is wider and shallower than the male's. Consequently, there is more space in the true pelvis of the female, especially in the pelvic inlet and pelvic outlet, to accommodate the passage of the infant's head at birth. Other significant structural differences between the pelves of females and males are listed and illustrated in Table 8.1.

CHECKPOINT

6. How is the female pelvis adapted for pregnancy and childbirth?
7. Using Table 8.1 as a guide; select the three ways that are easiest for you to distinguish a female from a male pelvis.

TABLE 8.1

Comparison of Female and Male Pelves

POINT OF COMPARISON	FEMALE	MALE
General structure	Light and thin.	Heavy and thick.
False (greater) pelvis	Shallow.	Deep.
Pelvic brim (inlet)	Wide and more oval.	Narrow and heart-shaped.
Acetabulum	Small and faces anteriorly.	Large and faces laterally.
Obturator foramen	Oval.	Round.
Pubic arch	Greater than 90° angle.	Less than 90° angle.

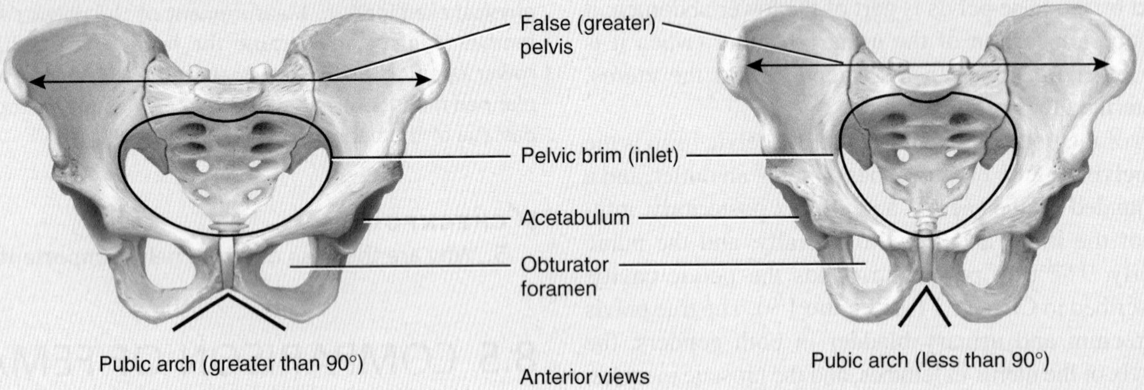
False (greater) pelvis
Pelvic brim (inlet)
Acetabulum
Obturator foramen
Pubic arch (greater than 90°)
Pubic arch (less than 90°)
Anterior views

Iliac crest	Less curved.	More curved.
Ilium	Less vertical.	More vertical.
Greater sciatic notch	Wide (almost 90°).	Narrow (about 70°; inverted V).
Coccyx	More movable and more curved anteriorly.	Less movable and less curved anteriorly.
Sacrum	Shorter, wider (see anterior views), and less curved anteriorly.	Longer, narrower (see anterior views), and more curved anteriorly.

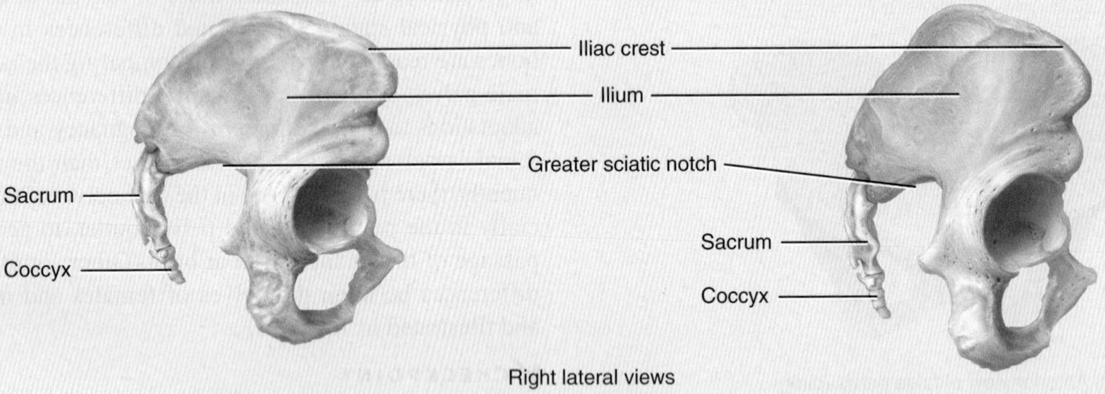
Iliac crest
Ilium
Greater sciatic notch
Sacrum
Coccyx
Sacrum
Coccyx
Right lateral views

Pelvic outlet	Wider.	Narrower.
Ischial tuberosity	Shorter, farther apart, and more medially projecting.	Longer, closer together, and more laterally projecting.

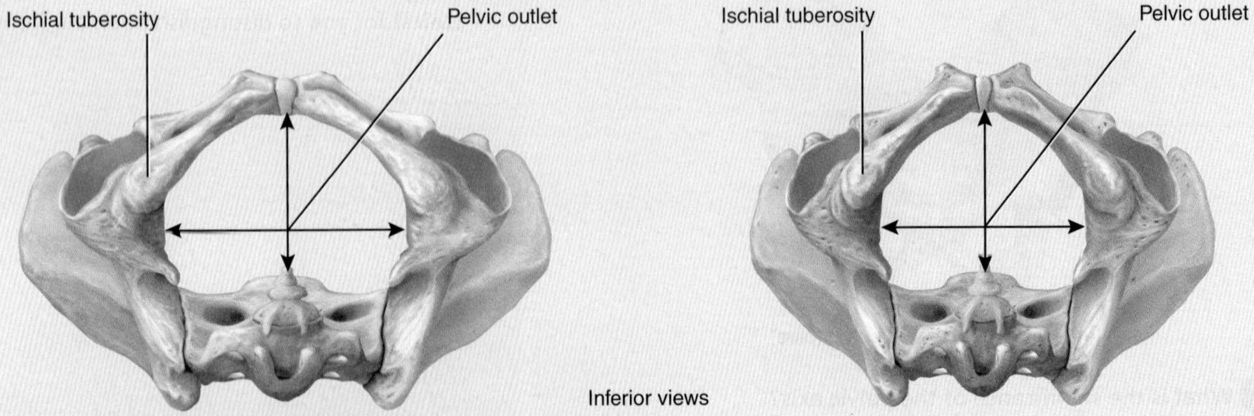
Ischial tuberosity
Pelvic outlet
Ischial tuberosity
Pelvic outlet
Inferior views

8.6 LOWER LIMB (EXTREMITY)

OBJECTIVE

- Identify the bones of the lower limb and their principal markings.

Each **lower limb (lower extremity)** has 30 bones in four locations—(1) the femur in the thigh; (2) the patella (kneecap); (3) the tibia and fibula in the leg; and (4) the 7 tarsals in the tarsus (ankle), the 5 metatarsals in the metatarsus, and the 14 phalanges (bones of the digits) in the foot (Figure 8.12).

The bones of the lower limb are discussed in Exhibits 8.G through 8.I.

CHECKPOINT

8. Name the bones that form the lower limb, from proximal to distal.
9. What are the names and functions of the arches of the foot?

Figure 8.12 **Right lower limb.**

Each lower limb includes a femur, a patella (kneecap), a tibia, a fibula, tarsals (ankle bones), metatarsals, and phalanges (bones of the digits).

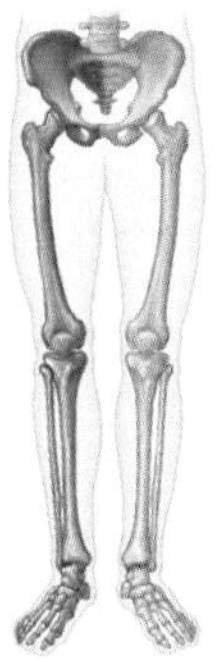

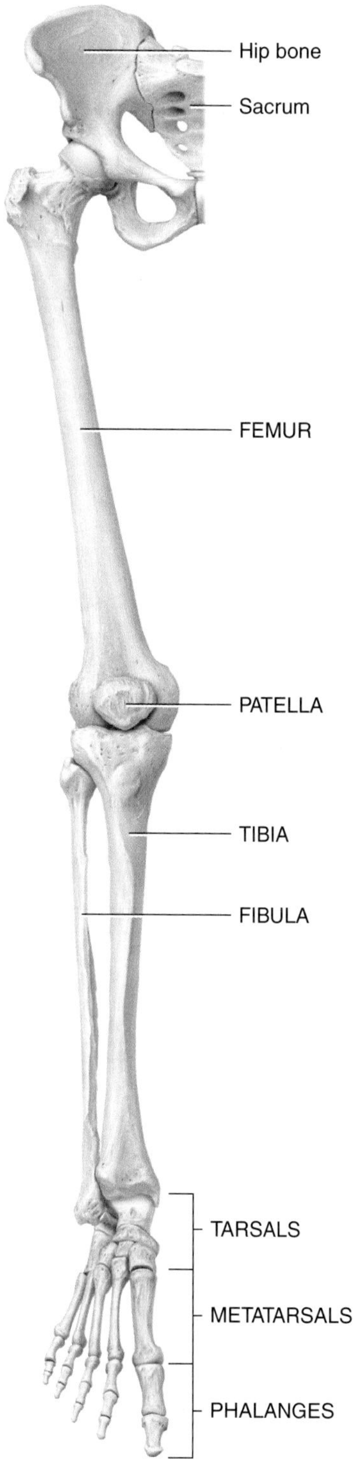
Hip bone
Sacrum
FEMUR
PATELLA
TIBIA
FIBULA
TARSALS
METATARSALS
PHALANGES

Anterior view of free lower limb

How many bones make up each lower limb?

EXHIBIT 8.G Skeleton of the Thigh—Femur and Patella *(Figures 8.13 and 8.14)*

OBJECTIVE

- Identify the location and surface features of the femur and patella.

Femur

The **femur,** or thigh bone, is the longest, heaviest, and strongest bone in the body (Figure 8.13). Its proximal end articulates with the acetabulum of the hip bone. Its distal end articulates with the tibia and patella. The

Figure 8.13 Right femur in relation to the hip bone, patella, tibia, and fibula.

The acetabulum of the hip bone and head of the femur articulate to form the hip joint.

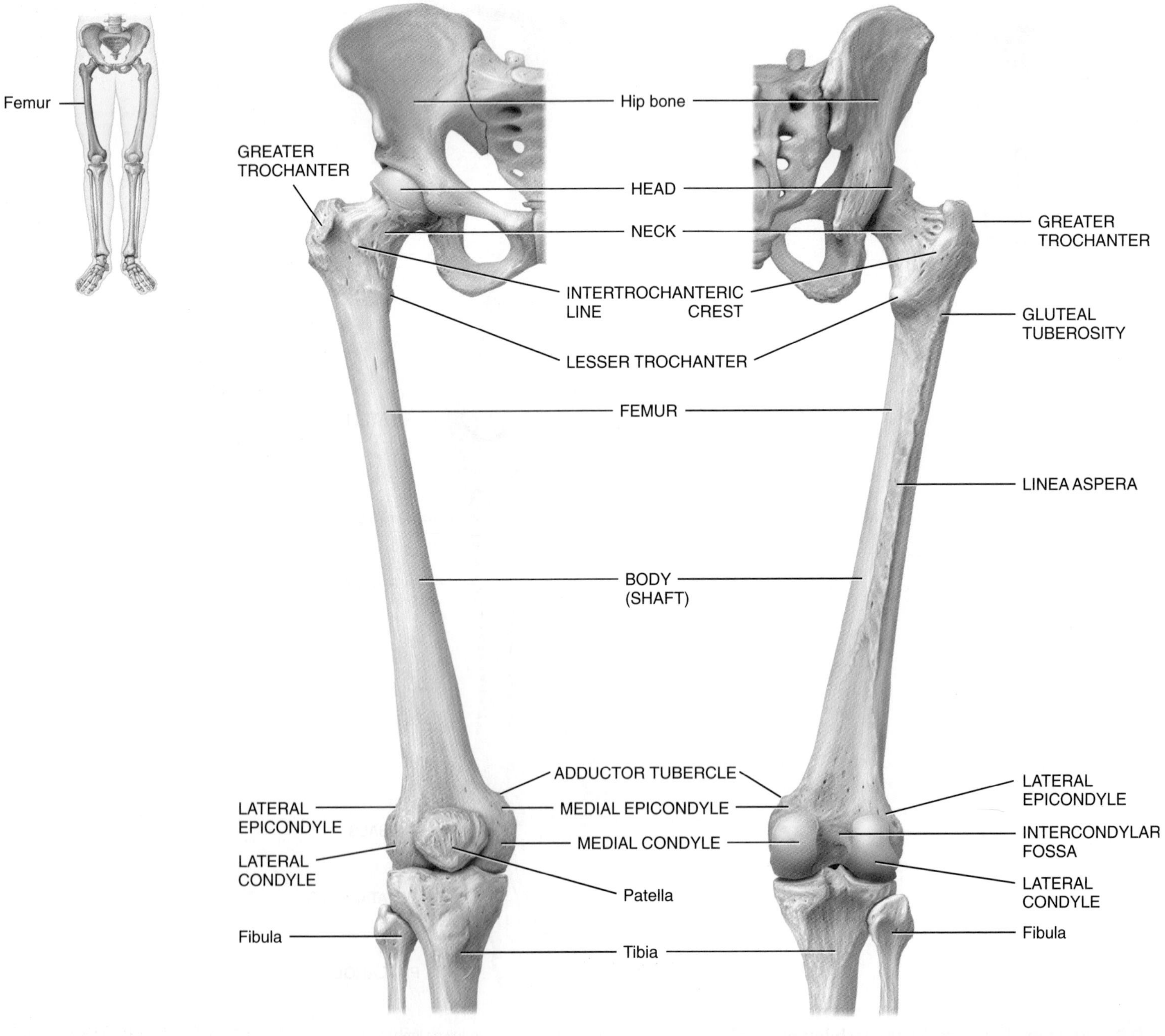

(a) Anterior view

(b) Posterior view

body (shaft) of the femur angles medially and, as a result, the knee joints are closer to the midline than the hip joints. This angle of the femoral shaft *(angle of convergence)* is greater in females because the female pelvis is broader.

The proximal end of the femur consists of a rounded *head* that articulates with the acetabulum of the hip bone to form the *hip (coxal) joint.* The head contains a small central depression (pit) called the *fovea capitis* (FŌ-vē-a CAP-i-tis; *fovea* = pit, *capitis* = of the head). The ligament of the head of the femur connects the fovea capitis of the femur to the acetabulum of the hip bone. The *neck* of the femur is a constricted region distal to the head. A "broken hip" is more often associated with a fracture in the neck of the femur than with fractures of the hip bones. The *greater trochanter* (trō-KAN-ter) and *lesser trochanter* are projections from the junction of the neck and shaft that serve as points of attachment for the tendons of some of the thigh and buttock muscles. The greater trochanter is the prominence felt and seen anterior to the hollow on the side of the hip. It is a landmark commonly used to locate the site for intramuscular injections into the lateral surface of the thigh. The lesser trochanter is inferior and medial to the greater trochanter. Between the anterior surfaces of the trochanters is a narrow *intertrochanteric line* (Figure 8.13a). A ridge called the *intertrochanteric crest* appears between the posterior surfaces of the trochanters (Figure 8.13b).

Inferior to the intertrochanteric crest on the posterior surface of the body of the femur is a vertical ridge called the *gluteal tuberosity.* It blends into another vertical ridge called the *linea aspera* (LIN-ē-a AS-per-a; *asper* = rough). Both ridges serve as attachment points for the tendons of several thigh muscles.

The expanded distal end of the femur includes the *medial condyle* (= knuckle) and the *lateral condyle.* These articulate with the medial and lateral condyles of the tibia. Superior to the condyles are the *medial epicondyle* and the *lateral epicondyle,* to which ligaments of the knee joint attach. A depressed area between the condyles on the posterior surface is called the *intercondylar fossa* (in-ter-KON-di-lar). The *patellar surface* is located between the condyles on the anterior surface. Just superior to the medial epicondyle is the *adductor tubercle,* a roughened projection that is a site of attachment for the adductor magnus muscle.

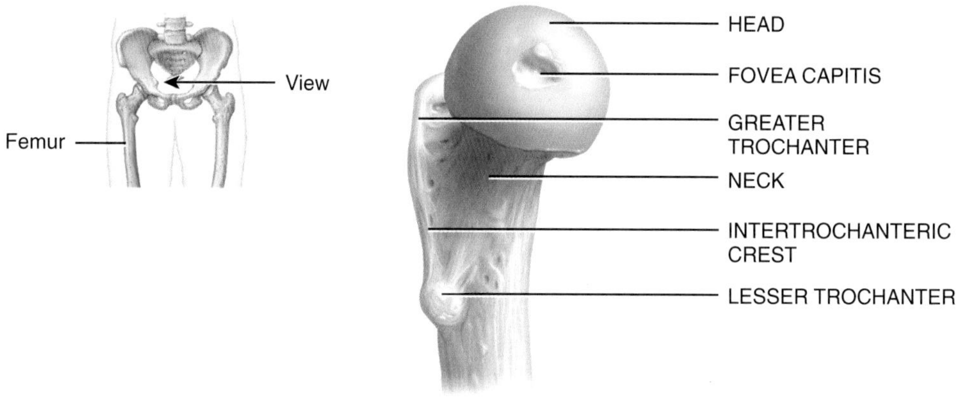
View
Femur
HEAD
FOVEA CAPITIS
GREATER
TROCHANTER
NECK
INTERTROCHANTERIC
CREST
LESSER TROCHANTER

(c) Medial view of proximal end of femur

Why is the angle of convergence of the femurs greater in females than males?

EXHIBIT 8.G CONTINUES

Patella

The **patella** (= little dish), or kneecap, is a small, triangular bone located anterior to the knee joint (Figure 8.14). The broad proximal end of this sesamoid bone, which develops in the tendon of the quadriceps femoris muscle, is called the *base;* the pointed distal end is referred to as the *apex.* The posterior surface contains two *articular facets,* one for the medial condyle of the femur and another for the lateral condyle of the femur. The patellar ligament attaches the patella to the tibial tuberosity. The *patellofemoral joint,* between the posterior surface of the patella and the patellar surface of the femur, is the intermediate component of the *tibiofemoral (knee) joint.* The patella increases the leverage of the tendon of the quadriceps femoris muscle, maintains the position of the tendon when the knee is bent (flexed), and protects the knee joint.

CLINICAL CONNECTION | *Patellofemoral Stress Syndrome*

Patellofemoral stress syndrome (runner's knee) is one of the most common problems runners experience. During normal flexion and extension of the knee, the patella tracks (glides) superiorly and inferiorly in the groove between the femoral condyles. In patellofemoral stress syndrome, normal tracking does not occur; instead, the patella tracks laterally as well as superiorly and inferiorly, and the increased pressure on the joint causes aching or tenderness around or under the patella. The pain typically occurs after a person has been sitting for a while, especially after exercise. It is worsened by squatting or walking down stairs. One cause of runner's knee is constantly walking, running, or jogging on the same side of the road. Other predisposing factors include running on hills, running long distances, and an anatomical deformity called genu valgum, or knock-knee (see Medical Terminology at the end of the chapter). •

✔CHECKPOINT

What is the clinical importance of the greater trochanter?

Which joints are formed by the femur with other bones?

Figure 8.14 Right patella.

The patella articulates with the lateral and medial condyles of the femur.

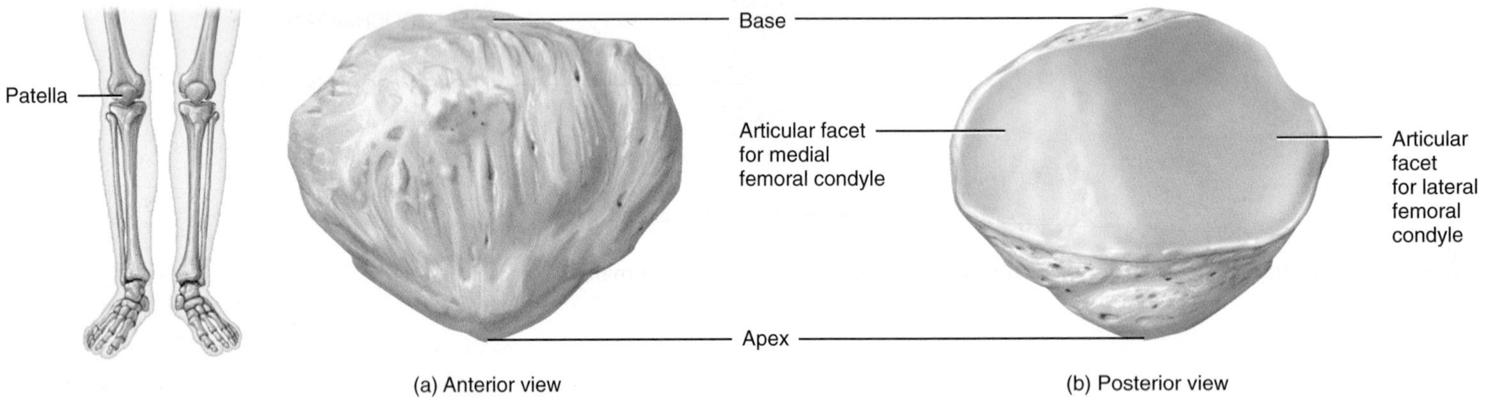
Patella
Base
Articular facet
for medial
femoral condyle
Articular
facet
for lateral
femoral
condyle
Apex
(a) Anterior view
(b) Posterior view

? The patella is classified as which type of bone? Why?

EXHIBIT 8.H Skeleton of the Leg—Tibia and Fibula *(Figure 8.15)*

OBJECTIVE

- Identify the location and surface features of the tibia and fibula.

Tibia

The **tibia,** or shin bone, is the larger, medial, weight-bearing bone of the leg (Figure 8.15). The term *tibia* means flute, because the tibial bones of birds were used in ancient times to make musical instruments. The tibia

Figure 8.15 Right tibia and fibula in relation to the femur, patella, and talus.

The tibia articulates with the femur and fibula proximally, and with the fibula and talus distally.

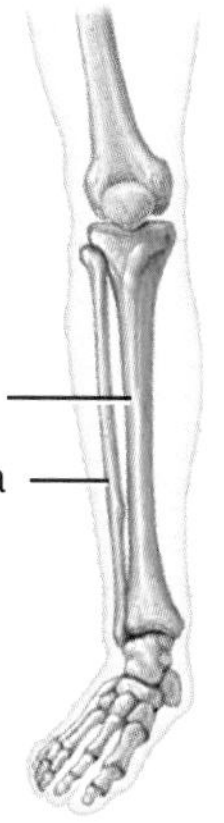

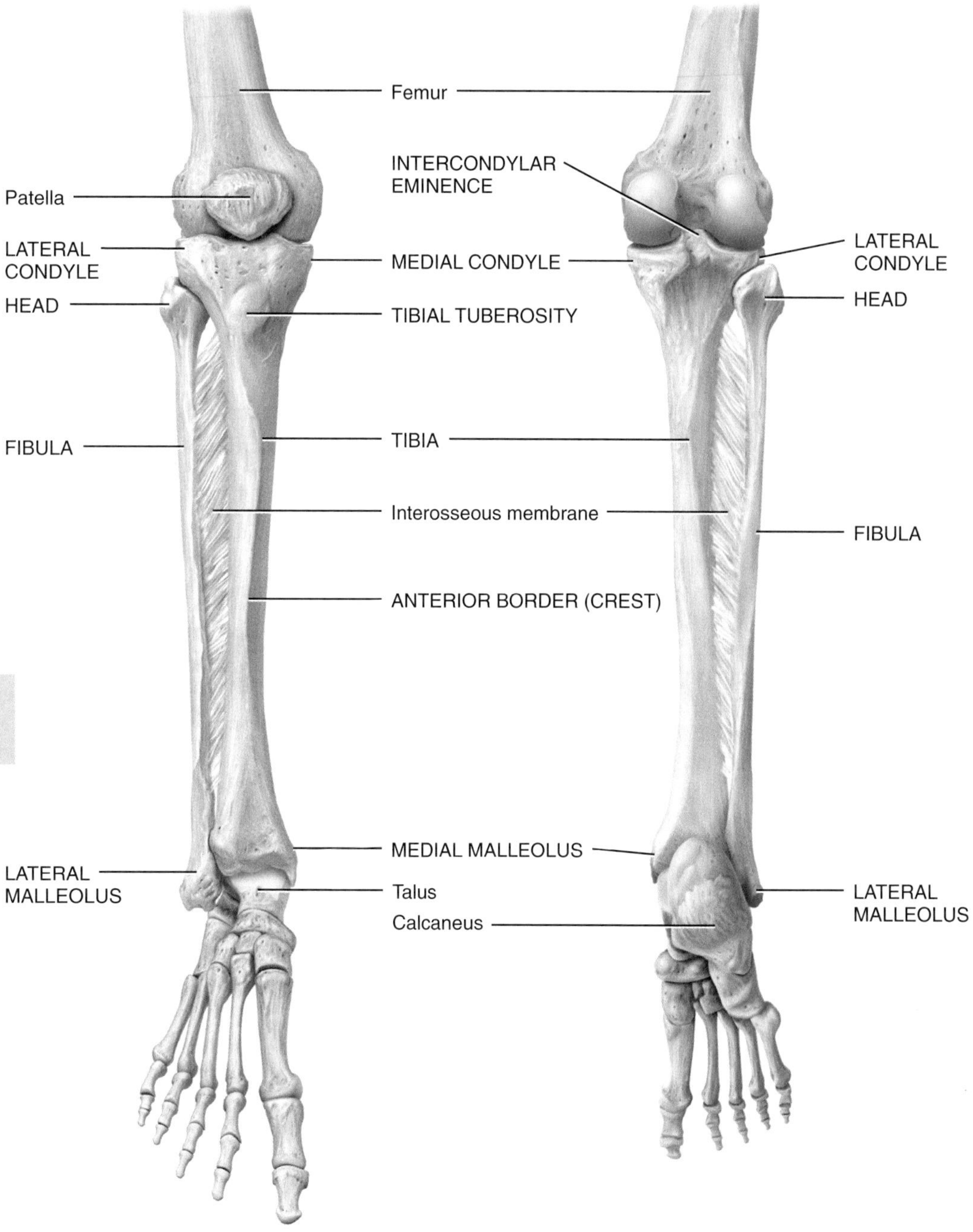

(a) Anterior view

(b) Posterior view

MNEMONIC for location of tibia and fibula:
The fibu**LA** is **LA**teral.

EXHIBIT 8.H CONTINUES

FIGURE 8.15 CONTINUED

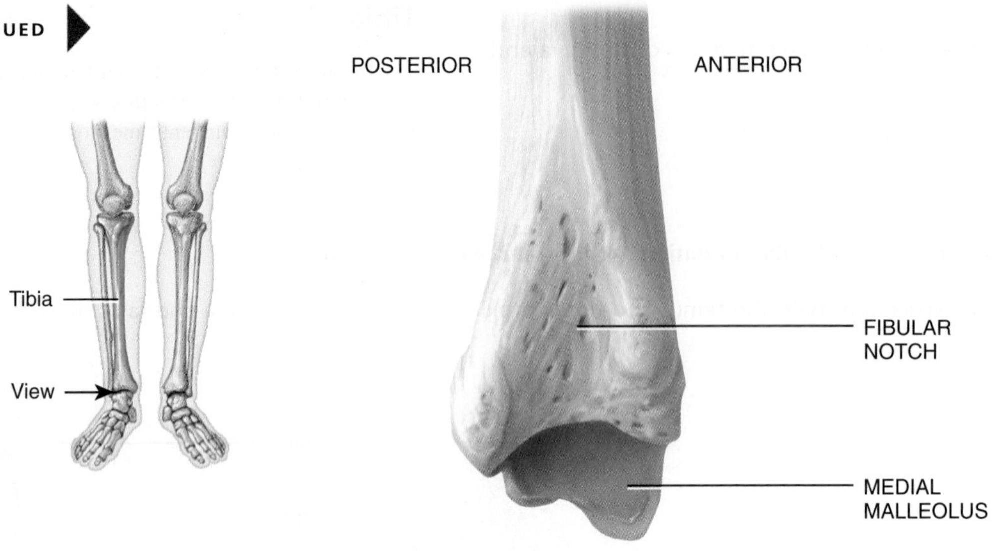
POSTERIOR
ANTERIOR
Tibia
View
FIBULAR NOTCH
MEDIAL MALLEOLUS

(c) Lateral view of distal end of tibia

Which leg bone bears the weight of the body?

articulates at its proximal end with the femur and fibula, and at its distal end with the fibula and the talus bone of the ankle. The tibia and fibula, like the ulna and radius, are connected by an interosseous membrane.

The proximal end of the tibia is expanded into a *lateral condyle* and a *medial condyle.* These articulate with the condyles of the femur to form the lateral and medial *tibiofemoral (knee) joints.* The inferior surface of the lateral condyle articulates with the head of the fibula. The slightly concave condyles are separated by an upward projection called the *intercondylar eminence* (Figure 8.15b). The *tibial tuberosity* on the anterior surface is a point of attachment for the patellar ligament. Inferior to and continuous with the tibial tuberosity is a sharp ridge that can be felt below the skin, known as the *anterior border (crest)* or *shin.*

The medial surface of the distal end of the tibia forms the *medial malleolus* (mal-LĒ-ō-lus = hammer). This structure articulates with the talus of the ankle and forms the prominence that can be felt on the medial surface of the ankle. The *fibular notch* (Figure 8.15c) articulates with the distal end of the fibula to form the *distal tibiofibular joint.* Of all the long bones of the body, the tibia is the most frequently fractured and is also the most frequent site of an open (compound) fracture.

Fibula

The **fibula** is parallel and lateral to the tibia, but it is considerably smaller. (See Figure 8.15 for a mnemonic describing the relative positions of the tibia and fibula.) Unlike the tibia, the fibula does not articulate with the femur, but it does help stabilize the ankle joint.

The *head* of the fibula, the proximal end, articulates with the inferior surface of the lateral condyle of the tibia below the level of the knee joint to form the *proximal tibiofibular joint.* The distal end is more arrowhead-shaped and has a projection called the *lateral malleolus* that articulates with the talus of the ankle. This forms the prominence on the lateral surface of the ankle. As noted previously, the fibula also articulates with the tibia at the fibular notch to form the distal tibiofibular joint.

CLINICAL CONNECTION | *Bone Grafting*

Bone grafting generally consists of taking a piece of bone, along with its periosteum and nutrient artery, from one part of the body to replace missing bone in another part of the body. The transplanted bone restores the blood supply to the transplanted site, and healing occurs as in a fracture. The fibula is a common source of bone for grafting because even after a piece of the fibula has been removed, walking, running, and jumping can be normal. Recall that the tibia is the weight-bearing bone of the leg. •

✔CHECKPOINT

Which structures form the medial and lateral prominences of the ankle? Which joints are formed by the tibia and fibula with other bones?

EXHIBIT 8.I Skeleton of the Foot *(Figures 8.16 and 8.17)*

OBJECTIVE

- Identify the location and surface features of the bones of the foot.

Tarsals, Metatarsals, and Phalanges

The **tarsus** (ankle) is the proximal region of the foot and consists of seven **tarsal bones** (Figure 8.16). They include the **talus** (TĀ-lus = ankle bone) and **calcaneus** (kal-KĀ-nē-us = heel), located in the posterior part of the foot. The calcaneus is the largest and strongest tarsal bone. The anterior tarsal bones are the **navicular** (na-VIK-ū-lar = like a little boat), three **cuneiform bones** (KYOO-nē-i-form = wedge-shaped) called the **third (lateral), second (intermediate),** and **first (medial) cuneiforms,** and the **cuboid** (KŪ-boyd = cube-shaped). (A mnemonic to help you remember the names of the tarsal bones is included in Figure 8.16.) Joints between tarsal bones are called *intertarsal joints.* The talus, the most superior tarsal bone, is the only bone of the foot that articulates with the fibula and tibia. It articulates on one side with the medial malleolus of the tibia and on the other side with the lateral malleolus of the fibula. These articulations form the *talocrural (ankle) joint.*

Figure 8.16 Right foot.

The skeleton of the foot consists of the proximal tarsals, the intermediate metatarsals, and the distal phalanges.

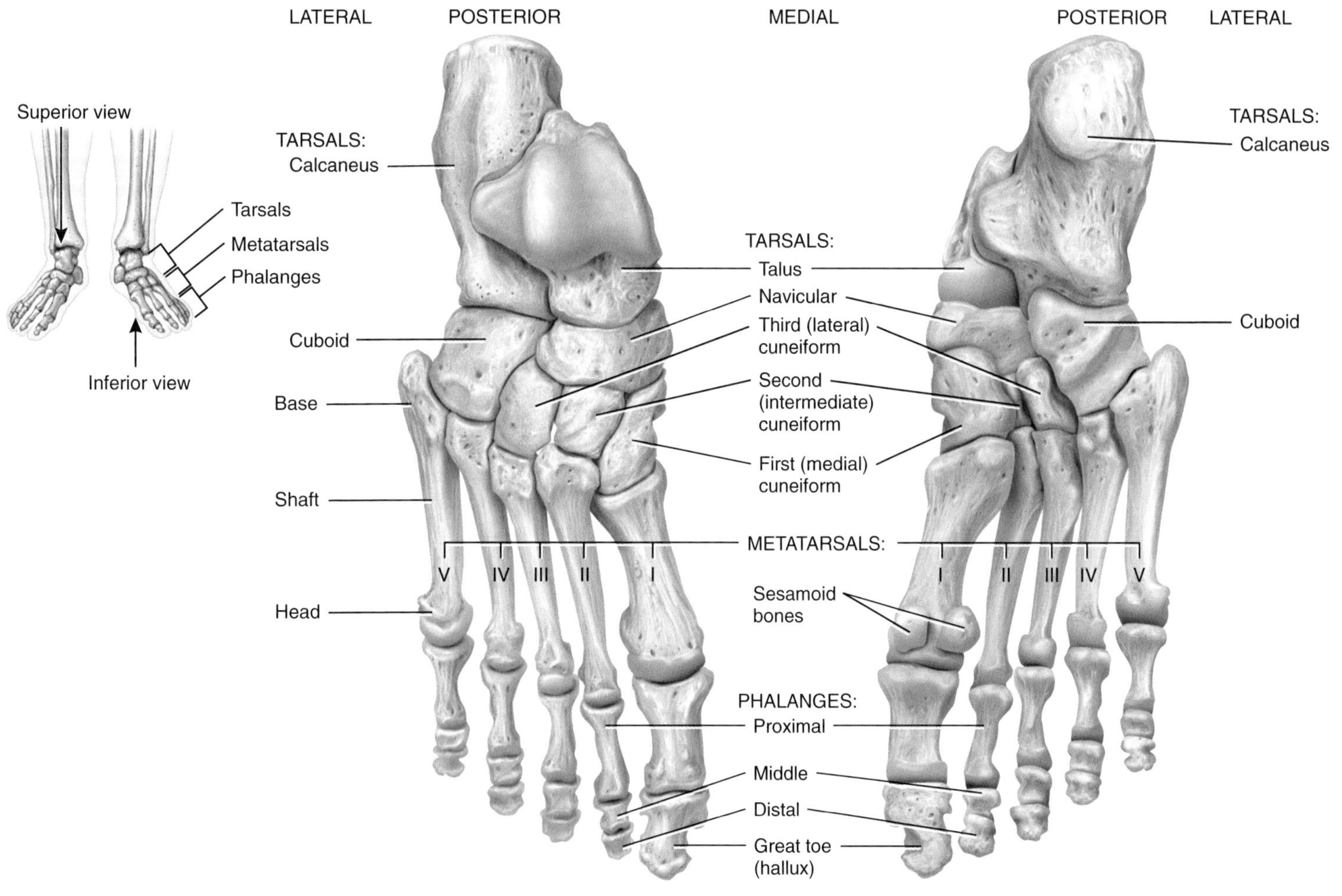

(a) Superior view (b) Inferior view

MNEMONIC for tarsals:

Tall	Centers	Never	Take	Shots	From	Corners.
Talus	Calcaneus	Navicular	Third cuneiform	Second cuneiform	First cuneiform	Cuboid

Which tarsal bone articulates with the tibia and fibula?

EXHIBIT 8.I CONTINUES

During walking, the talus transmits about half the weight of the body to the calcaneus. The remainder is transmitted to the other tarsal bones.

The **metatarsus,** the intermediate region of the foot, consists of five **metatarsal bones** numbered I to V (or 1–5) from the medial to lateral position (Figure 8.16). Like the metacarpals of the palm of the hand, each metatarsal consists of a proximal *base,* an intermediate *shaft,* and a distal *head.* The metatarsals articulate proximally with the first, second, and third cuneiform bones and with the cuboid to form the *tarsometatarsal joints.* Distally, they articulate with the proximal row of phalanges to form the *metatarsophalangeal joints.* The first metatarsal is thicker than the others because it bears more weight.

CLINICAL CONNECTION | *Fractures of the Metatarsals*

Fractures of the metatarsals occur when a heavy object falls on the foot or when a heavy object rolls over the foot. Such fractures are also common among dancers, especially ballet dancers. If a ballet dancer is on the tip of her toes and loses her balance, the full body weight is placed on the metatarsals, causing one or more of them to fracture. •

The **phalanges** comprise the distal component of the foot and resemble those of the hand both in number and arrangement. The toes are numbered I to V (or 1–5) beginning with the great toe, from medial to lateral. Each *phalanx* (singular) consists of a proximal *base,* an intermediate *shaft,* and a distal *head.* The great or big toe (*hallux;* HAL-eks) has two large, heavy phalanges called proximal and distal phalanges. The other four toes each have three phalanges—proximal, middle, and distal. Joints between phalanges of the foot, like those of the hand, are called *interphalangeal joints.*

Arches of the Foot

The bones of the foot are arranged in two **arches** that are held in position by ligaments and tendons (Figure 8.17). The arches enable the foot to support the weight of the body, provide an ideal distribution of body weight over the soft and hard tissues of the foot, and provide leverage while walking. The arches are not rigid; they yield as weight is applied and spring back when the weight is lifted, thus storing energy for the next step and helping to absorb shocks. Usually, the arches are fully developed by age 12 or 13.

The **longitudinal arch** has two parts, both of which consist of tarsal and metatarsal bones arranged to form an arch from the anterior to the posterior part of the foot. The *medial part* of the longitudinal arch, which originates at the calcaneus, rises to the talus and descends through the navicular, the three cuneiforms, and the heads of the three medial metatarsals. The *lateral part* of the longitudinal arch also begins at the calcaneus. It rises at the cuboid and descends to the heads of the two lateral metatarsals. The medial portion of the longitudinal arch is so high that the medial portion of the foot between the ball and heel does not touch the ground when you walk on a hard surface.

The **transverse arch** is found between the medial and lateral aspects of the foot and is formed by the navicular, three cuneiforms, and the bases of the five metatarsals.

As noted earlier, one function of the arches is to distribute body weight over the soft and hard tissues of the body. Normally, the ball of the foot carries about 40% of the weight and the heel carries about 60%. The ball of the foot is the padded portion of the sole superficial to the heads of the metatarsals. When a person wears high-heeled shoes, however, the distribution of weight changes so that the ball of the foot may carry up to 80% and the heel 20%. As a result, the fat pads at the ball of the foot are damaged, joint pain develops, and structural changes in bones may occur.

CLINICAL CONNECTION | *Flatfoot and Clawfoot*

The bones composing the arches of the foot are held in position by ligaments and tendons. If these ligaments and tendons are weakened, the height of the medial longitudinal arch may decrease or "fall." The result is **flatfoot,** the causes of which include excessive weight, postural abnormalities, weakened supporting tissues, and genetic predisposition. Fallen arches may lead to inflammation of the fascia of the sole (plantar fasciitis), Achilles tendinitis, shin splints, stress fractures, bunions, and calluses. A custom-designed arch support often is prescribed to treat flatfoot.

Clawfoot is a condition in which the medial longitudinal arch is abnormally elevated. It is often caused by muscle deformities, such as may occur in diabetics whose neurological lesions lead to atrophy of muscles of the foot. •

✔ **CHECKPOINT**

Which tarsal bone articulates with both the tibia and fibula?

What are the names and functions of the arches of the foot?

Figure 8.17 **Arches of the right foot.**

Arches help the foot support and distribute the weight of the body and provide leverage during walking.

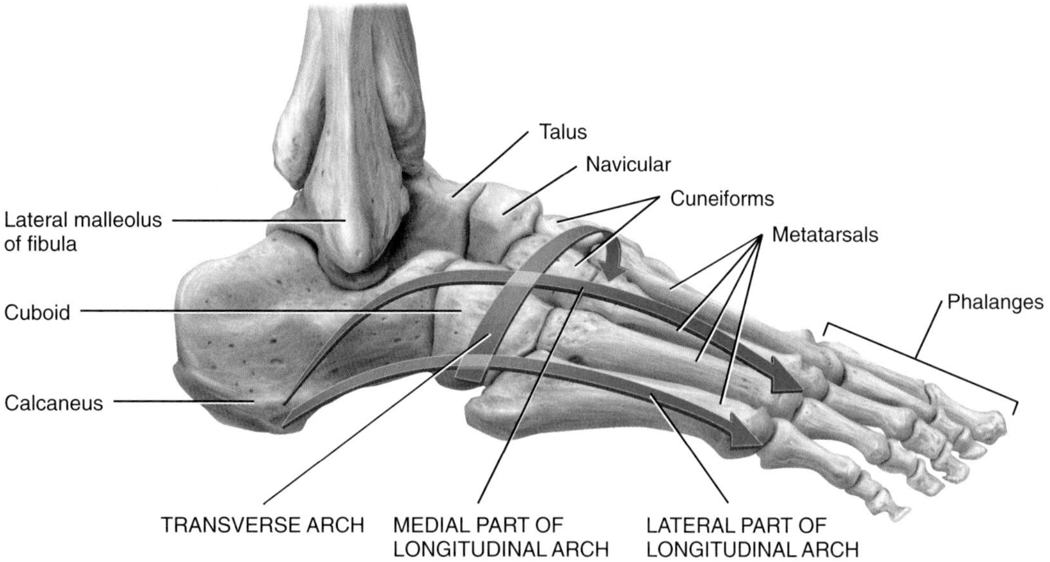

Lateral view of arches

What structural feature of the arches allows them to absorb shocks?

8.7 DEVELOPMENT OF THE SKELETAL SYSTEM

OBJECTIVE

- Describe the development of the skeletal system.

Most skeletal tissue arises from *mesenchymal cells,* connective tissue cells derived from **mesoderm.** However, much of the skeleton of the skull arises from **ectoderm.** The mesenchymal cells condense and form models of bones in areas where the bones themselves will ultimately form. In some cases, the bones form directly within the mesenchyme (intramembranous ossification; see Figure 6.5). In other cases, the bones form within hyaline cartilage that develops from mesenchyme (endochondral ossification; see Figure 6.6).

The *skull* begins development during the fourth week after fertilization. It develops from mesenchyme around the developing brain and consists of two major portions: **neurocranium** (mesodermal in origin), which forms the bones of the skull, and **viscerocranium** (ectodermal in origin), which forms the bones of the face (Figure 8.18a). The neurocranium is divided into two parts:

1. The **cartilaginous neurocranium** consists of hyaline cartilage developed from mesenchyme at the base of the developing skull. It later undergoes endochondral ossification to form the *bones at the base of the skull.*
2. The **membranous neurocranium** consists of mesenchyme and later undergoes intramembranous ossification to form the *flat bones that make up the roof and sides of the skull.*

During fetal life and infancy the flat bones are separated by membrane-filled spaces called fontanels (see Figure 7.14).

The viscerocranium, like the neurocranium, is divided into two parts:

1. The **cartilaginous viscerocranium** is derived from the cartilage of the first two pharyngeal (branchial) arches (see Figure 29.13). Endochondral ossification of these cartilages forms the *ear bones* and *hyoid bone.*
2. The **membranous viscerocranium** is derived from mesenchyme in the first pharyngeal arch and, following intramembranous ossification, forms the facial bones.

Vertebrae and *ribs* are derived from portions of cube-shaped masses of mesoderm called somites (see Figure 10.17). Mesenchymal cells from these regions surround the notochord (see Figure 10.17) at about 4 weeks after fertilization. The **notochord** is a solid cylinder of mesodermal cells that induces (stimulates) the mesenchymal cells to form the *vertebral bodies, costal (rib) centers,* and *vertebral arch centers.* Between the vertebral bodies, the notochord induces mesenchymal cells to form the *nucleus pulposus* of an intervertebral disc and surrounding mesenchymal cells form the *annulus fibrosus* of an intervertebral disc. As development continues, other parts of a vertebra form and the *vertebral arch* surrounds the spinal cord (failure of the vertebral arch to develop properly results in a condition called spina bifida; see Disorders: Homeostatic Imbalances in Chapter 7). In the thoracic region, processes from the vertebrae develop into the *ribs.* The *sternum* develops from mesoderm in the anterior body wall.

Figure 8.18 Development of the skeletal system. Bones that develop from the cartilaginous neurocranium are indicated in light blue; from the cartilaginous viscerocranium in dark blue; from the membranous neurocranium in dark red; and from the membranous viscerocranium in light red.

After the limb buds develop, endochondral ossification of the limb bones begins by the end of the eighth embryonic week.

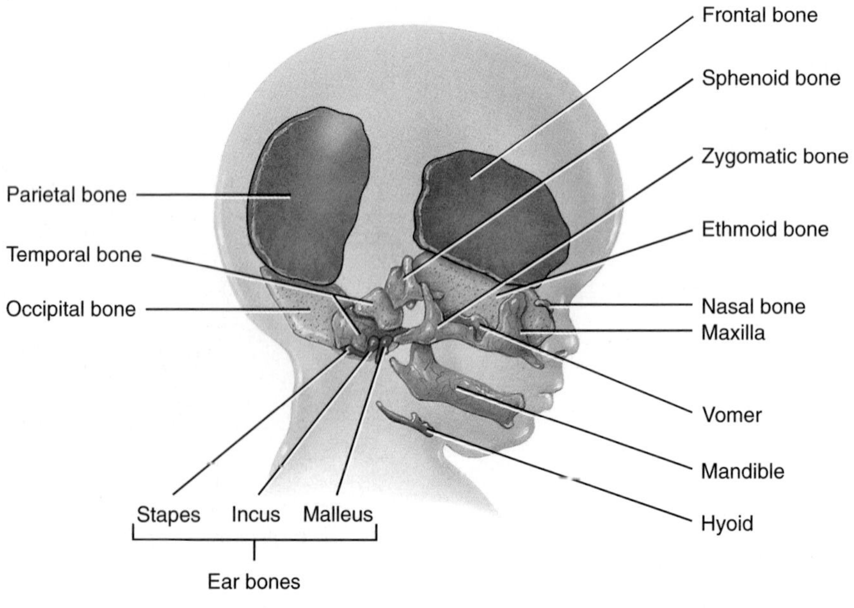

(a) Development of the skull

The *skeleton of the limbs* is derived from mesoderm. During the middle of the fourth week after fertilization, the upper limbs appear as small elevations at the sides of the trunk called **upper limb buds** (Figure 8.18b). About 2 days later, the **lower limb buds** appear. The limb buds consist of **mesenchyme** covered by **ectoderm.** At this point, a mesenchymal skeleton exists in the limbs; some of the masses of mesoderm surrounding the developing bones will become the skeletal muscles of the limbs.

By the sixth week, the limb buds develop a constriction around the middle portion. The constriction produces flattened distal segments of the upper buds called **hand plates** and distal segments of the lower buds called **foot plates** (Figure 8.18c). These plates represent the beginnings of the hands and feet, respectively. At this stage of limb development, a cartilaginous skeleton formed from mesenchyme is present. By the seventh week (Figure 8.18d), the *arm, forearm,* and *hand* are evident in the upper limb bud, and the *thigh, leg,* and *foot* appear in the lower limb bud. By the eighth week (Figure 8.18e), as the shoulder, elbow, and wrist areas become apparent, the upper limb bud is appropriately called the upper limb, and the free lower limb bud is now the free lower limb.

Endochondral ossification of the limb bones begins by the end of the eighth week after fertilization. By the twelfth week, primary ossification centers are present in most of the limb bones. Most secondary ossification centers appear after birth.

✔ **CHECKPOINT**

10. When and how do the limbs develop?

• • •

To appreciate the skeletal system's contributions to homeostasis of other body systems, examine *Focus on Homeostasis: The Skeletal System.* Next, in Chapter 9, we will see how joints both hold the skeleton together and permit it to participate in movements.

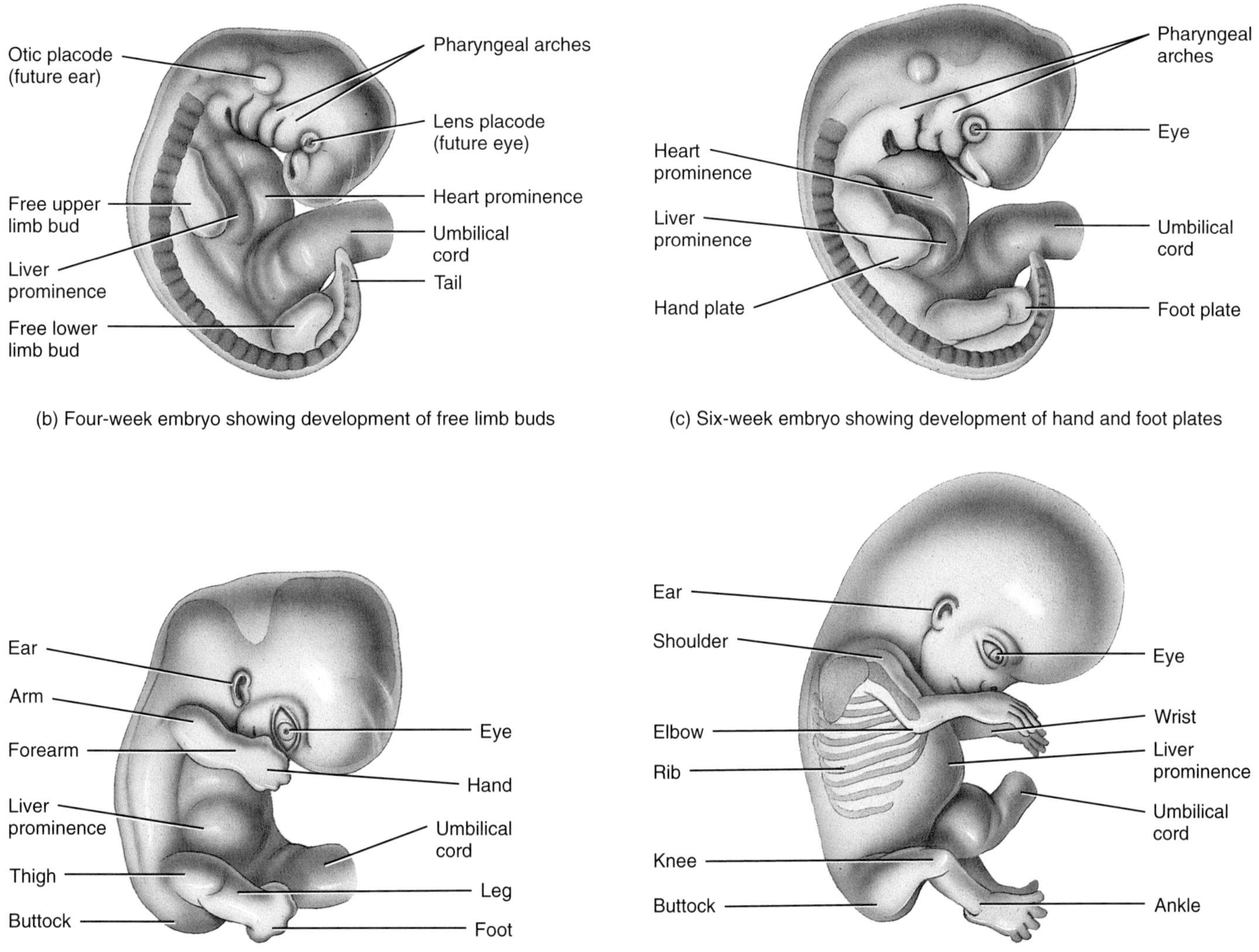

(b) Four-week embryo showing development of free limb buds

(c) Six-week embryo showing development of hand and foot plates

(d) Seven-week embryo showing development of arm, forearm, and hand in free upper limb bud and thigh, leg, and foot in free lower limb bud

(e) Eight-week embryo in which free limb buds have developed into free upper and lower limbs

? Which of the three basic embryonic tissues—ectoderm, mesoderm, and endoderm—gives rise to most of the skeletal system?

Focus on Homeostasis

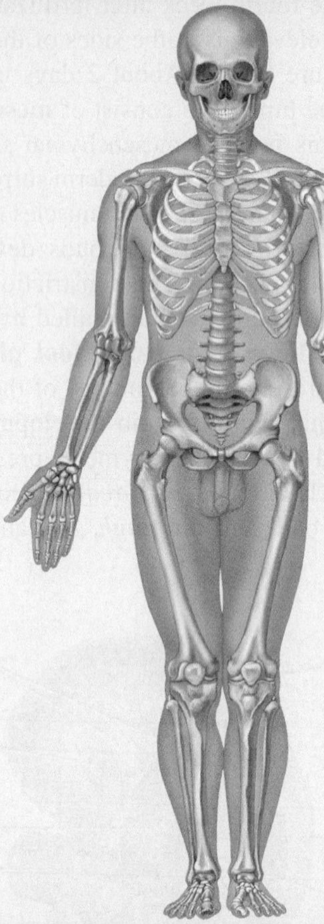

THE SKELETAL SYSTEM

BODY SYSTEM	CONTRIBUTION OF THE SKELETAL SYSTEM
For all body systems 	Bones provide support and protection for internal organs; bones store and release calcium, which is needed for proper functioning of most body tissues.
Integumentary system 	Bones provide strong support for overlying muscles and skin.
Muscular system 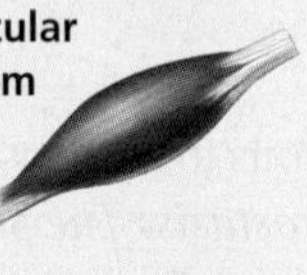	Bones provide attachment points for muscles and leverage for muscles to bring about body movements; contraction of skeletal muscle requires calcium ions.
Nervous system 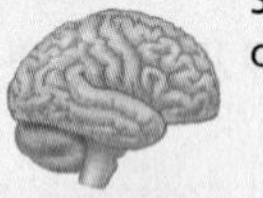	Skull and vertebrae protect brain and spinal cord; normal blood level of calcium is needed for normal functioning of neurons and neuroglia.
Endocrine system 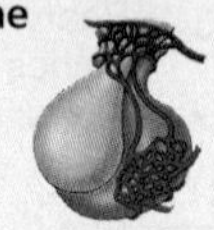	Bones store and release calcium, needed during exocytosis of hormone-filled vesicles and for normal actions of many hormones.
Cardiovascular system 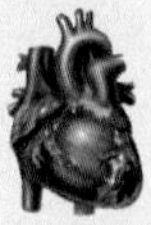	Red bone marrow carries out hemopoiesis (blood cell formation); rhythmic beating of the heart requires calcium ions.
Lymphatic system and immunity 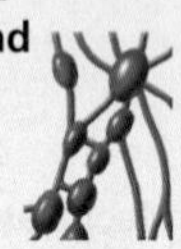	Red bone marrow produces lymphocytes, white blood cells that are involved in immune responses.
Respiratory system 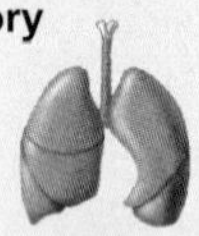	Axial skeleton of thorax protects lungs; rib movements assist in breathing; some muscles used for breathing attach to bones via tendons.
Digestive system 	Teeth masticate (chew) food; rib cage protects esophagus, stomach, and liver; pelvis protects portions of the intestines.
Urinary system 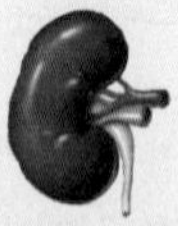	Ribs partially protect kidneys; pelvis protects urinary bladder and urethra.
Reproductive systems 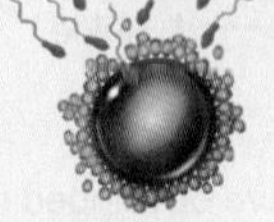	Pelvis protects ovaries, uterine (fallopian) tubes, and uterus in females and part of ductus (vas) deferens and accessory glands in males; bones are an important source of calcium needed for milk synthesis during lactation.

DISORDERS: HOMEOSTATIC IMBALANCES

Hip Fracture

Although any region of the hip girdle may fracture, the term **hip fracture** most commonly applies to a break in the bones associated with the hip joint—the head, neck, or trochanteric regions of the femur, or the bones that form the acetabulum. In the United States, 300,000 to 500,000 people sustain hip fractures each year. The incidence of hip fractures is increasing, due in part to longer life spans. Decreases in bone mass due to osteoporosis (which occurs more often in females), along with an increased tendency to fall, predispose elderly people to hip fractures.

Hip fractures often require surgical treatment, the goal of which is to repair and stabilize the fracture, increase mobility, and decrease pain. Sometimes the repair is accomplished by using surgical pins, screws, nails, and plates to secure the head of the femur. In severe hip fractures, the femoral head or the acetabulum of the hip bone may be replaced by prostheses (artificial devices). The procedure of replacing either the femoral head or the acetabulum is *hemiarthroplasty* (hem-ē-AR-thrō-plas-tē; *hemi-* = one-half; *-arthro-* = joint; *-plasty* = molding). Replacement of both the femoral head and acetabulum is *total hip arthroplasty.* The acetabular prosthesis is made of plastic, and the femoral prosthesis is metal; both are designed to withstand a high degree of stress. The prostheses are attached to healthy portions of bone with acrylic cement and screws (see Figure 9.16).

MEDICAL TERMINOLOGY

Clubfoot or talipes equinovarus (TA-li-pēz ĒK-wīn-ō-VA-rus; *-pes* = foot; *equino-* = horse) An inherited deformity in which the foot is twisted inferiorly and medially, and the angle of the arch is increased; occurs in 1 of every 1000 births. Treatment consists of manipulating the arch to a normal curvature by casts or adhesive tape, usually soon after birth. Corrective shoes or surgery may also be required.

Genu valgum (JĒ-noo VAL-gum; *genu* = knee; *valgum* = bent outward) A deformity in which the knees are abnormally close together and the space between the ankles is increased due to a lateral angulation of the tibia in relation to the femur. Also called **knock-knee.**

Genu varum (JĒ-noo VAR-um; *varum* = bent toward the midline) A deformity in which the knees are abnormally separated and there is a medial angulation of the tibia in relation to the femur and the lower limbs are bowed laterally. Also called **bowleg.**

Hallux valgus (HAL-uks VAL-gus; *hallux* = great toe) Angulation of the great toe away from the midline of the body, typically caused by wearing tightly fitting shoes. When the great toe angles toward the next toe, there is a bony protrusion at the base of the great toe. Also called a **bunion.**

CHAPTER REVIEW AND RESOURCE SUMMARY

WILEY PLUS

Review	Resource
8.1 Pectoral (Shoulder) Girdle 1. Each of the body's two pectoral (shoulder) girdles consists of a clavicle and scapula. 2. Each pectoral girdle attaches an upper limb to the axial skeleton (see Exhibits 8.A and 8.B).	Anatomy Overview - The Appendicular Skeleton: The Pectoral Girdle Figure 8.1 - Right Pectoral (Shoulder) Girdle
8.2 Upper Limb (Extremity) 1. Each of the two upper limbs (extremities) contains 30 bones. 2. The bones of each upper limb include the humerus, ulna, radius, carpals, metacarpals, and phalanges (see Exhibits 8.C through 8.E).	Anatomy Overview - The Upper Limb Figure 8.5 - Right Humerus
8.3 Pelvic (Hip) Girdle 1. The pelvic (hip) girdle consists of two hip bones. 2. Each hip bone consists of three parts: the ilium, pubis, and ischium (see Exhibit 8.F). 3. The hip bones, sacrum, and pubic symphysis form the bony pelvis. It supports the vertebral column and pelvic viscera and attaches the free lower limbs to the axial skeleton.	Anatomy Overview - The Pelvic Girdle
8.4 False and True Pelves 1. The false pelvis is separated from the true pelvis by the pelvic brim. 2. The true pelvis surrounds the pelvic cavity and houses the rectum and urinary bladder in both genders, the vagina and cervix of the uterus in females, and the prostate in males. 3. The false pelvis is the lower portion of the abdomen that is situated superior to the pelvic brim. It contains the superior portion of the urinary bladder (when full) and the lower intestines in both genders and the uterus, uterine tubes, and ovaries in the female.	Anatomy Overview - The Pelvic Girdle

Review	Resource

8.5 Comparison of Female and Male Pelves

Resource: Anatomy Overview - The Pelvic Girdle

1. Bones of the male skeleton are generally larger and heavier than bones of the female skeleton. They also have more prominent markings for muscle attachments.
2. The female pelvis is adapted for pregnancy and childbirth. Sex-related differences in pelvic structure are listed and illustrated in Table 8.1.

8.6 Lower Limb (Extremity)

Resource: Anatomy Overview - The Lower Limb; Figure 8.13 - Right Femur

1. Each of the two lower limbs (extremities) contains 30 bones.
2. The bones of each lower limb include the femur, the patella, the tibia, the fibula, the tarsals, the metatarsals, and the phalanges (see Exhibits 8.G through 8.I).
3. The bones of the foot are arranged in two arches, the longitudinal arch and the transverse arch, to provide support and leverage.

8.7 Development of the Skeletal System

1. Most bones form from mesoderm by intramembranous or endochondral ossification; much of the skeleton of the skull arises from ectoderm.
2. Bones of the limbs develop from limb buds, which consist of mesoderm and ectoderm.

SELF-QUIZ QUESTIONS

Fill in the blanks in the following statements.

1. The bones that make up the palm are the ______.
2. List the three bones that fuse to form a hip (coxal) bone: ______, ______, ______.
3. The portion of the bony pelvis that is inferior to the pelvic brim is the ______ pelvis; the portion that is superior to the pelvic brim is the ______ pelvis.

Indicate whether the following statements are true or false.

4. The largest carpal bone is the lunate.
5. The anterior joint formed by the two coxal (hip) bones is the pubic symphysis.

Choose the one best answer to the following questions.

6. Which of the following statements are *true?* (1) The pectoral girdle consists of the scapula, the clavicle, and the sternum. (2) Although the joints of the pectoral girdle are not very stable, they allow free movement in many directions. (3) The anterior component of the pectoral girdle is the scapula. (4) The pectoral girdle articulates directly with the vertebral column. (5) The posterior component of the pectoral girdle is the sternum.
   (a) 1, 2, and 3 (b) 2 only (c) 4 only
   (d) 2, 3, and 5 (e) 3, 4, and 5
7. Which of the following are *true* concerning the elbow joint? (1) When the forearm is extended, the olecranon fossa receives the olecranon. (2) When the forearm is flexed, the radial fossa receives the coronoid process. (3) The head of the radius articulates with the capitulum. (4) The trochlea articulates with the trochlear notch. (5) The head of the ulna articulates with the ulnar notch of the radius.
   (a) 1, 2, 3, 4, and 5 (b) 1, 3, and 4 (c) 1, 3, 4, and 5
   (d) 1, 2, 3, and 4 (e) 2, 3, and 4
8. Which of the following is the most superior of the tarsals and articulates with the distal end of the tibia?
   (a) calcaneus (b) navicular (c) cuboid
   (d) cuneiform (e) talus
9. Which is (are) *not true* concerning the scapula? (1) The lateral border is also known as the axillary border. (2) The scapular notch accommodates the head of the humerus. (3) The scapula is also known as the collarbone. (4) The acromion process articulates with the clavicle. (5) The coracoid process is utilized for muscle attachment.
   (a) 1, 2, and 3 (b) 3 only (c) 2 and 3
   (d) 3 and 4 (e) 2, 3, and 5
10. Which of the following is *false?*
    (a) A decrease in the height of the medial longitudinal arch creates a condition known as clawfoot.
    (b) The transverse arch is formed by the navicular, cuneiforms, and bases of the five metatarsals.
    (c) The longitudinal arch has medial and lateral parts, both of which originate at the calcaneus.
    (d) Arches help to absorb shocks.
    (e) Arches enable the foot to support the body's weight.
11. Which of the following is (are) involved in the knee joint? Choose the one best answer. (1) fibular notch of the tibia (2) lateral condyle of tibia (3) head of fibula (4) greater trochanter of femur (5) medial condyle of femur.
    (a) 1 and 2 (b) 2 and 5
    (c) 1, 2, 3, and 5 (d) 3 and 4
    (e) 4 only
12. The greater sciatic notch is located on the
    (a) ilium (b) ischium (c) femur
    (d) pubis (e) sacrum

13. Match the following:

____ (a) a large, triangular, flat bone found in the posterior part of the thorax
____ (b) an S-shaped bone lying horizontally in the superior and anterior part of the thorax
____ (c) articulates proximally with the scapula and distally with the radius and ulna
____ (d) located on the medial aspect of the forearm
____ (e) located on the lateral aspect of the forearm
____ (f) the longest, heaviest, and strongest bone of the body
____ (g) the larger, medial bone of the leg
____ (h) the smaller, lateral bone of the leg
____ (i) heel bone
____ (j) sesamoid bone that articulates with the femur and tibia

(1) calcaneus
(2) scapula
(3) patella
(4) radius
(5) femur
(6) clavicle
(7) ulna
(8) tibia
(9) humerus
(10) fibula

14. Match the following:

____ (a) largest and strongest tarsal bone
____ (b) most medial bone in the distal row of carpals; has a hook-shaped projection on anterior surface
____ (c) most medial, pea-shaped bone located in the proximal row of carpals
____ (d) articulate with metatarsals I–III and cuboid
____ (e) located in the proximal row of carpals; its name means "moon-shaped"
____ (f) most lateral bone in the distal row of carpals
____ (g) largest carpal bone
____ (h) generally classified as proximal, middle, and distal
____ (i) most lateral bone in the proximal row of carpals
____ (j) articulates with the tibia and fibula
____ (k) located in the proximal row of carpals; its name indicates that it is "three-cornered"
____ (l) lateral bone that articulates with the calcaneus and metatarsals; IV–V
____ (m) articulates with metacarpal II
____ (n) boat-shaped bone that articulates with the talus

(1) cuboid
(2) triquetrum
(3) calcaneus
(4) pisiform
(5) capitate
(6) phalanges
(7) trapezoid
(8) hamate
(9) lunate
(10) scaphoid
(11) cuneiforms
(12) navicular
(13) trapezium
(14) talus

15. Match the following (some answers will be used more than once):

____ (a) olecranon
____ (b) olecranon fossa
____ (c) trochlea
____ (d) greater trochanter
____ (e) medial malleolus
____ (f) acromial end
____ (g) capitulum
____ (h) acromion
____ (i) radial tuberosity
____ (j) acetabulum
____ (k) lateral malleolus
____ (l) glenoid cavity
____ (m) coronoid process
____ (n) linea aspera
____ (o) anterior border
____ (p) anterior superior iliac spine
____ (q) fovea capitis
____ (r) greater tubercle
____ (s) trochlear notch
____ (t) obturator foramen
____ (u) styloid process

(1) clavicle
(2) scapula
(3) humerus
(4) ulna
(5) radius
(6) femur
(7) tibia
(8) fibula
(9) hip bone

CRITICAL THINKING QUESTIONS

1. Mr. Smith's dog Rover dug up a complete set of human bones in the woods near his house. After examining the scene, the local police collected the bones and transported them to the coroner's office for identification. Later, Mr. Smith read in the newspaper that the bones belonged to an elderly female. How was this determined?

2. A proud dad holds his 5-month-old baby girl upright on her feet while supporting her under her arms. He states that she can never be a dancer because her feet are too flat. Is this true? Why or why not?

3. The local newspaper reported that Farmer White caught his hand in a piece of machinery last Tuesday. He lost the lateral two fingers of his left hand. His daughter, who is taking high school science, reports that Farmer White has three remaining phalanges. Is she correct, or does she need a refresher course in anatomy? Support your answer.

? ANSWERS TO FIGURE QUESTIONS

8.1 The pectoral girdles attach the upper limbs to the axial skeleton.

8.2 The weakest part of the clavicle is its midregion at the junction of the two curves.

8.3 The acromion of the scapula forms the high point of the shoulder.

8.4 Each upper limb has 30 bones.

8.5 The radius articulates at the elbow with the capitulum and radial fossa of the humerus. The ulna articulates at the elbow with the trochlea, coronoid fossa, and olecranon fossa of the humerus.

8.6 The olecranon is the "elbow" part of the ulna.

8.7 The radius and ulna form the proximal and distal radioulnar joints. Their shafts are also connected by the interosseous membrane.

8.8 The scaphoid is the most frequently fractured wrist bone.

8.9 The bony pelvis attaches the lower limbs to the axial skeleton and supports the backbone and pelvic viscera.

8.10 The femur articulates with the acetabulum of the hip bone; the sacrum articulates with the auricular surface of the hip bone.

8.11 The pelvic axis is the course taken by a baby's head as it descends through the pelvis during childbirth.

8.12 Each lower limb has 30 bones.

8.13 The angle of convergence of the femurs is greater in females than males because the female pelvis is broader.

8.14 The patella is classified as a sesamoid bone because it develops in a tendon (the tendon of the quadriceps femoris muscle of the thigh).

8.15 The tibia is the weight-bearing bone of the leg.

8.16 The talus is the only tarsal bone that articulates with the tibia and the fibula.

8.17 Because the arches are not rigid, they yield when weight is applied and spring back when weight is lifted, allowing them to absorb the shock of walking.

8.18 Most of the skeletal system arises from embryonic mesoderm.

9 JOINTS

JOINTS AND HOMEOSTASIS *The joints of the skeletal system contribute to homeostasis by holding bones together in ways that allow for movement and flexibility.*

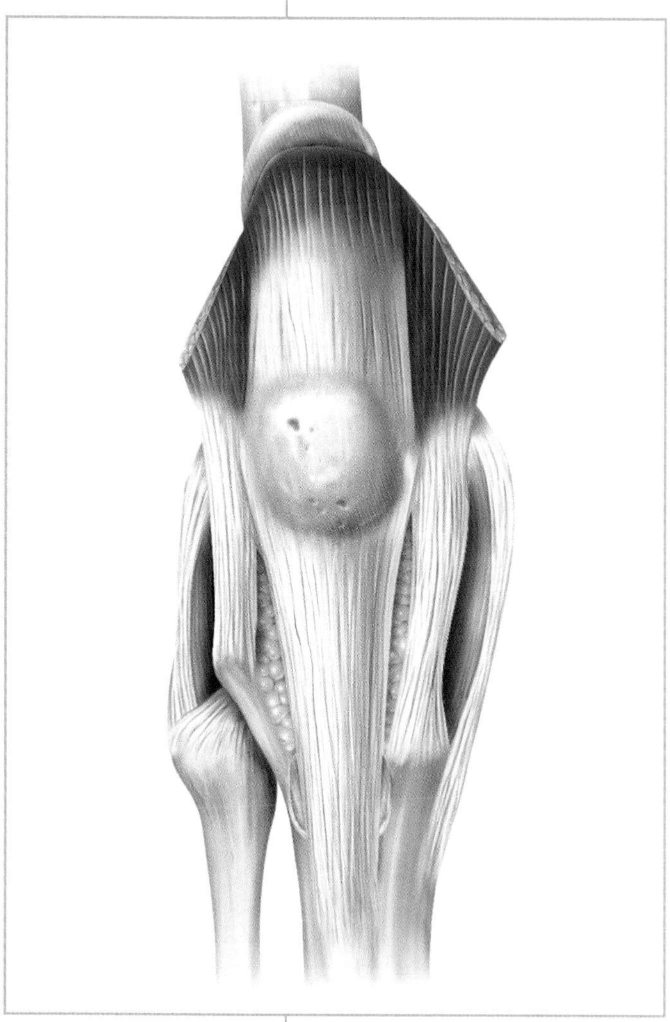

Bones are too rigid to bend without being damaged. Fortunately, flexible connective tissues form joints that hold bones together while still permitting, in most cases, some degree of movement. A **joint,** also called an **articulation** (ar-tik-ū-LĀ-shun) or **arthrosis** (ar-THRŌ-sis), is a point of contact between two bones, between bone and cartilage, or between bone and teeth. When we say one bone *articulates* with another bone, we mean that the bones form a joint. You can appreciate the importance of joints if you have ever had a cast over your knee joint, which makes walking difficult, or a splint on your finger, which limits your ability to manipulate small objects. The scientific study of joints is termed **arthrology** (ar-THROL-ō-jē; *arthr-* = joint; *-logy* = study of). The study of motion of the human body is called **kinesiology** (ki-nē-sē′-OL-ō-jē; *kinesi* = movement).

? ***Did you ever wonder why pitchers so often require rotator cuff surgery?***

9.1 JOINT CLASSIFICATIONS

OBJECTIVE

- Describe the structural and functional classifications of joints.

Joints are classified structurally, based on their anatomical characteristics, and functionally, based on the type of movement they permit.

The structural classification of joints is based on two criteria: (1) the presence or absence of a space between the articulating bones, called a synovial cavity, and (2) the type of connective tissue that binds the bones together. Structurally, joints are classified as one of the following types:

- **Fibrous joints** (FĪ-brus): There is no synovial cavity, and the bones are held together by dense irregular connective tissue that is rich in collagen fibers.
- **Cartilaginous joints** (kar′-ti-LAJ-i-nus): There is no synovial cavity, and the bones are held together by cartilage.
- **Synovial joints** (si-NŌ-vē-al): The bones forming the joint have a synovial cavity and are united by the dense irregular connective tissue of an articular capsule, and often by accessory ligaments.

The functional classification of joints relates to the degree of movement they permit. Functionally, joints are classified as one of the following types:

- **Synarthrosis** (sin′-ar-THRŌ-sis; *syn-* = together): An immovable joint. The plural is *synarthroses.*
- **Amphiarthrosis** (am′-fē-ar-THRŌ-sis; *amphi-* = on both sides): A slightly movable joint. The plural is *amphiarthroses.*
- **Diarthrosis** (dī-ar-THRŌ-sis = movable joint): A freely movable joint. The plural is *diarthroses.* All diarthroses are synovial joints. They have a variety of shapes and permit several different types of movements.

The following sections present the joints of the body according to their structural classifications. As we examine the structure of each type of joint, we will also outline its functions.

CHECKPOINT

1. On what basis are joints classified?

9.2 FIBROUS JOINTS

OBJECTIVE

- Describe the structure and functions of the three types of fibrous joints.

As previously noted, **fibrous joints** lack a synovial cavity, and the articulating bones are held very closely together by dense irregular connective tissue. Fibrous joints permit little or no movement. The three types of fibrous joints are sutures, syndesmoses, and interosseous membranes.

Sutures

A **suture** (SOO-chur; *sutur* = seam) is a fibrous joint composed of a thin layer of dense irregular connective tissue; sutures occur only between bones of the skull. An example is the coronal suture

Figure 9.1 Fibrous joints.

At a fibrous joint the bones are held together by dense irregular connective tissue.

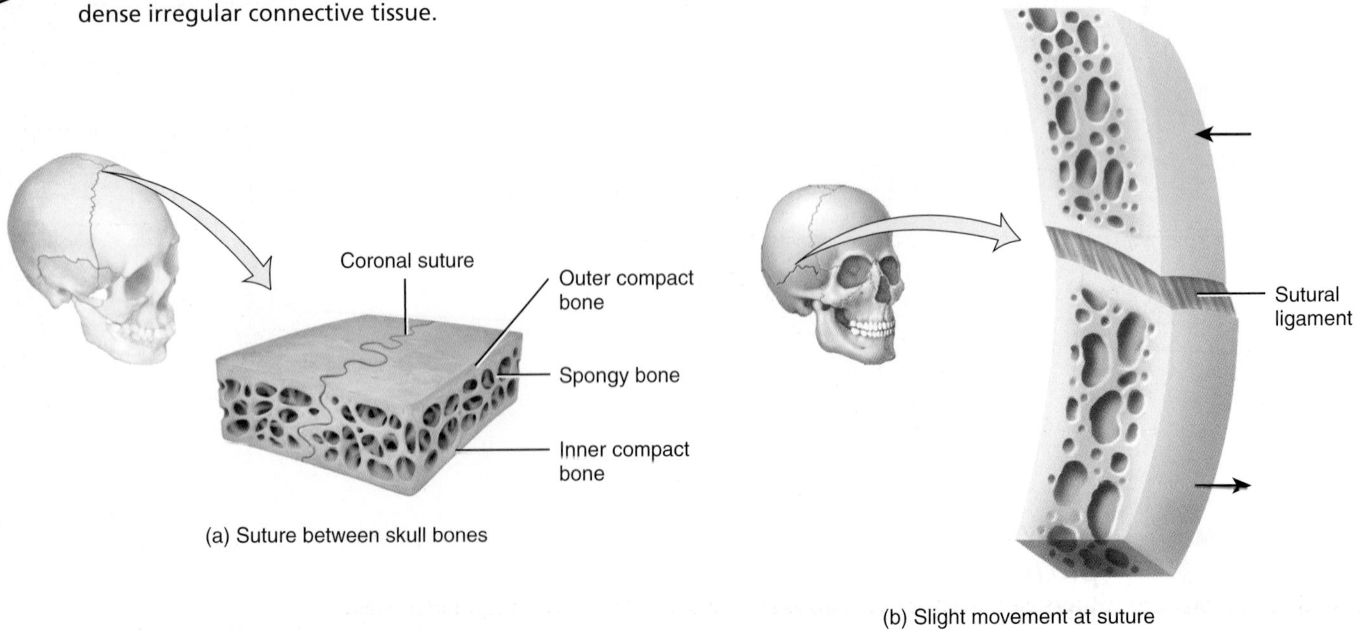
Coronal suture
Outer compact bone
Spongy bone
Inner compact bone
(a) Suture between skull bones
Sutural ligament
(b) Slight movement at suture

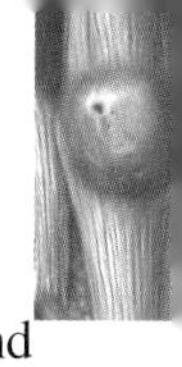

between the parietal and frontal bones (Figure 9.1a). The irregular, interlocking edges of sutures give them added strength and decrease their chance of fracturing. Sutures are joints that form as the numerous bones of the skull come in contact during development. They are immovable or slightly movable. In older individuals, sutures are immovable (synarthroses), but in infants and children they are slightly movable (amphiarthroses) (Figure 9.1b). Sutures play important roles in shock absorption in the skull.

Some sutures, although present during growth of the skull, are replaced by bone in the adult. Such a suture is called a **synostosis** (sin′-os-TŌ-sis; *os-* = bone), or bony joint—a joint in which there is a complete fusion of two separate bones into one. For example, the frontal bone grows in halves that join together across a suture line. Usually they are completely fused by age 6 and the suture becomes obscure. If the suture persists beyond age 6, it is called a **frontal** or **metopic suture** (me-TŌ-pik; *metopon* = forehead). A synostosis is classified as a synarthrosis because it is immovable.

Syndesmoses

A **syndesmosis** (sin′-dez-MŌ-sis; *syndesmo-* = band or ligament) is a fibrous joint in which there is a greater distance between the articulating surfaces and more dense irregular connective tissue than in a suture. The dense irregular connective tissue is typically arranged as a bundle (ligament), allowing the joint to permit limited movement. One example of a syndesmosis is the distal tibiofibular joint, where the anterior tibiofibular ligament connects the tibia and fibula (Figure 9.1c, left). It permits slight movement (amphiarthrosis). Another example of a syndesmosis is called a **gomphosis** (gom-FŌ-sis; *gompbo-* = bolt or nail) or *dentoalveolar joint,* in which a cone-shaped peg fits into a socket. The only examples of gomphoses in the human body are the articulations between the roots of the teeth and their sockets (alveoli) in the maxillae and mandible (Figure 9.1c, right). The dense irregular connective tissue between a tooth and its socket is the thin periodontal ligament (membrane). A healthy gomphosis permits no movement (synarthrosis). Inflammation and degeneration of the gums, periodontal ligament, and bone is called *periodontal disease.*

Interosseous Membranes

The final category of fibrous joint is the **interosseous membrane** (in′-ter-OS-ē-us), which is a substantial sheet of dense irregular connective tissue that binds neighboring long bones and permits slight movement (amphiarthrosis). There are two principal interosseous membrane joints in the human body. One occurs between the radius and ulna in the forearm (see Figure 8.6) and the other occurs between the tibia and fibula in the leg (Figure 9.1d).

✔ CHECKPOINT

2. Which fibrous joints are synarthroses? Which are amphiarthroses?

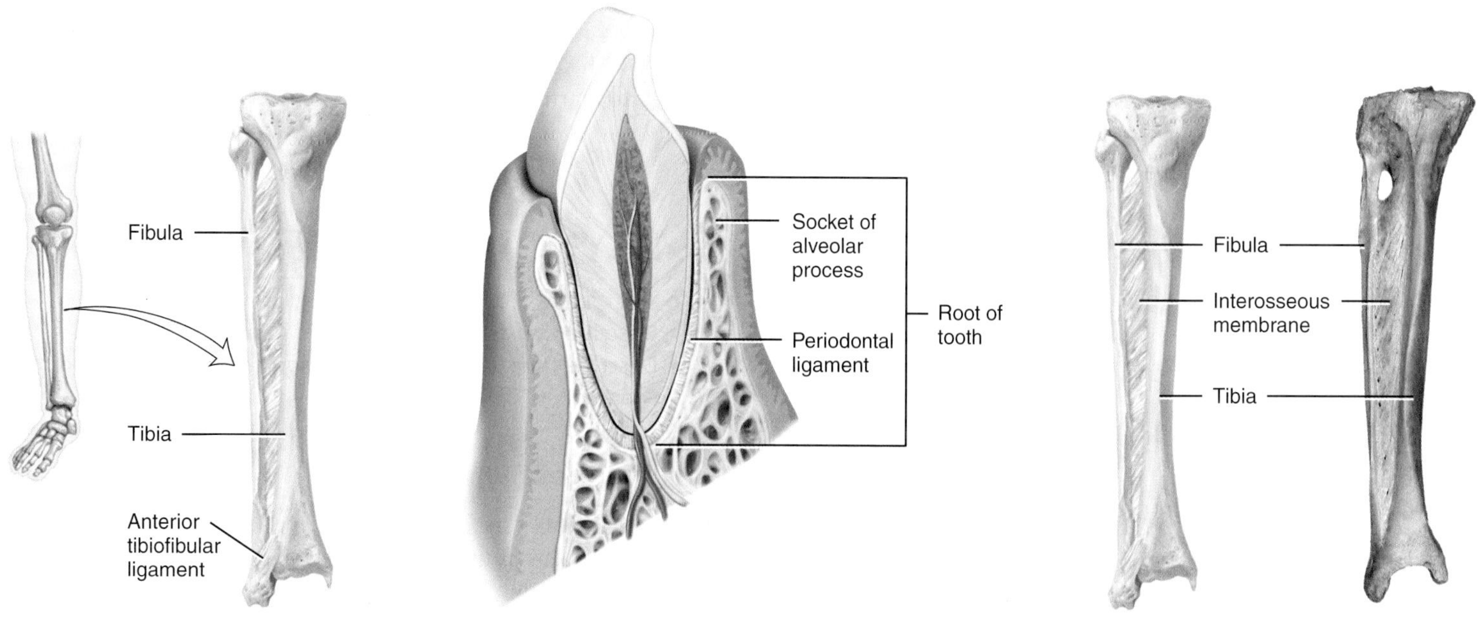

Syndesmosis between tibia and fibula at distal tibiofibular joint

Syndesmosis between tooth and socket of alveolar process (gomphosis)

(c) Syndesmosis

(d) Interosseous membrane between diaphyses of tibia and fibula

? Functionally, why are sutures classified as synarthroses, and syndesmoses as amphiarthroses?

9.3 CARTILAGINOUS JOINTS

OBJECTIVE

- Describe the structure and functions of the two types of cartilaginous joints.

Like a fibrous joint, a **cartilaginous joint** (kar′-ti-LAJ-i-nus) lacks a synovial cavity and allows little or no movement. Here the articulating bones are tightly connected by either hyaline cartilage or fibrocartilage (see Table 4.6). The two types of cartilaginous joints are synchondroses and symphyses.

Synchondroses

A **synchondrosis** (sin′-kon-DRŌ-sis; *chondro-* = cartilage) is a cartilaginous joint in which the connecting material is hyaline cartilage. An example of a synchondrosis is the epiphyseal (growth) plate that connects the epiphysis and diaphysis of a growing bone (Figure 9.2a). A photomicrograph of the epiphyseal plate is shown in Figure 6.7b. Functionally, a synchondrosis is immovable (*synarthrosis*). When bone elongation ceases, bone replaces the hyaline cartilage, and the synchondrosis becomes a synostosis, a bony joint. Another example of a synchondrosis is the joint between the first rib and the manubrium of the sternum, which also ossifies during adult life and becomes an immovable (synarthrosis) synostosis, or bony joint (see Figure 7.22b).

In an x-ray of a young person's skeleton, the synchondroses are easily seen as thin dark areas between the white-appearing bone tissue (see Figure 6.7a). This is how a medical professional can view an x-ray and determine that a person still has some degree of growth remaining. Breaks in a bone that extend into the epiphyseal plate and damage the cartilage of the synchondrosis can affect further growth of the bone, leading to abbreviated development of the length of the bone that leads to a bone of shortened length.

Figure 9.2 **Cartilaginous joints.**

At a cartilaginous joint the bones are held together by cartilage.

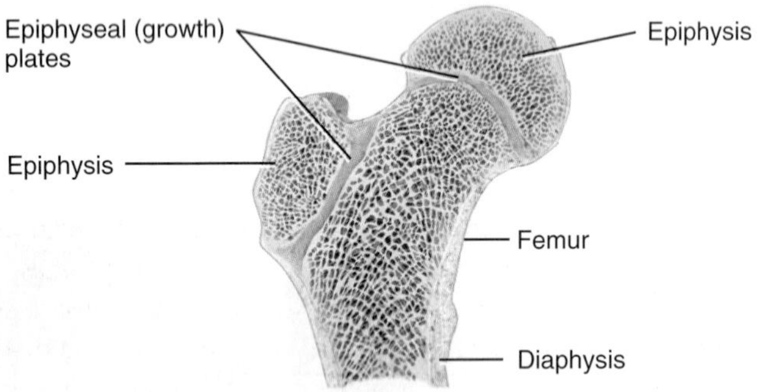
Epiphyseal (growth) plates
Epiphysis
Epiphysis
Femur
Diaphysis

(a) Synchondrosis

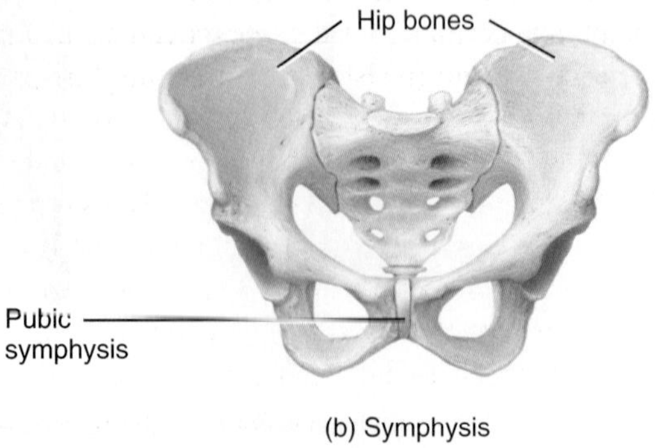
Hip bones
Pubic symphysis

(b) Symphysis

What is the structural difference between a synchondrosis and a symphysis?

Symphyses

A **symphysis** (SIM-fi-sis = growing together) is a cartilaginous joint in which the ends of the articulating bones are covered with hyaline cartilage, but a broad, flat disc of fibrocartilage connects the bones. All symphyses (plural) occur in the midline of the body. The pubic symphysis between the anterior surfaces of the hip bones is one example of a symphysis (Figure 9.2b). This type of joint is also found at the junction of the manubrium and body of the sternum (see Figure 7.22) and at the intervertebral joints between the bodies of vertebrae (see Figure 7.20a). A portion of the intervertebral disc is composed of fibrocartilage. A symphysis is a slightly movable joint (amphiarthrosis).

CHECKPOINT

3. Which cartilaginous joints are synarthroses? Which are amphiarthroses?

9.4 SYNOVIAL JOINTS

OBJECTIVES

- Describe the structure of synovial joints.
- Describe the structure and function of bursae and tendon sheaths.

Structure of Synovial Joints

Synovial joints (si-NŌ-vē-al) have certain characteristics that distinguish them from other joints. The unique characteristic of a synovial joint is the presence of a space called a **synovial (joint) cavity** between the articulating bones (Figure 9.3). Because the synovial cavity allows considerable movement at a joint, all synovial joints are classified functionally as freely movable (diarthroses). The bones at a synovial joint are covered by a layer of hyaline cartilage called **articular cartilage.** The cartilage covers the articulating surface of the bones with a smooth, slippery

Figure 9.3 **Structure of a typical synovial joint.** Note the two layers of the articular capsule—the fibrous membrane and the synovial membrane. Synovial fluid lubricates the synovial cavity, which is located between the synovial membrane and the articular cartilage.

The distinguishing feature of a synovial joint is the synovial cavity between the articulating bones.

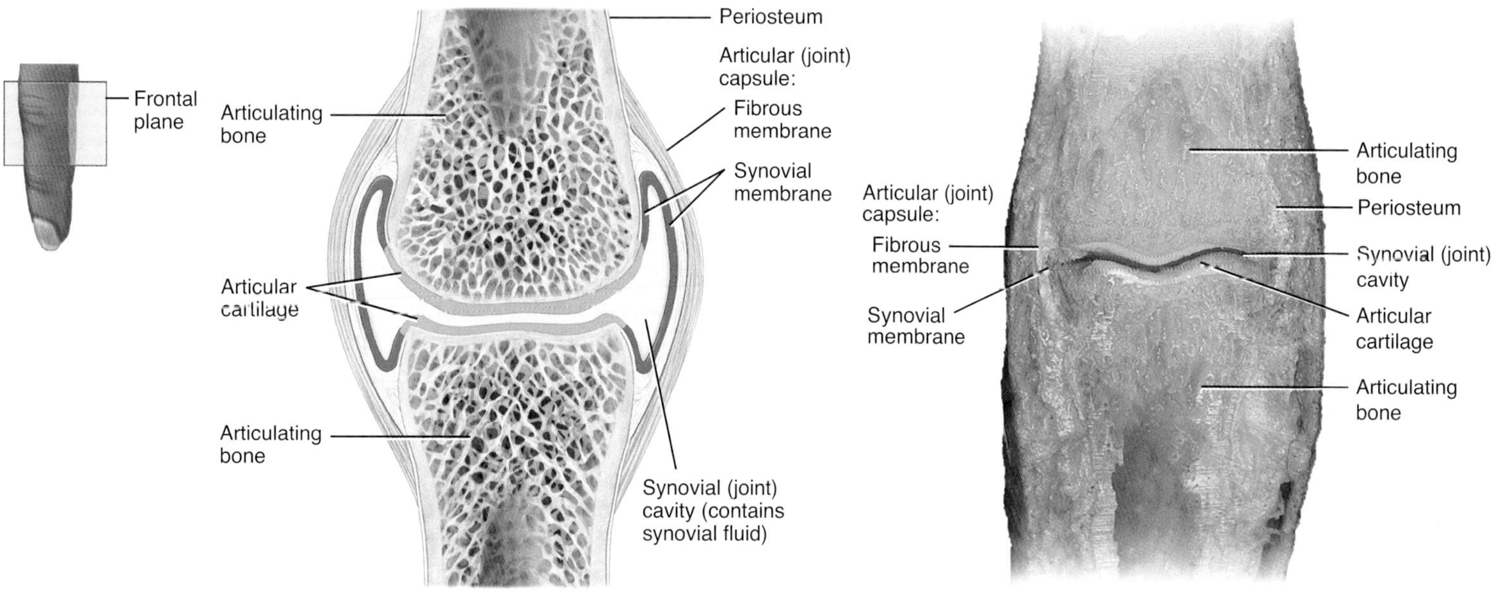

(a) Frontal section

(b) Frontal section

What is the functional classification of synovial joints?

surface but does not bind them together. Articular cartilage reduces friction between bones in the joint during movement and helps to absorb shock.

CLINICAL CONNECTION | *Autologous Chondrocyte Implantation*

When there is damage to articular cartilage in the knee joint, especially involving the femur, there is an alternative to partial or total knee replacement (see Section 9.10) called **autologous chondrocyte implantation** (aw-TOL-o-gus), or **ACI.** Candidates for ACI have cartilage damage due to acute or repetitive trauma, not arthritis. In the procedure, healthy chondrocytes (cartilage cells) are taken from an area of the femoral condyle that is not weight-bearing and sent to a laboratory, where they are cultured for 4 to 5 weeks to generate between 5 million and 10 million cells. When the cultured cells are available, the implantation takes place. The damaged area is prepared by removing dead cartilage from the defect, which is covered by a piece of periosteum, usually taken from the tibia. Then the cultured chondrocytes are injected under the periosteum, where they will grow and mature over time. The patient can put the full weight of the body on the knee in about 10 to 12 weeks. •

Articular Capsule

A sleevelike **articular (joint) capsule** surrounds a synovial joint, encloses the synovial cavity, and unites the articulating bones. The articular capsule is composed of two layers, an outer fibrous membrane and an inner synovial membrane (Figure 9.3a). The **fibrous membrane** usually consists of dense irregular connective tissue (mostly collagen fibers) that attaches to the periosteum of the articulating bones. In fact, the fibrous membrane is literally a thickened continuation of the periosteum between the bones. The flexibility of the fibrous membrane permits considerable movement at a joint, while its great tensile strength (resistance to stretching) helps prevent the bones from dislocating. The fibers of some fibrous membranes are arranged as parallel bundles of dense regular connective tissue that are highly adapted for resisting strains. The strength of these fiber bundles, called **ligaments** (*liga-* = bound or tied), is one of the principal mechanical factors that hold bones close together in a synovial joint. Ligaments are often designated by individual names. The inner layer of the articular capsule, the **synovial membrane,** is composed of areolar connective tissue with elastic fibers. At many synovial joints the synovial membrane includes accumulations of adipose tissue, called **articular fat pads.** An example is the infrapatellar fat pad in the knee (see Figure 9.15c).

A **"double-jointed"** person does not really have extra joints. Individuals who are double-jointed have greater flexibility in their articular capsules and ligaments; the resulting increase in range of motion allows them to entertain fellow partygoers with activities such as touching their thumbs to their wrists and putting their ankles or elbows behind their necks. Unfortunately, such flexible joints are less structurally stable and are more easily dislocated.

Synovial Fluid

The synovial membrane secretes **synovial fluid** (*ov-* = egg), a viscous, clear or pale yellow fluid named for its similarity in appearance and consistency to uncooked egg white. Synovial fluid consists of hyaluronic acid secreted by fibroblastlike cells in the synovial membrane and interstitial fluid filtered from blood plasma. It forms a thin film over the surfaces within the articular capsule. Its functions include reducing friction by lubricating the joint, absorbing shocks, and supplying oxygen and nutrients to and removing carbon dioxide and metabolic wastes from the chondrocytes within articular cartilage. (Recall that cartilage is an avascular tissue, so it does not have blood vessels to perform the latter function.) Synovial fluid also contains phagocytic cells that remove microbes and the debris that results from normal wear and tear in the joint. When a synovial joint is immobile for a time, the fluid becomes quite viscous (gel-like), but as joint movement increases, the fluid becomes less viscous. One of the benefits of warming up before exercise is that it stimulates the production and secretion of synovial fluid; more fluid means less stress on the joints during exercise.

We are all familiar with the cracking sounds heard as certain joints move, or the popping sounds that arise when a person pulls on his fingers to crack his knuckles. According to one theory, when the synovial cavity expands, the pressure inside the synovial cavity decreases, creating a partial vacuum. The suction draws carbon dioxide and oxygen out of blood vessels in the synovial membrane, forming bubbles in the fluid. When the fingers are flexed (bent) the volume of the cavity decreases and the pressure increases; this bursts the bubbles and creates cracking or popping sounds as the gases are driven back into solution.

CLINICAL CONNECTION | Aspiration of Synovial Fluid

As a result of various injuries or diseases, there may be an excessive buildup of synovial fluid in a joint cavity, resulting in pain and decreased mobility. In order to relieve the pressure and ease the pain, **aspiration of synovial fluid** (as′-pi-RĀ-shun) may be necessary. In this procedure, a needle is inserted into the joint cavity and the fluid is withdrawn into a syringe; the fluid is then analyzed for diagnostic purposes. For example, the fluid may contain bacteria, which confirms a diagnosis of infection, or it may contain urea crystals, which confirms a diagnosis of gout. It is also possible to inject medication into a joint cavity; anti-inflammatory drugs such as cortisol are also used to decrease joint swelling caused by retention of fluid from the blood that entered the joint. •

Accessory Ligaments, Articular Discs, and Labra

Many synovial joints also contain **accessory ligaments** called extracapsular ligaments and intracapsular ligaments (see Figure 9.15d). *Extracapsular ligaments* lie outside the articular capsule. Examples are the fibular and tibial collateral ligaments of the knee joint. *Intracapsular ligaments* occur within the articular capsule but are excluded from the synovial cavity by folds of the synovial membrane. Examples are the anterior and posterior cruciate ligaments of the knee joint.

Inside some synovial joints, such as the knee, crescent-shaped pads of fibrocartilage lie between the articular surfaces of the bones and are attached to the fibrous capsule. These pads are called **articular discs** or **menisci** (me-NIS-sī or me-NIS-kī; singular is *meniscus*). Figure 9.15d depicts the lateral and medial menisci in the knee joint. The discs bind strongly to the inside of the fibrous membrane and usually subdivide the synovial cavity into two spaces, allowing separate movements to occur in each space. As you will see later, separate movements also occur in the respective compartments of the temporomandibular joint (TMJ) (see Exhibit 9.A). The functions of the menisci are not completely understood but are known to include the following: (1) shock absorption; (2) a better fit between articulating bony surfaces; (3) providing adaptable surfaces for combined movements; (4) weight distribution over a greater contact surface; and (5) distribution of synovial lubricant across the articular surfaces of the joint.

A **labrum** (LĀ-brum; the plural is *labra*), prominent in the ball-and-socket joints of the shoulder and hip (see Figures 9.12c, d; 9.14c), is the fibrocartilaginous lip that extends from the edge of the joint socket. The labrum helps deepen the joint socket and increases the area of contact between the socket over the ball-like surface of the head of the humerus or femur.

CLINICAL CONNECTION | Torn Cartilage and Arthroscopy

The tearing of menisci in the knee, commonly called **torn cartilage,** occurs often among athletes. Such damaged cartilage will begin to wear and may cause arthritis to develop unless the damaged cartilage is treated surgically. Years ago, if a patient had torn cartilage, the entire meniscus was removed by a procedure called a *meniscectomy* (men′-i-SEK-tō-mē). The problem was that over time the articular cartilage was worn away more quickly. Currently, surgeons perform a partial meniscectomy, in which only the torn segment of the meniscus is removed. Surgical repair of the torn cartilage may be assisted by **arthroscopy** (ar-THROS-kō-pē; *-scopy* = observation). This minimally invasive procedure involves examination of the interior of a joint, usually the knee, with an arthroscope, a lighted, pencil-thin fiber optic camera used for visualizing the nature and extent of damage. Arthroscopy is also used to monitor the progression of disease and the effects of therapy. The insertion of surgical instruments through other incisions also enables a physician to remove torn cartilage and repair damaged cruciate ligaments in the knee; obtain tissue samples for analysis; and perform surgery on other joints, such as the shoulder, elbow, ankle, and wrist. •

Nerve and Blood Supply

The nerves that supply a joint are the same as those that supply the skeletal muscles that move the joint. Synovial joints contain many nerve endings that are distributed to the articular capsule and associated ligaments. Some of the nerve endings convey information about pain from the joint to the spinal cord and brain for processing. Other nerve endings respond to the degree of movement and stretch at a joint, such as when a physician strikes the tendon below your kneecap to test your reflexes. The spinal cord and brain respond by sending impulses through different nerves to the muscles to adjust body movements.

Although many of the components of synovial joints are avascular, arteries in the vicinity send out numerous branches that penetrate the ligaments and articular capsule to deliver oxygen and nutrients. Veins remove carbon dioxide and wastes from the joints. The arterial branches from several different arteries typically merge around a joint before penetrating the articular capsule. The chondrocytes in the articular cartilage of a synovial joint receive oxygen and nutrients from synovial fluid derived from blood; all other joint tissues are supplied directly by capillaries. Carbon dioxide and wastes pass from chondrocytes of articular cartilage into synovial fluid and then into veins; carbon dioxide and wastes from all other joint structures pass directly into veins.

CLINICAL CONNECTION | *Sprain and Strain*

A **sprain** is the forcible wrenching or twisting of a joint that stretches or tears its ligaments but does not dislocate the bones. It occurs when the ligaments are stressed beyond their normal capacity. Severe sprains may be so painful that the joint cannot be moved. There is considerable swelling, which results from chemicals released by the damaged cells and hemorrhage of ruptured blood vessels. The lateral ankle joint is most often sprained; the wrist is another area that is frequently sprained. A **strain** is a stretched or partially torn muscle or muscle and tendon. It often occurs when a muscle contracts suddenly and powerfully—such as the leg muscles of sprinters when they spring from the blocks.

Initially sprains should be treated with ***PRICE:*** protection, rest, ice, compression, and elevation. PRICE therapy may be used on muscle strains, joint inflammation, suspected fractures, and bruises. The five components of PRICE therapy are

- ***Protection*** means protecting the injury from further damage; for example, stop the activity and use padding and protection, and use splints or a sling, or crutches, if necessary.
- ***Rest*** the injured area to avoid further damage to the tissues. Stop the activity immediately. Avoid exercise or other activities that cause pain or swelling to the injured area. Rest is needed for repair. Exercising before an injury has healed may increase the probability of re-injury.
- ***Ice*** the injured area as soon as possible. Applying ice slows blood flow to the area, reduces swelling, and relieves pain. Ice works effectively when applied for 20 minutes, off for 40 minutes, back on for 20 minutes, and so on.
- ***Compression*** by wrap or bandage helps to reduce swelling. Care must be taken to compress the injured area but not to block blood flow.
- ***Elevation*** of the injured area above the level of the heart, when possible, will reduce potential swelling. •

Bursae and Tendon Sheaths

The various movements of the body create friction between moving parts. Saclike structures called **bursae** (BER-sē = purses; singular is *bursa*) are strategically situated to alleviate friction in some joints, such as the shoulder and knee joints (see Figures 9.12 and 9.15c). Bursae are not strictly part of synovial joints, but they do resemble joint capsules because their walls consist of an outer fibrous membrane of thin, dense connective tissue lined by a synovial membrane. They are filled with a small amount of fluid that is similar to synovial fluid. Bursae can be located between the skin and bones, tendons and bones, muscles and bones, or ligaments and bones. The fluid-filled bursal sacs cushion the movement of these body parts against one another.

CLINICAL CONNECTION | *Bursitis*

An acute or chronic inflammation of a bursa, called **bursitis** (bur-SĪ-tis), is usually caused by irritation from repeated, excessive exertion of a joint. The condition may also be caused by trauma, by an acute or chronic infection (including syphilis and tuberculosis), or by rheumatoid arthritis (described in Disorders: Homeostatic Imbalances at the end of this chapter). Symptoms include pain, swelling, tenderness, and limited movement. Treatment may include oral anti-inflammatory agents and injections of cortisol-like steroids. •

Structures called tendon sheaths also reduce friction at joints. **Tendon (synovial) sheaths** are tubelike bursae that wrap around certain tendons that experience considerable friction as they pass through tunnels formed by connective tissue and bone. The inner layer of a tendon sheath, the *visceral layer,* is attached to the surface of the tendon. The outer layer, known as the *parietal layer,* is attached to bone (see Figure 11.18a). Between the layers is a cavity that contains a film of synovial fluid. A tendon sheath protects all sides of a tendon from friction as the tendon slides back and forth. Tendon sheaths are found where tendons pass through synovial cavities, such as the tendon of the biceps brachii muscle at the shoulder joint (see Figure 9.12c). Tendon sheaths are also found at the wrist and ankle, where many tendons come together in a confined space (see Figure 11.18a), and in the fingers and toes, where there is a great deal of movement (see Figure 11.18).

CLINICAL CONNECTION | *Tenosynovitis*

Tenosynovitis (ten′-o-sīn-ō-VĪ-tis) is an inflammation of the tendons, tendon sheaths, and synovial membranes surrounding certain joints. The tendons most often affected are at the wrists, shoulders, elbows (resulting in *tennis elbow*), finger joints (resulting in *trigger finger*), ankles, and feet. The affected sheaths sometimes become visibly swollen because of fluid accumulation. Tenderness and pain are frequently associated with movement of the body part. The condition often follows trauma, strain, or excessive exercise. Tenosynovitis of the dorsum of the foot may be caused by tying shoelaces too tightly. Gymnasts are prone to developing the condition as a result of chronic, repetitive, and maximum hyperextension at the wrists. Other repetitive movements involving activities such as typing, haircutting, carpentry, and assembly line work can also result in tenosynovitis. •

✔ **CHECKPOINT**

4. How does the structure of synovial joints classify them as diarthroses?
5. What are the functions of articular cartilage, synovial fluid, and articular discs?
6. What types of sensations are perceived at joints, and from what sources do joints receive nourishment?
7. In what ways are bursae similar to joint capsules? How do they differ?

9.5 TYPES OF MOVEMENTS AT SYNOVIAL JOINTS

OBJECTIVE

- Describe the types of movements that can occur at synovial joints.

Anatomists, physical therapists, and kinesiologists (professionals who study the science of human movement and look for ways to improve the efficiency and performance of the human body at work, in sports, and in daily activities) use specific terminology to designate the movements that can occur at synovial joints. These precise terms may indicate the form of motion, the direction of movement, or the relationship of one body part to another during movement. Movements at synovial joints are grouped into four main categories: (1) gliding, (2) angular movements, (3) rotation, and (4) special movements, which occur only at certain joints.

Gliding

Gliding is a simple movement in which relatively flat bone surfaces move back-and-forth and from side-to-side with respect to one another (Figure 9.4). There is no significant alteration of the angle between the bones. Gliding movements are limited in range due to the structure of the articular capsule and associated ligaments and bones; however, these sliding movements can also be combined with rotation. The intercarpal and intertarsal joints are examples of articulations where gliding movements occur.

Angular Movements

In **angular movements,** there is an increase or a decrease in the angle between articulating bones. The major angular movements are flexion, extension, lateral flexion, hyperextension, abduction, adduction, and circumduction. These movements are discussed with respect to the body in the anatomical position (see Figure 1.5).

Flexion, Extension, Lateral Flexion, and Hyperextension

Flexion and extension are opposite movements. In **flexion** (FLEK-shun; *flex-* = to bend) there is a decrease in the angle between articulating bones; in **extension** (eks-TEN-shun; *exten-* = to stretch out) there is an increase in the angle between articulating bones, often to restore a part of the body to the anatomical position after it has been flexed (Figure 9.5). Both movements usually occur along the sagittal plane. All of the following are examples of flexion (as you have probably already guessed, extension is simply the reverse of these movements):

- Bending the head toward the chest at the atlanto-occipital joint between the atlas (the first vertebra) and the occipital bone of the skull, and at the cervical intervertebral joints between the cervical vertebrae (Figure 9.5a)
- Bending the trunk forward at the intervertebral joints
- Moving the humerus forward at the shoulder joint, as in swinging the arms forward while walking (Figure 9.5b)
- Moving the forearm toward the arm at the elbow joint between the humerus, ulna, and radius (Figure 9.5c)
- Moving the palm toward the forearm at the wrist or radiocarpal joint between the radius and carpals, as in the upward movement when doing wrist curls (Figure 9.5d)
- Bending the digits of the hand or feet at the interphalangeal joints between phalanges
- Moving the femur forward at the hip joint between the femur and hip bone, as in walking (Figure 9.5e)

Figure 9.4 Gliding movements at synovial joints.

Gliding movements consist of side-to-side and back-and-forth motions.

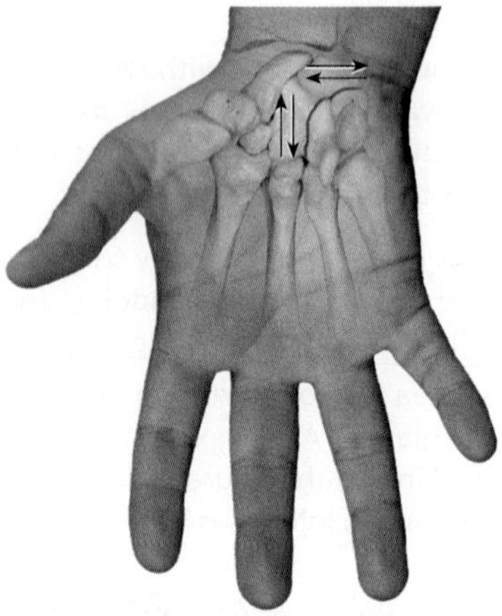

Gliding between intercarpals (arrows)

What are two examples of joints that permit gliding movements?

Figure 9.5 **Angular movements at synovial joints—flexion, extension, hyperextension, and lateral flexion.**

In angular movements, there is an increase or decrease in the angle between articulating bones.

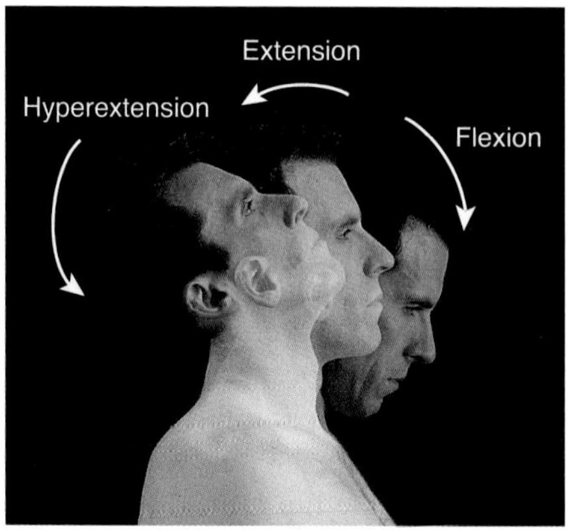

(a) Atlanto-occipital and cervical intervertebral joints

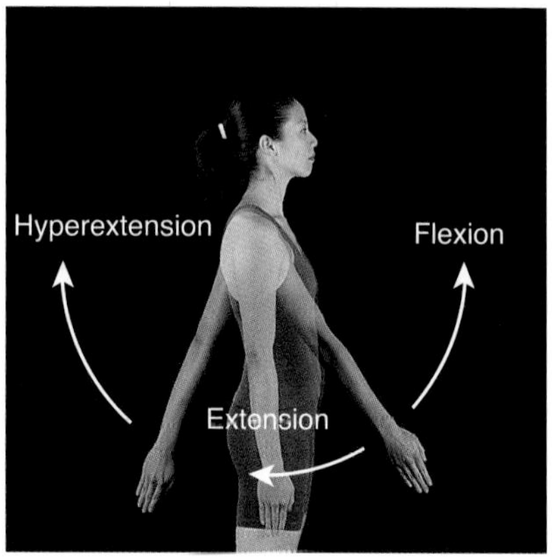

(b) Shoulder joint

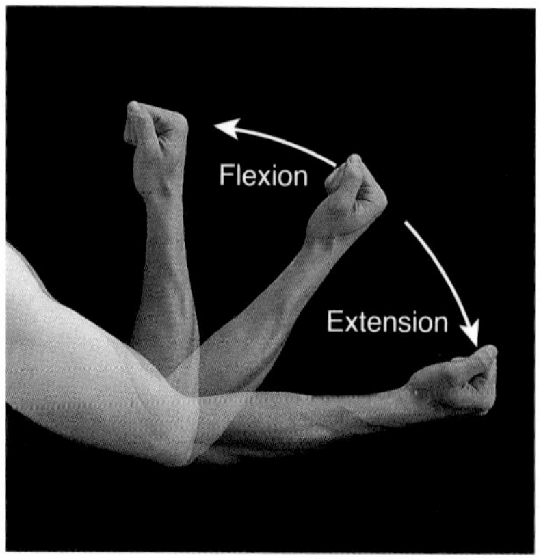

(c) Elbow joint

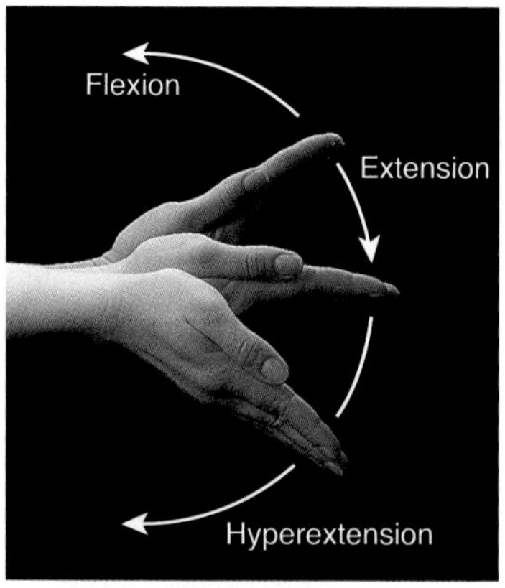

(d) Wrist joint

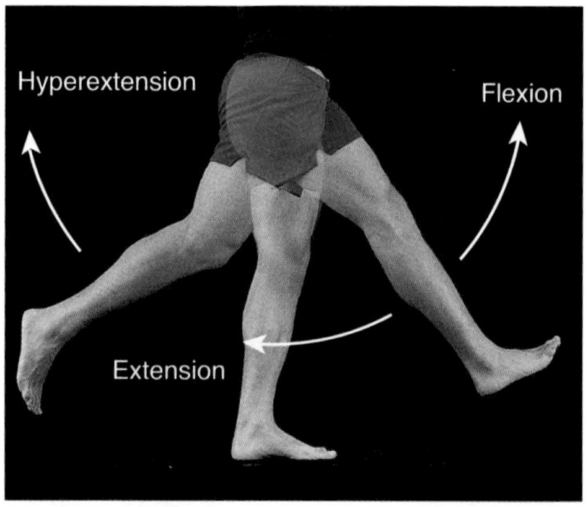

(e) Hip joint

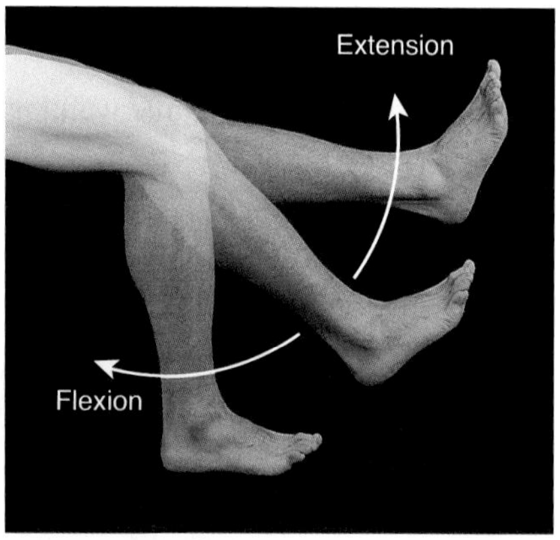

(f) Knee joint

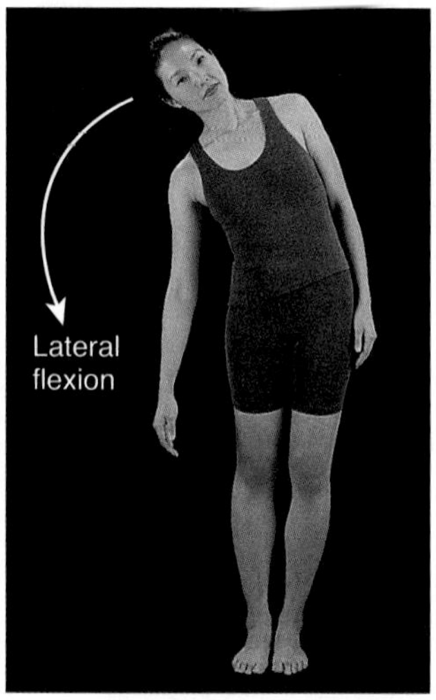

(g) Intervertebral joints

What are two examples of flexion that do not occur along the sagittal plane?

- Moving the heel toward the buttock at the tibiofemoral joint between the tibia, femur, and patella, as occurs when bending the knee (Figure 9.5f)

Although flexion and extension usually occur along the sagittal plane, there are a few exceptions. For example, flexion of the thumb involves movement of the thumb medially across the palm at the carpometacarpal joint between the trapezium and metacarpal of the thumb, as when you touch your thumb to the opposite side of your palm (see Figure 11.18g). Another example is movement of the trunk sideways to the right or left at the waist. This movement, which occurs along the frontal plane and involves the intervertebral joints, is called **lateral flexion** (Figure 9.5g).

Continuation of extension beyond the anatomical position is called **hyperextension** (hī-per-ek-STEN-shun; *hyper-* = beyond or excessive). Examples of hyperextension include:

- Bending the head backward at the atlanto-occipital and cervical intervertebral joints (Figure 9.5a)
- Bending the trunk backward at the intervertebral joints
- Moving the humerus backward at the shoulder joint, as in swinging the arms backward while walking (Figure 9.5b)
- Moving the palm backward at the wrist joint (Figure 9.5d)
- Moving the femur backward at the hip joint, as in walking (Figure 9.5e)

Hyperextension of hinge joints, such as the elbow, interphalangeal, and knee joints, is usually prevented by the arrangement of ligaments and the anatomical alignment of the bones.

Abduction, Adduction, and Circumduction

Abduction (ab-DUK-shun; *ab-* = away; *-duct* = to lead) is the movement of a bone away from the midline; **adduction** (ad-DUK-shun; *ad-* = toward) is the movement of a bone toward the midline. Both movements usually occur along the frontal plane. Examples of abduction include moving the humerus laterally at the shoulder joint, moving the palm laterally at the wrist joint, and moving the femur laterally at the hip joint (Figure 9.6a–c). The movement that returns each of these body parts to the anatomical position is adduction (Figure 9.6a–c).

The midline of the body is *not* used as a point of reference for abduction and adduction of the digits. In abduction of the fingers (but not the thumb), an imaginary line is drawn through the longitudinal axis of the middle (longest) finger, and the fingers move

Figure 9.6 Angular movements at synovial joints—abduction and adduction.

Abduction and adduction usually occur along the frontal plane.

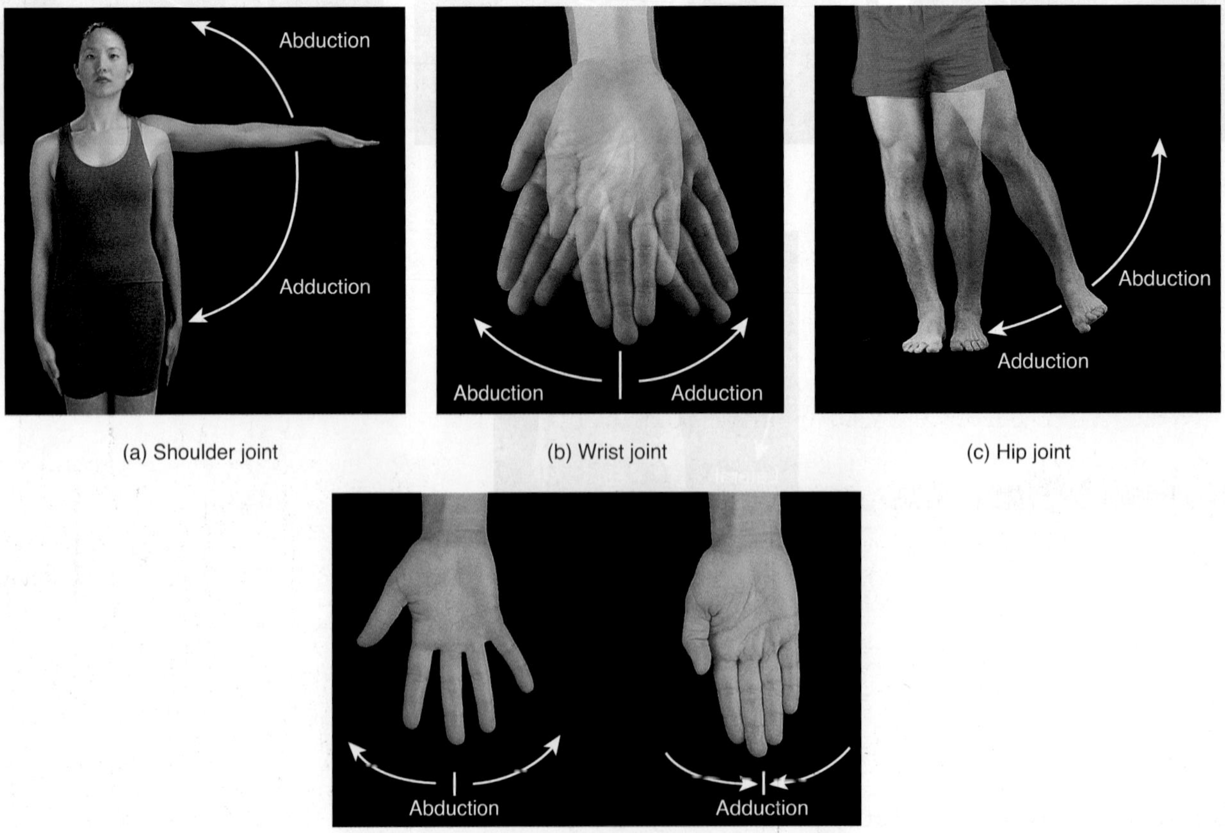

(a) Shoulder joint (b) Wrist joint (c) Hip joint

(d) Metacarpophalangeal joints of the fingers (not the thumb)

In what way is considering adduction as "adding your limb to your trunk" an effective learning device?

Figure 9.7 Angular movements at synovial joints—circumduction.

Circumduction is the movement of the distal end of a body part in a circle.

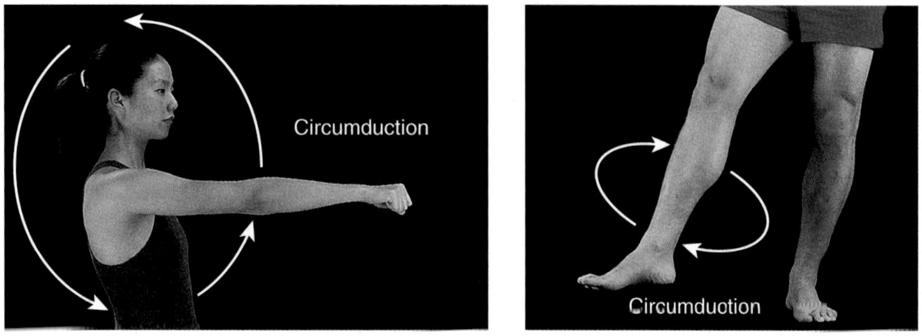

(a) Shoulder joint (b) Hip joint

Which movements in continuous sequence produce circumduction?

away (spread out) from the middle finger (Figure 9.6d). In abduction of the thumb, the thumb moves away from the palm in the sagittal plane (see Figure 11.18g). Abduction of the toes is relative to an imaginary line drawn through the second toe. Adduction of the fingers and toes returns them to the anatomical position. Adduction of the thumb moves the thumb toward the palm in the sagittal plane (see Figure 11.18g).

Circumduction (ser-kum-DUK-shun; *circ-* = circle) is movement of the distal end of a body part in a circle (Figure 9.7). Circumduction is not an isolated movement by itself but rather a continuous sequence of flexion, abduction, extension, adduction, and rotation of the joint (or in the opposite order). It does not occur along a separate axis or plane of movement. Examples of circumduction are moving the humerus in a circle at the shoulder joint (Figure 9.7a), moving the hand in a circle at the wrist joint, moving the thumb in a circle at the carpometacarpal joint, moving the fingers in a circle at the metacarpophalangeal joints (between the metacarpals and phalanges), and moving the femur in a circle at the hip joint (Figure 9.7b). Both the shoulder and hip joints permit circumduction. Flexion, abduction, extension, and adduction are more limited in the hip joints than in the shoulder joints due to the tension on certain ligaments and muscles and the depth of the acetabulum in the hip joint (see Exhibits 9.B and 9.D).

Rotation

In **rotation** (rō-TĀ-shun; *rota-* = revolve), a bone revolves around its own longitudinal axis. One example is turning the head from side to side at the atlanto-axial joint (between the atlas and axis), as when you shake your head "no" (Figure 9.8a). Another is turning the trunk from side to side at the intervertebral joints while keeping the hips and lower limbs in the anatomical position. In the limbs, rotation is defined relative to the midline, and specific qualifying terms are used. If the anterior surface of a bone of the limb is turned toward the midline, the movement is called *medial (internal) rotation.* You can medially rotate the humerus at the

Figure 9.8 Rotation at synovial joints.

In rotation, a bone revolves around its own longitudinal axis.

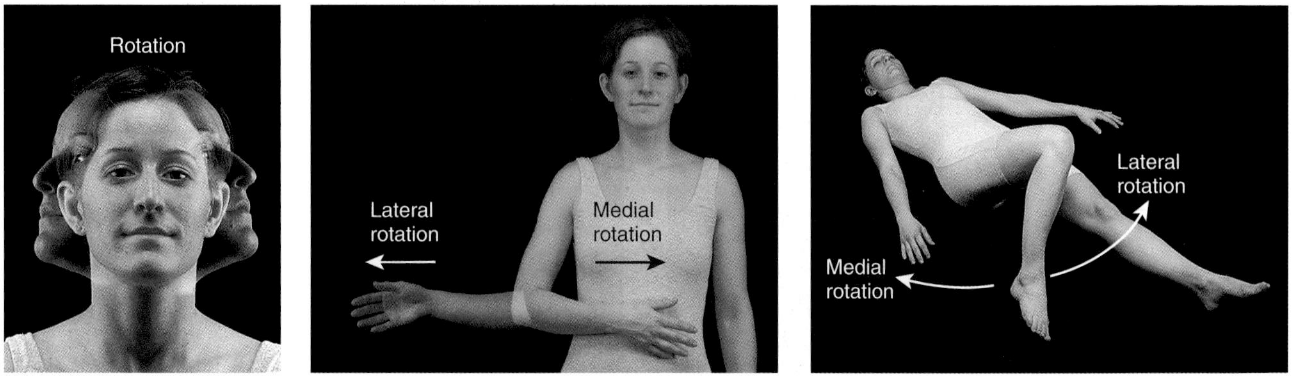

(a) Atlanto-axial joint (b) Shoulder joint (c) Hip joint

How do medial and lateral rotation differ?

shoulder joint as follows: Starting in the anatomical position, flex your elbow and then move your palm across the chest (Figure 9.8b). You can medially rotate the femur at the hip joint as follows: Lie on your back, bend your knee, and then move your leg and foot laterally from the midline. Although you are moving your leg and foot laterally, the femur is rotating medially (Figure 9.8c). Medial rotation of the leg at the knee joint can be produced by sitting on a chair, bending your knee, raising your lower limb off the floor, and turning your toes medially. If the anterior surface of the bone of a limb is turned away from the midline, the movement is called *lateral (external) rotation* (see Figure 9.8b, c).

Special Movements

Special movements occur only at certain joints. They include elevation, depression, protraction, retraction, inversion, eversion, dorsiflexion, plantar flexion, supination, pronation, and opposition (Figure 9.9):

- **Elevation** (el-e-VĀ-shun = to lift up) is a superior movement of a part of the body, such as closing the mouth at the temporomandibular joint (between the mandible and temporal bone) to elevate the mandible (Figure 9.9a) or shrugging the shoulders at the acromioclavicular joint to elevate the scapula and clavicle. Its opposing movement is depression. Other bones that may be elevated (or depressed) include the hyoid and ribs.
- **Depression** (de-PRESH-un = to press down) is an inferior movement of a part of the body, such as opening the mouth to depress the mandible (Figure 9.9b) or returning shrugged shoulders to the anatomical position to depress the scapula and clavicle.
- **Protraction** (prō-TRAK-shun = to draw forth) is a movement of a part of the body anteriorly in the transverse plane. Its

Figure 9.9 Special movements at synovial joints.

Special movements occur only at certain synovial joints.

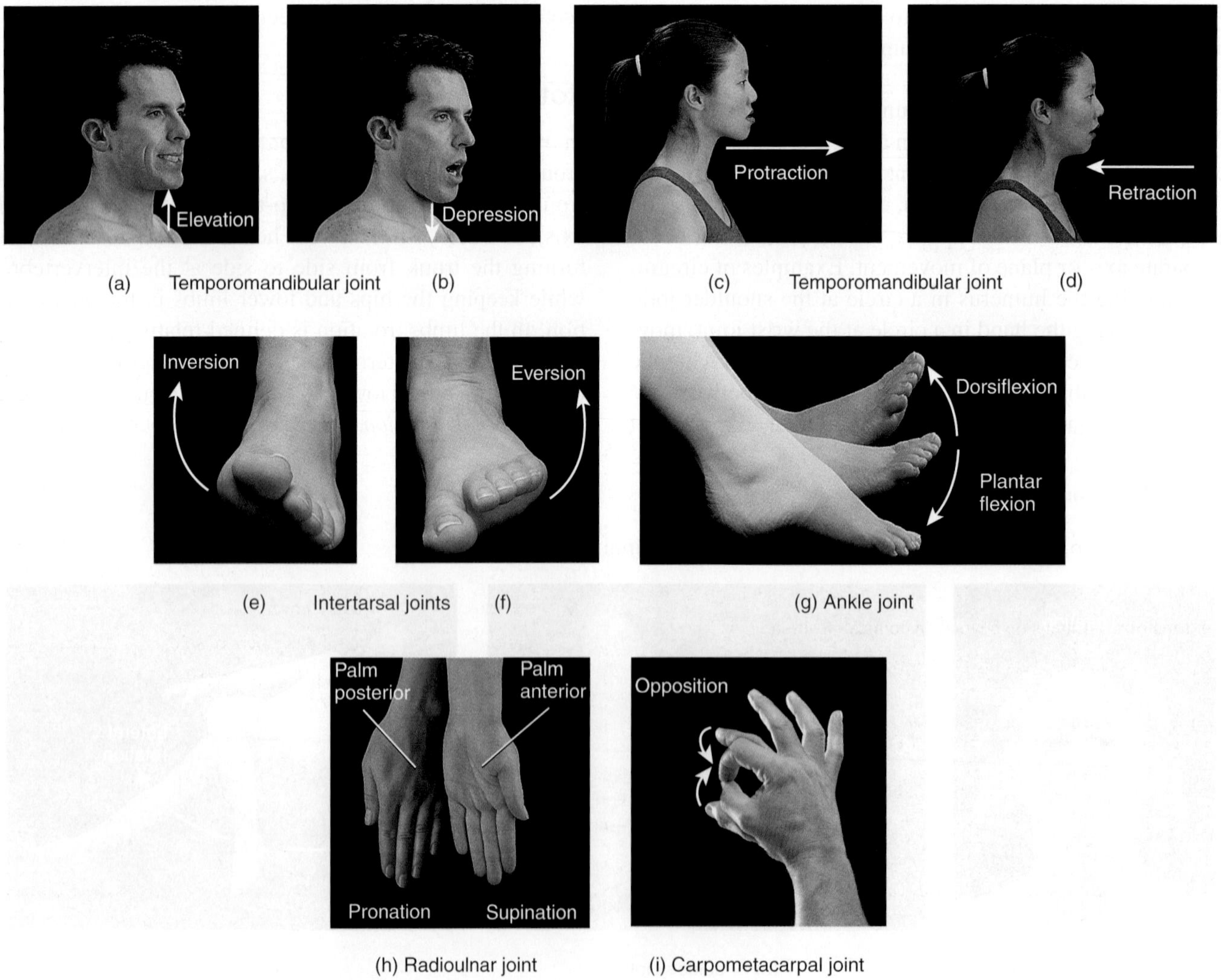

What movement of the shoulder girdle occurs when you bring your arms forward until the elbows touch?

opposing movement is retraction. You can protract your mandible at the temporomandibular joint by thrusting it outward (Figure 9.9c) or protract your clavicles at the acromioclavicular and sternoclavicular joints by crossing your arms.

- **Retraction** (rē-TRAK-shun = to draw back) is a movement of a protracted part of the body back to the anatomical position (Figure 9.9d).
- **Inversion** (in-VER-zhun = to turn inward) is movement of the sole medially at the intertarsal joints (between the tarsals) (Figure 9.9e). Its opposing movement is eversion. Physical therapists also refer to inversion of the feet as *supination.*
- **Eversion** (ē-VER-zhun = to turn outward) is a movement of the sole laterally at the intertarsal joints (Figure 9.9f). Physical therapists also refer to eversion of the feet as *pronation.*
- **Dorsiflexion** (dor-si-FLEK-shun) refers to bending of the foot at the ankle or talocrural joint (between the tibia, fibula, and talus) in the direction of the dorsum (superior surface) (Figure 9.9g). Dorsiflexion occurs when you stand on your heels. Its opposing movement is plantar flexion.
- **Plantar flexion** (PLAN-tar) involves bending of the foot at the ankle joint in the direction of the plantar or inferior surface (see Figure 9.9g), as when you elevate your body by standing on your toes.
- **Supination** (soo-pi-NĀ-shun) is a movement of the forearm at the proximal and distal radioulnar joints in which the palm is turned anteriorly (Figure 9.9h). This position of the palms is one of the defining features of the anatomical position. Its opposing movement is pronation.
- **Pronation** (prō-NĀ-shun) is a movement of the forearm at the proximal and distal radioulnar joints in which the distal end of the radius crosses over the distal end of the ulna and the palm is turned posteriorly (Figure 9.9h).
- **Opposition** (op-ō-ZISH-un) is the movement of the thumb at the carpometacarpal joint (between the trapezium and metacarpal of the thumb) in which the thumb moves across the palm to touch the tips of the fingers on the same hand (Figure 9.9i). These "opposable thumbs" allow the distinctive digital movement that gives humans and other primates the ability to grasp and manipulate objects very precisely.

A summary of the movements that occur at synovial joints is presented in Table 9.1.

✔CHECKPOINT

8. What are the four major categories of movements that occur at synovial joints?
9. On yourself or with a partner, demonstrate each movement listed in Table 9.1.

TABLE 9.1

Summary of Movements at Synovial Joints

MOVEMENT	DESCRIPTION
Gliding	Movement of relatively flat bone surfaces back and forth and side to side over one another; little change in angle between bones.
Angular	Increase or decrease in angle between bones.
Flexion	Decrease in angle between articulating bones, usually in sagittal plane.
Lateral flexion	Movement of trunk in frontal plane.
Extension	Increase in angle between articulating bones, usually in sagittal plane.
Hyperextension	Extension beyond anatomical position.
Abduction	Movement of bone away from midline, usually in frontal plane.
Adduction	Movement of bone toward midline, usually in frontal plane.
Circumduction	Flexion, abduction, extension, adduction, and rotation in succession (or in the opposite order); distal end of body part moves in circle.
Rotation	Movement of bone around longitudinal axis; in limbs, may be medial (toward midline) or lateral (away from midline).
Special	Occurs at specific joints.
Elevation	Superior movement of body part.
Depression	Inferior movement of body part.
Protraction	Anterior movement of body part in transverse plane.
Retraction	Posterior movement of body part in transverse plane.
Inversion	Medial movement of sole.
Eversion	Lateral movement of sole.
Dorsiflexion	Bending foot in direction of dorsum (superior surface).
Plantar flexion	Bending foot in direction of plantar surface (sole).
Supination	Movement of forearm that turns palm anteriorly.
Pronation	Movement of forearm that turns palm posteriorly.
Opposition	Movement of thumb across palm to touch fingertips on same hand.

9.6 TYPES OF SYNOVIAL JOINTS

OBJECTIVE

- Describe the six subtypes of synovial joints.

Although all synovial joints share many characteristics in common, the shapes of the articulating surfaces vary; thus, many types of movements are possible. Synovial joints are divided into six categories based on type of movement: plane, hinge, pivot, condyloid, saddle, and ball-and-socket.

Plane Joints

The articulating surfaces of bones in a **plane joint** (PLĀN), also called a *planar joint* (PLĀ-nar), are flat or slightly curved (Figure 9.10a). Plane joints primarily permit back-and-forth and side-to-side movements between the flat surfaces of bones, but they may also rotate against one another. Many plane joints are *biaxial,* meaning that they permit movement in two axes. An *axis* is a straight line around which a bone rotates (revolves) or slides. If plane joints rotate in addition to sliding, then they are *triaxial (multiaxial),* permitting movement in three axes. Some examples of plane joints are the intercarpal joints (between carpal bones at the wrist), intertarsal joints (between tarsal bones at the ankle), sternoclavicular joints (between the manubrium of the sternum and the clavicle), acromioclavicular joints (between the acromion of the scapula and the clavicle), sternocostal joints (between the sternum and ends of the costal cartilages at the tips of the second through seventh pairs of ribs), and vertebrocostal joints (between the heads and tubercles of ribs and transverse processes of thoracic vertebrae).

Hinge Joints

In a **hinge joint,** or *ginglymus joint* (JIN-gli-mus), the convex surface of one bone fits into the concave surface of another bone (Figure 9.10b). As the name implies, hinge joints produce an angular, opening-and-closing motion like that of a hinged door. In most joint movements, one bone remains in a fixed position while the other moves around an axis. Hinge joints are *uniaxial (monaxial)* because they typically allow motion around a single axis. Hinge joints permit only flexion and extension. Examples of hinge joints are the knee (actually a modified hinge joint, which will be described later), elbow, ankle, and interphalangeal joints (between the phalanges of the fingers and toes).

Pivot Joints

In a **pivot joint,** or *trochoid* (TRŌ-koyd) *joint,* the rounded or pointed surface of one bone articulates with a ring formed partly by another bone and partly by a ligament (Figure 9.10c). A pivot joint is *uniaxial* because it allows rotation only around its own longitudinal axis. Examples of pivot joints are the atlanto-axial joint, in which the atlas rotates around the axis and permits the head to turn from side to side as when you shake your head "no" (see Figure 9.8a), and the radioulnar joints that enable the palms to turn anteriorly and posteriorly as the head of the radius pivots around its long axis in the radial notch of the ulna (see Figure 9.9h).

Condyloid Joints

In a **condyloid joint** (KON-di-loyd; *condyl-* = knuckle) or *ellipsoidal joint,* the convex oval-shaped projection of one bone fits into the oval-shaped depression of another bone (Figure 9.10d). A condyloid joint is *biaxial* because the movement it permits is around two axes (flexion–extension and abduction–adduction), plus limited circumduction (remember that circumduction is not an isolated movement). Examples of condyloid joints are the radiocarpal (wrist) and metacarpophalangeal joints (between the metacarpals and proximal phalanges) of the second through fifth digits.

Saddle Joints

In a **saddle joint** or *sellar joint* (SEL-ar), the articular surface of one bone is saddle-shaped, and the articular surface of the other bone fits into the "saddle" as a sitting rider would sit (Figure 9.10e). The movements at a saddle joint are the same as those at a condyloid joint: *biaxial* (flexion–extension and abduction–adduction) plus limited circumduction. An example of a saddle joint is the carpometacarpal joint between the trapezium of the carpus and metacarpal of the thumb.

Ball-and-Socket Joints

A **ball-and-socket joint** or *spheroid* (SFĒ-royd) *joint* consists of the ball-like surface of one bone fitting into a cuplike depression of another bone (Figure 9.10f). Such joints are *triaxial (multiaxial),* permitting movements around three axes (flexion–extension, abduction–adduction, and rotation). Examples of ball-and-socket joints are the shoulder and hip joints. At the shoulder joint, the head of the humerus fits into the glenoid cavity of the scapula. At the hip joint, the head of the femur fits into the acetabulum of the hip (coxal) bone.

Table 9.2 summarizes the structural and functional categories of joints.

✔CHECKPOINT

10. Which types of joints are uniaxial, biaxial, and triaxial?

Figure 9.10 Types of synovial joints. For each type, a drawing of the actual joint and a simplified diagram are shown.

Synovial joints are classified into six principal types based on the shapes of the articulating bone surfaces.

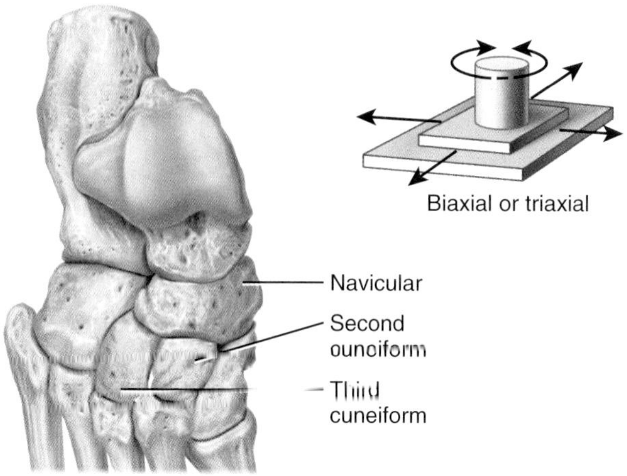

(a) Plane joint between navicular and second and third cuneiforms of tarsus in foot

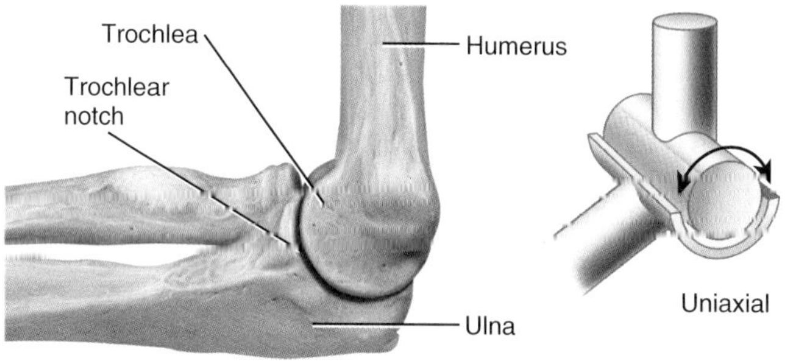

(b) Hinge joint between trochlea of humerus and trochlear notch of ulna at the elbow

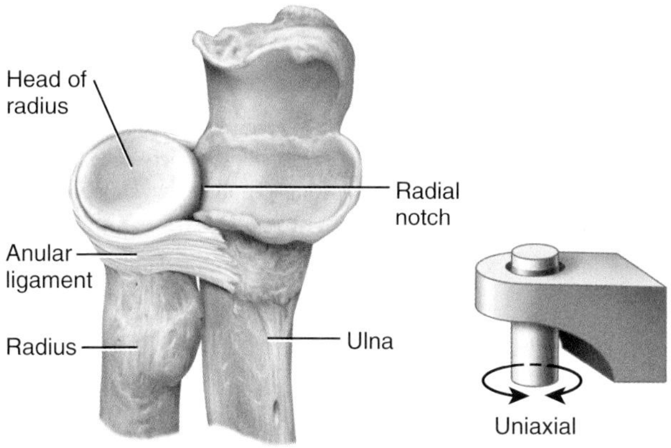

(c) Pivot joint between head of radius and radial notch of ulna

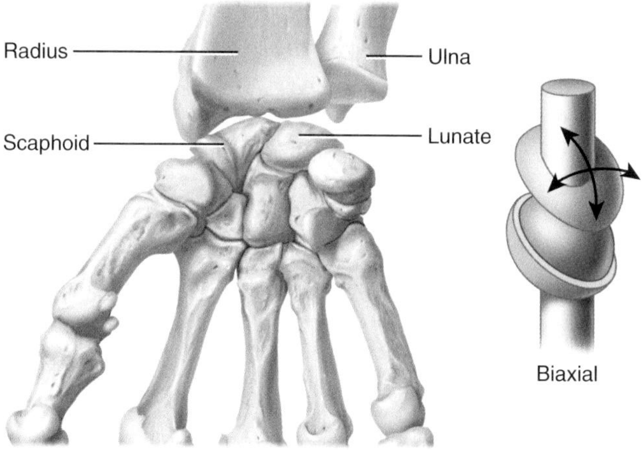

(d) Condyloid joint between radius and scaphoid and lunate bones of carpus (wrist)

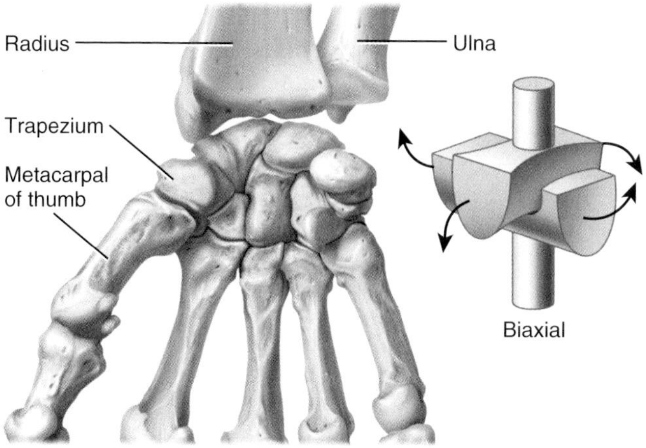

(e) Saddle joint between trapezium of carpus (wrist) and metacarpal of thumb

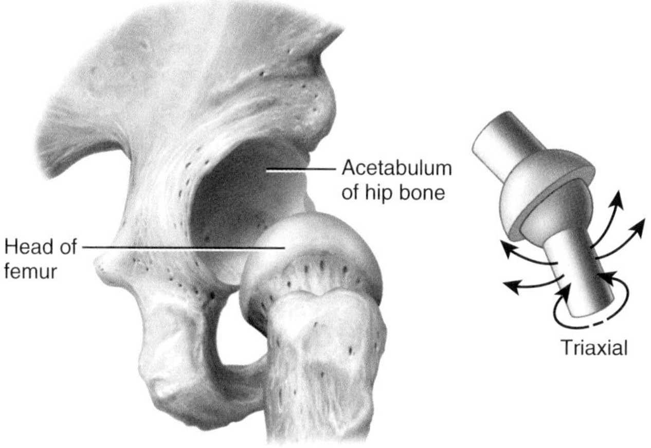

(f) Ball-and-socket joint between head of femur and acetabulum of hip bone

Which of the joints shown here are biaxial?

TABLE 9.2

Summary of Structural and Functional Classifications of Joints

STRUCTURAL CLASSIFICATION	DESCRIPTION	FUNCTIONAL CLASSIFICATION	EXAMPLE
FIBROUS No Synovial Cavity; Articulating Bones Held Together by Fibrous Connective Tissue.			
Suture	Articulating bones united by thin layer of dense irregular connective tissue, found between skull bones; with age, some sutures replaced by synostosis (separate cranial bones fuse into single bone).	Synarthrosis (immovable) and amphiarthrosis (slightly movable).	Coronal suture.
Syndesmosis	Articulating bones united by more dense irregular connective tissue, usually a ligament.	Amphiarthrosis (slightly movable).	Distal tibiofibular joint.
Interosseous membrane	Articulating bones united by substantial sheet of dense irregular connective tissue.	Amphiarthrosis (slightly movable).	Between tibia and fibula.
CARTILAGINOUS No Synovial Cavity; Articulating Bones United by Hyaline Cartilage or Fibrocartilage.			
Synchondrosis	Connecting material: hyaline cartilage; becomes synostosis when bone elongation ceases.	Synarthrosis (immovable).	Epiphyseal plate between diaphysis and epiphysis of long bone.
Symphysis	Connecting material: broad, flat disc of fibrocartilage.	Amphiarthrosis (slightly movable).	Pubic symphysis and intervertebral joints.
SYNOVIAL Characterized by Synovial Cavity, Articular Cartilage, and Articular (Joint) Capsule; May Contain Accessory Ligaments, Articular Discs, and Bursae.			
Plane	Articulated surfaces flat or slightly curved.	Many biaxial diarthroses (freely movable): back-and-forth and side-to-side movements. Some triaxial diarthrosis: back-and-forth, side-to-side, rotation.	Intercarpal, intertarsal, sternocostal (between sternum and second to seventh pairs of ribs), and vertebrocostal joints.
Hinge	Convex surface fits into concave surface.	Uniaxial diarthrosis: flexion–extension.	Knee (modified hinge), elbow, ankle, and interphalangeal joints.
Pivot	Rounded or pointed surface fits into ring formed partly by bone and partly by ligament.	Uniaxial diarthrosis: rotation.	Atlanto-axial and radioulnar joints.
Condyloid	Oval-shaped projection fits into oval-shaped depression.	Biaxial diarthrosis: flexion–extension, abduction–adduction.	Radiocarpal and metacarpophalangeal joints.
Saddle	Articular surface of one bone is saddle-shaped; articular surface of other bone "sits" in saddle.	Biaxial diarthrosis: flexion–extension, abduction–adduction.	Carpometacarpal joint between trapezium and metacarpal of thumb.
Ball-and-socket	Ball-like surface fits into cuplike depression.	Triaxial diarthrosis: flexion–extension, abduction–adduction, rotation.	Shoulder and hip joints.

9.7 FACTORS AFFECTING CONTACT AND RANGE OF MOTION AT SYNOVIAL JOINTS

OBJECTIVE

- Describe six factors that influence the type of movement and range of motion possible at a synovial joint.

The articular surfaces of synovial joints contact one another and determine the type and possible range of motion. **Range of motion (ROM)** refers to the range, measured in degrees of a circle, through which the bones of a joint can be moved. The following factors contribute to keeping the articular surfaces in contact and affect range of motion:

1. ***Structure or shape of the articulating bones.*** The structure or shape of the articulating bones determines how closely they can fit together. The articular surfaces of some bones have a complementary relationship. This spatial relationship is very obvious at the hip joint, where the head of the femur articulates with the acetabulum of the hip bone. An interlocking fit allows rotational movement.
2. ***Strength and tension (tautness) of the joint ligaments.*** The different components of a fibrous capsule are tense or taut only when the joint is in certain positions. Tense ligaments not only restrict the range of motion but also direct the movement of the articulating bones with respect to each other. In the knee joint, for example, the anterior cruciate ligament is taut and the posterior cruciate ligament is loose when the knee is straightened, and the reverse occurs when the knee is bent. In the hip joint, certain ligaments become taut when standing and more firmly attach the head of the femur to the acetabulum of the hip bone.
3. ***Arrangement and tension of the muscles.*** Muscle tension reinforces the restraint placed on a joint by its ligaments, and thus restricts movement. A good example of the effect of muscle tension on a joint is seen at the hip joint. When the thigh is flexed with the knee extended, the flexion of the hip joint is restricted by the tension of the hamstring muscles on the posterior surface of the thigh, so most of us can't raise a straightened leg more than a 90-degree angle from the floor. But if the knee is also flexed, the tension on the hamstring muscles is lessened, and the thigh can be raised farther, allowing you to raise your thigh to touch your chest.
4. ***Contact of soft parts.*** The point at which one body surface contacts another may limit mobility. For example, if you bend your arm at the elbow, it can move no farther after the anterior surface of the forearm meets with and presses against the biceps brachii muscle of the arm. Joint movement may also be restricted by the presence of adipose tissue.
5. ***Hormones.*** Joint flexibility may also be affected by hormones. For example, relaxin, a hormone produced by the placenta and ovaries, increases the flexibility of the fibrocartilage of the pubic symphysis and loosens the ligaments between the sacrum, hip bone, and coccyx toward the end of pregnancy. These changes permit expansion of the pelvic outlet, which assists in delivery of the baby.
6. ***Disuse.*** Movement at a joint may be restricted if a joint has not been used for an extended period. For example, if an elbow joint is immobilized by a cast, range of motion at the joint may be limited for a time after the cast is removed. Disuse may also result in decreased amounts of synovial fluid, diminished flexibility of ligaments and tendons, and *muscular atrophy,* a reduction in size or wasting of a muscle.

✔ CHECKPOINT

11. How do the strength and tension of ligaments determine range of motion?

9.8 SELECTED JOINTS OF THE BODY

OBJECTIVE

- Identify the major joints of the body by location, classification, and movements.

In Chapters 7 and 8, we discussed the major bones and their markings. In this chapter we have examined how joints are classified according to their structure and function, and we have introduced the movements that occur at joints. Table 9.3 (selected joints of the axial skeleton) and Table 9.4 (selected joints of the appendicular skeleton) will help you integrate the information you have learned in all three chapters. These tables list some of the major joints of the body according to their articular components (the bones that enter into their formation), their structural and functional classification, and the type(s) of movement that occur(s) at each joint.

Next we examine in detail several selected joints of the body in a series of exhibits. Each exhibit considers a specific synovial joint and contains (1) a definition—a description of the type of joint and the bones that form the joint; (2) the anatomical components—a description of the major connecting ligaments, articular disc (if present), articular capsule, and other distinguishing features of the joint; and (3) the joint's possible movements. Each exhibit also refers you to a figure that illustrates the joint. The joints described are the temporomandibular joint (TMJ), shoulder (humeroscapular or glenohumeral) joint, elbow joint, hip (coxal) joint, and knee (tibiofemoral) joint. Because these joints are described in Exhibits 9.A–9.E (Figures 9.11–9.15), they are not included in Tables 9.3 and 9.4.

✔ CHECKPOINT

12. Using Tables 9.3 and 9.4 as a guide, identify only the cartilaginous joints.

TABLE 9.3

Selected Joints of the Axial Skeleton

JOINT	ARTICULAR COMPONENTS	CLASSIFICATION	MOVEMENTS
Suture	Between skull bones.	*Structural:* fibrous. *Functional:* amphiarthrosis and synarthrosis.	None.
Atlanto-occipital	Between superior articular facets of atlas and occipital condyles of occipital bone.	*Structural:* synovial (condyloid). *Functional:* diarthrosis.	Flexion and extension of head; slight lateral flexion of head to either side.
Atlanto-axial	(1) Between dens of axis and anterior arch of atlas; (2) between lateral masses of atlas and axis.	*Structural:* synovial (pivot) between dens and anterior arch; synovial (planar) between lateral masses. *Functional:* diarthrosis.	Rotation of head.
Intervertebral	(1) Between vertebral bodies; (2) between vertebral arches.	*Structural:* cartilaginous (symphysis) between vertebral bodies; synovial (planar) between vertebral arches. *Functional:* amphiarthrosis between vertebral bodies; diarthrosis between vertebral arches.	Flexion, extension, lateral flexion, and rotation of vertebral column.
Vertebrocostal	(1) Between facets of heads of ribs and facets of bodies of adjacent thoracic vertebrae and intervertebral discs between them; (2) between articular part of tubercles of ribs and facets of transverse processes of thoracic vertebrae.	*Structural:* synovial (planar). *Functional:* diarthrosis.	Slight gliding.
Sternocostal	Between sternum and first seven pairs of ribs.	*Structural:* cartilaginous (synchondrosis) between sternum and first pair of ribs; synovial (plane) between sternum and second through seventh pairs of ribs. *Functional:* synarthrosis between sternum and first pair of ribs; diarthrosis between sternum and second through seventh pairs of ribs.	None between sternum and first pair of ribs; slight gliding between sternum and second through seventh pairs of ribs.
Lumbosacral	(1) Between body of fifth lumbar vertebra and base of sacrum; (2) between inferior articular facets of fifth lumbar vertebra and superior articular facets of first vertebra of sacrum.	*Structural:* cartilaginous (symphysis) between body and base; synovial (planar) between articular facets. *Functional:* amphiarthrosis between body and base; diarthrosis between articular facets.	Flexion, extension, lateral flexion, and rotation of vertebral column.

TABLE 9.4

Selected Joints of the Appendicular Skeleton

JOINT	ARTICULAR COMPONENTS	CLASSIFICATION	MOVEMENTS
Sternoclavicular	Between sternal end of clavicle, manubrium of sternum, and first costal cartilage.	*Structural:* synovial (plane, pivot). *Functional:* diarthrosis.	Gliding, with limited movements in nearly every direction.
Acromioclavicular	Between acromion of scapula and acromial end of clavicle.	*Structural:* synovial (plane). *Functional:* diarthrosis.	Gliding and rotation of scapula on clavicle.
Radioulnar	Proximal radioulnar joint between head of radius and radial notch of ulna; distal radioulnar joint between ulnar notch of radius and head of ulna.	*Structural:* synovial (pivot). *Functional:* diarthrosis.	Rotation of forearm.
Wrist (radiocarpal)	Between distal end of radius and scaphoid, lunate, and triquetrum of carpus.	*Structural:* synovial (condyloid). *Functional:* diarthrosis.	Flexion, extension, abduction, adduction, circumduction, and slight hyperextension of wrist.
Intercarpal	Between proximal row of carpal bones, distal row of carpal bones, and between both rows of carpal bones (midcarpal joints).	*Structural:* synovial (plane), except for hamate, scaphoid, and lunate (midcarpal) joint, which is synovial (saddle). *Functional:* diarthrosis.	Gliding plus flexion, extension, abduction, adduction, and slight rotation at midcarpal joints.
Carpometacarpal	Carpometacarpal joint of thumb between trapezium of carpus and first metacarpal; carpometacarpal joints of remaining digits formed between carpus and second through fifth metacarpals.	*Structural:* synovial (saddle) at thumb; synovial (plane) at remaining digits. *Functional:* diarthrosis.	Flexion, extension, abduction, adduction, and circumduction at thumb; gliding at remaining digits.
Metacarpophalangeal and metatarsophalangeal	Between heads of metacarpals (or metatarsals) and bases of proximal phalanges.	*Structural:* synovial (condyloid). *Functional:* diarthrosis.	Flexion, extension, abduction, adduction, and circumduction of phalanges.
Interphalangeal	Between heads of phalanges and bases of more distal phalanges.	*Structural:* synovial (hinge). *Functional:* diarthrosis.	Flexion and extension of phalanges.
Sacroiliac	Between auricular surfaces of sacrum and ilia of hip bones.	*Structural:* synovial (plane). *Functional:* diarthrosis.	Slight gliding (even more so during pregnancy).
Pubic symphysis	Between anterior surfaces of hip bones.	*Structural:* cartilaginous (symphysis). *Functional:* amphiarthrosis.	Slight movements (even more so during pregnancy).
Tibiofibular	Proximal tibiofibular joint between lateral condyle of tibia and head of fibula; distal tibiofibular joint between distal end of fibula and fibular notch of tibia.	*Structural:* synovial (plane) at proximal joint; fibrous (syndesmosis) at distal joint. *Functional:* diarthrosis at proximal joint; amphiarthrosis at distal joint.	Slight gliding at proximal joint; slight rotation of fibula during dorsiflexion of foot.
Ankle (talocrural)	(1) Between distal end of tibia and its medial malleolus and talus; (2) between lateral malleolus of fibula and talus.	*Structural:* synovial (hinge). *Functional:* diarthrosis.	Dorsiflexion and plantar flexion of foot.
Intertarsal	Subtalar joint between talus and calcaneus of tarsus; talocalcaneonavicular joint between talus and calcaneus and navicular of tarsus; calcaneocuboid joint between calcaneus and cuboid of tarsus.	*Structural:* synovial (plane) at subtalar and calcaneocuboid joints; synovial (saddle) at talocalcaneonavicular joint. *Functional:* diarthrosis.	Inversion and eversion of foot.
Tarsometatarsal	Between three cuneiforms of tarsus and bases of five metatarsal bones.	*Structural:* synovial (plane). *Functional:* diarthrosis.	Slight gliding.

EXHIBIT 9.A Temporomandibular Joint *(Figure 9.11)*

OBJECTIVE

- Describe the anatomical components of the temporomandibular joint and explain the movements that can occur at this joint.

Definition

The **temporomandibular joint (TMJ)** (tem′-po-rō-man-DIB-ū-lar) is a combined hinge and planer joint formed by the condylar process of the mandible and the mandibular fossa and articular tubercle of the temporal bone. The temporomandibular joint is the only freely movable joint between skull bones (with the exception of the ear ossicles); all other skull joints are sutures and therefore immovable or slightly movable.

Anatomical Components

1. ***Articular disc (meniscus).*** Fibrocartilage disc that separates the synovial cavity into superior and inferior compartments, each with a synovial membrane (Figure 9.11c).
2. ***Articular capsule.*** Thin, fairly loose envelope around the circumference of the joint (Figure 9.11a, b).
3. ***Lateral ligament.*** Two short bands on the lateral surface of the articular capsule that extend inferiorly and posteriorly from the inferior border and tubercle of the zygomatic process of the temporal bone to the lateral and posterior aspect of the neck of the mandible. The lateral ligament is covered by the parotid gland and helps strengthen the joint laterally and prevent displacement of the mandible (Figure 9.11a).
4. ***Sphenomandibular ligament*** (sfē-nō-man-DIB-ū-lar). Thin band that extends inferiorly and anteriorly from the spine of the sphenoid bone to the ramus of the mandible (Figure 9.11b). It does not contribute significantly to the strength of the joint.
5. ***Stylomandibular ligament*** (stī-lō-man-DIB-ū-lar). Thickened band of deep cervical fascia that extends from the styloid process of the temporal bone to the inferior and posterior border of the ramus of the mandible. This ligament separates the parotid gland from the submandibular gland and limits movement of the mandible at the TMJ (Figure 9.11a, b).

Movements

In the temporomandibular joint, only the mandible moves because the temporal bone is firmly anchored to other bones of the skull by sutures. Accordingly, the mandible may function in depression (jaw opening) and elevation (jaw closing), which occurs in the inferior compartment, and protraction, retraction, lateral displacement, and slight rotation, which occur in the superior compartment (see Figure 9.9a–d).

CLINICAL CONNECTION | *Dislocated Mandible*

A **dislocation** (dis′-lō-KĀ-shun; *dis-* = apart) or *luxation* (luks-Ā-shun; *luxatio* = dislocation) is the displacement of a bone from a joint with tearing of ligaments, tendons, and articular capsules. A **dislocated mandible** can occur in several ways. *Anterior displacements* are the most common and occur when the condylar processes of the mandible pass anterior to the articular tubercles. Common causes are extreme mouth opening, as in yawning or taking a large bite, dental procedures, or general anesthesia. *Posterior displacement* can be caused by a direct blow to the chin. *Superior displacements* are typically caused by a direct blow to a partially opened mouth. *Lateral dislocations* are usually associated with mandibular fractures. •

✔**CHECKPOINT**

What distinguishes the temporomandibular joint from the other joints of the skull?

Figure 9.11 **Right temporomandibular joint (TMJ).**

The TMJ is the only movable joint between skull bones.

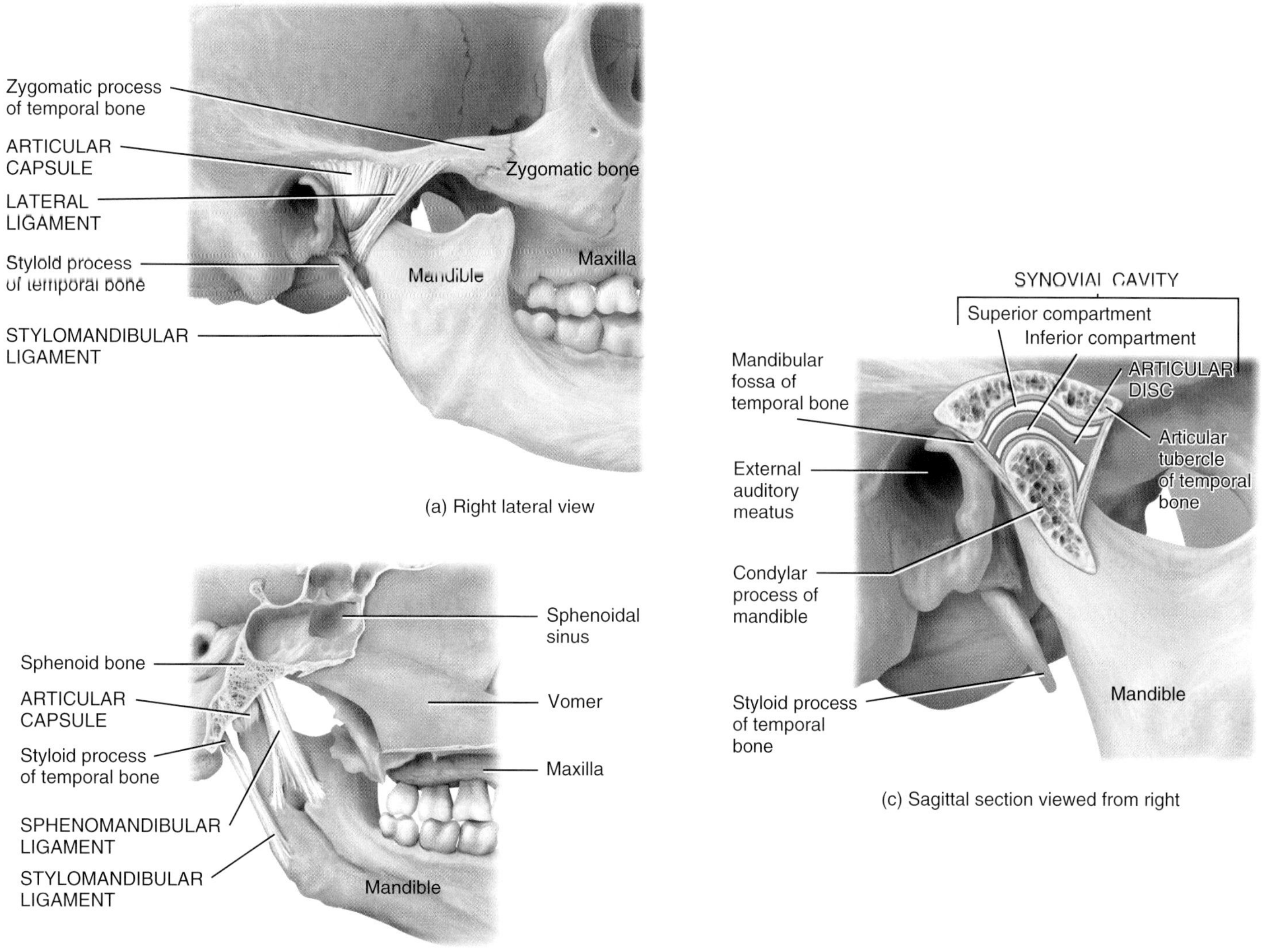

Zygomatic process of temporal bone
ARTICULAR CAPSULE
LATERAL LIGAMENT
Styloid process of temporal bone
STYLOMANDIBULAR LIGAMENT
Zygomatic bone
Maxilla
Mandible
Sphenoid bone
ARTICULAR CAPSULE
Styloid process of temporal bone
SPHENOMANDIBULAR LIGAMENT
STYLOMANDIBULAR LIGAMENT
Sphenoidal sinus
Vomer
Maxilla
Mandible
SYNOVIAL CAVITY
Superior compartment
Inferior compartment
ARTICULAR DISC
Mandibular fossa of temporal bone
Articular tubercle of temporal bone
External auditory meatus
Condylar process of mandible
Styloid process of temporal bone
Mandible

(a) Right lateral view

(b) Left medial view

(c) Sagittal section viewed from right

Which ligament prevents displacement of the mandible?

Shoulder Joint *(Figure 9.12)*

OBJECTIVE

- Describe the anatomical components of the shoulder joint and the movements that can occur at this joint.

Definition

The **shoulder joint** is a ball-and-socket joint formed by the head of the humerus and the glenoid cavity of the scapula. It also is referred to as the *humeroscapular* or *glenohumeral joint.*

Anatomical Components

1. ***Articular capsule.*** Thin, loose sac that completely envelops the joint and extends from the glenoid cavity to the anatomical neck of the humerus. The inferior part of the capsule is its weakest area (Figure 9.12).
2. ***Coracohumeral ligament*** (kor′-a-kō-Ū-mer-al). Strong, broad ligament that strengthens the superior part of the articular capsule and extends from the coracoid process of the scapula to the greater tubercle of the humerus (Figure 9.12a, b). The ligament strengthens the superior part of the articular capsule and reinforces the anterior aspect of the articular capsule.
3. ***Glenohumeral ligaments*** (glē-no-Ū-mer-al). Three thickenings of the articular capsule over the anterior surface of the joint that extend from the glenoid cavity to the lesser tubercle and anatomical neck of the humerus. These ligaments are often indistinct or absent and provide only minimal strength (Figure 9.12a, b). They play a role in joint stabilization when the humerus approaches or exceeds its limits of motion.
4. ***Transverse humeral ligament.*** Narrow sheet extending from the greater tubercle to the lesser tubercle of the humerus (Figure 9.12a). The ligament functions as a retinaculum (retaining band of connective tissue) to hold the long head of the biceps brachii muscle.
5. ***Glenoid labrum.*** Narrow rim of fibrocartilage around the edge of the glenoid cavity that slightly deepens and enlarges the glenoid cavity (Figure 9.12b, c).
6. ***Bursae.*** Four *bursae* (see Section 9.4) are associated with the shoulder joint. They are the *subscapular bursa* (Figure 9.12a), *subdeltoid bursa, subacromial bursa* (Figure 9.12a–c), and *subcoracoid bursa.*

Figure 9.12 **Right shoulder (humeroscapular or glenohumeral) joint.**

Most of the stability of the shoulder joint results from the arrangement of the rotator cuff muscles.

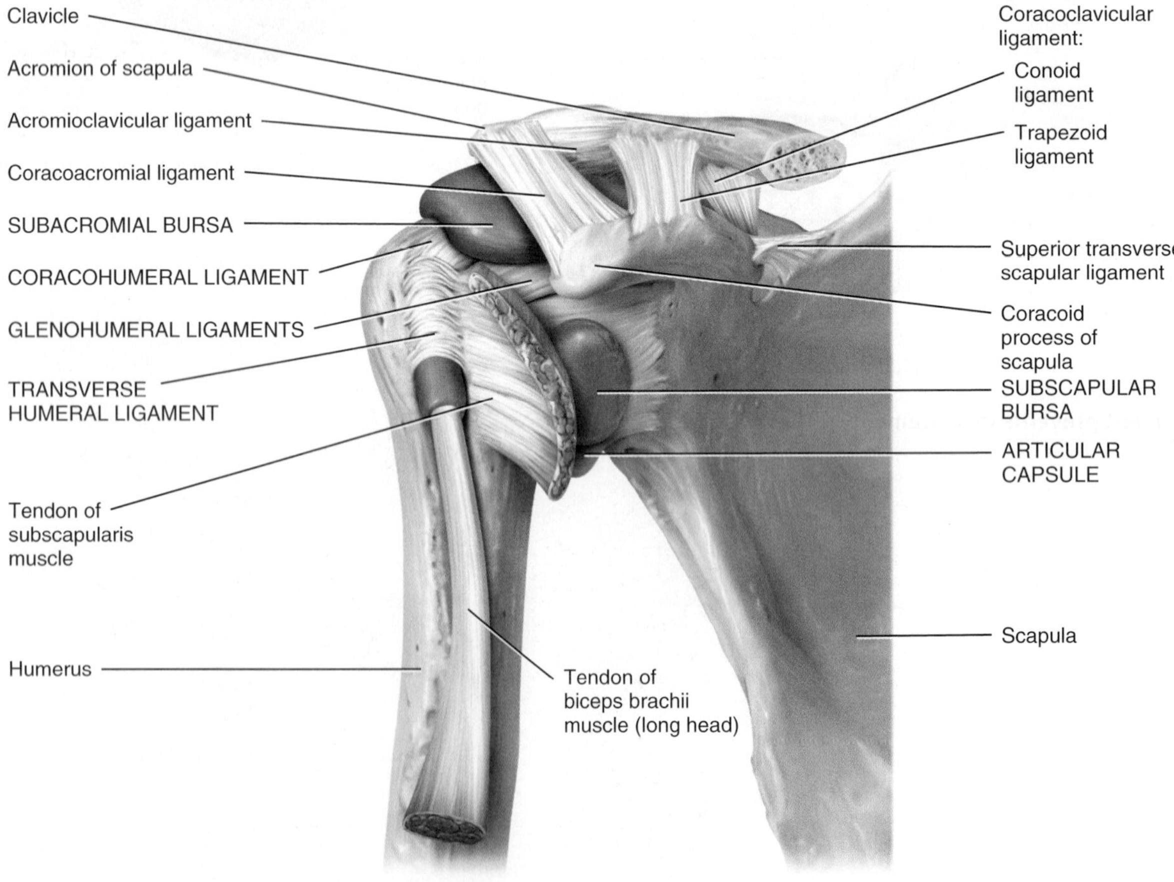
Clavicle
Acromion of scapula
Acromioclavicular ligament
Coracoacromial ligament
SUBACROMIAL BURSA
CORACOHUMERAL LIGAMENT
GLENOHUMERAL LIGAMENTS
TRANSVERSE HUMERAL LIGAMENT
Tendon of subscapularis muscle
Humerus
Tendon of biceps brachii muscle (long head)
Coracoclavicular ligament:
Conoid ligament
Trapezoid ligament
Superior transverse scapular ligament
Coracoid process of scapula
SUBSCAPULAR BURSA
ARTICULAR CAPSULE
Scapula

(a) Anterior view

Movements

The shoulder joint allows flexion, extension, hyperextension, abduction, adduction, medial rotation, lateral rotation, and circumduction of the arm (see Figures 9.5–9.8). It has more freedom of movement than any other joint of the body. This freedom results from the looseness of the articular capsule and the shallowness of the glenoid cavity in relation to the large size of the head of the humerus.

Although the ligaments of the shoulder joint strengthen it to some extent, most of the strength results from the muscles that surround the joint,

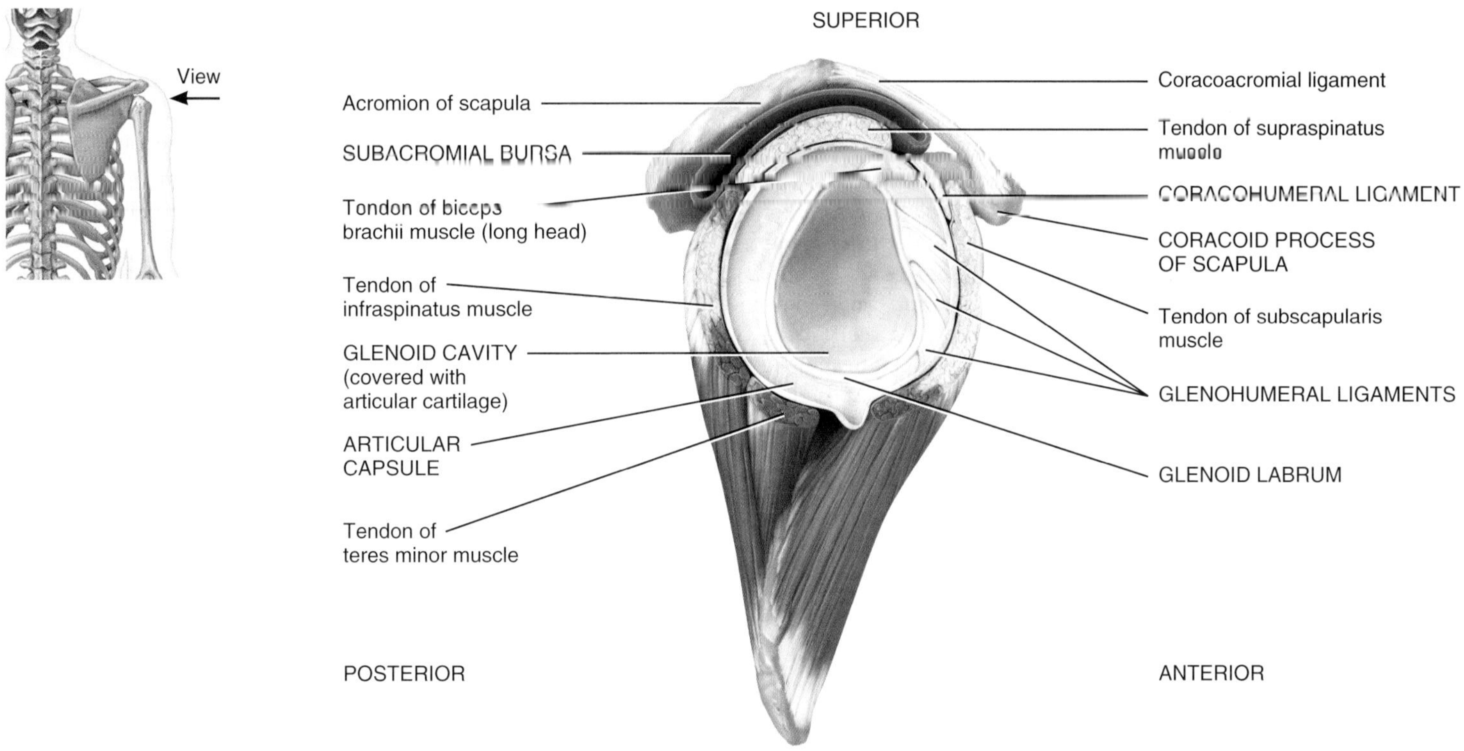

View
SUPERIOR
Acromion of scapula
SUBACROMIAL BURSA
Tendon of biceps brachii muscle (long head)
Tendon of infraspinatus muscle
GLENOID CAVITY (covered with articular cartilage)
ARTICULAR CAPSULE
Tendon of teres minor muscle
POSTERIOR
Coracoacromial ligament
Tendon of supraspinatus muscle
CORACOHUMERAL LIGAMENT
CORACOID PROCESS OF SCAPULA
Tendon of subscapularis muscle
GLENOHUMERAL LIGAMENTS
GLENOID LABRUM
ANTERIOR

(b) Lateral view (opened)

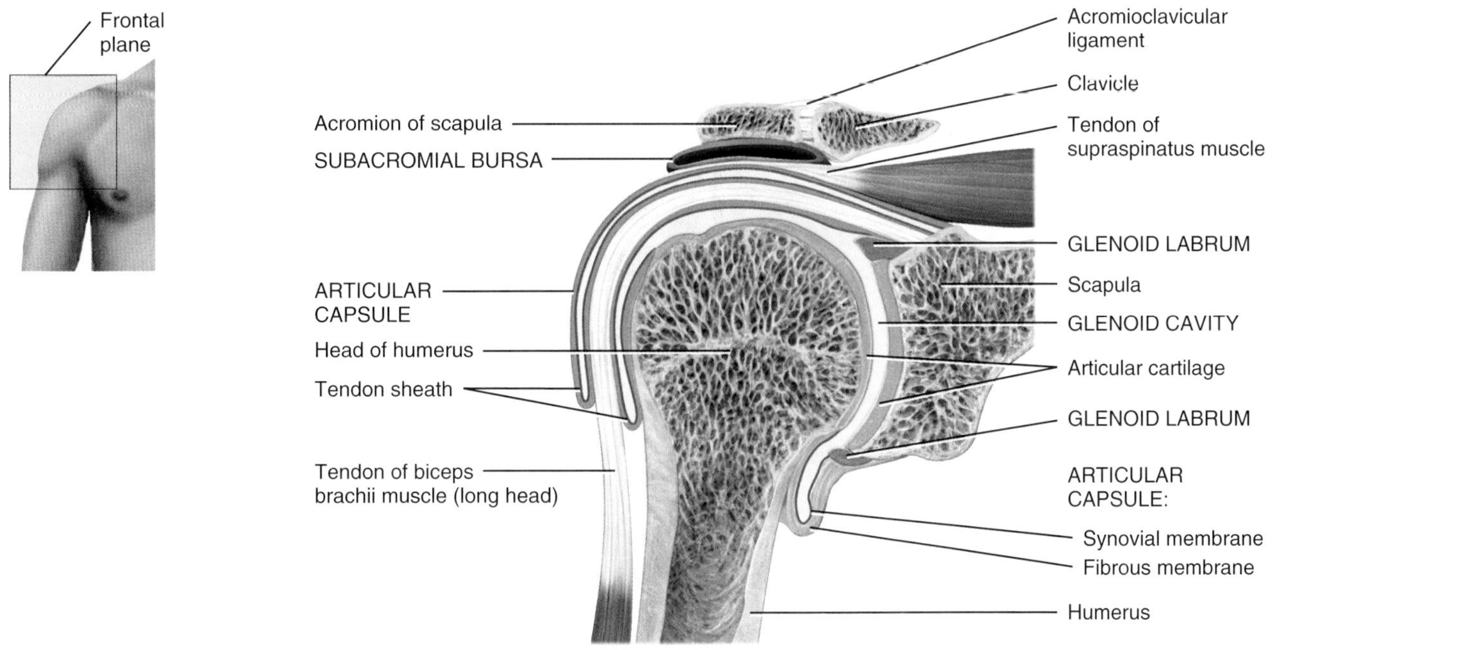

Frontal plane
Acromion of scapula
SUBACROMIAL BURSA
ARTICULAR CAPSULE
Head of humerus
Tendon sheath
Tendon of biceps brachii muscle (long head)
Acromioclavicular ligament
Clavicle
Tendon of supraspinatus muscle
GLENOID LABRUM
Scapula
GLENOID CAVITY
Articular cartilage
GLENOID LABRUM
ARTICULAR CAPSULE:
Synovial membrane
Fibrous membrane
Humerus

(c) Frontal section

EXHIBIT 9.B CONTINUES

especially the *rotator cuff muscles.* These muscles (supraspinatus, infraspinatus, teres minor, and subscapularis) join the scapula to the humerus (see also Figure 11.15). The tendons of the rotator cuff muscles encircle the joint (except for the inferior portion) and intimately surround the articular capsule. The rotator cuff muscles work as a group to hold the head of the humerus in the glenoid cavity.

CLINICAL CONNECTION | *Rotator Cuff Injury, Dislocated and Separated Shoulder, and Torn Glenoid Labrum*

Rotator cuff injury is a strain or tear in the rotator cuff muscles and is a common injury among baseball pitchers, volleyball players, racket sports players, swimmers, and violinists, due to shoulder movements that involve vigorous circumduction. It also occurs as a result of wear and tear, aging, trauma, poor posture, improper lifting, and repetitive motions in certain jobs, such as placing items on a shelf above your head. Most often, there is tearing of the supraspinatus muscle tendon of the rotator cuff. This tendon is especially predisposed to wear and tear because of its location between the head of the humerus and acromion of the scapula, which compresses the tendon during shoulder movements. Poor posture and poor body mechanics also increase compression of the supraspinatus muscle tendon.

The joint most commonly dislocated in adults is the shoulder joint because its socket is quite shallow and the bones are held together by supporting muscles. Usually in a **dislocated shoulder,** the head of the humerus becomes displaced inferiorly, where the articular capsule is least protected. Dislocations of the mandible, elbow, fingers, knee, or hip are less common. Dislocations are treated with rest, ice, pain relievers, manual manipulation, or surgery followed by use of a sling and physical therapy.

A **separated shoulder** actually refers to an injury not to the shoulder joint but to the acromioclavicular joint, a joint formed by the acromion of the scapula and the acromial end of the clavicle. This condition is usually the result of forceful trauma to the joint, as when the shoulder strikes the ground in a fall. Treatment options are similar to those for treating a dislocated shoulder, although surgery is rarely needed.

In a **torn glenoid labrum,** the fibrocartilaginous labrum may tear away from the glenoid cavity. This causes the joint to catch or feel like it's slipping out of place. The shoulder may indeed become dislocated as a result. A torn labrum is reattached to the glenoid surgically with anchors and sutures. The repaired joint is more stable. •

✔CHECKPOINT

Which tendons at the shoulder joint of a baseball pitcher are most likely to be torn due to excessive circumduction?

FIGURE 9.12 CONTINUED

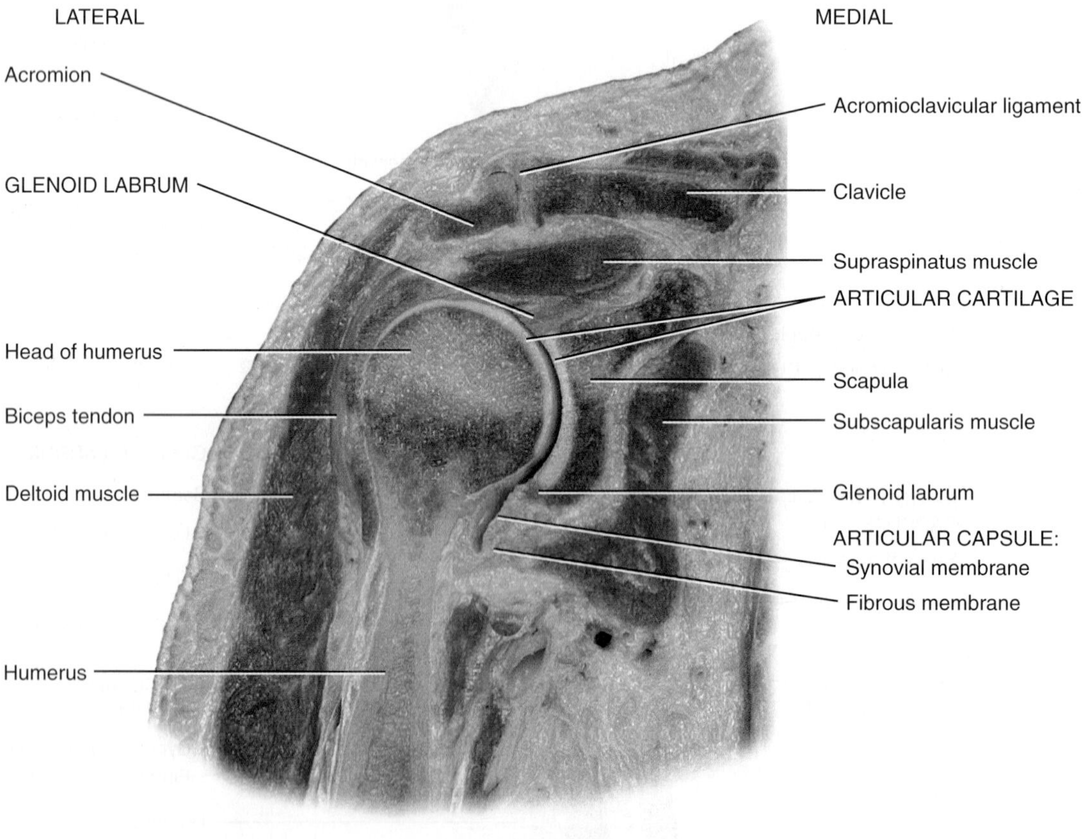
LATERAL
MEDIAL
Acromion
GLENOID LABRUM
Head of humerus
Biceps tendon
Deltoid muscle
Humerus
Acromioclavicular ligament
Clavicle
Supraspinatus muscle
ARTICULAR CARTILAGE
Scapula
Subscapularis muscle
Glenoid labrum
ARTICULAR CAPSULE:
Synovial membrane
Fibrous membrane

(d) Frontal section

Why does the shoulder joint have more freedom of movement than any other joint of the body?

Elbow Joint *(Figure 9.13)*

OBJECTIVE

- Describe the anatomical components of the elbow joint and the movements that can occur at this joint.

Definition

The **elbow joint** is a hinge joint formed by the trochlea and capitulum of the humerus, the trochlear notch of the ulna, and the head of the radius.

Anatomical Components

1. ***Articular capsule.*** The anterior part of the articular capsule covers the anterior part of the elbow joint, from the radial and coronoid fossae of the humerus to the coronoid process of the ulna and the anular ligament of the radius. The posterior part extends from the capitulum, olecranon fossa, and lateral epicondyle of the humerus to the anular ligament of the radius, the olecranon of the ulna, and the ulna posterior to the radial notch (Figure 9.13a, b).
2. ***Ulnar collateral ligament.*** Thick, triangular ligament that extends from the medial epicondyle of the humerus to the coronoid process and olecranon of the ulna (Figure 9.13a). Part of this ligament deepens the socket for the trochlea of the humerus.
3. ***Radial collateral ligament.*** Strong, triangular ligament that extends from the lateral epicondyle of the humerus to the anular ligament of the radius and the radial notch of the ulna (Figure 9.13b).
4. ***Anular ligament of the radius.*** Strong band that encircles the head of the radius. It holds the head of the radius in the radial notch of the ulna (Figure 9.13a, b).

Movements

The elbow joint allows flexion and extension of the forearm (see Figure 9.5c).

CLINICAL CONNECTION | *Tennis Elbow, Little-League Elbow, and Dislocation of the Radial Head*

Tennis elbow most commonly refers to pain at or near the lateral epicondyle of the humerus, usually caused by an improperly executed backhand. The extensor muscles strain or sprain, resulting in pain. **Little-League elbow**, inflammation of the medial epicondyle, typically develops as a result of a heavy pitching schedule and/or a schedule that involves throwing curve balls, especially among youngsters. In this disorder, the elbow joint may enlarge, fragment, or separate.

A **dislocation of the radial head** is the most common upper limb dislocation in children. In this injury, the head of the radius slides past or ruptures the radial anular ligament, a ligament that forms a collar around the head of the radius at the proximal radioulnar joint. Dislocation is most apt to occur when a strong pull is applied to the forearm while it is extended and supinated, for instance, while swinging a child around with outstretched arms. •

CHECKPOINT

At the elbow joint, which ligaments connect (a) the humerus and the ulna, and (b) the humerus and the radius?

Figure 9.13 **Right elbow joint.**

The elbow joint is formed by parts of three bones: humerus, ulna, and radius.

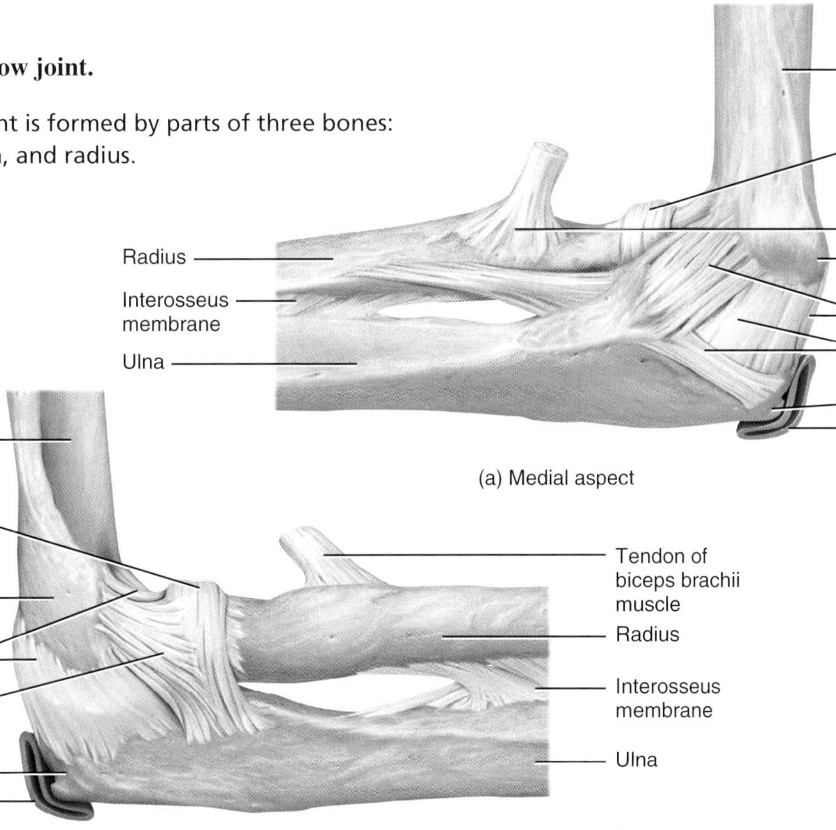

Which movements are possible at a hinge joint?

Hip Joint *(Figure 9.14)*

OBJECTIVE

- Describe the anatomical components of the hip joint and the movements that can occur at this joint.

Definition

The **hip joint** *(coxal joint)* is a ball-and-socket joint formed by the head of the femur and the acetabulum of the hip bone.

Anatomical Components

1. ***Articular capsule.*** Very dense and strong capsule that extends from the rim of the acetabulum to the neck of the femur (Figure 9.14c). With its accessory ligaments, this is one of the strongest structures of the body. The articular capsule consists of circular and longitudinal fibers. The circular fibers, called the zona orbicularis, form a collar around the neck of the femur. Accessory ligaments known as the

Figure 9.14 Right hip (coxal) joint.

The articular capsule of the hip joint is one of the strongest structures in the body.

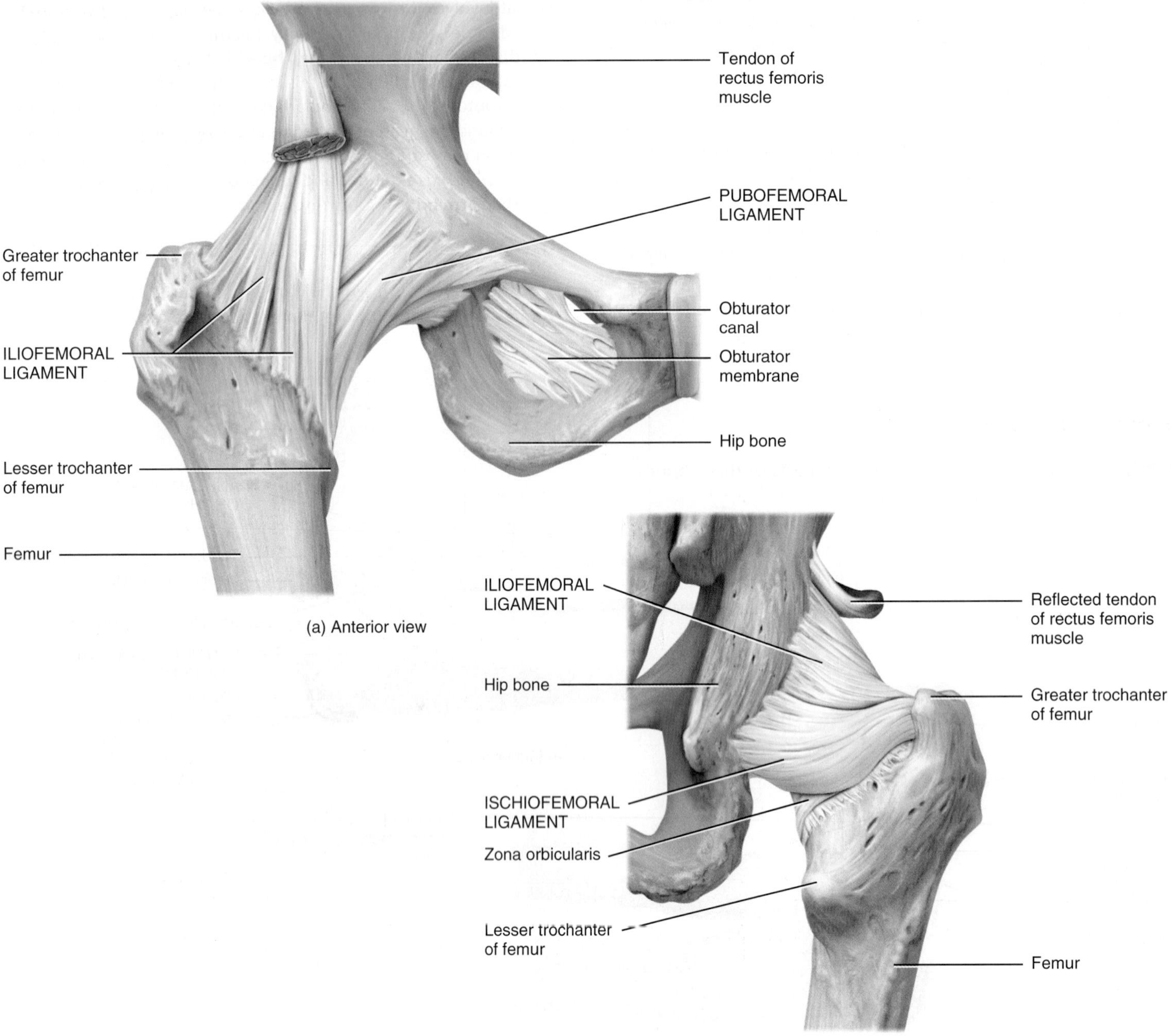

(a) Anterior view

(b) Posterior view

iliofemoral ligament, pubofemoral ligament, and *ischiofemoral ligament* reinforce the longitudinal fibers of the articular capsule.

2. ***Iliofemoral ligament*** (il′-ē-ō-FEM-ō-ral). Thickened portion of the articular capsule that extends from the anterior inferior iliac spine of the hip bone to the intertrochanteric line of the femur (Figure 9.14a, b). This ligament is said to be the body's strongest and prevents hyperextension of the femur at the hip joint during standing.
3. ***Pubofemoral ligament*** (pū′-bō-FEM-ō-ral). Thickened portion of the articular capsule that extends from the pubic part of the rim of the acetabulum to the neck of the femur (Figure 9.14a). This ligament prevents overabduction of the femur at the hip joint and strengthens the articular capsule.
4. ***Ischiofemoral ligament*** (is′-kē-ō-FEM-ō-ral). Thickened portion of the articular capsule that extends from the ischial wall of the acetabulum to the neck of the femur (Figure 9.14b). This ligament slackens during adduction, tenses during abduction, and strengthens the articular capsule.
5. ***Ligament of the head of the femur.*** Flat, triangular band (primarily a synovial fold) that extends from the fossa of the acetabulum to the fovea capitis of the head of the femur (Figure 9.14c). The ligament usually contains a small artery that supplies the head of the femur.
6. ***Acetabular labrum*** (as-e-TAB-ū-lar LĀ-brum). Fibrocartilage rim attached to the margin of the acetabulum that enhances the depth of the acetabulum. As a result, dislocation of the femur is rare (Figure 9.14c).
7. ***Transverse ligament of the acetabulum.*** Strong ligament that crosses over the acetabular notch. It supports part of the acetabular labrum and is connected with the ligament of the head of the femur and the articular capsule (Figure 9.14c).

Movements

The hip joint allows flexion, extension, abduction, adduction, circumduction, medial rotation, and lateral rotation of the thigh (see Figures 9.5–9.8). The extreme stability of the hip joint is related to the very strong articular capsule and its accessory ligaments, the manner in which the femur fits into the acetabulum, and the muscles surrounding the joint. Although the shoulder and hip joints are both ball-and-socket joints, the hip joints do not have as wide a range of motion. Flexion is limited by the anterior surface of the thigh coming into contact with the anterior abdominal wall when the knee is flexed and by tension of the hamstring muscles when the knee is extended. Extension is limited by tension of the iliofemoral, pubofemoral, and ischiofemoral ligaments. Abduction is limited by the tension of the pubofemoral ligament, and adduction is limited by contact with the opposite limb and tension in the ligament of the head of the femur. Medial rotation is limited by the tension in the ischiofemoral ligament, and lateral rotation is limited by tension in the iliofemoral and pubofemoral ligaments.

✔ CHECKPOINT

What factors limit the degree of flexion and abduction at the hip joint?

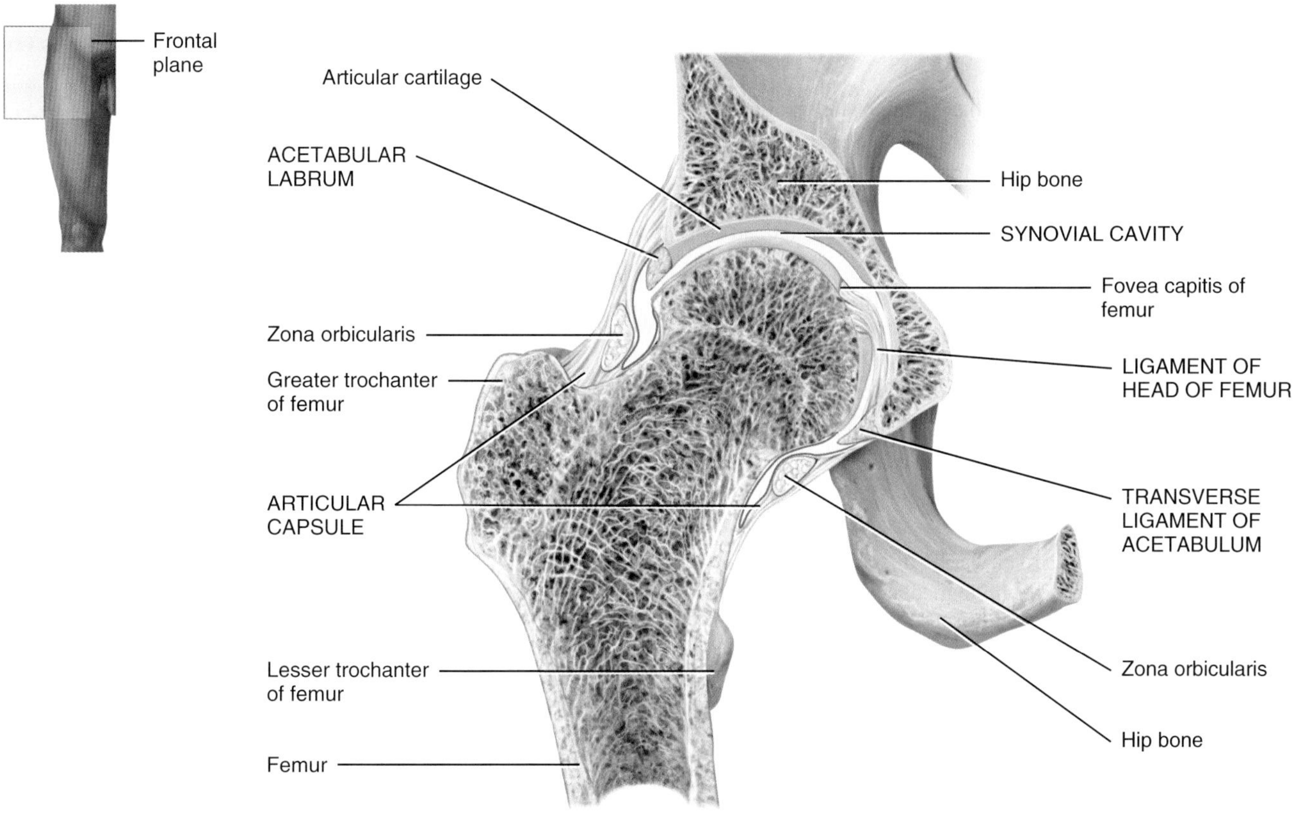

Frontal plane
Articular cartilage
ACETABULAR LABRUM
Zona orbicularis
Greater trochanter of femur
ARTICULAR CAPSULE
Lesser trochanter of femur
Femur
Hip bone
SYNOVIAL CAVITY
Fovea capitis of femur
LIGAMENT OF HEAD OF FEMUR
TRANSVERSE LIGAMENT OF ACETABULUM
Zona orbicularis
Hip bone

(c) Frontal section

? Which ligaments limit the degree of extension that is possible at the hip joint?

EXHIBIT 9.E Knee Joint *(Figure 9.15)*

OBJECTIVE

- Describe the main anatomical components of the knee joint and explain the movements that can occur at this joint.

Definition

The **knee joint** *(tibiofemoral joint)* is the largest and most complex joint of the body (Figure 9.15). It is a modified hinge joint (because its primary movement is a uniaxial hinge movement) that consists of three joints within a single synovial cavity:

1. Laterally is a *tibiofemoral joint,* between the lateral condyle of the femur, lateral meniscus, and lateral condyle of the tibia, which is the weight-bearing bone of the leg.
2. Medially is another *tibiofemoral joint,* between the medial condyle of the femur, medial meniscus, and medial condyle of the tibia.
3. An intermediate *patellofemoral joint* is between the patella and the patellar surface of the femur.

Anatomical Components

1. ***Articular capsule.*** No complete, independent capsule unites the bones of the knee joint. The ligamentous sheath surrounding the joint consists mostly of muscle tendons or their expansions (Figure 9.15e–g). There are, however, some capsular fibers connecting the articulating bones.
2. ***Medial and lateral patellar retinacula*** (ret′-i-NAK-ū-la). Fused tendons of insertion of the quadriceps femoris muscle and the fascia lata (deep fascia of thigh) that strengthen the anterior surface of the joint (Figure 9.15e).
3. ***Patellar ligament.*** Continuation of the common tendon of insertion of the quadriceps femoris muscle that extends from the patella to the tibial tuberosity. This ligament also strengthens the anterior surface of the joint. The posterior surface of the ligament is separated from the synovial membrane of the joint by an infrapatellar fat pad (Figure 9.15c–e).
4. ***Oblique popliteal ligament*** (pop-LIT-ē-al). Broad, flat ligament that extends from the intercondylar fossa and lateral condyle of the femur to the head and medial condyle of the tibia (Figure 9.15f, h). The ligament strengthens the posterior surface of the joint.
5. ***Arcuate popliteal ligament.*** Extends from the lateral condyle of the femur to the styloid process of the head of the fibula. It strengthens the lower lateral part of the posterior surface of the joint (Figure 9.15f).
6. ***Tibial collateral ligament.*** Broad, flat ligament on the medial surface of the joint that extends from the medial condyle of the femur to the medial condyle of the tibia (Figure 9.15a, e–h). Tendons of the sartorius, gracilis, and semitendinosus muscles, all of which strengthen the medial aspect of the joint, cross the ligament. The tibial collateral ligament is firmly attached to the medial meniscus.
7. ***Fibular collateral ligament.*** Strong, rounded ligament on the lateral surface of the joint that extends from the lateral condyle of the femur to the lateral side of the head of the fibula (Figure 9.15a, e–h). It strengthens the lateral aspect of the joint. The ligament is covered by the tendon of the biceps femoris muscle. The tendon of the popliteal muscle is deep to the ligament.
8. ***Intracapsular ligaments*** (in′-tra-KAP-sū-lar). Ligaments within the capsule that connect the tibia and femur. The anterior and posterior **cruciate ligaments** (KROO-shē-āt = like a cross) are named based on their origins relative to the intercondylar area of the tibia. From their origins, they cross on their way to their destinations on the femur.
   a. ***Anterior cruciate ligament (ACL).*** Extends posteriorly and laterally from a point anterior to the intercondylar area of the tibia to the posterior part of the medial surface of the lateral condyle of the femur (Figure 9.15a, b, h). The ACL limits hyperextension of the knee (which normally does not occur at this joint) and prevents the anterior sliding of the tibia on the femur. This ligament is stretched or torn in about 70% of all serious knee injuries.
   ACL injuries are much more common in females than males, perhaps as much as 3 to 6 times. The reasons are unclear but may be related to less space between the femoral condyle in females so that the space for ACL movement is limited; the wider pelvis of females that creates a greater angle between the femur and tibia and increases the risk for an ACL tear; female hormones that allow for greater flexibility of ligaments, muscles, and tendons but which do not permit them to absorb the stresses put on them and thus the stresses are transferred to the ACL; and females having less muscle strength so that they rely less on muscles and more on the ACL to hold the knee in place.
   b. ***Posterior cruciate ligament (PCL).*** Extends anteriorly and medially from a depression on the posterior intercondylar area of the tibia and lateral meniscus to the anterior part of the lateral surface of the medial condyle of the femur (Figure 9.15a, b, h). The PCL prevents the posterior sliding of the tibia (and anterior sliding of the femur) when the knee is flexed. This is very important when walking down stairs or a steep incline.
9. ***Articular discs (menisci).*** Two fibrocartilage discs between the tibial and femoral condyles help compensate for the irregular shapes of the bones and circulate synovial fluid.
   a. ***Medial meniscus.*** Semicircular piece of fibrocartilage (C-shaped). Its anterior end is attached to the anterior intercondylar fossa of the tibia, anterior to the anterior cruciate ligament. Its posterior end is attached to the posterior intercondylar fossa of the tibia between the attachments of the posterior cruciate ligament and lateral meniscus (Figure 9.15a, b, d, h).
   b. ***Lateral meniscus.*** Nearly circular piece of fibrocartilage (approaches an incomplete O in shape) (Figure 9.15a, b, d, h). Its anterior end is attached anteriorly to the intercondylar eminence of the tibia, and laterally and posteriorly to the anterior cruciate ligament. Its posterior end is attached posteriorly to the intercondylar eminence of the tibia, and anteriorly to the posterior end of the medial meniscus. The anterior surfaces of the medial and lateral menisci are connected to each other by the *transverse ligament of the knee* (Figure 9.15a) and to the margins of the head of the tibia by the *coronary ligaments* (not illustrated).
10. The more important bursae of the knee include the following.
    a. *Prepatellar bursa* between the patella and skin (Figure 9.15c, d).
    b. *Infrapatellar bursa* between superior part of tibia and patellar ligament (Figure 9.15c–e).
    c. *Suprapatellar bursa* between inferior part of femur and deep surface of quadriceps femoris muscle (Figure 9.15c–e).

Figure 9.15 Right knee (tibiofemoral) joint.

The knee joint is the largest and most complex joint in the body.

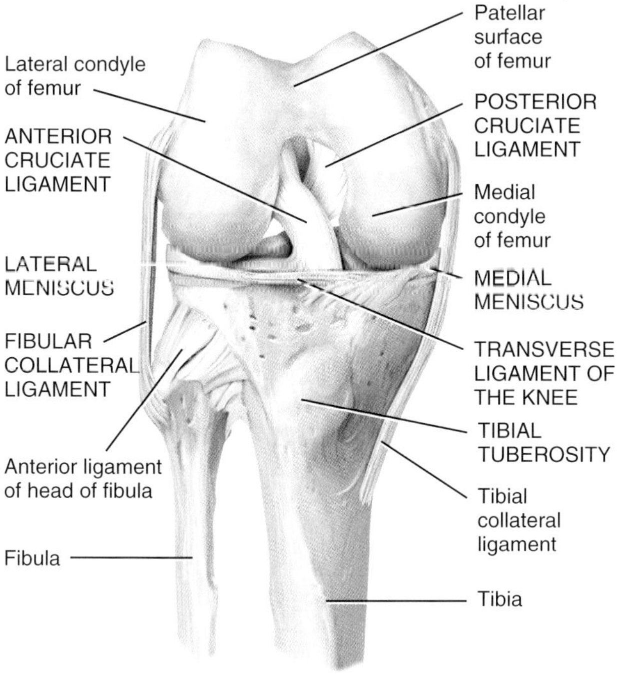

(a) Anterior deep view

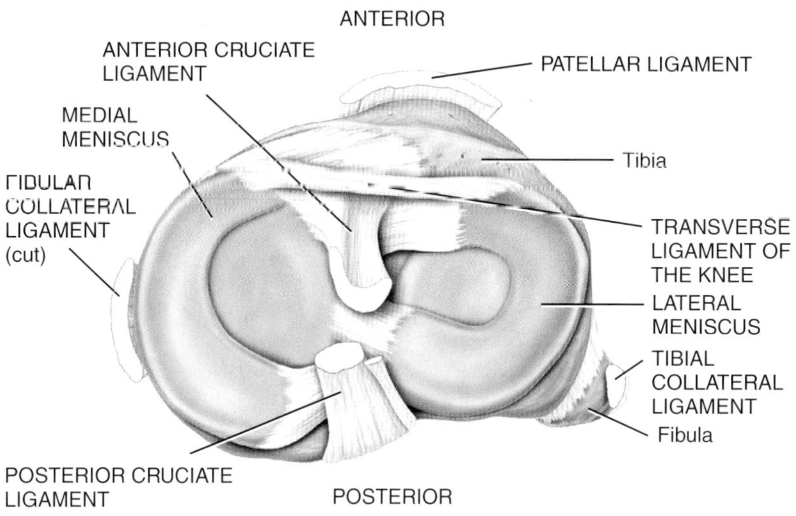

(b) Superior view of menisci

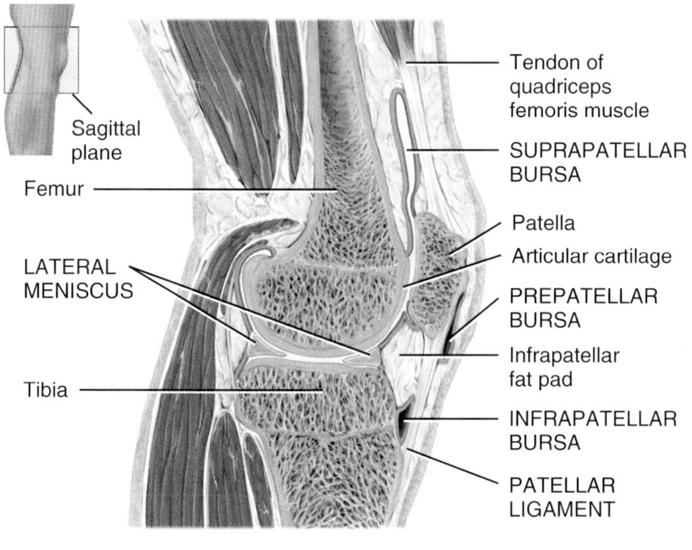

(c) Sagittal section

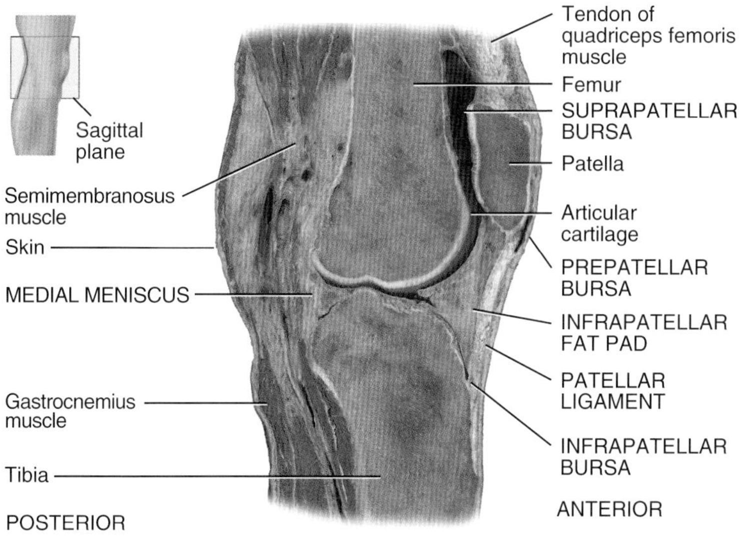

(d) Sagittal section

EXHIBIT 9.E CONTINUES

Movements

The knee joint allows flexion, extension, slight medial rotation, and lateral rotation of the leg in the flexed position (see Figures 9.5f and 9.8c).

FIGURE 9.15 CONTINUED

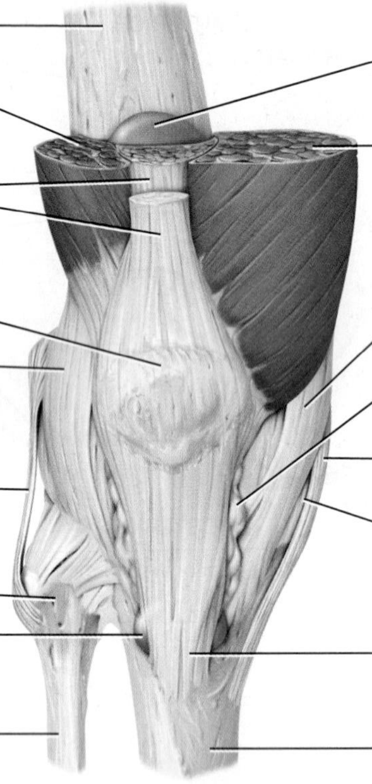

(e) Anterior superficial view

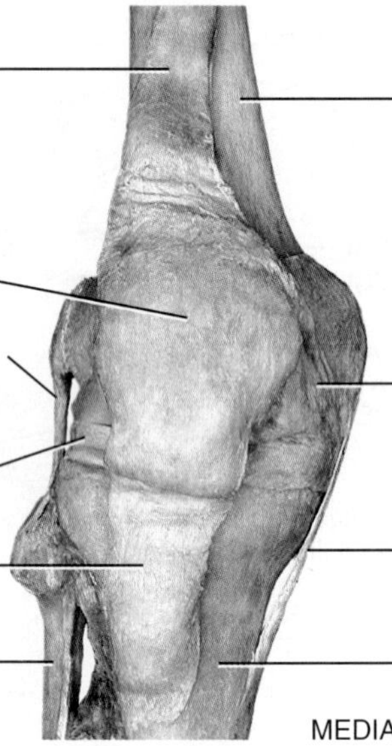

(g) Anterior view

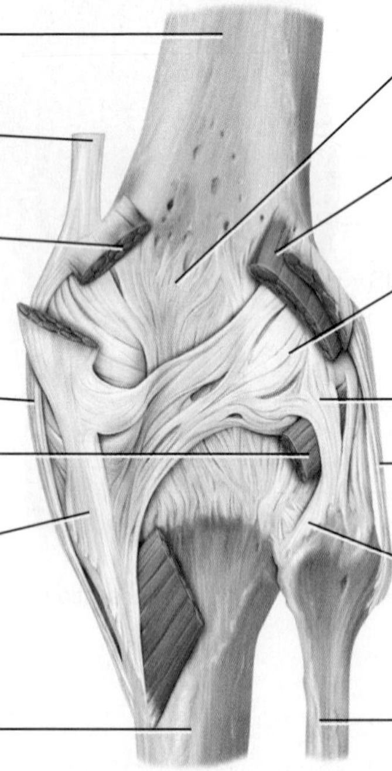

(f) Posterior deep view

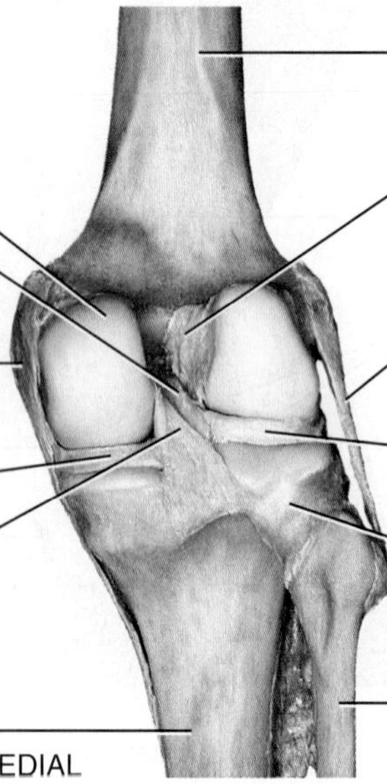

(h) Posterior view

CLINICAL CONNECTION | *Knee Injuries*

The knee joint is the joint most vulnerable to damage because it is a mobile, weight-bearing joint and its stability depends almost entirely on its associated ligaments and muscles. Further, there is no complementary fit between the surfaces of the articulating bones. Following are several kinds of **knee injuries.** A **swollen knee** may occur immediately or hours after an injury. The initial swelling is due to escape of blood from damaged blood vessels adjacent to areas of injury, including rupture of the anterior cruciate ligament, damage to synovial membranes, torn menisci, fractures, or collateral ligament sprains. Delayed swelling is due to excessive production of synovial fluid, a condition commonly referred to as "water on the knee."

The firm attachment of the tibial collateral ligament to the medial meniscus is clinically significant because tearing of the ligament typically also results in tearing of the meniscus. Such an injury may occur in sports such as football and rugby when the knee receives a blow from the lateral side while the foot is fixed on the ground. The force of the blow may also tear the anterior cruciate ligament, which is also connected to the medial meniscus. The term **"unhappy triad"** is applied to a knee injury that involves damage to the three components of the knee at the same time: the tibial collateral ligament, medial meniscus, and anterior cruciate ligament.

A **dislocated knee** refers to the displacement of the tibia relative to the femur. The most common type is dislocation anteriorly, resulting from hyperextension of the knee. A frequent consequence of a dislocated knee is damage to the popliteal artery.

If no surgery is required, treatment of knee injuries involves PRICE (protection, rest, ice, compression, and elevation) with some strengthening exercises and perhaps physical therapy. •

✔ CHECKPOINT

What are the opposing functions of the anterior and posterior cruciate ligaments?

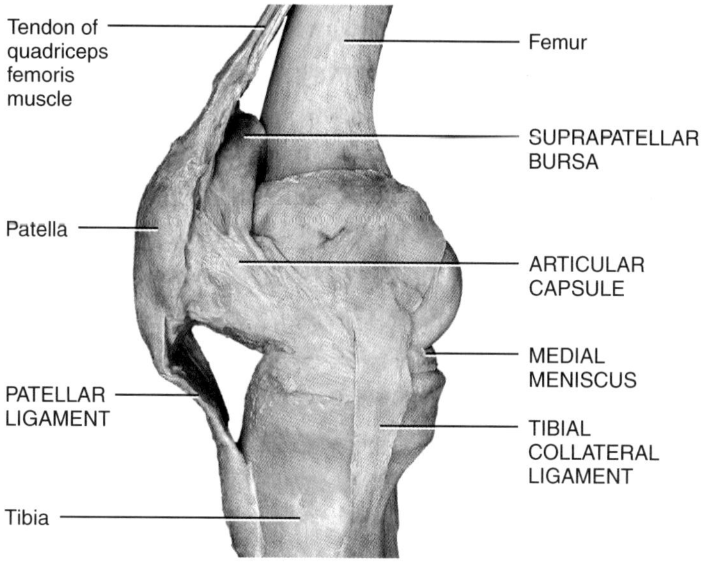

(i) Medial view

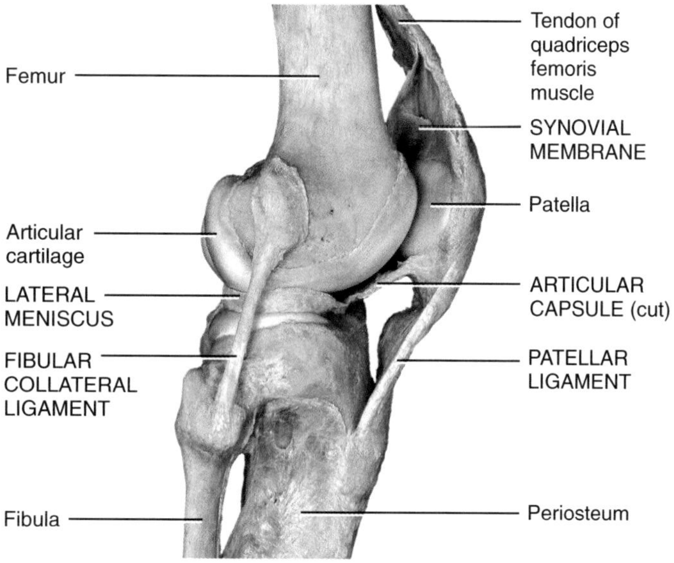

(j) Lateral view

What movement occurs at the knee joint when the quadriceps femoris (anterior thigh) muscles contract?

9.9 AGING AND JOINTS

OBJECTIVE

- Explain the effects of aging on joints.

Aging usually results in decreased production of synovial fluid in joints. In addition, the articular cartilage becomes thinner with age, and ligaments shorten and lose some of their flexibility. The effects of aging on joints are influenced by genetic factors and by wear and tear, and vary considerably from one person to another. Although degenerative changes in joints may begin as early as age 20, most changes do not occur until much later. By age 80, almost everyone develops some type of degeneration in the knees, elbows, hips, and shoulders. It is also common for elderly individuals to develop degenerative changes in the vertebral column, resulting in a hunched-over posture and pressure on nerve roots. One type of arthritis, called osteoarthritis (see Disorders: Homeostatic Imbalances at the end of this chapter), is at least partially age-related. Nearly everyone over age 70 has evidence of some osteoarthritic changes. Stretching and aerobic exercises that attempt to maintain full range of motion are helpful in minimizing the effects of aging. They help to maintain the effective functioning of ligaments, tendons, muscles, synovial fluid, and articular cartilage.

✔CHECKPOINT

13. Which joints show evidence of degeneration in nearly all individuals as aging progresses?

9.10 ARTHROPLASTY

OBJECTIVE

- Explain the procedures involved in arthroplasty, and describe how a total hip replacement is performed.

Joints that have been severely damaged by diseases such as arthritis, or by injury, may be replaced surgically with artificial joints in a procedure referred to as **arthroplasty** (AR-thrō-plas′-tē; *arthr-* = joint; *-plasty* = plastic repair of). Although most joints in the body can be repaired by arthroplasty, the ones most commonly replaced are the hips, knees, and shoulders. About 400,000 hip replacements and about 300,000 knee replacements are performed annually in the United States. During the procedure, the ends of the damaged bones are removed and metal, ceramic, or plastic components are fixed in place. The goals of arthroplasty are to relieve pain and increase range of motion.

Hip Replacements

Partial hip replacements involve only the femur. **Total hip replacements** involve both the acetabulum and head of the femur (Figure 9.16a–c). The damaged portions of the acetabulum and the head of the femur are replaced by prefabricated prostheses (artificial devices). The acetabulum is shaped to accept the new socket, the head of the femur is removed, and the center of the femur is shaped to fit the femoral component. The acetabular component consists of a plastic such as polyethylene, and the femoral component is composed of a metal such as cobalt-chrome, titanium alloys, or stainless steel. These materials are designed to withstand a high degree of stress and to prevent a response by the immune system. Once the appropriate acetabular and femoral components are selected, they are attached to the healthy portion of bone with acrylic cement, which forms an interlocking mechanical bond.

Knee Replacements

Knee replacements are actually a resurfacing of cartilage and, like hip replacements, may be partial or total. In a **total knee replacement,** the damaged cartilage is removed from the distal end of the femur, the proximal end of the tibia, and the back surface of the patella (if the back surface of the patella is not badly damaged, it may be left intact) (Figure 9.16d–f). The femur is reshaped and fitted with a metal femoral component and cemented in place. The tibia is reshaped and fitted with a plastic tibial component that is cemented in place. If the back surface of the patella is badly damaged, it is replaced with a plastic patellar implant.

In a **partial knee replacement,** also called a **unicompartmental knee replacement,** only one side of the knee joint is replaced. Once the damaged cartilage is removed from the distal end of the femur, the femur is reshaped and a metal femoral component is cemented in place. Then the damaged cartilage from the proximal end of the tibia is removed, along with the meniscus. The tibia is reshaped and fitted with a plastic tibial component that is cemented into place. If the back surface of the patella is badly damaged, it is replaced with a plastic patellar component.

Researchers are continually seeking to improve the strength of the cement and devise ways to stimulate bone growth around the implanted area. Potential complications of arthroplasty include infection, blood clots, loosening or dislocation of the replacement components, and nerve injury.

With increasing sensitivity of metal detectors at airports and other public areas, it is possible that metal point replacements may activate metal detectors.

✔CHECKPOINT

14. Which joints of the body most commonly undergo arthroplasty?

Figure 9.16 **Total hip and knee replacement.**

In a total hip replacement, damaged portions of the acetabulum and the head of the femur are replaced by prostheses.

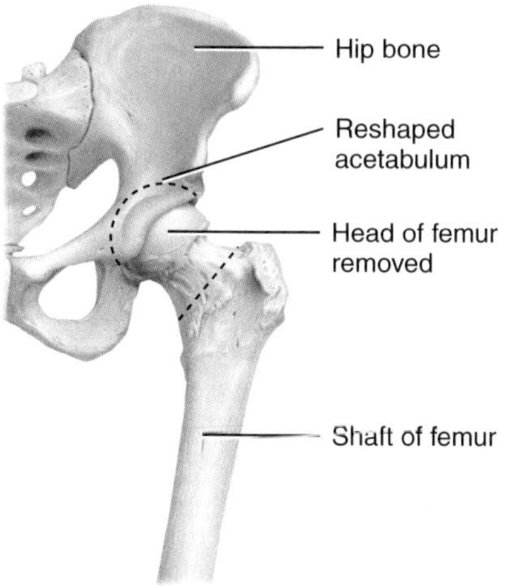
Hip bone
Reshaped acetabulum
Head of femur removed
Shaft of femur

(a) Preparation for total hip replacement

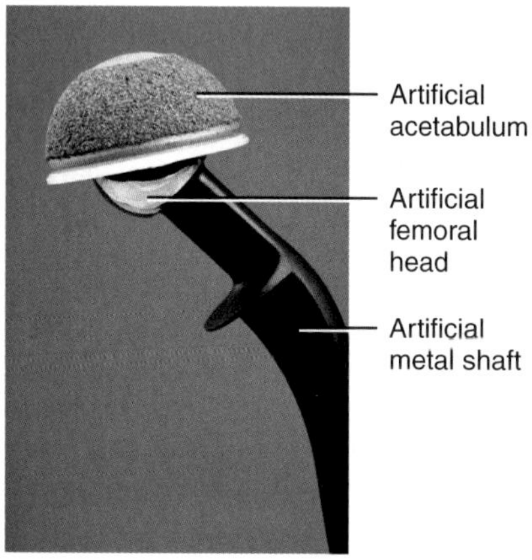
Artificial acetabulum
Artificial femoral head
Artificial metal shaft

(b) Components of an artificial hip joint prior to implantation

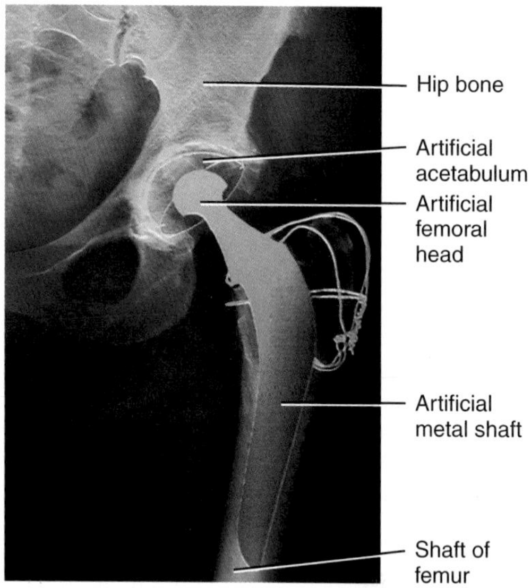
Hip bone
Artificial acetabulum
Artificial femoral head
Artificial metal shaft
Shaft of femur

(c) Radiograph of an artificial hip joint

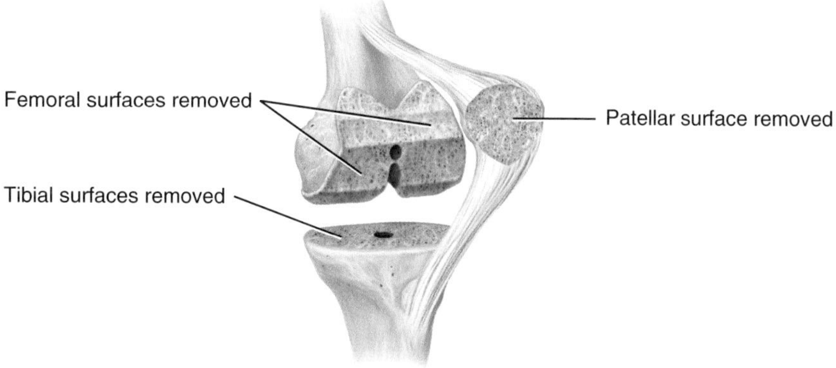
Femoral surfaces removed
Tibial surfaces removed
Patellar surface removed

(d) Preparation for total knee replacement

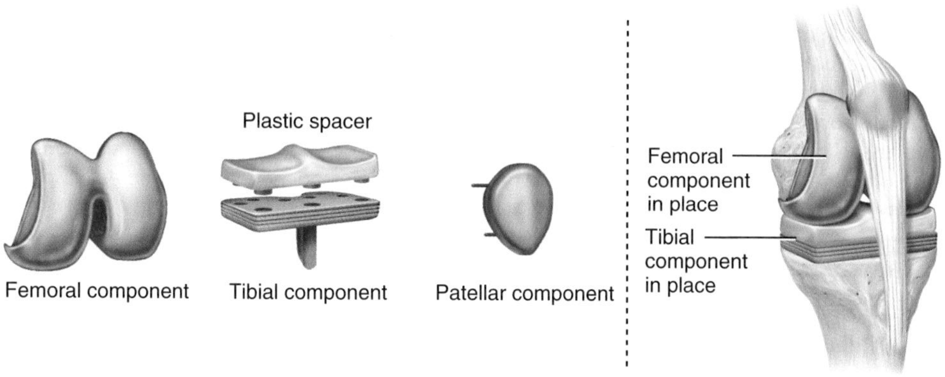
Plastic spacer
Femoral component
Tibial component
Patellar component
Femoral component in place
Tibial component in place

(e) Components of artificial knee joint prior to implantation (left) and implanted (right)

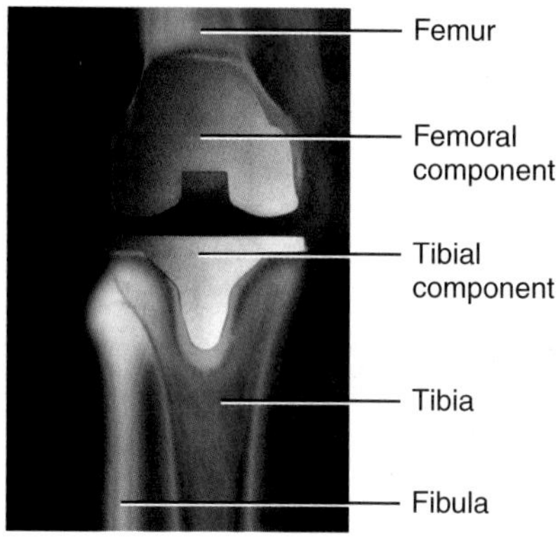
Femur
Femoral component
Tibial component
Tibia
Fibula

(f) Radiograph of a total knee replacement

What is the purpose of arthroplasty?

DISORDERS: HOMEOSTATIC IMBALANCES

Rheumatism and Arthritis

Rheumatism (ROO-ma-tizm) is any painful disorder of the supporting structures of the body—bones, ligaments, tendons, or muscles—that is not caused by infection or injury. **Arthritis** is a form of rheumatism in which the joints are swollen, stiff, and painful. It afflicts about 45 million people in the United States and is the leading cause of physical disability among adults over age 65.

Osteoarthritis

Osteoarthritis (OA) (os′-tē-ō-ar-THRĪ-tis) is a degenerative joint disease in which joint cartilage is gradually lost. It results from a combination of aging, obesity, irritation of the joints, muscle weakness, and wear and abrasion. Commonly known as "wear-and-tear" arthritis, osteoarthritis is the most common type of arthritis.

Osteoarthritis is a progressive disorder of synovial joints, particularly weight-bearing joints. Articular cartilage deteriorates and new bone forms in the subchondral areas and at the margins of the joint. The cartilage slowly degenerates, and as the bone ends become exposed, spurs (small bumps) of new osseous tissue are deposited on them in a misguided effort by the body to protect against increased friction. These spurs decrease the space of the joint cavity and restrict joint movement. Unlike rheumatoid arthritis (described next), osteoarthritis affects mainly the articular cartilage, although the synovial membrane often becomes inflamed late in the disease. Two major distinctions between osteoarthritis and rheumatoid arthritis are that osteoarthritis first afflicts the larger joints (knees, hips) and is due to wear and tear, whereas rheumatoid arthritis first strikes smaller joints and is an active attack on the cartilage. Osteoarthritis is the most common reason for hip- and knee-replacement surgery.

Rheumatoid Arthritis

Rheumatoid arthritis (RA) is an autoimmune disease in which the immune system of the body attacks its own tissues—in this case, its own cartilage and joint linings. RA is characterized by inflammation of the joint, which causes swelling, pain, and loss of function. Usually, this form of arthritis occurs bilaterally: If one wrist is affected, the other is also likely to be affected, although they are often not affected to the same degree.

The primary symptom of RA is inflammation of the synovial membrane. If untreated, the membrane thickens, and synovial fluid accumulates. The resulting pressure causes pain and tenderness. The membrane then produces an abnormal granulation tissue, called pannus, that adheres to the surface of the articular cartilage and sometimes erodes the cartilage completely. When the cartilage is destroyed, fibrous tissue joins the exposed bone ends. The fibrous tissue ossifies and fuses the joint so that it becomes immovable—the ultimate crippling effect of RA. Growth of the granulation tissue causes the distortion of the fingers that characterizes hands of RA sufferers.

Gouty Arthritis

Uric acid (a substance that gives urine its name) is a waste product produced during the metabolism of nucleic acid (DNA and RNA) subunits. A person who suffers from **gout** (GOWT) either produces excessive amounts of uric acid or is not able to excrete as much as normal. The result is a buildup of uric acid in the blood. This excess acid then reacts with sodium to form a salt called sodium urate. Crystals of this salt accumulate in soft tissues such as the kidneys and in the cartilage of the ears and joints.

In **gouty arthritis,** sodium urate crystals are deposited in the soft tissues of the joints. Gout most often affects the joints of the feet, especially at the base of the big toe. The crystals irritate and erode the cartilage, causing inflammation, swelling, and acute pain. Eventually, the crystals destroy all joint tissues. If the disorder is untreated, the ends of the articulating bones fuse, and the joint becomes immovable. Treatment consists of pain relief (ibuprofen, naproxen, colchicine, and cortisone) followed by administration of allopurinol to keep uric acid levels low so that crystals do not form.

Lyme Disease

A spiral-shaped bacterium called *Borrelia burgdorferi* causes **Lyme disease,** named for the town of Lyme, Connecticut, where it was first reported in 1975. The bacteria are transmitted to humans mainly by deer ticks (*Ixodes dammini*). These ticks are so small that their bites often go unnoticed. Within a few weeks of the tick bite, a rash may appear at the site. Although the rash often resembles a bull's-eye target, there are many variations, and some people never develop a rash. Other symptoms include joint stiffness, fever and chills, headache, stiff neck, nausea, and low back pain. In advanced stages of the disease, arthritis is the main complication. It usually afflicts the larger joints such as the knee, ankle, hip, elbow, or wrist. Antibiotics are generally effective against Lyme disease, especially if they are given promptly. However, some symptoms may linger for years.

Ankylosing Spondylitis

Ankylosing spondylitis (ang′-ki-LŌ-sing spon′-di-LĪ-tis; *ankyle* = stiff; *spondyl* = vertebra) is an inflammatory disease of unknown origin that affects joints between vertebrae (intervertebral) and between the sacrum and hip bone (sacroiliac joint). The disease, which is more common in males, sets in between ages 20 and 40. It is characterized by pain and stiffness in the hips and lower back that progress upward along the backbone. Inflammation can lead to *ankylosis* (severe or complete loss of movement at a joint) and *kyphosis* (hunchback). Treatment consists of anti-inflammatory drugs, heat, massage, and supervised exercise.

MEDICAL TERMINOLOGY

Arthralgia (ar-THRAL-jē-a; *arthr-* = joint; *-algia* = pain) Pain in a joint.

Bursectomy (bur-SEK-tō-mē; *-ectomy* = removal of) Removal of a bursa.

Chondritis (kon-DRĪ-tis; *chondr-* = cartilage) Inflammation of cartilage.

Subluxation (sub-luks-Ā-shun) A partial or incomplete dislocation.

Synovitis (sin′-ō-VĪ-tis) Inflammation of a synovial membrane in a joint.

CHAPTER REVIEW AND RESOURCE SUMMARY

WILEY PLUS

Review | **Resource**

Introduction

1. A joint (articulation or arthrosis) is a point of contact between two bones, between bone and cartilage, or between bone and teeth.
2. A joint's structure may permit no movement, slight movement, or free movement.

Resource: Anatomy Overview - Joints

9.1 Joint Classifications

1. Structural classification is based on the presence or absence of a synovial cavity and the type of connective tissue. Structurally, joints are classified as fibrous, cartilaginous, or synovial.
2. Functional classification of joints is based on the degree of movement permitted. Joints may be synarthroses (immovable), amphiarthroses (slightly movable), or diarthroses (freely movable).

Resource: Anatomy Overview - Joints

9.2 Fibrous Joints

1. The bones of fibrous joints are held together by dense irregular connective tissue.
2. These joints include immovable or slightly movable sutures (found between skull bones), immovable to slightly movable syndesmoses (such as roots of teeth in the sockets in the mandible and maxilla and the distal tibiofibular joint), and slightly movable interosseous membranes (found between the radius and ulna in the forearm and the tibia and fibula in the leg).

Resource: Anatomy Overview - Fibrous Joints

9.3 Cartilaginous Joints

1. The bones of cartilaginous joints are held together by cartilage.
2. These joints include immovable synchondroses united by hyaline cartilage (epiphyseal plates between diaphyses and epiphyses) and slightly movable symphyses united by fibrocartilage (pubic symphysis).

Resource: Anatomy Overview - Cartilaginous Joints

9.4 Synovial Joints

1. Synovial joints contain a space between bones called the synovial cavity. All synovial joints are diarthroses.
2. Other characteristics of synovial joints are the presence of articular cartilage and an articular capsule, made up of a fibrous membrane and a synovial membrane.
3. The synovial membrane secretes synovial fluid, which forms a thin, viscous film over the surfaces within the articular capsule.
4. Many synovial joints also contain accessory ligaments (extracapsular and intracapsular) and articular discs (menisci).
5. Synovial joints contain an extensive nerve and blood supply. The nerves convey information about pain, joint movements, and the degree of stretch at a joint. Blood vessels penetrate the articular capsule and ligaments.
6. Bursae are saclike structures, similar in structure to joint capsules, that alleviate friction in joints such as the shoulder and knee joints.
7. Tendon sheaths are tubelike bursae that wrap around tendons where there is considerable friction.

Resource:
- Anatomy Overview - Synovial Joints
- Anatomy Overview - Structural Subtypes of Synovial Joints
- Figure 9.3 - Structure of a Typical Synovial Joint
- Exercise - Judge That Joint

9.5 Types of Movements at Synovial Joints

1. In a gliding movement, the nearly flat surfaces of bones move back and forth and side to side.
2. In angular movements, a change in the angle between bones occurs. Examples are flexion–extension, lateral flexion, hyperextension, and abduction–adduction. Circumduction refers to the movement of the distal end of a body part in a circle and involves a continuous sequence of flexion, abduction, extension, adduction, and rotation of the joint (or in the opposite direction).
3. In rotation, a bone moves around its own longitudinal axis.
4. Special movements occur at specific synovial joints. Examples are elevation–depression, protraction–retraction, inversion–eversion, dorsiflexion–plantar flexion, supination–pronation, and opposition.
5. Table 9.1 summarizes the various types of movements at synovial joints.

Resource:
- Anatomy Overview - Movements Produced at Synovial Joints
- Figure 9.5 - Angular Movements at Synovial Joints
- Figure 9.6 - Angular Movements at Synovial Joints— Abduction and Adduction
- Figure 9.7 - Angular Movements at Synovial Joints—Circumduction
- Figure 9.8 - Rotation at Synovial Joints
- Figure 9.9 - Special Movements at Synovial Joints
- Exercise - Manage Movements

9.6 Types of Synovial Joints

1. Types of synovial joints are plane, hinge, pivot, condyloid, saddle, and ball-and-socket.
2. In a plane joint the articulating surfaces are flat, and the bones primarily glide back and forth and side to side (many are biaxial); they may also permit rotation (triaxial); examples are joints between carpals and tarsals.

Resource: Concepts and Connections - Joint Classifications

Review | **Resource**

3. In a hinge joint, the convex surface of one bone fits into the concave surface of another, and the motion is angular around one axis (uniaxial); examples are the elbow, knee (a modified hinge joint), and ankle joints.
4. In a pivot joint, a round or pointed surface of one bone fits into a ring formed by another bone and a ligament, and movement is rotational (uniaxial); examples are the atlanto-axial and radioulnar joints.
5. In a condyloid joint, an oval projection of one bone fits into an oval cavity of another, and motion is angular around two axes (biaxial); examples include the wrist joint and metacarpophalangeal joints of the second through fifth digits.
6. In a saddle joint, the articular surface of one bone is shaped like a saddle and the other bone fits into the saddle like a sitting rider; movement is biaxial. An example is the carpometacarpal joint between the trapezium and the metacarpal of the thumb.
7. In a ball-and-socket joint, the ball-shaped surface of one bone fits into the cuplike depression of another; motion is around three axes (triaxial). Examples include the shoulder and hip joints.
8. Table 9.2 summarizes the structural and functional categories of joints.

9.7 Factors Affecting Contact and Range of Motion at Synovial Joints

1. The ways that articular surfaces of synovial joints contact one another determine the type of movement that is possible.
2. Factors that contribute to keeping the surfaces in contact and affect range of motion are structure or shape of the articulating bones, strength and tension of the joint ligaments, arrangement and tension of the muscles, apposition of soft parts, hormones, and disuse.

9.8 Selected Joints of the Body

1. A summary of selected joints of the body, including articular components, structural and functional classifications, and movements, is presented in Tables 9.3 and 9.4.
2. The temporomandibular joint (TMJ) is between the condyle of the mandible and mandibular fossa and articular tubercle of the temporal bone (Exhibit 9.1).
3. The shoulder (humeroscapular or glenohumeral) joint is between the head of the humerus and glenoid cavity of the scapula (Exhibit 9.2).
4. The elbow joint is between the trochlea of the humerus, the trochlear notch of the ulna, and the head of the radius (Exhibit 9.3).
5. The hip (coxal) joint is between the head of the femur and acetabulum of the hip bone (Exhibit 9.4).
6. The knee (tibiofemoral) joint is between the patella and patellar surface of the femur; the lateral condyle of the femur, the lateral meniscus, and the lateral condyle of the tibia; and the medial condyle of the femur, the medial meniscus, and the medial condyle of the tibia (Exhibit 9.5).

Resource:
Anatomy Overview - Specific Synovial Joints
Anatomy Overview - Specific Synovial Diarthrotic Joints
Figure 9.11 - Right Temporomandibular Joint
Figure 9.15 - Right Knee Joint

9.9 Aging and Joints

1. With aging, a decrease in synovial fluid, thinning of articular cartilage, and decreased flexibility of ligaments occur.
2. Most individuals experience some degeneration in the knees, elbows, hips, and shoulders due to the aging process.

Resource:
Homeostatic Imbalances - The Case of the Painful Hands

9.10 Arthroplasty

1. Arthroplasty refers to the surgical replacement of joints.
2. The most commonly replaced joints are the hips, knees, and shoulders.

SELF-QUIZ QUESTIONS

Fill in the blanks in the following statements.

1. A point of contact between two bones, between bone and cartilage, or between bone and teeth is called a(n) ______.

2. The surgical procedure in which a severely damaged joint is replaced with an artificial joint is known as ______.

Indicate whether the following statements are true or false.

3. Menisci are fluid-filled sacs located outside of the joint cavity to ease friction between bones and softer tissue.

4. Shrugging your shoulders involves flexion and extension.

5. Synovial fluid becomes more viscous (thicker) as movement at the joint increases.

Choose the one best answer to the following questions.

6. Which of the following are structural classifications of joints? (1) amphiarthrosis, (2) cartilaginous, (3) synovial, (4) synarthrosis, (5) fibrous.
(a) 1, 2, 3, 4, and 5 (b) 2 and 5 (c) 1 and 4
(d) 1, 2, 4, and 5 (e) 2, 3, and 5

7. Which of the following joints could be classified functionally as synarthroses? (1) syndesmosis, (2) symphysis, (3) synovial, (4) gomphosis, (5) suture.
(a) 1 and 2 (b) 3 and 5 (c) 1, 2, and 3
(d) 4 and 5 (e) 5 only

8. The most common degenerative joint disease in the elderly, often caused by wear-and-tear, is
(a) rheumatoid arthritis. (b) osteoarthritis. (c) rheumatism.
(d) gouty arthritis. (e) ankylosing spondylitis.

9. Chewing your food involves (1) flexion, (2) extension, (3) hyperextension, (4) elevation, (5) depression.
(a) 1 and 2 (b) 1 and 3 (c) 4 and 5
(d) 3 and 5 (e) 1 and 4

10. Synovial fluid functions to (1) absorb shocks at joints, (2) lubricate joints, (3) form a blood clot in a joint injury, (4) supply oxygen and nutrients to chondrocytes, (5) provide phagocytes to remove debris from joints.
(a) 1, 2, 4, and 5 (b) 1, 2, 3, 4, and 5 (c) 1, 2, and 4
(d) 3 and 4 (e) 2, 4, and 5

11. Which of the following statements are *true* concerning a synovial joint? (1) The bones at a synovial joint are covered by a mucous membrane. (2) The articular capsule surrounds a synovial joint, encloses the synovial cavity, and unites the articulating bones. (3) The fibrous membrane of the articular capsule permits considerable movement at a joint. (4) The tensile strength of the fibrous membrane helps prevent bones from disarticulating. (5) All synovial joints contain a fibrous membrane.
(a) 1, 2, 3, and 4 (b) 2, 3, 4, and 5 (c) 2, 3, and 4
(d) 1, 2, and 3 (e) 2, 4, and 5

12. Which of the following keep the articular surfaces of synovial joints in contact and affect range of motion? (1) structure or shape of the articulating bones, (2) strength and tension of the joint ligaments, (3) arrangement and tension of muscles, (4) lack of use, (5) contact of soft parts.
(a) 1, 2, 3, and 5 (b) 2, 3, 4, and 5 (c) 1, 3, 4, and 5
(d) 1, 3, and 5 (e) 1, 2, 3, 4, and 5

13. Match the following:
____ (a) a fibrous joint that unites the bones of the skull; a synarthrosis
____ (b) a fibrous joint between the tibia and fibula; an amphiarthrosis
____ (c) the articulation between bone and teeth
____ (d) the epiphyseal (growth) plate
____ (e) joint between the two pubic bones
____ (f) joint with a cavity between the bones; diarthrosis
____ (g) a bony joint

(1) synostosis
(2) synchondrosis
(3) syndesmosis
(4) synovial
(5) suture
(6) symphysis
(7) gomphosis

14. Match the following:
____ (a) rounded or pointed surface of one bone articulates with a ring formed by another bone and a ligament; allows rotation around its own axis
____ (b) articulating bone surfaces are flat or slightly curved; permit gliding movement
____ (c) convex, oval projection of one bone fits into oval depression of another bone; permits movement in two axes
____ (d) convex surface of one bone articulates with concave surface of another bone; permits flexion and extension
____ (e) ball-shaped surface of one bone articulates with cuplike depression of another bone; permits largest degree of movement in three axes
____ (f) modified condyloid joint where articulating bones resemble a rider sitting in a saddle

(1) hinge joint
(2) saddle joint
(3) ball-and-socket joint
(4) plane joint
(5) condyloid joint
(6) pivot joint

15. Match the following:

____ (a) upward movement of a body part	(1) pronation
____ (b) downward movement of a body part	(2) plantar flexion
____ (c) movement of bone toward midline	(3) eversion
____ (d) movement in which relatively flat bone surfaces move back-and-forth and side-to-side with respect to one another	(4) abduction
____ (e) movement of a body part anteriorly in the transverse plane	(5) rotation
____ (f) decrease in angle between bones	(6) retraction
____ (g) movement of an anteriorly projected body part back to the anatomical position	(7) opposition
____ (h) movement of the sole medially	(8) elevation
____ (i) movement of the sole laterally	(9) flexion
____ (j) movement of bone away from midline	(10) adduction
____ (k) action that occurs when you stand on your heels	(11) depression
____ (l) action that occurs when you stand on your toes	(12) inversion
____ (m) movement of the forearm to turn the palm anteriorly	(13) gliding
____ (n) movement of the forearm to turn the palm posteriorly	(14) extension
____ (o) movement of thumb across the palm to touch the tips of the fingers of the same hand	(15) protraction
____ (p) increase in angle between bones	(16) dorsiflexion
____ (q) movement of distal end of a part of the body in a circle	(17) circumduction
____ (r) bone revolves around its own longitudinal axis	(18) supination

CRITICAL THINKING QUESTIONS

1. Katie loves pretending that she's a human cannonball. As she jumps off the diving board, she assumes the proper position before she pounds into the water: head and thighs tucked against her chest; back rounded; arms pressed against her sides while her forearms, crossed in front of her shins, hold her legs tightly folded against her chest. Use the proper anatomical terms to describe the position of Katie's back, head, and free limbs.

2. During football practice, Jeremiah was tackled and twisted his lower leg. There was a sharp pain, followed immediately by swelling of the knee joint. The pain and swelling worsened throughout the remainder of the afternoon until Jeremiah could barely walk. The coach told Jeremiah to see a doctor who might want to "drain the water off his knee." What was the coach referring to and what specifically do you think happened to Jeremiah's knee joint to cause these symptoms?

3. After lunch, during a particularly long and dull class video, Antonio became sleepy and yawned. To his dismay, he was then unable to close his mouth. Explain what happened and what should be done to correct this problem.

? ANSWERS TO FIGURE QUESTIONS

9.1 Functionally, sutures are classified as synarthroses because they are immovable; syndesmoses are classified as amphiarthroses because they are slightly movable.

9.2 The structural difference between a synchondrosis and a symphysis is the type of cartilage that holds the joint together: hyaline cartilage in a synchondrosis and fibrocartilage in a symphysis.

9.3 Functionally, synovial joints are diarthroses, freely movable joints.

9.4 Gliding movements occur at intercarpal joints and at intertarsal joints.

9.5 Two examples of flexion that do not occur along the sagittal plane are flexion of the thumb and lateral flexion of the trunk.

9.6 When you adduct your arm or leg, you bring it closer to the midline of the body, thus "adding" it to the trunk.

9.7 Circumduction involves flexion, abduction, extension, adduction, and rotation in a continuous sequence (or in the opposite order).

9.8 The anterior surface of a bone or limb rotates toward the midline in medial rotation, and away from the midline in lateral rotation.

9.9 Bringing your arms forward until the elbows touch is an example of protraction.

9.10 Many plane, saddle, and condyloid joints are biaxial joints.

9.11 The lateral ligament prevents displacement of the mandible.

9.12 The shoulder joint is the most freely movable joint in the body because of the looseness of its articular capsule and the shallowness of the glenoid cavity in relation to the size of the head of the humerus.

9.13 A hinge joint permits flexion and extension.

9.14 Tension in three ligaments—iliofemoral, pubofemoral, and ischiofemoral—limits the degree of extension at the hip joint.

9.15 Contraction of the quadriceps femoris muscle causes extension at the knee joint.

9.16 The purpose of arthroplasty is to relieve joint pain and permit greater range of motion.

10 MUSCULAR TISSUE

MUSCULAR TISSUE AND HOMEOSTASIS *Muscular tissue contributes to homeostasis by producing body movements, moving substances through the body, and producing heat to maintain normal body temperature.*

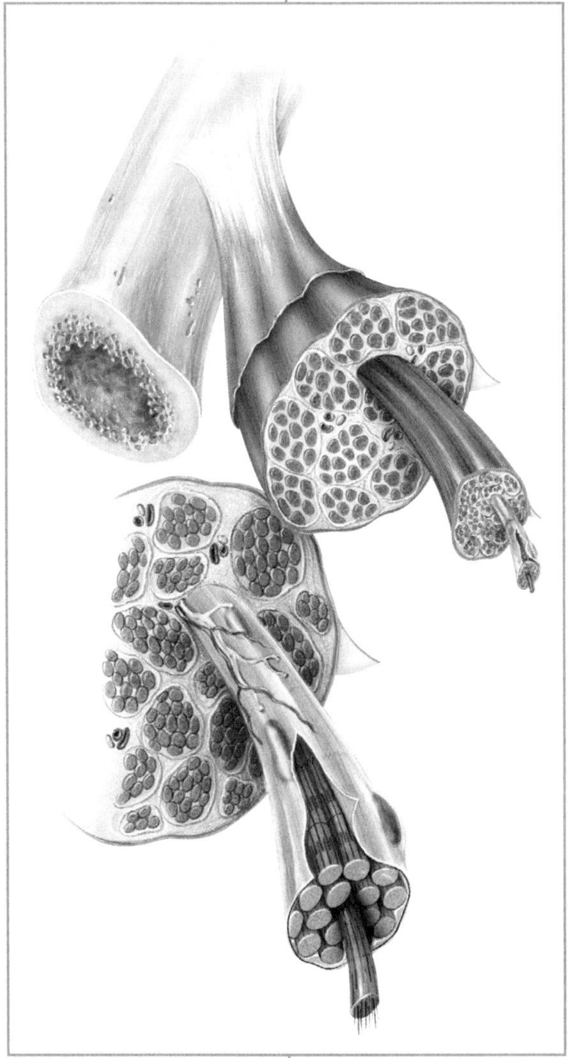

Although bones provide leverage and form the framework of the body, they cannot move body parts by themselves. Motion results from the alternating contraction and relaxation of muscles, which make up 40–50% of total adult body weight (depending on the percentage of body fat, gender, and exercise regimen). Your muscular strength reflects the primary function of muscle—the transformation of chemical energy into mechanical energy to generate force, perform work, and produce movement. In addition, muscle tissues stabilize body position, regulate organ volume, generate heat, and propel fluids and food matter through various body systems. The scientific study of muscles is known as **myology** (mī-OL-ō-jē; *myo-* = muscle; *-logy* = study of).

? ***Did you ever wonder what causes rigor mortis?***

10.1 OVERVIEW OF MUSCULAR TISSUE

OBJECTIVES

- Explain the structural differences between the three types of muscular tissue.
- Compare the functions and special properties of the three types of muscular tissue.

Types of Muscular Tissue

The three types of muscular tissue—skeletal, cardiac, and smooth—were introduced in Chapter 4 (see Table 4.9). Although the different types of muscular tissue share some properties, they differ from one another in their microscopic anatomy, location, and how they are controlled by the nervous and endocrine systems.

Skeletal muscle tissue is so named because most skeletal muscles move bones of the skeleton. (A few skeletal muscles attach to and move the skin or other skeletal muscles.) Skeletal muscle tissue is *striated:* Alternating light and dark protein bands (*striations*) are seen when the tissue is examined with a microscope (see Table 4.9). Skeletal muscle tissue works mainly in a *voluntary* manner. Its activity can be consciously controlled by neurons (nerve cells) that are part of the somatic (voluntary) division of the nervous system. (Figure 12.10 depicts the divisions of the nervous system.) Most skeletal muscles also are controlled subconsciously to some extent. For example, your diaphragm continues to alternately contract and relax without conscious control so that you don't stop breathing. Also, you do not need to consciously think about contracting the skeletal muscles that maintain your posture or stabilize body positions.

Only the heart contains **cardiac muscle tissue,** which forms most of the heart wall. Cardiac muscle is also *striated,* but its action is *involuntary.* The alternating contraction and relaxation of the heart is not consciously controlled. Rather, the heart beats because it has a pacemaker that initiates each contraction. This built-in rhythm is termed **autorhythmicity** (aw′-tō-rith-MISS-i-tē). Several hormones and neurotransmitters can adjust heart rate by speeding or slowing the pacemaker.

Smooth muscle tissue is located in the walls of hollow internal structures, such as blood vessels, airways, and most organs in the abdominopelvic cavity. It is also found in the skin, attached to hair follicles. Under a microscope, this tissue lacks the striations of skeletal and cardiac muscle tissue. For this reason, it looks *nonstriated,* which is why it is referred to as *smooth.* The action of smooth muscle is usually *involuntary,* and some smooth muscle tissue, such as the muscles that propel food through your gastrointestinal tract, has autorhythmicity. Both cardiac muscle and smooth muscle are regulated by neurons that are part of the autonomic (involuntary) division of the nervous system and by hormones released by endocrine glands.

Functions of Muscular Tissue

Through sustained contraction or alternating contraction and relaxation, muscular tissue has four key functions: producing body movements, stabilizing body positions, storing and moving substances within the body, and generating heat.

1. ***Producing body movements.*** Movements of the whole body such as walking and running, and localized movements such as grasping a pencil, keyboarding, or nodding the head as a result of muscular contractions, rely on the integrated functioning of skeletal muscles, bones, and joints.
2. ***Stabilizing body positions.*** Skeletal muscle contractions stabilize joints and help maintain body positions, such as standing or sitting. Postural muscles contract continuously when you are awake; for example, sustained contractions of your neck muscles hold your head upright when you are listening intently to your anatomy and physiology lecture.
3. ***Storing and moving substances within the body.*** Storage is accomplished by sustained contractions of ringlike bands of smooth muscle called *sphincters,* which prevent outflow of the contents of a hollow organ. Temporary storage of food in the stomach or urine in the urinary bladder is possible because smooth muscle sphincters close off the outlets of these organs. Cardiac muscle contractions of the heart pump blood through the blood vessels of the body. Contraction and relaxation of smooth muscle in the walls of blood vessels help adjust blood vessel diameter and thus regulate the rate of blood flow. Smooth muscle contractions also move food and substances such as bile and enzymes through the gastrointestinal tract, push gametes (sperm and oocytes) through the passageways of the reproductive systems, and propel urine through the urinary system. Skeletal muscle contractions promote the flow of lymph and aid the return of blood in veins to the heart.
4. ***Generating heat.*** As muscular tissue contracts, it produces heat, a process known as **thermogenesis** (ther′-mō-JEN-e-sis). Much of the heat generated by muscle is used to maintain normal body temperature. Involuntary contractions of skeletal muscles, known as *shivering,* can increase the rate of heat production.

Properties of Muscular Tissue

Muscular tissue has four special properties that enable it to function and contribute to homeostasis:

1. **Electrical excitability** (ek-sīt′-a-BIL-i-tē), a property of both muscle and nerve cells that was introduced in Chapter 4, is the ability to respond to certain stimuli by producing electrical signals called **action potentials (impulses).** Action potentials in muscles are referred to as *muscle action potentials;* those in nerve cells are called *nerve action potentials.* Chapter 12 provides more detail about how action potentials arise (see Section 12.3). For muscle cells, two main types of stimuli trigger action potentials. One is autorhythmic *electrical signals* arising in the muscular tissue itself, as in the heart's pacemaker. The other is *chemical stimuli,* such as neurotransmitters released by neurons, hormones distributed by the blood, or even local changes in pH.

2. **Contractility** (kon′-trak-TIL-i-tē) is the ability of muscular tissue to contract forcefully when stimulated by an action potential. When a skeletal muscle contracts, it generates tension (force of contraction) while pulling on its attachment points. In some muscle contractions, the muscle develops tension (force of contraction) but does not shorten. An example is holding this book in an outstretched hand. In other muscle contractions, the tension generated is great enough to overcome the load (resistance) of the object being moved so the muscle shortens and movement occurs. An example is lifting a book off a table.
3. **Extensibility** (ek-sten′-si-BIL-i-tē) is the ability of muscular tissue to stretch, within limits, without being damaged. The connective tissue within the muscle limits the range of extensibility and keeps it within the contractile range of the muscle cells. Normally, smooth muscle is subject to the greatest amount of stretching. For example, each time your stomach fills with food, the smooth muscle in the wall is stretched. Cardiac muscle also is stretched each time the heart fills with blood.
4. **Elasticity** (e-las-TIS-i-tē) is the ability of muscular tissue to return to its original length and shape after contraction or extension.

Skeletal muscle is the focus of much of this chapter. Cardiac muscle and smooth muscle are described briefly here. Cardiac muscle is discussed in more detail in Chapter 20 (the heart), and smooth muscle is included in Chapter 15 (the autonomic nervous system), as well as in discussions of the various organs containing smooth muscle.

✔ CHECKPOINT

1. What features distinguish the three types of muscular tissue?
2. List the general functions of muscular tissue.
3. Describe the four properties of muscular tissue.

10.2 SKELETAL MUSCLE TISSUE

OBJECTIVES

- Explain the importance of connective tissue components, blood vessels, and nerves to skeletal muscles.
- Describe the microscopic anatomy of a skeletal muscle fiber.
- Distinguish thick filaments from thin filaments.
- Describe the functions of skeletal muscle proteins.

Each of your skeletal muscles is a separate organ composed of hundreds to thousands of cells, which are called **muscle fibers** because of their elongated shapes. Thus, *muscle cell* and *muscle fiber* are two terms for the same structure. Skeletal muscle also contains connective tissues surrounding muscle fibers and whole muscles, and blood vessels and nerves (Figure 10.1). To understand how contraction of skeletal muscle can generate tension, you must first understand its gross and microscopic anatomy.

Connective Tissue Components

Connective tissue surrounds and protects muscular tissue. The **subcutaneous layer** or **hypodermis,** which separates muscle from skin (see Figure 11.21), is composed of areolar connective tissue and adipose tissue. It provides a pathway for nerves, blood vessels, and lymphatic vessels to enter and exit muscles. The adipose tissue of the subcutaneous layer stores most of the body's triglycerides, serves as an insulating layer that reduces heat loss, and protects muscles from physical trauma. **Fascia** (FASH-ē-a = bandage) is a dense sheet or broad band of irregular connective tissue that lines the body wall and limbs and supports and surrounds muscles and other organs of the body. As you will see, fascia holds muscles with similar functions together (see Figure 11.21). Fascia allows free movement of muscles; carries nerves, blood vessels, and lymphatic vessels; and fills spaces between muscles.

Three layers of connective tissue extend from the fascia to protect and strengthen skeletal muscle (Figure 10.1):

- The outermost layer of dense, irregular connective tissue, encircling the entire muscle, is the **epimysium** (ep-i-MĪZ-ē-um; *epi-* = upon).
- **Perimysium** (per-i-MĪZ-ē-um; *peri-* = around) is also a layer of dense, irregular connective tissue, but it surrounds groups of 10 to 100 or more muscle fibers, separating them into bundles called **fascicles** (FAS-i-kuls = little bundles). Many fascicles are large enough to be seen with the naked eye. They give a cut of meat its characteristic "grain"; if you tear a piece of meat, it rips apart along the fascicles.
- **Endomysium** (en′-dō-MĪZ-ē-um; *endo-* = within) penetrates the interior of each fascicle and separates individual muscle fibers from one another. The endomysium is mostly reticular fibers.

The epimysium, perimysium, and endomysium all are continuous with the connective tissue that attaches skeletal muscle to other structures, such as bone or another muscle. For example, all three connective tissue layers may extend beyond the muscle fibers to form a ropelike **tendon** that attaches a muscle to the periosteum of a bone. An example is the *calcaneal (Achilles) tendon* of the gastrocnemius (calf) muscle, which attaches the muscle to the calcaneus (heel bone) (shown in Figure 11.22c). When the connective tissue elements extend as a broad, flat sheet, it is called an **aponeurosis** (ap-ō-noo-RŌ-sis; *apo-* = from; *-neur-* = a sinew). An example is the *epicranial aponeurosis* on top of the skull between the frontal and occipital bellies of the occipitofrontalis muscle (shown in Figure 11.4a, c).

CLINICAL CONNECTION | *Fibromyalgia*

Fibromyalgia (fī-brō-mī-AL-jē-a; *algia* = painful condition) is a chronic, painful, nonarticular rheumatic disorder that affects the fibrous connective tissue components of muscles, tendons, and ligaments. A striking sign is pain that results from gentle pressure at specific "tender points." Even without pressure, there is pain, tenderness, and stiffness of muscles, tendons, and surrounding soft tissues. Besides muscle pain, those with fibromyalgia report severe fatigue, poor sleep, headaches, depression, irritable bowel syndrome, and inability to carry out their daily activities. There is no specific identifiable cause. Treatment consists of stress reduction, regular exercise, application of heat, gentle massage, physical therapy, medication for pain, and a low-dose antidepressant to help improve sleep. •

Figure 10.1 **Organization of skeletal muscle and its connective tissue coverings.**

A skeletal muscle consists of individual muscle fibers (cells) bundled into fascicles and surrounded by three connective tissue layers that are extensions of the fascia.

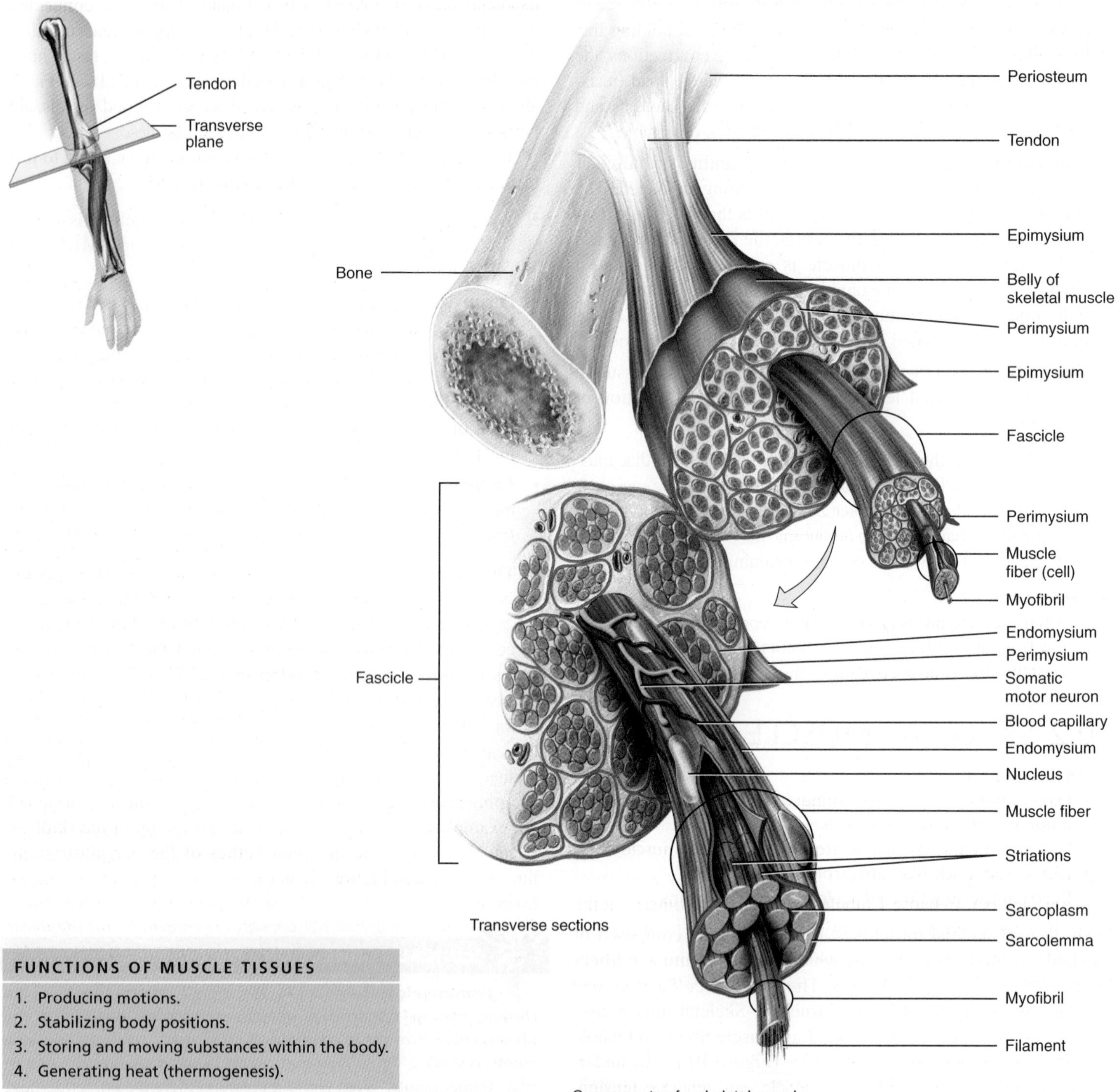

Tendon
Transverse plane
Periosteum
Tendon
Epimysium
Bone
Belly of skeletal muscle
Perimysium
Epimysium
Fascicle
Perimysium
Muscle fiber (cell)
Myofibril
Endomysium
Perimysium
Fascicle
Somatic motor neuron
Blood capillary
Endomysium
Nucleus
Muscle fiber
Striations
Sarcoplasm
Sarcolemma
Transverse sections
Myofibril
Filament

Components of a skeletal muscle

FUNCTIONS OF MUSCLE TISSUES
1. Producing motions.
2. Stabilizing body positions.
3. Storing and moving substances within the body.
4. Generating heat (thermogenesis).

Which connective tissue coat surrounds groups of muscle fibers, separating them into fascicles?

Nerve and Blood Supply

Skeletal muscles are well supplied with nerves and blood vessels. Generally, an artery and one or two veins accompany each nerve that penetrates a skeletal muscle. The neurons that stimulate skeletal muscle to contract are *somatic motor neurons.* Each somatic motor neuron has a threadlike axon that extends from the brain or spinal cord to a group of skeletal muscle fibers (see Figure 10.9d). The axon of a somatic motor neuron typically branches many times, each branch extending to a different skeletal muscle fiber.

Microscopic blood vessels called capillaries are plentiful in muscular tissue; each muscle fiber is in close contact with one or more capillaries (see Figure 10.9d). The blood capillaries bring in oxygen and nutrients and remove heat and the waste products of muscle metabolism. Especially during contraction, a muscle fiber synthesizes and uses considerable ATP (adenosine triphosphate). These reactions, which you will learn more about later on, require oxygen, glucose, fatty acids, and other substances that are delivered to the muscle fiber in the blood.

Microscopic Anatomy of a Skeletal Muscle Fiber

The most important components of a skeletal muscle are the muscle fibers themselves. The diameter of a mature skeletal muscle fiber ranges from 10 to 100 μm.* The typical length of a mature skeletal muscle fiber is about 10 cm (4 in.), although some are as long as 30 cm (12 in.). Because each skeletal muscle fiber arises during embryonic development from the fusion of a hundred or more small mesodermal cells called *myoblasts* (MĪ-ō-blasts) (Figure 10.2a), each mature skeletal muscle fiber has a hundred or more nuclei. Once fusion has occurred, the muscle fiber loses its ability to undergo cell division. Thus, the number of skeletal muscle fibers is set before you are born, and most of these cells last a lifetime.

Sarcolemma, Transverse Tubules, and Sarcoplasm

The multiple nuclei of a skeletal muscle fiber are located just beneath the **sarcolemma** (sar′-kō-LEM-ma; *sarc-* = flesh; *-lemma* = sheath), the plasma membrane of a muscle cell (Figure 10.2b, c). Thousands of tiny invaginations of the sarcolemma, called **transverse (T) tubules,** tunnel in from the surface toward the center of each muscle fiber. Because T tubules are open to the outside of the fiber, they are filled with interstitial fluid. Muscle action potentials travel along the sarcolemma and through the T tubules, quickly spreading throughout the muscle fiber. This arrangement ensures that an action potential excites all parts of the muscle fiber at essentially the same instant.

Within the sarcolemma is the **sarcoplasm** (SAR-kō-plazm), the cytoplasm of a muscle fiber. Sarcoplasm includes a substantial amount of glycogen, which is a large molecule composed of many glucose molecules (see Figure 2.16). Glycogen can be used for synthesis of ATP. In addition, the sarcoplasm contains a red-colored protein called **myoglobin** (mī-ō-GLŌB-in). This protein, found only in muscle, binds oxygen molecules that diffuse into muscle fibers from interstitial fluid. Myoglobin releases oxygen when it is needed by the mitochondria for ATP production. The mitochondria lie in rows throughout the muscle fiber, strategically close to the contractile muscle proteins that use ATP during contraction so that ATP can be produced quickly as needed (Figure 10.2c).

*One micrometer (μm) is 10^{-6} meter (1/25,000 in.).

Myofibrils and Sarcoplasmic Reticulum

At high magnification, the sarcoplasm appears stuffed with little threads. These small structures are the **myofibrils** (mī-ō-FĪ-brils; *myo-* = muscle; *-fibrilla* = little fiber), the contractile organelles of skeletal muscle (Figure 10.2c). Myofibrils are about 2 μm in diameter and extend the entire length of a muscle fiber. Their prominent striations make the entire skeletal muscle fiber appear striped (striated).

A fluid-filled system of membranous sacs called the **sarcoplasmic reticulum** (sar′-kō-PLAZ-mik re-TIK-ū-lum) or **SR** encircles each myofibril (Figure 10.2c). This elaborate system is similar to smooth endoplasmic reticulum in nonmuscular cells. Dilated end sacs of the sarcoplasmic reticulum called **terminal cisterns** (SIS-terns = reservoirs) butt against the T tubule from

CLINICAL CONNECTION | *Muscular Hypertrophy, Fibrosis, and Muscular Atrophy*

The muscle growth that occurs after birth occurs by enlargement of existing muscle fibers, called **muscular hypertrophy** (hī-PER-trō-fē; *hyper-* = above or excessive; *-trophy* = nourishment), rather than by **muscular hyperplasia** (hī-per-PLĀ-zē-a; *-plasis* = molding), an increase in the number of fibers. Muscular hypertrophy is due to increased production of myofibrils, mitochondria, sarcoplasmic reticulum, and other organelles. It results from very forceful, repetitive muscular activity, such as strength training. Because hypertrophied muscles contain more myofibrils, they are capable of more forceful contractions. During childhood, human growth hormone and other hormones stimulate an increase in the size of skeletal muscle fibers. The hormone testosterone (from the testes in males and in small amounts from other tissues such as the ovaries in females) promotes further enlargement of muscle fibers.

A few myoblasts do persist in mature skeletal muscle as *satellite cells* (Figure 10.2a, b). Satellite cells retain the capacity to fuse with one another or with damaged muscle fibers to regenerate functional muscle fibers. However, the number of new skeletal muscle fibers that can be formed by satellite cells is not enough to compensate for significant skeletal muscle damage or degeneration. In such cases, skeletal muscle tissue undergoes **fibrosis,** the replacement of muscle fibers by fibrous scar tissue. Thus, regeneration of skeletal muscle tissue is limited.

Muscular atrophy (AT-rō-fē; *a-* = without, *-trophy* = nourishment) is a wasting away of muscles. Individual muscle fibers decrease in size as a result of progressive loss of myofibrils. Muscular atrophy that occurs because muscles are not used is termed *disuse atrophy.* Bedridden individuals and people with casts experience disuse atrophy because the flow of nerve impulses (nerve action potentials) to inactive skeletal muscle is greatly reduced, but the condition is reversible. If instead the nerve supply to a muscle is disrupted or cut, the muscle undergoes *denervation atrophy.* Over a period of 6 months to 2 years, the muscle shrinks to about one-fourth its original size, and the muscle fibers are irreversibly replaced by fibrous connective tissue. •

Figure 10.2 Microscopic organization of skeletal muscle. (a) During embryonic development, many myoblasts fuse to form one skeletal muscle fiber. Once fusion has occurred, a skeletal muscle fiber loses the ability to undergo cell division, but satellite cells retain this ability. (b–d) The sarcolemma of the fiber encloses sarcoplasm and myofibrils, which are striated. Sarcoplasmic reticulum wraps around each myofibril. Thousands of transverse tubules, filled with interstitial fluid, invaginate from the sarcolemma toward the center of the muscle fiber. A triad is a transverse tubule and the two terminal cisterns of the sarcoplasmic reticulum on either side of it. A photomicrograph of skeletal muscle tissue is shown in Table 4.9.

6 The contractile elements of muscle fibers, the myofibrils, contain overlapping thick and thin filaments.

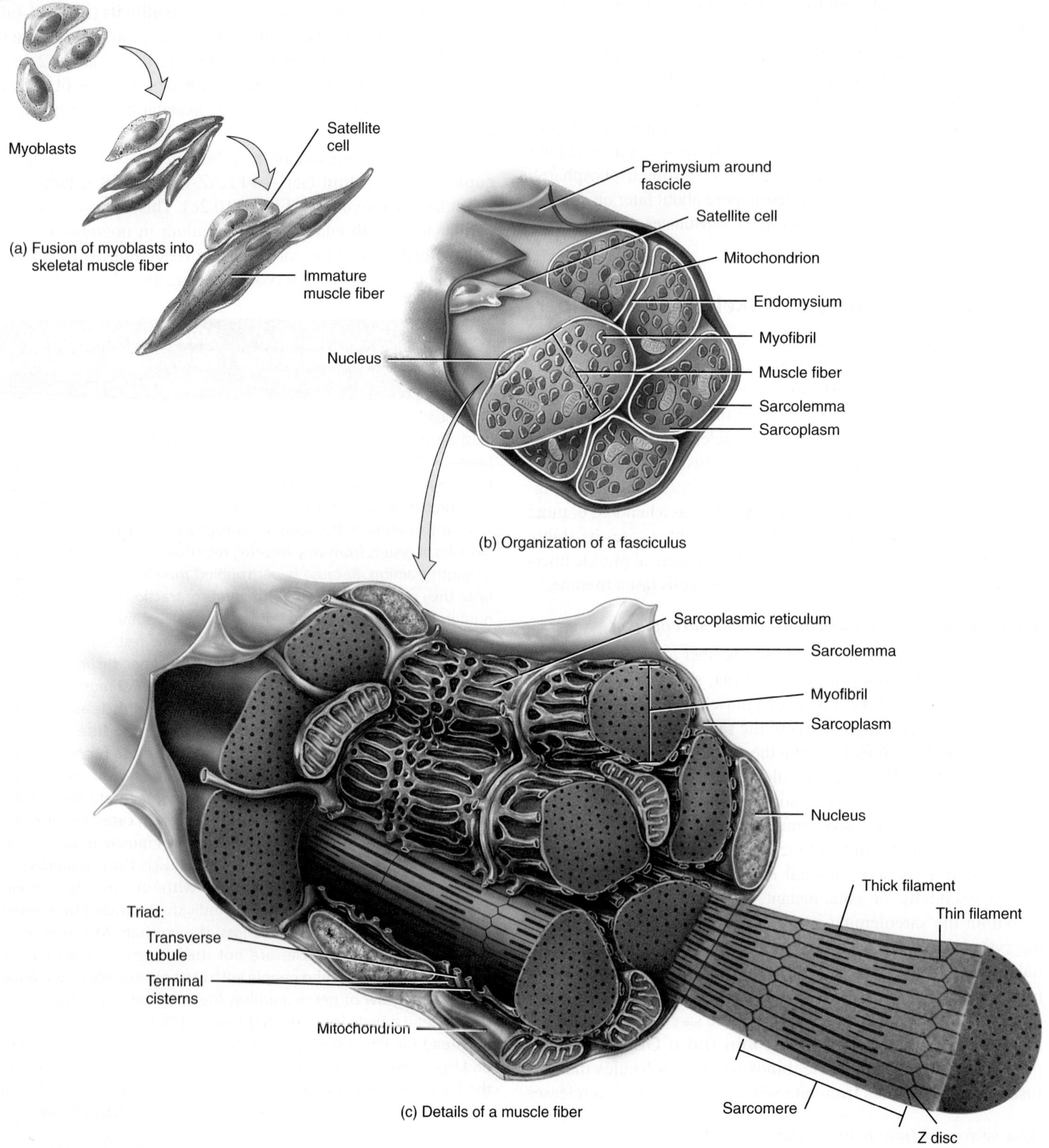
Myoblasts
Satellite cell
(a) Fusion of myoblasts into skeletal muscle fiber
Immature muscle fiber
Perimysium around fascicle
Satellite cell
Mitochondrion
Endomysium
Myofibril
Nucleus
Muscle fiber
Sarcolemma
Sarcoplasm
(b) Organization of a fasciculus
Sarcoplasmic reticulum
Sarcolemma
Myofibril
Sarcoplasm
Nucleus
Thick filament
Thin filament
Triad:
Transverse tubule
Terminal cisterns
Mitochondrion
(c) Details of a muscle fiber
Sarcomere
Z disc

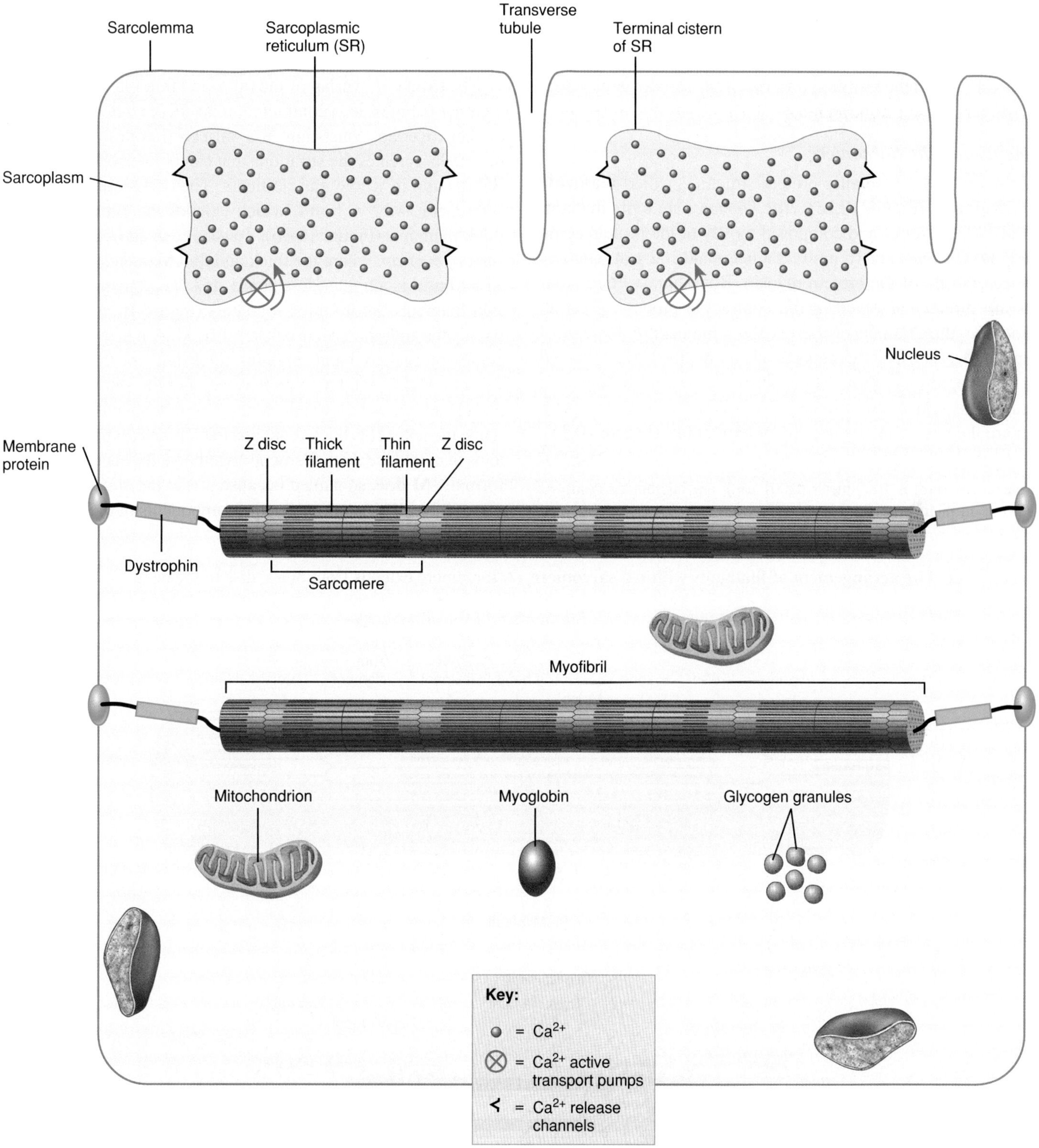
Sarcolemma
Sarcoplasmic reticulum (SR)
Transverse tubule
Terminal cistern of SR
Sarcoplasm
Nucleus
Membrane protein
Z disc
Thick filament
Thin filament
Z disc
Dystrophin
Sarcomere
Myofibril
Mitochondrion
Myoglobin
Glycogen granules
Key:
= Ca2+
= Ca2+ active transport pumps
= Ca2+ release channels

(d) Simplistic representation of a muscle fiber

Which structure shown here releases calcium ions to trigger muscle contraction?

both sides. A transverse tubule and the two terminal cisterns on either side of it form a **triad** (*tri-* = three). In a relaxed muscle fiber, the sarcoplasmic reticulum stores calcium ions (Ca^{2+}). Release of Ca^{2+} from the terminal cisterns of the sarcoplasmic reticulum triggers muscle contraction.

Filaments and the Sarcomere

Within myofibrils are smaller protein structures called **filaments** or **myofilaments** (Figure 10.2c). *Thin filaments* are 8 nm in diameter and 1–2 μm long† and composed mostly of the protein actin, while *thick filaments* are 16 nm in diameter and 1–2 μm long and composed mostly of the protein myosin. Both thin and thick filaments are directly involved in the contractile process. Overall, there are two thin filaments for every thick filament in the regions of filament overlap. The filaments inside a myofibril do not extend the entire length of a muscle fiber. Instead, they are arranged in compartments called **sarcomeres** (SAR-kō-mērs; *-mere* = part), which are the basic functional units of a myofibril (Figure 10.3a).

†One nanometer (nm) is 10^{-9} meter (0.001 μm); one micrometer (μm) = 1/25,000 of an inch.

Narrow, plate-shaped regions of dense protein material called **Z discs** separate one sarcomere from the next. Thus, a sarcomere extends from one Z disc to the next Z disc.

The extent of overlap of the thick and thin filaments depends on whether the muscle is contracted, relaxed, or stretched. The pattern of their overlap, consisting of a variety of zones and bands (Figure 10.3b), creates the striations that can be seen both in single myofibrils and in whole muscle fibers. The darker middle part of the sarcomere is the **A band,** which extends the entire length of the thick filaments (Figure 10.3b). Toward each end of the A band is a *zone of overlap,* where the thick and thin filaments lie side by side. The **I band** is a lighter, less dense area that contains the rest of the thin filaments but no thick filaments (Figure 10.3b), and a Z disc passes through the center of each I band. A narrow **H zone** in the center of each A band contains thick but not thin filaments. A mnemonic that will help you to remember the composition of the I and H bands is as follows: the letter I is thin (contains thin filaments), while the letter H is thick (contains thick filaments). Supporting proteins that hold the thick filaments together at the center of the H zone form the **M line,** so named because it is at the *middle* of the sarcomere. Table 10.1 summarizes the components of the sarcomere.

Figure 10.3 The arrangement of filaments within a sarcomere. A sarcomere extends from one Z disc to the next.

Myofibrils contain two types of filaments: thick filaments and thin filaments.

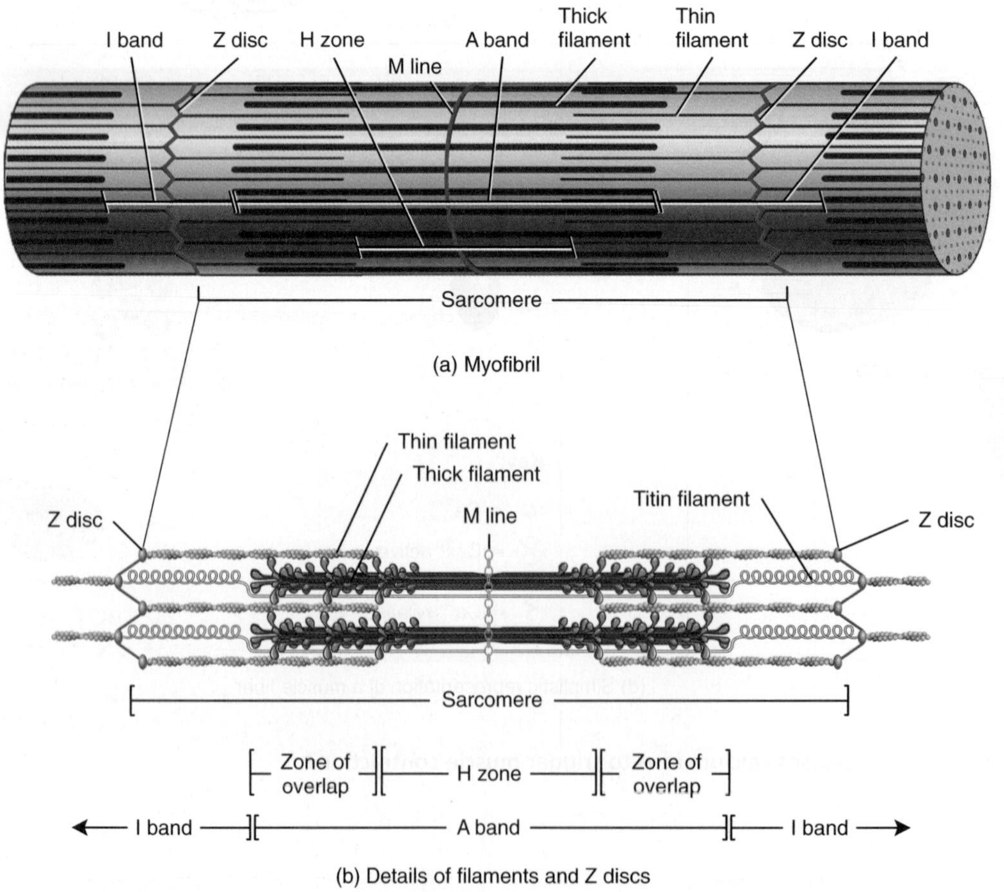

Which of the following is the smallest: muscle fiber, thick filament, or myofibril? Which is largest?

TABLE 10.1

Components of a Sarcomere

COMPONENT	DESCRIPTION
Z discs	Narrow, plate-shaped regions of dense material that separate one sarcomere from the next.
A band	Dark, middle part of sarcomere that extends entire length of thick filaments and includes those parts of thin filaments that overlap thick filaments.
I band	Lighter, less dense area of sarcomere that contains remainder of thin filaments but no thick filaments. A Z disc passes through center of each I band.
H zone	Narrow region in center of each A band that contains thick filaments but no thin filaments.
M line	Region in center of H zone that contains proteins that hold thick filaments together at center of sarcomere.

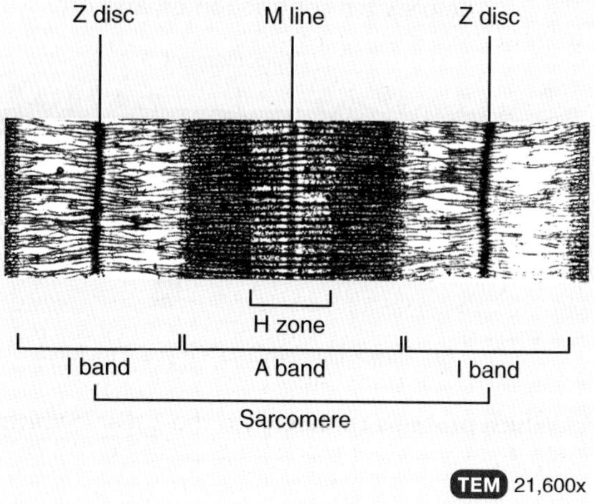

CLINICAL CONNECTION | *Exercise-Induced Muscle Damage*

Comparison of electron micrographs of muscle tissue taken from athletes before and after intense exercise reveals considerable **exercise-induced muscle damage,** including torn sarcolemmas in some muscle fibers, damaged myofibrils, and disrupted Z discs. Microscopic muscle damage after exercise also is indicated by increases in blood levels of proteins, such as myoglobin and the enzyme creatine kinase, that are normally confined within muscle fibers. From 12 to 48 hours after a period of strenuous exercise, skeletal muscles often become sore. Such **delayed onset muscle soreness (DOMS)** is accompanied by stiffness, tenderness, and swelling. Although the causes of DOMS are not completely understood, microscopic muscle damage appears to be a major factor. In response to exercise-induced muscle damage, muscle fibers undergo repair: New regions of sarcolemma are formed to replace torn sarcolemmas, and more muscle proteins (including those of the myofibrils) are synthesized in the sarcoplasm of the muscle fibers. •

Muscle Proteins

Myofibrils are built from three kinds of proteins: (1) contractile proteins, which generate force during contraction; (2) regulatory proteins, which help switch the contraction process on and off; and (3) structural proteins, which keep the thick and thin filaments in the proper alignment, give the myofibril elasticity and extensibility, and link the myofibrils to the sarcolemma and extracellular matrix.

The two *contractile proteins* in muscle are myosin and actin, components of thick and thin filaments, respectively. **Myosin** (MĪ-ō-sin) is the main component of thick filaments and functions as a motor protein in all three types of muscle tissue. *Motor proteins* pull various cellular structures to achieve movement by converting the chemical energy in ATP to the mechanical energy of motion, that is, the production of force. In skeletal muscle, about 300 molecules of myosin form a single thick filament. Each myosin molecule is shaped like two golf clubs twisted together (Figure 10.4a). The *myosin tail* (twisted golf club handles) points toward the M line in the center of the sarcomere. Tails of neighboring myosin molecules lie parallel to one another, forming the shaft of the thick filament. The two projections of each myosin molecule (golf club heads) are called *myosin heads.* The heads project outward from the shaft in a spiraling fashion, each extending toward one of the six thin filaments that surround each thick filament.

Thin filaments are anchored to Z discs (see Figure 10.3b). Their main component is the protein **actin** (AK-tin). Individual actin molecules join to form an actin filament that is twisted into a helix (Figure 10.4b). On each actin molecule is a *myosin-binding site,* where a myosin head can attach.

Smaller amounts of two *regulatory proteins*—**tropomyosin** (trō-pō-MĪ-ō-sin) and **troponin** (TRŌ-pō-nin)—are also part of the thin filament. In relaxed muscle, myosin is blocked from binding to actin because strands of tropomyosin cover the myosin-binding sites on actin. The tropomyosin strands in turn are held in place by troponin molecules. You will soon learn that when calcium ions (Ca^{2+}) bind to troponin, it undergoes a change in shape; this change moves tropomyosin away from myosin-binding sites on actin and muscle contraction subsequently begins as myosin binds to actin.

Besides contractile and regulatory proteins, muscle contains about a dozen *structural proteins,* which contribute to the alignment, stability, elasticity, and extensibility of myofibrils. Several key structural proteins are titin, α-actinin, myomesin, nebulin, and dystrophin. **Titin** (*titan* = gigantic) is the third most plentiful protein in skeletal muscle (after actin and myosin). This molecule's name reflects its huge size. With a molecular mass of about 3 million daltons, titin is 50 times larger than an average-sized protein. Each titin molecule spans half a sarcomere, from a Z disc to an M line (see Figure 10.3b), a distance of 1 to 1.2 μm in relaxed muscle. Each titin molecule connects a Z disc to the M line of the sarcomere, thereby

Figure 10.4 Structure of thick and thin filaments. (a) A thick filament contains about 300 myosin molecules, one of which is shown enlarged. The myosin tails form the shaft of the thick filament, and the myosin heads project outward toward the surrounding thin filaments. (b) Thin filaments contain actin, troponin, and tropomyosin.

Contractile proteins (myosin and actin) generate force during contraction; regulatory proteins (troponin and tropomyosin) help switch contraction on and off.

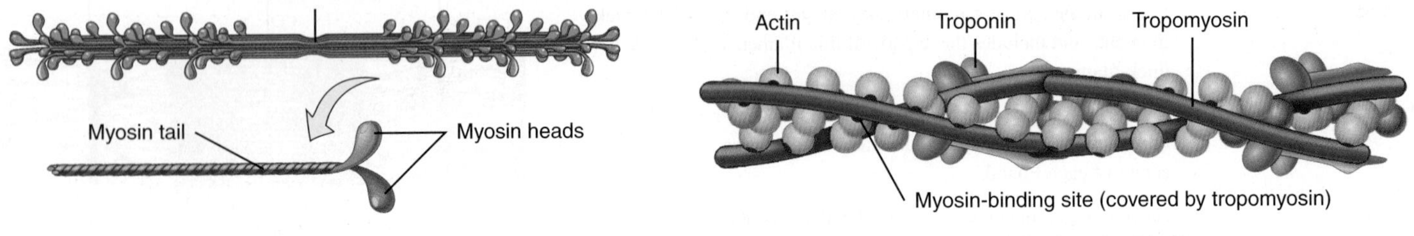

(a) A thick filament and a myosin molecule

(b) Portion of a thin filament

Which proteins connect into the Z disc? Which proteins are present in the A band? In the I band?

helping stabilize the position of the thick filament. The part of the titin molecule that extends from the Z disc is very elastic. Because it can stretch to at least four times its resting length and then spring back unharmed, titin accounts for much of the elasticity and extensibility of myofibrils. Titin probably helps the sarcomere return to its resting length after a muscle has contracted or been stretched, may help prevent overextension of sarcomeres, and maintains the central location of the A bands.

The dense material of the Z discs contains molecules of **α-actinin,** which bind to actin molecules of the thin filament and to titin. Molecules of the protein **myomesin** (mī-ō-MĒ-sin) form the M line. The M line proteins bind to titin and connect adjacent thick filaments to one another. Myosin holds the thick filaments in alignment at the M line. **Nebulin** (NEB-ū-lin) is a long, nonelastic protein wrapped around the entire length of each thin filament. It helps anchor the thin filaments to the Z discs and regulates the

TABLE 10.2

Summary of Skeletal Muscle Fiber Proteins

TYPE OF PROTEIN	DESCRIPTION
Contractile proteins	Proteins that generate force during muscle contractions.
Myosin	Contractile protein that makes up thick filament; molecule consists of a tail and two myosin heads, which bind to myosin-binding sites on actin molecules of thin filament during muscle contraction.
Actin	Contractile protein that is the main component of thin filament; each actin molecule has a myosin-binding site where myosin head of thick filament binds during muscle contraction.
Regulatory proteins	Proteins that help switch muscle contraction process on and off.
Tropomyosin	Regulatory protein that is a component of thin filament; when skeletal muscle fiber is relaxed, tropomyosin covers myosin-binding sites on actin molecules, thereby preventing myosin from binding to actin.
Troponin	Regulatory protein that is a component of thin filament; when calcium ions (Ca^{2+}) bind to troponin, it changes shape; this conformational change moves tropomyosin away from myosin-binding sites on actin molecules, and muscle contraction subsequently begins as myosin binds to actin.
Structural proteins	Proteins that keep thick and thin filaments of myofibrils in proper alignment, give myofibrils elasticity and extensibility, and link myofibrils to sarcolemma and extracellular matrix.
Titin	Structural protein that connects Z disc to M line of sarcomere, thereby helping to stabilize thick filament position; can stretch and then spring back unharmed, and thus accounts for much of the elasticity and extensibility of myofibrils.
α-Actinin	Structural protein of Z discs that attaches to actin molecules of thin filaments and to titin molecules.
Myomesin	Structural protein that forms M line of sarcomere; binds to titin molecules and connects adjacent thick filaments to one another
Nebulin	Structural protein that wraps around entire length of each thin filament; helps anchor thin filaments to Z discs and regulates length of thin filaments during development.
Dystrophin	Structural protein that links thin filaments of sarcomere to integral membrane proteins in sarcolemma, which are attached in turn to proteins in connective tissue matrix that surrounds muscle fibers; is thought to help reinforce sarcolemma and help transmit tension generated by sarcomeres to tendons.

length of thin filaments during development. **Dystrophin** (dis-TRŌ-fin) links thin filaments of the sarcomere to integral membrane proteins of the sarcolemma, which are attached in turn to proteins in the connective tissue extracellular matrix that surrounds muscle fibers (see Figure 10.2d). Dystrophin and its associated proteins are thought to reinforce the sarcolemma and help transmit the tension generated by the sarcomeres to the tendons. The relationship of dystrophin to muscular dystrophy is discussed in Disorders: Homeostatic Imbalances at the end of the chapter.

Table 10.2 summarizes the different types of skeletal muscle fiber proteins, and Table 10.3 summarizes the levels of organization within a skeletal muscle.

TABLE 10.3

Levels of Organization within a Skeletal Muscle

LEVEL	DESCRIPTION
Skeletal muscle	Organ made up of fascicles that contain muscle fibers (cells), blood vessels, and nerves; wrapped in epimysium.

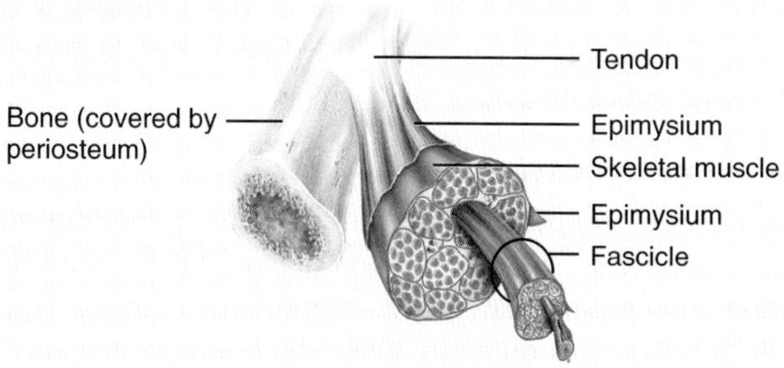

LEVEL	DESCRIPTION
Fascicle	Bundle of muscle fibers wrapped in perimysium.

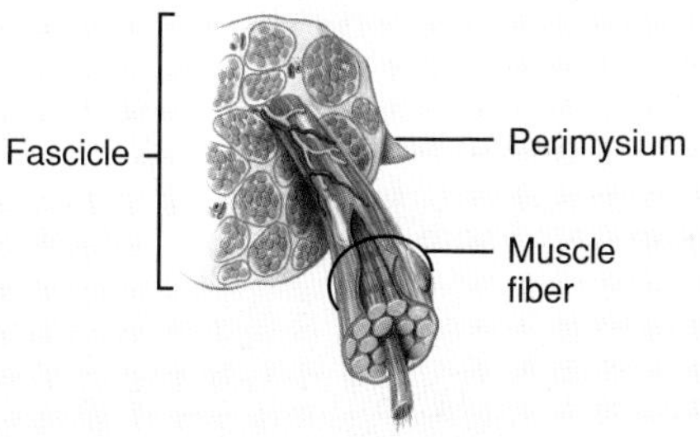

LEVEL	DESCRIPTION
Muscle fiber (cell)	Long cylindrical cell covered by endomysium and sarcolemma; contains sarcoplasm, myofibrils, many peripherally located nuclei, mitochondria, transverse tubules, sarcoplasmic reticulum, and terminal cisterns. The fiber has a striated appearance.

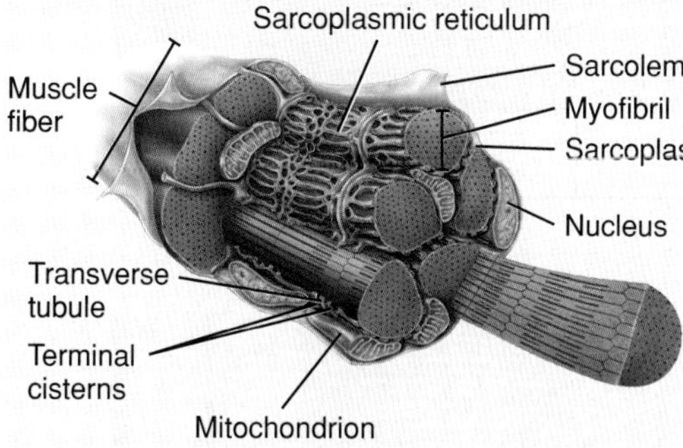

LEVEL	DESCRIPTION
Myofibril	Threadlike contractile elements within sarcoplasm of muscle fiber that extend entire length of fiber; composed of filaments.

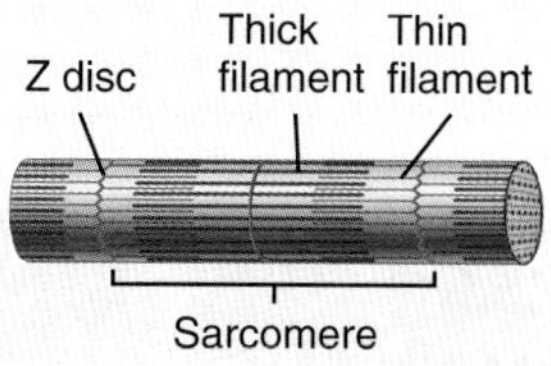

LEVEL	DESCRIPTION
Filaments (myofilaments)	Contractile proteins within myofibrils that are of two types: thick filaments composed of myosin and thin filaments composed of actin, tropomyosin, and troponin; sliding of thin filaments past thick filaments produces muscle shortening.

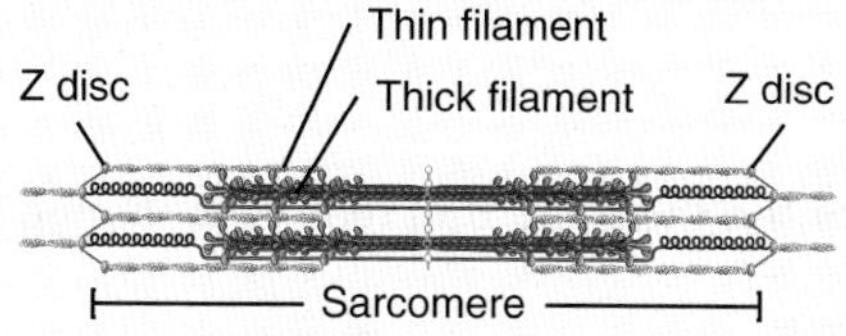

CHECKPOINT

4. What types of fascia cover skeletal muscles?
5. Why is a rich blood supply important for muscle contraction?
6. How are the structures of thin and thick filaments different?

10.3 CONTRACTION AND RELAXATION OF SKELETAL MUSCLE FIBERS

OBJECTIVES

- Outline the steps involved in the sliding filament mechanism of muscle contraction.
- Describe how muscle action potentials arise at the neuromuscular junction.

When scientists examined the first electron micrographs of skeletal muscle in the mid-1950s, they were surprised to see that the lengths of the thick and thin filaments were the same in both relaxed and contracted muscle. It had been thought that muscle contraction must be a folding process, somewhat like closing an accordion. Instead, researchers discovered that skeletal muscle shortens during contraction because the thick and thin filaments slide past one another. The model describing this process is known as the **sliding filament mechanism.**

The Sliding Filament Mechanism

Muscle contraction occurs because myosin heads attach to and "walk" along the thin filaments at both ends of a sarcomere, progressively pulling the thin filaments toward the M line (Figure 10.5). As a result, the thin filaments slide inward and meet at the center of a sarcomere. They may even move so far inward that their ends overlap (Figure 10.5c). As the thin filaments slide inward, the Z discs come closer together, and the sarcomere shortens. However, the lengths of the individual thick and thin filaments do not change. Shortening of the sarcomeres causes shortening of the whole muscle fiber, which in turn leads to shortening of the entire muscle.

The Contraction Cycle

At the onset of contraction, the sarcoplasmic reticulum releases calcium ions (Ca^{2+}) into the sarcoplasm. There, they bind to troponin. Troponin then moves tropomyosin away from the myosin-binding sites on actin. Once the binding sites are "free," the **contraction cycle**—the repeating sequence of events that

Figure 10.5 **Sliding filament mechanism of muscle contraction, as it occurs in two adjacent sarcomeres.**

During muscle contractions, thin filaments move toward the M line of each sarcomere.

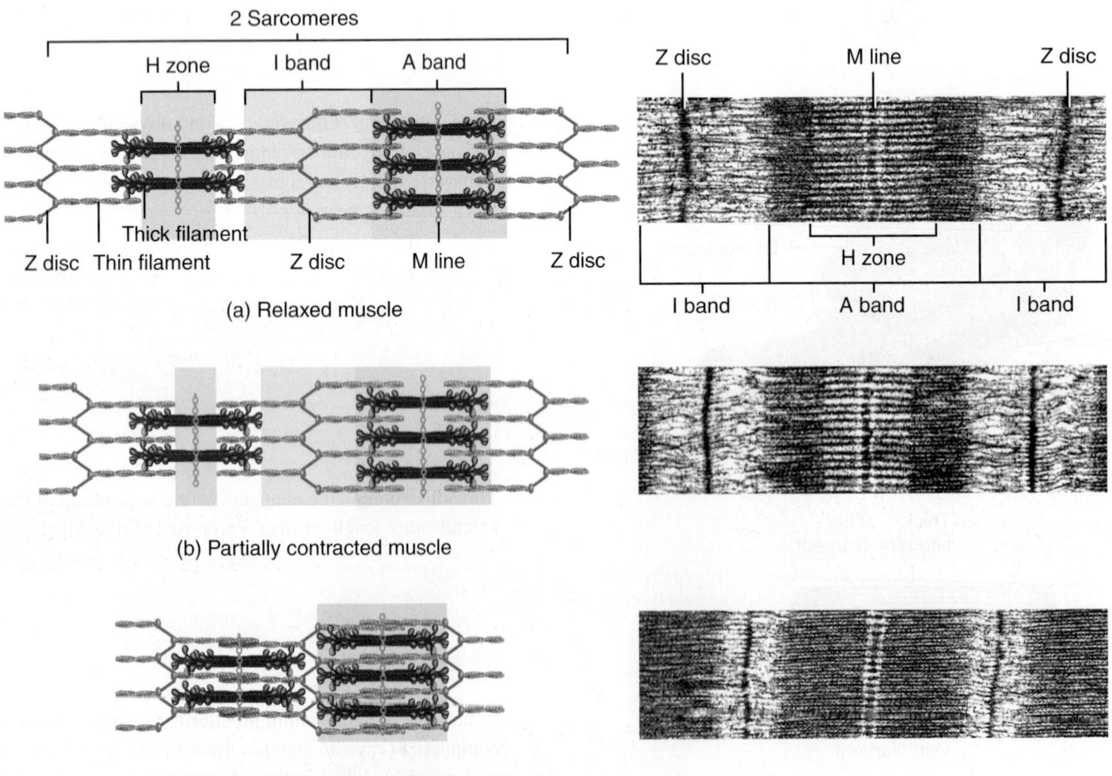

(a) Relaxed muscle

(b) Partially contracted muscle

(c) Maximally contracted muscle

What happens to the I band and H zone as muscle contracts? Do the lengths of the thick and thin filaments change?

causes the filaments to slide—begins. The contraction cycle consists of four steps (Figure 10.6):

1. ***ATP hydrolysis.*** The myosin head includes an ATP-binding site and an ATPase, an enzyme that hydrolyzes ATP into ADP (adenosine diphosphate) and a phosphate group. This hydrolysis reaction reorients and energizes the myosin head. Notice that the products of ATP hydrolysis—ADP and a phosphate group—are still attached to the myosin head.
2. ***Attachment of myosin to actin to form cross-bridges.*** The energized myosin head attaches to the myosin-binding site on actin and releases the previously hydrolyzed phosphate group. When the myosin heads attach to actin during contraction, they are referred to as **cross-bridges.**
3. ***Power stroke.*** After the cross-bridges form, the power stroke occurs. During the power stroke, the site on the cross-bridge where ADP is still bound opens. As a result, the cross-bridge rotates and releases the ADP. The cross-bridge generates force as it rotates toward the center of the sarcomere, sliding the thin filament past the thick filament toward the M line.
4. ***Detachment of myosin from actin.*** At the end of the power stroke, the cross-bridge remains firmly attached to actin until it binds another molecule of ATP. As ATP binds to the ATP-binding site on the myosin head, the myosin head detaches from actin.

The contraction cycle repeats as the myosin ATPase hydrolyzes the newly bound molecule of ATP, and continues as long as ATP is available and the Ca^{2+} level near the thin filament is sufficiently high. The cross-bridges keep rotating back and forth with each power stroke, pulling the thin filaments toward the M line. Each of the 600 cross-bridges in one thick filament attaches and detaches about five times per second. At any one instant, some of the myosin heads are attached to actin, forming cross-bridges and generating force, and other myosin heads are detached from actin, getting ready to bind again.

As the contraction cycle continues, movement of cross-bridges applies the force that draws the Z discs toward each other, and the sarcomere shortens. During a maximal muscle contraction, the distance between two Z discs can decrease to half the resting length. The Z discs in turn pull on neighboring sarcomeres, and the whole muscle fiber shortens. Some of the components of a muscle are elastic: They stretch slightly before they transfer the tension generated by the sliding filaments. The elastic components include titin molecules, connective tissue around the muscle fibers (endomysium, perimysium, and epimysium), and tendons

Figure 10.6 The contraction cycle. Sarcomeres exert force and shorten through repeated cycles during which the myosin heads attach to actin (cross-bridges), rotate, and detach.

During the power stroke of contraction, cross-bridges rotate and move the thin filaments past the thick filaments toward the center of the sarcomere.

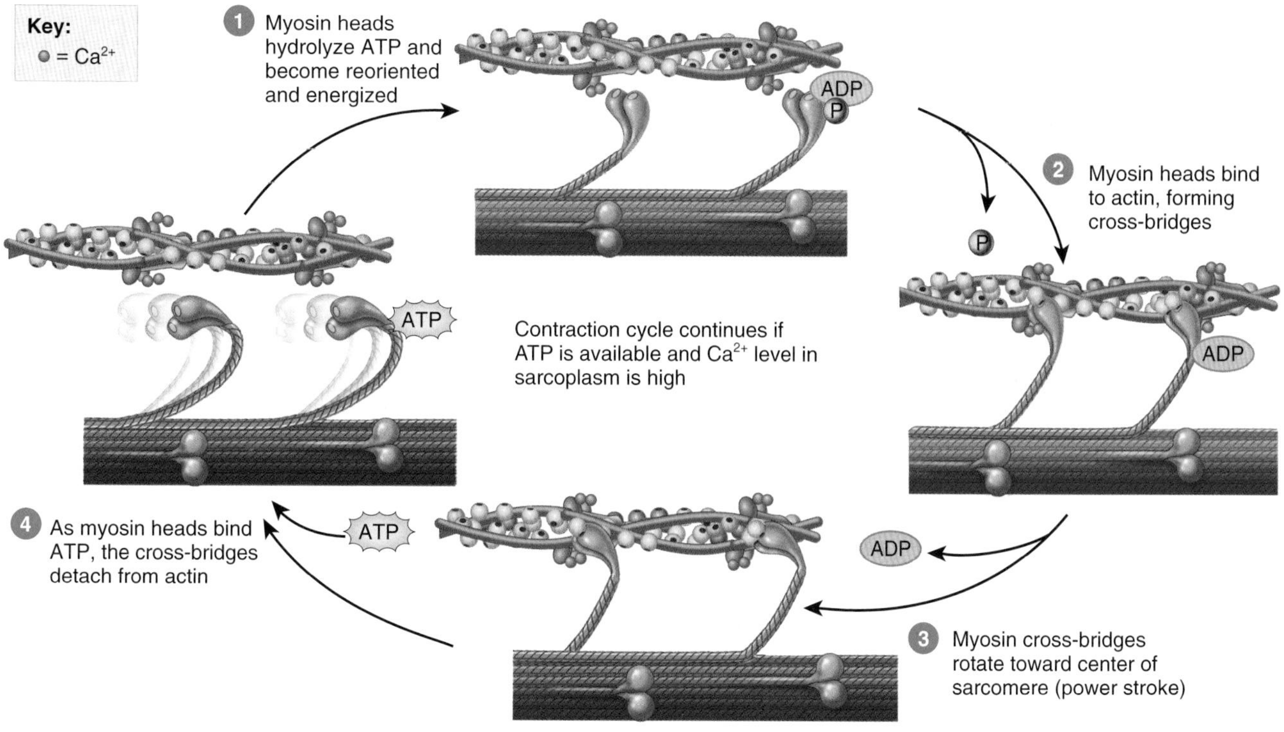
Key:
= Ca^{2+}
1 Myosin heads hydrolyze ATP and become reoriented and energized
ADP
P
2 Myosin heads bind to actin, forming cross-bridges
P
ATP
Contraction cycle continues if ATP is available and Ca^{2+} level in sarcoplasm is high
ADP
ATP
ADP
4 As myosin heads bind ATP, the cross-bridges detach from actin
3 Myosin cross-bridges rotate toward center of sarcomere (power stroke)

What would happen if ATP suddenly were not available after the sarcomere had started to shorten?

that attach muscle to bone. As the cells of a skeletal muscle start to shorten, they first pull on their connective tissue coverings and tendons. The coverings and tendons stretch and then become taut, and the tension passed through the tendons pulls on the bones to which they are attached. The result is movement of a part of the body. You will soon learn, however, that the contraction cycle does not always result in shortening of the muscle fibers and the whole muscle. In some contractions, the cross-bridges rotate and generate tension, but the thin filaments cannot slide inward because the tension they generate is not large enough to move the load on the muscle (such as trying to lift a whole box of books with one hand).

Excitation–Contraction Coupling

An increase in Ca^{2+} concentration in the sarcoplasm starts muscle contraction, and a decrease stops it. When a muscle fiber is relaxed, the concentration of Ca^{2+} in its sarcoplasm is very low, only about 0.1 micromole per liter (0.1 μmol/L). However, a huge amount of Ca^{2+} is stored inside the sarcoplasmic reticulum (Figure 10.7a). As a muscle action potential propagates along the sarcolemma and into the T tubules, it causes **Ca^{2+} release channels** in the SR membrane to open (Figure 10.7b). When these channels open, Ca^{2+} flows out of the SR into the sarcoplasm around the thick and thin filaments. As a result, the Ca^{2+} concentration in the sarcoplasm rises tenfold or more. The released calcium ions combine with troponin, causing it to change shape. This conformational change moves tropomyosin away from the myosin-binding sites on actin. Once these binding sites are free, myosin heads bind to them to form cross-bridges, and the contraction cycle begins. The events just described are referred to collectively as **excitation–contraction coupling,** as they are the steps that connect excitation (a muscle action potential propagating along the sarcolemma and into the T tubules) to contraction (sliding of the filaments).

The sarcoplasmic reticulum membrane also contains **Ca^{2+} active transport pumps** that use ATP to move Ca^{2+} constantly

Figure 10.7 The role of Ca^{2+} in the regulation of contraction by troponin and tropomyosin. (a) During relaxation, the level of Ca^{2+} in the sarcoplasm is low, only 0.1 μM (0.0001 mM), because calcium ions are pumped into the sarcoplasmic reticulum by Ca^{2+} active transport pumps. (b) A muscle action potential propagating along a transverse tubule opens Ca^{2+} release channels in the sarcoplasmic reticulum, calcium ions flow into the cytosol, and contraction begins.

An increase in the Ca^{2+} level in the sarcoplasm starts the sliding of thin filaments. When the level of Ca^{2+} in the sarcoplasm declines, sliding stops.

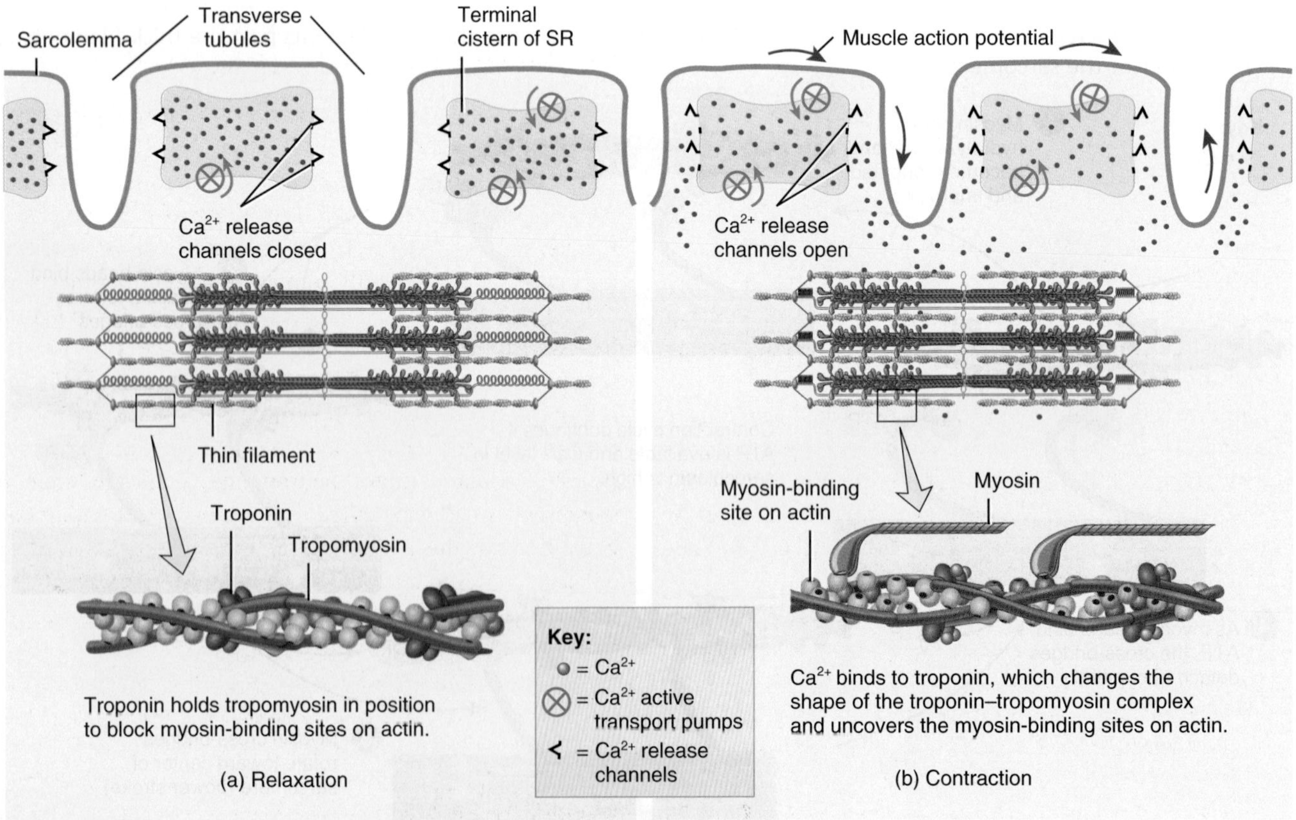
Sarcolemma
Transverse tubules
Terminal cistern of SR
Muscle action potential
Ca2+ release channels closed
Ca2+ release channels open
Thin filament
Troponin
Tropomyosin
Myosin-binding site on actin
Myosin
Troponin holds tropomyosin in position to block myosin-binding sites on actin.
(a) Relaxation
Key:
= Ca2+
= Ca2+ active transport pumps
= Ca2+ release channels
Ca2+ binds to troponin, which changes the shape of the troponin–tropomyosin complex and uncovers the myosin-binding sites on actin.
(b) Contraction

What are three functions of ATP in muscle contraction?

from the sarcoplasm into the SR (Figure 10.7). While muscle action potentials continue to propagate through the T tubules, the Ca^{2+} release channels are open. Calcium ions flow into the sarcoplasm more rapidly than they are transported back by the pumps. After the last action potential has propagated throughout the T tubules, the Ca^{2+} release channels close. As the pumps move Ca^{2+} back into the SR, the concentration of calcium ions in the sarcoplasm quickly decreases. Inside the SR, molecules of a calcium-binding protein, appropriately called **calsequestrin** (kal′-se-KWES-trin), bind to the Ca^{2+}, enabling even more Ca^{2+} to be sequestered or stored within the SR. As a result, the concentration of Ca^{2+} is 10,000 times higher in the SR than in the cytosol in a relaxed muscle fiber. As the Ca^{2+} level in the cytosol drops, tropomyosin covers the myosin-binding sites, and the muscle fiber relaxes.

CLINICAL CONNECTION | *Rigor Mortis*

After death, cellular membranes become leaky. Calcium ions leak out of the sarcoplasmic reticulum into the sarcoplasm and allow myosin heads to bind to actin. ATP synthesis ceases shortly after breathing stops, however, so the cross-bridges cannot detach from actin. The resulting condition, in which muscles are in a state of rigidity (cannot contract or stretch), is called **rigor mortis** (rigidity of death). Rigor mortis begins 3–4 hours after death and lasts about 24 hours; then it disappears as proteolytic enzymes from lysosomes digest the cross-bridges. •

Length–Tension Relationship

Figure 10.8 shows the **length–tension relationship** for skeletal muscle, which indicates how the forcefulness of muscle contraction depends on the length of the sarcomeres within a muscle *before contraction begins.* At a sarcomere length of about 2.0–2.4 μm (which is very close to the resting length in most muscles), the zone of overlap in each sarcomere is optimal, and the muscle fiber can develop maximum tension. Notice in Figure 10.8 that maximum tension (100%) occurs when the zone of overlap between a thick and thin filament extends from the edge of the H zone to one end of a thick filament.

As the sarcomeres of a muscle fiber are stretched to a longer length, the zone of overlap shortens, and fewer myosin heads can make contact with thin filaments. Therefore, the tension the fiber can produce decreases. When a skeletal muscle fiber is stretched to 170% of its optimal length, there is no overlap between the thick and thin filaments. Because none of the myosin heads can bind to thin filaments, the muscle fiber cannot contract, and tension is zero. As sarcomere lengths become increasingly shorter than the optimum, the tension that can develop again decreases. This is because thick filaments crumple as they are compressed by the Z discs, resulting in fewer myosin heads making contact with thin filaments. Normally, resting muscle fiber length is held very close to the optimum by firm attachments of skeletal muscle to bones (via their tendons) and to other inelastic tissues.

Figure 10.8 Length–tension relationship in a skeletal muscle fiber. Maximum tension during contraction occurs when the resting sarcomere length is 2.0–2.4 μm.

A muscle fiber develops its greatest tension when there is an optimal zone of overlap between thick and thin filaments.

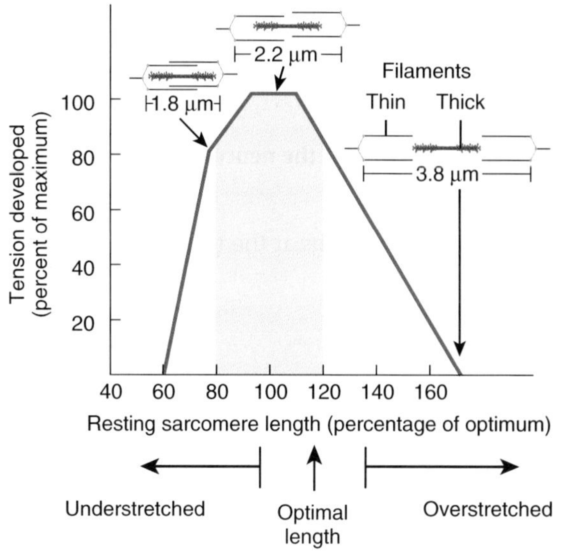

Why is tension maximal at a sarcomere length of 2.2 μm?

The Neuromuscular Junction

As noted earlier in the chapter, the neurons that stimulate skeletal muscle fibers to contract are called **somatic motor neurons.** Each somatic motor neuron has a threadlike axon that extends from the brain or spinal cord to a group of skeletal muscle fibers. A muscle fiber contracts in response to one or more action potentials propagating along its sarcolemma and through its system of T tubules. Muscle action potentials arise at the **neuromuscular junction (NMJ)** (noo-rō-MUS-kū-lar), the synapse between a somatic motor neuron and a skeletal muscle fiber (Figure 10.9a). A **synapse** is a region where communication occurs between two neurons, or between a neuron and a target cell—in this case, between a somatic motor neuron and a muscle fiber. At most synapses a small gap, called the **synaptic cleft,** separates the two cells. Because the cells do not physically touch, the action potential cannot "jump the gap" from one cell to another. Instead, the first cell communicates with the second by releasing a chemical called a **neurotransmitter.**

At the NMJ, the end of the motor neuron, called the **axon terminal,** divides into a cluster of **synaptic end bulbs** (Figure 10.9a, b), the neural part of the NMJ. Suspended in the cytosol within each synaptic end bulb are hundreds of membrane-enclosed sacs called **synaptic vesicles.** Inside each synaptic vesicle are thousands of molecules of **acetylcholine** (as′-ē-til-KŌ-lēn), abbreviated **ACh,** the neurotransmitter released at the NMJ.

The region of the sarcolemma opposite the synaptic end bulbs, called the **motor end plate** (Figure 10.9b, c), is the muscle fiber part of the NMJ. Within each motor end plate are 30 million to 40 million **acetylcholine receptors,** integral transmembrane proteins to which ACh specifically binds. These receptors are abundant in **junctional folds,** deep grooves in the motor end plate that provide a large surface area for ACh. As you will see, the ACh receptors are ligand-gated ion channels. A neuromuscular junction thus includes all the synaptic end bulbs on one side of the synaptic cleft, plus the motor end plate of the muscle fiber on the other side.

A nerve impulse (nerve action potential) elicits a muscle action potential in the following way (Figure 10.9c):

1. ***Release of acetylcholine.*** Arrival of the nerve impulse at the synaptic end bulbs stimulates voltage-gated channels to open. Because calcium ions are more concentrated in the extracellular fluid, Ca^{2+} flows inward through the open

Figure 10.9 Structure of the neuromuscular junction (NMJ), the synapse between a somatic motor neuron and a skeletal muscle fiber.

Synaptic end bulbs at the tips of axon terminals contain synaptic vesicles filled with acetylcholine (ACh).

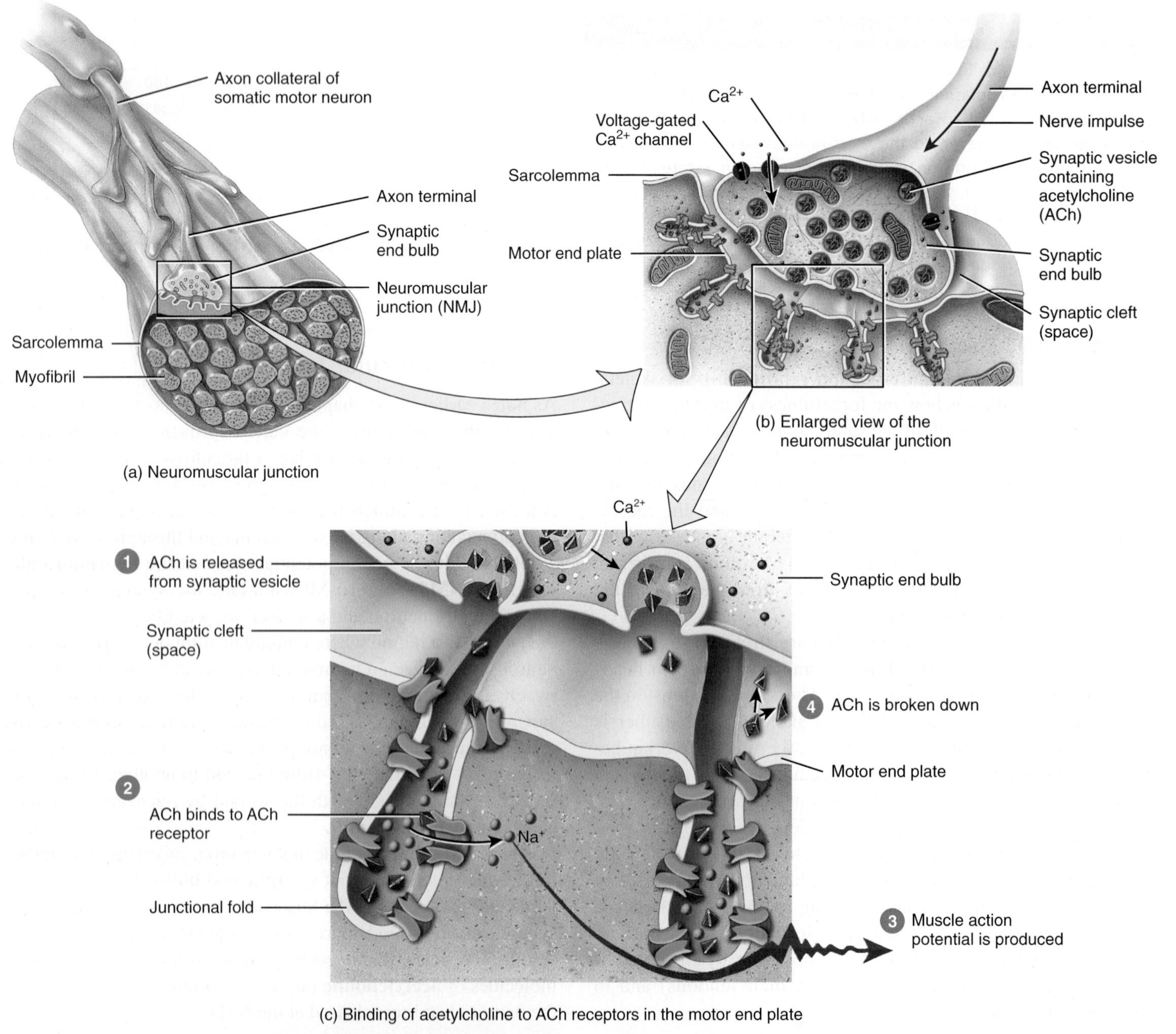

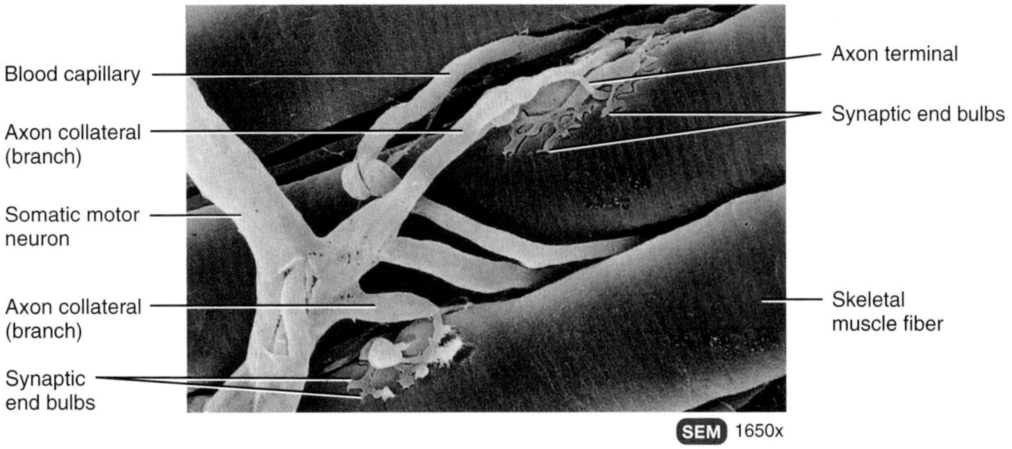
Blood capillary
Axon collateral (branch)
Somatic motor neuron
Axon collateral (branch)
Synaptic end bulbs
Axon terminal
Synaptic end bulbs
Skeletal muscle fiber
SEM 1650x

(d) Neuromuscular junction

What part of the sarcolemma contains acetylcholine receptors?

channels. The entering Ca^{2+} in turn stimulates the synaptic vesicles to undergo exocytosis. During exocytosis, the synaptic vesicles fuse with the motor neuron's plasma membrane, liberating ACh into the synaptic cleft. The ACh then diffuses across the synaptic cleft between the motor neuron and the motor end plate.

2. ***Activation of ACh receptors.*** Binding of two molecules of ACh to the receptor on the motor end plate opens an ion channel in the ACh receptor. Once the channel is open, small cations, most importantly Na^+, can flow across the membrane.
3. ***Production of muscle action potential.*** The inflow of Na^+ (down its electrochemical gradient) makes the inside of the muscle fiber more positively charged. This change in the membrane potential triggers a muscle action potential. Each nerve impulse normally elicits one muscle action potential. The muscle action potential then propagates along the sarcolemma into the system of T tubules. This causes the sarcoplasmic reticulum to release its stored Ca^{2+} into the sarcoplasm and the muscle fiber subsequently contracts.
4. ***Termination of ACh activity.*** The effect of ACh binding lasts only briefly because ACh is rapidly broken down by an enzyme called **acetylcholinesterase,** or **AChE** (as′-ē-til-kō′-lin-ES-ter-ās). This enzyme is attached to collagen fibers in the extracellular matrix of the synaptic cleft. AChE breaks down ACh into acetyl and choline, products that cannot activate the ACh receptor.

If another nerve impulse releases more acetylcholine, steps 2 and 3 repeat. When action potentials in the motor neuron cease, ACh is no longer released, and AChE rapidly breaks down the ACh already present in the synaptic cleft. This ends the production of muscle action potentials, the Ca^{2+} moves from the sarcoplasm of the muscle fiber back into the sarcoplasmic reticulum, and the Ca^{2+} release channels in the sarcoplasmic reticulum membrane close.

The NMJ usually is near the midpoint of a skeletal muscle fiber. Muscle action potentials that arise at the NMJ propagate toward both ends of the fiber. This arrangement permits nearly simultaneous activation (and thus contraction) of all parts of the muscle fiber. Figure 10.10 summarizes the events that occur during contraction and relaxation of a skeletal muscle fiber.

Several plant products and drugs selectively block certain events at the NMJ. *Botulinum toxin* (bot-u-LĪN-um), produced by the bacterium *Clostridium botulinum,* blocks exocytosis of synaptic vesicles at the NMJ. As a result, ACh is not released, and muscle contraction does not occur. The bacteria proliferate in improperly canned foods, and their toxin is one of the most lethal chemicals known. A tiny amount can cause death by paralyzing skeletal muscles. Breathing stops due to paralysis of respiratory muscles, including the diaphragm. Yet it is also the first bacterial toxin to be used as a medicine (Botox®). Injections of Botox into the affected muscles can help patients who have strabismus (crossed eyes), blepharospasm (uncontrollable blinking), or spasms of the vocal cords that interfere with speech. It is also used to alleviate chronic back pain due to muscle spasms in the lumbar region and as a cosmetic treatment to relax muscles that cause facial wrinkles.

The plant derivative *curare,* a poison used by South American Indians on arrows and blowgun darts, causes muscle paralysis by binding to and blocking ACh receptors. In the presence of curare, the ion channels do not open. Curare-like drugs are often used during surgery to relax skeletal muscles.

A family of chemicals called *anticholinesterase agents* have the property of slowing the enzymatic activity of acetylcholinesterase, thus slowing removal of ACh from the synaptic cleft. At low doses, these agents can strengthen weak muscle contractions. One example is neostigmine, which is used to treat patients with myasthenia gravis (see Disorders: Homeostatic Imbalances at the end of this chapter). Neostigmine is also used as an antidote for curare poisoning and to terminate the effects of curare-like drugs after surgery.

Figure 10.10 Summary of the events of contraction and relaxation in a skeletal muscle fiber.

Acetylcholine released at the neuromuscular junction triggers a muscle action potential, which leads to muscle contraction.

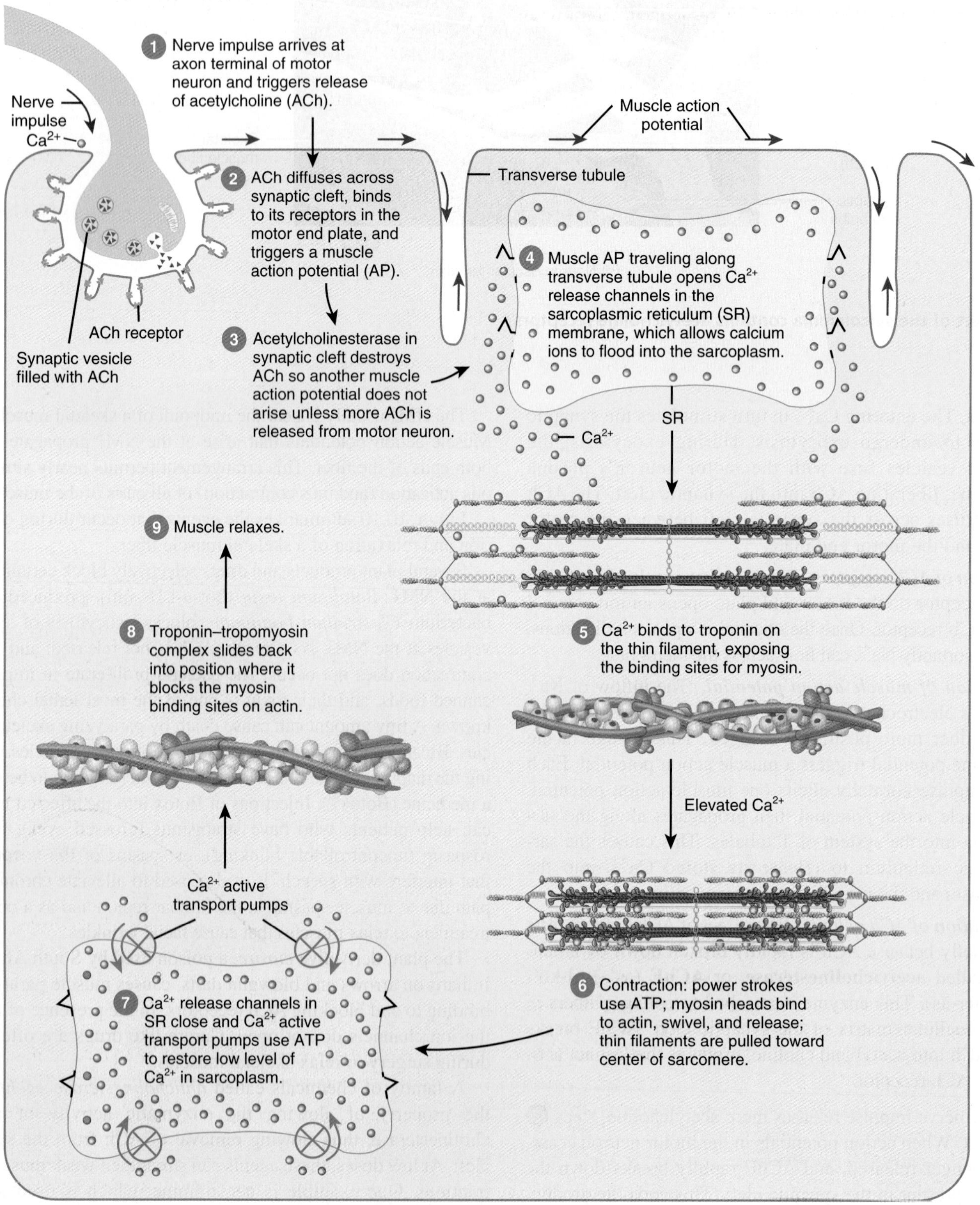

1 Nerve impulse arrives at axon terminal of motor neuron and triggers release of acetylcholine (ACh).
Nerve impulse
Ca2+
Muscle action potential
2 ACh diffuses across synaptic cleft, binds to its receptors in the motor end plate, and triggers a muscle action potential (AP).
Transverse tubule
4 Muscle AP traveling along transverse tubule opens Ca2+ release channels in the sarcoplasmic reticulum (SR) membrane, which allows calcium ions to flood into the sarcoplasm.
ACh receptor
Synaptic vesicle filled with ACh
3 Acetylcholinesterase in synaptic cleft destroys ACh so another muscle action potential does not arise unless more ACh is released from motor neuron.
SR
Ca2+
9 Muscle relaxes.
8 Troponin–tropomyosin complex slides back into position where it blocks the myosin binding sites on actin.
5 Ca2+ binds to troponin on the thin filament, exposing the binding sites for myosin.
Elevated Ca2+
Ca2+ active transport pumps
7 Ca2+ release channels in SR close and Ca2+ active transport pumps use ATP to restore low level of Ca2+ in sarcoplasm.
6 Contraction: power strokes use ATP; myosin heads bind to actin, swivel, and release; thin filaments are pulled toward center of sarcomere.

Which numbered steps in this figure are part of excitation–contraction coupling?

CLINICAL CONNECTION | *Electromyography*

Electromyography (e-lek′-trō-mī-OG-ra-fē; *electro-* = electricity; *-myo-* = muscle; *-graph* = to write) or **EMG** is a test that measures the electrical activity (muscle action potentials) in resting and contracting muscles. Normally, resting muscle produces no electrical activity; a slight contraction produces some electrical activity; and a more forceful contraction produces increased electrical activity. In the procedure, a ground electrode is placed over the muscle to be tested to eliminate background electrical activity. Then, a fine needle attached by wires to a recording instrument is inserted into the muscle. The electrical activity of the muscle is displayed as waves on an oscilloscope and heard through a loudspeaker.

EMG helps to determine if muscle weakness or paralysis is due to a malfunction of the muscle itself or the nerves supplying the muscle. EMG is also used to diagnose certain muscle disorders, such as muscular dystrophy, and to understand which muscles function during complex movements. •

✔CHECKPOINT

7. What roles do contractile, regulatory, and structural proteins play in muscle contraction and relaxation?
8. How do calcium ions and ATP contribute to muscle contraction and relaxation?
9. How does sarcomere length influence the maximum tension that is possible during muscle contraction?
10. How is the motor end plate different from other parts of the sarcolemma?

10.4 MUSCLE METABOLISM

OBJECTIVES

- Describe the reactions by which muscle fibers produce ATP.
- Distinguish between anaerobic and aerobic cellular respiration.
- Describe the factors that contribute to muscle fatigue.

Production of ATP in Muscle Fibers

Unlike most cells of the body, skeletal muscle fibers often switch between a low level of activity, when they are relaxed and using only a modest amount of ATP, and a high level of activity, when they are contracting and using ATP at a rapid pace. A huge amount of ATP is needed to power the contraction cycle, to pump Ca^{2+} into the sarcoplasmic reticulum, and for other metabolic reactions involved in muscle contraction. However, the ATP present inside muscle fibers is enough to power contraction for only a few seconds. If muscle contractions continue past that time, the muscle fibers must make more ATP. Muscle fibers have three ways to produce ATP: (1) from creatine phosphate, (2) by anaerobic cellular respiration, and (3) by aerobic cellular respiration (Figure 10.11). The use of creatine phosphate for ATP production is unique to muscle fibers, but all body cells make ATP by the reactions of anaerobic and aerobic cellular respiration. We consider the events of cellular respiration briefly here and then in detail in Chapter 25.

Creatine Phosphate

While muscle fibers are relaxed, they produce more ATP than they need for resting metabolism. Most of the excess ATP is used to synthesize **creatine phosphate** (KRĒ-a-tēn), an energy-rich molecule that is found in muscle fibers (Figure 10.11a). The enzyme creatine kinase (CK) catalyzes the transfer of one of the high-energy phosphate groups from ATP to creatine, forming creatine phosphate and ADP. **Creatine** is a small, amino acid–like molecule that is synthesized in the liver, kidneys, and pancreas and then transported to muscle fibers. Creatine phosphate is three to six times more plentiful than ATP in the sarcoplasm of a relaxed muscle fiber. When contraction begins and the ADP level starts to rise, CK catalyzes the transfer of a high-energy phosphate group from creatine phosphate back to ADP. This direct phosphorylation reaction quickly generates new ATP molecules. Since the formation of ATP from creatine phosphate occurs very rapidly, creatine phosphate is the first source of energy when muscle contraction begins. The other energy-generating mechanisms in a muscle fiber (the pathways of anaerobic and aerobic cellular respiration) take a relatively longer period of time to produce ATP compared to creatine phosphate. Together, stores of creatine phosphate and ATP provide enough energy for muscles to contract maximally for about 15 seconds.

CLINICAL CONNECTION | *Creatine Supplementation*

Creatine is both synthesized in the body and derived from foods such as milk, red meat, and some fish. Adults need to synthesize and ingest a total of about 2 grams of creatine daily to make up for the urinary loss of creatinine, the breakdown product of creatine. Some studies have demonstrated improved performance from **creatine supplementation** during explosive movements, such as sprinting. Other studies, however, have failed to find a performance-enhancing effect of creatine supplementation. Moreover, ingesting extra creatine decreases the body's own synthesis of creatine, and it is not known whether natural synthesis recovers after long-term creatine supplementation. In addition, creatine supplementation can cause dehydration and may cause kidney dysfunction. Further research is needed to determine both the long-term safety and the value of creatine supplementation. •

Anaerobic Cellular Respiration

Anaerobic cellular respiration (an′-ar-Ō-bik) is a series of ATP-producing reactions that do not require oxygen. When muscle activity continues and the supply of creatine phosphate within the muscle fibers is depleted, glucose is catabolized to generate ATP. Glucose easily passes from the blood into contracting muscle fibers via facilitated diffusion, and it is also produced by the breakdown of glycogen within muscle fibers (Figure 10.11b). Then, a series of reactions known as *glycolysis* quickly breaks down each glucose molecule into two molecules of pyruvic acid. Glycolysis occurs in the cytosol and produces a net gain of two molecules of ATP.

Figure 10.11 Production of ATP for muscle contraction. (a) Creatine phosphate, formed from ATP while the muscle is relaxed, transfers a high-energy phosphate group to ADP, forming ATP, during muscle contraction. (b) Breakdown of muscle glycogen into glucose and production of pyruvic acid from glucose via glycolysis produce both ATP and lactic acid. Because no oxygen is needed, this is an anaerobic pathway. (c) Within mitochondria, pyruvic acid, fatty acids, and amino acids are used to produce ATP via aerobic cellular respiration, an oxygen-requiring set of reactions.

During a long-term event such as a marathon race, most ATP is produced aerobically.

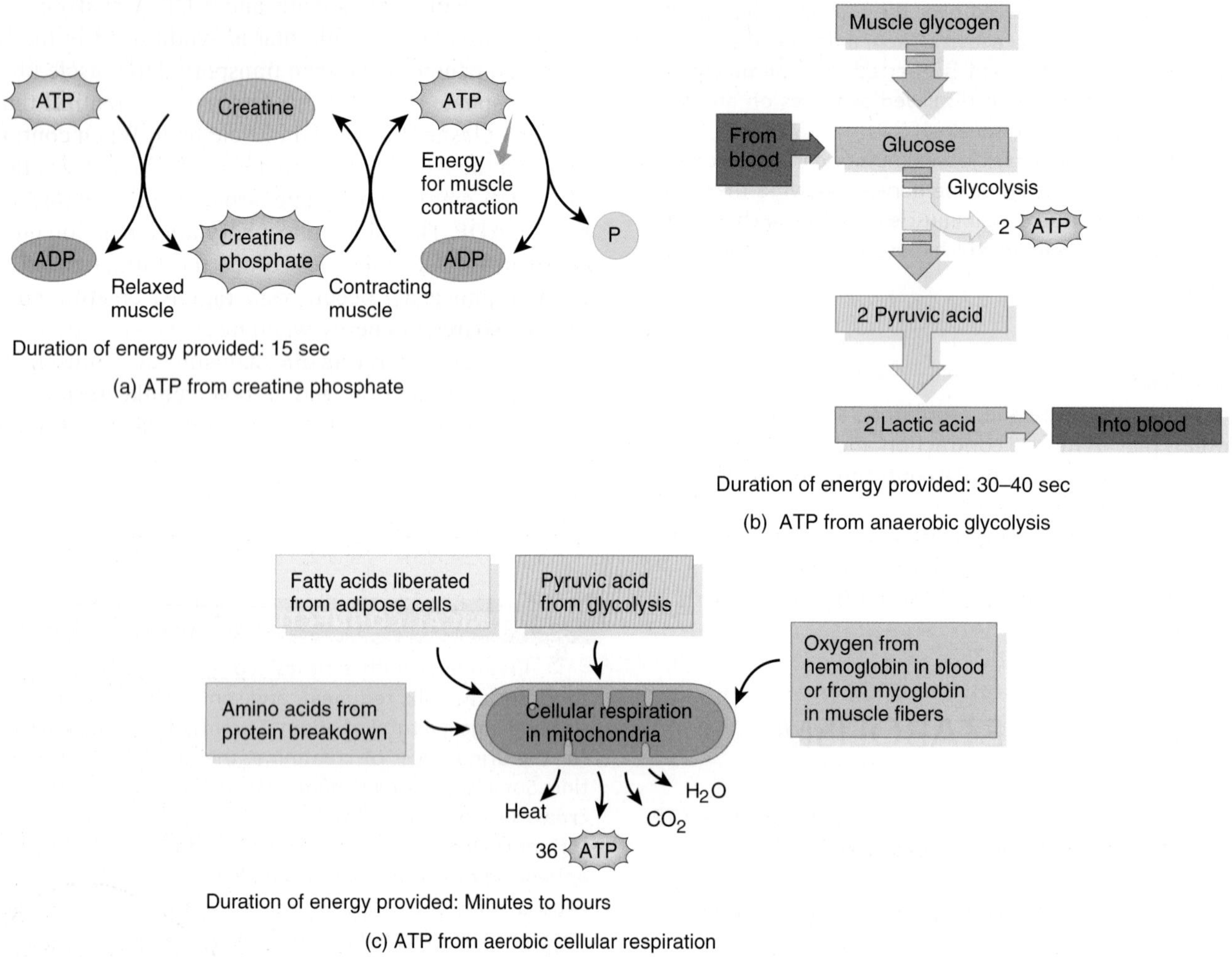

? Where inside a skeletal muscle fiber are the events shown here occurring?

Ordinarily, the pyruvic acid formed by glycolysis enters the mitochondria, where it undergoes a series of oxygen-requiring reactions called aerobic cellular respiration (described next) that produce a large amount of ATP. However, during periods of heavy exercise, not enough oxygen is available to skeletal muscle fibers. When this occurs, anaerobic reactions convert most of the pyruvic acid to lactic acid in the cytosol. About 80% of the lactic acid produced in this way diffuses out of the skeletal muscle fibers into the blood. Liver cells can convert some of the lactic acid back to glucose. In addition to providing new glucose molecules, this conversion reduces acidity of the blood. Anaerobic cellular respiration can provide enough energy for about 30 to 40 seconds of maximal muscle activity.

Aerobic Cellular Respiration

During periods of rest or light to moderate exercise, a sufficient amount of oxygen is available to skeletal muscle fibers. In such cases, ATP used for muscular activity is produced from a series of oxygen-requiring reactions called **aerobic cellular respiration** (ār-Ō-bik). During this process, pyruvic acid enters the mitochondria, where it is completely oxidized in reactions that generate ATP, carbon dioxide, water, and heat (Figure 10.11c). Although aerobic cellular respiration is slower than glycolysis, it yields much more ATP. One molecule of glucose yields about 36 molecules of ATP via aerobic cellular respiration, but yields only 2 molecules of ATP via anaerobic glycolysis.

Muscle tissue has two sources of oxygen: (1) oxygen that diffuses into muscle fibers from the blood and (2) oxygen released by myoglobin within muscle fibers. Both myoglobin (found only in muscle cells) and hemoglobin (found only in red blood cells) are oxygen-binding proteins. They bind oxygen when it is plentiful and release oxygen when it is scarce.

Aerobic cellular respiration supplies enough ATP for prolonged activity provided sufficient oxygen and nutrients are available. These nutrients include the pyruvic acid obtained from the glycolysis of glucose, fatty acids from the breakdown of triglycerides in adipose cells, and amino acids from the breakdown of proteins. In activities that last more than 10 minutes, the aerobic system provides more than 90% of the needed ATP. At the end of an endurance event such as a marathon race, nearly 100% of the ATP is being produced by aerobic cellular respiration.

Muscle Fatigue

The inability of a muscle to maintain force of contraction after prolonged activity is called **muscle fatigue** (fa-TĒG). Fatigue results mainly from changes within muscle fibers. Even before actual muscle fatigue occurs, a person may have feelings of tiredness and the desire to cease activity; this response, called *central fatigue,* is caused by changes in the central nervous system (brain and spinal cord). Although its exact mechanism is unknown, it may be a protective mechanism to stop a person from exercising before muscles become damaged. As you will see, certain types of skeletal muscle fibers fatigue more quickly than others.

Although the precise mechanisms that cause muscle fatigue are still not clear, several factors are thought to contribute. One is inadequate release of calcium ions from the SR, resulting in a decline of Ca^{2+} concentration in the sarcoplasm. Depletion of creatine phosphate also is associated with fatigue, but surprisingly, the ATP levels in fatigued muscle often are not much lower than those in resting muscle. Other factors that contribute to muscle fatigue include insufficient oxygen, depletion of glycogen and other nutrients, buildup of lactic acid and ADP, and failure of action potentials in the motor neuron to release enough acetylcholine.

Oxygen Consumption after Exercise

During prolonged periods of muscle contraction, increases in breathing rate and blood flow enhance oxygen delivery to muscle tissue. After muscle contraction has stopped, heavy breathing continues for a while, and oxygen consumption remains above the resting level. Depending on the intensity of the exercise, the recovery period may be just a few minutes, or it may last as long as several hours. The term **oxygen debt** refers to the added oxygen, over and above the resting oxygen consumption, that is taken into the body after exercise. This extra oxygen is used to "pay back" or restore metabolic conditions to the resting level in three ways: (1) to convert lactic acid back into glycogen stores in the liver, (2) to resynthesize creatine phosphate and ATP in muscle fibers, and (3) to replace the oxygen removed from myoglobin.

The metabolic changes that occur *during exercise* can account for only some of the extra oxygen used *after exercise.* Only a small amount of glycogen resynthesis occurs from lactic acid. Instead, most glycogen is made much later from dietary carbohydrates. Much of the lactic acid that remains after exercise is converted back to pyruvic acid and used for ATP production via aerobic cellular respiration in the heart, liver, kidneys, and skeletal muscle. Oxygen use after exercise also is boosted by ongoing changes. First, the elevated body temperature after strenuous exercise increases the rate of chemical reactions throughout the body. Faster reactions use ATP more rapidly, and more oxygen is needed to produce the ATP. Second, the heart and the muscles used in breathing are still working harder than they were at rest, and thus they consume more ATP. Third, tissue repair processes are occurring at an increased pace. For these reasons, **recovery oxygen uptake** is a better term than oxygen debt for the elevated use of oxygen after exercise.

✔ CHECKPOINT

11. Which ATP-producing reactions are aerobic and which are anaerobic?
12. Which sources provide ATP during a marathon race?
13. What factors contribute to muscle fatigue?
14. Why is the term recovery oxygen uptake more accurate than oxygen debt?

10.5 CONTROL OF MUSCLE TENSION

OBJECTIVES

- Describe the structure and function of a motor unit, and define motor unit recruitment.
- Explain the phases of a twitch contraction.
- Describe how frequency of stimulation affects muscle tension, and how muscle tone is produced.
- Distinguish between isotonic and isometric contractions.

A single nerve impulse in a somatic motor neuron elicits a single muscle action potential in all thc skeletal muscle fibers with which it forms synapses. Action potentials always have the same size in a given neuron or muscle fiber. In contrast, the force of muscle fiber contraction does vary; a muscle fiber is capable of producing a much greater force than the one that results from a single action potential. The total force or tension that a single muscle fiber can produce depends mainly on the rate at which nerve impulses arrive at the neuromuscular junction. The number of impulses per second is the *frequency of stimulation.* Maximum tension is also affected by the amount of stretch before contraction (see Figure 10.8) and by nutrient and oxygen availability. The total tension a whole muscle can produce depends on the number of muscle fibers that are contracting in unison.

Motor Units

Even though each skeletal muscle fiber has only a single neuromuscular junction, the axon of a somatic motor neuron branches out and forms neuromuscular junctions with many different muscle fibers. A **motor unit** consists of a somatic motor neuron plus all the skeletal muscle fibers it stimulates (Figure 10.12). A single somatic motor neuron makes contact with an average of 150 skeletal muscle fibers, and all of the muscle fibers in one motor unit contract in unison. Typically, the muscle fibers of a motor unit are dispersed throughout a muscle rather than clustered together.

Whole muscles that control precise movements consist of many small motor units. For instance, muscles of the larynx (voice box) that control voice production have as few as two or three muscle fibers per motor unit, and muscles controlling eye movements may have 10 to 20 muscle fibers per motor unit. In contrast, skeletal muscles responsible for large-scale and powerful movements, such as the biceps brachii muscle in the arm and the gastrocnemius muscle in the calf of the leg, have as many as 2000 to 3000 muscle fibers in some motor units. Because all the muscle fibers of a motor unit contract and relax together, the total strength of a contraction depends, in part, on the size of the motor units and the number that are activated at a given time.

Twitch Contraction

A **twitch contraction** is the brief contraction of all the muscle fibers in a motor unit in response to a single action potential in its motor neuron. In the laboratory, a twitch can be produced by direct electrical stimulation of a motor neuron or its muscle fibers. The record of a muscle contraction, called a **myogram** (MĪ-ō-gram), is shown in Figure 10.13. Twitches of skeletal muscle fibers last anywhere from 20 to 200 msec. This is very long compared to the brief 1–2 msec‡ that a muscle action potential lasts.

Note that a brief delay occurs between application of the stimulus (time zero on the graph) and the beginning of contraction. The delay, which lasts about 2 msec, is termed the **latent period.** During the latent period, the muscle action potential sweeps over the sarcolemma and calcium ions are released from the sarcoplasmic reticulum. The second phase, the **contraction period,** lasts 10–100 msec. During this time, Ca^{2+} binds to troponin, myosin-binding sites on actin are exposed, and cross-bridges form. Peak tension develops in the muscle fiber. During the third phase, the **relaxation period,** also lasting 10–100 msec, Ca^{2+} is actively transported back into the sarcoplasmic reticulum, myosin-binding sites are covered by tropomyosin, myosin heads detach from actin, and tension in the muscle fiber decreases. The actual duration of these periods depends on the type of skeletal muscle fiber. Some fibers, such as the fast-twitch fibers that move the eyes (described shortly), have contraction periods as brief as 10 msec and equally brief relaxation periods. Others, such as the slow-twitch fibers that move the legs, have contraction and relaxation periods of about 100 msec each.

If two stimuli are applied, one immediately after the other, the muscle will respond to the first stimulus but not to the second. When a muscle fiber receives enough stimulation to contract, it temporarily loses its excitability and cannot respond for a time. The period of lost excitability, called the **refractory period** (rē-FRAK-tō-rē), is a characteristic of all muscle and nerve cells. The duration of the refractory period varies with the muscle involved. Skeletal muscle has a short refractory period of about 5 msec; cardiac muscle has a longer refractory period of about 300 msec.

Frequency of Stimulation

When a second stimulus occurs after the refractory period of the first stimulus is over, but before the skeletal muscle fiber has relaxed, the second contraction will actually be stronger than the first (Figure 10.14b). This phenomenon, in which stimuli arriving at different times cause larger contractions, is

‡One millisecond (msec) is 10^{-3} second (0.001 sec).

Figure 10.12 **Motor units.** Two somatic motor neurons (one purple and one green) are shown, each supplying the muscle fibers of its motor unit.

A motor unit consists of a somatic motor neuron plus all the muscle fibers it stimulates.

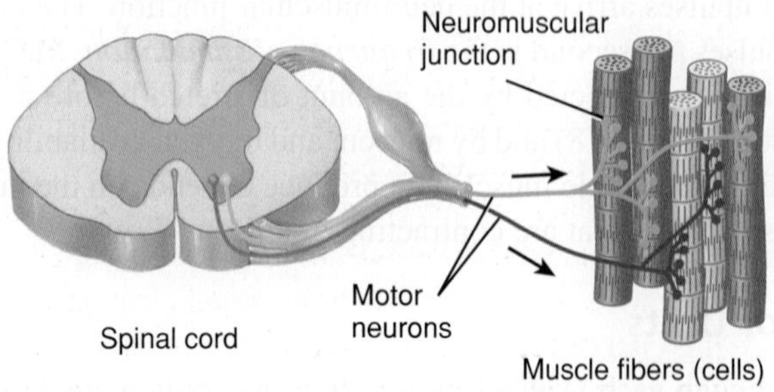

What is the effect of the size of a motor unit on its strength of contraction? (Assume that each muscle fiber can generate about the same amount of tension.)

Figure 10.13 **Myogram of a twitch contraction.** The arrow indicates the time at which the stimulus occurred.

A myogram is a record of a muscle contraction.

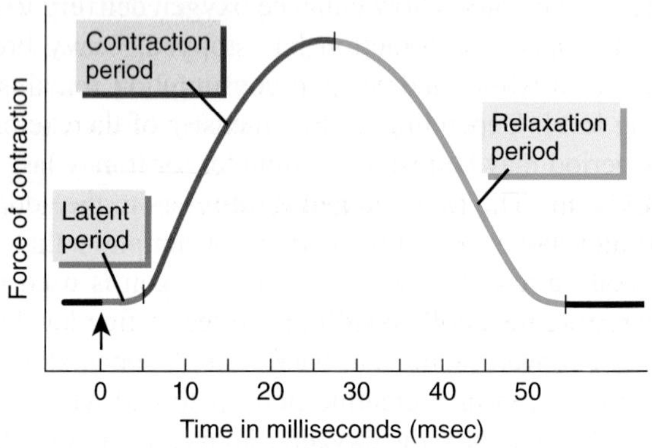

What events occur during the latent period?

Figure 10.14 Myograms showing the effects of different frequencies of stimulation. (a) Single twitch. (b) When a second stimulus occurs before the muscle fiber has relaxed, the second contraction is stronger than the first, a phenomenon called wave summation. (The solid black line indicates the force of contraction expected in a single twitch.) (c) Unfused tetanus produces a jagged curve due to partial relaxation of the muscle fiber between stimuli. (d) In fused tetanus, which occurs when there are 80–100 stimuli per second, the myogram line, like the contraction force, is steady and sustained.

Due to wave summation, the tension produced during a sustained contraction is greater than that produced by a single twitch.

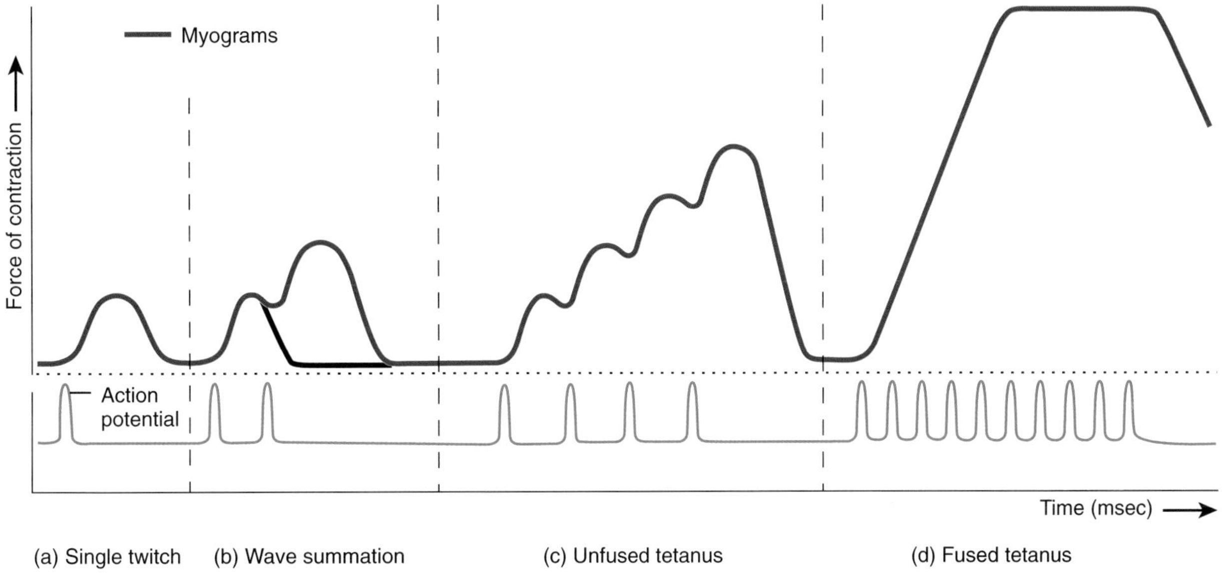

Would the peak force of the second contraction in part b be larger or smaller if the second stimulus were applied a few milliseconds later?

called **wave summation.** When a skeletal muscle fiber is stimulated at a rate of 20 to 30 times per second, it can only partially relax between stimuli. The result is a sustained but wavering contraction called **unfused (incomplete) tetanus** (*tetan-* = rigid, tense; Figure 10.14c). When a skeletal muscle fiber is stimulated at a higher rate of 80 to 100 times per second, it does not relax at all. The result is **fused (complete) tetanus,** a sustained contraction in which individual twitches cannot be detected (Figure 10.14d).

Wave summation and both kinds of tetanus occur when additional Ca^{2+} is released from the sarcoplasmic reticulum by subsequent stimuli while the levels of Ca^{2+} in the sarcoplasm are still elevated from the first stimulus. Because of the buildup in the Ca^{2+} level, the peak tension generated during fused tetanus is 5 to 10 times larger than the peak tension produced during a single twitch. Even so, smooth, sustained voluntary muscle contractions are achieved mainly by out-of-synchrony unfused tetanus in different motor units.

The stretch of elastic components, such as tendons and connective tissues around muscle fibers, also affects wave summation. During wave summation, elastic components are not given much time to spring back between contractions, and thus remain taut. While in this state, the elastic components do not require very much stretching before the beginning of the next muscular contraction. The combination of the tautness of the elastic components and the partially contracted state of the filaments enables the force of another contraction to be greater than the one before.

Motor Unit Recruitment

The process in which the number of active motor units increases is called **motor unit recruitment.** Typically, the different motor units of an entire muscle are not stimulated to contract in unison. While some motor units are contracting, others are relaxed. This pattern of motor unit activity delays muscle fatigue and allows contraction of a whole muscle to be sustained for long periods. The weakest motor units are recruited first, with progressively stronger motor units added if the task requires more force.

Recruitment is one factor responsible for producing smooth movements rather than a series of jerks. As mentioned, the number of muscle fibers innervated by one motor neuron varies greatly. Precise movements are brought about by small changes in muscle contraction. Therefore, the small muscles that produce precise movements are made up of small motor units. For this reason, when a motor unit is recruited or turned off, only slight changes occur in muscle tension. By contrast, large motor units are active when a large amount of tension is needed and precision is less important.

CLINICAL CONNECTION | *Aerobic Training versus Strength Training*

Regular, repeated activities such as jogging or aerobic dancing increase the supply of oxygen-rich blood available to skeletal muscles for aerobic cellular respiration. By contrast, activities such as weight lifting rely more on anaerobic production of ATP through glycolysis. Such anaerobic activities stimulate synthesis of muscle proteins and result, over time, in increased muscle size (muscle hypertrophy). Athletes who engage in anaerobic training should have a diet that includes an adequate amount of proteins. This protein intake will allow the body to synthesize muscle proteins and to increase muscle mass. As a result, aerobic training builds endurance for prolonged activities; in contrast, anaerobic training builds muscle strength for short-term feats. **Interval training** is a workout regimen that incorporates both types of training—for example, alternating sprints with jogging. •

Muscle Tone

Even at rest, a skeletal muscle exhibits **muscle tone** (*tonos* = tension), a small amount of tautness or tension in the muscle due to weak, involuntary contractions of its motor units. Recall that skeletal muscle contracts only after it is activated by acetylcholine released by nerve impulses in its motor neurons. Hence, muscle tone is established by neurons in the brain and spinal cord that excite the muscle's motor neurons. When the motor neurons serving a skeletal muscle are damaged or cut, the muscle becomes **flaccid** (FLAK-sid or FLAS-sid = flabby), a state of limpness in which muscle tone is lost. To sustain muscle tone, small groups of motor units are alternatively active and inactive in a constantly shifting pattern. Muscle tone keeps skeletal muscles firm, but it does not result in a force strong enough to produce movement. For example, when you are awake, the muscles in the back of the neck are in normal tonic contraction; they keep the head upright and prevent it from slumping forward on the chest. Muscle tone also is important in smooth muscle tissues, such as those found in the gastrointestinal tract, where the walls of the digestive organs maintain a steady pressure on their contents. The tone of smooth muscle fibers in the walls of blood vessels plays a crucial role in maintaining blood pressure.

CLINICAL CONNECTION | *Hypotonia and Hypertonia*

Hypotonia (hī′-pō-TŌ-nē-a; *hypo-* = below) refers to decreased or lost muscle tone. Such muscles are said to be flaccid. Flaccid muscles are loose and appear flattened rather than rounded. Certain disorders of the nervous system and disruptions in the balance of electrolytes (especially sodium, calcium, and, to a lesser extent, magnesium) may result in **flaccid paralysis** (pa-RAL-i-sis), which is characterized by loss of muscle tone, loss or reduction of tendon reflexes, and atrophy (wasting away) and degeneration of muscles.

Hypertonia (hī′-per-TŌ-nē-a; *hyper-* = above) refers to increased muscle tone and is expressed in two ways: spasticity or rigidity. **Spasticity** (spas-TIS-i-tē) is characterized by increased muscle tone (stiffness) associated with an increase in tendon reflexes and pathological reflexes (such as the Babinski sign, in which the great toe extends with or without fanning of the other toes in response to stroking the outer margin of the sole). Certain disorders of the nervous system and electrolyte disturbances such as those previously noted may result in **spastic paralysis,** partial paralysis in which the muscles exhibit spasticity. **Rigidity** refers to increased muscle tone in which reflexes are not affected, as occurs in tetanus. Tetanus is a disease caused by a bacterium, *Clostridium tetani,* that enters the body thorugh exposed wounds. It leads to muscle stiffness and spasms that can make breathing difficult and can become life threatening as a result. The bacteria produce a toxin that interferes with the nerves controlling the muscles. The first signs are typically spasms and stiffness in the muscles of the face and jaws. •

Isotonic and Isometric Contractions

Muscle contractions may be either isotonic or isometric. In an **isotonic contraction** (ī′-sō-TON-ik; *iso-* = equal; *-tonic* = tension), the *tension* (force of contraction) developed in the muscle remains almost constant while the muscle changes its length. Isotonic contractions are used for body movements and for moving objects. The two types of isotonic contractions are concentric and eccentric. In a **concentric isotonic contraction** (kon-SEN-trik), if the tension generated is great enough to overcome the resistance of the object to be moved, the muscle shortens and pulls on another structure, such as a tendon, to produce movement and to reduce the angle at a joint. Picking a book up off a table involves concentric isotonic contractions of the biceps brachii muscle in the arm (Figure 10.15a). By contrast, as you lower the book to place it back on the table, the previously shortened biceps lengthens in a controlled manner while it continues to contract. When the length of a muscle increases during a contraction, the contraction is an **eccentric isotonic contraction** (ek-SEN-trik) (Figure 10.15b). During an eccentric contraction, the tension exerted by the myosin cross-bridges resists movement of a load (the book, in this case) and slows the lengthening process. For reasons that are not well understood, repeated eccentric isotonic contractions (for example, walking downhill) produce more muscle damage and more delayed-onset muscle soreness than do concentric isotonic contractions.

In an **isometric contraction** (ī′-sō-MET-rik; *metro* = measure or length), the tension generated is not enough to exceed the resistance of the object to be moved, and the muscle does not change its length. An example would be holding a book steady using an outstretched arm (Figure 10.15c). These contractions are important for maintaining posture and for supporting objects in a fixed position. Although isometric contractions do not result in body movement, energy is still expended. The book pulls the arm downward, stretching the shoulder and arm muscles. The isometric contraction of the shoulder and arm muscles counteracts the stretch. Isometric contractions are important because they stabilize some joints as others are moved. Most activities include both isotonic and isometric contractions.

Figure 10.15 Comparison between isotonic (concentric and eccentric) and isometric contractions. (a and b) Isotonic contractions of the biceps brachii muscle in the arm. (c) Isometric contraction of shoulder and arm muscles.

6 In an isotonic contraction, tension remains constant as muscle length decreases or increases; in an isometric contraction, tension increases greatly without a change in muscle length.

(a) Concentric contraction while picking up a book

(b) Eccentric contraction while lowering a book

(c) Isometric contraction while holding a book steady

? What type of contraction occurs in your neck muscles while you are walking?

✔ **CHECKPOINT**

15. How are the sizes of motor units related to the degree of muscular control they allow?
16. What is motor unit recruitment?
17. Why is muscle tone important?
18. Define each of the following terms: concentric isotonic contraction, eccentric isotonic contraction, and isometric contraction.
19. Demonstrate an isotonic contraction. How does it feel? What do you think causes the physical discomfort you are experiencing?

10.6 TYPES OF SKELETAL MUSCLE FIBERS

OBJECTIVE

- Compare the structure and function of the three types of skeletal muscle fibers.

Skeletal muscle fibers are not all alike in composition and function. For example, muscle fibers vary in their content of myoglobin, the red-colored protein that binds oxygen in muscle fibers. Skeletal muscle fibers that have a high myoglobin content are termed *red muscle fibers* and appear darker (the dark meat in chicken legs and thighs); those that have a low content of myoglobin are called *white muscle fibers* and appear lighter (the white meat in chicken breasts). Red muscle fibers also contain more mitochondria and are supplied by more blood capillaries.

Skeletal muscle fibers also contract and relax at different speeds, and vary in which metabolic reactions they use to generate ATP and in how quickly they fatigue. For example, a fiber is categorized as either slow or fast depending on how rapidly the ATPase in its myosin heads hydrolyzes ATP. Based on all these structural and functional characteristics, skeletal muscle fibers are classified into three main types: (1) slow oxidative fibers, (2) fast oxidative–glycolytic fibers, and (3) fast glycolytic fibers.

Slow Oxidative Fibers

Slow oxidative (SO) fibers are smallest in diameter and thus are the least powerful type of muscle fibers. They appear dark red because they contain large amounts of myoglobin and many blood capillaries. Because they have many large mitochondria, SO fibers generate ATP mainly by aerobic cellular respiration, which is why they are called oxidative fibers. These fibers are said to be "slow" because the ATPase in the myosin heads hydrolyzes ATP relatively slowly and the contraction cycle proceeds at a slower pace than in "fast" fibers. As a result, SO fibers have a slow speed of contraction. Their twitch contractions last from 100 to 200 msec, and they take longer to reach peak tension. However, slow fibers are very resistant to fatigue and are capable of prolonged, sustained contractions for many hours. These slow-twitch, fatigue-resistant fibers are adapted for maintaining posture and for aerobic, endurance-type activities such as running a marathon.

Fast Oxidative–Glycolytic Fibers

Fast oxidative–glycolytic (FOG) fibers are intermediate in diameter between the other two types of fibers. Like slow oxidative fibers, they contain large amounts of myoglobin and many blood capillaries. Thus, they also have a dark red appearance. FOG fibers can generate considerable ATP by aerobic cellular respiration, which gives them a moderately high resistance to fatigue. Because their intracellular glycogen level is high, they also generate ATP by anaerobic glycolysis. FOG fibers are "fast" because the ATPase in their myosin heads hydrolyzes ATP three to five times faster than the myosin ATPase in SO fibers, which makes their speed of contraction faster. Thus, twitches of FOG fibers reach peak tension more quickly than those of SO fibers but are briefer in duration—less than 100 msec. FOG fibers contribute to activities such as walking and sprinting.

Fast Glycolytic Fibers

Fast glycolytic (FG) fibers are largest in diameter and contain the most myofibrils. Hence, they can generate the most powerful contractions. FG fibers have low myoglobin content, relatively few blood capillaries, and few mitochondria, and appear white in color. They contain large amounts of glycogen and generate ATP mainly by glycolysis. Due to their large size and their ability to hydrolyze ATP rapidly, FG fibers contract strongly and quickly. These fast-twitch fibers are adapted for intense anaerobic movements of short duration, such as weight lifting or throwing a ball, but they fatigue quickly. Strength training programs that engage a person in activities requiring great strength for short times increase the size, strength, and glycogen content of fast glycolytic fibers. The FG fibers of a weight lifter may be 50% larger than those of a sedentary person or an endurance athlete because of increased synthesis of muscle proteins. The overall result is muscle enlargement due to hypertrophy of the FG fibers.

Distribution and Recruitment of Different Types of Fibers

Most skeletal muscles are a mixture of all three types of skeletal muscle fibers; about half the fibers in a typical skeletal muscle are SO fibers. However, the proportions vary somewhat, depending on the action of the muscle, the person's training regimen, and genetic factors. For example, the continually active postural muscles of the neck, back, and legs have a high proportion of SO fibers. Muscles of the shoulders and arms, in contrast, are not constantly active but are used briefly now and then to produce large amounts of tension, such as in lifting and throwing. These muscles have a high proportion of FG fibers. Leg muscles, which not only support the body but are also used for walking and running, have large numbers of both SO and FOG fibers.

Within a particular motor unit, all of the skeletal muscle fibers are of the same type. The different motor units in a muscle are recruited in a specific order, depending on need. For example, if weak contractions suffice to perform a task, only SO motor units are activated. If more force is needed, the motor units of FOG fibers are also recruited. Finally, if maximal force is required, motor units of FG fibers are also called into action with the other two types. Activation of various motor units is controlled by the brain and spinal cord.

Table 10.4 summarizes the characteristics of the three types of skeletal muscle fibers.

✔ **CHECKPOINT**

20. Why are some skeletal muscle fibers classified as "fast" and others are said to be "slow"?
21. In what order are the various types of skeletal muscle fibers recruited when you sprint to make it to the bus stop?

10.7 EXERCISE AND SKELETAL MUSCLE TISSUE

OBJECTIVE

- Describe the effects of exercise on different types of skeletal muscle fibers.

The relative ratio of fast glycolytic (FG) and slow oxidative (SO) fibers in each muscle is genetically determined and helps account for individual differences in physical performance. For example, people with a higher proportion of FG fibers (see Table 10.4) often excel in activities that require periods of intense activity, such as weight lifting or sprinting. People with higher percentages of SO fibers are better at activities that require endurance, such as long-distance running.

Although the total number of skeletal muscle fibers usually does not increase with exercise, the characteristics of those present can change to some extent. Various types of exercises can induce changes in the fibers in a skeletal muscle. Endurance-type (aerobic) exercises, such as running or swimming, cause a gradual transformation of some FG fibers into fast oxidative–glycolytic (FOG) fibers. The transformed muscle fibers show slight increases in diameter, number of mitochondria, blood supply, and strength. Endurance exercises also result in cardiovascular and respiratory changes that cause skeletal muscles to receive better supplies of oxygen and nutrients but do not increase muscle mass. By contrast, exercises that require great strength for short periods produce an increase in the size and strength of FG fibers. The increase in size is due to increased synthesis of thick and thin filaments. The overall result is muscle enlargement (hypertrophy), as evidenced by the bulging muscles of body builders.

A certain degree of elasticity is an important attribute of skeletal muscles and their connective tissue attachments. Greater elasticity contributes to a greater degree of flexibility, increasing the range of motion of a joint. When a relaxed muscle is physically stretched, its ability to lengthen is limited by connective tissue structures, such as fasciae. Regular stretching gradually lengthens these structures, but the process occurs very slowly. To see an improvement in flexibility, stretching exercises must be performed regularly—daily, if possible—for many weeks.

Effective Stretching

Stretching cold muscles does not increase flexibility and may cause injury. Tissues stretch best when slow, gentle force is

TABLE 10.4

Characteristics of the Three Types of Skeletal Muscle Fibers

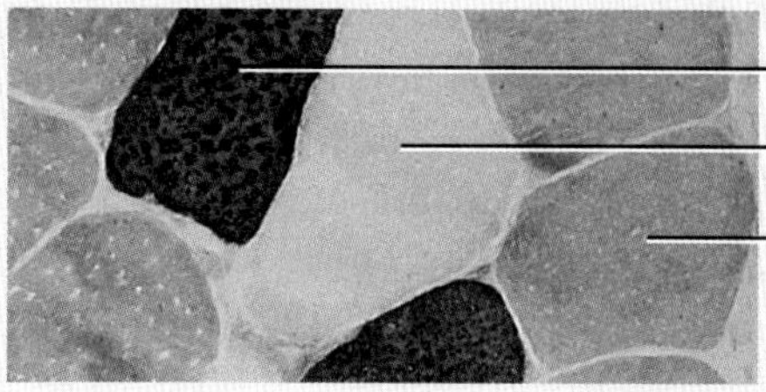

Transverse section of three types of skeletal muscle fibers

	SLOW OXIDATIVE (SO) FIBERS	FAST OXIDATIVE–GLYCOLYTIC (FOG) FIBERS	FAST GLYCOLYTIC (FG) FIBERS
STRUCTURAL CHARACTERISTIC			
Fiber diameter	Smallest.	Intermediate.	Largest.
Myoglobin content	Large amount.	Large amount.	Small amount.
Mitochondria	Many.	Many.	Few.
Capillaries	Many.	Many.	Few.
Color	Red.	Red-pink.	White (pale).
FUNCTIONAL CHARACTERISTIC			
Capacity for generating ATP and method used	High, by aerobic cellular respiration.	Intermediate, by both aerobic cellular respiration and anaerobic cellular respiration (glycolysis).	Low, by anaerobic cellular respiration (glycolysis).
Rate of ATP hydrolysis by myosin ATPase	Slow.	Fast.	Fast.
Contraction velocity	Slow.	Fast.	Fast.
Fatigue resistance	High.	Intermediate.	Low.
Creatine kinase	Lowest amount.	Intermediate amount.	Highest amount.
Glycogen stores	Low.	Intermediate.	High.
Order of recruitment	First.	Second.	Third.
Location where fibers are abundant	Postural muscles such as those of neck.	Lower limb muscles.	Upper limb muscles.
Primary functions of fibers	Maintaining posture and aerobic endurance activities.	Walking, sprinting.	Rapid, intense movements of short duration.

applied at elevated tissue temperatures. An external source of heat, such as hot packs or ultrasound, may be used, but 10 or more minutes of muscular contraction is also a good way to raise muscle temperature. Exercise heats muscle more deeply and thoroughly than external measures. That's where the term "warm-up" comes from. Many people stretch before they engage in exercise, but it's important to warm up (for example, walking, jogging, easy swimming, or easy aerobics) *before* stretching to avoid injury.

Strength Training

Strength training refers to the process of exercising with progressively heavier resistance for the purpose of strengthening the musculoskeletal system. This activity results not only in stronger muscles, but in many other health benefits as well. Strength training also helps to increase bone strength by increasing the deposition of bone minerals in young adults and helping to prevent, or at least slow, their loss in later life. By increasing muscle mass, strength training raises resting metabolic rate, the amount of energy expended at rest, so a person can eat more food without gaining weight. Strength training helps to prevent back injury and other injuries from participation in sports and other physical activities. Psychological benefits include reductions in feelings of stress and fatigue. As repeated training builds exercise tolerance, it takes increasingly longer before lactic acid is produced in the muscle, resulting in a reduced probability of muscle spasms.

CLINICAL CONNECTION | *Anabolic Steroids*

The use of **anabolic** (to build up proteins) **steroids,** also called **"roids,"** by athletes has received widespread attention. These steroid hormones, similar to testosterone, are taken to increase muscle size by increasing the synthesis of proteins in muscle and thus increasing strength during athletic contests. However, the large doses needed to produce an effect have damaging, sometimes even devastating side effects, including liver cancer, kidney damage, increased risk of heart disease, stunted growth, wide mood swings, increased acne, and increased irritability and aggression. Additionally, females who take anabolic steroids may experience atrophy of the breasts and uterus, menstrual irregularities, sterility, facial hair growth, and deepening of the voice. Males may experience diminished testosterone secretion, atrophy of the testes, sterility, and baldness. •

✔ **CHECKPOINT**

22. On a cellular level, what causes muscle hypertrophy?

10.8 CARDIAC MUSCLE TISSUE

OBJECTIVE

- Describe the main structural and functional characteristics of cardiac muscle tissue.

The principal tissue in the heart wall is **cardiac muscle tissue** (described in more detail in Chapter 20 and illustrated in Figure 20.9). Between the layers of **cardiac muscle fibers,** the contractile cells of the heart, are sheets of connective tissue that contain blood vessels, nerves, and the conduction system of the heart. Cardiac muscle fibers have the same arrangement of actin and myosin and the same bands, zones, and Z discs as skeletal muscle fibers. However, *intercalated discs* (in-TER-ka-lāt-ed; *intercal-* = to insert between) are unique to cardiac muscle fibers. These microscopic structures are irregular transverse thickenings of the sarcolemma that connect the ends of cardiac muscle fibers to one another. The discs contain *desmosomes,* which hold the fibers together, and *gap junctions,* which allow muscle action potentials to spread from one cardiac muscle fiber to another (see Figure 4.2e). Cardiac muscle tissue has an endomysium and perimysium, but lacks an epimysium.

In response to a single action potential, cardiac muscle tissue remains contracted 10 to 15 times longer than skeletal muscle tissue (see Figure 20.11). The long contraction is due to prolonged delivery of Ca^{2+} into the sarcoplasm. In cardiac muscle fibers, Ca^{2+} enters the sarcoplasm both from the sarcoplasmic reticulum (as in skeletal muscle fibers) and from the interstitial fluid that bathes the fibers. Because the channels that allow inflow of Ca^{2+} from interstitial fluid stay open for a relatively long time, a cardiac muscle contraction lasts much longer than a skeletal muscle twitch.

We have seen that skeletal muscle tissue contracts only when stimulated by acetylcholine released by a nerve impulse in a motor neuron. In contrast, cardiac muscle tissue contracts when stimulated by its own autorhythmic muscle fibers. Under normal resting conditions, cardiac muscle tissue contracts and relaxes about 75 times a minute. This continuous, rhythmic activity is a major physiological difference between cardiac and skeletal muscle tissue. The mitochondria in cardiac muscle fibers are larger and more numerous than in skeletal muscle fibers. This structural feature correctly suggests that cardiac muscle depends largely on aerobic cellular respiration to generate ATP, and thus requires a constant supply of oxygen. Cardiac muscle fibers can also use lactic acid produced by skeletal muscle fibers to make ATP, a benefit during exercise. Like skeletal muscle, cardiac muscle fibers can undergo hypertrophy in response to an increased workload. This is called a *physiological enlarged heart* and it is why many athletes have enlarged hearts. By contrast, a *pathological enlarged heart* is related to significant heart disease.

✔ **CHECKPOINT**

23. What are the similarities among and differences between skeletal and cardiac muscle?

10.9 SMOOTH MUSCLE TISSUE

OBJECTIVE

- Describe the main structural and functional characteristics of smooth muscle tissue.

Like cardiac muscle tissue, **smooth muscle tissue** is usually activated involuntarily. Of the two types of smooth muscle tissue, the more common type is **visceral (single-unit) smooth muscle tissue** (Figure 10.16a). It is found in the skin and in tubular arrangements that form part of the walls of small arteries and veins and of hollow organs such as the stomach, intestines, uterus, and urinary bladder. Like cardiac muscle, visceral smooth muscle is autorhythmic. The fibers connect to one another by gap junctions, forming a network through which muscle action potentials can spread. When a neurotransmitter, hormone, or autorhythmic signal stimulates one fiber, the muscle action potential is transmitted to neighboring fibers, which then contract in unison, as a single unit.

The second type of smooth muscle tissue, **multiunit smooth muscle tissue** (Figure 10.16b), consists of individual fibers, each with its own motor neuron terminals and with few gap junctions between neighboring fibers. Stimulation of one visceral muscle fiber causes contraction of many adjacent fibers, but stimulation of one multiunit fiber causes contraction of that fiber only. Multiunit smooth muscle tissue is found in the walls of large arteries, in airways to the lungs, in the arrector pili muscles that attach to hair follicles, in the muscles of the iris that adjust pupil diameter, and in the ciliary body that adjusts focus of the lens in the eye.

Figure 10.16 Smooth muscle tissue. (a) One autonomic motor neuron synapses with several visceral smooth muscle fibers, and action potentials spread to neighboring fibers through gap junctions. (b) Three autonomic motor neurons synapse with individual multiunit smooth muscle fibers; stimulation of one multiunit fiber causes contraction of that fiber only. (c) Relaxed and contracted smooth muscle fiber. A photomicrograph of smooth muscle tissue is shown in Table 4.9.

Visceral smooth muscle fibers connect to one another by gap junctions and contract as a single unit. Multiunit smooth muscle fibers lack gap junctions and contract independently.

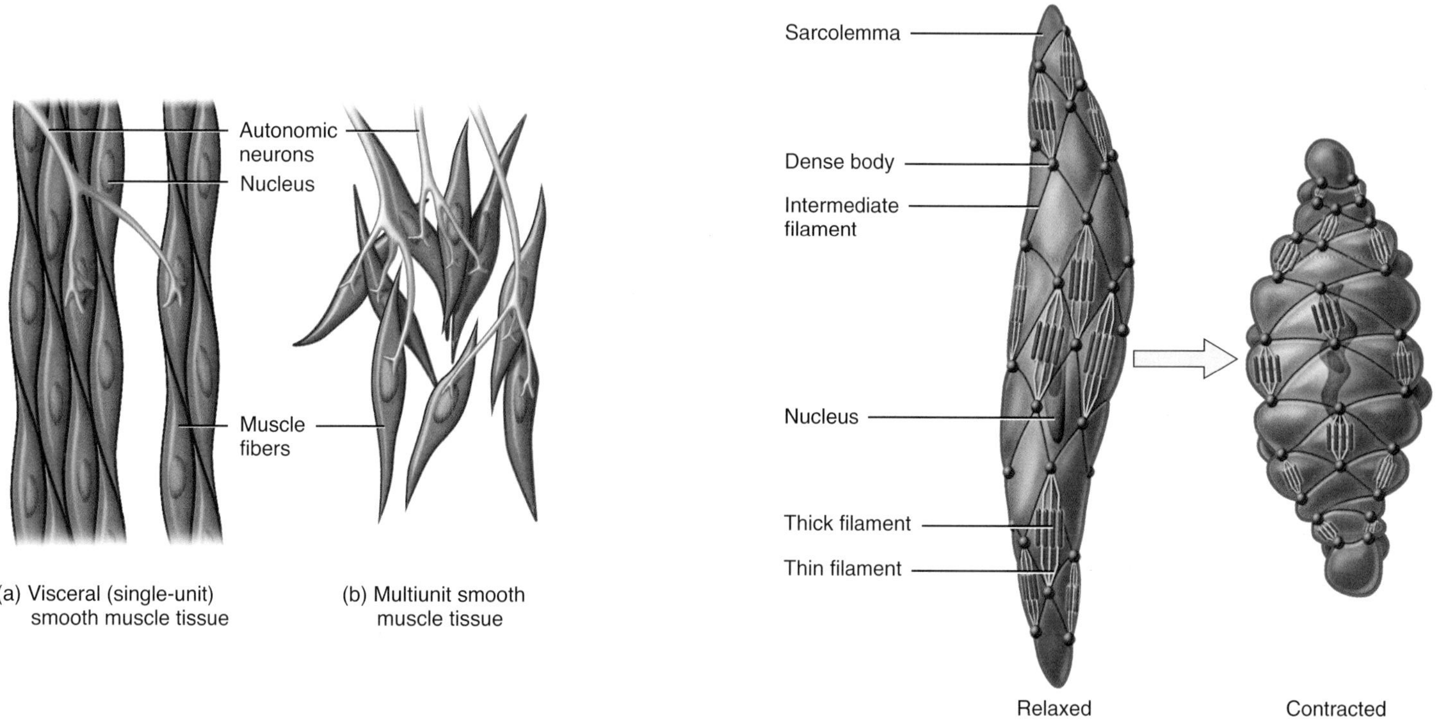

(a) Visceral (single-unit) smooth muscle tissue

(b) Multiunit smooth muscle tissue

(c) Microscopic anatomy of a relaxed and contracted smooth muscle fiber

? Which type of smooth muscle is more like cardiac muscle than skeletal muscle, with respect to both its structure and function?

Microscopic Anatomy of Smooth Muscle

A single relaxed smooth muscle fiber is 30–200 μm long. It is thickest in the middle (3–8 μm) and tapers at each end (Figure 10.16c). Within each fiber is a single, oval, centrally located nucleus. The sarcoplasm of smooth muscle fibers contains both *thick filaments* and *thin filaments,* in ratios between 1:10 and 1:15, but they are not arranged in orderly sarcomeres as in striated muscle. Smooth muscle fibers also contain *intermediate filaments.* Because the various filaments have no regular pattern of overlap, smooth muscle fibers do not exhibit striations (see Table 4.9), causing a smooth appearance. Smooth muscle fibers also lack transverse tubules and have only a small amount of sarcoplasmic reticulum for storage of Ca^{2+}. Although there are no transverse tubules in smooth muscle tissue, there are small pouchlike invaginations of the plasma membrane called **caveolae** (kav′-ē-Ō-lē; *cavus* = space) that contain extracellular Ca^{2+} that can be used for muscular contraction.

In smooth muscle fibers, the thin filaments attach to structures called **dense bodies,** which are functionally similar to Z discs in striated muscle fibers. Some dense bodies are dispersed throughout the sarcoplasm; others are attached to the sarcolemma. Bundles of intermediate filaments also attach to dense bodies and stretch from one dense body to another (Figure 10.16c). During contraction, the sliding filament mechanism involving thick and thin filaments generates tension that is transmitted to intermediate filaments. These in turn pull on the dense bodies attached to the sarcolemma, causing a lengthwise shortening of the muscle fiber. As a smooth muscle fiber contracts, it rotates as a corkscrew turns. The fiber twists in a helix as it contracts, and rotates in the opposite direction as it relaxes.

Physiology of Smooth Muscle

Although the principles of contraction are similar, smooth muscle tissue exhibits some important physiological differences from cardiac and skeletal muscle tissue. Contraction in a smooth muscle fiber starts more slowly and lasts much longer than skeletal muscle fiber contraction. Another difference is that smooth muscle can both shorten and stretch to a greater extent than the other muscle types.

An increase in the concentration of Ca^{2+} in the cytosol of a smooth muscle fiber initiates contraction, just as in striated muscle. Sarcoplasmic reticulum (the reservoir for Ca^{2+} in striated muscle) is found in small amounts in smooth muscle. Calcium ions flow into smooth muscle cytosol from both the interstitial fluid and sarcoplasmic reticulum. Because there are no transverse tubules in smooth muscle fibers (there are caveolae instead), it takes longer for Ca^{2+} to reach the filaments in the center of the fiber and trigger the contractile process. This accounts, in part, for the slow onset of contraction of smooth muscle.

Several mechanisms regulate contraction and relaxation of smooth muscle cells. In one such mechanism, a regulatory protein called **calmodulin** (cal-MOD-ū-lin) binds to Ca^{2+} in the cytosol. (Recall that troponin takes this role in striated muscle fibers.) After binding to Ca^{2+}, calmodulin activates an enzyme called *myosin light chain kinase.* This enzyme uses ATP to add a phosphate group to a portion of the myosin head. Once the phosphate group is attached, the myosin head can bind to actin, and contraction can occur. Because myosin light chain kinase works rather slowly, it contributes to the slowness of smooth muscle contraction.

Not only do calcium ions enter smooth muscle fibers slowly, they also move slowly out of the muscle fiber, which delays relaxation. The prolonged presence of Ca^{2+} in the cytosol provides for **smooth muscle tone,** a state of continued partial contraction. Smooth muscle tissue can thus sustain long-term tone, which is important in the gastrointestinal tract, where the walls maintain a steady pressure on the contents of the tract, and in the walls of blood vessels called *arterioles,* which maintain a steady pressure on blood.

Most smooth muscle fibers contract or relax in response to action potentials from the autonomic nervous system. In addition, many smooth muscle fibers contract or relax in response to stretching, hormones, or local factors such as changes in pH, oxygen and carbon dioxide levels, temperature, and ion concentrations. For example, the hormone epinephrine, released by the adrenal medulla, causes relaxation of smooth muscle in the airways and in some blood vessel walls (those that have so-called β_2 receptors; see Table 15.2).

Unlike striated muscle fibers, smooth muscle fibers can stretch considerably and still maintain their contractile function. When smooth muscle fibers are stretched, they initially contract, developing increased tension. Within a minute or so, the tension decreases. This phenomenon, which is called the **stress–relaxation response,** allows smooth muscle to undergo great changes in length while retaining the ability to contract effectively. Thus, even though smooth muscle in the walls of blood vessels and hollow organs such as the stomach, intestines, and urinary bladder can stretch, the pressure on the contents within them changes very little. After the organ empties, the smooth muscle in the wall rebounds, and the wall retains its firmness.

✔ **CHECKPOINT**

24. What are the differences between visceral and multiunit smooth muscle?
25. How are skeletal and smooth muscle similar? How do they differ?

10.10 REGENERATION OF MUSCULAR TISSUE

OBJECTIVE

- Explain how muscle fibers regenerate.

Because mature skeletal muscle fibers have lost the ability to undergo cell division, growth of skeletal muscle after birth is due mainly to **hypertrophy** (hī-PER-trō-fē), the enlargement of existing cells, rather than to **hyperplasia** (hī-per-PLĀ-zē-a), an increase in the number of fibers. Satellite cells divide slowly and fuse with existing fibers to assist both in muscle growth and in repair of damaged fibers. Thus, skeletal muscle tissue can regenerate only to a limited extent.

Until recently it was believed that damaged cardiac muscle fibers could not be replaced and that healing took place exclusively by fibrosis, the formation of scar tissue. New research described in Chapter 20 indicates that, under certain circumstances, cardiac muscle tissue can regenerate. In addition, cardiac muscle fibers can undergo hypertrophy in response to increased workload. Hence, many athletes have enlarged hearts.

Smooth muscle tissue, like skeletal and cardiac muscle tissue, can undergo hypertrophy. In addition, certain smooth muscle fibers, such as those in the uterus, retain their capacity for division and thus can grow by hyperplasia. Also, new smooth muscle fibers can arise from cells called *pericytes,* stem cells found in association with blood capillaries and small veins. Smooth muscle fibers can also proliferate in certain pathological conditions, such as occur in the development of atherosclerosis (see Disorders: Homeostatic Imbalances in Chapter 20). Compared with the other two types of muscle tissue, smooth muscle tissue has considerably greater powers of regeneration. Such powers are still limited when compared with other tissues, such as epithelium.

Table 10.5 summarizes the major characteristics of the three types of muscular tissue.

✔ **CHECKPOINT**

26. Which type of muscular tissue has the highest capacity for regeneration?

10.11 DEVELOPMENT OF MUSCLE

OBJECTIVE

- Describe the development of muscles.

Except for muscles such as those of the iris of the eyes and the arrector pili muscles attached to hairs, all muscles of the body are derived from **mesoderm.** As the mesoderm develops, part of it

TABLE 10.5

Summary of the Major Features of the Three Types of Muscular Tissue

CHARACTERISTIC	SKELETAL MUSCLE	CARDIAC MUSCLE	SMOOTH MUSCLE
Microscopic appearance and features	Long cylindrical fiber with many peripherally located nuclei; unbranched; striated. 	Branched cylindrical fiber with one centrally located nucleus; intercalated discs join neighboring fibers; striated. 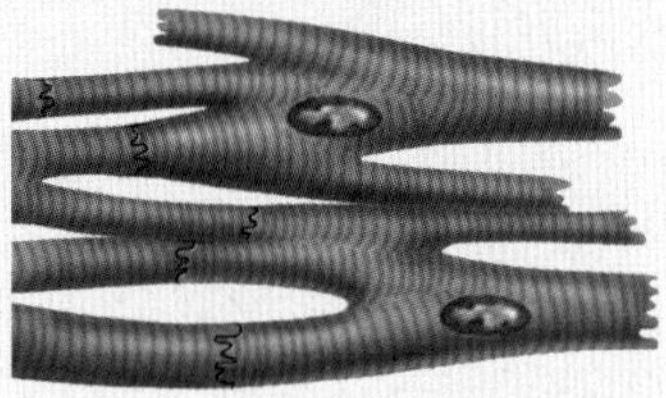	Fiber thickest in middle, tapered at each end, and with one centrally positioned nucleus; not striated. 
Location	Most commonly attached by tendons to bones.	Heart.	Walls of hollow viscera, airways, blood vessels, iris and ciliary body of eye, arrector pili muscles of hair follicles.
Fiber diameter	Very large (10–100 μm).	Large (10–20 μm).	Small (3–8 μm).
Connective tissue components	Endomysium, perimysium, and epimysium.	Endomysium and perimysium.	Endomysium.
Fiber length	Very large (100 μm–30 cm = 12 in.).	Large (50–100 μm).	Intermediate (30–200 μm).
Contractile proteins organized into sarcomeres	Yes.	Yes.	No.
Sarcoplasmic reticulum	Abundant.	Some.	Very little.
Transverse tubules present	Yes, aligned with each A–I band junction.	Yes, aligned with each Z disc.	No.
Junctions between fibers	None.	Intercalated discs contain gap junctions and desmosomes.	Gap junctions in visceral smooth muscle; none in multiunit smooth muscle.
Autorhythmicity	No.	Yes.	Yes, in visceral smooth muscle.
Source of Ca^{2+} for contraction	Sarcoplasmic reticulum.	Sarcoplasmic reticulum and interstitial fluid.	Sarcoplasmic reticulum and interstitial fluid.
Regulator proteins for contraction	Troponin and tropomyosin.	Troponin and tropomyosin.	Calmodulin and myosin light chain kinase.
Speed of contraction	Fast.	Moderate.	Slow.
Nervous control	Voluntary (somatic nervous system).	Involuntary (autonomic nervous system).	Involuntary (autonomic nervous system).
Contraction regulation	Acetylcholine released by somatic motor neurons.	Acetylcholine and norepinephrine released by autonomic motor neurons; several hormones.	Acetylcholine and norepinephrine released by autonomic motor neurons; several hormones; local chemical changes; stretching.
Capacity for regeneration	Limited, via satellite cells.	Limited, under certain conditions.	Considerable (compared with other muscle tissues, but limited compared with epithelium), via pericytes.

becomes arranged in dense columns on either side of the developing nervous system. These columns of mesoderm undergo segmentation into a series of cube-shaped structures called **somites** (SŌ-mīts) (Figure 10.17a). The first pair of somites appears on the 20th day of embryonic development. Eventually, 42 to 44 pairs of somites are formed by the end of the fifth week. The number of somites can be correlated to the approximate age of the embryo.

The cells of a somite differentiate into three regions: (1) a **myotome** (MĪ-ō-tōm), which, as the name suggests, forms the skeletal muscles of the head, neck, and limbs; (2) a **dermatome** (DER-ma-tōm), which forms the connective tissues, including the

Figure 10.17 **Location and structure of somites, key structures in the development of the muscular system.**

Most muscles are derived from mesoderm.

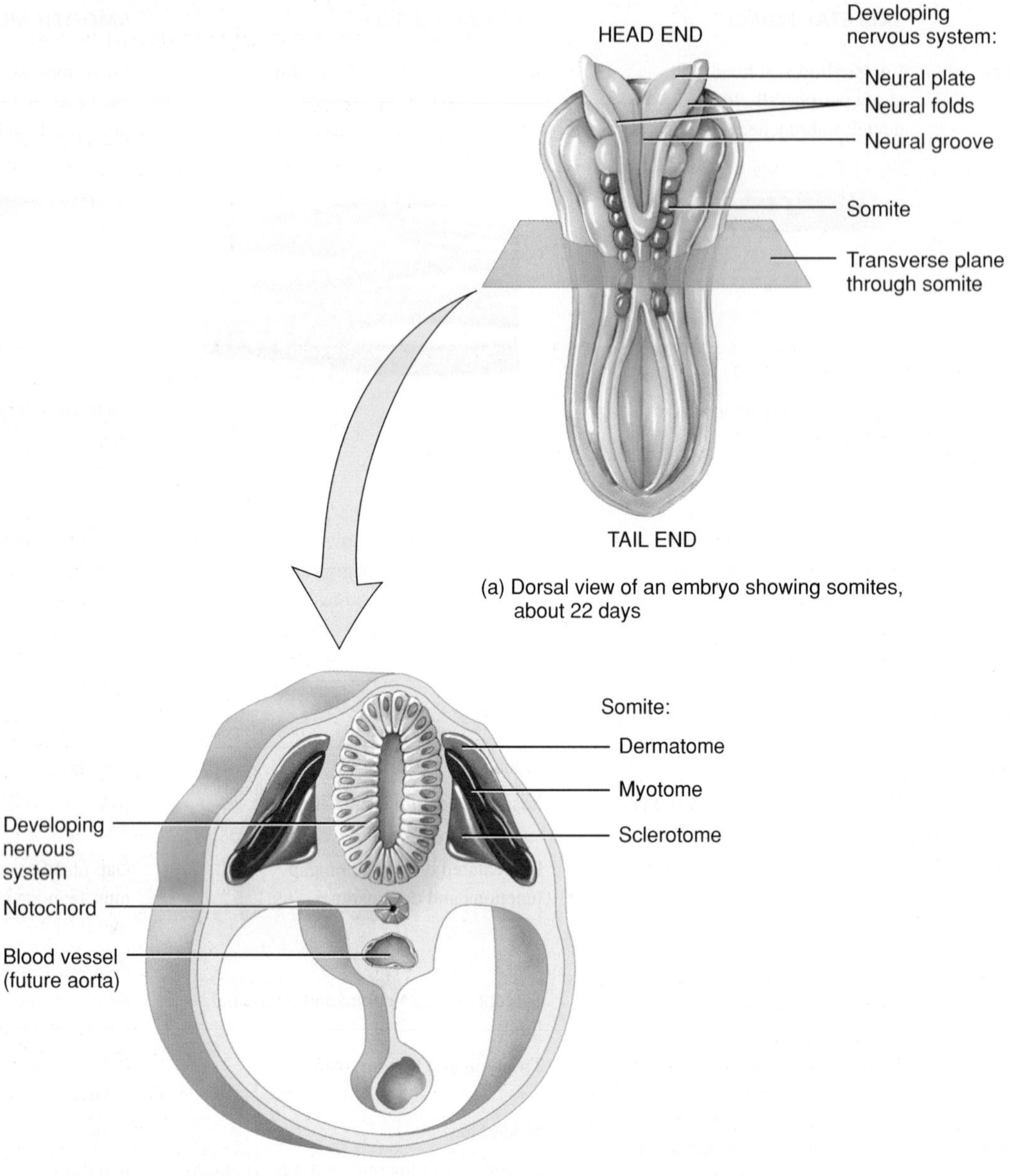

(a) Dorsal view of an embryo showing somites, about 22 days

(b) Transverse section through a somite showing its subdivisions

Which part of a somite differentiates into skeletal muscle?

dermis of the skin; and (3) a **sclerotome** (SKLE-rō-tōm), which gives rise to the vertebrae (Figure 10.17b, c).

Cardiac muscle develops from **mesodermal cells** that migrate to and envelop the developing heart while it is still in the form of endocardial heart tubes (see Figure 20.19).

Smooth muscle develops from **mesodermal cells** that migrate to and envelop the developing gastrointestinal tract and viscera.

CHECKPOINT

27. Which structures develop from myotomes, dermatomes, and sclerotomes?

10.12 AGING AND MUSCULAR TISSUE

OBJECTIVE

- Explain the effects of aging on skeletal muscle.

Between the ages of 30 and 50, humans undergo a slow, progressive loss of skeletal muscle mass that is replaced largely by fibrous connective tissue and adipose tissue. An estimated 10% of muscle mass is lost during these years. In part, this decline may be due to decreased levels of physical activity. Accompanying the loss of muscle mass is a decrease in maximal strength, a slowing of muscle reflexes, and a loss of flexibility. With aging, the relative number of slow oxidative (SO) fibers appears to increase. This could be due to either the atrophy of the other fiber types or their conversion into slow oxidative fibers. Another 40% of muscle is typically lost between the ages of 50 and 80. Loss of muscle strength is usually not perceived by persons until they reach the age of 60 to 65. At that point it is most common for muscles of the lower limbs to weaken before those of the upper limbs. Thus the independence of the elderly may be affected when it becomes difficult to climb stairs or get up from a seated position.

Assuming that there is not a chronic medical condition for which exercise is contraindicated, exercise has been shown to be effective at any age. Aerobic activities and strength training programs are effective in older people and can slow or even reverse the age-associated decline in muscular performance.

✔ **CHECKPOINT**

28. Why does muscle strength decrease with aging?
29. Why do you think a healthy 30-year-old can lift a 25-lb load much more comfortably than an 80-year-old?

DISORDERS: HOMEOSTATIC IMBALANCES

Abnormalities of skeletal muscle function may be due to disease or damage of any of the components of a motor unit: somatic motor neurons, neuromuscular junctions, or muscle fibers. The term **neuromuscular disease** encompasses problems at all three sites; the term **myopathy** (mī-OP-a-thē; *-pathy* = disease) signifies a disease or disorder of the skeletal muscle tissue itself.

Myasthenia Gravis

Myasthenia gravis (mī-as-THĒ-nē-a GRAV-is; *mys-* = muscle; *-aisthesis* = sensation) is an autoimmune disease that causes chronic, progressive damage of the neuromuscular junction. The immune system inappropriately produces antibodies that bind to and block some ACh receptors, thereby decreasing the number of functional ACh receptors at the motor end plates of skeletal muscles (see Figure 10.9). Because 75% of patients with myasthenia gravis have hyperplasia or tumors of the thymus, it is thought that thymic abnormalities cause the disorder. As the disease progresses, more ACh receptors are lost. Thus, muscles become increasingly weaker, fatigue more easily, and may eventually cease to function.

Myasthenia gravis occurs in about 1 in 10,000 people and is more common in women, typically ages 20 to 40 at onset; men usually are ages 50 to 60 at onset. The muscles of the face and neck are most often affected. Initial symptoms include weakness of the eye muscles, which may produce double vision, and weakness of the throat muscles that may produce difficulty in swallowing. Later, the person has difficulty chewing and talking. Eventually the muscles of the limbs may become involved. Death may result from paralysis of the respiratory muscles, but often the disorder does not progress to that stage.

Anticholinesterase drugs such as pyridostigmine (Mestinon) or neostigmine, the first line of treatment, act as inhibitors of acetylcholinesterase, the enzyme that breaks down ACh. Thus, the inhibitors raise the level of ACh that is available to bind with still-functional receptors. More recently, steroid drugs such as prednisone have been used with success to reduce antibody levels. Another treatment is plasmapheresis, a procedure that removes the antibodies from the blood. Often, surgical removal of the thymus (thymectomy) is helpful.

Muscular Dystrophy

The term **muscular dystrophy** (DIS-trō-fē′; *dys-* = difficult; *-trophy* = nourishment) refers to a group of inherited muscle-destroying diseases that cause progressive degeneration of skeletal muscle fibers. The most common form of muscular dystrophy is *Duchenne muscular dystrophy* (doo-SHĀN) or *DMD.* Because the mutated gene is on the X chromosome, and because males have only one X chromosome, DMD strikes boys almost exclusively. (Sex-linked inheritance is described in Chapter 29.) Worldwide, about 1 in every 3500 male babies—about 21,000 in all—are born with DMD each year. The disorder usually becomes apparent between the ages of 2 and 5, when parents notice the child falls often and has difficulty running, jumping, and hopping. By age 12 most boys with DMD are unable to walk. Respiratory or cardiac failure usually causes death by age 20.

In DMD, the gene that codes for the protein dystrophin is mutated, so little or no dystrophin is present in the sarcolemma. Without the reinforcing effect of dystrophin, the sarcolemma tears easily during muscle contraction, causing muscle fibers to rupture and die. The dystrophin gene was discovered in 1987, and by 1990 the first attempts were made to treat DMD patients with gene therapy. The muscles of three boys with DMD were injected with myoblasts bearing functional dystrophin genes, but only a few muscle fibers gained the ability to produce dystrophin. Similar clinical trials with additional patients have also failed. An alternative approach to the problem is to find a way to induce muscle fibers to produce the protein *utrophin,* which is similar to dystrophin. Experiments with dystrophin-deficient mice suggest this approach may work.

Abnormal Contractions of Skeletal Muscle

One kind of abnormal muscular contraction is a **spasm,** a sudden involuntary contraction of a single muscle in a large group of muscles. A painful spasmodic contraction is known as a **cramp.** Cramps may be caused by inadequate blood flow to muscles, overuse of a muscle, dehydration, injury, holding a position for prolonged periods, and low blood levels of electrolytes, such as potassium. A **tic** is a spasmodic twitching made involuntarily by muscles that are ordinarily under voluntary control. Twitching of the eyelid and facial muscles are examples of tics. A **tremor** is a rhythmic, involuntary, purposeless contraction that produces a quivering or shaking movement. A **fasciculation** (fa-sik-ū-LĀ-shun) is an involuntary, brief twitch of an entire motor unit that is visible under the skin; it occurs irregularly and is not associated with movement of the affected muscle. Fasciculations may be seen in multiple sclerosis (see Disorders: Homeostatic Imbalances in Chapter 12) or in amyotrophic lateral sclerosis (Lou Gehrig's disease, see Clinical Connection: Amyotrophic Lateral Sclerosis in Chapter 16). A **fibrillation** (fi-bri-LĀ-shun) is a spontaneous contraction of a single muscle fiber that is not visible under the skin but can be recorded by electromyography. Fibrillations may signal destruction of motor neurons.

MEDICAL TERMINOLOGY

Myalgia (mī-AL-jē-a; *-algia* = painful condition) Pain in or associated with muscles.

Myoma (mī-Ō-ma; *-oma* = tumor) A tumor consisting of muscle tissue.

Myomalacia (mī′-ō-ma-LĀ-shē-a; *-malacia* = soft) Pathological softening of muscle tissue.

Myositis (mī′-ō-SĪ-tis; *-itis* = inflammation of) Inflammation of muscle fibers (cells).

Myotonia (mī′-ō-TŌ-nē-a; *-tonia* = tension) Increased muscular excitability and contractility, with decreased power of relaxation; tonic spasm of the muscle.

Volkmann's contracture (FŌLK-manz kon-TRAK-chur; *contra-* = against) Permanent shortening (contracture) of a muscle due to replacement of destroyed muscle fibers by fibrous connective tissue, which lacks extensibility. Typically occurs in forearm flexor muscles. Destruction of muscle fibers may occur from interference with circulation caused by a tight bandage, a piece of elastic, or a cast.

CHAPTER REVIEW AND RESOURCE SUMMARY

WILEY PLUS

Review	Resource
Introduction	
1. Motion results from alternating contraction and relaxation of muscles, which constitute 40–50% of total body weight.	
2. The prime function of muscle is changing chemical energy into mechanical energy to perform work.	
10.1 Overview of Muscular Tissue	Anatomy Overview - Muscular Tissue
1. The three types of muscular tissue are skeletal, cardiac, and smooth. Skeletal muscle tissue is primarily attached to bones; it is striated and voluntary. Cardiac muscle tissue forms the wall of the heart; it is striated and involuntary. Smooth muscle tissue is located primarily in internal organs; it is nonstriated (smooth) and involuntary.	
2. Through contraction and relaxation, muscular tissue performs four important functions: producing body movements; stabilizing body positions; moving substances within the body and regulating organ volume; and producing heat.	
3. Four special properties of muscular tissues are (1) electrical excitability, the property of responding to stimuli by producing action potentials; (2) contractility, the ability to generate tension to do work; (3) extensibility, the ability to be extended (stretched); and (4) elasticity, the ability to return to original shape after contraction or extension.	
10.2 Skeletal Muscle Tissue	Anatomy Overview - Skeletal Muscle Anatomy Overview - Cross-section of Skeletal Muscle Animation - Muscle Cell Structures
1. The subcutaneous layer separates skin from muscles, provides a pathway for blood vessels and nerves to enter and exit muscles, and protects muscles from physical trauma. Fascia lines the body wall and limbs that surround and support muscles, allows free movement of muscles, carries nerves and blood vessels, and fills space between muscles.	
2. Tendons and aponeuroses are extensions of connective tissue beyond muscle fibers that attach the muscle to bone or to other muscle. A tendon is generally ropelike in shape; an aponeurosis is wide and flat.	
3. Skeletal muscles are well supplied with nerves and blood vessels. Generally, an artery and one or two veins accompany each nerve that penetrates a skeletal muscle.	
4. Somatic motor neurons provide the nerve impulses that stimulate skeletal muscle to contract.	
5. Blood capillaries bring in oxygen and nutrients and remove heat and waste products of muscle metabolism.	
6. The major cells of skeletal muscle tissue are termed skeletal muscle fibers. Each muscle fiber has 100 or more nuclei because it arises from the fusion of many myoblasts. Satellite cells are myoblasts that persist after birth. The sarcolemma is a muscle fiber's plasma membrane; it surrounds the sarcoplasm. Transverse tubules are invaginations of the sarcolemma.	
7. Each muscle fiber (cell) contains hundreds of myofibrils, the contractile elements of skeletal muscle. Sarcoplasmic reticulum (SR) surrounds each myofibril. Within a myofibril are thin and thick filaments, arranged in compartments called sarcomeres.	

Review | **Resource**

8. The overlapping of thick and thin filaments produces striations. Darker A bands alternate with lighter I bands. Table 10.1 summarizes the components of the sarcomere.
9. Myofibrils are composed of three types of proteins: contractile, regulatory, and structural. The contractile proteins are myosin (thick filament) and actin (thin filament). Regulatory proteins are tropomyosin and troponin, both of which are part of the thin filament. Structural proteins include titin (links Z disc to M line and stabilizes thick filament), myomesin (forms M line), nebulin (anchors thin filaments to Z discs and regulates length of thin filaments during development), and dystrophin (links thin filaments to sarcolemma). Table 10.2 summarizes the different types of skeletal muscle fiber proteins. Table 10.3 summarizes the levels of organization within a skeletal muscle.
10. Projecting myosin heads contain actin-binding and ATP-binding sites and are the motor proteins that power muscle contraction.

10.3 Contraction and Relaxation of Skeletal Muscle Fibers

1. Muscle contraction occurs because cross-bridges attach to and "walk" along the thin filaments at both ends of a sarcomere, progressively pulling the thin filaments toward the center of a sarcomere. As the thin filaments slide inward, the Z discs come closer together, and the sarcomere shortens.
2. The contraction cycle is the repeating sequence of events that causes sliding of the filaments: (1) Myosin ATPase hydrolyzes ATP and becomes energized; (2) the myosin head attaches to actin, forming a cross-bridge; (3) the cross-bridge generates force as it rotates toward the center of the sarcomere (power stroke); and (4) binding of ATP to the myosin head detaches it from actin. The myosin head again hydrolyzes the ATP, returns to its original position, and binds to a new site on actin as the cycle continues.
3. An increase in Ca^{2+} concentration in the cytosol starts filament sliding; a decrease turns off the sliding process.
4. The muscle action potential propagating into the T tubule system causes opening of Ca^{2+} release channels in the SR membrane. Calcium ions diffuse from the SR into the sarcoplasm and combine with troponin. This binding causes tropomyosin to move away from the myosin-binding sites on actin.
5. Ca^{2+} active transport pumps continually remove Ca^{2+} from the sarcoplasm into the SR. When the concentration of calcium ions in the sarcoplasm decreases, tropomyosin slides back over and blocks the myosin-binding sites, and the muscle fiber relaxes.
6. A muscle fiber develops its greatest tension when there is an optimal zone of overlap between thick and thin filaments. This dependency is the length–tension relationship.
7. The neuromuscular junction (NMJ) is the synapse between a somatic motor neuron and a skeletal muscle fiber. The NMJ includes the axon terminals and synaptic end bulbs of a motor neuron, plus the adjacent motor end plate of the muscle fiber sarcolemma.
8. When a nerve impulse reaches the synaptic end bulbs of a somatic motor neuron, it triggers exocytosis of the synaptic vesicles, which releases acetylcholine (ACh). ACh diffuses across the synaptic cleft and binds to ACh receptors, initiating a muscle action potential. Acetylcholinesterase then quickly breaks down ACh into its component parts.

Resource:
- Animation - Contraction and Movement
- Animation - Neuromuscular Junctions
- Animation - Contraction of a Sarcomere
- Animation - Control of Muscle Tension
- Figure 10.5 - Sliding Filament Mechanism of Muscle Contraction
- Figure 10.6 - The Contraction Cycle
- Figure 10.10 - Summary of Events of Contraction and Relaxation
- Exercise - Contraction Connections
- Exercise - Fueling Contraction and Recovery
- Exercise - Muscle Car
- Concepts and Connections - Events at the Neuromuscular Junction
- Concepts and Connections - Excitation of Skeletal Muscle
- Concepts and Connections - Skeletal Muscle Contraction Cycle

10.4 Muscle Metabolism

1. Muscle fibers have three sources for ATP production: creatine, anaerobic cellular respiration, and aerobic cellular respiration.
2. Creatine kinase catalyzes the transfer of a high-energy phosphate group from creatine phosphate to ADP to form new ATP molecules. Together, creatine phosphate and ATP provide enough energy for muscles to contract maximally for about 15 seconds.
3. Glucose is converted to pyruvic acid in the reactions of glycolysis, which yield two ATPs without using oxygen. Such anaerobic cellular respiration can provide enough energy for 30–40 seconds of maximal muscle activity.
4. Muscular activity that occurs over a prolonged period of time depends on aerobic cellular respiration, mitochondrial reactions that require oxygen to produce ATP.
5. The inability of a muscle to contract forcefully after prolonged activity is muscle fatigue.
6. Elevated oxygen use after exercise is called recovery oxygen uptake.

Resource:
- Animation - Muscle Metabolism
- Concepts and Connections - Muscle Metabolism
- Figure 10.11 - Production of ATP for Muscle Contraction

10.5 Control of Muscle Tension

1. A motor neuron and the muscle fibers it stimulates form a motor unit. A single motor unit may contain as few as 2 or as many as 3000 muscle fibers.

Resource:
- Animation - Control of Muscle Tension
- Exercise - Increase Muscle Tension

Review | **Resource**

2. Recruitment is the process of increasing the number of active motor units.
3. A twitch contraction is a brief contraction of all the muscle fibers in a motor unit in response to a single action potential.
4. A record of a contraction is called a myogram. It consists of a latent period, a contraction period, and a relaxation period.
5. Wave summation is the increased strength of a contraction that occurs when a second stimulus arrives before the muscle fiber has completely relaxed after a previous stimulus.
6. Repeated stimuli can produce unfused (incomplete) tetanus, a sustained muscle contraction with partial relaxation between stimuli. More rapidly repeating stimuli produce fused (complete) tetanus, a sustained contraction without partial relaxation between stimuli.
7. Continuous involuntary activation of a small number of motor units produces muscle tone, which is essential for maintaining posture.
8. In a concentric isotonic contraction, the muscle shortens to produce movement and to reduce the angle at a joint. During an eccentric isotonic contraction, the muscle lengthens.
9. Isometric contractions, in which tension is generated without muscle changing its length, are important because they stabilize some joints as others are moved.

10.6 Types of Skeletal Muscle Fibers

1. On the basis of their structure and function, skeletal muscle fibers are classified as slow oxidative (SO), fast oxidative–glycolytic (FOG), and fast glycolytic (FG) fibers.
2. Most skeletal muscles contain a mixture of all three fiber types. Their proportions vary with the typical action of the muscle.
3. The motor units of a muscle are recruited in the following order: first SO fibers, then FOG fibers, and finally FG fibers.
4. Table 10.4 summarizes the three types of skeletal muscle fibers.

Resource: Anatomy Overview - Skeletal Muscle; Animation - Muscle Cell Structures

10.7 Exercise and Skeletal Muscle Tissue

1. Various types of exercises can induce changes in the fibers in a skeletal muscle. Endurance-type (aerobic) exercises cause a gradual transformation of some fast glycolytic (FG) fibers into fast oxidative–glycolytic (FOG) fibers.
2. Exercises that require great strength for short periods produce an increase in the size and strength of fast glycolytic (FG) fibers. The increase in size is due to increased synthesis of thick and thin filaments.

10.8 Cardiac Muscle Tissue

1. Cardiac muscle is found only in the heart. Cardiac muscle fibers have the same arrangement of actin and myosin and the same bands, zones, and Z discs as skeletal muscle fibers. The fibers connect to one another through intercalated discs, which contain both desmosomes and gap junctions.
2. Cardiac muscle tissue remains contracted 10 to 15 times longer than skeletal muscle tissue due to prolonged delivery of Ca^{2+} into the sarcoplasm.
3. Cardiac muscle tissue contracts when stimulated by its own autorhythmic fibers. Due to its continuous, rhythmic activity, cardiac muscle depends greatly on aerobic cellular respiration to generate ATP.

Resource: Anatomy Overview - Cardiac Muscle; Anatomy Overview - Muscle Tissue

10.9 Smooth Muscle Tissue

1. Smooth muscle is nonstriated and involuntary.
2. Smooth muscle fibers contain intermediate filaments and dense bodies; the function of dense bodies is similar to that of the Z discs in striated muscle.
3. Visceral (single-unit) smooth muscle is found in the walls of hollow viscera and of small blood vessels. Many fibers form a network that contracts in unison.
4. Multiunit smooth muscle is found in large blood vessels, large airways to the lungs, arrector pili muscles, and the eye, where it adjusts pupil diameter and lens focus. The fibers operate independently rather than in unison.
5. The duration of contraction and relaxation of smooth muscle is longer than in skeletal muscle since it takes longer for Ca^{2+} to reach the filaments.
6. Smooth muscle fibers contract in response to nerve impulses, hormones, and local factors.
7. Smooth muscle fibers can stretch considerably and still maintain their contractile function.

Resource: Anatomy Overview - Smooth Muscle Tissue; Anatomy Overview - Muscle Tissue; Figure 10.16 - Smooth Muscle Tissue

Review **Resource**

10.10 Regeneration of Muscular Tissue

1. Skeletal muscle fibers cannot divide and have limited powers of regeneration; cardiac muscle fibers can regenerate under limited circumstances; and smooth muscle fibers have the best capacity for division and regeneration.
2. Table 10.5 summarizes the major characteristics of the three types of muscular tissue.

10.11 Development of Muscle

1. With few exceptions, muscles develop from mesoderm.
2. Skeletal muscles of the head and limbs develop from general mesoderm. Other skeletal muscles develop from the mesoderm of somites.

10.12 Aging and Muscular Tissue

1. With aging, there is a slow, progressive loss of skeletal muscle mass, which is replaced by fibrous connective tissue and fat.
2. Aging also results in a decrease in muscle strength, slower muscle reflexes, and loss of flexibility.

SELF-QUIZ QUESTIONS

Fill in the blanks in the following statements.

1. A single somatic motor neuron and all of the muscle fibers it stimulates is known as a _____.

2. The wasting away of muscle due to lack of use is known as _____ while the replacement of skeletal muscle fibers with scar tissue is known as _____.

3. The synaptic end bulbs of somatic motor neurons contain synaptic vesicles filled with the neurotransmitter _____.

Indicate whether the following statements are True *or* False.

4. The ability of muscle cells to respond to stimuli and produce electrical signals is known as excitability.

5. The sequence of events resulting in skeletal muscle contraction are
(a) generation of a nerve impulse;
(b) release of the neurotransmitter acetylcholine;
(c) generation of a muscle action potential;
(d) release of calcium ions from the sarcoplasmic reticulum;
(e) calcium ion binding to troponin;
(f) power stroke with actin and myosin binding and release.

Choose the one best answer to the following questions.

6. In muscle physiology, the latent period refers to
(a) the period of lost excitability that occurs when two stimuli are applied immediately one after the other.
(b) the brief contraction of a motor unit.
(c) the period of elevated oxygen use after exercise.
(d) an inability of a muscle to contract forcefully after prolonged activity.
(e) a brief delay that occurs between application of a stimulus and the beginning of contraction.

7. Which of the following muscle proteins and their descriptions are mismatched?
(a) titin: regulatory protein that holds troponin in place
(b) myosin: contractile motor protein
(c) tropomyosin: regulatory protein that blocks myosin-binding sites
(d) actin: contractile protein that contains myosin-binding sites
(e) calsequestrin: calcium-binding protein

8. During muscle contraction all of the following occur *except*
(a) cross-bridges are formed when the energized myosin head attaches to actin's myosin-binding site.
(b) ATP undergoes hydrolysis.
(c) the thick filaments slide inward toward the M line.
(d) calcium concentration in the cytosol increases.
(e) the Z discs are drawn toward each other.

9. Which of the following is *not* true concerning sarcomeres (before contraction begins) and muscle fiber length–tension relationships?
(a) If sarcomeres are stretched, the tension in the fiber decreases.
(b) If a muscle cell is stretched so that there is no overlap of the filaments, no tension is generated.
(c) Extremely compressed sarcomeres result in less muscle tension.
(d) Maximum tension occurs when the zone of overlap between a thick and thin filament extends from the edge of the H zone to one end of a thick filament.
(e) If sarcomeres shorten, the tension in them increases.

10. Which of the following are sources of ATP for muscle contraction? (1) creatine phosphate, (2) glycolysis, (3) anaerobic cellular respiration, (4) aerobic cellular respiration, (5) acetylcholine.
(a) 1, 2, and 3 (b) 2, 3, and 4 (c) 2, 3, and 5
(d) 1, 2, 3, and 4 (e) 2, 3, 4, and 5

11. What would happen if ATP were suddenly unavailable after the sarcomere had begun to shorten?
(a) Nothing. The contraction would proceed normally.
(b) The myosin heads would be unable to detach from actin.
(c) Troponin would bind with the myosin heads.
(d) Actin and myosin filaments would separate completely and be unable to recombine.
(e) The myosin heads would detach completely from actin.

12. Match the following:

____ (a) a sheath of areolar connective tissue that wraps around individual skeletal muscle fibers
____ (b) dense irregular connective tissue that separates a muscle into groups of individual muscle fibers
____ (c) bundles of muscle fibers
____ (d) the outermost connective tissue layer that encircles an entire skeletal muscle
____ (e) dense irregular connective tissue that lines the body wall and limbs and holds functional muscle units together
____ (f) a cord of dense regular connective tissue that attaches muscle to the periosteum of bone
____ (g) muscle cell
____ (h) areolar and adipose connective tissue that separates muscle from skin
____ (i) connective tissue elements extended as a broad, flat layer
____ (j) a two-layer tube of fibrous connective tissue enclosing certain tendons

(1) aponeurosis
(2) fascia
(3) subcutaneous layer
(4) tendon
(5) endomysium
(6) perimysium
(7) epimysium
(8) tendon (synovial) sheath
(9) fascicles
(10) muscle fiber

13. Match the following:

____ (a) synapse between a motor neuron and a muscle fiber
____ (b) invaginations of the sarcolemma from the surface toward the center of the muscle fiber
____ (c) myoblasts that persist in mature skeletal muscle
____ (d) plasma membrane of a muscle fiber
____ (e) oxygen-binding protein found only in muscle fibers
____ (f) Ca^{2+}-storing tubular system similar to smooth endoplasmic reticulum
____ (g) contracting unit of a skeletal muscle fiber
____ (h) middle area in the sarcomere where thick and thin filaments are found
____ (i) area in the sarcomere where only thin filaments are present but thick filaments are not
____ (j) separates the sarcomeres from each other
____ (k) area of only thick filaments
____ (l) cytoplasm of a muscle fiber
____ (m) composed of supporting proteins holding thick filaments together at the H zone

(1) A band
(2) I band
(3) Z disc
(4) H zone
(5) M line
(6) sarcomere
(7) neuromuscular junction
(8) myoglobin
(9) satellite cells
(10) transverse tubules
(11) sarcoplasmic reticulum
(12) sarcolemma
(13) sarcoplasm

14. Match the following (some questions will have more than one answer):

____ (a) has fibers joined by intercalated discs
____ (b) thick and thin filaments are not arranged as orderly sarcomeres
____ (c) uses satellite cells to repair damaged muscle fibers
____ (d) striated
____ (e) contraction begins slowly but lasts for long periods
____ (f) has an extended contraction due to prolonged calcium delivery from both the sarcoplasmic reticulum and the interstitial fluid
____ (g) does not exhibit autorhythmicity
____ (h) uses pericytes to repair damaged muscle fibers
____ (i) uses troponin as a regulatory protein
____ (j) can be classified as single-unit or multiunit
____ (k) can be autorhythmic
____ (l) uses calmodulin as a regulatory protein

(1) skeletal muscle
(2) cardiac muscle
(3) smooth muscle

15. Match the following:

____ (a) the smooth muscle action that allows the fibers to maintain their contractile function even when stretched
____ (b) a brief contraction of all the muscle fibers in a motor unit of a muscle in response to a single action potential in its motor neuron
____ (c) sustained contraction of a muscle, with no relaxation between stimuli
____ (d) larger contractions resulting from stimuli arriving at different times
____ (e) process of increasing the number of activated motor units
____ (f) contraction in which the muscle shortens
____ (g) inability of a muscle to maintain its strength of contraction or tension during prolonged activity
____ (h) sustained, but wavering contraction with partial relaxation between stimuli
____ (i) produced by the continual involuntary activation of a small number of skeletal muscle motor units; results in firmness in skeletal muscle
____ (j) contraction in which muscle tension is generated without shortening of the muscle
____ (k) amount of oxygen needed to restore the body's metabolic conditions back to resting levels after exercise
____ (l) contraction in which a muscle lengthens

(1) muscle fatigue
(2) twitch contraction
(3) wave summation
(4) fused (complete) tetanus
(5) concentric isotonic contraction
(6) motor unit recruitment
(7) muscle tone
(8) eccentric isotonic contraction
(9) isometric contraction
(10) stress–relaxation response
(11) recovery oxygen uptake
(12) unfused (incomplete) tetanus

CRITICAL THINKING QUESTIONS

1. Weight lifter Jamal has been practicing many hours a day, and his muscles have gotten noticeably bigger. He tells you that his muscle cells are "multiplying like crazy and making him get stronger and stronger." Do you believe his explanation? Why or why not?

2. Chicken breasts are composed of "white meat" while chicken legs are composed of "dark meat." The breasts and legs of migrating ducks are dark meat. The breasts of both chickens and ducks are used in flying. How can you explain the differences in the color of the meat (muscles)? How are they adapted for their particular functions?

3. Polio is a disease caused by a virus that can attack the somatic motor neurons in the central nervous system. Individuals who suffer from polio can develop muscle weakness and atrophy. In a certain percentage of cases, the individuals may die due to respiratory paralysis. Relate your knowledge of how muscle fibers function to the symptoms exhibited by infected individuals.

? ANSWERS TO FIGURE QUESTIONS

10.1 Perimysium bundles groups of muscle fibers into fascicles.

10.2 The sarcoplasmic reticulum releases calcium ions to trigger muscle contraction.

10.3 The following are arranged from smallest to largest: thick filament, myofibril, muscle fiber.

10.4 Actin and titin anchor into the Z disc. A bands contain myosin, actin, troponin, tropomyosin, and titin; I bands contain actin, troponin, tropomyosin, and titin.

10.5 The I bands and H zones disappear during muscle contraction; the lengths of the thin and thick filaments do not change.

10.6 If ATP were not available, the cross-bridges would not be able to detach from actin. The muscles would remain in a state of rigidity, as occurs in rigor mortis.

10.7 Three functions of ATP in muscle contraction are the following: (1) Its hydrolysis by an ATPase activates the myosin head so it can bind to actin and rotate; (2) its binding to myosin causes detachment from actin after the power stroke; and (3) it powers the pumps that transport Ca^{2+} from the cytosol back into the sarcoplasmic reticulum.

10.8 A sarcomere length of 2.2 μm gives a generous zone of overlap between the parts of the thick filaments that have myosin heads and the thin filaments without the overlap's being so extensive that sarcomere shortening is limited.

10.9 The part of the sarcolemma that contains acetylcholine receptors is the motor end plate.

10.10 Steps 4 through 6 are part of excitation–contraction coupling (muscle action potential through binding of myosin heads to actin).

10.11 Glycolysis, exchange of phosphate between creatine phosphate and ADP, and glycogen breakdown occur in the cytosol. Oxidation of pyruvic acid, amino acids, and fatty acids (aerobic cellular respiration) occurs in mitochondria.

10.12 Motor units having many muscle fibers are capable of more forceful contractions than those having only a few fibers.

10.13 During the latent period, the muscle action potential sweeps over the sarcolemma and calcium ions are released from the sarcoplasmic reticulum.

10.14 If the second stimulus were applied a little later, the second contraction would be smaller than the one illustrated in part b.

10.15 Holding your head upright without movement involves mainly isometric contractions.

10.16 Visceral smooth muscle is more like cardiac muscle; both contain gap junctions, which allow action potentials to spread from each cell to its neighbors.

10.17 The myotome of a somite differentiates into skeletal muscle.

11 THE MUSCULAR SYSTEM

THE MUSCULAR SYSTEM AND HOMEOSTASIS *The muscular system and muscular tissue of your body contribute to homeostasis by stabilizing body position, producing movements, regulating organ volume, moving substances within the body, and producing heat.*

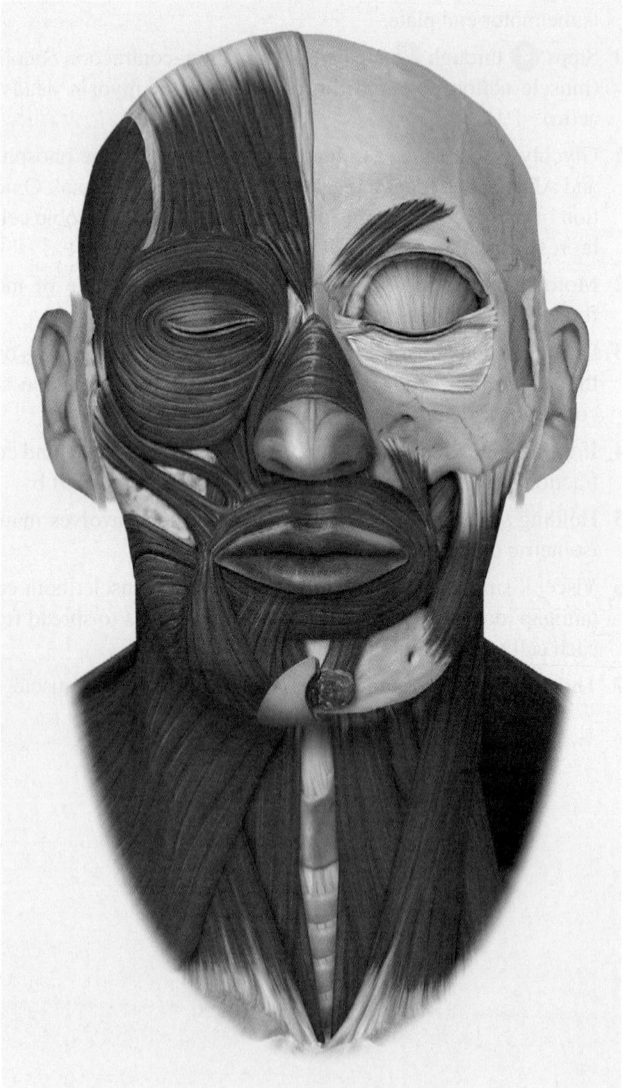

Together, the voluntarily controlled muscles of your body compose the **muscular system.** Almost all of the 700 individual muscles that make up the muscular system, such as the biceps brachii muscle, include both skeletal muscle tissue and connective tissue. The function of most muscles is to produce movements of body parts. A few muscles function mainly to stabilize bones so that other skeletal muscles can execute a movement more effectively. This chapter presents many of the major skeletal muscles in the body, most of which are found on both the right and left sides. We will identify the attachment sites and innervation (the nerve or nerves that stimulate contraction) of each muscle described. Developing a working knowledge of these key aspects of skeletal muscle anatomy will enable you to understand how normal movements occur. This knowledge is especially crucial for professionals, such as those in the allied health and physical rehabilitation fields, who work with patients whose normal patterns of movement and physical mobility have been disrupted by physical trauma, surgery, or muscular paralysis.

? ***Did you ever wonder how carpal tunnel syndrome occurs?***

11.1 HOW SKELETAL MUSCLES PRODUCE MOVEMENTS

OBJECTIVES

- Describe the relationship between bones and skeletal muscles in producing body movements.
- Define lever and fulcrum, and compare the three types of levers based on location of the fulcrum, effort, and load.
- Identify the types of fascicle arrangements in a skeletal muscle, and relate the arrangements to strength of contraction and range of motion.
- Explain how the prime mover, antagonist, synergist, and fixator in a muscle group work together to produce movement.

Muscle Attachment Sites: Origin and Insertion

Those skeletal muscles that produce movements do so by exerting force on tendons, which in turn pull on bones or other structures (such as skin). Most muscles cross at least one joint and are usually attached to articulating bones that form the joint (Figure 11.1a).

When a skeletal muscle contracts, it pulls one of the articulating bones toward the other. The two articulating bones usually do not move equally in response to contraction. One bone remains stationary or near its original position, either because other muscles stabilize that bone by contracting and pulling it in the opposite direction or because its structure makes it less movable. Ordinarily, the attachment of a muscle's tendon to the stationary bone is called the **origin** (OR-i-jin); the attachment of the muscle's other tendon to the movable bone is called the **insertion** (in-SER-shun). A good analogy is a spring on a door. In this example, the part of the spring attached to the frame is the origin; the part attached to the door represents the insertion. A useful rule of thumb is that the origin is usually proximal and the insertion distal; the insertion is usually pulled toward the origin. The fleshy portion of the muscle between the tendons is called the **belly (body),** the coiled middle portion of the spring in our example. The **actions** of

Figure 11.1 Relationship of skeletal muscles to bones. Muscles are attached to bones by tendons at their origin and insertion. Skeletal muscles produce movements by pulling on bones. Bones serve as levers, and joints act as fulcrums for the levers. Here the lever–fulcrum principle is illustrated by the movement of the forearm. Note where the load (resistance) and effort are applied in (b).

In the limbs, the origin of a muscle is usually proximal and the insertion is usually distal.

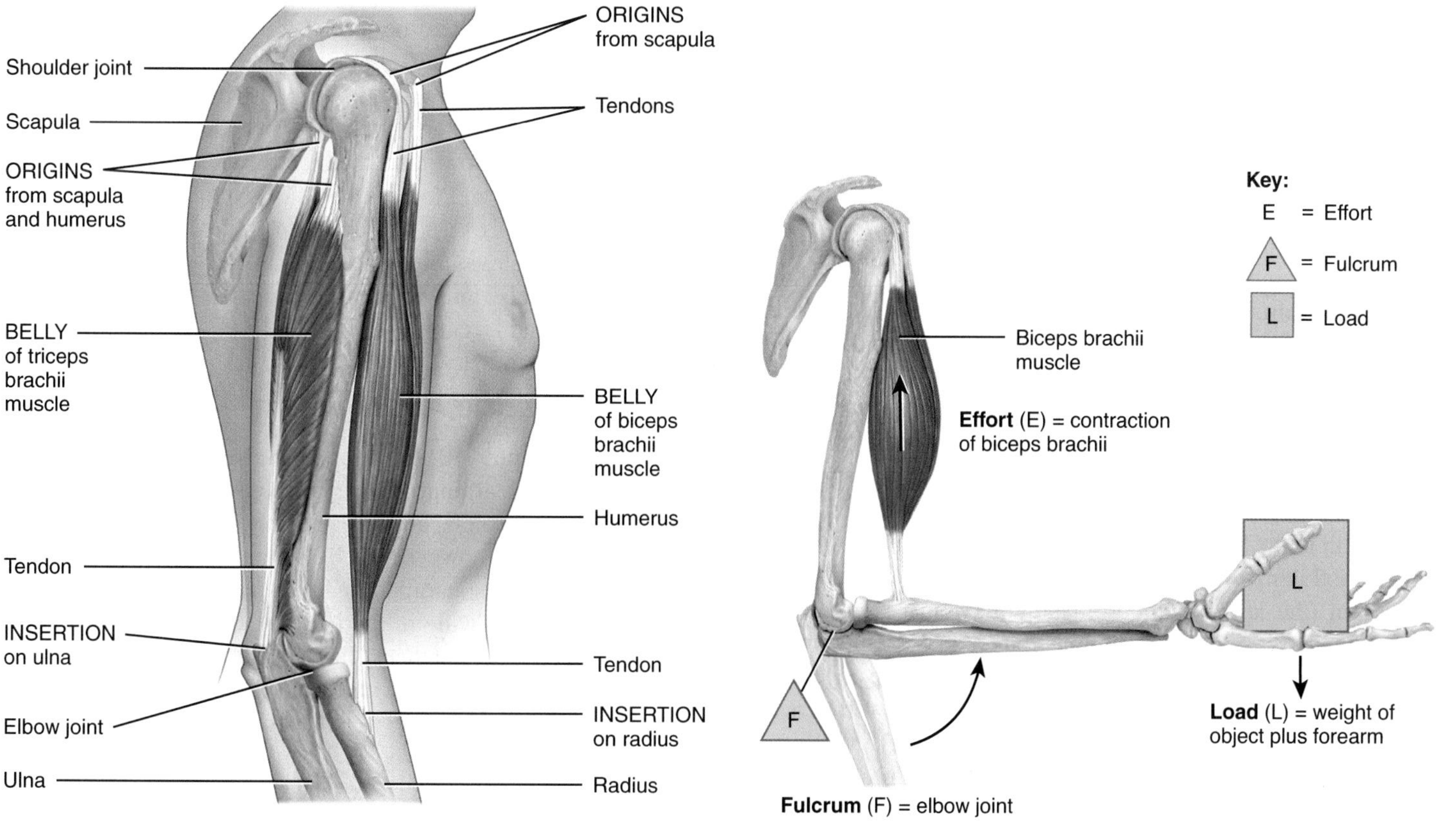

(a) Origin and insertion of a skeletal muscle

(b) Movement of the forearm lifting a weight

Where is the belly of the muscle that extends the forearm located?

a muscle are the main movements that occur when the muscle contracts. In our spring example, this would be the closing of the door. Certain muscles are also capable of **reverse muscle action (RMA).** This means that during specific movements of the body the actions are reversed and therefore the positions of the origin and insertion of a specific muscle are switched.

Muscles that move a body part often do not cover the moving part. Figure 11.1b shows that although one of the functions of the biceps brachii muscle is to move the forearm, the belly of the muscle lies over the humerus, not over the forearm. You will also see that muscles that cross two joints, such as the rectus femoris and sartorius of the thigh, have more complex actions than muscles that cross only one joint.

Lever Systems and Leverage

In producing movement, bones act as levers, and joints function as the fulcrums of these levers. A **lever** is a rigid structure that can move around a fixed point called a **fulcrum,** symbolized by F. A lever is acted on at two different points by two different forces: the **effort** (E), which causes movement, and the **load** L or **resistance,** which opposes movement. The effort is the force exerted by muscular contraction; the load is typically the weight of the body part that is moved or some resistance that the moving body part is trying to overcome (such as the weight of a book you might be picking up). Motion occurs when the effort applied to the bone at the insertion exceeds the load. Consider the biceps brachii flexing the forearm at the elbow as an object is lifted (Figure 11.1b). When the forearm is raised, the elbow is the fulcrum. The weight of the forearm plus the weight of the object in the hand is the load. The force of contraction of the biceps brachii pulling the forearm up is the effort.

The relative distance between the fulcrum and load and the point at which the effort is applied determine whether a given lever operates at a mechanical advantage or a mechanical disadvantage. For example, if the load is closer to the fulcrum and the effort farther from the fulcrum, then only a relatively small effort is required to move a large load over a small distance. This is called a **mechanical advantage.** If, instead, the load is farther from the fulcrum and the effort is applied closer to the fulcrum, then a relatively large effort is required to move a small load (but at greater speed). This is called a **mechanical disadvantage.** Compare chewing something hard (the load) with your front teeth and the teeth in the back of your mouth. It is much easier to crush the hard food item with the back teeth because they are closer to the fulcrum (the jaw or temporomandibular joint) than are the front teeth. Here is one more example you can try. Straighten out a paper clip. Now get a pair of scissors and try to cut the paper clip with the tip of the scissors (mechanical disadvantage) versus near the pivot point of the scissors (mechanical advantage).

Levers are categorized into three types according to the positions of the fulcrum, the effort, and the load:

1. The fulcrum is between the effort and the load in **first-class levers** (Figure 11.2a). (Think E*F*L.) Scissors and seesaws are examples of first-class levers. A first-class lever can produce either a mechanical advantage or mechanical disadvantage depending on whether the effort or the load is closer to the fulcrum. (Think of an adult and a child on a seesaw.) As we've seen in the preceding examples, if the effort (child) is farther from the fulcrum than the load (adult), a heavy load can be moved, but not very far or fast. If the effort is closer to the fulcrum than the load, only a lighter load can be moved, but it moves far and fast. There are few first-class levers in the body. One example is the lever formed by the head resting on the vertebral column (Figure 11.2a). When the head is raised, the contraction of the posterior neck muscles provides the effort E, the joint between the atlas and the occipital bone (atlanto-occipital joint) forms the fulcrum F, and the weight of the anterior portion of the skull is the load L.
2. The load is between the fulcrum and the effort in **second-class levers** (Figure 11.2b). (Think E*L*F.) Second-class levers operate like a wheelbarrow. They always produce a mechanical advantage because the load is always closer to the fulcrum than the effort. This arrangement sacrifices speed and range of motion for force; this type of lever produces the most force. This class of lever is uncommon in the human body. An example is standing up on your toes. The fulcrum F is the ball of the foot. The load is the weight of the body. The effort (E) is the contraction of the muscles of the calf, which raise the heel off the ground.
3. The effort is between the fulcrum and the load in **third-class levers** (Figure 11.2c). (Think F*E*L.) These levers operate like a pair of forceps and are the most common levers in the body. Third-class levers always produce a mechanical disadvantage because the effort is always closer to the fulcrum than the load. In the body, this arrangement favors speed and range of motion over force. The elbow joint, the biceps brachii muscle, and the bones of the arm and forearm are one example of a third-class lever (Figure 11.2c). As we have seen, in flexing the forearm at the elbow, the elbow joint is the fulcrum F, the contraction of the biceps brachii muscle provides the effort E, and the weight of the hand and forearm is the load L.

Effects of Fascicle Arrangement

Recall from Chapter 10 that the skeletal muscle fibers (cells) within a muscle are arranged in bundles known as **fascicles** (FAS-i-kuls). Within a fascicle, all muscle fibers are parallel to one another. The fascicles, however, may form one of five patterns with respect to the tendons: parallel, fusiform (spindle-shaped, narrow toward the ends and wide in the middle), circular, triangular, or pennate (shaped like a feather) (Table 11.1).

Fascicular arrangement affects a muscle's power and range of motion. As a muscle fiber contracts, it shortens to about 70% of its resting length. The longer the fibers in a muscle, the greater the range of motion it can produce. However, the power of a muscle depends not on length but on its total cross-sectional area, because a short fiber can contract as forcefully as a long one. So the more fibers per unit of cross-sectional area a muscle has, the more power it can produce. Fascicular arrangement often represents a compromise

Figure 11.2 Lever structure and types of levers.

Levers are divided into three types based on the placement of the fulcrum, effort, and load (resistance).

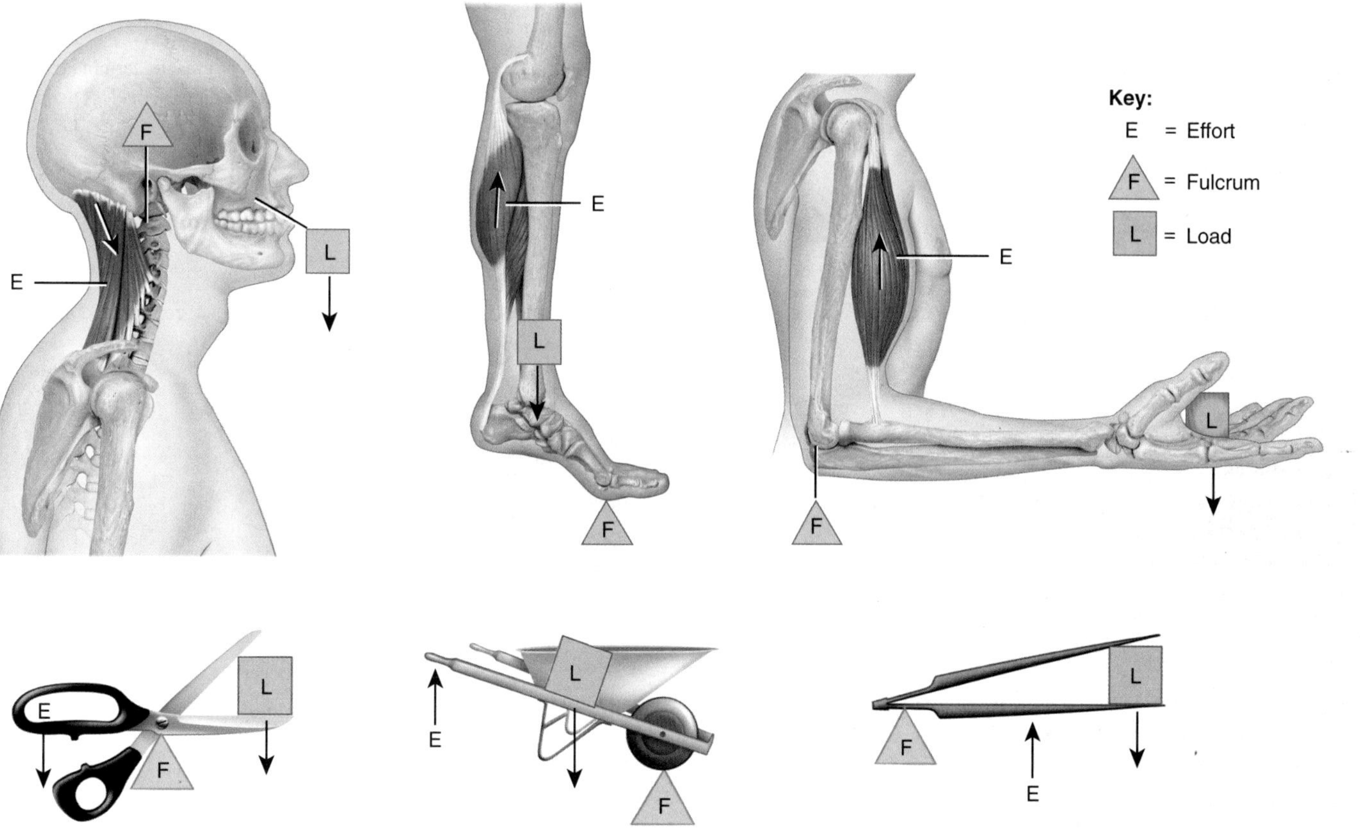
Key:
E = Effort
F = Fulcrum
L = Load
E
F
L

(a) First-class lever

(b) Second-class lever

(c) Third-class lever

Which type of lever produces the most force?

between power and range of motion. Pennate muscles, for instance, have a large number of short-fibered fascicles distributed over their tendons, giving them greater power but a smaller range of motion. In contrast, parallel muscles have comparatively fewer fascicles, but they have long fibers that extend the length of the muscle, so they have a greater range of motion but less power.

CLINICAL CONNECTION | *Intramuscular Injections*

An **intramuscular (IM) injection** penetrates the skin and subcutaneous layer to enter the muscle itself. Intramuscular injections are preferred when prompt absorption is desired, when larger doses than can be given subcutaneously are indicated, or when the drug is too irritating to give subcutaneously. The common sites for intramuscular injections include the gluteus medius muscle of the buttock (see Figure 11.3b), lateral side of the thigh in the midportion of the vastus lateralis muscle (see Figure 11.3a), and the deltoid muscle of the shoulder (see Figure 11.3b). Muscles in these areas, especially the gluteal muscles in the buttock, are fairly thick, and absorption is promoted by their extensive blood supply. To avoid injury, intramuscular injections are given deep within the muscle, away from major nerves and blood vessels. Intramuscular injections have a faster speed of delivery than oral medications but are slower than intravenous infusions. •

Coordination among Muscles

Movements often are the result of several skeletal muscles acting as a group. Most skeletal muscles are arranged in opposing (antagonistic) pairs at joints—that is, flexors–extensors, abductors–adductors, and so on. Within opposing pairs, one muscle, called the **prime mover** or **agonist** (= leader), contracts to cause an action while the other muscle, the **antagonist** (*anti-* = against), stretches and yields to the effects of the prime mover. In the process of flexing the forearm at the elbow, for instance, the biceps brachii is the prime mover, and the triceps brachii is the antagonist (see Figure 11.1a). The antagonist and prime mover are usually located on opposite sides of the bone or joint, as is the case in this example.

With an opposing pair of muscles, the roles of the prime mover and antagonist can switch for different movements. For example, while extending the forearm at the elbow against resistance (i.e., lowering the load shown in Figure 11.2c), the triceps brachii becomes the prime mover, and the biceps brachii is the antagonist. If a prime mover and its antagonist contract at the same time with equal force, there will be no movement.

Sometimes a prime mover crosses other joints before it reaches the joint at which its primary action occurs. The biceps brachii, for

TABLE 11.1

Arrangement of Fascicles

PARALLEL

Fascicles parallel to longitudinal axis of muscle; terminate at either end in flat tendons.

Example: Sternohyoid muscle (see Figure 11.8a)

FUSIFORM

Fascicles nearly parallel to longitudinal axis of muscle; terminate in flat tendons; muscle tapers toward tendons, where diameter is less than at belly.

Example: Digastric muscle (see Figure 11.8a)

CIRCULAR

Fascicles in concentric circular arrangements form sphincter muscles that enclose an orifice (opening).

Example: Orbicularis oculi muscle (see Figure 11.4a)

TRIANGULAR

Fascicles spread over broad area converge at thick central tendon; gives muscle a triangular appearance.

Example: Pectoralis major muscle (see Figure 11.3a)

PENNATE

Short fascicles in relation to total muscle length; tendon extends nearly entire length of muscle.

UNIPENNATE

Fascicles arranged on only one side of tendon.

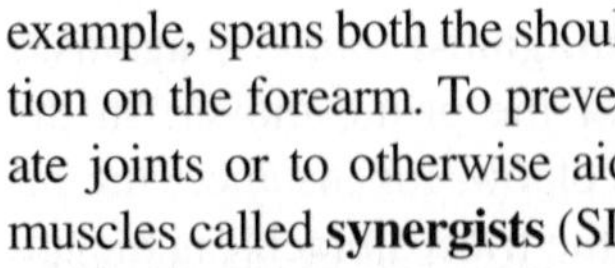

Example: Extensor digitorum longus muscle (see Figure 11.22b)

BIPENNATE

Fascicles arranged on both sides of centrally positioned tendons.

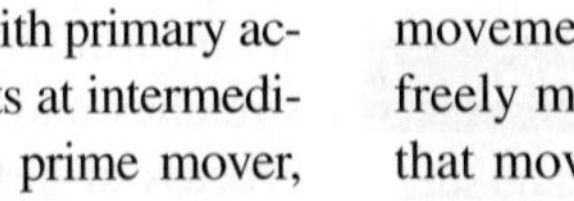

Example: Rectus femoris muscle (see Figure 11.20a)

MULTIPENNATE

Fascicles attach obliquely from many directions to several tendons.

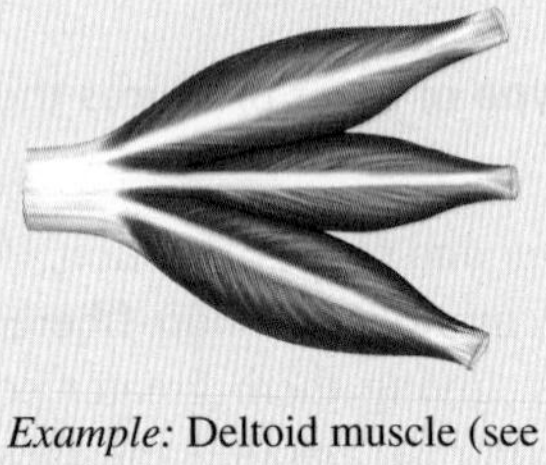

Example: Deltoid muscle (see Figure 11.10a)

example, spans both the shoulder and elbow joints, with primary action on the forearm. To prevent unwanted movements at intermediate joints or to otherwise aid the movement of the prime mover, muscles called **synergists** (SIN-er-jists; *syn-* = together; *-ergon* = work) contract and stabilize the intermediate joints. As an example, muscles that flex the fingers (prime movers) cross the intercarpal and radiocarpal joints (intermediate joints). If movement at these intermediate joints were unrestrained, you would not be able to flex your fingers without flexing the wrist at the same time. Synergistic contraction of the wrist extensor muscles stabilizes the wrist joint and prevents unwanted movement, while the flexor muscles of the fingers contract to bring about the primary action, efficient flexion of the fingers. Synergists are usually located close to the prime mover.

Some muscles in a group also act as **fixators,** stabilizing the origin of the prime mover so that the prime mover can act more efficiently. Fixators steady the proximal end of a limb while movements occur at the distal end. For example, the scapula is a freely movable bone that serves as the origin for several muscles that move the arm. When the arm muscles contract, the scapula must be held steady. In abduction of the arm, the deltoid muscle serves as the prime mover, and fixators (pectoralis minor, trapezius, subclavius, serratus anterior muscles, and others) hold the scapula firmly against the back of the chest (see Figure 11.14a, b). The insertion of the deltoid muscle pulls on the humerus to abduct the arm. Under different conditions—that is, for different movements—and at different times, many muscles may act as prime movers, antagonists, synergists, or fixators.

In the limbs, a **compartment** is a group of skeletal muscles, their associated blood vessels, and associated nerves, all of which have a common function. In the upper limbs, for example, flexor compartment muscles are anterior, and extensor compartment muscles are posterior.

CLINICAL CONNECTION | *Benefits of Stretching*

The overall goal of **stretching** is to achieve normal range of motion of joints and mobility of soft tissues surrounding the joints. For most individuals, the best stretching routine involves *static stretching,* that is, slow sustained stretching that holds a muscle in a lengthened position. The muscles should be stretched to the point of slight discomfort (not pain) and held for about 30 seconds. Stretching should be done after warming up to increase the range of motion most effectively.

1. ***Improved physical performance.*** A flexible joint has the ability to move through a greater range of motion, which improves performance.
2. ***Decreased risk of injury.*** Stretching decreases resistance in various soft tissues so there is less likelihood of exceeding maximum tissue extensibility during an activity (i.e., injuring the soft tissues).
3. ***Reduced muscle soreness.*** Stretching can reduce some of the muscle soreness that results after exercise.
4. ***Improved posture.*** Poor posture results from improper position of various parts of the body and the effects of gravity over a number of years. Stretching can help realign soft tissues to improve and maintain good posture. •

✔CHECKPOINT

1. Using the terms origin, insertion, and belly, describe how skeletal muscles produce body movements by pulling on bones.
2. List the three types of levers, and give an example of a first-, second-, and third-class lever found in the body.
3. Describe the various arrangements of fascicles.
4. Define the roles of the prime mover (agonist), antagonist, synergist, and fixator in producing various movements of the free upper limb.
5. What is a muscle compartment?

11.2 HOW SKELETAL MUSCLES ARE NAMED

OBJECTIVE

- Explain seven features used in naming skeletal muscles.

The names of most of the skeletal muscles contain combinations of the word roots of their distinctive features. This works two ways. You can learn the names of muscles by remembering the terms that refer to muscle features, such as the pattern of the muscle's fascicles; the size, shape, action, number of origins, and location of the muscle; and the sites of origin and insertion of the muscle. Knowing the names of a muscle will then give you clues about its features. Study Table 11.2 to become familiar with the terms used in muscle names.

✔CHECKPOINT

6. Select 10 muscles in Figure 11.3 and identify the features on which their names are based. (*Hint*: Use the prefix, suffix, and root of each muscle's name as a guide.)

11.3 PRINCIPAL SKELETAL MUSCLES

Exhibits 11.A–11.T will assist you in learning the names of the principal skeletal muscles in various regions of the body. The muscles in the exhibits are divided into groups according to the part of the body on which they act. As you study groups of muscles in the exhibits, refer to Figure 11.3 to see how each group is related to the others.

The exhibits contain the following elements:

- ***Objective.*** This statement describes what you should learn from the exhibit.
- ***Overview.*** These paragraphs provide a general introduction to the muscles under consideration and emphasize how the muscles are organized within various regions. The discussion also highlights any distinguishing features of the muscles.
- ***Muscle names.*** The word roots indicate how the muscles are named. As noted previously, once you have mastered the naming of the muscles, you can more easily understand their actions.
- ***Origins, insertions, and actions.*** You are also given the origin, insertion, and actions of each muscle.
- ***Relating muscles to movements.*** These exercises will help you organize the muscles in the body region under consideration according to the actions they produce.
- ***Innervation.*** This section lists the nerve or nerves that cause contraction of each muscle. In general, cranial nerves, which arise from the lower parts of the brain, serve muscles in the head region. Spinal nerves, which arise from the spinal cord within the vertebral column, innervate muscles in the rest of the body. Cranial nerves are designated by both a name and a Roman numeral: the facial (VII) nerve, for example. Spinal nerves are numbered in groups according to the part of the spinal cord from which they arise: C = cervical (neck region), T = thoracic (chest region), L = lumbar (lower-back region), and S = sacral (buttocks region). An example is T1, the first thoracic spinal nerve.
- ***Questions.*** These knowledge checkpoints relate specifically to information in each exhibit, and take the form of review, critical thinking, and/or application questions.
- ***Clinical Connections.*** Selected exhibits include clinical applications, which explore the clinical, professional, or everyday relevance of a particular muscle or its function through descriptions of disorders or clinical procedures.
- ***Figures.*** The figures in the exhibits may present superficial and deep, anterior and posterior, or medial and lateral views to show each muscle's position as clearly as possible. The muscle names in all capital letters are specifically referred to in the table part of the exhibit.

TABLE 11.2

Characteristics Used to Name Muscles

NAME	MEANING	EXAMPLE	FIGURE
DIRECTION: ORIENTATION OF MUSCLE FASCICLES RELATIVE TO THE BODY'S MIDLINE			
Rectus	Parallel to midline	Rectus abdominis	11.10c
Transverse	Perpendicular to midline	Transversus abdominis	11.10c
Oblique	Diagonal to midline	External oblique	11.10a
SIZE: RELATIVE SIZE OF THE MUSCLE			
Maximus	Largest	Gluteus maximus	11.20c
Minimus	Smallest	Gluteus minimus	11.20d
Longus	Long	Adductor longus	11.20a
Brevis	Short	Adductor brevis	11.20b
Latissimus	Widest	Latissimus dorsi	11.15b
Longissimus	Longest	Longissimus capitis	11.19a
Magnus	Large	Adductor magnus	11.20b
Major	Larger	Pectoralis major	11.10a
Minor	Smaller	Pectoralis minor	11.14a
Vastus	Huge	Vastus lateralis	11.20a
SHAPE: RELATIVE SHAPE OF THE MUSCLE			
Deltoid	Triangular	Deltoid	11.15b
Trapezius	Trapezoid	Trapezius	11.3b
Serratus	Saw-toothed	Serratus anterior	11.14b
Rhomboid	Diamond-shaped	Rhomboid major	11.15d
Orbicularis	Circular	Orbicularis oculi	11.4a
Pectinate	Comblike	Pectineus	11.20a
Piriformis	Pear-shaped	Piriformis	11.20d
Platys	Flat	Platysma	11.4c
Quadratus	Square, four-sided	Quadratus femoris	11.20d
Gracilis	Slender	Gracilis	11.20a
ACTION: PRINCIPAL ACTION OF THE MUSCLE			
Flexor	Decreases joint angle	Flexor carpi radialis	11.17a
Extensor	Increases joint angle	Extensor carpi ulnaris	11.17d
Abductor	Moves bone away from midline	Abductor pollicis longus	11.17e
Adductor	Moves bone closer to midline	Adductor longus	11.20a
Levator	Raises or elevates body part	Levator scapulae	11.14a
Depressor	Lowers or depresses body part	Depressor labii inferioris	11.4b
Supinator	Turns palm anteriorly	Supinator	11.17c
Pronator	Turns palm posteriorly	Pronator teres	11.17a
Sphincter	Decreases size of an opening	External anal sphincter	11.12
Tensor	Makes body part rigid	Tensor fasciae latae	11.20a
Rotator	Rotates bone around longitudinal axis	Rotatore	11.19b
NUMBER OF ORIGINS: NUMBER OF TENDONS OF ORIGIN			
Biceps	Two origins	Biceps brachii	11.16a
Triceps	Three origins	Triceps brachii	11.16b
Quadriceps	Four origins	Quadriceps femoris	11.20a
LOCATION: STRUCTURE NEAR WHICH A MUSCLE IS FOUND			
Example: Temporalis, muscle near temporal bone.			11.4c
ORIGIN AND INSERTION: SITES WHERE MUSCLE ORIGINATES AND INSERTS			
Example: Sternocleidomastoid, originating on sternum and clavicle and inserting on mastoid process of temporal bone.			11.3a

Figure 11.3 **Principal superficial skeletal muscles.**

Most movements require several skeletal muscles acting in groups rather than individually.

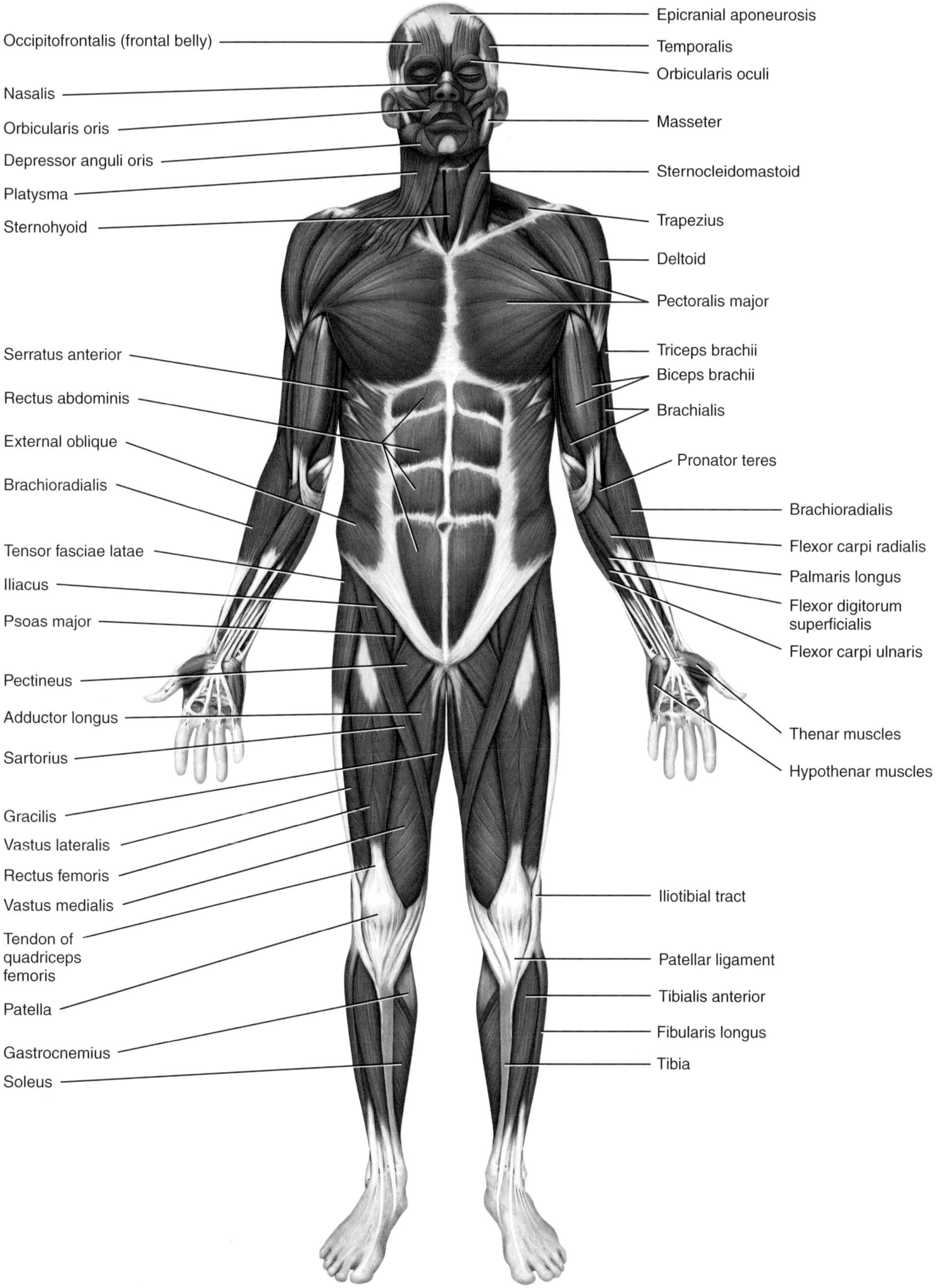
Epicranial aponeurosis
Occipitofrontalis (frontal belly)
Temporalis
Orbicularis oculi
Nasalis
Orbicularis oris
Masseter
Depressor anguli oris
Platysma
Sternocleidomastoid
Sternohyoid
Trapezius
Deltoid
Pectoralis major
Serratus anterior
Triceps brachii
Biceps brachii
Rectus abdominis
Brachialis
External oblique
Pronator teres
Brachioradialis
Brachioradialis
Flexor carpi radialis
Tensor fasciae latae
Palmaris longus
Iliacus
Flexor digitorum superficialis
Psoas major
Flexor carpi ulnaris
Pectineus
Adductor longus
Thenar muscles
Sartorius
Hypothenar muscles
Gracilis
Vastus lateralis
Rectus femoris
Vastus medialis
Iliotibial tract
Tendon of quadriceps femoris
Patellar ligament
Patella
Tibialis anterior
Fibularis longus
Gastrocnemius
Tibia
Soleus

(a) Anterior view

FIGURE 11.3 CONTINUES

FIGURE 11.3 CONTINUED

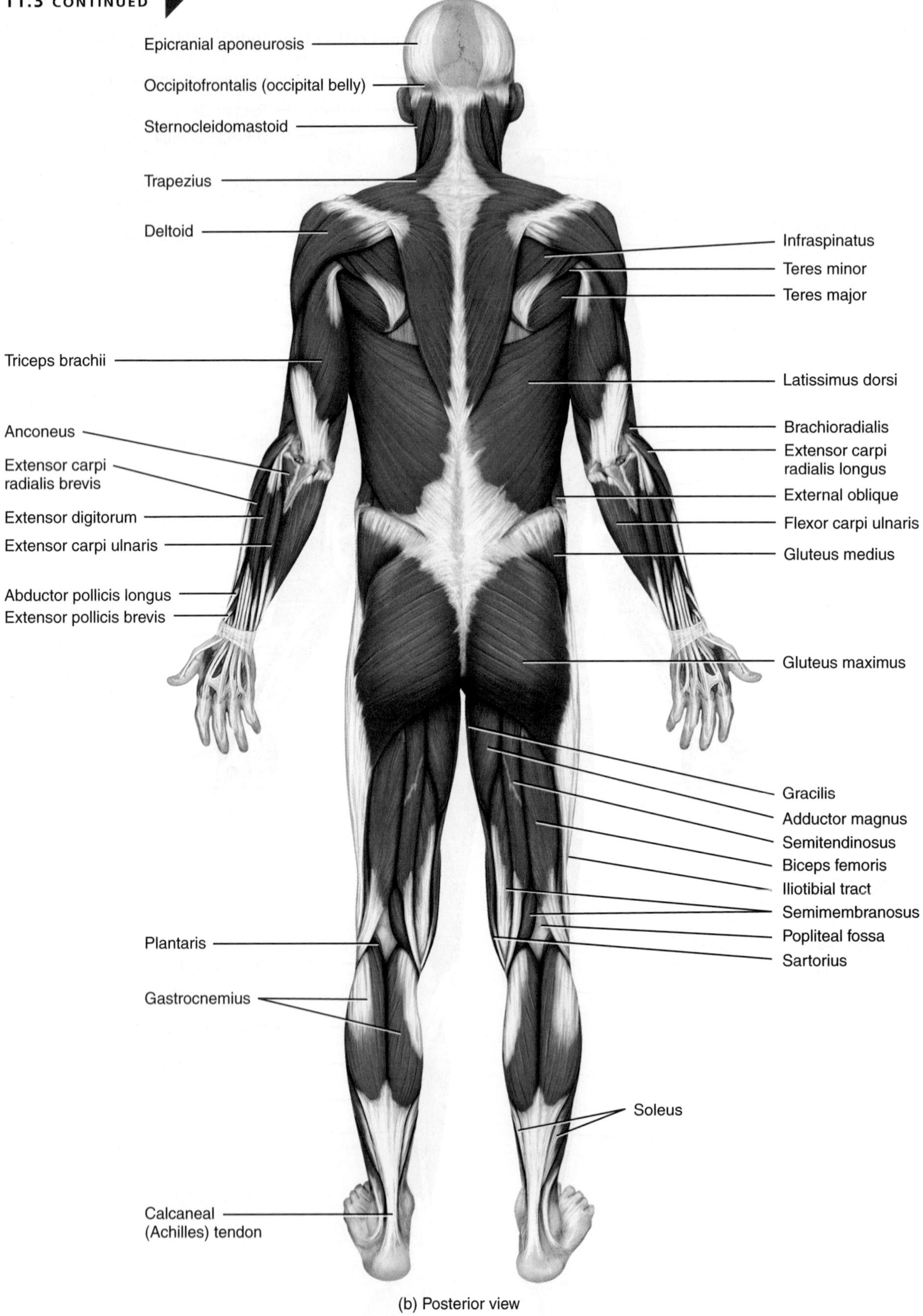
Epicranial aponeurosis
Occipitofrontalis (occipital belly)
Sternocleidomastoid
Trapezius
Deltoid
Infraspinatus
Teres minor
Teres major
Triceps brachii
Latissimus dorsi
Anconeus
Brachioradialis
Extensor carpi radialis brevis
Extensor carpi radialis longus
External oblique
Extensor digitorum
Flexor carpi ulnaris
Extensor carpi ulnaris
Gluteus medius
Abductor pollicis longus
Extensor pollicis brevis
Gluteus maximus
Gracilis
Adductor magnus
Semitendinosus
Biceps femoris
Iliotibial tract
Semimembranosus
Plantaris
Popliteal fossa
Sartorius
Gastrocnemius
Soleus
Calcaneal (Achilles) tendon

(b) Posterior view

Give an example of a muscle named for each of the following characteristics: direction of fibers, shape, action, size, origin and insertion, location, and number of tendons of origin.

EXHIBIT 11.A Muscles of the Head That Produce Facial Expressions *(Figure 11.4)*

OBJECTIVE

- Describe the origin, insertion, action, and innervation of the muscles of facial expression.

The muscles of facial expression, which provide us with the ability to express a wide variety of emotions, lie within the subcutaneous layer (Figure 11.4). They usually originate in the fascia or bones of the skull and insert into the skin. Because of their insertions, the muscles of facial expression move the skin rather than a joint when they contract.

Among the noteworthy muscles in this group are those surrounding the orifices (openings) of the head such as the eyes, nose, and mouth. These muscles function as *sphincters* (SFINGK-ters), which close the orifices, and *dilators* (DĪ-lā-tors), which dilate or open the orifices. For example, the **orbicularis oculi** muscle closes the eye, and the levator palpebrae superioris muscle opens it (discussed in Exhibit 11.B). The **occipitofrontalis** is an unusual muscle in this group because it is made up of two parts: an anterior part called the **frontal belly (frontalis),** which is superficial to the frontal bone, and a posterior part called the **occipital belly (occipitalis),** which is superficial to the occipital bone. The two muscular portions are held together by a strong **aponeurosis** (sheetlike tendon), the **epicranial aponeurosis** (ep-i-KRĀ-nē-al ap′-ō-noo-RŌ-sis), also called the **galea aponeurotica** (GĀ-lē-a ap′-ō-noo′-RŌ-ti-ka), that covers the superior and lateral surfaces of the skull. The **buccinator** muscle forms the major muscular portion of the cheek. The duct of the parotid gland (a salivary gland) passes through the buccinator muscle to reach the oral cavity. The buccinator muscle is so named because it compresses the cheeks (*bucc-* = cheek) during blowing—for example, when a musician plays a brass instrument such as a trumpet. It functions in whistling, blowing, and sucking and assists in chewing.

CLINICAL CONNECTION | *Bell's Palsy*

Bell's palsy, also known as **facial paralysis,** is a unilateral paralysis of the muscles of facial expression. It is due to damage or disease of the facial (VII) nerve. Possible causes include inflammation of the facial nerve due to an ear infection, ear surgery that damages the facial nerve, or infection by the herpes simplex virus. The paralysis causes the entire side of the face to droop in severe cases. The person cannot wrinkle the forehead, close the eye, or pucker the lips on the affected side. Drooling and difficulty in swallowing also occur. Eighty percent of patients recover completely within a few weeks to a few months. For other, paralysis is permanent. The symptoms of Bell's palsy mimic those of a stroke. •

RELATING MUSCLES TO MOVEMENTS

Arrange the muscles in this exhibit into two groups: (1) those that act on the mouth and (2) those that act on the eyes.

CHECKPOINT

Why do the muscles of facial expression move the skin rather than a joint?

EXHIBIT 11.A CONTINUES

Muscles of the Head That Produce Facial Expressions *(Figure 11.4)* CONTINUED

MUSCLE	ORIGIN	INSERTION	ACTION	INNERVATION
SCALP MUSCLES				
Occipitofrontalis (ok-sip′-i-tō-frun-TĀ-lis)				
Frontal belly	Epicranial aponeurosis.	Skin superior to supraorbital margin.	Draws scalp anteriorly, raises eyebrows, and wrinkles skin of forehead horizontally as in look of surprise.	Facial (VII) nerve.
Occipital belly (*occipit-* = back of the head)	Occipital bone and mastoid process of temporal bone.	Epicranial aponeurosis.	Draws scalp posteriorly.	Facial (VII) nerve.
MOUTH MUSCLES				
Orbicularis oris (or-bi′-kū-LAR-is OR-is; *orb-* = circular; *oris* = of the mouth)	Muscle fibers surrounding opening of mouth.	Skin at corner of mouth.	Closes and protrudes lips, as in kissing; compresses lips against teeth; and shapes lips during speech.	Facial (VII) nerve.
Zygomaticus major (zī-gō-MA-ti-kus; *zygomatic* = cheek bone; *major* = greater)	Zygomatic bone.	Skin at angle of mouth and orbicularis oris.	Draws angle of mouth superiorly and laterally, as in smiling.	Facial (VII) nerve.
Zygomaticus minor (*minor* = lesser)	Zygomatic bone.	Upper lip.	Raises (elevates) upper lip, exposing maxillary (upper) teeth.	Facial (VII) nerve.
Levator labii superioris (le-VĀ-tor LĀ-bē-ī soo-per′-ē-OR-is; *levator* = raises or elevates; *labii* = lip; *superioris* = upper)	Superior to infraorbital foramen of maxilla.	Skin at angle of mouth and orbicularis oris.	Raises upper lip.	Facial (VII) nerve.
Depressor labii inferioris (de-PRE-sor LĀ-bē-ī; *depressor* = depresses or lowers; *inferioris* = lower)	Mandible.	Skin of lower lip.	Depresses (lowers) lower lip.	Facial (VII) nerve.
Depressor anguli oris (ANG-ū-lī; *angul* = angle or corner; *oris* = mouth)	Mandible.	Angle of mouth.	Draws angle of mouth laterally and inferiorly, as in opening mouth.	Facial (VII) nerve.
Levator anguli oris	Inferior to infraorbital foramen.	Skin of lower lip and orbicularis oris.	Draws angle of mouth laterally and superiorly.	Facial (VII) nerve.
Buccinator (BUK-si-nā′-tor; *bucc-* = cheek)	Alveolar processes of maxilla and mandible and pterygomandibular raphe (fibrous band extending from pterygoid process of sphenoid bone to mandible).	Orbicularis oris.	Presses cheeks against teeth and lips, as in whistling, blowing, and sucking; draws corner of mouth laterally; and assists in mastication (chewing) by keeping food between the teeth (and not between teeth and cheeks).	Facial (VII) nerve.
Risorius (ri-ZOR-ē-us; *risor* = laughter)	Fascia over parotid (salivary) gland.	Skin at angle of mouth.	Draws angle of mouth laterally, as in grimacing.	Facial (VII) nerve.
Mentalis (men-TĀ-lis; *mental* = chin)	Mandible.	Skin of chin.	Elevates and protrudes lower lip and pulls skin of chin up, as in pouting.	Facial (VII) nerve.
NECK MUSCLES				
Platysma (pla-TIZ-ma; *platys* = flat, broad)	Fascia over deltoid and pectoralis major muscles.	Mandible, blends with muscles around angle of mouth, and skin of lower face.	Draws outer part of lower lip inferiorly and posteriorly as in pouting; depresses mandible.	Facial (VII) nerve.

MUSCLE	ORIGIN	INSERTION	ACTION	INNERVATION
ORBIT AND EYEBROW MUSCLES				
Orbicularis oculi (or-bi′-kū-LAR-is OK-ū-lī; *oculi* = eye)	Medial wall of orbit.	Circular path around orbit.	Closes eye.	Facial (VII) nerve.
Corrugator supercilii (KOR-a-gā′-tor soo-per-SIL-ē-ī; *corrugat* = wrinkle; *supercilii* = eyebrow)	Medial end of superciliary arch of frontal bone.	Skin of eyebrow.	Draws eyebrow inferiorly and wrinkles skin of forehead vertically as in frowning.	Facial (VII) nerve.

Figure 11.4 **Muscles of the head that produce facial expressions.**

When they contract, muscles of facial expression move the skin rather than a joint.

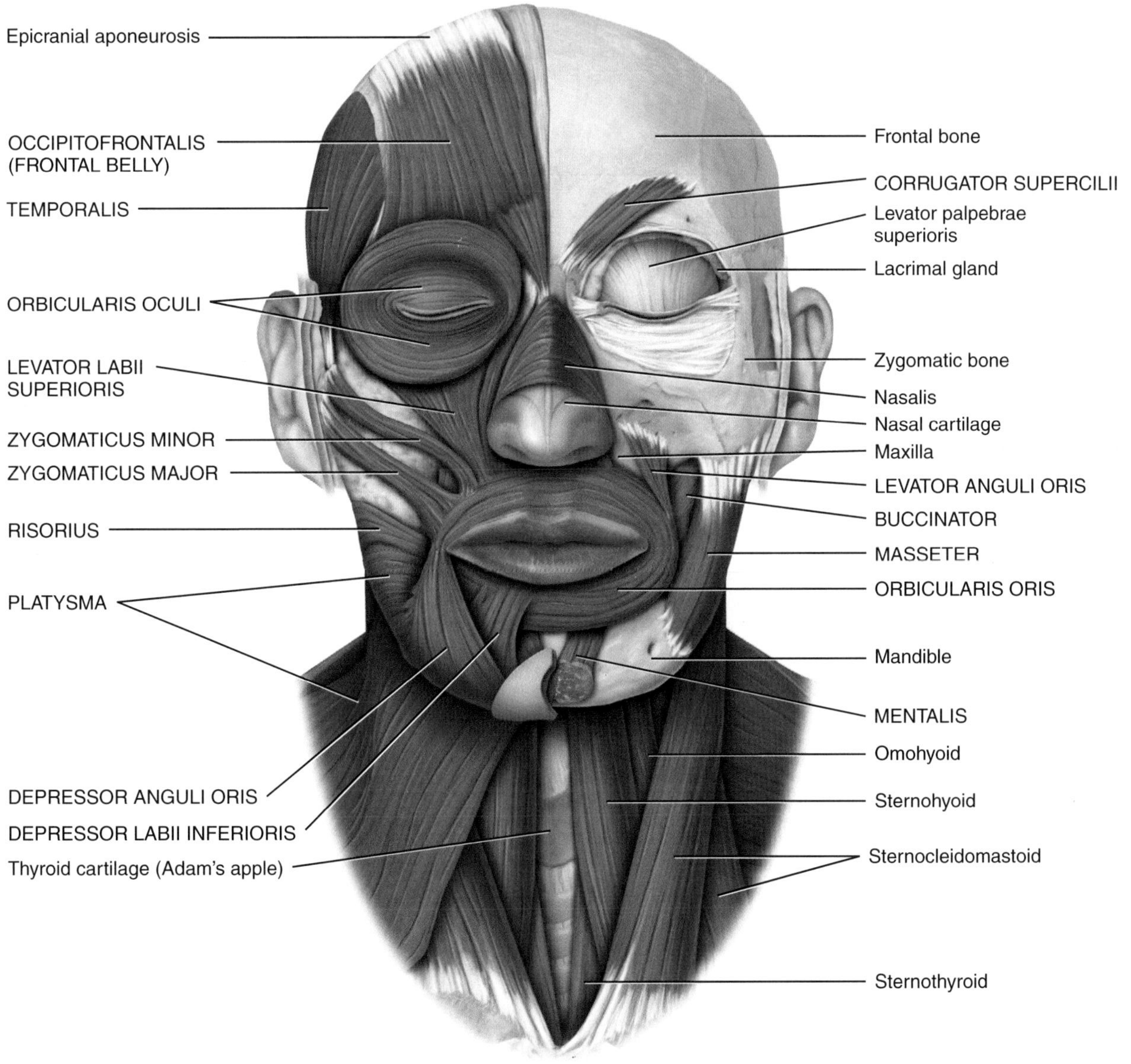

Epicranial aponeurosis
OCCIPITOFRONTALIS (FRONTAL BELLY)
TEMPORALIS
ORBICULARIS OCULI
LEVATOR LABII SUPERIORIS
ZYGOMATICUS MINOR
ZYGOMATICUS MAJOR
RISORIUS
PLATYSMA
DEPRESSOR ANGULI ORIS
DEPRESSOR LABII INFERIORIS
Thyroid cartilage (Adam's apple)
Frontal bone
CORRUGATOR SUPERCILII
Levator palpebrae superioris
Lacrimal gland
Zygomatic bone
Nasalis
Nasal cartilage
Maxilla
LEVATOR ANGULI ORIS
BUCCINATOR
MASSETER
ORBICULARIS ORIS
Mandible
MENTALIS
Omohyoid
Sternohyoid
Sternocleidomastoid
Sternothyroid

(a) Anterior superficial view (b) Anterior deep view

EXHIBIT 11.A CONTINUES

FIGURE 11.4 CONTINUED

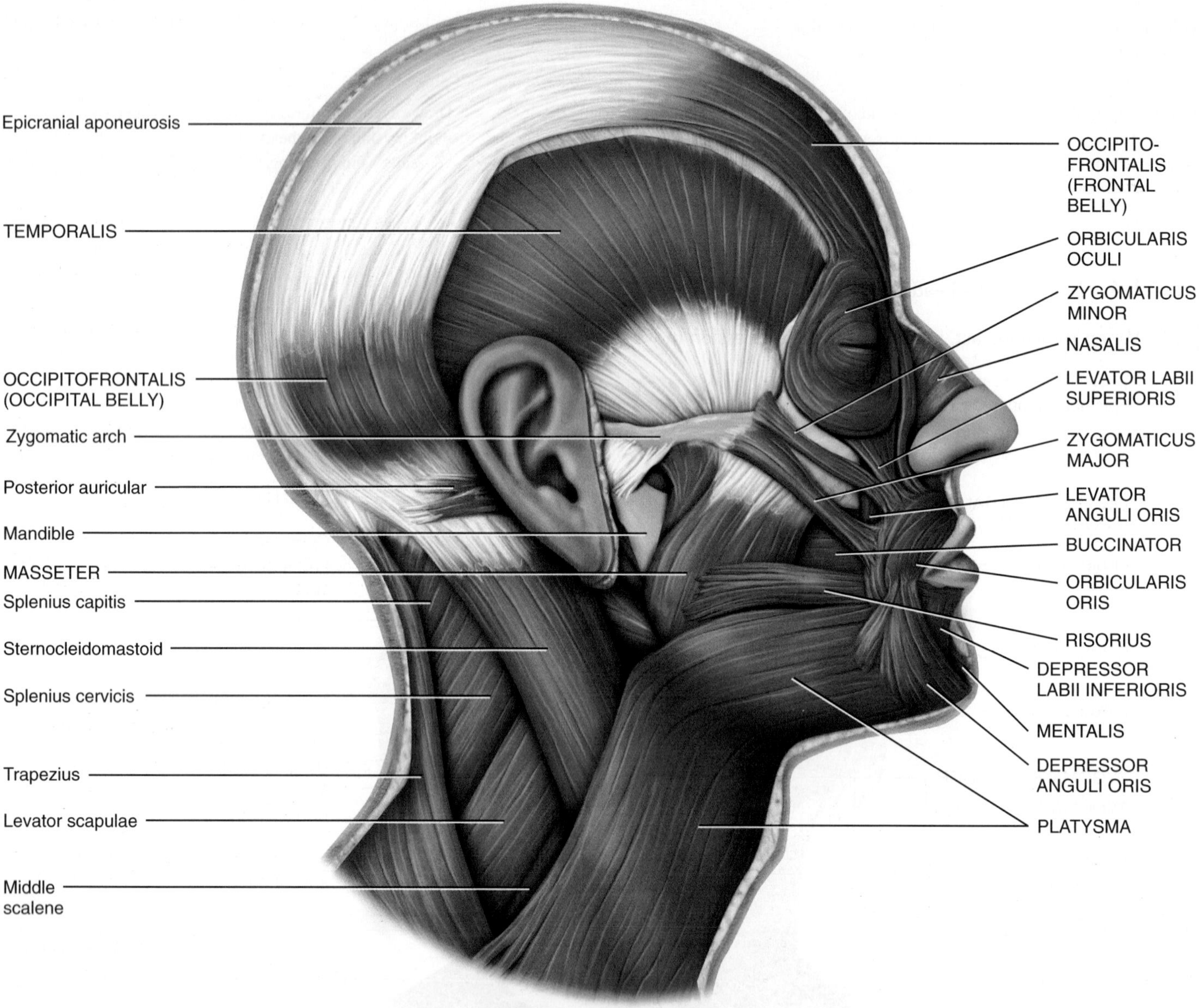
Epicranial aponeurosis
TEMPORALIS
OCCIPITOFRONTALIS (OCCIPITAL BELLY)
Zygomatic arch
Posterior auricular
Mandible
MASSETER
Splenius capitis
Sternocleidomastoid
Splenius cervicis
Trapezius
Levator scapulae
Middle scalene
OCCIPITO-FRONTALIS (FRONTAL BELLY)
ORBICULARIS OCULI
ZYGOMATICUS MINOR
NASALIS
LEVATOR LABII SUPERIORIS
ZYGOMATICUS MAJOR
LEVATOR ANGULI ORIS
BUCCINATOR
ORBICULARIS ORIS
RISORIUS
DEPRESSOR LABII INFERIORIS
MENTALIS
DEPRESSOR ANGULI ORIS
PLATYSMA

(c) Right lateral superficial view

Which muscles of facial expression cause frowning, smiling, pouting, and squinting?

EXHIBIT 11.B Muscles of the Head That Move the Eyeballs (Extrinsic Eye Muscles) and Upper Eyelids *(Figure 11.5)*

OBJECTIVE

- Describe the origin, insertion, action, and innervation of the extrinsic eye muscles that move the eyeballs and upper eyelids.

Muscles that move the eyeballs are called **extrinsic eye muscles** because they originate outside the eyeballs (in the orbit) and insert on the outer surface of the sclera ("white of the eye") (Figure 11.5). The extrinsic eye muscles are some of the fastest contracting and most precisely controlled skeletal muscles in the body.

Three pairs of extrinsic eye muscles control movements of the eyeballs: (1) superior and inferior recti, (2) lateral and medial recti, and (3) superior and inferior obliques. The four recti muscles (superior, inferior, lateral, and medial) arise from a tendinous ring in the orbit and insert into the sclera of the eye. As their names imply, the **superior** and **inferior recti** move the eyeballs superiorly and inferiorly; the **lateral** and **medial recti** move the eyeballs laterally and medially, respectively.

The actions of the oblique muscles cannot be deduced from their names. The **superior oblique** muscle originates posteriorly near the tendinous ring, then passes anteriorly superior to the medial rectus muscle, and ends in a round tendon. The tendon extends through a pulleylike loop of fibrocartilaginous tissue called the *trochlea* (= pulley) on the anterior and medial part of the roof of the orbit. Finally, the tendon turns and inserts on the posterolateral aspect of the eyeball. Accordingly, the superior oblique muscle moves the eyeballs inferiorly and laterally. The **inferior oblique** muscle originates on the maxilla at the anteromedial aspect of the floor of the orbit. It then passes posteriorly and laterally and inserts on the posterolateral aspect of the eyeball. Because of this arrangement, the inferior oblique muscle moves the eyeballs superiorly and laterally.

Unlike the recti and oblique muscles, the **levator palpebrae superioris** does not move the eyeballs, since its tendon passes the eyeball and inserts into the upper eyelid. Rather, it raises the upper eyelids, that is, opens the eyes. It is therefore an antagonist to the orbicularis oculi, which closes the eyes.

CLINICAL CONNECTION | *Strabismus*

Strabismus (stra-BIZ-mus; *strabismos* = squinting) is a condition in which the two eyeballs are not properly aligned. This can be hereditary or it can be due to birth injuries, poor attachments of the muscles, problems with the brain's control center, or localized disease. Strabismus can be constant or intermittent. In strabismus, each eye sends an image to a different area of the brain and because the brain usually ignores the messages sent by one of the eyes, the ignored eye becomes weaker; hence "lazy eye," or *amblyopia*, develops. *External strabismus* results when a lesion in the oculomotor (III) nerve causes the eyeball to move laterally when at rest, and results in an inability to move the eyeball medially and inferiorly. A lesion in the abducens (VI) nerve results in *internal strabismus*, a condition in which the eyeball moves medially when at rest and cannot move laterally.

Treatment options for strabismus depend on the specific type of problem and include surgery, visual therapy (retraining the brain's control center), and orthoptics (eye muscle training to straighten the eyes). •

MUSCLE	ORIGIN	INSERTION	ACTION	INNERVATION
Superior rectus (*rectus* = fascicles parallel to midline)	Common tendinous ring (attached to orbit around optic foramen).	Superior and central part of eyeballs.	Moves eyeballs superiorly (elevation) and medially (adduction), and rotates them medially.	Oculomotor (III) nerve.
Inferior rectus	Same as above.	Inferior and central part of eyeballs.	Moves eyeballs inferiorly (depression) and medially (adduction), and rotates them medially.	Oculomotor (III) nerve.
Lateral rectus	Same as above.	Lateral side of eyeballs.	Moves eyeballs laterally (abduction).	Abducens (VI) nerve.
Medial rectus	Same as above.	Medial side of eyeballs.	Moves eyeballs medially (adduction).	Oculomotor (III) nerve.
Superior oblique (*oblique* = fascicles diagonal to midline)	Sphenoid bone, superior and medial to common tendinous ring in orbit.	Eyeball between superior and lateral recti. Muscle inserts into superior and lateral surfaces of eyeball via tendon that passes through trochlea.	Moves eyeballs inferiorly (depression) and laterally (abduction), and rotates them medially.	Trochlear (IV) nerve.
Inferior oblique	Maxilla in floor of orbit.	Eyeballs between inferior and lateral recti.	Moves eyeballs superiorly (elevation) and laterally (abduction), and rotates them laterally.	Oculomotor (III) nerve.
Levator palpebrae superioris (le-VĀ-tor PAL-pe-brē soo-per′-ē-OR-is; *palpebrae* = eyelids)	Roof of orbit (lesser wing of sphenoid bone).	Skin and tarsal plate of upper eyelids. (opens eyes).	Elevates upper eyelids	Oculomotor (III) nerve.

EXHIBIT 11.B CONTINUES ▶

RELATING MUSCLES TO MOVEMENTS

Arrange the muscles in this exhibit according to their actions on the eyeballs: (1) elevation, (2) depression, (3) abduction, (4) adduction, (5) medial rotation, and (6) lateral rotation. The same muscle may be mentioned more than once.

✓CHECKPOINT

Which muscles that move the eyeballs contract and relax as you look to your left without moving your head?

Figure 11.5 Muscles of the head that move the eyeballs (extrinsic eye muscles) and upper eyelid.

The extrinsic muscles of the eyeball are among the fastest contracting and most precisely controlled skeletal muscles in the body.

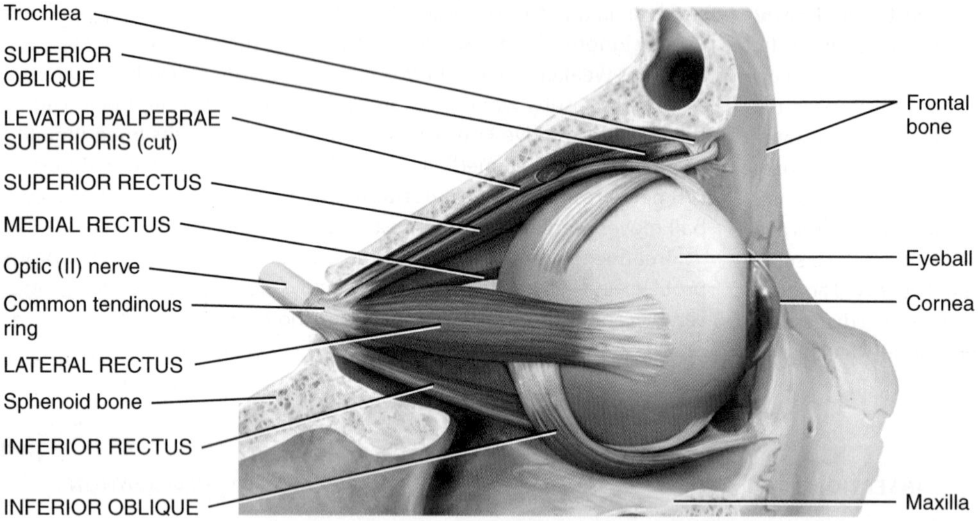

(a) Right lateral view of right eyeball

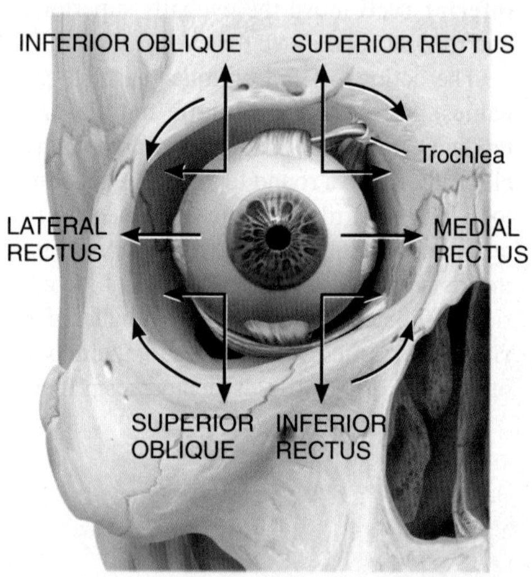

(b) Movements of right eyeball in response to contraction of extrinsic muscles

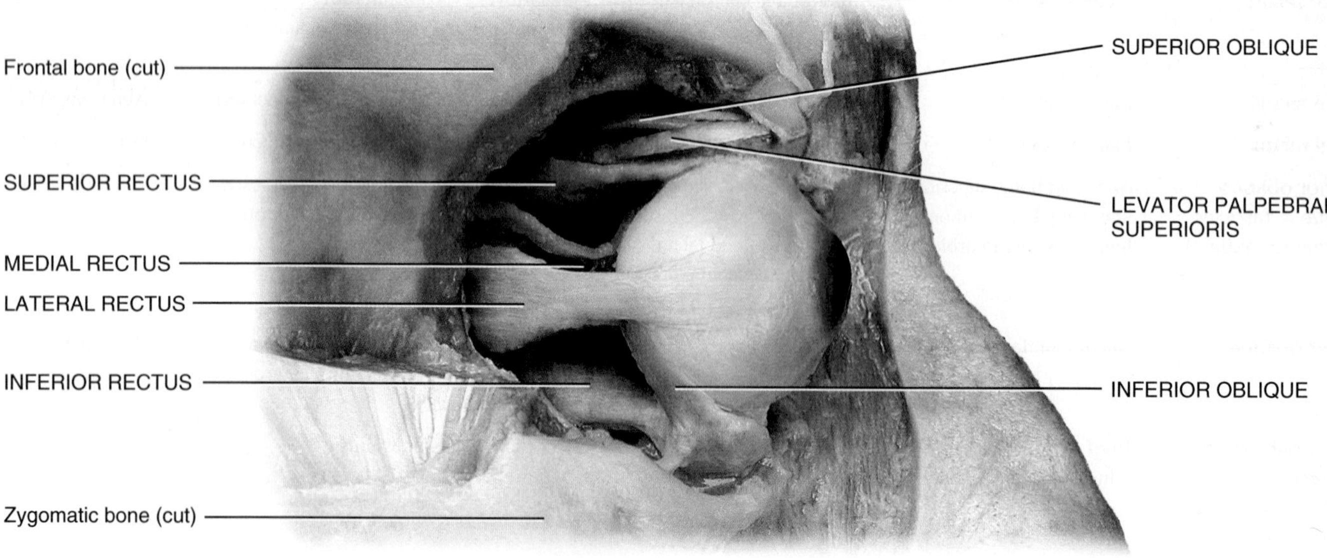

(c) Right lateral view of right eyeball

? How does the inferior oblique muscle move the eyeball superiorly and laterally?

EXHIBIT 11.C Muscles That Move the Mandible and Assist in Mastication and Speech *(Figure 11.6)*

OBJECTIVE

- Describe the origin, insertion, action, and innervation of the muscles that move the mandible and assist in mastication and speech.

The muscles that move the mandible (lower jawbone) at the temporomandibular joint (TMJ) are known as the **muscles of mastication (chewing)** (Figure 11.6). Of the four pairs of muscles involved in mastication, three are powerful closers of the jaw and account for the strength of the bite: **masseter, temporalis,** and **medial pterygoid.** Of these, the masseter is the strongest muscle of mastication. The **medial** and **lateral pterygoid** muscles assist in mastication by moving the mandible from side to side to help grind food. Additionally, the lateral pterygoid muscles protract (protrude) the mandible. The masseter muscle has been removed in Figure 11.6 to illustrate the deeper pterygoid muscles; the masseter can be seen in Figure 11.4c. Note the enormous bulk of the temporalis and masseter muscles compared to the smaller mass of the two pterygoid muscles.

Note: A mnemonic for muscles of mastication is **T**eeny **M**ice **M**ake **P**etite **L**ittle **P**rints = **T**emporalis, **M**asseter, **M**edial **P**terygoid, and **L**ateral **P**terygoid.

CLINICAL CONNECTION | *Gravity and the Mandible*

As just noted, three of the four muscles of mastication close the mandible and only the lateral pterygoid opens the mouth. The force of **gravity on the mandible** offsets this imbalance. When the masseter, temporalis, and medial pterygoid muscles relax, the mandible drops. Now you know why the mouth of many persons, particularly the elderly, is open while the person is asleep in a chair. In contrast, astronauts in zero gravity must work hard to open their mouths. •

RELATING MUSCLES TO MOVEMENTS

Arrange the muscles in this exhibit according to their actions on the mandible: (1) elevation, (2) depression, (3) retraction, (4) protraction, and (5) side-to-side movement. The same muscle may be mentioned more than once.

✔ CHECKPOINT

What would happen if you lost tone in the masseter and temporalis muscles?

MUSCLE	ORIGIN	INSERTION	ACTION	INNERVATION
Masseter (MA-se-ter = chewer) (see Figure 11.4c)	Maxilla and zygomatic arch.	Angle and ramus of mandible.	Elevates mandible, as in closing mouth.	Mandibular division of trigeminal (V) nerve.
Temporalis (tem′-pō-RĀ-lis; *tempor-* = time or temples)	Temporal bone.	Coronoid process and ramus of mandible.	Elevates and retracts mandible.	Mandibular division of trigeminal (V) nerve.
Medial pterygoid (TER-i-goyd; *medial* = closer to midline; *pterygoid* = winglike)	Medial surface of lateral portion of pterygoid process of sphenoid bone; maxilla.	Angle and ramus of mandible.	Elevates and protracts (protrudes) mandible and moves mandible from side to side.	Mandibular division of trigeminal (V) nerve.
Lateral pterygoid (TER-i-goyd; *lateral* = farther from midline)	Greater wing and lateral surface of lateral portion of pterygoid process of sphenoid bone.	Condyle of mandible; temporomandibular joint (TMJ).	Protracts mandible, depresses mandible as in opening mouth, and moves mandible from side to side.	Mandibular division of trigeminal (V) nerve.

EXHIBIT 11.C CONTINUES

Figure 11.6 Muscles that move the mandible (lower jawbone) and assist in mastication (chewing) and speech.

The muscles that move the mandible are also known as muscles of mastication.

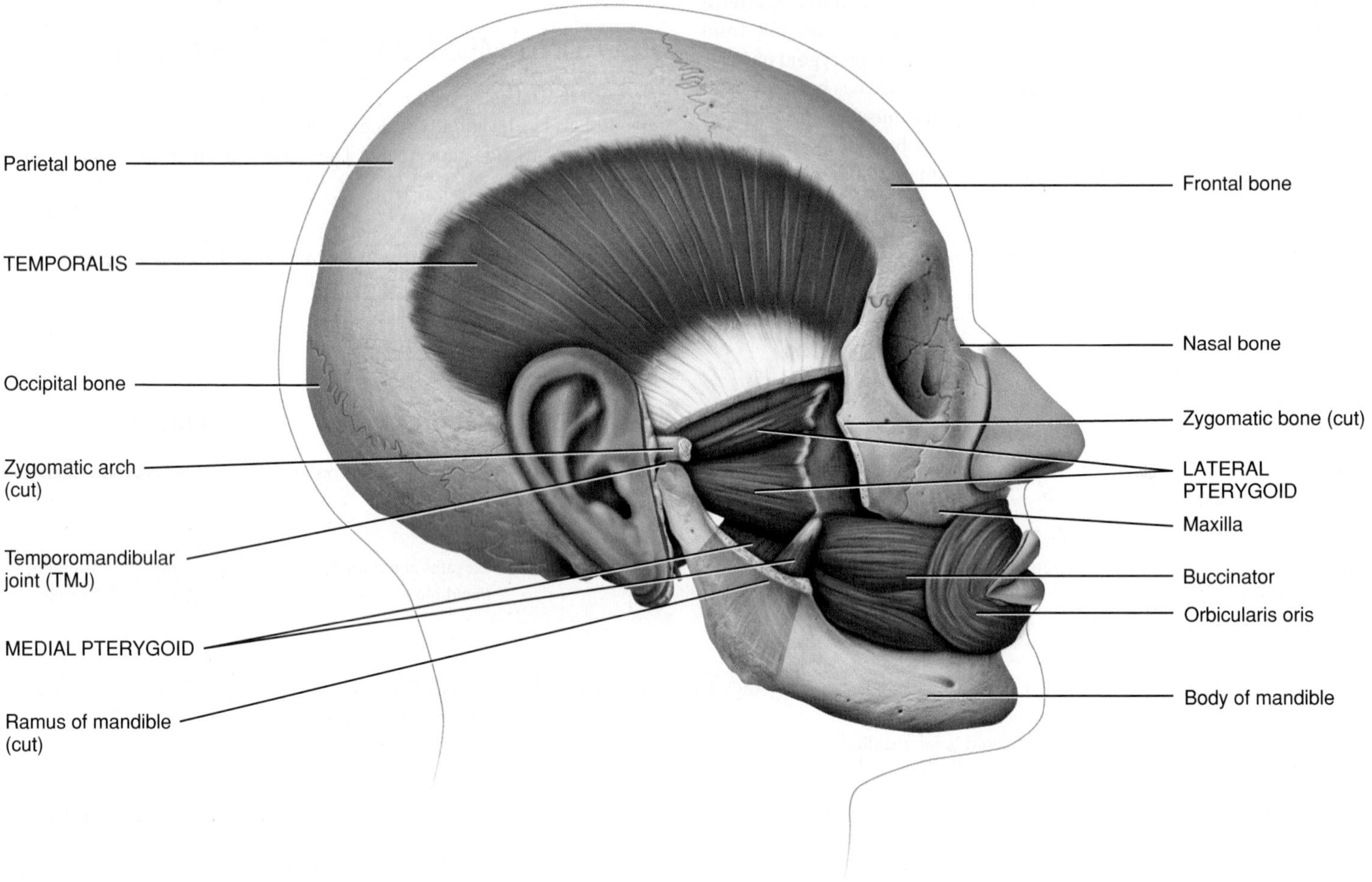

Right lateral superficial view

Which is the strongest muscle of mastication?

EXHIBIT 11.D Muscles of the Head That Move the Tongue and Assist in Mastication and Speech *(Figure 11.7)*

OBJECTIVE

- Describe the origin, insertion, action, and innervation of the muscles that move the tongue and assist in mastication and speech.

The tongue is a highly mobile structure that is vital to digestive functions such as *mastication* (chewing), detection of taste, and *deglutition* (swallowing). It is also important in speech. The tongue's mobility is greatly aided by its attachment to the mandible, styloid process of the temporal bone, and hyoid bone.

The tongue is divided into lateral halves by a median fibrous septum. The septum extends throughout the length of the tongue. Inferiorly, the septum attaches to the hyoid bone. Muscles of the tongue are of two principal types: extrinsic and intrinsic. **Extrinsic tongue muscles** originate outside the tongue and insert into it (Figure 11.7). They move the entire tongue in various directions, such as anteriorly, posteriorly, and laterally. **Intrinsic tongue muscles** originate and insert within the tongue. These muscles alter the shape of the tongue rather than moving the entire tongue. The extrinsic and intrinsic muscles of the tongue insert into both lateral halves of the tongue.

When you study the extrinsic tongue muscles, you will notice that all of their names end in *glossus,* meaning tongue. You will also notice that the actions of the muscles are obvious, considering the positions of the mandible, styloid process, hyoid bone, and soft palate, which serve as origins for these muscles. For example, the **genioglossus** (origin: the mandible) pulls the tongue downward and forward, the **styloglossus** (origin: the styloid process) pulls the tongue upward and backward, the **hyoglossus** (origin: the hyoid bone) pulls the tongue downward and flattens it, and the **palatoglossus** (origin: the soft palate) raises the back portion of the tongue.

MUSCLE	ORIGIN	INSERTION	ACTION	INNERVATION
Genioglossus (jē′-nē-ō-GLOS-us; *genio-* = chin; *-glossus* = tongue)	Mandible.	Undersurface of tongue and hyoid bone.	Depresses tongue and thrusts it anteriorly (protraction).	Hypoglossal (XII) nerve.
Styloglossus (stī′-lō-GLOS-us; *stylo* = stake or pole; styloid process of temporal bone)	Styloid process of temporal bone.	Side and undersurface of tongue.	Elevates tongue and draws it posteriorly (retraction).	Hypoglossal (XII) nerve.
Hyoglossus (hī′-ō-GLOS-us; *hyo* = U-shaped)	Greater horn and body of hyoid bone.	Side of tongue.	Depresses tongue and draws down its sides.	Hypoglossal (XII) nerve.
Palatoglossus (pal′-a-tō-GLOS-us; *palato* = roof of mouth or palate)	Anterior surface of soft palate.	Side of tongue.	Elevates posterior portion of tongue and draws soft palate down on tongue.	Pharyngeal plexus, which contains axons from the vagus (X) nerve.

EXHIBIT 11.D CONTINUES ▶

CLINICAL CONNECTION | *Intubation during Anesthesia*

When general anesthesia is administered during surgery, a total relaxation of the muscles results. Once the various types of drugs for anesthesia have been given (especially the paralytic agents), the patient's airway must be protected and the lungs ventilated because the muscles involved with respiration are among those paralyzed. Paralysis of the genioglossus muscle causes the tongue to fall posteriorly, which may obstruct the airway to the lungs. To avoid this, the mandible is either manually thrust forward and held in place (known as the "sniffing position"), or a tube is inserted from the lips through the laryngopharynx (inferior portion of the throat) into the trachea **(endotracheal intubation).** People can also be intubated nasally (through the nose). •

RELATING MUSCLES TO MOVEMENTS

Arrange the muscles in this exhibit according to the following actions on the tongue: (1) depression, (2) elevation, (3) protraction, and (4) retraction. The same muscle may be mentioned more than once.

✔ **CHECKPOINT**

When your physician says, "Open your mouth, stick out your tongue, and say ahh," to examine the inside of your mouth for possible signs of infection, which muscles do you contract?

Figure 11.7 **Muscles of the head that move the tongue and assist in mastication (chewing) and speech—extrinsic tongue muscles.**

The extrinsic and intrinsic muscles of the tongue are arranged in both lateral halves of the tongue.

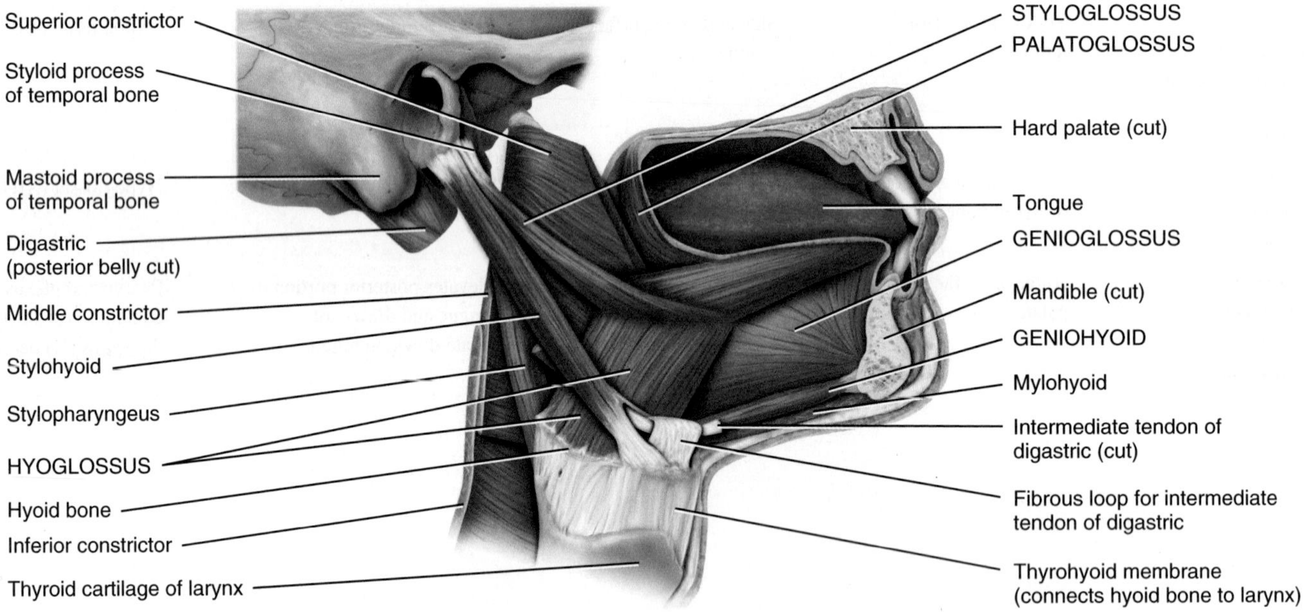
Superior constrictor
Styloid process of temporal bone
Mastoid process of temporal bone
Digastric (posterior belly cut)
Middle constrictor
Stylohyoid
Stylopharyngeus
HYOGLOSSUS
Hyoid bone
Inferior constrictor
Thyroid cartilage of larynx
STYLOGLOSSUS
PALATOGLOSSUS
Hard palate (cut)
Tongue
GENIOGLOSSUS
Mandible (cut)
GENIOHYOID
Mylohyoid
Intermediate tendon of digastric (cut)
Fibrous loop for intermediate tendon of digastric
Thyrohyoid membrane (connects hyoid bone to larynx)

Right side deep view

What are the functions of the tongue?

EXHIBIT 11.E Muscles of the Anterior Neck That Assist in Deglutition and Speech *(Figure 11.8)*

OBJECTIVE

- Describe the origin, insertion, action, and innervation of the muscles of the anterior neck that assist in deglutition and speech.

Two groups of muscles are associated with the anterior aspect of the neck: (1) the **suprahyoid muscles,** so called because they are located superior to the hyoid bone, and (2) the **infrahyoid muscles,** named for their position inferior to the hyoid bone (Figure 11.8). Both groups of muscles stabilize the hyoid bone, allowing it to serve as a firm base on which the tongue can move.

As a group, the suprahyoid muscles elevate the hyoid bone, floor of the oral cavity, and tongue during deglutition (swallowing). As its name suggests, the **digastric** muscle has two bellies, anterior and posterior, united by an intermediate tendon that is held in position by a fibrous loop. This muscle elevates the hyoid bone and larynx (voice box) during swallowing and speech. In a *reverse muscle action (RMA),* when the hyoid is stabilized, the digastric depresses the mandible and is therefore synergistic to the lateral pterygoid in the opening of the mouth. The **stylohyoid** muscle elevates and draws the hyoid bone posteriorly, thus elongating the floor of the oral cavity during swallowing. The **mylohyoid** muscle elevates the hyoid bone and helps press the tongue against the roof of the oral cavity during swallowing to move food from the oral cavity into the throat. The **geniohyoid** muscle (see Figure 11.7) elevates and draws the hyoid bone anteriorly to shorten the floor of the oral cavity and to widen the throat to receive food that is being swallowed. It also depresses the mandible.

The infrahyoid muscles are sometimes called "strap" muscles because of their ribbonlike appearance. Most of the infrahyoid muscles

MUSCLE	ORIGIN	INSERTION	ACTION	INNERVATION
SUPRAHYOID MUSCLES				
Digastric (dī′-GAS-trik; *di-* = two; *-gastr-* = belly)	Anterior belly from inner side of inferior border of mandible; posterior belly from temporal bone.	Body of hyoid bone via an intermediate tendon.	Elevates hyoid bone. RMA: Depresses mandible, as in opening mouth.	Anterior belly: mandibular division of trigeminal (V) nerve. Posterior belly: facial (VII) nerve.
Stylohyoid (stī′-lō-HĪ-oyd; *stylo-* = stake or pole, styloid process of temporal bone; *-hyo-* = U-shaped, pertaining to hyoid bone)	Styloid process of temporal bone.	Body of hyoid bone.	Elevates hyoid bone and draws it posteriorly.	Facial (VII) nerve.
Mylohyoid (mī′-lō-HĪ-oyd; *mylo* = mill)	Inner surface of mandible.	Body of hyoid bone.	Elevates hyoid bone and floor of mouth and depresses mandible.	Mandibular division of trigeminal (V) nerve.
Geniohyoid (jē′-nē-ō-HĪ-oyd; *genio* = chin) (see Figure 11.7)	Inner surface of mandible.	Body of hyoid bone.	Elevates hyoid bone, draws hyoid bone and tongue anteriorly. Depresses mandible.	First cervical spinal nerve (C1).
INFRAHYOID MUSCLES				
Omohyoid (ō-mō-HĪ-oyd; *omo* = relationship to shoulder)	Superior border of scapula and superior transverse ligament.	Body of hyoid bone.	Depresses hyoid bone.	Branches of spinal nerves C1–C3.
Sternohyoid (ster′-nō-HĪ-oyd; *sterno* = sternum)	Medial end of clavicle and manubrium of sternum.	Body of hyoid bone.	Depresses hyoid bone.	Branches of spinal nerves C1–C3.
Sternothyroid (ster′-nō-THĪ-royd; *thyro* = thyroid gland)	Manubrium of sternum.	Thyroid cartilage of larynx.	Depresses thyroid cartilage of larynx.	Branches of spinal nerves C1–C3.
Thyrohyoid (thī′-rō-HĪ-oyd)	Thyroid cartilage of larynx.	Greater horn of hyoid bone.	Depresses hyoid bone. RMA: Elevates thyroid cartilage.	Branches of spinal nerves C1–C2 and descending hypoglossal (XII) nerve.

EXHIBIT 11.E CONTINUES

depress the hyoid bone and some move the larynx during swallowing and speech. The **omohyoid** muscle, like the digastric muscle, is composed of two bellies connected by an intermediate tendon. In this case, however, the two bellies are referred to as *superior* and *inferior*, rather than anterior and posterior. Together, the omohyoid, **sternohyoid,** and **thyrohyoid** muscles depress the hyoid bone. In addition, the **sternothyroid** muscle depresses the thyroid cartilage (Adam's apple) of the larynx to produce low sounds; the RMA of the thyrohyoid muscle elevates the thyroid cartilage to produce high sounds.

CLINICAL CONNECTION | *Dysphagia*

Dysphagia (dis-FĀ-jē-a; *dys-* = abnormal; *-phagia* = to eat) is a clinical term for difficulty in swallowing. Some individuals are unable to swallow while others have difficulty swallowing liquids, foods, or saliva. Causes include nervous system disorders that weaken or damage muscles of deglutition (stroke, Parkinson disease, cerebral palsy); infections; cancer of the head, neck, or esophagus; and injuries to the head, neck, or chest. •

Figure 11.8 Muscles of the anterior neck that assist in deglutition (swallowing) and speech.

The suprahyoid muscles elevate the hyoid bone, the floor of the oral cavity, and the tongue during swallowing.

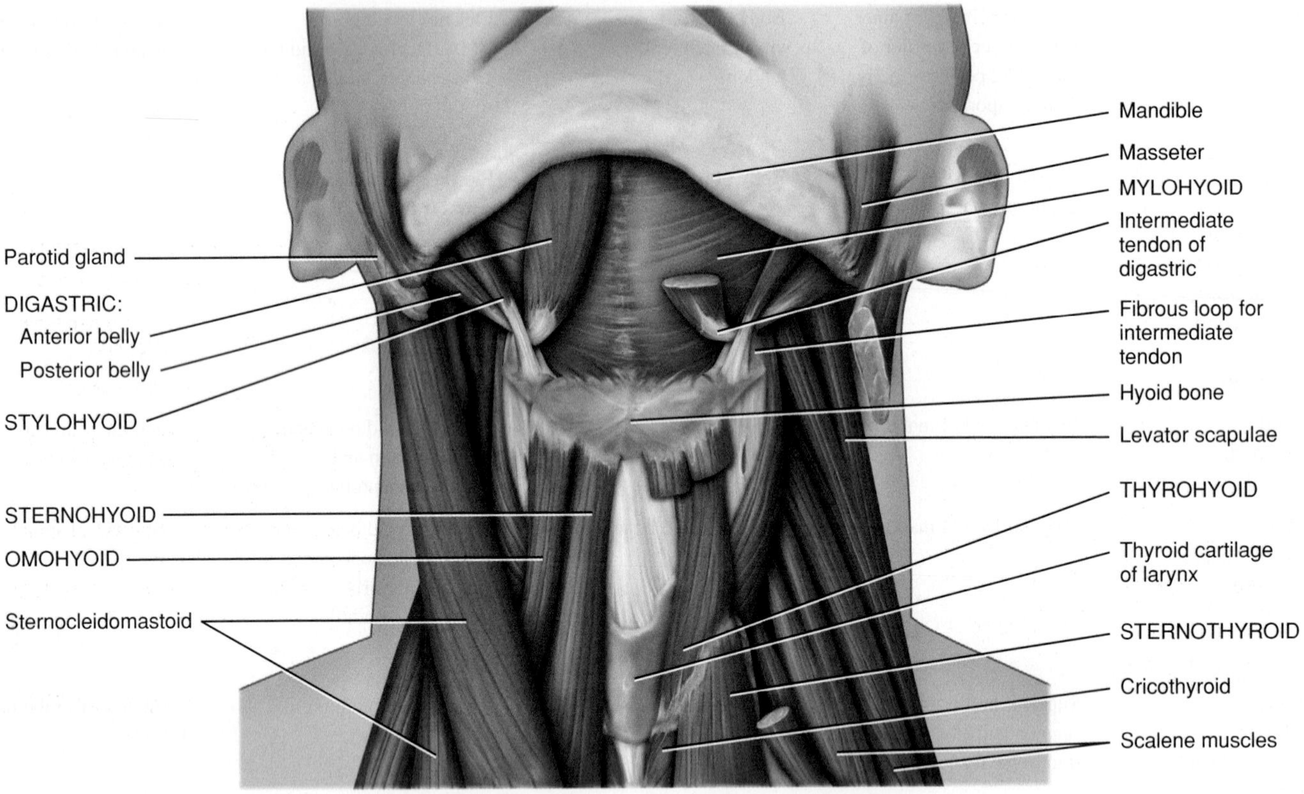
Mandible
Masseter
MYLOHYOID
Intermediate tendon of digastric
Parotid gland
DIGASTRIC:
Anterior belly
Posterior belly
Fibrous loop for intermediate tendon
Hyoid bone
STYLOHYOID
Levator scapulae
THYROHYOID
STERNOHYOID
OMOHYOID
Thyroid cartilage of larynx
Sternocleidomastoid
STERNOTHYROID
Cricothyroid
Scalene muscles

(a) Anterior superficial view (b) Anterior deep view

RELATING MUSCLES TO MOVEMENTS

Arrange the muscles in this exhibit according to the following actions on the hyoid bone: (1) elevating it, (2) drawing it anteriorly, (3) drawing it posteriorly, and (4) depressing it; and on the thyroid cartilage: (1) elevating it and (2) depressing it. The same muscle may be mentioned more than once.

✔ CHECKPOINT

Which tongue, facial, and mandibular muscles do you use for chewing?

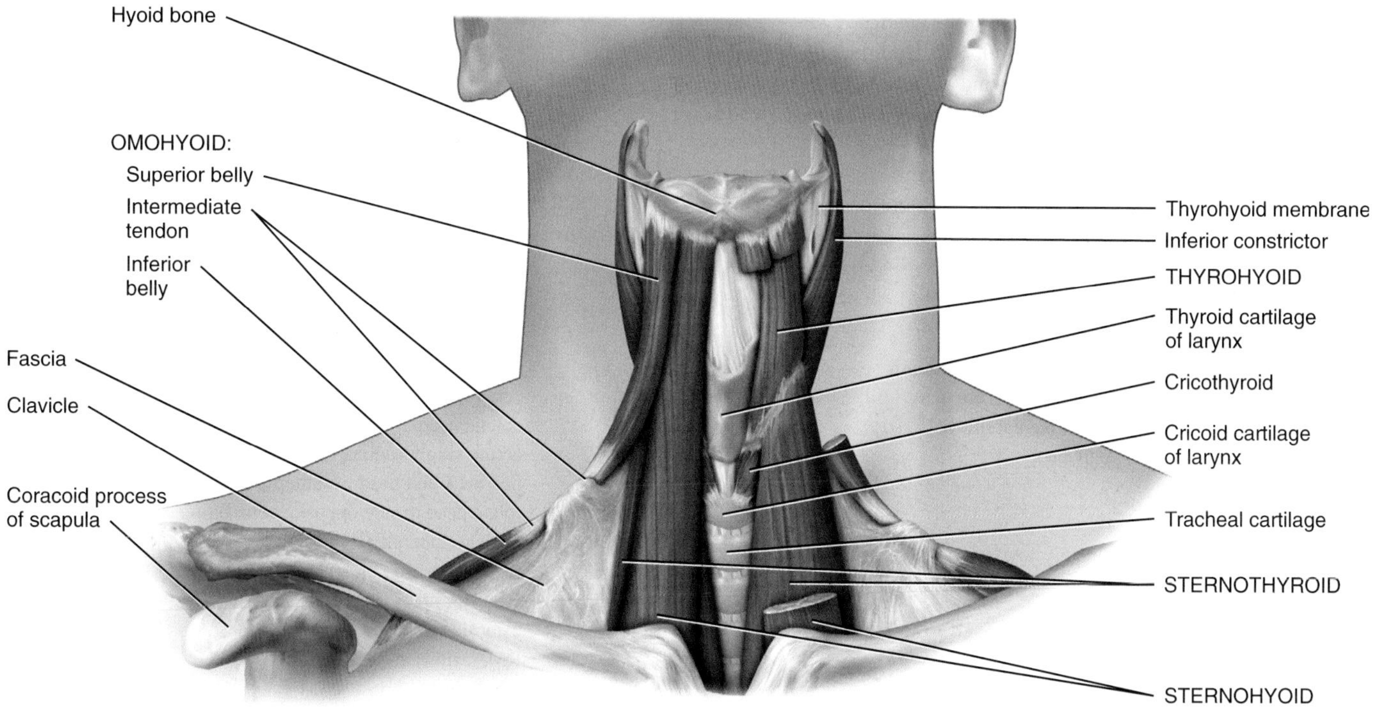

Hyoid bone
OMOHYOID:
Superior belly
Intermediate tendon
Inferior belly
Fascia
Clavicle
Coracoid process of scapula
Thyrohyoid membrane
Inferior constrictor
THYROHYOID
Thyroid cartilage of larynx
Cricothyroid
Cricoid cartilage of larynx
Tracheal cartilage
STERNOTHYROID
STERNOHYOID

Anterior superficial view (c) Anterior deep view

What is the combined action of the suprahyoid and infrahyoid muscles?

EXHIBIT 11.F Muscles of the Neck That Move the Head *(Figure 11.9)*

OBJECTIVE

- Describe the origin, insertion, action, and innervation of the muscles that move the head.

The head is attached to the vertebral column at the atlanto-occipital joints formed by the atlas and occipital bone. Balance and movement of the head on the vertebral column involves the action of several neck muscles. For example, acting together (bilaterally), contraction of the two **sternocleidomastoid (SCM)** muscles flexes the cervical portion of the vertebral column and flexes the head. Acting singly (unilaterally), each sternocleidomastoid muscle laterally flexes and rotates the head. Each SCM consists of two bellies (Figure 11.9c); they are more evident near the anterior attachments. The separation of the two bellies is variable and thus more evident in some persons than in others. The two bellies insert as the **sternal head** and the **clavicular head** of the SCM. The bellies also function differently; muscular spasm in the two bellies causes somewhat different symptoms. Bilateral contraction of the **semispinalis capitis, splenius capitis,** and **longissimus capitis** muscles extends the head (Figure 11.9a, b). However, when these same muscles contract unilaterally, their actions are quite different, involving primarily rotation of the head.

The sternocleidomastoid muscle is an important landmark that divides the neck into two major triangles: anterior and posterior (Figure 11.9c). The triangles are important anatomically and surgically because of the structures that lie within their boundaries.

The **anterior triangle** is bordered superiorly by the mandible, medially by the cervical midline, and laterally by the anterior border of the sternocleidomastoid muscle. It has its apex at the sternum (Figure 11.9c). The anterior triangle is subdivided into three paired triangles: *submandibular, carotid,* and *muscular*. An unpaired *submental triangle* is formed by the upper part of the combined right and left anterior triangles. The anterior triangle contains submental, submandibular, and deep cervical lymph nodes; the submandibular salivary gland and a portion of the parotid salivary gland; the facial artery and vein; carotid arteries and internal jugular vein; the thyroid gland and infrahyoid muscles; and the following cranial nerves: glossopharyngeal (IX), vagus (X), accessory (XI), and hypoglossal (XII).

The **posterior triangle** is bordered inferiorly by the clavicle, anteriorly by the posterior border of the sternocleidomastoid muscle, and posteriorly by the anterior border of the trapezius muscle (Figure 11.9c). The posterior triangle is subdivided into two triangles, *occipital* and *supraclavicular (omoclavicular),* by the inferior belly of the omohyoid muscle. The posterior triangle contains part of the subclavian artery, external jugular vein, cervical lymph nodes, brachial plexus, and the accessory (XI) nerve.

RELATING MUSCLES TO MOVEMENTS

Arrange the muscles in this exhibit according to the following actions on the head: (1) flexion, (2) lateral flexion, (3) extension, (4) rotation to side opposite contracting muscle, and (5) rotation to same side as contracting muscle. The same muscle may be mentioned more than once.

✔ CHECKPOINT

What muscles do you contract to signify "yes" and "no"?

MUSCLE	ORIGIN	INSERTION	ACTION	INNERVATION
Sternocleidomastoid (ster′-nō-klī′-dō-MAS-toyd; *sterno-* = breastbone; *-cleido-* = clavicle; *-mastoid* = mastoid process of temporal bone)	Sternal head: manubrium of sternum; clavicular head: medial third of clavicle.	Mastoid process of temporal bone and lateral half of superior nuchal line of occipital bone.	Acting together (bilaterally), flex cervical portion of vertebral column, extend head at atlanto-occipital joints; acting singly (unilaterally), laterally flex neck and head to same side and rotate head to side opposite contracting muscle. Laterally rotate and flex head to opposite side of contracting muscle. Posterior fibers of muscle can assist in extension of head. RMA: Elevate sternum during forced inhalation.	Accessory (XI) nerve, C2, and C3.
Semispinalis capitis (se′-mē-spi-NĀ-lis KAP-i-tis; *semi-* = half; *spine* = spinous process; *capit* = head)	Articular processes of C4–C6 and transverse processes of C7–T7.	Occipital bone between superior and inferior nuchal lines.	Acting together, extend head and vertebral column; acting singly, rotate head to side opposite contracting muscle.	Cervical spinal nerves.
Splenius capitis (SPLĒ-nē-us KAP-i-tis; *splenium* = bandage)	Ligamentum nuchae and spinous processes of C7–T4.	Occipital bone and mastoid process of temporal bone.	Acting together, extend head; acting singly, rotate head to side opposite contracting muscle.	Cervical spinal nerves.
Longissimus capitis (lon-JIS-i-mus KAP-i-tis; *longissimus* = longest)	Articular processes of T1–T4.	Mastoid process of temporal bone.	Acting together, extend head and vertebral column; acting singly, laterally flex and rotate head to same side as contracting muscle.	Cervical spinal nerves.
Spinalis capitis (spi-NĀ-lis KAP-i-tis; *spinal* = vertebral column)	Often absent or very small; arises with semispinalis capitis.	Occipital bone.	Extends head and vertebral column.	Cervical spinal nerves.

Figure 11.9 Muscles of the neck that move the head.

The sternocleidomastoid muscle divides the neck into two principal triangles: anterior and posterior.

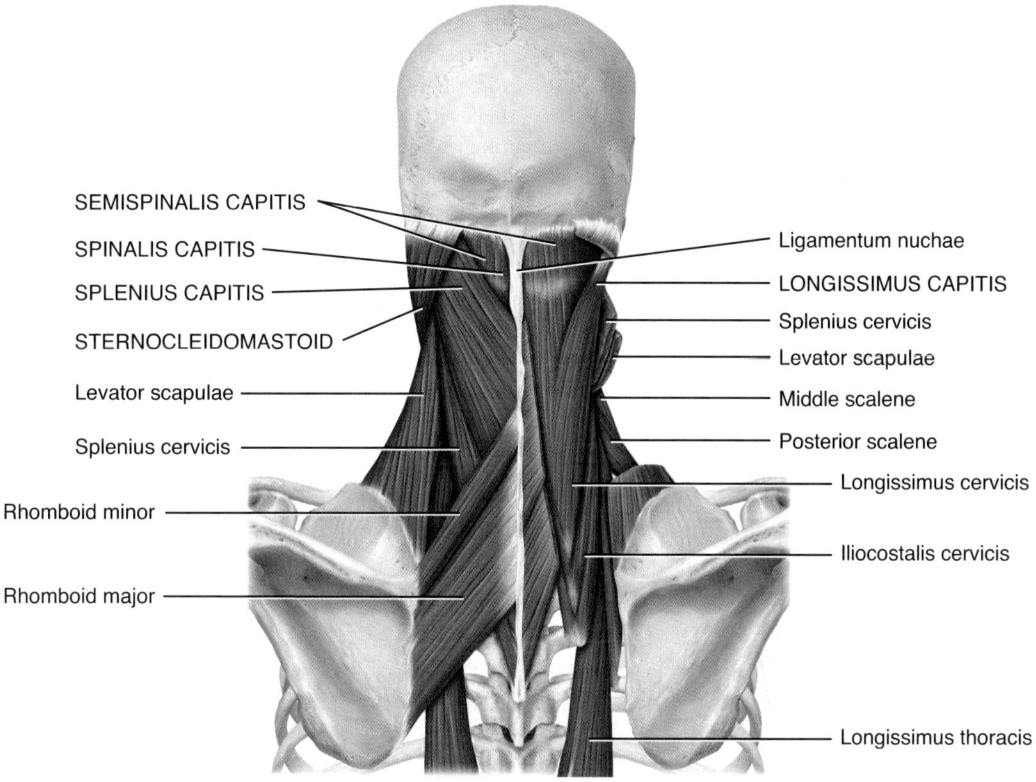

SEMISPINALIS CAPITIS
SPINALIS CAPITIS
SPLENIUS CAPITIS
STERNOCLEIDOMASTOID
Levator scapulae
Splenius cervicis
Rhomboid minor
Rhomboid major
Ligamentum nuchae
LONGISSIMUS CAPITIS
Splenius cervicis
Levator scapulae
Middle scalene
Posterior scalene
Longissimus cervicis
Iliocostalis cervicis
Longissimus thoracis

(a) Posterior superficial view (b) Posterior deep view

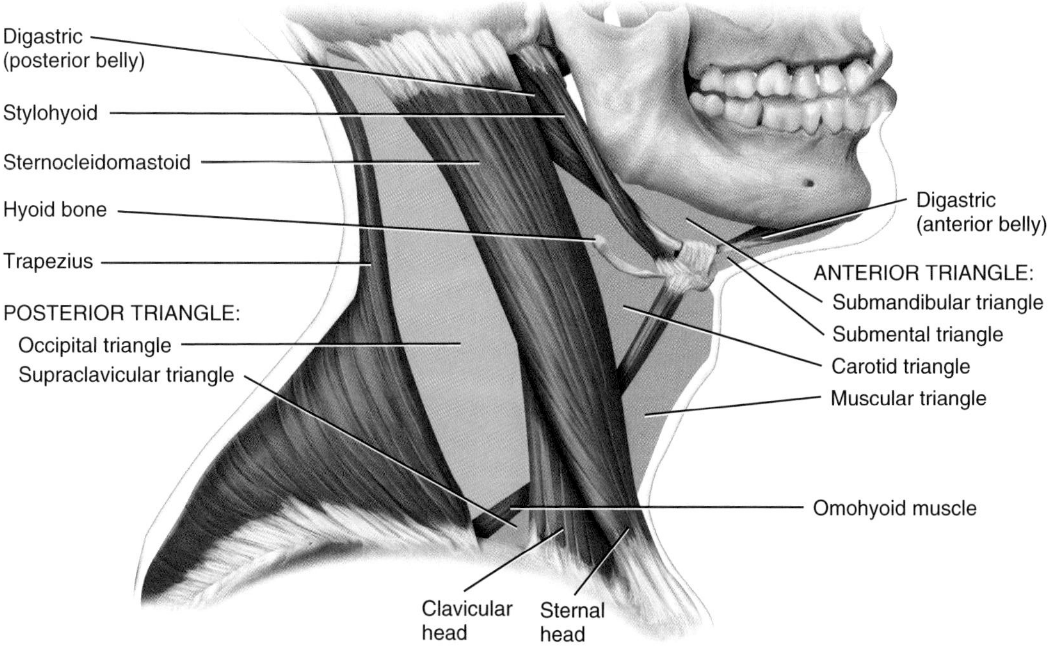

Digastric (posterior belly)
Stylohyoid
Sternocleidomastoid
Hyoid bone
Trapezius
POSTERIOR TRIANGLE:
Occipital triangle
Supraclavicular triangle
Digastric (anterior belly)
ANTERIOR TRIANGLE:
Submandibular triangle
Submental triangle
Carotid triangle
Muscular triangle
Omohyoid muscle
Clavicular head
Sternal head

(c) Right lateral view of triangles of neck

Why are triangles of the neck important?

Muscles of the Abdomen That Protect Abdominal Viscera and Move the Vertebral Column *(Figure 11.10)*

OBJECTIVE

- Describe the origin, insertion, action, and innervation of the muscles that protect the abdominal viscera and move the vertebral column.

The anterolateral abdominal wall is composed of skin, fascia, and four pairs of muscles: the external oblique, internal oblique, transversus abdominis, and rectus abdominis (Figure 11.10). The first three muscles named are arranged from superficial to deep.

The **external oblique** is the superficial muscle. Its fascicles extend inferiorly and medially. The **internal oblique** is the intermediate flat muscle. Its fascicles extend at right angles to those of the external oblique. The **transversus abdominis** is the deep muscle, with most of its fascicles directed transversely around the abdominal wall. Together, the external oblique, internal oblique, and transversus abdominis form three layers of muscle around the abdomen. In each layer, the muscle fascicles extend in a different direction. This is a structural arrangement that affords considerable protection to the abdominal viscera, especially when the muscles have good tone.

The **rectus abdominis** muscle is a long muscle that extends the entire length of the anterior abdominal wall, originating at the pubic crest and pubic symphysis and inserting on the cartilages of ribs 5–7 and the xiphoid process of the sternum. The anterior surface of the muscle is interrupted by three transverse fibrous bands of tissue called **tendinous intersections,** believed to be remnants of septa that separated myotomes during embryological development (see Figure 10.17). There are usually three tendinous intersections, one at the level of the umbilicus, one near the xiphoid process, and one midway between the other two. A fourth intersection is sometimes found below the level of the umbilicus. These tendinous intersections are fused with the anterior wall of the rectus sheath but have no connections to the posterior abdominal wall. Muscular persons may possess easily demonstrated intersections as the result of exercise and the ensuing hypertrophy of the rectus muscle. Hypertrophy of the muscle tissue, of course, has no effect on the connective tissue of the intersections. Body

Figure 11.10 Muscles of the abdomen that protect abdominal viscera and move the vertebral column (backbone).

The anterolateral abdominal muscles protect the abdominal viscera, move the vertebral column, and assist in forced exhalation, defecation, urination, and childbirth.

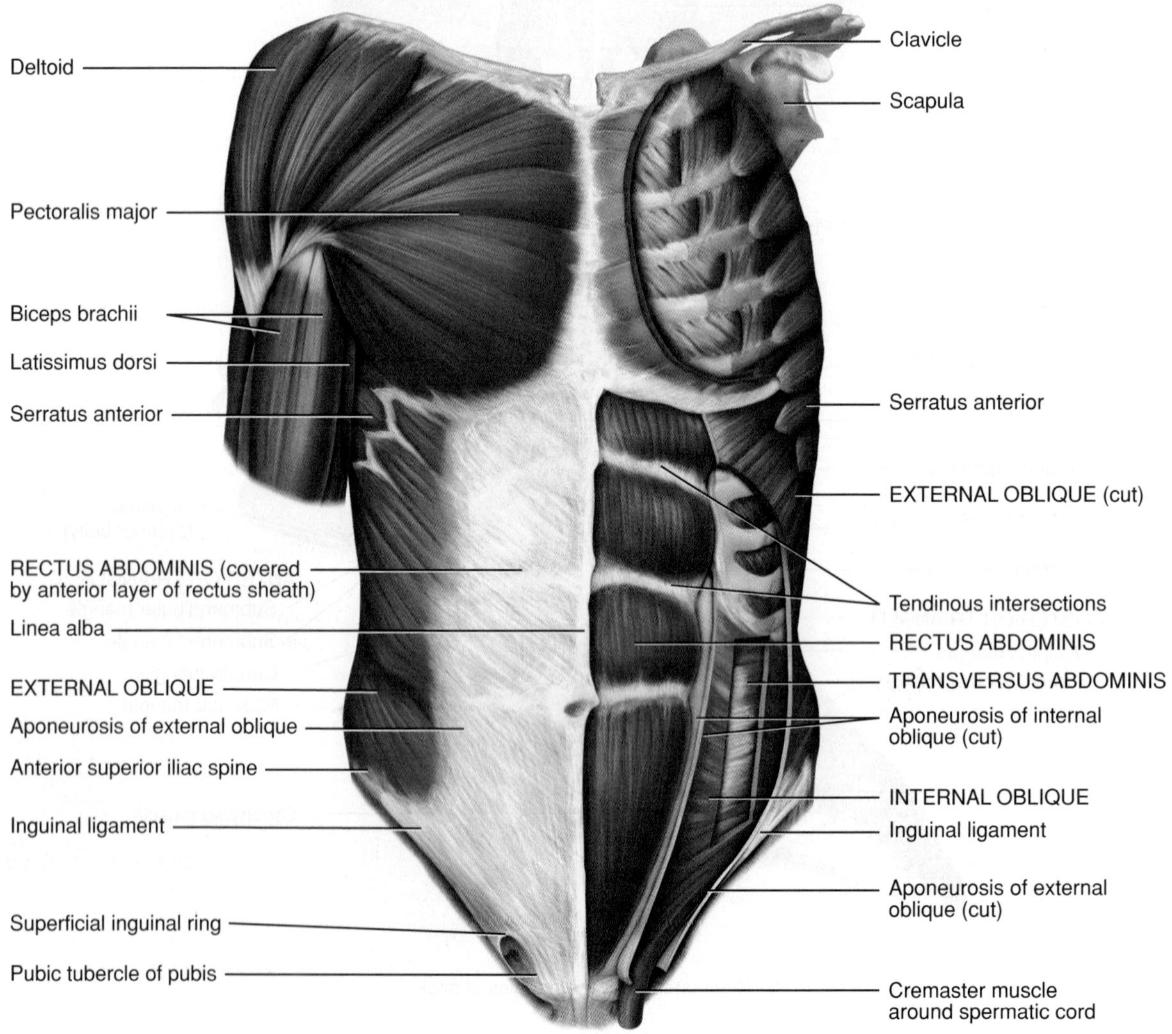

(a) Anterior superficial view (b) Anterior deep view

builders focus on the development of the **"six-pack"** effect of the abdomen. Small percentages of the population have a variant of the intersections and are able to develop an "eight-pack."

As a group, the muscles of the anterolateral abdominal wall help contain and protect the abdominal viscera; flex, laterally flex, and rotate the vertebral column (backbone) at the intervertebral joints; compress

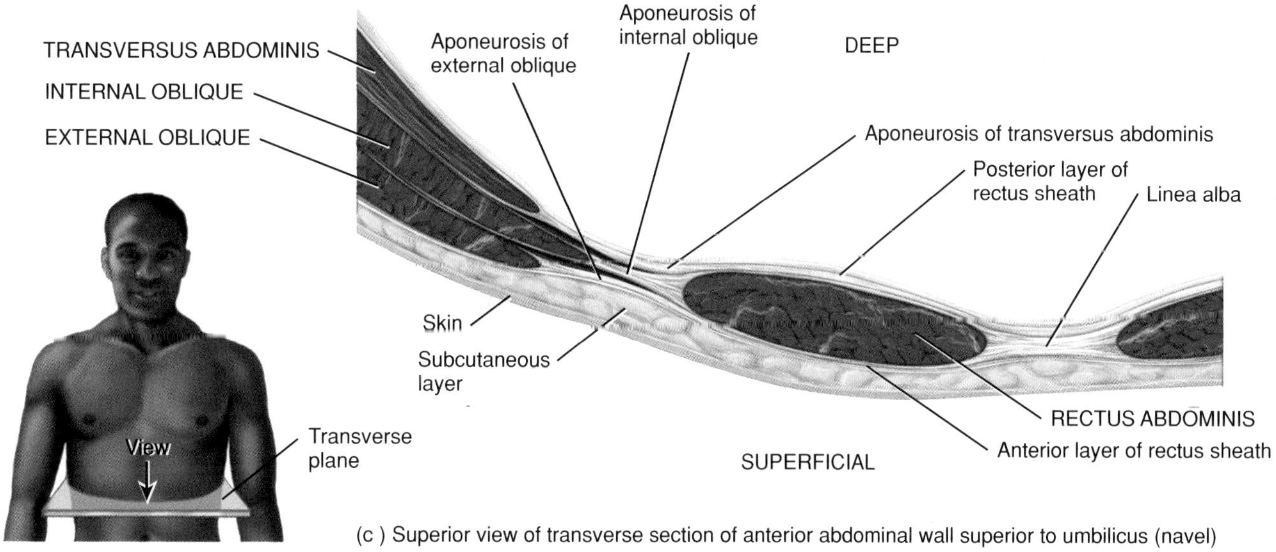

(c) Superior view of transverse section of anterior abdominal wall superior to umbilicus (navel)

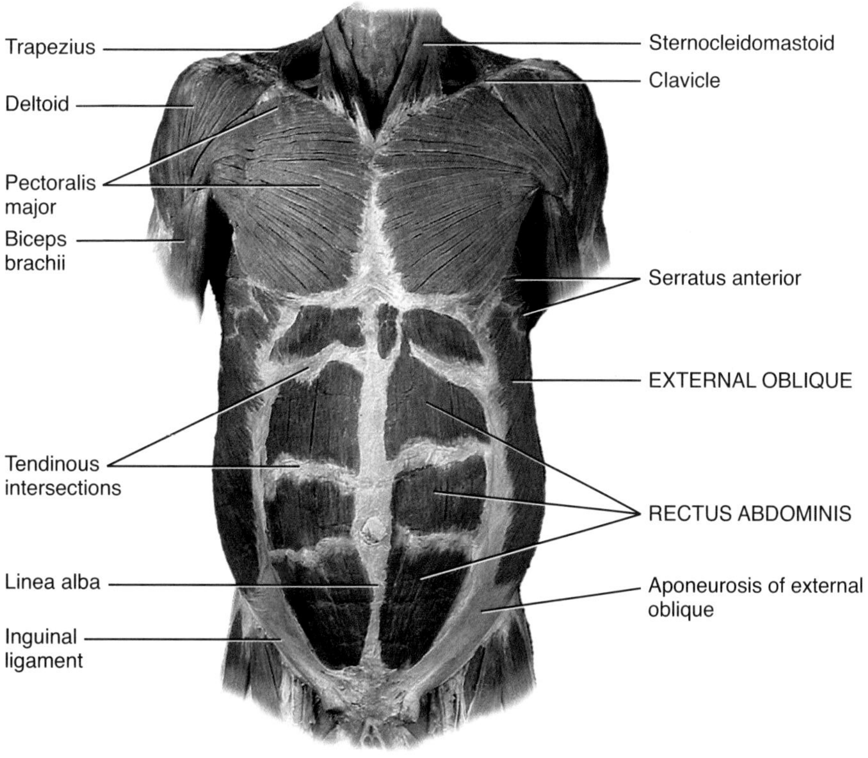

(d) Anterior view

Which abdominal muscle aids in urination?

EXHIBIT 11.G CONTINUES

the abdomen during forced exhalation; and produce the force required for defecation, urination, and childbirth.

The aponeuroses (sheathlike tendons) of the external oblique, internal oblique, and transversus abdominis muscles form the **rectus sheaths,** which enclose the rectus abdominis muscles. The sheaths meet at the midline to form the **linea alba** (= white line), a tough, fibrous band that extends from the xiphoid process of the sternum to the pubic symphysis. In the latter stages of pregnancy, the linea alba stretches to increase the distance between the rectus abdominis muscles. The inferior free border of the external oblique aponeurosis forms the **inguinal ligament,** which runs from the anterior superior iliac spine to the pubic tubercle (see Figure 11.20a). Just superior to the medial end of the inguinal ligament is a triangular slit in the aponeurosis referred to as the **superficial inguinal ring,** the outer opening of the inguinal canal (see Figure 28.2). The **inguinal canal** contains the spermatic cord and ilioinguinal nerve in males, and the round ligament of the uterus and ilioinguinal nerve in females.

The posterior abdominal wall is formed by the lumbar vertebrae, parts of the ilia of the hip bones, psoas major and iliacus muscles (described in Exhibit 11.Q), and quadratus lumborum muscle. The anterolateral abdominal wall can contract and distend; the posterior abdominal wall is bulky and stable by comparison.

CLINICAL CONNECTION | *Inguinal Hernia*

A **hernia** (HER-nē-a) is a protrusion of an organ through a structure that normally contains it, which creates a lump that can be seen or felt through the skin's surface. The inguinal region is a weak area in the abdominal wall. It is often the site of an **inguinal hernia,** a rupture or separation of a portion of the inguinal area of the abdominal wall resulting in the protrusion of a part of the small intestine. A hernia is much more common in males than in females because the inguinal canals in males are larger to accommodate the spermatic cord and ilioinguinal nerve. Treatment of hernias most often involves surgery. The organ that protrudes is "tucked" back into the abdominal cavity and the defect in the abdominal muscles is repaired. In addition, a mesh is often applied to reinforce the area of weakness. •

RELATING MUSCLES TO MOVEMENTS

Arrange the muscles in this exhibit according to the following actions on the vertebral column: (1) flexion, (2) lateral flexion, (3) extension, and (4) rotation. The same muscle may be mentioned more than once.

✔ **CHECKPOINT**

Which muscles do you contract when you "suck in your gut," thereby compressing the anterior abdominal wall?

MUSCLE	ORIGIN	INSERTION	ACTION	INNERVATION
Rectus abdominis (REK-tus ab-DOM-in-is; *rectus* = fascicles parallel to midline: *abdominis* = abdomen)	Pubic crest and pubic symphysis.	Cartilage of ribs 5–7 and xiphoid process.	Flexes vertebral column, especially lumbar portion, and compresses abdomen to aid in defecation, urination, forced exhalation, and childbirth. RMA: Flexes pelvis on the vertebral column.	Thoracic spinal nerves T7–T12.
External oblique (ō-BLĒK; *external* = closer to surface; *oblique* = fascicles diagonal to midline)	Ribs 5–12.	Iliac crest and linea alba.	Acting together (bilaterally), compress abdomen and flex vertebral column; acting singly (unilaterally), laterally flex vertebral column, especially lumbar portion, and rotate vertebral column.	Thoracic spinal nerves T7–T12 and the iliohypogastric nerve.
Internal oblique (ō-BLĒK; *internal* = farther from surface)	Iliac crest, inguinal ligament, and thoracolumbar fascia.	Cartilage of ribs 7–10 and linea alba.	Acting together, compress abdomen and flex vertebral column; acting singly, laterally flex vertebral column, especially lumbar portion, and rotate vertebral column.	Thoracic spinal nerves T8–T12, the iliohypogastric nerve, and ilioinguinal nerve.
Transversus abdominis (tranz-VER-sus ab-DOM-in-is; *transverse* = fascicles perpendicular to midline)	Iliac crest, inguinal ligament, lumbar fascia, and cartilages of ribs 5–10.	Xiphoid process, linea alba, and pubis.	Compresses abdomen.	Thoracic spinal nerves T8–T12, iliohypogastric nerve, and ilioinguinal nerve.
Quadratus lumborum (kwod-RĀ-tus lum-BŌR-um; *quad* = four; *lumbo* = lumbar region) (see Figure 11.11)	Iliac crest and iliolumbar ligament.	Inferior border of rib 12 and L1–L4.	Acting together, pull twelfth ribs inferiorly during forced exhalation, fix twelfth ribs to prevent their elevation during deep inhalation, and held extend lumbar portion of vertebral column; acting singly, laterally flex vertebral column, especially lumbar portion. RMA: Elevates hip bone, commonly on one side.	Thoracic spinal nerves T12 and lumbar spinal nerves L1–L3 or L1–L4.

EXHIBIT 11.H Muscles of the Thorax That Assist in Breathing *(Figure 11.11)*

OBJECTIVE

- Describe the origin, insertion, action, and innervation of the muscles of the thorax that assist in breathing.

The muscles of the thorax (chest) alter the size of the thoracic cavity so that breathing can occur. Inhalation (breathing in) occurs when the thoracic cavity increases in size, and exhalation (breathing out) occurs when the thoracic cavity decreases in size.

The dome-shaped **diaphragm** is the most important muscle that powers breathing. It also separates the thoracic and abdominal cavities. The diaphragm has a convex superior surface that forms the floor of the thoracic cavity (Figure 11.11b) and a concave, inferior surface that forms the roof of the abdominal cavity (Figure 11.11b). The **peripheral muscular portion** of the diaphragm originates on the xiphoid process of the sternum, the inferior six ribs and their costal cartilages, and the lumbar vertebrae and their intervertebral discs and the twelfth rib (Figure 11.11d). From their various origins, the fibers of the muscular portion converge and insert into the **central tendon,** a strong aponeurosis located near the center of the muscle (Figure 11.11a–d). The central tendon fuses with the inferior surface of the pericardium (covering of the heart) and the pleurae (coverings of the lungs).

The diaphragm has three major openings through which various structures pass between the thorax and abdomen. These structures include the aorta, along with the thoracic duct and azygous vein, which pass through the **aortic hiatus**; the esophagus with accompanying vagus (X) nerves, which pass through the **esophageal hiatus**; and the inferior vena cava, which passes through the **caval opening (foramen for the vena cava).** In a condition called a hiatus hernia, the stomach protrudes superiorly through the esophageal hiatus.

Movements of the diaphragm also help return venous blood passing through abdominal veins to the heart. Together with the anterolateral abdominal muscles, the diaphragm helps to increase intra-abdominal pressure to evacuate the pelvic contents during defecation, urination, and childbirth. This mechanism is further assisted when you take a deep breath and close the rima glottidis (the space between the vocal folds). The trapped air in the respiratory system prevents the diaphragm from elevating. The increase in intra-abdominal pressure also helps support the vertebral column and helps prevent flexion during weight lifting. This greatly assists the back muscles in lifting a heavy weight.

Other muscles involved in breathing, called **intercostal** muscles, span the intercostal spaces, the spaces between ribs. These muscles are arranged in three layers. The 11 pairs of **external intercostal** muscles occupy the superficial layer, and their fibers run in an oblique direction inferiorly and anteriorly from the rib above to the rib below. They elevate the ribs during inhalation to help expand the thoracic cavity. The 11 pairs of **internal intercostal** muscles occupy the intermediate layer of the intercostal spaces. The fibers of these muscles run at right angles to the external intercostals, in an oblique direction inferiorly and posteriorly from the inferior border of the rib above to the superior border of the rib below. They draw adjacent ribs together during forced exhalation to help decrease the size of the thoracic cavity.

Note: A mnemonic for the action of the intercostal muscles is singing "Old MacDonald had a farm, **E, I, E, I, O**" = **E**xternal **I**ntercostals **E**levate during **I**nhalation, **O**h!"

As you will see in Chapter 23, the diaphragm and external intercostal muscles are used during quiet inhalation and exhalation. However, during deep, forceful inhalation (during exercise or playing a wind instrument), the sternocleidomastoid, scalene, and pectoralis minor muscles are also used; during deep, forceful exhalation, the external oblique, internal oblique, transversus abdominis, rectus abdominis, and internal intercostals are also used.

MUSCLE	ORIGIN	INSERTION	ACTION	INNERVATION
Diaphragm (DĪ-a-fram; *dia-* = across; *-phragm* = wall)	Xiphoid process of sternum, costal cartilages and adjacent portions of inferior six ribs, lumbar vertebrae and their intervertebral discs.	Central tendon.	Contraction of diaphragm causes it to flatten and increases vertical dimension of thoracic cavity, resulting in inhalation; relaxation of diaphragm causes it to move superiorly and decreases vertical dimension of thoracic cavity, resulting in exhalation.	Phrenic nerve, which contains axons from cervical spinal nerves (C3–C5).
External intercostals (in′-ter-KOS-tals; *external* = closer to surface; *inter-* = between; *-costa* = rib)	Inferior border of rib above.	Superior border of rib below.	Contraction elevates ribs and increases anteroposterior and lateral dimensions of thoracic cavity, resulting in inhalation; relaxation depresses ribs and decreases anteroposterior and lateral dimensions of thoracic cavity, resulting in exhalation.	Thoracic spinal nerves T2–T12.
Internal intercostals (in′-ter-KOS-tals; *internal* = farther from surface)	Superior border of rib below.	Inferior border of rib above.	Contraction draws adjacent ribs together to further decrease anteroposterior and lateral dimensions of thoracic cavity during forced exhalation.	Thoracic spinal nerves T2–T12.

EXHIBIT 11.H CONTINUES

RELATING MUSCLES TO MOVEMENTS

Arrange the muscles in this exhibit according to the following actions: (1) increase in vertical length, (2) increase in lateral and anteroposterior dimensions, and (3) decrease in lateral and anteroposterior dimensions of the thorax.

✔CHECKPOINT

What are the names of the three openings in the diaphragm, and which structures pass through each?

Figure 11.11 **Muscles of the thorax (chest) that assist in breathing.**

Openings in the diaphragm permit the passage of the aorta, esophagus, and inferior vena cava.

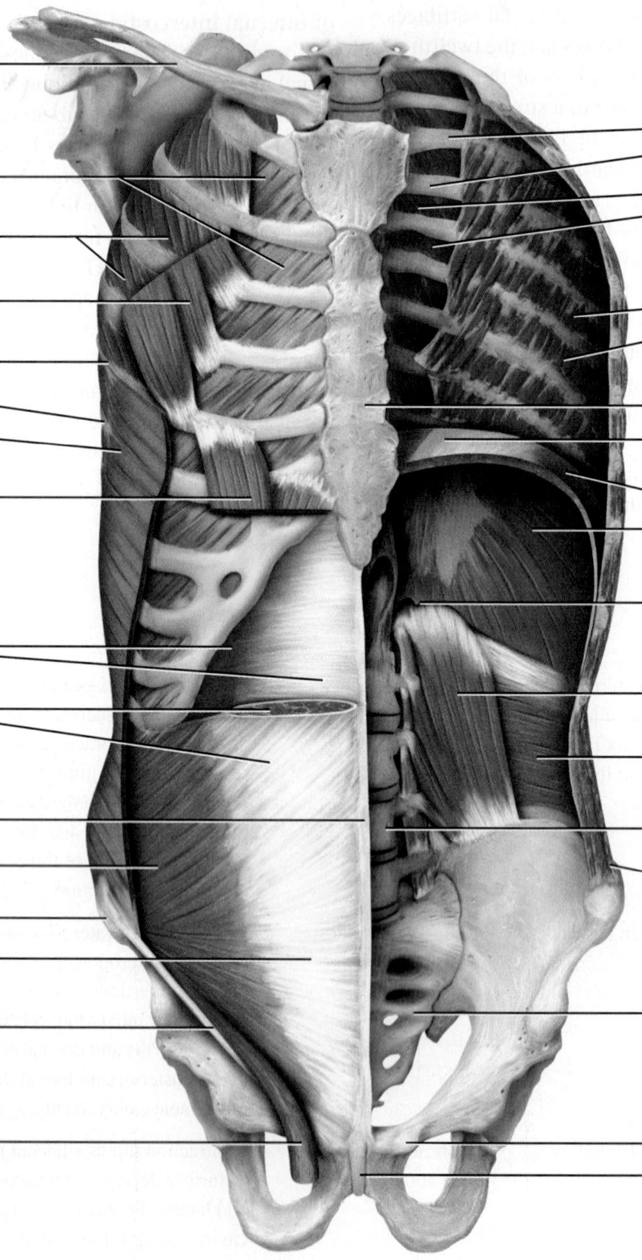

(a) Anterior superficial view (b) Anterior deep view

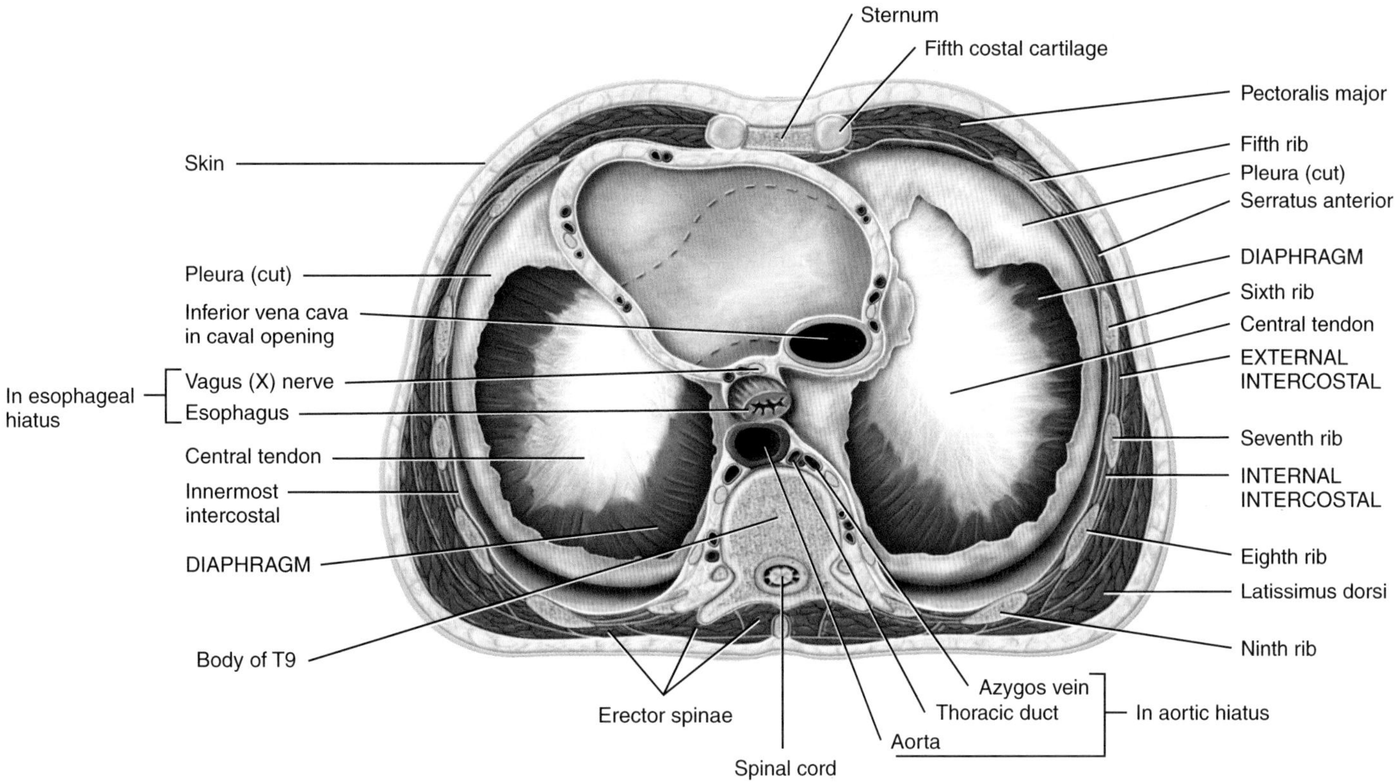

Sternum
Fifth costal cartilage
Pectoralis major
Fifth rib
Pleura (cut)
Serratus anterior
DIAPHRAGM
Sixth rib
Central tendon
EXTERNAL INTERCOSTAL
Seventh rib
INTERNAL INTERCOSTAL
Eighth rib
Latissimus dorsi
Ninth rib
Skin
Pleura (cut)
Inferior vena cava in caval opening
In esophageal hiatus
Vagus (X) nerve
Esophagus
Central tendon
Innermost intercostal
DIAPHRAGM
Body of T9
Erector spinae
Spinal cord
Azygos vein
Thoracic duct
Aorta
In aortic hiatus

(c) Superior view of diaphragm

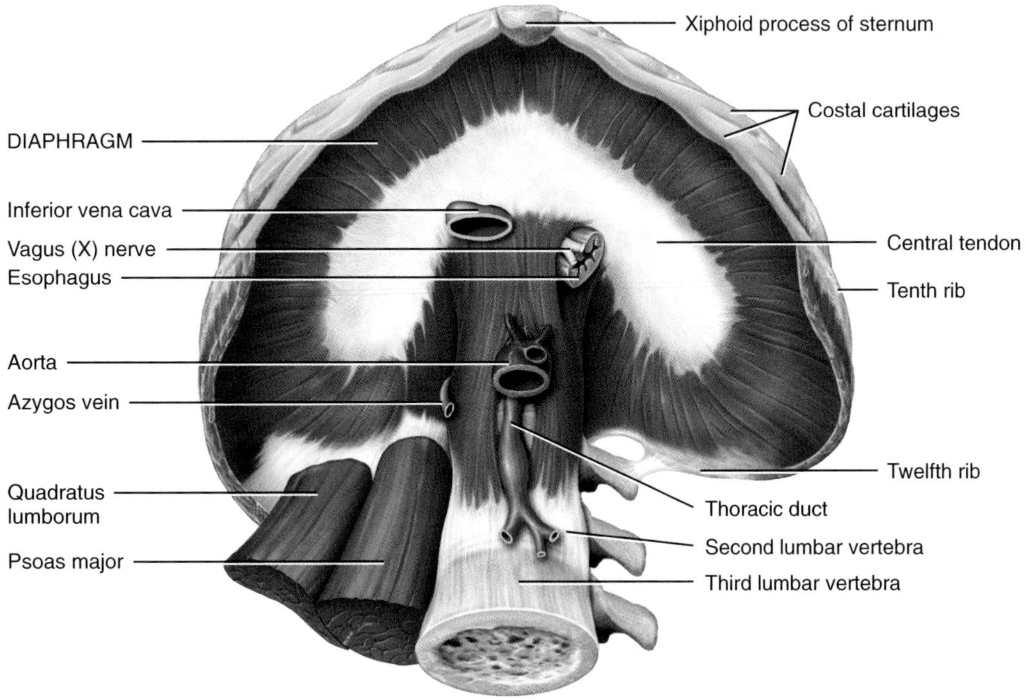

Xiphoid process of sternum
Costal cartilages
DIAPHRAGM
Inferior vena cava
Vagus (X) nerve
Esophagus
Central tendon
Tenth rib
Aorta
Azygos vein
Twelfth rib
Quadratus lumborum
Thoracic duct
Second lumbar vertebra
Psoas major
Third lumbar vertebra

(d) Inferior view of diaphragm

Which muscle associated with breathing is innervated by the phrenic nerve?

EXHIBIT 11.I Muscles of the Pelvic Floor That Support the Pelvic Viscera and Function as Sphincters *(Figure 11.12)*

OBJECTIVE

- Describe the origin, insertion, action, and innervation of the muscles of the pelvic floor that support the pelvic viscera and function as sphincters.

The muscles of the pelvic floor are the levator ani and ischiococcygeus. Together with the fascia covering their internal and external surfaces, these muscles are referred to as the **pelvic diaphragm,** which stretches from the pubis anteriorly to the coccyx posteriorly, and from one lateral wall of the pelvis to the other. This arrangement gives the pelvic diaphragm the appearance of a funnel suspended from its attachments. The pelvic diaphragm separates the pelvic cavity above from the perineum below (see Exhibit 11.J). The anal canal and urethra pierce the pelvic diaphragm in both sexes, and the vagina also goes through it in females.

The two components of the **levator ani** muscle are the **pubococcygeus** and **iliococcygeus.** Figure 11.12 shows these muscles in the female and Figure 11.13 in Exhibit 11.J illustrates them in the male. The levator ani is the largest and most important muscle of the pelvic floor. It supports the pelvic viscera and resists the inferior thrust that accompanies increases in intra-abdominal pressure during functions such as forced exhalation, coughing, vomiting, urination, and defecation. The muscle also functions as a sphincter at the anorectal junction, urethra, and vagina. In addition to assisting the levator ani, the **ischiococcygeus** pulls the coccyx anteriorly after it has been pushed posteriorly during defecation or childbirth.

MUSCLE	ORIGIN	INSERTION	ACTION	INNERVATION
Levator ani (le-VĀ-tor Ā-nē; *levator* = raises; *ani* = anus)	Muscle is divisible into two parts: pubococcygeus muscle and iliococcygeus muscle.			
Pubococcygeus (pū′-bō-kok-SIJ-ē-us; *pubo-* = pubis; *-coccygeus* = coccyx)	Pubis and ischial spine.	Coccyx, urethra, anal canal, perineal body of perineum (wedge-shaped mass of fibrous tissue in center of perineum), and anococcygeal ligament (narrow fibrous band that extends from anus to coccyx).	Supports and maintains position of pelvic viscera; resists increase in intra-abdominal pressure during forced exhalation, coughing, vomiting, urination, and defecation; constricts anus, urethra, and vagina.	Sacral spinal nerves S2–S4.
Iliococcygeus (il′-ē-ō-kok-SIJ-ē-us; *ilio* = ilium)	Ischial spine.	Coccyx.	As above.	Sacral spinal nerves S2–S4.
Ischiococcygeus (is′-kē-ō-kok-SIJ-ē-us; *ischio* = hip)	Ischial spine.	Lower sacrum and upper coccyx.	Supports and maintains position of pelvic viscera; resists increase in intra-abdominal pressure during forced exhalation, coughing, vomiting, urination, and defecation; pulls coccyx anteriorly following defecation or childbirth.	Sacral spinal nerves S4–S5.

CLINICAL CONNECTION | *Injury of Levator Ani and Urinary Stress Incontinence*

During childbirth, the levator ani muscle supports the head of the fetus, and the muscle may be injured during a difficult childbirth or traumatized during an *episiotomy* (a cut made with surgical scissors to prevent or direct tearing of the perineum during the birth of a baby). The consequence of such injuries may be **urinary stress incontinence,** that is, the leakage of urine whenever intra-abdominal pressure is increased—for example, during coughing. One way to treat urinary stress incontinence is to strengthen and tighten the muscles that support the pelvic viscera. This is accomplished by *Kegel exercises,* the alternate contraction and relaxation of muscles of the pelvic floor. To find the correct muscles, the person imagines that she is urinating and then contracts the muscles as if stopping in midstream. The muscles should be held for a count of three, then relaxed for a count of three. This should be done 5–10 times each hour—sitting, standing, and lying down. Kegel exercises are also encouraged during pregnancy to strengthen the muscles for delivery. •

RELATING MUSCLES TO MOVEMENTS

Arrange the muscles in this exhibit according to the following actions: (1) supporting and maintaining the position of the pelvic viscera; (2) resisting an increase in intra-abdominal pressure; and (3) constriction of the anus, urethra, and vagina. The same muscle may be mentioned more than once.

✔ **CHECKPOINT**

Which muscles are strengthened by Kegel exercises?

Figure 11.12 Muscles of the pelvic floor that support the pelvic viscera, assist in resisting intra-abdominal pressure, and function as sphincters.

The pelvic diaphragm supports the pelvic viscera.

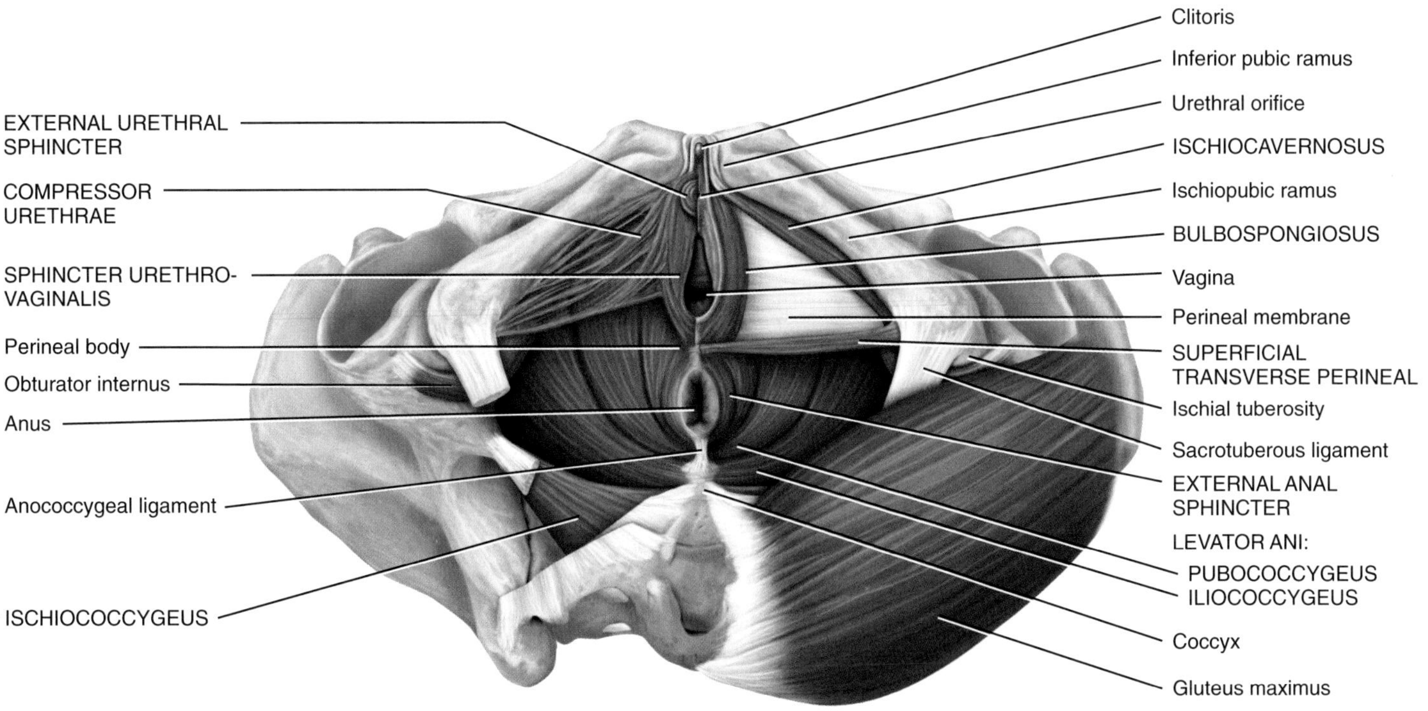

Inferior superficial view of a female perineum

What are the borders of the pelvic diaphragm?

EXHIBIT 11.J Muscles of the Perineum *(Figure 11.13)*

OBJECTIVE

- Describe the origin, insertion, action, and innervation of the muscles of the perineum.

The **perineum** is the region of the trunk inferior to the pelvic diaphragm. It is a diamond-shaped area that extends from the pubic symphysis anteriorly, to the coccyx posteriorly, and to the ischial tuberosities laterally. The female and the male perineums may be compared in Figures 11.12 and 11.13, respectively. A transverse line drawn between the ischial tuberosities divides the perineum into an **anterior urogenital triangle** that contains the external genitals and a posterior anal triangle that contains the anus (see Figure 28.21). Several perineal muscles insert into the perineal body of the perineum (described in Section 28.1). Clinically, the perineum is very important to physicians who care for women during pregnancy and treat disorders related to the female genital tract, urogenital organs, and the anorectal region.

The muscles of the perineum are arranged in two layers: **superficial** and **deep.** The muscles of the superficial layer are the **superficial transverse perineal** muscle, the **bulbospongiosus,** and the **ischiocavernosus** (Figures 11.12 and 11.13). The deep muscles of the male perineum are the **deep transverse perineal** muscle and **external urethral sphincter** (Figure 11.13). The deep muscles of the female perineum are the **compressor urethrae, sphincter urethrovaginalis,** and **external urethral sphincter** (see Figure 11.12). The deep muscles of the perineum assist in urination and ejaculation in males and urination and compression of the vagina in females. The **external anal sphincter** closely adheres to the skin around the margin of the anus and keeps the anal canal and anus closed except during defecation.

MUSCLE	ORIGIN	INSERTION	ACTION	INNERVATION
SUPERFICIAL PERINEAL MUSCLES				
Superficial transverse perineal (per-i-NĒ-al; *superficial* = closer to surface; *transverse* = across; *perineus* = perineum)	Ischial tuberosity.	Perineal body of perineum.	Stabilizes perineal body of perineum.	Perineal branch of pudendal nerve of sacral plexus.
Bulbospongiosus (bul′-bō-spon′-jē-Ō-sus; *bulb-* = bulb; *-spongio-* = sponge)	Perineal body of perineum.	Perineal membrane of deep muscles of perineum, corpus spongiosum of penis, and deep fascia on dorsum of penis in male; pubic arch and root and dorsum of clitoris in female.	Helps expel urine during urination, helps propel semen along urethra, assists in erection of penis in male; constricts vaginal orifice and assists in erection of clitoris in female.	Perineal branch of pudendal nerve of sacral plexus.
Ischiocavernosus (is′-kē-ō-ka′-ver-NŌ-sus; *ischio* = hip)	Ischial tuberosity and ischial and pubic rami.	Corpora cavernosa of penis in male and clitoris in female and pubic symphysis.	Maintains erection of penis in male and clitoris in female by decreasing urine drainage.	Perineal branch of pudendal nerve of sacral plexus.
DEEP PERINEAL MUSCLES				
Deep transverse perineal (per-i-NĒ-al; *deep* = farther from surface)	Ischial ramus.	Perineal body of perineum.	Helps expel last drops of urine and semen in male.	Perineal branch of pudendal nerve of sacral plexus.
External urethral sphincter (ū-RĒ-thral SFINGK-ter)	Ischial and pubic rami.	Median raphe in male and vaginal wall in female.	Helps expel last drops of urine and semen in male and urine in female.	Sacral spinal nerve S4 and inferior rectal branch of pudendal nerve.
Compressor urethrae (ū-RĒ-thrē) (see Figure 11.12)	Ischiopubic ramus.	Blends with partner on other side anterior to urethra.	Serves as accessory sphincter of urethra.	Perineal branch of pudendal nerve of sacral plexus.
Sphincter urethrovaginalis (ū-RĒ-thrō-vaj-i-NAL-is) (see Figure 11.12)	Perineal body.	Blends with partner on other side anterior to urethra.	Serves as accessory sphincter of urethra and facilitates closing of vagina.	Perineal branch of pudendal nerve of sacral plexus.
External anal sphincter (Ā-nal)	Anococcygeal ligament.	Perineal body of perineum.	Keeps anal canal and anus closed.	Sacral spinal nerve S4 and inferior rectal branch of pudendal nerve.

RELATING MUSCLES TO MOVEMENTS

Arrange the muscles in this exhibit according to the following actions; (1) expulsion of urine and semen, (2) erection of the clitoris and penis, (3) closure of the anal orifice, and (4) constriction of the vaginal orifice. The same muscle may be mentioned more than once.

✔CHECKPOINT

What are the borders and contents of the urogenital triangle and the anal triangle?

Figure 11.13 Muscles of the perineum.

The urogenital diaphragm assists in urination in females and males, plays a part in ejaculation in males, and helps strengthen the pelvic floor.

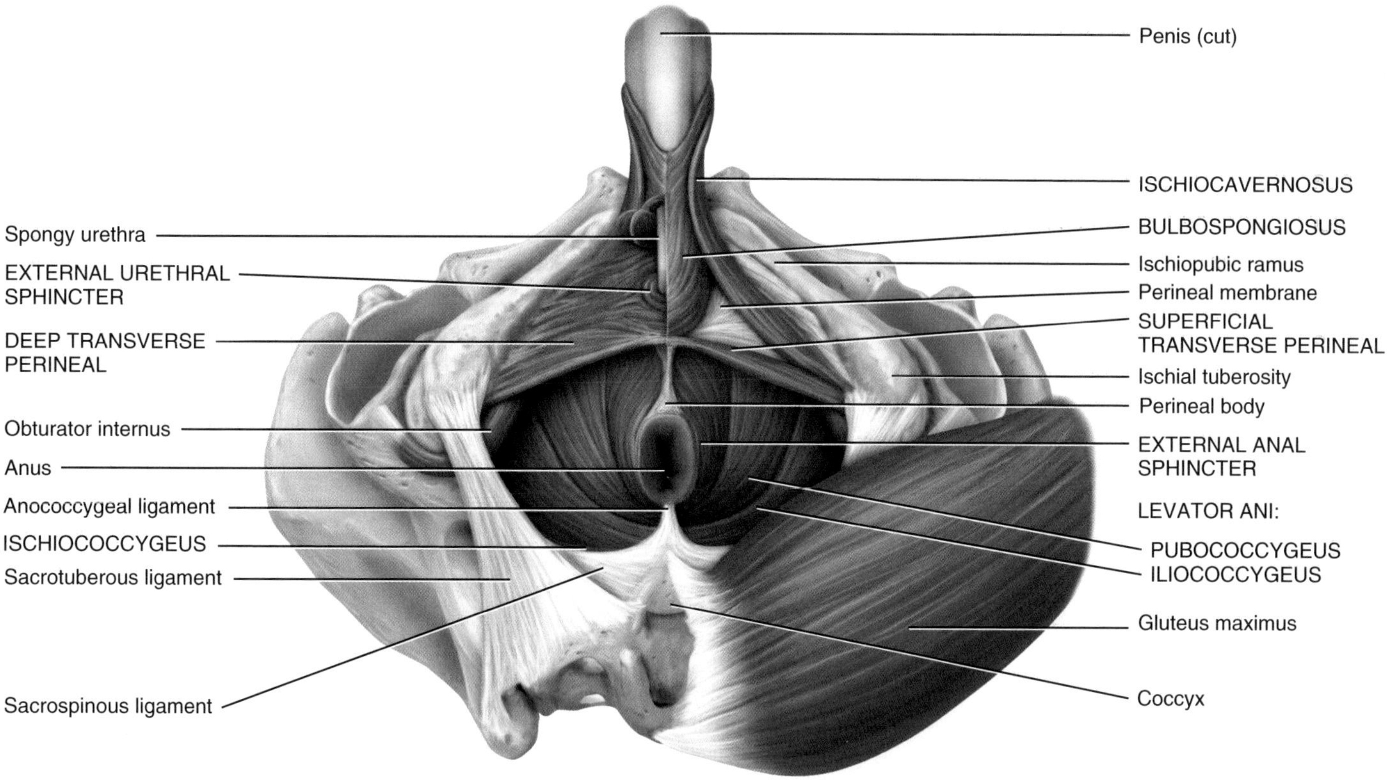

Inferior superficial view of male perineum

What are the borders of the perineum?

EXHIBIT 11.K Muscles of the Thorax That Move the Pectoral Girdle *(Figure 11.14)*

OBJECTIVE

- Describe the origin, insertion, action, and innervation of the muscles of the thorax that move the pectoral girdle.

The main action of the muscles that move the pectoral (shoulder) girdle (clavicle and scapula) is to stabilize the scapula so it can function as a steady origin for most of the muscles that move the humerus. Because scapular movements usually accompany humeral movements in the same direction, the muscles also move the scapula to increase the range of motion of the humerus. For example, it would not be possible to raise the arm above the head if the scapula did not move with the humerus. During abduction, the scapula follows the humerus by rotating upward.

Muscles that move the pectoral girdle can be classified into two groups based on their location in the thorax: **anterior** and **posterior thoracic muscles** (Figure 11.14). The anterior thoracic muscles are the subclavius, pectoralis minor, and serratus anterior. The **subclavius** is a small, cylindrical muscle under the clavicle that extends from the clavicle to the first rib. It steadies the clavicle during movements of the pectoral girdle. The **pectoralis minor** is a thin, flat, triangular muscle that is deep to the pectoralis major. Besides its role in movements of the scapula, the pectoralis minor muscle assists in forced inhalation. The **serratus anterior** is a large, flat, fan-shaped muscle between the ribs and scapula. It is so named because of the saw-toothed appearance of its origins on the ribs.

The posterior thoracic muscles are the trapezius, levator scapulae, rhomboid major, and rhomboid minor. The **trapezius** is a large, flat, triangular sheet of muscle extending from the skull and vertebral column medially to the pectoral girdle laterally. It is the most superficial back muscle and covers the posterior neck region and superior portion of the trunk. The two trapezius muscles form a trapezoid (diamond-shaped quadrangle)—hence its name. The **levator scapulae** is a narrow, elongated muscle in the posterior portion of the neck. It is deep to the sternocleidomastoid and trapezius muscles. As its name suggests, one of its actions is to elevate the scapula (see Figure 11.15c). The **rhomboid major** and **rhomboid minor** lie deep to the trapezius and are not always distinct from each other. They appear as parallel bands that pass inferiorly and laterally from the vertebrae to the scapula (see Figure 11.15c). Their names are based on their shape—that is, a rhomboid (an oblique parallelogram). The rhomboid major is about two times wider than the rhomboid minor. Both muscles are used when forcibly lowering the raised upper limbs, as in driving a stake with a sledgehammer.

MUSCLE	ORIGIN	INSERTION	ACTION	INNERVATION
ANTERIOR THORACIC MUSCLES				
Subclavius (sub-KLĀ-vē-us; *sub-* = under; *-clavius* = clavicle)	First rib.	Clavicle.	Depresses and moves clavicle anteriorly and helps stabilize pectoral girdle.	Subclavian nerve.
Pectoralis minor (pek′-tō-RĀ-lis; *pector* = breast, chest, thorax; *minor* = lesser)	Second through fifth, third through fifth, or second through fourth ribs.	Coracoid process of scapula.	Abducts scapula and rotates it downward. RMA: Elevates third through fifth ribs during forced inhalation when scapula is fixed.	Medial pectoral nerve.
Serratus anterior (ser-Ā-tus; *serratus* = saw-toothed; *anterior* = front)	Superior eight or nine ribs.	Vertebral border and inferior angle of scapula.	Abducts scapula and rotates it upward. RMA: Elevates ribs when scapula is stabilized. Known as "boxer's muscle" because it is important in horizontal arm movements such as punching and pushing.	Long thoracic nerve.
POSTERIOR THORACIC MUSCLES				
Trapezius (tra-PĒ-zē-us; *trapezi* = trapezoid-shaped)	Superior nuchal line of occipital bone, ligamentum nuchae, and spines of C7–T12.	Clavicle and acromion and spine of scapula.	Superior fibers upward rotate scapula; middle fibers adduct scapula; inferior fibers depress and upward rotate scapula; superior and inferior fibers together rotate scapula upward; stabilizes scapula. RMA: Superior fibers can help extend head.	Accessory (XI) nerve and cervical spinal nerves C3–C5.
Levator scapulae (le-VĀ-tor SKA-pū-lē; *levator* = raises; *scapulae* = scapula)	Transverse processes of C1–C4.	Superior vertebral border of scapula.	Elevates scapula and rotates it downward.	Dorsal scapular nerve and cervical spinal nerves C3–C5.
Rhomboid major (rom-BOYD; *rhomboid* = rhomboid or diamond-shaped) (see Figure 11.15c)	Spines of T2–T5.	Vertebral border of scapula inferior to spine.	Elevates and adducts scapula and rotates it downward; stabilizes scapula.	Dorsal scapular nerve.
Rhomboid minor (rom-BOYD) (see Figure 11.15c)	Spines of C7–T1.	Vertebral border of scapula superior to spine.	Elevates and adducts scapula and rotates it downward; stabilizes scapula.	Dorsal scapular nerve.

To understand the actions of muscles that move the scapula, it is first helpful to review the various movements of the scapula:

- **Elevation:** superior movement of the scapula, such as shrugging the shoulders or lifting a weight over the head.
- **Depression:** inferior movement of the scapula, as in pulling down on a rope attached to a pulley.
- **Abduction (protraction):** movement of the scapula laterally and anteriorly, as in doing a "push-up" or punching.
- **Adduction (retraction):** movement of the scapula medially and posteriorly, as in pulling the oars in a rowboat.
- **Upward rotation:** movement of the inferior angle of the scapula laterally so that the glenoid cavity is moved upward. This movement is required to move the humerus past the horizontal, as in raising the arms in a "jumping jack."
- **Downward rotation:** movement of the inferior angle of the scapula medially so that the glenoid cavity is moved downward. This movement is seen when a gymnast on parallel bars supports the weight of the body on the hands.

RELATING MUSCLES TO MOVEMENTS

Arrange the muscles in this exhibit according to the following actions on the scapula: (1) depression, (2) elevation, (3) abduction, (4) adduction, (5) upward rotation, and (6) downward rotation. The same muscle may be mentioned more than once.

✔ **CHECKPOINT**

What muscles in this exhibit are used to raise your shoulders, lower your shoulders, join your hands behind your back, and join your hands in front of your chest?

Figure 11.14 Muscles of the thorax (chest) that move the pectoral (shoulder) girdle (clavicle and scapula).

Muscles that move the pectoral girdle originate on the axial skeleton and insert on the clavicle or scapula.

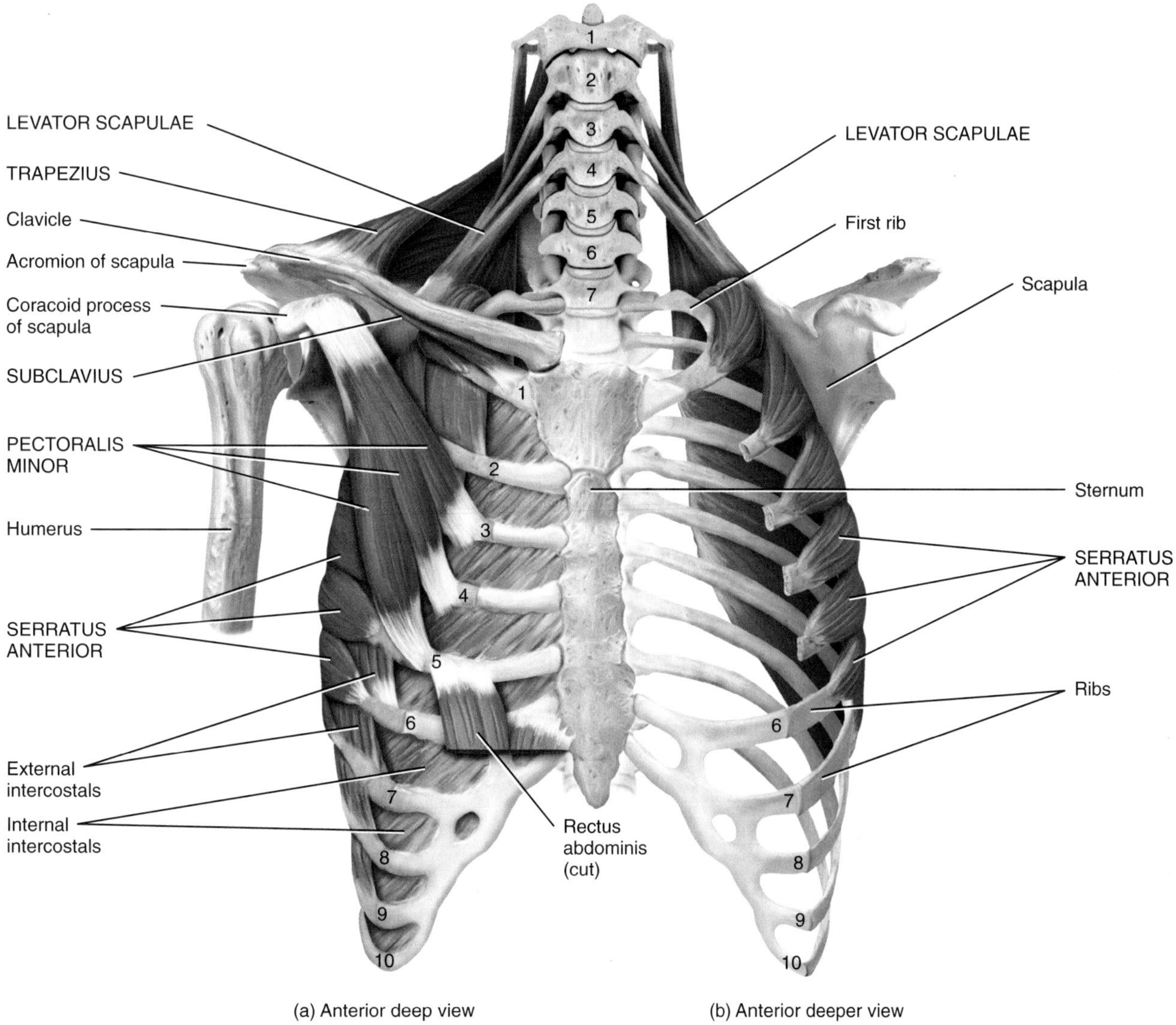

What is the main action of the muscles that move the pectoral girdle?

EXHIBIT 11.L Muscles of the Thorax and Shoulder That Move the Humerus *(Figure 11.15)*

OBJECTIVE

- Describe the origin, insertion, action, and innervation of the muscles of the thorax that move the humerus.

Of the nine muscles that cross the shoulder joint, all except the pectoralis major and latissimus dorsi originate on the scapula (shoulder blade). The pectoralis major and latissimus dorsi thus are called **axial muscles,** because they originate on the axial skeleton. The remaining seven muscles, the **scapular muscles,** arise from the scapula (Figure 11.15).

Of the two axial muscles that move the humerus (arm bone), the **pectoralis major** is a large, thick, fan-shaped muscle that covers the superior part of the thorax and forms the anterior fold of the thorax. It has two origins: a smaller clavicular head and a larger sternocostal head. The **latissimus dorsi** is a broad, triangular muscle located on the interior part of the back that forms most of the posterior wall of the axilla. The reverse muscle action (RMA) of the latissimus dorsi enables the vertebral column and torso to be elevated, as in doing a pullup. It is commonly called the "swimmer's muscle" because its many actions are used while swimming; consequently, many competitive swimmers have well-developed "lats."

Among the scapular muscles, the **deltoid** is a thick, powerful shoulder muscle that covers the shoulder joint and forms the rounded contour

MUSCLE	ORIGIN	INSERTION	ACTION	INNERVATION
AXIAL MUSCLES THAT MOVE THE HUMERUS				
Pectoralis major (pek′-tō-RĀ-lis; *pector* = chest; *major* = larger) (see also Figure 11.10a)	Clavicle (clavicular head), sternum, and costal cartilages of second to sixth ribs and sometimes first to seventh ribs (sternocostal head).	Greater tubercle and lateral lip of intertubercular sulcus of humerus.	As a whole, adducts and medially rotates arm at shoulder joint; clavicular head flexes arm, and sternocostal head extends flexed arm to side of trunk.	Medial and lateral pectoral nerves.
Latissimus dorsi (la-TIS-i-mus DOR-sī; *latissimus* = widest; *dorsi* = of the back)	Spines of T7–L5, lumbar vertebrae, crests of sacrum and ilium, inferior four ribs via thoracolumbar fascia.	Intertubercular sulcus of humerus.	Extends, adducts, and medially rotates arm at shoulder joint; draws arm inferiorly and posteriorly. RMA: Elevates vertebral column and torso.	Thoracodorsal nerve.
SCAPULAR MUSCLES THAT MOVE THE HUMERUS				
Deltoid (DEL-toyd = triangularly shaped)	Acromial extremity of clavicle (anterior fibers), acromion of scapula (lateral fibers), and spine of scapula (posterior fibers).	Deltoid tuberosity of humerus.	Lateral fibers abduct arm at shoulder joint; anterior fibers flex and medially rotate arm at shoulder joint; posterior fibers extend and laterally rotate arm at shoulder joint.	Axillary nerve.
Subscapularis (sub-scap′-ū-LĀ-ris; *sub-* = below; *-scapularis* = scapula)	Subscapular fossa of scapula.	Lesser tubercle of humerus.	Medially rotates arm at shoulder joint.	Upper and lower subscapular nerve.
Supraspinatus (soo-pra-spī-NĀ-tus; *supra-* = above; *-spina-* = spine [of the scapula])	Supraspinous fossa of scapula.	Greater tubercle of humerus.	Assists deltoid muscle in abducting arm at shoulder joint.	Suprascapular nerve.
Infraspinatus (in′-fra-spī-NĀ-tus; *infra* = below)	Infraspinous fossa of scapula.	Greater tubercle of humerus.	Laterally rotates arm at shoulder joint.	Suprascapular nerve.
Teres major (TE-rēz; *teres* = long and round)	Inferior angle of scapula.	Medial lip of intertubercular sulcus of humerus.	Extends arm at shoulder joint and assists in adduction and medial rotation of arm at shoulder joint.	Lower subscapular nerve.
Teres minor (TE-rēz)	Inferior lateral border of scapula.	Greater tubercle of humerus.	Laterally rotates and extends arm at shoulder joint.	Axillary nerve.
Coracobrachialis (kor′-a-kō-brā-kē-Ā-lis; *coraco-* = coracoid process [of the scapula]; *-brachi-* = arm)	Coracoid process of scapula.	Middle of medial surface of shaft of humerus.	Flexes and adducts arm at shoulder joint.	Musculocutaneous nerve.

of the shoulder. This muscle is a frequent site of intramuscular injections. As you study the deltoid, note that its fascicles originate from three different points and that each group of fascicles moves the humerus differently. The **subscapularis** is a large triangular muscle that fills the subscapular fossa of the scapula and forms a small part in the apex of the posterior wall of the axilla. The **supraspinatus,** a rounded muscle named for its location in the supraspinous fossa of the scapula, lies deep to the trapezius. The **infraspinatus** is a triangular muscle, also named for its location in the infraspinous fossa of the scapula. The **teres major** is a thick, flattened muscle inferior to the teres minor that also helps form part of the posterior wall of the axilla. The **teres minor** is a cylindrical, elongated muscle, often inseparable from the infraspinatus, which lies

Figure 11.15 **Muscles of the thorax (chest) and shoulder that move the humerus (arm bone).**

The strength and stability of the shoulder joint are provided by the tendons that form the rotator cuff.

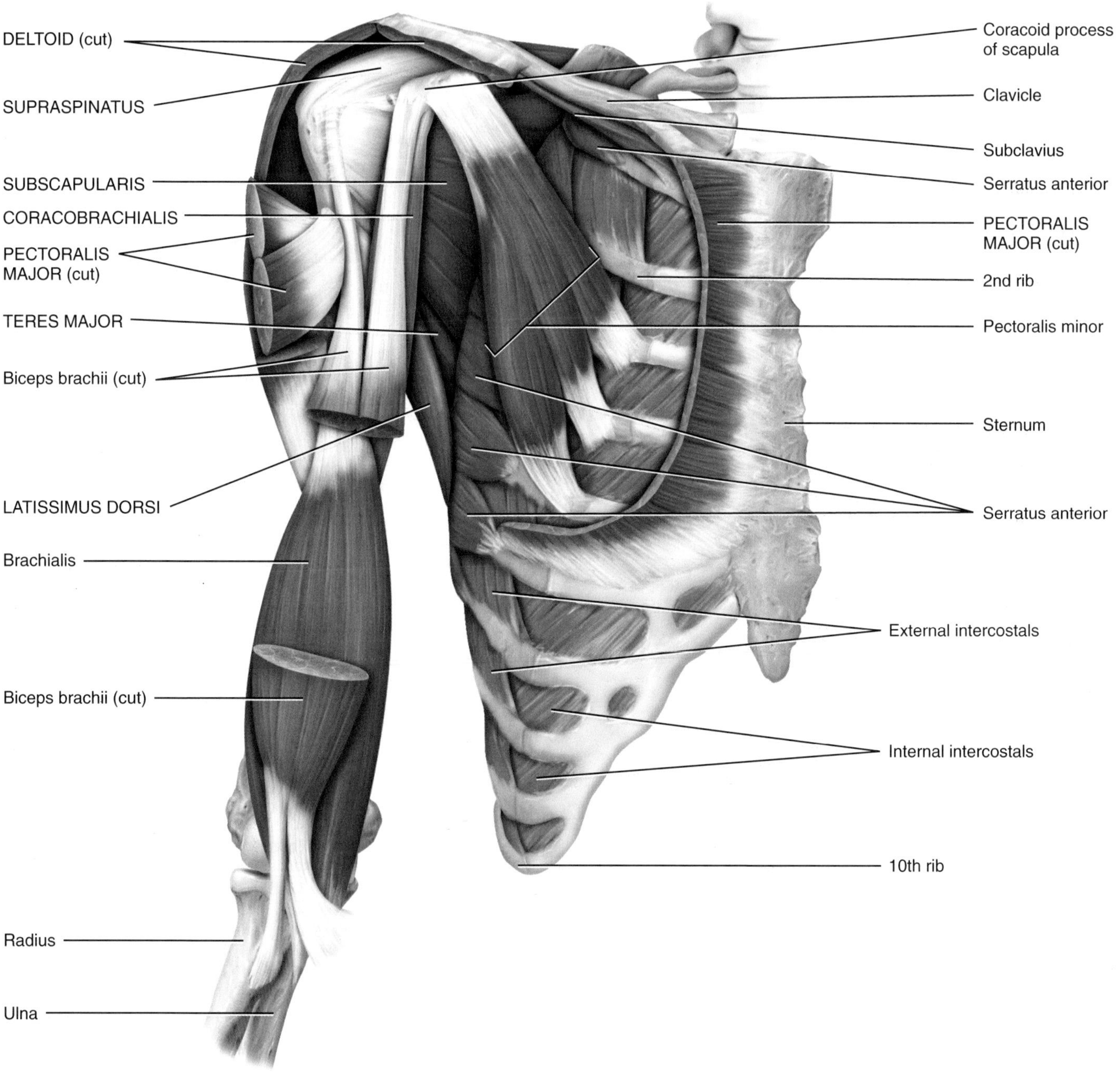

(a) Anterior deep view (the intact pectoralis major muscle is shown in Figure 11.3a)

EXHIBIT 11.L CONTINUES

Muscles of the Thorax and Shoulder That Move the Humerus *(Figure 11.15)* CONTINUED

along its superior border. The **coracobrachialis** is an elongated, narrow muscle in the arm.

Four deep muscles of the shoulder—subscapularis, supraspinatus, infraspinatus, and teres minor—strengthen and stabilize the shoulder joint. These muscles join the scapula to the humerus. Their flat tendons fuse together to form the **rotator (musculotendinous) cuff,** a nearly complete circle of tendons around the shoulder joint, like the cuff on a shirtsleeve. The supraspinatus muscle is especially subject to wear and tear because of its location between the head of the humerus and acromion of the scapula, which compress its tendon during shoulder movements, especially abduction of the arm. This is further aggravated by poor posture with slouched shoulders.

FIGURE 11.15 CONTINUED

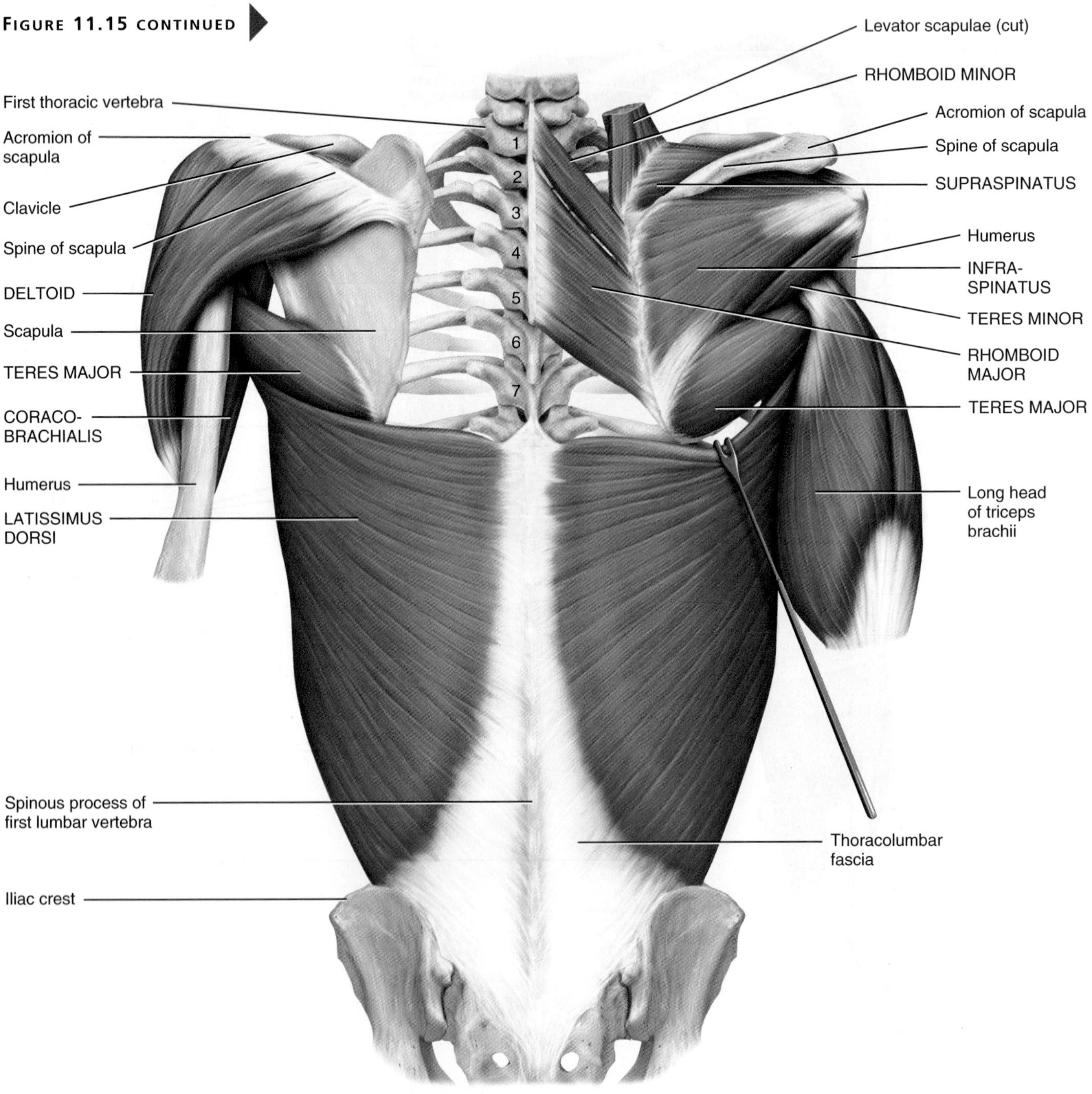

(b) Posterior view (c) Posterior view

CLINICAL CONNECTION | *Rotator Cuff Injury and Impingement Syndrome*

Rotator cuff injury is a strain or tear in the rotator cuff muscles and is common among baseball pitchers, volleyball players, racquet sports players, and swimmers due to shoulder movements that involve vigorous circumduction. It also occurs as a result of wear and tear, aging, trauma, poor posture, improper lifting, and repetitive motions in certain jobs, such as placing items on a shelf above your head. Most often, there is tearing of the supraspinatus muscle tendon or the rotator cuff. This tendon is especially predisposed to wear and tear because of its location between the head of the humerus and acromion of the scapula, which compresses the tendon during shoulder movements. Poor posture and poor body mechanics also increase compression of the supraspinatus muscle tendon.

One of the most common causes of shoulder pain and dysfunction in athletes is known as **impingement syndrome,** which is sometimes confused with another common complaint, compartment syndrome, discussed in Disorders: Homeostatic Imbalances at the end of this chapter. The repetitive movement of the arm over the head that is common in baseball, overhead racquet sports, lifting weights over the head, spiking a volleyball, and swimming puts these athletes at risk. Impingement syndrome may also be caused by a direct blow or stretch injury. Continual pinching of the supraspinatus tendon as a result of overhead motions causes it to become inflamed and results in pain. If movement is continued despite the pain, the tendon may degenerate near the attachment to the humerus and ultimately may tear away from the bone (rotator cuff injury). Treatment consists of resting the injured tendons, strengthening the shoulder through exercise, massage therapy, and surgery if the injury is particularly severe. During surgery, an inflamed bursa may be removed, bone may be trimmed, and/or the coracoacromial ligament may be detached. Torn rotator cuff tendons may be trimmed and then reattached with sutures, anchors, or surgical tacks. These steps make more space, thus relieving pressure and allowing the arm to move freely. •

RELATING MUSCLES TO MOVEMENTS

Arrange the muscles in this exhibit according to the following actions on the humerus at the shoulder joint: (1) flexion, (2) extension, (3) abduction, (4) adduction, (5) medial rotation, and (6) lateral rotation. The same muscle may be mentioned more than once.

✔ CHECKPOINT

Why are the two muscles that cross the shoulder joint called axial muscles, and the seven others called scapular muscles?

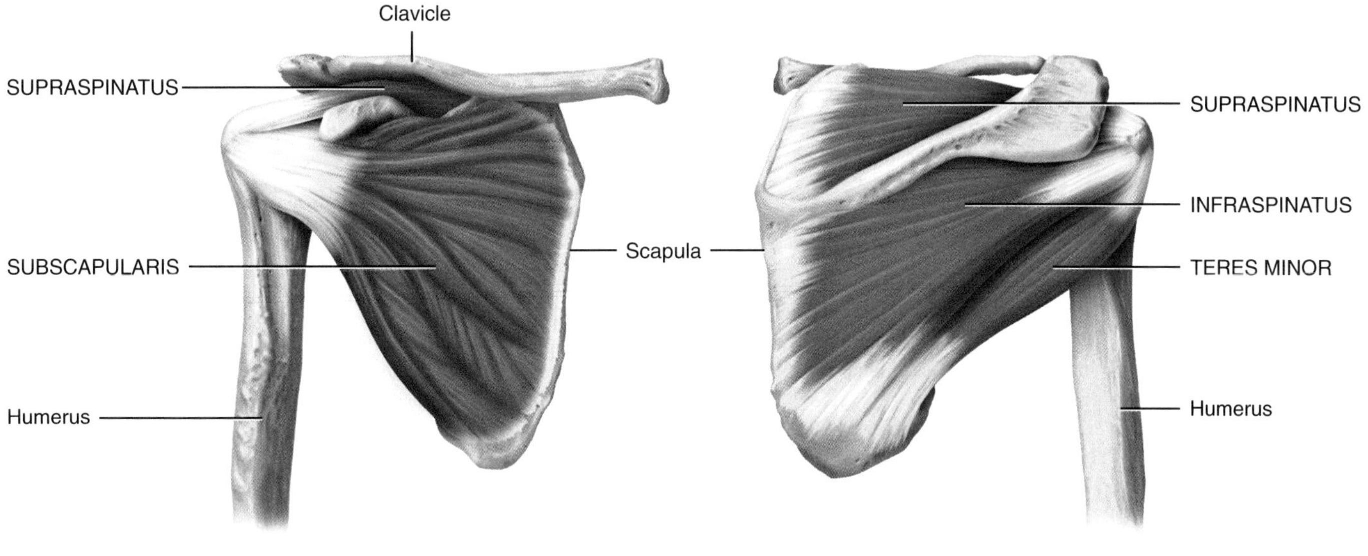

(d) Anterior deep view

(e) Posterior deep view

? Which tendons make up the rotator cuff?

EXHIBIT 11.M Muscles of the Arm That Move the Radius and Ulna *(Figure 11.16)*

OBJECTIVE

- Describe the origin, insertion, action, and innervation of the muscles of the arm that move the radius and ulna.

Most of the muscles that move the radius and ulna (forearm bones) cause flexion and extension at the elbow, which is a hinge joint. The biceps brachii, brachialis, and brachioradialis muscles are the flexor muscles. The extensor muscles are the triceps brachii and the anconeus (Figure 11.16).

The **biceps brachii** is the large muscle located on the anterior surface of the arm. As indicated by its name, it has two heads of origin (long and short), both from the scapula. The muscle spans both the shoulder and elbow joints. In addition to its role in flexing the forearm at the elbow joint, it also supinates the forearm at the radioulnar joints and flexes the arm at the shoulder joint. The **brachialis** is deep to the biceps brachii muscle. It is the most powerful flexor of the forearm at the elbow joint. For this reason, it is called the "workhorse" of the elbow flexors. The **brachioradi-**

MUSCLE	ORIGIN	INSERTION	ACTION	INNERVATION
FOREARM FLEXORS				
Biceps brachii (BĪ-seps BRĀ-kē-ī; *biceps* = two heads of origin; *brachii* = arm)	Long head originates from tubercle above glenoid cavity of scapula (supraglenoid tubercle). Short head originates from coracoid process of scapula.	Radial tuberosity of radius and bicipital aponeurosis.*	Flexes forearm at elbow joint, supinates forearm at radioulnar joints, and flexes arm at shoulder joint.	Musculocutaneous nerve.
Brachialis (brā-kē-Ā-lis)	Distal, anterior surface of humerus.	Ulnar tuberosity and coronoid process of ulna.	Flexes forearm at elbow joint.	Musculocutaneous and radial nerves.
Brachioradialis (brā′-kē-ō-rā-dē-Ā-lis; *radi* = radius)	Lateral border of distal end of humerus.	Superior to styloid process of radius.	Flexes forearm at elbow joint; supinates and pronates forearm at radioulnar joints to neutral position.	Radial nerve.
FOREARM EXTENSORS				
Triceps brachii (TRĪ-seps BRĀ-kē-ī; *triceps* = three heads of origin)	Long head originates from infraglenoid tubercle, a projection inferior to glenoid cavity of scapula. Lateral head originates from lateral and posterior surface of humerus. Medial head originates from entire posterior surface of humerus inferior to a groove for the radial nerve.	Olecranon of ulna.	Extends forearm at elbow joint and extends arm at shoulder joint.	Radial nerve.
Anconeus (an-KŌ-nē-us; *ancon* = elbow)	Lateral epicondyle of humerus.	Olecranon and superior portion of shaft of ulna.	Extends forearm at elbow joint.	Radial nerve.
FOREARM PRONATORS				
Pronator teres (PRŌ-nā-tor TE-rēz; *pronator* = turns palm posteriorly; *tero* = round and long) (see also Figure 11.17a)	Medial epicondyle of humerus and coronoid process of ulna.	Midlateral surface of radius.	Pronates forearm at radioulnar joints and weakly flexes forearm at elbow joint.	Median nerve.
Pronator quadratus (PRŌ-nā-tor kwod-RĀ-tus; *quadratus* = square, four-sided) (see also Figure 11.17a–c)	Distal portion of shaft of ulna.	Distal portion of shaft of radius.	Pronates forearm at radioulnar joints.	Median nerve.
FOREARM SUPINATOR				
Supinator (SOO-pi-nā-tor; *supinator* = turns palm anteriorly) (see also Figure 11.17b,c)	Lateral epicondyle of humerus and ridge near radial notch of ulna (supinator crest).	Lateral surface of proximal one-third of radius.	Supinates forearm at radioulnar joints.	Deep radial nerve.

*The **bicipital aponeurosis** is a broad aponeurosis from the tendon of insertion of the biceps brachii muscle that descends medially across the brachial artery and fuses with deep fascia over the forearm flexor muscles. It also helps to protect the median nerve and brachial artery.

alis flexes the forearm at the elbow joint, especially when a quick movement is required or when a weight is lifted slowly during flexion of the forearm.

The **triceps brachii** is the large muscle located on the posterior surface of the arm. It is the more powerful of the extensors of the forearm at the elbow joint. As its name implies, it has three heads of origin, one from the scapula (long head) and two from the humerus (lateral and medial heads). The long head crosses the shoulder joint; the other heads do not. The **anconeus** is a small muscle located on the lateral part of the posterior aspect of the elbow that assists the triceps brachii in extending the forearm at the elbow joint.

Some muscles that move the radius and ulna are involved in pronation and supination at the radioulnar joints. The pronators, as suggested by their names, are the **pronator teres** and **pronator quadratus** muscles. The supinator of the forearm is aptly named the **supinator** muscle. You use the powerful action of the supinator when you twist a corkscrew or turn a screw with a screwdriver.

In the limbs, functionally related skeletal muscles and their associated blood vessels and nerves are grouped together by fascia into regions called **compartments.** In the arm, the biceps brachii, brachialis, and coracobrachialis muscles compose the **anterior (flexor) compartment.** The triceps brachii muscle forms the **posterior (extensor) compartment.**

Figure 11.16 Muscles of the arm that move the radius and ulna (forearm bones).

The anterior arm muscles flex the forearm, and the posterior arm muscles extend it.

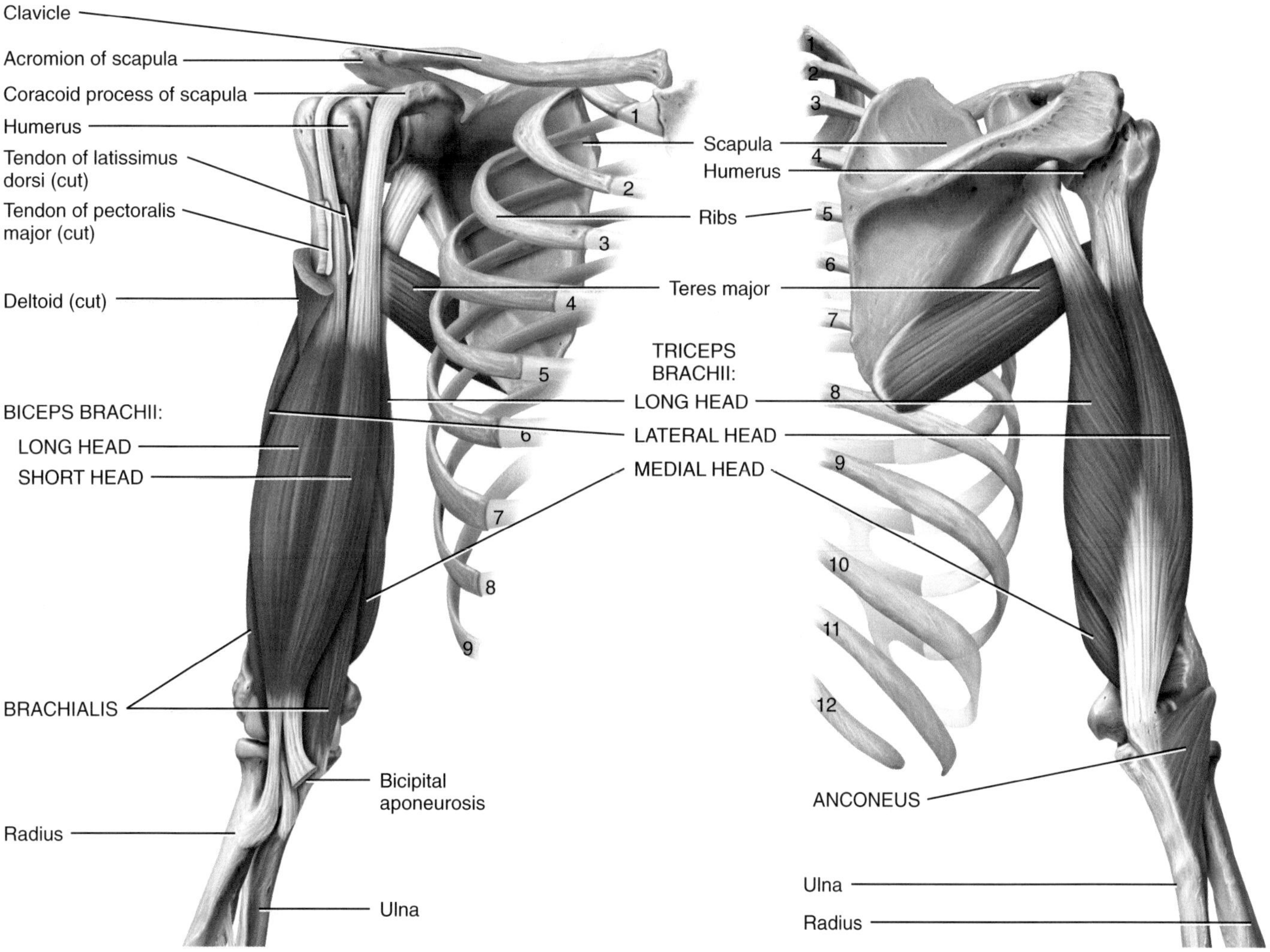

(a) Anterior view

(b) Posterior view

EXHIBIT 11.M CONTINUES

RELATING MUSCLES TO MOVEMENTS

Arrange the muscles in this exhibit according to the following actions on the elbow joint: (1) flexion and (2) extension; the following actions on the forearm at the radioulnar joints: (1) supination and (2) pronation; and the following actions on the humerus at the shoulder joint: (1) flexion and (2) extension. The same muscle may be mentioned more than once.

✔ **CHECKPOINT**

Flex your arm. Which group of muscles is contracting? Which group of muscles must relax so that you can flex your arm?

FIGURE 11.16 CONTINUED

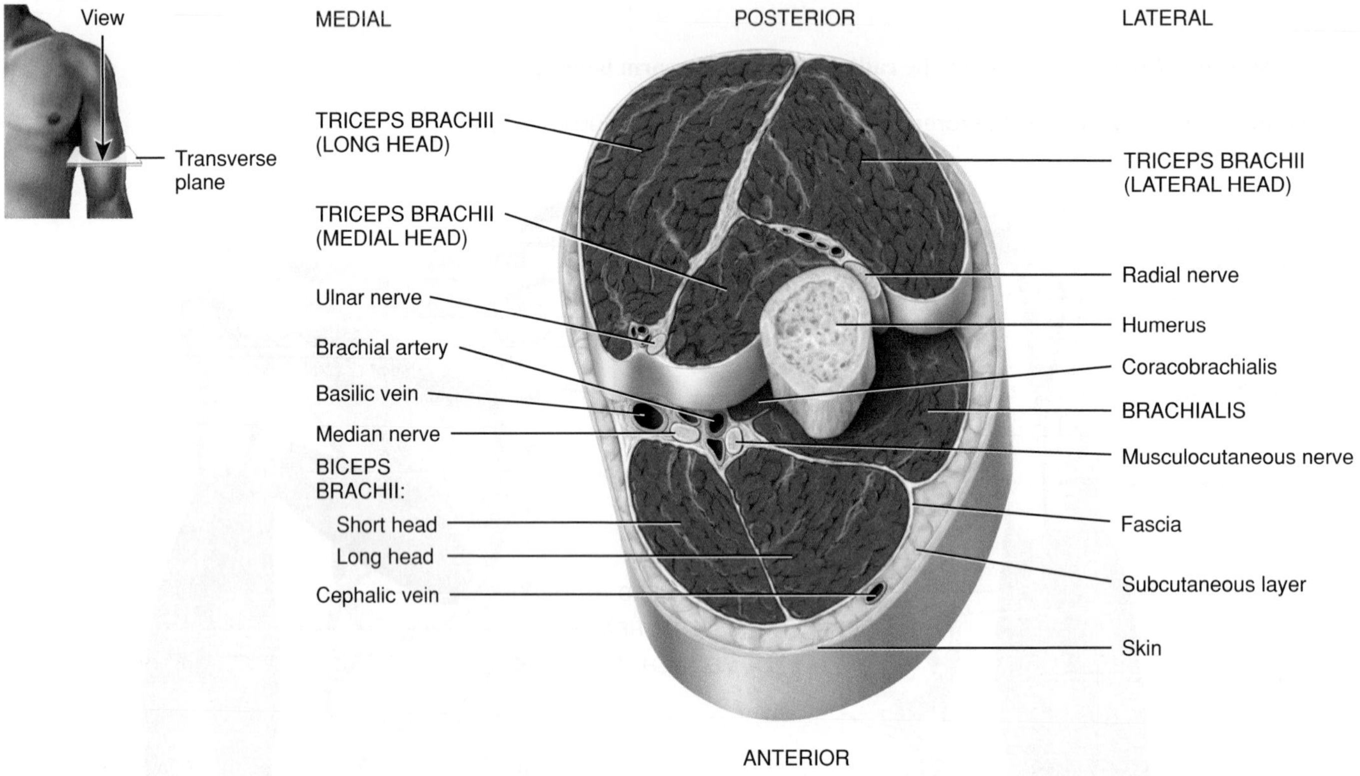
View
Transverse plane
MEDIAL
POSTERIOR
LATERAL
TRICEPS BRACHII (LONG HEAD)
TRICEPS BRACHII (MEDIAL HEAD)
Ulnar nerve
Brachial artery
Basilic vein
Median nerve
BICEPS BRACHII:
Short head
Long head
Cephalic vein
TRICEPS BRACHII (LATERAL HEAD)
Radial nerve
Humerus
Coracobrachialis
BRACHIALIS
Musculocutaneous nerve
Fascia
Subcutaneous layer
Skin
ANTERIOR

(c) Superior view of transverse section of arm

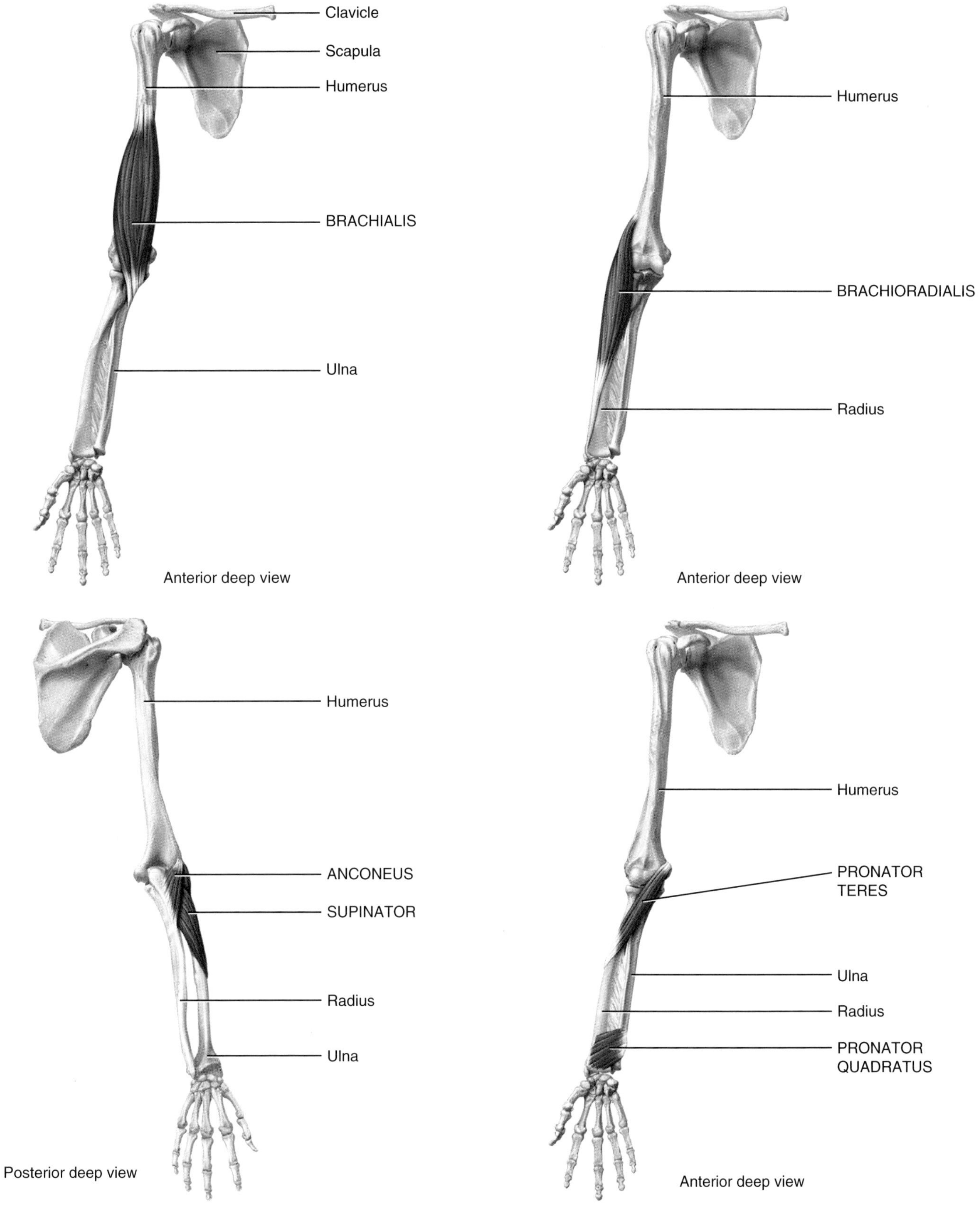
Clavicle
Scapula
Humerus
BRACHIALIS
Ulna
Anterior deep view
Humerus
BRACHIORADIALIS
Radius
Anterior deep view
Humerus
ANCONEUS
SUPINATOR
Radius
Ulna
Posterior deep view
Humerus
PRONATOR TERES
Ulna
Radius
PRONATOR QUADRATUS
Anterior deep view

(d) Isolated muscles

Which muscles are the most powerful flexor and the most powerful extensor of the forearm?

EXHIBIT 11.N Muscles of the Forearm That Move the Wrist, Hand, Thumb, and Digits *(Figure 11.17)*

OBJECTIVE

- Describe the origin, insertion, action, and innervation of the muscles of the forearm that move the wrist, hand, and digits.

Muscles of the forearm that move the wrist, hand, and digits are many and varied (Figure 11.17). Those in this group that act on the digits are known as **extrinsic muscles of the hand** (*ex* = outside) because they originate *outside* the hand and insert within it. As you will see, the names for the muscles that move the wrist, hand, and digits give some indication of their origin, insertion, or action. Based on location and function, the muscles of the forearm are divided into two groups: (1) anterior compartment muscles and (2) posterior compartment muscles. The **anterior (flexor) compartment** muscles of the forearm originate on the humerus; typically insert on the carpals, metacarpals, and phalanges; and function primarily as flexors. The bellies of these muscles form the bulk of the forearm. One of the muscles in the superficial anterior compartment, the palmaris longus muscle, is missing in about 10% of individuals (usually in the left forearm) and is commonly used for tendon repair. The **posterior (extensor) compartment** muscles of the forearm originate on the humerus, insert on the metacarpals and phalanges, and function as extensors. Within each compartment, the muscles are grouped as superficial or deep.

The **superficial anterior compartment** muscles are arranged in the following order from lateral to medial: **flexor carpi radialis, palmaris longus,** and **flexor carpi ulnaris** (the ulnar nerve and artery are just lateral to the tendon of this muscle at the wrist). The **flexor digitorum superficialis** muscle is deep to the other three muscles and is the largest superficial muscle in the forearm.

The **deep anterior compartment** muscles are arranged in the following order from lateral to medial: **flexor pollicis longus** (the only flexor of the distal phalanx of the thumb) and **flexor digitorum profundus** (ends in four tendons that insert into the distal phalanges of the fingers).

The **superficial posterior compartment** muscles are arranged in the following order from lateral to medial: **extensor carpi radialis longus, extensor carpi radialis brevis, extensor digitorum** (occupies most of the posterior surface of the forearm and divides into four tendons that insert into the middle and distal phalanges of the fingers), **extensor digiti minimi** (a slender muscle usually connected to the extensor digitorum), and **extensor carpi ulnaris.**

The **deep posterior compartment** muscles are arranged in the following order from lateral to medial: **abductor pollicis longus, extensor pollicis brevis, extensor pollicis longus,** and **extensor indicis.**

The tendons of the muscles of the forearm that attach to the wrist or continue into the hand, along with blood vessels and nerves, are held close to bones by strong fasciae. The tendons are also surrounded by tendon sheaths. At the wrist, the deep fascia is thickened into fibrous bands called **retinacula** (*retinacul* = holdfast). The **flexor retinaculum** is located over the palmar surface of the carpal bones. The long flexor tendons of the digits and wrist and the median nerve pass deep to the flexor retinaculum. The **extensor retinaculum** is located over the dorsal surface of the carpal bones. The extensor tendons of the wrist and digits pass deep to it.

MUSCLE	ORIGIN	INSERTION	ACTION	INNERVATION
SUPERFICIAL ANTERIOR (FLEXOR) COMPARTMENT OF THE FOREARM				
Flexor carpi radialis (FLEK-sor KAR-pē-rā′-dē-Ā-lis; *flexor* = decreases angle at joint; *carpi* = wrist; *radi* = radius)	Medial epicondyle of humerus.	Second and third metacarpals.	Flexes and abducts hand (radial deviation) at wrist joint.	Median nerve.
Palmaris longus (pal-MA-ris LON-gus; *palma* = palm; *longus* = long)	Medial epicondyle of humerus.	Flexor retinaculum and palmar aponeurosis (fascia in center of palm).	Weakly flexes hand at wrist joint.	Median nerve.
Flexor carpi ulnaris (FLEK-sor KAR-pē ūl-NAR-is; *ulnar* = ulna)	Medial epicondyle of humerus and superior posterior border of ulna.	Pisiform, hamate, and base of fifth metacarpal.	Flexes and adducts hand (ulnar deviation) at wrist joint.	Ulnar nerve.
Flexor digitorum superficialis (FLEK-sor di-ji-TOR-um soo′-per-fish′-ē-Ā-lis; *digit* = finger or toe; *superficialis* = closer to surface)	Medial epicondyle of humerus, coronoid process of ulna, and ridge along lateral margin or anterior surface (anterior oblique line) of radius.	Middle phalanx of each finger.*	Flexes middle phalanx of each finger at proximal interphalangeal joint, proximal phalanx of each finger at metacarpophalangeal joint, and hand at wrist joint.	Median nerve.
DEEP ANTERIOR (FLEXOR) COMPARTMENT OF THE FOREARM				
Flexor pollicis longus (FLEK-sor POL-li-sis LON-gus; *pollicis* = thumb)	Anterior surface of radius and interosseous membrane (sheet of fibrous tissue that holds shafts of ulna and radius together).	Base of distal phalanx of thumb.	Flexes distal phalanx of thumb at interphalangeal joint.	Median nerve.
Flexor digitorum profundus (FLEK-sor di′-ji-TOR-um prō-FUN-dus; *profundus* = deep)	Anterior medial surface of body of ulna.	Base of distal phalanx of each finger.	Flexes distal and middle phalanges of each finger at interphalangeal joints, proximal phalanx of each finger at metacarpophalangeal joint, and hand at wrist joint.	Median and ulnar nerves.

CLINICAL CONNECTION | *Golfer's Elbow*

Golfer's elbow is a condition that can be caused by strain of the flexor muscles, especially the flexor carpi radialis, as a result of repetitive movements such as swinging a golf club. Strain can, however, be caused by many actions. Pianists, violinists, movers, weight lifters, bikers, and those who use computers are among those who may develop pain near the medial epicondyle (*medial epicondylitis*). •

MUSCLE	ORIGIN	INSERTION	ACTION	INNERVATION
SUPERFICIAL POSTERIOR (EXTENSOR) COMPARTMENT OF THE FOREARM				
Extensor carpi radialis longus (eks-TEN-sor KAR-pē rā′-dē-Ā-lis LON-gus; *extensor* = increases angle at joint)	Lateral supracondylar ridge of humerus.	Second metacarpal.	Extends and abducts hand at wrist joint (ulnar deviation).	Radial nerve.
Extensor carpi radialis brevis (eks-TEN-sor KAR-pē rā-dē-Ā-lis BREV-is; *brevis* = short)	Lateral epicondyle of humerus.	Third metacarpal.	Extends and abducts hand at wrist joint.	Radial nerve.
Extensor digitorum (eks-TEN-sor di′-ji-TOR-um)	Lateral epicondyle of humerus.	Distal and middle phalanges of each finger.	Extends distal and middle phalanges of each finger at interphalangeal joints, proximal phalanx of each finger at metacarpophalangeal joint, and hand at wrist joint.	Radial nerve.
Extensor digiti minimi (eks-TEN-sor DIJ-i-tē MIN-i-mē; *digit* = finger or toe; *minimi* = smallest)	Lateral epicondyle of humerus.	Tendon of extensor digitorum on fifth phalanx.	Extends proximal phalanx of little finger at metacarpophalangeal joint and hand at wrist joint.	Deep radial nerve.
Extensor carpi ulnaris (eks-TEN-sor KAR-pē ul-NAR-is)	Lateral epicondyle of humerus and posterior border of ulna.	Fifth metacarpal.	Extends and adducts hand at wrist joint (ulnar deviation).	Deep radial nerve.
DEEP POSTERIOR (EXTENSOR) COMPARTMENT OF THE FOREARM				
Abductor pollicis longus (ab-DUK-tor POL-li-sis LON-gus; *abductor* = moves part away from midline)	Posterior surface of middle of radius and ulna and interosseous membrane.	First metacarpal.	Abducts and extends thumb at carpometacarpal joint and abducts hand at wrist joint.	Deep radial nerve.
Extensor pollicis brevis (eks-TEN-sor POL-li-sis BREV-is)	Posterior surface of middle of radius and interosseous membrane.	Base of proximal phalanx of thumb.	Extends proximal phalanx of thumb at metacarpophalangeal joint, first metacarpal of thumb at carpometacarpal joint, and hand at wrist joint.	Deep radial nerve.
Extensor pollicis longus (eks-TEN-sor POL-li-sis LON-gus)	Posterior surface of middle of ulna and interosseous membrane.	Base of distal phalanx of thumb.	Extends distal phalanx of thumb at interphalangeal joint, extends first metacarpal of thumb at carpometacarpal joint, and abducts hand at wrist joint.	Deep radial nerve.
Extensor indicis (eks-TEN-sor IN-di-kis; *indicis* = index)	Posterior surface of ulna and interosseous membrane.	Tendon of extensor digitorum of index finger.	Extends distal and middle phalanges of index finger at interphalangeal joints, proximal phalanx of index finger at metacarpophalangeal joint, and hand at wrist joint.	Deep radial nerve.

*Reminder: The thumb or pollex is the first digit and has two phalanges: proximal and distal. The remaining digits, the fingers, are numbered II–V (2–5), and each has three phalanges: proximal, middle, and distal.

EXHIBIT 11.N CONTINUES ▶

RELATING MUSCLES TO MOVEMENTS

Arrange the muscles in this exhibit according to the following actions on the wrist joint: (1) flexion, (2) extension, (3) abduction, and (4) adduction; the following actions on the fingers at the metacarpophalangeal joints: (1) flexion and (2) extension; the following actions on the fingers at the interphalangeal joints: (1) flexion and (2) extension; the following actions on the thumb at the carpometacarpal, metacarpophalangeal, and interphalangeal joints: (1) extension and (2) abduction; and the following action on the thumb at the interphalangeal joint: flexion. The same muscle may be mentioned more than once.

✔ **CHECKPOINT**

Which muscles and actions of the wrist, hand, thumb, and fingers are used when writing?

Figure 11.17 **Muscles of the forearm that move the wrist, hand, thumb, and digits.**

The anterior compartment muscles function as flexors, and the posterior compartment muscles function as extensors.

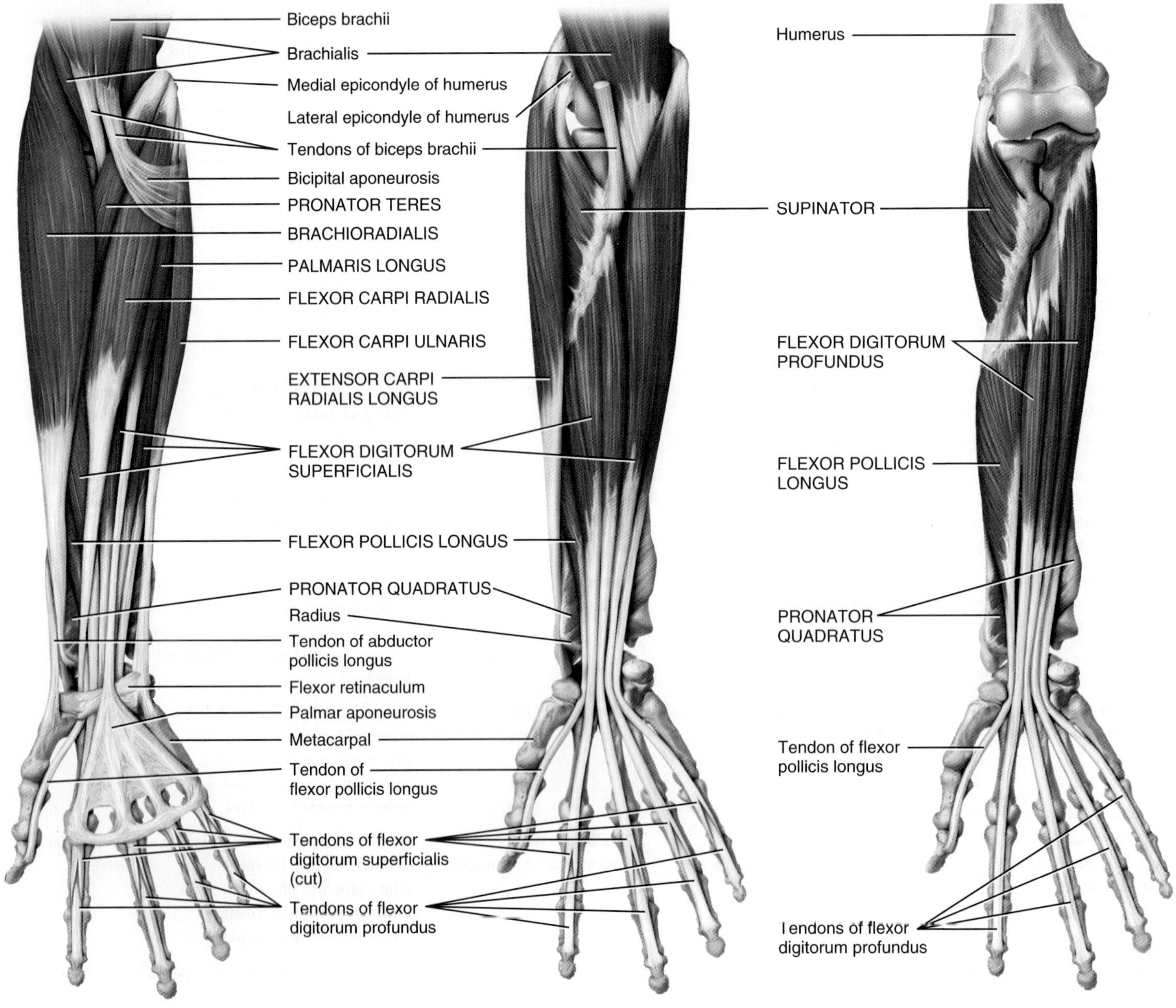

(a) Anterior superficial view (b) Anterior intermediate view (c) Anterior deep view

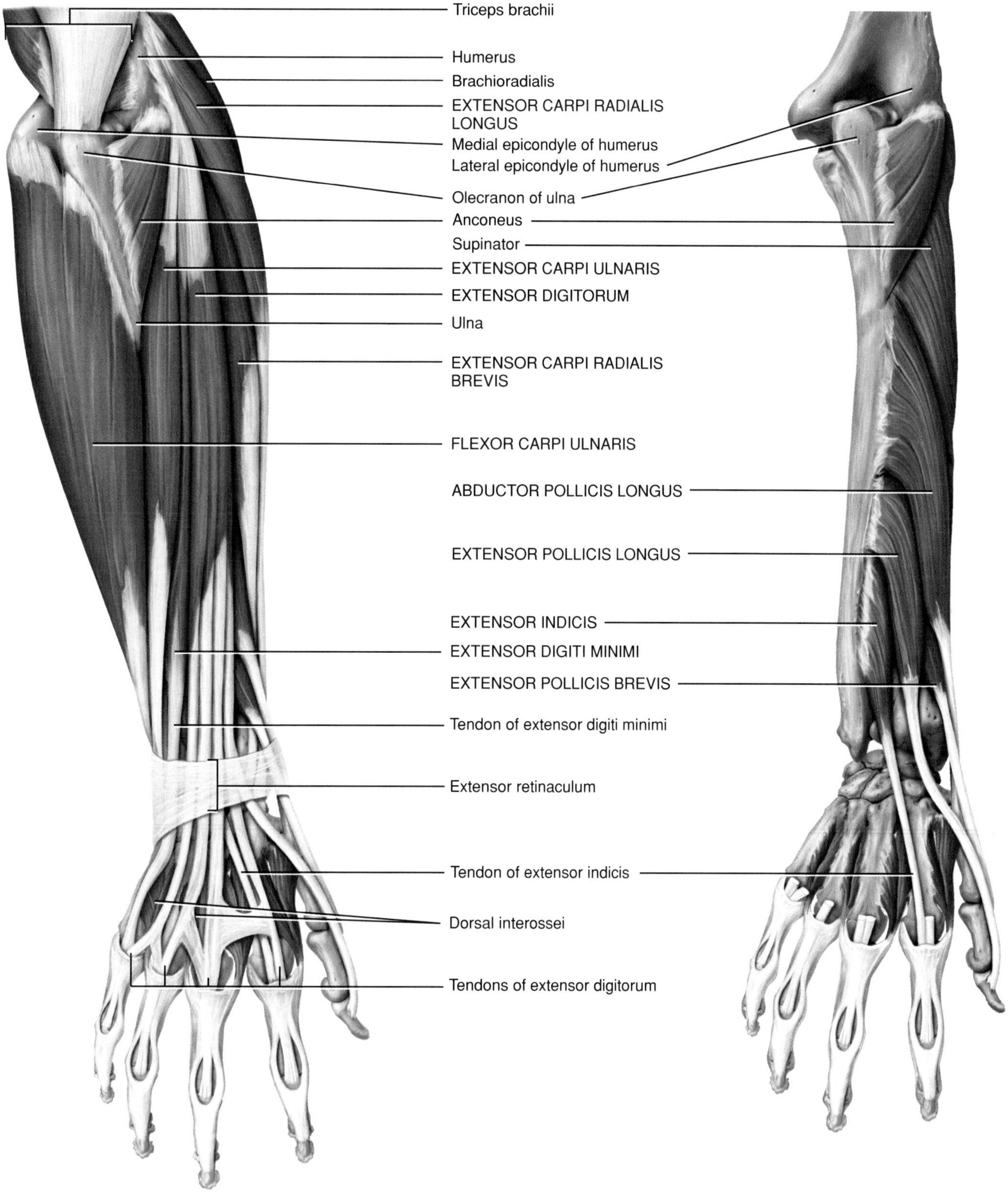
Triceps brachii
Humerus
Brachioradialis
EXTENSOR CARPI RADIALIS LONGUS
Medial epicondyle of humerus
Lateral epicondyle of humerus
Olecranon of ulna
Anconeus
Supinator
EXTENSOR CARPI ULNARIS
EXTENSOR DIGITORUM
Ulna
EXTENSOR CARPI RADIALIS BREVIS
FLEXOR CARPI ULNARIS
ABDUCTOR POLLICIS LONGUS
EXTENSOR POLLICIS LONGUS
EXTENSOR INDICIS
EXTENSOR DIGITI MINIMI
EXTENSOR POLLICIS BREVIS
Tendon of extensor digiti minimi
Extensor retinaculum
Tendon of extensor indicis
Dorsal interossei
Tendons of extensor digitorum

(d) Posterior superficial view

(e) Posterior deep view

What structures pass deep to the flexor retinaculum?

EXHIBIT 11.O Muscles of the Palm That Move the Digits—Intrinsic Muscles of the Hand *(Figure 11.18)*

OBJECTIVE

- Describe the origin, insertion, action, and innervation of the muscles of the palm that move the digits (the intrinsic muscles of the hand).

Several of the muscles discussed in Exhibit 11.N move the digits in various ways and are known as extrinsic muscles of the hand. They produce the powerful but crude movements of the digits. The **intrinsic muscles of the hand** in the palm produce the weak but intricate and precise movements of the digits that characterize the human hand (Figure 11.18). The muscles in this group are so named because their origins and insertions are *within* the hand.

The intrinsic muscles of the hand are divided into three groups: (1) **thenar,** (2) **hypothenar,** and (3) **intermediate.** The thenar muscles

MUSCLE	ORIGIN	INSERTION	ACTION	INNERVATION
THENAR (LATERAL ASPECT OF PALM)				
Abductor pollicis brevis (ab-DUK-tor POL-li-sis BREV-is; *abductor* = moves part away from middle; *pollicis* = thumb; *brevis* = short)	Flexor retinaculum, scaphoid, and trapezium.	Lateral side of proximal phalanx of thumb.	Abducts thumb at carpometacarpal joint.	Median nerve.
Opponens pollicis (op-PŌ-nenz POL-li-sis; *opponens* = opposes)	Flexor retinaculum and trapezium.	Lateral side of first metacarpal (thumb).	Moves thumb across palm to meet any finger (opposition) at carpometacarpal joint.	Median nerve.
Flexor pollicis brevis (FLEK-sor POL-li-sis BREV-is; *flexor* = decreases angle at joint)	Flexor retinaculum, trapezium, capitate, and trapezoid.	Lateral side of proximal phalanx of thumb.	Flexes thumb at carpometacarpal and metacarpophalangeal joints.	Median and ulnar nerves.
Adductor pollicis (ad-DUK-tor POL-li-sis; *adductor* = moves part toward midline)	Oblique head originates from capitate and second and third metacarpals. Transverse head originates from third metacarpal.	Medial side of proximal phalanx of thumb by tendon containing sesamoid bone.	Adducts thumb at carpometacarpal and metacarpophalangeal joints.	Ulnar nerve.
HYPOTHENAR (MEDIAL ASPECT OF PALM)				
Abductor digiti minimi (ab-DUK-tor DIJ-i-tē MIN-i-mē; *digit* = finger or toe; *minimi* = smallest)	Pisiform and tendon of flexor carpi ulnaris.	Medial side of proximal phalanx of little finger.	Abducts and flexes little finger at metacarpophalangeal joint.	Ulnar nerve.
Flexor digiti minimi brevis (FLEK-sor DIJ-i-tē MIN-i-mē BREV-is)	Flexor retinaculum and hamate.	Medial side of proximal phalanx of little finger.	Flexes little finger at carpometacarpal and metacarpophalangeal joints.	Ulnar nerve.
Opponens digiti minimi (op-PŌ-nenz DIJ-i-tē MIN-i-mē)	Flexor retinaculum and hamate.	Medial side of fifth metacarpal (little finger).	Moves little finger across palm to meet thumb (opposition) at carpometacarpal joint.	Ulnar nerve.
INTERMEDIATE (MIDPALMAR)				
Lumbricals (LUM-bri-kals; *lumbric* = earthworm) (four muscles)	Lateral sides of tendons and flexor digitorum profundus of each finger.	Lateral sides of tendons of extensor digitorum on proximal phalanges of each finger.	Flex each finger at metacarpophalangeal joints and extend each finger at interphalangeal joints.	Median and ulnar nerves.
Palmar interossei (PAL-mar in′-ter-OS-ē-ī; *palmar* = palm; *inter-* = between; *-ossei* = bones) (three muscles)	Sides of shafts of metacarpals of all digits (except middle one).	Sides of bases of proximal phalanges of all digits (except middle one).	Adduct and flex each finger (except middle finger) at metacarpophalangeal joints and extend these fingers at interphalangeal joints.	Ulnar nerve.
Dorsal interossei (DOR-sal in′-ter-OS-ē-ī; *dorsal* = back surface; *inter* = between; *ossei* = bones) (four muscles)	Adjacent sides of metacarpals.	Proximal phalanx of each finger.	Abduct fingers 2–4 at metacarpophalangeal joints, flex fingers 2–4 at metacarpophalangeal joints, and extend each finger at interphalangeal joints.	Ulnar nerve.

Figure 11.18 **Muscles of the palm that move the digits—intrinsic muscles of the hand.**

The intrinsic muscles of the hand produce the intricate and precise movements of the digits that characterize the human hand.

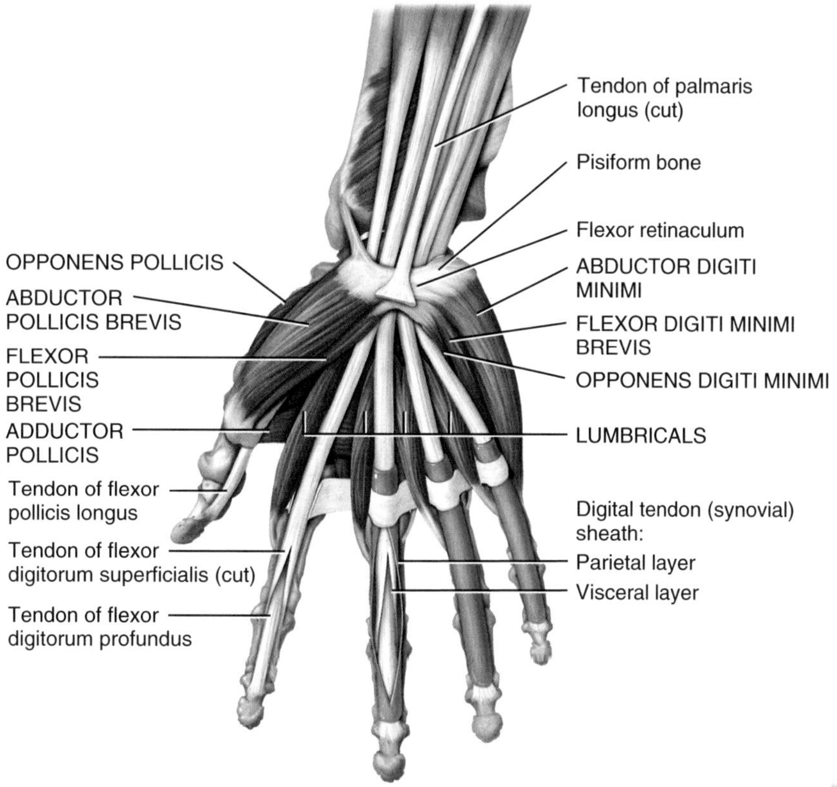

Tendon of palmaris longus (cut)
Pisiform bone
Flexor retinaculum
ABDUCTOR DIGITI MINIMI
FLEXOR DIGITI MINIMI BREVIS
OPPONENS DIGITI MINIMI
LUMBRICALS
Digital tendon (synovial) sheath:
Parietal layer
Visceral layer
OPPONENS POLLICIS
ABDUCTOR POLLICIS BREVIS
FLEXOR POLLICIS BREVIS
ADDUCTOR POLLICIS
Tendon of flexor pollicis longus
Tendon of flexor digitorum superficialis (cut)
Tendon of flexor digitorum profundus

(a) Anterior superficial view

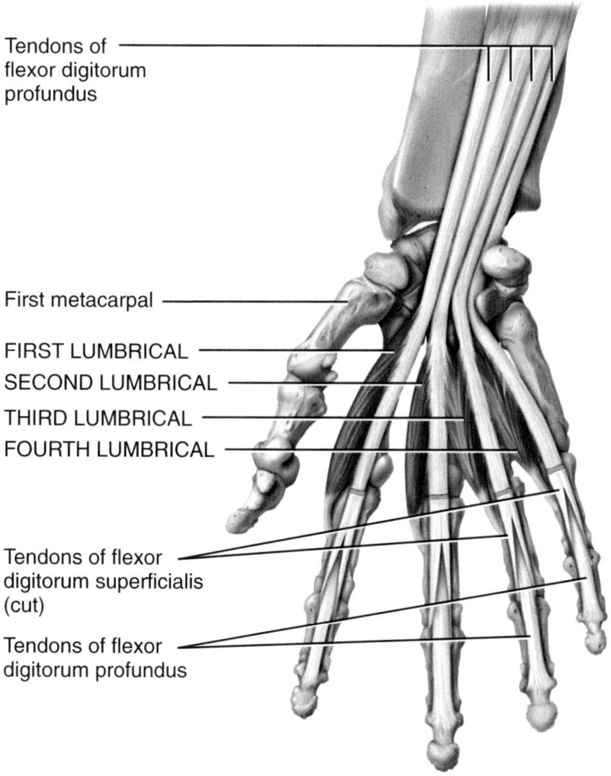

Tendons of flexor digitorum profundus
First metacarpal
FIRST LUMBRICAL
SECOND LUMBRICAL
THIRD LUMBRICAL
FOURTH LUMBRICAL
Tendons of flexor digitorum superficialis (cut)
Tendons of flexor digitorum profundus

(b) Anterior intermediate view of lumbricals

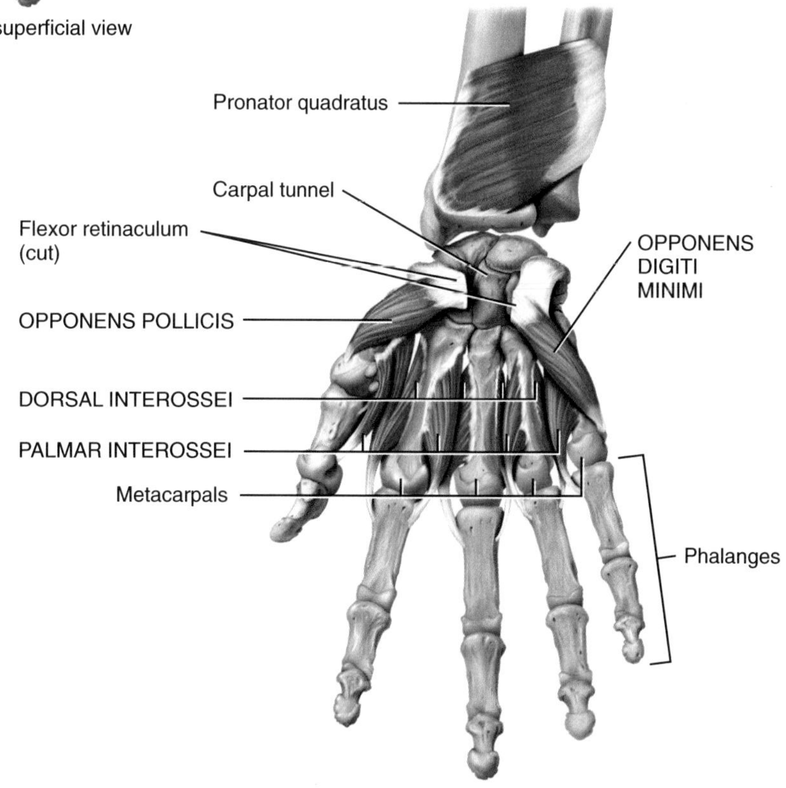

Pronator quadratus
Carpal tunnel
Flexor retinaculum (cut)
OPPONENS DIGITI MINIMI
OPPONENS POLLICIS
DORSAL INTEROSSEI
PALMAR INTEROSSEI
Metacarpals
Phalanges

(c) Anterior deep view

EXHIBIT 11.O CONTINUES ▶

include the abductor pollicis brevis, opponens pollicis, and flexor pollicis brevis. The **abductor pollicis brevis** is a thin, short, relatively broad superficial muscle on the lateral side of the thenar eminence. The **flexor pollicis brevis** is a short, wide muscle that is medial to the abductor pollicis brevis muscle. The **opponens pollicis** is a small, triangular muscle that is deep to the flexor pollicis brevis and abductor pollicis brevis muscles. The three thenar muscles plus the adductor pollicis form the **thenar eminence,** the lateral rounded contour on the palm that is also called the *ball of the thumb.* The **adductor pollicis** also acts on the thumb. The muscle is fan-shaped and has two heads (oblique and transverse) separated by a gap through which the radial artery passes.

The three hypothenar muscles act on the little finger and form the **hypothenar eminence,** the medial rounded contour on the palm that is also called the ball of the little finger. The hypothenar muscles are the abductor digiti minimi, flexor digiti minimi brevis, and opponens digiti minimi. The **abductor digiti minimi** is a short, wide muscle and is the most superficial of the hypothenar muscles. It is a powerful muscle that plays an important role in grasping an object with outspread fingers. The **flexor digiti minimi brevis** muscle is also short and wide and is lateral to the abductor digiti minimi muscle. The **opponens digiti minimi** muscle is triangular and deep to the other two hypothenar muscles.

The 11 intermediate (midpalmar) muscles include the lumbricals, palmar interossei, and dorsal interossei. The **lumbricals,** as their name indicates, are worm-shaped. They originate from and insert into the tendons of other muscles (flexor digitorum profundus and extensor digitorum). The **palmar interossei** are the smallest and more anterior of the interossei muscles. The **dorsal interossei** are the most posterior of this series of muscles. Both sets of interossei muscles are located between the metacarpals and are important in abduction, adduction, flexion, and extension of the fingers, and in movements in skilled activities such as writing, typing, and playing a piano.

The functional importance of the hand is readily apparent when you consider that certain hand injuries can result in permanent disability. Most of the dexterity of the hand depends on movements of the thumb. The general activities of the hand are free motion, power grip (forcible movement of the fingers and thumb against the palm, as in squeezing), precision handling (a change in position of a handled object that requires exact control of finger and thumb positions, as in winding a watch or threading a needle), and pinch (compression between the thumb and index finger or between the thumb and first two fingers).

Movements of the thumb are very important in the precise activities of the hand, and they are defined in different planes from comparable movements of other digits because the thumb is positioned at a right angle to the other digits. The five principal movements of the thumb are illustrated in Figure 11.18g and include *flexion* (movement of the thumb medially across the palm), *extension* (movement of the thumb laterally away from the palm), *abduction* (movement of the thumb in an anteroposterior plane away from the palm), *adduction* (movement of the thumb in an anteroposterior plane toward the palm), and *opposition* (movement of the thumb across the palm so that the tip of the thumb meets the tip of a finger). Opposition is the single most distinctive digital movement that gives humans and other primates the ability to grasp and manipulate objects precisely.

FIGURE 11.18 CONTINUED

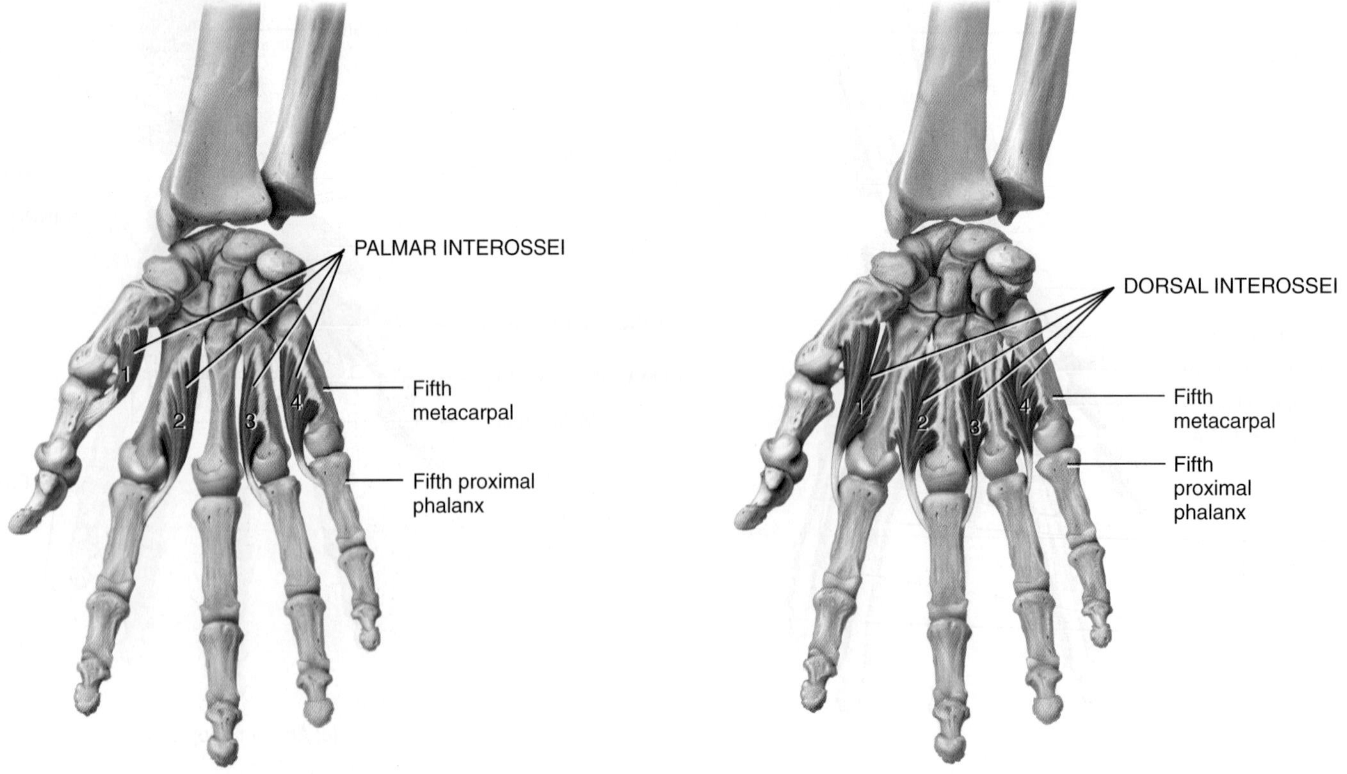

(d) Anterior deep view of palmar interossei

(e) Anterior deep view of dorsal interossei

CLINICAL CONNECTION | *Carpal Tunnel Syndrome*

The **carpal tunnel** is a narrow passageway formed anteriorly by the flexor retinaculum and posteriorly by the carpal bones. Through this tunnel pass the median nerve, the most superficial structure, and the long flexor tendons for the digits (see Figure 11.18f). Structures within the carpal tunnel, especially the median nerve, are vulnerable to compression, and the resulting condition is called **carpal tunnel syndrome.** Compression of the median nerve leads to sensory changes over the lateral side of the hand and muscle weakness in the thenar eminence. This results in pain, numbness, and tingling of the fingers. The condition may be caused by inflammation of the digital tendon sheaths, fluid retention, excessive exercise, infection, trauma, and/or repetitive activities that involve flexion of the wrist, such as keyboarding, cutting hair, or playing the piano. Treatment may involve the use of nonsteroidal anti-inflammatory drugs (such as ibuprofen or aspirin), wearing a wrist splint, corticosteroid injections, or surgery to cut the flexor retinaculum and release pressure on the median nerve. •

RELATING MUSCLES TO MOVEMENTS

Arrange the muscles in this exhibit according to the following actions on the thumb at the carpometacarpal and metacarpophalangeal joints: (1) abduction, (2) adduction, (3) flexion, and (4) opposition; and the following actions on the fingers at the metacarpophalangeal and interphalangeal joints: (1) abduction, (2) adduction, (3) flexion, and (4) extension. The same muscle may be mentioned more than once.

✔ **CHECKPOINT**

How do the actions of the extrinsic and intrinsic muscles of the hand differ?

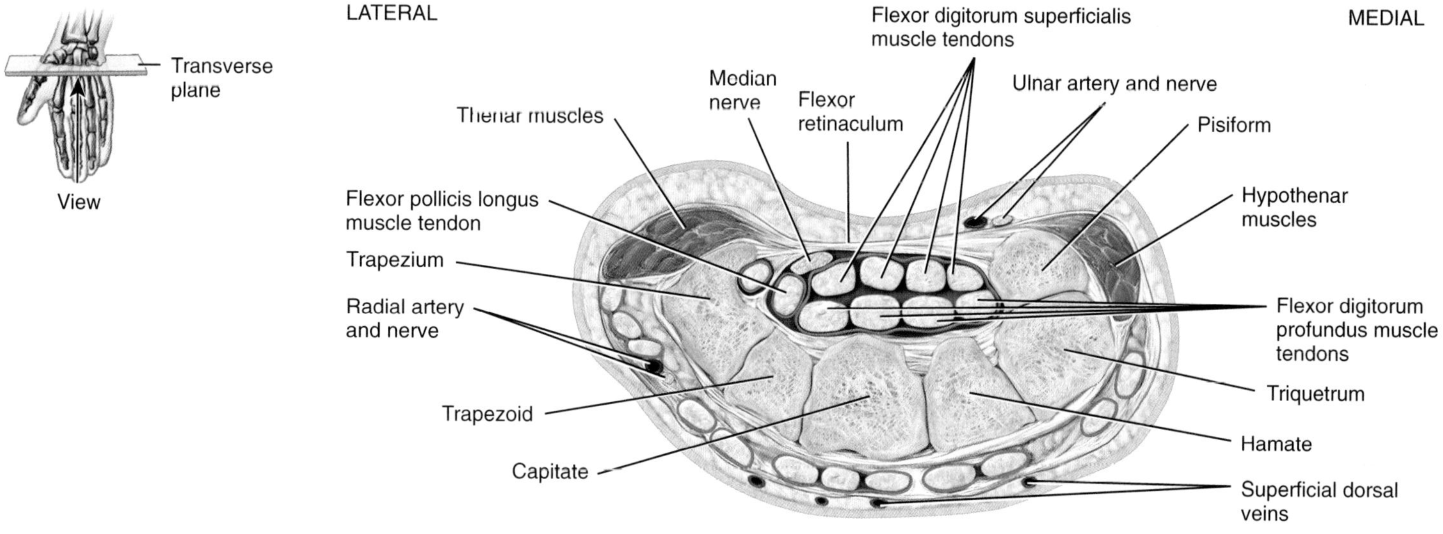

(f) Inferior view of transverse section

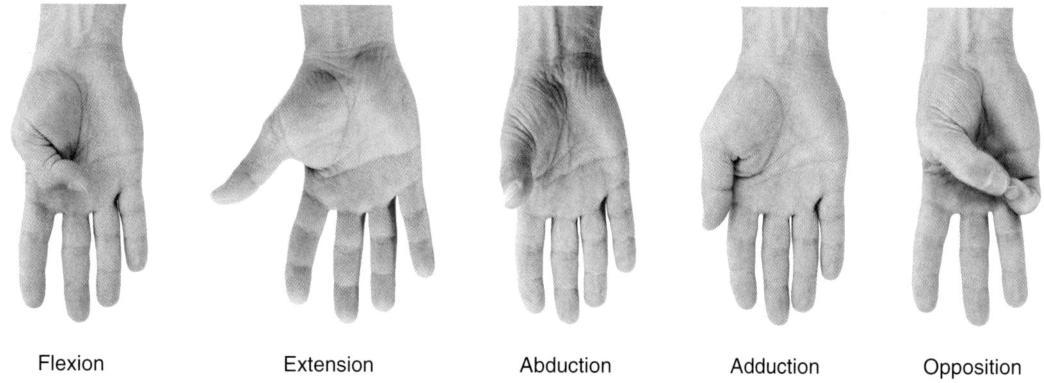

(g) Movements of the thumb

? Muscles of the thenar eminence act on which digit?

EXHIBIT 11.P

Muscles of the Neck and Back That Move the Vertebral Column *(Figure 11.19)*

OBJECTIVE

- Describe the origin, insertion, action, and innervation of the muscles that move the vertebral column.

The muscles that move the vertebral column (backbone) are quite complex because they have multiple origins and insertions and there is considerable overlap among them. One way to group the muscles is on the basis of the general direction of the muscle bundles and their approximate lengths. For example, the splenius muscles arise from the midline and extend laterally and superiorly to their insertions (Figure 11.19a). The erector spinae muscle group (consisting of the iliocostalis, longissimus, and spinalis muscles) arises either from the midline or more

MUSCLE	ORIGIN	INSERTION	ACTION	INNERVATION
SPLENIUS				
Splenius capitis (SPLĒ-nē-us KAP-i-tis; *splenium* = bandage; *capit* = head)	Ligamentum nuchae and spinous processes of C7–T4.	Occipital bone and mastoid process of temporal bone.	Acting together (bilaterally), extend head and extend vertebral column; acting singly (unilaterally), laterally flex and/or rotate head to same side as contracting muscle.	Middle cervical spinal nerves.
Splenius cervicis (SPLĒ-nē-us SER-vi-kis; *cervic* = neck)	Spinous processes of T3–T6.	Transverse processes of C1–C2 or C1–C4.	Acting together, extend head; acting singly, laterally flex and/or rotate head to same side as contracting muscle.	Inferior cervical spinal nerves.
ERECTOR SPINAE (e-REK-tor SPĪ-nē) Consists of iliocostalis muscles (lateral), longissimus muscles (intermediate), and spinalis muscles (medial).				
ILIOCOSTALIS GROUP (Lateral)				
Iliocostalis cervicis (il′-ē-ō-kos-TĀL-is SER-vi-kis; *ilio-* = flank; *-costa-* = rib)	Ribs 1–6.	Transverse processes of C4–C6.	Acting together, muscles of each region (cervical, thoracic, and lumbar) extend and maintain erect posture of vertebral column of their respective regions; acting singly, laterally flex vertebral column of their respective regions to same side as contracting muscle.	Cervical and thoracic spinal nerves.
Iliocostalis thoracis (il′-ē-ō-kos-TĀL-is thō-RĀ-sis; *thorac* = chest)	Ribs 7–12.	Ribs 1–6.		Thoracic spinal nerves.
Iliocostalis lumborum (il′-ē-ō-kos-TĀL-is lum-BOR-um)	Iliac crest.	Ribs 7–12.		Lumbar spinal nerves.
LONGISSIMUS GROUP (Intermediate)				
Longissimus capitis (lon-JIS-i-mus KAP-i-tis; *longissimus* = longest)	Articular processes of C4–C7 and transverse processes of T1–T4.	Mastoid process of temporal bone.	Acting together, both longissimus capitis muscles extend head and extend vertebral column; acting singly, rotate head to same side as contracting muscle.	Middle and inferior cervical spinal nerves.
Longissimus cervicis (lon-JIS-i-mus SER-vi-kis)	Transverse processes of T4–T5.	Transverse processes of C2–C6.	Acting together, longissimus cervicis and both longissimus thoracis muscles extend vertebral column of their respective regions; acting singly, laterally flex vertebral column of their respective regions.	Cervical and superior thoracic spinal nerves.
Longissimus thoracis (lon-JIS-i-mus thō-RĀ-sis)	Transverse processes of lumbar vertebrae.	Transverse processes of all thoracic and superior lumbar vertebrae and ribs 9 and 10.		Thoracic and lumbar spinal nerves.
SPINALIS GROUP (Medial)				
Spinalis capitis (spi-NĀ-lis KAP-i-tis; *spinal* = vertebral column)	Often absent or very small. Arises with semispinalis capitis.	Occipital bone.	Acting together, muscles of each region (cervical, thoracic, and lumbar) extend vertebral column of their respective regions and extend head.	Cervical spinal nerves.
Spinalis cervicis (spi-NĀ-lis SER-vi-kis)	Ligamentum nuchae and spinous process of C7.	Spinous process of axis.		Inferior cervical and thoracic spinal nerves.
Spinalis thoracis (spi-NĀ-lis thō-RĀ-sis)	Spinous processes of T10–L2.	Spinous processes of superior thoracic vertebrae.		Thoracic spinal nerves.

laterally but usually runs almost longitudinally, with neither a significant lateral nor medial direction as it is traced superiorly. The muscles of the transversospinalis group (semispinalis, multifidus, rotatores) arise laterally but extend toward the midline as they are traced superiorly. Deep to these three muscle groups are small segmental muscles that extend between spinous processes or transverse processes of vertebrae. Note in Exhibit 11.G that the rectus abdominis, external oblique, internal oblique, and quadratus lumborum muscles also play a role in moving the vertebral column.

The bandagelike **splenius** muscles are attached to the sides and back of the neck. The two muscles in this group are named on the basis of their superior attachments (insertions): **splenius capitis** (head region)

MUSCLE	ORIGIN	INSERTION	ACTION	INNERVATION
TRANSVERSOSPINALES (trans-ver-sō-spi-NĀ-lēz)				
Semispinalis capitis (sem′-ē-spi-NĀ-lis KAP-i-tis; *semi* = partially or one half)	Articular processes of C4–C6 and transverse processes of C7–T7.	Occipital bone between superior and inferior nuchal lines.	Acting together, extend head and vertebral column; acting singly, rotate head to side opposite contracting muscle.	Cervical and thoracic spinal nerves.
Semispinalis cervicis (sem′-ē-spi-NĀ-lis SER-vi-kis)	Transverse processes of T1–T5.	Spinous processes of C1–C5.	Acting together, both semispinalis cervicis and both semispinalis thoracis muscles extend vertebral column of their respective regions; acting singly, rotate head to side opposite contracting muscle.	Cervical and thoracic spinal nerves.
Semispinalis thoracis (sem′-ē-spi-NĀ-lis thō-RĀ-sis)	Transverse processes of T6–T10.	Spinous processes of C6–T4.		Thoracic spinal nerves.
Multifidus (mul-TIF-i-dus; *multi-* = many; *-fid-* = segmented)	Sacrum; ilium; transverse processes of lumbar, thoracic, and C4–C7.	Spinous process of a more superior vertebra.	Acting together, extend vertebral column; acting singly, weakly laterally flex vertebral column and weakly rotate vertebral column to side opposite contracting muscle.	Cervical, thoracic, and lumbar spinal nerves.
Rotatores (rō′-ta-TŌ-rēz; singular is **rotatore;** *rotatore* = to rotate)	Transverse processes of all vertebrae.	Spinous process of vertebra superior to the one of origin.	Acting together, weakly extend vertebral column; acting singly, weakly rotate vertebral column to side opposite contracting muscle.	Cervical, thoracic, and lumbar spinal nerves.
SEGMENTAL (seg-MEN-tal)				
Interspinales (in-ter-spī-NĀ-lēz; *inter* = between)	Superior surface of all spinous processes.	Inferior surface of spinous process of vertebra superior to the one of origin.	Acting together, weakly extend, vertebral column; acting singly, stabilize vertebral column during movement.	Cervical, thoracic, and lumbar spinal nerves.
Intertransversarii (in′-ter-trans-vers-AR-ē-ī; singular is **intertransversarius**)	Transverse processes of all vertebrae.	Transverse process of vertebra superior to the one of origin.	Acting together, weakly extend vertebral column; acting singly, weakly laterally flex vertebral column and stabilize it during movements.	Cervical, thoracic, and lumbar spinal nerves.
SCALENES (SKĀ-lēnz)				
Anterior scalene (SKĀ-lēn; *anterior* = front; *scalene* = uneven)	Transverse processes of C3–C6.	First rib.	Acting together, right and left anterior scalene and middle scalene muscles elevate first ribs during deep inhalation. RMA: Flex cervical vertebrae; acting singly, laterally flex and slightly rotate cervical vertebrae.	Cervical spinal nerves.
Middle scalene	Transverse processes of C2–C7.	First rib.		Cervical spinal nerves.
Posterior scalene	Transverse processes of C4–C6.	Second rib.	Acting together, right and left posterior scalene elevate second ribs during deep inhalation. RMA: Flex cervical vertebrae; acting singly, laterally flex and slightly rotate cervical vertebrae.	Cervical spinal nerves.

EXHIBIT 11.P CONTINUES

and **splenius cervicis** (cervical region). They extend the head and laterally flex and rotate the head.

The **erector spinae** is the largest muscle mass of the back, forming a prominent bulge on either side of the vertebral column. It is the chief extensor of the vertebral column. It is also important in controlling flexion, lateral flexion, and rotation of the vertebral column and in maintaining the lumbar curve. As noted above, it consists of three groups: iliocostalis (laterally placed), longissimus (intermedially placed), and spinalis (medially placed). These groups, in turn, consist of a series of overlapping muscles, and the muscles within the groups are named according to the regions of the body with which they are associated. The **iliocostalis group** consists of three muscles: the **iliocostalis cervicis** (cervical region), **iliocostalis thoracis** (thoracic region), and **iliocostalis lumborum** (lumbar region). The **longissimus group** resembles a herringbone and consists of three muscles: the **longissimus capitis** (head region), **longissimus cervicis** (cervical region), and **longissimus thoracis** (thoracic region). The **spinalis group** also consists of three muscles: the **spinalis capitis**, **spinalis cervicis**, and **spinalis thoracis.**

The **transversospinales** are so named because their fibers run from the transverse processes to the spinous processes of the vertebrae. The semispinalis muscles in this group are also named according to the region of the body with which they are associated: **semispinalis capitis** (head region), **semispinalis cervicis** (cervical region), and **semispinalis thoracis** (thoracic region). These muscles extend the vertebral column and rotate the head. The **multifidus** muscle in this group, as its name implies, is segmented into several bundles. It extends and laterally flexes the vertebral column. This muscle is large and thick in the lumbar region and is important in maintaining the lumbar curve. The **rotatores** muscles of this group are short and are found along the entire length of the vertebral column. These small muscles contribute little to vertebral movement but play important roles in monitoring the position of the vertebral column and providing proprioceptive feedback to the stronger vertebral muscles.

Within the segmental muscle group (Figure 11.19b), the **interspinales** and **intertransversarii** muscles unite the spinous and transverse processes of consecutive vertebrae. They function primarily in stabilizing the vertebral column during its movements, and providing proprioceptive feedback.

Within the **scalene group** (Figure 11.19c), the **anterior scalene** muscle is anterior to the middle scalene muscle, the **middle scalene** muscle is intermediate in placement and is the longest and largest of the scalene muscles, and the **posterior scalene** muscle is posterior to the middle scalene muscle and is the smallest of the scalene muscles. These muscles flex, laterally flex, and rotate the head and assist in deep inhalation.

CLINICAL CONNECTION | *Back Injuries and Heavy Lifting*

The four factors associated with increased risk of **back injury** are amount of force, repetition, posture, and stress applied to the backbone. Poor physical condition, poor posture, lack of exercise, and excessive body weight contribute to the number and severity of sprains and strains. Back pain caused by a muscle strain or ligament sprain will normally heal within a short time and may never cause further problems. However, if ligaments and muscles are weak, discs in the lower back can become weakened and may herniate (rupture) with excessive lifting or a sudden fall, causing considerable pain.

Full flexion at the waist, as in touching your toes, overstretches the erector spinae muscles. Muscles that are overstretched cannot contract effectively. Straightening up from such a position is therefore initiated by the hamstring muscles on the back of the thigh and the gluteus maximus muscles of the buttocks. The erector spinae muscles join in as the degree of flexion decreases. Improperly lifting a heavy weight, however, can strain the erector spinae muscles. The result can be painful muscle spasms, tearing of tendons and ligaments of the lower back, and herniating of intervertebral discs. The lumbar muscles are adapted for maintaining posture, not for lifting. This is why it is important to bend at the knees and use the powerful extensor muscles of the thighs and buttocks while lifting a heavy load. •

RELATING MUSCLES TO MOVEMENTS

Arrange the muscles in this exhibit according to the following actions on the head at the atlanto-occipital and intervertebral joints: (1) extension, (2) lateral flexion, (3) rotation to same side as contracting muscle, and (4) rotation to opposite side as contracting muscle; and arrange the muscles according to the following actions on the vertebral column at the intervertebral joints: (1) flexion, (2) extension, (3) lateral flexion, (4) rotation, and (5) stabilization. The same muscle may be mentioned more than once.

✔ **CHECKPOINT**

What is the largest muscle group of the back?

Figure 11.19 **Muscles of the neck and back that move the vertebral column (backbone).** The trapezius and occipitofrontalis muscles have been removed.

The erector spinae group (iliocostalis, longissimus, and spinalis muscles) is the largest muscular mass of the body and is the chief extensor of the vertebral column.

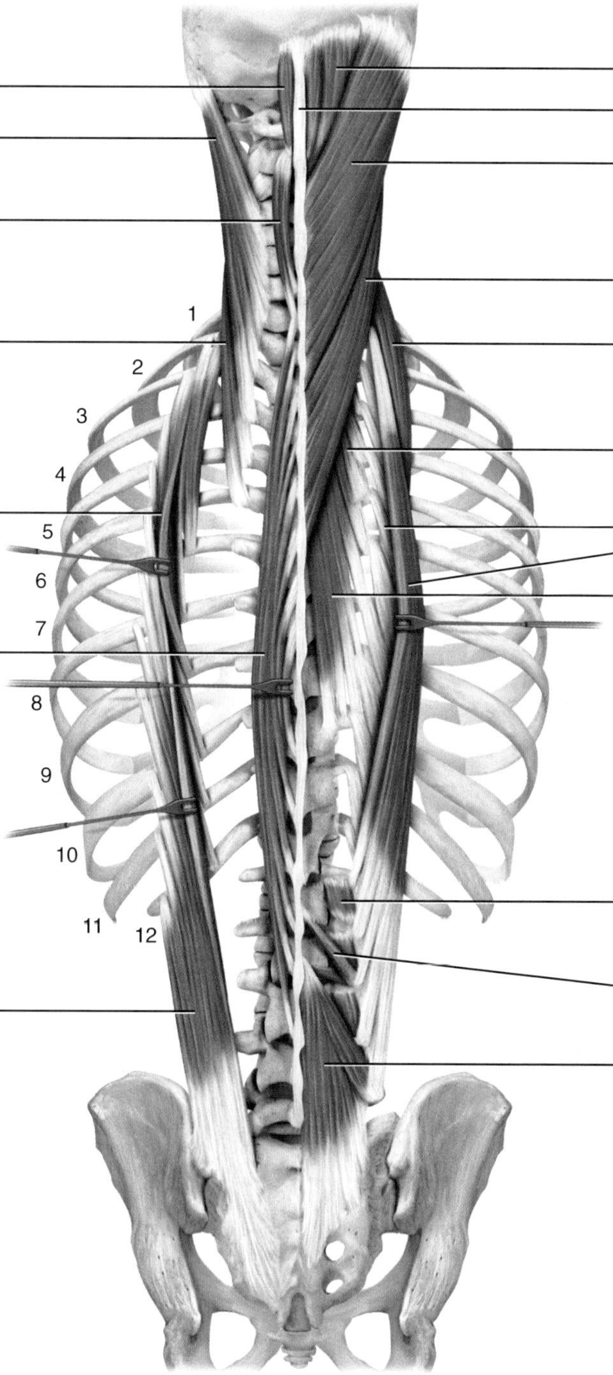

SPINALIS CAPITIS
LONGISSIMUS CAPITIS
SPINALIS CERVICIS
LONGISSIMUS CERVICIS
ILIOCOSTALIS THORACIS
SPINALIS THORACIS
ILIOCOSTALIS LUMBORUM
1
2
3
4
5
6
7
8
9
10
11
12
SEMISPINALIS CAPITIS
Ligamentum nuchae
SPLENIUS CAPITIS
SPLENIUS CERVICIS
ILIOCOSTALIS CERVICIS
SEMISPINALIS CERVICIS
LONGISSIMUS THORACIS
SEMISPINALIS THORACIS
INTERTRANSVERSARIUS
ROTATORE
MULTIFIDUS

(a) Posterior view

EXHIBIT 11.P CONTINUES

FIGURE 11.19 CONTINUED

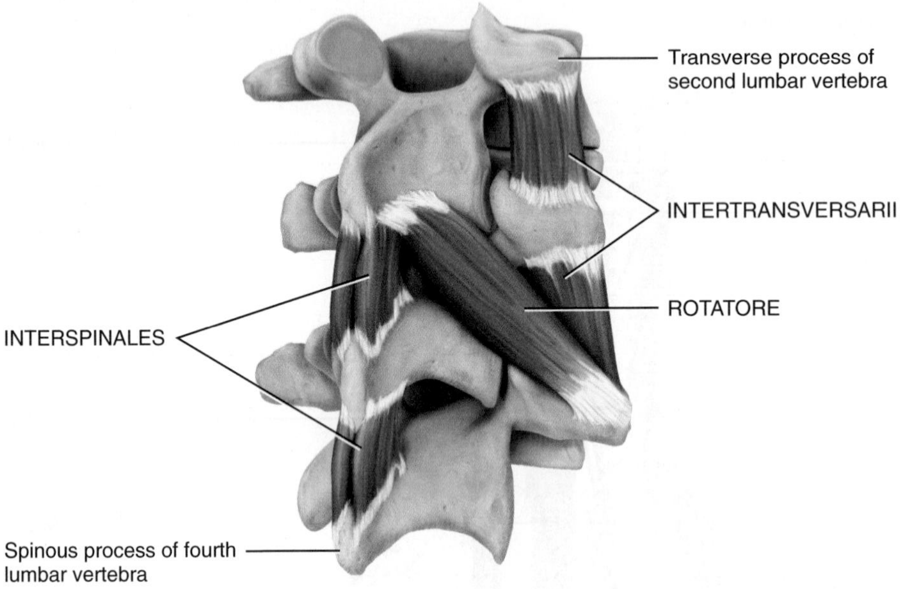
Transverse process of second lumbar vertebra
INTERTRANSVERSARII
ROTATORE
INTERSPINALES
Spinous process of fourth lumbar vertebra

(b) Posterolateral view

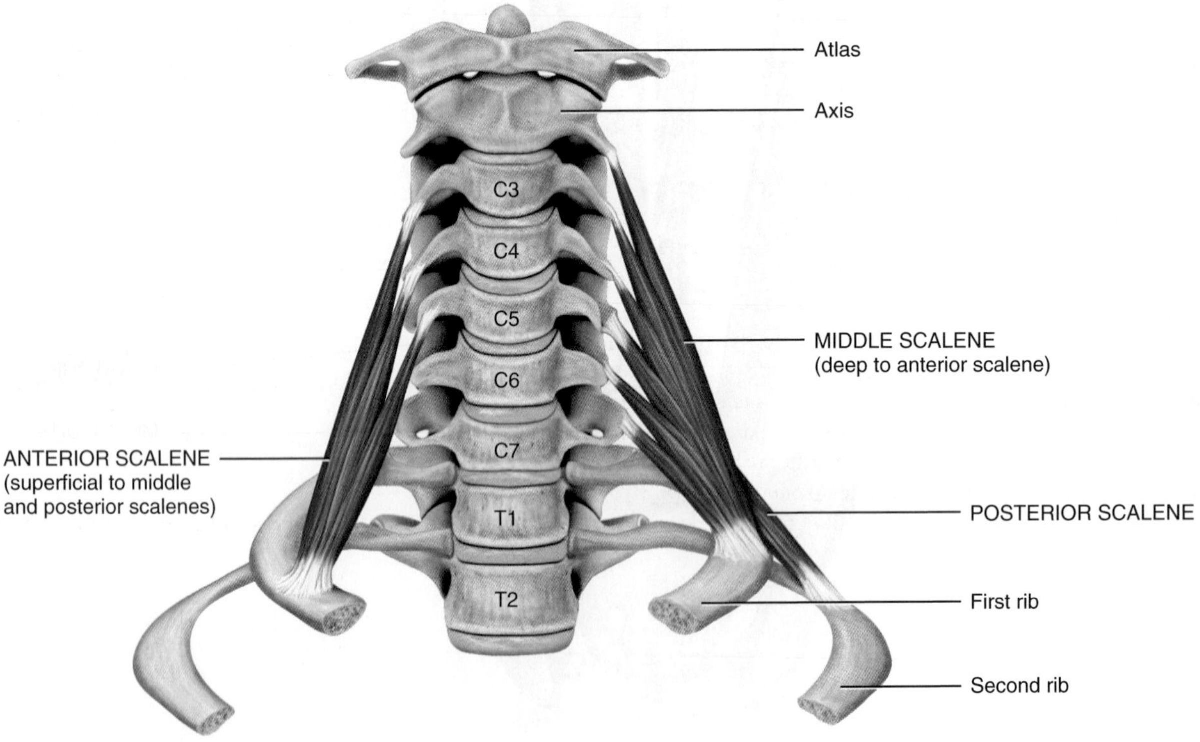
Atlas
Axis
C3
C4
C5
C6
C7
T1
T2
MIDDLE SCALENE (deep to anterior scalene)
ANTERIOR SCALENE (superficial to middle and posterior scalenes)
POSTERIOR SCALENE
First rib
Second rib

(c) Anterior view

Which muscles originate at the midline and extend laterally and upward to their insertion?

EXHIBIT 11.Q Muscles of the Gluteal Region That Move the Femur *(Figure 11.20)*

OBJECTIVE

- Describe the origin, insertion, action, and innervation of the muscles of the gluteal region that move the femur.

As you will see, muscles of the lower limbs are larger and more powerful than those of the upper limbs because of differences in function. While upper limb muscles are characterized by versatility of movement, lower limb muscles function in stability, locomotion, and maintenance of posture. In addition, muscles of the lower limbs often cross two joints and act equally on both.

The majority of muscles that move the femur (thigh bone) originate on the pelvic girdle and insert on the femur (Figure 11.20). The **psoas major** and **iliacus** muscles share a common insertion (lesser trochanter of femur) and are collectively known as the **iliopsoas** muscle. There are three gluteal muscles: gluteus maximus, gluteus medius, and gluteus minimus. The **gluteus maximus** is the largest and heaviest of the three muscles and is one of the largest muscles in the body. It is the chief extensor of the femur. In its reverse muscle action (RMA), it is a powerful extensor of the torso at the hip joint. The **gluteus medius** is mostly deep to the gluteus maximus and is a powerful abductor of the femur at the hip joint. It is a common site for intramuscular injection. The **gluteus minimus** is the smallest of the gluteal muscles and lies deep to the gluteus medius.

The **tensor fasciae latae** muscle is located on the lateral surface of the thigh. The *fascia lata* is a layer of deep fascia, composed of dense connective tissue, that encircles the entire thigh. It is well developed laterally where, together with the tendons of the tensor fasciae latae and gluteus maximus muscles, it forms a structure called the **iliotibial tract.** The tract inserts into the lateral condyle of the tibia.

The **piriformis, obturator internus, obturator externus, superior gemellus, inferior gemellus,** and **quadratus femoris** muscles are all deep to the gluteus maximus muscle and function as lateral rotators of the femur at the hip joint.

Three muscles on the medial aspect of the thigh are the **adductor longus, adductor brevis,** and **adductor magnus.** They originate on the pubic bone and insert on the femur. These three muscles adduct the thigh and are unique in their ability to both medially and laterally rotate the thigh. When the foot is on the ground, these muscles medially rotate the thigh, but when the foot is off the ground, they are lateral rotators of the thigh. This results from their oblique orientation from their anterior origin to their posterior insertion. In addition, the adductor longus flexes the thigh and the adductor magnus extends the thigh. The pectineus muscle also adducts and flexes the femur at the hip joint.

Technically, the adductor muscles and pectineus muscles are components of the medial compartment of the thigh and could be included in Exhibit 11.R. However, they are included here because they act on the femur.

At the junction between the trunk and lower limb is a space called the **femoral triangle.** The base is formed superiorly by the inguinal ligament, medially by the lateral border of the adductor longus muscle, and laterally by the medial border of the sartorius muscle. The apex is formed by the crossing of the adductor longus by the sartorius muscle (Figure 11.20a). The contents of the femoral triangle, from lateral to medial, are the femoral nerve and its branches, the femoral artery and several of its branches, the femoral vein and its proximal tributaries, and the deep inguinal lymph nodes. The femoral artery is easily accessible within the triangle and is the site for insertion of catheters that may extend into the aorta and ultimately into the coronary vessels of the heart. Such catheters are utilized during cardiac catheterization, coronary angiography, and other procedures involving the heart. Inguinal hernias frequently appear in this area.

CLINICAL CONNECTION | *Groin Pull*

The five major muscles of the inner thigh function to move the legs medially. This muscle group is important in activities such as sprinting, hurdling, and horseback riding. A rupture or tear of one or more of these muscles can cause a **groin pull.** Groin pulls most often occur during sprinting or twisting, or from kicking a solid, perhaps stationary object. Symptoms of a groin pull may be sudden or may not surface until the day after the injury; they include sharp pain in the inguinal region, swelling, bruising, or inability to contract the muscles. As with most strain injuries, treatment involves PRICE therapy, which stands for *p*rotection, *r*est, *i*ce, *c*ompression, and *e*levation. After the injured part is protected from further damage, ice should be applied immediately, and the injured part should be elevated and rested. An elastic bandage should be applied, if possible, to compress the injured tissue. •

RELATING MUSCLES TO MOVEMENTS

Arrange the muscles in this exhibit according to the following actions on the thigh at the hip joint: (1) flexion, (2) extension, (3) abduction, (4) adduction, (5) medial rotation, and (6) lateral rotation. The same muscle may be mentioned more than once.

✔CHECKPOINT

What is the origin of most muscles that move the femur?

EXHIBIT 11.Q CONTINUES ▶

Muscles of the Gluteal Region That Move the Femur *(Figure 11.20)* CONTINUED

MUSCLE	ORIGIN	INSERTION	ACTION	INNERVATION
Iliopsoas (il-ē-ō-SŌ-as) **Psoas major** (SŌ-as MĀ-jor; *psoa* = a muscle of the loin; *major* = larger)	Transverse processes and bodies of lumbar vertebrae.	With iliacus into lesser trochanter of femur.	Psoas major and iliacus muscles acting together flex thigh at hip joint, rotate thigh laterally, and flex trunk on hip as in sitting up from supine position.	Lumbar spinal nerves L2–L3.
Iliacus (il′-ē-A-cus; *iliac* = ilium)	Iliac fossa and sacrum.	With psoas major into lesser trochanter of femur.		Femoral nerve.
Gluteus maximus (GLOO-tē-us MAK-si-mus; *glute* = rump or buttock; *maximus* = largest)	Iliac crest, sacrum, coccyx, and aponeurosis of sacrospinalis.	Iliotibial tract of fascia lata and superior lateral part of linea aspera (gluteal tuberosity) under greater trochanter of femur.	Extends thigh at hip joint and laterally rotates thigh; helps lock knee in extension. RMA: Extends torso.	Inferior gluteal nerve.
Gluteus medius (GLOO-tē-us MĒ-dē-us; *medi* = middle)	Ilium.	Greater trochanter of femur.	Abducts thigh at hip joint and medially rotates thigh.	Superior gluteal nerve.
Gluteus minimus (GLOO-tē-us MIN-i-mus; *minimus* = smallest)	Ilium.	Greater trochanter of femur.	Abducts thigh at hip joint and medially rotates thigh.	Superior gluteal nerve.
Tensor fasciae latae (TEN-sor FA-shē-ē LĀ-tē; *tensor* = makes tense; *fasciae* = band; *lat* = wide)	Iliac crest.	Tibia by way of iliotibial tract.	Flexes and abducts thigh at hip joint.	Superior gluteal nerve.
Piriformis (pir-i-FOR-mis; *piri-* = pear; *-form-* = shape)	Anterior sacrum.	Superior border of greater trochanter of femur.	Laterally rotates and abducts thigh at hip joint.	Sacral spinal nerves S1 or S2, mainly S1.
Obturator internus (OB-too-rā′-tor in-TER-nus; *obturator* = obturator foramen; *intern* = inside)	Inner surface of obturator foramen, pubis, and ischium.	Medial surface of greater trochanter of femur.	Laterally rotates and abducts thigh at hip joint.	Nerve to obturator internus.
Obturator externus (OB-too-rā′-tor ex-TER-nus; *extern-* = outside)	Outer surface of obturator membrane.	Deep depression inferior to greater trochanter (trochanteric fossa) of femur.	Laterally rotates and abducts thigh at hip joint.	Obturator nerve.
Superior gemellus (jem-EL-lus; *superior* = above; *gemell* = twins)	Ischial spine.	Medial surface of greater trochanter of femur.	Laterally rotates and abducts thigh at hip joint.	Nerve to obturator internus.
Inferior gemellus (jem-EL-lus; *inferior* = below)	Ischial tuberosity.	Medial surface of greater trochanter of femur.	Laterally rotates and abducts thigh at hip joint.	Nerve to quadratus femoris.
Quadratus femoris (kwod-RĀ-tus FEM-or-is; *quad* = square, four-sided; *femoris* = femur)	Ischial tuberosity.	Elevation superior to mid-portion of intertrochanteric crest (quadrate tubercle) on posterior femur.	Laterally rotates and stabilizes hip joint.	Nerve to quadratus femoris.
Adductor longus (ad-DUK-tor LONG-us; *adductor* = moves part closer to midline; *longus* = long)	Pubic crest and pubic symphysis.	Linea aspera of femur.	Adducts and flexes thigh at hip joint and rotates thigh.* RMA: Extends thigh.	Obturator nerve.
Adductor brevis (BREV-is; *brevis* = short)	Inferior ramus of pubis.	Superior half of linea aspera of femur.	Adducts and flexes thigh at hip joint and rotates thigh.* RMA: Extends thigh.	Obturator nerve.
Adductor magnus (MAG-nus; *magnus* = large)	Inferior ramus of pubis and ischium to ischial tuberosity.	Linea aspera of femur.	Adducts thigh at hip joint and rotates thigh; anterior part flexes thigh at hip joint, and posterior part extends thigh at hip joint.*	Obturator and sciatic nerves.
Pectineus (pek-TIN-ē-us; *pectin* = comb)	Superior ramus of pubis.	Pectineal line of femur, between lesser trochanter and linea aspera.	Flexes and adducts thigh at hip joint.	Femoral nerve.

*All adductors are unique muscles that cross the thigh joint obliquely from an anterior origin to a posterior insertion. As a result they laterally rotate the hip joint when the foot is off the ground but medially rotate the hip joint when the foot is on the ground.

Figure 11.20 **Muscles of the gluteal region that move the femur (thigh bone).**

Most muscles that move the femur originate on the pelvic (hip) girdle and insert on the femur.

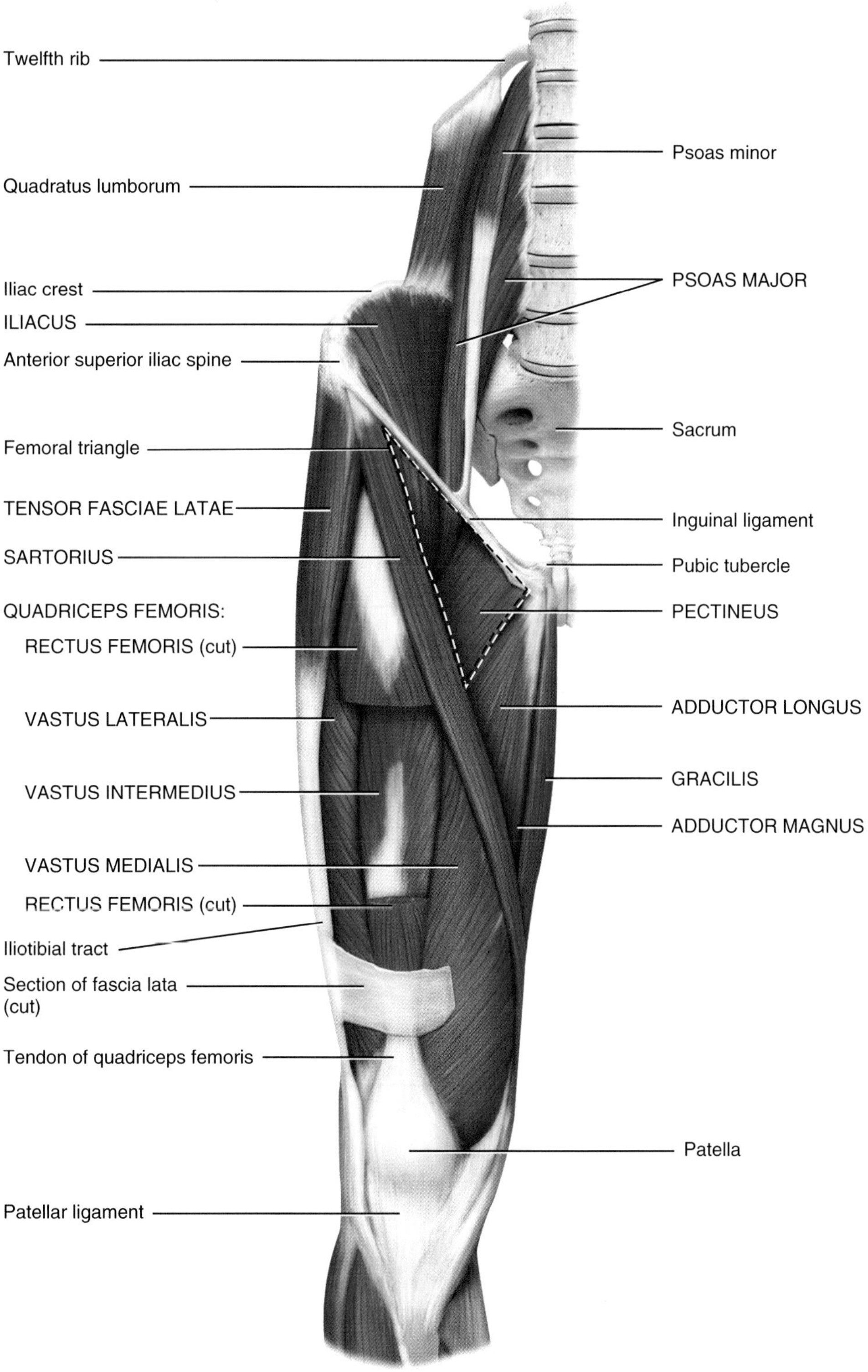

Twelfth rib
Quadratus lumborum
Iliac crest
ILIACUS
Anterior superior iliac spine
Femoral triangle
TENSOR FASCIAE LATAE
SARTORIUS
QUADRICEPS FEMORIS:
RECTUS FEMORIS (cut)
VASTUS LATERALIS
VASTUS INTERMEDIUS
VASTUS MEDIALIS
RECTUS FEMORIS (cut)
Iliotibial tract
Section of fascia lata (cut)
Tendon of quadriceps femoris
Patellar ligament
Psoas minor
PSOAS MAJOR
Sacrum
Inguinal ligament
Pubic tubercle
PECTINEUS
ADDUCTOR LONGUS
GRACILIS
ADDUCTOR MAGNUS
Patella

(a) Anterior superficial view (the femoral triangle is indicated by a dashed line)

EXHIBIT 11.Q CONTINUES

FIGURE 11.20 CONTINUED

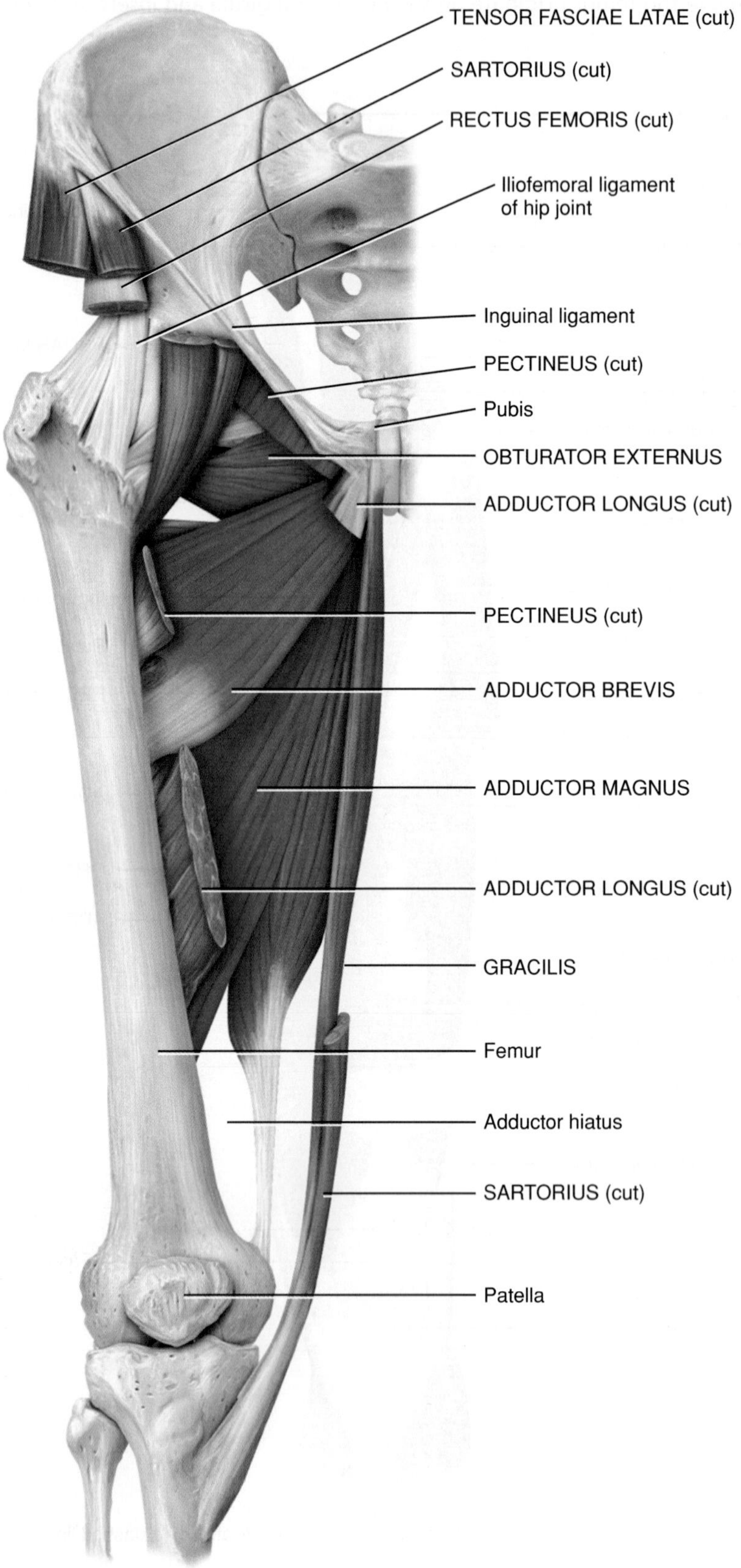
TENSOR FASCIAE LATAE (cut)
SARTORIUS (cut)
RECTUS FEMORIS (cut)
Iliofemoral ligament of hip joint
Inguinal ligament
PECTINEUS (cut)
Pubis
OBTURATOR EXTERNUS
ADDUCTOR LONGUS (cut)
PECTINEUS (cut)
ADDUCTOR BREVIS
ADDUCTOR MAGNUS
ADDUCTOR LONGUS (cut)
GRACILIS
Femur
Adductor hiatus
SARTORIUS (cut)
Patella

(b) Anterior deep view (femur rotated laterally)

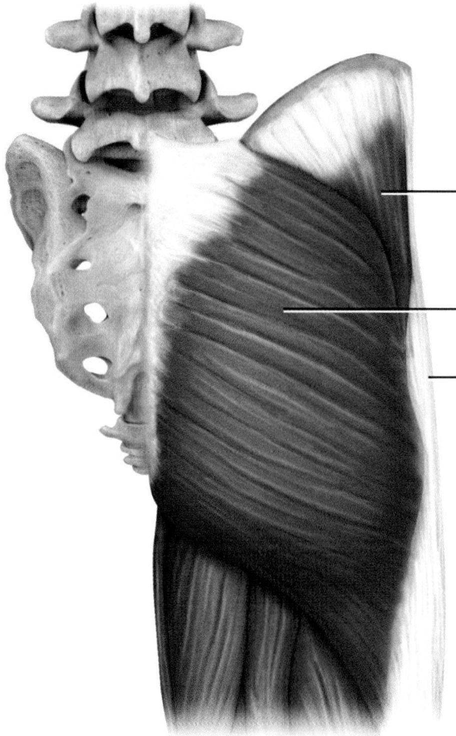
GLUTEUS MEDIUS
GLUTEUS MAXIMUS
TENSOR FASCIAE LATAE
(c) Posterior superficial view

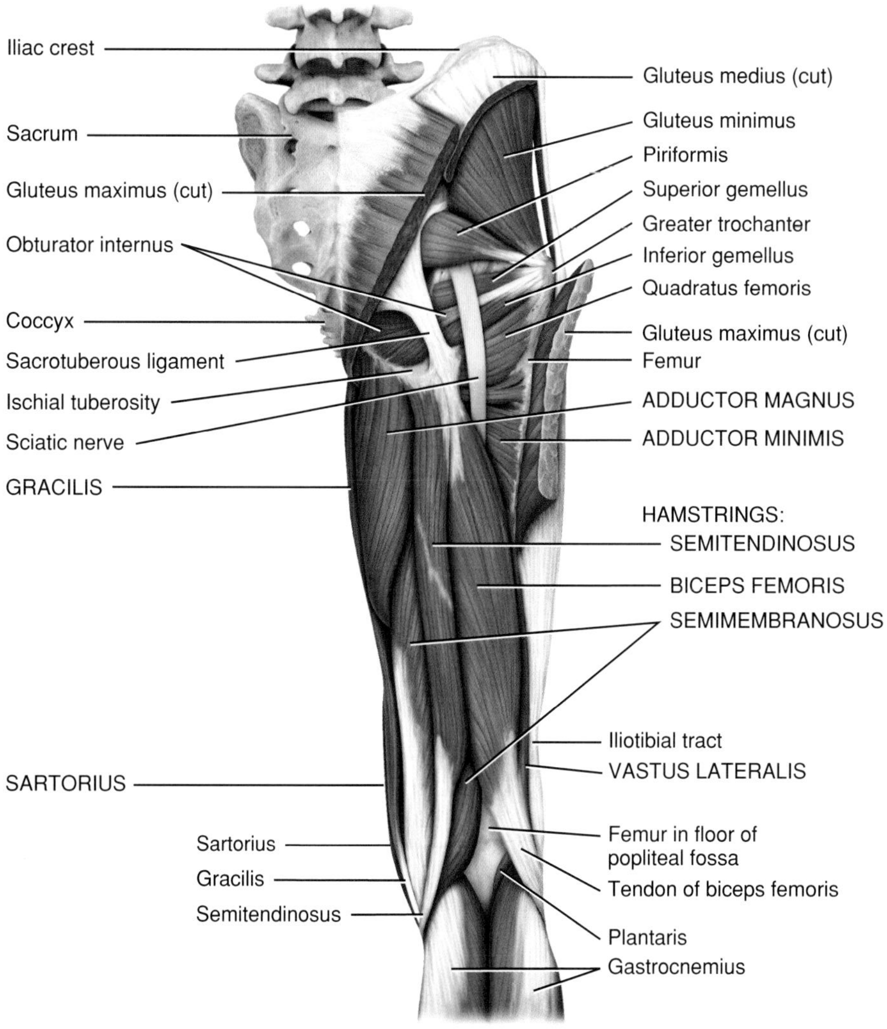
Iliac crest
Sacrum
Gluteus maximus (cut)
Obturator internus
Coccyx
Sacrotuberous ligament
Ischial tuberosity
Sciatic nerve
GRACILIS
SARTORIUS
Sartorius
Gracilis
Semitendinosus
Gluteus medius (cut)
Gluteus minimus
Piriformis
Superior gemellus
Greater trochanter
Inferior gemellus
Quadratus femoris
Gluteus maximus (cut)
Femur
ADDUCTOR MAGNUS
ADDUCTOR MINIMIS
HAMSTRINGS:
SEMITENDINOSUS
BICEPS FEMORIS
SEMIMEMBRANOSUS
Iliotibial tract
VASTUS LATERALIS
Femur in floor of popliteal fossa
Tendon of biceps femoris
Plantaris
Gastrocnemius
(d) Posterior superficial view of thigh and deep view of gluteal region

EXHIBIT 11.Q CONTINUES

EXHIBIT 11.Q Muscles of the Gluteal Region That Move the Femur *(Figure 11.20)* CONTINUED

FIGURE 11.20 CONTINUED

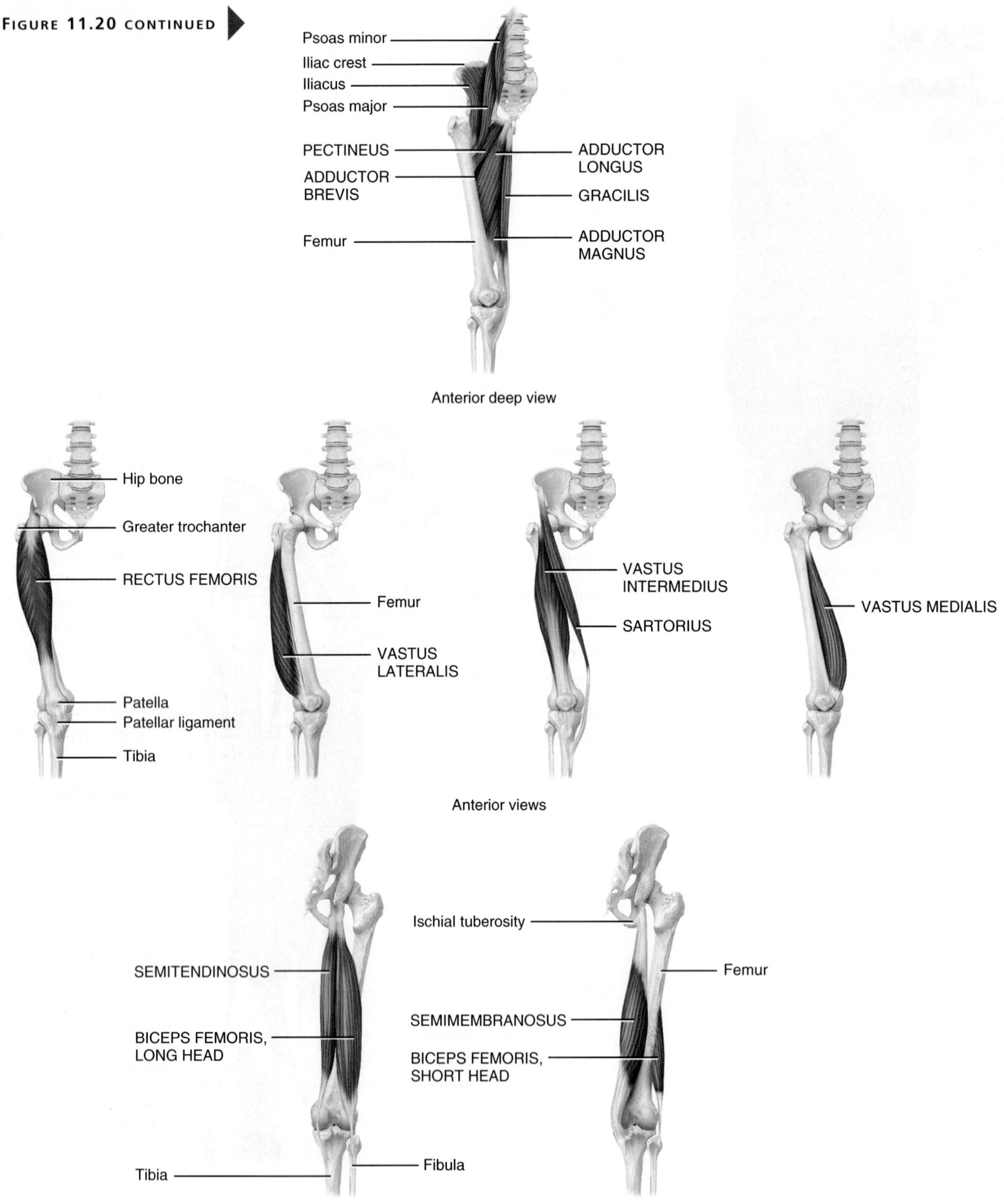

Posterior deep views

(e) Isolated muscles

What are the principal differences between the muscles of the free upper and lower limbs?

EXHIBIT 11.R Muscles of the Thigh That Move the Femur and Tibia and Fibula *(Figure 11.21)*

OBJECTIVE

- Describe the origin, insertion, action, and innervation of the muscles that move the femur, tibia, and fibula.

Deep fascia separates the muscles of the thigh that act on the femur (thigh bone) and tibia and fibula (leg bones) into medial, anterior, and posterior compartments (Figure 11.21). Most of the muscles of the **medial (adductor) compartment of the thigh** have a similar orientation and adduct the femur at the hip joint. (See the adductor magnus, adductor longus, adductor brevis, and pectineus, which are components of the medial compartment, in Exhibit 11.Q.) The **gracilis,** the other muscle in the medial compartment, is a long, straplike muscle on the medial aspect of the thigh and knee. This muscle not only adducts the thigh, but also medially rotates the thigh and flexes the leg at the knee joint. For this reason, it is discussed here.

The muscles of the **anterior (extensor) compartment of the thigh** extend the leg (and flex the thigh). This compartment contains the quadriceps femoris and sartorius muscles. The **quadriceps femoris** muscle is the largest muscle in the body, covering most of the anterior surface and sides of the thigh. The muscle is actually a composite muscle, usually described as four separate muscles: (1) **rectus femoris,** on the anterior aspect of the thigh; (2) **vastus lateralis,** on the lateral aspect of the thigh; (3) **vastus medialis,** on the medial aspect of the thigh; and (4) **vastus intermedius,** located deep to the rectus femoris between the vastus lateralis and vastus medialis. The common tendon for the four muscles is known as the **quadriceps tendon,** which inserts into the patella. The tendon continues below the patella as the **patellar ligament,** which attaches to the tibial tuberosity. The quadriceps femoris muscle is the great extensor muscle of the leg. The **sartorius** is a long, narrow muscle that forms a band across the thigh from the ilium of the hip bone to the medial side of the tibia. The various movements it produces (flexion of the leg at the knee joint and flexion, abduction, and lateral rotation at the hip joint) help effect the cross-legged sitting position in which the heel of one limb is placed on the knee of the opposite limb. Its name means *tailor's muscle;* it was so called because tailors often assume a cross-legged sitting position. (Because the major action of the sartorius muscle is to move the thigh rather than the leg, it could have been included in Exhibit 11.Q.)

The muscles of the **posterior (flexor) compartment of the thigh** flex the leg (and extend the thigh). This compartment is composed of three muscles collectively called the **hamstrings:** (1) biceps femoris, (2) semitendinosus, and (3) semimembranosus. The hamstrings are so named because their tendons are long and stringlike in the popliteal area. Because the hamstrings span two joints (hip and knee), they are both extensors of the thigh and flexors of the leg. The **popliteal fossa** is a diamond-shaped space on the posterior aspect of the knee bordered laterally by the tendons of the biceps femoris muscle and medially by the tendons of the semitendinosus and semimembranosus muscles.

CLINICAL CONNECTION | *Pulled Hamstrings and Charley Horse*

A strain or partial tear of the proximal hamstring muscles is referred to as **pulled hamstrings** or **hamstring strains.** Like pulled groins (see Exhibit 11.Q), they are common sports injuries in individuals who run very hard and/or are required to perform quick starts and stops. Sometimes the violent muscular exertion required to perform a feat tears away a part of the tendinous origins of the hamstrings, especially the biceps femoris, from the ischial tuberosity. This is usually accompanied by a contusion (bruising), tearing of some of the muscle fibers, and rupture of blood vessels, producing a hematoma (collection of blood) and sharp pain. Adequate training with good balance between the quadriceps femoris and hamstrings and stretching exercises before running or competing are important in preventing this injury.

The term **charley horse** (CHAR-lē HŌRS), a slang term, is a popular name for a cramp or stiffness of muscles due to tearing of the muscle, followed by bleeding into the area. It is a common sports injury due to trauma or excessive activity and frequently occurs in the quadriceps femoris muscle, especially among football players. •

RELATING MUSCLES TO MOVEMENTS

Arrange the muscles in this exhibit according to the following actions on the thigh at the hip joint: (1) abduction, (2) adduction, (3) lateral rotation, (4) flexion, and (5) extension; and according to the following actions on the leg at the knee joint: (1) flexion and (2) extension. The same muscle may be mentioned more than once.

✔ **CHECKPOINT**

Which muscles are part of the medial, anterior, and posterior compartments of the thigh?

EXHIBIT 11.R CONTINUES ▶

Muscles of the Thigh That Move the Femur and Tibia and Fibula *(Figure 11.21)* CONTINUED

MUSCLE	ORIGIN	INSERTION	ACTION	INNERVATION
MEDIAL (ADDUCTOR) COMPARTMENT OF THE THIGH				
Adductor magnus (ad-DUK-tor MAG-nus)	See Exhibit 11.Q			
Adductor longus (LONG-us)	See Exhibit 11.Q			
Adductor brevis (BREV-is)	See Exhibit 11.Q			
Pectineus (pek-TIN-ē-us)	See Exhibit 11.Q			
Gracilis (GRAS-i-lis = slender) (see also Figure 11.20a)	Body and inferior ramus of pubis.	Medial surface of body of tibia.	Adducts thigh at hip joint, medially rotates thigh, and flexes leg at knee joint.	Obturator nerve.
ANTERIOR (EXTENSOR) COMPARTMENT OF THE THIGH (see also Figure 11.20a)				
Quadriceps femoris (KWOD-ri-seps FEM-or-is; *quadriceps* = four heads [of origin]; *femoris* = femur)				
Rectus femoris (REK-tus FEM-or-is; *rectus* = fascicles parallel to midline)	Anterior inferior iliac spine.	Patella via quadriceps tendon and then tibial tuberosity via patellar ligament.	All four heads extend leg at knee joint; rectus femoris muscle acting alone also flexes thigh at hip joint.	Femoral nerve.
Vastus lateralis (VAS-tus lat′-e-RA-lis; *vast* = huge; *lateralis* = lateral)	Greater trochanter and linea aspera of femur.			
Vastus medialis (VAS-tus mē-dē-Ā-lis; *medialis* = medial)	Linea aspera of femur.			
Vastus intermedius (VAS-tus in′-ter-MĒ-dē-us; *intermedius* = middle)	Anterior and lateral surfaces of body of femur.			
Sartorius (sar-TOR-ē-us; *sartor* = tailor; longest muscle in body)	Anterior superior iliac spine.	Medial surface of body of tibia.	Weakly flexes leg at knee joint; weakly flexes, abducts, and laterally rotates thigh at hip joint.	Femoral nerve.
POSTERIOR (FLEXOR) COMPARTMENT OF THE THIGH (see also Figure 11.20d)				
Hamstrings A collective designation for three separate muscles.				
Biceps femoris (BĪ-seps FEM-or-is; *biceps* = two heads of origin)	Long head arises from ischial tuberosity; short head arises from linea aspera of femur.	Head of fibula and lateral condyle of tibia.	Flexes leg at knee joint and extends thigh at hip joint.	Tibial and fibular nerves from sciatic nerve.
Semitendinosus (sem′-ē-ten-di-NŌ-sus; *semi-* = half; *-tendo-* = tendon)	Ischial tuberosity.	Proximal part of medial surface of shaft of tibia.	Flexes leg at knee joint and extends thigh at hip joint.	Tibial nerve from sciatic nerve.
Semimembranosus (sem′-ē-mem-bra-NŌ-sus; *membran-* = membrane)	Ischial tuberosity.	Medial condyle of tibia.	Flexes leg at knee joint and extends thigh at hip joint.	Tibial nerve from sciatic nerve.

Figure 11.21 Muscles of the thigh that move the femur (thigh bone) and tibia and fibula (leg bones).

Muscles that act on the leg originate in the hip and thigh and are separated into compartments by deep fascia.

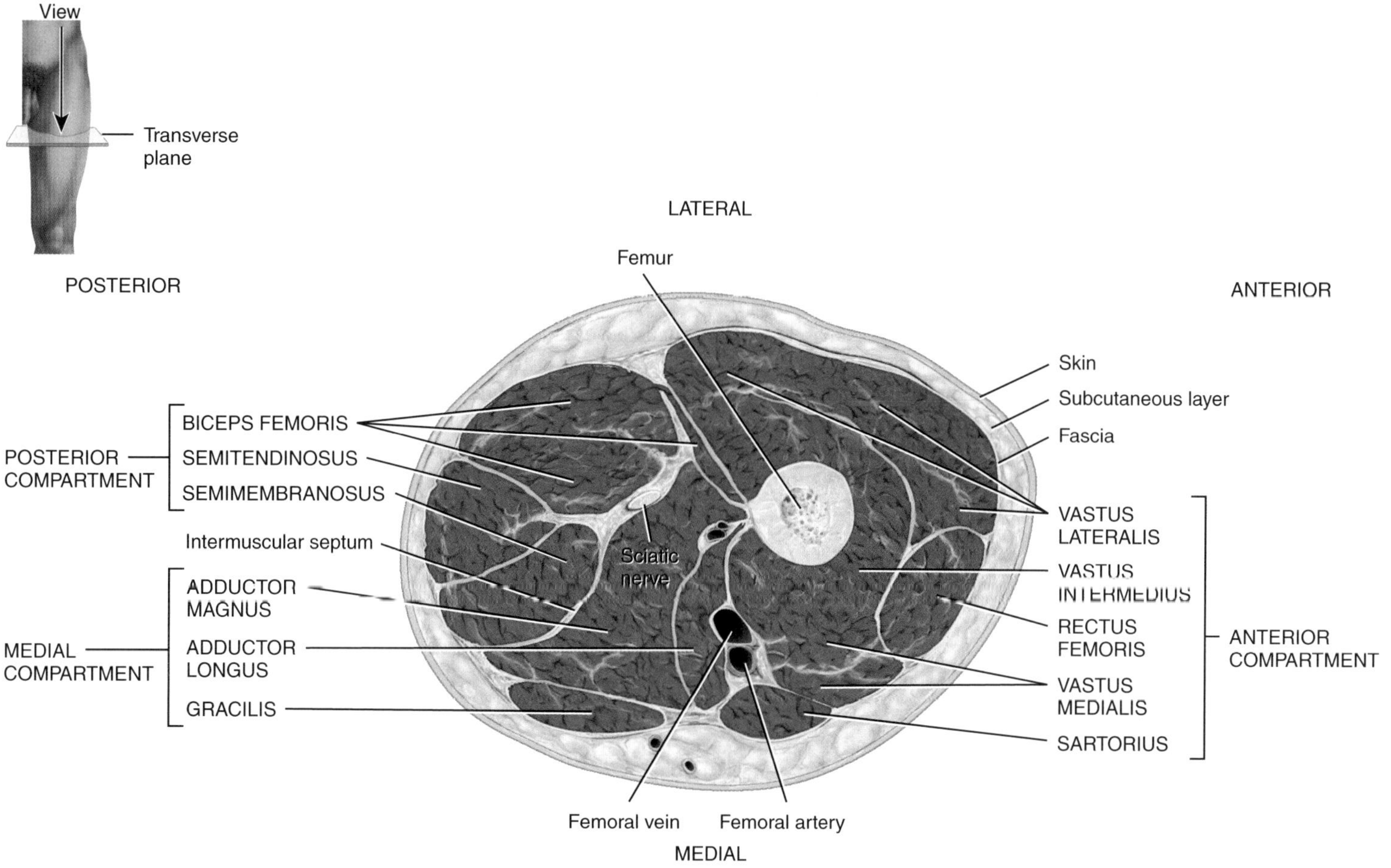

(a) Superior view of transverse section of thigh

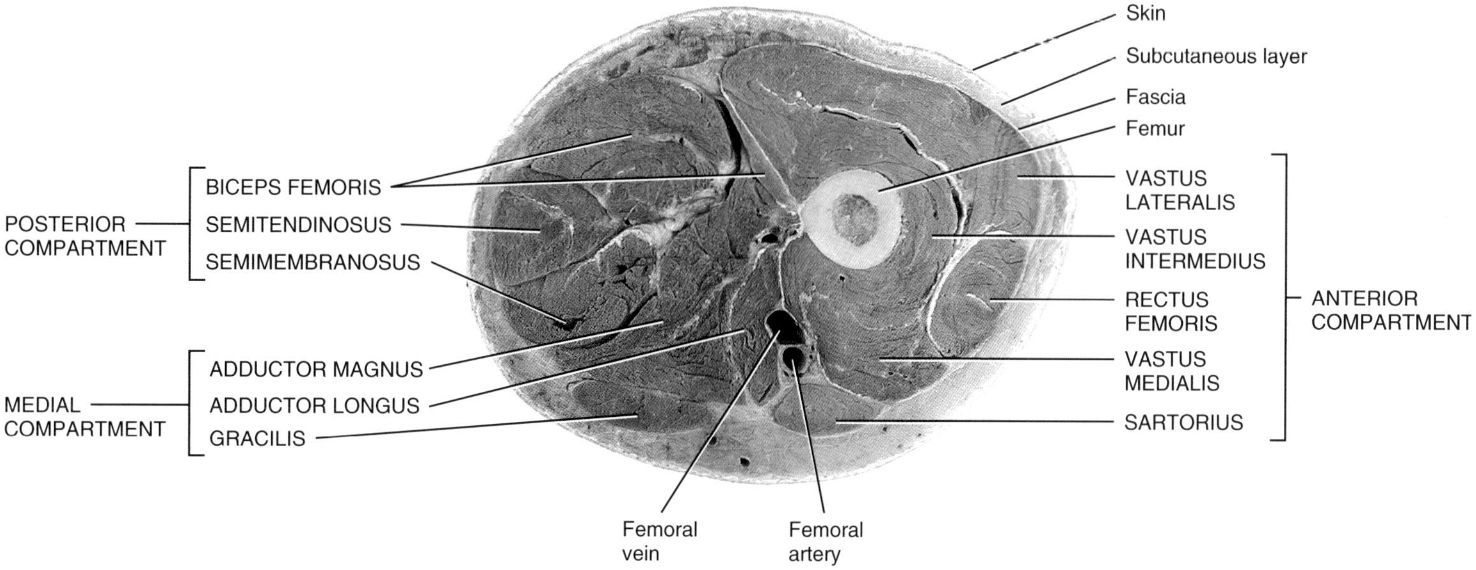

(b) Superior view of transverse section of thigh

Which muscles constitute the quadriceps femoris and hamstring muscles?

EXHIBIT 11.S

Muscles of the Leg That Move the Foot and Toes *(Figure 11.22)*

OBJECTIVE

- Describe the origin, insertion, action, and innervation of the muscles of the leg that move the foot and toes.

Muscles that move the foot and toes are located in the leg (Figure 11.22). The muscles of the leg, like those of the thigh, are divided by deep fascia into three compartments: anterior, lateral, and posterior. The **anterior compartment of the leg** consists of muscles that dorsiflex the foot. In a situation analogous to the wrist, the tendons of the muscles of the anterior compartment are held firmly to the ankle by thickenings of deep fascia called the **superior extensor retinaculum** (*transverse ligament of the ankle*) and **inferior extensor retinaculum** (*cruciate ligament of the ankle*).

Within the anterior compartment, the **tibialis anterior** is a long, thick muscle against the lateral surface of the tibia, where it is easy to palpate (feel). The **extensor hallucis longus** is a thin muscle between and partly deep to the tibialis anterior and **extensor digitorum longus** muscles. This featherlike muscle is lateral to the tibialis anterior muscle, where it can also be palpated easily. The **fibularis (peroneus) tertius** muscle is part of the extensor digitorum longus, with which it shares a common origin.

The **lateral (fibular) compartment of the leg** contains two muscles that plantar flex and evert the foot: the **fibularis (peroneus) longus** and **fibularis (peroneus) brevis.**

The **posterior compartment of the leg** consists of muscles in superficial and deep groups. The superficial muscles share a common tendon of insertion, the **calcaneal (Achilles) tendon,** the strongest tendon of the body. It inserts into the calcaneal bone of the ankle. The superficial and most of the deep muscles plantar flex the foot at the ankle joint. The superficial muscles of the posterior compartment are the gastrocnemius, soleus, and plantaris—the so-called calf muscles. The large size of these muscles is directly related to the characteristic upright stance of humans. The **gastrocnemius** is the most superficial muscle and forms the prominence of the calf. The **soleus,** which lies deep to the gastrocnemius, is broad and flat. It derives its name from its resemblance to a flat fish (sole). The **plantaris** is a small muscle that may be absent; conversely, sometimes there are two of them in each leg. It runs obliquely between the gastrocnemius and soleus muscles.

The deep muscles of the posterior compartment are the popliteus, tibialis posterior, flexor digitorum longus, and flexor hallucis longus. The **popliteus** is a triangular muscle that forms the floor of the popliteal fossa. The **tibialis posterior** is the deepest muscle in the posterior compartment. It lies between the flexor digitorum longus and flexor hallucis longus muscles. The **flexor digitorum longus** is smaller than the **flexor hallucis longus,** even though the former flexes four toes and the latter flexes only the great toe at the interphalangeal joint.

MUSCLE	ORIGIN	INSERTION	ACTION	INNERVATION
ANTERIOR COMPARTMENT OF THE LEG				
Tibialis anterior (tib′-ē-Ā-lis an-TĒR-ē-or; *tibialis* = tibia; *anterior* = front)	Lateral condyle and body of tibia and interosseous membrane (sheet of fibrous tissue that holds shafts of tibia and fibula together).	First metatarsal and first (medial) cuneiform.	Dorsiflexes foot at ankle joint and inverts (supinates) foot at intertarsal joints.	Deep fibular (peroneal) nerve.
Extensor hallucis longus (eks-TEN-sor HAL-ū-sis LON-gus; *extensor* = increases angle at joint; *hallucis* = hallux or great toe; *longus* = long)	Anterior surface of middle third of fibula and interosseous membrane.	Distal phalanx of great toe.	Dorsiflexes foot at ankle joint and extends proximal phalanx of great toe at metatarsophalangeal joint.	Deep fibular (peroneal) nerve.
Extensor digitorum longus (di′-ji-TOR-um LON-gus; *digit-* = finger or toe)	Lateral condyle of tibia, anterior surface of fibula, and interosseous membrane.	Middle and distal phalanges of toes 2–5.*	Dorsiflexes foot at ankle joint and extends distal and middle phalanges of each toe at interphalangeal joints and proximal phalanx of each toe at metatarsophalangeal joint.	Deep fibular (peroneal) nerve.
Fibularis (peroneus) tertius (fib-ū-LĀ-ris; per′-Ō-nē-us; *person* = fibula; TER-shus; *tertius* = third)	Distal third of fibula and interosseous membrane.	Base of fifth metatarsal.	Dorsiflexes foot at ankle joint and everts (pronates) foot at intertarsal joints.	Deep fibular (peroneal) nerve.
LATERAL (FIBULAR) COMPARTMENT OF THE LEG				
Fibularis (peroneus) longus (LON-gus)	Head and body of fibula.	First metatarsal and first cuneiform.	Plantar flexes foot at ankle joint and everts (pronates) foot at intertarsal joints.	Superficial fibular (peroneal) nerve.
Fibularis (peroneus) brevis (BREV-is; *brevis* = short)	Distal half of body of fibula.	Base of fifth metatarsal.	Plantar flexes foot at ankle joint and everts (pronates) foot at intertarsal joints.	Superficial fibular (peroneal) nerve.

CLINICAL CONNECTION | *Shin Splint Syndrome*

Shin splint syndrome, or simply **shin splints,** refers to pain or soreness along the tibia, specifically the medial, distal two-thirds. It may be caused by tendinitis of the anterior compartment muscles, especially the tibialis anterior muscle, inflammation of the periosteum (periostitis) around the tibia, or stress fractures of the tibia. The tendinitis usually occurs when poorly conditioned runners run on hard or banked surfaces with poorly supportive running shoes. The condition may also occur with vigorous activity of the legs following a period of relative inactivity or running in cold weather without proper warmup. The muscles in the anterior compartment (mainly the tibialis anterior) can be strengthened to balance the stronger posterior compartment muscles. •

RELATING MUSCLES TO MOVEMENTS

Arrange the muscles in this exhibit according to the following actions on the foot at the ankle joint: (1) dorsiflexion and (2) plantar flexion; according to the following actions on the foot at the intertarsal joints: (1) inversion and (2) eversion; and according to the following actions on the toes at the metatarsophalangeal and interphalangeal joints: (1) flexion and (2) extension. The same muscle may be mentioned more than once.

✔ **CHECKPOINT**

What are the superior extensor retinaculum and inferior extensor retinaculum?

MUSCLE	ORIGIN	INSERTION	ACTION	INNERVATION
SUPERFICIAL POSTERIOR COMPARTMENT OF THE LEG				
Gastrocnemius (gas′-trok-NĒ-mē-us; *gastro-* = belly; *cnem-* = leg)	Lateral and medial condyles of femur and capsule of knee	Calcaneus by way of calcaneal (Achilles) tendon.	Plantar flexes foot at ankle joint and flexes leg at knee joint.	Tibial nerve.
Soleus (SŌ-lē-us; *sole* = type of flat fish)	Head of fibula and medial border of tibia.	Calcaneus by way of calcaneal (Achilles) tendon.	Plantar flexes foot at ankle joint.	Tibial nerve.
Plantaris (plan-TĀR-is = sole)	Lateral epicondyle of femur.	Calcaneus by way of calcaneal (Achilles) tendon.	Plantar flexes foot at ankle joint and flexes leg at knee joint.	Tibial nerve.
DEEP POSTERIOR COMPARTMENT OF THE LEG				
Popliteus (pop-LIT-ē-us; *poplit* = back of knee)	Lateral condyle of femur.	Proximal tibia.	Flexes leg at knee joint and medially rotates tibia to unlock the extended knee.	Tibial nerve.
Tibialis posterior (tib′-ē-Ā-lis; *posterior* = back)	Proximal tibia, fibula, and interosseous membrane.	Second, third, and fourth metatarsals; navicular; and all three cuneiforms.	Plantar flexes foot at ankle joint and inverts (supinates) foot at intertarsal joints.	Tibial nerve.
Flexor digitorum longus (FLEK-sor di′-ji-TOR-um LON-gus; *flexor* = decreases angle at point; *digit* = finger or toe)	Middle third of posterior surface of tibia.	Distal phalanges of toes 2–5.	Plantar flexes foot at ankle joint; flexes distal and middle proximal phalanx of each toe phalanges of each toe at interphalangeal joints and at metatarsophalangeal joint.	Tibial nerve.
Flexor hallucis longus (HAL-ū-sis)	Inferior two-thirds of posterior portion of fibula.	Distal phalanx of great toe.	Plantar flexes foot at ankle joint; flexes distal phalanx of great toe at interphalangeal joint and proximal phalanx of great toe at metatarsophalangeal joint.	Tibial nerve.

*Reminder: The great toe or hallux is the first toe and has two phalanges: proximal and distal. The remaining toes are numbered II–V (2–5), and each has three phalanges: proximal, middle, and distal.

EXHIBIT 11.S CONTINUES ▶

Muscles of the Leg That Move the Foot and Toes *(Figure 11.22)* CONTINUED

Figure 11.22 Muscles of the leg that move the foot and toes.

The superficial muscles of the posterior compartment share a common tendon of insertion, the calcaneal (Achilles) tendon, which inserts into the calcaneal bone of the ankle.

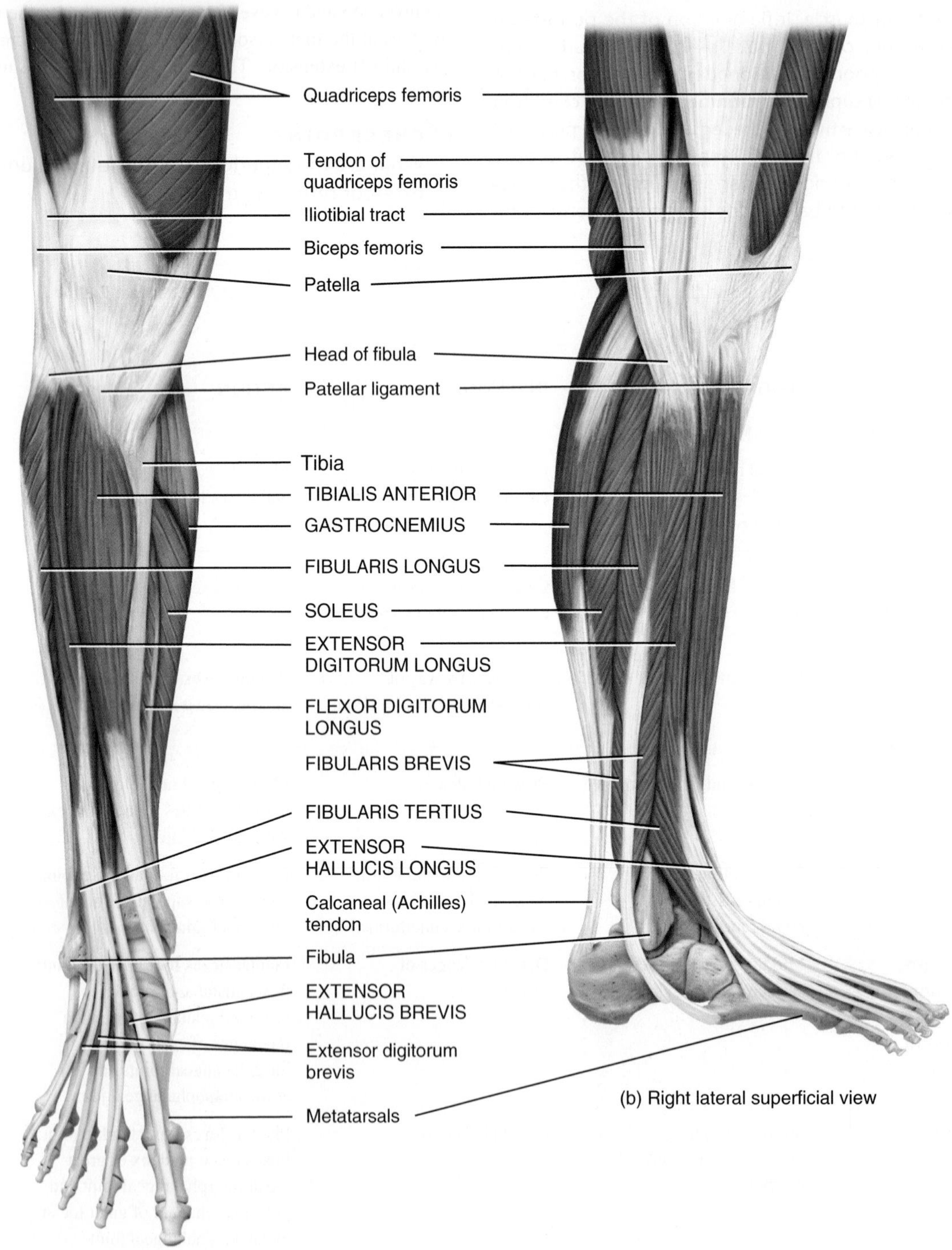

Quadriceps femoris
Tendon of quadriceps femoris
Iliotibial tract
Biceps femoris
Patella
Head of fibula
Patellar ligament
Tibia
TIBIALIS ANTERIOR
GASTROCNEMIUS
FIBULARIS LONGUS
SOLEUS
EXTENSOR DIGITORUM LONGUS
FLEXOR DIGITORUM LONGUS
FIBULARIS BREVIS
FIBULARIS TERTIUS
EXTENSOR HALLUCIS LONGUS
Calcaneal (Achilles) tendon
Fibula
EXTENSOR HALLUCIS BREVIS
Extensor digitorum brevis
Metatarsals

(a) Anterior superficial view

(b) Right lateral superficial view

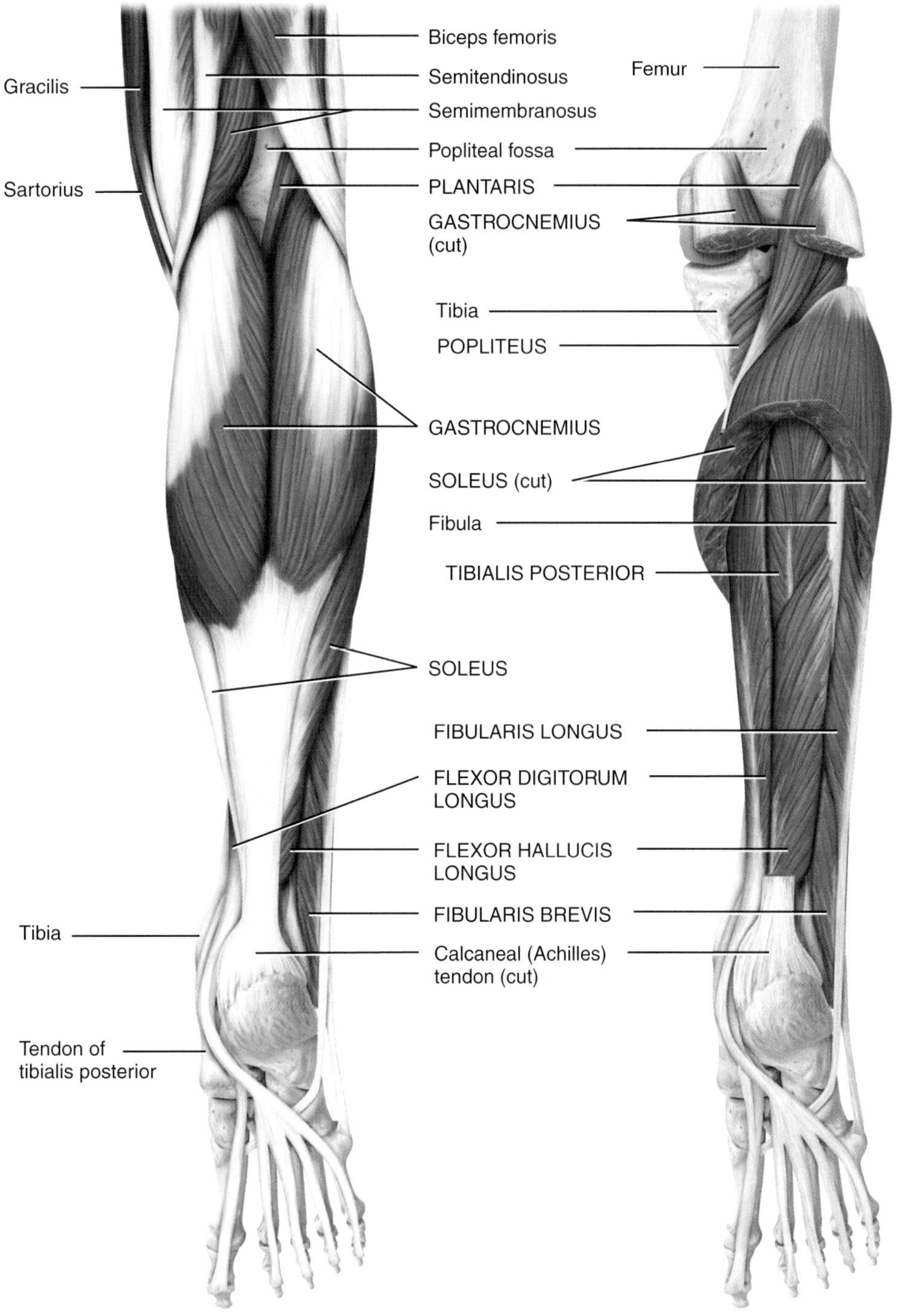
Biceps femoris
Gracilis
Semitendinosus
Semimembranosus
Popliteal fossa
Sartorius
PLANTARIS
GASTROCNEMIUS
(cut)
Femur
Tibia
POPLITEUS
GASTROCNEMIUS
SOLEUS (cut)
Fibula
TIBIALIS POSTERIOR
SOLEUS
FIBULARIS LONGUS
FLEXOR DIGITORUM
LONGUS
FLEXOR HALLUCIS
LONGUS
FIBULARIS BREVIS
Tibia
Calcaneal (Achilles)
tendon (cut)
Tendon of
tibialis posterior

(c) Posterior superficial view

(d) Posterior deep view

EXHIBIT 11.S CONTINUES

Muscles of the Leg That Move the Foot and Toes *(Figure 11.22)* CONTINUED

FIGURE 11.22 CONTINUED

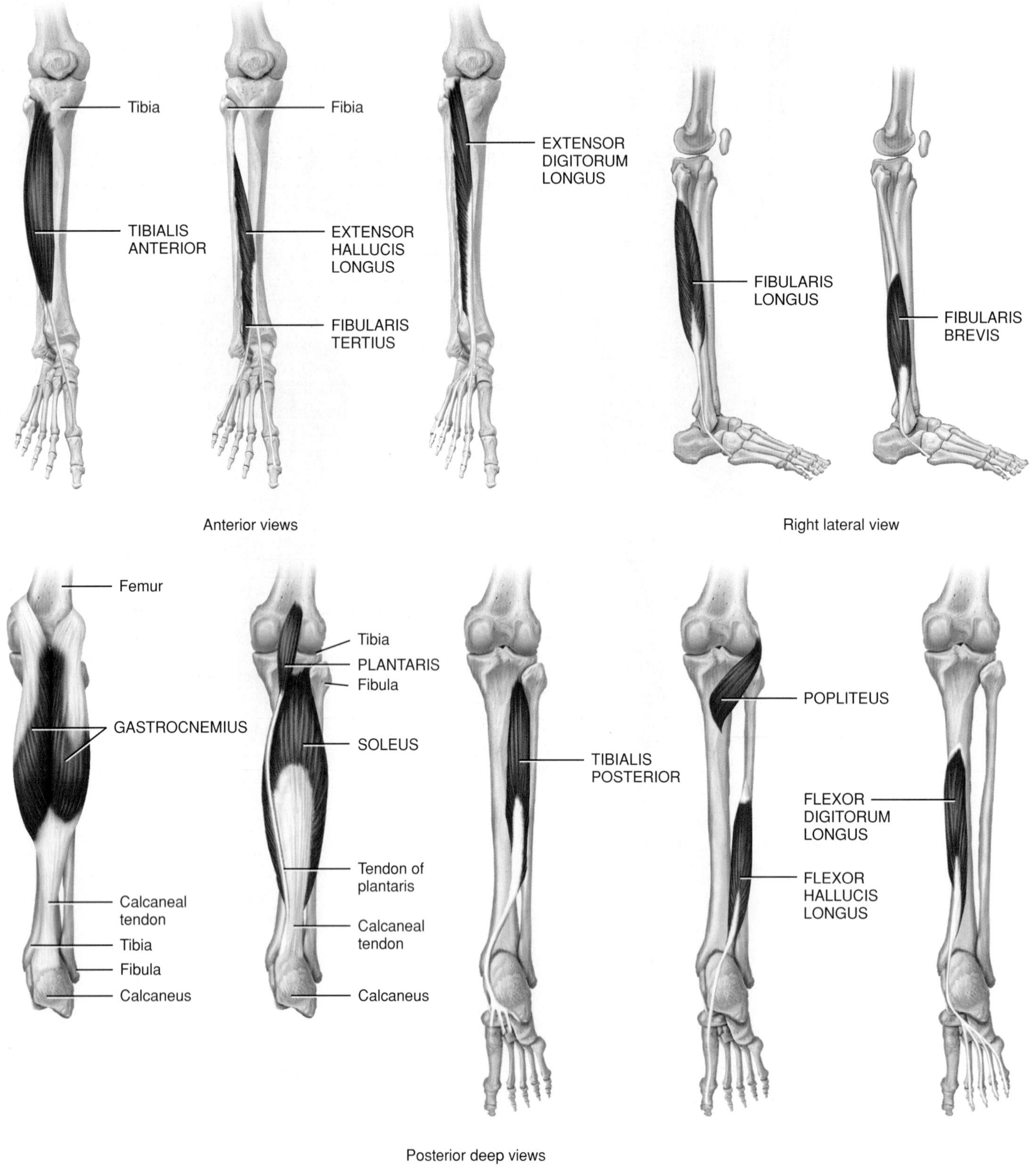
Tibia
TIBIALIS ANTERIOR
Fibia
EXTENSOR HALLUCIS LONGUS
FIBULARIS TERTIUS
EXTENSOR DIGITORUM LONGUS
FIBULARIS LONGUS
FIBULARIS BREVIS
Anterior views
Right lateral view
Femur
GASTROCNEMIUS
Calcaneal tendon
Tibia
Fibula
Calcaneus
Tibia
PLANTARIS
Fibula
SOLEUS
Tendon of plantaris
Calcaneal tendon
Calcaneus
TIBIALIS POSTERIOR
POPLITEUS
FLEXOR DIGITORUM LONGUS
FLEXOR HALLUCIS LONGUS
Posterior deep views

(e) Isolated muscles

What structures firmly hold the tendons of the anterior compartment muscles to the ankle?

EXHIBIT 11.T Intrinsic Muscles of the Foot That Move the Toes *(Figure 11.23)*

OBJECTIVE

- Describe the origin, insertion, action, and innervation of the intrinsic muscles of the foot that move the toes.

The muscles in this exhibit are termed **intrinsic muscles of the foot** because they originate and insert *within* the foot (Figure 11.23). The muscles of the hand are specialized for precise and intricate movements, but those of the foot are limited to support and locomotion. The deep fascia of the foot forms the **plantar aponeurosis (fascia)** that extends from the calcaneus bone to the phalanges of the toes. The aponeurosis supports the longitudinal arch of the foot and encloses the flexor tendons of the foot.

The intrinsic muscles of the foot are divided into two groups: **dorsal** and **plantar.** There are two dorsal muscles, the **extensor hallucis brevis** and the **extensor digitorum brevis.** The latter is a four-part muscle deep to the tendons of the extensor digitorum longus muscle, which extends toes 2–5 at the metatarsophalangeal joints.

The plantar muscles are arranged in four layers. The most superficial layer is called the first layer. Three muscles are in the first layer. The **abductor hallucis,** which lies along the medial border of the sole and is comparable to the abductor pollicis brevis in the hand, abducts the great toe at the metatarsophalangeal joint. The **flexor digitorum brevis,** which lies in the middle of the sole, flexes toes 2–5 at the interphalangeal and metatarsophalangeal joints. The **abductor digiti minimi,** which lies along the lateral border of the sole and is comparable to the same muscle in the hand, abducts the little toe.

The second layer consists of the **quadratus plantae,** a rectangular muscle that arises by two heads and flexes toes 2–5 at the metatarsophalangeal joints, and the **lumbricals,** four small muscles that are similar to the lumbricals in the hands. They flex the proximal phalanges and extend the distal phalanges of toes 2–5.

Three muscles compose the third layer. The **flexor hallucis brevis,** which lies adjacent to the plantar surface of the metatarsal of the great toe and is comparable to the same muscle in the hand, flexes the great toe. The **adductor hallucis,** which has an oblique and transverse head like the adductor pollicis in the hand, adducts the great toe. The **flexor digiti minimi brevis,** which lies superficial to the metatarsal of the little toe and is comparable to the same muscle in the hand, flexes the little toe.

The fourth layer is the deepest and consists of two muscle groups. The **dorsal interossei** are four muscles that abduct toes 2–4, flex the proximal phalanges, and extend the distal phalanges. The three **plantar interossei** abduct toes 3–5, flex the proximal phalanges, and extend the distal phalanges. The interossei of the feet are similar to those of the hand. However, their actions are relative to the midline of the second digit rather than the third digit as in the hand.

CLINICAL CONNECTION | *Plantar Fasciitis*

Plantar fasciitis (fas-ē-Ī-tis) or **painful heel syndrome** is an inflammatory reaction due to chronic irritation of the plantar aponeurosis (fascia) at its origin on the calcaneus (heel bone). The aponeurosis becomes less elastic with age. This condition is also related to weight-bearing activities (walking, jogging, lifting heavy objects), improperly constructed or fitting shoes, excess weight (puts pressure on the feet), and poor biomechanics (flat feet, high arches, and abnormalities in gait may cause uneven distribution of weight on the feet). Plantar fasciitis is the most common cause of heel pain in runners and arises in response to the repeated impact of running. Treatments include ice, deep heat, stretching exercises, weight loss, prosthetics (such as shoe inserts or heel lifts), steroid injections, and surgery. •

RELATING MUSCLES TO MOVEMENTS

Arrange the muscles in this exhibit according to the following actions on the great toe at the metatarsophalangeal joint: (1) flexion, (2) extension, (3) abduction, and (4) adduction; and according to the following actions on toes 2–5 at the metatarsophalangeal and interphalangeal joints: (1) flexion, (2) extension, (3) abduction, and (4) adduction. The same muscle may be mentioned more than once.

✔ **CHECKPOINT**

How do the intrinsic muscles of the hand and foot differ in function?

EXHIBIT 11.T CONTINUES ▶

Intrinsic Muscles of the Foot That Move the Toes *(Figure 11.23)* CONTINUED

MUSCLE	ORIGIN	INSERTION	ACTION	INNERVATION
DORSAL				
Extensor hallucis brevis (eks-TEN-sor HAL-ū-sis BREV-is; *extensor* = increases angle at joint; *hallucis* = hallux or great toe; *brevis* = short) (see Figure 11.22a)	Calcaneus and inferior extensor retinaculum.	Proximal phalanx of great toe.	Extends great toe at metatarsophalangeal joint.	Deep fibular (peroneal) nerve.
Extensor digitorum brevis *digit* = finger or toe; *brevis* = short) (see Figure 11.22a, b)	Calcaneus and inferior extensor retinaculum.	Middle phalanges of toes 2–4.	Extends toes 2–4 at interphalangeal joints.	Deep fibular (peroneal) nerve.
PLANTAR				
First layer (most superficial)				
Abductor hallucis (*abductor* = moves part away from midline; *hallucis* = hallux or great toe)	Calcaneus, plantar aponeurosis, and flexor retinaculum.	Medial side of proximal phalanx of great toe with the tendon of flexor hallucis brevis.	Abducts and flexes great toe at metatarsophalangeal joint.	Medial plantar nerve.
Flexor digitorum brevis (*flexor* = decreases angle at joint)	Calcaneus, plantar aponeurosis, and flexor retinaculum.	Sides of middle phalanx of toes 2–5.	Flexes toes 2–5 at proximal interphalangeal and metatarsophalangeal joints.	Medial plantar nerve.
Abductor digiti minimi (DIJ-i-tē MIN-i-mē; *minimi* = smallest)	Calcaneus, plantar aponeurosis, and flexor retinaculum.	Lateral side of proximal phalanx of little toe with tendon of flexor digiti minimi brevis.	Abducts and flexes little toe at metatarsophalangeal joint.	Lateral plantar nerve.
Second layer				
Quadratus plantae (kwod-RĀ-tus PLAN-tē; *quad* = square, four-sided; *planta* = sole)	Calcaneus.	Tendon of flexor digitorum longus.	Assists flexor digitorum longus to only flex toes 2–5 at interphalangeal and metatarsophalangeal joints.	Lateral plantar nerve.
Lumbricals (LUM-bri-kals; *lumbric* = earthworm)	Tendons of flexor digitorum longus.	Tendons of extensor digitorum longus on proximal phalanges of toes 2–5.	Extend toes 2–5 at interphalangeal joints and flex toes 2–5 at metatarsophalangeal joints.	Medial and lateral plantar nerves.
Third layer				
Flexor hallucis brevis	Cuboid and third (lateral) cuneiform.	Medial and lateral sides of proximal phalanx of great toe via tendon containing sesamoid bone.	Flexes great toe at metatarsophalangeal joint.	Medial plantar nerve.
Adductor hallucis (ad-DUK-tor = moves part closer to midline)	Metatarsals 2–4, ligaments of 3–5 metatarsophalangeal joints, and tendon of fibularis (peroneus) longus.	Lateral side of proximal phalanx of great toe.	Adducts and flexes great toe at metatarsophalangeal joint.	Lateral plantar nerve.
Flexor digiti minimi brevis	Metatarsal 5 and tendon of fibularis (peroneus) longus.	Lateral side of proximal phalanx of little toe.	Flexes little toe at metatarsophalangeal joint.	Lateral plantar nerve.
Fourth layer (deepest)				
Dorsal interossei (in-ter-OS-ē-ī) (not illustrated)	Adjacent side of all metatarsals.	Proximal phalanges: both sides of toe 2 and lateral side of toes 3 and 4.	Abduct and flex toes 2–4 at metatarsophalangeal joints and extend toes at interphalangeal joints.	Lateral plantar nerve.
Plantar interossei (PLAN-tar)	Metatarsals 3–5.	Medial side of proximal phalanges of toes 3–5.	Adduct and flex proximal metatarsophalangeal joints and extend toes at interphalangeal joints.	Lateral plantar nerve.

Figure 11.23 **Intrinsic muscles of the foot that move the toes.**

The muscles of the hand are specialized for precise and intricate movements; those of the foot are limited to support and movement.

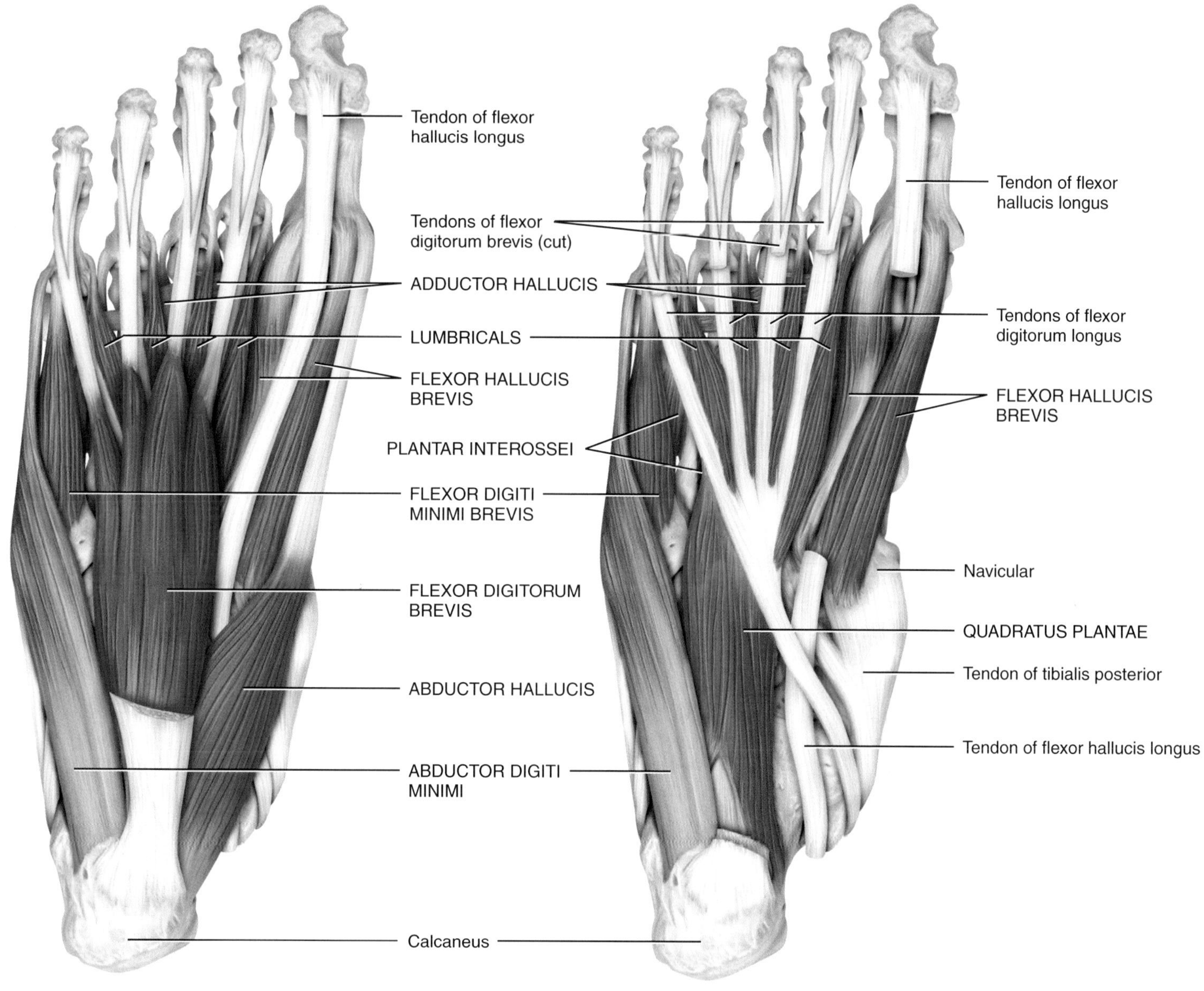

(a) Plantar superficial and deep view

(b) Plantar deep view

EXHIBIT 11.T CONTINUES

Intrinsic Muscles of the Foot That Move the Toes *(Figure 11.23)* CONTINUED

FIGURE 11.23 CONTINUED

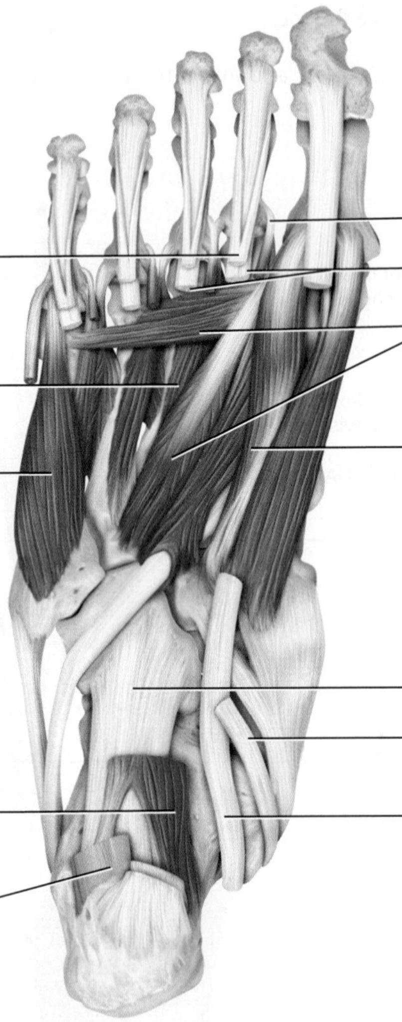
Tendon of lumbrical (cut)
Tendon of flexor digitorum brevis (cut)
Tendons of flexor digitorum longus (cut)
ADDUCTOR HALLUCIS
PLANTAR INTEROSSEI
FLEXOR HALLUCIS BREVIS
FLEXOR DIGITI MINIMI BREVIS
Long plantar ligament
Tendon of flexor digitorum longus (cut)
QUADRATUS PLANTAE (cut)
Tendon of flexor hallucis longus
ABDUCTOR DIGITI MINIMI (cut)

(c) Plantar deeper view

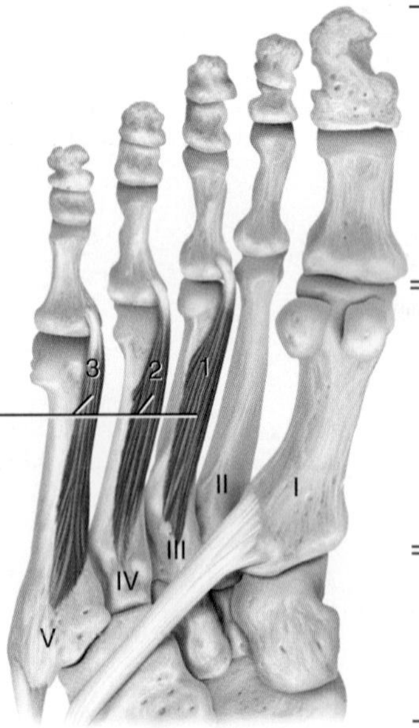
Phalanges
Metatarsals
Tarsals
PLANTAR INTEROSSEI
3
2
1
I
II
III
IV
V

(d) Plantar view

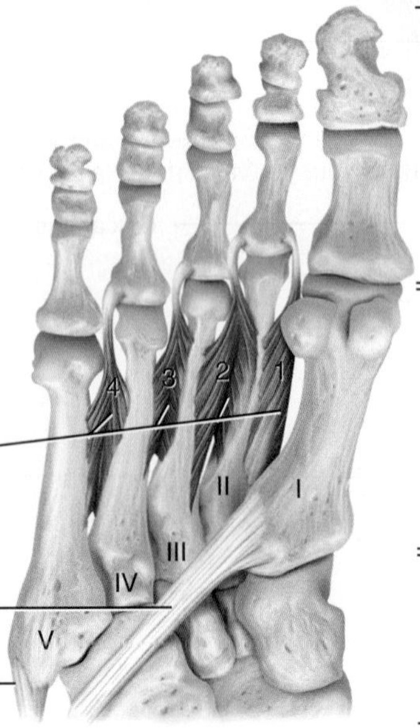
Phalanges
Metatarsals
Tarsals
DORSAL INTEROSSEI
4
3
2
1
I
II
III
IV
V
Tendon of fibularis longus
Tendon of fibularis brevis

(e) Plantar view

What structure supports the longitudinal arch and encloses the flexor tendons of the foot?

Focus on Homeostasis

BODY SYSTEM	CONTRIBUTION OF THE MUSCULAR SYSTEM
For all body systems 	The muscular system and muscle tissues produce body movements, stabilize body positions, move substances within the body, and produce heat that helps maintain normal body temperature.
Integumentary system 	Pull of skeletal muscles on attachments to skin of face causes facial expressions; muscular exercise increases skin blood flow.
Skeletal system 	Skeletal muscle causes movement of body parts by pulling on attachments to bones; skeletal muscle provides stability for bones and joints.
Nervous system 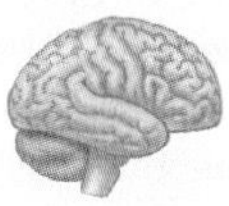	Smooth, cardiac, and skeletal muscles carry out commands for the nervous system; shivering—involuntary contraction of skeletal muscles that is regulated by the brain—generates heat to raise body temperature.
Endocrine system 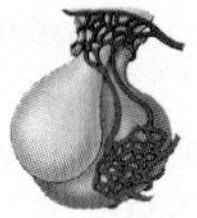	Regular activity of skeletal muscles (exercise) improves action and signaling mechanisms of some hormones, such as insulin; muscles protect some endocrine glands.
Cardiovascular system 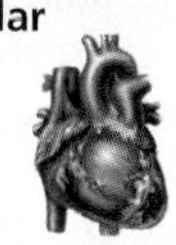	Cardiac muscle powers pumping action of heart; contraction and relaxation of smooth muscle in blood vessel walls help adjust the amount of blood flowing through various body tissues; contraction of skeletal muscles in the legs assists return of blood to the heart; regular exercise causes cardiac hypertrophy (enlargement) and increases heart's pumping efficiency; lactic acid produced by active skeletal muscles may be used for ATP production by the heart.
Lymphatic system and immunity 	Skeletal muscles protect some lymph nodes and lymphatic vessels and promote flow of lymph inside lymphatic vessels; exercise may increase or decrease some immune responses.
Respiratory system 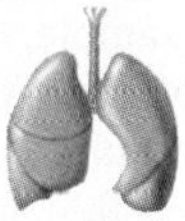	Skeletal muscles involved with breathing cause air to flow into and out of the lungs; smooth muscle fibers adjust size of airways; vibrations in skeletal muscles of larynx control air flowing past vocal cords, regulating voice production; coughing and sneezing, due to skeletal muscle contractions, help clear airways; regular exercise improves efficiency of breathing.
Digestive system 	Skeletal muscles protect and support organs in the abdominal cavity; alternating contraction and relaxation of skeletal muscles power chewing and initiate swallowing; smooth muscle sphincters control volume of organs of the gastrointestinal (GI) tract; smooth muscles in walls of GI tract mix and move its contents through the tract.
Urinary system 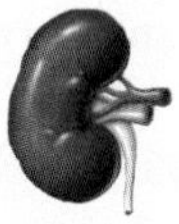	Skeletal and smooth muscle sphincters and smooth muscle in wall of urinary bladder control whether urine is stored in the urinary bladder or voided (urination).
Reproductive systems 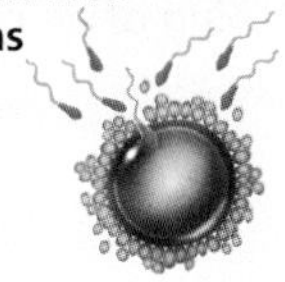	Skeletal and smooth muscle contractions eject semen from male; smooth muscle contractions propel oocyte along uterine tube, help regulate flow of menstrual blood from uterus, and force baby from uterus during childbirth; during intercourse, skeletal muscle contractions are associated with orgasm and pleasurable sensations in both sexes.

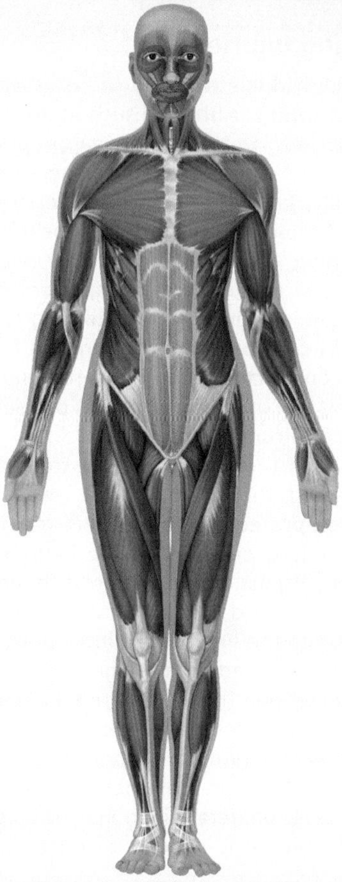

THE MUSCULAR SYSTEM

DISORDERS: HOMEOSTATIC IMBALANCES

Running Injuries

Many individuals who jog or run sustain some type of **running-related injury.** Although such injuries may be minor, some can be quite serious. Untreated or inappropriately treated minor injuries may become chronic. Among runners, common sites of injury include the ankle, knee, calcaneal (Achilles) tendon, hip, groin, foot, and back. Of these, the knee often is the most severely injured area.

Running injuries are frequently related to faulty training techniques. This may involve improper or lack of sufficient warm-up routines, running too much, or running too soon after an injury. Or it might involve extended running on hard and/or uneven surfaces. Poorly constructed or worn-out running shoes can also contribute to injury, as can any biomechanical problem (such as a fallen arch) aggravated by running.

Most sports injuries should be treated initially with PRICE therapy, which stands for *p*rotection, *r*est, *i*ce, *c*ompression, and *e*levation. Immediately protect the injured part, rest, and apply ice immediately after the injury, and elevate the injured part. Then apply an elastic bandage, if possible, to compress the injured tissue. Continue using PRICE for 2 to 3 days, and resist the temptation to apply heat, which may worsen the swelling. Follow-up treatment may include alternating moist heat and ice massage to enhance blood flow in the injured area. Sometimes it is helpful to take nonsteroidal anti-inflammatory drugs (NSAIDs) or to have local injections of corticosteroids. During the recovery period, it is important to keep active, using an alternative fitness program that does not worsen the original injury. This activity should be determined in consultation with a physician. Finally, careful exercise is needed to rehabilitate the injured area itself. Massage therapy may also be used to prevent or treat many sports injuries.

Compartment Syndrome

As noted earlier in this chapter, skeletal muscles in the limbs are organized into functional units called *compartments.* In a disorder called **compartment syndrome,** some external or internal pressure constricts the structures within a compartment, resulting in damaged blood vessels and subsequent reduction of the blood supply (ischemia) to the structures within the compartment. Symptoms include pain, burning, pressure, pale skin, and paralysis. Common causes of compartment syndrome include crushing and penetrating injuries, contusion (damage to subcutaneous tissues without the skin being broken), muscle strain (overstretching of a muscle), or an improperly fitted cast. The pressure increase in the compartment can have serious consequences, such as hemorrhage, tissue injury, and edema (buildup of interstitial fluid). Because deep fasciae (connective tissue coverings) that enclose the compartments are very strong, accumulated blood and interstitial fluid cannot escape, and the increased pressure can literally choke off the blood flow and deprive nearby muscles and nerves of oxygen. One treatment option is **fasciotomy** (fash-ē-OT-ō-mē), a surgical procedure in which muscle fascia is cut to relieve the pressure. Without intervention, nerves can suffer damage, and muscles can develop scar tissue that results in permanent shortening of the muscles, a condition called *contracture.* If left untreated, tissues may die and the limb may no longer be able to function. Once the syndrome has reached this stage, amputation may be the only treatment option.

MEDICAL TERMINOLOGY

Muscle strain Tearing of fibers in a skeletal muscle or its tendon that attaches the muscle to bone. The tearing can also damage small blood vessels, causing local bleeding (bruising) and pain (caused by irritation of nerve endings in the region). Muscle strains usually occur when a muscle is stretched beyond its limit, for example, in response to sudden, quick heavy lifting; during sports activities; or while performing work tasks. A similar injury occurs if there is a direct blow to a muscle. Factors that contribute to muscle strain include muscle tightness due to insufficient stretching, not working antagonistic muscles equally, poor conditioning, muscle fatigue, and insufficient warm-up. The condition is treated by PRICE therapy: protection (P), rest (R), ice immediately after the injury (I), compression via a supportive wrap (C), and elevation of the limb (E). Also called **muscle pull** or **muscle tear.**

Paralysis (pa-RAL-i-sis; *para-* = departure from normal; *-lysis* = loosening) Loss of muscle function (voluntary movement) through injury, disease, or damage to its nerve supply. Most paralysis is due to stroke or spinal cord injury.

Repetitive strain or motion injuries (RSIs) Conditions resulting from overuse of equipment, poor posture, poor body mechanics, or activity that requires repeated movements; for example, various conditions of assembly line workers. Examples of overuse of equipment include overuse of a computer, hammer, guitar, or piano, to name a few.

Rhabdomyosarcoma (rab′-dō-mī′-ō-sar-KŌ-ma; *rhab-* = rod-shaped; *-myo-* = muscle; *-sarc-* = flesh; *-oma* = tumor) A tumor of skeletal muscle. Usually occurs in children and is highly malignant, with rapid metastasis.

Torticollis (tor-ti-KOL-is; *tortus-* = twisted; *-column* = neck) A contraction or shortening of the sternocleidomastoid muscle that causes the head to tilt toward the affected side and the chin to rotate toward the opposite side. It may be acquired or congenital. Also called **wryneck.**

Tic Spasmodic twitching made involuntarily by muscles that are usually under conscious control, for example, twitching of an eyelid.

CHAPTER REVIEW AND RESOURCE SUMMARY

WILEY PLUS

Review

11.1 How Skeletal Muscles Produce Movements

1. Skeletal muscles that produce movement do so by pulling on bones.
2. The attachment to the more stationary bone is the origin; the attachment to the more movable bone is the insertion.
3. Bones serve as levers, and joints serve as fulcrums. Two different forces act on the lever: load (resistance) and effort.
4. Levers are categorized into three types—first-class, second-class, and third-class (most common)—according to the positions of the fulcrum, the effort, and the load on the lever.
5. Fascicular arrangements include parallel, fusiform, circular, triangular, and pennate. Fascicular arrangement affects a muscle's power and range of motion.
6. A prime mover produces the desired action; an antagonist produces an opposite action. Synergists assist a prime mover by reducing unnecessary movement. Fixators stabilize the origin of a prime mover so that it can act more efficiently.

Resource

Anatomy Overview - Skeletal Muscle
Animation - Contraction and Movement
Figure 11.1 - Relationship of Skeletal Muscles to Bones

11.2 How Skeletal Muscles Are Named

1. Distinctive features of different skeletal muscles include direction of muscle fascicles; size, shape, action, number of origins (or heads), and location of the muscle; and sites of origin and insertion of the muscle.
2. Most skeletal muscles are named based on combinations of features.

Resource

Anatomy Overview - The Muscular System

11.3 Principal Skeletal Muscles

1. Muscles of the head that produce facial expression move the skin rather than a joint when they contract, and they permit us to express a wide variety of emotions (see Exhibit 11.A). The muscles of the head that move the eyeballs are among the fastest contracting and most precisely controlled skeletal muscles in the body. They permit us to elevate, depress, abduct, adduct, and medially and laterally rotate the eyeballs. The muscles that move the eyelids open the eyes (see Exhibit 11.B).
2. Muscles that move the mandible play major roles in mastication (chewing) and speech (see Exhibit 11.C). The muscles of the head that move the tongue are important in mastication and speech, as well as deglutition (swallowing) (Exhibit 11.D).
3. Muscles of the anterior neck that assist in deglutition and speech, called suprahyoid muscles, are located above the hyoid bone (see Exhibit 11.E).
4. Muscles of the neck that move the head alter its position and help balance it on the vertebral column (see Exhibit 11.F).
5. Muscles of the abdomen help contain and protect the abdominal viscera, move the vertebral column, compress the abdomen, and produce the force required for defecation, urination, vomiting, and childbirth (see Exhibit 11.G).
6. Muscles of the thorax used in breathing alter the size of the thoracic cavity so that inhalation and exhalation can occur and assist in venous return of blood to the heart (see Exhibit 11.H).
7. Muscles of the pelvic floor support the pelvic viscera, resist the thrust that accompanies increases in intra-abdominal pressure, and function as sphincters at the anorectal junction, urethra, and vagina (see Exhibit 11.I). Muscles of the perineum assist in urination, erection of the penis and clitoris, ejaculation, and defecation (see Exhibit 11.J).
8. Muscles of the thorax that move the pectoral (shoulder) girdle stabilize the scapula so it can function as a stable point of origin for most of the muscles that move the humerus (see Exhibit 11.K).
9. Muscles of the thorax that move the humerus originate for the most part on the scapula (scapular muscles); the remaining muscles originate on the axial skeleton (axial muscles) (see Exhibit 11.L). Muscles of the arm that move the radius and ulna are involved in flexion and extension at the elbow joint and are organized into flexor and extensor compartments (see Exhibit 11.M).
10. Muscles of the forearm that move the wrist, hand, thumb, and digits (fingers) are many and varied; those muscles that act on the digits are called extrinsic muscles (see Exhibit 11.N). The muscles of the palm that move the digits (intrinsic muscles) are important in skilled activities and provide humans with the ability to grasp and manipulate objects precisely (see Exhibit 11.O).
11. Muscles of the neck and back that move the vertebral column are quite complex because they have multiple origins and insertions and because there is considerable overlap among them (see Exhibit 11.P).

Resource

Anatomy Overview - Muscles of Facial Expression
Anatomy Overview - Muscles Moving Eyeballs
Anatomy Overview - Muscles for Speech, Swallowing, and Chewing
Anatomy Overview - Muscles That Move the Head
Anatomy Overview - Muscles of the Torso
Anatomy Overview - Muscles for Breathing
Anatomy Overview - Muscles That Move the Pectoral Girdle
Anatomy Overview - Muscles That Move the Arm
Anatomy Overview - Muscles That Move the Forearm
Anatomy Overview - Muscles That Move the Hand
Anatomy Overview - Muscles That Move the Thigh
Anatomy Overview - Muscles of the Foot
Figure 11.3 - Principal Superficial Skeletal Muscles
Figure 11.11 - Muscles of the Thorax That Assist in Breathing
Figure 11.17 - Muscles of the Forearm That Move the Wrist, Hand, Thumb, and Digits
Exercise - Shoulder Movements Target Practice
Exercise - Lower Limb Movement Target Practice
Exercise - Manage Those Muscles

Review	Resource
12. Muscles of the gluteal region that move the femur originate for the most part on the pelvic girdle and insert on the femur; these muscles are larger and more powerful than comparable muscles in the upper limb (see Exhibit 11.Q). Muscles of the thigh that move the femur and the tibia and fibula are separated into medial (adductor), anterior (extensor), and posterior (flexor) compartments (see Exhibit 11.R).	
13. Muscles of the leg that move the foot and toes are divided into anterior, lateral, and posterior compartments (see Exhibit 11.S).	
14. Muscles of the foot that move the toes (intrinsic muscles), unlike those of the hand, are limited to the functions of support and locomotion (see Exhibit 11.T).	

SELF-QUIZ QUESTIONS

Fill in the blanks in the following statements.

1. The muscle that forms the major portion of the cheek is the ______.

2. The three superficial posterior plantar flexors of the leg are the ______, ______, and ______. All three of these muscles insert on the ______ by way of the Achilles tendon.

Indicate whether the following statements are true or false.

3. Longer fibers in a muscle result in a greater range of motion.

4. When flexing the forearm, the biceps brachii acts as the prime mover and the triceps brachii acts as the antagonist.

Choose the one best answer to the following questions.

5. Which of the following muscles does *not* flex the thigh?
(a) rectus femoris
(b) gracilis
(c) sartorius
(d) iliacus
(e) tensor fascia latae

6. The iliotibial tract is composed of the tendon of the gluteus maximus muscle, the deep fascia that encircles the thigh, and the tendon of which of the following muscles?
(a) iliacus
(b) gluteus minimus
(c) tensor fascia latae
(d) adductor longus
(e) vastus lateralis

7. In order for movement to occur, (1) muscles generally need to cross a joint, (2) contraction of the muscle will pull on the origin, (3) muscles that move a body part cannot cover the moving part, (4) muscles need to exert force on tendons that pull on bones, (5) the insertion must act to stabilize the joint.
(a) 1, 2, 3, 4, and 5 (b) 1, 2, 3, and 4
(c) 1, 2, and 4 (d) 1, 3, and 4
(e) 3 and 4

8. Because you did not do well on your recent anatomy and physiology exam, you leave the classroom pouting. Which one of these muscles are you using?
(a) mentalis
(b) orbicularis oris
(c) risorius
(d) levator labii superioris
(e) zygomaticus minor

9. The rectus femoris has fascicles arranged on both sides of a centrally positioned tendon. This pattern of fascicle arrangement is
(a) unipennate.
(b) fusiform.
(c) multipennate.
(d) parallel.
(e) bipennate.

10. Which of the following muscle names and their naming descriptors are mismatched?
(a) adductor brevis: short muscle that moves a bone closer to the midline
(b) rectus abdominis: muscle with fibers parallel to the midline of the abdomen
(c) levator scapula: muscle that raises the scapula
(d) sternohyoid: muscle attached to the sternum and hyoid
(e) serratus anterior: comblike muscle located on the body's anterior surface

11. Match the following:

____ (a) muscle that stabilizes the origin of the prime mover	(1) compartment
____ (b) site of muscle attachment to a stationary bone	(2) origin
____ (c) muscle that stretches to allow desired motion	(3) insertion
____ (d) muscle that contracts to stabilize intermediate joints	(4) belly
____ (e) site of muscle attachment to a movable bone	(5) synergist
____ (f) group of muscles, along with their blood and nerves, that have a common function	(6) fixator
____ (g) contracting muscle that produces the desired motion	(7) prime mover (agonist)
____ (h) fleshy part of the muscle	(8) antagonist

12. Match the following:

____ (a) compression of median nerve resulting in pain and numbness and tingling in the fingers
____ (b) tendinitis of the anterior compartment muscles of the leg; inflammation of the tibial periosteum
____ (c) improperly aligned eyeballs due to lesions in either the oculomotor or abducens nerves
____ (d) stretching or tearing of distal attachments of adductor muscles
____ (e) rupture of a portion of the inguinal area of the abdominal wall resulting in protrusion of part of the small intestine
____ (f) caused by repetitive movement of the arm over the head that results in inflammation of the supraspinatus tendon
____ (g) inflammation due to chronic irritation of the plantar aponeurosis at its origin on the calcaneus; most common cause of heel pain in runners
____ (h) painful inflammation of tendons, tendon sheaths, and synovial membranes of joints
____ (i) paralysis of facial muscles as a result of damage to the facial nerve
____ (j) common in individuals who perform quick starts and stops; tearing away of part of the tendinous origins from the ischial tuberosity
____ (k) permanent shortening of a muscle due to nerve damage and scar tissue development
____ (l) may occur as a result of injury to levator ani muscle
____ (m) external or internal pressure constricts structures in a compartment, causing a reduction of blood supply to the structures

(1) tenosynovitis
(2) Bell's palsy
(3) inguinal hernia
(4) urinary stress incontinence
(5) compartment syndrome
(6) groin pull
(7) pulled hamstrings
(8) strabismus
(9) shin splints
(10) plantar fasciitis
(11) impingement syndrome
(12) contracture
(13) carpal tunnel syndrome

13. Match the following:

____ (a) rectus femoris, vastus lateralis, vastus medialis, vastus intermedius
____ (b) biceps femoris, semitendinosus, semimembranosus
____ (c) erector spinae; includes iliocostalis, longissimus, and spinalis groups
____ (d) thenar, hypothenar, intermediate
____ (e) biceps brachii, brachialis, coracobrachialis
____ (f) latissimus dorsi
____ (g) subscapularis, supraspinatus, infraspinatus, teres minor
____ (h) diaphragm, external intercostals, internal intercostals
____ (i) trapezius, levator scapulae, rhomboid major, rhomboid minor

(1) breathing muscles
(2) constitute flexor compartment of the arm
(3) hamstrings
(4) intrinsic muscle groups of the hand
(5) muscles that strengthen and stabilize the shoulder joint; the rotator cuff
(6) quadriceps femoris muscle
(7) largest muscle mass of the back
(8) posterior thoracic muscles
(9) swimmer's muscle

14. Match the following (some answers may be used more than once):

____ (a) trapezius
____ (b) orbicularis oculi
____ (c) levator ani
____ (d) rectus abdominis
____ (e) triceps brachii
____ (f) gastrocnemius
____ (g) temporalis
____ (h) external anal sphincter
____ (i) external oblique
____ (j) iliocostalis thoracis
____ (k) digastric
____ (l) styloglossus
____ (m) masseter
____ (n) adductor longus
____ (o) zygomaticus major
____ (p) latissimus dorsi
____ (q) flexor carpi radialis
____ (r) pronator teres
____ (s) sternocleidomastoid
____ (t) quadriceps femoris
____ (u) deltoid
____ (v) tibialis anterior
____ (w) sartorius
____ (x) gluteus maximus
____ (y) superior rectus
____ (z) trapezius

(1) muscle of facial expression
(2) muscle of mastication
(3) muscle that moves the eyeballs
(4) extrinsic muscle that moves the tongue
(5) suprahyoid muscle
(6) muscle of the perineum
(7) muscle that moves the head
(8) abdominal wall muscle
(9) pelvic floor muscle
(10) pectoral girdle muscle
(11) muscle that moves the humerus
(12) muscle that moves the radius and ulna
(13) muscle that moves the wrist, hand, and digits
(14) muscle that moves the vertebral column
(15) muscle that moves the femur
(16) muscle that acts on the femur, tibia, and fibula
(17) muscle that moves the foot and toes

15. Match the following (some answers may be used more than once):

____ (a) most common lever in the body
____ (b) lever formed by the head resting on the vertebral column
____ (c) always produces a mechanical advantage
____ (d) E*F*L
____ (e) F*L*E
____ (f) F*E*L
____ (g) adduction of the thigh

(1) first-class lever
(2) second-class lever
(3) third-class lever

CRITICAL THINKING QUESTIONS

1. During a facelift, the cosmetic surgeon accidentally severs the facial nerve on the right side of the face. What are some of the effects this would have on the patient, and what muscles are involved?

2. While taking the bus to the supermarket, eleven-year-old Desmond informs his mother that he has to "go to the bathroom" (urinate). His mother tells him he must "hold it" until they arrive at the store. What muscles must remain contracted in order for him to prevent urination?

3. Minor-league pitcher José has been throwing a hundred pitches a day in order to perfect his curve ball. Lately he has experienced pain in his pitching arm. The doctor diagnosed a torn rotator cuff. José was confused because he thought cuffs were only found on shirt sleeves, not inside his shoulder. Explain to José what the doctor means and how this injury could affect his arm movement.

? ANSWERS TO FIGURE QUESTIONS

11.1 The belly of the muscle that extends the forearm, the triceps brachii, is located posterior to the humerus.

11.2 Second-class levers produce the most force.

11.3 For muscles named after their various characteristics, here are possible correct responses (for others, see Table 11.2): direction of fibers: external oblique; shape: deltoid; action: extensor digitorum; size: gluteus maximus; origin and insertion: sternocleidomastoid; location: tibialis anterior; number of tendons of origin: biceps brachii.

11.4 The corrugator supercilii muscle is involved in frowning; the zygomaticus major muscle contracts when you smile; the mentalis muscle contributes to pouting; the orbicularis oculi muscle contributes to squinting.

11.5 The inferior oblique muscle moves the eyeball superiorly and laterally because it originates at the anteromedial aspect of the floor of the orbit and inserts on the posterolateral aspect of the eyeball.

11.6 The masseter is the strongest muscle of mastication.

11.7 Functions of the tongue include chewing, detection of taste, swallowing, and speech.

11.8 The suprahyoid and infrahyoid muscles stabilize the hyoid bone to assist in tongue movements.

11.9 The triangles in the neck formed by the sternocleidomastoid muscles are important anatomically and surgically because of the structures that lie within their boundaries.

11.10 The rectus abdominis muscle aids in urination.

11.11 The diaphragm is innervated by the phrenic nerve.

11.12 The borders of the pelvic diaphragm are the pubic symphysis anteriorly, the coccyx posteriorly, and the walls of the pelvis laterally.

11.13 The borders of the perineum are the pubic symphysis anteriorly, the coccyx posteriorly, and the ischial tuberosities laterally.

11.14 The main action of the muscles that move the pectoral girdle is to stabilize the scapula to assist in movements of the humerus.

11.15 The rotator cuff consists of the flat tendons of the subscapularis, supraspinatus, infraspinatus, and teres minor muscles that form a nearly complete circle around the shoulder joint.

11.16 The brachialis is the most powerful forearm flexor; the triceps brachii is the most powerful forearm extensor.

11.17 Flexor tendons of the digits and wrist and the median nerve pass deep to the flexor retinaculum.

11.18 Muscles of the thenar eminence act on the thumb (pollex).

11.19 The splenius muscles arise from the midline and extend laterally and superiorly to their insertion.

11.20 Upper limb muscles exhibit diversity of movement; lower limb muscles function in stability, locomotion, and maintenance of posture. In addition, lower limb muscles usually cross two joints and act equally on both.

11.21 The quadriceps femoris consists of the rectus femoris, vastus lateralis, vastus medialis, and vastus intermedius; the hamstrings consist of the biceps femoris, semitendinosus, and semimembranosus.

11.22 The superior and inferior extensor retinacula firmly hold the tendons of the anterior compartment muscles to the ankle.

11.23 The plantar aponeurosis (fascia) supports the longitudinal arch and encloses the flexor tendons of the foot.

12 NERVOUS TISSUE

NERVOUS TISSUE AND HOMEOSTASIS *The excitable characteristic of nervous tissue allows for the generation of nerve impulses (action potentials) that provide communication with and regulation of most body tissues.*

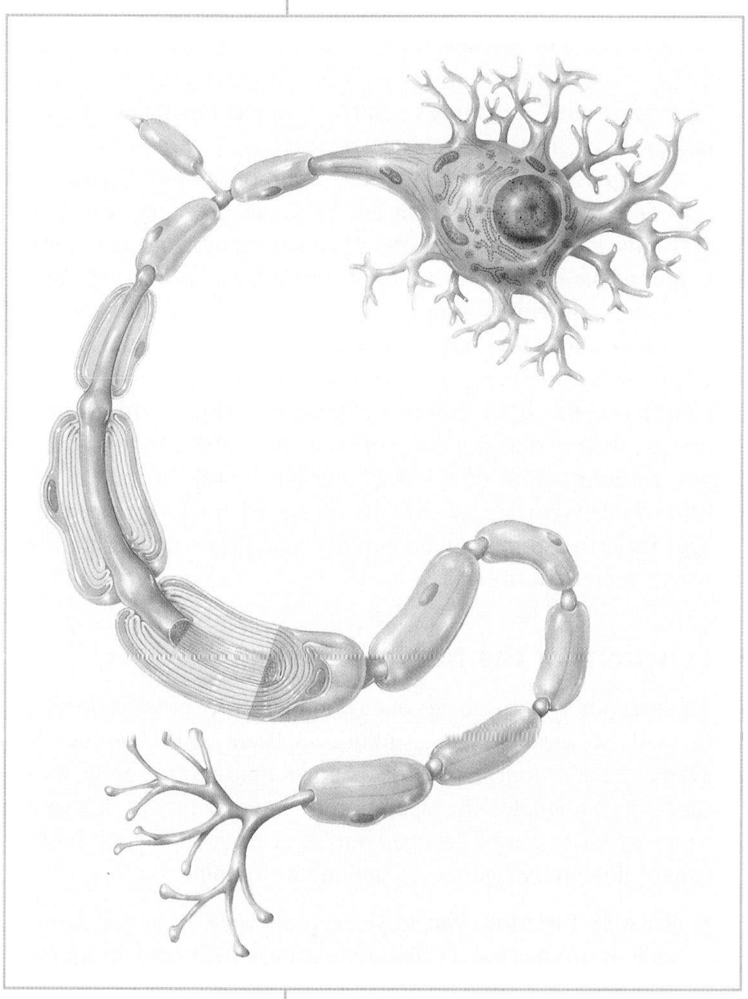

Both the nervous and endocrine systems have the same objective: to keep controlled conditions within limits that maintain life. The nervous system regulates body activities by responding rapidly using nerve impulses; the endocrine system responds by releasing hormones. Chapter 18 compares the roles of both systems in maintaining homeostasis.

The nervous system is also responsible for our perceptions, behaviors, and memories, and it initiates all voluntary movements. Because this system is quite complex, we discuss its structure and function in several chapters. This chapter focuses on the organization of the nervous system and the properties of neurons (nerve cells) and neuroglia (cells that support the activities of neurons). We then examine the structure and functions of the spinal cord and spinal nerves (Chapter 13), and of the brain and cranial nerves (Chapter 14). The autonomic nervous system, which operates without voluntary control, will be covered in Chapter 15. Chapter 16 will discuss the somatic senses—touch, pressure, warmth, cold, pain, and others—and their sensory and motor pathways to show how nerve impulses pass into the spinal cord and brain or from the spinal cord and brain to muscles and glands. Exploration of the nervous system concludes with a discussion of the special senses: smell, taste, vision, hearing, and equilibrium (Chapter 17).

Neurology deals with normal functioning and disorders of the nervous system. A **neurologist** (noo-ROL-ō-jist) is a physician who diagnoses and treats disorders of the nervous system.

? ***Did you ever wonder how the human nervous system coordinates and integrates all body systems so rapidly and efficiently?***

12.1 OVERVIEW OF THE NERVOUS SYSTEM

OBJECTIVES

- Describe the organization of the nervous system.
- Describe the three basic functions of the nervous system.

Organization of the Nervous System

With a mass of only 2 kg (4.5 lb), about 3% of the total body weight, the **nervous system** is one of the smallest and yet the most complex of the 11 body systems. This intricate network of billions of neurons and even more neuroglia is organized into two main subdivisions: the central nervous system and the peripheral nervous system.

Central Nervous System

The **central nervous system (CNS)** consists of the brain and spinal cord (Figure 12.1a). The brain is the part of the CNS that is located in the skull and contains about 100 billion (10^{11}) neurons. The spinal cord is connected to the brain through the foramen magnum of the occipital bone and is encircled by the bones of the vertebral column. The spinal cord contains about 100 million neurons. The CNS processes many different kinds of incoming sensory information. It is also the source of thoughts, emotions, and memories. Most signals that stimulate muscles to contract and glands to secrete originate in the CNS.

Peripheral Nervous System

The **peripheral nervous system (PNS)** (pe-RIF-e-ral) consists of all nervous tissue outside the CNS (Figure 12.1a). Components of the PNS include nerves, ganglia, enteric plexuses, and sensory receptors. A **nerve** is a bundle of hundreds to thousands of axons plus associated connective tissue and blood vessels that lies outside the brain and spinal cord. Twelve pairs of **cranial nerves** emerge from the brain and thirty-one pairs of **spinal nerves** emerge from the spinal cord. Each nerve follows a defined path and serves a specific region of the body. **Ganglia** (GANG-glē-a = swelling or knot; singular is *ganglion*) are small masses of nervous tissue, consisting primarily of neuron cell bodies, that are located outside of the brain and spinal cord. Ganglia are closely associated with cranial and spinal nerves. **Enteric plexuses** (PLEK-sus-ez) are extensive networks of neurons located in the walls of organs of the gastrointestinal tract. The neurons of these plexuses help regulate the digestive system (see Section 24.3). The term **sensory receptor** refers to a structure of the nervous system that monitors changes in the external or internal environment. Examples of sensory receptors include touch receptors in the skin, photoreceptors in the eye, and olfactory receptors in the nose.

The PNS is divided into a **somatic nervous system (SNS)** (sō-MAT-ik; *soma* = body), an **autonomic nervous system (ANS)** (aw′-tō-NOM-ik; *auto-* = self; *-nomic* = law), and an **enteric nervous system (ENS)** (en-TER-ik; *enteron* = intestines). The SNS consists of (1) sensory neurons that convey information from somatic receptors in the head, body wall, and limbs and from receptors for the special senses of vision, hearing, taste, and smell to the CNS and (2) motor neurons that conduct impulses from the CNS to *skeletal muscles* only. Because these motor responses can be consciously controlled, the action of this part of the PNS is *voluntary*.

The ANS consists of (1) sensory neurons that convey information from autonomic sensory receptors, located primarily in visceral organs such as the stomach and lungs, to the CNS and (2) motor neurons that conduct nerve impulses from the CNS to *smooth muscle, cardiac muscle,* and *glands*. Because its motor responses are not normally under conscious control, the action of the ANS is involuntary. The motor part of the ANS consists of two branches, the **sympathetic division** and the **parasympathetic division.** With a few exceptions, effectors receive nerves from both divisions, and usually the two divisions have opposing actions. For example, sympathetic neurons increase heart rate, and parasympathetic neurons slow it down. In general, the sympathetic division helps support exercise of emergency actions, the "fight-or-flight" responses, and the parasympathetic division takes care of "rest-and-digest" activities.

The operation of the ENS, the "brain of the gut," is involuntary. Once considered part of the ANS, the ENS consists of over 100 million neurons in enteric plexuses that extend most of the length of the gastrointestinal (GI) tract. Many of the neurons of the enteric plexuses function independently of the ANS and CNS to some extent, although they also communicate with the CNS via sympathetic and parasympathetic neurons. Sensory neurons of the ENS monitor chemical changes within the GI tract as well as the stretching of its walls. Enteric motor neurons govern contraction of GI tract smooth muscle to propel food through the GI tract, secretions of the GI tract organs such as acid from the stomach, and activity of GI tract endocrine cells, which secrete hormones.

Functions of the Nervous System

The nervous system carries out a complex array of tasks. It allows us to sense various smells, produce speech, and remember past events; in addition, it provides signals that control body movements and regulates the operation of internal organs. These diverse activities can be grouped into three basic functions: sensory (input), integrative (process), and motor (output).

- **Sensory function.** Sensory receptors *detect* internal stimuli, such as an increase in blood pressure, or external stimuli (for example, a raindrop landing on your arm). This sensory information is then carried into the brain and spinal cord through cranial and spinal nerves.
- **Integrative function.** The nervous system *processes* sensory information by analyzing it and making decisions for appropriate responses—an activity known as **integration.**
- **Motor function.** Once sensory information is integrated, the nervous system *may elicit an appropriate motor response* by activating **effectors** (muscles and glands) through cranial and spinal nerves. Stimulation of the effectors causes muscles to contract and glands to secrete.

Figure 12.1 Organization of the nervous system. (a) Subdivisions of the nervous system. (b) Nervous system organizational chart; blue boxes represent sensory components of the peripheral nervous system, red boxes represent motor components of the PNS, and green boxes represent effectors (muscles and glands).

The two main subdivisions of the nervous system are (1) the central nervous system (CNS), which consists of the brain and spinal cord, and (2) the peripheral nervous system (PNS), which consists of all nervous tissue outside the CNS.

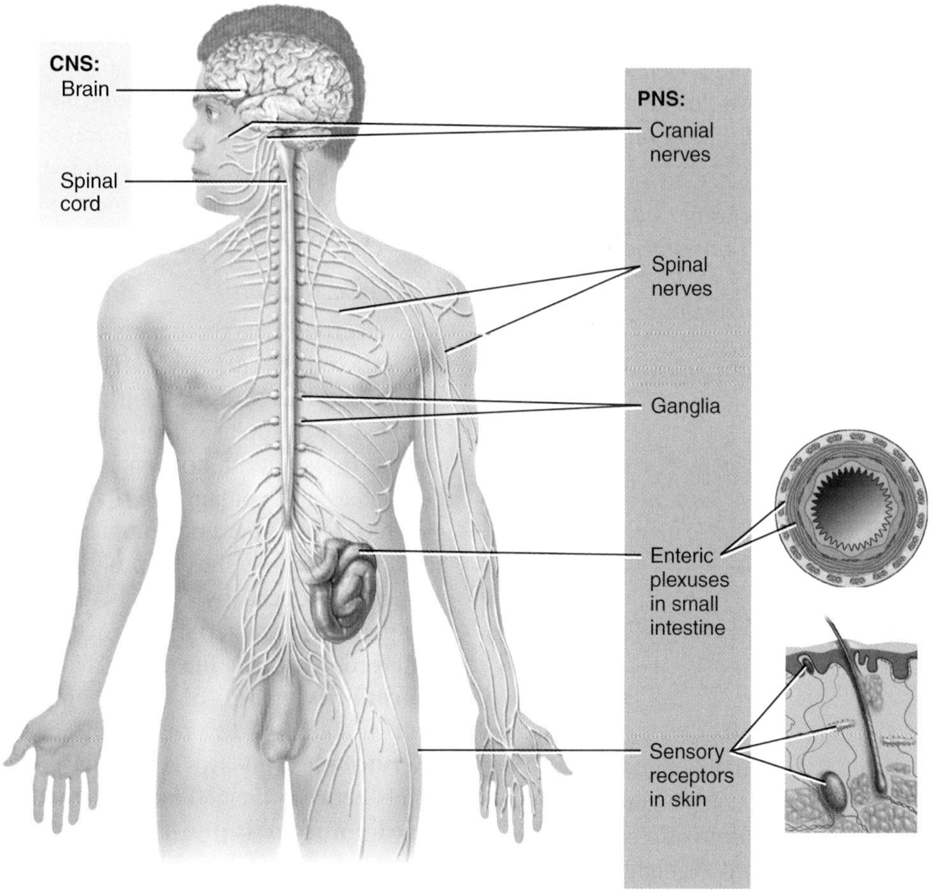

(a)

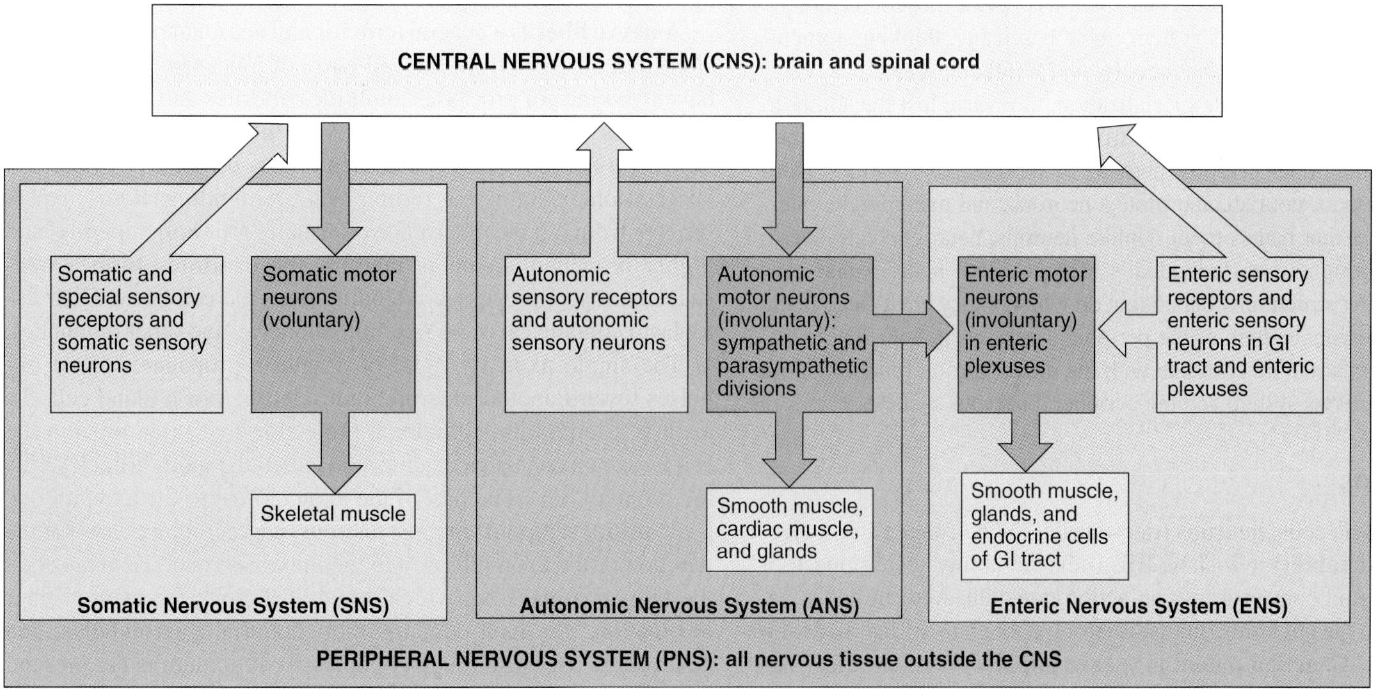

(b)

What are some of the functions of the CNS?

The three basic functions of the nervous system occur, for example, when you answer your cell phone after hearing it ring. The sound of the ringing cell phone stimulates sensory receptors in your ears (sensory function). This auditory information is subsequently relayed into your brain where it is processed and the decision to answer the phone is made (integrative function). The brain then stimulates the contraction of specific muscles that will allow you to grab the phone and press the appropriate button to answer it (motor function).

✔ **CHECKPOINT**

1. What is the purpose of a sensory receptor?
2. What are the components and functions of the SNS, ANS, and ENS?
3. Which subdivisions of the PNS control voluntary actions? Involuntary actions?
4. Explain the concept of integration and provide an example.

12.2 HISTOLOGY OF NERVOUS TISSUE

OBJECTIVES

- Contrast the histological characteristics and the functions of neurons and neuroglia.
- Distinguish between gray matter and white matter.

Nervous tissue comprises two types of cells—*neurons* and *neuroglia.* These cells combine in a variety of ways in different regions of the nervous system. In addition to forming the complex processing networks within the brain and spinal cord, neurons also connect all regions of the body to the brain and spinal cord. As highly specialized cells capable of reaching great lengths and making extremely intricate connections with other cells, neurons provide most of the unique functions of the nervous system, such as sensing, thinking, remembering, controlling muscle activity, and regulating glandular secretions. As a result of their specialization, they have lost the ability to undergo mitotic divisions. Neuroglia are smaller cells but they greatly outnumber neurons, perhaps by as much as 25 times. Neuroglia support, nourish, and protect neurons, and maintain the interstitial fluid that bathes them. Unlike neurons, neuroglia continue to divide throughout an individual's lifetime. Both neurons and neuroglia differ structurally depending on whether they are located in the central nervous system or the peripheral nervous system. These differences in structure correlate with the differences in function of the central nervous system and the peripheral nervous system.

Neurons

Like muscle cells, **neurons (nerve cells)** (NOO-rons) possess **electrical excitability** (ek-sīt′-a-BIL-i-tē), the ability to respond to a stimulus and convert it into an action potential. A **stimulus** is any change in the environment that is strong enough to initiate an action potential. An **action potential (nerve impulse)** is an electrical signal that propagates (travels) along the surface of the membrane of a neuron. It begins and travels due to the movement of ions (such as sodium and potassium) between interstitial fluid and the inside of a neuron through specific ion channels in its plasma membrane. Once begun, a nerve impulse travels rapidly and at a constant strength.

Some neurons are tiny and propagate impulses over a short distance (less than 1 mm) within the CNS. Others are the longest cells in the body. The neurons that enable you to wiggle your toes, for example, extend from the lumbar region of your spinal cord (just above waist level) to the muscles in your foot. Some neurons are even longer. Those that allow you to feel a feather tickling your toes stretch all the way from your foot to the lower portion of your brain. Nerve impulses travel these great distances at speeds ranging from 0.5 to 130 meters per second (1 to 290 mi/hr).

Parts of a Neuron

Most neurons have three parts: (1) a cell body, (2) dendrites, and (3) an axon (Figure 12.2). The **cell body,** also known as the *perikaryon* (per′-i-KAR-ē-on) or *soma,* contains a nucleus surrounded by cytoplasm that includes typical cellular organelles such as lysosomes, mitochondria, and a Golgi complex. Neuronal cell bodies also contain free ribosomes and prominent clusters of rough endoplasmic reticulum, termed **Nissl bodies** (NIS-el). The ribosomes are the sites of protein synthesis. Newly synthesized proteins produced by Nissl bodies are used to replace cellular components, as material for growth of neurons, and to regenerate damaged axons in the PNS. The cytoskeleton includes both **neurofibrils** (noo-rō-FĪ-brils), composed of bundles of intermediate filaments that provide the cell shape and support, and **microtubules** (mī-krō-TOO-būls′), which assist in moving materials between the cell body and axon. Aging neurons also contain **lipofuscin** (līp′-o-FYŪS-īn), a pigment that occurs as clumps of yellowish brown granules in the cytoplasm. Lipofuscin is a product of neuronal lysosomes that accumulates as the neuron ages, but does not seem to harm the neuron as it ages.

A **nerve fiber** is a general term for any neuronal process (extension) that emerges from the cell body of a neuron. Most neurons have two kinds of processes: multiple dendrites and a single axon. **Dendrites** (DEN-drīts = little trees) are the receiving or input portions of a neuron. The plasma membranes of dendrites (and cell bodies) contain numerous receptor sites for binding chemical messengers from other cells. Dendrites usually are short, tapering, and highly branched. In many neurons the dendrites form a tree-shaped array of processes extending from the cell body. Their cytoplasm contains Nissl bodies, mitochondria, and other organelles.

The single **axon** (= axis) of a neuron propagates nerve impulses toward another neuron, a muscle fiber, or a gland cell. An axon is a long, thin, cylindrical projection that often joins to the cell body at a cone-shaped elevation called the **axon hillock** (HIL-lok = small hill). The part of the axon closest to the axon hillock is the **initial segment.** In most neurons, nerve impulses arise at the junction of the axon hillock and the initial segment, an area called the **trigger zone,** from which they travel along the axon to their destination. An axon contains mitochondria, microtubules, and neurofibrils. Because rough endoplasmic reticulum is not present, protein synthesis does not occur in the axon. The cytoplasm of an axon, called **axoplasm,** is surrounded by a plasma membrane known as the **axolemma** (*lemma* = sheath or husk). Along the

Figure 12.2 **Structure of a multipolar neuron (a neuron with a large cell body, several short dendrites, and a single long axon).** Arrows indicate the direction of information flow: dendrites → cell body → axon → axon terminals.

The basic parts of a neuron are dendrites, a cell body, and an axon.

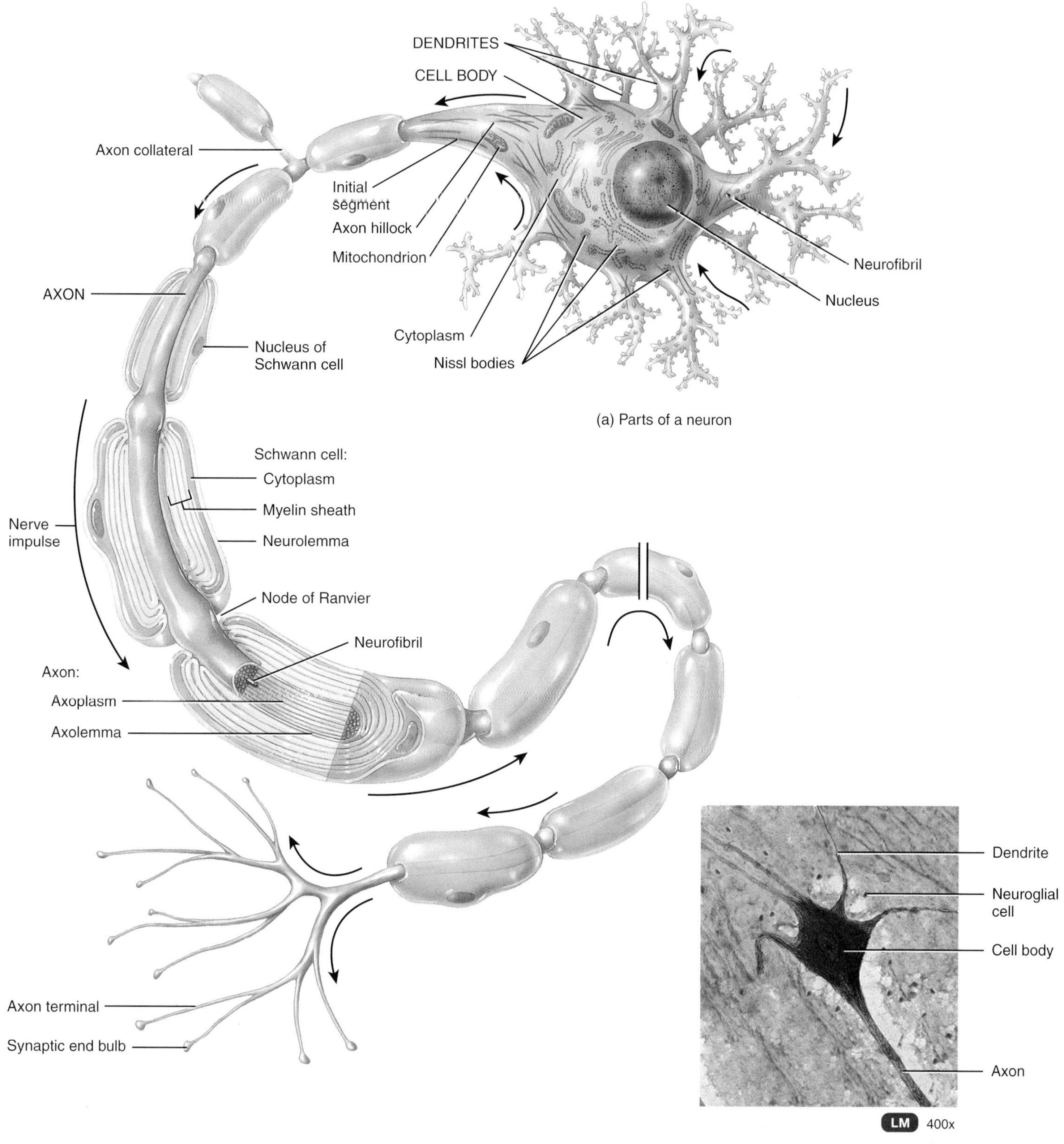

(a) Parts of a neuron

(b) Motor neuron

What roles do the dendrites, cell body, and axon play in communication of signals?

length of an axon, side branches called **axon collaterals** may branch off, typically at a right angle to the axon. The axon and its collaterals end by dividing into many fine processes called **axon terminals (telodendria)** (tēl′-ō-DEN-drē-a).

The site of communication between two neurons or between a neuron and an effector cell is called a **synapse** (SIN-aps). The tips of some axon terminals swell into bulb-shaped structures called **synaptic end bulbs;** others exhibit a string of swollen bumps called **varicosities** (var′-i-KOS-i-tēz). Both synaptic end bulbs and varicosities contain many tiny membrane-enclosed sacs called **synaptic vesicles** that store a chemical called a **neurotransmitter** (noo′-rō-trans′-MIT-ter). A neurotransmitter is a molecule released from a synaptic vesicle that excites or inhibits another neuron, muscle fiber, or gland cell. Many neurons contain two or even three types of neurotransmitters, each with different effects on the postsynaptic cell.

Because some substances synthesized or recycled in the neuron cell body are needed in the axon or at the axon terminals, two types of transport systems carry materials from the cell body to the axon terminals and back. The slower system, which moves materials about 1–5 mm per day, is called **slow axonal transport.** It conveys axoplasm in one direction only—from the cell body toward the axon terminals. Slow axonal transport supplies new axoplasm to developing or regenerating axons and replenishes axoplasm in growing and mature axons.

Fast axonal transport, which is capable of moving materials a distance of 200–400 mm per day, uses proteins that function as "motors" to move materials along the surfaces of microtubules of the neuron's cytoskeleton. Fast axonal transport moves materials in both directions—away from and toward the cell body. Fast axonal transport that occurs in an **anterograde** (forward) direction moves organelles and synaptic vesicles from the cell body to the axon terminals. Fast axonal transport that occurs in a **retrograde** (backward) direction moves membrane vesicles and other cellular materials from the axon terminals to the cell body to be degraded or recycled. Substances that enter the neuron at the axon terminals are also moved to the cell body by fast retrograde transport. These substances include trophic chemicals such as nerve growth factor and harmful agents such as tetanus toxin and the viruses that cause rabies, herpes simplex, and polio.

Structural Diversity in Neurons

Neurons display great diversity in size and shape. For example, their cell bodies range in diameter from 5 micrometers (μm) (slightly smaller than a red blood cell) up to 135 μm (barely large enough to see with the unaided eye). The pattern of dendritic branching is varied and distinctive for neurons in different parts of the nervous system. A few small neurons lack an axon, and many others have very short axons. As we have already discussed, the longest axons are almost as long as a person is tall, extending from the toes to the lowest part of the brain.

Classification of Neurons

Both structural and functional features are used to classify the various neurons in the body.

STRUCTURAL CLASSIFICATION Structurally, neurons are classified according to the number of processes extending from the cell body (Figure 12.3):

Figure 12.3 Structural classification of neurons. Breaks indicate that axons are longer than shown.

A multipolar neuron has many processes extending from the cell body, a bipolar neuron has two, and a unipolar neuron has one.

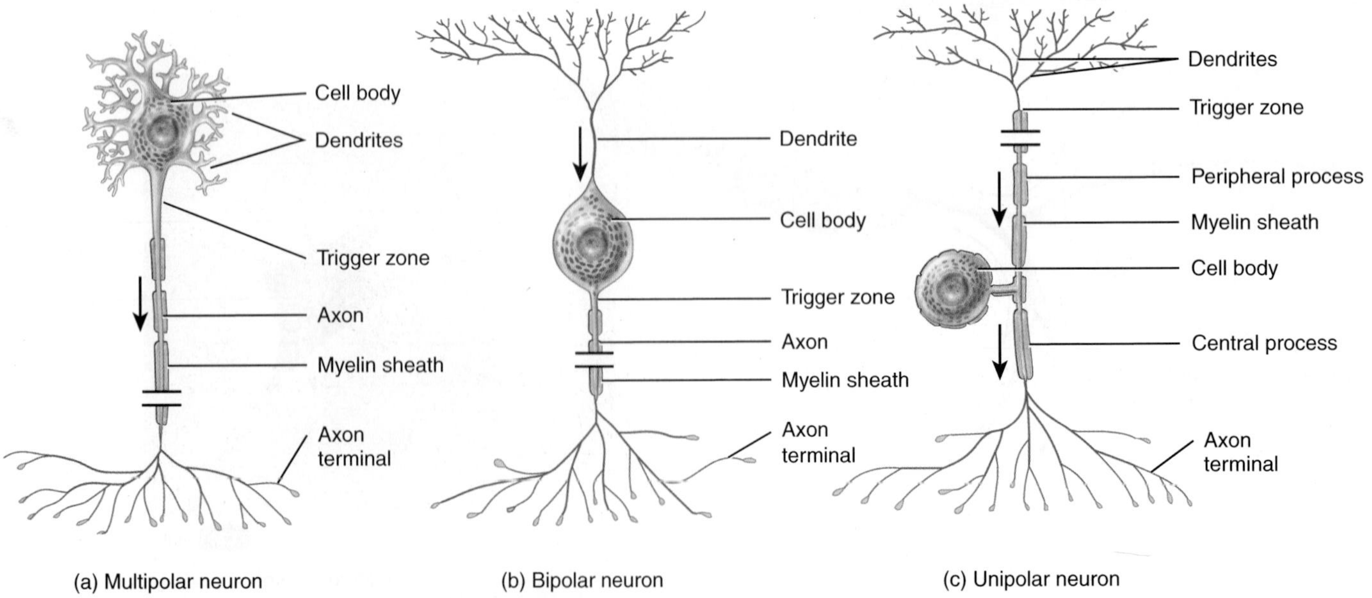

Which type of neuron shown in this figure is the most abundant type of neuron in the CNS?

1. **Multipolar neurons** usually have several dendrites and one axon (Figure 12.3a). Most neurons in the brain and spinal cord are of this type, as well as all motor neurons (described shortly).
2. **Bipolar neurons** have one main dendrite and one axon (Figure 12.3b). They are found in the retina of the eye, the inner ear, and the olfactory (*olfact* = to smell) area of the brain.
3. **Unipolar neurons** have dendrites and one axon that are fused together to form a continuous process that emerges from the cell body (Figure 12.3c). These neurons are more appropriately called **pseudounipolar neurons** (soo′-dō-ū′-ni-PŌ-lar) because they begin in the embryo as bipolar neurons. During development, the dendrites and axon fuse together and become a single process. The dendrites of most unipolar neurons function as **sensory receptors** that detect a sensory stimulus such as touch, pressure, pain, or thermal stimuli (see Figure 12.10). The trigger zone for nerve impulses in a unipolar neuron is at the junction of the dendrites and axon (Figure 12.3c). The impulses then propagate toward the synaptic end bulbs. The cell bodies of most unipolar neurons are located in the ganglia of spinal and cranial nerves. Several examples of sensory receptors that are dendrites of unipolar neurons are illustrated in Figure 12.4.

In addition to the structural classification scheme just described, some neurons are named for the histologist who first described them or for an aspect of their shape or appearance; examples include **Purkinje cells** (pur-KIN-jē) in the cerebellum and **pyramidal cells** (pi-RAM-i-dal), found in the cerebral cortex of the brain, which have pyramid-shaped cell bodies (Figure 12.5).

Figure 12.4 **Examples of sensory receptors that are dendrites of unipolar neurons.** Breaks indicate that axons are longer than shown. (a) A Meissner corpuscle is a touch receptor that consists of a mass of dendrites enclosed by a capsule of connective tissue. (b) A Merkel disc is a touch receptor that consists of free nerve endings (bare dendrites) that make contact with Merkel cells of the stratum basale of the skin. (c) A pacinian corpuscle is a pressure receptor composed of a multilayered connective tissue capsule that encloses a dendrite. (d) A nociceptor is a pain receptor that consists of free nerve endings (bare dendrites). Thermoreceptors (which detect thermal sensations), itch receptors, and tickle receptors resemble nociceptors in that they are unipolar neurons with free nerve endings that serve as sensory receptors.

The type of sensory receptor found in the dendrites of a unipolar neuron determines the type of sensory stimulus that the unipolar neuron can detect.

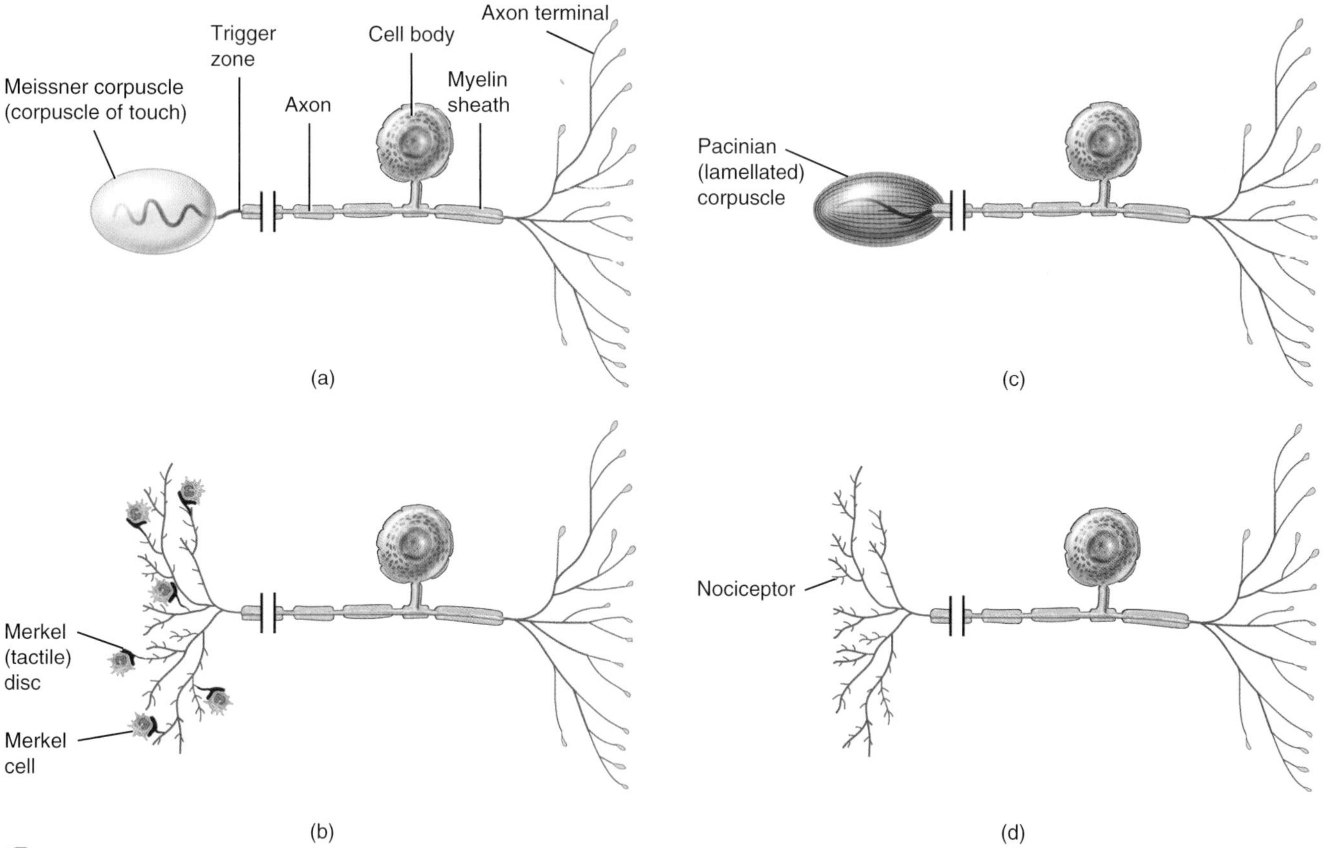

Where are most unipolar neurons located in the body?

Figure 12.5 Two examples of CNS neurons. Arrows indicate the direction of information flow.

The dendritic branching pattern often is distinctive for a particular type of neuron.

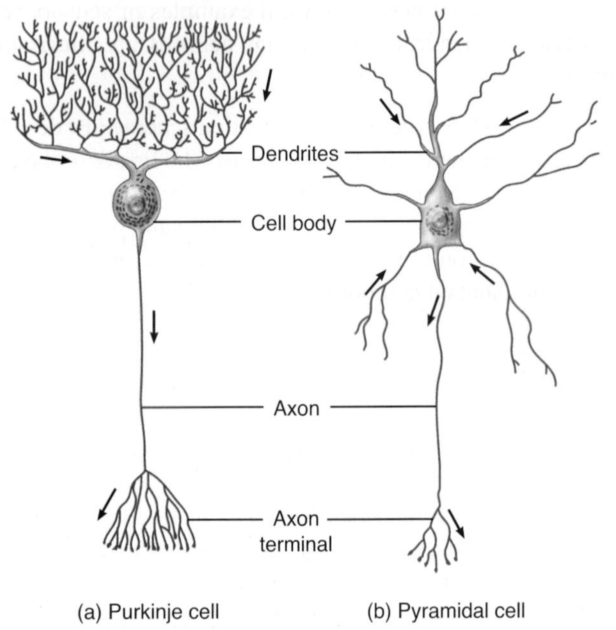
Dendrites
Cell body
Axon
Axon terminal
(a) Purkinje cell
(b) Pyramidal cell

How did the pyramidal cells get their name?

Functional Classification Functionally, neurons are classified according to the direction in which the nerve impulse (action potential) is conveyed with respect to the CNS.

1. **Sensory** or **afferent neurons** (AF-er-ent NOO-ronz; *af-* = toward; *-ferrent* = carried) either contain sensory receptors at their distal ends (dendrites) (see Figure 12.10) or are located just after sensory receptors that are separate cells. Once an appropriate stimulus activates a sensory receptor, the sensory neuron forms an action potential in its axon and the action potential is conveyed *into* the CNS through cranial or spinal nerves. Most sensory neurons are unipolar in structure.
2. **Motor** or **efferent neurons** (EF-e-rent; *ef-* = away from) convey action potentials *away* from the CNS to **effectors** (muscles and glands) in the periphery (PNS) through cranial or spinal nerves (see Figure 12.10). Motor neurons are multipolar in structure.
3. **Interneurons** or **association neurons** are mainly located within the CNS between sensory and motor neurons (see Figure 12.10). Interneurons integrate (process) incoming sensory information from sensory neurons and then elicit a motor response by activating the appropriate motor neurons. Most interneurons are multipolar in structure.

Neuroglia

Neuroglia (noo-RŌG-lē-a; *-glia* = glue) or **glia** (GLĒ-a) make up about half the volume of the CNS. Their name derives from the idea of early histologists that they were the "glue" that held nervous tissue together. We now know that neuroglia are not merely passive bystanders but rather actively participate in the activities of nervous tissue. Generally, neuroglia are smaller than neurons, and they are 5 to 25 times more numerous. In contrast to neurons, glia do not generate or propagate action potentials, and they can multiply and divide in the mature nervous system. In cases of injury or disease, neuroglia multiply to fill in the spaces formerly occupied by neurons. Brain tumors derived from glia, called **gliomas** (glē-Ō-mas), tend to be highly malignant and to grow rapidly. Of the six types of neuroglia, four—astrocytes, oligodendrocytes, microglia, and ependymal cells—are found only in the CNS. The remaining two types—Schwann cells and satellite cells—are present in the PNS.

Neuroglia of the CNS

Neuroglia of the CNS can be classified on the basis of size, cytoplasmic processes, and intracellular organization into four types: astrocytes, oligodendrocytes, microglia, and ependymal cells (Figure 12.6).

Astrocytes (AS-trō-sīts; *astro-* = star; *-cyte* = cell) These star-shaped cells have many processes and are the largest and most numerous of the neuroglia. There are two types of **astrocytes.** *Protoplasmic astrocytes* have many short branching processes and are found in gray matter (described shortly). *Fibrous astrocytes* have many long unbranched processes and are located mainly in white matter (also described shortly). The processes of astrocytes make contact with blood capillaries, neurons, and the pia mater (a thin membrane around the brain and spinal cord).

The functions of astrocytes include the following: (1) Astrocytes contain microfilaments that give them considerable strength, which enables them to support neurons. (2) Processes of astrocytes wrapped around blood capillaries isolate neurons of the CNS from various potentially harmful substances in blood by secreting chemicals that maintain the unique selective permeability characteristics of the endothelial cells of the capillaries. In effect, the endothelial cells create a *blood–brain barrier,* which restricts the movement of substances between the blood and interstitial fluid of the CNS. Details of the blood–brain barrier are discussed in Chapter 14. (3) In the embryo, astrocytes secrete chemicals that appear to regulate the growth, migration, and interconnection among neurons in the brain. (4) Astrocytes help to maintain the appropriate chemical environment for the generation of nerve impulses. For example, they regulate the concentration of important ions such as K^+; take up excess neurotransmitters; and serve as a conduit for the passage of nutrients and other substances between blood capillaries and neurons. (5) Astrocytes may also play a role in learning and memory by influencing the formation of neural synapses (see Section 16.5).

Oligodendrocytes (OL-i-gō-den′-drō-sīts; *oligo-* = few; *-dendro-* = tree) These resemble astrocytes but are smaller and contain fewer processes. **Oligodendrocyte** processes are responsible for forming and maintaining the myelin sheath around CNS axons. As you will see shortly, the **myelin sheath** is a multilayered lipid and protein covering around some axons that insulates them and increases the speed of nerve impulse conduction. Such axons are said to be **myelinated** (mī-e-li-NĀ-ted).

MICROGLIA (mī-KROG-lē-a; *micro-* = small) These neuroglia are small cells with slender processes that give off numerous spinelike projections. **Microglia** function as phagocytes. Like tissue macrophages, they remove cellular debris formed during normal development of the nervous system and phagocytize microbes and damaged nervous tissue.

EPENDYMAL CELLS (ep-EN-de-mal; *epen-* = above; *-dym-* = garment) **Ependymal cells** are cuboidal to columnar cells arranged in a single layer that possess microvilli and cilia. These cells line the ventricles of the brain and central canal of the spinal cord (spaces filled with cerebrospinal fluid, which protects and nourishes the brain and spinal cord). Functionally, ependymal cells produce, possibly monitor, and assist in the circulation of cerebrospinal fluid. They also form the blood–cerebrospinal fluid barrier, which is discussed in Chapter 14.

Figure 12.6 Neuroglia of the central nervous system.

Neuroglia of the CNS are distinguished on the basis of size, cytoplasmic processes, and intracellular organization.

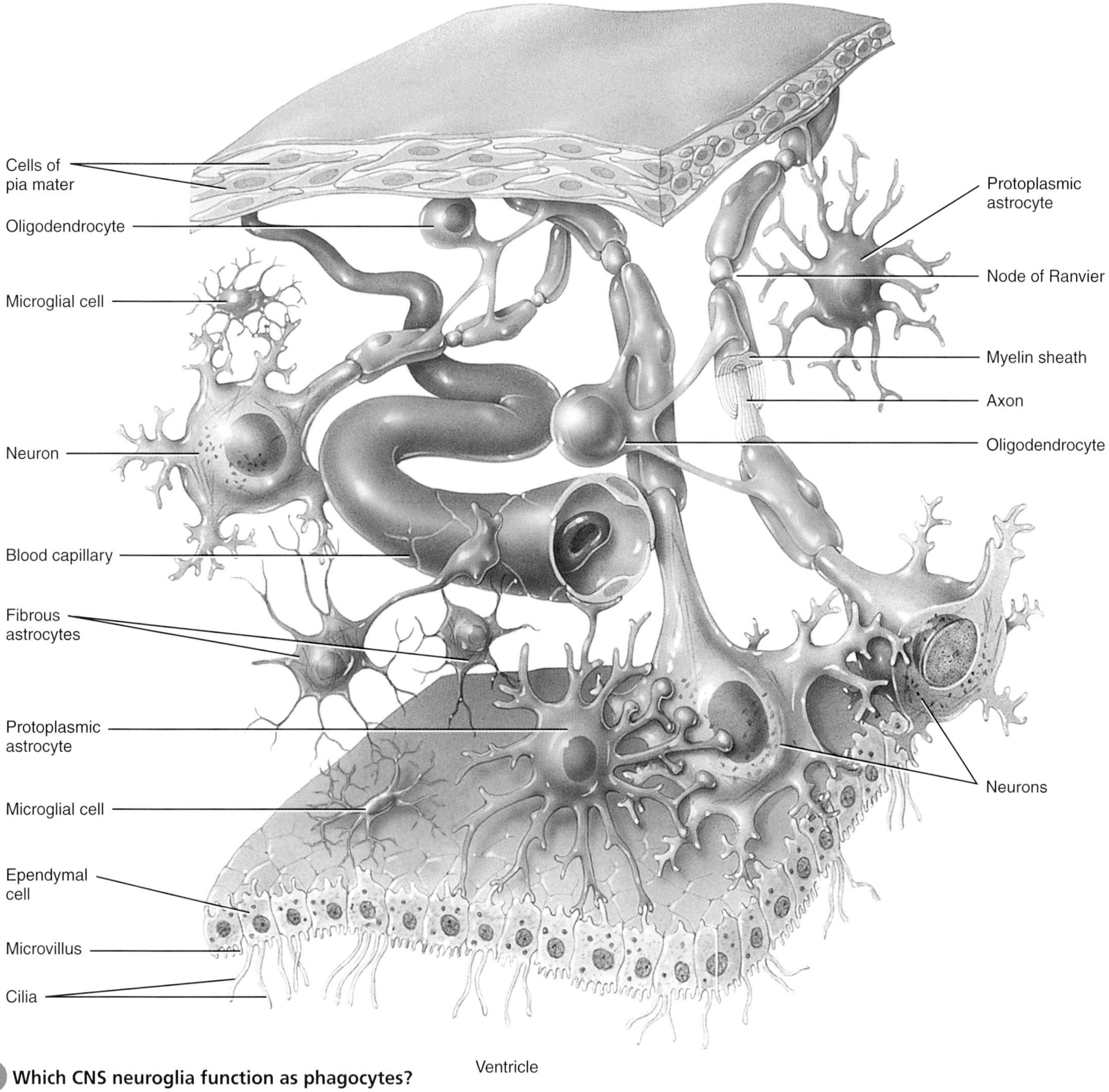

Which CNS neuroglia function as phagocytes?

Neuroglia of the PNS

Neuroglia of the PNS completely surround axons and cell bodies. The two types of glial cells in the PNS are Schwann cells and satellite cells (Figure 12.7).

SCHWANN CELLS (SCHVON or SCHWON) These cells encircle PNS axons. Like oligodendrocytes, they form the myelin sheath around axons. However, a single oligodendrocyte myelinates several axons, but each **Schwann cell** myelinates a single axon (Figure 12.7a; see also Figure 12.8a, c). A single Schwann cell can also enclose as many as 20 or more unmyelinated axons (axons that lack a myelin sheath) (Figure 12.7b). Schwann cells participate in axon regeneration, which is more easily accomplished in the PNS than in the CNS.

SATELLITE CELLS (SAT-i-līt) These flat cells surround the cell bodies of neurons of PNS ganglia (Figure 12.7c). Besides providing structural support, **satellite cells** regulate the exchanges of materials between neuronal cell bodies and interstitial fluid.

Myelination

As you have already learned, axons surrounded by a multilayered lipid and protein covering, called the myelin sheath, are said to be **myelinated** (Figure 12.8a). The sheath electrically insulates the axon of a neuron and increases the speed of nerve impulse conduction. Axons without such a covering are said to be **unmyelinated** (Figure 12.8b).

Two types of neuroglia produce myelin sheaths: Schwann cells (in the PNS) and oligodendrocytes (in the CNS). Schwann cells begin to form myelin sheaths around axons during fetal development. Each Schwann cell wraps about 1 millimeter (1 mm = 0.04 in.) of a single axon's length by spiraling many times around the axon (Figure 12.8a). Eventually, multiple layers of glial plasma membrane surround the axon, with the Schwann cell's cytoplasm and nucleus forming the outermost layer. The inner portion, consisting of up to 100 layers of Schwann cell membrane, is the myelin sheath. The outer nucleated cytoplasmic layer of the Schwann cell, which encloses the myelin sheath, is the **neurolemma (sheath of Schwann)** (noo′-rō-LEM-ma). A neurolemma is found only around axons in the PNS. When an axon is injured, the neurolemma aids regeneration by forming a regeneration tube that guides and stimulates regrowth of the axon. Gaps in the myelin sheath, called **nodes of Ranvier** (RON-vē-ā), appear at intervals along the axon (Figure 12.8; see also Figure 12.2). Each Schwann cell wraps one axon segment between two nodes.

In the CNS, an oligodendrocyte myelinates parts of several axons. Each oligodendrocyte puts forth about 15 broad, flat processes that spiral around CNS axons, forming a myelin sheath. A neurolemma is not present, however, because the oligodendrocyte cell body and nucleus do not envelop the axon. Nodes of Ranvier are present, but they are fewer in number. Axons in the CNS display little regrowth after injury. This is thought to be due, in part, to the absence of a neurolemma, and in part to an inhibitory influence exerted by the oligodendrocytes on axon regrowth.

Figure 12.7 Neuroglia of the peripheral nervous system.

Neuroglia of the PNS completely surround axons and cell bodies of neurons.

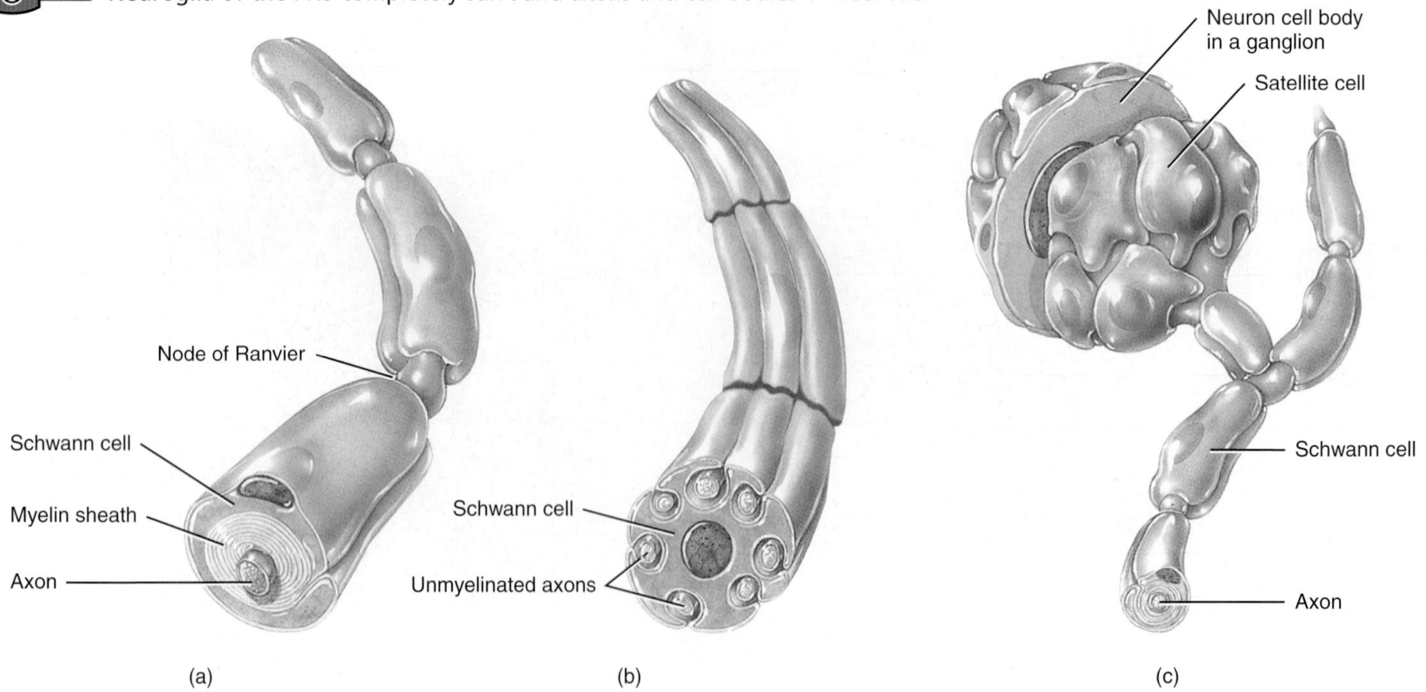
Node of Ranvier
Schwann cell
Myelin sheath
Axon
(a)
Schwann cell
Unmyelinated axons
(b)
Neuron cell body in a ganglion
Satellite cell
Schwann cell
Axon
(c)

How do Schwann cells and oligodendrocytes differ with respect to the number of axons they myelinate?

The amount of myelin increases from birth to maturity, and its presence greatly increases the speed of nerve impulse conduction. An infant's responses to stimuli are neither as rapid nor as coordinated as those of an older child or an adult, in part because myelination is still in progress during infancy.

Collections of Nervous Tissue

The components of nervous tissue are grouped together in a variety of ways. Neuronal cell bodies are often grouped together in clusters. The axons of neurons are usually grouped together in

Figure 12.8 Myelinated and unmyelinated axons. Notice that one layer of Schwann cell plasma membrane surrounds unmyelinated axons.

Axons surrounded by a myelin sheath produced either by Schwann cells in the PNS or by oligodendrocytes in the CNS are said to be myelinated.

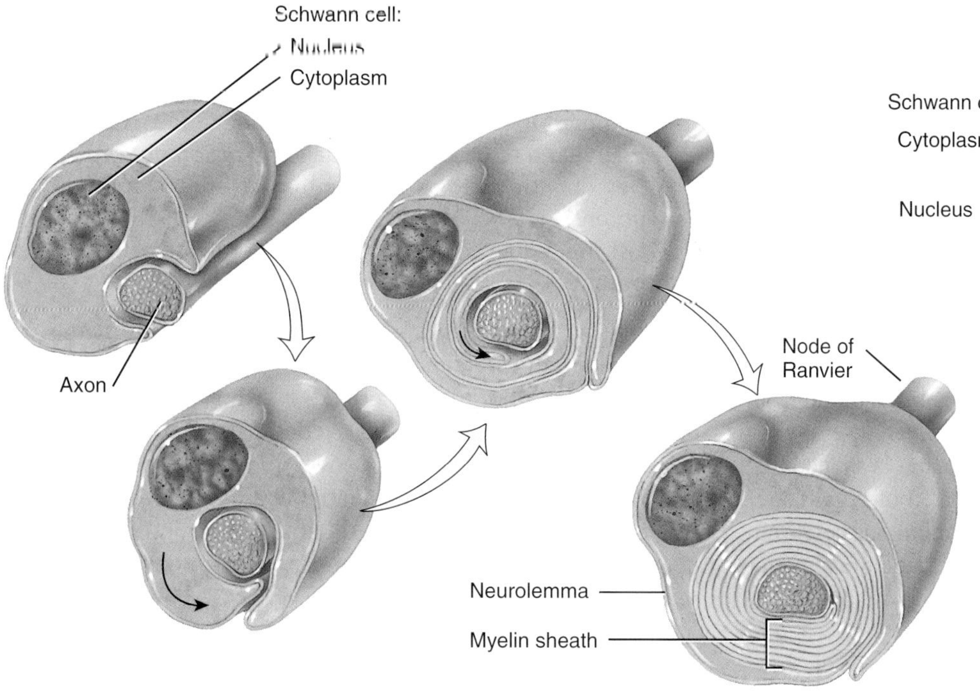
Schwann cell:
Nucleus
Cytoplasm
Axon
Node of Ranvier
Neurolemma
Myelin sheath

(a) Transverse sections of stages in the formation of a myelin sheath

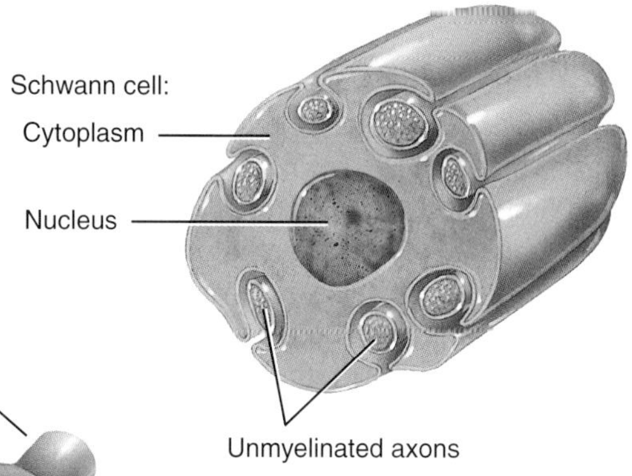
Schwann cell:
Cytoplasm
Nucleus
Unmyelinated axons

(b) Transverse section of unmyelinated axons

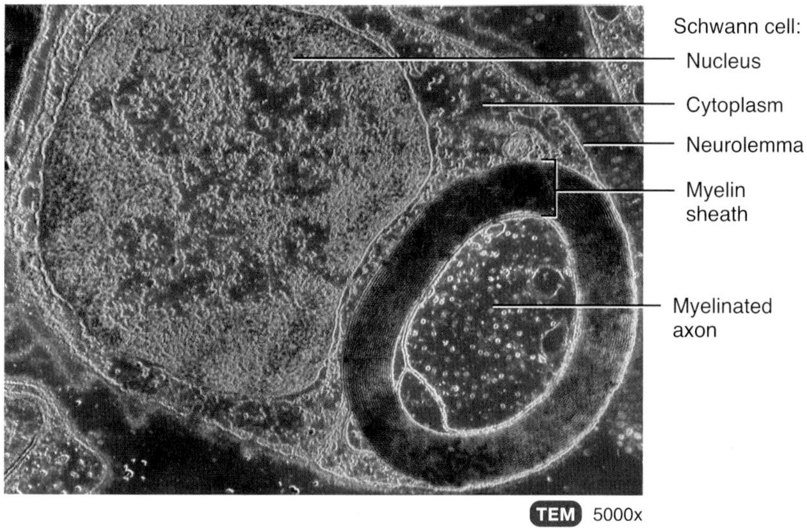
Schwann cell:
Nucleus
Cytoplasm
Neurolemma
Myelin sheath
Myelinated axon
TEM 5000x

(c) Transverse section of myelinated axon

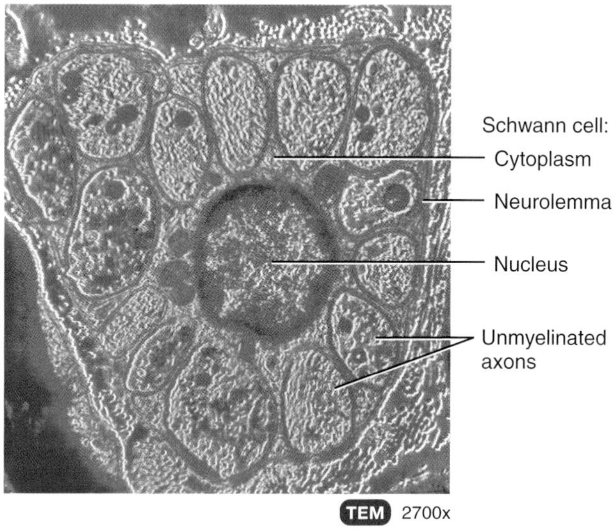
Schwann cell:
Cytoplasm
Neurolemma
Nucleus
Unmyelinated axons
TEM 2700x

(d) Transverse section of unmyelinated axons

What is the functional advantage of myelination?

bundles. In addition, widespread regions of nervous tissue are grouped together as either gray matter or white matter.

Clusters of Neuronal Cell Bodies

Recall that a **ganglion** (plural is *ganglia*) refers to a cluster of neuronal cell bodies located in the *PNS*. As mentioned earlier, ganglia are closely associated with cranial and spinal nerves. By contrast, a **nucleus** is a cluster of neuronal cell bodies located in the *CNS*.

Bundles of Axons

Recall that a **nerve** is a bundle of axons that is located in the *PNS*. Cranial nerves connect the brain to the periphery, whereas spinal nerves connect the spinal cord to the periphery. A **tract** is a bundle of axons that is located in the *CNS*. Tracts interconnect neurons in the spinal cord and brain.

Gray and White Matter

In a freshly dissected section of the brain or spinal cord, some regions look white and glistening, and others appear gray (Figure 12.9). **White matter** is composed primarily of myelinated axons. The whitish color of myelin gives white matter its name. The **gray matter** of the nervous system contains neuronal cell bodies, dendrites, unmyelinated axons, axon terminals, and neuroglia. It appears grayish, rather than white, because the Nissl bodies impart a gray color and there is little or no myelin in these areas. Blood vessels are present in both white and gray matter. In the spinal cord, the white matter surrounds an inner core of gray matter that, depending on how imaginative you are, is shaped like a butterfly or the letter H; in the brain, a thin shell of gray matter covers the surface of the largest portions of the brain, the cerebrum and cerebellum (Figure 12.9). The arrangement of gray matter and white matter in the spinal cord and brain is discussed more extensively in Chapters 13 and 14, respectively.

✔CHECKPOINT

5. Describe the parts of a neuron and the functions of each.
6. Give several examples of the structural and functional classifications of neurons.
7. What is a neurolemma and why is it important?
8. With reference to the nervous system, what is a nucleus?

12.3 ELECTRICAL SIGNALS IN NEURONS

OBJECTIVES

- Describe the cellular properties that permit communication among neurons and effectors.
- Compare the basic types of ion channels, and explain how they relate to graded potentials and action potentials.
- Describe the factors that maintain a resting membrane potential.
- List the sequence of events that generate an action potential.

Like muscle fibers, neurons are electrically excitable. They communicate with one another using two types of electrical signals: (1) **Graded potentials** (described shortly) are used for short-distance communication only. (2) **Action potentials** (also described shortly) allow communication over long distances within the body. Recall that an action potential in a muscle fiber is called a **muscle action potential.** When an action potential occurs in a neuron (nerve cell), it is called a **nerve action potential (nerve**

Figure 12.9 Distribution of gray matter and white matter in the spinal cord and brain.

White matter primarily consists of myelinated axons of many neurons. Gray matter consists of neuron cell bodies, dendrites, unmyelinated axons, axon terminals, and neuroglia.

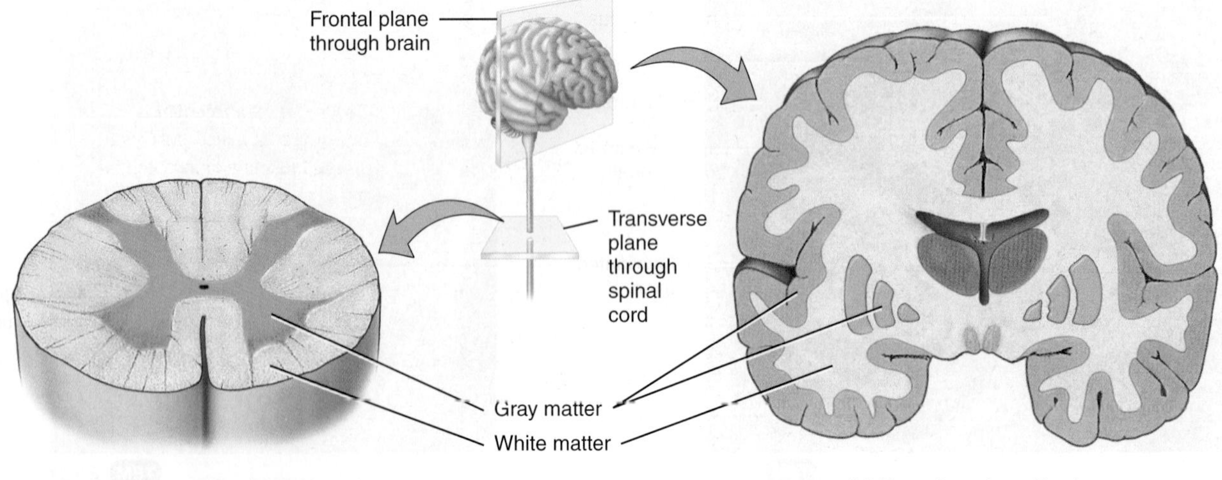

Frontal plane through brain
Transverse plane through spinal cord
Gray matter
White matter

(a) Transverse section of spinal cord

(b) Frontal section of brain

What is responsible for the white appearance of white matter?

impulse). To understand the functions of graded potentials and action potentials, consider how the nervous system allows you to feel the smooth surface of a pen that you have picked up off of a table (Figure 12.10):

1. As you touch the pen, a graded potential develops in a sensory receptor in the skin of the fingers.
2. The graded potential triggers the axon of the sensory neuron to form a nerve action potential, which travels along the axon into the CNS and ultimately causes the release of neurotransmitter at a synapse with an interneuron.

Figure 12.10 Overview of nervous system functions.

Graded potentials and nerve and muscle action potentials are involved in the relay of sensory stimuli, integrative functions such as perception, and motor activities.

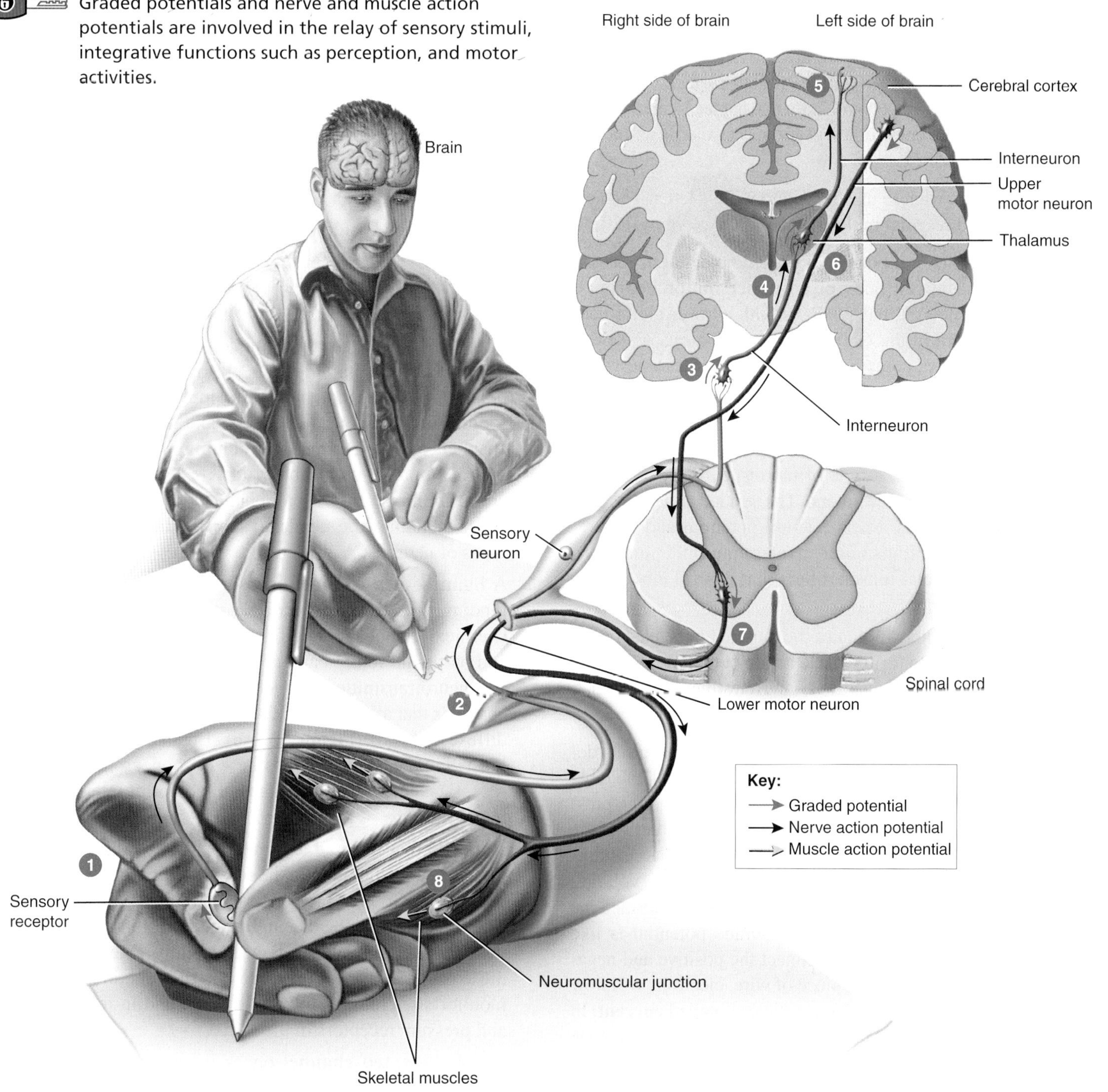

In which region of the brain does perception primarily occur?

3 The neurotransmitter stimulates the interneuron to form a graded potential in its dendrites and cell body.

4 In response to the graded potential, the axon of the interneuron forms a nerve action potential. The nerve action potential travels along the axon, which results in neurotransmitter release at the next synapse with another interneuron.

5 This process of neurotransmitter release at a synapse followed by the formation of a graded potential and then a nerve action potential occurs over and over as interneurons in higher parts of the brain (such as the thalamus and cerebral cortex) are activated. Once interneurons in the **cerebral cortex,** the outer part of the brain, are activated, perception occurs and you are able to feel the smooth surface of the pen touch your fingers. As you will learn in Chapter 14, perception, the conscious awareness of a sensation, is primarily a function of the cerebral cortex.

Suppose that you want to use the pen to write a letter. The nervous system would respond in the following way (Figure 12.10):

6 A stimulus in the brain causes a graded potential to form in the dendrites and cell body of an **upper motor neuron,** a type of motor neuron that synapses with a lower motor neuron farther down in the CNS in order to contract a skeletal muscle. The graded potential subsequently causes a nerve action potential to occur in the axon of the upper motor neuron, followed by neurotransmitter release.

7 The neurotransmitter generates a graded potential in a **lower motor neuron,** a type of motor neuron that directly supplies skeletal muscle fibers. The graded potential triggers the formation of a nerve action potential and then release of the neurotransmitter at neuromuscular junctions formed with skeletal muscle fibers that control movements of the fingers.

8 The neurotransmitter stimulates the muscle fibers that control finger movements to form muscle action potentials. The muscle action potentials cause these muscle fibers to contract, which allows you to write with the pen.

The production of graded potentials and action potentials depends on two basic features of the plasma membrane of excitable cells: the existence of a resting membrane potential and the presence of specific types of ion channels. Like most other cells in the body, the plasma membrane of excitable cells exhibits a **membrane potential,** an electrical potential difference (voltage) across the membrane. In excitable cells, this voltage is termed the **resting membrane potential.** The membrane potential is like voltage stored in a battery. If you connect the positive and negative terminals of a battery with a piece of wire, electrons will flow along the wire. This flow of charged particles is called **current.** In living cells, the flow of ions (rather than electrons) constitutes the electrical current.

Graded potentials and action potentials occur because the membranes of neurons contain many different kinds of ion channels that open or close in response to specific stimuli. Because the lipid bilayer of the plasma membrane is a good electrical insulator, the main paths for current to flow across the membrane are through the ion channels.

Ion Channels

When ion channels are open, they allow specific ions to move across the plasma membrane, down their **electrochemical gradient**—a concentration (chemical) difference plus an electrical difference. Recall that ions move from areas of higher concentration to areas of lower concentration (the chemical part of the gradient). Also, positively charged cations move toward a negatively charged area, and negatively charged anions move toward a positively charged area (the electrical aspect of the gradient). As ions move, they create a flow of electrical current that can change the membrane potential.

Ion channels open and close due to the presence of "gates." The gate is a part of the channel protein that can seal the channel pore shut or move aside to open the pore (see Figure 3.6). The electrical signals produced by neurons and muscle fibers rely on four types of ion channels: leak channels, ligand-gated channels, mechanically gated channels, and voltage-gated channels:

1. The gates of **leak channels** randomly alternate between open and closed positions (Figure 12.11a). Typically, plasma membranes have many more potassium ion (K^+) leak channels than sodium ion (Na^+) leak channels, and the potassium ion leak channels are leakier than the sodium ion leak channels. Thus, the membrane's permeability to K^+ is much higher than its permeability to Na^+. Leak channels are found in nearly all cells, including the dendrites, cell bodies, and axons of all types of neurons.
2. A **ligand-gated channel** opens and closes in response to the binding of a ligand (chemical) stimulus. A wide variety of chemical ligands—including neurotransmitters, hormones, and particular ions—can open or close ligand-gated channels. The neurotransmitter acetylcholine, for example, opens cation channels that allow Na^+ and Ca^{2+} to diffuse inward and K^+ to diffuse outward (Figure 12.11b). Ligand-gated channels are located in the dendrites of some sensory neurons, such as pain receptors, and in dendrites and cell bodies of interneurons and motor neurons.
3. A **mechanically gated channel** opens or closes in response to mechanical stimulation in the form of vibration (such as sound waves), touch, pressure, or tissue stretching (Figure 12.11c). The force distorts the channel from its resting position, opening the gate. Examples of mechanically gated channels are those found in auditory receptors in the ears, in receptors that monitor stretching of internal organs, and in touch receptors and pressure receptors in the skin.
4. A **voltage-gated channel** opens in response to a change in membrane potential (voltage) (Figure 12.11d). Voltage-gated channels participate in the generation and conduction of action potentials in the axons of all types of neurons.

Figure 12.11 Ion channels in the plasma membrane. (a) Leak channels randomly open and close. (b) A chemical stimulus—here, the neurotransmitter acetylcholine—opens a ligand-gated channel. (c) A mechanical stimulus opens a mechanically gated channel. (d) A change in membrane potential opens voltage-gated K^+ channels during an action potential.

The electrical signals produced by neurons and muscle fibers rely on four types of ion channels: leak channels, ligand-gated channels, mechanically gated channels, and voltage-gated channels.

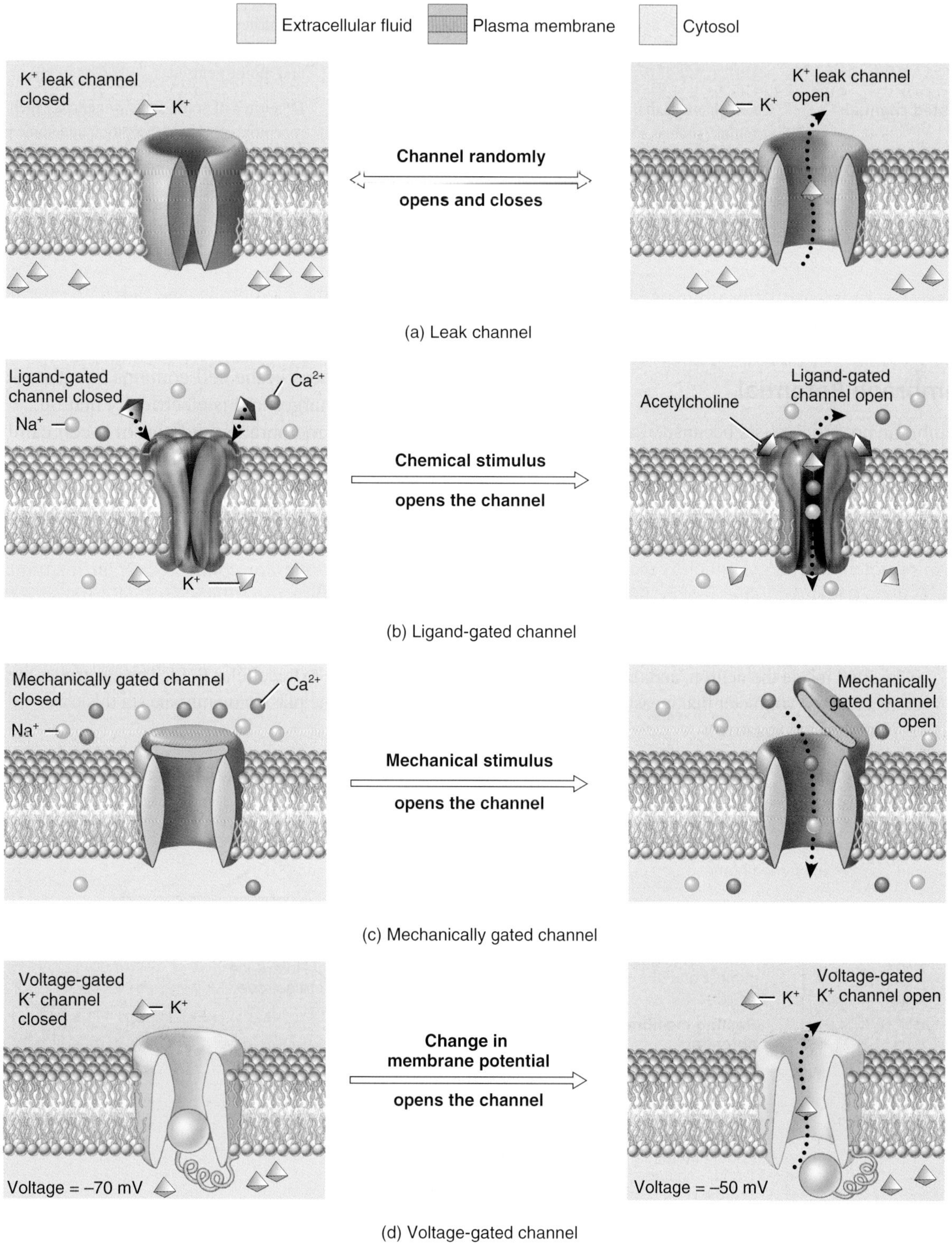

What type of gated channel is activated by a touch on the arm?

TABLE 12.1

Ion Channels in Neurons

TYPE OF ION CHANNEL	DESCRIPTION	LOCATION
Leak channels	Gated channels that randomly open and close.	Found in nearly all cells, including dendrites, cell bodies, and axons of all types of neurons.
Ligand-gated channels	Gated channels that open in response to binding of ligand (chemical) stimulus.	Dendrites of some sensory neurons such as pain receptors and dendrites and cell bodies of interneurons and motor neurons.
Mechanically gated channels	Gated channels that open in response to mechanical stimulus (such as touch, pressure, vibration, or tissue stretching).	Dendrites of some sensory neurons such as touch receptors, pressure receptors, and some pain receptors.
Voltage-gated channels	Gated channels that open in response to voltage stimulus (change in membrane potential).	Axons of all types of neurons.

Table 12.1 presents a summary of the four major types of ion channels in neurons.

Resting Membrane Potential

The resting membrane potential exists because of a small buildup of negative ions in the cytosol along the inside of the membrane, and an equal buildup of positive ions in the extracellular fluid (ECF) along the outside surface of the membrane (Figure 12.12a). Such a separation of positive and negative electrical charges is a form of potential energy, which is measured in volts or millivolts (1 mV = 0.001 V). The greater the difference in charge across the membrane, the larger the membrane potential (voltage). Notice in Figure 12.12a that the buildup of charge occurs only very close to the membrane. The cytosol or extracellular fluid elsewhere in the cell contains equal numbers of positive and negative charges and is electrically neutral.

The resting membrane potential of a cell can be measured in the following way: The tip of a recording microelectrode is inserted inside the cell, and a reference electrode is placed outside the cell in the extracellular fluid. *Electrodes* are devices that conduct electrical charges. The recording microelectrode and the reference electrode are connected to an instrument known as a *voltmeter,* which detects the electrical difference (voltage) across

Figure 12.12 Resting membrane potential. To measure resting membrane potential, the tip of the recording microelectrode is inserted inside the neuron, and the reference electrode is placed in the extracellular fluid. The electrodes are connected to a voltmeter that measures the difference in charge across the plasma membrane (in this case −70 mV inside negative).

The resting membrane potential is an electrical potential difference (voltage) that exists across the plasma membrane of an excitable cell under resting conditions.

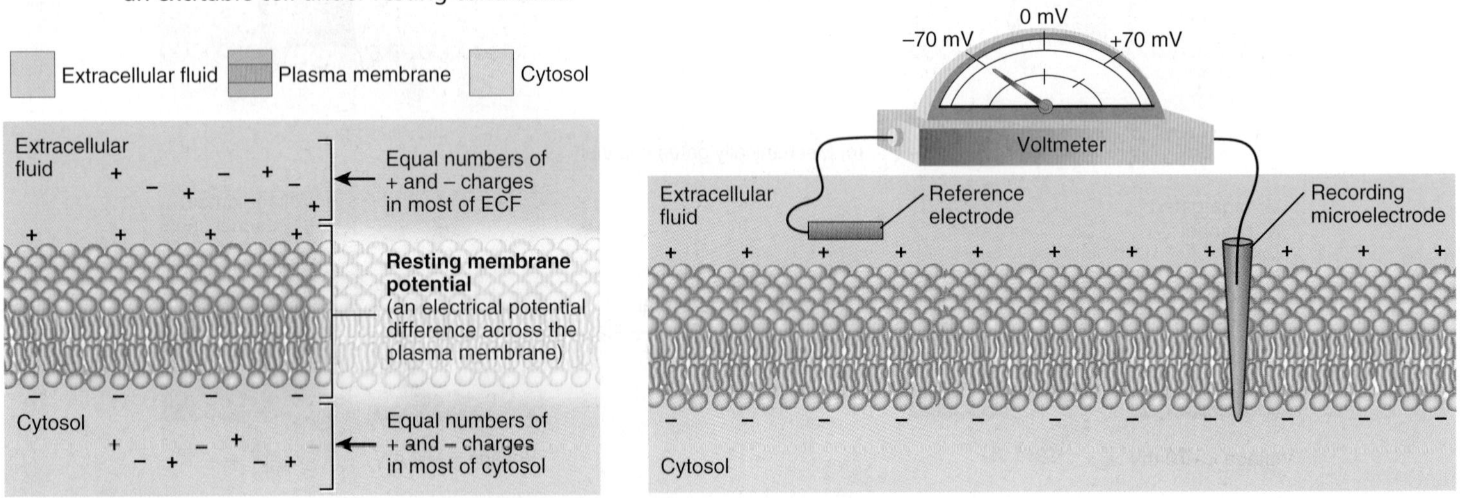

(a) Distribution of charges that produce the resting membrane potential of a neuron

(b) Measurement of the resting membrane potential of a neuron

The resting membrane potential of a neuron typically is −70 mV. What does this mean?

the plasma membrane (Figure 12.12b). In neurons, the resting membrane potential ranges from −40 to −90 mV. A typical value is −70 mV. The minus sign indicates that the inside of the cell is negative relative to the outside. A cell that exhibits a membrane potential is said to be **polarized.** Most body cells are polarized; the membrane potential varies from +5 mV to −100 mV in different types of cells.

The resting membrane potential arises from three major factors:

1. ***Unequal distribution of ions in the ECF and cytosol.*** A major factor that contributes to the resting membrane potential is the unequal distribution of various ions in extracellular fluid and cytosol (Figure 12.13). Extracellular fluid is rich in Na^+ and chloride ions (Cl^-). In cytosol, however, the main cation is K^+, and the two dominant anions are phosphates attached to molecules, such as the three phosphates in ATP, and amino acids in proteins. Because the plasma membrane typically has more K^+ leak channels than Na^+ leak channels, the number of potassium ions that diffuse down their concentration gradient out of the cell into the ECF is greater than the number of sodium ions that diffuse down their concentration gradient from the ECF into the cell. As more and more positive potassium ions exit, the inside of the membrane becomes increasingly negative, and the outside of the membrane becomes increasingly positive.

Figure 12.13 Factors that contribute to the resting membrane potential. A major factor that contributes to the inside-negative resting membrane potential of a cell is the unequal distribution of ions in the ECF and cytosol. The ECF is rich in Na^+ and Cl^- ions and the cytosol is rich in K^+ ions and anions such as phosphates in ATP and amino acids in proteins. Because the plasma membrane has more K^+ leak channels than Na^+ leak channels, the number of K^+ ions that leave the cell is greater than the number of Na^+ ions that enter the cell. As more and more K^+ ions leave the cell, the inside of the membrane becomes increasingly negative and the outside of the membrane becomes increasingly positive. The inability of most anions to leave the cell also contributes to the negativity of the resting membrane potential. These trapped anions cannot follow K^+ out of the cell because they are attached to nondiffusible molecules such as ATP and large proteins. The activity of the Na^+/K^+ ATPases is another factor that contributes to the resting membrane potential. The Na^+/K^+ ATPase expels 3 Na^+ ions for every 2 K^+ ions imported. Since the Na^+/K^+ ATPase removes more positive charges from the cell than it brings into the cell, it is electrogenic and helps establish the inside-negative resting membrane potential.

The resting membrane potential is determined by three major factors: (1) unequal distribution of ions in the ECF and cytosol, (2) inability of most anions to leave the cell, and (3) the electrogenic nature of the Na^+/K^+ ATPases.

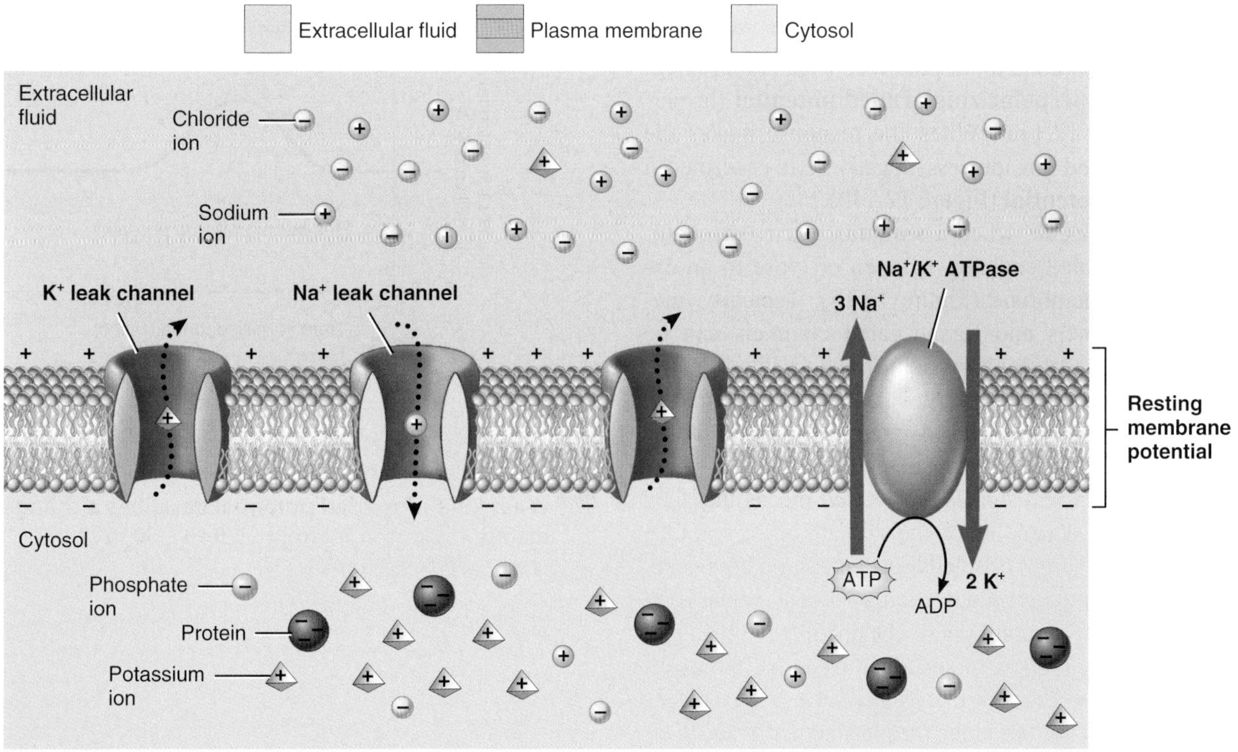

Suppose that the plasma membrane of a neuron has more Na^+ leak channels than K^+ leak channels. What effect would this have on the resting membrane potential?

2. ***Inability of most anions to leave the cell.*** Another factor contributes to the inside-negative resting membrane potential: Most anions inside the cell are not free to leave (Figure 12.13). They cannot follow the K^+ out of the cell because they are attached to nondiffusible molecules such as ATP and large proteins.
3. ***Electrogenic nature of the Na^+/K^+ ATPases.*** Membrane permeability to Na^+ is very low because there are only a few sodium leak channels. Nevertheless, sodium ions do slowly diffuse inward, down their concentration gradient. Left unchecked, such inward leakage of Na^+ would eventually destroy the resting membrane potential. The small inward Na^+ leak and outward K^+ leak are offset by the Na^+/K^+ ATPases (sodium–potassium pumps) (Figure 12.13). These pumps help maintain the resting membrane potential by pumping out Na^+ as fast as it leaks in. At the same time, the Na^+/K^+ ATPases bring in K^+. However, the potassium ions eventually leak back out of the cell as they move down their concentration gradient. Recall that the Na^+/K^+ ATPases expel three Na^+ for each two K^+ imported (see Figure 3.10). Since these pumps remove more positive charges from the cell than they bring into the cell, they are *electrogenic,* which means they contribute to the negativity of the resting membrane potential. Their total contribution, however, is very small: only -3 mV of the total -70 mV resting membrane potential in a typical neuron.

Graded Potentials

A **graded potential** is a small deviation from the membrane potential that makes the membrane either more polarized (inside more negative) or less polarized (inside less negative). When the response makes the membrane more polarized (inside more negative), it is termed a **hyperpolarizing graded potential** (hī-per-PŌ-lar-ī′-zing) (Figure 12.14a). When the response makes the membrane less polarized (inside less negative), it is termed a **depolarizing graded potential** (Figure 12.14b).

A graded potential occurs when a stimulus causes mechanically gated or ligand-gated channels to open or close in an excitable cell's plasma membrane (Figure 12.15). Typically, mechanically gated channels and ligand-gated channels can be present in the dendrites of sensory neurons, and ligand-gated channels are numerous in the dendrites and cell bodies of interneurons and motor neurons. Hence, graded potentials occur mainly in the dendrites and cell body of a neuron.

To say that these electrical signals are *graded* means that they vary in amplitude (size), depending on the strength of the stimulus (Figure 12.16). They are larger or smaller depending on how many ligand-gated or mechanically gated channels have opened (or closed) and how long each remains open. The opening or closing of these ion channels alters the flow of specific ions across the membrane, producing a flow of current that is *localized,* which means that it spreads to adjacent regions along the plasma membrane in either direction from the stimulus source for a short distance and then gradually dies out as the charges are lost across the membrane

Figure 12.14 Graded potentials. Most graded potentials occur in the dendrites and cell body (areas colored blue).

During a hyperpolarizing graded potential, the membrane potential is inside more negative than the resting level. During a depolarizing graded potential, the membrane potential is inside less negative than the resting level.

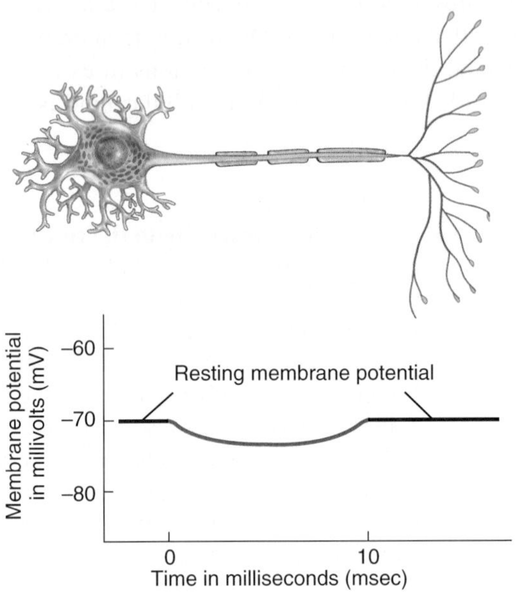

(a) Hyperpolarizing graded potential

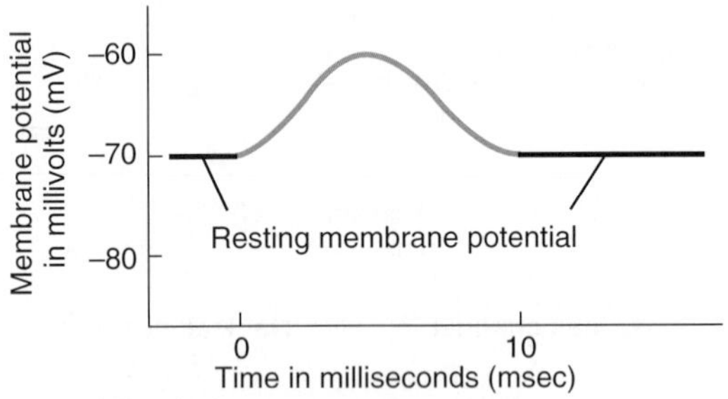

(b) Depolarizing graded potential

What kind of graded potential describes a change in membrane potential from -70 to -60 mV? From -70 to -80 mV?

Figure 12.15 Generation of graded potentials in response to the opening of mechanically gated channels or ligand-gated channels. (a) A mechanical stimulus (pressure) opens a mechanically gated channel that allows passage of cations (mainly Na^+ and Ca^{2+}) into the cell; a depolarizing graded potential occurs because the membrane potential becomes inside less negative than at rest. (b) The neurotransmitter acetylcholine (a ligand stimulus) opens a cation channel that allows passage of Na^+, K^+, and Ca^{2+}; Na^+ inflow is greater than either Ca^{2+} inflow or K^+ outflow, and a depolarizing graded potential occurs because the membrane potential becomes inside less negative than at rest. (c) The neurotransmitter glycine (a ligand stimulus) opens a Cl^- channel that allows passage of Cl^- ions into the cell; a hyperpolarizing graded potential occurs because the membrane potential becomes inside more negative than at rest.

A graded potential forms in response to the opening of mechanically gated channels or ligand-gated channels.

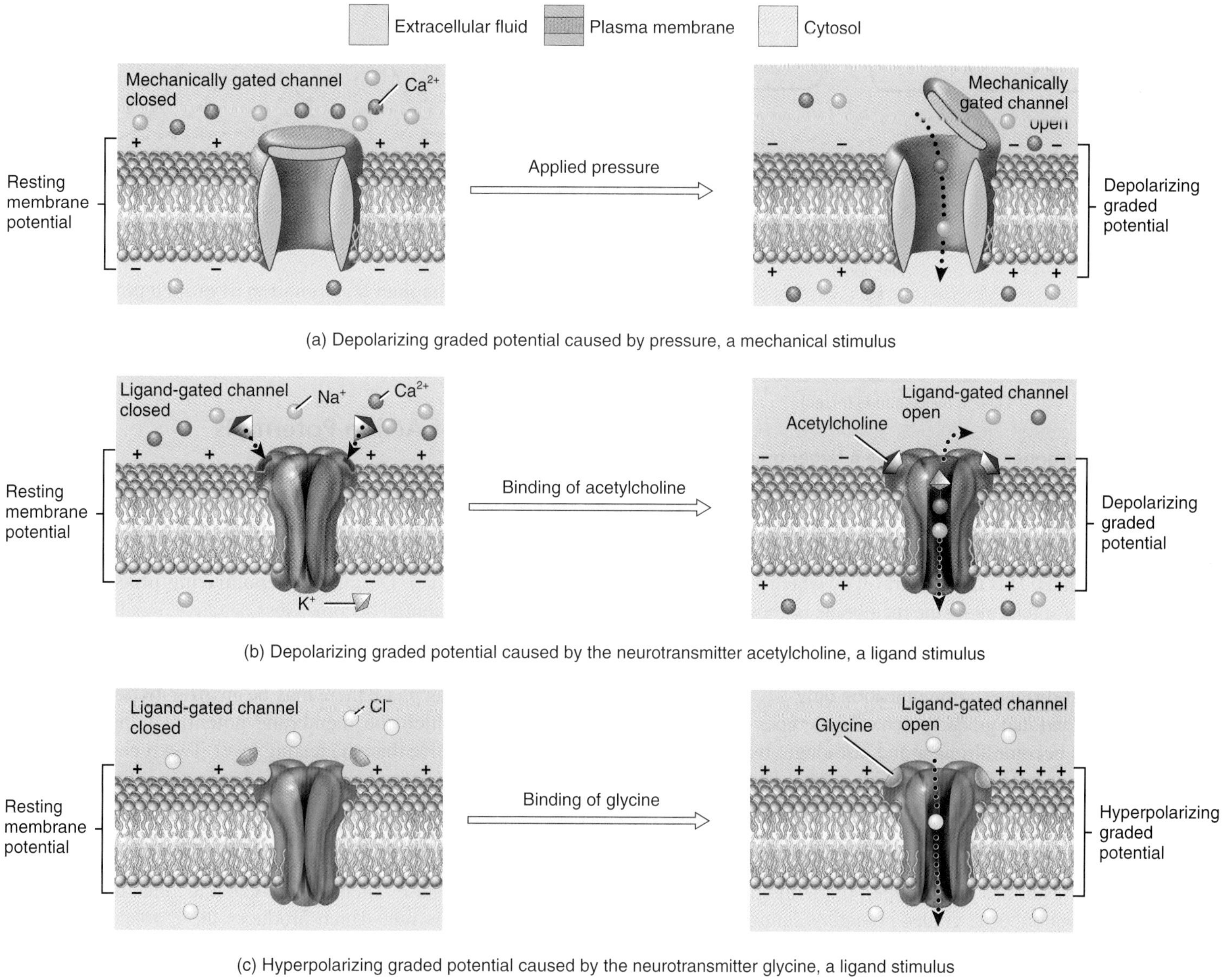

(a) Depolarizing graded potential caused by pressure, a mechanical stimulus

(b) Depolarizing graded potential caused by the neurotransmitter acetylcholine, a ligand stimulus

(c) Hyperpolarizing graded potential caused by the neurotransmitter glycine, a ligand stimulus

Which parts of a neuron contain mechanically gated channels? Ligand-gated channels?

Figure 12.16 The graded nature of graded potentials. As stimulus strength increases (stimuli 1, 2, and 3), the amplitude (size) of each resulting depolarizing graded potential increases. Although not shown, a similar relationship exists between stimulus strength and the amplitude of a hyperpolarizing graded potential.

The amplitude of a graded potential depends on the stimulus strength. The greater the stimulus strength, the larger the amplitude of the graded potential.

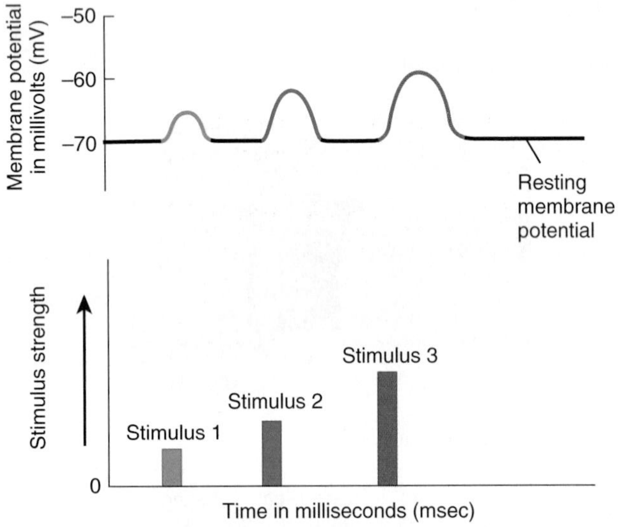

Why does a stronger stimulus cause a larger graded potential than a weaker stimulus?

Figure 12.17 Summation of graded potentials. Summation of two depolarizing graded potentials happens in response to two stimuli of the same strength that occur very close together in time. The dotted lines represent the individual depolarizing graded potentials that would form if summation did not occur.

Summation occurs when two or more graded potentials add together to become larger in amplitude.

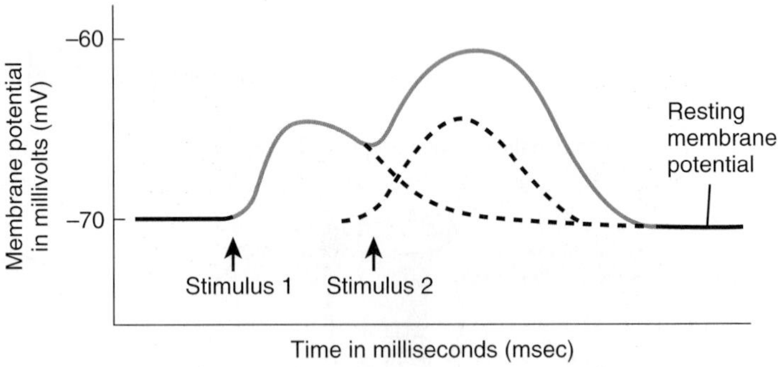

What would happen if summation of graded potentials in a neuron did not occur?

through leak channels. This mode of travel by which graded potentials die out as they spread along the membrane is known as **decremental conduction** (dek-re-MENT-al). Because they die out within a few millimeters of their point of origin, graded potentials are useful for short-distance communication only.

Although an individual graded potential undergoes decremental conduction, it can become stronger and last longer by summating with other graded potentials. **Summation** is the process by which graded potentials add together. If two depolarizing graded potentials summate, the net result is a larger depolarizing graded potential (Figure 12.17). If two hyperpolarizing graded potentials summate, the net result is a larger hyperpolarizing graded potential. If two equal but opposite graded potentials summate (one depolarizing and the other hyperpolarizing), then they cancel each other out and the overall graded potential disappears. You will learn more about the process of summation later in this chapter.

Graded potentials have different names depending on which type of stimulus causes them and where they occur. For example, when a graded potential occurs in the dendrites or cell body of a neuron in response to a neurotransmitter, it is called a *postsynaptic potential* (explained shortly). On the other hand, the graded potentials that occur in sensory receptors and sensory neurons are termed *receptor potentials* and *generator potentials* (explained in Chapter 16).

Generation of Action Potentials

An **action potential (AP)** or **impulse** is a sequence of rapidly occurring events that decrease and reverse the membrane potential and then eventually restore it to the resting state. An action potential has two main phases: a depolarizing phase and a repolarizing phase (Figure 12.18). During the **depolarizing phase,** the negative membrane potential becomes less negative, reaches zero, and then becomes positive. During the **repolarizing phase,** the membrane potential is restored to the resting state of −70 mV. Following the repolarizing phase there may be an **after-hyperpolarizing phase,** during which the membrane potential temporarily becomes more negative than the resting level. Two types of voltage-gated channels open and then close during an action potential. These channels are present mainly in the axon plasma membrane and axon terminals. The first channels that open, the voltage-gated Na^+ channels, allow Na^+ to rush into the cell, which causes the depolarizing phase. Then voltage-gated K^+ channels open, allowing K^+ to flow out, which produces the repolarizing phase. The after-hyperpolarizing phase occurs when the voltage-gated K^+ channels remain open after the repolarizing phase ends.

An action potential occurs in the membrane of the axon of a neuron when depolarization reaches a certain level termed the **threshold** (about −55 mV in many neurons). Different neurons may have different thresholds for generation of an action potential, but the threshold in a particular neuron usually is constant. The generation of an action potential depends on whether a particular stimulus is able to bring the membrane potential to threshold (Figure 12.19). An action potential will not occur in response to a **subthreshold stimulus,** a stimulus that is a weak depolarization

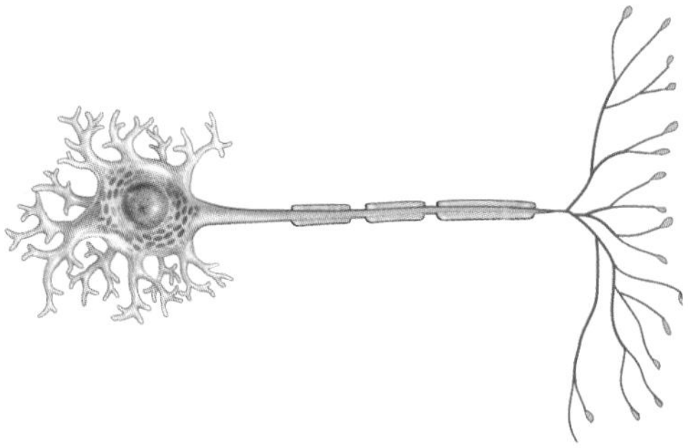

Figure 12.18 Action potential (AP) or impulse. When a stimulus depolarizes the membrane to threshold (−55 mV), an AP is generated. The action potential arises at the trigger zone (here, at the junction of the axon hillock and the initial segment) and then propagates along the axon to the axon terminals. The green-colored regions of the neuron indicate parts that typically have voltage-gated Na^+ and K^+ channels (axon plasma membrane and axon terminals).

An action potential consists of a depolarizing phase and a repolarizing phase, which may be followed by an after-hyperpolarizing phase.

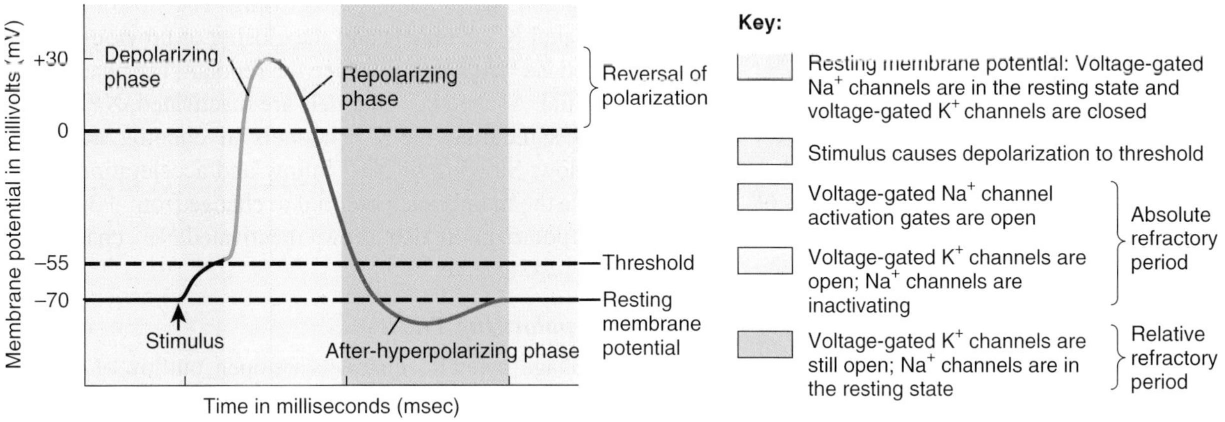

Which channels are open during the depolarizing phase? During the repolarizing phase?

Figure 12.19 Stimulus strength and action potential generation. A subthreshold stimulus does not cause an action potential because it does not bring the membrane potential to threshold. However, an action potential does occur in response to a threshold stimulus because a threshold stimulus is just strong enough to depolarize the membrane to threshold. Several action potentials form in response to a suprathreshold stimulus, which depolarizes the membrane above threshold. Each of the action potentials caused by the suprathreshold stimulus has the same amplitude (size) as the action potential caused by the threshold stimulus. For simplicity, the after-hyperpolarizing phase of the action potential is not shown.

An action potential will only occur once the membrane potential reaches threshold.

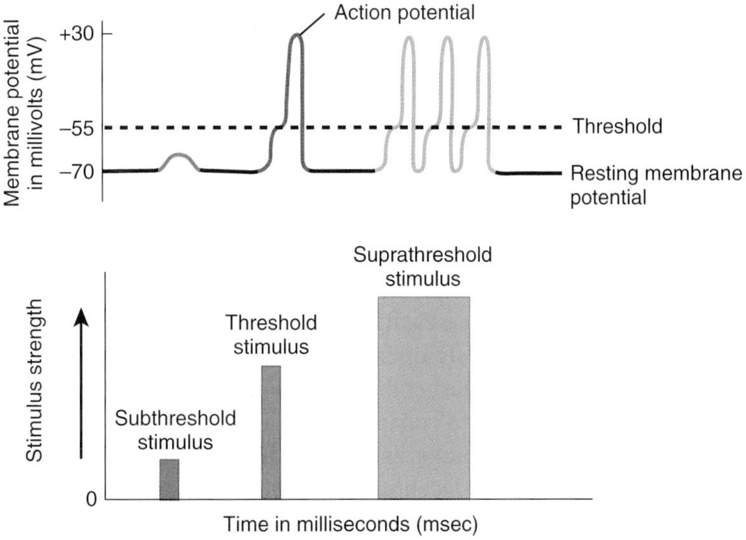

Will an action potential occur in response to a hyperpolarizing graded potential that spreads from the dendrites or cell body to the trigger zone of the axon of a neuron? Why or why not?

that cannot bring the membrane potential to threshold. However, an action potential will occur in response to a **threshold stimulus,** a stimulus that is just strong enough to depolarize the membrane to threshold. Several action potentials will form in response to a **suprathreshold stimulus,** a stimulus that is strong enough to depolarize the membrane *above* threshold. Each of the action potentials caused by a suprathreshold stimulus has the same amplitude (size) as an action potential caused by a threshold stimulus. Therefore, once an action potential is generated, the amplitude of an action potential is always the same and does not depend on stimulus intensity. Instead, the greater the stimulus strength above threshold, the greater the frequency of the action potentials until a maximum frequency is reached as determined by the absolute refractory period (described shortly).

As you have just learned, an action potential is generated in response to a threshold stimulus but does not form when there is a subthreshold stimulus. In other words, an action potential either occurs completely or it does not occur at all. This characteristic of an action potential is known as the **all-or-none principle.** The all-or-none principle of the action potential is similar to pushing the first domino in a long row of standing dominoes. When the push on the first domino is strong enough (when depolarization reaches threshold), that domino falls against the second domino, and the *entire* row topples (an action potential occurs). Stronger pushes on the first domino produce the identical effect—toppling of the entire row. Thus, pushing on the first domino produces an all-or-none event: The dominoes all fall or none fall.

Depolarizing Phase

When a depolarizing graded potential or some other stimulus causes the membrane of the axon to depolarize to threshold, voltage-gated Na^+ channels open rapidly. Both the electrical and the chemical gradients favor inward movement of Na^+, and the resulting inrush of Na^+ causes the depolarizing phase of the action potential (see Figure 12.18). The inflow of Na^+ changes the membrane potential from -55 mV to $+30$ mV. At the peak of the action potential, the inside of the membrane is 30 mV more positive than the outside.

Each voltage-gated Na^+ channel has two separate gates, an *activation gate* and an *inactivation gate.* In the *resting state* of a voltage-gated Na^+ channel, the inactivation gate is open, but the activation gate is closed (step 1 in Figure 12.20). As a result, Na^+ cannot move into the cell through these channels. At threshold, voltage-gated Na^+ channels are activated. In the *activated state* of a voltage-gated Na^+ channel, both the activation and inactivation gates in the channel are open and Na^+ inflow begins (step 2 in Figure 12.20). As more channels open, Na^+ inflow increases, the membrane depolarizes further, and more Na^+ channels open. This is an example of a positive feedback mechanism. During the few ten-thousandths of a second that the voltage-gated Na^+ channel is open, about 20,000 Na^+ flow across the membrane and change the membrane potential considerably. But the concentration of Na^+ hardly changes because of the millions of Na^+ present in the extracellular fluid. The sodium–potassium pumps easily bail out the 20,000 or so Na^+ that enter the cell during a single action potential and maintain the low concentration of Na^+ inside the cell.

Repolarizing Phase

Shortly after the activation gates of the voltage-gated Na^+ channels open, the inactivation gates close (step 3 in Figure 12.20). Now the voltage-gated Na^+ channel is in an *inactivated state.* In addition to opening voltage-gated Na^+ channels, a threshold-level depolarization also opens voltage-gated K^+ channels (steps 3 and 4 in Figure 12.20). Because the voltage-gated K^+ channels open more slowly, their opening occurs at about the same time the voltage-gated Na^+ channels are closing. The slower opening of voltage-gated K^+ channels and the closing of previously open voltage-gated Na^+ channels produce the repolarizing phase of the action potential. As the Na^+ channels are inactivated, Na^+ inflow slows. At the same time, the K^+ channels are opening, accelerating K^+ outflow. Slowing of Na^+ inflow and acceleration of K^+ outflow cause the membrane potential to change from $+30$ mV to -70 mV. Repolarization also allows inactivated Na^+ channels to revert to the resting state.

After-hyperpolarizing Phase

While the voltage-gated K^+ channels are open, outflow of K^+ may be large enough to cause an after-hyperpolarizing phase of the action potential (see Figure 12.18). During this phase, the voltage-gated K^+ channels remain open and the membrane potential becomes even more negative (about -90 mV). As the voltage-gated K^+ channels close, the membrane potential returns to the resting level of -70 mV. Unlike voltage-gated Na^+ channels, most voltage-gated K^+ channels do not exhibit an inactivated state. Instead, they alternate between closed (resting) and open (activated) states.

Refractory Period

The period of time after an action potential begins during which an excitable cell cannot generate another action potential in response to a *normal* threshold stimulus is called the **refractory period** (rē-FRAK-tor-ē) (see key in Figure 12.18). During the **absolute refractory period,** even a very strong stimulus cannot initiate a second action potential. This period coincides with the period of Na^+ channel activation and inactivation (steps 2–4 in Figure 12.20). Inactivated Na^+ channels cannot reopen; they first must return to the resting state (step 1 in Figure 12.20). In contrast to action potentials, graded potentials do not exhibit a refractory period.

Large-diameter axons have a larger surface area and have a brief absolute refractory period of about 0.4 msec. Because a second nerve impulse can arise very quickly, up to 1000 impulses per second are possible. Small-diameter axons have absolute refractory periods as long as 4 msec, enabling them to transmit a maximum of 250 impulses per second. Under normal body conditions, the maximum frequency of nerve impulses in different axons ranges between 10 and 1000 per second.

The **relative refractory period** is the period of time during which a second action potential can be initiated, but only by a larger-than-normal stimulus. It coincides with the period when

Figure 12.20 Changes in ion flow through voltage-gated channels during the depolarizing and repolarizing phases of an action potential. Leak channels and sodium–potassium pumps are not shown.

Inflow of sodium ions (Na^+) causes the depolarizing phase, and outflow of potassium ions (K^+) causes the repolarizing phase of an action potential.

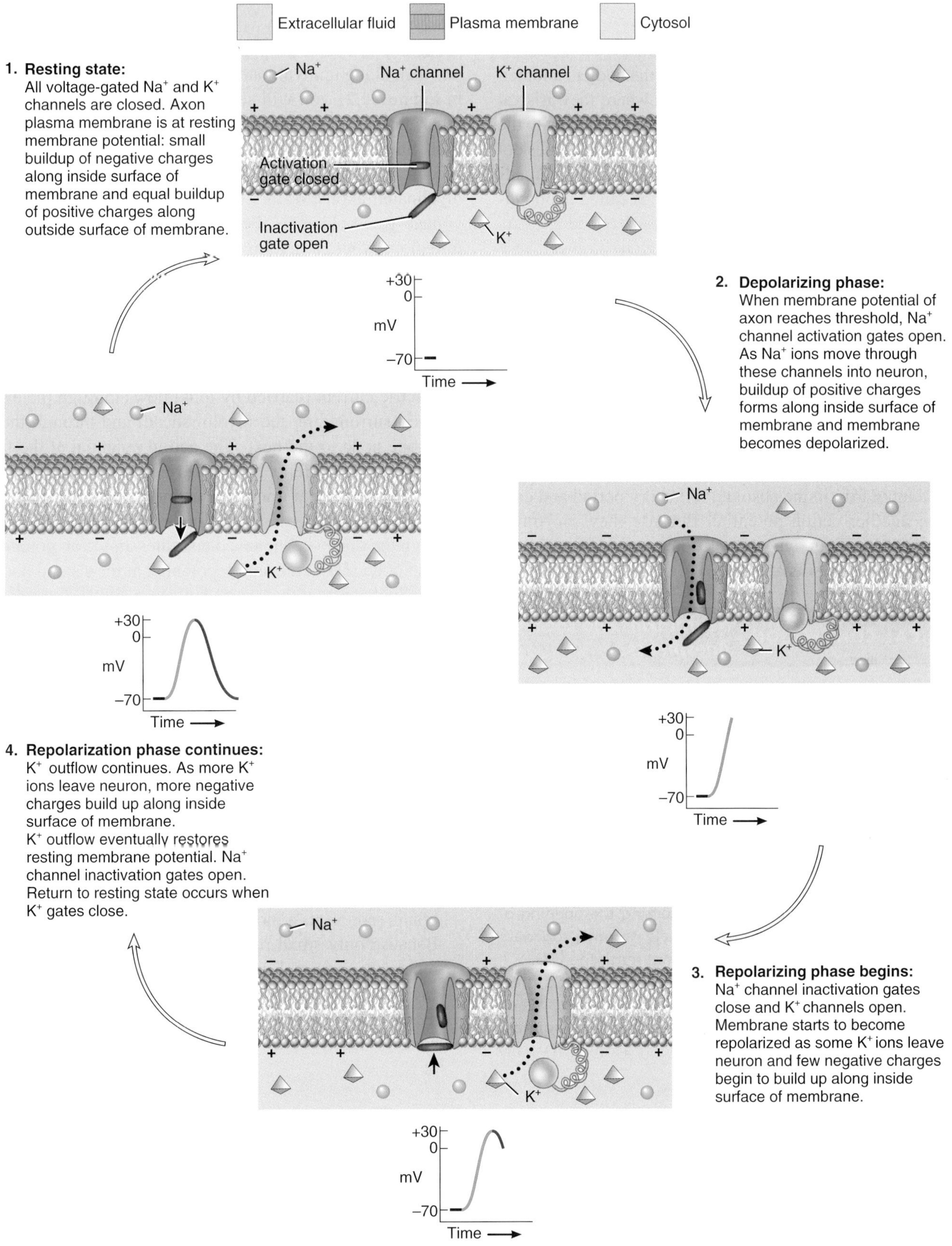

Extracellular fluid
Plasma membrane
Cytosol
1. Resting state:
All voltage-gated Na+ and K+ channels are closed. Axon plasma membrane is at resting membrane potential: small buildup of negative charges along inside surface of membrane and equal buildup of positive charges along outside surface of membrane.
Na+
Na+ channel
K+ channel
Activation gate closed
Inactivation gate open
K+
+30
0
mV
–70
Time
2. Depolarizing phase:
When membrane potential of axon reaches threshold, Na+ channel activation gates open. As Na+ ions move through these channels into neuron, buildup of positive charges forms along inside surface of membrane and membrane becomes depolarized.
3. Repolarizing phase begins:
Na+ channel inactivation gates close and K+ channels open. Membrane starts to become repolarized as some K+ ions leave neuron and few negative charges begin to build up along inside surface of membrane.
4. Repolarization phase continues:
K+ outflow continues. As more K+ ions leave neuron, more negative charges build up along inside surface of membrane.
K+ outflow eventually restores resting membrane potential. Na+ channel inactivation gates open. Return to resting state occurs when K+ gates close.

Given the existence of leak channels for both K^+ and Na^+, could the membrane repolarize if the voltage-gated K^+ channels did not exist?

the voltage-gated K^+ channels are still open after inactivated Na^+ channels have returned to their resting state (see Figure 12.18).

Propagation of Action Potentials

To communicate information from one part of the body to another, action potentials in a neuron must travel from where they arise at the trigger zone of the axon to the axon terminals. In contrast to the graded potential, an action potential is not decremental (it does not die out). Instead, an action potential keeps its strength as it spreads along the membrane. This mode of conduction is called **propagation** (prop′-a-GĀ-shun), and it depends on positive feedback. As you have already learned, when sodium ions flow in, they cause voltage-gated Na^+ channels in adjacent segments of the membrane to open. Thus, the action potential travels along the membrane rather like the activity of that long row of dominoes. In actuality, it is not the same action potential that propagates along the entire axon. Instead, the action potential regenerates over and over at adjacent regions of membrane from the trigger zone to the axon terminals. In a neuron, an action potential can propagate in this direction only—it cannot propagate back toward the cell body because any region of membrane that has just undergone an action potential is temporarily in the absolute refractory period and cannot generate another action potential. Because they can travel along a membrane without dying out, action potentials function in communication over long distances.

CLINICAL CONNECTION | Neurotoxins and Local Anesthetics

Certain shellfish and other organisms contain **neurotoxins** (noo′-rō-TOK-sins), substances that produce their poisonous effects by acting on the nervous system. One particularly lethal neurotoxin is tetrodotoxin (TTX), present in the viscera of Japanese pufferfish. TTX effectively blocks action potentials by inserting itself into voltage-gated Na^+ channels so they cannot open.

Local anesthetics are drugs that block pain and other somatic sensations. Examples include procaine (Novocaine®) and lidocaine, which may be used to produce anesthesia in the skin during suturing of a gash, in the mouth during dental work, or in the lower body during childbirth. Like TTX, these drugs act by blocking the opening of voltage-gated Na^+ channels. Action potentials cannot propagate past the obstructed region, so pain signals do not reach the CNS.

Localized cooling of a nerve can also produce an anesthetic effect because axons propagate action potentials at lower speeds when cooled. The application of ice to injured tissue can reduce pain because propagation of the pain sensations along axons is partially blocked. •

Continuous and Saltatory Conduction

There are two types of propagation: continuous conduction and saltatory conduction. The type of action potential propagation described so far is **continuous conduction,** which involves step-by-step depolarization and repolarization of each adjacent segment of the plasma membrane (Figure 12.21a). In continuous conduction, ions flow through their voltage-gated channels in each adjacent segment of the membrane. Note that the action potential propagates only a relatively short distance in a few milliseconds. Continuous conduction occurs in unmyelinated axons and in muscle fibers.

Action potentials propagate more rapidly along myelinated axons than along unmyelinated axons. If you compare parts a and b in Figure 12.21 you will see that the action potential propagates much farther along the myelinated axon in the same period of time. **Saltatory conduction** (SAL-ta-tō-rē; *saltat-* = leaping), the special mode of action potential propagation that occurs along myelinated axons, occurs because of the uneven distribution of voltage-gated channels. Few voltage-gated channels are present in regions where a myelin sheath covers the axolemma. By contrast, at the nodes of Ranvier (where there is no myelin sheath), the axolemma has many voltage-gated channels. Hence, current carried by Na^+ and K^+ flows across the membrane mainly at the nodes.

When an action potential propagates along a myelinated axon, an electric current (carried by ions) flows through the extracellular fluid surrounding the myelin sheath and through the cytosol from one node to the next. The action potential at the first node generates ionic currents in the cytosol and extracellular fluid that depolarize the membrane to threshold, opening voltage-gated Na^+ channels at the second node. The resulting ionic flow through the opened channels constitutes an action potential at the second node. Then, the action potential at the second node generates an ionic current that opens voltage-gated Na^+ channels at the third node, and so on. Each node repolarizes after it depolarizes.

The flow of current across the membrane only at the nodes of Ranvier has two consequences:

1. The action potential appears to "leap" from node to node as each nodal area depolarizes to threshold, thus the name "saltatory." Because an action potential leaps across long segments of the myelinated axolemma as current flows from one node to the next, it travels much faster than it would in an unmyelinated axon of the same diameter.
2. Opening a smaller number of channels only at the nodes, rather than many channels in each adjacent segment of membrane, represents a more energy-efficient mode of conduction. Because only small regions of the membrane depolarize and repolarize, minimal inflow of Na^+ and outflow of K^+ occurs each time an action potential passes by. Thus, less ATP is used by sodium–potassium pumps to maintain the low intracellular concentration of Na^+ and the low extracellular concentration of K^+.

Factors That Affect the Speed of Propagation

The speed of propagation of an action potential is affected by three major factors: amount of myelination, axon diameter, and temperature.

1. ***Amount of myelination.*** As you have just learned, action potentials propagate more rapidly along myelinated axons than along unmyelinated axons.

Figure 12.21 Propagation of an action potential in a neuron after it arises at the trigger zone. Dotted lines indicate ionic current flow. The insets show the path of current flow. (a) In continuous conduction along an unmyelinated axon, ionic currents flow across each adjacent segment of the membrane. (b) In saltatory conduction along a myelinated axon, the action potential (nerve impulse) at the first node generates ionic currents in the cytosol and interstitial fluid that open voltage-gated Na^+ channels at the second node, and so on at each subsequent node.

Unmyelinated axons exhibit continuous conduction; myelinated axons exhibit saltatory conduction.

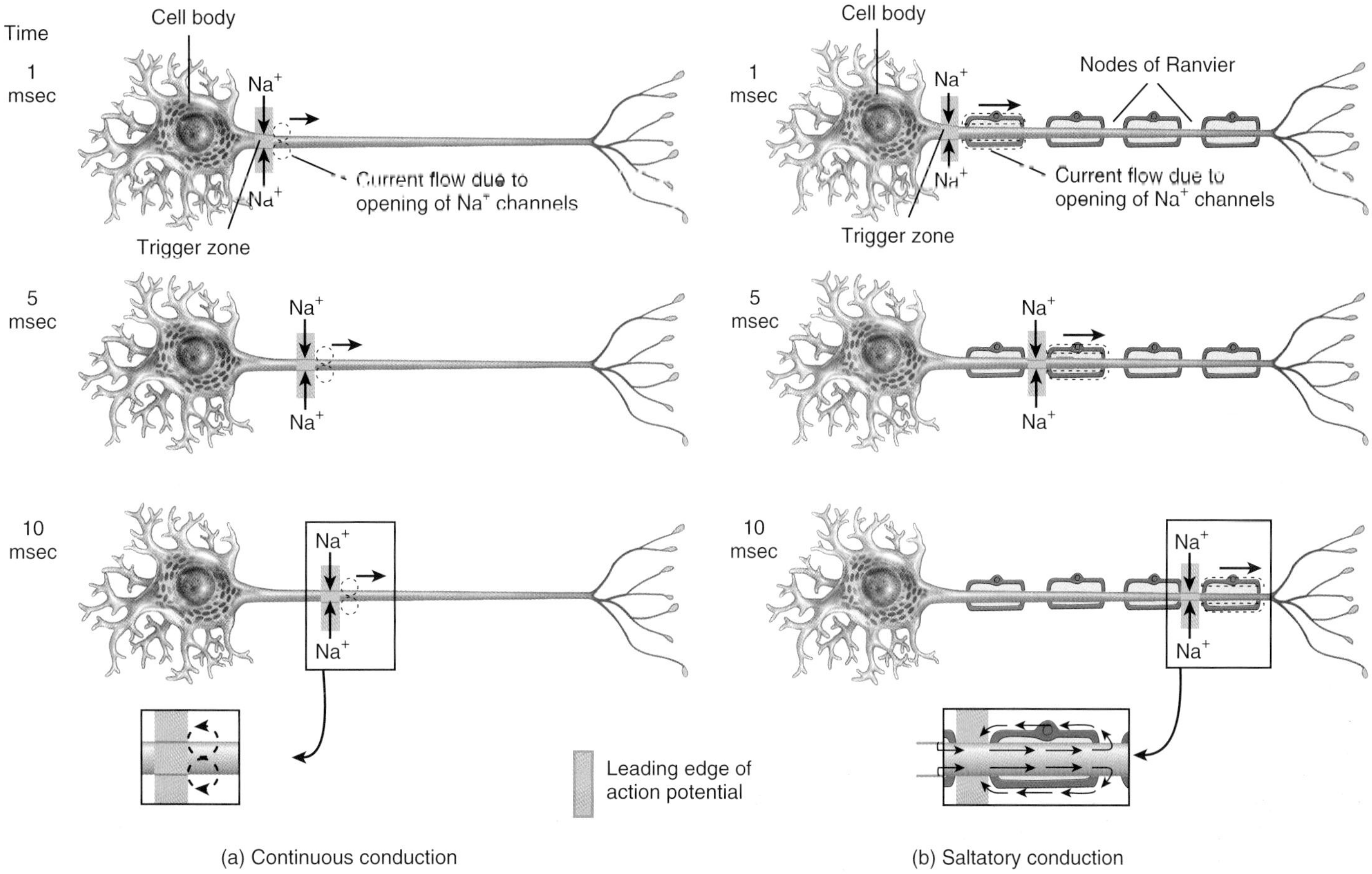

What factors determine the speed of propagation of an action potential?

2. ***Axon diameter.*** Larger-diameter axons propagate action potentials faster than smaller ones due to their larger surface areas.
3. ***Temperature.*** Axons propagate action potentials at lower speeds when cooled.

Classification of Nerve Fibers

Axons can be classified into three major groups based on the amount of myelination, their diameters, and their propagation speeds:

- **A fibers** are the largest-diameter axons (5–20 μm) and are myelinated. A fibers have a brief absolute refractory period and conduct nerve impulses (action potentials) at speeds of 12 to 130 m/sec (27–290 mi/hr). The axons of sensory neurons that propagate impulses associated with touch, pressure, position of joints, and some thermal and pain sensations are A fibers, as are the axons of motor neurons that conduct impulses to skeletal muscles.
- **B fibers** are axons with diameters of 2–3 μm. Like A fibers, B fibers are myelinated and exhibit saltatory conduction at speeds up to 15 m/sec (34 mi/hr). B fibers have a somewhat longer absolute refractory period than A fibers. B fibers conduct sensory nerve impulses from the viscera to the brain and spinal cord. They also constitute all the axons of the autonomic motor neurons that extend from the brain and spinal cord to the ANS relay stations called autonomic ganglia.
- **C fibers** are the smallest-diameter axons (0.5–1.5 μm) and all are unmyelinated. Nerve impulse propagation along a C fiber ranges from 0.5 to 2 m/sec (1–4 mi/hr). C fibers exhibit the longest absolute refractory periods. These unmyelinated axons

conduct some sensory impulses for pain, touch, pressure, heat, and cold from the skin, and pain impulses from the viscera. Autonomic motor fibers that extend from autonomic ganglia to stimulate the heart, smooth muscle, and glands are C fibers. Examples of motor functions of B and C fibers are constricting and dilating the pupils, increasing and decreasing the heart rate, and contracting and relaxing the urinary bladder.

Encoding of Stimulus Intensity

How can your sensory systems detect stimuli of differing intensities if all nerve impulses are the same size? Why does a light touch feel different from firmer pressure? The main answer to this question is the *frequency of action potentials*—how often they are generated at the trigger zone. A light touch generates a low frequency of action potentials. A firmer pressure elicits action potentials that pass down the axon at a higher frequency. In addition to this "frequency code," a second factor is the number of sensory neurons recruited (activated) by the stimulus. A firm pressure stimulates a larger number of pressure-sensitive neurons than does a light touch.

Comparison of Electrical Signals Produced by Excitable Cells

We have seen that excitable cells—neurons and muscle fibers—produce two types of electrical signals: graded potentials and action potentials (impulses). One obvious difference between them is that the propagation of action potentials permits communication over long distances, but graded potentials can function only in short-distance communication because they are not propagated. Table 12.2 presents a summary of the differences between graded potentials and action potentials.

As we discussed in Chapter 10, propagation of a muscle action potential along the sarcolemma and into the T tubule system initiates the events of muscle contraction. Although action potentials in muscle fibers and in neurons are similar, there are some notable differences. The typical resting membrane potential of a neuron is −70 mV, but it is closer to −90 mV in skeletal and cardiac muscle fibers. The duration of a nerve impulse is 0.5–2 msec, but a muscle action potential is considerably longer—about 1.0–5.0 msec for skeletal muscle fibers and 10–300 msec for cardiac and smooth muscle fibers. Finally, the propagation speed of action potentials along the largest-diameter, myelinated axons is about 18 times faster than the propagation speed along the sarcolemma of a skeletal muscle fiber.

✔CHECKPOINT

9. Define the terms resting membrane potential, depolarization, repolarization, nerve impulse, and refractory period and identify the factors responsible for each.
10. How is saltatory conduction different from continuous conduction?
11. What effect does myelination have on the speed of propagation of an action potential?
12. How can you tell the difference between a stroke on the cheek and a slap across the face?

TABLE 12.2

Comparison of Graded Potentials and Action Potentials in Neurons

CHARACTERISTIC	GRADED POTENTIALS	ACTION POTENTIALS
Origin	Arise mainly in dendrites and cell body.	Arise at trigger zones and propagate along axon.
Types of channels	Ligand-gated or mechanically gated ion channels.	Voltage-gated channels for Na^+ and K^+.
Conduction	Decremental (not propagated); permit communication over short distances.	Propagate and thus permit communication over longer distances.
Amplitude (size)	Depending on strength of stimulus, varies from less than 1 mV to more than 50 mV.	All or none; typically about 100 mV.
Duration	Typically longer, ranging from several milliseconds to several minutes.	Shorter, ranging from 0.5 to 2 msec.
Polarity	May be hyperpolarizing (inhibitory to generation of action potential) or depolarizing (excitatory to generation of action potential).	Always consist of depolarizing followed by repolarizing phase and return to resting membrane potential.
Refractory period	Not present, thus summation can occur.	Present, thus summation cannot occur.

12.4 SIGNAL TRANSMISSION AT SYNAPSES

OBJECTIVES

- Explain the events of signal transmission at electrical and chemical synapses.
- Distinguish between spatial and temporal summation.
- Give examples of excitatory and inhibitory neurotransmitters, and describe how they act.

Recall from Chapter 10 that a synapse (SIN-aps) is a region where communication occurs between two neurons or between a neuron and an effector cell (muscle cell or glandular cell). The term **presynaptic neuron** (*pre-* = before) refers to a nerve cell that carries a nerve impulse toward a synapse. It is the cell that sends a signal. A **postsynaptic cell** is the cell that receives a signal. It may be a nerve cell called a **postsynaptic neuron** (*post-* = after) that carries a nerve impulse away from a synapse or an effector cell that responds to the impulse at the synapse.

Most synapses are either **axodendritic** (ak′-so-den-DRIT-ik = from axon to dendrite), while others are **axosomatic** (ak′-sō-sō-MAT-ik = from axon to cell body) or **axoaxonic** (ak′-so-ak-SON-ik = from axon to axon). In addition, synapses may be electrical or chemical and they differ both structurally and functionally.

In Chapter 10 we described the events occurring at one type of synapse, the neuromuscular junction. Our focus in this chapter is on synaptic communication among the billions of neurons in the nervous system. Synapses are essential for homeostasis because they allow information to be filtered and integrated. During learning, the structure and function of particular synapses change. The changes may allow some signals to be transmitted while others are blocked. For example, the changes in your synapses from studying will determine how well you do on your anatomy and physiology tests! Synapses are also important because some diseases and neurological disorders result from disruptions of synaptic communication, and many therapeutic and addictive chemicals affect the body at these junctions.

Electrical Synapses

At an **electrical synapse,** action potentials (impulses) conduct directly between the plasma membranes of adjacent neurons through structures called **gap junctions.** Each gap junction contains a hundred or so tubular *connexons,* which act like tunnels to connect the cytosol of the two cells directly (see Figure 4.2e). As ions flow from one cell to the next through the connexons, the action potential spreads from cell to cell. Gap junctions are common in visceral smooth muscle, cardiac muscle, and the developing embryo. They also occur in the brain.

Electrical synapses have two main advantages:

1. ***Faster communication.*** Because action potentials conduct directly through gap junctions, electrical synapses are faster than chemical synapses. At an electrical synapse, the action potential passes directly from the presynaptic cell to the postsynaptic cell. The events that occur at a chemical synapse take some time and delay communication slightly.
2. ***Synchronization.*** Electrical synapses can synchronize (coordinate) the activity of a group of neurons or muscle fibers. In other words, a large number of neurons or muscle fibers can produce action potentials in unison if they are connected by gap junctions. The value of synchronized action potentials in the heart or in visceral smooth muscle is coordinated contraction of these fibers to produce a heartbeat or move food through the gastrointestinal tract.

Chemical Synapses

Although the plasma membranes of presynaptic and postsynaptic neurons in a **chemical synapse** are close, they do not touch. They are separated by the **synaptic cleft,** a space of 20–50 nm* that is filled with interstitial fluid. Nerve impulses cannot conduct across the synaptic cleft, so an alternative, indirect form of communication occurs. In response to a nerve impulse, the presynaptic neuron releases a neurotransmitter that diffuses through the fluid in the synaptic cleft and binds to receptors in the plasma membrane of the postsynaptic neuron. The postsynaptic neuron receives the chemical signal and in turn produces a **postsynaptic potential,** a type of graded potential. Thus, the presynaptic neuron converts an electrical signal (nerve impulse) into a chemical signal (released neurotransmitter). The postsynaptic neuron receives the chemical signal and in turn generates an electrical signal (postsynaptic potential). The time required for these processes at a chemical synapse, a **synaptic delay** of about 0.5 msec, is the reason that chemical synapses relay signals more slowly than electrical synapses.

A typical chemical synapse transmits a signal as follows (Figure 12.22):

1. A nerve impulse arrives at a synaptic end bulb (or at a varicosity) of a presynaptic axon.
2. The depolarizing phase of the nerve impulse opens **voltage-gated Ca^{2+} channels,** which are present in the membrane of synaptic end bulbs. Because calcium ions are more concentrated in the extracellular fluid, Ca^{2+} flows inward through the opened channels.
3. An increase in the concentration of Ca^{2+} inside the presynaptic neuron serves as a signal that triggers exocytosis of the synaptic vesicles. As vesicle membranes merge with the plasma membrane, neurotransmitter molecules within the vesicles are released into the synaptic cleft. Each synaptic vesicle contains several thousand molecules of neurotransmitter.
4. The neurotransmitter molecules diffuse across the synaptic cleft and bind to **neurotransmitter receptors** in the postsynaptic neuron's plasma membrane. The receptor shown in Figure 12.22 is part of a ligand-gated channel (see

*1 nanometer (nm) = 10^{-9} (0.000000001) meter.

Figure 12.22 Signal transmission at a chemical synapse. Through exocytosis of synaptic vesicles, a presynaptic neuron releases neurotransmitter molecules. After diffusing across the synaptic cleft, the neurotransmitter binds to receptors in the plasma membrane of the postsynaptic neuron and produces a postsynaptic potential.

At a chemical synapse, a presynaptic neuron converts an electrical signal (nerve impulse) into a chemical signal (neurotransmitter release). The postsynaptic neuron then converts the chemical signal back into an electrical signal (postsynaptic potential).

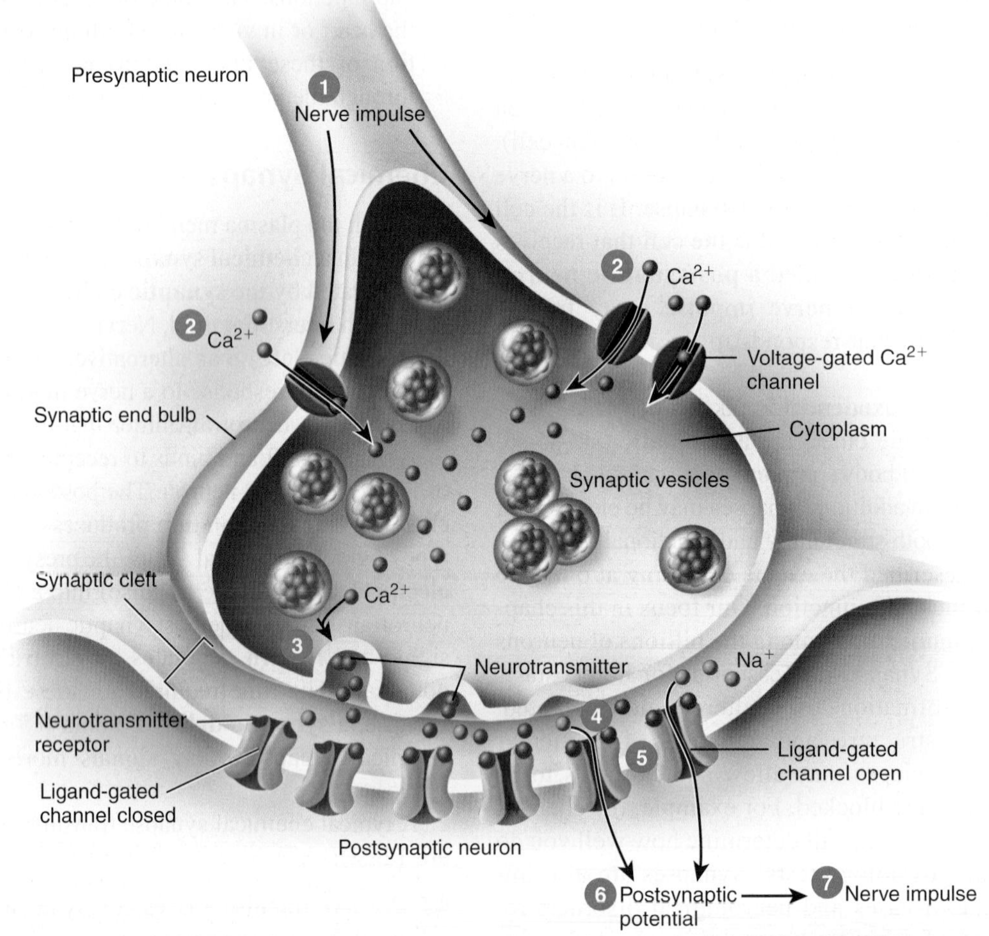

Why may electrical synapses work in two directions, but chemical synapses can transmit a signal in only one direction?

Figure 12.11b); you will soon learn that this type of neurotransmitter receptor is called an *ionotropic receptor*. Not all neurotransmitters bind to ionotropic receptors; some bind to *metabotropic receptors* (described shortly).

5 Binding of neurotransmitter molecules to their receptors on ligand-gated channels opens the channels and allows particular ions to flow across the membrane.

6 As ions flow through the opened channels, the voltage across the membrane changes. This change in membrane voltage is a **postsynaptic potential.** Depending on which ions the channels admit, the postsynaptic potential may be a depolarization (excitation) or a hyperpolarization (inhibition). For example, opening of Na^+ channels allows inflow of Na^+, which causes depolarization. However, opening of Cl^- or K^+ channels causes hyperpolarization. Opening Cl^- channels permits Cl^- to move into the cell, while opening the K^+ channels allows K^+ to move out—in either event, the inside of the cell becomes more negative.

7 When a depolarizing postsynaptic potential reaches threshold, it triggers an action potential in the axon of the postsynaptic neuron.

At most chemical synapses, only *one-way information transfer* can occur—from a presynaptic neuron to a postsynaptic neuron or an effector, such as a muscle fiber or a gland cell. For example, synaptic transmission at a neuromuscular junction (NMJ) proceeds from a somatic motor neuron to a skeletal muscle fiber (but not in the opposite direction). Only synaptic end bulbs of presynaptic neurons can release neurotransmitter, and only the postsynaptic neuron's membrane has the receptor proteins that can recognize and bind that neurotransmitter. As a result, action potentials move in one direction.

Excitatory and Inhibitory Postsynaptic Potentials

A neurotransmitter causes either an excitatory or an inhibitory graded potential. A neurotransmitter that causes *depolarization* of the postsynaptic membrane is excitatory because it brings the membrane closer to threshold (see Figure 12.14b). A depolarizing postsynaptic potential is called an **excitatory postsynaptic potential (EPSP).** Although a single EPSP normally does not initiate a nerve impulse, the postsynaptic cell does become more excitable. Because it is partially depolarized, it is more likely to reach threshold when the next EPSP occurs.

A neurotransmitter that causes *hyperpolarization* of the postsynaptic membrane (see Figure 12.14a) is inhibitory. During hyperpolarization, generation of an action potential is more difficult than usual because the membrane potential becomes inside more negative and thus even farther from threshold than in its resting state. A hyperpolarizing postsynaptic potential is termed an **inhibitory postsynaptic potential (IPSP).**

Structure of Neurotransmitter Receptors

As you have already learned, neurotransmitters released from a presynaptic neuron bind to **neurotransmitter receptors** in the plasma membrane of a postsynaptic cell. Each type of neurotransmitter receptor has one or more neurotransmitter binding sites where its specific neurotransmitter binds. When a neurotransmitter binds to the correct neurotransmitter receptor, an ion channel opens and a postsynaptic potential (either an EPSP or IPSP) forms in the membrane of the postsynaptic cell. Neurotransmitter receptors are classified as either ionotropic receptors or metabotropic receptors based on whether the neurotransmitter binding site and the ion channel are components of the same protein or are components of different proteins.

Ionotropic Receptors

An **ionotropic receptor** (ī-on-ō-TROP-ik) is a type of neurotransmitter receptor that contains a neurotransmitter binding site and an ion channel. In other words, the neurotransmitter binding site and the ion channel are components of the *same* protein. An ionotropic receptor is a type of ligand-gated channel (see Figure 12.11b). In the absence of neurotransmitter (the ligand), the ion channel component of the ionotropic receptor is closed. When the correct neurotransmitter binds to the ionotropic receptor, the ion channel opens, and an EPSP or IPSP occurs in the postsynaptic cell.

Many excitatory neurotransmitters bind to ionotropic receptors that contain cation channels (Figure 12.23a). EPSPs result from opening these cation channels. When cation channels open, they allow passage of the three most plentiful cations (Na^+, K^+, and Ca^{2+}) through the postsynaptic cell membrane, but Na^+ inflow is greater than either Ca^{2+} inflow or K^+ outflow and the inside of the postsynaptic cell becomes less negative (depolarized).

Many inhibitory neurotransmitters bind to ionotropic receptors that contain chloride channels (Figure 12.23b). IPSPs result from opening these Cl^- channels. When Cl^- channels open, a larger number of chloride ions diffuse inward. The inward flow of Cl^- ions causes the inside of the postsynaptic cell to become more negative (hyperpolarized).

Metabotropic Receptors

A **metabotropic receptor** (me-tab′-ō-TRO-pik) is a type of neurotransmitter receptor that contains a neurotransmitter binding site but lacks an ion channel as part of its structure. However, a metabotropic receptor is coupled to a separate ion channel by a type of membrane protein called a *G protein.* When a neurotransmitter binds to a metabotropic receptor, the G protein either directly opens (or closes) the ion channel or it may act indirectly by activating another molecule, a "second messenger," in the cytosol, which in turn opens (or closes) the ion channel (see Chapter 18 for a detailed discussion of G proteins). Thus, a metabotropic receptor differs from an ionotropic receptor in that the neurotransmitter binding site and the ion channel are components of *different* proteins.

Some inhibitory neurotransmitters bind to metabotropic receptors that are linked to K^+ channels (Figure 12.23c). IPSPs result from the opening of these K^+ channels. When K^+ channels open, a larger number of potassium ions diffuses outward. The outward flow of K^+ ions causes the inside of the postsynaptic cell to become more negative (hyperpolarized).

Different Postsynaptic Effects for the Same Neurotransmitter

The same neurotransmitter can be excitatory at some synapses and inhibitory at other synapses, depending on the structure of the neurotransmitter receptor to which it binds. For example, at some excitatory synapses acetylcholine (ACh) binds to ionotropic receptors that contain cation channels that open and subsequently generate EPSPs in the postsynaptic cell (Figure 12.23a). By contrast, at some inhibitory synapses ACh binds to metabotropic receptors coupled to G proteins that open K^+ channels, resulting in the formation of IPSPs in the postsynaptic cell (Figure 12.23c).

Removal of Neurotransmitter

Removal of the neurotransmitter from the synaptic cleft is essential for normal synaptic function. If a neurotransmitter could linger in the synaptic cleft, it would influence the postsynaptic neuron, muscle fiber, or gland cell indefinitely. Neurotransmitter is removed in three ways:

1. ***Diffusion.*** Some of the released neurotransmitter molecules diffuse away from the synaptic cleft. Once a neurotransmitter molecule is out of reach of its receptors, it can no longer exert an effect.
2. ***Enzymatic degradation.*** Certain neurotransmitters are inactivated through enzymatic degradation. For example, the enzyme acetylcholinesterase breaks down acetylcholine in the synaptic cleft.
3. ***Uptake by cells.*** Many neurotransmitters are actively transported back into the neuron that released them (reuptake).

Figure 12.23 Ionotropic and metabotropic neurotransmitter receptors. (a) The ionotropic acetylcholine (ACh) receptor contains two binding sites for the neurotransmitter ACh and a cation channel. Binding of ACh to this receptor causes the cation channel to open. Opening the cation channel allows passage of the three most plentiful cations (Na^+, K^+, and Ca^{2+}) through the postsynaptic cell membrane, but Na^+ inflow is greater than either Ca^{2+} inflow or K^+ outflow and an excitatory postsynaptic potential (EPSP) is generated. (b) The ionotropic gamma aminobutyric acid (GABA) receptor contains two binding sites for the neurotransmitter GABA and a Cl^- channel. Binding of GABA to this receptor causes the Cl^- channel to open. Opening the Cl^- channel allows a larger number of chloride ions to diffuse inward and an inhibitory postsynaptic potential (IPSP) is generated. (c) The metabotropic acetylcholine (ACh) receptor contains a binding site for the neurotransmitter ACh. Binding of ACh to this receptor activates a G protein, which in turn opens a K^+ channel. Opening the K^+ channel allows a larger number of potassium ions to diffuse out of the cell and an IPSP forms.

An ionotropic receptor is a type of neurotransmitter receptor that contains a neurotransmitter binding site and an ion channel; a metabotropic receptor is a type of neurotransmitter receptor that contains a neurotransmitter binding site and is coupled to a separate ion channel by a G protein.

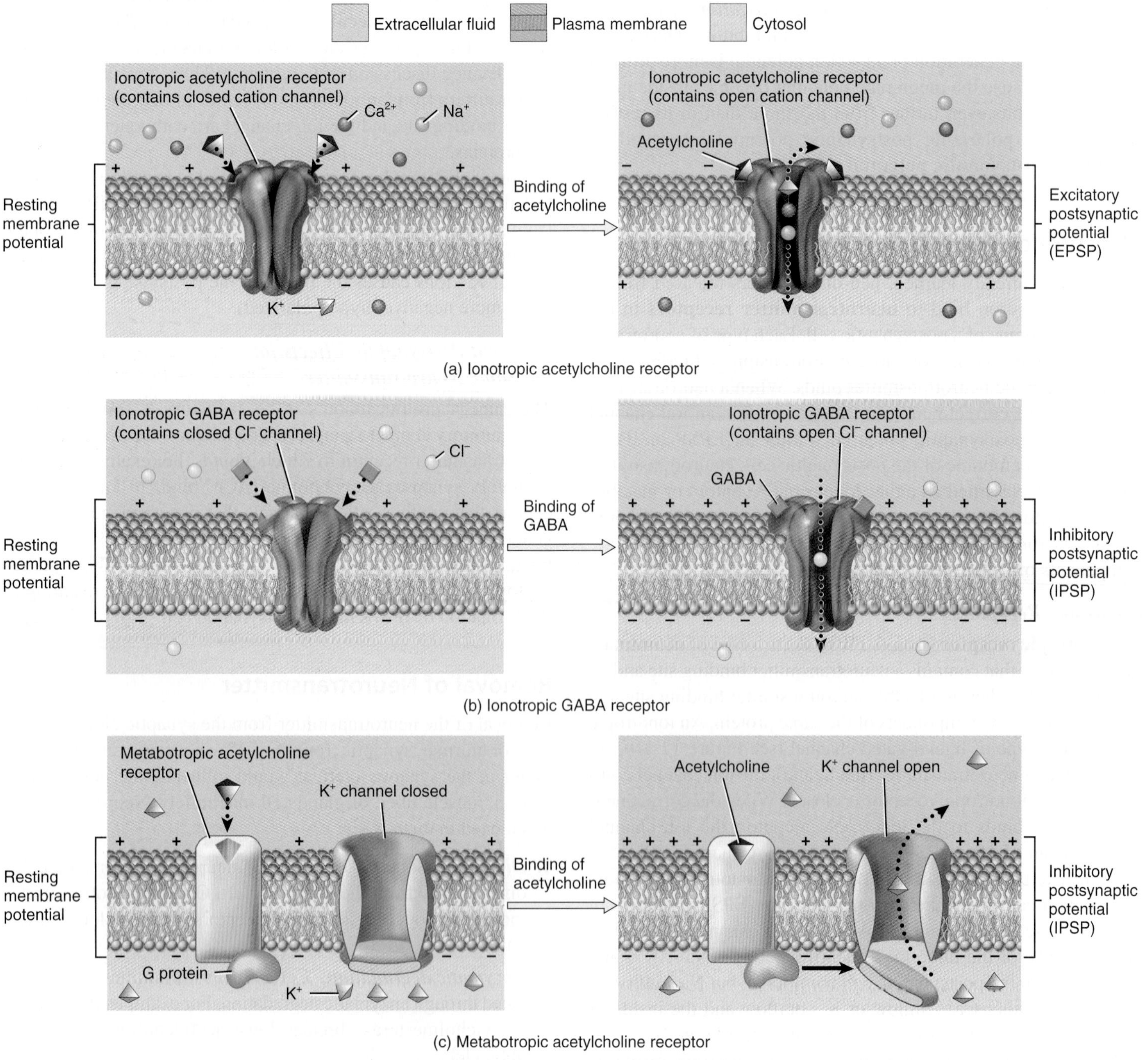

The neurotransmitter acetylcholine (ACh) is excitatory at some synapses and inhibitory at other synapses. How is this possible?

Others are transported into neighboring neuroglia (uptake). The neurons that release norepinephrine, for example, rapidly take up the norepinephrine and recycle it into new synaptic vesicles. The membrane proteins that accomplish such uptake are called *neurotransmitter transporters*.

Spatial and Temporal Summation of Postsynaptic Potentials

A typical neuron in the CNS receives input from 1000 to 10,000 synapses. Integration of these inputs involves summation of the postsynaptic potentials that form in the postsynaptic neuron. Recall that summation is the process by which graded potentials add together. The greater the summation of EPSPs, the greater the chance that threshold will be reached. At threshold, one or more nerve impulses (action potentials) arise.

There are two types of summation: spatial summation and temporal summation. **Spatial summation** is summation of postsynaptic potentials in response to stimuli that occur at different *locations* in the membrane of a postsynaptic cell at the same time. For example, spatial summation results from the buildup of neurotransmitter released simultaneously by *several* presynaptic end bulbs (Figure 12.24a). **Temporal summation** is summation of postsynaptic potentials in response to stimuli that occur at the same location in the membrane of the postsynaptic cell but at different *times*. For example, temporal summation results from buildup of neurotransmitter released by a *single* presynaptic end bulb two or more times in rapid succession (Figure 12.24b). Because a typical EPSP lasts about 15 msec, the second (and subsequent) release of neurotransmitter must occur soon after the first one if temporal summation is to occur. Summation is rather like a vote on the Internet. Many people voting "yes" or "no" on an issue at the same time can be compared to spatial summation. One person voting repeatedly and rapidly is like temporal summation. Most of the time, spatial and temporal summations are acting together to influence the chance that a neuron fires an action potential.

A single postsynaptic neuron receives input from many presynaptic neurons, some of which release excitatory neurotransmitters and some of which release inhibitory neurotransmitters

Figure 12.24 **Spatial and temporal summation.** (a) When presynaptic neurons 1 and 2 separately cause EPSPs (arrows) in postsynaptic neuron 3, the threshold level is not reached in neuron 3. Spatial summation occurs only when neurons 1 and 2 act simultaneously on neuron 3; their EPSPs sum to reach the threshold level and trigger a nerve impulse (action potential). (b) Temporal summation occurs when stimuli applied to the same axon in rapid succession (arrows) cause overlapping EPSPs that sum. When depolarization reaches the threshold level, a nerve impulse is triggered.

Spatial summation results from the buildup of neurotransmitter released simultaneously by several presynaptic end bulbs; temporal summation results from the buildup of neurotransmitter released by a single presynaptic end bulb two or more times in rapid succession.

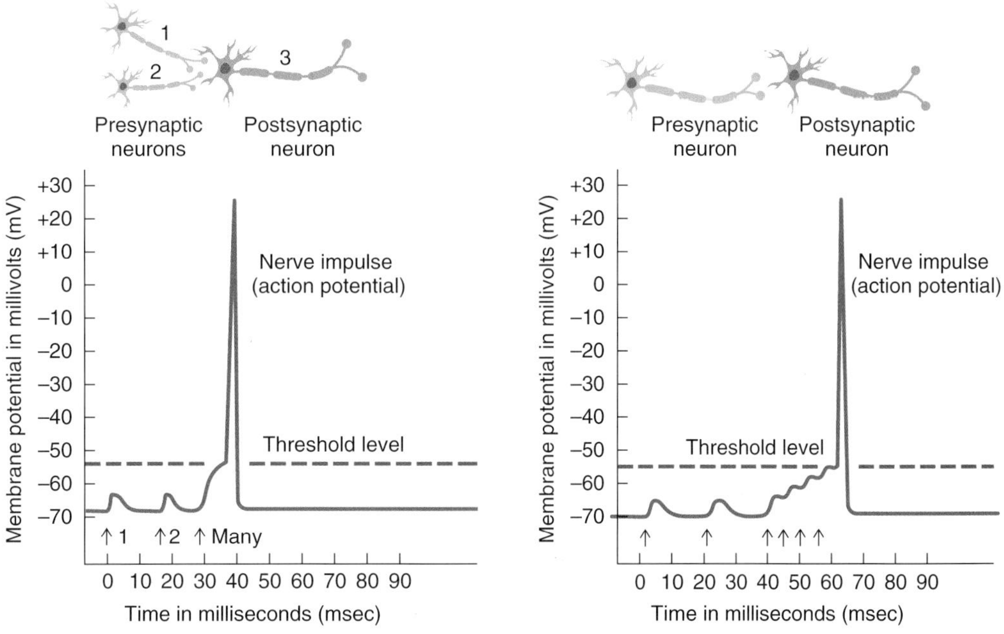

(a) Spatial summation

(b) Temporal summation

Suppose that EPSPs summate in a postsynaptic neuron in response to simultaneous stimulation by the neurotransmitters glutamate, serotonin, and acetylcholine released by three separate presynaptic neurons. Is this an example of spatial or temporal summation?

(Figure 12.25). The sum of all the excitatory and inhibitory effects at any given time determines the effect on the postsynaptic neuron, which may respond in the following ways:

1. ***EPSP.*** If the total excitatory effects are greater than the total inhibitory effects but less than the threshold level of stimulation, the result is an EPSP that does not reach threshold. Fol-

Figure 12.25 Summation of postsynaptic potentials at the trigger zone of a postsynaptic neuron. Presynaptic neurons 1, 3, and 5 release excitatory neurotransmitters (red dots) that generate excitatory postsynaptic potentials (EPSPs) (red arrows) in the membrane of a postsynaptic neuron. Presynaptic neurons 2 and 4 release inhibitory neurotransmitters (purple dots) that generate inhibitory postsynaptic potentials (IPSPs) (purple arrows) in the membrane of the postsynaptic neuron. The net summation of these EPSPs and IPSPs determines whether an action potential will be generated at the trigger zone of the postsynaptic neuron.

If the net summation of EPSPs and IPSPs is a depolarization that reaches threshold, then an action potential will occur at the trigger zone of a postsynaptic neuron.

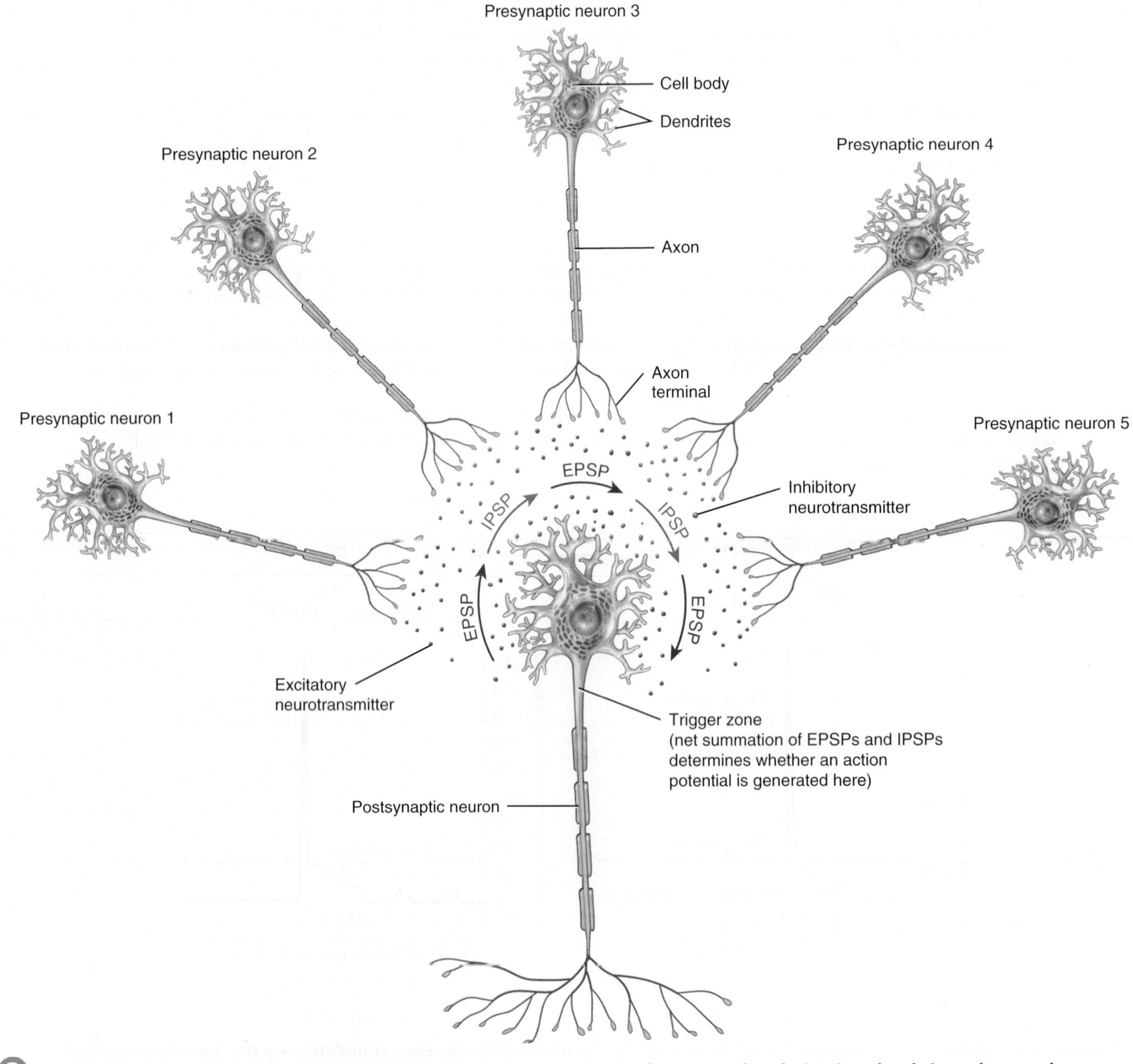

Suppose that the net summation of the EPSPs and IPSPs shown in this figure is a depolarization that brings the membrane potential of the trigger zone of the postsynaptic neuron to −60 mV. Will an action potential occur in the postsynaptic neuron?

CLINICAL CONNECTION | *Strychnine Poisoning*

The importance of inhibitory neurons can be appreciated by observing what happens when their activity is blocked. Normally, inhibitory neurons in the spinal cord called *Renshaw cells* release the neurotransmitter glycine at inhibitory synapses with somatic motor neurons. This inhibitory input to their motor neurons prevents excessive contraction of skeletal muscles. **Strychnine** (STRIK-nīn) is a lethal poison that is mainly used as a pesticide to control rats, moles, gophers, and coyotes. When ingested, it binds to and blocks glycine receptors. The normal, delicate balance between excitation and inhibition in the CNS is disturbed, and motor neurons generate nerve impulses without restraint. All skeletal muscles, including the diaphragm, contract fully and remain contracted. Because the diaphragm cannot relax, the victim cannot inhale, and suffocation results. •

lowing an EPSP, subsequent stimuli can more easily generate a nerve impulse through summation because the neuron is partially depolarized.

2. ***Nerve impulse(s).*** If the total excitatory effects are greater than the total inhibitory effects and threshold is reached, one or more nerve impulses (action potentials) will be triggered. Impulses continue to be generated as long as the EPSP is at or above the threshold level.
3. ***IPSP.*** If the total inhibitory effects are greater than the excitatory effects, the membrane hyperpolarizes (IPSP). The result is inhibition of the postsynaptic neuron and an inability to generate a nerve impulse.

Table 12.3 summarizes the structural and functional elements of a neuron.

✔ CHECKPOINT

13. How is neurotransmitter removed from the synaptic cleft?
14. How are excitatory and inhibitory postsynaptic potentials similar and different?
15. Why are action potentials said to be "all-or-none," and EPSPs and IPSPs are described as "graded"?

TABLE 12.3

Summary of Neuronal Structure and Function

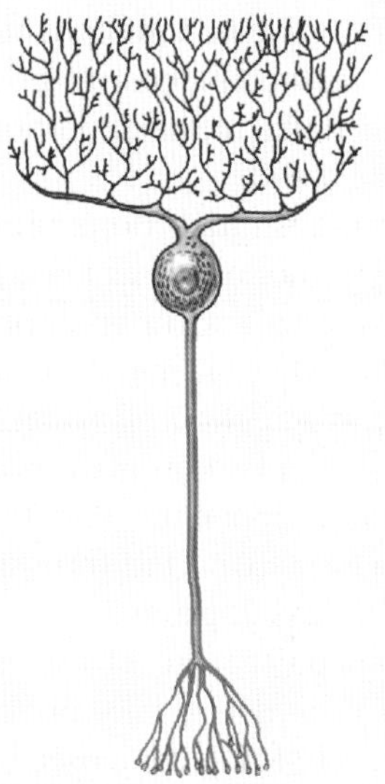

STRUCTURE	FUNCTIONS
Dendrites	Receive stimuli through activation of ligand-gated or mechanically gated ion channels; in sensory neurons, produce generator or receptor potentials; in motor neurons and interneurons, produce excitatory and inhibitory postsynaptic potentials (EPSPs and IPSPs).
Cell body	Receives stimuli and produces EPSPs and IPSPs through activation of ligand-gated ion channels.
Junction of axon hillock and initial segment of axon	Trigger zone in many neurons; integrates EPSPs and IPSPs and, if sum is depolarization that reaches threshold, initiates action potential (nerve impulse).
Axon	Propagates nerve impulses from initial segment (or from dendrites of sensory neurons) to axon terminals in self-regenerating manner; impulse amplitude does not change as it propagates along axon.
Axon terminals and synaptic end bulbs (or varicosities)	Inflow of Ca^{2+} caused by depolarizing phase of nerve impulse triggers exocytosis of neurotransmitter from synaptic vesicles.

Key:

- Plasma membrane includes chemically gated channels
- Plasma membrane includes voltage-gated Na^+ and K^+ channels
- Plasma membrane includes voltage-gated Ca^{2+} channels

12.5 NEUROTRANSMITTERS

OBJECTIVE

- Describe the classes and functions of neurotransmitters.

About 100 substances are either known or suspected neurotransmitters. Some neurotransmitters bind to their receptors and act quickly to open or close ion channels in the membrane. Others act more slowly via second-messenger systems to influence chemical reactions inside cells. The result of either process can be excitation or inhibition of postsynaptic neurons. Many neurotransmitters are also hormones released into the bloodstream by endocrine cells in organs throughout the body. Within the brain, certain neurons, called **neurosecretory cells,** also secrete hormones. Neurotransmitters can be divided into two classes based on size: small-molecule neurotransmitters and neuropeptides (Figure 12.26).

Small-Molecule Neurotransmitters

The small-molecule neurotransmitters include acetylcholine, amino acids, biogenic amines, ATP and other purines, nitric oxide, and carbon monoxide.

Acetylcholine

The best-studied neurotransmitter is **acetylcholine (ACh)** (a-sē′-til-KŌ-lēn), which is released by many PNS neurons and by some CNS neurons. ACh is an excitatory neurotransmitter at some synapses, such as the neuromuscular junction, where the binding of ACh to ionotropic receptors opens cation channels (see Figure 12.23a). It is also an inhibitory neurotransmitter at other synapses, where it binds to metabotropic receptors coupled to G proteins that open K^+ channels (see Figure 12.23c). For example, ACh slows heart rate at inhibitory synapses made by parasympathetic neurons of the vagus (X) nerve. The enzyme *acetylcholinesterase (AChE)* (a-sē′-til-kō′-lin-ES-ter-ās) inactivates ACh by splitting it into acetate and choline fragments.

Amino Acids

Several amino acids are neurotransmitters in the CNS. **Glutamate** (gloo-TA-māt) (glutamic acid) and **aspartate** (as-PAR-tāt) (aspartic acid) have powerful excitatory effects. Most excitatory neurons in the CNS and perhaps half of the synapses in the brain communicate via glutamate. At some glutamate synapses, binding of the neurotransmitter to ionotropic receptors opens cation channels. The consequent inflow of cations (mainly Na^+ ions) produces an EPSP. Inactivation of glutamate occurs via reuptake. Glutamate transporters actively transport glutamate back into the synaptic end bulbs and neighboring neuroglia.

Gamma aminobutyric (GAM-ma am-i-nō-bu-TIR-ik) **acid (GABA)** and **glycine** are important inhibitory neurotransmitters. At many synapses, the binding of GABA to ionotropic receptors opens Cl^- channels (see Figure 12.23b). GABA is found only in the CNS, where it is the most common inhibitory neurotransmitter. As many as one-third of all brain synapses use GABA. Antianxiety drugs such as diazepam (Valium®) enhance the action of GABA. Like GABA, the binding of glycine to ionotropic receptors opens Cl^- channels. About half of the inhibitory synapses in the spinal cord use the amino acid glycine; the rest use GABA.

CLINICAL CONNECTION | *Excitotoxicity*

A high level of glutamate in the interstitial fluid of the CNS causes **excitotoxicity** (ek-sī′-tō-tok-SIS-i-tē)—destruction of neurons through prolonged activation of excitatory synaptic transmission. The most common cause of excitotoxicity is oxygen deprivation of the brain due to ischemia (inadequate blood flow), as happens during a stroke. Lack of oxygen causes the glutamate transporters to fail, and glutamate accumulates in the interstitial spaces between neurons and glia, literally stimulating the neurons to death. Clinical trials are underway to see whether antiglutamate drugs administered after a stroke can offer some protection from excitotoxicity. •

Biogenic Amines

Certain amino acids are modified and decarboxylated (carboxyl group removed) to produce biogenic amines. Those that are prevalent in the nervous system include norepinephrine, epinephrine, dopamine, and serotonin. Most biogenic amines bind to metabotropic receptors; there are many different types of metabotropic receptors for each biogenic amine. Biogenic amines may cause either excitation or inhibition, depending on the type of metabotropic receptor at the synapse.

Norepinephrine (NE) (nor′-ep-i-NEF-rin) plays roles in arousal (awakening from deep sleep), dreaming, and regulating mood. A smaller number of neurons in the brain use **epinephrine** as a neurotransmitter. Both epinephrine and norepinephrine also serve as hormones. Cells of the adrenal medulla, the inner portion of the adrenal gland, release them into the blood.

Brain neurons containing the neurotransmitter **dopamine (DA)** (DŌ-pa-mēn) are active during emotional responses, addictive behaviors, and pleasurable experiences. In addition, dopamine-releasing neurons help regulate skeletal muscle tone and some aspects of movement due to contraction of skeletal muscles. The muscular stiffness that occurs in Parkinson disease is due to degeneration of neurons that release dopamine (see Disorders: Homeostatic Imbalances in Chapter 16). One form of schizophrenia is due to accumulation of excess dopamine.

Norepinephrine, dopamine, and epinephrine are classified chemically as **catecholamines** (kat-e-KŌL-a-mēns). They all include an amino group ($—NH_2$) and a catechol ring composed of six carbons and two adjacent hydroxyl (—OH) groups. Catecholamines are synthesized from the amino acid tyrosine. Inactivation of catecholamines occurs via reuptake into synaptic end bulbs. Then they are either recycled back into the synaptic vesicles or destroyed by enzymes. The two enzymes that break down catecholamines are **catechol-*O*-methyltransferase** (kat′-e-kōl-ō-meth-il-TRANS-fer-ās), or **COMT,** and **monoamine oxidase** (mon-ō-AM-īn OK-si-dās), or **MAO.**

Serotonin (ser′-ō-TŌ-nin), which is also known as **5-hydroxytryptamine (5-HT),** is concentrated in the neurons in a part of the brain called the raphe nucleus. It is thought to be involved in

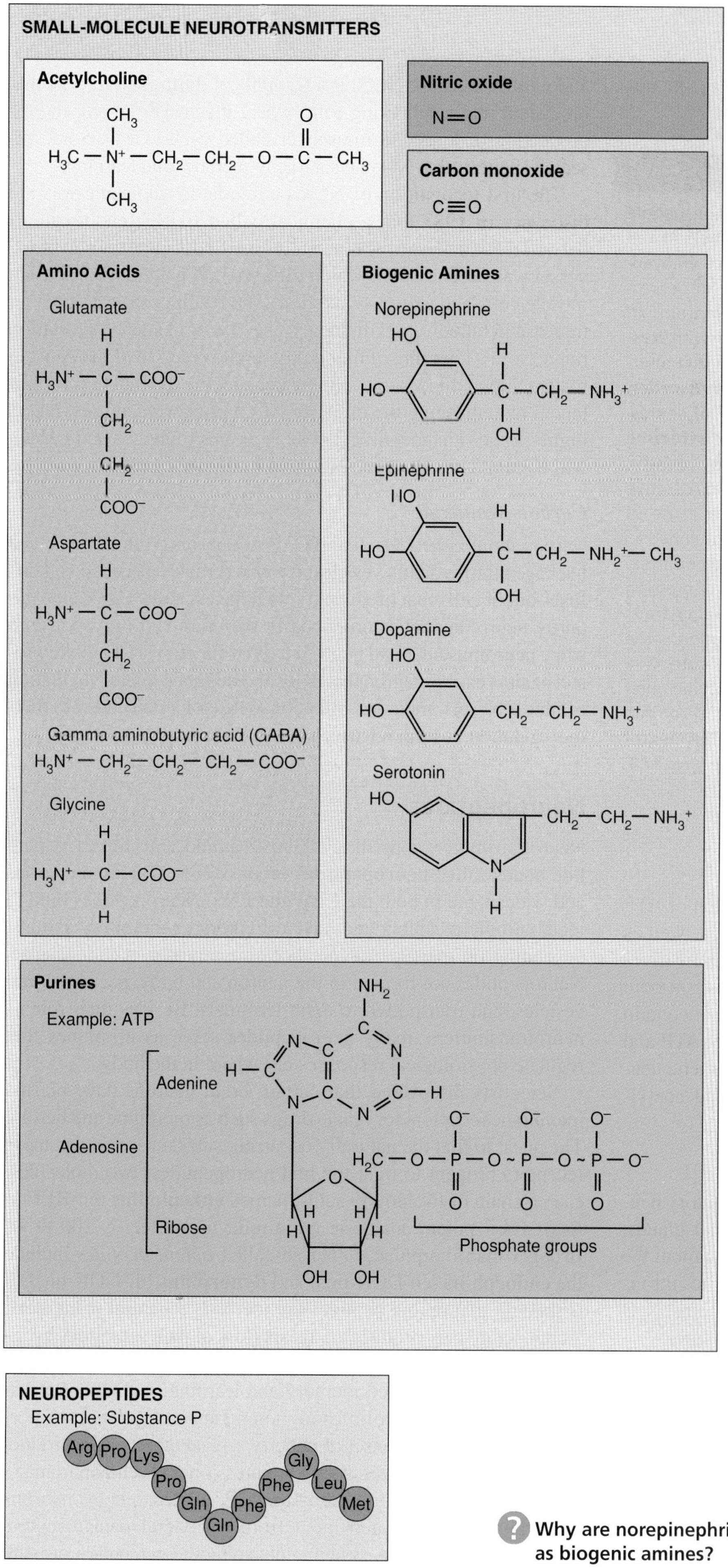
SMALL-MOLECULE NEUROTRANSMITTERS
Acetylcholine
Nitric oxide
N=O
Carbon monoxide
C≡O
Amino Acids
Glutamate
Aspartate
Gamma aminobutyric acid (GABA)
Glycine
Biogenic Amines
Norepinephrine
Epinephrine
Dopamine
Serotonin
Purines
Example: ATP
Adenine
Adenosine
Ribose
Phosphate groups
NEUROPEPTIDES
Example: Substance P
Arg
Pro
Lys
Pro
Gln
Gln
Phe
Phe
Gly
Leu
Met

Figure 12.26 Neurotransmitters. Neurotransmitters are divided into two major classes based on size: small-molecule neurotransmitters and neuropeptides. Small-molecule neurotransmitters include acetylcholine, amino acids, biogenic amines, purines, nitric oxide, and carbon monoxide. The neuropeptide shown is substance P, which consists of 11 amino acids linked by peptide bonds in the following order: arginine (Arg), proline (Pro), lysine (Lys), proline, glutamine (Gln), glutamine, phenylalanine (Phe), phenylalanine, glycine (Gly), leucine (Leu), and methionine (Met).

Neurotransmitters are chemical substances that neurons use to communicate with other neurons, muscle fibers, and glands.

Why are norepinephrine, epinephrine, dopamine, and serotonin classified as biogenic amines?

sensory perception, temperature regulation, control of mood, appetite, and the induction of sleep.

CLINICAL CONNECTION | *Depression*

Depression is a disorder that affects over 18 million people each year in the United States. People who are depressed feel sad and helpless, have a lack of interest in activities that they once enjoyed, and experience suicidal thoughts. There are several types of depression. A person with **major depression** experiences symptoms of depression that last for more than two weeks. A person with **dysthymia** (dis-THĪ-mē-a) experiences episodes of depression that alternate with periods of feeling normal. A person with **bipolar disorder,** or *manic-depressive illness,* experiences recurrent episodes of depression and extreme elation (mania). A person with **seasonal affective disorder (SAD)** experiences depression during the winter months, when day length is short (see Clinical Connection: Seasonal Affective Disorder and Jet Lag in Chapter 18). Although the exact cause of depression is unknown, research suggests that depression is linked to an imbalance of the neurotransmitters serotonin, norepinephrine, and dopamine in the brain. Factors that may contribute to depression include heredity, stress, chronic illnesses, certain personality traits (such as low self-esteem), and hormonal changes. Medication is the most common treatment for depression. For example, **selective serotonin reuptake inhibitors (SSRIs)** are drugs that provide relief from some forms of depression. By inhibiting reuptake of serotonin by serotonin transporters, SSRIs prolong the activity of this neurotransmitter at synapses in the brain. SSRIs include fluoxetine (Prozac®), paroxetine (Paxil®), and sertraline (Zoloft®). •

ATP and Other Purines

The characteristic ring structure of the adenosine portion of ATP (Figure 12.26) is called a purine ring. Adenosine itself, as well as its triphosphate, diphosphate, and monophosphate derivatives (ATP, ADP, and AMP), is an excitatory neurotransmitter in both the CNS and the PNS. Most of the synaptic vesicles that contain ATP also contain another neurotransmitter. In the PNS, ATP and norepinephrine are released together from some sympathetic neurons; some parasympathetic neurons release ATP and acetylcholine in the same vesicles.

Nitric Oxide

The simple gas **nitric oxide (NO)** is an important excitatory neurotransmitter secreted in the brain, spinal cord, adrenal glands, and nerves to the penis and has widespread effects throughout the body. NO contains a single nitrogen atom, in contrast to nitrous oxide (N_2O), or laughing gas, which has two nitrogen atoms. N_2O is sometimes used as an anesthetic during dental procedures.

The enzyme **nitric oxide synthase (NOS)** catalyzes formation of NO from the amino acid arginine. Based on the presence of NOS, it is estimated that more than 2% of the neurons in the brain produce NO. Unlike all previously known neurotransmitters, NO is not synthesized in advance and packaged into synaptic vesicles. Rather, it is formed on demand and acts immediately. Its action is brief because NO is a highly reactive free radical. It exists for less than 10 seconds before it combines with oxygen and water to form inactive nitrates and nitrites. Because NO is lipid-soluble, it diffuses from cells that produce it into neighboring cells, where it activates an enzyme for production of a second messenger called cyclic GMP. Some research suggests that NO plays a role in memory and learning.

The first recognition of NO as a regulatory molecule was the discovery in 1987 that a chemical called EDRF (endothelium-derived relaxing factor) was actually NO. Endothelial cells in blood vessel walls release NO, which diffuses into neighboring smooth muscle cells and causes relaxation. The result is vasodilation, an increase in blood vessel diameter. The effects of such vasodilation range from a lowering of blood pressure to erection of the penis in males. Sildenafil (Viagra®) alleviates erectile dysfunction (impotence) by enhancing the effect of NO. In larger quantities, NO is highly toxic. Phagocytic cells, such as macrophages and certain white blood cells, produce NO to kill microbes and tumor cells.

Carbon Monoxide

Carbon monoxide (CO), like NO, is not produced in advance and packaged into synaptic vesicles. It too is formed as needed and diffuses out of cells that produce it into adjacent cells. CO is an excitatory neurotransmitter produced in the brain and in response to some neuromuscular and neuroglandular functions. CO might protect against excess neuronal activity and might be related to dilation of blood vessels, memory, olfaction (sense of smell), vision, thermoregulation, insulin release, and anti-inflammatory activity.

Neuropeptides

Neurotransmitters consisting of 3 to 40 amino acids linked by peptide bonds called **neuropeptides** (noor-ō-PEP-tīds) are numerous and widespread in both the CNS and PNS. Neuropeptides bind to metabotropic receptors and have excitatory or inhibitory actions, depending on the type of metabotropic receptor at the synapse. Neuropeptides are formed in the neuron cell body, packaged into vesicles, and transported to axon terminals. Besides their role as neurotransmitters, many neuropeptides serve as hormones that regulate physiological responses elsewhere in the body.

Scientists discovered that certain brain neurons have plasma membrane receptors for opiate drugs such as morphine and heroin. The quest to find the naturally occurring substances that use these receptors brought to light the first neuropeptides: two molecules, each a chain of five amino acids, named **enkephalins** (en-KEF-a-lins). Their potent analgesic (pain-relieving) effect is 200 times stronger than morphine. Other so-called *opioid peptides* include the **endorphins** (en-DOR-fins) and **dynorphins** (dī-NOR-fins). It is thought that opioid peptides are the body's natural painkillers. Acupuncture may produce analgesia (loss of pain sensation) by increasing the release of opioids. These neuropeptides have also been linked to improved memory and learning; feelings of pleasure or euphoria; control of body temperature; regulation of hormones that affect the onset of puberty, sexual drive, and reproduction; and mental illnesses such as depression and schizophrenia.

Another neuropeptide, **substance P,** is released by neurons that transmit pain-related input from peripheral pain receptors into the central nervous system, enhancing the perception of pain.

Enkephalin and endorphin suppress the release of substance P, thus decreasing the number of nerve impulses being relayed to the brain for pain sensations. Substance P has also been shown to counter the effects of certain nerve-damaging chemicals, prompting speculation that it might prove useful as a treatment for nerve degeneration.

Table 12.4 provides brief descriptions of these neuropeptides, as well as others that will be discussed in later chapters.

CLINICAL CONNECTION | *Modifying the Effects of Neurotransmitters*

Substances naturally present in the body as well as drugs and toxins can **modify the effects of neurotransmitters** in several ways:

1. Neurotransmitter synthesis can be stimulated or inhibited. For instance, many patients with Parkinson disease (see Disorders: Homeostatic Imbalances in Chapter 16) receive benefit from the drug L-dopa because it is a precursor of dopamine. For a limited period of time, taking L-dopa boosts dopamine production in affected brain areas.
2. Neurotransmitter release can be enhanced or blocked. Amphetamines promote release of dopamine and norepinephrine. Botulinum toxin causes paralysis by blocking release of acetylcholine from somatic motor neurons.
3. The neurotransmitter receptors can be activated or blocked. An agent that binds to receptors and enhances or mimics the effect of a natural neurotransmitter is an **agonist** (AG-ōn-ist). Isoproterenol (Isuprel®) is a powerful agonist of epinephrine and norepinephrine. It can be used to dilate the airways during an asthma attack. An agent that binds to and blocks neurotransmitter receptors is an **antagonist** (an-TAG-ō-nist). Zyprexa®, a drug prescribed for schizophrenia, is an antagonist of serotonin and dopamine.
4. Neurotransmitter removal can be stimulated or inhibited. For example, cocaine produces euphoria—intensely pleasurable feelings—by blocking transporters for dopamine reuptake. This action allows dopamine to linger longer in synaptic clefts, producing excessive stimulation of certain brain regions. •

✓ **CHECKPOINT**

16. Which neurotransmitters are excitatory and which are inhibitory? How do they exert their effects?
17. In what ways is nitric oxide different from all previously known neurotransmitters?

12.6 NEURAL CIRCUITS

OBJECTIVE

- Identify the various types of neural circuits in the nervous system

The CNS contains billions of neurons organized into complicated networks called **neural circuits,** functional groups of neurons that process specific types of information. In a **simple series circuit,** a presynaptic neuron stimulates a single postsynaptic neuron. The second neuron then stimulates another, and so on. However, most neural circuits are more complex.

A single presynaptic neuron may synapse with several postsynaptic neurons. Such an arrangement, called **divergence,** permits one presynaptic neuron to influence several postsynaptic neurons (or several muscle fibers or gland cells) at the same time. In a **diverging circuit,** the nerve impulse from a single presynaptic neuron causes the stimulation of increasing numbers of cells along the circuit (Figure 12.27a). For example, a small number of neurons in the brain that govern a particular body movement stimulate a much larger number of neurons in the spinal cord. Sensory signals are also arranged in diverging circuits, allowing a sensory impulse to be relayed to several regions of the brain. This arrangement amplifies the signal.

In another arrangement, called **convergence,** several presynaptic neurons synapse with a single postsynaptic neuron. This arrangement permits more effective stimulation or inhibition of the postsynaptic neuron. In a **converging circuit** (Figure 12.27b), the

TABLE 12.4

Neuropeptides

SUBSTANCE	DESCRIPTION
Substance P	Found in sensory neurons, spinal cord pathways, and parts of brain associated with pain; enhances perception of pain.
Enkephalins	Inhibit pain impulses by suppressing release of substance P; may have role in memory and learning, control of body temperature, sexual activity, and mental illness.
Endorphins	Inhibit pain by blocking release of substance P; may have role in memory and learning, sexual activity, control of body temperature, and mental illness.
Dynorphins	May be related to controlling pain and registering emotions.
Hypothalamic releasing and inhibiting hormones	Produced by hypothalamus; regulate release of hormones by anterior pituitary.
Angiotensin II	Stimulates thirst; may regulate blood pressure in brain. As a hormone, causes vasoconstriction and promotes release of aldosterone, which increases rate of salt and water reabsorption by kidneys.
Cholecystokinin (CCK)	Found in brain and small intestine; may regulate feeding as a "stop eating" signal. As a hormone, regulates pancreatic enzyme secretion during digestion, and contraction of smooth muscle in gastrointestinal tract.

Figure 12.27 Examples of neural circuits.

A neural circuit is a functional group of neurons that processes a specific kind of information.

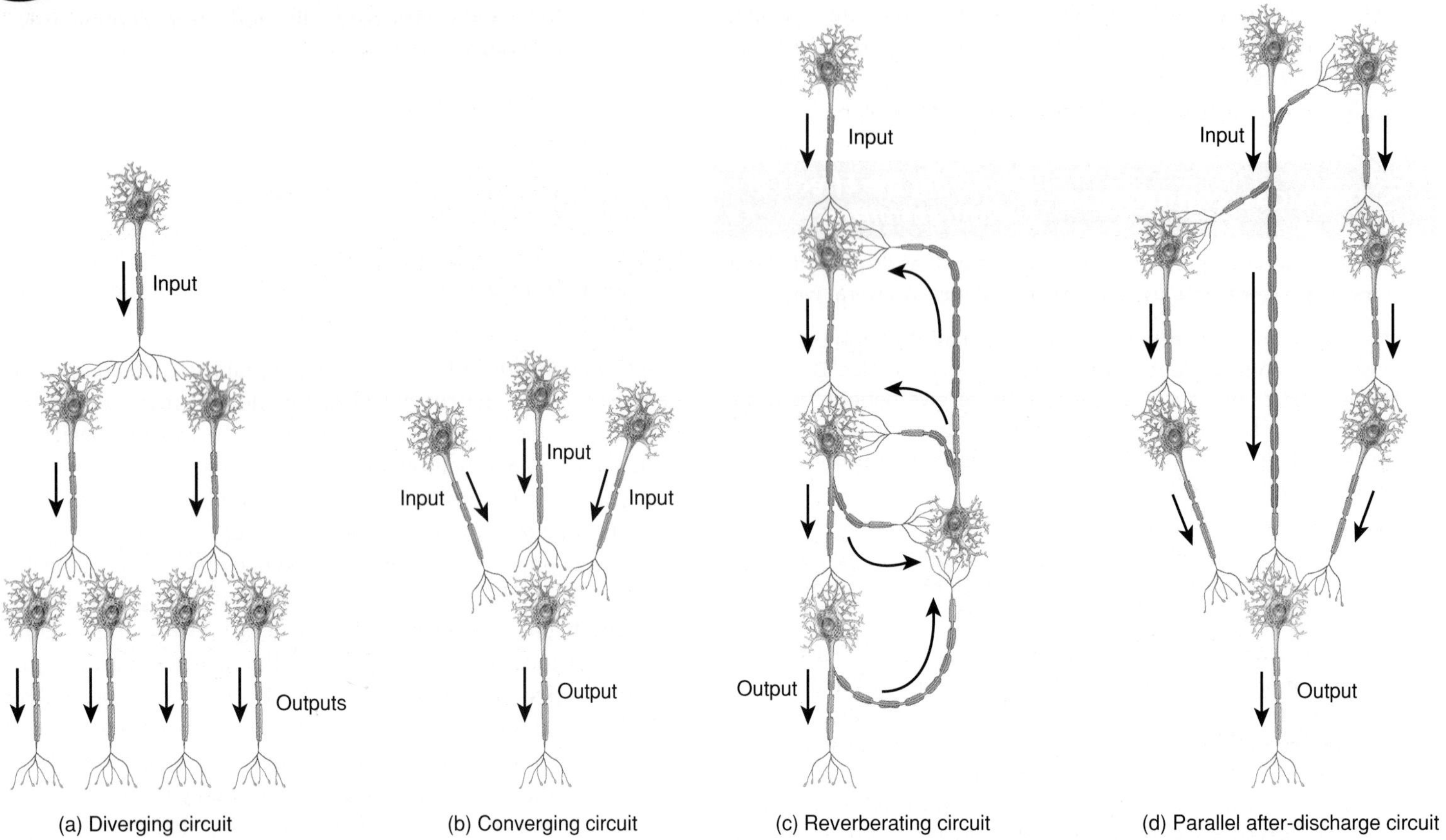

A motor neuron in the spinal cord typically receives input from neurons that originate in several different regions of the brain. Is this an example of convergence or divergence?

postsynaptic neuron receives nerve impulses from several different sources. For example, a single motor neuron that synapses with skeletal muscle fibers at neuromuscular junctions receives input from several pathways that originate in different brain regions.

Some circuits are organized so that stimulation of the presynaptic cell causes the postsynaptic cell to transmit a series of nerve impulses. One such circuit is called a **reverberating circuit** (Figure 12.27c). In this pattern, the incoming impulse stimulates the first neuron, which stimulates the second, which stimulates the third, and so on. Branches from later neurons synapse with earlier ones. This arrangement sends impulses back through the circuit again and again. The output signal may last from a few seconds to many hours, depending on the number of synapses and the arrangement of neurons in the circuit. Inhibitory neurons may turn off a reverberating circuit after a period of time. Among the body responses thought to be the result of output signals from reverberating circuits are breathing, coordinated muscular activities, waking up, and short-term memory.

A fourth type of circuit is the **parallel after-discharge circuit** (Figure 12.27d). In this circuit, a single presynaptic cell stimulates a group of neurons, each of which synapses with a common postsynaptic cell. A differing number of synapses between the first and last neurons imposes varying synaptic delays, so that the last neuron exhibits multiple EPSPs or IPSPs. If the input is excitatory, the postsynaptic neuron then can send out a stream of impulses in quick succession. Parallel after-discharge circuits may be involved in precise activities such as mathematical calculations.

CHECKPOINT

18. What is a neural circuit?
19. What are the functions of diverging, converging, reverberating, and parallel after-discharge circuits?

12.7 REGENERATION AND REPAIR OF NERVOUS TISSUE

OBJECTIVES

- Define plasticity and neurogenesis.
- Describe the events involved in damage and repair of peripheral nerves.

Throughout your life, your nervous system exhibits **plasticity** (plas-TIS-i-tē), the capability to change based on experience. At the level of individual neurons, the changes that can occur include the sprouting of new dendrites, synthesis of new proteins, and changes in synaptic contacts with other neurons. Undoubtedly, both chemical and electrical signals drive the changes that occur. Despite this plasticity, mammalian neurons have very limited powers of **regeneration,** the capability to replicate or repair themselves. In the PNS, damage to dendrites and myelinated axons may be repaired if the cell body remains intact and if the Schwann cells that produce myelination remain active. In the CNS, little or no repair of damage to neurons occurs. Even when the cell body remains intact, a severed axon cannot be repaired or regrown.

Neurogenesis in the CNS

Neurogenesis (noo′-rō-JEN-e-sis)—the birth of new neurons from undifferentiated stem cells—occurs regularly in some animals. For example, new neurons appear and disappear every year in some songbirds. Until recently, the dogma in humans and other primates was "no new neurons" in the adult brain. Then, in 1992, Canadian researchers published their unexpected finding that **epidermal growth factor (EGF)** stimulated cells taken from the brains of adult mice to proliferate into both neurons and astrocytes. Previously, EGF was known to trigger mitosis in a variety of nonneuronal cells and to promote wound healing and tissue regeneration. In 1998, scientists discovered that significant numbers of new neurons do arise in the adult human hippocampus, an area of the brain that is crucial for learning.

The nearly complete lack of neurogenesis in other regions of the brain and spinal cord seems to result from two factors: (1) inhibitory influences from neuroglia, particularly oligodendrocytes, and (2) absence of growth-stimulating cues that were present during fetal development. Axons in the CNS are myelinated by oligodendrocytes rather than Schwann cells, and this CNS myelin is one of the factors inhibiting regeneration of neurons. Perhaps this same mechanism stops axonal growth once a target region has been reached during development. Also, after axonal damage, nearby astrocytes proliferate rapidly, forming a type of scar tissue that acts as a physical barrier to regeneration. Thus, injury of the brain or spinal cord usually is permanent. Ongoing research seeks ways to improve the environment for existing spinal cord axons to bridge the injury gap. Scientists also are trying to find ways to stimulate dormant stem cells to replace neurons lost through damage or disease and to develop tissue-cultured neurons that can be used for transplantation purposes.

Figure 12.28 **Damage and repair of a neuron in the PNS.**

Myelinated axons in the peripheral nervous system may be repaired if the cell body remains intact and if Schwann cells remain active.

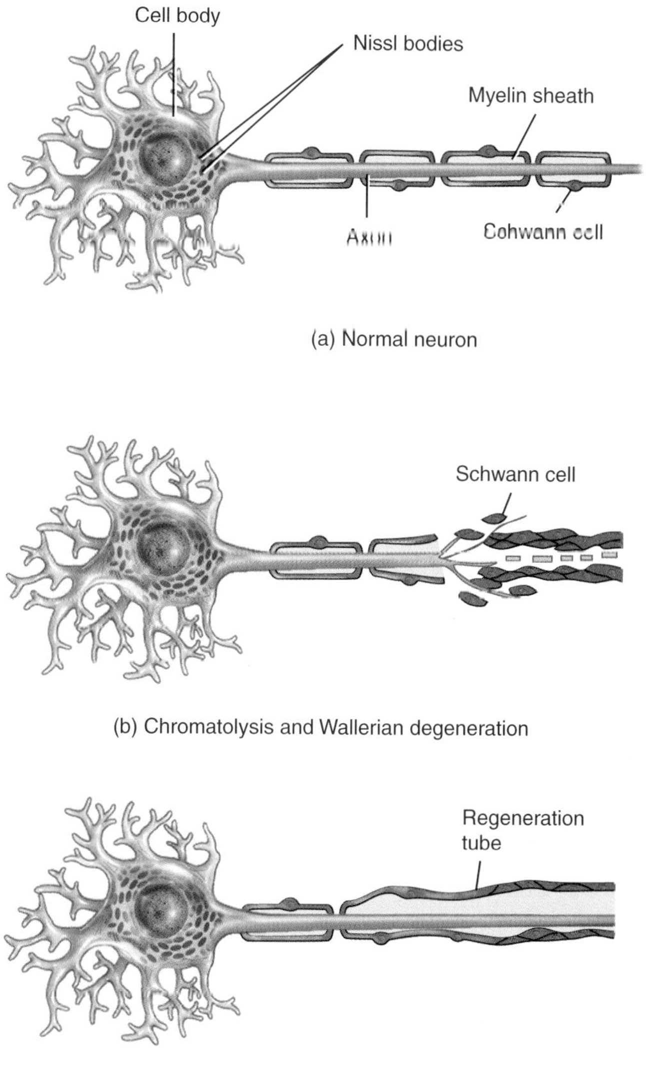

What is the role of the neurolemma in regeneration?

Damage and Repair in the PNS

Axons and dendrites that are associated with a neurolemma may undergo repair if the cell body is intact, if the Schwann cells are functional, and if scar tissue formation does not occur too rapidly (Figure 12.28). Most nerves in the PNS consist of processes that are covered with a neurolemma. A person who injures axons of a nerve in an upper limb, for example, has a good chance of regaining nerve function.

When there is damage to an axon, changes usually occur both in the cell body of the affected neuron and in the portion of the axon distal to the site of injury. Changes also may occur in the portion of the axon proximal to the site of injury.

About 24 to 48 hours after injury to a process of a normal peripheral neuron (Figure 12.28a), the Nissl bodies break up into fine granular masses. This alteration is called **chromatolysis** (krō′-ma-TOL-i-sis; *chromato-* = color; *-lysis* = destruction). By the third to fifth day, the part of the axon distal to the damaged region becomes slightly swollen and then breaks up into fragments; the myelin sheath also deteriorates (Figure 12.28b). Even though the axon and myelin sheath degenerate, the neurolemma remains. Degeneration of the distal portion of the axon and myelin sheath is called **Wallerian degeneration** (waw-LE-rē′-an).

Following chromatolysis, signs of recovery in the cell body become evident. Macrophages phagocytize the debris. Synthesis of RNA and protein accelerates, which favors rebuilding or **regeneration** of the axon. The Schwann cells on either side of the injured site multiply by mitosis, grow toward each other, and may form a **regeneration tube** across the injured area (Figure 12.28c). The tube guides growth of a new axon from the proximal area across the injured area into the distal area previously occupied by the original axon. However, new axons cannot grow if the gap at the site of injury is too large or if the gap becomes filled with collagen fibers.

During the first few days following damage, buds of regenerating axons begin to invade the tube formed by the Schwann cells (Figure 12.28b). Axons from the proximal area grow at a rate of about 1.5 mm (0.06 in.) per day across the area of damage, find their way into the distal regeneration tubes, and grow toward the distally located receptors and effectors. Thus, some sensory and motor connections are reestablished and some functions restored. In time, the Schwann cells form a new myelin sheath.

✔CHECKPOINT

20. What factors contribute to a lack of neurogenesis in most parts of the brain?
21. What is the function of the regeneration tube in repair of neurons?

DISORDERS: HOMEOSTATIC IMBALANCES

Multiple Sclerosis

Multiple sclerosis (MS) is a disease that causes a progressive destruction of myelin sheaths surrounding neurons in the CNS. It afflicts about 350,000 people in the United States and 2 million people worldwide. It usually appears between the ages of 20 and 40, affecting females twice as often as males. MS is most common in whites, less common in blacks, and rare in Asians. MS is an autoimmune disease—the body's own immune system spearheads the attack. The condition's name describes the anatomical pathology: In *multiple* regions the myelin sheaths deteriorate to *scleroses,* which are hardened scars or plaques. Magnetic resonance imaging (MRI) studies reveal numerous plaques in the white matter of the brain and spinal cord. The destruction of myelin sheaths slows and then short-circuits propagation of nerve impulses.

The most common form of the condition is relapsing–remitting MS, which usually appears in early adulthood. The first symptoms may include a feeling of heaviness or weakness in the muscles, abnormal sensations, or double vision. An attack is followed by a period of remission during which the symptoms temporarily disappear. One attack follows another over the years, usually every year or two. The result is a progressive loss of function interspersed with remission periods, during which symptoms abate.

Although the cause of MS is unclear, both genetic susceptibility and exposure to some environmental factor (perhaps a herpes virus) appear to contribute. Since 1993, many patients with relapsing–remitting MS have been treated with injections of beta interferon. This treatment lengthens the time between relapses, decreases the severity of relapses, and slows formation of new lesions in some cases. Unfortunately, not all MS patients can tolerate beta interferon, and therapy becomes less effective as the disease progresses.

Epilepsy

Epilepsy (ep′-i-LEP-sē) is characterized by short, recurrent attacks of motor, sensory, or psychological malfunction, although it almost never affects intelligence. The attacks, called *epileptic seizures,* afflict about 1% of the world's population. They are initiated by abnormal, synchronous electrical discharges from millions of neurons in the brain, perhaps resulting from abnormal reverberating circuits. The discharges stimulate many of the neurons to send nerve impulses over their conduction pathways. As a result, lights, noise, or smells may be sensed when the eyes, ears, and nose have not been stimulated. Moreover, the skeletal muscles of a person having a seizure may contract involuntarily. *Partial seizures* begin in a small area on one side of the brain and produce milder symptoms; *generalized seizures* involve larger areas on both sides of the brain and loss of consciousness.

Epilepsy has many causes, including brain damage at birth (the most common cause); metabolic disturbances (hypoglycemia, hypocalcemia, uremia, hypoxia); infections (encephalitis or meningitis); toxins (alcohol, tranquilizers, hallucinogens); vascular disturbances (hemorrhage, hypotension); head injuries; and tumors and abscesses of the brain. Seizures associated with fever are most common in children under the age of two. However, most epileptic seizures have no demonstrable cause.

Epileptic seizures often can be eliminated or alleviated by antiepileptic drugs, such as phenytoin, carbamazepine, and valproate sodium. An implantable device that stimulates the vagus (X) nerve has produced dramatic results in reducing seizures in some patients whose epilepsy was not well-controlled by drugs. In very severe cases, surgical intervention may be an option.

MEDICAL TERMINOLOGY

Guillain-Barré syndrome (GBS) (GHE-an ba-RĀ) An acute demyelinating disorder in which macrophages strip myelin from axons in the PNS. It is the most common cause of acute paralysis in North America and Europe and may result from the immune system's response to a bacterial infection. Most patients recover completely or partially, but about 15% remain paralyzed.

Neuroblastoma (noor-ō-blas-TŌ-ma) A malignant tumor that consists of immature nerve cells (neuroblasts); occurs most commonly in the abdomen and most frequently in the adrenal glands. Although rare, it is the most common tumor in infants.

Neuropathy (noo-ROP-a-thē; *neuro-* = a nerve; *-pathy* = disease) Any disorder that affects the nervous system but particularly a disorder of a cranial or spinal nerve. An example is *facial neuropathy* (Bell's palsy), a disorder of the facial (VII) nerve.

Rabies (RĀ-bēz; *rabi-* = mad, raving) A fatal disease caused by a virus that reaches the CNS via fast axonal transport. It is usually transmitted by the bite of an infected dog or other meat-eating animal. The symptoms are excitement, aggressiveness, and madness, followed by paralysis and death.

CHAPTER REVIEW AND RESOURCE SUMMARY

Review

12.1 Overview of the Nervous System

1. The central nervous system (CNS) consists of the brain and spinal cord.
2. The peripheral nervous system (PNS) consists of all nervous tissue outside the CNS. Components of the PNS include nerves, ganglia, enteric plexuses, and sensory receptors.
3. The PNS is divided into a somatic nervous system (SNS), autonomic nervous system (ANS), and enteric nervous system (ENS).
4. The SNS consists of sensory neurons that conduct impulses from somatic and special sense receptors to the CNS and motor neurons from the CNS to skeletal muscles.
5. The ANS contains sensory neurons from visceral organs and motor neurons that convey impulses from the CNS to smooth muscle tissue, cardiac muscle tissue, and glands.
6. The ENS consists of neurons in enteric plexuses in the gastrointestinal (GI) tract that function somewhat independently of the ANS and CNS. The ENS monitors chemical changes within the GI tract and stretching of its walls; the ENS also controls contraction of GI tract smooth muscle.
7. The nervous system helps maintain homeostasis and integrates all body activities by sensing changes (sensory function), interpreting them (integrative function), and reacting to them (motor function).

Resource

Anatomy Overview - Nervous System
Anatomy Overview - Nerve
Animation - Structure and Function of the Nervous System: System Organization

12.2 Histology of Nervous Tissue

1. Nervous tissue consists of neurons (nerve cells) and neuroglia. Neurons have the property of electrical excitability and are responsible for most unique functions of the nervous system: sensing, thinking, remembering, controlling muscle activity, and regulating glandular secretions.
2. Most neurons have three parts. The dendrites are the main receiving or input region. Integration occurs in the cell body, which includes typical cellular organelles. The output part typically is a single axon, which propagates nerve impulses toward another neuron, a muscle fiber, or a gland cell.
3. Synapses are the site of functional contact between two excitable cells. Axon terminals contain synaptic vesicles filled with neurotransmitter molecules.
4. Slow axonal transport and fast axonal transport are systems for conveying materials to and from the cell body and axon terminals.
5. On the basis of their structure, neurons are classified as multipolar, bipolar, or unipolar.
6. Neurons are functionally classified as sensory (afferent) neurons, motor (efferent) neurons, and interneurons. Sensory neurons carry sensory information into the CNS. Motor neurons carry information out of the CNS to effectors (muscles and glands). Interneurons are located within the CNS between sensory and motor neurons.
7. Neuroglia support, nurture, and protect neurons and maintain the interstitial fluid that bathes them. Neuroglia in the CNS include astrocytes, oligodendrocytes, microglia, and ependymal cells. Neuroglia in the PNS include Schwann cells and satellite cells.
8. Two types of neuroglia produce myelin sheaths: Oligodendrocytes myelinate axons in the CNS, and Schwann cells myelinate axons in the PNS.
9. White matter consists of aggregates of myelinated axons; gray matter contains cell bodies, dendrites, and axon terminals of neurons, unmyelinated axons, and neuroglia.
10. In the spinal cord, gray matter forms an H-shaped inner core that is surrounded by white matter. In the brain, a thin, superficial shell of gray matter covers the cerebral and cerebellar hemispheres.

Resource

Anatomy Overview - Nervous Tissue: Neuron
Anatomy Overview - Nervous Tissue: Neuroglia
Animation - Neuron Structure and Function
Animation - Factors That Affect Conduction Rates: Myelination
Figure 12.6 - Neuroglia of the Central Nervous System
Exercise - Paint a Neuron

12.3 Electrical Signals in Neurons

1. Neurons communicate with one another using graded potentials, which are used for short-distance communication only, and action potentials, which allow communication over long distances within the body.
2. The electrical signals produced by neurons and muscle fibers rely on four kinds of ion channels: leak channels, ligand-gated channels, mechanically gated channels, and voltage-gated channels. Table 12.1 summarizes the different types of ion channels in neurons.
3. A resting membrane potential exists across the plasma membrane of excitable cells that are unstimulated (at rest). The resting membrane potential exists because of a small buildup of negative ions in the cytosol along the inside surface of the membrane, and an equal buildup of positive ions in the extracellular fluid (ECF) along the outside surface of the membrane. A typical value for the resting membrane potential of a neuron is -70 mV. A cell that exhibits a membrane potential is polarized. The resting membrane potential is determined by three major factors: (1) unequal distribution of ions

Resource

Animation - Introduction to Membrane Potentials
Animation - Resting Membrane Potentials
Animation - Action Potentials
Animation - Graded Potentials
Figure 12.12 - Resting Membrane Potential
Figure 12.18 - Action Potential or Impulse
Exercise - Ruling the Gated Channels
Exercise - Keep the Resting Potential
Exercise - Nervous Race

Review | **Resource**

in the ECF and cytosol; (2) inability of most cytosolic anions to leave the cell; and (3) electrogenic nature of the Na^+/K^+ ATPases.

4. A graded potential is a small deviation from the resting membrane potential that occurs because ligand-gated or mechanically gated channels open or close. A hyperpolarizing graded potential makes the membrane potential more negative (more polarized); a depolarizing graded potential makes the membrane potential less negative (less polarized). The amplitude of a graded potential varies, depending on the strength of the stimulus.
5. According to the all-or-none principle, if a stimulus is strong enough to generate an action potential, the impulse generated is of a constant size. A stronger stimulus does not generate a larger action potential. Instead, the greater the stimulus strength above threshold, the greater the frequency of the action potentials.
6. During an action potential, voltage-gated Na^+ and K^+ channels open and close in sequence. This results first in depolarization, the reversal of membrane polarization (from -70 mV to $+30$ mV). Then repolarization, the recovery of the resting membrane potential (from $+30$ mV to -70 mV), occurs.
7. During the first part of the refractory period (RP), another impulse cannot be generated at all (absolute RP); a little later, it can be triggered only by a larger-than-normal stimulus (relative RP).
8. Because an action potential travels from point to point along the membrane without getting smaller, it is useful for long-distance communication. Nerve impulse propagation in which the impulse "leaps" from one node of Ranvier to the next along a myelinated axon is saltatory conduction. Saltatory conduction is faster than continuous conduction.
9. Axons with larger diameters conduct impulses at higher speeds than do axons with smaller diameters.
10. The intensity of a stimulus is encoded in the frequency of action potentials and in the number of sensory neurons that are recruited. Table 12.2 compares graded potentials and action potentials.

Exercise - Define Stages of the Action Potential
Exercise - Investigate Types of Impulse Conduction
Concepts and Connections - Membrane Potentials

12.4 Signal Transmission at Synapses

1. A synapse is the functional junction between one neuron and another, or between a neuron and an effector such as a muscle or a gland. The two types of synapses are electrical and chemical.
2. A chemical synapse produces one-way information transfer—from a presynaptic neuron to a postsynaptic neuron.
3. An excitatory neurotransmitter is one that can depolarize the postsynaptic neuron's membrane, bringing the membrane potential closer to threshold. An inhibitory neurotransmitter hyperpolarizes the membrane of the postsynaptic neuron, moving it further from threshold.
4. There are two major types of neurotransmitter receptors: ionotropic receptors and metabotropic receptors. An ionotropic receptor contains a neurotransmitter binding site and an ion channel. A metabotropic receptor contains a neurotransmitter binding site and is coupled to a separate ion channel by a G protein.
5. Neurotransmitter is removed from the synaptic cleft in three ways: diffusion, enzymatic degradation, and uptake by cells (neurons and neuroglia).
6. If several presynaptic end bulbs release their neurotransmitter at about the same time, the combined effect may generate a nerve impulse, due to summation. Summation may be spatial or temporal.
7. The postsynaptic neuron is an integrator. It receives excitatory and inhibitory signals, integrates them, and then responds accordingly.
8. Table 12.3 summarizes the structural and functional elements of a neuron.

Animation - Events at the Synapse
Exercise - Are Your Synapses Working?
Figure 12.22 - Signal Transmission at a Chemical Synapse

12.5 Neurotransmitters

1. Both excitatory and inhibitory neurotransmitters are present in the CNS and the PNS. A given neurotransmitter may be excitatory in some locations and inhibitory in others.
2. Neurotransmitters can be divided into two classes based on size: (1) small-molecule neurotransmitters (acetylcholine, amino acids, biogenic amines, ATP and other purines, nitric oxide, and carbon monoxide), and (2) neuropeptides, which are composed of 3 to 40 amino acids.
3. Chemical synaptic transmission may be modified by affecting synthesis, release, or removal of a neurotransmitter or by blocking or stimulating neurotransmitter receptors.
4. Table 12.4 describes several important neuropeptides.

Anatomy Overview - Neurotransmitters

12.6 Neural Circuits

1. Neurons in the central nervous system are organized into networks called neural circuits.
2. Neural circuits include simple series, diverging, converging, reverberating, and parallel after-discharge circuits.

Animation - Introduction to Membrane Potentials
Figure 12.27 - Neural Circuits

Review	Resource

12.7 Regeneration and Repair of Nervous Tissue

1. The nervous system exhibits plasticity (the capability to change based on experience), but it has very limited powers of regeneration (the capability to replicate or repair damaged neurons).
2. Neurogenesis, the birth of new neurons from undifferentiated stem cells, is normally very limited. Repair of damaged axons does not occur in most regions of the CNS.
3. Axons and dendrites that are associated with a neurolemma in the PNS may undergo repair if the cell body is intact, the Schwann cells are functional, and scar tissue formation does not occur too rapidly.

SELF-QUIZ QUESTIONS

Fill in the blanks in the following statements.

1. The subdivisions of the PNS are the ____, ____, and ____.

2. The two divisions of the autonomic nervous system are the ____ division and the ____ division.

Indicate whether the following statements are true or false.

3. At a chemical synapse between two neurons, the neuron receiving the signal is called the presynaptic neuron, and the neuron sending the signal is called the postsynaptic neuron.

4. Neurons in the PNS are always capable of repair while those in the CNS are not.

Choose the one best answer to the following questions.

5. Which of the following statements are *true*? (1) The sensory function of the nervous system involves sensory receptors sensing certain changes in the internal and external environments. (2) Sensory neurons receive electrical signals from sensory receptors. (3) The integrative function of the nervous system involves analyzing sensory information, storing some of it, and making decisions regarding appropriate responses. (4) Interneurons are located primarily in the PNS. (5) Motor function involves the activation of effectors (muscles and glands).
(a) 1, 2, 3, and 4 (b) 2, 4, and 5 (c) 1, 2, 3, and 5
(d) 1, 2, and 4 (e) 2, 3, 4, and 5

6. A neuron's resting membrane potential is established and maintained by (1) a high concentration of K^+ in the extracellular fluid and a high concentration of Na^+ in the cytosol, (2) the plasma membrane's higher permeability to Na^+ because of the presence of numerous Na^+ leakage channels, (3) differences in both ion concentrations and electrical gradients, (4) the fact that there are numerous large, nondiffusible anions in the cytosol, (5) sodium–potassium pumps that help to maintain the proper distribution of sodium and potassium.
(a) 1, 2, and 5 (b) 1, 2, and 3 (c) 2, 3, and 4
(d) 3, 4, and 5 (e) 1, 2, 3, 4, and 5

7. Place the following events in a chemical synapse in the correct order: (1) release of neurotransmitters into the synaptic cleft, (2) arrival of nerve impulse at the presynaptic neuron's synaptic end bulb (or varicosity), (3) either depolarization or hyperpolarization of postsynaptic membrane, (4) inward flow of Ca^{2+} through activated voltage-gated Ca^{2+} channels in the synaptic end bulb membrane, (5) exocytosis of synaptic vesicles, (6) opening of ligand-gated channels on the postsynaptic plasma membrane, (7) binding of neurotransmitters to receptors in the postsynaptic neuron's plasma membrane.
(a) 2, 1, 5, 4, 7, 6, 3 (b) 1, 2, 4, 5, 7, 6, 3
(c) 2, 4, 5, 1, 7, 6, 3 (d) 4, 5, 1, 7, 6, 3, 2
(e) 2, 5, 1, 4, 6, 7, 3

8. Several neurons in the brain sending impulses to a single motor neuron that terminates at a neuromuscular junction is an example of a ____ circuit.
(a) reverberating (b) simple series
(c) parallel after-discharge (d) diverging
(e) converging

9. Which of the following statements are *true*? (1) If the excitatory effect is greater than the inhibitory effect but less than the threshold of stimulation, the result is a subthreshold EPSP. (2) If the excitatory effect is greater than the inhibitory effect and reaches or surpasses the threshold level of stimulation, the result is a threshold or suprathreshold EPSP and one or more nerve impulses. (3) If the inhibitory effect is greater than the excitatory effect, the membrane hyperpolarizes, resulting in inhibition of the postsynaptic neuron and the inability of the neuron to generate a nerve impulse. (4) The greater the summation of hyperpolarizations, the more likely a nerve impulse will be initiated.
(a) 1 and 4 (b) 2 and 4 (c) 1, 3, and 4
(d) 2, 3, and 4 (e) 1, 2, and 3

10. Which of the following statements are *true*? (1) The basic types of ion channels are gated, leakage, and electrical. (2) Ion channels allow for the development of graded potentials and action potentials. (3) Voltage gated channels open in response to changes in membrane potential. (4) Ligand-gated channels open due to the presence of specific chemicals. (5) A graded potential is useful for communication over long distances.
(a) 1, 2, and 3 (b) 2, 3, and 4 (c) 2, 3, and 5
(d) 2, 3, 4, and 5 (e) 1, 3, and 5

11. Which of the following statements are *true*? (1) The frequency of impulses and number of activated sensory neurons encode differences in stimuli intensity. (2) Larger-diameter axons conduct nerve impulses faster than smaller-diameter ones. (3) Continuous conduction is faster than saltatory conduction. (4) The presence or absence of a myelin sheath is an important factor that determines the speed of nerve impulse propagation. (5) Action potentials are localized, but graded potentials are propagated.
(a) 1, 3, and 5 (b) 3 and 4 (c) 2, 4, and 5
(d) 2 and 4 (e) 1, 2, and 4

12. Neurotransmitters are removed from the synaptic cleft by (1) axonal transport, (2) diffusion away from the cleft, (3) neurosecretory cells, (4) enzymatic breakdown, (5) cellular uptake.
(a) 1, 2, 3, and 4 (b) 2, 4, and 5 (c) 2, 3, and 4
(d) 1, 4, and 5 (e) 1, 2, 3, 4, and 5

13. Match the following:

____ (a) neurons with just one process extending from the cell body; are always sensory neurons	(1) astrocytes
____ (b) small phagocytic neuroglia	(2) oligodendrocytes
____ (c) help maintain an appropriate chemical environment for generation of action potentials by neurons; part of the blood–brain barrier	(3) ganglion
____ (d) provide myelin sheath for CNS axons	(4) ependymal cells
____ (e) contains neuronal cell bodies, dendrites, axon terminals, unmyelinated axons, and neuroglia	(5) satellite cells
____ (f) a cluster of cell bodies within the CNS	(6) unipolar neurons
____ (g) form CSF and assist in its circulation; form blood–cerebrospinal barrier	(7) bipolar neurons
____ (h) neurons having several dendrites and one axon; most common neuronal type	(8) multipolar neurons
____ (i) neurons with one main dendrite and one axon; found in the retina of the eye	(9) gray matter
____ (j) provide myelin sheath for PNS axons	(10) white matter
____ (k) support neurons in PNS ganglia	(11) enteric plexuses
____ (l) a cluster of neuronal cell bodies located outside the brain and spinal cord	(12) microglia
____ (m) composed primarily of myelinated axons	(13) Schwann cells
____ (n) bundles of axons and associated connective tissue and blood vessels in the PNS	(14) nucleus
____ (o) extensive neuronal networks that help regulate the digestive system	(15) nerves

14. Match the following:

____ (a) a sequence of rapidly occurring events that decreases and eventually reverses the membrane potential and then restores it to the resting state; a nerve impulse	(1) graded potential
____ (b) a small deviation from the resting membrane potential that makes the membrane either more or less polarized	(2) action potential
____ (c) period of time when a second action potential can be initiated with a very strong stimulus	(3) excitatory postsynaptic potential
____ (d) the minimum level of depolarization required for a nerve impulse to be generated	(4) inhibitory postsynaptic potential
____ (e) the recovery of the resting membrane potential	(5) absolute refractory period
____ (f) a neurotransmitter-caused depolarization of the postsynaptic membrane	(6) repolarization
____ (g) a neurotransmitter-caused hyperpolarization of the postsynaptic membrane	(7) after-hyperpolarizing phase
____ (h) time during which a neuron cannot produce an action potential even with a very strong stimulus	(8) spatial summation
____ (i) polarization that is less negative than the resting level	(9) threshold
____ (j) results from the buildup of neurotransmitter released simultaneously by several presynaptic end bulbs	(10) relative refractory period
____ (k) the hyperpolarization that occurs after the repolarizing phase of an action potential	(11) temporal summation
____ (l) polarization that is more negative than the resting level	(12) depolarizing graded potential
____ (m) results from the buildup of neurotransmitter from the rapid, successive release by a single presynaptic end bulb	(13) hyperpolarizing graded potential

15. Match the following:

____ (a) the part of the neuron that contains the nucleus and organelles	(1) myelin sheath
____ (b) rough endoplasmic reticulum in neurons; site of protein synthesis	(2) neurolemma
____ (c) store neurotransmitter	(3) nodes of Ranvier
____ (d) the process that propagates nerve impulses toward another neuron, muscle fiber, or gland cell	(4) cell body
____ (e) the highly branched receiving or input portions of a neuron	(5) Nissl bodies
____ (f) a multilayered lipid and protein covering for axons produced by neuroglia	(6) neurofibrils
____ (g) the outer nucleated cytoplasmic layer of the Schwann cell	(7) dendrites
____ (h) first portion of the axon, closest to the axon hillock	(8) axon
____ (i) site of communication between two neurons or between a neuron and an effector cell	(9) axon hillock
____ (j) form the cytoskeleton of a neuron	(10) initial segment
____ (k) gaps in the myelin sheath of an axon	(11) trigger zone
____ (l) general term for any neuronal process	(12) synaptic cleft
____ (m) area where the axon joins the cell body	(13) nerve fiber
____ (n) area where nerve impulses arise	(14) axon terminals
____ (o) the numerous fine processes at the ends of an axon and its collaterals	(15) synapse
____ (p) interstitial fluid-filled space separating two neurons	(16) synaptic vesicles

CRITICAL THINKING QUESTIONS

1. The buzzing of the alarm clock woke Carrie. She stretched, yawned, and started to salivate as she smelled the brewing coffee. She could feel her stomach rumble. List the divisions of the nervous system that are involved in each of these actions.

2. Baby Ming is learning to crawl. He also likes to pull himself onto window sills, gnawing on the painted wood of his century-old home as he looks out the windows. Lately his mother, an anatomy and physiology student, has noticed some odd behavior and took Ming to the pediatrician. Blood work determined that Ming had a high level of lead in his blood, ingested from the old leaded paint on the window sill. The doctor indicated that lead poisoning is a type of demyelination disorder. Why should Ming's mother be concerned?

3. As a torture procedure for his enemies, mad scientist Dr. Moro is trying to develop a drug that will enhance the effects of substance P. What cellular mechanisms could he enlist to design such a drug?

? ANSWERS TO FIGURE QUESTIONS

12.1 The CNS processes many different kinds of sensory information; is the source of thoughts, emotions, and memories; and gives rise to signals that stimulate muscles to contract and glands to secrete.

12.2 Dendrites and the cell body receive input; the axon conducts nerve impulses (action potentials) and transmits the message to another neuron or effector cell by releasing a neurotransmitter at its synaptic end bulbs.

12.3 Most neurons in the CNS are multipolar neurons.

12.4 Most unipolar neurons are located in the PNS; the cell bodies of these neurons are mainly in the ganglia of cranial and spinal nerves.

12.5 The cell body of a pyramidal cell is shaped like a pyramid.

12.6 Microglia function as phagocytes in the central nervous system.

12.7 One Schwann cell myelinates a single axon; one oligodendrocyte myelinates several axons.

12.8 Myelination increases the speed of nerve impulse conduction.

12.9 Myelin makes white matter look shiny and white.

12.10 Perception primarily occurs in the cerebral cortex.

12.11 A touch on the arm activates mechanically gated channels.

12.12 The -70 mV resting membrane potential of a neuron means that the inside of the neuron is 70 mV more negative than the outside when the neuron is at rest (not excited by a stimulus).

12.13 More Na^+ ions would leak into the cell and fewer K^+ ions would leak out of the cell, which would make the resting membrane potential more inside-positive.

12.14 A change in membrane potential from -70 to -60 mV is a depolarizing graded potential since the membrane potential is inside less negative than at rest. A change in membrane potential from -70 to -80 mV is a hyperpolarizing graded potential since the membrane potential is inside more negative than at rest.

12.15 Ligand-gated channels and mechanically gated channels can be present in the dendrites of sensory neurons, and ligand-gated channels are numerous in the dendrites and cell bodies of interneurons and motor neurons.

12.16 A stronger stimulus opens more mechanically gated channels or ligand-gated channels than a weaker stimulus.

12.17 Since individual graded potentials undergo decremental conduction, they would die out as they spread through the dendrites and cell body if summation did not occur and an action potential would not be generated at the trigger zone of the axon.

12.18 Voltage-gated Na^+ channels are open during the depolarizing phase, and voltage-gated K^+ channels are open during the repolarizing phase.

12.19 An action potential will not occur in response to a hyperpolarizing graded potential because a hyperpolarizing graded potential causes the membrane potential to become inside more negative and, therefore, farther away from threshold (-55 mV).

12.20 Yes, because the leak channels would still allow K^+ to exit more rapidly than Na^+ could enter the axon. Some mammalian myelinated axons have only a few voltage-gated K^+ channels.

12.21 The diameter of an axon, presence or absence of a myelin sheath, and temperature determine the speed of propagation of an action potential.

12.22 In some electrical synapses (gap junctions), ions may flow equally well in either direction, so either neuron may be the presynaptic one. At a chemical synapse, one neuron releases neurotransmitter and the other neuron has receptors that bind this chemical. Thus, the signal can proceed in only one direction.

12.23 At some excitatory synapses, ACh binds to ionotropic receptors with cation channels that open and subsequently generate EPSPs in the postsynaptic cell. At some inhibitory synapses, ACh binds to metabotropic receptors coupled to G proteins that open K^+ channels, resulting in the formation of IPSPs in the postsynaptic cell.

12.24 This is an example of spatial summation since the summation results from the buildup of neurotransmitter released simultaneously by several presynaptic end bulbs.

12.25 Since -60 mV is below threshold, an action potential will not occur in the postsynaptic neuron.

12.26 Norepinephrine, epinephrine, dopamine, and serotonin are classified as biogenic amines because they are derived from amino acids that have been chemically modified.

12.27 A motor neuron receiving input from several other neurons is an example of convergence.

12.28 The neurolemma provides a regeneration tube that guides regrowth of a severed axon.

13 THE SPINAL CORD AND SPINAL NERVES

THE SPINAL CORD AND SPINAL NERVES AND HOMEOSTASIS *The spinal cord and spinal nerves contribute to homeostasis by providing quick, reflexive responses to many stimuli. The spinal cord is the pathway for sensory input to the brain and motor output from the brain.*

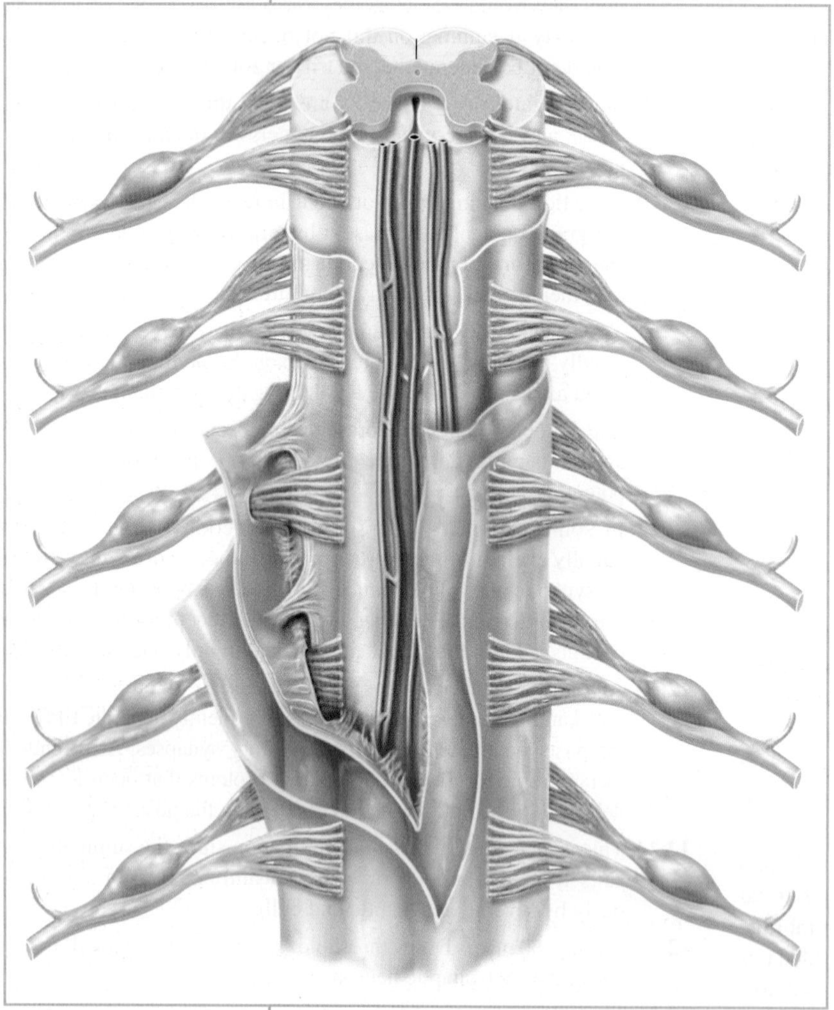

About 100 million neurons and even more neuroglia compose the **spinal cord,** the part of the central nervous system that extends from the brain. The spinal cord and its associated spinal nerves contain neural circuits that control some of your most rapid reactions to environmental changes. If you pick up something hot, the grasping muscles may relax and you may drop the hot object even before you are consciously aware of the extreme heat or pain. This is an example of a spinal cord reflex—a quick, automatic response to certain kinds of stimuli that involves neurons only in the spinal nerves and spinal cord. Besides processing reflexes, the gray matter of the spinal cord also is a site for integration (summing) of excitatory postsynaptic potentials (EPSPs) and inhibitory postsynaptic potentials (IPSPs), which you learned about in Chapter 12. These graded potentials arise as neurotransmitter molecules interact with their receptors at synapses in the spinal cord. The white matter of the spinal cord contains a dozen major sensory and motor tracts, which function as the "highways" along which sensory input travels to the brain and motor output travels from the brain to skeletal muscles and other effectors. Recall that the spinal cord is continuous with the brain and that together they make up the central nervous system (CNS).

? ***Did you ever wonder why spinal cord injuries can have such widespread effects on the body?***

13.1 SPINAL CORD ANATOMY

OBJECTIVES

- Describe the protective structures and the gross anatomical features of the spinal cord.
- Describe how spinal nerves are connected to the spinal cord.

Protective Structures

Recall from the previous chapter that the nervous tissue of the central nervous system is very delicate and does not respond well to injury or damage. Accordingly, nervous tissue requires considerable protection. The first layer of protection for the central nervous system is the hard bony skull and vertebral column. The skull encases the brain and the vertebral column surrounds the spinal cord, providing strong protective defenses against damaging blows or bumps. The second protective layer is the meninges, three membranes that lie between the bony encasement and the nervous tissue in both the brain and spinal cord. Finally, a space between two of the meningeal membranes contains cerebrospinal fluid, a buoyant liquid that suspends the central nervous tissue in a weightless environment while surrounding it with a shock-absorbing, hydraulic cushion.

Vertebral Column

The spinal cord is located within the vertebral canal of the vertebral column. As you learned in Chapter 7, the vertebral foramina of all the vertebrae, stacked one on top of the other, form the vertebral canal. The surrounding vertebrae provide a sturdy shelter for the enclosed spinal cord (see Figure 13.1c). The vertebral ligaments, meninges, and cerebrospinal fluid provide additional protection.

Meninges

The **meninges** (me-NIN-jēz; singular is **meninx** [MĒ-ninks]) are three protective, connective tissue coverings that encircle the spinal cord and brain. From superficial to deep they are the (1) dura mater, (2) arachnoid mater, and (3) pia mater. The **spinal meninges** surround the spinal cord (Figure 13.1a) and are continuous with the **cranial meninges,** which encircle the brain (shown in Figure 14.2a). All three spinal meninges cover the spinal nerves up to the point where they exit the spinal column through the intervertebral foramina. The spinal cord is also protected by a cushion of fat and connective tissue located in the **epidural space** (ep′-i-DOO-ral), a space between the dura mater and the wall of the vertebral canal (Figure 13.1c). Following is a description of each meningeal layer.

1. **Dura mater** (DOO-ra MĀ-ter = tough mother). The most superficial of the three spinal meninges is a thick strong layer composed of dense, irregular connective tissue. The dura mater forms a sac from the level of the foramen magnum in the occipital bone, where it is continuous with the meningeal dura mater of the brain, to the second sacral vertebra. The dura mater is also continuous with the epineurium, the outer covering of spinal and cranial nerves.
2. **Arachnoid mater** (a-RAK-noyd MĀ-ter; *arachn-* = spider; *-oid* = similar to). This layer, the middle of the meningeal membranes, is a thin, avascular covering comprised of cells and thin, loosely arranged collagen and elastic fibers. It is called the arachnoid mater because of its spider's web arrangement of delicate collagen fibers and some elastic fibers. It is deep to the dura mater and is continuous through the foramen magnum with the arachnoid mater of the brain. Between the dura mater and the arachnoid mater is a thin **subdural space,** which contains interstitial fluid.
3. **Pia mater** (PĒ-a MĀ-ter; *pia* = delicate). This innermost meninx is a thin transparent connective tissue layer that adheres to the surface of the spinal cord and brain. It consists of thin squamous to cuboidal cells within interlacing bundles of collagen fibers and some fine elastic fibers. Within the pia mater are many blood vessels that supply oxygen and nutrients to the spinal cord. Triangular-shaped membranous extensions of the pia mater suspend the spinal cord in the middle of its dural sheath. These extensions, called **denticulate ligaments** (den-TIK-ū-lāt = small tooth), are thickenings of the pia mater. They project laterally and fuse with the arachnoid mater and inner surface of the dura mater between the anterior and posterior nerve roots of spinal nerves on either side (Figure 13.1a, b). Extending along the entire length of the spinal cord, the denticulate ligaments protect the spinal cord against sudden displacement that could result in shock. Between the arachnoid mater and pia mater is a space, the **subarachnoid space,** which contains shock-absorbing cerebrospinal fluid.

CLINICAL CONNECTION | *Spinal Tap*

In a **spinal tap (lumbar puncture)**, a local anesthetic is given, and a long hollow needle is inserted into the subarachnoid space to withdraw cerebrospinal fluid (CSF) for diagnostic purposes; to introduce antibiotics, contrast media for myelography, or anesthetics; to administer chemotherapy; to measure CSF pressure; and/or to evaluate the effects of treatment for diseases such as meningitis. During this procedure, the patient lies on his or her side with the vertebral column flexed. Flexion of the vertebral column increases the distance between the spinous processes of the vertebrae, which allows easy access to the subarachnoid space. The spinal cord ends around the second lumbar vertebra (L2); however, the spinal meninges and circulating cerebrospinal fluid extend to the second sacral vertebra (S2). Between vertebrae L2 and S2 the spinal meninges are present, but the spinal cord is absent. Consequently, a spinal tap is normally performed in adults between the L3 and L4 or L4 and L5 lumbar vertebrae because this region provides safe access to the subarachnoid space without the risk of damaging the spinal cord. (A line drawn across the highest points of the iliac crests, called the *supracristal line*, passes through the spinous process of the fourth lumbar vertebra and is used as a landmark for administering a spinal tap.) •

External Anatomy of the Spinal Cord

The **spinal cord** is roughly oval in shape, being flattened slightly anteriorly and posteriorly. In adults, it extends from the medulla oblongata, the inferior part of the brain, to the superior border of the second lumbar vertebra (Figure 13.2). In newborn infants, it extends to the third or fourth lumbar vertebra. During early childhood, both the spinal cord and the vertebral column grow longer as part of overall body growth. Elongation of the spinal cord stops around age 4 or 5, but growth of the vertebral column continues. Thus, the spinal cord does not extend the entire length of the adult vertebral column. The length of the adult spinal cord ranges from 42 to 45 cm (16–18 in.). Its diameter is about 2 cm (0.75 in.) in the midthoracic region, somewhat larger in the lower cervical and midlumbar regions, and smallest at the inferior tip.

When the spinal cord is viewed externally, two conspicuous enlargements can be seen. The superior enlargement, the **cervical enlargement,** extends from the fourth cervical vertebra (C4) to the first thoracic vertebra (T1). Nerves to and from the upper limbs arise from the cervical enlargement. The inferior enlargement, called the **lumbar enlargement,** extends from the ninth to the twelfth thoracic vertebra. Nerves to and from the lower limbs arise from the lumbar enlargement.

Inferior to the lumbar enlargement, the spinal cord terminates as a tapering, conical structure called the **conus medullaris** (KŌ-nus med-ū-LAR-is; *conus* = cone), which ends at the level of the intervertebral disc between the first and second lumbar vertebrae (L1–L2) in adults. Arising from the conus medullaris is the **filum terminale** (FĪ-lum ter-mi-NAL-ē = terminal filament), an extension of the pia mater that extends inferiorly and fuses with the arachnoid mater and dura mater and anchors the spinal cord to the coccyx.

Spinal nerves are the paths of communication between the spinal cord and specific regions of the body. The spinal cord appears to be segmented because the 31 pairs of spinal nerves emerge at regular intervals from intervertebral foramina (Figure 13.2). Indeed, each pair of spinal nerves is said to arise from a *spinal segment*. Within the spinal cord there is no obvious segmentation but, for convenience, the naming of spinal nerves is based on the segment in

Figure 13.1 **Gross anatomy of the spinal cord.** The spinal meninges are evident in parts (a) and (c).

Meninges are connective tissue coverings that surround the spinal cord and brain.

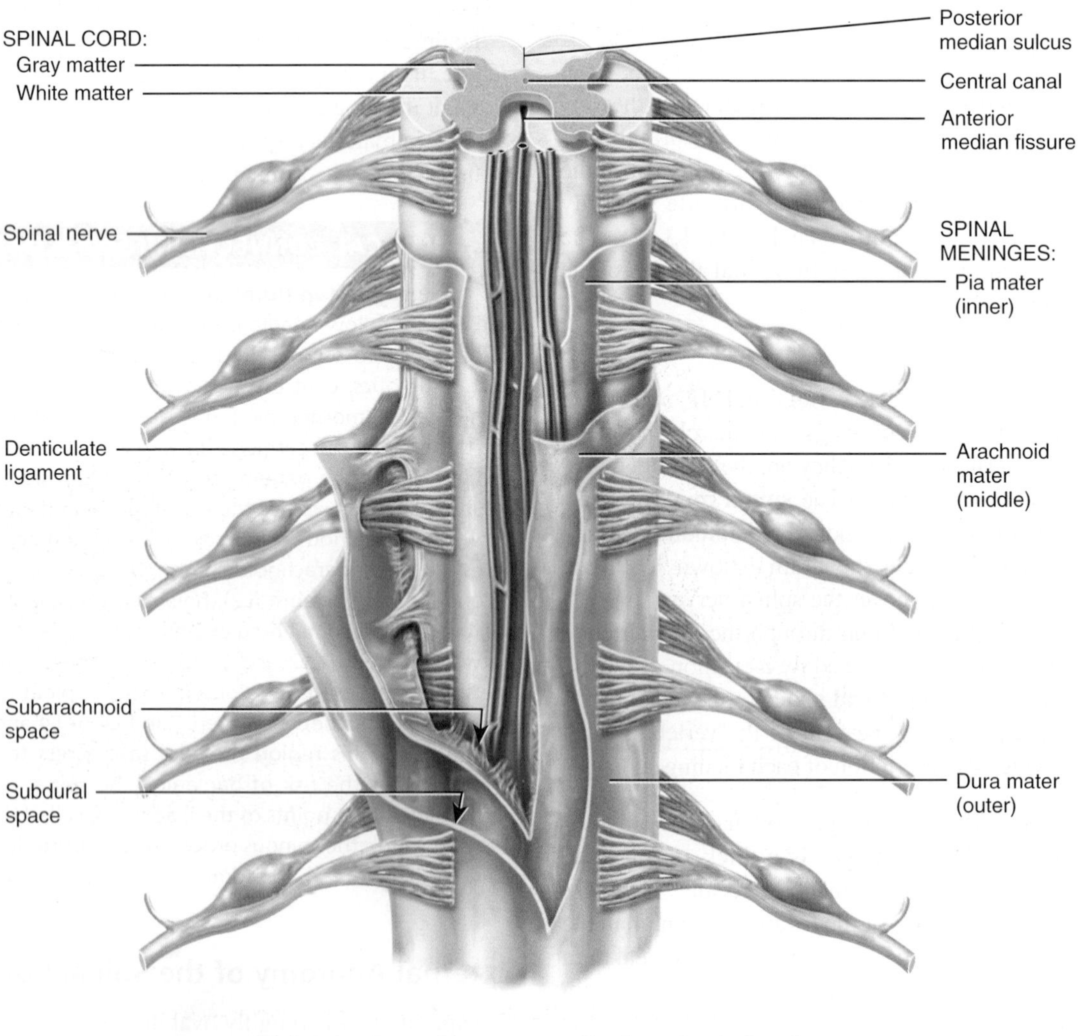

SPINAL CORD:
Gray matter
White matter
Posterior median sulcus
Central canal
Anterior median fissure
Spinal nerve
SPINAL MENINGES:
Pia mater (inner)
Denticulate ligament
Arachnoid mater (middle)
Subarachnoid space
Subdural space
Dura mater (outer)

(a) Anterior view and transverse section through spinal cord

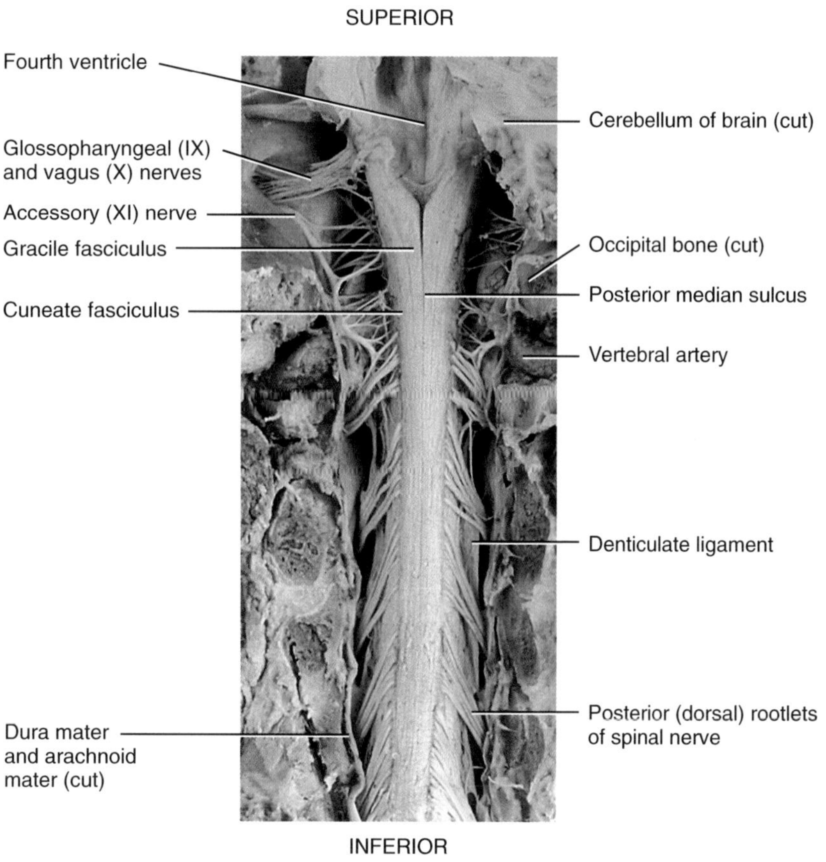
SUPERIOR
Fourth ventricle
Glossopharyngeal (IX) and vagus (X) nerves
Accessory (XI) nerve
Gracile fasciculus
Cuneate fasciculus
Dura mater and arachnoid mater (cut)
Cerebellum of brain (cut)
Occipital bone (cut)
Posterior median sulcus
Vertebral artery
Denticulate ligament
Posterior (dorsal) rootlets of spinal nerve
INFERIOR

(b) Posterior view of cervical region of spinal cord

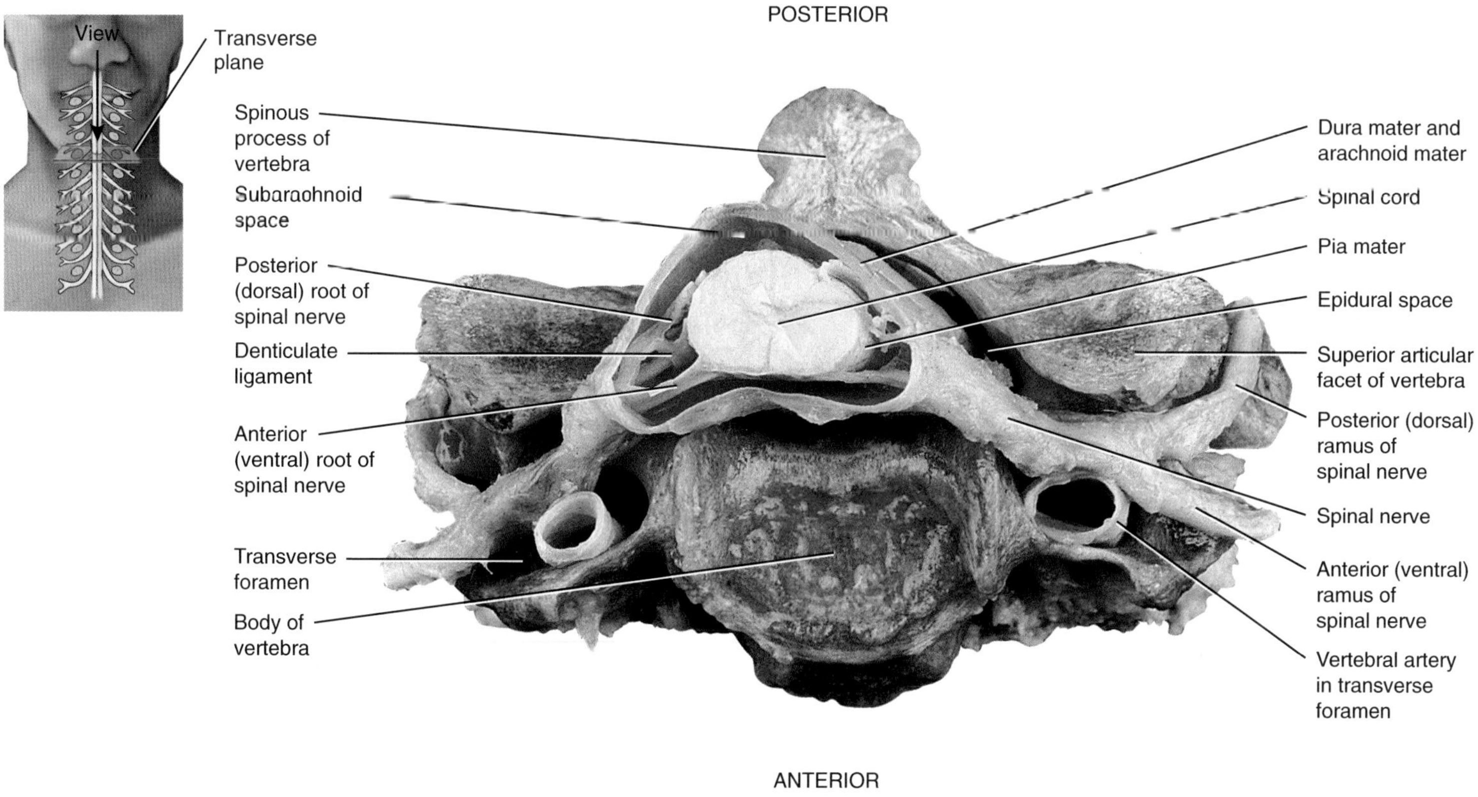
POSTERIOR
View
Transverse plane
Spinous process of vertebra
Subarachnoid space
Posterior (dorsal) root of spinal nerve
Denticulate ligament
Anterior (ventral) root of spinal nerve
Transverse foramen
Body of vertebra
Dura mater and arachnoid mater
Spinal cord
Pia mater
Epidural space
Superior articular facet of vertebra
Posterior (dorsal) ramus of spinal nerve
Spinal nerve
Anterior (ventral) ramus of spinal nerve
Vertebral artery in transverse foramen
ANTERIOR

(c) Transverse section of the spinal cord within a cervical vertebra

? What are the superior and inferior boundaries of the spinal dura mater?

Figure 13.2 **External anatomy of the spinal cord and the spinal nerves.**

The spinal cord extends from the medulla oblongata of the brain to the superior border of the second lumbar vertebra.

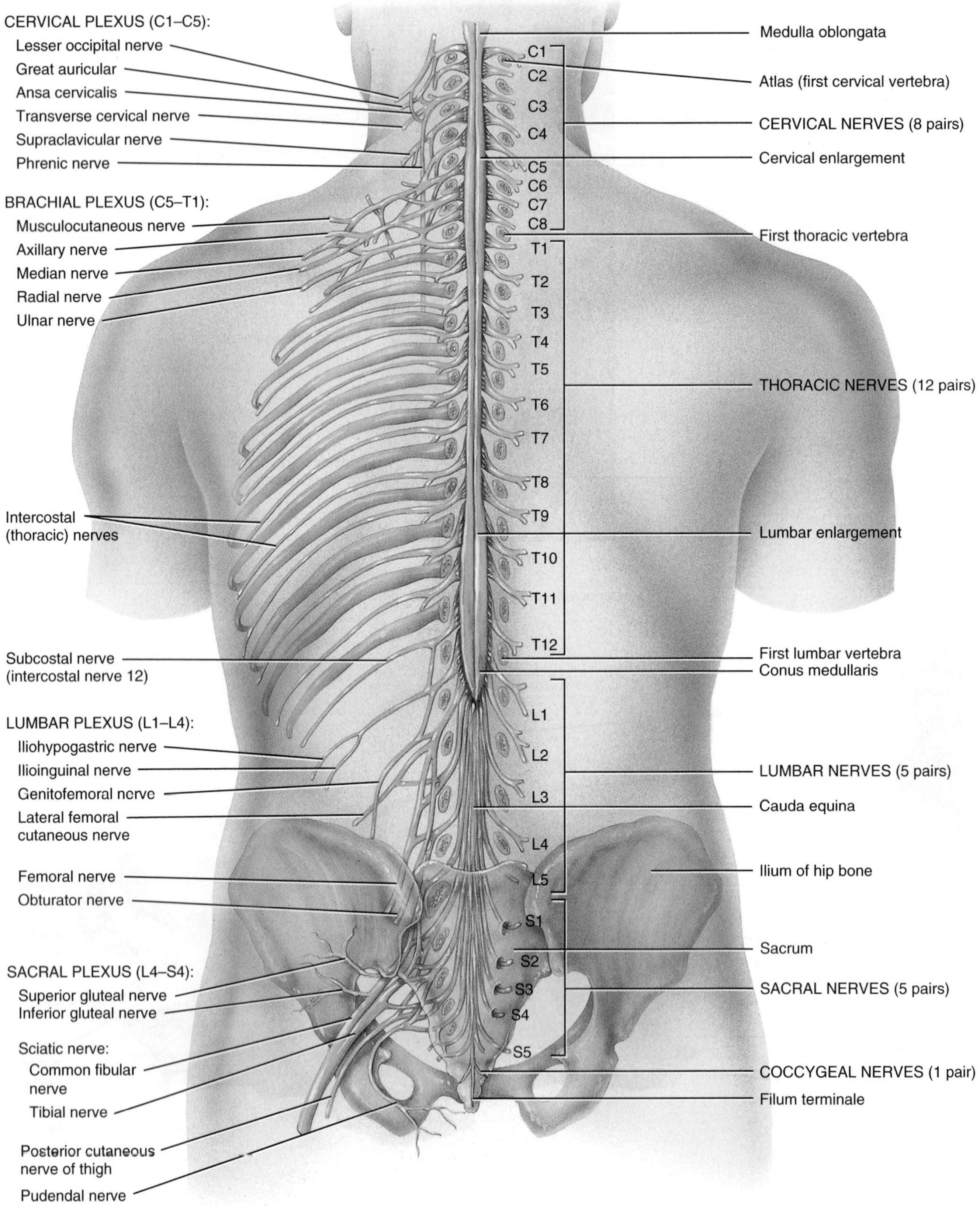

CERVICAL PLEXUS (C1–C5):
Lesser occipital nerve
Great auricular
Ansa cervicalis
Transverse cervical nerve
Supraclavicular nerve
Phrenic nerve
BRACHIAL PLEXUS (C5–T1):
Musculocutaneous nerve
Axillary nerve
Median nerve
Radial nerve
Ulnar nerve
Intercostal (thoracic) nerves
Subcostal nerve (intercostal nerve 12)
LUMBAR PLEXUS (L1–L4):
Iliohypogastric nerve
Ilioinguinal nerve
Genitofemoral nerve
Lateral femoral cutaneous nerve
Femoral nerve
Obturator nerve
SACRAL PLEXUS (L4–S4):
Superior gluteal nerve
Inferior gluteal nerve
Sciatic nerve:
Common fibular nerve
Tibial nerve
Posterior cutaneous nerve of thigh
Pudendal nerve
C1
C2
C3
C4
C5
C6
C7
C8
T1
T2
T3
T4
T5
T6
T7
T8
T9
T10
T11
T12
L1
L2
L3
L4
L5
S1
S2
S3
S4
S5
Medulla oblongata
Atlas (first cervical vertebra)
CERVICAL NERVES (8 pairs)
Cervical enlargement
First thoracic vertebra
THORACIC NERVES (12 pairs)
Lumbar enlargement
First lumbar vertebra
Conus medullaris
LUMBAR NERVES (5 pairs)
Cauda equina
Ilium of hip bone
Sacrum
SACRAL NERVES (5 pairs)
COCCYGEAL NERVES (1 pair)
Filum terminale

Posterior view of entire spinal cord and portions of spinal nerves

What portion of the spinal cord connects with nerves of the upper limbs?

which they are located. There are 8 pairs of *cervical nerves* (represented in Figure 13.2 as C1–C8), 12 pairs of *thoracic nerves* (T1–T12), 5 pairs of *lumbar nerves* (L1–L5), 5 pairs of *sacral nerves* (S1–S5), and 1 pair of *coccygeal nerves* (Co1).

Two bundles of axons, called **roots,** connect each spinal nerve to a segment of the cord by even smaller bundles of axons called **rootlets** (see Figure 13.3a). The **posterior (dorsal) root** and rootlets contain only sensory axons, which conduct nerve

Figure 13.3 Internal anatomy of the spinal cord: the organization of gray matter and white matter. For simplicity, dendrites are not shown in this and several other illustrations of transverse sections of the spinal cord. Blue, red, and green arrows indicate the direction of nerve impulse propagation.

The posterior gray horn contains axons of sensory neurons and cell bodies of interneurons; the lateral gray horn contains cell bodies of autonomic motor neurons; and the anterior gray horn contains cell bodies of somatic motor neurons.

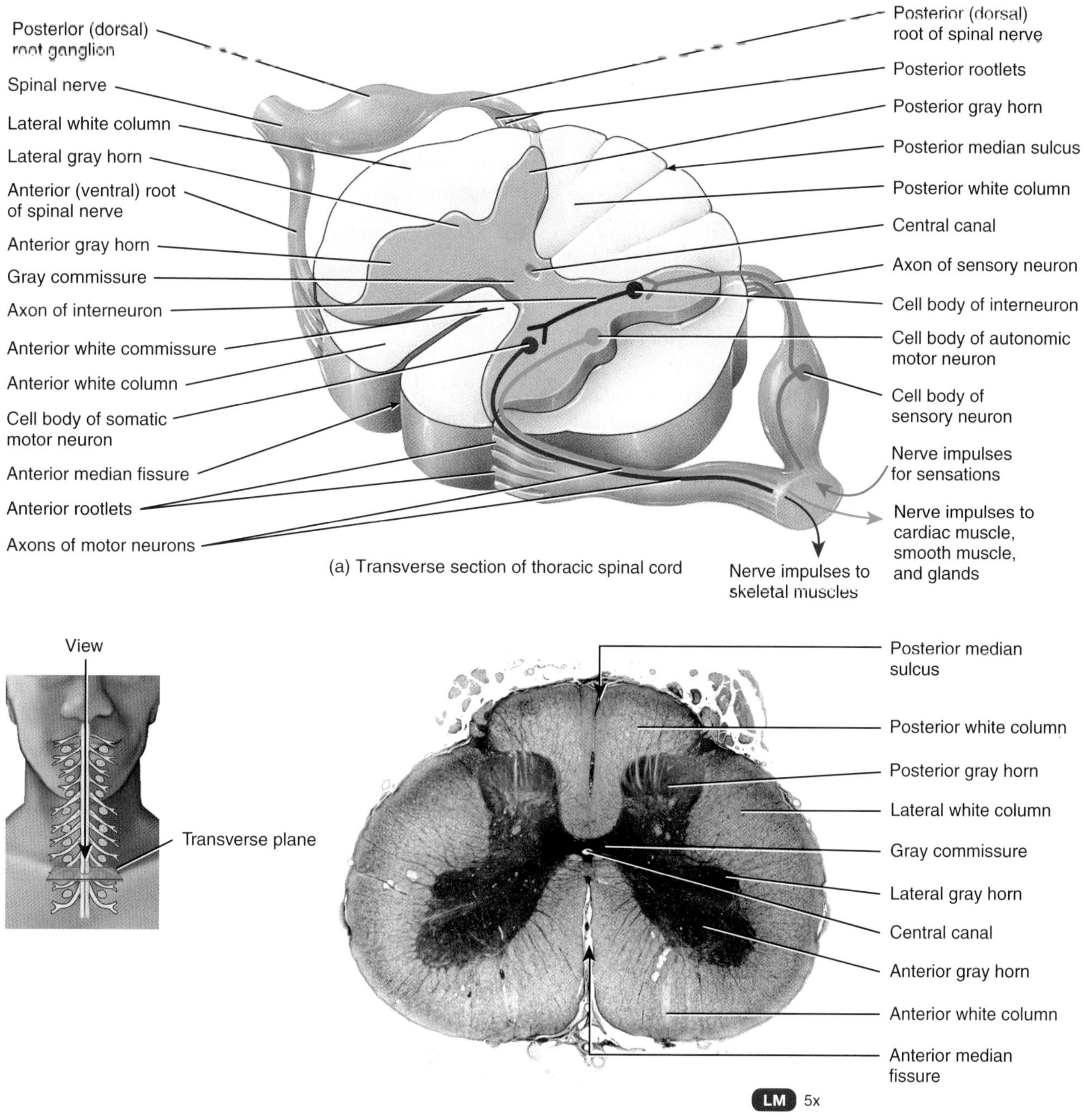

(a) Transverse section of thoracic spinal cord

(b) Transverse section of thoracic spinal cord

What is the difference between a horn and a column in the spinal cord?

impulses from sensory receptors in the skin, muscles, and internal organs into the central nervous system. Each posterior root has a swelling, the **posterior (dorsal) root ganglion,** which contains the cell bodies of sensory neurons. The **anterior (ventral) root** and rootlets contain axons of motor neurons, which conduct nerve impulses from the CNS to effectors (muscles and glands).

As spinal nerves branch from the spinal cord, they pass laterally to exit the vertebral canal through the intervertebral foramina between adjacent vertebrae. However, because the spinal cord is shorter than the vertebral column, nerves that arise from the lumbar, sacral, and coccygeal regions of the spinal cord do not leave the vertebral column at the same level they exit the cord. The roots of these lower spinal nerves angle inferiorly alongside the filum terminale in the vertebral canal like wisps of hair. Accordingly, the roots of these nerves are collectively named the **cauda equina** (KAW-da ē-KWĪ-na), meaning "horse's tail" (Figure 13.2).

Internal Anatomy of the Spinal Cord

A transverse section of the spinal cord reveals regions of white matter that surround an inner core of gray matter (Figure 13.3). The white matter of the spinal cord consists primarily of bundles of myelinated axons of neurons. Two grooves penetrate the white matter of the spinal cord and divide it into right and left sides. The **anterior median fissure** is a wide groove on the anterior (ventral) side. The **posterior median sulcus** is a narrow furrow on the posterior (dorsal) side. The gray matter of the spinal cord is shaped like the letter H or a butterfly; it consists of dendrites and cell bodies of neurons, unmyelinated axons, and neuroglia. The **gray commissure** (KOM-mi-shur) forms the crossbar of the H. In the center of the gray commissure is a small space called the **central canal;** it extends the entire length of the spinal cord and is filled with cerebrospinal fluid. At its superior end, the central canal is continuous with the fourth ventricle (a space that contains cerebrospinal fluid) in the medulla oblongata of the brain. Anterior to the gray commissure is the **anterior (ventral) white commissure,** which connects the white matter of the right and left sides of the spinal cord.

In the gray matter of the spinal cord and brain, clusters of neuronal cell bodies form functional groups called **nuclei.** *Sensory nuclei* receive input from receptors via sensory neurons, and *motor nuclei* provide output to effector tissues via motor neurons. The gray matter on each side of the spinal cord is subdivided into regions called **horns** (Figure 13.3). The **posterior (dorsal) gray horns** contain cell bodies and axons of interneurons as well as axons of incoming sensory neurons. Recall that cell bodies of sensory neurons are located in the posterior (dorsal) root ganglion of a spinal nerve. The **anterior (ventral) gray horns** contain *somatic motor nuclei,* which are clusters of cell bodies of somatic motor neurons that provide nerve impulses for contraction of skeletal muscles. Between the posterior and anterior gray horns are the **lateral gray horns,** which are present only in thoracic and upper lumbar segments of the spinal cord. The lateral gray horns contain *autonomic motor nuclei,* which are clusters of cell bodies of autonomic motor neurons that regulate the activity of cardiac muscle, smooth muscle, and glands.

The white matter of the spinal cord, like the gray matter, is organized into regions. The anterior and posterior gray horns divide the white matter on each side into three broad areas called **columns:** (1) **anterior (ventral) white columns,** (2) **posterior (dorsal) white columns,** and (3) **lateral white columns** (Figure 13.3). Each column in turn contains distinct bundles of axons having a common origin or destination and carrying similar information. These bundles, which may extend long distances up or down the spinal cord, are called **tracts.** Recall that tracts are bundles of axons in the CNS, whereas nerves are bundles of axons in the PNS. **Sensory (ascending) tracts** consist of axons that conduct nerve impulses toward the brain. Tracts consisting of axons that carry nerve impulses from the brain are called **motor (descending) tracts.** Sensory and motor tracts of the spinal cord are continuous with sensory and motor tracts in the brain.

The internal organization of the spinal cord allows sensory input and motor output to be processed by the spinal cord in the following way (Figure 13.4):

1. Sensory receptors detect a sensory stimulus.
2. Sensory neurons convey this sensory input in the form of nerve impulses along their axons, which extend from sensory receptors into the spinal nerve and then into the posterior root. From the posterior root, axons of sensory neurons may proceed along three possible paths (see steps 3, 4, and 5).
3. Axons of sensory neurons may extend into the white matter of the spinal cord and ascend to the brain as part of a sensory tract.
4. Axons of sensory neurons may enter the posterior gray horn and synapse with interneurons whose axons extend into the white matter of the spinal cord and then ascend to the brain as part of a sensory tract.
5. Axons of sensory neurons may enter the posterior gray horn and synapse with interneurons that in turn synapse with somatic motor neurons that are involved in spinal reflex pathways. Spinal cord reflexes are described in more detail later in this chapter.
6. Motor output from the spinal cord to skeletal muscles involves somatic motor neurons of the anterior gray horn. Many somatic motor neurons are regulated by the brain. Axons from higher brain centers form motor tracts that descend from the brain into the white matter of the spinal cord. There they synapse with somatic motor neurons either directly or indirectly by first synapsing with interneurons that in turn synapse with somatic motor neurons.
7. When activated, somatic motor neurons convey motor output in the form of nerve impulses along their axons, which sequentially pass through the anterior gray horn and anterior root to enter the spinal nerve. From the spinal nerve, axons of somatic motor neurons extend to skeletal muscles of the body.
8. Motor output from the spinal cord to cardiac muscle, smooth muscle, and glands involves autonomic motor neurons of the

Figure 13.4 Processing of sensory input and motor output by the spinal cord.

Sensory input is conveyed from sensory receptors to the posterior gray horns of the spinal cord, whereas motor output is conveyed from the anterior and lateral gray horns of the spinal cord to effectors (muscles and glands).

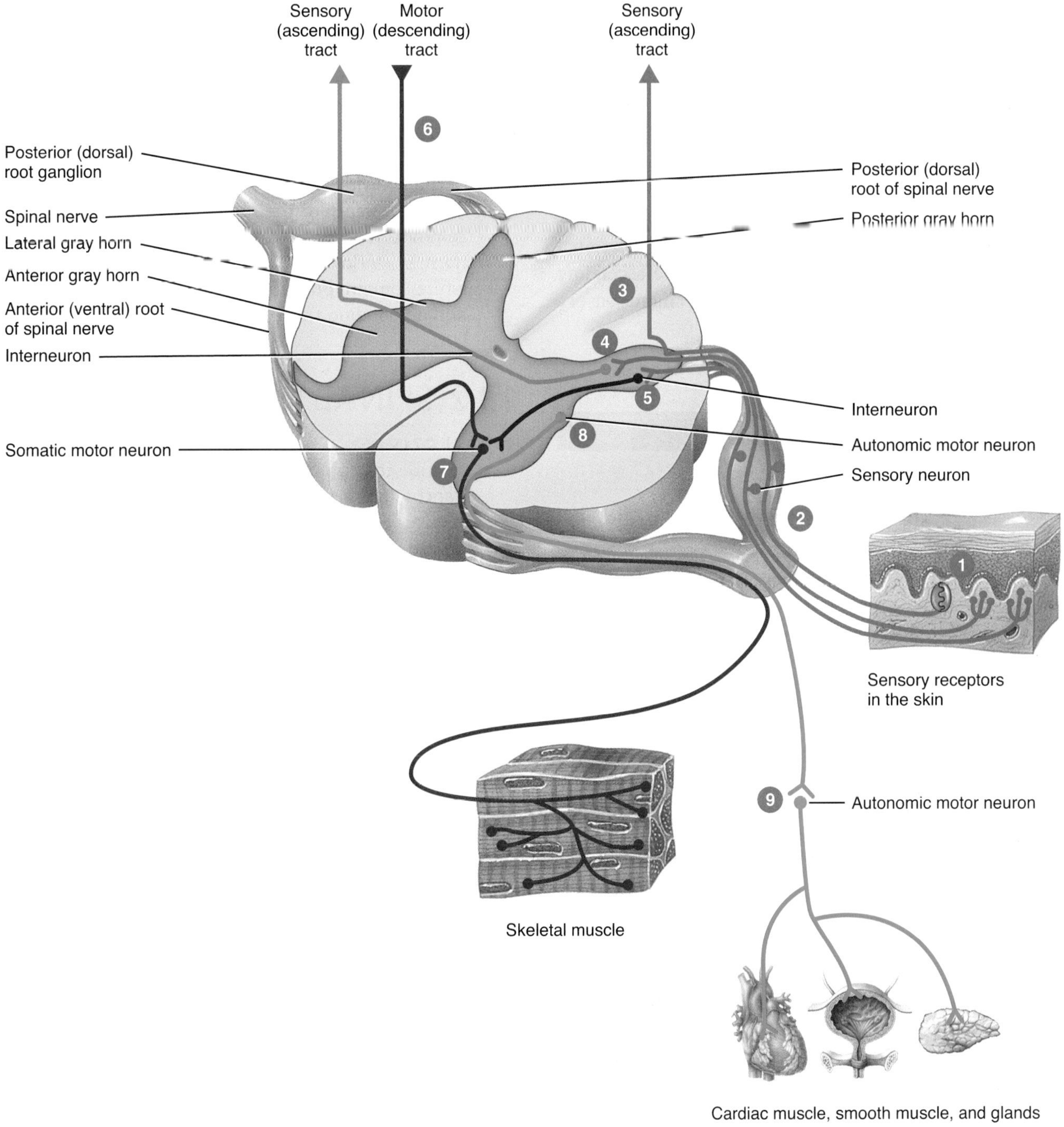

Lateral gray horns are found in which segments of the spinal cord?

lateral gray horn. When activated, autonomic motor neurons convey motor output in the form of nerve impulses along their axons, which sequentially pass through the lateral gray horn, anterior gray horn, and anterior root to enter the spinal nerve.

9 From the spinal nerve, axons of autonomic motor neurons from the spinal cord synapse with another group of autonomic motor neurons located in the peripheral nervous system (PNS). The axons of this second group of autonomic motor

neurons in turn synapse with cardiac muscle, smooth muscle, and glands. You will learn more about autonomic motor neurons when the autonomic nervous system is described in Chapter 15.

The various spinal cord segments vary in size, shape, relative amounts of gray and white matter, and distribution and shape of gray matter. For example, the amount of gray matter is largest in the cervical and lumbar segments of the spinal cord because these segments are responsible for sensory and motor innervation of the limbs. In addition, more sensory and motor tracts are present in the upper segments of the spinal cord than in the lower segments. Therefore, the amount of white matter decreases from cervical to sacral segments of the spinal cord. There are two major reasons for this variation in spinal cord white matter: (1) As the spinal cord ascends from sacral to cervical segments, more ascending axons are added to spinal cord white matter to form more sensory tracts. (2) As the spinal cord descends from cervical to sacral segments, the motor tracts decrease in thickness as more descending axons leave the motor tracts to synapse with neurons in the gray matter of the spinal cord. Table 13.1 summarizes the variations in spinal cord segments.

✔CHECKPOINT

1. Where are the spinal meninges located? Where are the epidural, subdural, and subarachnoid spaces located?
2. What are the cervical and lumbar enlargements?
3. Define conus medullaris, filum terminale, and cauda equina. What is a spinal segment? How is the spinal cord partially divided into right and left sides?
4. What does each of the following terms mean? Gray commissure, central canal, anterior gray horn, lateral gray horn, posterior gray horn, anterior white column, lateral white column, posterior white column, ascending tract, and descending tract.

TABLE 13.1

Comparison of Various Spinal Cord Segments

SEGMENT	DISTINGUISHING CHARACTERISTICS
Cervical 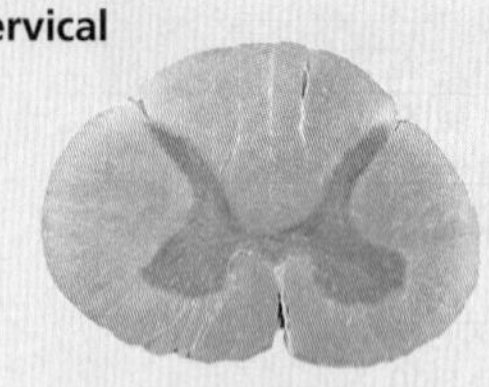 (Segment C6)	Relatively large diameter, relatively large amounts of white matter, oval; in upper cervical segments (C1–C4), posterior gray horn is large but anterior gray horn is relatively small; in lower cervical segments (C5 and below), posterior gray horns are enlarged and anterior gray horns are well developed.
Thoracic 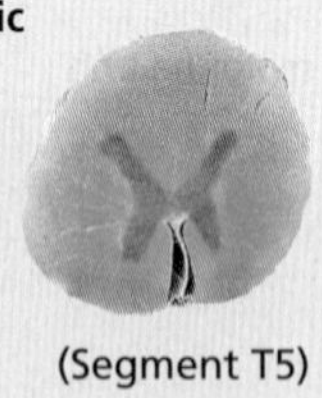 (Segment T5)	Small diameter due to relatively small amounts of gray matter; except for first thoracic segment, anterior and posterior gray horns are relatively small; small lateral gray horn is present.
Lumbar 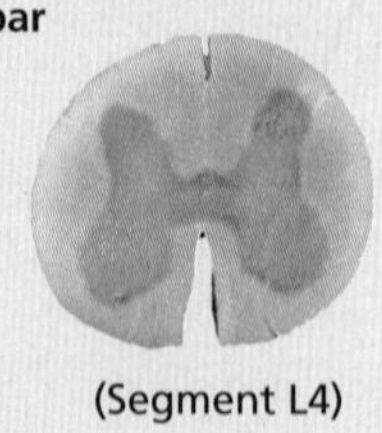 (Segment L4)	Nearly circular; very large anterior and posterior gray horns; small lateral gray horn is present in upper segments; relatively less white matter than cervical segments.
Sacral 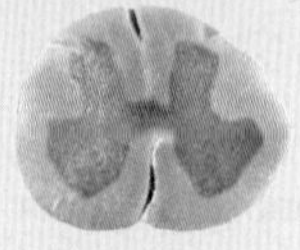 (Segment S3)	Relatively small, but relatively large amounts of gray matter; relatively small amounts of white matter; anterior and posterior gray horns are large and thick.
Coccygeal	Resembles lower sacral spinal segments, but much smaller.

13.2 SPINAL NERVES

OBJECTIVES

- Describe the components, connective tissue coverings, and branching of a spinal nerve.
- Define plexus, and identify the distribution of nerves of the cervical, brachial, lumbar, and sacral plexuses.
- Describe the clinical significance of dermatomes.

Spinal nerves are associated with the spinal cord and, like all nerves of the peripheral nervous system (PNS), are parallel bundles of axons and their associated neuroglial cells wrapped in several layers of connective tissue. Spinal nerves connect the CNS to sensory receptors, muscles, and glands in all parts of the body. The 31 pairs of spinal nerves are named and numbered according to the region and level of the vertebral column from which they emerge (see Figure 13.2). Not all spinal cord segments are aligned with their corresponding vertebrae. Recall that the spinal cord ends near the level of the superior border of the second lumbar vertebra (L2), and that the roots of the lumbar, sacral, and coccygeal nerves descend at an angle to reach their respective foramina before emerging from the vertebral column. This arrangement constitutes the cauda equina.

The first cervical pair of spinal nerves emerges from the spinal cord between the occipital bone and the atlas (first cervical vertebra, or C1). Most of the remaining spinal nerves emerge from the spinal cord through the intervertebral foramina between adjoining vertebrae. Spinal nerves C1–C7 exit the vertebral canal *above* their corresponding vertebrae. Spinal nerve C8 exits the vertebral canal between vertebrae C7 and T1. Spinal nerves T1–L5 exit the vertebral canal *below* their corresponding vertebrae. From the spinal cord, the roots of the sacral spinal nerves (S1–S5) and the coccygeal spinal nerves (Co1) enter the sacral canal, the part of the vertebral canal in the sacrum (see Figure 7.21). Subsequently, spinal nerves S1–S4 exit the sacral canal via the four pairs of anterior and posterior sacral foramina, and spinal nerves S5 and Co1 exit the sacral canal via the sacral hiatus.

As noted earlier, a typical spinal nerve has two connections to the cord: a posterior root and an anterior root (see Figure 13.3a). The posterior and anterior roots unite to form a spinal nerve at the intervertebral foramen. Because the posterior root contains sensory axons and the anterior root contains motor axons, a spinal nerve is classified as a **mixed nerve.** The posterior root contains a posterior root ganglion in which cell bodies of sensory neurons are located.

Connective Tissue Coverings of Spinal Nerves

Each spinal nerve and cranial nerve consists of many individual axons and contains layers of protective connective tissue coverings (Figure 13.5). Individual axons within a nerve, whether myelinated or unmyelinated, are wrapped in **endoneurium** (en′-dō NOO-rē-um; *endo-* = within or inner; *-neurium* = nerve), the innermost layer. The endoneurium consists of a mesh of collagen fibers, fibroblasts, and macrophages. Groups of axons with their endoneurium are held together in bundles called **fascicles,** each of which is wrapped in **perineurium** (per′-i-NOO-rē-um; *peri-* = around), the middle layer. The perineurium is a thicker layer of connective tissue. It consists of up to 15 layers of fibroblasts within a network of collagen fibers. The outermost covering over the entire nerve is the **epineurium** (ep′-i-NOO-rē-um; *epi-* = over). It consists of fibroblasts and thick collagen fibers. Extensions of the epineurium also fill the spaces between fascicles. The dura mater of the spinal meninges fuses with the epineurium as the nerve passes through the intervertebral foramen. Note the presence of many blood vessels, which nourish the spinal meninges (Figure 13.5b). You may recall from Chapter 10 that the connective tissue coverings of skeletal muscles—endomysium, perimysium, and epimysium—are similar in organization to those of nerves.

Distribution of Spinal Nerves

Branches

A short distance after passing through its intervertebral foramen, a spinal nerve divides into several branches (Figure 13.6). These branches are known as **rami** (RĀ-mī = branches). The **posterior (dorsal) ramus** (RĀ-mus; singular form) serves the deep muscles and skin of the posterior surface of the trunk. The **anterior (ventral) ramus** serves the muscles and structures of the upper and lower limbs and the skin of the lateral and anterior surfaces of the trunk. In addition to posterior and anterior rami, spinal nerves also give off a **meningeal branch** (me-NIN-jē′-al). This branch reenters the vertebral cavity through the intervertebral foramen and supplies the vertebrae, vertebral ligaments, blood vessels of the spinal cord, and meninges. Other branches of a spinal nerve

Figure 13.5 Organization and connective tissue coverings of a spinal nerve. (Part b © Dr. Richard Kessel and Dr. Randy Kardon/Visuals Unlimited.)

Three layers of connective tissue wrappings protect axons: Endoneurium surrounds individual axons, perineurium surrounds bundles of axons (fascicles), and epineurium surrounds an entire nerve.

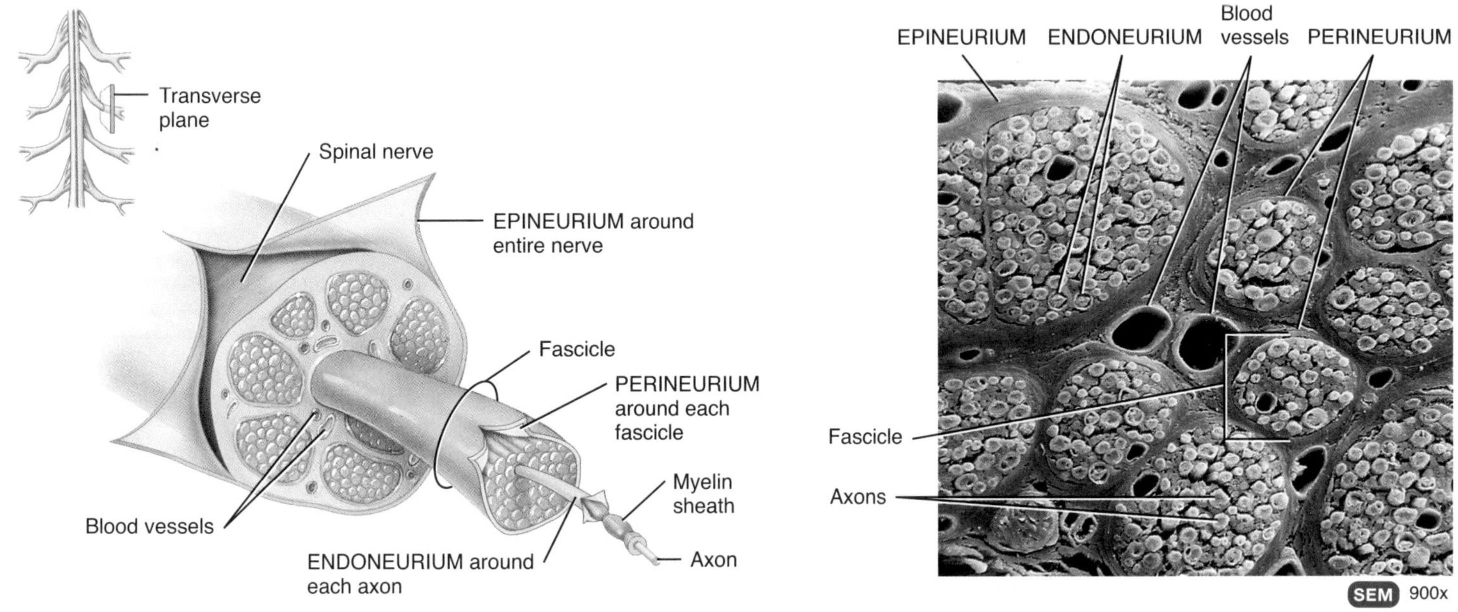

(a) Transverse section showing the coverings of a spinal nerve

(b) Transverse section of 12 nerve fascicles

Why are all spinal nerves classified as mixed nerves?

Figure 13.6 **Branches of a typical spinal nerve, shown in transverse section through the thoracic portion of the spinal cord.** (See also Figure 13.1c.)

The branches of a spinal nerve are the posterior ramus, the anterior ramus, the meningeal branch, and the rami communicantes.

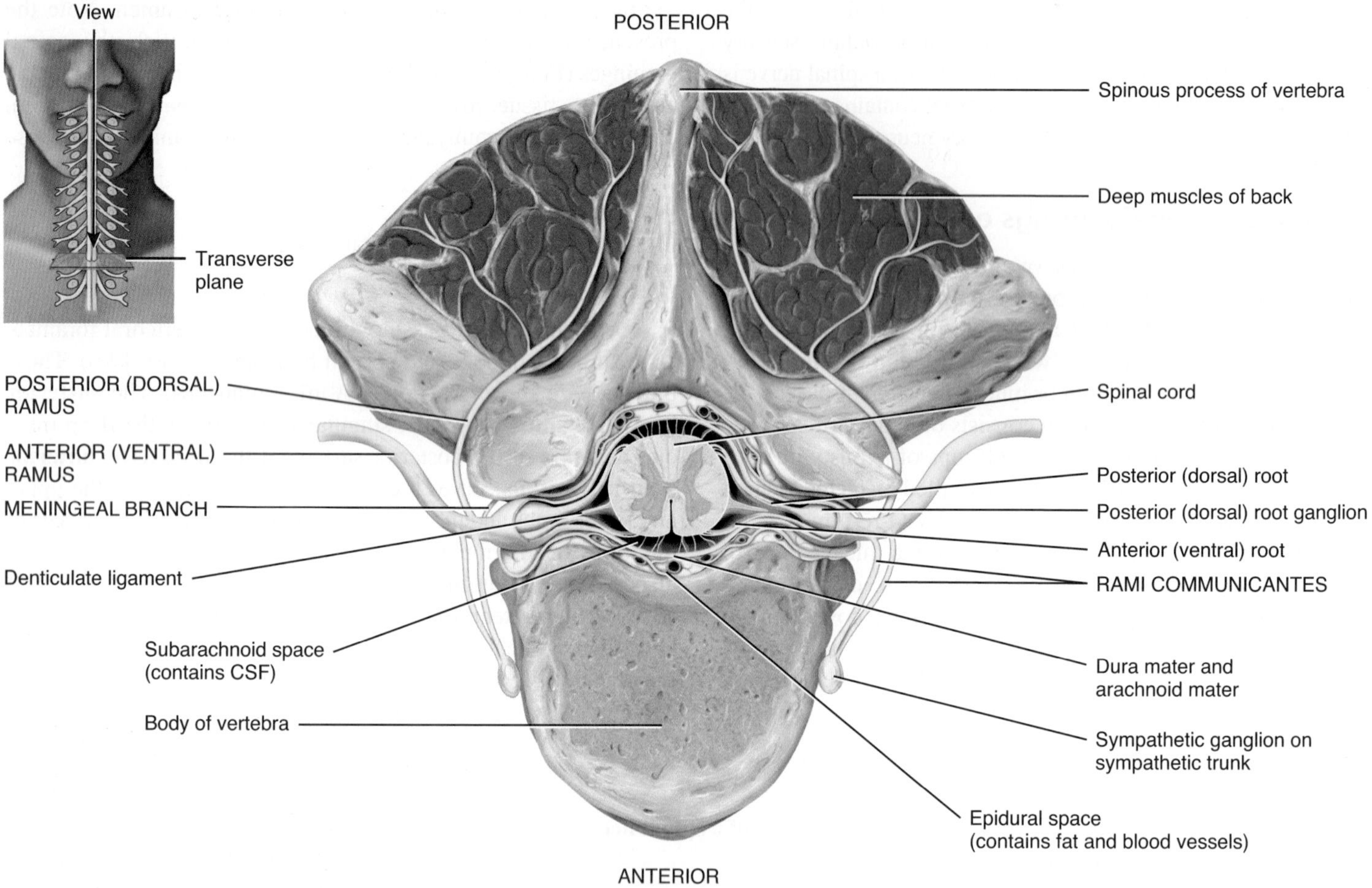

View
Transverse plane
POSTERIOR
Spinous process of vertebra
Deep muscles of back
POSTERIOR (DORSAL) RAMUS
ANTERIOR (VENTRAL) RAMUS
MENINGEAL BRANCH
Denticulate ligament
Subarachnoid space (contains CSF)
Body of vertebra
Spinal cord
Posterior (dorsal) root
Posterior (dorsal) root ganglion
Anterior (ventral) root
RAMI COMMUNICANTES
Dura mater and arachnoid mater
Sympathetic ganglion on sympathetic trunk
Epidural space (contains fat and blood vessels)
ANTERIOR

(a) Superior view

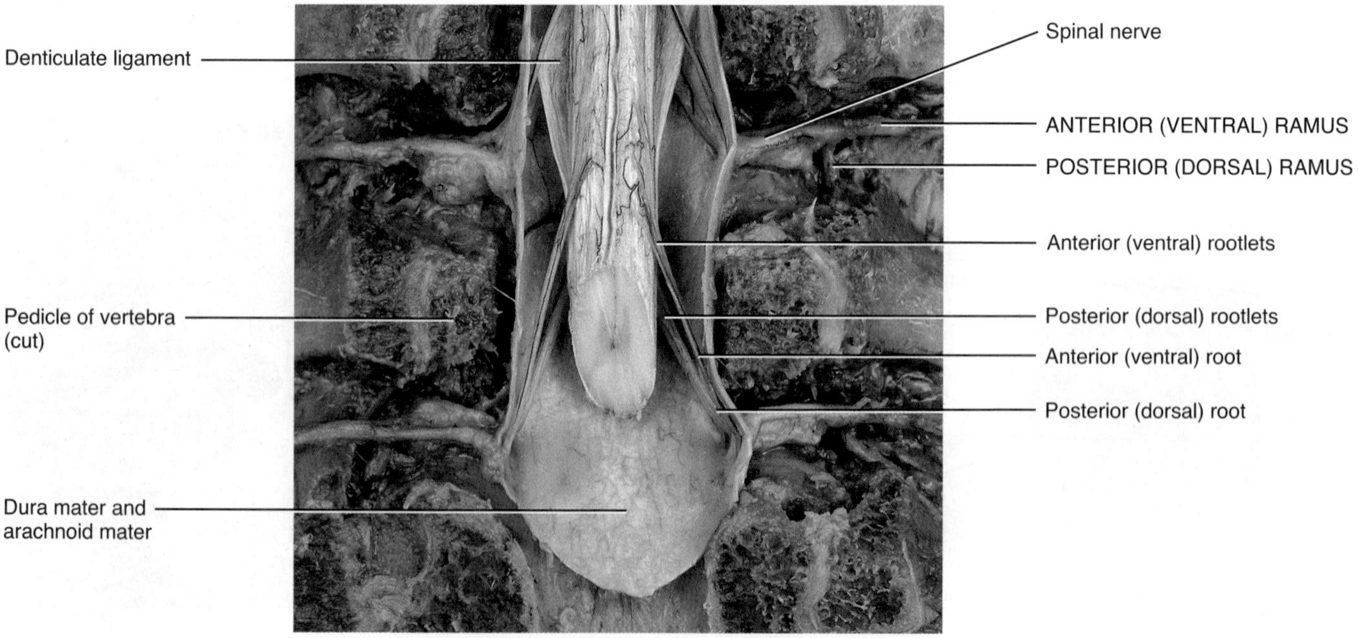

Denticulate ligament
Pedicle of vertebra (cut)
Dura mater and arachnoid mater
Spinal nerve
ANTERIOR (VENTRAL) RAMUS
POSTERIOR (DORSAL) RAMUS
Anterior (ventral) rootlets
Posterior (dorsal) rootlets
Anterior (ventral) root
Posterior (dorsal) root

(b) Anterior view and oblique section of spinal cord

Which spinal nerve branches serve the upper and lower limbs?

are the **rami communicantes** (kō-mū-ni-KAN-tēz), components of the autonomic nervous system that will be discussed in Chapter 15.

Plexuses

Axons from the anterior rami of spinal nerves, except for thoracic nerves T2–T12, do not go directly to the body structures they supply. Instead, they form networks on both the left and right sides of the body by joining with various numbers of axons from anterior rami of adjacent nerves. Such a network of axons is called a **plexus** (PLEK-sus = braid or network). The principal plexuses are the **cervical plexus, brachial plexus, lumbar plexus,** and **sacral plexus.** A smaller **coccygeal plexus** is also present. Refer to Figure 13.2 to see their relationships to one another. Emerging from the plexuses are nerves bearing names that are often descriptive of the general regions they serve or the course they take. Each of the nerves in turn may have several branches named for the specific structures they innervate.

Exhibits 13.A–13.D (Figures 13.7–13.10) summarize the principal plexuses. The anterior rami of spinal nerves T2–T12 are called intercostal nerves and will be discussed next.

Intercostal Nerves

The anterior rami of spinal nerves T2–T12 do not enter into the formation of plexuses and are known as **intercostal** or **thoracic nerves.** These nerves directly connect to the structures they supply in the intercostal spaces. After leaving its intervertebral foramen, the anterior ramus of nerve T2 innervates the intercostal muscles of the second intercostal space and supplies the skin of the axilla and posteromedial aspect of the arm. Nerves T3–T6 extend along the costal grooves of the ribs and then to the intercostal muscles and skin of the anterior and lateral chest wall. Nerves T7–T12 supply the intercostal muscles and abdominal muscles, along with the overlying skin. The posterior rami of the intercostal nerves supply the deep back muscles and skin of the posterior aspect of the thorax.

EXHIBIT 13.A Cervical Plexus *(Figure 13.7)*

OBJECTIVE

- Describe the origin and distribution of the cervical plexus.

The **cervical plexus** (SER-vi-kul) is formed by the roots (anterior rami) of the first four cervical nerves (C1–C4), with contributions from C5 (Figure 13.7). There is one on each side of the neck alongside the first four cervical vertebrae.

The cervical plexus supplies the skin and muscles of the head, neck, and superior part of the shoulders and chest. The phrenic nerves arise from the cervical plexuses and supply motor fibers to the diaphragm. Branches of the cervical plexus also run parallel to two cranial nerves, the accessory (XI) nerve and hypoglossal (XII) nerve.

NERVE	ORIGIN	DISTRIBUTION
SUPERFICIAL (SENSORY) BRANCHES		
Lesser occipital	C2	Skin of scalp posterior and superior to ear.
Great auricular (aw-RIK-ū-lar)	C2–C3	Skin anterior, inferior, and over ear, and over parotid glands.
Transverse cervical	C2–C3	Skin over anterior aspect of neck.
Supraclavicular	C3–C4	Skin over superior portion of chest and shoulder.
DEEP (LARGELY MOTOR) BRANCHES		
Ansa cervicalis (AN-sa ser-vi-KAL-is)		Divides into superior and inferior roots.
Superior root	C1	Infrahyoid and geniohyoid muscles of neck.
Inferior root	C2–C3	Infrahyoid muscles of neck.
Phrenic (FREN-ik)	C3–C5	Diaphragm.
Segmental branches	C1–C5	Prevertebral (deep) muscles of neck, levator scapulae, and middle scalene muscles.

EXHIBIT 13.A CONTINUES

CLINICAL CONNECTION | *Injuries to the Phrenic Nerves*

The phrenic nerves originate from C3, C4, and C5 and supply the diaphragm. Complete severing of the spinal cord above the origin of the phrenic nerves (C3, C4, and C5) causes respiratory arrest. In **injuries to the phrenic nerves,** breathing stops because the phrenic nerves no longer send nerve impulses to the diaphragm. The phrenic nerves may also be damaged due to pressure from malignant tracheal or esophageal tumors in the mediastinum. •

✔ **CHECKPOINT**

Which nerve from the cervical plexus causes contraction of the diaphragm?

Figure 13.7 Cervical plexus in anterior view.

The cervical plexus supplies the skin and muscles of the head, neck, superior portion of the shoulders and chest, and diaphragm.

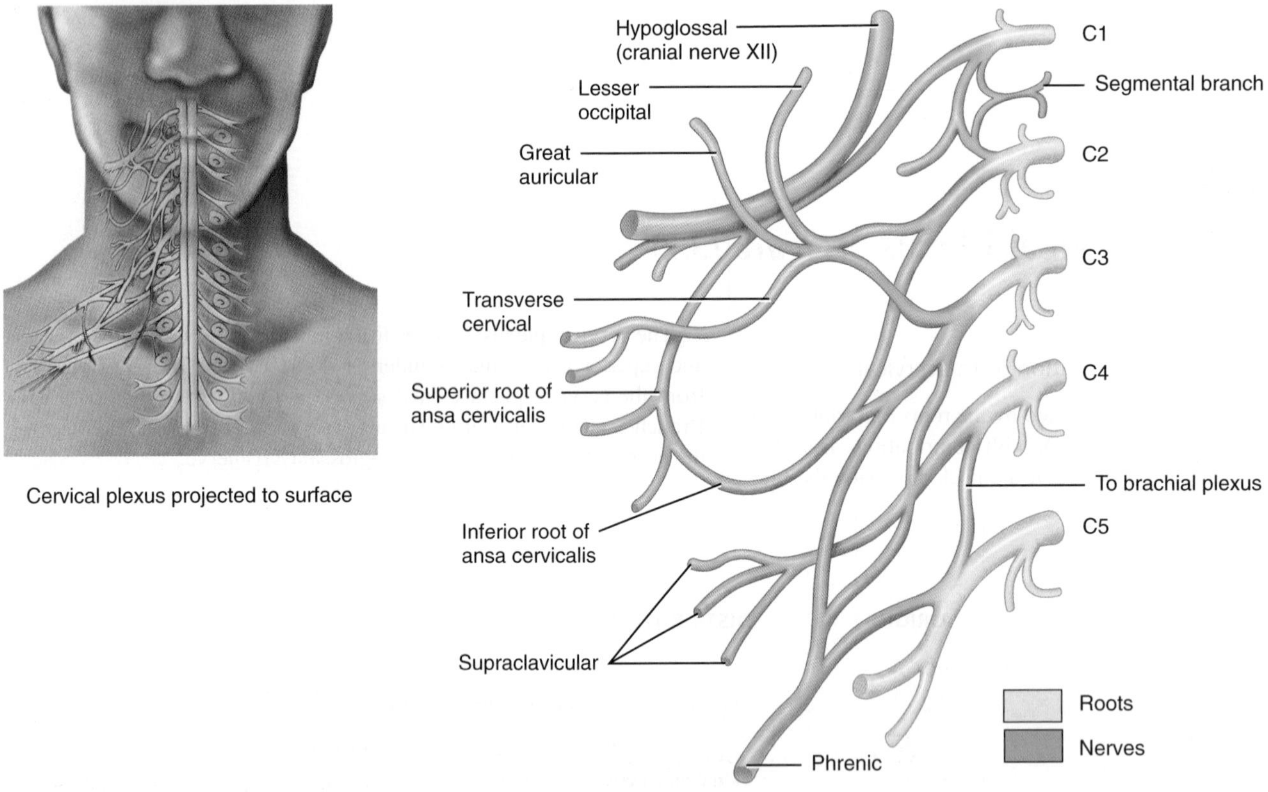

Cervical plexus projected to surface

Origin of cervical plexus

Why does complete severing of the spinal cord at level C2 cause respiratory arrest?

EXHIBIT 13.B Brachial Plexus *(Figure 13.8)*

OBJECTIVE

- Describe the origin, distribution, and effects of damage to the brachial plexus.

The roots (anterior rami) of spinal nerves C5–C8 and T1 form the **brachial plexus** (BRĀ-kē-al), which extends inferiorly and laterally on either side of the last four cervical and first thoracic vertebrae (Figure 13.8a). It passes above the first rib posterior to the clavicle and then enters the axilla.

Since the brachial plexus is so complex, an explanation of its various parts is helpful. As with the cervical and other plexuses, the **roots** are the anterior rami of the spinal nerves. The roots of several spinal nerves unite to form **trunks** in the inferior part of the neck. These are the *superior, middle,* and *inferior trunks.* Posterior to the clavicles, the trunks divide into **divisions,** called the *anterior* and *posterior divisions.* In the axillae, the divisions unite to form **cords** called the *lateral, medial,* and *posterior cords.* The cords are named for their relationship to the axillary artery, a large artery that supplies blood to the upper limb. The principal **nerves** of the brachial plexus branch from the cords.

The brachial plexus provides almost the entire nerve supply of the shoulders and upper limbs (Figure 13.8b). Five important nerves arise from the brachial plexus: (1) The **axillary nerve** supplies the deltoid and teres minor muscles. (2) The **musculocutaneous nerve** supplies the anterior muscles of the arm. (3) The **radial nerve** supplies the muscles on the posterior aspect of the arm and forearm. (4) The **median nerve** supplies most of the muscles of the anterior forearm and some of the muscles of the hand. (5) The **ulnar nerve** supplies the anteromedial muscles of the forearm and most of the muscles of the hand.

CLINICAL CONNECTION | *Injuries to Nerves Emerging from the Brachial Plexus*

Injury to the superior roots of the brachial plexus (C5–C6) may result from forceful pulling away of the head from the shoulder, as might occur from a heavy fall on the shoulder or excessive stretching of an infant's neck during childbirth. The presentation of this injury is characterized by an upper limb in which the shoulder is adducted, the arm is medially rotated, the elbow is extended, the forearm is pronated, and the wrist is flexed (Figure 13.8c). This condition is called **Erb-Duchenne palsy** or **waiter's tip position.** There is loss of sensation along the lateral side of the arm.

Injury to the radial (and axillary) **nerve** can be caused by improperly administered intramuscular injections into the deltoid muscle. The radial nerve may also be injured when a cast is applied too tightly around the mid-humerus. Radial nerve injury is indicated by **wrist drop,** the inability to extend the wrist and fingers (Figure 13.8c). Sensory loss is minimal due to the overlap of sensory innervation by adjacent nerves.

NERVE	ORIGIN	DISTRIBUTION
Dorsal scapular (SKAP-ū-lar)	C5	Levator scapulae, rhomboid major, and rhomboid minor muscles.
Long thoracic (thor-RAS-ik)	C5–C7	Serratus anterior muscle.
Nerve to subclavius (sub-KLĀ-vē-us)	C5–C6	Subclavius muscle.
Suprascapular	C5–C6	Supraspinatus and infraspinatus muscles.
Musculocutaneous (mus′-kū-lō-kū-TĀN-ē-us)	C5–C7	Coracobrachialis, biceps brachii, and brachialis muscles.
Lateral pectoral (PEK-to-ral)	C5–C7	Pectoralis major muscle.
Upper subscapular	C5–C6	Subscapularis muscle.
Thoracodorsal (tho-RĀ-kō-dor-sal)	C6–C8	Latissimus dorsi muscle.
Lower subscapular	C5–C6	Subscapularis and teres major muscles.
Axillary (AK-si-lar-ē)	C5–C6	Deltoid and teres minor muscles; skin over deltoid and superior posterior aspect of arm.
Median	C5–T1	Flexors of forearm, except flexor carpi ulnaris; ulnar half of flexor digitorum profundus, and some muscles of hand (lateral palm); skin of lateral two-thirds of palm of hand and fingers.
Radial		Triceps brachii, anconeus, and extensor muscles of forearm; skin of posterior arm and forearm, lateral two-thirds of dorsum of hand, and fingers over proximal and middle phalanges.
Medial pectoral	C8–T1	Pectoralis major and pectoralis minor muscles.
Medial cutaneous nerve of arm (kū′-TĀ-nē-us)	C8–T1	Skin of medial and posterior aspects of distal third of arm.
Medial cutaneous nerve of forearm	C8–T1	Skin of medial and posterior aspects of forearm.
Ulnar	C8–T1	Flexor carpi ulnaris, ulnar half of flexor digitorum profundus, and most muscles of hand; skin of medial side of hand, little finger, and medial half of ring finger.

EXHIBIT 13.B CONTINUES ▶

Injury to the median nerve may result in **median nerve palsy,** which is indicated by numbness, tingling, and pain in the palm and fingers. There is also inability to pronate the forearm and flex the proximal interphalangeal joints of all digits and the distal interphalangeal joints of the second and third digits (Figure 13.8c). In addition, wrist flexion is weak and is accompanied by adduction, and thumb movements are weak.

Injury to the ulnar nerve may result in **ulnar nerve palsy,** which is indicated by an inability to abduct or adduct the fingers, atrophy of the interosseous muscles of the hand, hyperextension of the metacarpophalangeal joints, and flexion of the interphalangeal joints, a condition called **clawhand** (Figure 13.8c). There is also loss of sensation over the little finger.

Injury to the long thoracic nerve results in paralysis of the serratus anterior muscle. The medial border of the scapula protrudes, giving it the appearance of a wing. When the arm is raised, the vertebral border and inferior angle of the scapula pull away from the thoracic wall and protrude outward, causing the medial border of the scapula to protrude; because the scapula looks like a wing, this condition is called **winged scapula** (Figure 13.8c). The arm cannot be abducted beyond the horizontal position.

Compression of the brachial plexus on one or more of its nerves is sometimes known as **thoracic outlet syndrome.** The subclavian artery and subclavian vein may also be compressed. The compression may result from spasm of the scalene or pectoralis minor muscles, the presence of a cervical rib (an embryological anomaly), or misaligned ribs. The patient may experience pain, numbness, weakness, or tingling in the upper limb, across the upper thoracic area, and over the scapula on the affected side. The symptoms of thoracic outlet syndrome are exaggerated during physical or emotional stress because the added stress increases the contraction of the involved muscles. •

✔ **CHECKPOINT**

Injury of which nerve could cause paralysis of the serratus anterior muscle?

Figure 13.8 **Brachial plexus in anterior view.**

The brachial plexus supplies the shoulders and upper limbs.

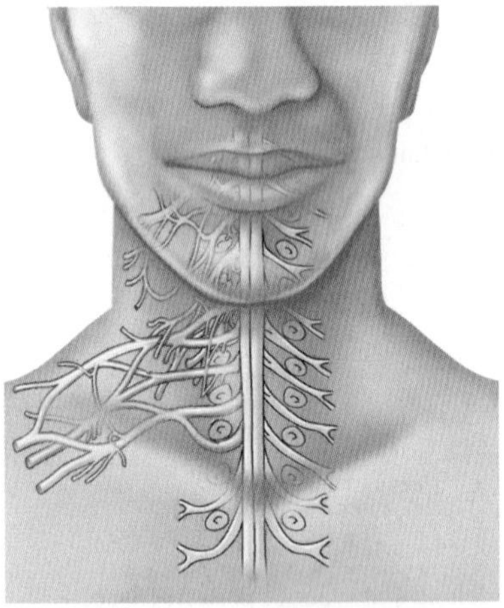

Brachial plexus projected to surface

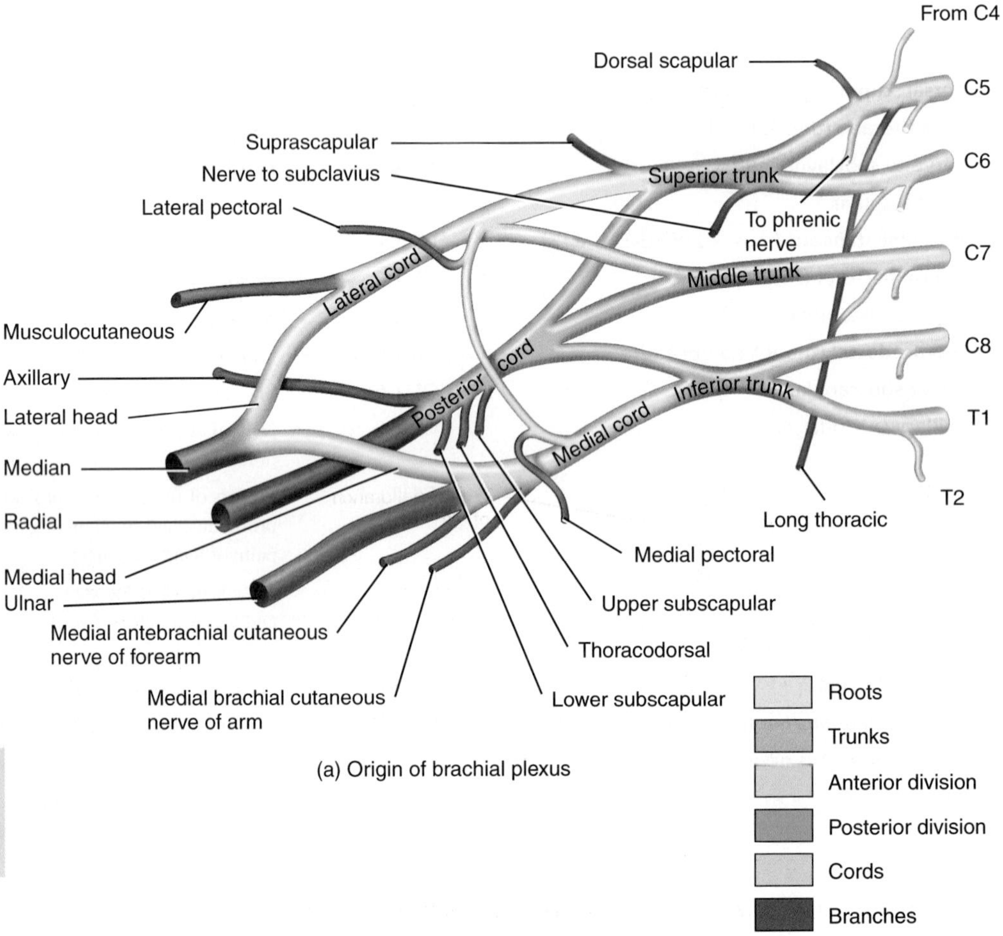
From C4
C5
C6
C7
C8
T1
T2
Dorsal scapular
Suprascapular
Nerve to subclavius
Lateral pectoral
Superior trunk
To phrenic nerve
Middle trunk
Lateral cord
Posterior cord
Inferior trunk
Medial cord
Musculocutaneous
Axillary
Lateral head
Median
Radial
Medial head
Ulnar
Medial antebrachial cutaneous nerve of forearm
Medial brachial cutaneous nerve of arm
Long thoracic
Medial pectoral
Upper subscapular
Thoracodorsal
Lower subscapular
Roots
Trunks
Anterior division
Posterior division
Cords
Branches

(a) Origin of brachial plexus

MNEMONIC for subunits of the brachial plexus:
Risk **T**akers **D**on't **C**autiously **B**ehave.
Roots, **T**runks, **D**ivisions, **C**ords, **B**ranches

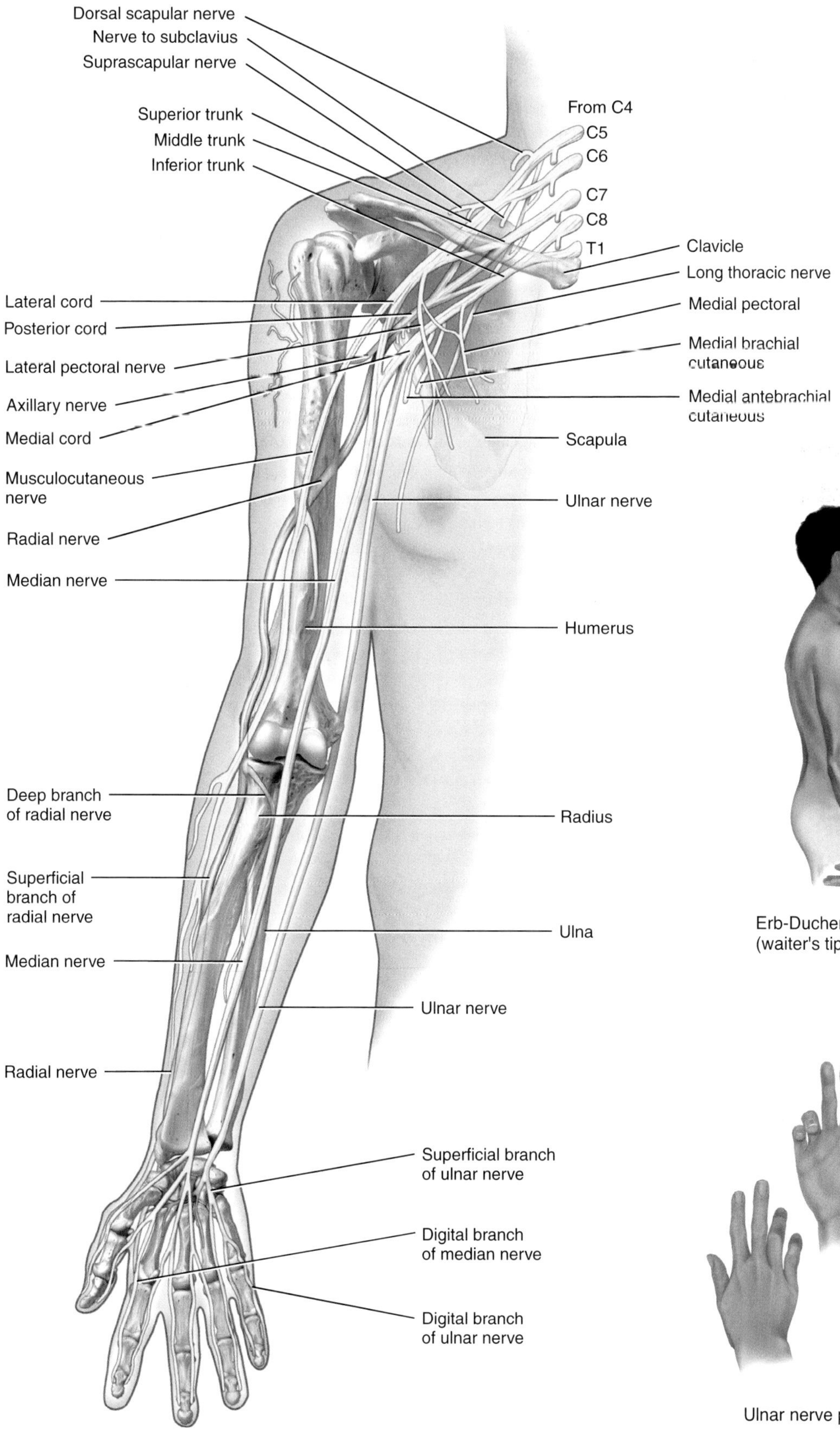
Dorsal scapular nerve
Nerve to subclavius
Suprascapular nerve
Superior trunk
Middle trunk
Inferior trunk
From C4
C5
C6
C7
C8
T1
Clavicle
Long thoracic nerve
Medial pectoral
Medial brachial cutaneous
Medial antebrachial cutaneous
Lateral cord
Posterior cord
Lateral pectoral nerve
Axillary nerve
Medial cord
Scapula
Musculocutaneous nerve
Ulnar nerve
Radial nerve
Median nerve
Humerus
Deep branch of radial nerve
Radius
Superficial branch of radial nerve
Ulna
Median nerve
Ulnar nerve
Radial nerve
Superficial branch of ulnar nerve
Digital branch of median nerve
Digital branch of ulnar nerve

(b) Distribution of nerves from the brachial plexus

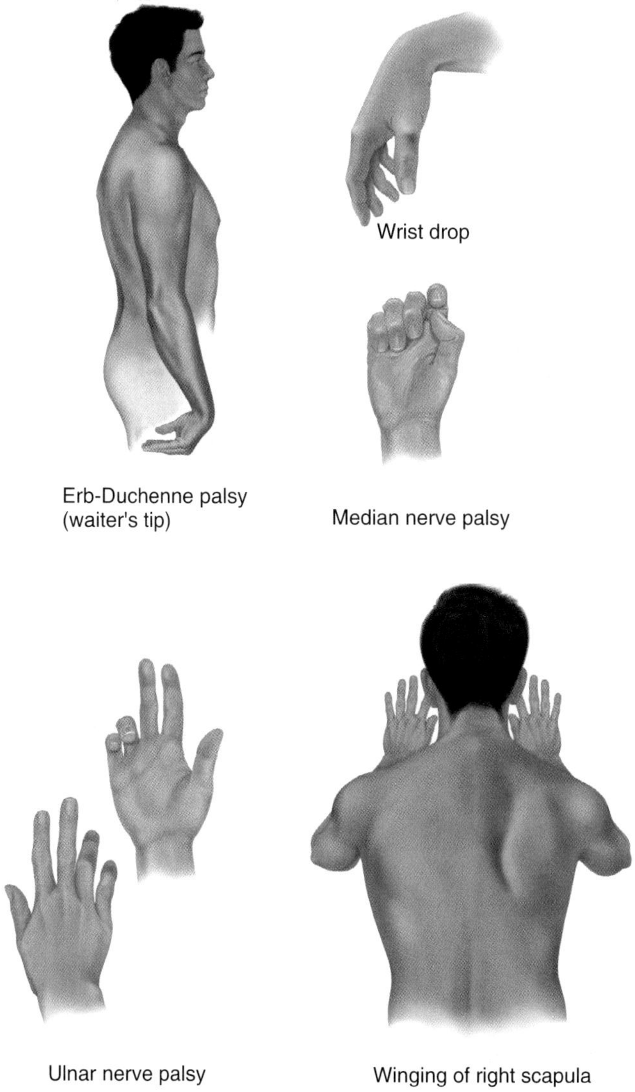
Wrist drop
Erb-Duchenne palsy (waiter's tip)
Median nerve palsy
Ulnar nerve palsy
Winging of right scapula

(c) Injuries to the brachial plexus

What five important nerves arise from the brachial plexus?

Lumbar Plexus *(Figure 13.9)*

OBJECTIVE

- Describe the origin and distribution of the lumbar plexus.

The roots (anterior rami) of spinal nerves L1–L4 form the **lumbar plexus** (LUM-bar) (Figure 13.9). Unlike the brachial plexus, there is minimal intermingling of fibers in the lumbar plexus. On either side of the first four lumbar vertebrae, the lumbar plexus passes obliquely outward, between the superficial and deep heads of the psoas major muscle and anterior to the quadratus lumborum muscle. It then gives rise to its peripheral nerves.

The lumbar plexus supplies the anterolateral abdominal wall, external genitals, and part of the lower limbs.

NERVE	ORIGIN	DISTRIBUTION
Iliohypogastric (il′-ē-ō-hī-pō-GAS-trik)	L1	Muscles of anterolateral abdominal wall; skin of inferior abdomen and buttock.
Ilioinguinal (il′-ē-ō-IN-gwi-nal)	L1	Muscles of anterolateral abdominal wall; skin of superior and medial aspect of thigh, root of penis and scrotum in male, and labia majora and mons pubis in female.
Genitofemoral (jen′-i-tō-FEM-or-al)	L1–L2	Cremaster muscle; skin over middle anterior surface of thigh, scrotum in male, and labia majora in female.
Lateral cutaneous nerve of thigh	L2–L3	Skin over lateral, anterior, and posterior aspects of thigh.
Femoral	L2–L4	Largest nerve arising from lumbar plexus; distributed to flexor muscles of hip joint and extensor muscles of knee joint, skin over anterior and medial aspect of thigh and medial side of leg and foot.
Obturator (OB-too-rā-tor)	L2–L4	Adductor muscles of hip joint; skin over medial aspect of thigh.

Figure 13.9 Lumbar plexus in anterior view.

The lumbar plexus supplies the anterolateral abdominal wall, external genitals, and part of the lower limbs.

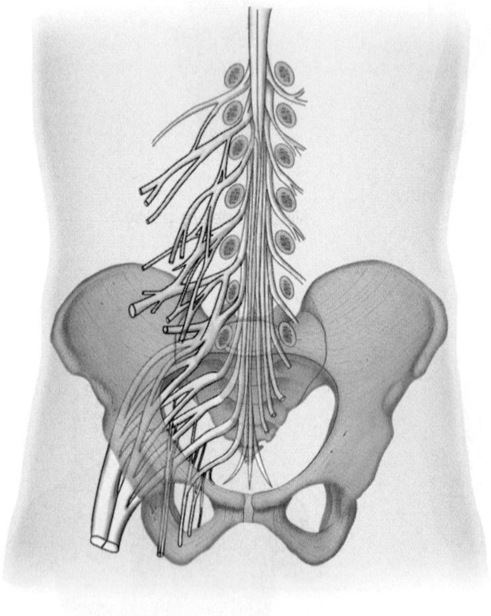

Lumbar plexus

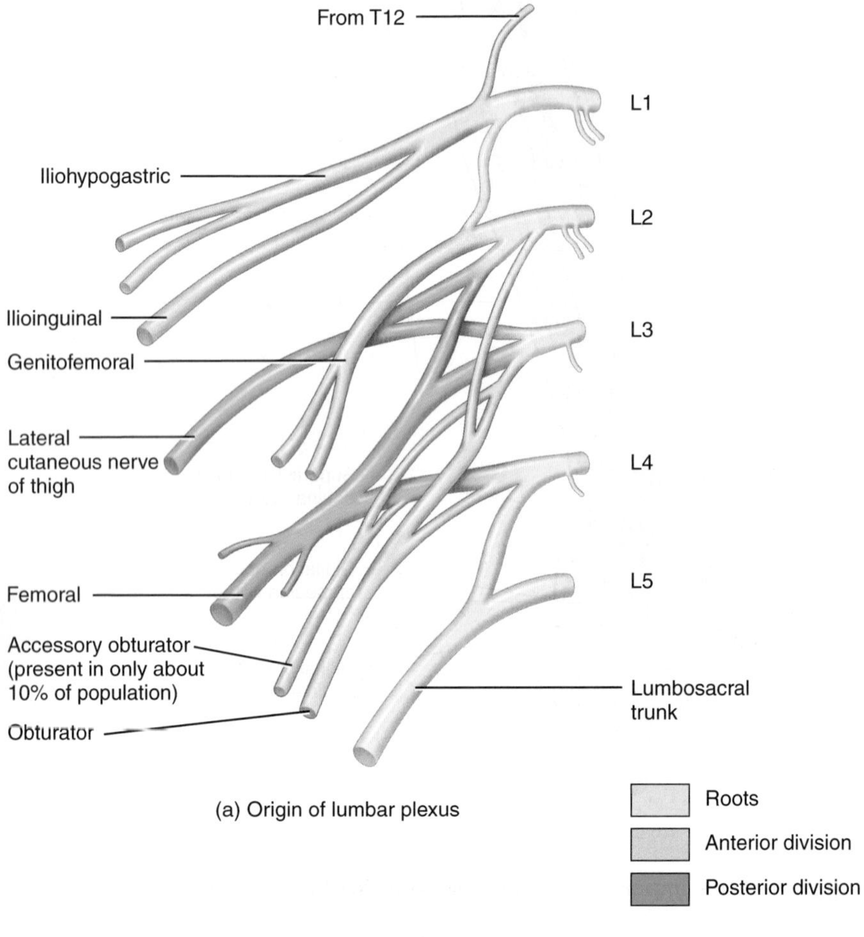
From T12
L1
L2
L3
L4
L5
Iliohypogastric
Ilioinguinal
Genitofemoral
Lateral cutaneous nerve of thigh
Femoral
Accessory obturator (present in only about 10% of population)
Obturator
Lumbosacral trunk
Roots
Anterior division
Posterior division

(a) Origin of lumbar plexus

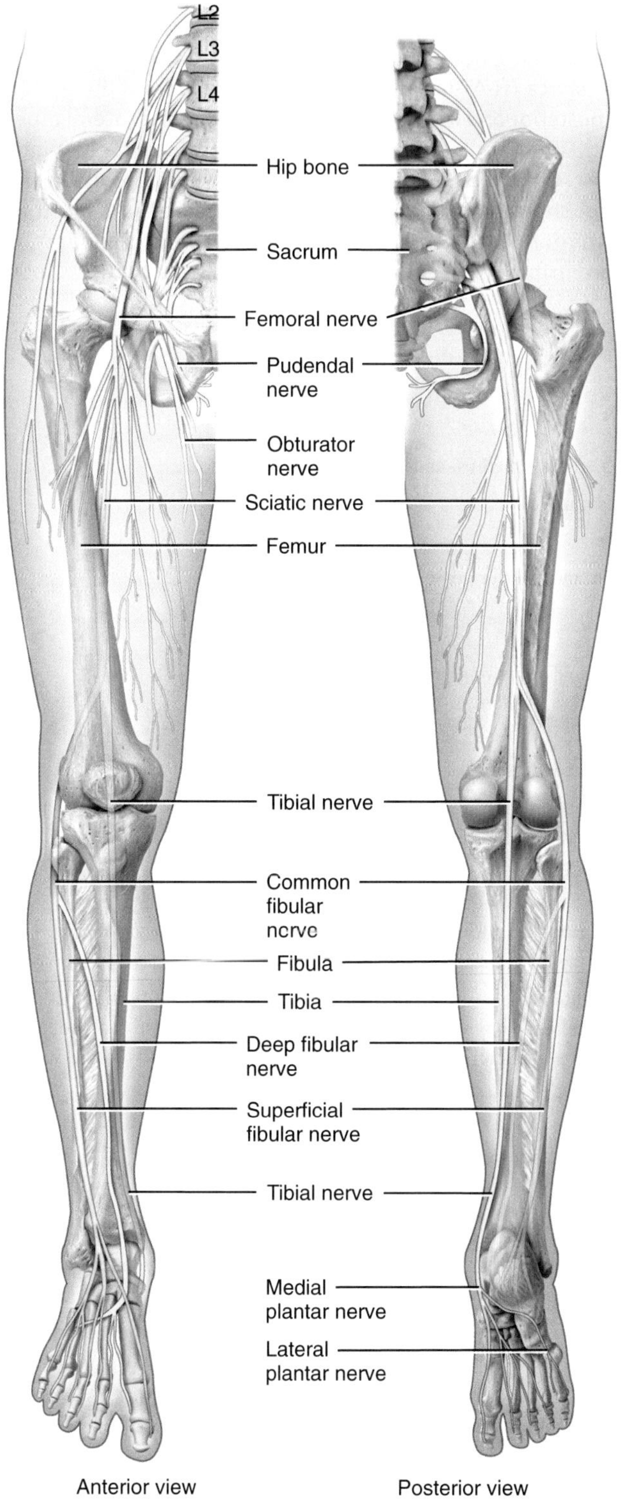
L2
L3
L4
Hip bone
Sacrum
Femoral nerve
Pudendal nerve
Obturator nerve
Sciatic nerve
Femur
Tibial nerve
Common fibular nerve
Fibula
Tibia
Deep fibular nerve
Superficial fibular nerve
Tibial nerve
Medial plantar nerve
Lateral plantar nerve
Anterior view
Posterior view

(b) Distribution of nerves from the lumbar and sacral plexuses

? What are the signs of femoral nerve injury?

CLINICAL CONNECTION | *Injuries to the Lumbar Plexus*

The largest nerve arising from the lumbar plexus is the femoral nerve. **Injuries to the femoral nerve,** which can occur in stab or gunshot wounds, are indicated by an inability to extend the leg and by loss of sensation in the skin over the anteromedial aspect of the thigh.

Injuries to the obturator nerve result in paralysis of the adductor muscles of the thigh and loss of sensation over the medial aspect of the thigh. It may result from pressure on the nerve by the fetal head during pregnancy. •

✔ CHECKPOINT

What is the largest nerve arising from the lumbar plexus?

EXHIBIT 13.D Sacral and Coccygeal Plexuses *(Figure 13.10)*

OBJECTIVE

- Describe the origin and distribution of the sacral and coccygeal plexuses.

The roots (anterior rami) of spinal nerves L4–L5 and S1–S4 form the **sacral plexus** (SĀ-kral) (Figure 13.10). This plexus is situated largely anterior to the sacrum. The sacral plexus supplies the buttocks, perineum, and lower limbs. The largest nerve in the body—the sciatic nerve—arises from the sacral plexus.

The roots (anterior rami) of spinal nerves S4–S5 and the coccygeal nerves form a small **coccygeal plexus** (kok-SIG-ē-al), which supplies a small area of skin in the coccygeal region.

CLINICAL CONNECTION | *Injury to the Sciatic Nerve*

The most common form of back pain is caused by compression or irritation of the sciatic nerve, the longest nerve in the human body. The sciatic nerve is actually two nerves—tibial and common fibular—bound together by a common sheath of connective tissue. It splits into its two divisions, usually at the knee. **Injury to the sciatic nerve** results in **sciatica** (sī-AT-i-ka), pain that may extend from the buttock down the posterior and lateral aspect of the leg and the lateral aspect of the foot. The sciatic nerve may be injured because of a herniated (slipped) disc, dislocated hip, osteoarthritis of the lumbosacral spine, pathological shortening of the lateral rotator muscles of the thigh (especially piriformis), pressure from the uterus during pregnancy, inflammation, irritation, or an improperly administered gluteal intra-

NERVE	ORIGIN	DISTRIBUTION
Superior gluteal (GLOO-tē-al)	L4–L5 and S1	Gluteus minimus, gluteus medius, and tensor fasciae latae muscles.
Inferior gluteal	L5–S2	Gluteus maximus muscle.
Nerve to piriformis (pir-i-FORM-is)	S1–S2	Piriformis muscle.
Nerve to quadratus femoris (quad-RĀ-tus FEM-or-is) and **inferior gemellus** (jem-EL-us)	L4–L5 and S1	Quadratus femoris and inferior gemellus muscles.
Nerve to obturator internus (OB-too-rā′-tor in-TER-nus) and **superior gemellus**	L5–S2	Obturator internus and superior gemellus muscles.
Perforating cutaneous (kū′-TĀ-nē-us)	S2–S3	Skin over inferior medial aspect of buttock.
Posterior cutaneous nerve of thigh	S1–S3	Skin over anal region, inferior lateral aspect of buttock, superior posterior aspect of thigh, superior part of calf, scrotum in male, and labia majora in female.
Sciatic (sī-AT-ik)	L4–S3	Actually two nerves—tibial and common fibular—bound together by common sheath of connective tissue; splits into its two divisions, usually at the knee. (See below for distributions.) As sciatic nerve descends through thigh, it sends branches to hamstring muscles and adductor magnus.
Tibial (TIB-ē-al)	L4–S3	Gastrocnemius, plantaris, soleus, popliteus, tibialis posterior, flexor digitorum longus, and flexor hallucis longus muscles. Branches of tibial nerve in foot are medial plantar nerve and lateral plantar nerve.
Medial plantar (PLAN-tar) (see Figure 13.9b)		Abductor hallucis, flexor digitorum brevis, and flexor hallucis brevis muscles; skin over medial two-thirds of plantar surface of foot.
Lateral plantar (see Figure 13.9b)		Remaining muscles of foot not supplied by medial plantar nerve; skin over lateral third of plantar surface of foot.
Common fibular (FIB-ū-lar)	L4–S2	Divides into superficial fibular and deep fibular branch.
Superficial fibular		Fibularis longus and fibularis brevis muscles; skin over distal third of anterior aspect of leg and dorsum of foot.
Deep fibular		Tibialis anterior, extensor hallucis longus, fibularis tertius, and extensor digitorum longus and extensor digitorum brevis muscles; skin on adjacent sides of great and second toes.
Pudendal (pū-DEN-dal)	S2–S4	Muscles of perineum; skin of penis and scrotum in male and clitoris, labia majora, labia minora, and vagina in female.

muscular injection. In addition, sitting on a wallet or other object for a long period of time can compress the nerve and induce pain.

In many sciatic nerve injuries, the common fibular portion is the most affected, frequently from fractures of the fibula or by pressure from casts or splints over the thigh or leg. Damage to the common fibular nerve causes the foot to be plantar flexed, a condition called **foot drop**, and inverted, a condition called **equinovarus** (e-KWĪ-nō-va-rus). There is also loss of function along the anterolateral apects of the leg and dorsum of the foot and toes. Injury to the tibial portion of the sciatic nerve results in dorsiflexion of the foot plus eversion, a condition called **calcaneovalgus** (kal-KĀ-nē-ō-val′-gus). Loss of sensation on the sole also occurs. Treatments for sciatica are similar to those for a herniated (slipped) disc—rest, pain medications, exercises, ice or heat, and massage. •

✔ **CHECKPOINT**

Injury of which nerve causes footdrop?

Figure 13.10 Sacral and coccygeal plexuses in anterior view. The distribution of the nerves of the sacral plexus is shown in Figure 13.9b.

The sacral plexus supplies the buttocks, perineum, and lower limbs.

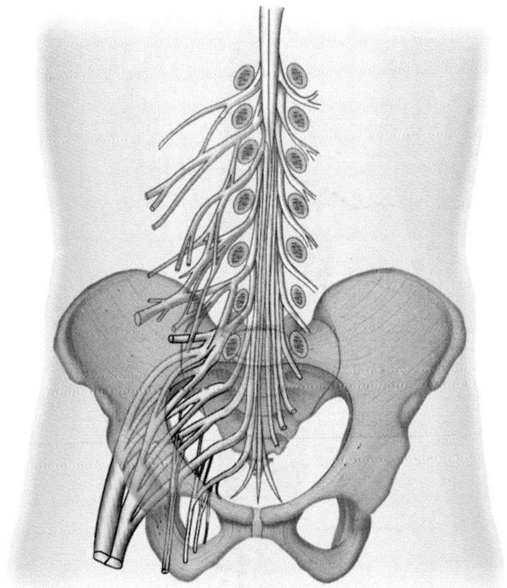

Sacral and coccygeal plexuses

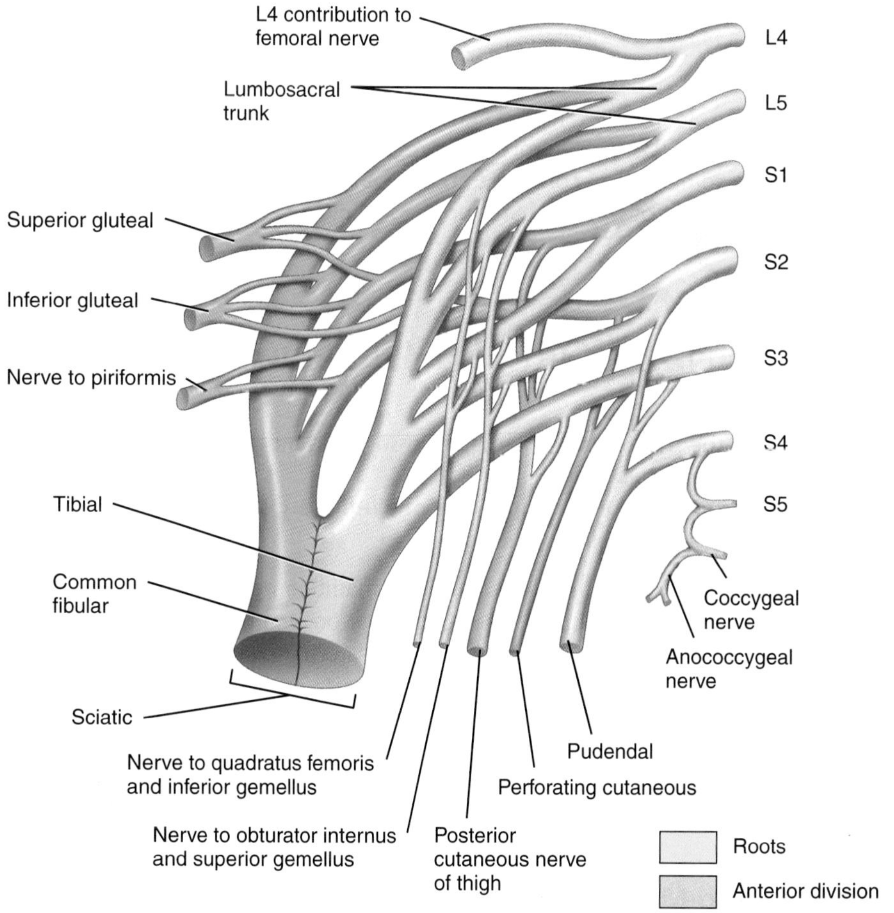

Origin of sacral plexus

What is the origin of the sacral plexus?

Dermatomes

The skin over the entire body is supplied by somatic sensory neurons that carry nerve impulses from the skin into the spinal cord and brain. Each spinal nerve contains sensory neurons that serve a specific, predictable segment of the body. One of the cranial nerves, the trigeminal (V) nerve, serves most of the skin of the face and scalp. The area of the skin that provides sensory input to the CNS via one pair of spinal nerves or the trigeminal (V) nerve is called a **dermatome** (DER-ma-tōm; *derma-* = skin; *-tome* = thin segment) (Figure 13.11). The nerve supply in adjacent dermatomes overlaps somewhat. Knowing which spinal cord segments supply each dermatome makes it possible to locate damaged regions of the spinal cord. If the skin in a particular region is stimulated but the sensation is not perceived, the nerves supplying that dermatome are probably damaged. In regions where the overlap is considerable, little loss of sensation may result if only one of the nerves supplying the dermatome is damaged. Information about the innervation patterns of spinal nerves can also be used therapeutically. Cutting posterior roots or infusing local anesthetics can block pain either permanently or transiently. Because dermatomes overlap, deliberate production of a region of complete anesthesia may require that at least three adjacent spinal nerves be cut or blocked by an anesthetic drug.

✔ **CHECKPOINT**

5. How are spinal nerves named and numbered? Why are all spinal nerves classified as mixed nerves?
6. How do spinal nerves connect to the spinal cord?
7. Which regions of the body are supplied by plexuses and by intercostal nerves?

13.3 SPINAL CORD PHYSIOLOGY

OBJECTIVES

- Describe the functions of the major sensory and motor tracts of the spinal cord.
- Describe the functional components of a reflex arc and the ways reflexes maintain homeostasis.

The spinal cord has two principal functions in maintaining homeostasis: nerve impulse propagation and integration of information. The *white matter tracts* in the spinal cord are highways for nerve impulse propagation. Sensory input travels along these tracts toward the brain, and motor output travels from the brain along these tracts toward skeletal muscles and other effector tissues. The *gray matter* of the spinal cord receives and integrates incoming and outgoing information.

Sensory and Motor Tracts

As noted previously, one of the ways the spinal cord promotes homeostasis is by conducting nerve impulses along tracts. Often, the name of a tract indicates its position in the white matter and where it begins and ends. For example, the anterior corticospinal tract is located in the *anterior* white column; it begins in the *cerebral*

Figure 13.11 Distribution of dermatomes.

A dermatome is an area of skin that provides sensory input to the CNS via the posterior roots of one pair of spinal nerves or via the trigeminal (V) nerve.

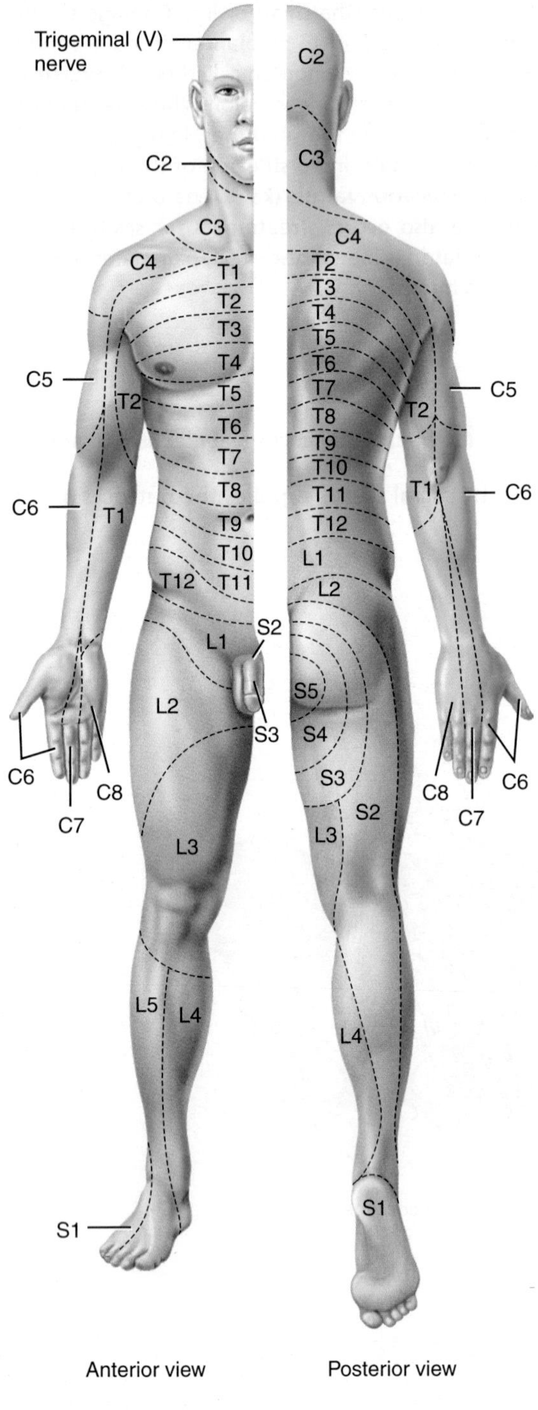

Which is the only spinal nerve that does not have a corresponding dermatome?

cortex (superficial gray matter of the cerebrum of the brain) and ends in the *spinal cord*. Notice that the location of the axon terminals comes last in the name. This regularity in naming allows you to determine the direction of information flow along any tract named according to this convention. Because the anterior corticospinal tract conveys nerve impulses from the brain toward the spinal cord, it is a motor (descending) tract. Figure 13.12 highlights the major sensory and motor tracts in the spinal cord. These tracts are described in detail in Chapter 16 and summarized in Tables 16.3 and 16.4.

Nerve impulses from sensory receptors propagate up the spinal cord to the brain along two main routes on each side: the spinothalamic tract and the posterior column. The **spinothalamic tract** (spī′-nō-tha-LAM-ik) conveys nerve impulses for sensing pain, warmth, coolness, itching, tickling, deep pressure, and crude touch. The **posterior column** consists of two tracts: the **gracile fasciculus** (GRAS-īl fa-SIK-ū-lus) and the **cuneate fasciculus** (KŪ-nē-āt). The posterior column tracts convey nerve impulses for discriminative touch, light pressure, vibration, and conscious proprioception (the awareness of the positions and movements of muscles, tendons, and joints).

The sensory systems keep the CNS informed of changes in the external and internal environments. The sensory information is integrated (processed) by interneurons in the spinal cord and brain. Responses to the integrative decisions are brought about by motor activities (muscular contractions and glandular secretions). The cerebral cortex, the outer part of the brain, plays a major role in controlling precise voluntary muscular movements. Other brain regions provide important integration for regulation of automatic movements. Motor output to skeletal muscles travels down

Figure 13.12 Locations of major sensory and motor tracts, shown in a transverse section of the spinal cord. Sensory tracts are indicated on one half and motor tracts on the other half of the cord, but actually all tracts are present on both sides.

The name of a tract often indicates its location in the white matter and where it begins and ends.

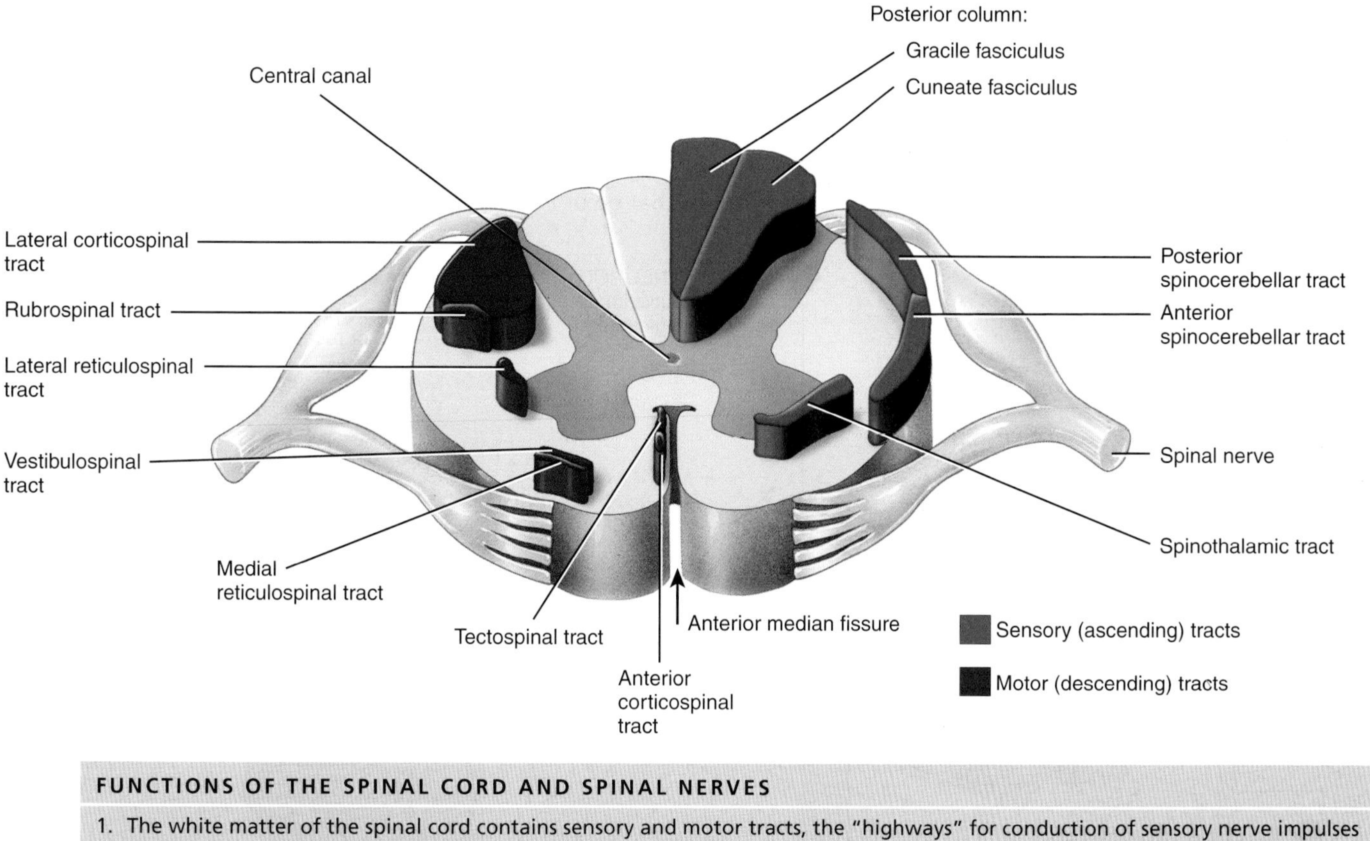

FUNCTIONS OF THE SPINAL CORD AND SPINAL NERVES

1. The white matter of the spinal cord contains sensory and motor tracts, the "highways" for conduction of sensory nerve impulses toward the brain and motor nerve impulses from the brain toward effector tissues.
2. The spinal cord gray matter is a site for integration (summing) of excitatory postsynaptic potentials (EPSPs) and inhibitory postsynaptic potentials (IPSPs).
3. Spinal nerves and the nerves that branch from them connect the CNS to the sensory receptors, muscles, and glands in all parts of the body.

Based on its name, list the origin and destination of the spinothalamic tract. Is this a sensory or a motor tract?

the spinal cord in two types of descending pathways: direct and indirect. The **direct pathways** include the **lateral corticospinal** (kor′-ti-kō-SPĪ-nal), **anterior corticospinal,** and **corticobulbar tracts** (kor′-ti-kō-BUL-bar). They convey nerve impulses that originate in the cerebral cortex and are destined to cause *voluntary* movements of skeletal muscles. **Indirect pathways** include the **rubrospinal** (ROO-brō-spī-nal), **tectospinal** (TEK-tō-spī-nal), **vestibulospinal** (ves-TIB-ū-lō-spī-nal), **lateral reticulospinal** (re-TIK-ū-lō-spī-nal), and **medial reticulospinal tracts.** These tracts convey nerve impulses from the brain stem to cause *automatic movements* and help coordinate body movements with visual stimuli. Indirect pathways also maintain skeletal muscle tone, sustain contraction of postural muscles, and play a major role in equilibrium by regulating muscle tone in response to movements of the head.

Reflexes and Reflex Arcs

The second way the spinal cord promotes homeostasis is by serving as an integrating center for some reflexes. A **reflex** is a fast, involuntary, unplanned sequence of actions that occurs in response to a particular stimulus. Some reflexes are inborn, such as pulling your hand away from a hot surface before you even feel that it is hot. Other reflexes are learned or acquired. For instance, you learn many reflexes while acquiring driving expertise. Slamming on the brakes in an emergency is one example. When integration takes place in the spinal cord gray matter, the reflex is a **spinal reflex.** An example is the familiar patellar reflex (knee jerk). If integration occurs in the brain stem rather than the spinal cord, the reflex is called a **cranial reflex.** An example is the tracking movements of your eyes as you read this sentence. You are probably most aware of **somatic reflexes,** which involve contraction of skeletal muscles. Equally important, however, are the **autonomic (visceral) reflexes,** which generally are not consciously perceived. They involve responses of smooth muscle, cardiac muscle, and glands. As you will see in Chapter 15, body functions such as heart rate, digestion, urination, and defecation are controlled by the autonomic nervous system through autonomic reflexes.

Nerve impulses propagating into, through, and out of the CNS follow specific pathways, depending on the kind of information, its origin, and its destination. The pathway followed by nerve impulses that produce a reflex is a **reflex arc (reflex circuit).** A reflex arc includes the following five functional components (Figure 13.13):

1. **Sensory receptor.** The distal end of a sensory neuron (dendrite) or an associated sensory structure serves as a sensory

Figure 13.13 General components of a reflex arc. Arrows show the direction of nerve impulse propagation.

A reflex is a fast, predictable sequence of involuntary actions that occur in response to certain changes in the environment.

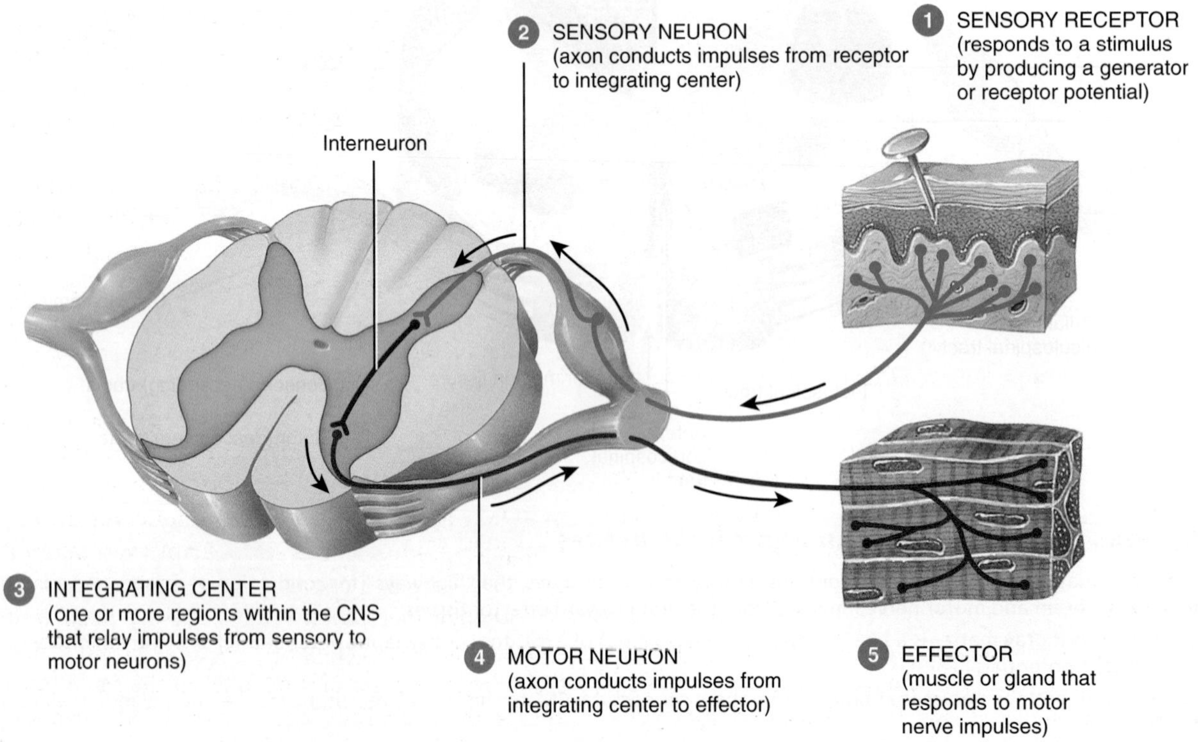
2 SENSORY NEURON
(axon conducts impulses from receptor to integrating center)
1 SENSORY RECEPTOR
(responds to a stimulus by producing a generator or receptor potential)
Interneuron
3 INTEGRATING CENTER
(one or more regions within the CNS that relay impulses from sensory to motor neurons)
4 MOTOR NEURON
(axon conducts impulses from integrating center to effector)
5 EFFECTOR
(muscle or gland that responds to motor nerve impulses)

What initiates a nerve impulse in a sensory neuron? Which branch of the nervous system includes all integrating centers for reflexes?

receptor. It responds to a specific **stimulus**—a change in the internal or external environment—by producing a graded potential called a generator (or receptor) potential (described in Section 16.1). If a generator potential reaches the threshold level of depolarization, it will trigger one or more nerve impulses in the sensory neuron.

2. **Sensory neuron.** The nerve impulses propagate from the sensory receptor along the axon of the sensory neuron to the axon terminals, which are located in the gray matter of the spinal cord or brain stem. From here, relay neurons send nerve impulses to the area of the brain that allows conscious awareness that the reflex has occurred.
3. **Integrating center.** One or more regions of gray matter within the CNS act as an integrating center. In the simplest type of reflex, the integrating center is a single synapse between a sensory neuron and a motor neuron. A reflex pathway having only one synapse in the CNS is termed a **monosynaptic reflex arc** (mon′-ō-si-NAP-tik; *mono-* = one). More often, the integrating center consists of one or more interneurons, which may relay impulses to other interneurons as well as to a motor neuron. A **polysynaptic reflex arc** (*poly-* = many) involves more than two types of neurons and more than one CNS synapse.
4. **Motor neuron.** Impulses triggered by the integrating center propagate out of the CNS along a motor neuron to the part of the body that will respond.
5. **Effector.** The part of the body that responds to the motor nerve impulse, such as a muscle or gland, is the effector. Its action is called a reflex. If the effector is skeletal muscle, the reflex is a **somatic reflex.** If the effector is smooth muscle, cardiac muscle, or a gland, the reflex is an **autonomic (visceral) reflex.**

Because reflexes are normally so predictable, they provide useful information about the health of the nervous system and can greatly aid diagnosis of disease. Damage or disease anywhere along its reflex arc can cause a reflex to be absent or abnormal. For example, tapping the patellar ligament normally causes reflex extension of the knee joint. Absence of the patellar reflex could indicate damage of the sensory or motor neurons, or a spinal cord injury in the lumbar region. Somatic reflexes generally can be tested simply by tapping or stroking the body surface.

Next, we examine four important somatic spinal reflexes: the stretch reflex, the tendon reflex, the flexor (withdrawal) reflex, and the crossed extensor reflex.

The Stretch Reflex

A **stretch reflex** causes contraction of a skeletal muscle (the effector) in response to stretching of the muscle. This type of reflex occurs via a monosynaptic reflex arc. The reflex can occur by activation of a single sensory neuron that forms one synapse in the CNS with a single motor neuron. Stretch reflexes can be elicited by tapping on tendons attached to muscles at the elbow, wrist, knee, and ankle joints. An example of a stretch reflex is the patellar reflex (knee jerk), which is described in Clinical Connection: Reflexes and Diagnosis later in the chapter.

A stretch reflex operates as follows (Figure 13.14):

1. Slight stretching of a muscle stimulates sensory receptors in the muscle called **muscle spindles** (shown in more detail in Figure 16.4). The spindles monitor changes in the length of the muscle.
2. In response to being stretched, a muscle spindle generates one or more nerve impulses that propagate along a somatic sensory neuron through the posterior root of the spinal nerve and into the spinal cord.
3. In the spinal cord (integrating center), the sensory neuron makes an excitatory synapse with and thereby activates a motor neuron in the anterior gray horn.
4. If the excitation is strong enough, one or more nerve impulses arise in the motor neuron and propagate along its axon, which extends from the spinal cord into the anterior root and through peripheral nerves to the stimulated muscle. The axon terminals of the motor neuron form neuromuscular junctions (NMJs) with skeletal muscle fibers of the stretched muscle.
5. Acetylcholine released by nerve impulses at the NMJs triggers one or more muscle action potentials in the stretched muscle (effector), and the muscle contracts. Thus, muscle stretch is followed by muscle contraction, which relieves the stretching.

In the reflex arc just described, sensory nerve impulses enter the spinal cord on the same side from which motor nerve impulses leave it. This arrangement is called an **ipsilateral reflex** (ip′-si-LAT-er-al = same side). All monosynaptic reflexes are ipsilateral.

In addition to the large-diameter motor neurons that innervate typical skeletal muscle fibers, smaller-diameter motor neurons innervate smaller, specialized muscle fibers within the muscle spindles themselves. The brain regulates muscle spindle sensitivity through pathways to these smaller motor neurons. This regulation ensures proper muscle spindle signaling over a wide range of muscle lengths during voluntary and reflex contractions. By adjusting how vigorously a muscle spindle responds to stretching, the brain sets an overall level of **muscle tone,** which is the small degree of contraction present while the muscle is at rest. Because the stimulus for the stretch reflex is stretching of muscle, this reflex helps avert injury by preventing overstretching of muscles.

Although the stretch reflex pathway itself is monosynaptic (just two neurons and one synapse), a polysynaptic reflex arc to the antagonistic muscles operates at the same time. This arc involves three neurons and two synapses. An axon collateral (branch) from the muscle spindle sensory neuron also synapses with an inhibitory interneuron in the integrating center. In turn, the interneuron synapses with and inhibits a motor neuron that

Figure 13.14 **Stretch reflex.** This monosynaptic reflex arc has only one synapse in the CNS—between a single sensory neuron and a single motor neuron. A polysynaptic reflex arc to antagonistic muscles that includes two synapses in the CNS and one interneuron is also illustrated. Plus signs (+) indicate excitatory synapses; the minus sign (−) indicates an inhibitory synapse.

The stretch reflex causes contraction of a muscle that has been stretched.

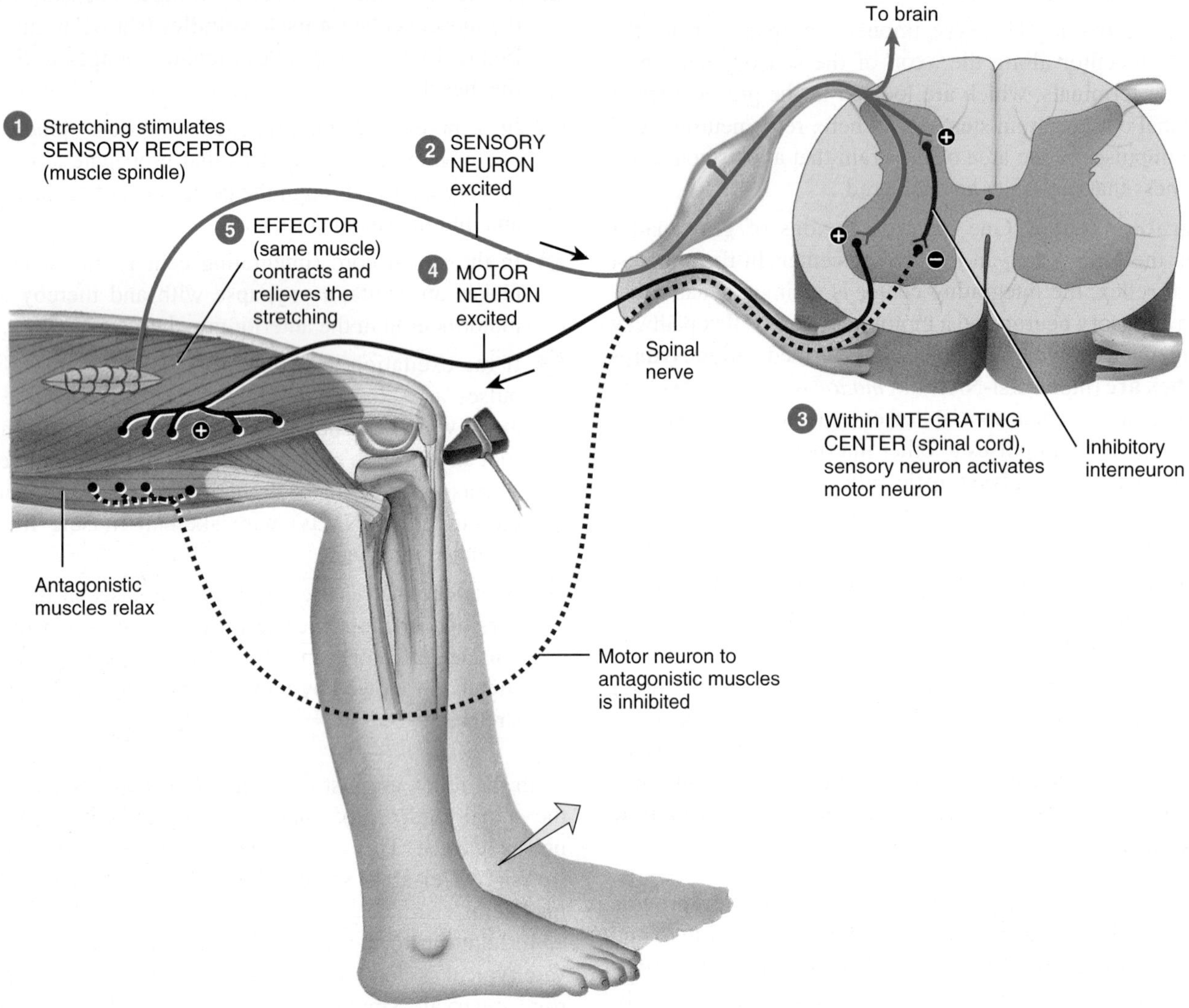
To brain
1 Stretching stimulates SENSORY RECEPTOR (muscle spindle)
2 SENSORY NEURON excited
5 EFFECTOR (same muscle) contracts and relieves the stretching
4 MOTOR NEURON excited
Spinal nerve
3 Within INTEGRATING CENTER (spinal cord), sensory neuron activates motor neuron
Inhibitory interneuron
Antagonistic muscles relax
Motor neuron to antagonistic muscles is inhibited

What makes this an ipsilateral reflex?

normally excites the antagonistic muscles (Figure 13.14). Thus, when the stretched muscle contracts during a stretch reflex, antagonistic muscles that oppose the contraction relax. This type of arrangement, in which the components of a neural circuit simultaneously cause contraction of one muscle and relaxation of its antagonists, is termed **reciprocal innervation** (rē-SIP-ro′-kal in′-er-VĀ-shun). Reciprocal innervation prevents conflict between opposing muscles and is vital in coordinating body movements.

Axon collaterals of the muscle spindle sensory neuron also relay nerve impulses to the brain over specific ascending pathways. In this way, the brain receives input about the state of stretch or contraction of skeletal muscles, enabling it to coordinate muscular movements. The nerve impulses that pass to the brain also allow conscious awareness that the reflex has occurred.

The stretch reflex can also help maintain posture. For example, if a standing person begins to lean forward, the gastrocnemius and other calf muscles are stretched. Consequently, stretch reflexes are initiated in these muscles, which cause them to contract and reestablish the body's upright posture. Similar types of stretch reflexes occur in the muscles of the shin when a standing person begins to lean backward.

The Tendon Reflex

The stretch reflex operates as a feedback mechanism to control muscle *length* by causing muscle contraction. In contrast, the **tendon reflex** operates as a feedback mechanism to control muscle *tension* by causing muscle relaxation before muscle force becomes so great that tendons might be torn. Although the tendon reflex is less sensitive than the stretch reflex, it can override the stretch reflex when tension is great, making you drop a very heavy weight, for example. Like the stretch reflex, the tendon reflex is ipsilateral. The sensory receptors for this reflex are called **tendon (Golgi tendon) organs** (shown in more detail in Figure 16.4), which lie within a tendon near its junction with a muscle. In contrast to muscle spindles, which are sensitive to changes in muscle length, tendon organs detect and respond to changes in muscle tension that are caused by passive stretch or muscular contraction.

A tendon reflex operates as follows (Figure 13.15):

1. As the tension applied to a tendon increases, the tendon organ (sensory receptor) is stimulated (depolarized to threshold).
2. Nerve impulses arise and propagate into the spinal cord along a sensory neuron.
3. Within the spinal cord (integrating center), the sensory neuron activates an inhibitory interneuron that synapses with a motor neuron.
4. The inhibitory neurotransmitter inhibits (hyperpolarizes) the motor neuron, which then generates fewer nerve impulses.
5. The muscle relaxes and relieves excess tension.

Thus, as tension on the tendon organ increases, the frequency of inhibitory impulses increases; inhibition of the motor neurons to the muscle developing excess tension (effector) causes relaxation

Figure 13.15 Tendon reflex. This reflex arc is polysynaptic—more than one CNS synapse and more than two different neurons are involved in the pathway. The sensory neuron synapses with two interneurons. An inhibitory interneuron causes relaxation of the effector, and a stimulatory interneuron causes contraction of the antagonistic muscle. Plus signs (+) indicate excitatory synapses; the minus sign (−) indicates an inhibitory synapse.

The tendon reflex causes relaxation of the muscle attached to the stimulated tendon organ.

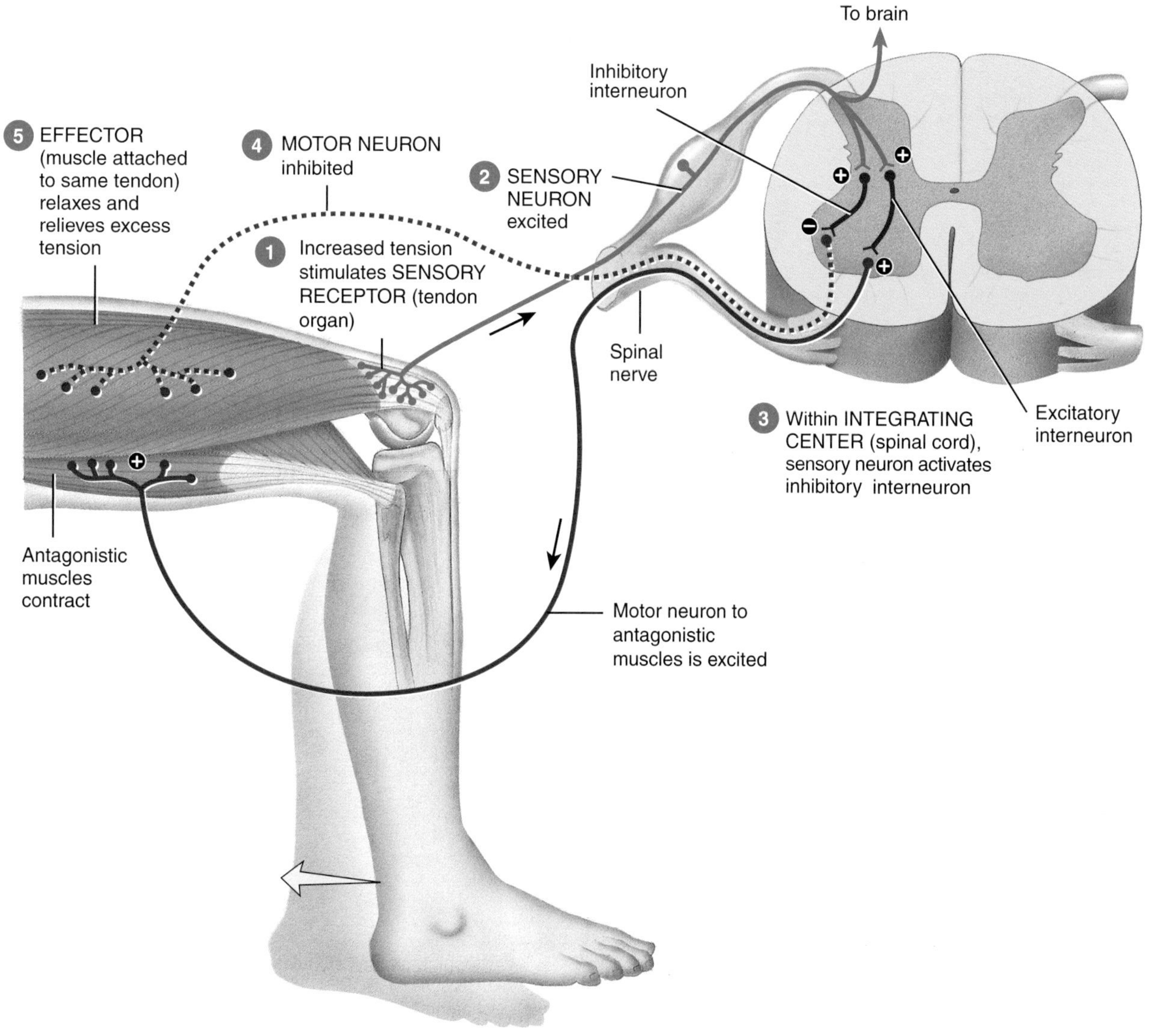

What is reciprocal innervation?

of the muscle. In this way, the tendon reflex protects the tendon and muscle from damage due to excessive tension.

Note in Figure 13.15 that the sensory neuron from the tendon organ also synapses with an excitatory interneuron in the spinal cord. The excitatory interneuron in turn synapses with motor neurons controlling antagonistic muscles. Thus, while the tendon reflex brings about relaxation of the muscle attached to the tendon organ, it also triggers contraction of antagonists. Here we have another example of reciprocal innervation. The sensory neuron also relays nerve impulses to the brain by way of sensory tracts, thus informing the brain about the state of muscle tension throughout the body.

The Flexor and Crossed Extensor Reflexes

Another reflex involving a polysynaptic reflex arc results when, for instance, you step on a tack. In response to such a painful stimulus, you immediately withdraw your leg. This reflex, called the **flexor** or **withdrawal reflex,** operates as follows (Figure 13.16):

1. Stepping on a tack stimulates the dendrites (sensory receptor) of a pain-sensitive neuron.
2. This sensory neuron then generates nerve impulses, which propagate into the spinal cord.

Figure 13.16 Flexor (withdrawal) reflex. Plus signs (+) indicate excitatory synapses.

The flexor reflex causes withdrawal of a part of the body in response to a painful stimulus.

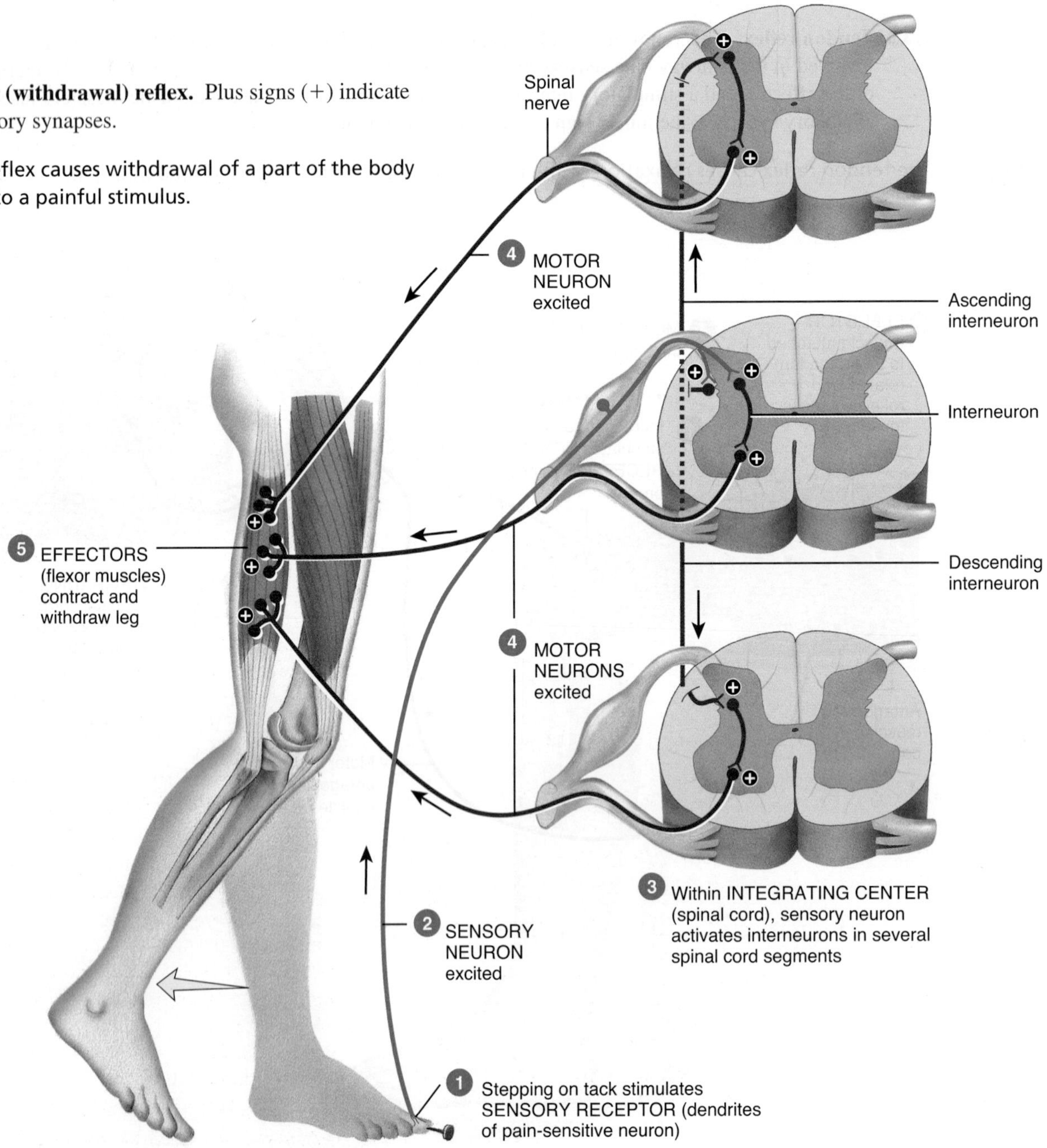
Spinal nerve
4 MOTOR NEURON excited
Ascending interneuron
Interneuron
Descending interneuron
5 EFFECTORS (flexor muscles) contract and withdraw leg
4 MOTOR NEURONS excited
3 Within INTEGRATING CENTER (spinal cord), sensory neuron activates interneurons in several spinal cord segments
2 SENSORY NEURON excited
1 Stepping on tack stimulates SENSORY RECEPTOR (dendrites of pain-sensitive neuron)

? Why is the flexor reflex classified as an intersegmental reflex arc?

3. Within the spinal cord (integrating center), the sensory neuron activates interneurons that extend to several spinal cord segments.
4. The interneurons activate motor neurons in several spinal cord segments. As a result, the motor neurons generate nerve impulses, which propagate toward the axon terminals.
5. Acetylcholine released by the motor neurons causes the flexor muscles in the thigh (effectors) to contract, producing withdrawal of the leg. This reflex is protective because contraction of flexor muscles moves a limb away from the source of a possibly damaging stimulus.

The flexor reflex, like the stretch reflex, is ipsilateral—the incoming and outgoing impulses propagate into and out of the same side of the spinal cord. The flexor reflex also illustrates another feature of polysynaptic reflex arcs. Moving your entire lower or upper limb away from a painful stimulus involves contraction of more than one muscle group. Hence, several motor neurons must simultaneously convey impulses to several limb muscles. Because nerve impulses from one sensory neuron ascend and descend in the spinal cord and activate interneurons in several segments of the spinal cord, this type of reflex is called an **intersegmental reflex arc** (in′-ter-seg-MEN-tal; *inter-* = between). Through intersegmental reflex arcs, a single sensory neuron can activate several motor neurons, thereby stimulating more than one effector. The monosynaptic stretch reflex, in contrast, involves muscles receiving nerve impulses from one spinal cord segment only.

Something else may happen when you step on a tack: You may start to lose your balance as your body weight shifts to the other foot. Besides initiating the flexor reflex that causes you to withdraw the limb, the pain impulses from stepping on the tack also initiate a **crossed extensor reflex** to help you maintain your balance; it operates as follows (Figure 13.17):

1. Stepping on a tack stimulates the sensory receptor of a pain-sensitive neuron in the right foot.
2. This sensory neuron then generates nerve impulses, which propagate into the spinal cord.
3. Within the spinal cord (integrating center), the sensory neuron activates several interneurons that synapse with motor neurons on the left side of the spinal cord in several spinal cord segments. Thus, incoming pain signals cross to the opposite side through interneurons at that level, and at several levels above and below the point of entry into the spinal cord.
4. The interneurons excite motor neurons in several spinal cord segments that innervate extensor muscles. The motor neurons in turn generate more nerve impulses, which propagate toward the axon terminals.
5. Acetylcholine released by the motor neurons causes extensor muscles in the thigh (effectors) of the unstimulated left limb to contract, producing extension of the left leg. In this way, weight can be placed on the foot that must now support the entire body. A comparable reflex occurs with painful stimulation of the left lower limb or either upper limb.

Unlike the flexor reflex, which is an ipsilateral reflex, the crossed extensor reflex involves a **contralateral reflex arc** (kon′-tra-LAT-er-al = opposite side): Sensory impulses enter one side of the spinal cord and motor impulses exit on the opposite side. Thus, a crossed extensor reflex synchronizes the extension of the contralateral limb with the withdrawal (flexion) of the stimulated limb. Reciprocal innervation also occurs in both the flexor reflex and the crossed extensor reflex. In the flexor reflex, when the flexor muscles of a painfully stimulated lower limb are contracting, the extensor muscles of the same limb are relaxing to some degree. If both sets of muscles contracted at the same time, the two sets of muscles would pull on the bones in opposite directions, which might immobilize the limb. Because of reciprocal innervation, one set of muscles contracts while the other relaxes.

CLINICAL CONNECTION | *Reflexes and Diagnosis*

Reflexes are often used for diagnosing disorders of the nervous system and locating injured tissue. If a reflex ceases to function or functions abnormally, the physician may suspect that the damage lies somewhere along a particular conduction pathway. Many somatic reflexes can be tested simply by tapping or stroking the body. Among the somatic reflexes of clinical significance are the following:

- **Patellar reflex (knee jerk).** This stretch reflex involves extension of the leg at the knee joint by contraction of the quadriceps femoris muscle in response to tapping the patellar ligament (see Figure 13.14). This reflex is blocked by damage to the sensory or motor nerves supplying the muscle or to the integrating centers in the second, third, or fourth lumbar segments of the spinal cord. It is often absent in people with chronic diabetes mellitus or neurosyphilis, both of which cause degeneration of nerves. It is exaggerated in disease or injury involving certain motor tracts descending from the higher centers of the brain to the spinal cord.
- **Achilles reflex (a-KIL-ēz) (ankle jerk).** This stretch reflex involves plantar flexion of the foot by contraction of the gastrocnemius and soleus muscles in response to tapping the calcaneal (Achilles) tendon. Absence of the Achilles reflex indicates damage to the nerves supplying the posterior leg muscles or to neurons in the lumbosacral region of the spinal cord. This reflex may also disappear in people with chronic diabetes, neurosyphilis, alcoholism, and subarachnoid hemorrhages. An exaggerated Achilles reflex indicates cervical cord compression or a lesion of the motor tracts of the first or second sacral segments of the cord.
- **Babinski sign (ba-BIN-skē).** This reflex results from gentle stroking of the lateral outer margin of the sole. The great toe extends, with or without a lateral fanning of the other toes. This phenomenon normally occurs in children under $1\frac{1}{2}$ years of age and is due to incomplete myelination of fibers in the corticospinal tract. A positive Babinski sign after age $1\frac{1}{2}$ is abnormal and indicates an interruption of the corticospinal tract as the result of a lesion of the tract, usually in the upper portion. The normal response after age $1\frac{1}{2}$ is the **plantar flexion reflex,** or **negative Babinski**—a curling under of all the toes.
- **Abdominal reflex.** This reflex involves contraction of the muscles that compress the abdominal wall in response to stroking the side

Figure 13.17 Crossed extensor reflex. The flexor reflex arc is shown (at left) for comparison with the crossed extensor reflex arc. Plus signs (+) indicate excitatory synapses.

A crossed extensor reflex causes contraction of muscles that extend joints in the limb opposite a painful stimulus.

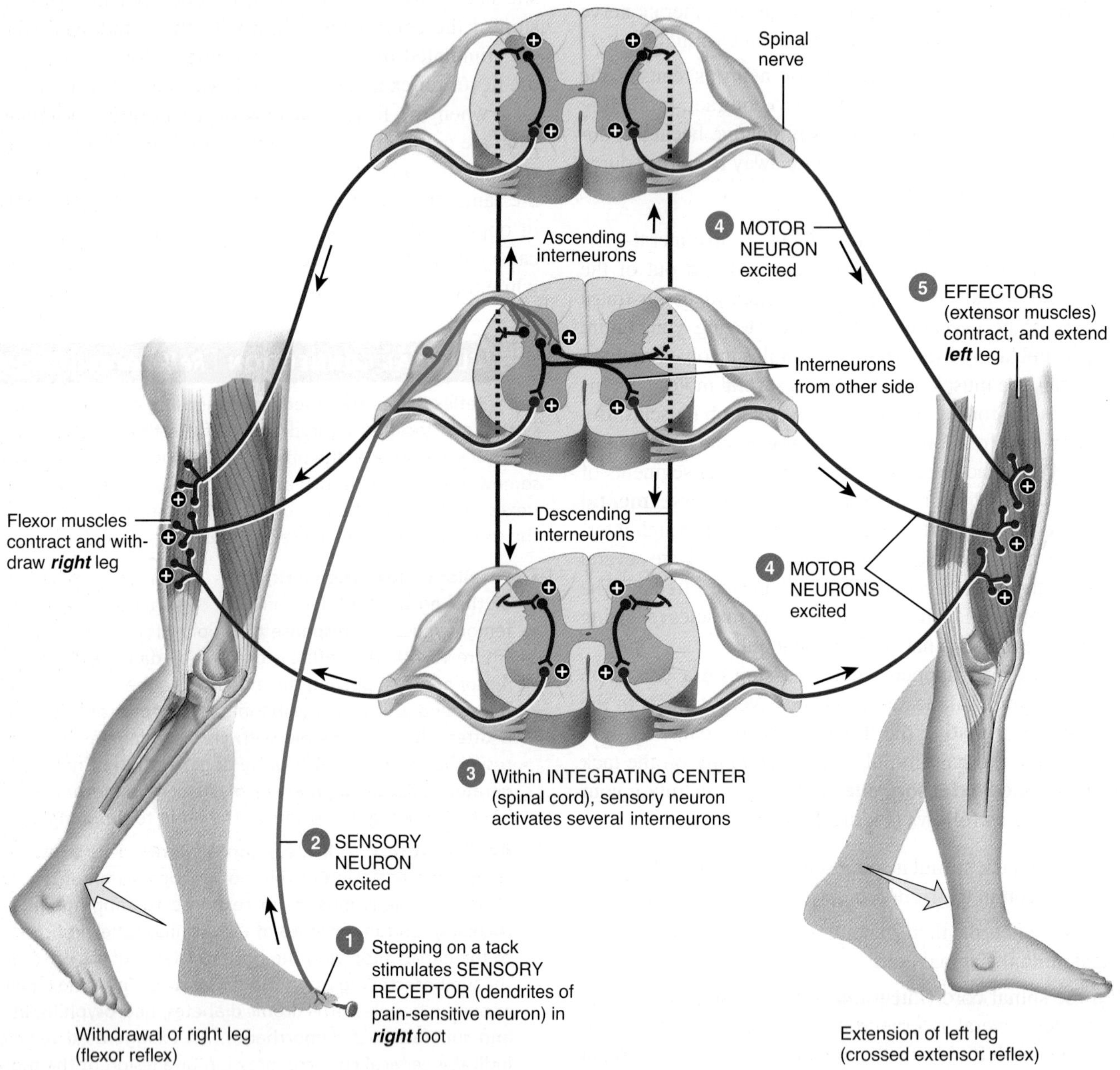
Spinal nerve
Ascending interneurons
4 MOTOR NEURON excited
5 EFFECTORS (extensor muscles) contract, and extend left leg
Interneurons from other side
Flexor muscles contract and withdraw right leg
Descending interneurons
4 MOTOR NEURONS excited
3 Within INTEGRATING CENTER (spinal cord), sensory neuron activates several interneurons
2 SENSORY NEURON excited
1 Stepping on a tack stimulates SENSORY RECEPTOR (dendrites of pain-sensitive neuron) in right foot
Withdrawal of right leg (flexor reflex)
Extension of left leg (crossed extensor reflex)

Why is the crossed extensor reflex classified as a contralateral reflex arc?

of the abdomen. The response is an abdominal muscle contraction that causes the umbilicus to move in the direction of the stimulus. Absence of this reflex is associated with lesions of the corticospinal tracts. It may also be absent because of lesions of the peripheral nerves, lesions of integrating centers in the thoracic part of the cord, or multiple sclerosis.

Most autonomic reflexes are not practical diagnostic tools because it is difficult to stimulate visceral effectors, which are deep inside the body. An exception is the pupillary light reflex, in which the pupils of both eyes decrease in diameter when either eye is exposed to light. Because the reflex arc includes synapses in lower parts of the brain, the **absence of a normal pupillary light reflex** may indicate brain damage or injury. •

✔CHECKPOINT

8. Which spinal cord tracts are ascending tracts? Which are descending tracts?
9. How are somatic and autonomic reflexes similar and different?
10. Describe the mechanism and function of a stretch reflex, tendon reflex, flexor (withdrawal) reflex, and crossed extensor reflex.
11. What does each of the following terms mean in relation to reflex arcs? Monosynaptic, ipsilateral, polysynaptic, intersegmental, contralateral, and reciprocal innervation.

DISORDERS: HOMEOSTATIC IMBALANCES

The spinal cord can be damaged in several ways. Outcomes range from little or no long-term neurological deficits to severe deficits and even death.

Traumatic Injuries

Most **spinal cord injuries** are due to trauma as a result of factors such as automobile accidents, falls, contact sports, diving, and acts of violence. The effects of the injury depend on the extent of direct trauma to the spinal cord or compression of the cord by fractured or displaced vertebrae or blood clots. Although any segment of the spinal cord may be involved, the most common sites of injury are in the cervical, lower thoracic, and upper lumbar regions. Depending on the location and extent of spinal cord damage, paralysis may occur. **Monoplegia** (mon′-ō-PLĒ-jē-a; *mono-* = one; *-plegia* = blow or strike) is paralysis of one limb only. **Diplegia** (*di-* = two) is paralysis of both upper limbs or both lower limbs. **Paraplegia** (*para-* = beyond) is paralysis of both lower limbs. **Hemiplegia** (*hemi-* = half) is paralysis of the upper limb, trunk, and lower limb on one side of the body, and **quadriplegia** (*quad-* = four) is paralysis of all four limbs.

Complete transection (tran-SEK-shun; *trans-* = across; *-section* = a cut) of the spinal cord means that the cord is severed from one side to the other, thus cutting all sensory and motor tracts. It results in a loss of all sensations and voluntary movement below the level of the transection. A person will have permanent loss of all sensations in dermatomes below the injury because ascending nerve impulses cannot propagate past the transection to reach the brain. A **dermatome** (DER-ma-tōm; *derma* = skin, *-tome* = thin segment) is an area of skin that provides sensory input to the CNS via a pair of spinal nerves. At the same time, all voluntary muscle contractions will be lost below the transection because nerve impulses descending from the brain also cannot pass. The extent of paralysis of skeletal muscles depends on the level of injury. The closer the injury is to the head, the greater the area of the body that may be affected. The following list outlines which muscle functions may be *retained* at progressively lower levels of spinal cord transection. (These are spinal cord levels and not vertebral column levels. Recall that spinal cord levels differ from vertebral column levels because of the differential growth of the cord versus the column, especially as you progress inferiorly.)

- C1–C3: no function maintained from the neck down; ventilator needed for breathing; electric wheelchair with breath, head, or shoulder-controlled device required (see Figure A)
- C4–C5: diaphragm, which allows breathing
- C6–C7: some arm and chest muscles, which allows feeding, some dressing, and manual wheelchair required (see Figure B)
- T1–T3: intact arm function
- T4–T9: control of trunk above the umbilicus
- T10–L1: most thigh muscles, which allows walking with long leg braces (see Figure C)
- L1–L2: most leg muscles, which allows walking with short leg braces (see Figure D)

Hemisection is a partial transection of the cord on either the right or the left side. After hemisection, three main symptoms, known together as *Brown-Séquard syndrome* (sē-KAR), occur below the level of the injury: (1) Damage of the posterior column (sensory tracts)

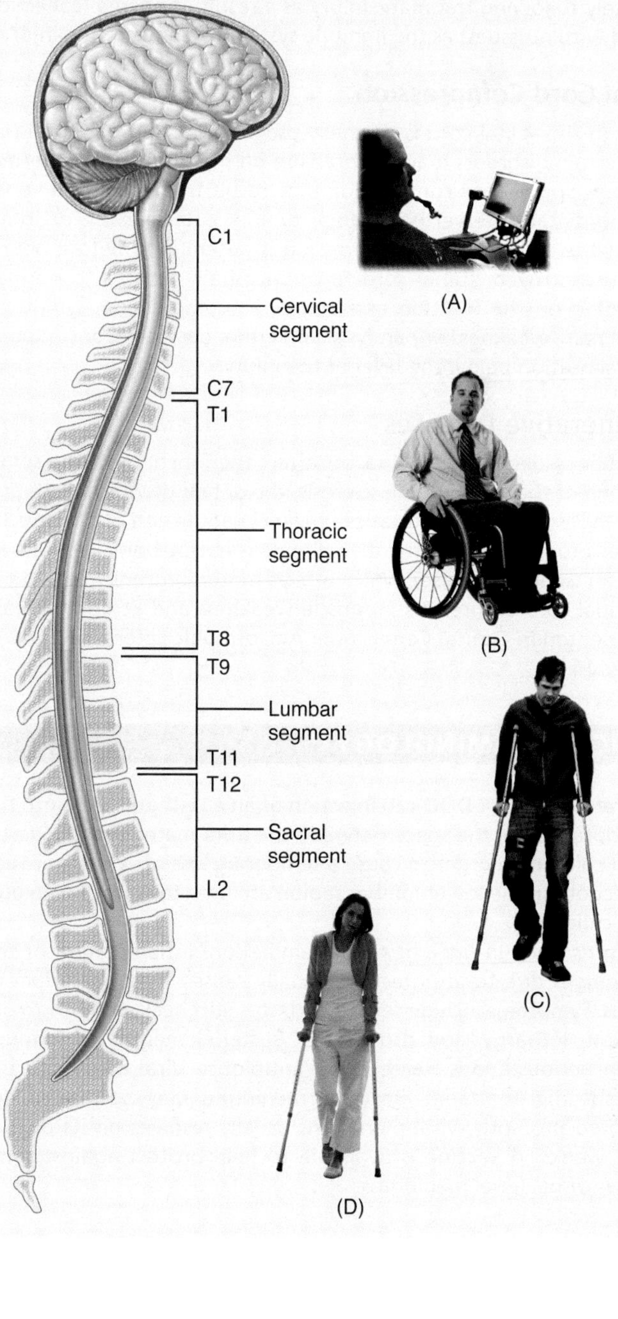
C1
Cervical segment
C7
T1
Thoracic segment
T8
T9
Lumbar segment
T11
T12
Sacral segment
L2
(A)
(B)
(C)
(D)

causes loss of proprioception and fine touch sensations on the *ipsilateral* (same) side as the injury. (2) Damage of the lateral corticospinal tract (motor tract) causes ipsilateral paralysis. (3) Damage of the spinothalamic tracts (sensory tracts) causes loss of pain and temperature sensations on the *contralateral* (opposite) side.

Following complete transection, and to varying degrees after hemisection, spinal shock occurs. **Spinal shock** is an immediate response to spinal cord injury characterized by temporary **areflexia** (a′-re-FLEK-sē-a), loss of reflex function. The areflexia occurs in parts of the body served by spinal nerves below the level of the injury. Signs of acute spinal shock include slow heart rate, low blood pressure, flaccid paralysis of skeletal muscles, loss of somatic sensations, and urinary bladder dysfunction. Spinal shock may begin within 1 hour after injury and may last from several minutes to several months, after which reflex activity gradually returns.

In many cases of traumatic injury of the spinal cord, the patient may have an improved outcome if an anti-inflammatory corticosteroid drug called methylprednisolone is given within 8 hours of the injury. This is because the degree of neurologic deficit is greatest immediately following traumatic injury as a result of *edema* (collection of fluid within tissues) as the immune system responds to the injury.

Spinal Cord Compression

Although the spinal cord is normally protected by the vertebral column, certain disorders may put pressure on it and disrupt its normal functions. Spinal cord compression may result from fractured vertebrae, herniated intervertebral discs, tumors, osteoporosis, or infections. If the source of the compression is determined before neural tissue is destroyed, spinal cord function usually returns to normal. Depending on the location and degree of compression, symptoms include pain, weakness or paralysis, and either decreased or complete loss of sensation below the level of the injury.

Degenerative Diseases

A number of degenerative diseases affect the functions of the spinal cord. One of these is multiple sclerosis, the details of which were presented in Disorders: Homeostatic Imbalances at the end of Chapter 12. Another progressive degenerative disease is amyotrophic lateral sclerosis (Lou Gehrig's disease), which affects motor neurons of the brain and spinal cord and results in muscle weakness and atrophy. Details are presented in Clinical Connection: Amyotrophic Lateral Sclerosis in Chapter 16.

Shingles

Shingles is an acute infection of the peripheral nervous system caused by herpes zoster (HER-pēz ZOS-ter), the virus that also causes chickenpox. After a person recovers from chickenpox, the virus retreats to a posterior root ganglion. If the virus is reactivated, the immune system usually prevents it from spreading. From time to time, however, the reactivated virus overcomes a weakened immune system, leaves the ganglion, and travels down sensory neurons of the skin by fast axonal transport (described in Section 12.2). The result is pain, discoloration of the skin, and a characteristic line of skin blisters. The line of blisters marks the distribution (dermatome) of the particular cutaneous sensory nerve belonging to the infected posterior root ganglion.

Poliomyelitis

Poliomyelitis (pō′-lē-ō-mī-e-LĪ-tus), or simply **polio,** is caused by a virus called poliovirus. The onset of the disease is marked by fever, severe headache, a stiff neck and back, deep muscle pain and weakness, and loss of certain somatic reflexes. In its most serious form, the virus produces paralysis by destroying cell bodies of motor neurons, specifically those in the anterior horns of the spinal cord and in the nuclei of the cranial nerves. Polio can cause death from respiratory or heart failure if the virus invades neurons in vital centers that control breathing and heart functions in the brain stem. Even though polio vaccines have virtually eradicated polio in the United States, outbreaks of polio continue throughout the world. Due to international travel, polio could be easily reintroduced into North America if individuals are not vaccinated appropriately.

Several decades after suffering a severe attack of polio and following their recovery from it, some individuals develop a condition called **post-polio syndrome.** This neurological disorder is characterized by progressive muscle weakness, extreme fatigue, loss of function, and pain, especially in muscles and joints. Post-polio syndrome seems to involve a slow degeneration of motor neurons that innervate muscle fibers. Triggering factors appear to be a fall, a minor accident, surgery, or prolonged bed rest. Possible causes include overuse of surviving motor neurons over time, smaller motor neurons because of the initial infection by the virus, reactivation of dormant polio viral particles, immune-mediated responses, hormone deficiencies, and environmental toxins. Treatment consists of muscle-strengthening exercises, administration of pyridostigmine to enhance the action of acetylcholine in stimulating muscle contraction, and administration of nerve growth factors to stimulate both nerve and muscle growth.

MEDICAL TERMINOLOGY

Epidural block (ep′-i-DOO-ral) Injection of an anesthetic drug into the epidural space, the space between the dura mater and the vertebral column, in order to cause a temporary loss of sensation. Such injections in the lower lumbar region are used to control pain during childbirth.

Meningitis (men-in-JĪ-tis; *-itis* = inflammation) Inflammation of the meninges due to an infection, usually caused by a bacterium or virus. Symptoms include fever, headache, stiff neck, vomiting, confusion, lethargy, and drowsiness. Bacterial meningitis is much more serious and is treated with antibiotics. Viral meningitis has no specific treatment. Bacterial meningitis may be fatal if not treated promptly; viral meningitis usually resolves on its own in 1–2 weeks. A vaccine is available to help protect against some types of bacterial meningitis.

Myelitis (mī-e-LĪ-tis; *myel-* = spinal cord) Inflammation of the spinal cord.

Nerve block Loss of sensation in a region due to injection of a local anesthetic; an example is local dental anesthesia.

Neuralgia (noo-RAL-jē-a; *neur-* = nerve; *-algia* = pain) Attacks of pain along the entire course or a branch of a sensory nerve.

Neuritis (*neur-* = nerve; *-itis* = inflammation) Inflammation of one or several nerves that may result from irritation to the nerve produced by direct blows, bone fractures, contusions, or penetrating injuries. Additional causes include infections, vitamin deficiency (usually thiamine), and poisons such as carbon monoxide, carbon tetrachloride, heavy metals, and some drugs.

Paresthesia (par-es-THĒ-zē-a; *par-* = departure from normal; *-esthesia* = sensation) An abnormal sensation such as burning, pricking, tickling, or tingling resulting from a disorder of a sensory nerve.

CHAPTER REVIEW AND RESOURCE SUMMARY

Review | **Resource**

13.1 Spinal Cord Anatomy

1. The spinal cord is protected by the vertebral column, the meninges, cerebrospinal fluid, and denticulate ligaments.
2. The three meninges are coverings that run continuously around the spinal cord and brain. They are the dura mater, arachnoid mater, and pia mater.
3. The spinal cord begins as a continuation of the medulla oblongata and ends at about the second lumbar vertebra in an adult.
4. The spinal cord contains cervical and lumbar enlargements that serve as points of origin for nerves to the limbs.
5. The tapered inferior portion of the spinal cord is the conus medullaris, from which arise the filum terminale and cauda equina.
6. Spinal nerves connect to each segment of the spinal cord by two roots. The posterior or dorsal root contains sensory axons, and the anterior or ventral root contains motor neuron axons.
7. The anterior median fissure and the posterior median sulcus partially divide the spinal cord into right and left sides.
8. The gray matter in the spinal cord is divided into horns, and the white matter into columns. In the center of the spinal cord is the central canal, which runs the length of the spinal cord.
9. Parts of the spinal cord observed in transverse section are the gray commissure; central canal; anterior, posterior, and lateral gray horns; and anterior, posterior, and lateral white columns, which contain ascending and descending tracts. Each part has specific functions.
10. The spinal cord conveys sensory and motor information by way of ascending and descending tracts, respectively.

Resource:
Anatomy Overview - The Spinal Cord
Figure 13.1 - Gross Anatomy of the Spinal Cord
Figure 13.2 - External Anatomy of the Spinal Cord and Spinal Nerves
Figure 13.4 - Processing of Sensory and Motor Input

13.2 Spinal Nerves

1. The 31 pairs of spinal nerves are named and numbered according to the region and level of the spinal cord from which they emerge. There are 8 pairs of cervical, 12 pairs of thoracic, 5 pairs of lumbar, 5 pairs of sacral, and 1 pair of coccygeal nerves.
2. Spinal nerves typically are connected with the spinal cord by a posterior root and an anterior root. All spinal nerves contain both sensory and motor axons (are mixed nerves).
3. Three connective tissue coverings associated with spinal nerves are the endoneurium, perineurium, and epineurium.
4. Branches of a spinal nerve include the posterior ramus, anterior ramus, meningeal branch, and rami communicantes.
5. The anterior rami of spinal nerves, except for T2–T12, form networks of nerves called plexuses.
6. Emerging from the plexuses are nerves bearing names that typically describe the general regions they supply or the route they follow.
7. Nerves of the cervical plexus supply the skin and muscles of the head, neck, and upper part of the shoulders; they connect with some cranial nerves and innervate the diaphragm. Nerves of the brachial plexus supply the upper limbs and several neck and shoulder muscles. Nerves of the lumbar plexus supply the anterolateral abdominal wall, external genitals, and part of the lower limbs. Nerves of the sacral plexus supply the buttocks, perineum, and part of the lower limbs. Nerves of the coccygeal plexus supply the skin of the coccygeal region.
8. Anterior rami of nerves T2–T12 do not form plexuses and are called intercostal (thoracic) nerves. They are distributed directly to the structures they supply in intercostal spaces.
9. Sensory neurons within spinal nerves and the trigeminal (V) nerve serve specific, constant segments of the skin called dermatomes.
10. Knowledge of dermatomes helps health care providers determine which segment of the spinal cord or which spinal nerve is damaged.

Resource:
Anatomy Overview - Spinal Nerves

13.3 Spinal Cord Physiology

1. The white matter tracts in the spinal cord are highways for nerve impulse propagation. Along these tracts, sensory input travels toward the brain, and motor output travels from the brain toward skeletal muscles and other effector tissues. Sensory input travels along two main routes in the white matter of the spinal cord: the posterior column and the spinothalamic tract. Motor output travels along two main routes in the white matter of the spinal cord: direct pathways and indirect pathways.

Resource:
Animation - Somatic Sensory and Motor Pathways
Animation - Reflex Arcs
Animation - Reflexes
Figure 13.12 - Locations of Major Sensory and Motor Tracts

Review

2. A second major function of the spinal cord is to serve as an integrating center for spinal reflexes. This integration occurs in the gray matter.
3. A reflex is a fast, predictable sequence of involuntary actions, such as muscle contractions or glandular secretions, which occurs in response to certain changes in the environment. Reflexes may be spinal or cranial and somatic or autonomic (visceral).
4. The components of a reflex arc are sensory receptor, sensory neuron, integrating center, motor neuron, and effector.
5. Somatic spinal reflexes include the stretch reflex, the tendon reflex, the flexor (withdrawal) reflex, and the crossed extensor reflex; all exhibit reciprocal innervation.
6. A two-neuron or monosynaptic reflex arc consists of one sensory neuron and one motor neuron. A stretch reflex, such as the patellar reflex, is an example.
7. The stretch reflex is ipsilateral and is important in maintaining muscle tone.
8. A polysynaptic reflex arc contains sensory neurons, interneurons, and motor neurons. The tendon reflex, flexor (withdrawal) reflex, and crossed extensor reflexes are examples.
9. The tendon reflex is ipsilateral and prevents damage to muscles and tendons when muscle force becomes too extreme. The flexor reflex is ipsilateral and moves a limb away from the source of a painful stimulus. The crossed extensor reflex extends the limb contralateral to a painfully stimulated limb, allowing the weight of the body to shift when a supporting limb is withdrawn.
10. Several important somatic reflexes are used to diagnose various disorders. These include the patellar reflex, Achilles reflex, Babinski sign, and abdominal reflex.

Resource

Figure 13.17 - Crossed Extensor Reflex
Exercise - Assemble an Arc
Exercise - Stretch Reflex

SELF-QUIZ QUESTIONS

Fill in the blanks in the following statements.

1. Because they contain both sensory and motor axons, spinal nerves are considered to be _____ nerves.
2. The five components of a reflex arc, in order from the beginning to the end, are (1) _____, (2) _____, (3) _____, (4) _____, and (5) _____.

Indicate whether the following statements are true or false.

3. Gray matter of the spinal cord contains somatic motor and sensory nuclei, autonomic motor and sensory nuclei, and functions to receive and integrate both incoming and outgoing information.
4. The epidural space is located between the wall of the vertebral canal and the pia mater.

Choose the one best answer to the following questions.

5. Which of the following is *not* true? (1) Dermatomes are areas of the body that are stimulated by motor neurons exiting specific spinal nerves. (2) The stretch reflex helps to maintain muscle tone. (3) The Achilles reflex is an example of a stretch reflex. (4) The abdominal reflex is used to diagnose problems with autonomic reflexes. (5) Spinal nerves T2–T12 do not enter into the formation of a plexus.
   (a) 1, 2, and 4 (b) 2 and 5 (c) 1 and 4
   (d) 1, 3, and 5 (e) 1, 3, and 4
6. While identifying and labeling cadaver muscles, your lab partner accidentally pokes your finger with a pin. Place the following steps in the correct order from beginning to end of your body's response. (1) Impulses travel through anterior (ventral) root of spinal nerve(s). (2) Sensory neuron relays impulse to spinal cord. (3) Motor impulses reach muscles, causing withdrawal of the affected limb. (4) Integrating centers interpret sensory impulses, and then generate motor impulses. (5) Sensory receptor activated by stimulus. (6) Impulse travels through posterior (dorsal) root of spinal nerve.
   (a) 5, 3, 6, 4, 1, 2 (b) 5, 2, 1, 4, 6, 3 (c) 5, 2, 6, 4, 1, 3
   (d) 3, 5, 1, 2, 4, 6 (e) 2, 1, 5, 4, 6, 3
7. The connective tissue surrounding each individual axon is
   (a) endoneurium. (b) epineurium. (c) perineurium.
   (d) fascicle. (e) arachnoid mater.
8. The tracts of the posterior column are involved in (1) conscious proprioception. (2) touch, (3) pain, (4) thermal sensations, (5) pressure, (6) vibration.
   (a) 1, 2, 4, and 5 (b) 2, 4, and 6 (c) 1, 2, 5, and 6
   (d) 3, 4, 5, and 6 (e) 1, 3, 5, and 6
9. Which of the following is a motor tract?
   (a) posterior spinocerebellar (b) lateral spinothalamic
   (c) anterior spinocerebellar (d) lateral corticospinal
   (e) posterior column
10. Cutting the posterior root of a spinal nerve would
    (a) interfere with the circulation of cerebrospinal fluid.
    (b) impair motor control of skeletal muscles.
    (c) interfere with the ability of the brain to transmit motor impulses.
    (d) impair motor control of organs.
    (e) interfere with the flow of sensory impulses.
11. Which of the following statements is *false*?
    (a) The two main spinal cord sensory paths are the spinothalamic and anterior columns.
    (b) The spinothalamic tracts convey impulses for sensing pain, temperature, itching, and tickling.
    (c) Direct pathways convey nerve impulses destined to cause precise, voluntary movements of skeletal muscles.
    (d) Indirect pathways convey nerve impulses that program automatic movements, help coordinate body movements with visual stimuli, maintain skeletal muscle tone and posture, and contribute to equilibrium.
    (e) The direct pathways are motor pathways.

12. Which of the following are *true*?
(1) The anterior (ventral) gray horns contain cell bodies of neurons that cause skeletal muscle contraction. (2) The gray commissure connects the white matter of the right and left sides of the spinal cord. (3) Cell bodies of autonomic motor neurons are located in the lateral gray horns. (4) Sensory (ascending) tracts conduct motor impulses down the spinal cord. (5) Gray matter in the spinal cord consists of cell bodies of neurons, neuroglia, unmyelinated axons, and dendrites of interneurons and motor neurons.
(a) 1, 2, 3, and 5 (b) 2 and 4 (c) 2, 3, 4, and 5
(d) 1, 3, and 5 (e) 1, 2, 3, and 4

13. Match the following (some answers may be used more than once):

____ (a) a reflex resulting in the contraction of a skeletal muscle when it is stretched
____ (b) receptors that monitor changes in muscle length
____ (c) a balance-maintaining reflex
____ (d) operates as a feedback mechanism to control muscle tension by causing muscle relaxation when muscle force becomes too extreme
____ (e) reflex arc that consists of one sensory and one motor neuron
____ (f) acts as a feedback mechanism to control muscle length by causing muscle contraction
____ (g) sensory impulses enter on one side of the spinal cord and motor impulses exit on the opposite side
____ (h) occurs when sensory nerve impulse travels up and down the spinal cord, thereby activating several motor neurons and more than one effector
____ (i) polysynaptic reflex initiated in response to a painful stimulus
____ (j) receptors that monitor changes in muscle tension
____ (k) maintains proper muscle tone
____ (l) reflex pathway that contains sensory neurons, interneurons, and motor neurons
____ (m) motor nerve impulses exit the spinal cord on the same side that sensory impulses entered the spinal cord
____ (n) protects the tendon and muscle from damage due to excessive tension
____ (o) a neural circuit that coordinates body movements by causing contraction of one muscle and relaxation of antagonistic muscles or relaxation of a muscle and contraction of the antagonists

(1) stretch reflex
(2) tendon reflex
(3) flexor (withdrawal) reflex
(4) crossed extensor reflex
(5) intersegmental reflex arc
(6) contralateral reflex arc
(7) ipsilateral reflex arc
(8) muscle spindles
(9) tendon (Golgi tendon) organs
(10) reciprocal innervation
(11) monosynaptic reflex
(12) polysynaptic reflex

14. Match the following:

____ (a) the joining together of the anterior rami of adjacent nerves
____ (b) spinal nerve branches that serve the deep muscles and skin of the posterior surface of the trunk
____ (c) spinal nerve branches that serve the muscles and structures of the upper and lower limbs and the lateral and ventral trunk
____ (d) area of the spinal cord from which nerves to and from the upper limbs arise
____ (e) area of the spinal cord from which nerves to and from the lower limbs arise
____ (f) the roots form the nerves that arise from the inferior part of the spinal cord but do not leave the vertebral column at the same level as they exit the cord
____ (g) contains motor neuron axons and conducts impulses from the spinal cord to the peripheral organs and cells
____ (h) avascular covering of spinal cord composed of delicate collagen fibers and some elastic fibers
____ (i) contains sensory neuron axons and conducts impulses from the peripheral receptors into the spinal cord
____ (j) superficial spinal cord covering of dense, irregular connective tissue
____ (k) an extension of the pia mater that anchors the spinal cord to the coccyx
____ (l) extending the length of the spinal cord, these pia mater thickenings fuse with the arachnoid mater and dura mater and help to protect the spinal cord from shock and sudden displacement
____ (m) thin transparent connective tissue composed of interlacing bundles of collagen fibers and some elastic fibers adhering to the spinal cord's surface
____ (n) space within the spinal cord filled with cerebrospinal fluid
____ (o) spinal nerve branch that supplies vertebrae, vertebral ligaments, blood vessels of the spinal cord, and meninges

(1) cervical enlargement
(2) lumbar enlargement
(3) central canal
(4) denticulate ligaments
(5) cauda equina
(6) meningeal branch
(7) pia mater
(8) arachnoid mater
(9) dura mater
(10) posterior (dorsal) root
(11) anterior (ventral) root
(12) posterior (dorsal) ramus
(13) anterior (ventral) ramus
(14) plexus
(15) filum terminale

15. Match the following:

____ (a) provides the entire nerve supply of the shoulders and upper limbs
____ (b) provides the nerve supply of the skin and muscles of the head, neck, and superior part of the shoulders and chest
____ (c) provides the nerve supply of the anterolateral abdominal wall, external genitals, and part of the lower limbs
____ (d) supplies the buttocks, perineum, and lower limbs
____ (e) formed by the anterior rami of C1–C4 with some contribution by C5
____ (f) formed by anterior rami of S4–S5 and coccygeal nerves
____ (g) formed by the anterior rami of L1–L4
____ (h) formed by the anterior rami of C5–C8 and T1
____ (i) formed by the anterior rami of L4–L5 and S1–S4
____ (j) phrenic nerve arises from this plexus
____ (k) median nerve arises from this plexus
____ (l) sciatic nerve arises from this plexus
____ (m) femoral nerve arises from this plexus
____ (n) supplies a small area of skin in coccygeal region
____ (o) injury to this plexus can affect breathing

(1) cervical plexus
(2) brachial plexus
(3) lumbar plexus
(4) sacral plexus
(5) coccygeal plexus

CRITICAL THINKING QUESTIONS

1. Evalina's severe headaches and other symptoms were suggestive of meningitis, so her physician ordered a spinal tap. List the structures that the needle will pierce from the most superficial to the deepest. Why would the physician order a test in the spinal region to check a problem in Evalina's head?

2. Sunil has developed an infection that is destroying cells in the anterior gray horns in the lower cervical region of the spinal cord. What kinds of symptoms would you expect to occur?

3. Allyson is in a car accident and suffers spinal cord compression in the lower spinal cord. Although she is in pain, she cannot distinguish when the doctor is touching her calf or her toes and she is having trouble telling how her lower limbs are positioned. What part of the spinal cord has been affected by the accident?

? ANSWERS TO FIGURE QUESTIONS

13.1 The superior boundary of the spinal dura mater is the foramen magnum of the occipital bone. The inferior boundary is the second sacral vertebra.

13.2 The cervical enlargement connects with sensory and motor nerves of the upper limbs.

13.3 A horn is an area of gray matter, and a column is a region of white matter in the spinal cord.

13.4 Lateral gray horns are found in the thoracic and upper lumbar segments of the spinal cord.

13.5 All spinal nerves are classified as mixed because their posterior roots contain sensory axons and their anterior roots contain motor axons.

13.6 The anterior rami serve the upper and lower limbs.

13.7 Severing the spinal cord at level C2 causes respiratory arrest because it prevents descending nerve impulses from reaching the phrenic nerve, which stimulates contraction of the diaphragm, the main muscle needed for breathing.

13.8 The axillary, musculocutaneous, radial, median, and ulnar nerves are five important nerves that arise from the brachial plexus.

13.9 Signs of femoral nerve injury include inability to extend the leg and loss of sensation in the skin over the anterolateral aspect of the thigh.

13.10 The origin of the sacral plexus is the anterior rami of spinal nerves L4–L5 and S1–S4.

13.11 The only spinal nerve without a corresponding dermatome is C1.

13.12 The spinothalamic tract originates in the spinal cord and ends in the thalamus (a region of the brain). Because "spinal" comes first in the name, you know it contains ascending axons and thus is a sensory tract.

13.13 A sensory receptor produces a generator potential, which triggers a nerve impulse if the generator potential reaches threshold. Reflex integrating centers are in the CNS.

13.14 In an ipsilateral reflex, the sensory and motor neurons are on the same side of the spinal cord.

13.15 Reciprocal innervation is a type of arrangement of a neural circuit involving simultaneous contraction of one muscle and relaxation of its antagonist.

13.16 The flexor reflex is intersegmental because impulses go out over motor neurons located in several spinal nerves, each arising from a different segment of the spinal cord.

13.17 The crossed extensor reflex is a contralateral reflex arc because the motor impulses leave the spinal cord on the side opposite the entry of sensory impulses.

14 THE BRAIN AND CRANIAL NERVES

THE BRAIN, CRANIAL NERVES, AND HOMEOSTASIS *Your brain contributes to homeostasis by receiving sensory input, integrating new and stored information, making decisions, and executing responses through motor activities.*

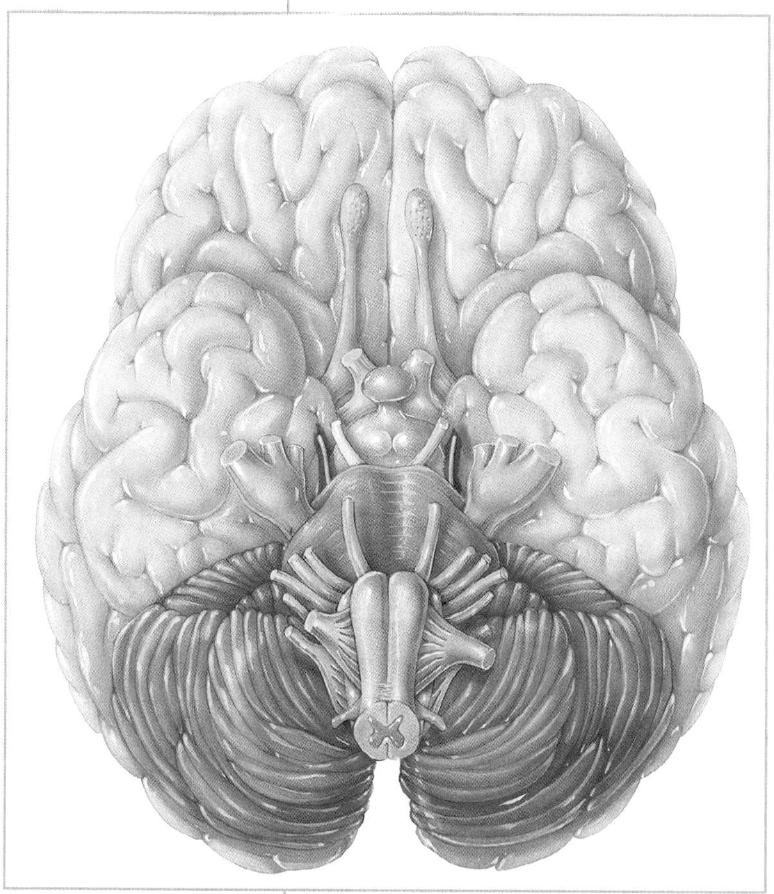

Solving an equation, feeling hungry, laughing—the neural processes needed for each of these activities occur in different regions of the **brain,** that portion of the central nervous system contained within the cranium. About 100 billion neurons and 10 trillion to 50 trillion neuroglia make up the brain, which has a mass of about 1300 g (almost 3 lb) in adults. On average, each neuron forms 1000 synapses with other neurons. Thus, the total number of synapses, about a thousand trillion or 10^{15}, is larger than the number of stars in our galaxy.

The brain is the control center for registering sensations, correlating them with one another and with stored information, making decisions, and taking actions. It also is the center for the intellect, emotions, behavior, and memory. But the brain encompasses yet a larger domain: It directs our behavior toward others. With ideas that excite, artistry that dazzles, or rhetoric that mesmerizes, one person's thoughts and actions may influence and shape the lives of many others. As you will see shortly, different regions of the brain are specialized for different functions. Different parts of the brain also work together to accomplish certain shared functions. This chapter explores how the brain is protected and nourished, what functions occur in the major regions of the brain, and how the spinal cord and the 12 pairs of cranial nerves connect with the brain to form the control center of the human body.

? ***Did you ever wonder how cerebrovascular accidents (strokes) occur and how they are treated?***

14.1 BRAIN ORGANIZATION, PROTECTION, AND BLOOD SUPPLY

OBJECTIVES

- Identify the major parts of the brain.
- Describe how the brain is protected.
- Describe the blood supply of the brain.

In order to understand the terminology used for the principal parts of the adult brain, it will be helpful to know how the brain develops. The brain and spinal cord develop from the ectodermal **neural tube** (see Figure 14.27). The anterior part of the neural tube expands, along with the associated neural crest tissue. Constrictions in this expanded tube soon appear, creating three regions called **primary brain vesicles:** *prosencephalon, mesencephalon,* and *rhombencephalon* (see Figure 14.28). Both the prosencephalon and rhombencephalon subdivide further, forming **secondary brain vesicles.** The *prosencephalon* (PROS-en-sef′-a-lon), or forebrain, gives rise to the telencephalon and diencephalon, and the *rhombencephalon* (ROM-ben-sef′-a-lon), or hindbrain, develops into the metencephalon and myelencephalon. The various brain vesicles give rise to the following adult structures:

- The **telencephalon** develops into the *cerebrum* and *lateral ventricles.*
- The **diencephalon** forms the *thalamus, hypothalamus, epithalamus,* and *third ventricle.*
- The **metencephalon** becomes the *pons, cerebellum,* and *upper part of the fourth ventricle.*
- The **myelencephalon** forms the *medulla oblongata* and *lower part of the fourth ventricle.*
- The **mesencephalon** (mes′-en-SEF-a-lon), or midbrain, gives rise to the *midbrain* and *aqueduct of the midbrain (cerebral aqueduct).*

The walls of these brain regions develop into the nervous tissue of the brain, while the hollow interior of the tube is transformed into the various ventricles (fluid-filled spaces) of the brain. The expanded neural crest tissue becomes prominent in head development. Most of the protective structures of the brain—that is, most of the bones of the skull, associated connective tissues, and meningeal membranes—arise from this expanded neural crest tissue.

These relationships are summarized in Table 14.1.

Major Parts of the Brain

The adult brain consists of four major parts: brain stem, cerebellum, diencephalon, and cerebrum (Figure 14.1). The **brain stem** is continuous with the spinal cord and consists of the medulla oblongata, pons, and midbrain. Posterior to the brain stem is the **cerebellum** (ser′-e-BEL-um = little brain). Superior to the brain stem is the **diencephalon** (dī′-en-SEF-a-lon; *di-* = through; *-encephalon* = brain), which consists of the thalamus, hypothalamus, and epithalamus. Supported on the diencephalon and brain stem is the **cerebrum** (se-RĒ-brum = brain), the largest part of the brain.

Protective Coverings of the Brain

The cranium (see Figure 7.4) and the cranial meninges surround and protect the brain. The **cranial meninges** (me-NIN-jēz) are continuous with the spinal meninges, have the same basic structure, and bear the same names: the outer **dura mater** (DOO-ra MĀ-ter), the middle **arachnoid mater** (a-RAK-noyd), and the inner **pia mater** (PĪ-a or PĒ-a) (Figure 14.2). However, the cranial

TABLE 14.1

Development of the Brain

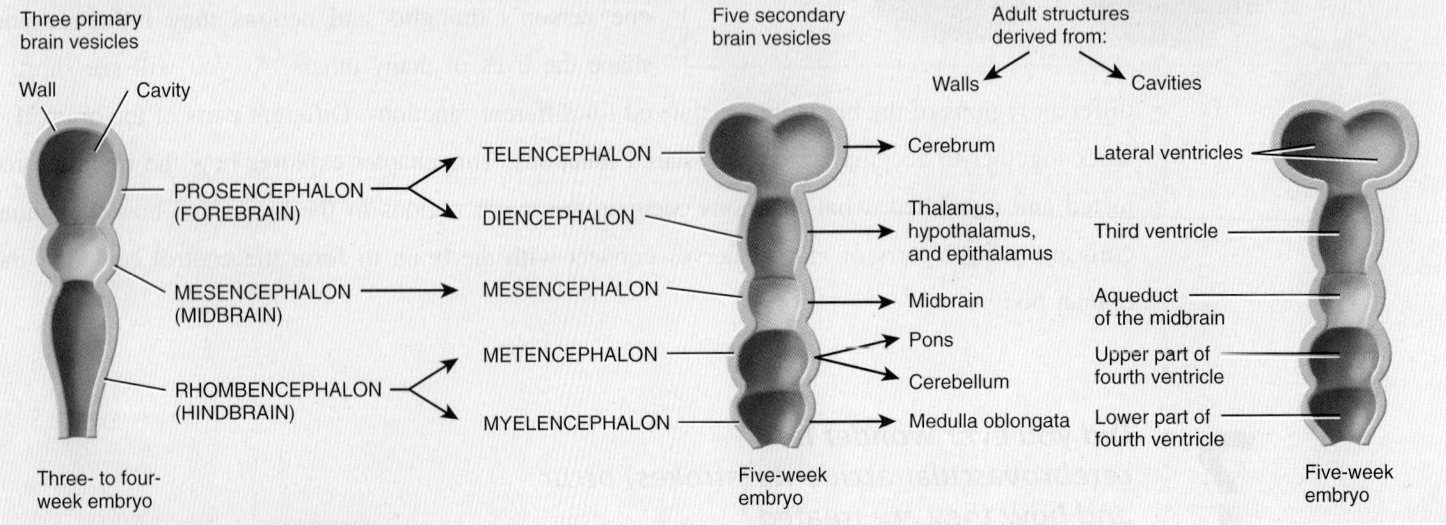

Figure 14.1 The brain. The pituitary gland is discussed with the endocrine system in Chapter 18.

6 The four principal parts of the brain are the brain stem, cerebellum, diencephalon, and cerebrum.

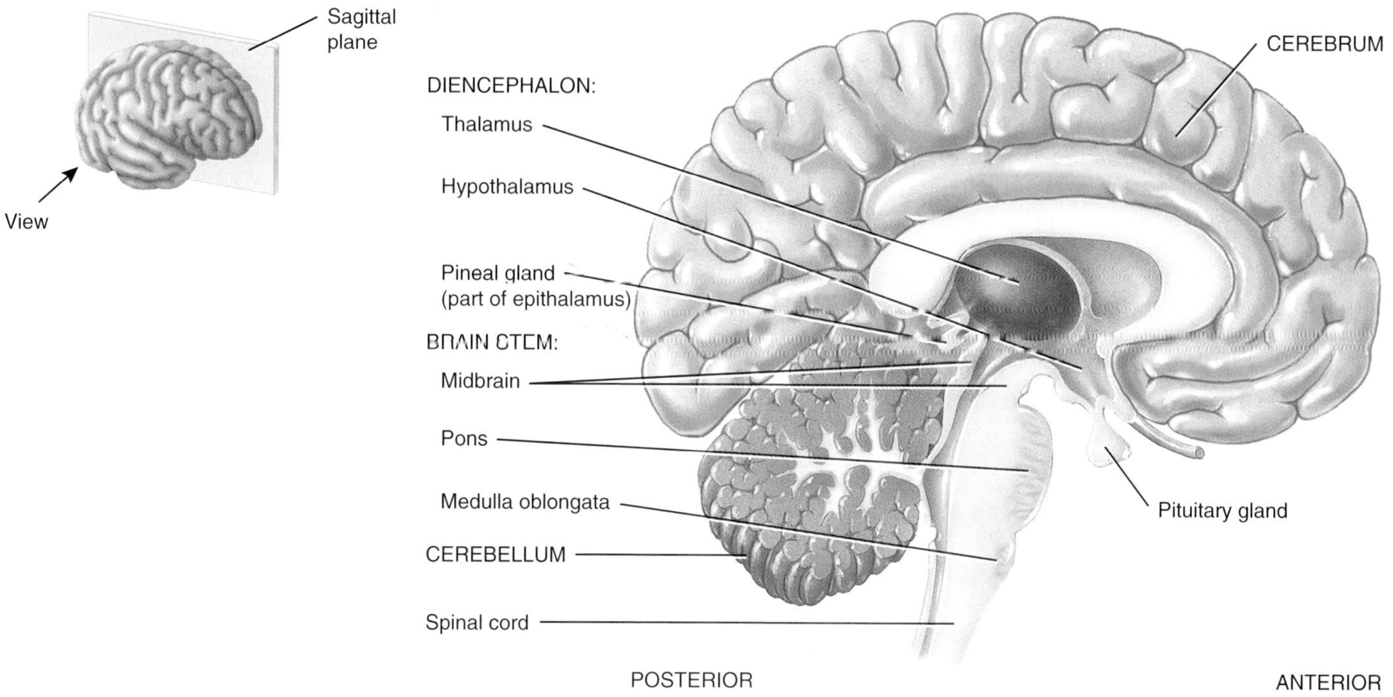
Sagittal plane
View
CEREBRUM
DIENCEPHALON:
Thalamus
Hypothalamus
Pineal gland (part of epithalamus)
BRAIN STEM:
Midbrain
Pons
Medulla oblongata
CEREBELLUM
Spinal cord
Pituitary gland
POSTERIOR
ANTERIOR

(a) Sagittal section, medial view

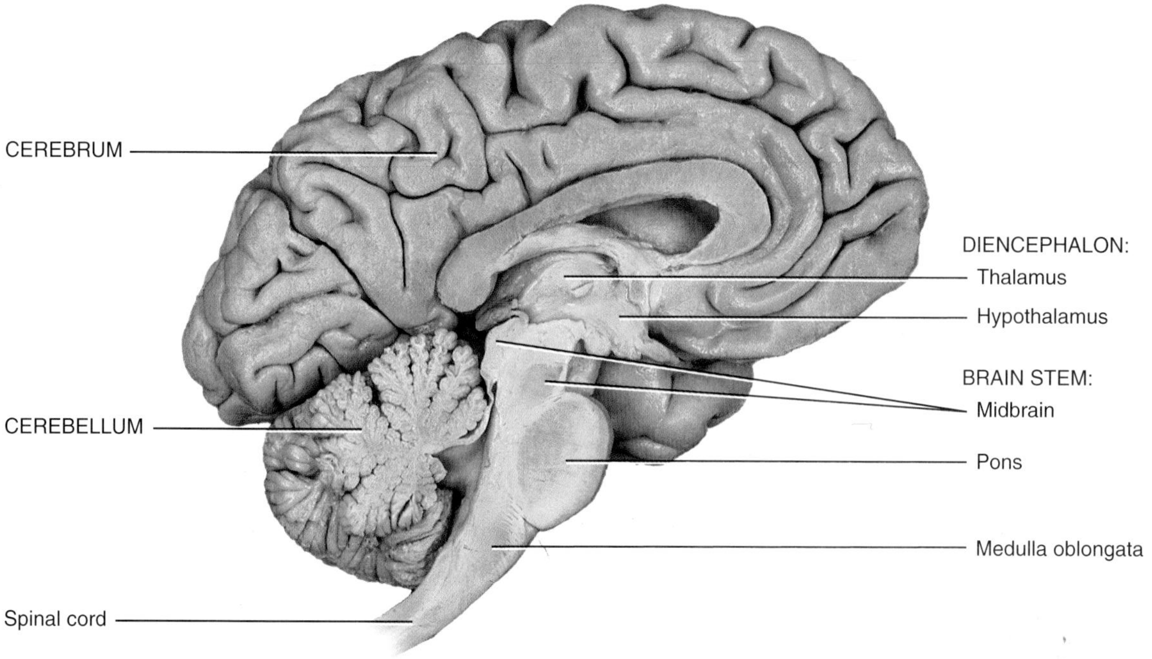
CEREBRUM
CEREBELLUM
Spinal cord
DIENCEPHALON:
Thalamus
Hypothalamus
BRAIN STEM:
Midbrain
Pons
Medulla oblongata

(b) Sagittal section, medial view

? Which part of the brain is the largest?

Figure 14.2 The protective coverings of the brain.

Cranial bones and cranial meninges protect the brain.

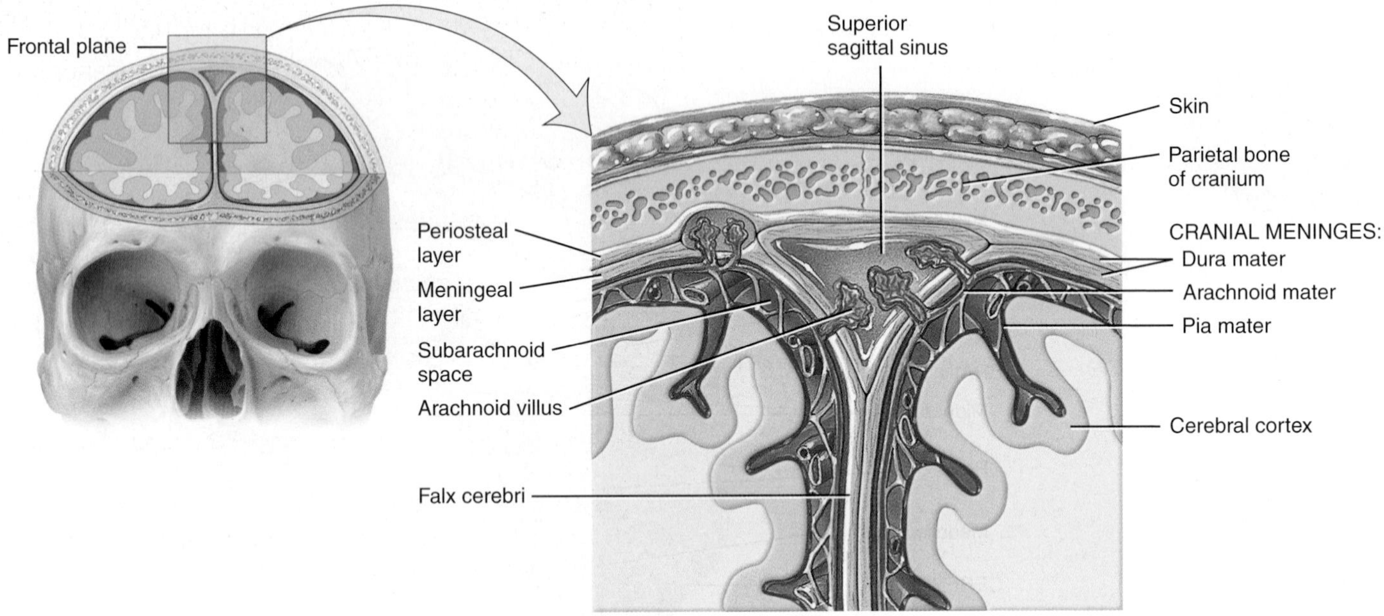

(a) Anterior view of frontal section through skull showing the cranial meninges

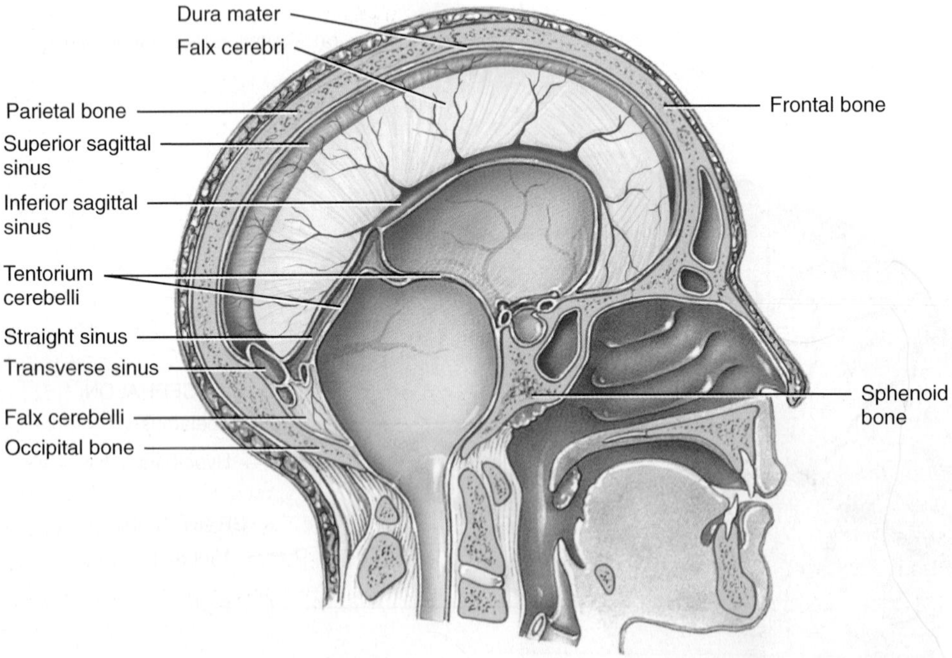

(b) Sagittal section of extensions of the dura mater

What are the three layers of the cranial meninges, from superficial to deep?

dura mater has two layers; the spinal dura mater has only one. The two dural layers are called the *periosteal layer* (which is external) and the *meningeal layer* (which is internal). The dural layers around the brain are fused together except where they separate to enclose the dural venous sinuses (endothelial-lined venous channels) that drain venous blood from the brain and deliver it into the internal jugular veins. Also, there is no epidural space around the brain. Blood vessels that enter brain tissue pass along the surface of the brain, and as they penetrate inward, they are sheathed by a loose-fitting sleeve of pia mater. Three extensions of the dura mater separate parts of the brain: (1) The **falx cerebri** (FALKS ser-i-BRĒ; *falx* = sickle-shaped) separates the two hemispheres (sides) of the cerebrum. (2) The **falx cerebelli** (ser′-e-BEL-ī) separates the two hemispheres of the cerebellum. (3) The **tentorium cerebelli** (ten-TŌ-re-um = tent) separates the cerebrum from the cerebellum.

Brain Blood Flow and the Blood–Brain Barrier

Blood flows to the brain mainly via the internal carotid and vertebral arteries (see Figure 21.19); the dural venous sinuses drain into the internal jugular veins to return blood from the head to the heart (see Figure 21.24).

In an adult, the brain represents only 2% of total body weight, but it consumes about 20% of the oxygen and glucose used by the body, even when you are resting. Neurons synthesize ATP almost exclusively from glucose via reactions that use oxygen. When the activity of neurons and neuroglia increases in a particular region of the brain, blood flow to that area also increases. Even a brief slowing of brain blood flow may cause disorientation or a lack of consciousness, such as when you stand up too quickly after sitting for a long period of time. Typically, an interruption in blood flow for 1 or 2 minutes impairs neuronal function, and total deprivation of oxygen for about 4 minutes causes permanent injury. Because virtually no glucose is stored in the brain, the supply of glucose also must be continuous. If blood entering the brain has a low level of glucose, mental confusion, dizziness, convulsions, and loss of consciousness may occur. People with diabetes must be vigilant about their blood sugar levels because these levels can drop quickly, leading to diabetic shock, which is characterized by seizure, coma, and possibly death.

The **blood–brain barrier (BBB)** consists mainly of tight junctions that seal together the endothelial cells of brain blood capillaries and a thick basement membrane that surrounds the capillaries. As you learned in Chapter 12, astrocytes are one type of neuroglia; the processes of many astrocytes press up against the capillaries and secrete chemicals that maintain the permeability characteristics of the tight junctions. A few water-soluble substances, such as glucose, cross the BBB by active transport. Other substances, such as creatinine, urea, and most ions, cross the BBB very slowly. Still other substances—proteins and most antibiotic drugs—do not pass at all from the blood into brain tissue. However, lipid-soluble substances, such as oxygen, carbon dioxide, alcohol, and most anesthetic agents, are able to access brain tissue freely. Trauma, certain toxins, and inflammation can cause a breakdown of the blood–brain barrier.

CLINICAL CONNECTION | *Breaching the Blood–Brain Barrier*

Because it is so effective, the BBB prevents the passage of helpful substances as well as those that are potentially harmful. Researchers are exploring ways to move drugs that could be therapeutic for brain cancer or other CNS disorders past the BBB. In one method, the drug is injected in a concentrated sugar solution. The high osmotic pressure of the sugar solution causes the endothelial cells of the capillaries to shrink, which opens gaps between their tight junctions, making the BBB more leaky and allowing the drug to enter the brain tissue. •

✔CHECKPOINT

1. Compare the sizes and locations of the cerebrum and cerebellum.
2. Describe the locations of the cranial meninges.
3. Explain the blood supply to the brain and the importance of the blood–brain barrier.

14.2 CEREBROSPINAL FLUID

OBJECTIVE

- Explain the formation and circulation of cerebrospinal fluid.

Cerebrospinal fluid (CSF) is a clear, colorless liquid composed primarily of water that protects the brain and spinal cord from chemical and physical injuries. It also carries small amounts of oxygen, glucose, and other needed chemicals from the blood to neurons and neuroglia. CSF continuously circulates through cavities in the brain and spinal cord and around the brain and spinal cord in the subarachnoid space (the space between the arachnoid mater and pia mater). The total volume of CSF is 80 to 150 mL (3 to 5 oz) in an adult. CSF contains small amounts of glucose, proteins, lactic acid, urea, cations (Na^+, K^+, Ca^{2+}, Mg^{2+}), and anions (Cl^- and HCO_3^-); it also contains some white blood cells.

Figure 14.3 shows the four CSF-filled cavities within the brain, which are called **ventricles** (VEN-tri-kuls = little cavities). There is one **lateral ventricle** in each hemisphere of the cerebrum. (Think of them as ventricles 1 and 2.) Anteriorly, the lateral ventricles are separated by a thin membrane, the **septum pellucidum** (SEP-tum pe-LOO-si-dum; *pellucid* = transparent). The **third ventricle** is a narrow slitlike cavity along the midline superior to the hypothalamus and between the right and left halves of the thalamus. The **fourth ventricle** lies between the brain stem and the cerebellum.

Functions of CSF

The CSF has three basic functions:

1. ***Mechanical protection.*** CSF serves as a shock-absorbing medium that protects the delicate tissues of the brain and spinal

Figure 14.3 **Locations of ventricles within a "transparent" brain.** One interventricular foramen on each side connects a lateral ventricle to the third ventricle, and the aqueduct of the midbrain connects the third ventricle to the fourth ventricle.

Ventricles are cavities within the brain that are filled with cerebrospinal fluid.

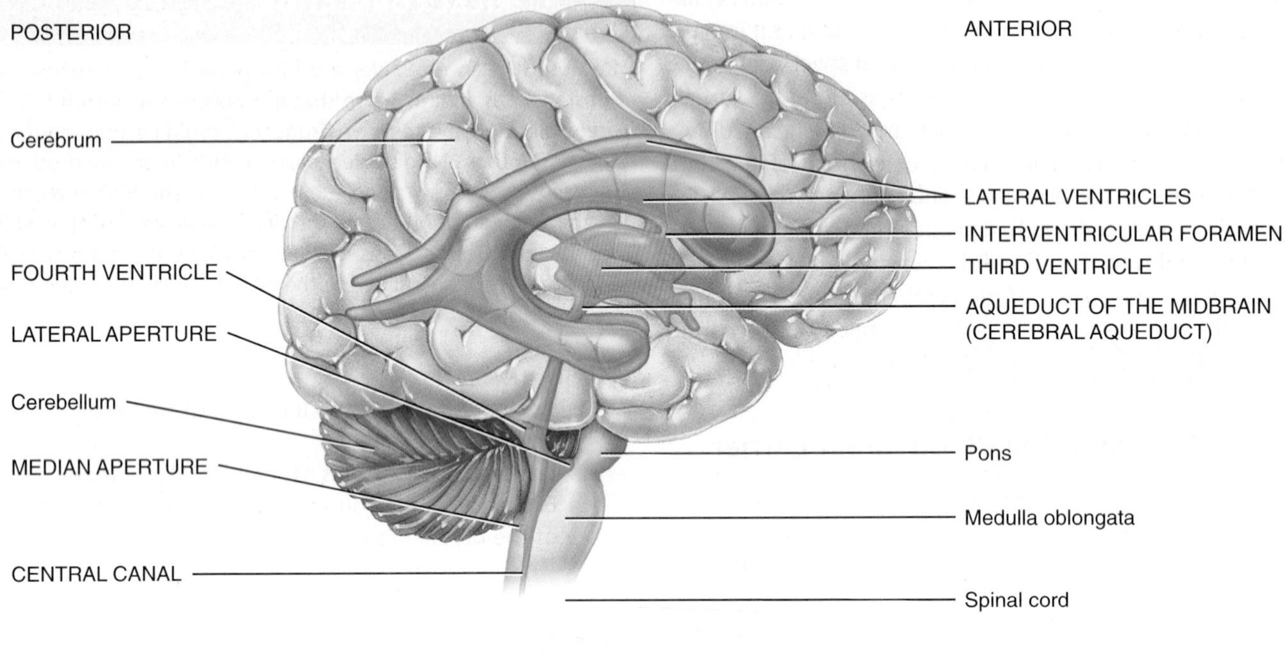
POSTERIOR
ANTERIOR
Cerebrum
LATERAL VENTRICLES
INTERVENTRICULAR FORAMEN
THIRD VENTRICLE
FOURTH VENTRICLE
AQUEDUCT OF THE MIDBRAIN (CEREBRAL AQUEDUCT)
LATERAL APERTURE
Cerebellum
Pons
MEDIAN APERTURE
Medulla oblongata
CENTRAL CANAL
Spinal cord

Right lateral view of brain

Which brain region is anterior to the fourth ventricle? Which is posterior to it?

cord from jolts that would otherwise cause them to hit the bony walls of the cranial cavity and vertebral canal. The fluid also buoys the brain so that it "floats" in the cranial cavity.

2. ***Homeostatic function.*** The pH of the CSF affects pulmonary ventilation and cerebral blood flow, which is important in maintaining homeostatic controls for brain tissue. CSF also serves as a transport system for polypeptide hormones secreted by hypothalamic neurons that act at remote sites in the brain.
3. ***Circulation.*** CSF is a medium for minor exchange of nutrients and waste products between the blood and adjacent nervous tissue.

Formation of CSF in the Ventricles

The majority of CSF production is from the **choroid plexuses** (KŌ-royd = membranelike), networks of blood capillaries in the walls of the ventricles (Figure 14.4a). Ependymal cells joined by tight junctions cover the capillaries of the choroid plexuses. Selected substances (mostly water) from the blood plasma, which are filtered from the capillaries, are secreted by the ependymal cells to produce the cerebrospinal fluid. This secretory capacity is bidirectional and accounts for continuous production of CSF and transport of metabolites from the nervous tissue back to the blood. Because of the tight junctions between ependymal cells, materials entering CSF from choroid capillaries cannot leak between these cells; instead, they must pass through the ependymal cells. This **blood–cerebrospinal fluid barrier** permits certain substances to enter the CSF but excludes others, protecting the brain and spinal cord from potentially harmful blood-borne substances. In contrast to the blood–brain barrier, which is formed mainly by tight junctions of brain capillary endothelial cells, the blood–cerebrospinal fluid barrier is formed by tight junctions of ependymal cells.

Circulation of CSF

The CSF formed in the choroid plexuses of each lateral ventricle flows into the third ventricle through two narrow, oval openings, the **interventricular foramina** (in′-ter-ven-TRIK-ū-lar) (singular is *foramen;* Figure 14.4b). More CSF is added by the choroid plexus in the roof of the third ventricle. The fluid then flows through the **aqueduct of the midbrain (cerebral aqueduct)** (AK-we-dukt), which passes through the midbrain, into the fourth ventricle. The choroid plexus of the fourth ventricle contributes more fluid. CSF enters the subarachnoid space through three openings in the roof of the fourth ventricle: a single **median aperture** (AP-er-chur) and paired **lateral apertures,** one on each side. CSF then circulates in the central canal of the spinal cord and in the subarachnoid space around the surface of the brain and spinal cord.

CSF is gradually reabsorbed into the blood through **arachnoid villi,** fingerlike extensions of the arachnoid that project into the dural venous sinuses, especially the **superior sagittal sinus** (see Figure 14.2). (A cluster of arachnoid villi is called an **arachnoid granulation.**) Normally, CSF is reabsorbed as rapidly as it is formed by the choroid plexuses, at a rate of about 20 mL/hr (480 mL/day). Because the rates of formation and reabsorption are the same, the pressure of CSF normally is constant. For the same reason, the volume of CSF remains constant. Figure 14.4d summarizes the production and flow of CSF.

CLINICAL CONNECTION | *Hydrocephalus*

Abnormalities in the brain—tumors, inflammation, or developmental malformations—can interfere with the circulation of CSF from the ventricles into the subarachnoid space. When excess CSF accumulates in the ventricles, the CSF pressure rises. Elevated CSF pressure causes a condition called **hydrocephalus** (hī′-drō-SEF-a-lus; *hydro-* = water; *-cephal-* = head). The abnormal accumulation of CSF may be due to an obstruction to CSF flow or an abnormal rate of CSF production and/or reabsorption. In a baby whose fontanels have not yet closed, the head bulges due to the increased pressure. If the condition persists, the fluid buildup compresses and damages the delicate nervous tissue. Hydrocephalus is relieved by draining the excess CSF. In one procedure, called *endoscopic third ventriculostomy (ETV),* a neurosurgeon makes a hole in the floor of the third ventricle and the CSF drains directly into the subarachnoid space. In adults, hydrocephalus may occur after head injury, meningitis, or subarachnoid hemorrhage. Because the adult skull bones are fused together, this condition can quickly become life-threatening and requires immediate intervention. •

✔CHECKPOINT

4. What structures produce CSF, and where are they located?
5. What is the difference between the blood–brain barrier and the blood–cerebrospinal fluid barrier?

Figure 14.4 Pathways of circulating cerebrospinal fluid.

CSF is formed from blood plasma by ependymal cells that cover the choroid plexuses of the ventricles.

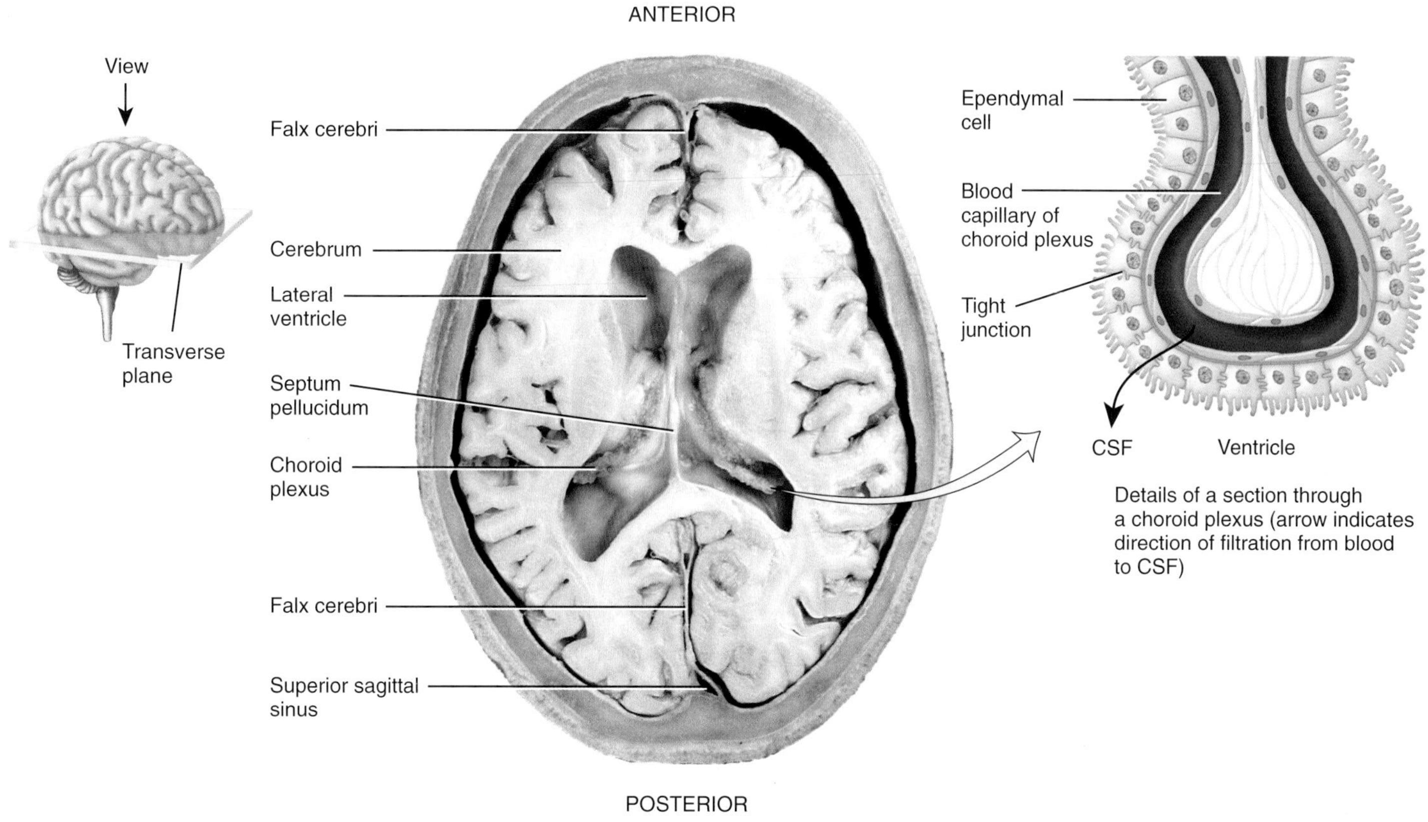

(a) Superior view of transverse section of brain showing choroid plexuses

FIGURE 14.4 CONTINUES ▶

FIGURE 14.4 CONTINUED

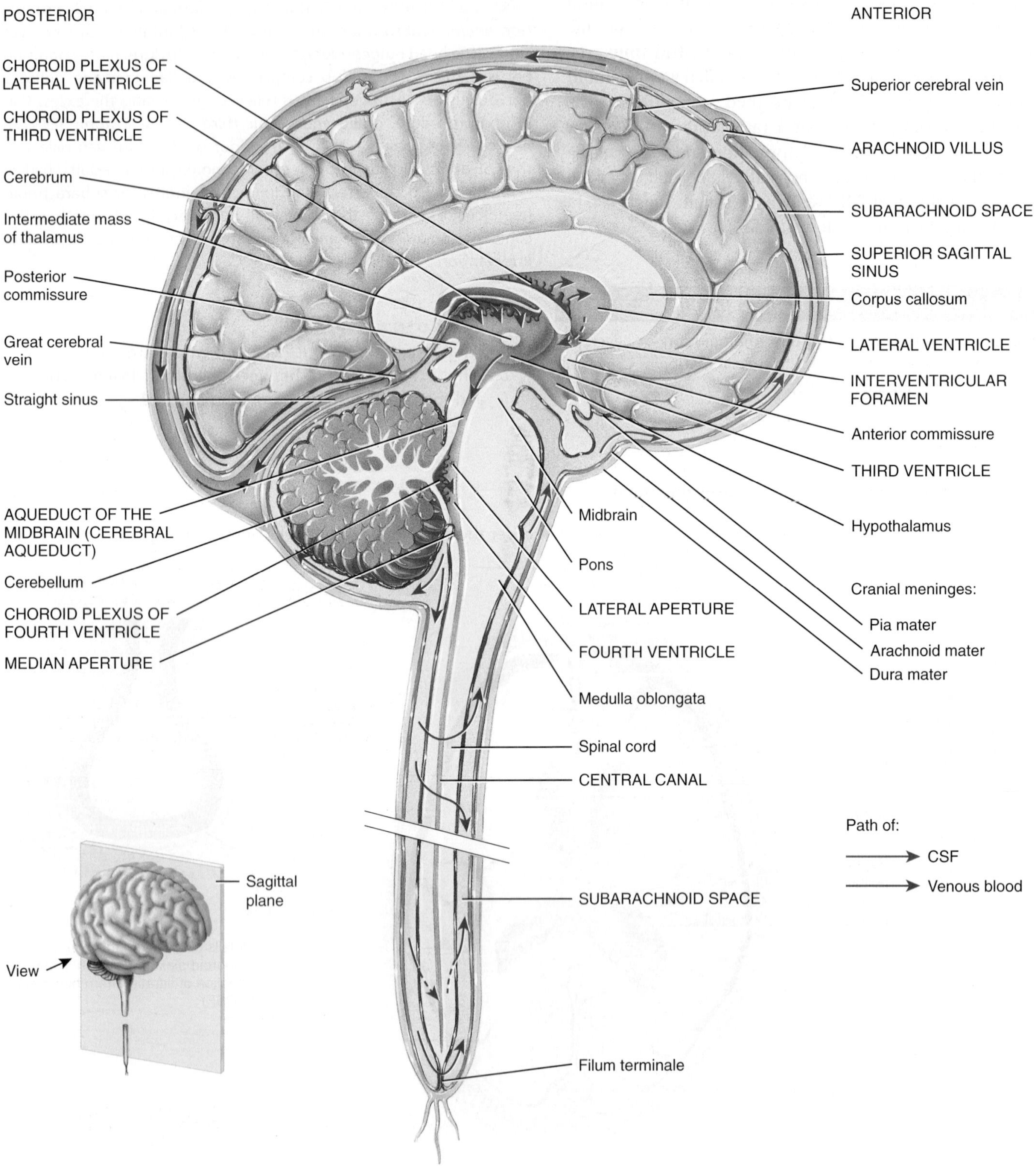
POSTERIOR
ANTERIOR
CHOROID PLEXUS OF LATERAL VENTRICLE
CHOROID PLEXUS OF THIRD VENTRICLE
Cerebrum
Intermediate mass of thalamus
Posterior commissure
Great cerebral vein
Straight sinus
AQUEDUCT OF THE MIDBRAIN (CEREBRAL AQUEDUCT)
Cerebellum
CHOROID PLEXUS OF FOURTH VENTRICLE
MEDIAN APERTURE
Superior cerebral vein
ARACHNOID VILLUS
SUBARACHNOID SPACE
SUPERIOR SAGITTAL SINUS
Corpus callosum
LATERAL VENTRICLE
INTERVENTRICULAR FORAMEN
Anterior commissure
THIRD VENTRICLE
Hypothalamus
Cranial meninges:
Pia mater
Arachnoid mater
Dura mater
Midbrain
Pons
LATERAL APERTURE
FOURTH VENTRICLE
Medulla oblongata
Spinal cord
CENTRAL CANAL
SUBARACHNOID SPACE
Filum terminale
Path of:
CSF
Venous blood
Sagittal plane
View

(b) Sagittal section of brain and spinal cord

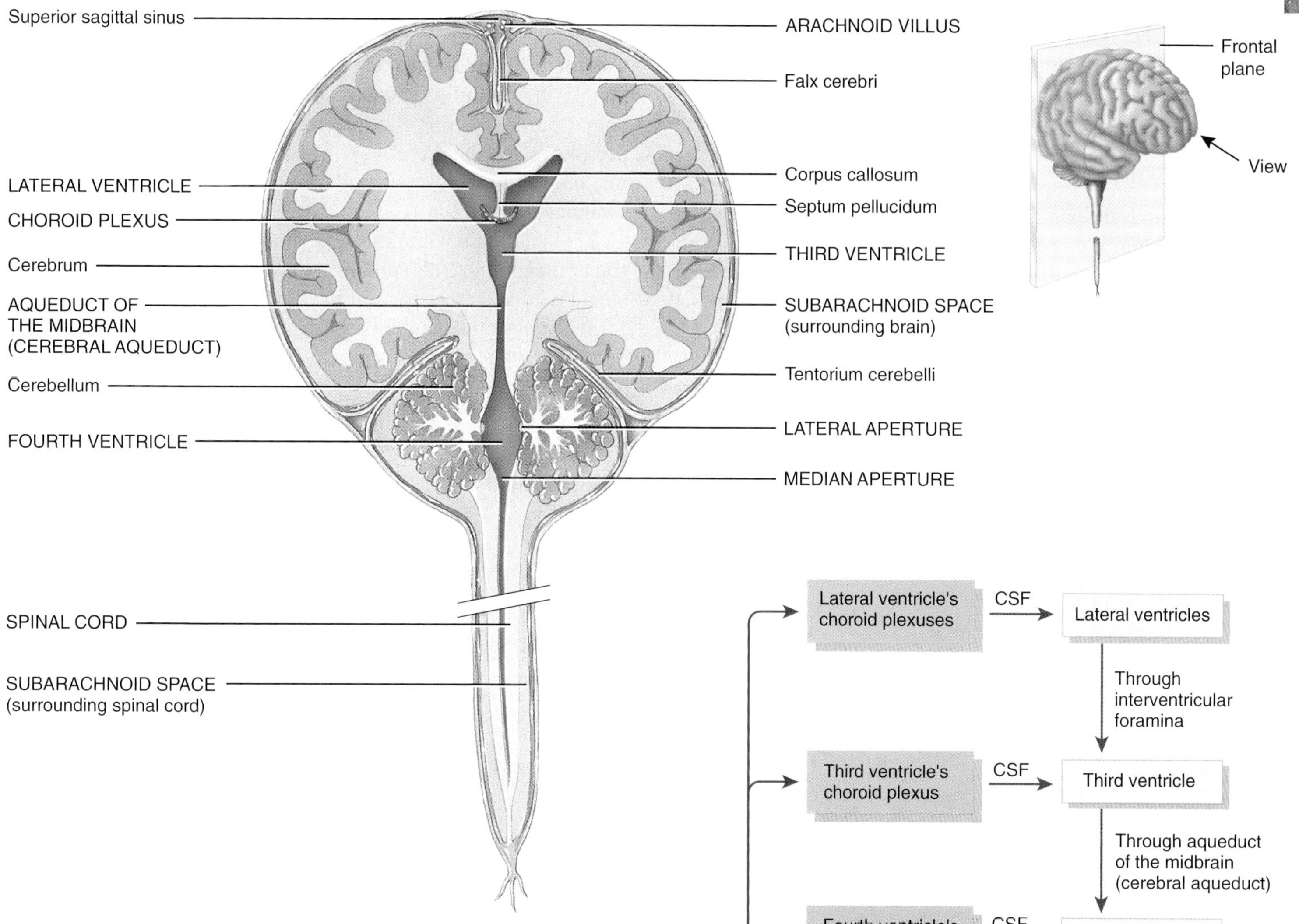

(c) Frontal section of brain and spinal cord

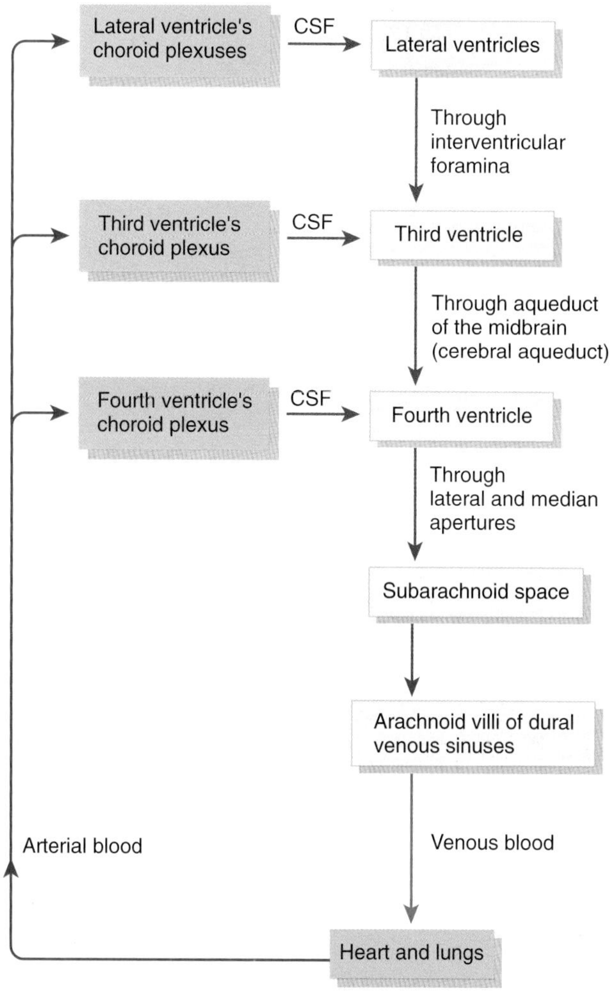

(d) Summary of the formation, circulation, and absorption of cerebrospinal fluid (CSF)

Where is CSF reabsorbed?

14.3 THE BRAIN STEM AND RETICULAR FORMATION

OBJECTIVE

- Describe the structures and functions of the brain stem and reticular formation.

The brain stem is the part of the brain between the spinal cord and the diencephalon. It consists of three structures: (1) medulla oblongata, (2) pons, and (3) midbrain. Extending through the brain stem is the reticular formation, a netlike region of interspersed gray and white matter.

Medulla Oblongata

The **medulla oblongata** (me-DOOL-la ob′-long-GA-ta), or more simply the **medulla,** is continuous with the superior part of the spinal cord; it forms the inferior part of the brain stem (Figure 14.5; see also Figure 14.1). The medulla begins at the foramen magnum and extends to the inferior border of the pons, a distance of about 3 cm (1.2 in.).

The medulla's white matter contains all sensory (ascending) tracts and motor (descending) tracts that extend between the spinal cord and other parts of the brain. Some of the white matter forms bulges on the anterior aspect of the medulla. These protru-

Figure 14.5 Medulla oblongata in relation to the rest of the brain stem.

The brain stem consists of the medulla oblongata, pons, and midbrain.

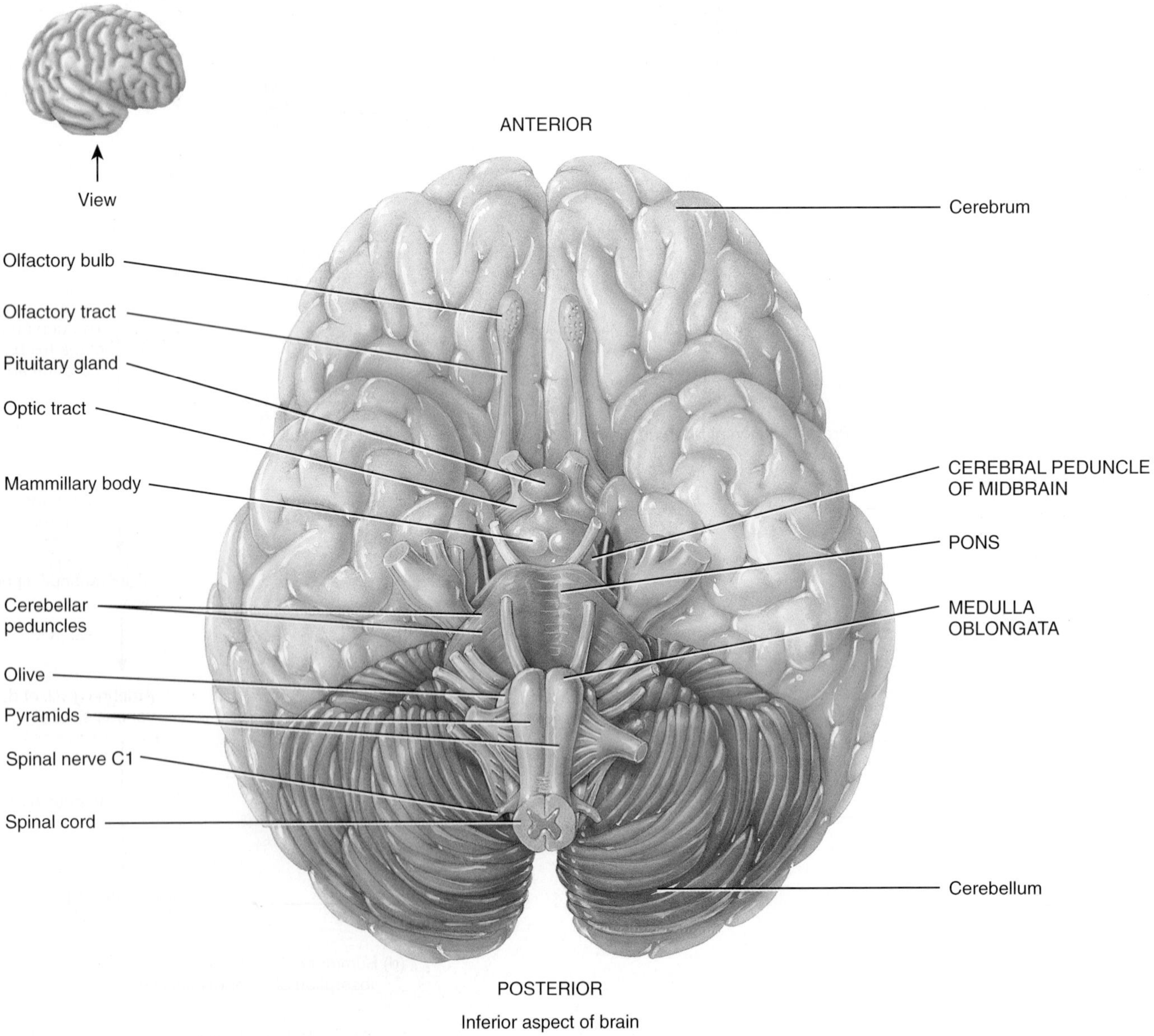

What part of the brain stem contains the pyramids? The cerebral peduncles? Literally means "bridge"?

sions, called the **pyramids** (Figure 14.6; see also Figure 14.5), are formed by the large corticospinal tracts that pass from the cerebrum to the spinal cord. The corticospinal tracts control voluntary movements of the limbs and trunk (see Figure 16.10). Just superior to the junction of the medulla with the spinal cord, 90% of the axons in the left pyramid cross to the right side, and 90% of the axons in the right pyramid cross to the left side. This crossing is called the **decussation of pyramids** (dē′-ku-SĀ-shun; *decuss* = crossing) and explains why each side of the brain controls voluntary movements on the opposite side of the body.

The medulla also contains several **nuclei.** (Recall that a nucleus is a collection of neuronal cell bodies within the CNS.) Some of these nuclei control vital body functions. Examples of nuclei in the medulla that regulate vital activities include the cardiovascular center and the medullary rhythmicity area. The **cardiovascular center** regulates the rate and force of the heartbeat and the diameter of blood vessels (see Figure 21.13). The **medullary rhythmicity area** (rith-MIS-i-tē) of the **respiratory center** adjusts the basic rhythm of breathing (see Figure 23.25).

Besides regulating heartbeat, blood vessel diameter, and the normal breathing rhythm, nuclei in the medulla also control reflexes for vomiting, swallowing, sneezing, coughing, and hiccupping. The **vomiting center** of the medulla causes **vomiting,** the forcible expulsion of the contents of the upper gastrointestinal (GI) tract through the mouth (see Section 24.8). The **deglutition center** (dē-gloo-TISH-un) of the medulla promotes **deglutition** (swallowing) of a mass of food that has moved from the oral cavity of the mouth into the pharynx (throat) (see Section 24.8). **Sneezing** involves spasmodic contraction of breathing muscles that forcefully expel air through the nose and mouth. **Coughing** involves a long-drawn and deep inhalation and then a strong exhalation that suddenly sends a blast of air through the upper respiratory passages. **Hiccupping** is caused by spasmodic contractions of the diaphragm (a muscle of breathing) that ultimately result in the production of a sharp sound on inhalation. Sneezing, coughing, and hiccupping are described in more detail in Table 23.1.

Just lateral to each pyramid is an oval-shaped swelling called an **olive** (see Figures 14.5, 14.6). Within the olive is the **inferior olivary nucleus,** which receives input from the cerebral cortex, red nucleus of the midbrain, and spinal cord. Neurons of the inferior olivary nucleus extend their axons into the cerebellum, where they regulate the activity of cerebellar neurons. By influencing cerebellar neuron activity, the inferior olivary nucleus provides instructions that the cerebellum uses to make adjustments to muscle activity as you learn new motor skills.

Nuclei associated with sensations of touch, pressure, vibration, and conscious proprioception are located in the posterior part of

Figure 14.6 Internal anatomy of the medulla oblongata.

The pyramids of the medulla contain the large motor tracts that run from the cerebrum to the spinal cord.

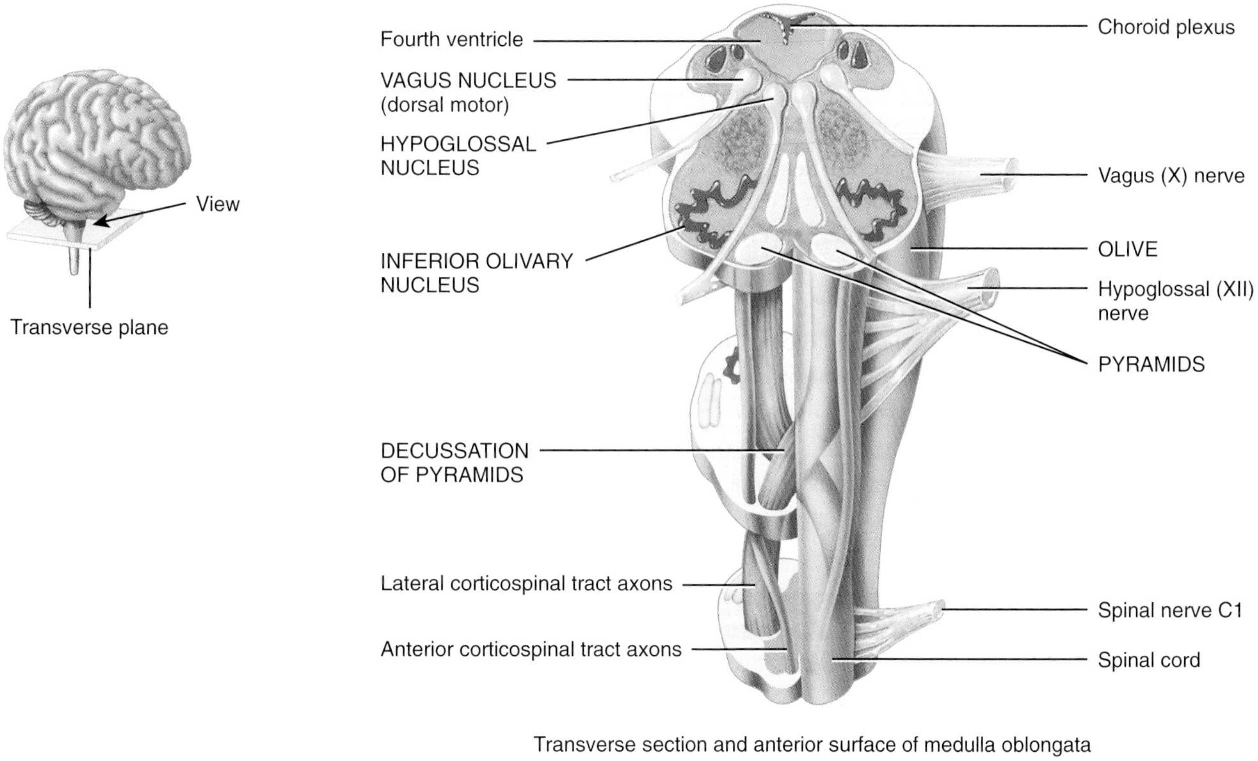

View
Transverse plane
Fourth ventricle
VAGUS NUCLEUS (dorsal motor)
HYPOGLOSSAL NUCLEUS
INFERIOR OLIVARY NUCLEUS
DECUSSATION OF PYRAMIDS
Lateral corticospinal tract axons
Anterior corticospinal tract axons
Choroid plexus
Vagus (X) nerve
OLIVE
Hypoglossal (XII) nerve
PYRAMIDS
Spinal nerve C1
Spinal cord

Transverse section and anterior surface of medulla oblongata

What does decussation mean? What is the functional consequence of decussation of the pyramids?

the medulla. These nuclei are the right and left **gracile nucleus** (GRAS-il = slender) and **cuneate nucleus** (KŪ-nē-āt = wedge). Ascending sensory axons of the **gracile fasciculus** (fa-SIK-ū-lus) and the **cuneate fasciculus,** which are two tracts in the posterior columns of the spinal cord, form synapses in these nuclei (see Figure 16.5). Postsynaptic neurons then relay the sensory information to the thalamus on the opposite side of the brain. The axons ascend to the thalamus in a band of white matter called the **medial lemniscus** (lem-NIS-kus = ribbon), which extends through the medulla, pons, and midbrain (see Figure 14.7b). The tracts of the posterior columns and the axons of the medial lemniscus are collectively known as the **posterior column–medial lemniscus pathway.**

The medulla also contains nuclei that are components of sensory pathways for gustation (taste), audition (hearing), and equilibrium (balance). The **gustatory nucleus** (GUS-ta-tō′-rē) of the medulla is part of the gustatory pathway from the tongue to the brain; it receives gustatory input from the taste buds of the tongue (see Figure 17.3e). The **cochlear nuclei** (KOK-lē-ar) of the medulla are part of the auditory pathway from the inner ear to the brain; they receive auditory input from the cochlea of the inner ear (see Figure 17.23). The **vestibular nuclei** (ves-TIB-ū-lar) of the medulla and pons are components of the equilibrium pathway from the inner ear to the brain; they receive sensory information associated with equilibrium from proprioceptors in the vestibular apparatus of the inner ear (see Figure 17.26).

Finally, the medulla contains nuclei associated with the following five pairs of cranial nerves (see Figure 14.5):

1. **Vestibulocochlear (VIII) nerves.** Several nuclei in the medulla receive sensory input from and provide motor output to the cochlea of the internal ear via the vestibulocochlear nerves. These nerves convey impulses related to hearing.
2. **Glossopharyngeal (IX) nerves.** Nuclei in the medulla relay sensory and motor impulses related to taste, swallowing, and salivation via the glossopharyngeal nerves.
3. **Vagus (X) nerves.** Nuclei in the medulla receive sensory impulses from and provide motor impulses to the pharynx and larynx and many thoracic and abdominal viscera via the vagus nerves.
4. **Accessory (XI) nerves (cranial portion).** These fibers are actually part of the vagus (X) nerves. Nuclei in the medulla are the origin for nerve impulses that control swallowing via the vagus nerves (cranial portion of the accessory nerves).
5. **Hypoglossal (XII) nerves.** Nuclei in the medulla are the origin for nerve impulses that control tongue movements during speech and swallowing via the hypoglossal nerves.

CLINICAL CONNECTION | *Injury of the Medulla*

Given the vital activities controlled by the medulla, it is not surprising that **injury to the medulla** from a hard blow to the back of the head or upper neck such as falling back on ice can be fatal. Damage to the medullary rhythmicity area is particularly serious and can rapidly lead to death. Symptoms of nonfatal injury to the medulla may include cranial nerve malfunctions on the same side of the body as the injury, paralysis and loss of sensation on the opposite side of the body, and irregularities in breathing or heart rhythm. Alcohol overdose also suppresses the medullary rhythmicity area and may result in death. •

Pons

The **pons** (= bridge) lies directly superior to the medulla and anterior to the cerebellum and is about 2.5 cm (1 in.) long (see Figures 14.1, 14.5). Like the medulla, the pons consists of both nuclei and tracts. As its name implies, the pons is a bridge that connects parts of the brain with one another. These connections are provided by bundles of axons. Some axons of the pons connect the right and left sides of the cerebellum. Others are part of ascending sensory tracts and descending motor tracts.

The pons has two major structural components: a ventral region and a dorsal region. The ventral region of the pons forms a large synaptic relay station consisting of scattered gray centers called the **pontine nuclei** (PON-tīn). Entering and exiting these nuclei are numerous white matter tracts, each of which provides a connection between the cortex (outer layer) of a cerebral hemisphere and that of the opposite hemisphere of the cerebellum. This complex circuitry plays an essential role in coordinating and maximizing the efficiency of voluntary motor output throughout the body. The dorsal region of the pons is more like the other regions of the brain stem, the medulla and midbrain. It contains ascending and descending tracts along with the nuclei of cranial nerves.

Other nuclei located in the pons are the **pneumotaxic area** (noo-mō-TAK-sik) and the **apneustic area** (ap-NOO-stik), shown in Figure 23.24. Together with the medullary rhythmicity area, the pneumotaxic and apneustic areas help control breathing.

The pons also contains nuclei associated with the following four pairs of cranial nerves (see Figure 14.5):

1. **Trigeminal (V) nerves.** Nuclei in the pons receive sensory impulses for somatic sensations from the head and face and provide motor impulses that govern chewing via the trigeminal nerves.
2. **Abducens (VI) nerves.** Nuclei in the pons provide motor impulses that control eyeball movement via the abducens nerves.
3. **Facial (VII) nerves.** Nuclei in the pons receive sensory impulses for taste and provide motor impulses to regulate secretion of saliva and tears and contraction of muscles of facial expression via the facial nerves.
4. **Vestibulocochlear (VIII) nerves.** Nuclei in the pons receive sensory impulses from and provide motor impulses to the vestibular apparatus via the vestibulocochlear nerves. These nerves convey impulses related to balance and equilibrium.

Midbrain

The **midbrain** or **mesencephalon** extends from the pons to the diencephalon (see Figures 14.1, 14.5) and is about 2.5 cm (1 in.)

long. The aqueduct of the midbrain (cerebral aqueduct) passes through the midbrain, connecting the third ventricle above with the fourth ventricle below. Like the medulla and the pons, the midbrain contains both nuclei and tracts (Figure 14.7).

The anterior part of the midbrain contains paired bundles of axons known as the **cerebral peduncles** (pe-DUNK-kuls or PĒ-dung-kuls = little feet; see Figures 14.5, 14.7b). The cerebral peduncles consist of axons of the corticospinal, corticobulbar, and corticopontine tracts, which conduct nerve impulses from motor areas in the cerebral cortex to the spinal cord, medulla, and pons, respectively.

The posterior part of the midbrain, called the **tectum** (TEK-tum = roof), contains four rounded elevations (Figure 14.7a). The two superior elevations, nuclei known as the **superior colliculi** (ko-LIK-ū-lī = little hills; singular is *colliculus*), serve as reflex centers for certain visual activities. Through neural circuits from the retina of the eye to the superior colliculi to the extrinsic eye muscles, visual stimuli elicit eye movements for tracking moving images (such as a moving car) and scanning stationary images (as you are doing to read this sentence). The superior colliculi are also responsible for reflexes that govern movements of the head, eyes, and trunk in response to visual stimuli. The two inferior elevations, the **inferior colliculi,** are part of the auditory pathway, relaying impulses from the receptors for hearing in the inner ear to the brain. These two nuclei are also reflex centers for the *startle reflex,* sudden movements of the head, eyes, and trunk that occur when you are surprised by a loud noise such as a gunshot.

The midbrain contains several other nuclei, including the left and right **substantia nigra** (sub-STAN-shē-a = substance; NĪ-gra = black), which are large and darkly pigmented (Figure 14.7b). Neurons that release dopamine, extending from the substantia nigra to the basal nuclei, help control subconscious muscle activities. Loss of these neurons is associated with Parkinson disease (see Disorders: Homeostatic Imbalances at the end of Chapter 16). Also present are the left and right **red nuclei,** which look reddish due to their rich blood supply and an iron-containing pigment in their neuronal cell bodies. Axons from the cerebellum and cerebral cortex form synapses in the red nuclei, which help control muscular movements.

Still other nuclei in the midbrain are associated with two pairs of cranial nerves (see Figure 14.5):

1. **Oculomotor (III) nerves.** Nuclei in the midbrain provide motor impulses that control movements of the eyeball, while

Figure 14.7 **Midbrain.**

The midbrain connects the pons to the diencephalon.

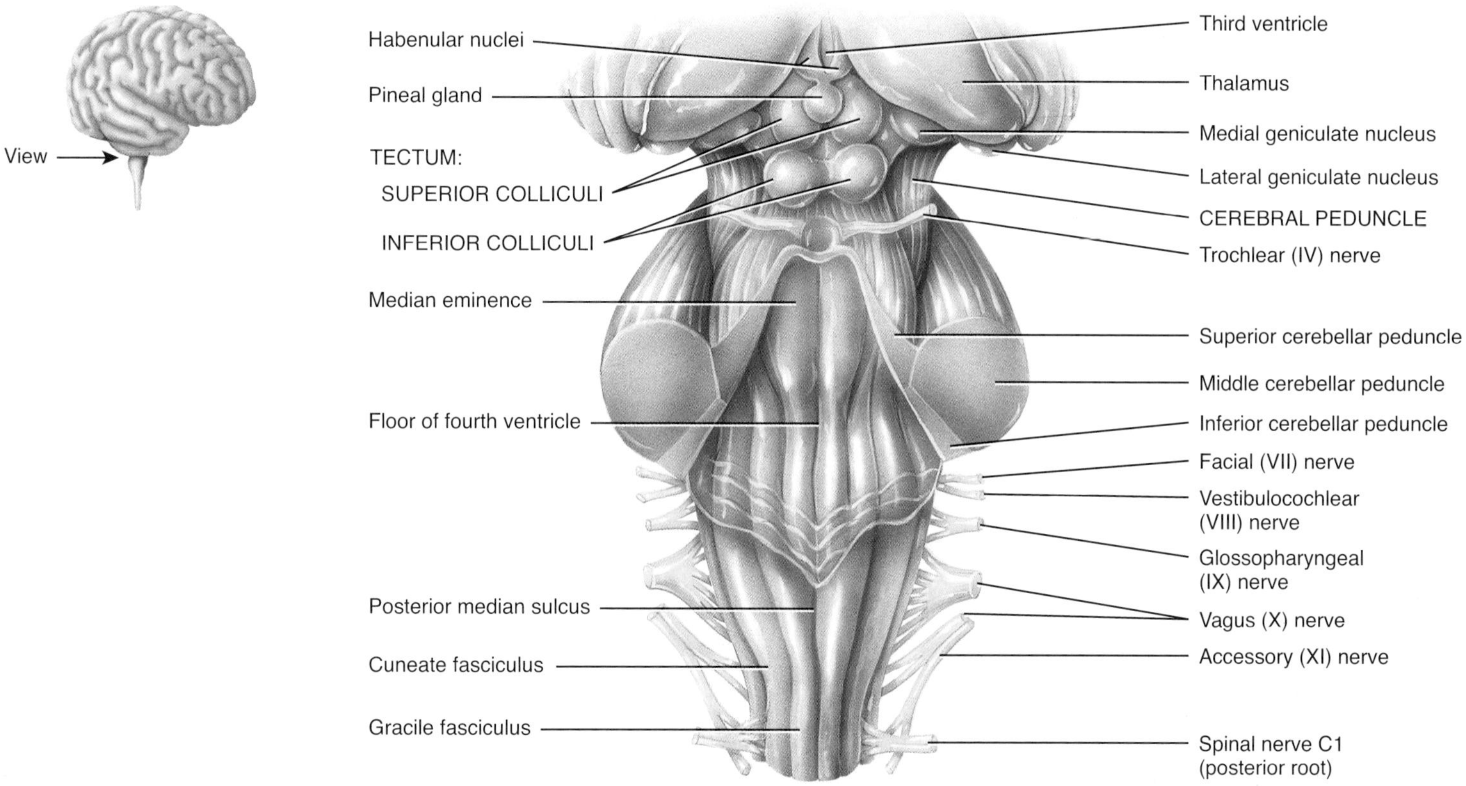
View
Habenular nuclei
Pineal gland
TECTUM:
SUPERIOR COLLICULI
INFERIOR COLLICULI
Median eminence
Floor of fourth ventricle
Posterior median sulcus
Cuneate fasciculus
Gracile fasciculus
Third ventricle
Thalamus
Medial geniculate nucleus
Lateral geniculate nucleus
CEREBRAL PEDUNCLE
Trochlear (IV) nerve
Superior cerebellar peduncle
Middle cerebellar peduncle
Inferior cerebellar peduncle
Facial (VII) nerve
Vestibulocochlear
(VIII) nerve
Glossopharyngeal
(IX) nerve
Vagus (X) nerve
Accessory (XI) nerve
Spinal nerve C1
(posterior root)

(a) Posterior view of midbrain in relation to brain stem

FIGURE 14.7 CONTINUES

FIGURE 14.7 CONTINUED

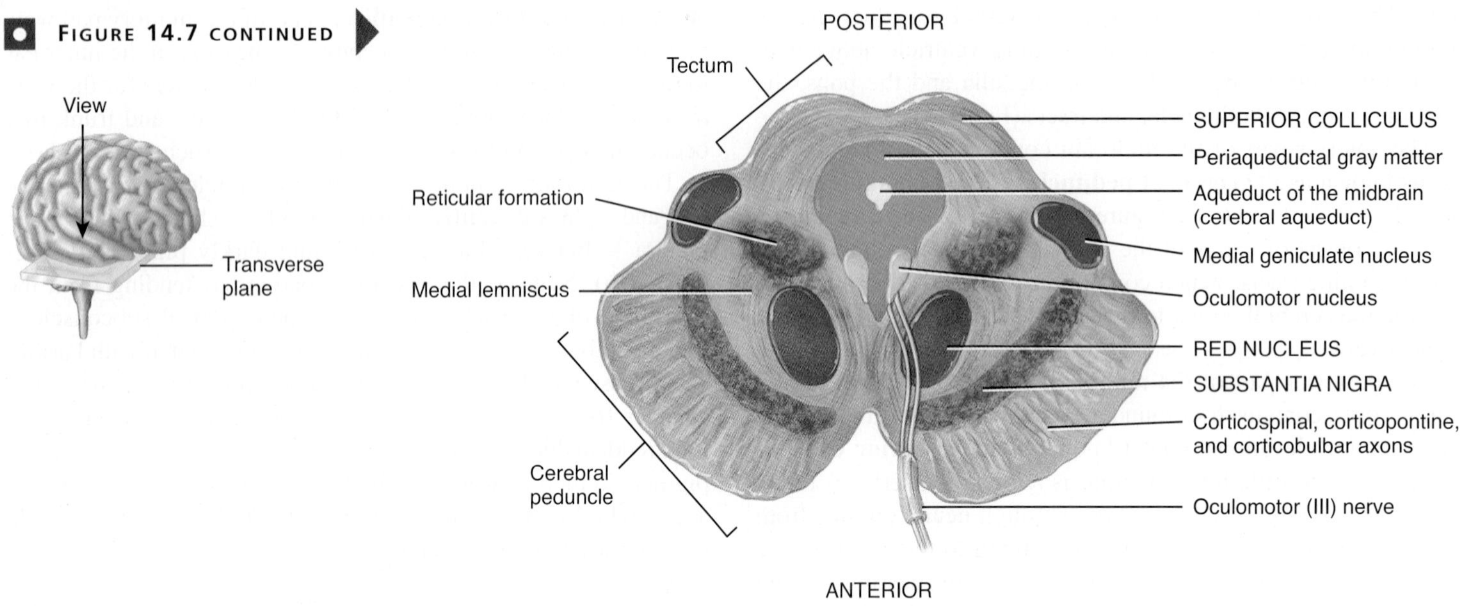
View
Transverse plane
POSTERIOR
Tectum
Reticular formation
Medial lemniscus
Cerebral peduncle
SUPERIOR COLLICULUS
Periaqueductal gray matter
Aqueduct of the midbrain (cerebral aqueduct)
Medial geniculate nucleus
Oculomotor nucleus
RED NUCLEUS
SUBSTANTIA NIGRA
Corticospinal, corticopontine, and corticobulbar axons
Oculomotor (III) nerve
ANTERIOR

(b) Transverse section of midbrain

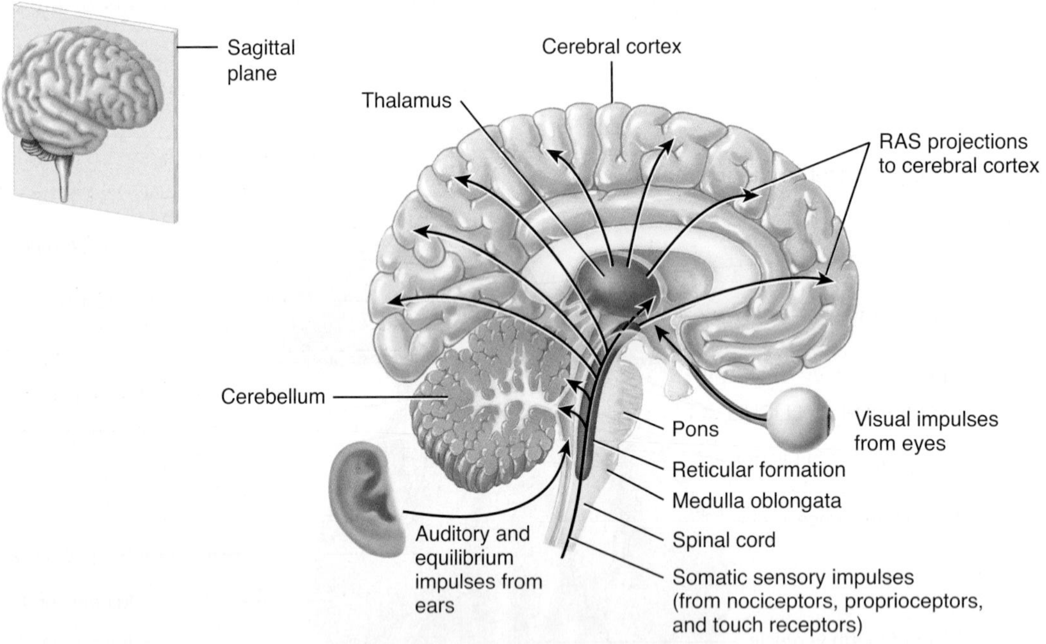
Sagittal plane
Cerebral cortex
Thalamus
RAS projections to cerebral cortex
Cerebellum
Pons
Visual impulses from eyes
Reticular formation
Medulla oblongata
Spinal cord
Auditory and equilibrium impulses from ears
Somatic sensory impulses (from nociceptors, proprioceptors, and touch receptors)

(c) Sagittal section through brain and spinal cord showing the reticular formation

What is the importance of the cerebral peduncles?

accessory oculomotor nuclei provide motor control to the smooth muscles that regulate constriction of the pupil and changes in shape of the lens via the oculomotor nerves.

2. **Trochlear (IV) nerves.** Nuclei in the midbrain provide motor impulses that control movements of the eyeball via the trochlear nerves.

Reticular Formation

In addition to the well-defined nuclei already described, much of the brain stem consists of small clusters of neuronal cell bodies (gray matter) interspersed among small bundles of myelinated axons (white matter). The broad region where white matter and gray matter exhibit a netlike arrangement is known as the **reticular**

formation (re-TIK-ū-lar; *ret-* = net; Figure 14.7c). It extends from the superior part of the spinal cord, throughout the brain stem, and into the inferior part of the diencephalon. Neurons within the reticular formation have both ascending (sensory) and descending (motor) functions.

The ascending portion of the reticular formation is called the **reticular activating system (RAS),** which consists of sensory axons that project to the cerebral cortex, both directly and through the thalamus. Many sensory stimuli can activate the ascending portion of the RAS. Among these are visual and auditory stimuli; mental activities; stimuli from pain, touch, and pressure receptors; and receptors in our limbs and head that keep us aware of the position of our body parts. Perhaps the most important function of the RAS is **consciousness,** a state of wakefulness in which an individual is fully alert, aware, and oriented. Visual and auditory stimuli and mental activities can stimulate the RAS to help maintain consciousness. The RAS is also active during **arousal,** or awakening from sleep. Another function of the RAS is to help maintain **attention** and *alertness.* The RAS also prevents **sensory overload** by filtering out insignificant information so that it does not reach consciousness. For example, while waiting in the hallway for your anatomy class to begin, you may be unaware of all the noise around you while reviewing your notes for class. Inactivation of the RAS produces **sleep,** a state of partial consciousness from which an individual can be aroused. Damage to the RAS, on the other hand, results in **coma,** a state of unconsciousness from which an individual cannot be aroused. In the lightest stages of coma, brain stem and spinal cord reflexes persist, but in the deepest states even those reflexes are lost, and if respiratory and cardiovascular controls are lost, the patient dies. Drugs such as melatonin affect the RAS by helping to induce sleep, and general anesthetics turn off consciousness via the RAS. The descending portion of the RAS has connections to the cerebellum and spinal cord and helps regulate **muscle tone,** the slight degree of involuntary contraction in normal resting skeletal muscles. This portion of the RAS also assists in the regulation of heart rate, blood pressure, and respiratory rate.

Even though the RAS receives input from the eyes, ears, and other sensory receptors, there is no input from receptors for the sense of smell; even strong odors may fail to cause arousal. People who die in house fires usually succumb to smoke inhalation without awakening. For this reason, all sleeping areas should have a nearby smoke detector that emits a loud alarm. A vibrating pillow or flashing light can serve the same purpose for those who are hearing impaired.

The functions of the brain stem are summarized in Table 14.2.

✔ **CHECKPOINT**

6. Where are the medulla, pons, and midbrain located relative to one another?
7. What body functions are governed by nuclei in the brain stem?
8. List the functions of the reticular formation.

14.4 THE CEREBELLUM

OBJECTIVE

- Describe the structure and functions of the cerebellum.

The **cerebellum** (ser′-e-BEL-um), second only to the cerebrum in size, occupies the inferior and posterior aspects of the cranial cavity. Like the cerebrum, the cerebellum has a highly folded surface that greatly increases the surface area of its outer gray matter cortex, allowing for a greater number of neurons. The cerebellum accounts for about a tenth of the brain mass yet contains nearly half of the neurons in the brain. The cerebellum is posterior to the medulla and pons and inferior to the posterior portion of the cerebrum (see Figure 14.1). A deep groove known as the **transverse fissure,** along with the **tentorium cerebelli** (ten-TŌ-rē-um), which supports the posterior part of the cerebrum, separates the cerebellum from the cerebrum (see Figures 14.2b, 14.11b).

In superior or inferior views, the shape of the cerebellum resembles a butterfly. The central constricted area is the **vermis** (= worm), and the lateral "wings" or lobes are the **cerebellar hemispheres** (Figure 14.8a, b). Each hemisphere consists of lobes separated by deep and distinct fissures. The **anterior lobe** and **posterior lobe** govern subconscious aspects of skeletal muscle movements. The **flocculonodular lobe** (flok-ū-lō-NOD-ū-lar; *flocculo-* = wool-like tuft) on the inferior surface contributes to equilibrium and balance.

The superficial layer of the cerebellum, called the **cerebellar cortex,** consists of gray matter in a series of slender, parallel folds called **folia** (= leaves). Deep to the gray matter are tracts of white matter called **arbor vitae** (AR-bor VĪ-tē = tree of life) that resemble branches of a tree. Even deeper, within the white matter, are the **cerebellar nuclei,** regions of gray matter that give rise to axons carrying impulses from the cerebellum to other brain centers.

Three paired **cerebellar peduncles** (pe-DUNG-kuls) attach the cerebellum to the brain stem (see Figures 14.7a, 14.8b). These bundles of white matter consist of axons that conduct impulses between the cerebellum and other parts of the brain. The **superior cerebellar peduncles** contain axons that extend from the cerebellum to the red nuclei of the midbrain and to several nuclei of the thalamus. The **middle cerebellar peduncles** are the largest peduncles; their axons carry impulses for voluntary movements from the pontine nuclei (which receive input from motor areas of the cerebral cortex) into the cerebellum. The **inferior cerebellar peduncles** consist of (1) axons of the spinocerebellar tracts that carry sensory information into the cerebellum from proprioceptors in the trunk and limbs; (2) axons from the vestibular apparatus of the inner ear and from the vestibular nuclei of the medulla and pons that carry sensory information into the cerebellum from proprioceptors in the head; (3) axons from the inferior olivary nucleus of the medulla that enter the cerebellum and regulate the activity of cerebellar neurons; (4) axons that extend from the cerebellum to the vestibular nuclei of the medulla and pons; and (5) axons that extend from the cerebellum to the reticular formation.

Figure 14.8 Cerebellum.

The cerebellum coordinates skilled movements and regulates posture and balance.

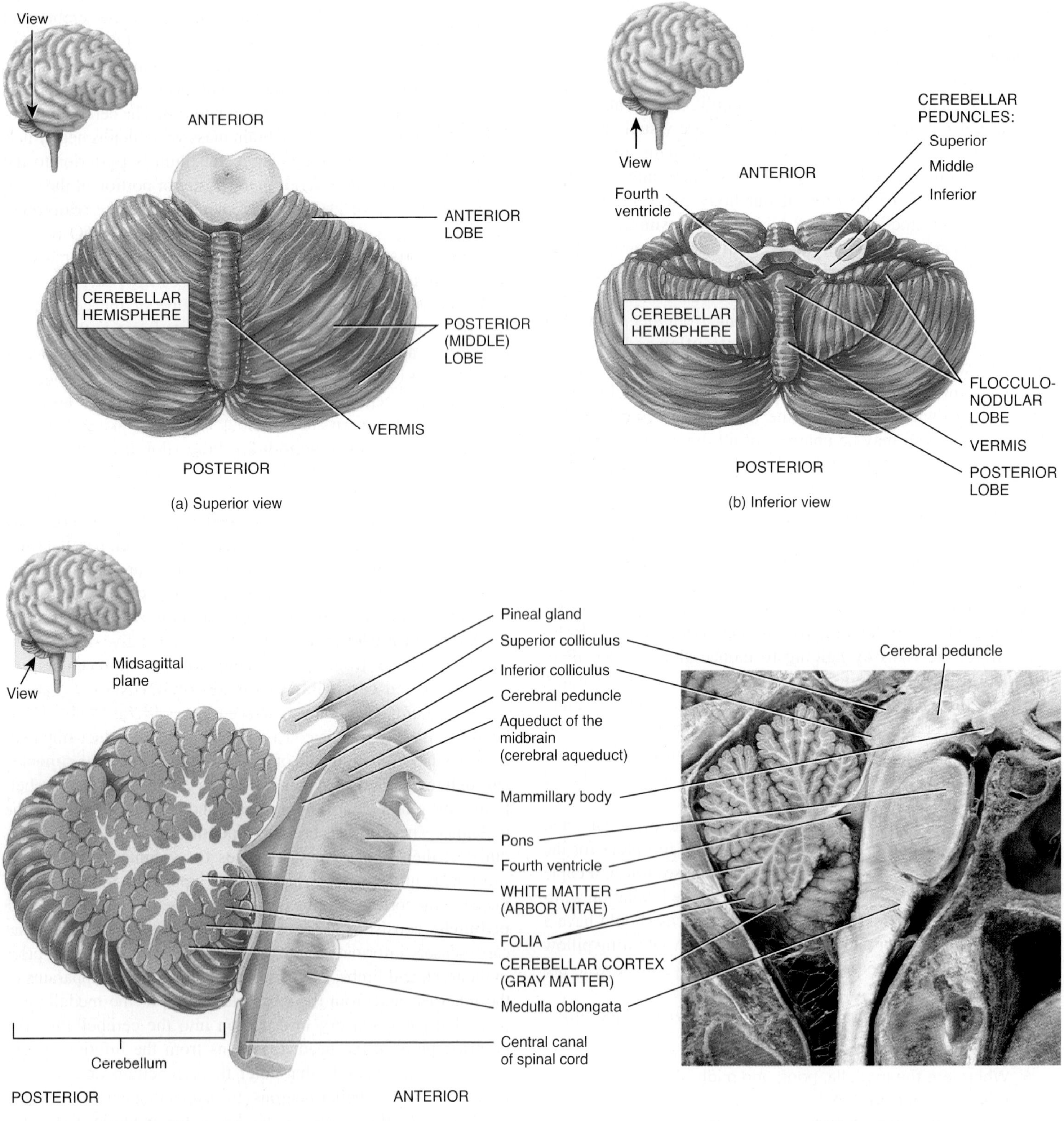

(a) Superior view

(b) Inferior view

(c) Midsagittal section of cerebellum and brain stem

(d) Midsagittal section

Which structures contain the axons that carry information into and out of the cerebellum?

The primary function of the cerebellum is to evaluate how well movements initiated by motor areas in the cerebrum are actually being carried out. When movements initiated by the cerebral motor areas are not being carried out correctly, the cerebellum detects the discrepancies. It then sends feedback signals to motor areas of the cerebral cortex, via its connections to the thalamus. The feedback signals help correct the errors, smooth the movements, and coordinate complex sequences of skeletal muscle contractions. Aside from this coordination of skilled movements, the cerebellum is the main brain region that regulates posture and balance. These aspects of cerebellar function make possible all skilled muscular activities, from catching a baseball to dancing to speaking. The presence of reciprocal connections between the cerebellum and association areas of the cerebral cortex suggests that the cerebellum may also have nonmotor functions such as cognition (acquisition of knowledge) and language processing. This view is supported by imaging studies using MRI and PET. Studies also suggest that the cerebellum may play a role in processing sensory information.

The functions of the cerebellum are summarized in Table 14.2.

CLINICAL CONNECTION | *Ataxia*

Damage to the cerebellum can result in a loss of ability to coordinate muscular movements, a condition called **ataxia** (a-TAK-sē-a; *a-* = without; *-taxia* = order). Blindfolded people with ataxia cannot touch the tip of their nose with a finger because they cannot coordinate movement with their sense of where a body part is located. Another sign of ataxia is a changed speech pattern due to uncoordinated speech muscles. Cerebellar damage may also result in staggering or abnormal walking movements. People who consume too much alcohol show signs of ataxia because alcohol inhibits activity of the cerebellum. Such individuals have difficulty in passing sobriety tests. Ataxia can also occur as a result of degenerative diseases (multiple sclerosis and Parkinson disease), trauma, brain tumors, and genetic factors, and as a side effect of medication prescribed for bipolar disorder. •

✔CHECKPOINT

9. Describe the location and principal parts of the cerebellum.
10. Where do the axons of each of the three pairs of cerebellar peduncles begin and end? What are their functions?

14.5 THE DIENCEPHALON

OBJECTIVE

- Describe the components and functions of the diencephalon (thalamus, hypothalamus, and epithalamus).

The **diencephalon** forms a central core of brain tissue just superior to the midbrain. It is almost completely surrounded by the cerebral hemispheres and contains numerous nuclei involved in a wide variety of sensory and motor processing between higher and lower brain centers. The diencephalon extends from the brain stem to the cerebrum and surrounds the third ventricle; it includes the thalamus, hypothalamus, and epithalamus. Projecting from the hypothalamus is the hypophysis, or pituitary gland. Portions of the diencephalon in the wall of the third ventricle are called circumventricular organs and will be discussed shortly. The optic tracts carrying neurons from the retina enter the diencephalon.

Thalamus

The **thalamus** (THAL-a-mus = inner chamber), which measures about 3 cm (1.2 in.) in length and makes up 80% of the diencephalon, consists of paired oval masses of gray matter organized into nuclei with interspersed tracts of white matter (Figure 14.9). A bridge of gray matter called the **intermediate mass (interthalamic adhesion)** joins the right and left halves of the thalamus in about 70% of human brains. A vertical Y-shaped sheet of white matter called the **internal medullary lamina** divides the gray matter of the right and left sides of the thalamus (Figure 14.9c). It consists of myelinated axons that enter and leave the various thalamic nuclei. Axons that connect the thalamus and cerebral cortex pass through the **internal capsule,** a thick band of white matter lateral to the thalamus (see Figure 14.13b).

The thalamus is the major relay station for most sensory impulses that reach the primary sensory areas of the cerebral cortex from the spinal cord and brain stem. In addition, the thalamus contributes to motor functions by transmitting information from the cerebellum and basal nuclei to the primary motor area of the cerebral cortex. The thalamus also relays nerve impulses between different areas of the cerebrum and plays a role in the maintenance of consciousness.

Based on their positions and functions, there are seven major groups of nuclei on each side of the thalamus (Figure 14.9c, d):

1. The **anterior nucleus** receives input from the hypothalamus and sends output to the limbic system (described in Section 14.6). It functions in emotions and memory.
2. The **medial nuclei** receive input from the limbic system and basal nuclei and send output to the cerebral cortex. They function in emotions, learning, memory, and cognition (thinking and knowing).
3. Nuclei in the **lateral group** receive input from the limbic system, superior colliculi, and cerebral cortex and send output to the cerebral cortex. The **lateral dorsal nucleus** functions in the expression of emotions. The **lateral posterior nucleus** and **pulvinar nucleus** help integrate sensory information.
4. Five nuclei are part of the **ventral group.** The **ventral anterior nucleus** receives input from the basal nuclei and sends output to motor areas of the cerebral cortex; it plays a role in movement control. The **ventral lateral nucleus** receives input from the cerebellum and basal nuclei and sends output to motor areas of the cerebral cortex; it also plays a role in movement control. The **ventral posterior nucleus** relays impulses for somatic sensations such as touch, pressure, vibration, itch,

Figure 14.9 Thalamus. Note the position of the thalamus in the lateral view (a) and in the medial view (b). The various thalamic nuclei shown in (c) and (d) are correlated by color to the cortical regions to which they project in (a) and (b).

The thalamus is the principal relay station for sensory impulses that reach the cerebral cortex from other parts of the brain and the spinal cord.

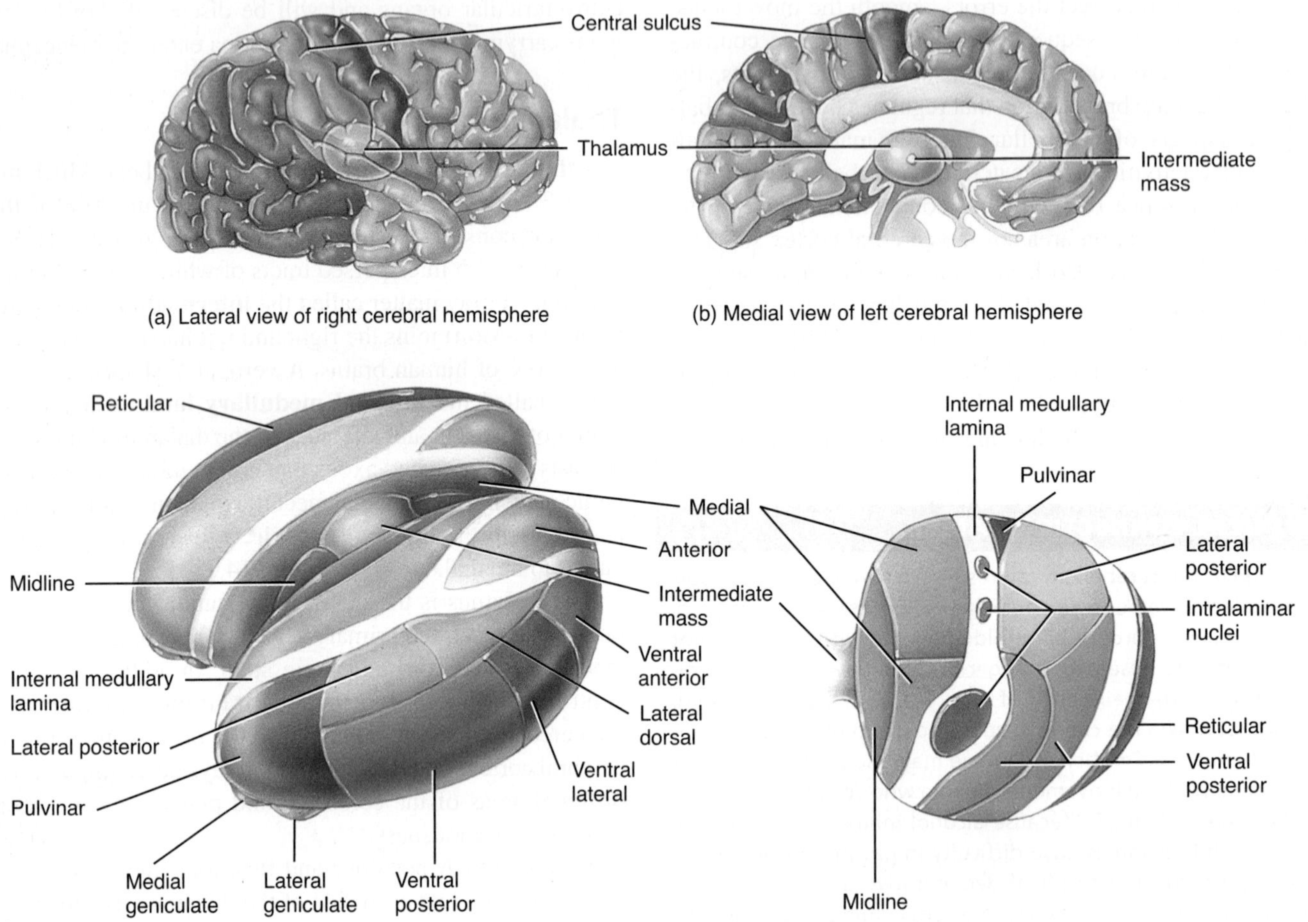

(c) Superolateral view of thalamus showing locations of thalamic nuclei (reticular nucleus is shown on the left side only; all other nuclei are shown on the right side)

(d) Transverse section of right side of thalamus showing locations of thalamic nuclei

What structure usually connects the right and left halves of the thalamus?

tickle, temperature, pain, and proprioception from the face and body to the cerebral cortex. The **lateral geniculate nucleus** (je-NIK-ū-lat = bent like a knee) relays visual impulses for sight from the retina to the primary visual area of the cerebral cortex. The **medial geniculate nucleus** relays auditory impulses for hearing from the ear to the primary auditory area of the cerebral cortex.

5. **Intralaminar nuclei** (in′-tra-LA-mi′-nar) lie within the internal medullary lamina and make connections with the reticular formation, cerebellum, basal nuclei, and wide areas of the cerebral cortex. They function in arousal (activation of the cerebral cortex from the brain stem reticular formation) and integration of sensory and motor information.
6. The **midline nucleus** forms a thin band adjacent to the third ventricle and has a presumed function in memory and olfaction.
7. The **reticular nucleus** surrounds the lateral aspect of the thalamus, next to the internal capsule. This nucleus monitors, filters, and integrates activities of other thalamic nuclei.

Hypothalamus

The **hypothalamus** (hī′-pō-THAL-a-mus; *hypo-* = under) is a small part of the diencephalon located inferior to the thalamus. It is composed of a dozen or so nuclei in four major regions:

1. The **mammillary region** (MAM-i-ler-ē; *mammill-* = nipple-shaped), adjacent to the midbrain, is the most posterior part of

the hypothalamus. It includes the *mammillary bodies* and *posterior hypothalamic nuclei* (Figure 14.10). The mammillary bodies are two small, rounded projections that serve as relay stations for reflexes related to the sense of smell.

2. The **tuberal region** (TOO-ber-al), the widest part of the hypothalamus, includes the *dorsomedial nucleus, ventromedial nucleus,* and *arcuate nucleus* (AR-kū-āt), plus the stalklike **infundibulum** (in-fun-DIB-ū-lum = funnel), which connects the pituitary gland to the hypothalamus (Figure 14.10). The **median eminence** is a slightly raised region that encircles the infundibulum (see Figure 14.7a).
3. The **supraoptic region** (*supra-* = above; *-optic* = eye) lies superior to the optic chiasm (point of crossing of optic nerves) and contains the *paraventricular nucleus, supraoptic nucleus, anterior hypothalamic nucleus,* and *suprachiasmatic nucleus* (soo′-pra-kī′-az-MA-tik) (Figure 14.10). Axons from the paraventricular and supraoptic nuclei form the hypothalamohypophyseal tract (hī′-pō-thal′-a-mō-hī-pō-FIZ-ē-al), which extends through the infundibulum to the posterior lobe of the pituitary (see Figure 18.8).
4. The **preoptic region** anterior to the supraoptic region is usually considered part of the hypothalamus because it participates with the hypothalamus in regulating certain autonomic activities. The preoptic region contains the *medial* and *lateral preoptic nuclei* (Figure 14.10).

The hypothalamus controls many body activities and is one of the major regulators of homeostasis. Sensory impulses related to both somatic and visceral senses arrive at the hypothalamus, as do impulses from receptors for vision, taste, and smell. Other receptors within the hypothalamus itself continually monitor osmotic pressure, blood glucose level, certain hormone concentrations, and the temperature of blood. The hypothalamus has several very important connections with the pituitary gland and produces a variety of hormones, which are described in more detail in Chapter 18. Some functions can be attributed to specific hypothalamic nuclei, but others are not so precisely localized. Important functions of the hypothalamus include the following:

- ***Control of the ANS.*** The hypothalamus controls and integrates activities of the autonomic nervous system, which regulates contraction of smooth muscle and cardiac muscle and the secretions of many glands. Axons extend from the hypothalamus to parasympathetic and sympathetic nuclei in the brain stem and spinal cord. Through the ANS, the hypothalamus is a major regulator of visceral activities, including regulation of heart rate, movement of food through the gastrointestinal tract, and contraction of the urinary bladder.

Figure 14.10 Hypothalamus. Selected portions of the hypothalamus and a three-dimensional representation of hypothalamic nuclei are shown (after Netter).

6 The hypothalamus controls many body activities and is an important regulator of homeostasis.

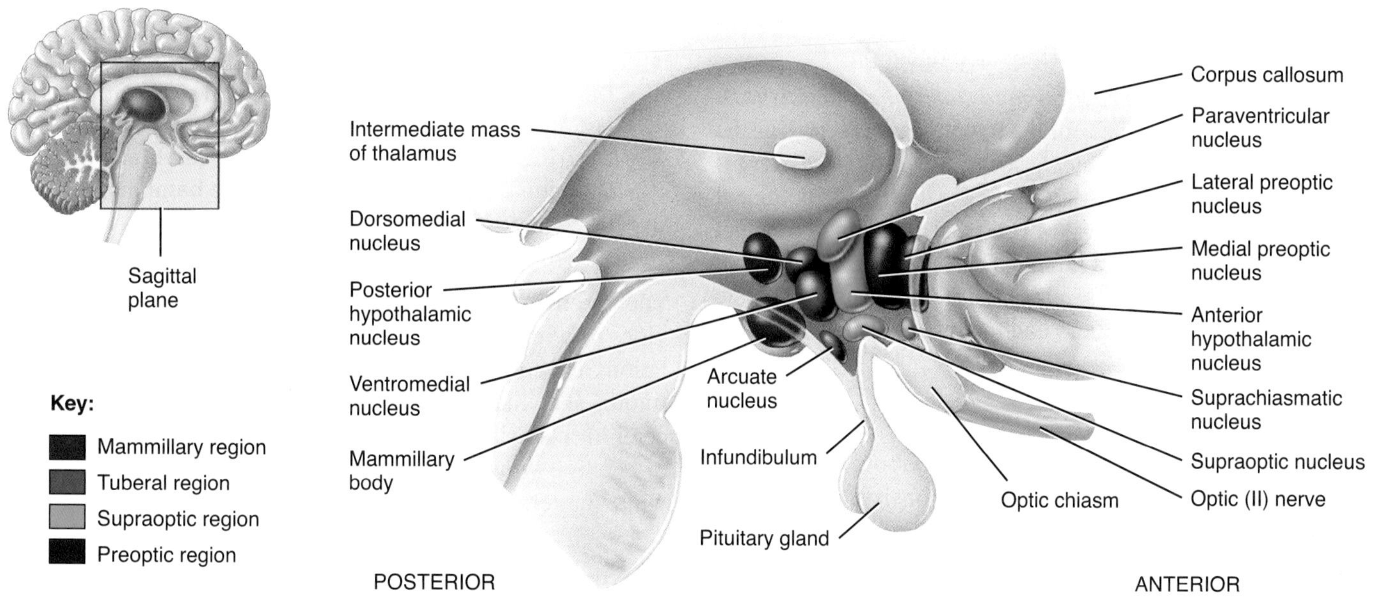

Sagittal section of brain showing hypothalamic nuclei

? What are the four major regions of the hypothalamus, from posterior to anterior?

- ***Production of hormones.*** The hypothalamus produces several hormones and has two types of important connections with the pituitary gland, an endocrine gland located inferior to the hypothalamus (see Figure 14.1). First, hypothalamic hormones known as *releasing hormones* and *inhibiting hormones* are released into capillary networks in the median eminence (see Figure 18.5). The bloodstream carries these hormones directly to the anterior lobe of the pituitary, where they stimulate or inhibit secretion of anterior pituitary hormones. Second, axons extend from the paraventricular and supraoptic nuclei through the infundibulum into the posterior lobe of the pituitary (see Figure 18.8). The cell bodies of these neurons make one of two hormones (*oxytocin* or *antidiuretic hormone*). Their axons transport the hormones to the posterior pituitary, where they are released.
- ***Regulation of emotional and behavioral patterns.*** Together with the limbic system (described shortly), the hypothalamus participates in expressions of rage, aggression, pain, and pleasure, and the behavioral patterns related to sexual arousal.
- ***Regulation of eating and drinking.*** The hypothalamus regulates food intake. It contains a **feeding center,** which promotes eating, and a **satiety center,** which causes a sensation of fullness and cessation of eating. The hypothalamus also contains a **thirst center.** When certain cells in the hypothalamus are stimulated by rising osmotic pressure of the extracellular fluid, they cause the sensation of thirst. The intake of water by drinking restores the osmotic pressure to normal, removing the stimulation and relieving the thirst.
- ***Control of body temperature.*** The hypothalamus also functions as the body's **thermostat.** If the temperature of blood flowing through the hypothalamus is above normal, the hypothalamus directs the autonomic nervous system to stimulate activities that promote heat loss. When blood temperature is below normal, by contrast, the hypothalamus generates impulses that promote heat production and retention.
- ***Regulation of circadian rhythms and states of consciousness.*** The suprachiasmatic nucleus of the hypothalamus serves as the body's internal biological clock because it establishes **circadian (daily) rhythms** (ser-KĀ-dē-an), patterns of biological activity (such as the sleep–wake cycle) that occur on a circadian schedule (cycle of about 24 hours). This nucleus receives input from the eyes (retina) and sends output to other hypothalamic nuclei, the reticular formation, and the pineal gland.

Epithalamus

The **epithalamus** (ep′-i-THAL-a-mus; *epi-* = above), a small region superior and posterior to the thalamus, consists of the pineal gland and habenular nuclei. The **pineal gland** (PĪN-ē-al = pineconelike) is about the size of a small pea and protrudes from the posterior midline of the third ventricle (see Figure 14.1). The pineal gland is part of the endocrine system because it secretes the hormone **melatonin.** As more melatonin is liberated during darkness than in light, this hormone is thought to promote sleepiness. When taken orally, melatonin also appears to contribute to the setting of the body's biological clock by inducing sleep and helping the body to adjust to jet lag. The **habenular nuclei** (ha-BEN-ū-lar), shown in Figure 14.7a, are involved in olfaction, especially emotional responses to odors such as a loved one's cologne or Mom's chocolate chip cookies baking in the oven.

The functions of the three parts of the diencephalon are summarized in Table 14.2.

Circumventricular Organs

Parts of the diencephalon, called **circumventricular** (ser′-kum-ven-TRIK-ū-lar) **organs (CVOs)** because they lie in the wall of the third ventricle, can monitor chemical changes in the blood because they lack a blood–brain barrier. CVOs include part of the hypothalamus, the pineal gland, the pituitary gland, and a few other nearby structures. Functionally, these regions coordinate homeostatic activities of the endocrine and nervous systems, such as the regulation of blood pressure, fluid balance, hunger, and thirst. CVOs are also thought to be the sites of entry into the brain of HIV, the virus that causes AIDS. Once in the brain, HIV may cause dementia (irreversible deterioration of mental state) and other neurological disorders.

✔ **CHECKPOINT**

11. Why is the thalamus considered a "relay station" in the brain?
12. Why is the hypothalamus considered part of both the nervous system and the endocrine system?
13. What are the functions of the epithalamus?
14. Define a circumventricular organ.

14.6 THE CEREBRUM

OBJECTIVES

- Describe the cortex, gyri, fissures, and sulci of the cerebrum.
- List and locate the lobes of the cerebrum.
- Describe the tracts that compose the cerebral white matter.
- Describe the nuclei that compose the basal nuclei.
- List the structures and describe the functions of the limbic system.

The **cerebrum** (se-RĒ-brum) is the "seat of intelligence." It provides us with the ability to read, write, and speak; to make calculations and compose music; and to remember the past, plan for the future, and imagine things that have never existed before. The cerebrum consists of an outer cerebral cortex, an internal region of cerebral white matter, and gray matter nuclei deep within the white matter.

Cerebral Cortex

The **cerebral cortex** (*cortex* = rind or bark) is a region of gray matter that forms the outer rim of the cerebrum (Figure 14.11a). Although only 2–4 mm (0.08–0.16 in.) thick, the cerebral cortex contains billions of neurons arranged in layers. During embryonic development, when brain size increases rapidly, the gray matter of the cortex enlarges much faster than the deeper white matter. As a result, the cortical region rolls and folds on itself. The folds are called **gyri** (JĪ-rī = circles; singular is *gyrus*) or **convolutions** (kon′-vō-LOO-shuns) (Figure 14.11a, b). The deepest grooves between folds are known as **fissures;** the shallower grooves between folds are termed **sulci** (SUL-sī = grooves; singular is *sulcus*). The most prominent fissure, the **longitudinal fissure,** separates the cerebrum into right and left halves called **cerebral**

Figure 14.11 Cerebrum. Because the insula cannot be seen externally, it has been projected to the surface in (b).

The cerebrum is the "seat of intelligence"; it provides us with the ability to read, write, and speak; to make calculations and compose music; to remember the past and plan for the future; and to create.

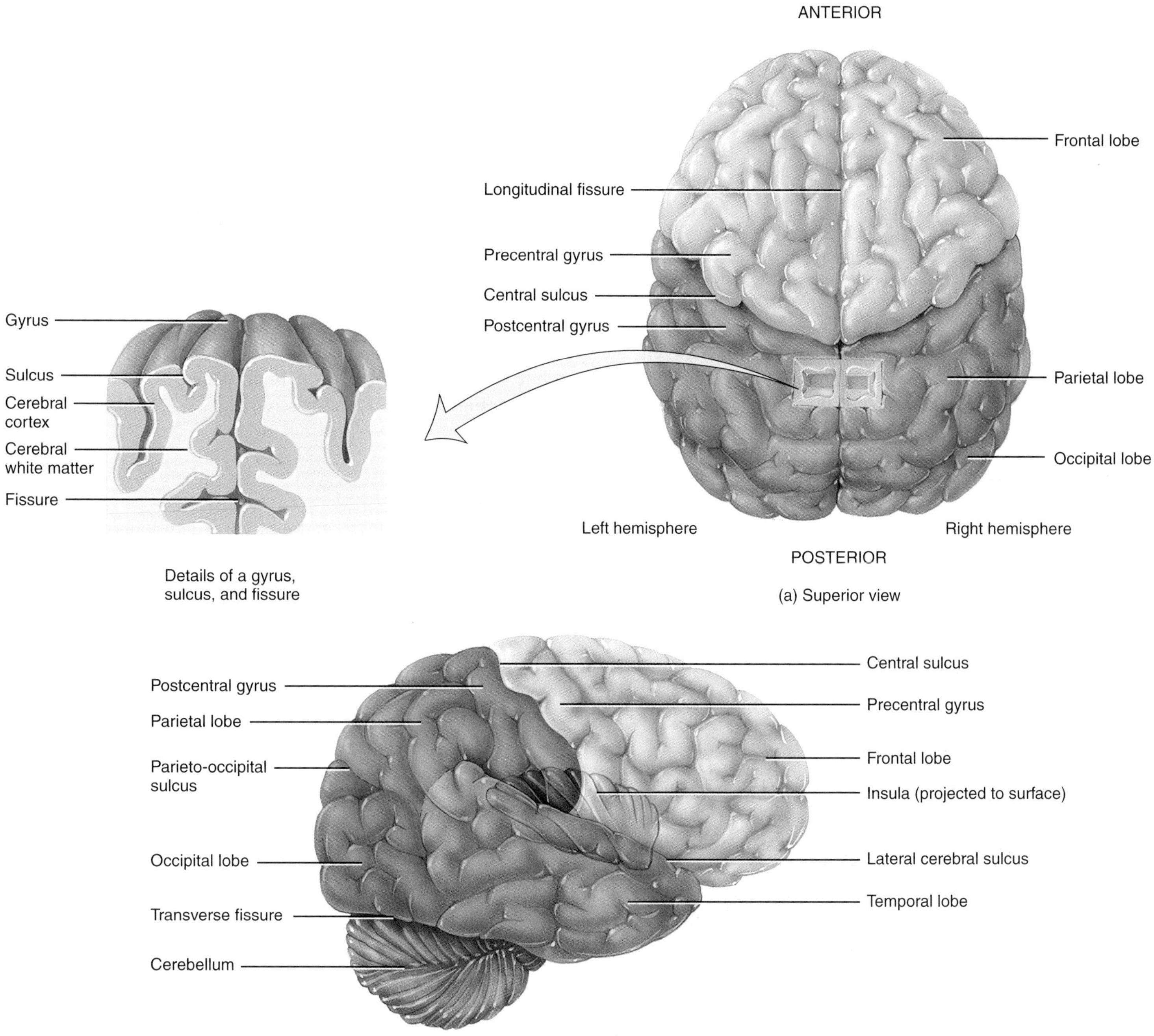
ANTERIOR
Frontal lobe
Longitudinal fissure
Precentral gyrus
Central sulcus
Postcentral gyrus
Parietal lobe
Occipital lobe
Left hemisphere
Right hemisphere
POSTERIOR
Gyrus
Sulcus
Cerebral cortex
Cerebral white matter
Fissure
Details of a gyrus, sulcus, and fissure
Postcentral gyrus
Parietal lobe
Parieto-occipital sulcus
Occipital lobe
Transverse fissure
Cerebellum
Central sulcus
Precentral gyrus
Frontal lobe
Insula (projected to surface)
Lateral cerebral sulcus
Temporal lobe

(a) Superior view

(b) Right lateral view

FIGURE 14.11 CONTINUES

FIGURE 14.11 CONTINUED

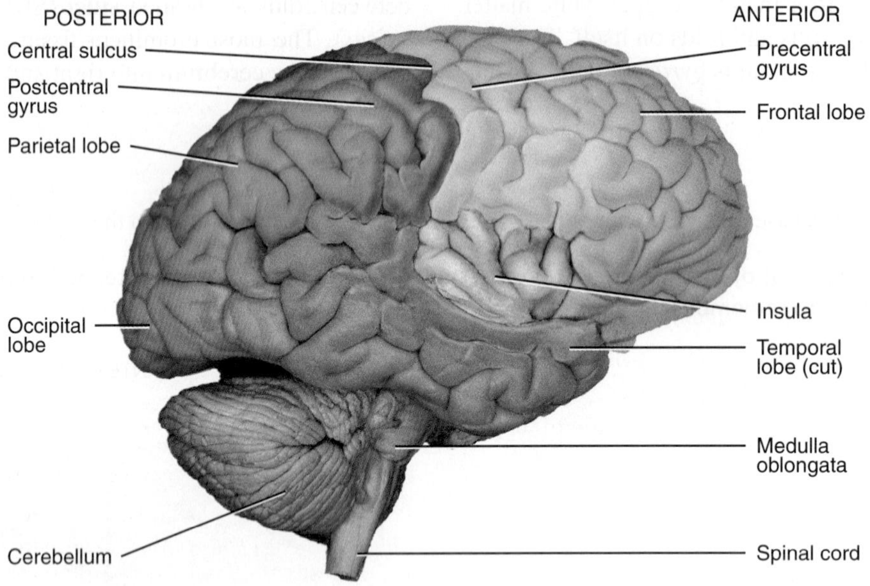
POSTERIOR
ANTERIOR
Central sulcus
Postcentral gyrus
Parietal lobe
Occipital lobe
Cerebellum
Precentral gyrus
Frontal lobe
Insula
Temporal lobe (cut)
Medulla oblongata
Spinal cord

(c) Right lateral view with temporal lobe cut away

During development, does the gray matter or the white matter enlarge more rapidly? What are the brain folds, shallow grooves, and deep grooves called?

hemispheres. Within the longitudinal fissure between the cerebral hemispheres is the falx cerebri. The cerebral hemispheres are connected internally by the **corpus callosum** (kal-LŌ-sum; *corpus* = body; *callosum* = hard), a broad band of white matter containing axons that extend between the hemispheres (Figure 14.12).

Figure 14.12 Organization of white matter tracts of the left cerebral hemisphere.

Association tracts, commissural tracts, and projection tracts form white matter tracts in the cerebral hemispheres.

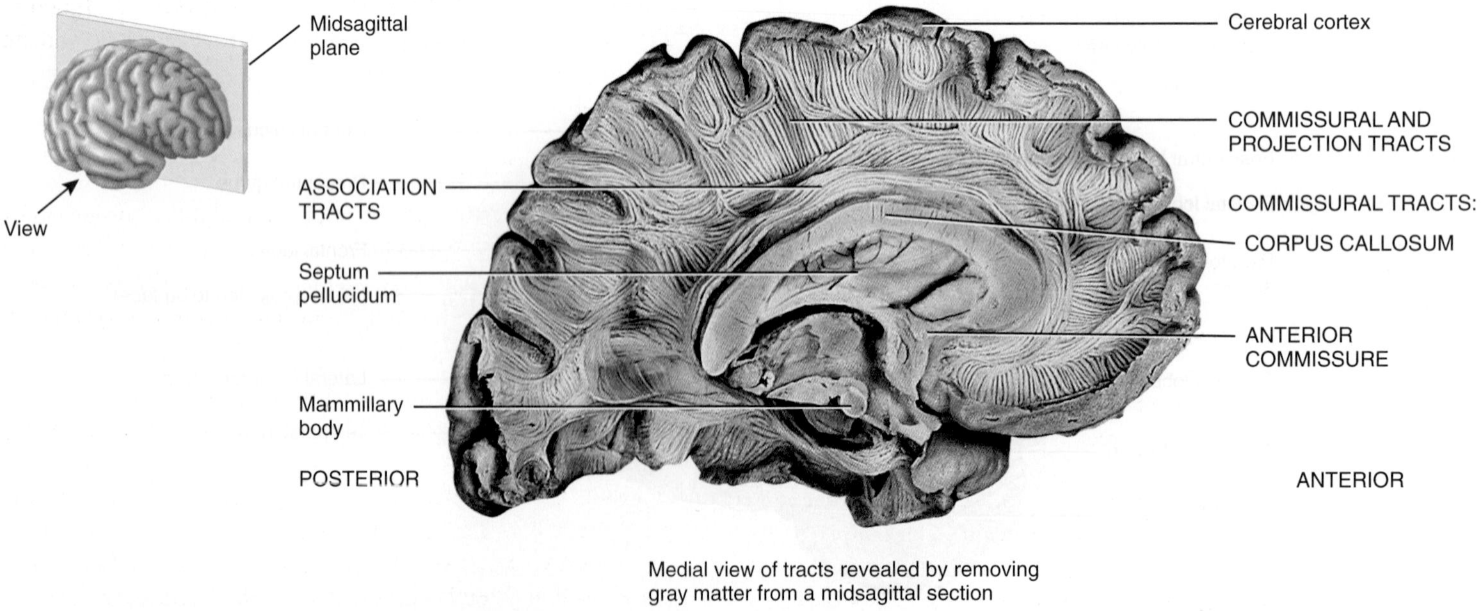
Midsagittal plane
View
ASSOCIATION TRACTS
Septum pellucidum
Mammillary body
POSTERIOR
Cerebral cortex
COMMISSURAL AND PROJECTION TRACTS
COMMISSURAL TRACTS:
CORPUS CALLOSUM
ANTERIOR COMMISSURE
ANTERIOR

Medial view of tracts revealed by removing gray matter from a midsagittal section

Which tracts carry impulses between gyri of the same hemisphere? Between gyri in opposite hemispheres? Between the cerebrum and thalamus, brain stem, and spinal cord?

Lobes of the Cerebrum

Each cerebral hemisphere can be further subdivided into several lobes. The lobes are named after the bones that cover them: frontal, parietal, temporal, and occipital lobes (see Figure 14.11a, b). The **central sulcus** (SUL-kus) separates the **frontal lobe** from the **parietal lobe.** A major gyrus, the **precentral gyrus**—located immediately anterior to the central sulcus—contains the primary motor area of the cerebral cortex. Another major gyrus, the **postcentral gyrus,** which is located immediately posterior to the central sulcus, contains the primary somatosensory area of the cerebral cortex. The **lateral cerebral sulcus (fissure)** separates the **frontal lobe** from the **temporal lobe.** The **parieto-occipital sulcus** separates the **parietal lobe** from the **occipital lobe.** A fifth part of the cerebrum, the **insula,** cannot be seen at the surface of the brain because it lies within the lateral cerebral sulcus, deep to the parietal, frontal, and temporal lobes (Figure 14.11b).

Cerebral White Matter

The **cerebral white matter** consists primarily of myelinated axons in three types of tracts (Figure 14.12):

1. **Association tracts** contain axons that conduct nerve impulses between gyri in the same hemisphere.
2. **Commissural tracts** (kom′-i-SYUR-al) contain axons that conduct nerve impulses from gyri in one cerebral hemisphere to corresponding gyri in the other cerebral hemisphere. Three important groups of commissural tracts are the **corpus callosum** (the largest fiber bundle in the brain, containing about 300 million fibers), **anterior commissure,** and **posterior commissure.**
3. **Projection tracts** contain axons that conduct nerve impulses from the cerebrum to lower parts of the CNS (thalamus, brain stem, or spinal cord) or from lower parts of the CNS to the cerebrum. An example is the **internal capsule,** a thick band of white matter that contains both ascending and descending axons (see Figure 14.13).

Basal Nuclei

Deep within each cerebral hemisphere are three nuclei (masses of gray matter) that are collectively termed the **basal nuclei** (Figure 14.13). (Historically, these nuclei have been called the basal *ganglia.* However, this is a misnomer because a *ganglion* is an aggregate of neuronal cell bodies in the peripheral nervous system. While both terms still appear in the literature, we use *nuclei,* as this is the correct term as determined by the *Terminologia Anatomica,* the final say on correct anatomical terminology.)

Two of the basal nuclei are side-by-side, just lateral to the thalamus. They are the **globus pallidus** (GLŌ-bus PAL-i-dus; *globus* = ball; *pallidus* = pale), which is closer to the thalamus, and the **putamen** (pū-TĀ-men = shell), which is closer to the cerebral cortex. Together, the globus pallidus and putamen are referred to as the **lentiform nucleus** (LEN-ti-form = shaped like a lens). The third of the basal nuclei is the **caudate nucleus** (KAW-dāt; *caud-* = tail), which has a large "head" connected to a smaller "tail" by a long comma-shaped "body." Together, the lentiform and caudate nuclei are known as the **corpus striatum** (strī-Ā-tum; *corpus* = body; *striatum* = striated). The term corpus striatum refers to the striated (striped) appearance of the internal capsule as it passes among the basal nuclei. Nearby structures that are functionally linked to the basal nuclei are the *substantia nigra* of the midbrain and the *subthalamic nuclei* of the diencephalon (see Figures 14.7b, 14.13b). Axons from the substantia nigra terminate in the caudate nucleus and putamen. The subthalamic nuclei interconnect with the globus pallidus.

The **claustrum** (KLAWS-trum) is a thin sheet of gray matter situated lateral to the putamen. It is considered by some to be a subdivision of the basal nuclei. The function of the claustrum in humans has not been clearly defined, but it may be involved in visual attention.

The basal nuclei receive input from the cerebral cortex and provide output to motor parts of the cortex via the medial and ventral group nuclei of the thalamus. In addition, the basal nuclei have extensive connections with one another. A major function of the basal nuclei is to help regulate initiation and termination of movements. Activity of neurons in the putamen precedes or anticipates body movements; activity of neurons in the caudate nucleus occurs prior to eye movements. The globus pallidus helps regulate the muscle tone required for specific body movements. The basal nuclei also control subconscious contractions of skeletal muscles. Examples include automatic arm swings while walking and true laughter in response to a joke (not the kind you consciously initiate to humor your A&P instructor).

In addition to influencing motor functions, the basal nuclei have other roles. They help initiate and terminate some cognitive processes, such as attention, memory, and planning, and may act with the limbic system to regulate emotional behaviors. Disorders such as Parkinson disease, obsessive–compulsive disorder, schizophrenia, and chronic anxiety are thought to involve dysfunction of circuits between the basal nuclei and the limbic system and are described in more detail in Chapter 16.

The Limbic System

Encircling the upper part of the brain stem and the corpus callosum is a ring of structures on the inner border of the cerebrum and floor of the diencephalon that constitutes the **limbic system** (*limbic* = border). The main components of the limbic system are as follows (Figure 14.14):

- The so-called **limbic lobe** is a rim of cerebral cortex on the medial surface of each hemisphere. It includes the **cingulate gyrus** (SIN-gyu-lat; *cingul-* = belt), which lies above the corpus callosum, and the **parahippocampal gyrus** (par′-a-hip-ō-KAM-pal), which is in the temporal lobe below. The **hippocampus** (hip′-ō-KAM-pus = seahorse) is a portion of the parahippocampal gyrus that extends into the floor of the lateral ventricle.
- The **dentate gyrus** (*dentate* = toothed) lies between the hippocampus and parahippocampal gyrus.

Figure 14.13 Basal nuclei. In (a) the basal nuclei have been projected to the surface; in both (a) and (b) they are shown in purple.

The basal nuclei help initiate and terminate movements, suppress unwanted movements, and regulate muscle tone.

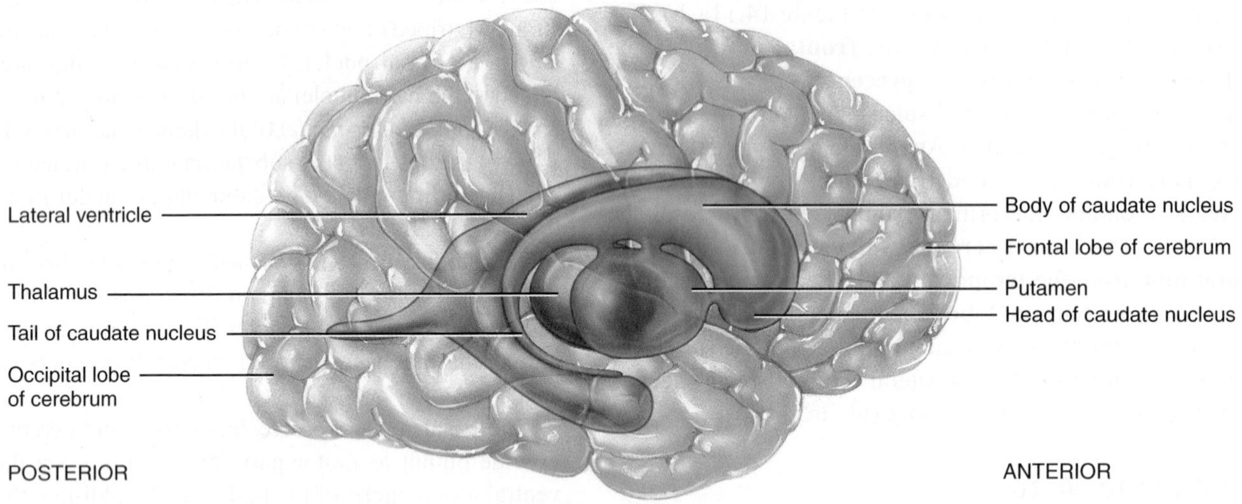

Lateral ventricle
Thalamus
Tail of caudate nucleus
Occipital lobe of cerebrum
Body of caudate nucleus
Frontal lobe of cerebrum
Putamen
Head of caudate nucleus
POSTERIOR
ANTERIOR

(a) Lateral view of right side of brain

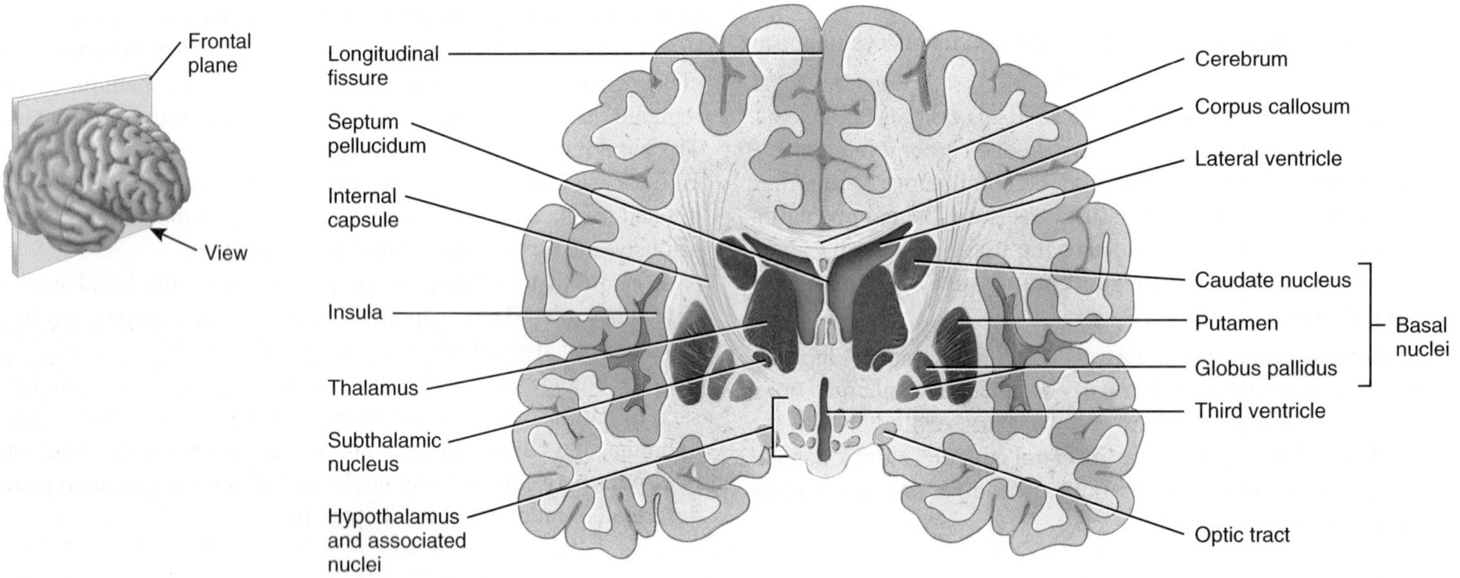

Frontal plane
View
Longitudinal fissure
Septum pellucidum
Internal capsule
Insula
Thalamus
Subthalamic nucleus
Hypothalamus and associated nuclei
Cerebrum
Corpus callosum
Lateral ventricle
Caudate nucleus
Putamen
Globus pallidus
Basal nuclei
Third ventricle
Optic tract

(b) Anterior view of frontal section

Where are the basal nuclei located relative to the thalamus?

- The **amygdala** (a-MIG-da-la; *amygda-* = almond-shaped) is composed of several groups of neurons located close to the tail of the caudate nucleus.
- The **septal nuclei** are located within the septal area formed by the regions under the corpus callosum and the paraterminal gyrus (a cerebral gyrus).
- The **mammillary bodies** of the hypothalamus are two round masses close to the midline near the cerebral peduncles.
- Two nuclei of the thalamus, the **anterior nucleus** and the **medial nucleus,** participate in limbic circuits (see Figure 14.9c, d).
- The **olfactory bulbs** are flattened bodies of the olfactory pathway that rest on the cribriform plate.
- The **fornix, stria terminalis, stria medullaris, medial forebrain bundle,** and **mammillothalamic tract** (mam-i-lō-tha-LAM-ik) are linked by bundles of interconnecting myelinated axons.

The limbic system is sometimes called the "emotional brain" because it plays a primary role in a range of emotions, including pain, pleasure, docility, affection, and anger. It also is involved in olfaction (smell) and memory. Experiments have shown that when different areas of animals' limbic systems are stimulated, the animals' reactions indicate that they are experiencing intense pain or extreme pleasure. Stimulation of other limbic system areas in animals produces tameness and signs of affection.

Figure 14.14 Components of the limbic system (shaded green) and surrounding structures.

The limbic system governs emotional aspects of behavior.

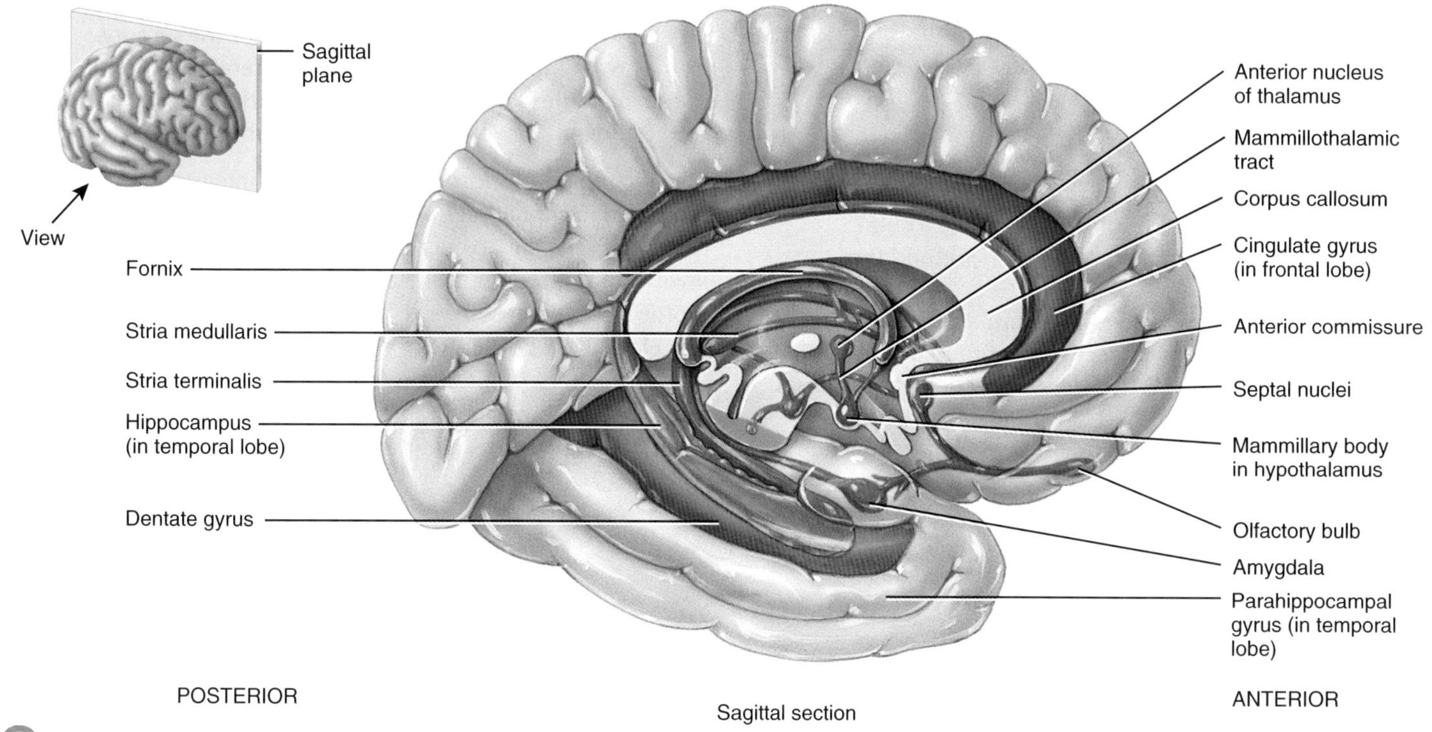
Sagittal plane
View
Fornix
Stria medullaris
Stria terminalis
Hippocampus (in temporal lobe)
Dentate gyrus
POSTERIOR
Sagittal section
Anterior nucleus of thalamus
Mammillothalamic tract
Corpus callosum
Cingulate gyrus (in frontal lobe)
Anterior commissure
Septal nuclei
Mammillary body in hypothalamus
Olfactory bulb
Amygdala
Parahippocampal gyrus (in temporal lobe)
ANTERIOR

Which part of the limbic system functions with the cerebrum in memory?

Stimulation of a cat's amygdala or certain nuclei of the hypothalamus produces a behavioral pattern called rage—the cat extends its claws, raises its tail, opens its eyes wide, hisses, and spits. By contrast, removal of the amygdala produces an animal that lacks fear and aggression. Likewise, a person whose amygdala is damaged fails to recognize fearful expressions in others or to express fear in situations where this emotion would normally be appropriate, for example, while being attacked by an animal.

Together with parts of the cerebrum, the limbic system also functions in memory; damage to the limbic system causes memory impairment. One portion of the limbic system, the hippocampus, is seemingly unique among structures of the central nervous system—it has cells reported to be capable of mitosis. Thus, the portion of the brain that is responsible for some aspects of memory may develop new neurons, even in the elderly.

The functions of the cerebrum are summarized in Table 14.2.

CLINICAL CONNECTION | *Brain Injuries*

Brain injuries are commonly associated with head trauma and result in part from displacement and distortion of neural tissue at the moment of impact. Additional tissue damage may occur when normal blood flow is restored after a period of ischemia (reduced blood flow). The sudden increase in oxygen level produces large numbers of oxygen free radicals (charged oxygen molecules with an unpaired electron). Brain cells recovering from the effects of a stroke or cardiac arrest also release free radicals. Free radicals cause damage by disrupting cellular DNA and enzymes and by altering plasma membrane permeability. Brain injuries can also result from hypoxia (cellular oxygen deficiency).

Various degrees of brain injury are described by specific terms. A **concussion** (kon-KUSH-un) is an injury characterized by an abrupt, but temporary, loss of consciousness (from seconds to hours), disturbances of vision, and problems with equilibrium. It is caused by a blow to the head or the sudden stopping of a moving head (as in an automobile accident) and is the most common brain injury. A concussion produces no obvious bruising of the brain. Signs of a concussion are headache, drowsiness, nausea and/or vomiting, lack of concentration, confusion, or post-traumatic amnesia (memory loss).

A brain **contusion** (kon-TOO-zhun) is bruising due to trauma and includes the leakage of blood from microscopic vessels. It is usually associated with a concussion. In a contusion, the pia mater may be torn, allowing blood to enter the subarachnoid space. The area most commonly affected is the frontal lobe. A contusion usually results in an immediate loss of consciousness (generally lasting no longer than 5 minutes), loss of reflexes, transient cessation of respiration, and decreased blood pressure. Vital signs typically stabilize in a few seconds.

A **laceration** (las-er-Ā-shun) is a tear of the brain, usually from a skull fracture or a gunshot wound. A laceration results in rupture of large blood vessels, with bleeding into the brain and subarachnoid space. Consequences include cerebral hematoma (localized pool of blood, usually clotted, that swells against the brain tissue), edema, and increased intracranial pressure. If the blood clot is small enough, it may pose no major threat and may be absorbed. If the blood clot is large, it may require surgical removal. Swelling infringes on the limited space that the brain occupies in the cranial cavity. Swelling causes excruciating headaches. Brain tissue can also undergo *necrosis* (cellular death) due to the swelling; if the swelling is severe enough, the brain can herniate through the foramen magnum, resulting in death. •

✔CHECKPOINT

15. List and locate the lobes of the cerebrum. How are they separated from one another? What is the insula?
16. Distinguish between the precentral gyrus and the postcentral gyrus.
17. Describe the organization of cerebral white matter and indicate the function of each major group of fibers.
18. List the basal nuclei. What are the functions of the basal nuclei?
19. Define the limbic system and list several of its functions.

14.7 FUNCTIONAL ORGANIZATION OF THE CEREBRAL CORTEX

OBJECTIVES

- Describe the locations and functions of the sensory, association, and motor areas of the cerebral cortex.
- Explain the significance of hemispheric lateralization.
- Define brain waves and indicate their significance.

Specific types of sensory, motor, and integrative signals are processed in certain regions of the cerebral cortex (Figure 14.15). Generally, **sensory areas** receive sensory information and are involved in **perception,** the conscious awareness of a sensation; **motor areas** control the execution of voluntary movements; and **association areas** deal with more complex integrative functions such as memory, emotions, reasoning, will, judgment, personality traits, and intelligence. In this section we will also discuss hemispheric lateralization and brain waves.

Figure 14.15 **Functional areas of the cerebrum.** Broca's speech area and Wernicke's area are in the left cerebral hemisphere of most people; they are shown here to indicate their relative locations. The numbers, still used today, are from K. Brodmann's map of the cerebral cortex, first published in 1909.

Particular areas of the cerebral cortex process sensory, motor, and integrative signals.

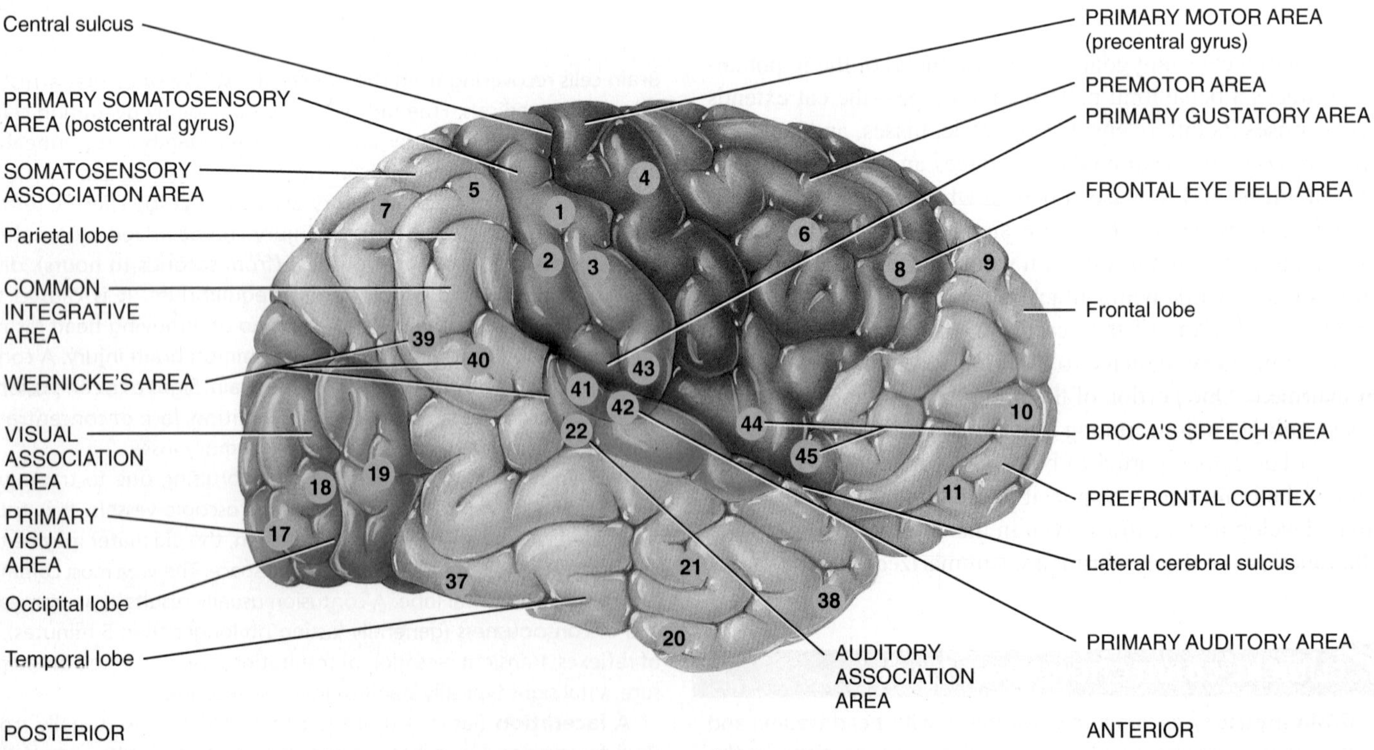

Lateral view of right cerebral hemisphere

? **What area(s) of the cerebrum integrate(s) interpretation of visual, auditory, and somatic sensations? Translates thoughts into speech? Controls skilled muscular movements? Interprets sensations related to taste? Interprets pitch and rhythm? Interprets shape, color, and movement of objects? Controls voluntary scanning movements of the eyes?**

Sensory Areas

Sensory impulses arrive mainly in the posterior half of both cerebral hemispheres, in regions behind the central sulci. In the cerebral cortex, primary sensory areas receive sensory information that has been relayed from peripheral sensory receptors through lower regions of the brain. Sensory association areas often are adjacent to the primary areas. They usually receive input both from the primary areas and from other brain regions. Sensory association areas integrate sensory experiences to generate meaningful patterns of recognition and awareness. For example, a person with damage in the *primary* visual area would be blind in at least part of his visual field, but a person with damage to a visual *association* area might see normally yet be unable to recognize ordinary objects such as a lamp or a toothbrush just by looking at them.

The following are some important sensory areas (Figure 14.15; the significance of the numbers in parentheses is explained in the figure caption):

- The **primary somatosensory area** (areas 1, 2, and 3) is located directly posterior to the central sulcus of each cerebral hemisphere in the postcentral gyrus of each parietal lobe. It extends from the lateral cerebral sulcus, along the lateral surface of the parietal lobe to the longitudinal fissure, and then along the medial surface of the parietal lobe within the longitudinal fissure. The primary somatosensory area receives nerve impulses for touch, pressure, vibration, itch, tickle, temperature (coldness and warmth), pain, and proprioception (joint and muscle position) and is involved in the perception of these somatic sensations. A "map" of the entire body is present in the primary somatosensory area: Each point within the area receives impulses from a specific part of the body (see Figure 16.8a). The size of the cortical area receiving impulses from a particular part of the body depends on the number of receptors present there rather than on the size of the body part. For example, a larger region of the somatosensory area receives impulses from the lips and fingertips than from the thorax or hip. This distorted somatic sensory map of the body is known as the **sensory homunculus** (*homunculus* = little man). The primary somatosensory area allows you to pinpoint where somatic sensations originate, so that you know exactly where on your body to swat that mosquito.
- The **primary visual area** (area 17), located at the posterior tip of the occipital lobe mainly on the medial surface (next to the longitudinal fissure), receives visual information and is involved in visual perception.
- The **primary auditory area** (areas 41 and 42), located in the superior part of the temporal lobe near the lateral cerebral sulcus, receives information for sound and is involved in auditory perception.
- The **primary gustatory area** (area 43), located at the base of the postcentral gyrus superior to the lateral cerebral sulcus in the parietal cortex, receives impulses for taste and is involved in gustatory perception and taste discrimination.
- The **primary olfactory area** (area 28), located in the temporal lobe on the medial aspect (and thus not visible in Figure 14.15), receives impulses for smell and is involved in olfactory perception.

Motor Areas

Motor output from the cerebral cortex flows mainly from the anterior part of each hemisphere. Among the most important motor areas are the following (Figure 14.15):

- The **primary motor area** (area 4) is located in the precentral gyrus of the frontal lobe. As is true for the primary somatosensory area, a "map" of the entire body is present in the primary motor area: Each region within the area controls voluntary contractions of specific muscles or groups of muscles (see Figure 16.8b). Electrical stimulation of any point in the primary motor area causes contraction of specific skeletal muscle fibers on the opposite side of the body. Different muscles are represented unequally in the primary motor area. More cortical area is devoted to those muscles involved in skilled, complex, or delicate movement. For instance, the cortical region devoted to muscles that move the fingers is much larger than the region for muscles that move the toes. This distorted muscle map of the body is called the **motor homunculus.**
- **Broca's speech area** (BRŌ-kaz) (areas 44 and 45) is located in the frontal lobe close to the lateral cerebral sulcus. Speaking and understanding language are complex activities that involve several sensory, association, and motor areas of the cortex. In about 97% of the population, these language areas are localized in the *left* hemisphere. The planning and production of speech occur in the *left* frontal lobe in most people. From Broca's speech area, nerve impulses pass to the premotor regions that control the muscles of the larynx, pharynx, and mouth. The impulses from the premotor area result in specific, coordinated muscle contractions. Simultaneously, impulses propagate from Broca's speech area to the primary motor area. From here, impulses also control the breathing muscles to regulate the proper flow of air past the vocal cords. The coordinated contractions of your speech and breathing muscles enable you to speak your thoughts. People who suffer a cerebrovascular accident (CVA) or stroke in this area can still have clear thoughts but are unable to form words, a phenomenon referred to as *nonfluent aphasia*; see the next Clinical Connection.

Association Areas

The association areas of the cerebrum consist of large areas of the occipital, parietal, and temporal lobes and of the frontal lobes anterior to the motor areas. Association areas are connected with one another by association tracts and include the following (Figure 14.15):

- The **somatosensory association area** (areas 5 and 7) is just posterior to and receives input from the primary somatosensory area, as well as from the thalamus and other parts of the brain. This area permits you to determine the exact shape and texture of an object by feeling it, to determine the orientation of one object with respect to another as they are felt, and to sense the relationship of one body part to another. Another role of the somatosensory association area is the storage of memories of past somatic sensory experiences, enabling you to compare current sensations with previous experiences. For example, the somatosensory association area allows you to recognize objects such as a pencil and a paperclip simply by touching them.

- The **visual association area** (areas 18 and 19), located in the occipital lobe, receives sensory impulses from the primary visual area and the thalamus. It relates present and past visual experiences and is essential for recognizing and evaluating what is seen. For example, the visual association area allows you to recognize an object such as a spoon simply by looking at it.
- The **facial recognition area,** corresponding roughly to areas 20, 21, and 37 in the inferior temporal lobe, receives nerve impulses from the visual association area. This area stores information about faces, and it allows you to recognize people by their faces. The facial recognition area in the *right* hemisphere is usually more dominant than the corresponding region in the left hemisphere.
- The **auditory association area** (area 22), located inferior and posterior to the primary auditory area in the temporal cortex, allows you to recognize a particular sound as speech, music, or noise.
- The **orbitofrontal cortex,** corresponding roughly to area 11 along the lateral part of the frontal lobe, receives sensory impulses from the primary olfactory area. This area allows you to identify odors and to discriminate among different odors. During olfactory processing, the orbitofrontal cortex of the *right* hemisphere exhibits greater activity than the corresponding region in the left hemisphere.
- **Wernicke's (posterior language) area** (VER-ni-kēz) (area 22, and possibly areas 39 and 40), a broad region in the *left* temporal and parietal lobes, interprets the meaning of speech by recognizing spoken words. It is active as you translate words into thoughts. The regions in the *right* hemisphere that correspond to Broca's and Wernicke's areas in the left hemisphere also contribute to verbal communication by adding emotional content, such as anger or joy, to spoken words. Unlike those who have CVAs in Broca's area, people who suffer strokes in Wernicke's area can still speak, but cannot arrange words in a coherent fashion (fluent aphasia, or "word salad"; see the Clinical Connection on this page).
- The **common integrative area** (areas 5, 7, 39, and 40) is bordered by somatosensory, visual, and auditory association areas. It receives nerve impulses from these areas and from the primary gustatory area, the primary olfactory area, the thalamus, and parts of the brain stem. This area integrates sensory interpretations from the association areas and impulses from other areas, allowing the formation of thoughts based on a variety of sensory inputs. It then transmits signals to other parts of the brain for the appropriate response to the sensory signals it has interpreted.
- The **prefrontal cortex (frontal association area)** is an extensive area in the anterior portion of the frontal lobe that is well developed in primates, especially humans (areas 9, 10, 11, and 12; area 12 is not illustrated since it can be seen only in a medial view). This area has numerous connections with other areas of the cerebral cortex, thalamus, hypothalamus, limbic system, and cerebellum. The prefrontal cortex is concerned with the makeup of a person's personality, intellect, complex learning abilities, recall of information, initiative, judgment, foresight, reasoning, conscience, intuition, mood, planning for the future, and development of abstract ideas. A person with bilateral damage to the prefrontal cortices typically becomes rude, inconsiderate, incapable of accepting advice, moody, inattentive, less creative, unable to plan for the future, and incapable of anticipating the consequences of rash or reckless words or behavior.
- The **premotor area** (area 6) is a motor association area that is immediately anterior to the primary motor area. Neurons in this area communicate with the primary motor cortex, the sensory association areas in the parietal lobe, the basal nuclei, and the thalamus. The premotor area deals with learned motor activities of a complex and sequential nature. It generates nerve impulses that cause specific groups of muscles to contract in a specific sequence, as when you write your name. The premotor area also serves as a memory bank for such movements.
- The **frontal eye field area** (area 8) in the frontal cortex is sometimes included in the premotor area. It controls voluntary scanning movements of the eyes—like those you just used in reading this sentence.

CLINICAL CONNECTION | *Aphasia*

Much of what we know about language areas comes from studies of patients with language or speech disturbances that have resulted from brain damage. Broca's speech area, Wernicke's area, and other language areas are located in the left cerebral hemisphere of most people, regardless of whether they are left-handed or right-handed. Injury to language areas of the cerebral cortex results in **aphasia** (a-FĀ-zē-a; *a-* = without; *-phasia* = speech), an inability to use or comprehend words. Damage to Broca's speech area results in *nonfluent aphasia,* an inability to properly articulate or form words; people with nonfluent aphasia know what they wish to say but cannot speak. Damage to Wernicke's area, the common integrative area, or auditory association area results in *fluent aphasia,* characterized by faulty understanding of spoken or written words. A person experiencing this type of aphasia may fluently produce strings of words that have no meaning ("word salad"). For example, someone with fluent aphasia might say, "I rang car porch dinner light river pencil." The underlying deficit may be **word deafness** (an inability to understand spoken words), **word blindness** (an inability to understand written words), or both. •

The functions of the various parts of the brain are summarized in Table 14.2.

TABLE 14.2

Summary of Functions of Principal Parts of the Brain

PART	FUNCTION
BRAIN STEM	
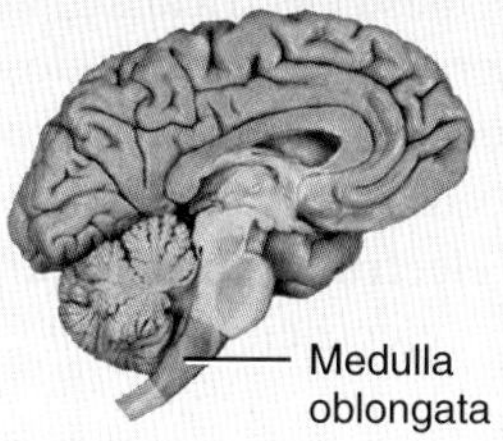	***Medulla oblongata:*** Contains sensory (ascending) and motor (descending) tracts. Cardiovascular center regulates heartbeat and blood vessel diameter. Medullary rhythmicity area (together with pons) regulates breathing. Contains gracile nucleus, cuneate nucleus, gustatory nucleus, cochlear nuclei, and vestibular nuclei (components of sensory pathways to brain). Inferior olivary nucleus provides instructions that cerebellum uses to adjust muscle activity when learning new motor skills. Other nuclei coordinate vomiting, swallowing, sneezing, coughing, and hiccupping. Contains nuclei of origin for cranial nerves VIII, IX, X, XI, and XII. Reticular formation (also in pons, midbrain, and diencephalon) functions in consciousness and arousal.
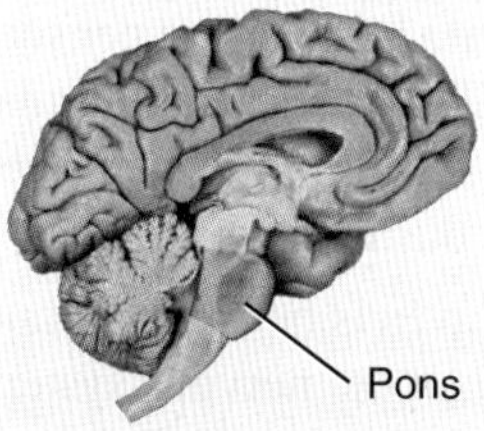	***Pons:*** Contains sensory and motor tracts. Pontine nuclei relay nerve impulses from motor areas of cerebral cortex to cerebellum. Contains vestibular nuclei (along with medulla) that are part of equilibrium pathway to brain. Pneumotaxic area and apneustic area (together with the medulla) help control breathing. Contains nuclei of origin for cranial nerves V, VI, VII, and VIII.
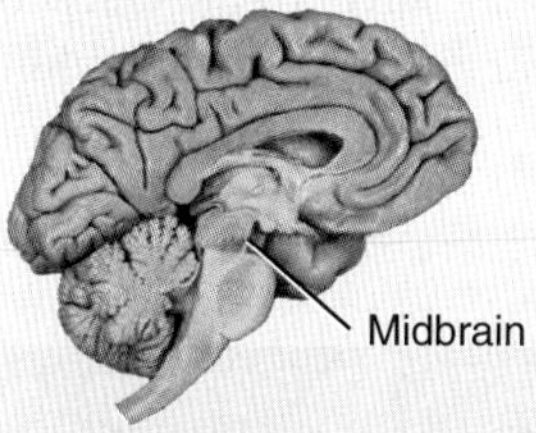	***Midbrain:*** Contains sensory and motor tracts. Superior colliculi coordinate movements of head, eyes, and trunk in response to visual stimuli. Inferior colliculi coordinate movements of head, eyes, and trunk in response to auditory stimuli. Substantia nigra and red nucleus contribute to control of movement. Contains nuclei of origin for cranial nerves III and IV.
CEREBELLUM	
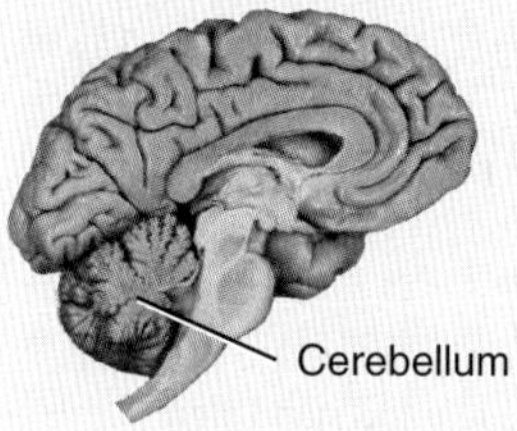	Smooths and coordinates contractions of skeletal muscles. Regulates posture and balance. May have role in cognition and language processing.
DIENCEPHALON	
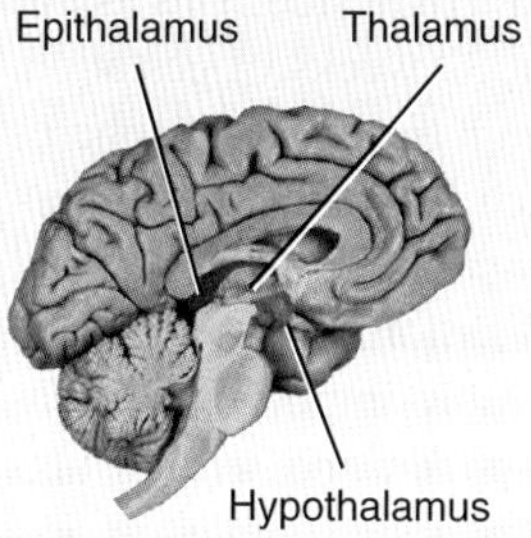	***Thalamus:*** Relays almost all sensory input to cerebral cortex. Contributes to motor functions by transmitting information from cerebellum and basal nuclei to primary motor area of cerebral cortex. Plays role in maintenance of consciousness. ***Hypothalamus:*** Controls and integrates activities of autonomic nervous system. Produces hormones, including releasing hormones, inhibiting hormones, oxytocin, and antidiuretic hormone (ADH). Regulates emotional and behavioral patterns (together with limbic system). Contains feeding and satiety centers (regulate eating), thirst center (regulates drinking), and suprachiasmatic nucleus (regulates circadian rhythms). Controls body temperature by serving as body's thermostat. ***Epithalamus:*** Consists of pineal gland (secretes melatonin) and habenular nuclei.
CEREBRUM	
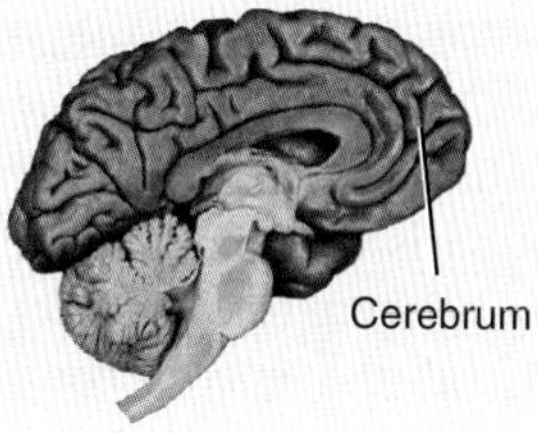	Sensory areas of cerebral cortex are involved in perception of sensory information; motor areas control execution of voluntary movements; association areas deal with more complex integrative functions such as memory, personality traits, and intelligence. Basal nuclei help initiate and terminate movements, suppress unwanted movements, and regulate muscle tone. Limbic system promotes range of emotions, including pleasure, pain, docility, affection, fear, and anger.

Hemispheric Lateralization

Although the brain is almost symmetrical on its right and left sides, subtle anatomical differences between the two hemispheres exist. For example, in about two-thirds of the population, the planum temporale, a region of the temporal lobe that includes Wernicke's area, is 50% larger on the left side than on the right side. This asymmetry appears in the human fetus at about 30 weeks of gestation. Physiological differences also exist; although the two hemispheres share performance of many functions, each hemisphere also specializes in performing certain unique functions. This functional asymmetry is termed **hemispheric lateralization.**

In most people the left hemisphere is more important for reasoning, numerical and scientific skills, spoken and written language, and the ability to use and understand sign language. Patients with damage in the left hemisphere often exhibit aphasia. Conversely, the right hemisphere is more specialized for musical and artistic awareness; spatial and pattern perception; recognition of faces and emotional content of language; discrimination of different smells; and generating mental images of sight, sound, touch, taste, and smell to compare relationships among them. Patients with damage in right hemisphere regions that correspond to Broca's and Wernicke's areas in the left hemisphere speak in a monotonous voice, having lost the ability to impart emotional inflection to what they say.

Despite some dramatic differences in functions of the two hemispheres, there is considerable variation from one person to another. Also, lateralization seems less pronounced in females than in males, both for language (left hemisphere) and for visual and spatial skills (right hemisphere). For instance, females are less likely than males to suffer aphasia after damage to the left hemisphere. A possibly related observation is that the anterior commissure is 12% larger and the corpus callosum has a broader posterior portion in females. Recall that both the anterior commissure and the corpus callosum are commissural tracts that provide communication between the two hemispheres.

Table 14.3 summarizes some of the functional differences between the two cerebral hemispheres.

Brain Waves

At any instant, brain neurons are generating millions of nerve impulses (action potentials). Taken together, these electrical signals are called **brain waves.** Brain waves generated by neurons close to the brain surface, mainly neurons in the cerebral cortex, can be detected by sensors called electrodes placed on the forehead and scalp. A record of such waves is called an **electroencephalogram** (e-lek′-trō-en-SEF-a-lō-gram; *electro-* = electricity; *-gram* = recording) or **EEG.** Electroencephalograms are useful both in studying normal brain functions, such as changes that occur during sleep, and in diagnosing a variety of brain disorders, such as epilepsy, tumors, trauma, hematomas, metabolic abnormalities, sites of trauma, and degenerative diseases. The EEG is also utilized to determine if "life" is present, that is, to establish or confirm that brain death has occurred.

Patterns of activation of brain neurons produce four types of brain waves (Figure 14.16):

1. **Alpha waves.** These rhythmic waves occur at a frequency of about 8–13 cycles per second. (The unit commonly used to express frequency is the hertz, Hz. One hertz is one cycle per second.) Alpha waves are present in the EEGs of nearly all normal individuals when they are awake and resting with their eyes closed. These waves disappear entirely during sleep.
2. **Beta waves.** The frequency of these waves is between 14 and 30 Hz. Beta waves generally appear when the nervous system is active—that is, during periods of sensory input and mental activity.
3. **Theta waves.** Theta waves (THĀ-ta) have frequencies of 4–7 Hz. These waves normally occur in children and adults experiencing emotional stress. They also occur in many disorders of the brain.

TABLE 14.3

Functional Differences between Right and Left Hemispheres

RIGHT HEMISPHERE FUNCTIONS	LEFT HEMISPHERE FUNCTIONS
Receives somatic sensory signals from, and controls muscles on, left side of body.	Receives somatic sensory signals from, and controls muscles on, right side of body.
Musical and artistic awareness.	Reasoning.
Space and pattern perception.	Numerical and scientific skills.
Recognition of faces and emotional content of facial expressions.	Ability to use and understand sign language.
Generating emotional content of language.	Spoken and written language.
Generating mental images to compare spatial relationships.	
Identifying and discriminating among odors.	

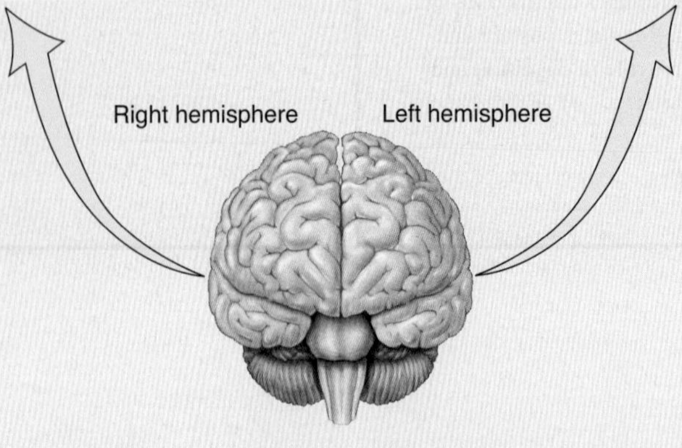

Figure 14.16 **Types of brain waves recorded in an electroencephalogram (EEG).**

Brain waves indicate electrical activity of the cerebral cortex.

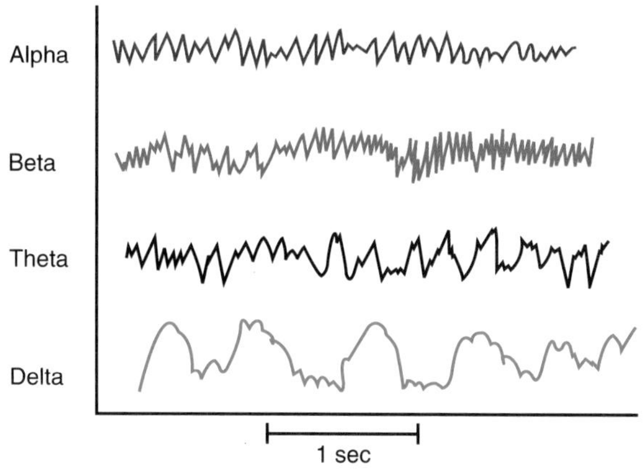
Alpha
Beta
Theta
Delta
1 sec

Which type of brain wave indicates emotional stress?

4. **Delta waves.** The frequency of these waves is 1–5 Hz. Delta waves occur during deep sleep in adults, but they are normal in awake infants. When produced by an awake adult, they indicate brain damage.

CHECKPOINT

20. Compare the functions of the sensory, motor, and association areas of the cerebral cortex.
21. What is hemispheric lateralization?
22. What is the diagnostic value of an EEG?

14.8 CRANIAL NERVES

OBJECTIVE

- Identify the cranial nerves by name, number, and type, and give the function of each.

The 12 pairs of **cranial nerves** are so named because they pass through various foramina in the bones of the cranium and arise from the brain inside the cranial cavity. Like the 31 pairs of spinal nerves, they are part of the peripheral nervous system (PNS). Each cranial nerve has both a number, designated by a roman numeral, and a name. The numbers indicate the order, from anterior to posterior, in which the nerves arise from the brain. The names designate a nerve's distribution or function.

Three cranial nerves (I, II, and VIII) carry axons of sensory neurons and thus are called **special sensory nerves.** These nerves are unique to the head and are associated with the special senses of smelling, seeing, and hearing. The cell bodies of most sensory neurons are located in ganglia outside the brain.

Five cranial nerves (III, IV, VI, XI, and XII) are classified as **motor nerves** because they contain only axons of motor neurons as they leave the brain stem. The cell bodies of motor neurons lie in nuclei within the brain. Motor axons that innervate skeletal muscles are of two types:

1. *Branchial motor axons* innervate skeletal muscles that develop from the pharyngeal (branchial) arches (see Figure 14.28). These neurons leave the brain through the mixed cranial nerves and the accessory nerve.
2. *Somatic motor axons* innervate skeletal muscles that develop from head somites (eye muscles and tongue muscles). These neurons exit the brain through five motor cranial nerves (III, IV, VI, XI, and XII). Motor axons that innervate smooth muscle, cardiac muscle, and glands are called *autonomic motor axons* and are part of the parasympathetic division.

The remaining four cranial nerves (V, VII, IX, and X) are **mixed nerves**—they contain axons of both sensory neurons entering the brain stem and motor neurons leaving the brain stem.

Each cranial nerve is covered in detail in Exhibits 14.A through 14.J. Although the cranial nerves are mentioned singly in the exhibits with regard to their type, location, and function, remember that they are paired structures.

Table 14.4 presents a summary of the components and principal functions of the cranial nerves, including a mnemonic to help you remember their names.

CLINICAL CONNECTION | *Dental Anesthesia*

The inferior alveolar nerve, a branch of the mandibular nerve, supplies all the teeth in one half of the mandible; it is often anesthetized in dental procedures. The same procedure will anesthetize the lower lip because the mental nerve is a branch of the inferior alveolar nerve. Because the lingual nerve runs very close to the inferior alveolar nerve near the mental foramen, it too is often anesthetized at the same time. For anesthesia to the upper teeth, the superior alveolar nerve endings, which are branches of the maxillary nerve, are blocked by inserting the needle beneath the mucous membrane. The anesthetic solution is then infiltrated slowly throughout the area of the roots of the teeth to be treated. •

CHECKPOINT

23. How are cranial nerves named and numbered?
24. What is the difference between a special sensory, motor, and mixed cranial nerve?
25. Briefly list the function for each cranial nerve.

EXHIBIT 14.A Olfactory (I) Nerve *(Figure 14.17)*

OBJECTIVE

- Identify the termination of the olfactory nerve in the brain, the foramen through which it passes, and its function.

The **olfactory (I) nerve** (ōl-FAK-tō-rē; *olfact-* = to smell) is entirely sensory; it contains axons that conduct nerve impulses for olfaction, the sense of smell (Figure 14.17). The olfactory epithelium occupies the superior part of the nasal cavity, covering the inferior surface of the cribriform plate and extending down along the superior nasal concha. The olfactory receptors within the olfactory epithelium are bipolar neurons. Each has a single odor-sensitive, knob-shaped dendrite projecting from one side of the cell body and an unmyelinated axon extending from the other side. Bundles of axons of olfactory receptors extend through about 20 olfactory foramina in the cribriform plate of the ethmoid bone on each side of the nose. These 40 or so bundles of axons collectively form the right and left olfactory nerves.

Olfactory nerves end in the brain in paired masses of gray matter called the **olfactory bulbs,** two extensions of the brain that rest on the cribriform plate. Within the olfactory bulbs, the axon terminals of olfactory receptors form synapses with the dendrites and cell bodies of the next neurons in the olfactory pathway. The axons of these neurons make up the **olfactory tracts,** which extend posteriorly from the olfactory bulbs (Figure 14.17). Axons in the olfactory tracts end in the primary olfactory area in the temporal lobe of the cerebral cortex.

✔ CHECKPOINT

Where is the olfactory epithelium located?

Figure 14.17 Olfactory (I) nerve.

The olfactory epithelium is located on the inferior surface of the cribriform plate and superior nasal conchae.

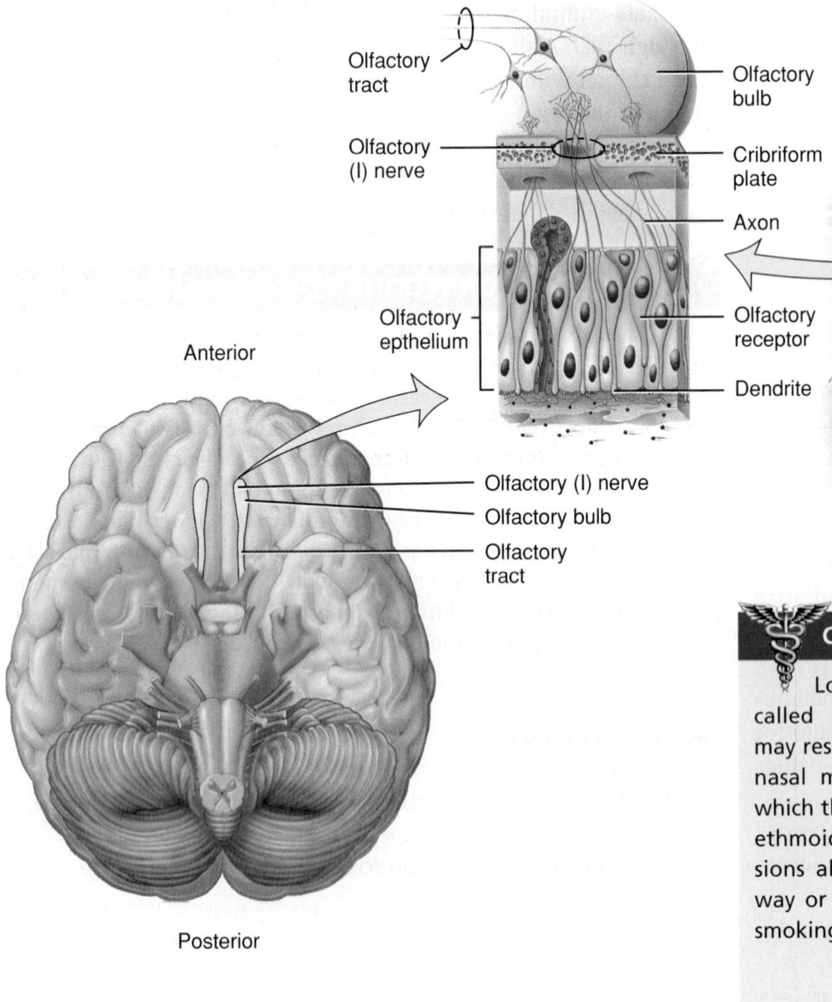

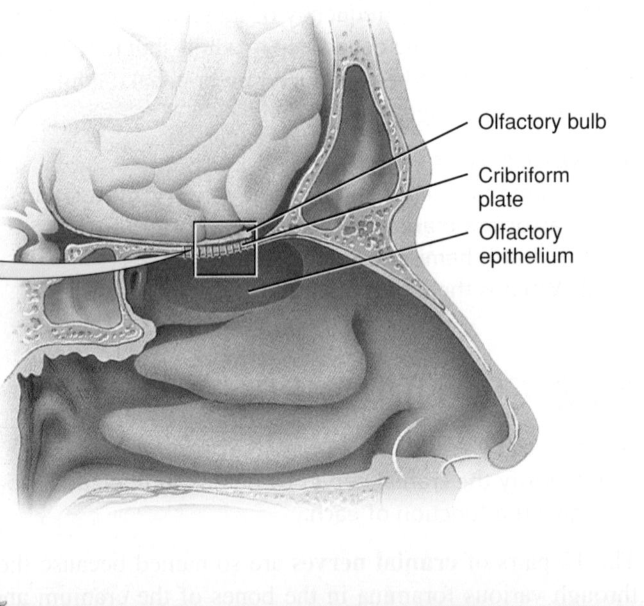

CLINICAL CONNECTION | *Anosmia*

Loss of the sense of smell, called **anosmia** (an-OZ-mē-a), may result from infections of the nasal mucosa, head injuries in which the cribriform plate of the ethmoid bone is fractured, lesions along the olfactory pathway or in the brain, meningitis, smoking, or cocaine use. •

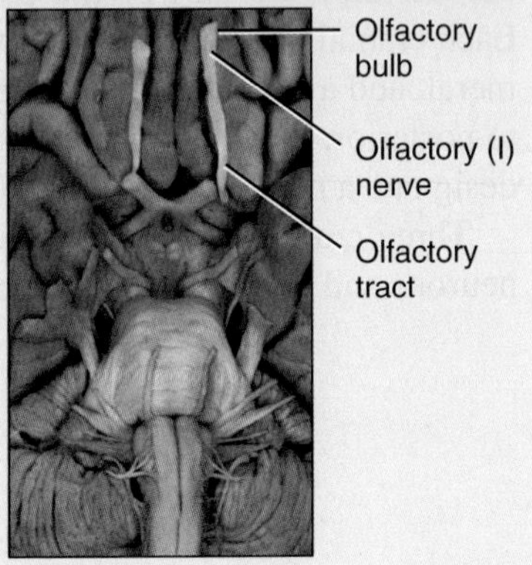

? **Where do axons in the olfactory tracts terminate?**

EXHIBIT 14.B Optic (II) Nerve *(Figure 14.18)*

OBJECTIVE

- Identify the termination of the optic nerve in the brain, the foramen through which it exits the skull, and its function.

The **optic (II) nerve** (OP-tik; *opti-* = the eye, vision) is entirely sensory; it contains axons that conduct nerve impulses for vision (Figure 14.18). In the retina, rods and cones initiate visual signals and relay them to bipolar cells, which transmit the signals to ganglion cells. Axons of all the ganglion cells in the retina of each eye join to form an optic nerve, which passes through the optic foramen. About 10 mm (0.4 in.) posterior to the eyeball, the two optic nerves merge to form the **optic chiasm** (KĪ-azm = a crossover, as in the letter X). Within the chiasm, axons from the medial half of each eye cross to the opposite side; axons from the lateral half remain on the same side. Posterior to the chiasm, the regrouped axons, some from each eye, form the **optic tracts.** Most axons in the optic tracts end in the lateral geniculate nucleus of the thalamus. There they synapse with neurons whose axons extend to the primary visual area in the occipital lobe of the cerebral cortex (area 17 in Figure 14.15). A few axons pass through the lateral geniculate nucleus and then extend to the superior colliculi of the midbrain. They synapse with motor neurons that control the extrinsic and intrinsic eye muscles.

CHECKPOINT

Trace the sequence of nerve cells that process visual impulses within the retina.

Figure 14.18 **Optic (II) nerve.**

In sequence, visual signals are relayed from rods and cones to bipolar cells to ganglion cells.

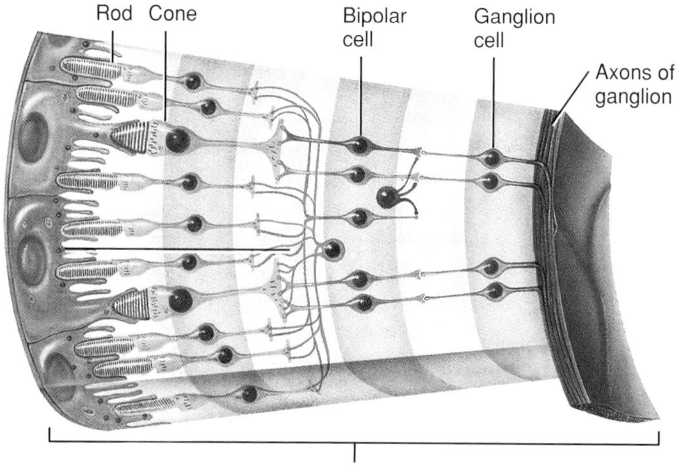

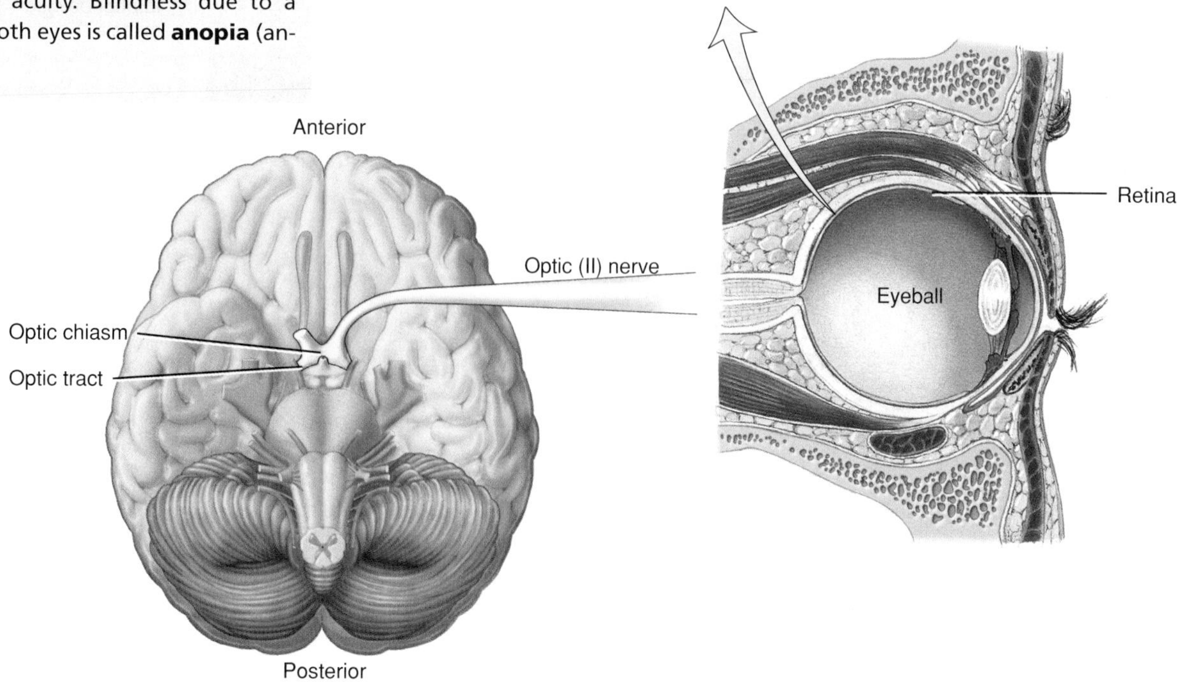

CLINICAL CONNECTION | *Anopia*

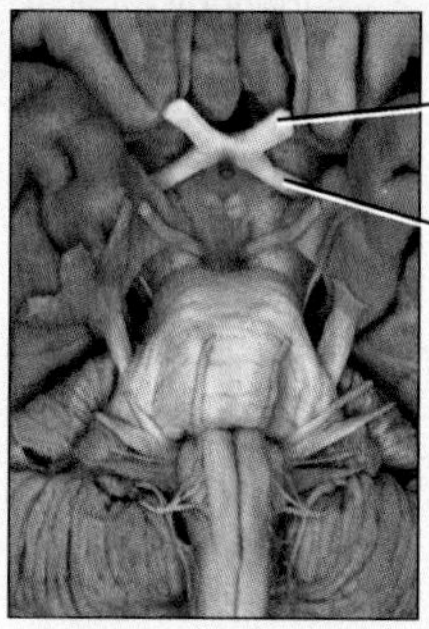

Fractures in the orbit, brain lesions, damage along the visual pathway, diseases of the nervous system (such as multiple sclerosis), pituitary gland tumors, or cerebral aneurysms (enlargements of blood vessels due to weakening of their walls) may result in visual field defects and loss of visual acuity. Blindness due to a defect in or loss of one or both eyes is called **anopia** (an-Ō-pē-a). •

Where do most axons in the optic tracts terminate?

EXHIBIT 14.C Oculomotor (III), Trochlear (IV), and Abducens (VI) Nerves *(Figure 14.19)*

OBJECTIVE

- Identify the origins of the oculomotor, trochlear, and abducens nerves in the brain, the foramen through which each exits the skull, and their functions.

The oculomotor, trochlear, and abducens nerves are the cranial nerves that control the muscles that move the eyeballs. They are all motor nerves that contain only motor axons as they exit the brain stem. Sensory axons from the extrinsic eyeball muscles begin their course toward the brain in each of these nerves, but eventually these sensory axons leave these nerves to join the ophthalmic branch of the trigeminal nerve. The sensory axons *do not* return to the brain in the oculomotor, trochlear, or abducens nerves. The cell bodies of the unipolar sensory neurons reside in the mesencephalic nucleus and they enter the midbrain via the trigeminal nerve. These axons convey nerve impulses from the extrinsic eyeball muscles for *proprioception,* the perception of the movements and position of the body independent of vision.

The **oculomotor (III) nerve** (ok′-ū-lō-MŌ-tor; *oculo-* = eye; *-motor* = a mover) has its motor nucleus in the anterior part of the midbrain. The oculomotor nerve extends anteriorly and divides into superior and inferior branches, both of which pass through the superior orbital fissure into the orbit (Figure 14.19a). Axons in the superior branch innervate the superior rectus (an extrinsic eyeball muscle) and the levator palpebrae superioris (the muscle of the upper eyelid). Axons in the inferior branch supply the medial rectus, inferior rectus, and inferior oblique muscles—all extrinsic eyeball muscles. These somatic motor neurons control movements of the eyeball and upper eyelid.

Figure 14.19 Oculomotor (III), trochlear (IV), and abducens (VI) nerves.

6 The oculomotor nerve has the widest distribution among extrinsic eye muscles.

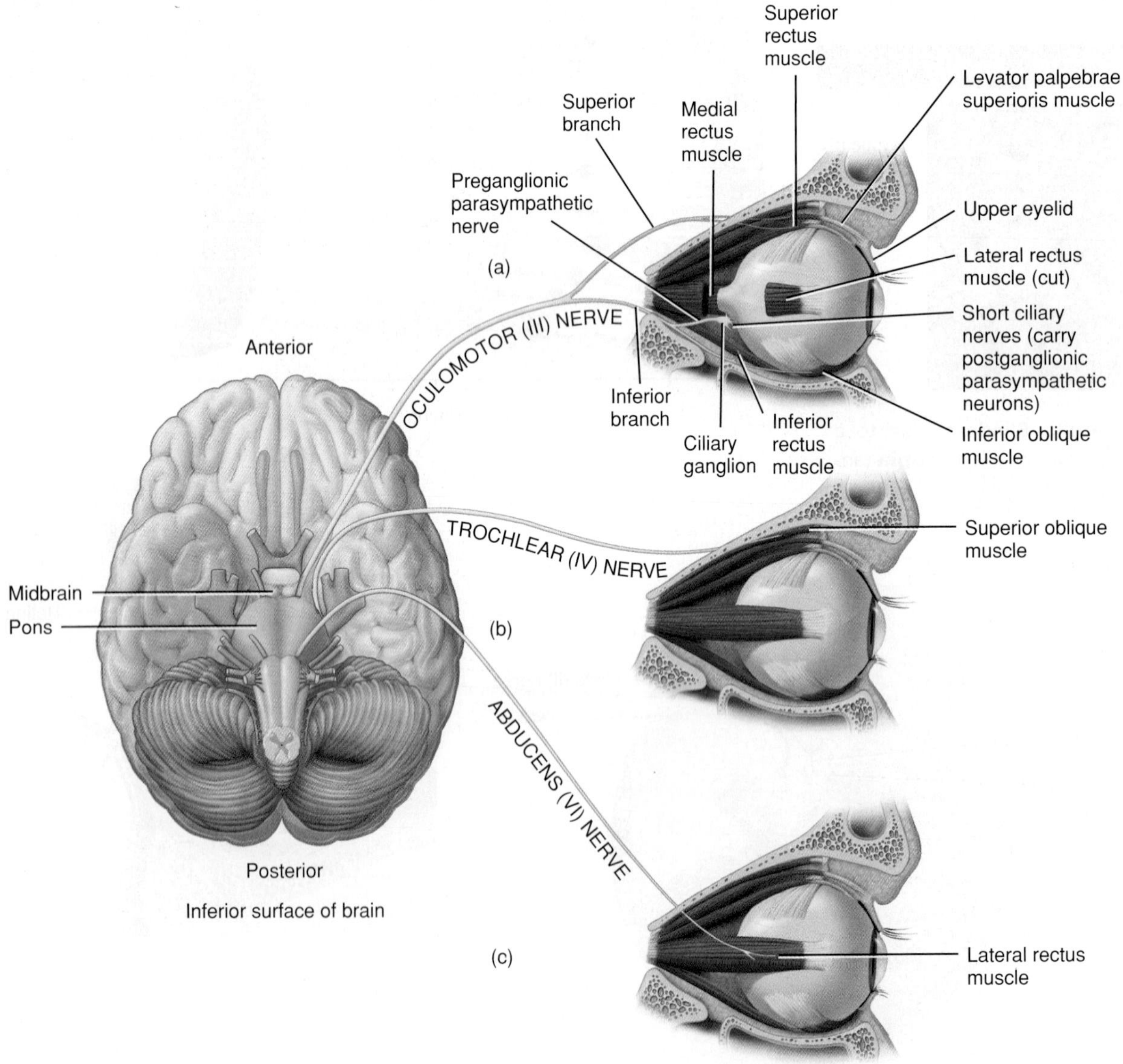

Which branch of the oculomotor nerve is distributed to the superior rectus muscle? Which is the smallest cranial nerve?

The inferior branch of the oculomotor nerve also supplies autonomic motor axons to intrinsic eyeball muscles, which consist of smooth muscle. They include the ciliary muscle of the eyeball and the circular muscles (sphincter pupillae) of the iris. Autonomic impulses propagate from a nucleus in the midbrain (accessory oculomotor) to the **ciliary ganglion,** a synaptic relay center for the two motor neurons of the autonomic nervous system. From the ciliary ganglion, autonomic motor axons extend to the ciliary muscle, which adjusts the lens for near vision *(accommodation).* Other autonomic motor axons stimulate the circular muscles of the iris to contract when bright light stimulates the eye, causing a decrease in the size of the pupil *(constriction).*

The **trochlear (IV) nerve** (TRŌK-lē-ar; *trochle-* = a pulley) is the smallest of the 12 cranial nerves and is the only one that arises from the posterior aspect of the brain stem. The somatic motor neurons originate in a nucleus in the midbrain (trochlear nucleus), and axons from the nucleus cross to the opposite side as they exit the brain on its posterior aspect. The nerve then wraps around the pons and exits through the superior orbital fissure into the orbit. These somatic motor axons innervate the superior oblique muscle of the eyeball, another extrinsic eyeball muscle that controls movement of the eyeball (Figure 14.19b).

Neurons of the **abducens (VI) nerve** (ab-DOO-senz; *ab-* = away; *-ducens* = to lead) originate from a nucleus in the pons (abducens nucleus). Somatic motor axons extend from the nucleus to the lateral rectus muscle of the eyeball, an extrinsic eyeball muscle, through the superior orbital fissure of the orbit (Figure 14.19c). The abducens nerve is so named because nerve impulses cause abduction (lateral rotation) of the eyeball.

✔CHECKPOINT

How are the oculomotor, trochlear, and abducens nerves related functionally?

CLINICAL CONNECTION | *Strabismus, Ptosis, and Diplopia*

Damage to the oculomotor (III) nerve causes **strabismus** (stra-BIZ-mus) (a condition in which both eyes do not fix on the same object, since one or both eyes may turn inward or outward), **ptosis** (TŌ-sis) (drooping) of the upper eyelid, dilation of the pupil, movement of the eyeball downward and outward on the damaged side, loss of accommodation for near vision, and **diplopia** (di-PLŌ-pē-a) (double vision).

Trochlear (IV) nerve damage can also result in strabismus and diplopia.

With damage to the abducens (VI) nerve, the affected eyeball cannot move laterally beyond the midpoint, and the eyeball usually is directed medially. This leads to strabismus and diplopia.

Causes of damage to the oculomotor, trochlear, and abducens nerves include trauma to the skull or brain, compression resulting from aneurysms, and lesions of the superior orbital fissure. Individuals with damage to these nerves are forced to tilt their heads in various directions to help bring the affected eyeball into the correct frontal plane. •

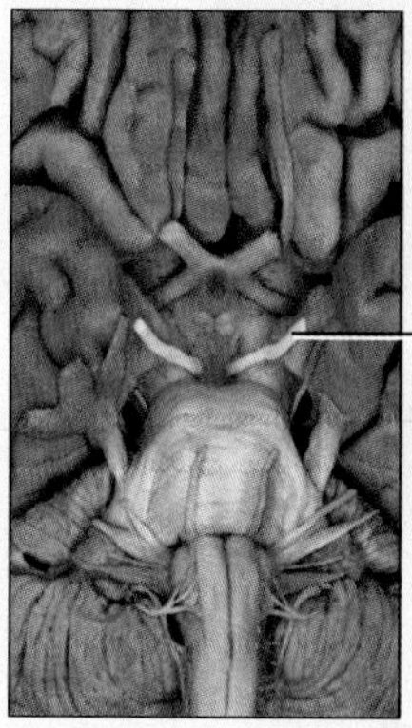

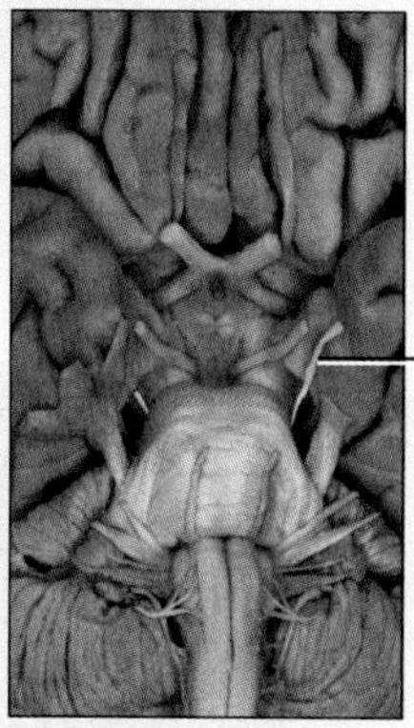

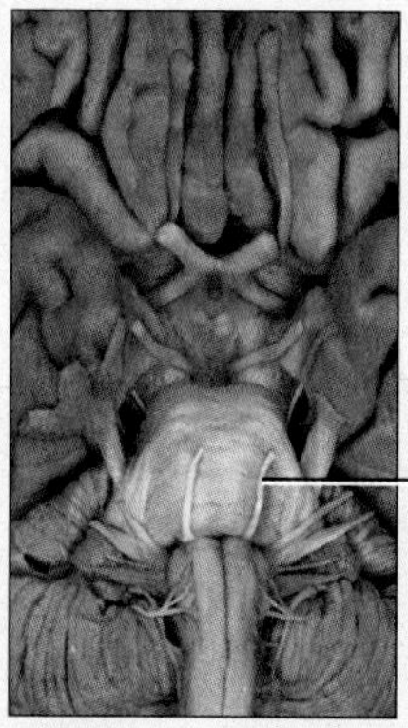

EXHIBIT 14.D Trigeminal (V) Nerve *(Figure 14.20)*

OBJECTIVE

- Identify the origin of the trigeminal nerve in the brain, describe the foramina through which each of its three major branches exits the skull, and explain the function of each branch.

The **trigeminal (V) nerve** (trī-JEM-i-nal = triple, for its three branches) is a mixed cranial nerve and the largest of the cranial nerves. The trigeminal nerve emerges from two roots on the anterolateral surface of the pons. The large sensory root has a swelling called the **trigeminal (semilunar) ganglion,** which is located in a fossa on the inner surface of the petrous portion of the temporal bone. The ganglion contains cell bodies of most of the primary sensory neurons. Neurons of the smaller motor root originate in a nucleus in the pons.

As indicated by its name, the trigeminal nerve has three branches: ophthalmic, maxillary, and mandibular (Figure 14.20). The **ophthalmic nerve** (of-THAL-mik; *ophthalm-* = the eye), the smallest branch, passes into the orbit via the superior orbital fissure. The **maxillary nerve** (*maxilla* = upper jawbone) is intermediate in size between the ophthalmic and mandibular nerves and passes through the foramen rotundum. The **mandibular nerve** (*mandibula* = lower jawbone), the largest branch, passes through the foramen ovale.

Figure 14.20 Trigeminal (V) nerve.

The three branches of the trigeminal nerve leave the cranium through the superior orbital fissure, foramen rotundum, and foramen ovale.

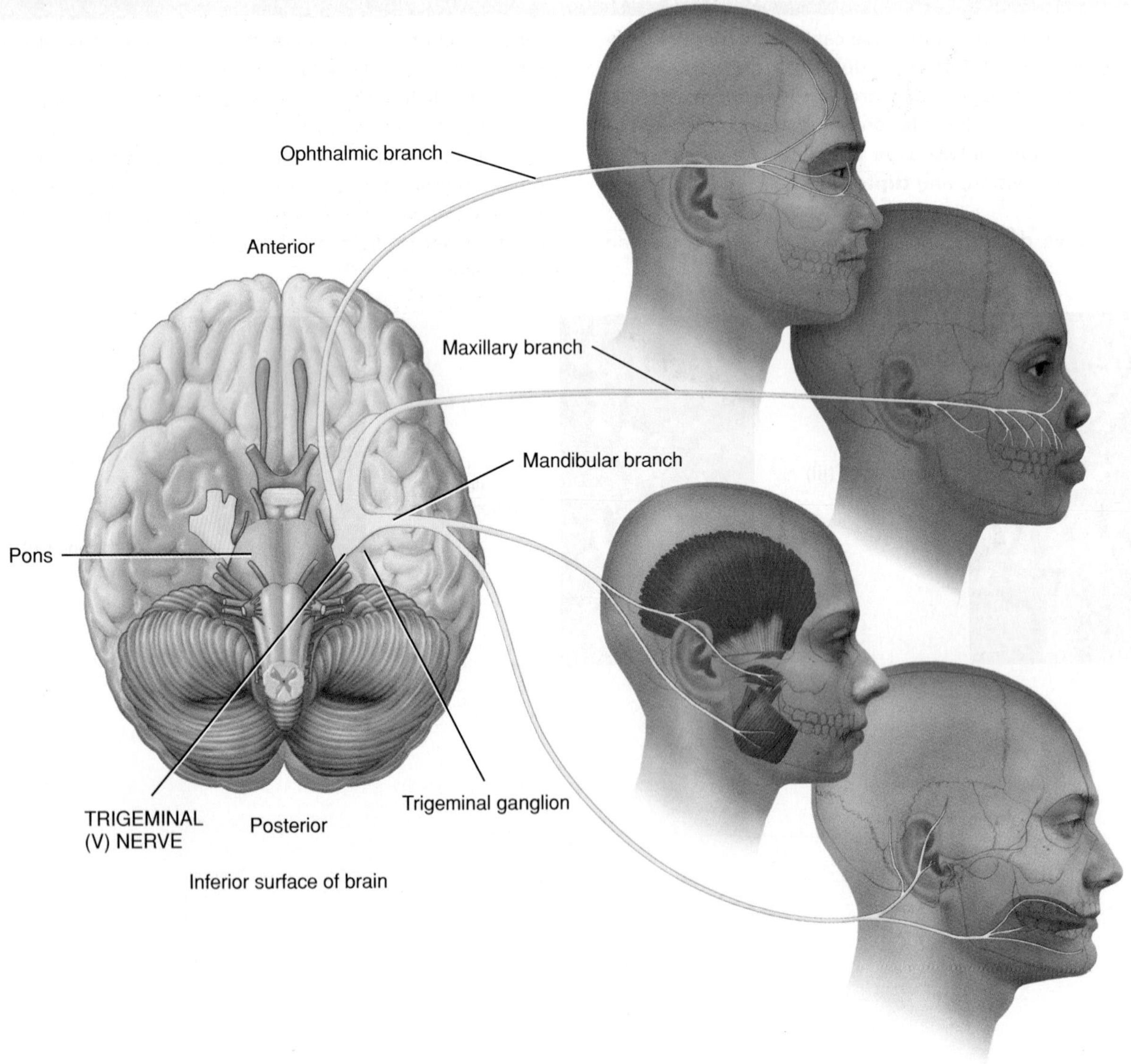
Ophthalmic branch
Anterior
Maxillary branch
Mandibular branch
Pons
TRIGEMINAL
(V) NERVE
Posterior
Trigeminal ganglion
Inferior surface of brain

How does the trigeminal nerve compare in size with the other cranial nerves?

Sensory axons in the trigeminal nerve carry nerve impulses for touch, pain, and thermal sensations (heat and cold). The ophthalmic nerve contains sensory axons from the skin over the upper eyelid, cornea, lacrimal glands, upper part of the nasal cavity, side of the nose, forehead, and anterior half of the scalp. The maxillary nerve includes sensory axons from the mucosa of the nose, palate, part of the pharynx, upper teeth, upper lip, and lower eyelid. The mandibular nerve contains sensory axons from the anterior two-thirds of the tongue (not taste), cheek and mucosa deep to it, lower teeth, skin over the mandible and side of the head anterior to the ear, and mucosa of the floor of the mouth. The sensory axons from the three branches enter the trigeminal ganglion, where their cell bodies are located, and terminate in nuclei in the pons. The trigeminal nerve also contains sensory axons from proprioceptors (receptors that provide information regarding body position and movements) located in the muscles of mastication and extrinsic muscles of the eyeball, but the cell bodies of these neurons are located in the mesencephalic nucleus.

Branchial motor neurons of the trigeminal nerve are part of the mandibular nerve and supply muscles of mastication (masseter, temporalis, medial pterygoid, lateral pterygoid, anterior belly of digastric, and mylohyoid muscles, as well as the tensor veli palatini muscle in the soft palate and tensor tympani muscle in the middle ear). These motor neurons mainly control chewing movements.

✔ **CHECKPOINT**

What are the three branches of the trigeminal nerve, and which branch is the largest?

CLINICAL CONNECTION | *Trigeminal Neuralgia*

Neuralgia (pain) relayed via one or more branches of the trigeminal (V) nerve, caused by conditions such as inflammation or lesions, is called **trigeminal neuralgia (tic douloureux).** This is a sharp cutting or tearing pain that lasts for a few seconds to a minute and is caused by anything that presses on the trigeminal nerve or its branches. It occurs almost exclusively in people over 60 and can be the first sign of a disease, such as multiple sclerosis or diabetes, or lack of vitamin B_{12}, which damage the nerves. Injury of the mandibular nerve may cause paralysis of the chewing muscles and a loss of the sensations of touch, temperature, and proprioception in the lower part of the face. •

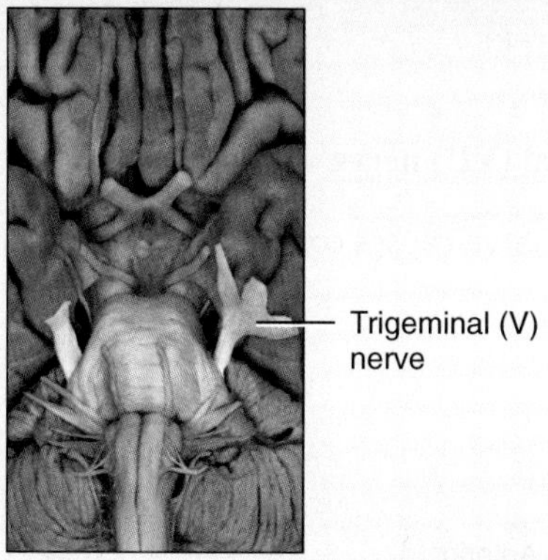

EXHIBIT 14.E Facial (VII) Nerve *(Figure 14.21)*

OBJECTIVE

- Identify the origins of the facial nerve in the brain, the foramen through which it exits the skull, and its function.

The **facial (VII) nerve** (FĀ-shal = face) is a mixed cranial nerve. Its sensory axons extend from the taste buds of the anterior two-thirds of the tongue, which enter the temporal bone to join the facial nerve. From here the sensory axons pass to the **geniculate ganglion** (je-NIK-ū-lat), a cluster of cell bodies of sensory neurons of the facial nerve within the temporal bone, and ends in the pons. From the pons, axons extend to the thalamus, and then to the gustatory areas of the cerebral cortex (Figure 14.21). The sensory portion of the facial nerve also contains axons from skin in the ear canal that relay touch, pain, and thermal sensations. Additionally, proprioceptors from muscles of the face and scalp relay information through their cell bodies in a nucleus in the midbrain (mesencephalic nucleus).

Axons of branchial motor neurons arise from a nucleus in the pons and exit the stylomastoid foramen to innervate middle ear, facial, scalp, and neck muscles. Nerve impulses propagating along these axons cause contraction of the muscles of facial expression plus the stylohyoid muscle, the posterior belly of the digastric muscle, and the stapedius muscle. The facial nerve innervates more named muscles than any other nerve in the body.

Axons of the autonomic motor neurons run in branches of the facial nerve and end in two ganglia: the **pterygopalatine ganglion** (ter′-i-gō-PAL-a-tīn) and the **submandibular ganglion,** respectively. From synaptic relays in the two ganglia, postganglionic autonomic motor axons extend to lacrimal glands (which secrete tears), nasal glands, palatine glands, and saliva-producing sublingual and submandibular glands.

✔ **CHECKPOINT**

Why is the facial nerve considered the major motor nerve of the head?

Figure 14.21 **Facial (VII) nerve.**

The facial nerve causes contraction of the muscles of facial expression.

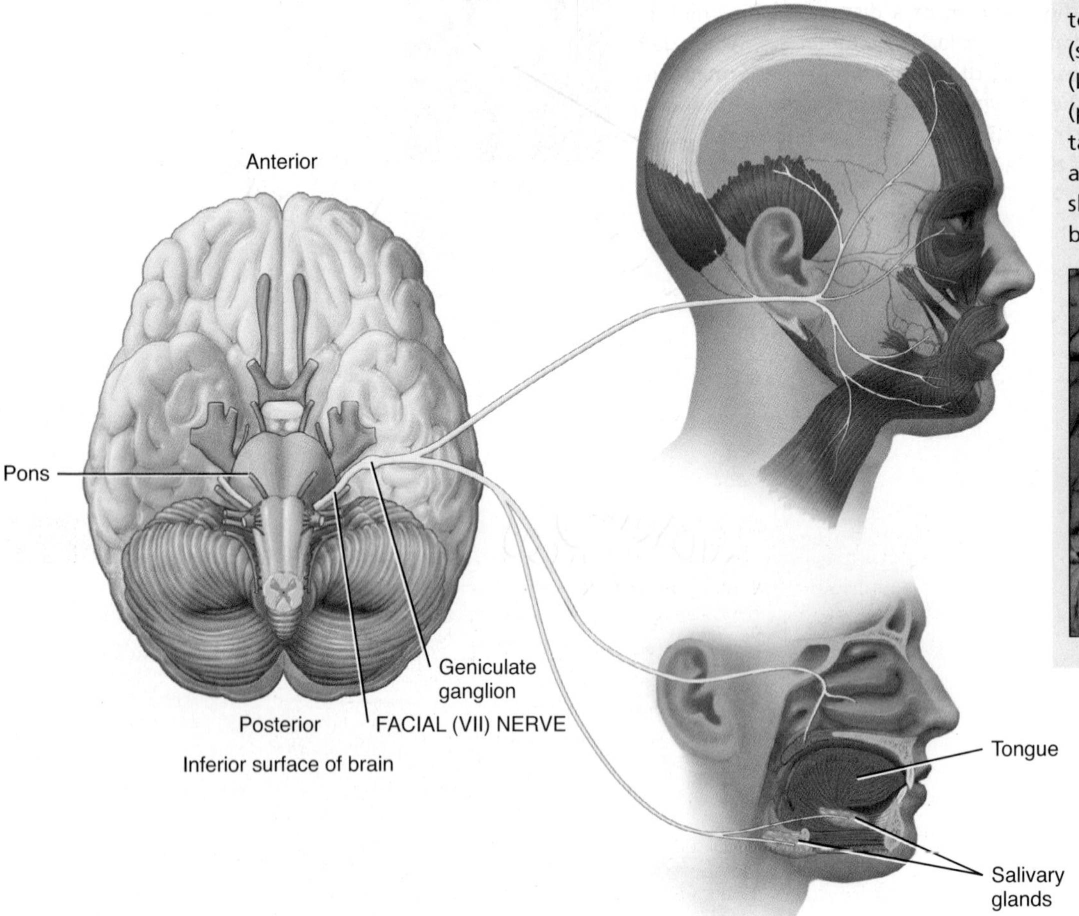
Anterior
Pons
Geniculate ganglion
Posterior
FACIAL (VII) NERVE
Inferior surface of brain
Tongue
Salivary glands

CLINICAL CONNECTION | *Bell's Palsy*

Damage to the facial (VII) nerve due to conditions such as viral infection (shingles) or a bacterial infection (Lyme disease) produces **Bell's palsy** (paralysis of the facial muscles), loss of taste, decreased salivation, and loss of ability to close the eyes, even during sleep. The nerve can also be damaged by trauma, tumors, and stroke. •

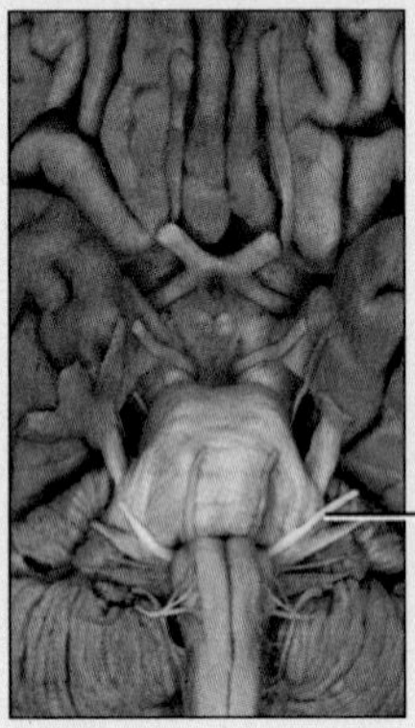
Facial (VII) nerve

Where do the motor axons of the facial nerve originate?

EXHIBIT 14.F Vestibulocochlear (VIII) Nerve *(Figure 14.22)*

OBJECTIVE

- Identify the origin of the vestibulocochlear nerve in the brain, the foramen through which it exits the skull, and the functions of each of its branches.

The **vestibulocochlear (VIII) nerve** (ves-tib-ū-lō-KOK-lē-ar; *vestibulo-* = small cavity; *-cochlear* = spiral, snail-like) was formerly known as the *acoustic* or *auditory nerve*. It is a sensory cranial nerve and has two branches, the vestibular branch and the cochlear branch (Figure 14.22). The **vestibular branch** carries impulses for equilibrium and the **cochlear branch** carries impulses for hearing.

Sensory axons in the vestibular branch extend from the semicircular canals, the saccule, and the utricle of the inner ear to the **vestibular ganglion,** where the cell bodies of the neurons are located (see Figure 17.21b), and end in vestibular nuclei in the pons and cerebellum. Some sensory axons also enter the cerebellum via the inferior cerebellar peduncle.

Sensory axons in the cochlear branch arise in the spiral organ (organ of Corti) in the cochlea of the internal ear. The cell bodies of cochlear branch sensory neurons are located in the **spiral ganglion** of the cochlea (see Figure 17.21b). From there, axons extend to nuclei in the medulla oblongata and end in the thalamus.

The nerve contains some motor fibers, but they do not innervate muscle tissue. Instead, they modulate the hair cells in the inner ear.

CHECKPOINT

What are the functions of each of the two branches of the vestibulocochlear nerve?

Figure 14.22 Vestibulocochlear (VIII) nerve.

The vestibular branch of the vestibulocochlear nerve carries impulses for equilibrium, while the cochlear branch carries impulses for hearing.

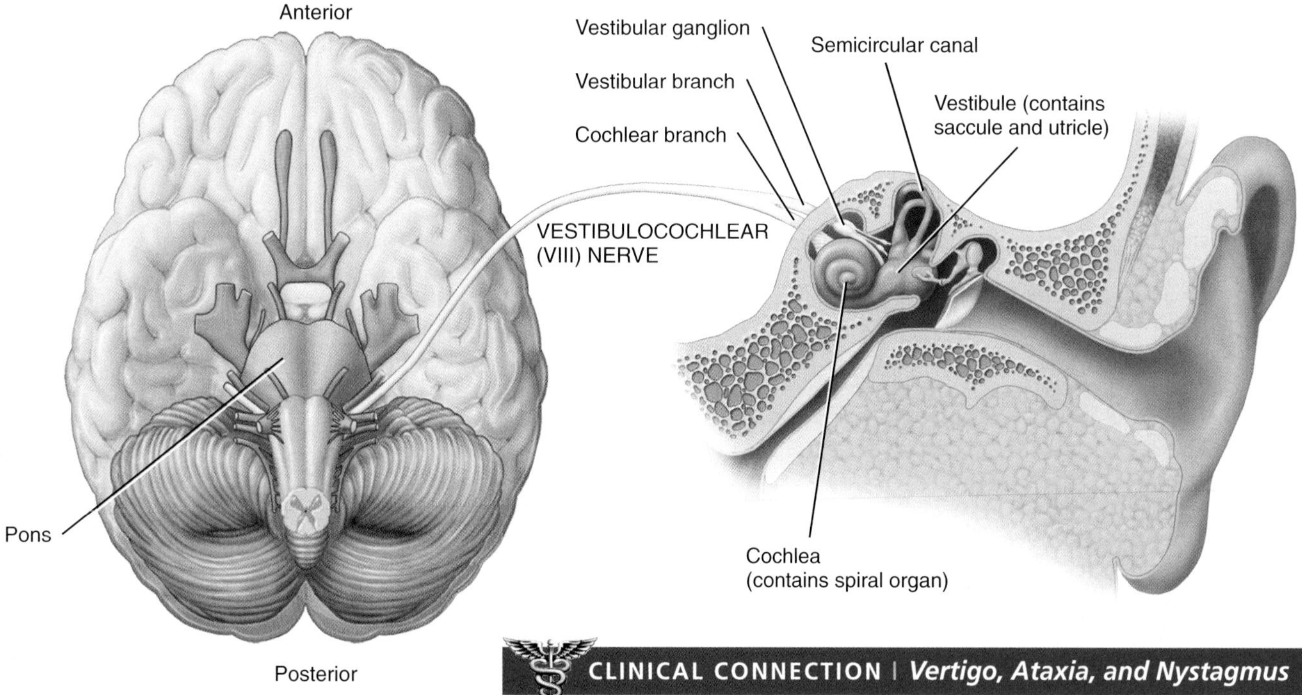

CLINICAL CONNECTION | *Vertigo, Ataxia, and Nystagmus*

Injury to the vestibular branch of the vestibulocochlear (VIII) nerve may cause **vertigo** (ver-TI-gō) (a subjective feeling that one's own body or the environment is rotating), **ataxia** (a-TAK-sē-a) (muscular incoordination), and **nystagmus** (nis-TAG-mus) (involuntary rapid movement of the eyeball). Injury to the cochlear branch may cause **tinnitus** (ringing in the ears) or deafness. The vestibulocochlear nerve may be injured as a result of conditions such as trauma, lesions, or middle ear infections. •

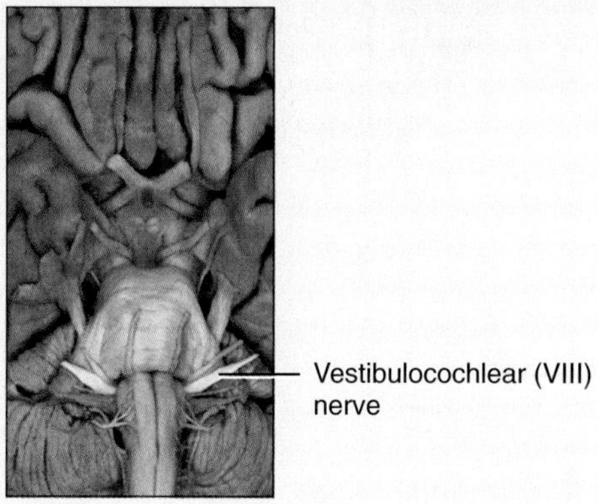

What structures are found in the vestibular and spiral ganglia?

EXHIBIT 14.G Glossopharyngeal (IX) Nerve *(Figure 14.23)*

OBJECTIVE

- Identify the origin of the glossopharyngeal nerve in the brain, the foramen through which it exits the skull, and its function.

The **glossopharyngeal (IX) nerve** (glos′-ō-fa-RIN-jē-al; *glosso-* = tongue; *-pharyngeal* = throat) is a mixed cranial nerve (Figure 14.23). Sensory axons of the glossopharyngeal nerve arise from (1) taste buds on the posterior one-third of the tongue, (2) proprioceptors from some swallowing muscles supplied by the motor portion, (3) baroreceptors (pressure-monitoring receptors) in the carotid sinus that monitor blood pressure, (4) chemoreceptors (receptors that monitor blood levels of oxygen and carbon dioxide) in the carotid bodies near the carotid arteries (see Figure 23.26) and aortic bodies near the arch of the aorta (see Figure 23.26), and (5) the external ear to convey touch, pain, and thermal (heat and cold) sensations. The cell bodies of these sensory neurons are located in the **superior** and **inferior ganglia.** From the ganglia, sensory axons pass through the jugular foramen and end in the medulla.

Axons of motor neurons in the glossopharyngeal nerve arise in nuclei of the medulla and exit the skull through the jugular foramen. Branchial motor neurons innervate the stylopharyngeus muscle, which assists in swallowing, and axons of autonomic motor neurons stimulate the parotid gland to secrete saliva. The postganglionic cell bodies of autonomic motor neurons are located in the **otic ganglion.**

CHECKPOINT

Which other cranial nerves are also distributed to the tongue?

Figure 14.23 **Glossopharyngeal (IX) nerve.**

Sensory axons in the glossopharyngeal nerve supply the taste buds.

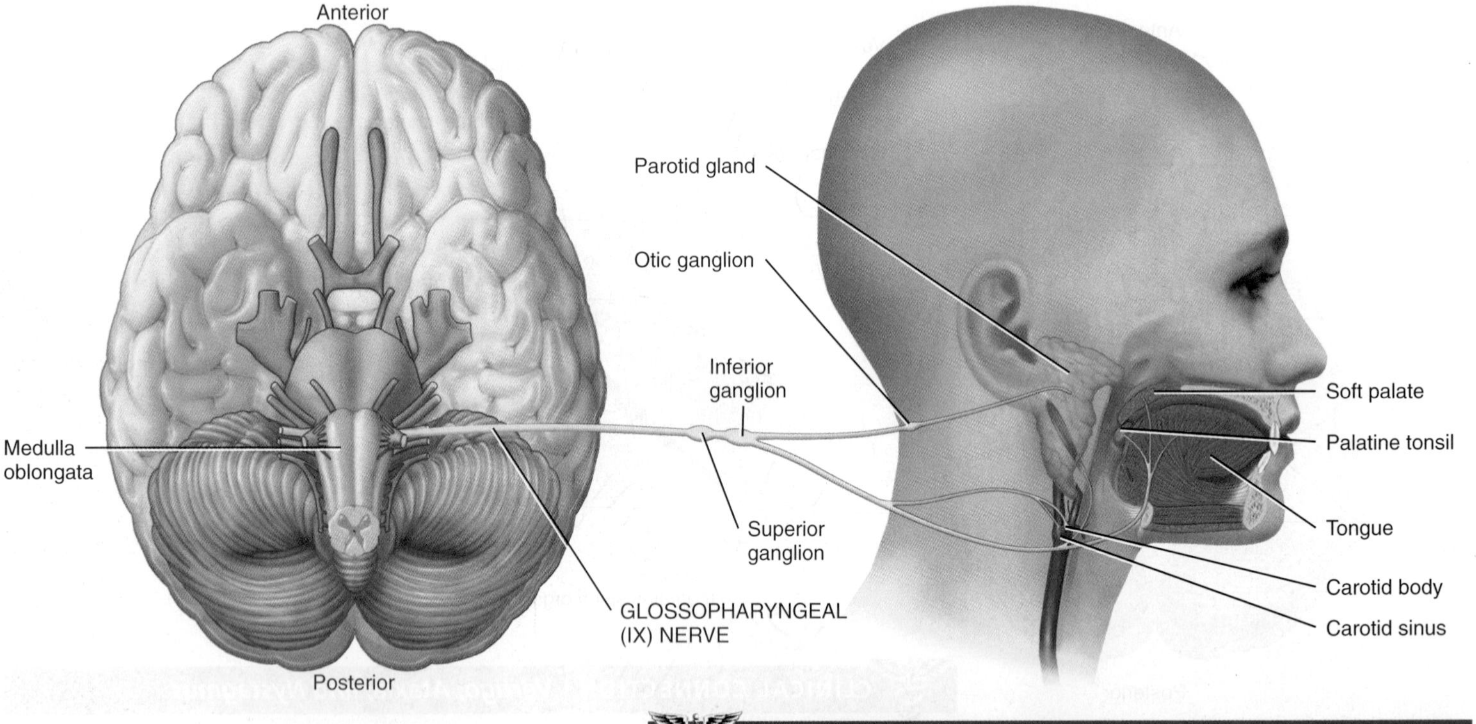

Inferior surface of brain

CLINICAL CONNECTION | *Dysphagia, Aptyalia, and Ageusia*

Injury to the glossopharyngeal (IX) nerve causes **dysphagia** (dis-FĀ-gē-a), or difficulty in swallowing; **aptyalia** (ap-tē-Ā-lē-a), or reduced secretion of saliva; loss of sensation in the throat; and **ageusia** (a-GOO-sē-a), or loss of taste sensation. The glossopharyngeal nerve may be injured as a result of conditions such as trauma or lesions. •

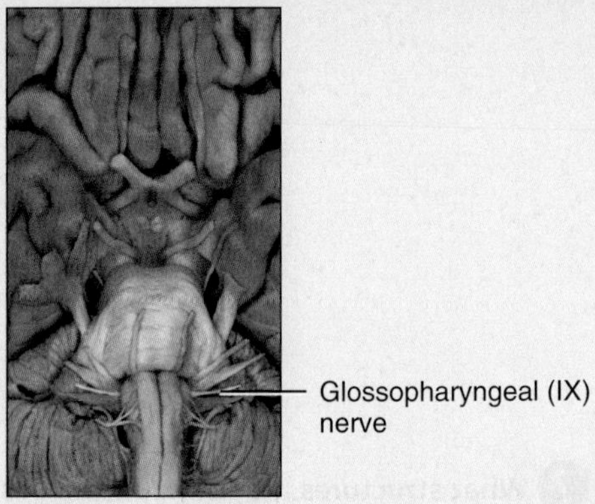

Through which foramen does the glossopharyngeal nerve exit the skull?

EXHIBIT 14.H Vagus (X) Nerve *(Figure 14.24)*

OBJECTIVE

- Identify the origin of the vagus nerve in the brain, the foramen through which it exits the skull, and its function.

The **vagus (X) nerve** (VĀ-gus = vagrant or wandering) is a mixed cranial nerve that is distributed from the head and neck into the thorax and abdomen (Figure 14.24). The nerve derives its name from its wide distribution. In the neck, it lies medial and posterior to the internal jugular vein and common carotid artery.

Sensory axons in the vagus nerve arise from the skin of the external ear for touch, pain, and thermal sensations; a few taste buds in the epiglottis and pharynx; and proprioceptors in muscles of the neck and throat. Also, sensory axons come from baroreceptors in the carotid sinus and chemoreceptors in the carotid and aortic bodies. The majority of sensory neurons come from visceral sensory receptors in most organs of the thoracic and abdominal cavities that convey sensations (such as hunger, fullness, and discomfort) from these organs. The sensory neurons have cell bodies in the **superior** and **inferior ganglia** and then pass through the jugular foramen to end in the medulla and pons.

The branchial motor neurons, which run briefly with the accessory nerve, arise from nuclei in the medulla oblongata and supply muscles of the pharynx, larynx, and soft palate that are used in swallowing, vocalization, and coughing. Historically these motor neurons have been called the cranial accessory nerve, but these fibers actually belong to the vagus (X) nerve.

Axons of autonomic motor neurons in the vagus nerve originate in nuclei of the medulla and supply the lungs, heart, glands of the gastrointestinal (GI) tract, and smooth muscle of the respiratory passageways, esophagus, stomach, gallbladder, small intestine, and most of the large intestine (see Figure 15.3). Autonomic motor axons initiate smooth muscle contractions in the gastrointestinal tract to aid motility and stimulate secretion by digestive glands; activate smooth muscle to constrict respiratory passageways; and decrease heart rate.

✔CHECKPOINT

On what basis is the vagus nerve named?

Figure 14.24 Vagus (X) nerve.

The vagus nerve is widely distributed in the head, neck, thorax, and abdomen.

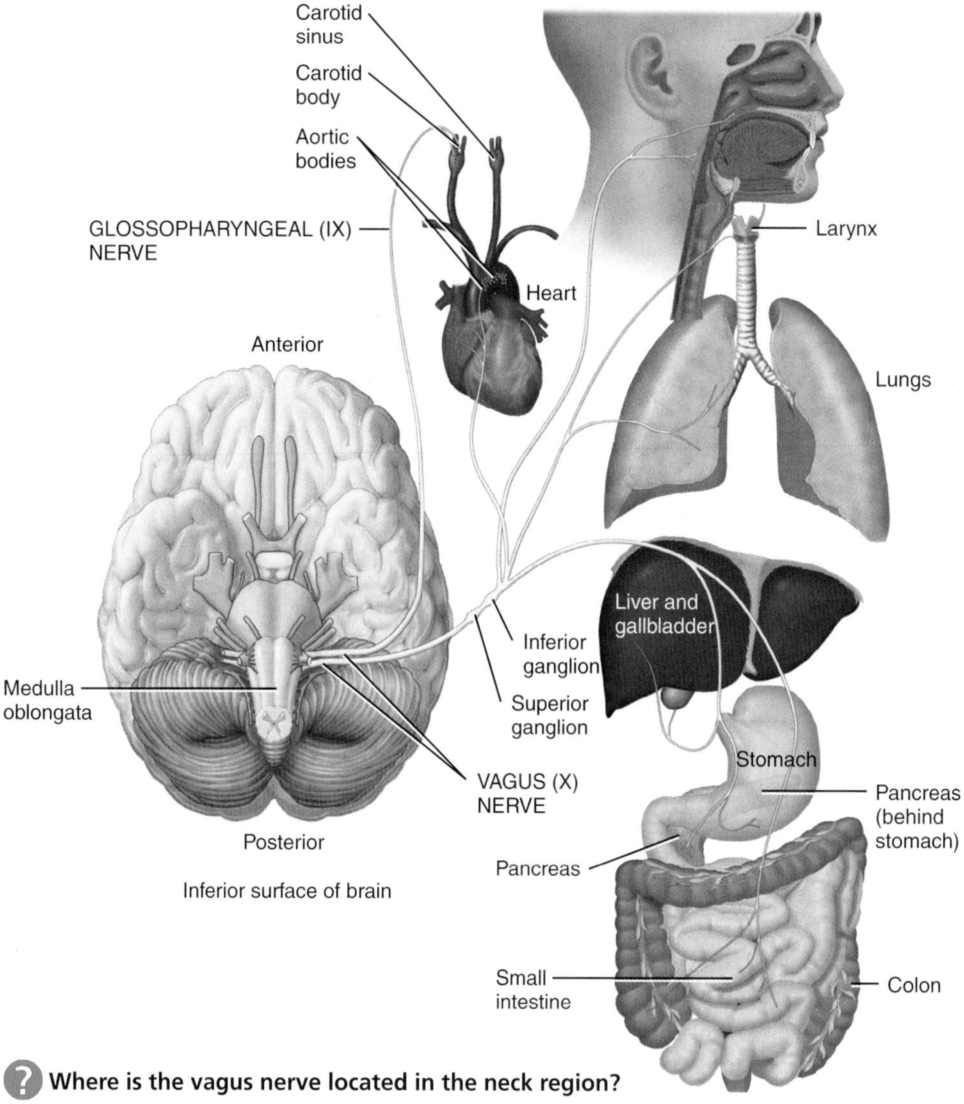

Carotid sinus
Carotid body
Aortic bodies
GLOSSOPHARYNGEAL (IX) NERVE
Heart
Larynx
Lungs
Anterior
Posterior
Inferior surface of brain
Medulla oblongata
Inferior ganglion
Superior ganglion
VAGUS (X) NERVE
Liver and gallbladder
Stomach
Pancreas (behind stomach)
Pancreas
Small intestine
Colon

? **Where is the vagus nerve located in the neck region?**

CLINICAL CONNECTION | *Vagal Paralysis, Dysphagia, and Tachycardia*

Injury to the vagus (X) nerve due to conditions such as trauma or lesions causes **vagal paralysis,** or interruptions of sensations from many organs in the thoracic and abdominal cavities; **dysphagia** (dis-FĀ-gē-a), or difficulty in swallowing; and **tachycardia** (tak′-i-KAR-dē-a), or increased heart rate. •

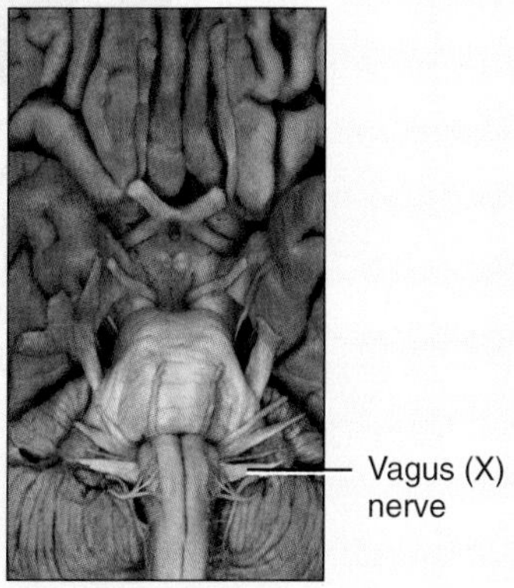

Vagus (X) nerve

EXHIBIT 14.I Accessory (XI) Nerve *(Figure 14.25)*

OBJECTIVE

- Identify the origin of the accessory nerve in the spinal cord, the foramina through which it first enters and then exits the skull, and its function.

The **accessory (XI) nerve** (ak-SES-ō-rē = assisting) is a motor cranial nerve (Figure 14.25). Historically it has been divided into two parts, a cranial accessory nerve and a spinal accessory nerve. The cranial accessory nerve actually is part of the vagus (X) nerve (see Exhibit 14.H). The "old" spinal accessory nerve is the accessory nerve we discuss in this exhibit. Its motor axons arise in the anterior gray horn of the first five segments of the cervical portion of the spinal cord. The axons from the segments exit the spinal cord laterally and come together, ascend through the foramen magnum, and then exit through the jugular foramen along with the vagus and glossopharyngeal nerves. The accessory nerve conveys motor impulses to the sternocleidomastoid and trapezius muscles to coordinate head movements. Sensory axons in the accessory nerve, which originate from proprioceptors in the sternocleidomastoid and trapezius muscles, begin their course toward the brain in the accessory nerve, but eventually leave the nerve to join nerves of the cervical plexus. From the cervical plexus they enter the spinal cord via the posterior roots of cervical spinal nerves; their cell bodies are located in the posterior root ganglia of those nerves. In the spinal cord the axons ascend to nuclei in the medulla oblongata.

✔ CHECKPOINT

Where do the motor axons of the accessory nerve originate?

Figure 14.25 **Accessory (XI) nerve.**

The accessory nerve exits the cranium through the jugular foramen.

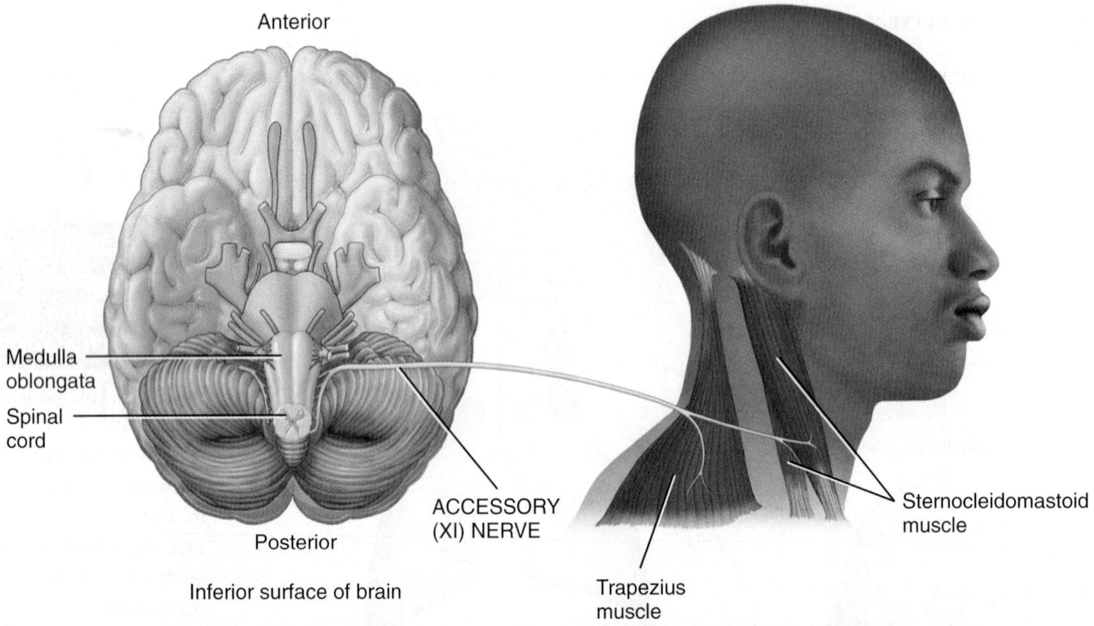

CLINICAL CONNECTION | *Paralysis of the Sternocleidomastoid and Trapezius Muscles*

If the accessory (XI) nerve is damaged due to conditions such as trauma, lesions, or stroke, the result is **paralysis of the sternocleidomastoid and trapezius muscles** so that the person is unable to raise the shoulders and has difficulty in turning the head. •

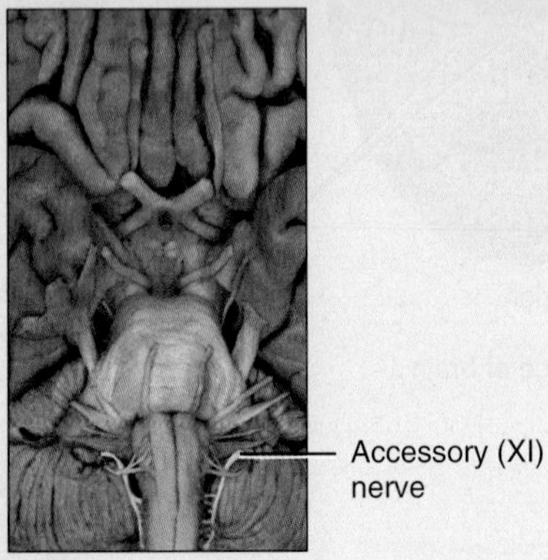

? How does the accessory nerve differ from the other cranial nerves?

EXHIBIT 14.J Hypoglossal (XII) Nerve *(Figure 14.26)*

OBJECTIVE

- Identify the origin of the hypoglossal nerve in the brain, the foramen through which it exits the skull, and its function.

The **hypoglossal (XII) nerve** (hī′-pō-GLOS-al; *hypo-* = below; *-glossal* = tongue) is a motor cranial nerve. The somatic motor axons originate in a nucleus in the medulla oblongata (hypoglossal nucleus), exit the medulla on its anterior surface, and pass through the hypoglossal canal to supply the muscles of the tongue (Figure 14.26). These axons conduct nerve impulses for speech and swallowing. The sensory axons do not return to the brain in the hypoglossal nerve. Instead, sensory axons that originate from proprioceptors in the tongue muscles begin their course toward the brain in the hypoglossal nerve but leave the nerve to join cervical spinal nerves and end in the medulla oblongata, again entering the central nervous system via posterior roots of cervical spinal nerves.

✔CHECKPOINT

In what portion of the brain does the hypoglossal nucleus originate?

Figure 14.26 Hypoglossal (XII) nerve.

The hypoglossal nerve exits the cranium through the hypoglossal canal.

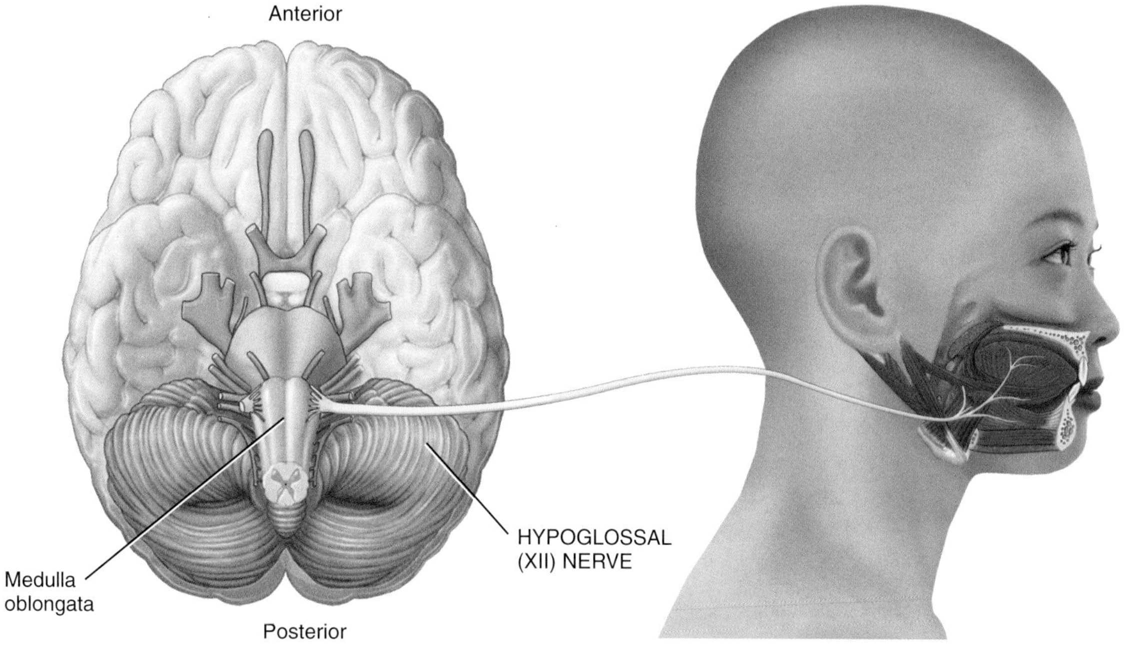

Inferior surface of brain

CLINICAL CONNECTION | *Dysarthria and Dysphagia*

Injury to the hypoglossal (XII) nerve results in difficulty in chewing; **dysarthria** (dis-AR-thrē-a), or difficulty in speaking; and **dysphagia** (dis-FĀ-gē-a), or difficulty in swallowing. The tongue, when protruded, curls toward the affected side, and the affected side atrophies. The hypoglossal nerve may be injured as a result of conditions such as trauma, lesions, stroke, amyotrophic lateral sclerosis (Lou Gehrig's disease), or infections in the brain stem. •

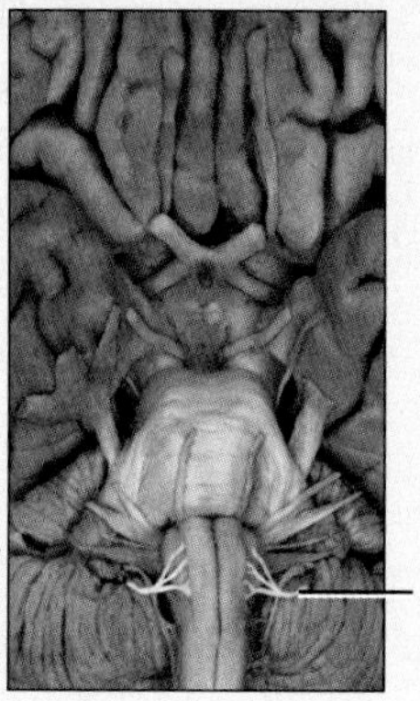

? What important motor functions are related to the hypoglossal nerve?

TABLE 14.4

Summary of Cranial Nerves*

CRANIAL NERVE	COMPONENTS	PRINCIPAL FUNCTIONS
Olfactory (I)	***Special sensory***	Olfaction (smell).
Optic (II)	***Special sensory***	Vision (sight).
Oculomotor (III)	***Motor***	
	Somatic	Movement of eyeballs and upper eyelid.
	Motor (autonomic)	Adjusts lens for near vision (accommodation).
		Constriction of pupil.
Trochlear (IV)	***Motor***	
	Somatic	Movement of eyeballs.
Trigeminal (V)	***Mixed***	
	Sensory	Touch, pain, and thermal sensations from scalp, face, and oral cavity (including teeth and anterior two-thirds of tongue).
	Motor (branchial)	Chewing and controls middle ear muscle.
Abducens (VI)	***Motor***	
	Somatic	Movement of eyeballs.
Facial (VII)	***Mixed***	
	Sensory	Taste from anterior two-thirds of tongue.
		Touch, pain, and thermal sensations from skin in external ear canal.
	Motor (branchial)	Control of muscles of facial expression and middle ear muscle.
	Motor (autonomic)	Secretion of tears and saliva.
Vestibulocochlear (VIII)	***Special sensory***	Hearing and equilibrium.
Glossopharyngeal (IX)	***Mixed***	
	Sensory	Taste from posterior one-third of tongue.
		Proprioception in some swallowing muscles.
		Monitors blood pressure and oxygen and carbon dioxide levels in blood.
		Touch, pain, and thermal sensations from skin of external ear and upper pharynx.
	Motor (branchial)	Assists in swallowing.
	Motor (autonomic)	Secretion of saliva.
Vagus (X)	***Mixed***	
	Sensory	Taste from epiglottis.
		Proprioception from throat and voice box muscles.
		Monitors blood pressure and oxygen and carbon dioxide levels in blood.
		Touch, pain, and thermal sensations from skin of external ear.
		Sensations from thoracic and abdominal organs.
	Motor (branchial)	Swallowing, vocalization, and coughing.
	Motor (autonomic)	Motility and secretion of gastrointestinal organs.
		Constriction of respiratory passageways.
		Decreases heart rate.
Accessory (XI)	***Motor***	
	Branchial	Movement of head and pectoral girdle.
Hypoglossal (XII)	***Motor***	
	Somatic	Speech, manipulation of food, and swallowing.

***MNEMONIC FOR CRANIAL NERVES:**

Oh	Oh	Oh	To	Touch	And	Feel	Very	Green	Vegetables	AH!	
Olfactory	Optic	Oculomotor	Trochlear	Trigeminal	Abducens	Facial	Vestibulocochlear	Glossopharyngeal	Vagus	Accessory	Hypoglossal

14.9 DEVELOPMENT OF THE NERVOUS SYSTEM

OBJECTIVE

- Describe how the parts of the brain develop.

Development of the nervous system begins in the third week of gestation with a thickening of the **ectoderm** called the **neural plate** (Figure 14.27). The plate folds inward and forms a longitudinal groove, the **neural groove.** The raised edges of the neural plate are called **neural folds.** As development continues, the neural folds increase in height and meet to form a tube called the **neural tube.**

Three layers of cells differentiate from the wall that encloses the neural tube. The outer or **marginal layer** cells develop into the *white matter* of the nervous system. The middle or **mantle layer** cells develop into the *gray matter.* The inner or **ependymal layer** (e-PEN-di-mal) cells eventually form the *lining of the central canal of the spinal cord* and *ventricles of the brain.*

The **neural crest** is a mass of tissue between the neural tube and the skin ectoderm (Figure 14.27b). It differentiates and eventually forms the *posterior (dorsal) root ganglia of spinal nerves, spinal nerves, ganglia of cranial nerves, cranial nerves, ganglia of the autonomic nervous system, adrenal medulla,* and *meninges.*

Figure 14.27 **Origin of the nervous system.** (a) Dorsal view of an embryo in which the neural folds have partially united, forming the early neural tube. (b) Transverse sections through the embryo showing the formation of the neural tube.

6 The nervous system begins developing in the third week from a thickening of ectoderm called the neural plate.

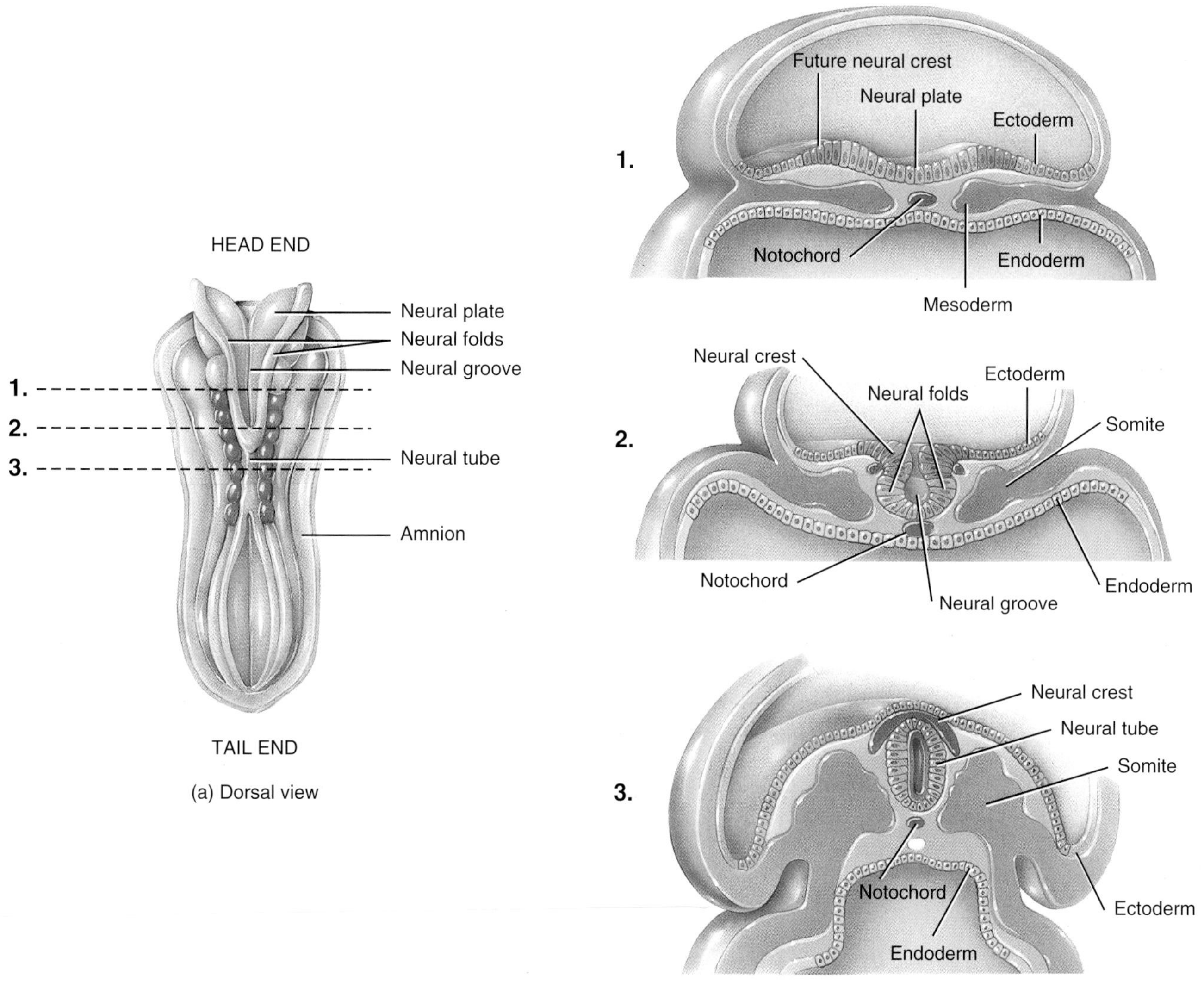
HEAD END
Neural plate
Neural folds
Neural groove
1.
2.
3.
Neural tube
Amnion
TAIL END
Future neural crest
Neural plate
Ectoderm
Notochord
Endoderm
Mesoderm
Neural crest
Neural folds
Ectoderm
Somite
Notochord
Neural groove
Endoderm
Neural crest
Neural tube
Somite
Notochord
Ectoderm
Endoderm

(a) Dorsal view

(b) Transverse sections

What is the origin of the gray matter of the nervous system?

As discussed at the beginning of this chapter, during the third to fourth week of embryonic development, the anterior part of the neural tube develops into three enlarged areas called **primary brain vesicles** that are named for their relative positions. These are the **prosencephalon** (pros′-en-SEF-a-lon; *pros-* = before) or forebrain, **mesencephalon** (mes′-en-SEF-a-lon; *mes-* = middle) or midbrain, and **rhombencephalon** (rom′-ben-SEF-a-lon; *rhomb-* = behind) or hindbrain (Figure 14.28a; see also Table 14.1). During the fifth week of development, **secondary brain vesicles** begin to develop. The prosencephalon develops into two secondary brain vesicles called the **telencephalon** (tel′-en-SEF-a-lon; *tel-* = distant) and the **diencephalon** (dī-en-SEF-a-lon; *di-* = through) (Figure 14.28b). The rhombencephalon also develops into two secondary brain vesicles called the **metencephalon** (met′-en-SEF-a-lon; *met-* = after) and the **myelencephalon** (mī-el-en-SEF-a-lon; *myel-* = marrow). The area of the neural tube inferior to the myelencephalon gives rise to the *spinal cord.*

The brain vesicles continue to develop as follows (Figure 14.28c, d; see also Table 14.1):

- The telencephalon develops into the *cerebral hemispheres,* including the *basal nuclei,* and houses the paired *lateral ventricles.*

Figure 14.28 Development of the brain and spinal cord.

⑥ The various parts of the brain develop from the primary brain vesicles.

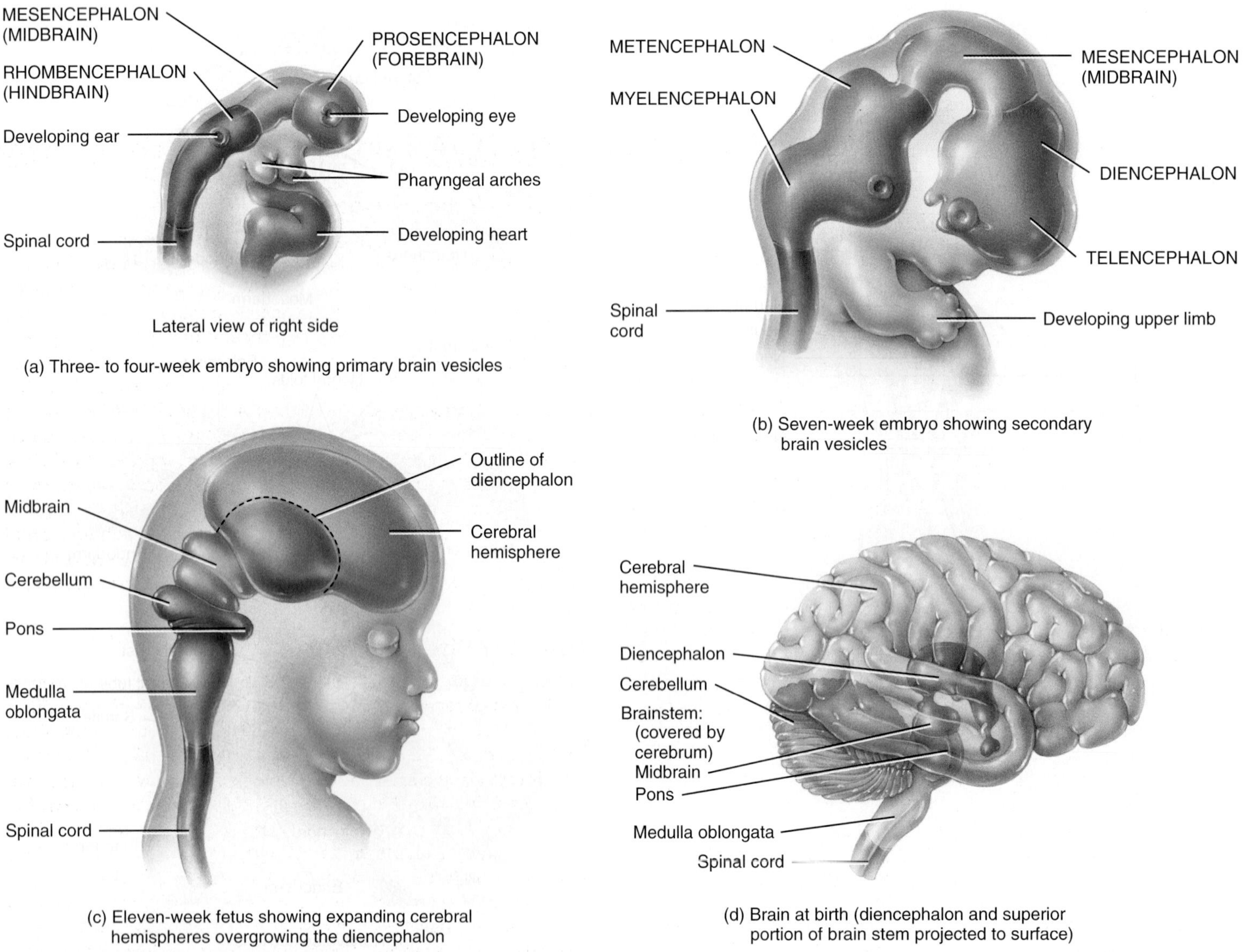

(a) Three- to four-week embryo showing primary brain vesicles

(b) Seven-week embryo showing secondary brain vesicles

(c) Eleven-week fetus showing expanding cerebral hemispheres overgrowing the diencephalon

(d) Brain at birth (diencephalon and superior portion of brain stem projected to surface)

Which primary brain vesicle does not develop into a secondary brain vesicle?

- The diencephalon develops into the *thalamus, hypothalamus,* and *epithalamus.*
- The mesencephalon develops into the *midbrain,* which surrounds the *aqueduct of the midbrain (cerebral aqueduct).*
- The metencephalon becomes the *pons* and *cerebellum* and houses part of the *fourth ventricle.*
- The myelencephalon develops into the *medulla oblongata* and houses the remainder of the *fourth ventricle.*

Two **neural tube defects**—spina bifida (see Disorders: Homeostatic Imbalances at the end of Chapter 7) and anencephaly (absence of the skull and cerebral hemispheres, discussed in Section 29.1)—are associated with low levels of folic acid (folate), one of the B vitamins, in the first few weeks of development. Many foods, especially grain products such as cereals and bread, are now fortified with folic acid; however, the incidence of both disorders is greatly reduced when women who are or may become pregnant take folic acid supplements.

✔ CHECKPOINT

26. What parts of the brain develop from each primary brain vesicle?

14.10 AGING AND THE NERVOUS SYSTEM

OBJECTIVE

- Describe the effects of aging on the nervous system.

The brain grows rapidly during the first few years of life. Growth is due mainly to an increase in the size of neurons already present, the proliferation and growth of neuroglia, the development of dendritic branches and synaptic contacts, and continuing myelination of axons. From early adulthood onward, brain mass declines. By the time a person reaches 80, the brain weighs about 7% less than it did in young adulthood. Although the number of neurons present does not decrease very much, the number of synaptic contacts declines. Associated with the decrease in brain mass is a decreased capacity for sending nerve impulses to and from the brain. As a result, processing of information diminishes. Conduction velocity decreases, voluntary motor movements slow down, and reflex times increase.

✔ CHECKPOINT

27. How is brain mass related to age?

DISORDERS: HOMEOSTATIC IMBALANCES

Cerebrovascular Accident

The most common brain disorder is a **cerebrovascular accident (CVA),** also called a **stroke** or **brain attack.** CVAs affect 500,000 people each year in the United States and represent the third leading cause of death, behind heart attacks and cancer. A CVA is characterized by abrupt onset of persisting neurological symptoms, such as paralysis or loss of sensation, that arise from destruction of brain tissue. Common causes of CVAs are intracerebral hemorrhage (bleeding from a blood vessel in the pia mater or brain), emboli (blood clots), and atherosclerosis (formation of cholesterol-containing plaques that block blood flow) of the cerebral arteries.

Among the risk factors implicated in CVAs are high blood pressure, high blood cholesterol, heart disease, narrowed carotid arteries, transient ischemic attacks (TIAs; discussed next), diabetes, smoking, obesity, and excessive alcohol intake.

A clot-dissolving drug called *tissue plasminogen activator (t-PA)* is used to open up blocked blood vessels in the brain. The drug is most effective when administered *within three hours* of the onset of the CVA, however, and is helpful only for CVAs due to a blood clot *(ischemic CVAs).* Use of t-PA can decrease the permanent disability associated with these types of CVAs by 50%. However, t-PA should not be administered to individuals with strokes caused by hemorrhaging *(hemorrhagic CVAs)* since it can cause further injury or even death. The distinction between the types of CVA is made on the basis of a CT scan.

New studies show that "cold therapy" might be successful in limiting the amount of residual damage from a CVA. States of hypothermia, such as those experienced by cold-water drowning victims, seem to trigger a survival response in which the body requires less oxygen; application of this principle to stroke victims has showed promise. Some commercial companies now provide "CVA survival kits" including cooling blankets that can be kept in the home.

Transient Ischemic Attacks

A **transient ischemic attack (TIA)** (is-KĒ-mik) is an episode of temporary cerebral dysfunction caused by impaired blood flow to part of the brain. Symptoms include dizziness, weakness, numbness, or paralysis in a limb or on one side of the body; drooping of one side of the face; headache; slurred speech or difficulty understanding speech; and/or a partial loss of vision or double vision. Sometimes nausea or vomiting also occurs. The onset of symptoms is sudden and reaches maximum intensity almost immediately. A TIA usually persists for 5 to 10 minutes and only rarely lasts as long as 24 hours. It leaves no permanent neurological deficits. The causes of TIA include blood clots, atherosclerosis, and certain blood disorders. About one-third of patients who experience a TIA will have a CVA eventually. Therapy for TIAs includes drugs such as aspirin, which blocks the aggregation of blood platelets, and anticoagulants; cerebral artery bypass grafting; and carotid endarterectomy (removal of the cholesterol-containing plaques and inner lining of an artery).

Alzheimer Disease

Alzheimer disease (ALTZ-hī-mer) or **AD** is a disabling senile dementia, the loss of reasoning and ability to care for oneself, that afflicts about 11% of the population over age 65. In the United States, about 4 million people suffer from AD. Claiming over 100,000 lives a year, AD is the fourth leading cause of death among the elderly, after heart disease, cancer, and stroke. The cause of most AD cases is still unknown, but evidence suggests it is due to a combination of genetic factors, environmental or lifestyle factors, and the aging process. Mutations in three different genes (coding for presenilin-1, presenilin-2, and amyloid precursor protein) lead to early-onset forms of AD in afflicted families but account for less than 1% of all cases. An environmental risk factor for developing AD is a history of head injury. A similar dementia occurs in boxers, probably caused by repeated blows to the head.

Individuals with AD initially have trouble remembering recent events. They then become confused and forgetful, often repeating questions or getting lost while traveling to familiar places. Disorientation grows, memories of past events disappear, and episodes of paranoia, hallucination, or violent changes in mood may occur. As their minds continue to deteriorate, they lose their ability to read, write, talk, eat, or walk. The disease culminates in dementia. A person with AD usually dies of some complication that afflicts bedridden patients, such as pneumonia.

At autopsy, brains of AD victims show three distinct structural abnormalities:

1. ***Loss of neurons that liberate acetylcholine.*** A major center of neurons that liberate ACh is the *nucleus basalis*, a group of large cells below the globus pallidus. Axons of these neurons project widely throughout the cerebral cortex and limbic system. Their destruction is a hallmark of Alzheimer disease.
2. ***Beta-amyloid plaques*** (bā′-ta-AM-i-loyd), clusters of abnormal proteins deposited outside neurons.
3. ***Neurofibrillary tangles*** (noo-rō-FĪ-bril-ler′-ē), abnormal bundles of filaments inside neurons in affected brain regions. These filaments consist of a protein called *tau* (TAW) that has been *hyperphosphorylated* (hī-per-fos-FOR-i-lā-ted, meaning too many phosphate groups have been added to it).

Drugs that inhibit acetylcholinesterase (AChE), the enzyme that inactivates ACh, improve alertness and behavior in about 5% of AD patients. Tacrine®, the first anticholinesterase inhibitor approved for treatment of AD in the United States, has significant side effects and requires dosing four times a day. Donepezil®, approved in 1998, is less toxic to the liver and has the advantage of once-a-day dosing. Some evidence suggests that vitamin E (an antioxidant), estrogen, ibuprofen, and ginkgo biloba extract may have slight beneficial effects in AD patients. In addition, researchers are currently exploring ways to develop drugs that will prevent beta-amyloid plaque formation by inhibiting the enzymes involved in beta-amyloid synthesis and by increasing the activity of the enzymes involved in beta-amyloid degradation. Researchers are also trying to develop drugs that will reduce the formation of neurofibrillary tangles by inhibiting the enzymes that hyperphosphorylate tau.

Brain Tumors

A **brain tumor** is an abnormal growth of tissue in the brain that may be malignant or benign. Unlike most other tumors in the body, malignant and benign brain tumors may be equally serious, compressing adjacent tissues and causing a buildup of pressure in the skull. The most common malignant tumors are secondary tumors that metastasize from other cancers in the body, such as those in the lungs, breasts, skin (malignant melanoma), blood (leukemia), and lymphatic organs (lymphoma). Most primary brain tumors (those that originate within the brain) are gliomas, which develop in neuroglia. The symptoms of a brain tumor depend on its size, location, and rate of growth. Among the symptoms are headache, poor balance and coordination, dizziness, double vision, slurred speech, nausea and vomiting, fever, abnormal pulse and breathing rates, personality changes, numbness and weakness of the limbs, and seizures. Treatment options for brain tumors vary with their size, location, and type and may include surgery, radiation therapy, and/or chemotherapy. Unfortunately, chemotherapeutic agents do not readily cross the blood–brain barrier.

Attention Deficit Hyperactivity Disorder

Attention deficit hyperactivity disorder (ADHD) is a learning disorder characterized by poor or short attention span, a consistent level of hyperactivity, and a level of impulsiveness inappropriate for the child's age. ADHD is believed to affect about 5% of children and is diagnosed 10 times more often in boys than in girls. The condition typically begins in childhood and continues into adolescence and adulthood. Symptoms of ADHD develop in early childhood, often before age four, and include difficulty in organizing and finishing tasks, lack of attention to details, short attention span and inability to concentrate, difficulty following instructions, talking excessively and frequently interrupting others, frequent running or excessive climbing, inability to play quietly alone, and difficulty waiting or taking turns.

The causes of ADHD are not fully understood, but it does have a strong genetic component. Some evidence also suggests that ADHD is related to problems with neurotransmitters. In addition, recent imaging studies have demonstrated that people with ADHD have less nervous tissue in specific regions of the brain such as the frontal and temporal lobes, caudate nucleus, and cerebellum. Treatment may involve remedial education, behavioral modification techniques, restructuring routines, and drugs that calm the child and help focus attention.

MEDICAL TERMINOLOGY

Agnosia (ag-NŌ-zē-a; *a-* = without; *-gnosia* = knowledge) Inability to recognize the significance of sensory stimuli such as sounds, sights, smells, tastes, and touch.

Apraxia (a-PRAK-sē-a; *-praxia* = coordinated) Inability to carry out purposeful movements in the absence of paralysis.

Consciousness (KON-shus-nes) A state of wakefulness in which an individual is fully alert, aware, and oriented, partly as a result of feedback between the cerebral cortex and reticular activating system.

Delirium (dē-LIR-ē-um = off the track) A transient disorder of abnormal cognition and disordered attention accompanied by disturbances of the sleep–wake cycle and psychomotor behavior (hyperactivity or hypoactivity of movements and speech). Also called **acute confusional state (ACS).**

Dementia (de-MEN-shē-a; *de-* = away from; *-mentia* = mind) Permanent or progressive general loss of intellectual abilities, including impairment of memory, judgment, and abstract thinking and changes in personality.

Encephalitis (en′-sef-a-LĪ-tis) An acute inflammation of the brain caused by either a direct attack by any of several viruses or an allergic reaction to any of the many viruses that are normally harmless to the central nervous system. If the virus affects the spinal cord as well, the condition is called **encephalomyelitis.**

Encephalopathy (en-sef′-a-LOP-a-thē; *encephalo-* = brain; *-pathos* = disease) Any disorder of the brain.

Lethargy (LETH-ar-jē) A condition of functional sluggishness.

Microcephaly (mī-krō-SEF-a-lē; *micro-* = small; *-cephal* = head) A congenital condition that involves the development of a small brain and skull and frequently results in mental retardation.

Prosopagnosia (pros′-ō-pag-NŌ-sē-a; *a-* = without; *-gnosia* = knowledge) Inability to recognize faces, usually caused by damage to the facial recognition area in the inferior temporal lobe of both cerebral hemispheres.

Reye's syndrome (RĪZ) Occurs after a viral infection, particularly chickenpox or influenza, most often in children or teens who have taken aspirin; characterized by vomiting and brain dysfunction (disorientation, lethargy, and personality changes) that may progress to coma and death.

Stupor (STOO-por) Unresponsiveness from which a patient can be aroused only briefly and only by vigorous and repeated stimulation.

CHAPTER REVIEW AND RESOURCE SUMMARY

WILEY PLUS

Review

Resource

14.1 Brain Organization, Protection, and Blood Supply

1. The major parts of the brain are the brain stem, cerebellum, diencephalon, and cerebrum.
2. The brain is protected by cranial bones and the cranial meninges.
3. The cranial meninges are continuous with the spinal meninges. From superficial to deep they are the dura mater, arachnoid mater, and pia mater.
4. Blood flow to the brain is mainly via the internal carotid and vertebral arteries.
5. Any interruption of the oxygen or glucose supply to the brain can result in weakening of, permanent damage to, or death of brain cells.
6. The blood–brain barrier (BBB) causes different substances to move between the blood and the brain tissue at different rates and prevents the movement of some substances from blood into the brain.

Anatomy Overview - Nervous System
Anatomy Overview - The Brain
Figure 14.1 - The Brain
Exercise - Concentrate on Your Brain

14.2 Cerebrospinal Fluid

1. Cerebrospinal fluid (CSF) is formed in the choroid plexuses and circulates through the lateral ventricles, third ventricle, fourth ventricle, subarachnoid space, and central canal. Most of the fluid is absorbed into the blood across the arachnoid villi of the superior sagittal sinus.
2. Cerebrospinal fluid provides mechanical protection, chemical protection, and circulation of nutrients.

14.3 The Brain Stem and Reticular Formation

1. The medulla oblongata is continuous with the superior part of the spinal cord and contains both sensory tracts and motor tracts. It contains a cardiovascular center, which regulates heart rate and blood vessel diameter, and a medullary rhythmicity area, which helps control breathing. It also contains the gracile nucleus, cuneate nucleus, gustatory nucleus, cochlear nuclei, and vestibular nuclei, which are components of sensory pathways to the brain. Also present in the medulla is the inferior olivary nucleus, which provides instructions that the cerebellum uses to adjust muscle activity when you learn new motor skills. Other nuclei of the medulla coordinate vomiting, swallowing, sneezing, coughing, and hiccupping. The medulla also contains nuclei associated with cranial nerves VIII–XII.
2. The pons is superior to the medulla. It contains both sensory tracts and motor tracts. Pontine nuclei relay nerve impulses related to voluntary skeletal movements from the cerebral cortex to the cerebellum. The pons also contains the pneumotaxic and apneustic centers, which help control breathing. Vestibular nuclei, which are present in the pons and medulla, are part of the equilibrium pathway to the brain. Also present in the pons are nuclei associated with cranial nerves V–VIII.
3. The midbrain connects the pons and diencephalon and surrounds the cerebral aqueduct. It contains both sensory tracts and motor tracts. The superior colliculi coordinate movements of the head, eye, and trunk in response to visual stimuli; the inferior colliculi coordinate movements of the head, eyes, and trunk in response to auditory stimuli. The midbrain also contains nuclei associated with cranial nerves III and IV.
4. A large part of the brain stem consists of small areas of gray matter and white matter called the reticular formation, which helps maintain consciousness, causes awakening from sleep, and contributes to regulating muscle tone.

Anatomy Overview - Brain Stem
Figure 14.5 - Medulla Oblongata in Relation to the Rest of the Brain Stem

14.4 The Cerebellum

1. The cerebellum occupies the inferior and posterior aspects of the cranial cavity. It consists of two lateral hemispheres and a medial, constricted vermis.
2. It connects to the brain stem by three pairs of cerebellar peduncles.
3. The cerebellum smooths and coordinates the contractions of skeletal muscles. It also maintains posture and balance.

Anatomy Overview - Cerebellum

Review

Resource

14.5 The Diencephalon

Anatomy Overview - Diencephalon
Figure 14.10 - Hypothalamus

1. The diencephalon surrounds the third ventricle and consists of the thalamus, hypothalamus, and epithalamus.
2. The thalamus is superior to the midbrain and contains nuclei that serve as relay stations for most sensory input to the cerebral cortex. It also contributes to motor functions by transmitting information from the cerebellum and basal nuclei to the primary motor area of the cerebral cortex. In addition, the thalamus plays a role in maintenance of consciousness.
3. The hypothalamus is inferior to the thalamus. It controls the autonomic nervous system, produces hormones, and regulates emotional and behavioral patterns (along with the limbic system). The hypothalamus also contains a feeding center and satiety center, which regulate eating, and a thirst center, which regulates drinking. In addition, the hypothalamus controls body temperature by serving as the body's thermostat. Also present in the hypothalamus is the suprachiasmatic nucleus, which regulates circadian rhythms and functions as the body's internal biological clock.
4. The epithalamus consists of the pineal gland and the habenular nuclei. The pineal gland secretes melatonin, which is thought to promote sleep and to help set the body's biological clock.
5. Circumventricular organs (CVOs) can monitor chemical changes in the blood because they lack the blood–brain barrier.

14.6 The Cerebrum

Anatomy Overview - Cerebrum
Figure 14.11 - Cerebrum

1. The cerebrum is the largest part of the brain. Its cortex contains gyri (convolutions), fissures, and sulci.
2. The cerebral hemispheres are divided into four lobes: frontal, parietal, temporal, and occipital.
3. The white matter of the cerebrum is deep to the cortex and consists primarily of myelinated axons extending to other regions as association, commissural, and projection fibers.
4. The basal nuclei are several groups of nuclei in each cerebral hemisphere. They help initiate and terminate movements, suppress unwanted movements, and regulate muscle tone.
5. The limbic system encircles the upper part of the brain stem and the corpus callosum. It functions in emotional aspects of behavior and memory.
6. Table 14.2 summarizes the functions of various parts of the brain.

14.7 Functional Organization of the Cerebral Cortex

1. The sensory areas of the cerebral cortex allow perception of sensory information. The motor areas control the execution of voluntary movements. The association areas are concerned with more complex integrative functions such as memory, personality traits, and intelligence.
2. The primary somatosensory area (areas 1, 2, and 3) receives nerve impulses from somatic sensory receptors for touch, pressure, vibration, itch, tickle, temperature, pain, and proprioception and is involved in the perception of these sensations. Each point within the area receives impulses from a specific part of the face or body. The primary visual area (area 17) receives visual information and is involved in visual perception. The primary auditory area (areas 41 and 42) receives information for sound and is involved in auditory perception. The primary gustatory area (area 43) receives impulses for taste and is involved in gustatory perception and taste discrimination. The primary olfactory area (area 28) receives impulses for smell and is involved in olfactory perception.
3. Motor areas include the primary motor area (area 4), which controls voluntary contractions of specific muscles or groups of muscles, and Broca's speech area (areas 44 and 45), which controls production of speech.
4. The somatosensory association area (areas 5 and 7) permits you to determine the exact shape and texture of an object simply by touching it and to sense the relationship of one body part to another. It also stores memories of past somatic sensory experiences.
5. The visual association area (areas 18 and 19) relates present to past visual experiences and is essential for recognizing and evaluating what is seen. The facial recognition area (areas 20, 21, and 37) stores information about faces and allows you to recognize people by their faces. The auditory association area (area 22) allows you to recognize a particular sound as speech, music, or noise.
6. The orbitofrontal cortex (area 11) allows you to identify odors and discriminate among different odors. Wernicke's area (area 22 and possibly 39 and 40) interprets the meaning of speech by translating words into thoughts. The common integrative area (areas 5, 7, 39, and 40) integrates sensory interpretations from the association areas and impulses from other areas, allowing thoughts based on sensory inputs.
7. The prefrontal cortex (areas 9, 10, 11, and 12) is concerned with personality, intellect, complex learning abilities, judgment, reasoning, conscience, intuition, and development of abstract ideas. The premotor area (area 6) generates nerve impulses that cause specific groups of muscles to contract in

Review	Resource
specific sequences. It also serves as a memory bank for complex movements. The frontal eye field area (area 8) controls voluntary scanning movements of the eyes.	
8. Subtle anatomical differences exist between the two hemispheres, and each has unique functions. Each hemisphere receives sensory signals from and controls movements of the opposite side of the body. The left hemisphere is more important for language, numerical and scientific skills, and reasoning. The right hemisphere is more important for musical and artistic awareness, spatial and pattern perception, recognition of faces, emotional content of language, identifying odors, and generating mental images of sight, sound, touch, taste, and smell.	
9. Brain waves generated by the cerebral cortex are recorded from the surface of the head in an electroencephalogram (EEG). The EEG may be used to diagnose epilepsy, infections, and tumors.	
14.8 Cranial Nerves	Anatomy Overview - Cranial Nerves Exercise - Cranial Nerve Target Practice: Motor Functions Exercise - Cranial Nerve Target Practice: Sensory Functions
1. Twelve pairs of cranial nerves originate from the nose, eyes, inner ear, brain stem, and spinal cord. 2. They are named primarily based on their distribution and are numbered I–XII in order of attachment to the brain. Exhibits 14.A–14.J and Table 14.4 summarize the types, locations, functions, and disorders of the cranial nerves.	
14.9 Development of the Nervous System	
1. The development of the nervous system begins with a thickening of a region of the ectoderm called the neural plate. 2. During embryological development, primary brain vesicles form from the neural tube and serve as forerunners of various parts of the brain. 3. The telencephalon forms the cerebrum, the diencephalon develops into the thalamus and hypothalamus, the mesencephalon develops into the midbrain, the metencephalon develops into the pons and cerebellum, and the myelencephalon forms the medulla.	
14.10 Aging and the Nervous System	
1. The brain grows rapidly during the first few years of life. 2. Age-related effects involve loss of brain mass and decreased capacity for sending nerve impulses.	

SELF-QUIZ QUESTIONS

Fill in the blanks in the following statements.

1. The cerebral hemispheres are connected internally by a broad band of white matter known as the ______.

2. List the five lobes of the cerebrum: ______, ______, ______, ______, ______.

3. The _______ separates the cerebrum into right and left halves.

Indicate whether the following statements are true or false.

4. The brain stem consists of the medulla oblongata, pons, and diencephalon.

5. You are the greatest student of anatomy and physiology, and you are well-prepared for your exam on the brain. As you confidently answer the questions, your brain is exhibiting beta waves.

Choose the one best answer to the following questions.

6. Which of the following is *not* a function of the thalamus?
   (a) relaying information from the cerebellum and basal nuclei to primary motor areas of the cerebral cortex
   (b) helping maintain consciousness
   (c) playing a role in emotions and memory
   (d) regulating body temperature
   (e) relaying sensory impulses to the cerebral cortex

7. Which of the following statements is *false*?
   (a) The blood supply to the brain is provided mainly by the internal carotid and vertebral arteries.
   (b) Neurons in the brain rely almost exclusively on aerobic respiration to produce ATP.
   (c) An interruption of blood flow to the brain for even 20 seconds may impair brain function.
   (d) Glucose supply to the brain must be continuous.
   (e) Low levels of glucose in the blood to the brain may result in unconsciousness.

8. In which of the following ways does cerebrospinal fluid contribute to homeostasis? (1) mechanical protection, (2) chemical protection, (3) electrical protection, (4) circulation, (5) immunity.
   (a) 1, 2, and 3 (b) 2, 3, and 4 (c) 3, 4, and 5
   (d) 1, 2, and 4 (e) 2, 4, and 5

9. Which of the following are functions of the hypothalamus? (1) control of the ANS, (2) production of hormones, (3) regulation of emotional and behavioral patterns, (4) regulation of eating and drinking, (5) control of body temperature, (6) regulation of circadian rhythms.
   (a) 1, 2, 4, and 6 (b) 2, 3, 5, and 6 (c) 1, 3, 5, and 6
   (d) 1, 4, 5, and 6 (e) 1, 2, 3, 4, 5, and 6

10. Which of the following statements is *false*?
(a) Association tracts transmit nerve impulses between gyri in the same hemisphere.
(b) Commissural tracts transmit impulses from the gyri in one cerebral hemisphere to the corresponding gyri in the other hemisphere.
(c) Projection tracts form descending and ascending tracts that transmit impulses from the cerebrum and other parts of the brain to the spinal cord, or from the spinal cord to the brain.
(d) The internal capsule is an example of a commissural tract.
(e) The corpus callosum is an example of a commissural tract.

11. Which of the following statements is *true*?
(a) The right and left hemispheres of the cerebrum are completely symmetrical.
(b) The left hemisphere controls the left side of the body.
(c) The right hemisphere is more important for spoken and written language.
(d) The left hemisphere is more important for musical and artistic awareness.
(e) Hemispheric lateralization is more pronounced in males than in females.

12. Match the following (some answers will be used more than once):

____ (a) oculomotor
____ (b) trigeminal
____ (c) abducens
____ (d) vestibulocochlear
____ (e) accessory
____ (f) vagus
____ (g) facial
____ (h) glossopharyngeal
____ (i) olfactory
____ (j) trochlear
____ (k) optic
____ (l) hypoglossal
____ (m) functions in sense of smell
____ (n) functions in hearing and equilibrium
____ (o) functions in chewing
____ (p) functions in facial expression and secretion of saliva and tears
____ (q) functions in movement of tongue during speech and swallowing
____ (r) functions in secretion of digestive fluids
____ (s) functions in secretion of saliva, taste, regulation of blood pressure, and muscle sense
____ (t) sensory only
____ (u) functions in eye movement by controlling extrinsic eye muscles
____ (v) functions in swallowing and head movements

(1) cranial nerve I
(2) cranial nerve II
(3) cranial nerve III
(4) cranial nerve IV
(5) cranial nerve V
(6) cranial nerve VI
(7) cranial nerve VII
(8) cranial nerve VIII
(9) cranial nerve IX
(10) cranial nerve X
(11) cranial nerve XI
(12) cranial nerve XII

13. Match the following (some answers may be used more than once):

____ (a) emotional brain; involved in olfaction and memory
____ (b) bridge connecting parts of the brain with each other
____ (c) sensory relay area
____ (d) alerts the cerebral cortex to incoming sensory signals
____ (e) regulates posture and balance
____ (f) lacks a blood-brain barrier; can monitor chemical changes in the blood
____ (g) site of decussation of pyramids
____ (h) site of pneumotaxic and apneustic areas
____ (i) secretes melatonin
____ (j) contains sensory, motor, and association areas
____ (k) responsible for maintaining consciousness and awakening from sleep
____ (l) controls ANS
____ (m) contains reflex centers for movements of the eyes, head, and neck in response to visual and other stimuli, and reflex center for movements of the head and trunk in response to auditory stimuli
____ (n) plays an essential role in awareness and in the acquisition of knowledge; cognition
____ (o) several groups of nuclei that control large autonomic movements of skeletal muscles and help regulate muscle tone required for specific body movements
____ (p) produces hormones that regulate endocrine gland function
____ (q) contains the vital cardiovascular center and medullary rhythmicity center

(1) medulla oblongata
(2) pons
(3) midbrain
(4) cerebellum
(5) pineal gland
(6) thalamus
(7) hypothalamus
(8) cerebrum
(9) limbic system
(10) reticular formation
(11) circumventricular organs
(12) reticular activating system
(13) basal nuclei

14. Match the following:

____ (a) protrusions in the medulla formed by the large corticospinal tracts
____ (b) dura mater extension that separates the two cerebral hemispheres
____ (c) fingerlike extensions of arachnoid mater where CSF is reabsorbed
____ (d) dura mater extension that separates the two cerebellar hemispheres
____ (e) located in the hypothalamus; relay stations for reflexes related to smell
____ (f) folds in the cerebral cortex
____ (g) shallow grooves in the cerebral cortex
____ (h) bundles of white matter that relay information between the cerebellum and other parts of the brain
____ (i) a thick band of sensory and motor tracts that connect the cerebral cortex with the brain stem and spinal cord
____ (j) dura mater extension that separates the cerebrum from the cerebellum
____ (k) thin membranous partition between the lateral ventricles

(1) gyri
(2) internal capsule
(3) mammillary bodies
(4) tentorium cerebelli
(5) pyramids
(6) falx cerebelli
(7) septum pellucidum
(8) cerebellar peduncles
(9) falx cerebri
(10) sulci
(11) arachnoid villi

15. Match the following:

____ (a) allows planning and production of speech
____ (b) receives impulses for sound
____ (c) controls voluntary contraction of muscles
____ (d) allows recognition and evaluation of visual experiences
____ (e) integration and interpretation of somatic sensations; comparison of past to present sensations
____ (f) receives impulses for touch, proprioception, pain, and temperature
____ (g) receives impulses for taste
____ (h) interpretation of sounds as speech, music, or noise
____ (i) receives impulses from many sensory and association areas as well as the thalamus and brain stem; allows formation of thoughts so appropriate action can occur
____ (j) translates words into thoughts
____ (k) receives impulses for smell
____ (l) allows awareness of shape, color, and movement
____ (m) coordinates muscle movement for complex, learned sequential motor activities
____ (n) involved in scanning eye movements
____ (o) allows discrimination among different odors

(1) primary visual area
(2) primary auditory area
(3) primary gustatory area
(4) primary olfactory area
(5) primary somatosensory area
(6) primary motor area
(7) somatosensory association area
(8) visual association area
(9) frontal eye field
(10) Broca's area
(11) auditory association area
(12) premotor area
(13) Wernicke's area
(14) common integrative area
(15) orbitofrontal cortex

CRITICAL THINKING QUESTIONS

1. An elderly relative suffered a CVA (stroke) and now has difficulty moving her right arm, and she also has speech problems. What areas of the brain were damaged by the stroke?

2. Nicky has recently had a viral infection and now she cannot move the muscles on the right side of her face. In addition, she is experiencing a loss of taste and a dry mouth, and she cannot close her right eye. What cranial nerve has been affected by the viral infection?

3. You have been hired by a pharmaceutical company to develop a drug to regulate a specific brain disorder. What is a major physiological roadblock to developing such a drug, and how can you design a drug to bypass that roadblock so that the drug is delivered to the brain where it is needed?

? ANSWERS TO FIGURE QUESTIONS

14.1 The largest part of the brain is the cerebrum.

14.2 From superficial to deep, the three cranial meninges are the dura mater, arachnoid, and pia mater.

14.3 The brain stem is anterior to the fourth ventricle, and the cerebellum is posterior to it.

14.4 Cerebrospinal fluid is reabsorbed by the arachnoid villi that project into the dural venous sinuses.

14.5 The medulla oblongata contains the pyramids; the midbrain contains the cerebral peduncles; *pons* means "bridge."

14.6 Decussation means crossing to the opposite side. The functional consequence of decussation of the pyramids is that each side of the cerebrum controls muscles on the opposite side of the body.

14.7 The cerebral peduncles are the main sites through which tracts extend and nerve impulses are conducted between the superior parts of the brain and the inferior parts of the brain and the spinal cord.

14.8 The cerebellar peduncles carry information into and out of the cerebellum.

14.9 In about 70% of human brains, the intermediate mass connects the right and left halves of the thalamus.

14.10 From posterior to anterior, the four major regions of the hypothalamus are the mammillary, tuberal, supraoptic, and preoptic regions.

14.11 The gray matter enlarges more rapidly during development, in the process producing convolutions or gyri (folds), sulci (shallow grooves), and fissures (deep grooves).

14.12 Association tracts connect gyri of the same hemisphere; commissural tracts connect gyri in opposite hemispheres; projection tracts connect the cerebrum with the thalamus, brain stem, and spinal cord.

14.13 The basal nuclei are lateral, superior, and inferior to the thalamus.

14.14 The hippocampus is the part of the limbic system that functions with the cerebrum in memory.

14.15 The common integrative association area integrates interpretation of visual, auditory, and somatic sensations; Broca's speech area translates thoughts into speech; the premotor area controls skilled muscular movements; the primary gustatory area interprets sensations related to taste; the primary auditory area allows you to interpret pitch and rhythm; the primary visual area allows you to interpret shape, color, and movement of objects; and the frontal eye field area controls voluntary scanning movements of the eyes.

14.16 In an EEG, theta waves indicate emotional stress.

14.17 Axons in the olfactory tracts terminate in the primary olfactory area in the temporal lobe of the cerebral cortex.

14.18 Most axons in the optic tracts terminate in the lateral geniculate nucleus of the thalamus.

14.19 The superior branch of the oculomotor nerve is distributed to the superior rectus muscle; the trochlear nerve is the smallest cranial nerve.

14.20 The trigeminal nerve is the largest cranial nerve.

14.21 Motor axons of the facial nerve originate in the pons.

14.22 The vestibular ganglion contains cell bodies from sensory axons that arise in the semicircular canals, saccule, and utricle; the spiral ganglion contains cell bodies from axons that arise in the spiral organ of the cochlea.

14.23 The glossopharyngeal nerve exits the skull through the jugular foramen.

14.24 The vagus nerve is located medial and posterior to the internal jugular vein and common carotid artery in the neck.

14.25 The accessory nerve is the only cranial nerve that originates from both the brain and spinal cord.

14.26 Two important motor functions of the hypoglossal nerve are speech and swallowing.

14.27 The gray matter of the nervous system derives from the mantle layer cells of the neural tube.

14.28 The mesencephalon does not develop into a secondary brain vesicle.

15 THE AUTONOMIC NERVOUS SYSTEM

THE AUTONOMIC NERVOUS SYSTEM AND HOMEOSTASIS *The autonomic nervous system contributes to homeostasis by responding to subconscious visceral sensations and exciting or inhibiting smooth muscle, cardiac muscle, and glands.*

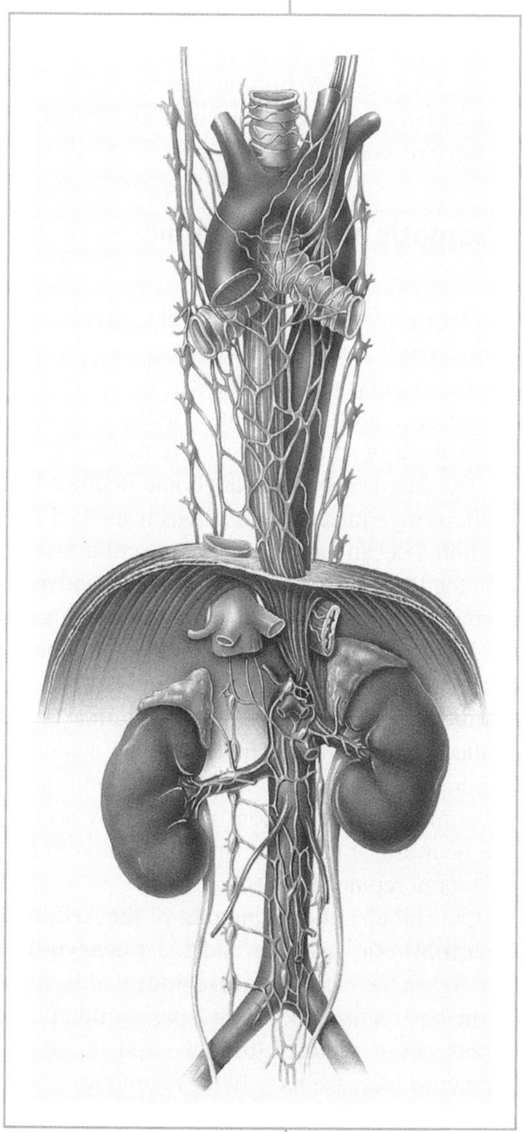

As you learned in Chapter 12, the peripheral nervous system (PNS) includes cranial and spinal nerves and is divided into a somatic nervous system (SNS), autonomic nervous system (ANS), and enteric nervous system (ENS). Like the somatic nervous system, the **autonomic nervous system (ANS)** (aw′-tō-NOM-ik) operates via reflex arcs. (The derivation is *auto-* = self; *-nomic* = law.) Structurally, the ANS includes autonomic sensory neurons, integrating centers in the central nervous system (CNS), autonomic motor neurons, and the enteric division. A continual flow of nerve impulses from (1) *autonomic sensory neurons* in visceral organs and blood vessels propagate into (2) *integrating centers* in the CNS. Then, impulses in (3) *autonomic motor neurons* propagate to various effector tissues, thereby regulating the activity of smooth muscle, cardiac muscle, and many glands. (4) The *enteric division* is a specialized network of nerves and ganglia forming an independent nerve network within the wall of the gastrointestinal (GI) tract. The ANS usually operates without conscious control. However, centers in the hypothalamus and brain stem do regulate ANS reflexes. In this chapter, we compare structural and functional features of the somatic and autonomic nervous systems. Then we discuss the anatomy of the motor portion of the ANS and compare the organization and actions of its two major parts, the sympathetic and parasympathetic divisions.

? ***Did you ever wonder how some blood pressure medications exert their effects through the autonomic nervous system?***

15.1 COMPARISON OF SOMATIC AND AUTONOMIC NERVOUS SYSTEMS

OBJECTIVE

- Compare the structural and functional differences between the somatic and autonomic parts of the nervous system.

Somatic Nervous System

The somatic nervous system includes both sensory and motor neurons. Sensory neurons convey input from receptors for somatic senses (tactile, thermal, pain, and proprioceptive sensations; see Chapter 16) and from receptors for the special senses (sight, hearing, taste, smell, and equilibrium; see Chapter 17). All these sensations normally are consciously perceived. In turn, somatic motor neurons innervate skeletal muscles—the effectors of the somatic nervous system—and produce both reflexive and voluntary movements. When a somatic motor neuron stimulates the muscle, it contracts; the effect always is excitation. If somatic motor neurons cease to stimulate a muscle, the result is a paralyzed, limp muscle that has no muscle tone. Although we are generally not conscious of breathing, the muscles that generate respiratory movements are also skeletal muscles controlled by somatic motor neurons. If the respiratory motor neurons become inactive, breathing stops. A few skeletal muscles, such as those in the middle ear, are controlled by reflexes and cannot be contracted voluntarily.

Autonomic Nervous System

The main input to the ANS comes from **autonomic (visceral) sensory neurons.** Mostly, these neurons are associated with **interoceptors** (IN-ter-ō-sep′-tors), which are sensory receptors located in blood vessels, visceral organs, muscles, and the nervous system that monitor conditions in the *internal* environment. Examples of interoceptors are chemoreceptors that monitor blood CO_2 level and mechanoreceptors that detect the degree of stretch in the walls of organs or blood vessels. Unlike those triggered by a flower's perfume, a beautiful painting, or a delicious meal, these sensory signals are not consciously perceived most of the time, although intense activation of interoceptors may produce conscious sensations. Two examples of perceived visceral sensations are pain sensations from damaged viscera and angina pectoris (chest pain) from inadequate blood flow to the heart. Input that influences the ANS also includes some sensations monitored by somatic sensory and special sensory neurons. For example, pain can produce dramatic changes in some autonomic activities.

Autonomic motor neurons regulate visceral activities by either increasing (exciting) or decreasing (inhibiting) ongoing activities in their effector tissues (cardiac muscle, smooth muscle, and glands). Changes in the diameter of the pupils, dilation and constriction of blood vessels, and adjustment of the rate and force of the heartbeat are examples of autonomic motor responses. Unlike skeletal muscle, tissues innervated by the ANS often function to some extent even if their nerve supply is damaged. For example, the heart continues to beat when it is removed for transplantation into another person, smooth muscle in the lining of the gastrointestinal tract contracts rhythmically on its own, and glands produce some secretions in the absence of ANS control.

Most autonomic responses cannot be consciously altered to any great degree. You probably cannot voluntarily slow your heartbeat to half its normal rate. For this reason, some autonomic responses are the basis for polygraph ("lie detector") tests. However, practitioners of yoga or other techniques of meditation may learn how to regulate at least some of their autonomic activities through long practice. **Biofeedback,** in which monitoring devices display information about a body function such as heart rate or blood pressure, enhances the ability to learn such conscious control. Signals from the general somatic and special senses, acting via the limbic system, also influence responses of autonomic motor neurons. Seeing a bike about to hit you, hearing squealing brakes of a nearby car, or being grabbed by an attacker would all increase the rate and force of your heartbeat.

Comparison of Somatic and Autonomic Motor Neurons

Recall from Chapter 10 that the axon of a single, myelinated somatic motor neuron extends from the central nervous system (CNS) all the way to the skeletal muscle fibers in its motor unit (Figure 15.1a). By contrast, most autonomic motor pathways consist of two motor neurons in series, that is, one following the other (Figure 15.1b). The first neuron (preganglionic neuron) has its cell body in the CNS; its myelinated axon extends from the CNS to an **autonomic ganglion.** (Recall that a *ganglion* is a collection of neuronal cell bodies in the PNS.) The cell body of the second neuron (postganglionic neuron) is also in that same autonomic ganglion; its unmyelinated axon extends directly from the ganglion to the effector (smooth muscle, cardiac muscle, or a gland). Alternatively, in some autonomic pathways, the first motor neuron extends to specialized cells called *chromaffin cells* in the adrenal medullae (inner portions of the adrenal glands) rather than an autonomic ganglion. In addition, all somatic motor neurons release only acetylcholine (ACh) as their neurotransmitter, but autonomic motor neurons release either ACh or norepinephrine (NE).

Unlike somatic output (motor), the output part of the ANS has two divisions: the **sympathetic division** and the **parasympathetic division.** Most organs have **dual innervation;** that is, they receive impulses from both sympathetic and parasympathetic neurons. In some organs, nerve impulses from one division of the ANS stimulate the organ to increase its activity (excitation), and impulses from the other division decrease the organ's activity (inhibition). For example, an increased rate of nerve impulses from the *sympathetic* division increases heart rate, and an increased rate of nerve impulses from the *parasympathetic* division decreases heart rate. The sympathetic division is often called the *fight-or-flight division.* Sympathetic activities result in increased alertness and metabolic activities in order to prepare the body for an emergency situation. Responses to such situations, which may

Figure 15.1 Motor neuron pathways in the (a) somatic nervous system and (b) autonomic nervous system (ANS). Note that autonomic motor neurons release either acetylcholine (ACh) or norepinephrine (NE); somatic motor neurons release ACh.

Somatic nervous system stimulation always excites its effectors (skeletal muscle fibers); stimulation by the autonomic nervous system either excites or inhibits visceral effectors.

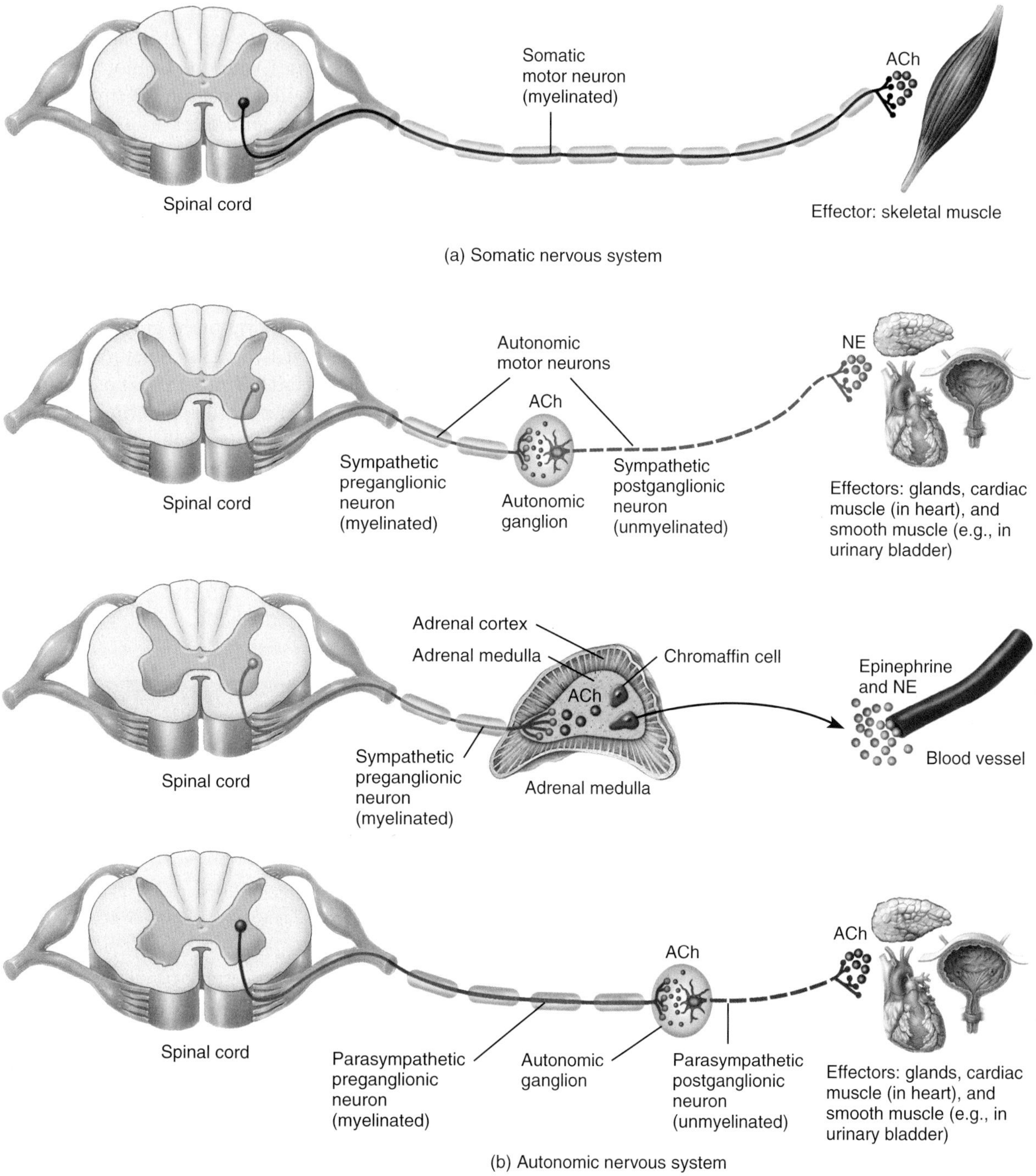

What does dual innervation mean?

occur during physical activity or emotional stress, include a rapid heart rate, faster breathing rate, dilation of the pupils, dry mouth, sweaty but cool skin, dilation of blood vessels to organs involved in combating stress (such as the heart and skeletal muscles), constriction of blood vessels to organs not involved in combating stress (for example, the gastrointestinal tract and kidneys), and release of glucose from the liver.

The parasympathetic division is often referred to as the *rest-and-digest division* because its activities conserve and restore body energy during times of rest or digesting a meal; the majority of its output is directed to the smooth muscle and glandular tissue of the gastrointestinal and respiratory tracts. The parasympathetic division conserves energy and replenishes nutrient stores. Although both the sympathetic and parasympathetic divisions are concerned with maintaining homeostasis, they do so in dramatically different ways.

Table 15.1 compares the somatic and autonomic nervous systems.

✔ **CHECKPOINT**

1. How do the autonomic nervous system and somatic nervous system compare in structure and function?
2. What are the main input and output components of the autonomic nervous system?

15.2 ANATOMY OF AUTONOMIC MOTOR PATHWAYS

OBJECTIVES

- Describe preganglionic and postganglionic neurons of the autonomic nervous system.
- Compare the anatomical components of the sympathetic and parasympathetic divisions of the autonomic nervous system.

Anatomical Components

Each division of the ANS has two motor neurons. The first of the two motor neurons in any autonomic motor pathway is called a **preganglionic neuron** (Figure 15.1b). Its cell body is in the brain or spinal cord; its axon exits the CNS as part of a cranial or spinal nerve. The axon of a preganglionic neuron is a small-diameter, myelinated type B fiber that usually extends to an autonomic ganglion, where it synapses with a **postganglionic neuron,** the second neuron in the autonomic motor pathway. Note that the postganglionic neuron lies entirely outside the CNS in the PNS. Its cell body and dendrites are located in an **autonomic ganglion,** where it forms synapses with one or more preganglionic axons. The axon of a postganglionic neuron is a small-diameter, unmyelinated type C fiber that terminates in a visceral effector. Thus, preganglionic neurons convey nerve impulses from the CNS to autonomic ganglia, and postganglionic neurons relay the impulses from autonomic ganglia to visceral effectors.

Preganglionic Neurons

In the sympathetic division, the preganglionic neurons have their cell bodies in the lateral horns of the gray matter in the 12 thoracic segments and the first two (and sometimes three) lumbar segments of the spinal cord (Figure 15.2). For this reason, the sympathetic division is also called the **thoracolumbar division** (thōr′-a-kō-LUM-bar), and the axons of the sympathetic preganglionic neurons are known as the **thoracolumbar outflow.**

Cell bodies of preganglionic neurons of the parasympathetic division are located in the nuclei of four cranial nerves in the brain stem (III, VII, IX, and X) and in the lateral gray matter of the second through fourth sacral segments of the spinal cord (Figure 15.3). Hence, the parasympathetic division is also known

TABLE 15.1

Comparison of the Somatic and Autonomic Nervous Systems

	SOMATIC NERVOUS SYSTEM	AUTONOMIC NERVOUS SYSTEM
Sensory input	From somatic senses and special senses.	Mainly from interoceptors; some from somatic senses and special senses.
Control of motor output	Voluntary control from cerebral cortex, with contributions from basal ganglia, cerebellum, brain stem, and spinal cord.	Involuntary control from hypothalamus, limbic system, brain stem, and spinal cord; limited control from cerebral cortex.
Motor neuron pathway	One-neuron pathway: Somatic motor neurons extending from CNS synapse directly with effector.	Usually two-neuron pathway: Preganglionic neurons extending from CNS synapse with postganglionic neurons in autonomic ganglion, and postganglionic neurons extending from ganglion synapse with visceral effector. Alternatively, preganglionic neurons may extend from CNS to synapse with chromaffin cells of adrenal medullae.
Neurotransmitters and hormones	All somatic motor neurons release ACh.	All sympathetic and parasympathetic preganglionic neurons release ACh. Most sympathetic postganglionic neurons release NE; those to most sweat glands release ACh. All parasympathetic postganglionic neurons release ACh. Chromaffin cells of adrenal medullae release epinephrine and norepinephrine.
Effectors	Skeletal muscle.	Smooth muscle, cardiac muscle, and glands.
Responses	Contraction of skeletal muscle.	Contraction or relaxation of smooth muscle; increased or decreased rate and force of contraction of cardiac muscle; increased or decreased secretions of glands.

Figure 15.2 Structure of the sympathetic division of the autonomic nervous system. Solid lines represent preganglionic axons; dashed lines represent postganglionic axons. Although the innervated structures are shown for only one side of the body for diagrammatic purposes, the sympathetic division actually innervates tissues and organs on both sides.

Cell bodies of sympathetic preganglionic neurons are located in the lateral horns of gray matter in the 12 thoracic and first two lumbar segments of the spinal cord.

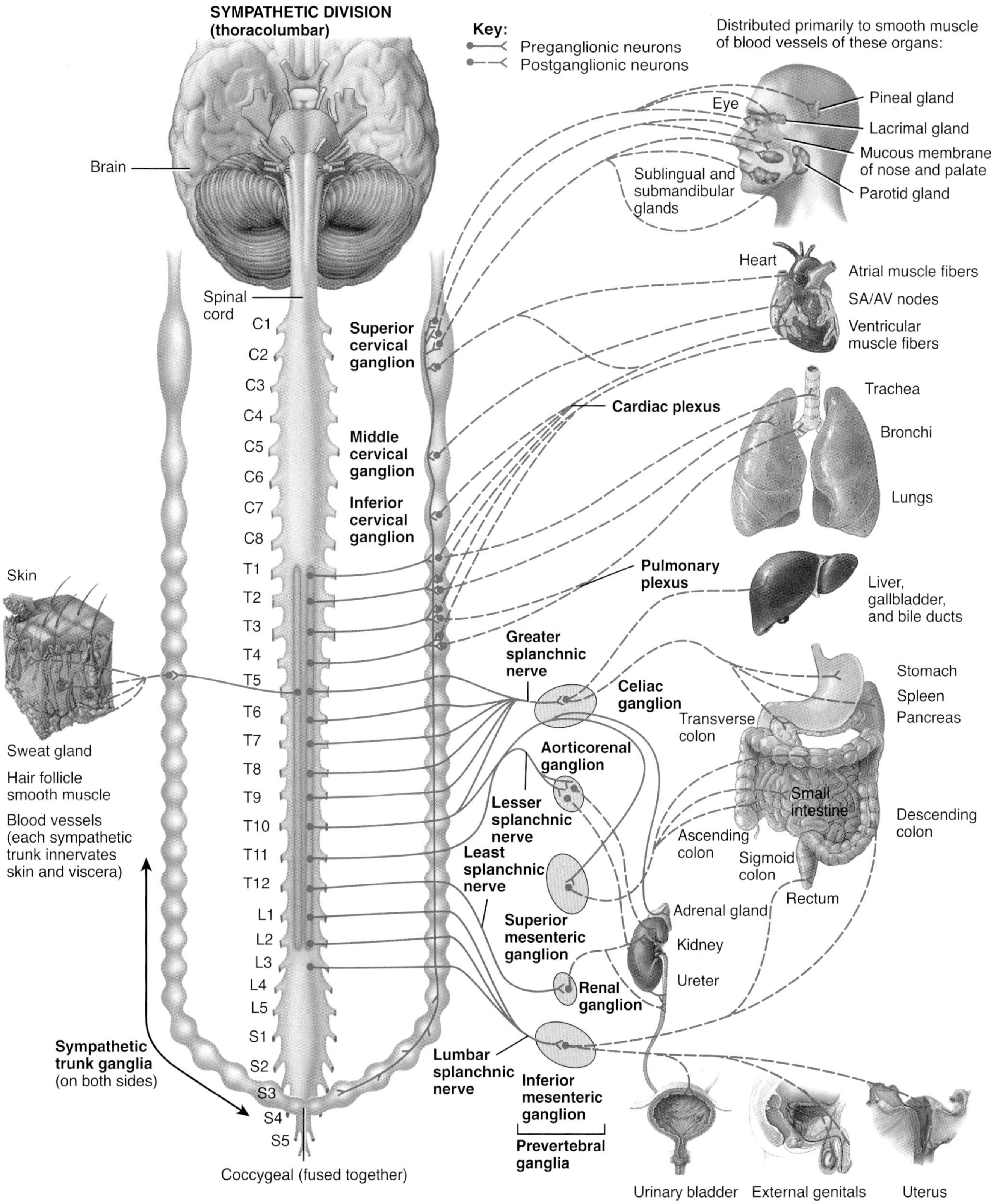

Which division, sympathetic or parasympathetic, has longer preganglionic axons? Why?

Figure 15.3 Structure of the parasympathetic division of the autonomic nervous system. Solid lines represent preganglionic axons; dashed lines represent postganglionic axons. Although the innervated structures are shown only for one side of the body for diagrammatic purposes, the parasympathetic division actually innervates tissues and organs on both sides.

Cell bodies of parasympathetic preganglionic neurons are located in brain stem nuclei and in the lateral gray matter in the second through fourth sacral segments of the spinal cord.

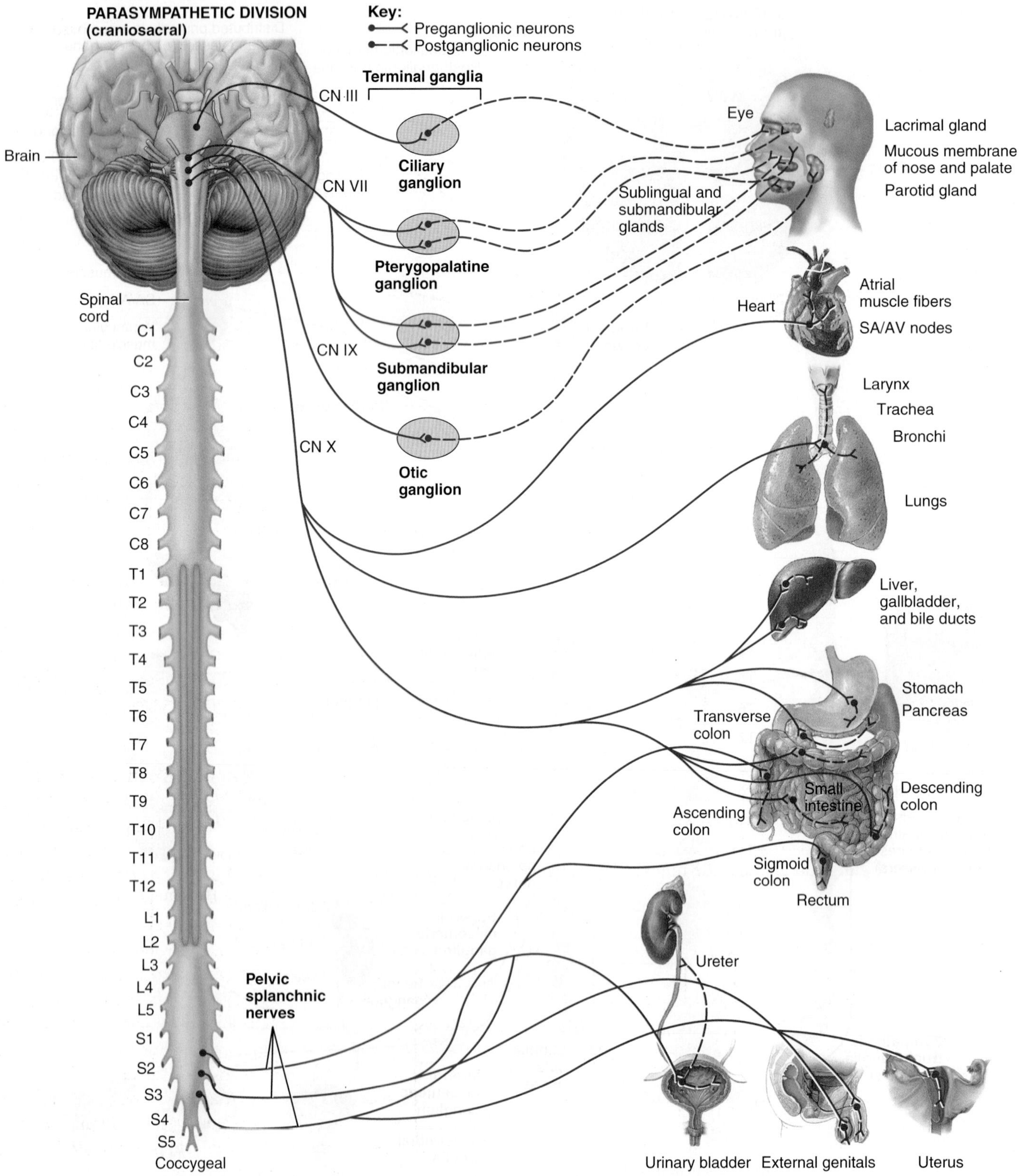

Which ganglia are associated with the parasympathetic division? Sympathetic division?

as the **craniosacral division** (krā′-nē-ō-SĀK-ral), and the axons of the parasympathetic preganglionic neurons are referred to as the **craniosacral outflow.**

Autonomic Ganglia

There are two major groups of autonomic ganglia: (1) sympathetic ganglia, which are components of the sympathetic division of the ANS, and (2) parasympathetic ganglia, which are components of the parasympathetic division of the ANS.

SYMPATHETIC GANGLIA The sympathetic ganglia are the sites of synapses between sympathetic preganglionic and postganglionic neurons. There are two major types of sympathetic ganglia: sympathetic trunk ganglia and prevertebral ganglia. **Sympathetic trunk ganglia** (also called *vertebral chain ganglia* or *paravertebral ganglia*) lie in a vertical row on either side of the vertebral column. These ganglia extend from the base of the skull to the coccyx (Figure 15.2). Postganglionic axons from sympathetic trunk ganglia primarily innervate organs above the diaphragm, such as the head, neck, shoulders, and heart. Sympathetic trunk ganglia in the neck have specific names. They are the **superior, middle,** and **inferior cervical ganglia.** The remaining sympathetic trunk ganglia do not have individual names. Because the sympathetic trunk ganglia are near the spinal cord, most sympathetic preganglionic axons are short and most sympathetic postganglionic axons are long.

The second group of sympathetic ganglia, the **prevertebral (collateral) ganglia,** lies anterior to the vertebral column and close to the large abdominal arteries. In general, postganglionic axons from prevertebral ganglia innervate organs below the diaphragm. There are five major prevertebral ganglia (Figure 15.2; see also Figure 15.5): (1) The **celiac ganglion** (SĒ-lē-ak) is on either side of the celiac trunk, an artery that is just inferior to the diaphragm. (2) The **superior mesenteric ganglion** (MEZ-en-ter′-ik) is near the beginning of the superior mesenteric artery in the upper abdomen. (3) The **inferior mesenteric ganglion** is near the beginning of the inferior mesenteric artery in the middle of the abdomen. (4) The **aorticorenal ganglion** (ā-or′-ti-kō-RĒ-nal) and (5) the **renal ganglion** are near the renal artery of each kidney.

PARASYMPATHETIC GANGLIA Preganglionic axons of the parasympathetic division synapse with postganglionic neurons in **terminal (intramural) ganglia.** Most of these ganglia are located close to or actually within the wall of a visceral organ. Terminal ganglia in the head have specific names. They are the **ciliary ganglion, pterygopalatine ganglion** (ter′-i-gō-PAL-a-tīn), **submandibular ganglion,** and **otic ganglion** (Figure 15.3). The remaining terminal ganglia do not have specific names. Because terminal ganglia are located either close to or in the wall of the visceral organ, parasympathetic preganglionic axons are long, in contrast to parasympathetic postganglionic axons, which are short.

Postganglionic Neurons

Once axons of sympathetic preganglionic neurons pass to sympathetic trunk ganglia, they may connect with postganglionic neurons in one of the following ways (Figure 15.4):

1. An axon may synapse with postganglionic neurons in the ganglion it first reaches.
2. An axon may ascend or descend to a higher or lower ganglion before synapsing with postganglionic neurons. The axons of incoming sympathetic preganglionic neurons that pass up or down the sympathetic trunk collectively form the **sympathetic chains,** the fibers on which the ganglia are strung.
3. An axon may continue, without synapsing, through the sympathetic trunk ganglion to end at a prevertebral ganglion and synapse with postganglionic neurons there.
4. An axon may also pass, without synapsing, through the sympathetic trunk ganglion and a prevertebral ganglion and then extend to chromaffin cells of the adrenal medullae that are functionally similar to sympathetic postganglionic neurons.

A single sympathetic preganglionic fiber has many axon collaterals (branches) and may synapse with 20 or more postganglionic neurons. This pattern of projection is an example of divergence and helps explain why many sympathetic responses affect almost the entire body simultaneously. After exiting their ganglia, the postganglionic axons typically terminate in several visceral effectors (see Figure 15.2).

Axons of preganglionic neurons of the parasympathetic division pass to terminal ganglia near or within a visceral effector (see Figure 15.3). In the ganglion, the presynaptic neuron usually synapses with only four or five postsynaptic neurons, all of which supply a single visceral effector, allowing parasympathetic responses to be localized to a single effector.

Autonomic Plexuses

In the thorax, abdomen, and pelvis, axons of both sympathetic and parasympathetic neurons form tangled networks called **autonomic plexuses,** many of which lie along major arteries. The autonomic plexuses also may contain sympathetic ganglia and axons of autonomic sensory neurons. The major plexuses in the thorax are the **cardiac plexus,** which supplies the heart, and the **pulmonary plexus,** which supplies the bronchial tree (Figure 15.5).

The abdomen and pelvis also contain major autonomic plexuses (Figure 15.5), and often the plexuses are named after the artery along which they are distributed. The **celiac (solar) plexus** is the largest autonomic plexus and surrounds the celiac trunk. It contains two large celiac ganglia, two aorticorenal ganglia, and a dense network of autonomic axons and is distributed to the stomach, spleen, pancreas, liver, gallbladder, kidneys, adrenal medullae, testes, and ovaries. The **superior mesenteric plexus** contains the superior mesenteric ganglion and supplies the small and large intestines. The **inferior mesenteric plexus** contains the inferior mesenteric ganglion, which innervates the large intestine. Axons of some sympathetic postganglionic neurons from the inferior

Figure 15.4 Types of connections between ganglia and postganglionic neurons in the sympathetic division of the ANS. Numbers correspond to descriptions in the text. Also illustrated are the gray and white rami communicantes.

Sympathetic ganglia lie in two chains on either side of the vertebral column (sympathetic trunk ganglia) and near large abdominal arteries anterior to the vertebral column (prevertebral ganglia).

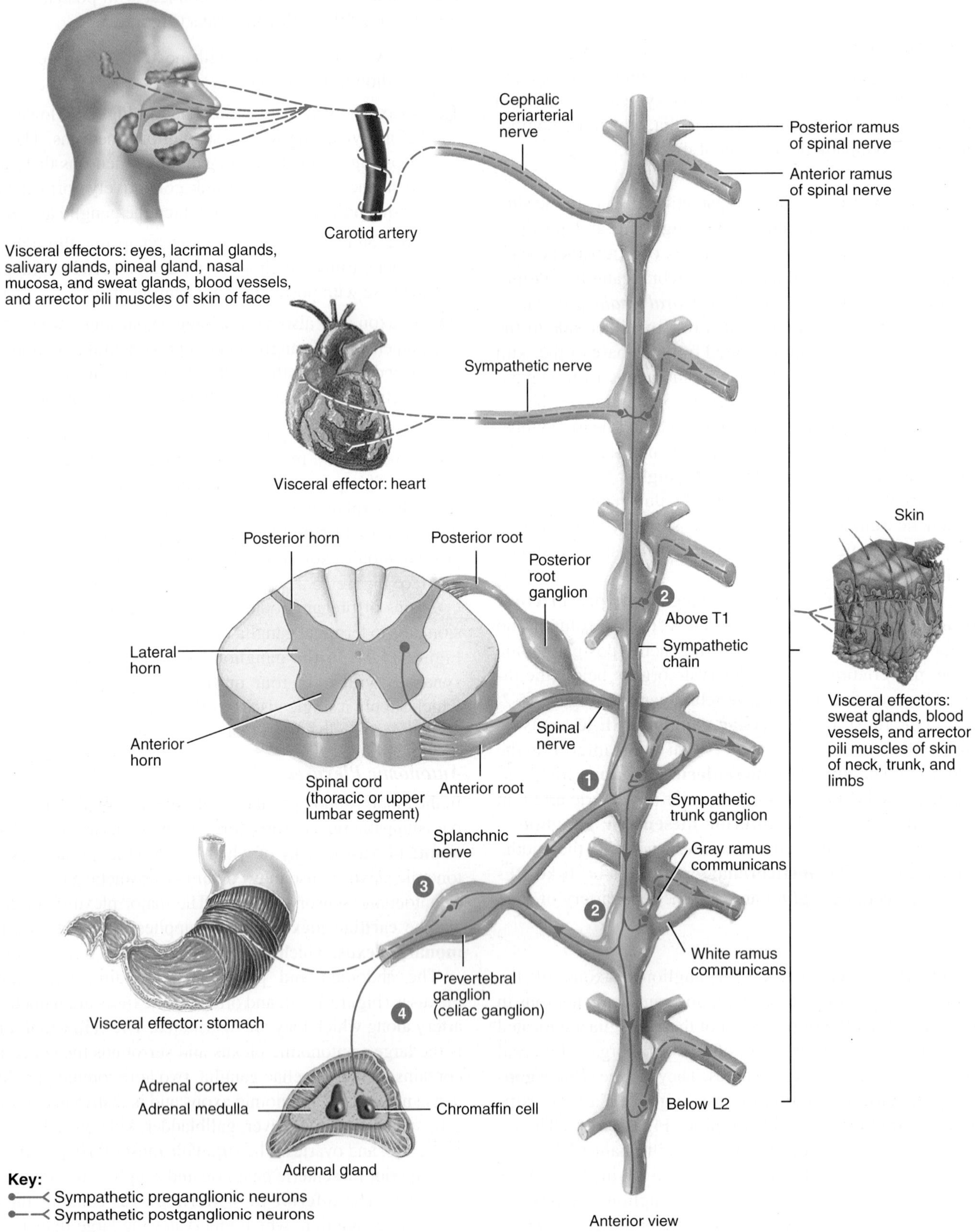

Cephalic periarterial nerve
Posterior ramus of spinal nerve
Anterior ramus of spinal nerve
Carotid artery
Visceral effectors: eyes, lacrimal glands, salivary glands, pineal gland, nasal mucosa, and sweat glands, blood vessels, and arrector pili muscles of skin of face
Sympathetic nerve
Visceral effector: heart
Skin
Posterior horn
Posterior root
Posterior root ganglion
2
Above T1
Sympathetic chain
Lateral horn
Visceral effectors: sweat glands, blood vessels, and arrector pili muscles of skin of neck, trunk, and limbs
Spinal nerve
Anterior horn
1
Spinal cord (thoracic or upper lumbar segment)
Anterior root
Sympathetic trunk ganglion
Splanchnic nerve
Gray ramus communicans
3
2
White ramus communicans
Prevertebral ganglion (celiac ganglion)
4
Visceral effector: stomach
Adrenal cortex
Adrenal medulla
Chromaffin cell
Below L2
Adrenal gland
Key:
Sympathetic preganglionic neurons
Sympathetic postganglionic neurons
Anterior view

What is the significance of the sympathetic trunk ganglia?

Figure 15.5 Autonomic plexuses in the thorax, abdomen, and pelvis.

An autonomic plexus is a network of sympathetic and parasympathetic axons that sometimes also includes autonomic sensory axons and sympathetic ganglia.

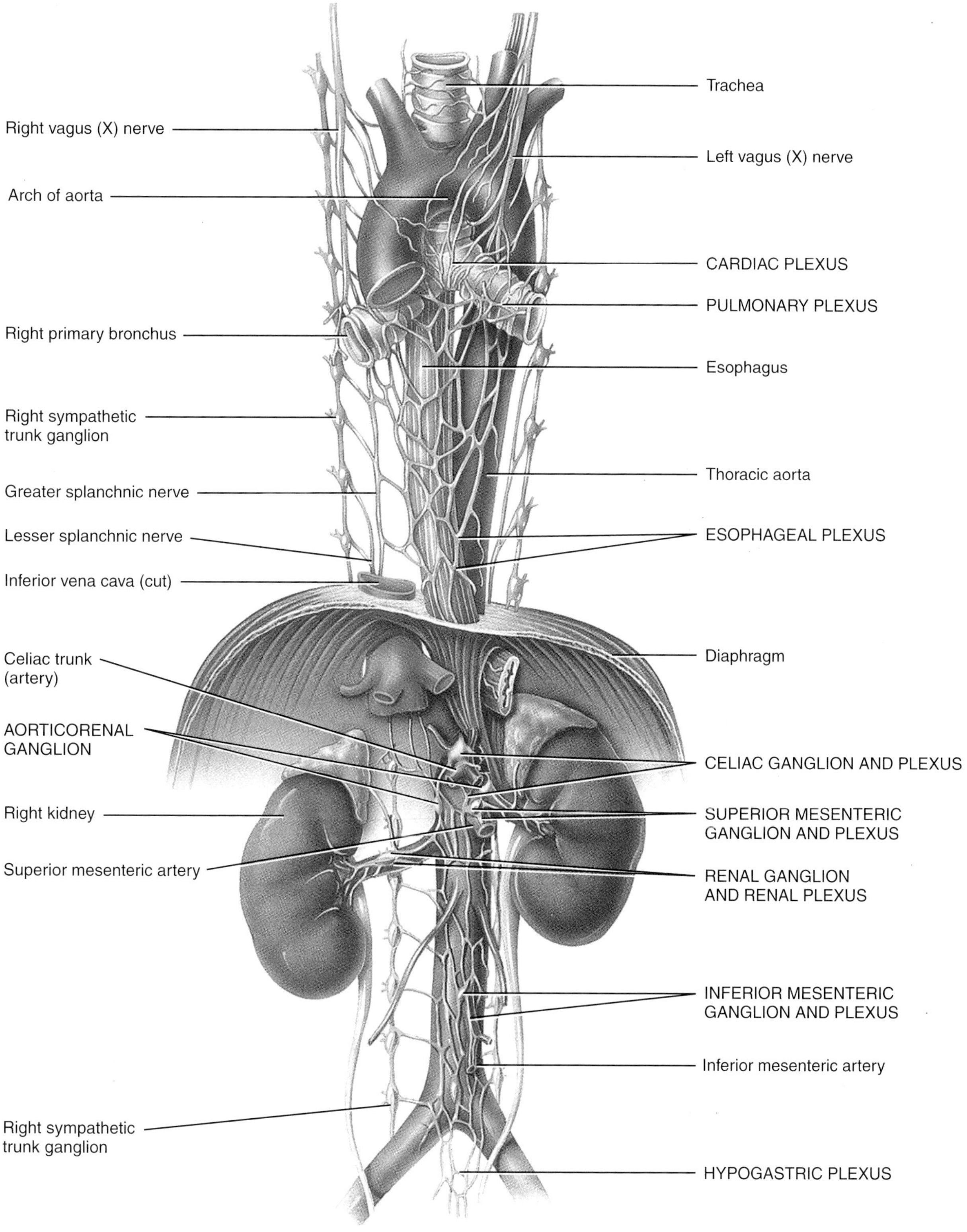

Anterior view

Which is the largest autonomic plexus?

mesenteric ganglion also extend through the **hypogastric plexus,** which is anterior to the fifth lumbar vertebra, to supply the pelvic viscera. The **renal plexus** contains the renal ganglion and supplies the renal arteries within the kidneys and ureters.

With this background in mind, we can now examine some of the specific structural features of the sympathetic and parasympathetic divisions of the ANS in more detail.

Structure of the Sympathetic Division

Pathway from Spinal Cord to Sympathetic Trunk Ganglia

Cell bodies of sympathetic preganglionic neurons are part of the lateral gray horns of all thoracic segments and of the first two lumbar segments of the spinal cord (see Figure 15.2). The preganglionic axons leave the spinal cord along with the somatic motor neurons at the same segmental level. After exiting through the intervertebral foramina, the myelinated preganglionic sympathetic axons pass into the anterior root of a spinal nerve and enter a short pathway called a **white ramus** (RĀ-mus) before passing to the nearest sympathetic trunk ganglion on the same side (see Figure 15.4). Collectively, the white rami are called the **white rami communicantes** (kō-mū-ni-KAN-tēz; singular is **ramus communicans**). Thus, white rami communicantes are structures containing sympathetic preganglionic axons that connect the anterior ramus of the spinal nerve with the ganglia of the sympathetic trunk. The "white" in their name indicates that they contain myelinated axons. Only the thoracic and first two or three lumbar nerves have white rami communicantes.

Organization of Sympathetic Trunk Ganglia

The paired **sympathetic trunk ganglia** are arranged anterior and lateral to the vertebral column, one on either side. Typically, there are 3 cervical, 11 or 12 thoracic, 4 or 5 lumbar, 4 or 5 sacral sympathetic trunk ganglia, and 1 coccygeal ganglion. The right and left coccygeal ganglia are fused together and usually lie at the midline. Although the sympathetic trunk ganglia extend inferiorly from the neck, chest, and abdomen to the coccyx, they receive preganglionic axons only from the thoracic and lumbar segments of the spinal cord (see Figure 15.2).

The cervical portion of each sympathetic trunk is located in the neck and is subdivided into superior, middle, and inferior ganglia (see Figure 15.2). Postganglionic neurons leaving the **superior cervical ganglion** serve the head and heart. They are distributed to the sweat glands, smooth muscle of the eye, blood vessels of the face, lacrimal glands, pineal gland, nasal mucosa, salivary glands (which include the submandibular, sublingual, and parotid glands), and heart. Postganglionic neurons leaving the **middle cervical ganglion** and the **inferior cervical ganglion** innervate the heart.

The thoracic portion of each sympathetic trunk lies anterior to the necks of the corresponding ribs. This region of the sympathetic trunk receives most of the sympathetic preganglionic axons. Postganglionic neurons from the thoracic sympathetic trunk innervate the heart, lungs, bronchi, and other thoracic viscera. In the skin, these neurons also innervate sweat glands, blood vessels, and arrector pili muscles of hair follicles. The lumbar portion of each sympathetic trunk lies lateral to the corresponding lumbar vertebrae. The sacral region of the sympathetic trunk lies in the pelvic cavity on the medial side of the anterior sacral foramina.

Pathways from Sympathetic Trunk Ganglia to Visceral Effectors

Axons leave the sympathetic trunk in four possible ways: (1) They can enter spinal nerves; (2) they can form cephalic periarterial nerves; (3) they can form sympathetic nerves; and (4) they can form splanchnic nerves.

SPINAL NERVES Recall that some of the incoming sympathetic preganglionic neurons synapse with postganglionic neurons in the sympathetic trunk, either in the ganglion at the level of entry or in a ganglion farther up or down the sympathetic trunk. The axons of some of these postganglionic neurons leave the sympathetic trunk by entering a short pathway called a **gray ramus** and then merge with the anterior ramus of a spinal nerve. Therefore, **gray rami communicantes** are structures containing sympathetic postganglionic axons that connect the ganglia of the sympathetic trunk to spinal nerves (see Figure 15.4). The "gray" in their name indicates that they contain unmyelinated axons. Gray rami communicantes outnumber the white rami because there is a gray ramus leading to each of the 31 pairs of spinal nerves. The axons of the postganglionic neurons that leave the sympathetic trunk to enter spinal nerves provide sympathetic innervation to the visceral effectors in the skin of the neck, trunk, and limbs, including sweat glands, smooth muscle in blood vessels, and arrector pili muscles of hair follicles.

CEPHALIC PERIARTERIAL NERVES Some sympathetic preganglionic neurons that enter the sympathetic trunk ascend to the superior cervical ganglion, where they synapse with postganglionic neurons. The axons of some of these postganglionic neurons leave the sympathetic trunk by forming **cephalic periarterial nerves** (per′-ē-ar-TĒ-rē-al), nerves that extend to the head by wrapping around and following the course of various arteries (such as the carotid arteries) that pass from the neck to the head (see Figure 15.4). Cephalic periarterial nerves provide sympathetic innervation to visceral effectors in the skin of the face (sweat glands, smooth muscle of blood vessels, and arrector pili muscles of hair follicles), as well as other visceral effectors of the head (smooth muscle of the eye, lacrimal glands, pineal gland, nasal mucosa, and salivary glands).

SYMPATHETIC NERVES Some of the incoming sympathetic preganglionic neurons synapse with postganglionic neurons in one or more ganglia of the sympathetic trunk. Then, the axons of the postganglionic neurons leave the trunk by forming **sympathetic nerves** that extend to visceral effectors in the thoracic cavity (Figure 15.4). Sympathetic nerves provide sympathetic innervation to the heart and lungs.

- ***Sympathetic nerves to the heart.*** Sympathetic innervation of the heart consists of axons of preganglionic neurons that enter the sympathetic trunk and then form synapses with postganglionic neurons in the superior, middle, and inferior cervical ganglia and first through fourth thoracic ganglia (T1–T4). From these ganglia, axons of postganglionic neurons exit the sympathetic trunk by forming sympathetic nerves that enter the cardiac plexus to supply the heart (see Figure 15.2).

- ***Sympathetic nerves to the lungs.*** Sympathetic innervation of the lungs consists of axons of preganglionic neurons that enter the sympathetic trunk and then form synapses with postganglionic neurons in the second through fourth thoracic ganglia (T2–T4). From these ganglia, axons of sympathetic postganglionic neurons exit the trunk by forming sympathetic nerves that enter the pulmonary plexus to supply the smooth muscle of the bronchi and bronchioles of the lungs (see Figure 15.2).

SPLANCHNIC NERVES Recall that some sympathetic preganglionic axons pass through the sympathetic trunk without terminating in it. Beyond the trunk, they form nerves known as **splanchnic nerves** (SPLANK-nik; see Figures 15.2 and 15.4), which extend to outlying prevertebral ganglia.

- ***Splanchnic nerves to abdominopelvic organs.*** Most sympathetic preganglionic axons that enter splanchnic nerves are destined to synapse with sympathetic postganglionic neurons in the prevertebral ganglia that supply the organs of the abdominopelvic cavity. Preganglionic axons from the fifth through ninth or tenth thoracic ganglia (T5–T9 or T10) form the **greater splanchnic nerve.** It pierces the diaphragm and enters the **celiac ganglion** of the celiac plexus. From there, postganglionic neurons follow and innervate blood vessels to the stomach, spleen, liver, kidneys, and small intestine. Preganglionic axons from the tenth and eleventh thoracic ganglia (T10–T11) form the **lesser splanchnic nerve.** It pierces the diaphragm and passes through the celiac plexus to enter the aorticorenal ganglion and superior mesenteric ganglion of the superior mesenteric plexus. Postganglionic neurons from the superior mesenteric ganglion follow and innervate blood vessels of the small intestine and proximal colon. The **least (lowest) splanchnic nerve,** which is not always present, is formed by preganglionic axons from the twelfth thoracic ganglia (T12) or a branch of the lesser splanchnic nerve. It pierces the diaphragm and enters the renal plexus near the kidney. Postganglionic neurons from the renal plexus supply kidney arterioles and the ureters. Preganglionic axons that form the **lumbar splanchnic nerve** from the first through fourth lumbar ganglia (L1–L4) enter the inferior mesenteric plexus and terminate in the **inferior mesenteric ganglion,** where they synapse with postganglionic neurons. Axons of postganglionic neurons extend through the inferior mesenteric plexus to supply the distal colon and rectum; they also extend through the hypogastric plexus to supply blood vessels of the distal colon, rectum, urinary bladder, and genital organs. Postganglionic axons leaving the prevertebral ganglia follow the course of various arteries to abdominal and pelvic visceral effectors.

- ***Splanchnic nerves to the adrenal medulla.*** Some sympathetic preganglionic axons pass, without synapsing, through the sympathetic trunk, greater splanchnic nerves, and celiac ganglion, and then extend to **chromaffin cells** in the adrenal medullae of the adrenal glands (see Figures 15.2 and 15.4). Developmentally, the adrenal medullae and sympathetic ganglia are derived from the same tissue, the neural crest (see Figure 14.27). The adrenal medullae are modified sympathetic ganglia, and the chromaffin cells are similar to sympathetic postganglionic neurons except they lack dendrites and axons. Rather than extending to another organ, however, these cells release hormones into the blood. Upon stimulation by sympathetic preganglionic neurons, the chromaffin cells of the adrenal medullae release a mixture of catecholamine hormones—about 80% **epinephrine,** 20% **norepinephrine,** and a trace amount of **dopamine.** These hormones circulate throughout the body and intensify responses elicited by sympathetic postganglionic neurons.

CLINICAL CONNECTION | *Horner's Syndrome*

In **Horner's syndrome,** the sympathetic innervation to one side of the face is lost due to an inherited mutation, an injury, or a disease that affects sympathetic outflow through the superior cervical ganglion. Symptoms occur on the affected side and include ptosis (drooping of the upper eyelid), miosis (constricted pupil), and anhidrosis (lack of sweating). •

Structure of the Parasympathetic Division

Cell bodies of parasympathetic preganglionic neurons are found in nuclei in the brain stem and in the lateral gray matter of the second through fourth sacral segments of the spinal cord (see Figure 15.3). Their axons emerge as part of a cranial nerve or as part of the anterior root of a spinal nerve. The **cranial parasympathetic outflow** consists of preganglionic axons that extend from the brain stem in four cranial nerves. The **sacral parasympathetic outflow** consists of preganglionic axons in anterior roots of the second through fourth sacral spinal nerves. The preganglionic axons of both the cranial and sacral outflows end in terminal ganglia, where they synapse with postganglionic neurons.

The cranial outflow has four pairs of ganglia and the ganglia associated with the vagus (X) nerve. The four pairs of cranial parasympathetic ganglia innervate structures in the head and are located close to the organs they innervate (see Figure 15.3).

1. The **ciliary ganglia** lie lateral to each optic (II) nerve near the posterior aspect of the orbit. Preganglionic axons pass with the oculomotor (III) nerves to the ciliary ganglia. Postganglionic axons from the ganglia innervate smooth muscle fibers in the eyeball.
2. The **pterygopalatine ganglia** (ter′-i-gō-PAL-a-tīn) are located lateral to the sphenopalatine foramen, between the sphenoid and palatine bones. They receive preganglionic axons from the facial (VII) nerve and send postganglionic axons to the nasal mucosa, palate, pharynx, and lacrimal glands.

3. The **submandibular ganglia** are found near the ducts of the submandibular salivary glands. They receive preganglionic axons from the facial nerves and send postganglionic axons to the submandibular and sublingual salivary glands.
4. The **otic ganglia** are situated just inferior to each foramen ovale. They receive preganglionic axons from the glossopharyngeal (IX) nerves and send postganglionic axons to the parotid salivary glands.

Preganglionic axons that leave the brain as part of the vagus (X) nerves carry nearly 80% of the total craniosacral outflow. Vagal axons extend to many terminal ganglia in the thorax and abdomen. As the vagus nerve passes through the thorax, it sends axons to the heart and the airways of the lungs. In the abdomen, it supplies the liver, gallbladder, stomach, pancreas, small intestine, and part of the large intestine.

The sacral parasympathetic outflow consists of preganglionic axons from the anterior roots of the second through fourth sacral spinal nerves (S2–S4). As the preganglionic axons course through the sacral spinal nerves, they branch off these nerves to form **pelvic splanchnic nerves** (Figure 15.6). Pelvic splanchnic nerves synapse with parasympathetic postganglionic neurons located in terminal ganglia in the walls of the innervated viscera. From the terminal ganglia, parasympathetic postganglionic axons innervate

Figure 15.6 Pelvic splanchnic nerves.

Through pelvic splanchnic nerves, axons of parasympathetic preganglionic neurons extend to parasympathetic postganglionic neurons in terminal ganglia in the walls of the colon, ureters, urinary bladder, and reproductive organs.

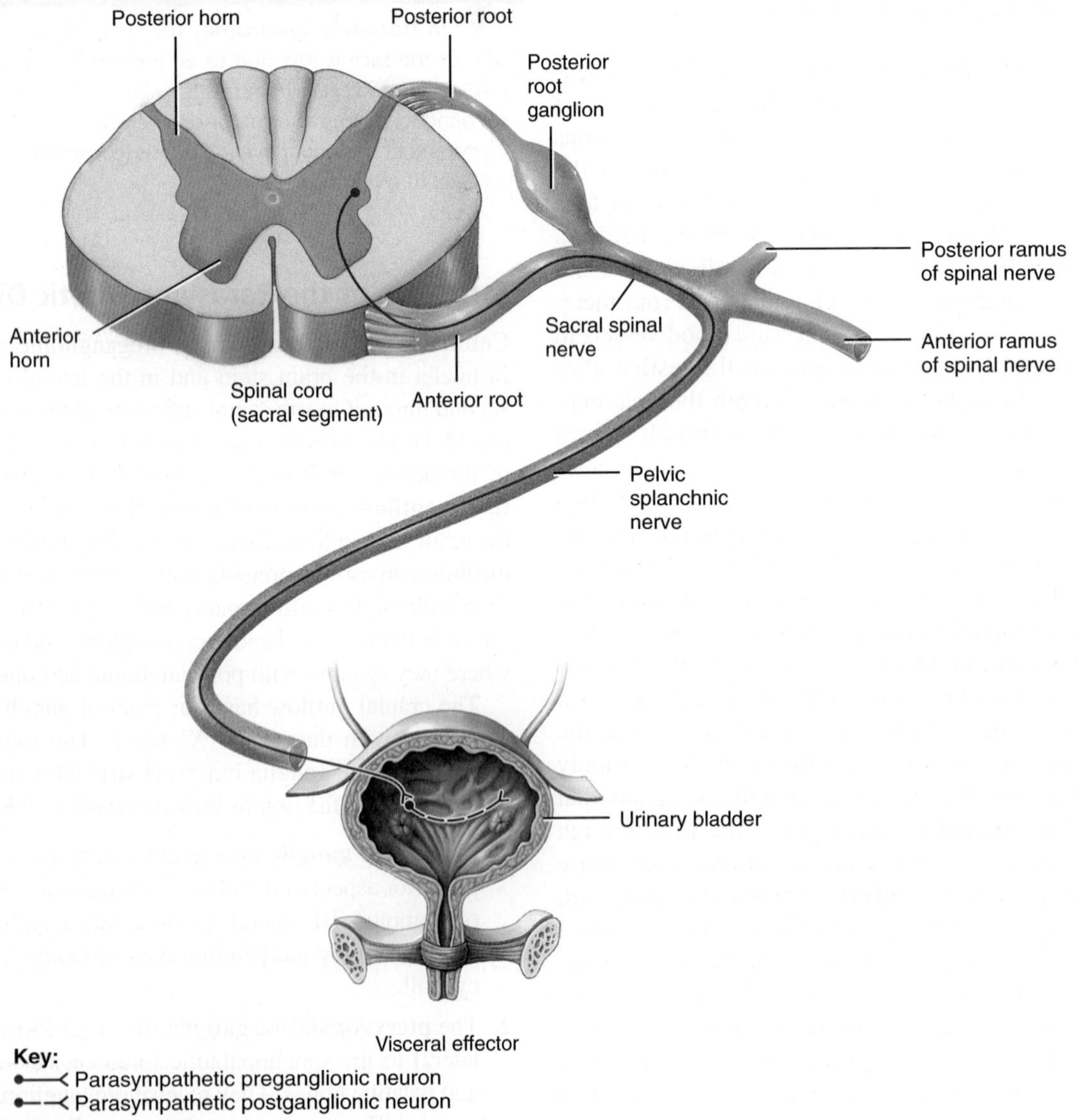

? Pelvic splanchnic nerves branch from which spinal nerves?

smooth muscle and glands in the walls of the colon, ureters, urinary bladder, and reproductive organs.

Structure of the Enteric Division

To understand and appreciate the **enteric division** (en-TER-ik) of the autonomic nervous system, it is important to realize that the gastrointestinal tract, like the surface of the body, forms an extensive area of contact with the environment. Although this gastrointestinal environment is inside the body, it is still considered part of the external environment. Just as the surface of the body must respond to important environmental stimuli in order to function properly, the surface of the gastrointestinal tract must respond to surrounding stimuli to generate proper homeostatic controls. In fact, these responses and controls are so important that the gastrointestinal tract has its own nervous system with intrinsic input, processing, and output. The enteric division can and does function independently of central nervous system activity, but it can also receive controlling input from the central nervous system.

The enteric division is the specialized collection of nerves and ganglia forming a complex, integrated neuronal network within the wall of the gastrointestinal tract, pancreas, and gallbladder. This incredible nerve network contains in the neighborhood of 100 million neurons, approximately the same number as the spinal cord, and is capable of continued function without input from the central nervous system. The enteric network of nerves and ganglia contains sensory neurons capable of monitoring tension in the intestinal wall and accessing the composition of the intestinal contents. These sensory neurons relay their input signals to interneurons within the enteric ganglia. The interneurons establish an integrative network that processes the incoming signals and generates regulatory output signals to motor neurons throughout plexuses within the wall of the digestive organs. The motor neurons carry the output signals to the smooth muscle and glands of the gastrointestinal tract to exert control over its motility and secretory activities.

Most of the nerve fibers that innervate the digestive organs arise from two plexuses within the enteric system. The largest, the **myenteric plexus** (mī-en-TER-ik), is positioned between the outer longitudinal and circular muscle layers from the upper esophagus to the anus. The myenteric plexus communicates extensively with a somewhat smaller plexus, the **submucosal plexus,** which occupies the gut wall between the circular muscle layer and the muscularis mucosae (see Section 24.3) and runs from the stomach to the anus. Neurons emerge from the ganglia of these two plexuses to form smaller plexuses around blood vessels and within the muscle layers and mucosa of the gut wall. It is this system of nerves that makes possible the normal motility and secretory functions of the gastrointestinal tract.

✔ CHECKPOINT

3. Why is the sympathetic division called the thoracolumbar division even though its ganglia extend from the cervical to the sacral region?
4. List the organs served by each sympathetic and parasympathetic ganglion.
5. Describe the locations of sympathetic trunk ganglia, prevertebral ganglia, and terminal ganglia. Which types of autonomic neurons synapse in each type of ganglion?
6. Why does the sympathetic division produce simultaneous effects throughout the body, in contrast to parasympathetic effects, which typically are localized to specific organs?
7. What are the functions of the enteric division of the ANS?

15.3 ANS NEUROTRANSMITTERS AND RECEPTORS

OBJECTIVE

- Describe the neurotransmitters and receptors involved in autonomic responses.

Based on the neurotransmitter they produce and release, autonomic neurons are classified as either cholinergic or adrenergic. The receptors for the neurotransmitters are integral membrane proteins located in the plasma membrane of the postsynaptic neuron or effector cell.

Cholinergic Neurons and Receptors

Cholinergic neurons (kō′-lin-ER-jik) release the neurotransmitter **acetylcholine (ACh).** In the ANS, the cholinergic neurons include (1) all sympathetic and parasympathetic preganglionic neurons, (2) sympathetic postganglionic neurons that innervate most sweat glands, and (3) all parasympathetic postganglionic neurons (Figure 15.7).

ACh is stored in synaptic vesicles and released by exocytosis. It then diffuses across the synaptic cleft and binds with specific **cholinergic receptors,** integral membrane proteins in the *postsynaptic* plasma membrane. The two types of cholinergic receptors, both of which bind ACh, are nicotinic receptors and muscarinic receptors. **Nicotinic receptors** (nik′-ō-TIN-ik) are present in the plasma membrane of dendrites and cell bodies of both sympathetic and parasympathetic postganglionic neurons (Figure 15.7a, b), the plasma membranes of chromaffin cells of the adrenal medullae, and in the motor end plate at the neuromuscular junction. They are so named because nicotine mimics the action of ACh by binding to these receptors. (Nicotine, a natural substance in tobacco leaves, is not a naturally occurring substance in humans and is not normally present in nonsmokers.) **Muscarinic receptors** (mus′-ka-RIN-ik) are present in the plasma membranes of all effectors (smooth muscle, cardiac muscle, and glands) innervated by parasympathetic postganglionic axons. In addition, most sweat glands receive their innervation from *cholinergic* sympathetic postganglionic neurons and possess muscarinic receptors (see Figure 15.7b). These receptors are so named because a mushroom poison called muscarine mimics the actions of ACh by binding to them. Nicotine does not activate muscarinic receptors, and muscarine does not activate nicotinic receptors, but ACh does activate both types of cholinergic receptors.

Figure 15.7 Cholinergic neurons and adrenergic neurons in the sympathetic and parasympathetic divisions.

Cholinergic neurons release acetylcholine; adrenergic neurons release norepinephrine. Cholinergic receptors (nicotinic or muscarinic) and adrenergic receptors are integral membrane proteins located in the plasma membrane of a postsynaptic neuron or an effector cell.

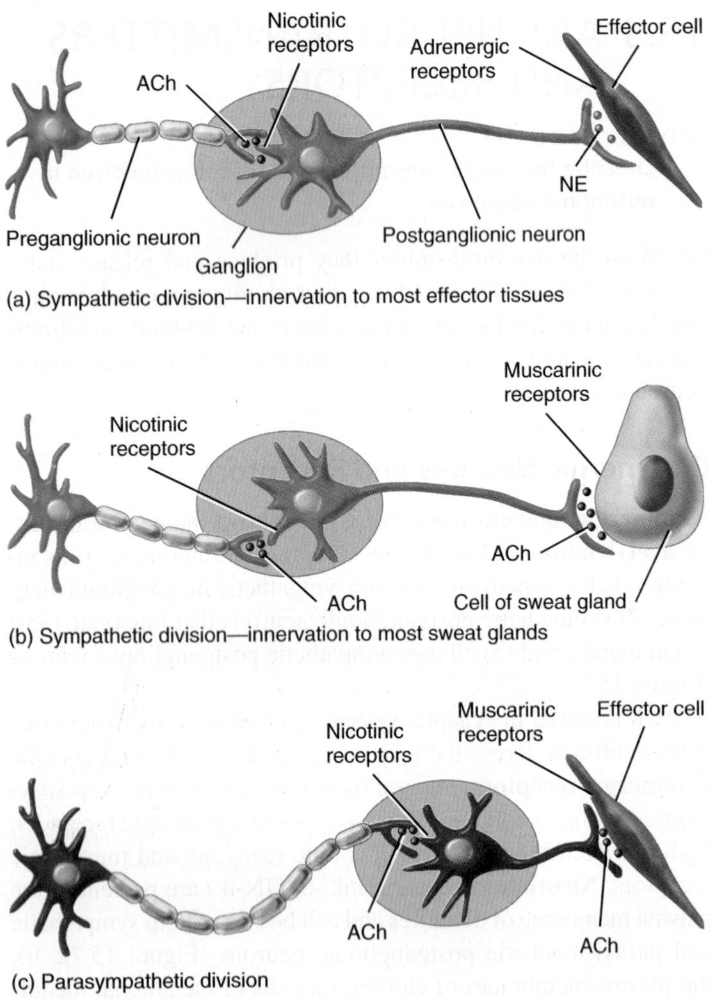

Which ANS neurons are adrenergic? What types of effector tissues contain muscarinic receptors?

Activation of nicotinic receptors by ACh causes depolarization and thus excitation of the postsynaptic cell, which can be a postganglionic neuron, an autonomic effector, or a skeletal muscle fiber. Activation of muscarinic receptors by ACh sometimes causes depolarization (excitation) and sometimes causes hyperpolarization (inhibition), depending on which particular cell bears the muscarinic receptors. For example, binding of ACh to muscarinic receptors inhibits (relaxes) smooth muscle sphincters in the gastrointestinal tract. By contrast, ACh excites muscarinic receptors in smooth muscle fibers in the circular muscles of the iris of the eye, causing them to contract. Because acetylcholine is quickly inactivated by the enzyme **acetylcholinesterase (AChE),** effects triggered by cholinergic neurons are brief.

Adrenergic Neurons and Receptors

In the ANS, **adrenergic neurons** (ad′-ren-ER-jik) release **norepinephrine (NE),** also known as **noradrenalin** (Figure 15.7a). Most sympathetic postganglionic neurons are adrenergic. Like ACh, NE is stored in synaptic vesicles and released by exocytosis. Molecules of NE diffuse across the synaptic cleft and bind to specific adrenergic receptors on the postsynaptic membrane, causing either excitation or inhibition of the effector cell.

Adrenergic receptors bind both norepinephrine and epinephrine. The norepinephrine can be either released as a neurotransmitter by sympathetic postganglionic neurons or released as a hormone into the blood by chromaffin cells of the adrenal medullae; epinephrine is released as a hormone. The two main types of adrenergic receptors are **alpha (α) receptors** and **beta (β) receptors,** which are found on visceral effectors innervated by most sympathetic postganglionic axons. These receptors are further classified into subtypes—α_1, α_2, β_1, β_2, and β_3—based on the specific responses they elicit and by their selective binding of drugs that activate or block them. Although there are some exceptions, activation of α_1 and β_1 receptors generally produces excitation, and activation of α_2 and β_2 receptors causes inhibition of effector tissues. β_3 receptors are present only on cells of brown adipose tissue, where their activation causes *thermogenesis* (heat production). Cells of most effectors contain either alpha or beta receptors; some visceral effector cells contain both. Norepinephrine stimulates alpha receptors more strongly than beta receptors; epinephrine is a potent stimulator of both alpha and beta receptors.

The activity of norepinephrine at a synapse is terminated either when the NE is taken up by the axon that released it or when the NE is enzymatically inactivated by either **catechol-*O*-methyltransferase (COMT)** (kat′-e-kol′-ō-meth′-il-TRANS-fer-ās) or **monoamine oxidase (MAO)** (mon′-ō-AM-ēn OK-si-dās′). Compared to ACh, norepinephrine lingers in the synaptic cleft for a longer time. Thus, effects triggered by adrenergic neurons typically are longer lasting than those triggered by cholinergic neurons.

Table 15.2 describes the locations of cholinergic and adrenergic receptors and summarizes the responses that occur when each type of receptor is activated.

Receptor Agonists and Antagonists

A large variety of drugs and natural products can selectively activate or block specific cholinergic or adrenergic receptors. An **agonist** (*agon* = a contest) is a substance that binds to and activates a receptor, in the process mimicking the effect of a natural

TABLE 15.2

Location and Responses of Adrenergic and Cholinergic Receptors

TYPE OF RECEPTOR	MAJOR LOCATIONS	EFFECTS OF RECEPTOR ACTIVATION
CHOLINERGIC	Integral proteins in postsynaptic plasma membranes; activated by the neurotransmitter acetylcholine.	
Nicotinic	Plasma membrane of postganglionic sympathetic and parasympathetic neurons.	Excitation → impulses in postganglionic neurons.
	Chromaffin cells of adrenal medullae.	Epinephrine and norepinephrine secretion.
	Sarcolemma of skeletal muscle fibers (motor end plate).	Excitation → contraction.
Muscarinic	Effectors innervated by parasympathetic postganglionic neurons.	In some receptors, excitation; in others, inhibition.
	Sweat glands innervated by cholinergic sympathetic postganglionic neurons.	Increased sweating.
	Skeletal muscle blood vessels innervated by cholinergic sympathetic postganglionic neurons.	Inhibition → relaxation → vasodilation.
ADRENERGIC	Integral proteins in postsynaptic plasma membranes; activated by the neurotransmitter norepinephrine and the hormones norepinephrine and epinephrine.	
α_1	Smooth muscle fibers in blood vessels that serve salivary glands, skin, mucosal membranes, kidneys, and abdominal viscera; radial muscle in iris of eye; sphincter muscles of stomach and urinary bladder.	Excitation → contraction, which causes vasoconstriction, dilation of pupil, and closing of sphincters.
	Salivary gland cells.	Secretion of K^+ and water.
	Sweat glands on palms and soles.	Increased sweating.
α_2	Smooth muscle fibers in some blood vessels.	Inhibition → relaxation → vasodilation.
	Cells of pancreatic islets that secrete the hormone insulin (beta cells).	Decreased insulin secretion.
	Pancreatic acinar cells.	Inhibition of digestive enzyme secretion.
	Platelets in blood.	Aggregation to form platelet plug.
β_1	Cardiac muscle fibers.	Excitation → increased force and rate of contraction.
	Juxtaglomerular cells of kidneys.	Renin secretion.
	Posterior pituitary.	Antidiuretic hormone (ADH) secretion.
	Adipose cells.	Breakdown of triglycerides → release of fatty acids into blood.
β_2	Smooth muscle in walls of airways; in blood vessels that serve heart, skeletal muscle, adipose tissue, and liver; and in walls of visceral organs, such as urinary bladder.	Inhibition → relaxation, which causes dilation of airways, vasodilation, and relaxation of organ walls.
	Ciliary muscle in eye.	Inhibition → relaxation.
	Hepatocytes in liver.	Glycogenolysis (breakdown of glycogen into glucose).
β_3	Brown adipose tissue.	Thermogenesis (heat production).

neurotransmitter or hormone. Phenylephrine, an adrenergic agonist at α_1 receptors, is a common ingredient in cold and sinus medications. Because it constricts blood vessels in the nasal mucosa, phenylephrine reduces production of mucus, thus relieving nasal congestion. An **antagonist** (*anti* = against) is a substance that binds to and blocks a receptor, thereby preventing a natural neurotransmitter or hormone from exerting its effect. For example, atropine blocks muscarinic ACh receptors, dilates the pupils, reduces glandular secretions, and relaxes smooth muscle in the gastrointestinal tract. As a result, it is used to dilate the pupils during eye examinations, in the treatment of smooth muscle disorders such as iritis and intestinal hypermotility, and as an antidote for chemical warfare agents that inactivate acetylcholinesterase.

Propranolol (Inderal®) often is prescribed for patients with hypertension (high blood pressure). It is a nonselective beta blocker, meaning it binds to all types of beta receptors and prevents their activation by epinephrine and norepinephrine. The desired effects of propranolol are due to its *blockade* of β_1 receptors—namely, decreased heart rate and force of contraction and a consequent decrease in blood pressure. Undesired effects due to blockade of β_2 receptors may include hypoglycemia (low blood glucose), resulting from decreased glycogen breakdown and decreased gluconeogenesis (the conversion of a noncarbohydrate into glucose in the liver), and mild bronchoconstriction (narrowing of the airways). If these side effects pose a threat to the patient, a selective β_1 blocker (which binds only to specific beta receptors) such as metoprolol (Lopressor®) can be prescribed instead of propranolol.

✔CHECKPOINT

8. Why are cholinergic and adrenergic neurons so named?
9. What neurotransmitters and hormones bind to adrenergic receptors?
10. What do the terms agonist and antagonist mean?

15.4 PHYSIOLOGY OF THE ANS

OBJECTIVE

- Describe the major responses of the body to stimulation by the sympathetic and parasympathetic divisions of the ANS.

Autonomic Tone

As noted earlier, most body organs receive innervation from both divisions of the ANS, which typically work in opposition to one another. The balance between sympathetic and parasympathetic activity, called **autonomic tone,** is regulated by the hypothalamus. Typically, the hypothalamus turns up sympathetic tone at the same time it turns down parasympathetic tone, and vice versa. The two divisions can affect body organs differently because their postganglionic neurons release different neurotransmitters and because the effector organs possess different adrenergic and cholinergic receptors. A few structures receive only sympathetic innervation—sweat glands, arrector pili muscles attached to hair follicles in the skin, the kidneys, the spleen, most blood vessels, and the adrenal medullae (see Figure 15.2). In these structures there is no opposition from the parasympathetic division. Still, an increase in sympathetic tone has one effect, and a decrease in sympathetic tone produces the opposite effect.

Sympathetic Responses

During physical or emotional stress, the sympathetic division dominates the parasympathetic division. High sympathetic tone favors body functions that can support vigorous physical activity and rapid production of ATP. At the same time, the sympathetic division reduces body functions that favor the storage of energy. Besides physical exertion, various emotions—such as fear, embarrassment, or rage—stimulate the sympathetic division. Visualizing body changes that occur during "E situations" such as exercise, emergency, excitement, and embarrassment will help you remember most of the sympathetic responses. Activation of the sympathetic division and release of hormones by the adrenal medullae set in motion a series of physiological responses collectively called the **fight-or-flight response,** which includes the following effects:

- The pupils of the eyes dilate.
- Heart rate, force of heart contraction, and blood pressure increase.
- The airways dilate, allowing faster movement of air into and out of the lungs.
- The blood vessels that supply the kidneys and gastrointestinal tract constrict, which decreases blood flow through these tissues. The result is a slowing of urine formation and digestive activities, which are not essential during exercise.
- Blood vessels that supply organs involved in exercise or fighting off danger—skeletal muscles, cardiac muscle, liver, and adipose tissue—dilate, allowing greater blood flow through these tissues.
- Liver cells perform glycogenolysis (breakdown of glycogen to glucose), and adipose tissue cells perform lipolysis (breakdown of triglycerides to fatty acids and glycerol).
- Release of glucose by the liver increases blood glucose level.
- Processes that are not essential for meeting the stressful situation are inhibited. For example, muscular movements of the gastrointestinal tract and digestive secretions slow down or even stop.

The effects of sympathetic stimulation are longer lasting and more widespread than the effects of parasympathetic stimulation for three reasons: (1) Sympathetic postganglionic axons diverge more extensively; as a result, many tissues are activated simultaneously. (2) Acetylcholinesterase quickly inactivates acetylcholine, but norepinephrine lingers in the synaptic cleft for a longer period. (3) Epinephrine and norepinephrine secreted into the blood from the adrenal medullae intensify and prolong the responses caused by NE liberated from sympathetic postganglionic axons. These blood-borne hormones circulate throughout the body, affecting all tissues that have alpha and beta receptors. In time, blood-borne NE and epinephrine are destroyed by enzymes in the liver.

Parasympathetic Responses

In contrast to the fight-or-flight activities of the sympathetic division, the parasympathetic division enhances **rest-and-digest** activities. Parasympathetic responses support body functions that conserve and restore body energy during times of rest and recovery. In the quiet intervals between periods of exercise, parasympathetic impulses to the digestive glands and the smooth muscle of the gastrointestinal tract predominate over sympathetic impulses. This allows energy-supplying food to be digested and absorbed. At the same time, parasympathetic responses reduce body functions that support physical activity.

The acronym *SLUDD* can be helpful in remembering five parasympathetic responses. It stands for salivation (S), lacrimation (L), urination (U), digestion (D), and defecation (D). All of these activities are stimulated mainly by the parasympathetic division. Besides the increasing SLUDD responses, other important parasympathetic responses are "three decreases": decreased heart rate, decreased diameter of airways (bronchoconstriction), and decreased diameter (constriction) of the pupils.

Table 15.3 compares the structural and functional features of the sympathetic and parasympathetic divisions of the ANS. Table 15.4 lists the responses of glands, cardiac muscle, and smooth muscle to stimulation by the sympathetic and parasympathetic divisions of the ANS.

✓CHECKPOINT

11. Define autonomic tone.
12. What are some examples of the antagonistic effects of the sympathetic and parasympathetic divisions of the autonomic nervous system?
13. What happens during the fight-or-flight response?
14. Why is the parasympathetic division of the ANS called an energy conservation/restoration system?
15. Describe the sympathetic response in a frightening situation for each of the following body parts: hair follicles, iris of eye, lungs, spleen, adrenal medullae, urinary bladder, stomach, intestines, gallbladder, liver, heart, arterioles of the abdominal viscera, and arterioles of skeletal muscles.

TABLE 15.3

Comparison of Sympathetic and Parasympathetic Divisions of the ANS

	SYMPATHETIC (THORACOLUMBAR)	PARASYMPATHETIC (CRANIOSACRAL)
Distribution	Wide regions of body: skin, sweat glands, arrector pili muscles of hair follicles, adipose tissue, smooth muscle of blood vessels.	Limited mainly to head and to viscera of thorax, abdomen, and pelvis; some blood vessels.
Location of preganglionic neuron cell bodies and site of outflow	Lateral gray horns of spinal cord segments T1–L2. Axons of preganglionic neurons constitute thoracolumbar outflow.	Nuclei of cranial nerves III, VII, IX, and X and lateral gray matter of spinal cord segments S2–S4. Axons of preganglionic neurons constitute craniosacral outflow.
Associated ganglia	Sympathetic trunk ganglia and prevertebral ganglia.	Terminal ganglia.
Ganglia locations	Close to CNS and distant from visceral effectors.	Typically near or within wall of visceral effectors.
Axon length and divergence	Preganglionic neurons with short axons synapse with many postganglionic neurons with long axons that pass to many visceral effectors.	Preganglionic neurons with long axons usually synapse with four to five postganglionic neurons with short axons that pass to single visceral effector.
White and gray rami communicantes	Both present; white rami communicantes contain myelinated preganglionic axons; gray rami communicantes contain unmyelinated postganglionic axons.	Neither present.
Neurotransmitters	Preganglionic neurons release acetylcholine (ACh), which is excitatory and stimulates postganglionic neurons; most postganglionic neurons release norepinephrine (NE); postganglionic neurons that innervate most sweat glands and some blood vessels in skeletal muscle release ACh.	Preganglionic neurons release ACh, which is excitatory and stimulates postganglionic neurons; postganglionic neurons release ACh.
Physiological effects	Fight-or-flight responses.	Rest-and-digest activities.

TABLE 15.4

Effects of Sympathetic and Parasympathetic Divisions of the ANS

VISCERAL EFFECTOR	EFFECT OF SYMPATHETIC STIMULATION (α OR β ADRENERGIC RECEPTORS, EXCEPT AS NOTED)*	EFFECT OF PARASYMPATHETIC STIMULATION (MUSCARINIC ACh RECEPTORS)
GLANDS		
Adrenal medullae	Secretion of epinephrine and norepinephrine (nicotinic ACh receptors).	No known effect.
Lacrimal (tear)	Slight secretion of tears (α).	Secretion of tears.
Pancreas	Inhibits secretion of digestive enzymes and the hormone insulin (α_2); promotes secretion of the hormone glucagon (β_2).	Secretion of digestive enzymes and the hormone insulin.
Posterior pituitary	Secretion of antidiuretic hormone (ADH) (β_1).	No known effect.
Pineal	Increases synthesis and release of melatonin (β).	No known effect.
Sweat	Increases sweating in most body regions (muscarinic ACh receptors); sweating on palms and soles (α_1).	No known effect.
Adipose tissue†	Lipolysis (breakdown of triglycerides into fatty acids and glycerol) (β_1); release of fatty acids into blood (β_1 and β_3).	No known effect.
Liver†	Glycogenolysis (conversion of glycogen into glucose); gluconeogenesis (conversion of noncarbohydrates into glucose); decreased bile secretion (α and β_2).	Glycogen synthesis; increased bile secretion.
Kidney, juxtaglomerular cells†	Secretion of renin (β_1).	No known effect.
CARDIAC (HEART) MUSCLE		
	Increased heart rate and force of atrial and ventricular contractions (β_1).	Decreased heart rate; decreased force of atrial contraction.
SMOOTH MUSCLE		
Iris, radial muscle	Contraction → dilation of pupil (α_1).	No known effect.
Iris, circular muscle	No known effect.	Contraction → constriction of pupil.
Ciliary muscle of eye	Relaxation to adjust shape of lens for distant vision (β_2).	Contraction for close vision.
Lungs, bronchial muscle	Relaxation → airway dilation (β_2).	Contraction → airway constriction.
Gallbladder and ducts	Relaxation to facilitate storage of bile in the gallbladder ($\beta 2$).	Contraction → release of bile into small intestine.
Stomach and intestines	Decreased motility and tone (α_1, α_2, β_2); contraction of sphincters (α_1).	Increased motility and tone; relaxation of sphincters.
Spleen	Contraction and discharge of stored blood into general circulation (α_1).	No known effect.
Ureter	Increases motility (α_1).	Increases motility (?).
Urinary bladder	Relaxation of muscular wall (β_2); contraction of internal urethral sphincter (α_1).	Contraction of muscular wall; relaxation of internal urethral sphincter.
Uterus	Inhibits contraction in nonpregnant women (β_2); promotes contraction in pregnant women (α_1).	Minimal effect.
Sex organs	In males: contraction of smooth muscle of ductus (vas) deferens, prostate, and seminal vesicle resulting in ejaculation (α_1).	Vasodilation; erection of clitoris (females) and penis (males).
Hair follicles, arrector pili muscle	Contraction → erection of hairs resulting in goosebumps (α_1).	No known effect.

TABLE 15.4 CONTINUED

Effects of Sympathetic and Parasympathetic Divisions of the ANS

VISCERAL EFFECTOR	EFFECT OF SYMPATHETIC STIMULATION (α OR β ADRENERGIC RECEPTORS, EXCEPT AS NOTED)*	EFFECT OF PARASYMPATHETIC STIMULATION (MUSCARINIC ACh RECEPTORS)
VASCULAR SMOOTH MUSCLE		
Salivary gland arterioles	Vasoconstriction, which decreases secretion of saliva (α_1).	Vasodilation, which increases secretion of saliva.
Gastric gland arterioles	Vasoconstriction, which inhibits secretion (α_1).	Secretion of gastric juice.
Intestinal gland arterioles	Vasoconstriction, which inhibits secretion (α_1).	Secretion of intestinal juice.
Coronary (heart) arterioles	Relaxation → vasodilation (β_2); contraction → vasoconstriction (α_1, α_2); contraction → vasoconstriction (muscarinic ACh receptors).	Contraction → vasoconstriction.
Skin and mucosal arterioles	Contraction → vasoconstriction (α_1).	Vasodilation, which may not be physiologically significant.
Skeletal muscle arterioles	Contraction → vasoconstriction (α_1); relaxation → vasodilation (β_2); relaxation → vasodilation (muscarinic ACh receptors).	No known effect.
Abdominal viscera arterioles	Contraction → vasoconstriction (α_1, β_2).	No known effect.
Brain arterioles	Slight contraction → vasoconstriction (α_1).	No known effect.
Kidney arterioles	Constriction of blood vessels → decreased urine volume (α_1).	No known effect.
Systemic veins	Contraction → constriction (α_1); relaxation → dilation (β_2).	No known effect.

*Subcategories of α and β receptors are listed if known.

†Grouped with glands because they release substances into the blood.

15.5 INTEGRATION AND CONTROL OF AUTONOMIC FUNCTIONS

OBJECTIVES

- Describe the components of an autonomic reflex.
- Explain the relationship of the hypothalamus to the ANS.

Autonomic Reflexes

Autonomic reflexes are responses that occur when nerve impulses pass through an autonomic reflex arc. These reflexes play a key role in regulating controlled conditions in the body, such as *blood pressure,* by adjusting heart rate, force of ventricular contraction, and blood vessel diameter; *digestion,* by adjusting the motility (movement) and muscle tone of the gastrointestinal tract; and *defecation* and *urination,* by regulating the opening and closing of sphincters.

The components of an autonomic reflex arc are as follows:

- **Receptor.** Like the receptor in a somatic reflex arc (see Figure 13.14), the receptor in an autonomic reflex arc is the distal end of a sensory neuron, which responds to a stimulus and produces a change that will ultimately trigger nerve impulses. Autonomic sensory receptors are mostly associated with interoceptors.
- **Sensory neuron.** Conducts nerve impulses from receptors to the CNS.
- **Integrating center.** Interneurons within the CNS relay signals from sensory neurons to motor neurons. The main integrating centers for most autonomic reflexes are located in the hypothalamus and brain stem. Some autonomic reflexes, such as those for urination and defecation, have integrating centers in the spinal cord.
- **Motor neurons.** Nerve impulses triggered by the integrating center propagate out of the CNS along motor neurons to an effector. In an autonomic reflex arc, two motor neurons connect the CNS to an effector: The preganglionic neuron conducts motor impulses from the CNS to an autonomic ganglion, and the postganglionic neuron conducts motor impulses from an autonomic ganglion to an effector (see Figure 15.1).
- **Effector.** In an autonomic reflex arc, the effectors are smooth muscle, cardiac muscle, and glands, and the reflex is called an autonomic reflex.

Focus on Homeostasis

BODY SYSTEM	CONTRIBUTION OF THE NERVOUS SYSTEM
For all body systems 	Together with hormones from the endocrine system, nerve impulses provide communication and regulation of most body tissues.
Integumentary system 	Sympathetic nerves of the autonomic nervous system (ANS) control contraction of smooth muscles attached to hair follicles and secretion of perspiration from sweat glands.
Skeletal system 	Pain receptors in bone tissue warn of bone trauma or damage.
Muscular system 	Somatic motor neurons receive instructions from motor areas of the brain and stimulate contraction of skeletal muscles to bring about body movements; basal nuclei and reticular formation set level of muscle tone; cerebellum coordinates skilled movements.
Endocrine system 	Hypothalamus regulates secretion of hormones from anterior and posterior pituitary; ANS regulates secretion of hormones from adrenal medulla and pancreas.
Cardiovascular system 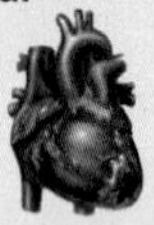	Cardiovascular center in the medulla oblongata provides nerve impulses to ANS that govern heart rate and the forcefulness of the heartbeat; nerve impulses from ANS also regulate blood pressure and blood flow through blood vessels.
Lymphatic system and immunity 	Certain neurotransmitters help regulate immune responses; activity in nervous system may increase or decrease immune responses.
Respiratory system 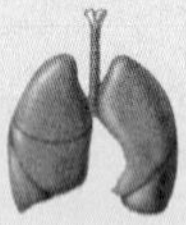	Respiratory areas in brain stem control breathing rate and depth; ANS helps regulate diameter of airways.
Digestive system 	ANS and enteric nervous system (ENS) help regulate digestion; parasympathetic division of ANS stimulates many digestive processes.
Urinary system 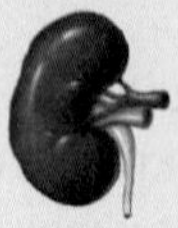	ANS helps regulate blood flow to kidneys, thereby influencing the rate of urine formation; brain and spinal cord centers govern emptying of the urinary bladder.
Reproductive systems 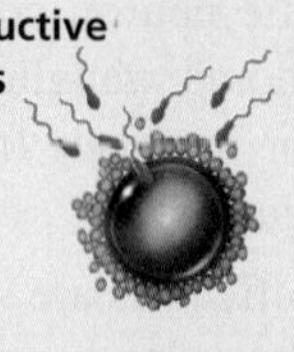	Hypothalamus and limbic system govern a variety of sexual behaviors; ANS brings about erection of penis in males and clitoris in females and ejaculation of semen in males; hypothalamus regulates release of anterior pituitary hormones that control gonads (ovaries and testes); nerve impulses elicited by touch stimuli from suckling infant cause release of oxytocin and milk ejection in nursing mothers.

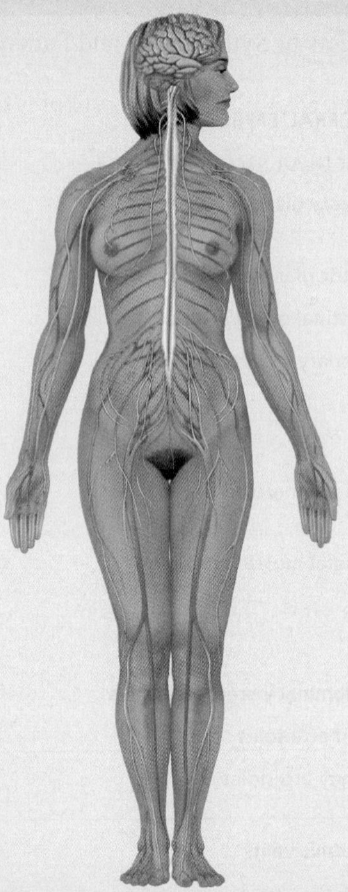

THE NERVOUS SYSTEM

Autonomic Control by Higher Centers

Normally, we are not aware of muscular contractions of our digestive organs, our heartbeat, changes in the diameter of our blood vessels, and pupil dilation and constriction because the integrating centers for these autonomic responses are in the spinal cord or the lower regions of the brain. Somatic or autonomic sensory neurons deliver input to these centers, and autonomic motor neurons provide output that adjusts activity in the visceral effector, usually without our conscious perception.

The hypothalamus is the major control and integration center of the ANS. The hypothalamus receives sensory input related to visceral functions, olfaction (smell), and gustation (taste), as well as changes in temperature, osmolarity, and levels of various substances in blood. It also receives input relating to emotions from the limbic system. Output from the hypothalamus influences autonomic centers in both the brain stem (such as the cardiovascular, salivation, swallowing, and vomiting centers) and the spinal cord (such as the defecation and urination reflex centers in the sacral spinal cord).

Anatomically, the hypothalamus is connected to both the sympathetic and parasympathetic divisions of the ANS by axons of neurons with dendrites and cell bodies in various hypothalamic nuclei. The axons form tracts from the hypothalamus to parasympathetic and sympathetic nuclei in the brain stem and spinal cord through relays in the reticular formation. The posterior and lateral parts of the hypothalamus control the sympathetic division. Stimulation of these areas produces an increase in heart rate and force of contraction, a rise in blood pressure due to constriction of blood vessels, an increase in body temperature, dilation of the pupils, and inhibition of the gastrointestinal tract. In contrast, the anterior and medial parts of the hypothalamus control the parasympathetic division. Stimulation of these areas results in a decrease in heart rate, lowering of blood pressure, constriction of the pupils, and increased secretion and motility of the gastrointestinal tract.

✔CHECKPOINT

16. Give three examples of controlled conditions in the body that are kept in homeostatic balance by autonomic reflexes.
17. How does an autonomic reflex arc differ from a somatic reflex arc?

• • •

Now that we have discussed the structure and function of the nervous system, you can appreciate the many ways that this system contributes to homeostasis of other body systems by examining *Focus on Homeostasis: The Nervous System.*

DISORDERS: HOMEOSTATIC IMBALANCES

Autonomic Dysreflexia

Autonomic dysreflexia (dis′-rē-FLEKS-sē-a) is an exaggerated response of the sympathetic division of the ANS that occurs in about 85% of individuals with spinal cord injury at or above the level of T6. The condition is seen after recovery from spinal shock (see Disorders: Homeostatic Imbalances in Chapter 13) and occurs due to interruption of the control of ANS neurons by higher centers. When certain sensory impulses, such as those resulting from stretching of a full urinary bladder, are unable to ascend the spinal cord, mass stimulation of the sympathetic nerves inferior to the level of injury occurs. Other triggers include stimulation of pain receptors and the visceral contractions resulting from sexual stimulation, labor/delivery, and bowel stimulation. Among the effects of increased sympathetic activity is severe vasoconstriction, which elevates blood pressure. In response, the cardiovascular center in the medulla oblongata (1) increases parasympathetic output via the vagus (X) nerve, which decreases heart rate, and (2) decreases sympathetic output, which causes dilation of blood vessels superior to the level of the injury.

The condition is characterized by a pounding headache; hypertension; flushed, warm skin with profuse sweating above the injury level; pale, cold, dry skin below the injury level; and anxiety. This emergency condition requires immediate intervention. The first approach is to quickly identify and remove the problematic stimulus. If this does not relieve the symptoms, an antihypertensive drug such as clonidine or nitroglycerin can be administered. Untreated autonomic dysreflexia can cause seizures, stroke, or heart attack.

Raynaud Phenomenon

In **Raynaud phenomenon** (rā-NŌ) the digits (fingers and toes) become ischemic (lack blood) after exposure to cold or with emotional stress. The condition is due to excessive sympathetic stimulation of smooth muscle in the arterioles of the digits and a heightened response to stimuli that cause vasoconstriction. When arterioles in the digits vasoconstrict in response to sympathetic stimulation, blood flow is greatly diminished. As a result, the digits may blanch (look white due to blockage of blood flow) or become cyanotic (look blue due to deoxygenated blood in capillaries). In extreme cases, the digits may become necrotic from lack of oxygen and nutrients. With rewarming after cold exposure, the arterioles may dilate, causing the fingers and toes to look red. Many patients with Raynaud phenomenon have low blood pressure. Some have increased numbers of alpha adrenergic receptors. Raynaud is most common in young women and occurs more often in cold climates. Patients with Raynaud phenomenon should avoid exposure to cold, wear warm clothing, and keep the hands and feet warm. Drugs used to treat Raynaud include nifedipine, a calcium channel blocker that relaxes vascular smooth muscle, and prazosin, which relaxes smooth muscle by blocking alpha receptors. Smoking and the use of alcohol or illicit drugs can exacerbate the symptoms of this condition.

MEDICAL TERMINOLOGY

Autonomic nerve neuropathy (noo-ROP-a-thē) A *neuropathy* (disorder of a cranial or spinal nerve) that affects one or more autonomic nerves, with multiple effects on the autonomic nervous system, including constipation, urinary incontinence, impotence, and fainting and low blood pressure when standing (*orthostatic hypotension*) due to decreased sympathetic control of the cardiovascular system. Often caused by long-term diabetes mellitus (*diabetic neuropathy*).

Biofeedback A technique in which an individual is provided with information regarding an autonomic response such as heart rate, blood pressure, or skin temperature. Various electronic monitoring devices provide visual or auditory signals about the autonomic responses. By concentrating on positive thoughts, individuals learn to alter autonomic responses. For example, biofeedback has been used to decrease heart rate and blood pressure and increase skin temperature in order to decrease the severity of migraine headaches.

Dysautonomia (dis-aw-tō-NŌ-mē-a; *dys-* = difficult; *-autonomia* = self-governing) An inherited disorder in which the autonomic nervous system functions abnormally, resulting in reduced tear gland secretions, poor vasomotor control, motor incoordination, skin blotching, absence of pain sensation, difficulty in swallowing, hyporeflexia, excessive vomiting, and emotional instability.

Hyperhidrosis (hī′-per-hī-DRŌ-sis; *hyper-* = above or too much; *-hidrosis* = sweat) Excessive or profuse sweating due to intense stimulation of sweat glands.

Mass reflex In cases of severe spinal cord injury above the level of the sixth thoracic vertebra, stimulation of the skin or overfilling of a visceral organ (such as the urinary bladder or colon) below the level of the injury results in intense activation of autonomic and somatic output from the spinal cord as reflex activity returns. The exaggerated response occurs because there is no inhibitory input from the brain. The mass reflex consists of flexor spasms of the lower limbs, evacuation of the urinary bladder and colon, and profuse sweating below the level of the lesion.

Megacolon (*mega-* = big) An abnormally large colon. In congenital megacolon, parasympathetic nerves to the distal segment of the colon do not develop properly. Loss of motor function in the segment causes massive dilation of the normal proximal colon. The condition results in extreme constipation, abdominal distension, and occasionally, vomiting. Surgical removal of the affected segment of the colon corrects the disorder.

Reflex sympathetic dystrophy (RSD) A syndrome that includes spontaneous pain, painful hypersensitivity to stimuli such as light touch, and excessive coldness and sweating in the involved body part. The disorder frequently involves the forearms, hands, knees, and feet. It appears that activation of the sympathetic division of the autonomic nervous system due to traumatized nociceptors as a result of trauma or surgery on bones or joints is involved. Treatment consists of anesthetics and physical therapy. Recent clinical studies also suggest that the drug baclofen can be used to reduce pain and restore normal function to the affected body part. Also called **complex regional pain syndrome type 1.**

Vagotomy (vā-GOT-ō-mē; *-tome* = incision) Cutting the vagus (X) nerve. It is frequently done to decrease the production of hydrochloric acid in persons with ulcers.

CHAPTER REVIEW AND RESOURCE SUMMARY

Review	Resource
15.1 Comparison of Somatic and Autonomic Nervous Systems	Anatomy Overview - The Nervous System: Overview
1. The somatic nervous system operates under conscious control; the ANS usually operates without conscious control.	Anatomy Overview - Organization of the ANS
2. Sensory input for the somatic nervous system is mainly from the somatic senses and special senses; sensory input for the ANS is from interoceptors, in addition to somatic senses and special senses.	Animation - Somatic Sensory and Motor Pathways
3. The axons of somatic motor neurons extend from the CNS and synapse directly with an effector. Autonomic motor pathways consist of two motor neurons in series. The axon of the first motor neuron extends from the CNS and synapses in an autonomic ganglion with the second motor neuron; the second neuron synapses with an effector.	Figure 15.1 - Motor Neuron Pathways in the Somatic and Autonomic Nervous Systems 
4. The output (motor) portion of the ANS has two major divisions: sympathetic and parasympathetic. Most body organs receive dual innervation; usually one ANS division causes excitation and the other causes inhibition. The enteric division consists of nerves and ganglia within the wall of the GI tract.	Exercise - Assemble the Structure of the ANS Exercise - Sort ANS Functions Exercise - What Is Your ANS Status?
5. Somatic nervous system effectors are skeletal muscles; ANS effectors include cardiac muscle, smooth muscle, and glands.	
6. Table 15.1 compares the somatic and autonomic nervous systems.	
15.2 Anatomy of Autonomic Motor Pathways	Animation - ANS: Motor Pathways Anatomy Overview - Visceral Receptors Anatomy Overview - Visceral Effectors
1. A preganglionic neuron is the first of the two motor neurons in any autonomic motor pathway; the axon of the preganglionic neuron extends to an autonomic ganglion, where it synapses with a postganglionic neuron, the second neuron in the autonomic motor pathway. Preganglionic neurons are myelinated; postganglionic neurons are unmyelinated.	
2. The cell bodies of sympathetic preganglionic neurons are in the lateral gray horns of the 12 thoracic and the first two or three lumbar segments of the spinal cord; the cell bodies of parasympathetic	

Review | **Resource**

preganglionic neurons are in four cranial nerve nuclei (III, VII, IX, and X) in the brain stem and lateral gray matter of the second through fourth sacral segments of the spinal cord.

3. There are two major groups of autonomic ganglia: sympathetic ganglia and parasympathetic ganglia. Sympathetic ganglia include sympathetic trunk ganglia (on both sides of vertebral column) and prevertebral ganglia (anterior to vertebral column). Parasympathetic ganglia are known as terminal ganglia (near or inside visceral effectors).
4. Sympathetic preganglionic neurons synapse with postganglionic neurons in ganglia of the sympathetic trunk or in prevertebral ganglia; parasympathetic preganglionic neurons synapse with postganglionic neurons in terminal ganglia.

15.3 ANS Neurotransmitters and Receptors

1. Cholinergic neurons release acetylcholine. In the ANS, cholinergic neurons include all sympathetic and parasympathetic preganglionic neurons, sympathetic postganglionic neurons that innervate most sweat glands, and all parasympathetic postganglionic neurons.
2. Acetylcholine binds to cholinergic receptors. The two types of cholinergic receptors, both of which bind acetylcholine, are nicotinic receptors and muscarinic receptors. Nicotinic receptors are present in the plasma membranes of dendrites and cell bodies of both sympathetic and parasympathetic postganglionic neurons, in the plasma membranes of chromaffin cells of the adrenal medullae, and in the motor end plate at the neuromuscular junction. Muscarinic receptors are present in the plasma membranes of all effectors innervated by parasympathetic postganglionic neurons and in most sweat glands innervated by cholinergic sympathetic postganglionic neurons.
3. In the ANS, adrenergic neurons release norepinephrine. Most sympathetic postganglionic neurons are adrenergic.
4. Both epinephrine and norepinephrine bind to adrenergic receptors, which are found on visceral effectors innervated by most sympathetic postganglionic neurons. The two main types of adrenergic receptors are alpha receptors and beta receptors.
5. Table 15.2 summarizes the types of cholinergic and adrenergic receptors.
6. An agonist is a substance that binds to and activates a receptor, mimicking the effect of a natural neurotransmitter or hormone. An antagonist is a substance that binds to and blocks a receptor, thereby preventing a natural neurotransmitter or hormone from exerting its effect.

Resource:
Anatomy Overview - Neurotransmitters
Animation - The ANS: Types of Neurotransmitters and Neurons
Figure 15.7 - Cholinergic and Adrenergic Neurons in the Sympathetic and Parasympathetic Divisions

15.4 Physiology of the ANS

1. The sympathetic division favors body functions that can support vigorous physical activity and rapid production of ATP (fight-or-flight response); the parasympathetic division regulates activities that conserve and restore body energy.
2. The effects of sympathetic stimulation are longer lasting and more widespread than the effects of parasympathetic stimulation.
3. Table 15.3 compares structural and functional features of the sympathetic and parasympathetic divisions.
4. Table 15.4 lists sympathetic and parasympathetic responses.

Resource:
Anatomy Overview - Effectors
Animation - Physiological Effects of the ANS
Animation - The Alarm Reaction

15.5 Integration and Control of Autonomic Functions

1. An autonomic reflex adjusts the activities of smooth muscle, cardiac muscle, and glands.
2. An autonomic reflex arc consists of a receptor, a sensory neuron, an integrating center, two autonomic motor neurons, and a visceral effector.
3. The hypothalamus is the major control and integration center of the ANS. It is connected to both the sympathetic and the parasympathetic divisions.

Resource:
Anatomy Overview - The ANS Control Centers
Animation - Reflexes

SELF-QUIZ QUESTIONS

Fill in the blanks in the following statements.

1. Cholinergic neurons release _____ and adrenergic neurons release _____.
2. Because of the location of the preganglionic cell bodies, the sympathetic division of the ANS is also called the _____ division; the parasympathetic division is also called the _____ division.

Indicate whether the following statements are true or false.

3. The vagus nerves transmit 80% of the outflow of the parasympathetic preganglionic axons.
4. Organs that receive both sympathetic and parasympathetic motor impulses are said to have dual innervation.

Choose the one best answer to the following questions.

5. Which of the following statements is *false*?
   (a) A single sympathetic preganglionic fiber may synapse with 20 or more postganglionic fibers, which partly explains why sympathetic responses are widespread throughout the body.
   (b) Parasympathetic effects tend to be localized because parasympathetic neurons usually synapse in the terminal ganglia with only four or five postsynaptic neurons (all of which supply a single effector).
   (c) Some sympathetic preganglionic neurons extend to and terminate in the adrenal medullae.
   (d) The parasympathetic preganglionic neurons synapse with the postganglionic axons in the prevertebral ganglia.
   (e) Parasympathetic preganglionic neurons emerge from the CNS as part of a cranial nerve or anterior root of a spinal nerve.

6. Which autonomic plexus supplies the large intestine? (1) renal, (2) inferior mesenteric, (3) hypogastric, (4) superior mesenteric, (5) celiac.
   (a) 2, 3, and 4 (b) 1, 2, 3, 4, and 5 (c) 3 and 4
   (d) 4 and 5 (e) 2 and 4

7. Which of the following statements are *true*? (1) The somatic nervous system and the ANS both include sensory and motor neurons. (2) Somatic motor neurons release the neurotransmitter norepinephrine. (3) The effect of an autonomic motor neuron is either excitation or inhibition, but that of a somatic motor neuron is always excitation. (4) Autonomic sensory neurons are mostly associated with interoceptors. (5) Autonomic motor pathways consist of two motor neurons in series. (6) Somatic motor pathways consist of two motor neurons in series.
   (a) 1, 2, 3, 4, and 5 (b) 1, 3, 4, and 5 (c) 2, 3, 5, and 6
   (d) 1, 3, 5, and 6 (e) 2, 4, 5, and 6

8. Which of the following statements is *false*?
   (a) The first neuron in an autonomic pathway is the preganglionic neuron.
   (b) The axons of preganglionic neurons are located in spinal or cranial nerves.
   (c) The postganglionic neuron's cell body is within the CNS.
   (d) Postganglionic neurons relay impulses from autonomic ganglia to visceral effectors.
   (e) All somatic motor neurons release acetylcholine.

9. Which of the following are *true*? (1) Monoamine oxidase enzymatically breaks down norepinephrine. (2) Activation of α_2 and β_2 receptors generally produces excitation in the effectors. (3) A beta blocker works by preventing activation of β receptors by epinephrine and norepinephrine. (4) An agonist is a substance that binds to a receptor and prevents the natural neurotransmitter from exerting its effect. (5) Activation of nicotinic receptors always causes excitation of the postsynaptic cell.
   (a) 2 and 3 (b) 1, 2, and 3 (c) 2, 4, and 5
   (d) 1, 2, 3, 4, and 5 (e) 1, 3, and 5

10. Which of the following are cholinergic neurons? (1) all sympathetic preganglionic neurons, (2) all parasympathetic preganglionic neurons, (3) all parasympathetic postganglionic neurons, (4) all sympathetic postganglionic neurons, (5) some sympathetic postganglionic neurons.
    (a) 1, 2, 3, and 5 (b) 1, 2, 3, and 4 (c) 2, 3, and 5
    (d) 2 and 5 (e) 1, 3, and 5

11. Which of the following statements are *true*? (1) Most sympathetic postganglionic axons are adrenergic. (2) Cholinergic receptors are classified as nicotinic and muscarinic. (3) Adrenergic receptors are classified as alpha and beta. (4) Muscarinic receptors are present on all effectors innervated by parasympathetic postganglionic axons. (5) In general, norepinephrine stimulates alpha receptors more vigorously than beta receptors; epinephrine is a potent stimulator of both alpha and beta receptors.
    (a) 1, 2, 3, 4, and 5 (b) 2, 3, 4 and 5 (c) 1, 3, 4, and 5
    (d) 3, 4, and 5 (e) 1, 2, 3, and 4

12. Which of the following are reasons why the effects of sympathetic stimulation are longer lasting and more widespread than those of parasympathetic stimulation? (1) There is greater divergence of sympathetic postganglionic fibers. (2) There is less divergence of sympathetic postganglionic fibers. (3) Acetylcholinesterase quickly inactivates ACh, but norepinephrine lingers in the synaptic cleft for a longer time. (4) Norepinephrine and epinephrine secreted into the blood by the adrenal medullae intensify the actions of the sympathetic division. (5) ACh remains in the synaptic cleft until norepinephrine is produced.
    (a) 1 and 3 (b) 1, 3, and 5 (c) 1, 3, and 4
    (d) 2, 3, and 4 (e) 2, 3, and 5

13. Place the following components of an autonomic reflex arc in the correct order from beginning to end.
    (a) postganglionic neuron (b) sensory neuron
    (c) effector (d) autonomic ganglion
    (e) receptor (f) preganglionic neuron
    (g) integrating center

14. Match the following:
    ____ (a) also known as intramural ganglia
    ____ (b) includes the celiac, superior mesenteric, and inferior mesenteric ganglia
    ____ (c) also called vertebral chain or paravertebral ganglia
    ____ (d) lie in a vertical row on either side of the vertebral column
    ____ (e) postganglionic fibers, in general, innervate organs below the diaphragm
    ____ (f) ganglia located at the end of an autonomic motor pathway close to or actually within the wall of a visceral organ
    ____ (g) includes ciliary, pterygopalatine, submandibular, and otic ganglia
    ____ (h) extend from base of the skull to the coccyx
    ____ (i) myelinated preganglionic fibers that connect the anterior rami of spinal nerves with the ganglia of the sympathetic trunk
    ____ (j) also known as collateral ganglia
    ____ (k) unmyelinated postganglionic axons that connect the ganglia of the sympathetic trunk to spinal nerves

    (1) sympathetic trunk ganglia
    (2) prevertebral ganglia
    (3) terminal ganglia
    (4) white rami communicantes
    (5) gray rami communicantes

15. Match the following:

____ (a) stimulates urination and defecation
____ (b) prepares the body for emergency situations
____ (c) fight-or-flight response
____ (d) promotes digestion and absorption of food
____ (e) concerned primarily with processes involving the expenditure of energy
____ (f) controlled by the posterior and lateral portions of the hypothalamus
____ (g) controlled by the anterior and medial portions of the hypothalamus
____ (h) causes a decrease in heart rate

(1) increased activity of the sympathetic division of the ANS
(2) increased activity of the parasympathetic division of the ANS

CRITICAL THINKING QUESTIONS

1. You've been to the "all-you-can-eat" buffet and have consumed large amounts of food. After returning home, you recline on the couch to watch television. Which division of the nervous system will be handling your body's after-dinner activities? List several organs involved, the major nerve supply to each organ, and the effects of the nervous system on their functions.

2. Ciara is driving home from school, listening to her favorite music, when a dog darts into the street in front of her car. She manages to swerve to avoid hitting the dog. As she continues on her way, she notices her heart is racing, she has "goose-bumps," and her hands are sweaty. Why is she experiencing these effects?

3. Mrs. Young is experiencing a bout of diarrhea that is keeping her house-bound. She would like to attend a birthday party for her brother but is afraid to attend because of her diarrhea. What type of drug, related to the autonomic nervous system function, could she take to help relieve her diarrhea?

? ANSWERS TO FIGURE QUESTIONS

15.1 Dual innervation means that a body organ receives neural innervation from both sympathetic and parasympathetic neurons of the ANS.

15.2 Most parasympathetic preganglionic axons are longer than most sympathetic preganglionic axons because most parasympathetic ganglia are in the walls of visceral organs, but most sympathetic ganglia are close to the spinal cord in the sympathetic trunk.

15.3 Terminal ganglia are associated with the parasympathetic division; sympathetic trunk and prevertebral ganglia are associated with the sympathetic division.

15.4 Sympathetic trunk ganglia contain sympathetic postganglionic neurons that lie in a vertical row on either side of the vertebral column.

15.5 The largest autonomic plexus is the celiac (solar) plexus.

15.6 Pelvic splanchnic nerves branch from the second through fourth sacral spinal nerves.

15.7 Most (but not all) sympathetic postganglionic neurons are adrenergic. Muscarinic receptors are present in the plasma membranes of all effectors (smooth muscle, cardiac muscle, and glands) innervated by parasympathetic postganglionic neurons and in sweat glands innervated by cholinergic sympathetic postganglionic neurons.

16 SENSORY, MOTOR, AND INTEGRATIVE SYSTEMS

SENSORY, MOTOR, AND INTEGRATIVE SYSTEMS AND HOMEOSTASIS *The sensory and motor pathways of the body provide routes for input into the brain and spinal cord and for output to targeted organs for responses, such as muscle contraction.*

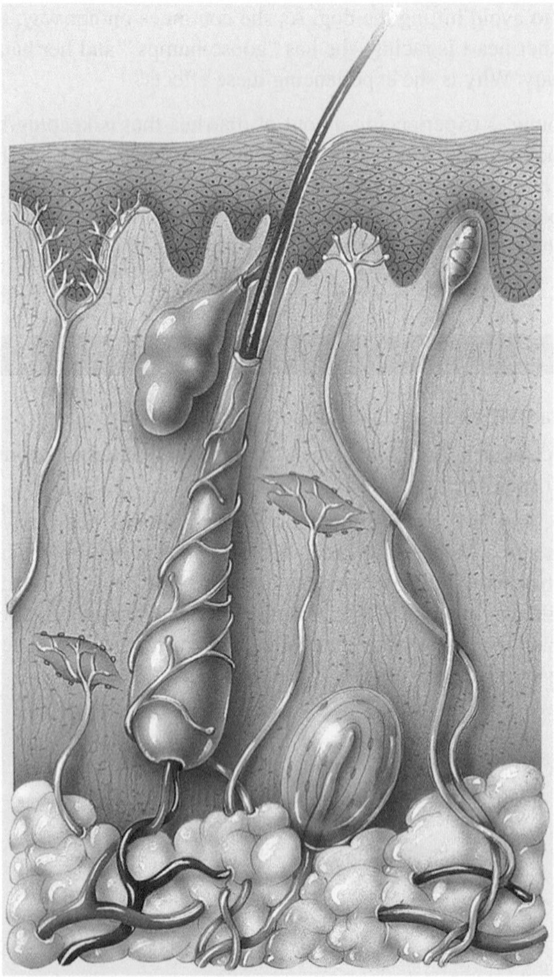

In the previous four chapters we described the organization of the nervous system. In this chapter, we explore the levels and components of sensation. We also examine the pathways that convey somatic sensory nerve impulses from the body to the brain, and the pathways that carry impulses from the brain to skeletal muscles to produce movements. As sensory impulses reach the CNS, they become part of a large pool of sensory input. However, not every nerve impulse transmitted to the CNS elicits a response. Rather, each piece of incoming information is combined with other arriving and previously stored information in a process called *integration*. Integration occurs at many places along pathways in the CNS, such as the spinal cord, brain stem, cerebellum, basal nuclei, and cerebral cortex. You will also learn how the motor responses that govern muscle contraction are modified at several of these levels. To conclude this chapter, we introduce two complex integrative functions of the brain: (1) wakefulness and sleep and (2) learning and memory.

? ***Did you ever wonder how drugs such as aspirin and ibuprofen relieve pain?***

16.1 SENSATION

OBJECTIVES

- Define sensation, and discuss the components of sensation.
- Describe the different ways to classify sensory receptors.

In its broadest definition, **sensation** is the conscious or subconscious awareness of changes in the external or internal environment. The nature of the sensation and the type of reaction generated vary according to the ultimate destination of nerve impulses that convey sensory information to the CNS. Sensory impulses that reach the spinal cord may serve as input for spinal reflexes, such as the stretch reflex you learned about in Chapter 13. Sensory impulses that reach the lower brain stem elicit more complex reflexes, such as changes in heart rate or breathing rate. When sensory impulses reach the cerebral cortex, we become consciously aware of the sensory stimuli and can precisely locate and identify specific sensations such as touch, pain, hearing, or taste. As you learned in Chapter 14, **perception** is the conscious awareness and interpretation of sensations and is primarily a function of the cerebral cortex. We have no perception of some sensory information because it never reaches the cerebral cortex. For example, certain sensory receptors constantly monitor the pressure of blood in blood vessels. Because the nerve impulses conveying blood pressure information propagate to the cardiovascular center in the medulla oblongata rather than to the cerebral cortex, blood pressure is not consciously perceived.

Sensory Modalities

Each unique type of sensation—such as touch, pain, vision, or hearing—is called a **sensory modality** (mō-DAL-i-tē). A given sensory neuron carries information for only one sensory modality. Neurons relaying impulses for touch to the somatosensory area of the cerebral cortex do not transmit impulses for pain. Likewise, nerve impulses from the eyes are perceived as sight, and those from the ears are perceived as sounds.

The different sensory modalities can be grouped into two classes: general senses and special senses.

1. The **general senses** refer to both somatic senses and visceral senses. **Somatic senses** (*somat-* = of the body) include tactile sensations (touch, pressure, vibration, itch, and tickle), thermal sensations (warm and cold), pain sensations, and proprioceptive sensations. Proprioceptive sensations allow perception of both the static (nonmoving) positions of limbs and body parts (joint and muscle position sense) and movements of the limbs and head. **Visceral senses** provide information about conditions within internal organs, for example, pressure, stretch, chemicals, nausea, hunger, and temperature.
2. The **special senses** include the sensory modalities of smell, taste, vision, hearing, and equilibrium or balance.

In this chapter we discuss the somatic senses and visceral pain. The special senses are the focus of Chapter 17. Visceral senses were discussed in Chapter 15 and will be discussed in association with individual organs in later chapters.

The Process of Sensation

The process of sensation begins in a **sensory receptor,** which can be either a specialized cell or the dendrites of a sensory neuron. As previously noted, a given sensory receptor responds vigorously to one particular kind of **stimulus,** a change in the environment that can activate certain sensory receptors. A sensory receptor responds only weakly or not at all to other stimuli. This characteristic of sensory receptors is known as **selectivity.**

For a sensation to arise, the following four events typically occur:

1. ***Stimulation of the sensory receptor.*** An appropriate stimulus must occur within the sensory receptor's *receptive field,* that is, the body region where stimulation activates the receptor and produces a response.
2. ***Transduction of the stimulus.*** A sensory receptor *transduces* (converts) energy in a stimulus into a graded potential. Recall that graded potentials vary in amplitude (size), depending on the strength of the stimulus that causes them, and are not propagated. (See Section 12.3 to review the differences between action potentials and graded potentials.) Each type of sensory receptor exhibits selectivity: It can transduce only one kind of stimulus. For example, odorant molecules in the air stimulate olfactory (smell) receptors in the nose, which transduce the molecules' chemical energy into electrical energy in the form of a graded potential.
3. ***Generation of nerve impulses.*** When a graded potential in a sensory neuron reaches threshold, it triggers one or more nerve impulses, which then propagate toward the CNS. Sensory neurons that conduct impulses from the PNS into the CNS are called **first-order neurons.**
4. ***Integration of sensory input.*** A particular region of the CNS receives and integrates the sensory nerve impulses. Conscious sensations or perceptions are integrated in the cerebral cortex. You seem to see with your eyes, hear with your ears, and feel pain in an injured part of your body because sensory impulses from each part of the body arrive in a specific region of the cerebral cortex, which interprets the sensation as coming from the stimulated sensory receptors.

Sensory Receptors

Types of Sensory Receptors

Several structural and functional characteristics of sensory receptors can be used to group them into different classes. These include (1) microscopic structure, (2) location of the receptors and the origin of stimuli that activate them, and (3) type of stimulus detected.

MICROSCOPIC STRUCTURE On a microscopic level, sensory receptors may be (1) free nerve endings of first-order sensory neurons, (2) encapsulated nerve endings of first-order sensory neurons, or (3) separate cells that synapse with first-order sensory neurons. **Free nerve endings** are bare dendrites; they lack any structural specializations that can be seen under a light microscope (Figure 16.1a). Receptors for pain, temperature, tickle, itch, and some touch sensations are free nerve endings. Receptors for other somatic and visceral sensations, such as pressure, vibration, and some touch sensations, are **encapsulated nerve endings.** Their dendrites are enclosed in a connective tissue capsule that has a distinctive microscopic structure—for example, Pacinian corpuscles (Figure 16.1b). The different types of capsules enhance the sensitivity or specificity of the receptor. Sensory receptors for some special senses are specialized, **separate cells** that synapse with sensory neurons. These include *hair cells* for hearing and equilibrium in the inner ear, *gustatory receptor cells* in taste buds (Figure 16.1c), and *photoreceptors* in the retina of the eye for vision; you will learn more about separate cells in Chapter 17.

Sensory receptors produce two different kinds of graded potentials—generator potentials and receptor potentials—in response to a stimulus. When stimulated, the dendrites of free nerve endings, encapsulated nerve endings, and the receptive part of olfactory receptors produce a **generator potential** (Figure 16.1a, b). When a generator potential is large enough to reach threshold, it triggers one or more nerve impulses in the axon of a first-order sensory neuron. The resulting nerve impulse propagates along the axon into the CNS. Thus, generator potentials generate action potentials.

By contrast, sensory receptors that are separate cells produce graded potentials termed **receptor potentials.** Receptor potentials trigger release of neurotransmitter through exocytosis of

Figure 16.1 **Types of sensory receptors and their relationship to first-order sensory neurons.** (a) Free nerve endings, in this case, a cold-sensitive receptor. These endings are bare dendrites of first-order neurons with no apparent structural specialization. (b) An encapsulated nerve ending, in this case a pressure-sensitive receptor. Encapsulated nerve endings are dendrites of first-order neurons. (c) A separate receptor cell—here, a gustatory (taste) receptor—and its synapse with a first-order neuron.

Free nerve endings and encapsulated nerve endings produce generator potentials that trigger nerve impulses in first-order neurons. Separate sensory receptors produce a receptor potential that causes the release of a neurotransmitter. The neurotransmitter then triggers nerve impulses in a first-order neuron.

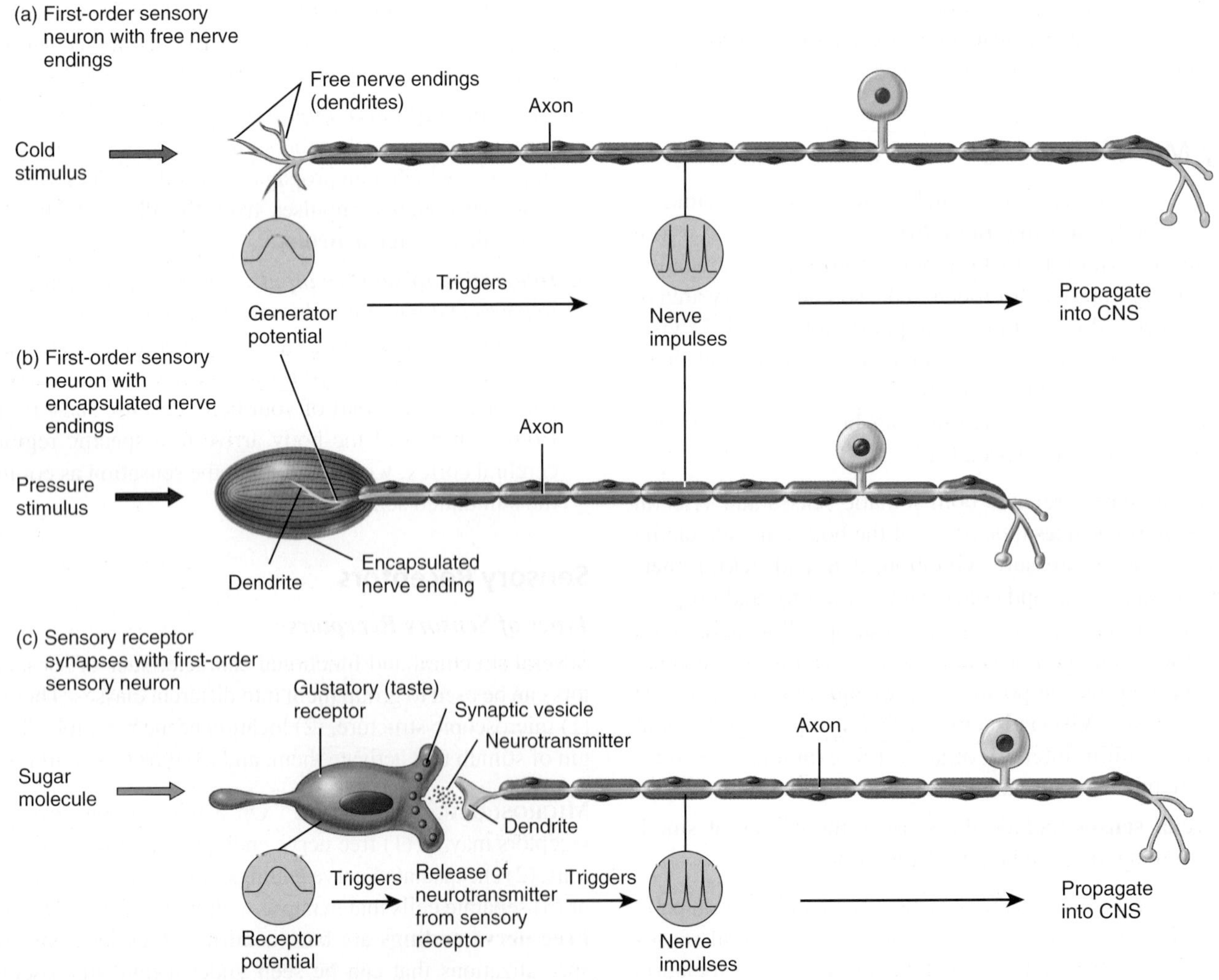

Which senses are served by receptors that are separate cells?

synaptic vesicles (Figure 16.1c). The neurotransmitter molecules liberated from synaptic vesicles diffuse across the synaptic cleft and produce a postsynaptic potential (PSP) in the first-order neuron. In turn, the PSPs may trigger one or more nerve impulses, which propagate along the axon into the CNS.

The amplitude of both generator potentials and receptor potentials varies with the intensity of the stimulus, with an intense stimulus producing a large potential and a weak stimulus eliciting a small one. Similarly, large generator potentials or receptor potentials trigger nerve impulses at high frequencies in the first-order neuron, in contrast to small generator potentials or receptor potentials, which trigger nerve impulses at lower frequencies.

Location of Receptors and Origin of Activating Stimuli Another way to group sensory receptors is based on the location of the receptors and the origin of the stimuli that activate them.

- **Exteroceptors** (EKS-ter-ō-sep′-tors) are located at or near the external surface of the body; they are sensitive to stimuli originating outside the body and provide information about the *external* environment. The sensations of hearing, vision, smell, taste, touch, pressure, vibration, temperature, and pain are conveyed by exteroceptors.
- **Interoceptors** (IN-ter-ō-sep′-tors) or **visceroceptors** are located in blood vessels, visceral organs, muscles, and the nervous system and monitor conditions in the *internal* environment. The nerve impulses produced by interoceptors usually are not consciously perceived; occasionally, however, activation of interoceptors by strong stimuli may be felt as pain or pressure.
- **Proprioceptors** (PRŌ-prē-ō-sep′-tors; *proprio-* = one's own) are located in muscles, tendons, joints, and the inner ear. They provide information about body position, muscle length and tension, and the position and movement of your joints.

Type of Stimulus Detected A third way to group sensory receptors is according to the type of stimulus they detect. Most stimuli are in the form of mechanical energy, such as sound waves or pressure changes; electromagnetic energy, such as light or heat; or chemical energy, such as in a molecule of glucose.

- **Mechanoreceptors** are sensitive to mechanical stimuli such as the deformation, stretching, or bending of cells. Mechanoreceptors provide sensations of touch, pressure, vibration, proprioception, and hearing and equilibrium. They also monitor the stretching of blood vessels and internal organs.
- **Thermoreceptors** detect changes in temperature.
- **Nociceptors** (nō′-sē-SEP-tors) respond to painful stimuli resulting from physical or chemical damage to tissue.
- **Photoreceptors** detect light that strikes the retina of the eye.
- **Chemoreceptors** detect chemicals in the mouth (taste), nose (smell), and body fluids.
- **Osmoreceptors** detect the osmotic pressure of body fluids.

Table 16.1 summarizes the classification of sensory receptors.

Adaptation in Sensory Receptors

A characteristic of most sensory receptors is **adaptation,** in which the generator potential or receptor potential decreases in amplitude during a maintained, constant stimulus. As you may already have guessed, this causes the frequency of nerve impulses in the first-order neuron to decrease. Because of adaptation, the perception of a sensation may fade or disappear even though the stimulus persists. For example, when you first step into a hot shower, the water may feel very hot, but soon the sensation decreases to one of comfortable warmth even though the stimulus (the high temperature of the water) does not change.

TABLE 16.1

Classification of Sensory Receptors

BASIS OF CLASSIFICATION	DESCRIPTION
MICROSCOPIC STRUCTURE	
Free nerve endings	Bare dendrites associated with pain, thermal, tickle, itch, and some touch sensations.
Encapsulated nerve endings	Dendrites enclosed in connective tissue capsule for pressure, vibration, and some touch sensations.
Separate cells	Receptor cells synapse with first-order sensory neurons; located in retina of eye (photoreceptors), inner ear (hair cells), and taste buds of tongue (gustatory receptor cells).
RECEPTOR LOCATION AND ACTIVATING STIMULI	
Exteroceptors	Located at or near body surface; sensitive to stimuli originating outside body; provide information about external environment; convey visual, smell, taste, touch, pressure, vibration, thermal, and pain sensations.
Interoceptors	Located in blood vessels, visceral organs, and nervous system; provide information about internal environment; impulses usually are not consciously perceived but occasionally may be felt as pain or pressure.
Proprioceptors	Located in muscles, tendons, joints, and inner ear; provide information about body position, muscle length and tension, position and motion of joints, and equilibrium (balance).
TYPE OF STIMULUS DETECTED	
Mechanoreceptors	Detect mechanical stimuli; provide sensations of touch, pressure, vibration, proprioception, and hearing and equilibrium; also monitor stretching of blood vessels and internal organs.
Thermoreceptors	Detect changes in temperature.
Nociceptors	Respond to painful stimuli resulting from physical or chemical damage to tissue.
Photoreceptors	Detect light that strikes the retina of the eye.
Chemoreceptors	Detect chemicals in mouth (taste), nose (smell), and body fluids.
Osmoreceptors	Sense osmotic pressure of body fluids.

Receptors vary in how quickly they adapt. **Rapidly adapting receptors** adapt very quickly. They are specialized for signaling *changes* in a stimulus. Receptors associated with pressure, touch, and smell are rapidly adapting. **Slowly adapting receptors,** by contrast, adapt slowly and continue to trigger nerve impulses as long as the stimulus persists. Slowly adapting receptors monitor stimuli associated with pain, body position, and chemical composition of the blood.

✔CHECKPOINT

1. How is sensation different from perception?
2. What is a sensory modality?
3. How are generator potentials and receptor potentials similar? How are they different?
4. What is the difference between rapidly adapting and slowly adapting receptors?

16.2 SOMATIC SENSATIONS

OBJECTIVES

- Describe the location and function of the somatic sensory receptors for tactile, thermal, and pain sensations.
- Identify the receptors for proprioception and describe their functions.

Somatic sensations arise from stimulation of sensory receptors embedded in the skin or subcutaneous layer; in mucous membranes of the mouth, vagina, and anus; in muscles, tendons, and joints; and in the inner ear. The sensory receptors for somatic sensations are distributed unevenly—some parts of the body surface are densely populated with receptors, and others contain only a few. The areas with the highest density of somatic sensory receptors are the tip of the tongue, the lips, and the fingertips. Somatic sensations that arise from stimulating the skin surface are **cutaneous sensations** (kū-TĀ-nē-us; *cutane-* = skin). There are four modalities of somatic sensation: tactile, thermal, pain, and proprioceptive.

Tactile Sensations

The **tactile sensations** (TAK-tīl; *tact-* = touch) include touch, pressure, vibration, itch, and tickle. Although we perceive differences among these sensations, they arise by activation of some of the same types of receptors. Several types of encapsulated mechanoreceptors attached to large-diameter myelinated A fibers mediate sensations of touch, pressure, and vibration. Other tactile sensations, such as itch and tickle sensations, are detected by free nerve endings attached to small-diameter, unmyelinated C fibers. Recall that larger-diameter, myelinated axons propagate nerve impulses more rapidly than do smaller-diameter, unmyelinated axons. Tactile receptors in the skin or subcutaneous layer include Meissner corpuscles, hair root plexuses, Merkel discs, Ruffini corpuscles, Pacinian corpuscles, and free nerve endings (Figure 16.2).

Touch

Sensations of **touch** generally result from stimulation of tactile receptors in the skin or subcutaneous layer. There are two types of rapidly adapting touch receptors. **Meissner corpuscles** (MĪS-ner) or *corpuscles of touch* are touch receptors that are located in the dermal papillae of hairless skin. Each corpuscle is an egg-shaped mass of dendrites enclosed by a capsule of connective tissue. Because Meissner corpuscles are rapidly adapting receptors, they generate nerve impulses mainly at the onset of a touch. They are abundant in the fingertips, hands, eyelids, tip of the tongue, lips, nipples, soles, clitoris, and tip of the penis. **Hair root plexuses** are rapidly adapting touch receptors found in hairy skin; they consist of free nerve endings wrapped around hair follicles. Hair root plexuses detect movements on the skin surface that disturb hairs. For example, an insect landing on a hair causes movement of the hair shaft that stimulates the free nerve endings.

There are also two types of slowly adapting touch receptors. **Merkel discs,** also known as *tactile discs* or *type I cutaneous mechanoreceptors,* are saucer-shaped, flattened free nerve endings that make contact with Merkel cells of the stratum basale (see Figure 5.2d). These touch receptors are plentiful in the fingertips, hands, lips, and external genitalia. **Ruffini corpuscles** or *type II cutaneous mechanoreceptors* are elongated, encapsulated receptors located deep in the dermis, and in ligaments and tendons. Present in the hands and abundant on the soles, they are most sensitive to stretching that occurs as digits or limbs are moved.

Pressure

Pressure, a sustained sensation that is felt over a larger area than touch, occurs with deformation of deeper tissues. Receptors that contribute to sensations of pressure include Meissner corpuscles, Merkel discs, and Pacinian corpuscles. A **Pacinian** (pa-SIN-ē-an), or **lamellated, corpuscle** is a large oval structure composed of a multilayered connective tissue capsule that encloses a dendrite. Like Meissner corpuscles, Pacinian corpuscles adapt rapidly. They are widely distributed in the body: in the dermis and subcutaneous layer; in submucosal tissues that underlie mucous and serous membranes; around joints, tendons, and muscles; in the periosteum; and in the mammary glands, external genitalia, and certain viscera, such as the pancreas and urinary bladder.

Vibration

Sensations of **vibration,** such as using an electric knife to carve a turkey, result from rapidly repetitive sensory signals from tactile receptors. The receptors for vibration sensations are Meissner corpuscles and Pacinian corpuscles. Meissner corpuscles can detect lower-frequency vibrations, and Pacinian corpuscles detect higher-frequency vibrations.

Itch

The **itch** sensation results from stimulation of free nerve endings by certain chemicals, such as bradykinin or antigens in mosquito saliva injected from a bite, often because of a local inflammatory response (bradykinin, a kinin, is a potent vasodilator).

Tickle

Free nerve endings are thought to mediate the **tickle** sensation. This intriguing sensation typically arises only when someone else touches you, not when you touch yourself. The solution to this puzzle seems to lie in the impulses that conduct to and from the cerebellum when you are moving your fingers and touching yourself that don't occur when someone else is tickling you.

CLINICAL CONNECTION | *Phantom Limb Sensation*

Patients who have had a limb amputated may still experience sensations such as itching, pressure, tingling, or pain as if the limb were still there. This phenomenon is called **phantom limb sensation.** Although the limb has been removed, severed endings of sensory axons are still present in the remaining stump. If these severed endings are activated, the cerebral cortex interprets the sensation as coming from the sensory receptors in the nonexisting (phantom) limb. Another explanation for phantom limb sensation is that the area of the cerebral cortex that previously received sensory input from the missing limb undergoes extensive functional reorganization that allows it to respond to stimuli from another body part. The remodeling of this cortical area is thought to give rise to false sensory perceptions from the missing limb. Phantom limb pain can be very distressing to an amputee. Many report that the pain is severe or extremely intense, and that it often does not respond to traditional pain medication therapy. In such cases, alternative treatments may include electrical nerve stimulation, acupuncture, and biofeedback. •

Thermal Sensations

Thermoreceptors are free nerve endings that have receptive fields about 1 mm in diameter on the skin surface. Two distinct **thermal sensations**—coldness and warmth—are detected by different receptors. **Cold receptors** are located in the stratum basale of the epidermis and are attached to medium-diameter, myelinated A fibers, although a few connect to small-diameter, unmyelinated C fibers. Temperatures between 10° and 40°C (50–105°F) activate cold receptors. **Warm receptors,** which are not as abundant as cold receptors, are located in the dermis and are attached to small-diameter, unmyelinated C fibers; they are activated by temperatures between 32° and 48°C (90–118°F). Cold and warm receptors both adapt rapidly at the onset of a stimulus, but as noted earlier in the chapter they continue to generate impulses at a lower frequency throughout a prolonged stimulus. Temperatures below 10°C and above 48°C primarily stimulate pain receptors, rather than thermoreceptors, producing painful sensations, which we discuss next.

Pain Sensations

Pain is indispensable for survival. It serves a protective function by signaling the presence of noxious, tissue-damaging conditions.

Figure 16.2 Structure and location of sensory receptors in the skin and subcutaneous layer.

The somatic sensations of touch, pressure, vibration, warmth, cold, and pain arise from sensory receptors in the skin, subcutaneous layer, and mucous membranes.

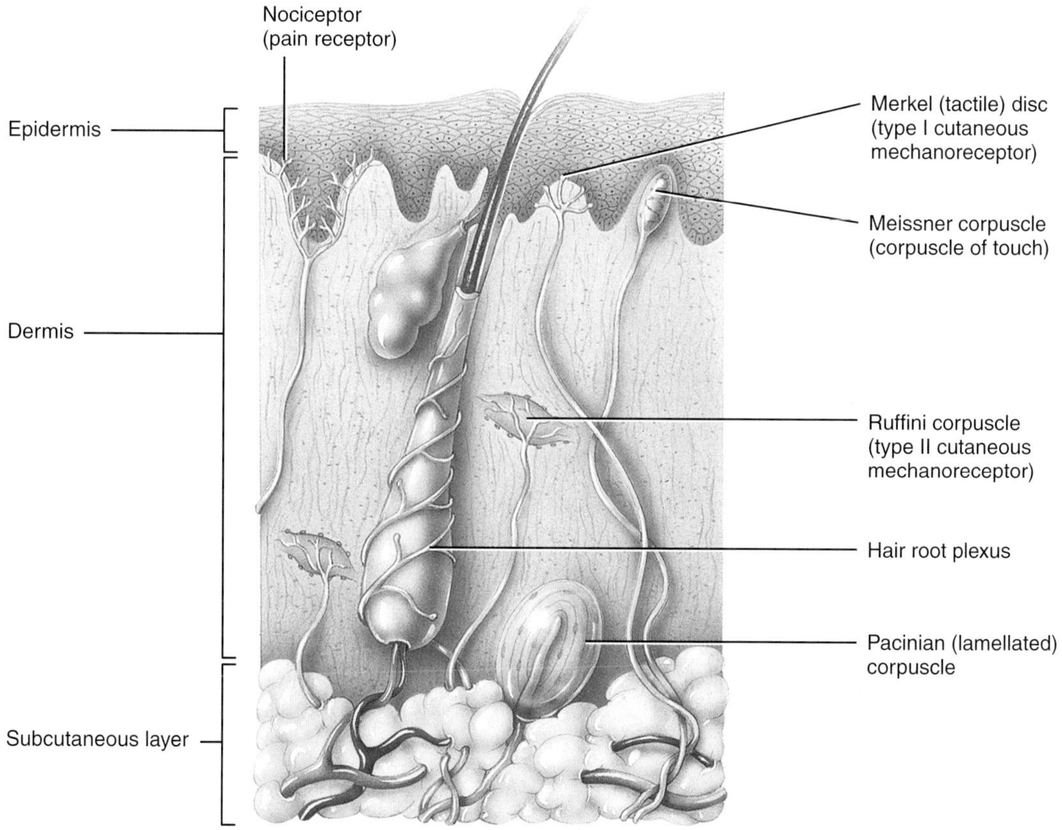

Which sensations can arise when free nerve endings are stimulated?

From a medical standpoint, the subjective description and indication of the location of pain may help pinpoint the underlying cause of disease.

Nociceptors (*noci-* = harmful), the receptors for pain, are free nerve endings found in every tissue of the body except the brain (Figure 16.2). Intense thermal, mechanical, or chemical stimuli can activate nociceptors. Tissue irritation or injury releases chemicals such as prostaglandins, kinins, and potassium ions (K^+) that stimulate nociceptors. Pain may persist even after a pain-producing stimulus is removed because pain-mediating chemicals linger, and because nociceptors exhibit very little adaptation. Conditions that elicit pain include excessive distention (stretching) of a structure, prolonged muscular contractions, muscle spasms, or ischemia (inadequate blood flow to an organ).

Types of Pain

There are two types of pain: fast and slow. The perception of **fast pain** occurs very rapidly, usually within 0.1 second after a stimulus is applied, because the nerve impulses propagate along medium-diameter, myelinated A fibers. This type of pain is also known as acute, sharp, or pricking pain. The pain felt from a needle puncture or knife cut to the skin is fast pain. Fast pain is not felt in deeper tissues of the body. The perception of **slow pain,** by contrast, begins a second or more after a stimulus is applied. It then gradually increases in intensity over a period of several seconds or minutes. Impulses for slow pain conduct along small-diameter, unmyelinated C fibers. This type of pain, which may be excruciating, is also referred to as chronic, burning, aching, or throbbing pain. Slow pain can occur both in the skin and in deeper tissues or internal organs. An example is the pain associated with a toothache. You can perceive the difference in onset of these two types of pain best when you injure a body part that is far from the brain because the conduction distance is long. When you stub your toe, for example, you first feel the sharp sensation of fast pain and then feel the slower, aching sensation of slow pain.

Pain that arises from stimulation of receptors in the skin is called **superficial somatic pain;** stimulation of receptors in skeletal muscles, joints, tendons, and fascia causes **deep somatic pain. Visceral pain** results from stimulation of nociceptors in visceral organs. If stimulation is *diffuse* (involves large areas), visceral pain can be severe. Diffuse stimulation of visceral nociceptors might result from distention or ischemia of an internal organ. For example, a kidney stone or a gallstone might cause severe pain by obstructing and distending a ureter or bile duct.

Localization of Pain

Fast pain is very precisely localized to the stimulated area. For example, if someone pricks you with a pin, you know exactly which part of your body was stimulated. Somatic slow pain also is well localized but more diffuse (involves large areas); it usually appears to come from a larger area of the skin. In some instances of visceral slow pain, the affected area is where the pain is felt. If the pleural membranes around the lungs are inflamed, for example, you experience chest pain.

However, in many instances of visceral pain, the pain is felt in or just deep to the skin that overlies the stimulated organ, or in a surface area far from the stimulated organ. This phenomenon is called **referred pain.** Figure 16.3 shows skin regions to which

Figure 16.3 Distribution of referred pain. Colored parts indicate skin areas to which visceral pain is referred.

Nociceptors are present in almost every tissue of the body.

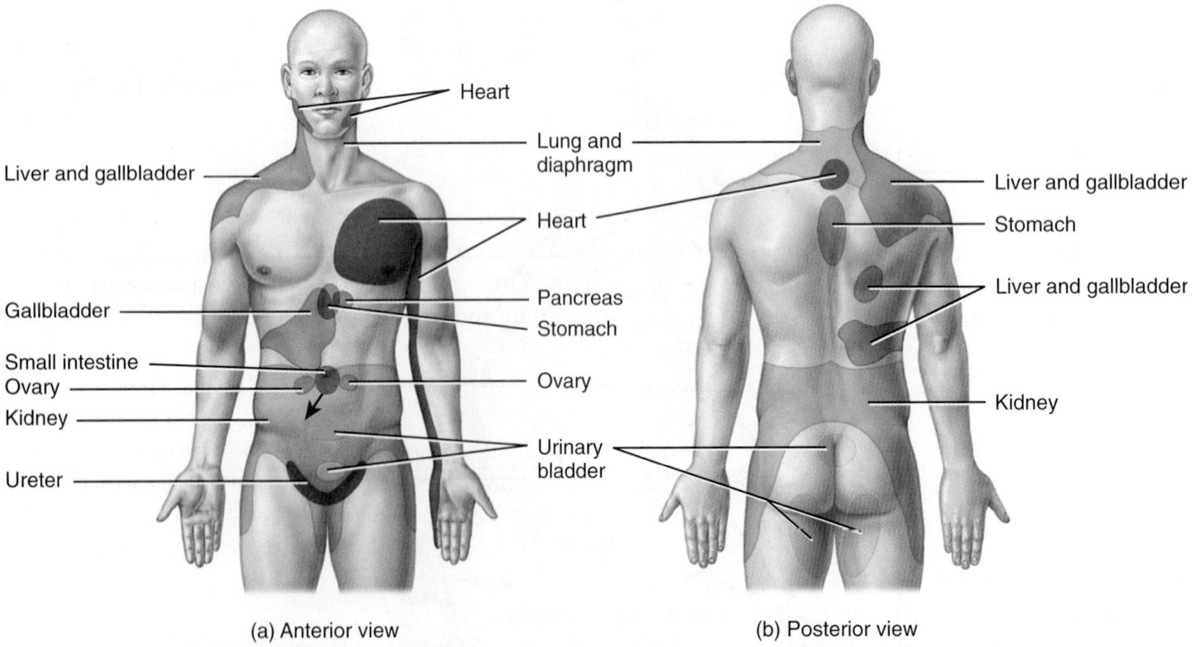

Which visceral organ has the broadest area for referred pain?

visceral pain may be referred. In general, the visceral organ involved and the area to which the pain is referred are served by the same segment of the spinal cord. For example, sensory fibers from the heart, the skin superficial to the heart, and the skin along the medial aspect of the left arm enter spinal cord segments T1 to T5. Thus, the pain of a heart attack typically is felt in the skin over the heart and along the left arm.

CLINICAL CONNECTION | *Analgesia: Relief from Pain*

Pain sensations sometimes occur out of proportion to minor damage, persist chronically due to an injury, or even appear for no obvious reason. In such cases, **analgesia** (an-al-JĒ-zē-a; *an-* = without; *-algesia* = pain) or pain relief is needed. Analgesic drugs such as aspirin and ibuprofen (for example, Advil® or Motrin®) block formation of prostaglandins, which stimulate nociceptors. Local anesthetics, such as Novocaine®, provide short-term pain relief by blocking conduction of nerve impulses along the axons of first-order pain neurons. Morphine and other opiate drugs alter the quality of pain perception in the brain; pain is still sensed but it is no longer perceived as being so noxious. Many pain clinics use anticonvulsant and antidepressant medications to treat those suffering from chronic pain. •

Proprioceptive Sensations

Proprioceptive sensations allow us to know where our head and limbs are located and how they are moving even if we are not looking at them, so that we can walk, type, or dress without using our eyes. **Kinesthesia** (kin′-es-THĒ-zē-a; *kin-* = motion; *-esthesia* = perception) is the perception of body movements. Proprioceptive sensations arise in receptors termed **proprioceptors.** Those proprioceptors embedded in muscles (especially postural muscles) and tendons inform us of the degree to which muscles are contracted, the amount of tension on tendons, and the positions of joints. Hair cells of the inner ear monitor the orientation of the head relative to the ground and head position during movements. The way they provide information for maintaining balance and equilibrium will be described in Chapter 17. Because proprioceptors adapt slowly and only slightly, the brain continually receives nerve impulses related to the position of different body parts and makes adjustments to ensure coordination.

Proprioceptors also allow **weight discrimination,** the ability to assess the weight of an object. This type of information helps you to determine the muscular effort necessary to perform a task. For example, as you pick up a shopping bag, you quickly realize whether it contains books or feathers, and you then exert the correct amount of effort needed to lift it.

Here we discuss three types of proprioceptors: muscle spindles within skeletal muscles, tendon organs within tendons, and joint kinesthetic receptors within synovial joint capsules.

Muscle Spindles

Muscle spindles are the proprioceptors in skeletal muscles that monitor changes in the length of skeletal muscles and participate in stretch reflexes (shown in Figure 13.15). By adjusting how vigorously a muscle spindle responds to stretching of a skeletal muscle, the brain sets an overall level of **muscle tone,** the small degree of contraction that is present while the muscle is at rest.

Each **muscle spindle** consists of several slowly adapting sensory nerve endings that wrap around 3 to 10 specialized muscle fibers, called **intrafusal muscle fibers** (in′-tra-FŪ-sal = within a spindle). A connective tissue capsule encloses the sensory nerve endings and intrafusal fibers and anchors the spindle to the endomysium and perimysium (Figure 16.4). Muscle spindles are interspersed among most skeletal muscle fibers and aligned parallel to them. In muscles that produce finely controlled movements, such as those of the fingers or eyes as you read music and play a musical instrument, muscle spindles are plentiful. Muscles involved in coarser but more forceful movements, like the quadriceps femoris and hamstring muscles of the thigh, have fewer muscle spindles. The only skeletal muscles that lack spindles are the tiny muscles of the middle ear.

The main function of muscle spindles is to measure *muscle length*—how much a muscle is being stretched. Either sudden or prolonged stretching of the central areas of the intrafusal muscle fibers stimulates the sensory nerve endings. The resulting nerve impulses propagate into the CNS. Information from muscle spindles arrives quickly at the somatic sensory areas of the cerebral cortex, which allows conscious perception of limb positions and movements. At the same time, impulses from muscle spindles pass to the cerebellum, where the input is used to coordinate muscle contractions.

In addition to their sensory nerve endings near the middle of intrafusal fibers, muscle spindles contain motor neurons called **gamma motor neurons.** These motor neurons terminate near both ends of the intrafusal fibers and adjust the tension in a muscle spindle to variations in the length of the muscle. For example, when your biceps muscle shortens in response to lifting a weight, gamma motor neurons stimulate the ends of the intrafusal fibers to contract slightly. This keeps the intrafusal fibers taut even though the contractile muscle fibers surrounding the spindle are reducing spindle tension. This maintains the sensitivity of the muscle spindle to stretching of the muscle. As the frequency of impulses in its gamma motor neuron increases, a muscle spindle becomes more sensitive to stretching of its midregion.

Surrounding muscle spindles are ordinary skeletal muscle fibers, called **extrafusal muscle fibers** (*extrafusal* = outside a spindle), which are supplied by large-diameter A fibers called **alpha motor neurons.** The cell bodies of both gamma and alpha motor neurons are located in the anterior gray horn of the spinal cord (or in the brain stem for muscles in the head). During the stretch reflex, impulses in muscle spindle sensory axons propagate into the spinal cord and brain stem and activate alpha motor neurons that connect to extrafusal muscle fibers in the same muscle. In this way, activation of its muscle spindles causes contraction of a skeletal muscle, which relieves the stretching.

Tendon Organs

Tendon organs are located at the junction of a tendon and a muscle. By initiating tendon reflexes (see Figure 13.16), tendon

Figure 16.4 Two types of proprioceptors: a muscle spindle and a tendon organ. In muscle spindles, which monitor changes in skeletal muscle length, sensory nerve endings wrap around the central portion of intrafusal muscle fibers. In tendon organs, which monitor the force of muscle contraction, sensory nerve endings are activated by increasing tension on a tendon. If you examine Figure 13.15 you can see the relationship of a muscle spindle to the spinal cord as a component of a stretch reflex. In Figure 13.16, you can see the relationship of a tendon organ to the spinal cord as a component of a tendon reflex.

Proprioceptors provide information about body position and movement.

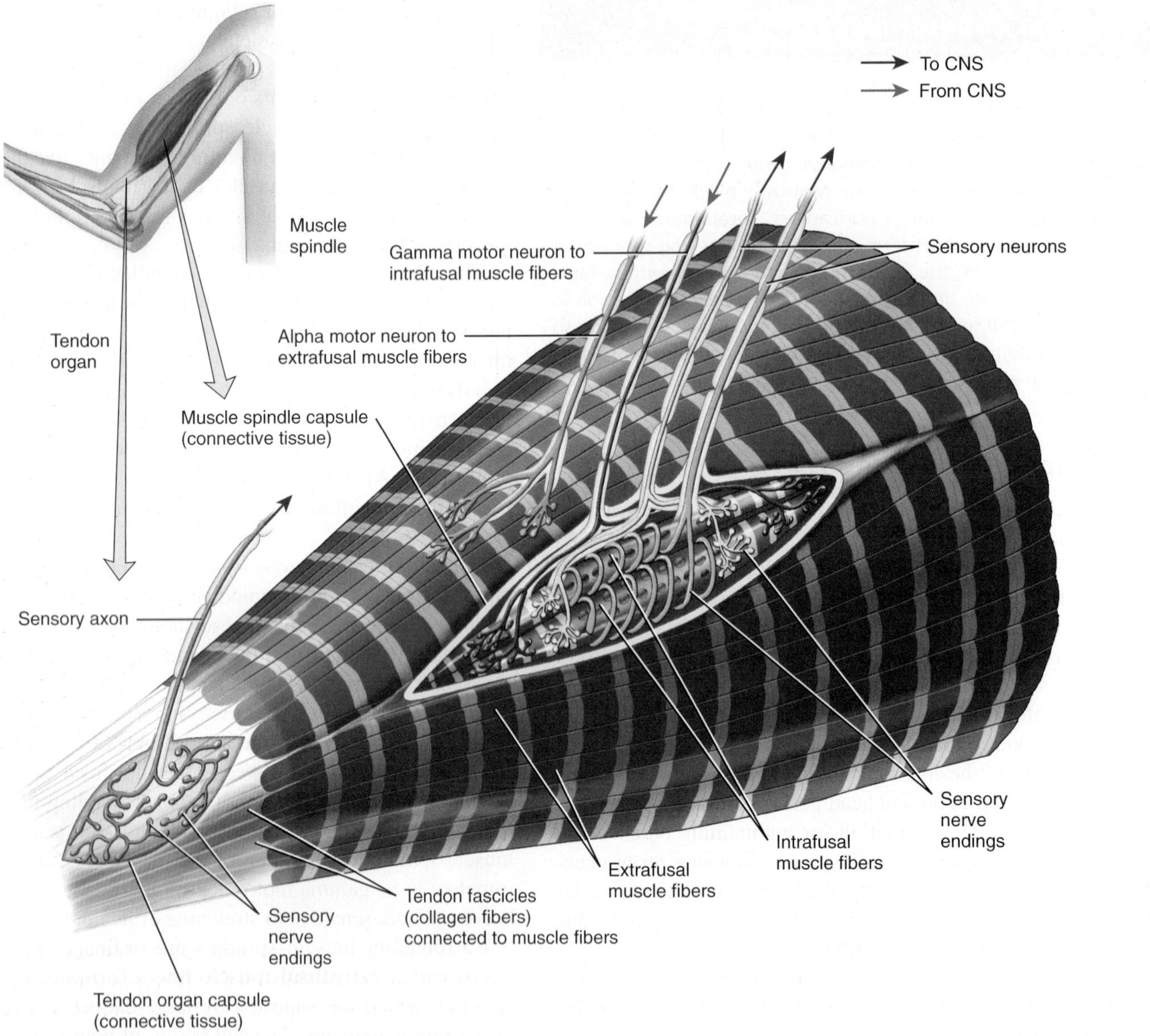

How is a muscle spindle activated?

organs protect tendons and their associated muscles from damage due to excessive tension. (When a muscle contracts, it exerts a force that pulls the points of attachment of the muscle at either end toward each other. This force is the muscle tension.) Each tendon organ consists of a thin capsule of connective tissue that encloses a few tendon fascicles (bundles of collagen fibers) (Figure 16.4). Penetrating the capsule are one or more sensory nerve endings that entwine among and around the collagen fibers of the tendon. When tension is applied to a muscle, the tendon organs generate nerve impulses that propagate into the CNS, providing information about changes in muscle tension. The resulting tendon reflexes decrease muscle tension by causing muscle relaxation.

Joint Kinesthetic Receptors

Several types of **joint kinesthetic receptors** (kin′-es-THET-ik) are present within and around the articular capsules of synovial joints. Free nerve endings and Ruffini corpuscles in the capsules

TABLE 16.2

Summary of Receptors for Somatic Sensations

RECEPTOR TYPE	RECEPTOR STRUCTURE AND LOCATION	SENSATIONS	ADAPTATION RATE
TACTILE RECEPTORS			
Meissner corpuscles (corpuscles of touch)	Capsule surrounds mass of dendrites in dermal papillae of hairless skin.	Touch, pressure, and slow vibrations.	Rapid.
Hair root plexuses	Free nerve endings wrapped around hair follicles in skin.	Touch.	Rapid.
Merkel (tactile) discs (type I cutaneous mechanoreceptors)	Saucer-shaped free nerve endings make contact with Merkel cells in epidermis.	Touch and pressure.	Slow.
Ruffini corpuscles (type II cutaneous mechanoreceptors)	Elongated capsule surrounds dendrites deep in dermis and in ligaments and tendons.	Stretching of skin.	Slow.
Pacinian (lamellated) corpuscles	Oval, layered capsule surrounds dendrites; present in dermis and subcutaneous layer, submucosal tissues, joints, periosteum, and some viscera.	Pressure and fast vibrations.	Rapid.
Itch and tickle receptors	Free nerve endings in skin and mucous membranes.	Itching and tickling.	Both slow and rapid.
THERMORECEPTORS			
Warm receptors and cold receptors	Free nerve endings in skin and mucous membranes of mouth, vagina, and anus.	Warmth or cold.	Initially rapid, then slow.
PAIN RECEPTORS			
Nociceptors	Free nerve endings in every body tissue except brain.	Pain.	Slow.
PROPRIOCEPTORS			
Muscle spindles	Sensory nerve endings wrap around central area of encapsulated intrafusal muscle fibers within most skeletal muscles.	Muscle length.	Slow.
Tendon organs	Capsule encloses collagen fibers and sensory nerve endings at junction of tendon and muscle.	Muscle tension.	Slow.
Joint kinesthetic receptors	Pacinian corpuscles, Ruffini corpuscles, tendon organs, and free nerve endings.	Joint position and movement.	Rapid.

of joints respond to pressure. Small Pacinian corpuscles in the connective tissue outside articular capsules respond to acceleration and deceleration of joints during movement. Joint ligaments contain receptors similar to tendon organs that adjust reflex inhibition of the adjacent muscles when excessive strain is placed on the joint.

Table 16.2 summarizes the types of somatic sensory receptors and the sensations they convey.

✔CHECKPOINT

5. Which somatic sensory receptors are encapsulated?
6. Why do some receptors adapt slowly, and others adapt rapidly?
7. Which somatic sensory receptors mediate touch sensations?
8. How does fast pain differ from slow pain?
9. What is referred pain, and how is it useful in diagnosing internal disorders?
10. What aspects of muscle function are monitored by muscle spindles and tendon organs?

16.3 SOMATIC SENSORY PATHWAYS

OBJECTIVE

- Describe the neuronal components and functions of the posterior column–medial lemniscus pathway, the anterolateral pathway, and the spinocerebellar pathway.

Somatic sensory pathways relay information from the somatic sensory receptors just described to the primary somatosensory area in the cerebral cortex and to the cerebellum. The pathways to the cerebral cortex consist of thousands of sets of three neurons: a first-order neuron, a second-order neuron, and a third-order neuron.

1. **First-order neurons** conduct impulses from somatic receptors into the brain stem or spinal cord. From the face, mouth, teeth, and eyes, somatic sensory impulses propagate along *cranial nerves* into the brain stem. From the neck, trunk, limbs, and posterior aspect of the head, somatic sensory impulses propagate along *spinal nerves* into the spinal cord.

2. **Second-order neurons** conduct impulses from the brain stem and spinal cord to the thalamus. Axons of second-order neurons *decussate* (cross over to the opposite side) in the brain stem or spinal cord before ascending to the ventral posterior nucleus of the thalamus. Thus, all somatic sensory information from one side of the body reaches the thalamus on the opposite side.
3. **Third-order neurons** conduct impulses from the thalamus to the primary somatosensory area of the cortex on the same side.

Regions within the CNS where neurons synapse with other neurons that are a part of a particular sensory or motor pathway are known as **relay stations** because neural signals are being relayed from one region of the CNS to another. For example, the neurons of many sensory pathways synapse with neurons in the thalamus; therefore the thalamus functions as a major relay station. In addition to the thalamus, many other regions of the CNS, including the spinal cord and brain stem, can function as relay stations.

Somatic sensory impulses ascend to the cerebral cortex via three general pathways: (1) the posterior column–medial lemniscus pathway, (2) the anterolateral (spinothalamic) pathway, and (3) the trigeminothalamic pathway. Somatic sensory impulses reach the cerebellum via the spinocerebellar tracts.

Posterior Column–Medial Lemniscus Pathway to the Cortex

Nerve impulses for touch, pressure, vibration, and conscious proprioception from the limbs, trunk, neck, and posterior head ascend to the cerebral cortex along the **posterior column–medial lemniscus pathway** (lem-NIS-kus = ribbon) (Figure 16.5). The name of the pathway comes from the names of two white-matter tracts that convey the impulses: the posterior column of the spinal cord and the medial lemniscus of the brain stem.

First-order neurons in the posterior column–medial lemniscus pathway extend from sensory receptors in the limbs, trunk, neck, and posterior head into the spinal cord and ascend to the medulla oblongata on the same side of the body. The cell bodies of these first-order neurons are in the posterior (dorsal) root ganglia of spinal nerves. In the spinal cord, their axons form the **posterior (dorsal) columns,** which consist of two parts: the **gracile fasciculus** (GRAS-īl fa-SIK-ū-lus) and the **cuneate fasciculus** (KŪ-nē-āt). The axons synapse with the dendrites of second-order neurons whose cell bodies are located in the gracile nucleus or cuneate nucleus of the medulla. Nerve impulses for touch, pressure, vibration, and conscious proprioception from the upper limbs, upper trunk, neck, and posterior head propagate along axons in the cuneate fasciculus and arrive at the cuneate nucleus. Nerve impulses for touch, pressure, and vibration from the lower limbs and lower trunk propagate along axons in the gracile fasciculus and arrive at the gracile nucleus.

The axons of the second-order neurons cross to the opposite side of the medulla and enter the **medial lemniscus,** a thin ribbonlike projection tract that extends from the medulla to the ventral posterior nucleus of the thalamus. In the thalamus, the axon terminals of second-order neurons synapse with third-order neurons,

Figure 16.5 The posterior column–medial lemniscus pathway.

The posterior column–medial lemniscus pathway conveys nerve impulses for touch, pressure, vibration, and conscious proprioception from the limbs, trunk, neck, and posterior head to the cerebral cortex.

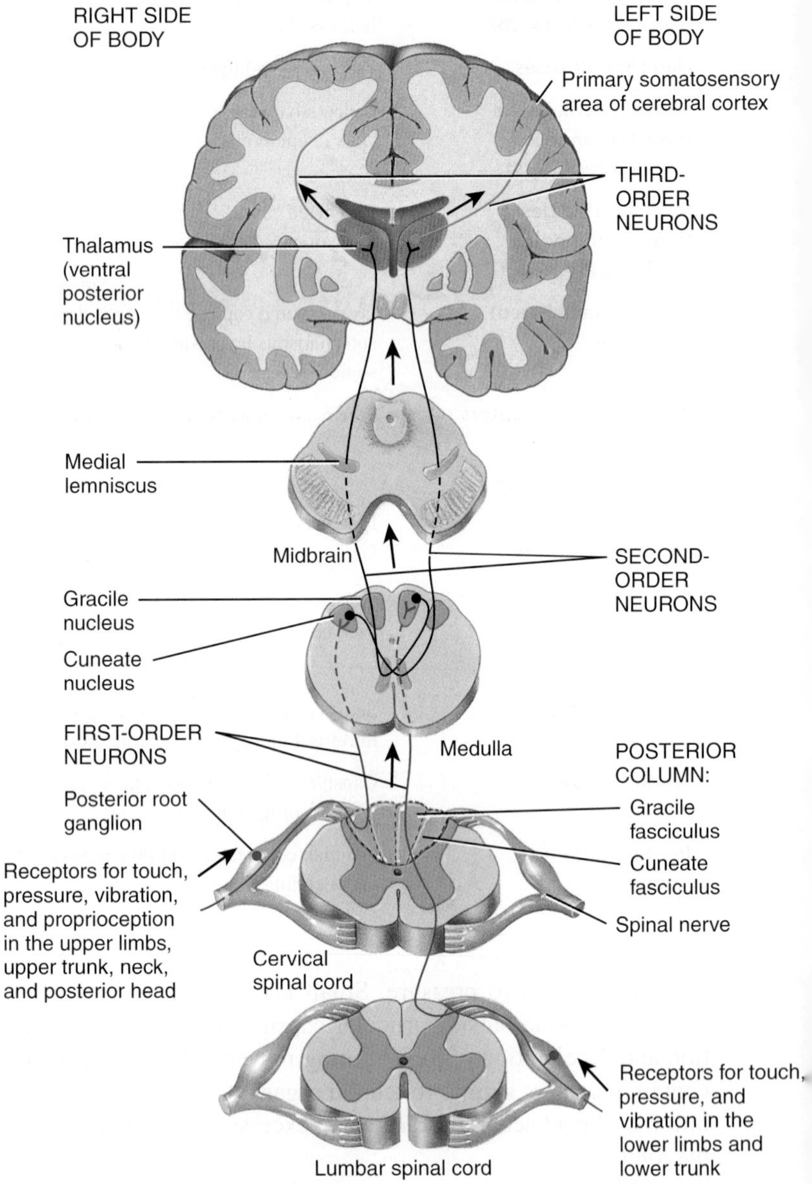

What are the two major tracts that form the posterior columns?

which project their axons to the primary somatosensory area of the cerebral cortex.

Anterolateral Pathway to the Cortex

Nerve impulses for pain, temperature, itch, and tickle from the limbs, trunk, neck, and posterior head ascend to the cerebral cortex along the **anterolateral** or **spinothalamic pathway** (spī-nō-tha-

LAM-ik). Like the posterior column–medial lemniscus pathway, the anterolateral pathway is composed of three-neuron sets (Figure 16.6). The first-order neurons connect a receptor of the limbs, trunk, neck, or posterior head with the spinal cord. The cell bodies of the first-order neurons are in the posterior root ganglion. The axon terminals of the first-order neurons synapse with second-order neurons, whose cell bodies are located in the posterior gray horn of the spinal cord. The axons of the second-order neurons cross to the opposite side of the spinal cord. Then, they pass upward to the brain stem as the **spinothalamic tract.** The axons of the second-order neurons end in the ventral posterior nucleus of the thalamus, where they synapse with the third-order neurons. The axons of the third-order neurons project to the primary somatosensory area on the same side of the cerebral cortex as the thalamus.

Figure 16.6 The anterolateral (spinothalamic) pathway.

The anterolateral pathway conveys nerve impulses for pain, cold, warmth, itch, and tickle from the limbs, trunk, neck, and posterior head to the cerebral cortex.

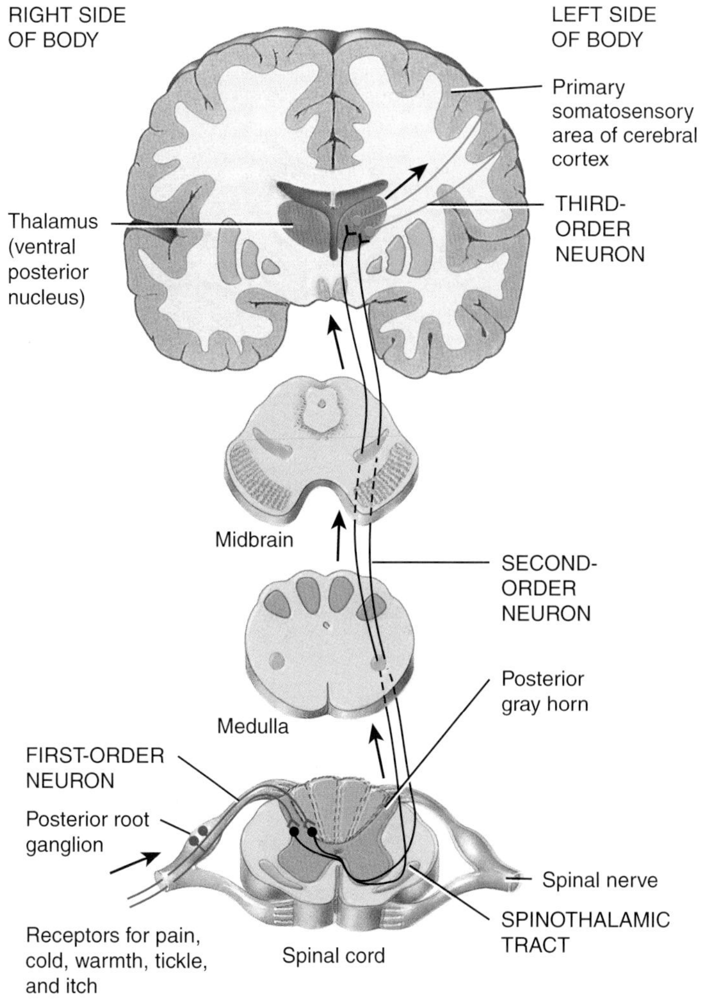

What types of sensory deficits could be produced by damage to the right spinothalamic tract?

Trigeminothalamic Pathway to the Cortex

Nerve impulses for most somatic sensations (tactile, thermal, and pain) from the face, nasal cavity, oral cavity, and teeth ascend to the cerebral cortex along the **trigeminothalamic pathway** (trī-JEM-i-nō-tha-LAM-ik). Like the other somatosensory pathways just described, the trigeminothalamic pathway consists of three-neuron sets (Figure 16.7). First-order neurons extend from somatic sensory receptors in the face, nasal cavity, oral cavity, and teeth into the pons through the trigeminal (V) nerves. The cell

Figure 16.7 The trigeminothalamic pathway.

The trigeminothalamic pathway conveys nerve impulses for most somatic sensations (tactile, thermal, pain, and proprioceptive) from the face, nasal cavity, oral cavity, and teeth to the cerebral cortex.

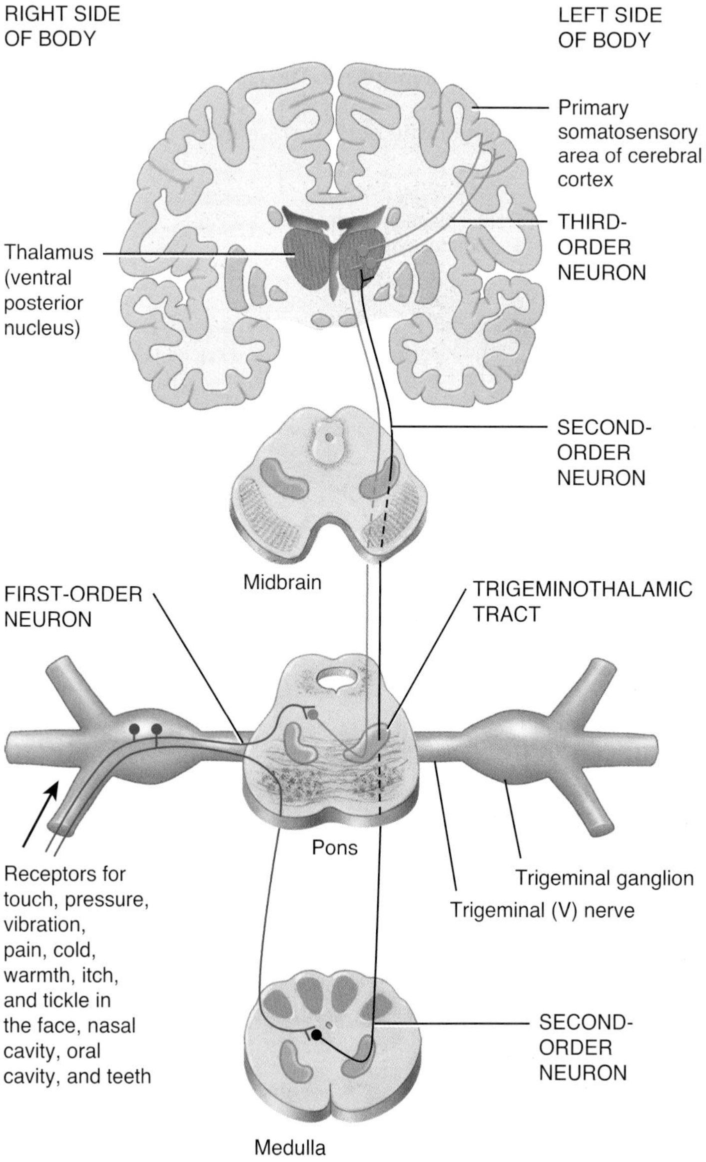

Which cranial nerve conveys impulses for most somatic sensations from the left side of the face into the pons?

bodies of these first-order neurons are in the trigeminal ganglion. The axon terminals of some first-order neurons synapse with second-order neurons in the pons. The axons of other first-order neurons descend into the medulla to synapse with second-order neurons. The axons of the second-order neurons cross to the opposite side of the pons and medulla and then ascend as the **trigeminothalamic tract** to the ventral posterior nucleus of the thalamus. In the thalamus, the axon terminals of the second-order neurons synapse with third-order neurons, which project their axons to the primary somatosensory area on the same side of the cerebral cortex as the thalamus.

Mapping the Primary Somatosensory Area

Specific areas of the cerebral cortex receive somatic sensory input from particular parts of the body. Other areas of the cerebral cortex provide output in the form of instructions for movement of particular parts of the body. The *somatic sensory map* and the *somatic motor map* relate body parts to these cortical areas.

Precise localization of somatic sensations occurs when nerve impulses arrive at the **primary somatosensory area** (areas 1, 2, and 3 in Figure 14.15), which occupies the postcentral gyri of the parietal lobes of the cerebral cortex. Each region in this area receives sensory input from a different part of the body. Figure 16.8a maps the destination of somatic sensory signals from different parts of the left side of the body in the somatosensory area of the right cerebral hemisphere. The left cerebral hemisphere has a similar primary somatosensory area that receives sensory input from the right side of the body.

Note that some parts of the body—chiefly the lips, face, tongue, and hand—provide input to large regions in the somatosensory area. Other parts of the body, such as the trunk and lower limbs,

Figure 16.8 Somatic sensory and somatic motor maps in the cerebral cortex, right hemisphere. (a) Primary somatosensory area (postcentral gyrus) and (b) primary motor area (precentral gyrus) of the right cerebral hemisphere. The left hemisphere has similar representation. (After Penfield and Rasmussen.)

Each point on the body surface maps to a specific region in both the primary somatosensory area and the primary motor area.

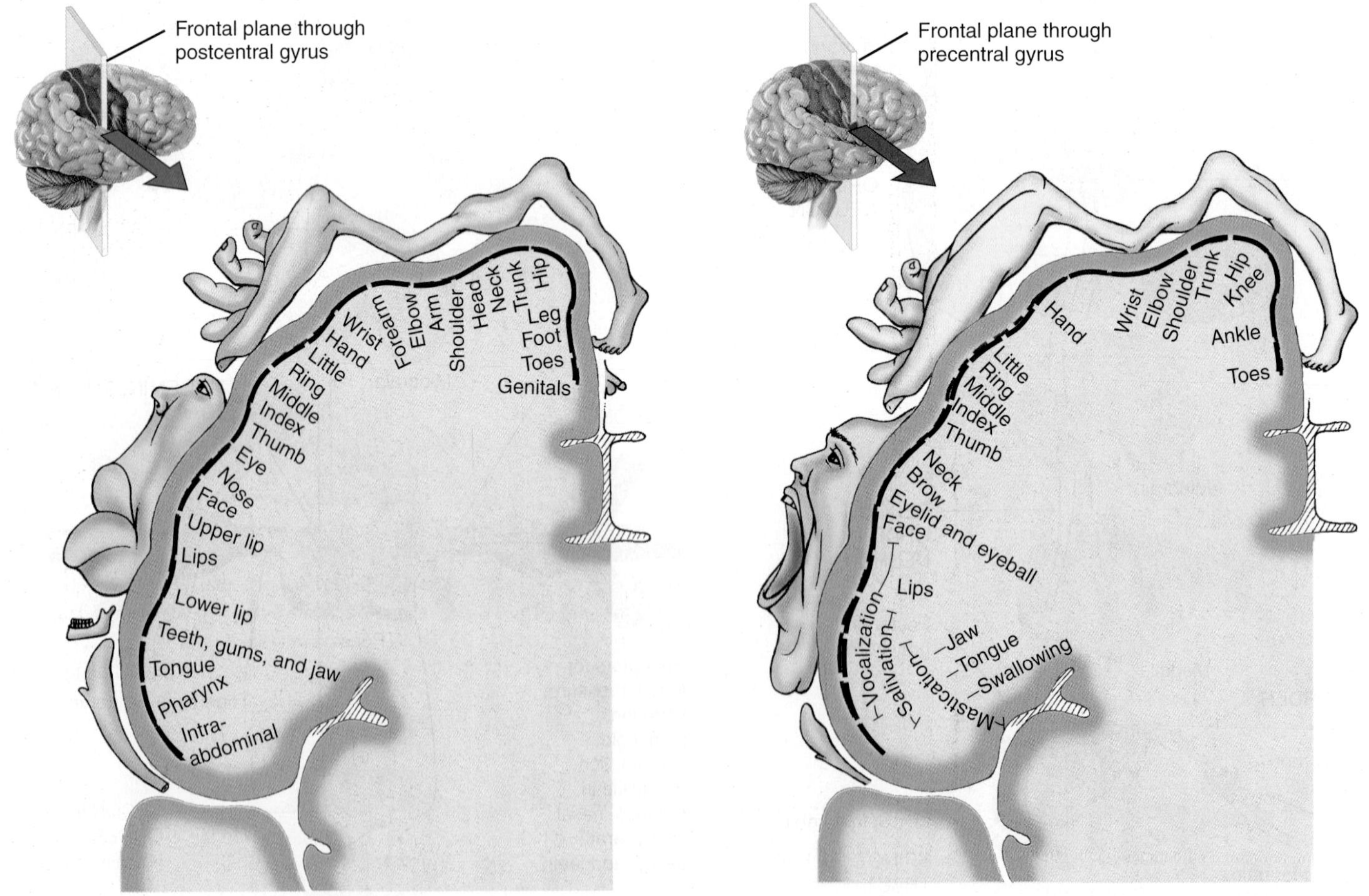

(a) Frontal section of primary somatosensory area in right cerebral hemisphere

(b) Frontal section of primary motor area in right cerebral hemisphere

How do the somatosensory and motor representations compare for the hand, and what does this difference imply?

project to much smaller cortical regions. The relative sizes of these regions in the somatosensory area are proportional to the number of specialized sensory receptors within the corresponding part of the body. For example, there are many sensory receptors in the skin of the lips but few in the skin of the trunk. This distorted somatic sensory map of the body is known as the **sensory homunculus.** The size of the cortical region that represents a body part may expand or shrink somewhat, depending on the quantity of sensory impulses received from that body part. For example, people who learn to read Braille eventually have a larger cortical region in the somatosensory area to represent the fingertips.

Somatic Sensory Pathways to the Cerebellum

Two tracts in the spinal cord—the **posterior spinocerebellar tract** (spī-nō-ser-e-BEL-ar) and the **anterior spinocerebellar tract**—are the major routes proprioceptive impulses take to reach the cerebellum. Although they are not consciously perceived, sensory impulses conveyed to the cerebellum along these two pathways are critical for posture, balance, and coordination of skilled movements.

Table 16.3 summarizes the major somatic sensory tracts and pathways.

TABLE 16.3

Major Somatic Sensory Tracts and Pathways

TRACTS AND LOCATIONS	PATHWAY FUNCTIONS
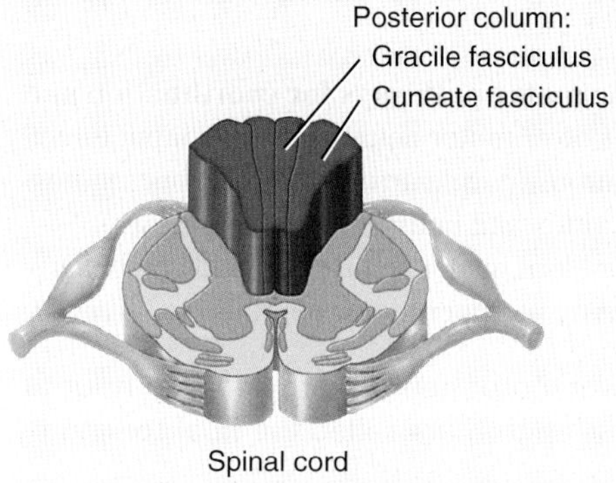	**Posterior column:** Consists of two tracts: (1) **cuneate fasciculus,** which conveys nerve impulses for touch, pressure, vibration, and conscious proprioception from upper limbs, upper trunk, neck, and posterior head, and (2) **gracile fasciculus,** which conveys nerve impulses for touch, pressure, and vibration from lower limbs and lower trunk. Axons of first-order neurons from one side of body form posterior column on same side and end in medulla, where they synapse with dendrites and cell bodies of second-order neurons. Axons of second-order neurons decussate, enter medial lemniscus on opposite side, and extend to thalamus. Third-order neurons transmit nerve impulses from thalamus to primary somatosensory cortex on side opposite the site of stimulation.
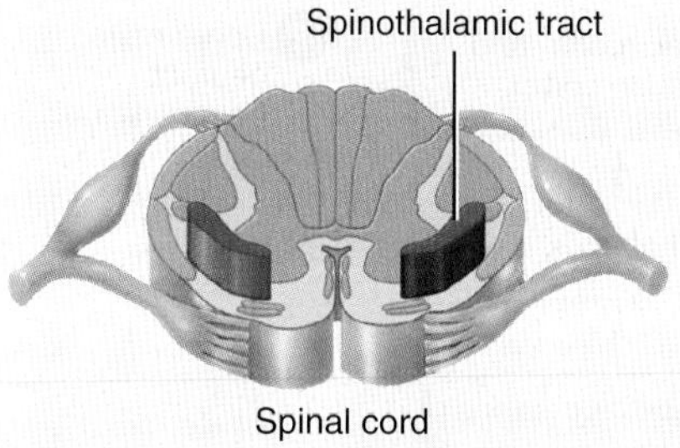	**Spinothalamic:** Conveys nerve impulses for pain, cold, warmth, itch, and tickle from limbs, trunk, neck, and posterior head. Axons of first-order neurons from one side of body synapse with dendrites and cell bodies of second-order neurons in posterior gray horn on same side of body. Axons of second-order neurons decussate, enter spinothalamic tract on opposite side, and extend to thalamus. Third-order neurons transmit nerve impulses from thalamus to primary somatosensory cortex on side opposite the site of stimulation.
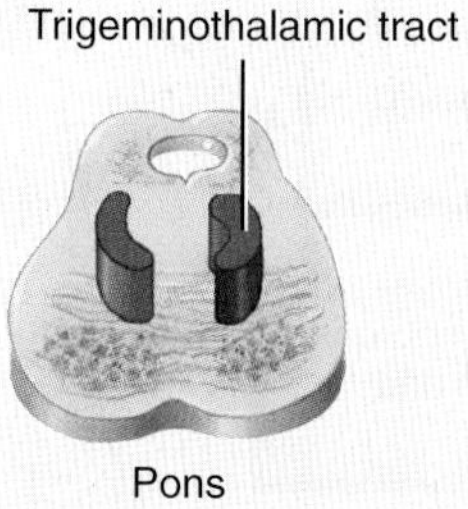	**Trigeminothalamic:** Conveys nerve impulses for touch, pressure, vibration, pain, cold, warmth, itch, and tickle from face, nasal cavity, oral cavity, and teeth. Axons of first-order neurons from one side of head synapse with dendrites and cell bodies of second-order neurons in pons and medulla on same side of head. Axons of second-order neurons decussate, enter trigeminothalamic tract on opposite side, and extend to thalamus. Third-order neurons transmit nerve impulses from thalamus to primary somatosensory cortex on side opposite the site of stimulation.
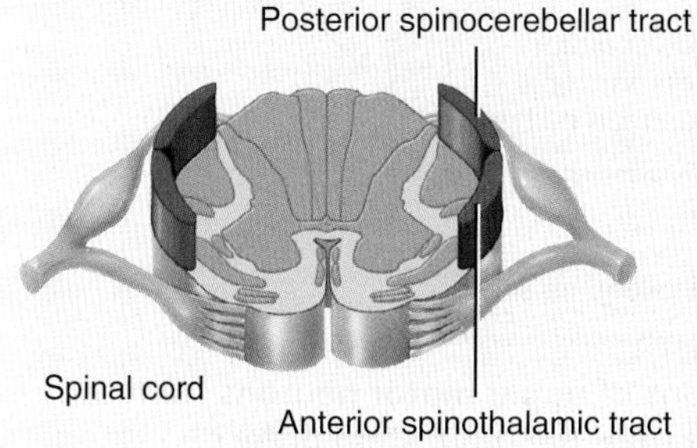	**Anterior and posterior spinocerebellar:** Convey nerve impulses from proprioceptors in trunk and lower limb of one side of body to same side of cerebellum. Proprioceptive input informs cerebellum of actual movements, allowing it to coordinate, smooth, and refine skilled movements and maintain posture and balance.

CLINICAL CONNECTION | *Syphilis*

Syphilis is a sexually transmitted disease caused by the bacterium *Treponema pallidum.* Because it is a bacterial infection, it can be treated with antibiotics. However, if the infection is not treated, the third stage of syphilis typically causes debilitating neurological symptoms. A common outcome is progressive degeneration of the posterior portions of the spinal cord, including the posterior columns, posterior spinocerebellar tracts, and posterior roots. Somatic sensations are lost, and the person's gait becomes uncoordinated and jerky because proprioceptive impulses fail to reach the cerebellum. •

✔CHECKPOINT

11. What are the functional differences between the posterior column–medial lemniscus pathway, the anterolateral pathway, and the trigeminothalamic pathway?
12. Which body parts have the largest representation in the primary somatosensory area?
13. What type of sensory information is carried in the spinocerebellar tracts, and what is its function?

16.4 SOMATIC MOTOR PATHWAYS

OBJECTIVES

- Identify the locations and functions of the different types of neurons in the somatic motor pathways.
- Compare the locations and functions of the direct and indirect motor pathways.
- Explain how the basal nuclei and cerebellum contribute to movements.

Neural circuits in the brain and spinal cord orchestrate all voluntary and involuntary movements. Ultimately, all excitatory and inhibitory signals that control movement converge on the motor neurons that extend out of the brain stem and spinal cord to innervate skeletal muscles in the body. These neurons, also known as **lower motor neurons (LMNs),** have their cell bodies in the brain stem and spinal cord. From the brain stem, axons of LMNs extend through *cranial nerves* to innervate skeletal muscles of the face and head. From the spinal cord, axons of LMNs extend through *spinal nerves* to innervate skeletal muscles of the limbs and trunk. Only LMNs provide output from the CNS to skeletal muscle fibers. For this reason, they are also called the *final common pathway.*

Neurons in four distinct but highly interactive neural circuits, collectively termed the **somatic motor pathways,** participate in control of movement by providing input to lower motor neurons (Figure 16.9):

1. ***Local circuit neurons.*** Input arrives at lower motor neurons from nearby interneurons called **local circuit neurons.** These neurons are located close to the lower motor neuron cell bodies in the brain stem and spinal cord. Local circuit neurons receive input from somatic sensory receptors, such as nociceptors and muscle spindles, as well as from higher centers in the brain. They help coordinate rhythmic activity in specific muscle groups, such as alternating flexion and extension of the lower limbs during walking.
2. ***Upper motor neurons.*** Both local circuit neurons and lower motor neurons receive input from **upper motor neurons (UMNs).** Most upper motor neurons synapse with local circuit neurons, which in turn synapse with lower motor neurons. (A few upper motor neurons synapse directly with lower motor neurons.) UMNs from the cerebral cortex are essential for the execution of voluntary movements of the body. Other UMNs originate in motor centers of the brain stem: the red nucleus, the vestibular nucleus, the superior colliculus, and the reticular formation. UMNs from the brain stem regulate muscle tone, control postural muscles, and help maintain balance and orientation of the head and body. Both the basal nuclei and cerebellum exert influence on upper motor neurons.

Figure 16.9 Somatic motor pathways for coordination and control of movement. Lower motor neurons receive input directly from (1) local circuit neurons (purple arrow) and (2) upper motor neurons in the cerebral cortex and brain stem (green arrows). Neural circuits involving basal nuclei neurons (3) and cerebellar neurons (4) regulate activity of upper motor neurons (red arrows).

Because lower motor neurons provide all output to skeletal muscles, they are called the final common pathway.

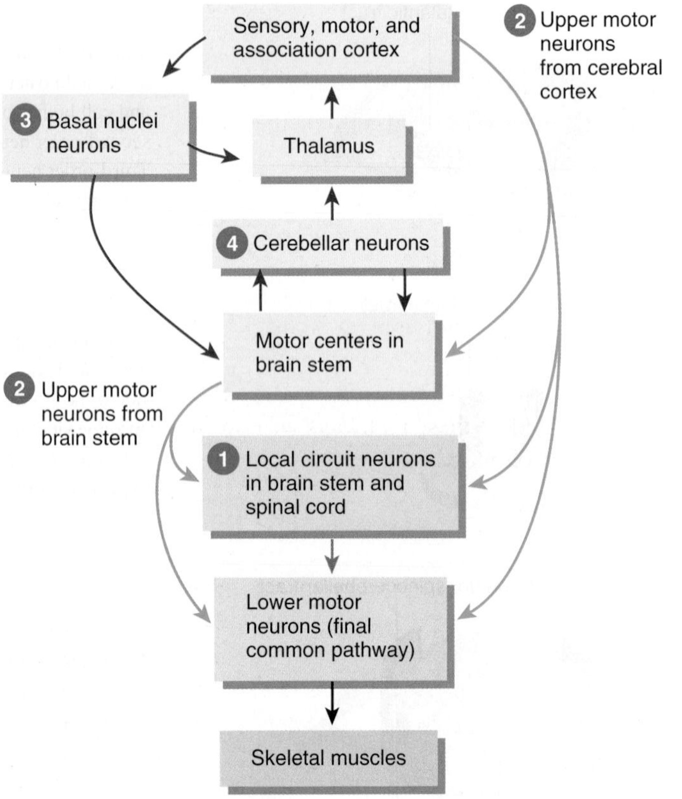

How do the functions of upper motor neurons from the cerebral cortex and from the brain stem differ?

3. ***Basal nuclei neurons.*** Basal nuclei neurons assist movement by providing input to upper motor neurons. Neural circuits interconnect the basal nuclei with motor areas of the cerebral cortex (via the thalamus) and the brain stem. These circuits help initiate and terminate movements, suppress unwanted movements, and establish a normal level of muscle tone.

4. ***Cerebellar neurons.*** Cerebellar neurons also aid movement by controlling the activity of upper motor neurons. Neural circuits interconnect the cerebellum with motor areas of the cerebral cortex (via the thalamus) and the brain stem. A prime function of the cerebellum is to monitor differences between intended movements and movements actually performed. Then, it issues commands to upper motor neurons to reduce errors in movement. The cerebellum thus coordinates body movements and helps maintain normal posture and balance.

CLINICAL CONNECTION | *Paralysis*

Damage or disease of *lower* motor neurons produces **flaccid paralysis** (FLAS-id or FLAK-sid) of muscles on the same side of the body. There is neither voluntary nor reflex action of the innervated muscle fibers, muscle tone is decreased or lost, and the muscle remains limp or flaccid. Injury or disease of *upper* motor neurons in the cerebral cortex removes inhibitory influences that some of these neurons have on lower motor neurons, which causes **spastic paralysis** of muscles on the opposite side of the body. In this condition muscle tone is increased, reflexes are exaggerated, and pathological reflexes such as the Babinski sign appear (see Clinical Connection: Reflexes and Diagnosis in Chapter 13). •

Organization of Upper Motor Neuron Pathways

The axons of upper motor neurons extend from the brain to lower motor neurons via two types of somatic motor pathways—direct and indirect. **Direct motor pathways** provide input to lower motor neurons via axons that extend directly from the cerebral cortex. **Indirect motor pathways** provide input to lower motor neurons from motor centers in the basal nuclei, cerebellum, and cerebral cortex. Direct and indirect pathways both govern generation of nerve impulses in the lower motor neurons, the neurons that stimulate contraction of skeletal muscles.

Before we examine these pathways we consider the role of the motor cortex in voluntary movement.

Mapping the Motor Areas

Control of body movements occurs via neural circuits in several regions of the brain. The **primary motor area** (area 4 in Figure 14.15), located in the precentral gyrus of the frontal lobe (see Figure 16.8b) of the cerebral cortex, is a major control region for the execution of voluntary movements. The adjacent **premotor area** (area 6) also contributes axons to the descending motor pathways. As is true for somatic sensory representation in the somatosensory area, different muscles are represented unequally in the primary motor area. More cortical area is devoted to those muscles involved in skilled, complex, or delicate movements. Muscles in the thumb, fingers, lips, tongue, and vocal cords have large representations; the trunk has a much smaller representation. This distorted muscle map of the body is referred to as the **motor homunculus.** By comparing Figures 16.8a and 16.8b, you can see that somatosensory and somatic motor representations are similar but not identical for most parts of the body.

Direct Motor Pathways

Nerve impulses for voluntary movements propagate from the cerebral cortex to lower motor neurons via the direct motor pathways. The direct motor pathways, which are also known as the *pyramidal pathways,* consist of axons that descend from pyramidal cells. *Pyramidal cells* are upper motor neurons with pyramid-shaped cell bodies (see Figure 12.5b) located in the primary motor area and the premotor area of the cerebral cortex (areas 4 and 6, respectively, in Figure 14.15). The direct motor pathways consist of the corticospinal pathways and the corticobulbar pathway.

CORTICOSPINAL PATHWAYS The **corticospinal pathways** (kor′-ti-kō-SPĪ-nal) conduct impulses for the control of muscles of the limbs and trunk. Axons of upper motor neurons in the cerebral cortex form the **corticospinal tracts,** which descend through the internal capsule of the cerebrum and the cerebral peduncle of the midbrain. In the medulla oblongata, the axon bundles of the corticospinal tracts form the ventral bulges known as the *pyramids.* About 90% of the corticospinal axons *decussate* (cross over) to the *contralateral* (opposite) side in the medulla oblongata and then descend into the spinal cord where they synapse with a local circuit neuron or a lower motor neuron. The 10% that remain on the *ipsilateral* (same) side eventually decussate at the spinal cord levels where they synapse with a local circuit neuron or lower motor neuron. Thus, the right cerebral cortex controls most of the muscles on the left side of the body, and the left cerebral cortex controls most of the muscles on the right side of the body. There are two types of corticospinal tracts: the lateral corticospinal tract and the anterior corticospinal tract.

1. ***Lateral corticospinal tract.*** Corticospinal axons that decussate in the medulla form the **lateral corticospinal tract** in the lateral white column of the spinal cord (Figure 16.10a). These axons synapse with local circuit neurons or lower motor neurons in the anterior gray horn of the spinal cord. Axons of these lower motor neurons exit the cord in the anterior roots of spinal nerves and terminate in skeletal muscles that control movements of the distal parts of the limbs. The distal muscles are responsible for precise, agile, and highly skilled movements of the hands and feet. Examples include the movements needed to button a shirt or play the piano.
2. ***Anterior corticospinal tract.*** Corticospinal axons that do not decussate in the medulla form the **anterior corticospinal tract** in the anterior white column of the spinal cord (Figure 16.10b). At each spinal cord level, some of these axons decussate via the anterior white commissure. Then, they synapse with local circuit neurons or lower motor neurons in the anterior gray horn. Axons of these lower motor neurons exit the cord in the

Figure 16.10 The corticospinal pathways.

The corticospinal pathways conduct nerve impulses for the control of muscles of the limbs and trunk.

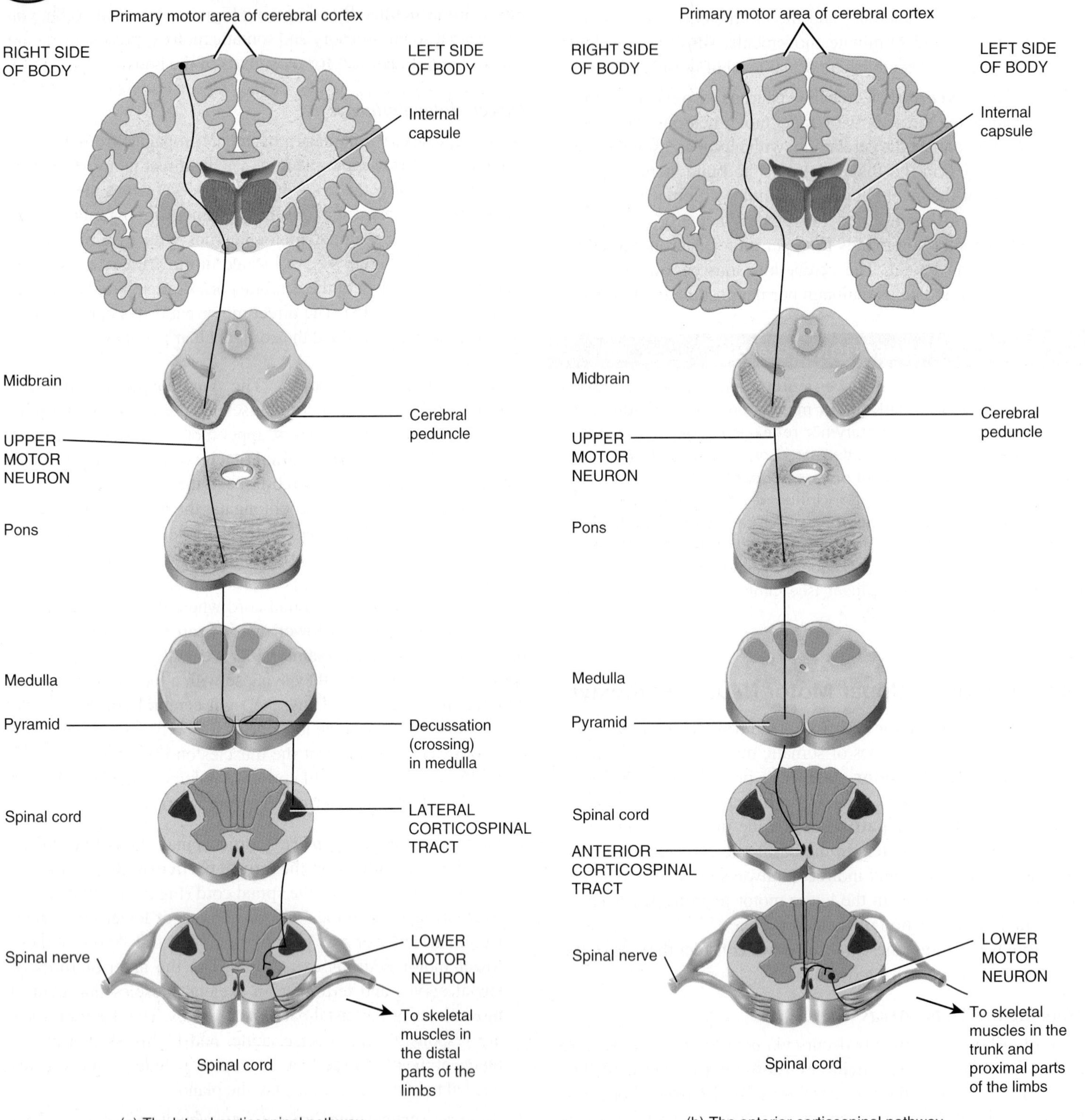

(a) The lateral corticospinal pathway

(b) The anterior corticospinal pathway

Which tract conveys nerve impulses that result in contractions of muscles in the distal parts of the limbs?

anterior roots of spinal nerves. They terminate in skeletal muscles that control movements of the trunk and proximal parts of the limbs.

Corticobulbar Pathway The **corticobulbar pathway** (kor′-ti-kō-BUL-bar) conducts impulses for the control of skeletal muscles in the head. Axons of upper motor neurons from the cerebral cortex form the **corticobulbar tract**, which descends along with the corticospinal tracts through the internal capsule of the cerebrum and cerebral peduncle of the midbrain (Figure 16.11). Some of the axons of the corticobulbar tract decussate; others do not. The axons terminate in the motor nuclei of nine pairs of cranial nerves in the brain stem: the oculomotor (III), trochlear (IV), trigeminal (V), abducens (VI), facial (VII), glossopharyngeal (IX),

Figure 16.11 **The corticobulbar pathway.** For simplicity, only two cranial nerves are illustrated.

The corticobulbar pathway conducts nerve impulses for the control of skeletal muscles in the head.

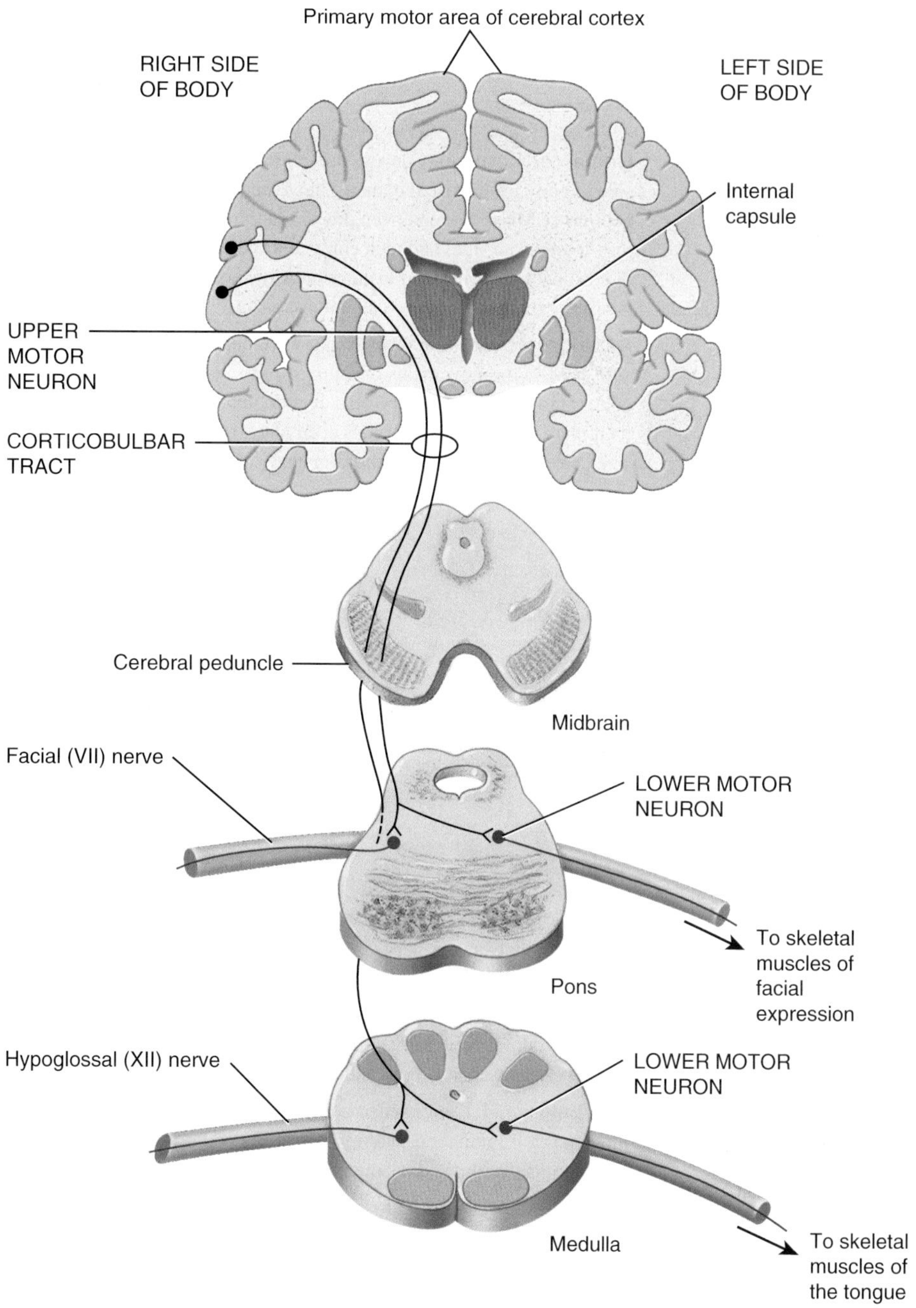

? The axons of the corticobulbar tract terminate in the motor nuclei of which cranial nerves?

vagus (X), accessory (XI), and hypoglossal (XII). The lower motor neurons of the cranial nerves convey impulses that control precise, voluntary movements of the eyes, tongue, and neck, plus chewing, facial expression, speech, and swallowing.

Table 16.4 summarizes the major somatic motor tracts and pathways.

CLINICAL CONNECTION | *Amyotrophic Lateral Sclerosis*

Amyotrophic lateral sclerosis (ALS) (ā′-mī-ō-TRŌF-ik; *a-* = without; *-myo-* = muscle; *-trophic* = nourishment) is a progressive degenerative disease that attacks motor areas of the cerebral cortex, axons of upper motor neurons in the lateral white columns (corticospinal and rubrospinal tracts), and lower motor neuron cell bodies. It causes progressive muscle weakness and atrophy. ALS often begins in sections of the spinal cord that serve the hands and arms but rapidly spreads to involve the whole body and face, without affecting intellect or

TABLE 16.4

Major Somatic Motor Tracts and Pathways

TRACTS AND LOCATIONS | **PATHWAY FUNCTIONS**

DIRECT (PYRAMIDAL) PATHWAYS

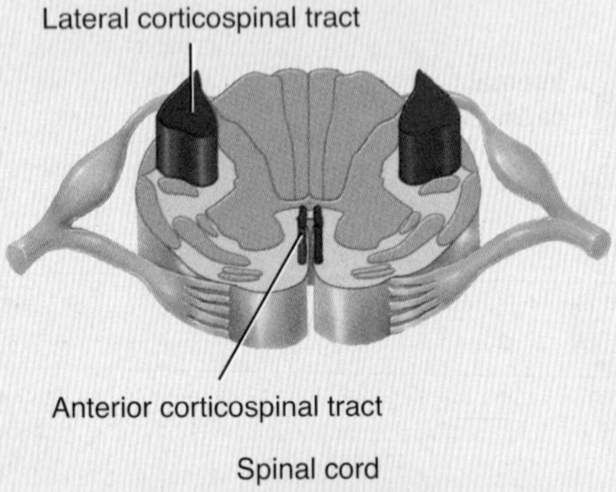

Lateral corticospinal: Conveys nerve impulses from motor cortex to skeletal muscles on opposite side of body for precise, voluntary movements of distal parts of limbs. Axons of upper motor neurons (UMNs) descend from precentral gyrus of cortex into medulla. Here 90% decussate (cross over to opposite side) and then enter contralateral side of spinal cord to form this tract. At their level of termination, UMNs end in anterior gray horn on same side. They provide input to lower motor neurons, which innervate skeletal muscles.

Anterior corticospinal: Conveys nerve impulses from motor cortex to skeletal muscles on opposite side of body for movements of trunk and proximal parts of limbs. Axons of UMNs descend from cortex into medulla. Here the 10% that do not decussate enter spinal cord and form this tract. At their level of termination, UMNs decussate and end in anterior gray horn on opposite side of body. They provide input to lower motor neurons, which innervate skeletal muscles.

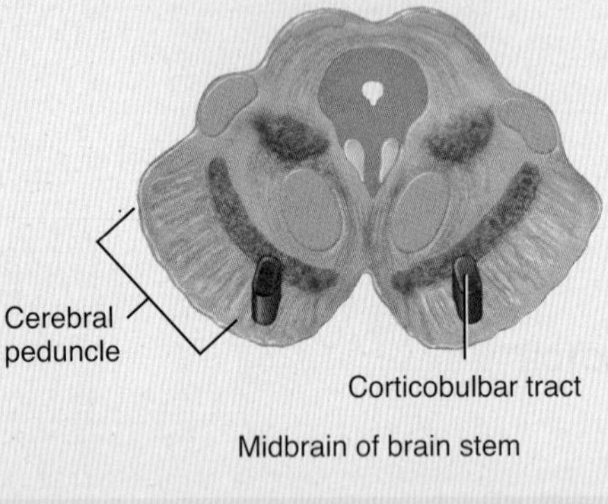

Corticobulbar: Conveys nerve impulses from motor cortex to skeletal muscles of head and neck to coordinate precise, voluntary movements. Axons of UMNs descend from cortex into brain stem, where some decussate and others do not. They provide input to lower motor neurons in nuclei of cranial nerves III, IV, V, VI, VII, IX, X, XI, and XII, which control voluntary movements of eyes, tongue, and neck; chewing; facial expression; and speech.

INDIRECT (EXTRAPYRAMIDAL) PATHWAYS

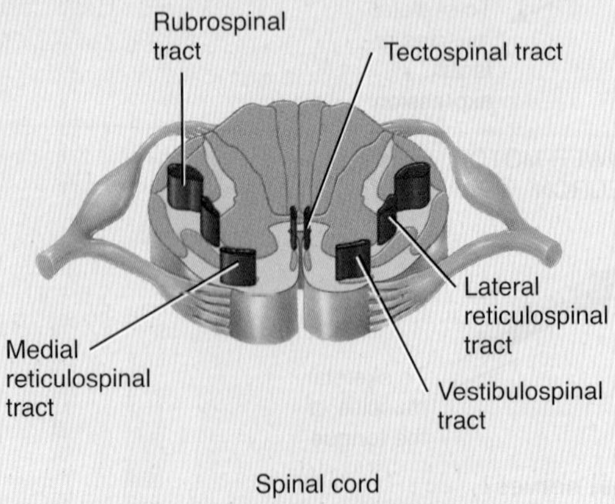

Rubrospinal: Conveys nerve impulses from red nucleus (which receives input from cerebral cortex and cerebellum) to contralateral skeletal muscles that govern precise, voluntary movements of distal parts of upper limbs.

Tectospinal: Conveys nerve impulses from superior colliculus to contralateral skeletal muscles that reflexively move head, eyes, and trunk in response to visual or auditory stimuli.

Vestibulospinal: Conveys nerve impulses from vestibular nucleus (which receives input about head movements from inner ear) to ipsilateral skeletal muscles of trunk and proximal parts of limbs for maintaining posture and balance in response to head movements.

Medial and lateral reticulospinal: Conveys nerve impulses from reticular formation to ipsilateral skeletal muscles of trunk and proximal parts of limbs for maintaining posture and regulating muscle tone in response to ongoing body movements.

sensations. Death typically occurs in 2 to 5 years. ALS is commonly known as *Lou Gehrig's disease* after the New York Yankees baseball player who died from it at age 37 in 1941.

Inherited mutations account for about 15% of all cases of ALS (familial ALS). Noninherited (sporadic) cases of ALS appear to have several implicating factors. According to one theory there is a buildup in the synaptic cleft of the neurotransmitter glutamate released by motor neurons due to a mutation of the protein that normally deactivates and recycles the neurotransmitter. The excess glutamate causes motor neurons to malfunction and eventually die. The drug riluzole, which is used to treat ALS, reduces damage to motor neurons by decreasing the release of glutamate. Other factors may include damage to motor neurons by free radicals, autoimmune responses, viral infections, deficiency of nerve growth factor, apoptosis (programmed cell death), environmental toxins, and trauma.

In addition to riluzole, ALS is treated with drugs that relieve symptoms such as fatigue, muscle pain and spasticity, excessive saliva, and difficulty sleeping. The only other treatment is supportive care provided by physical, occupational, and speech therapists; nutritionists; social workers; and home care and hospice nurses. •

Indirect Motor Pathways

The **indirect motor pathways** or **extrapyramidal pathways** include all somatic motor tracts other than the corticospinal and corticobulbar tracts. Axons of upper motor neurons that give rise to the indirect motor pathways descend from various nuclei of the brain stem into five major tracts of the spinal cord and terminate on local circuit neurons or lower motor neurons. These tracts are the **rubrospinal** (ROO-brō-spī-nal), **tectospinal** (TEK-tō-spī-nal), **vestibulospinal** (ves-TIB-ū-lō-spī-nal), **lateral reticulospinal** (re-TIK-ū-lō-spī-nal), and **medial reticulospinal tracts.**

Table 16.4 summarizes the tracts of the indirect motor pathways.

Roles of the Basal Nuclei

As previously noted, the basal nuclei and cerebellum influence movement through their effects on upper motor neurons. The functions of the basal nuclei include the following:

1. The basal nuclei play a major role in the initiation and termination of movements. Two parts of the basal nuclei, the caudate nucleus and the putamen, receive input from sensory, association, and motor areas of the cerebral cortex and from the substantia nigra. Output from the basal nuclei comes from the globus pallidus and substantia nigra, which send feedback signals to the upper motor cortex by way of the thalamus. (Figure 14.13b shows these parts of the basal nuclei.) This circuit—from cortex to basal nuclei to thalamus to cortex—appears to function in initiating and terminating movements. Neurons in the putamen generate impulses just before body movements occur, and neurons in the caudate nucleus generate impulses just before eye movements occur.
2. The basal nuclei suppress unwanted movements by their inhibitory effects on the thalamus and superior colliculus.
3. The basal nuclei influence muscle tone. The globus pallidus sends impulses into the reticular formation that reduce muscle tone. Damage or destruction of some basal nuclei connections causes a generalized increase in muscle tone.
4. The basal nuclei influence many aspects of cortical function, including sensory, limbic, cognitive, and linguistic functions. For example, the basal nuclei help initiate and terminate some cognitive processes, such as attention, memory, and planning. In addition, the basal nuclei may act with the limbic system to regulate emotional behaviors.

CLINICAL CONNECTION | *Disorders of the Basal Nuclei*

Disorders of the basal nuclei can affect body movements, cognition, and behavior. Uncontrollable shaking (tremor) and muscle rigidity (stiffness) are hallmark signs of **Parkinson disease (PD)** (see Disorders: Homeostatic Imbalances at the end of this chapter). In this disorder, dopamine-releasing neurons that extend from the substantia nigra to the putamen and caudate nucleus degenerate.

Huntington disease (HD) is an inherited disorder in which the caudate nucleus and putamen degenerate, with loss of neurons that normally release GABA or acetylcholine. A key sign of HD is **chorea** (KŌ-rē-a = a dance), in which rapid, jerky movements occur involuntarily and without purpose. Progressive mental deterioration also occurs. Symptoms of HD often do not appear until age 30 or 40. Death occurs 10 to 20 years after symptoms first appear.

Tourette syndrome is a disorder that is characterized by involuntary body movements (motor tics) and the use of inappropriate or unnecessary sounds or words (vocal tics). Although the cause is unknown, research suggests that this disorder involves a dysfunction of the cognitive neural circuits between the basal nuclei and the prefrontal cortex.

Some psychiatric disorders, such as schizophrenia and obsessive–compulsive disorder, are thought to involve dysfunction of the behavioral neural circuits between the basal nuclei and the limbic system. In **schizophrenia,** excess dopamine activity in the brain causes a person to experience delusions, distortions of reality, paranoia, and hallucinations. People who have **obsessive–compulsive disorder (OCD)** experience repetitive thoughts (obsessions) that cause repetitive behaviors (compulsions) that they feel obligated to perform. For example, a person with OCD might have repetitive thoughts about someone breaking into the house; these thoughts might drive that person to check the doors of the house over and over again (for minutes or hours at a time) to make sure that they are locked. •

Modulation of Movement by the Cerebellum

In addition to maintaining proper posture and balance, the cerebellum is active in both learning and performing rapid, coordinated, highly skilled movements such as hitting a golf ball, speaking, and swimming. Cerebellar function involves four activities (Figure 16.12):

(1) **Monitoring intentions for movement.** The cerebellum receives impulses from the motor cortex and basal nuclei via

Figure 16.12 Input to and output from the cerebellum.

The cerebellum coordinates and smooths contractions of skeletal muscles during skilled movements and helps maintain posture and balance.

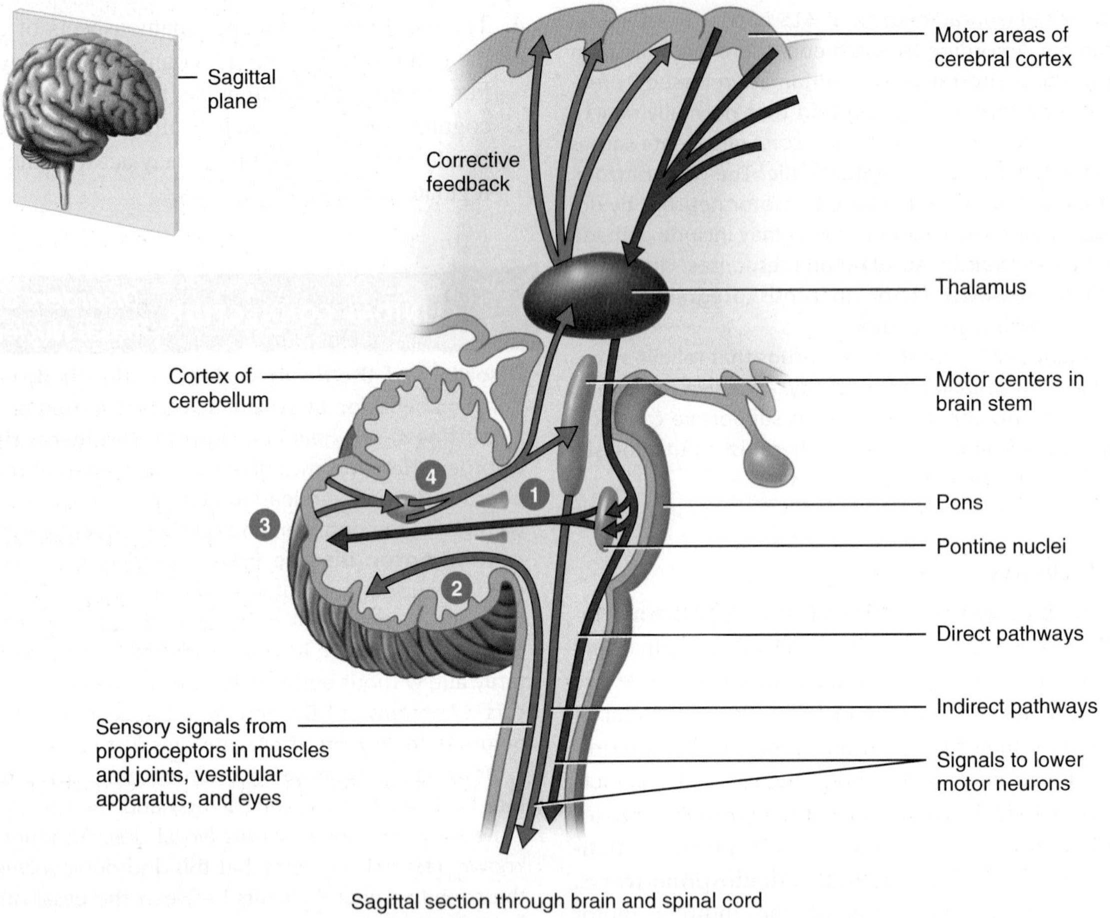
Sagittal plane
Motor areas of cerebral cortex
Corrective feedback
Thalamus
Cortex of cerebellum
Motor centers in brain stem
4
1
3
Pons
Pontine nuclei
2
Direct pathways
Indirect pathways
Sensory signals from proprioceptors in muscles and joints, vestibular apparatus, and eyes
Signals to lower motor neurons
Sagittal section through brain and spinal cord

Which tracts carry information from proprioceptors in joints and muscles to the cerebellum?

the pontine nuclei in the pons regarding what movements are planned (red arrows).

2. **Monitoring actual movement.** The cerebellum receives input from proprioceptors in joints and muscles that reveals what actually is happening. These nerve impulses travel in the anterior and posterior spinocerebellar tracts. Nerve impulses from the vestibular (equilibrium-sensing) apparatus in the inner ear and from the eyes also enter the cerebellum.
3. **Comparing command signals with sensory information.** The cerebellum compares intentions for movement with the actual movement performed.
4. **Sending out corrective feedback.** If there is a discrepancy between intended and actual movement, the cerebellum sends feedback to upper motor neurons. This information travels via the thalamus to UMNs in the cerebral cortex but goes directly to UMNs in brain stem motor centers (green arrows). As movements occur, the cerebellum continuously provides error corrections to upper motor neurons, which decreases errors and smoothes the motion. It also contributes over longer periods to the learning of new motor skills.

Skilled activities such as tennis or volleyball provide good examples of the contribution of the cerebellum to movement. To make a good serve or to block a spike, you must bring your racket or arms forward just far enough to make solid contact. How do you stop at exactly the right point? Before you even hit the ball, the cerebellum has sent nerve impulses to the cerebral cortex and basal nuclei informing them where your swing must stop. In response to impulses from the cerebellum, the cortex and basal nuclei transmit motor impulses to opposing body muscles to stop the swing.

✔CHECKPOINT

14. Trace the path of a motor impulse from the upper motor neurons through the final common pathway.
15. Which parts of the body have the largest representation in the motor cortex? Which have the smallest?
16. Explain why the two main somatic motor pathways are called "direct" and "indirect."
17. Explain the role of the cerebellum in performing rapid, coordinated, highly skilled movements.

16.5 INTEGRATIVE FUNCTIONS OF THE CEREBRUM

OBJECTIVES

- Compare the integrative cerebral functions of wakefulness and sleep, and learning and memory.
- Describe the four stages of sleep.
- Explain the factors that contribute to memory.

We turn now to a fascinating, though incompletely understood, function of the cerebrum: integration, the processing of sensory information by analyzing and storing it and making decisions for various responses. The **integrative functions** include cerebral activities such as sleep and wakefulness, learning and memory, and emotional responses. (The role of the limbic system in emotional behavior was discussed in Chapter 14.)

Wakefulness and Sleep

Humans sleep and awaken in a 24-hour cycle called a **circadian rhythm** (ser-KĀ-dē-an; *circa-* = about; *-dia* = a day) that is established by the suprachiasmatic nucleus of the hypothalamus (see Figure 14.10). A person who is awake is in a state of readiness and is able to react consciously to various stimuli. EEG recordings show that the cerebral cortex is very active during wakefulness; fewer impulses arise during most stages of sleep.

The Role of the Reticular Activating System in Awakening

How does your nervous system make the transition between these two states? Because stimulation of some of its parts increases activity of the cerebral cortex, a portion of the reticular formation is known as the **reticular activating system (RAS)** (Figure 16.13). When this area is active, many nerve impulses are transmitted to widespread areas of the cerebral cortex, both directly and via the thalamus. The effect is a generalized increase in cortical activity.

Arousal, or awakening from sleep, also involves increased activity in the RAS. For arousal to occur, the RAS must be stimulated. Many sensory stimuli can activate the RAS: painful stimuli detected by nociceptors, touch and pressure on the skin, movement of the limbs, bright light, or the buzz of an alarm clock. Once the RAS is activated, the cerebral cortex is also activated, and arousal occurs. The result is a state of wakefulness called **consciousness** (KON-shus-nes). Notice in Figure 16.13 that even though the RAS receives input from somatic sensory receptors, the eyes, and the ears, there is no input from olfactory receptors; even strong odors may fail to cause arousal. People who die in house fires usually succumb to smoke inhalation without awakening. For this reason, all sleeping areas should have a nearby

Figure 16.13 The reticular activating system (RAS) consists of neurons whose axons project from the reticular formation through the thalamus to the cerebral cortex.

Increased activity of the RAS causes awakening from sleep (arousal).

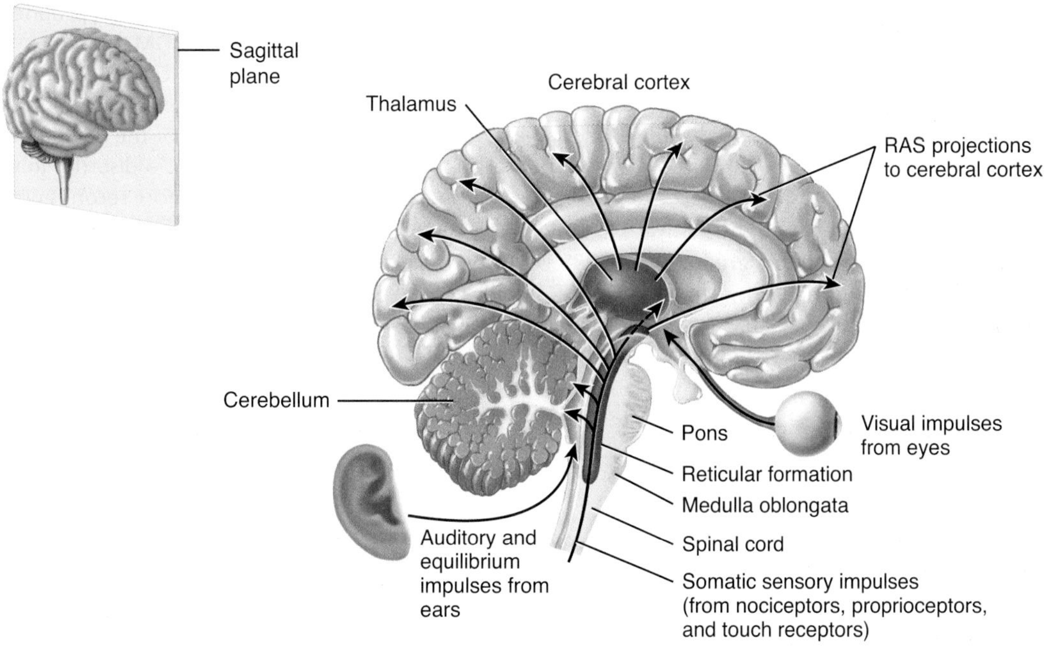

Sagittal section through brain and spinal cord showing the reticular formation

Why should every sleeping room have a smoke detector?

smoke detector that emits a loud alarm. A vibrating pillow or flashing light can serve the same purpose for those who are hearing impaired.

Sleep

Sleep is a state of altered consciousness or partial unconsciousness from which an individual can be aroused. Although it is essential, the exact functions of sleep are still unclear. Sleep deprivation impairs attention, learning, and performance. Normal sleep consists of two components: non-rapid eye movement (NREM) sleep and rapid eye movement (REM) sleep. **NREM sleep** consists of four gradually merging stages:

1. *Stage 1* is a transition stage between wakefulness and sleep that normally lasts 1–7 minutes. The person is relaxed with eyes closed and has fleeting thoughts. People awakened during this stage often say they have not been sleeping.
2. *Stage 2* or *light sleep* is the first stage of true sleep. In it, a person is a little more difficult to awaken. Fragments of dreams may be experienced, and the eyes may slowly roll from side to side.
3. *Stage 3* is a period of moderately deep sleep. Body temperature and blood pressure decrease, and it is difficult to awaken the person. This stage occurs about 20 minutes after falling asleep.
4. *Stage 4* is the deepest level of sleep. Although brain metabolism decreases significantly and body temperature drops slightly at this time, most reflexes are intact, and muscle tone is decreased only slightly. When sleepwalking occurs, it does so during this stage.

Typically, a person goes from stage 1 to stage 4 of NREM sleep in less than an hour. During a typical 7- or 8-hour sleep period, there are three to five episodes of **REM sleep,** during which the eyes move rapidly back and forth under closed eyelids. The person may rapidly ascend through stages 3 and 2 before entering REM sleep. The first episode of REM sleep lasts 10–20 minutes. Then, another interval of NREM sleep follows.

REM and NREM sleep alternate throughout the night. REM periods, which occur approximately every 90 minutes, gradually lengthen, until the final one lasts about 50 minutes. In adults, REM sleep totals 90–120 minutes during a typical sleep period. As a person ages, the average total time spent sleeping decreases, and the percentage of REM sleep declines. As much as 50% of an infant's sleep is REM sleep, as opposed to 35% for 2-year-olds and 25% for adults. Although we do not yet understand the function of REM sleep, the high percentage of REM sleep in infants and children is thought to be important for the maturation of the brain. Neuronal activity is high during REM sleep—brain blood flow and oxygen use are higher during REM sleep than during intense mental or physical activity while awake.

Different parts of the brain mediate NREM and REM sleep. Neurons in the preoptic area of the hypothalamus, the basal forebrain, and the medulla oblongata govern NREM sleep; neurons in the pons and midbrain turn REM sleep on and off. Several lines of evidence suggest the existence of sleep-inducing chemicals in the brain. One apparent sleep-inducer is adenosine, which accumulates during periods of high ATP (adenosine triphosphate) use by the nervous system. Adenosine binds to specific receptors, called A1 receptors, and inhibits certain cholinergic (acetylcholine-releasing) neurons of the RAS that participate in arousal. Thus, activity in the RAS during sleep is low due to the inhibitory effect of adenosine. Caffeine (in coffee) and theophylline (in tea)—substances known for their ability to maintain wakefulness—bind to and block the A1 receptors, preventing adenosine from binding and inducing sleep.

Several physiological changes occur during sleep. Most dreaming occurs during REM sleep, and the EEG readings are similar to those of a person who is awake. With the exception of motor neurons that govern breathing and eye movements, most somatic motor neurons are inhibited during REM sleep, which decreases muscle tone and even paralyzes the skeletal muscles. Many people experience a momentary feeling of paralysis if they are awakened during REM sleep. During sleep, activity in the parasympathetic division of the autonomic nervous system (ANS) increases while sympathetic activity decreases. Heart rate and blood pressure decrease during NREM sleep and decrease further during REM sleep. Increased parasympathetic activity during REM sleep sometimes causes erection of the penis, even when dream content is not sexual. The presence of penile erections during REM sleep in a man with erectile dysfunction (inability to attain an erection while awake) indicates that his problem has a psychological, rather than a physical, cause.

CLINICAL CONNECTION | *Sleep Disorders*

Sleep disorders affect over 70 million Americans each year. Common sleep disorders include insomnia, sleep apnea, and narcolepsy. A person with **insomnia** (in-SOM-nē-a) has difficulty in falling asleep or staying asleep. Possible causes of insomnia include stress, excessive caffeine intake, disruption of circadian rhythms (for example, working the night shift instead of the day shift at your job), and depression. **Sleep apnea** (AP-nē-a) is a disorder in which a person repeatedly stops breathing for 10 or more seconds while sleeping. Most often, it occurs because a loss of muscle tone in pharyngeal muscles allows the airway to collapse. **Narcolepsy** (NAR-kō-lep-sē) is a condition in which REM sleep cannot be inhibited during waking periods. As a result, involuntary periods of sleep that last about 15 minutes occur throughout the day. Recent studies have revealed that people with narcolepsy have a deficiency of the neuropeptide *orexin*, which is also known as **hypocretin.** Orexin is released from certain neurons of the hypothalamus and has a role in promoting wakefulness. •

Learning and Memory

Without memory, we would repeat mistakes and be unable to learn. Similarly, we would not be able to repeat our successes or accomplishments, except by chance. Although both learning and memory have been studied extensively, we still have no completely satisfactory explanation for how we recall information or how we remember events. However, we do know something about how information is acquired and stored, and it is clear that there are different categories of memory.

Learning is the ability to acquire new information or skills through instruction or experience. **Memory** is the process by which information acquired through learning is stored and retrieved. For an experience to become part of memory, it must produce persistent structural and functional changes that represent the experience in the brain. This capability for change associated with learning is termed **plasticity** (plas-TIS-i-tē). Nervous system plasticity underlies our ability to change our behavior in response to stimuli from the external and internal environments. It involves changes in individual neurons—for example, synthesis of different proteins or sprouting of new dendrites—as well as changes in the strengths of synaptic connections among neurons. The parts of the brain known to be involved with memory include the association areas of the frontal, parietal, occipital, and temporal lobes; parts of the limbic system, especially the hippocampus and amygdala; and the diencephalon. The primary somatosensory and primary motor areas in the brain also exhibit plasticity. If a particular body part is used more intensively or in a newly learned activity, such as reading Braille, the cortical areas devoted to that body part gradually expand.

Memory occurs in stages over a period of time. **Immediate memory** is the ability to recall ongoing experiences for a few seconds. It provides a perspective to the present time that allows us to know where we are and what we are doing. **Short-term memory** is the temporary ability to recall a few pieces of information for seconds to minutes. One example is when you look up an unfamiliar telephone number, cross the room to the phone, and then dial the new number. If the number has no special significance, it is usually forgotten within a few seconds. Brain areas involved in immediate and short-term memory include the hippocampus, the mammillary bodies, and two nuclei of the thalamus (anterior and medial nuclei). Some evidence supports the notion that short-term memory depends more on electrical and chemical events in the brain than on structural changes, such as the formation of new synapses.

Information in short-term memory may later be transformed into a more permanent type of memory, called **long-term memory,** which lasts from days to years. If you use that new telephone number often enough, it becomes part of long-term memory. Information in long-term memory usually can be retrieved for use whenever needed. The reinforcement that results from the frequent retrieval of a piece of information is called **memory consolidation.** Long-term memories for information that can be expressed by language, such as a telephone number, apparently are stored in wide regions of the cerebral cortex. Memories for motor skills, such as how to serve a tennis ball, are stored in the basal nuclei and cerebellum as well as in the cerebral cortex.

Although the brain receives many stimuli, we pay attention to only a few of them at a time. It has been estimated that only 1% of all the information that comes to our consciousness is stored as long-term memory. Moreover, much of what goes into long-term memory is eventually forgotten. Memory does not record every detail as if it were magnetic tape. Even when details are lost, we can often explain the idea or concept using our own words and ways of viewing things.

CLINICAL CONNECTION | *Amnesia*

Amnesia (am-NĒ-zē-a = forgetfulness) refers to the lack or loss of memory. It is a total or partial inability to remember past experiences. In *anterograde amnesia,* there is memory loss for events that occur *after* the trauma or disease that caused the condition. In other words, it is an inability to form new memories. In *retrograde amnesia,* there is a memory loss for events that occurred *before* the trauma or disease that caused the condition. In other words, it is an inability to recall past events. •

Several conditions that inhibit the electrical activity of the brain, such as anesthesia, coma, electroconvulsive therapy (ECT), and ischemia of the brain, disrupt retention of recently acquired information without altering previously established long-term memories. People who suffer retrograde amnesia cannot remember anything that occurred during the 30 minutes or so before the amnesia developed. As a person recovers from amnesia, the most recent memories return last.

Anatomical changes occur in neurons when they are stimulated. For example, electron micrographs of neurons subjected to prolonged, intense activity reveal an increase in the number of presynaptic terminals and enlargement of synaptic end bulbs in presynaptic neurons, as well as an increase in the number of dendritic branches in postsynaptic neurons. Moreover, neurons grow new synaptic end bulbs with increasing age, presumably because of increased use. Opposite changes occur when neurons are inactive. For example, the cerebral cortex in the visual area of animals that have lost their eyesight becomes thinner.

A phenomenon called **long-term potentiation (LTP)** (pō-ten′-shē-Ā-shun) is believed to underlie some aspects of memory; transmission at some synapses within the hippocampus is enhanced (potentiated) for hours or weeks after a brief period of high-frequency stimulation. The neurotransmitter released is glutamate, which acts on NMDA* glutamate receptors on the postsynaptic neurons. In some cases, induction of LTP depends on the release of nitric oxide (NO) from the postsynaptic neurons after they have been activated by glutamate. The NO in turn diffuses into the presynaptic neurons and causes LTP.

✔ CHECKPOINT

18. Describe how sleep and wakefulness are related to the reticular activating system (RAS).
19. What are the four stages of non-rapid eye movement (NREM) sleep? How is NREM sleep distinguished from rapid eye movement (REM) sleep?
20. Define memory. What are the three kinds of memory? What is memory consolidation?
21. What is long-term potentiation?

*Named after the chemical *N*-methyl-D-aspartate, which is used to detect this type of glutamate receptor.

DISORDERS: HOMEOSTATIC IMBALANCES

Parkinson Disease

Parkinson disease (PD) is a progressive disorder of the CNS that typically affects its victims around age 60. Neurons that extend from the substantia nigra to the putamen and caudate nucleus, where they release the neurotransmitter dopamine (DA), degenerate in PD. The caudate nucleus of the basal nuclei contains neurons that liberate the neurotransmitter acetylcholine (ACh). Although the level of ACh does not change as the level of DA declines, the imbalance of neurotransmitter activity—too little DA and too much ACh—is thought to cause most of the symptoms. The cause of PD is unknown, but toxic environmental chemicals, such as pesticides, herbicides, and carbon monoxide, are suspected contributing agents. Only 5% of PD patients have a family history of the disease.

In PD patients, involuntary skeletal muscle contractions often interfere with voluntary movement. For instance, the muscles of the upper limb may alternately contract and relax, causing the hand to shake. This shaking, called **tremor,** is the most common symptom of PD. Also, muscle tone may increase greatly, causing rigidity of the involved body part. Rigidity of the facial muscles gives the face a masklike appearance. The expression is characterized by a wide-eyed, unblinking stare and a slightly open mouth with uncontrolled drooling.

Motor performance is also impaired by **bradykinesia** (*brady-* = slow), slowness of movements. Activities such as shaving, cutting food, and buttoning a shirt take longer and become increasingly more difficult as the disease progresses. Muscular movements also exhibit **hypokinesia** (*hypo-* = under), decreasing range of motion. For example, words are written smaller, letters are poorly formed, and eventually handwriting becomes illegible. Often, walking is impaired; steps become shorter and shuffling, and arm swing diminishes. Even speech may be affected.

Treatment of PD is directed toward increasing levels of DA and decreasing levels of ACh. Although people with PD do not manufacture enough dopamine, taking it orally is useless because DA cannot cross the blood–brain barrier. Even though symptoms are partially relieved by a drug developed in the 1960s called levodopa (L-dopa), a precursor of DA, the drug does not slow the progression of the disease. As more and more affected brain cells die, the drug becomes useless. Another drug, called selegiline (Deprenyl®), is used to inhibit monoamine oxidase, an enzyme that degrades dopamine. This drug slows progression of PD and may be used together with levodopa. Anticholinergic drugs such as benztropine and trihexyphenidyl can also be used to block the effects of ACh at some of the synapses between basal nuclei neurons. This helps to restore the balance between ACh and DA. Anticholinergic drugs effectively reduce symptomatic tremor, rigidity, and drooling.

For more than a decade, surgeons have sought to reverse the effects of Parkinson disease by transplanting dopamine-rich fetal nervous tissue into the basal nuclei (usually the putamen) of patients with severe PD. Only a few postsurgical patients have shown any degree of improvement, such as less rigidity and improved quickness of motion. Another surgical technique that has produced improvement for some patients is *pallidotomy,* in which a part of the globus pallidus that generates tremors and produces muscle rigidity is destroyed. In addition, some patients are being treated with a surgical procedure called *deep-brain stimulation (DBS),* which involves the implantation of electrodes into the subthalamic nucleus. The electrical currents released by the implanted electrodes reduce many of the symptoms of PD.

MEDICAL TERMINOLOGY

Acupuncture (ak-ū-PUNK-chur) The use of fine needles (lasers, ultrasound, or electricity) inserted into specific exterior body locations (acupoints) and manipulated to relieve pain and provide therapy for various conditions. The placement of needles may cause the release of neurotransmitters such as endorphins, painkillers that may inhibit pain pathways.

Cerebral palsy (CP) A motor disorder that results in the loss of muscle control and coordination; caused by damage of the motor areas of the brain during fetal life, birth, or infancy. Radiation during fetal life, temporary lack of oxygen during birth, and hydrocephalus during infancy may also cause cerebral palsy.

Coma (KŌ-ma) A state of unconsciousness in which a person's responses to stimuli are reduced or absent. In a *light coma,* an individual may respond to certain stimuli, such as sound, touch, or light, and move his or her eyes, cough, and even murmur. In a *deep coma,* a person does not respond to any stimuli and does not make any movements. Causes of coma include head injuries, cardiac arrest, stroke, brain tumors, infections (encephalitis and meningitis), seizures, alcoholic intoxication, drug overdose, severe lung disorders (chronic obstructive pulmonary disease, pulmonary edema, pulmonary embolism), inhalation of large amounts of carbon monoxide, liver or kidney failure, low or high blood sugar or sodium levels, and low or high body temperature. If brain damage is minor or reversible, a person may come out of a coma and recover fully; if brain damage is severe and irreversible, recovery is unlikely.

Pain threshold The smallest intensity of a painful stimulus at which a person perceives pain. All individuals have the same pain threshold.

Pain tolerance The greatest intensity of painful stimulation that a person is able to tolerate. Individuals vary in their tolerance to pain.

Synesthesia (sin-es-THĒ-zē-a; *syn-* = together; *-aisthesis* = sensation) A condition in which sensations of two or more modalities accompany one another. In some cases, a stimulus for one sensation is perceived as a stimulus for another; for example, a sound produces a sensation of color. In other cases, a stimulus from one part of the body is experienced as coming from a different part.

CHAPTER REVIEW AND RESOURCE SUMMARY

Review | **Resource**

16.1 Sensation

1. Sensation is the conscious or subconscious awareness of changes in the external or internal environment. Perception is the conscious awareness and interpretation of sensations and is primarily a function of the cerebral cortex.
2. The nature of a sensation and the type of reaction generated vary according to the destination of sensory impulses in the CNS.
3. Each different type of sensation is a sensory modality; usually, a given sensory neuron serves only one modality.
4. General senses include somatic senses (touch, pressure, vibration, warmth, cold, pain, itch, tickle, and proprioception) and visceral senses; special senses include the modalities of smell, taste, vision, hearing, and equilibrium.
5. For a sensation to arise, four events typically occur: stimulation, transduction, generation of impulses, and integration.
6. Simple receptors, consisting of free nerve endings and encapsulated nerve endings, are associated with the general senses; complex receptors are associated with the special senses.
7. Sensory receptors respond to stimuli by producing receptor or generator potentials.
8. Table 16.1 summarizes the classification of sensory receptors.
9. Adaptation is a decrease in sensitivity during a long-lasting stimulus. Receptors are either rapidly adapting or slowly adapting.

Resource:
Anatomy Overview - Chemoreceptors
Anatomy Overview - Baroreceptors
Anatomy Overview - Proprioceptors

16.2 Somatic Sensations

1. Somatic sensations include tactile sensations (touch, pressure, vibration, itch, and tickle), thermal sensations (warmth and cold), pain, and proprioception.
2. Receptors for tactile, thermal, and pain sensations are located in the skin, subcutaneous layer, and mucous membranes of the mouth, vagina, and anus.
3. Receptors for touch are (a) Meissner corpuscles and hair root plexuses, which are rapidly adapting, and (b) slowly adapting Merkel discs. Ruffini corpuscles, which are slowly adapting, are sensitive to stretching.
4. Receptors for pressure include Meissner corpuscles, Merkel discs, and Pacinian corpuscles.
5. Receptors for vibration are Meissner corpuscles and Pacinian corpuscles.
6. Itch receptors, tickle receptors, and thermoreceptors are free nerve endings. Cold receptors are located in the stratum basale of the epidermis; warm receptors are located in the dermis.
7. Pain receptors (nociceptors) are free nerve endings that are located in nearly every body tissue.
8. Nerve impulses for fast pain propagate along medium-diameter, myelinated A fibers, whereas those for slow pain conduct along small-diameter, unmyelinated C fibers.
9. Receptors for proprioceptive sensations (position and movement of body parts) are located in muscles, tendons, joints, and the inner ear. Proprioceptors include muscle spindles, tendon organs, joint kinesthetic receptors, and hair cells of the inner ear.
10. Table 16.2 summarizes the somatic sensory receptors and the sensations they convey.

Resource:
Anatomy Overview - Sensory Receptors of the Skin
Anatomy Overview - Proprioceptors
Figure 16.3 - Referred Pain
Figure 16.4 - Two Types of Proprioceptors

16.3 Somatic Sensory Pathways

1. Somatic sensory pathways from receptors to the cerebral cortex involve three-neuron sets: first-order, second-order, and third-order neurons.
2. Axon collaterals (branches) of somatic sensory neurons simultaneously carry signals into the cerebellum and the reticular formation of the brain stem.
3. Nerve impulses for touch, pressure, vibration, and conscious proprioception in the limbs, trunk, neck, and posterior head ascend to the cerebral cortex along the posterior column–medial lemniscus pathway.
4. Nerve impulses for pain, temperature, itch, and tickle from the limbs, trunk, neck, and posterior head ascend to the cerebral cortex along the anterolateral (spinothalamic) pathway.
5. Nerve impulses for most somatic sensations (tactile, thermal, pain, and proprioceptive) from the face, nasal cavity, oral cavity, and teeth ascend to the cerebral cortex along the trigeminothalamic pathway.
6. Specific regions of the primary somatosensory area (postcentral gyrus) of the cerebral cortex receive somatic sensory input from different parts of the body.
7. The neural pathways to the cerebellum are the anterior and posterior spinocerebellar tracts, which transmit impulses for subconscious proprioception from the trunk and lower limbs.
8. Table 16.3 summarizes the major somatic sensory pathways.

Resource:
Anatomy Overview - Cerebrum
Animation - Somatic Sensory Pathways
Figure 16.6 - The Anterolateral Pathway
Figure 16.8 - Somatic Sensory and Somatic Motor Maps in the Cerebral Cortex

Review	Resource
16.4 Somatic Motor Pathways	Animation - Somatic Motor Pathways Figure 16.10 - The Corticospinal Pathways
1. All excitatory and inhibitory signals that control movement converge on motor neurons, also known as lower motor neurons (LMNs) or the final common pathway.	
2. Neurons in four neural circuits, collectively termed the somatic motor pathways, participate in control of movement by providing input to lower motor neurons: local circuit neurons, upper motor neurons, basal nuclei neurons, and cerebellar neurons.	
3. The primary motor area (precentral gyrus) of the cortex is a major control region for executing voluntary movements.	
4. The axons of upper motor neurons (UMNs) extend from the brain to lower motor neurons via direct and indirect motor pathways.	
5. The direct (pyramidal) pathways include the corticospinal pathways and the corticobulbar pathway. The corticospinal pathways convey nerve impulses from the motor cortex to skeletal muscles in the limbs and trunk. The corticobulbar pathway conveys nerve impulses from the motor cortex to skeletal muscles in the head.	
6. Indirect (extrapyramidal) pathways extend from several motor centers of the brain stem into the spinal cord. Indirect pathways include the rubrospinal, tectospinal, vestibulospinal, and medial and lateral reticulospinal tracts.	
7. Table 16.4 summarizes the major somatic motor pathways.	
8. Neurons of the basal nuclei assist movement by providing input to the upper motor neurons. They help initiate and terminate movements, suppress unwanted movements, and establish a normal level of muscle tone.	
9. The cerebellum is active in learning and performing rapid, coordinated, highly skilled movements. It also contributes to maintaining balance and posture.	
16.5 Integrative Functions of the Cerebrum	
1. Sleep and wakefulness are integrative functions that are controlled by the suprachiasmatic nucleus and the reticular activating system (RAS).	
2. Non-rapid eye movement (NREM) sleep consists of four stages.	
3. Most dreaming occurs during rapid eye movement (REM) sleep.	
4. Memory, the ability to store and recall thoughts, involves persistent changes in the brain, a capability called plasticity. Three types are immediate, short-term, and long-term memory.	

SELF-QUIZ QUESTIONS

Fill in the blanks in the following statements.

1. _____ is the conscious or subconscious awareness of external or internal stimuli; _____ is the conscious awareness and interpretation of sensory input.

2. The term used to describe the crossing over of axons from one side of the brain or spinal cord to the other side is _____.

Indicate whether the following statements are true or false.

3. Touch, pressure, and pain are all classified as tactile sensations.

4. Awakening from sleep involves increased activity in the reticular activating system.

Choose the one best answer to the following questions.

5. A nurse touches the lower back of a patient, but the patient does not feel the sensation. Which of the following could explain the lack of sensation? (1) The stimulus was not in the receptive field. (2) The generator potential has not reached threshold. (3) There is damage to the somatosensory region of the cerebral cortex. (4) The nurse was stimulating a proprioceptor. (5) A slowly adapting receptor has been stimulated.
(a) 1, 3 and 5 (b) 3, 4, and 5 (c) 1, 2, and 3
(d) 2, 3, and 4 (e) 1 only

6. Which of the following statements is *false*? (1) Upper motor neurons transmit impulses from the CNS to skeletal muscle fibers. (2) Lower motor neurons have their cell bodies in the brain stem and spinal cord. (3) Local circuit neurons receive input from somatic sensory receptors and help coordinate rhythmic activity in specific muscle groups. (4) The activity of upper motor neurons is influenced by both the basal ganglia and cerebellum. (5) The cerebellum helps to monitor differences between intended movements and actual movements for coordination, posture, and balance.

7. Which of the following statements are *true*? (1) Slow pain is a result of impulse propagation along myelinated A nerve fibers. (2) Visceral pain occurs when nociceptors in the skin are stimulated. (3) Referred pain is pain felt in an area far from the stimulated organ. (4) Nociceptors exhibit very little adaptation. (5) Nociceptors are located in every body tissue.
(a) 1, 3, 4, and 5 (b) 2, 3, and 5 (c) 1 and 5
(d) 3 and 4 (e) 3, 4, and 5

8. You cannot "hear" with your eyes because
(a) hearing is a somatic sense and vision is a special sense.
(b) the sensory neurons for sight carry information only for the modality of vision.

(c) the impulses for hearing are transmitted to the somatosensory area of the cerebral cortex.
(d) hearing receptors are selective and vision receptors are not.
(e) hearing receptors produce a generator potential and vision receptors produce a receptor potential.

9. Which of the following statements is *false*?
(a) First-order sensory neurons carry signals from the somatic receptors into either the brain stem or spinal cord.
(b) Second-order neurons carry signals from the spinal cord and brain stem to the thalamus.
(c) Third-order neurons project to the primary somatosensory area of the cortex, where conscious perception of the sensation results.
(d) The somatic sensory pathways to the cerebellum are the posterior column–medial lemniscus pathway and the anterolateral pathway.
(e) Axons of second-order neurons decussate (cross) in the spinal cord or brain stem before ascending to the thalamus.

10. Which of the following is not an aspect of cerebellar function?
(a) monitoring movement intention
(b) monitoring actual movement
(c) comparing intent with actual performance
(d) sending out corrective signals
(e) directing sensory input to effectors

11. During REM sleep (1) neuronal activity in the pons and midbrain is high, (2) most somatic motor neurons are inhibited, (3) most dreaming occurs, (4) sleepwalking can occur, (5) there is an increase in heart rate and blood pressure.
(a) 1, 2, 4, and 5 (b) 2, 3, and 5 (c) 1, 2, 3, 4, and 5
(d) 2, 3, and 4 (e) 1, 2, and 3

12. Which of the following statements is *incorrect*?
(a) The graded potentials produced by receptors that serve the senses of touch, pressure, stretching, vibration, pain, proprioception, and smell are generator potentials.
(b) The graded potentials produced by receptors that serve the special senses of vision, hearing, equilibrium, and taste are receptor potentials.
(c) When a generator potential is large enough to reach threshold, it generates one or more nerve impulses in its first-order sensory neuron.
(d) A receptor potential generates nerve impulses in a second-order neuron.
(e) The amplitude of both generator and receptor potentials varies with the intensity of the stimulus.

13. Match the following (some answers may be used more than once):

____ (a) receptors located in muscles, tendons, joints, and the inner ear
____ (b) receptors located in blood vessels, visceral organs, muscles, and the nervous system
____ (c) receptors that detect temperature changes
____ (d) receptors that detect light that strikes the retina of the eye
____ (e) receptors located at or near the external surface of the body
____ (f) bare dendrites associated with pain, thermal, tickle, itch, and some touch sensations
____ (g) receptors that provide information about body position, muscle tension, and position and activity of joints
____ (h) receptors that sense osmotic pressures of body fluids
____ (i) receptors that detect chemicals in the mouth, nose, and body fluids
____ (j) receptors that detect mechanical pressure or stretching
____ (k) receptors that respond to stimuli resulting from physical or chemical damage to tissues
____ (l) dendrites enclosed in a connective tissue capsule

(1) exteroceptors
(2) interoceptors
(3) proprioceptors
(4) mechanoreceptors
(5) thermoreceptors
(6) nociceptors
(7) photoreceptors
(8) chemoreceptors
(9) free nerve endings
(10) encapsulated nerve endings
(11) osmoreceptors

14. Match the following:

____ (a) specialized groupings of muscle fibers interspersed among regular skeletal muscle fibers and oriented parallel to them; monitor changes in the length of a skeletal muscle
____ (b) inform the CNS about changes in muscle tension
____ (c) widely distributed free nerve ending receptors for pain
____ (d) encapsulated receptors for touch located in the dermal papillae; found in hairless skin, eyelids, tip of the tongue, and lips
____ (e) lamellated corpuscles that detect pressure
____ (f) type II cutaneous mechanoreceptors; most sensitive to stretching that occurs as digits or limbs are moved
____ (g) located in the stratum basale and activated by low temperatures
____ (h) located in the dermis and activated by high temperatures
____ (i) found within and around the articular capsules of synovial joints; respond to pressure and acceleration and deceleration of joints
____ (j) type I cutaneous mechanoreceptors that function in touch

(1) Meissner corpuscles
(2) Merkel discs
(3) Ruffini corpuscles
(4) Pacinian corpuscles
(5) cold receptors
(6) warm receptors
(7) nociceptors
(8) tendon organs
(9) joint kinesthetic receptors
(10) muscle spindles

15. Match the following:

____ (a) located in the precentral gyrus, this is the major control region of the cerebral cortex for initiation of voluntary movements

____ (b) direct pathways conveying impulses from the cerebral cortex to the spinal cord that result in precise, voluntary movements

____ (c) contains motor neurons that control skilled movements of the hands and feet

____ (d) tracts include rubrospinal, tectospinal, vestibulospinal, lateral reticulospinal, and medial reticulospinal

____ (e) contain neurons that help initiate and terminate movements; can suppress unwanted movements; influence muscle tone

____ (f) carries impulses for pain, temperature, tickle, and itch

____ (g) the major routes relaying proprioceptive input to the cerebellum; critical for posture, balance, and coordination of skilled movements

____ (h) composed of axons of first-order neurons; include the gracile fasciculus and cuneate fasciculus

____ (i) contains motor neurons that coordinate movements of the axial skeleton

____ (j) contains axons that convey impulses for precise, voluntary movements of the eyes, tongue, and neck, plus chewing, facial expression, and speech

____ (k) conveys sensations of touch, conscious proprioception, pressure, and vibration to the cerebral cortex

____ (l) carries impulses for most somatic sensations from the face, nasal cavity, oral cavity, and teeth

(1) posterior column
(2) anterolateral (spinothalamic) pathway
(3) spinocerebellar tracts
(4) lateral corticospinal tract
(5) anterior corticospinal tract
(6) corticobulbar tracts
(7) extrapyramidal pathways
(8) pyramidal pathways
(9) primary motor area
(10) basal nuclei
(11) posterior column–medial lemniscus pathway
(12) trigeminothalamic pathway

CRITICAL THINKING QUESTIONS

1. When Joni first stepped onto the sailboat, she smelled the tangy sea air and felt the motion of water beneath her feet. After a few minutes, she no longer noticed the smell, but unfortunately she was aware of the rolling motion for hours. What types of receptors are involved in smell and detection of motion? Why did her sensation of smell fade but the rolling sensation remain?

2. Monique sticks her left hand into a hot tub heated to about 43°C (110°F) in order to decide if she wants to enter the hot tub. Trace the pathway involved in transmitting the sensation of heat from her left hand to the somatosensory area in the cerebral cortex.

3. Marvin has had trouble sleeping. Last night his mother found him sleepwalking and gently led him back to his bed. When Marvin was awakened by his alarm clock the next day, he had no recollection of sleepwalking and, in fact, told his mother about the vivid dreams he had. What specific stages of sleep did Marvin undergo during the night? What neurological mechanism awakened Marvin in the morning?

? ANSWERS TO FIGURE QUESTIONS

16.1 The special senses of vision, taste, hearing, and equilibrium are served by separate sensory cells.

16.2 Pain, thermal sensations, and tickle and itch arise with activation of different free nerve endings.

16.3 The kidneys have the broadest area for referred pain.

16.4 Muscle spindles are activated when the central areas of the intrafusal fibers are stretched.

16.5 The posterior columns consist of the cuneate fasciculus and the gracile fasciculus.

16.6 Damage to the right spinothalamic tract could result in loss of pain, thermal, itch, and tickle sensations on the left side of the body.

16.7 The left trigeminal (V) nerve conveys nerve impulses for most somatic sensations from the left side of the face into the pons.

16.8 The hand has a larger representation in the motor area than in the somatosensory area, which implies greater precision in the hand's movement control than fine ability in its sensation.

16.9 Cerebral cortex UMNs are essential for the execution of voluntary movements of the body. Brain stem UMNs regulate muscle tone, control postural muscles, and help maintain balance and orientation of the head and body.

16.10 The lateral corticospinal tract conducts impulses that result in contractions of the muscles in the distal parts of the limbs.

16.11 The axons of the corticobulbar tract terminate in the motor nuclei of the following cranial nerves: oculomotor (III), trochlear (IV), trigeminal (V), abducens (VI), facial (VII), glossopharyngeal (IX), vagus (X), accessory (XI), and hypoglossal (XII).

16.12 The anterior and posterior spinocerebellar tracts carry information from proprioceptors in joints and muscles to the cerebellum.

16.13 Olfactory input does not stimulate the RAS; a smoke detector responds to smoke by sounding a loud bell or buzzer, which awakens sleepers by providing auditory input that stimulates the RAS.

17 THE SPECIAL SENSES

THE SPECIAL SENSES AND HOMEOSTASIS *Sensory organs have special receptors that allow us to smell, taste, see, hear, and maintain equilibrium or balance. Information conveyed from these receptors to the central nervous system is used to help maintain homeostasis.*

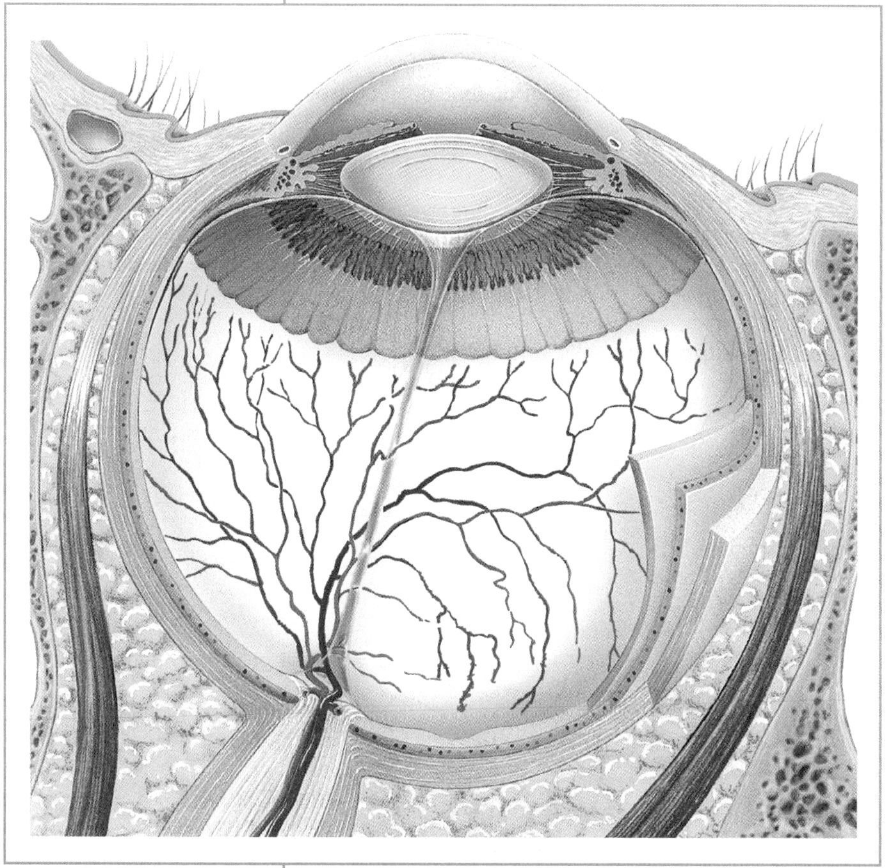

Recall from Chapter 16 that the general senses include somatic senses (tactile, thermal, pain, and proprioceptive) and visceral sensations. As you learned in that chapter, receptors for the general senses are scattered throughout the body and are relatively simple in structure. Receptors for the special senses—smell, taste, vision, hearing, and equilibrium—are anatomically distinct from one another and are concentrated in specific locations in the head. They are usually embedded in the epithelial tissue within complex sensory organs such as the eyes and ears. Neural pathways for the special senses are also more complex than those for the general senses.

In this chapter we examine the structure and function of the special sense organs, and the pathways involved in conveying their information to the central nervous system. **Ophthalmology** (of-thal-MOL-ō-jē; *ophthalmo-* = eye; *-logy* = study of) is the science that deals with the eyes and their disorders. The other special senses are, in large part, the concern of **otorhinolaryngology** (ō-tō-rī′-nō-lar-in-GOL-ō-jē; *oto-* = ear; *-rhino-* = nose; *-laryngo-* = larynx), the science that deals with the ears, nose, pharynx (throat), and larynx (voice box) and their disorders.

? ***Did you ever wonder how LASIK is performed?***

17.1 OLFACTION: SENSE OF SMELL

OBJECTIVES

- Describe the structure of the olfactory receptors and other cells involved in olfaction.
- Outline the neural pathway for olfaction.

Last night as you were studying anatomy in the lounge, all of a sudden you were surrounded by the smell of freshly baked brownies. When you followed your nose and begged for one, biting into the moist, flavorful treat transported you back 10 years into your mother's kitchen. Both smell and taste are chemical senses; the sensations arise from the interaction of molecules with smell or taste receptors. To be detected by either sense, the stimulating molecules must be dissolved. Because impulses for smell and taste propagate to the limbic system (and to higher cortical areas as well), certain odors and tastes can evoke strong emotional responses or a flood of memories.

Anatomy of Olfactory Receptors

It is estimated that humans can recognize about 10,000 different odors. To make this possible, the nose contains 10 million to 100 million receptors for the sense of smell or **olfaction** (ōl-FAK-shun; *olfact-* = smell), contained within a region called the olfactory epithelium. With a total area of 5 cm^2 (a little less than 1 $in.^2$), the **olfactory epithelium** (ōl-FAK-tō-rē) occupies the superior part of the nasal cavity, covering the inferior surface of the cribriform plate and extending along the superior nasal concha (Figure 17.1a). The olfactory epithelium consists of three kinds of cells: olfactory receptors, supporting cells, and basal cells (Figure 17.1b).

Olfactory receptors are the first-order neurons of the olfactory pathway. Each olfactory receptor is a bipolar neuron with an exposed knob-shaped dendrite and an axon projecting through the cribriform plate and ending in the olfactory bulb. The parts of the olfactory receptors that respond to inhaled chemicals are the **olfactory hairs,** cilia that project from the dendrite. (Recall that

Figure 17.1 Olfactory epithelium and olfactory pathway. (a) Location of olfactory epithelium in nasal cavity. (b) Anatomy of olfactory receptors, consisting of first-order neurons whose axons extend through the cribriform plate and terminate in the olfactory bulb. (c) Histology of the olfactory epithelium. (d) Olfactory pathway.

The olfactory epithelium consists of olfactory receptors, supporting cells, and basal cells.

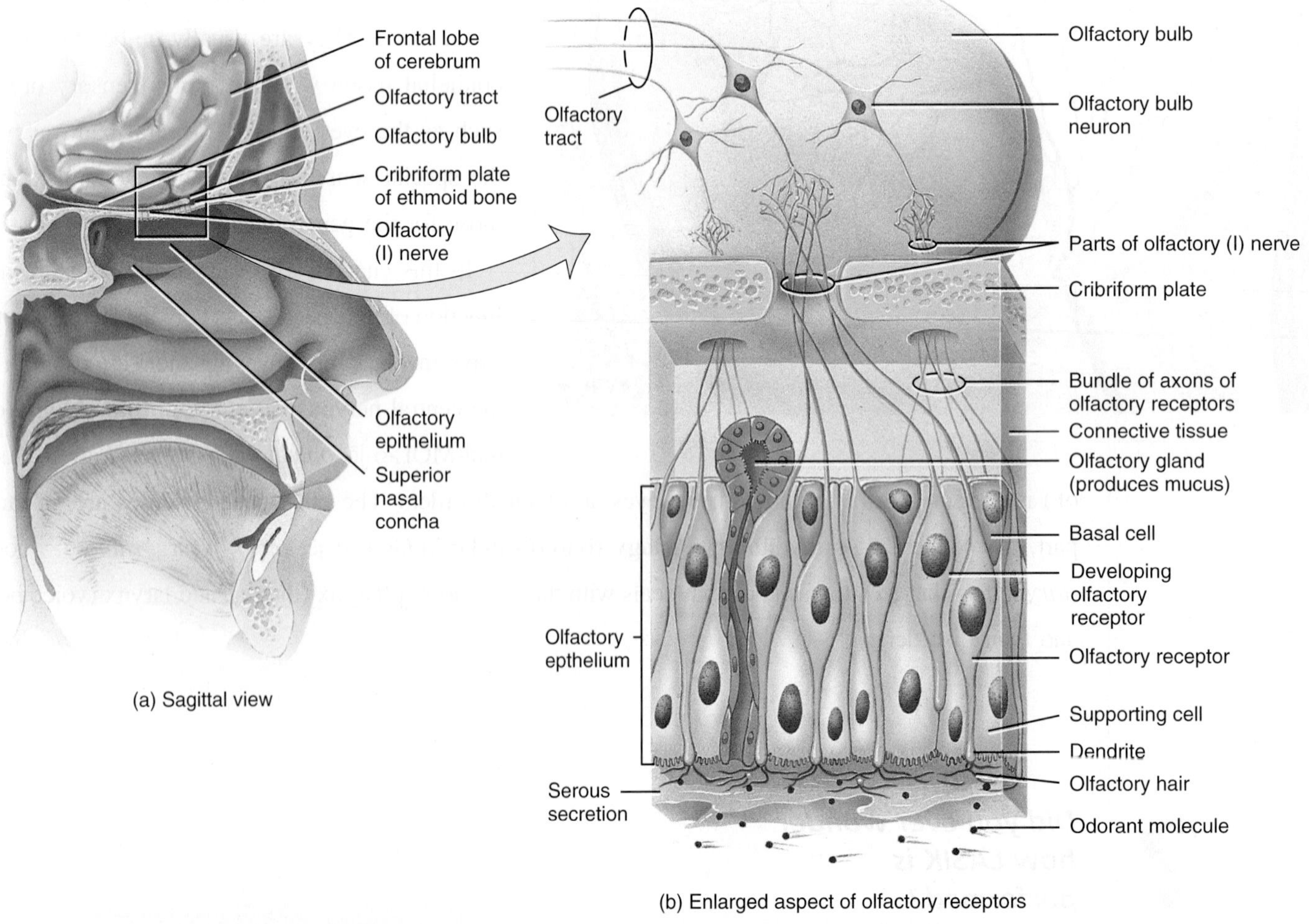
Frontal lobe of cerebrum
Olfactory tract
Olfactory bulb
Cribriform plate of ethmoid bone
Olfactory (I) nerve
Olfactory epithelium
Superior nasal concha
(a) Sagittal view
Olfactory tract
Olfactory bulb
Olfactory bulb neuron
Parts of olfactory (I) nerve
Cribriform plate
Bundle of axons of olfactory receptors
Connective tissue
Olfactory gland (produces mucus)
Basal cell
Developing olfactory receptor
Olfactory receptor
Supporting cell
Dendrite
Olfactory hair
Odorant molecule
Olfactory epthelium
Serous secretion
(b) Enlarged aspect of olfactory receptors

transduction is the conversion of stimulus energy into a graded potential in a sensory receptor.) Chemicals that have an odor and can therefore stimulate the olfactory hairs are called **odorants.** Olfactory receptors respond to the chemical stimulation of an odorant molecule by producing a generator potential, thus initiating the olfactory response.

Supporting cells are columnar epithelial cells of the mucous membrane lining the nose. They provide physical support, nourishment, and electrical insulation for the olfactory receptors, and they help detoxify chemicals that come in contact with the olfactory epithelium. **Basal cells** are stem cells located between the bases of the supporting cells. They continually undergo cell division to produce new olfactory receptors, which live for only a month or so before being replaced. This process is remarkable considering that olfactory receptors are neurons, and as you have already learned, mature neurons are generally not replaced.

Within the connective tissue that supports the olfactory epithelium are **olfactory (Bowman's) glands,** which produce mucus that is carried to the surface of the epithelium by ducts. The secretion moistens the surface of the olfactory epithelium and dissolves odorants so that transduction can occur. Both supporting cells of the nasal epithelium and olfactory glands are innervated by autonomic neurons within branches of the facial (VII) nerve, which can be stimulated by certain chemicals. Impulses in these nerves in turn stimulate the lacrimal glands in the eyes and nasal mucous glands. The result is tears and a runny nose after inhaling substances such as pepper or the vapors of household ammonia.

Physiology of Olfaction

Many attempts have been made to distinguish among and classify "primary" sensations of smell. Genetic evidence now suggests the existence of hundreds of primary odors. Our ability to recognize about 10,000 different odors probably depends on patterns of activity in the brain that arise from activation of many different combinations of olfactory receptors.

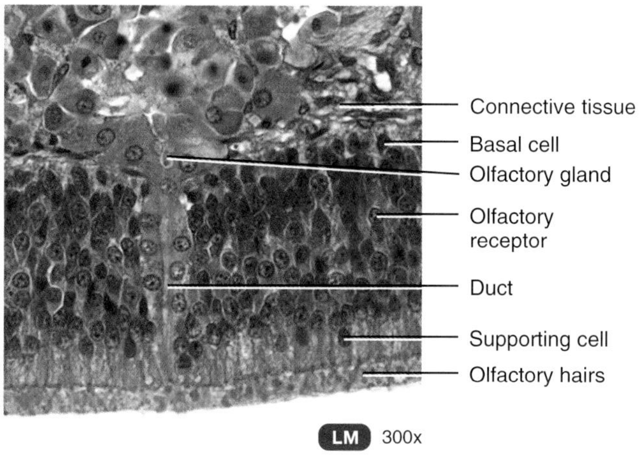

(c) Histology of olfactory epithelium

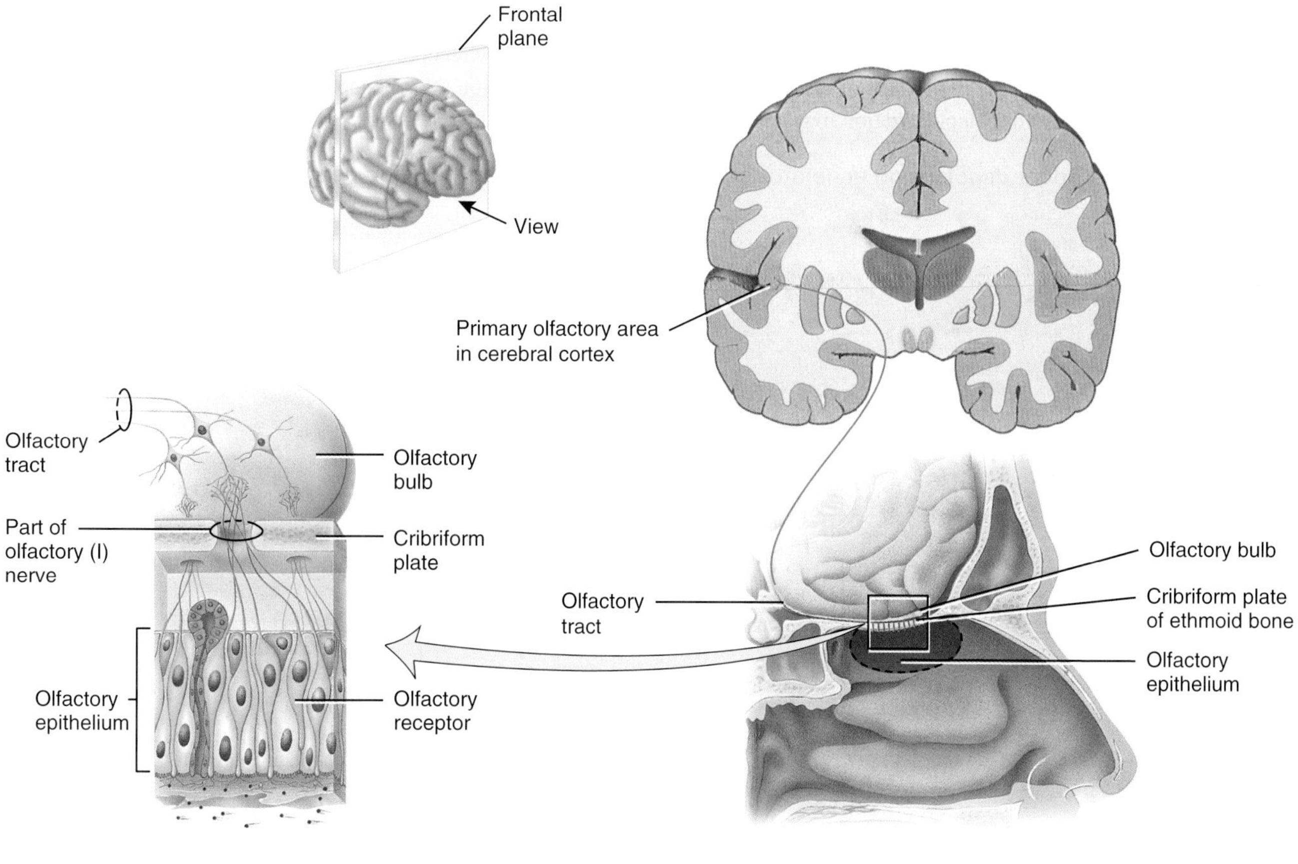

(d) Olfactory pathway

What is the life span of an olfactory receptor?

Olfactory receptors react to odorant molecules in the same way that most sensory receptors react to their specific stimuli: A generator potential (depolarization) develops and triggers one or more nerve impulses. In some cases, an odorant binds to an *olfactory receptor protein* in the plasma membrane of an olfactory hair (Figure 17.2). The olfactory receptor protein is coupled to a membrane protein called a G protein, which in turn activates the enzyme adenylate cyclase (see Section 18.4). The result is the following chain of events: production of cyclic adenosine monophosphate (cAMP) → opening of sodium ion (Na^+) channels → inflow of Na^+ → depolarizing generator potential → generation of nerve impulse and propagation along an axon of the olfactory receptor.

Odor Thresholds and Adaptation

Olfaction, like all the special senses, has a low threshold. Only a few molecules of certain substances need be present in air to be perceived as an odor. A good example is the chemical methyl mercaptan, which smells like rotten cabbage and can be detected in concentrations as low as 1/25 billionth of a milligram per milliliter of air. Because the natural gas used for cooking and heating is odorless but lethal and potentially explosive if it accumulates, a small amount of methyl mercaptan is added to natural gas to provide olfactory warning of gas leaks.

Adaptation (decreasing sensitivity) to odors occurs rapidly. Olfactory receptors adapt by about 50% in the first second or so after stimulation but adapt very slowly thereafter. Still, complete insensitivity to certain strong odors occurs about a minute after exposure. Apparently, reduced sensitivity involves an adaptation process in the central nervous system as well.

The Olfactory Pathway

On each side of the nose, about 40 bundles of the slender, unmyelinated axons of olfactory receptors extend through about 20 olfactory foramina in the cribriform plate of the ethmoid bone (see Figure 17.1b). These 40 or so bundles of axons collectively form the right and left **olfactory (I) nerves.** The olfactory nerves terminate in the brain in paired masses of gray matter called the **olfactory bulbs,** which are located below the frontal lobes of the cerebrum and lateral to the crista galli of the ethmoid bone. Within the olfactory bulbs, the axon terminals of olfactory receptors form synapses with the dendrites and cell bodies of olfactory bulb neurons in the olfactory pathway.

Axons of olfactory bulb neurons extend posteriorly and form the **olfactory tract** (see Figure 17.1b). Some of the axons of the olfactory tract project to the **primary olfactory area** of the cerebral cortex; located at the inferior and medial surface of the temporal lobe, the primary olfactory area is where conscious awareness

Figure 17.2 Olfactory transduction. Binding of an odorant molecule to an olfactory receptor protein activates a G protein and adenylate cyclase, resulting in the production of cAMP. Cyclic AMP opens sodium ion (Na^+) channels, and Na^+ ions enter the olfactory receptor. The resulting depolarization may generate an action potential, which propagates along the axon of the olfactory receptor.

6 Odorants can produce depolarizing generator potentials, which can lead to action potentials.

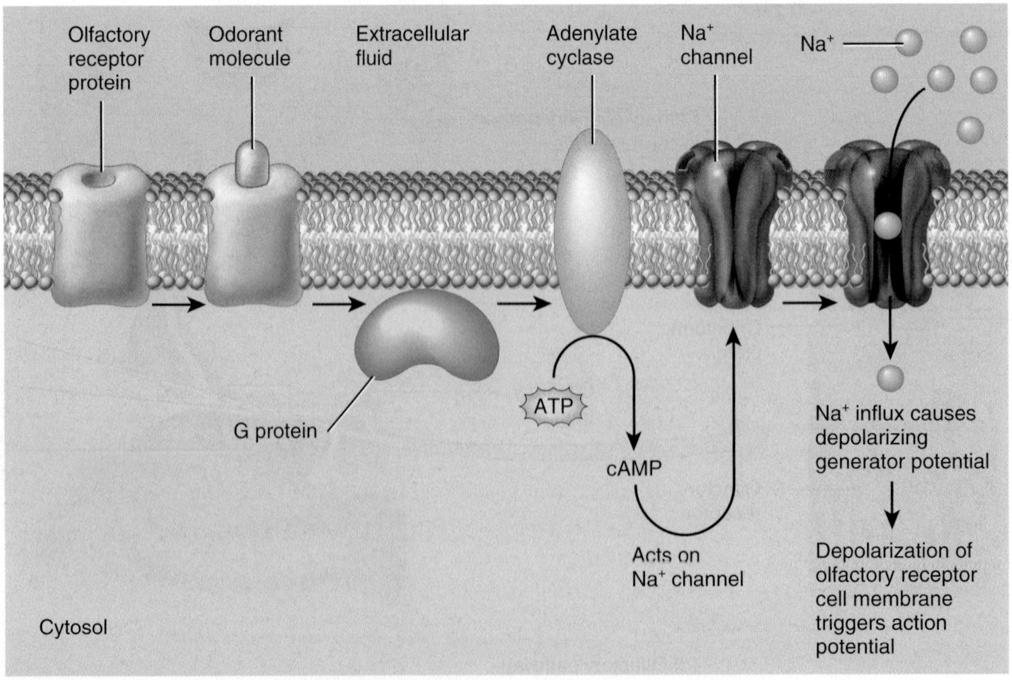

In which part of an olfactory receptor does olfactory transduction occur?

of smell begins (see Figure 17.1d). Olfactory sensations are the only sensations that reach the cerebral cortex without first synapsing in the thalamus. Other axons of the olfactory tract project to the limbic system and hypothalamus; these connections account for our emotional and memory-evoked responses to odors. Examples include sexual excitement on smelling a certain perfume, nausea on smelling a food that once made you violently ill, or an odor-evoked memory of a childhood experience.

From the primary olfactory area, pathways also extend to the **frontal lobe.** An important region for odor identification and discrimination is the **orbitofrontal area** (area 11 in Figure 14.15). People who suffer damage in this area have difficulty identifying different odors. Positron emission tomography (PET) studies suggest some degree of hemispheric lateralization: The orbitofrontal area of the *right* hemisphere exhibits greater activity during olfactory processing.

CLINICAL CONNECTION | *Hyposmia*

Women often have a keener sense of smell than men do, especially at the time of ovulation. Smoking seriously impairs the sense of smell in the short term and may cause long-term damage to olfactory receptors. With aging the sense of smell deteriorates. **Hyposmia** (hī-POZ-mē-a; *osmi* = smell, odor), a reduced ability to smell, affects half of those over age 65 and 75% of those over age 80. Hyposmia also can be caused by neurological changes, such as a head injury, Alzheimer disease, or Parkinson disease; certain drugs, such as antihistamines, analgesics, or steroids; and the damaging effects of smoking. •

✔CHECKPOINT

1. How do basal cells contribute to olfaction?
2. What is the sequence of events from the binding of an odorant molecule to an olfactory hair to the arrival of a nerve impulse in the orbitofrontal area?

17.2 GUSTATION: SENSE OF TASTE

OBJECTIVE

- Describe the structure of the gustatory receptors and the neural pathway for gustation.

Taste or **gustation** (gus-TĀ-shun; *gust-* = taste), like olfaction, is a chemical sense. However, it is much simpler than olfaction in that only five primary tastes can be distinguished: *sour, sweet, bitter, salty,* and *umami* (ū-MAM-ē). The umami taste, more recently discovered than the other tastes, was first reported by Japanese scientists and is described as "meaty" or "savory." Umami is believed to arise from taste receptors that are stimulated by monosodium glutamate (MSG), a substance naturally present in many foods and added to others as a flavor enhancer. All other flavors, such as chocolate, pepper, and coffee, are combinations of the five primary tastes, plus accompanying olfactory and tactile (touch) sensations. Odors from food can pass upward from the mouth into the nasal cavity, where they stimulate olfactory receptors. Because olfaction is much more sensitive than taste, a given concentration of a food substance may stimulate the olfactory system thousands of times more strongly than it stimulates the gustatory system. When you have a cold or are suffering from allergies and cannot taste your food, it is actually olfaction that is blocked, not taste.

Anatomy of Taste Buds and Papillae

The receptors for sensations of taste are located in the taste buds (Figure 17.3). Most of the nearly 10,000 taste buds of a young adult are on the tongue, but some are found on the soft palate (posterior portion of the roof of the mouth), pharynx (throat), and epiglottis (cartilage lid over voice box). The number of taste buds declines with age. Each **taste bud** is an oval body consisting of three kinds of epithelial cells: supporting cells, gustatory receptor cells, and basal cells (see Figure 17.3c). The **supporting cells** surround about 50 **gustatory receptor cells** (GUS-ta-tōr-ē) in each taste bud. A single, long microvillus, called a **gustatory hair,** projects from each gustatory receptor cell to the external surface through the **taste pore,** an opening in the taste bud. **Basal cells,** stem cells found at the periphery of the taste bud near the connective tissue layer, produce supporting cells, which then develop into gustatory receptor cells. Each gustatory receptor cell has a life span of about 10 days. This is why it doesn't take taste sensors on the tongue too long to recover from being burned by that too hot cup of coffee or cocoa. At their base, the gustatory receptor cells synapse with dendrites of the first-order neurons that form the first part of the gustatory pathway. The dendrites of each first-order neuron branch profusely and contact many gustatory receptor cells in several taste buds.

Taste buds are found in elevations on the tongue called **papillae** (pa-PIL-ē; singular is *papilla*), which increase the surface area and provide a rough texture to the upper surface of the tongue (Figure 17.3a, b). Three types of papillae contain taste buds:

1. About 12 very large, circular **vallate (circumvallate) papillae** (VAL-āt = wall-like) form an inverted V-shaped row at the back of the tongue. Each of these papillae houses 100–300 taste buds.
2. **Fungiform papillae** (FUN-ji-form = mushroomlike) are mushroom-shaped elevations scattered over the entire surface of the tongue that contain about five taste buds each.
3. **Foliate papillae** (FO-lē-āt = leaflike) are located in small trenches on the lateral margins of the tongue, but most of their taste buds degenerate in early childhood.

In addition, the entire surface of the tongue has **filiform papillae** (FIL-i-form = threadlike). These pointed, threadlike structures contain tactile receptors but no taste buds. They increase friction between the tongue and food, making it easier for the tongue to move food in the oral cavity.

Physiology of Gustation

Chemicals that stimulate gustatory receptor cells are known as **tastants.** Once a tastant is dissolved in saliva, it can make contact with the plasma membrane of the gustatory hairs, which are the

Figure 17.3 The relationship of gustatory receptor cells in taste buds to tongue papillae.

6 Gustatory receptor cells are located in taste buds.

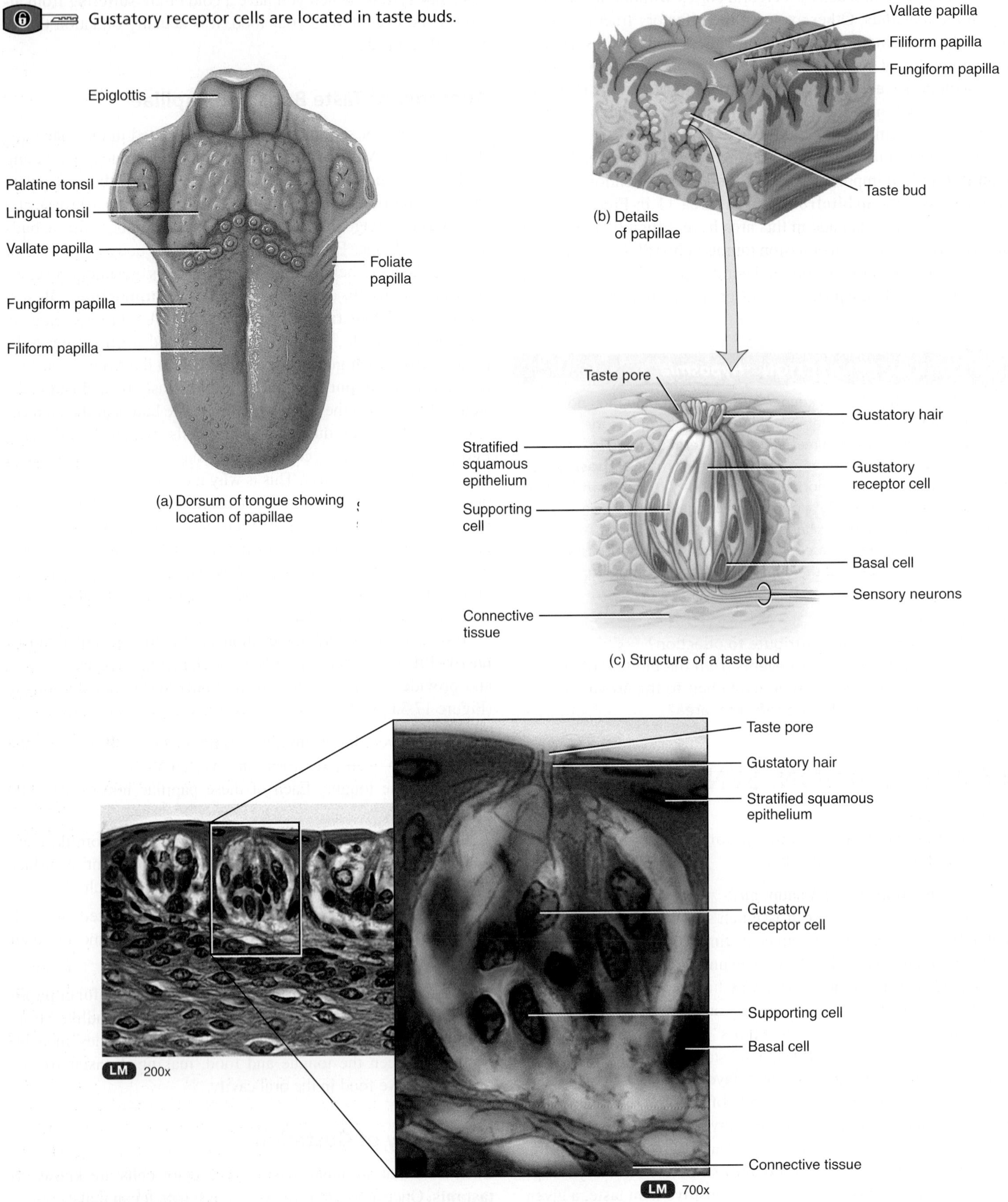

Epiglottis
Palatine tonsil
Lingual tonsil
Vallate papilla
Foliate papilla
Fungiform papilla
Filiform papilla
Vallate papilla
Filiform papilla
Fungiform papilla
Taste bud
Taste pore
Gustatory hair
Stratified squamous epithelium
Gustatory receptor cell
Supporting cell
Basal cell
Sensory neurons
Connective tissue
LM 200x
Taste pore
Gustatory hair
Stratified squamous epithelium
Gustatory receptor cell
Supporting cell
Basal cell
Connective tissue
LM 700x

(a) Dorsum of tongue showing location of papillae

(b) Details of papillae

(c) Structure of a taste bud

(d) Histology of a taste bud from a vallate papilla

sites of taste transduction. The result is a receptor potential that stimulates exocytosis of synaptic vesicles from the gustatory receptor cell. In turn, the liberated neurotransmitter molecules trigger nerve impulses in the first-order sensory neurons that synapse with gustatory receptor cells.

The receptor potential arises differently for different tastants. The sodium ions (Na^+) in a salty food enter gustatory receptor cells via Na^+ channels in the plasma membrane. The accumulation of Na^+ inside causes depolarization that leads to release of neurotransmitter. The hydrogen ions (H^+) in sour tastants may flow into gustatory receptor cells via H^+ channels. They also influence opening and closing of other types of ion channels. Again, the result is depolarization and the liberation of neurotransmitter.

Other tastants, responsible for stimulating sweet, bitter, and umami tastes, do not themselves enter gustatory receptor cells. Rather, they bind to receptors on the plasma membrane that are linked to G proteins. The G proteins then activate several different chemicals known as second messengers inside the gustatory receptor cell. Different second messengers cause depolarization in different ways, but the result is the same—release of neurotransmitter.

If all tastants cause release of neurotransmitter from many gustatory receptor cells, why do foods taste different? The answer to this question is thought to lie in the patterns of nerve impulses in groups of first-order taste neurons that synapse with the gustatory receptor cells. Different tastes arise from activation of different groups of taste neurons. In addition, although each individual gustatory receptor cell responds to more than one of the five primary tastes, it may respond more strongly to some tastants than to others.

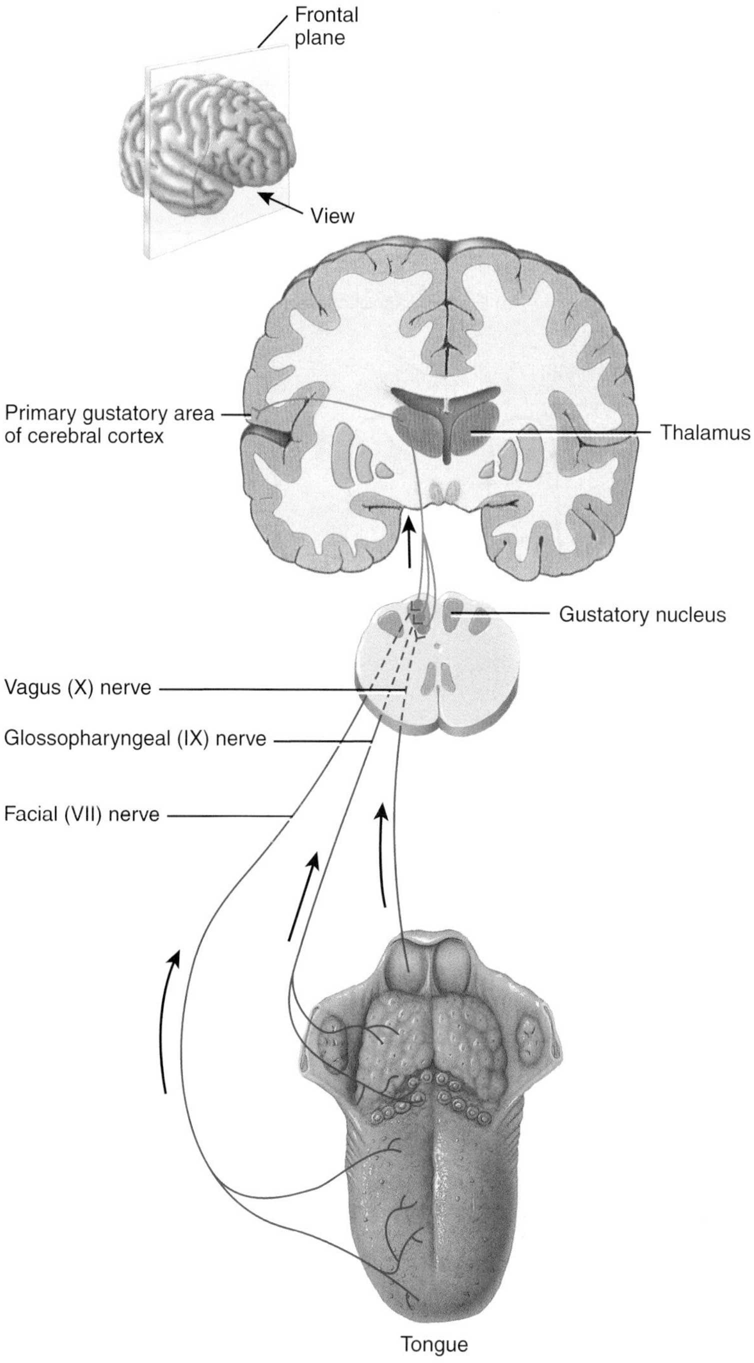

(e) Gustatory pathway

What role do basal cells play in taste buds?

Taste Thresholds and Adaptation

The threshold for taste varies for each of the primary tastes. The threshold for bitter substances, such as quinine, is lowest. Because poisonous substances often are bitter, the low threshold (or high sensitivity) may have a protective function. The threshold for sour substances, such as lemon, as measured by using hydrochloric acid, is somewhat higher. The thresholds for salty substances, represented by sodium chloride, and for sweet substances, as measured by using sucrose, are similar, and are higher than those for bitter or sour substances.

Complete adaptation to a specific taste can occur in 1–5 minutes of continuous stimulation. Taste adaptation is due to changes that occur in the taste receptors, in olfactory receptors, and in neurons of the gustatory pathway in the CNS.

The Gustatory Pathway

Three cranial nerves contain axons of the first-order gustatory neurons that innervate the taste buds. The **facial (VII) nerve** serves taste buds in the anterior two-thirds of the tongue; the **glossopharyngeal (IX) nerve** serves taste buds in the posterior one-third of the tongue; and the **vagus (X) nerve** serves taste buds in the throat and epiglottis (Figure 17.3e). From the taste buds, nerve impulses propagate along these cranial nerves to the **gustatory nucleus** in the medulla oblongata. From the medulla, some axons carrying taste signals project to the **limbic system** and the **hypothalamus**; others project to the **thalamus.** Taste signals that project from the thalamus to the **primary gustatory area** in the parietal lobe of the cerebral cortex (see area 43 in Figure 14.15) give rise to the conscious perception of taste.

CLINICAL CONNECTION | *Taste Aversion*

Probably because of taste projections to the hypothalamus and limbic system, there is a strong link between taste and pleasant or unpleasant emotions. Sweet foods evoke reactions of pleasure while bitter ones cause expressions of disgust even in newborn babies. This phenomenon is the basis for **taste aversion,** in which people and animals quickly learn to avoid a food if it upsets the digestive system. The advantage of avoiding foods that cause such illness is longer survival. However, the drugs and radiation treatments used to combat cancer often cause nausea and gastrointestinal upset regardless of what foods are consumed. Thus, cancer patients may lose their appetite because they develop taste aversions for most foods. •

✔CHECKPOINT

3. How do olfactory receptors and gustatory receptor cells differ in structure and function?
4. Trace the path of a gustatory stimulus from contact of a tastant with saliva to the primary gustatory area in the cerebral cortex.
5. Compare the olfactory and gustatory pathways.

17.3 VISION

OBJECTIVES

- List and describe the accessory structures of the eye and the structural components of the eyeball.
- Discuss image formation by describing refraction, accommodation, and constriction of the pupil.
- Describe the processing of visual signals in the retina and the neural pathway for vision.

Vision, the act of seeing, is extremely important to human survival. More than half the sensory receptors in the human body are located in the eyes, and a large part of the cerebral cortex is devoted to processing visual information. In this section, we examine electromagnetic radiation, the accessory structures of the eye, the eyeball itself, the formation of visual images, the physiology of vision, and the visual pathway from the eye to the brain.

Electromagnetic Radiation

Electromagnetic radiation (e-lek′-trō-mag′-NET-ik) is energy in the form of waves that radiates from the sun. There are many types of electromagnetic radiation, including gamma rays, x-rays, UV rays, visible light, infrared radiation, microwaves, and radio waves. This range of electromagnetic radiation is known as the **electromagnetic spectrum** (Figure 17.4). The distance between two consecutive peaks of an electromagnetic wave is the *wavelength.* Wavelengths range from short to long; for example, gamma rays have wavelengths smaller than a nanometer, and most radio waves have wavelengths greater than a meter.

The eyes are responsible for the detection of **visible light,** the part of the electromagnetic spectrum with wavelengths ranging from about 400 to 700 nm. Visible light exhibits colors: The color

Figure 17.4 The electromagnetic spectrum.

Visible light is the part of the electromagnetic spectrum with wavelengths ranging from about 400 to 700 nm.

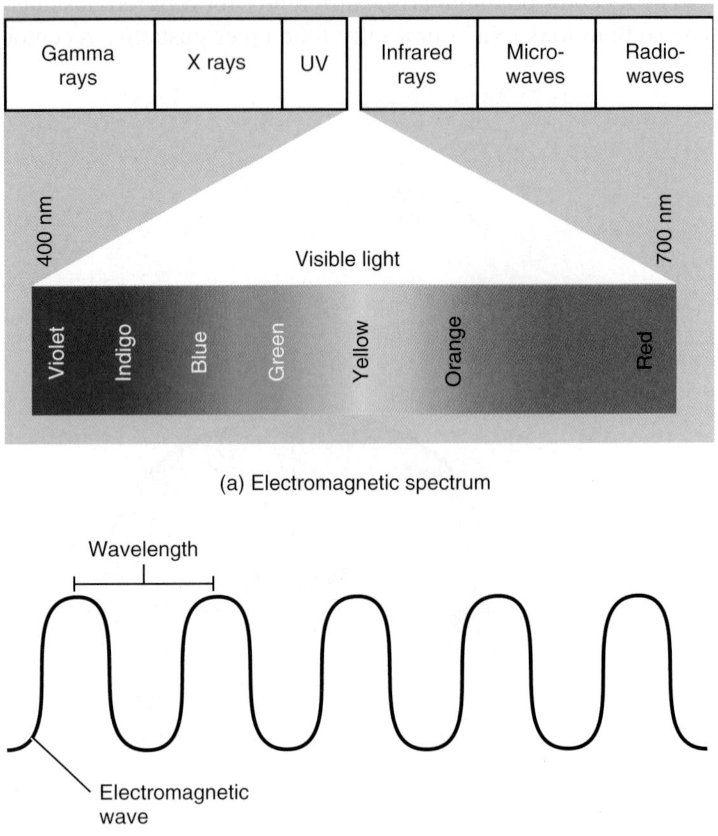

(a) Electromagnetic spectrum

(b) An electromagnetic wave

? Visible light that has a wavelength of 700 nm is what color?

of visible light depends on its wavelength. For example, light that has a wavelength of 400 nm is violet, and light that has a wavelength of 700 nm is red. An object can absorb certain wavelengths of visible light and reflect others; the object will appear the color of the wavelength that is reflected. For example, a green apple appears green because it reflects mostly green light and absorbs most other wavelengths of visible light. An object appears white because it reflects all wavelengths of visible light. An object appears black because it absorbs all wavelengths of visible light.

Accessory Structures of the Eye

The **accessory structures** of the eye include the eyelids, eyelashes, eyebrows, the lacrimal (tear-producing) apparatus, and extrinsic eye muscles.

Eyelids

The upper and lower **eyelids,** or **palpebrae** (PAL-pe-brē; singular is *palpebra*), shade the eyes during sleep, protect the eyes from

Figure 17.5 Surface anatomy of the right eye.

The palpebral fissure is the space between the upper and lower eyelids that exposes the eyeball.

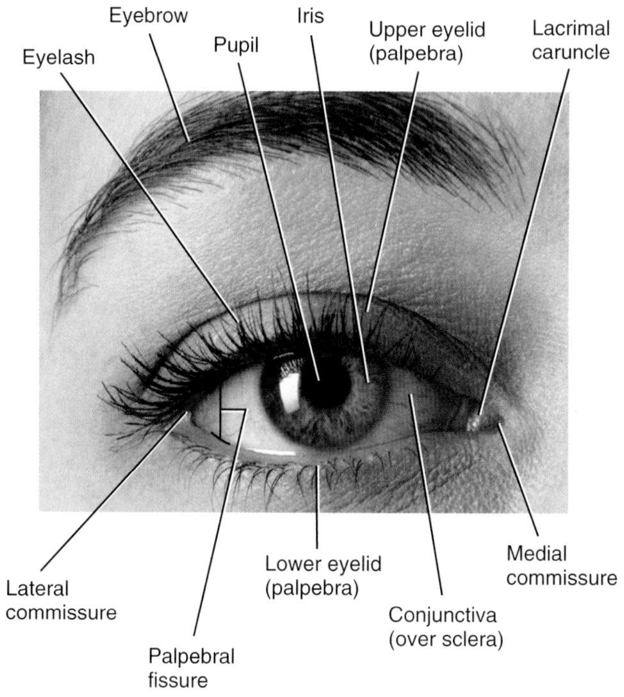
Eyebrow
Iris
Eyelash
Pupil
Upper eyelid
(palpebra)
Lacrimal
caruncle
Lateral
commissure
Palpebral
fissure
Lower eyelid
(palpebra)
Conjunctiva
(over sclera)
Medial
commissure

Which structure shown here is continuous with the inner lining of the eyelids?

excessive light and foreign objects, and spread lubricating secretions over the eyeballs (Figure 17.5). The upper eyelid is more movable than the lower and contains in its superior region the **levator palpebrae superioris muscle** (see Figure 17.6a). Sometimes a person may experience an annoying *twitch* in an eyelid, an involuntary quivering similar to muscle twitches in the hand, forearm, leg, or foot. Twitches are almost always harmless and usually last for only a few seconds. They are often associated with stress and fatigue. The space between the upper and lower eyelids that exposes the eyeball is the **palpebral fissure** (PAL-pe-bral). Its angles are known as the **lateral commissure** (KOM-i-shur), which is narrower and closer to the temporal bone, and the **medial commissure,** which is broader and nearer the nasal bone. In the medial commissure is a small, reddish elevation, the **lacrimal caruncle** (KAR-ung-kul), which contains sebaceous (oil) glands and sudoriferous (sweat) glands. The whitish material that sometimes collects in the medial commissure comes from these glands.

From superficial to deep, each eyelid consists of epidermis, dermis, subcutaneous tissue, fibers of the orbicularis oculi muscle, a tarsal plate, tarsal glands, and conjunctiva (Figure 17.6a). The **tarsal plate** is a thick fold of connective tissue that gives form and support to the eyelids. Embedded in each tarsal plate is a row of elongated modified sebaceous glands, known as **tarsal** or **Meibomian glands** (mī-BŌ-mē-an), that secrete a fluid that helps keep the eyelids from adhering to each other. Infection of the tarsal glands produces a tumor or cyst on the eyelid called a **chalazion** (ka-LĀ-zē-on = small bump). The **conjunctiva** (kon′-junk-TĪ-va) is a thin, protective mucous membrane composed of nonkeratinized stratified squamous epithelium with numerous goblet cells that is supported by areolar connective tissue. The **palpebral conjunctiva** lines the inner aspect of the eyelids, and the **bulbar conjunctiva** passes from the eyelids onto the surface of the eyeball, where it covers the sclera (the "white" of the eye) but not the cornea, which is a transparent region that forms the outer anterior surface of the eyeball. Over the sclera, the conjunctiva is vascular. Both the sclera and the cornea will be discussed in more detail shortly. Dilation and congestion of the blood vessels of the bulbar conjunctiva due to local irritation or infection are the cause of **bloodshot eyes.**

Eyelashes and Eyebrows

The **eyelashes,** which project from the border of each eyelid, and the **eyebrows,** which arch transversely above the upper eyelids, help protect the eyeballs from foreign objects, perspiration, and the direct rays of the sun. Sebaceous glands at the base of the hair follicles of the eyelashes, called **sebaceous ciliary glands,** release a lubricating fluid into the follicles. Infection of these glands, usually by bacteria, causes a painful, pus-filled swelling called a **sty.**

The Lacrimal Apparatus

The **lacrimal apparatus** (LAK-ri-mal; *lacrim-* = tears) is a group of structures that produces and drains **lacrimal fluid** or **tears.** The **lacrimal glands,** each about the size and shape of an almond, secrete lacrimal fluid, which drains into 6–12 **excretory lacrimal ducts** that empty tears onto the surface of the conjunctiva of the upper lid (Figure 17.6b). From here the tears pass medially over the anterior surface of the eyeball to enter two small openings called **lacrimal puncta** (singular is *punctum*). Tears then pass into two ducts, the **lacrimal canals,** which lead into the **lacrimal sac** and then into the **nasolacrimal duct.** This duct carries the lacrimal fluid into the nasal cavity just inferior to the inferior nasal concha. An infection of the lacrimal sacs is called **dacryocystitis** (dak′-rē-ō-sis-TĪ-tis; *dacryo-* = lacrimal sac; *-itis* = inflammation of). It is usually caused by a bacterial infection and results in blockage of the nasolacrimal ducts.

The lacrimal glands are supplied by parasympathetic fibers of the facial (VII) nerves. The lacrimal fluid produced by these glands is a watery solution containing salts, some mucus, and **lysozyme** (LĪ-sō-zīm), a protective bactericidal enzyme. The fluid protects, cleans, lubricates, and moistens the eyeball. After being secreted from the lacrimal gland, lacrimal fluid is spread medially over the surface of the eyeball by the blinking of the eyelids. Each gland produces about 1 mL of lacrimal fluid per day.

Normally, tears are cleared away as fast as they are produced, either by evaporation or by passing into the lacrimal canals and then into the nasal cavity. If an irritating substance makes contact with the conjunctiva, however, the lacrimal glands are stimulated

Figure 17.6 Accessory structures of the eye.

Accessory structures of the eye include the eyelids, eyelashes, eyebrows, lacrimal apparatus, and extrinsic eye muscles.

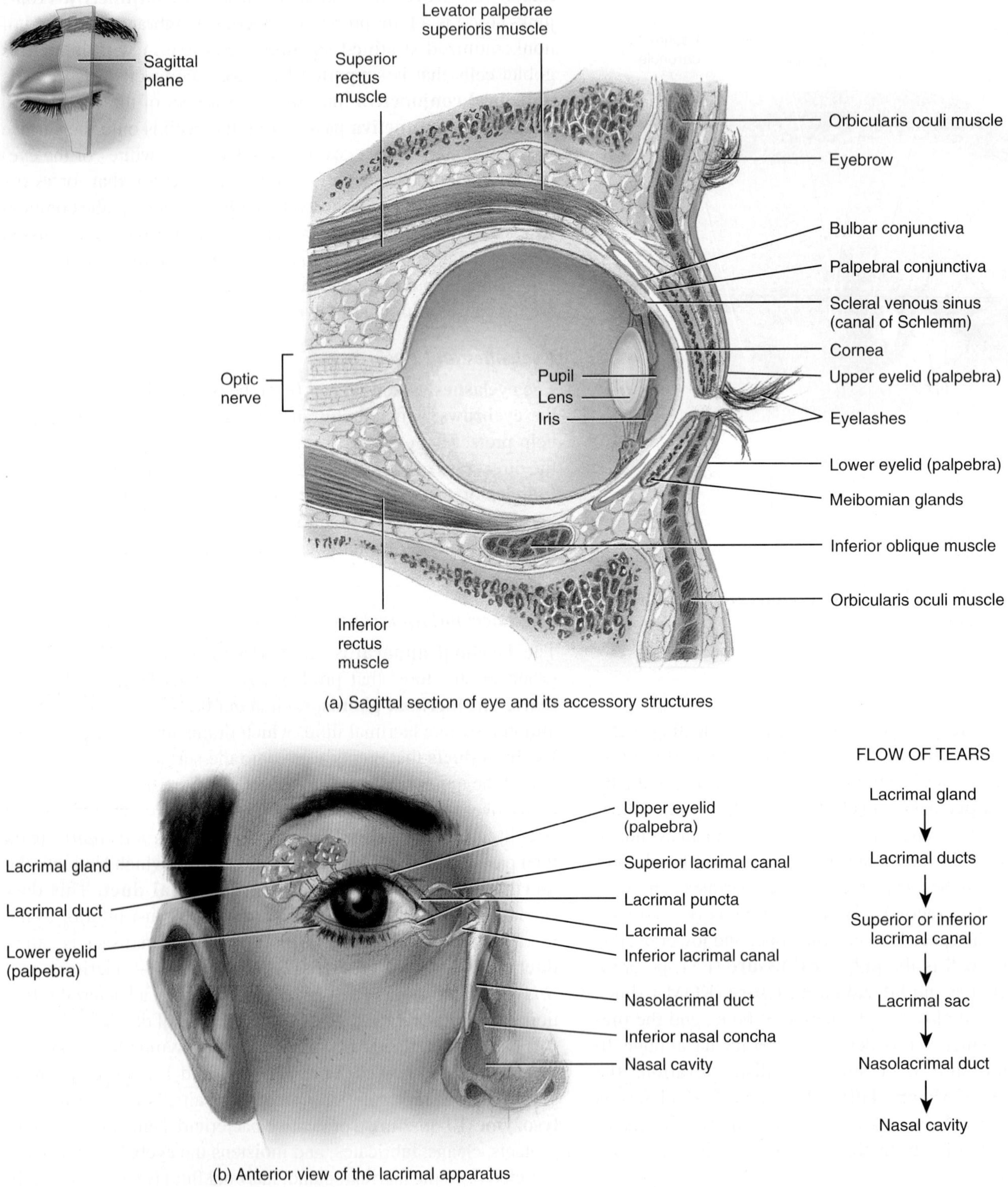

(b) Anterior view of the lacrimal apparatus

What is lacrimal fluid, and what are its functions?

to oversecrete, and tears accumulate (watery eyes). Lacrimation is a protective mechanism, as the tears dilute and wash away the irritating substance. Watery eyes also occur when an inflammation of the nasal mucosa, such as occurs with a cold, obstructs the nasolacrimal ducts and blocks drainage of tears. Only humans express emotions, both happiness and sadness, by **crying.** In response to parasympathetic stimulation, the lacrimal glands produce excessive lacrimal fluid that may spill over the edges of the eyelids and even fill the nasal cavity with fluid. This is how crying produces a runny nose.

Extrinsic Eye Muscles

The eyes sit in the bony depressions of the skull called the *orbits.* The orbits help protect the eyes, stabilize them in three-dimensional space, and anchor them to the muscles that produce their essential movements. The extrinsic eye muscles extend from the walls of the bony orbit to the sclera (white) of the eye and are surrounded in the orbit by a significant quantity of **periorbital fat** (per′-ē-OR-bi-tal). These muscles are capable of moving the eye in almost any direction. Six extrinsic eye muscles move each eye: the **superior rectus, inferior rectus, lateral rectus, medial rectus, superior oblique,** and **inferior oblique** (Figures 17.6a and 17.7). They are supplied by cranial nerves III, IV, or VI. In general, the motor units in these muscles are small. Some motor neurons serve only two or three muscle fibers—fewer than in any other part of the body except the larynx (voice box). Such small motor units permit smooth, precise, and rapid movement of the eyes. As indicated in Exhibit 11.B, the extrinsic eye muscles move the eyeball laterally, medially, superiorly, and inferiorly. For example, looking to the right requires simultaneous contraction of the right lateral rectus and left medial rectus muscles of the eyeball and relaxation of the left lateral rectus and right medial rectus of the

Figure 17.7 Anatomy of the eyeball.

The wall of the eyeball consists of three layers: the fibrous tunic, the vascular tunic, and the retina.

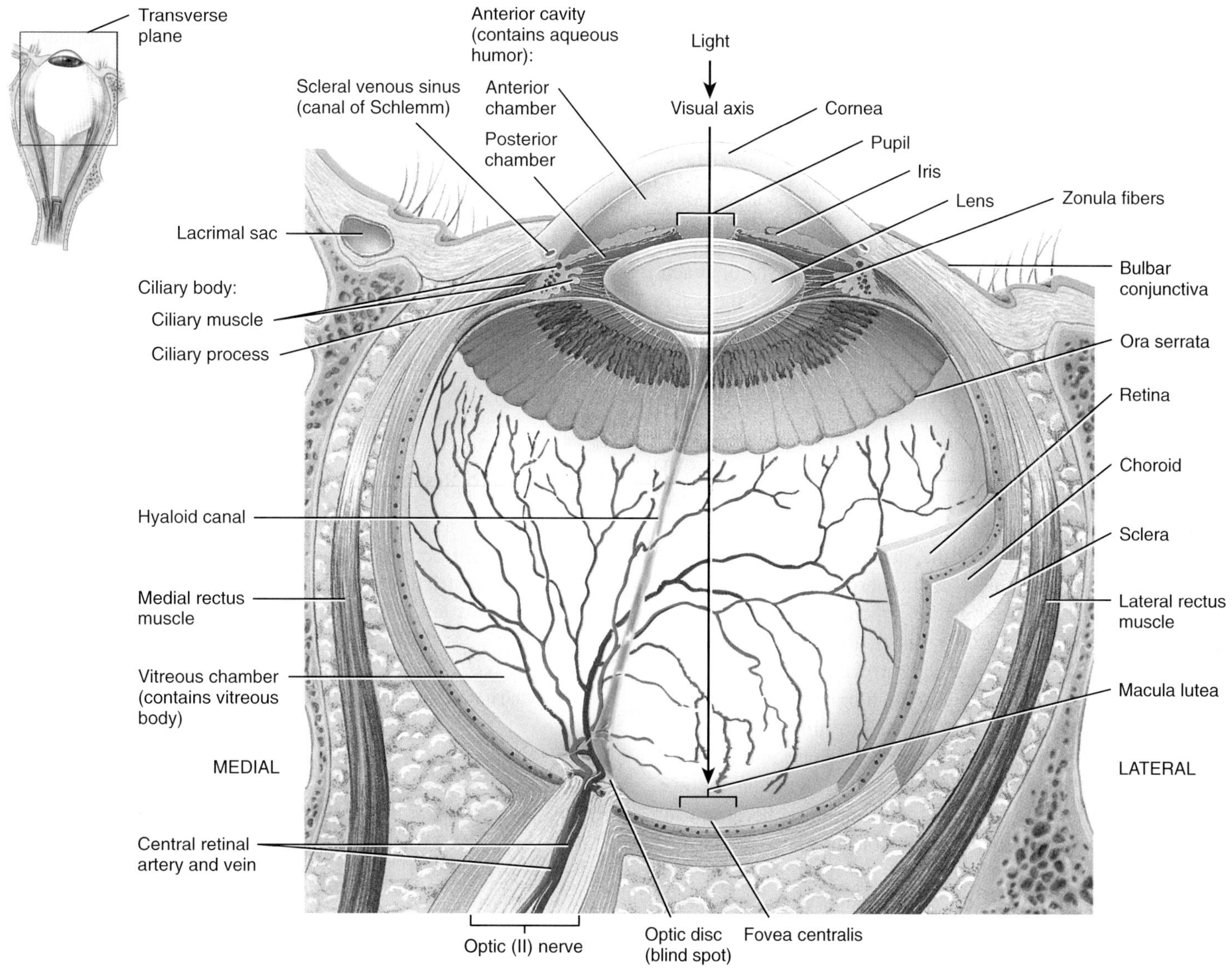

Superior view of transverse section of right eyeball

What are the components of the fibrous tunic and vascular tunic?

eyeball. The oblique muscles preserve rotational stability of the eyeball. Neural circuits in the brain stem and cerebellum coordinate and synchronize the movements of the eyes.

Anatomy of the Eyeball

The adult **eyeball** measures about 2.5 cm (1 in.) in diameter. Of its total surface area, only the anterior one-sixth is exposed; the remainder is recessed and protected by the orbit, into which it fits. Anatomically, the wall of the eyeball consists of three layers: (1) fibrous tunic, (2) vascular tunic, and (3) retina.

Fibrous Tunic

The **fibrous tunic** (TOO-nik) is the superficial layer of the eyeball and consists of the anterior cornea and posterior sclera (Figure 17.7). The **cornea** (KOR-nē-a) is a transparent coat that covers the colored iris. Because it is curved, the cornea helps focus light onto the retina. Its outer surface consists of nonkeratinized stratified squamous epithelium. The middle coat of the cornea consists of collagen fibers and fibroblasts, and the inner surface is simple squamous epithelium. Since the central part of the cornea receives oxygen from the outside air, contact lenses that are worn for long periods of time must be permeable to permit oxygen to pass through them. The **sclera** (SKLE-ra; *scler-* = hard), the "white" of the eye, is a layer of dense connective tissue made up mostly of collagen fibers and fibroblasts. The sclera covers the entire eyeball except the cornea; it gives shape to the eyeball, makes it more rigid, protects its inner parts, and serves as a site of attachment for the extrinsic eye muscles. At the junction of the sclera and cornea is an opening known as the **scleral venous sinus (canal of Schlemm).** A fluid called *aqueous humor,* which will be described later, drains into this sinus (Figure 17.7).

Vascular Tunic

The **vascular tunic** or **uvea** (Ū-ve-a) is the middle layer of the eyeball. It is composed of three parts: choroid, ciliary body, and iris (Figure 17.7). The highly vascularized **choroid** (KŌ-royd), which is the posterior portion of the vascular tunic, lines most of the internal surface of the sclera. Its numerous blood vessels provide nutrients to the posterior surface of the retina. The choroid also contains melanocytes that produce the pigment melanin, which causes this layer to appear dark brown in color. Melanin in the choroid absorbs stray light rays, which prevents reflection and scattering of light within the eyeball. As a result, the image cast on the retina by the cornea and lens remains sharp and clear. Albinos lack melanin in all parts of the body, including the eye. They often need to wear sunglasses, even indoors, because even moderately bright light is perceived as bright glare due to light scattering.

In the anterior portion of the vascular tunic, the choroid becomes the **ciliary body** (SIL-ē-ar′-ē). It extends from the **ora serrata** (Ō-ra ser-RĀ-ta), the jagged anterior margin of the retina, to a point just posterior to the junction of the sclera and cornea. Like the choroid, the ciliary body appears dark brown in color because it contains melanin-producing melanocytes. In addition, the ciliary body consists of ciliary processes and ciliary muscle. The **ciliary processes** are protrusions or folds on the internal surface of the ciliary body. They contain blood capillaries that secrete aqueous humor. Extending from the ciliary process are **zonular fibers (suspensory ligaments)** that attach to the lens. The fibers consist of thin, hollow fibrils that resemble elastic connective tissue fibers. The **ciliary muscle** is a circular band of smooth muscle. Contraction or relaxation of the ciliary muscle changes the tightness of the zonular fibers, which alters the shape of the lens, adapting it for near or far vision.

The **iris** (= rainbow), the colored portion of the eyeball, is shaped like a flattened donut. It is suspended between the cornea and the lens and is attached at its outer margin to the ciliary processes. It consists of melanocytes and circular and radial smooth muscle fibers. The amount of melanin in the iris determines the eye color. The eyes appear brown to black when the iris contains a large amount of melanin, blue when its melanin concentration is very low, and green when its melanin concentration is moderate.

A principal function of the iris is to regulate the amount of light entering the eyeball through the **pupil** (*pupil* = little person; because this is where you see a reflection of yourself when looking into someone's eyes), the hole in the center of the iris. The pupil appears black because, as you look through the lens, you see the heavily pigmented back of the eye (choroid and retina). However, if bright light is directed into the pupil, the reflected light is red because of the blood vessels on the surface of the retina. It is for this reason that a person's eyes appear red in a photograph ("red eye") when the flash is directed into the pupil. Autonomic reflexes regulate pupil diameter in response to light levels (Figure 17.8). When bright light stimulates the eye, parasympathetic

Figure 17.8 Responses of the pupil to light of varying brightness.

⑥ Contraction of the circular muscles causes constriction of the pupil; contraction of the radial muscles causes dilation of the pupil.

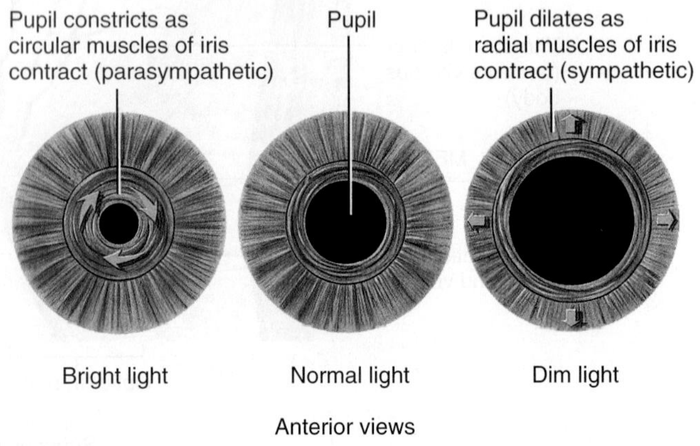
Pupil constricts as circular muscles of iris contract (parasympathetic)
Pupil
Pupil dilates as radial muscles of iris contract (sympathetic)
Bright light
Normal light
Dim light
Anterior views

? Which division of the autonomic nervous system causes pupillary constriction? Which causes pupillary dilation?

fibers of the oculomotor (III) nerve stimulate the **circular muscles** or **sphincter pupillae** (pu-PIL-ē) of the iris to contract, causing a decrease in the size of the pupil (constriction). In dim light, sympathetic neurons stimulate the **radial muscles** or **dilator pupillae** of the iris to contract, causing an increase in the pupil's size (dilation).

Retina

The third and inner layer of the eyeball, the **retina,** lines the posterior three-quarters of the eyeball and is the beginning of the visual pathway (see Figure 17.7). This layer's anatomy can be viewed with an *ophthalmoscope* (of-THAL-mō-skōp; *ophthalmos-* = eye; *-skopeo* = to examine), an instrument that shines light into the eye and allows an observer to peer through the pupil, providing a magnified image of the retina and its blood vessels as well as the optic (II) nerve (Figure 17.9). The surface of the retina is the only place in the body where blood vessels can be viewed directly and examined for pathological changes, such as those that occur with hypertension, diabetes mellitus, cataracts, and age-related macular disease. Several landmarks are visible through an ophthalmoscope. The **optic disc** is the site where the optic (II) nerve exits the eyeball. Bundled together with the optic nerve are the **central retinal artery,** a branch of the ophthalmic artery, and the **central retinal vein** (see Figure 17.7). Branches of the central retinal artery fan out to nourish the anterior surface of the retina; the central retinal vein drains blood from the retina through the optic disc. Also visible are the macula lutea and fovea centralis, which are described shortly.

The retina consists of a pigmented layer and a neural layer. The **pigmented layer** is a sheet of melanin-containing epithelial cells located between the choroid and the neural part of the retina. The melanin in the pigmented layer of the retina, as in the choroid, also helps to absorb stray light rays. The **neural (sensory) layer** of the retina is a multilayered outgrowth of the brain that processes visual data extensively before sending nerve impulses into axons that form the optic nerve. Three distinct layers of retinal neurons—the **photoreceptor layer,** the **bipolar cell layer,** and the **ganglion cell layer**—are separated by two zones, the *outer* and *inner synaptic layers,* where synaptic contacts are made (Figure 17.10). Note that light passes through the ganglion and bipolar cell layers and both synaptic layers before it reaches the photoreceptor layer. Two other types of cells present in the bipolar cell layer of the retina are called **horizontal cells** and **amacrine cells** (AM-a-krin). These cells form laterally directed neural circuits that modify the signals being transmitted along the pathway from photoreceptors to bipolar cells to ganglion cells.

Figure 17.9 A normal retina, as seen through an ophthalmoscope. Blood vessels in the retina can be viewed directly and examined for pathological changes.

6 The optic disc is the site where the optic nerve exits the eyeball. The fovea centralis is the area of highest visual acuity.

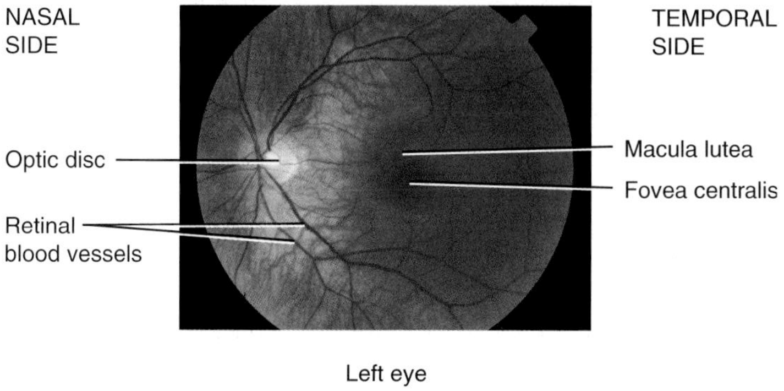
NASAL SIDE
TEMPORAL SIDE
Optic disc
Retinal blood vessels
Macula lutea
Fovea centralis
Left eye

? **Evidence of what diseases may be seen through an ophthalmoscope?**

CLINICAL CONNECTION | *Detached Retina*

A **detached retina** may occur due to trauma, such as a blow to the head, in various eye disorders, or as a result of age-related degeneration. The detachment occurs between the neural portion of the retina and the pigmented epithelium. Fluid accumulates between these layers, forcing the thin, pliable retina to billow outward. The result is distorted vision and blindness in the corresponding field of vision. The retina may be reattached by laser surgery or cryosurgery (localized application of extreme cold), and reattachment must be accomplished quickly to avoid permanent damage to the retina. •

Photoreceptors are specialized cells that begin the process by which light rays are ultimately converted to nerve impulses. There are two types of photoreceptors: rods and cones. Each retina has about 6 million cones and 120 million rods. **Rods** allow us to see in dim light, such as moonlight. Because rods do not provide color vision, in dim light we can see only black, white, and all shades of gray in between. Brighter lights stimulate **cones,** which produce color vision. Three types of cones are present in the retina: (1) *blue cones,* which are sensitive to blue light, (2) *green cones,* which are sensitive to green light, and (3) *red cones,* which are sensitive to red light. Color vision results from the stimulation of various combinations of these three types of cones. Most of our experiences are mediated by the cone system, the loss of which produces legal blindness. A person who loses rod vision mainly has difficulty seeing in dim light and thus should not drive at night.

From photoreceptors, information flows through the outer synaptic layer to bipolar cells and then from bipolar cells through the inner synaptic layer to ganglion cells. The axons of ganglion cells extend posteriorly to the optic disc and exit the eyeball as the optic (II) nerve. The optic disc is also called the **blind spot.** Because it contains no rods or cones, we cannot see images that strike the blind spot. Normally, you are not aware of having a blind spot, but you can easily demonstrate its presence. Hold this page about 20 in. from your face with the cross shown at the end

Figure 17.10 **Microscopic structure of the retina.** The downward blue arrow at left indicates the direction of the signals passing through the neural layer of the retina. Eventually, nerve impulses arise in ganglion cells and propagate along their axons, which make up the optic (II) nerve.

In the retina, visual signals pass from photoreceptors to bipolar cells to ganglion cells.

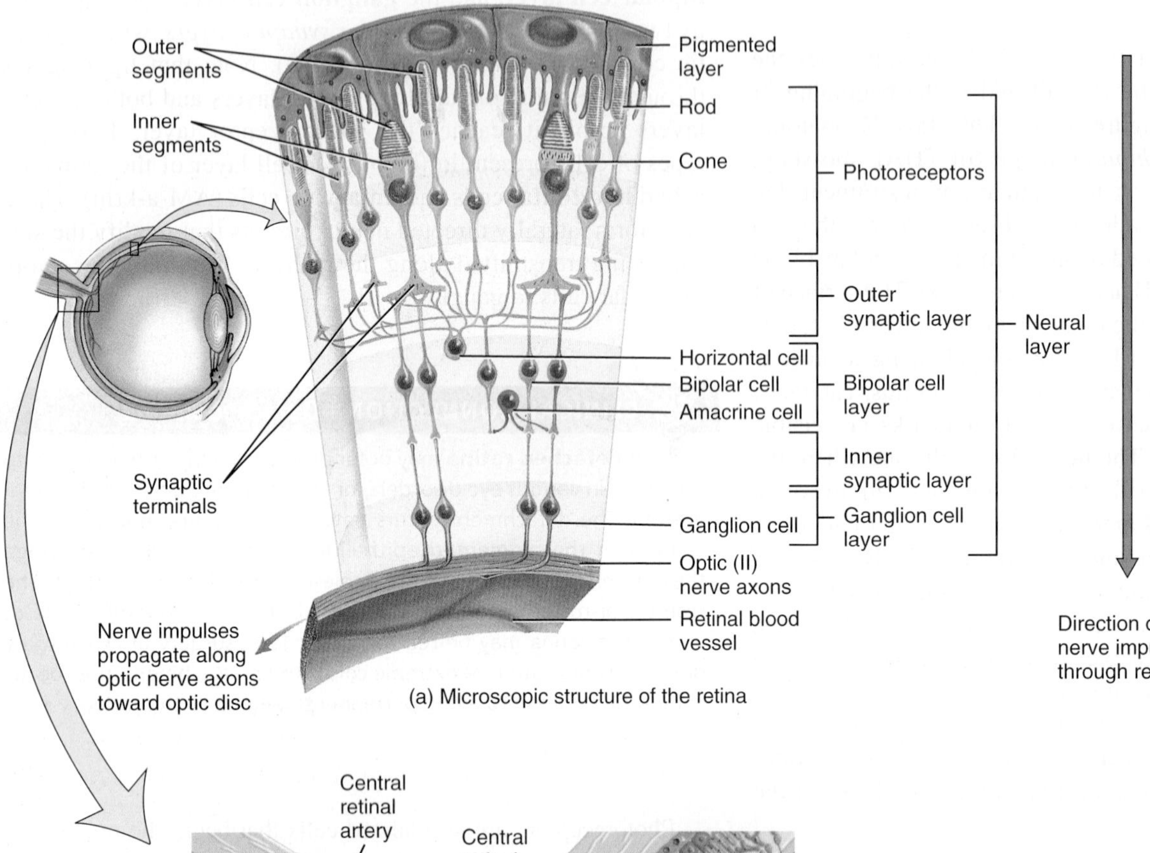

(a) Microscopic structure of the retina

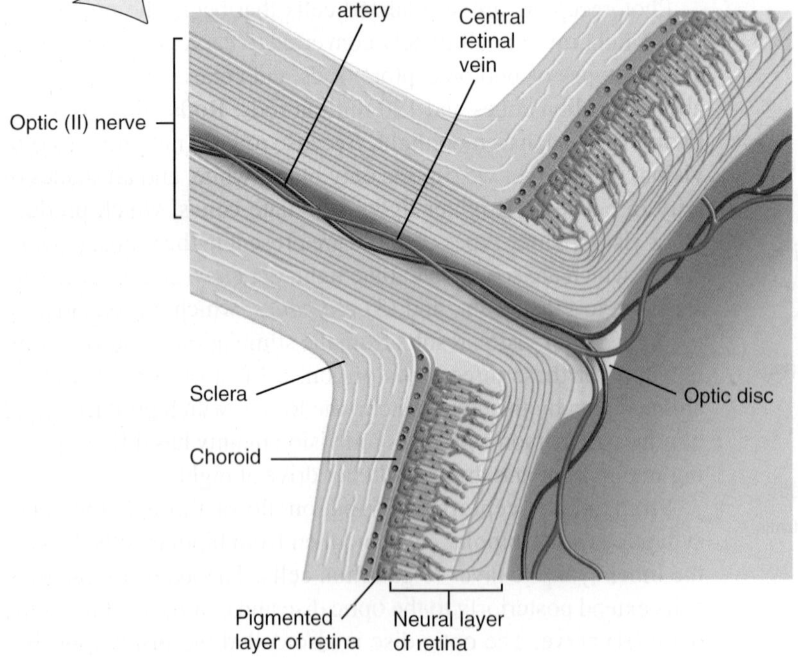

(b) Transverse section of posterior eyeball at optic disc

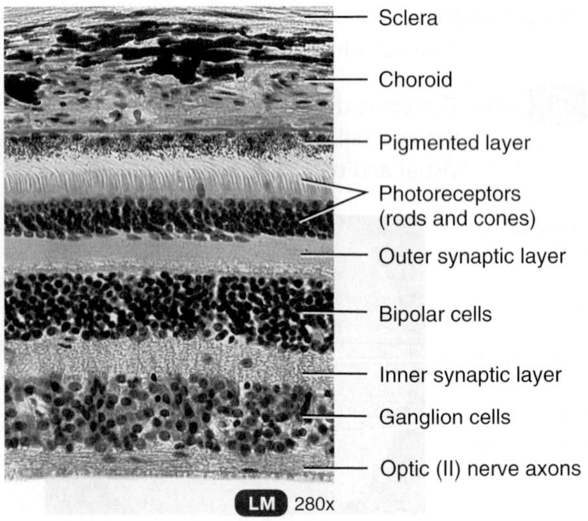

(c) Histology of a portion of the retina

What are the two types of photoreceptors, and how do their functions differ?

of this paragraph directly in front of your right eye. You should be able to see the cross and the square when you close your left eye. Now, keeping the left eye closed, slowly bring the page closer to your face while keeping the right eye on the cross. At a certain distance the square will disappear from your field of vision because its image falls on the blind spot.

+ ■

The **macula lutea** (MAK-ū-la LOO-tē-a; *macula* = a small, flat spot; *lute-* = yellowish) is in the exact center of the posterior portion of the retina, at the visual axis of the eye (see Figure 17.9). The **fovea centralis** (FŌ-vē-a) (see Figures 17.7 and 17.9), a small depression in the center of the macula lutea, contains only cones. In addition, the layers of bipolar and ganglion cells, which scatter light to some extent, do not cover the cones here; these layers are displaced to the periphery of the fovea centralis. As a result, the fovea centralis is the area of highest **visual acuity** (a-KU-i-tē) or **resolution** (sharpness of vision). A main reason that you move your head and eyes while looking at something is to place images of interest on your fovea centralis—as you do to read the words in this sentence! Rods are absent from the fovea centralis and are more plentiful toward the periphery of the retina. Because rod vision is more sensitive than cone vision, you can see a faint object (such as a dim star) better if you gaze slightly to one side rather than looking directly at it.

CLINICAL CONNECTION | *Age-related Macular Disease*

Age-related macular disease (AMD), also known as *macular degeneration,* is a degenerative disorder of the retina in persons 50 years of age and older. In AMD, abnormalities occur in the region of the macula lutea, which is ordinarily the area of most acute vision. Victims of advanced AMD retain their peripheral vision but lose the ability to see straight ahead. For instance, they cannot see facial features to identify a person in front of them. AMD is the leading cause of blindness in those over age 75, afflicting 13 million Americans, and is 2.5 times more common in pack-a-day smokers than in nonsmokers. Initially, a person may experience blurring and distortion at the center of the visual field. In "dry" AMD, central vision gradually diminishes because the pigmented layer atrophies and degenerates. There is no effective treatment. In about 10% of cases, dry AMD progresses to "wet" AMD, in which new blood vessels form in the choroid and leak plasma or blood under the retina. Vision loss can be slowed by using laser surgery to destroy the leaking blood vessels. •

Lens

Behind the pupil and iris, within the cavity of the eyeball, is the **lens** (see Figure 17.7). Within the cells of the lens, proteins called **crystallins** (KRIS-ta-lins), arranged like the layers of an onion, make up the refractive media of the lens, which normally is perfectly transparent and lacks blood vessels. It is enclosed by a clear connective tissue capsule and held in position by encircling zonular fibers, which attach to the ciliary processes. The lens helps focus images on the retina to facilitate clear vision.

Interior of the Eyeball

The lens divides the interior of the eyeball into two cavities: the anterior cavity and vitreous chamber. The **anterior cavity**—the space anterior to the lens—consists of two chambers. The **anterior chamber** lies between the cornea and the iris. The **posterior chamber** lies behind the iris and in front of the zonular fibers and lens (Figure 17.11). Both chambers of the anterior cavity are filled with **aqueous humor** (AK-wē-us HŪ-mer; *aqua* = water), a transparent watery fluid that nourishes the lens and cornea. Aqueous humor continually filters out of blood capillaries in the ciliary processes of the ciliary body and enters the posterior chamber. It then flows forward between the iris and the lens, through the pupil, and into the anterior chamber. From the anterior chamber, aqueous humor drains into the scleral venous sinus (canal of Schlemm) and then into the blood. Normally, aqueous humor is completely replaced about every 90 minutes.

The larger posterior cavity of the eyeball is the **vitreous chamber** (VIT-rē-us), which lies between the lens and the retina. Within the vitreous chamber is the **vitreous body,** a transparent jellylike substance that holds the retina flush against the choroid, giving the retina an even surface for the reception of clear images. It occupies about four-fifths of the eyeball. Unlike the aqueous humor, the vitreous body does not undergo constant replacement. It is formed during embryonic life and consists of mostly water plus collagen fibers and hyaluronic acid. The vitreous body also contains phagocytic cells that remove debris, keeping this part of the eye clear for unobstructed vision. Occasionally, collections of debris may cast a shadow on the retina and create the appearance of specks that dart in and out of the field of vision. These *vitreal floaters,* which are more common in older individuals, are usually harmless and do not require treatment. The **hyaloid canal** (HĪ-a-loyd) is a narrow channel that is inconspicuous in adults and runs through the vitreous body from the optic disc to the posterior aspect of the lens. In the fetus, it is occupied by the hyaloid artery (see Figure 17.27d).

The pressure in the eye, called **intraocular pressure,** is produced mainly by the aqueous humor and partly by the vitreous body; normally it is about 16 mmHg (millimeters of mercury). The intraocular pressure maintains the shape of the eyeball and prevents it from collapsing. Puncture wounds to the eyeball may cause the loss of aqueous humor and the vitreous body. This in turn causes a decrease in intraocular pressure, a detached retina, and in some cases blindness.

Table 17.1 summarizes the structures associated with the eyeball.

Image Formation

In some ways the eye is like a camera: Its optical elements focus an image of some object on a light-sensitive "film"—the retina—while ensuring the correct amount of light to make the proper "exposure." To understand how the eye forms clear images of objects on the retina, we must examine three processes: (1) the refraction or bending of light by the lens and cornea; (2) accommodation, the change in shape of the lens; and (3) constriction or narrowing of the pupil.

Figure 17.11 The iris separating the anterior and posterior chambers of the eye. The section is through the anterior portion of the eyeball at the junction of the cornea and sclera. Arrows indicate the flow of aqueous humor.

The lens separates the posterior chamber of the anterior cavity from the vitreous chamber.

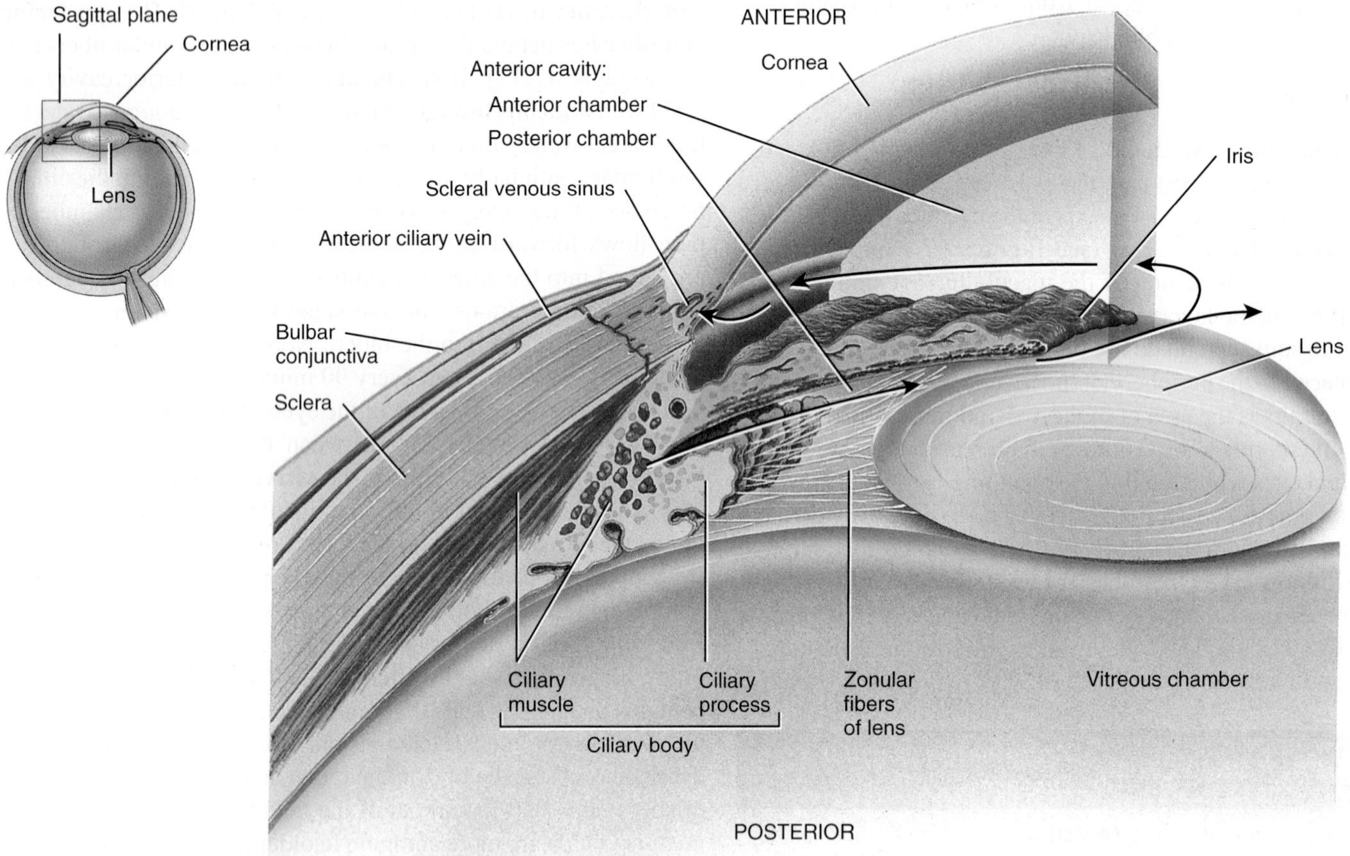

Where is aqueous humor produced, what is its circulation path, and where does it drain from the eyeball?

Refraction of Light Rays

When light rays traveling through a transparent substance (such as air) pass into a second transparent substance with a different density (such as water), they bend at the junction between the two substances. This bending is called **refraction** (re-FRAK-shun) (Figure 17.12a). As light rays enter the eye, they are refracted at the anterior and posterior surfaces of the cornea. Both surfaces of the lens of the eye further refract the light rays so they come into exact focus on the retina.

Images focused on the retina are inverted (upside down) (Figure 17.12b, c). They also undergo right-to-left reversal; that is, light from the right side of an object strikes the left side of the retina, and vice versa. The reason the world does not look inverted and reversed is that the brain "learns" early in life to coordinate visual images with the orientations of objects. The brain stores the inverted and reversed images we acquired when we first reached for and touched objects and interprets those visual images as being correctly oriented in space.

About 75% of the total refraction of light occurs at the cornea. The lens provides the remaining 25% of focusing power and also changes the focus to view near or distant objects. When an object is 6 m (20 ft) or more away from the viewer, the light rays reflected from the object are nearly parallel to one another (Figure 17.12b). The lens must bend these parallel rays just enough so that they fall exactly focused on the central fovea, where vision is sharpest. Because light rays that are reflected from objects closer than 6 m (20 ft) are divergent rather than parallel (Figure 17.12c), the rays must be refracted more if they are to be focused on the retina. This additional refraction is accomplished through a process called accommodation.

Accommodation and the Near Point of Vision

A surface that curves outward, like the surface of a ball, is said to be *convex.* When the surface of a lens is convex, that lens will refract incoming light rays toward each other, so that they eventually intersect. If the surface of a lens curves inward, like the inside of a hollow ball, the lens is said to be *concave* and causes light rays to refract away from each other. The lens of the eye is convex on both its anterior and posterior surfaces, and its focusing power increases as its curvature becomes greater. When the eye is focusing on a close object, the lens becomes more curved, causing greater refraction of the light rays. This increase in the curvature

TABLE 17.1

Summary of the Structures of the Eyeball

STRUCTURE	FUNCTION
Fibrous tunic 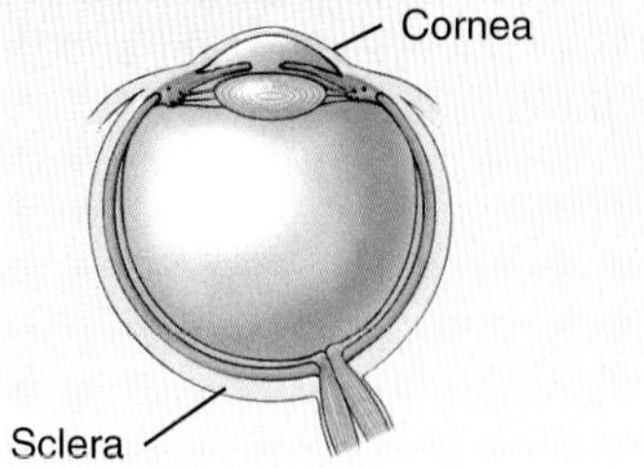	***Cornea:*** Admits and refracts (bends) light. ***Sclera:*** Provides shape and protects inner parts.
Vascular tunic 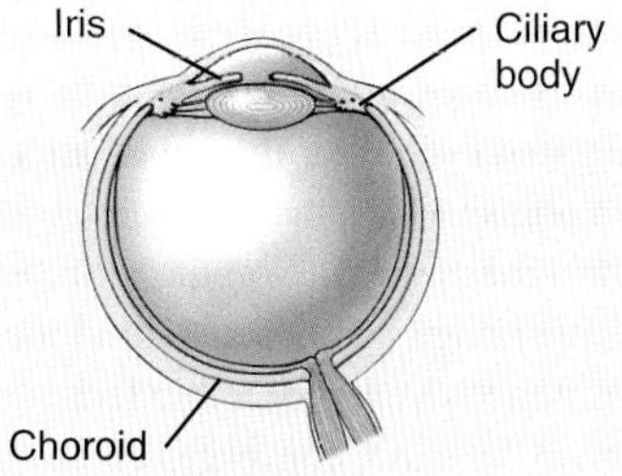	***Iris:*** Regulates amount of light that enters eyeball. ***Ciliary body:*** Secretes aqueous humor and alters shape of lens for near or far vision (accommodation). ***Choroid:*** Provides blood supply and absorbs scattered light.
Retina 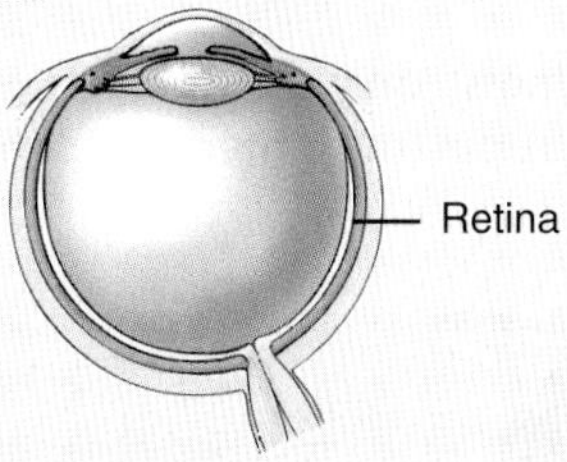	Receives light and converts it into receptor potentials and nerve impulses. Output to brain via axons of ganglion cells, which form optic (II) nerve.
Lens 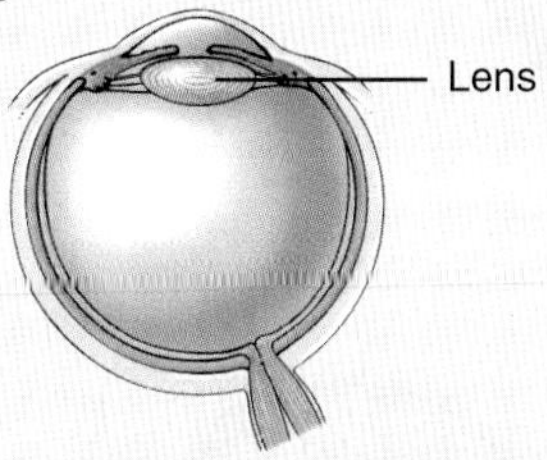	Refracts light.
Anterior cavity 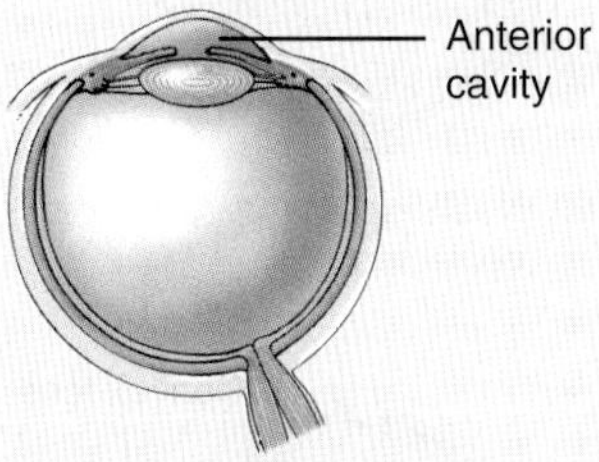	Contains aqueous humor that helps maintain shape of eyeball and supplies oxygen and nutrients to lens and cornea.
Vitreous chamber 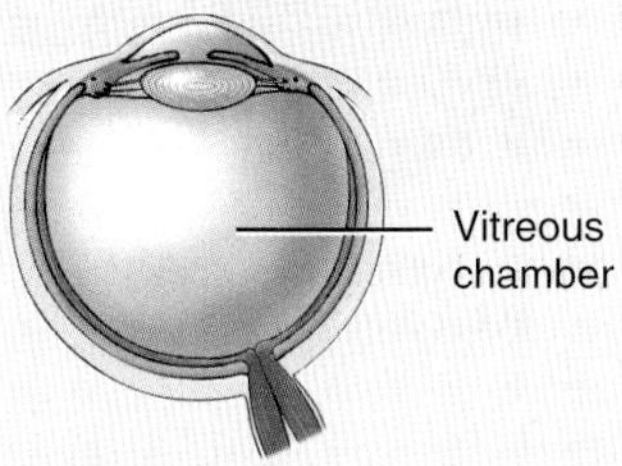	Contains vitreous body that helps maintain shape of eyeball and keeps retina attached to choroid.

Figure 17.12 Refraction of light rays. (a) Refraction is the bending of light rays at the junction of two transparent substances with different densities. (b) The cornea and lens refract light rays from distant objects so the image is focused on the retina. (c) In accommodation, the lens becomes more spherical, which increases the refraction of light.

Images focused on the retina are inverted and left-to-right reversed.

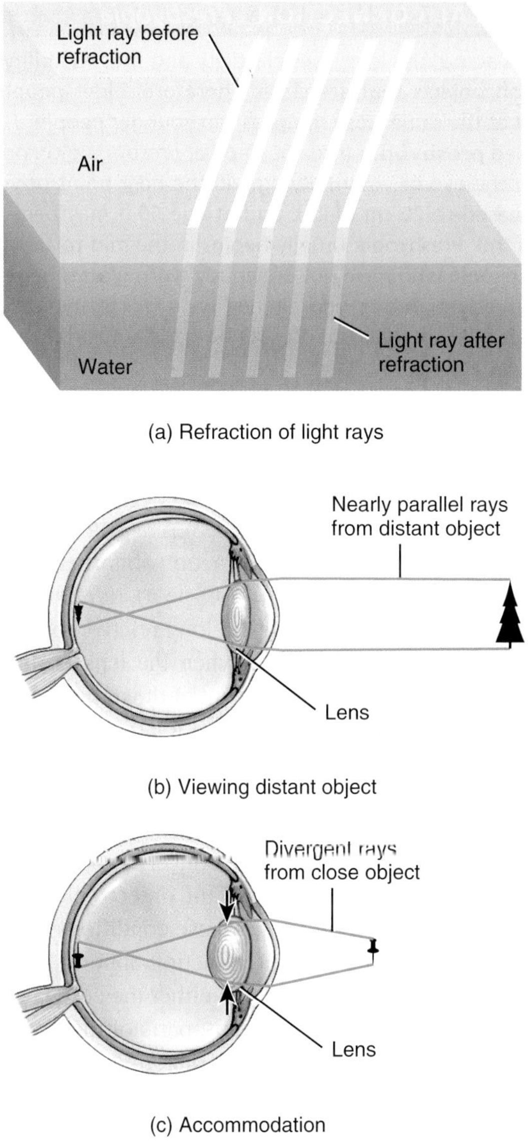

What sequence of events occurs during accommodation?

of the lens for near vision is called **accommodation** (a-kom-a-DĀ-shun) (Figure 17.12c). The **near point of vision** is the minimum distance from the eye that an object can be clearly focused with maximum accommodation. This distance is about 10 cm (4 in.) in a young adult.

How does accommodation occur? When you are viewing distant objects, the ciliary muscle of the ciliary body is relaxed and the lens is flatter because it is stretched in all directions by taut zonular fibers. When you view a close object, the ciliary muscle contracts, which pulls the ciliary process and choroid forward toward the lens. This action releases tension on the lens and zonular fibers. Because it is elastic, the lens becomes more spherical (more convex), which increases its focusing power and causes greater convergence of the light rays. Parasympathetic fibers of the oculomotor (III) nerve innervate the ciliary muscle of the ciliary body and, therefore, mediate the process of accommodation.

CLINICAL CONNECTION | *Presbyopia*

With aging, the lens loses elasticity and thus its ability to curve to focus on objects that are close. Therefore, older people cannot read print at the same close range as can younger people. This condition is called **presbyopia** (prez-bē-Ō-pē-a; *presby-* = old; *-opia* = pertaining to the eye or vision). By age 40 the near point of vision may have increased to 20 cm (8 in.), and at age 60 it may be as much as 80 cm (31 in.). Presbyopia usually begins in the mid-forties. At about that age, people who have not previously worn glasses begin to need them for reading. Those who already wear glasses typically start to need bifocals, lenses that can focus for both distant and close vision. •

Refraction Abnormalities

The normal eye, known as an **emmetropic eye** (em′-e-TROP-ik), can sufficiently refract light rays from an object 6 m (20 ft) away so that a clear image is focused on the retina. Many people, however, lack this ability because of refraction abnormalities. Among these abnormalities are **myopia** (mī-Ō-pē-a), or *nearsightedness,* which occurs when the eyeball is too long relative to the focusing power of the cornea and lens, or when the lens is thicker than normal, so an image converges in front of the retina. Myopic individuals can see close objects clearly, but not distant objects. In **hyperopia** (hī-per-Ō-pē-a) or *farsightedness,* also known as **hypermetropia** (hī′-per-me-TRŌ-pē-a), the eyeball length is short relative to the focusing power of the cornea and lens, or the lens is thinner than normal, so an image converges behind the retina. Hyperopic individuals can see distant objects clearly, but not close ones. Figure 17.13 illustrates these conditions and explains how they are corrected. Another refraction abnormality is **astigmatism** (a-STIG-ma-tizm), in which either the cornea or the lens has an irregular curvature. As a result, parts of the image are out of focus, and thus vision is blurred or distorted.

Most errors of vision can be corrected by eyeglasses, contact lenses, or surgical procedures. A contact lens floats on a film of tears over the cornea. The anterior outer surface of the contact lens corrects the visual defect, and its posterior surface matches the curvature of the cornea. LASIK involves reshaping the cornea to permanently correct refraction abnormalities.

CLINICAL CONNECTION | *LASIK*

An increasingly popular alternative to wearing glasses or contact lenses is refractive surgery to correct the curvature of the cornea for conditions such as farsightedness, nearsightedness, and astigmatism. The most common type of refractive surgery is **LASIK** (laser-assisted in-situ keratomileusis). After anesthetic drops are placed in the eye, a circular flap of tissue from the center of the cornea is cut. The flap is folded out of the way, and the underlying layer of cornea is reshaped with a laser, one microscopic layer at a time. A computer assists the physician in removing very precise layers of the cornea. After the sculpting is complete, the corneal flap is repositioned over the treated area. A patch is placed over the eye overnight and the flap quickly reattaches to the rest of the cornea. •

Figure 17.13 Refraction abnormalities in the eyeball and their correction. (a) Normal (emmetropic) eye. (b) In the nearsighted or myopic eye, the image is focused in front of the retina. The condition may result from an elongated eyeball or thickened lens. (c) Correction of myopia is by use of a concave lens that diverges entering light rays so that they come into focus directly on the retina. (d) In the farsighted or hyperopic eye, the image is focused behind the retina. The condition results from a shortened eyeball or a thin lens. (e) Correction of hyperopia is by a convex lens that converges entering light rays so that they focus directly on the retina.

In myopia (nearsightedness), only close objects can be seen clearly; in hyperopia (farsightedness), only distant objects can be seen clearly.

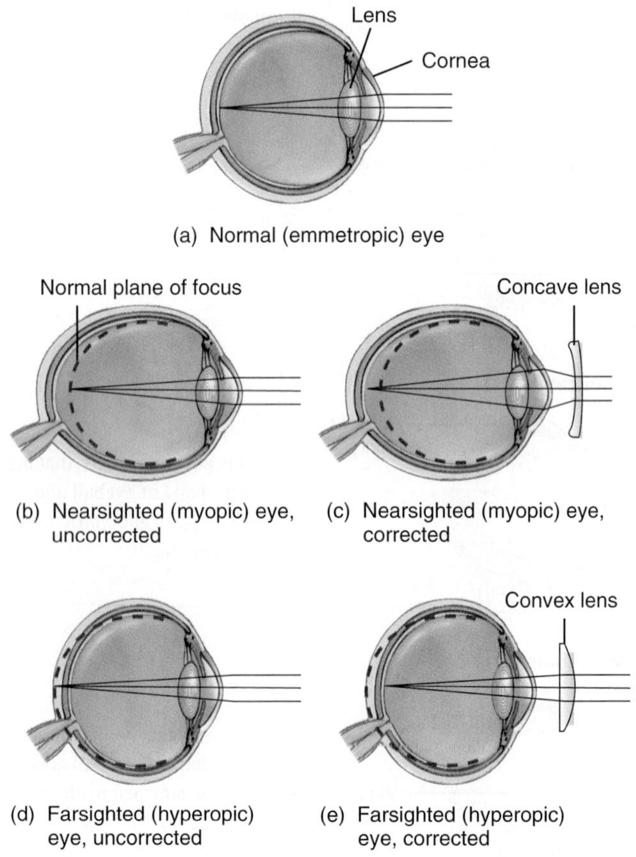
Lens
Cornea
(a) Normal (emmetropic) eye
Normal plane of focus
Concave lens
(b) Nearsighted (myopic) eye, uncorrected
(c) Nearsighted (myopic) eye, corrected
Convex lens
(d) Farsighted (hyperopic) eye, uncorrected
(e) Farsighted (hyperopic) eye, corrected

What is presbyopia?

Constriction of the Pupil

The circular muscle fibers of the iris also have a role in the formation of clear retinal images. As you have already learned, **constriction of the pupil** is a narrowing of the diameter of the hole through which light enters the eye due to the contraction of the circular muscles of the iris. This autonomic reflex occurs simultaneously with accommodation and prevents light rays from entering the eye through the periphery of the lens. Light rays entering at the periphery would not be brought to focus on the retina and would result in blurred vision. The pupil, as noted earlier, also constricts in bright light.

Convergence

Because of the position of their eyes in their heads, many animals, such as horses and goats, see one set of objects off to the left through one eye, and an entirely different set of objects off to the right through the other. In humans, both eyes focus on only one set of objects—a characteristic called **binocular vision.** This feature of our visual system allows the perception of depth and an appreciation of the three-dimensional nature of objects.

Binocular vision occurs when light rays from an object strike corresponding points on the two retinas. When we stare straight ahead at a distant object, the incoming light rays are aimed directly at both pupils and are refracted to comparable spots on the retinas of both eyes. As we move closer to an object, however, the eyes must rotate medially if the light rays from the object are to strike the same points on both retinas. The term **convergence** refers to this medial movement of the two eyeballs so that both are directed toward the object being viewed, for example, tracking a pencil moving toward your eyes. The nearer the object, the greater the degree of convergence needed to maintain binocular vision. The coordinated action of the extrinsic eye muscles brings about convergence.

Physiology of Vision

Photoreceptors and Photopigments

Rods and cones were named for the different appearance of the *outer segment*—the distal end next to the pigmented layer—of each of these types of photoreceptors. The outer segments of rods are cylindrical or rod-shaped; those of cones are tapered or cone-shaped (Figure 17.14). Transduction of light energy into a receptor potential occurs in the outer segment of both rods and cones. The photopigments are integral proteins in the plasma membrane of the outer segment. In cones the plasma membrane is folded back and forth in a pleated fashion; in rods the pleats pinch off from the plasma membrane to form discs. The outer segment of each rod contains a stack of about 1000 discs, piled up like coins inside a wrapper.

Photoreceptor outer segments are renewed at an astonishingly rapid pace. In rods, one to three new discs are added to the base of the outer segment every hour while old discs slough off at the tip and are phagocytized by pigment epithelial cells. The *inner segment* contains the cell nucleus, Golgi complex, and many mitochondria. At its proximal end, the photoreceptor expands into bulblike synaptic terminals filled with synaptic vesicles.

Figure 17.14 Structure of rod and cone photoreceptors. The inner segments contain the metabolic machinery for synthesis of photopigments and production of ATP. The photopigments are embedded in the membrane discs or folds of the outer segments. New discs in rods and new folds in cones form at the base of the outer segment. Pigmented epithelial cells phagocytize the old discs and folds that slough off the distal tip of the outer segments.

6 Transduction of light energy into a receptor potential occurs in the outer segments of rods and cones.

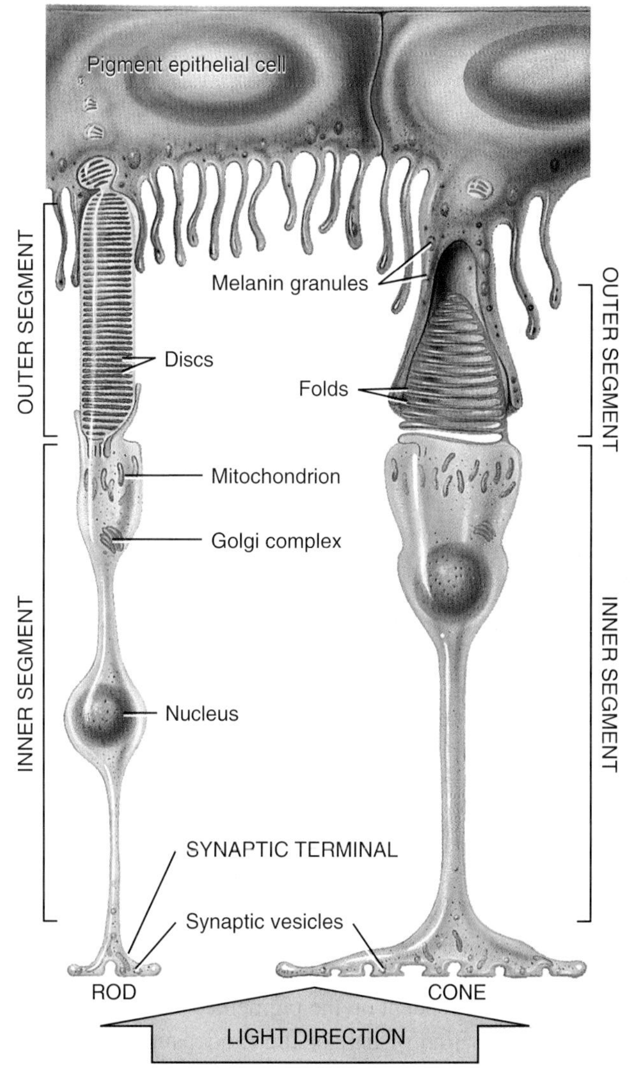

? What are the functional similarities between rods and cones?

The first step in visual transduction is absorption of light by a **photopigment,** a colored protein that undergoes structural changes when it absorbs light, in the outer segment of a photoreceptor. Light absorption initiates the events that lead to the production of a receptor potential. The single type of photopigment in rods is **rhodopsin** (rō-DOP-sin; *rhod-* = rose; *-opsin* = related to vision). Three different **cone photopigments** are present in the retina, one in each of the three types of cones. Color vision results from different colors of light selectively activating the different cone photopigments.

All photopigments associated with vision contain two parts: a glycoprotein known as **opsin** and a derivative of vitamin A called **retinal.** Vitamin A derivatives are formed from carotene, the plant pigment that gives carrots their orange color. Good vision depends on adequate dietary intake of carotene-rich vegetables such as carrots, spinach, broccoli, and yellow squash, or foods that contain vitamin A, such as liver.

Retinal is the light-absorbing part of all visual photopigments. In the human retina, there are four different opsins, three in the cones and one in the rods (rhodopsin). Small variations in the amino acid sequences of the different opsins permit the rods and cones to absorb different colors (wavelengths) of incoming light.

Photopigments respond to light in the following cyclical process (Figure 17.15):

1. In darkness, retinal has a bent shape, called *cis*-retinal, which fits snugly into the opsin portion of the photopigment. When *cis*-retinal absorbs a photon of light, it straightens out to a shape called *trans*-retinal. This cis-to-trans conversion is called **isomerization** (ī-som′-er-i-ZĀ-shun) and is the first step in visual transduction. After retinal isomerizes, several unstable chemical intermediates form and disappear. These chemical changes lead to production of a receptor potential (see Figure 17.16).
2. In about a minute, *trans*-retinal completely separates from opsin. The final products look colorless, so this part of the cycle is termed **bleaching** of photopigment.
3. An enzyme called **retinal isomerase** converts *trans*-retinal back to *cis*-retinal.
4. The *cis*-retinal then can bind to opsin, reforming a functional photopigment. This part of the cycle—resynthesis of a photopigment—is called **regeneration.**

The pigmented layer of the retina adjacent to the photoreceptors stores a large quantity of vitamin A and contributes to the regeneration process in rods. The extent of rhodopsin regeneration decreases drastically if the retina detaches from the pigmented layer. Cone photopigments regenerate much more quickly than the rhodopsin in rods and are less dependent on the pigmented layer. After complete bleaching, regeneration of half of the rhodopsin takes 5 minutes; half of the cone photopigments regenerate in only 90 seconds. Full regeneration of bleached rhodopsin takes 30 to 40 minutes.

Light and Dark Adaptation

When you emerge from dark surroundings (say, a tunnel) into the sunshine, **light adaptation** occurs—your visual system adjusts in seconds to the brighter environment by decreasing its sensitivity. On the other hand, when you enter a darkened room such as a theater, your visual system undergoes **dark adaptation**—its sensitivity increases slowly over many minutes. The difference in the rates of bleaching and regeneration of the photopigments in the rods and cones accounts for some (but not all) of the sensitivity changes during light and dark adaptation.

Figure 17.15 The cyclical bleaching and regeneration of photopigment. Blue arrows indicate bleaching steps; black arrows indicate regeneration steps.

Retinal, a derivative of vitamin A, is the light-absorbing part of all visual photopigments.

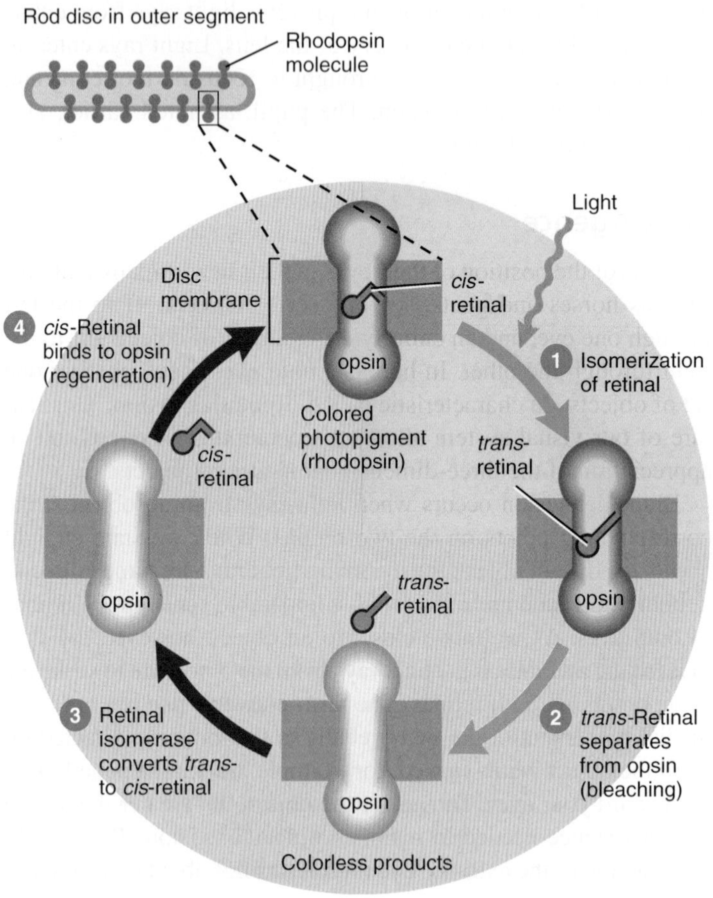

What is the conversion of *cis*-retinal to *trans*-retinal called?

As the light level increases, more and more photopigment is bleached. While light is bleaching some photopigment molecules, however, others are being regenerated. In daylight, regeneration of rhodopsin cannot keep up with the bleaching process, so rods contribute little to daylight vision. In contrast, cone photopigments regenerate rapidly enough that some of the cis form is always present, even in very bright light.

If the light level decreases abruptly, sensitivity increases rapidly at first and then more slowly. In complete darkness, full regeneration of cone photopigments occurs during the first 8 minutes of dark adaptation. During this time, a threshold (barely perceptible) light flash is seen as having color. Rhodopsin regenerates more slowly, and our visual sensitivity increases until even a single photon (the smallest unit of light) can be detected. In that situation, although much dimmer light can be detected, threshold flashes appear gray-white, regardless of their color. At very low light levels, such as starlight, objects appear as shades of gray because only the rods are functioning.

Figure 17.16 Operation of rod photoreceptors.

Light causes a hyperpolarizing receptor potential in photoreceptors, which decreases release of an inhibitory neurotransmitter (glutamate).

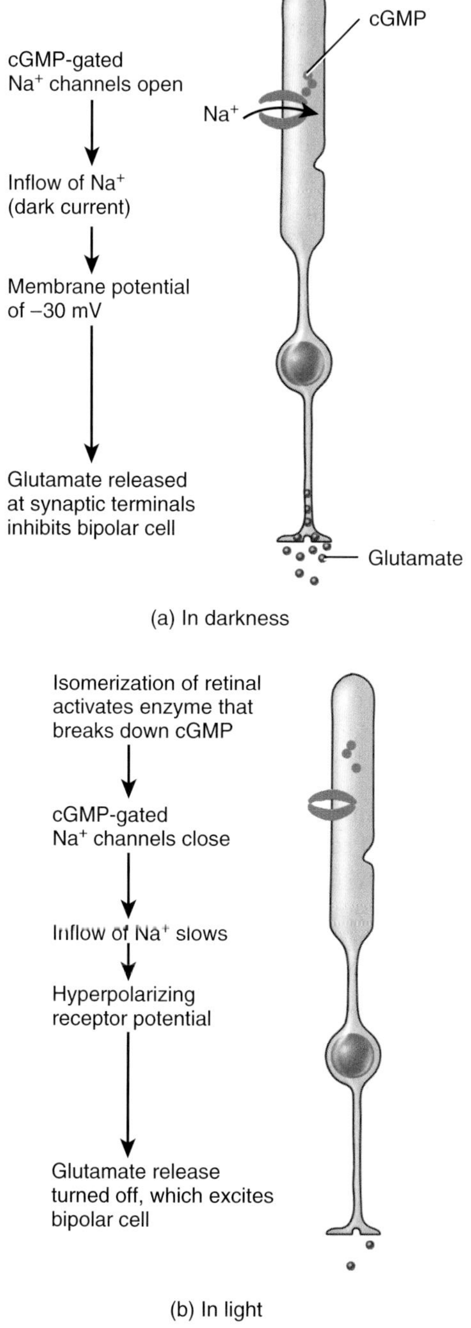
cGMP
cGMP-gated
Na+ channels open
Na+
Inflow of Na+
(dark current)
Membrane potential
of –30 mV
Glutamate released
at synaptic terminals
inhibits bipolar cell
Glutamate
(a) In darkness
Isomerization of retinal
activates enzyme that
breaks down cGMP
cGMP-gated
Na+ channels close
Inflow of Na+ slows
Hyperpolarizing
receptor potential
Glutamate release
turned off, which excites
bipolar cell
(b) In light

What is the function of cyclic GMP in photoreceptors?

Release of Neurotransmitter by Photoreceptors

As mentioned previously, the absorption of light and isomerization of retinal initiates chemical changes in the photoreceptor outer segment that lead to production of a receptor potential. To understand how the receptor potential arises, however, we first need to examine the operation of photoreceptors in the absence of light. In darkness, sodium ions (Na^+) flow into photoreceptor outer segments through ligand-gated Na^+ channels (Figure 17.16a). The ligand that holds these channels open is **cyclic GMP (guanosine monophosphate)** or **cGMP.** The inflow of Na^+, called the "dark current," partially depolarizes the photoreceptor. As a result, in darkness the membrane potential of a photoreceptor is about -30 mV. This is much closer to zero than a typical neuron's resting membrane potential of -70 mV. The partial depolarization during darkness triggers continual release of neurotransmitter at the synaptic terminals. The neurotransmitter in rods, and perhaps in cones, is the amino acid glutamate (glutamic acid). At synapses between rods and some bipolar cells, glutamate is an inhibitory neurotransmitter: It triggers inhibitory postsynaptic potentials (IPSPs) that hyperpolarize the bipolar cells and prevent them from sending signals on to the ganglion cells.

When light strikes the retina and *cis*-retinal undergoes isomerization, enzymes are activated that break down cGMP. As a result, some cGMP-gated Na^+ channels close, Na^+ inflow decreases, and the membrane potential becomes more negative, approaching -70 mV (Figure 17.16b). This sequence of events produces a hyperpolarizing receptor potential that decreases the release of glutamate. Dim lights cause small and brief receptor potentials that partially turn off glutamate release; brighter lights elicit larger and longer receptor potentials that more completely shut down neurotransmitter release. Thus, light excites the bipolar cells that synapse with rods by turning off the release of an inhibitory neurotransmitter. The excited bipolar cells subsequently stimulate the ganglion cells to form action potentials in their axons.

CLINICAL CONNECTION | *Color Blindness and Night Blindness*

Most forms of **color blindness,** an inherited inability to distinguish between certain colors, result from the absence or deficiency of one of the three types of cones. The most common type is *red-green color blindness,* in which red cones or green cones are missing. As a result, the person cannot distinguish between red and green. Prolonged vitamin A deficiency and the resulting below-normal amount of rhodopsin may cause **night blindness** or **nyctalopia** (nik′-ta-LŌ-pē-a), an inability to see well at low light levels. •

The Visual Pathway

Visual signals in the retina undergo considerable processing at synapses among the various types of neurons in the retina (horizontal cells, bipolar cells, and amacrine cells; see Figure 17.10). Then, the axons of retinal ganglion cells provide output from the retina to the brain, exiting the eyeball as the **optic (II) nerve.**

Processing of Visual Input in the Retina

Within the neural layer of the retina, certain features of visual input are enhanced while other features may be discarded. Input from several cells may either converge on a smaller number of postsynaptic neurons or diverge to a large number. Overall, convergence predominates: There are only 1 million ganglion cells, but 126 million photoreceptors in the human eye.

Once receptor potentials arise in the outer segments of rods and cones, they spread through the inner segments to the synaptic terminals. Neurotransmitter molecules released by rods and cones induce local graded potentials in both bipolar cells and horizontal cells. Between 6 and 600 rods synapse with a single bipolar cell in the outer synaptic layer of the retina; a cone more often synapses with a single bipolar cell. The convergence of many rods onto a single bipolar cell increases the light sensitivity of rod vision but slightly blurs the image that is perceived. Cone vision, although less sensitive, is sharper because of the one-to-one synapses between cones and their bipolar cells. Stimulation of rods by light excites bipolar cells; cone bipolar cells may be either excited or inhibited when a light is turned on.

Horizontal cells transmit inhibitory signals to bipolar cells in the areas lateral to excited rods and cones. This lateral inhibition enhances contrasts in the visual scene between areas of the retina that are strongly stimulated and adjacent areas that are more weakly stimulated. Horizontal cells also assist in the differentiation of various colors. Amacrine cells, which are excited by bipolar cells, synapse with ganglion cells and transmit information to them that signals a change in the level of illumination of the retina. When bipolar or amacrine cells transmit excitatory signals to ganglion cells, the ganglion cells become depolarized and initiate nerve impulses.

Brain Pathway and Visual Fields

The axons within the **optic (II) nerve** pass through the **optic chiasm** (kī-AZ-em = a crossover, as in the letter X), a crossing point of the optic nerves (Figure 17.17a, b). Some axons cross to the opposite side, but others remain uncrossed. After passing through the optic chiasm, the axons, now part of the **optic tract,** enter the brain and most of them terminate in the **lateral geniculate nucleus** of the thalamus. Here they synapse with neurons whose axons form the **optic radiations,** which project to the **primary visual areas** in the occipital lobes of the cerebral cortex (area 17 in Figure 14.15), and visual perception begins. Some of the fibers in the optic tracts terminate in the **superior colliculi,** which control the extrinsic eye muscles, and the **pretectal nuclei,** which control pupillary and accommodation reflexes.

Everything that can be seen by one eye is that eye's **visual field.** As noted earlier, because our eyes are located anteriorly in our heads, the visual fields overlap considerably (Figure 17.17b). We have binocular vision due to the large region where the visual fields of the two eyes overlap—the **binocular visual field.** The visual field of each eye is divided into two regions: the **nasal** or **central half** and the **temporal** or **peripheral half.** For each eye, light rays from an object in the nasal half of the visual field fall on the temporal half of the retina, and light rays from an object in the temporal half of the visual field fall on the nasal half of the retina. Visual information from the *right* half of each visual field is conveyed to the *left* side of the brain, and visual information from the *left* half of each visual field is conveyed to the *right* side of the brain, as follows (Figure 17.17c, d):

1. The axons of all retinal ganglion cells in one eye exit the eyeball at the optic disc and form the optic nerve on that side.
2. At the optic chiasm, axons from the temporal half of each retina do not cross but continue directly to the lateral geniculate nucleus of the thalamus on the same side.
3. In contrast, axons from the nasal half of each retina cross the optic chiasm and continue to the opposite thalamus.
4. Each optic tract consists of crossed and uncrossed axons that project from the optic chiasm to the thalamus on one side.
5. Axon collaterals (branches) of the retinal ganglion cells project to the midbrain, where they participate in neural circuits that govern constriction of the pupils in response to light and coordination of head and eye movements. Collaterals also extend to the suprachiasmatic nucleus of the hypothalamus, which establishes patterns of sleep and other activities that occur on a circadian or daily schedule in response to intervals of light and darkness.
6. The axons of thalamic neurons form the optic radiations as they project from the thalamus to the primary visual area of the cortex on the same side.

Although we have just described the visual pathway as a single pathway, visual signals are thought to be processed by at least three separate systems in the cerebral cortex, each with its own function. One system processes information related to the shape of objects, another system processes information regarding color of objects, and a third system processes information about movement, location, and spatial organization.

✔**CHECKPOINT**

6. What is the function of the lacrimal apparatus?
7. What types of cells make up the neural layer and the pigmented layer of the retina?
8. How do photopigments respond to light and recover in darkness?
9. How do receptor potentials arise in photoreceptors?
10. By what pathway do nerve impulses triggered by an object in the nasal half of the visual field of the left eye reach the primary visual area of the cortex?

17.4 HEARING AND EQUILIBRIUM

OBJECTIVES

- Describe the anatomy of the structures in the three main regions of the ear.
- List the major events in the physiology of hearing.
- Identify the receptor organs for equilibrium, and describe how they function.
- Describe the auditory and equilibrium pathways.

Hearing is the ability to perceive sounds. The ear is an engineering marvel because its sensory receptors can transduce sound vibrations with amplitudes as small as the diameter of an atom of gold (0.3 nm) into electrical signals 1000 times faster than photoreceptors can respond to light. The ear also contains receptors

Figure 17.17 The visual pathway. (a) Partial dissection of the brain reveals the optic radiations (axons extending from the thalamus to the occipital lobe). (b) An object in the binocular visual field can be seen with both eyes. In (c) and (d), note that information from the right side of the visual field of each eye projects to the left side of the brain, and information from the left side of the visual field of each eye projects to the right side of the brain.

6 The axons of ganglion cells in the temporal half of each retina extend to the thalamus on the same side; the axons of ganglion cells in the nasal half of each retina extend to the thalamus on the opposite side.

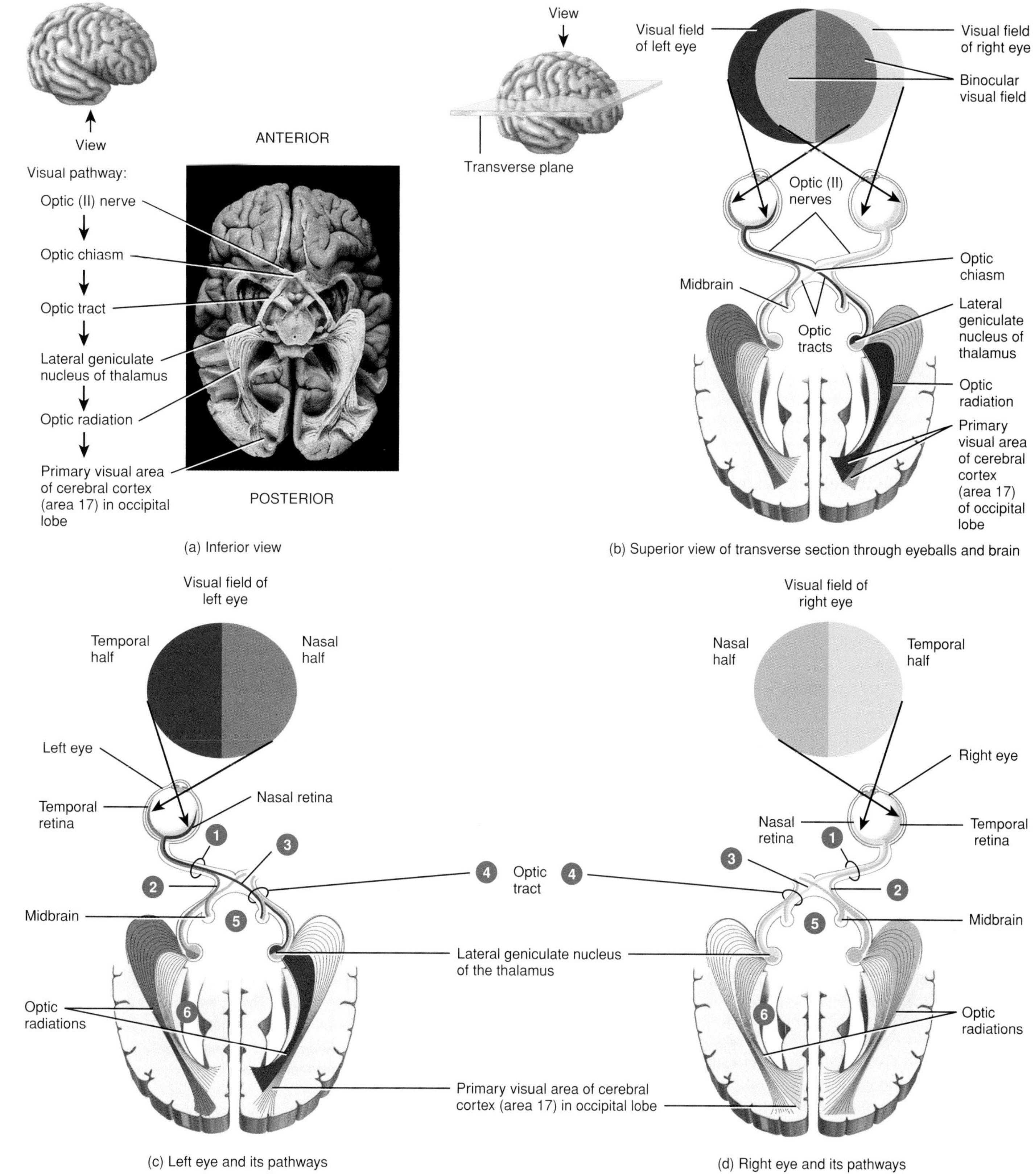

? Light rays from an object in the temporal half of the visual field strike which half of the retina?

for equilibrium, the sense that helps you maintain your balance and be aware of your orientation in space.

Anatomy of the Ear

The ear is divided into three main regions: (1) the external ear, which collects sound waves and channels them inward; (2) the middle ear, which conveys sound vibrations to the oval window; and (3) the internal ear, which houses the receptors for hearing and equilibrium.

External (Outer) Ear

The **external (outer) ear** consists of the auricle, external auditory canal, and eardrum (Figure 17.18). The **auricle** (AW-ri-kul) or **pinna** is a flap of elastic cartilage shaped like the flared end of a trumpet and covered by skin. The rim of the auricle is the **helix;** the inferior portion is the **lobule.** Ligaments and muscles attach the auricle to the head. The **external auditory canal** (*audit-* = hearing) is a curved tube about 2.5 cm (1 in.) long that lies in the temporal bone and leads to the eardrum. The **tympanic membrane** (tim-PAN-ik; *tympan-* = a drum) or **eardrum** is a thin, semitransparent partition between the external auditory canal and middle ear. The tympanic membrane is covered by epidermis and lined by simple cuboidal epithelium. Between the epithelial layers is connective tissue composed of collagen, elastic fibers, and fibroblasts. Tearing of the tympanic membrane is called a **perforated eardrum.** It may be due to pressure from a cotton swab, trauma, or a middle ear infection, and usually heals within a month. The tympanic membrane may be examined directly by an **otoscope** (Ō-tō-skōp; *oto-* = ear; *-skopeo* = to view), a viewing instrument that illuminates and magnifies the external auditory canal and tympanic membrane.

Near the exterior opening, the external auditory canal contains a few hairs and specialized sweat glands called **ceruminous glands** (se-RŪ-mi-nus) that secrete **earwax** or **cerumen** (se-RŪ-men). The combination of hairs and cerumen helps prevent dust and foreign objects from entering the ear. Cerumen also prevents damage to the delicate skin of the external ear canal by water and insects. Cerumen usually dries up and falls out of the ear canal. However, some people produce a large amount of cerumen, which can become

Figure 17.18 Anatomy of the ear.

The ear has three principal regions: the external (outer) ear, the middle ear, and the internal (inner) ear.

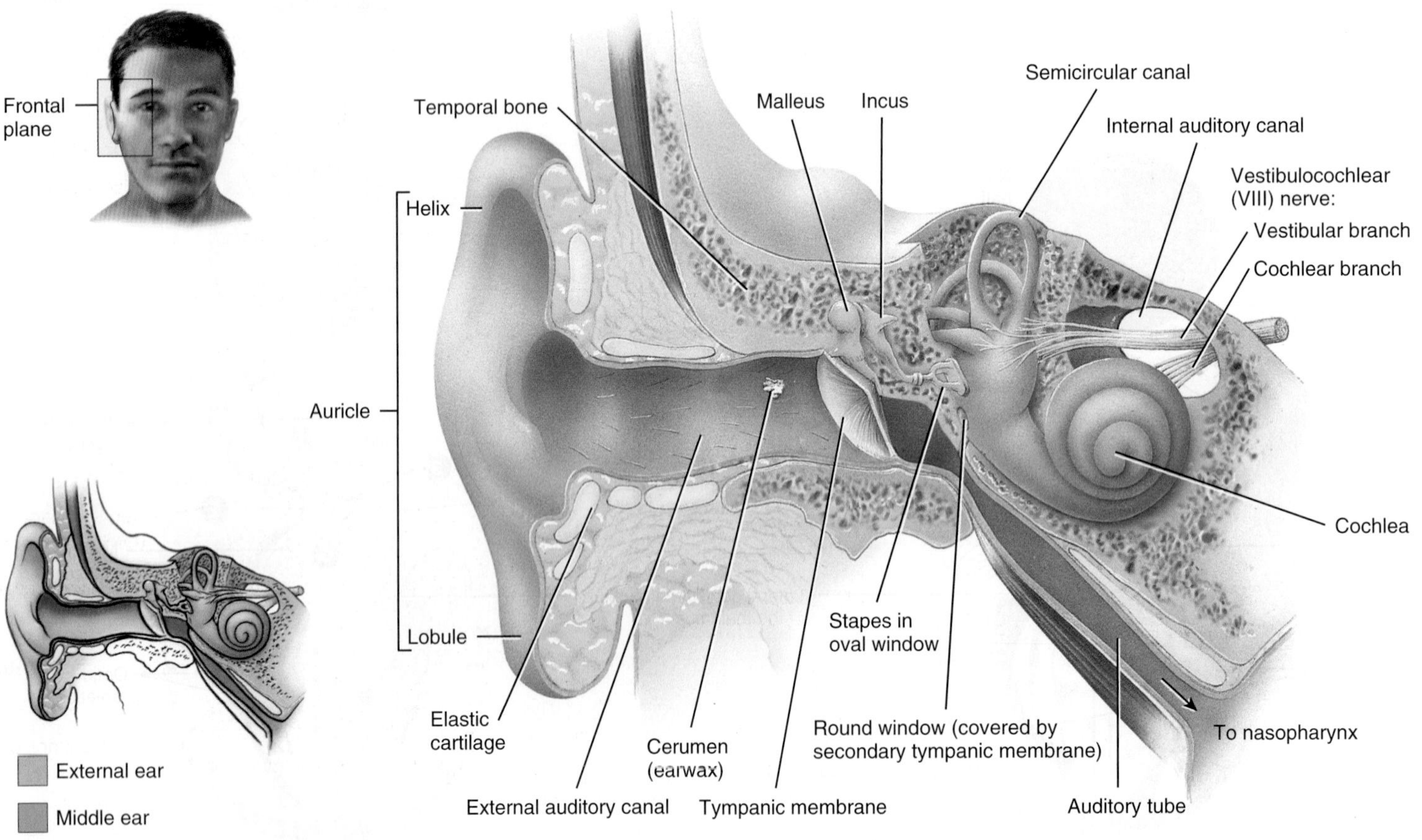
Frontal plane
Temporal bone
Malleus
Incus
Semicircular canal
Internal auditory canal
Vestibulocochlear (VIII) nerve:
Vestibular branch
Cochlear branch
Helix
Auricle
Cochlea
Lobule
Stapes in oval window
Elastic cartilage
Cerumen (earwax)
Round window (covered by secondary tympanic membrane)
To nasopharynx
External auditory canal
Tympanic membrane
Auditory tube
External ear
Middle ear
Internal ear

Frontal section through right side of skull showing three principal regions of ear

To which structure of the external ear does the malleus of the middle ear attach?

impacted and can muffle incoming sounds. The treatment for **impacted cerumen** is usually periodic ear irrigation or removal of wax with a blunt instrument by trained medical personnel.

Middle Ear

The **middle ear** is a small, air-filled cavity in the petrous portion of the temporal bone that is lined by epithelium (Figure 17.19). It is separated from the external ear by the tympanic membrane and from the internal ear by a thin bony partition that contains two small membrane-covered openings: the oval window and the round window. Extending across the middle ear and attached to it by ligaments are the three smallest bones in the body, the **auditory ossicles** (OS-si-kuls), which are connected by synovial joints. The bones, named for their shapes, are the malleus, incus, and stapes—commonly called the hammer, anvil, and stirrup, respectively. The "handle" of the **malleus** (MAL-ē-us) attaches to the internal surface of the tympanic membrane. The head of the malleus articulates with the body of the incus. The **incus** (ING-kus), the middle bone in the series, articulates with the head of the stapes. The base or footplate of the **stapes** (STĀ-pēz) fits into the **oval window.** Directly below the oval window is another opening, the **round window,** which is enclosed by a membrane called the **secondary tympanic membrane.**

Besides the ligaments, two tiny skeletal muscles also attach to the ossicles (Figure 17.19). The **tensor tympani** (TIM-pan-ē) **muscle,** which is supplied by the mandibular branch of the trigeminal (V) nerve, limits movement and increases tension on the eardrum to prevent damage to the inner ear from loud noises. The **stapedius** (sta-PĒ-de-us) **muscle,** which is supplied by the facial (VII) nerve, is the smallest skeletal muscle in the human body. By dampening large vibrations of the stapes due to loud noises, it protects the oval window, but it also decreases the sensitivity of hearing. For this reason, paralysis of the stapedius muscle is associated with **hyperacusia** (hī′-per-a-KŪ-sē-a), which is abnormally sensitive hearing. Because it takes a fraction of a second for the tensor tympani and stapedius muscles to contract, they can protect the inner ear from prolonged loud noises but not from brief ones such as a gunshot.

The anterior wall of the middle ear contains an opening that leads directly into the **auditory (pharyngotympanic) tube,**

Figure 17.19 The right middle ear and the auditory ossicles.

Common names for the malleus, incus, and stapes are the hammer, anvil, and stirrup, respectively.

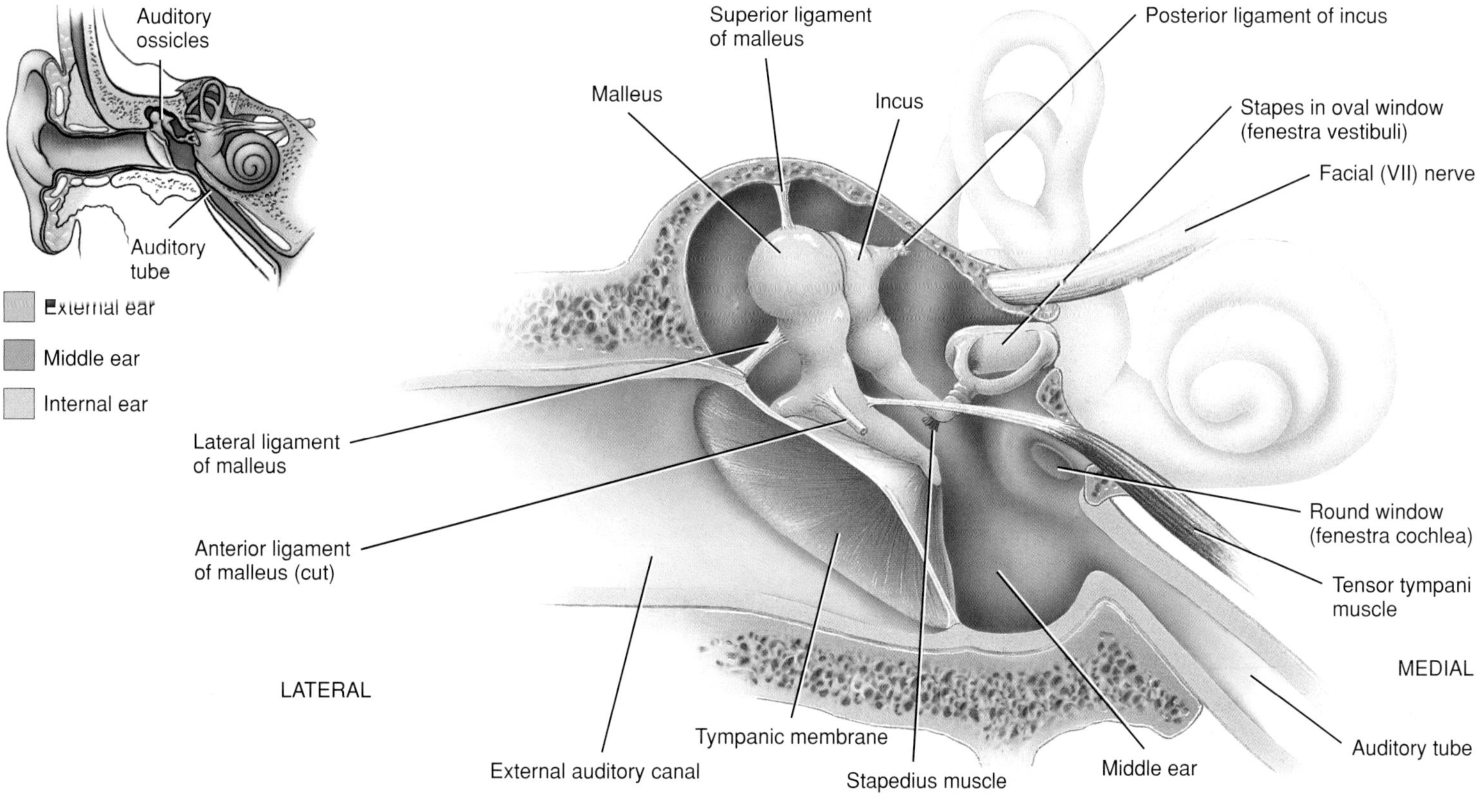

Frontal section showing location of auditory ossicles in the middle ear

What structures separate the middle ear from the internal ear?

commonly known as the **eustachian tube** (ū′-STĀ-kē-an, ū-STĀ-shun). The auditory tube, which consists of both bone and elastic cartilage, connects the middle ear with the nasopharynx (superior portion of the throat). It is normally closed at its medial (pharyngeal) end. During swallowing and yawning, it opens, allowing air to enter or leave the middle ear until the pressure in the middle ear equals the atmospheric pressure. Most of us have experienced our ears popping as the pressures equalize. When the pressures are balanced, the tympanic membrane vibrates freely as sound waves strike it. If the pressure is not equalized, intense pain, hearing impairment, ringing in the ears, and vertigo could develop. The auditory tube also is a route for pathogens to travel from the nose and throat to the middle ear, causing the most common type of ear infection (see *otitis media* in Disorders: Homeostatic Imbalances at the end of this chapter).

Internal (Inner) Ear

The **internal (inner) ear** is also called the **labyrinth** (LAB-i-rinth) because of its complicated series of canals (Figure 17.20). Structurally, it consists of two main divisions: an outer bony labyrinth that encloses an inner membranous labyrinth. It is like long balloons put inside a rigid tube. The **bony labyrinth** is a series of cavities in the petrous portion of the temporal bone divided into three areas: (1) the semicircular canals, (2) the vestibule, and (3) the cochlea. The bony labyrinth is lined with periosteum and contains **perilymph.** This fluid, which is chemically similar to cerebrospinal fluid, surrounds the **membranous labyrinth,** a series of epithelial sacs and tubes inside the bony labyrinth that have the same general form as the bony labyrinth and house the receptors for hearing and equilibrium. The epithelial membranous labyrinth contains **endolymph.** The level of potassium ions (K^+) in endolymph is unusually high for an extracellular fluid, and potassium ions play a role in the generation of auditory signals (described shortly).

The **vestibule** (VES-ti-būl) is the oval central portion of the bony labyrinth. The membranous labyrinth in the vestibule consists of two sacs called the **utricle** (Ū-tri-kul = little bag) and the **saccule** (SAK-ūl = little sac), which are connected by a small duct. Projecting superiorly and posteriorly from the vestibule are the three bony **semicircular canals,** each of which lies at approximately right angles to the other two. Based on their positions,

Figure 17.20 The right internal ear. The outer, cream-colored area is part of the bony labyrinth; the inner, pink-colored area is the membranous labyrinth.

6 The bony labyrinth contains perilymph, and the membranous labyrinth contains endolymph.

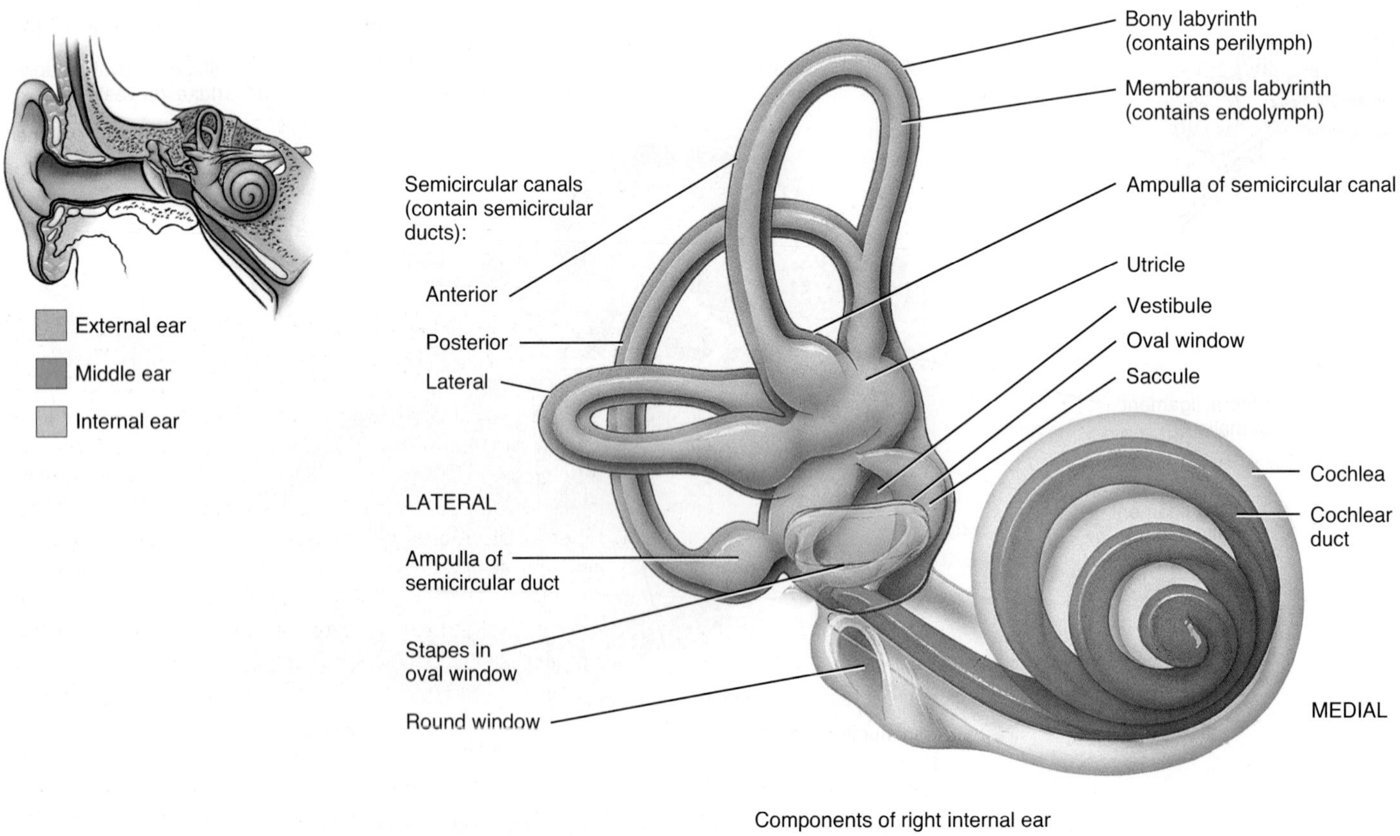

Components of right internal ear

What are the names of the two sacs that lie in the membranous labyrinth of the vestibule?

they are named the anterior, posterior, and lateral semicircular canals. The anterior and posterior semicircular canals are vertically oriented; the lateral one is horizontally oriented. At one end of each canal is a swollen enlargement called the **ampulla** (am-PUL-la = saclike duct). The portions of the membranous labyrinth that lie inside the bony semicircular canals are called the **semicircular ducts.** These structures connect with the utricle of the vestibule.

The vestibular branch of the vestibulocochlear (VIII) nerve consists of *ampullary, utricular,* and *saccular nerves.* These nerves contain both first-order sensory neurons and motor neurons that synapse with receptors for equilibrium. The first-order sensory neurons carry sensory information from the receptors, and the motor neurons carry feedback signals to the receptors, apparently to modify their sensitivity. Cell bodies of the sensory neurons are located in the **vestibular ganglia** (ves-TIB-ū-lar) (see Figure 17.21b).

Anterior to the vestibule is the **cochlea** (KOK-lē-a = snail-shaped), a bony spiral canal (Figure 17.21a) that resembles a snail's shell and makes almost three turns around a central bony core called the **modiolus** (mō-DĪ-ō′-lus; Figure 17.21b). Sections through the cochlea reveal that it is divided into three channels: cochlear duct, scala vestibuli, and scala tympani (Figure 17.21a–c). The **cochlear duct (scala media)** (KOK-lē-ar) is a continuation of the membranous labyrinth into the cochlea; it is filled with endolymph. The channel above the cochlear duct is the **scala vestibuli,** which ends at the oval window. The channel below is the **scala tympani,** which ends at the round window. Both the scala vestibuli and scala tympani are part of the bony labyrinth of the cochlea; therefore, these chambers are filled with perilymph. The scala vestibuli and scala tympani are completely separated by the cochlear duct, except for an opening at the apex of the cochlea, the **helicotrema** (hel-i-kō-TRĒ-ma; Figure 17.22). The cochlea adjoins the wall of the vestibule, into which the scala vestibuli opens. The perilymph in the vestibule is continuous with that of the scala vestibuli.

The **vestibular membrane** separates the cochlear duct from the scala vestibuli, and the **basilar membrane** (BĀS-i-lar) separates the cochlear duct from the scala tympani. Resting on the basilar membrane is the **spiral organ** or **organ of Corti** (KOR-tē) (Figure 17.21c, d). The spiral organ is a coiled sheet of epithelial cells, including supporting cells and about 16,000 **hair cells,** which are the receptors for hearing. There are two groups of hair cells: The *inner hair cells* are arranged in a single row, whereas the *outer hair cells* are arranged in three rows. At the apical tip of each hair cell are 40–80 stereocilia that extend into the endolymph of the cochlear duct. Despite their name, stereocilia are actually long, hairlike microvilli arranged in several rows of graded height.

At their basal ends, inner and outer hair cells synapse both with first-order sensory neurons and with motor neurons from the cochlear branch of the vestibulocochlear (VIII) nerve. Cell bodies of the sensory neurons are located in the **spiral ganglion** (Figure 17.21b, c). Although outer hair cells outnumber them by 3 to 1, the inner hair cells synapse with 90–95% of the first-order sensory neurons in the cochlear nerve that relay auditory information to the brain. By contrast, 90% of the motor neurons in the cochlear nerve synapse with outer hair cells. The **tectorial membrane** (tek-TŌ-rē-al; *tector-* = covering), a flexible gelatinous membrane, covers the hair cells of the spiral organ (Figure 17.21d). In fact, the ends of the stereocilia of the hair cells are embedded in the tectorial membrane while the bodies of the hair cells rest on the basilar membrane.

The Nature of Sound Waves

In order to understand the physiology of hearing, it is necessary to learn something about its input, which occurs in the form of sound waves. **Sound waves** are alternating high- and low-pressure regions traveling in the same direction through some medium (such as air). They originate from a vibrating object in much the same way that ripples arise and travel over the surface of a pond when you toss a stone into it. The *frequency* of a sound vibration is its *pitch.* The higher the frequency of vibration, the higher is the pitch. The sounds heard most acutely by the human ear are those from sources that vibrate at frequencies between 500 and 5000 hertz (Hz; 1 Hz = 1 cycle per second). The entire audible range extends from 20 to 20,000 Hz. Sounds of speech primarily contain frequencies between 100 and 3000 Hz, and the "high C" sung by a soprano has a dominant frequency at 1048 Hz. The sounds from a jet plane several miles away range from 20 to 100 Hz.

The larger the *intensity* (size or amplitude) of the vibration, the *louder* is the sound. Sound intensity is measured in units called **decibels (dB).** An increase of one decibel represents a tenfold increase in sound intensity. The hearing threshold—the point at which an average young adult can just distinguish sound from silence—is defined as 0 dB at 1000 Hz. Rustling leaves have a decibel level of 15; whispered speech, 30; normal conversation, 60; a vacuum cleaner, 75; shouting, 80; and a nearby motorcycle or jackhammer, 90. Sound becomes uncomfortable to a normal ear at about 120 dB, and painful above 140 dB.

CLINICAL CONNECTION | Loud Sounds and Hair Cell Damage

Exposure to loud music and the engine roar of jet planes, revved-up motorcycles, lawn mowers, and vacuum cleaners damages hair cells of the cochlea. Because prolonged noise exposure causes hearing loss, employers in the United States must require workers to use hearing protectors when occupational noise levels exceed 90 dB. Rock concerts and even inexpensive headphones can easily produce sounds over 110 dB. Continued exposure to high-intensity sounds is one cause of **deafness,** a significant or total hearing loss. The louder the sounds, the more rapid is the hearing loss. Deafness usually begins with loss of sensitivity for high-pitched sounds. If you are listening to music through ear buds and bystanders can hear it, the decibel level is in the damaging range. Most people fail to notice their progressive hearing loss until destruction is extensive and they begin having difficulty understanding speech. Wearing earplugs with a noise-reduction rating of 30 dB while engaging in noisy activities can protect the sensitivity of your ears. •

Figure 17.21 Semicircular canals, vestibule, and cochlea of the right ear. Note that the cochlea makes nearly three complete turns.

The three channels in the cochlea are the scala vestibuli, the scala tympani, and the cochlear duct.

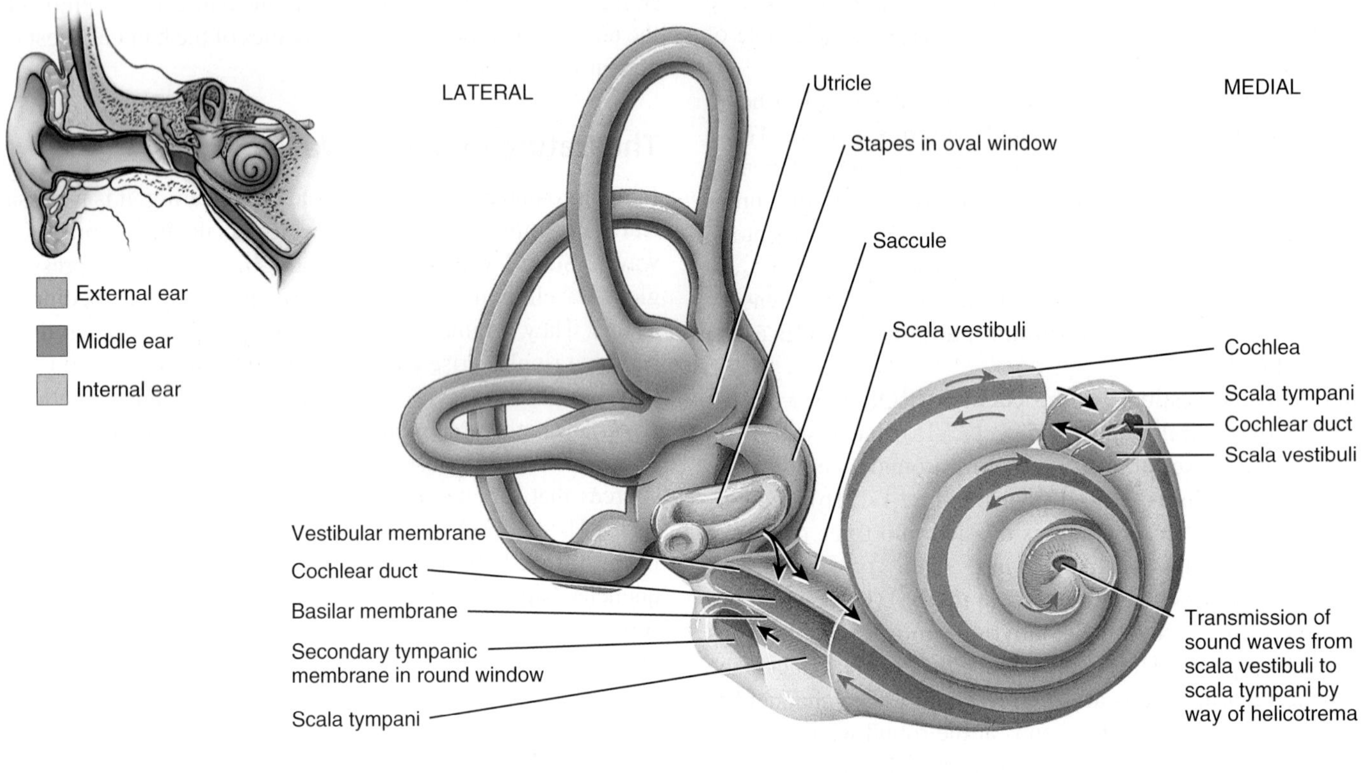

(a) Sections through cochlea

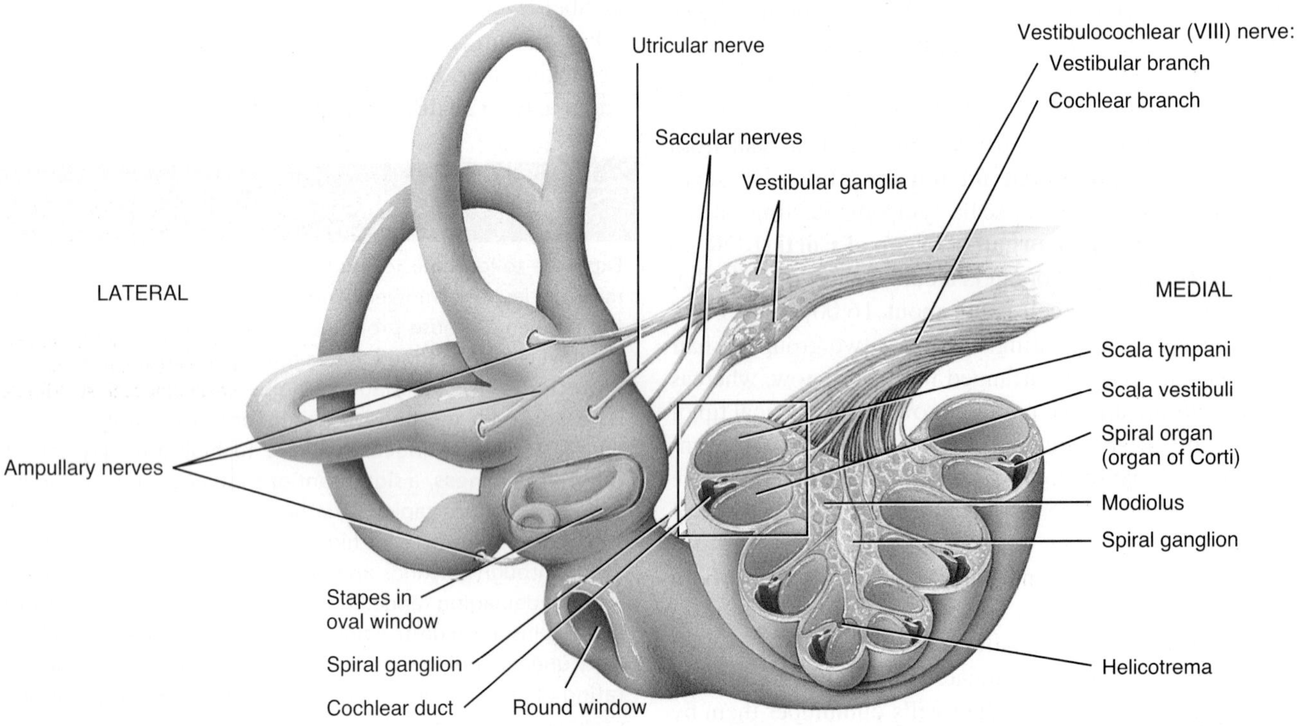

(b) Components of vestibulocochlear nerve (cranial nerve VIII)

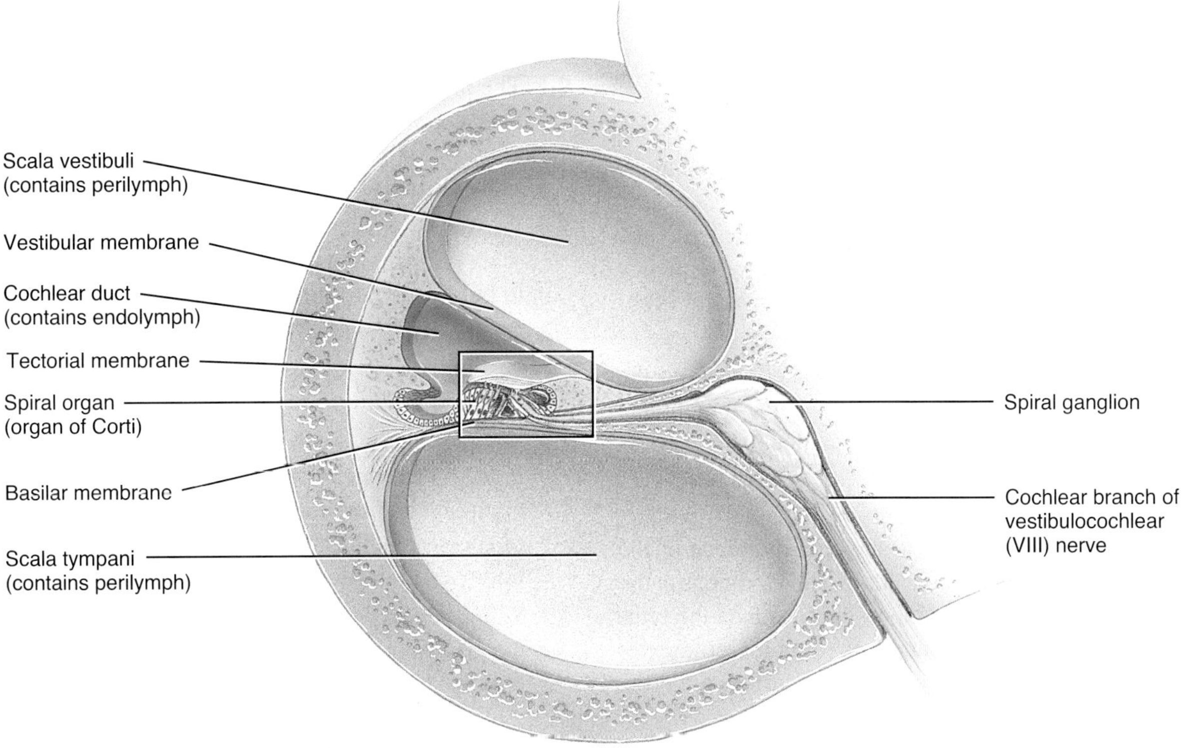
Scala vestibuli
(contains perilymph)
Vestibular membrane
Cochlear duct
(contains endolymph)
Tectorial membrane
Spiral organ
(organ of Corti)
Basilar membrane
Scala tympani
(contains perilymph)
Spiral ganglion
Cochlear branch of
vestibulocochlear
(VIII) nerve

(c) Section through one turn of cochlea

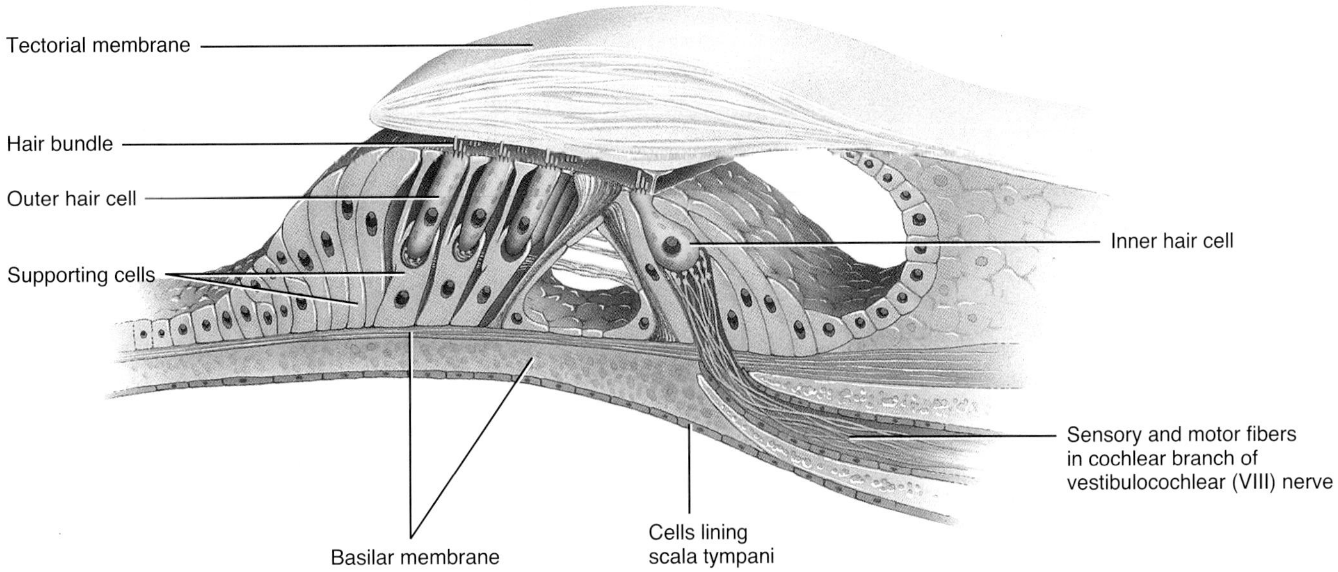
Tectorial membrane
Hair bundle
Outer hair cell
Supporting cells
Inner hair cell
Sensory and motor fibers
in cochlear branch of
vestibulocochlear (VIII) nerve
Basilar membrane
Cells lining
scala tympani

(d) Enlargement of spiral organ (organ of Corti)

What are the three subdivisions of the bony labyrinth?

Physiology of Hearing

The following events are involved in hearing (Figure 17.22):

1. The auricle directs sound waves into the external auditory canal.
2. When sound waves strike the tympanic membrane, the alternating waves of high and low pressure in the air cause the tympanic membrane to vibrate back and forth. The tympanic membrane vibrates slowly in response to low-frequency (low-pitched) sounds and rapidly in response to high-frequency (high-pitched) sounds.
3. The central area of the tympanic membrane connects to the malleus, which vibrates along with the tympanic membrane. This vibration is transmitted from the malleus to the incus and then to the stapes.
4. As the stapes moves back and forth, its oval-shaped footplate, which is attached via a ligament to the circumference of the oval window, vibrates in the oval window. The vibrations at the oval window are about 20 times more vigorous than the tympanic membrane because the auditory ossicles efficiently transmit small vibrations spread over a large surface area (the tympanic membrane) into larger vibrations at a smaller surface (the oval window).
5. The movement of the stapes at the oval window sets up fluid pressure waves in the perilymph of the cochlea. As the oval window bulges inward, it pushes on the perilymph of the scala vestibuli.
6. Pressure waves are transmitted from the scala vestibuli to the scala tympani and eventually to the round window, causing it to bulge outward into the middle ear. (See 9 in the figure.)
7. The pressure waves travel through the perilymph of the scala vestibuli, then the vestibular membrane, and then move into the endolymph inside the cochlear duct.
8. The pressure waves in the endolymph cause the basilar membrane to vibrate, which moves the hair cells of the spiral organ against the tectorial membrane. This leads to bending of the stereocilia and ultimately to the generation of nerve impulses in first-order neurons in cochlear nerve fibers.
9. Sound waves of various frequencies cause certain regions of the basilar membrane to vibrate more intensely than other regions. Each segment of the basilar membrane is "tuned" for

Figure 17.22 Events in the stimulation of auditory receptors in the right ear. The numbers correspond to the events listed in the text. The cochlea has been uncoiled to more easily visualize the transmission of sound waves and their distortion of the vestibular and basilar membranes of the cochlear duct.

Hair cells of the spiral organ (organ of Corti) convert a mechanical vibration (stimulus) into an electrical signal (receptor potential).

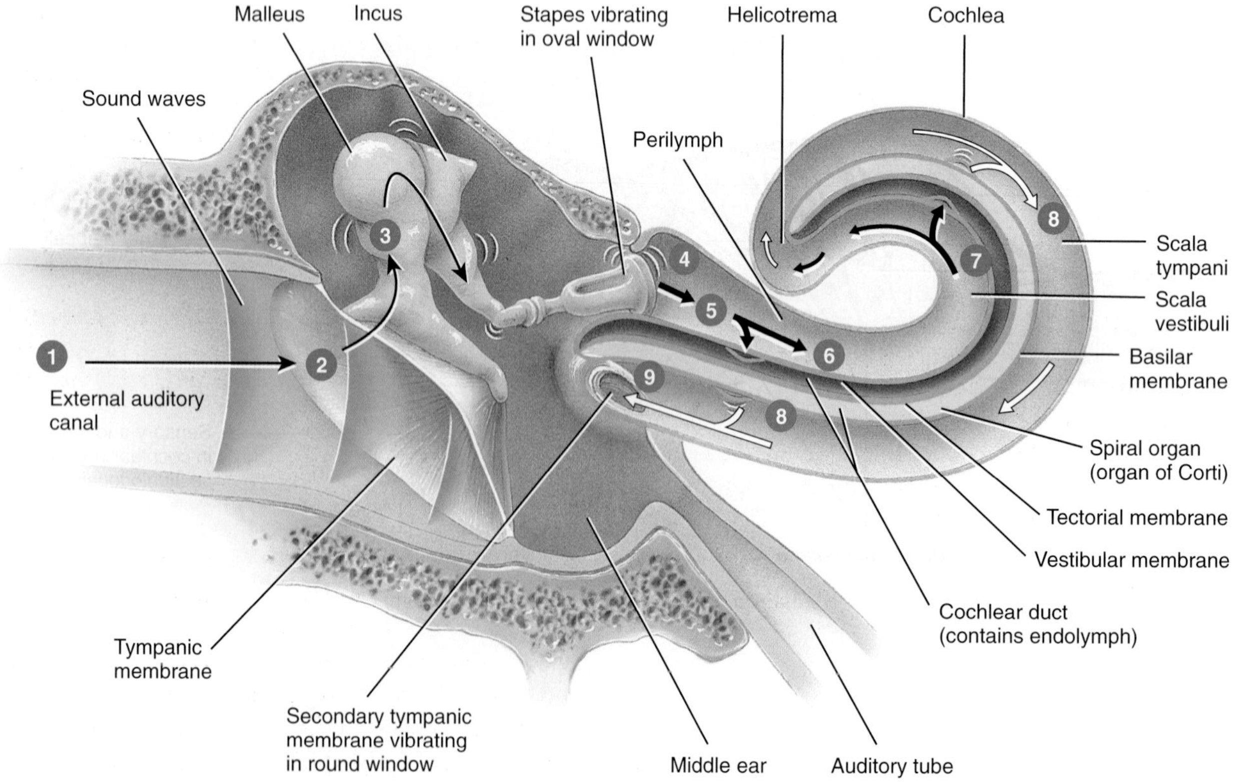

Which part of the basilar membrane vibrates most vigorously in response to high-frequency (high-pitched) sounds?

a particular pitch. Because the membrane is narrower and stiffer at the base of the cochlea (closer to the oval window), high-frequency (high-pitched) sounds induce maximal vibrations in this region. Toward the apex of the cochlea, the basilar membrane is wider and more flexible; low-frequency (low-pitched) sounds cause maximal vibration of the basilar membrane there. Loudness is determined by the intensity of sound waves. High-intensity sound waves cause larger vibrations of the basilar membrane, which leads to a higher frequency of nerve impulses reaching the brain. Louder sounds also may stimulate a larger number of hair cells.

The hair cells transduce mechanical vibrations into electrical signals. As the basilar membrane vibrates, the hair bundles at the apex of the hair cell bend back and forth and slide against one another. A *tip link* protein connects the tip of each stereocilium to a mechanically gated ion channel called the **transduction channel** in its taller stereocilium neighbor. As the stereocilia bend in the direction of the taller stereocilia, the tip links tug on the transduction channels and open them. These channels allow cations in the endolymph, primarily K^+, to enter the hair cell cytosol. As cations enter, they produce a depolarizing receptor potential. Depolarization quickly spreads along the plasma membrane and opens voltage-gated Ca^{2+} channels in the base of the hair cell. The resulting inflow of Ca^{2+} triggers exocytosis of synaptic vesicles containing a neurotransmitter, which is probably glutamate. As more neurotransmitter is released, the frequency of nerve impulses in the first-order sensory neurons that synapse with the base of the hair cell increases. Bending of the stereocilia in the opposite direction closes the transduction channels, allows hyperpolarization to occur, and reduces neurotransmitter release from the hair cells. This decreases the frequency of nerve impulses in the sensory neurons.

Besides its role in detecting sounds, the cochlea has the surprising ability to produce sounds. These usually inaudible sounds, called **otoacoustic emissions** (ō-tō-a-KOO-stik), can be picked up by placing a sensitive microphone next to the eardrum. They are caused by vibrations of the outer hair cells that occur in response to sound waves and to signals from motor neurons. As they depolarize and repolarize, the outer hair cells rapidly shorten and lengthen. This vibratory behavior appears to change the stiffness of the tectorial membrane and is thought to enhance the movement of the basilar membrane, which amplifies the responses of the inner hair cells. At the same time, the outer hair cell vibrations set up a traveling wave that goes back toward the stapes and leaves the ear as an otoacoustic emission. Detection of these inner ear–produced sounds is a fast, inexpensive, and noninvasive way to screen newborns for hearing defects. In deaf babies, otoacoustic emissions are not produced or are greatly reduced in size.

The Auditory Pathway

Bending of the stereocilia of the hair cells of the spiral organ causes the release of a neurotransmitter (probably glutamate), which generates nerve impulses in the sensory neurons that innervate the hair cells. The cell bodies of the sensory neurons are located in the **spiral ganglia.** Nerve impulses pass along the axons of these neurons, which form the cochlear branch of the vestibulocochlear (VIII) nerve (Figure 17.23). These axons synapse with neurons in the **cochlear nuclei** in the medulla oblongata on the same side. Some of the axons from the cochlear nuclei decussate (cross over) in the medulla, ascend in a tract called the **lateral meniscus** on the opposite side, and terminate in the **inferior colliculus** in the midbrain. Other axons from the cochlear nuclei end in the **superior olivary nucleus** in the pons on each side. Slight differences in the timing of nerve impulses arriving from the two ears at the superior olivary nuclei allow us to locate the source of a sound. Axons from the superior olivary nuclei also ascend in the lateral meniscus tracts on both sides and end in the inferior colliculi. From each inferior colliculus, nerve impulses are conveyed to the **medial geniculate nucleus** in the thalamus and finally to the **primary auditory area** of the cerebral cortex in the temporal lobe of the cerebrum (see areas 41 and 42 in Figure 14.15). Because many auditory axons decussate in the medulla while others remain on the same side, the right and left primary auditory areas receive nerve impulses from both ears.

CLINICAL CONNECTION | *Cochlear Implants*

A **cochlear implant** is a device that translates sounds into electrical signals that can be interpreted by the brain. Such a device is useful for people with deafness that is caused by damage to hair cells in the cochlea. The external parts of a cochlear implant consist of (1) a *microphone* worn around the ear that picks up sound waves, (2) a *sound processor,* which may be placed in a shirt pocket, that converts sound waves into electrical signals, and (3) a *transmitter,* worn behind the ear, which receives signals from the sound processor and passes them to an internal receiver. The internal parts of a cochlear implant are the (1) *internal receiver,* which relays signals to (2) *electrodes* implanted in the cochlea, where they trigger nerve impulses in sensory neurons in the cochlear branch of the vestibulocochlear (VIII) nerve. These artificially induced nerve impulses propagate over their normal pathways to the brain. The perceived sounds are crude compared to normal hearing, but they provide a sense of rhythm and loudness; information about certain noises, such as those made by telephones and automobiles; and the pitch and cadence of speech. Some patients hear well enough with a cochlear implant to use the telephone. •

Physiology of Equilibrium

There are two types of **equilibrium** (ē-kwi-LIB-rē-um) or balance. **Static equilibrium** refers to the maintenance of the position of the body (mainly the head) relative to the force of gravity. Body movements that stimulate the receptors for static equilibrium include tilting the head and *linear* acceleration or deceleration, such as when the body is being moved in an elevator or in a car that is speeding up or slowing down. **Dynamic equilibrium** is the maintenance of body position (mainly the head) in response to sudden movements such as *rotational* acceleration or deceleration.

Figure 17.23 The auditory pathway.

From hair cells of the cochlea, auditory information is conveyed along the cochlear branch of the vestibulocochlear (VIII) nerve and then to the brain stem, thalamus, and cerebral cortex.

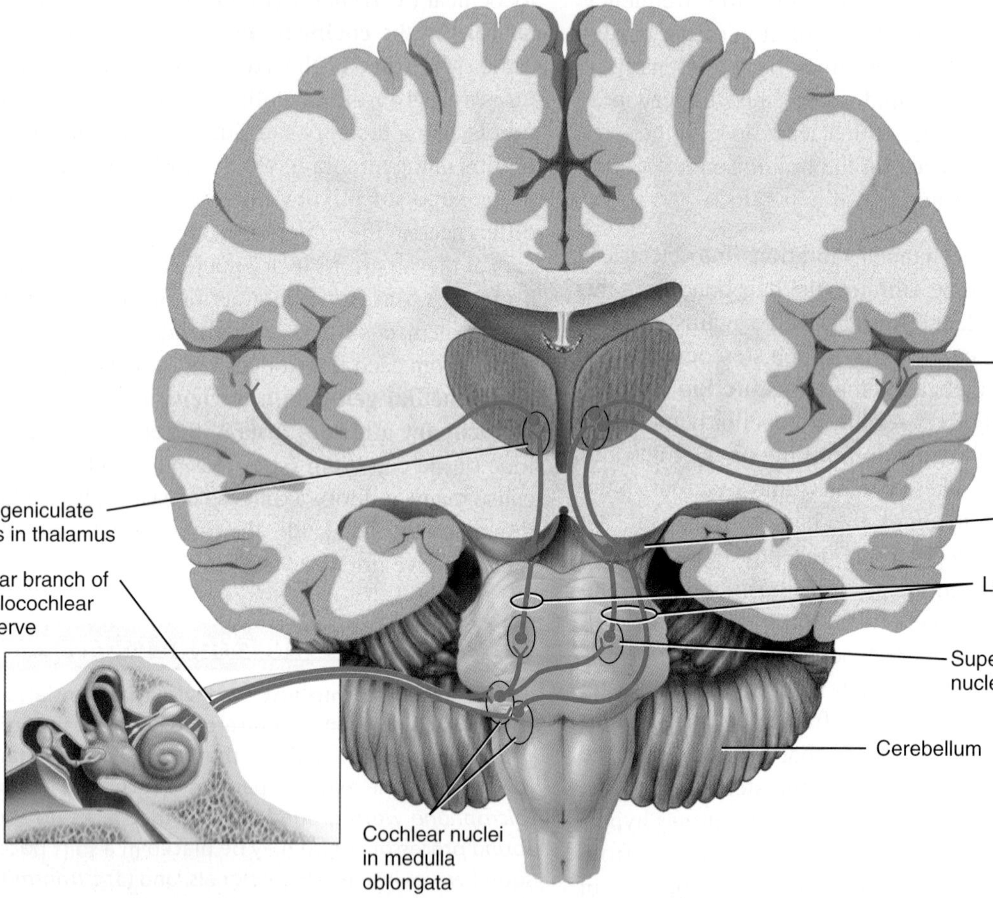

What is the function of the superior olivary nucleus of the pons?

Collectively, the receptor organs for equilibrium are called the **vestibular apparatus** (ves-TIB-ū-lar); these include the saccule, utricle, and semicircular ducts.

Otolithic Organs: Saccule and Utricle

The walls of both the utricle and the saccule contain a small, thickened region called a **macula** (MAK-ū-la; Figure 17.24). The two *maculae* (plural) (MAK-u-lē), which are perpendicular to one another, are the receptors for static equilibrium. They provide sensory information on the position of the head in space and are essential for maintaining appropriate posture and balance. The maculae also detect linear acceleration and deceleration—for example, the sensations you feel while in an elevator or car that is speeding up or slowing down.

The maculae consist of two kinds of cells: **hair cells,** which are the sensory receptors, and **supporting cells.** Hair cells have on their surface 40–80 stereocilia (which are actually microvilli) of graduated height, plus one *kinocilium,* a conventional cilium anchored firmly to its basal body and extending beyond the longest stereocilium. As in the cochlea, the stereocilia are connected by tip links. Collectively, the stereocilia and kinocilium are called a **hair bundle.** Scattered among the hair cells are columnar supporting cells that probably secrete the thick, gelatinous, glycoprotein layer, called the **otolithic membrane** (ō-tō-LITH-ik), that rests on the hair cells. A layer of dense calcium carbonate crystals, called **otoliths** (Ō-tō-liths; *oto-* = ear; *-liths* = stones), extends over the entire surface of the otolithic membrane.

Because the otolithic membrane sits on top of the macula, if you tilt your head forward, the otolithic membrane (along with the otoliths as well) is pulled by gravity. It slides "downhill" over the hair cells in the direction of the tilt, bending the hair bundles. However, if you are sitting upright in a car that suddenly jerks forward, the otolithic membrane lags behind the head movement, pulls on the hair bundles, and makes them bend in the other direction. Bending of the hair bundles in one direction stretches the tip links, which pull open transduction channels, producing depolarizing receptor potentials; bending in the opposite direction closes the transduction channels and produces hyperpolarization.

As the hair cells depolarize and repolarize, they release neurotransmitter at a faster or slower rate. The hair cells synapse with

Figure 17.24 Location and structure of receptors in the maculae of the right ear. Both first-order sensory neurons (blue) and motor neurons (red) synapse with the hair cells.

The movement of stereocilia initiates depolarizing receptor potentials.

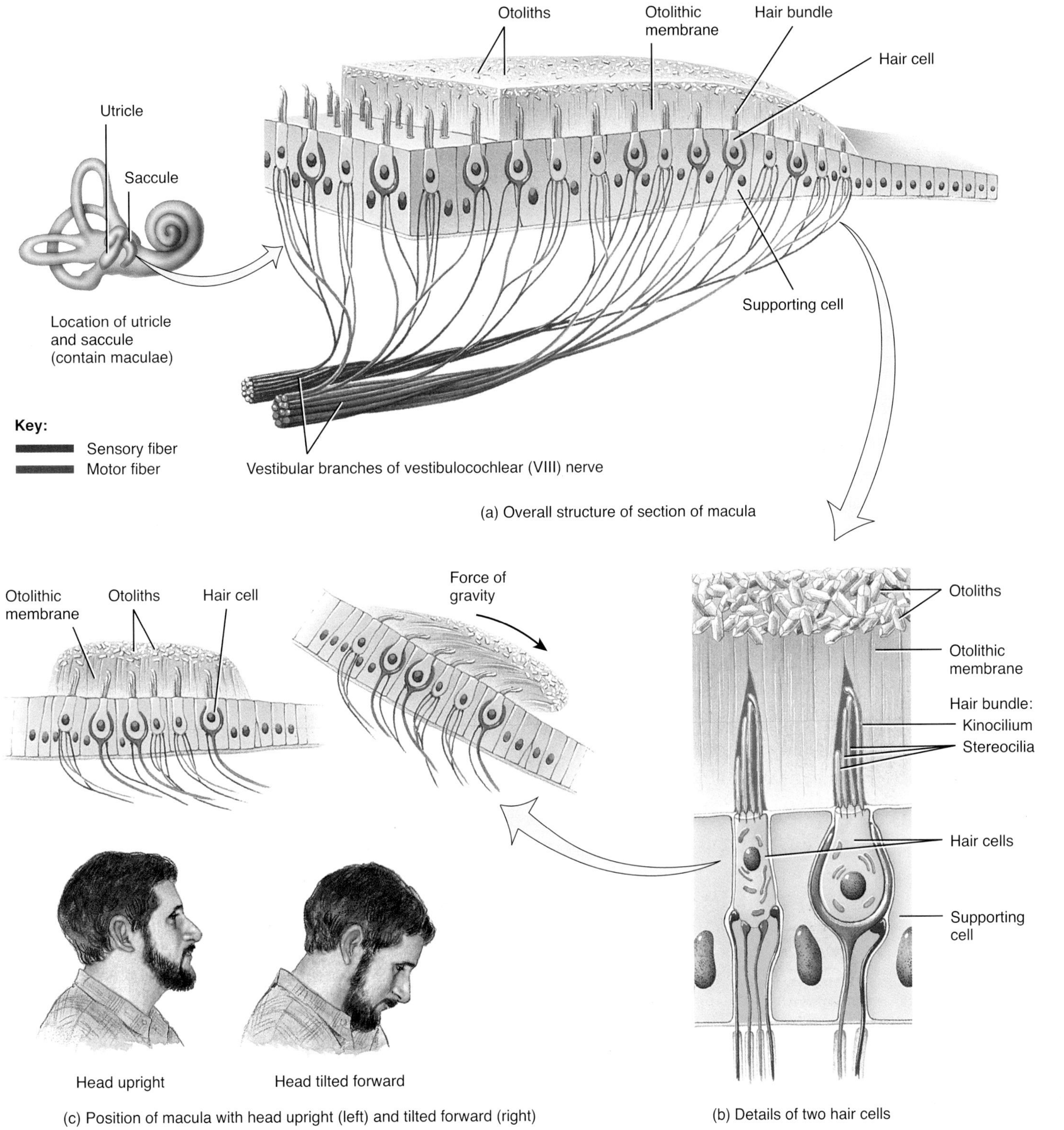

With which type of equilibrium are the maculae associated?

Figure 17.25 Location and structure of the semicircular ducts of the right ear. Both first-order sensory neurons (blue) and motor neurons (red) synapse with the hair cells. The ampullary nerves are branches of the vestibular division of the vestibulocochlear (VIII) nerve.

The positions of the semicircular ducts permit detection of rotational movements.

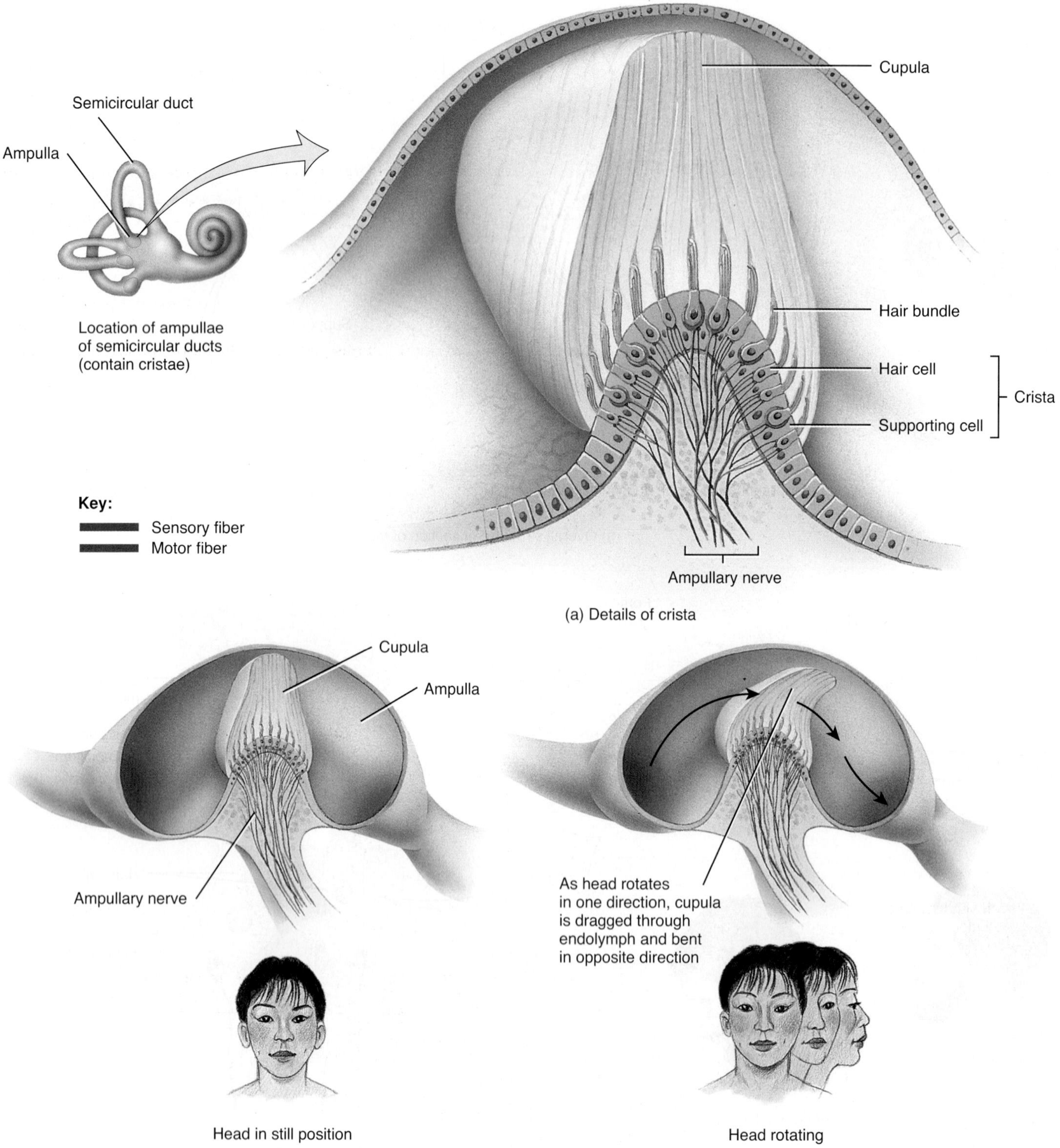
Semicircular duct
Ampulla
Location of ampullae
of semicircular ducts
(contain cristae)
Key:
Sensory fiber
Motor fiber
Cupula
Hair bundle
Hair cell
Supporting cell
Crista
Ampullary nerve
(a) Details of crista
Cupula
Ampulla
Ampullary nerve
As head rotates
in one direction, cupula
is dragged through
endolymph and bent
in opposite direction
Head in still position
Head rotating

(b) Position of cupula with head in still position (left) and when head rotates (right)

With which type of equilibrium are the semicircular ducts associated?

first-order sensory neurons in the vestibular branch of the vestibulocochlear (VIII) nerve (see Figure 17.21b). These neurons fire impulses at a slow or rapid pace depending on the amount of neurotransmitter present. Motor neurons also synapse with the hair cells and sensory neurons. Evidently, the motor neurons regulate the sensitivity of the hair cells and sensory neurons.

Semicircular Ducts

The three semicircular ducts function in dynamic equilibrium. The ducts lie at right angles to one another in three planes (Figure 17.25): The two vertical ducts are the anterior and posterior semicircular ducts, and the horizontal one is the lateral semicircular duct (see also Figure 17.20). This positioning permits detection of rotational acceleration or deceleration. In the **ampulla,** the dilated portion of each duct, is a small elevation called the **crista** (KRIS-ta = crest; plural is *cristae*). Each crista contains a group of hair cells and supporting cells. Covering the crista is a mass of gelatinous material called the **cupula** (KU-pū-la). When you move your head, the attached semicircular ducts and hair cells move with it. The endolymph within the ampulla, however, is not attached and lags behind. As the moving hair cells drag along the stationary endolymph, the hair bundles bend. Bending of the hair bundles produces receptor potentials. In turn, the receptor potentials lead to nerve impulses that pass along the vestibular branch of the vestibulocochlear (VIII) nerve.

Equilibrium Pathways

Bending of **hair bundles of the hair cells** in the semicircular ducts, utricle, or saccule causes the release of a neurotransmitter (probably glutamate), which generates nerve impulses in the sensory neurons that innervate the hair cells. The cell bodies of sensory neurons are located in the **vestibular ganglia.** Nerve impulses pass along the axons of these neurons, which form the **vestibular branch of the vestibulocochlear (VIII) nerve** (Figure 17.26). Most of these axons synapse with sensory neurons in **vestibular nuclei,** the major integrating centers for equilibrium, in the medulla oblongata and pons. The vestibular nuclei also receive input from the eyes and proprioceptors, especially proprioceptors in the neck and limb muscles that indicate the position of the head

Figure 17.26 The equilibrium pathway.

From hair cells of the semicircular ducts, utricle, and saccule, vestibular information is conveyed along the vestibular branch of the vestibulocochlear (VIII) nerve and then to the brain stem, cerebellum, thalamus, and cerebral cortex.

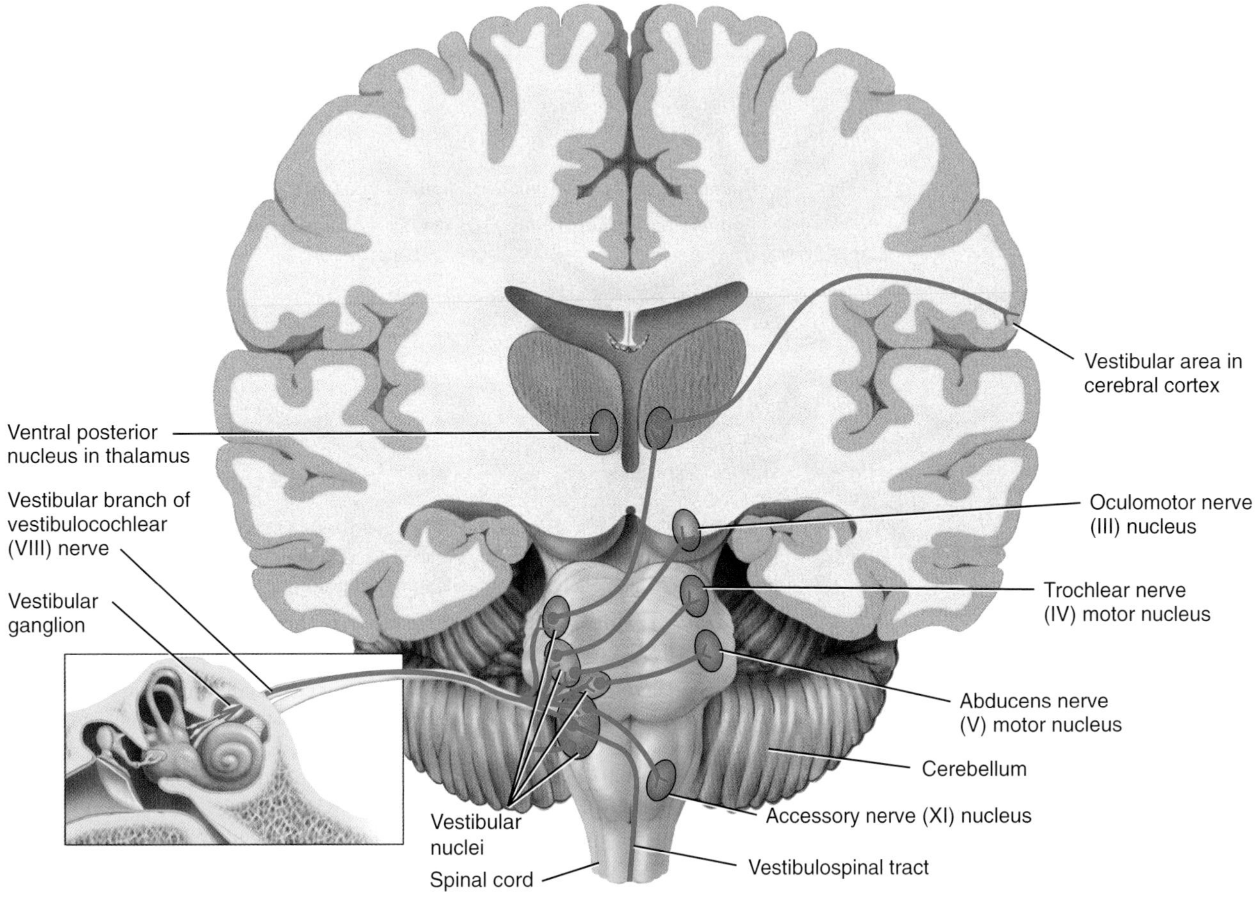

Where are the vestibular nuclei located?

and limbs. The remaining axons enter the cerebellum through the **inferior cerebellar peduncles** (see Figure 14.8b). Bidirectional pathways connect the cerebellum and vestibular nuclei.

The vestibular nuclei integrate information from vestibular, visual, and somatic receptors and then send commands to (1) the **nuclei of cranial nerves**—oculomotor (III), trochlear (IV), and abducens (VI)—that control coupled movements of the eyes with those of the head to help maintain focus on the visual field; (2) **nuclei of the accessory (XI) nerves** to help control head and neck movements to assist in maintaining equilibrium; (3) the **vestibulospinal tract,** which conveys impulses down the spinal cord to maintain muscle tone in skeletal muscles to help maintain equilibrium; and (4) the **ventral posterior nucleus** in the thalamus and then to the **vestibular area** in the parietal lobe of the cerebral cortex (which is part of the primary somatosensory area; see areas 1, 2, and 3 in Figure 14.15) to provide us with the conscious awareness of the position and movements of the head and limbs.

Table 17.2 summarizes the structures of the ear related to hearing and equilibrium.

TABLE 17.2

Summary of Structures of the Ear

<table>
<tr><th>REGIONS OF THE EAR AND KEY STRUCTURES</th><th>FUNCTION</th></tr>
<tr><td>**External (outer) ear**
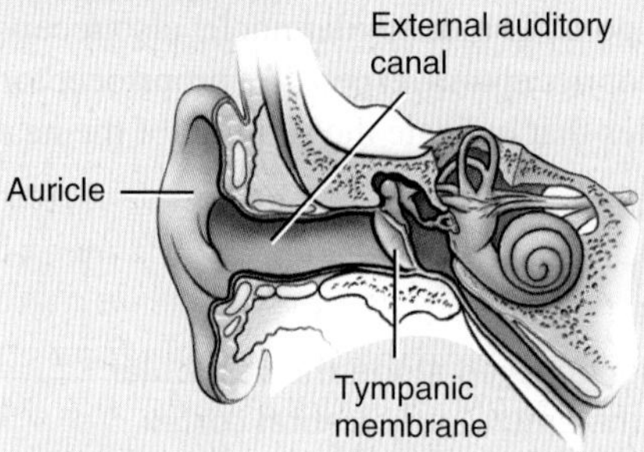
</td><td>*Auricle (pinna):* Collects sound waves.
External auditory canal (meatus): Directs sound waves to eardrum.
Tympanic membrane (eardrum): Sound waves cause it to vibrate, which in turn causes malleus to vibrate.</td></tr>
<tr><td>**Middle ear**
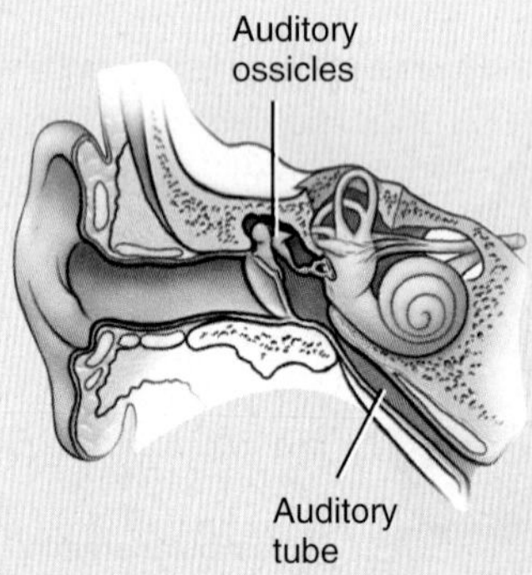
</td><td>*Auditory ossicles:* Transmit and amplify vibrations from tympanic membrane to oval window.
Auditory (eustachian) tube: Equalizes air pressure on both sides of tympanic membrane.</td></tr>
<tr><td>**Internal (inner) ear**
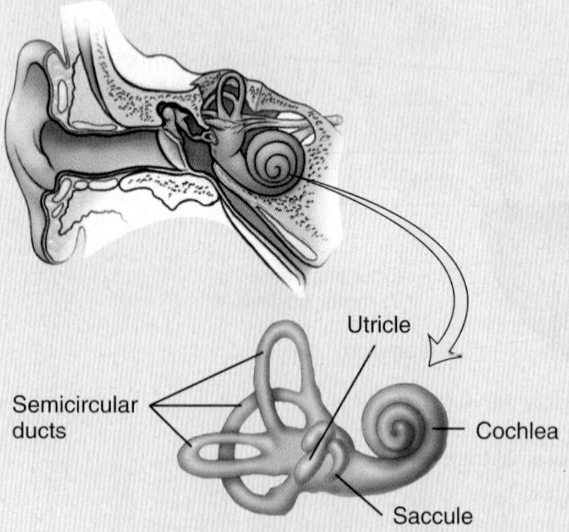
</td><td>*Cochlea:* Contains a series of fluids, channels, and membranes that transmit vibrations to spiral organ (organ of Corti), the organ of hearing; hair cells in spiral organ produce receptor potentials, which elicit nerve impulses in cochlear branch of vestibulocochlear (VIII) nerve.
Vestibular apparatus: Includes semicircular ducts, utricle, and saccule, which generate nerve impulses that propagate along vestibular branch of vestibulocochlear (VIII) nerve.
Semicircular ducts: Contain cristae, site of hair cells for dynamic equilibrium (maintenance of body position, mainly the head, in response to rotational acceleration and deceleration movements).
Utricle: Contains macula, site of hair cells for static equilibrium (maintenance of body position, mainly the head, relative to force of gravity).
Saccule: Contains macula, site of hair cells for static equilibrium (maintenance of body position, mainly the head, relative to force of gravity).</td></tr>
</table>

CLINICAL CONNECTION | *Motion Sickness*

Motion sickness is a condition that results when there is a conflict among the senses with regard to motion. For example, the vestibular apparatus senses angular and vertical motion, while the eyes and proprioceptors in muscles and joints determine the position of the body in space. If you are in the cabin of a moving ship, your vestibular apparatus informs the brain that there is movement from waves. But your eyes don't see any movement. This leads to the conflict among the senses. Motion sickness can also be experienced in other situations that involve movement, for example, in a car or airplane or on a train or amusement park ride.

Symptoms of motion sickness include paleness, restlessness, excess salivation, nausea, dizziness, cold sweats, headache, and malaise that may progress to vomiting. Once the motion is stopped, the symptoms disappear. If it is not possible to stop the motion, you might try sitting in the front seat of a car, the forward car of a train, the upper deck on a boat, or the wing seats in a plane. Looking at the horizon and not reading also help. Medications for motion sickness are usually taken in advance of travel and include scopolamine in time-release patches or tablets, dimenhydrinate (Dramamine®), and meclizine (Bonine®). •

✔CHECKPOINT

11. How are sound waves transmitted from the auricle to the spiral organ of Corti?
12. How do hair cells in the cochlea and vestibular apparatus transduce mechanical vibrations into electrical signals?
13. What is the pathway for auditory impulses from the cochlea to the cerebral cortex?
14. Compare the function of the maculae in maintaining static equilibrium with the role of the cristae in maintaining dynamic equilibrium.
15. What is the role of vestibular input to the cerebellum?
16. Describe the equilibrium pathways.

17.5 DEVELOPMENT OF THE EYES AND EARS

OBJECTIVE

- Describe the development of the eyes and the ears.

Eyes

The *eyes* begin to develop about 22 days after fertilization when the **ectoderm** of the lateral walls of the prosencephalon (forebrain) bulges out to form a pair of shallow grooves called the **optic grooves.** Within a few days, as the neural tube is closing, the optic grooves enlarge and grow toward the surface ectoderm and become known as the **optic vesicles.** When the optic vesicles reach the surface ectoderm, the surface ectoderm thickens to form the **lens placodes** (PLAK-ods). In addition, the distal portions of the optic vesicles invaginate, forming the **optic cups;** they remain attached to the prosencephalon by narrow, hollow proximal structures called **optic stalks.** Figure 17.27 shows the stages in the development of the eyes.

The lens placodes also invaginate and develop into lens vesicles that sit in the optic cups. The lens vesicles eventually develop into the *lenses.* Blood is supplied to the developing lenses (and retina) by the hyaloid arteries. These arteries gain access to the developing eyes through a groove on the inferior surface of the optic cup and optic stalk called the **choroid fissure.** As the lenses mature, part of the hyaloid arteries that pass through the vitreous chamber degenerate; the remaining portions of the hyaloid arteries become the *central retinal arteries.*

The inner wall of the optic cup forms the *neural layer* of the retina, while the outer layer forms the *pigmented layer* of the retina. Axons from the neural layer grow through the optic stalk to the brain, converting the optic stalk to the *optic (II) nerve.* Although myelination of the optic nerves begins late in fetal life, it is not completed until the tenth week after birth.

The anterior portion of the optic cup forms the epithelium of the *ciliary body, iris,* and *circular and radial muscles* of the iris. The connective tissue of the ciliary body, *ciliary muscle,* and *zonular fibers* of the lens develop from **mesenchyme** around the anterior portion of the optic cup.

Mesenchyme surrounding the optic cup and optic stalk differentiates into an inner layer that gives rise to the *choroid* and an outer layer that develops into the *sclera* and part of the *cornea.* The remainder of the cornea is derived from surface ectoderm.

The *anterior chamber* develops from a cavity that forms in the mesenchyme between the iris and cornea; the *posterior chamber* develops from a cavity that forms in the mesenchyme between the iris and lens.

Some mesenchyme around the developing eye enters the optic cup through the choroid fissure. This mesenchyme occupies the space between the lens and retina and differentiates into a delicate network of fibers. Later the spaces between the fibers fill with a jellylike substance, thus forming the *vitreous body* in the vitreous chamber.

The *eyelids* form from surface ectoderm and mesenchyme. The upper and lower eyelids meet and fuse at about eight weeks of development and remain closed until about 26 weeks of development.

Ears

The first portion of the ear to develop is the *internal ear.* It begins to form about 22 days after fertilization as a thickening of the surface ectoderm, called **otic placodes** (Figure 17.28a), that appears on either side of the rhombencephalon (hindbrain). The otic placodes invaginate quickly (Figure 17.28b) to form the **otic pits** (Figure 17.28c). Next, the otic pits pinch off from the surface ectoderm to form the **otic vesicles** within the mesenchyme of the head (Figure 17.28d). During later development, the otic vesicles will form the structures associated with the *membranous labyrinth* of the internal ear. Mesenchyme around the otic vesicles produces cartilage that later ossifies to form the bone associated with the *bony labyrinth* of the internal ear.

The *middle ear* develops from a structure called the first **pharyngeal (branchial) pouch,** an **endoderm**-lined outgrowth of the primitive pharynx (see Figure 18.21a). The pharyngeal

Figure 17.27 Development of the eyes.

The eyes begin to develop about 22 days after fertilization from ectoderm of the prosencephalon.

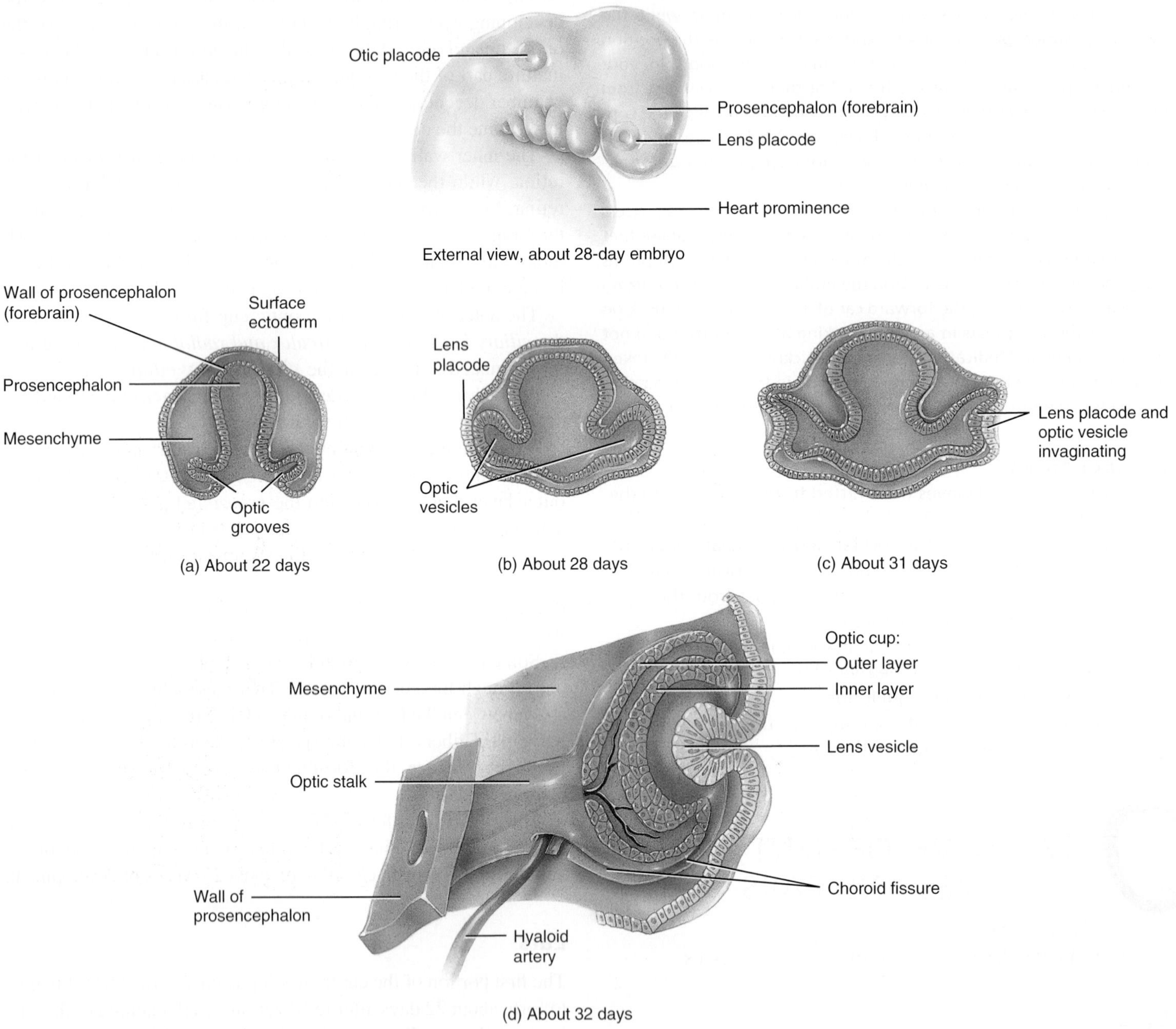

Which structure gives rise to the neural and pigmented layers of the retina?

pouches are discussed in detail in Section 29.1. The *auditory ossicles* develop from the first and second pharyngeal arches.

The *external ear* develops from the first **pharyngeal cleft,** an endoderm-lined groove between the first and second pharyngeal arches (see the inset in Figure 17.28). The pharyngeal clefts are discussed in detail in Section 29.1.

CHECKPOINT

17. How do the origins of the eyes and ears differ?

Figure 17.28 Development of the ears.

The first parts of the ears to develop are the internal ears, which begin to form about 22 days after fertilization as thickenings of surface ectoderm.

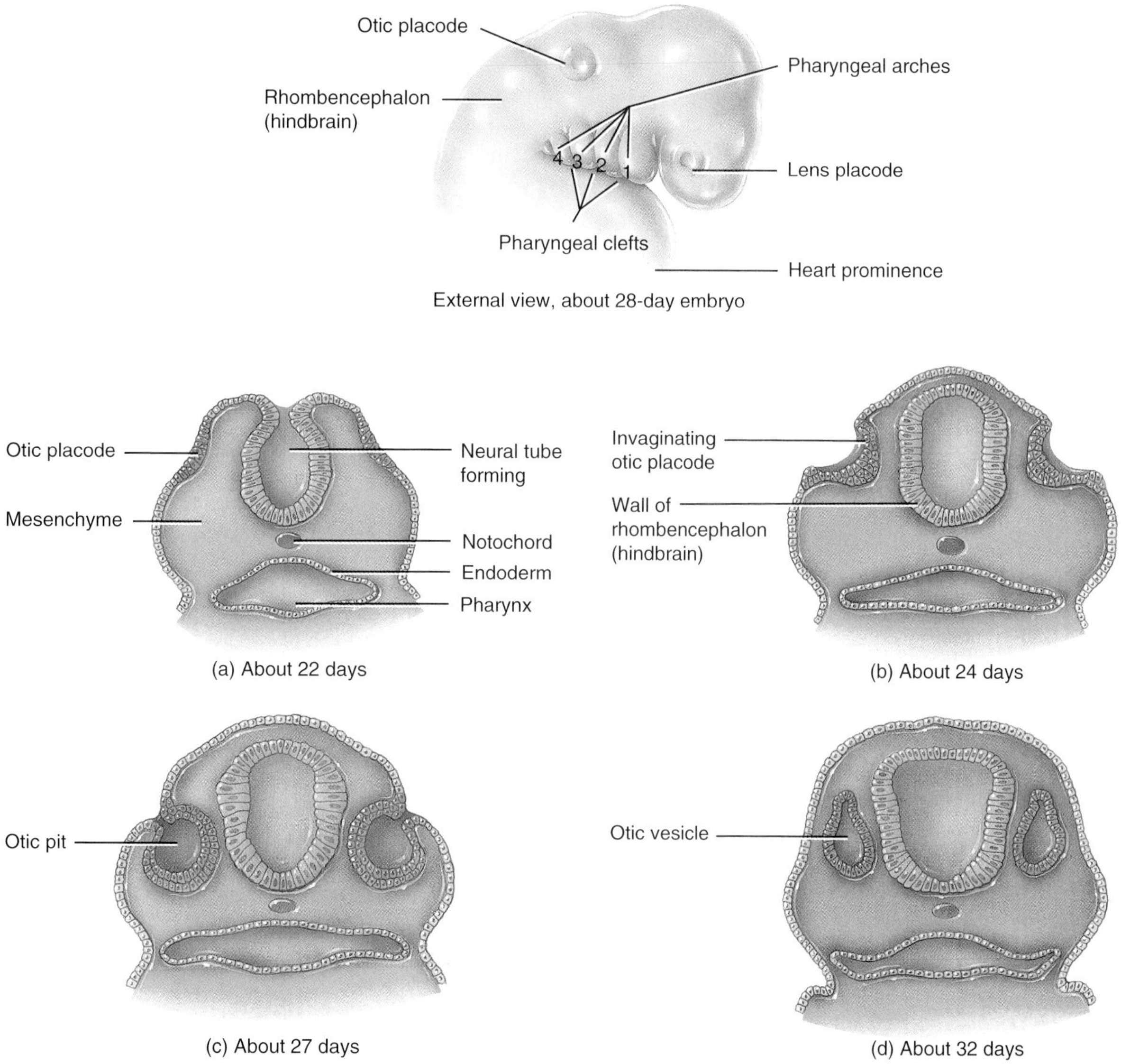

How do the three parts of the ear differ in origin?

17.6 AGING AND THE SPECIAL SENSES

OBJECTIVE

- Describe the age-related changes that occur in the eyes and ears.

Most people do not experience any problems with the senses of smell and taste until about age 50. This is due to a gradual loss of olfactory receptors and gustatory receptor cells coupled with their slower rate of replacement as we age.

Several age-related changes occur in the eyes. As noted earlier, the lens loses some of its elasticity and thus cannot change shape as easily, resulting in presbyopia (see Section 17.3). Cataracts (loss of transparency of the lenses) also occur with aging (see Disorders: Homeostatic Imbalances at the end of the chapter). In old age, the sclera ("white" of the eye) becomes thick and rigid and develops a yellowish or brownish coloration due to many years of exposure to ultraviolet light, wind, and dust. The sclera may also develop random splotches of pigment, especially in people with dark complexions. The iris fades or develops irregular pigment. The muscles that regulate the size of the pupil weaken with age and the pupils become smaller, react more slowly to light, and dilate more slowly in the dark. For these reasons, elderly people find that objects are not as bright, their eyes may adjust more slowly when going outdoors, and they have problems going from brightly lit to darkly lit places. Some diseases of the retina are more likely to occur in old age, including age-related macular disease and detached retina (see the Clinical Connections in Section 17.3). A disorder called glaucoma (see below) develops in the eyes of aging people as a result of the buildup of aqueous humor. Tear production and the number of mucous cells in the conjunctiva may decrease with age, resulting in dry eyes. The eyelids lose their elasticity, becoming baggy and wrinkled. The amount of fat around the orbits may decrease, causing the eyeballs to sink into the orbits. Finally, as we age the sharpness of vision decreases, color and depth perception are reduced, and "vitreal floaters" increase.

By about age 60, around 25% of individuals experience a noticeable hearing loss, especially for higher-pitched sounds. The age-associated progressive loss of hearing in both ears is called **presbycusis** (pres′-bī-KŪ-sis; *presby-* = old; *-acou-* = hearing; *-sis* = condition). It may be related to damaged and lost hair cells in the spiral organ or degeneration of the nerve pathway for hearing. Tinnitus (ringing in the ears) and vestibular imbalance also occur more frequently in the elderly.

✓CHECKPOINT

18. What changes in the eyes and ears are related to the aging process, and how do they take place?

DISORDERS: HOMEOSTATIC IMBALANCES

Cataracts

A common cause of blindness is a loss of transparency of the lens known as a **cataract** (KAT-a-rakt = waterfall). The lens becomes cloudy (less transparent) due to changes in the structure of the lens proteins. Cataracts often occur with aging but may also be caused by injury, excessive exposure to ultraviolet rays, certain medications (such as long-term use of steroids), or complications of other diseases (for example, diabetes). People who smoke also have increased risk of developing cataracts. Fortunately, sight can usually be restored by surgical removal of the old lens and implantation of a new artificial one.

Glaucoma

Glaucoma (glaw-KŌ-ma) is the most common cause of blindness in the United States, afflicting about 2 percent of the population over age 40. In many cases, glaucoma is due to an abnormally high intraocular pressure as a result of a buildup of aqueous humor within the anterior cavity. The fluid compresses the lens into the vitreous body and puts pressure on the neurons of the retina. Persistent pressure results in a progression from mild visual impairment to irreversible destruction of neurons of the retina, damage to the optic nerve, and blindness. Glaucoma is painless, and the other eye compensates largely, so a person may experience considerable retinal damage and loss of vision before the condition is diagnosed. Because glaucoma occurs more often with advancing age, regular measurement of intraocular pressure is an increasingly important part of an eye exam as people grow older. Risk factors include race (blacks are more susceptible), increasing age, family history, and past eye injuries and disorders.

Some individuals have another form of glaucoma called **normal-tension (low-tension) glaucoma.** In this condition, there is damage to the optic nerve with a corresponding loss of vision, even though intraocular pressure is normal. Although the cause is unknown, it appears to be related to a fragile optic nerve, vasospasm of blood vessels around the optic nerve, and ischemia due to narrowed or obstructed blood vessels around the optic nerve. The incidence of normal-tension glaucoma is higher among Japanese and Koreans and females.

Deafness

Deafness is significant or total hearing loss. **Sensorineural deafness** (sen′-so-rē-NOO-ral) is caused by either impairment of hair cells in the cochlea or damage of the cochlear branch of the vestibulocochlear (VIII) nerve. This type of deafness may be caused by atherosclerosis, which reduces blood supply to the ears; by repeated exposure to loud noise, which destroys hair cells of the spiral organ; by certain drugs such as aspirin and streptomycin; and/or by genetic factors. **Conduction deafness** is caused by impairment of the external and middle ear mechanisms for transmitting sounds to the cochlea. Causes of conduction deafness include otosclerosis, the deposition of new bone around the oval window; impacted cerumen; injury to the eardrum; and aging, which often results in thickening of the eardrum and stiffening of the joints of the auditory ossicles. A hearing test called *Weber's test* is used to distinguish between sensorineural and conduction deafness. In the test, the stem of a vibrating fork is held to the forehead. In people with normal hearing, the sound is heard equally in both ears. If the sound is heard best in the affected ear, the deafness is probably of the conduction type; if the sound is heard best in the normal ear, it is probably of the sensorineural type.

Ménière's Disease

Ménière's disease (men′-ē-ĀRZ) results from an increased amount of endolymph that enlarges the membranous labyrinth. Among the symptoms are fluctuating hearing loss (caused by distortion of the basilar membrane of the cochlea) and roaring tinnitus (ringing). Spinning or whirling vertigo (dizziness) is characteristic of Ménière's disease. Almost total destruction of hearing may occur over a period of years.

Otitis Media

Otitis media (ō-TĪ-tis MĒ-dē-a) is an acute infection of the middle ear caused mainly by bacteria and associated with infections of the nose and throat. Symptoms include pain, malaise, fever, and a reddening and outward bulging of the eardrum, which may rupture unless prompt treatment is received. (This may involve draining pus from the middle ear.) Bacteria passing into the auditory tube from the nasopharynx are the primary cause of middle ear infections. Children are more susceptible than adults to middle ear infections because their auditory tubes are almost horizontal, which decreases drainage. If otitis media occurs frequently, a surgical procedure called **tympanotomy** (tim′-pa-NOT-ō-mē; *tympano-* = drum; *-tome* = incision) is often employed. This consists of the insertion of a small tube into the eardrum to provide a pathway for the drainage of fluid from the middle ear.

MEDICAL TERMINOLOGY

Ageusia (a-GOO-sē-a; *a-* = without; *-geusis* = taste) Loss of the sense of taste.

Amblyopia (am′-blē-Ō-pē-a; *ambly-* = dull or dim) Term used to describe the loss of vision in an otherwise normal eye that, because of muscle imbalance, cannot focus in synchrony with the other eye. Sometimes called "wandering eyeball" or a "lazy eye."

Anosmia (an-OZ-mē-a; *a-* = without; *-osmi* = smell, odor) Total lack of the sense of smell.

Barotrauma (bar′-ō-TRAW-ma; *baros-* = weight) Damage or pain, mainly affecting the middle ear, as a result of pressure changes. It occurs when pressure on the outer side of the tympanic membrane is higher than on the inner side, for example, when flying in an airplane or diving. Swallowing or holding your nose and exhaling with your mouth closed usually opens the auditory tubes, allowing air into the middle ear to equalize the pressure.

Blepharitis (blef-a-RĪ-tis; *blephar-* = eyelid; *-itis* = inflammation of) An inflammation of the eyelid.

Conjunctivitis (pinkeye) An inflammation of the conjunctiva; when caused by bacteria such as pneumococci, staphylococci, or *Haemophilus influenzae,* it is very contagious and more common in children. Conjunctivitis may also be caused by irritants, such as dust, smoke, or pollutants in the air, in which case it is not contagious.

Corneal abrasion (KOR-nē-al a-BRĀ-zhun) A scratch on the surface of the cornea, for example, from a speck of dirt or damaged contact lenses. Symptoms include pain, redness, watering, blurry vision, sensitivity to bright light, and frequent blinking.

Corneal transplant A procedure in which a defective cornea is removed and a donor cornea of similar diameter is sewn in. It is the most common and most successful transplant operation. Since the cornea is avascular, antibodies in the blood that might cause rejection do not enter the transplanted tissue, and rejection rarely occurs. The shortage of donor corneas has been partially overcome by the development of artificial corneas made of plastic.

Diabetic retinopathy (ret-i-NOP-a-thē; *retino-* = retina; *-pathos* = suffering) Degenerative disease of the retina due to diabetes mellitus, in which blood vessels in the retina are damaged or new ones grow and interfere with vision.

Exotropia (ek′-sō-TRŌ-pē-a; *ex-* = out; *-tropia* = turning) Turning outward of the eyes.

Keratitis (ker′-a-TĪ-tis; *kerat-* = cornea) An inflammation or infection of the cornea.

Miosis (mī-Ō-sis) Constriction of the pupil.

Mydriasis (mi-DRĪ-a-sis) Dilation of the pupil.

Nystagmus (nis-TAG-mus; *nystagm-* = nodding or drowsy) A rapid involuntary movement of the eyeballs, possibly caused by a disease of the central nervous system. It is associated with conditions that cause vertigo.

Otalgia (ō-TAL-jē-a; *oto-* = ear; *-algia* = pain) Earache.

Photophobia (fō′-tō-FŌ-bē-a; *photo-* = light; *-phobia* = fear) Abnormal visual intolerance to light.

Ptosis (TŌ-sis = fall) Falling or drooping of the eyelid (or slippage of any organ below its normal position).

Retinoblastoma (ret-i-nō-blas-TŌ-ma; *-oma* = tumor) A tumor arising from immature retinal cells; it accounts for 2% of childhood cancers.

Scotoma (skō-TŌ-ma = darkness) An area of reduced or lost vision in the visual field.

Strabismus (stra-BIZ-mus; *strabismos* = squinting) Misalignment of the eyeballs so that the eyes do not move in unison when viewing an object; the affected eye turns either medially or laterally with respect to the normal eye and the result is double vision (diplopia). It may be caused by physical trauma, vascular injuries, or tumors of the extrinsic eye muscle or the oculomotor (III), trochlear (IV), or abducens (VI) cranial nerves.

Tinnitus (ti-NĪ-tus) A ringing, roaring, or clicking in the ears.

Tonometer (tō-NOM-ē-ter; *tono-* = tension or pressure; *-metron* = measure) An instrument for measuring pressure, especially intraocular pressure.

Trachoma (tra-KŌ-ma) A serious form of conjunctivitis and the greatest single cause of blindness in the world. It is caused by the bacterium *Chlamydia trachomatis.* The disease produces an excessive growth of subconjunctival tissue and invasion of blood vessels into the cornea, which progresses until the entire cornea is opaque.

Vertigo (VER-ti-gō = dizziness) A sensation of spinning or movement in which the world seems to revolve or the person seems to revolve in space, often associated with nausea and, in some cases, vomiting. It may be caused by arthritis of the neck or an infection of the vestibular apparatus.

CHAPTER REVIEW AND RESOURCE SUMMARY

WILEY PLUS

Review	Resource
17.1 Olfaction: Sense of Smell 1. The receptors for olfaction, which are bipolar neurons, are in the nasal epithelium along with olfactory glands, which produce mucus that dissolves odorants. 2. In olfactory reception, a generator potential develops and triggers one or more nerve impulses. 3. The threshold of smell is low, and adaptation to odors occurs quickly. 4. Axons of olfactory receptors form the olfactory (I) nerves, which convey nerve impulses to the olfactory bulbs, olfactory tracts, limbic system, and cerebral cortex (temporal and frontal lobes).	Anatomy Overview - The Special Senses Anatomy Overview - Chemoreceptors Anatomy Overview - Olfactory Nerve
17.2 Gustation: Sensation of Taste 1. The receptors for gustation, the gustatory receptor cells, are located in taste buds. 2. Dissolved chemicals, called tastants, stimulate gustatory receptor cells by flowing through ion channels in the plasma membrane or by binding to receptors attached to G proteins in the membrane. 3. Receptor potentials developed in gustatory receptor cells cause the release of neurotransmitter, which can generate nerve impulses in first-order sensory neurons.	Anatomy Overview - Chemoreceptors Anatomy Overview - Facial Nerve Anatomy Overview - Glossopharyngeal Nerve Anatomy Overview - Vagus Nerve

Review | **Resource**

4. The threshold varies with the taste involved, and adaptation to taste occurs quickly.
5. Gustatory receptor cells trigger nerve impulses in cranial nerves VII, IX, and X. Taste signals then pass to the medulla oblongata, thalamus, and cerebral cortex (parietal lobe).

17.3 Vision

1. Accessory structures of the eyes include the eyebrows, eyelids, eyelashes, lacrimal apparatus, and extrinsic eye muscles. The lacrimal apparatus consists of structures that produce and drain tears.
2. The eye is constructed of three layers: (a) fibrous tunic (sclera and cornea), (b) vascular tunic (choroid, ciliary body, and iris), and (c) retina.
3. The retina consists of a pigmented layer and a neural layer that includes a photoreceptor layer, bipolar cell layer, ganglion cell layer, horizontal cells, and amacrine cells.
4. The anterior cavity contains aqueous humor; the vitreous chamber contains the vitreous body.
5. Image formation on the retina involves refraction of light rays by the cornea and lens, which focus an inverted image on the fovea centralis of the retina. For viewing close objects, the lens increases its curvature (accommodation) and the pupil constricts to prevent light rays from entering the eye through the periphery of the lens.
6. The near point of vision is the minimum distance from the eye at which an object can be clearly focused with maximum accommodation.
7. In convergence, the eyeballs move medially so they are both directed toward an object being viewed.
8. The first step in vision is the absorption of light by photopigments in rods and cones and isomerization of *cis*-retinal. Receptor potentials in rods and cones decrease the release of inhibitory neurotransmitter, which induces graded potentials in bipolar cells and horizontal cells.
9. Horizontal cells transmit inhibitory signals to bipolar cells; bipolar or amacrine cells transmit excitatory signals to ganglion cells, which depolarize and initiate nerve impulses.
10. Impulses from ganglion cells are conveyed into the optic (II) nerve, through the optic chiasm and optic tract, to the thalamus. From the thalamus, impulses for vision propagate to the cerebral cortex (occipital lobe). Axon collaterals of retinal ganglion cells extend to the midbrain and hypothalamus.

Resource:
Anatomy Overview - Oculomotor Nerve
Anatomy Overview - Trochlear Nerve
Anatomy Overview - Abducens Nerve
Anatomy Overview - Optic Nerve
Figure 17.7 - Anatomy of the Eyeball
Figure 17.17 - The Visual Pathway

17.4 Hearing and Equilibrium

1. The external (outer) ear consists of the auricle, external auditory canal, and tympanic membrane (eardrum).
2. The middle ear consists of the auditory tube, ossicles, oval window, and round window.
3. The internal (inner) ear consists of the bony labyrinth and membranous labyrinth. The internal ear contains the spiral organ (organ of Corti), the organ of hearing.
4. Sound waves enter the external auditory canal, strike the tympanic membrane, pass through the ossicles, strike the oval window, set up waves in the perilymph, strike the vestibular membrane and scala tympani, increase pressure in the endolymph, vibrate the basilar membrane, and stimulate hair bundles on the spiral organ (organ of Corti).
5. Hair cells convert mechanical vibrations into a receptor potential, which releases neurotransmitter that can initiate nerve impulses in first-order sensory neurons.
6. Sensory axons in the cochlear branch of the vestibulocochlear (VIII) nerve terminate in the medulla oblongata. Auditory signals then pass to the inferior colliculus, thalamus, and temporal lobes of the cerebral cortex.
7. Static equilibrium is the orientation of the body relative to the pull of gravity. The maculae of the utricle and saccule are the sense organs of static equilibrium. Body movements that stimulate the receptors for static equilibrium include tilting the head and linear acceleration or deceleration.
8. Dynamic equilibrium is the maintenance of body position in response to rotational acceleration or deceleration. The cristae in the semicircular ducts are the main sense organs of dynamic equilibrium.
9. Most vestibular branch axons of the vestibulocochlear nerve enter the brain stem and terminate in the medulla and pons; other axons enter the cerebellum.

Resource:
Anatomy Overview - Vestibulocochlear Nerve
Figure 17.19 - The Right Middle Ear and Auditory Ossicles
Figure 17.20 - The Right Internal Ear
Figure 17.22 - Stimulation of Auditory Receptors in the Right Ear

17.5 Development of the Eyes and Ears

1. The eyes begin their development about 22 days after fertilization from ectoderm of the lateral walls of the prosencephalon (forebrain).
2. The ears begin their development about 22 days after fertilization from a thickening of ectoderm on either side of the rhombencephalon (hindbrain). The sequence of development of the ear is internal ear, middle ear, and external ear.

Review **Resource**

17.6 Aging and the Special Senses

1. Most people do not experience problems with the senses of smell and taste until about age 50.
2. Among the age-related changes to the eyes are presbyopia, cataracts, difficulty adjusting to light, macular disease, glaucoma, dry eyes, and decreased sharpness of vision.
3. With age there is a progressive loss of hearing, and tinnitus occurs more frequently.

SELF-QUIZ QUESTIONS

Fill in the blanks in the following statements.

1. The five primary taste sensations are _____, _____, _____, _____, and _____.

2. _____ equilibrium refers to the maintenance of the position of the body relative to the force of gravity; _____ equilibrium refers to the maintenance of body position in response to rotational acceleration or deceleration.

Indicate whether the following statements are true or false.

3. Of all of the special senses, only smell and taste sensations project both to higher cortical areas and to the limbic system.

4. The ability to change the curvature of the lens for near vision is convergence.

Choose the one best answer to the following questions.

5. Which of the following are *true*? (1) The sites of olfactory transduction are the olfactory hairs. (2) The olfactory bulbs transmit impulses to the temporal lobe of the cerebral cortex. (3) The axons of olfactory receptors pass through the olfactory foramina in the cribriform plate of the ethmoid bone. (4) The olfactory nerves are bundles of axons that terminate in the olfactory tracts. (5) Within the olfactory bulbs, the first-order neurons synapse with the second-order neurons.
(a) 1, 2, and 4 (b) 2, 3, 4, and 5 (c) 1, 2, 3, 4, and 5
(d) 1, 3, and 5 (e) 1, 2, 3, and 5

6. Which of the following statements is *incorrect*?
(a) Olfactory receptors respond to the chemical stimulation of an odorant molecule by producing a receptor potential.
(b) Basal stem cells continually produce new olfactory receptors.
(c) Adaptation to odors is rapid and occurs in both olfactory receptors and the CNS.
(d) Production of nasal mucus by olfactory glands serves to moisten the olfactory epithelium and dissolve odorants.
(e) The orbitofrontal area is an important region for odor identification and discrimination.

7. Which of the following statements is *incorrect*?
(a) Taste is a chemical sense.
(b) The receptors for taste sensations are found in taste buds located on the tongue, the soft palate, the pharynx, and the epiglottis.
(c) Gustatory hairs are the sites of taste transduction.
(d) The threshold for bitter substances is the highest.
(e) Complete adaptation to taste can occur in 1 to 5 minutes.

8. When viewing an object close to your eyes, which of the following are required for proper image formation on the retina? (1) increased curvature of the lens, (2) contraction of the ciliary muscle, (3) divergence of the eyeballs, (4) refraction of light at the anterior and posterior surfaces of the cornea, (5) constriction of the pupil by contraction of the extrinsic eye muscles.
(a) 1, 2, 3, 4, and 5 (b) 1, 2, and 4 (c) 1, 2, 3, and 4
(d) 2, 4, and 5 (e) 2, 3, and 4

9. Which of the following are *mismatched*?
(a) fungiform papillae: scattered over the entire tongue's surface.
(b) filiform papillae: contain taste buds in early childhood.
(c) vallate papillae: each houses 100–300 taste buds.
(d) foliate papillae: located in trenches on the lateral margins of the tongue.
(e) fungiform papillae: each houses about five taste buds.

10. Place in order the structures involved in the visual pathway.
(a) optic tract (b) ganglion cells (c) cornea
(d) lens (e) bipolar cells (f) optic nerve
(g) visual cortex (h) vitreous body (i) optic chiasm
(j) aqueous humor (k) pupil (l) photoreceptors
(m) thalamus

11. Which of the following statements is *incorrect*?
(a) Retinal is the light-absorbing portion of all visual photopigments.
(b) The only photopigment in rods is rhodopsin, but three different cone photopigments are present in the retina.
(c) Retinal is a derivative of vitamin C.
(d) Color vision results from different colors of light selectively activating different cone photopigments.
(e) Bleaching and regeneration of the photopigments account for much but not all of the sensitivity changes during light and dark adaptation.

12. Which of the following is the *correct* sequence for the auditory pathway?
(a) external auditory canal, tympanic membrane, auditory ossicles, oval window, cochlea and spiral organ
(b) tympanic membrane, external auditory canal, auditory ossicles, cochlea and spiral organ, round window
(c) auditory ossicles, tympanic membrane, cochlea and spiral organ, round window, oval window, external auditory canal
(d) auricle, tympanic membrane, round window, cochlea and spiral organ, oval window
(e) external auditory canal, tympanic membrane, auditory ossicles, internal auditory canal, spiral organ, oval window

13. Match the following:

____ (a) upper and lower eyelids; shade the eyes during sleep, spread lubricating secretions over the eyeballs
____ (b) produces and drains tears
____ (c) arch transversely above the eyeballs and help protect the eyeballs from foreign objects, perspiration, and the direct rays of the sun
____ (d) move the eyeball medially, laterally, superiorly, or inferiorly
____ (e) a thick fold of connective tissue that gives form and support to the eyelids
____ (f) modified sebaceous glands; secretion helps keep eyelids from adhering to one another
____ (g) project from the border of each eyelid; help protect the eyeballs from foreign objects, perspiration, and direct rays of the sun
____ (h) a thin, protective mucous membrane that lines the inner aspect of the eyelids and passes from the eyelids onto the surface of the eyeball, where it covers the sclera

(1) palpebrae
(2) tarsal or Meibomian glands
(3) conjunctiva
(4) eyelashes
(5) lacrimal apparatus
(6) extrinsic eye muscles
(7) eyebrows
(8) tarsal plate

14. Match the following:

____ (a) colored portion of the eyeball; regulates the amount of light entering the posterior part of the eyeball
____ (b) innermost layer of the eyeball; beginning of the visual pathway; contains rods and cones
____ (c) biconvex transparent structure that fine tunes focusing of light rays for clear vision
____ (d) circular band of smooth muscle that alters the shape of the lens for near or far vision
____ (e) watery fluid in the anterior cavity that helps nourish the lens and cornea; helps maintain shape of the eyeball
____ (f) the hole in the center of the iris
____ (g) white of the eye; gives shape to the eyeball, makes it more rigid, protects its inner parts
____ (h) avascular superficial layer of the eyeball; includes cornea and sclera
____ (i) middle, vascularized layer of the eyeball; includes choroid, ciliary body, and iris

(1) sclera
(2) ciliary muscle
(3) iris
(4) pupil
(5) retina
(6) fibrous tunic
(7) aqueous humor
(8) lens
(9) vascular tunic

15. Match the following:

____ (a) partition between external auditory canal and middle ear
____ (b) oval central portion of the bony labyrinth; contains utricle and saccule
____ (c) receptor for static equilibrium; also contributes to some aspects of dynamic equilibrium; consists of hair cells and supporting cells
____ (d) contains hair cells which are the receptors for hearing
____ (e) ear bones: malleus, incus, stapes
____ (f) the pressure equalization tube that connects the middle ear to the nasopharynx
____ (g) contains the spiral organ
____ (h) fluid found within the membranous labyrinth; pressure waves in this fluid cause vibration of the basilar membrane
____ (i) receptor organs for equilibrium; the saccule, utricle, and semicircular canals
____ (j) swollen enlargement in semicircular canals; contains structures involved in dynamic equilibrium
____ (k) opening between the middle ear and internal ear; is enclosed by a membrane called the secondary tympanic membrane
____ (l) the flap of elastic cartilage covered by skin that captures sound waves; the pinna
____ (m) fluid found inside bony labyrinth; bulging of the oval window causes pressure waves in this fluid
____ (n) opening between the middle and inner ear; receives base of stapes

(1) auricle
(2) tympanic membrane
(3) auditory ossicles
(4) vestibular apparatus
(5) ampulla
(6) cochlea
(7) perilymph
(8) oval window
(9) round window
(10) auditory or eustachian tube
(11) vestibule
(12) endolymph
(13) spiral organ
(14) macula

CRITICAL THINKING QUESTIONS

1. Mario has experienced damage to his facial nerve. How would this affect his special senses?

2. The shift nurse brings ailing eighty-year-old Gertrude her dinner. As Gertrude eats a small amount of her food, she comments that she isn't hungry and that "hospital food just doesn't taste good!" The nurse gives Gertrude a menu so she can choose her morning breakfast. Gertrude complains that she is having trouble reading the menu and asks the nurse to read it to her. As the nurse begins to read, Gertrude loudly asks her to "speak up and turn off the buzzing." What does the nurse know about aging and the special senses that help to explain Gertrude's comments?

3. As you help your neighbor put drops in her six-year-old daughter's eyes, the daughter states, "That medicine tastes bad." How do you explain to the neighbor how her daughter can "taste" the eyedrops?

? ANSWERS TO FIGURE QUESTIONS

17.1 An olfactory receptor has a life span of about one month.

17.2 Olfactory transduction occurs in the olfactory hairs of an olfactory receptor.

17.3 Basal cells develop into gustatory receptor cells.

17.4 Visible light that has a wavelength of 700 nm is red.

17.5 The conjunctiva is continuous with the inner lining of the eyelids.

17.6 Lacrimal fluid, or tears, is a watery solution containing salts, some mucus, and lysozyme that protects, cleans, lubricates, and moistens the eyeball.

17.7 The fibrous tunic consists of the cornea and sclera; the vascular tunic consists of the choroid, ciliary body, and iris.

17.8 The parasympathetic division of the ANS causes pupillary constriction; the sympathetic division causes pupillary dilation.

17.9 An ophthalmoscopic examination of the blood vessels of the eye can reveal evidence of hypertension, diabetes mellitus, cataracts, and age-related macular disease.

17.10 The two types of photoreceptors are rods and cones. Rods provide black-and-white vision in dim light; cones provide high visual acuity and color vision in bright light.

17.11 After its secretion by the ciliary process, aqueous humor flows into the posterior chamber, around the iris, into the anterior chamber, and out of the eyeball through the scleral venous sinus.

17.12 During accommodation the ciliary muscle contracts, causing the zonular fibers to slacken. The lens then becomes more convex, increasing its focusing power.

17.13 Presbyopia is the loss of lens elasticity that occurs with aging.

17.14 Both rods and cones transduce light into receptor potentials, use a photopigment embedded in outer segment discs or folds, and release neurotransmitter at synapses with bipolar cells and horizontal cells.

17.15 The conversion of *cis*-retinal to *trans*-retinal is called isomerization.

17.16 Cyclic GMP is the ligand that opens Na^+ channels in photoreceptors, causing the dark current to flow.

17.17 Light rays from an object in the temporal half of the visual field fall on the nasal half of the retina.

17.18 The malleus of the middle ear is attached to the eardrum, which is part of the external ear.

17.19 The oval and round windows separate the middle ear from the internal ear.

17.20 The two sacs in the membranous labyrinth of the vestibule are the utricle and saccule.

17.21 The three subdivisions of the bony labyrinth are the semicircular canals, vestibule, and cochlea.

17.22 The region of the basilar membrane close to the oval and round windows vibrates most vigorously in response to high-frequency sounds.

17.23 The superior olivary nucleus of the pons is the part of the auditory pathway that allows a person to locate the source of a sound.

17.24 The maculae are associated with static equilibrium; they provide sensory information about the position of the head in space.

17.25 The semicircular ducts are associated with dynamic equilibrium.

17.26 The vestibular nuclei are located in the medulla and the pons.

17.27 The optic cup forms the neural and pigmented layers of the retina.

17.28 The internal ear develops from surface ectoderm, the middle ear develops from pharyngeal pouches, and the external ear develops from a pharyngeal cleft.

18 THE ENDOCRINE SYSTEM

THE ENDOCRINE SYSTEM AND HOMEOSTASIS *Circulating or local hormones of the endocrine system contribute to homeostasis by regulating the activity and growth of target cells in your body. Hormones also regulate your metabolism.*

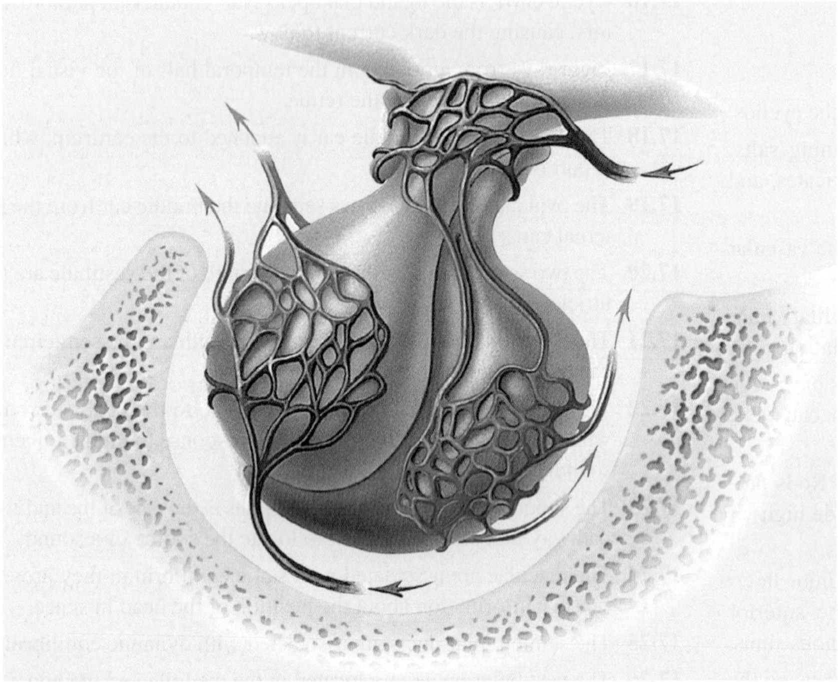

As girls and boys enter puberty, they start to develop striking differences in physical appearance and behavior. Perhaps no other period in life so dramatically shows the impact of the endocrine system in directing development and regulating body functions. In girls, estrogens promote accumulation of adipose tissue in the breasts and hips, sculpting a feminine shape. At the same time or a little later, increasing levels of testosterone in boys begin to help build muscle mass and enlarge the vocal cords, producing a lower-pitched voice. These changes are just a few examples of the powerful influence of endocrine secretions. Less dramatically, perhaps, multitudes of hormones help maintain homeostasis on a daily basis. They regulate the activity of smooth muscle, cardiac muscle, and some glands; alter metabolism; spur growth and development; influence reproductive processes; and participate in circadian (daily) rhythms established by the suprachiasmatic nucleus of the hypothalamus.

? ***Did you ever wonder why thyroid gland disorders affect all major body systems?***

18.1 COMPARISON OF CONTROL BY THE NERVOUS AND ENDOCRINE SYSTEMS

OBJECTIVE

- Compare control of body functions by the nervous system and endocrine system.

The nervous and endocrine systems act together to coordinate functions of all body systems. Recall that the nervous system acts through nerve impulses (action potentials) conducted along axons of neurons. At synapses, nerve impulses trigger the release of mediator (messenger) molecules called *neurotransmitters* (shown in Figure 12.22). The endocrine system also controls body activities by releasing mediators, called *hormones,* but the means of control of the two systems are very different.

A **hormone** (*hormon* = to excite or get moving) is a mediator molecule that is released in one part of the body but regulates the activity of cells in other parts of the body. Most hormones enter interstitial fluid and then the bloodstream. The circulating blood delivers hormones to cells throughout the body. Both neurotransmitters and hormones exert their effects by binding to receptors on or in their "target" cells. Several mediators act as both neurotransmitters and hormones. One familiar example is norepinephrine, which is released as a neurotransmitter by sympathetic postganglionic neurons and as a hormone by chromaffin cells of the adrenal medullae.

Responses of the endocrine system often are slower than responses of the nervous system; although some hormones act within seconds, most take several minutes or more to cause a response. The effects of nervous system activation are generally briefer than those of the endocrine system. The nervous system acts on specific muscles and glands. The influence of the endocrine system is much broader; it helps regulate virtually all types of body cells.

We will also have several opportunities to see how the nervous and endocrine systems function together as an interlocking "supersystem." For example, certain parts of the nervous system stimulate or inhibit the release of hormones by the endocrine system.

Table 18.1 compares the characteristics of the nervous and endocrine systems. In this chapter, we focus on the major endocrine glands and hormone-producing tissues and examine how their hormones govern body activities.

✔CHECKPOINT

1. List the similarities among and differences between the nervous and endocrine systems with regard to the control of homeostasis.

18.2 ENDOCRINE GLANDS

OBJECTIVE

- Distinguish between exocrine and endocrine glands.

Recall from Chapter 4 that the body contains two kinds of glands: exocrine glands and endocrine glands. **Exocrine glands** (EK-sō-krin; *exo-* = outside) secrete their products into ducts that carry the secretions into body cavities, into the lumen of an organ, or to the outer surface of the body. Exocrine glands include sudoriferous (sweat), sebaceous (oil), mucous, and digestive glands. **Endocrine glands** (EN-dō-krin; *endo-* = within) secrete their products (hormones) into the interstitial fluid surrounding the secretory cells rather than into ducts. From the interstitial fluid, hormones diffuse into blood capillaries and blood carries them to target cells throughout the body. Because of their dependence on the cardiovascular system to distribute their products, endocrine glands are some of the most vascular tissues of the body. Considering that most hormones are required in very small amounts, circulating levels typically are low.

The endocrine glands include the pituitary, thyroid, parathyroid, adrenal, and pineal glands (Figure 18.1). In addition, several organs and tissues are not exclusively classified as endocrine glands but contain cells that secrete hormones. These include the hypothalamus, thymus, pancreas, ovaries, testes, kidneys, stomach, liver, small intestine, skin, heart, adipose tissue, and placenta. Taken together, all endocrine glands and hormone-secreting cells constitute the **endocrine system.** The science of the structure and function of

TABLE 18.1

Comparison of Control by the Nervous and Endocrine Systems

CHARACTERISTIC	NERVOUS SYSTEM	ENDOCRINE SYSTEM
Mediator molecules	Neurotransmitters released locally in response to nerve impulses.	Hormones delivered to tissues throughout body by blood.
Site of mediator action	Close to site of release, at synapse; binds to receptors in postsynaptic membrane.	Far from site of release (usually); binds to receptors on or in target cells.
Types of target cells	Muscle (smooth, cardiac, and skeletal) cells, gland cells, other neurons.	Cells throughout body.
Time to onset of action	Typically within milliseconds (thousandths of a second).	Seconds to hours or days.
Duration of action	Generally briefer (milliseconds).	Generally longer (seconds to days).

Figure 18.1 **Location of many endocrine glands.** Also shown are other organs that contain endocrine cells and associated structures.

Endocrine glands secrete hormones, which circulating blood delivers to target tissues.

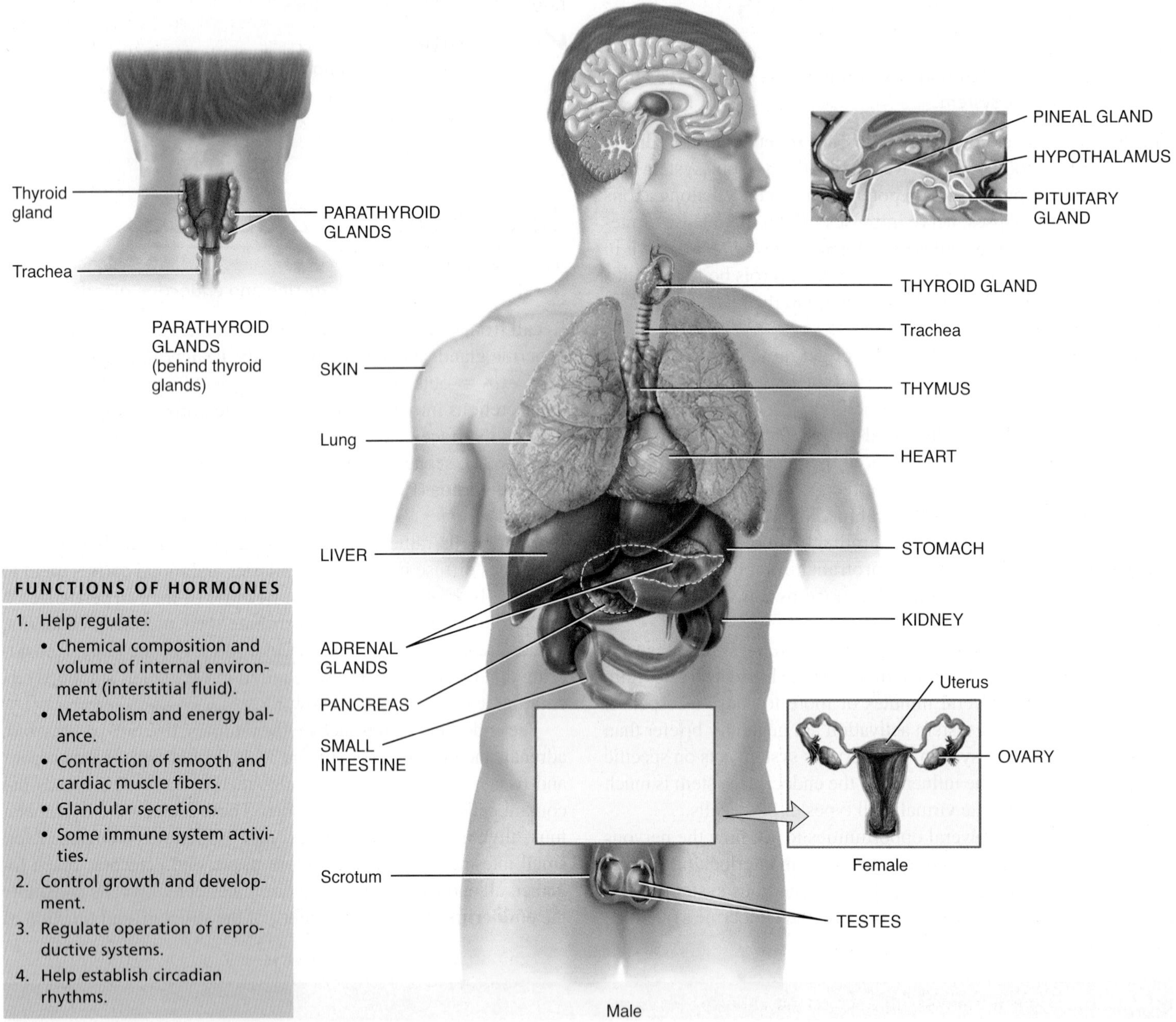

FUNCTIONS OF HORMONES

1. Help regulate:
   - Chemical composition and volume of internal environment (interstitial fluid).
   - Metabolism and energy balance.
   - Contraction of smooth and cardiac muscle fibers.
   - Glandular secretions.
   - Some immune system activities.
2. Control growth and development.
3. Regulate operation of reproductive systems.
4. Help establish circadian rhythms.

What is the basic difference between endocrine glands and exocrine glands?

the endocrine glands and the diagnosis and treatment of disorders of the endocrine system is **endocrinology** (en′-dō-kri-NOL-ō-jē; *endo-* = within; *-crino-* = to secrete; *-logy* = study of).

CHECKPOINT

2. List three organs or tissues that are not exclusively classified as endocrine glands but contain cells that secrete hormones.

18.3 HORMONE ACTIVITY

OBJECTIVES

- Describe how hormones interact with target-cell receptors.
- Compare the two chemical classes of hormones based on their solubility.

The Role of Hormone Receptors

Although a given hormone travels throughout the body in the blood, it affects only specific target cells. Hormones, like neurotransmitters, influence their target cells by chemically binding to specific protein **receptors.** Only the target cells for a given hormone have receptors that bind and recognize that hormone. For example, thyroid-stimulating hormone (TSH) binds to receptors on cells of the thyroid gland, but it does not bind to cells of the ovaries because ovarian cells do not have TSH receptors.

Receptors, like other cellular proteins, are constantly being synthesized and broken down. Generally, a target cell has 2000 to 100,000 receptors for a particular hormone. If a hormone is present in excess, the number of target-cell receptors may decrease—an effect called **down-regulation.** For example, when certain cells of the testes are exposed to a high concentration of luteinizing hormone (LH), the number of LH receptors decreases. Down-regulation makes a target cell *less sensitive* to a hormone. In contrast, when a hormone is deficient, the number of receptors may increase. This phenomenon, known as **up-regulation,** makes a target cell *more sensitive* to a hormone.

CLINICAL CONNECTION | *Blocking Hormone Receptors*

Synthetic hormones that **block the receptors** for some naturally occurring hormones are available as drugs. For example, RU486 (mifepristone), which is used to induce abortion, binds to the receptors for progesterone (a female sex hormone) and prevents progesterone from exerting its normal effect, in this case preparing the lining of the uterus for implantation. When RU486 is given to a pregnant woman, the uterine conditions needed for nurturing an embryo are not maintained, embryonic development stops, and the embryo is sloughed off along with the uterine lining. This example illustrates an important endocrine principle: If a hormone is prevented from interacting with its receptors, the hormone cannot perform its normal functions. •

Circulating and Local Hormones

Most endocrine hormones are **circulating hormones**—they pass from the secretory cells that make them into interstitial fluid and then into the blood (Figure 18.2a). Other hormones, termed **local hormones,** act locally on neighboring cells or on the same cell that secreted them without first entering the bloodstream (Figure 18.2b). Local hormones that act on neighboring cells are called **paracrines** (PAR-a-krins; *para-* = beside or near), and those that act on the same cell that secreted them are called **autocrines** (AW-tō-krins; *auto-* = self). One example of a local hormone is interleukin-2 (IL-2), which is released by helper T cells (a type of white blood cell) during immune responses (see Chapter 22). IL-2 helps activate other nearby immune cells, a paracrine effect. But it also acts as an autocrine by stimulating the same cell that released it to proliferate. This action generates more helper T cells that can secrete even more IL-2 and thus strengthen the immune response. Another example of a local hormone is the gas nitric oxide (NO), which is released by endothelial cells lining blood vessels. NO causes relaxation of nearby smooth muscle fibers in blood vessels, which in turn causes vasodilation (increase in blood vessel diameter). The effects of such vasodilation range from a lowering of blood pressure to erection of the penis in males. The drug *Viagra*® (sildenafil) enhances the effects stimulated by nitric oxide in the penis.

Local hormones usually are inactivated quickly; circulating hormones may linger in the blood and exert their effects for a few minutes or occasionally for a few hours. In time, circulating hormones are inactivated by the liver and excreted by the kidneys. In cases of kidney or liver failure, excessive levels of hormones may build up in the blood.

Figure 18.2 Comparison between circulating hormones and local hormones (autocrines and paracrines).

Circulating hormones are carried through the bloodstream to act on distant target cells. Paracrines act on neighboring cells, and autocrines act on the same cells that produced them.

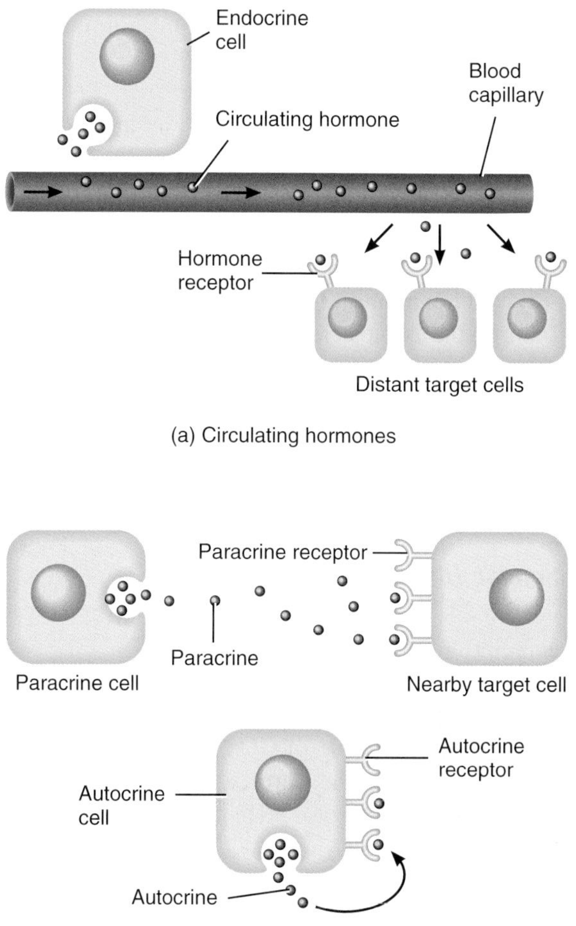

(b) Local hormones (paracrines and autocrines)

In the stomach, one stimulus for secretion of hydrochloric acid by parietal cells is the release of histamine by neighboring mast cells. Is histamine an autocrine or a paracrine in this situation?

Chemical Classes of Hormones

Chemically, hormones can be divided into two broad classes: those that are soluble in lipids, and those that are soluble in water. This chemical classification is also useful functionally because the two classes exert their effects differently.

Lipid-Soluble Hormones

The lipid-soluble hormones include steroid hormones, thyroid hormones, and nitric oxide.

1. **Steroid hormones** are derived from cholesterol. Each steroid hormone is unique due to the presence of different chemical groups attached at various sites on the four rings at the core of its structure. These small differences allow for a large diversity of functions.
2. Two **thyroid hormones** (T_3 and T_4) are synthesized by attaching iodine to the amino acid tyrosine. The presence of two benzene rings within a T_3 or T_4 molecule makes these molecules very lipid-soluble.
3. The gas **nitric oxide (NO)** is both a hormone and a neurotransmitter. Its synthesis is catalyzed by the enzyme nitric oxide synthase.

Water-Soluble Hormones

The water-soluble hormones include amine hormones, peptide and protein hormones, and eicosanoid hormones.

1. **Amine hormones** (a-MĒN) are synthesized by decarboxylating (removing a molecule of CO_2) and otherwise modifying certain amino acids. They are called amines because they retain an amino group ($—NH_3^+$). The catecholamines—epinephrine, norepinephrine, and dopamine—are synthesized by modifying the amino acid tyrosine. Histamine is synthesized from the amino acid histidine by mast cells and platelets. Serotonin and melatonin are derived from tryptophan.
2. **Peptide hormones** and **protein hormones** are amino acid polymers. The smaller peptide hormones consist of chains of 3 to 49 amino acids; the larger protein hormones include 50 to 200 amino acids. Examples of peptide hormones are antidiuretic hormone and oxytocin; protein hormones include human growth hormone and insulin. Several of the protein hormones, such as thyroid-stimulating hormone, have attached carbohydrate groups and thus are **glycoprotein hormones.**
3. The **eicosanoid hormones** (ī-KŌ-sa-noyd; *eicos-* = twenty forms; *-oid* = resembling) are derived from arachidonic acid, a 20-carbon fatty acid. The two major types of eicosanoids are **prostaglandins** and **leukotrienes.** The eicosanoids are important local hormones, and they may act as circulating hormones as well.

Table 18.2 summarizes the classes of lipid-soluble and water-soluble hormones and provides an overview of the major hormones and their sites of secretion.

Hormone Transport in the Blood

Most water-soluble hormone molecules circulate in the watery blood plasma in a "free" form (not attached to other molecules), but most lipid-soluble hormone molecules are bound to **transport proteins.** The transport proteins, which are synthesized by cells in the liver, have three functions:

1. They make lipid-soluble hormones temporarily water-soluble, thus increasing their solubility in blood.
2. They retard passage of small hormone molecules through the filtering mechanism in the kidneys, thus slowing the rate of hormone loss in the urine.
3. They provide a ready reserve of hormone, already present in the bloodstream.

In general, 0.1–10% of the molecules of a lipid-soluble hormone are not bound to a transport protein. This **free fraction** diffuses out of capillaries, binds to receptors, and triggers responses. As free hormone molecules leave the blood and bind to their receptors, transport proteins release new ones to replenish the free fraction.

CLINICAL CONNECTION | *Administering Hormones*

Both steroid hormones and thyroid hormones are effective when taken by mouth. They are not split apart during digestion and easily cross the intestinal lining because they are lipid-soluble. By contrast, peptide and protein hormones, such as insulin, are not effective oral medications because digestive enzymes destroy them by breaking their peptide bonds. This is why people who need insulin must take it by injection. •

✔ CHECKPOINT

3. What is the difference between down-regulation and up-regulation?
4. Identify the chemical classes of hormones, and give an example of each.
5. How are hormones transported in the blood?

18.4 MECHANISMS OF HORMONE ACTION

OBJECTIVE

- Describe the two general mechanisms of hormone action.

The response to a hormone depends on both the hormone and the target cell. Various target cells respond differently to the same hormone. Insulin, for example, stimulates synthesis of glycogen in liver cells and synthesis of triglycerides in adipose cells.

The response to a hormone is not always the synthesis of new molecules, as is the case for insulin. Other hormonal effects include changing the permeability of the plasma mem-

TABLE 18.2

Summary of Hormones by Chemical Class

CHEMICAL CLASS	HORMONES	SITE OF SECRETION
LIPID-SOLUBLE		
Steroid hormones	Aldosterone, cortisol, androgens.	Adrenal cortex.
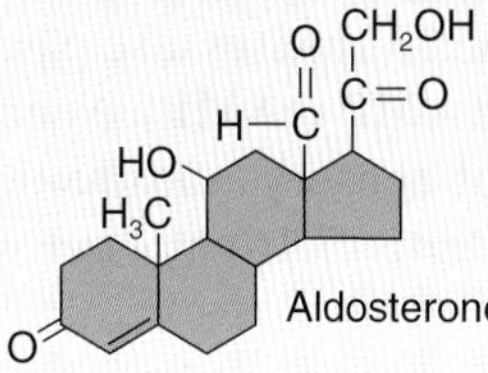 Aldosterone	Calcitriol.	Kidneys.
	Testosterone.	Testes.
	Estrogens, progesterone.	Ovaries.
Thyroid hormones 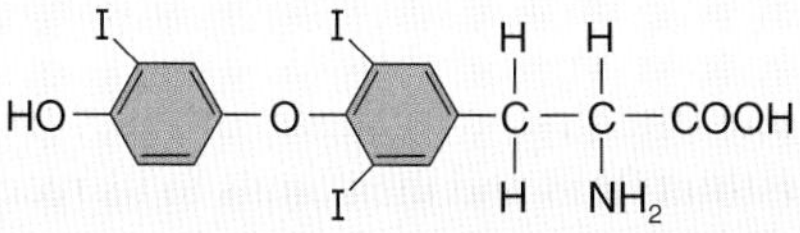 Triiodothyronine (T_3)	T_3 (triiodothyronine), T_4 (thyroxine).	Thyroid gland (follicular cells).
Gas	Nitric oxide (NO).	Endothelial cells lining blood vessels.
WATER-SOLUBLE		
Amines	Epinephrine, norepinephrine (catecholamines).	Adrenal medulla.
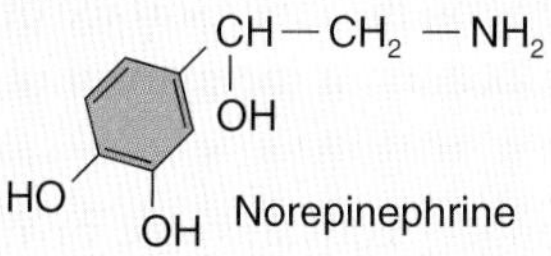 Norepinephrine	Melatonin.	Pineal gland.
	Histamine.	Mast cells in connective tissues.
	Serotonin.	Platelets in blood.
Peptides and proteins	All hypothalamic releasing and inhibiting hormones.	Hypothalamus.
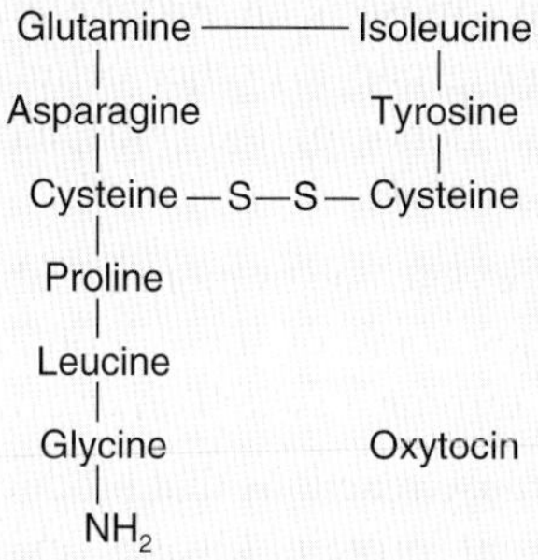 Oxytocin	Oxytocin, antidiuretic hormone.	Posterior pituitary.
	Human growth hormone, thyroid-stimulating hormone, adrenocorticotropic hormone, follicle-stimulating hormone, luteinizing hormone, prolactin, melanocyte-stimulating hormone.	Anterior pituitary.
	Insulin, glucagon, somatostatin, pancreatic polypeptide.	Pancreas.
	Parathyroid hormone.	Parathyroid glands.
	Calcitonin.	Thyroid gland (parafollicular cells).
	Gastrin, secretin, cholecystokinin, GIP (glucose-dependent insulinotropic peptide).	Stomach and small intestine (enteroendocrine cells).
	Erythropoietin.	Kidneys.
	Leptin.	Adipose tissue.
Eicosanoids 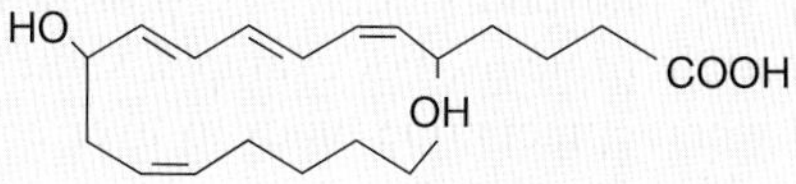 A leukotriene (LTB_4)	Prostaglandins, leukotrienes.	All cells except red blood cells.

brane, stimulating transport of a substance into or out of the target cells, altering the rate of specific metabolic reactions, or causing contraction of smooth muscle or cardiac muscle. In part, these varied effects of hormones are possible because a single hormone can set in motion several different cellular responses. However, a hormone must first "announce its arrival" to a target cell by binding to its receptors. The receptors for lipid-soluble hormones are located inside target cells. The receptors for water-soluble hormones are part of the plasma membrane of target cells.

Action of Lipid-Soluble Hormones

As you just learned, lipid-soluble hormones, including steroid hormones and thyroid hormones, bind to receptors within target cells. Their mechanism of action is as follows (Figure 18.3):

1. A free lipid-soluble hormone molecule diffuses from the blood, through interstitial fluid, and through the lipid bilayer of the plasma membrane into a cell.
2. If the cell is a target cell, the hormone binds to and activates receptors located within the cytosol or nucleus. The activated receptor–hormone complex then alters gene expression: It turns specific genes of the nuclear DNA on or off.
3. As the DNA is transcribed, new messenger RNA (mRNA) forms, leaves the nucleus, and enters the cytosol. There, it directs synthesis of a new protein, often an enzyme, on the ribosomes.
4. The new proteins alter the cell's activity and cause the responses typical of that hormone.

Action of Water-Soluble Hormones

Because amine, peptide, protein, and eicosanoid hormones are not lipid-soluble, they cannot diffuse through the lipid bilayer of the plasma membrane and bind to receptors inside target cells. Instead, water-soluble hormones bind to receptors that protrude from the target-cell surface. The receptors are integral transmembrane proteins in the plasma membrane. When a water-soluble hormone binds to its receptor at the outer surface of the plasma membrane, it acts as the **first messenger.** The first messenger (the hormone) then causes production of a **second messenger** inside the cell, where specific hormone-stimulated responses take place. One common second messenger is **cyclic AMP (cAMP).** Neurotransmitters, neuropeptides, and several sensory transduction mechanisms (for example, vision; see Figure 17.16) also act via second-messenger systems.

The action of a typical water-soluble hormone occurs as follows (Figure 18.4):

1. A water-soluble hormone (the first messenger) diffuses from the blood through interstitial fluid and then binds to its receptor at the exterior surface of a target cell's plasma membrane. The hormone–receptor complex activates a membrane protein called a **G protein.** The activated G protein in turn activates **adenylate cyclase** (a-DEN-i-lāt SĪ-klās).
2. Adenylate cyclase converts ATP into cyclic AMP (cAMP). Because the enzyme's active site is on the inner surface of the plasma membrane, this reaction occurs in the cytosol of the cell.
3. Cyclic AMP (the second messenger) activates one or more protein kinases, which may be free in the cytosol or bound to the plasma membrane. A **protein kinase** is an enzyme that phosphorylates (adds a phosphate group to) other cellular proteins (such as enzymes). The donor of the phosphate group is ATP, which is converted to ADP.
4. Activated protein kinases phosphorylate one or more cellular proteins. Phosphorylation activates some of these proteins and inactivates others, rather like turning a switch on or off.
5. Phosphorylated proteins in turn cause reactions that produce physiological responses. Different protein kinases exist within different target cells and within different organelles of the same target cell. Thus, one protein kinase might trigger glycogen synthesis, a second might cause the breakdown of triglyceride, a third may promote protein synthesis, and so forth. As noted in step 4, phosphorylation by a protein kinase can also inhibit certain proteins. For example, some of the kinases unleashed when epinephrine binds to liver cells inactivate an enzyme needed for glycogen synthesis.
6. After a brief period, an enzyme called **phosphodiesterase** (fos′-fō-dī-ES-ter′-ās) inactivates cAMP. Thus, the cell's response is turned off unless new hormone molecules continue to bind to their receptors in the plasma membrane.

The binding of a hormone to its receptor activates many G-protein molecules, which in turn activate molecules of adenylate cyclase (as noted in step 1). Unless they are further stimulated by the binding of more hormone molecules to receptors, G proteins

Figure 18.3 Mechanism of action of the lipid-soluble steroid hormones and thyroid hormones.

6 Lipid-soluble hormones bind to receptors inside target cells.

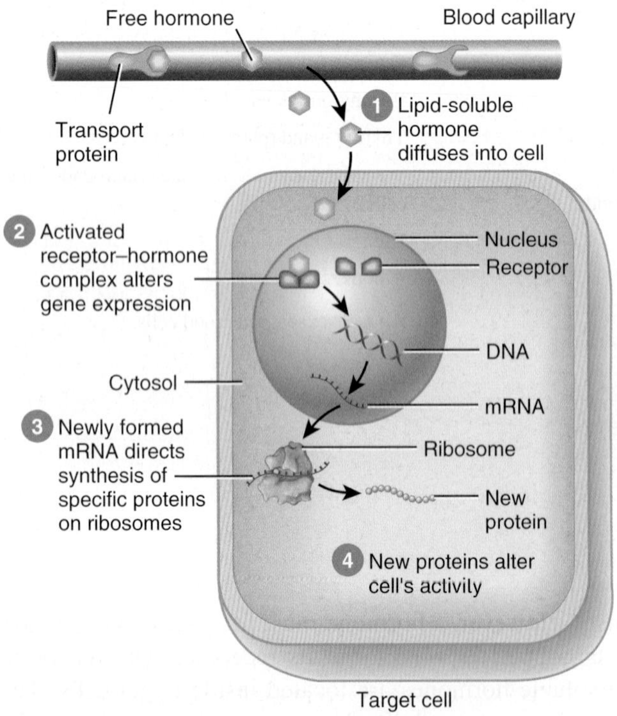
Free hormone
Blood capillary
Transport protein
1 Lipid-soluble hormone diffuses into cell
2 Activated receptor–hormone complex alters gene expression
Nucleus
Receptor
DNA
Cytosol
mRNA
3 Newly formed mRNA directs synthesis of specific proteins on ribosomes
Ribosome
New protein
4 New proteins alter cell's activity
Target cell

? What is the action of the receptor–hormone complex?

slowly inactivate, thus decreasing the activity of adenylate cyclase and helping to stop the hormone response. G proteins are a common feature of most second-messenger systems.

Many hormones exert at least some of their physiological effects through the *increased* synthesis of cAMP. Examples include antidiuretic hormone (ADH), thyroid-stimulating hormone (TSH), adrenocorticotropic hormone (ACTH), glucagon, epinephrine, and hypothalamic releasing hormones. In other cases, such as growth hormone–inhibiting hormone (GHIH), the level of cyclic AMP *decreases* in response to the binding of a hormone to its receptor. Besides cAMP, other second messengers include calcium ions (Ca^{2+}), cGMP (cyclic guanosine monophosphate, a cyclic nucleotide similar to cAMP), inositol trisphosphate (IP_3), and diacylglycerol (DAG). A given hormone may use different second messengers in different target cells.

Figure 18.4 **Mechanism of action of the water-soluble hormones (amines, peptides, proteins, and eicosanoids).**

Water-soluble hormones bind to receptors embedded in the plasma membranes of target cells.

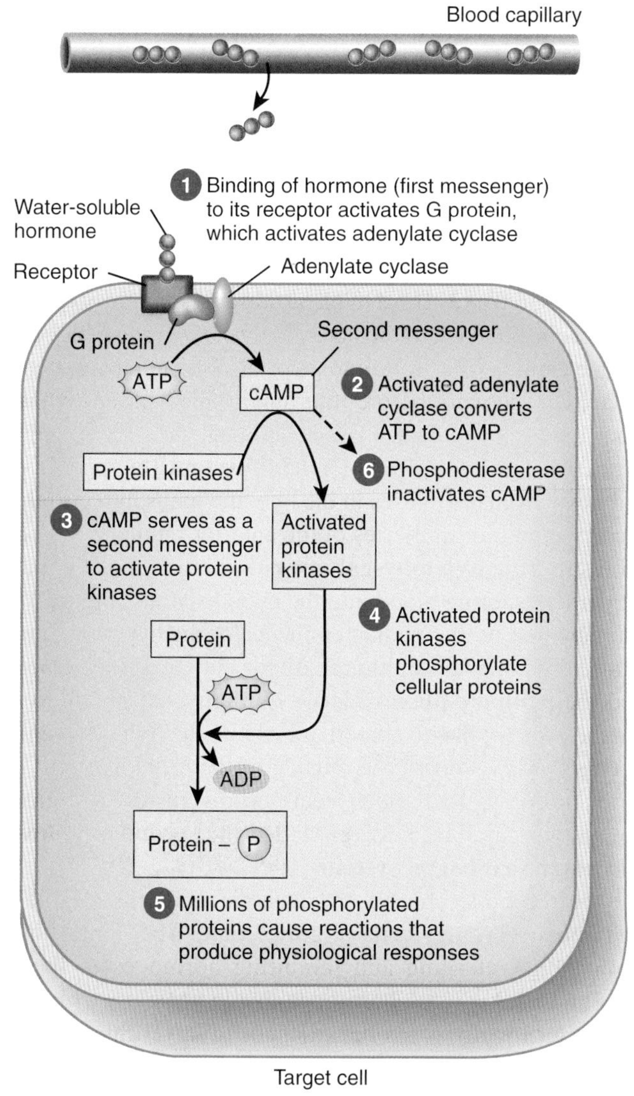
Blood capillary
Water-soluble hormone
1 Binding of hormone (first messenger) to its receptor activates G protein, which activates adenylate cyclase
Receptor
Adenylate cyclase
G protein
Second messenger
ATP
cAMP
2 Activated adenylate cyclase converts ATP to cAMP
Protein kinases
6 Phosphodiesterase inactivates cAMP
3 cAMP serves as a second messenger to activate protein kinases
Activated protein kinases
4 Activated protein kinases phosphorylate cellular proteins
Protein
ATP
ADP
Protein – P
5 Millions of phosphorylated proteins cause reactions that produce physiological responses
Target cell

Why is cAMP called a "second messenger"?

Hormones that bind to plasma membrane receptors can induce their effects at very low concentrations because they initiate a cascade or chain reaction, each step of which multiplies or amplifies the initial effect. For example, the binding of a single molecule of epinephrine to its receptor on a liver cell may activate a hundred or so G proteins, each of which activates an adenylate cyclase molecule. If each adenylate cyclase produces even 1000 cAMP, then 100,000 of these second messengers will be liberated inside the cell. Each cAMP may activate a protein kinase, which in turn can act on hundreds or thousands of substrate molecules. Some of the kinases phosphorylate and activate a key enzyme needed for glycogen breakdown. The end result of the binding of a single molecule of epinephrine to its receptor is the breakdown of millions of glycogen molecules into glucose monomers.

Hormone Interactions

The responsiveness of a target cell to a hormone depends on (1) the hormone's concentration in the blood, (2) the abundance of the target cell's hormone receptors, and (3) influences exerted by other hormones. A target cell responds more vigorously when the level of a hormone rises or when it has more receptors (up-regulation). In addition, the actions of some hormones on target cells require a simultaneous or recent exposure to a second hormone. In such cases, the second hormone is said to have a **permissive effect.** For example, epinephrine alone only weakly stimulates lipolysis (the breakdown of triglycerides), but when small amounts of thyroid hormones (T_3 and T_4) are present, the same amount of epinephrine stimulates lipolysis much more powerfully. Sometimes the permissive hormone increases the number of receptors for the other hormone, and sometimes it promotes the synthesis of an enzyme required for the expression of the other hormone's effects.

When the effect of two hormones acting together is greater or more extensive than the effect of each hormone acting alone, the two hormones are said to have a **synergistic effect.** For example, normal development of oocytes in the ovaries requires both follicle-stimulating hormone from the anterior pituitary and estrogens from the ovaries. Neither hormone alone is sufficient.

When one hormone opposes the actions of another hormone, the two hormones are said to have **antagonistic effects.** An example of an antagonistic pair of hormones is insulin, which promotes synthesis of glycogen by liver cells, and glucagon, which stimulates breakdown of glycogen in the liver.

CHECKPOINT

6. What factors determine the responsiveness of a target cell to a hormone?
7. What are the differences among permissive effects, synergistic effects, and antagonistic effects of hormones?

18.5 CONTROL OF HORMONE SECRETION

OBJECTIVE

- Describe the mechanisms of control of hormone secretion.

The release of most hormones occurs in short bursts, with little or no secretion between bursts. When stimulated, an endocrine gland will release its hormone in more frequent bursts, increasing the concentration of the hormone in the blood. In the absence of stimulation, the blood level of the hormone decreases. Regulation of secretion normally prevents overproduction or underproduction of any given hormone.

Hormone secretion is regulated by (1) signals from the nervous system, (2) chemical changes in the blood, and (3) other hormones. For example, nerve impulses to the adrenal medullae regulate the release of epinephrine; blood Ca^{2+} level regulates the secretion of parathyroid hormone; and a hormone from the anterior pituitary (adrenocorticotropic hormone) stimulates the release of cortisol by the adrenal cortex. Most hormonal regulatory systems work via negative feedback (see Figure 1.3), but a few operate via positive feedback (see Figure 1.4). For example, during childbirth, the hormone oxytocin stimulates contractions of the uterus, and uterine contractions in turn stimulate more oxytocin release, a positive feedback effect.

Now that you have a general understanding of the roles of hormones in the endocrine system, we turn to discussions of the various endocrine glands and the hormones they secrete.

✔CHECKPOINT

8. What three types of signals control hormone secretion?

18.6 HYPOTHALAMUS AND PITUITARY GLAND

OBJECTIVES

- Describe the locations of and relationships between the hypothalamus and pituitary gland.
- Describe the location, histology, hormones, and functions of the anterior and posterior pituitary.

For many years, the **pituitary gland** (pi-TOO-i-tār-ē) or **hypophysis** (hī-POF-i-sis) was called the "master" endocrine gland because it secretes several hormones that control other endocrine glands. We now know that the pituitary gland itself has a master—the **hypothalamus.** This small region of the brain below the thalamus is the major link between the nervous and endocrine systems. Cells in the hypothalamus synthesize at least nine different hormones, and the pituitary gland secretes seven. Together, these hormones play important roles in the regulation of virtually all aspects of growth, development, metabolism, and homeostasis.

The pituitary gland is a pea-shaped structure that measures 1–1.5 cm (0.5 in.) in diameter and lies in the hypophyseal fossa of the sella turcica of the sphenoid bone. It attaches to the hypothalamus by a stalk, the **infundibulum** (in′-fun-DIB-ū-lum = a funnel; Figure 18.5), and has two anatomically and functionally separate portions: the anterior pituitary and the posterior pituitary. The **anterior pituitary (anterior lobe),** also called the **adenohypophysis,** accounts for about 75% of the total weight of the gland and is composed of epithelial tissue. The anterior pituitary consists of two parts in an adult: The **pars distalis** is the larger portion, and the **pars tuberalis** (PARS too-be′-RAL-is) forms a sheath around the infundibulum. The **posterior pituitary (posterior lobe),** also called the **neurohypophysis,** is composed of neural tissue. It also consists of two parts: the **pars nervosa** (ner-VŌ-sa), the larger bulbar portion, and the infundibulum. A third region of the pituitary gland called the **pars intermedia** atrophies during human fetal development and ceases to exist as a separate lobe in adults (see Figure 18.21b). However, some of its cells migrate into adjacent parts of the anterior pituitary, where they persist.

Anterior Pituitary

The **anterior pituitary** or **adenohypophysis** (ad′-e-nō-hī-POF-i-sis; *adeno-* = gland; *-hypophysis* = undergrowth) secretes hormones that regulate a wide range of bodily activities, from growth to reproduction. Release of anterior pituitary hormones is stimulated by **releasing hormones** and suppressed by **inhibiting hormones** from the hypothalamus. Thus, the hypothalamic hormones are an important link between the nervous and endocrine systems.

Hypophyseal Portal System

Hypothalamic hormones that release or inhibit anterior pituitary hormones reach the anterior pituitary through a portal system. Usually, blood passes from the heart through an artery to a capillary to a vein and back to the heart. In a *portal system,* blood flows from one capillary network into a portal vein, and then into a second capillary network before returning to the heart. The name of the portal system indicates the location of the second capillary network. In the **hypophyseal portal system** (hī′-pō-FIZ-ē-al), blood flows from capillaries in the hypothalamus into portal veins that carry blood to capillaries of the anterior pituitary.

The **superior hypophyseal arteries,** branches of the internal carotid arteries, bring blood into the hypothalamus (Figure 18.5a). At the junction of the median eminence of the hypothalamus and the infundibulum, these arteries divide into a capillary network called the **primary plexus of the hypophyseal portal system.** From the primary plexus, blood drains into the **hypophyseal portal veins** that pass down the outside of the infundibulum. In the anterior pituitary, the hypophyseal portal veins divide again and form another capillary network called the **secondary plexus of the hypophyseal portal system.**

Above the optic chiasm are clusters of specialized neurons called **neurosecretory cells** (Figure 18.5b). They synthesize the hypothalamic releasing and inhibiting hormones in their cell bodies and package the hormones inside vesicles, which reach the axon terminals by axonal transport. Nerve impulses stimulate the vesicles to undergo exocytosis. The hormones then diffuse into the primary plexus of the hypophyseal portal system. Quickly, the hypothalamic hormones flow with the blood

Figure 18.5 Hypothalamus and pituitary gland, and their blood supply. Releasing and inhibiting hormones synthesized by hypothalamic neurosecretory cells are transported within axons and released at the axon terminals. The hormones diffuse into capillaries of the primary plexus of the hypophyseal portal system and are carried by the hypophyseal portal veins to the secondary plexus of the hypophyseal portal system for distribution to target cells in the anterior pituitary.

Hypothalamic hormones are an important link between the nervous and endocrine systems.

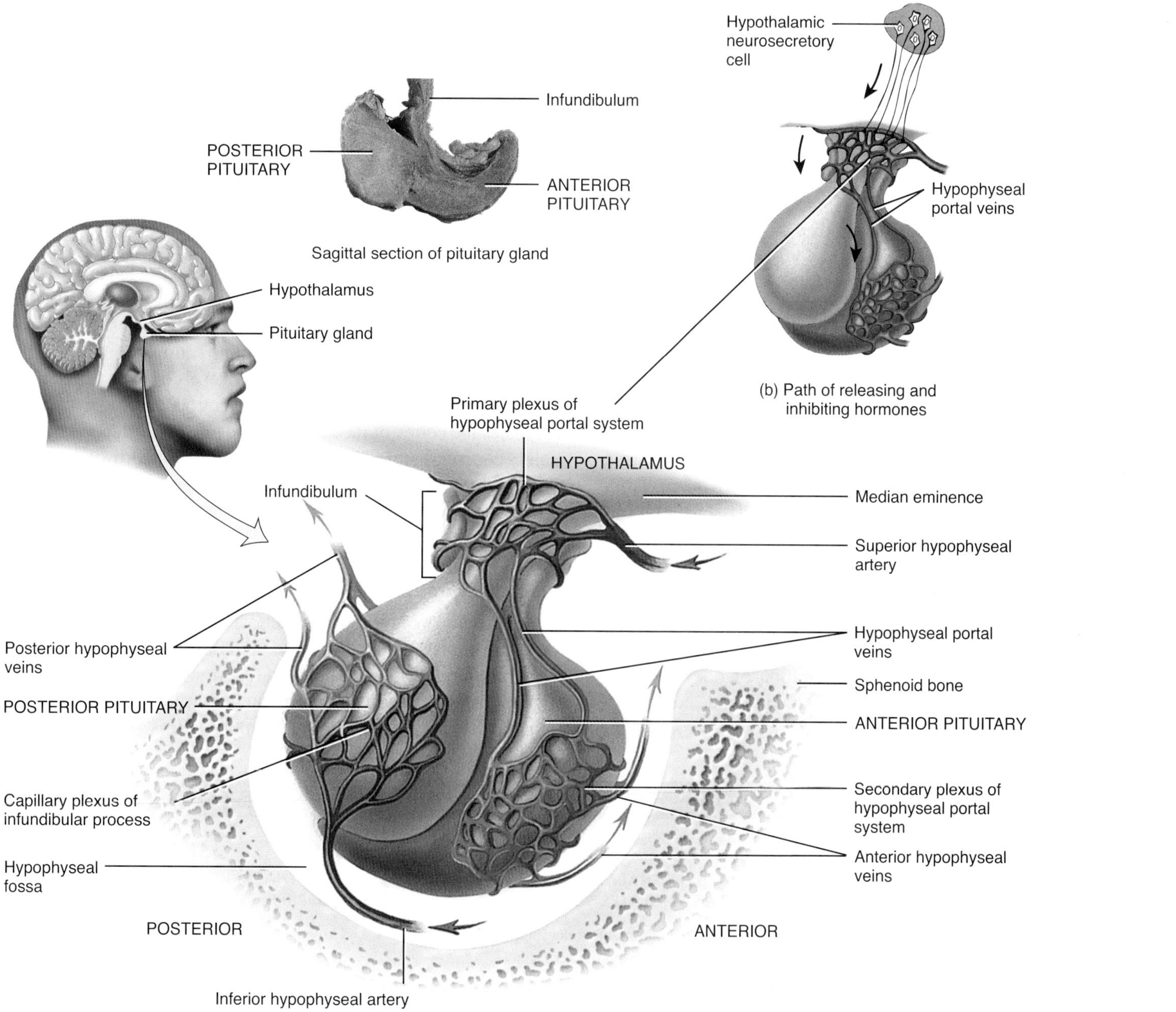

(a) Relationship of hypothalamus to pituitary gland

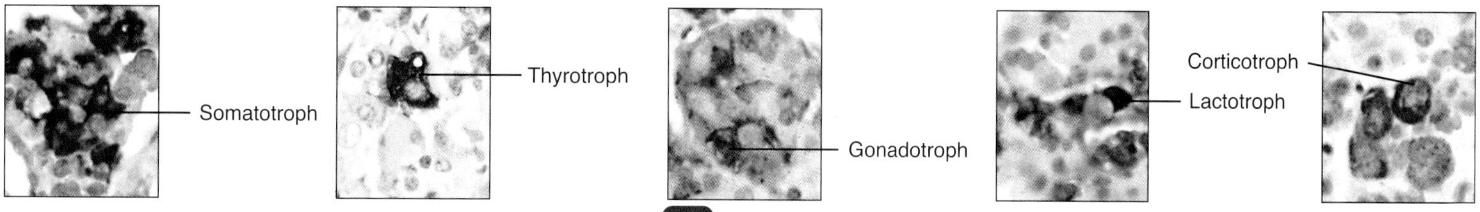

(c) Histology of anterior pituitary

What is the functional importance of the hypophyseal portal veins?

through the portal veins and into the secondary plexus. This direct route permits hypothalamic hormones to act immediately on anterior pituitary cells, before the hormones are diluted or destroyed in the general circulation. Hormones secreted by anterior pituitary cells pass into the secondary plexus capillaries, which drain into the anterior hypophyseal veins and out into the general circulation. Anterior pituitary hormones then travel to target tissues throughout the body. Those anterior pituitary hormones that act on other endocrine glands are called **tropic hormones** (TRŌ-pik) or tropins.

Types of Anterior Pituitary Cells and Their Hormones

Five types of anterior pituitary cells—somatotrophs, thyrotrophs, gonadotrophs, lactotrophs, and corticotrophs—secrete seven hormones (Table 18.3):

1. **Somatotrophs** (sō-MAT-ō-trōfs) secrete **human growth hormone (hGH),** also known as **somatotropin** (sō′-ma-tō-TRŌ-pin; *somato-* = body; *-tropin* = change). Human growth hormone in turn stimulates several tissues to secrete **insulinlike growth factors,** hormones that stimulate general body growth and regulate aspects of metabolism.
2. **Thyrotrophs** (THĪ-rō-trōfs) secrete **thyroid-stimulating hormone (TSH),** also known as **thyrotropin** (thī-rō-TRŌ-pin; *thyro-* = pertaining to the thyroid gland). TSH controls the secretions and other activities of the thyroid gland.
3. **Gonadotrophs** (gō-NAD-ō-trōfs; *gonado-* = seed) secrete two **gonadotropins: follicle-stimulating hormone (FSH)** and **luteinizing hormone (LH)** (LOO-tē-in′-īz-ing). FSH and LH both act on the gonads. They stimulate secretion of estrogens and progesterone and the maturation of oocytes in the ovaries, and they stimulate sperm production and secretion of testosterone in the testes.
4. **Lactotrophs** (LAK-tō-trōfs; *lacto-* = milk) secrete **prolactin (PRL),** which initiates milk production in the mammary glands.
5. **Corticotrophs** (KOR-ti-kō-trōfs) secrete **adrenocorticotropic hormone (ACTH),** also known as **corticotropin** (kor′-ti-kō-TRŌ-pin; *cortico-* = rind or bark), which stimulates the adrenal cortex to secrete glucocorticoids such as cortisol. Some corticotrophs, remnants of the pars intermedia, also secrete **melanocyte-stimulating hormone (MSH).**

Control of Secretion by the Anterior Pituitary

Secretion of anterior pituitary hormones is regulated in two ways. First, neurosecretory cells in the hypothalamus secrete five releasing hormones, which stimulate secretion of anterior pituitary hormones, and two inhibiting hormones, which suppress secretion of anterior pituitary hormones (Table 18.3). Second, negative feedback in the form of hormones released by target glands decreases secretions of three types of anterior pituitary cells (Figure 18.6). In such negative feedback loops, the secretory activity of thyrotrophs, gonadotrophs, and corticotrophs decreases when blood levels of their target gland hormones rise. For example, adrenocorticotropic hormone (ACTH) stimulates the cortex of the adrenal gland to secrete glucocorticoids, mainly cortisol. In turn, an elevated blood level of cortisol decreases secretion of both corticotropin and corticotropin-releasing hormone (CRH) by suppressing the activity of the anterior pituitary corticotrophs and hypothalamic neurosecretory cells.

Human Growth Hormone and Insulinlike Growth Factors

Somatotrophs are the most numerous cells in the anterior pituitary, and human growth hormone (hGH) is the most plentiful anterior pituitary hormone. The main function of hGH is to promote synthesis and secretion of small protein hormones called **insulinlike growth factors (IGFs)** or **somatomedins** (sō′-ma-tō-MĒ-

TABLE 18.3

Hormones of the Anterior Pituitary

HORMONE	SECRETED BY	HYPOTHALAMIC-RELEASING HORMONE (STIMULATES SECRETION)	HYPOTHALAMIC-INHIBITING HORMONE (SUPPRESSES SECRETION)
Human growth hormone (hGH), also known as somatotropin	Somatotrophs.	Growth hormone–releasing hormone (GHRH), also known as somatocrinin.	Growth hormone–inhibiting hormone (GHIH), also known as somatostatin.
Thyroid-stimulating hormone (TSH), also known as thyrotropin	Thyrotrophs.	Thyrotropin-releasing hormone (TRH).	Growth hormone–inhibiting hormone (GHIH).
Follicle-stimulating hormone (FSH)	Gonadotrophs.	Gonadotropin-releasing hormone (GnRH).	—
Luteinizing hormone (LH)	Gonadotrophs.	Gonadotropin-releasing hormone (GnRH).	—
Prolactin (PRL)	Lactotrophs.	Prolactin-releasing hormone (PRH).*	Prolactin-inhibiting hormone (PIH), which is dopamine.
Adrenocorticotropic hormone (ACTH), also known as corticotropin	Corticotrophs.	Corticotropin-releasing hormone (CRH).	—
Melanocyte-stimulating hormone (MSH)	Corticotrophs.	Corticotropin-releasing hormone (CRH).	Dopamine.

*Thought to exist, but exact nature is uncertain.

Figure 18.6 Negative feedback regulation of hypothalamic neurosecretory cells and anterior pituitary corticotrophs. Solid arrows indicate stimulation of secretions; dashed arrows indicate inhibition of secretion via negative feedback.

⑥ Cortisol secreted by the adrenal cortex suppresses secretion of CRH and ACTH.

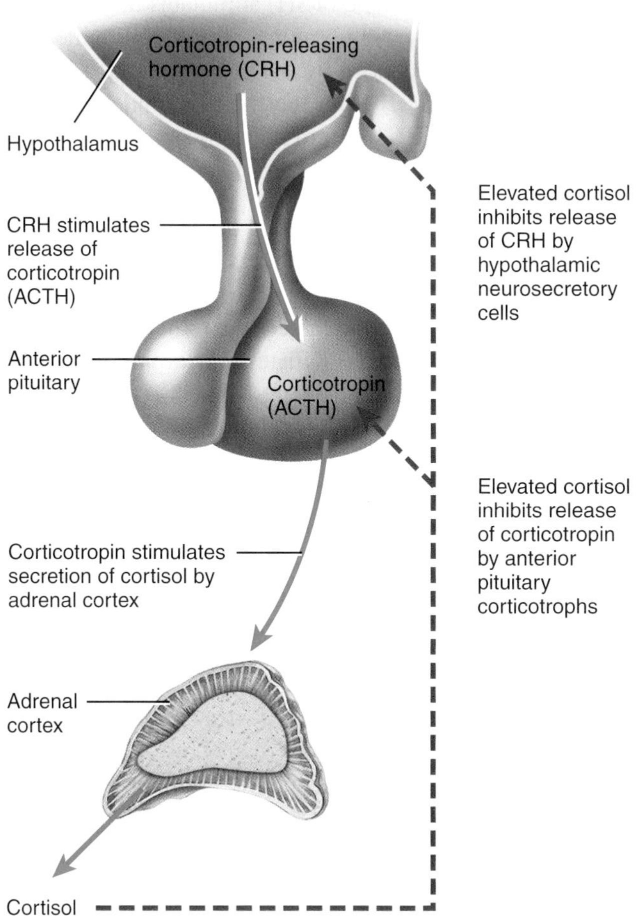

? Which other target gland hormones suppress secretion of hypothalamic and anterior pituitary hormones by negative feedback?

dins). In response to human growth hormone, cells in the liver, skeletal muscles, cartilage, bones, and other tissues secrete IGFs, which may either enter the bloodstream from the liver or act locally in other tissues as autocrines or paracrines.

The functions of IGFs include the following:

1. IGFs cause cells to grow and multiply by increasing uptake of amino acids into cells and accelerating protein synthesis. IGFs also decrease the breakdown of proteins and the use of amino acids for ATP production. Due to these effects of the IGFs, human growth hormone increases the growth rate of the skeleton and skeletal muscles during childhood and the teenage years. In adults, human growth hormone and IGFs help maintain the mass of muscles and bones and promote healing of injuries and tissue repair.
2. IGFs also enhance lipolysis in adipose tissue, which results in increased use of the released fatty acids for ATP production by body cells.
3. In addition to affecting protein and lipid metabolism, human growth hormone and IGFs influence carbohydrate metabolism by decreasing glucose uptake, which decreases the use of glucose for ATP production by most body cells. This action spares glucose so that it is available to neurons for ATP production in times of glucose scarcity. IGFs and human growth hormone may also stimulate liver cells to release glucose into the blood.

Somatotrophs in the anterior pituitary release bursts of human growth hormone every few hours, especially during sleep. Their secretory activity is controlled mainly by two hypothalamic hormones: (1) growth hormone–releasing hormone (GHRH) promotes secretion of human growth hormone, and (2) growth hormone–inhibiting hormone (GHIH) suppresses it. A major regulator of GHRH and GHIH secretion is the blood glucose level (Figure 18.7):

1. **Hypoglycemia** (hī′-pō-glī-SĒ-mē-a), an abnormally low blood glucose concentration, stimulates the hypothalamus to secrete GHRH, which flows toward the anterior pituitary in the hypophyseal portal veins.
2. On reaching the anterior pituitary, GHRH stimulates somatotrophs to release human growth hormone.
3. Human growth hormone stimulates secretion of insulinlike growth factors, which speed breakdown of liver glycogen into glucose, causing glucose to enter the blood more rapidly.
4. As a result, blood glucose rises to the normal level (about 90 mg/100 mL of blood plasma).
5. An increase in blood glucose above the normal level inhibits release of GHRH.
6. **Hyperglycemia** (hī′-per-glī-SĒ-mē-a), an abnormally high blood glucose concentration, stimulates the hypothalamus to secrete GHIH (while inhibiting the secretion of GHRH).
7. On reaching the anterior pituitary in portal blood, GHIH inhibits secretion of human growth hormone by somatotrophs.
8. A low level of human growth hormone and IGFs slows breakdown of glycogen in the liver, and glucose is released into the blood more slowly.
9. Blood glucose falls to the normal level.
10. A decrease in blood glucose below the normal level (hypoglycemia) inhibits release of GHIH.

Other stimuli that promote secretion of human growth hormone include decreased fatty acids and increased amino acids in the blood; deep sleep (stages 3 and 4 of non-rapid eye movement sleep); increased activity of the sympathetic division of the autonomic nervous system, such as might occur with stress or vigorous physical exercise; and other hormones, including glucagon, estrogens, cortisol, and insulin. Factors that inhibit human growth hormone secretion are increased levels of fatty acids and decreased levels of amino acids in the blood; rapid eye movement sleep;

Figure 18.7 Effects of human growth hormone (hGH) and insulinlike growth factors (IGFs). Solid arrows indicate stimulation of secretion; dashed arrows indicate inhibition of secretion via negative feedback.

Secretion of hGH is stimulated by growth hormone–releasing hormone (GHRH) and inhibited by growth hormone–inhibiting hormone (GHIH).

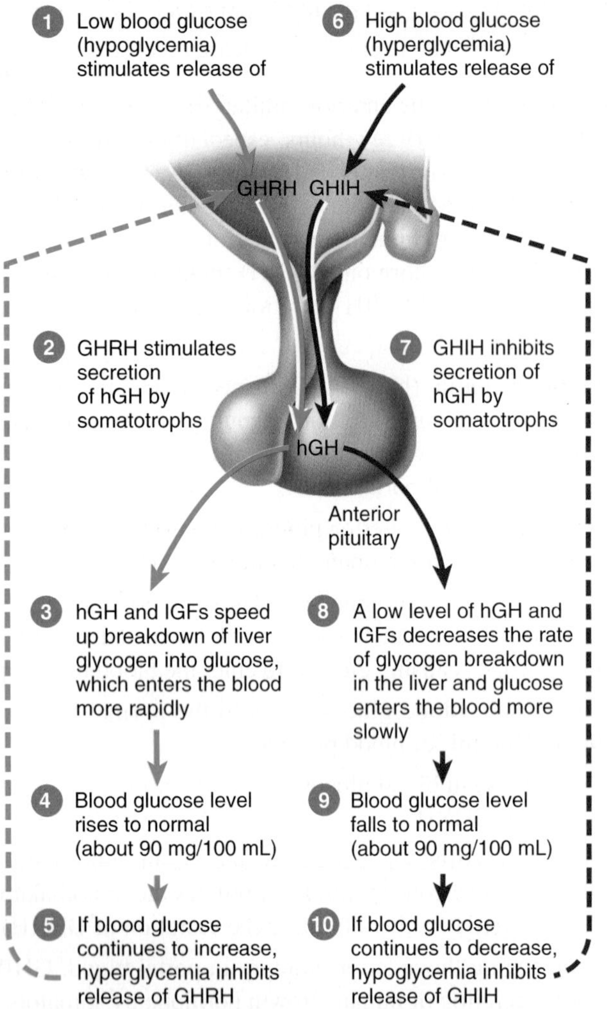
1 Low blood glucose (hypoglycemia) stimulates release of
6 High blood glucose (hyperglycemia) stimulates release of
GHRH
GHIH
2 GHRH stimulates secretion of hGH by somatotrophs
7 GHIH inhibits secretion of hGH by somatotrophs
hGH
Anterior pituitary
3 hGH and IGFs speed up breakdown of liver glycogen into glucose, which enters the blood more rapidly
8 A low level of hGH and IGFs decreases the rate of glycogen breakdown in the liver and glucose enters the blood more slowly
4 Blood glucose level rises to normal (about 90 mg/100 mL)
9 Blood glucose level falls to normal (about 90 mg/100 mL)
5 If blood glucose continues to increase, hyperglycemia inhibits release of GHRH
10 If blood glucose continues to decrease, hypoglycemia inhibits release of GHIH

If a person has a pituitary tumor that secretes a large amount of hGH and the tumor cells are not responsive to regulation by GHRH and GHIH, will hyperglycemia or hypoglycemia be more likely?

emotional deprivation; obesity; low levels of thyroid hormones; and human growth hormone itself (through negative feedback). Growth hormone–inhibiting hormone (GHIH), alternatively known as somatostatin, also inhibits the secretion of human growth hormone.

CLINICAL CONNECTION | *Diabetogenic Effect of hGH*

One symptom of excess human growth hormone (hGH) is hyperglycemia. Persistent hyperglycemia in turn stimulates the pancreas to secrete insulin continually. Such excessive stimulation, if it lasts for weeks or months, may cause "beta-cell burnout," a greatly decreased capacity of pancreatic beta cells to synthesize and secrete insulin. Thus, excess secretion of human growth hormone may have a **diabetogenic effect** (dī′-a-bet′-o-JEN-ik); that is, it causes diabetes mellitus (lack of insulin activity). •

Thyroid-Stimulating Hormone

Thyroid-stimulating hormone (TSH) stimulates the synthesis and secretion of the two thyroid hormones, triiodothyronine (T_3) and thyroxine (T_4), both produced by the thyroid gland. Thyrotropin-releasing hormone (TRH) from the hypothalamus controls TSH secretion. Release of TRH in turn depends on blood levels of T_3 and T_4; high levels of T_3 and T_4 inhibit secretion of TRH via negative feedback. There is no thyrotropin-inhibiting hormone. The release of TRH is explained later in the chapter (see Figure 18.12).

Follicle-Stimulating Hormone

In females, the ovaries are the targets for follicle-stimulating hormone (FSH). Each month FSH initiates the development of several ovarian follicles, saclike arrangements of secretory cells that surround a developing oocyte. FSH also stimulates follicular cells to secrete estrogens (female sex hormones). In males, FSH stimulates sperm production in the testes. Gonadotropin-releasing hormone (GnRH) from the hypothalamus stimulates FSH release. Release of GnRH and FSH is suppressed by estrogens in females and by testosterone (the principal male sex hormone) in males through negative feedback systems. There is no gonadotropin-inhibiting hormone.

Luteinizing Hormone

In females, luteinizing hormone (LH) triggers **ovulation,** the release of a secondary oocyte (future ovum) by an ovary. LH stimulates formation of the corpus luteum (structure formed after ovulation) in the ovary and the secretion of progesterone (another female sex hormone) by the corpus luteum. Together, FSH and LH also stimulate secretion of estrogens by ovarian cells. Estrogens and progesterone prepare the uterus for implantation of a fertilized ovum and help prepare the mammary glands for milk secretion. In males, LH stimulates cells in the testes to secrete testosterone. Secretion of LH, like that of FSH, is controlled by gonadotropin-releasing hormone (GnRH).

Prolactin

Prolactin (PRL), together with other hormones, initiates and maintains milk production by the mammary glands. By itself, prolactin has only a weak effect. Only after the mammary glands have been primed by estrogens, progesterone, glucocorticoids, human growth hormone, thyroxine, and insulin, which exert permissive effects, does PRL bring about milk production. Ejection of milk from the mammary glands depends on the hormone oxytocin, which is released from the posterior pituitary. Together, milk production and ejection constitute *lactation.*

The hypothalamus secretes both inhibitory and excitatory hormones that regulate prolactin secretion. In females, prolactin-inhibiting hormone (PIH), which is dopamine, inhibits the release of prolactin from the anterior pituitary most of the time. Each month, just before menstruation begins, the secretion of PIH diminishes and the blood level of prolactin rises, but not enough to stimulate milk production. Breast tenderness just before menstruation may be caused by elevated prolactin. As the menstrual cycle begins anew, PIH is again secreted and the prolactin level drops. During pregnancy, the prolactin level rises, stimulated by prolactin-releasing hormone (PRH) from the hypothalamus. The sucking action of a nursing infant causes a reduction in hypothalamic secretion of PIH.

The function of prolactin is not known in males, but its hypersecretion causes erectile dysfunction (impotence, the inability to have an erection of the penis). In females, hypersecretion of prolactin causes galactorrhea (inappropriate lactation) and amenorrhea (absence of menstrual cycles).

Adrenocorticotropic Hormone

Corticotrophs secrete mainly adrenocorticotropic hormone (ACTH). ACTH controls the production and secretion of cortisol and other glucocorticoids by the cortex (outer portion) of the adrenal glands. Corticotropin-releasing hormone (CRH) from the hypothalamus stimulates secretion of ACTH by corticotrophs. Stress-related stimuli, such as low blood glucose or physical trauma, and interleukin-1, a substance produced by macrophages, also stimulate release of ACTH. Glucocorticoids inhibit CRH and ACTH release via negative feedback.

Melanocyte-Stimulating Hormone

Melanocyte-stimulating hormone (MSH) increases skin pigmentation in amphibians by stimulating the dispersion of melanin granules in melanocytes. Its exact role in humans is unknown, but the presence of MSH receptors in the brain suggests it may influence brain activity. There is little circulating MSH in humans. However, continued administration of MSH for several days does produce a darkening of the skin. Excessive levels of corticotropin-releasing hormone (CRH) can stimulate MSH release; dopamine inhibits MSH release.

Table 18.4 summarizes the principal actions of the anterior pituitary hormones.

TABLE 18.4

Summary of the Principal Actions of Anterior Pituitary Hormones

HORMONE	TARGET TISSUES	PRINCIPAL ACTIONS
Human growth hormone (hGH), also known as somatotropin	 Liver (and other tissues)	Stimulates liver, muscle, cartilage, bone, and other tissues to synthesize and secrete insulinlike growth factors (IGFs); IGFs promote growth of body cells, protein synthesis, tissue repair, lipolysis, and elevation of blood glucose concentration.
Thyroid-stimulating hormone (TSH), also known as thyrotropin	 Thyroid gland	Stimulates synthesis and secretion of thyroid hormones by thyroid gland.
Follicle-stimulating hormone (FSH)	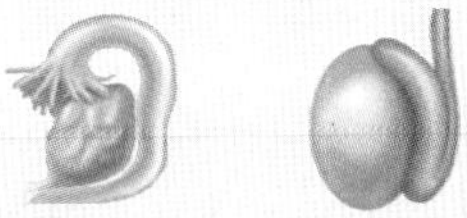 Ovaries Testes	In females, initiates development of oocytes and induces ovarian secretion of estrogens. In males, stimulates testes to produce sperm.
Luteinizing hormone (LH)	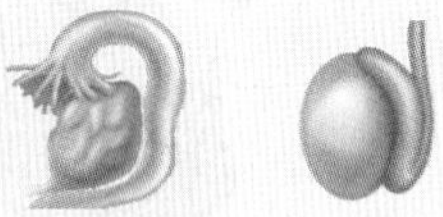 Ovaries Testes	In females, stimulates secretion of estrogens and progesterone, ovulation, and formation of corpus luteum. In males, stimulates testes to produce testosterone.
Prolactin (PRL)	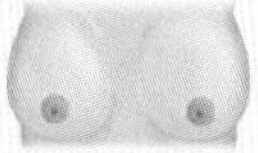 Mammary glands	Together with other hormones, promotes milk production by mammary glands.
Adrenocorticotropic hormone (ACTH), also known as corticotropin	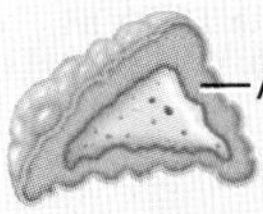	Stimulates secretion of glucocorticoids (mainly cortisol) by adrenal cortex.
Melanocyte-stimulating hormone (MSH)	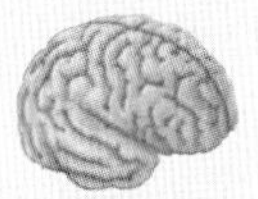 Brain	Exact role in humans is unknown but may influence brain activity; when present in excess, can cause darkening of skin.

Posterior Pituitary

Although the **posterior pituitary** or **neurohypophysis** (noo′-rō-hī-POF-i-sis; *neuro-* = nerve; *-hypophysis* = undergrowth) does not *synthesize* hormones, it does *store* and *release* two hormones. It consists of axons and axon terminals of more than 10,000 hypothalamic neurosecretory cells. The cell bodies of the neurosecretory cells are in the paraventricular and supraoptic nuclei of the hypothalamus; their axons form the **hypothalamohypophyseal tract** (hī′-pō-thal′-a-mō-hī-pō-FIZ-ē-al). This tract begins in the hypothalamus and ends near blood capillaries in the posterior pituitary (Figure 18.8). The neuronal cell bodies in the paraventricular nucleus synthesize the hormone **oxytocin** (**OT;** ok′-sē-TŌ-sin; *okytoc* = quick birth) and the neuronal cell bodies in the supraoptic nucleus produce **antidiuretic hormone** (**ADH**)**,** also called **vasopressin.** The axon terminals in the posterior pituitary are associated with specialized neuroglia called **pituicytes** (pi-TOO-i-sītz). These cells have a supporting role similar to that of astrocytes (see Chapter 12).

After their production in the cell bodies of neurosecretory cells, oxytocin and antidiuretic hormone are packaged into secretory vesicles, which move by fast axonal transport (described in Section 12.2) to the axon terminals in the posterior pituitary, where they are stored until nerve impulses trigger exocytosis and release of the hormone.

Blood is supplied to the posterior pituitary by the **inferior hypophyseal arteries,** which branch from the internal carotid arteries. In the posterior pituitary, the inferior hypophyseal arteries drain into the **capillary plexus of the infundibular process,** a capillary network that receives secreted oxytocin and antidiuretic hormone (see Figure 18.5). From this plexus, hormones pass into the **posterior hypophyseal veins** for distribution to target cells in other tissues.

Oxytocin

During and after delivery of a baby, oxytocin affects two target tissues: the mother's uterus and breasts. During delivery, stretching of the cervix of the uterus stimulates the release of oxytocin which, in turn, enhances contraction of smooth muscle cells in the wall of the uterus (see Figure 1.4); after delivery, it stimulates milk ejection ("letdown") from the mammary glands in response to the mechanical stimulus provided by a suckling infant. The

Figure 18.8 Axons of hypothalamic neurosecretory cells form the hypothalamohypophyseal tract, which extends from the paraventricular and supraoptic nuclei to the posterior pituitary. Hormone molecules synthesized in the cell body of a neurosecretory cell are packaged into secretory vesicles that move down to the axon terminals. Nerve impulses trigger exocytosis of the vesicles, thereby releasing the hormone.

Oxytocin and antidiuretic hormone are synthesized in the hypothalamus and released into the capillary plexus of the infundibular process in the posterior pituitary.

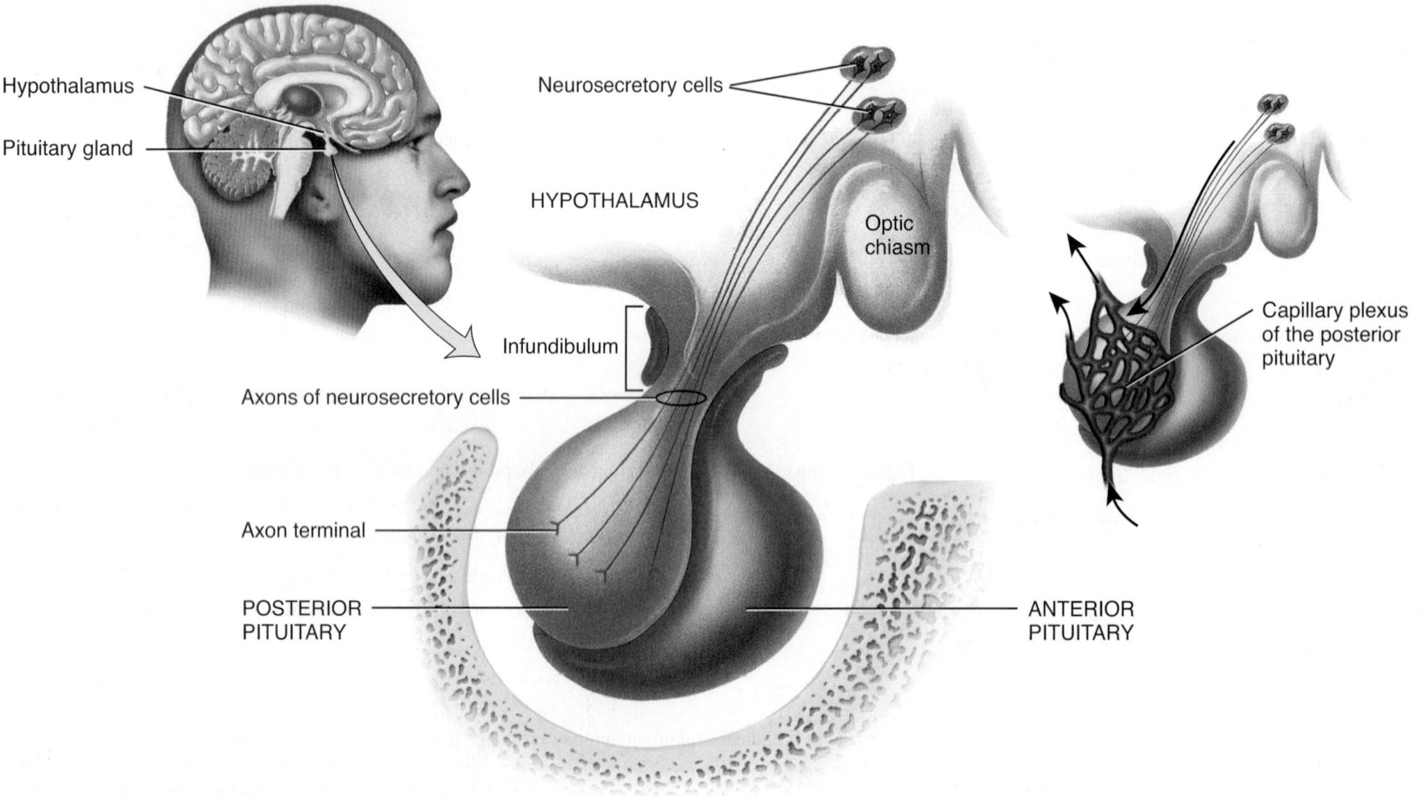

Functionally, how are the hypothalamohypophyseal tract and the hypophyseal portal veins similar? Structurally, how are they different?

function of oxytocin in males and in nonpregnant females is not clear. Experiments with animals have suggested that it has actions within the brain that foster parental caretaking behavior toward young offspring. It may also be responsible, in part, for the feelings of sexual pleasure during and after intercourse.

CLINICAL CONNECTION | *Oxytocin and Childbirth*

Years before oxytocin was discovered, it was common practice in midwifery to let a first-born twin nurse at the mother's breast to speed the birth of the second child. Now we know why this practice is helpful—it stimulates the release of oxytocin. Even after a single birth, nursing promotes expulsion of the placenta (afterbirth) and helps the uterus regain its smaller size. Synthetic oxytocin (Pitocin) often is given to induce labor or to increase uterine tone and control hemorrhage just after giving birth. •

Antidiuretic Hormone

As its name implies, an **antidiuretic** (an-ti-dī-ū-RET-ik; *anti-* = against; *-dia-* = throughout; *-ouresis* = urination) is a substance that decreases urine production. ADH causes the kidneys to return more water to the blood, thus decreasing urine volume. In the absence of ADH, urine output increases more than tenfold, from the normal 1 to 2 liters to about 20 liters a day. Drinking alcohol often causes frequent and copious urination because alcohol inhibits secretion of ADH. ADH also decreases the water lost through sweating and causes constriction of arterioles, which increases blood pressure. This hormone's other name, **vasopressin** (vā-sō-PRESS-in; *vaso-* = blood; *-pressus* = to press), reflects this effect on blood pressure.

The amount of ADH secreted varies with blood osmotic pressure and blood volume. Figure 18.9 shows regulation of ADH secretion and the actions of ADH:

1. High blood osmotic pressure (or decreased blood volume)—due to dehydration or a decline in blood volume because of hemorrhage, diarrhea, or excessive sweating—stimulates **osmoreceptors,** neurons in the hypothalamus that monitor blood osmotic pressure. Elevated blood osmotic pressure activates the osmoreceptors directly; they also receive excitatory input from other brain areas when blood volume decreases.
2. Osmoreceptors activate the hypothalamic neurosecretory cells that synthesize and release ADH.
3. When neurosecretory cells receive excitatory input from the osmoreceptors, they generate nerve impulses that cause exocytosis of ADH-containing vesicles from their axon terminals in the posterior pituitary. This liberates ADH, which diffuses into blood capillaries of the posterior pituitary.
4. The blood carries ADH to three target tissues: the kidneys, sudoriferous (sweat) glands, and smooth muscle in blood vessel walls. The kidneys respond by retaining more water, which decreases urine output. Secretory activity of sweat glands decreases, which lowers the rate of water loss by perspiration from the skin. Smooth muscle in the walls of arterioles (small arteries) contracts in response to high levels of ADH, which constricts (narrows) the lumen of these blood vessels and increases blood pressure.
5. Low osmotic pressure of blood (or increased blood volume) inhibits the osmoreceptors.
6. Inhibition of osmoreceptors reduces or stops ADH secretion. The kidneys then retain less water by forming a larger volume of urine, secretory activity of sweat glands increases, and arterioles dilate. The blood volume and osmotic pressure of body fluids return to normal.

Figure 18.9 **Regulation of secretion and actions of antidiuretic hormone (ADH).**

ADH acts to retain body water and increase blood pressure.

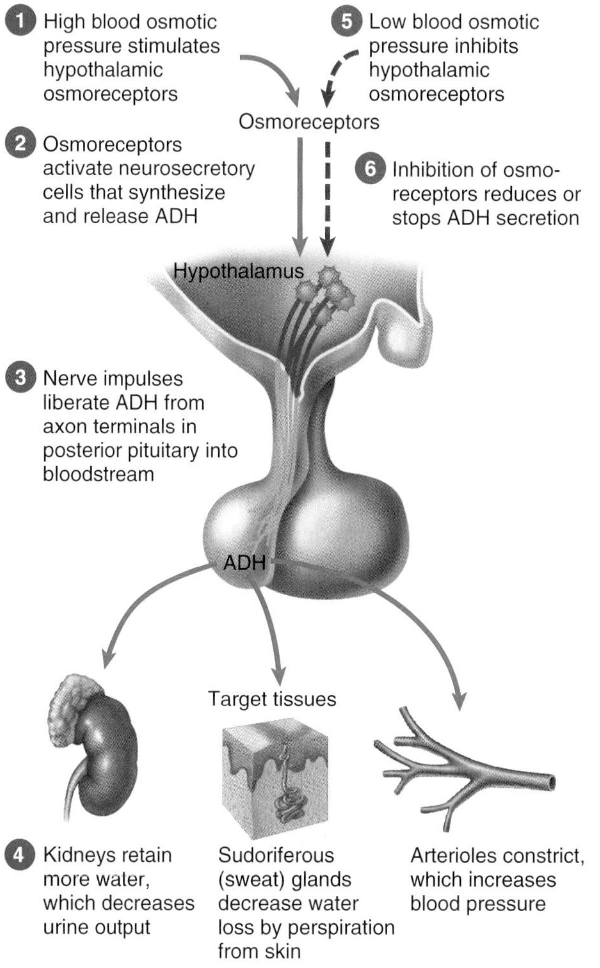
1 High blood osmotic pressure stimulates hypothalamic osmoreceptors
5 Low blood osmotic pressure inhibits hypothalamic osmoreceptors
Osmoreceptors
2 Osmoreceptors activate neurosecretory cells that synthesize and release ADH
6 Inhibition of osmoreceptors reduces or stops ADH secretion
Hypothalamus
3 Nerve impulses liberate ADH from axon terminals in posterior pituitary into bloodstream
ADH
Target tissues
4 Kidneys retain more water, which decreases urine output
Sudoriferous (sweat) glands decrease water loss by perspiration from skin
Arterioles constrict, which increases blood pressure

If you drank a liter of water, what effect would this have on the osmotic pressure of your blood, and how would the level of ADH change in your blood?

Secretion of ADH can also be altered in other ways. Pain, stress, trauma, anxiety, acetylcholine, nicotine, and drugs such as morphine, tranquilizers, and some anesthetics stimulate ADH secretion. The dehydrating effect of alcohol, which has already been mentioned, may cause both the thirst and the headache typical of a hangover. Hyposecretion of ADH or nonfunctional ADH receptors cause diabetes insipidus (see Disorders: Homeostatic Imbalances at the end of the chapter).

Table 18.5 lists the posterior pituitary hormones, control of their secretion, and their principal actions.

✔CHECKPOINT

9. In what respect is the pituitary gland actually two glands?
10. How do hypothalamic releasing and inhibiting hormones influence secretions of the anterior pituitary?
11. Describe the structure and importance of the hypothalamohypophyseal tract.

18.7 THYROID GLAND

OBJECTIVE

- Describe the location, histology, hormones, and functions of the thyroid gland.

The butterfly-shaped **thyroid gland** is located just inferior to the larynx (voice box). It is composed of right and left **lateral lobes,** one on either side of the trachea, that are connected by an **isthmus** (IS-mus = a narrow passage) anterior to the trachea (Figure 18.10a). About 50% of thyroid glands have a small third lobe, called the *pyramidal lobe.* It extends superiorly from the isthmus. The normal mass of the thyroid is about 30 g (1 oz).

Microscopic spherical sacs called **thyroid follicles** (Figure 18.10b) make up most of the thyroid gland. The wall of each follicle consists primarily of cells called **follicular cells** (fo-LIK-u-lar), most of which extend to the lumen (internal space) of the follicle. A **basement membrane** surrounds each follicle. When the follicular cells are inactive, their shape is low cuboidal to squamous, but under the influence of TSH they become active in secretion and range from cuboidal to low columnar in shape. The follicular cells produce two hormones: **thyroxine** (thī-ROK-sēn), which is also called **tetraiodothyronine** (tet-ra-ī-ō-dō-THĪ-rō-nēn) or **T_4** because it contains four atoms of iodine, and **triiodothyronine** (trī-ī′-ō-dō-THĪ-rō-nēn) or **T_3,** which contains three atoms of iodine. T_3 and T_4 together are also known as **thyroid hormones.** A few cells called **parafollicular cells** (par′-a-fo-LIK-u-lar) or **C cells** lie between follicles. They produce the hormone **calcitonin** (kal-si-TŌ-nin), which helps regulate calcium homeostasis.

Formation, Storage, and Release of Thyroid Hormones

The thyroid gland is the only endocrine gland that stores its secretory product in large quantities—normally about a 100-day supply. Synthesis and secretion of T_3 and T_4 occurs as follows (Figure 18.11):

1. ***Iodide trapping.*** Thyroid follicular cells trap iodide ions (I^-) by actively transporting them from the blood into the cytosol. As a result, the thyroid gland normally contains most of the iodide in the body.
2. ***Synthesis of thyroglobulin.*** While the follicular cells are trapping I^-, they are also synthesizing **thyroglobulin (TGB)**

TABLE 18.5

Summary of Posterior Pituitary Hormones

HORMONE AND TARGET TISSUES	CONTROL OF SECRETION	PRINCIPAL ACTIONS
Oxytocin (OT) 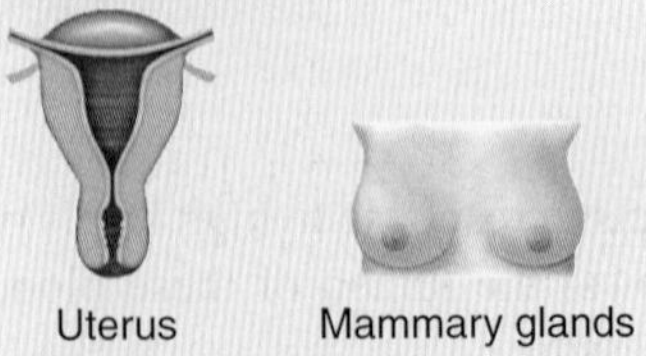	Neurosecretory cells of hypothalamus secrete OT in response to uterine distention and stimulation of nipples.	Stimulates contraction of smooth muscle cells of uterus during childbirth; stimulates contraction of myoepithelial cells in mammary glands to cause milk ejection.
Antidiuretic hormone (ADH) or vasopressin 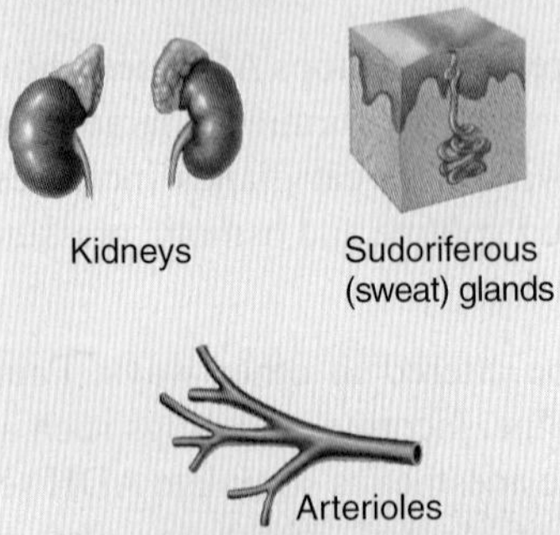	Neurosecretory cells of hypothalamus secrete ADH in response to elevated blood osmotic pressure, dehydration, loss of blood volume, pain, or stress; inhibitors of ADH secretion include low blood osmotic pressure, high blood volume, and alcohol.	Conserves body water by decreasing urine volume; decreases water loss through perspiration; raises blood pressure by constricting arterioles.

Figure 18.10 Location, blood supply, and histology of the thyroid gland.

Thyroid hormones regulate (1) oxygen use and basal metabolic rate, (2) cellular metabolism, and (3) growth and development.

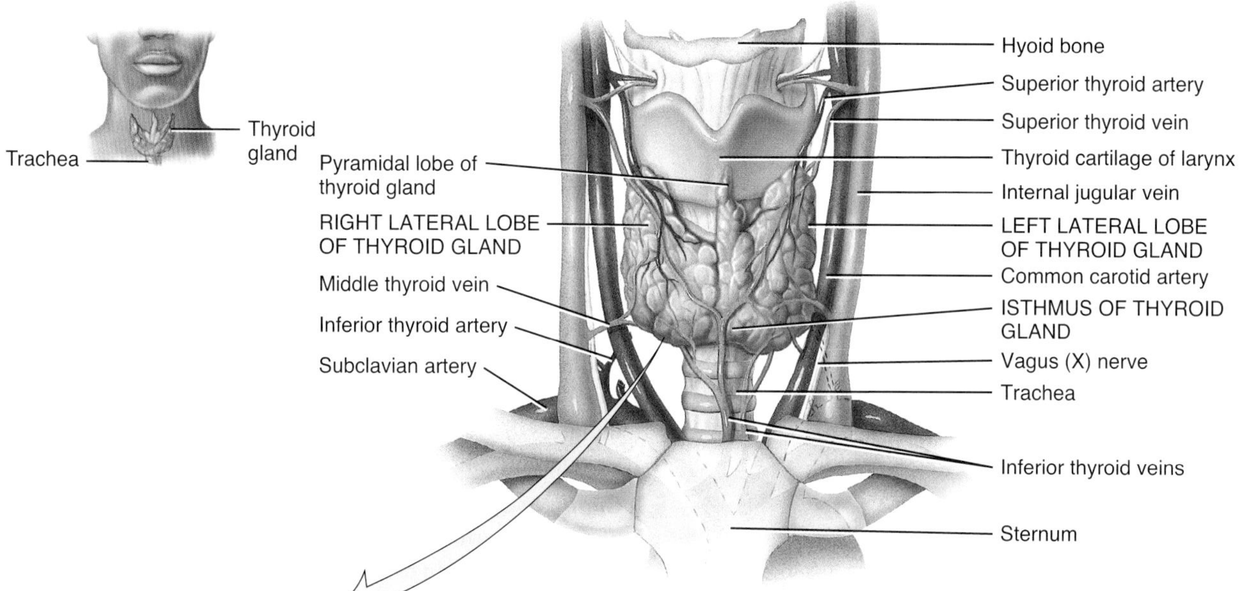

(a) Anterior view of thyroid gland

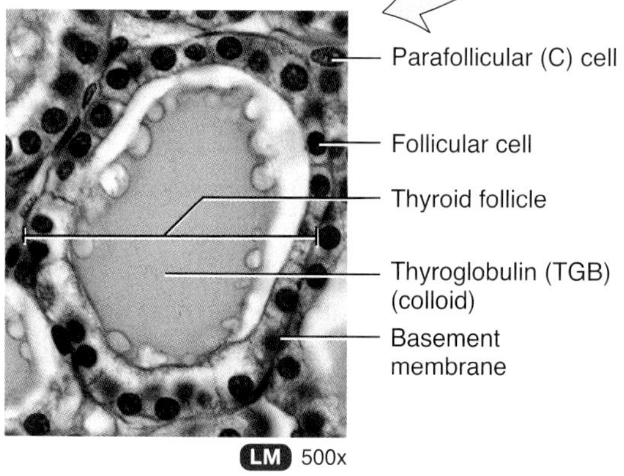

(b) Thyroid follicles

(c) Anterior view of thyroid gland

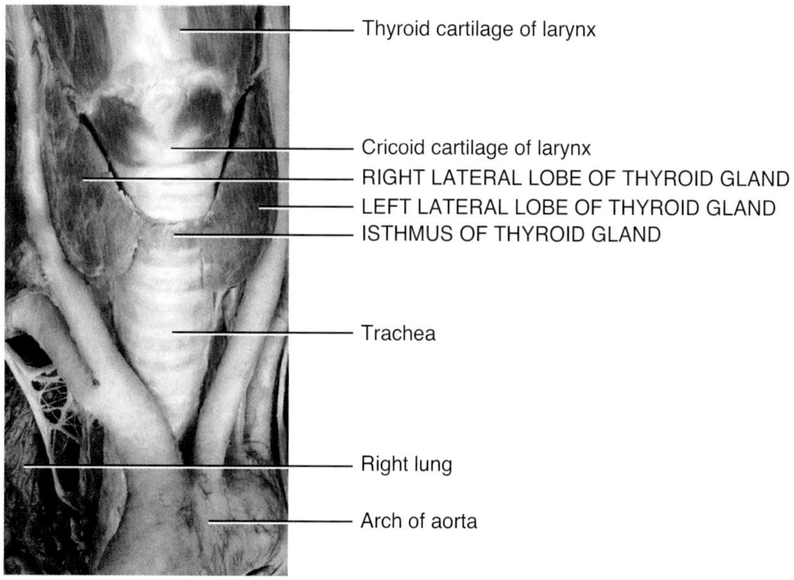

(d) Anterior view

Which cells secrete T_3 and T_4? Which secrete calcitonin? Which of these hormones are also called thyroid hormones?

Figure 18.11 Steps in the synthesis and secretion of thyroid hormones.

Thyroid hormones are synthesized by attaching iodine atoms to the amino acid tyrosine.

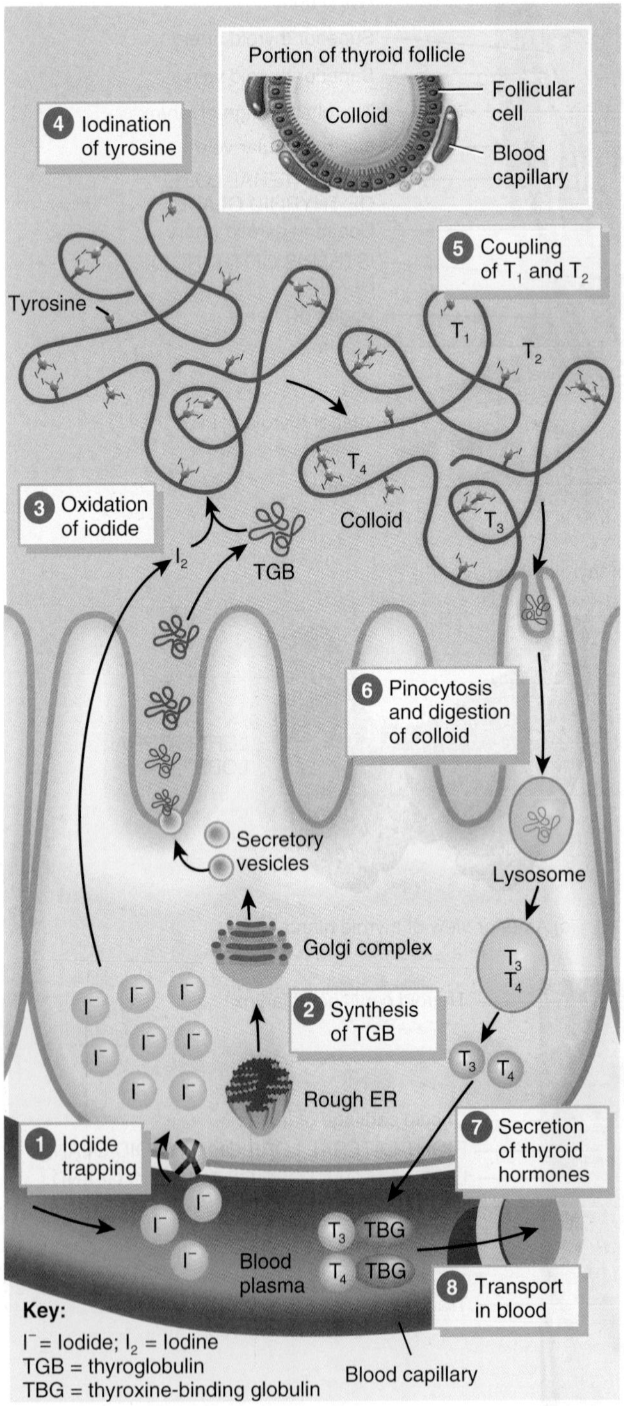
Portion of thyroid follicle
Colloid
Follicular cell
Blood capillary
4 Iodination of tyrosine
5 Coupling of T1 and T2
Tyrosine
T1
T2
T4
T3
3 Oxidation of iodide
I2
TGB
Colloid
6 Pinocytosis and digestion of colloid
Secretory vesicles
Lysosome
Golgi complex
2 Synthesis of TGB
Rough ER
1 Iodide trapping
7 Secretion of thyroid hormones
TBG
Blood plasma
8 Transport in blood
Key:
I⁻ = Iodide; I2 = Iodine
TGB = thyroglobulin
TBG = thyroxine-binding globulin
Blood capillary

What is the storage form of thyroid hormones?

(thī′-rō-GLOB-u-lin), a large glycoprotein that is produced in the rough endoplasmic reticulum, modified in the Golgi complex, and packaged into secretory vesicles. The vesicles then undergo exocytosis, which releases TGB into the lumen of the follicle.

3. ***Oxidation of iodide.*** Some of the amino acids in TGB are tyrosines that will become iodinated. However, negatively charged iodide ions cannot bind to tyrosine until they undergo oxidation (removal of electrons) to iodine: $2\ I^- \rightarrow I_2$. As the iodide ions are being oxidized, they pass through the membrane into the lumen of the follicle.

4. ***Iodination of tyrosine.*** As iodine molecules (I_2) form, they react with tyrosines that are part of thyroglobulin molecules. Binding of one iodine atom yields monoiodotyrosine (T_1), and a second iodination produces diiodotyrosine (T_2). The TGB with attached iodine atoms, a sticky material that accumulates and is stored in the lumen of the thyroid follicle, is termed **colloid.**

5. ***Coupling of T_1 and T_2.*** During the last step in the synthesis of thyroid hormone, two T_2 molecules join to form T_4, or one T_1 and one T_2 join to form T_3.

6. ***Pinocytosis and digestion of colloid.*** Droplets of colloid reenter follicular cells by pinocytosis and merge with lysosomes. Digestive enzymes in the lysosomes break down TGB, cleaving off molecules of T_3 and T_4.

7. ***Secretion of thyroid hormones.*** Because T_3 and T_4 are lipid-soluble, they diffuse through the plasma membrane into interstitial fluid and then into the blood. T_4 normally is secreted in greater quantity than T_3, but T_3 is several times more potent. Moreover, after T_4 enters a body cell, most of it is converted to T_3 by removal of one iodine.

8. ***Transport in the blood.*** More than 99% of both the T_3 and the T_4 combine with transport proteins in the blood, mainly **thyroxine-binding globulin (TBG).**

Actions of Thyroid Hormones

Because most body cells have receptors for thyroid hormones, T_3 and T_4 exert their effects throughout the body.

1. Thyroid hormones increase **basal metabolic rate (BMR),** the rate of oxygen consumption under standard or basal conditions (awake, at rest, and fasting), by stimulating the use of cellular oxygen to produce ATP. When the basal metabolic rate increases, cellular metabolism of carbohydrates, lipids, and proteins increases.
2. A second major effect of thyroid hormones is to stimulate synthesis of additional sodium–potassium pumps (Na^+/K^+ ATPase), which use large amounts of ATP to continually eject sodium ions (Na^+) from the cytosol into the extracellular fluid and potassium ions (K^+) from the extracellular fluid into the cytosol. As cells produce and use more ATP, more heat is given off, and body temperature rises. This phenomenon is called the **calorigenic effect** (ka-lor′-i-JEN-ik). In this way,

thyroid hormones play an important role in the maintenance of normal body temperature. Normal mammals can survive in freezing temperatures, but those whose thyroid glands have been removed cannot.

3. In the regulation of metabolism, the thyroid hormones stimulate protein synthesis and increase the use of glucose and fatty acids for ATP production. They also increase lipolysis and enhance cholesterol excretion, thus reducing blood cholesterol level.
4. The thyroid hormones enhance some actions of the catecholamines (norepinephrine and epinephrine) because they up-regulate beta (β) receptors. For this reason, symptoms of hyperthyroidism include increased heart rate, more forceful heartbeats, and increased blood pressure.
5. Together with human growth hormone and insulin, thyroid hormones accelerate body growth, particularly the growth of the nervous and skeletal systems. Deficiency of thyroid hormones during fetal development, infancy, or childhood causes severe mental retardation and stunted bone growth.

Control of Thyroid Hormone Secretion

Thyrotropin-releasing hormone (TRH) from the hypothalamus and thyroid-stimulating hormone (TSH) from the anterior pituitary stimulate synthesis and release of thyroid hormones, as shown in Figure 18.12:

1. Low blood levels of T_3 and T_4 or low metabolic rate stimulates the hypothalamus to secrete TRH.
2. TRH enters the hypophyseal portal veins and flows to the anterior pituitary, where it stimulates thyrotrophs to secrete TSH.
3. TSH stimulates virtually all aspects of thyroid follicular cell activity, including iodide trapping (1 in Figure 18.11), hormone synthesis and secretion (2 and 7 in Figure 18.11), and growth of the follicular cells.
4. The thyroid follicular cells release T_3 and T_4 into the blood until the metabolic rate returns to normal.
5. An elevated level of T_3 inhibits release of TRH and TSH (negative feedback inhibition).

Conditions that increase ATP demand—a cold environment, hypoglycemia, high altitude, and pregnancy—also increase the secretion of the thyroid hormones.

Calcitonin

The hormone produced by the **parafollicular cells** of the thyroid gland (see Figure 18.10b) is **calcitonin (CT).** CT can decrease the level of calcium in the blood by inhibiting the action of osteoclasts, the cells that break down bone extracellular matrix. The secretion of CT is controlled by a negative feedback system (see Figure 18.14).

When its blood level is high, calcitonin lowers the amount of blood calcium and phosphates by inhibiting bone resorption (breakdown of bone extracellular matrix) by osteoclasts and by accelerating uptake of calcium and phosphates into bone extracellular

Figure 18.12 Regulation of secretion and actions of thyroid hormones. TRH = thyrotropin-releasing hormone, TSH = thyroid-stimulating hormone, T_3 = triiodothyronine, and T_4 = thyroxine (tetraiodothyronine).

TSH promotes release of thyroid hormones (T_3 and T_4) by the thyroid gland.

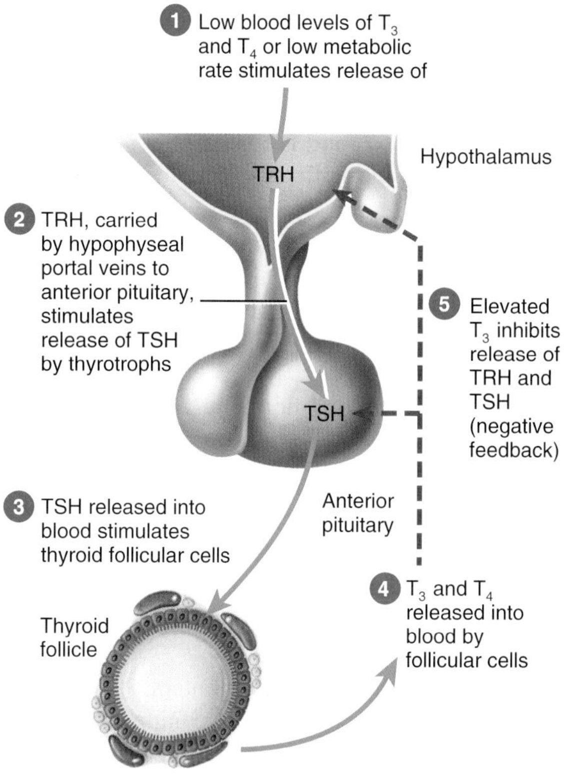

Actions of Thyroid Hormones:

Increase basal metabolic rate
Stimulate synthesis of Na^+/K^+ ATPase
Increase body temperature (calorigenic effect)
Stimulate protein synthesis
Increase the use of glucose and fatty acids for ATP production
Stimulate lipolysis
Enhance some actions of catecholamines
Regulate development and growth of nervous tissue and bones

? How could an iodine-deficient diet lead to goiter, which is an enlargement of the thyroid gland?

matrix. Miacalcin, a calcitonin extract derived from salmon that is 10 times more potent than human calcitonin, is prescribed to treat osteoporosis.

Table 18.6 summarizes the hormones produced by the thyroid gland, control of their secretion, and their principal actions.

✔ CHECKPOINT

12. Explain how blood levels of T_3/T_4, TSH, and TRH would change in a laboratory animal that has undergone a thyroidectomy (complete removal of its thyroid gland).
13. How are the thyroid hormones synthesized, stored, and secreted?
14. How is the secretion of T_3 and T_4 regulated?
15. What are the physiological effects of the thyroid hormones?

18.8 PARATHYROID GLANDS

OBJECTIVE

- Describe the location, histology, hormone, and functions of the parathyroid glands.

Partially embedded in the posterior surface of the lateral lobes of the thyroid gland are several small, round masses of tissue called the **parathyroid glands** (*para-* = beside). Each has a mass of about 40 mg (0.04 g). Usually, one superior and one inferior parathyroid gland are attached to each lateral thyroid lobe (Figure 18.13a), for a total of four.

Microscopically, the parathyroid glands contain two kinds of epithelial cells (Figure 18.13b, c). The more numerous cells, called **chief (principal) cells,** produce **parathyroid hormone (PTH),** also called **parathormone.** The function of the other kind of cell, called an **oxyphil cell,** is not known in a normal parathyroid gland. However, its presence clearly helps to identify the parathyroid gland histologically due to its unique staining characteristics. Furthermore, in a cancer of the parathyroid glands, oxyphil cells secrete excess PTH.

Parathyroid Hormone

Parathyroid hormone is the major regulator of the levels of calcium (Ca^{2+}), magnesium (Mg^{2+}), and phosphate ($HPO_4^{\,2-}$) ions in the blood. The specific action of PTH is to increase the number and activity of osteoclasts. The result is elevated bone *resorption,* which releases ionic calcium (Ca^{2+}) and phosphates ($HPO_4^{\,2-}$) into the blood. PTH also acts on the kidneys. First, it slows the rate at which Ca^{2+} and Mg^{2+} are lost from blood into the urine. Second, it increases loss of $HPO_4^{\,2-}$ from blood into the urine. Because more $HPO_4^{\,2-}$ is lost in the urine than is gained from the bones, PTH decreases blood $HPO_4^{\,2-}$ level and increases blood Ca^{2+} and Mg^{2+} levels. A third effect of PTH on the kidneys is to promote formation of the hormone **calcitriol** (kal′-si-TRĪ-ol), the active form of vitamin D. Calcitriol, also known as *1,25-dihydroxyvitamin* D_3, increases the rate of Ca^{2+}, $HPO_4^{\,2-}$, and Mg^{2+} *absorption* from the gastrointestinal tract into the blood.

The blood calcium level directly controls the secretion of both calcitonin and parathyroid hormone via negative feedback loops that do not involve the pituitary gland (Figure 18.14):

1. A higher-than-normal level of calcium ions (Ca^{2+}) in the blood stimulates parafollicular cells of the thyroid gland to release more calcitonin.
2. Calcitonin inhibits the activity of osteoclasts, thereby decreasing the blood Ca^{2+} level.

TABLE 18.6

Summary of Thyroid Gland Hormones

HORMONE AND SOURCE	CONTROL OF SECRETION	PRINCIPAL ACTIONS
T_3 (triiodothyronine) and **T_4 (thyroxine)** or thyroid hormones from follicular cells 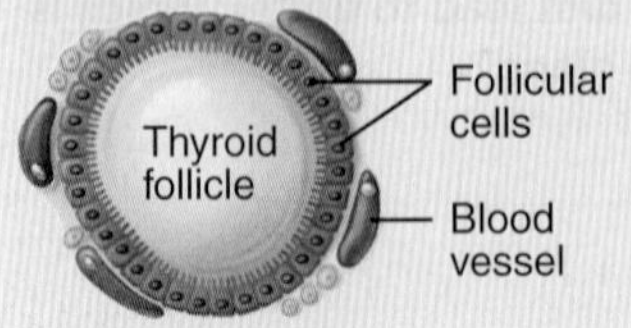	Secretion is increased by thyrotropin-releasing hormone (TRH), which stimulates release of thyroid-stimulating hormone (TSH) in response to low thyroid hormone levels, low metabolic rate, cold, pregnancy, and high altitudes; TRH and TSH secretions are inhibited in response to high thyroid hormone levels; high iodine level suppresses T_3/T_4 secretion.	Increase basal metabolic rate; stimulate synthesis of proteins; increase use of glucose and fatty acids for ATP production; increase lipolysis; enhance cholesterol excretion; accelerate body growth; contribute to development of nervous system.
Calcitonin (CT) from parafollicular cells 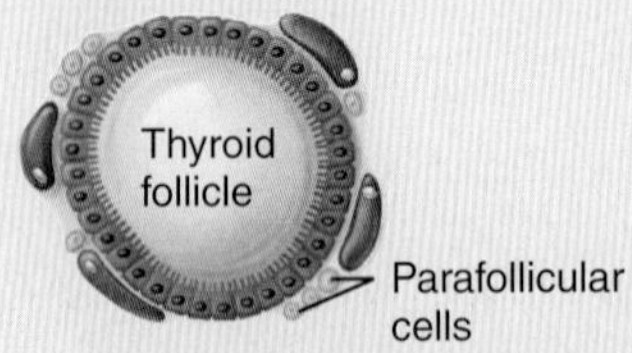	High blood Ca^{2+} levels stimulate secretion; low blood Ca^{2+} levels inhibit secretion.	Lowers blood levels of Ca^{2+} and $HPO_4^{\,2-}$ by inhibiting bone resorption by osteoclasts and by accelerating uptake of calcium and phosphates into bone extracellular matrix.

Figure 18.13 Location, blood supply, and histology of the parathyroid glands.

The parathyroid glands, normally four in number, are embedded in the posterior surface of the thyroid gland.

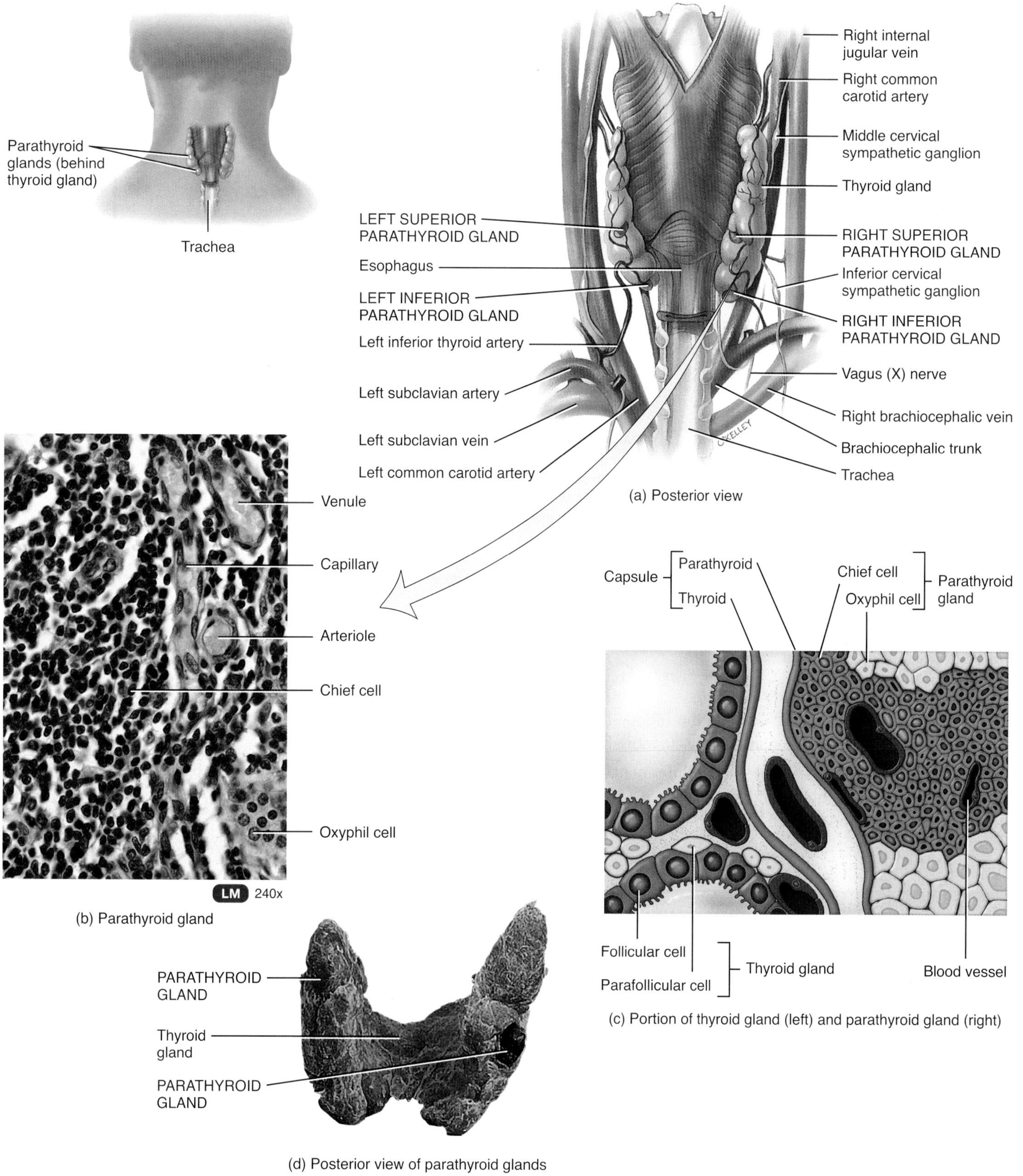
Parathyroid glands (behind thyroid gland)
Trachea
Right internal jugular vein
Right common carotid artery
Middle cervical sympathetic ganglion
Thyroid gland
LEFT SUPERIOR PARATHYROID GLAND
Esophagus
LEFT INFERIOR PARATHYROID GLAND
Left inferior thyroid artery
Left subclavian artery
Left subclavian vein
Left common carotid artery
RIGHT SUPERIOR PARATHYROID GLAND
Inferior cervical sympathetic ganglion
RIGHT INFERIOR PARATHYROID GLAND
Vagus (X) nerve
Right brachiocephalic vein
Brachiocephalic trunk
Trachea
Venule
Capillary
Arteriole
Chief cell
Oxyphil cell
LM 240x
Capsule
Parathyroid
Thyroid
Chief cell
Oxyphil cell
Parathyroid gland
Follicular cell
Parafollicular cell
Thyroid gland
Blood vessel
PARATHYROID GLAND
Thyroid gland
PARATHYROID GLAND

(a) Posterior view

(b) Parathyroid gland

(c) Portion of thyroid gland (left) and parathyroid gland (right)

(d) Posterior view of parathyroid glands

What are the secretory products of (1) parafollicular cells of the thyroid gland and (2) chief (principal) cells of the parathyroid glands?

Figure 18.14 **The roles of calcitonin (green arrows), parathyroid hormone (blue arrows), and calcitriol (orange arrows) in calcium homeostasis.**

With respect to regulation of blood Ca^{2+} level, calcitonin and PTH are antagonists.

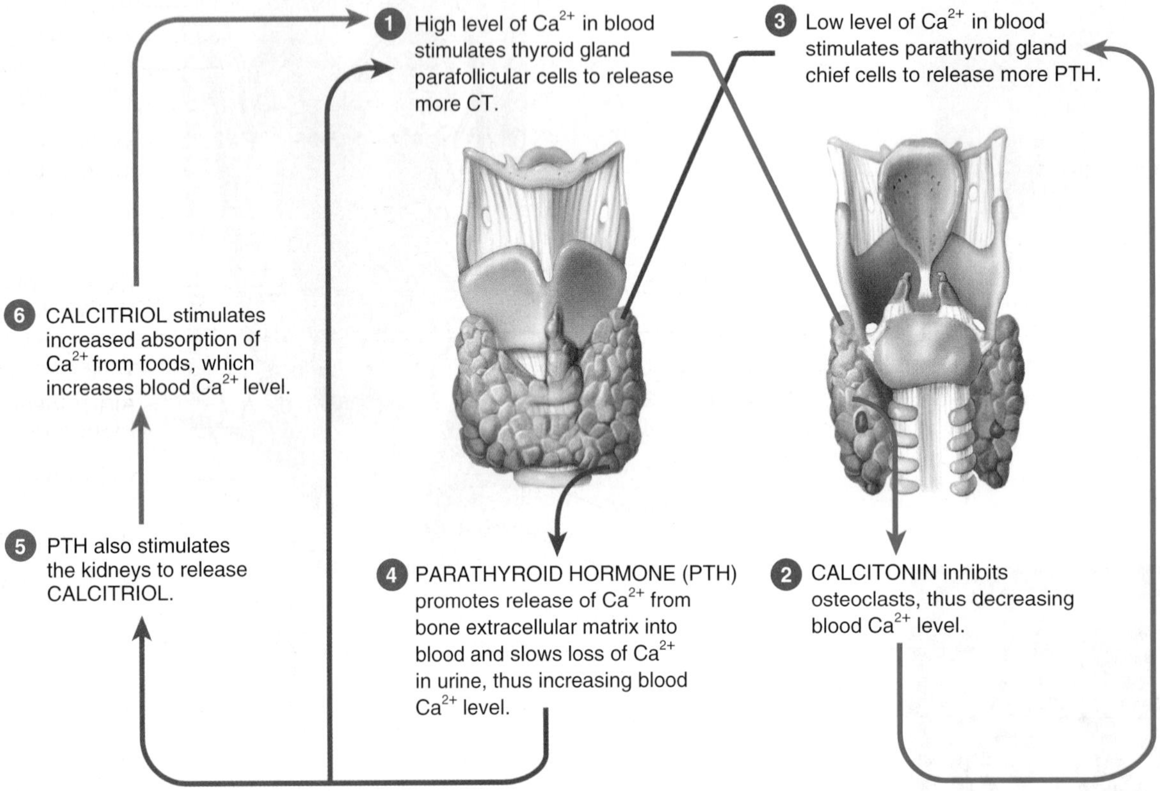
1 High level of Ca2+ in blood stimulates thyroid gland parafollicular cells to release more CT.
3 Low level of Ca2+ in blood stimulates parathyroid gland chief cells to release more PTH.
6 CALCITRIOL stimulates increased absorption of Ca2+ from foods, which increases blood Ca2+ level.
5 PTH also stimulates the kidneys to release CALCITRIOL.
4 PARATHYROID HORMONE (PTH) promotes release of Ca2+ from bone extracellular matrix into blood and slows loss of Ca2+ in urine, thus increasing blood Ca2+ level.
2 CALCITONIN inhibits osteoclasts, thus decreasing blood Ca2+ level.

? What are the primary target tissues for PTH, CT, and calcitriol?

3. A lower-than-normal level of Ca^{2+} in the blood stimulates chief cells of the parathyroid gland to release more PTH.
4. PTH promotes resorption of bone extracellular matrix, which releases Ca^{2+} into the blood and slows loss of Ca^{2+} in the urine, raising the blood level of Ca^{2+}.
5. PTH also stimulates the kidneys to synthesize calcitriol, the active form of vitamin D.
6. Calcitriol stimulates increased absorption of Ca^{2+} from foods in the gastrointestinal tract, which helps increase the blood level of Ca^{2+}.

TABLE 18.7

Summary of Parathyroid Gland Hormone

HORMONE AND SOURCE	CONTROL OF SECRETION	PRINCIPAL ACTIONS
Parathyroid hormone (PTH) from chief cells 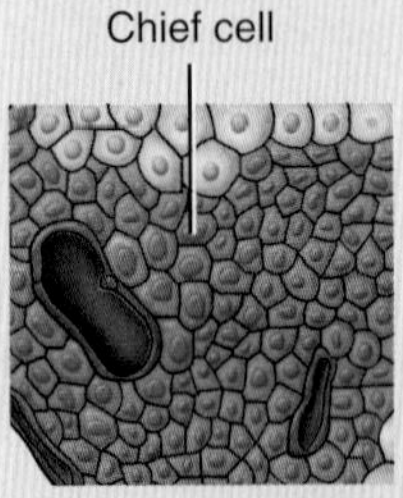 Chief cell	Low blood Ca^{2+} levels stimulate secretion; high blood Ca^{2+} levels inhibit secretion.	Increases blood Ca^{2+} and Mg^{2+} levels and decreases blood HPO_4^{2-} level; increases bone resorption by osteoclasts; increases Ca^{2+} reabsorption and HPO_4^{2-} excretion by kidneys; promotes formation of calcitriol (active form of vitamin D), which increases rate of dietary Ca^{2+} and Mg^{2+} absorption.

Table 18.7 summarizes control of secretion and the principal actions of parathyroid hormone.

✔ **CHECKPOINT**

16. How is secretion of parathyroid hormone regulated?
17. In what ways are the actions of PTH and calcitriol similar? How are they different?

18.9 ADRENAL GLANDS

OBJECTIVE

- Describe the location, histology, hormones, and functions of the adrenal glands.

The paired **adrenal (suprarenal) glands,** one of which lies superior to each kidney in the retroperitoneal space (Figure 18.15a), have a flattened pyramidal shape. In an adult, each adrenal gland is 3–5 cm in height, 2–3 cm in width, and a little less than 1 cm thick, with a mass of 3.5–5 g, only half its size at birth. During embryonic development, the adrenal glands differentiate into two structurally and functionally distinct regions: a large, peripherally located **adrenal cortex,** comprising 80–90% of the gland, and a small, centrally located **adrenal medulla** (Figure 18.15b). A connective tissue capsule covers the gland. The adrenal glands, like the thyroid gland, are highly vascularized.

The adrenal cortex produces steroid hormones that are essential for life. Complete loss of adrenocortical hormones leads to death due to dehydration and electrolyte imbalances in a few days to a week, unless hormone replacement therapy begins promptly. The adrenal medulla produces three catecholamine hormones—norepinephrine, epinephrine, and a small amount of dopamine.

Adrenal Cortex

The adrenal cortex is subdivided into three zones, each of which secretes different hormones (Figure 18.15d). The outer zone, just deep to the connective tissue capsule, is the **zona glomerulosa** (glo-mer′-ū-LŌ-sa; *zona* = belt; *glomerul-* = little ball). Its cells, which are closely packed and arranged in spherical clusters and arched columns, secrete hormones called **mineralocorticoids** (min′-er-al-ō-KOR-ti-koyds) because they affect mineral homeostasis. The middle zone, or **zona fasciculata** (fa-sik′-ū-LA-ta; *fascicul-* = little bundle), is the widest of the three zones and consists of cells arranged in long, straight columns. The cells of the

Figure 18.15 Location, blood supply, and histology of the adrenal (suprarenal) glands.

The adrenal cortex secretes steroid hormones that are essential for life; the adrenal medulla secretes norepinephrine and epinephrine.

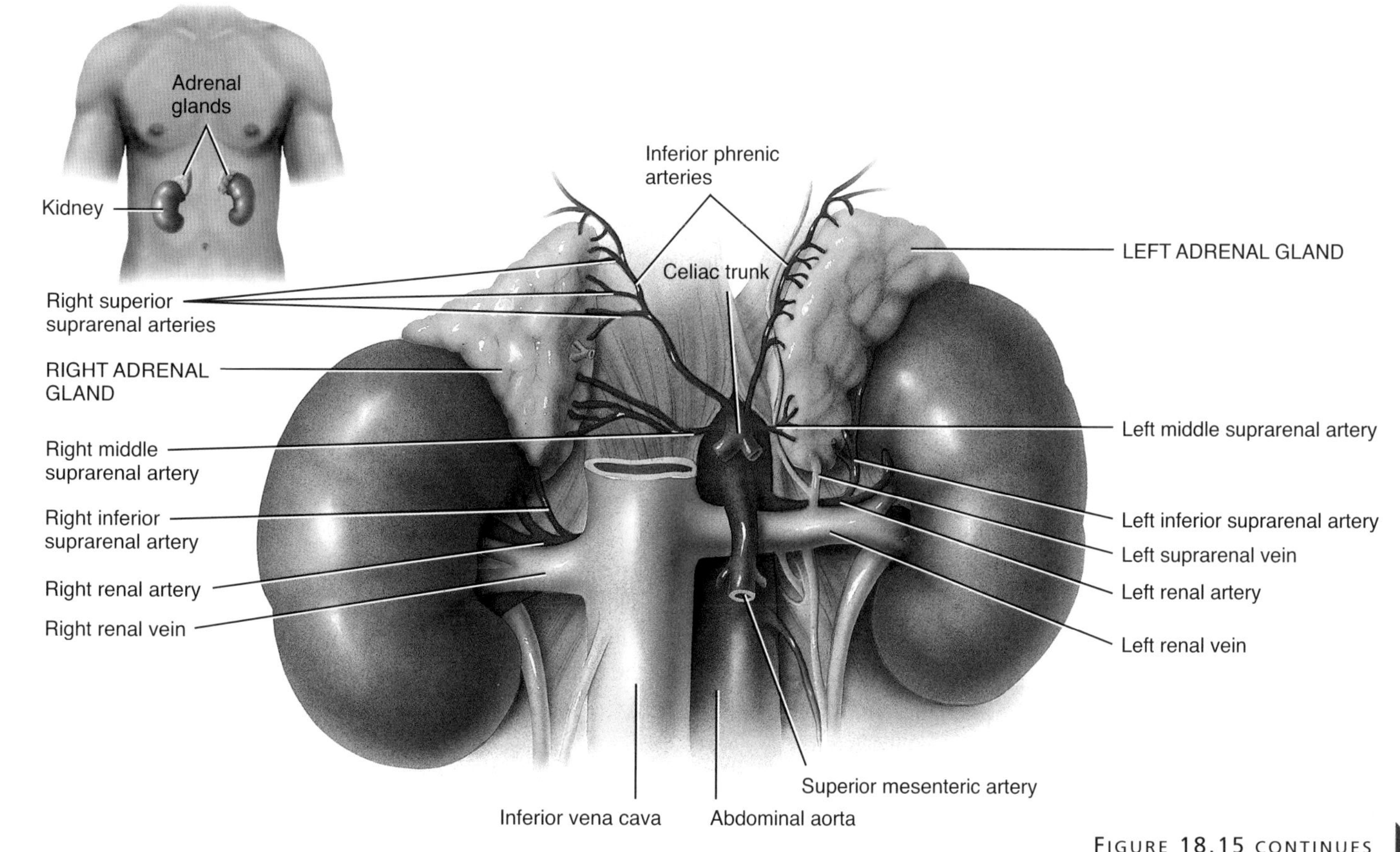

(a) Anterior view

FIGURE 18.15 CONTINUES ▶

FIGURE 18.15 CONTINUED

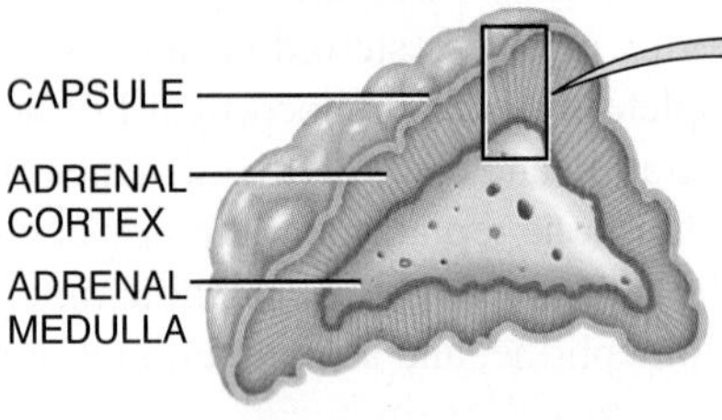
CAPSULE
ADRENAL CORTEX
ADRENAL MEDULLA

(b) Section through left adrenal gland

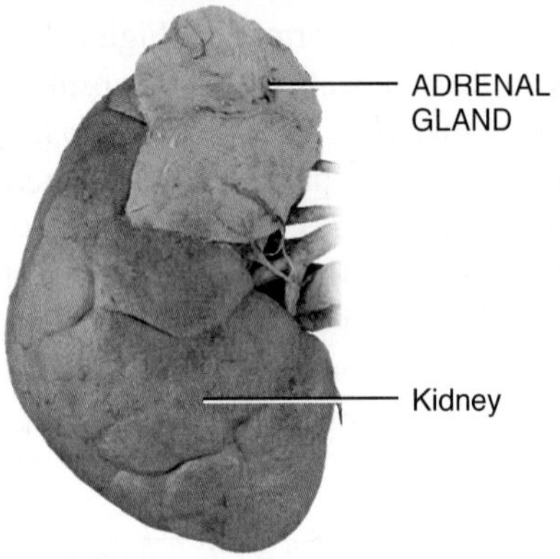
ADRENAL GLAND
Kidney

(c) Anterior view of adrenal gland and kidney

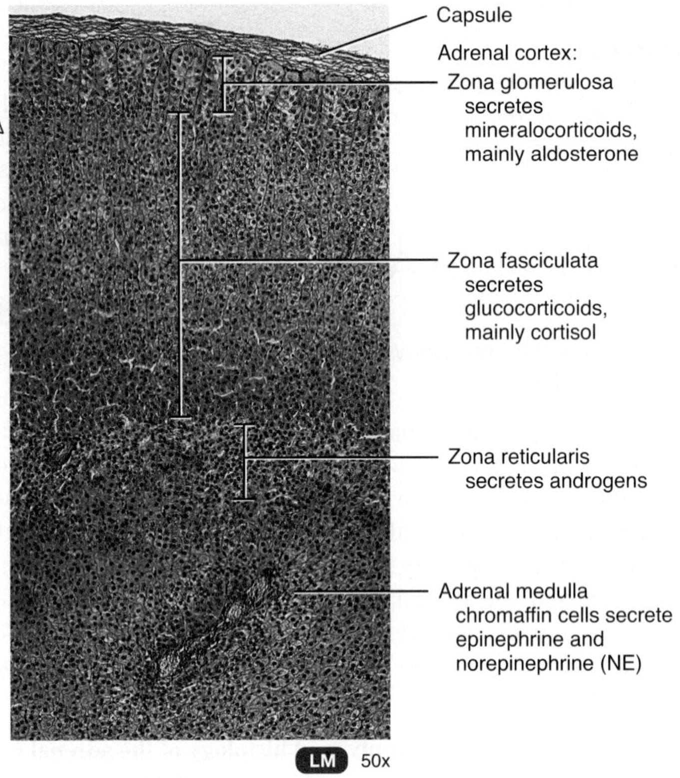
Capsule
Adrenal cortex:
Zona glomerulosa secretes mineralocorticoids, mainly aldosterone
Zona fasciculata secretes glucocorticoids, mainly cortisol
Zona reticularis secretes androgens
Adrenal medulla chromaffin cells secrete epinephrine and norepinephrine (NE)
LM 50x

(d) Subdivisions of adrenal gland

What is the position of the adrenal glands relative to the kidneys?

zona fasciculata secrete mainly **glucocorticoids** (gloo′-kō-KOR-ti-koyds), primarily cortisol, so named because they affect glucose homeostasis. The cells of the inner zone, the **zona reticularis** (re-tik′-ū-LAR-is; *reticul-* = network), are arranged in branching cords. They synthesize small amounts of weak **androgens** (*andro-* = a man), steroid hormones that have masculinizing effects.

Mineralocorticoids

Aldosterone (al-DOS-ter-ōn) is the major mineralocorticoid. It regulates homeostasis of two mineral ions—namely, sodium ions (Na^+) and potassium ions (K^+)—and helps adjust blood pressure and blood volume. Aldosterone also promotes excretion of H^+ in the urine; this removal of acids from the body can help prevent acidosis (blood pH below 7.35), which is discussed in Chapter 27.

The **renin–angiotensin–aldosterone** or **RAA pathway** (RĒ-nin an′-jē-ō-TEN-sin) controls secretion of aldosterone (Figure 18.16):

1. Stimuli that initiate the renin–angiotensin–aldosterone pathway include dehydration, Na^+ deficiency, or hemorrhage.
2. These conditions cause a decrease in blood volume.
3. Decreased blood volume leads to decreased blood pressure.
4. Lowered blood pressure stimulates certain cells of the kidneys, called juxtaglomerular cells, to secrete the enzyme **renin.**
5. The level of renin in the blood increases.
6. Renin converts **angiotensinogen** (an′-jē-ō-ten-SIN-ō-jen), a plasma protein produced by the liver, into **angiotensin I.**
7. Blood containing increased levels of angiotensin I circulates in the body.
8. As blood flows through capillaries, particularly those of the lungs, the enzyme **angiotensin-converting enzyme (ACE)** converts angiotensin I into the hormone **angiotensin II.**
9. Blood level of angiotensin II increases.
10. Angiotensin II stimulates the adrenal cortex to secrete aldosterone.
11. Blood containing increased levels of aldosterone circulates to the kidneys.
12. In the kidneys, aldosterone increases reabsorption of Na^+, which in turn causes reabsorption of water by osmosis. As a result, less water is lost in the urine. Aldosterone also stimulates the kidneys to increase secretion of K^+ and H^+ into the urine.
13. With increased water reabsorption by the kidneys, blood volume increases.
14. As blood volume increases, blood pressure increases to normal.
15. Angiotensin II also stimulates contraction of smooth muscle in the walls of arterioles. The resulting vasoconstriction of

Figure 18.16 Regulation of aldosterone secretion by the renin–angiotensin–aldosterone (RAA) pathway.

Aldosterone helps regulate blood volume, blood pressure, and levels of Na^+, K^+, and H^+ in the blood.

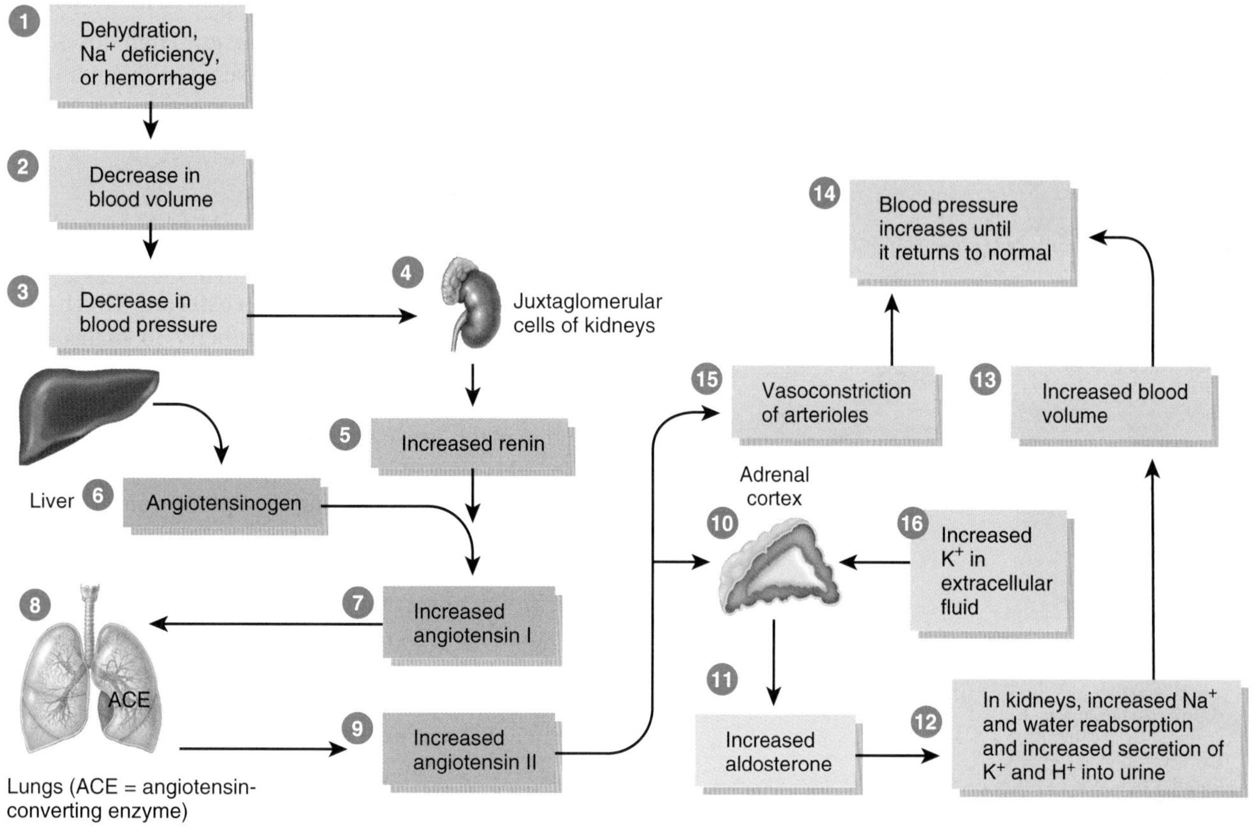

In what two ways can angiotensin II increase blood pressure, and what are its target tissues in each case?

the arterioles increases blood pressure and thus helps raise blood pressure to normal.

16 Besides angiotensin II, a second stimulator of aldosterone secretion is an increase in the K^+ concentration of blood (or interstitial fluid). A decrease in the blood K^+ level has the opposite effect.

Glucocorticoids

The glucocorticoids, which regulate metabolism and resistance to stress, include **cortisol** (KOR-ti-sol; also called **hydrocortisone**), **corticosterone** (kor′-ti-KOS-ter-ōn), and **cortisone** (KOR-ti-sōn). Of these three hormones secreted by the zona fasciculata, cortisol is the most abundant, accounting for about 95% of glucocorticoid activity.

Control of glucocorticoid secretion occurs via a typical negative feedback system (Figure 18.17). Low blood levels of glucocorticoids, mainly cortisol, stimulate neurosecretory cells in the hypothalamus to secrete **corticotropin-releasing hormone (CRH).** CRH (together with a low level of cortisol) promotes the release of ACTH from the anterior pituitary. ACTH flows in the blood to the adrenal cortex, where it stimulates glucocorticoid secretion. (To a much smaller extent, ACTH also stimulates secretion of aldosterone.) The discussion of stress at the end of the chapter describes how the hypothalamus also increases CRH release in response to a variety of physical and emotional stresses (see Section 18.15).

Glucocorticoids have the following effects:

1. ***Protein breakdown.*** Glucocorticoids increase the rate of protein breakdown, mainly in muscle fibers, and thus increase the liberation of amino acids into the bloodstream. The amino acids may be used by body cells for synthesis of new proteins or for ATP production.
2. ***Glucose formation.*** On stimulation by glucocorticoids, liver cells may convert certain amino acids or lactic acid to glucose, which neurons and other cells can use for ATP production. Such conversion of a substance other than glycogen or another monosaccharide into glucose is called **gluconeogenesis** (gloo′-ko-nē′-ō-JEN-e-sis).
3. ***Lipolysis.*** Glucocorticoids stimulate **lipolysis** (lī-POL-i-sis), the breakdown of triglycerides and release of fatty acids from adipose tissue into the blood.
4. ***Resistance to stress.*** Glucocorticoids work in many ways to provide resistance to stress. The additional glucose supplied by the liver cells provides tissues with a ready source of ATP to combat a range of stresses, including exercise, fasting, fright, temperature extremes, high altitude, bleeding, infection, surgery, trauma, and

Figure 18.17 Negative feedback regulation of glucocorticoid secretion.

A high level of CRH and a low level of glucocorticoids promote the release of ACTH, which stimulates glucocorticoid secretion by the adrenal cortex.

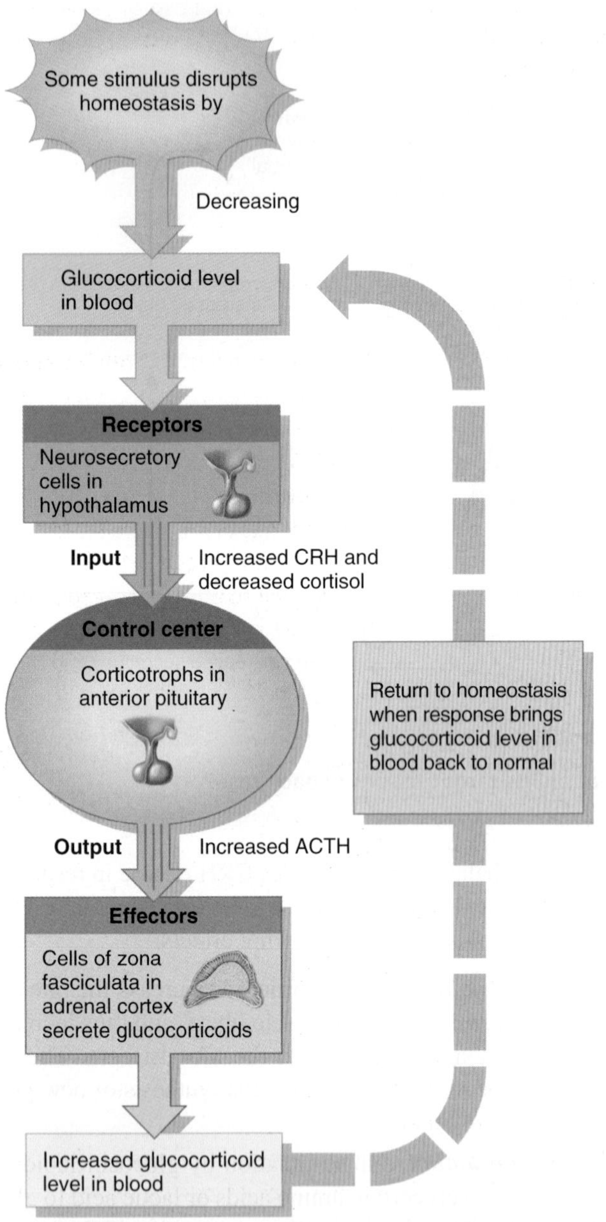

Some stimulus disrupts homeostasis by
Decreasing
Glucocorticoid level in blood
Receptors
Neurosecretory cells in hypothalamus
Input
Increased CRH and decreased cortisol
Control center
Corticotrophs in anterior pituitary
Return to homeostasis when response brings glucocorticoid level in blood back to normal
Output
Increased ACTH
Effectors
Cells of zona fasciculata in adrenal cortex secrete glucocorticoids
Increased glucocorticoid level in blood

If a heart transplant patient receives prednisone (a glucocorticoid) to help prevent rejection of the transplanted tissue, will blood levels of ACTH and CRH be high or low? Explain.

disease. Because glucocorticoids make blood vessels more sensitive to other hormones that cause vasoconstriction, they raise blood pressure. This effect would be an advantage in cases of severe blood loss, which causes blood pressure to drop.

5. ***Anti-inflammatory effects.*** Glucocorticoids inhibit white blood cells that participate in inflammatory responses. Unfortunately, glucocorticoids also retard tissue repair, and as a result, they slow wound healing. Although high doses can cause severe mental disturbances, glucocorticoids are very useful in the treatment of chronic inflammatory disorders such as rheumatoid arthritis.
6. ***Depression of immune responses.*** High doses of glucocorticoids depress immune responses. For this reason, glucocorticoids are prescribed for organ transplant recipients to retard tissue rejection by the immune system.

Androgens

In both males and females, the adrenal cortex secretes small amounts of weak androgens. The major androgen secreted by the adrenal gland is **dehydroepiandrosterone (DHEA)** (dē-hī-drō-ep′-ē-an-DROS-ter-ōn). After puberty in males, the androgen testosterone is also released in much greater quantity by the testes. Thus, the amount of androgens secreted by the adrenal gland in males is usually so low that their effects are insignificant. In females, however, adrenal androgens play important roles. They promote libido (sex drive) and are converted into estrogens (feminizing sex steroids) by other body tissues. After menopause, when ovarian secretion of estrogens ceases, all female estrogens come from conversion of adrenal androgens. Adrenal androgens also stimulate growth of axillary and pubic hair in boys and girls and contribute to the prepubertal growth spurt. Although control of adrenal androgen secretion is not fully understood, the main hormone that stimulates its secretion is ACTH.

CLINICAL CONNECTION | *Congenital Adrenal Hyperplasia*

Congenital adrenal hyperplasia (CAH) (hī-per-PLĀ-zhē-a) is a genetic disorder in which one or more enzymes needed for synthesis of cortisol are absent. Because the cortisol level is low, secretion of ACTH by the anterior pituitary is high due to lack of negative feedback inhibition. ACTH in turn stimulates growth and secretory activity of the adrenal cortex. As a result, both adrenal glands are enlarged. However, certain steps leading to synthesis of cortisol are blocked. Thus, precursor molecules accumulate, and some of these are weak androgens that can undergo conversion to testosterone. The result is **virilism** (VIR-i-lizm), or masculinization. In a female, virile characteristics include growth of a beard, development of a much deeper voice and a masculine distribution of body hair, growth of the clitoris so it may resemble a penis, atrophy of the breasts, and increased muscularity that produces a masculine physique. In prepubertal males, the syndrome causes the same characteristics as in females, plus rapid development of the male sexual organs and emergence of male sexual desires. In adult males, the virilizing effects of CAH are usually completely obscured by the normal virilizing effects of the testosterone secreted by the testes. As a result, CAH is often difficult to diagnose in adult males. Treatment involves cortisol therapy, which inhibits ACTH secretion and thus reduces production of adrenal androgens. •

Adrenal Medulla

The inner region of the adrenal gland, the **adrenal medulla,** is a modified sympathetic ganglion of the autonomic nervous system (ANS). It develops from the same embryonic tissue as all other sympathetic ganglia, but its cells, which lack axons, form clusters around large blood vessels. Rather than releasing a neurotransmitter, the cells of the adrenal medulla secrete hormones. The hormone-producing cells, called **chromaffin cells** (KRŌ-maf-in; *chrom-* = color; *-affin* = affinity for; see Figure 18.15d), are innervated by sympathetic preganglionic neurons of the ANS. Because the ANS exerts direct control over the chromaffin cells, hormone release can occur very quickly.

The two major hormones synthesized by the adrenal medulla are **epinephrine** (ep′-i-NEF-rin) and **norepinephrine (NE),** also called *adrenaline* and *noradrenaline,* respectively. The chromaffin cells of the adrenal medulla secrete an unequal amount of these hormones—about 80% epinephrine and 20% norepinephrine. The hormones of the adrenal medulla intensify sympathetic responses that occur in other parts of the body.

In stressful situations and during exercise, impulses from the hypothalamus stimulate sympathetic preganglionic neurons, which in turn stimulate the chromaffin cells to secrete epinephrine and norepinephrine. These two hormones greatly augment the fight-or-flight response that you learned about in Chapter 15. By increasing heart rate and force of contraction, epinephrine and norepinephrine increase the output of the heart, which increases blood pressure. They also increase blood flow to the heart, liver, skeletal muscles, and adipose tissue; dilate airways to the lungs; and increase blood levels of glucose and fatty acids.

Table 18.8 summarizes the hormones produced by the adrenal glands, control of their secretion, and their principal actions.

✔ **CHECKPOINT**

18. How do the adrenal cortex and adrenal medulla compare with regard to location and histology?
19. How is secretion of adrenal cortex hormones regulated?
20. How is the adrenal medulla related to the autonomic nervous system?

18.10 PANCREATIC ISLETS

OBJECTIVE

- Describe the location, histology, hormones, and functions of the pancreatic islets.

The **pancreas** (*pan-* = all; *-creas* = flesh) is both an endocrine gland and an exocrine gland. We discuss its endocrine functions here and include its exocrine functions in Chapter 24 in the coverage of the digestive system. A flattened organ that measures about 12.5–15 cm (5–6 in.) in length, the pancreas is located in the curve of the duodenum, the first part of the small intestine, and consists of a head, a body, and a tail (Figure 18.18a). Roughly 99% of the exocrine cells of the pancreas are arranged in clusters called **acini** (AS-i-nī). The acini produce digestive enzymes,

TABLE 18.8

Summary of Adrenal Gland Hormones

HORMONES AND SOURCE	CONTROL OF SECRETION	PRINCIPAL ACTIONS
ADRENAL CORTEX HORMONES		
Mineralocorticoids (mainly aldosterone) from zona glomerulosa cells	Increased blood K^+ level and angiotensin II stimulate secretion.	Increase blood levels of Na^+ and water; decrease blood level of K^+.
Glucocorticoids (mainly cortisol) from zona fasciculata cells	ACTH stimulates release; corticotropin-releasing hormone (CRH) promotes ACTH secretion in response to stress and low blood levels of glucocorticoids.	Increase protein breakdown (except in liver), stimulate gluconeogenesis and lipolysis, provide resistance to stress, dampen inflammation, depress immune responses.
Androgens (mainly dehydroepiandrosterone, or DHEA) from zona reticularis cells 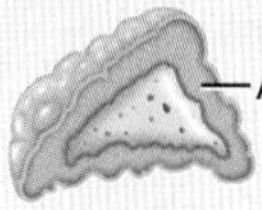	ACTH stimulates secretion.	Assist in early growth of axillary and pubic hair in both sexes; in females, contribute to libido and are source of estrogens after menopause.
ADRENAL MEDULLA HORMONES		
Epinephrine and **norepinephrine** from chromaffin cells 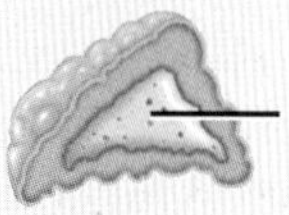	Sympathetic preganglionic neurons release acetylcholine, which stimulates secretion.	Enhance effects of sympathetic division of autonomic nervous system (ANS) during stress.

Figure 18.18 Location, blood supply, and histology of the pancreas.

Pancreatic hormones regulate blood glucose level.

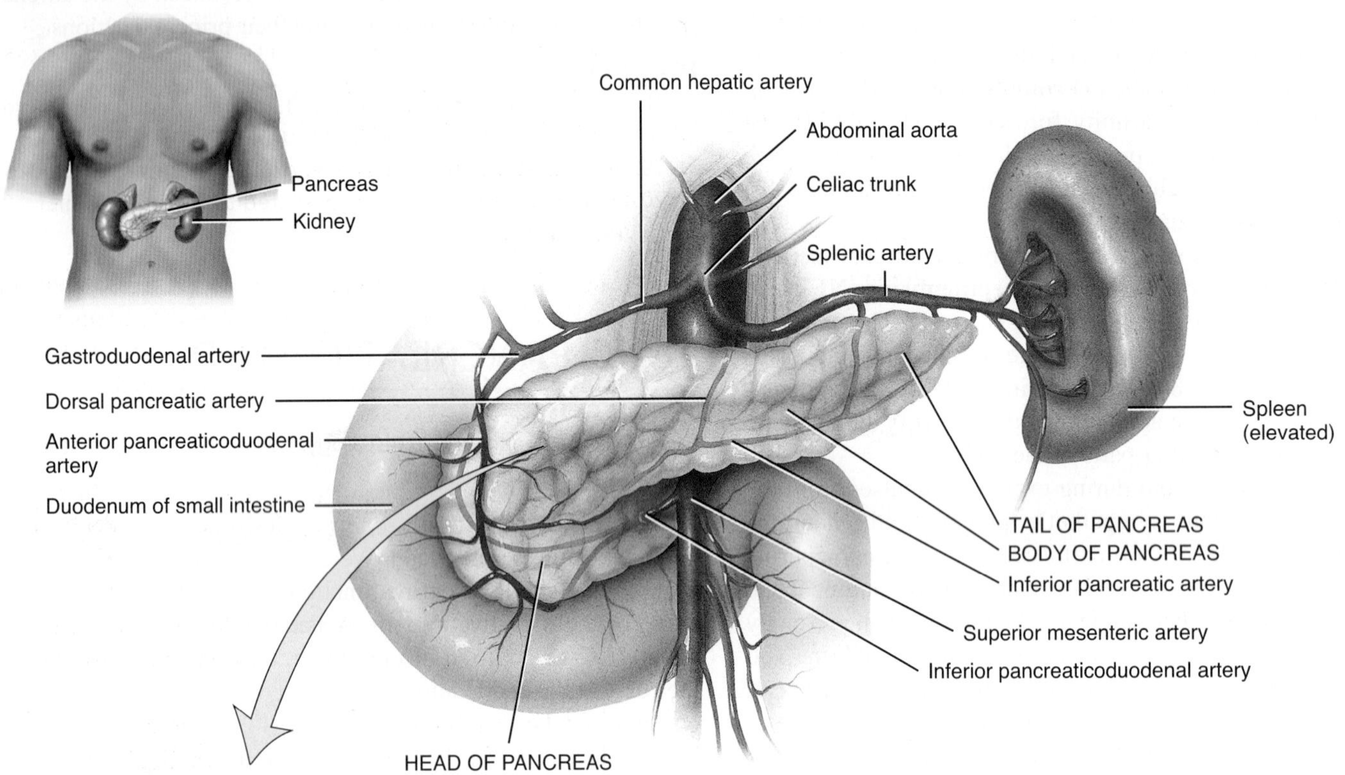
Common hepatic artery
Abdominal aorta
Celiac trunk
Splenic artery
Pancreas
Kidney
Gastroduodenal artery
Dorsal pancreatic artery
Anterior pancreaticoduodenal artery
Duodenum of small intestine
Spleen (elevated)
TAIL OF PANCREAS
BODY OF PANCREAS
Inferior pancreatic artery
Superior mesenteric artery
Inferior pancreaticoduodenal artery
HEAD OF PANCREAS

(a) Anterior view

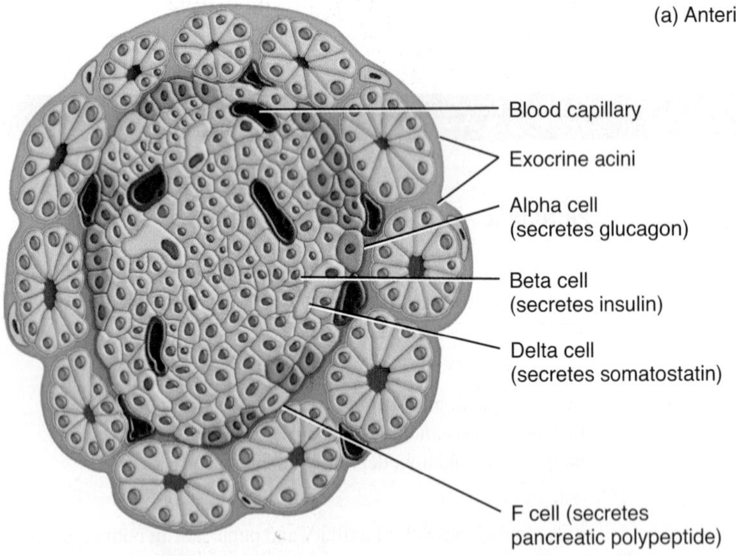
Blood capillary
Exocrine acini
Alpha cell (secretes glucagon)
Beta cell (secretes insulin)
Delta cell (secretes somatostatin)
F cell (secretes pancreatic polypeptide)

(b) Pancreatic islet and surrounding acini

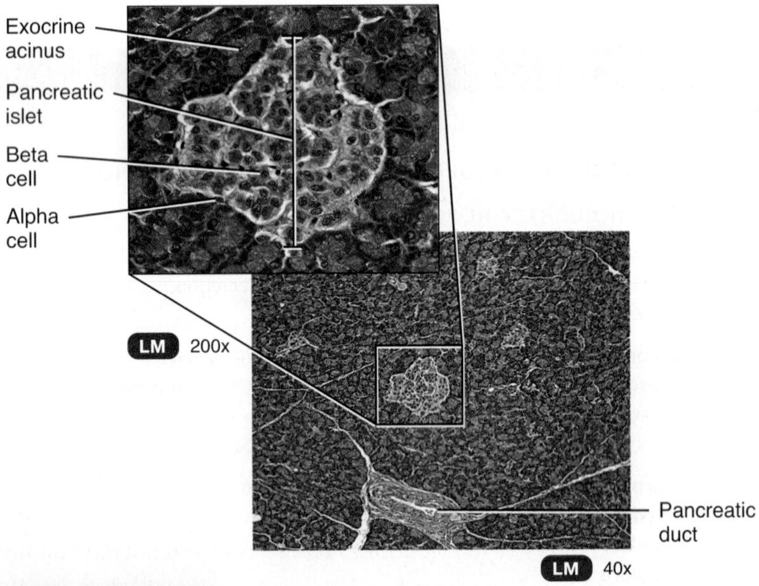
Exocrine acinus
Pancreatic islet
Beta cell
Alpha cell
LM 200x
Pancreatic duct
LM 40x

(c) Pancreatic islet and surrounding acini

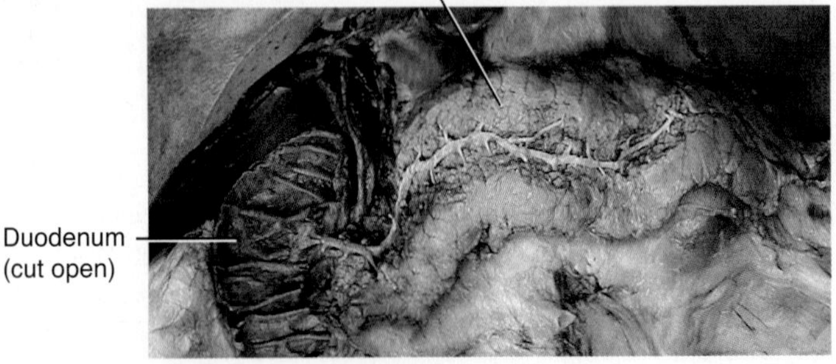
Pancreas
Duodenum (cut open)

(d) Anterior view of pancreas dissected to reveal pancreatic duct

Is the pancreas an exocrine gland or an endocrine gland?

which flow into the gastrointestinal tract through a network of ducts. Scattered among the exocrine acini are 1–2 million tiny clusters of endocrine tissue called **pancreatic islets** (Ī-lets) or **islets of Langerhans** (LAHNG-er-hanz; Figure 18.18b, c). Abundant capillaries serve both the exocrine and endocrine portions of the pancreas.

Cell Types in the Pancreatic Islets

Each pancreatic islet includes four types of hormone-secreting cells:

1. **Alpha** or **A cells** constitute about 17% of pancreatic islet cells and secrete **glucagon** (GLOO-ka-gon).
2. **Beta** or **B cells** constitute about 70% of pancreatic islet cells and secrete **insulin** (IN-soo-lin).
3. **Delta** or **D cells** constitute about 7% of pancreatic islet cells and secrete **somatostatin** (sō-ma-to-STAT-in).
4. **F cells** constitute the remainder of pancreatic islet cells and secrete **pancreatic polypeptide.**

The interactions of the four pancreatic hormones are complex and not completely understood. We do know that glucagon raises blood glucose level, and insulin lowers it. Somatostatin acts in a paracrine manner to inhibit both insulin and glucagon release from neighboring beta and alpha cells. It may also act as a circulating hormone to slow absorption of nutrients from the gastrointestinal tract. In addition, somatostatin inhibits the secretion of growth hormone. Pancreatic polypeptide inhibits somatostatin secretion, gallbladder contraction, and secretion of digestive enzymes by the pancreas.

Regulation of Glucagon and Insulin Secretion

The principal action of glucagon is to increase blood glucose level when it falls below normal. Insulin, on the other hand, helps lower blood glucose level when it is too high. The level of blood glucose controls secretion of glucagon and insulin via negative feedback (Figure 18.19):

1. Low blood glucose level (hypoglycemia) stimulates secretion of glucagon from alpha cells of the pancreatic islets.
2. Glucagon acts on hepatocytes (liver cells) to accelerate the conversion of glycogen into glucose (glycogenolysis) and to promote formation of glucose from lactic acid and certain amino acids (gluconeogenesis).
3. As a result, hepatocytes release glucose into the blood more rapidly, and blood glucose level rises.
4. If blood glucose continues to rise, high blood glucose level (hyperglycemia) inhibits release of glucagon (negative feedback).
5. High blood glucose (hyperglycemia) stimulates secretion of insulin by beta cells of the pancreatic islets.
6. Insulin acts on various cells in the body to accelerate facilitated diffusion of glucose into cells; to speed conversion of glucose into glycogen (glycogenesis); to increase uptake of amino acids by cells and to increase protein synthesis; to speed synthesis of fatty acids (lipogenesis); to slow the conversion of glycogen to glucose (glycogenolysis); and to slow the formation of glucose from lactic acid and amino acids (gluconeogenesis).
7. As a result, blood glucose level falls.
8. If blood glucose level drops below normal, low blood glucose inhibits release of insulin (negative feedback) and stimulates release of glucagon.

Although blood glucose level is the most important regulator of insulin and glucagon, several hormones and neurotransmitters also regulate the release of these two hormones. In addition to the responses to blood glucose level just described, glucagon

Figure 18.19 Negative feedback regulation of the secretion of glucagon (blue arrows) and insulin (orange arrows).

Low blood glucose stimulates release of glucagon; high blood glucose stimulates secretion of insulin.

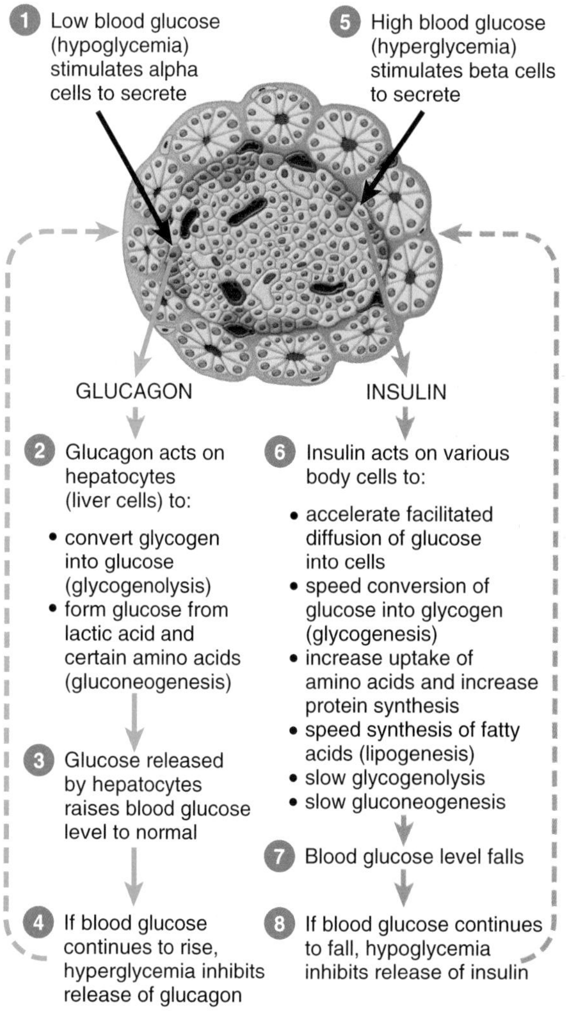

? Does glycogenolysis increase or decrease blood glucose level?

stimulates insulin release directly; insulin has the opposite effect, suppressing glucagon secretion. As blood glucose level declines and less insulin is secreted, the alpha cells of the pancreas are released from the inhibitory effect of insulin so they can secrete more glucagon. Indirectly, human growth hormone (hGH) and adrenocorticotropic hormone (ACTH) stimulate secretion of insulin because they act to elevate blood glucose.

Insulin secretion is also stimulated by:

- Acetylcholine, the neurotransmitter liberated from axon terminals of parasympathetic vagus nerve fibers that innervate the pancreatic islets
- The amino acids arginine and leucine, which would be present in the blood at higher levels after a protein-containing meal
- Glucose-dependent insulinotropic peptide (GIP),* a hormone released by enteroendocrine cells of the small intestine in response to the presence of glucose in the gastrointestinal tract

Thus, digestion and absorption of food containing both carbohydrates and proteins provide strong stimulation for insulin release.

Glucagon secretion is stimulated by:

- Increased activity of the sympathetic division of the ANS, as occurs during exercise

*GIP—previously called gastric inhibitory peptide—was renamed because at physiological concentration its inhibitory effect on stomach function is negligible.

TABLE 18.9

Summary of Pancreatic Islet Hormones

HORMONE AND SOURCE	CONTROL OF SECRETION	PRINCIPAL ACTIONS
Glucagon from alpha cells of pancreatic islets 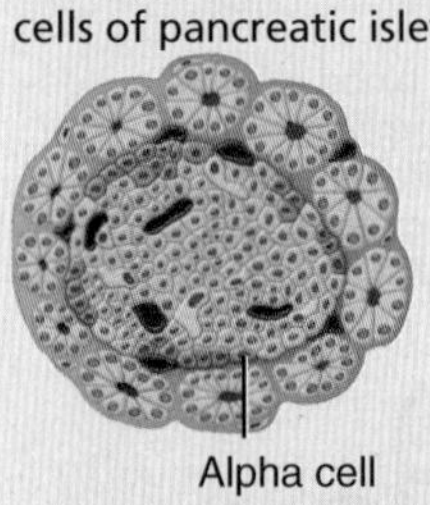	Decreased blood level of glucose, exercise, and mainly protein meals stimulate secretion; somatostatin and insulin inhibit secretion.	Raises blood glucose level by accelerating breakdown of glycogen into glucose in liver (glycogenolysis), converting other nutrients into glucose in liver (gluconeogenesis), and releasing glucose into blood.
Insulin from beta cells of pancreatic islets 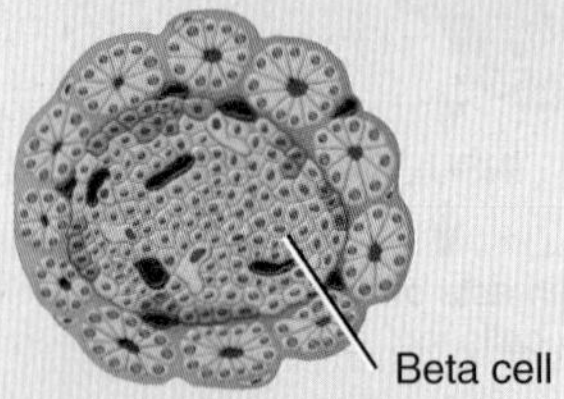	Increased blood level of glucose, acetylcholine (released by parasympathetic vagus nerve fibers), arginine and leucine (two amino acids), glucagon GIP, hGH, and ACTH stimulate secretion; somatostatin inhibits secretion.	Lowers blood glucose level by accelerating transport of glucose into cells, converting glucose into glycogen (glycogenesis), and decreasing glycogenolysis and gluconeogenesis; increases lipogenesis and stimulates protein synthesis.
Somatostatin from delta cells of pancreatic islets 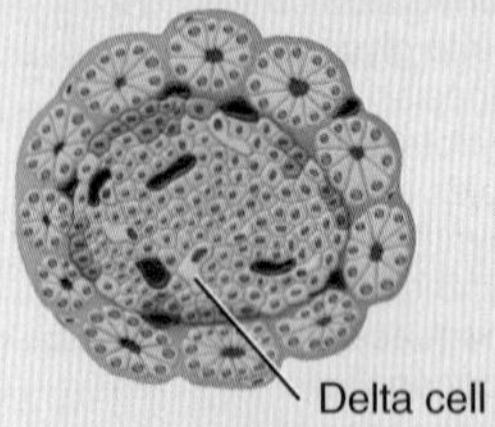	Pancreatic polypeptide inhibits secretion.	Inhibits secretion of insulin and glucagon; slows absorption of nutrients from gastrointestinal tract.
Pancreatic polypeptide from F cells of pancreatic islets 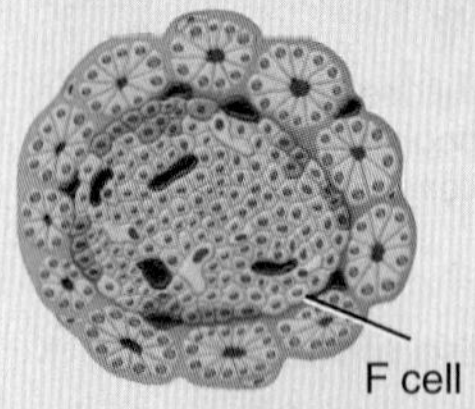	Meals containing protein, fasting, exercise, and acute hypoglycemia stimulate secretion; somatostatin and elevated blood glucose level inhibit secretion.	Inhibits somatostatin secretion, gallbladder contraction, and secretion of pancreatic digestive enzymes.

- A rise in blood amino acids if blood glucose level is low, which could occur after a meal that contained mainly protein

Table 18.9 summarizes the hormones produced by the pancreas, control of their secretion, and their principal actions.

✔CHECKPOINT

21. How are blood levels of glucagon and insulin controlled?
22. What are the effects on secretion of insulin and glucagon of exercise versus eating a carbohydrate- and protein-rich meal?

18.11 OVARIES AND TESTES

OBJECTIVE

- Describe the location, hormones, and functions of the male and female gonads.

Gonads are the organs that produce gametes—sperm in males and oocytes in females. In addition to their reproductive function, the gonads secrete hormones. The **ovaries,** paired oval bodies located in the female pelvic cavity, produce several steroid hormones including two **estrogens** (estradiol and estrone) and **progesterone.** These female sex hormones, along with FSH and LH from the anterior pituitary, regulate the menstrual cycle, maintain pregnancy, and prepare the mammary glands for lactation. They also promote enlargement of the breasts and widening of the hips at puberty, and help maintain these female secondary sex characteristics. The ovaries also produce **inhibin,** a protein hormone that inhibits secretion of follicle-stimulating hormone (FSH). During pregnancy, the ovaries and placenta produce a peptide hormone called **relaxin (RLX),** which increases the flexibility of the pubic symphysis during pregnancy and helps dilate the uterine cervix during labor and delivery. These actions help ease the baby's passage by enlarging the birth canal.

The male gonads, the **testes,** are oval glands that lie in the scrotum. The main hormone produced and secreted by the testes is **testosterone,** an **androgen** or male sex hormone. Testosterone stimulates descent of the testes before birth, regulates production of sperm, and stimulates the development and maintenance of male secondary sex characteristics, such as beard growth and deepening of the voice. The testes also produce inhibin, which inhibits secretion of FSH. The detailed structure of the ovaries and testes and the specific roles of sex hormones are discussed in Chapter 28.

Table 18.10 summarizes the hormones produced by the ovaries and testes and their principal actions.

✔CHECKPOINT

23. Why are the ovaries and testes classified as endocrine glands as well as reproductive organs?

18.12 PINEAL GLAND AND THYMUS

OBJECTIVE

- Describe the location, histology, hormone, and functions of the pineal gland.
- Describe the role of the thymus in immunity.

TABLE 18.10

Summary of Hormones of the Ovaries and Testes

HORMONE	PRINCIPAL ACTIONS
OVARIAN HORMONES	
Estrogens and progesterone  Ovaries	Together with gonadotropic hormones of anterior pituitary, regulate female reproductive cycle and oogenesis, maintain pregnancy, prepare mammary glands for lactation, and promote development and maintenance of female secondary sex characteristics.
Relaxin	Increases flexibility of pubic symphysis during pregnancy; helps dilate uterine cervix during labor and delivery.
Inhibin	Inhibits secretion of FSH from anterior pituitary.
TESTICULAR HORMONES	
Testosterone 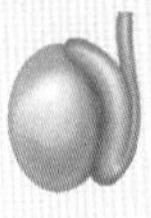 Testes	Stimulates descent of testes before birth; regulates spermatogenesis; promotes development and maintenance of male secondary sex characteristics.
Inhibin	Inhibits secretion of FSH from anterior pituitary.

The **pineal gland** (PĪN-ē-al = pinecone shape) is a small endocrine gland attached to the roof of the third ventricle of the brain at the midline (see Figure 18.1). Part of the epithalamus, it is positioned between the two superior colliculi, has a mass of 0.1–0.2 g, and is covered by a capsule formed by the pia mater. The gland consists of masses of neuroglia and secretory cells called **pinealocytes** (pīn-ē-AL-ō-sītz).

The pineal gland secretes **melatonin,** an amine hormone derived from serotonin. Melatonin appears to contribute to the setting of the body's biological clock, which is controlled by the suprachiasmatic nucleus of the hypothalamus. As more melatonin is liberated during darkness than in light, this hormone is thought to promote sleepiness. In response to visual input from the eyes (retina), the suprachiasmatic nucleus stimulates sympathetic postganglionic neurons of the superior cervical ganglion, which in turn stimulate the pinealocytes of the pineal gland to secrete melatonin in a rhythmic pattern, with low levels of melatonin secreted during the day and significantly higher levels secreted at night. During sleep, plasma levels of melatonin increase tenfold and then decline to a low level again before awakening. Small doses of melatonin given orally can induce sleep and reset daily rhythms, which might benefit workers whose shifts alternate between daylight and nighttime hours. Melatonin also is a potent antioxidant that may provide some protection against damaging oxygen free radicals.

In animals that breed during specific seasons, melatonin inhibits reproductive functions, but it is unclear whether melatonin influences human reproductive function. Melatonin levels are higher in children and decline with age into adulthood, but there is no evidence that changes in melatonin secretion correlate with the onset of puberty and sexual maturation. Nevertheless, because melatonin causes atrophy of the gonads in several animal species, the possibility of adverse effects on human reproduction must be studied before its use to reset daily rhythms can be recommended.

CLINICAL CONNECTION | *Seasonal Affective Disorder and Jet Lag*

Seasonal affective disorder (SAD) is a type of depression that afflicts some people during the winter months, when day length is short. It is thought to be due, in part, to overproduction of melatonin. Full-spectrum bright-light therapy—repeated doses of several hours of exposure to artificial light as bright as sunlight—provides relief for some people. Three to six hours of exposure to bright light also appears to speed recovery from jet lag, the fatigue suffered by travelers who quickly cross several time zones. •

The **thymus** is located behind the sternum between the lungs. Because of the role of the thymus in immunity, the details of its structure and functions are discussed in Chapter 22. The hormones produced by the thymus—**thymosin, thymic humoral factor (THF), thymic factor (TF),** and **thymopoietin** (thī-mō-poy-Ē-tin)—promote the maturation of T cells (a type of white blood cell that destroys microbes and foreign substances) and may retard the aging process.

✔**CHECKPOINT**

24. What is the relationship between melatonin and sleep?
25. Which thymic hormones play a role in immunity?

18.13 OTHER ENDOCRINE TISSUES AND ORGANS, EICOSANOIDS, AND GROWTH FACTORS

OBJECTIVES

- List the hormones secreted by cells in tissues and organs other than endocrine glands, and describe their functions.
- Describe the actions of eicosanoids and growth factors.

Hormones from Other Endocrine Tissues and Organs

As you learned at the beginning of this chapter, cells in organs other than those usually classified as endocrine glands have an endocrine function and secrete hormones. You learned about several of these in this chapter: the hypothalamus, thymus, pancreas, ovaries, and testes. Table 18.11 provides an overview of these organs and tissues and their hormones and actions.

Eicosanoids

Two families of eicosanoid molecules—the **prostaglandins** (pros′-ta-GLAN-dins), or **PGs,** and the **leukotrienes** (loo-kō-TRĪ-ēns), or **LTs**—are found in virtually all body cells except red blood cells, where they act as local hormones (paracrines or autocrines) in response to chemical or mechanical stimuli. They are synthesized by clipping a 20-carbon fatty acid called **arachidonic acid** (a-rak-i-DON-ik) from membrane phospholipid molecules. From arachidonic acid, different enzymatic reactions produce PGs or LTs. **Thromboxane (TX)** (throm-BOK-sān) is a modified PG that constricts blood vessels and promotes platelet activation.

TABLE 18.11

Summary of Hormones Produced by Other Organs and Tissues That Contain Endocrine Cells

HORMONE	PRINCIPAL ACTIONS
GASTROINTESTINAL TRACT	
Gastrin	Promotes secretion of gastric juice; increases movements of the stomach.
Glucose-dependent insulinotropic peptide (GIP)	Stimulates release of insulin by pancreatic beta cells.
Secretin	Stimulates secretion of pancreatic juice and bile.
Cholecystokinin (CCK)	Stimulates secretion of pancreatic juice; regulates release of bile from gallbladder; causes feeling of fullness after eating.
PLACENTA	
Human chorionic gonadotropin (hCG)	Stimulates corpus luteum in ovary to continue production of estrogens and progesterone to maintain pregnancy.
Estrogens and progesterone	Maintain pregnancy; help prepare mammary glands to secrete milk.
Human chorionic somatomammotropin (hCS)	Stimulates development of mammary glands for lactation.
KIDNEYS	
Renin	Part of reaction sequence that raises blood pressure by bringing about vasoconstriction and secretion of aldosterone.
Erythropoietin (EPO)	Increases rate of red blood cell formation.
Calcitriol* (active form of vitamin D)	Aids in absorption of dietary calcium and phosphorus.
HEART	
Atrial natriuretic peptide (ANP)	Decreases blood pressure.
ADIPOSE TISSUE	
Leptin	Suppresses appetite; may increase FSH and LH activity.

*Synthesis begins in the skin, continues in the liver, and ends in the kidneys.

Appearing in the blood in minute quantities, eicosanoids are present only briefly due to rapid inactivation.

To exert their effects, eicosanoids bind to receptors on target-cell plasma membranes and stimulate or inhibit the synthesis of second messengers such as cyclic AMP. Leukotrienes stimulate chemotaxis (attraction to a chemical stimulus) of white blood cells and mediate inflammation. The prostaglandins alter smooth muscle contraction, glandular secretions, blood flow, reproductive processes, platelet function, respiration, nerve impulse transmission, lipid metabolism, and immune responses. They also have roles in promoting inflammation and fever, and in intensifying pain.

CLINICAL CONNECTION | *Nonsteroidal Anti-inflammatory Drugs*

In 1971, scientists solved the long-standing puzzle of how aspirin works. Aspirin and related **nonsteroidal anti-inflammatory drugs (NSAIDs),** such as ibuprofen (Motrin®), inhibit *cyclooxygenase (COX),* a key enzyme involved in prostaglandin synthesis. NSAIDs are used to treat a wide variety of inflammatory disorders, from rheumatoid arthritis to tennis elbow. The success of NSAIDs in reducing fever, pain, and inflammation shows how prostaglandins contribute to these woes. •

Growth Factors

Several of the hormones we have described—insulinlike growth factor, thymosin, insulin, thyroid hormones, human growth hormone, and prolactin—stimulate cell growth and division. In addition, several more recently discovered hormones called **growth factors** play important roles in tissue development, growth, and repair. Growth factors are *mitogenic* substances—they cause growth by stimulating cell division. Many growth factors act locally, as autocrines or paracrines.

A summary of sources and actions of six important growth factors is presented in Table 18.12.

TABLE 18.12

Summary of Selected Growth Factors

GROWTH FACTOR	COMMENT
Epidermal growth factor (EGF)	Produced in submaxillary (salivary) glands; stimulates proliferation of epithelial cells, fibroblasts, neurons, and astrocytes; suppresses some cancer cells and secretion of gastric juice by stomach.
Platelet-derived growth factor (PDGF)	Produced in blood platelets; stimulates proliferation of neuroglia, smooth muscle fibers, and fibroblasts; appears to have role in wound healing; may contribute to atherosclerosis development.
Fibroblast growth factor (FGF)	Found in pituitary gland and brain; stimulates proliferation of many cells derived from embryonic mesoderm (fibroblasts, adrenocortical cells, smooth muscle fibers, chondrocytes, and endothelial cells); stimulates formation of new blood vessels (angiogenesis).
Nerve growth factor (NGF)	Produced in submandibular (salivary) glands and hippocampus of brain; stimulates growth of ganglia in embryo; maintains sympathetic nervous system; stimulates hypertrophy and differentiation of neurons.
Tumor angiogenesis factors (TAFs)	Produced by normal and tumor cells; stimulate growth of new capillaries, organ regeneration, and wound healing.
Transforming growth factors (TGFs)	Produced by various cells as separate molecules: TGF-alpha has activities similar to epidermal growth factor; TGF-beta inhibits proliferation of many cell types.

✔CHECKPOINT

26. What hormones are secreted by the gastrointestinal tract, placenta, kidneys, skin, adipose tissue, and heart?
27. What are some functions of prostaglandins, leukotrienes, and growth factors?

18.14 THE STRESS RESPONSE

OBJECTIVE

- Describe how the body responds to stress.

It is impossible to remove all of the stress from our everyday lives. Some stress, called **eustress,** prepares us to meet certain challenges and thus is helpful. Other stress, called **distress,** is harmful. Any stimulus that produces a stress response is called a **stressor.** A stressor may be almost any disturbance of the human body—heat or cold, environmental poisons, toxins given off by bacteria, heavy bleeding from a wound or surgery, or a strong emotional reaction. The responses to stressors may be pleasant or unpleasant, and they vary among people and even within the same person at different times.

Your body's homeostatic mechanisms attempt to counteract stress. When they are successful, the internal environment remains within normal physiological limits. If stress is extreme, unusual, or long lasting, the normal mechanisms may not be enough. In 1936, Hans Selye, a pioneer in stress research, showed that a variety of stressful conditions or noxious agents elicit a similar sequence of bodily changes. These changes, called the **stress response** or **general adaptation syndrome (GAS),** are controlled mainly by the hypothalamus. The stress response occurs in three stages: (1) an initial fight-or-flight response, (2) a slower resistance reaction, and eventually (3) exhaustion.

The Fight-or-Flight Response

The **fight-or-flight response,** initiated by nerve impulses from the hypothalamus to the sympathetic division of the autonomic nervous system (ANS), including the adrenal medulla, quickly mobilizes the body's resources for immediate physical activity

(Figure 18.20a). It brings huge amounts of glucose and oxygen to the organs that are most active in warding off danger: the brain, which must become highly alert; the skeletal muscles, which may have to fight off an attacker or flee; and the heart, which must work vigorously to pump enough blood to the brain and muscles. During the fight-or-flight response, nonessential body functions such as digestive, urinary, and reproductive activities are inhibited. Reduction of blood flow to the kidneys promotes release of renin, which sets into motion the renin–angiotensin–aldosterone pathway (see Figure 18.16). Aldosterone causes the kidneys to retain Na^+, which leads to water retention and elevated blood pressure. Water retention also helps preserve body fluid volume in the case of severe bleeding.

The Resistance Reaction

The second stage in the stress response is the **resistance reaction** (Figure 18.20b). Unlike the short-lived fight-or-flight response, which is initiated by nerve impulses from the hypothalamus, the resistance reaction is initiated in large part by hypothalamic releasing hormones and is a longer-lasting response. The hormones involved are corticotropin-releasing hormone (CRH), growth hormone–releasing hormone (GHRH), and thyrotropin-releasing hormone (TRH).

CRH stimulates the anterior pituitary to secrete ACTH, which in turn stimulates the adrenal cortex to increase release of cortisol. Cortisol then stimulates gluconeogenesis by liver cells, breakdown of triglycerides into fatty acids (lipolysis), and catabolism of proteins into amino acids. Tissues throughout the body can use the resulting glucose, fatty acids, and amino acids to produce ATP or to repair damaged cells. Cortisol also reduces inflammation.

A second hypothalamic releasing hormone, GHRH, causes the anterior pituitary to secrete human growth hormone (hGH). Acting via insulinlike growth factors, hGH stimulates lipolysis and glycogenolysis, the breakdown of glycogen to glucose, in the liver. A third hypothalamic releasing hormone, TRH, stimulates the anterior pituitary to secrete thyroid-stimulating hormone (TSH). TSH promotes secretion of thyroid hormones, which stimulate the increased use of glucose for ATP production. The combined actions of hGH and TSH supply additional ATP for metabolically active cells throughout the body.

The resistance stage helps the body continue fighting a stressor long after the fight-or-flight response dissipates. This is why your heart continues to pound for several minutes even after the stressor is removed. Generally, it is successful in seeing us through a stressful episode, and our bodies then return to normal. Occasionally, however, the resistance stage fails to combat the stressor, and the body moves into the state of exhaustion.

Exhaustion

The resources of the body may eventually become so depleted that they cannot sustain the resistance stage, and **exhaustion** ensues. Prolonged exposure to high levels of cortisol and other hormones involved in the resistance reaction causes wasting of muscle, suppression of the immune system, ulceration of the gastrointestinal tract, and failure of pancreatic beta cells. In addition, pathological changes may occur because resistance reactions persist after the stressor has been removed.

Stress and Disease

Although the exact role of stress in human diseases is not known, it is clear that stress can lead to particular diseases by temporarily inhibiting certain components of the immune system. Stress-related disorders include gastritis, ulcerative colitis, irritable bowel syndrome, hypertension, asthma, rheumatoid arthritis (RA), migraine headaches, anxiety, and depression. People under stress are at a greater risk of developing chronic disease or dying prematurely.

Interleukin-1, a substance secreted by macrophages of the immune system (see the discussion of ACTH in Section 18.6), is an important link between stress and immunity. One action of interleukin-1 is to stimulate secretion of ACTH, which in turn stimulates the production of cortisol. Not only does cortisol provide resistance to stress and inflammation, but it also suppresses further production of interleukin-1. Thus, the immune system turns on the stress response, and the resulting cortisol then turns off one immune system mediator. This negative feedback system keeps the immune response in check once it has accomplished its goal. Because of this activity, cortisol and other glucocorticoids are used as immunosuppressive drugs for organ transplant recipients.

CLINICAL CONNECTION | *Posttraumatic Stress Disorder*

Posttraumatic stress disorder (PTSD) is an anxiety disorder that may develop in an individual who has experienced, witnessed, or learned about a physically or psychologically distressing event. The immediate cause of PTSD appears to be the specific stressors associated with the events. Among the stressors are terrorism, hostage taking, imprisonment, military duty, serious accidents, torture, sexual or physical abuse, violent crimes, school shootings, massacres, and natural disasters. In the United States, PTSD affects 10% of females and 5% of males. Symptoms of PTSD include reexperiencing the event through nightmares or flashbacks; avoidance of any activity, person, place, or event associated with the stressors; loss of interest and lack of motivation; poor concentration; irritability; and insomnia. Treatment may include the use of antidepressants, mood stabilizers, and antianxiety and antipsychotic agents. •

✔CHECKPOINT

28. What is the central role of the hypothalamus during stress?
29. What body reactions occur during the fight-or-flight response, the resistance reaction, and exhaustion?
30. What is the relationship between stress and immunity?

Figure 18.20 **Responses to stressors during the stress response.** Red arrows (hormonal responses) and green arrows (neural responses) in (a) indicate immediate fight-or-flight reactions; black arrows in (b) indicate long-term resistance reactions.

Stressors stimulate the hypothalamus to initiate the stress response through the fight-or-flight response and the resistance reaction.

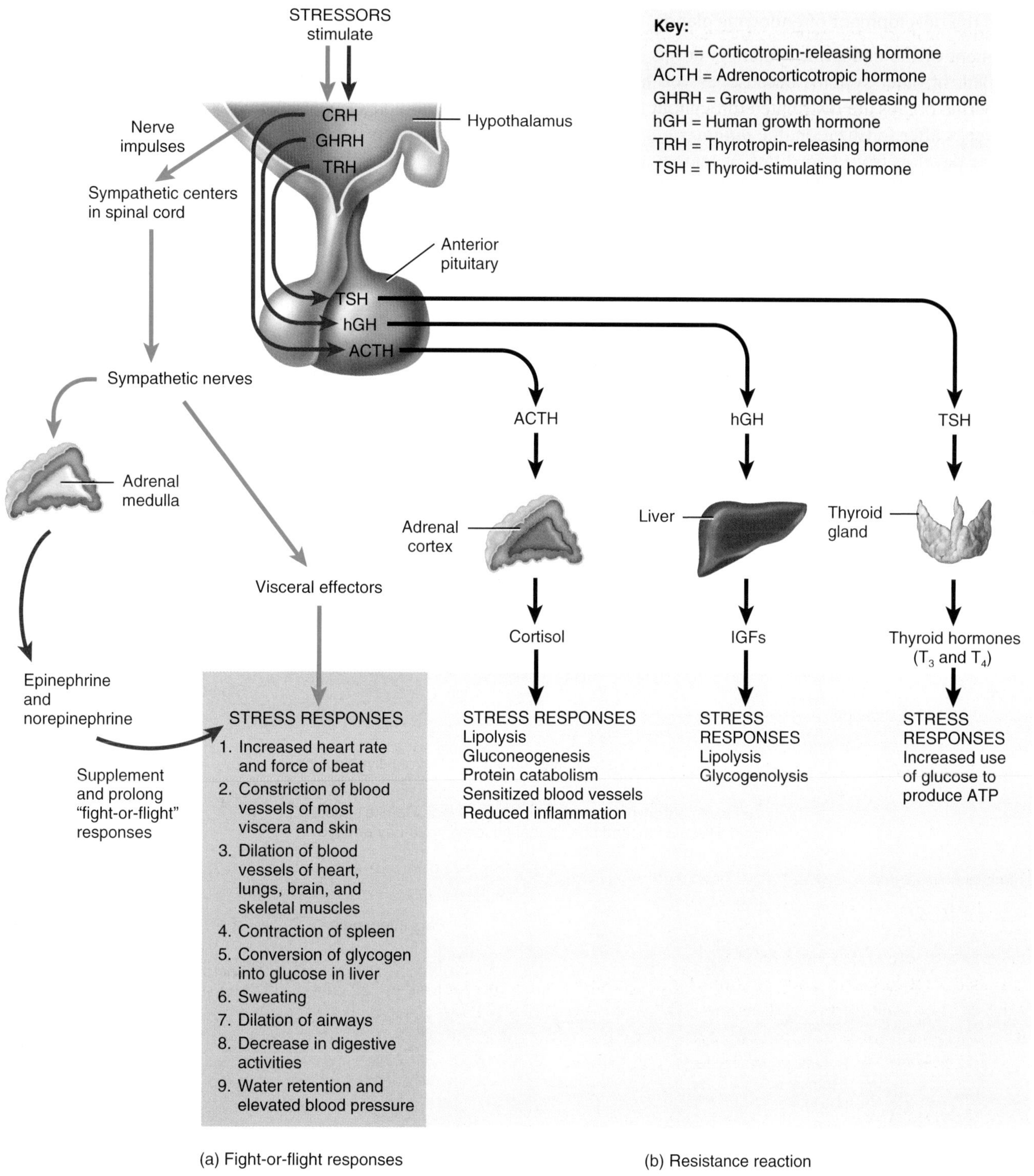

What is the basic difference between the stress response and homeostasis?

18.15 DEVELOPMENT OF THE ENDOCRINE SYSTEM

OBJECTIVE

- Describe the development of endocrine glands.

The development of the endocrine system is not as localized as the development of other systems because, as you have already learned, endocrine organs are distributed throughout the body.

About 3 weeks after fertilization, the *pituitary gland (hypophysis)* begins to develop from two different regions of the **ectoderm.** The *posterior pituitary (neurohypophysis)* is derived from an outgrowth of ectoderm called the **neurohypophyseal bud** (noor′-ō-hī-pō-FIZ-ē-al), located on the floor of the hypothalamus (Figure 18.21). The *infundibulum,* also an outgrowth of the neurohypophyseal bud, connects the posterior pituitary to the hypothalamus. The *anterior pituitary (adenohypophysis)* is derived from an outgrowth of ectoderm from the roof of the mouth called the **hypophyseal (Rathke's) pouch.** The pouch grows toward the neurohypophyseal bud and eventually loses its connection with the roof of the mouth.

The *thyroid gland* develops during the fourth week as a midventral outgrowth of **endoderm,** called the **thyroid diverticulum** (dī′-ver-TIK-ū-lum), from the floor of the pharynx at the level of the second pair of pharyngeal pouches (Figure 18.21a). The outgrowth projects inferiorly and differentiates into the right and left lateral lobes and the isthmus of the gland.

The *parathyroid glands* develop during the fourth week from **endoderm** as outgrowths from the third and fourth **pharyngeal pouches** (fa-RIN-jē-al), which help to form structures of the head and neck.

Figure 18.21 **Development of the endocrine system.**

Glands of the endocrine system develop from all three primary germ layers: ectoderm, mesoderm, and endoderm.

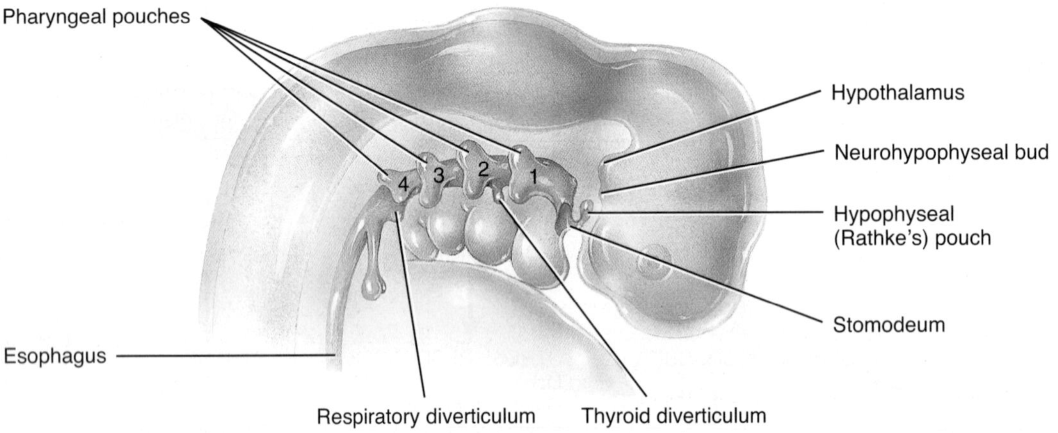

(a) Location of neurohypophyseal bud, hypophyseal (Rathke's) pouch, thyroid diverticulum, and pharyngeal pouches in 28-day embryo

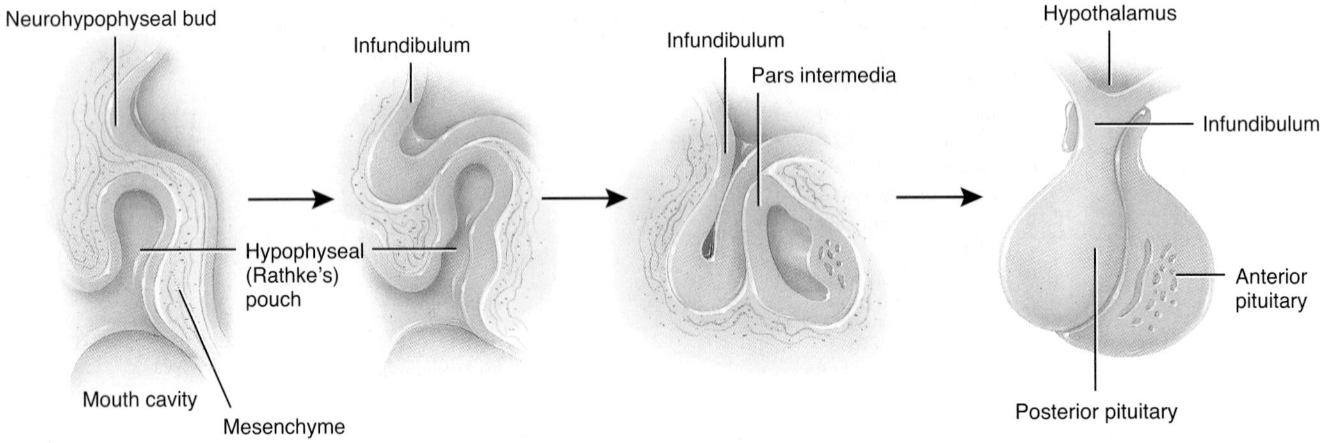

(b) Development of pituitary gland between 5 and 16 weeks

Which endocrine gland develops from tissues with two different embryological origins?

The adrenal cortex and adrenal medulla develop during the fifth week and have completely different embryological origins. The *adrenal cortex* is derived from the same region of **mesoderm** that produces the gonads. Endocrine tissues that secrete steroid hormones all are derived from mesoderm. The *adrenal medulla* is derived from **ectoderm** from **neural crest** cells that migrate to the superior pole of the kidney. Recall that neural crest cells also give rise to sympathetic ganglia and other structures of the nervous system (see Figure 14.27b).

The *pancreas* develops during the fifth through seventh weeks from two outgrowths of **endoderm** from the part of the **foregut** that later becomes the duodenum (see Figure 29.12c). The two outgrowths eventually fuse to form the pancreas. The origin of the ovaries and testes is discussed in Section 28.5.

The *pineal gland* arises during the seventh week as an outgrowth between the thalamus and colliculi of the midbrain from **ectoderm** associated with the **diencephalon** (see Figure 14.28).

The *thymus* arises during the fifth week from **endoderm** of the third pharyngeal pouches.

✔ **CHECKPOINT**

31. Compare the origins of the adrenal cortex and adrenal medulla.

18.16 AGING AND THE ENDOCRINE SYSTEM

OBJECTIVE

- Describe the effects of aging on the endocrine system.

Although some endocrine glands shrink as we get older, their performance may or may not be compromised. Production of human growth hormone by the anterior pituitary decreases, which is one cause of muscle atrophy as aging proceeds. The thyroid gland often decreases its output of thyroid hormones with age, causing a decrease in metabolic rate, an increase in body fat, and hypothyroidism, which is seen more often in older people. Because there is less negative feedback (lower levels of thyroid hormones), the level of thyroid-stimulating hormone increases with age (see Figure 18.12).

With aging, the blood level of PTH rises, perhaps due to inadequate dietary intake of calcium. In a study of older women who took 2400 mg/day of supplemental calcium, blood levels of PTH were as low as those of younger women. Both calcitriol and calcitonin levels are lower in older persons. Together, the rise in PTH and the fall in calcitonin level heighten the age-related decrease in bone mass that leads to osteoporosis and increased risk of fractures (see Figure 18.14).

The adrenal glands contain increasingly more fibrous tissue and produce less cortisol and aldosterone with advancing age. However, production of epinephrine and norepinephrine remains normal. The pancreas releases insulin more slowly with age, and receptor sensitivity to glucose declines. As a result, blood glucose levels in older people increase faster and return to normal more slowly than in younger individuals.

The thymus is largest in infancy. After puberty, its size begins to decrease, and thymic tissue is replaced by adipose and areolar connective tissue. In older adults, the thymus has atrophied significantly. However, it still produces new T cells for immune responses.

The ovaries decrease in size with age, and they no longer respond to gonadotropins. The resultant decreased output of estrogens leads to conditions such as osteoporosis, high blood cholesterol, and atherosclerosis. FSH and LH levels are high due to less negative feedback inhibition of estrogens. Although testosterone production by the testes decreases with age, the effects are not usually apparent until very old age; and many elderly males can still produce active sperm in normal numbers, even though there are higher numbers of morphologically abnormal sperm and decreased sperm motility.

✔ **CHECKPOINT**

32. Which hormone is related to the muscle atrophy that occurs with aging?

• • •

To appreciate the many ways the endocrine system contributes to homeostasis of other body systems, examine *Focus on Homeostasis: The Endocrine System*. Next, in Chapter 19, we will begin to explore the cardiovascular system, starting with a description of the composition and functions of blood.

Focus on Homeostasis

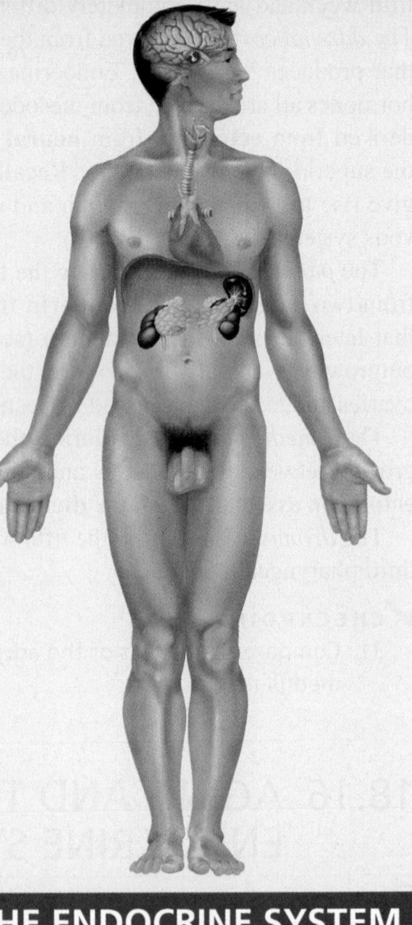

THE ENDOCRINE SYSTEM

BODY SYSTEM	CONTRIBUTION OF THE ENDOCRINE SYSTEM
For all body systems 	Together with the nervous system, circulating and local hormones of the endocrine system regulate activity and growth of target cells throughout the body; several hormones regulate metabolism, uptake of glucose, and molecules used for ATP production by body cells.
Integumentary system 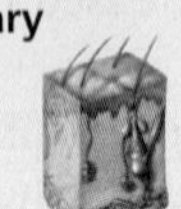	Androgens stimulate growth of axillary and pubic hair and activation of sebaceous glands; excess melanocyte-stimulating hormone (MSH) causes darkening of skin.
Skeletal system 	Human growth hormone (hGH) and insulinlike growth factors (IGFs) stimulate bone growth; estrogens cause closure of the epiphyseal plates at the end of puberty and help maintain bone mass in adults; parathyroid hormone (PTH) and calcitonin regulate levels of calcium and other minerals in bone matrix and blood; thyroid hormones are needed for normal development and growth of the skeleton.
Muscular system 	Epinephrine and norepinephrine help increase blood flow to exercising muscle; PTH maintains proper level of Ca^{2+}, needed for muscle contraction; glucagon, insulin, and other hormones regulate metabolism in muscle fibers; hGH, IGFs, and thyroid hormones help maintain muscle mass.
Nervous system 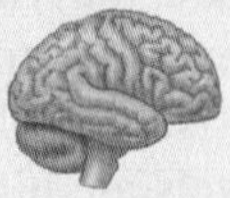	Several hormones, especially thyroid hormones, insulin, and growth hormone, influence growth and development of the nervous system; PTH maintains proper level of Ca^{2+}, needed for generation and conduction of nerve impulses.
Cardiovascular system 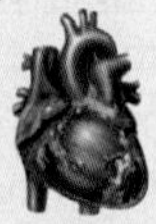	Erythropoietin (EPO) promotes formation of red blood cells; aldosterone and antidiuretic hormone (ADH) increase blood volume; epinephrine and norepinephrine increase heart rate and force of contraction; several hormones elevate blood pressure during exercise and other stresses.
Lymphatic system and immunity 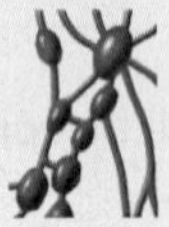	Glucocorticoids such as cortisol depress inflammation and immune responses; thymic hormones promote maturation of T cells (a type of white blood cell).
Respiratory system 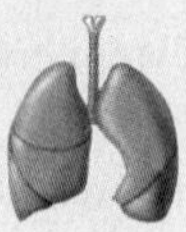	Epinephrine and norepinephrine dilate (widen) airways during exercise and other stresses; erythropoietin regulates amount of oxygen carried in blood by adjusting number of red blood cells.
Digestive system 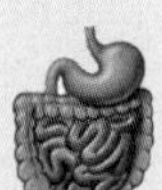	Epinephrine and norepinephrine depress activity of the digestive system; gastrin, cholecystokinin, secretin, and GIP help regulate digestion; calcitriol promotes absorption of dietary calcium; leptin suppresses appetite.
Urinary system 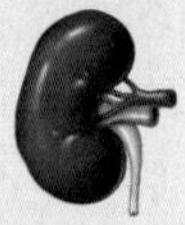	ADH, aldosterone, and atrial natriuretic peptide (ANP) adjust the rate of loss of water and ions in the urine, thereby regulating blood volume and ion content of the blood.
Reproductive systems 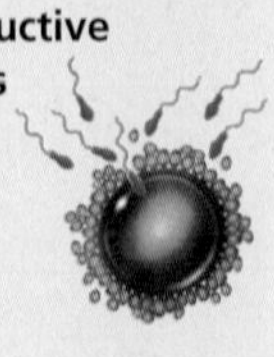	Hypothalamic releasing and inhibiting hormones, follicle-stimulating hormone (FSH), and luteinizing hormone (LH) regulate development, growth, and secretions of the gonads (ovaries and testes); estrogens and testosterone contribute to development of oocytes and sperm and stimulate development of secondary sex characteristics; prolactin promotes milk secretion in mammary glands; oxytocin causes contraction of the uterus and ejection of milk from the mammary glands.

DISORDERS: HOMEOSTATIC IMBALANCES

Disorders of the endocrine system often involve either **hyposecretion** (*hypo-* = too little or under), inadequate release of a hormone, or **hypersecretion** (*hyper-* = too much or above), excessive release of a hormone. In other cases, the problem is faulty hormone receptors, an inadequate number of receptors, or defects in second-messenger systems. Because hormones are distributed in the blood to target tissues throughout the body, problems associated with endocrine dysfunction may also be widespread.

Pituitary Gland Disorders

Pituitary Dwarfism, Giantism, and Acromegaly

Several disorders of the anterior pituitary involve human growth hormone (hGH). Hyposecretion of hGH during the growth years slows bone growth, and the epiphyseal plates close before normal height is reached. This condition is called **pituitary dwarfism** (see Clinical Connection: Hormonal Abnormalities That Affect Height in Chapter 6). Other organs of the body also fail to grow, and the body proportions are childlike. Treatment requires administration of hGH during childhood, before the epiphyseal plates close.

Hypersecretion of hGH during childhood causes **giantism,** an abnormal increase in the length of long bones. The person grows to be very tall, but body proportions are about normal. Figure 18.22a shows identical twins; one brother developed giantism due to a pituitary tumor. Hypersecretion of hGH during adulthood is called **acromegaly** (ak′-rō-MEG-a-lē). Although hGH cannot produce further lengthening of the long bones because the epiphyseal plates are already closed, the bones of the hands, feet, cheeks, and jaws thicken and other tissues enlarge. In addition, the eyelids, lips, tongue, and nose enlarge, and the skin thickens and develops furrows, especially on the forehead and soles (Figure 18.22b).

Diabetes Insipidus

The most common abnormality associated with dysfunction of the posterior pituitary is **diabetes insipidus** (dī-a-BĒ-tēs in-SIP-i-dus;

Figure 18.22 Various endocrine disorders.

Disorders of the endocrine system often involve hyposecretion or hypersecretion of hormones.

(a) A 22-year-old man with pituitary giantism shown beside his identical twin

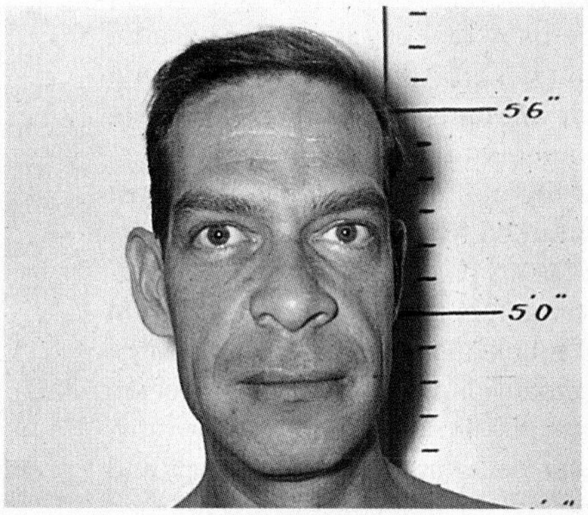

(b) Acromegaly (excess hGH during adulthood)

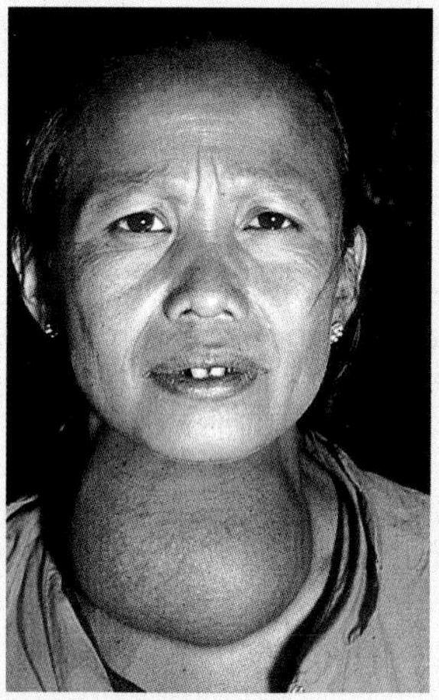

(c) Goiter (enlargement of thyroid gland)

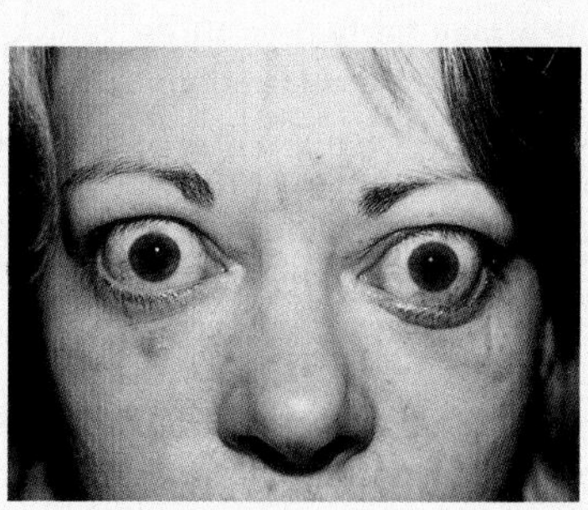

(d) Exophthalmos (excess thyroid hormones, as in Graves' disease)

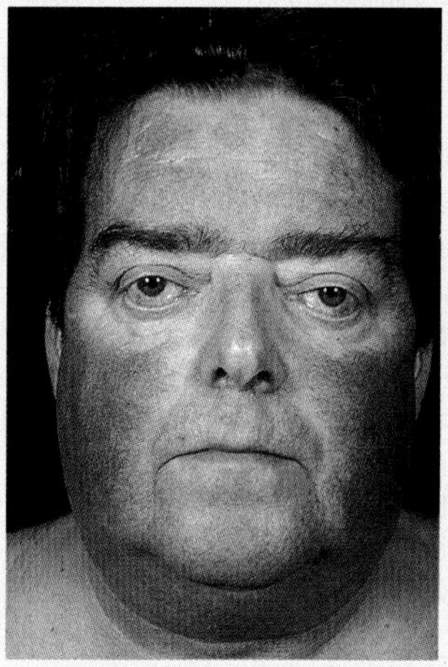

(e) Cushing's syndrome (excess glucocorticoids)

Which endocrine disorder is due to antibodies that mimic the action of TSH?

diabetes = overflow; *insipidus* = tasteless) or **DI.** This disorder is due to defects in antidiuretic hormone (ADH) receptors or an inability to secrete ADH. *Neurogenic diabetes insipidus* results from hyposecretion of ADH, usually caused by a brain tumor, head trauma, or brain surgery that damages the posterior pituitary or the hypothalamus. In *nephrogenic diabetes insipidus,* the kidneys do not respond to ADH. The ADH receptors may be nonfunctional, or the kidneys may be damaged. A common symptom of both forms of DI is excretion of large volumes of urine, with resulting dehydration and thirst. Bed-wetting is common in afflicted children. Because so much water is lost in the urine, a person with DI may die of dehydration if deprived of water for only a day or so.

Treatment of neurogenic diabetes insipidus involves hormone replacement, usually for life. Either subcutaneous injection or nasal spray application of ADH analogs is effective. Treatment of nephrogenic diabetes insipidus is more complex and depends on the nature of the kidney dysfunction. Restriction of salt in the diet and, paradoxically, the use of certain diuretic drugs, are helpful.

Thyroid Gland Disorders

Thyroid gland disorders affect all major body systems and are among the most common endocrine disorders. **Congenital hypothyroidism,** hyposecretion of thyroid hormones that is present at birth, has devastating consequences if not treated promptly. Previously termed *cretinism,* this condition causes severe mental retardation and stunted bone growth. At birth, the baby typically is normal because lipid-soluble maternal thyroid hormones crossed the placenta during pregnancy and allowed normal development. Most states require testing of all newborns to ensure adequate thyroid function. If congenital hypothyroidism exists, oral thyroid hormone treatment must be started soon after birth and continued for life.

Hypothyroidism during the adult years produces **myxedema** (mix-e-DĒ-ma), which occurs about five times more often in females than in males. A hallmark of this disorder is edema (accumulation of interstitial fluid) that causes the facial tissues to swell and look puffy. A person with myxedema has a slow heart rate, low body temperature, sensitivity to cold, dry hair and skin, muscular weakness, general lethargy, and a tendency to gain weight easily. Because the brain has already reached maturity, mental retardation does not occur, but the person may be less alert. Oral thyroid hormones reduce the symptoms.

The most common form of hyperthyroidism is **Graves' disease,** which also occurs seven to ten times more often in females than in males, usually before age 40. Graves' disease is an autoimmune disorder in which the person produces antibodies that mimic the action of thyroid-stimulating hormone (TSH). The antibodies continually stimulate the thyroid gland to grow and produce thyroid hormones. A primary sign is an enlarged thyroid, which may be two to three times its normal size. Graves' patients often have a peculiar edema behind the eyes, called **exophthalmos** (ek'-sof-THAL-mos), which causes the eyes to protrude (Figure 18.22d). Treatment may include surgical removal of part or all of the thyroid gland (thyroidectomy), the use of radioactive iodine (^{131}I) to selectively destroy thyroid tissue, and the use of antithyroid drugs to block synthesis of thyroid hormones.

A **goiter** (GOY-ter; *guttur* = throat) is simply an enlarged thyroid gland. It may be associated with hyperthyroidism, hypothyroidism, or **euthyroidism** (ū-THĪ-royd-izm; *eu* = good), which means normal secretion of thyroid hormone. In some places in the world, dietary iodine intake is inadequate; the resultant low level of thyroid hormone in the blood stimulates secretion of TSH, which causes thyroid gland enlargement (Figure 18.22c).

Parathyroid Gland Disorders

Hypoparathyroidism (hī-pō-par'-a-THĪ-royd-izm)—too little parathyroid hormone—leads to a deficiency of blood Ca^{2+}, which causes neurons and muscle fibers to depolarize and produce action potentials spontaneously. This leads to twitches, spasms, and **tetany** (maintained contraction) of skeletal muscle. The leading cause of hypoparathyroidism is accidental damage to the parathyroid glands or to their blood supply during thyroidectomy surgery.

Hyperparathyroidism, an elevated level of parathyroid hormone, most often is due to a tumor of one of the parathyroid glands. An elevated level of PTH causes excessive resorption of bone matrix, raising the blood levels of calcium and phosphate ions and causing bones to become soft and easily fractured. High blood calcium level promotes formation of kidney stones. Fatigue, personality changes, and lethargy are also seen in patients with hyperparathyroidism.

Adrenal Gland Disorders

Cushing's Syndrome

Hypersecretion of cortisol by the adrenal cortex produces **Cushing's syndrome** (Figure 18.22e). Causes include a tumor of the adrenal gland that secretes cortisol, or a tumor elsewhere that secretes adrenocorticotropic hormone (ACTH), which in turn stimulates excessive secretion of cortisol. The condition is characterized by breakdown of muscle proteins and redistribution of body fat, resulting in spindly arms and legs accompanied by a rounded "moon face," "buffalo hump" on the back, and pendulous (hanging) abdomen. Facial skin is flushed, and the skin covering the abdomen develops stretch marks. The person also bruises easily, and wound healing is poor. The elevated level of cortisol causes hyperglycemia, osteoporosis, weakness, hypertension, increased susceptibility to infection, decreased resistance to stress, and mood swings. People who need long-term glucocorticoid therapy—for instance, to prevent rejection of a transplanted organ—may develop a cushingoid appearance.

Addison's Disease

Hyposecretion of glucocorticoids and aldosterone causes **Addison's disease (chronic adrenocortical insufficiency).** The majority of cases are autoimmune disorders in which antibodies cause adrenal cortex destruction or block binding of ACTH to its receptors. Pathogens, such as the bacterium that causes tuberculosis, also may trigger adrenal cortex destruction. Symptoms, which typically do not appear until 90% of the adrenal cortex has been destroyed, include mental lethargy, anorexia, nausea and vomiting, weight loss, hypoglycemia, and muscular weakness. Loss of aldosterone leads to elevated potassium and decreased sodium in the blood, low blood pressure, dehydration, decreased cardiac output, arrhythmias, and even cardiac arrest. The skin may have a "bronzed" appearance that often is mistaken for a suntan. Such was true in the case of President John F. Kennedy, whose Addison's disease was known to only a few while he was alive. Treatment consists of replacing glucocorticoids and mineralocorticoids and increasing sodium in the diet.

Pheochromocytomas

Usually benign tumors of the chromaffin cells of the adrenal medulla, called **pheochromocytomas** (fē-ō-krō'-mō-si-TŌ-mas; *pheo-* = dusky; *-chromo-* = color; *-cyto-* = cell), cause hypersecretion of epinephrine and norepinephrine. The result is a prolonged version of the fight-or-flight response: rapid heart rate, high blood pressure, high levels of glucose in blood and urine, an elevated basal metabolic rate

(BMR), flushed face, nervousness, sweating, and decreased gastrointestinal motility. Treatment is surgical removal of the tumor.

Pancreatic Islet Disorders

The most common endocrine disorder is **diabetes mellitus** (MEL-i-tus; *melli-* = honey sweetened), caused by an inability to produce or use insulin. Diabetes mellitus is the fourth leading cause of death by disease in the United States, primarily because of its damage to the cardiovascular system. Because insulin is unavailable to aid transport of glucose into body cells, blood glucose level is high and glucose "spills" into the urine (glucosuria). Hallmarks of diabetes mellitus are the three "polys": *polyuria,* excessive urine production due to an inability of the kidneys to reabsorb water; *polydipsia,* excessive thirst; and *polyphagia,* excessive eating.

Both genetic and environmental factors contribute to onset of the two types of diabetes mellitus—type 1 and type 2—but the exact mechanisms are still unknown. **Type 1 diabetes,** previously known as **insulin-dependent diabetes mellitus (IDDM),** occurs because the person's immune system destroys the pancreatic beta cells. As a result, the pancreas produces little or no insulin. Type 1 diabetes usually develops in people younger than age 20 and it persists throughout life. By the time symptoms of type 1 diabetes arise, 80–90% of the islet beta cells have been destroyed. Type 1 diabetes is most common in northern Europe, especially in Finland, where nearly 1% of the population develops type 1 diabetes by 15 years of age. In the United States, type 1 diabetes is 1.5–2.0 times more common in whites than in African American or Asian populations.

The cellular metabolism of an untreated type 1 diabetic is similar to that of a starving person. Because insulin is not present to aid the entry of glucose into body cells, most cells use fatty acids to produce ATP. Stores of triglycerides in adipose tissue are catabolized to yield fatty acids and glycerol. The by-products of fatty acid breakdown—organic acids called ketones or ketone bodies—accumulate. Buildup of ketones causes blood pH to fall, a condition known as **ketoacidosis** (kē′-tō-as-i-DŌ-sis). Unless treated quickly, ketoacidosis can cause death.

The breakdown of stored triglycerides also causes weight loss. As lipids are transported by the blood from storage depots to cells, lipid particles are deposited on the walls of blood vessels, leading to atherosclerosis and a multitude of cardiovascular problems, including cerebrovascular insufficiency, ischemic heart disease, peripheral vascular disease, and gangrene. A major complication of diabetes is loss of vision due either to cataracts (excessive glucose attaches to lens proteins, causing cloudiness) or to damage to blood vessels of the retina. Severe kidney problems also may result from damage to renal blood vessels.

Type 1 diabetes is treated through self-monitoring of blood glucose level (up to 7 times daily), regular meals containing 45–50% carbohydrates and less than 30% fats, exercise, and periodic insulin injections (up to 3 times a day). Several implantable pumps are available to provide insulin without the need for repeated injections. Because they lack a reliable glucose sensor, however, the person must self-monitor blood glucose level to determine insulin doses. It is also possible to successfully transplant a pancreas, but immunosuppressive drugs must then be taken for life. Another promising approach under investigation is transplantation of isolated islets in semipermeable hollow tubes. The tubes allow glucose and insulin to enter and leave but prevent entry of immune system cells that might attack the islet cells.

Type 2 diabetes, formerly known as **non-insulin-dependent diabetes mellitus (NIDDM),** is much more common than type 1, representing more than 90% of all cases. Type 2 diabetes most often occurs in obese people who are over age 35. However, the number of obese children and teenagers with type 2 diabetes is increasing. Clinical symptoms are mild, and the high glucose levels in the blood often can be controlled by diet, exercise, and weight loss. Sometimes, drugs such as *glyburide* (DiaBeta®) and metformin (Fortamet®) are used to stimulate secretion of insulin by pancreatic beta cells. Although some type 2 diabetics need insulin, many have a sufficient amount (or even a surplus) of insulin in the blood. For these people, diabetes arises not from a shortage of insulin but because target cells become less sensitive to it due to down-regulation of insulin receptors.

Hyperinsulinism most often results when a diabetic injects too much insulin. The main symptom is **hypoglycemia,** decreased blood glucose level, which occurs because the excess insulin stimulates too much uptake of glucose by body cells. The resulting hypoglycemia stimulates the secretion of epinephrine, glucagon, and human growth hormone. As a consequence, anxiety, sweating, tremor, increased heart rate, hunger, and weakness occur. When blood glucose falls, brain cells are deprived of the steady supply of glucose they need to function effectively. Severe hypoglycemia leads to mental disorientation, convulsions, unconsciousness, and shock. Shock due to an insulin overdose is termed **insulin shock.** Death can occur quickly unless blood glucose is restored to normal levels. From a clinical standpoint, a diabetic suffering from either a hyperglycemia or a hypoglycemia crisis can have very similar symptoms—mental changes, coma, seizures, and so on. It is important to quickly and correctly identify the cause of the underlying symptoms and treat them appropriately.

MEDICAL TERMINOLOGY

Gynecomastia (gī-ne′-kō-MAS-tē-a; *gyneco-* = woman; *-mast-* = breast) Excessive development of mammary glands in a male. Sometimes a tumor of the adrenal gland may secrete sufficient amounts of estrogen to cause the condition.

Hirsutism (HER-soo-tizm; *hirsut-* = shaggy) Presence of excessive body and facial hair in a male pattern, especially in women; may be due to excess androgen production due to tumors or drugs.

Thyroid crisis (storm) A severe state of hyperthyroidism that can be life-threatening. It is characterized by high body temperature, rapid heart rate, high blood pressure, gastrointestinal symptoms (abdominal pain, vomiting, diarrhea), agitation, tremors, confusion, seizures, and possibly coma.

Virilizing adenoma (*aden-* = gland; *-oma* = tumor) Tumor of the adrenal gland that liberates excessive androgens, causing virilism (masculinization) in females. Occasionally, adrenal tumor cells liberate estrogens to the extent that a male patient develops gynecomastia. Such a tumor is called a **feminizing adenoma.**

42

CHAPTER REVIEW AND RESOURCE SUMMARY

Review	Resource
Introduction 1. Hormones regulate the activity of smooth muscle, cardiac muscle, and some glands; alter metabolism; spur growth and development; influence reproductive processes; and participate in circadian (daily) rhythms.	Anatomy Overview - The Endocrine System
18.1 Comparison of Control by the Nervous and Endocrine Systems 1. The nervous system controls homeostasis through nerve impulses and neurotransmitters, which act locally and quickly. The endocrine system uses hormones, which act more slowly in distant parts of the body. (See Table 18.1.) 2. The nervous system controls neurons, muscle cells, and glandular cells; the endocrine system regulates virtually all body cells.	Anatomy Overview - The Nervous System
18.2 Endocrine Glands 1. Exocrine glands (sudoriferous, sebaceous, mucous, and digestive) secrete their products through ducts into body cavities or onto body surfaces. Endocrine glands secrete hormones into interstitial fluid. Then, the hormones diffuse into the blood. 2. The endocrine system consists of endocrine glands (pituitary, thyroid, parathyroid, adrenal, and pineal glands) and other hormone-secreting tissues (hypothalamus, thymus, pancreas, ovaries, testes, kidneys, stomach, liver, small intestine, skin, heart, adipose tissue, and placenta).	Anatomy Overview - The Endocrine System
18.3 Hormone Activity 1. Hormones affect only specific target cells that have receptors to recognize (bind) a given hormone. The number of hormone receptors may decrease (down-regulation) or increase (up-regulation). 2. Circulating hormones enter the bloodstream; local hormones (paracrines and autocrines) act locally on neighboring cells. 3. Chemically, hormones are either lipid-soluble (steroids, thyroid hormones, and nitric oxide) or water-soluble (amines; peptides, proteins, and glycoproteins; and eicosanoids). (See Table 18.2.) 4. Water-soluble hormone molecules circulate in the watery blood plasma in a "free" form (not attached to plasma proteins); most lipid-soluble hormones are bound to transport proteins synthesized by the liver.	Anatomy Overview - Hormones Anatomy Overview - Lipid-Soluble Hormones Anatomy Overview - Water-Soluble Hormones Anatomy Overview - Local Hormones
18.4 Mechanisms of Hormone Action 1. Lipid-soluble steroid hormones and thyroid hormones affect cell function by altering gene expression. 2. Water-soluble hormones alter cell function by activating plasma membrane receptors, which elicit production of a second messenger that activates various enzymes inside the cell. 3. Hormonal interactions can have three types of effects: permissive, synergistic, or antagonistic.	Animation - Introduction to Hormonal Regulation, Secretion, and Concentration Animation - Mechanisms of Hormone Action Exercise - Hormone Actions
18.5 Control of Hormone Secretion 1. Hormone secretion is controlled by signals from the nervous system, chemical changes in blood, and other hormones. 2. Negative feedback systems regulate the secretion of many hormones.	Animation - Introduction to Hormone Feedback Loops Exercise - Produce That Hormone
18.6 Hypothalamus and Pituitary Gland 1. The hypothalamus is the major integrating link between the nervous and endocrine systems. The hypothalamus and pituitary gland regulate virtually all aspects of growth, development, metabolism, and homeostasis. The pituitary gland is located in the hypophyseal fossa and is divided into two main portions: the anterior pituitary (glandular portion) and the posterior pituitary (nervous portion). 2. Secretion of anterior pituitary hormones is stimulated by releasing hormones and suppressed by inhibiting hormones from the hypothalamus. 3. The blood supply to the anterior pituitary is from the superior hypophyseal arteries. Hypothalamic releasing and inhibiting hormones enter the primary plexus and flow to the secondary plexus in the anterior pituitary by the hypophyseal portal veins. 4. The anterior pituitary consists of somatotrophs that produce human growth hormone (hGH), lactotrophs that produce prolactin (PRL), corticotrophs that secrete adrenocorticotropic hormone (ACTH) and	Anatomy Overview - The Hypothalamus and Pituitary Gland Anatomy Overview - Hormones of the Anterior Pituitary Gland Anatomy Overview - The Hypothalamus Anatomy Overview - Hormones of the Hypothalamus Animation - hGH: Growth and Development Animation - GHRH/hGR

Review | **Resource**

melanocyte-stimulating hormone (MSH), thyrotrophs that secrete thyroid-stimulating hormone (TSH), and gonadotrophs that synthesize follicle-stimulating hormone (FSH) and luteinizing hormone (LH). (See Tables 18.3 and 18.4.)

5. Human growth hormone (hGH) stimulates body growth through insulinlike growth factors (IGFs). Secretion of hGH is inhibited by GHIH (growth hormone–inhibiting hormone, or somatostatin) and promoted by GHRH (growth hormone–releasing hormone).
6. TSH regulates thyroid gland activities. Its secretion is stimulated by TRH (thyrotropin-releasing hormone) and suppressed by GHIH.
7. FSH and LH regulate the activities of the gonads—ovaries and testes. Their secretion is controlled by GnRH (gonadotropin-releasing hormone).
8. Prolactin (PRL) helps initiate milk secretion. Prolactin-inhibiting hormone (PIH) suppresses secretion of PRL; prolactin-releasing hormone (PRH) stimulates PRL secretion.
9. ACTH regulates the activities of the adrenal cortex and is controlled by CRH (corticotropin-releasing hormone). Dopamine inhibits secretion of MSH.
10. The posterior pituitary contains axon terminals of neurosecretory cells whose cell bodies are in the hypothalamus. Hormones made by the hypothalamus and stored in the posterior pituitary are oxytocin (OT), which stimulates contraction of the uterus and ejection of milk from the breasts, and antidiuretic hormone (ADH), which stimulates water reabsorption by the kidneys and constriction of arterioles. (See Table 18.5.) Oxytocin secretion is stimulated by uterine stretching and suckling during nursing; ADH secretion is controlled by osmotic pressure of the blood and blood volume.

Animation - ACTH/Cortisol: Glycogenolysis
Animation - TRH/TSH: Production
Animation - hGH: Glycogenolysis and Lipolysis
Animation - Antidiuretic Hormone
Figure 18.5 - Hypothalamus and Pituitary Gland and Their Blood Supply

18.7 Thyroid Gland

1. The thyroid gland is located inferior to the larynx.
2. It consists of thyroid follicles composed of follicular cells, which secrete the thyroid hormones thyroxine (T_4) and triiodothyronine (T_3), and parafollicular cells, which secrete calcitonin (CT).
3. Thyroid hormones are synthesized from iodine and tyrosine within thyroglobulin (TGB). They are transported in the blood bound to plasma proteins, mostly thyroxine-binding globulin (TBG).
4. Secretion is controlled by TRH from the hypothalamus and thyroid-stimulating hormone (TSH) from the anterior pituitary.
5. Thyroid hormones regulate oxygen use and metabolic rate, cellular metabolism, and growth and development.
6. Calcitonin (CT) can lower the blood level of calcium ions (Ca^{2+}) and promote deposition of Ca^{2+} into bone matrix. Secretion of CT is controlled by the Ca^{2+} level in the blood. (See Table 18.6.)

Anatomy Overview - The Thyroid
Animation - Thyroid Hormones and Glucose and Lipid Catabolism
Animation - TRH/TSH
Animation - Calcitonin
Figure 18.11 - Steps in the Synthesis and Secretion of Thyroid Hormones
Exercise - Calcium Homeostasis

18.8 Parathyroid Glands

1. The parathyroid glands are embedded in the posterior surfaces of the lateral lobes of the thyroid gland. They consist of chief cells and oxyphil cells.
2. Parathyroid hormone (PTH) regulates the homeostasis of calcium, magnesium, and phosphate ions by increasing blood calcium and magnesium levels and decreasing blood phosphate levels. PTH secretion is controlled by the level of calcium in the blood. (See Table 18.7.)

Anatomy Overview - Parathyroid Glands
Anatomy Overview - Hormones of the Parathyroid
Animation - Parathyroid Hormone

18.9 Adrenal Glands

1. The adrenal glands are located superior to the kidneys. They consist of an outer adrenal cortex and inner adrenal medulla.
2. The adrenal cortex is divided into a zona glomerulosa, a zona fasciculata, and a zona reticularis; the adrenal medulla consists of chromaffin cells and large blood vessels.
3. Cortical secretions include mineralocorticoids, glucocorticoids, and androgens.
4. Mineralocorticoids (mainly aldosterone) increase sodium and water reabsorption and decrease potassium reabsorption. Secretion is controlled by the renin–angiotensin–aldosterone (RAA) pathway and by K^+ level in the blood.
5. Glucocorticoids (mainly cortisol) promote protein breakdown, gluconeogenesis, and lipolysis; help resist stress; and serve as anti-inflammatory substances. Their secretion is controlled by ACTH.
6. Androgens secreted by the adrenal cortex stimulate growth of axillary and pubic hair, aid the prepubertal growth spurt, and contribute to libido.
7. The adrenal medulla secretes epinephrine and norepinephrine (NE), which are released during stress and produce effects similar to sympathetic responses. (See Table 18.8.)

Anatomy Overview - Adrenal Glands
Anatomy Overview - Hormones of the Adrenal Glands
Animation - ACTH/Cortisol
Animation - Epinephrine/NE
Figure 18.15 - The Adrenal Glands
Figure 18.16 - Regulation of Aldosterone Secretion

Review | **Resource**

18.10 Pancreatic Islets

1. The pancreas lies in the curve of the duodenum. It has both endocrine and exocrine functions.
2. The endocrine portion consists of pancreatic islets or islets of Langerhans, made up of four types of cells: alpha, beta, delta, and F cells.
3. Alpha cells secrete glucagon, beta cells secrete insulin, delta cells secrete somatostatin, and F cells secrete pancreatic polypeptide.
4. Glucagon increases blood glucose level; insulin decreases blood glucose level. Secretion of both hormones is controlled by the level of glucose in the blood. (See Table 18.9.)

Resource:
Anatomy Overviews: The Pancreas; Hormones of the Pancreas
Animations: Glucagon; Insulin
Figure 18.18 - The Pancreas
Figure 18.19 - Secretion of Glucagon and Insulin
Exercise - Glucose Regulation Feedback Loop
Concepts and Connections - Blood Glucose Regulation

18.11 Ovaries and Testes

1. The ovaries are located in the pelvic cavity and produce estrogens, progesterone, and inhibin. These sex hormones govern the development and maintenance of female secondary sex characteristics, reproductive cycles, pregnancy, lactation, and normal female reproductive functions. (See Table 18.10.)
2. The testes lie inside the scrotum and produce testosterone and inhibin. These sex hormones govern the development and maintenance of male secondary sex characteristics and normal male reproductive functions. (See Table 18.10.)

Resource:
Anatomy Overviews: The Ovaries; Ovarian Hormones; The Testes; Testicular Hormones
Animations: Hormonal Control of Male Reproductive Function; Hormonal Regulation of Female Reproductive System
Exercises: Match the Female Hormones; Match the Male Hormones

18.12 Pineal Gland and Thymus

1. The pineal gland is attached to the roof of the third ventricle of the brain. It consists of secretory cells called pinealocytes, neuroglia, and endings of sympathetic postganglionic axons.
2. The pineal gland secretes melatonin, which contributes to setting the body's biological clock (controlled in the suprachiasmatic nucleus). During sleep, plasma levels of melatonin increase.
3. The thymus secretes several hormones related to immunity.
4. Thymosin, thymic humoral factor (THF), thymic factor (TF), and thymopoietin promote the maturation of T cells.

Resource:
Anatomy Overview - The Pineal Gland

18.13 Other Endocrine Tissues and Organs, Eicosanoids, and Growth Factors

1. Body tissues other than those normally classified as endocrine glands contain endocrine tissue and secrete hormones; they include the gastrointestinal tract, placenta, kidneys, skin, and heart. (See Table 18.11.)
2. Prostaglandins and leukotrienes are eicosanoids that act as local hormones in most body tissues.
3. Growth factors are local hormones that stimulate cell growth and division. (See Table 18.12.)

Resource:
Animations: Hormonal Control of Digestive Activities; Hormonal Regulation of Pregnancy
Exercise - Concentrate on Endocrine Activities

18.14 The Stress Response

1. Productive stress is termed eustress, and harmful stress is termed distress.
2. If stress is extreme, it triggers the stress response (general adaptation syndrome), which occurs in three stages: the fight-or-flight response, resistance reaction, and exhaustion.
3. The stimuli that produce the stress response are called stressors. Stressors include surgery, poisons, infections, fever, and strong emotional responses.
4. The fight-or-flight response is initiated by nerve impulses from the hypothalamus to the sympathetic division of the autonomic nervous system and the adrenal medulla. This response rapidly increases circulation, promotes ATP production, and decreases nonessential activities.
5. The resistance reaction is initiated by releasing hormones secreted by the hypothalamus, most importantly CRH, TRH, and GHRH. Resistance reactions are longer lasting and accelerate breakdown reactions to provide ATP for counteracting stress.
6. Exhaustion results from depletion of body resources during the resistance stage.
7. Stress may trigger certain diseases by inhibiting the immune system. An important link between stress and immunity is interleukin-l, produced by macrophages; it stimulates secretion of ACTH.

Resource:
Animation - General Adaptation Syndrome
Concepts and Connections - General Adaptation Syndrome

18.15 Development of the Endocrine System

1. The development of the endocrine system is not as localized as in other systems because endocrine organs develop in widely separated parts of the embryo.
2. The pituitary gland, adrenal medulla, and pineal gland develop from ectoderm; the adrenal cortex develops from mesoderm; and the thyroid gland, parathyroid glands, pancreas, and thymus develop from endoderm.

Review	Resource
18.16 Aging and the Endocrine System	
1. Although some endocrine glands shrink as we get older, their performance may or may not be compromised. 2. Production of human growth hormone, thyroid hormones, cortisol, aldosterone, and estrogens decreases with advancing age. 3. With aging, the blood levels of TSH, LH, FSH, and PTH rise. 4. The pancreas releases insulin more slowly with age, and receptor sensitivity to glucose declines. 5. After puberty, thymus size begins to decrease, and thymic tissue is replaced by adipose and areolar connective tissue.	

SELF-QUIZ QUESTIONS

Fill in the blanks in the following statements.

1. The three stages of the stress response or general adaptation syndrome, in order of occurrence, are ______, ______ , and ______.

2. The ______ is the major integrating link between the nervous and endocrine systems, acts as an endocrine gland itself, and helps control the stress response.

3. Down-regulation makes a target cell ______ sensitive to a hormone while up-regulation makes a target cell ______ sensitive to a hormone.

Indicate whether the following statements are true or false.

4. If the effect of two or more hormones acting together is greater than the sum of each acting alone, then the two hormones are said to have a permissive effect.

5. In the direct gene activation method of hormone action, the hormone enters the target cell and binds to an intracellular receptor. The activated receptor–hormone complex then alters gene expression to produce the protein that causes the physiological responses that are characteristic of the hormone.

Choose the one best answer to the following questions.

6. Which of the following comparisons are *true*? (1) Nerve impulses produce their effects quickly; hormonal responses generally are slower. (2) Nervous system effects are brief; endocrine system effects are longer lasting. (3) The nervous system controls homeostasis through mediator molecules called neurotransmitters; the endocrine system works through mediator molecules called hormones. (4) The nervous system can stimulate or inhibit the release of hormones; some hormones are released by neurons as neurotransmitters. (5) Neurotransmitters transmit impulses directly; hormones must bind to receptors on or in target cells in order to exert their effects.
(a) 1, 2, 3, 4, and 5 (b) 1, 2, 3, and 4 (c) 2, 3, 4, and 5
(d) 2, 4, and 5 (e) 1, 4, and 5

7. Insulin and thyroxine arrive at an organ at the same time. Thyroxine causes an effect on the organ but insulin does not. Why?
(a) Insulin is a lipid-soluble hormone and thyroxine is not.
(b) The target cells in the organ have up-regulated for thyroxine.
(c) Thyroxine is a local hormone and insulin is a circulating hormone.
(d) Thyroxine inhibits the action of insulin.
(e) The organ's cells have receptors for thyroxine but not for insulin.

8. Which of the following is not a category of water-soluble hormones?
(a) peptides (b) amines (c) eicosanoids
(d) steroids (e) proteins

9. Place in correct order the action of a water-soluble hormone on its target cell. (1) Adenylate cyclase is activated, catalyzing the conversion of ATP to cAMP. (2) Enzymes catalyze reactions that produce a physiological response attributed to the hormone. (3) The hormone binds to a membrane receptor. (4) Activated protein kinases phosphorylate cellular proteins. (5) The hormone–receptor complex activates G proteins. (6) Cyclic AMP activates protein kinases.
(a) 3, 5, 1, 6, 4, 2 (b) 3, 1, 5, 6, 4, 2 (c) 5, 1, 4, 2, 3, 6
(d) 3, 4, 5, 1, 6, 2 (e) 6, 3, 5, 1, 4, 2

10. Hormones (1) generally utilize negative feedback mechanisms to regulate their secretion; (2) will only affect target cells far removed from the hormone-producing secretory cells; (3) must bind to transport proteins in order to circulate in the blood; (4) may be released in low concentrations but can produce large effects in the target cells because of amplification; (5) can regulate the responsiveness of the target tissue by controlling the number of receptor sites for the hormone.
(a) 1, 2, and 3 (b) 1, 2, 4, and 5 (c) 2, 3, and 4
(d) 2, 3, 4, and 5 (e) 1, 4, and 5

11. The pituitary gland (1) is located in the cribriform plate of the ethmoid bone, (2) is linked to the hypothalamus by the infundibulum, (3) has a posterior portion that contains axon terminals from hypothalamic neurosecretory cells, (4) produces releasing and inhibiting hormones, (5) has a vascular connection with the hypothalamus known as the hypophyseal portal system.
(a) 1, 2, and 4 (b) 2, 3, 4, and 5 (c) 2, 3, and 5
(d) 1, 2, 3, 4, and 5 (e) 2, 4, and 5

12. The class of adrenal gland hormones that provides resistance to stress, produces anti-inflammatory effects, and promotes normal metabolism to ensure adequate quantities of ATP is
(a) glucocorticoids.
(b) mineralocorticoids.
(c) androgens.
(d) catecholamines.
(e) gonadocorticoids.

13. Match the following:

____ (a) increases blood Ca^{2+} level
____ (b) increases blood glucose level
____ (c) decreases blood Ca^{2+} level
____ (d) decreases blood glucose level
____ (e) initiates and maintains milk secretion by the mammary glands
____ (f) regulates the body's biological clock
____ (g) stimulates sex hormone production; triggers ovulation
____ (h) augments fight-or-flight responses
____ (i) regulates metabolism and resistance to stress
____ (j) helps control water and electrolyte homeostasis
____ (k) suppresses release of FSH
____ (l) stimulates growth of axillary and pubic hair
____ (m) promotes T cell maturation
____ (n) regulates oxygen use, basal metabolic rate, cellular metabolism, and growth and development
____ (o) stimulates protein synthesis, inhibits protein breakdown, stimulates lipolysis, and retards use of glucose for ATP production
____ (p) inhibits water loss through the kidneys
____ (q) stimulates egg and sperm formation
____ (r) enhances uterine contractions during labor; stimulates milk ejection
____ (s) stimulates and inhibits secretion of anterior pituitary hormones
____ (t) increases skin pigmentation when present in excess
____ (u) stimulates synthesis and release of T_3 and T_4
____ (v) local hormones involved in inflammation, smooth muscle contraction, blood flow, and inflammation

(1) insulin
(2) glucagon
(3) inhibin
(4) follicle-stimulating hormone
(5) luteinizing hormone
(6) thyroxine and triiodothyronine
(7) calcitonin
(8) parathormone
(9) melanocyte-stimulating hormone
(10) oxytocin
(11) antidiuretic hormone
(12) prolactin
(13) human growth hormone
(14) hypothalamic regulating hormones
(15) aldosterone
(16) thyroid-stimulating hormone
(17) androgens
(18) epinephrine and norepinephrine
(19) prostaglandins
(20) melatonin
(21) thymosin
(22) cortisol

14. Match the following hormone-secreting cells to the hormone(s) they release:

____ (a) ACTH and MSH
____ (b) TSH
____ (c) glucagon
____ (d) PTH
____ (e) glucocorticoids
____ (f) calcitonin
____ (g) insulin
____ (h) androgens
____ (i) progesterone
____ (j) FSH and LH
____ (k) epinephrine and norepinephrine
____ (l) hGH
____ (m) testosterone
____ (n) mineralocorticoids
____ (o) thyroxine and triiodothyronine
____ (p) PRL

(1) beta cells of pancreatic islet
(2) alpha cells of pancreatic islet
(3) follicular cells of thyroid gland
(4) parafollicular cells of thyroid gland
(5) testes
(6) ovary
(7) somatotrophs
(8) thyrotrophs
(9) gonadotrophs
(10) corticotrophs
(11) lactotrophs
(12) chief cells
(13) chromaffin cells
(14) zona glomerulosa cells
(15) zona fasciculata cells
(16) zona reticularis cells

15. Match the endocrine disorder to the problem that produced the disorder:

____ (a) hyposecretion of insulin or down-regulation of insulin receptors
____ (b) hypersecretion of hGH before closure of epiphyseal plates
____ (c) hyposecretion of thyroid hormone that is present at birth
____ (d) hypersecretion of glucocorticoids
____ (e) hyposecretion of hGH before closure of epiphyseal plates
____ (f) hypersecretion of epinephrine and norepinephrine
____ (g) hypersecretion of hGH after closure of epiphyseal plates
____ (h) hyposecretion of glucocorticoids and aldosterone
____ (i) hyposecretion of ADH
____ (j) hypersecretion of melatonin
____ (k) hyposecretion of thyroid hormone in adults
____ (l) hyperthyroidism, an autoimmune disease

(1) giantism
(2) acromegaly
(3) pituitary dwarfism
(4) diabetes insipidus
(5) myxedema
(6) Graves' disease
(7) Cushing's syndrome
(8) seasonal affective disorder
(9) Addison's disease
(10) pheochromocytomas
(11) congenital hypothyroidism
(12) diabetes mellitus

CRITICAL THINKING QUESTIONS

1. Amanda hates her new student ID photo. Her hair looks dry, the extra weight that she's gained shows, and her neck looks fat. In fact, there's an odd butterfly-shaped swelling across the front of her neck, under her chin. Amanda's also been feeling very tired and mentally "dull" lately, but she figures all new A&P students feel that way. Should she visit the clinic or just wear turtlenecks?

2. Amanda (from question 1 above) goes to the clinic and blood is drawn. The results show that her T_4 levels are low and her TSH levels are low. Later she is given a TSH stimulation test in which TSH is injected and the T_4 levels are monitored. After TSH injection, her T_4 levels rise. Does Amanda have problems with her pituitary or with her thyroid gland? How did you come to your conclusion?

3. Mr. Hernandez visited his doctor complaining that he is constantly thirsty and is "in the bathroom day and night" relieving his bladder. The doctor ordered blood and urine tests to check for glucose and ketones, which were all negative. What is the doctor's diagnosis of Mr. Hernandez, and what gland(s) or organ(s) is(are) involved?

? ANSWERS TO FIGURE QUESTIONS

18.1 Secretions of endocrine glands diffuse into interstitial fluid and then into the blood; exocrine secretions flow into ducts that lead into body cavities or to the body surface.

18.2 In the stomach, histamine is a paracrine because it acts on nearby parietal cells without entering the blood.

18.3 The receptor–hormone complex alters gene expression by turning specific genes of nuclear DNA on or off.

18.4 Cyclic AMP is termed a second messenger because it translates the presence of the first messenger, the water-soluble hormone, into a response inside the cell.

18.5 The hypophyseal portal veins carry blood from the median eminence of the hypothalamus, where hypothalamic releasing and inhibiting hormones are secreted, to the anterior pituitary, where these hormones act.

18.6 Thyroid hormones suppress secretion of TSH by thyrotrophs and of TRH by hypothalamic neurosecretory cells; gonadal hormones suppress secretion of FSH and LH by gonadotrophs and of GnRH by hypothalamic neurosecretory cells.

18.7 Excess levels of hGH would cause hyperglycemia.

18.8 Functionally, both the hypothalamohypophyseal tract and the hypophyseal portal veins carry hypothalamic hormones to the pituitary gland. Structurally, the tract is composed of axons of neurons that extend from the hypothalamus to the posterior pituitary; the portal veins are blood vessels that extend from the hypothalamus to the anterior pituitary.

18.9 Absorption of a liter of water in the intestines would decrease the osmotic pressure of your blood plasma, turning off secretion of ADH and decreasing the ADH level in your blood.

18.10 Follicular cells secrete T_3 and T_4, also known as thyroid hormones. Parafollicular cells secrete calcitonin.

18.11 The storage form of thyroid hormones is thyroglobulin.

18.12 Lack of iodine in the diet → diminished production of T_3 and T_4 → increased release of TSH → growth (enlargement) of the thyroid gland → goiter.

18.13 Parafollicular cells of the thyroid gland secrete calcitonin; chief (principal) cells of the parathyroid gland secrete PTH.

18.14 Target tissues for PTH are bones and the kidneys; target tissue for CT is bone; target tissue for calcitriol is the GI tract.

18.15 The adrenal glands are superior to the kidneys in the retroperitoneal space.

18.16 Angiotensin II acts to constrict blood vessels by causing contraction of vascular smooth muscle, and it stimulates secretion of aldosterone (by zona glomerulosa cells of the adrenal cortex), which in turn causes the kidneys to conserve water and thereby increase blood volume.

18.17 A transplant recipient who takes prednisone will have low blood levels of ACTH and CRH due to negative feedback suppression of the anterior pituitary and hypothalamus by the prednisone.

18.18 The pancreas is both an endocrine and an exocrine gland.

18.19 Glycogenolysis is the conversion of glycogen into glucose and therefore it increases blood glucose level.

18.20 Homeostasis maintains controlled conditions typical of a normal internal environment; the stress response resets controlled conditions at a different level to cope with various stressors.

18.21 The adrenal cortex of the adrenal gland is derived from mesoderm, while the adrenal medulla of the adrenal gland is derived from ectoderm.

18.22 Antibodies that mimic the action of TSH are produced in Graves' disease.

Appendix A

MEASUREMENTS

U.S. Customary System

PARAMETER	UNIT	RELATION TO OTHER U.S. UNITS	SI (METRIC) EQUIVALENT
Length	inch	1/12 foot	2.54 centimeters
foot	12 inches	0.305 meter	
yard	36 inches	0.914 meter	
mile	5,280 feet	1.609 kilometers	
Mass	grain	1/1,000 pound	64.799 milligrams
dram	1/16 ounce	1.772 grams	
ounce	16 drams	28.350 grams	
pound	16 ounces	453.6 grams	
ton	2,000 pounds	907.18 kilograms	
Volume (Liquid)	ounce	1/16 pint	29.574 milliliters
pint	16 ounces	0.473 liter	
quart	2 pints	0.946 liter	
gallon	4 quarts	3.785 liters	
Volume (Dry)	pint	1/2 quart	0.551 liter
quart	2 pints	1.101 liters	
peck	8 quarts	8.810 liters	
bushel	4 pecks	35.239 liters	

International System (SI)

BASE UNITS			PREFIXES		
UNIT	QUANTITY	SYMBOL	PREFIX	MULTIPLIER	SYMBOL
meter	length	m	tera-	$10^{12} = 1{,}000{,}000{,}000{,}000$	T
kilogram	mass	kg	giga-	$10^{9} = 1{,}000{,}000{,}000$	G
second	time	s	mega-	$10^{6} = 1{,}000{,}000$	M
liter	volume	L	kilo-	$10^{3} = 1{,}000$	k
mole	amount of matter	mol	hecto-	$10^{2} = 100$	h
			deca-	$10^{1} = 10$	da
			deci-	$10^{-1} = 0.1$	d
			centi-	$10^{-2} = 0.01$	c
			milli-	$10^{-3} = 0.001$	m
			micro-	$10^{-6} = 0.000{,}001$	μ
			nano-	$10^{-9} = 0.000{,}000{,}001$	n
			pico-	$10^{-12} = 0.000{,}000{,}000{,}001$	p

Temperature Conversion

FAHRENHEIT (F) TO CELSIUS (C)

$$°C = (°F - 32) \div 1.8$$

CELSIUS (C) TO FAHRENHEIT (F)

$$°F = (°C \times 1.8) + 32$$

U.S. to SI (Metric) Conversion

WHEN YOU KNOW	MULTIPLY BY	TO FIND
inches	2.54	centimeters
feet	30.48	centimeters
yards	0.91	meters
miles	1.61	kilometers
ounces	28.35	grams
pounds	0.45	kilograms
tons	0.91	metric tons
fluid ounces	29.57	milliliters
pints	0.47	liters
quarts	0.95	liters
gallons	3.79	liters

SI (Metric) to U.S. Conversion

WHEN YOU KNOW	MULTIPLY BY	TO FIND
millimeters	0.04	inches
centimeters	0.39	inches
meters	3.28	feet
kilometers	0.62	miles
liters	1.06	quarts
cubic meters	35.32	cubic feet
grams	0.035	ounces
kilograms	2.21	pounds

PERIODIC TABLE

The periodic table lists the known **chemical elements,** the basic units of matter. The elements in the table are arranged left-to-right in rows in order of their **atomic number,** the number of protons in the nucleus. Each horizontal row, numbered from 1 to 7, is a **period.** All elements in a given period have the same number of electron shells as their period number. For example, an atom of hydrogen or helium each has one electron shell, while an atom of potassium or calcium each has four electron shells. The elements in each column, or **group,** share chemical properties. For example, the elements in column IA are very chemically reactive, whereas the elements in column VIIIA have full electron shells and thus are chemically inert.

Scientists now recognize 117 different elements; 92 occur naturally on Earth, and the rest are produced from the natural elements using particle accelerators or nuclear reactors. Elements are designated by **chemical symbols,** which are the first one or two letters of the element's name in English, Latin, or another language.

Twenty-six of the 92 naturally occurring elements normally are present in your body. Of these, just four elements—oxygen (O), carbon (C), hydrogen (H), and nitrogen (N) (coded blue)—constitute about 96% of the body's mass. Eight others—calcium (Ca), phosphorus (P), potassium (K), sulfur (S), sodium (Na), chlorine (Cl), magnesium (Mg), and iron (Fe) (coded pink)—contribute 3.8% of the body's mass. An additional 14 elements, called **trace elements** because they are present in tiny amounts, account for the remaining 0.2% of the body's mass. The trace elements are aluminum, boron, chromium, cobalt, copper, fluorine, iodine, manganese, molybdenum, selenium, silicon, tin, vanadium, and zinc (coded yellow). Table 2.1 on page 30 provides information about the main chemical elements in the body.

23 ← Atomic number
V ← Chemical symbol
50.942 ← Atomic mass (weight)

Percentage of body mass
- 96% (4 elements)
- 3.8% (8 elements)
- 0.2% (14 elements)

	IA	IIA	IIIB	IVB	VB	VIB	VIIB	VIIIB			IB	IIB	IIIA	IVA	VA	VIA	VIIA	VIIIA
1	1 Hydrogen **H** 1.0079																	2 Helium **He** 4.003
2	3 Lithium **Li** 6.941	4 Beryllium **Be** 9.012											5 Boron **B** 10.811	6 Carbon **C** 12.011	7 Nitrogen **N** 14.007	8 Oxygen **O** 15.999	9 Fluorine **F** 18.998	10 Neon **Ne** 20.180
3	11 Sodium **Na** 22.989	12 Magnesium **Mg** 24.305											13 Aluminum **Al** 26.9815	14 Silicon **Si** 28.086	15 Phosphorus **P** 30.974	16 Sulfur **S** 32.066	17 Chlorine **Cl** 35.453	18 Argon **Ar** 39.948
4	19 Potassium **K** 39.098	20 Calcium **Ca** 40.08	21 Scandium **Sc** 44.956	22 Titanium **Ti** 47.87	23 Vanadium **V** 50.942	24 Chromium **Cr** 51.996	25 Manganese **Mn** 54.938	26 Iron **Fe** 55.845	27 Cobalt **Co** 58.933	28 Nickel **Ni** 58.69	29 Copper **Cu** 63.546	30 Zinc **Zn** 65.38	31 Gallium **Ga** 69.723	32 Germanium **Ge** 72.59	33 Arsenic **As** 74.992	34 Selenium **Se** 78.96	35 Bromine **Br** 79.904	36 Krypton **Kr** 83.80
5	37 Rubidium **Rb** 85.468	38 Strontium **Sr** 87.62	39 Yttrium **Y** 88.905	40 Zirconium **Zr** 91.22	41 Niobium **Nb** 92.906	42 Molybdenum **Mo** 95.94	43 Technetium **Tc** (99)	44 Ruthenium **Ru** 101.07	45 Rhodium **Rh** 102.905	46 Palladium **Pd** 106.42	47 Silver **Ag** 107.868	48 Cadmium **Cd** 112.40	49 Indium **In** 114.82	50 Tin **Sn** 118.69	51 Antimony **Sb** 121.75	52 Tellurium **Te** 127.60	53 Iodine **I** 126.904	54 Xenon **Xe** 131.30
6	55 Cesium **Cs** 132.905	56 Barium **Ba** 137.33		72 Hafnium **Hf** 178.49	73 Tantalum **Ta** 180.948	74 Tungsten **W** 183.85	75 Rhenium **Re** 186.2	76 Osmium **Os** 190.2	77 Iridium **Ir** 192.22	78 Platinum **Pt** 195.08	79 Gold **Au** 196.967	80 Mercury **Hg** 200.59	81 Thallium **Tl** 204.38	82 Lead **Pb** 207.19	83 Bismuth **Bi** 208.980	84 Polonium **Po** (209)	85 Astatine **At** (210)	86 Radon **Rn** (222)
7	87 Francium **Fr** (223)	88 Radium **Ra** (226)		104 Rutherfordium **Rf** (267)	105 Dubnium **Db** (268)	106 Seaborgium **Sg** (271)	107 Bohrium **Bh** (272)	108 Hassium **Hs** (270)	109 Meitnerium **Mt** (276)	110 **Ds** (281)	111 **Rg** (280)	112 **Uub** (285)	113 **Uut** (284)	114 **Uuq** (289)	115 **Uup** (288)	116 **Uuh** (293)		118 **Uuo** (294)

57–71, Lanthanides

57 Lanthanum **La** 138.91	58 Cerium **Ce** 140.12	59 Praseodymium **Pr** 140.907	60 Neodymium **Nd** 144.24	61 Promethium **Pm** 144.913	62 Samarium **Sm** 150.35	63 Europium **Eu** 151.96	64 Gadolinium **Gd** 157.25	65 Terbium **Tb** 158.925	66 Dysprosium **Dy** 162.50	67 Holmium **Ho** 164.930	68 Erbium **Er** 167.26	69 Thulium **Tm** 168.934	70 Ytterbium **Yb** 173.04	71 Lutetium **Lu** 174.97

89–103, Actinides

89 Actinium **Ac** (227)	90 Thorium **Th** 232.038	91 Protactinium **Pa** (231)	92 Uranium **U** 238.03	93 Neptunium **Np** (237)	94 Plutonium **Pu** 244.064	95 Americium **Am** (243)	96 Curium **Cm** (247)	97 Berkelium **Bk** (247)	98 Californium **Cf** 242.058	99 Einsteinium **Es** (254)	100 Fermium **Fm** 257.095	101 Mendelevium **Md** 258.10	102 Nobelium **No** 259.10	103 Lawrencium **Lr** 260.105

NORMAL VALUES FOR SELECTED BLOOD TESTS

The system of international (SI) units (Système Internationale d'Unités) is used in most countries and in many medical and scientific journals. Clinical laboratories in the United States, by contrast, usually report values for blood and urine tests in conventional units. The laboratory values in this Appendix give conventional units first, followed by SI equivalents in parentheses. Values listed for various blood tests should be viewed as reference values rather than absolute "normal" values for all well people. Values may vary due to age, gender, diet, and environment of the subject or the equipment, methods, and standards of the lab performing the measurement.

KEY TO SYMBOLS

g = gram	mL = milliliter
mg = milligram = 10^{-3} gram	μL = microliter
μg = microgram = 10^{-6} gram	mEq/L = milliequivalents per liter
U = units	mmol/L = millimoles per liter
L = liter	μmol/L = micromoles per liter
dL = deciliter	> = greater than; < = less than

Blood Tests

TEST (SPECIMEN)	U.S. REFERENCE VALUES (SI UNITS)	VALUES INCREASE IN	VALUES DECREASE IN
Aminotransferases (serum)			
Alanine aminotransferase (ALT)	0–35 U/L (same)	Liver disease or liver damage due to toxic drugs.	
Aspartate aminotransferase (AST)	0–35 U/L (same)	Myocardial infarction, liver disease, trauma to skeletal muscles, severe burns.	Beriberi, uncontrolled diabetes mellitus with acidosis, pregnancy.
Ammonia (plasma)	20–120 μg/dL (12–55 μmol/L)	Liver disease, heart failure, emphysema, pneumonia, hemolytic disease of newborn.	Hypertension.
Bilirubin (serum)	Conjugated: <0.5 mg/dL (<5.0 μmol/L)	Conjugated bilirubin: liver dysfunction or gallstones.	
	Unconjugated: 0.2–1.0 mg/dL (18–20 μmol/L) Newborn: 1.0–12.0 mg/dL (<200 μmol/L)	Unconjugated bilirubin: excessive hemolysis of red blood cells.	
Blood urea nitrogen (BUN) (serum)	8–26 mg/dL (2.9–9.3 mmol/L)	Kidney disease, urinary tract obstruction, shock, diabetes, burns, dehydration, myocardial infarction.	Liver failure, malnutrition, overhydration, pregnancy.
Carbon dioxide content (bicarbonate + dissolved CO_2) (whole blood)	Arterial: 19–24 mEq/L (19–24 mmol/L) Venous: 22–26 mEq/L (22–26 mmol/L)	Severe diarrhea, severe vomiting, starvation, emphysema, aldosteronism.	Renal failure, diabetic ketoacidosis, shock.

Blood Tests Continued

TEST (SPECIMEN)	U.S. REFERENCE VALUES (SI UNITS)	VALUES INCREASE IN	VALUES DECREASE IN
Cholesterol, total (plasma) **HDL cholesterol** (plasma) **LDL cholesterol** (plasma)	<200 mg/dL (<5.2 mmol/L) is desirable >40 mg/dL (>1.0 mmol/L) is desirable <130 mg/dL (<3.2 mmol/L) is desirable	Hypercholesterolemia, uncontrolled diabetes mellitus, hypothyroidism, hypertension, atherosclerosis, nephrosis.	Liver disease, hyperthyroidism, fat malabsorption, pernicious or hemolytic anemia, severe infections.
Creatine (serum)	Males: 0.15–0.5 mg/dL (10–40 μmol/L) Females: 0.35–0.9 mg/dL (30–70 μmol/L)	Muscular dystrophy, damage to muscle tissue, electric shock, chronic alcoholism.	
Creatine kinase (CK), also known as **creatine phosphokinase (CPK)** (serum)	0–130 U/L (same)	Myocardial infarction, progressive muscular dystrophy, hypothyroidism, pulmonary edema.	
Creatinine (serum)	0.5–1.2 mg/dL (45–105 μmol/L)	Impaired renal function, urinary tract obstruction, giantism, acromegaly.	Decreased muscle mass, as occurs in muscular dystrophy or myasthenia gravis.
Electrolytes (plasma)	See Table 27.2 on page 1119		
Gamma-glutamyl transferase (GGT) (serum)	0–30 U/L (same)	Bile duct obstruction, cirrhosis, alcoholism, metastatic liver cancer, congestive heart failure.	
Glucose (plasma)	70–110 mg/dL (3.9–6.1 mmol/L)	Diabetes mellitus, acute stress, hyperthyroidism, chronic liver disease, Cushing's disease.	Addison's disease, hypothyroidism, hyperinsulinism.
Hemoglobin (whole blood)	Males: 14–18 g/100 mL (140–180 g/L) Females: 12–16 g/100 mL (120–160 g/L) Newborns: 14–20 g/100 mL (140–200 g/L)	Polycythemia, congestive heart failure, chronic obstructive pulmonary disease, living at high altitude.	Anemia, severe hemorrhage, cancer, hemolysis, Hodgkin's disease, nutritional deficiency of vitamin B_{12}, systemic lupus erythematosus, kidney disease.
Iron, total (serum)	Males: 80–180 μg/dL (14–32 μmol/L) Females: 60–160 μg/dL (11–29 μmol/L)	Liver disease, hemolytic anemia, iron poisoning.	Iron-deficiency anemia, chronic blood loss, pregnancy (late), chronic heavy menstruation.
Lactic dehydrogenase (LDH) (serum)	71–207 U/L (same)	Myocardial infarction, liver disease, skeletal muscle necrosis, extensive cancer.	
Lipids (serum) **Total** **Triglycerides**	 400–850 mg/dL (4.0–8.5 g/L) 10–190 mg/dL (0.1–1.9 g/L)	Hyperlipidemia, diabetes mellitus.	Fat malabsorption, hypothyroidism.
Platelet (thrombocyte) count (whole blood)	150,000–400,000/μL	Cancer, trauma, leukemia, cirrhosis.	Anemias, allergic conditions, hemorrhage.
Protein (serum) **Total** **Albumin** **Globulin**	 6–8 g/dL (60–80 g/L) 4–6 g/dL (40–60 g/L) 2.3–3.5 g/dL (23–35 g/L)	Dehydration, shock, chronic infections.	Liver disease, poor protein intake, hemorrhage, diarrhea, malabsorption, chronic renal failure, severe burns.
Red blood cell (erythrocyte) count (whole blood)	Males: 4.5–6.5 million/μL Females: 3.9–5.6 million/μL	Polycythemia, dehydration, living at high altitude.	Hemorrhage, hemolysis, anemias, cancer, overhydration.
Uric acid (urate) (serum)	2.0–7.0 mg/dL (120–420) μmol/L	Impaired renal function, gout, metastatic cancer, shock, starvation.	
White blood cell (leukocyte) count, total (whole blood)	5,000–10,000/μL (See Table 19.3 on page 742 for relative percentages of different types of WBCs.)	Acute infections, trauma, malignant diseases, cardiovascular diseases. (See also Table 19.2 on page 741.)	Diabetes mellitus, anemia. (See also Table 19.2 on page 741.)

Appendix D

NORMAL VALUES FOR SELECTED URINE TESTS

Urine Tests

TEST (SPECIMEN)	U.S. REFERENCE VALUES (SI UNITS)	CLINICAL IMPLICATIONS
Amylase (2 hour)	35–260 Somogyi units/hr (6.5–48.1 units/hr)	Values increase in inflammation of the pancreas (pancreatitis) or salivary glands, obstruction of the pancreatic duct, and perforated peptic ulcer.
Bilirubin* (random)	Negative	Values increase in liver disease and obstructive biliary disease.
Blood* (random)	Negative	Values increase in renal disease, extensive burns, transfusion reactions, and hemolytic anemia.
Calcium (Ca^{2+}) (random)	10 mg/dL (2.5 mmol/liter); up to 300 mg/24 hr (7.5 mmol/24 hr)	Amount depends on dietary intake; values increase in hyperparathyroidism, metastatic malignancies, and primary cancer of breasts and lungs; values decrease in hypoparathyroidism and vitamin D deficiency.
Casts (24 hour)		
Epithelial	Occasional	Values increase in nephrosis and heavy metal poisoning.
Granular	Occasional	Values increase in nephritis and pyelonephritis.
Hyaline	Occasional	Values increase in kidney infections.
Red blood cell	Occasional	Values increase in glomerular membrane damage and fever.
White blood cell	Occasional	Values increase in pyelonephritis, kidney stones, and cystitis.
Chloride (Cl^-) (24 hour)	140–250 mEq/24 hr (140–250 mmol/24 hr)	Amount depends on dietary salt intake; values increase in Addison's disease, dehydration, and starvation; values decrease in pyloric obstruction, diarrhea, and emphysema.
Color (random)	Yellow, straw, amber	Varies with many disease states, hydration, and diet.
Creatinine (24 hour)	Males: 1.0–2.0 g/24 hr (9–18 mmol/24 hr) Females: 0.8–1.8 g/24 hr (7–16 mmol/24 hr)	Values increase in infections; values decrease in muscular atrophy, anemia, and kidney diseases.
Glucose*	Negative	Values increase in diabetes mellitus, brain injury, and myocardial infarction.
Hydroxycorticosteroids (17-hydroxysteroids) (24 hour)	Males: 5–15 mg/24 hr (13–41 μmol/24 hr) Females: 2–13 mg/24 hr (5–36 μmol/24 hr)	Values increase in Cushing's syndrome, burns, and infections; values decrease in Addison's disease.
Ketone bodies* (random)	Negative	Values increase in diabetic acidosis, fever, anorexia, fasting, and starvation.
17-ketosteroids (24 hour)	Males: 8–25 mg/24 hr (28–87 μmol/24 hr) Females: 5–15 mg/24 hr (17–53 μmol/24 hr)	Values decrease in surgery, burns, infections, adrenogenital syndrome, and Cushing's syndrome.

TEST (SPECIMEN)	U.S. REFERENCE VALUES (SI UNITS)	CLINICAL IMPLICATIONS
Odor (random)	Aromatic	Becomes acetonelike in diabetic ketosis.
Osmolality (24 hour)	500–1400 mOsm/kg water (500–1400 mmol/kg water)	Values increase in cirrhosis, congestive heart failure (CHF), and high-protein diets; values decrease in aldosteronism, diabetes insipidus, and hypokalemia.
pH* (random)	4.6–8.0	Values increase in urinary tract infections and severe alkalosis; values decrease in acidosis, emphysema, starvation, and dehydration.
Phenylpyruvic acid (random)	Negative	Values increase in phenylketonuria (PKU).
Potassium (K^+) (24 hour)	40–80 mEq/24 hr (40–80 mmol/24 hr)	Values increase in chronic renal failure, dehydration, starvation, and Cushing's syndrome; values decrease in diarrhea, malabsorption syndrome, and adrenal cortical insufficiency.
Protein* (albumin) (random)	Negative	Values increase in nephritis, fever, severe anemias, trauma, and hyperthyroidism.
Sodium (Na^+) (24 hour)	75–200 mg/24 hr (75–200 mmol/24 hr)	Amount depends on dietary salt intake; values increase in dehydration, starvation, and diabetic acidosis; values decrease in diarrhea, acute renal failure, emphysema, and Cushing's syndrome.
Specific gravity* (random)	1.001–1.035 (same)	Values increase in diabetes mellitus and excessive water loss; values decrease in absence of antidiuretic hormone (ADH) and severe renal damage.
Urea (random)	25–35 g/24 hr (420–580 mmol/24 hr)	Values increase in response to increased protein intake; values decrease in impaired renal function.
Uric acid (24 hour)	0.4–1.0 g/24 hr (1.5–4.0 mmol/24 hr)	Values increase in gout, leukemia, and liver disease; values decrease in kidney disease.
Urobilinogen* (2 hour)	0.3–1.0 Ehrlich units (1.7–6.0 μmol/24 hr)	Values increase in anemias, hepatitis A (infectious), biliary disease, and cirrhosis; values decrease in cholelithiasis and renal insufficiency.
Volume, total (24 hour)	1000–2000 mL/24 hr (1.0–2.0 liters/24 hr)	Varies with many factors.

* Test often performed using a **dipstick,** a plastic strip impregnated with chemicals that is dipped into a urine specimen to detect particular substances. Certain colors indicate the presence or absence of a substance and sometimes give a rough estimate of the amount(s) present.

Appendix E

ANSWERS

Answers to Self-Quiz Questions

Chapter 1 1. tissue 2. metabolism, anabolism, catabolism 3. intracellular fluid (ICF), extracellular fluid (ECF) 4. true 5. false 6. false 7. e 8. d 9. a 10. c 11. c 12. (a) 4, (b) 6, (c) 8, (d) 1, (e) 9, (f) 5, (g) 2, (h) 7, (i) 3, (j) 10 13. (a) 1, (b) 12, (c) 1, 6, (d) 6, (e) , (f) 8, (g) 7 (h) 3, (i) 2, (j) 10 14. (a) 4, (b) 1, (c) 3, (d) 6, (e) 5, (f) 7, (g) 2 15. (a) 6, (b) 1, (c) 11, (d) 5, (e) 10, (f) 8, (g) 7, (h) 9, (i) 4, (j) 3, (k) 2

Chapter 2 1. 8 2. solid, liquid, gas 3. monosaccharides, amino acids 4. true 5. false 6. true 7. c 8. a 9. d 10. b 11. e 12. a 13. e 14. (a) 1, (b) 2, (c) 1, (d) 4, (e) 3 15. (a) 11, (b) 1, (c) 8, (d) 3, (e) 7, (f) 4, (g) 5, (h) 9, (i) 10, (j) 12, (k) 6, (l) 2

Chapter 3 1. plasma membrane, cytoplasm, nucleus 2. apoptosis, necrosis 3. Telomeres 4. UAG 5. false 6. true 7. true 8. e 9. c 10. c, g, i, b, d, k, f, j, a, e, h 11. a 12. c 13. (a) 2, (b) 3, (c) 5, (d) 7, (e) 6, (f) 8, (g) 1, (h) 4 14. (a) 2, (b) 9, (c) 3, (d) 5, (e) 11, (f) 8, (g) 1, (h) 6, (i) 10, (j) 7, (k) 13, (l) 4, (m) 12 15. (a) 3, (b) 9, (c) 1, (d) 5, (e) 11, (f) 4, (g) 8, (h) 7, (i) 2, (j) 10, (k) 6

Chapter 4 1. epithelial, connective, muscle, nervous 2. arrangement of cells in layers, cell shape 3. true 4. true 5. e 6. b 7. a 8. c 9. e 10. b 11. d 12. c 13. (a) 4, (b) 8, (c) 5, (d) 2, (e) 6, (f) 3, (g) 1, (h) 7 14. (a) C, (b) M, (c) N, (d) E, (e) C, (f) E, (g) M, (h) E, (i) C, (j) M, (k) N, (l) E, (m) C, (n) E, (o) M and N 15. (a) 3, (b) 5, (c) 8, (d) 13, (e) 9, (f) 7, (g) 11, (h) 6, (i) 2, (j) 4, (k) 10, (l) 12, (m) 1

Chapter 5 1. stratum lucidum 2. eccrine, ceruminous, apocrine 3. false 4. true 5. c 6. e 7. a 8. c 9. b 10. e 11. a 12. c 13. (a) 3, (b) 5, (c) 4, (d) 1, (e) 6, (f) 11, (g) 2, (h) 8, (i) 9, (j) 10, (k) 7 14. (a) 3, (b) 4, (c) 1, (d) 2 15. (a) 4, (b) 3, (c) 2, (d) 1, inflammatory, migratory, proliferative, maturation

Chapter 6 1. interstitial, appositional 2. hardness, tensile strength 3. true 4. true 5. true 6. d 7. a 8. e 9. c 10. a 11. (a) 1, (b) 4, (c) 3, (d) 2 12. (a) 2, (b) 6, (c) 4, (d) 5, (e) 7, (f) 3, (g) 1 13. (a) 3, (b) 9, (c) 8, (d) 1, (e) 5, (f) 4, (g) 6, (h) 7, (i) 12, (j) 2, (k) 11 (l) 10 14. (a) 3, (b) 7, (c) 6, (d) 1, (e) 4, (f) 2, (g) 5, (h) 9, (i) 8, (j) 10 15. (a) 12, (b) 4, (c) 8, (d) 6, (e) 3, (f) 9, (g) 13, (h) 10, (i) 7, (j) 5, (k) 2, (l) 11, (m) 1

Chapter 7 1. fontanels 2. pituitary gland 3. sacrum, coccyx 4. false 5. false 6. c 7. c 8. a 9. e 10. d 11. e 12. (a) 4, (b) 9, (c) 7, (d) 5, (e) 3, (f) 1, (g) 2, (h) 8, (i) 6 13. (a) 7, (b) 5, (c) 1, (d) 6, (e) 2, (f) 4, (g) 8, (h) 9, (i) 3, (j) 10, (k) 11, (l) 13, (m) 12 14. (a) 2, (b) 3, (c) 5, (d) 6, (e) 4, (f) 1, (g) 5, (h) 4, (i) 2, (j) 4, (k) 3 15. (a) 3, (b) 1, (c) 6, (d) 9, (e) 13, (f) 12, (g) 2, (h) 4, (i) 5, (j) 7, (k) 10, (l) 15, (m) 8, (n) 11, (o) 14

Chapter 8 1. metacarpals 2. ilium, ischium, pubis 3. true (lesser), false (greater) 4. false 5. true 6. b 7. c 8. e 9. c 10. a 11. b and e 12. a 13. (a) 2, (b) 6, (c) 9, (d) 7, (e) 4, (f) 5, (g) 8, (h) 10, (i) 1, (j) 3 14. (a) 3, (b) 8, (c) 4, (d) 11, (e) 9, (f) 13, (g) 5, (h) 6, (i) 10, (j) 14, (k) 2, (l) 1, (m) 7, (n) 12 15. (a) 4, (b) 3, (c) 3, (d) 6, (e) 7, (f) 1, (g) 3, (h) 2, (i) 5, (j) 9, (k) 8, (l) 2, (m) 4, (n) 6, (o) 7, (p) 9, (q) 6, (r) 3, (s) 4, (t) 9, (u) 4 and 5

Chapter 9 1. joint, articulation or arthrosis 2. arthroplasty 3. false 4. false 5. false 6. e 7. d 8. b 9. c 10. a 11. c 12. e 13. (a) 5, (b) 3, (c) 7, (d) 2, (e) 6, (f) 4, (g) 1 14. (a) 6, (b) 4, (c) 5, (d) 1, (e) 3, (f) 2 15. (a) 8, (b) 11, (c) 10, (d) 13, (e) 15, (f) 9, (g) 6, (h) 12, (i) 3, (j) 4, (k) 16, (l) 2, (m) 18, (n) 1, (o) 7, (p) 14, (q) 17, (r) 5

Chapter 10 1. motor unit 2. muscular atrophy, fibrosis 3. acetylcholine 4. true 5. true 6. e 7. a 8. c 9. e 10. d 11. b 12. (a) 5, (b) 6, (c) 9, (d) 7, (e) 2, (f) 4, (g) 10, (h) 3, (i) 1, (j) 8 13. (a) 7, (b) 10, (c) 9, (d) 12, (e) 8, (f) 11, (g) 6, (h) 1, (i) 2, (j) 3, (k) 4, (l) 13, (m) 5 14. (a) 2, (b) 3, (c) 1, (d) 1 and 2, (e) 3, (f) 2, (g) 1, (h) 3, (i) 1 and 2, (j) 3, (k) 2 and 3, (l) 3 15. (a) 10, (b) 2, (c) 4, (d) 3, (e) 6, (f) 5, (g) 1, (h) 12, (i) 7, (j) 9, (k) 11, (l) 8

Chapter 11 1. buccinator 2. gastrocnemius, soleus, plantaris, calcaneus 3. true 4. true 5. b 6. c 7. c 8. a 9. e 10. e 11. (a) 6, (b) 2, (c) 8, (d) 5, (e) 3, (f) 1, (g) 7, (h) 4 12. (a) 13, (b) 9, (c) 8, (d) 6, (e) 3, (f) 11, (g) 10, (h) 1, (i) 2, (j) 7, (k) 12, (l) 4, (m) 5 13. (a) 6, (b) 3, (c) 7, (d) 4, (e) 2, (f) 9, (g) 5, (h) 1, (i) 8 14. (a) 10, (b) 1, (c) 9, (d) 8, (e) 12, (f) 17, (g) 2, (h) 6, (i) 8, (j) 14, (k) 5, (l) 4, (m) 2, (n) 15, (o) 1, (p) 11, (q) 13, (r) 12, (s) 7, (t) 16, (u) 11, (v) 17, (w) 16, (x) 15, (y) 3, (z) 10 15. (a) 3, (b) 1, (c) 2, (d) 1, (e) 2, (f) 3, (g) 3

Chapter 12 1. somatic, autonomic, enteric 2. sympathetic, parasympathetic 3. false 4. false 5. c 6. d 7. c 8. e 9. e 10. b 11. e 12. b 13. (a) 6, (b) 12, (c) 1, (d) 2, (e) 9, (f) 14, (g) 4, (h) 8, (i) 7, (j) 13, (k) 5, (l) 3, (m) 10, (n) 15, (o) 11 14. (a) 2, (b) 1, (c) 10, (d) 9, (e) 6, (f) 3, (g) 4, (h) 5, (i) 12, (j) 8, (k) 7, (l) 13, (m) 11 15. (a) 4, (b) 5, (c) 16, (d) 8, (e) 7, (f) 1, (g) 2, (h) 10, (i) 15, (j) 6, (k) 3, (l) 13, (m) 9, (n) 11, (o) 14, (p) 12

Chapter 13 1. mixed 2. sensory receptor, sensory neuron, integrating center, motor neuron, effector 3. true 4. false 5. c 6. c 7. a 8. c 9. d 10. e 11. a 12. d 13. (a) 1, (b) 8, (c) 4, (d) 2, (e) 11, (f) 1, (g) 6, (h) 5, (i) 3, (j) 9, (k) 1, (l) 12, (m) 7, (n) 2, (o) 10 14. (a) 14, (b) 12, (c) 13, (d) 1, (e) 2, (f) 5, (g) 11, (h) 8, (i) 10, (j) 9, (k) 15, (l) 4, (m) 7, (n) 3, (o) 6 15. (a) 2, (b) 1, (c) 3, (d) 4, (e) 1, (f) 5, (g) 3, (h) 2, (i) 4, (j) 1, (k) 2, (l) 4, (m) 3, (n) 5, (o) 1

Chapter 14 1. corpus callosum 2. frontal, temporal, parietal, occipital, insula 3. longitudinal fissure 4. false 5. true 6. d 7. c 8. d 9. e 10. d 11. e 12. (a) 3, (b) 5, (c) 6, (d) 8, (e) 11, (f) 10, (g) 7, (h) 9, (i) 1, (j) 4, (k) 2, (l) 12, (m) 1, (n) 8, (o) 5, (p) 7, (q) 12, (r) 10, (s) 9, (t) 1 and 2, (u) 3, 4, and 6, (v) 11 13. (a) 9, (b) 2, (c) 6, (d) 10, (e) 4, (f) 11, (g) 1, (h) 2, (i) 5, (j) 8, (k) 12, (l) 7, (m) 3, (n) 6 and 8, (o) 13, (p) 7, (q) 1 14. (a) 5, (b) 9, (c) 11, (d) 6, (e) 3, (f) 1, (g) 10, (h) 8, (i) 2, (j) 4, (k) 7 15. (a) 10, (b) 2, (c) 6, (d) 8, (e) 7, (f) 5, (g) 3, (h) 11, (i) 14, (j) 13, (k) 4, (l) 1, (m) 12, (n) 9, (o) 15

Chapter 15 1. acetylcholine, epinephrine or norepinephrine 2. thoracolumbar, craniosacral 3. true 4. true 5. d 6. e 7. b 8. c 9. e 10. a 11. a 12. c 13. e, b, g, f, d, a, c 14. (a) 3, (b) 2, (c) 1, (d) 1, (e) 2, (f) 3, (g) 3, (h) 1, (i) 4, (j) 2, (k) 5 15. (a) 2, (b) 1, (c) 1, (d) 2, (e) 1, (f) 1, (g) 2, (h) 2

Chapter 16 1. sensation, perception 2. decussation 3. false 4. true 5. c 6. 1 7. d 8. b 9. d 10. e 11. e 12. d 13. (a) 3, (b) 2, (c) 5, (d) 7, (e) 1, (f) 9, (g) 3, (h) 11, (i) 8, (j) 4, (k) 6, (l) 10 14. (a) 10, (b) 8, (c) 7, (d) 1, (e) 4, (f) 3, (g) 5, (h) 6, (i) 9, (j) 2 15. (a) 9, (b) 8, (c) 4, (d) 7, (e) 10, (f) 2, (g) 3, (h) 1, (i) 5, (j) 6, (k) 11, (l) 12

Chapter 17 1. sweet, sour, salty, bitter, umami 2. static, dynamic 3. true 4. false 5. d 6. a 7. d 8. b 9. b 10. c, j, k, d, h, l, e, b, f, i, a, m, g 11. c 12. a 13. (a) 1, (b) 5, (c) 7, (d) 6, (e) 8, (f) 2, (g) 4, (h) 3 14. (a) 3, (b) 5, (c) 8, (d) 2, (e) 7, (f) 4, (g) 1, (h) 6, (i) 9 15. (a) 2, (b) 11, (c) 14, (d) 13, (e) 3, (f) 10, (g) 6, (h) 12, (i) 4, (j) 5, (k) 9, (l) 1, (m) 7, (n) 8

Chapter 18 1. fight-or-flight response, resistance reaction, exhaustion 2. hypothalamus 3. less, more 4. false 5. true 6. b 7. e 8. d 9. a 10. e 11. c 12. a 13. (a) 8, (b) 2 (and 13, 18, 22), (c) 7, (d) 1, (e) 12, (f) 20, (g) 5, (h) 18, (i) 22, (j) 15 (and 11), (k) 3, (l) 17, (m) 21, (n) 6, (o) 13, (p) 11 (and 15), (q) 4, (r) 10, (s) 14, (t) 9, (u) 16, (v) 19 14. (a) 10, (b) 8, (c) 2, (d) 12, (e) 15, (f) 4, (g) 1, (h) 16, (i) 6, (j) 9, (k) 13, (l) 7, (m) 5, (n) 14, (o) 3, (p) 11 15. (a) 12, (b) 1, (c) 11, (d) 7, (e) 3, (f) 10, (g) 2, (h) 9, (i) 4, (j) 8, (k) 5, (l) 6

Chapter 19 1. serum 2. clot retraction 3. true 4. true 5. e 6. a 7. b 8. c 9. d 10. a 11. d 12. e 13. (a) 4, (b) 7, (c) 6, (d) 1, (e) 3, (f) 5, (g) 2 14. (a) 4, (b) 6, (c) 2, (d) 7, (e) 1, (f) 5, (g) 3 15. (a) 4, (b) 6, (c) 1, (d) 3, (e) 5, (f) 2

Chapter 20 1. left ventricle 2. systole, diastole 3. false 4. true 5. a 6. c 7. d 8. b 9. b 10. e 11. c 12. (a) 3, (b) 6, (c) 1, (d) 5, (e) 2, (f) 4 13. (a) 8, (b) 4, (c) 11, (d) 5, (e) 1, (f) 9, (g) 7, (h) 2, (i) 10, (j) 6, (k) 3 14. (a) 3, (b) 2, (c) 9, (d) 14, (e) 8, (f) 7, (g) 11, (h) 12, (i) 15, (j) 4, (k) 5, (l) 1, (m) 6, (n) 21, (o) 22, (p) 19, (q) 17, (r) 18, (s) 20, (t) 16, (u) 13, (v) 10 15. (a) 3, (b) 7, (c) 2, (d) 5, (e) 1, (f) 6, (g) 4 and 7

Chapter 21 1. carotid sinus, aortic 2. skeletal muscle pump, respiratory pump 3. true 4. true 5. b 6. a 7. c 8. e 9. a 10. d 11. (a) D, (b) C, (c) C, (d) D, (e) D, (f) C, (g) C, (h) C, (i) D, (j) D, (k) C 12. (a) 2, (b) 5, (c) 1, (d) 4, (e) 3 13. (a) 11, (b) 1, (c) 4, (d) 9, (e) 3, (f) 8, (g) 6, (h) 2, (i) 7, (j) 5, (k) 10, (l) 12, (m) 13 14. (a) 2, (b) 6, (c) 4, (d) 1, (e) 3, (f) 5 15. (a) 5, (b) 3, (c) 1, (d) 4, (e) 2, (f) 4, (g) 1, (h) 5, (i) 3

Chapter 22 1. skin, mucous membranes, antimicrobial proteins, natural killer cells, phagocytes 2. antigens 3. true 4. true 5. c 6. d 7. e 8. d 9. e 10. b 11. c 12. e, h, b, f, a, d, g, i, c 13. (a) 3, (b) 1, (c) 7, (d) 4, (e) 2, (f) 5, (g) 6 14. (a) 2, (b) 3, (c) 4, (d) 7, (e) 1, (f) 6, (g) 5

Chapter 23 1. oxyhemoglobin; dissolved CO_2, carbamino compounds (primarily carbaminohemoglobin), and bicarbonate ion 2. $CO_2 + H_2O \rightarrow H_2CO_3 \rightarrow H^+ + HCO_3^-$ 3. false 4. true 5. c 6. e 7. b 8. d 9. a 10. e 11. e, g, b, h, a, d, f, c 12. (a) 3, (b) 8, (c) 5, (d) 9, (e) 2, (f) 7, (g) 10, (h) 1, (i) 4, (j) 6 13. (a) 2, (b) 11, (c) 3, (d) 9, (e) 1, (f) 12, (g) 10, (h) 5, (i) 13, (j) 8, (k) 4, (l) 6, (m) 7 14. (a) 7, (b) 8, (c) 1, (d) 5, (e) 6, (f) 9, (g) 2, (h) 3, (i) 4 15. (a) 9, (b) 11, (c) 3, (d) 4, (e) 6, (f) 1, (g) 5, (h) 10, (i) 7, (j) 8, (k) 2

Chapter 24 1. monosaccharides; amino acids; monoglycerides, fatty acids; pentoses, phosphates, nitrogenous bases 2. diffusion, facilitated diffusion, osmosis, active transport 3. true 4. true 5. b 6. d 7. e 8. c 9. a 10. b 11. d 12. a 13. (a) 5, (b) 13, (c) 8, (d) 11, (e) 9, (f) 1, (g) 7, (h) 12, (i) 14, (j) 2, (k) 4, (l) 10, (m) 6, (n) 3 14. (a) 4, (b) 6, (c) 7, (d) 1, (e) 5, (f) 3, (g) 9, (h) 2, (i) 11, (j) 10, (k) 8

Chapter 25 1. hypothalamus 2. glucose 6-phosphate, pyruvic acid, acetyl coenzyme A 3. true 4. false 5. e 6. c 7. b 8. d 9. a 10. b 11. e 12. a 13. (a) 2 and 3, (b) 1, (c) 3 and 5, (d) 2, 4, 5, and 6, (e) 1, (f) 2, (g) 1, 4, and 6 14. (a) 9, (b) 12, (c) 11, (d) 10, (e) 4, (f) 13, (g) 3, (h) 5, (i) 8, (j) 1, (k) 6, (l) 2, (m) 7 15. (a) 17, (b) 15, (c) 8, (d) 19, (e) 1, (f) 7, (g) 4, (h) 10, (i) 16, (j) 14, (k) 11, (l) 2, (m) 13, (n) 6, (o) 20, (p) 9, (q) 5, (r) 18, (s) 3, (t) 12

Chapter 26 1. glomerulus, glomerular (Bowman's) capsule 2. micturition 3. false 4. true 5. d 6. e 7. c 8. b 9. e 10. a 11. g, a, h, c, o, n, l, j, i, b, d, e, m, k, f 12. g, f, b, d, h, c, a, e, i 13. (a) 8, (b) 2, (c) 10, (d) 5, (e) 3, (f) 1, (g) 7, (h) 4, (i) 11, (j) 9, (k) 6 14. (a) 4, (b) 3, (c) 7, (d) 6, (e) 2, (f) 8, (g) 1, (h) 5 15. (a) 5, (b) 4, (c) 6, (d) 8, (e) 1, (f) 2, (g) 7, (h) 3, (i) 9

Chapter 27 1. metabolic 2. bicarbonate ion, carbonic acid 3. true 4. false 5. d 6. e 7. a 8. b 9. (a) 8, (b) 9, (c) 7, (d) 1, (e) 6, (f) 2, (g) 4, (h) 5, (i) 3 10. (a) 8, (b) 12, (c) 7, (d) 5, (e) 10, (f) 6, (g) 9, (h) 13, (i) 1, (j) 11, (k) 3, (l) 4, (m) 2

Chapter 28 1. puberty, menarche, menopause 2. true 3. true 4. e 5. c 6. a 7. c 8. b 9. a 10. e 11. d 12. (a) 4, (b) 2, (c) 1, (d) 6, (e) 5, (f) 3 13. (a) 13, (b) 10, (c) 12, (d) 1, (e) 5, (f) 2, (g) 4, (h) 6, (i) 14, (j) 11, (k) 8, (l) 3, (m) 9, (n) 7 14. (a) 6, (b) 4, (c) 1, (d) 12, (e) 8, (f) 5, (g) 7, (h) 13, (i) 11, (j) 3, (k) 14, (l) 2, (m) 10, (n) 15, (o) 9

Chapter 29 1. dilation, expulsion, placental 2. corpus luteum, human chorionic gonadotropin 3. mesoderm, ectoderm, endoderm 4. false 5. a 6. d 7. b 8. e 9. b 10. c 11. (a) 6, (b) 2, (c) 11, (d) 5, (e) 10, (f) 1, (g) 3, (h) 8, (i) 9, (j) 4, (k) 7 12. (a) 3, (b) 6, (c) 5, (d) 2, (e) 1, (f) 4 13. (a) 3, (b) 4, (c) 5, (d) 1, (e) 2, (f) 6, (g) 7, (h) 10, (i) 8, (j) 9 14. (a) 7, (b) 3, (c) 14, (d) 6, (e) 4, (f) 10, (g) 8, (h) 13, (i) 5, (j) 2, (k) 1, (l) 9, (m) 16, (n) 11, (o) 15, (p) 12 15. (a) 3, (b) 6, (c) 4, (d) 1, (e) 9, (f) 2, (g) 5, (h) 8, (i) 7

Answers to Critical Thinking Questions

Chapter 1

1. No. Computed tomography is used to look at differences in tissue density. To assess activity in an organ such as the brain, a positron emission tomography (PET) scan or a single-photo-emission computerized tomography (SPECT) scan would provide a colorized visual assessment of brain activity.

2. Stem cells are undifferentiated cells. Research using stem cells has shown that these undifferentiated cells may be prompted to differentiate into the specific cells needed to replace those which are damaged or malfunctioning.

3. Homeostasis is the relative constancy (or dynamic equilibrium) of the body's internal environment. Homeostasis is maintained as the body changes in response to shifting external and internal conditions, including those of temperature, pressure, fluid, electrolytes, and other chemicals.

Chapter 2

1. Neither butter nor margarine is a particularly good choice for frying eggs. Butter contains saturated fats that are associated with heart disease. However, many margarines contain hydrogenated or partially hydrogenated trans-fatty acids that also increase the risk of heart disease. An alternative would be frying the eggs in any of the mono- or polyunsaturated fats such as olive oil, peanut oil, or corn oil. Boiling or poaching eggs instead of frying them would reduce the fat content of his breakfast, as would eating only the egg whites (not the high-fat yolks).

2. High body temperatures can be life-threatening, especially in infants. The increased temperature can cause denaturing of structural proteins and vital enzymes. When this happens, the proteins become nonfunctional. If the denatured enzymes are required for reactions that are necessary for life, then the infant could die.

3. Simply adding water to the table sugar does not cause it to break apart into monosaccharides. The water acts as a solvent, dissolving the sucrose and forming a sugar-water solution. To complete the breakdown of table sugar to glucose and fructose would require the presence of the enzyme sucrase.

Chapter 3

1. Synthesis of mucin by ribosomes on rough endoplasmic reticulum, to transport vesicle, to entry face of Golgi complex, to transfer vesicle, to medial cisternae where protein is modified, to transfer vesicle, to exit face, to secretory vesicle, to plasma membrane where it undergoes exocytosis.

2. Since smooth ER inactivates or detoxifies drugs, and peroxisomes also destroy harmful substances such as alcohol, we would expect to see increased numbers of these organelles in Sebastian's liver cells.

3. In order to restore water balance to the cells, the runners need to consume hypotonic solutions. The water in the hypotonic solution will move from the blood, into the interstitial fluid, and then into the cells. Plain water works well; sports drinks contain water and some electrolytes (which may have been lost due to sweating) but will still be hypotonic in relation to the body cells.

Chapter 4

1. There are many possible adaptations, including: more adipose tissue for insulation; thicker bones for support; more red blood cells for oxygen transport; increased thickness of skin to prevent water loss; etc.

2. Infants tend to have a high proportion of brown fat, which contains many mitochondria and is highly vascularized. When broken down, brown fat produces heat that helps to maintain infants' body temperatures. This heat can also warm the blood, which then distributes the heat throughout the body.

3. Your bread-and-water diet is not providing you with the necessary nutrients to encourage tissue repair. You need proper amounts of many essential vitamins, especially vitamin C, which is required for repair of the matrix and blood vessels. Vitamin A is needed to help properly maintain epithelial tissue. Adequate protein is also needed in order to synthesize the structural proteins of the damaged tissue.

Chapter 5

1. The dust particles are primarily keratinocytes that are shed from the stratum corneum of the skin.

2. Tattoos are created by depositing ink into the dermis, which does not undergo shedding as the epidermis does. Although the tattoo will fade due to exposure to sunlight and the flushing away of ink particles by the lymphatic system, the tattoo is indeed permanent.

3. Chef Eduardo has damaged the nail matrix—the part of the nail that produces growth. Because the damaged area has not regrown properly, the nail matrix may be permanently damaged.

Chapter 6

1. Due to the strenuous, repetitive activity, Taryn has probably developed a stress fracture of her right tibia (lower leg bone). Stress fractures are due to repeated stress on a bone that causes microscopic breaks in the bone without any evidence of injury to other tissue. An x-ray would not reveal the stress fracture, but a bone scan would. Thus the bone scan would either confirm or negate the physician's diagnosis.

2. When Marcus broke his arm as a child, he injured his epiphyseal (growth) plate. Damage to the cartilage in the epiphyseal plate resulted in premature closure of the plate, which interfered with the lengthwise growth of the arm bone.

3. Exercise causes mechanical stress on bones, but because there is effectively zero gravity in space, the pull of gravity on bones is missing. The lack of stress from gravity results in bone demineralization and weakness.

Chapter 7

1. Inability to open mouth—damage to the mandible, probably at temporomandibular joint; black eye—trauma to the ridge over the supraorbital margin; broken nose—probably damage to the nasal septum (includes the vomer, septal cartilage, and perpendicular plate of the ethmoid) and possibly the nasal bones; broken cheek—fracture of zygomatic bone; broken upper jaw—fracture of maxilla; damaged eye socket—fracture of parts of the sphenoid, frontal, ethmoid, palatine, zygomatic, lacrimal and maxilla (all compose the eye socket); punctured lung—damage to the thoracic vertebrae, which have punctured the lung.

2. Due to the repeated and extensive tension on his bone surfaces, Bubba would experience deposition of new bone tissue. His arm bones would be thicker and with increased raised areas (projections) where the tendons attach his muscles to bone.

3. The "soft area" being referred to is the anterior fontanel, located between the parietal and frontal bones. This is one of several areas of fibrous connective tissue in the skull that has not ossified; it should complete its ossification at 18–24 months after birth. Fontanels allow flexibility of the skull for childbirth and for brain growth after birth. The connective tissue will not allow passage of water; thus no brain damage will occur through simply washing the baby's hair.

Chapter 8

1. There are several characteristics of the bony pelves that can be used to differentiate male from female: (1) The pelvis in the female is wider and more shallow than the male's; (2) the pelvic brim of the female is larger and more oval; (3) the pubic arch has an angle greater than 90º; (4) the pelvic outlet is wider than in a male; (5) the female's iliac crest is less curved and the ilium less vertical. Table 8.1 provides additional differences between female and male pelves. Age of the skeleton can be determined by the size of the bones, the presence or absence of epiphyseal plates, the degree of demineralization of the bones, and the general appearance of the "bumps" and ridges of bones.

2. Infants do have "flat feet" because their arches have not yet developed. As they begin to stand and walk, the arches should begin to develop in order to accommodate and support their body weight. The arches are usually fully developed by age 12 or 13, so Dad doesn't need to worry yet!

3. There are 14 phalanges in each hand: two bones in the thumb and three in each of the other fingers. Farmer White has lost five phalanges on his left hand (two in his thumb and three in his index finger), so he has nine remaining on his left and 14 remaining on his right for a total of 23.

Chapter 9

1. Katie's vertebral column, head, thighs, lower legs, lower arms, and fingers are flexed. Her forearms and shoulders are medially rotated. Her thighs and arms are adducted.

2. The knee joint is commonly injured, especially among athletes. The twisting of Jeremiah's leg could have resulted in a multitude of internal injuries to the knee joint but often football players suffer tearing of the anterior cruciate ligament and medial meniscus. The immediate swelling is due to blood from damaged blood vessels, damaged synovial membranes, and the torn meniscus. Continued swelling is a result of a buildup of synovial fluid, which can result in pain and decreased mobility. Jeremiah's doctor may aspirate some of the fluid ("draining the water off his knee") and might want to perform arthroscopy to check for the extent of the knee damage.

3. The condylar processes of the mandible passed anteriorly to the articular tubercles of the temporal bones, and this dislocated Antonio's mandible. It could be corrected by pressing the thumbs downward on the lower molar teeth and pushing the mandible backward.

Chapter 10

1. Muscle cells lose their ability to undergo cell division after birth. Therefore, the increase in size is not due to an increase in the number of muscle cells but rather is due to enlargement of the existing muscle fibers (hypertrophy). This enlargement can occur from forceful, repetitive muscular activity. It will cause the muscle fibers to increase their production of internal structures such as mitochondria and myofibrils and produce an increase in the muscle fiber diameter.

2. The "dark meat" of both chickens and ducks is composed primarily of slow oxidative (SO) muscle fibers. These fibers contain large amounts of myoglobin and capillaries, which accounts for their dark color. In addition, these fibers contain large numbers of mitochondria and generate ATP by aerobic respiration. SO fibers are resistant to fatigue and can produce sustained contractions for many hours. The legs of chickens and ducks are used for support, walking, and swimming (in ducks), all activities in which endurance is needed. In addition, migrating ducks require SO fibers in their breasts to enable them to have enough energy to fly for extremely long distances while migrating. There may be some fast oxidative–glycolytic (FOG) fibers in the dark meat. FOG fibers also contain large amounts of myoglobin and capillaries, contributing to the dark color. They can use aerobic or anaerobic cellular respiration to generate ATP and have high-to-moderate resistance to fatigue. These fibers would be good for the occasional "sprint" that ducks and chickens undergo to escape dangerous situations. In contrast, the white meat of a chicken breast is composed primarily of fast glycolytic (FG) fibers. FG fibers have lower amounts of myoglobin and capillaries that give the meat its white color. There are also few mitochondria in FG fibers, so these fibers generate ATP mainly by glycolysis. These fibers contract strongly and quickly and are adapted for intense anaerobic movements of short duration. Chickens occasionally use their breasts for flying extremely short distances, usually to escape prey or perceived danger, so FG fibers are appropriate for their breast muscle.

3. Destruction of the somatic motor neurons to skeletal muscle fibers will result in a loss of stimulation to the skeletal muscles. When not stimulated on a regular basis, a muscle begins to lose its muscle tone. Through lack of use, the muscle fibers will weaken, begin to decrease in size, and can be replaced by fibrous connective tissue, resulting in a type of denervation atrophy. A lack of stimulation of the breathing muscles (especially the diaphragm) from motor neurons can result in inability of the breathing muscles to contract, thus causing respiratory paralysis and possibly death of the individual from respiratory failure.

Chapter 11

1. All of the following could occur on the affected (right) side of the face: (1) drooping of eyelid—levator palpebrae superioris; (2) drooping of the mouth, drooling, keeping food in mouth—orbicularis oris, buccinator; (3) uneven smile—zygomaticus major, levator labii superioris, risorius; (4) inability to wrinkle forehead—occipitofrontalis; (5) trouble sucking through a straw—buccinator.

2. Bulbospongiosus, external urethral sphincter, and deep transverse perineal.

3. The rotator cuff is formed by a combination of the tendons of four deep muscles of the shoulder—subscapularis, supraspinatus, infraspinatus, and teres minor. These muscles add strength and stability to the shoulder joint. Although any of the muscles' tendons can be injured, the subscapularis is most often damaged. Dependent upon the injured muscle, Jose may have trouble medially rotating his arm (subscapularis), abducting his arm (supraspinatus), laterally rotating his arm (infraspinatus, teres minor), or extending his arm (teres minor).

Chapter 12

1. Smelling the coffee and hearing the alarm are somatic sensory, stretching and yawning are somatic motor, salivating is autonomic (parasympathetic) motor, stomach rumble is enteric motor.

2. Demyelinaton or destruction of the myelin sheath can lead to multiple problems, especially in infants and children whose myelin sheaths are still in the process of developing. The affected axons deteriorate, which will interfere with function in both the CNS and PNS. There will be lack of sensation and loss of motor control with less rapid and less coordinated body responses. Damage to the axons in the CNS can be permanent and Ming's brain development may be irreversibly affected.

3. Dr. Moro could develop a drug that: (1) is an agonist of substance P; (2) blocks the breakdown of substance P; (3) blocks the reuptake of substance P; (4) promotes the release of substance P; (5) suppresses the release of enkephalins.

Chapter 13

1. The needles will pierce the epidermis, dermis, and subcutaneous layer and then go between the vertebrae through the epidural space, the dura mater, the subdural space, the arachnoid mater, and into the CSF in the subarchnoid space. CSF is produced in the brain, and the spinal meninges are continuous with the cranial meninges.

2. The anterior gray horns contain cell bodies of somatic motor neurons and motor nuclei that are responsible for the nerve impulses for skeletal muscle contraction. Because the lower cervical region is affected (brachial plexus, C5–C8), you would expect that Sunil may have trouble with movement in his shoulder, arm, and hand on the affected side.

3. Allyson has damaged her posterior columns in the lower (lumbar) region of the spinal cord. The posterior columns are responsible for transmitting nerve impulses responsible for awareness of muscle position (proprioception) and touch—which are affected in Allyson—as well as other functions such as light pressure sensations and vibration sensations. Relating Allyson's symptoms to the distribution of dermatomes, it is likely that regions L4, L5, and S1 of her spinal cord were compressed.

Chapter 14

1. Movement of the right arm is controlled by the left hemisphere's primary motor area, located in the precentral gyrus. Speech is controlled by Broca's area in the left hemisphere's frontal lobe just superior to the lateral cerebral sulcus.

2. Nicky's right facial nerve has been affected; she is suffering from Bell's palsy due to the viral infection. The facial nerve controls contraction of skeletal muscles of the face, tear gland and salivary gland secretion, as well as conveying sensory impulses from many of the taste buds on the tongue.

3. You will need to design a drug that can get through the brain's blood–brain barrier (BBB). The drug should be lipid- or water-soluble. If the drug can open a gap between the tight junctions of the endothelial cells of the brain capillaries, it would be more likely to pass through the BBB. Targeting the drug to enter the brain in certain areas near the third ventricle (the circumventricular organs) might be an option as the BBB is entirely absent in those areas and the capillary endothelium is more permeable, allowing the blood-borne drug to more readily enter the brain tissue.

Chapter 15

1. Digestion and relaxation are controlled by increased stimulation of the parasympathetic division of the ANS. The salivary glands, pancreas, and liver will show increased secretion; the stomach and intestines will have increased activity; the gallbladder will have increased contractions; heart contractions will have decreased force and rate. Following is the nerve supply to each listed organ: salivary glands—facial nerves (cranial nerve VII) and glossopharyngeal nerves (cranial nerve IX); pancreas, liver, stomach, gallbladder, intestines and heart—vagus nerves (cranial nerve X).

2. Ciara experienced one of the "E situations" (emergency in her case), which has activated the fight-or-flight response. Some noticeable effects of increased sympathetic activity include an increase in heart rate, sweating on the palms, and contraction of the arrector pili muscles, which causes the goose flesh. Secretion of epinephrine and norepinphrine from the adrenal medullae will intensify and prolong the responses.

3. Mrs. Young needs to slow down the activity of her digestive system, which seems to be experiencing increased parasympathetic response. A parasympathetic blocking agent is needed. Because the stomach and intestines have muscarinic receptors, she needs to be provided with a muscarinic blocking agent (such as atropine), which will result in decreased motility in the stomach and intestines.

Chapter 16

1. Chemoreceptors in the nose detect odors. Proprioceptors detect body position and are involved in equilibrium. The chemoreceptors in the nose are rapidly adapting, whereas proprioceptors are slowly adapting. Thus the smell faded while the sensation of motion remained.

2. Thermal (heat) receptors in her left hand detect the stimulus. A nerve impulse is transmitted to the spinal cord through first-order neurons with cell bodies in posterior root ganglia. The impulses travel into the spinal cord, where the first-order neurons synapse with second-order neurons, whose cell bodies are located in the posterior gray horn of the spinal cord. The axons of the second-order neurons decussate to the right side in the spinal cord and then the impulses ascend through the lateral spinothalamic tract. The axons of the second-order neurons end in the ventral posterior nucleus of the right side of the thalamus, where they synapse with the third-order neurons. Axons of the third-order neurons transmit impulses to the specific primary somatosensory areas in the postcentral gyrus of the right parietal lobe.

3. When Marvin settled down for the night, he passed through Stages 1–3 of NREM sleep. Sleepwalking occurred when he was in Stage 4 (slow-wave sleep). Because this is the deepest stage of sleep, his mother was able to return him to his bed without awakening him. Marvin then cycled through REM and NREM sleep. His dreaming occurred during the REM phases of sleep. The noise of the alarm clock provided the sensory stimulus that stimulated the reticular activating system. Activation of this system sends numerous nerve impulses to widespread areas of the cerebral cortex, both directly and via the thalamus. The result is the state of wakefulness.

Chapter 17

1. Damage to the facial nerve would affect smell, taste, and hearing. Within the nasal epithelium and connective tissue, both the supporting cells and olfactory glands are innervated by branches of the facial nerve. Without input from the facial nerve, there will be a lack of mucus production required to dissolve odorants. The facial nerve also serves taste buds in the anterior two-thirds of the tongue, so damage can affect taste sensations. Hearing will be affected by a damaged facial nerve because the stapedius muscle, which is attached to the stapes, is innervated by the facial nerve. Contraction of the stapedius muscle helps to protect the inner ear from prolonged loud noises. Damage to the facial nerve will result in sounds that are excessively loud, resulting in more susceptibility to damage by prolonged loud noises.

2. With age, Gertrude has lost much of her sense of smell and taste due to a decline in olfactory and gustatory receptors. Since smell and taste are intimately linked, food no longer smells nor tastes as good to Gertrude. Gertrude has presbyopia, a loss of lens elasticity, which makes it difficult to read. She may also be experiencing age-related loss of sharpness of vision and depth perception. Gertrude's hearing difficulties could be a result of damage to hair cells in the organ of Corti or degeneration of the nerve pathway for hearing. The "buzzing" Gertrude hears may be tinnitus, which also occurs more frequently in the elderly.

3. Some of the eyedrops placed in the eye may pass through the nasolacrimal duct into the nasal cavity where olfactory receptors are stimulated. Because most "tastes" are actually smells, the child will "taste" the medicine from her eye.

Chapter 18

1. Yes, Amanda should visit the clinic, as these are serious signs and symptoms. She has an enlarged thyroid gland, or goiter. The goiter is probably due to hypothyroidism, which is causing the weight gain, fatigue, mental dullness, and other symptoms.

2. Amanda's problem is her pituitary gland, which is not secreting normal levels of TSH. Rising thyroxine (T_4) levels after the TSH injection indicates that her thyroid is functioning normally and able to respond to the increased TSH levels. If the thyroxine levels had not risen, then the problem would have been her thyroid gland.

3. Mr. Hernandez has diabetes insipidus caused by either insufficient production or release of ADH due to hypothalamus or posterior pituitary damage. He also could have defective ADH receptors in the kidneys. Diabetes insipidus is characterized by production of large volumes of urine, dehydration, and increased thirst, but with no glucose or ketones present in the urine (which would be indicative of diabetes mellitus rather than diabetes insipidus).

Chapter 19

1. The broad spectrum antibiotics may have destroyed the bacteria that caused Shilpa's bladder infection but also destroyed the naturally occurring large intestine bacteria that produce vitamin K. Vitamin K is required for the synthesis of four clotting factors (II, VII, IX, and X). Without these clotting factors present in normal amounts, Shilpa will experience clotting problems until the intestinal bacteria reach normal levels and produce additional vitamin K.

2. Mrs. Brown's kidney failure is interfering with her ability to produce erythropoietin (EPO). Her physician can prescribe Epoetin alfa, a recombinant EPO, which is very effective in treating the decline in RBC production with renal failure.

3. A primary problem Thomas may experience is with clotting. Clotting time becomes longer because the liver is responsible for producing many of the clotting factors and clotting proteins such as fibrinogen. Thrombopoietin, which stimulates the formation of platelets, is also produced in the liver. In addition, the liver is responsible for eliminating bilirubin, produced from the breakdown of RBCs. With a malfunctioning liver, the bilirubin will accumulate, resulting in jaundice. In addition, there can be decreased concentrations of the plasma protein albumin, which can affect blood pressure.

Chapter 20

1. The dental procedures introduced bacteria into Gerald's blood. The bacteria colonized his endocardium and heart valves, resulting in bacterial endocarditis. Gerald may have had a previously undetected heart murmur, or the heart murmur may have resulted from his endocarditis. His physician will want to monitor his heart to watch for further damage to the valve.

2. Extremely rapid heart rates can result in a decreased stroke volume due to insufficient ventricular filling. As a result, the cardiac output will decline to the point where there may not be enough blood reaching the central nervous system. She initially may experience light-headedness but could lose consciousness if the cardiac output declines dramatically.

3. Mr. Perkins is suffering from angina pectoris and has several risk factors for coronary artery disease such as smoking, obesity, sedentary lifestyle, and male gender. Cardiac angiography involves the use of a cardiac catheter to inject a radiopaque medium into the heart and its vessels. The angiogram may reveal blockages such as atherosclerotic plaques in his coronary arteries.

Chapter 21

1. The hole in the heart was the foramen ovale, which is an opening between the right and left atria. In fetal circulation it allows blood to bypass the right ventricle, enter the left atrium, and join systemic circulation. The "hole" should close shortly after birth to become the fossa ovalis. Closure of the foramen ovale after birth will allow the deoxgenated blood from the right atrium to enter pulmonary circulation so that the blood can become oxygenated prior to entering systemic circulation. If closure doesn't occur, surgery may be required.

2. Michael is suffering from hypovolemic shock due to the loss of blood. The low blood pressure is a result of low blood volume and a subsequent decrease in cardiac output. His rapid, weak pulse is an attempt of the heart to compensate for the decrease in cardiac output through sympathetic stimulation of the heart and increased blood levels of epinephrine and norepinephrine. His pale, cool, and clammy skin is a result of sympathetic constriction of the blood vessels of the skin and sympathetic stimulation of sweat glands. The lack of urine production is due to increased secretion of aldosterone and ADH, both of which are produced to increase blood volume in order to compensate for Michael's hypotension. The fluid loss from his bleeding results in activation of the thirst center in the hypothalamus. His confusion and disorientation are caused by a reduced oxygen supply to the brain from the decreased cardiac output.

3. Maureen has varicose veins, a condition in which the venous valves become leaky. The leaking valves allow the backflow of blood and an increased pressure that distends the veins and allows fluid to leak into the surrounding tissue. Standing on hard surfaces for long periods of time can cause varicosities to develop. Maureen needs to elevate her legs when possible to counteract the effects of gravity on the blood flow in the lower legs. She could also utilize support hose, which add external support for the superficial veins, much like skeletal muscle does for deeper veins. If the varices become severe, Maureen may require more extensive treatment such as sclerotherapy, radiofrequency endovenous occlusion, laser occlusion, or stripping.

Chapter 22

1. The influenza vaccination introduces a weakened or killed virus (which will not cause the disease) to the body. The immune system recognizes the antigen and mounts a primary immune response. Upon exposure to the same flu virus that was in the vaccine, the body will produce a secondary response, which will usually prevent a case of the flu. This is artificially acquired active immunity.

2. Mrs. Franco's lymph nodes were removed because metastasis of cancerous cells can occur through the lymph nodes and lymphatic vessels. Mrs. Franco's swelling is a lymphedema that is occurring due to the buildup of interstitial fluid from interference with drainage in the lymph vessels.

3. Tariq's physician would need to perform an antibody titer, which is a measure of the amount of antibody in the serum. If Tariq has previously been exposed to mumps (or been vaccinated for mumps), his blood should have elevated levels of IgG antibodies after this exposure from his sister. His immune system would be experiencing a secondary response. If he has not previously had mumps and has contracted mumps from his sister, his immune system would initiate a primary response. In that case, his blood would show an elevated titer of IgM antibodies, which are secreted by plasma cells after an initial exposure to the mumps antigen.

Chapter 23

1. Aretha's excess mucus production is causing blockage of the paranasal sinuses, which are used as hollow resonating chambers for singing and speech. In addition, her sore throat could be due to inflammation of the pharynx and larynx, which will affect their normal functions. Normally, the pharynx also acts as a resonating chamber and the true vocal cords, located in the larynx, vibrate for speech and singing. Inflammation of the true vocal cords (laryngitis) interferes with their ability to freely vibrate, which will affect both singing and speech.

2. In emphysema, there is destruction of the alveolar walls, producing abnormally large air spaces that remain filled with air during exhalation. The destruction of alveoli decreases the surface area for gas exchange across the respiratory membrane, resulting in a decreased blood O_2 level. Damage to the alveolar walls also causes a loss of elasticity, making exhalation more difficult. This can result in a buildup of CO_2. Cigarette smoke contains nicotine, carbon monoxide, and a variety of irritants, all of which affect the lungs. Nicotine constricts terminal bronchioles, decreasing the air flow into and out of the lungs; carbon monoxide binds to hemoglobin, reducing its ability to carry oxygen; irritants such as tar and fine particulate matter destroy cilia and increase mucus secretion, interfering with the ability of the respiratory passages to cleanse themselves.

3. The squirrel's nest blocked the passage of exhaust gas from the furnace, causing an accumulation of carbon monoxide (CO), a colorless, odorless gas, in the home. As they were sleeping, their blood was saturated with CO, which has a stronger affinity for hemoglobin than oxygen. As a result, the Robinsons became oxygen deficient. Without adequate oxygenation of the brain, the Robinsons died during their sleep.

Chapter 24

1. HCl has several important roles in digestion. HCl stimulates the secretion of hormones that promote the flow of bile and pancreatic juice. The presence of HCl destroys certain microbes that may have been ingested with food. HCl begins denaturing proteins in food, and provides the proper chemical environment for activating pepsinogen into pepsin, which breaks apart certain peptide bonds in proteins. It also helps in the action of gastric lipase, which splits triglycerides in fat molecules found in milk into fatty acids and monoglycerides.

2. Blockage of the pancreatic and bile ducts prevents pancreatic digestive enzymes and bile from reaching the duodenum. As a consequence, there will be problems digesting carbohydrates, proteins, nucleic acids, and lipids. Of particular concern is lipid digestion since the pancreatic juices contain the primary lipid-digesting enzyme. Fats will not be adequately digested, and Trey's feces will contain larger than normal amounts of lipids. In addition, the lack of bile salts will affect the body's ability to emulsify lipids and to form micelles required for absorption of fatty acids and monoglycerides (from lipid breakdown).

When lipids are not absorbed properly, then there will be malabsorption of the lipid-soluble vitamins (A, D, E, and K).

3. Antonio experienced gastroesophageal reflux. The stomach contents backed up (refluxed) into Antonio's esophagus due to a failure of the lower esophageal sphincter to fully close. The HCl from the stomach irritated the esophageal wall, which resulted in the burning sensation he felt; this is commonly known as "heartburn," even though it is not related to the heart. Antonio's recent meal worsened the problem. Alcohol and smoking both can cause the sphincter to relax, while certain foods such tomatoes, chocolate, and coffee can stimulate stomach acid secretion. In addition, lying down immediately after a meal can exacerbate the problem.

Chapter 25

1. Ingestion of cyanide affects cellular respiration. The cyanide binds to the cytochrome oxidase complex in the inner membrane of mitochondria. Blocking this complex interferes with the last step in electron transport in aerobic ATP production. Jane Doe's body would quickly run out of energy to perform vital functions, resulting in her death.

2. Glenn's total cholesterol and LDL levels are very high, while his HDL levels are low. Total cholesterol above 239 mg/dL and LDL above 159 mg/dL are considered high. The ratio of total cholesterol (TC) to HDL-cholesterol is a predictor of the risk of developing coronary artery disease. Glenn's TC to HDL is 15; a ratio above 4 is undesirable. His ratio places him at high risk of developing coronary artery disease. In addition, for every 50 mg/dL TC over 200 mg/dL, the risk of a heart attack doubles. Glenn needs to reduce his TC and LDL-cholesterol while raising his HDL-cholesterol levels. LDLs contribute to fatty plaque formation on coronary artery walls. On the other hand, HDLs help remove excess cholesterol from the blood, which helps decrease the risk of coronary artery disease. Glenn will need to reduce his dietary intake of total fat, saturated fats, and cholesterol, all of which contribute to raising LDL levels. Exercise will raise HDL levels. If those changes are not successful, drug therapy may be required.

3. The goal of weight loss programs is to reduce caloric intake so that the body utilizes stored lipids as an energy source. As part of that desired lipid metabolism, ketone bodies are produced. Some of these ketone bodies will be excreted in the urine. If no ketones are present, then Marissa's body is not breaking down lipids. Only through using fewer calories than needed will her body break down the stored fat and release ketones. Thus, she must be eating more calories than needed to support her daily activities—she is "cheating."

Chapter 26

1. Without reabsorption, initially 105–125 mL of filtrate would be lost per minute, assuming normal glomerular filtration rate. Fluid loss from the blood would cause a decrease in blood pressure, and therefore a decrease in GBHP. When GBHP dropped below 45 mmHg, filtration would stop (assuming normal CHP and BCOP) because NFP would be zero.

2. a. Although normally pale yellow, urine color can vary based upon concentration, diet, drugs, and disease. A dark yellow color would not necessarily indicate a problem, but further investigation may be needed. Turbidity or cloudiness can be caused by urine that has been standing for a period of time, from certain foods, or from bacterial infections. Further investigation is needed. b. Ammonia-like odor occurs when the urine sample is allowed to stand. c. Albumin should not be present in urine (or be present only in very small amounts) because it is too large to pass through the filtration membranes. The presence of high levels of albumin is cause for concern as it indicates damage to the filtration membranes. d. Casts are hardened masses of material that are flushed out in the urine. The presence of casts is not normal and indicates a pathology. e. The pH of normal urine ranges from 4.8 to 8.0. A pH of 5.5 is in normal range. f. Hematuria is the presence of red blood cells in the urine. It can occur with certain pathological conditions or from kidney trauma. Hematuria may occur if the urine sample was contaminated with menstrual blood.

3. Bruce has developed renal calculi (kidney stones), which are blocking his ureters and interfering with the flow of urine from the kidneys to the bladder. The rhythmic pains are a result of the peristaltic contractions of the ureters as they attempt to move the stones toward the bladder. Bruce can wait for the stones to pass, can have them surgically removed, or can use shock-wave lithotripsy to break apart the stones into smaller fragments that can be eliminated with urine. To prevent future episodes, Bruce needs to watch his diet (limit calcium) and drink fluids, and may need drug intervention.

Chapter 27

1. The loss of stomach acids can result in metabolic alkalosis. Robin's HCO_3^- levels would be higher than normal. She would be hypoventilating in order to decrease her pH by slowing the loss of CO_2. Excessive vomiting can result in hyponatremia, hypokalemia, and hypochloremia. Both hyponatremia and hypokalemia can cause mental confusion.

2. (Step 1) pH = 7.30 indicates slight acidosis, which could be caused by elevated PCO_2 or lowered HCO_3^-. (Step 2) The HCO_3^- is lower than normal (20 mEq/liter), so (Step 3) the cause is metabolic. (Step 4) The PCO_2 is lower than normal (32 mmHg), so hyperventilation is providing some compensation. Diagnosis: Henry has partially compensated metabolic acidosis. A possible cause is kidney damage that resulted from interruption of blood flow during the heart attack.

3. Sam will experience increased fluid loss through increased evaporation from the skin and water vapor from the respiratory system through his increased respiratory rate. His insensible water loss will also increase (loss of water from mucous membranes of the mouth and respiratory system). Sam will have a decrease in urine formation.

Chapter 28

1. Monica's excessive training has resulted in an abnormally low amount of body fat. A certain amount of body fat is needed in order to produce the hormones required for the ovarian cycle. Several hormones are involved. Her amenorrhea is due to a lack of gonadotropin-releasing hormone, which in turn reduces the release of LH and FSH. Her follicles with their enclosed ova fail to develop and ovulation will not occur. In addition, synthesis of estrogens and progesterone declines from the lack of hormonal feedback. Usually a gain of weight will allow normal hormonal feedback mechanisms to return.

2. Along with estrogens, progesterone helps to prepare the endometrium for possible implantation of a zygote by promoting growth of the endometrium. The endometrial glands secrete glycogen, which will help sustain an embryo if one should implant. If implantation occurs, progesterone helps maintain the endometrium for the developing fetus. In addition, it helps prepare mammary glands to secrete milk. It inhibits the release of GnRH and LH, which stops a new ovarian cycle from occurring.

3. The ductus deferens is cut and tied in a vasectomy. This stops the release of sperm into the ejaculatory duct and urethra. Mark will still produce the secretions from his accessory glands (prostate, seminal vesicles, bulbourethral glands) in his ejaculate. In addition, a vasectomy does not affect sexual performance; he will be able to achieve erection and ejaculation, as those events are nervous system responses.

Chapter 29

1. As part of the feedback mechanism for lactation, oxytocin is released from the posterior pituitary. It is carried to the mammary glands where it causes milk to be released into the mammary ducts (milk ejection). The oxytocin is also transported in the blood to the uterus, which contains oxytocin receptors on the myometrium. The oxytocin causes contraction of the myometrium, resulting in the painful sensations that Kathy is experiencing. The uterine contractions can help return the uterus back to its prepregnancy size.

2. Sex-linked genetic traits, such as hemophilia, are present on the X chromosomes but not on the Y chromosomes. In males, the X chromosome is always inherited from the mother, and the Y chromosome from the father. Thus, Jack's hemophilia gene was inherited from his mother on his X chromosome. The gene for hemophilia is a recessive gene. His mother would need two recessive genes, one on each of her X chromosomes, to be hemophiliac. His father must carry the dominant (nonhemophiliac) gene on his X chromosome, so he also would not have hemophilia.

3. The cord blood is a source of pluripotent stem cells, which are unspecialized cells that have the potential to specialize into cells with specific functions. The hope is that stem cells can be used to generate cells and tissues to treat a variety of disorders. It is assumed that the tissues would not be rejected since they would contain the same genetic material as the patient—in this case Alisa's baby.

GLOSSARY

Pronunciation Key

1. The most strongly accented syllable appears in capital letters, for example, bilateral (bī-LAT-er-al) and diagnosis (dī-ag-NŌ-sis).

2. If there is a secondary accent, it is noted by a prime (′), for example, constitution (kon′-sti-TOO-shun) and physiology (fiz′-ē-OL-ō-jē). Any additional secondary accents are also noted by a prime, for example, decarboxylation (dē′-kar-bok′-si-LĀ-shun).

3. Vowels marked by a line above the letter are pronounced with the long sound, as in the following common words:

ā as in *māke*	*ō* as in *pōle*
ē as in *bē*	*ū* as in *cūte*
ī as in *īvy*	

4. Vowels not marked by a line above the letter are pronounced with the short sound, as in the following words:

a as in *above* or *at*	*o* as in *not*
e as in *bet*	*u* as in *bud*
i as in *sip*	

5. Other vowel sounds are indicated as follows:

oy as in *oil*
oo as in *root*

6. Consonant sounds are pronounced as in the following words:

b as in *bat*	*m* as in *mother*
ch as in *chair*	*n* as in *no*
d as in *dog*	*p* as in *pick*
f as in *father*	*r* as in *rib*
g as in *get*	*s* as in *so*
h as in *hat*	*t* as in *tea*
j as in *jump*	*v* as in *very*
k as in *can*	*w* as in *welcome*
ks as in *tax*	*z* as in *zero*
kw as in *quit*	*zh* as in *lesion*
l as in *let*	

A

Abdomen (ab-DŌ-men *or* AB-dō-men) The area between the diaphragm and pelvis.

Abdominal (ab-DŌM-i-nal) **cavity** Superior portion of the abdominopelvic cavity that contains the stomach, spleen, liver, gallbladder, pancreas, kidneys, small intestine, and part of the large intestine.

Abdominal thrust manoeuvre A first-aid procedure for choking. Employs a quick, upward thrust against the diaphragm that forces air out of the lungs with sufficient force to eject any lodged material.

Abdominopelvic (ab-dom′-i-nō-PEL-vik) **cavity** Inferior to the diaphragm and subdivided into a superior abdominal cavity and an inferior pelvic cavity.

Abduction (ab-DUK-shun) Movement away from the midline of the body.

Abortion (a-BOR-shun) The premature loss **(spontaneous)** or removal **(induced)** of the embryo or nonviable fetus; miscarriage due to a failure in the normal process of developing or maturing.

Abscess (AB-ses) A localized collection of pus and liquefied tissue in a cavity.

Absorption (ab-SORP-shun) Intake of fluids or other substances by cells of the skin or mucous membranes; the passage of digested foods from the gastrointestinal tract into blood or lymph.

Accessory duct A duct of the pancreas that empties into the duodenum about 2.5 cm (1 in.) superior to the ampulla of Vater (hepatopancreatic ampulla). Also called the **duct of Santorini** (san′-tō-RĒ-nē).

Acetabulum (as′-e-TAB-ū-lum) The rounded cavity on the external surface of the hip bone that receives the head of the femur.

Acetylcholine (as′-ē-til-KŌ-lēn) **(ACh)** A neurotransmitter liberated by many peripheral nervous system neurons and some central nervous system neurons. It is excitatory at neuromuscular junctions but inhibitory at some other synapses (for example, it slows heart rate).

Achalasia (ak′-a-LĀ-zē-a) A condition, caused by malfunction of the myenteric plexus, in which the lower esophageal sphincter fails to relax normally as food approaches. A whole meal may become lodged in the oesophagus and enter the stomach very slowly. Distension of the oesophagus results in chest pain that is often confused with pain originating from the heart.

Achilles tendon *See* **Calcaneal tendon.**

Acini (AS-i-nē) Groups of cells in the pancreas that secrete digestive enzymes. *Singular* is **acinus** (AS-i-nus).

Acoustic (a-KOOS-tik) Pertaining to sound or the sense of hearing.

Acquired immunodeficiency syndrome (AIDS) A fatal disease caused by the human immunodeficiency virus (HIV). Characterized by a positive HIV-antibody test, low helper T cell count, and certain indicator diseases (for example, Kaposi's sarcoma, pneumocystis carinii pneumonia, tuberculosis, fungal diseases). Other symptoms include fever or night sweats, coughing, sore throat, fatigue, body aches, weight loss, and enlarged lymph nodes.

Acrosome (AK-rō-sōm) A lysosomelike organelle in the head of a sperm cell containing enzymes that facilitate the penetration of a sperm cell into a secondary oocyte.

Actin (AK-tin) A contractile protein that is part of thin filaments in muscle fibers.

Action potential (AP) An electrical signal that propagates along the membrane of a neuron or muscle fiber (cell); a rapid change in membrane potential that involves a depolarization followed by a repolarization. Also called a **nerve action potential** or **nerve impulse** as it relates to a neuron, and a **muscle action potential** as it relates to a muscle fiber.

Activation (ak′-ti-VĀ-shun) **energy** The minimum amount of energy required for a chemical reaction to occur.

Active transport The movement of substances across cell membranes against a concentration gradient, requiring the expenditure of cellular energy (ATP).

Acute (a-KŪT) Having rapid onset, severe symptoms, and a short course; not chronic.

Adaptation (ad′-ap-TĀ-shun) The adjustment of the pupil of the eye to changes in light intensity. The property by which a sensory neuron relays a decreased frequency of action potentials from a receptor, even though the strength of the stimulus remains constant; the decrease in perception of a sensation over time while the stimulus is still present.

Adduction (ad-DUK-shun) Movement toward the midline of the body.

Adenoids (AD-e-noyds) The pharyngeal tonsils.

Adenosine triphosphate (a-DEN-ō-sēn trī-FOS-fāt) **(ATP)** The main energy currency in living cells; used to transfer the chemical energy needed for metabolic reactions. ATP consists of the purine base *adenine* and the five-carbon sugar *ribose,* to which are added, in linear array, three *phosphate* groups.

Adhesion (ad-HĒ-zhun) Abnormal joining of parts to each other.

Adipocyte (AD-i-pō-sīt) Fat cell, derived from a fibroblast. Also called **fat cell** or **adipose cell.**

Adipose (AD-i-pōz) **tissue** Tissue composed of adipocytes specialized for triglyceride storage and present in the form of soft pads between various organs for support, protection, and insulation.

Adrenal cortex (a-DRĒ-nal KOR-teks) The outer portion of an adrenal gland, divided into three zones; the zona glomerulosa secretes mineralocorticoids, the zona fasciculata secretes glucocorticoids, and the zona reticularis secretes androgens.

Adrenal glands Two glands located superior to each kidney. Also called the **suprarenal** (soo′-pra-RĒ-nal) **glands.**

Adrenal medulla (me-DUL-a) The inner part of an adrenal gland, consisting of cells that secrete epinephrine, norepinephrine, and a small amount of dopamine in response to stimulation by sympathetic preganglionic neurons.

Adrenergic (ad′-ren-ER-jik) **neuron** A neuron that releases epinephrine (adrenaline) or norepinephrine (noradrenaline) as its neurotransmitter.

Adrenocorticotropic (ad-rē′-nō-kor-ti-kō-TRŌP-ik) **hormone (ACTH)** A hormone produced by the anterior pituitary that influences the production and secretion of certain hormones of the adrenal cortex. Also called **corticotropin** (kor′-ti-kō-TRŌ-pin).

Adventitia (ad-ven-TISH-a) The outermost covering of a structure or organ.

Aerobic (air-Ō-bik) Requiring molecular oxygen.

Aerobic (ār-Ō-bik) **cellular respiration** The production of ATP (36 molecules) from the complete oxidation of pyruvic acid in mitochondria. Carbon dioxide, water, and heat are also produced.

Afferent arteriole (AF-er-ent ar-TĒ-rē-ōl) A blood vessel of a kidney that divides into the capillary network called a glomerulus; there is one afferent arteriole for each glomerulus.

Agglutination (a-gloo-ti-NĀ-shun) Clumping of microorganisms or blood cells, typically due to an antigen–antibody reaction.

Aggregated lymphatic follicles Clusters of lymph nodules that are most numerous in the ileum. Also called **Peyer's** (PĪ-erz) **patches.**

Albinism (AL-bin-izm) Abnormal, nonpathological, partial, or total absence of pigment in skin, hair, and eyes.

Aldosterone (al-DOS-ter-ōn) A mineralocorticoid produced by the adrenal cortex that promotes sodium and water reabsorption by the kidneys and potassium excretion in urine.

All-or-none principle If a stimulus depolarizes a neuron to threshold, the neuron fires at its maximum voltage (all); if threshold is not reached, the neuron does not fire at all (none). Given above threshold, stronger stimuli do not produce stronger action potentials.

Allantois (a-LAN-tō-is) A small, vascularized outpouching of the yolk sac that serves as an early site for blood formation and development of the urinary bladder.

Alleles (a-LĒLZ) Alternate forms of a single gene that control the same inherited trait (such as type A blood) and are located at the same position on homologous chromosomes.

Allergen (AL-er-jen) An antigen that evokes a hypersensitivity reaction.

Alopecia (al′-ō-PĒ-shē-a) The partial or complete lack of hair as a result of factors such as genetics, aging, endocrine disorders, chemotherapy, and skin diseases.

Alpha (AL-fa) **cell** A type of cell in the pancreatic islets (islets of Langerhans) that secretes the hormone glucagon. Also termed an **A cell.**

Alpha (α) receptor A type of receptor for norepinephrine and epinephrine; present on visceral effectors innervated by sympathetic postganglionic neurons.

Alveolar duct Branch of a respiratory bronchiole around which alveoli and alveolar sacs are arranged.

Alveolar macrophage (MAK-rō-fāj) Highly phagocytic cell found in the alveolar walls of the lungs. Also called a **dust cell.**

Alveolar sac A cluster of alveoli that share a common opening.

Alveolus (al-VĒ-ō-lus) A small hollow or cavity; an air sac in the lungs; milk-secreting portion of a mammary gland. *Plural* is **alveoli** (al-VĒ-ol-ī).

Alzheimer's (ALTZ-hī-mer) **disease (AD)** Disabling neurological disorder characterized by dysfunction and death of specific cerebral neurons, resulting in widespread intellectual impairment, personality changes, and fluctuations in alertness.

Amenorrhoea (ā-men-ō-RĒ-a) Absence of menstruation.

Amnesia (am-NĒ-zē-a) A lack or loss of memory.

Amnion (AM-nē-on) A thin, protective fetal membrane that develops from the epiblast; holds the fetus suspended in amniotic fluid. Also called the **"bag of waters."**

Amniotic (am′-nē-OT-ik) **fluid** Fluid in the amniotic cavity, the space between the developing embryo (or fetus) and amnion; the fluid is initially produced as a filtrate from maternal blood and later includes fetal urine. It functions as a shock absorber, helps regulate fetal body temperature, and helps prevent desiccation.

Amphiarthrosis (am′-fē-ar-THRŌ-sis) A slightly movable joint, in which the articulating bony surfaces are separated by fibrous connective tissue or fibrocartilage to which both are attached; types are syndesmosis and symphysis.

Ampulla (am-PUL-la) A saclike dilation of a canal or duct.

Ampulla of Vater *See* **Hepatopancreatic ampulla.**

Anabolism (a-NAB-ō-lizm) Synthetic, energy-requiring reactions whereby small molecules are built up into larger ones.

Anaemia (a-NĒ-mē-a) Condition of the blood in which the number of functional red blood cells or their haemoglobin content is below normal.

Anaerobic (an-ar-Ō-bik) Not requiring oxygen.

Anaesthesia (an′-es-THĒ-zē-a) A total or partial loss of feeling or sensation; may be general or local.

Anal (Ā-nal) **canal** The last 2 or 3 cm (1 in.) of the rectum; opens to the exterior through the anus.

Anal column A longitudinal fold in the mucous membrane of the anal canal that contains a network of arteries and veins.

Anal triangle The subdivision of the female or male perineum that contains the anus.

Analgesia (an-al-JĒ-zē-a) Pain relief; absence of the sensation of pain.

Anaphase (AN-a-fāz) The third stage of mitosis in which the chromatids that have separated at the centromeres move to opposite poles of the cell.

Anaphylaxis (an′-a-fi-LAK-sis) A hypersensitivity (allergic) reaction in which IgE antibodies attach to mast cells and basophils, causing them to produce mediators of anaphylaxis (histamine, leukotrienes, kinins, and prostaglandins) that bring about increased blood permeability, increased smooth muscle contraction, and increased mucus production. Examples are hay fever, hives, and anaphylactic shock.

Anastomosis (a-nas′-tō-MŌ-sis) An end-to-end union or joining of blood vessels, lymphatic vessels, or nerves.

Anatomic dead space Spaces of the nose, pharynx, larynx, trachea, bronchi, and bronchioles totaling about 150 mL of the 500 mL in a quiet breath (tidal volume); air in the anatomic dead space does not reach the alveoli to participate in gas exchange.

Anatomical (an′-a-TOM-i-kal) **position** A position of the body universally used in anatomical descriptions in which the body is erect, the head is level, the eyes face forward, the upper limbs are at the sides, the palms face forward, and the feet are flat on the floor.

Anatomy (a-NAT-ō-mē) The structure or study of the structure of the body and the relation of its parts to each other.

Androgens (AN-drō-jenz) Masculinizing sex hormones produced by the testes in males and the adrenal cortex in both sexes; responsible for libido (sexual desire); the two main androgens are testosterone and dihydrotestosterone.

Aneurysm (AN-ū-rizm) A saclike enlargement of a blood vessel caused by a weakening of its wall.

Angina pectoris (an-JĪ-na *or* AN-ji-na PEK-tō-ris) A pain in the chest related to reduced coronary circulation due to coronary artery disease (CAD) or spasms of vascular smooth muscle in coronary arteries.

Angiogenesis (an′-jē-ō-JEN-e-sis) The formation of blood vessels in the extraembryonic mesoderm of the yolk sac, connecting stalk, and chorion at the beginning of the third week of development.

Ankylosis (ang′-ki-LŌ-sis) Severe or complete loss of movement at a joint as the result of a disease process.

Antagonist (an-TAG-ō-nist) A muscle that has an action opposite that of the prime mover (agonist) and yields to the movement of the prime mover.

Antagonistic (an-tag-o-NIST-ik) **effect** A hormonal interaction in which the effect of one hormone on a target cell is opposed by another hormone. For example, calcitonin (CT) lowers blood calcium level, whereas parathyroid hormone (PTH) raises it.

Anterior (an-TĒR-ē-or) Nearer to or at the front of the body. Equivalent to **ventral** in bipeds.

Anterior pituitary Anterior lobe of the pituitary gland. Also called the **adenohypophysis** (ad′-e-nō-hī-POF-i-sis).

Anterior root The structure composed of axons of motor (efferent) neurons that emerges from the anterior aspect of the spinal cord and extends laterally to join a posterior root, forming a spinal nerve. Also called a **ventral root.**

Anterolateral (an′-ter-ō-LAT-er-al) **pathway** Sensory pathway that conveys information related to pain, temperature, tickle, and itch. Also called **spinothalamic pathway.**

Antibody (AN-ti-bod′-ē) **(Ab)** A protein produced by plasma cells in response to a specific antigen; the antibody combines with that antigen to neutralize, inhibit, or destroy it. Also called an **immunoglobulin** (im-ū-nō-GLOB-ū-lin) or **Ig.**

Anticoagulant (an-tī-cō-AG-ū-lant) A substance that can delay, suppress, or prevent the clotting of blood.

Antidiuretic (an′-ti-dī-ū-RET-ik) Substance that inhibits urine formation.

Antidiuretic hormone (ADH) Hormone produced by neurosecretory cells in the paraventricular and supraoptic nuclei of the hypothalamus that stimulates water reabsorption from kidney tubule cells into the blood and vasoconstriction of arterioles. Also called **vasopressin** (vāz-ō-PRES-in).

Antigen (AN-ti-jen) **(Ag)** A substance that has immunogenicity (the ability to provoke an immune response) and reactivity (the ability to react with the antibodies or cells that result from the immune response); contraction of *anti*body *gen*erator. Also termed a **complete antigen.**

Antigen-presenting cell (APC) Special class of migratory cell that processes and presents antigens to T cells during an immune response; APCs include macrophages, B cells, and dendritic cells, which are present in the skin, mucous membranes, and lymph nodes.

Antioxidant A substance that inactivates oxygen-derived free radicals. Examples are selenium, zinc, beta carotene, and vitamins C and E.

Antrum (AN-trum) Any nearly closed cavity or chamber, especially one within a bone, such as a sinus.

Anuria (an-Ū-rē-a) Absence of urine formation or daily urine output of less than 50 mL.

Anus (Ā-nus) The distal end and outlet of the rectum.

Aorta (ā-OR-ta) The main systemic trunk of the arterial system of the body that emerges from the left ventricle.

Aortic (ā-OR-tik) **body** Cluster of chemoreceptors on or near the arch of the aorta that respond to changes in blood levels of oxygen, carbon dioxide, and hydrogen ions (H^+).

Aortic reflex A reflex that helps maintain normal systemic blood pressure; initiated by baroreceptors in the wall of the ascending aorta and arch of the aorta. Nerve impulses from aortic baroreceptors reach the cardiovascular center via sensory axons of the vagus (X) nerves.

Apex (Ā-peks) The pointed end of a conical structure, such as the apex of the heart.

Aphasia (a-FĀ-zē-a) Loss of ability to express oneself properly through speech or loss of verbal comprehension.

Apneustic (ap-NOO-stik) **area** A part of the respiratory center in the pons that sends stimulatory nerve impulses to the inspiratory area that activate and prolong inhalation and inhibit exhalation.

Apnoea (AP-nē-a) Temporary cessation of breathing.

Apocrine (AP-ō-krin) **gland** A type of gland in which the secretory products gather at the free end of the secreting cell and are pinched off, along with some of the cytoplasm, to become the secretion, as in mammary glands.

Aponeurosis (ap′-ō-noo-RŌ-sis) A sheetlike tendon joining one muscle with another or with bone.

Apoptosis (ap′-ōp-TŌ-sis *or* ap-ō-TŌ-sis) Programmed cell death; a normal type of cell death that removes unneeded cells during embryological development, regulates the number of cells in tissues, and eliminates many potentially dangerous cells such as cancer cells. During apoptosis, the DNA fragments, the nucleus condenses, mitochondria cease to function, and the cytoplasm shrinks, but the plasma membrane remains intact. Phagocytes engulf and digest the apoptotic cells, and an inflammatory response does not occur.

Appositional (a-pō-ZISH-o-nal) **growth** Growth due to surface deposition of material, as in the growth in diameter of cartilage and bone. Also called **exogenous** (eks-OJ-e-nus) **growth.**

Aqueous humour (A-kwē-us HŪ-mor) The watery fluid, similar in composition to cerebrospinal fluid, that fills the anterior cavity of the eye.

Arachnoid (a-RAK-noyd) **mater** The middle of the three meninges (coverings) of the brain and spinal cord. Also termed the **arachnoid.**

Arachnoid villus (VIL-us) Berrylike tuft of the arachnoid mater that protrudes into the superior sagittal sinus and through which cerebrospinal fluid is reabsorbed into the bloodstream.

Arbor vitae (AR-bor VĪ-tē) The white matter tracts of the cerebellum, which have a treelike appearance when seen in midsagittal section.

Arch of the aorta The most superior portion of the aorta, lying between the ascending and descending segments of the aorta.

Areola (a-RĒ-ō-la) Any tiny space in a tissue. The pigmented ring around the nipple of the breast.

Arm The part of the upper limb from the shoulder to the elbow.

Arousal (a-ROW-zal) Awakening from sleep, a response due to stimulation of the reticular activating system (RAS).

Arrector pili (a-REK-tor PĪ-lē) Smooth muscles attached to hairs; contraction pulls the hairs into a vertical position, resulting in "goose bumps."

Arrhythmia (a-RITH-mē-a) An irregular heart rhythm. Also called a **dysrhythmia.**

Arteriole (ar-TĒ-rē-ōl) A small, almost microscopic, artery that delivers blood to a capillary.

Arteriosclerosis (ar-tē-rē-ō-skle-RŌ-sis) Group of diseases characterized by thickening of the walls of arteries and loss of elasticity.

Artery (AR-ter-ē) A blood vessel that carries blood away from the heart.

Arthritis (ar-THRĪ-tis) Inflammation of a joint.

Arthrology (ar-THROL-ō-jē) The study or description of joints.

Arthroplasty (AR-thrō-plas′-tē) Surgical replacement of joints, for example, the hip and knee joints.

Arthroscopy (ar-THROS-kō-pē) A procedure for examining the interior of a joint, usually the knee, by inserting an arthroscope into a small incision; used to determine extent of damage, remove torn cartilage, repair cruciate ligaments, and obtain samples for analysis.

Arthrosis (ar-THRŌ-sis) A joint or articulation.

Articular (ar-TIK-ū-lar) **capsule** Sleevelike structure around a synovial joint composed of a fibrous capsule and a synovial membrane.

Articular cartilage (KAR-ti-lij) Hyaline cartilage attached to articular bone surfaces.

Articular disc Fibrocartilage pad between articular surfaces of bones of some synovial joints. Also called a **meniscus** (men-IS-kus).

Articulation (ar-tik-ū-LĀ-shun) A joint; a point of contact between bones, cartilage and bones, or teeth and bones.

Arytenoid (ar′-i-TĒ-noyd) **cartilages** A pair of small, pyramidal cartilages of the larynx that attach to the vocal folds and intrinsic pharyngeal muscles and can move the vocal folds.

Ascending colon (KŌ-lon) The part of the large intestine that passes superiorly from the caecum to the inferior border of the liver, where it bends at the right colic (hepatic) flexure to become the transverse colon.

Ascites (a-SĪ-tēz) Abnormal accumulation of serous fluid in the peritoneal cavity.

Association areas Large cortical regions on the lateral surfaces of the occipital, parietal, and temporal lobes and on the frontal lobes anterior to the motor areas connected by many motor and sensory axons to other parts of the cortex. The association areas are concerned with motor patterns, memory, concepts of word-hearing and word-seeing, reasoning, will, judgment, and personality traits.

Asthma (AZ-ma) Usually allergic reaction characterized by smooth muscle spasms in bronchi resulting in wheezing and difficult breathing. Also called **bronchial asthma.**

Astigmatism (a-STIG-ma-tizm) An irregularity of the lens or cornea of the eye causing the image to be out of focus and producing faulty vision.

Astrocyte (AS-trō-sīt) A neuroglial cell having a star shape that participates in brain development and the metabolism of neurotransmitters, helps form the blood–brain barrier, helps maintain the proper balance of K^+ for generation of nerve impulses, and provides a link between neurons and blood vessels.

Ataxia (a-TAK-sē-a) A lack of muscular coordination, lack of precision.

Atherosclerosis (ath-er-ō-skle-RŌ-sis) A progressive disease characterized by the formation in the walls of large and medium-sized arteries of lesions called atherosclerotic plaques.

Atherosclerotic (ath′-er-ō-skle-RO-tik) **plaque** (PLAK) A lesion that results from accumulated cholesterol and smooth muscle fibers (cells) of the tunica media of an artery; may become obstructive.

Atom Unit of matter that makes up a chemical element; consists of a nucleus (containing positively charged protons and uncharged neutrons) and negatively charged electrons that orbit the nucleus.

Atresia (a-TRĒ-zē-a) Degeneration and reabsorption of an ovarian follicle before it fully matures and ruptures; abnormal closure of a passage, or absence of a normal body opening.

Atrial fibrillation (Ā-trē-al fib-ri-LĀ-shun) **(AF)** Asynchronous contraction of cardiac muscle fibers in the atria that results in the cessation of atrial pumping.

Atrial natriuretic (nā′-trē-ū-RET-ik) **peptide (ANP)** Peptide hormone, produced by the atria of the heart in response to stretching, that inhibits aldosterone production and thus lowers blood pressure; causes natriuresis, increased urinary excretion of sodium.

Atrioventricular (AV) (ā′-trē-ō-ven-TRIK-ū-lar) **bundle** The part of the conduction system of the heart that begins at the atrioventricular (AV) node, passes through the cardiac skeleton separating the atria and the ventricles, then extends a short distance down the interventricular septum before splitting into right and left bundle branches. Also called the **bundle of His** (HISS).

Atrioventricular (AV) node The part of the conduction system of the heart made up of a compact mass of conducting cells located in the septum between the two atria.

Atrioventricular (AV) valve A heart valve made up of membranous flaps or cusps that allows blood to flow in one direction only, from an atrium into a ventricle.

Atrium (Ā-trē-um) A superior chamber of the heart.

Atrophy (AT-rō-fē) Wasting away or decrease in size of a part, due to a failure, abnormality of nutrition, or lack of use.

Auditory ossicle (AW-di-tō-rē OS-si-kul) One of the three small bones of the middle ear called the malleus, incus, and stapes.

Auditory tube The tube that connects the middle ear with the nose and nasopharynx region of the throat. Also called the **eustachian** (ū-STĀ-kē-an *or* ū-STĀ-shun) **tube** or **pharyngotympanic tube.**

Auscultation (aws-kul-TĀ-shun) Examination by listening to sounds in the body.

Autoimmunity An immunological response against a person's own tissues.

Autolysis (aw-TOL-i-sis) Self destruction of cells by their own lysosomal digestive enzymes after death or in a pathological process.

Autonomic ganglion (aw′-tō-NOM-ik GANG-lē-on) A cluster of cell bodies of sympathetic or parasympathetic neurons located outside the central nervous system.

Autonomic nervous system (ANS) Visceral sensory (afferent) and visceral motor (efferent) neurons. Autonomic motor neurons, both sympathetic and parasympathetic, conduct nerve impulses from the central nervous system to smooth muscle, cardiac muscle, and glands. So named because this part of the nervous system was thought to be self-governing or spontaneous.

Autonomic plexus (PLEK-sus) A network of sympathetic and parasympathetic axons; examples are the cardiac, celiac, and pelvic plexuses, which are located in the thorax, abdomen, and pelvis, respectively.

Autophagy (aw-TOF-a-jē) Process by which worn-out organelles are digested within lysosomes.

Autopsy (AW-top-sē) The examination of the body after death.

Autorhythmic (aw′-tō-RITH-mik) **cells** Cardiac or smooth muscle fibers that are self-excitable (generate impulses without an external stimulus); act as the heart's pacemaker and conduct the pacing impulse through the conduction system of the heart; self-excitable neurons in the central nervous system, as in the inspiratory area of the brain stem.

Autosome (AW-tō-sōm) Any chromosome other than the X and Y chromosomes (sex chromosomes).

Axilla (ak-SIL-a) The small hollow beneath the arm where it joins the body at the shoulders. Also called the **armpit.**

Axon (AK-son) The usually single, long process of a nerve cell that propagates a nerve impulse toward the axon terminals.

Axon terminal Terminal branch of an axon where synaptic vesicles undergo exocytosis to release neurotransmitter molecules. Also called **telodendria** (tel′-o-DEN-drea).

B

B cell A lymphocyte that can develop into a clone of antibody-producing plasma cells or memory cells when properly stimulated by a specific antigen.

Babinski (ba-BIN-skē) **sign** Extension of the great toe, with or without fanning of the other toes, in response to stimulation of the outer margin of the sole; normal up to 18 months of age and indicative of damage to descending motor pathways such as the corticospinal tracts after that.

Back The posterior part of the body; the dorsum.

Ball-and-socket joint A synovial joint in which the rounded surface of one bone moves within a cup-shaped depression or socket of another bone, as in the shoulder or hip joint. Also called a **spheroid** (SFĒ-royd) **joint.**

Baroreceptor (bar′-ō-re-SEP-tor) Neuron capable of responding to changes in blood or **stretch receptor** air or fluid pressure. Also called a **pressoreceptor.**

Basal nuclei Paired clusters of grey matter deep in each cerebral hemisphere including the globus pallidus, putamen, and caudate nucleus. Together, the caudate nucleus and putamen are known as the corpus striatum. Nearby structures that are functionally linked to the basal nuclei are the substantia nigra of the midbrain and the subthalamic nuclei of the diencephalon.

Basement membrane Thin, extracellular layer between epithelium and connective tissue consisting of a basal lamina and a reticular lamina.

Basilar (BĀS-i-lar) **membrane** A membrane in the cochlea of the internal ear that separates the cochlear duct from the scala tympani and on which the spiral organ (organ of Corti) rests.

Basophil (BĀ-sō-fil) A type of white blood cell characterized by a pale nucleus and large granules that stain blue-purple with basic dyes.

Belly The abdomen. The gaster or prominent, fleshy part of a skeletal muscle.

Beta (BĀ-ta) **cell** A type of cell in the pancreatic islets (islets of Langerhans) in the pancreas that secretes the hormone insulin. Also called a **B cell.**

Beta (β) receptor A type of adrenergic receptor for epinephrine and norepinephrine; found on visceral effectors innervated by sympathetic postganglionic neurons.

Bicuspid (bī-KUS-pid) **valve** Atrioventricular (AV) valve on the left side of the heart. Also called the **mitral valve.**

Bilateral (bī-LAT-er-al) Pertaining to two sides of the body.

Bile (BĪL) A secretion of the liver consisting of water, bile salts, bile pigments, cholesterol, lecithin, and several ions; it emulsifies lipids prior to their digestion.

Bilirubin (bil-ē-ROO-bin) An orange pigment that is one of the end products of haemoglobin breakdown in the hepatocytes and is excreted as a waste material in bile.

Biopsy (BĪ-op-sē) The removal of a sample of living tissue to help diagnose a disorder, for example, cancer.

Blastocyst (BLAS-tō-sist) In the development of an embryo, a hollow ball of cells that consists of a blastocele (the internal cavity), trophoblast (outer cells), and embryoblast (inner cell mass).

Blastocyst (BLAS-tō-sist) **cavity** The fluid-filled cavity within the blastocyst. Also called the **blastocele.**

Blastomere (BLAS-tō-mēr) One of the cells resulting from the cleavage of a fertilized ovum.

Blastula (BLAS-tyū-la) An early stage in the development of a zygote.

Blind spot Area in the retina at the end of the optic (II) nerve in which there are no photoreceptors. Also called **optic disc.**

Blood The fluid that circulates through the heart, arteries, capillaries, and veins and that constitutes the chief means of transport within the body.

Blood–brain barrier (BBB) A barrier consisting of specialized brain capillaries and astrocytes that prevents the passage of materials from the blood to the cerebrospinal fluid and brain.

Blood clot A gel that consists of the formed elements of blood trapped in a network of insoluble protein fibers.

Blood island Isolated mass of mesoderm derived from angioblasts and from which blood vessels develop.

Blood pressure (BP) Force exerted by blood against the walls of blood vessels due to contraction of the heart and influenced by the elasticity of the vessel walls; clinically, a measure of the pressure in arteries during ventricular systole and ventricular diastole.

Blood reservoir (REZ-er-vwar) Systemic veins and venules that contain large amounts of blood that can be moved quickly to parts of the body requiring the blood.

Blood–testis barrier (BTB) A barrier formed by Sertoli cells that prevents an immune response against antigens produced by spermatogenic cells by isolating the cells from the blood.

Body cavity A space within the body that contains various internal organs.

Bolus (BŌ-lus) A soft, rounded mass, usually food, that is swallowed.

Bone remodelling Replacement of old bone by new bone tissue.

Bony labyrinth (LAB-i-rinth) A series of cavities within the petrous portion of the temporal bone forming the vestibule, cochlea, and semicircular canals of the inner ear.

Bowman's capsule *See* **Glomerular capsule.**

Brachial plexus (BRĀ-kē-al PLEK-sus) A network of nerve axons of the ventral rami of spinal nerves C5, C6, C7, C8, and T1. The nerves that emerge from the brachial plexus supply the upper limb.

Bradycardia (brād′-i-KAR-dē-a) A slow resting heart or pulse rate (under 50 beats per minute).

Brain The part of the central nervous system contained within the cranial cavity.

Brain stem The portion of the brain immediately superior to the spinal cord, made up of the medulla oblongata, pons, and midbrain.

Brain waves Electrical signals that can be recorded from the skin of the head due to electrical activity of brain neurons.

Broad ligament A double fold of parietal peritoneum attaching the uterus to the side of the pelvic cavity.

Broca's (BRŌ-kaz) **speech area** Motor area of the brain in the frontal lobe that translates thoughts into speech. Also called the **motor speech area.**

Bronchi (BRON-kī) Branches of the respiratory passageway including primary bronchi (the two divisions of the trachea), secondary or lobar bronchi (divisions of the primary bronchi that are distributed to the lobes of the lung), and tertiary or segmental bronchi (divisions of the secondary bronchi that are distributed to bronchopulmonary segments of the lung). *Singular* is **bronchus.**

Bronchial (BRON-kē-al) **tree** The trachea, bronchi, and their branching structures up to and including the terminal bronchioles.

Bronchiole (BRONG-kē-ōl) Branch of a tertiary bronchus further dividing into terminal bronchioles (distributed to lobules of the lung), which divide into respiratory bronchioles (distributed to alveolar sacs).

Bronchitis (brong-KĪ-tis) Inflammation of the mucous membrane of the bronchial tree; characterized by hypertrophy and hyperplasia of seromucous glands and goblet cells that line the bronchi which results in a productive cough.

Bronchopulmonary (brong′-kō-PUL-mō-ner-ē) **segment** One of the smaller divisions of a lobe of a lung supplied by its own branches of a bronchus.

Brunner's gland *See* **Duodenal gland.**

Buccal (BUK-al) Pertaining to the cheek or mouth.

Buffer system A weak acid and the salt of that acid (that functions as a weak base). Buffers prevent drastic changes in pH by converting strong acids and bases to weak acids and bases.

Bulb of penis Expanded portion of the base of the corpus spongiosum penis.

Bulbourethral (bul′-bō-ū-RĒ-thral) **gland** One of a pair of glands located inferior to the prostate on either side of the urethra that secretes an alkaline fluid into the cavernous urethra. Also called a **Cowper's** (KOW-perz) **gland.**

Bulimia (boo-LIM-ē-a *or* boo-LĒ-mē-a) A disorder characterized by overeating at least twice a week followed by purging by self-induced vomiting, strict dieting or fasting, vigorous exercise, or use of laxatives or diuretics. Also called **binge–purge syndrome.**

Bulk-phase endocytosis A process by which most body cells can ingest membrane-surrounded droplets of interstitial fluid.

Bundle branch One of the two branches of the atrioventricular (AV) bundle made up of specialized muscle fibers (cells) that transmit electrical impulses to the ventricles.

Bundle of His *See* **Atrioventricular (AV) bundle.**

Burn Tissue damage caused by excessive heat, electricity, radioactivity, or corrosive chemicals that denature (break down) proteins in the skin.

Bursa (BUR-sa) A sac or pouch of synovial fluid located at friction points, especially about joints.

Bursitis (bur-SĪ-tis) Inflammation of a bursa.

Buttocks (BUT-oks) The two fleshy masses on the posterior aspect of the inferior trunk, formed by the gluteal muscles.

C

Caecum (SĒ-kum) A blind pouch at the proximal end of the large intestine that attaches to the ileum.

Calcaneal (kal-KĀ-nē-al) **tendon** The tendon of the soleus, gastrocnemius, and plantaris muscles at the back of the heel. Also called the **Achilles** (a-KIL-ēz) **tendon.**

Calcification (kal′-si-fi-KĀ-shun) Deposition of mineral salts, primarily hydroxyapatite, in a framework formed by collagen fibers in which the tissue hardens. Also called **mineralization** (min′-e-ral-i-ZĀ-shun).

Calcitonin (kal-si-TŌ-nin) **(CT)** A hormone produced by the parafollicular cells of the thyroid gland that can lower the amount of blood calcium and phosphates by inhibiting bone resorption (breakdown of bone extracellular matrix) and by accelerating uptake of calcium and phosphates into bone matrix.

Calculus (KAL-kū-lus) A stone, or insoluble mass of crystallized salts or other material, formed within the body, as in the gallbladder, kidney, or urinary bladder.

Callous (KAL-lus) A growth of new bone tissue in and around a fractured area, ultimately replaced by mature bone. An acquired, localized thickening.

Calyx (KĀL-iks) Any cuplike division of the kidney pelvis. *Plural* is **calyces** (KĀ-li-sēz).

Canal (ka-NAL) A narrow tube, channel, or passageway.

Canaliculus (kan′-a-LIK-ū-lus) A small channel or canal, as in bones, where they connect lacunae. *Plural* is **canaliculi** (kan′-a-LIK-ū-lī).

Canal of Schlemm *See* **Scleral venous sinus.**

Cancer A group of diseases characterized by uncontrolled or abnormal cell division.

Capacitation (ka-pas′-i-TĀ-shun) The functional changes that sperm undergo in the female reproductive tract that allow them to fertilize a secondary oocyte.

Capillary (KAP-i-lar′-ē) A microscopic blood vessel located between an arteriole and venule through which materials are exchanged between blood and interstitial fluid.

Carbohydrate Organic compound consisting of carbon, hydrogen, and oxygen; the ratio of hydrogen to oxygen atoms is usually 2:1. Examples include sugars, glycogen, starches, and glucose.

Carcinogen (kar-SIN-ō-jen) A chemical substance or radiation that causes cancer.

Cardiac (KAR-dē-ak) **arrest** Cessation of an effective heartbeat in which the heart is completely stopped or in ventricular fibrillation.

Cardiac cycle A complete heartbeat consisting of systole (contraction) and diastole (relaxation) of both atria plus systole and diastole of both ventricles.

Cardiac muscle Striated muscle fibers (cells) that form the wall of the heart; stimulated by an intrinsic conduction system and autonomic motor neurons.

Cardiac notch An angular notch in the anterior border of the left lung into which part of the heart fits.

Cardiac output (CO) The volume of blood ejected from the left ventricle (or the right ventricle) into the aorta (or pulmonary trunk) each minute.

Cardinal ligament A ligament of the uterus, extending laterally from the cervix and vagina as a continuation of the broad ligament.

Cardiogenic area (kar-dē-ō-JEN-ik) A group of mesodermal cells in the head end of an embryo that gives rise to the heart.

Cardiology (kar-dē-OL-ō-jē) The study of the heart and diseases associated with it.

Cardiovascular (kar-dē-ō-VAS-kū-lar) **(CV) center** Groups of neurons scattered within the medulla oblongata that regulate heart rate, force of contraction, and blood vessel diameter.

Cardiovascular physiology Study of the functions of the heart and blood vessels.

Cardiovascular system System that consists of blood, the heart, and blood vessels.

Carotene (KAR-ō-tēn) Antioxidant precursor of vitamin A, which is needed for synthesis of photopigments; yellow-orange pigment present in the stratum corneum of the epidermis. Accounts for the yellowish coloration of skin. Also termed **beta carotene.**

Carotid (ka-ROT-id) **body** Cluster of chemoreceptors on or near the carotid sinus that respond to changes in blood levels of oxygen, carbon dioxide, and hydrogen ions.

Carotid sinus A dilated region of the internal carotid artery just superior to where it branches from the common carotid artery; it contains baroreceptors that monitor blood pressure.

Carpal bones The eight bones of the wrist. Also called **carpals.**

Carpus (KAR-pus) A collective term for the eight bones of the wrist.

Cartilage (KAR-ti-lij) A type of connective tissue consisting of chondrocytes in lacunae embedded in a dense network of collagen and elastic fibers and an extracellular matrix of chondroitin sulfate.

Cartilaginous (kar′-ti-LAJ-i-nus) **joint** A joint without a synovial (joint) cavity where the articulating bones are held tightly together by cartilage, allowing little or no movement.

Catabolism (ka-TAB-ō-lizm) Chemical reactions that break down complex organic compounds into simple ones, with the net release of energy.

Cataract (KAT-a-rakt) Loss of transparency of the lens of the eye or its capsule or both.

Cauda equina (KAW-da ē-KWĪ-na) A tail-like array of roots of spinal nerves at the inferior end of the spinal cord.

Caudal (KAW-dal) Pertaining to any tail-like structure; inferior in position.

Coeliac plexus (SĒ-lē-ak PLEK-sus) A large mass of autonomic ganglia and axons located at the level of the superior part of the first lumbar vertebra. Also called the **solar plexus.**

Cell The basic structural and functional unit of all organisms; the smallest structure capable of performing all the activities vital to life.

Cell biology The study of cellular structure and function. Also called **cytology.**

Cell cycle Growth and division of a single cell into two identical cells; consists of interphase and cell division.

Cell division Process by which a cell reproduces itself that consists of a nuclear division (mitosis) and a cytoplasmic division (cytokinesis); types include somatic and reproductive cell division.

Cell junction Point of contact between plasma membranes of tissue cells.

Cellular respiration The oxidation of glucose to produce ATP that involves glycolysis, acetyl coenzyme A formation, the Krebs cycle, and the electron transport chain.

Cementum (se-MEN-tum) Calcified tissue covering the root of a tooth.

Central canal A microscopic tube running the length of the spinal cord in the grey commissure. A circular channel running longitudinally in the center of an osteon (haversian system) of mature compact bone, containing blood and lymphatic vessels and nerves. Also called a **haversian** (ha-VER-shun) **canal.**

Central fovea (FŌ-ve-a) A depression in the center of the macula lutea of the retina, containing cones only and lacking blood vessels; the area of highest visual acuity (sharpness of vision).

Central nervous system (CNS) That portion of the nervous system that consists of the brain and spinal cord.

Centrioles (SEN-trē-ōlz) Paired, cylindrical structures of a centrosome, each consisting of a ring of microtubules and arranged at right angles to each other.

Centromere (SEN-trō-mēr) The constricted portion of a chromosome where the two chromatids are joined; serves as the point of attachment for the microtubules that pull chromatids during anaphase of cell division.

Centrosome (SEN-trō-sōm) A dense network of small protein fibers near the nucleus of a cell, containing a pair of centrioles and pericentriolar material.

Cephalic (se-FAL-ik) Pertaining to the head; superior in position.

Cerebellar peduncle (ser-e-BEL-ar pe-DUNG-kul) A bundle of nerve axons connecting the cerebellum with the brain stem.

Cerebellum (ser′-e-BEL-um) The part of the brain lying posterior to the medulla oblongata and pons; governs balance and coordinates skilled movements.

Cerebral aqueduct (SER-ē-bral AK-we-dukt) A channel through the midbrain connecting the third and fourth ventricles and containing cerebrospinal fluid. Also termed the **aqueduct of the midbrain.**

Cerebral arterial circle A ring of arteries forming an anastomosis at the base of the brain between the internal carotid and basilar arteries and arteries supplying the cerebral cortex. Also called the **circle of Willis.**

Cerebral cortex The surface of the cerebral hemispheres, 2–4 mm thick, consisting of grey matter; arranged in six layers of neuronal cell bodies in most areas.

Cerebral peduncle (pe-DUNG-kul or PĒ-dung-kul) One of a pair of nerve axon bundles located on the anterior surface of the midbrain, conducting nerve impulses between the pons and the cerebral hemispheres.

Cerebrospinal (se-rē′-brō-SPĪ-nal) **fluid (CSF)** A fluid produced by ependymal cells that cover choroid plexuses in the ventricles of the brain; the fluid circulates in the ventricles, the central canal, and the subarachnoid space around the brain and spinal cord.

Cerebrovascular (se- rē′-brō-VAS-kū-lar) **accident (CVA)** Destruction of brain tissue (infarction) resulting from obstruction or rupture of blood vessels that supply the brain. Also called a **stroke** or **brain attack.**

Cerebrum (se-RĒ-brum *or* SER-e-brum) The two hemispheres of the forebrain (derived from the telencephalon), making up the largest part of the brain.

Cerumen (se-ROO-men) Waxlike secretion produced by ceruminous glands in the external auditory meatus (ear canal). Also termed **earwax.**

Ceruminous (se-RŪ-mi-nus) **gland** A modified sudoriferous (sweat) gland in the external auditory meatus that secretes cerumen (ear wax).

Cervical ganglion (SER-vi-kul GANG-glē-on) A cluster of cell bodies of postganglionic sympathetic neurons located in the neck, near the vertebral column.

Cervical plexus (PLEK-sus) A network formed by nerve axons from the ventral rami of the first four cervical nerves and receiving gray rami communicantes from the superior cervical ganglion.

Cervix (SER-viks) Neck; any constricted portion of an organ, such as the inferior cylindrical part of the uterus.

Chemical reaction The formation of new chemical bonds or the breaking of old chemical bonds between atoms.

Chemistry (KEM-is-trē) The science of the structure and interactions of matter.

Chemoreceptor (kē′-mō-rē-SEP-tor) Sensory receptor that detects the presence of a specific chemical.

Chiasm (KĪ-azm) A crossing; especially the crossing of axons in the optic (II) nerve.

Chief cell The secreting cell of a gastric gland that produces pepsinogen, the precursor of the enzyme pepsin, and the enzyme gastric lipase. Also called a **zymogenic** (zī′-mō-JEN-ik) **cell.** Cell in the parathyroid glands that secretes parathyroid hormone (PTH). Also called a **principal cell.**

Cholecystectomy (kō′-lē-sis-TEK-tō-mē) Surgical removal of the gallbladder.

Cholecystitis (kō′-lē-sis-TĪ-tis) Inflammation of the gallbladder.

Cholesterol (kō-LES-te-rol) Classified as a lipid, the most abundant steroid in animal tissues; located in cell membranes and used for the synthesis of steroid hormones and bile salts.

Cholinergic (kō′-lin-ER-jik) **neuron** A neuron that liberates acetylcholine as its neurotransmitter.

Chondrocyte (KON-drō-sīt) Cell of mature cartilage.

Chondroitin (kon-DROY-tin) **sulfate** An amorphous extracellular matrix material found outside connective tissue cells.

Chordae tendineae (KOR-dē TEN-di-nē-ē) Tendonlike, fibrous cords that connect atrioventricular valves of the heart with papillary muscles.

Chorion (KŌ-rē-on) The most superficial fetal membrane that becomes the principal embryonic portion of the placenta; serves a protective and nutritive function.

Chorionic villi (kō-rē-ON-ik VIL-li) Fingerlike projections of the chorion that grow into the decidua basalis of the endometrium and contain fetal blood vessels.

Chorionic villi sampling (CVS) The removal of a sample of chorionic villus tissue by means of a catheter to analyze the tissue for prenatal genetic defects.

Choroid (KŌ-royd) One of the vascular coats of the eyeball.

Choroid plexus (PLEK-sus) A network of capillaries located in the roof of each of the four ventricles of the brain; ependymal cells around choroid plexuses produce cerebrospinal fluid.

Chromaffin (KRŌ-maf-in) **cell** Cell that has an affinity for chrome salts, due in part to the presence of the precursors of the neurotransmitter epinephrine; found, among other places, in the adrenal medulla.

Chromatid (KRŌ-ma-tid) One of a pair of identical connected nucleoprotein strands that are joined at the centromere and separate during cell division, each becoming a chromosome of one of the two daughter cells.

Chromatin (KRŌ-ma-tin) The threadlike mass of genetic material, consisting of DNA and histone proteins, that is present in the nucleus of a nondividing or interphase cell.

Chromatolysis (krō′-ma-TOL-i-sis) The breakdown of Nissl bodies into finely granular masses in the cell body of a neuron whose axon has been damaged.

Chromosome (KRŌ-mō-sōm) One of the small, threadlike structures in the nucleus of a cell, normally 46 in a human diploid cell, that bears the genetic material; composed of DNA and proteins (histones) that form a delicate chromatin thread during interphase; becomes packaged into compact rodlike structures that are visible under the light microscope during cell division.

Chronic (KRON-ik) Long term or frequently recurring; applied to a disease that is not acute.

Chronic obstructive pulmonary disease (COPD) A disease, such as bronchitis or emphysema, in which there is some degree of obstruction of airways and consequent increase in airway resistance.

Chyle (KĪL) The milky-appearing fluid found in the lacteals of the small intestine after absorption of lipids in food.

Chyme (KĪM) The semifluid mixture of partly digested food and digestive secretions found in the stomach and small intestine during digestion of a meal.

Ciliary (SIL-ē-ar′-ē) **body** One of the three parts of the vascular tunic of the eyeball, the others being the choroid and the iris; includes the ciliary muscle and the ciliary processes.

Ciliary ganglion (GANG-glē-on) A very small parasympathetic ganglion whose preganglionic axons come from the oculomotor (III) nerve and whose postganglionic axons carry nerve impulses to the ciliary muscle and the sphincter muscle of the iris.

Cilium (SIL-ē-um) A hair or hairlike process projecting from a cell that may be used to move the entire cell or to move substances along the surface of the cell. *Plural* is **cilia.**

Circadian (ser-KĀ-dē-an) **rhythm** The pattern of biological activity on a 24-hour cycle, such as the sleep–wake cycle.

Circle of Willis *See* **Cerebral arterial circle.**

Circular folds Permanent, deep, transverse folds in the mucosa and submucosa of the small intestine that increase the surface area for absorption. Also called **plicae circulares** (PLĪ-kē SER-kū-lar-ēs).

Circulation time The time required for a drop of blood to pass from the right atrium, through pulmonary circulation, back to the left atrium, through systemic circulation down to the foot, and back again to the right atrium.

Circumduction (ser-kum-DUK-shun) A movement at a synovial joint in which the distal end of a bone moves in a circle while the proximal end remains relatively stable.

Cirrhosis (si-RŌ-sis) A liver disorder in which the parenchymal cells are destroyed and replaced by connective tissue.

Cisterna chyli (sis-TER-na KĪ-lē) The origin of the thoracic duct.

Cleavage (KLĒV-ij) The rapid mitotic divisions following the fertilisation of a secondary oocyte, resulting in an increased number of progressively smaller cells, called blastomeres.

Clitoris (KLI-to-ris) An erectile organ of the female, located at the anterior junction of the labia minora, that is homologous to the male penis.

Clone (KLŌN) A population of identical cells.

Coarctation (kō′-ark-TĀ-shun) **of the aorta** A congenital heart defect in which a segment of the aorta is too narrow. As a result, the flow of oxygenated blood to the body is reduced, the left ventricle is forced to pump harder, and high blood pressure develops.

Coccyx (KOK-siks) The fused bones at the inferior end of the vertebral column.

Cochlea (KOK-lē-a) A winding, cone-shaped tube forming a portion of the inner ear and containing the spiral organ (organ of Corti).

Cochlear duct The membranous cochlea consisting of a spirally arranged tube enclosed in the bony cochlea and lying along its outer wall. Also called the **scala media** (SCA-la MĒ-dē-a).

Collagen (KOL-a-jen) A protein that is the main organic constituent of connective tissue.

Collateral circuit The alternate route taken by blood through an anastomosis.

Colliculus (ko-LIK-ū-lus) A small elevation.

Colon The portion of the large intestine consisting of ascending, transverse, descending, and sigmoid portions.

Colony-stimulating factor (CSF) One of a group of molecules that stimulates development of white blood cells. Examples are macrophage CSF and granulocyte CSF.

Colostrum (kō-LOS-trum) A thin, cloudy fluid secreted by the mammary glands a few days prior to or after delivery before true milk is produced.

Column (KOL-um) Group of white matter tracts in the spinal cord.

Common bile duct A tube formed by the union of the common hepatic duct and the cystic duct that empties bile into the duodenum at the hepatopancreatic ampulla (ampulla of Vater).

Compact (dense) bone tissue Bone tissue that contains few spaces between osteons (haversian systems); forms the external portion of all bones and the bulk of the diaphysis (shaft) of long bones; is found immediately deep to the periosteum and external to spongy bone.

Concha (KON-ka) A scroll-like bone found in the skull. *Plural* is **conchae** (KON-kē).

Concussion (kon-KUSH-un) Traumatic injury to the brain that produces no visible bruising but may result in abrupt, temporary loss of consciousness.

Conduction system A group of autorhythmic cardiac muscle fibers that generates and distributes electrical impulses to stimulate coordinated contraction of the heart chambers; includes the sinoatrial (SA) node, the atrioventricular (AV) node, the atrioventricular (AV) bundle, the right and left bundle branches, and the Purkinje fibers.

Condyloid (KON-di-loyd) **joint** A synovial joint structured so that an oval-shaped condyle of one bone fits into an elliptical cavity of another bone, permitting side-to-side and back-and-forth movements, such as the joint at the wrist between the radius and carpals. Also called an **ellipsoidal** (ē-lip-SOYD-al) **joint.**

Cone (KŌN) The type of photoreceptor in the retina that is specialized for highly acute color vision in bright light.

Congenital (kon-JEN-i-tal) Present at the time of birth.

Conjunctiva (kon′-junk-TĪ-va) The delicate membrane covering the eyeball and lining the eyes.

Connective tissue One of the most abundant of the four basic tissue types in the body, performing the functions of binding and supporting; consists of relatively few cells in a generous extracellular matrix (the ground substance and fibers between the cells).

Consciousness (KON-shus-nes) A state of wakefulness in which an individual is fully alert, aware, and oriented, partly as a result of feedback between the cerebral cortex and reticular activating system.

Continuous conduction (kon-DUK-shun) Propagation of an action potential (nerve impulse) in a step-by-step depolarization of each adjacent area of an axon membrane.

Contraception (kon′-tra-SEP-shun) The prevention of fertilisation or impregnation without destroying fertility.

Contractility (kon′-trak-TIL-i-tē) The ability of cells or parts of cells to actively generate force to undergo shortening for movements. Muscle fibers (cells) exhibit a high degree of contractility.

Contralateral (KON-tra-lat-er-al) On the opposite side; affecting the opposite side of the body.

Conus medullaris (KŌ-nus med-ū-LAR-is) The tapered portion of the spinal cord inferior to the lumbar enlargement.

Convergence (con-VER-jens) A synaptic arrangement in which the synaptic end bulbs of several presynaptic neurons terminate on one postsynaptic neuron. The medial movement of the two eyeballs so that both are directed toward a near object being viewed in order to produce a single image.

Cornea (KOR-nē-a) The nonvascular, transparent fibrous coat through which the iris of the eye can be seen.

Corona (kō-RŌ-na) Margin of the glans penis.

Corona radiata The innermost layer of granulosa cells that is firmly attached to the zona pellucida around a secondary oocyte.

Coronary artery disease (CAD) A condition such as atherosclerosis that causes narrowing of coronary arteries so that blood flow to the heart is reduced. The result is **coronary heart disease (CHD),** in which the heart muscle receives inadequate blood flow due to an interruption of its blood supply.

Coronary circulation The pathway followed by the blood from the ascending aorta through the blood vessels supplying the heart and returning to the right atrium. Also called **cardiac circulation.**

Coronary sinus (SĪ-nus) A wide venous channel on the posterior surface of the heart that collects the blood from the coronary circulation and returns it to the right atrium.

Corpus albicans (KOR-pus AL-bi-kanz) A white fibrous patch in the ovary that forms after the corpus luteum regresses.

Corpus callosum (kal-LŌ-sum) The great commissure of the brain between the cerebral hemispheres.

Corpuscle of touch See **Meissner corpuscle.**

Corpus luteum (LOO-tē-um) A yellowish body in the ovary formed when a follicle has discharged its secondary oocyte; secretes oestrogens, progesterone, relaxin, and inhibin.

Corpus striatum (strī-Ā-tum) An area in the interior of each cerebral hemisphere composed of the caudate and putamen of the basal ganglia and white matter of the internal capsule, arranged in a striated manner.

Cortex (KOR-teks) An outer layer of an organ. The convoluted layer of grey matter covering each cerebral hemisphere.

Costal (KOS-tal) Pertaining to a rib.

Cramp A spasmodic, usually painful contraction of a muscle.

Cranial (KRĀ-nē-al) **cavity** A body cavity formed by the cranial bones and containing the brain.

Cranial nerve One of 12 pairs of nerves that leave the brain; pass through foramina in the skull; and supply sensory and motor neurons to the head, neck, part of the trunk, and viscera of the thorax and abdomen. Each is designated by a Roman numeral and a name.

Craniosacral (krā-nē-ō-SĀK-ral) **outflow** The axons of parasympathetic preganglionic neurons, which have their cell bodies located in nuclei in the brain stem and in the lateral grey matter of the sacral portion of the spinal cord.

Cranium (KRĀ-nē-um) The skeleton of the skull that protects the brain and the organs of sight, hearing, and balance; includes the frontal, parietal, temporal, occipital, sphenoid, and ethmoid bones.

Crista (KRIS-ta) A crest or ridged structure. A small elevation in the ampulla of each semicircular duct that contains receptors for dynamic equilibrium. *Plural* is **cristae.**

Crossing-over The exchange of a portion of one chromatid with another during meiosis. It permits an exchange of genes among chromatids and is one factor that results in genetic variation of progeny.

Crus (KRUS) **of penis** Separated, tapered portion of the corpora cavernosa penis. *Plural* is **crura** (KROO-ra).

Crypt of Lieberkühn *See* **Intestinal gland.**

Cryptorchidism (krip-TOR-ki-dizm) The condition of undescended testes.

Cuneate (KŪ-nē-āt) **nucleus** A group of neurons in the inferior part of the medulla oblongata in which axons of the cuneate fasciculus terminate.

Cupula (KU-pū-la) A mass of gelatinous material covering the hair cells of a crista; a sensory receptor in the ampulla of a semicircular canal stimulated when the head moves.

Cushing's syndrome Condition caused by a hypersecretion of glucocorticoids characterized by spindly legs, "moon face," "buffalo hump," pendulous abdomen, flushed facial skin, poor wound healing, hyperglycemia, osteoporosis, hypertension, and increased susceptibility to disease.

Cutaneous (kū-TĀ-nē-us) Pertaining to the skin.

Cyanosis (sī-a-NŌ-sis) A blue or dark purple discoloration, most easily seen in nail beds and mucous membranes, that results from an increased concentration of deoxygenated (reduced) haemoglobin (more than 5 g/dL).

Cyst (SIST) A sac with a distinct connective tissue wall, containing a fluid or other material.

Cystic (SIS-tik) **duct** The duct that carries bile from the gallbladder to the common bile duct.

Cystitis (sis-TĪ-tis) Inflammation of the urinary bladder.

Cytokinesis (sī′-tō-ki-NĒ-sis) Distribution of the cytoplasm into two separate cells during cell division; coordinated with nuclear division (mitosis).

Cytolysis (sī-TOL-i-sis) The rupture of living cells in which the contents leak out.

Cytoplasm (SĪ-tō-plasm) Cytosol plus all organelles except the nucleus.

Cytoskeleton Complex internal structure of cytoplasm consisting of microfilaments, microtubules, and intermediate filaments.

Cytosol (SĪ-tō-sol) Semifluid portion of cytoplasm in which organelles and inclusions are suspended and solutes are dissolved. Also called **intracellular fluid.**

D

Dartos (DAR-tōs) The contractile smooth muscular tissue deep to the skin of the scrotum.

Decidua (dē-SID-ū-a) That portion of the endometrium of the uterus (all but the deepest layer) that is modified during pregnancy and shed after childbirth.

Deciduous (dē-SID-ū-us) Falling off or being shed seasonally or at a particular stage of development. In the body, referring to the first set of teeth.

Decussation (dē′-ku-SĀ-shun) A crossing-over to the opposite (contralateral) side; an example is the crossing of 90% of the axons in the large motor tracts to opposite sides in the medullary pyramids.

Decussation (dē′-ku-SĀ-shun) **of pyramids** The crossing of most axons (90%) in the left pyramid of the medulla to the right side and the crossing of most axons (90%) in the right pyramid to the left side.

Deep Away from the surface of the body or an organ.

Deep abdominal inguinal (IN-gwi-nal) **ring** A slit-like opening in the aponeurosis of the transversus abdominis muscle that represents the origin of the inguinal canal.

Deep-vein thrombosis (usu.) The presence of a thrombus in a vein, usually a deep vein of the lower limbs.

Defaecation (def-e-KĀ-shun) The discharge of faeces from the rectum.

Deglutition (dē-gloo-TISH-un) The act of swallowing.

Dehydration (dē-hī-DRĀ-shun) Excessive loss of water from the body or its parts.

Delta cell A cell in the pancreatic islets (islets of Langerhans) that secretes somatostatin. Also termed a **D cell.**

Demineralisation (dē-min′-er-al-i-ZĀ-shun) Loss of calcium and phosphorus from bones.

Dendrite (DEN-drīt) A neuronal process that carries electrical signals, usually graded potentials, toward the cell body.

Dendritic (den-DRIT-ik) **cell** One type of antigen-presenting cell with long branchlike projections that commonly is present in mucosal linings such as the vagina, in the skin (Langerhans cells in the epidermis), and in lymph nodes (follicular dendritic cells).

Dental caries (KA-rēz) Gradual demineralisation of the enamel and dentin of a tooth that may invade the pulp and alveolar bone. Also called **tooth decay.**

Denticulate (den-TIK-ū-lāt) Finely toothed or serrated; characterized by a series of small, pointed projections.

Dentin (DEN-tin) The bony tissues of a tooth enclosing the pulp cavity.

Dentition (den-TI-shun) The eruption of teeth. The number, shape, and arrangement of teeth.

Deoxyribonucleic (dē-ok′-sē-rī-bō-nū-KLĒ-ik) **acid (DNA)** A nucleic acid constructed of nucleotides consisting of one of four bases (adenine, cytosine, guanine, or thymine), deoxyribose, and a phosphate group; encoded in the nucleotides is genetic information.

Depression (de-PRESH-un) Movement in which a part of the body moves inferiorly.

Dermal papilla (pa-PIL-a) Fingerlike projection of the papillary region of the dermis that may contain blood capillaries or corpuscles of touch (Meissner corpuscles).

Dermatology (der′-ma-TOL-ō-jē) The medical specialty dealing with diseases of the skin.

Dermatome (DER-ma-tōm) The cutaneous area developed from one embryonic spinal cord segment and receiving most of its sensory innervation from one spinal nerve. An instrument for incising the skin or cutting thin transplants of skin.

Dermis (DER-mis) A layer of dense irregular connective tissue lying deep to the epidermis.

Descending colon (KŌ-lon) The part of the large intestine descending from the left colic (splenic) flexure to the level of the left iliac crest.

Detrusor (de-TROO-ser) **muscle** Smooth muscle that forms the wall of the urinary bladder.

Developmental biology The study of development from the fertilized egg to the adult form.

Deviated nasal septum A nasal septum that does not run along the midline of the nasal cavity. It deviates (bends) to one side.

Diabetes mellitus (dī-a-BĒ-tēz MEL-i-tus) An endocrine disorder caused by an inability to produce or use insulin. It is characterized by the three "polys": polyuria (excessive urine production), polydipsia (excessive thirst), and polyphagia (excess eating).

Diagnosis (dī′-ag-NŌ-sis) Distinguishing one disease from another or determining the nature of a disease from signs and symptoms by inspection, palpation, laboratory tests, and other means.

Dialysis (dī-AL-i-sis) The removal of waste products from blood by diffusion through a selectively permeable membrane.

Diaphragm (DĪ-a-fram) Any partition that separates one area from another, especially the dome-shaped skeletal muscle between the thoracic and abdominal cavities. Also a dome-shaped device that is placed over the cervix, usually with a spermicide, to prevent conception.

Diaphysis (dī-AF-i-sis) The shaft of a long bone.

Diarrhoea (dī-a-RĒ-a) Frequent defecation of liquid faeces caused by increased motility of the intestines.

Diarthrosis (dī-ar-THRŌ-sis) A freely movable joint; types are gliding, hinge, pivot, condyloid, saddle, and ball-and-socket.

Diastole (dī-AS-tō-lē) In the cardiac cycle, the phase of relaxation or dilation of the heart muscle, especially of the ventricles.

Diastolic (dī-as-TOL-ik) **blood pressure (DBP)** The force exerted by blood on arterial walls during ventricular relaxation; the lowest blood pressure measured in the large arteries, normally about 80 mmHg in a young adult.

Diencephalon (DĪ-en-sef′-a-lon) A part of the brain consisting of the thalamus, hypothalamus, and epithalamus.

Differentiation (dif′-er-en-shē-Ā-shun) Development of a cell from an unspecialized to a specialized one.

Diffusion (di-FŪ-zhun) A passive process in which there is a net or greater movement of molecules or ions from a region of high concentration to a region of low concentration until equilibrium is reached.

Digestion (dī-JES-chun) The mechanical and chemical breakdown of food to simple molecules that can be absorbed and used by body cells.

Digestive system A system that consists of the gastrointestinal tract (mouth, pharynx, oesophagus, stomach, small intestine, and large intestine) and accessory digestive organs (teeth, tongue, salivary glands, liver, gallbladder, and pancreas). Its function is to break down foods into small molecules that can be used by body cells.

Dilate (DĪ-lāt) To expand or swell.

Diploid (DIP-loyd) **cell** Having the number of chromosomes characteristically found in the somatic cells of an organism; having two haploid sets of chromosomes, one each from the mother and father. Symbolized 2*n*.

Direct motor pathways Collections of upper motor neurons with cell bodies in the motor cortex that project axons into the spinal cord, where they synapse with lower motor neurons or interneurons in the anterior horns. Also called the **pyramidal pathways.**

Disease Any change from a state of health.

Disease Refers to an illness characterized by a recognizable set of signs and symptoms.

Dislocation (dis′-lō-KĀ-shun) Displacement of a bone from a joint with tearing of ligaments, tendons, and articular capsules. Also called **luxation** (luks-Ā-shun).

Dissect (di-SEKT) To separate tissues and parts of a cadaver or an organ for anatomical study.

Distal (DIS-tal) Farther from the attachment of a limb to the trunk; farther from the point of origin or attachment.

Diuretic (dī-ū-RET-ik) A chemical that increases urine volume by decreasing reabsorption of water, usually by inhibiting sodium reabsorption.

Divergence (dī-VER-jens) A synaptic arrangement in which the synaptic end bulbs of one presynaptic neuron terminate on several postsynaptic neurons.

Diverticulum (dī′-ver-TIK-ū-lum) A sac or pouch in the wall of a canal or organ, especially in the colon.

Dorsal ramus (RĀ-mus) A branch of a spinal nerve containing motor and sensory axons supplying the muscles, skin, and bones of the posterior part of the head, neck, and trunk.

Dorsiflexion (dor-si-FLEK-shun) Bending the foot in the direction of the dorsum (upper surface).

Down-regulation Phenomenon in which there is a decrease in the number of receptors in response to an excess of a hormone or neurotransmitter.

Dual innervation The concept by which most organs of the body receive impulses from sympathetic and parasympathetic neurons.

Duct of Santorini *See* **Accessory duct.**

Duct of Wirsung *See* **Pancreatic duct.**

Ductus arteriosus (DUK-tus ar-tē-rē-O-sus) A small vessel connecting the pulmonary trunk with the aorta; found only in the fetus.

Ductus (vas) deferens (DEF-er-ens) The duct that carries sperm from the epididymis to the ejaculatory duct. Also called the **seminal duct.**

Ductus epididymis (ep′-i-DID-i-mis) A tightly coiled tube inside the epididymis, distinguished into a head, body, and tail, in which sperm undergo maturation.

Ductus venosus (ve-NŌ-sus) A small vessel in the fetus that helps the circulation bypass the liver.

Duodenal (doo-ō-DĒ-nal) **gland** Gland in the submucosa of the duodenum that secretes an alkaline mucus to protect the lining of the small intestine from the action of enzymes and to help neutralize the acid in chyme. Also called a **Brunner's** (BRUN-erz) **gland.**

Duodenal papilla (pa-PIL-a) An elevation on the duodenal mucosa that receives the hepatopancreatic ampulla (ampulla of Vater).

Duodenum (doo′-ō-DĒ-num *or* doo-OD-e-num) The first 25 cm (10 in.) of the small intestine, which connects the stomach and the ileum.

Dura mater (DOO-ra MĀ-ter) The outermost of the three meninges (coverings) of the brain and spinal cord.

Dynamic equilibrium (ē-kwi-LIB-rē-um) The maintenance of body position, mainly the head, in response to sudden movements such as rotation.

Dysmenorrhoea (dis′-men-ō-RĒ-a) Painful menstruation.

Dysplasia (dis-PLĀ-zē-a) Change in the size, shape, and organization of cells due to chronic irritation or inflammation; may either revert to normal if stress is removed or progress to neoplasia.

Dyspnoea (DISP-nē-a) Shortness of breath; painful or laboured breathing.

E

Ectoderm The primary germ layer that gives rise to the nervous system and the epidermis of skin and its derivatives.

Ectopic (ek-TOP-ik) Out of the normal location, as in ectopic pregnancy.

Effector (e-FEK-tor) An organ of the body, either a muscle or a gland, that is innervated by somatic or autonomic motor neurons.

Efferent arteriole (EF-er-ent ar-TĒ-rē-ōl) A vessel of the renal vascular system that carries blood from a glomerulus to a peritubular capillary.

Efferent (EF-er-ent) **ducts** A series of coiled tubes that transport sperm from the rete testis to the epididymis.

Ejaculation (ē-jak-ū-LĀ-shun) The reflex ejection or expulsion of semen from the penis.

Ejaculatory (ē-JAK-ū-la-tō-rē) **duct** A tube that transports sperm from the ductus (vas) deferens to the prostatic urethra.

Elasticity (e-las-TIS-i-tē) The ability of tissue to return to its original shape after contraction or extension.

Electrocardiogram (e-lek′-trō-KAR-dē-ō-gram) **(ECG** or **EKG)** A recording of the electrical changes that accompany the cardiac cycle that can be detected at the surface of the body; may be resting, stress, or ambulatory.

Elevation (el-e-VĀ-shun) Movement in which a part of the body moves superiorly.

Embolus (EM-bō-lus) A blood clot, bubble of air or fat from broken bones, mass of bacteria, or other debris or foreign material transported by the blood.

Embryo (EM-brē-ō) The young of any organism in an early stage of development; in humans, the developing organism from fertilisation to the end of the eighth week of development.

Embryoblast (EM-brē-ō-blast) A region of cells of a blastocyst that differentiates into the three primary germ layers—ectoderm, mesoderm, and endoderm—from which all tissues and organs develop; also called an **inner cell mass.**

Embryology (em′-brē-OL-ō-jē) The study of development from the fertilized egg to the end of the eighth week of development.

Emesis (EM-e-sis) Vomiting.

Emigration (em′-i-GRĀ-shun) Process whereby white blood cells (WBCs) leave the bloodstream by rolling along the endothelium, sticking to it, and squeezing between the endothelial cells. Adhesion molecules help WBCs stick to the endothelium. Also known as **migration** or **extravasation.**

Emission (ē-MISH-un) Propulsion of sperm into the urethra due to peristaltic contractions of the ducts of the testes, epididymides, and ductus (vas) deferens as a result of sympathetic stimulation.

Emphysema (em-fi-SĒ-ma) A lung disorder in which alveolar walls disintegrate, producing abnormally large air spaces and loss of elasticity in the lungs; typically caused by exposure to cigarette smoke.

Emulsification (e-mul-si-fi-KĀ-shun) The dispersion of large lipid globules into smaller, uniformly distributed particles in the presence of bile.

Enamel (e-NAM-el) The hard, white substance covering the crown of a tooth.

Endocardium (en-dō-KAR-dē-um) The layer of the heart wall, composed of endothelium and smooth muscle, that lines the inside of the heart and covers the valves and tendons that hold the valves open.

Endochondral (en′-dō-KON-dral) **ossification** Bone formation within hyaline cartilage that develops from mesenchyme.

Endocrine (EN-dō-krin) **gland** A gland that secretes hormones into interstitial fluid and then the blood; a ductless gland.

Endocrine system (EN-dō-krin) All endocrine glands and hormone-secreting cells.

Endocrinology (en′-dō-kri-NOL-ō-jē) The science concerned with the structure and functions of endocrine glands and the diagnosis and treatment of disorders of the endocrine system.

Endocytosis (en′-dō-sī-TŌ-sis) The uptake into a cell of large molecules and particles in which a segment of plasma membrane surrounds the substance, encloses it, and brings it in; includes phagocytosis, pinocytosis, and receptor-mediated endocytosis.

Endoderm (EN-dō-derm) A primary germ layer of the developing embryo; gives rise to the gastrointestinal tract, urinary bladder, urethra, and respiratory tract.

Endodontics (en′-dō-DON-tiks) The branch of dentistry concerned with the prevention, diagnosis, and treatment of diseases that affect the pulp, root, periodontal ligament, and alveolar bone.

Endolymph (EN-dō-limf′) The fluid within the membranous labyrinth of the internal ear.

Endometriosis (en′-dō-me′-trē-Ō-sis) The growth of endometrial tissue outside the uterus.

Endometrium (en′-dō-MĒ-trē-um) The mucous membrane lining the uterus.

Endomysium (en′-dō-MĪZ-ē-um) Invagination of the perimysium separating each individual muscle fiber (cell).

Endoneurium (en′-dō-NOO-rē-um) Connective tissue wrapping around individual nerve axons.

Endoplasmic reticulum (en′-dō-PLAS-mik re-TIK-ū-lum) **(ER)** A network of channels running through the cytoplasm of a cell that serves in intracellular transportation, support, storage, synthesis, and packaging of molecules. Portions of ER where ribosomes are attached to the outer surface are called **rough ER;** portions that have no ribosomes are called **smooth ER.**

End organ of Ruffini *See* **Type II cutaneous mechanoreceptor.**

Endosteum (end-OS-tē-um) The membrane that lines the medullary (marrow) cavity of bones, consisting of osteogenic cells and scattered osteoclasts.

Endothelium (en′-dō-THĒ-lē-um) The layer of simple squamous epithelium that lines the cavities of the heart, blood vessels, and lymphatic vessels.

Enteric (en-TER-ik) **nervous system** A portion of the autonomic nervous system within the wall of the gastrointestinal tract, pancreas, and gallbladder. Its sensory neurons monitor tension in the intestinal wall and assess the composition of intestinal contents; its motor neurons exert control over the motility and secretions of the gastrointestinal tract.

Enteroendocrine (en-ter-ō-EN-dō-krin) **cell** A cell of the mucosa of the gastrointestinal tract that secretes a hormone that governs function of the GI tract; hormones secreted include gastrin, cholecystokinin, glucose-dependent insulinotropic peptide (GIP), and secretin.

Enzyme (EN-zīm) A substance that accelerates chemical reactions; an organic catalyst, usually a protein.

Eosinophil (ē-ō-SIN-ō-fil) A type of white blood cell characterized by granules that stain red or pink with acid dyes.

Ependymal (ep-EN-de-mal) **cells** Neuroglial cells that cover choroid plexuses and produce cerebrospinal fluid (CSF); they also line the ventricles of the brain and probably assist in the circulation of CSF.

Epicardium (ep′-i-KAR-dē-um) The thin outer layer of the heart wall, composed of serous tissue and mesothelium. Also called the **visceral pericardium.**

Epidemiology (ep′-i-dē-mē-OL-ō-jē) Study of the occurrence and transmission of diseases and disorders in human populations.

Epidermis (ep′-i-DERM-is) The superficial, thinner layer of skin, composed of keratinized stratified squamous epithelium.

Epididymis (ep′-i-DID-i-mis) A comma-shaped organ that lies along the posterior border of the testis and contains the ductus epididymis, in which sperm undergo maturation. *Plural* is **epididymides** (ep′-i-di-DIM-i-dēz).

Epidural (ep′-i-DOO-ral) **space** A space between the spinal dura mater and the vertebral canal, containing areolar connective tissue and a plexus of veins.

Epiglottis (ep′-i-GLOT-is) A large, leaf-shaped piece of cartilage lying on top of the larynx, attached to the thyroid cartilage; its unattached portion is free to move up and down to cover the glottis (vocal folds and rima glottidis) during swallowing.

Epimysium (ep-i-MĪZ-ē-um) Fibrous connective tissue around muscles.

Epinephrine (ep-ē-NEF-rin) Hormone secreted by the adrenal medulla that produces actions similar to those that result from sympathetic stimulation. Also called **adrenaline** (a-DREN-a-lin).

Epineurium (ep′-i-NOO-rē-um) The superficial connective tissue covering around an entire nerve.

Epiphyseal (ep′-i-FIZ-ē-al) **line** The remnant of the epiphyseal plate in the metaphysis of a long bone.

Epiphyseal plate The hyaline cartilage plate in the metaphysis of a long bone; site of lengthwise growth of long bones.

Epiphysis (e-PIF-i-sis) The end of a long bone, usually larger in diameter than the shaft (diaphysis).

Epiphysis cerebri (se-RĒ-brē) Pineal gland.

Episiotomy (e-piz′-ē-OT-ō-mē) A cut made with surgical scissors to avoid tearing of the perineum at the end of the second stage of labour.

Epistaxis (ep′-i-STAK-sis) Loss of blood from the nose due to trauma, infection, allergy, neoplasm, and bleeding disorders. Also called **nosebleed.**

Epithalamus (ep′-i-THAL-a-mus) Part of the diencephalon superior and posterior to the thalamus, comprising the pineal gland and associated structures.

Epithelial (ep-i-THĒ-lē-al) **tissue** The tissue that forms the innermost and outermost surfaces of body structures and forms glands.

Eponychium (ep′-o-NIK-ē-um) Narrow band of stratum corneum at the proximal border of a nail that extends from the margin of the nail wall. Also called the **cuticle.**

Equilibrium (ē-kwi-LIB-rē-um) The state of being balanced.

Erectile dysfunction (ED) Failure to maintain an erection long enough for sexual intercourse. Previously known as **impotence** (IM-pō-tens).

Erection (ē-REK-shun) The enlarged and stiff state of the penis or clitoris resulting from the engorgement of the spongy erectile tissue with blood.

Eructation (e-ruk′-TĀ-shun) The forceful expulsion of gas from the stomach. Also called **belching.**

Erythema (er-e-THĒ-ma) Skin redness usually caused by dilation of the capillaries.

Erythrocyte (e-RITH-rō-sīt) A mature red blood cell.

Erythropoietin (e-rith′-rō-POY-e-tin) **(EPO)** A hormone released by the juxtaglomerular cells of the kidneys that stimulates red blood cell production.

Eupnea (ŪP-nē-ă) Normal quiet breathing.

Eustachian tube *See* **Auditory tube.**

Eversion (ē-VER-zhun) The movement of the sole laterally at the ankle joint or of an atrioventricular valve into an atrium during ventricular contraction.

Excitability (ek-sīt′-a-BIL-i-tē) The ability of muscle fibers to receive and respond to stimuli; the ability of neurons to respond to stimuli and generate nerve impulses.

Excretion (eks-KRĒ-shun) The process of eliminating waste products from the body; also the products excreted.

Exercise physiology Study of the changes in cell and organ function due to muscular activity.

Exhalation (eks-ha-LĀ-shun) Breathing out; expelling air from the lungs into the atmosphere. Also called **expiration.**

Exocrine (EK-sō-krin) **gland** A gland that secretes its products into ducts that carry the secretions into body cavities, into the lumen of an organ, or to the outer surface of the body.

Exocytosis (ek-sō-sī-TŌ-sis) A process in which membrane-enclosed secretory vesicles form inside the cell, fuse with the plasma membrane, and release their contents into the interstitial fluid; achieves secretion of materials from a cell.

Extensibility (ek-sten′-si-BIL-i-tē) The ability of muscle tissue to stretch when it is pulled.

Extension (eks-TEN-shun) An increase in the angle between two bones; restoring a body part to its anatomical position after flexion.

External Located on or near the surface.

External auditory (AW-di-tōr-ē) **canal** or **meatus** (mē-Ā-tus) A curved tube in the temporal bone that leads to the middle ear.

External ear The **outer ear,** consisting of the pinna, external auditory canal, and tympanic membrane (eardrum).

External nares (NĀ-rez) The openings into the nasal cavity on the exterior of the body. Also called the **nostrils.**

External respiration The exchange of respiratory gases between the lungs and blood. Also called **pulmonary respiration** or **pulmonary gas exchange.**

Exteroceptor (EKS-ter-ō-sep′-tor) A sensory receptor adapted for the reception of stimuli from outside the body.

Extracellular fluid (ECF) Fluid outside body cells, such as interstitial fluid and plasma.

Extracellular matrix (MĀ-triks) The ground substance and fibers between cells in a connective tissue.

Eyebrow The hairy ridge superior to the eye.

F

F cell A cell in the pancreatic islets (islets of Langerhans) that secretes pancreatic polypeptide.

Face The anterior aspect of the head.

Faeces (FĒ-sēz) Material discharged from the rectum and made up of bacteria, excretions, and food residue. Also called **stool.**

Falciform ligament (FAL-si-form LIG-a-ment) A sheet of parietal peritoneum between the two principal lobes of the liver. The ligamentum teres, or remnant of the umbilical vein, lies within its fold.

Falx cerebelli (FALKS′ ser-e-BEL-lī) A small triangular process of the dura mater attached to the occipital bone in the posterior cranial fossa and projecting inward between the two cerebellar hemispheres.

Falx cerebri (FALKS SER-e-brē) A fold of the dura mater extending deep into the longitudinal fissure between the two cerebral hemispheres.

Fascia (FASH-ē-a) Large connective tissue sheet that wraps around groups of muscles.

Fascicle (FAS-i-kul) A small bundle or cluster, especially of nerve or muscle fibers (cells). Also called a **fasciculus** (fa-SIK-ū-lus). *Plural* is **fasciculi** (fa-SIK-yoo-lī).

Fasciculation (fa-sik-ū-LĀ-shun) Abnormal, spontaneous twitch of all skeletal muscle fibers in one motor unit that is visible at the skin surface; not associated with movement of the affected muscle; present in progressive diseases of motor neurons, for example, poliomyelitis.

Fat A triglyceride that is a solid at room temperature.

Fatty acid A simple lipid that consists of a carboxyl group and a hydrocarbon chain; used to synthesize triglyceride and phospholipids.

Fauces (FAW-sēs) The opening from the mouth into the pharynx.

Feedback system (loop) A cycle of events in which the status of a body condition is monitored, evaluated, changed, remonitored, reevaluated, and so on.

Female reproductive cycle General term for the ovarian and uterine cycles, the hormonal changes that accompany them, and cyclic changes in the breasts and cervix; includes changes in the endometrium of a nonpregnant female that prepares the lining of the uterus to receive a fertilized ovum. Less correctly termed the **menstrual cycle.**

Fertilisation (fer-til-i-ZĀ-shun) Penetration of a secondary oocyte by a sperm cell, meiotic division of a secondary oocyte to form an ovum, and subsequent union of the nuclei of the gametes.

Fetal circulation The cardiovascular system of the fetus, including the placenta and special blood vessels involved in the exchange of materials between fetus and mother.

Fetus (usu., also foetus, informal) In humans, the developing organism *in utero* from the beginning of the third month to birth.

Fever An elevation in body temperature above the normal temperature of 37°C (98.6°F) due to a resetting of the hypothalamic thermostat.

Fibroblast (FĪ-brō-blast) A large, flat cell that secretes most of the extracellular matrix of areolar and dense connective tissues.

Fibrosis The process by which fibroblasts synthesize collagen fibers and other extracellular matrix materials that aggregate to form scar tissue.

Fibrous (FĪ-brus) **joint** A joint that allows little or no movement, such as a suture or a syndesmosis.

Fibrous tunic (TOO-nik) The superficial coat of the eyeball, made up of the posterior sclera and the anterior cornea.

Fight-or-flight response The effects produced on stimulation of the sympathetic division of the autonomic nervous system.

Filiform papilla (FIL-i-form pa-PIL-a) One of the conical projections that are distributed in parallel rows over the anterior two-thirds of the tongue and lack taste buds.

Filtration (fil-TRĀ-shun) The flow of a liquid through a filter (or membrane that acts like a filter) due to a hydrostatic pressure; occurs in capillaries due to blood pressure.

Filum terminale (FĪ-lum ter-mi-NAL-ē) Non-nervous fibrous tissue of the spinal cord that extends inferiorly from the conus medullaris to the coccyx.

Fimbriae (FIM-brē-ē) Fingerlike structures, especially the lateral ends of the uterine (fallopian) tubes.

Fissure (FISH-ur) A groove, fold, or slit that may be normal or abnormal.

Fixator A muscle that stabilizes the origin of the prime mover so that the prime mover can act more efficiently.

Fixed macrophage (MAK-rō-fāj) Stationary phagocytic cell found in the liver, lungs, brain, spleen, lymph nodes, subcutaneous tissue, and red bone marrow. Also called a **tissue macrophage** or **histiocyte** (HIS-tē-ō-sīt).

Flaccid (FLASS-id) Relaxed, flabby, or soft; lacking muscle tone.

Flagellum (fla-JEL-um) A hairlike, motile process on the extremity of a bacterium, protozoan, or sperm cell. *Plural* is **flagella** (fla-JEL-a).

Flatus (FLĀ-tus) Gas in the stomach or intestines; commonly used to denote expulsion of gas through the anus.

Flexion (FLEK-shun) Movement in which there is a decrease in the angle between two bones.

Follicle (FOL-i-kul) A small secretory sac or cavity; the group of cells that contains a developing oocyte in the ovaries.

Follicle-stimulating hormone (FSH) Hormone secreted by the anterior pituitary; it initiates development of ova and stimulates the ovaries to secrete oestrogens in females, and initiates sperm production in males.

Fontanel (fon-ta-NEL) A mesenchyme-filled space where bone formation is not yet complete, especially between the cranial bones of an infant's skull.

Foot The terminal part of the lower limb, from the ankle to the toes.

Foramen (fō-RĀ-men) A passage or opening; a communication between two cavities of an organ, or a hole in a bone for passage of vessels or nerves. *Plural* is **foramina** (fō-RAM-i-na).

Foramen ovale (fō-RĀ-men ō-VAL-ē) An opening in the fetal heart in the septum between the right and left atria. A hole in the greater wing of the sphenoid bone that transmits the mandibular branch of the trigeminal (V) nerve.

Forearm (FOR-arm) The part of the upper limb between the elbow and the wrist.

Fornix (FOR-niks) An arch or fold; a tract in the brain made up of association fibers, connecting the hippocampus with the mammillary bodies; a recess around the cervix of the uterus where it protrudes into the vagina.

Fossa (FOS-a) A furrow or shallow depression.

Fourth ventricle (VEN-tri-kul) A cavity filled with cerebrospinal fluid within the brain lying between the cerebellum and the medulla oblongata and pons.

Fracture (FRAK-choor) Any break in a bone.

Free radical An atom or group of atoms with an unpaired electron in the outermost shell. It is unstable, highly reactive, and destroys nearby molecules.

Frontal plane A plane at a right angle to a midsagittal plane that divides the body or organs into anterior and posterior portions. Also called a **coronal** (kō-RŌ-nal) **plane.**

Fundus (FUN-dus) The part of a hollow organ farthest from the opening.

Fungiform papilla (FUN-ji-form pa-PIL-a) A mushroomlike elevation on the upper surface of the tongue appearing as a red dot; most contain taste buds.

Furuncle (FŪ-rung-kul) A boil; painful nodule caused by bacterial infection and inflammation of a hair follicle or sebaceous (oil) gland.

G

Gallbladder A small pouch, located inferior to the liver, that stores bile and empties by means of the cystic duct.

Gallstone A solid mass, usually containing cholesterol, in the gallbladder or a bile-containing duct; formed anywhere between bile canaliculi in the liver and the hepatopancreatic ampulla (ampulla of Vater), where bile enters the duodenum. Also called a **biliary calculus.**

Gamete (GAM-ēt) A male or female reproductive cell; a sperm cell or secondary oocyte.

Ganglion (GANG-glē-on) Usually, a group of neuronal cell bodies lying outside the central nervous system (CNS). *Plural* is **ganglia** (GANG-glē-a).

Gastric (GAS-trik) **glands** Glands in the mucosa of the stomach composed of cells that empty their secretions into narrow channels called gastric pits. Types of cells are chief cells (secrete pepsinogen), parietal cells (secrete hydrochloric acid and intrinsic factor), surface mucous and mucous neck cells (secrete mucus), and G cells (secrete gastrin).

Gastroenterology (gas′-trō-en′-ter-OL-ō-jē) The medical specialty that deals with the structure, function, diagnosis, and treatment of diseases of the stomach and intestines.

Gastrointestinal (gas-trō-in-TES-ti-nal) **(GI) tract** A continuous tube running through the ventral body cavity extending from the mouth to the anus. Also called the **alimentary** (al′-i-MEN-tar-ē) **canal.**

Gastrulation (gas-troo-LĀ-shun) The migration of groups of cells from the epiblast that transform a bilaminar embryonic disc into a trilaminar embryonic disc with three primary germ layers; transformation of the blastula into the gastrula.

Gene (JĒN) Biological unit of heredity; a segment of DNA located in a definite position on a particular chromosome; a sequence of DNA that codes for a particular mRNA, rRNA, or tRNA.

Genetic engineering The manufacture and manipulation of genetic material.

Genetics The study of genes and heredity.

Genome (JĒ-nōm) The complete set of genes of an organism.

Genotype (JĒ-nō-tīp) The genetic makeup of an individual; the combination of alleles present at one or more chromosomal locations, as distinguished from the appearance, or phenotype, that results from those alleles.

Geriatrics (jer′-ē-AT-riks) The branch of medicine devoted to the medical problems and care of elderly persons.

Gestation (jes-TĀ-shun) The period of development from fertilisation to birth.

Gingivae (jin-JI-vē) Gums. They cover the alveolar processes of the mandible and maxilla and extend slightly into each socket.

Gland Specialized epithelial cell or cells that secrete substances; may be exocrine or endocrine.

Glans penis (glanz PĒ-nis) The slightly enlarged region at the distal end of the penis.

Glaucoma (glaw-KŌ-ma) An eye disorder in which there is increased intraocular pressure due to an excess of aqueous humour.

Glomerular (glō-MER-ū-lar) **capsule** A double-walled globe at the proximal end of a nephron that encloses the glomerular capillaries. Also called **Bowman's** (BŌ-manz) **capsule.**

Glomerular filtrate (glō-MER-ū-lar FIL-trāt) The fluid produced when blood is filtered by the filtration membrane in the glomeruli of the kidneys.

Glomerular filtration The first step in urine formation in which substances in blood pass through the filtration membrane and the filtrate enters the proximal convoluted tubule of a nephron.

Glomerular filtration rate The amount of filtrate formed in all renal corpuscles per minute. It averages 125 mL/min in males and 105 mL/min in females.

Glomerulus (glō-MER-ū-lus) A rounded mass of nerves or blood vessels, especially the microscopic tuft of capillaries that is surrounded by the glomerular (Bowman's) capsule of each kidney tubule. *Plural* is **glomeruli** (glō-MER-ū-li).

Glottis (GLOT-is) The vocal folds (true vocal cords) in the larynx plus the space between them (rima glottidis).

Glucagon (GLOO-ka-gon) A hormone produced by the alpha cells of the pancreatic islets (islets of Langerhans) that increases blood glucose level.

Glucocorticoids (gloo′-kō-KOR-ti-koyds) Hormones secreted by the cortex of the adrenal gland, especially cortisol, that influence glucose metabolism.

Glucose (GLOO-kōs) A hexose (six-carbon sugar), $C_6H_{12}O_6$, that is a major energy source for the production of ATP by body cells.

Glucosuria (gloo′-kō-SOO-rē-a) The presence of glucose in the urine; may be temporary or pathological. Also called **glycosuria.**

Glycogen (GLĪ-kō-jen) A highly branched polymer of glucose containing thousands of subunits; functions as a compact store of glucose molecules in liver and muscle fibers (cells).

Goblet cell A goblet-shaped unicellular gland that secretes mucus; present in epithelium of the airways and intestines.

Goiter (GOY-ter) An enlarged thyroid gland.

Golgi (GOL-jē) **complex** An organelle in the cytoplasm of cells consisting of four to six flattened sacs (cisternae), stacked on one another, with expanded areas at their ends; functions in processing, sorting, packaging, and delivering proteins and lipids to the plasma membrane, lysosomes, and secretory vesicles.

Golgi tendon organ *See* **Tendon organ.**

Gomphosis (gom-FŌ-sis) A fibrous joint in which a cone-shaped peg fits into a socket.

Gonad (GŌ-nad) A gland that produces gametes and hormones; the ovary in the female and the testis in the male.

Gonadotropic hormone Anterior pituitary hormone that affects the gonads.

Gout (GOWT) Hereditary condition associated with excessive uric acid in the blood; the acid crystallizes and deposits in joints, kidneys, and soft tissue.

Graafian follicle *See* **Mature follicle.**

Gracile (GRAS-īl) **nucleus** A group of nerve cells in the inferior part of the medulla oblongata in which axons of the gracile fasciculus terminate.

Greater oementum (ō-MEN-tum) A large fold in the serosa of the stomach that hangs down like an apron anterior to the intestines.

Greater vestibular (ves-TIB-ū-lar) **glands** A pair of glands on either side of the vaginal orifice that open by a duct into the space between the hymen and the labia minora. Also called **Bartholin's** (BAR-to-linz) **glands.**

Grey commisure (KOM-mi-shur) A narrow strip of grey matter connecting the two lateral gray masses within the spinal cord.

Grey matter Areas in the central nervous system and ganglia containing neuronal cell bodies, dendrites, unmyelinated axons, axon terminals, and neuroglia; Nissl bodies impart a gray color and there is little or no myelin in grey matter.

Grey ramus communicans (RĀ-mus kō-MŪ-ni-kans) A short nerve containing axons of sympathetic postganglionic neurons; the cell bodies of the neurons are in a sympathetic chain ganglion, and the unmyelinated axons extend via the gray ramus to a spinal nerve and then to the periphery to supply smooth muscle in blood vessels, arrector pili muscles, and sweat glands. *Plural* is **rami communicantes** (RĀ-mē kō-mū-ni-KAN-tēz).

Groin (GROYN) The depression between the thigh and the trunk; the inguinal region.

Gross anatomy The branch of anatomy that deals with structures that can be studied without using a microscope. Also called **macroscopic anatomy.**

Growth An increase in size due to an increase in (1) the number of cells, (2) the size of existing cells as internal components increase in size, or (3) the size of intercellular substances.

Gustation (gus-TĀ-shun). The sense of taste.

Gustatory (GUS-ta-tō′-rē) Pertaining to taste.

Gynaecology (gī′-ne-KOL-ō-jē) The branch of medicine dealing with the study and treatment of disorders of the female reproductive system.

Gynaecomastia (gīn′-e-kō-MAS-tē-a) Excessive growth (benign) of the male mammary glands due to secretion of oestrogens by an adrenal gland tumor (feminizing adenoma).

Gyrus (JI-rus) One of the folds of the cerebral cortex of the brain. *Plural* is **gyri** (JĪ-rī). Also called a **convolution.**

H

Haemangioblast (hē-MAN-jē-ō-blast) A precursor mesodermal cell that develops into blood and blood vessels.

Haematocrit (he-MAT-ō-krit) **(Hct)** The percentage of blood made up of red blood cells. Usually measured by centrifuging a blood sample in a graduated tube and then reading the volume of red blood cells and dividing it by the total volume of blood in the sample.

Haematology (hēm-a-TOL-ō-jē) The study of blood.

Haematoma (hē′-ma-TŌ-ma) A tumor or swelling filled with blood.

Haemodialysis (hē-mō-dī-AL-i-sis) Direct filtration of blood by removing wastes and excess electrolytes and fluid and then returning the cleansed blood.

Haemoglobin (hē′-mō-GLŌ-bin) **(Hb)** A substance in red blood cells consisting of the protein globin and the iron-containing red pigment heme that transports most of the oxygen and some carbon dioxide in blood.

Haemolysis (hē-MOL-i-sis) The escape of haemoglobin from the interior of a red blood cell into the surrounding medium; results from disruption of the cell membrane by toxins or drugs, freezing or thawing, or hypotonic solutions.

Haemolytic disease of the newborn (HDN) A hemolytic anaemia of a newborn child that results from the destruction of the infant's erythrocytes (red blood cells) by antibodies produced by the mother; usually the antibodies are due to an Rh blood type incompatibility. Also called **erythroblastosis fetalis** (e-rith′-rō-blas-TŌ-sis fe-TAL-is).

Haemophilia (hē′-mō-FIL-ē-a) A hereditary blood disorder where there is a deficient production of certain factors involved in blood clotting, resulting in excessive bleeding into joints, deep tissues, and elsewhere.

Haemopoiesis (hēm-ō-poy-Ē-sis) Blood cell production, which occurs in red bone marrow after birth. Also called **hematopoiesis** (hem′-a-tō-poy-Ē-sis).

Haemorrhage (HEM-o-rij) Bleeding; the escape of blood from blood vessels, especially when the loss is profuse.

Haemorrhoid (HEM-ō-royds) Dilated or varicosed blood vessels (usually veins) in the anal region. Also called **piles.**

Hair A threadlike structure produced by hair follicles that develops in the dermis. Also called a **pilus** (PĪ-lus).

Hair follicle (FOL-li-kul) Structure composed of epithelium and surrounding the root of a hair from which hair develops.

Hair root plexus (PLEK-sus) A network of dendrites arranged around the root of a hair as free or naked nerve endings that are stimulated when a hair shaft is moved.

Hand The terminal portion of an upper limb, including the carpals, metacarpals, and phalanges.

Haploid (HAP-loyd) **cell** Having half the number of chromosomes characteristically found in the somatic cells of an organism; characteristic of mature gametes. Symbolized *n*.

Hard palate (PAL-at) The anterior portion of the roof of the mouth, formed by the maxillae and palatine bones and lined by mucous membrane.

Haustra (HAWS-tra) A series of pouches that characterize the colon; caused by tonic contractions of the teniae coli. *Singular* is **haustrum.**

Haversian canal *See* **Central canal.**

Haversian system *See* **Osteon.**

Head The superior part of a human, cephalic to the neck. The superior or proximal part of a structure.

Hearing The ability to perceive sound.

Heart A hollow muscular organ lying slightly to the left of the midline of the chest that pumps the blood through the cardiovascular system.

Heart block An arrhythmia (dysrhythmia) of the heart in which the atria and ventricles contract independently because of a blocking of electrical impulses through the heart at some point in the conduction system.

Heart murmur (MER-mer) An abnormal sound that consists of a flow noise that is heard before, between, or after the normal heart sounds, or that may mask normal heart sounds.

Hemiplegia (hem-i-PLĒ-jē-a) Paralysis of the upper limb, trunk, and lower limb on one side of the body.

Hemodynamics (hē-mō-dī-NAM-iks) The forces involved in circulating blood throughout the body.

Hepatic (he-PAT-ik) Refers to the liver.

Hepatic duct A duct that receives bile from the bile capillaries. Small hepatic ducts merge to form the larger right and left hepatic ducts that unite to leave the liver as the common hepatic duct.

Hepatic portal circulation The flow of blood from the gastrointestinal organs to the liver before returning to the heart.

Hepatocyte (he-PAT-ō-cyte) A liver cell.

Hepatopancreatic (hep′-a-tō-pan′-krē-A-tik) **ampulla** A small, raised area in the duodenum where the combined common bile duct and main pancreatic duct empty into the duodenum. Also called the **ampulla of Vater** (FAH-ter).

Hernia (HER-nē-a) The protrusion or projection of an organ or part of an organ through a membrane or cavity wall, usually the abdominal cavity.

Herniated (HER-nē-ā′-ted) **disc** A rupture of an intervertebral disc so that the nucleus pulposus protrudes into the vertebral cavity. Also called a **slipped disc.**

Hiatus (hī-Ā-tus) An opening; a foramen.

Hilum (HĪ-lum) An area, depression, or pit where blood vessels and nerves enter or leave an organ. Also called a **hilus.**

Hinge joint A synovial joint in which a convex surface of one bone fits into a concave surface of another bone, such as the elbow, knee, ankle, and interphalangeal joints. Also called a **ginglymus** (JIN-gli-mus) **joint.**

Hirsutism (HER-soo-tizm) An excessive growth of hair in females and children, with a distribution similar to that in adult males, due to the conversion of vellus hairs into large terminal hairs in response to higher-than-normal levels of androgens.

Histamine (HISS-ta-mēn) Substance found in many cells, especially mast cells, basophils, and platelets, that is released when the cells are injured; results in vasodilation, increased permeability of blood vessels, and constriction of bronchioles.

Histology (his′-TOL-ō-jē) Microscopic study of the structure of tissues.

Holocrine (HŌ-lō-krin) **gland** A type of gland in which entire secretory cells, along with their accumulated secretions, make up the secretory product of the gland, as in the sebaceous (oil) glands.

Homoeostasis (hō′-mē-ō-STĀ-sis) The condition in which the body's internal environment remains relatively constant within physiological limits.

Homologous (hō-MOL-ō-gus) **chromosomes** Two chromosomes that belong to a pair. Also called **homologs.**

Hormone (HOR-mōn) A secretion of endocrine cells that alters the physiological activity of target cells of the body.

Horn An area of grey matter (anterior, lateral, or posterior) in the spinal cord.

Human chorionic gonadotropin (kō-rē-ON-ik gō-nad-ō-TRŌ-pin) **(hCG)** A hormone produced by the developing placenta that maintains the corpus luteum.

Human chorionic somatomammotropin (sō-mat-ō-mam-ō-TRŌ-pin) **(hCS)** Hormone produced by the chorion of the placenta that stimulates breast tissue for lactation, enhances body growth, and regulates metabolism. Also called **human placental lactogen (hPL).**

Human growth hormone (hGH) Hormone secreted by the anterior pituitary that stimulates growth of body tissues, especially skeletal and muscular tissues. Also known as **somatotropin** (sō′-ma-tō-TRŌ-pin) and **somatotropic hormone (STH).**

Hyaluronic (hī′-a-loo-RON-ik) **acid** A viscous, amorphous extracellular material that binds cells together, lubricates joints, and maintains the shape of the eyeballs.

Hymen (HĪ-men) A thin fold of vascularized mucous membrane at the vaginal orifice.

Hyperextension (hī′-per-ek-STEN-shun) Continuation of extension beyond the anatomical position, as in bending the head backward.

Hyperplasia (hī-per-PLĀ-zē-a) An abnormal increase in the number of normal cells in a tissue or organ, increasing its size.

Hypersecretion (hī′-per-se-KRĒ-shun) Overactivity of glands resulting in excessive secretion.

Hypersensitivity (hī′-per-sen-si-TI-vi-tē) Overreaction to an allergen that results in pathological changes in tissues. Also called **allergy.**

Hypertension (hī′-per-TEN-shun) High blood pressure.

Hyperthermia (hī′-per-THERM-ē-a) An elevated body temperature.

Hypertonia (hī′-per-TŌ-nē-a) Increased muscle tone that is expressed as spasticity or rigidity.

Hypertonic (hī′-per-TON-ik) Solution that causes cells to shrink due to loss of water by osmosis.

Hypertrophy (hī-PER-trō-fē) An excessive enlargement or overgrowth of tissue without cell division.

Hyperventilation (hī′-per-ven-ti-LĀ-shun) A rate of inhalation and exhalation higher than that required to maintain a normal partial pressure of carbon dioxide in the blood.

Hyponychium (hī′-pō-NIK-ē-um) Free edge of the fingernail.

Hypophyseal fossa (hī′-pō-FIZ-ē-al FOS-a) A depression on the superior surface of the sphenoid bone that houses the pituitary gland.

Hypophyseal (hī′-pō-FIZ-ē-al) **pouch** An outgrowth of ectoderm from the roof of the mouth from which the anterior pituitary develops.

Hypophysis (hī-POF-i-sis) Pituitary gland.

Hyposecretion (hī′-pō-se-KRĒ-shun) Underactivity of glands resulting in diminished secretion.

Hypothalamohypophyseal (hī′-pō-thal′-a-mō-hī-pō-FIZ-ē-al) **tract** A bundle of axons containing secretory vesicles filled with oxytocin or antidiuretic hormone that extend from the hypothalamus to the posterior pituitary.

Hypothalamus (hī′-pō-THAL-a-mus) A portion of the diencephalon, lying beneath the thalamus and forming the floor and part of the wall of the third ventricle.

Hypothermia (hī′-pō-THER-mē-a) Lowering of body temperature below 35°C (95°F); in surgical procedures, it refers to deliberate cooling of the body to slow down metabolism and reduce oxygen needs of tissues.

Hypotonia (hī′-pō-TŌ-nē-a) Decreased or lost muscle tone in which muscles appear flaccid.

Hypotonic (hī′-pō-TON-ik) Solution that causes cells to swell and perhaps rupture due to gain of water by osmosis.

Hypoventilation (hī-pō-ven-ti-LĀ-shun) A rate of inhalation and exhalation lower than that required to maintain a normal partial pressure of carbon dioxide in plasma.

Hypoxia (hī-POKS-ē-a) Lack of adequate oxygen at the tissue level.

Hysterectomy (hiss-te-REK-tō-mē) The surgical removal of the uterus.

I

Ileocaecal (il-ē-ō-SĒ-kal) **sphincter** A fold of mucous membrane that guards the opening from the ileum into the large intestine. Also called the **ileocecal valve.**

Ileum (IL-ē-um) The terminal part of the small intestine.

Immunity (i-MŪ-ni-tē) The state of being resistant to injury, particularly by poisons, foreign proteins, and invading pathogens.

Immunoglobulin (im-ū-nō-GLOB-ū-lin) **(Ig)** An antibody synthesized by plasma cells derived from B lymphocytes in response to the introduction of an antigen. Immunoglobulins are divided into five kinds (IgG, IgM, IgA, IgD, IgE).

Immunology (im′-ū-NOL-ō-jē) The study of the responses of the body when challenged by antigens.

Imperforate (im′-PER-fō-rāt) Abnormally closed.

Implantation (im′-plan-TĀ-shun) The insertion of a tissue or a part into the body. The attachment of the blastocyst to the stratum basalis of the endometrium about 6 days after fertilisation.

Incontinence (in-KON-ti-nens) Inability to retain urine, semen, or faeces through loss of sphincter control.

Indirect motor pathways Motor tracts that convey information from the brain down the spinal cord for automatic movements, coordination of body movements with visual stimuli, skeletal muscle tone and posture, and balance. Also known as **extrapyramidal pathways.**

Induction (in-DUK-shun) The process by which one tissue (inducting tissue) stimulates the development of an adjacent unspecialized tissue (responding tissue) into a specialized one.

Infarction (in-FARK-shun) A localized area of necrotic tissue, produced by inadequate oxygenation of the tissue.

Infection (in-FEK-shun) Invasion and multiplication of microorganisms in body tissues, which may be inapparent or characterized by cellular injury.

Inferior (in-FĒR-ē-or) Away from the head or toward the lower part of a structure. Also called **caudal** (KAW-dal).

Inferior vena cava (VĒ-na KĀ-va) **(IVC)** Large vein that collects blood from parts of the body inferior to the heart and returns it to the right atrium.

Infertility Inability to conceive or to cause conception. Also called **sterility** in males.

Inflammation (in′-fla-MĀ-shun) Localized, protective response to tissue injury designed to destroy, dilute, or wall off the infecting agent or injured tissue; characterized by redness, pain, heat, swelling, and sometimes loss of function.

Infundibulum (in-fun-DIB-ū-lum) The stalklike structure that attaches the pituitary gland to the hypothalamus of the brain. The funnel-shaped, open, distal end of the uterine (fallopian) tube.

Ingestion (in-JES-chun) The taking in of food, liquids, or drugs, by mouth.

Inguinal (IN-gwin-al) Pertaining to the groin.

Inguinal canal An oblique passageway in the anterior abdominal wall just superior and parallel to the medial half of the inguinal ligament that transmits the spermatic cord and ilioinguinal nerve in the male and round ligament of the uterus and ilioinguinal nerve in the female.

Inhalation (in-ha-LĀ-shun) The act of drawing air into the lungs. Also termed **inspiration.**

Inheritance The acquisition of body traits by transmission of genetic information from parents to offspring.

Inhibin A hormone secreted by the gonads that inhibits release of follicle-stimulating hormone (FSH) by the anterior pituitary.

Inhibiting hormone Hormone secreted by the hypothalamus that can suppress secretion of hormones by the anterior pituitary.

Insertion (in-SER-shun) The attachment of a muscle tendon to a movable bone or the end opposite the origin.

Insula (IN-soo-la) A triangular area of the cerebral cortex that lies deep within the lateral cerebral fissure, under the parietal, frontal, and temporal lobes.

Insulin (IN-soo-lin) A hormone produced by the beta cells of a pancreatic islet (islet of Langerhans) that decreases the blood glucose level.

Integrins (IN-te-grinz) A family of transmembrane glycoproteins in plasma membranes that function in cell adhesion; they are present in hemidesmosomes, which anchor cells to a basement membrane, and they mediate adhesion of neutrophils to endothelial cells during emigration.

Integumentary (in-teg-ū-MEN-tar-ē) Relating to the skin.

Integumentary (in-teg-ū-MEN-tar-ē) **system** A system composed of organs such as the skin, hair, oil and sweat glands, nails, and sensory receptors.

Intercalated (in-TER-ka-lāt-ed) **disc** An irregular transverse thickening of sarcolemma that contains desmosomes, which hold cardiac muscle fibers (cells) together, and gap junctions, which aid in conduction of muscle action potentials from one fiber to the next.

Intercostal (in′-ter-KOS-tal) **nerve** A nerve supplying a muscle located between the ribs. Also called **thoracic nerve.**

Intermediate (in′-ter-MĒ-dē-at) Between two structures, one of which is medial and one of which is lateral.

Intermediate filament Protein filament, ranging from 8 to 12 nm in diameter, that may provide structural reinforcement, hold organelles in place, and give shape to a cell.

Internal Away from the surface of the body.

Internal capsule A large tract of projection fibers lateral to the thalamus that is the major connection between the cerebral cortex and the brain stem and spinal cord; contains axons of sensory neurons carrying auditory, visual, and somatic sensory signals to the cerebral cortex plus axons of motor neurons descending from the cerebral cortex to the thalamus, subthalamus, brain stem, and spinal cord.

Internal ear The inner ear or labyrinth, lying inside the temporal bone, containing the organs of hearing and balance.

Internal nares (NĀ-rez) The two openings posterior to the nasal cavities opening into the nasopharynx. Also called the **choanae** (kō-Ā-nē).

Internal respiration The exchange of respiratory gases between blood and body cells. Also called **tissue respiration** or **systemic gas exchange.**

Interneurons (in′-ter-NOO-ronz) Neurons whose axons extend only for a short distance and contact nearby neurons in the brain, spinal cord, or a ganglion; they comprise the vast majority of neurons in the body. Also called **association neurons.**

Interoceptor (IN-ter-ō-sep′-tor) Sensory receptor located in blood vessels and viscera that provides information about the body's internal environment. Also called **visceroceptor.**

Interphase (IN-ter-fāz) The period of the cell cycle between cell divisions, consisting of the G_0 phase; G_1 (gap or growth) phase, when the cell is engaged in growth, metabolism, and production of substances required for division; S (synthesis) phase, during which chromosomes are replicated; and G_2 phase.

Interstitial cell of Leydig *See* **Interstitial endocrinocyte.**

Interstitial (in′-ter-STISH-al) **endocrinocyte** A cell that is located in the connective tissue between seminiferous tubules in a mature testis that secretes testosterone. Also called an **interstitial cell of Leydig** (LĪ-dig).

Interstitial (in′-ter-STISH-al) **fluid** The portion of extracellular fluid that fills the microscopic spaces between the cells of tissues; the internal environment of the body. Also called **intercellular** or **tissue fluid.**

Interstitial growth Growth from within, as in the growth of cartilage. Also called **endogenous** (en-DOJ-e-nus) **growth.**

Interventricular (in′-ter-ven-TRIK-ū-lar) **foramen** A narrow, oval opening through which the lateral ventricles of the brain communicate with the third ventricle. Also called the **foramen of Monro.**

Intervertebral (in′-ter-VER-te-bral) **disc** A pad of fibrocartilage located between the bodies of two vertebrae.

Intestinal gland A gland that opens onto the surface of the intestinal mucosa and secretes digestive enzymes. Also called a **crypt of Lieberkühn** (LĒ-ber-kūn).

Intracellular (in′-tra-SEL-yū-lar) **fluid** (**ICF**) Fluid located within cells.

Intrafusal (in′-tra-FŪ-sal) **fibers** Three to ten specialized muscle fibers (cells), partially enclosed in a spindle-shaped connective tissue capsule, that make up a muscle spindle.

Intramembranous (in′-tra-MEM-bra-nus) **ossification** Bone formation within mesenchyme arranged in sheetlike layers that resemble membranes.

Intramuscular injection An injection that penetrates the skin and subcutaneous layer to enter a skeletal muscle. Common sites are the deltoid, gluteus medius, and vastus lateralis muscles.

Intraocular (in′-tra-OK-ū-lar) **pressure (IOP)** Pressure in the eyeball, produced mainly by aqueous humour.

Intrinsic factor (IF) A glycoprotein, synthesized and secreted by the parietal cells of the gastric mucosa, that facilitates vitamin B_{12} absorption in the small intestine.

Invagination (in-vaj′-i-NĀ-shun) The pushing of the wall of a cavity into the cavity itself.

Inversion (in-VER-zhun) The movement of the sole medially at the ankle joint.

In vitro (VĒ-trō) Literally, in glass; outside the living body and in an artificial environment such as a laboratory test tube.

Ipsilateral (ip-si-LAT-er-al) On the same side, affecting the same side of the body.

Iris The colored portion of the vascular tunic of the eyeball seen through the cornea that contains circular and radial smooth muscle; the hole in the center of the iris is the pupil.

Irritable bowel syndrome (IBS) Disease of the entire gastrointestinal tract in which a person reacts to stress by developing symptoms (such as cramping and abdominal pain) associated with alternating patterns of diarrhoea and constipation. Excessive amounts of mucus may appear in faeces, and other symptoms include flatulence, nausea, and loss of appetite. Also known as **irritable colon** or **spastic colitis.**

Ischaemia (is-KĒ-mē-a) A lack of sufficient blood to a body part due to obstruction or constriction of a blood vessel.

Islet of Langerhans *See* **Pancreatic islet.**

Isotonic (ī′-sō-TON-ik) Having equal tension or tone. A solution having the same concentration of impermeable solutes as cytosol.

Isthmus (IS-mus) A narrow strip of tissue or narrow passage connecting two larger parts.

J

Jaundice (JON-dis) A condition characterized by yellowness of the skin, the white of the eyes, mucous membranes, and body fluids because of a buildup of bilirubin.

Jejunum (je-JOO-num) The middle part of the small intestine.

Joint A point of contact between two bones, between bone and cartilage, or between bone and teeth. Also called an **articulation** or **arthrosis**.

Joint kinesthetic (kin′-es-THET-ik) **receptor** A proprioceptive receptor located in a joint, stimulated by joint movement.

Juxtaglomerular (juks-ta-glō-MER-ū-lar) **apparatus (JGA)** Consists of the macula densa (cells of the distal convoluted tubule adjacent to the afferent and efferent arteriole) and juxtaglomerular cells (modified cells of the afferent and sometimes efferent arteriole); secretes renin when blood pressure starts to fall.

K

Keratin (KER-a-tin) An insoluble protein found in the hair, nails, and other keratinized tissues of the epidermis.

Keratinocyte (ker-a-TIN-ō-sīt) The most numerous of the epidermal cells; produces keratin.

Kidney (KID-nē) One of the paired reddish organs located in the lumbar region that regulates the composition, volume, and pressure of blood and produces urine.

Kidney stone A solid mass, usually consisting of calcium oxalate, uric acid, or calcium phosphate crystals, that may form in any portion of the urinary tract. Also called a **renal calculus.**

Kinaesthesia (kin′-es-THĒ-zē-a) The perception of the extent and direction of movement of body parts; this sense is possible due to nerve impulses generated by proprioceptors.

Kinesiology (ki-nē-sē′-OL-ō-jē) The study of the movement of body parts.

Kinetochore (ki-NET-ō-kor) Protein complex attached to the outside of a centromere to which kinetochore microtubules attach.

Kupffer's cell *See* **Stellate reticuloendothelial cell.**

Kyphosis (kī-FŌ-sis) An exaggeration of the thoracic curve of the vertebral column, resulting in a "round-shouldered" appearance. Also called **hunchback.**

L

Labial frenulum (LĀ-bē-al FREN-ū-lum) A medial fold of mucous membrane between the inner surface of the lip and the gums.

Labia majora (LĀ-bē-a ma-JŌ-ra) Two longitudinal folds of skin extending downward and backward from the mons pubis of the female.

Labia minora (min-OR-a) Two small folds of mucous membrane lying medial to the labia majora of the female.

Labium (LĀ-bē-um) A lip. A liplike structure. *Plural* is **labia** (LA-bē-a).

Labour The process of giving birth in which a fetus is expelled from the uterus through the vagina.

Labyrinth (LAB-i-rinth) Intricate communicating passageway, especially in the internal ear.

Lacrimal canal A duct, one on each eyelid, beginning at the punctum at the medial margin of an eyelid and conveying tears medially into the nasolacrimal sac.

Lacrimal gland Secretory cells, located at the superior anterolateral portion of each orbit, that secrete tears into excretory ducts that open onto the surface of the conjunctiva.

Lacrimal sac The superior expanded portion of the nasolacrimal duct that receives the tears from a lacrimal canal.

Lactation (lak-TĀ-shun) The secretion and ejection of milk by the mammary glands.

Lacteal (LAK-tē-al) One of many lymphatic vessels in villi of the intestines that absorb triglycerides and other lipids from digested food.

Lacuna (la-KOO-na) A small, hollow space, such as that found in bones in which the osteocytes lie. *Plural* is **lacunae** (la-KOO-nē).

Lambdoid (LAM-doyd) **suture** The joint in the skull between the parietal bones and the occipital bone; sometimes contains sutural (Wormian) bones.

Lamellae (la-MEL-ē) Concentric rings of hard, calcified extracellular matrix found in compact bone.

Lamellated corpuscle *See* **Pacinian corpuscle.**

Lamina (LAM-i-na) A thin, flat layer or membrane, as the flattened part of either side of the arch of a vertebra. *Plural* is **laminae** (LAM-i-nē).

Lamina propria (PRŌ-prē-a) The connective tissue layer of a mucosa.

Langerhans (LANG-er-hans) **cell** Epidermal dendritic cell that functions as an antigen-presenting cell (APC) during an immune response.

Lanugo (la-NOO-gō) Fine downy hairs that cover the fetus.

Large intestine The portion of the gastrointestinal tract extending from the ileum of the small intestine to the anus, divided structurally into the caecum, colon, rectum, and anal canal.

Laryngopharynx (la-rin′-gō-FAIR-inks) The inferior portion of the pharynx, extending downward from the level of the hyoid bone that divides posteriorly into the oesophagus and anteriorly into the larynx. Also called the **hypopharynx.**

Larynx (LAIR-inks) The **voice box,** a short passageway that connects the pharynx with the trachea.

Lateral (LAT-er-al) Farther from the midline of the body or a structure.

Lateral ventricle (VEN-tri-kul) A cavity within a cerebral hemisphere that communicates with the lateral ventricle in the other cerebral hemisphere and with the third ventricle by way of the interventricular foramen.

Leg The part of the lower limb between the knee and the ankle.

Lens A transparent organ constructed of proteins (crystallins) lying posterior to the pupil and iris of the eyeball and anterior to the vitreous body.

Lesion (LĒ-zhun) Any localized, abnormal change in a body tissue.

Lesser oementum (ō-MEN-tum) A fold of the peritoneum that extends from the liver to the lesser curvature of the stomach and the first part of the duodenum.

Lesser vestibular (ves-TIB-ū-lar) **gland** One of the paired mucus-secreting glands with ducts that open on either side of the urethral orifice in the vestibule of the female.

Leukaemia (loo-KĒ-mē-a) A malignant disease of the blood-forming tissues characterized by either uncontrolled production and accumulation of immature leukocytes in which many cells fail to reach maturity (acute) or an accumulation of mature leukocytes in the blood because they do not die at the end of their normal life span (chronic).

Leukocyte (LOO-kō-sīt) A white blood cell.

Leydig (LĪ-dig) **cell** A type of cell that secretes testosterone; located in the connective tissue between seminiferous tubules in a mature testis. Also known as **interstitial cell of Leydig** or **interstitial endocrinocyte.**

Ligament (LIG-a-ment) Dense regular connective tissue that attaches bone to bone.

Ligand (LĪ-gand) A chemical substance that binds to a specific receptor.

Limbic system A part of the forebrain, sometimes termed the visceral brain, concerned with various aspects of emotion and behavior; includes the limbic lobe, dentate gyrus, amygdala, septal nuclei, mammillary bodies, anterior thalamic nucleus, olfactory bulbs, and bundles of myelinated axons.

Lingual frenulum (LIN-gwal FREN-ū-lum) A fold of mucous membrane that connects the tongue to the floor of the mouth.

Lipase An enzyme that splits fatty acids from triglycerides and phospholipids.

Lipid (LIP-id) An organic compound composed of carbon, hydrogen, and oxygen that is usually insoluble in water, but soluble in alcohol, ether, and chloroform; examples include triglycerides (fats and oils), phospholipids, steroids, and eicosanoids.

Lipid bilayer Arrangement of phospholipid, glycolipid, and cholesterol molecules in two parallel sheets in which the hydrophilic "heads" face outward and the hydrophobic "tails" face inward; found in cellular membranes.

Lipoprotein (lip′-ō-PRŌ-tēn) One of several types of particles containing lipids (cholesterol and triglycerides) and proteins that make it water soluble for transport in the blood; high levels of **low-density lipoproteins (LDLs)** are associated with increased risk of atherosclerosis, whereas high levels of **high-density lipoproteins (HDLs)** are associated with decreased risk of atherosclerosis.

Liver Large organ under the diaphragm that occupies most of the right hypochondriac region and part of the epigastric region. Functionally, it produces bile and synthesizes most plasma proteins; interconverts nutrients; detoxifies substances; stores glycogen, iron, and vitamins; carries on phagocytosis of worn-out blood cells and bacteria; and helps synthesize the active form of vitamin D.

Long-term potentiation (po-ten′-shē-Ā-shun) **(LTP)** Prolonged, enhanced synaptic transmission that occurs at certain synapses within the hippocampus of the brain; believed to underlie some aspects of memory.

Lordosis (lor-DŌ-sis) An exaggeration of the lumbar curve of the vertebral column. Also called **hollow back.**

Lower limb The appendage attached at the pelvic (hip) girdle, consisting of the thigh, knee, leg, ankle, foot, and toes. Also called the **lower extremity** or **lower appendage.**

Lumbar (LUM-bar) Region of the back and side between the ribs and pelvis; loin.

Lumbar plexus (PLEK-sus) A network formed by the anterior (ventral) branches of spinal nerves L1 through L4.

Lumen (LOO-men) The space within an artery, vein, intestine, renal tubule, or other tubular structure.

Lungs Main organs of respiration that lie on either side of the heart in the thoracic cavity.

Lunula (LOO-noo-la) The moon-shaped white area at the base of a nail.

Luteinising (LOO-tē-in′-īz-ing) **hormone (LH)** A hormone secreted by the anterior pituitary that stimulates ovulation, stimulates progesterone secretion by the corpus luteum, and readies the mammary glands for milk secretion in females; stimulates testosterone secretion by the testes in males.

Lymph (LIMF) Fluid confined in lymphatic vessels and flowing through the lymphatic system until it is returned to the blood.

Lymph node An oval or bean-shaped structure located along lymphatic vessels.

Lymphatic (lim-FAT-ik) **capillary** Closed-ended microscopic lymphatic vessel that begins in spaces between cells and converges with other lymphatic capillaries to form lymphatic vessels.

Lymphatic system (lim-FAT-ik) A system consisting of a fluid called lymph, vessels called lymphatics that transport lymph, a number of organs containing lymphatic tissue (lymphocytes within a filtering tissue), and red bone marrow.

Lymphatic tissue A specialized form of reticular tissue that contains large numbers of lymphocytes.

Lymphatic vessel A large vessel that collects lymph from lymphatic capillaries and converges with other lymphatic vessels to form the thoracic and right lymphatic ducts.

Lymphocyte (LIM-fō-sīt) A type of white blood cell that helps carry out cell-mediated and antibody-mediated immune responses; found in blood and in lymphatic tissues.

Lysosome (LĪ-sō-sōm) An organelle in the cytoplasm of a cell, enclosed by a single membrane and containing powerful digestive enzymes.

Lysosyme (LĪ-sō-zīm) A bactericidal enzyme found in tears, saliva, and perspiration.

M

Macrophage (MAK-rō-fāj) Phagocytic cell derived from a monocyte; may be fixed or wandering.

Macula (MAK-ū-la) A discolored spot or a colored area. A small, thickened region on the wall of the utricle and saccule that contains receptors for static equilibrium.

Macula lutea (LOO-tē-a) The yellow spot in the center of the retina.

Major histocompatibility (MHC) antigens Surface proteins on white blood cells and other nucleated cells that are unique for each person (except for identical siblings); used to type tissues and help prevent rejection of transplanted tissues. Also known as **human leukocyte antigens (HLA).**

Malignant (ma-LIG-nant) Referring to diseases that tend to become worse and cause death, especially the invasion and spreading of cancer.

Mammary (MAM-ar-ē) **gland** Modified sudoriferous (sweat) gland of the female that produces milk for the nourishment of the young.

Mammillary (MAM-i-ler-ē) **bodies** Two small rounded bodies on the inferior aspect of the hypothalamus that are involved in reflexes related to the sense of smell.

Marrow (MAR-ō) Soft, spongelike material in the cavities of bone. *Red bone marrow* produces blood cells; *yellow bone marrow* contains adipose tissue that stores triglycerides.

Mast cell A cell found in areolar connective tissue that releases histamine, a dilator of small blood vessels, during inflammation.

Mastication (mas′-ti-KĀ-shun) Chewing.

Mature follicle A large, fluid-filled follicle containing a secondary oocyte and surrounding granulosa cells that secrete oestrogens. Also called a **graafian** (GRAF-ē-an) **follicle.**

Meatus (mē-Ā-tus) A passage or opening, especially the external portion of a canal.

Mechanoreceptor (me-KAN-ō-rē-sep-tor) Sensory receptor that detects mechanical deformation of the receptor itself or adjacent cells; stimuli so detected include those related to touch, pressure, vibration, proprioception, hearing, equilibrium, and blood pressure.

Medial (MĒ-dē-al) Nearer the midline of the body or a structure.

Medial lemniscus (lem-NIS-kus) A white matter tract that originates in the gracile and cuneate nuclei of the medulla oblongata and extends to the thalamus on the same side; sensory axons in this tract conduct nerve impulses for the sensations of proprioception, fine touch, vibration, hearing, and equilibrium.

Median aperture (AP-er-choor) One of the three openings in the roof of the fourth ventricle through which cerebrospinal fluid enters the subarachnoid space of the brain and cord. Also called the **foramen of Magendie.**

Median plane A vertical plane dividing the body into right and left halves. Situated in the middle.

Mediastinum (mē′-dē-as-TĪ-num) The anatomical region on the thoracic cavity between the pleurae of the lungs that extends from the sternum to the vertebral column and from the first rib to the diaphragm.

Medulla (me-DOOL-la) An inner layer of an organ, such as the medulla of the kidneys.

Medulla oblongata (me-DOOL-la ob′-long-GA-ta) The most inferior part of the brain stem. Also termed the **medulla.**

Medullary (MED-ū-lar′-ē) **cavity** The space within the diaphysis of a bone that contains yellow bone marrow. Also called the **marrow cavity.**

Medullary rhythmicity (rith-MIS-i-tē) **area** The neurons of the respiratory center in the medulla oblongata that control the basic rhythm of respiration.

Meibomian gland *See* **Tarsal gland.**

Meiosis (mī-Ō-sis) A type of cell division that occurs during production of gametes, involving two successive nuclear divisions that result in cells with the haploid *(n)* number of chromosomes.

Meissner (MĪS-ner) **corpuscle** A sensory receptor for touch; found in dermal papillae, especially in the palms and soles. Also called a **corpuscle of touch.**

Melanin (MEL-a-nin) A dark black, brown, or yellow pigment found in some parts of the body such as the skin, hair, and pigmented layer of the retina.

Melanocyte (MEL-a-nō-sīt′) A pigmented cell, located between or beneath cells of the deepest layer of the epidermis, that synthesizes melanin.

Melanocyte-stimulating hormone (MSH) A hormone secreted by the anterior pituitary that stimulates the dispersion of melanin granules in melanocytes in amphibians; continued administration produces darkening of skin in humans.

Melatonin (mel-a-TŌN-in) A hormone secreted by the pineal gland that helps set the timing of the body's biological clock.

Membrane A thin, flexible sheet of tissue composed of an epithelial layer and an underlying connective tissue layer, as in an epithelial membrane, or of areolar connective tissue only, as in a synovial membrane.

Membranous labyrinth (mem-BRA-nus LAB-i-rinth) The part of the labyrinth of the internal ear that is located inside the bony labyrinth and separated from it by the perilymph; made up of the semicircular ducts, the saccule and utricle, and the cochlear duct.

Memory The ability to recall thoughts; commonly classifed as short-term (activated) and long-term.

Menarche (me-NAR-kē) The first menses (menstrual flow) and beginning of ovarian and uterine cycles.

Meninges (me-NIN-jēz) Three membranes covering the brain and spinal cord, called the dura mater, arachnoid mater, and pia mater. *Singular* is **meninx** (MEN-inks).

Menopause (MEN-ō-pawz) The termination of the menstrual cycles.

Menstruation (men′-stroo-Ā-shun) Periodic discharge of blood, tissue fluid, mucus, and epithelial cells that usually lasts for 5 days; caused by a sudden reduction in oestrogens and progesterone. Also called the **menstrual phase** or **menses.**

Merkel (MER-kel) **cell** Type of cell in the epidermis of hairless skin that makes contact with a Merkel (tactile) disc, which functions in touch.

Merkel disc Saucer-shaped free nerve endings that make contact with Merkel cells in the epidermis and function as touch receptors. Also called a **tactile disc.**

Merocrine (MER-ō-krin) **gland** Gland made up of secretory cells that remain intact throughout the process of formation and discharge of the secretory product, as in the salivary and pancreatic glands.

Mesenchyme (MEZ-en-kīm) An embryonic connective tissue from which all other connective tissues arise.

Mesentery (MEZ-en-ter′-ē) A fold of peritoneum attaching the small intestine to the posterior abdominal wall.

Mesocolon (mez′-ō-KŌ-lon) A fold of peritoneum attaching the colon to the posterior abdominal wall.

Mesoderm The middle primary germ layer that gives rise to connective tissues, blood and blood vessels, and muscles.

Mesothelium (mez′-ō-THĒ-lē-um) The layer of simple squamous epithelium that lines serous membranes.

Mesovarium (mez′-ō-VAR-ē-um) A short fold of peritoneum that attaches an ovary to the broad ligament of the uterus.

Metabolism (me-TAB-ō-lizm) All the biochemical reactions that occur within an organism, including the synthetic (anabolic) reactions and decomposition (catabolic) reactions.

Metacarpus (met′-a-KAR-pus) A collective term for the five bones that make up the palm.

Metaphase (MET-a-fāz) The second stage of mitosis, in which chromatid pairs line up on the metaphase plate of the cell.

Metaphysis (me-TAF-i-sis) Region of a long bone between the diaphysis and epiphysis that contains the epiphyseal plate in a growing bone.

Metarteriole (met′-ar-TĒ-rē-ōl) A blood vessel that emerges from an arteriole, traverses a capillary network, and empties into a venule.

Metastasis (me-TAS-ta-sis) The spread of cancer to surrounding tissues (local) or to other body sites (distant).

Metatarsus (met′-a-TAR-sus) A collective term for the five bones located in the foot between the tarsals and the phalanges.

Microfilament (mī-krō-FIL-a-ment) Rodlike protein filament about 6 nm in diameter; constitutes contractile units in muscle fibers (cells) and provides support, shape, and movement in nonmuscle cells.

Microglia (mī-KROG-lē-a) Neuroglial cells that carry on phagocytosis.

Microtubule (mī-krō-TOO-būl) Cylindrical protein filament, from 18 to 30 nm in diameter, consisting of the protein tubulin; provides support, structure, and transportation.

Microvilli (mī-krō-VIL-ī) Microscopic, fingerlike projections of the plasma membranes of cells that increase surface area for absorption, especially in the small intestine and proximal convoluted tubules of the kidneys.

Micturition (mik′-choo-RISH-un) The act of expelling urine from the urinary bladder. Also called **urination** (ū-ri-NĀ-shun).

Midbrain The part of the brain between the pons and the diencephalon. Also called the **mesencephalon** (mes′-en-SEF-a-lon).

Middle ear A small, epithelial-lined cavity hollowed out of the temporal bone, separated from the external ear by the eardrum and from the internal ear by a thin bony partition containing the oval and round windows; extending across the middle ear are the three auditory ossicles. Also called the **tympanic** (tim-PAN-ik) **cavity.**

Midline An imaginary vertical line that divides the body into equal left and right sides.

Midsagittal plane A vertical plane through the midline of the body that divides the body or organs into *equal* right and left sides. Also called a **median plane.**

Mineralocorticoids (min′-er-al-ō-KOR-ti-koyds) A group of hormones of the adrenal cortex that help regulate sodium and potassium balance.

Mitochondrion (mī-tō-KON-drē-on) A double-membraned organelle that plays a central role in the production of ATP; known as the "powerhouse" of the cell. *Plural* is **mitochondria.**

Mitosis (mī-TŌ-sis) The orderly division of the nucleus of a cell that ensures that each new nucleus has the same number and kind of chromosomes as the original nucleus. The process includes the replication of chromosomes and the distribution of the two sets of chromosomes into two separate and equal nuclei.

Mitotic spindle Collective term for a football-shaped assembly of microtubules (nonkinetochore, kinetochore, and aster) that is responsible for the movement of chromosomes during cell division.

Modality (mō-DAL-i-tē) Any of the specific sensory entities, such as vision, smell, taste, or touch.

Modiolus (mō-DĪ-ō′-lus) The central pillar or column of the cochlea.

Molecule (mol′-e-KŪL) A substance composed of two or more atoms chemically combined.

Monocyte (MON-ō-sīt′) The largest type of white blood cell, characterized by agranular cytoplasm.

Monounsaturated fat A fatty acid that contains one double covalent bond between its carbon atoms; it is not completely saturated with hydrogen atoms. Plentiful in triglycerides of olive and peanut oils.

Mons pubis (MONZ PŪ-bis) The rounded, fatty prominence over the pubic symphysis, covered by coarse pubic hair.

Morula (MOR-ū-la) A solid sphere of cells produced by successive cleavages of a fertilized ovum about 4 days after fertilisation.

Motor area The region of the cerebral cortex that governs muscular movement, particularly the precentral gyrus of the frontal lobe.

Motor end plate Region of the sarcolemma of a muscle fiber (cell) that includes acetylcholine (ACh) receptors, which bind ACh released by synaptic end bulbs of somatic motor neurons.

Motor neurons (NOO-ronz) Neurons that conduct impulses from the brain toward the spinal cord or out of the brain and spinal cord into cranial or spinal nerves to effectors that may be either muscles or glands. Also called **efferent neurons.**

Motor unit A motor neuron together with the muscle fibers (cells) it stimulates.

Mucosa-associated lymphatic tissue (MALT) Lymphatic nodules scattered throughout the lamina propria (connective tissue) of mucous membranes lining the gastrointestinal tract, respiratory airways, urinary tract, and reproductive tract.

Mucous (MŪ-kus) **cell** A unicellular gland that secretes mucus. Two types are mucous neck cells and surface mucous cells in the stomach.

Mucous membrane A membrane that lines a body cavity that opens to the exterior. Also called the **mucosa** (mū-KŌ-sa).

Mucus The thick fluid secretion of goblet cells, mucous cells, mucous glands, and mucous membranes.

Muscarinic (mus′-ka-RIN-ik) **receptor** Receptor for the neurotransmitter acetylcholine found on all effectors innervated by parasympathetic postganglionic axons and on sweat glands innervated by cholinergic sympathetic postganglionic axons; so named because muscarine activates these receptors but does not activate nicotinic receptors for acetylcholine.

Muscle An organ composed of one of three types of muscle tissue (skeletal, cardiac, or smooth), specialized for contraction to produce voluntary or involuntary movement of parts of the body.

Muscle action potential A stimulating impulse that propagates along the sarcolemma and transverse tubules; in skeletal muscle, it is generated by acetylcholine, which increases the permeability of the sarcolemma to cations, especially sodium ions (Na^+).

Muscle fatigue (fa-TĒG) Inability of a muscle to maintain its strength of contraction or tension; may be related to insufficient oxygen, depletion of glycogen, and/or lactic acid buildup.

Muscle spindle An encapsulated proprioceptor in a skeletal muscle, consisting of specialized intrafusal muscle fibers and nerve endings; stimulated by changes in length or tension of muscle fibers.

Muscle strain Tearing of skeletal muscle fibers or tendon. Also called a **muscle pull** or **muscle tear.**

Muscle tone A sustained, partial contraction of portions of a skeletal or smooth muscle in response to activation of stretch receptors or a baseline level of action potentials in the innervating motor neurons.

Muscular dystrophies (DIS-trō-fēz) Inherited muscle-destroying diseases, characterized by degeneration of muscle fibers (cells), which causes progressive atrophy of the skeletal muscle.

Muscularis (MUS-kū-la′-ris) A muscular layer (coat or tunic) of an organ.

Muscularis mucosae (mū-KŌ-sē) A thin layer of smooth muscle fibers that underlie the lamina propria of the mucosa of the gastrointestinal tract.

Muscular system Usually refers to the approximately 100 voluntary muscles of the body that are composed of skeletal muscle tissue.

Muscular system Usually refers to the voluntary muscles of the body that are composed of skeletal muscle tissue.

Muscular tissue A tissue specialized to produce motion in response to muscle action potentials by its qualities of contractility, extensibility, elasticity, and excitability; types include skeletal, cardiac, and smooth.

Musculoskeletal (mus′-kyū-lō-skel-ETAL) **system** An integrated body system consisting of bones, joints, and muscles.

Mutation (mū-TĀ-shun) Any change in the sequence of bases in a DNA molecule resulting in a permanent alteration in some inheritable trait.

Myasthenia (mī-as-THĒ-nē-a) **gravis** Weakness and fatigue of skeletal muscles caused by antibodies directed against acetylcholine receptors.

Myelin (MĪ-e-lin) **sheath** Multilayered lipid and protein covering, formed by Schwann cells and oligodendrocytes, around axons of many peripheral and central nervous system neurons.

Myenteric (mī-en-TER-ik) **plexus** A network of autonomic axons and postganglionic cell bodies located in the muscularis of the gastrointestinal tract. Also called the **plexus of Auerbach** (OW-er-bak).

Myocardial infarction (mī′-ō-KAR-dē-al in-FARK-shun) **(MI)** Gross necrosis of myocardial tissue due to interrupted blood supply. Also called a **heart attack.**

Myocardium (mī′-ō-KAR-dē-um) The middle layer of the heart wall, made up of cardiac muscle tissue, lying between the epicardium and the endocardium and constituting the bulk of the heart.

Myofibril (mī-ō-FĪ-bril) A threadlike structure, extending longitudinally through a muscle fiber (cell), consisting mainly of thick filaments (myosin) and thin filaments (actin, troponin, and tropomyosin).

Myoglobin (mī-ō-GLŌB-in) The oxygen-binding, iron-containing protein present in the sarcoplasm of muscle fibers (cells); contributes the red color to muscle.

Myogram (MĪ-ō-gram) The record or tracing produced by a myograph, an apparatus that measures and records the force of muscular contractions.

Myology (mī-OL-ō-jē) The study of muscles.

Myometrium (mī′-ō-MĒ-trē-um) The smooth muscle layer of the uterus.

Myopathy (mī-OP-a-thē) Any abnormal condition or disease of muscle tissue.

Myopia (mī-Ō-pē-a) Defect in vision in which objects can be seen distinctly only when very close to the eyes; nearsightedness.

Myosin (MĪ-ō-sin) The contractile protein that makes up the thick filaments of muscle fibers.

Myotome (MĪ-ō-tōm) A group of muscles innervated by the motor neurons of a single spinal segment. In an embryo, the portion of a somite that develops into some skeletal muscles.

N

Nail A hard plate, composed largely of keratin, that develops from the epidermis of the skin to form a protective covering on the dorsal surface of the distal phalanges of the fingers and toes.

Nail matrix (MĀ-triks) The part of the nail beneath the body and root from which the nail is produced.

Nasal (NĀ-zal) **cavity** A mucosa-lined cavity on either side of the nasal septum that opens onto the face at the external nares and into the nasopharynx at the internal nares.

Nasal septum (SEP-tum) A vertical partition composed of bone (perpendicular plate of ethmoid and vomer) and cartilage, covered with a mucous membrane, separating the nasal cavity into left and right sides.

Nasolacrimal (nā′-zō-LAK-ri-mal) **duct** A canal that transports the lacrimal secretion (tears) from the nasolacrimal sac into the nose.

Nasopharynx (nā′-zō-FAR-inks) The superior portion of the pharynx, lying posterior to the nose and extending inferiorly to the soft palate.

Neck The part of the body connecting the head and the trunk. A constricted portion of an organ, such as the neck of the femur or uterus.

Necrosis (ne-KRŌ-sis) A pathological type of cell death that results from disease, injury, or lack of blood supply in which many adjacent cells swell, burst, and spill their contents into the interstitial fluid, triggering an inflammatory response.

Negative feedback system A feedback cycle that reverses a change in a controlled condition.

Neonatal (nē-ō-NĀ-tal) Pertaining to the first 4 weeks after birth.

Neoplasm (NĒ-ō-plazm) A new growth that may be benign or malignant.

Nephron (NEF-ron) The functional unit of the kidney.

Nerve A cordlike bundle of neuronal axons and/or dendrites and associated connective tissue coursing together outside the central nervous system.

Nerve fiber General term for any process (axon or dendrite) projecting from the cell body of a neuron.

Nerve impulse A wave of depolarization and repolarization that self-propagates along the plasma membrane of a neuron; also called a **nerve action potential.**

Nervous system A network of billions of neurons and even more neuroglia that is organized into two main divisions. Central nervous system (brain and spinal cord) and peripheral nervous system (nerves, ganglia, enteric plexuses, and sensory receptors outside the central nervous system).

Nervous tissue Tissue containing neurons that initiate and conduct nerve impulses to coordinate homoeostasis, and neuroglia that provide support and nourishment to neurons.

Neuralgia (noo-RAL-jē-a) Attacks of pain along the entire course or branch of a peripheral sensory nerve.

Neural plate A thickening of ectoderm, induced by the notochord, that forms early in the third week of development and represents the beginning of the development of the nervous system.

Neural tube defect (NTD) A developmental abnormality in which the neural tube does not close properly. Examples are spina bifida and anencephaly.

Neuritis (noo-RĪ-tis) Inflammation of one or more nerves.

Neurofibral node *See* **Node of Ranvier.**

Neuroglia (noo-RŌG-lē-a) Cells of the nervous system that perform various supportive functions. The neuroglia of the central nervous system are the astrocytes, oligodendrocytes, microglia, and ependymal cells; neuroglia of the peripheral nervous system include Schwann cells and satellite cells. Also called **glial** (GLĒ-al) **cells.**

Neurohypophyseal (noo′-rō-hī′-pō-FIZ-ē-al) **bud** An outgrowth of ectoderm located on the floor of the hypothalamus that gives rise to the posterior pituitary.

Neurolemma (noo-rō-LEM-ma) The peripheral, nucleated cytoplasmic layer of the Schwann cell. Also called **sheath of Schwann** (SCHWON).

Neurology (noo-ROL-ō-jē) The study of the normal functioning and disorders of the nervous system.

Neuromuscular (noo-rō-MUS-kū-lar) **junction (NMJ)** A synapse between the axon terminals of a motor neuron and the sarcolemma of a muscle fiber (cell).

Neuron (NOO-ron) A nerve cell, consisting of a cell body, dendrites, and an axon.

Neurophysiology (NOOR-ō-fiz-ē-ol′-ō-jē) Study of the functional properties of nerves.

Neurosecretory (noo-rō-SĒK-re-tō-rē) **cell** A neuron that secretes a hypothalamic releasing hormone or inhibiting hormone into blood capillaries of the hypothalmus; a neuron that secretes oxytocin or antidiuretic hormone into blood capillaries of the posterior pituitary.

Neurotransmitter (noo′-rō-trans′-MIT-er) One of a variety of molecules within axon terminals that are released into the synaptic cleft in response to a nerve impulse and that change the membrane potential of the postsynaptic neuron.

Neurulation (noor-oo-LĀ-shun) The process by which the neural plate, neural folds, and neural tube develop.

Neutrophil (NOO-trō-fil) A type of white blood cell characterized by granules that stain pale lilac with a combination of acidic and basic dyes.

Nicotinic (nik′-ō-TIN-ik) **receptor** Receptor for the neurotransmitter acetylcholine found on both sympathetic and parasympathetic postganglionic neurons and on skeletal muscle in the motor end plate; so named because nicotine activates these receptors but does not activate muscarinic receptors for acetylcholine.

Nipple A pigmented, wrinkled projection on the surface of the breast that is the location of the openings of the lactiferous ducts for milk release.

Nociceptor (nō′-sē-SEP-tor) A free (naked) nerve ending that detects painful stimuli.

Node of Ranvier (RON-vē-ā) A space along a myelinated axon between the individual Schwann cells that form the myelin sheath and the neurolemma. Also called a **neurofibral node.**

Norepinephrine (nor′-ep-ē-NEF-rin) **(NE)** A hormone secreted by the adrenal medulla that produces actions similar to those that result from sympathetic stimulation. Also called **noradrenaline** (nor-a-DREN-a-lin).

Notochord (NŌ-tō-cord) A flexible rod of mesodermal tissue that lies where the future vertebral column will develop and plays a role in induction.

Nucleic (noo-KLĒ-ik) **acid** An organic compound that is a long polymer of nucleotides, with each nucleotide containing a pentose sugar, a phosphate group, and one of four possible nitrogenous bases (adenine, cytosine, guanine, and thymine or uracil).

Nucleolus (noo′-KLĒ-ō-lus) Spherical body within a cell nucleus composed of protein, DNA, and RNA that is the site of the assembly of small and large ribosomal subunits. *Plural* is **nucleoli.**

Nucleosome (NOO-klē-ō-sōm) Structural subunit of a chromosome consisting of histones and DNA.

Nucleus (NOO-klē-us) A spherical or oval organelle of a cell that contains the hereditary factors of the cell, called genes. A cluster of unmyelinated nerve cell bodies in the central nervous system. The central part of an atom made up of protons and neutrons.

Nucleus pulposus (pul-PŌ-sus) A soft, pulpy, highly elastic substance in the center of an intervertebral disc; a remnant of the notochord.

Nutrient A chemical substance in food that provides energy, forms new body components, or assists in various body functions.

O

Obesity (ō-BĒS-i-tē) Body weight more than 20% above a desirable standard due to excessive accumulation of fat.

Oblique (ō-BLĒK) **plane** A plane that passes through the body or an organ at an angle between the transverse plane and either the midsagittal, parasagittal, or frontal plane.

Obstetrics (ob-STET-riks) The specialized branch of medicine that deals with pregnancy, labour, and the period of time immediately after delivery (about 6 weeks).

Oedema (usu., also edema) An abnormal accumulation of interstitial fluid.

Oesophagus (e-SOF-a-gus) The hollow muscular tube that connects the pharynx and the stomach.

Oestrogens (ES-tro-jenz) Feminizing sex hormones produced by the ovaries; govern development of oocytes, maintenance of female reproductive structures, and appearance of secondary sex characteristics; also affect fluid and electrolyte balance, and protein anabolism. Examples are β-estradiol, estrone, and estriol.

Olfaction (ōl-FAK-shun) The sense of smell.

Olfactory (ōl-FAK-tō-rē) Pertaining to smell.

Olfactory bulb A mass of grey matter containing cell bodies of neurons that form synapses with neurons of the olfactory (I) nerve, lying inferior to the frontal lobe of the cerebrum on either side of the crista galli of the ethmoid bone.

Olfactory receptor A bipolar neuron with its cell body lying between supporting cells located in the mucous membrane lining the superior portion of each nasal cavity; transduces odors into neural signals.

Olfactory tract A bundle of axons that extends from the olfactory bulb posteriorly to olfactory regions of the cerebral cortex.

Oligodendrocyte (OL-i-gō-den′-drō-sīt) A neuroglial cell that supports neurons and produces a myelin sheath around axons of neurons of the central nervous system.

Oliguria (ol′-i-GŪ-rē-a) Daily urinary output usually less than 250 mL.

Olive A prominent oval mass on each lateral surface of the superior part of the medulla oblongata.

Oncogene (ON-kō-jēn) Cancer-causing gene; it derives from a normal gene, termed a proto-oncogene, that encodes proteins involved in cell growth or cell regulation but has the ability to transform a normal cell into a cancerous cell when it is mutated or inappropriately activated. One example is *p53*.

Oncology (on-KOL-ō-jē) The study of tumors.

Oogenesis (ō-ō-JEN-e-sis) Formation and development of female gametes (oocytes).

Oophorectomy (ō′-of-ō-REK-tō-me) Surgical removal of the ovaries.

Ophthalmic (of-THAL-mik) Pertaining to the eye.

Ophthalmologist (of′-thal-MOL-ō-jist) A physician who specializes in the diagnosis and treatment of eye disorders using drugs, surgery, and corrective lenses.

Ophthalmology (of-thal-MOL-ō-jē) The study of the structure, function, and diseases of the eye.

Optic (OP-tik) Refers to the eye, vision, or properties of light.

Optic chiasm (kī-AZM) A crossing point of the two branches of the optic (II) nerve, anterior to the pituitary gland. Also called **optic chiasma.**

Optic disc A small area of the retina containing openings through which the axons of the ganglion cells emerge as the optic (II) nerve. Also called the **blind spot.**

Optic tract A bundle of axons that carry nerve impulses from the retina of the eye between the optic chiasm and the thalamus.

Ora serrata (Ō-ra ser-RĀ-ta) The irregular margin of the retina lying internal and slightly posterior to the junction of the choroid and ciliary body.

Orbit (OR-bit) The bony, pyramidal-shaped cavity of the skull that holds the eyeball.

Organ A structure composed of two or more different kinds of tissues with a specific function and usually a recognizable shape.

Organelle (or-ga-NEL) A permanent structure within a cell with characteristic morphology that is specialized to serve a specific function in cellular activities.

Organism (OR-ga-nizm) A total living form; one individual.

Organogenesis (or′-ga-nō-JEN-e-sis) The formation of body organs and systems. By the end of the eighth week of development, all major body systems have begun to develop.

Orifice (OR-i-fis) Any aperture or opening.

Origin (OR-i-jin) The attachment of a muscle tendon to a stationary bone or the end opposite the insertion.

Oropharynx (or′-ō-FAR-inks) The intermediate portion of the pharynx, lying posterior to the mouth and extending from the soft palate to the hyoid bone.

Orthopaedics (or′-thō-PĒ-diks) The branch of medicine that deals with the preservation and restoration of the skeletal system, articulations, and associated structures.

Osmoreceptor (oz′-mō-rē-CEP-tor) Receptor in the hypothalamus that is sensitive to changes in blood osmolarity and, in response to high osmolarity (low water concentration), stimulates synthesis and release of antidiuretic hormone (ADH).

Osmosis (oz-MŌ-sis) The net movement of water molecules through a selectively permeable membrane from an area of higher water concentration to an area of lower water concentration until equilibrium is reached.

Osseous (OS-ē-us) Bony.

Ossicle (OS-si-kul) One of the small bones of the middle ear (malleus, incus, stapes).

Ossification (os′-i-fi-KĀ-shun) Formation of bone. Also called **osteogenesis.**

Ossification (os′-i-fi-KĀ-shun) **center** An area in the cartilage model of a future bone where the cartilage cells hypertrophy, secrete enzymes that calcify their extracellular matrix, and die, and the area they occupied is invaded by osteoblasts that then lay down bone.

Osteoblast (OS-tē-ō-blast′) Cell formed from an osteogenic cell that participates in bone formation by secreting some organic components and inorganic salts.

Osteoclast (OS-tē-ō-klast′) A large, multinuclear cell that resorbs (destroys) bone matrix.

Osteocyte (OS-tē-ō-sīt′) A mature bone cell that maintains the daily activities of bone tissue.

Osteogenic (os′-tē-ō-JEN-ik) **cell** Stem cell derived from mesenchyme that has mitotic potential and the ability to differentiate into an osteoblast.

Osteogenic layer The inner layer of the periosteum that contains cells responsible for forming new bone during growth and repair.

Osteology (os-tē-OL-ō-jē) The study of bones.

Osteon (OS-tē-on) The basic unit of structure in adult compact bone, consisting of a central (haversian) ca nal with its concentrically arranged lamellae, lacunae, osteocytes, and canaliculi. Also called a **haversian** (ha-VER-shun) **system.**

Osteoporosis (os′-tē-ō-pō-RŌ-sis) Age-related disorder characterized by decreased bone mass and increased susceptibility to fractures, often as a result of decreased levels of oestrogens.

Otic (Ō-tik) Pertaining to the ear.

Otolith (Ō-tō-lith) A particle of calcium carbonate embedded in the otolithic membrane that functions in maintaining static equilibrium.

Otolithic (ō-tō-LITH-ik) **membrane** Thick, gelatinous, glycoprotein layer located directly over hair cells of the macula in the saccule and utricle of the internal ear.

Otorhinolaryngology (ō-tō-rī′-nō-lar-in-GOL-ō-jē) The branch of medicine that deals with the diagnosis and treatment of diseases of the ears, nose, and throat.

Oval window A small, membrane-covered opening between the middle ear and inner ear into which the footplate of the stapes fits.

Ovarian (ō-VAR-ē-an) **cycle** A monthly series of events in the ovary associated with the maturation of a secondary oocyte.

Ovarian follicle (FOL-i-kul) A general name for oocytes (immature ova) in any stage of development, along with their surrounding epithelial cells.

Ovarian ligament (LIG-a-ment) A rounded cord of connective tissue that attaches the ovary to the uterus.

Ovary (Ō-var-ē) Female gonad that produces oocytes and the estrogen, progesterone, inhibin, and relaxin hormones.

Ovulation (ov′-ū-LĀ-shun) The rupture of a mature ovarian (graafian) follicle with discharge of a secondary oocyte into the pelvic cavity.

Ovum (Ō-vum) The female reproductive or germ cell; an egg cell; arises through completion of meiosis in a secondary oocyte after penetration by a sperm.

Oxyhemoglobin (ok′-sē-HĒ-mō-glō-bin) **(Hb—O$_2$)** Haemoglobin combined with oxygen.

Oxytocin (ok′-sē-TŌ-sin) **(OT)** A hormone secreted by neurosecretory cells in the paraventricular and supraoptic nuclei of the hypothalamus that stimulates contraction of smooth muscle in the pregnant uterus and myoepithelial cells around the ducts of mammary glands.

P

P wave The deflection wave of an electrocardiogram that signifies atrial depolarization.

Pacinian corpuscle (pa-SIN-ē-an) Oval-shaped pressure receptor located in the dermis or subcutaneous tissue and consisting of concentric layers of a connective tissue wrapped around the dendrites of a sensory neuron. Also called a **lamellated corpuscle.**

Palate (PAL-at) The horizontal structure separating the oral and the nasal cavities; the roof of the mouth.

Palpate (PAL-pāt) To examine by touch; to feel.

Pancreas (PAN-krē-as) A soft, oblong organ lying along the greater curvature of the stomach and connected by a duct to the duodenum. It is both an exocrine gland (secreting pancreatic juice) and an endocrine gland (secreting insulin, glucagon, somatostatin, and pancreatic polypeptide).

Pancreatic (pan′-krē-AT-ik) **duct** A single large tube that unites with the common bile duct from the liver and gallbladder and drains pancreatic juice into the duodenum at the hepatopancreatic ampulla (ampulla of Vater). Also called the **duct of Wirsung** (VĒR-sung).

Pancreatic islet (Ī-let) Cluster of endocrine gland cells in the pancreas that secretes insulin, glucagon, somatostatin, and pancreatic polypeptide. Also called an **islet of Langerhans** (LAHNG-er-hanz).

Papanicolaou (pa-pa-NI-kō-lō) **test** A cytological staining test for the detection and diagnosis of premalignant and malignant conditions of the female genital tract. Cells scraped from the epithelium of the cervix of the uterus are examined microscopically. Also called a **Pap test** or **Pap smear.**

Papilla (pa-PIL-a) A small nipple-shaped projection or elevation.

Paralysis (pa-RAL-a-sis) Loss or impairment of motor function due to a lesion of nervous or muscular origin.

Paranasal sinus (par′-a-NĀ-zal SĪ-nus) A mucus-lined air cavity in a skull bone that communicates with the nasal cavity. Paranasal sinuses are located in the frontal, maxillary, ethmoid, and sphenoid bones.

Paraplegia (par-a-PLĒ-jē-a) Paralysis of both lower limbs.

Parasagittal plane (par-a-SAJ-i-tal) A vertical plane that does not pass through the midline and that divides the body or organs into *unequal* left and right portions.

Parasympathetic (par′-a-sim-pa-THET-ik) **division** One of the two subdivisions of the autonomic nervous system, having cell bodies of preganglionic neurons in nuclei in the brain stem and in the lateral gray horn of the sacral portion of the spinal cord; primarily concerned with activities that conserve and restore body energy.

Parathyroid (par′-a-THĪ-royd) **gland** One of usually four small endocrine glands embedded in the posterior surfaces of the lateral lobes of the thyroid gland.

Parathyroid hormone (PTH) A hormone secreted by the chief (principal) cells of the parathyroid glands that increases blood calcium level and decreases blood phosphate level. Also called **parathormone.**

Paraurethral (par′-a-ū-RĒ-thral) **gland** Gland embedded in the wall of the urethra whose duct opens on either side of the urethral orifice and secretes mucus. Also called **Skene's** (SKĒNZ) **gland.**

Parenchyma (pa-RENG-ki-ma) The functional parts of any organ, as opposed to tissue that forms its stroma or framework.

Parietal (pa-RĪ-e-tal) Pertaining to or forming the outer wall of a body cavity.

Parietal cell A type of secretory cell in gastric glands that produces hydrochloric acid and intrinsic factor. Also called an **oxyntic cell.**

Parietal pleura (PLOO-ra) The outer layer of the serous pleural membrane that encloses and protects the lungs; the layer that is attached to the wall of the pleural cavity.

Parkinson's disease (PD) Progressive degeneration of the basal nuclei and substantia nigra of the cerebrum resulting in decreased production of dopamine (DA) that leads to tremor, slowing of voluntary movements, and muscle weakness.

Parotid (pa-ROT-id) **gland** One of the paired salivary glands located inferior and anterior to the ears and connected to the oral cavity via a duct (Stensen's) that opens into the inside of the cheek opposite the maxillary (upper) second molar tooth.

Pars intermedia A small avascular zone between the anterior and posterior pituitary glands.

Parturition (par-toor-ISH-un) Act of giving birth to young; childbirth, delivery.

Patent (PĀ-tent) **ductus arteriosus (PDA)** A congenital heart defect in which the ductus arteriosus remains open. As a result, aortic blood flows into the lower-pressure pulmonary trunk, increasing pulmonary trunk pressure and overworking both ventricles.

Pathogen (PATH-ō-jen) A disease-producing microbe.

Pathological (path′-ō-LOJ-i-kal) **anatomy** The study of structural changes caused by disease.

Pathophysiology (PATH-ō-fez-ē-ol-ō-jē) Study of functional changes associated with disease and aging.

Pectinate (PEK-ti-nāt) **muscles** Projecting muscle bundles of the anterior atrial walls and the lining of the auricles.

Pectoral (PEK-tō-ral) Pertaining to the chest or breast.

Pedicel (PED-i-sel) Footlike structure, as on podocytes of a glomerulus.

Pelvic (PEL-vik) **cavity** Inferior portion of the abdominopelvic cavity that contains the urinary bladder, sigmoid colon, rectum, and internal female and male reproductive structures.

Pelvic splanchnic (PEL-vik SPLANGK-nik) **nerves** Consist of preganglionic parasympathetic axons from the levels of S2, S3, and S4 that supply the urinary bladder, reproductive organs, and the descending and sigmoid colon and rectum.

Pelvis The basinlike structure formed by the two hip bones, the sacrum, and the coccyx. The expanded, proximal portion of the ureter, lying within the kidney and into which the major calyces open.

Penis (PĒ-nis) The organ of urination and copulation in males; used to deposit semen into the female vagina.

Pepsin Protein-digesting enzyme secreted by chief cells of the stomach in the inactive form pepsinogen, which is converted to active pepsin by hydrochloric acid.

Peptic ulcer An ulcer that develops in areas of the gastrointestinal tract exposed to hydrochloric acid; classified as a gastric ulcer if in the lesser curvature of the stomach and as a duodenal ulcer if in the first part of the duodenum.

Percussion (pur-KUSH-un) The act of striking (percussing) an underlying part of the body with short, sharp taps as an aid in diagnosing the part by the quality of the sound produced.

Perforating canal A minute passageway by means of which blood vessels and nerves from the periosteum penetrate into compact bone. Also called **Volkmann's** (FŌLK-mans) **canal.**

Pericardial (per′-i-KAR-dē-al) **cavity** Small potential space between the visceral and parietal layers of the serous pericardium that contains pericardial fluid.

Pericardium (per′-i-KAR-dē-um) A loose-fitting membrane that encloses the heart, consisting of a superficial fibrous layer and a deep serous layer.

Perichondrium (per′-i-KON-drē-um) A covering of dense irregular connective tissue that surrounds the surface of most cartilage.

Perilymph (PER-i-limf) The fluid contained between the bony and membranous labyrinths of the inner ear.

Perimetrium (per′-i-MĒ-trē-um) The serosa of the uterus.

Perimysium (per-i-MĪZ-ē-um) Invagination of the epimysium that divides muscles into bundles.

Perineum (per′-i-NĒ-um) The pelvic floor; the space between the anus and the scrotum in the male and between the anus and the vulva in the female.

Perineurium (per′-i-NOO-rē-um) Connective tissue wrapping around fascicles in a nerve.

Periodontal (per′-ē-ō-DON-tal) **disease** A collective term for conditions characterized by degeneration of gingivae, alveolar bone, periodontal ligament, and cementum.

Periodontal ligament The periosteum lining the alveoli (sockets) for the teeth in the alveolar processes of the mandible and maxillae.

Periosteum (per′-ē-OS-tē-um) The covering of a bone that consists of connective tissue, osteogenic cells, and osteoblasts; is essential for bone growth, repair, and nutrition.

Peripheral (pe-RIF-er-al) Located on the outer part or a surface of the body.

Peripheral nervous system (PNS) The part of the nervous system that lies outside the central nervous system, consisting of nerves and ganglia.

Peristalsis (per′-i-STAL-sis) Successive muscular contractions along the wall of a hollow muscular structure.

Peritoneum (per′-i-tō-NĒ-um) The largest serous membrane of the body that lines the abdominal cavity and covers the viscera within it.

Peritonitis (per′-i-tō-NĪ-tis) Inflammation of the peritoneum.

Peroxisome (pe-ROKS-i-sōm) Organelle similar in structure to a lysosome that contains enzymes that use molecular oxygen to oxidize various organic compounds; such reactions produce hydrogen peroxide; abundant in liver cells.

Perspiration Sweat; produced by sudoriferous (sweat) glands and containing water, salts, urea, uric acid, amino acids, ammonia, sugar, lactic acid, and ascorbic acid. Helps maintain body temperature and eliminate wastes.

Peyer's patches *See* **Aggregated lymphatic follicles.**

pH A measure of the concentration of hydrogen ions (H^+) in a solution. The **pH scale** extends from 0 to 14, with a value of 7 expressing neutrality, values lower than 7 expressing increasing acidity, and values higher than 7 expressing increasing alkalinity.

Phagocytosis (fag′-ō-sī-TŌ-sis) The process by which phagocytes ingest and destroy microbes, cell debris, and other foreign matter.

Phalanx (FĀ-lanks) The bone of a finger or toe. *Plural* is **phalanges** (fa-LAN-jēz).

Pharmacology (far′-ma-KOL-ō-jē) The science of the effects and uses of drugs in the treatment of disease.

Pharynx (FAR-inks) The throat; a tube that starts at the internal nares and runs partway down the neck, where it opens into the oesophagus posteriorly and the larynx anteriorly.

Phenotype (FĒ-nō-tīp) The observable expression of genotype; physical characteristics of an organism determined by genetic makeup and influenced by interaction between genes and internal and external environmental factors.

Phlebitis (fle-BĪ-tis) Inflammation of a vein, usually in a lower limb.

Photopigment A substance that can absorb light and undergo structural changes that can lead to the development of a receptor potential. An example is rhodopsin. In the eye, also called **visual pigment.**

Photoreceptor Receptor that detects light shining on the retina of the eye.

Physiology (fiz′-ē-OL-ō-jē) Science that deals with the functions of an organism or its parts.

Pia mater (PĪ-a MĀ-ter *or* PĒ-a MA-ter) The innermost of the three meninges (coverings) of the brain and spinal cord.

Pineal (PĪN-ē-al) **gland** A cone-shaped gland located in the roof of the third ventricle that secretes melatonin. Also called the **epiphysis cerebri** (ē-PIF-i-sis se-RĒ-brē).

Pinealocyte (pin-ē-AL-ō-sīt) Secretory cell of the pineal gland that releases melatonin.

Pinna (PIN-na) The projecting part of the external ear composed of elastic cartilage and covered by skin and shaped like the flared end of a trumpet. Also called the **auricle** (OR-i-kul).

Pituicyte (pi-TOO-i-sīt) Supporting cell of the posterior pituitary.

Pituitary (pi-TOO-i-tār-ē) **gland** A small endocrine gland occupying the hypophyseal fossa of the sphenoid bone and attached to the hypothalamus by the infundibulum. Also called the **hypophysis** (hī-POF-i-sis).

Pivot joint A synovial joint in which a rounded, pointed, or conical surface of one bone articulates with a ring formed partly by another bone and partly by a ligament, as in the joint between the atlas and axis and between the proximal ends of the radius and ulna. Also called a **trochoid** (TRŌ-koyd) **joint.**

Placenta (pla-SEN-ta) The special structure through which the exchange of materials between fetal and maternal circulations occurs. Also called the **afterbirth.**

Plane joint A synovial joint having articulating surfaces that are usually flat, permitting only side-to-side and back-and-forth movements, as between carpal bones, tarsal bones, and the scapula and clavicle. Also called an **arthrodial** (ar-THRŌ-dē-al) **joint.**

Plantar flexion (PLAN-tar FLEK-shun) Bending the foot in the direction of the plantar surface (sole).

Plaque (PLAK) A layer of dense proteins on the inside of a plasma membrane in adherens junctions and desmosomes. A mass of bacterial cells, dextran (polysaccharide), and other debris that adheres to teeth (dental plaque). *See also* **Atherosclerotic plaque.**

Plasma (PLAZ-ma) The extracellular fluid found in blood vessels; blood minus the formed elements.

Plasma cell Cell that develops from a B cell (lymphocyte) and produces antibodies.

Plasma (cell) membrane Outer, limiting membrane that separates the cell's internal parts from extracellular fluid or the external environment.

Platelet (PLĀT-let) A fragment of cytoplasm enclosed in a cell membrane and lacking a nucleus; found in the circulating blood; plays a role in hemostasis. Also called a **thrombocyte** (THROM-bō-sīt).

Platelet plug Aggregation of platelets (thrombocytes) at a site where a blood vessel is damaged that helps stop or slow blood loss.

Pleura (PLOO-ra) The serous membrane that covers the lungs and lines the walls of the chest and the diaphragm.

Pleural cavity Small potential space between the visceral and parietal pleurae.

Plexus (PLEK-sus) A network of nerves, veins, or lymphatic vessels.

Plexus of Auerbach *See* **Myenteric plexus.**

Plexus of Meissner *See* **Submucosal plexus.**

Pluripotent (ploo-RI-pō-tent) **stem cell** Immature stem cell in red bone marrow that gives rise to precursors of all the different mature blood cells.

Pneumotaxic (noo-mō-TAK-sik) **area** A part of the respiratory center in the pons that continually sends inhibitory nerve impulses to the inspiratory area, limiting inhalation and facilitating exhalation.

Polycythaemia (pol′-ē-sī-THĒ-mē-a) Disorder characterized by an above-normal haematocrit (above 55%) in which hypertension, thrombosis, and haemorrhage can occur.

Polyunsaturated fat A fatty acid that contains more than one double covalent bond between its carbon atoms; abundant in triglycerides of corn oil, safflower oil, and cottonseed oil.

Polyuria (pol′-ē-Ū-rē-a) An excessive production of urine.

Pons (PONZ) The part of the brain stem that forms a "bridge" between the medulla oblongata and the midbrain, anterior to the cerebellum.

Portal system The circulation of blood from one capillary network into another through a vein.

Positive feedback system A feedback cycle that strengthens or reinforces a change in a controlled condition.

Postcentral gyrus Gyrus of cerebral cortex located immediately posterior to the central sulcus; contains the primary somatosensory area.

Posterior (pos-TĒR-ē-or) Nearer to or at the back of the body. Equivalent to **dorsal** in bipeds.

Posterior column–medial lemniscus pathway Sensory pathway that carries information related to proprioception, fine touch, two-point discrimination, pressure, and vibration. First-order neurons project from the spinal cord to the ipsilateral medulla in the posterior columns (gracile fasciculus and cuneate fasciculus). Second-order neurons project from the medulla to the contralateral thalamus in the medial lemniscus. Third-order neurons project from the thalamus to the somatosensory cortex (postcentral gyrus) on the same side.

Posterior pituitary Posterior lobe of the pituitary gland. Also called the **neurohypophysis** (noo-rō-hī-POF-i-sis).

Posterior root The structure composed of sensory axons lying between a spinal nerve and the dorsolateral aspect of the spinal cord. Also called the **dorsal (sensory) root.**

Posterior root ganglion (GANG-glē-on) A group of cell bodies of sensory neurons and their supporting cells located along the posterior root of a spinal nerve. Also called a **dorsal (sensory) root ganglion.**

Postganglionic neuron (pōst′-gang-lē-ON-ik NOO-ron) The second autonomic motor neuron in an autonomic pathway, having its cell body and dendrites located in an autonomic ganglion and its unmyelinated axon ending at cardiac muscle, smooth muscle, or a gland.

Postsynaptic (pōst-sin-AP-tik) **neuron** The nerve cell that is activated by the release of a neurotransmitter from another neuron and carries nerve impulses away from the synapse.

Pouch of Douglas *See* **Rectouterine pouch.**

Precapillary sphincter (SFINGK-ter) The distal-most muscle fiber (cell) at the metarteriole–capillary junction that regulates blood flow into capillaries.

Precentral gyrus Gyrus of cerebral cortex located immediately anterior to the central sulcus; contains the primary motor area.

Preganglionic (pre′-gang-lē-ON-ik) **neuron** The first autonomic motor neuron in an autonomic pathway, with its cell body and dendrites in the brain or spinal cord and its myelinated axon ending at an autonomic ganglion, where it synapses with a postganglionic neuron.

Pregnancy Sequence of events that normally includes fertilisation, implantation, embryonic growth, and fetal growth and terminates in birth.

Premenstrual syndrome (PMS) Severe physical and emotional stress ocurring late in the postovulatory phase of the menstrual cycle and sometimes overlapping with menstruation.

Prepuce (PRĒ-poos) The loose-fitting skin covering the glans of the penis and clitoris. Also called the **foreskin.**

Presbyopia (prez-bē-Ō-pē-a) A loss of elasticity of the lens of the eye due to advancing age with resulting inability to focus clearly on near objects.

Presynaptic (prē-sin-AP-tik) **neuron** A neuron that propagates nerve impulses toward a synapse.

Prevertebral ganglion (prē-VER-te-bral GANG-glē-on) A cluster of cell bodies of postganglionic sympathetic neurons anterior to the spinal column and close to large abdominal arteries. Also called a **collateral ganglion.**

Primary germ layer One of three layers of embryonic tissue, called ectoderm, mesoderm, and endoderm, that give rise to all tissues and organs of the body.

Primary motor area A region of the cerebral cortex in the precentral gyrus of the frontal lobe of the cerebrum that controls specific muscles or groups of muscles.

Primary somatosensory area A region of the cerebral cortex posterior to the central sulcus in the postcentral gyrus of the parietal lobe of the cerebrum that localizes exactly the points of the body where somatic sensations originate.

Prime mover The muscle directly responsible for producing a desired motion. Also called an **agonist** (AG-ō-nist).

Primitive gut Embryonic structure formed from the dorsal part of the yolk sac that gives rise to most of the gastrointestinal tract.

Primordial (prī-MŌR-dē-al) Existing first; especially primordial egg cells in the ovary.

Principal cell Cell type in the distal convoluted tubules and collecting ducts of the kidneys that is stimulated by aldosterone and antidiuretic hormone.

Proctology (prok-TOL-ō-jē) The branch of medicine concerned with the rectum and its disorders.

Progeny (PROJ-e-nē) Offspring or descendants.

Progesterone (prō-JES-te-rōn) A female sex hormone produced by the ovaries that helps prepare the endometrium of the uterus for implantation of a fertilized ovum and the mammary glands for milk secretion.

Prognosis (prog-NŌ-sis) A forecast of the probable results of a disorder; the outlook for recovery.

Prolactin (prō-LAK-tin) **(PRL)** A hormone secreted by the anterior pituitary that initiates and maintains milk secretion by the mammary glands.

Prolapse (PRŌ-laps) A dropping or falling down of an organ, especially the uterus or rectum.

Proliferation (prō-lif′-er-Ā-shun) Rapid and repeated reproduction of new parts, especially cells.

Pronation (prō-NĀ-shun) A movement of the forearm in which the palm is turned posteriorly.

Prophase (PRŌ-fāz) The first stage of mitosis during which chromatid pairs are formed and aggregate around the metaphase plate of the cell.

Proprioception (prō-prē-ō-SEP-shun) The perception of the position of body parts, especially the limbs, independent of vision; this sense is possible due to nerve impulses generated by proprioceptors.

Proprioceptor (PRŌ-prē-ō-sep′-tor) A receptor located in muscles, tendons, joints, or the internal ear (muscle spindles, tendon organs, joint kinesthetic receptors, and hair cells of the vestibular apparatus) that provides information about body position and movements.

Prostaglandin (pros′-ta-GLAN-din) **(PG)** A membrane-associated lipid; released in small quantities and acts as a local hormone.

Prostate (PROS-tāt) A doughnut-shaped gland inferior to the urinary bladder that surrounds the superior portion of the male urethra and secretes a slightly acidic solution that contributes to sperm motility and viability.

Proteasome (PRŌ-tē-a-sōm) Tiny cellular organelle in cytosol and nucleus containing proteases that destroy unneeded, damaged, or faulty proteins.

Protein An organic compound consisting of carbon, hydrogen, oxygen, nitrogen, and sometimes sulfur and phosphorus; synthesized on ribosomes and made up of amino acids linked by peptide bonds.

Prothrombin (prō-THROM-bin) An inactive blood-clotting factor synthesized by the liver, released into the blood, and converted to active thrombin in the process of blood clotting by the activated enzyme prothrombinase.

Proto-oncogene (prō′-tō-ON-kō-jēn) Gene responsible for some aspect of normal growth and development; it may transform into an oncogene, a gene capable of causing cancer.

Protraction (prō-TRAK-shun) The movement of the mandible or shoulder girdle forward on a plane parallel with the ground.

Proximal (PROK-si-mal) Nearer the attachment of a limb to the trunk; nearer to the point of origin or attachment.

Pseudopod (SOO-dō-pod) Temporary protrusion of the leading edge of a migrating cell; cellular projection that surrounds a particle undergoing phagocytosis.

Pterygopalatine ganglion (ter′-i-gō-PAL-a-tīn GANG-glē-on) A cluster of cell bodies of parasympathetic postganglionic neurons ending at the lacrimal and nasal glands.

Ptosis (TŌ-sis) Drooping, as of the eyelid or the kidney.

Puberty (PŪ-ber-tē) The time of life during which the secondary sex characteristics begin to appear and the capability for sexual reproduction is possible; usually occurs between the ages of 10 and 17.

Pubic symphysis A slightly movable cartilaginous joint between the anterior surfaces of the hip bones.

Puerperium (pū-er-PER-ē-um) The period immediately after childbirth, usually 4–6 weeks.

Pulmonary (PUL-mo-ner′-ē) Concerning or affected by the lungs.

Pulmonary circulation The flow of deoxygenated blood from the right ventricle to the lungs and the return of oxygenated blood from the lungs to the left atrium.

Pulmonary embolism (EM-bō-lizm) **(PE)** The presence of a blood clot or a foreign substance in a pulmonary arterial blood vessel that obstructs circulation to lung tissue.

Pulmonary oedema (usu., also edema) An abnormal accumulation of interstitial fluid in the tissue spaces and alveoli of the lungs due to increased pulmonary capillary permeability or increased pulmonary capillary pressure.

Pulmonary ventilation The inflow (inhalation) and outflow (exhalation) of air between the atmosphere and the lungs. Also called **breathing.**

Pulp cavity A cavity within the crown and neck of a tooth, which is filled with pulp, a connective tissue containing blood vessels, nerves, and lymphatic vessels.

Pulse (PULS) The rhythmic expansion and elastic recoil of a systemic artery after each contraction of the left ventricle.

Pupil The hole in the center of the iris, the area through which light enters the posterior cavity of the eyeball.

Purkinje (pur-KIN-jē) **fiber** Muscle fiber (cell) in the ventricular tissue of the heart specialized for conducting an action potential to the myocardium; part of the conduction system of the heart.

Pus The liquid product of inflammation containing leukocytes or their remains and debris of dead cells.

Pyloric (pī-LOR-ik) **sphincter** A thickened ring of smooth muscle through which the pylorus of the stomach communicates with the duodenum. Also called the **pyloric valve.**

Pyorrhoea (pī-ō-RĒ-a) A discharge or flow of pus, especially in the alveoli (sockets) and the tissues of the gums.

Pyramid (PIR-a-mid) A pointed or cone-shaped structure. One of two roughly triangular structures on the anterior aspect of the medulla oblongata composed of the largest motor tracts that run from the cerebral cortex to the spinal cord. A triangular structure in the renal medulla.

Pyramidal (pi-RAM-i-dal) **tracts (pathways)** *See* **Direct motor pathways.**

Q

QRS complex The deflection waves of an electrocardiogram that represent onset of ventricular depolarization.

Quadrant (KWOD-rant) One of four parts.

Quadriplegia (kwod′-ri-PLĒ-jē-a) Paralysis of four limbs: two upper and two lower.

R

Radiographic (rā′-dē-ō-GRAF-ik) **anatomy** Diagnostic branch of anatomy that includes the use of x rays.

Rami communicantes (RĀ-mē kō-mū-ni-KAN-tēz) Branches of a spinal nerve. *Singular* is **ramus communicans** (RĀ-mus kō-MŪ-ni-kans).

Rathke's pouch *See* **Hypophyseal pouch.**

Receptor A specialized cell or a distal portion of a neuron that responds to a specific sensory modality, such as touch, pressure, cold, light, or sound, and converts it to an electrical signal (generator or receptor potential). A specific molecule or cluster of molecules that recognizes and binds a particular ligand.

Receptor-mediated endocytosis A highly selective process whereby cells take up specific ligands, which usually are large molecules or particles, by enveloping them within a sac of plasma membrane. Ligands are eventually broken down by enzymes in lysosomes.

Recombinant DNA Synthetic DNA, formed by joining a fragment of DNA from one source to a portion of DNA from another.

Rectouterine (rek-tō-Ū-ter-in) **pouch** A pocket formed by the parietal peritoneum as it moves posteriorly from the surface of the uterus and is reflected onto the rectum; the most inferior point in the pelvic cavity. Also called the **pouch of Douglas.**

Rectum (REK-tum) The last 20 cm (8 in.) of the gastrointestinal tract, from the sigmoid colon to the anus.

Recumbent (re-KUM-bent) Lying down.

Red bone marrow A highly vascularized connective tissue located in microscopic spaces between trabeculae of spongy bone tissue.

Red nucleus A cluster of cell bodies in the midbrain, occupying a large part of the tectum from which axons extend into the rubroreticular and rubrospinal tracts.

Red pulp That portion of the spleen that consists of venous sinuses filled with blood and thin plates of splenic tissue called splenic (Billroth's) cords.

Referred pain Pain that is felt at a site remote from the place of origin.

Reflex Fast response to a change (stimulus) in the internal or external environment that attempts to restore homoeostasis.

Reflex arc The most basic conduction pathway through the nervous system, connecting a receptor and an effector and consisting of a receptor, a sensory neuron, an integrating center in the central nervous system, a motor neuron, and an effector. Also called **reflex circuit.**

Regional anatomy The division of anatomy dealing with a specific region of the body, such as the head, neck, chest, or abdomen.

Regurgitation (rē-gur′-ji-TĀ-shun) Return of solids or fluids to the mouth from the stomach; backward flow of blood through incompletely closed heart valves.

Relaxin (RLX) A female hormone produced by the ovaries and placenta that increases flexibility of the pubic symphysis and helps dilate the uterine cervix to ease delivery of a baby.

Releasing hormone Hormone secreted by the hypothalamus that can stimulate secretion of hormones of the anterior pituitary.

Renal (RĒ-nal) Pertaining to the kidneys.

Renal corpuscle (KOR-pus-l) A glomerular (Bowman's) capsule and its enclosed glomerulus.

Renal pelvis A cavity in the center of the kidney formed by the expanded, proximal portion of the ureter, lying within the kidney, and into which the major calyces open.

Renal physiology Study of the functions of the kidneys.

Renal pyramid A triangular structure in the renal medulla containing the straight segments of renal tubules and the vasa recta.

Reproduction (rē-prō-DUK-shun) The formation of new cells for growth, repair, or replacement; the production of a new individual.

Reproductive cell division Type of cell division in which gametes (sperm and oocytes) are produced; consists of meiosis and cytokinesis.

Respiration (res-pi-RĀ-shun) Overall exchange of gases between the atmosphere, blood, and body cells consisting of pulmonary ventilation, external respiration, and internal respiration.

Respiratory center Neurons in the pons and medulla oblongata of the brain stem that regulate the rate and depth of pulmonary ventilation.

Respiratory (res-PEER-a-tōr-ē) **physiology** Study of the functions of the air passageways and lungs.

Respiratory (res-PEER-a-tōr-ē) **system** System composed of the nose, pharynx, larynx, trachea, bronchi, and lungs that obtains oxygen for body cells and eliminates carbon dioxide from them.

Retention (rē-TEN-shun) A failure to void urine due to obstruction, nervous contraction of the urethra, or absence of sensation of desire to urinate.

Rete (RĒ-tē) **testis** The network of ducts in the testes.

Reticular (re-TIK-ū-lar) **activating system (RAS)** A portion of the reticular formation that has many ascending connections with the cerebral cortex; when this area of the brain stem is active, nerve impulses pass to the thalamus and widespread areas of the cerebral cortex, resulting in generalized alertness or arousal from sleep.

Reticular formation A network of small groups of neuronal cell bodies scattered among bundles of axons (mixed gray and white matter) beginning in the medulla oblongata and extending superiorly through the central part of the brain stem.

Reticulocyte (re-TIK-ū-lō-sīt) An immature red blood cell.

Reticulum (re-TIK-ū-lum) A network.

Retina (RET-i-na) The deep coat of the posterior portion of the eyeball consisting of nervous tissue (where the process of vision begins) and a pigmented layer of epithelial cells that contact the choroid.

Retinaculum (ret-i-NAK-ū-lum) A thickening of fascia that holds structures in place, for example, the superior and inferior retinacula of the ankle.

Retraction (rē-TRAK-shun) The movement of a protracted part of the body posteriorly on a plane parallel to the ground, as in pulling the lower jaw back in line with the upper jaw.

Retroperitoneal (re′-trō-per-i-tō-NĒ-al) External to the peritoneal lining of the abdominal cavity.

Rh factor An inherited antigen on the surface of red blood cells in Rh^+ individuals; not present in Rh^- individuals.

Rhinology (rī-NOL-ō-jē) The study of the nose and its disorders.

Ribonucleic (rī-bō-noo-KLĒ-ik) **acid (RNA)** A single-stranded nucleic acid made up of nucleotides, each consisting of a nitrogenous base (adenine, cytosine, guanine, or uracil), ribose, and a phosphate group; major types are messenger RNA (mRNA), transfer RNA (tRNA), and ribosomal RNA (rRNA), each of which has a specific role during protein synthesis.

Ribosome (RĪ-bō-sōm) A cellular structure in the cytoplasm of cells, composed of a small subunit and a large subunit that contain ribosomal RNA and ribosomal proteins; the site of protein synthesis.

Right lymphatic (lim-FAT-ik) **duct** A vessel of the lymphatic system that drains lymph from the upper right side of the body and empties it into the right subclavian vein.

Rigidity (ri-JID-i-tē) Hypertonia characterized by increased muscle tone, but reflexes are not affected.

Rigor mortis State of partial contraction of muscles after death due to lack of ATP; myosin heads (cross-bridges) remain attached to actin, thus preventing relaxation.

Rod One of two types of photoreceptors in the retina of the eye; specialized for vision in dim light.

Root canal A narrow extension of the pulp cavity lying within the root of a tooth.

Root of penis Attached portion of penis that consists of the bulb and crura.

Rotation (rō-TĀ-shun) Moving a bone around its own axis, with no other movement.

Rotator cuff Refers to the tendons of four deep shoulder muscles (subscapularis, supraspinatus, infraspinatus, and teres minor) that form a complete circle around the shoulder; they strengthen and stabilize the shoulder joint.

Round ligament (LIG-a-ment) A band of fibrous connective tissue enclosed between the folds of the broad ligament of the uterus, emerging from the uterus just inferior to the uterine tube, extending laterally along the pelvic wall and through the deep inguinal ring to end in the labia majora.

Round window A small opening between the middle and internal ear, directly inferior to the oval window, covered by the secondary tympanic membrane.

Ruffini corpuscle A sensory receptor embedded deeply in the dermis and deeper tissues that detects the stretching of the skin.

Rugae (ROO-gē) Large folds in the mucosa of an empty hollow organ, such as the stomach or vagina.

S

Saccule (SAK-ūl) The inferior and smaller of the two chambers in the membranous labyrinth inside the vestibule of the internal ear containing a receptor organ for static equilibrium.

Sacral plexus (SĀ-kral PLEK-sus) A network formed by the ventral branches of spinal nerves L4 through S3.

Sacral promontory (PROM-on-tor′-ē) The superior surface of the body of the first sacral vertebra that projects anteriorly into the pelvic cavity; a line from the sacral promontory to the superior border of the pubic symphysis divides the abdominal and pelvic cavities.

Saddle joint A synovial joint in which the articular surface of one bone is saddle-shaped and the articular surface of the other bone is shaped like the legs of the rider sitting in the saddle, as in the joint between the trapezium and the metacarpal of the thumb.

Sagittal (SAJ-i-tal) **plane** A plane that divides the body or organs into left and right portions. Such a plane may be **midsagittal (median),** in which the divisions are equal, or **parasagittal,** in which the divisions are unequal.

Saliva (sa-LĪ-va) A clear, alkaline, somewhat viscous secretion produced mostly by the three pairs of salivary glands; contains various salts, mucin, lysosyme, salivary amylase, and lingual lipase (produced by glands in the tongue).

Salivary amylase (SAL-i-ver-ē AM-i-lās) An enzyme in saliva that initiates the chemical breakdown of starch.

Salivary gland One of three pairs of glands that lie external to the mouth and pour their secretory product (saliva) into ducts that empty into the oral cavity; the parotid, submandibular, and sublingual glands.

Sarcolemma (sar′-kō-LEM-ma) The cell membrane of a muscle fiber (cell), especially of a skeletal muscle fiber.

Sarcomere (SAR-kō-mēr) A contractile unit in a striated muscle fiber (cell) extending from one Z disc to the next Z disc.

Sarcoplasm (SAR-kō-plazm) The cytoplasm of a muscle fiber (cell).

Sarcoplasmic reticulum (sar′-kō-PLAZ-mik re-TIK-ū-lum) **(SR)** A network of saccules and tubes surrounding myofibrils of a muscle fiber (cell), comparable to endoplasmic reticulum; functions to reabsorb calcium ions during relaxation and to release them to cause contraction.

Satellite (SAT-i-līt) **cell** Flat neuroglial cells that surround cell bodies of peripheral nervous system ganglia to provide structural support and regulate the exchange of material between a neuronal cell body and interstitial fluid.

Saturated fat A fatty acid that contains only single bonds (no double bonds) between its carbon atoms; all carbon atoms are bonded to the maximum number of hydrogen atoms; prevalent in triglycerides of animal products such as meat, milk, milk products, and eggs.

Scala tympani (SKA-la TIM-pan-ē) The inferior spiral-shaped channel of the bony cochlea, filled with perilymph.

Scala vestibuli (ves-TIB-ū-lē) The superior spiral-shaped channel of the bony cochlea, filled with perilymph.

Schwann (SCHVON or SCHWON) **cell** A neuroglial cell of the peripheral nervous system that forms the myelin sheath and neurolemma around a nerve axon by wrapping around the axon in a jellyroll fashion.

Sciatica (sī-AT-i-ka) Inflammation and pain along the sciatic nerve; felt along the posterior aspect of the thigh extending down the inside of the leg.

Sclera (SKLE-ra) The white coat of fibrous tissue that forms the superficial protective covering over the eyeball except in the most anterior portion; the posterior portion of the fibrous tunic.

Scleral venous sinus A circular venous sinus located at the junction of the sclera and the cornea through which aqueous humour drains from the anterior chamber of the eyeball into the blood. Also called the **canal of Schlemm** (SHLEM).

Sclerosis (skle-RŌ-sis) A hardening with loss of elasticity of tissues.

Scoliosis (skō-lē-Ō-sis) An abnormal lateral curvature from the normal vertical line of the backbone.

Scrotum (SKRŌ-tum) A skin-covered pouch that contains the testes and their accessory structures.

Sebaceous (se-BĀ-shus) **gland** An exocrine gland in the dermis of the skin, almost always associated with a hair follicle, that secretes sebum. Also called an **oil gland.**

Sebum (SĒ-bum) Secretion of sebaceous (oil) glands.

Secondary sex characteristic A characteristic of the male or female body that develops at puberty under the influence of sex hormones but is not directly involved in sexual reproduction; examples are distribution of body hair, voice pitch, body shape, and muscle development.

Secretion (se-KRĒ-shun) Production and release from a cell or a gland of a physiologically active substance.

Selective permeability (per′-mē-a-BIL-i-tē) The property of a membrane by which it permits the passage of certain substances but restricts the passage of others.

Semen (SĒ-men) A fluid discharged at ejaculation by a male that consists of a mixture of sperm and the secretions of the seminiferous tubules, seminal vesicles, prostate, and bulbourethral (Cowper's) glands.

Semicircular canals Three bony channels (anterior, posterior, lateral), filled with perilymph, in which lie the membranous semicircular canals filled with endolymph. They contain receptors for equilibrium.

Semicircular ducts The membranous semicircular canals filled with endolymph and floating in the perilymph of the bony semicircular canals; they contain cristae that are concerned with dynamic equilibrium.

Semilunar (sem′-ē-LOO-nar) **(SL) valve** A valve between the aorta or the pulmonary trunk and a ventricle of the heart.

Seminal vesicle (SEM-i-nal VES-i-kul) One of a pair of convoluted, pouchlike structures, lying posterior and inferior to the urinary bladder and anterior to the rectum, that secrete a component of semen into the ejaculatory ducts. Also termed **seminal gland.**

Seminiferous tubule (sem′-i-NI-fer-us TOO-būl) A tightly coiled duct, located in the testis, where sperm are produced.

Sensation A state of awareness of external or internal conditions of the body.

Sensory area A region of the cerebral cortex concerned with the interpretation of sensory impulses.

Sensory neurons (NOO-ronz) Neurons that carry sensory information from cranial and spinal nerves into the brain and spinal cord or from a lower to a higher level in the spinal cord and brain. Also called **afferent neurons.**

Septal defect An opening in the atrial septum (atrial septal defect) because the foramen ovale fails to close, or the ventricular septum (ventricular septal defect) due to incomplete development of the ventricular septum.

Septum (SEP-tum) A wall dividing two cavities.

Serous (SĒR-us) **membrane** A membrane that lines a body cavity that does not open to the exterior. The external layer of an organ formed by a serous membrane. The membrane that lines the pleural, pericardial, and peritoneal cavities. Also called a **serosa** (se-RŌ-sa).

Sertoli (ser-TŌ-lē) **cell** A supporting cell in the seminiferous tubules that secretes fluid for supplying nutrients to sperm and the hormone inhibin, removes excess cytoplasm from spermatogenic cells, and mediates the effects of FSH and testosterone on spermatogenesis. Also called a **sustentacular** (sus′-ten-TAK-ū-lar) **cell.**

Serum Blood plasma minus its clotting proteins.

Sesamoid (SES-a-moyd) **bones** Small bones usually found in tendons.

Sex chromosomes The twenty-third pair of chromosomes, designated X and Y, which determines the genetic sex of an individual; in males, the pair is XY; in females, XX.

Sexual intercourse The insertion of the erect penis of a male into the vagina of a female. Also called **coitus** (KŌ-i-tus).

Sheath of Schwann *See* **Neurolemma.**

Shock Failure of the cardiovascular system to deliver adequate amounts of oxygen and nutrients to meet the metabolic needs of the body due to inadequate cardiac output. It is characterized by hypotension; clammy, cool, and pale skin; sweating; reduced urine formation; altered mental state; acidosis; tachycardia; weak, rapid pulse; and thirst. Types include hypovolemic, cardiogenic, vascular, and obstructive.

Shoulder joint A synovial joint where the humerus articulates with the scapula.

Sigmoid colon (SIG-moyd KŌ-lon) The S-shaped part of the large intestine that begins at the level of the left iliac crest, projects medially, and terminates at the rectum at about the level of the third sacral vertebra.

Sign Any objective evidence of disease that can be observed or measured, such as a lesion, swelling, or fever.

Sinoatrial (si-nō-Ā-trē-al) **(SA) node** A small mass of cardiac muscle fibers (cells) located in the right atrium inferior to the opening of the superior vena cava that spontaneously depolarize and generate a cardiac action potential about 100 times per minute. Also called the natural **pacemaker.**

Sinus (SĪ-nus) A hollow in a bone (paranasal sinus) or other tissue; a channel for blood (vascular sinus); any cavity having a narrow opening.

Sinusoid (SĪ-nū-soyd) A large, thin-walled, and leaky type of capillary, having large intercellular clefts that may allow proteins and blood cells to pass from a tissue into the bloodstream; present in the liver, spleen, anterior pituitary, parathyroid glands, and red bone marrow.

Skeletal muscle An organ specialized for contraction, composed of striated muscle fibers (cells), supported by connective tissue, attached to a bone by a tendon or an aponeurosis, and stimulated by somatic motor neurons.

Skeletal system Framework of bones and their associated cartilages, ligaments, and tendons.

Skene's gland *See* **Paraurethral gland.**

Skin The external covering of the body that consists of a superficial, thinner epidermis (epithelial tissue) and a deep, thicker dermis (connective tissue) that is anchored to the subcutaneous layer. Also called **cutaneous membrane.**

Skin graft The transfer of a patch of healthy skin taken from a donor site to cover a wound.

Skull The skeleton of the head consisting of the cranial and facial bones.

Sleep A state of partial unconsciousness from which a person can be aroused; associated with a low level of activity in the reticular activating system.

Sliding filament mechanism A model that describes muscle contraction in which thin filaments slide past thick ones so that the filaments overlap, causing shortening of a sarcomere, and thus shortening of muscle fibers and alternately shortening of the entire muscle.

Small intestine A long tube of the gastrointestinal tract that begins at the pyloric sphincter of the stomach, coils through the central and inferior part of the abdominal cavity, and ends at the large intestine; divided into three segments: duodenum, jejunum, and ileum.

Smooth muscle A tissue specialized for contraction, composed of smooth muscle fibers (cells), located in the walls of hollow internal organs, and innervated by autonomic motor neurons.

Sodium–potassium ATPase An active transport pump located in the plasma membrane that transports sodium ions out of the cell and potassium ions into the cell at the expense of cellular ATP. It functions to keep the ionic concentrations of these ions at physiological levels. Also called the **sodium–potassium pump.**

Soft palate (PAL-at) The posterior portion of the roof of the mouth, extending from the palatine bones to the uvula. It is a muscular partition lined with mucous membrane.

Somatic (sō-MAT-ik) **cell division** Type of cell division in which a single starting cell duplicates itself to produce two identical cells; consists of mitosis and cytokinesis.

Somatic motor pathway Pathway that carries information from the cerebral cortex, basal nuclei, and cerebellum that stimulates contraction of skeletal muscles.

Somatic nervous system (SNS) The portion of the peripheral nervous system consisting of somatic sensory (afferent) neurons and somatic motor (efferent) neurons.

Somatic sensory pathway Pathway that carries information from somatic sensory receptor to the primary somatosensory area in the cerebral cortex and cerebellum.

Somite (SŌ-mīt) Block of mesodermal cells in a developing embryo that is distinguished into a myotome (which forms most of the skeletal muscles), dermatome (which forms connective tissues), and sclerotome (which forms the vertebrae).

Spasm (SPAZM) A sudden, involuntary contraction of large groups of muscles.

Spasticity (spas-TIS-i-tē) Hypertonia characterized by increased muscle tone, increased tendon reflexes, and pathological reflexes (Babinski sign).

Spermatic (sper-MAT-ik) **cord** A supporting structure of the male reproductive system, extending from a testis to the deep inguinal ring, that includes the ductus (vas) deferens, arteries, veins, lymphatic vessels, nerves, cremaster muscle, and connective tissue.

Spermatogenesis (sper′-ma-tō-JEN-e-sis) The formation and development of sperm in the seminiferous tubules of the testes.

Sperm cell A mature male gamete. Also termed **spermatozoon** (sper′-ma-tō-ZŌ-on).

Spermiogenesis (sper′-mē-ō-JEN-e-sis) The maturation of spermatids into sperm.

Sphincter (SFINGK-ter) A circular muscle that constricts an opening.

Sphincter of Oddi *See* **Sphincter of the hepatopancreatic ampulla.**

Sphincter of the hepatopancreatic ampulla A circular muscle at the opening of the common bile and main pancreatic ducts in the duodenum. Also called the **sphincter of Oddi** (OD-ē).

Spinal (SPĪ-nal) **cord** A mass of nerve tissue located in the vertebral canal from which 31 pairs of spinal nerves originate.

Spinal nerve One of the 31 pairs of nerves that originate on the spinal cord from posterior and anterior roots.

Spinal shock A period from several days to several weeks following transection of the spinal cord that is characterized by the abolition of all reflex activity.

Spinothalamic (spī-nō-tha-LAM-ik) **tract** Sensory (ascending) tract that conveys information up the spinal cord to the thalamus for sensations of pain, temperature, itch, and tickle.

Spinous (SPĪ-nus) **process** A sharp or thornlike process or projection. Also called a **spine.** A sharp ridge running diagonally across the posterior surface of the scapula.

Spiral organ The organ of hearing, consisting of supporting cells and hair cells that rest on the basilar membrane and extend into the endolymph of the cochlear duct. Also called the **organ of Corti** (KOR-tē).

Splanchnic (SPLANK-nik) Pertaining to the viscera.

Spleen (SPLĒN) Large mass of lymphatic tissue between the fundus of the stomach and the diaphragm that functions in formation of blood cells during early fetal development, phagocytosis of ruptured blood cells, and proliferation of B cells during immune responses.

Spongy (cancellous) bone tissue Bone tissue that consists of an irregular latticework of thin plates of bone called trabeculae; spaces between trabeculae of some bones are filled with red bone marrow; found inside short, flat, and irregular bones and in the epiphyses (ends) of long bones.

Sprain Forcible wrenching or twisting of a joint with partial rupture or other injury to its attachments without dislocation.

Squamous (SKWĀ-mus) Flat or scalelike.

Starvation (star-VĀ-shun) The loss of energy stores in the form of glycogen, triglycerides, and proteins due to inadequate intake of nutrients or inability to digest, absorb, or metabolize ingested nutrients.

Static equilibrium (ē-kwi-LIB-rē-um) The maintenance of posture in response to changes in the orientation of the body, mainly the head, relative to the ground.

Stellate reticuloendothelial (STEL-āt re-tik′-ū-lō-en′-dō-THĒ-lē-al) **cell** Phagocytic cell bordering a sinusoid of the liver. Also called a **Kupffer** (KOOP-fer) **cell.**

Stem cell An unspecialized cell that has the ability to divide for indefinite periods and give rise to a specialized cell.

Stenosis (sten-Ō-sis) An abnormal narrowing or constriction of a duct or opening.

Stereocilia (ste′-rē-ō-SIL-ē-a) Groups of extremely long, slender, nonmotile microvilli projecting from epithelial cells lining the epididymis.

Sterile (STE-ril) Free from any living microorganisms. Unable to conceive or produce offspring.

Sterilisation (ster′-i-li-ZĀ-shun) Elimination of all living microorganisms. Any procedure that renders an individual incapable of reproduction (for example, castration, vasectomy, hysterectomy, or oophorectomy).

Stimulus Any stress that changes a controlled condition; any change in the internal or external environment that excites a sensory receptor, a neuron, or a muscle fiber.

Stomach The J-shaped enlargement of the gastrointestinal tract directly inferior to the diaphragm in the epigastric, umbilical, and left hypochondriac regions of the abdomen, between the oesophagus and small intestine.

Straight tubule (TOO-būl) A duct in a testis leading from a convoluted seminiferous tubule to the rete testis.

Stratum (STRĀ-tum) A layer.

Stratum basalis (ba-SAL-is) The layer of the endometrium next to the myometrium that is maintained during menstruation and gestation and produces a new stratum functionalis following menstruation or parturition.

Stratum functionalis (funk′-shun-AL-is) The layer of the endometrium next to the uterine cavity that is shed during menstruation and that forms the maternal portion of the placenta during gestation.

Stretch receptor Receptor in the walls of blood vessels, airways, or organs that monitors the amount of stretching. Also termed **baroreceptor.**

Striae (STRĪ-ē) Internal scarring due to overstretching of the skin in which collagen fibers and blood vessels in the dermis are damaged. Also called **stretch marks.**

Stroma (STRŌ-ma) The tissue that forms the ground substance, foundation, or framework of an organ, as opposed to its functional parts (parenchyma).

Subarachnoid (sub′-a-RAK-noyd) **space** A space between the arachnoid mater and the pia mater that surrounds the brain and spinal cord and through which cerebrospinal fluid circulates.

Subcutaneous (sub′-kū-TĀ-nē-us) Beneath the skin. Also called **hypodermic** (hī-pō-DER-mik).

Subcutaneous (subQ) layer A continuous sheet of areolar connective tissue and adipose tissue between the dermis of the skin and the deep fascia of the muscles. Also called the **hypodermis.**

Subdural (sub-DOO-ral) **space** A space between the dura mater and the arachnoid mater of the brain and spinal cord that contains a small amount of fluid.

Sublingual (sub-LING-gwal) **gland** One of a pair of salivary glands situated in the floor of the mouth deep to the mucous membrane and to the side of the lingual frenulum, with a duct (Rivinus') that opens into the floor of the mouth.

Submandibular (sub′-man-DIB-ū-lar) **gland** One of a pair of salivary glands found inferior to the base of the tongue deep to the mucous membrane in the posterior part of the floor of the mouth, posterior to the sublingual glands, with a duct (Wharton's) situated to the side of the lingual frenulum. Also called the **submaxillary** (sub′-MAK-si-ler-ē) **gland.**

Submucosa (sub-mū-KŌ-sa) A layer of connective tissue located deep to a mucous membrane, as in the gastrointestinal tract or the urinary bladder; the submucosa connects the mucosa to the muscularis layer.

Submucosal plexus A network of autonomic nerve fibers located in the superficial part of the submucous layer of the small intestine. Also called the **plexus of Meissner** (MĪZ-ner).

Substrate A molecule on which an enzyme acts.

Subthalamus (sub-THAL-a-mus) Part of the diencephalon inferior to the thalamus; the substantia nigra and red nucleus extend from the midbrain into the subthalamus.

Sudoriferous (soo′-dor-IF-er-us) **gland** An apocrine or eccrine exocrine gland in the dermis or subcutaneous layer that produces perspiration. Also called a **sweat gland.**

Sulcus (SUL-kus) A groove or depression between parts, especially between the convolutions of the brain. *Plural* is **sulci** (SUL-sī).

Superficial (soo′-per-FISH-al) Located on or near the surface of the body or an organ. Also called **external.**

Superficial subcutaneous inguinal (IN-gwi-nal) **ring** A triangular opening in the aponeurosis of the external oblique muscle that represents the termination of the inguinal canal.

Superior (soo-PĒR-ē-or) Toward the head or upper part of a structure. Also called **cephalic** or **cranial.**

Superior vena cava (VĒ-na KĀ-va) **(SVC)** Large vein that collects blood from parts of the body superior to the heart and returns it to the right atrium.

Supination (soo-pi-NĀ-shun) A movement of the forearm in which the palm is turned anteriorly.

Surface anatomy The study of the structures that can be identified from the outside of the body.

Surfactant (sur-FAK-tant) Complex mixture of phospholipids and lipoproteins, produced by type II alveolar (septal) cells in the lungs, that decreases surface tension.

Suspensory ligament (sus-PEN-so-rē LIG-a-ment) A fold of peritoneum extending laterally from the surface of the ovary to the pelvic wall.

Sutural (SOO-chur-al) **bone** A small bone located within a suture between certain cranial bones. Also called **Wormian** (WER-mē-an) **bone.**

Suture (SOO-chur) An immovable fibrous joint that joins skull bones.

Sympathetic (sim′-pa-THET-ik) **division** One of the two subdivisions of the autonomic nervous system, having cell bodies of preganglionic neurons in the lateral gray columns of the thoracic segment and the first two or three lumbar segments of the spinal cord; primarily concerned with processes involving the expenditure of energy.

Sympathetic trunk ganglion (GANG-glē-on) A cluster of cell bodies of sympathetic postganglionic neurons lateral to the vertebral column, close to the body of a vertebra. These ganglia extend inferiorly through the neck, thorax, and abdomen to the coccyx on both sides of the vertebral column and are connected to one another to form a chain on each side of the vertebral column. Also called **sympathetic chain, vertebral chain ganglia,** or **paravertebral ganglia.**

Symphysis (SIM-fi-sis) A line of union. A slightly movable cartilaginous joint such as the pubic symphysis.

Symptom (SIMP-tum) A subjective change in body function not apparent to an observer, such as pain or nausea, that indicates the presence of a disease or disorder of the body.

Synapse (SIN-aps) The functional junction between two neurons or between a neuron and an effector, such as a muscle or gland; may be electrical or chemical.

Synapsis (sin-AP-sis) The pairing of homologous chromosomes during prophase I of meiosis.

Synaptic (sin-AP-tik) **cleft** The narrow gap at a chemical synapse that separates the axon terminal of one neuron from another neuron or muscle fiber (cell) and across which a neurotransmitter diffuses to affect the postsynaptic cell.

Synaptic end bulb Expanded distal end of an axon terminal that contains synaptic vesicles. Also called a **synaptic knob.**

Synaptic vesicle Membrane-enclosed sac in a synaptic end bulb that stores neurotransmitters.

Synarthrosis (sin′-ar-THRŌ-sis) An immovable joint such as a suture, gomphosis, or synchondrosis.

Synchondrosis (sin′-kon-DRŌ-sis) A cartilaginous joint in which the connecting material is hyaline cartilage.

Syndesmosis (sin′-dez-MŌ-sis) A slightly movable joint in which articulating bones are united by fibrous connective tissue.

Synergist (SIN-er-jist) A muscle that assists the prime mover by reducing undesired action or unnecessary movement.

Synergistic (syn-er-JIS-tik) **effect** A hormonal interaction in which the effects of two or more hormones acting together is greater or more extensive than the sum of each hormone acting alone.

Synostosis (sin′-os-TŌ-sis) A joint in which the dense fibrous connective tissue that unites bones at a suture has been replaced by bone, resulting in a complete fusion across the suture line.

Synovial (si-NŌ-vē-al) **cavity** The space between the articulating bones of a synovial joint, filled with synovial fluid. Also called a **joint cavity.**

Synovial fluid Secretion of synovial membranes that lubricates joints and nourishes articular cartilage.

Synovial joint A fully movable or diarthrotic joint in which a synovial (joint) cavity is present between the two articulating bones.

Synovial membrane The deeper of the two layers of the articular capsule of a synovial joint, composed of areolar connective tissue that secretes synovial fluid into the synovial (joint) cavity.

System An association of organs that have a common function.

Systemic (sis-TEM-ik) Affecting the whole body; generalized.

Systemic anatomy The anatomical study of particular systems of the body, such as the skeletal, muscular, nervous, cardiovascular, or urinary systems.

Systemic circulation The routes through which oxygenated blood flows from the left ventricle through the aorta to all the organs of the body and deoxygenated blood returns to the right atrium.

Systole (SIS-tō-lē) In the cardiac cycle, the phase of contraction of the heart muscle, especially of the ventricles.

Systolic (sis-TOL-ik) **blood pressure (SBP)** The force exerted by blood on arterial walls during ventricular contraction; the highest pressure measured in the large arteries, about 120 mmHg under normal conditions for a young adult.

T

T cell A lymphocyte that becomes immunocompetent in the thymus and can differentiate into a helper T cell or a cytotoxic T cell, both of which function in cell-mediated immunity.

T wave The deflection wave of an electrocardiogram that represents ventricular repolarization.

Tachycardia (tak′-i-KAR-dē-a) An abnormally rapid resting heartbeat or pulse rate (over 100 beats per minute).

Tactile (TAK-tīl) Pertaining to the sense of touch.

Tactile disc *See* **Merkel disc.**

Target cell A cell whose activity is affected by a particular hormone.

Tarsal bones The seven bones of the ankle. Also called **tarsals.**

Tarsal gland Sebaceous (oil) gland that opens on the edge of each eyelid. Also called a **Meibomian** (mī-BŌ-mē-an) **gland.**

Tarsal plate A thin, elongated sheet of connective tissue, one in each eyelid, giving the eyelid form and support. The aponeurosis of the levator palpebrae superioris is attached to the tarsal plate of the superior eyelid.

Tarsus (TAR-sus) A collective term for the seven bones of the ankle.

Tectorial (tek-TŌ-rē-al) **membrane** A gelatinous membrane projecting over and in contact with the hair cells of the spiral organ (organ of Corti) in the cochlear duct.

Teeth (TĒTH) Accessory structures of digestion, composed of calcified connective tissue and embedded in bony sockets of the mandible and maxilla, that cut, shred, crush, and grind food. Also called **dentes** (DEN-tēz).

Telophase (TEL-ō-fāz) The final stage of mitosis.

Tendon (TEN-don) A white fibrous cord of dense regular connective tissue that attaches muscle to bone.

Tendon organ A proprioceptive receptor, sensitive to changes in muscle tension and force of contraction, found chiefly near the junctions of tendons and muscles. Also called a **Golgi** (GOL-jē) **tendon organ.**

Tendon reflex A polysynaptic, ipsilateral reflex that protects tendons and their associated muscles from damage that might be brought about by excessive tension. The receptors involved are called tendon organs (Golgi tendon organs).

Teniae coli (TĒ-nē-ē KŌ-lī) The three flat bands of thickened, longitudinal smooth muscle running the length of the large intestine, except in the rectum. *Singular* is **tenia coli.**

Tentorium cerebelli (ten-TŌ-rē-um ser′-e-BEL-ī) A transverse shelf of dura mater that forms a partition between the occipital lobe of the cerebral hemispheres and the cerebellum and that covers the cerebellum.

Teratogen (TER-a-tō-jen) Any agent or factor that causes physical defects in a developing embryo.

Terminal ganglion (TER-min-al GANG-glē-on) A cluster of cell bodies of parasympathetic postganglionic neurons either lying very close to the visceral effectors or located within the walls of the visceral effectors supplied by the postganglionic neurons. Also called **intramural ganglion.**

Testis (TES-tis) Male gonad that produces sperm and the hormones testosterone and inhibin. Also called a **testicle.**

Testosterone (tes-TOS-te-rōn) A male sex hormone (androgen) secreted by interstitial (Leydig) cells of a mature testis; needed for development of sperm; together with a second androgen termed **dihydrotestosterone** (dī-hī-drō-tes-TOS-ter-ōn) **(DHT),** controls the growth and development of male reproductive organs, secondary sex characteristics, and body growth.

Tetralogy of Fallot (tet-RAL-ō-jē of fal-Ō) A combination of four congenital heart defects: (1) constricted pulmonary semilunar valve, (2) interventricular septal opening, (3) emergence of the aorta from both ventricles instead of from the left only, and (4) enlarged right ventricle.

Thalamus (THAL-a-mus) A large, oval structure located bilaterally on either side of the third ventricle, consisting of two masses of grey matter organized into nuclei; main relay center for sensory impulses ascending to the cerebral cortex.

Thermoreceptor (THER-mō-rē-sep-tor) Sensory receptor that detects changes in temperature.

Thermoregulation Homeostatic regulation of body temperature through sweating and adjustment of blood flow in the dermis.

Thigh The portion of the lower limb between the hip and the knee.

Third ventricle (VEN-tri-kul) A slitlike cavity between the right and left halves of the thalamus and between the lateral ventricles of the brain.

Thoracic (thor-AS-ik) **cavity** Cavity superior to the diaphragm that contains two pleural cavities, the mediastinum, and the pericardial cavity.

Thoracic duct A lymphatic vessel that begins as a dilation called the cisterna chyli, receives lymph from the left side of the head, neck, and chest, left arm, and the entire body below the ribs, and empties into the junction between the internal jugular and left subclavian veins. Also called the **left lymphatic** (lim-FAT-ik) **duct.**

Thoracolumbar (thōr′-a-kō-LUM-bar) **outflow** The axons of sympathetic preganglionic neurons, which have their cell bodies in the lateral gray columns of the thoracic segments and first two or three lumbar segments of the spinal cord.

Thorax (THŌ-raks) The chest.

Thrombosis (THROM-BŌ-sis) The formation of a clot in an unbroken blood vessel, usually a vein.

Thrombus (THROM-bus) A stationary clot formed in an unbroken blood vessel, usually a vein.

Thymus (THĪ-mus) A bilobed organ, located in the superior mediastinum posterior to the sternum and between the lungs, in which T cells develop immunocompetence.

Thyroid cartilage (THĪ-royd KAR-ti-lij) The largest single cartilage of the larynx, consisting of two fused plates that form the anterior wall of the larynx. Also called **Adam's apple.**

Thyroid follicle (FOL-i-kul) Spherical sac that forms the parenchyma of the thyroid gland and consists of follicular cells that produce thyroxine (T_4) and triiodothyronine (T_3).

Thyroid gland An endocrine gland with right and left lateral lobes on either side of the trachea connected by an isthmus; located anterior to the trachea just inferior to the cricoid cartilage; secretes thyroxine (T_4), triiodothyronine (T_3), and calcitonin.

Thyroid-stimulating hormone (TSH) A hormone secreted by the anterior pituitary that stimulates the synthesis and secretion of thyroxine (T_4) and triiodothyronine (T_3). Also called **thyrotropin** (THĪ-rō-TRŌ-pin).

Thyroxine (thī-ROK-sēn) **(T_4)** A hormone secreted by the thyroid gland that regulates metabolism, growth and development, and the activity of the nervous system. Also called **tetraiodothyronine** (tet-ra-ī-ō-dō-THĪ-rō-nēn).

Tic Spasmodic, involuntary twitching of muscles that are normally under voluntary control.

Tissue A group of similar cells and their intercellular substance joined together to perform a specific function.

Tissue rejection Phenomenon by which the body recognizes the protein (HLA antigens) in transplanted tissues or organs as foreign and produces antibodies against them.

Tongue A large skeletal muscle covered by a mucous membrane located on the floor of the oral cavity.

Tonsil (TON-sil) An aggregation of large lymphatic nodules embedded in the mucous membrane of the throat.

Topical (TOP-i-kal) Applied to the surface rather than ingested or injected.

Torn cartilage A tearing of an articular disc (meniscus) in the knee.

Trabecula (tra-BEK-ū-la) Irregular latticework of thin plates of spongy bone tissue. Fibrous cord of connective tissue serving as supporting fiber by forming a septum extending into an organ from its wall or capsule. *Plural* is **trabeculae** (tra-BEK-ū-lē).

Trabeculae carneae (KAR-nē-ē) Ridges and folds of the myocardium in the ventricles.

Trachea (TRĀ-kē-a) Tubular air passageway extending from the larynx to the fifth thoracic vertebra. Also called the **windpipe.**

Tract A bundle of nerve axons in the central nervous system.

Transplantation (tranz-plan-TĀ-shun) The transfer of living cells, tissues, or organs from a donor to a recipient or from one part of the body to another in order to restore a lost function.

Transverse colon (trans-VERS KŌ-lon) The portion of the large intestine extending across the abdomen from the right colic (hepatic) flexure to the left colic (splenic) flexure.

Transverse fissure (FISH-er) The deep cleft that separates the cerebrum from the cerebellum.

Transverse plane A plane that divides the body or organs into superior and inferior portions. Also called a **cross-sectional** or **horizontal plane.**

Transverse tubules (TOO-būls) **(T tubules)** Small, cylindrical invaginations of the sarcolemma of striated muscle fibers (cells) that conduct muscle action potentials toward the center of the muscle fiber.

Tremor (TREM-or) Rhythmic, involuntary, purposeless contraction of opposing muscle groups.

Triad (TRĪ-ad) A complex of three units in a muscle fiber composed of a transverse tubule and the sarcoplasmic reticulum terminal cisterns on both sides of it.

Tricuspid (trī-KUS-pid) **valve** Atrioventricular (AV) valve on the right side of the heart.

Triglyceride A lipid that consists of one glycerol molecule and three fatty acid molecules.

Triglyceride (trī-GLI-ser-īd) A lipid formed from one molecule of glycerol and three molecules of fatty acids that may be either solid (fats) or liquid (oils) at room temperature; the body's most highly concentrated source of chemical potential energy. Found mainly within adipocytes. Also called a **neutral fat** or a **triacylglycerol.**

Trigone (TRĪ-gōn) A triangular region at the base of the urinary bladder.

Triiodothyronine (trī-ī-ō-dō-THĪ-rō-nēn) **(T_3)** A hormone produced by the thyroid gland that regulates metabolism, growth and development, and the activity of the nervous system.

Trophoblast (TRŌF-ō-blast) The superficial covering of cells of the blastocyst.

Tropic (TRŌ-pik) **hormone** A hormone whose target is another endocrine gland.

Trunk The part of the body to which the upper and lower limbs are attached.

Tubal ligation (lī-GĀ-shun) A sterilisation procedure in which the uterine (fallopian) tubes are tied and cut.

Tubular reabsorption The movement of filtrate from renal tubules back into blood in response to the body's specific needs.

Tubular secretion The movement of substances in blood into renal tubular fluid in response to the body's specific needs.

Tumor-suppressor gene A gene coding for a protein that normally inhibits cell division; loss or alteration of a tumor suppressor gene called *p53* is the most common genetic change in a wide variety of cancer cells.

Tunica albuginea (TOO-ni-ka al′-bū-JIN-ē-a) A dense white fibrous capsule covering a testis or deep to the surface of an ovary.

Tunica externa (eks-TER-na) The superficial coat of an artery or vein, composed mostly of elastic and collagen fibers. Also called the **adventitia.**

Tunica interna (in-TER-na) The deep coat of an artery or vein, consisting of a lining of endothelium, basement membrane, and internal elastic lamina. Also called the **tunica intima** (IN-ti-ma).

Tunica media (MĒ-dē-a) The intermediate coat of an artery or vein, composed of smooth muscle and elastic fibers.

Tympanic antrum (tim-PAN-ik AN-trum) An air space in the middle ear that leads into the mastoid air cells or sinus.

Tympanic (tim-PAN-ik) **membrane** A thin, semitransparent partition of fibrous connective tissue between the external auditory meatus and the middle ear. Also called the **eardrum.**

U

Umbilical (um-BIL-i-kul) **cord** The long, ropelike structure containing the umbilical arteries and vein that connect the fetus to the placenta.

Umbilicus (um-BIL-i-kus *or* um-bi-LĪ-kus) A small scar on the abdomen that marks the former attachment of the umbilical cord to the fetus. Also called the **navel.**

Upper limb The appendage attached at the shoulder girdle, consisting of the arm, forearm, wrist, hand, and fingers. Also called **upper extremity** or **upper appendage.**

Uraemia (ū-RĒ-mē-a) Accumulation of toxic levels of urea and other nitrogenous waste products in the blood, usually resulting from severe kidney malfunction.

Ureter (Ū-rē-ter) One of two tubes that connect the kidney with the urinary bladder.

Urethra (ū-RĒ-thra) The duct from the urinary bladder to the exterior of the body that conveys urine in females and urine and semen in males.

Urinalysis (ū-ri-NAL-i-sis) An analysis of the volume and physical, chemical, and microscopic properties of urine.

Urinary (Ū-ri-ner-ē) **bladder** A hollow, muscular organ situated in the pelvic cavity posterior to the pubic symphysis; receives urine via two ureters and stores urine until it is excreted through the urethra.

Urinary system A system that consists of the kidneys, ureters, urinary bladder, and urethra. The system regulates the ionic composition, pH, volume, pressure, and osmolarity of blood.

Urine The fluid produced by the kidneys that contains wastes and excess materials; excreted from the body through the urethra.

Urogenital (ū′-rō-JEN-i-tal) **triangle** The region of the pelvic floor inferior to the pubic symphysis, bounded by the pubic symphysis and the ischial tuberosities, and containing the external genitalia.

Urology (ū-ROL-ō-jē) The specialized branch of medicine that deals with the structure, function, and diseases of the male and female urinary systems and the male reproductive system.

Uterine cycle A series of changes in the endometrium of a nonpregnant female that prepares the lining of the uterus to receive a fertilized ovum. Also called **menstrual cycle.**

Uterine (Ū-ter-in) **tube** Duct that transports ova from the ovary to the uterus. Also called the **fallopian** (fal-LŌ-pē-an) **tube** or **oviduct.**

Uterosacral ligament (ū-ter-ō-SĀ-kral LIG-a-ment) A fibrous band of tissue extending from the cervix of the uterus laterally to the sacrum.

Uterus (Ū-te-rus) The hollow, muscular organ in females that is the site of menstruation, implantation, development of the fetus, and labour. Also called the **womb.**

Utricle (Ū-tri-kul) The larger of the two divisions of the membranous labyrinth located inside the vestibule of the inner ear, containing a receptor organ for static equilibrium.

Uvea (Ū-vē-a) The three structures that together make up the vascular tunic of the eye.

Uvula (Ū-vū-la) A soft, fleshy mass, especially the U-shaped pendant part, descending from the soft palate.

V

Vagina (va-JĪ-na) A muscular, tubular organ that leads from the uterus to the vestibule, situated between the urinary bladder and the rectum of the female.

Vallate papilla (VAL-āt pa-PIL-a) One of the circular projections that is arranged in an inverted V-shaped row at the back of the tongue; the largest of the elevations on the upper surface of the tongue containing taste buds. Also called **circumvallate papilla.**

Varicocele (VAR-i-kō-sēl) A twisted vein; especially, the accumulation of blood in the veins of the spermatic cord.

Varicose (VAR-i-kōs) Pertaining to an unnatural swelling, as in the case of a varicose vein.

Vas A vessel or duct.

Vasa recta (VĀ-sa REK-ta) Extensions of the efferent arteriole of a juxtamedullary nephron that run alongside the loop of the nephron (loop of Henle) in the medullary region of the kidney.

Vasa vasorum (va-SŌ-rum) Blood vessels that supply nutrients to the larger arteries and veins.

Vascular (VAS-kū-lar) Pertaining to or containing many blood vessels.

Vascular (venous) sinus A vein with a thin endothelial wall that lacks a tunica media and externa and is supported by surrounding tissue.

Vascular spasm Contraction of the smooth muscle in the wall of a damaged blood vessel to prevent blood loss.

Vascular tunic (TOO-nik) The middle layer of the eyeball, composed of the choroid, ciliary body, and iris. Also called the **uvea** (Ū-vē-a).

Vasectomy (va-SEK-tō-mē) A means of sterilisation of males in which a portion of each ductus (vas) deferens is removed.

Vasoconstriction (vāz-ō-kon-STRIK-shun) A decrease in the size of the lumen of a blood vessel caused by contraction of the smooth muscle in the wall of the vessel.

Vasodilation (vāz′-ō-dī-LĀ-shun) An increase in the size of the lumen of a blood vessel caused by relaxation of the smooth muscle in the wall of the vessel.

Vein A blood vessel that conveys blood from tissues back to the heart.

Vena cava (VĒ-na KĀ-va) One of two large veins that open into the right atrium, returning to the heart all of the deoxygenated blood from the systemic circulation except from the coronary circulation.

Ventral (VEN-tral) Pertaining to the anterior or front side of the body; opposite of dorsal.

Ventral ramus (RĀ-mus) The anterior branch of a spinal nerve, containing sensory and motor fibers to the muscles and skin of the anterior surface of the head, neck, trunk, and the limbs.

Ventricle (VEN-tri-kul) A cavity in the brain filled with cerebrospinal fluid. An inferior chamber of the heart.

Ventricular fibrillation (ven-TRIK-ū-lar fib-ri-LĀ-shun) **(VF** or **V-fib)** Asynchronous ventricular contractions; unless reversed by defibrillation, results in heart failure.

Venule (VEN-ūl) A small vein that collects blood from capillaries and delivers it to a vein.

Vermiform appendix (VER-mi-form a-PEN-diks) A twisted, coiled tube attached to the caecum.

Vermis (VER-mis) The central constricted area of the cerebellum that separates the two cerebellar hemispheres.

Vertebrae (VER-te-brē) Bones that make up the vertebral column.

Vertebral (VER-te-bral) **canal** A cavity within the vertebral column formed by the vertebral foramina of all the vertebrae and containing the spinal cord. Also called the **spinal canal.**

Vertebral column The 26 vertebrae of an adult and 33 vertebrae of a child; encloses and protects the spinal cord and serves as a point of attachment for the ribs and back muscles. Also called the **backbone, spine,** or **spinal column.**

Vesicle (VES-i-kul) A small bladder or sac containing liquid.

Vesicouterine (ves′-ik-ō-Ū-ter-in) **pouch** A shallow pouch formed by the reflection of the peritoneum from the anterior surface of the uterus, at the junction of the cervix and the body, to the posterior surface of the urinary bladder.

Vestibular (ves-TIB-ū-lar) **apparatus** Collective term for the organs of equilibrium, which include the saccule, utricle, and semicircular ducts.

Vestibular membrane The membrane that separates the cochlear duct from the scala vestibuli.

Vestibule (VES-ti-būl) A small space or cavity at the beginning of a canal, especially the inner ear, larynx, mouth, nose, and vagina.

Villus (VIL-lus) A projection of the intestinal mucosal cells containing connective tissue, blood vessels, and a lymphatic vessel; functions in the absorption of the end products of digestion. *Plural* is **villi** (VIL-ī).

Viscera (VIS-er-a) The organs inside the ventral body cavity. *Singular* is **viscus** (VIS-kus).

Visceral (VIS-er-al) Pertaining to the organs or to the covering of an organ.

Visceral effectors (e-FEK-torz) Organs of the ventral body cavity that respond to neural stimulation, including cardiac muscle, smooth muscle, and glands.

Vision The act of seeing.

Vitamin An organic molecule necessary in trace amounts that acts as a catalyst in normal metabolic processes in the body.

Vitreous (VIT-rē-us) **body** A soft, jellylike substance that fills the vitreous chamber of the eyeball, lying between the lens and the retina.

Vocal folds Pair of mucous membrane folds below the ventricular folds that function in voice production. Also called **true vocal cords.**

Volkmann's canal *See* **Perforating canal.**

Vulva (VUL-va) Collective designation for the external genitalia of the female. Also called the **pudendum** (poo-DEN-dum).

W

Wallerian (wal-LE-rē-an) **degeneration** Degeneration of the portion of the axon and myelin sheath of a neuron distal to the site of injury.

Wandering macrophage (MAK-rō-fāj) Phagocytic cell that develops from a monocyte, leaves the blood, and migrates to infected tissues.

White matter Aggregations or bundles of myelinated and unmyelinated axons located in the brain and spinal cord.

White pulp The regions of the spleen composed of lymphatic tissue, mostly B lymphocytes.

White ramus communicans (RĀ-mus kō-MŪ-ni-kans) The portion of a preganglionic sympathetic axon that branches from the anterior ramus of a spinal nerve to enter the nearest sympathetic trunk ganglion.

X

Xiphoid (ZĪ-foyd) Sword-shaped.

Xiphoid (ZĪ-foyd) **process** The inferior portion of the sternum **(breastbone).**

Y

Yolk sac An extraembryonic membrane composed of the exocoelomic membrane and hypoblast. It transfers nutrients to the embryo, is a source of blood cells, contains primordial germ cells that migrate into the gonads to form primitive germ cells, forms part of the gut, and helps prevent desiccation of the embryo.

Z

Zona fasciculata (ZŌ-na fa-sik′-ū-LA-ta) The middle zone of the adrenal cortex consisting of cells arranged in long, straight cords that secrete glucocorticoid hormones, mainly cortisol.

Zona glomerulosa (glo-mer′-ū-LŌ-sa) The outer zone of the adrenal cortex, directly under the connective tissue covering, consisting of cells arranged in arched loops or round balls that secrete mineralocorticoid hormones, mainly aldosterone.

Zona pellucida (pe-LOO-si-da) Clear glycoprotein layer between a secondary oocyte and the surrounding granulosa cells of the corona radiata.

Zona reticularis (ret-ik′-ū-LAR-is) The inner zone of the adrenal cortex, consisting of cords of branching cells that secrete sex hormones, chiefly androgens.

Zygote (ZĪ-gōt) The single cell resulting from the union of male and female gametes; the fertilized ovum.

CREDITS

Illustration Credits

Chapter 1 1.1, 1.11, 1.12: Kevin Somerville. 1.2–1.4, 1.8, 1.9: Imagineering. 1.5: Molly Borman. 1.6: Kevin Somerville/Imagineering. 1.7, Table 1.2: DNA Illustrations.

Chapter 2 2.1–2.25: Imagineering.

Chapter 3 3.1, 3.2, 3.16–3.20, 3.22–3.24, Table 3.2: Tomo Narashima. 3.3, 3.5–3.15, 3.21, 3.25–3.35: Imagineering.

Chapter 4 4.1, 4.9, Table 4.1–4.6: Kevin Somerville/Imagineering. 4.2, 4.4, 4.7, 4.8: Imagineering. 4.5, 4.6: Kevin Somerville.

Chapter 5 5.1a, 5.2–5.7: Kevin Somerville. 5.9: Kevin Somerville /Imagineering. 5.10: Imagineering.

Chapter 6 6.1, 6.4, 6.6a, Table 6.1: John Gibb. 6.2, 6.3, 6.5, 6.9: Kevin Somerville. 6.8, 6.10: Imagineering.

Chapter 7 7.1–7.24, Table 7.1, Table 7.4: John Gibb. 7.25: Imagineering.

Chapter 8 8.1–8.17, Table 8.1: John Gibb. 8.18: Kevin Somerville.

Chapter 9 9.1, 9.2, 9.3a, 9.10–9.16: John Gibb.

Chapter 10 10.1, 10.2, 10.17, 10.18: Kevin Somerville. 10.2d, 10.3, 10.4–10.8, 10.10–10.14: Imagineering. 10.9a-c: Kevin Somerville/Imagineering. 10.16: Imagineering.

Chapter 11 11.1, 11.2: John Gibb/Imagineering. 11.3–11.23: John Gibb.

Chapter 12 12.1a, 12.2a, 12.6–12.8: Kevin Somerville. 12.3–12.5, 12.9–12.28, Table 12.3: Imagineering.

Chapter 13 13.1–13.3, 13.5, 13.8b, 13.13: Kevin Somerville. 13.6, 13.10b: John Gibb. 13.7, 13.8a, 13.10a, 13.11: Steve Oh. 13.12 Imagineering. 13.14–13.18: Leonard Dank/Imagineering.

Chapter 14 14.1–14.3, 14.4b-c, 14.5–14.15, 14.27, 14.28, Table 14.3: Kevin Somerville. 14.4a, 14.4d, 14.16 Table 14.1: Imagineering. 14.17–14.26: Kevin Somerville/Richard Combs/Imagineering.

Chapter 15 15.1, 15.4, 15.6: Kevin Somerville/ Imagineering. 15.2, 15.3, 15.7: Imagineering. 15.5: Kevin Somerville

Chapter 16 16.1, 16.3, 16.4, 16.8, 16.9, 16.12: Imagineering. 16.2, 16.13: Kevin Somerville. 16.5–16.7, 16.10–16.12: Kevin Somerville/ Imagineering.

Chapter 17 17.1, 17.7, 17.14, 17.21b-d, 17.22: Tomo Narashima. 17.2, 17.10, 17.17, 17.23, 17.26, Table 17.1, Table 17.2: Imagineering. 17.3: Molly Borman. 17.6: Sharon Ellis. 17.12, 17.13, 17.15, 17.16: Imagineering. 17.18–17.21a, 17.25a: Tomo Narashima/Imagineering. 17.24, 17.25b: Tomo Narashima/Sharon Ellis/Imagineering. 17.27, 17.28: Kevin Somerville/Imagineering.

Chapter 18 18.1, 18.21: Kevin Somerville. 18.5, 18.8, 18.10, 18.13a, 18.15, 18.18: Lynn O'Kelley/ Imagineering. 18.2–18.13c, 18.14, 18.16, 18.17, 18.19, 18.20: Imagineering.

Chapter 19 19.1, 19.3–19.6, 19.8, 19.9, 19.11–19.13, Table 19.2: Imagineering.

Chapter 20 20.1–20.5, 20.18–20.20, Table 20.1: Kevin Somerville. 20.7, 20.9, 20.14–20.16, 20.22, 20.23: Kevin Somerville/Imagineering. 20.10–20.13, 20.24: Imagineering

Chapter 21 21.1, 21.9, 21.12, 21.13, 21.17, 21.18a, 21.21, 21.23, 21.28, 21.29a, 21.30, 21.31: Kevin Somerville. 21.2, 20.5–20.8, 21.10, 21.11, 21.14–21.16 21.29b: Imagineering. 21.3, 21.4: Kevin Somerville/Imagineering. 21.18b, 21.19–21.22, 21.24–21.27: John Gibb. Exhibits: Imagineering/ John Gibb.

Chapter 22 22.1: Kevin Somerville/Imagineering/Richard Coombs. 22.2, 22.4, 22.9, 22.11–22.18, 22.20, 22.23: Imagineering. 22.3: John Gibb. 22.5, 22.7: Steve Oh. 22.6, 22.10: Kevin Somerville/ Imagineering. 22.8: Kevin Somerville. 22.19: Jared Schneidman Design.

Chapter 23 23.1, 23.2a, 23.10, 23.11: Kevin Somerville. 23.2b, 23.4a-c, 23.5, 23.6, 23.9: Molly Borman. 23.4d, 23.14–23.18, 23.23, 23.24: Imagineering. 23.13: John Gibb/Imagineering. 23.26: Kevin Somerville/Imagineering. 23.27: Imagineering.

Chapter 24 24.1, 24.2, 24.12, 24.16a-b, 24.18, 24.19, 24.24: Kevin Somerville. 24.4, 24.13, 24.16d-e, 24.21, 24.22, 24.25: Imagineering. 24.5, 24.7, 24.8, 24.10: Nadine Sokol. 24.6: DNA Illustrations. 24.11, 24.15: Steve Oh. 24.23: Molly Borman

Chapter 25 25.1–25.20: Imagineering.

Chapter 26 26.1, 26.2, 26.6, 26.8, 26.9, 26.23: Kevin Somerville. 26.3, 26.21: Steve Oh/Imagineering. 26.5, 26.10–26.20, Table 26.1: Imagineering. 26.22: Kevin Somerville/Imagineering.

Chapter 27 27.1–27.8: Imagineering

Chapter 28 28.1–28.4, 28.6, 28.9–28.14, 28.16, 28.19–28.22, 28.27, 28.28: Kevin Somerville. 28.5, 28.7, 28.8, 28.15, 28.23–28.26: Imagineering.

Chapter 29 29.1–29.13, 29.17, Table 29.2: Kevin Somerville. 29.15, 29.18: Kevin Somerville/ Imagineering. 29.16, 29.19–29.26: Imagineering.

Photo Credits

Chapter 1 Fig. 1.1 Rubberball Productions/ Getty Images. Fig. 1.8a Dissection Shawn Miller; Photograph Mark Nielsen. Fig. 1.8b Dissection Shawn Miller; Photograph Mark Nielsen. Fig. 1.8c Dissection Shawn Miller; Photograph Mark Nielsen. Fig. 1.12a Andy Washnik. Page 21 (left) Biophoto Associates/Photo Researchers. Page 21 (center) Breast Cancer Unit, Kings College Hospital, London/Photo Researchers, Inc. Page 21 (right) Zephyr/Photo Researchers, Inc. Page 22 (top left) Cardio-Thoracic Centre, Freeman Hospital, Newcastle-Upon-Tyne/ Photo Researchers, Inc. Page 22 (top center) CNRI/ Science Photo Library/Photo Researchers, Inc. Page 22 (top right) Science Photo Library/Photo Researchers, Inc. Page 22 (bottom left) Scott Camazine/Photo Researchers, Inc. Page 22 (bottom right) ISM/Phototake. Page 23 (center left) Simon Fraser/Photo Researchers, Inc. Page 23 (center right) ISM/Phototake. Page 23 (bottom) Courtesy Andrew Joseph Tortora and Damaris Soler. Page 24 (center left) Department of Nuclear Medicine, Charing Cross Hospital /Photo Researchers, Inc. Page 24 (center) SIU/Visuals Unlimited. Page 24 (center right) Dept. of Nuclear Medicine, Charing Cross Hospital/Photo Researchers, Inc. Page 24 (bottom right) ©Camal/Phototake.

Chapter 2 Fig. 2.23 Courtesy T.A. Steitz, Yale University.

Chapter 3 Fig. 3.4 Andy Washnik. Fig. 3.9 David Phillips/Photo Researchers, Inc. Fig. 3.13 ©Omikron/ Photo Researchers. Fig. 3.15a Albert Tousson/ Phototake. Fig. 3.15b Albert Tousson/Phototake. Fig. 3.15c Alexey Khodjakov/Photo Researchers, Inc. Fig. 3.16c David M. Phillips/Visuals Unlimited. Fig. 3.17b P. Motta/Photo Researchers, Inc. Fig. 3.17c David M. Phillips/Visuals Unlimited. Fig. 3.19b D. W. Fawcett/Photo Researchers, Inc. Fig. 3.20b Biophoto Associates/Photo Researchers, Inc. Fig. 3.22b David M. Phillips/Visuals Unlimited. Fig. 3.23b Don W. Fawcett/Visuals Unlimited. Fig. 3.24 D.W. Fawcett/Photo Researchers, Inc. Fig. 3.25b Andrew Syred/Photo Researchers, Inc. Fig. 3.32 Courtesy Michael Ross, University of Florida. Page 104 Steve Gschmeissner/Science Photo Library/Photo Researchers, Inc.

Chapter 4 All photos by Mark Nielsen except Fig 4.9a, which was taken by Michael Ross, University of Florida.

Chapter 5 Fig. 5.1b Courtesy Michael Ross, University of Florida. Fig. 5.1c David Becker/Photo Researchers, Inc. Fig. 5.1d Courtesy Andrew J. Kuntzman. Fig. 5.3b Mark Nielsen. Fig. 5.4b VVG/Science Photo Library/Photo Researchers, Inc. Fig. 5.8a Alain Dex/Photo Researchers, Inc. Fig. 5.8b

Biophoto Associates/Photo Researchers, Inc. Fig. 5.9a Sheila Terry/Science Photo Library/Photo Researchers, Inc. Fig. 5.9b St. Stephen's Hospital/Science Photo Library/Photo Researchers, Inc. Fig. 5.9c St. Stephen's Hospital/Science Photo Library/Photo Researchers, Inc. Fig. 5.11 Dr. P. Marazzi/Science Photo Library/Photo Researchers, Inc.

Chapter 6 Fig. 6.1b Mark Nielsen. Fig. 6.2 (left) CNRI/Photo Researchers, Inc. Fig. 6.2 (center) Dr. Richard Kessel and Randy Kardon/Tissues & Organs/Visuals Unlimited. Fig. 6.2 (right) Dr. Richard Kessel and Randy Kardon/Tissues & Organs/Visuals Unlimited. Fig. 6.6 Ralph Hutchings/Visuals Unlimited. Fig. 6.7a The Bergman Collection. Fig. 6.7b Mark Nielsen. Fig. 6.11a P. Motta/Photo Researchers, Inc. Fig. 6.11b P. Motta/Photo Researchers, Inc. Table 6.1a Courtesy Dr. Brent Layton. Table 6.1b Courtesy Per Amundson, M.D. Table 6.1c Courtesy Dr. Brent Layton. Table 6.1d Courtesy Dr. Brent Layton. Table 6.1e Courtesy Dr. Brent Layton. Table 6.1f Watney Collection/Phototake.

Chapter 7 Fig. 7.3b Mark Nielsen. Fig. 7.4c Mark Nielsen. Fig. 7.18b Mark Nielsen. Fig. 7.18c Mark Nielsen. Fig. 7.18d Mark Nielsen. Fig. 7.18e Mark Nielsen. Fig. 7.20b Mark Nielsen. Fig. 7.20c Mark Nielsen. Fig. 7.19b Mark Nielsen. Fig. 7.19c Mark Nielsen. Fig. 7.22b Mark Nielsen. Fig. 7.25a Princess Margaret Rose Orthopaedic Hospital/Photo Researchers, Inc. Fig. 7.25b Dr. P. Marazzi/Photo Researchers, Inc. Fig. 7.25c Custom Medical Stock Photo, Inc. Fig. 7.26 Center for Disease Control.

Chapter 9 Fig. 9.3b Mark Nielsen. Fig. 9.4 Mark Nielsen. Fig. 9.5 John Wilson White. Fig. 9.6 John Wilson White. Fig. 9.7 John Wilson White. Fig. 9.8 John Wilson White. Fig. 9.9 John Wilson White. Fig. 9.12d Dissection Shawn Miller, Photograph Mark Nielsen. Fig. 9.15d Dissection Shawn Miller, Photograph Mark Nielsen. Fig. 9.15g Dissection Shawn Miller, Photograph Mark Nielsen. Fig. 9.15h Dissection Shawn Miller, Photograph Mark Nielsen. Fig. 9.15i Dissection Shawn Miller, Photograph Mark Nielsen. Fig. 9.15j Dissection Shawn Miller, Photograph Mark Nielsen. Fig. 9.16b SIU/Visuals Unlimited. Fig. 9.16c ISM/Phototake. Fig. 9.16f Charles McRae/Visuals Unlimited.

Chapter 10 Fig. 10.5 Courtesy Hiroyouki Sasaki, Yale E. Goldman and Clara Franzini-Armstrong. Fig. 10.9d Don Fawcett/Photo Researchers, Inc. Fig. 10.15 John Wiley & Sons. Table 10.1 Courtesy Denah Appelt and Clara Franzini-Armstrong. Table 10.4 Biophoto Associates/Photo Researchers, Inc.

Chapter 11 Fig. 11.5c Dissection Shawn Miller, Photograph Mark Nielsen. Fig. 11.10d Dissection Nathan Mortensen and Shawn Miller; Photograph Mark Nielsen. Fig. 11.18g Andy Washnik. Fig. 11.21b Dissection Shawn Miller, Photograph Mark Nielsen.

Chapter 12 Fig. 12.2b Mark Nielsen. Fig. 12.8c Dennis Kunkel/Phototake. Fig. 12.8d Martin Rotker/Phototake.

Chapter 13 Fig. 13.1b Dissection Chris Roach; Photograph Mark Nielsen. Fig. 13.1c Dissection Shawn Miller, Photograph Mark Nielsen. Fig. 13.3b Courtesy Michael Ross, University of Florida. Fig. 13.5b Dr. Richard Kessel and Randy Kardon/Tissues & Organs/Visuals Unlimited. Fig. 13.6b Dissection Shawn Miller, Photograph Mark Nielsen. Table 13.1 Dissection Shawn Miller, Photograph Mark Nielsen. Table 13.1 Dissection Shawn Miller, Photograph Mark Nielsen. Table 13.1 Mark Nielsen and Shawn Miller. Table 13.1 Dissection Shawn Miller, Photograph Mark Nielsen. Page 529 (top) Bold Stock/Age Fotostock America, Inc. Page 529 (just below top) Masterfile. Page 529 (center) Andersen Ross/Getty Images, Inc. Page 529 (bottom) Masterfile.

Chapter 14 Fig. 14.1b Dissection Shawn Miller, Photograph Mark Nielsen. Fig. 14.4a Dissection Shawn Miller, Photograph Mark Nielsen. Fig. 14.8d Dissection Shawn Miller, Photograph Mark Nielsen. Fig. 14.11c Dissection Shawn Miller, Photograph Mark Nielsen. Fig. 14.12 N. Gluhbegovic and T.H. Williams, The Human Brain: A Photographic Guide, Harper and Row, Publishers, Inc. Hagerstown, MD, 1980. Table 14.2 Dissection Shawn Miller, Photograph Mark Nielsen. Table 14.2 Dissection Shawn Miller, Photograph Mark Nielsen. Table 14.2 Dissection Shawn Miller, Photograph Mark Nielsen. Table 14.2 Dissection Shawn Miller, Photograph Mark Nielsen. Table 14.2 Dissection Shawn Miller, Photograph Mark Nielsen. Table 14.2 Dissection Shawn Miller, Photograph Mark Nielsen. Fig. 14.17 Dissection Shawn Miller, Photograph Mark Nielsen. Fig. 14.18 Dissection Shawn Miller, Photograph Mark Nielsen. Fig. 14.19 Dissection Shawn Miller, Photograph Mark Nielsen. Fig. 14.19 Dissection Shawn Miller, Photograph Mark Nielsen. Fig. 14.19 Dissection Shawn Miller, Photograph Mark Nielsen. Fig. 14.20 Dissection Shawn Miller, Photograph Mark Nielsen. Fig. 14.21 Dissection Shawn Miller, Photograph Mark Nielsen. Fig. 14.22 Dissection Shawn Miller, Photograph Mark Nielsen. Fig. 14.23 Dissection Shawn Miller, Photograph Mark Nielsen. Fig. 14.24 Dissection Shawn Miller, Photograph Mark Nielsen. Fig. 14.25 Dissection Shawn Miller, Photograph Mark Nielsen. Fig. 14.26 Dissection Shawn Miller, Photograph Mark Nielsen.

Chapter 17 Fig. 17.1c Courtesy Michael Ross, University of Florida. Fig. 17.3d Mark Nielsen. Fig. 17.5 Geirge Diebold/Getty Images, Inc. Fig. 17.9 Paul Parker/Photo Researchers, Inc. Fig. 17.10c Mark Nielsen. Fig. 17.17 N. Gluhbegovic and T. H. Williams, The Human Brain: A Photographic Guide, Harper and Row, Publishers, Inc., Hagerstown, MD, 1980.

Chapter 18 Fig. 18.5 (top) Dissection Shawn Miller, Photograph Mark Nielsen. Fig. 18.5d Courtesy James Lowe, University of Nottingham, Nottingham, United Kingdom. Fig. 18.10b Mark Nielsen. Fig. 18.10c Dissection Shawn Miller, Photograph Mark Nielsen. Fig. 18.10d Dissection Shawn Miller, Photograph Mark Nielsen. Fig. 18.13b Mark Nielsen . Fig. 18.13d Dissection Shawn Miller, Photograph Mark Nielsen. Fig. 18.15c Dissection Shawn Miller, Photograph Mark Nielsen. Fig. 18.15d Mark Nielsen. Fig. 18.18 Mark Nielsen. Fig. 18.22a From New England Journal of Medicine, February 18, 1999, vol. 340, No. 7, page 524. Photo provided courtesy of Robert Gagel, Department of Internal Medicine, University of Texas M.D. Anderson Cancer Center, Houston Texas Fig. 18.22b Lester Bergman/The Bergman Collection. Fig. 18.22c Lester Bergman/The Bergman Collection. Fig. 18.22d Lester Bergman/The Bergman Collection. Fig. 18.22e Biophoto Associates/Photo Researchers, Inc.

Chapter 19 Opener Juergen Berger/Photo Researchers, Inc. Fig. 19.2a Juergen Berger/Photo Researchers, Inc. Fig. 19.2b Mark Nielsen. Fig. 19.7a Courtesy Michael Ross, University of Florida. Fig. 19.7b Courtesy Michael Ross, University of Florida. Fig. 19.7c Courtesy Michael Ross, University of Florida. Fig. 19.7d Courtesy Michael Ross, University of Florida. Fig. 19.7e Courtesy Michael Ross, University of Florida. Fig. 19.10a Dennis Kunkel/Visuals Unlimited. Fig. 19.10b Steve Gschmeissner/Photo Researchers, Inc. Fig. 19.10c David Phillips/Visuals Unlimited. Fig. 19.10d Dennis Kunkel/Phototake. Fig. 19.14 Jean Claude Revy/Phototake. Fig. 19.15 Stanley Fleger/Visuals Unlimited.

Chapter 20 Fig. 20.1c Dissection Shawn Miller, Photograph Mark Nielsen. Fig. 20.3b Dissection Shawn Miller, Photograph Mark Nielsen. Fig. 20.4b Mark Nielsen. Fig. 20.4c Mark Nielsen. Fig. 20.6c Dissection Shawn Miller, Photograph Mark Nielsen. Fig. 20.6f Dissection Shawn Miller, Photograph Mark Nielsen. Fig. 20.6g Dissection Shawn Miller, Photograph Mark Nielsen. Fig. 20.8c Dissection Shawn Miller, Photograph Mark Nielsen. Fig. 20.17 Jim cummins/Taxi/Getty Images, Inc. Fig. 20.21 ©Vu/Cabisco/Visuals Unlimited. Fig. 20.21 W. Ober/Visuals Unlimited. Fig. 20.22d ©ISM/Phototake.

Chapter 21 Fig. 21.1d Dennis Strete. Fig. 21.1e Courtesy Michael Ross, University of Florida. Fig. 21.5(top) Dissection Shawn Miller, Photograph Mark Nielsen. Fig. 21.5(bottom) Dissection Shawn Miller, Photograph Mark Nielsen. Exhibit 21.B Dissection Shawn Miller, Photograph Mark Nielsen. Fig. 21.19d Dissection Shawn Miller, Photograph Mark Nielsen. Fig. 21.21d Dissection Shawn Miller, Photograph Mark Nielsen. Fig. 21.25d Dissection Shawn Miller, Photograph Mark Nielsen.

Chapter 22 Fig. 22.5c Courtesy Michael Ross, University of Florida. Fig. 22.5b Courtesy Michael Ross, University of Florida. Fig. 22.6b Mark Nielsen. Fig. 22.6c Dissection Shawn Miller, Photograph Mark Nielsen. Fig. 22.7c Mark Nielsen. Fig. 22.9b Science Photo Library/Photo Researchers, Inc.

Chapter 23 Fig. 23.1b Dissection Shawn Miller, Photograph Mark Nielsen. Fig. 23.2c Dissection Shawn Miller, Photograph Mark Nielsen. Fig. 23.3 Courtesy Lynne Marie Barghesi. Fig. 23.5c CNRI/Photo Researchers, Inc. Fig. 23.6 Dissection Shawn Miller, Photograph Mark Nielsen. Fig. 23.8 Dissection Shawn Miller, Photograph Mark Nielsen. Fig. 23.10b Biophoto Associates/Photo Researchers, Inc. Fig. 23.11c Biophoto Associates/Photo Researchers, Inc.

Chapter 24 Fig. 24.1b Dissection Shawn Miller, Photograph Mark Nielsen. Fig. 24.4e Dissection

Shawn Miller, Photograph Mark Nielsen. Fig. 24.6b Mark Nielsen. Fig. 24.9 Mark Nielsen. Fig. 24.11b Dissection Shawn Miller, Photograph Mark Nielsen. Fig. 24.12 (inset) Hessler/Vu/Visuals Unlimited. Fig. 24.12c Mark Nielsen. Fig. 24.15d Dissection Shawn Miller, Photograph Mark Nielsen. Fig. 24.15e Dissection Shawn Miller, Photograph Mark Nielsen. Fig. 24.16c (top inset) Mark Nielsen. Fig. 24.16c Mark Nielsen. Fig. 24.18b Dissection Shawn Miller, Photograph Mark Nielsen. Fig. 24.19 Dissection Shawn Miller, Photograph Mark Nielsen. Fig. 24.20a Fred E. Hossler/Visuals Unlimited. Fig. 24.20b G. W. Willis/ Visuals Unlimited. Fig. 24.20c Courtesy Michael Ross, University of Florida. Fig. 24.20d Courtesy Michael Ross, University of Florida. Fig. 24.24c Courtesy Michael Ross, University of Florida. Fig. 24.24d Courtesy Michael Ross, University of Florida.

Chapter 26 Fig. 26.1b Dissection Shawn Miller, Photograph Mark Nielsen. Fig. 26.3b Mark Nielsen and Shawn Miller. Fig. 26.6b Dennis Strete. Fig. 26.8b Courtesy Michael Ross, University of Florida.

Chapter 28 Fig. 28.1b Dissection Shawn Miller, Photograph Mark Nielsen. Fig. 28.3b Dissection Shawn Miller, Photograph Mark Nielsen. Fig. 28.3c Dissection Shawn Miller, Photograph Mark Nielsen. Fig. 28.4a Mark Nielsen. Fig. 28.9b Dissection Shawn Miller, Photograph Mark Nielsen. Fig. 28.10c Dissection Shawn Miller, Photograph Mark Nielsen. Fig. 28.11b Dissection Shawn Miller, Photograph Mark Nielsen. Fig. 28.13b Courtesy Michael Ross, University of Florida. Fig. 28.13c Claude Edelmann/Photo Researchers, Inc. Fig. 28.14c Mark Nielsen. Fig. 28.14f Mark Nielsen. Fig. 28.17a Mark Nielsen. Fig. 28.17b P. Motta/Photo Researchers, Inc. Fig. 28.18a Mark Nielsen. Fig. 28.18b Courtesy Michael Ross, University of Florida. Fig. 28.20b Courtesy Michael Ross, University of Florida. Fig. 28.20c Mark Nielsen.

Chapter 29 Fig. 29.1b Don W. Fawcett/Photo Researchers, Inc. Fig. 29.1c Myriam Wharman/ Phototake. Fig. 29.11b Preparation and photography Mark Nielsen. Fig. 29.14a Photo provided courtesy of Kohei Shiota, Congenital Anomaly Research Center, Kyoto University, Graduate School of Medicine. Fig. 29.14b Courtesy National Museum of Health and Medicine, Armed Forces Institute of Pathology. Fig. 29.14c Courtesy National Museum of Health and Medicine, Armed Forces Institute of Pathology. Fig. 29.14d Courtesy National Museum of Health and Medicine, Armed Forces Institute of Pathology. Fig. 29.14e Courtesy National Museum of Health and Medicine, Armed Forces Institute of Pathology. Fig. 29.14f Photo by Lennart Nilsson/ Albert Bonniers Fîrlag AB, A Child is Born, Dell Publishing Company. Reproduced with permission. Fig. 29.14g Photo provided courtesy of Kohei Shiota, Congenital Anomaly Research Center, Kyoto University, Graduate School of Medicine. Fig. 29.14h Photo provided courtesy of Kohei Shiota, Congenital Anomaly Research Center, Kyoto University, Graduate School of Medicine.

INDEX

O

EPONYMS USED IN THIS TEXT

In the life sciences, an eponym is the name of a structure, drug, or disease that is based on the name of a person. For example, you may be more familiar with the Achilles tendon than you are with its more anatomically descriptive term, the calcaneal tendon. Because eponyms remain in frequent use, this listing correlates common eponyms with their anatomical terms.

EPONYM	ANATOMICAL TERM
Achilles tendon	calcaneal tendon
Adam's apple	thyroid cartilage
ampulla of Vater (VA-ter)	hepatopancreatic ampulla
Bartholin's (BAR-tō-linz) gland	greater vestibular gland
Billroth's (BIL-rōtz) cord	splenic cord
Bowman's (BŌ-manz) capsule	glomerular capsule
Bowman's (BŌ-manz) gland	olfactory gland
Broca's (BRŌ-kaz) area	motor speech area
Brunner's (BRUN-erz) gland	duodenal gland
bundle of His (HISS)	artrioventricular (AV) bundle
canal of Schlemm (SHLEM)	scleral venous sinus
circle of Willis (WIL-is)	cerebral arterial circle
Cooper's (KOO-perz) ligament	suspensory ligament of the breast
Cowper's (KOW-perz) gland	bulbourethral gland
crypt of Lieberkühn (LE-ber-kyūn)	intestinal gland
duct of Santorini (san′-tō-RĒ-nē)	accessory duct
duct of Wirsung (VĒR-sung)	pancreatic duct
Eustachian (yoo-STĀ-kē-an)	auditory tube
Fallopian (fal-LŌ-pē-an) tube	uterine tube
gland of Littré (LĒ-tra)	urethral gland
Golgi (GOL-jē) tendon organ	tendon organ
Graafian (GRAF-ē-an) follicle	mature ovarian follicle
Hassall's (HAS-alz) corpuscle	thymic corpuscle
Haversian (ha-VĒR-shun) canal	central canal
Haversian (ha-VĒR-shun) system	osteon
Heimlich (HĪM-lik) maneuver	abdomial thrust maneuver
islet of Langerhans (LANG-er-hanz)	pancreatic islet
Kupffer (KOOP-fer) cell	stellate reticuloendothelial cell
Leydig (LĪ-dig) cell	interstitial endocrinocyte
loop of Henle (HEN-lē)	loop of the nephron
Luschka's (LUSH-kaz) aperture	lateral aperture
Magendie's (ma-JEN-dēz) aperture	median aperture
Meibomian (mi-BŌ-mē-an) gland	tarsal gland
Meissner (MĪS-ner) corpuscle	corpuscle of touch
Merkel (MER-kel) disc	tactile disc
Müllerian (mil-E rē-an) duct	paramesonephric duct
organ of Corti (KOR-tē)	spiral organ
Pacinian (pa-SIN-ē-an) corpuscle	lamellated corpuscle
Peyer's (PĪ-erz) patch	aggregated lymphatic follicle
plexus of Auerbach (OW-er-bak)	myenteric plexus
plexus of Meissner (MĪS-ner)	submucosal plexus
pouch of Douglas	rectouterine pouch
Purkinje (pur-KIN-jē) fiber	conduction myofiber
Rathke's (rath-KĒZ) pouch	hypophyseal pouch
Ruffini (roo-FĒ-nē) corpuscle	type II cutaneous mechanoreceptor
Sertoli (ser-TŌ-lē) cell	sustentacular cell
Skene's (SKĒNZ) gland	paraurethral gland
sphincter of Oddi (OD-dē)	sphincter of the hepatopancreatic ampulla
Volkmann's (FŌLK-manz) canal	perforating canal
Wernicke's (VER-ni-kēz) area	auditory association area
Wharton's (HWAR-tunz) jelly	mucous connective tissue
Wolffian duct	mesonephric duct
Wormian (WER-mē-an) bone	sutural bone

COMBINING FORMS, WORD ROOTS, PREFIXES, AND SUFFIXES

Many of the terms used in anatomy and phsiology are compound words; that is, they are made up of word roots and one or more prefixes or suffixes. For example, *leukocyte* is formed from the word roots *leuk-* meaning "white", a connecting vowel (o), and *cyte* meaning "cell." Thus, a leukocyte is a white blood cell. The following list includes some of the most commonly used combining forms, word roots, prefixes, ad suffixes used in the study of anatomy and physiology. Each entry includes a usage example. Learning the meanings of these fundamental word parts will help you remember terms that, at first glance, may seem long or complicated.

COMBINING FORMS AND WORD ROOTS

Acous-, Acu- hearing Acoustics.
Acr- extremity Acromegaly.
Aden- gland Adenoma.
Alg-, Algia- pain Neuralgia.
Angi- vessel Angiocardiography.
Anthr- joint Arthropathy.
Aut-, Auto- self Autolysis.
Audit- hearing Auditory canal.

Bio- life, living Biopsy.
Blast- germ, bud Blastula.
Blephar- eyelid Blepharitis.
Brachi- arm Brachial plexus.
Bronch- trachea, windpipe Bronchoscopy.
Bucc- cheek Buccal.

Capit- head Decapitate.
Carcin- cancer Carcinogenic.
Cardi-, Cardia-, Cardio- heart Cardiogram.
Cephal- head Hydrocephalus.
Cerebro- brain Cerebrospinal fluid.
Chole- bile, gall Cholecystogram.
Chondr-, cartilage Chondrocyte.
Cor-, Coron- heart Coronary.
Cost- rib Costal.
Crani- skull Craniotomy.
Cut- skin Subcutaneous.
Cyst- sac, bladder Cystoscope.

Derma-, Dermato- skin Dermatosis.
Dura- hard Dura mater.

Enter- intestine Enteritis.
Erythr- red Erythrocyte.

Gastr- stomach Gastrointestinal.
Gloss- tongue Hypoglossal.
Glyco- sugar Glycogen.
Gyn-, Gynec- female, woman Gynecology.

Hem-, Hemat- blood Hematoma.
Hepar-, Hepat- liver Hepatitis.
Hist-, Histio- tissue Histology.
Hydr- water Dehydration.
Hyster- uterus Hysterectomy.

Ischi- hip, hip joint Ischium.

Kines- motion Kinesiology.

Labi- lip Labial.
Lacri- tears Lacrimal glands.
Laparo- loin, flank, abdomen Laparoscopy.
Leuko- white Leukocyte.
Lingu- tongue Sublingual glands.
Lip- fat Lipid.
Lumb- lower back, loin Lumbar.

Macul- spot, blotch Macula.
Malign- bad, harmful Malignant.
Mamm-, Mast- breast Mammography, Mastitis.
Meningo- membrane Meningitis.
Myel- marrow, spinal cord Myeloblast.
My-, Myo- muscle Myocardium.

Necro- corpse, dead Necrosis.
Nephro- kidney Nephron.
Neuro- nerve Neurotransmitter.

Ocul- eye Binocular.
Odont- tooth Orthodontic
Onco- mass, tumor Oncology.
Oo- egg Oocyte.
Opthalm- eye Ophthalmology.
Or- mouth Oral.
Osm- odor, sense of small Anosmia.
Os-, Osseo-, Osteo- bone Osteocyte.
Ot- ear Otitus media.

Palpebr- eyelid Palpebra.
Patho- disease Pathogen.
Pelv- basin Renal pelvis.
Phag- to eat Phagocytosis.
Phleb- vein Phlebitis.
Phren- diaphragm Phrenic.
Pilo- hair Depilatory.
Pneumo- lung, air Pneumothorax.
Pod- foot Podocyte.
Procto- anus, rectum Proctology.
Pulmon- lung Pulmonary.

Ren- kidneys Renal artery.
Rhin- nose Rhinitis.

Scler-, Sclero- hard Atherosclerosis.
Sep-, Spetic- toxic condition due to micoorganisms Septicemia.
Soma-, Somato- body Somatotropin.
Sten- narrow Stenosis.
Stasis-, Stat- stand still Homeostasis.

Tegument- skin, covering Integumentary.
Therm- heat Thermogenesis.
Thromb- clot, lump Thrombus.

Vas- vessel, duct Vasoconstriction.

Zyg- joined Zygote.

PREFIXES

A-, An- without, lack of, deficient Anesthesia.
Ab- away from, from Abnormal.
Ad-, Af- to, toward Adduction, Afferent neuron.
Alb- white Albino.
Alveol- cavity, socket Alveolus.
Andro- male, masculine Androgen.
Ante- before Antebrachial vein.
Anti- against Anticoagulant.

Bas- base, foundation Basal ganglia.
Bi- two, double Biceps.
Brady- slow Bradycardia.

Cata- down, lower, under Catabolism.
Circum- around Circumduction.
Cirrh- yellow Cirrhosis of the liver.
Co-, Con-, Com with, together Congenital.
Contra- against, opposite Contraception.
Crypt- hidden, concealed Cryptorchidism.
Cyano- blue Cyanosis.

De- down, from Deciduous.
Demi-, hemi- half Hemiplegia.
Di-, Diplo- two Diploid.
Dis- separation, apart, away from Dissection.
Dys- painful, difficult Dyspnea.

E-, Ec-, Ef- out from, out of Efferent neuron.
Ecto-, Exo- outside Ectopic pregnancy.
Em-, En- in, on Emmetropia.
End-, Endo- within, inside Endocardium.
Epi- upon, on, above Epidermis.
Eu- good, easy, normal Eupnea.
Ex-, Exo- outside, beyond Exocrine gland.
Extra- outside, beyond, in addition to Extracellular fluid.

Fore- before, in front of Forehead.

Gen- originate, produce, form Genitalia.
Gingiv- gum Gingivitis.

Hemi- half Hemiplegia.
Heter-, Hetero- other, different Heterozygous.
Homeo-, Homo- unchanging, the same, steady Homeostasis.
Hyper- over, above, excessive Hyperglycemia.
Hypo- under, beneath, deficient Hypothalamus.

In-, Im- in, inside, not Incontinent.
Infra- beneath Infraorbital.
Inter- among, between Intercostal.
Intra- within, inside Intracellular fluid.
Ipsi- same Ipsilateral.
Iso- equal, like Isotonic.

Juxta- near to Juxtaglomerular apparatus.

Later- side Lateral.

Macro- large, great Macrophage.
Mal- bad, abnormal Malnutrition.
Medi-, Meso- middle Medial.
Mega-, Megalo- great, large Magakaryocyte.
Melan- black Melanin.
Meta- after, beyond Metacarpus.
Micro- small Microfilament.
Mono- one Monounsaturated fat.

Neo- new Neonatal.

Oligo- **small, few** Oliguria.
Ortho- **straight, normal** Orthopedics.
Para- **near, beyond, beside** Paranasal sinus.
Peri- **around** Pericardium.
Poly- **much, many, too much** Polycythemia.
Post- **after, beyond** Postnatal.
Pre-, Pro- **before, in front of** Presynaptic.
Pseudo- **false** Pseudostratified.
Retro- **backward, behind** Retroperitoneal.
Semi- **half** Semicircular canals.
Sub- **under, beneath, below** Submucosa.
Super- **above, beyond** Superficial.
Supra- **above, over** Suprarenal.
Sym-, Syn- **with, together** Symphysis.
Tachy- **rapid** Tachycardia.
Trans- **across, through, beyond** Transudation.
Tri- **three** Trigone.

SUFFIXES

-able **capable of, having ability** to Viable.
-ac, -al **pertaining to** Cardiac.
-algia **painful condition** Myalgia.
-an, -ian **pertaining** to Circadian.
-ant **having the characteristic of** Malignant.
-ary **connected with** Ciliary.
-asis, -asia, -esis, -osis **condition or state of** Hemostasis.
-asthenia **weakness** Myasthenia.
-ation **process, action, condition** Inhalation.
-centesis **puncture, usually for drainage** Amniocentesis.
-cid, -cide, -cis, **cut, kill destroy** Spermicide.
-ectomize, -ectomy **excision of, removal of** Thyroidectomy.
-emia **condition of blood** Anemia.
-esthesia **sensation, feeling** Anesthesia.
-fer carry Efferent arteriole.
-gen **agent that produces or originates** Pathogen.
-genic **producing** Pyogenic.
-gram **record** Electrocardiogram.
-graph **instrument for recording** Electroencephalograph.
-ia **state, condition** Hypermetropia.
-ician **person associated with** Pediatrician.
-ics **art of, science of** Optics.
-ism **condition, state** Rheumatism.
-itis **inflammation** Neuritis.
-logy **the study or science of** Physiology.
-lysis **dissolution, loosening, destruction** Hemolysis.
-malacia **softening** Osteomalacia.
-megaly **enlarged** Cardiomegaly.
-mers, -meres **parts** Polymers.
-oma **tumor** Fibroma.
-osis **condition, disease** Necrosis.
-ostomy **create an opening** Colostomy.
-otomy **surgical incision** Tracheotomy.
-pathy **disease** Myopathy.
-penia **deficiency** Thrombocytopenia
-philic **to like, have an affinity for** Hydrophilic.
-phobe, -phobia **fear of, aversion to** Photophobia.
-plasia, -plasty **forming, molding** Rhinoplasty.
-pnea **breath** Apnea.
-poiesis **making** Hemopoiesis.
-ptosis **falling, sagging** Blepharoptosis.
-rrhage **bursting forth, abnormal discharge** Hemorrhage.
-rrhea **flow, discharge** Diarrhea.
-scope **instrument for viewing** Bronchoscope.
-stomy **creation of a mouth or artificial opening** Tracheostomy.
-tomy **cutting into, incision into** Laparotomy.
-tripsy **crushing** Lithotripsy.
-trophy **relating to nutrition or growth** Atrophy.
-uria **urine** Polyuria.